ENCYCLOPEDIA OF
BIOINFORMATICS AND
COMPUTATIONAL BIOLOGY

ENCYCLOPEDIA OF BIOINFORMATICS AND COMPUTATIONAL BIOLOGY

EDITORS IN CHIEF

Shoba Ranganathan

Macquarie University, Sydney, NSW, Australia

Michael Gribskov

Purdue University, West Lafayette, IN, United States

Kenta Nakai

The University of Tokyo, Tokyo, Japan

Christian Schönbach

Nazarbayev University, School of Science and Technology, Department of Biology, Astana, Kazakhstan

VOLUME 2
Topics

Bruno Gaeta

University of New South Wales, Sydney, Australia

ELSEVIER

AMSTERDAM • BOSTON • HEIDELBERG • LONDON • NEW YORK • OXFORD
PARIS • SAN DIEGO • SAN FRANCISCO • SINGAPORE • SYDNEY • TOKYO

Elsevier
Radarweg 29, PO Box 211, 1000 AE Amsterdam, Netherlands
The Boulevard, Langford Lane, Kidlington, Oxford OX5 1GB, United Kingdom
50 Hampshire Street, 5th Floor, Cambridge MA 02139, United States

Library of Congress Cataloging-in-Publication Data
A catalog record for this book is available from the Library of Congress

British Library Cataloguing-in-Publication Data
A catalogue record for this book is available from the British Library

ISBN 978-0-12-811414-8

For information on all publications visit our
website at http://store.elsevier.com

www.elsevier.com • www.bookaid.org

Working together
to grow libraries in
developing countries

Publisher: Oliver Walter
Acquisition Editor: Sam Crowe
Content Project Manager: Paula Davies
Associate Content Project Manager: Ebin Clinton Rozario
Designer: Greg Harris

EDITORS IN CHIEF

Shoba Ranganathan holds a Chair in Bioinformatics at Macquarie University since 2004. She has held research and academic positions in India, USA, Singapore and Australia as well as a consultancy in industry. She hosted the Macquarie Node of the ARC Centre of Excellence in Bioinformatics (2008–2013). She was elected the first Australian Board Director of the International Society for Computational Biology (ISCB; 2003–2005); President of Asia-Pacific Bioinformatics Network (2005–2016) and Steering Committee Member (2007–2012) of Bioinformatics Australia. She initiated the Workshops on Education in Bioinformatics (WEB) as an ISMB2001 Special Interest Group meeting and also served as Chair of ICSB's Educaton Committee. Shoba currently serves as Co-Chair of the Computational Mass Spectrometry (CompMS) initiative of the Human Proteome Organization (HuPO), ISCB and Metabolomics Society and as Board Director, APBioNet Ltd.

Shoba's research addresses several key areas of bioinformatics to understand biological systems using computational approaches. Her group has achieved both experience and expertise in different aspects of computational biology, ranging from metabolites and small molecules to biochemical networks, pathway analysis and computational systems biology. She has authored as well as edited several books as well as articles for the 2013 Encyclopedia of Systems Biology. She is currently an Editor-in-Chief of the Encyclopedia of Bioinformatics and Computational Biology and the Bioinformatics Section Editor of the Reference Module in Life Science as well as an editorial board member of several bioinformatics journals.

Dr. Gribskov graduated from Oregon State University in 1979 with a Bachelors of Science degree (with Honors) in Biochemistry and Biophysics. He then moved to the University of Wisconsin-Madison for graduate studies focused on the structure and function of the sigma subunit of *E. coli* RNA polymerase, receiving his Ph.D. in 1985. Dr. Gribskov studied X-ray crystallography as an American Cancer Society post-doctoral fellow at UCLA in the laboratory of David Eisenberg, and followed this with both crystallographic and computational studies at the National Cancer Institute. In 1992, Dr. Gribskov moved to the San Diego Supercomputer Center at the University of California, San Diego where he was lead scientist in the area of computational biology and an adjunct associate professor in the department of Biology. From 2003 to 2007, Dr. Gribskov was the president of the International Society for Computational Biology, the largest professional society devoted to bioinformatics and computational biology. In 2004, Dr. Gribskov moved to Purdue University where he holds an appointment as a full professor in the Biological Sciences and Computer Science departments (by courtesy). Dr. Gribskov's interests include genomic and transcriptomic analysis of model and non-model organisms, the application of pattern recognition and machine learning techniques to biomolecules, the design and implementation of biological databases to support molecular and systems biology, development of methods to study RNA structural patterns, and systems biology studies of human disease.

Kenta Nakai received the PhD degree on the prediction of subcellular localization sites of proteins from Kyoto University in 1992. From 1989, he has worked at Kyoto University, National Institute of Basic Biology, and Osaka University. From 1999 to 2003, he was an Associate Professor at the Human Genome Center, the Institute of Medical Science, the University of Tokyo, Japan. Since 2003, he has been a full Professor at the same institute. His main research interest is to develop computational ways for interpreting biological information, especially that of transcriptional regulation, from genome sequence data. He has published more than 150 papers, some of which have been cited more than 1,000 times.

Christian Schönbach is currently Department Chair and Professor at Department of Biology, School of Science and Technology, Nazarbayev University, Kazakhstan and Visiting Professor at International Research Center for Medical Sciences at Kumamoto University, Japan. He is a bioinformatics practitioner interfacing genetics, immunology and informatics conducting research on major histocompatibility complex, immune responses following virus infection, biomedical knowledge discovery, peroxisomal diseases, and autism spectrum disorder that resulted in more than 80 publications. His previous academic appointments included Professor at Kumamoto University (2016–2017), Nazarbayev University (2013–2016), Kazakhstan, Kyushu Institute of Technology (2009–2013) Japan, Associate Professor at Nanyang Technological University (2006–2009), Singapore, and Team Leader at RIKEN Genomic Sciences Center (2002–2006), Japan. Other prior positions included Principal Investigator at Kent Ridge Digital Labs, Singapore and Research Scientist at Chugai Institute for Molecular Medicine, Inc., Japan. In 2018 he became a member of International Society for Computational Biology (ISCB) Board of Directors. Since 2010 he is serving Asia-Pacific Bioinformatics Network (APBioNet) as Vice-President (Conferences 2010–2016) and President (2016–2018).

VOLUME EDITORS

Mario Cannataro is a Full Professor of Computer Engineering and Bioinformatics at University "Magna Graecia" of Catanzaro, Italy. He is the director of the Data Analytics research center and the chair of the Bioinformatics Laboratory at University "Magna Graecia" of Catanzaro. His current research interests include bioinformatics, medical informatics, data analytics, parallel and distributed computing. He is a Member of the editorial boards of Briefings in Bioinformatics, High-Throughput, Encyclopedia of Bioinformatics and Computational Biology, Encyclopedia of Systems Biology. He was guest editor of several special issues on bioinformatics and he is serving as a program committee member of several conferences. He published three books and more than 200 papers in international journals and conference proceedings. Prof. Cannataro is a Senior Member of IEEE, ACM and BITS, and a member of the Board of Directors for ACM SIGBIO.

Bruno Gaeta is Senior Lecturer and Director of Studies in Bioinformatics in the School of Computer Science and Engineering at UNSW Australia. His research interests cover multiple areas of bioinformatics including gene regulation and protein structure, currently with a focus on the immune system, antibody genes and the generation of antibody diversity. He is a pioneer of bioinformatics education and has trained thousands of biologists and trainee bioinformaticians in the use of computational tools for biological research through courses, workshops as well as a book series. He has worked both in academia and in the bioinformatics industry, and currently coordinates the largest bioinformatics undergraduate program in Australia.

Mohammad Asif Khan, PhD, is an associate professor and the Dean of the School of Data Sciences, as well as the Director of the Centre for Bioinformatics at Perdana University, Malaysia. He is also a visiting scientist at the Department of Pharmacology and Molecular Sciences, Johns Hopkins University School of Medicine (JHUSOM), USA. His research interests are in the area of biological data warehousing and applications of bioinformatics to the study of immune responses, vaccines, inhibitory drugs, venom toxins, and disease biomarkers. He has published in these areas, been involved in the development of several novel bioinformatics methodologies, tools, and specialized databases, and currently has three patent applications granted. He has also led the curriculum development of a Postgraduate Diploma in Bioinformatics programme and an MSc (Bioinformatics) programme at Perdana University. He is an elected ExCo member of the Asia-Pacific Bioinformatics Network (APBioNET) since 2010 and is currently the President of Association for Medical and Bio-Informatics, Singapore (AMBIS). He has donned various important roles in the organization of many local and international bioinformatics conferences, meetings and workshops.

CONTENTS OF VOLUME 2

LIST OF CONTRIBUTORS FOR VOLUME 2

Sanne Abeln
*Center for Integrative Bioinformatics VU (IBIVU),
The Netherlands*

Luciano A Abriata
*Swiss Federal Institute of Technology in Lausanne,
Lausanne, Switzerland; and Swiss Institute of
Bioinformatics, Lausanne, Switzerland*

Josep F. Abril
*University of Barcelona (UB), Barcelona, Spain;
and Institute of Biomedicine of the University of
Barcelona (IBUB), Barcelona, Spain*

Arpah Abu
University of Malaya, Kuala Lumpur, Malaysia

Nikhil Agrawal
University of KwaZulu-Natal, Durban, South Africa

Shandar Ahmad
Jawaharlal Nehru University, New Delhi, India

Amr A. Alhossary
Nanyang Technological University, Singapore

Mohd N.M. Ali
University of Malaya, Kuala Lumpur, Malaysia

Hamid Alinejad-Rokny
University of New South Wales, Sydney, NSW, Australia

Guruprasad Ananda
*The Jackson Laboratory, Farmington, CT,
United States*

Miguel Arenas
University of Vigo, Vigo, Spain

David R. Armstrong
*Protein Data Bank in Europe, EMBL-EBI, Hinxton,
United Kingdom*

Kiyoshi Asai
*University of Tokyo, Kashiwa, Japan; and Artificial
Intelligence, Research Center, Koto-ku, Japan*

Haitham Ashoor
*The Jackson Laboratory, Farmington, CT,
United States*

Laura Astolfi
*Department of Computer, Control and Management
Engineering "A. Ruberti", Sapienza University of Rome,
Italy; and Neuroelectrical Imaging and BCI Lab,
Fondazione Santa Lucia IRCSS, Rome, Italy*

Teresa K. Attwood
*The University of Manchester, Manchester, United
Kingdom*

Vasco Azevedo
*Federal University of Minas Gerais, Belo Horizonte,
Brazil*

Chidambaram S. Babu
JSS University, Mysuru, India

Shaarmini Balakrishnan
*Manipal International University, Nilai, Negeri
Sembilan, Malaysia*

Priyanka Baloni
Institute for Systems Biology, Seattle, WA, United States

Debmalya Barh
*Institute of Integrative Omics and Applied Biotechnology,
Purba Medinipur, India*

Avner Bar-Hen
CNAM, Paris, France

John M. Berrisford
*Protein Data Bank in Europe, EMBL-EBI, Hinxton,
United Kingdom*

Aivett Bilbao
*Pacific Northwest National Laboratory, Richland, WA,
United States*

Tobias Brinkjost
TU Dortmund University, Dortmund, Germany

Conrad J. Burden
*Australian National University, Canberra, ACT,
Australia*

William S. Bush
*Case Western Reserve University, Cleveland, OH,
United States*

Rita Casadio
University of Bologna, Bologna, Italy

Sergi Castellano
*Great Ormond Street Institute of Child Health (ICH),
University College London (UCL), London, United
Kingdom; and UCL Genomics, University College
London (UCL), London, United Kingdom*

Filippo Castiglione
National Research Council of Italy, Rome, Italy

Ramesh Chandra
University of Delhi, Delhi, India

Balakumar Chandrasekaran
University of KwaZulu-Natal, Durban, South Africa

Nai-Wen Chang
National Taiwan University, Taipei city, Taiwan, Republic of China; and National Yang-Ming University, Taipei, Taiwan, Republic of China

Michael A. Charleston
University of Tasmania, Hobart, TAS, Australia

Smriti Chawla
Indraprastha Institute of Information Technology, Delhi, India

Liang Chen
University of Macau, Macau, China

Lijun Cheng
The Ohio State University, Columbus, OH, United States

Sun Sook Chung
King's College, London, UK

Alice R. Clark
Protein Data Bank in Europe, EMBL-EBI, Hinxton, United Kingdom

Christopher T. Clarkson
University of Essex, Colchester, United Kingdom

Brian S. Cole
University of Pennsylvania, Philadelphia, PA, United States

Matthew J. Conroy
Protein Data Bank in Europe, EMBL-EBI, Hinxton, United Kingdom

Marco Da Costa Schulze
Federal University of Paraíba, João Pessoa, Brazil

Camila E. Da Paz Barbosa
Federal University of Paraíba, João Pessoa, Brazil

Hong-Jie Dai
National Taitung University, Taipei, Taiwan, Republic of China

Sayoni Das
Institute of Structural and Molecular Biology, University College London, London, United Kingdom

Hamed Dashti
Sharif University of Technology, Tehran, Iran

Wayne K. Dawson
CeNT, University of Warsaw, Warsaw, Poland; and BiLab, The University of Tokyo, Tokyo, Japan

Natalie L. Dawson
Institute of Structural and Molecular Biology, University College London, London, United Kingdom

Saravanan Dayalan
The University of Melbourne, Parkville, VIC, Australia

Marc Delord
Université Paris Diderot-Paris 7, Sorbone Paris Cité, Paris, France

Jaspreet Kaur Dhanja
Indian Institute of Technology Delhi, New Delhi, India

Sarinder K. Dhillon
University of Malaya, Kuala Lumpur, Malaysia

Socrates Dokos
University of New South Wales, Sydney, NSW, Australia

Richard J. Edwards
University of New South Wales, Sydney, NSW, Australia

Christiane Ehrt
TU Dortmund University, Dortmund, Germany

Christoph Endrullat
University of Applied Sciences, Wildau, Germany

Christine L.P. Eng
National University of Singapore, Singapore

Kiyoshi Ezawa
Kyushu Institute of Technology, Fukuoka, Japan

Lorenzo Farina
Department of Computer, Control and Management Engineering "A. Ruberti", Sapienza University of Rome, Italy; and Neuroelectrical Imaging and BCI Lab, Fondazione Santa Lucia IRCSS, Rome, Italy

Klaas Anton Feenstra
Center for Integrative Bioinformatics VU (IBIVU), The Netherlands

Edson L. Folador
Federal University of Paraíba, João Pessoa, Brazil

Franca Fraternali
King's College, London, UK

Marcus Frohme
University of Applied Sciences, Wildau, Germany

Marta Gómez Perosanz
Complutense University of Madrid, Madrid, Spain

Pascale Gaudet
SIB Swiss Institute of Bioinformatics, Geneva, Switzerland

Joshy George
The Jackson Laboratory, Farmington, CT, United States

Samik Ghosh
The Systems Biology Institute, Tokyo, Japan

Martin Golebiewski
Heidelberg Institute for Theoretical Studies (HITS gGmbH), Heidelberg, Germany

Madhu Goyal
University of Technology Sydney, Sydney, NSW, Australia

Michael Gribskov
Purdue University, West Lafayette, IN, United States

M. Michael Gromiha
Tokyo Institute of Technology, Tokyo, Japan

Deepti Gupta
Protein Data Bank in Europe, EMBL-EBI, Hinxton, United Kingdom

Molly A. Hall
The Pennsylvania State University, University Park, PA, United States

Ari Hardianto
Universitas Padjadjaran, Jatinangor, Indonesia

Sabeeha Hasnain
Jawaharlal Nehru University, New Delhi, India

Jaap Heringa
Vrije Universiteit Amsterdam, Amsterdam, The Netherlands

Andreas Hildebrandt
Institut für Informatik, Mainz, Germany

Joshua W.K. Ho
Victor Chang Cardiac Research Institute, Darlinghurst, NSW, Australia; and University of New South Wales, Sydney, NSW, Australia

Simon Y.W. Ho
University of Sydney, Sydney NSW, Australia

Barbara R. Holland
University of Tasmania, Hobart, TAS, Australia

Amy Y. Then Hui
University of Malaya, Kuala Lumpur, Malaysia

Lina Humbeck
TU, Dortmund, Germany

Midori Iida
Kyushu Institute of Technology, Fukuoka, Japan

Kenichiro Imai
Advanced Industrial Science and Technology, Tokyo, Japan

Apichart Intarapanich
National Electronics and Computer Technology Center, Pathum Thani, Thailand

Junichi Iwakiri
University of Tokyo, Kashiwa, Japan

Shingo Iwami
Kyushu University, Fukuoka, Japan; and Japan Science and Technology Agency, Kawaguchi, Japan

Shoya Iwanami
Kyushu University, Fukuoka, Japan

Katherine J.L. Jackson
Garvan Institute of Medical Research, Darlinghurst, NSW, Australia

Syed B. Jamal
Federal University of Minas Gerais, Belo Horizonte, Brazil

Abdul Salam Jarrah
American University of Sharjah, Sharjah, UAE

Mannu Jayakanthan
Tamil Nadu Agricultural University, Coimbatore, India

Jitendra Jonnagaddala
University of New South Wales, Sydney, NSW, Australia

Kho S. Jye
University of Malaya, Kuala Lumpur, Malaysia

Saowaluck Kaewkamnerd
National Electronics and Computer Technology Center, Pathum Thani, Thailand

Ulykbek Kairov
Nazarbayev University, Astana, Kazakhstan

Kazunari Kaizu
RIKEN Quantitative Biology Center (QBiC), Osaka, Japan

Yerbol Kalykhbergenov
Nazarbayev University, Astana, Kazakhstan

R. Krishna Murty Karuturi
The Jackson Laboratory, Farmington, CT, United States

Shalini Kaushik
Indian Institute of Technology Roorkee, India

Sandeep Kaushik
European Institute of Excellence on Tissue Engineering and Regenerative Medicine, Guimaraes, Portugal; and University of Minho, Braga, Portugal

Takeshi Kawashima
National Institute of Genetics, Mishima, Japan

Alexander Kel
geneXplain GmbH, Wolfenbüttel, Germany

Anil K. Kesarwani
The Jackson Laboratory, Farmington, CT, United States

Rupesh K. Kesharwani
The Jackson Laboratory, Farmington, CT, United States

Haris A. Khan
University of Malaya, Kuala Lumpur, Malaysia

Varun Khanna
Flinders University, Adelaide, SA, Australia; and Vaxine Pty Ltd., Adelaide, SA, Australia

Sunghwan Kim
National Institutes of Health, Bethesda, MD, United States

Hirohisa Kishino
University of Tokyo, Tokyo, Japan

Oliver Koch
Technical University of, Dortmund, Germany

Roman Kogay
Nazarbayev University, Astana, Republic of Kazakhstan

Nils M. Kriege
TU, Dortmund, Germany

Mathias Krull
geneXplain GmbH, Wolfenbüttel, Germany

Parveen Kumar
The Jackson Laboratory, Farmington, CT, United States

Subhas C Kundu
Headquarters of the European Institute of Excellence on Tissue Engineering and Regenerative Medicine, Guimaraes, Portugal

Chee-Keong Kwoh
Nanyang Technological University, Singapore

Anna Laddach
King's College, London, UK

Michal Lazniewski
CeNT, University of Warsaw, Warsaw, Poland; and Medical University of Warsaw, Warsaw, Poland

Jonathan G. Lees
Institute of Structural and Molecular Biology, University College London, London, United Kingdom

Jeremy Leipzig
Drexel University, Philadelphia, PA, United States

Emmanuelle Lerat
Université Lyon, CNRS, Villeurbanne, France

Lang Li
The Ohio State University, Columbus, OH, United States

Yi Li
The Jackson Laboratory, Farmington, CT, United States

David A. Liberles
Temple University, Philadelphia, PA, United States

Frédérique Lisacek
SIB Swiss Institute of Bioinformatics, Geneva, Switzerland; and University of Geneva, Geneva, Switzerland

Enze Liu
Indiana University School of Medicine, Indianapolis, IN, United States; and Indiana University, Indianapolis, IN, United States

Pier Luigi Martelli
University of Bologna, Bologna, Italy

Jitender Madan
Chandigarh College of Pharmacy Landran, Mohali, India

Sorayya Malek
University of Malaya, Kuala Lumpur, Malaysia

Ankit Malhotra
The Jackson Laboratory, Farmington, CT, United States

Vidhi Malik
Indian Institute of Technology Delhi, New Delhi, India

Emiliano Mancini
University of Amsterdam, Amsterdam, Netherlands

Mahendra Mariadassou
INRA, Paris, France

Venkata S.K. Mattaparthi
Tezpur University, Tezpur, India

Ramit Mehr
Bar-Ilan University, Ramt Gan, Israel

Charu K. Midha
Institute for Systems Biology, Seattle, WA, United States

Pozi Milow
University of Malaya, Kuala Lumpur, Malaysia

Alex L. Mitchell
EMBL-EBI, Hinxton, United Kingdom

Askhat Molkenov
Nazarbayev University, Astana, Kazakhstan

Jason H. Moore
University of Pennsylvania, Philadelphia, PA, United States

Laurentino Quiroga Moreno
Birkbeck College, University of London, United Kingdom

Mogeeb Mosleh
University of Taiz, Taiz, Yemen

Yuguang Mu
Nanyang Technological University, Singapore

Muchtaridi Muchtaridi
Universitas Padjadjaran, Jatinangor, Indonesia

Abhik Mukhopadhyay
Protein Data Bank in Europe, EMBL-EBI, Hinxton, United Kingdom

Saule Mussurova
Nazarbayev University, Astana, Kazakhstan

Raju Nagarajan
Indian Institute of Technology Madras, Chennai, India; and Vanderbilt University Medical Center, Nashville, TN, United States

Kenta Nakai
The University of Tokyo, Tokyo, Japan

Rina F. Nuwarda
Universitas Padjadjaran, Jatinangor, Indonesia

Mina Obuca
Institute of Molecular Genetics of the ASCR, Prague, Czech Republic

Christine Orengo
Institute of Structural and Molecular Biology, University College London, London, United Kingdom

Paola Paci
CNR National Research Council, IASI Institute of Systems Analysis and Computer Science "Antonio Ruberti", Rome, Italy

Sucheendra K. Palaniappan
The Systems Biology Institute, Tokyo, Japan

Preeti Pandey
Jawaharlal Nehru University, New Delhi, India

Chi N.I. Pang
The University of New South Wales, Sydney, NSW, Australia

Francesco Pappalardo
University of Catania, Catania, Italy

Ashwini Patil
The University of Tokyo, Tokyo, Japan

Josch K. Pauling
Humboldt University of Berlin, Berlin, Germany

Marco Pedicini
Roma Tre University, Rome, Italy

Marzio Pennisi
University of Catania, Catania, Italy

Swathik Clarancia Peter
Tamil Nadu Agricultural University, Coimbatore, India

Nikolai Petrovsky
Flinders University, Adelaide, SA, Australia; and Vaxine Pty Ltd., Adelaide, SA, Australia

Manuela Petti
Department of Computer, Control and Management Engineering "A. Ruberti", Sapienza University of Rome, Italy; and Neuroelectrical Imaging and BCI Lab, Fondazione Santa Lucia IRCSS, Rome, Italy

Russell Pickford
University of New South Wales, Kensington, NSW, Australia

Dariusz Plewczynski
CeNT, University of Warsaw, Warsaw, Poland; and MINI, Warsaw University of Technology, Warsaw, Poland

Pazit Polak
Bar-Ilan University, Ramt Gan, Israel

Jaroslaw Polanski
University of Silesia Institute of Chemistry, Katowice, Poland

Sarita Poonia
Indraprastha Institute of Information Technology, Delhi, India

Joram M. Posma
Imperial College London, London, United Kingdom

Heike Pospisil
University of Applied Sciences, Wildau, Germany

Chakrawarti Prasun
Indian Institute of Technology Roorkee, India

Giuseppe Profiti
University of Bologna, Bologna, Italy

Navaneethan Radhakrishnan
Indian Institute of Technology Delhi, New Delhi, India

Karthic Rajamanickam
University of Saskatchewan, Saskatoon, SK, Canada

Shoba Ranganathan
Macquarie University, Sydney, NSW, Australia

Soumya Lipsa Rath
Nagoya University, Nagoya, Japan

Hukam C. Rawal
ICAR – National Research Centre on Plant Biotechnology, New Delhi, India

Pedro A. Reche
Complutense University of Madrid, Madrid, Spain

Jonas Reeb
Technical University of Munich (TUM), Garching/ Munich, Germany

Rui L Reis
Headquarters of the European Institute of Excellence on Tissue Engineering and Regenerative Medicine, Guimaraes, Portugal

Thibault Robin
University of Geneva, CUI, Carouge, Switzerland

Christian Rockmann
University of Applied Sciences, Wildau, Germany

Ute Roessner
The University of Melbourne, Parkville, VIC, Australia

Burkhard Rost
Technical University of Munich (TUM), Garching/ Munich, Germany

Giulia Russo
University of Catania, Catania, Italy

Meena K. Sakharkar
University of Saskatchewan, Saskatoon, SK, Canada

Reza Salek
European Bioinformatics Institute (EMBL-EBI), Hinxton, United Kingdom

Jose Luis Sanchez-Trincado Lopez
Complutense University of Madrid, Madrid, Spain

Airy Sanjeev
Tezpur University, Tezpur, Assam, India

Aliya Sarkytbayeva
Nazarbayev University, Astana, Kazakhstan

Vishal K. Sarsani
The Jackson Laboratory, Farmington, CT, United States

Castrense Savojardo
University of Bologna, Bologna, Italy

Christian Schönbach
Nazarbayev University, Astana, Kazakhstan

Bertil Schmidt
Institut für Informatik, Mainz, Germany

Daniela Schuster
Paracelsus Medical University of Salzburg, Salzburg, Austria

Jean-Marc Schwartz
University of Manchester, Manchester, United Kingdom

Samuel Selvaraj
Bharathidasan University, Tiruchirappalli, India

Debarka Sengupta
Indraprastha Institute of Information Technology, Delhi, India

Deepak Sharma
Indian Institute of Technology Roorkee, India

Hagit Shatkay
University of Delaware, Newark, DE, United States

Adrian Shepherd
University of London, London, United Kingdom

Tsuyoshi Shirai
Nagahama Institute of Bio-Science and Technology, Nagahama, Japan

Onkar Singh
Institute of Information Science, Academia Sinica, Taipei, Taiwan, Republic of China; and National Yang-Ming University, Taipei, Taiwan, Republic of China

Zita Soons
Maastricht University, Maastricht, The Netherlands

Dean Southwood
Macquarie University, Sydney, NSW, Australia

Rachel A. Spicer
European Bioinformatics Institute (EMBL-EBI), Hinxton, United Kingdom

Anuj Srivastava
The Jackson Laboratory, Farmington, CT, United States

Toto Subroto
Universitas Padjadjaran, Bandung, West Java, Indonesia

Xiangying Sun
Purdue University, West Lafayette, IN, United States

Durai Sundar
Indian Institute of Technology Delhi, New Delhi, India

Wing-Kin Sung
National University of Singapore, Singapore; and Genome Institute of Singapore, Singapore

Nur A. Taib
University of Malaya, Kuala Lumpur, Malaysia

Kazuhiro Takemoto
Kyushu Institute of Technology, Fukuoka, Japan

Tin Wee Tan
National University of Singapore, Singapore

Vladimir B. Teif
University of Essex, Colchester, United Kingdom

Prasoon K. Thakur
Institute of Molecular Genetics of the ASCR, Prague, Czech Republic

Paolo Tieri
CNR National Research Council, IAC Institute for Applied Computing "Mauro Picone", Rome, Italy

Sandeep Tiwari
Federal University of Minas Gerais, Belo Horizonte, Brazil

Hiroyuki Toh
Kwansei Gakuin University, Sanda, Japan

Kentaro Tomii
Artificial Intelligent Research Center (AIRC), National Institute of Advanced Industrial Science and Technology (AIST), Tokyo, Japan

Joo Chuan Tong
National University of Singapore, Singapore; and Institute of High Performance Computing, Singapore

Sissades Tongsima
National Center for Genetic Engineering and Biotechnology, Pathum Thani, Thailand

Fatemeh Vafaee
University of New South Wales, Sydney, NSW, Australia

Cara Van Der Wal
University of Sydney, Sydney, NSW, Australia; and Australian Museum Research Institute, Australian Museum, Sydney, NSW, Australia

Marc R. Wilkins
The University of New South Wales, Sydney, NSW, Australia

Kelly L. Williams
Macquarie University, Sydney, NSW, Australia

Edgar Wingender
geneXplain GmbH, Wolfenbüttel, Germany; and University Medical Center Göttingen, Göttingen, Germany

Garry Wong
University of Macau, Macau, China

Jianguo Xia
McGill University, Sainte Anne de Bellevue, QC, Canada; and McGill University, Montreal, QC, Canada

Junfeng Xia
Anhui University, Hefei, China

Gur Yaari
Bar-Ilan University, Ramt Gan, Israel

Ayako Yachie-Kinoshita
The Systems Biology Institute, Tokyo, Japan; and University of Toronto, Toronto, ON, Canada

Tetsushi Yada
Kyushu Institute of Technology, Fukuoka, Japan

Jian Yang
University of Saskatchewan, Saskatoon, SK, Canada

Xiaoxin Ye
Victor Chang Cardiac Research Institute, Darlinghurst, NSW, Australia

Akbar Yermekov
Nazarbayev University, Astana, Kazakhstan

Hong Yung Yip
University of Malaya, Kuala Lumpur, Malaysia

Muhammad Yusuf
Universitas Padjadjaran, Jatinangor, Indonesia

Lina A.M. Zalani
University of Malaya, Kuala Lumpur, Malaysia

Zhongming Zhao
The University of Texas Health Science Center at Houston, Houston, TX, United States

Altyn Zhelambayeva
Nazarbayev University, Astana, Kazakhstan

Liangzhen Zheng
Nanyang Technological University, Singapore

Zhaxybay Zhumadilov
Nazarbayev University, Astana, Kazakhstan

PREFACE

Bioinformatics and Computational Biology (BCB) combine elements of computer science, information technology, mathematics, statistics, and biotechnology, providing the methodology and *in silico* solutions to mine biological data and processes, for knowledge discovery. In the era of molecular diagnostics, targeted drug design and Big Data for personalized or even precision medicine, computational methods for data analysis are essential for biochemistry, biology, biotechnology, pharmacology, biomedical science, and mathematics and statistics. Bioinformatics and Computational Biology are essential for making sense of the molecular data from many modern high-throughput studies of mice and men, as well as key model organisms and pathogens. This Encyclopedia spans basics to cutting-edge methodologies, authored by leaders in the field, providing an invaluable resource to students as well as scientists, in academia and research institutes as well as biotechnology, biomedical and pharmaceutical industries.

Navigating the maze of confusing and often contradictory jargon combined with a plethora of software tools is often confusing for students and researchers alike. This comprehensive and unique resource provides up-to-date theory and application content to address molecular data analysis requirements, with precise definition of terminology, and lucid explanations by experts.

No single authoritative entity exists in this area, providing a comprehensive definition of the myriad of computer science, information technology, mathematics, statistics, and biotechnology terms used by scientists working in bioinformatics and computational biology. Current books available in this area as well as existing publications address parts of a problem or provide chapters on the topic, essentially addressing practicing bioinformaticists or computational biologists. Newcomers to this area depend on Google searches leading to published literature as well as several textbooks, to collect the relevant information.

Although curricula have been developed for Bioinformatics education for two decades now (Altman, 1998), offering education in bioinformatics continues to remain challenging from the multidisciplinary perspective, and is perhaps an NP-hard problem (Ranganathan, 2005). A minimum Bioinformatics skill set for university graduates has been suggested (Tan *et al.*, 2009). The Bioinformatics section of the Reference Module in Life Sciences (Ranganathan, 2017) commenced by addressing the paucity of a comprehensive reference book, leading to the development of this Encyclopedia. This compilation aims to fill the "gap" for readers with succinct and authoritative descriptions of current and cutting-edge bioinformatics areas, supplemented with the theoretical concepts underpinning these topics.

This Encyclopedia comprises three sections, covering Methods, Topics and Applications. The theoretical methodology underpinning BCB are described in the Methods section, with Topics covering traditional areas such as phylogeny, as well as more recent areas such as translational bioinformatics, cheminformatics and computational systems biology. Additionally, Applications will provide guidance for commonly asked "how to" questions on scientific areas described in the Topics section, using the methodology set out in the Methods section. Throughout this Encyclopedia, we have endeavored to keep the content as lucid as possible, making the text "... as simple as possible, but not simpler," attributed to Albert Einstein. Comprehensive chapters provide overviews while details are provided by shorter, encyclopedic chapters.

During the planning phase of this Encyclopedia, the encouragement of Elsevier's Priscilla Braglia and the constructive comments from no less than ten reviewers lead our small preliminary editorial team (Christian Schönbach, Kenta Nakai and myself) to embark on this massive project. We then welcomed one more Editor-in-Chief, Michael Gribskov and three section editors, Mario Cannataro, Bruno Gaeta and Asif Khan, whose toils have results in gathering most of the current content, with all editors reviewing the submissions. Throughout the production phase, we have received invaluable support and guidance as well as milestone reminders from Paula Davies, for which we remain extremely grateful.

Finally we would like to acknowledge all our authors, from around the world, who dedicated their valuable time to share their knowledge and expertise to provide educational guidance for our readers, as well as leave a lasting legacy of their work.

We hope the readers will enjoy this Encyclopedia as much as the editorial team have, in compiling this as an ABC of bioinformatics, suitable for naïve as well as experienced scientists and as an essential reference and invaluable teaching guide for students, post-doctoral scientists, senior scientists, academics in universities and research institutes as well as pharmaceutical, biomedical and biotechnological industries. Nobel laureate Walter Gilbert predicted in 1990 that "In the year 2020 you will be able to go into the drug store, have your DNA sequence read in an hour or so, and given back to you on a compact disk so you can analyze it." While technology may have already arrived at this milestone, we are confident one of the readers of this Encyclopedia will be ready to extract valuable biological data by computational analysis, resulting in biomedical and therapeutic solutions, using bioinformatics to "measure" health for early diagnosis of "disease."

References

Altman, R.B., 1998. A curriculum for bioinformatics: the time is ripe. Bioinformatics. 14 (7), 549–550.
Ranganathan, S., 2005. Bioinformatics education–perspectives and challenges. PLoS Comput Biol 1 (6), e52.
Tan, T.W., Lim, S.J., Khan, A.M., Ranganathan, S., 2009. A proposed minimum skill set for university graduates to meet the informatics needs and challenges of the "-omics" era. BMC Genomics. 10 (Suppl 3), S36.
Ranganathan, S., 2017. Bioinformatics. Reference Module in Life Sciences. Oxford: Elsevier.

Shoba Ranganathan

The Gene Ontology

Pascale Gaudet, SIB Swiss Institute of Bioinformatics, Geneva, Switzerland

Background and Context

Up until the 1990s, data generated in biological and biomedical research were mostly shared via research articles. The development and democratisation of computers, on one hand, as well as the creation of network of this informatics infrastructure, on the other, opened a completely new way to share information and provided new possibilities to structure and integrate data that free text research articles cannot achieve. Concomitantly, high throughput technologies with applications for the biomedical sciences started to emerge, which allowed to analyze large numbers of biological samples (for example, microarrays) and to sequence ever larger and longer nucleic acid sequences (DNA and RNA). This gave rise to the new fields of proteomics, genomics, and transcriptomics. Repositories storing this data have been growing at exponential rates. One of the most important sequence repositories, GenBank, contains nucleotide sequences for 370,000 different species, with yearly rates of increase in certain data types such as transcriptomics of nearly 50% (Benson et al., 2017). Biological research had entered a new era, in which data processing and analysis require informatics tools, leading to the emergence of the field of bioinformatics.

Gene function analysis in the genomics era. The Gene Ontology project was started at the beginning of the genomics era, when the rapidly increasing number of sequenced genomes made it necessary to have a way to describe gene products' functions that would be applicable to all realms of life. Up until that point, in each species the vast majority of the genes had not been identified nor studied. The relatively few genes that had been identified had some amount of knowledge associated with it (even if very small): Either a mutant and its associated phenotype, an enzymatic activity, a genetic interaction, etc. The availability of entire genome sequences, and moreover, the sequences of genomes of multiple species, increased by many orders of magnitude the number of different genes for which we at least knew the existence. However, there was no systematic way to describe gene functions, with different biology communities having developed their own, often organism-specific nomenclatures. For example, the term 'budding' has different meaning when describing plant propagation, tooth formation initiation, or asexual reproduction that occurs for example in *Saccharomyces cerevisiae*. The Gene Ontology was created to meet that need. The specific question that got the project started was the need, formulated by Prof Michael Ashburner, to classify all genes from *Drosophila*, the model system he was studying and whose genome had just been sequenced, and have the same classification scheme for other model organisms so that gene functions (and hence, genomes), could be compared (Lewis, 2017). The ability to compare homologous and orthologous genes allows formulating research hypotheses for uncharacterized genes, or when certain biological modules are well conserved, providing a model of a gene's function.

Why an Ontology?

In information sciences, an ontology is a computational structure that describe entities and relationships of a domain of interest in a structured, and hence computable format. Ontologies are composed of a set of entities, also called classes, arranged into a hierarchy from the general to the specific (Hastings, 2017). This structure is powerful because it is amenable to computational analysis using logical reasoning. Moreover, linking ontology classes with biological entities provides a structured classification of gene function which itself can be further used for analysis of experimental data.

The Gene Ontology

Organisation of GO. The overarching goal of the GO project is to provide life science researchers with current functional information concerning gene functions and the cellular context under which these occur. GO is a structured, controlled vocabulary that formally represents biological knowledge. In the GO system, the terms are subdivided in three non-overlapping ontologies (also known as aspects): Molecular Function (MF), which describe the molecular activities of gene products; Biological Process (BP), the pathways and larger processes made up of the activities of multiple gene products, and Cellular Component (CC), that illustrate where gene products are active (Ashburner et al., 2000). These aspects are non-redundant and share a common space of identifiers and a well-specified syntax.

Classes and definitions. In GO, each term defines a *class*. The class itself is defined in two different ways: Textually, with a human-readable definition, and logically by relationships to other terms in the ontology and which can be computed over. The most common relation is is_a, which represents the ontological definition of a class (*i.e.*, it describes what the term *is* relative to others named classes in the ontology). Other commonly used relations in GO include *part_of*, representing partonomy between classes. For example, the brain is *part_of* the body.

Molecular functions are generally described as *types* of other functions, using the *is_a* relationship. For example, *DNA helicase activity* and *lyase activity* are both types of catalytic activity, and hence are related to *catalytic activity* by *is_a* relationships.

Biological process has three main types of relations, *is_a*, *part_of*, and *regulates*. The *is_a* relation represents types of a process, for example *mitotic DNA replication* (GO:1902969) is a particular type of *nuclear DNA replication* (GO:0033260). On the other hand, sub-processes are represented by *part_of* relations: mitotic DNA replication is a step of the *cell cycle* (GO:0000278). Biological regulation is a type of Biological Process, represented by the relation *regulates* and two more specific relations, *negatively regulates* and *positively regulates*.

Each class must be defined as a *is_a* class of another class, to ensure that the ontology is completely representing all classes logically. In addition, each class may have multiple relationships to other classes' terms. This allows for more expressivity than a simple hierarchy. This structure makes GO a directed graph with terms as nodes and relationships as edges. **Fig. 1** graphically illustrates the structure of GO.

An increasingly larger number of terms are being defined with equivalence axioms or logical definitions (Gene Ontology Consortium, 2015). These use explicitly defined relations from the Relations Ontology to produce a description for terms that is amenable to logical reasoning. An example is shown in **Fig. 2**. The equivalence axiom makes the class unique relative to all other classes in the ontology.

As of November 2017, the full ontology contained 49,968 terms: 4146 cellular components, 29,669 biological processes and 11,140 molecular functions. The ontology also contains 73,776 explicitly encoded *is_a* relationships, 30,879 explicitly encoded *part_of* relationships, and 20,564 explicitly encoded *regulates*, *negatively_regulates* or *positively_regulates* relationships.

GO uses external ontologies extensively to represent other types of terms such as chemicals (ChEBI ontology (de Matos *et al.*, 2010)), vertebrate anatomical parts (Uberon (Mungall *et al.*, 2012)), plant parts (Cooper *et al.*, 2013), cell types (Bard *et al.*, 2005), and sequence features (Eilbeck *et al.*, 2005).

Managing changes in the ontology. GO undergoes frequent revisions: changes of relations between terms, addition of new terms, or term removal. Terms are never deleted from the ontology, but their status changes to "obsolete" and all relationships to other terms are removed (Rhee *et al.*, 2008). Term obsoletion may lead to loss of annotations if the responsible group does not re-house the annotations under a more appropriate term. Other types of changes such as changes to the relationships between terms and term merges do not impact annotations, because annotations always refer to specific terms, not their position within the GO. However, these changes can affect the analyses done using the ontology.

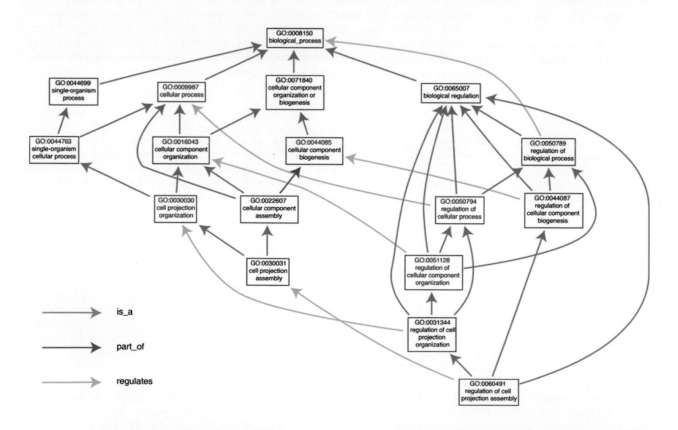

Fig. 1 *Illustration of the Gene Ontology structure.* Paths of the term *regulation of cell projection assembly*, GO:0060491 to its root term. GO is a directed graph with terms as nodes and relationships as edges; these relationships are either *is_a*, *part_of*, *has_part*, or *regulates*. In its basic representation, there should be no cycles in this graph, and we can therefore establish parent (more general) and child (more specific) terms. Reproduced from PMID:27812933.

name: L-glutamate import across plasma membrane

equivalentTo:

transport that ('has target start location' some 'extracellular region')

and ('has target end location' some cytosol)

and (imports some 'L-glutamate')

and ('results in transport across' some 'plasma membrane')

inferred classifications:

'import across plasma membrane'

'L-alpha-amino acid transmembrane transport'

'L-glutamate import into cell'

Fig. 2 Logical definition for the class L-glutamate import across plasma membrane.

GO Annotations

Annotation objects. Ideally, gene ontology terms are associated to the biological entity having the function being captured, ie the gene product. In practice, database entries for *genes* are often used as a proxy for gene products. Gene products can be either a protein, a non-protein-coding RNA, or any other gene product, including isoforms and post-translationally-modified proteins, which can be captured using the Protein Ontology (Natale *et al.*, 2017). Moreover, GO allows capturing isoform-specific data when appropriate. Identifiers are for example UniProtKB isoform identifiers, such as P12345-2.

Semantics of a GO annotation. The association of a GO term and a gene product means that the gene product inherently possesses an activity (or, more precisely ontologically, the potential for that activity) or a molecular role (MF term), directly participates in a process (BP), and that its function occurs in a specific cellular localization (CC) (Ashburner *et al.*, 2000). Therefore, transient localizations such as endoplasmic reticulum and Golgi apparatus for secreted proteins are not in the scope of GO. Dynamic localizations such as the movement to the mitotic spindle during cytokinesis should be captured, as these correspond to locations at which the gene product is active. A thorough overview of the meaning of biological function in GO has been written by Thomas (2017).

A gene product can be annotated using as many GO terms as necessary to completely describe its function, and the GO terms can be at varying levels in the hierarchy, depending on the state of knowledge when the annotation was made. If a gene product is annotated to any particular term, then the annotation also holds for all the is-a and part-of parent terms. Annotations to more granular terms carry more information.

Certain annotations additionally contain *qualifiers: contributes_to, colocalizes_with* and *NOT*. The qualifier can profoundly change the meaning of an annotation, in particular the *NOT* qualifier, which means that a gene protein does not have a particular function, or does not participate in a given process, or is not localized to the a given cellular component. *Contributes_to* is used for molecular function annotations, and means that while the gene product has some role in the molecular function, on its own it does not possess the function. This usually applies to a protein complex: for example, each RNA polymerase subunit contributes to RNA polymerase activity, but none of the individual subunits has that function on its own. *Colocalizes_with* is used to capture a localization near an organelle, but when the gene product is not an intrinsic part of the organelle. Details about the qualifiers used in GO and their definitions can be found on the GO website, (see "Relevant Websites" section).

Data provenance. The supporting evidence for each annotation is captured using evidence codes and references. Evidence Codes represent the type of information from which the annotation is derived, and are grouped in five broad categories: experimental evidence codes, computational analysis evidence codes, author statement evidence codes, curatorial statement evidence codes, and automatically assigned evidence codes (**Fig. 3**). Experimental evidence codes represent data captured from experimental data. The data can be derived from a direct assay (*IDA*), from a mutant (*IMP* and *IGI*), from a physical interaction (IPI), or from the expression pattern of a gene product (*IEP*). The GOC has recently introduced new evidence codes to label data derived from high throughput experiments: *High throughput direct assay evidence (HDA), high throughput mutant phenotype evidence (HMP), high throughput genetic interaction evidence (HGI),* and *high throughput expression pattern evidence (HEP).* The largest fraction of annotations is attributed automatically, e.g., annotations via the InterPro resource (Burge *et al.*, 2012).

Manual annotation of Panther protein families (Mi *et al.*, 2017) based on the phylogenetic relationships between the members of these protein families (PMID:21873635) are assigned the evidence code *Inferred from Biological Aspect of Ancestor* (*IBA*). Curators also manually assign annotations derived from the known functions of similar sequences (*ISS* and related methods *ISA, ISM* and *ISO*). Curators can also capture annotations based on statements that can be traced to an original publication (*TAS*) and they can make statements based on well-known biological knowledge using the Inferred from Curator (IC) code (for example, a protein

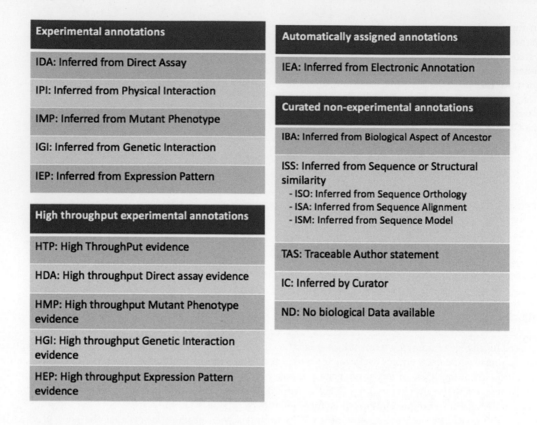

Fig. 3 The main evidence codes used by the GOC project.

with transcription factor activity is expected to be active in the nucleus of eukaryotic cells). Finally, when no data are available for an aspect of GO for a given protein, a curator can annotate the root term of the ontology (*molecular_function*, *biological_process*, or *cellular_component*), and use the evidence code *ND*. More details about the use of evidence codes and how they are used in GO can be found in Gaudet *et al.* (2017); Chibucos *et al.* (2017).

The Reference represents the source of the annotation. For annotations based on experimental evidence, the Reference contains the PubMed (PMID or PMC) accession ID. When the evidence code denotes an automatically assigned annotation, the reference will contain one of the GO_REF codes maintained by the GO Consortium (see "Relevant Websites" section) that details the automated assignment methods.

Contributing groups. Various contribute annotations, including UniProt (UniProt Consortium, 2014), Mouse Genome Informatics (Drabkin *et al.*, 2012), Saccharomyces Genome Database (Hong *et al.*, 2008), Wormbase (Harris *et al.*, 2010), Flybase (dos Santos *et al.*, 2015), dictyBase (Gaudet *et al.*, 2011), EcoCyc (Keseler *et al.*, 2009) and the Functional Gene Annotation group at University College London (Buchan *et al.*, 2010). While most funding for these efforts come from the US National Institutes of Health, many smaller funding streams have enabled a wide consortium of groups to contribute to GO resources.

In addition to the above-mentioned groups who have dedicated resources to produce GO annotations, the GOC is fortunate to have fostered collaborations with research groups that can improve the ontology, the annotations, or both. Major revisions or expansions of a specific GO domain are often undertaken in consultation with domain experts. Notable successful collaborative projects include that of the immune system (Diehl *et al.*, 2007), heart development (Khodiyar *et al.*, 2011), kidney development (Alam-Faruque *et al.*, 2014), fetal development (Wick *et al.*, 2014), muscle processes and cellular components (Feltrin *et al.*, 2009), transcription (Tripathi *et al.*, 2013), and cilium biology (van Dam *et al.*, 2013). Some model organism databases invite direct contributions from researchers, such as PomBase for *Schizosaccharomyces pombe*, who has developed curation tool, CANTO (Rutherford *et al.*, 2014).

Members of the broader community are welcome to suggest improvements to the ontology and annotations. The GO project is now managed via GitHub (see "Relevant Websites" section). Users are welcome to submit their feedback via the tracker. Details about contributing to GO are available on the GOC website (see "Relevant Websites" section).

Main Applications of the Gene Ontology

One of the main uses of GO is to analyze the results of high-throughput experiments. Two types of analyses can be done to interpret this type of data. One is to group the location or function of genes that are over- or under-expressed (enrichment analysis). In functional profiling, GO is used to determine which processes are different between sets of genes. This is done by using a likelihood-ratio test to determine if GO terms are represented differently between the gene sets (Mi *et al.*, 2013).

GO can also be used to infer the function of unannotated genes. Gene prediction with significant similarity to annotated genes can be assigned one or several of the functions of the characterized genes (Conesa and Götz, 2008; Meehan *et al.*, 2013). Other methods such as the presence of specific protein domains can also be used to assign GO terms (Burge *et al.*, 2012; Pedruzzi *et al.*, 2015).

Another usage of GO is as a dictionary to support text mining of the biomedical literature. In this case domain-specific concepts derived from GO are used to classify the literature or attach concepts to documents (Ruch, 2017).

Accessing Gene Ontology Project Data

Ontology: There are three different editions of the GO, in increasing order of complexity: *Go-basic*, *go*, and *go-plus*.

Go-basic: The basic edition of the GO, filtered such that annotations can be propagated up the graph (*i.e.*, it does not contain any cycles). The relations included are is a, *part_of*, *regulates*, *negatively_regulates*, and *positively_regulates*. Moreover, go-basic excludes relationships that cross different aspect of the ontology. This version is available in Open Biological Ontology (OBO) format only.

Go: This core edition of the GO includes additional relationship types, including some that span the three GO aspects, such as *has_part* and *occurs_in*, connecting the otherwise disjoint hierarchies found in go-basic. This version of the GO ontology is available in OBO format and in OWL-RDF/XML.

Go-plus: This is the most expressive edition of the GO; it includes more relationships than the *go* file, as well as connections to external ontologies, including the Chemical Entities of Biological Interest ontology (ChEBI; (Hastings *et al.*, 2013)), the Uberon anatomy ontology (Mungall *et al.*, 2012), and the Plant Ontology for plant structure/stage (PO (Cooper and Jaiswal, 2016)). It also includes import modules that are minimal subsets of those ontologies. This version of the GO ontology is available in OWL-RDF/XML.

See "Relevant Websites" section for instructions to download those files.

Ontology subsets. Gene Ontology subsets (also known as *slims*) are reduced versions of the ontology containing fewer terms and usually designed to summarize and aggregate the available data for specific application; for e.g., species-specific subsets or more generic subsets with informative terms in various categories). Subsets are particularly useful for giving a high level summary of GO annotations. Further information is available from the GO website (see "Relevant Websites" section).

Annotations. Annotations are available in Gene Association File format (GAF) (see Relevant Websites Section). A newer format, GPAD, is designed to be more normalized than GAF, and is intended to work in conjunction with a separate format for exchanging gene product information (GPI). Details about the contents of the GAF format can be found in Gaudet *et al.* (2017).

Web interfaces. Association of GO terms with gene products is made by different groups who contribute the data to a central database. That data is disseminated by the GO consortium via the AmiGO platform (see "Relevant Websites" section), the official web-based open-source tool for querying, browsing, and visualizing the Gene Ontology and annotations. AmiGO supports basic searching, browsing, the ability to download custom data sets, and a common question wizard interface.

Best practices for using GO data. GO and its annotations are updated and released daily. Therefore, for reproducibility and correct reporting of methodology, researchers should provide the data at which files used for a particular analysis were downloaded (Blake, 2013).

Future Directions

The expressivity of GO annotations is limited by the fact that the current paradigm associates a gene product with a GO class, but there are no links between independent annotations. One example is a steroid hormone receptor, which act as a hormone receptor when localized in the cytosol, and upon binding to its ligand, translocates to the nucleus where it acts as a transcription factor. In GO, the gene encoding that receptor would be annotated to both molecular functions (transcription factor activity and receptor activity) and two cellular components (cytosol and nucleus), but the connections between these annotations are unfortunately not part of the data model. The GO Consortium is working towards the elaboration of a new data model, the GO-Causal Activity Model (GO-CAM), which links different annotations to each other as appropriate, to produce a more complete model of biology. Improving the expressivity of GO annotations will enable researchers working on logic-based models in systems biology to use the rich GO data to enhance their models.

Closing Remarks

GO is among the most significant informatics resources in the biomedical research community, providing scientists a way to interpret genomic, proteomic and transcriptomic data, and discover gene function and build genomic profiles. Understanding the design and data production in GO is essential for researchers to make the best use of this extensive resource.

Acknowledgements

The GO project is funded by the NIH HG002273–17 to the multiple PI team of Judith A. Blake, J. M. Cherry, Suzanna E. Lewis, Paul Sternberg, and Paul D. Thomas.

See also: Biological and Medical Ontologies: GO and GOA. Biological and Medical Ontologies: Introduction. Natural Language Processing Approaches in Bioinformatics. Ontology in Bioinformatics. Ontology: Introduction

References

Alam-Faruque, Y., Hill, D.P., Dimmer, E.C., *et al.*, 2014. Representing kidney development using the gene ontology. PLOS ONE 9, e99864.
Ashburner, M., Ball, C.A., Blake, J.A., *et al.*, 2000. Gene ontology: Tool for the unification of biology. The gene ontology consortium. Nat. Genet. 25, 25–29.
Bard, J., Rhee, S.Y., Ashburner, M., 2005. An ontology for cell types. Genome Biol. 6, R21.
Benson, D.A., Cavanaugh, M., Clark, K., *et al.*, 2017. GenBank. Nucleic Acids Res. 45, D37–D42.
Blake, J.A., 2013. Ten quick tips for using the gene ontology. PLOS Comput. Biol. 9, e1003343.
Buchan, D.W.A., Ward, S.M., Lobley, A.E., *et al.*, 2010. Protein annotation and modelling servers at University College London. Nucleic Acids Res. 38, W563–W568.
Burge, S., Kelly, E., Lonsdale, D., *et al.*, 2012. Manual GO annotation of predictive protein signatures: The InterPro approach to GO curation. Database J. Biol. Databases Curation 2012, bar068.
Chibucos, M.C., Siegele, D.A., Hu, J.C., Giglio, M., 2017. The evidence and conclusion ontology (eco): Supporting go annotations. Methods Mol. Biol. Clifton NJ 1446, 245–259.
Conesa, A., Götz, S., 2008. Blast2GO: A comprehensive suite for functional analysis in plant genomics. Int. J. Plant Genomics 2008, 619832.
Cooper, L., Jaiswal, P., 2016. The plant ontology: A tool for plant genomics. Methods Mol. Biol. Clifton NJ 1374, 89–114.
Cooper, L., Walls, R.L., Elser, J., *et al.*, 2013. The plant ontology as a tool for comparative plant anatomy and genomic analyses. Plant Cell Physiol. 54, e1.
de Matos, P., Alcántara, R., Dekker, A., *et al.*, 2010. Chemical Entities of Biological Interest: An update. Nucleic Acids Res. 38, D249–D254.
Diehl, A.D., Lee, J.A., Scheuermann, R.H., Blake, J.A., 2007. Ontology development for biological systems: Immunology. Bioinform. Oxf. Engl. 23, 913–915.
dos Santos, G., Schroeder, A.J., Goodman, J.L., *et al.*, 2015. FlyBase: Introduction of the Drosophila melanogaster release 6 reference genome assembly and large-scale migration of genome annotations. Nucleic Acids Res. 43, D690–D697.
Drabkin, H.J., Blake, J.A., Mouse Genome Informatics Database, 2012. Manual Gene Ontology annotation workflow at the Mouse Genome Informatics Database. Database J. Biol. Databases Curation 2012, bas045.
Eilbeck, K., Lewis, S.E., Mungall, C.J., *et al.*, 2005. The sequence ontology: A tool for the unification of genome annotations. Genome Biol. 6, R44.
Feltrin, E., Campanaro, S., Diehl, A.D., *et al.*, 2009. Muscle research and gene ontology: New standards for improved data integration. BMC Med. Genomics 2, 6.
Gaudet, P., Fey, P., Basu, S., *et al.*, 2011. DictyBase update 2011: Web 2.0 functionality and the initial steps towards a genome portal for the Amoebozoa. Nucleic Acids Res. 39, D620–D624.
Gaudet, P., Škunca, N., Hu, J.C., Dessimoz, C., 2017. Primer on the gene ontology. Methods Mol. Biol. Clifton NJ 1446, 25–37.
Gene Ontology Consortium, 2015. Gene ontology consortium: Going forward. Nucleic Acids Res. 43, D1049–D1056.
Harris, T.W., Antoshechkin, I., Bieri, T., *et al.*, 2010. WormBase: A comprehensive resource for nematode research. Nucleic Acids Res. 38, D463–D467.
Hastings, J., 2017. Primer on Ontologies. Methods Mol. Biol. Clifton NJ 1446, 3–13.
Hastings, J., de Matos, P., Dekker, A., *et al.*, 2013. The ChEBI reference database and ontology for biologically relevant chemistry: Enhancements for 2013. Nucleic Acids Res. 41, D456–D463.
Hong, E.L., Balakrishnan, R., Dong, Q., *et al.*, 2008. Gene Ontology annotations at SGD: New data sources and annotation methods. Nucleic Acids Res. 36, D577–D581.
Keseler, I.M., Bonavides-Martínez, C., Collado-Vides, J., *et al.*, 2009. EcoCyc: A comprehensive view of Escherichia coli biology. Nucleic Acids Res. 37, D464–D470.
Khodiyar, V.K., Hill, D.P., Howe, D., *et al.*, 2011. The representation of heart development in the gene ontology. Dev. Biol. 354, 9–17.
Lewis, S.E., 2017. The vision and challenges of the gene ontology. Methods Mol. Biol. Clifton NJ 1446, 291–302.
Meehan, T.F., Vasilevsky, N.A., Mungall, C.J., *et al.*, 2013. Ontology based molecular signatures for immune cell types via gene expression analysis. BMC Bioinform. 14, 263.
Mi, H., Huang, X., Muruganujan, A., *et al.*, 2017. PANTHER version 11: Expanded annotation data from gene ontology and reactome pathways, and data analysis tool enhancements. Nucleic Acids Res. 45, D183–D189.
Mi, H., Muruganujan, A., Casagrande, J.T., Thomas, P.D., 2013. Large-scale gene function analysis with the PANTHER classification system. Nat. Protoc. 8, 1551–1566.
Mungall, C.J., Torniai, C., Gkoutos, G.V., Lewis, S.E., Haendel, M.A., 2012. Uberon, an integrative multi-species anatomy ontology. Genome Biol. 13, R5.
Natale, D.A., Arighi, C.N., Blake, J.A., *et al.*, 2017. Protein Ontology (PRO): Enhancing and scaling up the representation of protein entities. Nucleic Acids Res. 45, D339–D346.
Pedruzzi, I., Rivoire, C., Auchincloss, A.H., *et al.*, 2015. HAMAP in 2015: Updates to the protein family classification and annotation system. Nucleic Acids Res. 43, D1064–D1070.
Rhee, S.Y., Wood, V., Dolinski, K., Draghici, S., 2008. Use and misuse of the gene ontology annotations. Nat. Rev. Genet. 9, 509–515.
Ruch, P., 2017. Text mining to support gene ontology curation and vice versa. Methods Mol. Biol. Clifton NJ 1446, 69–84.
Rutherford, K.M., Harris, M.A., Lock, A., Oliver, S.G., Wood, V., 2014. Canto: An online tool for community literature curation. Bioinform. Oxf. Engl. 30, 1791–1792.
Thomas, P.D., 2017. The gene ontology and the meaning of biological function. Methods Mol. Biol. Clifton NJ 1446, 15–24.
Tripathi, S., Christie, K.R., Balakrishnan, R., *et al.*, 2013. Gene ontology annotation of sequence-specific DNA binding transcription factors: Setting the stage for a large-scale curation effort. Database J. Biol. Databases Curation 2013, bat062.
UniProt Consortium, 2014. Activities at the Universal Protein Resource (UniProt). Nucleic Acids Res. 42, D191–D198.
van Dam, T.J., Wheway, G., Slaats, G.G., SYSCILIA Study GroupHuynen, M.A., Giles, R.H., 2013. The SYSCILIA gold standard (SCGSv1) of known ciliary components and its applications within a systems biology consortium. Cilia 2, 7.
Wick, H.C., Drabkin, H., Ngu, H., *et al.*, 2014. DFLAT: Functional annotation for human development. BMC Bioinform. 15, 45.

Further Reading

Gaudet, P., Dessimoz, C., 2017. Gene ontology: Pitfalls, biases, and remedies. In: Proceedings of the Gene Ontology Handbook, pp. 189–206. Methods in Molecular Biology, ISSN 1940–6029.

Gaudet, P., Škunca, N., Hu, J., Dessimoz, C., 2017. Primer on the gene ontology. In: Proceedings of the Gene Ontology Handbook, pp. 25–40. Methods in Molecular Biology, ISSN 1940–6029.
Hastings, J., 2017. Primer on ontologies. In: Proceedings of the Gene Ontology Handbook, pp. 3–14. Methods in Molecular Biology, ISSN 1940–6029.
Lewis, S.E., 2017. The vision and challenges of the gene ontology. In: Proceedings of the Gene Ontology Handbook, pp. 291–302. Methods in Molecular Biology, ISSN 1940–6029.
Lovering, R.C., 2017. How does the scientific community contribute to Gene Ontology? In: Proceedings of the Gene Ontology Handbook, pp. 85–96. ISSN 1940–6029.
Monoz-Torres, M., Carbon, S., 2017. Get GO! retrieving GO data using AmiGO, QuickGO, API, files, and tools. In: Proceedings of the Gene Ontology Handbook, pp. 149–160. Methods in Molecular Biology, ISSN 1940–6029.
Poux, S., Gaudet, P., 2017. Best practices. In: Proceedings of the Manual Annotation With the Gene Ontology pp. 41–54. Methods in Molecular Biology, ISSN 1940–6029.
Thomas, P.D., 2017. The gene ontology and the meaning of biological function. In: Proceedings of the Gene Ontology Handbook, pp. 15–24. ISSN 1940–6029.

Relevant Websites

http://amigo.geneontology.org
AmiGO, the Gene Ontology Consortium's web-based set of tools for searching and browsing the Gene Ontology database.
http://geneontology.org/page/contributing-go
Contributing to GO.
http://geneontology.org/page/download-annotations
Download GO annotations.
http://geneontology.org/page/download-ontology
Download Ontology.
www.geneontology.org
Gene Ontology Consortium.
http://www.geneontology.org/cgi-bin/references.cgi
GO Reference Collection.
http://geneontology.org/page/go-slim-and-subset-guide
GO Slim and Subset Guide.
https://www.ebi.ac.uk/QuickGO/
QuickGO - EMBL-EBI.
https://github.com/geneontology/
The Gene Ontology Consortium software and issue tracker on GitHub.
https://en.wikipedia.org/wiki/Web_Ontology_Language
Web Ontology Language.

Biographical Sketch

Pascale Gaudet works in the field of biocuration since 2003, working on various projects, including the Model Organism Database dictyBase, the neXtProt database on human proteins of the Swiss Institute of Bioinformatics (SIB), and the Gene Ontology project. Since 2017 Pascale Gaudet is the Project Manager of the GO project. Additionally, Pascale Gaudet is one of the founding members and the first chairperson of the International Society for Biocuration (ISB). The ISB seeks to connect biocurators, developers, and researchers with an interest in biocuration, both amongst themselves and with the users and funders of biological informatics resources.

Protein Functional Annotation

Pier Luigi Martelli, Giuseppe Profiti, and Rita Casadio, University of Bologna, Bologna, Italy

Introduction

Functional annotation is one of the most crucial activities of a Bioinformatician. As a matter of fact, as long as a sequence of characters, no matter of nucleic acids or protein residues, is a string, biologists can make little usage of it. It is therefore mandatory to try to endow with structural and functional features any string of biological origin. In the following, in spite of being deeply aware of the relevance of the annotation of the whole genome in different species, we will focus into the problem of protein sequence annotation that starts immediately after the recognition of a coding region in the DNA sequence. Functional annotation is a prerequisite for the inclusion of the sequence in the public archive/database.

UniProt

Today, the largest public repository of protein sequences is the Universal Protein Resource (UniProt) (http://www.uniprot.org/help/about). UniProtKB (Uniprot Knowledgebase, UniProtKB/Swiss-Prot 2017_02) comprises two databases, TrEMBL and SwissProt. The difference consists in the quality of the annotation. In SwissProt (including 553,655 sequences), protein annotation is manually curated with the aid of literature and computational analysis; in TrEMBL (including 77,483,538 sequences) annotation is automatically computed and not reviewed. More than 95% of the protein sequences come from the translations of coding sequences (CDS) submitted to the EMBL-Bank/GenBank/DDBJ nucleotide sequence resources of the International Nucleotide Sequence Database Collaboration (INSDC). The problem of protein annotation starts here and it is extremely relevant, since all our molecular knowledge is based on what the database contains.

How many proteins do we know? For each record, Uniprot includes indications relative to the protein existence. The protein existence can derive from wet experiments (e.g., 3D structure, biochemical characterization, protein–protein interaction). In this case, the record lists "Evidence at protein level." Alternatively, the protein existence may be associated with the existence of a transcript ("Evidence at transcript level"). It is quite evident from **Table 1** that proteins with experimental evidence and evidence at the transcript level are only 1.5% of the total database. About 23% of the proteins are inferred by sequence similarity with existing orthologues in evolutionarily related species. The remaining 75% are "predicted proteins," without any evidence of the above qualifiers (http://www.uniprot.org/docs/pe_criteria). The percentage of predicted proteins is much less in the SwissProt database (2.5%). Interestingly, SwissProt contains some 0.3% of "Uncertain" proteins, so to say proteins that may be products of pseudogenes, dubious Coding DNA Sequences and genes, but that have been manually curated. Protein allocation in TrEMBL or SwissProt is indeed entirely dependent on the manual curation or automatic procedure of the annotation (http://www.uniprot.org/help/about). From the statistics, one may conclude that we know with some experimental evidence only 1.7% of the proteins listed in Uniprot KB (78,037,193).

The Gene Ontology

The Gene Ontology project started in 1998 with the goal of providing a controlled vocabulary of terms for describing gene product functional characteristics in a computationally feasible way, by associating words, actions and sentences suited to characterize the complete function of a gene with numbered terms (GO terms). GO terms have three major routes: biological process (BP), molecular function (MF) and cellular components (CC). Each route is hierarchically organized from more general (less informative) to more specific terms (leaves). Since its foundations, the Gene Ontology database has developed formal ontologies and constantly has been revised to include new discoveries. Information was derived from experiments described in some 100,000 peer-reviewed scientific papers (http://www.geneontology.org/). AmiGO 2 (http://amigo.geneontology.org/amigo) is a tool that allows the gene-GO term association search in the database. The release (March 2017) of GO includes 29,385 BP, 10,543 MF and

Table 1   Some statistics of the latest release of UniProtKB (release 2017_02).

Protein existence	#UNIPROT/TRrEMBL entries	#UNIPROT/SwissProt entries
Evidence at protein level	125,718	94,730
Evidence at transcript level	1,066,148	57,857
Inferred from homology	17,797,259	385,421
Predicted	58,494,413	13,697
Uncertain	0	1950

4045 CC terms. A GO term has experimental origins when the gene/gene product has a physical characterization in a paper that describes the corresponding association. Experimental Evidence codes, as well as Computational Analysis and Author Statement evidence codes are listed at http://www.geneontology.org/page/guide-go-evidence-codes

Foundations of Functional Annotations

Our information on the biological complexity can be dissected into three major levels of knowledge: sequence, structure, and function. Accordingly, we may think of the relationships among different spaces (universes). **Fig. 1** shows the relationships among the sequence and the structure space. Information is maximal when we know the structure of the protein, stored in the Protein Data Bank (PDB, http://www.rcsb.org/pdb/home/home.do). The structure of the protein is mainly derived from its electron density determination, and this allows us to determine even its function based on quantum molecular computations. Problem is that we know the structure of some 124,114 proteins (PDB, March 2017); however, only 40,463 UniProt entries are linked to nearly all the PDB entries. This is mostly due to redundancy of structures with respect to the same reference sequence. In conclusions, we have information based on the structure for only 0.05% of the total number of sequences. These proteins are associated with about 23,000 experimental GO terms (**Table 2**). Ideally, we can then connect the space of functions with that of structures and sequences. The question poses as to whether it is possible to infer functions from sequences, taking advantage of what it can be derived in terms of basic association rules from known structures and their associated sequences and functions. Evolutionary relationships can help: we know from biological observations that functions are conserved through species, that orthologous proteins conserve functions, have similar structures and similar sequences. We also know that even when alignment methods fail in retrieving significant scoring below 30% sequence identity, proteins may conserve structures and functions, being included in what we call remote homologs. All this biological information, related to the general concept of evolution, has been promoting in the last decades an enormous effort in developing computational methods. Although different and based on different strategy, different methods try to address the problem of how is it possible to fill the gap between the deluge of protein sequences and their structural and functional annotation.

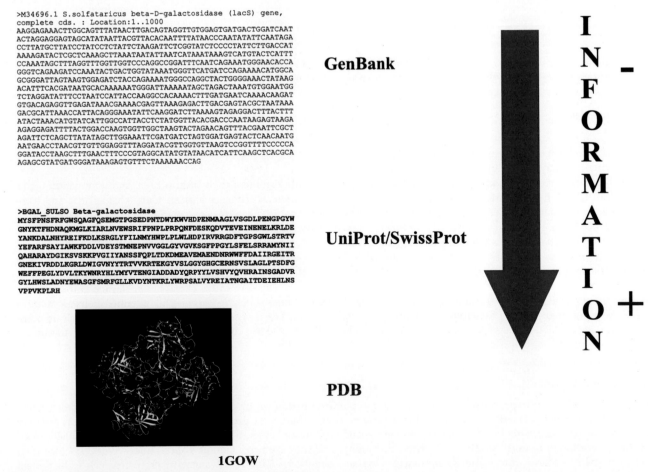

Fig. 1 Information content increases from genes to protein structures.

Table 2   Links between sequence, structure and function in SwissProt

Protein existence	#UNIPROT/SwissProt entries	#Distinct Protein Data Banks (PDBs)	#Distinct experimental GO terms[a]	#Distinct nonexperimental GO terms[a]
Evidence at protein level	94730	93843[a]	22861	20135
With PDB and exp GO	13056	61214	15018	12802
With PDB and no exp GO	11227	35535	0	6546
With exp GO and no PDB	34513	0	19676	15885
No PDB and no exp GO	35934	0	0	11532
Evidence at transcript level	57857	77	9506	14534
With PDB and exp GO	18	22	121	203
With PDB and no exp GO	53	55	0	205
With exp GO and no PDB	11029	0	9492	9084
No PDB and no exp GO	46757	0	0	12949
Inferred from homology	385421	92	3376	10895
With PDB and exp GO	6	6	14	17
With PDB and no exp GO	80	86	0	233
With exp GO and no PDB	4619	0	3372	4384
No PDB and no exp GO	380716	0	0	10293
Predicted	13697	5	679	1556
With PDB and no exp GO	4	5	0	8
With exp GO and no PDB	906	0	679	698
No PDB and no exp GO	12787	0	0	1228
Uncertain	1950	1	71	849
With PDB and no exp GO	1	1	0	4
With exp GO and no PDB	72	0	71	177
No PDB and no exp GO	1877	0	0	744

[a]Counting is performed considering experimental and electronic labels, as detailed at http://www.geneontology.org/page/guide-go-evidence-codes

The Annotation Process in UniProt

UniProt performs both expertly curated manual and automatic annotation (a flow diagram is reported at http://www.uniprot.org/help/biocuration) and highlights also the interplay between manual and automatic annotation.

Manual Annotation

Manual annotation in UniProt is a multistep-curated process that includes similarity search, feature prediction, and validation (http://www.uniprot.org/help/manual_curation). Details are described in the standard operating procedure (SOP) for UniProt manual curation (http://www.uniprot.org/help/biocuration).

Briefly, a sequence is aligned with other well-known sequences with alignment tools such as Blast, and information for its annotation derives from a variety of resources, including scientific literature, sequence analysis tools, phylogenetic and comparative genomics databases (Ensembl Compara, http://www.ensembl.org/info/genome/compara/index.html, and other species-specific collections). A protein is practically annotated with InterPro (https://www.ebi.ac.uk/interpro/about.html), a cluster of different resources that classifies sequences at superfamily, family and subfamily levels and comprises predictive models of protein functions from several databases. High-quality automated and manual annotation of proteins (HAMAP, http://hamap.expasy.org/) is part of the InterPro cluster for family classification. For association of functional GO terms to structural and functional features, InterPro integrates InterPro2GO that maps GO term to InterPro entries. The core of the classification scheme is the software package InterProScan that allows scanning sequences against InterPro features (https://www.ebi.ac.uk/interpro/interproscan.html) and integrating prediction of functional domains and important sites. For inclusion of distinguished features such as transmembrane regions and signal peptides, manual annotation follows a set of expert rules (http://www.uniprot.org/help/transmem; http://www.uniprot.org/help/signal).

Automatic Annotation

The automatic annotation process of UniProt is a multistep process performed by a pipeline that is based again on the protein classification schemes of InterPro. Once the structural and functional features are expertly determined, two prediction systems are active for the automatic annotation: UniRule and the Statistical Automatic Annotation System (SAAS). Both are essentially based on the application of rules (if conditions–>then annotation), differently extracted. UniRule collects rules (some 5557 Identifiers (ID)) derived by curators from manually annotated entries and templates, and routinely annotates properties such as the protein name, functions (GO terms), catalytic activity, pathway membership, and subcellular location, along with sequence specific information, such as the positions of posttranslational modifications and active sites. Similarly, SAAS generates automatic rules for functional annotation from expertly annotated entries in UniProtKB. The algorithm uses machine learning to find the most concise rule for an annotation based on the properties of sequence length, InterPro group membership and taxonomy. SAAS includes 13,674 IDs and can annotate protein properties such as function (GO term), catalytic activity, pathway membership, and subcellular location, with the exclusion of protein names and feature predictions. A sequence analysis method (SAM) is adopted to enrich records with extra sequence-specific information. Predictions of sequence features such as Signal, Transmembrane and Coil regions are computed by other prediction methods, including THMM, Phobius, and SignalP (http://www.uniprot.org/help/sam).

Our Present Knowledge

When addressing the problem of functional annotations, we should cope with what we know. In SwissProt, some 94,730 reviewed sequences with evidence at the protein level have links with some 93,843 PDB files and only 22,861 GO terms with experimental labels (**Table 2**). Some other 20,135 GO terms have only electronic indexes. The number of protein sequences with experimentally detected functions increases when we include proteins with evidence at the transcript level. Therefore, the actual situation is not particularly sound in terms of putative subsets to be adopted for training/testing specific computational methods to predict function from sequence. In the following, we will briefly review the actual state of the art methods (some of them have been already mentioned, being part of the InterPro consortium).

Computational methods have been implemented to associate uncharacterized proteins (targets) to terms describing the function, mostly complying with the Gene Ontology. In general, these methods compare a target to a set of annotated proteins in order to extract and transfer the functional terms. Different approaches are available, in relation to the information available for the target protein at the sequence and/or structure levels and to the similarity with the functionally characterized proteins (Chothia and Lesk, 1986). In the following methods are clustered according to their basic principles of implementation (**Table 3**).

Methods Based on Structure Similarity

When the target protein is endowed with experimental 3D structure, methods based on structure superimposition can be applied, relying on the observation that a high level of structural similarity among proteins generally implies functional similarity. This approach can be extended to proteins with modeled 3D structure, with limitations that strongly depend on the reliability of the model.

Methods for large-scale comparison of protein structures are therefore adopted to scan the target protein against the Protein Data Bank to retrieve annotation terms. Among them, CE (Shindyalov and Bourne, 2001), PDB-eFOLD (Krissinel and Henrick, 2004), DALI (Holm and Rosenström, 2010), VAST (Madej *et al.*, 2014). They adopt different strategies to compare structures at the level of the backbone atoms. However, the conservation of the overall fold is not sufficient to ensure function similarity. A comprehensive structural classification of proteins domains is available at CATH (http://www.cathdb.info/, Sillitoe *et al.*, 2015) and it clearly shows a high heterogeneity of functional terms for some of the folds, including TIM barrel fold, ferredoxin fold, and Rossmann fold.

Table 3 Methods for function prediction

Name	Input	WEB address
CE	Structure	http://source.rcsb.org/jfatcatserver/
DALIStructure	Structure	http://ekhidna.biocenter.helsinki.fi/dali_server/
PDB-eFOLD	Structure	http://www.ebi.ac.uk/msd-srv/ssm/
VAST	Structure	https://structure.ncbi.nlm.nih.gov/Structure/VAST
ProFunc	Structure	http://www.ebi.ac.uk/thornton-srv/databases/ProFunc/
ProBis	Structure	https://probis.nih.gov/
Blast2GO	Sequence	https://www.blast2go.com/
Argot2	Sequence	http://www.medcomp.medicina.unipd.it/Argot2/
SIFTER	Sequence	http://sifter.berkeley.edu/
BAR3	Sequence	http://bar.biocomp.unibo.it/
COG	Sequence	https://www.ncbi.nlm.nih.gov/COG/
SMART	Sequence	http://smart.embl-heidelberg.de/
PFAM	Sequence	http://pfam.xfam.org/
FunFams	Sequence	http://www.cathdb.info/
CCD-SPARCLE	Sequence	https://www.ncbi.nlm.nih.gov/Structure/cdd/cdd.shtml
PROSITE	Sequence	http://prosite.expasy.org/
PRINTS	Sequence	http://130.88.97.239/PRINTS/index.php
INGA	Sequence	http://protein.bio.unipd.it/inga
FFPred	Sequence	http://bioinf.cs.ucl.ac.uk/web_servers/ffpred/ffpred_help/
CombFunc	Sequence	http://www.sbg.bio.ic.ac.uk/~mwass/combfunc/

The reliable association between structure and function requires the detailed conservation of the protein sites important for catalytic activity, substrate binding, and intermolecular interaction. Repertoires of 3D templates of enzyme catalytic sites are available at Catalytic Site Atlas (http://www.ebi.ac.uk/thornton-srv/databases/CSA/, Furnham *et al.*, 2014.) and Pocketome (http://www.pocketome.org/, Kufareva *et al.*, 2012).

When global structural alignments of the target do not retrieve significant results in the PDB, the recognition of structural motifs can assist the functional annotation. Methods implementing local structure alignment, such as ProFunc (Laskowski *et al.*, 2005) can identify local regions, functionally relevant and conserved across several globally divergent structures. Putative active site pockets on the protein surface can be determined with geometric-based approaches like Ligsite (Huang and Schroeder, 2006) and ProBIS (Konc *et al.*, 2015).

Methods Based on Sequence Similarity

When the 3D structure of the target proteins is not available, protein annotation is routinely performed by aligning the residue sequence against the annotated proteins available in sequence databases. The inference of functional similarity requires a higher sequence identity with respect to structural similarity. Generally, the more specific the function to be transferred, the higher the level of identity. Rost (2002) showed that less than 30% of the enzyme sequences sharing more than 50% identity have entirely identical EC numbers and enzymes with different specific functions can align with BLAST E-values below 10^{-50}. Moreover, multidomain proteins increase the complexity of the protein sequence–structure–function relationship and require a careful check of the extent of alignment length with respect to the target sequence (coverage).

BLAST (https://blast.ncbi.nlm.nih.gov/Blast.cgi) is nowadays considered as the baseline method for transferring function annotations upon similarity search against a database of annotated sequence. Due to the complex patterns of association between sequences and function, statistical procedures are routinely applied on the set of retrieved terms to strengthen the reliability of the transferred annotations. For example, the widely adopted Blast2GO (Götz *et al.*, 2008) weights each term by considering the number of retrieved sequences it is associated to and the evidence codes linked to the annotations in the database. Argot2 (Falda *et al.*, 2012) weights GO terms according to their semantic similarity relations and the alignment scores of the retrieved sequences.

Phylogenetic analysis has been also adopted to support the transfer of annotation (e.g., SIFTER, Sahraeian *et al.*, 2015), under the assumption that function evolves parsimoniously within a phylogeny, mainly by duplication events; the position of the query sequence in a phylogenetic tree is more informative than the mere sequence similarity. An alternative strategy to perform a reliable transfer of annotation is based on the preliminary similarity-based clustering of the protein universe into groups that can be functionally characterized on the basis of the existing annotations. COG (Galperin *et al.*, 2015) contains orthology-based clusters of microbial sequences and endows them with manually curated annotation. A complete nonhierarchical clustering of UniProt is at the basis of BAR3 (Profiti *et al.*, 2017): pairs of sequences sharing identity >40% with an alignment coverage >90% are linked to form a graph, whose connected components are extracted to form the clusters. Functional terms associated to clustered proteins are statistically tested for overrepresentation and significant annotation are assigned to the whole cluster. A new sequence falling in the cluster upon alignment inherits the validated terms.

Methods Based on Sequence Profiles and Motifs

Sequences sharing low level of identity with already characterized proteins present the hardest challenges to the process of annotation. Procedures based on sequence profiles are able to extend the range of application of annotation transfer. Sequence profiles are built starting from the multiple alignment of sequences or structures of protein domains already characterized at the functional level. Profiles are routinely represented with either position specific scoring matrices (PSSMs) or Hidden Markov Models (HMMs). Uncharacterized sequences can be aligned against a collection of profiles to retrieve remote similarities and transfer annotation. Profile representation allows increasing the sensitivity of searches by scoring in a different and family-dependent way substitutions and indels in each position of the alignment. The main difference among methods lies in the strategy adopted to build the collection.

Popular repositories of HMMs obtained after manual curation of multiple sequence alignments are PFAM (Finn *et al.*, 2016) and SMART (Letunic *et al.*, 2015), containing models for more than 16,000 and 1200 protein domains, respectively.

The NCBI Conserved Domain Database (CDD) collects PSSMs for curated domains, integrating information from 3D-structure, SMART, PFAM, COG and other external source databases (Marchler-Bauer *et al.*, 2017). On top of CDD, SPARCLE improves the annotation by analyzing the sequential order of the different domains associated to a protein (domain architecture) (Marchler-Bauer *et al.*, 2017).

CATH-Gene3D Functional Families (FunFams) is a compilation of 92,882 HMMs obtained starting from multiple structural alignments of domains belonging to the same CATH family and sharing the same function (Sillitoe *et al.*, 2013).

When also remote similarity search with sequence profiles fails, function annotation can exploit the presence of sequence motifs or fingerprints. They generally represent fragments of the sequence that, in the folded protein, are part of relevant active or functional sites, as highlighted in multiple sequence alignments. Motifs and fingerprints are usually represented with regular expressions or small sequence profiles. PROSITE (Sigrist *et al.*, 2013) and PRINTS (Attwood *et al.*, 2012) are two of the most popular resources, that are also included in the InterPro annotation pipeline.

Integrative and Machine Learning Based Methods

Function annotation of sequences sharing no detectable similarity with characterized proteins requires the development of integrative methods that try to bridge the gap by exploiting different sources of information, including the prediction of relevant features and the mining of interactomics and transcriptomics data. The information is integrated with different approaches, including the expert-driven definition of decision trees, the application of simple regression rules, and the implementation of machine-learning algorithms. The rules of associations between features and annotation are extracted with manual or automatic procedures by analyzing sets of characterized proteins available in datasets (training set). The unbiased estimation of the prediction performance on uncharacterized sequences requires to collect sets of annotated proteins (testing sets), separated and as different as possible from the training sets. This operation is sometimes difficult, in particular when different tools, trained on different training sets, are integrated. For that reason, critical assessment experiments are a useful way to estimate, compare and interpret the methods performance (see next session).

It is in the set of considered features that lies the major difference among the available tools.

FFPred (Minneci *et al.*, 2013) combines with support vector machines (SVM) information on residue composition and presence of low complexity regions and predictions of the following features: secondary structure, signal peptide, transmembrane regions, coiled-coil or disordered segments, phosphorylation and glycosylation sites.

The availability of experimental data on transcriptomics and interactomics leads to the so called "guilty by association" approach: coexpressed genes and proteins in physical or functional relation reliably share common functional features. For example, INGA (Piovesan *et al.*, 2015) combines information on sequence similarity, domain architecture and protein–protein interactions as derived from STRING. In MS-kNN (Lan *et al.*, 2013) a k-Nearest-Neighbor (kNN) system analyzes similar information, integrated with experimental data on gene expression. CombFunc (Wass *et al.*, 2012), combines with SVMs prediction of structural domains and ligand binding sites, sequence motifs, sequence similarity and conservation, protein–protein interaction data from STRING and MINT, coexpression data from CoexpressDB.

An alternative source of information to be integrated is the literature published in scientific papers: EVEX (Van Landeghem *et al.*, 2012.) analyzes the PubMed abstracts to mine information about relevant biomolecular events such as phosphorylation, regulation targets, binding partners, and several other, assigning confidence values to these events.

Critical Assessment of Function Annotation

As we mentioned in the previous section, the evaluation of the performance of methods for functional annotation requires the availability of testing dataset of annotated proteins not used, in any way, during the training phase of the method. Moreover, the comparison of different methods requires the adoption of uniform testing sets. In order to provide an independent and unbiased assessment of computational methods and to provide insights on the state-of-the-art of function annotation, in 2010 the first critical assessment of function annotation (CAFA, http://biofunctionprediction.org/cafa/), experiment was run, followed by two more editions in 2013 and 2016. CAFA was inspired by critical assessment of structure prediction (CASP) and critical assessment of prediction of interactions (CAPRI), devoted to the blind evaluation of computational methods for the prediction of the structure of proteins and of protein complexes, respectively. In CAFA experiments, organizers provide for the community a large set of sequences with unknown or incomplete function. The task consists in predicting Gene Ontology annotations for each of these sequences, before a fixed deadline. After the prediction deadline, the experiment enters the "annotation growth" period in which protein functions are expected to accumulate in public databases (6–12 months). The predictions of different methods are afterwards compared with the experimental annotation deposited in UniProt during the "annotation growth" phase.

The most updated assessment is the CAFA2, i.e., the 2013 edition (CAFA2, Jiang *et al.*, 2016). The benchmark dataset comprised 3681 sequences and 126 different prediction models were evaluated. Results show that the predictive performance is increasing with respect to the CAFA1 edition. Top-performing algorithms are specific for sub-ontology (molecular function, biological process, cellular component) and biological domain (prokaryota, eukaryota). The main index of evaluation is the F1-score, which is the harmonic mean of the precision (# of correctly predicted annotations / # of predicted annotations) and the recall (# of correctly predicted annotations / # annotations of the benchmark dataset). The best F1-scores reached on eukaryotic proteins are about 0.6, 0.4, and 0.5 for MF, BP, and CC, respectively. In the case of prokaryotes, F1-scores are higher for BP and CC (0.5 and 0.7, respectively). These results, although encouraging, show that there is plenty of room for improvement of the methods for automatic function annotation.

Acknowledgments

This work was partially supported by the Italian Ministry of Education, University and Research [PRIN 2010–2011 project 20108XYHJS] to P.L.M., [PON projects PON01_02249 and PAN Lab PONa3_00166] to R.C. and P.L.M; European Union RTD Framework Program [COST BMBS Action TD1101, Action BM1405] to R.C and University of Bologna [FARB 2012] to R.C. G.P. would like to thank also ELIXIR-IIB and ELIXIR Europe for the support.

See also: Functional Enrichment Analysis Methods. Natural Language Processing Approaches in Bioinformatics. Ontology-Based Annotation Methods. Semantic Similarity Functions and Measures. Tools for Semantic Analysis Based on Semantic Similarity

References

Attwood, T.K., Coletta, A., Muirhead, G., *et al.*, 2012. The PRINTS database: A fine-grained protein sequence annotation and analysis resource – Its status in 2012. Database 2012. doi:10.1093/database/bas019.

Chothia, C., Lesk, A.M., 1986. The relation between the divergence of sequence and structure in proteins. EMBO Journal 5, 823–826.

Falda, M., Toppo, S., Pescarolo, A., *et al.*, 2012. Argot2: A large scale function prediction tool relying on semantic similarity of weighted gene ontology terms. BMC Bioinformatics 13, S14.

Finn, R.D., Coggill, P., Eberhardt, R.Y., *et al.*, 2016. The Pfam protein families database: Towards a more sustainable future. Nucleic Acids Research 44, D279–D285.

Furnham, N., Holliday, G.L., de Beer, T.A., *et al.*, 2014. The catalytic site atlas 2.0: Cataloging catalytic sites and residues identified in enzymes. Nucleic Acids Research 42, D485–D489.

Galperin, M.Y., Makarova, K.S., Wolf, Y.I., Koonin, E.V., 2015. Expanded microbial genome coverage and improved protein family annotation in the COG database. Nucleic Acids Research 43, D261–D269.

Götz, S., García-Gómez, J.M., Terol, J., *et al.*, 2008. High-throughput functional annotation and data mining with the Blast2GO suite. Nucleic Acids Research 36, 3420–3435.

Holm, L., Rosenström, P., 2010. Dali server: Conservation mapping in 3D. Nucleic Acids Research 38, W545–W549.

Huang, B., Schroeder, M., 2006. LIGSITEcsc: Predicting ligand binding sites using the Connolly surface and degree of conservation. BMC Structural Biology 6, 19.

Jiang, Y., Oron, T.R., Clark, W.T., *et al.*, 2016. An expanded evaluation of protein function prediction methods shows an improvement in accuracy. Genome Biology 17, 184.

Konc, J., Miller, B.T., Štular, T., *et al.*, 2015. ProBiS-CHARMMing: Web interface for prediction and optimization of ligands in protein binding sites. Journal of Chemical Information and Modeling 55, 2308–2314.

Krissinel, E., Henrick, K., 2004. Secondary-structure matching (SSM), a new tool for fast protein structure alignment in three dimensions. Acta Crystallographica D 60, 2256–2268.

Kufareva, I., Ilatovskiy, A.V., Abagyan, R., 2012. Pocketome: An encyclopedia of small-molecule binding sites in 4D. Nucleic Acids Research 40, D535–D540.

Lan, L., Djuric, N., Guo, Y., Vucetic, S., 2013. MS-kNN: Protein function prediction by integrating multiple data sources. BMC Bioinformatics 14, S8.

Laskowski, R.A., Watson, J.D., Thornton, J.M., 2005. ProFunc: A server for predicting protein function from 3D structure. Nucleic Acids Research 33, W89–W93.

Letunic, I., Doerks, T., Bork, P., 2015. SMART: Recent updates, new developments and status in 2015. Nucleic Acids Research 43, D257–D260.

Madej, T., Lanczycki, C.J., Zhang, D., *et al.*, 2014. MMDB and VAST +: Tracking structural similarities between macromolecular complexes. Nucleic Acids Research 42, D297–D303.

Marchler-Bauer, A., Bo, Y., Han, L., *et al.*, 2017. CDD/SPARCLE: Functional classification of proteins via subfamily domain architectures. Nucleic Acids Research 45, D200–D203.

Minneci, F., Piovesan, D., Cozzetto, D., Jones, D.T., 2013. FFPred 2.0: Improved homology-independent prediction of gene ontology terms for eukaryotic protein sequences. PLOS ONE 8, e63754.

Piovesan, D., Giollo, M., Leonardi, E., *et al.*, 2015. INGA: Protein function prediction combining interaction networks, domain assignments and sequence similarity. Nucleic Acids Research 43, W134–W140.

Profiti, G., Martelli, P.L., Casadio, R., 2017. The Bologna Annotation Resource (BAR 3.0): Improving protein functional annotation. Nucleic Acids Research 45, W285–W290.

Rost, B., 2002. Enzyme function less conserved than anticipated. Journal of Molecular Biology 318, 595–608.

Sahraeian, S.M., Luo, K.R., Brenner, S.E., 2015. SIFTER search: A web server for accurate phylogeny-based protein function prediction. Nucleic Acids Research 43, W141–W147.

Shindyalov, I.N., Bourne, P.E., 2001. A database and tools for 3-D protein structure comparison and alignment using the combinatorial extension (CE) algorithm. Nucleic Acids Research 29, 228–229.

Sigrist, C.J., de Castro, E., Cerutti, L., *et al.*, 2013. New and continuing developments at PROSITE. Nucleic Acids Research 41, D344–D347.

Sillitoe, I., Cuff, A.L., Dessailly, B.H., *et al.*, 2013. New functional families (FunFams) in CATH to improve the mapping of conserved functional sites to 3D structures. Nucleic Acids Research 41, D490–D498.

Sillitoe, I., Lewis, T.E., Cuff, A., *et al.*, 2015. CATH: Comprehensive structural and functional annotations for genome sequences. Nucleic Acids Research 43, D376–D381.

Van Landeghem, S., Hakala, K., Rönnqvist, S., *et al.*, 2012. Exploring biomolecular literature with EVEX: Connecting genes through events, homology, and indirect associations. Advanced Bioinformatics 2012, 582765.

Wass, M.N., Barton, G., Sternberg, M.J., 2012. CombFunc: Predicting protein function using heterogeneous data sources. Nucleic Acids Research 40, W466–W470.

Protein Post-Translational Modification Prediction

Chi Nl Pang and Marc R Wilkins, The University of New South Wales, Sydney, NSW, Australia

Introduction

Post-translational modifications (PTMs) of proteins involve the covalent modification of amino acid side chains or of the protein's N- and C-termini. There are two main types of PTMs. The first involve the enzymatic covalent addition of small molecules (e.g., phosphoryl group for phosphorylation) or polypeptide (e.g., ubiquitin for ubiquitination) to the amino acids, whereas the second involve the proteolytic cleavage of the polypeptide chain. The most common types of PTM additions include phosphorylation, acetylation, methylation, ubiquitination, and glycosylation (**Fig. 1**), of the more than 450 different types of PTM documented in the Uniprot Database (Bateman *et al.*, 2017).

There are many biological functions for post-translational modifications. One way in which the addition of PTM regulates the activity of a protein is through changing the conformation of the target protein, a function well-documented for phosphorylation (Lin *et al.*, 1996; Cohen, 2000). Other post-translation modifications act as a subcellular localization signal (e.g., phosphorylation, methylation, ubiquitination) (Hunter, 2007; Scott and Pawson, 2009), while fatty-acid modifications can be used to anchor a protein to membranes. Post-translational modifications can also affect the half-life of a protein. For example, ubiquitination of a protein can mark it for degradation via the ubiquitin-proteasome pathway (Ciechanover, 1994), while N-terminal modifications are well known to regulate protein stability (Varshavsky, 2011). The presence of PTMs can also affect protein folding and solubility (Tokmakov *et al.*, 2012).

In addition to the above functional roles, PTMs are known to modulate protein-protein interactions. Enzymes can act as 'writers' or 'erasers' of modifications (Beltrao *et al.*, 2013), often through transient binding with the target (Stein *et al.*, 2009). Protein interaction domains, a number of which are known, can mediate direct binding to the target proteins and act as a 'reader' for specific

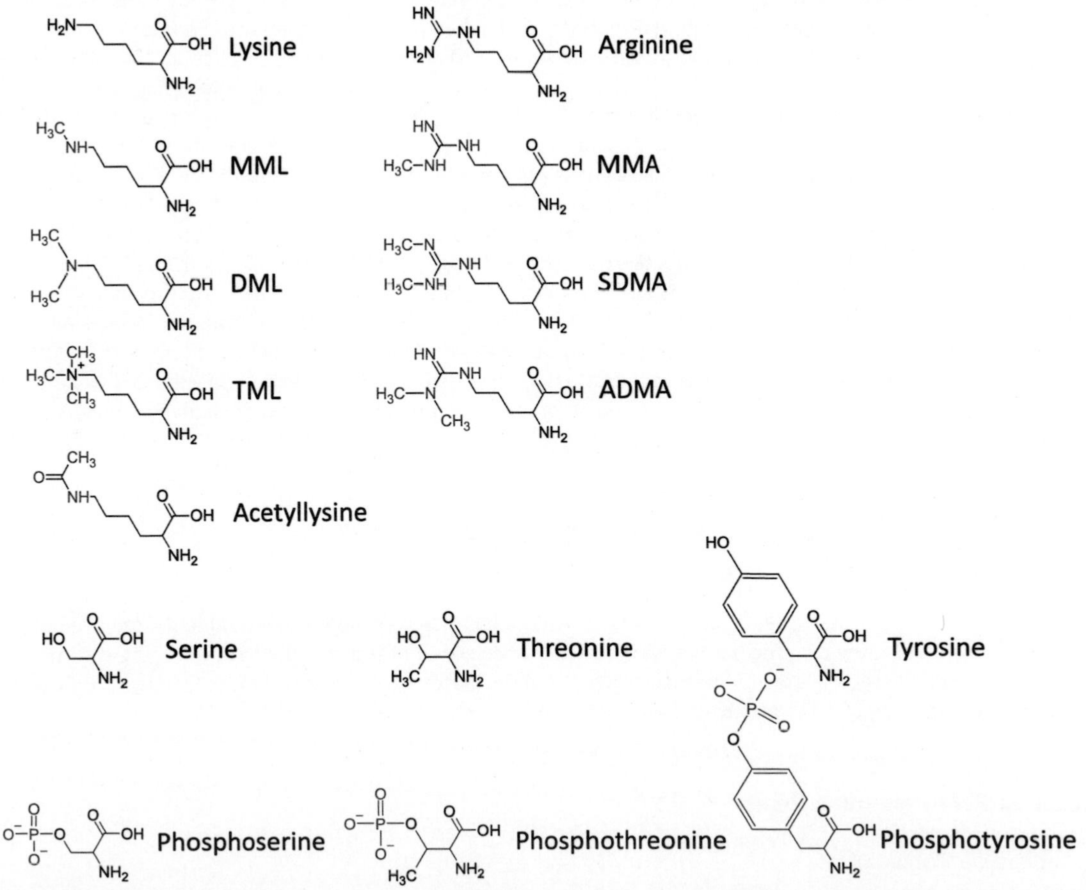

Fig. 1 Chemical structures of several common types of modified amino acids. This figure shows the chemical structures of amino acids (black) and their chemical modifications (red). These include acetylated lysine and phosphorylated serine, threonine, and tyrosine, and different types of methylated residues, including mono-methyllysine (MML), di-methyllysine (DML), tri-methyllysine (TML), mono-methylarginine (MMA), symmetrical di-methylarginine (SDMA), and asymmetrical di-methylarginine (ADMA).

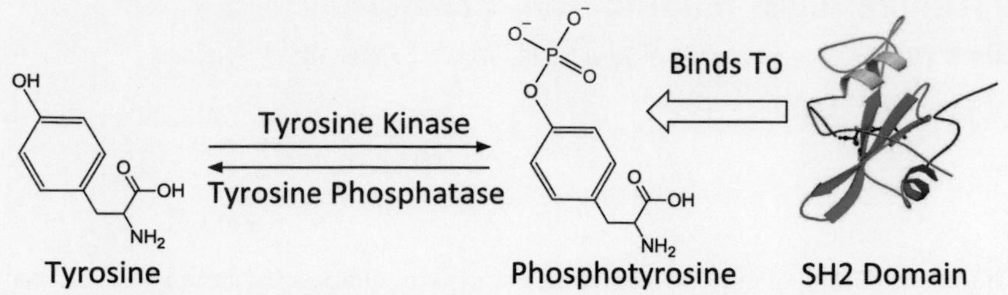

Fig. 2 Reversible tyrosine phosphorylation regulates the binding of SH2 domains to tyrosine phosphorylated peptides. A tyrosine kinase catalyzes the transfer of a phosphoryl moiety to a tyrosine residue, resulting in a phosphotyrosine residue. This reaction can be reversed by a tyrosine phosphatase. A phosphotyrosine residue in the peptide can be recognized and bound by SH2 domains (PDB: 1LKK; Tong *et al.*, 1996). The SH2 domain is shown binding to a tyrosine-phosphorylated peptide.

PTMs (reviewed in Seet *et al.*, 2006). For modifications which are reversible through the action of 'writer' and 'eraser' enzymes, the binding of the 'reader' to the modified protein can be dynamically regulated in response to signals upstream of the enzymes. An example of this is the binding of SH2 domains to phosphotyrosine residues, which can be reversed by a competing tyrosine phosphatase (**Fig. 2**; Seet *et al.*, 2006). In other cases, PTMs can modulate protein-protein interactions by changing the physio-chemical properties (e.g., hydrophobicity and charge) of a binding interface (Winter *et al.*, 2014) and/or causing conformational changes that exposes the interaction surface (Venne *et al.*, 2014). Recently, systematic analysis of protein interaction networks has identified that proteins with a high number of interaction partners, also called network 'hubs', are often enriched for PTMs. This observation is further evidence that PTMs have important roles in regulating protein-protein interactions (Duan and Walther, 2015).

The 'histone code' is a well-known process in which PTMs regulate the recruitment of transcription factors and their regulatory factors to active chromatin regions (Bannister and Kouzarides, 2011). The histone code involves multiple types of PTMs that are located at different sites along the tail of histone proteins (Bannister and Kouzarides, 2011). Different combinations of PTMs, involving 'PTM crosstalk', lead to the binding of different transcription factors to the histones. Recently, it has been suggested that histones may not be unique in having a 'modification code' and other proteins, including p53 (Gu and Zhu, 2012), RNA polymerase II (Eick and Geyer, 2013) and FoxO (Calnan and Brunet, 2008) were reported to have diverse PTMs which regulate their binding with interaction partners. This led to the general hypothesis that 'interaction codes' may be widespread in the proteome (Winter *et al.*, 2014). Interestingly, modification enzymes may also be subjected to regulation by PTMs, which adds further complexity to the analysis of the PTM code (Venne *et al.*, 2014). Deciphering this code would require the analysis of multiple types of PTMs, knowing the specificity of modifying enzymes, and knowing the role of PTMs in modulating protein folding, stability, localization and protein-protein interactions (Prabakaran *et al.*, 2012; Venne *et al.*, 2014).

Large-scale PTM datasets (Dinkel *et al.*, 2011; Zanzoni *et al.*, 2011; Huang *et al.*, 2016; Bateman *et al.*, 2017) have been generated mainly from proteomic studies (Hornbeck *et al.*, 2015; Minguez *et al.*, 2015). Such datasets have accelerated the development of PTM prediction tools. As for all software, vital steps for the development of these tools include defining the needs of the users, incorporating features that will address the needs, and ensuring that the tool is robust and achieves the desired goals. Recently, a comprehensive review of PTM prediction tools for 10 types of PTMs has been published elsewhere (Audagnotto and Dal Peraro, 2017). Accordingly, this article will not explore what all the existing PTM prediction tools are, but will explore the aims and hypothesis that are widely applicable to PTM predictions tools and will highlight a series of challenges involved in designing, engineering, and testing these prediction tools to achieve these aims.

What are the Criteria for Evaluating and Selecting a PTM Prediction Tool?

When evaluating PTM prediction tools, it is important to consider the number of steps that is involved in building and benchmarking the performance of the tool (**Fig. 3**), any of which could affect whether the tool is suitable for the need of the research project. The main steps involve the training dataset, the predictive of selection of features, the testing dataset, the machine learning algorithm, the evaluation dataset, the benchmarking approach and the performance metrics. Below, we illustrate the critical factors that influence the performance of PTM prediction tools through a number of case studies.

Case Studies on PTM Prediction Tools

Prediction of Phosphorylation Sites

PhosphoPredict
PhosphoPredict is a human kinase-specific phosphorylation site prediction tool developed by Song *et al.* (2017) (**Table 1**). This tool is of particular interest as it uses a combination of feature selection and protein-protein interaction to predict phosphorylation sites. The tool predicts phosphorylation sites for 12 human kinase families, including ATM, CDKs, GSK-3, MAPKs, PKA, PKB, PKC, and SRC. The

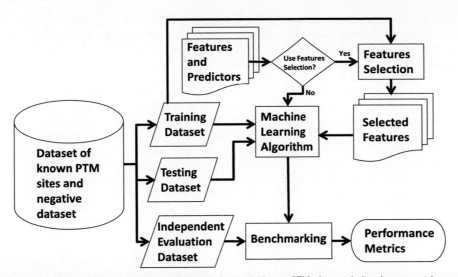

Fig. 3 Flow chart illustrating the process for developing a PTM prediction tool. Known PTM sites and sites known not be modified by the same type of PTM are collected. Datasets can be split into three, to generate a training dataset, the testing dataset, and an independent evaluation dataset. The features (also called predictors) of the model are defined (e.g., predicted surface accessibility and predicted intrinsically disordered regions). These features can be used directly for machine learning. Optionally, a subset of features which are least redundant and best describe the training dataset can be chosen using a feature selection algorithm. The training dataset, the testing dataset and the selected features can be used to train the machine learning algorithm and to develop the PTM prediction tool. The performance of the machine learning algorithm is benchmarked using the independent evaluation dataset and the results are then reported using various performance metrics (e.g., sensitivity, specificity, and accuracy).

Table 1   Case studies on post-translational modifications prediction tools

Name of tool and reference	Species specific?	Enzyme specific?	Dataset	Features[a]	Machine learning algorithm
Phosphorylation					
PhosphoPredict (Song et al., 2017)	Human only	Yes, 12 kinase families	Phospho.ELM	Amino Acid Type, PSS, PSA, PIDR, Gene Ontology, Protein Domains, KEGG Pathways, Protein-Protein Interactions from STRING. Maximum Relevance Minimum Redundancy features selection	Random forest
ELM (Dinkel et al., 2016)	Many organisms	Yes	Eukaryotic Linear Motif database and Phospho.ELM	Regular Expression Patterns, PSS, PSA, PIDR, and Protein Domains	Regular expression matching
Acetylation					
GPS-PAIL (Deng et al., 2016)	Seven eukaryotes including Human and Yeast	Yes, seven histone acetyl-transferase (HAT)	702 non-redundant HAT-specific acetyllysine sites for 205 proteins	Sequence +/− 30 amino acids surrounding acetylation sites included	Simulated annealing to train scoring matrix and position specific weights. Optimization of motif length.
Methylation					
GPS-MSP (Deng et al., 2017)	Many organisms	No, but specific to type of methylation	1521 meth-K from 962 proteins, and 3900 meth-R sites from 1751 protein	Sequence +/− 30 amino acids surrounding methylation sites included	Simulated annealing to train scoring matrix and position specific weights. Optimization of motif length.
Ubiquitination					
Nguyen et al. (2016)	Many organisms	No, but use clustering to predict putative motifs	2566 ubiquitination sites	Amino acid composition, amino acid pairwise composition, and PSSM	Support Vector Machine, Maximal Dependence Decomposition

[a]PSS: Predicted Secondary Structure; PSA: Predicted Solvent Accessibility; PIDR: Predicted Intrinsically-Disordered Region; PSSM: Position-specific scoring matrix.

human phosphorylation sites and their kinase-specificity were obtained from Phospho.ELM (Dinkel *et al.*, 2011). A wide-range of predictive features are utilized, which includes those that are sequence-based, including amino acid type, predicted secondary structure (Wagner *et al.*, 2005), predicted solvent accessibility (Wagner *et al.*, 2005) and predicted intrinsic-disorder (Ward *et al.*, 2004). In addition, functional features were also used for predictions, including GO annotations (Carbon *et al.*, 2017) and protein domains of the query substrate protein (Finn *et al.*, 2016, 2017), KEGG biochemical pathways (Kanehisa *et al.*, 2016) and protein-protein Interactions (Szklarczyk *et al.*, 2017). These functional features are represented as categorical labels with one feature for each category.

Song *et al.* (2017) used the feature selection tool called maximum Relevance Minimum Redundancy (mRMR) (Peng *et al.*, 2005) to identify the most important subset of features for each kinase, which led to increased prediction performance. There features were used in the Random Forest (Breiman, 2001; Liaw and Wiener, 2002) machine learning algorithm, trained using five-fold cross-validation. Independent datasets from PhosphoSitePlus (Hornbeck *et al.*, 2015) and Uniprot (Bateman *et al.*, 2017) were used to verify that the tools were not affected by overfitting. The benchmarking resulted in high specificity (88.3%–96.7%) but variable sensitivity (29.2%––80.5%) for different kinases. PhosphoPredict was found to have better or similar performance as compared to other tools, including KinasePhos (Huang *et al.*, 2005), PPSP (Xue *et al.*, 2006), GPS (Xue *et al.*, 2006), and Musite (Xue *et al.*, 2006). Together, this tool highlights how a careful and well-informed design, that considers many aspects of PTMs, can lead to high quality predictions.

Eukaryotic linear motif (ELM)

The ELM database contains kinase-specific phosphorylation motifs represented as regular expression patterns, which enable users to predict the presence of kinase-specific phosphorylation sites within query protein sequences (**Table 1**) (Gouw *et al.*, 2017). The ELM database currently has 20 kinase-specific phosphorylation motifs and 12 kinase binding motifs and has recently been updated to include new motifs for CDKs, MAPKs and Plks (Gouw *et al.*, 2017). For a user-defined sequence, the tool can identify matches to entries in the Phospho.ELM database (Dinkel *et al.*, 2011), including experimentally identified phosphorylation sites and their kinase specificity if known. Users can also search the companion Switches.ELM database, which contain a list of sites in which the presence or absence of phosphorylation is known to affect protein-protein interaction (Van Roey *et al.*, 2013). In addition to kinase-specific phosphorylation sites, other types of modifications with known conserved motifs in ELM includes N-glycosylation, N-myristoylation, O-fucose modification, O-glucose modification, S-palmitoylation, sumoylation, and enzyme-specific peptide cleavage sites. The ELM database also contains annotations for conserved motif regions, also called Short Linear Motifs (SLiMs). SLiMS are compact linear sequences on average six amino acids in length, preferentially found in intrinsically disordered regions and involved in PTMs, molecular targeting or protein-protein interaction interfaces (Davey *et al.*, 2012).

In ELM, users can apply filters to increase the specificity of their motif searches. These include filters for motifs present in specific taxonomic groups or cell compartments. The probability filter is based on a score calculated from multiplying the probabilities of occurrence of each amino acid within the regular expression pattern. This filter can be used to reduce matches to degenerate motifs. The globular domain filters identify structurally ordered domains, which include predicted protein domains from Pfam (Finn *et al.*, 2016) and SMART (Letunic and Bork, 2017) and predicted globular regions from GlobPlot (Letunic and Bork, 2017). Via *et al.* (2009) showed the utility of structural filters for phosphorylation, which can be used to prioritize matches to disordered regions. Predicted secondary structures from SMART are used as a proxy for surface accessible loops. The disordered region filters include GlobPlot and IUPred (Linding *et al.*, 2003). Unlike many other PTM prediction tools, which rely on machine learning algorithms whose predictive mechanisms are opaque, the ELM database provides clear information on why a site was predicted. The predictions made by ELM are based on known kinase-specific phosphorylation motifs and sequence and structural features. These can be visualized as sequence tracks, which enables users to inspect the structural environment, motifs or other PTMs surrounding the phosphorylation site, thus providing a rich functional and structural context.

Prediction of Acetylation Sites

GPS-PAIL

The **GPS-PAIL** tool (Deng *et al.*, 2016) predicts acetylation sites that are specific to seven histone acetyltransferases (HATs): CREBBP, EP300, HAT1, KAT2A, KAT2B, KAT5 and KAT8 (**Table 1**). Few tools can predict acetylation sites specific to single acetyltransferases (Li *et al.*, 2012; Wang *et al.*, 2012; Deng *et al.*, 2016) and thus GPS-PAIL is of particular interest. Since not all acetyltransferases are present ubiquitously, Deng *et al.* (2016) also checked for the presence of each acetyltransferase in the proteomes of seven eukaryotic species. The training dataset included 702 experimentally determined acetyltransferase-specific acetylation sites for 205 proteins from seven eukaryotes, with 544 sites for 160 human proteins. The method uses the BLOSUM62 amino acid substitution matrix to compare a query site with known enzyme-specific acetylation sites; predicted acetylation sites have a high score. Simulated annealing was used to train position-specific weights and to mutate the BLOSUM62 matrix to achieve better predictions. This is a complementary method to other commonly used machine learning tools, requiring a problem and domain-specific 'coding scheme' to describe the search space (Gouw *et al.*, 2017); in this case it was the BLOSUM62 matrix and the position specific weights. The use of simulated annealing is notable as it is an active research area. Optimal motif length in GPS-PAIL was identified through an exhaustive search. Analysis of the sequence motifs surrounding acetylation sites of each acetyl-transferase highlighted their sequence specificity, with K, R, G and A residues most often enriched in these motifs.

Performance evaluation of GPS-PAIL was carried out through leave-one-out cross-validation and N-fold cross-validation (Deng *et al.*, 2016). Among the seven acetyltransferases, the tool achieved a high specificity of 80%–95% but considerably lower sensitivity of 38%–57.5%. Of the 10,505 experimentally determined acetylation sites from 4488 human proteins, the tool predicted the acetyltransferase responsible for 1492 sites (14.2%) and 913 proteins (20.34%), consistent with low sensitivity. GPS-PAIL was found to be more efficient for predicting acetyltransferase-specific sites for mammalian and plants as compared to other proteomes. The majority of acetylation sites were found to be targeted by one acetyltransferase only, while no acetylation sites were the target of all seven acetyltransferases, which suggests acetylation sites tend to be acetyltransferase-specific.

Prediction of Methylation Sites

GPS-MSP

Methyltransferase-specific protein methylation motifs have been described (e.g., Hamey *et al.*, 2016) however, to the best of our knowledge, there is currently a lack of prediction tools for methyltransferase-specific protein methylation sites. Thus the tools predict sites, in a general sense, but not the possible methyltransferases that may be responsible. Deng *et al.* (2017) developed a suite of prediction tools (**GPS-MSP**) specific for different types of protein methylation, including mono-, di-, and tri-methyllysine, and mono-, symmetric di-, and asymmetric di-methylarginine. Separate prediction models for each type of methylation were shown to improve prediction accuracy. The same algorithm as GPS-PAIL (above) was used for GPS-MSP. The training dataset contained 1521 methyllysines from 962 proteins, and 3900 methylarginine sites from 1751 proteins. For different types of methylation sites, leave-one-out cross-validation and receiver operating characteristic analyses showed that the sensitivity ranged from ∼35% to 50% at a fixed specificity of 90%. Analysis of the sequence motif surrounding the arginine methylation sites identified the 'RGG' motif, confirming previous reports (Lischwe *et al.*, 1985; Kim *et al.*, 1997), while analysis of lysine methylation sites identified putative new sequence motifs, including the overrepresentation of leucine at the − 1 position of mono-methyl-lysine. This method shows how in the absence of enzyme-specific data, different types of methylation could be used as a proxy for enzyme-specificity to enhance prediction performance.

Prediction of ubiquitination sites

Protein ubiquitination is catalyzed by the E1 activating enzyme, the E2 conjugating enzyme and the E3 ubiquitin ligase that mediates substrate specificity (Amoutzias *et al.*, 2012). There are > 600 E3 ligases in human yet there is currently little information on the specificity of each E3 ligases (Amoutzias *et al.*, 2012). Therefore, current prediction tools are based on global prediction of ubiquitination sites, as reviewed by Chen *et al.* (2014). Nguyen *et al.* (2015) used unsupervised clustering of the ubiquitination sites and developed a prediction model for each cluster. Analysis of sequence motifs for each of the clusters showed that each have varying sequence specificity, suggesting that clustering and motif analysis could be used to identify putative E3 ligase recognition motifs (Nguyen *et al.*, 2015). Nguyen *et al.* (2017) later improved their prediction model by incorporating additional features, including amino acid composition, amino acid pairwise composition and a position-specific scoring matrix, into support vector machines (SVM). The SVM model for each cluster was combined into an ensemble model. This resulted in sensitivity of 67.6% and specificity of 58.8% from five-fold cross validation. Similar to protein methyltransferases and methylation, the E3 ligase-specificity for most ubiquitination sites are poorly defined. Nguyen *et al.* (2017) demonstrated that sequence-based clustering of ubiquitination sites, prior to training, could act as a viable proxy in the absence of enzyme-specific motifs. It is interesting to note that the same research group has also developed the UbiNet database of protein-protein interaction networks focusing on the interaction partners of the E1, E2 and E3 enzymes (Nguyen *et al.*, 2016). This ubiquitination enzyme protein-protein interaction database, in conjunction with their prediction algorithm, could be used to explore the relationship between specific E3 ligases and their potential substrates.

What are the Biological Questions Explored Through the Prediction of PTMs?

As noted above, prediction of post-translational modifications involves the use of simple motif-driven methods, through to machine learning and/or artificial intelligence approaches, to identify and also characterize PTMs in the proteome. Whilst it is possible to generate a list of predicted PTMs, a major challenge is to analyze the list of predicted modification sites to generate valuable biological insights. Here, we will highlight three main biological questions that potential users of PTM prediction tools may like to address and provide rationale on why these questions are important. (1) Where are the modification site(s) and how do these relate to the structure and function of the protein? (2) What are the modification enzyme(s) responsible for writing or erasing a PTM at a specific amino acid within a protein? (3) What is the functional role of the PTM in the proteome? These three questions are complementary and depend on whether the analysis is focused on the modification enzymes, the protein substrates, or the specific type of modification.

Where are the Modification Site(s) and How do These Relate to the Structure and Function of the Protein?

After obtaining a list of modification sites from a PTM prediction tool, one subsequent step can involve mapping modifications to the structural features of the target protein. This helps understand how the modifications on a specific protein may affect its function. These analyses often involve the use of information from three-dimensional protein structures, along with predicted structural properties (e.g., domains, disordered regions) from protein sequence analysis tools.

When the three-dimensional structure of a modified protein is available, the modification site can be mapped onto structure. This can provide accurate information on surface accessibility of the modification site (Pang *et al.*, 2007) and, for enzymes, whether the modification is in the vicinity of active sites and may affect catalytic activity (Zanzoni *et al.*, 2011; Craveur *et al.*, 2014). Molecular dynamic simulations of modified proteins can be used to predict whether the modifications may cause conformational change (reviewed in Audagnotto and Dal Peraro, 2017).

Structures of protein complexes can provide information on whether modification sites are in the vicinity of protein-DNA, protein-RNA and protein-protein interaction interfaces and therefore have the potential to modulate interactions (Beltrao *et al.*, 2012). There are large variety of known protein-protein interfaces (Winter, 2006; Zhao *et al.*, 2011; Garma *et al.*, 2012) and these include two main types, domain-domain and domain-motif interactions, which have been well characterized in the literature (Stein and Aloy, 2010; Mosca *et al.*, 2014). Domain-domain interactions are often more stable interactions while domain-motif interactions tend to mediate transient interactions (Mosca *et al.*, 2014). The 3DID database systematically curates domain-domain or domain-motif interaction interfaces from known 3-D structures of protein complexes (Mosca *et al.*, 2014). Using the 3DID database, it is possible to determine whether PTM sites are present at or adjacent to the interface residues of an interaction domain or motif. This can help understand whether a PTM could possibly mediate or block a protein-protein interaction, and thus act as a protein interaction switch (Akiva *et al.*, 2012; Beltrao *et al.*, 2012; Van Roey *et al.*, 2012).

Modifications on critical sites of a protein can have important effects on signaling. A well-known example involves receptor tyrosine kinases where, upon ligand binding, there is autophosphorylation which leads to conformational changes and homo-dimerization (Ullrich and Schlessinger, 1990). These autophosphorylation sites have a variety of known functions, which include maintaining kinase activity in the absence of bound ligand by disrupting autoinhibitory interaction (Rosen *et al.*, 1983; Lemmon and Schlessinger, 2010) and modulating interaction with specific downstream substrates as part of signal transduction (Kazlauskas and Cooper, 1989; Arvidsson *et al.*, 1994; Hanke and Mann, 2009). Analysis of signaling proteins, such as receptor tyrosine kinases, may involve a multi-faceted and integrative approach, including but not limited to the application of tools and concepts discussed above.

What are the Enzyme(s) Responsible for Writing or Erasing a PTM on a Specific Amino Acid Within a Protein?

This question is applicable to PTMs that can be catalyzed by multiple enzymes. This enquiry is suited to researchers who are interested in the regulatory role of modifying enzymes and their effect on one target protein of interest. Knowing the specific upstream modifying enzymes would then, in turn, enable the researcher to perform further investigative experiments. This may involve verifying the activity of an enzyme through the absence or decreased abundance of the modification site upon perturbation of the enzyme through knockout (CRISPR) (Hamey *et al.*, 2017), knockdown (RNA inhibition) (Weiss *et al.*, 2007), or targeting with an inhibitor (Weiss *et al.*, 2007). Identifying the enzyme responsible for the modification site also enables us to contextualize the modified protein within an enzyme-substrate interaction network (Linding *et al.*, 2008; Erce *et al.*, 2012). This enables researchers to analyze the function of the modification enzyme, and one or more substrates, as a group, including identifying enriched canonical pathways, biological functions, sub-cellular localization and coordinated changes in PTM abundance under different cellular contexts (Linding *et al.*, 2008; Yang *et al.*, 2015).

What is the Functional Role of a PTM in the Proteome?

It is important to gain a general perspective on the function of each type of PTM in the proteome. Functional enrichment analysis of proteins which contain a specific PTM type might help to highlight the role of a PTM within the cell, including the biological pathways that are overrepresented in the modified proteins. These analyses could be performed using tools for Gene Ontology (GO) enrichment (Maere *et al.*, 2005; Grossmann *et al.*, 2007; Bindea *et al.*, 2009; Eden *et al.*, 2009; Reimand *et al.*, 2016; Mi *et al.*, 2017) and Pathway Analysis (Krämer *et al.*, 2014; Fabregat *et al.*, 2016). Different types of PTMs are associated with diverse biological functions. The best known examples include serine, threonine and tyrosine phosphorylation, which are often involved in signaling (Linding *et al.*, 2008). By contrast, lysine acetylation is frequently observed on proteins involved in energy metabolism (Choudhary *et al.*, 2014), arginine methylation is associated with RNA process and transport (Bedford and Clarke, 2009; Blanc and phane Richard, 2017), lysine methylation is enriched among protein translation machineries (Lanouette *et al.*, 2014) and fatty acid modifications are frequently present on membrane-bound proteins and receptors (Schwenk *et al.*, 2010). Among eukaryotes, protein glycosylation is associated with secreted proteins, proteins with extracellular domains and protein quality control (Ben-ham, 2012; Xu and Ng, 2015). Interestingly, recent studies have also shown that glycosylation is also present among bacterial species (Szymanski and Wren, 2005; Lu *et al.*, 2015). This bacterial glycosylation has been suggested to modulate the function of virulence factors or factors involved in hijacking the host cellular machinery (Szymanski and Wren, 2005; Lu *et al.*, 2015).

Results and Discussion

Machine learning tools have a capacity to provide accurate predictions if the approach used to train, test and evaluate the tools are well-designed and the datasets are of high quality. However it must be acknowledged that PTM data remains sparse, will contain

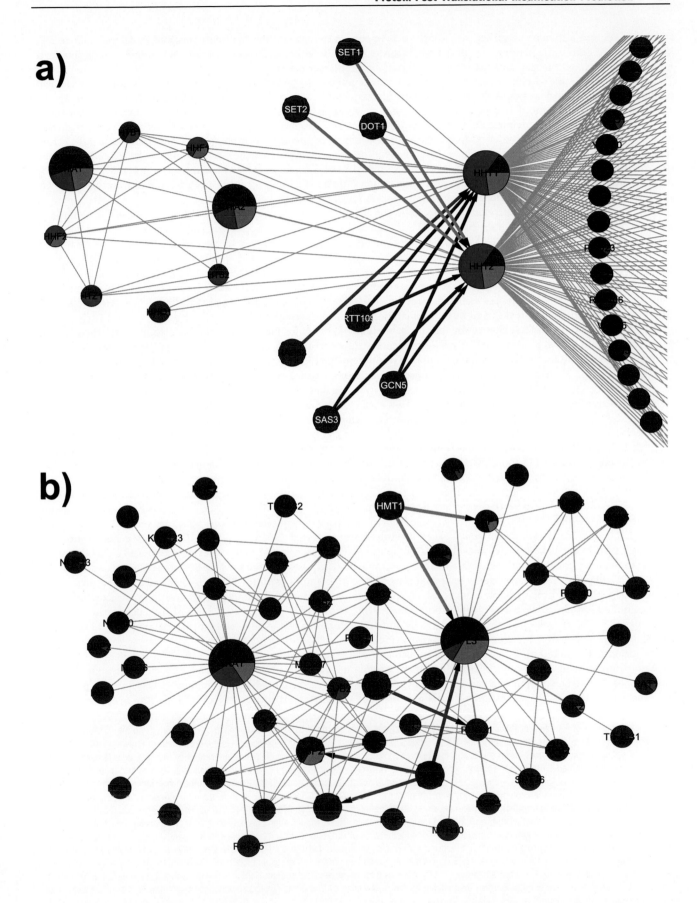

false positives and also false negatives. Accordingly, it is important to highlight the challenges involved in dataset curation and how these could affect the performance of the prediction tools. Major factors to consider include the enzymes responsible for a modification site, and the size, quality, composition of the modification datasets.

Incorporating the site specificity of PTM enzymes is likely to improve the performance of predictive tools. Modifications known to arise from specific enzymes can be analyzed for particular sequence motifs using tools such as pLogo (O'Shea *et al.*, 2013). The existence of a sequence motif is a good indication that enzyme-specific prediction would improve prediction accuracy, and conversely that using all available data without considering enzyme-specificity in such cases would lead to worse prediction performance. As described above, enzyme-specific prediction improved predictions of phosphorylation (Song *et al.*, 2017) and acetylation (Deng *et al.*, 2016) sties. A downstream benefit of enzyme-specific prediction is the ability to better understand the specific function of each PTM-catalyzing enzyme. For example, GO enrichment analysis of kinase-specific substrates predicted by PhosphoPredict show that while all kinase families were significantly enriched among diverse biological pathways, channel activity and ion channel activity were significantly overrepresented among MAPK only.

The taxonomic relevance of each enzyme should be considered when using enzyme-specific PTM prediction. For example, S. *cerevisiae* has three relevant acetyltransferases (HAT1, KAT2A and KAT6), and therefore predictions made by GPS-PAIL for other acetyltransferases should not be used for yeast (Deng *et al.*, 2016). However, in cases where a PTM is highly conserved across cellular organisms, such as N-glycosylation (Deshpande *et al.*, 2008; Aebi, 2013; Jarrell *et al.*, 2014; Pinho and Reis, 2015; Strasser, 2016), many enzymes may be conserved and taxonomic filters may be less useful (Beltrao *et al.*, 2013). Most enzyme-specific PTM data are focused on human and a few model organisms such as yeast. Databases that specialize in one type of PTM often include annotation for enzyme specificity. For example, both Phospho.ELM (Dinkel *et al.*, 2011) and PhosphoSitePlus (Hornbeck *et al.*, 2015) curates phosphorylation sites and include annotation for kinase-specific sites.

An adequate number of PTM sites in the training datasets are necessary for the machine learning algorithms to detect reliable patterns from the data and to use these patterns for predictions. However, it is difficult to know what actually represents an adequate number. It is important for tools to report the number of positive modification sites and negative sites considered, such that users can evaluate tools based on the size of the datasets. However there remains a strong possibility of false negatives being present and used. False negatives could be present due to a number of reasons, including difficulties in detecting low abundance proteins (Picotti *et al.*, 2009) or low stoichiometry modification sites (Amoutzias *et al.*, 2012) and challenges in the analysis of some membrane proteins, which are difficult to solubilize for analysis with mass spectrometry (Barrera and Robinson, 2011). To increase the coverage of the training datasets and to mitigate the effect of false negatives, data can be combined from multiple sources, including publically available PTM databases and those curated from literature. There are a number of publically available PTM databases, including Uniprot (Bateman *et al.*, 2017) and dbPTM (Huang *et al.*, 2016), which could be used for building collections of modification sites. However, there will typically be inconsistencies between different large-scale PTM resources which are difficult to reconcile.

Similar to false negatives, there are likely to be false positive modifications in experimental data and, when used for a tool's training set, these could lead to decreased prediction accuracy and specificity. This problem could be mitigated by filtering experimental data with a stringent false discovery rate (FDR), or selecting only modification sites which are verified by two or more independent experiments (Amoutzias *et al.*, 2012). Recently, Hart-Smith *et al.* (2016) showed that large-scale mass spectrometry analysis of protein methylation sites are subjected to high FDR. The reason is that many types of amino acid substitutions are isobaric to the mass change associated with a methyl group, and in many cases, it is not possible to distinguish between them. The study also highlighted that a target-decoy approach is ineffective for controlling for the FDR of protein methylation sites. Heavy methyl-group labelling through Stable Isotope Labelling with Amino Acid in Cell Culture (SILAC) was suggested as the validation criteria required for methylation sites identification. The MethylQuant tool (Tay *et al.*, 2017a,b) was recently developed to support heavy-methyl SILAC analyses. The above highlights the importance of using high quality experimental datasets, such as those that uses SILAC for methylation site identification, to avoid inclusion of false positive sites.

The ideal training and testing datasets should contain an equal proportion of positive and negative PTM site data. Yet there are often more unmodified than modified sites. One common solution is to train multiple machine learning models, e.g. an ensemble approach, each time using a random subset of the negative sites and averaging results from many models to increase prediction performance (Yang *et al.*, 2015). A further problem with modification datasets is the presence of inspection bias. For example,

---

Fig. 4   Visualization of modified proteins, protein-protein interactions, and enzyme-substrate interactions in *S. cerevisiae* using the PTMOracle app in Cytoscape. Each protein is represented as a node in the network. The pie chart for each protein represents the proportion of known modification sites that are phosphorylation (red), acetylation (blue) and methylation (grey). Protein-protein interactions are represented by thin grey lines, kinase-substrate interactions are represented by thick red arrows, acetyltransferase-substrate interactions are represented by thick blue arrows and methyltransferase-substrate interactions are represented by thick grey arrows. All enzyme-substrate interactions are directional, with the arrow indicating the substrate. (a) The interactions between histone proteins are shown in the network. The network focuses on the enzyme-substrate interactions that target histone proteins Hht1p and Hht2p (large nodes), and a subset of non-histone interaction partners for these two histone proteins are also shown. Hht1p is phosphorylated, acetylated and methylated, and the enzymes which catalyze these modifications are known and shown. Unlike Hht1p, the enzymes which are responsible for the phosphorylation, acetylation and methylation of Hht1p have not yet been experimentally identified. The methyltransferases Set1p, Set2p, and Dot1p form protein-protein interactions with Hht1p, predicting that these methyltransferases catalyze Hht1p methylation. (b) The protein-protein interaction network focusing on Npl3p and Yra1p (large nodes). The enzymes responsible for Npl3p methylation and phosphorylation are known and shown (arrows). The methyltransferase Hmt1p interacts with the methylated protein Nab2p, suggesting that it is responsible for Nab2p methylation. In contrast, the methyltransferase responsible for the methylation of Yra1p has not been confirmed, and Yra1p is not known to bind any existing protein modifying enzymes.

histone protein is the most extensively analyzed protein for protein methylation sites (Lothrop *et al.*, 2013) and there will likely be a high number of histone proteins in any methylation sites training set. This may potentially lead to bias. This could be resolved by identifying redundant protein sequences and redundant modification sites from the dataset, using tools such as CD-HIT (Huang *et al.*, 2010) and keeping only one representative example from each group of homologous sites (for example, Deng *et al.*, 2017; Nguyen *et al.*, 2017).

Future Directions

There remains scope for the improvement of PTM site prediction. Since functionally important modification sites are more likely to be evolutionarily conserved, others have used this as a means to rank and prioritize predicted PTM sites. Beltrao *et al.* (2012) developed the PTMfunc database which predicts the functional relevance of phosphorylation, acetylation and ubiquitination sites for 11 eukaryotic organisms. The PTM sites that are evolutionarily conserved are more likely to be biologically important. Users of PTM prediction tools could query the PTMcode database developed by Minguez *et al.* (2012) to identify single or pairs of co-evolving sites that are within the same domain or motifs sequences, thus highlighting putative important residues within domains or motifs, or co-evolving sites between a interacting proteins. This would help predict putative sites of greatest functional importance.

Downstream, it is important to decipher how PTMs regulate the dynamics of biological networks and participate in protein 'interaction codes' (Winter *et al.*, 2014). Bioinformatic tools are emerging that enable users to investigate cross-talk between PTMs, and explore enzyme-substrate relationships in the context of the interactome. Tay *et al.* (2017a,b) developed the PTMOracle app for the Cytoscape network software platform. This tool enables users to generate pie charts for each node in the network to represent the proportion of different types of PTMs on the protein (**Fig. 4**). More importantly, the tool can co-map and co-analyze PTMs, enzyme-substrate relationships along with other protein-protein interactions, and to generate and visualize a series of complex queries involving protein structural data, domains, motifs and interactions. Whilst the PTMOracle tool currently only uses known PTM sites in its underlying databases, predicted PTMs could also be used to enrich the networks generated and to help develop new hypotheses about PTM and protein function.

Closing Remarks

In conclusion, this article has highlighted several critical factors for developing robust and reliable PTM prediction tools. These factors include careful curation of PTM datasets, the use of enzyme-specific modification sites for tool training datasets, and the use of feature selection tools to identify non-redundant sets of features. We also discussed the biological questions that could be explored when analyzing a list of predicted PTMs. These questions include how a PTM relates to the structure and functions of the protein, the effect of PTMs on protein-protein interactions, the binding specificity of the modification enzymes and the biological processes that they regulate, and the function of different types of PTMs. It is hoped that these factors, together, will inspire the development of novel bioinformatics tools, to decipher the function of PTMs and to illuminate the roles of PTMs in protein-protein 'interaction codes'.

Acknowledgements

MRW acknowledges support from NCRIS, administered by Bioplatforms Australia, and from the New South Wales State Government RAAP Scheme. We thank Aidan Tay for kindly providing the figure that illustrates PTMOracle and Daniel Winter for kindly providing the figures that illustrate the chemical structures of modified amino acids.

See also: Algorithms for Strings and Sequences: Searching Motifs. Algorithms for Structure Comparison and Analysis: Docking. Computational Approaches for Modeling Signal Transduction Networks. Data Mining: Classification and Prediction. Data Mining: Prediction Methods. Functional Genomics. Gene Prioritization Tools. Identification of Sequence Patterns, Motifs and Domains. Natural Language Processing Approaches in Bioinformatics. Prediction of Protein Localization. Proteomics Mass Spectrometry Data Analysis Tools. The Evolution of Protein Family Databases. The Evolution of Protein Family Databases

References

Aebi, M., 2013. N-linked protein glycosylation in the ER. Biochimica et Biophysica Acta (BBA) – Molecular Cell Research 1833 (11), 2430–2437. doi:10.1016/j.bbamcr.2013.04.001.
Akiva, E., *et al.*, 2012. A dynamic view of domain-motif interactions. PLOS Computational Biology 8 (1), doi:10.1371/journal.pcbi.1002341.
Amoutzias, G.D., *et al.*, 2012. Evaluation and properties of the budding yeast phosphoproteome. Molecular & Cellular Proteomics : MCP 11 (6), doi:10.1074/mcp.M111.009555.
Arvidsson, A.K., *et al.*, 1994. Tyr-716 in the platelet-derived growth factor beta-receptor kinase insert is involved in GRB2 binding and Ras activation. Molecular and Cellular Biology 14 (10), 6715–6726. https://doi.org/10.1128/MCB.14.10.6715.
Audagnotto, M., Dal Peraro, M., 2017. Protein post-translational modifications: In silico prediction tools and molecular modeling. Computational and Structural Biotechnology Journal 15, 307–319. doi:10.1016/j.csbj.2017.03.004.

Bannister, A.J., Kouzarides, T., 2011. Regulation of chromatin by histone modifications. Cell Research 21 (3), 381–395. doi:10.1038/cr.2011.22.

Barrera, N.P., Robinson, C.V., 2011. Advances in the mass spectrometry of membrane proteins: From individual proteins to intact complexes. Annual Review of Biochemistry 80, 247–271. doi:10.1146/annurev-biochem-062309-093307.

Bateman, A., et al., 2017. UniProt: The universal protein knowledgebase. Nucleic Acids Research 45 (D1), D158–D169. doi:10.1093/nar/gkw1099.

Bedford, M.T., Clarke, S.G., 2009. Protein arginine methylation in mammals: Who, what, and why. Molecular Cell 33 (1), 1–13. https://doi.org/10.1016/j.molcel.2008.12.013.

Beltrao, P., et al., 2012. Systematic functional prioritization of protein posttranslational modifications. Cell 150 (2), 413–425. doi:10.1016/j.cell.2012.05.036.

Beltrao, P., et al., 2013. Evolution and functional cross-talk of protein post-translational modifications. Molecular Systems Biology 9 (1), 1–13. doi:10.1002/msb.201304521.

Benham, A.M., 2012. Protein secretion and the endoplasmic reticulum. Cold Spring Harbor Perspectives in Biology 4 (8), a012872. doi:10.1101/cshperspect.a012872.

Bindea, G., et al., 2009. ClueGO: A cytoscape plug-in to decipher functionally grouped gene ontology and pathway annotation networks. Bioinformatics 25 (8), 1091–1093. doi:10.1093/bioinformatics/btp101.

Blanc, R.S., phane Richard, S., 2017. Arginine methylation: The coming of age. Molecular Cell 65, 8–24. doi:10.1016/j.molcel.2016.11.003.

Breiman, L., 2001. Random forests. Machine Learning 45 (1), 5–32. doi:10.1023/A:1010933404324.

Calnan, D.R., Brunet, A., 2008. The FoxO code. Oncogene 27 (16), 2276–2288. doi:10.1038/onc.2008.21.

Carbon, S., et al., 2017. Expansion of the gene ontology knowledgebase and resources: The gene ontology consortium. Nucleic Acids Research 45 (D1), D331–D338. doi:10.1093/nar/gkw1108.

Chen, Z., et al., 2014. Towards more accurate prediction of ubiquitination sites: A comprehensive review of current methods, tools and features. Briefings in Bioinformatics 16 (4), 640–657. doi:10.1093/bib/bbu031.

Choudhary, C., et al., 2014. The growing landscape of lysine acetylation links metabolism and cell signalling. Nature Reviews Molecular Cell Biology 15 (8), 536–550. doi:10.1038/nrm3841.

Ciechanover, A., 1994. The ubiquitin-proteasome proteolytic pathway. Cell 79 (1), 13–21. doi:10.1016/0092-8674(94)90396-4.

Cohen, P., 2000. The regulation of protein function by multisite phosphorylation – A 25 year update. Trends in Biochemical Sciences 25 (12), 596–601. doi:10.1016/S0968-0004(00)01712-6.

Craveur, P., Rebehmed, J., De Brevern, A.G., 2014. PTM-SD: A database of structurally resolved and annotated posttranslational modifications in proteins. Database 2014, 1–9. doi:10.1093/database/bau041.

Davey, N.E., et al., 2012. Attributes of short linear motifs. Molecular BioSystems 8 (1), 268–281. doi:10.1039/C1MB05231D.

Deng, W., et al., 2016. GPS-PAIL: Prediction of lysine acetyltransferase-specific modification sites from protein sequences. Scientific Reports 6 (1), 39787. doi:10.1038/srep39787.

Deng, W., et al., 2017. Computational prediction of methylation types of covalently modified lysine and arginine residues in proteins. Briefings in Bioinformatics 18 (4), 647–658. doi:10.1093/bib/bbw041.

Deshpande, N., et al., 2008. Protein glycosylation pathways in filamentous fungi. Glycobiology 18 (8), 626–637. doi:10.1093/glycob/cwn044.

Dinkel, H., et al., 2011. Phospho.ELM: A database of phosphorylation sites-update 2011. Nucleic Acids Research 39 (Database issue), D261–D267. doi:10.1093/nar/gkq1104.

Dinkel, H., et al., 2016. ELM 2016–data update and new functionality of the eukaryotic linear motif resource. Nucleic Acids Research 44 (D1), D294–300. https://doi.org/10.1093/nar/gkv1291.

Duan, G., Walther, D., 2015. The roles of post-translational modifications in the context of protein interaction networks. PLOS Computational Biology 11 (2), 1–23. doi:10.1371/journal.pcbi.1004049.

Eden, E., et al., 2009. GOrilla: A tool for discovery and visualization of enriched GO terms in ranked gene lists. BMC Bioinformatics 10 (1), 48. doi:10.1186/1471-2105-10-48.

Eick, D., Geyer, M., 2013. The RNA polymerase II carboxy-terminal domain (CTD) code. Chemical Reviews 113 (11), 8456–8490. doi:10.1021/cr400071f.

Erce, M.A., et al., 2012. The methylproteome and the intracellular methylation network. Proteomics 12 (4–5), 564–586. doi:10.1002/pmic.201100397.

Fabregat, A., et al., 2016. The reactome pathway knowledgebase. Nucleic Acids Research 44 (D1), D481–D487. doi:10.1093/nar/gkv1351.

Finn, R.D., et al., 2016. The Pfam protein families database: Towards a more sustainable future. Nucleic Acids Research 44 (D1), D279–D285. doi:10.1093/nar/gkv1344.

Finn, R.D., et al., 2017. InterPro in 2017 – Beyond protein family and domain annotations. Nucleic Acids Research 45, D190–D199. doi:10.1093/nar/gkw1107.

Garma, L., et al., 2012. How many protein-protein interactions types exist in nature? PLOS ONE 7 (6), e38913. doi:10.1371/journal.pone.0038913.

Gouw, M., et al., 2017. The eukaryotic linear motif resource – 2018 update. Nucleic Acids Research. doi:10.1093/nar/gkx1077.

Grossmann, S., et al., 2007. Improved detection of overrepresentation of gene ontology annotations with parent-child analysis. Bioinformatics 23 (22), 3024–3031. doi:10.1093/bioinformatics/btm440.

Gu, B., Zhu, W.G., 2012. Surf the post-translational modification network of p53 regulation. International Journal of Biological Sciences 8 (5), 672–684. doi:10.7150/ijbs.4283.

Hamey, J.J., et al., 2016. The activity of a yeast Family 16 methyltransferase, Efm2, is affected by a conserved tryptophan and its N-terminal region. FEBS Open Bio 6 (12), 1320–1330. doi:10.1002/2211-5463.12153.

Hamey, J.J., et al., 2017. METTL21B is a novel human lysine methyltransferase of translation elongation factor 1A: Discovery by CRISPR/Cas9 knock out. Molecular & Cellular Proteomics : MCP. doi:10.1074/mcp.M116.066308. pii, p. mcp.M116.066308.

Hanke, S., Mann, M., 2009. The phosphotyrosine interactome of the insulin receptor family and its substrates IRS-1 and IRS-2. Molecular & Cellular Proteomics 8 (3), 519–534. doi:10.1074/mcp.M800407-MCP200.

Hart-Smith, G., et al., 2016. Large scale mass spectrometry-based identifications of enzyme-mediated protein methylation are subject to high false discovery rates. Molecular & Cellular Proteomics 15 (3), 989–1006. doi:10.1074/mcp.M115.055384.

Hornbeck, P.V., et al., 2015. PhosphoSitePlus, 2014: Mutations, PTMs and recalibrations. Nucleic Acids Research 43 (D1), D512–D520. doi:10.1093/nar/gku1267.

Huang, H.D., et al., 2005. KinasePhos: A web tool for identifying protein kinase-specific phosphorylation sites. Nucleic Acids Research 33 (Web Server issue), W226–W229. doi:10.1093/nar/gki471.

Huang, K.Y., et al., 2016. dbPTM 2016: 10-year anniversary of a resource for post-translational modification of proteins. Nucleic Acids Research 44 (D1), D435–D446. doi:10.1093/nar/gkv1240.

Huang, Y., et al., 2010. CD-HIT suite: A web server for clustering and comparing biological sequences. Bioinformatics 26 (5), 680–682. doi:10.1093/bioinformatics/btq003.

Hunter, T., 2007. The age of crosstalk: Phosphorylation, ubiquitination, and beyond. Molecular Cell 28 (5), 730–738. doi:10.1016/j.molcel.2007.11.019.

Jarrell, K.F., et al., 2014. N-linked glycosylation in archaea: A structural, functional, and genetic analysis. Microbiology and Molecular Biology Reviews 78 (2), 304–341. doi:10.1128/MMBR.00052-13.

Kanehisa, M., et al., 2016. KEGG as a reference resource for gene and protein annotation. Nucleic Acids Research 44 (D1), D457–D462. doi:10.1093/nar/gkv1070.

Kazlauskas, A., Cooper, J.A., 1989. Autophosphorylation of the PDGF receptor in the kinase insert region regulates interactions with cell proteins. Cell 58 (6), 1121–1133. https://doi.org/10.1016/0092-8674(89)90510-2.

Kim, S., et al., 1997. Identification of N(G)-methylarginine residues in human heterogeneous RNP protein A1: Phe/Gly-Gly-Gly-Arg-Gly-Gly-Gly/Phe is a preferred recognition motif. Biochemistry 36 (17), 5185–5192. doi:10.1021/bi9625509.

Krämer, A., et al., 2014. Causal analysis approaches in ingenuity pathway analysis. Bioinformatics 30 (4), 523–530. doi:10.1093/bioinformatics/btt703.

Lanouette, S., et al., 2014. The functional diversity of protein lysine methylation. Molecular Systems Biology. 724. doi:10.1002/msb.134974.

Lemmon, M.A., Schlessinger, J., 2010. Cell signaling by receptor tyrosine kinases. Cell 141 (7), 1117–1134. doi:10.1016/j.cell.2010.06.011.

Letunic, I., Bork, P., 2017. 20 years of the SMART protein domain annotation resource. Nucleic Acids Research. gkx922. doi:10.1093/nar/gkx922.

Liaw, A., Wiener, M., 2002. Classification and regression by random forest. R News 2 (3), 18–22. (Available at: http://cran.r-project.org/doc/Rnews/).

Lin, K., et al., 1996. A protein phosphorylation switch at the conserved allosteric site in GP. Science 273 (5281), 1539–1542. doi:10.1126/science.273.5281.1539.

Linding, R., et al., 2003. GlobPlot: Exploring protein sequences for globularity and disorder. Nucleic Acids Research 31 (13), 3701–3708, https://doi.org/10.1093/nar/gkg519.

Linding, R., *et al.*, 2008. NetworkKIN: A resource for exploring cellular phosphorylation networks. Nucleic Acids Research 36 (Database issue), D695–D699. doi:10.1093/nar/gkm902.

Lischwe, M.A., *et al.*, 1985. Clustering of glycine and NG,NG-dimethylarginine in nucleolar protein C23. Biochemistry 24 (22), 6025–6028, https://doi.org/10.1021/bi00343a001.

Li, T., *et al.*, 2012. Characterization and prediction of lysine (K)-acetyl-transferase specific acetylation sites. Molecular & Cellular proteomics: MCP 11 (1), doi:10.1074/mcp.M111.011080.

Lothrop, A.P., Torres, M.P., Fuchs, S.M., 2013. Deciphering post-translational modification codes. FEBS Letters 587 (8), 1247–1257. doi:10.1016/j.febslet.2013.01.047.

Lu, Q., Li, S., Shao, F., 2015. Sweet Talk: Protein glycosylation in bacterial interaction with the host. Trends in Microbiology 23 (10), 630–641. doi:10.1016/j.tim.2015.07.003.

Maere, S., Heymans, K., Kuiper, M., 2005. BiNGO: A cytoscape plugin to assess overrepresentation of Gene Ontology categories in biological networks. Bioinformatics 21 (16), 3448–3449. doi:10.1093/bioinformatics/bti551.

Mi, H., *et al.*, 2017. PANTHER version 11: Expanded annotation data from Gene Ontology and Reactome pathways, and data analysis tool enhancements. Nucleic Acids Research 45 (D1), D183–D189. doi:10.1093/nar/gkw1138.

Minguez, P., *et al.*, 2012. Deciphering a global network of functionally associated post-translational modifications. Molecular Systems Biology 8, 599. doi:10.1038/msb.2012.31.

Minguez, P., *et al.*, 2015. PTMcode v2: A resource for functional associations of post-translational modifications within and between proteins. Nucleic Acids Research 43 (D1), D494–D502. doi:10.1093/nar/gku1081.

Mosca, R., *et al.*, 2014. 3did: A catalog of domain-based interactions of known three-dimensional structure. Nucleic Acids Research 42 (Database issue), D374–D379. doi:10.1093/nar/gkt887.

Nguyen, V.-N., *et al.*, 2015. Characterization and identification of ubiquitin conjugation sites with E3 ligase recognition specificities. BMC Bioinformatics. BioMed Central 16 (Suppl. 1), S1. doi:10.1186/1471-2105-16-S1-S1.

Nguyen, V.N., *et al.*, 2016. UbiNet: An online resource for exploring the functional associations and regulatory networks of protein ubiquitylation. Database 2016, 1–14. doi:10.1093/database/baw054.

Nguyen, V.N., *et al.*, 2017. A new scheme to characterize and identify protein ubiquitination sites. IEEE/ACM Transactions on Computational Biology and Bioinformatics 14 (2), 393–403. doi:10.1109/TCBB.2016.2520939.

O'Shea, J.P., *et al.*, 2013. pLogo: A probabilistic approach to visualizing sequence motifs. Nature Methods 10 (12), 1211–1212. doi:10.1038/nmeth.2646.

Pang, C.N.I., Hayen, A., Wilkins, M.R., 2007. Surface accessibility of protein post-translational modifications. Journal of Proteome Research 6 (5), 1833–1845. doi:10.1021/pr060674u.

Peng, H., Long, F., Ding, C., 2005. Feature selection based on mutual information: Criteria of max-dependency, max-relevance, and min-redundancy. IEEE Transactions on Pattern Analysis and Machine Intelligence 27 (8), 1226–1238. doi:10.1109/TPAMI.2005.159.

Picotti, P., *et al.*, 2009. Full dynamic range proteome analysis of *S. cerevisiae* by targeted proteomics. Cell 138 (4), 795–806. https://doi.org/10.1016/j.cell.2009.05.051.

Pinho, S.S., Reis, C.A., 2015. Glycosylation in cancer: Mechanisms and clinical implications. Nature Reviews Cancer 15 (9), 540–555. doi:10.1038/nrc3982.

Prabakaran, S., *et al.*, 2012. Post-translational modification: Nature's escape from genetic imprisonment and the basis for dynamic information encoding. Wiley Interdisciplinary Reviews: Systems Biology and Medicine 4 (6), 565–583. doi:10.1002/wsbm.1185.

Reimand, J., *et al.*, 2016. g:Profiler – A web server for functional interpretation of gene lists (2016 update). Nucleic Acids Research 44 (W1), W83–W89. doi:10.1093/nar/gkw199.

Rosen, O.M., *et al.*, 1983. Phosphorylation activates the insulin receptor tyrosine protein kinase. Proceedings of the National Academy of Sciences of the United States of America 80 (11), 3237–3240. https://doi.org/10.1073/pnas.80.11.3237.

Schwenk, R.W., *et al.*, 2010. Fatty acid transport across the cell membrane: Regulation by fatty acid transporters. Prostaglandins Leukotrienes and Essential Fatty Acids 82 (4–6), 149–154. doi:10.1016/j.plefa.2010.02.029.

Scott, J.D., Pawson, T., 2009. Cell signaling in space and time: Where proteins come together and when they're apart. Science 326 (5957), 1220–1224. doi:10.1126/science.1175668.

Seet, B.T., *et al.*, 2006. Reading protein modifications with interaction domains. Nature Reviews Molecular Cell Biology 7 (7), 473–483. doi:10.1038/nrm1960.

Song, J., *et al.*, 2017. PhosphoPredict: A bioinformatics tool for prediction of human kinase-specific phosphorylation substrates and sites by integrating heterogeneous feature selection. Scientific Reports 7 (1), 6862. doi:10.1038/s41598-017-07199-4.

Stein, A., *et al.*, 2009. Dynamic interactions of proteins in complex networks: A more structured view. The FEBS Journal 276 (19), 5390–5405. doi:10.1111/j.1742-4658.2009.07251.x. 2009/08/29.

Stein, A., Aloy, P., 2010. Novel peptide-mediated interactions derived from high-resolution 3-dimensional structures. PLOS Computational Biology 6 (5), doi:10.1371/journal.pcbi.1000789.

Strasser, R., 2016. Plant protein glycosylation. Glycobiology 26 (9), 926–939. doi:10.1093/glycob/cww023.

Szklarczyk, D., *et al.*, 2017. The STRING database in 2017: Quality-controlled protein-protein association networks, made broadly accessible. Nucleic Acids Research 45 (D1), D362–D368. doi:10.1093/nar/gkw937.

Szymanski, C.M., Wren, B.W., 2005. Protein glycosylation in bacterial mucosal pathogens. Nature Reviews Microbiology 3 (3), 225–237. doi:10.1038/nrmicro1100.

Tay, A.P., *et al.*, 2017a. MethylQuant: A tool for sensitive validation of enzyme-mediated protein methylation sites from heavy-methyl SILAC data. Journal of Proteome Research. doi:10.1021/acs.jproteome.7b00601.

Tay, A.P., *et al.*, 2017b. PTMOracle: A Cytoscape app for covisualizing and coanalyzing post-translational modifications in protein interaction networks. Journal of Proteome Research 16 (5), 1988–2003. doi:10.1021/acs.jproteome.6b01052.

Tokmakov, A.A., *et al.*, 2012. Multiple post-translational modifications affect heterologous protein synthesis. The Journal of Biological Chemistry 287 (32), 27106–27116. doi:10.1074/jbc.M112.366351.

Tong, L., *et al.*, 1996. Crystal structures of the human p56lckSH2 domain in complex with two short phosphotyrosyl peptides at 1.0 Å and 1.8 Å resolution. Journal of Molecular Biology 256 (3), 601–610. doi:10.1006/jmbi.1996.0112.

Ullrich, A., Schlessinger, J., 1990. Signal transduction by receptors with tyrosine kinase activity. Cell 61 (2), 203–212. doi:10.1007/s00497-011-0177-9.

Van Roey, K., *et al.*, 2013. The switches.ELM Resource: A compendium of conditional regulatory interaction interfaces. Science Signaling 6 (269), https://doi.org/10.1126/scisignal.2003345.

Van Roey, K., Gibson, T.J., Davey, N.E., 2012. Motif switches: Decision-making in cell regulation. Current Opinion in Structural Biology 22 (3), 378–385. doi:10.1016/j.sbi.2012.03.004.

Varshavsky, A., 2011. The N-end rule pathway and regulation by proteolysis. Protein Science 20 (8), 1298–1345. doi:10.1002/pro.666.

Venne, A.S., Kollipara, L., Zahedi, R.P., 2014. The next level of complexity: Crosstalk of posttranslational modifications. Proteomics 14 (4–5), 513–524. doi:10.1002/pmic.201300344.

Via, A., *et al.*, 2009. A structure filter for the eukaryotic linear motif resource. BMC Bioinformatics 10, 351. https://doi.org/10.1186/1471-2105-10-351.

Wagner, M., *et al.*, 2005. Linear regression models for solvent accessibility prediction in proteins. Journal of Computational Biology 12 (3), 355–369. doi:10.1089/cmb.2005.12.355.

Wang, L., *et al.*, 2012. ASEB: A web server for KAT-specific acetylation site prediction. Nucleic Acids Research 40 (Web Server issue), W376–W379. doi:10.1093/nar/gks437.

Ward, J.J., *et al.*, 2004. Prediction and functional analysis of native disorder in proteins from the three kingdoms of life. Journal of Molecular Biology 337 (3), 635–645. doi:10.1016/j.jmb.2004.02.002.

Weiss, W.A., Taylor, S.S., Shokat, K.M., 2007. Recognizing and exploiting differences between RNAi and small-molecule inhibitors. Nature Chemical Biology 3 (12), 739–744. doi:10.1038/nchembio1207-739.

Winter, C., 2006. SCOPPI: A structural classification of protein-protein interfaces. Nucleic Acids Research 34 (90001), D310–D314. doi:10.1093/nar/gkj099.

Winter, D.L., Erce, M.A., Wilkins, M.R., 2014. A web of possibilities: Network-based discovery of protein interaction codes. Journal of Proteome Research 13 (12), 5333–5338. doi:10.1021/pr500585p.

Xu, C., Ng, D.T.W., 2015. Glycosylation-directed quality control of protein folding. Nature Reviews Molecular Cell Biology 16 (12), 742–752. doi:10.1038/nrm4073.

Xue, Y., *et al.*, 2006. PPSP: Prediction of PK-specific phosphorylation site with Bayesian decision theory. BMC Bioinformatics 7, 163. doi:10.1186/1471-2105-7-163.

Yang, P., *et al.*, 2015. Positive-unlabeled ensemble learning for kinase substrate prediction from dynamic phosphoproteomics data. Bioinformatics 32 (2), btv550. doi:10.1093/bioinformatics/btv550.

Zanzoni, A., *et al.*, 2011. Phospho3D 2.0: An enhanced database of three-dimensional structures of phosphorylation sites. Nucleic Acids Research 39 (Suppl. 1), doi:10.1093/nar/gkq936.

Zhao, N., *et al.*, 2011. Structural similarity and classification of protein interaction interfaces. PLOS ONE 6 (5), e19554. doi:10.1371/journal.pone.0019554.

Further Readings

Basu, A., Rose, K.L., Zhang, J., et al., 2009. Proteome-wide prediction of acetylation substrates. Proceedings of the National Academy of Sciences of the United States of America 106, 13785–13790.

Betts, M.J., Wichmann, O., Utz, M., et al., 2017. Systematic identification of phosphorylation-mediated protein interaction switches. PLOS Computational Biology 13, e1005462.

Chaudhuri, R., Yang, J., 2017. Cross-species PTM mapping from phosphoproteomic data. In: Wu, C.H., A., C.N., Ross, Karen E. (Eds.), Protein Bioinformatics: From Protein Modifications and Networks to Proteomics. New York: Humana Press, pp. 459–469.

Chauhan, J.S., Bhat, A.H., Raghava, G.P.S., Rao, A., 2012. GlycoPP: A webserver for prediction of N- and O-glycosites in prokaryotic protein sequences. PLOS ONE 7, e40155.

Chen, X., Shi, S.-P., Xu, H.-D., Suo, S.-B., Qiu, J.-D., 2016. A homology-based pipeline for global prediction of post-translational modification sites. Scientific Reports 6, 25801.

Chuang, G.-Y., Boyington, J.C., Joyce, M.G., et al., 2012. Computational prediction of N-linked glycosylation incorporating structural properties and patterns. Bioinformatics 28, 2249–2255.

Horn, H., Schoof, E.M., Kim, J., et al., 2014. KinomeXplorer: An integrated platform for kinome biology studies. Nature Methods 11, 603–604.

Huang, Y., Xu, B., Zhou, X., et al., 2015. Systematic characterization and prediction of post-translational modification cross-talk. Molecular & Cellular Proteomics 14, 761–770.

Jia, J., Zhang, L., Liu, Z., Xiao, X., Chou, K.-C., 2016. pSumo-CD: Predicting sumoylation sites in proteins with covariance discriminant algorithm by incorporating sequence-coupled effects into general PseAAC. Bioinformatics 32, 3133–3141.

Maurer-Stroh, S., Eisenhaber, B., Eisenhaber, F., 2002. N-terminal N-myristoylation of proteins: Prediction of substrate proteins from amino acid sequence. Journal of Molecular Biology 317, 541–557.

Maurer-Stroh, S., Eisenhaber, F., 2005. Refinement and prediction of protein prenylation motifs. Genome Biology 6, R55.

Shi, Y., Guo, Y., Hu, Y., Li, M., 2015. Position-specific prediction of methylation sites from sequence conservation based on information theory. Scientific Reports 5, 12403.

Wang, X.-B., Wu, L.-Y., Wang, Y.-C., Deng, N.-Y., 2009. Prediction of palmitoylation sites using the composition of k-spaced amino acid pairs. Protein Engineering, Design & Selection 22, 707–712.

Woodsmith, J., Kamburov, A., Stelzl, U., 2013. Dual coordination of post translational modifications in human protein networks. PLOS Computational Biology 9, e1002933.

Xu, Y., Ding, J., Wu, L.-Y., Chou, K.-C., 2013. iSNO-PseAAC: Predict cysteine S-nitrosylation sites in proteins by incorporating position specific amino acid propensity into pseudo amino acid composition. PlOS ONE 8, e55844.

Relevant Websites

msp.biocuckoo.org
    GPS-MSP – Methyl-group Specific Predictor 1.0.
pail.biocuckoo.org
    GPS-PAIL 2.0 – Prediction of Acetylation on Internal Lysines.
github.com/PengyiYang/KSP-PUEL
    KSP-PUEL – Positive-unlabeled ensemble learning for kinase-substrate prediction from dynamic phosphoproteomics data.
PhosphoPredictphosphopredict.erc.monash.edu.
    PhosphoPredict – Prediction of kinase-specific phosphorylation sites.
ptmcode.embl.de
    PTMCode 2 – Known and predicted PTM functional associations.
ptmfunc.com
    PTMfunc – Repository of functional predictions for protein PTMs.
apps.cytoscape.org/apps/ptmoracle
    PTMOracle – Co-visualisation and co-analysis of PTM and PPI data.
elm.eu.org
    The Eukaryotic Linear Motif (ELM) resource for Functional Sites in Proteins.
csb.cse.yzu.edu.tw/UbiNet
    UbiNet – A web resource for exploring ubiquitination networks.

Biographical Sketch

Dr. Chi Nam Ignatius Pang is a research associate currently working with Prof Wilkins. He completed his PhD in 2010 on the topic "The Dynamics of Protein Interaction Networks." In his thesis, he explored different mechanisms that govern the dynamics of protein-protein interactions – such as how changes in domain-domain interactions, post-translational modifications, and protein abundance can dynamically switch on/off protein-protein interactions. He also performed a large-scale screen of arginine and lysine methylation in the yeast proteome, and found that there are more methylated proteins than previously thought. In collaboration with Dr. Gene Hart-Smith, he is currently investigating how protein methylation affects protein-protein interactions. This involve identifying changes in protein complexes between wild type yeasts and methyltransferase knockout mutants using protein correlation profiling. Prior to his postdoc he had 2 years of experience in finance involving data analytics and fraud detection.

Prof Marc Wilkins is the Director of the Initiative. Professor Wilkins holds the chair of Systems Biology in the School of Biotechnology and Biomolecular Sciences. His research interests involve the proteomics of protein-protein interactions and systems biology. Specifically, Professor Wilkins' research focuses on the dynamics of protein-protein interaction networks, and the role that gene expression and protein post-translational modifications play in the control of protein interactions and thus delivery of cellular function in yeast and human cells.

Protein Properties

Kentaro Tomii, Artificial Intelligent Research Center (AIRC), National Institute of Advanced Industrial Science and Technology (AIST), Tokyo, Japan

Introduction

Protein molecules, the main constituents of living things, are organic compounds consisting mainly of carbon, nitrogen, oxygen, hydrogen, and sulfur. Protein behavior such as dynamics and interactions depends on the protein's physicochemical properties, even when a protein performs complex and exquisite roles *in vivo*. Therefore, knowing protein properties is expected to be useful for analyzing and understanding proteins.

A protein possesses its own unique properties to perform its function. Proteins are classifiable roughly into four types: globular proteins, membrane proteins, fibrous proteins, and intrinsically disordered proteins. Protein properties differ depending on the protein type. For instance, most membrane proteins show higher hydrophobicity than proteins of other types. This nature is strongly associated with their roles in a cell, i.e., the manner in which they localize and work at membranes. This characteristic, which is related to the protein solubility and other properties, is also useful to classify them using computational methods as introduced in the section below, although proteins of different types can be fused. For instance, a protein can have soluble domain(s) and transmembrane domain(s), or can have soluble domain(s) and intrinsically disordered regions. The function of the entire protein is based on the properties of individual domains.

Various properties of proteins or protein domains can be measured using physicochemical and biochemical experiments. However, indeed, precise measurements cannot be obtained easily. Therefore, conducting research on an unknown protein is expected to be effective to ascertain the approximate values of various protein properties using computational methods. This article briefly introduces widely various *in silico* methods and tools used for predicting and calculating protein properties such as molecular weight, isoelectric point, solubility, crystallizability, and hydrophobicity.

Amino Acids

A protein is a string of amino acids. The arrangement of that string (primary structure) determines a protein's structure and function. That is to say, amino acid properties are the origin of the structure, function (and also evolution) of proteins. In general, naturally occurring proteins comprise the 20 most common amino acids with different side-chains. Each type of amino acid possesses unique properties that affect the structure and function of proteins, depending on its side-chains. Based on the chemical properties of side-chains, the 20 amino acids are classified roughly into four groups: hydrophobic, polar, and charged residues, and Gly which only has H (hydrogen) as its side-chain. Already, numerous physicochemical and biochemical properties of amino acids have been measured and published.

A compendium of such properties is available. Based on the 188 amino acid properties collected and analyzed by Kidera and his colleagues (Kidera *et al.*, 1985), Nakai *et al.* (1988) launched the so-called AAindex database, which includes 222 amino acid indices: sets of 20 numerical values representing their various properties. Since then, the database has been developed and expanded. In 1996, Tomii and Kanehisa (1996) reformatted the database and released a collection of 402 amino acid indices, with a list of similar indices having correlation coefficients of 0.8 (-0.8) or more (less), and 42 amino acid substitution matrices representing similarity between amino acids. In a substitution matrix, higher (better) scores are assigned to pairs of residues with more similar properties. They are used for protein sequence analysis such as similarity search and phylogenetic inference. The ever-growing database currently (AAindex Release 9.2) contains 566 indices, 94 matrices, and 47 contact potentials (Kawashima *et al.*, 2008). The AAindex database is available at "see Relevant Websites section". According to the result of cluster analysis, most of the collected indices are roughly classified into several classes according to characteristics such as hydrophobicity, propensities related to secondary structures, volumes of amino acids, and amino acid composition.

ExPASy offers a tool called ProtScale to visualize a profile as a two-dimensional plot for a protein using 57 indices at "see Relevant Websites section" (Gasteiger *et al.*, 2005). EMBOSS also provides a program called pepwindow via the web browser interface (see "Relevant Websites section") and Web Services, for drawing a hydropathy plot for a protein based on the hydropathy index (**Fig. 1**) proposed in an earlier report (Kyte and Doolittle, 1982). It can also be used to draw various plots using various indices by substituting the index in 'datafile'.

Amino Acid Composition and Molecular Weight

As described above, naturally occurring proteins consist mainly of 20 types of amino acids. The 20 most common amino acids are known to be used unevenly. The frequency of use of individual amino acids is biased. On average, frequently occurring residues such as Ala and Leu account for approximately 9%, although rare residues such as Cys and Trp account for about 1%. Proteins are roughly classified into four types: water-soluble globular proteins, membrane proteins, fibrous proteins, and intrinsically disordered proteins. Amino acid compositions for those four types of proteins differ (see below). As one might expect, they differ for individual proteins. Especially, fibrous proteins such as collagen consist of a characteristic repeat such as Gly Pro-X and Gly X-hydroxyproline, which is a modified proline residue.

```
H KYTJ820101
D Hydropathy index (Kyte-Doolittle, 1982)
R PMID:7108955
A Kyte, J. and Doolittle, R.F.
T A simple method for displaying the hydropathic character of a protein
J J. Mol. Biol. 157, 105-132 (1982)
C JURD980101    0.996   CHOC760103    0.964   OLSK800101    0.942
  JANJ780102    0.922   NADH010102    0.920   NADH010101    0.918
  DESM900102    0.898   EISD860103    0.897   CHOC760104    0.889
  NADH010103    0.885   WOLR810101    0.885   RADA880101    0.884
  MANP780101    0.881   EISD840101    0.878   PONP800103    0.870
  WOLR790101    0.869   NADH920108    0.868   JANJ790101    0.867
  JANJ790102    0.866   BASU050103    0.863   PONP800102    0.861
  MEIH800103    0.856   NADH010104    0.856   PONP800101    0.851
  PONP800108    0.850   CORJ870101    0.848   WARP780101    0.845
  COWR900101    0.845   PONP930101    0.844   RADA880108    0.842
  ROSG850102    0.841   DESM900101    0.837   BLAS910101    0.836
  BIOV880101    0.829   RADA880107    0.828   BASU050101    0.826
  KANM800104    0.824   LIFS790102    0.824   CIDH920104    0.824
  MIYS850101    0.821   RADA880104    0.819   NAKH900111    0.817
  CORJ870104    0.812   NISK800101    0.812   FAUJ830101    0.811
  ROSM880105    0.806   ARGP820103    0.806   CORJ870103    0.806
  NADH010105    0.804   NAKH920105    0.803   ARGP820102    0.803
  CORJ870107    0.801   MIYS990104   -0.800   CORJ870108   -0.802
  KRIW790101   -0.805   MIYS990105   -0.818   MIYS990103   -0.833
  CHOC760102   -0.838   MIYS990101   -0.840   MIYS990102   -0.840
  MONM990101   -0.842   GUYH850101   -0.843   FASG890101   -0.844
  RACS770102   -0.844   ROSM880101   -0.845   JANJ780103   -0.845
  ENGD860101   -0.850   PRAM900101   -0.850   JANJ780101   -0.852
  GRAR740102   -0.859   PUNT030102   -0.862   GUYH850104   -0.869
  MEIH800102   -0.871   PUNT030101   -0.872   ROSM880102   -0.878
  KUHL950101   -0.883   GUYH850105   -0.883   OOBM770101   -0.899
I    A/L     R/K     N/M     D/F     C/P     Q/S     E/T     G/W     H/Y     I/V
     1.8    -4.5    -3.5    -3.5     2.5    -3.5    -3.5    -0.4    -3.2     4.5
     3.8    -3.9     1.9     2.8    -1.6    -0.8    -0.7    -0.9    -1.3     4.2
//
```

Fig. 1 The AAindex entry of the hydropathy index proposed by Kyte and Doolittle. Each entry in AAindex includes its accession number (denoted as 'H'), the name or short description of the index ('D'), the PubMed ID of literature which publishes the index ('R'), the author name(s) ('A'), the title of the literature ('T'), the bibliographic information of the literature ('J'), the list of similar indices and their correlation coefficients with the index ('C'), the index value for each amino acid ('I'), and the termination symbol ("//").

It is often argued that amino acid composition is associated with structural and biological characteristics of proteins. Actually, chain-length (=domain size) dependence of amino acid composition is also known to exist (White, 1992; Carugo, 2008). Therefore, many computational prediction methods have been proposed, such as the molar extinction coefficient, solubility and crystallizability of proteins, for folding types, structural classes, secondary structure contents, quaternary structural types, and subcellular localization, and for membrane proteins, thermophilic proteins, disordered proteins, based on amino acid and/or dipeptide composition.

A protein's molecular weight is the total of the molecular weights of the protein's constituent amino acids, in addition to the weights originated from posttranslational modifications. Knowing the molecular weight of a protein is helpful for setting and selecting methods for expression and purification systems of it. Although the molecular weight of a protein can be ascertained using analytical biochemistry, exact measurements are quite laborious. At present, as the genome sequences are accumulated, the molecular weights of constituent amino acids are often used to calculate the total by converting the decoded nucleotide sequence into an amino acid sequence. In doing so, it is noteworthy that the molecular weight of the water molecules desorbed along with peptide bond formation is subtracted from the sum of the constituent amino acid molecular weights of the proteins.

Isoelectric Point

Proteins contain ionizable residues including both acidic (aspartic acid and glutamic acid) and basic (arginine, lysine, and histidine) residues. "The isoelectric point (pI) is defined as the level of pH at which the protein/peptide has a net charge of zero

(Smoluch *et al.*, 2016)". Proteins show different pI depending on the type and number of constituent amino acids. For instance, pI of a protein such as aldolase is around 10, although the pI of a protein such as ornithine decarboxylase is around 4 (Dice and Goldberg, 1975). In proteomes, the distribution of pI values of proteins has been known to show a generally bimodal distribution. It has a low fraction of proteins with pI close to 7.4, which closely approximates the pH of the major compartments inside of most cells. A report of one study describes the relation between the length of proteins and their pI values (Kiraga *et al.*, 2007). It might have different charge states depending primarily on the surface charge and solution state. At the pI of a protein, as attractions between protein molecules are maximized, its solubility is minimized. Using such a difference in pI, proteins are separable. Therefore, it is important to calculate pI for each protein. Actually, the pI of a protein is calculable approximately from the protein's amino acid composition. However, generally, it is difficult to obtain the exact value of pI without experimental research.

Generally speaking, although accurate calculation is not easy, various approximate calculation methods have been proposed to estimate the pI of a protein. In brief, the pI of a protein is calculated based on pKa, the inverse logarithm of an acid dissociation constant, of charged residues. Many sets of pKa values for the ionizable groups of seven charged residues (Cys, Asp, Glu, His, Lys, Arg, and Tyr) have been published along with the terminal groups: amino and carboxyl groups. Recently, an isoelectric point calculator, called IPC (see "Relevant Websites section"), has been proposed based on the optimized sets of pKa values using a basin-hopping procedure (Kozlowski, 2016). To develop accurate methods for structure-based calculation of pKa values, a blind prediction challenge was organized: The pKa Cooperative (see "Relevant Websites section"). For this challenge, invited research groups have attempted to predict an unpublished set of experimental pKa values, especially for interior residues in proteins (Nielsen *et al.*, 2011). Actually, pKa calculations for those residues can be difficult.

The pI values of proteins can provide useful information to advance protein science. One report has described that "a relationship exists between a protein's pI and the pH under which it will crystallize" (Kirkwood *et al.*, 2015). One report describes that mammalian complexes of three or more unique proteins have greater homogeneity than the expected values in terms of predicted pI values (Wong *et al.*, 2008).

Solubility and Crystallizability

The solubility of proteins in aqueous solution is regarded as dependent mainly on the surface structure of proteins and the types of amino acids on their surfaces. Generally, proteins with more charged and polar residues on their accessible surface with water molecules show higher solubility. Therefore, the optimum pH differs for each protein. The solubility of a protein also depends on the salt concentration of the solution and the type of salt. Proteins such as membrane proteins, with many hydrophobic residues on their surface, are insoluble in most cases, although they can be solubilized in some cases using surfactants.

Early computational methods for protein solubility prediction from its sequence were simple and were based on the rate of charged residues contained in a protein. In recent years, with the progress of the structural genomics projects worldwide, large amounts of data related to the expression system of many proteins and their results have been accumulated. Therefore, most current prediction methods use such huge datasets as learning datasets. Furthermore, they are based on machine learning methods, which makes it possible to use not only the rate of charged residues but also various physicochemical properties such as those in AAindex for solubility prediction. The prediction accuracy has improved as a consequence. A useful interpretation and list of solubility prediction methods for protein production in *Escherichia coli* are available in a recent review (Chang *et al.*, 2014).

X-ray crystallographic analyses still play a major role in protein tertiary structural studies. About 90% of entries containing proteins in PDB were determined using X-ray crystallography as of December 2017. In X-ray crystallography, obtaining diffraction quality crystals of proteins is a central issue. As is the case with solubility prediction, prediction methods for protein crystallization have been developed for and along with the development of structural genomics projects. For advancing structural genomics projects, it is indispensable to select good target proteins and to search optimal crystallization conditions. Like solubility prediction methods, as structural genomics projects progress, large amounts of data related to protein crystallization are becoming available, leading to a vigorous construction of prediction methods using such huge data and machine learning methods. For these reasons, crystallization prediction methods have progressed in recent years by consideration of various protein characteristics related to their crystallization, although it remains controversial, which is the influential factor for protein crystallizability. An exhaustive and practical review of prediction for protein crystallizability has been published (Smialowski and Wong, 2016).

Hydrophobicity

Proteins usually include hydrophobic residues as well as charged and polar residues. The proportion of hydrophobic residues in a protein is useful for classifying proteins because membrane proteins tend to have more hydrophobic residues in their transmembrane region(s) than other proteins do. For instance, a GRAVY (GRand AVerage of hydropathY) score, which is calculated by summing the hydropathy values of all the amino acids and by dividing by the number of residues in a protein, has been developed and used for predicting membrane proteins (Kyte and Doolittle, 1982).

Because hydrophilicity and hydrophobicity are two sides of the same coin, Kyte and Doolitle used the term 'hydropathy', which means "strong feelings about water." They originally proposed their hydropathy index based on both the water-vapor transfer free energies and the interior–exterior distribution of the side-chains. Recently, an engineering measurement, contact angles of a water nanodroplet on a(n artificial) planar surface (of peptides), has been used for characterizing hydrophobicity of amino acids

(Zhu *et al.*, 2016). Including these, many different scales for representing hydrophobicity of amino acids have been published to date (Simm *et al.*, 2016). The variation of hydrophobicity scales can be attributed to their underlying formulations. A prominent difference is the way of dealing with amino acids. For instance, Cys residues can be hydrophobic when we consider their locations in proteins, although their hydrophilic nature emerges when we consider individual free amino acids.

Disordered Proteins

Soluble globular proteins, membrane proteins, and fibrous proteins must possess unique tertiary structures under physiological conditions to perform their functions. However, it is considered that intrinsically disordered proteins (IDPs) do not possess unique tertiary structures but instead exist as dynamic structural ensembles, although they have specific structures when they bind to other molecules under physiological conditions. Most IDPs are involved in important roles in cellular signalling and regulation. Many nucleic proteins are known to contain IDPs. Actually, it is considered that about 20%–30% of eukaryotic proteins are IDPs. Intrinsically disordered proteins (or regions) tend to have more abundant Gly, Ser, and Pro residues, and less hydrophobic residues than other proteins (or regions). Elucidating intrinsically disordered regions in a protein is helpful to ascertain the boundaries of protein domains suitable for X-ray crystallographic analysis. Numerous IDP (or region) prediction methods using this tendency and other features have been reported. A comprehensive summary of methods and databases for disorder prediction is available (Lieutaud *et al.*, 2016). Because IDP prediction can reflect difficulties of discrimination between ordered and disordered proteins (or regions) and because more information about IDPs has been accumulated than ever, machine learning methods such as support vector machine (SVM) and artificial neural network (ANN) have been applied for this purpose.

Quaternary Structure

A complex composed of multiple protein chains (subunits) is defined as a quaternary structure, also called a multimer, or biological assembly. Many proteins form their quaternary structures when they function *in vivo*. The quaternary structure of a protein can be ascertained by physicochemical and biochemical studies. However, accurate measurements are generally elaborate, and exhaustive measurements of protein quaternary structures are not easy. Actually, the estimated state(s) of protein quaternary structure measured in solution, those described by the author(s) in PDB, and the states estimated based on analysis of contacts between protein chains in the crystal in PDB are often different. Under these circumstances, development of an accurate method for estimating the quaternary structure of proteins is expected. However, even among homologous proteins, the quaternary structure of proteins is often not conserved as much as its three-dimensional structure is, and in not as many cases. However, even between distantly related proteins, predictions using the quaternary structure of homologous proteins can be effective in some cases (Nakamura *et al.*, 2017).

Useful Resources

Below, we introduce and summarize useful sites and tools other than those introduced into the papers and reviews described above.

	Name	*URL*	*Descriptions*
Amino acids	Amino acid explorer	https://www.ncbi.nlm.nih.gov/Class/Structure/aa/aa_explorer.cgi	Starting points for learning biochemical and physicochemical properties of amino acids
	Amino acid information finder	http://www.currentprotocols.com/WileyCDA/CurPro3Tool/toolId-1.html	
Molecular weight	DNA/RNA/Protein molecular weight calculator	http://www.currentprotocols.com/WileyCDA/CurPro3Tool/toolId-8.html	Molecular weight calculator for both nucleotide and amino acid sequences
Isoelectric point	Proteome-*pI*	http://isoelectricpointdb.org	Database of isoelectric points pre-calculated with 18 methods for 21,721,250 protein sequences from 5029 organisms (Kozlowski, 2017)
Solubility	Recombinant protein solubility prediction	http://www.biotech.ou.edu/	Prediction method for proteins overexpressed in *E. coli* using logistic regression (Diaz *et al.*, 2010)
	CamSol	http://www-mvsoftware.ch.cam.ac.uk/index.php/camsolhome	Prediction method for protein solubility and aggregation propensity (Sormanni *et al.*, 2015)
	Periscope	http://lightning.med.monash.edu/periscope/	Sequence-based predictor for soluble protein expression in the periplasm of *E. coli* (Chang *et al.*, 2016)

	Protein–Sol	https://protein-sol.manchester.ac.uk/	Sequence-based method for predicting protein solubility and lysine and arginine content for solubility design (Hebditch *et al.*, 2017)
Hydrophobicity	GRAVY calculator	http://www.gravy-calculator.de/	Tools for calculating the GRAVY score
	Protein GRAVY	http://www.bioinformatics.org/sms2/protein_gravy.html	
Disordered proteins/ regions List of approx. 40 methods for prediction of intrinsically disordered proteins (regions)	List of protein disorder predictors	http://iimcb.genesilico.pl/metadisorder/	list_of_protein_disorder_tools_programs.html
Quaternary structure	Online analysis tools – Protein quaternary structure	https://molbiol-tools.ca/Protein_quaternary_structure.htm	List of tools for predicting and analyzing quaternary structure(s) of proteins
Various properties	ProtParam	https://web.expasy.org/protparam/	Tool for computing various physicochemical properties, including amino acid composition, molecular weight, *pI*, GRAVY, and more (Gasteiger *et al.*, 2005)
	EMBOSS Pepstats	https://www.ebi.ac.uk/Tools/seqstats/emboss_pepstats	Tool for computing various protein properties such as molecular weight, *pI*, and improbability of expression in inclusion bodies
Portal sites	Current protocols tools and calculators	http://www.currentprotocols.com/WileyCDA/Section/id-810245.html	Large list of tools and calculators for life scientists
	ExPASy bioinformatics resource portal	https://www.expasy.org/	Extensive collection of databases and tools in widely diverse life sciences (Artimo *et al.*, 2012)
	Online analysis tools	https://molbiol-tools.ca	Well-organized list of computational tools for molecular biologist by Dr. Kropinski
	OMICtools	https://omictools.com/proteomics-category	Comprehensive collection of tools and databases in proteomics

Closing Remarks

Elucidation of protein properties is important and helpful for advancing protein research. Many computational methods for the broad range of protein properties have been developed and have been made available, especially along with the development of structural genomics projects as described above. These methods are expected to become more helpful and reliable for our research, according to their improvement and the accumulation of related data.

Acknowledgement

We thank Professor Kenta Nakai for the invitation to contribute this article.

See also: *Ab initio* Protein Structure Prediction. Algorithms for Strings and Sequences: Searching Motifs. Algorithms for Structure Comparison and Analysis: Docking. Algorithms for Structure Comparison and Analysis: Homology Modelling of Proteins. Algorithms for Structure Comparison and Analysis: Prediction of Tertiary Structures of Proteins. Epitope Predictions. Functional Genomics. Natural Language Processing Approaches in Bioinformatics. Prediction of Protein Localization. Prediction of Protein-Protein Interactions: Looking Through the Kaleidoscope. Protein Functional Annotation. Protein Post-Translational Modification Prediction. Protein Post-Translational Modification Prediction. Proteomics Data Representation and Databases. Proteomics Mass Spectrometry Data Analysis Tools. Secondary Structure Prediction. Sequence Analysis. Sequence Composition. The Evolution of Protein Family Databases. Transcription Factor Databases. Transmembrane Domain Prediction

References

Artimo, P., Jonnalagedda, M., Arnold, K., *et al.*, 2012. ExPASy: SIB bioinformatics resource portal. Nucleic Acids Research 40, W597–W603.
Carugo, O., 2008. Amino acid composition and protein dimension. Protein Science 17, 2187–2191.

Chang, C.C., Li, C., Webb, G.I., *et al.*, 2016. Periscope: Quantitative prediction of soluble protein expression in the periplasm of *Escherichia coli*. Scientific Reports 6, 21844.

Chang, C.C., Song, J., Tey, B.T., Ramanan, R.N., 2014. Bioinformatics approaches for improved recombinant protein production in *Escherichia coli*: Protein solubility prediction. Briefings in Bioinformatics 15, 953–962.

Diaz, A.A., Tomba, E., Lennarson, R., *et al.*, 2010. Prediction of protein solubility in *Escherichia coli* using logistic regression. Biotechnology and Bioengineering 105, 374–383.

Dice, J.F., Goldberg, A.L., 1975. Relationship between *in vivo* degradative rates and isoelectric points of proteins. Proceedings of the National Academy of Sciences of the United States of America 72, 3893–3897.

Gasteiger, E., Hoogland, C., Gattiker, A., 2005. Protein identification and analysis tools on the ExPASy server. In: Walker, J.M. (Ed.), The Proteomics Protocols Handbook. Humana Press, pp. 571–607.

Hebditch, M., Carballo-Amador, M.A., Charonis, S., Curtis, R., Warwicker, J., 2017. Protein-Sol: A web tool for predicting protein solubility from sequence. Bioinformatics 33, 3098–3100.

Kawashima, S., Pokarowski, P., Pokarowska, M., *et al.*, 2008. AAindex: Amino acid index database, progress report 2008. Nucleic Acids Research 36, D202–D205.

Kidera, A., Konishi, Y., Oka, M., Ooi, T., Scheraga, H.A., 1985. Statistical analysis of the physical properties of the 20 naturally occurring amino acids. Journal of Protein Chemistry 4, 23–55.

Kiraga, J., Mackiewicz, P., Mackiewicz, D., *et al.*, 2007. The relationships between the isoelectric point and length of proteins, taxonomy and ecology of organisms. BMC Genomics 8, 163.

Kirkwood, J., Hargreaves, D., O'Keefe, S., Wilson, J., 2015. Using isoelectric point to determine the pH for initial protein crystallization trials. Bioinformatics 31, 1444–1451.

Kozlowski, L.P., 2016. IPC – Isoelectric point calculator. Biology Direct 11, 55.

Kozlowski, L.P., 2017. Proteome-*pI*: Proteome isoelectric point database. Nucleic Acids Research 45, D1112–D1116.

Kyte, J., Doolittle, R.F., 1982. A simple method for displaying the hydropathic character of a protein. Journal of Molecular Biology 157, 105–132.

Lieutaud, P., Ferron, F., Uversky, A.V., *et al.*, 2016. How disordered is my protein and what is its disorder for? A guide through the "dark side" of the protein universe. Intrinsically Disordered Proteins 4, e1259708.

Nakai, K., Kidera, A., Kanehisa, M., 1988. Cluster analysis of amino acid indices for prediction of protein structure and function. Protein Engineering 2, 93–100.

Nakamura, T., Oda, T., Fukasawa, Y., Tomii, K., 2017. Template-based quaternary structure prediction of proteins using enhanced profile-profile alignments. Proteins.

Nielsen, J.E., Gunner, M.R., García-Moreno, B.E., 2011. The pKa Cooperative: A collaborative effort to advance structure-based calculations of pKa values and electrostatic effects in proteins. Proteins 79, 3249–3259.

Simm, S., Einloft, J., Mirus, O., Schleiff, E., 2016. 50 years of amino acid hydrophobicity scales: Revisiting the capacity for peptide classification. Biological Research 49, 31.

Smialowski, P., Wong, P., 2016. Protein crystallizability. Methods in Molecular Biology 1415, 341–370.

Smoluch, M., Mielczarek, P., Drabik, A., Silberring, J., 2016. Online and offline sample fractionation. In: Ciborowski, P., Silberring, J. (Eds.), Proteomic Profiling and Analytical Chemistry, second ed. Elsevier, pp. 63–99.

Sormanni, P., Aprile, F.A., Vendruscolo, M., 2015. The CamSol method of rational design of protein mutants with enhanced solubility. Journal of Molecular Biology 427, 478–490.

Tomii, K., Kanehisa, M., 1996. Analysis of amino acid indices and mutation matrices for sequence comparison and structure prediction of proteins. Protein Engineering 9, 27–36.

White, S.H., 1992. Amino acid preferences of small proteins. Implications for protein stability and evolution. Journal of Molecular Biology 227, 991–995.

Wong, P., Althammer, S., Hildebrand, A., *et al.*, 2008. An evolutionary and structural characterization of mammalian protein complex organization. BMC Genomics 9, 629.

Zhu, C., Gao, Y.A., Li, H., *et al.*, 2016. Characterizing hydrophobicity of amino acid side chains in a protein environment via measuring contact angle of a water nanodroplet on planar peptide network. Proceedings of the National Academy of Sciences of the United States of America 113, 12946–12951.

Relevant Websites

http://www.genome.jp/aaindex/
    AAindex.
https://www.ebi.ac.uk/Tools/seqstats/emboss_pepwindow/
    EMBOSS Pepwindow.
https://web.expasy.org/protscale/
    ExPASy
    ProtScale.
http://isoelectric.ovh.org
    IPC.
http://www.pkacoop.org
    pKa Cooperative.

The Evolution of Protein Family Databases

Teresa K Attwood, The University of Manchester, Manchester, United Kingdom
Alex L Mitchell, EMBL-EBI, Hinxton, United Kingdom

Introduction

Elucidation of protein function forms a major part of understanding fundamental natural processes, including the basis of diseases and the interaction of species with their environments. Characterising protein function using laboratory-based experiments is a high-cost, low-throughput process that cannot keep pace with the volumes of data being generated by genomic and environmental sequencing projects. The problem has become particularly acute in recent years, as sequencing technology has developed to the point where it is possible to sequence entire genomes, or to generate hundreds of millions of DNA sequences from environmental samples, in a single experiment. The pace of data generation now outstrips the rate of experimental characterisation by many orders of magnitude, and the disparity is likely to get worse as sequencing technologies improve. For the custodians of the data, charged with the task of assigning functions to the mass of raw sequences, the deluge has created almost unimaginable annotation burdens. To give an example, during the last 30 years, dedicated teams of curators have manually annotated ~555,000 Uni-ProtKB/Swiss-Prot sequences; this figure is roughly equal to the number of sequences now entering UniProtKB per week. To put these figures in context, the European Bioinformatics Institute Metagenomics peptide database now contains almost 50 million sequences, derived from 400 data-sets (Mitchell *et al.*, 2017). More than 15 million of these are predicted to be full length, but only ~1 million currently have counterparts in UniProtKB; and the database is expected to grow to 100s of millions or billions of sequences as proteins continue to be assimilated. The future implications for resources like UniProt (UniProt Consortium, 2017) are staggering. Inevitably, this impossible situation has mandated the use of computational approaches to complement and mitigate curators' Herculean annotation tasks.

Before the situation became as acute as it is today, an innovative strategy to try to address the problem focused on the use of protein family databases. Such databases have been part of the ecosystem of bioinformatics tools and resources for almost 30 years. Underpinned by information derived from multiple sequence alignments, they offer a more sensitive and *scalable* alternative to pairwise sequence analysis approaches (e.g., such as BLAST (Altschul *et al.*, 1990)) that compare a query sequence against all proteins in a target database (consider, for example, that Tara Oceans scientists have estimated that a BLASTx comparison of ~120 million of their eukaryotic sequences against UniRef90 (containing ~26 million sequences) and the many millions of sequences in the Marine Microbial Eukaryote Transcriptome Sequencing Project (MMETSP; Keeling *et al.*, 2014) would require ~9 million CPU hours on a high-performance computing facility). Nevertheless, the pivotal role that protein family databases now play in sequence analysis is probably under-appreciated: today, much of their work is conducted behind the scenes in support of the principal players – the primary sequence repositories, like UniProtKB.

The origins of protein family databases can be traced to a description – in *Of URFs and ORFs* (Doolittle, 1986) – of short amino acid 'patterns' that often characterise specific functional sites. In his book, Doolittle asserted that such patterns could be grouped and potentially used to search for similar sites in query sequences. Reading this inspired Bairoch (then a postgraduate student in Geneva working on developing the Swiss-Prot database) to write a program to do just this – he called the program PROSITE (Bairoch, 2000). At the same time, he scoured the literature for other sequence patterns, but was dismayed to discover that there were few of them; and those he did find weren't sufficiently diagnostic to be useful (Bairoch, 2000). He therefore decided to create some patterns of his own. In the sections that follow, we explore how Bairoch's pioneering work spawned the creation of numerous protein family databases, and how these became such critical tools for sequence annotation.

Protein Family Classification – Different Approaches

PROSITE

Seeking to create a collection of his own diagnostic 'signatures', Bairoch began by developing short patterns for various binding sites, active sites, sites of post-translational modification, etc. and/or particular family groupings. The process involved creating multiple alignments of representative sequences, inspecting the alignments for conserved residues, or residue groups, and then encoding the observed pattern of conservation in a regular expression-like syntax that would, ideally, provide a consensus for all sequences in the family or sharing the functional site. Importantly, for each, he provided a detailed abstract describing the biological entity (site, domain, family, etc.) encompassed by the expression. The first collection of these patterns – 58 in all – was bundled with the PROSITE search tool and released in March 1988 as part of the PC/Gene software package, which was then being commercialised by Intelligenetics.

At the time, PROSITE was very useful for diagnosing potential functional sites in individual sequences and/or for suggesting their likely family relationships; but it was also becoming clear that the resource had a much larger role in characterising entire sequence collections, such as those represented by Swiss-Prot (Bairoch and Boeckmann, 1991). Hence, before long, Bairoch was

Table 1   Sample of some of the first sequence patterns released in PROSITE

Pattern and documentation IDs	Functional site	Sequence Pattern[a]	Pattern length
PS00001; PDOC00001	*N-glycosylation site	N-{P}-[ST]-{P}	4
PS00004; PDOC00004	*cAMP- and cGMP-dependent protein kinase phosphorylation site	[RK](2)-x-[ST]	4
PS00005; PDOC00005	*Protein kinase C phosphorylation site	[ST]-x-[RK]	3
PS00006; PDOC00006	*Casein kinase II phosphorylation site	[ST]-x(2)-[DE]	4
PS00007; PDOC00007	*Tyrosine kinase phosphorylation site	[RK]-x(2,3)-[DE]-x(3)-Y	8–9
PS00008; PDOC00008	*N-myristoylation site	G-{EDRKHPFYW}-x(2)-[STAGCN]-{P}	6
PS00009; PDOC00009	*Amidation site	x-G-[RK]-[RK]	4
PS00010; PDOC00010	Aspartic acid and asparagine hydroxylation site	C-x-[DN]-x(4)-[FY]-x-C-x-C	12
PS00011; PDOC00011	Vitamin K-dependent carboxylation domain	E-x(2)-[ERK]-E-x-C-x(6)-[EDR]-x(10,11)-[FYA]-[YW]	26–27
PS00012; PDOC00012	Phosphopantetheine attachment site	[DEQGSTALMKRH]-[LIVMFYSTAC]-[GNQ]-[LIVMFYAG]-[DNEKHS]-S-[LIVMST]-{PCFY}-[STAGCPQLIVMF]-[LIVMATN]-[DENQGTAKRHLM]-[LIVMWSTA]-[LIVGSTACR]-{LPIY}-{VY}-[LIVMFA]	16

[a]Within these expressions, amino acid residues (denoted using the IUPAC single-letter notation) in square brackets are allowed at that position; those in curly brackets are forbidden at that position; x is a wild-card, denoting any amino acid; numbers in parentheses indicate how many times a residue, or residue group, can occur; and a residue on its own is strictly conserved.

Note: Asterisks denote patterns that have been shown to exhibit poor diagnostic performance.

under pressure to liberate PROSITE from its commercial shackles. Accordingly, in October 1989, he announced that he was making the resource publicly available, and released a new version (4.0) the following month, with 202 entries. **Table 1** shows some of the first patterns collected in PROSITE (Bairoch, 1991).

The first thing to notice in **Table 1** is that most of the expressions are very short and, in a sense, quite 'primitive' – some are just 3 or 4 residues long. In the late 1980s, when sequence databases were still relatively small (by comparison with the vast, genome-scale repositories of today), these and other PROSITE patterns were nevertheless quite useful indicators of potential functional sites or family relationships. However, the shorter a pattern and the more variable it is (i.e., the larger the allowed residue groups and the more wild-cards used), the poorer its diagnostic performance. Inevitably, as target sequence databases grew larger, this was to become increasingly problematic.

In 1994, the diagnostic performance of then current PROSITE patterns was analysed statistically with respect to Swiss-Prot 30.0 and to a randomised database derived from it by regional shuffling. The frequency of occurrence of the patterns depicted in **Table 1** is shown in **Table 2**. As can be seen, the most promiscuous patterns are those that are shortest and most variable. In fact, patterns PS00001 to PS00009 are so noisy that they're flagged as 'patterns with a high probability of occurrence' (this flag allows such promiscuous matchers to be ignored by modern PROSITE-scanning tools).

Interestingly, in **Table 1**, pattern identifiers (IDs) PS00002 and PS00003 are missing; the latter is also missing from **Table 2**. PS00002 encoded a glycosaminoglycan attachment site (S-G-x-G); this expression was qualified by an additional condition that there should be at least two acidic amino acids from −2 to −4 relative to the serine attachment site. Similarly, PS00003 encoded a tyrosine sulphation site, the consensus features of which were described in terms of the acidic, hydrophobic and polar residues within 1 and 7 residues N- and C-terminally to the tyrosine (specifically, E or D within two residues of the Y (typically at −1); at least three acidic residues from −5 to +5; no more than 1 basic residue and 3 hydrophobic from −5 to +5; at least one P or G from −7 to −2 and from +1 to +7; or at least two or three D, S or N from −7 to +7; absence of disulphide-bonded C residues from −7 to +7; and absence of N-linked glycans near the Y).

In early PROSITE releases, PS00002 and PS00003 were termed 'rules'. Rules were logical natural-language assertions used to complement or replace patterns that couldn't be described effectively by the formal consensus-expression syntax. Being written in free-text English, the only way to scan rules against a sequence database was to encode them directly inside the search program; many of the earliest PROSITE-scanning tools were unable to do this (Gattiker *et al.*, 2002).

As the mass of accumulating protein sequences grew, and their diversity widened, the diagnostic power of several PROSITE entries began to erode. Some of these were revised and updated in light of knowledge derived from new family members in Swiss-Prot (see **Table 3**); PS00002, PS00003 and other poorly performing rules and patterns were deleted. However, from the outset, the considerable divergence of some protein families (e.g., globins, immunoglobulins, heat-shock proteins), functional sites and domains (SH2, SH3, PH, Kringle domains, etc.) couldn't be encapsulated efficiently using sequence patterns. Part of the problem is

Table 2   Pattern frequencies based on Swiss-Prot 30.0 and the corresponding PROSITE release. Patterns PS00001 to PS00009, which are the shortest and most variable, are the most promiscuous

Pattern		Swiss-Prot 30.0		Randomised		Exp. hits	Quota
ID	Length	Matches	Sequences	Matches	Sequences	/1000 res.	%
PS00001	4	79,458	25,162	74,272	25,707	5.25E + 00	93.47
PS00002	4	6872	5357	6155	4895	4.35E − 01	89.57
PS00004	4	23,439	14,466	23,128	14,653	1.63E + 00	98.67
PS00005	3	185,672	35,425	190,343	35,449	1.35E + 01	102.52
PS00006	4	205,059	34,946	194,184	34,658	1.37E + 01	94.70
PS00007	8–9	21,117	13,635	20,903	13,618	1.48E + 00	98.99
PS00008	6	179,880	34,404	177,861	34,284	1.26E + 01	98.88
PS00009	4	13,478	10,339	12,080	9360	8.54E − 01	89.63
PS00010	12	299	67	4	4	2.83E − 04	1.34
PS00011	26–27	38	38	6	6	4.24E − 04	15.79
PS00012	16	46	46	9	9	6.36E − 04	19.57

Table 3   Evolution of the pattern for rhodopsin-like G protein-coupled receptors (GPCRs) since February 1991

Vsn., date	Pattern PS00237	FP	FN
6.1, Feb. 91	[LMRK]-[RKHQNSL]-x(3)-[NTHR]-[LIVMFYW](2)-[LIVM]-x-[SNH]-[LIV]-x(3)-[DEG]-[LIVMFYWA]	2	4
7.1, Jun. 91	[GSTALIVMC]-[GSTA]-[GSTALIVMF]-x(2)-[LIVMN]-x(2)-[LIVMFT]-[GSTAN]-[LIVMFYWAS]-[DEN]-**R**-[FYWCH]-x(2)-[LIVM]	0	2
8.0, Dec. 91	[GSTALIVMC]-[GSTAPDE]-{EDPKRH}-x(2)-[LIVMNG]-x(2)-[LIVMFT]-[GSTAN]-[LIVMFYWAS]-[DEN]-**R**-[FYWCH]-x(2)-[LIVM]	1	0
12.0, Jun. 94	[GSTALIVMYWC]-[GSTANCPDE]-{EDPKRH}-x(2)-[LIVMNGA]-x(2)-[LIVMFT]-[GSTANC]-[LIVMFYWSTAC]-[DENH]-**R**-[FYWCSH]-x(2)-[LIVM]	20	8
15.0, Jul. 98	[GSTALIVMFYWC]-[GSTANCPDE]-{EDPKRH}-x(2)-[LIVMNQGA]-x(2)-[LIVMFT]-[GSTANC]-[LIVMFYWSTAC]-[DENH]-**R**-[FYWCSH]-x(2)-[LIVM]	44	66
19.0, Apr. 05	[GSTALIVMFYWC]-[GSTANCPDE]-{EDPKRH}-x-{PQ}-[LIVMNQGA]-x(2)-[LIVMFT]-[GSTANC]-[LIVMFYWSTAC]-[DENH]-**R**-[FYWCSH]-{PE}-x-[LIVM]	76	282
20.0, Nov. 06	[GSTALIVMFYWC]-[GSTANCPDE]-{EDPKRH}-x-{PQ}-[LIVMNQGA]-{RK}-{RK}-[LIVMFT]-[GSTANC]-[LIVMFYWSTAC]-[DENH]-**R**-[FYWCSH]-{PE}-x-[LIVM]	74	387
2017_08 Aug, 17	[GSTALIVMFYWC]-[GSTANCPDE]-{EDPKRH}-x-{PQ}-[LIVMNQGA]-{RK}-{RK}-[LIVMFT]-[GSTANC]-[LIVMFYWSTAC]-[DENH]-**R**-[FYWCSH]-{PE}-x-[LIVM]	153	454

Note: To keep pace with the growth of Swiss-Prot, successive updates aimed to maximise the number of true-positive matches, and minimise the number of false-positives (FP) and missed true sequences (FN). Inevitably, attempts to accommodate the family's sequence diversity led the pattern to become less selective over time. The Arg residue, implicated in G protein coupling, is highlighted.

that matching a consensus expression is a binary event: i.e., a sequence either matches exactly, or not at all. Diagnostically, this approach is very brittle. Efforts to capture highly divergent sequence groups in this way tend to result in patterns that are either too permissive (i.e., make significant numbers of erroneous (false-positive) matches), or too selective (i.e., fail to match significant numbers of genuine family members – false negatives). Attempting to address this problem, two parallel initiatives emerged, complementing some of the diagnostic limitations of patterns and rules with more sophisticated pattern-recognition techniques: these were 'profiles' (Bairoch and Bucher, 1994) and protein 'fingerprints' (Attwood and Beck, 1994).

PROSITE's Profile Library

Profiles (or position-specific weight matrices) are tables of amino acid weights and gap penalties used to compute a similarity score, relative to a pre-defined cut-off value, when aligned with a protein sequence. They were introduced into PROSITE 12.0, in June 1994. The data structure used was a generalisation of that originally described by Gribskov and colleagues (Gribskov *et al.*, 1987, 1990), and modified by Lüthy and co-workers (Lüthy *et al.*, 1994).

Like patterns, profiles are built from sequence alignments; however, they are more tolerant of amino acid changes and of sequence length differences arising from insertions or deletions. In principle, this allows relationships between sequences to be modelled more 'realistically'. One advantage of this is that, by contrast with the specificity of patterns (which focus on short, highly similar regions), profiles can be used to characterise the full length of sequence alignments, whether they encode family-specific groups or family-agnostic domains. Consequently, by design, profiles can encapsulate both conserved and divergent regions. This gives rise to the possibility of unrelated sequences achieving significant scores with partially incorrect alignments. This situation is mitigated to some extent by the requirement that, in order to be accepted as a true match, a sequence must align correctly with residues known to have particular structural or functional properties (metal-ion binding, catalysis, protein-protein interaction, etc.) as verified by experimental data (Bairoch and Bucher, 1994). Part of a profile is shown in **Fig. 1**, illustrating the complexity of the scoring system relative to the simplicity of the consensus expressions in **Table 1**.

Broadly speaking, a profile is a scoring 'matrix' (the elements of which are denoted MA in **Fig. 1**) containing an alternating sequence of position-specific scores. When a sequence is aligned with a profile, its overall similarity score is derived by summing the scorable components, or 'states': specifically, the 'beginning' (B), 'match' (M), 'insert' (I), 'delete' (D) and 'end' (E) states. In addition, so-called 'state-transition' steps between consecutive positions are also scored: 16 state-transition scores, including transitions between identical states, are defined for all possible transitions between the set of states [B,M,I,D] and [M,I,D,E]: BI, BD, MI, MD, IM, etc. (state-transition scores are analogous to gap-opening penalties used in alignment algorithms).

```
ID   G_PROTEIN_RECEP_F1_2; MATRIX.
AC   PS50262;
DT   01-DEC-2001 CREATED; 01-OCT-2013 DATA UPDATE; 30-AUG-2017 INFO UPDATE.
DE   G-protein coupled receptors family 1 profile.
MA   /GENERAL_SPEC: ALPHABET='ABCDEFGHIKLMNPQRSTVWYZ'; LENGTH=259;
MA   /DISJOINT: DEFINITION=PROTECT; N1=6; N2=254;
MA   /NORMALIZATION: MODE=1; FUNCTION=LINEAR; R1=1.9359; R2=0.02006056; TEXT='NScore';
MA   /CUT_OFF: LEVEL=0; SCORE=328; N_SCORE=8.5; MODE=1; TEXT='!';
MA   /CUT_OFF: LEVEL=-1; SCORE=228; N_SCORE=6.5; MODE=1; TEXT='?';
MA   /DEFAULT: D=-20; I=-20; B1=-100; E1=-100; MI=-105; MD=-105; IM=-105; DM=-105; MM=1; M0=-10;
MA   /I: B1=0; BI=-105; BD=-105;
MA   /M: SY='G'; M=1,-11,-24,-13,-15,-19,30,-19,-20,-18,-15,-11,-5,-19,-16,-18,-2,-11,-15,-22,-20,-16;
MA   /M: SY='N'; M=-9,33,-19,15,-2,-18,-2,8,-18,-2,-26,-18,51,-20,-1,-2,9,1,-26,-38,-17,-2;
MA   /M: SY='I'; M=-1,-21,-16,-26,-21,0,-16,-22,10,-22,8,4,-17,-22,-19,-20,-8,-3,10,-21,-7,-20;
MA   /M: SY='L'; M=-6,-24,-17,-28,-21,8,-25,-20,14,-24,23,12,-21,-25,-20,-19,-17,-5,11,-17,0,-20;
MA   /M: SY='V'; M=-1,-19,-16,-23,-21,-2,-24,-14,15,-18,6,7,-16,-24,-18,-17,-7,-1,21,-26,-5,-20;
MA   /M: SY='I'; M=-8,-28,-16,-33,-25,6,-31,-24,26,-26,24,15,-24,-26,-21,-23,-20,-8,20,-20,-1,-24;
MA   /M: SY='I'; M=-6,-23,-19,-27,-21,5,-24,-16,8,-19,8,5,-20,-23,-17,-17,-15,-6,5,-2,7,-19;
MA   /M: SY='V'; M=4,-22,-14,-26,-21,-5,-23,-23,16,-18,7,6,-19,-22,-18,-19,-7,-1,20,-24,-8,-21;
MA   /M: SY='I'; M=-8,-25,-19,-30,-24,14,-29,-20,19,-24,19,11,-21,-25,-21,-20,-17,-4,16,-16,5,-23;
MA   /M: SY='F'; M=-3,-18,-12,-23,-18,4,-20,-16,2,-17,3,1,-14,-22,-16,-14,-7,0,3,-16,0,-17;
MA   /M: SY='R'; M=-9,-10,-22,-12,-6,-10,-18,-7,-15,7,-11,-5,-5,-17,-1,14,-6,-4,-11,-18,-5,-5;
MA   /M: SY='K'; M=-8,1,-21,-1,0,-14,-16,-1,-18,4,-17,-10,3,-12,-1,3,-1,-1,-15,-23,-5,-1;
MA   /M: SY='R'; M=-8,-7,-25,-7,0,-19,-16,-6,-19,11,-18,-7,-3,-5,2,13,-3,-4,-15,-23,-10,0;
MA   /M: SY='R'; M=-8,-7,-21,-9,-3,-16,-16,-4,-14,7,-12,-1,-3,-14,3,11,-4,-4,-11,-23,-8,-1;
MA   /I: I=-4; MD=-23;
MA   /M: SY='L'; M=-7,-17,-20,-19,-11,-2,-20,-11,1,-8,11,8,-14,-19,-8,-1,-14,-7,0,-18,-3,-10; D=-4;
MA   /I: I=-4; MI=0; MD=-23; IM=0; DM=-23;
MA   /M: SY='R'; M=-11,-5,-24,-7,-1,-18,-17,7,-20,8,-16,-6,1,-15,6,17,-5,-6,-17,-22,-6,0;
MA   /M: SY='T'; M=-2,3,-16,-3,-3,-17,-11,-7,-17,-1,-20,-13,9,-13,-1,0,15,16,-11,-31,-13,-2;
.
.
.
MA   /M: SY='P'; M=-9,-21,-32,-16,-7,-13,-21,-20,-14,-14,-20,-14,-20,57,-14,-20,-11,-8,-20,-22,-19,-14;
MA   /M: SY='I'; M=-10,-29,-22,-34,-27,18,-31,-24,23,-26,18,11,-24,-25,-22,-20,-9,17,-7,-26;
MA   /M: SY='I'; M=-7,-28,-20,-34,-27,4,-33,-26,32,-27,22,15,-23,-24,-22,-24,-19,-7,25,-21,-2,-26;
MA   /M: SY='Y'; M=-17,-20,-21,-22,-21,27,-29,10,-1,-13,1,0,-18,-29,-14,-12,-18,-8,-7,18,59,-20;
MA   /I: E1=0;
NR   /RELEASE=2017_08,555426;
NR   /TOTAL=2510(2508); /POSITIVE=2456(2454); /UNKNOWN=41(41);
NR   /FALSE_POS=13(13); /FALSE_NEG=12; /PARTIAL=3;
CC   /MATRIX_TYPE=protein_domain;
CC   /SCALING_DD=reversed;
CC   /AUTHOR=K_Hofmann;
CC   /TAXO-RANGE=??E?V; /MAX-REPEAT=2;
CC   /VERSION=3;
DR   O42385    , 5H1AA_TAKRU, T; O42384    , 5H1AB_TAKRU, T;
DR   Q6XXX9    , 5HT1A_CANLF, T; Q9N297    , 5HT1A_GORGO, T;
DR   Q0EAB6    , 5HT1A_HORSE, T; P08908    , 5HT1A_HUMAN, T;
DR   Q64264    , 5HT1A_MOUSE, T; Q9N298    , 5HT1A_PANTR, T;
DR   Q9N296    , 5HT1A_PONPY, T; P19327    , 5HT1A_RAT  , T;
.
.
.
DR   P59528    , TR123_MOUSE, ?; Q7M710    , TR125_MOUSE, ?;
DR   Q67ET4    , TR125_RAT  , ?; Q67ES0    , TR140_RAT  , ?;
DR   P16751    , UL78_HCMVA , ?;
DR   A1WHP5    , COXX_VEREI , F; Q6UUW6    , CRLC_DICDI , F;
DR   Q54QV5    , CRLD_DICDI , F; Q54GG1    , CRLF_DICDI , F;
DR   Q5UAW9    , GP157_HUMAN, F; Q8C206    , GP157_MOUSE, F;
DR   Q5FVG1    , GP157_RAT  , F; Q9VRN2    , MTH2_DROME , F;
DR   Q09344    , SRABE_CAEEL, F; Q19992    , SRD1_CAEEL , F;
DR   O01609    , SRD33_CAEEL, F; O17240    , SRD3_CAEEL , F;
DR   P46564    , SRV1_CAEEL , F;
3D   1F88; 1GZM; 1HLL; 1HOF; 1HZX; 1JFP; 1L9H; 1LN6; 1U19; 2G87; 2HPY; 2I35;
3D   2I36; 2I37; 2J4Y; 2KS9; 2KSA; 2KSB; 2LNL; 2PED; 2R4R; 2R4S; 2RH1; 2VT4;
3D   2X72; 2Y00; 2Y01; 2Y02; 2Y03; 2Y04; 2YCW; 2YCX; 2YCY; 2YCZ; 2YDO; 2YDV;
.
.
.
3D   4Z34; 4Z35; 4Z36; 4ZJ8; 4ZJC; 4ZUD; 4ZWJ; 5A8E; 5C1M; 5CXV; 5D5A; 5D5B;
3D   5D6L; 5DGY; 5DHG; 5DHH; 5DSG; 5DYS; 5EN0; 5F8U; 5G53; 5GLH; 5GLI; 5IU4;
3D   5IU7; 5IU8; 5IUA; 5IUB; 5JQH; 5K2A; 5K2B; 5K2C; 5K2D;
PR   PRU00521;
DO   PDOC00210;
//
```

Fig. 1 Example profile showing position-specific and state-transition scores for rhodopsin-like GPCRs. The 'ALPHABET' and 'LENGTH' parameters indicate that the profile spans a 259-residue portion of the protein. All true-positive (T) and partial (P) matches are listed in the database cross-references (DR); links to structures in the PDB (Rose *et al.*, 2017) are made in the 3D lines. The profile's diagnostic performance is shown in the numerical results (NR).

Fig. 1 shows excerpts from the PROSITE profile for rhodopsin-like GPCRs (G_PROTEIN_ RECEP_F1_2, PS50262). Of particular interest is the profile's diagnostic power, encoded in the NR lines. These reveal that the profile statistics were derived from a search of UniProtKB/Swiss-Prot 2017_08, which contained 555,426 sequences. The profile made 2510 matches in 2508 sequences (hence, some matches were repeats); there were 13 false-positive matches and 12 false negatives; three partial matches were also made. With 99.47% precision and 99.51% recall, the diagnostic performance of this profile is near perfect.

Alongside this profile, PROSITE also contains a 17-residue pattern for the family, as seen in **Table 3** (G_PROTEIN_ RECEP_F1_1, PS00237). Against the same version of UniProtKB/Swiss-Prot, the pattern made 2142 matches in 2138 sequences: 153 of these were false positive; and 31 were partial matches. In addition, 454 sequences were noted as false negatives. Thus, with 92.86% precision and 81.42% recall, the pattern doesn't match the diagnostic performance of the profile. In general, where both patterns and profiles are used to describe a particular protein family or domain, the profile generally has the greater diagnostic power.

As profiles continued to outperform patterns, a growing profile library began to augment PROSITE: by August 2012, the number of profiles exceeded one thousand (Sigrist *et al.*, 2013); by August 2017, they comprised almost half the database. However, creating patterns and profiles, regularly assessing and refining their diagnostic behaviour in response to the changing size and composition of Swiss-Prot, and maintaining their annotation accordingly, is extremely labour intensive. Thus, while the philosophy of providing comprehensive manual annotation for every database entry adds significant value to the database, and still contributes to its popularity, the time-consuming nature of the task is one of the hardest challenges for curators, and has placed an inevitable brake on its growth. By August 2017, PROSITE provided descriptions of 1788 sites, domains and families, diagnosed by 2500 patterns and profiles.

PRINTS

The second pattern-recognition technique devised in response to the diagnostic limitations of some of the original PROSITE patterns was 'fingerprinting' (Attwood *et al.*, 1994). An early idea behind the development of PROSITE was that a pattern, defining a single conserved region, or motif, within a sequence alignment was sufficient to be able to diagnose a complete protein family. However, most alignments contain several motifs, often separated by regions that are poorly conserved and/or gapped. The concept of fingerprints thus stems from the observation that groups of motifs, and their unique inter-relationships, can together create distinctive signatures for particular protein families or domains. From a diagnostic perspective, this offers greater flexibility than patterns, as it affords the possibility to tolerate mis-matches at the level of individual motifs, without compromising the discriminating power of the fingerprint as a whole, as shown in **Fig. 2**.

The first fingerprint was created in an attempt to capture all known rhodopsin-like GPCRs in Swiss-Prot (Attwood and Findlay, 1993, 1994). The distinguishing trait of GPCRs is the presence of seven conserved transmembrane (TM) domains; the fingerprint therefore encoded each of these motifs. The resulting discriminator proved to be more robust than the existing PROSITE pattern for this family, which was originally based on a conserved region within TM domain 2 (see **Table 3**). This led the pattern to be

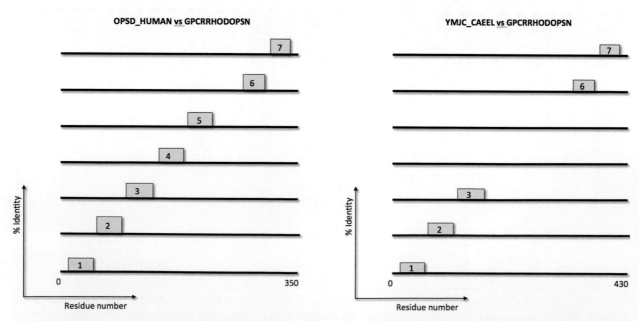

Fig. 2 Complete (human rhodopsin, OPSD_HUMAN) and partial (*C. elegans* putative GPCR, YMJC_CAEEL) matches with the rhodopsin-like GPCR fingerprint. The *C. elegans* protein has a clear relationship with this GPCR family, but lacks significant matches with the fourth and fifth TM domains.

revised (Attwood and Findlay, 1993, 1994), based instead on TM domain 3: this is more tightly conserved, and includes the G protein-interaction site at the N-terminal extremity of the second cytoplasmic loop; the revised pattern was also later complemented by a more potent profile.

In light of the success of the GPCR fingerprint, the technique was used to characterise a range of other protein families, several of which weren't yet available in PROSITE. The results were released in October 1991 in a new resource, initially known as the Features Database (Akrigg et al., 1992). Inspired by PROSITE, the Features Database adhered to the underlying philosophy of adding value to its contents through hand-crafted annotation. Given their similarities, and their growing annotation burdens, there were obvious advantages to unifying their formats (Bairoch, 2000). Nevertheless, a proposal to the European Commission to unite their contents, to create the world's first integrated protein family resource, proved unsuccessful (Attwood et al., 2011).

Despite this setback, both databases continued to evolve. The Features Database, which was growing at a steady rate of 200 fingerprints annually, was subsequently re-branded 'PRINTS' (Attwood and Beck, 1994; Attwood et al., 1994). However, without a dedicated team of curators, this growth rate proved unsustainable. The last release of the PRINTS database was made available in 2012, with 2156 fingerprints encoding 12,444 motifs (Attwood et al., 2012), and is still available for online searching.

Blocks

Following the development of PROSITE and PRINTS, similar attempts were made to characterise protein families, but avoiding heavy reliance on manual approaches to create the signatures and annotate their results. Amongst the first of these was a method that, like fingerprints, made use of multiple conserved motifs, or 'blocks', derived from sequence alignments (Henikoff and Henikoff, 1991). In this approach, the most conserved alignment regions were identified automatically using two independent motif-finding algorithms; only motifs identified by both algorithms were accepted in the final family-specific set of blocks, which were then deposited in the Blocks database.

Blocks' algorithmic approach meant that, for the same family of proteins, the identified blocks generally differed from motifs in the 'equivalent' fingerprint. Importantly also, the blocks-based approach didn't explicitly exploit the relationships between motifs, but treated each motif separately; the diagnostic behaviour of blocks and fingerprints designed to characterise the same families was therefore different. To afford users these different diagnostic perspectives, each release of Blocks was complemented with a copy of Blocks-format PRINTS, derived directly from the manually-selected motifs from the PRINTS database.

In the original versions of the Blocks database, blocks were derived from alignments of sequence families identified in PROSITE. However, as PROSITE's growth was curbed by its high level of manual processing, this was also limiting the growth of Blocks. By the late 1990s, information from a range of additional protein families had become available in complementary databases like PRINTS (Attwood and Beck, 1994), ProDom (Sonnhammer and Kahn, 1994), Pfam (Sonnhammer et al., 1997) and DOMO (Gracy and Argos, 1998). A more comprehensive, non-redundant version of the Blocks database was therefore devised – termed Blocks + – which included, in a stepwise fashion, i) all blocks derived from PROSITE, ii) blocks derived from PRINTS not present in PROSITE, iii) blocks derived from Pfam not present in PROSITE or PRINTS, and so on for Prodom and DOMO (Henikoff et al., 1999). Eventually, this multiple-database pipeline was replaced by a more streamlined approach utilising protein families from the integrated database, InterPro (Apweiler et al., 2001).

Even though Blocks releases were not reliant on manual annotation, database maintenance was still laborious, and production ceased in April 2007. The final release contained 29,068 blocks, representing 5900 sequence groups documented in InterPro 14.0. Although no longer maintained, Blocks and Blocks-format PRINTS are still available for searching online.

ProDom

A rather different analytical approach emerged from a study of protein modules. These are analogous to molecular building-blocks, discrete domains that have been combined in different ways during the course of evolution to create a range of multi-functional proteins. An automatic method was developed to recognise such modules and domains, and to cluster them into groups. The algorithm was applied to more than 21,000 full-length sequences in Swiss-Prot 21.0, and the resulting domain families were organised automatically into a database – this was ProDom (Sonnhammer and Kahn, 1994). Multiple alignments were also created for each group, from which consensus sequences were generated to facilitate database searches.

A particular challenge with this approach was that the clusters (and their alignments and consensus sequences) had to be re-calculated for each release of Swiss-Prot. Not only was this onerous, but the resultant domain 'families' were not consistent between computations on successive Swiss-Prot releases. This lack of stability made indexing ProDom's contents problematic, and led to integration issues with early releases of InterPro (Apweiler et al., 2001); moreover, as Swiss-Prot grew and TrEMBL (Bairoch and Apweiler, 1996) became part of the expanding target database (in 1996, TrEMBL 1.0 held ∼ 105,000 sequences, almost twice the size of Swiss-Prot 34.0, with which it was released in parallel), this brought significant maintenance overheads. The last version of ProDom (ProDom 2012.1), based on more than 21 million full-length sequences from the December 2012 release of Uni-ProtKB/Swiss-Prot and /TrEMBL (UniProt Consortium, 2013), included more than 1.7 million domain families containing at least two sequences.

Pfam

During the decade following the release of the first 58 PROSITE patterns, protein family databases proliferated, and database maintenance issues began to take centre stage. The principal tension was between the need to provide users with high-quality, reliable information, and their desire for rapid access to the most comprehensive data. Databases like PROSITE and PRINTS focused on the former, using manual approaches to craft quality alignments and detailed annotation; resources like Blocks and ProDom, on the other hand, made concerted efforts to address the latter, as automation was the only practical way for databases to become more 'complete'. A kind of hybrid approach was introduced with the advent of Pfam (Sonnhammer et al., 1997), a Hidden Markov Model (HMM)-based database that included both annotated and automatically derived components (denoted Pfam A and Pfam B, respectively).

Conceptually, HMMs are similar to profiles, which model sequence alignments as strings of match, insert and delete states. The main difference is that, rather than using absolute scores, HMMs are probabilistic: i.e., match states instead contain amino acid probabilities within a given alignment column; and transition states denote transition probabilities to and from insert and delete states, reflecting the propensity to insert or skip a residue at a particular position. Pfam A (release 1.0) contained 175 of the most widely populated sequence clusters, seeded by family information from sources like PROSITE, ProDom, etc.; Pfam B contained 11,929 clusters, based on alignments constructed from the portion of Swiss-Prot not already represented in Pfam A (Sonnhammer et al., 1997) – for Pfam 1.0, this was 81% of Swiss-Prot 33.0.

Although Pfam emerged almost ten years after PROSITE, the database nevertheless grew rapidly. In large part, this resulted from a shifting emphasis towards automation, as Pfam adapted to cope with the engorgement of TrEMBL (Finn et al., 2016). Annotating the growing numbers of sequence clusters had become the most burdensome task; it was therefore decided to adopt the Wikipedia model, effectively outsourcing annotation to the community. This meant that, from release 25.0 onwards, the notes traditionally assigned to Pfam entries were to be largely replaced by Wikipedia articles. By March 2017, Pfam 29.0 contained 16,712 entries, and Pfam B, its automatically generated supplement, had been removed.

InterPro

The TrEMBL database, containing translations of all coding sequences in the EMBL data library (Emmert et al., 1994), was released in 1996 as a computer-annotated supplement to Swiss-Prot (Bairoch and Apweiler, 1996). This provided rapid access to protein products of genomic data within a familiar Swiss-Prot-like framework, but in a relatively 'raw' state, with little annotation. In 1997, Bairoch, Attwood and Apweiler returned to the earlier discussion of the feasibility of uniting PROSITE and PRINTS. By then, however, the vision was more ambitious: the quest, to develop a resource to annotate TrEMBL entries, to add value to the deluge of uncharacterised genomic data pouring from sequencing projects, using signatures from several protein family resources, including ProDom and the newly announced Pfam database. This time, the case for integration was stronger, and the funding bid for the new project – termed InterPro – was successful. A beta release of InterPro was made in October 1999, characterising 2423 families and domains in Swiss-Prot 38.0 and TrEMBL 11.0: this 'pilot' release combined 1370 entries from PROSITE 16.0; 1157 fingerprints from PRINTS 23.1; 241 preliminary profiles from the Profile library; and 1465 entries from Pfam 4.0. It had thus taken seven years finally to realise the goal of an integrated protein family database (Apweiler et al., 2001).

Although ProDom was part of the original InterPro proposal, it was nevertheless excluded from initial releases because the nature of its clusters fluctuated between different versions of Swiss-Prot. Before it could be meaningfully integrated, some means of assigning stable accession numbers to ProDom entries therefore had to be found. Another important factor

Table 4   Composition of InterPro release 64.0, July 2017

Partner database	Version	Entries	# Integrated	% Integrated
Pfam*	31.0	16,712	16,119	96
PANTHER*	11.1	91,538	6,224	7
TIGRFAMs*	15.0	4,488	4,448	99
PIRSF*	3.02	3,285	3,201	97
CDD	3.16	12,805	2,165	17
HAMAP	201,701.18	2,160	2,160	100
PRINTS	42.0	2,106	1,980	94
SUPERFAMILY*	1.75	2,019	1,479	73
ProDom	2006.1	1,894	1,308	69
PROSITE patterns	20.132	1,309	1,298	99
SMART*	7.1	1,312	1,264	96
CATH-Gene3D*	4.1.0	2,737	1,235	45
PROSITE profiles	20.132	1,174	1,142	97
SFLD*	2	480	146	30

Note: Asterisks denote partner databases that contribute HMM-based signatures to the resource.

tempered the early development of InterPro (and remains a challenge today): although conceptually straightforward, amalgamating the contents of PROSITE, PRINTS and Pfam was a major undertaking. The databases use different analysis methods to identify sequence groups; importantly, they also use very different terminologies to describe them - at the time, there was no common definition of what constituted a 'family', a 'superfamily', a 'domain family', and so on. Thus, trying sensibly to merge apparently equivalent database entries, whose terminologies and whose sets of identified sequences were different, was extremely difficult. To facilitate the integration process, initial work therefore focused on partner databases that offered some level of annotation. Hence, ProDom didn't appear in InterPro until release 1.2 in June 2000, to which it contributed 540 entries.

During the next decade, additional resources were considered as potential InterPro partners (Blocks was amongst these; however, by 2001 it was taking source data directly from InterPro, so its inclusion in InterPro would have added a kind of circularity), resulting in seven more databases joining the InterPro consortium (Hunter *et al.*, 2009). By 2017, with 14

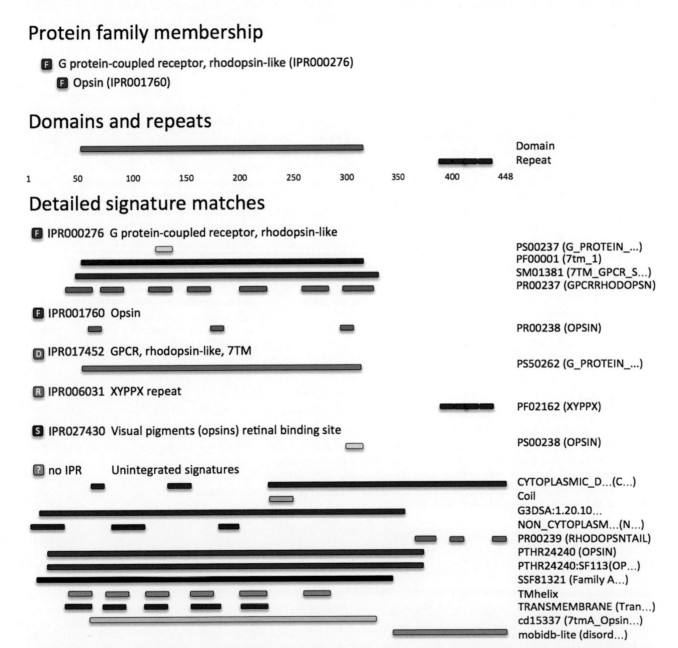

Fig. 3 Output from a search of InterPro with the sequence of rhodopsin from the Japanese flying squid (P31356, OPSD_TODPA). Matches to signatures within each partner database are colour-coded: PROSITE (patterns), yellow, (profiles) orange; PRINTS, green; Pfam, blue; PANTHER, brown; Gene3D, purple; SUPERFAMILY, black. The 'anatomy' of the sequence is revealed in terms of the overarching superfamily and specific family to which it belongs, the functional sites (G protein-coupling, retinal-binding) and C-terminal repeats it contains, the location of its 7TM-domain and the positions of its constituent TM regions.

contributing data sources, the annotation challenges had become immense, and now lend InterPro a level a complexity that was inconceivable 25 years ago, when integrating protein family databases was first discussed. **Table 4** gives a hint of this complexity: in it are listed the versions of the partner databases contributing to the July 2017 release, their sizes and the numbers of their entries included in the resource.

While most of the contents of each partner database are incorporated, aberrant signatures (i.e., those that match unrelated sequences outside their intended scope), or differences in and between them, cause integration of some entries to be deferred. Other concerns include i) how to reconcile sequence-based families with structure-based families (e.g., such as defined in CATH-Gene3D (Dawson *et al.*, 2017) and SUPERFAMILY (Oates *et al.*, 2015)), whose conceptual models of protein families differ considerably; and ii) how to manage and compare the contents of partner databases that contain vast amounts of (largely automatically generated) data relative to others (e.g., PANTHER (Mi *et al.*, 2017)), especially in cases where their signature collections change substantially from release to release.

The data complexity and semantic issues outlined above make efficient processing very difficult, and annotating the results a continual headache. In consequence, note the discrepancies between columns 3 and 4 of **Table 4** – while the integration level of the contents of some partner databases is 90% or more, for others it is less than 50%. Notwithstanding the challenges, with 30,876 entries in release 64.0, InterPro is the largest integrated protein family database, offering the most comprehensive opportunity for protein family characterisation in the world (Finn *et al.*, 2017). Helping to fulfil its original rationale, more than 80% of TrEMBL entries can now achieve at least some level of annotation using the resource.

InterPro's considerable diagnostic power, with its multiple-database perspective, is illustrated in **Fig. 3**, which shows the result of searching the database with the sequence of rhodopsin from the Japanese flying squid. The sequence is diagnosed as belonging to the rhodopsin-like GPCR superfamily, being specifically a member of the opsin family; it contains a G protein-coupling motif in TM region 3, and a retinal-binding site in TM region 7; the location of its 7TM-domain is identified, as are the positions of each of the constituent TM regions; and a repetitive region is confirmed to reside C-terminally to the 7TM-domain. No other single database can dissect, in such an extensive, fine-grained way, the 'anatomy' of a protein sequence.

As InterPro has evolved, its composition has changed markedly. Early releases provided a powerful balance of the diagnostic opportunities afforded by patterns, profiles, fingerprints and HMMs, yielding a resource whose diagnostic power was greater than the sum of its parts. Over the years, this balance has, to a large extent, been diluted as more HMM-based databases joined the consortium. **Table 4** reveals that the bulk of the contributions to InterPro are now HMM-based (Pfam, PANTHER, TIGRFAMs, PIRSF, etc.); a significant challenge for maintaining InterPro's health in future will therefore be to retain the broad scope and resilience of its original diagnostic diversity. Perhaps a more pressing and invidious threat concerns the financial viability of InterPro's partner resources. As mentioned earlier, several family databases have struggled to sustain regular release schedules; some have ceased production. Many of these resources began effectively as cottage industries, the 'pet' projects of enthusiastic individuals, shared publicly for the benefit of the community. However, without adequate funds, databases cannot survive. It is particularly ironic, then, that the value and importance of InterPro was recently recognised when it was designated an ELIXIR 'core resource'. InterPro's uniqueness derives from the contributions of its partner databases, yet many of these are neither part of ELIXIR, nor considered sufficiently financially viable to be eligible as core resources, presenting a kind of conundrum for InterPro's continuity (and a problem for its future diagnostic finesse).

Closing Remarks

Protein family databases have come a long way since Doolittle's suggestion, more than three decades ago, that sequence patterns could characterise the potential functions of unannotated sequences, and Bairoch created the first protein family database based on such patterns. Today, there are numerous publicly available databases that offer insights into protein functions, structures and family relationships, many more than have been integrated into InterPro. Nevertheless, InterPro stands apart, a union of many different partners, whose different philosophical and methodological approaches have been melded into a uniquely powerful whole, one that now provides the backbone of annotation for newly determined sequences entering TrEMBL, and one that continues to provide a versatile, high-quality diagnostic tool for researchers around the world.

Acknowledgment

We are extremely grateful to Amos Bairoch for the time he devoted to excavating his personal archives, retrieving 'ancient' pre-Web files and emails to verify and augment some of the PROSITE-related information presented here.

See also: *Ab initio* Protein Structure Prediction. Algorithms for Strings and Sequences: Searching Motifs. Algorithms for Structure Comparison and Analysis: Docking. Algorithms for Structure Comparison and Analysis: Homology Modelling of Proteins. Algorithms for Structure Comparison and Analysis: Prediction of Tertiary Structures of Proteins. Bioinformatics Data Models, Representation and Storage. Biological Database Searching. Data Storage and Representation. Functional Genomics. Hidden Markov Models. Hidden Markov Models. Homologous Protein Detection. Identification of Homologs. Identification of Sequence Patterns, Motifs and Domains. Information Retrieval in Life Sciences.

Information Retrieval in Life Sciences. Natural Language Processing Approaches in Bioinformatics. Protein Functional Annotation. Protein Structural Bioinformatics: An Overview. Protein Structure Classification. Standards and Models for Biological Data: FGED and HUPO. Supervised Learning: Classification. Transmembrane Domain Prediction. Transmembrane Domain Prediction

References

Akrigg, D.A., Attwood, T.K., Bleasby, A.J., *et al.*, 1992. SERPENT – An information storage and analysis resource for protein sequences. CABIOS 8 (3), 295–296.

Altschul, S.F., Gish, W., Miller, W., Myers, E.W., Lipman, D.J., 1990. Basic local alignment search tool. J. Mol. Biol. 215 (3), 403–410.

Apweiler, R., Attwood, T.K., Bairoch, A., *et al.*, 2001. The InterPro database, an integrated documentation resource for protein families, domains and functional sites. Nucleic Acids Res. 29 (1), 37–40.

Attwood, T.K., Beck, M.E., 1994. PRINTS – A protein motif fingerprint database. Protein Eng. 7 (7), 841–848.

Attwood, T.K., Beck, M.E., Bleasby, A.J., Parry-Smith, D.J., 1994. PRINTS – A database of protein motif fingerprints. Nucleic Acids Res. 22 (17), 3590–3596.

Attwood, T.K., Coletta, A., Muirhead, G., *et al.*, 2012. The PRINTS database: A fine-grained protein sequence annotation and analysis resource – Its status in 2012. Database. doi:10.1093/database/base019.

Attwood, T.K., Findlay, J.B.C., 1993. Design of a discriminating fingerprint for G-protein-coupled receptors. Protein Eng. 6 (2), 167–176.

Attwood, T.K., Findlay, J.B.C., 1994. Fingerprinting G-protein-coupled receptors. Protein Eng. 7 (2), 195–203.

Attwood, T.K., Gisel, A., Eriksson, N.-E., Bongcam-Rudloff, E., 2011. Concepts, historical milestones and the central place of bioinformatics in modern biology: A European perspective. In: Mahdavi, M.A. (Ed.), Bioinformatics – Trends and Methodologies. Intech Online Publishers. ISBN 978-953-307-282-1.

Bairoch, A., 1991. PROSITE: A dictionary of sites and patterns in proteins. Nucleic Acids Res. 19, 2241–2244.

Bairoch, A., 2000. Serendipity in bioinformatics, the tribulations of a Swiss bioinformatician through exciting times!. Bioinformatics 16 (1), 48–64.

Bairoch, A., Apweiler, R., 1996. The SWISS-PROT protein sequence data bank and its new supplement TREMBL. Nucleic Acids Res. 24 (1), 21–25.

Bairoch, A., Boeckmann, B., 1991. The SWISS-PROT protein sequence data bank. Nucleic Acids Res. 19, 2247–2249.

Bairoch, A., Bucher, P., 1994. PROSITE: Recent developments. Nucleic Acids Res. 22 (17), 3583–3589.

Dawson, N.L., Lewis, T.E., Das, S., *et al.*, 2017. CATH: An expanded resource to predict protein function through structure and sequence. Nucleic Acids Res. 45 (D1), D289–D295. doi:10.1093/nar/gkw1098.

Doolittle, R.F., 1986. Of Urfs and Orfs: A Primer on How to Analyze Derived Amino Acid Sequences. Mill Valley, CA: University Science Books.

Emmert, D.B., Stoehr, P.J., Stoesser, G., Cameron, G.N., 1994. The European Bioinformatics Institute (EBI) databases. Nucleic Acids Res. 22 (17), 3445–3449.

Finn, R.D., Attwood, T.K., Babbitt, P.C., *et al.*, 2017. InterPro in 2017 – Beyond protein family and domain annotations. Nucleic Acids Res. 45 (D1), D190–D199. doi:10.1093/nar/gkw1107.

Finn, R.D., Coggill, P., Eberhardt, R.Y., *et al.*, 2016. The Pfam protein families database: Towards a more sustainable future. Nucleic Acids Res. 44 (D1), D279–D285.

Gattiker, A., Gasteiger, E., Bairoch, A., 2002. ScanProsite: A reference implementation of a PROSITE scanning tool. Appl. Bioinform. 1 (2), 107–108.

Gracy, J., Argos, P., 1998. DOMO: A new database of aligned protein domains. Trends Biochem. Sci. 23 (12), 495–497.

Gribskov, M., Luethy, R., Eisenberg, D., 1990. Profile analysis. Method Enzymol. 183, 146–159.

Gribskov, M., McLachlan, A.D., Eisenberg, D., 1987. Profile analysis: Detection of distantly related proteins. Proc. Natl. Acad. Sci. USA 84 (13), 4355–4358.

Henikoff, S., Henikoff, J.G., 1991. Automated assembly of protein blocks for database searching. Nucleic Acids Res. 19 (23), 6565–6572.

Henikoff, S., Henikoff, J.G., Pietrokovski, S., 1999. Blocks + : A non-redundant database or protein alignment blocks derived from multiple compilations. Bioinformatics 15 (6), 471–479.

Hunter, S., Apweiler, R., Attwood, T.K., *et al.*, 2009. InterPro: The integrative protein signature database. Nucleic Acids Res. 37 (Database Issue), D211–D215.

Keeling, P.J., Burki, F., Wilcox, H.M., *et al.*, 2014. The Marine Microbial Eukaryote Transcriptome Sequencing Project (MMETSP): Illuminating the functional diversity of eukaryotic life in the oceans through transcriptome sequencing. PLOS Biology 12 (6), e1001889. Available at: https://doi.org/10.1371/journal.pbio.1001889

Lüthy, R., Xenarios, I., Bucher, P., 1994. Improving the sensitivity of the sequence profile method. Protein Sci. 3 (1), 139–146.

Mi, H., Huang, X., Muruganujan, A., *et al.*, 2017. PANTHER version 11: Expanded annotation data from Gene Ontology and Reactome pathways, and data analysis tool enhancements. Nucleic Acids Res. 45 (D1), D183–D189. doi:10.1093/nar/gkw1138.

Mitchell, A.L., Scheremetjew, M., Denise, H., *et al.*, 2017. EBI Metagenomics in 2017: Enriching the analysis of microbial communities, from sequence reads to assemblies. Nucleic Acids Res. doi:10.1093/nar/gkx967.

Oates, M.E., Stahlhacke, J., Vavoulis, D.V., *et al.*, 2015. The SUPERFAMILY 1.75 database in 2014: A doubling of data. Nucleic Acids Res. 43 (D1), D227–D233. doi:10.1093/nar/gku1041.

Rose, P.W., Prlić, A., Altunkaya, A., *et al.*, 2017. The RSCB protein data bank: Integrative view of protein, gene and 3D structural information. Nucleic Acids Res. 45 (D1), D158–D169. doi:10.1093/nar/gkw1099.

Sigrist, C.J.A., de Castro, E., Cerutti, L., *et al.*, 2013. New and continuing developments at PROSITE. Nucleic Acids Res. 41 (D1), D344–D347.

Sonnhammer, E.L., Eddy, S.R., Durbin, R., 1997. Pfam: A comprehensive database of protein domain families based on seed alignments. Proteins 28 (3), 405–420.

Sonnhammer, E.L., Kahn, D., 1994. Modular arrangement of proteins as inferred from analysis of homology. Protein Sci. 3 (3), 482–492.

The UniProt Consortium, 2013. Update on activities at the Universal Protein Resource (UniProt). Nucleic Acids Res. 41 (D1), D43–D47. doi:10.1093/nar/gks1068.

The UniProt Consortium, 2017. UniProt: The universal protein knowledgebase. Nucleic Acids Res. 45 (D1), D158–D169. doi:10.1093/nar/gkw1099.

Further Reading

Attwood, T.K., Miller, C.J., 2002. Progress in bioinformatics and the importance of being earnest. Biotechnol. Annu. Rev. 8, 1–54.

Attwood, T.K., Pettifer, S.R., Thorne, D., 2016. Bioinformatics Challenges at the Interface of Biology and Computer Science: Mind the Gap. John Wiley and Sons. ISBN: 978-0-470-03548-1.

Bork, P., Bairoch, A., 1996. Go hunting in sequence databases but watch out for the traps. Trends Genet. 12 (10), 425–427. Available at: https://doi.org/10.1016/0168-9525(96)60040-7.

Faria, D., Schlicker, A., Pesquita, C., *et al.*, 2012. Mining GO annotations for improving annotation consistency. PLOS ONE 7 (7), e40519. https://doi.org/10.1371/journal.pone.0040519.

Higgs, P., Attwood, T.K., 2005. Bioinformatics and Molecular Evolution. Blackwell Publishing. ISBN: 1-4051-0683-2.

Holliday, G.L., Davidson, R., Akiva, E., Babbitt, P.C., 2017. Evaluating functional annotations of enzymes using the Gene Ontology. Methods Mol. Biol. 1446, 111–132.

Ioannidis, J.P.A., 2005. Why most published research findings are false. PLOS Med. 2 (8), e124. Available at: https://doi.org/10.1371/journal.pmed.0020124.

Mushegian, A., 2011. Grand challenges in bioinformatics and computational biology. Front. Genet. 2, 60. doi:10.3389/fgene.2011.00060.

Poux, S., Magrane, M., Arighi, C.N., *et al.*, 2014. Expert curation in UniProtKB: A case study on dealing with conflicting and erroneous data. Database. Available at: https://doi.org/10.1093/database/bau016

Sangrador-Vegas, A., Mitchell, A.L., Chang, H.-Y., Yong, S.-Y., Finn, R.D., 2016. GO annotation in InterPro: Why stability does not indicate accuracy in a sea of changing annotations. Database. Available at: https://doi.org/10.1093/database/baw027

Schnoes, A.M., Brown, S.D., Dodevski, I., Babbitt, P.C., 2009. Annotation error in public databases: Misannotation of molecular function in enzyme superfamilies. PLOS Comput. Biol. 5 (12), e1000605. doi:10.1371/journal.pcbi.1000605.

Smith, T.F., 1990. The history of the genetic sequence databases. Genomics 6 (4), 701–707. Available at: https://doi.org/10.1016/0888-7543(90)90509-S.

Strasser, B.J., 2008. GenBank – Natural history in the 21st century? Science 322 (5901), 537–538. doi:10.1126/science.1163399.

Wheelan, S.J., Boguski, M.S., 1998. Late-night thoughts on the sequence annotation problem. Genome Res. 8 (3), 168–169.

Wong, W.-C., Maurer-Stroh, S., Eisenhaber, F., 2010. More than 1001 problems with protein domain databases: Transmembrane regions, signal peptides and the issue of sequence homology. PLOS Comput. Biol. 6 (7), e1000867. Available at: https://doi.org/10.1371/journal.pcbi.1000867.

Relevant Websites

http://blocks.fhcrc.org/blocks/blocks_search.html
 Blocks.
http://www.cathdb.info
 CATH/Gene3D.
https://elixir-europe.org/
 ELIXIR.
http://www.ebi.ac.uk/interpro/
 InterPro.
http://pfam.xfam.org/
 Pfam.
http://130.88.97.239/PRINTS/
 PRINTS.
http://prodom.prabi.fr/prodom/current/html/home.php
 ProDom.
http://prosite.expasy.org/
 PROSITE.
https://www.embl.de/tara-oceans/
 Tara Oceans.
https://www.uniprot.org/
 UniProt.

Biographical Sketch

Terri is professor of Bioinformatics at the University of Manchester. Her interests in protein sequence analysis led to the creation of various databases (e.g., PRINTS, InterPro) and software tools like Utopia Documents, which uses semantic technologies to integrate research data with scientific articles. She has written several bioinformatics textbooks and reference works. She currently Chairs the GOBLET Foundation, and leads the development of ELIXIR's Training e-Support System, TeSS.

Alex is coordinator for the InterPro and EMBL-EBI Metagenomics databases at the European Bioinformatics Institute. He obtained his DPhil in pharmacology from the University of Oxford, and was previously employed as a molecular biologist at the Institute of Psychiatry in London, before moving to the University of Manchester to work on protein family databases. He joined EMBL-EBI in 2011.

Transmembrane Domain Prediction

Pier Luigi Martelli and Castrense Savojardo, University of Bologna, Bologna, Italy
Giuseppe Profiti, University of Bologna, Bologna, Italy and National Research Council, Italy
Rita Casadio, University of Bologna, Bologna, Italy

Introduction

A fraction ranging from 20% to 30% of all genes in any organism encodes for transmembrane (TM) proteins (Krogh *et al.*, 2001). This class is involved in a wide range of functions, enabling almost all communications of matter, energy and information across the biological membranes enclosing cells and organelles. Indeed, only hydrophobic and small molecules can diffuse through membranes and cross them without need of proteins. Owing to their crucial role in physiological processes, TM proteins are the privileged target of drugs: in humans, TM proteins represent around 50% of known pharmaceutical targets (Bakheet and Doig, 2009).

TM proteins consist of three regions: the TM region spans the hydrophobic phase of the lipid bilayer and separate two domains (called intra-cellular and extra-cellular domains, in the case of proteins inserted in the plasma membrane) that are in contact with the aqueous solvent. In multi-pass TM proteins (having more than one membrane-spanning segment), the three regions comprise discontinuous protein segments that fold to form structural domains responsible of protein functions and interaction with other proteins and molecules. Specifically, TM domains can serve, for example, as anchors, channels (gated or not, nonspecific or selective), signal transducers, modifiers of physical and chemical properties of the membrane.

Two major structural classes of transmembrane proteins have been discriminated so far: all-α and β-barrel, whose membrane-spanning segments are α-helices or β-strands, respectively (**Fig. 1**). Indeed, to stabilize the interaction between the protein and the membrane, the membrane-spanning segments adopt a regular secondary structure: this allows to avoid contacts among the polar groups of the backbone and the hydrophobic chains of the lipid bilayer.

All-α TM proteins known to date comprise a number of helices ranging from 1 to 24 (as in the case of the voltage-gated sodium channel NavPaS from *Periplaneta americana*, PDB code: 5X0M) and have been found in all membrane types with the exception of the outer membrane of Gram-negative bacteria where only β-barrel TM proteins are present. TM β-barrel domains result from the interaction of different β-strands, ranging from 8 to 60 (as in the case of Type II secretion system GspD from *Escherichia coli*, PDB code: 5WQ7) and often including different protein chains. β-barrel TM proteins are also present together with all-α TM proteins in the outer membrane of mitochondria and chloroplasts.

The crystallization of TM proteins and the possibility to resolve their structure by means of X-ray diffractometry is hampered by the need of mimicking with detergents the tripartite bi-phasic environment where TM proteins are stable. Recently, improvements of cryo-electron microscopy technique, not requiring crystallization, opened a new way to resolve the structure of TM protein (Fernandez-Leiro and Scheres, 2016). However, experimental characterization of TM protein structure is still challenging. Computational methods are required for formulating reliable and testable hypotheses on conformation and functional role of TM proteins, whose covalent structure can be deduced from the translation of genomes sequenced with modern Next Generation Sequencing technologies at ever increasing throughput and decreasing cost.

The major computational problems related to TM domains are:

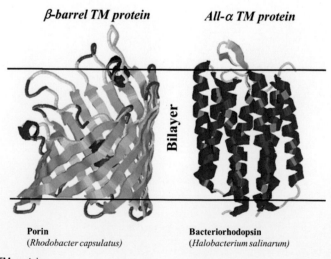

β-barrel TM protein   All-α TM protein

Bilayer

Porin
(*Rhodobacter capsulatus*)

Bacteriorhodopsin
(*Halobacterium salinarum*)

Fig. 1 Two structural classes of TM proteins.

Table 1   TM proteins in UniProtKB (Dec 2017)

Keyword	UniProtKB	With structure	SwissProt	With structure
Transmembrane	20,227,696	4,648	78,045	3,112
Transmembrane α-helix	20,079,319	4,496	76,687	2,990
Transmembrane β-strand	74,199	115	1,027	95

- The determination of the number of membrane-spanning segments and their position along the sequence.
- The determination of the orientation of the membrane protein with respect to the lipid bilayer (topology).
- The determination of the three-dimensional arrangement of the TM spanning segments.
- The determination of the function of TM proteins, including their subcellular localization.

Databases

UniprotKB

Table 1 shows the status of the 2017_12 release of UniProtKB (The UniProt Consortium, 2017): it contains 20,227,696 proteins labelled with the keyword "transmembrane" (KW-0812), out of 102,804,649 total sequences (19.7%). When restricting to the subset containing the manually curated entries of SwissProt, TM proteins represent 78,045 sequences out of 556,388 (14.0%). Only a 3112 SwissProt sequences (corresponding to 4.0%) are endowed with a three-dimensional structure, at least for one domain. The fraction is much lower when considering all UniProt (0.02%). Moreover, in most cases the experimentally resolved structure is partial and does not include the TM domain. For that reason, specialized datasets collecting proteins with structurally resolved TM domain are needed (see next section).

With few exceptions, TM proteins in UniProtKB are also marked with two more labels, identifying the type: "transmembrane helix" (KW-1133) and "transmembrane beta strand" (KW-1134). It is evident, from the figures in **Table 1**, that most of the known TM proteins belong to the all-α class.

Databases of Structurally Resolved TM Domains

The oldest collection of TM domains with well resolved structure is maintained at UC Irvine laboratory of Prof. White (see "Relevant Websites section") and, when excluding monotopic proteins not spanning the membrane, it counts 2372 coordinate files relative to 735 unique proteins (release: Jan 2018). The resource collects and expertly classifies the information deposited at the PDB about the available TM domain structures.

One major problem in the analysis of TM protein structure is that neither the exact position of the phospholipids nor the orientation of the protein with respect to the membrane plane are known. Different procedures have been implemented to infer the interaction between membranes and TM proteins, analyzing the physical-chemical properties of the protein surface as determined from the crystallographic structure. Some of them are listed below.

- PDBTM (see "Relevant Websites section", Kozma *et al.*, 2013) uses a geometrical approach to locate the most likely position of the lipid bilayer: accessible surfaces of hydrophobic (F, G, I, L, M, V, W and Y) and hydrophilic residues (A, C, D, E, H, K, N, P, Q, R, S and T) are computed; the membrane, represented as two parallel planes more than 15 Å apart, is positioned to minimize the exposure of hydrophobic residues with some constraint on the protein backbone geometry. PDBTM (release: Jun 2017) contains data from 3227 PDB chains, 2848 all-α and 366 β-barrel.
- OPM (Orientations of Proteins in Membranes, see "Relevant Websites section", Lomize *et al.*, 2006) adopts an energy based approach in which each protein is modeled as a rigid body with flexible side chains, while the bilayer is represented as a planar hydrophobic slab of adjustable thickness ranging between 21.1 Å and 43.8 Å, depending on the type of biological membrane; transfer energy is calculated at an all-atom level using atomic solvation parameters determined for the water-decadiene system, neglecting explicit electrostatic interactions and contributions of pore facing atoms; most probable protein arrangement is computed by minimizing its transfer energy from water to the membrane. OPM contains (Dec 2017) data from 1783 non-redundant PDB files, 1554 all-α and 229 β-barrel.
- MemProtMD (see "Relevant Websites section", Stansfeld *et al.*, 2015) performs a coarse-grained molecular dynamics simulation: groups of atoms are treated collectively as large particles, allowing to increase timescales and system dimensions; the dynamics of the system comprising protein, membrane lipids and solvent is simulated for 1 µs and the self-assembly of the system is followed. Different lipid species are used in different simulations. MemProtMD currently contains simulations for some 3000 proteins.

Computational Prediction of TM Segments and Topology

Computational characterization of TM protein sequences routinely starts with the prediction of the number of TM segments, of their location along the sequence and their orientation with respect to the lipid bilayer.

Different methods implement different strategies for detecting signals embodied in the sequence in relation to the interaction with the membrane phospholipids and the internal and external polar environments. Local sequence composition is the major source of signal, often complemented with the evolutionary information obtained from multiple sequence alignment of similar proteins. Depending on the TM protein type, the prediction must satisfy different constraints on the number of membrane-spanning segments and their length. In all-α TM proteins the number of membrane-spanning elements is not lower-bounded and their length ranges from 16 to more than 40 residues, depending on their tilt with respect to the membrane plane. Most helices have a clear hydrophobic character, but helices surrounding a pore or in reciprocal interaction routinely show a weaker hydrophobic signal and exhibit a more amphipathic quality. An important topogenic signal is the so-called "positive inside rule" stating that, owing to the unbalance of charge density between the two sides of plasma membranes, the loops facing the cytoplasmic phase (inner loops) are richer in positively charged residues than loops facing the extra-cellular environment (outer loops) (von Heijne, 1989). In eukaryotes, the rule can be applied also to proteins inserted in other membranes, but the definition of inner and outer loops must be redefined for each membrane type (Sharpe *et al.*, 2010).

The generation of the circular arrangement of β-barrels requires at least 8 segments and, so far, the largest resolved barrel consists of 60 strands. In some cases (e.g., TolC, α-hemolysin, Type II secretion system GspD), the functional protein is a multimeric complex in which strands from different subunits interact to form the complete barrel. The minimum number of strands per chain is equal to two. The minimum length of each strand is 6 residues, needed to span the membrane width with an extended conformation. The longest TM β-strand crystallized so far is 24 residue long. Since β-barrels usually delimitate water-filled pores, the strands consist of alternating hydrophobic and hydrophilic residues. Statistical analysis on available structures also reveals that in Gram-negative bacteria, loops exposed towards the extracellular space (outer loops) are, on average, longer than loops facing the periplasmic space (inner loops) (Martelli *et al.*, 2002).

Several confounding factors complicate the prediction of the correct topology. In particular, in the case of all-α membrane proteins the presence of a N-terminal signal peptide with a clear hydrophobic character can lead to the misprediction of an extra helix. Moreover, some all-α TM protein contain partially reentrant segments that span only a portion of the bilayer and strongly challenge the computational characterization of TM segments (Tsirigos *et al.*, 2018).

Scale-Based Methods

Early methods for the localization of TM segments, in particular α-helices, rely on the search of hydrophobic segments along the sequence. More than 20 different scales endowing each residue type with a hydrophobicity value have been proposed to this aim (see "Relevant Websites section"). Among them, the most adopted is the Kyte-Doolittle hydropathy scale (Kyte and Doolittle, 1982) that combines data on water-vapor transfer free energies of the different residues and their exposure distribution in a set of structurally resolved proteins.

Thermodynamic scales have been also generated on the basis of the free energy of transfer between water and hydrophobic liquid (octanol) and between water and phosphocholine interface (White and Wimley, 1999). More recently, a "biological" hydrophobicity scale has been compiled starting from the measurement of the propensity of each residue to be translocated in the endoplasmic reticulum membrane by translocon protein Sec61 (Hessa *et al.*, 2005). This scale has been proved to outperform the others because it probably better casts the physical-chemical constraints deriving from both the protein-membrane interaction and the insertion process. It has been implemented in the SCAMPI2 predictor (Peters *et al.*, 2016).

Hydrophobicity scales are routinely applied to the recognition of TM α-helices by averaging over a window and by searching for hydrophobic stretches at least 15 residue long. The presence of weakly hydrophobic or amphipathic helices strongly limits the performance of scale-based methods. An evolution of the scale-based approach is the first version of MEMSAT (Jones *et al.*, 1994) where 5 different scales (Helix inner end, Helix middle, Helix outer end, Inner loop, Outer loop) are derived from the analysis of 83 all-α TM proteins. During the prediction phase, the five signals computed along the sequence are integrated with a dynamic programming procedure casting the basic grammatical constraints of the topology of all-α TM proteins. Scale-based methods are of scarce relevance for the prediction of β-barrel membrane proteins.

Machine-Learning Based Methods

The best performing methods for the topology prediction of both all-α and β-barrel TM proteins are based on machine-learning approaches such as Neural Networks (NN), Support Vector Machines (SVM), Hidden Markov Models (HMM) and Conditional Random Fields (CRF). Machine learning methods require the availability of a possibly large dataset of proteins with known topology that is used to fix the trainable parameters of the different methods. If the training set is large and diverse enough, the training procedure allows to extract the general rules of the mapping between the sequence and the topology. After the training phase, the parameters can be applied to the prediction of uncharacterized protein sequences. The performance assessment of machine-learning based methods requires a careful validation on protein sets completely independent of the training set. The adoption of machine learning tools allows to analyze more complex input encoding than the simple sequence. In particular, it has been proved that the introduction of evolutionary information extracted from multiple sequence alignments of similar proteins and encoded with sequence profiles largely improves the prediction performance. This has been proved for the prediction of both all-α and β-barrel TM proteins using NNs (Rost *et al.*, 1995; Jacoboni *et al.*, 2001) and HMMs (Martelli *et al.*, 2002; Martelli *et al.*, 2003). While NNs and

SVMs are very efficient in analyzing the local context of a residue, graphical models such as HMMs and Grammatical Restrained Hidden CRFs cast in a more natural way the grammatical constraints that TM proteins must satisfy (schematically depicted in **Fig. 2**). The complementarity among methods makes it possible to improve the predictive performance by adopting ensemble methods (Martelli *et al.*, 2003; Bernsel *et al.*, 2009) or hybrid approaches (Savojardo *et al.*, 2013).

Table 2 lists some of the most adopted methods for the prediction of all-α TM proteins, specifying whether they are able to deal with reentrant regions (RR column) and with the possible presence of N-terminal signal peptides (SP columns).

Table 3 lists some of the most popular methods for the prediction of β-barrel TM proteins.

Three-Dimensional Reconstruction

The prediction of the TM protein topology is of great help for classifying new protein sequences, restricting the set of possible conformations they can assume and functions they can perform. In some cases, prediction of protein topology improves the search of a structural template in the PDB and assists the target-template alignment for comparative modelling, even at low sequence identity. However, due to the scarcity of structural information on TM proteins, methods allowing the template-independent reconstruction of the 3D structure of TM domains are needed. In most cases, these methods start from a topology prediction and predict the most probable contacts among TM-segments. On this basis, optimization procedures can be applied to search for the most favorable arrangement of the TM segments inside the lipid bilayer, after defining a score accounting for inter-segment and segment-membrane interactions.

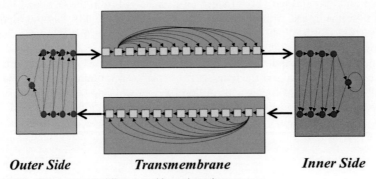

Outer Side   *Transmembrane*   *Inner Side*

Fig. 2   Generic model of TM protein. State represent different positions along the sequence.

Table 2   Available methods for the topology prediction of all-α TM proteins

	Method[a]	Input[b]	SP[c]	RH[d]	Reference/site
ENSEMBLE	HMM + NN	Prof	No	No	Martelli *et al.* (2002) mu2py.biocomp.unibo.it/mempype
HMMTOP	HMM	Seq	No	No	Tusnády and Simon (1998) www.enzim.hu/hmmtop/
MEMSAT-SVM	SVM	Prof	Yes	Yes	Nugent and Jones (2009) http://bioinf.cs.ucl.ac.uk/psipred/
PHDhtm	NN	Prof	No	No	Rost *et al.* (1996) www.predictprotein.org
PHILIUS	DBN	No	Yes	No	Reynolds *et al.* (2008) www.yeastrc.org/philius/
PolyPHOBIUS	HMM	MSA	Yes	No	Käll *et al.* (2005) phobius.sbc.su.se/poly.html
SCAMPI2	Scales	Seq	No	No	Peters *et al.* (2016) scampi.bioinfo.se/
SPOCTOPUS	NN + HMM	Prof	Yes	Yes	Viklund *et al.* (2008) octopus.cbr.su.se
TMHMM	HMM	Seq	No	No	Sonnhammer *et al.* (1998) www.cbs.dtu.dk/services/TMHMM/
TOPCONS	Consensus	Prof	Yes	Yes	Tsirigos *et al.* (2015) topcons.cbr.su.se/

[a]Method: NN = Neural Networks; HMM = Hidden Markov Models; SVM = Support Vector Machines; DBN = Dynamic Bayesian Networks.
[b]Input: Seq = Sequence; Prof = Sequence profile; MSA = Multiple Sequence Alignments.
[c]SP = prediction of Signal peptide.
[d]RR = prediction of reentrant regions.

Table 3 Available methods for the topology prediction of β-barrel TM proteins

	Method[a]	Input[b]	Reference/site
BETAWARE	NN + GRHCRF	Prof	Savojardo *et al.* (2013) betaware.biocomp.unibo.it
BOCTOPUS2	SVM + HMM	Prof	Hayat *et al.* (2016) boctopus.bioinfo.se/
PRED-TMBB	HMM	Seq	Bagos *et al.* (2004) bioinformatics.biol.uoa.gr/PRED-TMBB/
PredβTM	SVM + rules	Seq	Roy Choudhury and Novič (2015) http://transpred.ki.si/
PROF-TMB	HMM	Prof	Bigelow *et al.* (2004) www.predictprotein.org

[a]Method: NN = Neural Networks; HMM = Hidden Markov Models; SVM = Support Vector Machines; GRHCRF = Grammatical Restrained Hidden Conditional Random Fields.
[b]Input: Seq = Sequence; Prof = Sequence profile.

Available methods for contact predictions in all-α TM proteins include:

- MEMPACK (see "Relevant Websites section", Nugent and Jones, 2010), a method to predict lipid exposure, residue-residue contacts, helix-helix interactions and ultimately overall protein helical packing. Individual classifiers for lipid exposure and residue contacts are based on SVMs and have been trained on data derived from the OPM and PDBTM as well as molecular dynamics data extracted from the Coarse Grained Database (Chetwynd *et al.*, 2008) (for lipid exposure prediction).
- TMHcon (see "Relevant Websites section", Fuchs *et al.*, 2009), a method for predicting helix-helix contact in TM proteins based on NNs and different input features, including: evolutionary information, sequence distance and co-evolution measures of residue pairs as well as protein global features such as protein length and number of TM segments.
- FILM3 (see "Relevant Websites section", Nugent and Jones, 2013), a method combining PSICOV (Jones *et al.*, 2012), an advanced residue-residue contact predictor, with TM topology prediction (performed using MEMSAT-SVM) and fragment selection and assembly. The main novelty of FILM3 consists in the adoption of scoring energy functions (used for fragment selection and assembly) entirely based on distance constraint imposed by predicted TM segments and contacts predicted by PSICOV.
- MemBrain (see "Relevant Websites section", Yang and Shen, 2018), a recent approach to predict inter-helix contacts based on deep learning methods.

Functional Characterization

Structural characterization of TM proteins helps in formulating hypotheses about their role in the complexity of biological functions. However, the three-dimensional reconstruction is not always feasible and, in some cases, similar structures are associated to different specific functions or subcellular localizations. For this reason, methods for inferring protein function and localization from sequence have been implemented.

To date, the only prediction method for protein localization specific for TM proteins is MemLoci (Pierleoni *et al.*, 2011a; see "Relevant Websites section") that classifies protein sequences into three subcellular localizations (Plasma membrane, Organelle membranes, Internal membranes) with a SVM. The MemPype system (Pierleoni *et al.*, 2011b; see "Relevant Websites section") integrates MemLoci with similarity-based inference methods and tools for the prediction of topology, presence of signal peptide and GPI-anchor.

General-purpose systems for inferring protein functions can be applied to TM proteins (see also Martelli *et al.*, 2017). The basic idea of these methods is to extrapolate functional information present in databases in order to annotate sequences lacking experimental characterization. For example, the Bologna Annotation Resource (BAR3.0, see "Relevant Websites section") is a prediction tool that exploits large-scale clustering of UniProtKB sequences resulting from the application of strict similarity constraints (more than 40% sequence identity on an alignment coverage larger than 90%; Profiti *et al.*, 2017). Functional characterization of each cluster is derived from proteins carrying functional annotation (in terms of Gene Ontology, GO) in UniProtKB, after an analysis of statistical significance. This procedure allows to spread functional annotation to all the members of the clusters. For example, in the present version, BAR 3.0 contains 2294 protein sequences experimentally associated with the GO term "integral component of membrane" (GO:0016021) or its descendants, and the BAR3.0 structure allows spreading the annotation to 751,272 protein sequences.

See also: Algorithms for Strings and Sequences: Searching Motifs. Cell Modeling and Simulation. Data Mining: Classification and Prediction. Data Mining: Prediction Methods. Functional Genomics. Identification of Sequence Patterns, Motifs and Domains. Natural Language Processing Approaches in Bioinformatics. Prediction of Protein Localization. Protein Functional Annotation. Protein Structural Bioinformatics: An Overview. Secondary Structure Prediction. Sequence Analysis. Sequence Composition. The Evolution of Protein Family Databases

References

Bagos, P.G., Liakopoulos, T.D., Spyropoulos, I.C., Hamodrakas, S.J., 2004. PRED-TMBB: A web server for predicting the topology of beta-barrel outer membrane proteins. Nucleic Acids Res. 32, W400–W404.

Bakheet, T.M., Doig, A.J., 2009. Properties and identification of human protein drug targets. Bioinformatics 25, 451–457.

Bernsel, A., Viklund, H., Hennerdal, A., Elofsson, A., 2009. TOPCONS: Consensus predictionof membrane protein topology. Nucleic Acids Res. 37, W465–W468.

Bigelow, H.R., Petrey, D.S., Liu, J., et al., 2004. Predicting transmembrane beta-barrels in proteomes. Nucleic Acids Res. 32, 2566–2577.

Chetwynd, A.P., Scott, K.A., Mokrab, Y., Sansom, M.S., 2008. CGDB: A database of membrane protein/lipid interactions by coarse-grained molecular dynamics simulations. Mol. Membr. Biol. 225, 662–669.

Fernandez-Leiro, R., Scheres, S.H., 2016. Unravelling biological macromolecules with cryo-electron microscopy. Nature 537, 339–346.

Fuchs, A., Kirschner, A., Frishman, D., 2009. Prediction of helix-helix contacts and interacting helices in polytopic membrane proteins using neural networks. Proteins 74, 857–871.

Hayat, S., Peters, C., Shu, N., et al., 2016. Inclusion of dyad-repeat pattern improves topology prediction of transmembrane β-barrel proteins. Bioinformatics 32, 1571–1573.

Hessa, T., Kim, H., Bihlmaier, K., et al., 2005. Recognition of transmembrane helices by the endoplasmic reticulum translocon. Nature 433, 377–381.

Jacoboni, I., Martelli, P.L., Fariselli, P., et al., 2001. Prediction of the transmembrane regions of beta-barrel membrane proteins with a neural network-based predictor. Protein Sci. 10, 779–787.

Jones, D.T., Buchan, D.W., Cozzetto, D., Pontil, M., 2012. PSICOV: Precise structural contact prediction using sparse inverse covariance estimation on large multiple sequence alignments. Bioinformatics 28, 184–190.

Jones, D.T., Taylor, W.R., Thornton, J.M., 1994. A model recognition approach to the prediction of all-helical membrane protein structure and topology. Biochemistry 33, 3038–3049.

Käll, L., Krogh, A., Sonnhammer, E.L., 2005. An HMM posterior decoder for sequence feature prediction that includes homology information. Bioinformatics 21, i251–i257.

Kozma, D., Simon, I., Tusnády, G.E., 2013. PDBTM: Protein Data Bank of transmembrane proteins after 8 years. Nucleic Acids Res. 41, D524–D529.

Krogh, A., Larsson, B., von Heijne, G., Sonnhammer, E.L., 2001. Predicting transmembrane protein topology with a hidden Markov model: Application to complete genomes. J. Mol. Biol. 305, 567–580.

Kyte, J., Doolittle, R.F., 1982. A simple method for displaying the hydropathic character of a protein. J. Mol. Biol. 157, 105–132.

Lomize, M.A., Lomize, A.L., Pogozheva, I.D., Mosberg, H.I., 2006. OPM: Orientations of proteins in membranes database. Bioinformatics 22, 623–625.

Martelli, P.L., Fariselli, P., Casadio, R., 2003. An ENSEMBLE machine learning approach for the prediction of all-alpha membrane proteins. Bioinformatics 19, i205–i211.

Martelli, P.L., Fariselli, P., Krogh, A., Casadio, R., 2002. A sequence-profile-based HMM for predicting and discriminating beta barrel membrane proteins. Bioinformatics 18, S46–S53.

Martelli, P.L., Profiti, G., Casadio, R., 2017. Protein function prediction. Ref. Modul. Life Sci.

Nugent, T., Jones, D.T., 2009. Transmembrane protein topology prediction using support vector machines. BMC Bioinform. 10, 159.

Nugent, T., Jones, D.T., 2010. Predicting transmembrane helix packing arrangements using residue contacts and a force-directed algorithm. PLOS Comput. Biol. 6, e1000714.

Nugent, T., Jones, D.T., 2013. Accurate de novo structure prediction of large transmembrane protein domains using fragment-assembly and correlated mutation analysis. PNAS 109, E1540–E1547.

Peters, C., Tsirigos, K.D., Shu, N., Elofsson, A., 2016. Improved topology prediction using the terminal hydrophobic helices rule. Bioinformatics 32, 1158–1162.

Pierleoni, A., Martelli, P.L., Casadio, R., 2011a. MemLoci: Predicting subcellular localization of membrane proteins in eukaryotes. Bioinformatics 27, 1224–1230.

Pierleoni, A., Indio, V., Savojardo, C., et al., 2011b. MemPype: A pipeline for the annotation of eukaryotic membrane proteins. Nucleic Acids Res. 39, W375–W380.

Profiti, G., Martelli, P.L., Casadio, R., 2017. The Bologna Annotation Resource (BAR 3.0): Improving protein functional annotation. Nucleic Acids Res. 45, W285–W290.

Reynolds, S.M., Käll, L., Riffle, M.E., Bilmes, J.A., Noble, W.S., 2008. Transmembrane topology and signal peptide prediction using dynamic bayesian networks. PLOS Comput. Biol. 4, e1000213.

Rost, B., Casadio, R., Fariselli, P., Sander, C., 1995. Transmembrane helices predicted at 95% accuracy. Protein Sci. 4, 521–533.

Rost, B., Fariselli, P., Casadio, R., 1996. Topology prediction for helical transmembrane proteins at 86% accuracy. Protein Sci. 5, 1704–1718.

Roy Choudhury, A., Novič, M., 2015. PredβTM: A novel β-transmembrane region prediction algorithm. PLOS One 10, e0145564.

Savojardo, C., Fariselli, P., Casadio, R., 2013. BETAWARE: A machine-learning tool to detect and predict transmembrane beta-barrel proteins in prokaryotes. Bioinformatics 29, 504–505.

Sharpe, H.J., Stevens, T.J., Munro, S., 2010. A comprehensive comparison of transmembrane domains reveals organelle-specific properties. Cell 142, 158–169.

Sonnhammer, E.L., von Heijne, G., Krogh, A., 1998. A hidden Markov model for predicting transmembrane helices in protein sequences. Proc. Int. Conf. Intell. Syst. Mol. Biol. 6, 175–182.

Stansfeld, P.J., Goose, J.E., Caffrey, M., et al., 2015. MemProtMD: Automated insertion of membrane protein structures into explicit lipid membranes. Structure 23, 1350–1361.

The UniProt Consortium, 2017. UniProt the universal protein knowledgebase. Nucleic Acids Res. 45, D158–D169.

Tsirigos, K.D., Govindarajan, S., Bassot, C., et al., 2018. Topology of membrane proteins-predictions, limitations and variations. Curr. Opin. Struct. Biol. http://dx.doi.org/10.1016/j.sbi.2017.10.00 (in press).

Tsirigos, K.D., Peters, C., Shu, N., et al., 2015. The TOPCONS web server for consensus prediction of membrane protein topology and signal peptides. Nucleic Acids Res. 43, W401–W407.

Tusnády, G.E., Simon, I., 1998. Principles governing amino acid composition of integral membrane proteins: Application to topology prediction. J. Mol. Biol. 283, 489–506.

Viklund, H., Bernsel, A., Skwark, M., Elofsson, A., 2008. SPOCTOPUS: A combined predictor of signal peptides and membrane protein topology. Bioinformatics 24, 2928–2929.

von Heijne, G., 1989. Control of topology and mode of assembly of a polytopic membrane protein by positively charged residues. Nature 341, 456–458.

White, S.H., Wimley, W.C., 1999. Membrane protein folding and stability: Physical principles. Annu. Rev. Biophys. Biomol. Struct. 28, 319–365.

Yang, J., Shen, H.B., 2018. MemBrain-contact 2.0: A new two-stage machine learning model for the prediction enhancement of transmembrane protein residue contacts in the full chain. Bioinformatics. http://dx.doi.org/10.1093/bioinformatics/btx593 (in press).

Relevant Websites

https://bar.biocomp.unibo.it
    Biocomputing Group-University of Bologna.
www.csbio.sjtu.edu.cn/bioinf/MemBrain
    MemBrain.
http://blanco.biomol.uci.edu/mpstruc/
    Membrane Proteins of Known Structure.

https://mu2py.biocomp.unibo.it/memloci
 MemLoci-Biocomputing Group.
http://memprotmd.bioch.ox.ac.uk/
 MemProtMD.
https://mu2py.biocomp.unibo.it/mempype
 MemPype-Biocomputing Group.
http://opm.phar.umich.edu/
 Orientations of Proteins in Membranes.
http://pdbtm.enzim.hu/
 Protein Data Bank of Transmembrane Proteins.
http://web.expasy.org/protscale/
 ProtScale.
http://bioinf.cs.ucl.ac.uk/psipred/
 PSIPRED Protein Sequence Analysis Workbench.
http://webclu.bio.wzw.tum.de/tmhcon/
 TMHcon.
http://bioinf.cs.ucl.ac.uk/software_downloads/
 UCL-CS Bioinformatics: Software & Downloads.

Prediction of Protein Localization

Kenta Nakai, The University of Tokyo, Tokyo, Japan
Kenichiro Imai, Advanced Industrial Science and Technology, Tokyo, Japan

Introduction

In the body of multicellular organisms, there are many cell types, each of which contains a set of proteins, some of which are specific to it while some are ubiquitous to all cell types. With the development of single-cell transcriptomics technology, precise mapping of each protein to its expressed cells is ongoing. Such information will be undoubtedly useful in clarifying the cellular function of these proteins. Similarly, each protein is localized at a compartment(s) (organelle) within a cell (or excreted to outside of the cell). Such knowledge would be also useful in understanding the function of the protein because each compartment plays some specific roles within the cell. Especially, knowledge on whether a protein is secreted or not may have particular importance in pharmaceutical and medical sciences. And it should be noted that the information of protein subcellular localization is also important for unicellular organisms, such as bacteria. Attempts to systematically determine the subcellular localization within a cell have been performed in the field of proteomics. Nowadays, even the information of semi-quantitative localization frequency at multiple localization sites for a protein is accumulating.

It has been demonstrated that many (if not all) proteins are sorted to their final localization site after their synthesis in the cytosol according to the information that is encoded as part(s) of their own amino acid sequence (known as the signal hypothesis). Typically, such information is encoded as an N-terminal presequence and is called the targeting signal. In other words, some subcellular machinery recognizes and interprets such signals in the nascent proteins. Therefore, the prediction of protein subcellular localization from their amino acid sequence can have some extra biological rationale: If we can mimic the recognition process within the cell, we would be able to predict the localization, in principle. For practical purposes, however, additional information, such as the protein–protein interaction data, could be used. It should be also emphasized that the prediction problem has been used as a test bed for a variety of machine-learning algorithms probably because of its simplicity in scheme and its potential practical value. In this article, we will sketch a brief overview of this field, introducing some representative works. Finally, we will discuss its future directions.

Topics

Prediction of Subcellular Localization Sites

Basic prediction scheme

Since possible repertory of localization sites can be different between organisms, the basic input information is a set of a (precursor) amino acid sequence and its species origin. In eukaryotic cells, the category has been conventionally divided into animal cells and plant ones; unicellular eukaryotes, such as yeasts, can be treated as an animal or a special type of plant lacking chloroplasts. In many cases, the repertory for animal cells are: The cytosol, the nucleus, the mitochondrion, the endoplasmic reticulum (ER), the Golgi apparatus, the lysosome, the plasma membrane and the extracellular space (i.e., excreted). In plants, vacuoles are usually treated as equivalent with lysosomes in animals. In addition, plastids (typically chloroplasts) are added in plants. Sometimes, (integral) membrane proteins in compartments, such as mitochondria, are collectively treated as membrane proteins. It is known that some of the above compartments, such as mitochondria and plastids, have further precise structures because they are composed of multiple membranes. The nucleolus in the nucleus would be regarded as an independent site. However, attempts to include more precise predictions are rare mainly because of the difficulty in their data size: If a site contains only a small number of known proteins, its treatment in the machine-learning scheme is difficult. Similarly, in spite of their biological importance, peroxisomes are often not included in the repertory because of their relatively small sample size. Sometimes, cytoskeletal proteins are regarded as an independent site. The treatment of peripheral membrane proteins, that is, proteins that are weakly bound to the membrane mainly through protein–protein interaction, is quite difficult (except the cases when they are the lipid-anchored proteins). Theoretically, they can be treated as soluble proteins but both experimental data and practical expectation prefer them to be treated as membrane proteins. In bacterial predictions, the repertories have been classified into two, depending on the presence or the absence of the outer membrane: Gram-negative and Gram-positive bacteria. In Gram-positive bacteria, the repertory is the cytoplasm, the cell membrane, and the extracellular space while in Gram-negative ones, it is the cytoplasm, the inner membrane, the periplasm, the outer membrane, and the extracellular space. The cell membrane in Gram-positive bacteria is considered to be equivalent with the inner membrane in Gram-negative bacteria. Detailed attempts to predict archaebacterial proteins are few and they seem to have been practically treated as a kind of Gram-positive bacterial protein.

Since the N-terminal part of a protein can affect its localization significantly, the input amino acid sequence must be the accurate precursor sequence. However, it is often the case that the translation site of a protein is not experimentally determined. Moreover, it is known that the translation site can be switched in various conditions because of alternative choices of transcriptional start sites and/or alternative splicing. Thus, predicting the potential changes of localization sites depending on the choices of translational start sites would be a future direction in this field (see the concluding remarks).

In many cases, the most probable localization site or a list of possible sites with p values are provided in the prediction. Recently, there have also been many attempts to extend the scheme to treat with proteins that are localized at multiple sites. However, objective evaluation of their performances is difficult because both systematic and quantitative data of multiple localization have been quite rare, if any.

Conventionally, the prediction accuracy of a prediction method is assessed using the n-fold cross validation test, including the jack-knife test, where the number of testing data is one. As for the measure of prediction performance, standard measures in machine learning, such as the overall accuracy, the sensitivity/specificity (or precision/recall), and the Matthew's correlation coefficient are often used.

General prediction methods

Many prediction methods of subcellular localization sites have been developed and improved by combining biological or empirical sequence features correlated with subcellular localization with a variety of machine-learning algorithms, such as the k-nearest neighbor classifier, the support vector machine , and deep learning: PSORT (Nakai and Kanehisa, 1992), CELLO 2.5 (Yu *et al.*, 2006), WoLF PSORT (Horton *et al.*, 2007), MultiLoc2 (Blum *et al.*, 2009), SherLoc2 (Briesemeister *et al.*, 2009), YLoc (Briesemeister *et al.*, 2010), iLoc-Euk (Chou *et al.*, 2011), Loctree3 (Goldberg *et al.*, 2014), and DeepLoc (Almagro Armenteros *et al.*, 2017).

Prediction features generally used in the prediction of subcellular localization sites are roughly categorized into three types: Targeting signal features, sequence features, and annotation-based features. The information of targeting signals gives powerful prediction features and thus is detailed in the next section. The targeting signals are roughly classified into two classes: N-terminal targeting signals and non-N-terminal targeting signals. Representative N-terminal targeting signals are the signal sequence for the secretory pathway, the chloroplastic targeting signal, and the mitochondrial targeting signal. Whereas, the nuclear localization signal (NLS) and the nuclear export signal (NES) are usually located in the internal position of protein sequences, and peroxisomal targeting signal 1 (PTS1) is located at the C-terminus of proteins.

Sequence features are heavily used in localization prediction since some differences in the sequence features are empirically known to be correlated with different localization sites. Note that these sequence features may not be directly related to protein targeting itself. Representative sequence features are a series of expanded amino acid compositions of the entire sequence, such as the content of dipeptides, n-grams, k-mers, and the pseudo amino acid composition. Pseudo amino acid composition is reported to be more informative in terms of incorporating sequence-order information of a protein sequence (Chou, 2001). Functional motifs are also used in the prediction because some sequence motifs are directly related to the targeting signals and some functional information is closely related to the function of a localization site (e.g., the DNA-binding motif).

Typical annotation-based features are experimentally verified localization information found in UniProt or GO terms, in addition to the information of functional domains, protein–protein interaction and the text information from PubMed abstracts. Those annotation-based features are transferred from homologous proteins or the protein itself. In those features, the transfer of localization annotation from homologous proteins seems to be practically useful. Indeed, a simple homology-based inference outperforms methods based on machine learnings (Imai and Nakai, 2010) if a homologous protein with localization annotation is available.

Evaluation of prediction performance is a difficult issue in subcellular localization prediction. There is often some overlap between the training and the test sets of different methods since most methods use UniProt annotation for making these datasets. To evaluate the prediction performance with less bias, Salvatore *et al.* (2017) recently made a benchmark dataset from new data obtained by recent large-scale experimental studies in human cells and examined the performance of six state-of-the-art methods (CELLO 2.5, LocTree2, MultiLoc2, SherLoc2, WoLF PSORT and YLoc) (Salvatore *et al.*, 2017). In this assessment, CELLO 2.5, SherLoc2, LocTree2, and YLoc showed better performance (F_1 score$=0.70$), where F_1 score$=2((\text{precision} \times \text{recall})/(\text{precision} + \text{recall}))$. Also they developed an ensemble method, SubCons that combines four predictors (CELLO2.5, LocTree2, MultiLoc2 and SherLoc2) using a Random Forest classifier and it outperformed the six methods (F_1 score$=0.79$).

Targeting Signal Prediction

Prediction of signal sequence

Signal sequences (signal peptides) are the N-terminal sorting signal that targets the linked protein to the secretory pathway in eukaryotes and prokaryotes. About 10%–20% of eukaryotic proteome and 10% of bacterial proteome have been estimated to have the signal sequence (Kanapin *et al.*, 2003; Ivankov *et al.*, 2013). In eukaryotic cells, the signal recognition particle (SRP) cotranslationally recognizes the signal sequence upon its emergence from the ribosome and transfers it to the Sec61 translocon in the ER membrane (Nilsson *et al.*, 2015) via the SRP receptor. In bacterial cells, the SRP pathway is mainly used for targeting the membrane proteins with noncleaved signal sequence (signal anchors). Instead, secretory proteins containing the signal sequence are recognized by SecA, which drives the translocation through the SecYEG translocon (Kudva *et al.*, 2013). The signal peptidase cleaves signal sequences and the mature proteins are generated. Signal sequences do not share sequence similarity but share some characteristic features (von Heijne, 1990): Signal sequences are on average 16–30 amino acid residues in length and comprise characteristic tripartite architecture: A positively charged n-region, a central hydrophobic h-region and a c-region with the cleavage site for signal peptidases. The cleavage site has a consensus pattern known as the -1,-3 rule: Amino acids with small side chains are preferred at the − 1 position of signal sequences and no charged amino acid residues are preferred at their − 3 position.

For predicting the signal sequence and its cleavage site, many prediction methods, such as SignalP (Bendtsen et al., 2004; Petersen et al., 2011), Signal-CF (Chou and Shen, 2007), and Signal-BLAST (Frank and Sippl, 2008), have been developed with the characteristic features. According to an assessment study in 2009 (Choo et al., 2009), SignalP 3.0 achieves the best overall accuracy (0.872–0.914) in eukaryotic, Gram-positive, and Gram-negative bacterial data sets. Therefore, the discrimination between secretory and nonsecretory proteins (not including membrane proteins) based on the signal sequence prediction is the most successful in targeting signal predictions. However, those methods share a problem: Difficulty in the discrimination between the signal sequence and the transmembrane region. This problem will yield many false-positive predictions from N-terminal trans-membrane regions when the methods are applied in proteomic analyses. Recently, SignalP (SignalP 4.0) was updated to tackle the problem using two types of neural network-based methods, SignalP-TM and SignalP-noTM: SignalP-TM has been trained with sequences containing transmembrane segments in the dataset while SignalP-noTM has been trained without those sequences (Petersen et al., 2011). SignalP 4.0 shows better discrimination between signal peptides and transmembrane regions, and consequently achieves the best signal sequence prediction. On the other hand, there is still room for improvement on the cleavage site prediction: Precision and sensitivity of current methods hovers around ∼66% and ∼68%, respectively.

Prediction of mitochondrial targeting signal

Mitochondria have been estimated to host 1000–1500 distinct proteins, respectively (Meisinger et al., 2008). The vast majority of mitochondrial proteins are encoded in the nuclear genome and imported by the translocases in the mitochondrial outer and inner membranes. About a half of mitochondrial proteins possess an N-terminal cleavable targeting signal (presequence), while the remaining proteins are thought to have a noncleavable internal targeting signal (Chacinska et al., 2009). The mechanism of presequence-dependent targeting and the features of the presequence are well studied, however, the targeting mechanism with the internal targeting signal is still unclear. Thus, the main target of mitochondrial targeting signal prediction has been based on the presequences. Presequences-containing proteins are translocated by the outer and the inner membrane translocase, TOM and TIM23-PAM complexes (Schulz et al., 2015; Wiedemann and Pfanner, 2017). Presequences have a length of 20–60 amino acid residues (Calvo et al., 2017) and do not share sequence similarity. However, they exhibit a few characteristic features: A high composition of arginine and near absence of negatively charged residues (von Heijne, 1986; Schneider et al., 1998). They are also often capable of forming a local amphiphilic α-helical structure with hydrophobic residues on one face and positively charged residues on the opposite face (Chacinska et al., 2009; Fukasawa et al., 2015). The presequence is cleaved in the matrix by the mitochondrial processing peptidase (MPP) and some of them are subsequently further cleaved by the intermediate peptidases, such as Oct1 and Icp55 (Mossmann et al., 2012). The cleavage by MPP occurs at a position that is two amino acids C-terminal from an arginine (the R-2 motif). Icp55 and Oct1 subsequently cleave one amino acid and eight amino acids, respectively. Therefore, proteins cleaved by MPP + Icp55 have an arginine at position -3 (the R-3 motif) in the presequence while proteins cleaved by MPP + Oct1 have an arginine at position -10 (the R-10 motif).

Widely used presequence prediction methods, such as MitoProtII (Claros, 1995), TargetP (Emanuelsson et al., 2000), Predotar (Small et al., 2004), and MitoFates (Fukasawa et al., 2015) were developed using machine-learning techniques with these properties of the presequences. These tools are also capable of predicting the presequence with its cleavage site. MitoProtII and MitoFates are specific predictors for the (mitochondrial) presequence, while TargetP and Predotar integrate other N-terminal targeting signal predictions, such as the secretory signal sequence and the chloroplastic targeting signal. Among those tools, Mitofates performs better in predicting both the presequence existence and the cleavage site: Precision and recall of 0.79 and 0.80 in the presequence prediction and the sensitivity of cleavage site detection is 71% (Fukasawa et al., 2015). However, proteomic analyses of N-termini of mitochondrial proteins of mouse and yeast point out low accuracy (especially in the cleavage site prediction) of those presequence prediction tools (Vögtle et al., 2009; Calvo et al., 2017). A cluster analysis of yeast presequence data obtained from the proteomic analysis with features used in the presequence prediction suggests that presequences can be grouped into at least three clusters (Fukasawa et al., 2015). The largest cluster (60% of presequences) shows typical presequence features such as the enrichment of arginines, almost no negatively charged residues, positively charged amphiphilicity, and the MPP cleavage pattern. On the other hand, the two remaining clusters differ from the classical view of presequences: Many of them have low net-charge and their cleavage sites did not match to any known patterns. Presequences belonging to these clusters should be the reason of the low prediction accuracy. To further improve the presequence prediction, it will be necessary to better characterize these untypical presequences.

Prediction of chloroplastic targeting signal

Approximately 3000 different proteins are estimated to be needed to develop a chloroplast. Similar to mitochondria, the vast majority of those chloroplast proteins are nuclear-encoded and imported into the chloroplast. Most chloroplast precursor proteins have a cleavable N-terminal extension, the chloroplastic targeting signal (transit peptide) (Li and Chiu, 2010; Paila et al., 2015). Thus, existing chloroplastic targeting signal prediction tools deal with the cleavable N-terminal transit peptides. Precursor proteins containing the transit peptide are imported through the translocase in the outer and inner envelope membranes, the TOC and TIC complex, respectively. Similar to other N-terminal targeting signals, the transit peptide is cleaved off by the stroma processing peptidase during or after the translocation. Chloroplastic transit peptides exhibit a few characteristic features: A high content of hydroxylated amino acids (e.g., serines), lack of negatively charged residues, and a propensity to form α-helical structures in hydrophobic environments (Bruce, 2001; Jarvis, 2008). In this respect, transit peptides are similar to mitochondrial presequences. Their cleavage sites are characterized by the higher frequency of Ala, Ile, Cys, and Val residues. A loosely conserved consensus motif ([V,I]X[A,C] ↓ A, where ↓ represents the cleavage site) was previously found from a limited set of 32 cleavage sites (Gavel and von Heijne, 1990). A more recent 198 cleavage site set gives three motifs, [V,I][R,A] ↓ [A,C]AAE, S[V,I][R,S,V] ↓ [C,A]A, and [A,V]N ↓ A [A,M]AG[E,D] (Savojardo *et al.*, 2015).

A widely used prediction method for the transit peptide and its cleavage site is ChloroP (Emanuelsson *et al.*, 1999). However, predictors that are specialized for transit peptides are minor and the transit peptide prediction is usually integrated as a part of the prediction of N-terminal targeting signals: TargetP (Emanuelsson *et al.*, 2000), iPSORT (Bannai *et al.*, 2002), Predotar (Small *et al.*, 2004), and TPpred3 (Savojardo *et al.*, 2015) are capable of predicting transit peptides. Indeed, TargetP includes ChloroP in its algorithm. Among those methods, TPpred3 achieves better performance for transit peptide prediction (46% precision and 64% recall) (Savojardo *et al.*, 2015), although further improvement seems to be needed for practical purposes. Comparing with mitochondrial targeting signals, the data size of known transit peptides is smaller. Larger sets of established N-terminal proteomics data should be useful to further improve the prediction performance.

Prediction of nuclear localization signals and nuclear export signals

Nuclear proteins are carried into and out of the nuclei through the nuclear pore complex by nucleocytoplasmic transport receptors, that is, the importin (karyopherin)-β family. The human genome encodes 20 importin-β family proteins, of which 10 are nuclear import receptors (importin-β, transportin-1,-2, -SR, importin-4, -5 (RanBP5), -7, -8, -9, and -11), 7 are export receptors (exportin-1 (CRM1), -2(CAS/CSE1L), -5, -6, -7, -t, and RanBP17), 2 are bidirectional receptors (imporin-13 and exportin-4), and the function of RanBP6 is undetermined (Kimura and Imamoto, 2014). Those nucleocytoplasmic transport receptors are thought to recognize specific targeting signals on those cargo proteins. Nevertheless the targeting signals have been identified for only a few nucleocytoplasmic transport receptors, so far (Soniat and Chook, 2015): The classical nuclear localization signal (cNLS) for the importin-α adaptor, which links cargos and importin β (Lange *et al.*, 2007), PY-NLS for transportin-1 and -2 (Lee *et al.*, 2006), the NES for exportin-1/CRM1 (Hutten and Kehlenbach, 2007), and the SR-rich domain that binds to transportin-SR (Maertens *et al.*, 2014). Among those nuclear targeting signals, cNLS and NES are well characterized. Thus, recent prediction algorithms for NLSs and NESs mainly target the two types of the signals.

cNLSs include two types of signals, the monopartite and the bipartite NLSs: The monopartite NLS means a single cluster of basic amino acid residues (e.g., KR[K/R]R and K[K/R]RK) and the bipartite NLSs means two clusters of basic amino acids separated by a 10–12 amino acid linker (e.g., $KRX_{10-12}K[K/R][K/R]$) (Kosugi *et al.*, 2009). Nucpred (Brameier Krings and MacCallum, 2007), cNLSmapper (Kosugi *et al.*, 2009), NLStradamus (Ba *et al.*, 2009), NucImport (Mehdi *et al.*, 2011), and seqNLS (Lin and Hu, 2013) are widely used for the cNLS prediction. According to a recent assessment based on experimentally identified human NLSs (77% of them are classified as known cNLSs) (Lisitsyna *et al.*, 2017), the highest Matthews' correlation coefficient (∼0.3) was obtained from NucPred, seqNLS, and NLStradamus. However, those prediction methods correctly identified only ∼45 % of the human NLS data, suggesting the necessity for further improvement on the NLS prediction. Recent proteomic analysis for the identification of cargo proteins of 12 nucleocytoplasmic transport receptors (10 nuclear import receptors and 2 bidirectional receptors) (Kimura *et al.*, 2017) would be a valuable resource for further improvement of NLS prediction and/or for finding novel NLSs. Meanwhile, the proteomic analysis pointed out that about 30% of the identified cargos are shared by multiple receptors. Better characterization of such multiple recognitions may be fruitful for future improvements.

Nuclear export receptor exportin-1/CRM1 binds to 8–15 residue-long NESs in hundreds of distinct cargos. NES sequences are diverse and 11 patterns were defined by a peptide library-based study and structural analyses of exportin-1/CRM1-NES (Kosugi *et al.*, 2008; Fung *et al.*, 2015, 2017). Those NESs usually have 4–5 hydrophobic residues, which bind to the hydrophobic pocket of the receptors. The hydrophobic residues are arranged in various patterns: $\Phi X_3 \Phi X_2 \Phi X \Phi$, $\Phi X \Phi X_2 \Phi X \Phi$ and $\Phi X_2 \Phi X_3 \Phi X_2 \Phi$ (Φ represents Leu, Val, Ile, Phe, or Met). Several computational tools, such as NetNES (La Cour *et al.*, 2004), NESsential (Fu *et al.*, 2011), NESmapper (Kosugi *et al.*, 2014), LocNES (Xu *et al.*, 2014), and Wregex-based prediction (Prieto *et al.*, 2014) have been developed to predict NESs, based on the sequence profiles of NESs and some biophysical properties (e.g., disorder propensity and solvent accessibilities). However, the performance of those predictors is still not enough (precision is ∼50% with 20% recall). The diverse NES patterns and the low performance of predictors suggest that there may be no fixed hydrophobic pattern in NESs. There may be a limitation to predict NESs from only sequence. Combination with structure-based predictions might help to find a breakthrough.

Other signals

Besides the above signals the predictors of which are relatively well developed, there are also known signals that may be used for the prediction. Here, we briefly sketch several of them.

As noted above, peroxisomes are often not included in the prediction repertory because known peroxisomal proteins are not so many. However, two kinds of signals, namely PTS1 and PTS2, are known to exist. PTS1 is characterized by the presence of the SKL motif (the more precise consensus is [SAC][KRH][LM]) in the C-terminus of proteins and is recognized by the receptor Pex5, which is a modular protein containing a C-terminal tetratricopeptide repeat domain (Gould *et al.*, 1987, Kim and Hettema, 2015). Of course, its computational recognition is easy. It is also possible to include weaker conserved patterns near PTS1. However, the number of proteins that are sorted with PTS1 in vivo is not enough to make the prediction of the entire peroxisomal proteins reliable. The other signal, PTS2, is recognized by the receptor Pex7. The signal is characterized as an about nine-residue motif near the N-terminus; the consensus motif is R[LIVQ]XX[LIVQH][LSGA]X[HQ][LA] (Petriv *et al.*, 2004). But its reliable predictors have not been developed because of its small sample size, again.

It has been shown that some lysosomal proteins are recognized from their special modification with mannose-6-phosphate (M6P), allowing their recognition by M6P receptors in the Golgi complex and ensuing their transport to the endosomal/lysosomal system (Braulke and Bonifacino, 2009). However, their sequence determinant is still a mystery and thus there have not been effective predictors for lysosomal proteins, so far.

Similarly, some other posttranslational modifications can affect the localization of proteins significantly. Typical examples are lipid anchors, where attached lipid moiety is inserted into the membrane as an anchor. Three types of lipid anchors are known: N-myristoylation, glycosylphosphatidylinositol (GPI) attachment, and prenylation. In N-myristoylation, a 14-carbon fatty acid is attached to an N-terminal Gly residue (the consensus motif is MGXXX[ST]) (Johnson *et al.*, 1994). GPI is covalently attached to the C-terminus of proteins (Ferguson *et al.*, 2009). In prenylation, a 15-carbon (farnesylation), or a 20-carbon (geranylgeranylation) lipid is attached to a cysteine by farnesyltransferases or by protein geranlygeranyl transferases I, respectively (the consensus motif is CAAX at the C-terminus) (Fu and Casey, 1999). For each of them, prediction algorithms, such as NMT (Maurer-Stroh *et al.*, 2002), PredGPI (Pierleoni *et al.*, 2008), GPS-Lipid (Xie *et al.*, 2016), have been developed (Audagnotto and Dal Peraro, 2017). However, distinctions on which membrane the predicted lipid will be targeted might be necessary in the future when enough data for each membrane are accumulated.

For the prediction of the localization membrane for (integral) membrane proteins, the prediction of their membrane topology is also important because it is known that some important signals can exist on the so-called cytoplasmic tail (i.e., the terminal region that is protruded to the cytoplasmic side with a membrane) of membrane proteins. For the prediction of membrane topology itself, please refer to its specific article. Here, we would like to point out just two things: The prediction of membrane topology can be related to the prediction of the cleavage of N-terminal hydrophobic segment because N-terminal transmembrane segments are often wrongly predicted as cleavable signal peptides, as explained above. Second, the most important localization signal existing on the C-terminus of a protein, which also exists as a part of the cytoplasmic tail, is known as the KDEL signal, which works as a retention signal to the ER (Munro and Pelham, 1987). Again, its computational recognition is not so difficult, though there are some differences in the optimum motif between organisms, such as HDEL in yeasts. However, the distinction of ER membrane proteins from the others remains to be a very difficult prediction problem because the KDEL signal is not so universal.

For some of the proteins localized at relatively minor sites, such as the Golgi body and the cell wall (in yeasts and plants), there have been several reports on their potential signals (Banfield, 2011; Mao *et al.*, 2008) but their known examples are too few to be tested in the scheme of machine learning.

Concluding Remarks

When publishing a paper on a new prediction method, it is generally required that the method should outperform existing ones. Indeed, many recent papers in the field of protein localization prediction claim that their predictors are apparently quite reliable. However, as far as the amino acid sequence is used as its input, it is questionable if these predictors can be practically useful for totally new proteins. It seems to be a fundamental problem that we do not know how many nuclear DNA-encoded proteins are sorted with their own amino acid sequence signals. Thus, perhaps, a promising new direction in this field would be to use prediction tools to evaluate the possibility of whether each protein is sorted with a classical pathway or not. On the other hand, it is certain that the practical importance of subcellular localization prediction would become even less with the future accumulation of more systematic and precise proteomic data. Then, what new directions will be pursued? As mentioned above, one possibility is to develop a more precise predictor that can predict the semiquantitative distribution between multiple sites. To do this, more comprehensive data for localization are necessary. Although this is not the scope of this article, the prediction of subcellular localization for RNAs would also is an interesting frontier. Finally, we believe that the field should be extended to more general goals. Three possibilities in this direction would be (1)to explore the possibility of predicting differential localization of a protein between different cell types/conditions, incorporating the information of protein isoform expression between cells/conditions; (2) to explore the possibility of more general prediction of protein in vivo fates, including the predictions of more general post-translational modifications and degradations; and (3) to explore the possibility to develop predictors for artificial proteins, incorporating a vast amount of synthetic biological data. Such predictors will have some practical value for designing new drugs, for example, that are optimized to be targeted to a certain localization site.

See also: Algorithms for Strings and Sequences: Searching Motifs. Artificial Intelligence and Machine Learning in Bioinformatics. Artificial Intelligence. Cell Modeling and Simulation. Data Mining: Classification and Prediction. Data Mining: Prediction Methods. Epitope Predictions. Functional Genomics. Machine Learning in Bioinformatics. Natural Language Processing Approaches in Bioinformatics. Protein Functional Annotation. Protein Post-Translational Modification Prediction. Protein Properties. Protein–Protein Interaction Databases. Sequence Analysis. Sequence Composition. Supervised Learning: Classification. The Evolution of Protein Family Databases. Transmembrane Domain Prediction

References

Almagro Armenteros, J.J., Sønderby, C.K., Sønderby, S.K., Nielsen, H., Winther, O., 2017. DeepLoc: Prediction of protein subcellular localization using deep learning. Bioinformatics 33 (21), 3387–3395.

Audagnotto, M., Dal Peraro, M., 2017. Protein post-translational modifications: In silico prediction tools and molecular modeling. Computational and Structural Biotechnology Journal 15, 307–319.

Ba, A.N.N., Pogoutse, A., Provart, N., Moses, A.M., 2009. NLStradamus: A simple Hidden Markov Model for nuclear localization signal prediction. BMC Bioinformatics 10 (1), 202.

Banfield, D.K., 2011. Mechanisms of protein retention in the Golgi. Cold Spring Harbor Perspectives in Biology 3 (8), a005264.

Bannai, H., Tamada, Y., Maruyama, O., Nakai, K., Miyano, S., 2002. Extensive feature detection of N-terminal protein sorting signals. Bioinformatics 18 (2), 298–305.

Bendtsen, J.D., Nielsen, H., von Heijne, G., Brunak, S., 2004. Improved prediction of signal peptides: SignalP 3.0. Journal of Molecular Biology 340 (4), 783–795.

Blum, T., Briesemeister, S., Kohlbacher, O., 2009. MultiLoc2: Integrating phylogeny and Gene Ontology terms improves subcellular protein localization prediction. BMC Bioinformatics 10 (1), 274.

Brameier, M., Krings, A., MacCallum, R.M., 2007. NucPred – Predicting nuclear localization of proteins. Bioinformatics 23 (9), 1159–1160.

Braulke, T., Bonifacino, J.S., 2009. Sorting of lysosomal proteins. Biochimica et Biophysica Acta (BBA) - Molecular Cell Research 1793 (4), 605–614.

Briesemeister, S., Blum, T., Brady, S., et al., 2009. SherLoc2: A high-accuracy hybrid method for predicting subcellular localization of proteins. Journal of Proteome Research 8 (11), 5363–5366.

Briesemeister, S., Rahnenführer, J., Kohlbacher, O., 2010. YLoc – An interpretable web server for predicting subcellular localization. Nucleic Acids Research 38 (Suppl_2), W497–W502.

Bruce, B.D., 2001. The paradox of plastid transit peptides: Conservation of function despite divergence in primary structure. Biochimica et Biophysica Acta (BBA)-Molecular Cell Research 1541 (1–2), 2–21.

Calvo, S.E., Julien, O., Clauser, K.R., et al., 2017. Comparative analysis of mitochondrial N-termini from mouse, human, and yeast. Molecular and Cellular Proteomics 16 (4), 512–523.

Chacinska, A., Koehler, C.M., Milenkovic, D., Lithgow, T., Pfanner, N., 2009. Importing mitochondrial proteins: Machineries and mechanisms. Cell 138 (4), 628–644.

Choo, K.H., Tan, T.W., Ranganathan, S., 2009. A comprehensive assessment of N-terminal signal peptides prediction methods. BMC Bioinformatics 10 (Suppl. 5), S2.

Chou, K.C., 2001. Prediction of protein cellular attributes using pseudo-amino acid composition. Proteins: Structure, Function, and Bioinformatics 43 (3), 246–255.

Chou, K.C., Shen, H.B., 2007. Signal CF: A subsite-coupled and window-fusing approach for predicting signal peptides. Biochemical and Biophysical Research Communications 357 (3), 633–640.

Chou, K.C., Wu, Z.C., Xiao, X., 2011. iLoc-Euk: A multi-label classifier for predicting the subcellular localization of singleplex and multiplex eukaryotic proteins. PLOS ONE 6 (3), e18258.

Claros, M.G., 1995. MitoProt, a Macintosh application for studying mitochondrial proteins. Bioinformatics 11 (4), 441–447.

Emanuelsson, O., Nielsen, H., Brunak, S., Von Heijne, G., 2000. Predicting subcellular localization of proteins based on their N-terminal amino acid sequence. Journal of Molecular Biology 300 (4), 1005–1016.

Emanuelsson, O., Nielsen, H., Von Heijne, G., 1999. ChloroP, a neural network-based method for predicting chloroplast transit peptides and their cleavage sites. Protein Science 8 (5), 978–984.

Ferguson, M.A., Kinoshita, T., Hart, G.W., 2009. Glycosylphosphatidylinositol Anchors. NY, United States: Cold Spring Harbor Laboratory Press.

Frank, K., Sippl, M.J., 2008. High-performance signal peptide prediction based on sequence alignment techniques. Bioinformatics 24 (19), 2172–2176.

Fu, H.W., Casey, P.J., 1999. Enzymology and biology of CaaX protein prenylation. Recent Progress in Hormone Research 54, 315–342.

Fukasawa, Y., Tsuji, J., Fu, S.C., et al., 2015. MitoFates: Improved prediction of mitochondrial targeting sequences and their cleavage sites. Molecular & Cellular Proteomics 14 (4), 1113–1126.

Fung, H.Y.J., Fu, S.C., Brautigam, C.A., Chook, Y.M., 2015. Structural determinants of nuclear export signal orientation in binding to exportin CRM1. eLife 4, e10034.

Fung, H.Y.J., Fu, S.C., Chook, Y.M., 2017. Nuclear export receptor CRM1 recognizes diverse conformations in nuclear export signals. eLife 6. doi:10.7554/eLife.23961.

Fu, S.C., Imai, K., Horton, P., 2011. Prediction of leucine-rich nuclear export signal containing proteins with NESsential. Nucleic Acids Research 39 (16), e111.

Gavel, Y., von Heijne, G., 1990. A conserved cleavage-site motif in chloroplast transit peptides. FEBS Letters. 261 (2), 455–458.

Goldberg, T., Hecht, M., Hamp, T., et al., 2014. LocTree3 prediction of localization. Nucleic Acids Research 42 (W1), W350–W355.

Gould, S.G., Keller, G.A., Subramani, S., 1987. Identification of a peroxisomal targeting signal at the carboxy terminus of firefly luciferase. The Journal of cell biology 105 (6), 2923–2931.

Horton, P., Park, K.J., Obayashi, T., et al., 2007. WoLF PSORT: Protein localization predictor. Nucleic Acids Research 35 (Suppl_2), W585–W587.

Hutten, S., Kehlenbach, R.H., 2007. CRM1-mediated nuclear export: To the pore and beyond. Trends in Cell Biology 17 (4), 193–201.

Imai, K., Nakai, K., 2010. Prediction of subcellular locations of proteins: Where to proceed? Proteomics 10 (22), 3970–3983.

Ivankov, D.N., Payne, S.H., Galperin, M.Y., et al., 2013. How many signal peptides are there in bacteria? Environmental Microbiology 15 (4), 983–990.

Jarvis, P., 2008. Targeting of nucleus-encoded proteins to chloroplasts in plants. New Phytologist 179 (2), 257–285.

Johnson, D.R., Bhatnagar, R.S., Knoll, L.J., Gordon, J.I., 1994. Genetic and biochemical studies of protein N-myristoylation. Annual Review of Biochemistry 63 (1), 869–914.

Kanapin, A., Batalov, S., Davis, M.J., et al., 2003. Mouse proteome analysis. Genome Research. 13 (6b), 1335–1344.

Kim, P.K., Hettema, E.H., 2015. Multiple pathways for protein transport to peroxisomes. Journal of molecular biology 427 (6), 1176–1190.

Kimura, M., Imamoto, N., 2014. Biological significance of the importin-β family-dependent nucleocytoplasmic transport pathways. Traffic 15 (7), 727–748.

Kimura, M., Morinaka, Y., Imai, K., et al., 2017. Extensive cargo identification reveals distinct biological roles of the 12 importin pathways. eLife 6, e21184.

Kosugi, S., Hasebe, M., Matsumura, N., et al., 2009. Six classes of nuclear localization signals specific to different binding grooves of importin α. Journal of Biological Chemistry 284 (1), 478–485.

Kosugi, S., Hasebe, M., Tomita, M., Yanagawa, H., 2008. Nuclear export signal consensus sequences defined using a localization-based yeast selection system. Traffic 9 (12), 2053–2062.

Kosugi, S., Hasebe, M., Tomita, M., Yanagawa, H., 2009. Systematic identification of cell cycle-dependent yeast nucleocytoplasmic shuttling proteins by prediction of composite motifs. Proceedings of the National Academy of Sciences 106 (25), 10171–10176.

Kosugi, S., Yanagawa, H., Terauchi, R., Tabata, S., 2014. NESmapper: Accurate prediction of leucine-rich nuclear export signals using activity-based profiles. PLOS Computational Biology 10 (9), e1003841.

Kudva, R., Denks, K., Kuhn, P., et al., 2013. Protein translocation across the inner membrane of Gram-negative bacteria: The Sec and Tat dependent protein transport pathways. Research in Microbiology 164 (6), 505–534.

La Cour, T., Kiemer, L., Mølgaard, A., et al., 2004. Analysis and prediction of leucine-rich nuclear export signals. Protein Engineering Design and Selection 17 (6), 527–536.

Lange, A., Mills, R.E., Lange, C.J., et al., 2007. Classical nuclear localization signals: Definition, function, and interaction with importin α. Journal of Biological Chemistry 282 (8), 5101–5105.

Lee, B.J., Cansizoglu, A.E., Süel, K.E., et al., 2006. Rules for nuclear localization sequence recognition by karyopherinβ2. Cell 126 (3), 543–558.

Lin, J.R., Hu, J., 2013. SeqNLS: Nuclear localization signal prediction based on frequent pattern mining and linear motif scoring. PLOS ONE 8 (10), e76864.

Lisitsyna, O.M., Seplyarskiy, V.B., Sheval, E.V., 2017. Comparative analysis of nuclear localization signal (NLS) prediction methods. Biopolymers and Cell 33, 147–154.

Li, H.M., Chiu, C.C., 2010. Protein transport into chloroplasts. Annual Review of Plant Biology 61, 157–180.

Maertens, G.N., Cook, N.J., Wang, W., *et al.*, 2014. Structural basis for nuclear import of splicing factors by human Transportin 3. Proceedings of the National Academy of Sciences 111 (7), 2728–2733.

Mao, Y., Zhang, Z., Gast, C., Wong, B., 2008. C-terminal signals regulate targeting of glycosylphosphatidylinositol-anchored proteins to the cell wall or plasma membrane in Candida albicans. Eukaryotic Cell 7 (11), 1906–1915.

Maurer-Stroh, S., Eisenhaber, B., Eisenhaber, F., 2002. N-terminal N-myristoylation of proteins: Prediction of substrate proteins from amino acid sequence1. Journal of Molecular Biology 317 (4), 541–557.

Mehdi, A.M., Sehgal, M.S.B., Kobe, B., Bailey, T.L., Bodén, M., 2011. A probabilistic model of nuclear import of proteins. Bioinformatics 27 (9), 1239–1246.

Meisinger, C., Sickmann, A., Pfanner, N., 2008. The mitochondrial proteome: From inventory to function. Cell 134 (1), 22–24.

Mossmann, D., Meisinger, C., Vögtle, F.N., 2012. Processing of mitochondrial presequences. Biochimica et Biophysica Acta (BBA)-Gene Regulatory Mechanisms 1819 (9), 1098–1106.

Munro, S., Pelham, H.R., 1987. A C-terminal signal prevents secretion of luminal ER proteins. Cell 48 (5), 899–907.

Nakai, K., Kanehisa, M., 1992. A knowledge base for predicting protein localization sites in eukaryotic cells. Genomics 14 (4), 897–911.

Nilsson, I., Lara, P., Hessa, T., *et al.*, 2015. The code for directing proteins for translocation across ER membrane: SRP cotranslationally recognizes specific features of a signal sequence. Journal of Molecular Biology 427 (6), 1191–1201.

Paila, Y.D., Richardson, L.G., Schnell, D.J., 2015. New insights into the mechanism of chloroplast protein import and its integration with protein quality control, organelle biogenesis and development. Journal of Molecular Biology 427 (5), 1038–1060.

Petersen, T.N., Brunak, S., von Heijne, G., Nielsen, H., 2011. SignalP 4.0: Discriminating signal peptides from transmembrane regions. Nature Methods 8 (10), 785.

Petriv, I., Tang, L., Titorenko, V.I., *et al.*, 2004. A new definition for the consensus sequence of the peroxisome targeting signal type 2. Journal of molecular biology 341 (1), 119–134.

Pierleoni, A., Martelli, P.L., Casadio, R., 2008. PredGPI: A GPI-anchor predictor. BMC Bioinformatics 9 (1), 392.

Prieto, G., Fullaondo, A., Rodriguez, J.A., 2014. Prediction of nuclear export signals using weighted regular expressions (Wregex). Bioinformatics 30 (9), 1220–1227.

Salvatore, M., Warholm, P., Shu, N., Basile, W., Elofsson, A., 2017. SubCons: A new ensemble method for improved human subcellular localization predictions. Bioinformatics 33 (14), 2464–2470.

Savojardo, C., Martelli, P.L., Fariselli, P., Casadio, R., 2015. TPpred3 detects and discriminates mitochondrial and chloroplastic targeting peptides in eukaryotic proteins. Bioinformatics 31 (20), 3269–3275.

Schneider, G., Sjöling, S., Wallin, E., *et al.*, 1998. Feature-extraction from endopeptidase cleavage sites in mitochondrial targeting peptides. Proteins: Structure, Function, and Bioinformatics 30 (1), 49–60.

Schulz, C., Schendzielorz, A., Rehling, P., 2015. Unlocking the presequence import pathway. Trends in Cell Biology 25 (5), 265–275.

Small, I., Peeters, N., Legeai, F., Lurin, C., 2004. Predotar: A tool for rapidly screening proteomes for N-terminal targeting sequences. Proteomics 4 (6), 1581–1590.

Soniat, M., Chook, Y.M., 2015. Nuclear localization signals for four distinct karyopherin-β nuclear import systems. Biochemical Journal 468 (3), 353–362.

Vögtle, F.N., Wortelkamp, S., Zahedi, R.P., *et al.*, 2009. Global analysis of the mitochondrial N-proteome identifies a processing peptidase critical for protein stability. Cell 139 (2), 428–439.

von Heijne, G., 1986. Mitochondrial targeting sequences may form amphiphilic helices. The EMBO Journal 5 (6), 1335–1342.

von Heijne, G., 1990. The signal peptide. The Journal of Membrane Biology 115 (3), 195–201.

Wiedemann, N., Pfanner, N., 2017. Mitochondrial machineries for protein import and assembly. Annual Review of Biochemistry 86, 685–714.

Xie, Y., Zheng, Y., Li, H., *et al.*, 2016. GPS-Lipid: A robust tool for the prediction of multiple lipid modification sites. Scientific Reports 6, 28249.

Xu, D., Marquis, K., Pei, J., *et al.*, 2014. LocNES: A computational tool for locating classical NESs in CRM1 cargo proteins. Bioinformatics. 31 (9), 1357–1365.

Yu, C.S., Chen, Y.C., Lu, C.H., Hwang, J.K., 2006. Prediction of protein subcellular localization. Proteins: Structure, Function, and Bioinformatics 64 (3), 643–651.

Further Reading

Nakai, K., 2000. Protein sorting signals and prediction of subcellular localization. Advances in Protein Chemistry 54, 277–344.

Nielsen, H., 2017. Predicting subcellular localization of proteins by bioinformatic algorithms. Current Topics in Microbiology and Immunology 404, 129–158.

Proteome Informatics

Frédérique Lisacek, SIB Swiss Institute of Bioinformatics, Geneva, Switzerland and University of Geneva, Geneva, Switzerland

Introduction

Proteomics sets the ambitious challenge of the large-scale determination of gene and cellular function directly at the protein level. As suggested in a Nature Methods Commentary: "Sequencing the human genome was perhaps the easy part, and now making sense of the constantly moving and changing picture of the proteome will require a lot of time, effort and creativity" (Nilsson *et al.*, 2010). Proteomics has thus evolved as a multidisciplinary field relying on both traditional analytical methods and the development of new technologies. Considering that in complex mixtures proteins may span up to 12 orders of magnitude in relative abundance (Angel *et al.*, 2012), reliable profiling of protein expression is obviously not straightforward.

More generally, the elucidation of biological processes requires information on protein abundance, amino acid variations and modifications, as well as interacting partners. Proteomics large-scale studies undertaken to collect this information roughly span four main objectives: (1) identification of all proteins from a cell or tissue creating a catalogue of information; (2) analysis of differential protein expression associated with specific conditions, different cell states and sample treatments; (3) characterisation of proteins by discovering their function, cellular location, posttranslational modifications (PTMs), etc and (4) contribution to understanding protein interaction networks. When met, these are useful to design effective diagnostic techniques (Liu *et al.*, 2013) or means of monitoring plant growth (Baginsky, 2009).

The present article is introductory to Robin (2018) and Bilbao (2018) where details of the relevant proteome databases and tools are provided. This trilogy is intended as explanatory for the common terminology found in the dense and sometimes cryptic data analysis paragraphs of the Material and Methods section of published proteomics studies.

Proteomics Technical Issues Bearing on Data Processing

The major steps leading to the generation of proteomics data involve sample preparation, molecule separation and mass measurement. Despite its paramount importance in the outcome of a proteomics experiment sample preparation cannot be detailed or discussed here. This information is nonetheless accounted for in the "Minimum Information About a Proteomics Experiment" (MIAPE, (Taylor *et al.*, 2007)) standard with which any stored experimental dataset must comply in proteomics data repositories (see Robin, 2018). Techniques for molecule separation and mass measurement are now introduced briefly in relation to the questions raised by their automation.

Separation Techniques

Technical means to separate complex mixtures have been and still are a central concern in physics and chemistry. The physico-chemical properties of biological molecules account for the common use of electrophoresis and liquid chromatography in molecular biology.

The major difference in using one or the other techniques in the proteomics workflow is the position of the protein digestion step in the overall process. Proteins are separated by 2D-gel electrophoresis, excised from the gel and then digested so as to feed the mass spectrometer whereas proteins are digested prior to running the chromatographic separation that is coupled to the mass spectrometer.

2D-gel Electrophoresis and image processing

Under the influence of an electrical field, charged molecules migrate in the direction of the electrode bearing the opposite charge. Because of their varying charges and masses, different molecules will migrate at different speeds and will thus be separated into single fractions. Such a principle defines electrophoresis.

The electrophoretic mobility, which influences the speed of migration, is a significant and characteristic parameter of a charged molecule. The electrophoretic separation of samples is done in a buffer with a precise pH value and a constant ionic strength. Proteins are separated in two dimensions: mass and pI (isoelectric point). On gels they are not visible and various staining methods are used for detection. Proteins are revealed on a 2D-gel as spots of various sizes and intensity that correspond to protein abundance (darker spots for more abundant proteins).

An image capture of a 2D-gel can be analysed by software to assist spot picking by a robot and/or identify differentially expressed proteins. Gel suppliers often support 2D-gel analysis software most of which is licensed. The two commonly used tools are SameSpots (see "Relevant Websites section") and ImageMaster/Melanie (see "Relevant Websites section"). They have been developed over decades. The first version of SameSpots was called Phoretix 2D (Mahon and Dupree, 2001) and Melanie was popularised in Appel *et al.* (1991). Despite their respective good performance, both tools still stumble on the inherent ambiguity of faded spots. Other algorithmic issues hindering 2D-gel qualitative and quantitative comparison are well explained in Dowsey *et al.* (2010).

Digitalised 2D-gels have been annotated and stored in on-line databases for comparative and benchmarking purposes. Many of these were created in the 1990s and are no longer maintained. The reference database remains World 2D-page (see "Relevant Websites section") and more recently the GelMap portal was initiated for plant species (see "Relevant Websites section").

Although many hundreds and even thousands of proteins can be separated on a 2D-gel, this technique remains less popular in high-throughput settings despite notable attempts such as Dowsey et al. (2008). Nonetheless, 2D-electrophoresis cannot be ignored as a useful approach for instance to assess the extent of protein forms in a given sample and tends to be considered as a complement to the popular Liquid Chromatography (LC)-based proteomics (Rogowska-Wrzesinska et al., 2013).

Liquid chromatography and retention time prediction

Liquid Chromatography (LC) utilises a liquid mobile phase to separate the components of a peptide mixture. A liquid medium continuously flows through a chromatographic column and carries the analytes. The chromatographic separation process is based on the difference in the surface interactions of the analyte and eluent molecules. The velocity at which individual molecules in the mixtures move through the column is a function of the physical property of the analytes, the content of the column and the composition of the mobile phase. Chromatograms are output as plots of the retention time (duration from tip to bottom of the column) in the x-axis and compound abundance in the y-axis. The most popular technique in proteomics is the reverse-phase chromatography. Although reproducible, small variations are observable in these plots as a reflection of peptide conformational changes among others.

The hydrophobicity of peptides is highly correlated to retention time. Due to surface interactions, hydrophobic peptides tend to stay longer in the column. However the sole consideration of this property is not enough to accurately predict retention time from the peptide amino acid sequence. The long path to designing an efficient method is well described in Henneman and Palmblad (2013) where the many attempts with mixed results and performance are referred to in a timeline starting in 1951. Furthermore, the comparison of different methods for retention time prediction has often not led to a consensus in the metric to be used in the evaluation of the accuracy of prediction. As very nicely put by Henneman and Palmblad (2013), discussions centred on the (non)-linearity of the model and variations of the tolerance to error are pointless when the need for assigning a confidence interval to a predicted retention time is prevalent. Another excellent review of peptide retention time prediction methods can be found in Moruz and Käll (2017) where the contribution of prediction is shown (i) to improve peptide identification in simple mixtures by matching theoretical and observed masses and retention times (ii) to increase the throughput of proteomics experiments or (iii) to decrease the number of candidate peptides during database searching.

Interestingly, software designed for 2D-gel image analysis was transposed in the 1990s to LC-MS images defined as plots of mass versus retention time mainly for quantitative studies. Algorithmic aspects of the associated software are reviewed in Dowsey et al. (2010).

Ion mobility

Ion mobility (IM) is a fast and reproducible technique for separation of molecules based on their size and shape. Molecules having the same mass-to-charge ratio, but different conformational arrangements can be distinguished. IM has been and is used in combination with LC and mass spectrometry (see next section), providing an additional dimension of information with benefits for proteomics analyses (Baker et al., 2015). The following list of software supports IM data processing: Synapter (Bond et al., 2013), LC-IMS-MS FeatureFinder (Crowell et al., 2013), ISOQuant (Distler et al., 2014) and Skyline (Pino et al., 2017).

While IM has shown great utility when coupled with MS for analysis of complex samples, methods for processing the complex data generated still lag behind. In fact, the incorporation of this extra separation dimension requires upgrades and optimisation of existing computational pipelines and the development of new algorithmic strategies to fully exploit the advantages of the technology.

Mass Spectrometry

A mass spectrometer produces charged particles (ions) from the molecules to be analysed, in the case of proteomics, peptide mixtures. Various means are implemented in the spectrometer to exert forces on the charged particles and measure the mass or molecular weight of the analysed substance.

From raw data to peak lists

All mass spectrometers carry out three distinct functions:

1. *Ionisation* for charging peptides; the major two processes are Electrospray Ionisation (ESI) and Matrix-Assisted Laser Desorption Ionisation (MALDI). These techniques are detailed in all basic mass spectrometry (MS) manuals.
2. *Ion analysis* for separating ions by mass; the different analysers in proteomics are Time-Of-Flight (TOF), Quadrupole, Ion Trap, Fourier Transform Ion Cyclotron Resonance (FTICR). These techniques are detailed in all basic MS manuals.
3. *Ion detection* that requires the collection of ions of different masses. The charged ion beam generated by the analyser is directed sequentially (point) or in parallel (array) onto a point/array collector. This causes an electric current the flow of which marks the arrival of successive ions and the magnitude of which marks their abundance. The analogue electrical signal is digitised and processed by a computer.

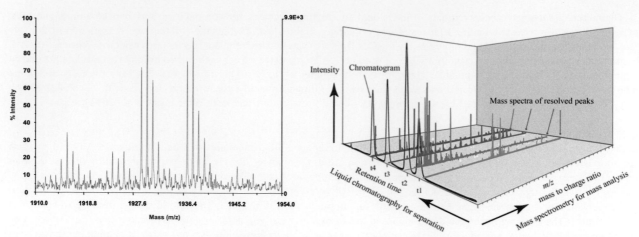

Fig. 1 (a) Typical MALDI-TOF spectrum, (b) Relationship between dimensions *m/z*, intensity, retention time. *Source*: https://en.wikipedia.org/wiki/File:Liquid_chromatography_MS_spectrum_3D_analysis.png.

A mass spectrometer outputs raw mass spectra featuring the mass-to-charge ratio (*m/z*) in the x-axis and the intensity in the y-axis (**Fig. 1**).

In theory, the mass of each peptide should correspond to a peak in the spectrum. In reality, the imperfection of the sample, its handling and processing, etc., introduce a background noise and cause variations in intensity. The selection of the relevant peaks, that is, those definitely corresponding to peptide masses, is not straightforward. The most intense peak in an isotopic group is not necessarily the monoisotopic mass of a peptide. The intensity ratio of all peaks in this group must be examined. Moreover, a spectrum may contain single- and double- or triple-charged species. Indeed, a population of variably charged ions is generated in the ionisation process. In a given population, peptides may have 1, 2 and 0 sites of protonation. In this case, the intensity of the peaks is a reflection of the population generated in the ionisation process. Multiple-charged ions are generally of low abundance in MALDI but frequent in ESI. At sufficient resolution, a single-charged ion will show isotopic peaks that differ by 1 mass unit, a double-charged ion will show peaks that differ by 0.5 mass units and so on.

Despite these objective difficulties, peak detection is automated. Many solutions are proprietary and integrated in software licensed with an instrument. OpenMS (see "Relevant Websites section") detailed in Bilbao (2018) is a popular software library implementing a wide range of useful functions for MS data analysis. It includes for instance, peak detection which main task is centroiding. According to the most recent IUPAC recommendations (Murray *et al.*, 2013) centroiding spectra entails calculating the centroid based on the average *m/z* value weighed by the intensity, and assigning *m/z* values based on a calibration file (Internal calibration consists in running a well-known sample, detect the signals, and use a correction function to associate the known values to the experimentally measured values. It is a recalibration step that corrects the experimental data with the use of information in the same spectrum. Once the correction is made, the instrument can update its hardware parameters accordingly. Regular calibration of the *m/z* scale is necessary to maintain accuracy in the instrument.). Only the centroid *m/z* value and the peak magnitude are stored. This is the basis of peak lists that are input in the various software tools that identify proteins from mass spectra. Another popular open software library for performing basic mass spectra processing converting raw data to peak lists is ProteoWizard (see "Relevant Websites section"). As OpenMS, its major advantage is that it is instrument independent and provides tools to convert vendor proprietary formats to standard and open formats (Robin, 2018; Bilbao, 2018).

Inherent Role of Bioinformatics in Proteomics

In 1993, five independent groups published on a method for protein identification using mass spectrometry (MS) referred to as Peptide Mass Fingerprinting, shortened as PMF (Henzel *et al.*, 1993; James *et al.*, 1993; Mann *et al.*, 1993; Pappin *et al.*, 1993; Yates *et al.*, 1993). Each of these was integrating software to match a list of experimentally observed peptide masses with theoretical peptide masses computed from protein sequences stored in a selected database. This computation is fairly simple given the known monoisotopic mass of each of the 20 amino acids. By definition, the mass of a peptide is the sum of masses of the amino acids it is made of.

This protein identification technology relies heavily on the fast and accurate mapping of mass spectra to protein database entries, and gave birth to a bioinformatics discipline focusing on the development of software tools for the processing of MS data (proteome informatics). Several software tools implementing this *Peptide Mass Fingerprinting* (PMF) identification were developed in the following years such as Perkins *et al.* (1999) the seed of the most popular Mascot software (see Bilbao, 2018).

Soon after, the first publication on another MS based approach for identifying proteins emerged, tandem MS or MS/MS (Eng *et al.*, 1994). In this setup the mass spectrometer performs sequential MS analyses. In a first stage intact peptides are analysed and a mass spectrum is produced. Next, a selection of these peptides is fragmented at the peptide bond by bombarding them with fast-moving atoms. The mass and intensity of each peptide fragment is detected producing a MS/MS spectrum. Each MS/MS spectrum can be mapped to its originating peptide using a similar approach to PMF, where the assembly of peptide fragment masses making up the mass spectrum is screened against theoretical peptide fragment masses. Technically, the workflow can be fully automated: digested peptide mixtures are run through LC, then upon elution from the chromatographic column the analytes are ionised and directly sprayed into a first mass spectrometer, where they are measured. The resulting spectrum provides the mass to charge (m/z) ratio of the intact analyte ions (precursor ions) that are coming off the LC. The ions of a precursor are selected, and fragment ions are created and measured in a second mass spectrometer to produce the MS/MS spectrum for the analyte. This approach was called *shotgun proteomics* (Wolters *et al.*, 2001) referring to earlier *shotgun genomics*, which led to sequencing the first human genome, as a means to boost the field to the high-throughput level.

Early automation: Shotgun proteomics

From the turn of the 21st century, mass spectrometry combined with liquid chromatography has quickly become a central analytical platform for the high-throughput investigation of complex protein samples (Aebersold and Mann, 2003). MS instrumentation with higher resolving power and mass accuracy, increased sensitivity, alternative fragmentation mechanisms and new data acquisition strategies has boosted the throughput, quality and depth of proteomics data. At the same time, the ability to quickly and consistently analyse the large amounts of experimental data produced in proteomics research has relied on software advances well reviewed in Nesvizhskii (2010) and illustrated in Bilbao (2018).

Over the years, bioinformatics has emerged as a crucial contributor with the development of algorithms, the connection of various tools into pipelines, as well as the deposition and dissemination of both software and data. Currently, the prominence of open source software and public data in reproducible research is emphasised and clearly details in both aspects of software in Bilbao (2018) and public data in Robin (2018).

Fragmentation techniques

Tandem mass spectrometry (abbreviated MS/MS) consists in determining the masses of selected ions. In MS/MS, a precursor or parent ion is mass-selected and fragmented and the masses of the resulting product or daughter ions are analysed. Some peptides are further analysed to determine their precise amino-acid content. The technique requires either two or more analysers in series or relies on the intrinsic characteristics of some spectrometers such as the MALDI-TOF. When several analysers are involved, precursor ions are separated according to their m/z value in the first analyser and selected for fragmentation. Fragments are analysed in a second analyser.

Different types of fragment ions observed in an MS/MS spectrum depend on the peptide primary sequence, the amount of internal energy, how the energy was introduced, the charge state, etc. Fragment ions are named according to an accepted nomenclature first proposed by Roepstorff and Fohlman (1984), and modified by Johnson *et al.* (1987). It is represented in the **Fig. 2**

The most common ions observed in peptide mass spectra are *a*, *b*, and *y*. The *a* type ions occur at a lower frequency and abundance compared to the *b* and *y* type ions. The *a-b* pairs are often observed in fragment MS/MS spectra. The presence of such pairs can then be used for the detection of *b* ions.

The most commonly used peptide fragmentation techniques are:

- Collision-Induced Dissociation (CID).
- Electron-Capture Dissociation (ECD).
- Electron-Transfer Dissociation (ETD).

In CID, molecular ions are fragmented in the gas phase (Mitchell Wells and McLuckey, 2005). In this case, the molecular ions are pushed to collide with neutral molecules such as helium, nitrogen or argon. Collision induces kinetic to internal energy transfer. The transfer of energy results in bond breakage so that ions are split into smaller fragments. CID most often leads to the generation of *b* and *y* type ions. Note that Higher-energy Collisional Dissociation (HCD) often cited in the literature is a CID technique specific to a particular type of mass spectrometer, namely the Orbitrap which principle was introduced in Makarov (2000).

In ECD, molecular ions are fragmented by the direct introduction of low energy electrons to trapped gas phase ions (Zubarev *et al.*, 1998). During ECD, the parent ions capture single electrons and fragment simultaneously. In such case, the

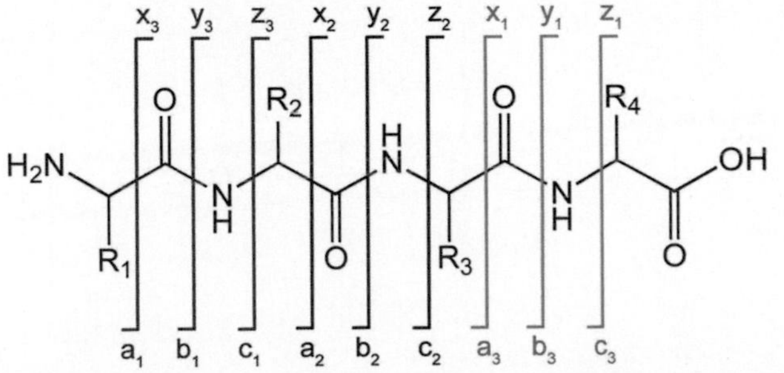

Fig. 2 The *a*,*b*,*c* ions extend from the N-terminus while *x*,*y*,*z* ions extend from the C-terminus. *Source:* https://commons.wikimedia.org/wiki/File: Peptide_fragmentation.gif

fragmentation occurs also in the positions that are not energetically favoured. This generates more complete ions series, most often *a*, *c* and *z* type ions.

ETD is another technique to fragment molecular ions in the gas phase (Syka *et al.*, 2004). Similar to ECD, ETD induces fragmentation of peptides by transferring electrons to them. ETD cleaves randomly along the peptide backbone while side chains are left intact. The technique works well for higher charge state ions ($> +2$) generating c and z type ions. ETD is advantageous for the fragmentation of longer peptides or even entire proteins.

Tandem MS provides structural information by establishing relationships between precursor ions and their fragmentation products.

Data Processing

A tandem (MS/MS or MS2) mass spectrum is composed of a precursor peptide and a combination of fragment peaks produced from the fragmentation the peptide sequence. The number of peaks may vary from ten to several hundreds. The peak of a MS/MS spectrum is characterised by the same two values as the MS1 spectrum: mass-to charge ratio (m/z) in its intensity (abundance).

Peptide identification relies on matching peaks from MS/MS spectra with either theoretical or reference experimental spectra. Software for performing this task is detailed in Bilbao (2018). Generally, it first entails the *in silico* digestion of protein sequences into peptides in accordance with the cleavage rules of the protease used in the sample preparation step of the experimental workflow. Then, for each peptide a theoretical spectrum is generated and the calculated peptide fragments are compared with the experimentally observed peaks. A peptide match score is derived which reflects the similarity between the MS/MS spectrum and the theoretical peptide spectrum. This aspect is detailed in the next section.

This type of analysis has been extensively reviewed in the literature where different algorithmic solutions are described in terms of finding the optimal Peptide Spectrum Matches (PSMs). The need to optimise arises from a variety of situations inherent to mass spectrometry. An MS/MS spectrum contains peaks that cannot be assigned to backbone fragments. Possible sources of these peaks include neutral losses, side chain fragments and ions due to multiple backbone fragmentations. Other challenges include ambiguous co-fragmented peptides and frequently missing fragment ions at the beginning or the end of a peptide sequence.

Manually or automatically annotated MS/MS spectra are commonly displaying the *ion ladders* reflecting the ion nomenclature as shown in the figure below where peaks matching an ion series are labelled (**Fig. 3**).

PSM Scoring and Validation of Protein Inference

A number of different scoring functions have been described in the literature including simple spectral correlation functions such as the dot-product (Liu *et al.*, 2007), more advanced cross correlation functions, scoring functions based on empirically observed rules or statistically derived fragmentation frequencies. The score reported for a given PSM can be on some arbitrary scale or converted to a statistical measure such as a p-value or an expectation value, E-value. The final list of identified peptides is compiled into a protein 'hit list', which is the output of a typical proteomics experiment

Several properties can contribute to scoring the similarity between an experimental spectrum and a reference. The precursor mass is mainly used to restrict the search space. The mass differences for both the matching fragments and the precursor are usually considered in the scoring. Other properties include the number and intensities of the experimental peaks, number of complementary ions, number of missed cleavages and number (and types) of modifications.

A great variety of scoring strategies have been proposed for peptide identification. Broadly speaking, they can be divided into heuristic and probabilistic schemes. Heuristic methods typically combine a set of scoring properties in a practical form reflecting expert knowledge. A product or a linear function are simple examples of how these properties can be combined into a scoring function. Probabilistic schemes compute the probability that a given sequence is the origin of the spectrum.

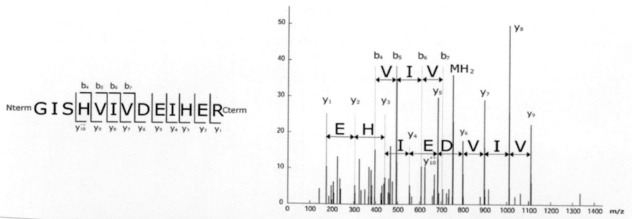

Fig. 3 Example of annotated spectrum that shows the ion ladders (y and b series) reconstituted for peptide GISHVIVDEIHER. Reproduced from Hernandez, P., *et al.*, 2006. Automated protein identification by tandem mass spectrometry: Issues and strategies. Mass Spectrom. Rev. 25 (2), 235–254.

The "target-decoy" approach is commonly used for estimating the false discovery rate (FDR) (Elias and Gygi, 2007). A target-decoy database consists of all sequences to be used in the identification process plus the same amount of decoy sequences. For each sequence entry in the database, a 'matching' decoy entry is produced by either reversing the original sequence or by shuffling the letter code. Seemingly counterintuitive, searching against a larger search space decreases the number of reliable identifications because of the higher probability of retrieving false positives.

De Novo Sequencing

De novo sequencing algorithms attempt to derive a peptide sequence directly from the MS/MS spectra by considering the mass difference between pairs of peaks and all possible amino acid combinations (longest ion ladders). The "peptide sequence tag" algorithm was a stepping-stone in the implementation of the identification process (Mann and Wilm, 1994). It is based on extracting short unambiguous amino acid sequences from the spectrum peak pattern by linking two peaks with a mass difference corresponding to the mass of an amino acid. In theory the full peptide sequence can be found using an algorithm based on tag-extraction by finding a tag spanning the full mass range of the spectrum. The approach seemed well suited to solving *de novo* sequencing. However, the accuracy of such algorithms hinges on very high quality spectral data and therefore remains limited to sequencing a small amount of peptides. In practise, sequence tags in combination with the spectrum precursor mass information have been used as a specific probe to identify the peptide origin of the MS/MS spectrum.

Many algorithms have been published whether dependent on dynamic programming, integer linear programming, machine learning and a few other modelling techniques. The coverage, the description and the evaluation of software solutions is provided in Muth and Renard (2017) where all methods that were published were tested. In the end, only a handful remains operational and usable with any type of fragmentation data. Only two of these are open (Frank and Pevzner, 2005; Ma, 2015). Interestingly, the most convincing application of this software so far is antibody sequencing (Tran *et al.*, 2016; Guthals *et al.*, 2017) which brings out sequence polymorphism independently of DNA sequencing. Such an alternative method is helpful since an individual antibody repertoire cannot be predicted from the genome alone.

Identification and Localisation of Post-Translational Modifications

At this point, it is probably important to note that only a minority of spectra are actually assigned to peptides when processing data. In fact, the vast majority, that is at least 60% of spectra produced by the mass spectrometer remain unassigned and this is sometimes called "proteomics dark matter" (Skinner and Kelleher, 2015). This substantial unidentifiable portion of the data is often thought to be attributable to our poor knowledge of Post-Translational Modifications (PTMs). PTMs are alteration of proteins during or after protein biosynthesis, resulting in changes of the protein physicochemical properties, which modulate and regulate cellular functions. Abnormal PTM abundances are involved in disease states.

By identifying the occurrence of regular mass shifts corresponding to the addition or removal of chemical groups of known masses in a peptide MS/MS spectrum, it is possible to identify a PTM and its position on the peptide sequence. When specified in the search parameters, search engines can consider relevant PTMs, such as phosphorylation. However, this number is usually limited because the search space is substantially expanded.

Exhaustive identification of PTMs in high-throughput MS-experiments has proven to be difficult for a number of reasons including the large number of possible PTMs, sub-stoichiometric amounts of modified proteins, and as peptides carrying certain PTMs display MS2 fragmentation patterns which can be difficult to interpret (Creasy and Cottrell, 2004). Therefore successful protein modification studies rely on carefully designed experimental setups applying extensive sample fractionation/enrichment protocols for the detection of low abundant protein species. MS data need to be accurate and contain information rich fragmentation patterns. Software for PTM automated detection is detailed in Bilbao (2018).

It was proposed to rename modified proteins into proteoforms (Smith *et al.*, 2013) or modforms (Prabakaran *et al.*, 2012) suggesting that protein-protein interactions should actually be described as *form-*form interactions as pointed out in Chichester *et al.* (2003).

Proteomics Workflows

In this section, the impact of the technical and experimental choices on data processing and analysis is reviewed.

Database Versus Spectral Library Search

As explained in Section "Proteomics Technical Issues Bearing on Data Processing", the fragmentation pattern encoded by a tandem (or MS/MS) spectrum supports the identification of the corresponding peptide sequence. Search engines annotate the MS/MS spectra with peptide sequences using various algorithms and scoring strategies (Nesvizhskii, 2010). Even though database search engines remain very popular, a significant number of articles promote the use of direct spectral matching and the construction of spectral libraries to support reliable peptide identification and quantification especially in high-throughput settings (Lam *et al.*, 2007). Computationally, using spectral libraries is also much more efficient as it skips the construction of theoretical spectra. The downside is of course, the potential incompleteness of the library and its subsequent limited coverage. Nonetheless, the constant improvement of instrument accuracy and sensitivity goes along with the potential for generating libraries with good coverage.

In database search, protein sequences are digested *in silico* into peptides. Considering knowledge about the type of fragmentation technique employed, theoretical spectra are generated from each peptide sequence. The search engine then scores the observed tandem mass spectra by matching against the predicted fragmentation, producing a list of peptide-to-spectrum matches (PSMs).

Spectral library searching offers a natural way of incorporating previous knowledge about observed peptides and their respective fragmentation patterns into a new search. The National Institute of Standards and Technology (NIST, see "Relevant Websites section") and the Institute for Systems Biology (ISB, see "Relevant Websites section") have invested effort in compiling high quality publicly available spectrum libraries for different organisms and instrument types including some specialised libraries rich in modified peptides. Spectral libraries generated from previously annotated spectra can be used as reference to further annotate new generated spectra, producing a list of spectrum-to-spectrum matches (SSMs).

To enable the exploration of these spectral libraries when analysing new experimental data, several tools have been developed (Griss, 2016). Firstly, peptide identification with these tools is fast as the experimental data is screened against small size databases and the time-consuming modelling of theoretical spectra can be skipped. Secondly, a library spectrum is rich in information (including fragmentation intensities of a wide range of fragment ions) relative to a theoretical peptide spectrum. Finally, library search tools typically employ simple and fast scoring algorithms and are expected to yield more discriminative match scores than sequence search tools. The main drawback remains the impossible identification of peptides, which are absent in the library. Note that decoy spectral library tools have been developed to evaluate the quality of spectral matching based on an FDR calculation. These are reviewed in Zhang *et al.* (2018).

A particular representation spectral data is in the form of spectral network in which spectral similarity is mapped (nodes are spectra and edges are similarity between spectra as measured by peak alignment). As noted in Guthals *et al.* (2012), the spectral network paradigm is founded on two core principles beyond mainstream approaches: (1) matching unidentified spectra to a spectral library or to other unidentified spectra is more efficient than to theoretical spectra based on reference sequences and (2) consensus interpretation of sets of related spectra is more reliable than identification of one spectrum at a time.

At this point in time, the consensus view when comparing the much more frequently used database search to spectral library matching is that each corresponding output complements each other. The combination of both strategies tends to increase the number of confidently identified peptides and proteins.

Data Dependent Acquisition (DDA) Versus Data Independent Acquisition (DIA)

In shotgun proteomics, the selection of precursor ions in mass spectra (MS1) for fragmentation that will produce tandem mass spectra (MS2) is defined upon intensity. In other words, the ten or twenty (parametered choice) most intense peaks designate which peptide will be fragmented. To make this choice less dependent on the data and somehow less arbitrary, a new approach called "Data Independent Acquisition" (DIA) was introduced contrasting to "Data Dependent Acquisition" (DDA) by means of co-selection and co-fragmentation. DIA is by design avoiding the detection and selection of individual precursor ions during LC-MS analysis.

An early DIA proof-of-principle was reported in Purvine *et al.* (2003). Then Venable *et al.* (2004) suggested the use of sequential isolation and fragmentation of several precursor ion windows (10u each) using a linear ion trap over a defined mass range, thereby introducing the term "data-independent acquisition". DIA integrates qualitative and quantitative information. Recorded spectra are a continuous map of multiplexed chromatographic profiles from all detectable peptides, which are eluted, ionized and fragmented. As a consequence of this parallel peptide analysis, one of the main challenges to perform identification in DIA spectra is the lack of a direct relationship between each precursor ion and its fragment ions, as opposed to DDA. The most straightforward way to process DIA spectra is the direct submission to DDA search engines. This topic is well covered in Bilbao (2018).

Overall, DDA performance declines as sample complexity increases because the semi-stochastic selection of precursor ions aggravates known limitations for both identification and quantification: limited reproducibility and dynamic range, bias toward high abundance peptides and under-sampling to name a few. In turn, DIA solves these issues while generating highly convoluted, dense and redundant datasets that remain challenging to analyse.

Top Down Versus Bottom up MS

Shotgun proteomics and other strategies that emerged with the increasing resolution power of mass spectrometers are still facing with challenges of protein information loss resulting from the ambiguity of the peptide reassembly process into proteins. To begin with, identified peptides are often not specific to an individual protein and even less to a protein form, which may lead to piecing together chimeric proteins. Secondly, significant portions of a protein may not be identified, for instance due to the absence of lysine or arginine and the oversized peptides generated by tryptic digestion. Consequently, potential meaningful PTMs or sequence variants occurring in these regions cannot be detected.

So far in the present article, all approaches have been based on breaking proteins into pieces (peptides and peptide fragments). The generic term for qualifying these is "bottom-up" proteomics. In recent years, "top-down" proteomics has emerged as an answer to the shortcomings just listed. In this set-up, the intact protein is introduced into the mass spectrometer where both intact and fragment ions masses are measured. Taking advantage of high resolution and mass accuracy, Top Down proteomics studies have largely been implemented using ESI coupled to either Fourier transform ion cyclotron resonance (FT-ICR) or Orbitrap mass analysers. These approaches are well reviewed in Catherman *et al.* (2014).

The most renowned software for the identification of intact proteins is ProSight which latest version is defined as light (Fellers *et al.*, 2015). It uses the precursor mass and a mass tolerance window to generate a possible list of candidates from a larger annotated database. The theoretical fragment ions from the candidates are then compared to the experimentally determined fragment ions within a fragment mass tolerance. A P-score is calculated for each hit, representing the probability that a random sequence could account for the matching ions. ProSight versions were gradually improved to include fixed (e.g., alkylation of cysteine residues) and terminal (e.g., N-terminal acetylation) modifications. An alternative to ProSight is mainly BIG Mascot or MascotTD, based on the popular Bottom Up software platform Mascot and extends the precursor mass cutoff from 16 kDa to 110 kDa (see "Relevant Websites section").

To be complete, the term "middle-down" proteomics should also be defined in this section. This approach generates larger size peptides than the regular bottom-up. The major challenge is often the selection of the enzyme that will digest proteins into peptides of the appropriate size.

Label-Free Versus Labelled Quantification

Broadly speaking, absolute quantification estimates intracellular protein concentrations on individual samples, and relative quantification compares the measured protein abundances in one sample relative to another. In terms of sample preparation, quantification methods can include stable isotope labels or label-free based.

Adding stable isotope labels significantly reduces experimental bias and improves accuracy, which requires specific but typically straightforward data processing approaches. Relative quantification relies on three main types of techniques in proteomics: metabolic, chemical or enzymatic. The metabolic approach is characterised by the Stable Isotope Labelling with Amino acids in Cell culture (SILAC) method that uses non-radioactive isotopic labelling (Ong *et al.*, 2002). The chemical methods include isotope-coded affinity tags (ICAT) (Gygi *et al.*, 1999), isobaric labelling tandem mass tags (TMT) (Thompson *et al.*, 2003) and isobaric tags for relative and absolute quantification (iTRAQ)(Ross *et al.*, 2004). The enzymatic approach is illustrated with the $^{16}O/^{18}O$ where the enzyme-catalyzed incorporation of oxygen is performed in the C-terminal carboxylic acids of peptides during the digestion procedure (Yao *et al.*, 2001). In the workflows using of these different methods, samples corresponding to distinct experimental conditions are labelled and combined, and all samples are compared in the same LC-MS analysis of the pooled samples. Labelling is performed on proteins (pre-digestion) or on peptides (post digestion). There is no computational challenge as each mass change caused by labels is known and only mass differences are sought. In the end, label-based methods increase costs (reagents are expensive) and sample preparation time, while limiting the number of different samples that can be compared.

In contrast, in label-free methods different samples are prepared and analysed individually, requiring computational methods to align retention time across multiple samples and to normalise intensity. As (Bilbao, 2018) is fairly detailed on software available for supporting peptide and protein quantification, this section will only mention image-based analysis. For example, MS1 "2D-images" plotting m/z and retention time (rt) values of peptides have been used and processed in much the same way 2D-gels images are scanned and compared for differential analysis. In this representation, retention time alignment is approached by superimposing 2D (m/z, rt) maps. Solutions to these image processing issues overlapping those raised by 2D-gel analysis are reviewed in (Dowsey *et al.*, 2010). Furthermore, the technique known as MALDI-imaging mass spectrometry offers yet another label-free technology. A tissue section is coated with matrix and irradiated with a laser. Molecules are then desorbed and ionised according to the MALDI technique. Spectra are acquired from each location (pixel) over the surface of the tissue and 2D ion density images are reconstructed from the spectra (Seeley and Caprioli, 2008). Image analysis software is listed on the website of the active MALDI-imaging community: https://ms-imaging.org.

A comprehensive coverage of quantification should include a mention of targeted proteomics usually opposed to discovery (shotgun) proteomics. Selected reaction monitoring (SRM) has become the MS method of choice to accurately and consistently quantify a selection of proteins across samples. It is typically used a means of validating the expression of a protein or a proteotypic (i.e., uniquely characterising a protein (Mallick *et al.*, 2007)) peptide. Note that so-called "directed proteomics" is an intermediary strategy that entails using a list of predetermined precursor ions, otherwise called "inclusion list" in order to focus on a particular segment of a global proteome and increase reproducibility. To conclude this section on quantification approaches, **Fig. 4** (inspired from **Fig. 1** in (Bensimon *et al.*, 2012)) highlights the requirements of the different label-free strategies and the corresponding depth of analysis achieved.

Some Omics Bridging Issues

Needless to say, proteomics as part of the omics landscape is integrated with many other omics. It is virtually impossible to cover all existing options and the following highlights a selection.

Proteogenomics

Genomic sequencing data can be used to improve protein identification through refining the construction of a protein sequence database. Proteomic data can in turn support large scale RNA and DNA sequencing initiatives by providing evidence of coding sequence variants, novel coding transcripts or refined gene boundaries. In fact support to complete genome annotation was suggested

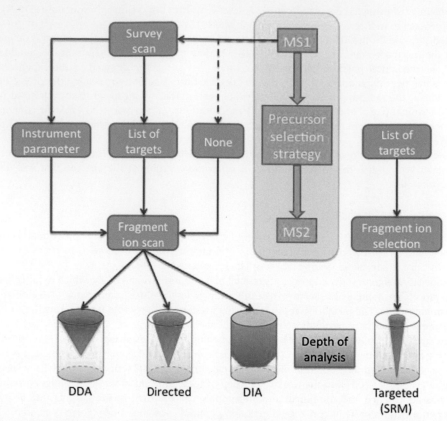

Fig. 4 The various protein identification and quantification strategies and corresponding options at the MS1 level and subsequent MS2 level following precursor ion selection (full red line, required; dashed red line, optional). DDA=Data-Dependent Acquisition (implemented in shotgun proteomics); DIA=Data-Independent Acquisition; "Directed" and "Targeted" are defined in the text. Results are represented in terms of coverage in red volumes to illustrate the depth of analysis.

very early on in proteomics with the prospect of revealing information not accessible in genomics (Küster *et al.*, 2001; Choudhary *et al.*, 2001). But it took longer than expected to design systematic approaches such as Branca *et al.* (2014) and the work that follows to shape the field of proteogenomics. In the meanwhile, isolated attempts led to contribute to plant gene/protein characterisation, e.g., identification of paralogue genes from multigenic families (Delalande *et al.*, 2005) as well as refining bacterial genome annotation, e.g., identification of precise translation start in proteins of *Mycobacterium tuberculosis* (Rison *et al.*, 2007).

Since 2013, efforts in proteogenomics are more structured. For instance it is playing an important role in the Chromosome-Centric Human Proteome Project (C-HPP) developed within the Human Proteome Organisation (HUPO). C-HPP aims at characterising all protein coding genes of the human proteome and the information is collected and centralised in the neXtprot database (Gaudet *et al.*, 2013). Proteogenomics is also gaining attention in bacterial genome sequencing and annotation. Beyond the resolution of ambiguous cases of translations start sites, generic solutions are now available (Omasits *et al.*, 2017) and make a difference in terms of coding sequence annotation in complete genomes.

Another aspect of proteogenomics is shown in the range of software was developed for the identification of sequence variants through building customised databases of protein sequences of known variant peptides. There are basically three options for the search engines used in this context: mass spectra are searched (i) once against a database of mixed of wild-type and variant peptides, (ii) twice against a first database of wild-type and a second database of variant peptides (iii) once against a database of variant peptides. These approaches are benchmarked on two large human and bacterial datasets with two different search engines in Ivanov *et al.* (2017). Despite a preference for the first strategy, results with others are close and possible gaps are influenced by the search database size or the size of the training data set used for validation.

The existence of small open reading frames (less than 300 nucleotides) undetected by gene prediction software (especially in the human and mouse genomes) has been debated prior the finding experimental means to produce evidence. A solution was brought by ribosome profiling coupled to MS-based protein and peptide identification (Koch *et al.*, 2014). This also led to the creation of a database of small ORFs (see "Relevant Websites section").

Ultimately, proteogenomics has unique information to contribute in the context of other omics. For example, proteogenomic techniques can help assess protein-level expression in viral infections as they alter gene expression and protein production when the virus highjacks cellular processes that allow for self-replication (Ruggles *et al.*, 2017).

Interactomics

The rising of network biology (Barabási and Oltvai, 2004) has significantly modified knowledge representation in molecular biology. In protein science, experimental techniques for studying protein-protein interactions and protein expression in cells and tissues have evolved in parallel for a while and finally met to benefit each other. (Bensimon *et al.*, 2012) provide a very clear account of this mutual influence and resulting integration mostly in the context of systems biology. In fact, proteomics has been used for solving protein-protein interactions in complexes since the publication of the Tandem Affinity Purification (TAP) method (Rigaut *et al.*, 1999). Ten years later, a full pipeline known as ProHits processes raw MS data files to produce fully annotated protein-protein interaction datasets and considered a reference (Liu *et al.*, 2010). The noisy nature of interactomics data led the authors to also release a contaminant database known as the Crapome (http://www.crapome.org).

Substantial progress in generating reliable MS-based interactome data is expected to also arise from work that has been and still is invested in detecting protein interactions by cross-link experiments (shortened as XL-MS). The identification of cross-linked peptides from mass spectra and the mapping of identified cross-links in a structural context naturally relies on efficient computational methods, as reviewed in (Yılmaz *et al.*, 2018). Cryo-EM (Electron Microscopy) data was shown to complement XL-MS (Snijder *et al.*, 2017) and such type of "hybrid" MS strategies (i.e., combined with other technologies) is very likely to significantly expand in the years to come.

PTM Detection and Mapping

In this section, two examples of the most frequent PTMs are briefly covered to highlight the productive contribution of bioinformatics in the past decade. Other related issues are discussed later in the open questions.

Phosphoproteomics

O-Phosphorylation, which corresponds to the addition of a phosphate residue on a serine, a threonine or a tyrosine, is a simple modification that plays a central role in signalling processes. A rough estimate of 60% phospho-proteins is given for the content in the human genome (Vlastaridis *et al.*, 2017).

Phosphoproteomics is massively contributing to populating databases such as PhosphoSite (see "Relevant Websites section"). Similar experimental techniques are commonly applied. In particular, metal-based phosphopeptide enrichment strategies methods including immobilized metal affinity chromatography (IMAC) and metal oxide affinity chromatography (MOAC) are prevalent in the field. Phosphopeptide search is often included in database search engines though the remaining challenge remains the positioning of the PTM. Many peptides contain several potential sites and scoring is necessary (Savitski *et al.*, 2011)

As pointed out in Riley and Coon (2016) phosphoproteomics studies tend to focus on O-phosphorylation (serine, threonine, and tyrosine) while alternative N-phosphorylation (lysine, arginine, and histidine) and S-phosphorylation (cysteine) are known to regulate/modulate signalling in bacterial systems, and possibly in eukaryotes as well Kee and Muir (2012). Conventional LC–MS/MS approaches need to be revised to accommodate the specificity of these modifications in higher throughput.

Glycoproteomics

Glycosylation entails the attachment of glycan molecules (oligosaccharides) on proteins. At least 60% of proteins are expected to be glycosylated in mammalian proteomes and many thousands of fully defined glycan structures have been published. Mass spectrometry is the most popular technique to characterise the structure of glycans and glycopeptides; NMR is a common alternative to MS for solving structures with a lesser throughput and higher amounts needed of sample material. A repository of glycan structures each one given a unique identifier was released in an effort to channel the diverse contributions of the glycoscience community (Tiemeyer *et al.*, 2017). In parallel, curated databases are also developed as summarised in Lisacek *et al.* (2017) **Fig. 5** illustrates the variety of glycan molecules attached to a specific site of human elastase..

To date, many studies are based on MS1 data and limited to identifying glycopeptides associated with glycan monosaccharide compositions as opposed to fully defined structures, e.g., Rombouts *et al.* (2016). The fragmentation pattern of glycans was analysed and formalised by Domon and Costello (1988). Multiple fragmentation energies may be required to gain enough information on a glycan structure. Then, standard CID fragmentation of glycopeptides tends to reveals mostly losses of glycans via glycosidic bond cleavage, and relatively little peptide bond cleavage. This lack of information means that determining the peptide identity will be difficult.

Technical steps of current glycoproteomics workflows are not evenly mature or optimised. Advanced methods are the result of applicable or transposable proteomics principles while open questions reflect issues specific to the field of glycosciences and still in need to be tackled. Throughput is increasing and in much the same way lessons learnt in proteomics techniques have benefitted those in glycoproteomics, ideas implemented in proteome informatics also deeply influence glycoproteomics data analysis software. For example, data pre-processing approaches are considered operational (handling of raw files, peak picking, spectral matching or deconvolution). In contrast, open questions such as site-specific localisation and profiling of glycosylation, integration of candidate peptide and glycan scores, target-decoy models and FDR calculation still require time to reach a consensus stage. Reviews such as Dallas *et al.* (2013), Hu *et al.* (2016) and Gaunitz *et al.* (2017) provide rich overviews of glycoproteome informatics with detailed analyses of the obstacles to full automation. Interestingly, undertaking the reanalysis of a dataset initially

← ★ **Elastase iiib**

UniProt: **P08861**

neXtProt: **NX_P08861**

GeneCards: **CELA3B**

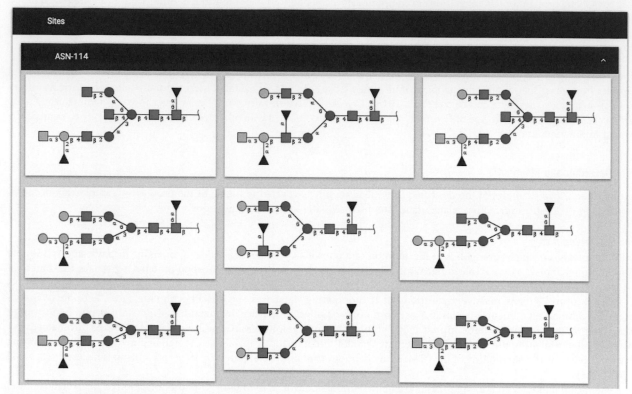

Fig. 5 Example of glycan molecules attached to the asparagine residue (N-linked) at position 114 of human elastase. The cartoon code is detailed in Varki, A., Cummings, R.D., Aebi, M., *et al.*, 2015. Symbol nomenclature for graphical representations of glycans. Glycobiology 25, pp. 1323–1324. Available at: https://doi.org/10.1093/glycob/cwv091. This information was found at https://glyconnect.expasy.org/.

released as global proteomic and phosphoproteomic analyses led to the identification of a large number of glycopeptides (Hu *et al.*, 2017) thereby demonstrating both the value of reanalysis and the wealth of information hidden in MS data.

Open Questions and Future Directions

Standards and Coverage

To link with the discussion in Robin (2018), the importance of complying with standards in order to support comparability and reproducibility of data is also stressed here. More effort is needed in that direction and not surprisingly new standards are emerging in proteogenomics (proBAM Menschaert *et al.*, 2018) or in glycoproteomics (MIRAGE Kolarich *et al.*, 2013)

The proBAM format is based on the BAM format originally designed to store alignments of short DNA or RNA reads to a reference genome. The main advantages of proBAM are (i) the connection of proteomics data to a range of bioinformatics tools that process BAM files, (ii) the option to import information about peptide identification to a genome browser, (iii) exchange and re-use of MS-based proteomics data made easier. The production of this format directly stems from recent progress in proteogenomics.

It can be safely assumed that further effort will be invested in developing the necessary tools in particular with the goal of increasing the peptide coverage in protein identification software and mapping the various sources of variations ranging from SNPs (Single Molecular Polymorphism) to PTMs.

Method Improvement and Further Integration With Other – Omics

As mentioned throughout this article, technological development plays a key role in the evolution of proteomics. MS-based proteomics methodologies have evolved rapidly over recent years along with a remarkable progress on specialised software. In fact, proteome informatics has to keep up with constant innovation such as moving from DDA to DIA technologies or tackling the challenges of top down proteomics. Furthermore, new approaches are nowadays used in combination with other omics such as genomics, transcriptomics and metabolomics (see related discussion Bilbao, 2018).

This increasing throughput and large-scale context is grounded in the big data era and requires robust and efficient MS software as well as advanced MS hardware for comprehensively exploiting the information and generating knowledge. In this sense, progress in instrumentation and analytical methods are expected, and importantly, development of new computational methods to cope with these advances will continue to be an active area of research. An increasing number of available tools including proprietary and open source can be expected.

It should also be noted that both in transcriptomics and proteomics, the unavailability of sequenced or well-annotated genomes and therefore the lack of reference templates limit the interpretation of data. This is particularly acute in plant biology (see for example Special Issue of Biochimica et Biophysica Acta (BBA) – Proteins and Proteomics Edited by Hans-Peter Mock Plant Proteomics: a bridge between fundamental processes and crop production, Volume 1864, Issue 8, Pages 881–1060, 2016) and with animals that are not considered as model organisms (see for example De Almeida et al., 2018). This gap is likely to be filled as genome-sequencing technology advances.

Definite Potential for Elucidating PTM Cross-Talk

As pointed in Venne et al. (2014), PTMs can be mutually exclusive such as the phosphorylation and the O-GlcNAcylation of serine and threonine of signalling proteins presented very early on as the "yin-yang hypothesis" (Hart et al., 1995) but they can also be cooperative as in the well-studied case of histones (Schwämmle et al., 2014). However, crosstalk appears to be species-dependent (Beltrao et al., 2012) making the concept of "PTM code" still difficult to define and circumscribe despite being referred to in many articles (e.g., in relation to chaperones (Cloutier and Coulombe, 2013) or tubulins (Janke, 2014)). To grasp an understanding of the scenario of protein modifications and better apprehend interplay, work should however not be limited to describing the modified protein "landscape" of a cell or a tissue. As a matter of fact, understanding also means entangling a tight bundle of interactions, which encompasses ascertaining the role, the origin and the purpose of enzymes that carry out PTMs. This task is likely to take a while.

The generalised use of mass spectrometry (MS) following top down, middle down or bottom-up strategies mentioned in Section "Proteomics Workflows", as well as the refinement of dedicated approaches described in Section "PTM Detection and Mapping", explain the rather recent accumulation of detected PTMs. The role of PTMs has an obvious bearing on the overall system of interactions. In fact, examples of protein-protein interaction networks integrating PTM knowledge are rare and devoted to mapping data collected in eukaryotes such the yeast methylome (Erce et al., 2012) or the phospho-tyrosine interaction network in human (Grossmann et al., 2015). Bioinformatics resources mapping modified protein interactions are still scarce. For example, with a focus on human and mouse data, TOPFind attempts to address this question with the PathFinder tool (Fortelny et al., 2015) mapping and modelling a protease interaction network.

Conclusion

This overview spans over two decades of bioinformatics development for the reliable identification of proteins in complex mixtures. Many more points and applications could be raised and the reader will find finer details regarding a broad range of topics in cited references. Complexity is unlikely to be reduced in future large-scale studies as hinted in Section "Open Questions and Future Directions" and confirmed for instance by recent effort to launch metaproteomics applications (proteomics of microbiota), which in much the same metagenomics was initiated, cannot be tackled without powerful software (Muth et al., 2015; Cheng et al., 2017). Proteome informatics can still prosper.

See also: Clinical Proteomics. Data-Information-Concept Continuum From a Text Mining Perspective. Experimental Platforms for Extracting Biological Data: Mass Spectrometry, Microarray, Next Generation Sequencing. Identification of Proteins from Proteomic Analysis. Identification of Sequence Patterns, Motifs and Domains. Natural Language Processing Approaches in Bioinformatics. Prediction of Protein Localization. Prediction of Protein-Protein Interactions: Looking Through the Kaleidoscope. Protein Post-Translational Modification Prediction. Protein Properties. Protein–Protein Interaction Databases. Proteomics Data Representation and Databases. Proteomics Mass Spectrometry Data Analysis Tools. Quantification of Proteins from Proteomic Analysis. Text Mining for Bioinformatics Using Biomedical Literature. The Evolution of Protein Family Databases. Transmembrane Domain Prediction. Utilising IPG-IEF to Identify Differentially-Expressed Proteins

References

Aebersold, R., Mann, M., 2003. Mass spectrometry-based proteomics. Nature 422, 198–207. Available at: https://doi.org/10.1038/nature01511.
De Almeida, A.M., Eckersall, D., Miller, I., 2018. Proteomics in Domestic Animals: From Farm to Systems Biology. Springer.

Angel, T.E., Aryal, U.K., Hengel, S.M., *et al.*, 2012. Mass spectrometry-based proteomics: Existing capabilities and future directions. Chem. Soc. Rev. 41, 3912. Available at: https://doi.org/10.1039/c2cs15331a.

Appel, R.D., Hochstrasser, D.F., Funk, M., *et al.*, 1991. The MELANIE project: From a biopsy to automatic protein map interpretation by computer. Electrophoresis 12, 722–735. Available at: https://doi.org/10.1002/elps.1150121006.

Baginsky, S., 2009. Plant proteomics: Concepts, applications, and novel strategies for data interpretation. Mass Spectrom. Rev. 28, 93–120. Available at: https://doi.org/10.1002/mas.20183.

Baker, E.S., Burnum-Johnson, K.E., Ibrahim, Y.M., *et al.*, 2015. Enhancing bottom-up and top-down proteomic measurements with ion mobility separations. Proteomics 15, 2766–2776. Available at: https://doi.org/10.1002/pmic.201500048.

Barabási, A.-L., Oltvai, Z.N., 2004. Network biology: Understanding the cell's functional organization. Nat. Rev. Genet. 5, 101–113. Available at: https://doi.org/10.1038/nrg1272.

Beltrao, P., Albanèse, V., Kenner, L.R., *et al.*, 2012. Systematic functional prioritization of protein posttranslational modifications. Cell 150, 413–425. Available at: https://doi.org/10.1016/j.cell.2012.05.036.

Bensimon, A., Heck, A.J.R., Aebersold, R., 2012. Mass spectrometry-based proteomics and network biology. Annu. Rev. Biochem. 81, 379–405. Available at: https://doi.org/10.1146/annurev-biochem-072909-100424.

Bilbao, A., 2018. encyclopedia of bioinformatics and computational biology: Proteomics mass spectrometry data analysis tools. In: Reference Module in Life Sciences. Elsevier. https://doi.org/10.1016/B978-0-12-809633-8.20274-4.

Bond, N.J., Shliaha, P.V., Lilley, K.S., Gatto, L., 2013. Improving qualitative and quantitative performance for MS ^{E} -based Label-free Proteomics. J. Proteome Res. 12, 2340–2353. Available at: https://doi.org/10.1021/pr300776t.

Branca, R.M.M., Orre, L.M., Johansson, H.J., *et al.*, 2014. HiRIEF LC-MS enables deep proteome coverage and unbiased proteogenomics. Nat. Methods 11, 59–62. Available at: https://doi.org/10.1038/nmeth.2732.

Catherman, A.D., Skinner, O.S., Kelleher, N.L., 2014. Top down proteomics: Facts and perspectives. Biochem. Biophys. Res. Commun. 445, 683–693. Available at: https://doi.org/10.1016/j.bbrc.2014.02.041.

Cheng, K., Ning, Z., Zhang, X., *et al.*, 2017. MetaLab: An automated pipeline for metaproteomic data analysis. Microbiome 5.Available at: https://doi.org/10.1186/s40168-017-0375-2.

Chichester, C., Nikitin, F., Ravarini, J.-C., Lisacek, F., 2003. Consistency checks for characterizing protein forms. Comput. Biol. Chem. 27, 29–35.

Choudhary, J.S., Blackstock, W.P., Creasy, D.M., Cottrell, J.S., 2001. Interrogating the human genome using uninterpreted mass spectrometry data. Proteomics 1, 651–667. Available at: https://doi.org/10.1002/1615-9861(200104)1:5651::AID-PROT6513.0.CO;2-N.

Cloutier, P., Coulombe, B., 2013. Regulation of molecular chaperones through post-translational modifications: Decrypting the chaperone code. Biochim. Biophys. Acta 1829, 443–454. Available at: https://doi.org/10.1016/j.bbagrm.2013.02.010.

Creasy, D.M., Cottrell, J.S., 2004. Unimod: Protein modifications for mass spectrometry. Proteomics 4, 1534–1536. Available at: https://doi.org/10.1002/pmic.200300744.

Crowell, K.L., Slysz, G.W., Baker, E.S., *et al.*, 2013. LC-IMS-MS Feature Finder: Detecting multidimensional liquid chromatography, ion mobility and mass spectrometry features in complex datasets. Bioinformatics 29, 2804–2805. Available at: https://doi.org/10.1093/bioinformatics/btt465.

Dallas, D.C., Martin, W.F., Hua, S., German, J.B., 2013. Automated glycopeptide analysis-review of current state and future directions. Brief. Bioinform. 14, 361–374. Available at: https://doi.org/10.1093/bib/bbs045.

Delalande, F., Carapito, C., Brizard, J.-P., Brugidou, C., Van Dorsselaer, A., 2005. Multigenic families and proteomics: Extended protein characterization as a tool for paralog gene identification. Proteomics 5, 450–460. Available at: https://doi.org/10.1002/pmic.200400954.

Distler, U., Kuharev, J., Navarro, P., *et al.*, 2014. Drift time-specific collision energies enable deep-coverage data-independent acquisition proteomics. Nat. Methods 11, 167–170. Available at: https://doi.org/10.1038/nmeth.2767.

Domon, B., Costello, C.E., 1988. A systematic nomenclature for carbohydrate fragmentations in FAB-MS/MS spectra of glycoconjugates. Glycoconj. J. 5, 397–409. Available at: https://doi.org/10.1007/BF01049915.

Dowsey, A.W., Dunn, M.J., Yang, G.-Z., 2008. Automated image alignment for 2D gel electrophoresis in a high-throughput proteomics pipeline. Bioinformatics 24, 950–957. Available at: https://doi.org/10.1093/bioinformatics/btn059.

Dowsey, A.W., English, J.A., Lisacek, F., *et al.*, 2010. Image analysis tools and emerging algorithms for expression proteomics. Proteomics 10, 4226–4257. Available at: https://doi.org/10.1002/pmic.200900635.

Elias, J.E., Gygi, S.P., 2007. Target-decoy search strategy for increased confidence in large-scale protein identifications by mass spectrometry. Nat. Methods 4, 207–214. Available at: https://doi.org/10.1038/nmeth1019.

Eng, J.K., McCormack, A.L., Yates, J.R., 1994. An approach to correlate tandem mass spectral data of peptides with amino acid sequences in a protein database. J. Am. Soc. Mass Spectrom. 5, 976–989. Available at: https://doi.org/10.1016/1044-0305(94)80016-2.

Erce, M.A., Pang, C.N.I., Hart-Smith, G., Wilkins, M.R., 2012. The methylproteome and the intracellular methylation network. Proteomics 12, 564–586. Available at: https://doi.org/10.1002/pmic.201100397.

Fellers, R.T., Greer, J.B., Early, B.P., *et al.*, 2015. ProSight Lite: Graphical software to analyze top-down mass spectrometry data. Proteomics 15, 1235–1238. Available at: https://doi.org/10.1002/pmic.201400313.

Fortelny, N., Yang, S., Pavlidis, P., Lange, P.F., Overall, C.M., 2015. Proteome TopFIND 3.0 with TopFINDer and PathFINDer: Database and analysis tools for the association of protein termini to pre- and post-translational events. Nucleic Acids Res. 43, D290–D297. Available at: https://doi.org/10.1093/nar/gku1012.

Frank, A., Pevzner, P., 2005. PepNovo: De novo peptide sequencing via probabilistic network modeling. Anal. Chem. 77, 964–973. Available at: https://doi.org/10.1021/ac048788h.

Gaudet, P., Argoud-Puy, G., Cusin, I., *et al.*, 2013. neXtProt: Organizing protein knowledge in the context of human proteome projects. J. Proteome Res. 12, 293–298. Available at: https://doi.org/10.1021/pr300830v.

Gaunitz, S., Nagy, G., Pohl, N.L.B., Novotny, M.V., 2017. Recent advances in the analysis of complex glycoproteins. Anal. Chem. 89, 389–413. Available at: https://doi.org/10.1021/acs.analchem.6b04343.

Griss, J., 2016. Spectral library searching in proteomics. Proteomics 16, 729–740. Available at: https://doi.org/10.1002/pmic.201500296.

Grossmann, A., Benlasfer, N., Birth, P., *et al.*, 2015. Phospho-tyrosine dependent protein-protein interaction network. Mol. Syst. Biol. 11 (3), 794. Available at: https://doi.org/10.15252/msb.20145968.

Guthals, A., Gan, Y., Murray, L., *et al.*, 2017. De novo MS/MS sequencing of native human antibodies. J. Proteome Res. 16, 45–54. Available at: https://doi.org/10.1021/acs.jproteome.6b00608.

Guthals, A., Watrous, J.D., Dorrestein, P.C., Bandeira, N., 2012. The spectral networks paradigm in high throughput mass spectrometry. Mol. Biosyst. 8 (10), 2535–2544. Available at: https://doi.org/10.1039/c2mb25085c.

Gygi, S.P., Rist, B., Gerber, S.A., *et al.*, 1999. Quantitative analysis of complex protein mixtures using isotope-coded affinity tags. Nat. Biotechnol. 17, 994–999. Available at: https://doi.org/10.1038/13690.

Hart, G.W., Greis, K.D., Dong, L.Y., *et al.*, 1995. O-linked N-acetylglucosamine: The "yin-yang" of Ser/Thr phosphorylation? Nuclear and cytoplasmic glycosylation. Adv. Exp. Med. Biol. 376, 115–123.

Henneman, A.A., Palmblad, M., 2013. Retention time prediction and protein identification. In: Matthiesen, R. (Ed.), Mass Spectrometry Data Analysis in Proteomics. Totowa, NJ: Humana Press, pp. 101–118. Available at: https://doi.org/10.1007/978-1-62703-392-3_4.

Henzel, W.J., Billeci, T.M., Stults, J.T., *et al.*, 1993. Identifying proteins from two-dimensional gels by molecular mass searching of peptide fragments in protein sequence databases. Proc. Natl. Acad. Sci. USA 90, 5011–5015.

Hu, H., Khatri, K., Zaia, J., 2017. Algorithms and design strategies towards automated glycoproteomics analysis: Algorithms and design strategies. Mass Spectrom. Rev. 36 (4), 475–498. Available at: https://doi.org/10.1002/mas.21487.

Hu, Y., Shah, P., Clark, D.J., Ao, M., Zhang, H., 2017. Reanalysis of global proteomic and phosphoproteomic data identified a large number of glycopeptides. Available at: https://doi.org/10.1101/233247.

Ivanov, M.V., Lobas, A.A., Karpov, D.S., Moshkovskii, S.A., Gorshkov, M.V., 2017. Comparison of false discovery rate control strategies for variant peptide identifications in shotgun proteogenomics. J. Proteome Res. 16, 1936–1943. Available at: https://doi.org/10.1021/acs.jproteome.6b01014.

James, P., Quadroni, M., Carafoli, E., Gonnet, G., 1993. Protein identification by mass profile fingerprinting. Biochem. Biophys. Res. Commun. 195, 58–64. Available at: https://doi.org/10.1006/bbrc.1993.2009.

Janke, C., 2014. The tubulin code: Molecular components, readout mechanisms, and functions. J. Cell Biol. 206, 461–472. Available at: https://doi.org/10.1083/jcb.201406055.

Johnson, R.S., Martin, S.A., Biemann, K., Stults, J.T., Watson, J.T., 1987. Novel fragmentation process of peptides by collision-induced decomposition in a tandem mass spectrometer: Differentiation of leucine and isoleucine. Anal. Chem. 59, 2621–2625. Available at: https://doi.org/10.1021/ac00148a019.

Kee, J.-M., Muir, T.W., 2012. Chasing phosphohistidine, an elusive sibling in the phosphoamino acid family. ACS Chem. Biol. 7, 44–51. Available at: https://doi.org/10.1021/cb200445w.

Koch, A., Gawron, D., Steyaert, S., et al., 2014. A proteogenomics approach integrating proteomics and ribosome profiling increases the efficiency of protein identification and enables the discovery of alternative translation start sites. Proteomics 14, 2688–2698. Available at: https://doi.org/10.1002/pmic.201400180.

Kolarich, D., Rapp, E., Struwe, W.B., et al., 2013. The minimum information required for a glycomics experiment (MIRAGE) project: Improving the Standards for reporting mass-spectrometry-based glycoanalytic data. Mol. Cell. Proteomics 12, 991–995. Available at: https://doi.org/10.1074/mcp.O112.026492.

Küster, B., Mortensen, P., Andersen, J.S., Mann, M., 2001. Mass spectrometry allows direct identification of proteins in large genomes. Proteomics 1, 641–650. Available at: https://doi.org/10.1002/1615-9861(200104)1:5<641::AID-PROT6413.0.CO;2-R.

Lam, H., Deutsch, E.W., Eddes, J.S., et al., 2007. Development and validation of a spectral library searching method for peptide identification from MS/MS. Proteomics 7, 655–667. Available at: https://doi.org/10.1002/pmic.200600625.

Lisacek, F., Mariethoz, J., Alocci, D., et al., 2017. Databases and associated tools for glycomics and glycoproteomics. In: Lauc, G., Wuhrer, M. (Eds.), High-Throughput Glycomics and Glycoproteomics. New York, NY: Springer New York, pp. 235–264. https://doi.org/10.1007/978-1-4939-6493-2_18.

Liu, J., Bell, A.W., Bergeron, J.J.M., et al., 2007. Methods for peptide identification by spectral comparison. Proteome Sci. 5, 3. Available at: https://doi.org/10.1186/1477-5956-5-3.

Liu, Y., Hüttenhain, R., Collins, B., Aebersold, R., 2013. Mass spectrometric protein maps for biomarker discovery and clinical research. Expert Rev. Mol. Diagn. 13, 811–825. Available at: https://doi.org/10.1586/14737159.2013.845089.

Liu, G., Zhang, J., Larsen, B., et al., 2010. ProHits: Integrated software for mass spectrometry–based interaction proteomics. Nat. Biotechnol. 28, 1015–1017. Available at: https://doi.org/10.1038/nbt1010-1015.

Ma, B., 2015. Novor: Real-Time peptide de novo sequencing software. J. Am. Soc. Mass Spectrom. 26, 1885–1894. Available at: https://doi.org/10.1007/s13361-015-1204-0.

Mahon, P., Dupree, P., 2001. Quantitative and reproducible two-dimensional gel analysis using Phoretix 2D Full. Electrophoresis 22, 2075–2085. Available at: https://doi.org/10.1002/1522-2683(200106)22:10<2075::AID-ELPS20753.0.CO;2-C.

Makarov, A., 2000. Electrostatic axially harmonic orbital trapping: A high-performance technique of mass analysis. Anal. Chem. 72, 1156–1162. Available at: https://doi.org/10.1021/ac991131p.

Mann, M., Højrup, P., Roepstorff, P., 1993. Use of mass spectrometric molecular weight information to identify proteins in sequence databases. Biol. Mass Spectrom. 22, 338–345. Available at: https://doi.org/10.1002/bms.1200220605.

Mallick, P., Schirle, M., Chen, S.S., et al., 2007. Computational prediction of proteotypic peptides for quantitative proteomics. Nat. Biotechnol. 25, 125–131. Available at: https://doi.org/10.1038/nbt1275.

Mann, M., Wilm, M., 1994. Error-tolerant identification of peptides in sequence databases by peptide sequence tags. Anal. Chem. 66, 4390–4399. Available at: https://doi.org/10.1021/ac00096a002.

Menschaert, G., Wang, X., Jones, A.R., et al., 2018. The proBAM and proBed standard formats: Enabling a seamless integration of genomics and proteomics data. Genome Biol. 19. Available at: https://doi.org/10.1186/s13059-017-1377-x.

Mitchell Wells, J., McLuckey, S.A., 2005. Collision-induced dissociation (CID) of peptides and proteins. In: Burlingame, A.L. (Ed.), Methods in Enzymology 402. Elsevier, pp. 148–185. https://doi.org/10.1016/S0076-6879(05)02005-7.

Moruz, L., Käll, L., 2017. Peptide retention time prediction. Mass Spectrom. Rev. 36, 615–623. Available at: https://doi.org/10.1002/mas.21488.

Murray, K.K., Boyd, R.K., Eberlin, M.N., et al., 2013. Definitions of terms relating to mass spectrometry (IUPAC recommendations 2013). Pure Appl. Chem. 85, 1515–1609. Available at: https://doi.org/10.1351/PAC-REC-06-04-06.

Muth, T., Behne, A., Heyer, R., et al., 2015. The MetaProteomeAnalyzer: A powerful open-source software suite for metaproteomics data analysis and interpretation. J. Proteome Res. 14, 1557–1565. Available at: https://doi.org/10.1021/pr501246w.

Muth, T., Renard, B.Y., 2017. Evaluating de novo sequencing in proteomics: Already an accurate alternative to database-driven peptide identification? Brief. Bioinform. Available at: https://doi.org/10.1093/bib/bbx033.

Nesvizhskii, A.I., 2010. A survey of computational methods and error rate estimation procedures for peptide and protein identification in shotgun proteomics. J. Proteomics 73, 2092–2123. Available at: https://doi.org/10.1016/j.jprot.2010.08.009.

Nilsson, T., Mann, M., Aebersold, R., et al., 2010. Mass spectrometry in high-throughput proteomics: Ready for the big time. Nat. Methods 7, 681–685. Available at: https://doi.org/10.1038/nmeth0910-681.

Omasits, U., Varadarajan, A.R., Schmid, M., et al., 2017. An integrative strategy to identify the entire protein coding potential of prokaryotic genomes by proteogenomics. Genome Res. 27, 2083–2095. Available at: https://doi.org/10.1101/gr.218255.116.

Ong, S.-E., Blagoev, B., Kratchmarova, I., et al., 2002. Stable isotope labeling by amino acids in cell culture, silac, as a simple and accurate approach to expression proteomics. Mol. Cell. Proteomics 1, 376–386. Available at: https://doi.org/10.1074/mcp.M200025-MCP200.

Pappin, D.J., Hojrup, P., Bleasby, A.J., 1993. Rapid identification of proteins by peptide-mass fingerprinting. Curr. Biol. 3, 327–332.

Perkins, D.N., Pappin, D.J.C., Creasy, D.M., Cottrell, J.S., 1999. Probability-based protein identification by searching sequence databases using mass spectrometry data. Electrophoresis 20, 3551–3567. Available at: https://doi.org/10.1002/(SICI)1522-2683(19991201)20:18<3551::AID-ELPS3551>3.0.CO;2-2.

Pino, L.K., Searle, B.C., Bollinger, J.G., et al., 2017. The skyline ecosystem: Informatics for quantitative mass spectrometry proteomics. Mass Spectrom. Rev. Available at: https://doi.org/10.1002/mas.21540.

Prabakaran, S., Lippens, G., Steen, H., Gunawardena, J., 2012. Post-translational modification: Nature's escape from genetic imprisonment and the basis for dynamic information encoding. Wiley Interdiscip. Rev. Syst. Biol. Med. 4, 565–583. Available at: https://doi.org/10.1002/wsbm.1185.

Purvine, S., Eppel*, J.-T., Yi, E.C., Goodlett, D.R., 2003. Shotgun collision-induced dissociation of peptides using a time of flight mass analyzer. Proteomics 3, 847–850. Available at: https://doi.org/10.1002/pmic.200300362.

Rigaut, G., Shevchenko, A., Rutz, B., et al., 1999. A generic protein purification method for protein complex characterization and proteome exploration. Nat. Biotechnol. 17, 1030–1032. Available at: https://doi.org/10.1038/13732.

Riley, N.M., Coon, J.J., 2016. Phosphoproteomics in the age of rapid and deep proteome profiling. Anal. Chem. 88, 74–94. Available at: https://doi.org/10.1021/acs.analchem.5b04123.

Rison, S.C.G., Mattow, J., Jungblut, P.R., Stoker, N.G., 2007. Experimental determination of translational starts using peptide mass mapping and tandem mass spectrometry within the proteome of Mycobacterium tuberculosis. Microbiology 153, 521–528. Available at: https://doi.org/10.1099/mic.0.2006/001537-0.

Robin, T., 2018. Proteomics Data Representation and Databases. In: Reference Module in Life Sciences. Elsevier. Available at: https://doi.org/10.1016/B978-0-12-809633-8.20273-2/B978-0-12-809633-8.20273-2.

Roepstorff, P., Fohlman, J., 1984. Letter to the editors. Biol. Mass Spectrom. 11 (11), 601. Available at: https://onlinelibrary.wiley.com/doi/abs/10.1002/bms.1200111109.

Rogowska-Wrzesinska, A., Le Bihan, M.-C., Thaysen-Andersen, M., Roepstorff, P., 2013. 2D gels still have a niche in proteomics. J. Proteomics 88, 4–13. Available at: https://doi.org/10.1016/j.jprot.2013.01.010.

Rombouts, Y., Jónasdóttir, H.S., Hipgrave Ederveen, A.L., et al., 2016. Acute phase inflammation is characterized by rapid changes in plasma/peritoneal fluid N-glycosylation in mice. Glycoconj. J. 33, 457–470. Available at: https://doi.org/10.1007/s10719-015-9648-9.

Ross, P.L., Huang, Y.N., Marchese, J.N., et al., 2004. Multiplexed protein quantitation in saccharomyces cerevisiae using amine-reactive isobaric tagging reagents. Mol. Cell. Proteomics 3, 1154–1169. Available at: https://doi.org/10.1074/mcp.M400129-MCP200.

Ruggles, K.V., Krug, K., Wang, X., et al., 2017. Methods, tools and current perspectives in proteogenomics. Mol. Cell. Proteomics 16, 959–981. Available at: https://doi.org/10.1074/mcp.MR117.000024.

Savitski, M.M., Lemeer, S., Boesche, M., et al., 2011. Confident phosphorylation site localization using the mascot delta score. Mol. Cell. Proteomics 10. Available at:. https://doi.org/10.1074/mcp.M110.003830.

Schwämmle, V., Aspalter, C.-M., Sidoli, S., Jensen, O.N., 2014. Large scale analysis of co-existing post-translational modifications in histone tails reveals global fine structure of cross-talk. Mol. Cell. Proteomics 13, 1855–1865. Available at: https://doi.org/10.1074/mcp.O113.036335.

Seeley, E.H., Caprioli, R.M., 2008. Molecular imaging of proteins in tissues by mass spectrometry. Proc. Natl. Acad. Sci. USA 105, 18126–18131. Available at: https://doi.org/10.1073/pnas.0801374105.

Skinner, O.S., Kelleher, N.L., 2015. Illuminating the dark matter of shotgun proteomics. Nat. Biotechnol. 33, 717–718. Available at: https://doi.org/10.1038/nbt.3287.

Smith, L.M., Kelleher, N.L., 2013. Proteoform: A single term describing protein complexity. Nat. Methods 10, 186–187. Available at: https://doi.org/10.1038/nmeth.2369.

Snijder, J., Schuller, J.M., Wiegard, A., et al., 2017. Structures of the cyanobacterial circadian oscillator frozen in a fully assembled state. Science 355, 1181–1184. Available at: https://doi.org/10.1126/science.aag3218.

Syka, J.E.P., Coon, J.J., Schroeder, M.J., Shabanowitz, J., Hunt, D.F., 2004. Peptide and protein sequence analysis by electron transfer dissociation mass spectrometry. Proc. Natl. Acad. Sci. 101, 9528–9533. Available at: https://doi.org/10.1073/pnas.0402700101.

Taylor, C.F., Paton, N.W., Lilley, K.S., et al., 2007. The minimum information about a proteomics experiment (MIAPE). Nat. Biotechnol. 25, 887–893. Available at: https://doi.org/10.1038/nbt1329.

Thompson, A., Schäfer, J., Kuhn, K., et al., 2003. Tandem mass tags: A novel quantification strategy for comparative analysis of complex protein mixtures by MS/MS. Anal. Chem. 75, 1895–1904. Available at: https://doi.org/10.1021/ac0262560.

Tiemeyer, M., Aoki, K., Paulson, J., et al., 2017. GlyTouCan: An accessible glycan structure repository. Glycobiology. 1–5. Available at: https://doi.org/10.1093/glycob/cwx066.

Tran, N.H., Rahman, M.Z., He, L., et al., 2016. Complete de novo assembly of monoclonal antibody sequences. Sci. Rep. 6. Available at: https://doi.org/10.1038/srep31730.

Venable, J.D., Dong, M.-Q., Wohlschlegel, J., Dillin, A., Yates, J.R., 2004. Automated approach for quantitative analysis of complex peptide mixtures from tandem mass spectra. Nat. Methods 1, 39–45. Available at: https://doi.org/10.1038/nmeth705.

Venne, A.S., Kollipara, L., Zahedi, R.P., 2014. The next level of complexity: Crosstalk of posttranslational modifications. Proteomics 14, 513–524. Available at: https://doi.org/10.1002/pmic.201300344.

Vlastaridis, P., Kyriakidou, P., Chaliotis, A., et al., 2017. Estimating the total number of phosphoproteins and phosphorylation sites in eukaryotic proteomes. GigaScience 6, 1–11. Available at: https://doi.org/10.1093/gigascience/giw015.

Wolters, D.A., Washburn, M.P., Yates, J.R., 2001. An automated multidimensional protein identification technology for shotgun proteomics. Anal. Chem. 73, 5683–5690. Available at: https://doi.org/10.1021/ac010617e.

Yao, X., Freas, A., Ramirez, J., Demirev, P.A., Fenselau, C., 2001. Proteolytic ^{18}O labeling for comparative proteomics: Model studies with two serotypes of adenovirus. Anal. Chem. 73, 2836–2842. Available at: https://doi.org/10.1021/ac001404c.

Yates, J.R., Speicher, S., Griffin, P.R., Hunkapiller, T., 1993. Peptide mass maps: A highly informative approach to protein identification. Anal. Biochem. 214, 397–408. Available at: https://doi.org/10.1006/abio.1993.1514.

Yılmaz, Ş., Shiferaw, G.A., Rayo, J., et al., 2018. Cross-linked peptide identification: A computational forest of algorithms. Mass Spectrom. Rev. Available at: https://doi.org/10.1002/mas.21559.

Zhang, Z., Burke, M., Mirokhin, Y.A., et al., 2018. Reverse and random decoy methods for false discovery rate estimation in high mass accuracy peptide spectral library searches. J. Proteome Res. 17, 846–857. Available at: https://doi.org/10.1021/acs.jproteome.7b00614.

Zubarev, R.A., Kelleher, N.L., McLafferty, F.W., 1998. Electron capture dissociation of multiply charged protein cations. A nonergodic process. J. Am. Chem. Soc. 120, 3265–3266. Available at: https://doi.org/10.1021/ja973478k.

Relevant Websites

http://totallab.com/2d-products
  2D Products - TotalLab - Supporting your research.
http://www.crapome.org
  Crapome.org.
https://www.gelmap.de
  Gelmap. Spot visualization by LUH.
http://www.matrixscience.com/help/top_down_help.html
  Mascot database search: Top-down searches - Matrix Science.
http://2d-gel-analysis.com
  Melanie 2D gel analysis software for protein expression profiling.
https://ms-imaging.org
  MS Imaging—Home of Mass Spectrometry Imaging.
https://www.openms.de
  OpenMS.
http://peptide.nist.gov/
  Peptide Mass Spectral Libraries.
https://www.phosphosite.org
  PhosphoSitePlus.
http://proteowizard.sourceforge.net
  ProteoWizard.

http://www.peptideatlas.org/speclib/
    Spectrum Library Central at PeptideAtlas.
http://www.sorfs.org
    sORFs.org: repository of small ORFs identified by ribosome profiling.
https://world-2dpage.expasy.org
    The World-2DPAGE Constellation - ExPASy.

Proteomics Data Representation and Databases

Thibault Robin, University of Geneva, CUI, Carouge, Switzerland

Introduction

Proteomics distinguishes itself from other life sciences by the variety and the complexity of the data generated. Compared to genomics, many biological events such as post-translational modifications, protein isoforms, and protein degradation add challenges to data analysis and interpretation (Altelaar *et al.*, 2012). Furthermore, proteins rarely act independently but rather interact with each other in large intricate, dynamic and plastic networks (Baker, 2012; Altelaar *et al.*, 2012).

In recent years, significant breakthroughs were achieved in proteomics technologies (Altelaar *et al.*, 2012). Mass spectrometry (MS) in particular has emerged as the reference approach for the high-throughput identification of peptides and proteins. This progress was made possible through improvements in sample preparation, instrumentation and computational tools (Altelaar *et al.*, 2012). The widespread adoption of next-generation mass spectrometers by the proteomics community yielded to a massive increase in the amount of data generated, raising concerns about their storage and sharing policies. Despite some initial reluctance, the need for open access data in proteomics is nowadays fully embraced by the community. Although data sharing was introduced a long time ago in other omics fields, it was not effortless in proteomics and required a collective input over the years to convince both researchers and instrument manufacturers (Prince *et al.*, 2004; Verheggen and Martens, 2015; Deutsch *et al.*, 2017).

To meet the demand, a large panel of databases with their respective underlying purpose and content specificities was developed over time. Despite the undeniable biological value provided, the resulting diversity can prove to be troublesome for users who have to grasp the extent of the various data formats for both input and output files. Another consequence of such fragmented information is that long-term database maintenance is not guaranteed. Some databases may be financially impacted when the only source of income lies in research grants and donations. A recent setback was the successive discontinuation of the Peptidome (2011) (Slotta *et al.*, 2009; Csordas *et al.*, 2013) and Tranche (2013) (Smith *et al.*, 2011; Science Signaling, 2013) databases due to the lack of funding, resulting in the partial loss of each dataset (Deutsch *et al.*, 2017). These incidents led the ProteomeXchange consortium to make data sustainability one of its priorities (Deutsch *et al.*, 2017).

This article will describe the proteomics field in regard to the available databases and the data they contain. First, the principal data types are presented along with the corresponding standard formats that were established through the years. Next, the main categories of databases are detailed and illustrated, with representative examples that have the greatest impact on the field today. Then, the ProteomeXchange consortium is detailed to present the significant contribution it brought to the proteomics field. Finally, the prospects of proteomics data and databases are examined.

Data Types and Standard Formats

The data generated in most MS-based protemics experiments can be divided into two main categories depending on their origin: raw data and processed data. Metadata is a separate though significant category, present throughout the analysis process. Its purpose is to store and track a wide range of technical and biological information, essential to proper data handling and interpretation. Metadata information is especially essential in the proteomics field, since a broad range of approaches and technologies are being used (**Table 1**).

Raw Data

In its primal form, proteomics raw data is mass spectra. A mass spectrum consists at its core in a collection of mass-to-charge ratios (m/z) plotted against their corresponding intensities (Aebersold and Mann, 2003). Supplementary information is usually provided, such as the precursor ion m/z value and charge along with the retention time when a chromatographic separation step was performed.

Proprietary formats

Mass spectrometers usually produce raw data in the form of proprietary binary files presenting various levels of compression (Martens *et al.*, 2005b). Most of the time, the exact specifications of the format encoding-schemes are not publicly available. Proprietary binary files cannot consequently be accessed by users without software supplied by the instrument vendor. Likewise, third-party software, such as MSConvert from the ProteoWizard suite (Kessner *et al.*, 2008), will require the implementation of specific proprietary libraries to decode the information contained in the data files. The data formats also change drastically from one vendor to another, or even sometimes between instruments, raising important standardization concerns. Despite all the restrictions, proprietary formats contain the greatest amount of information and present overall good performances when being processed by compatible software.

Table 1 Summary of the main open data formats in the proteomics field

Name	Supported Data	Organization	URL
mzData	Raw mass spectra	Proteomics Standards Initiative, Human Proteome Organization	http://www.psidev.info/mzdata
mzXML	Raw mass spectra	Seattle Proteome Center, Institute for Systems Biology	http://tools.proteomecenter.org/wiki/index.php?title=Formats:mzXML
mzML	Raw mass spectra	Proteomics Standards Initiative, Human Proteome Organization	http://www.psidev.info/mzml
mz5	Raw mass spectra	Proteomics Center, Boston Children's Hospital	http://software.steenlab.org/mz5
TraML	SRM transition lists	Proteomics Standards Initiative, Human Proteome Organization	http://www.psidev.info/traml
mzIdentML	Identification data	Proteomics Standards Initiative, Human Proteome Organization	http://www.psidev.info/mzidentml
mzQuantML	Quantification data	Proteomics Standards Initiative, Human Proteome Organization	http://www.psidev.info/mzquantml
pepXML	Identification and quantification data	Seattle Proteome Center, Institute for Systems Biology	http://tools.proteomecenter.org/wiki/index.php?title=Formats:pepXML
protXML	Identification and quantification data	Seattle Proteome Center, Institute for Systems Biology	http://tools.proteomecenter.org/wiki/index.php?title=Formats:protXML
mzTab	Identification and quantification data	Proteomics Standards Initiative, Human Proteome Organization	http://www.psidev.info/mztab

Peak list formats

The limitations of the proprietary formats favored the common practice of extracting the mass spectra from the binary files as peak lists stored in text files at the cost of an overall reduced amount of information. In addition to the almost complete loss of metadata information, the mass spectra usually go through denoising, deisotoping, centroiding and charge deconvolution steps during the format conversion (Martens *et al.*, 2005b). The resulting peak list files are therefore significantly more compact while becoming easily manageable for users. The lack of metadata represents the main shortcoming of this format, making it poorly adapted to data storing and sharing.

Many peak list formats were developed through the years, such as the Mascot generic format (MGF), the Micromass peak lists (PKL) and the SEQUEST files (DTA). Despite slight variations in the amount of information they contain, these formats are easily inter-convertible without significant data loss using publicly available software (Martens *et al.*, 2005b).

mzData and mzXML formats

In order to determine a middle solution between proprietary binary files and basic peak lists, new open data formats were proposed at the beginning of the 2000s owing to the expansion of XML (Extensible Markup Language) in bioinformatics. For several years, the mzData (HUPO-PSI, 2006) and mzXML (Pedrioli *et al.*, 2004) formats used to be the most common open data formats to store and share raw data information (Martens *et al.*, 2011). Although sharing the same XML architecture, the two formats mainly differed in regard to their ontology policies (Martens *et al.*, 2011).

On the one hand, the mzData format was developed by the HUPO Proteomics Standards Initiative (PSI) as an open format standard to share and archive data. Its metadata annotation was based on a controlled vocabulary that could be regularly updated (Martens *et al.*, 2011). On the other hand, the mzXML format was developed at the Institute for Systems Biology (ISB) as an open data format accompanied by a suite of tools (Pedrioli *et al.*, 2004). Its metadata annotation was based on a fixed schema that required modification along with the corresponding software to incorporate new annotations (Martens *et al.*, 2011). Although nowadays deprecated, the mzData and mzXML formats are still compatible with most current software and are being used at a smaller scale by the proteomics community.

mzML format

With the purpose of format unification, the mzML format was developed as part of a collective effort under the guidance of the HUPO-PSI to replace both the mzData and mzXML formats (Martens *et al.*, 2011). The coexistence of two open formats having similar functionality was perceived at the time as confusing for users. The goal was then to design a unique XML-based format supporting all past features while providing a consensus way to encode the information (Martens *et al.*, 2011). In addition to instrument support improvement, the mzML format features notably metadata annotation following a flexible controlled vocabulary that allows the inclusion of custom new terms by users without requiring the modification of the XML schema (Martens *et al.*, 2011). The mzML format quickly became compatible with most tools available in proteomics, strengthening its format unification goal.

mz5 format

More recently, the mz5 open data format was developed at the Proteomics Center of the Boston Children's Hospital with the aim of providing a more speedy and storage efficient alternative to mzML while preserving the same ontology (Wilhelm *et al.*, 2012).

Distinguishing itself from its predecessors, the mz5 format is designed based on the Hierarchical Data Format (HDF5) instead of the traditional XML architecture. Despite the partial base-64 encoding being used in mzML, XML was designed to remain easily manipulable by users and consequently struggles to handle massive data files. With data benchmarks, the mz5 format was shown to increase 3–4 times read and write performances while halving data file size in comparison to its XML-based counterparts (Wilhelm *et al.*, 2012). Despite its obvious benefits, the low amount of compatible software prevented the widespread adoption of the mz5 format by the proteomics community.

Processed Data

In most high-throughput proteomics experiments, peptide and protein identification is achieved through the database search approach (Nesvizhskii, 2006). In this method, the experimental spectra are aligned against theoretical spectra inferred from the fragmentation patterns of the peptides contained in a reference protein sequence database. The tools performing the search, referred to as search engines, additionally provide numerous statistical scores assessing the quality of the resulting peptide-spectrum matches (PSMs), allowing to rank the peptides matching a given experimental spectrum (Nesvizhskii, 2006). In a further step, the identified peptides are used to determine the proteins from which they originated. Protein inference can however prove to be challenging, especially in eukaryote species, because peptides can belong to several distinct proteins and multiple protein isoforms may coexist (Nesvizhskii, 2005). The notion of proteotypic peptide was then introduced to single out those peptides that are unique to a protein.

Protein identification is the main task in most MS data processing workflows. Nonetheless, protein quantification is also frequently performed. Quantification experiments usually have the aim of comparing protein expression across multiple samples prepared in distinct conditions, although absolute quantification of proteins can also be achieved. Two mutually exclusive strategies have been established through time for quantification in proteomics: stable isotope labeling and label-free quantification (Bantscheff *et al.*, 2007). In stable isotope labeling approaches such as SILAC (Ong *et al.*, 2002), a mass tag is inserted in proteins through a wide range of experimental protocols. The induced mass shift between the light and heavy forms of peptides is detectable by mass spectrometers, leading to the relative quantification of the proteins of interest through the comparison of the signal intensities (Bantscheff *et al.*, 2007). The label-free approaches expanded more recently with the emergence of more accurate instruments. Being usually based either on spectral counting or on precursor ion intensity, label-free approaches do not include any isotope-tagging step and consequently reduce the sample preparation complexity (Bantscheff *et al.*, 2007).

PepXML and protXML formats

The pepXML and protXML formats (Keller *et al.*, 2005) were originally developed with the purpose of improving and facilitating data transfer between the various tools being part of the Trans-Proteomic Pipeline (TPP) (Keller *et al.*, 2005). The pepXML format stores and shares peptide identification and quantification data while the protXML format focuses solely on the corresponding protein data (Deutsch *et al.*, 2015). Some search engines can directly output their results into files in the pepXML format to be further processed by the TPP. Converter tools are also available in the TPP for the other formats. Although the pepXML and protXML formats were not defined with the underlying aim of becoming official standard formats, many tools outside the TPP adopted them over the years (Deutsch *et al.*, 2015).

mzIdentML and mzQuantML formats

To provide a reliable open standard format for the distribution of peptide and protein identification data, mzIdentML was developed by the HUPO-PSI alongside with a suite of converters for most open and proprietary formats available (Jones *et al.*, 2012). As in the case of pepXML and protXML, search engines that evolve over the years offer the option to export directly their results in the mzIdentML format (Jones *et al.*, 2012).

Based on the success of mzIdentML, the mzQuantML format was launched shortly thereafter under the direction of the HUPO-PSI as a quantitative-focused counterpart (Walzer *et al.*, 2013). It was developed to answer the pressing need for a standard format in quantitative proteomics, especially since the techniques and approaches may diverge significantly between individual experiments (Walzer *et al.*, 2013). The mzQuantML format notably features the ability to cross-reference other XML-based open formats such as mzML and mzIdentML. These cross-links help to keep track of the numerous data and metadata produced throughout the analysis workflow (Walzer *et al.*, 2013). Both formats are based on the same controlled vocabulary that was introduced by the HUPO-PSI for mzML (Jones *et al.*, 2012; Walzer *et al.*, 2013), thus providing a robust yet flexible way to annotate metadata information.

mzTab format

The mzTab format was recently launched by the HUPO-PSI as a complement to both the mzIdentML and mzQuantML formats (Griss *et al.*, 2014). The underlying purpose was to design a simple tabular format summarizing identification and quantification data from proteomics and metabolomics experiments. As the strength of the format resides in the easy report of experimental results, it usually contains an overall lower amount of information compared to most XML-based formats (Griss *et al.*, 2014). Files in the mzTab format do not require specific parsing software since no special encoding is involved, which significantly simplifies their handling by users.

TraML format

Through the years, more specialized open formats also made their apparition in the proteomics field. The TraML (Deutsch *et al.*, 2012) standard format developed by the HUPO-PSI is such an example, capturing solely the transition lists resulting from selected reaction monitoring (SRM) experiments. In SRM approaches, a transition represents the pair of *m/z* values of the targeted peptide precursor ion and of one of its corresponding fragment ions. TraML is designed around the same principles that were established for the mzML and mzIdentML formats, using an XML-based architecture along with a controlled vocabulary encoding the metadata information (Deutsch *et al.*, 2012).

Databases and Repositories

Databases in the proteomics field differ strongly in terms of both their purpose and the data they contain. While some solely serve as repositories for the dissemination of experimental datasets, others have been developed with more advanced functions such as sequence annotation or targeted research goals.

Proteomics databases and repositories are still under substantial development in comparison to other life science fields. This delay can partially be explained by the initial skepticism of a part of the proteomics community in regard to sharing publicly experimental data due to data submission imposed by publishers. One of the common concerns was the possible third-party reanalysis of datasets prior to result release in a publication of the original submitter. To address this issue, most databases feature nowadays the ability to keep dataset submission private for the duration of the reviewing process (Vizcaíno *et al.*, 2014).

Needless to say, UniProt (The UniProt Consortium, 2017) remains arguably the most popular and internationally renowned resource for proteins, playing a central role in the proteomics database landscape as the ever-present reference for protein annotation. As UniProt is however a knowledge database that contains almost to no MS data, it will not be presented in details here (**Table 2**). Likewise, the species-specific resources will also not be discussed in this article.

GPMdb

The Global Proteome Machine database (Craig *et al.*, 2004) (GPMdb) was initially designed by Beavis Informatics for the purpose of storing the reprocessed data produced by the GPM data analysis servers. In this automated pipeline, the MS raw data submitted by users or contained in other databases are reanalyzed using the online version of the X!Tandem (Craig and Beavis, 2004) search engine. The input files submitted to the GPMdb have to be formatted either as peak list (DTA, PKL or MGF) or in a compatible open standard format (mzXML or mzData). In addition to a large panel of search parameters, the users can choose prior to the database search if they authorize the inclusion of the dataset to the GPMdb. With the accumulation of a large amount of data over the years, the GPMdb has diversified its functions and is today commonly used by researchers to compile spectral libraries and develop related search tools. This notably led to the inception of the X!Hunter (Craig *et al.*, 2006) online search engine, which can perform the search of experimental mass spectra against consensus spectral libraries built from the GPMdb.

PeptideAtlas

The PeptideAtlas database (Desiere *et al.*, 2006) was originally developed in 2004 at the Institute for Systems Biology (ISB) as a resource providing genome annotations from the automatic reprocessing of MS raw data by the TPP pipeline. PeptideAtlas presents the peculiarity of being composed of distinct builds according to the organism or tissue of interest (Deutsch *et al.*, 2008). Another noteworthy feature of PeptideAtlas consists in the computation of an observability score for each proteotypic peptide represented in the database (Deutsch *et al.*, 2008). This offers the option of determining the peptides that are the most likely to be detected in a MS experiment for a given protein. As in the case of the GPMdb, the PeptideAtlas database surpassed its initial annotation purpose and is nowadays frequently used as a research database for the compilation of spectral libraries.

Table 2   Summary of the main public databases and repositories in the proteomics field

Name	Supported Data	Organization	URL
PRIDE Archive	Raw and processed data	European Molecular Biology Laboratory, European Bioinformatics Institute	https://www.ebi.ac.uk/pride/archive
MassIVE	Raw and processed data	Center for Computational Mass Spectrometry, University of California San Diego	https://massive.ucsd.edu
jPOSTrepo	Raw and processed data	National Bioscience Data Center, Japan Science and Technology Agency	http://jpostdb.org
PASSEL	SRM raw data	Seattle Proteome Center, Institute for Systems Biology	http://www.peptideatlas.org/passel
GPMdb	Processed data	Beavis Informatics	http://gpmdb.thegpm.org
PeptideAtlas	Processed data	Seattle Proteome Center, Institute for Systems Biology	http://www.peptideatlas.org

PASSEL

The PeptideAtlas SRM Experiment Library (PASSEL) was developed in the framework of the PeptideAtlas project (Farrah *et al.*, 2012). It was designed as a receiving repository for data produced in SRM experiments. In addition, the raw data submitted to PASSEL are automatically reprocessed by specific tools such as mQuest from the mProphet suite (Reiter *et al.*, 2011). The results of this workflow are stored in a separate database that can be accessed through the PASSEL web interface. Among all the obtained transitions, those considered to be the best based on quality metrics are included in the SRMAtlas database (Picotti *et al.*, 2008).

PRIDE/PRIDE Archive

The Proteomics Identifications (PRIDE) database (Martens *et al.*, 2005a) was established in 2004 at the European Bioinformatics Institute (EBI) as a structured repository designed to propagate experimental proteomics data. The underlying intent was to provide a resource that would keep the submitted data intact, preserving it from any kind of control or alteration. PRIDE contains both the raw MS data and the resulting peptide and protein identifications, along with the corresponding biological and experimental metadata (Martens *et al.*, 2005a). As of 2014, the PRIDE database was entirely redesigned and replaced by PRIDE Archive (Vizcaíno *et al.*, 2016). The most significant structural changes brought by PRIDE Archive lie in a new data storage architecture and submission system along with a reworked web interface (Vizcaíno *et al.*, 2016).

The Peptidome database (Slotta *et al.*, 2009) was a public data repository developed in 2009 at the National Center for Biotechnology Information (NCBI) that aimed to allow the exchange of experimental proteomics data. Peptidome was thus closely related to PRIDE in terms of both its aim and features. After the discontinuation of Peptidome in 2011 due to lack of funds, a coordinated effort of both teams led to transfer most of its content to the PRIDE repository (Csordas *et al.*, 2013).

MassIVE

The Mass spectrometry Interactive Virtual Environment (MassIVE) repository was developed at the Center for Computational Mass Spectrometry (CCMS) as a public data repository for datasets produced in MS-based proteomics experiments. Compared to other repositories, MassIVE aims to enhance the interactivity of its content. A noteworthy feature is the ability to add comments and reanalyzes for a given submitted dataset. MassIVE also provides the possibility to directly reprocess the raw data that were submitted using several online workflows.

Tranche (Smith *et al.*, 2011) was a public data repository developed in 2005 as part of the Proteome Commons network that aimed to transfer proteomics raw datasets at a large scale. Since the repository had to handle a high data traffic in relation with the large file size, a special effort was made in the design of an efficient peer-to-peer infrastructure coupled with a reliable client-server architecture (Smith *et al.*, 2011). After the funding shortfall that led to the discontinuation of Proteome Commons and Tranche in 2013 (Science Signaling, 2013), as many data sets as possible were recovered and transferred to MassIVE (Deutsch *et al.*, 2017).

jPOST

The Japan Proteome Standard (jPOST) is an emerging resource that ultimately aims to provide both a repository and a database for the worldwide proteomics community. On the one hand, the jPOST repository (jPOSTrepo) (Okuda *et al.*, 2017) was launched at the National Bioscience Data Center (NBDC) in 2015 as a platform for the global exchange of datasets produced in MS-based proteomics experiments. The underlying aim was to offer an alternative for the Asia and Oceania regions to the existing proteomics repositories. Indeed, most public MS data resources are either hosted in Europe (PRIDE Archive) or in the United States (PASSEL, MassIVE). This can prove to be problematic in terms of data transfer speed for researchers who agree to share their experimental datasets but are located far away from these regions. A special effort was thus additionally undertaken to deploy an efficient file upload system, further facilitating data exchange (Okuda *et al.*, 2017).

On the other hand, the jPOST database (jPOSTdb) is still under development and has not been made publicly available yet. Eventually, jPOSTdb will include the automatic reprocessing of the MS raw data submitted to jPOSTrepo through a customized workflow using a combination of several search engines.

ProteomeXchange Consortium

The ProteomeXchange (PX) consortium (Vizcaíno *et al.*, 2014) was originally formed in 2006 and officially launched in 2011 as a collective effort to coordinate data exchange and submission between MS-based proteomics databases and repositories. Since the partner resources had been developed independently at distinct research institutions, it resulted in limited coordination in regard to their respective data submission guidelines and policies (Vizcaíno *et al.*, 2014). This proved to be troublesome for researchers willing to share their experimental data, struggling to determine where and in which form datasets should be submitted. Likewise, data comparison through the use of several sources could turn out to be challenging. The purpose of the PX consortium was to remove these obstacles by providing a regulated infrastructure and framework (Vizcaíno *et al.*, 2014). Such an initiative had already been successfully implemented in other omics fields, as for instance with the development in genomics of the International Nucleotide Sequence Database Collaboration (INSDC) (Cochrane *et al.*, 2016) to regulate the dissemination of DNA sequencing data.

In its first inception, the PX consortium had only two core members: PRIDE and PeptideAtlas (along with its PASSEL resource) (Vizcaíno *et al.*, 2014). PRIDE was used at the time as the unique entry point for data produced in shotgun proteomics experiments, whereas SRM data were redirected to PASSEL. In the pursuit of its unification role, more receiving repositories progressively joined the PX consortium in recent years. This led notably to the inclusion of the MassIVE repository in 2014, followed shortly by the addition of jPOST in 2016 (Deutsch *et al.*, 2017). More emerging resources are also expected to become members of the PX consortium in the future, further improving the state of data sharing in the proteomics field.

Data submission in the PX consortium can be subdivided in two distinct workflows: complete and partial (Deutsch *et al.*, 2017). In a complete submission, the format of all the submitted files can be recognized and parsed by the receiving repository. It is the recommended submission process in most cases, as it allows searching directly through the results with precise queries. A partial submission contains files that are in a non-compatible format, thus preventing the receiving repository to read them. As with the complete submissions, the files can however still be searched using the associated submission metadata. Despite its limitations, the partial submission process remains essential to support the inclusion of datasets coming from a broad range of proteomics experiments, including marginal or emerging techniques (Deutsch *et al.*, 2017).

Both complete and partial workflows require the submission of the raw data, the processed data and the corresponding biological and technical metadata information. The latter is encoded in an XML format that was specifically developed for the PX consortium (PX XML) (Vizcaíno *et al.*, 2014). After submission, each dataset is assigned a unique PX identifier. A receiving repository specific identifier is additionally assigned, except for PRIDE (Deutsch *et al.*, 2017). If a dataset was submitted as complete, a digital object identifier (DOI) is also created to allow proper referencing in case of reanalysis in another project (Vizcaíno *et al.*, 2014). Since data reuse is also one of the main focuses of the PX consortium, distinct identifiers are specifically generated to track reprocessed datasets and store corresponding results (Deutsch *et al.*, 2017).

The publicly available datasets stored in the resources of the PX consortium are accessed and queried through the ProteomeCentral online portal, which was deployed from the first version. ProteomeCentral notably features the ability to filter the datasets using metadata keywords, which can be for instance used to retrieve solely the set of results matching a given organism, tissue or MS instrument (Deutsch *et al.*, 2017). A RSS feed was also set up to push updates about new publicly available datasets that are released in the PX consortium, providing links to both the ProteomeCentral entry and the corresponding PX XML file storing the submission metadata (Vizcaíno *et al.*, 2014).

Despite the essential contribution brought by PX consortium to the proteomics field, there is still room for improvement. Some work remains for example to be done regarding the compatibly of complete submission with data produced by increasingly popular experimental approaches, such as the data independent acquisition, top-down proteomics or MS-based imaging techniques (Deutsch *et al.*, 2017). The MassIVE repository already started this process by becoming compatible with some DIA-based workflows (Deutsch *et al.*, 2017). The PX consortium is arguably the most impactful initiative that occurred in recent years for databases and repositories in proteomics, bringing together the different resources in the field and standardizing data submission, processing and dissemination.

Conclusion and Prospects

The successful outcome of collective efforts such as the ProteomeXchange initiative has significantly boosted data sharing in the proteomics field. An obvious consequence is the increasing number of journal editors who nowadays mandate the deposition of the original raw datasets to enhance the publication of scientific articles and allow for evidence checks. Further into the future, more integration of the different omics fields is to be expected. Progress has already been made in this direction with the recent development of the Omics Discovery Index (OmicsDI) portal (Perez-Riverol *et al.*, 2016). OmicsDI aims to provide an online platform for the dissemination and access of datasets from genomics, transcriptomics, metabolomics and proteomics resources. Tighter collaboration between proteomics and metabolomics is also likely, based in particular on the fact that both fields rely predominantly on MS-based approaches and techniques to produce their data. Some open standard formats such as mzTab have already started to become compatible with metabolomics workflows.

Acknowledgement

The author would like to thank the Dr. Frédérique Lisacek and Emma Ricart Altimiras for their support.

See also: Bioinformatics Data Models, Representation and Storage. Biological Database Searching. Clinical Proteomics. Data Storage and Representation. Experimental Platforms for Extracting Biological Data: Mass Spectrometry, Microarray, Next Generation Sequencing. Identification of Proteins from Proteomic Analysis. Information Retrieval in Life Sciences. Natural Language Processing Approaches in Bioinformatics. Protein Structure Databases. Proteomics Mass Spectrometry Data Analysis Tools. Quantification of Proteins from Proteomic Analysis. Standards and Models for Biological Data: FGED and HUPO. Supervised Learning: Classification. The Evolution of Protein Family Databases

References

Aebersold, R., Mann, M., 2003. Mass spectrometry-based proteomics. Nature 422 (6928), 198–207.

Altelaar, A.F.M., Munoz, J., Heck, A.J.R., 2012. Next-generation proteomics: Towards an integrative view of proteome dynamics. Nature Reviews Genetics 14 (1), 35–48.

Baker, M., 2012. Proteomics: The interaction map. Nature 484 (7393), 271–275.

Bantscheff, M., Schirle, M., Sweetman, G., Rick, J., Kuster, B., 2007. Quantitative mass spectrometry in proteomics: A critical review. Analytical and Bioanalytical Chemistry 389 (4), 1017–1031.

Cochrane, G., Karsch-Mizrachi, I., Takagi, T., International Nucleotide Sequence Database Collaboration, 2016. The international nucleotide sequence database collaboration. Nucleic Acids Research 44 (D1), D48–D50.

Craig, R., Beavis, R.C., 2004. TANDEM: Matching proteins with tandem mass spectra. Bioinformatics. 20 (9), 1466–1467.

Craig, R., Cortens, J.P., Beavis, R.C., 2004. Open source system for analyzing, validating, and storing protein identification data. Journal of Proteome Research 3 (6), 1234–1242.

Craig, R., Cortens, J.C., Fenyo, D., Beavis, R.C., 2006. Using annotated peptide mass spectrum libraries for protein identification. Journal of Proteome Research 5 (8), 1843–1849.

Csordas, A., Wang, R., Ríos, D., et al., 2013. From peptidome to PRIDE: Public proteomics data migration at a large scale. Proteomics 13 (10–11), 1692–1695.

Desiere, F., Deutsch, E.W., King, N.L., et al., 2006. The PeptideAtlas project. Nucleic Acids Research 34 (Database issue), D655–D658.

Deutsch, E.W., Chambers, M., Neumann, S., et al., 2012. TraML– A standard format for exchange of selected reaction monitoring transition lists. Molecular & Cellular Proteomics 11 (4), (R111.015040-R111.015040).

Deutsch, E.W., Csordas, A., Sun, Z., et al., 2017. The ProteomeXchange consortium in 2017: Supporting the cultural change in proteomics public data deposition. Nucleic Acids Research 45 (D1), D1100–D1106.

Deutsch, E.W., Lam, H., Aebersold, R., 2008. PeptideAtlas: A resource for target selection for emerging targeted proteomics workflows. EMBO Reports 9 (5), 429–434.

Deutsch, E.W., Mendoza, L., Shteynberg, D., et al., 2015. Trans-Proteomic Pipeline, a standardized data processing pipeline for large-scale reproducible proteomics informatics. PROTEOMICS – Clinical Applications 9 (7–8), 745–754.

Farrah, T., Deutsch, E.W., Kreisberg, R., et al., 2012. PASSEL: The PeptideAtlas SRM experiment library. Proteomics 12 (8), 1170–1175.

Griss, J., Jones, A.R., Sachsenberg, T., et al., 2014. The mzTab data exchange format: Communicating mass-spectrometry-based proteomics and metabolomics experimental results to a wider audience. Molecular & Cellular Proteomics 13 (10), 2765–2775.

HUPO-PSI, 2006. The mzData standard. Available at: http://www.psidev.info/mass-spectrometry#mzdata (accessed 27.03.17).

Jones, A.R., Eisenacher, M., Mayer, G., et al., 2012. The mzIdentML data standard for mass spectrometry-based proteomics results. Molecular & Cellular Proteomics 11 (7), (M111.014381-M111.014381).

Keller, A., Eng, J., Zhang, N., Li, X.K., Aebersold, R., 2005. A uniform proteomics MS/MS analysis platform utilizing open XML file formats. Molecular Systems Biology 1 (1), E1–E8.

Kessner, D., Chambers, M., Burke, R., Agus, D., Mallick, P., 2008. ProteoWizard: Open source software for rapid proteomics tools development. Bioinformatics 24 (21), 2534–2536.

Martens, L., Chambers, M., Sturm, M., et al., 2011. mzML – A community standard for mass spectrometry data. Molecular & Cellular Proteomics 10 (1), doi:10.1074/mcp.R110.000133.

Martens, L., Hermjakob, H., Jones, P., et al., 2005a. PRIDE: The proteomics identifications database. Proteomics 5 (13), 3537–3545.

Martens, L., Nesvizhskii, A.I., Hermjakob, H., et al., 2005b. Do we want our data raw? Including binary mass spectrometry data in public proteomics data repositories. Proteomics 5 (13), 3501–3505.

Nesvizhskii, A.I., 2005. Interpretation of shotgun proteomic data: The protein inference problem. Molecular & Cellular Proteomics 4 (10), 1419–1440.

Nesvizhskii, A.I., 2006. Protein identification by tandem mass spectrometry and sequence database searching. Mass Spectrometry Data Analysis in Proteomics. New Jersey, NJ: Humana Press, pp. 87–120.

Okuda, S., Watanabe, Y., Moriya, Y., et al., 2017. jPOSTrepo: An international standard data repository for proteomes. Nucleic Acids Research 45 (D1), D1107–D1111.

Ong, S.-E., Blagoev, B., Kratchmarova, I., et al., 2002. Stable isotope labeling by amino acids in cell culture, SILAC, as a simple and accurate approach to expression proteomics. Molecular & Cellular Proteomics 1 (5), 376–386.

Pedrioli, P.G.A., Eng, J.K., Hubley, R., et al., 2004. A common open representation of mass spectrometry data and its application to proteomics research. Nature Biotechnology 22 (11), 1459–1466.

Perez-Riverol, Y., Bai, M., Leprevost, F., Squizzato, S., et al., 2016. Omics Discovery Index–Discovering and Linking Public Omics Datasets.

Picotti, P., Lam, H., Campbell, D., et al., 2008. A database of mass spectrometric assays for the yeast proteome. Nature Methods 5 (11), 913–914.

Prince, J.T., Carlson, M.W., Wang, R., Lu, P., Marcotte, E.M., 2004. The need for a public proteomics repository. Nature Biotechnology 22 (4), 471–472.

Reiter, L., Rinner, O., Picotti, P., et al., 2011. mProphet: Automated data processing and statistical validation for large-scale SRM experiments. Nature Methods 8 (5), 430–435.

Science Signaling, 2013. ST NetWatch: Protein Databases. Available at: http://stke.sciencemag.org/resources/st-netwatch-archive/st-netwatch-protein-databases (accessed 23.03.17).

Slotta, D.J., Barrett, T., Edgar, R., 2009. NCBI peptidome: A new public repository for mass spectrometry peptide identifications. Nature Biotechnology 27 (7), 600–601.

Smith, B.E., Hill, J.A., Gjukich, M.A., Andrews, P.C., 2011. Tranche distributed repository and ProteomeCommons.org. In: Hamacher, M, Eisenacher, M, Stephan, C (Eds.), Data Mining in Proteomics. Totowa, NJ: Humana Press, pp. 123–145.

The UniProt Consortium, 2017. UniProt: The universal protein knowledgebase. Nucleic Acids Research 45 (D1), D158–D169.

Verheggen, K., Martens, L., 2015. Ten years of public proteomics data: How things have evolved, and where the next ten years should lead us. EuPA Open Proteomics 8, 28–35.

Vizcaíno, J.A., Csordas, A., del-Toro, N., et al., 2016. 2016 update of the PRIDE database and its related tools. Nucleic Acids Research 44 (D1), D447–D456.

Vizcaíno, J.A., Deutsch, E.W., Wang, R., et al., 2014. ProteomeXchange provides globally coordinated proteomics data submission and dissemination. Nature Biotechnology 32 (3), 223–226.

Walzer, M., Qi, D., Mayer, G., et al., 2013. The mzQuantML data standard for mass spectrometry-based quantitative studies in proteomics. Molecular & Cellular Proteomics 12 (8), 2332–2340.

Wilhelm, M., Kirchner, M., Steen, J.A.J., Steen, H., 2012. mz5: Space- and time-efficient storage of mass spectrometry data sets. Molecular & Cellular Proteomics 11 (1), [O111.011379-O111.011379].

Further Reading

Deutsch, E.W., 2012. File formats commonly used in mass spectrometry proteomics. Molecular & Cellular Proteomics 11 (12), 1612–1621.

Martens, L., 2011. Proteomics databases and repositories. In: Wu, C.H., Chen, C (Eds.), Bioinformatics for Comparative Proteomics. Totowa, NJ: Humana Press, pp. 213–227.

Martens, L., Vizcaíno, J.A., 2017. A golden age for working with public proteomics data. Trends in Biochemical Sciences 42 (5), 333–341.

Mayer, G., Montecchi-Palazzi, L., Ovelleiro, D., *et al.*, 2013. The HUPO proteomics standards initiative-mass spectrometry controlled vocabulary. Database 2013 (0), bat009.

Perez-Riverol, Y., Alpi, E., Wang, R., *et al.*, 2015. Making proteomics data accessible and reusable: Current state of proteomics databases and repositories. Proteomics 15 (5–6), 930–950.

Riffle, M., Eng, J.K., 2009. Proteomics data repositories. Proteomics 9 (20), 4653–4663.

Taylor, C.F., Paton, N.W., Lilley, K.S., *et al.*, 2007. The minimum information about a proteomics experiment (MIAPE). Nature Biotechnology 25 (8), 887–893.

Vaudel, M., Verheggen, K., Csordas, A., *et al.*, 2016. Exploring the potential of public proteomics data. Proteomics 16 (2), 214–225.

Verheggen, K., Martens, L., 2015. Ten years of public proteomics data: How things have evolved, and where the next ten years should lead us. EuPA Open Proteomics 8, 28–35.

Vizcaíno, J.A., Foster, J.M., Martens, L., 2010. Proteomics data repositories: Providing a safe haven for your data and acting as a springboard for further research. Journal of Proteomics 73 (11), 2136–2146.

Biographical Sketch

Thibault Robin is a Ph.D. student in bioinformatics at the Proteome Informatics Group (PIG) of the Swiss Institute of Bioinformatics (SIB) affiliated with the University of Geneva. After obtaining a Bachelor degree in biology in 2013, he developed a strong interest for informatics and programming. This led to his graduation with a Master degree in bioinformatics in 2015. He is currently working on the development of new proteomics tools, focusing in particular on post-translational modifications and sequence variants.

Proteomics Mass Spectrometry Data Analysis Tools

Aivett Bilbao, Pacific Northwest National Laboratory, Richland, WA, United States

Introduction

The overall purpose of proteomics is the large-scale study of genes and cellular functions at the phenotype level. Information on protein abundances, their variations and modifications, and their interacting networks are critical for understanding complex biological processes, which in turn, are useful to accomplish scientific developments such as effective diagnostic techniques and disease treatments (Aebersold and Mann, 2016; Szabo and Janaky, 2015). The ability to quickly and consistently analyze the large amounts of experimental data produced in proteomics research relies on software advances (Nesvizhskii, 2010; Yates, 2015). Bioinformatics have emerged as a crucial contributor with the development of novel algorithms, the connection of various tools into complex pipelines, as well as the deposition and dissemination of both software and data. In particular, the prominence of open source software and public data in reproducible research has been recently highlighted in several publications (Deutsch *et al.*, 2017; Jiménez *et al.*, 2017; Martens and Vizcaíno, 2017; da Veiga Leprevost *et al.*, 2017; Wilkinson *et al.*, 2016).

As a specialized section of "Proteome Informatics" (Lisacek), this article provides an up-to-date and focused literature review of the software for MS-based proteomics. In particular, freely available and open source tools are cited and discussed, providing a solid starting point with relevant notions for both non-specialized end-users and bioinformaticians.

Since understanding of the data structure is critical to comprehend the processing software methods and obtaining accurate results, the first section summarizes the MS data acquisition process and spectral types currently employed in discovery or untargeted proteomics. Next, the general workflows are discussed and the community efforts to facilitate reproducible research are highlighted. In the remaining sections, algorithmic concepts related to identification, statistical assessment of error rates and quantification tools are compared with an emphasis on the most popular free and open source tools. The review concludes with a discussion of pertinent guidelines and future directions in the field.

Understanding Mass Spectrometry Data From Proteomics Analyses

Discovery of the proteins present in a sample is typically achieved by liquid-chromatography (LC) coupled to mass spectrometry (MS). In the so-called shotgun or bottom-up approach, complex protein samples are digested into peptides. The most popular enzyme used for protein digestion is trypsin, which leads to peptides with C-terminally protonated amino acids, providing an advantage in subsequent MS-based peptide sequencing (Aebersold and Mann, 2016; Aebersold, 2003). As peptides elute from the LC column, they are ionized by electrospray (ESI) and subsequent ions are analyzed by the mass spectrometer. In the traditional data-dependent acquisition (DDA) mode, the MS instrument is operated to maximize the number of selected and fragmented peptides, as they elute from the LC column. Within each DDA cycle, ion signals are recorded in a MS1 or survey scan (precursor ion signals) and the top-*n* most abundant ions are then selected and serially isolated for fragmentation (MS/MS, MS2 or tandem MS) to generate structural information. Alternatively, the mass spectrometer can be operated in data-independent acquisition mode (DIA), where all ions from peptides eluting during LC-MS analysis are fragmented and analyzed in parallel, regardless of their intensities. The advantages and disadvantages as well as specific flavors of DIA with a focus on processing software strategies are discussed in Bilbao *et al.* (2015a).

The different type of generated spectra is illustrated in **Fig. 1**. Data analysis tools pertinent for each spectral type are discussed in the following sections.

The Data Processing Workflow in MS-Based Proteomics

A wide range of informatics methods comprises signal processing of raw MS data, identification and quantification approaches for protein characterization, assessment and control of error rates, as well as application of statistical and machine learning techniques to unravel relevant protein targets and interacting networks in complex processes. Development and software implementation of algorithms for every step is a laborious process that requires substantial amounts of time and effort. As a result, the complete proteomics data analysis is a composite procedure involving several specialized tasks which are interconnected as a software workflow or pipeline (**Fig. 2**). These specialized tools have been developed and improved over the years in a stepwise fashion (Noble and MacCoss, 2012), which has greatly promoted the field but also raised major challenges when moving from single and individual tools to complex and integrated workflows: software availability and reproducible results. The challenges have historically rendered bioinformatics as the most difficult area in MS-based proteomics research (Aebersold, 2009). Multiple tools normally require substantial effort and informatics skills for correct installation and configuration (da Veiga

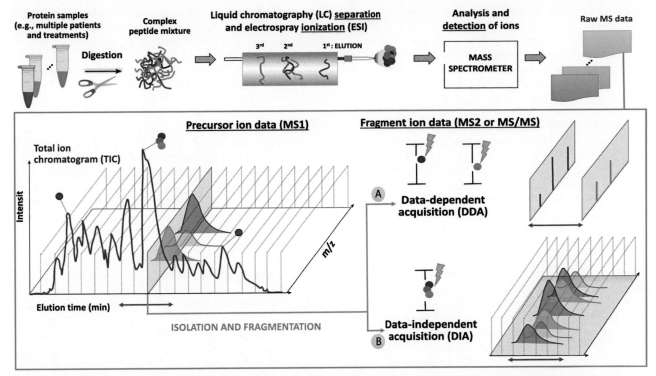

Fig. 1 Sample analysis and mass spectrometry data in discovery proteomics. Following sample preparation and enzymatic digestion, complex peptide mixtures are separated by LC and peptide ions are generated by ESI. The mass spectrometer analyses and detects the peptide ions as they elute. Files with raw MS data are generated from each sample. A file usually contains precursor (MS1) and fragment (MS2) ion spectra. Consecutive spectra in the analysis run are represented by dashed lines. A first representation of the data is the MS1 total ion chromatogram (TIC, blue trace), which is the sum of all ions detected at each time point or spectrum. Examples of MS2 spectra acquired for a specific time range are shown, depending on the MS operation mode used in the analysis run. In DDA mode, the mass spectrometer generates MS2 spectra after isolation and fragmentation of selected precursor ions. The top-2 most intense ions (red and blue LC-MS peaks) are selected to collect individual MS2 spectra at discrete time points and the low abundance one (green LC-MS peak) is undetected. Ions can be selected or not at any time point across the LC peak. In DIA mode, the mass spectrometer generates MS2 spectra of multiple co-eluting ions which are fragmented all together. Continuous fragment ion traces are recorded in convoluted or multiplexed MS2 spectra. Multiple LC-MS/MS peaks from the 3 fragmented peptide ions are detected, including the low abundance one (green LC-MS/MS peaks).

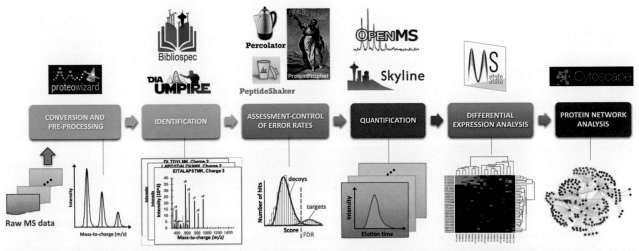

Fig. 2 General proteomics data processing workflow. Processing steps are shown as interconnected boxes. Above and below each step, available logos from example tools and visual representations are included, respectively. Some tools include a graphical user interface (GUI) or only command-line access. In the first step, raw MS data in proprietary formats from different instrument vendors are typically converted to open data formats. Pre-processing can involve centroid data generation, where multiple measurements per peak (black traces) are reduced to single centroid peaks (blue sticks). Next, identification followed by assessment and control of error rates are performed. Quantification information of confident identifications across multiple samples are used as input for downstream analyses and biological interpretations. For label-free data, quantification include a multi-sample alignment step.

Leprevost *et al.*, 2017), lack appropriate documentation and have poor or none graphical user interface (GUI) (Cappadona *et al.*, 2012).

Fortunately, software availability challenges are being recently addressed with the incorporation of advances in software engineering and the contributions from open-source and community-driven efforts.

The BioContainers framework (da Veiga Leprevost *et al.*, 2017) (see Relevant Websites Section) adopted container based technologies such as Docker or rkt to provide complete pipelines as easily installable, deployable and executable packages.

The ELIXIR Tools and Data Services Registry (see Relevant Websites Section) (Ison *et al.*, 2016), is a catalogue of bioinformatics software and resources with extensive metadata descriptions.

A useful list of free software for analysis of mass spectrometry data can be found in http://www.ms-utils.org.

Other efforts to provide well-documented software parameters and dependencies include workflow systems such as KNIME (Berthold *et al.*, 2009) and the Galaxy platform (Afgan *et al.*, 2016). Examples for proteomics are the OpenMS 2 tools (Aiche *et al.*, 2015) delivered as KNIME (see Relevant Websites Section), and proteomics workflows in Galaxy (see Relevant Websites Section) for OpenMS, the Trans-Proteomic Pipeline (TPP) and MaxQuant.

The other component for reproducible research, besides the software, is the data itself. Along the same lines of efforts, current requirements from scientific journals and funding agencies have promoted a change of mentality: public data sharing is a common acceptance as good scientific practice in the proteomics field, as embraced earlier in genomics and transcriptomics (Martens and Vizcaíno, 2017).

The set of principles referred to as the FAIR Data Principles that requires data to be Findable, Accessible, Interoperable, and Reusable in the long-term has been recently published as a guideline to support and enhance the reusability in contemporary e-Science (Wilkinson *et al.*, 2016).

The ProteomeXchange (PX) Consortium (see Relevant Websites Section) standardizes submission and dissemination of mass spectrometry proteomics data worldwide (Deutsch *et al.*, 2017).

Recent and detailed information about proteomics data representation and databases can be found in the complementing article "Proteomics data representation and databases".

As a remarkable outcome of the mentioned efforts, the majority of the software tools and related data discussed in the following sections are freely available to the community. All readers, from end-users to software developers, are invited to explore and download content from the cited resources for gaining a better understanding of the tools, familiarize with the data and optimizing parameters for better results.

It is important to note that there is a myriad of tools and that new ones regularly come out. Evaluating a tool can be time-consuming and demands patience. It is common to find many tools stating to have the needed functionality. However, a deep examination is often required to determine whether the statements are true or whether the tool fits the specific user needs. It is recommended to select a tool taking into account the input and output file formats, available documentation, tutorials and test data, and related publications including application studies. It is equally important to check if the tool is actively maintained and updated, which informatics skills are needed and how scalable it is. Regarding the later point, a few files should be tried first to become familiar with the tool, adjust parameters and evaluate results, and then more files could be processed according to a given study to evaluate if the tool is likely or not to handle all the data.

Preprocessing MS Data

After the mass spectra acquisition, the first step in the informatics pipeline involves several preprocessing methods. Since current MS instruments acquire high resolution data in profile mode, centroid data is computationally generated either during or post-acquisition (**Fig. 2**, conversion and pre-processing). Preprocessing methods include mass calibration, noise removal, peak picking and grouping of peaks into isotopic profiles with deconvolution and charge-state determination. Preprocessing allows significant data reduction and enhancement for further analysis. Format of raw MS files depends on the instrument vendor, as each of the major vendors uses its own proprietary binary format and continuously update it to support new features of their instruments. Therefore, open formats such as mzXML (Pedrioli *et al.*, 2004) and mzML (Martens *et al.*, 2011) have been created by the proteomics community and developers of analysis software to enable the exchange of data between tools. An overview of the most common file formats used in MS-based proteomics can be found in Deutsch (2012).

Raw MS data processing can be performed using proprietary software provided by the instrument vendors, which may produce better results because methods are customized to pair the characteristics of the data generated by the specific instrument. Alternatively, raw MS data can be parsed from the vendor format to open file formats – with or without preprocessing – to continue the processing pipeline using other software tools. The freely available ProteoWizard toolkit (Chambers *et al.*, 2012) provides software for converting proprietary files from most vendors to open file formats and includes several preprocessing algorithms both open source from the community and non-open source from the vendors. Other software platforms including preprocessing tools are OpenMS (Röst *et al.*, 2016; Sturm *et al.*, 2008), MaxQuant (Cox and Mann, 2008; Tyanova *et al.*, 2016) and MzJava (Horlacher *et al.*, 2015b).

A special case is the Skyline software tool for quantification (MacLean *et al.*, 2010; Pino *et al.*, 2017), where MS data from many instrument vendor formats can be imported directly, i.e., a conversion to open file formats is usually not required.

Identification by Tandem Mass Spectrometry

The fragmentation pattern encoded by a tandem or MS/MS spectrum yield identification of the corresponding peptide sequence. The so-called search engines (**Fig. 2**, identification) annotate the MS/MS spectra with peptide sequences using various algorithms and scoring strategies (Nesvizhskii, 2010). Peptide identification strategies, discussed in detail in (Lisacek), can be roughly classified into four categories: database search, spectral library search, de-novo sequencing and hybrid approaches such as sequence tag-assisted database search.

SEQUEST (Eng *et al.*, 1994) and MASCOT (Perkins *et al.*, 1999) proprietary database search engines have been traditionally and predominantly used. Nowadays, open source tools are becoming more popular, with increased robustness, usability and flexibility for customization or addition of new features. For instance, Comet is an academic version of SEQUEST. **Table 1** summarizes examples of free and/or open source peptide identification software currently available.

As **Table 1** shows, most of the identification tools do not have a GUI. Andromeda is available as standalone or integrated into MaxQuant. The SearchGUI (Vaudel *et al.*, 2011) (see Relevant Websites Section) provides an open source, user friendly framework to operate multiple command line search engines.

In terms of scoring, several properties can be taken into account to measure the similarity between an experimental spectrum and a reference. The precursor mass is mainly used to restrict the search space. However, the mass differences for both the matching fragments and precursor can be considered into the scoring. Other properties include the number and intensities of the experimental peaks, number of complementary ions, number of missed cleavages and number (and types) of modifications.

A great variety of scoring strategies have been proposed for peptide identification. Broadly speaking, they can be divided into heuristic and probabilistic schemes. Heuristic methods typically combine a set of scoring properties in a practical form reflecting expert knowledge. A product or a linear function are simple examples of how these properties can be combined into a scoring function. Probabilistic schemes compute the probability that a given sequence is the origin of the spectrum.

With the introduction of machine learning, algorithms that can automatically learn from training data have been applied to many fields, including proteomics. Specifically, MSGF+ is able to automatically derive scoring parameters from annotated spectra. More recently, DeepNovo implements a deep neural network model for de novo peptide sequencing.

Finally, the actual software implementation of the algorithm makes a great difference in terms of computing resources required by the tool. By using a fragment-ion indexing method, MSFragger enables a significant improvement in speed over most existing proteome database search tools.

Post-Translational Modifications

Post-translational modifications (PTMs) are alteration of proteins during or after protein biosynthesis, resulting in changes of the protein physicochemical properties which modulate and regulate cellular functions. Abnormal PTM abundances are involved in disease states.

Table 1   Selected list of peptide identification software freely available

Software	Programming language	Open source	GUI	Identification type	Scoring and modelling strategies	Reference and website
Comet	C++	Yes	No	Sequence DB	Heuristic. Cross-correlation	(Eng *et al.*, 2015, 2013) http://comet-ms.sourceforge.net
MSGF+	Java	Yes	No	Sequence DB	Probabilistic. Directed acyclic graph, dot product, generating function	(Kim and Pevzner, 2014) https://omics.pnl.gov
X!Tandem	C++	Yes	No[a]	Sequence DB	Heuristic. Hyperscore (hypergeometric distribution-based)	(Craig and Beavis, 2004) http://www.thegpm.org
Andromeda	C#	No	Yes	Sequence DB	Probabilistic. Binominal distribution-based	(Cox *et al.*, 2011) http://www.maxquant.org
MSFragger	Java	Yes	Yes	Sequence DB	Heuristic. Hyperscore (similar to X!Tandem)	(Kong *et al.*, 2017) http://www.nesvilab.org
SpectraST	C++	Yes	No[a]	Spectral library	Heuristic. Dot product	(Lam *et al.*, 2007) http://tools.proteomecenter.org
BiblioSpec	C++	Yes	No	Spectral library	Heuristic. Dot product	(Frewen *et al.*, 2006) https://skyline.ms
PepNovo+	C++	Yes	No	De-novo	Probabilistic. Directed acyclic graph	(Frank *et al.*, 2007) http://proteomics.ucsd.edu
DeepNovo	Python	Yes	No	De-novo	Probabilistic. Deep learning neural network	(Tran *et al.*, 2017) https://github.com/nh2tran/DeepNovo

[a]An alternative version is available through TPP GUI.

By identifying the occurrence of regular mass shifts corresponding to the addition or removal of chemical groups of known masses in a peptide MS/MS spectrum, it is possible to identify a PTM and its position on the peptide sequence. When specified in the search parameters, search engines can consider relevant PTMs, such as phosphorylation. However, this number is usually limited because the search space is substantially expanded. Consequently, an important fraction of MS/MS spectra usually remains unidentified, mainly due to the diversity of PTMs and other chemical modifications that are unaccounted. An attractive exception is the automatic PTMs detection algorithm in the proprietary search engine ProteinPilot (Shilov et al., 2007), where modifications, substitutions, and cleavage events are modeled with probabilities rather than by discrete user-controlled settings. This alleviates the user with the tedious task of optimizing parameters.

A range of tools have been developed to perform the so-called "open modification search", i.e., searching with no a priori PTM information (Ahrné et al., 2010). Approaches aiming to automatically discover unexpected modifications in high throughput proteome data include: MODa (Na et al., 2012), MzMod (Horlacher et al., 2015a) and MSFragger (Kong et al., 2017).

In addition, quantification of PTMs in large-scale studies using DIA has been reported with the tool called inference of peptidoforms (IPF), a fully automated algorithm that uses spectral libraries to query, validate and quantify peptidoforms in DIA datasets (Rosenberger et al., 2017b).

Recent advances in instrumentation and development of dedicated software are making possible the characterization of proteomic samples by limiting or circumventing the enzymatic protein digestion. These approaches, referred to as middle-down and top-down, provide more complete sequence data, as oppose to bottom-up proteomics. Middle-down proteomics are hybrid methods, employing limited digestion to create large peptides which retain a richer sequence and PTM information and can be analyzed relatively easily by MS (Sidoli et al., 2014). In top-down proteomics, high resolution MS/MS data is obtained directly from fragmentation of the intact protein (Smith and Kelleher, 2013; Zhang and Ge, 2011). Various software tools are available to analyze this type of complex spectra, adapted to the identification of proteoforms and PTMs, such as MS-Align+ (Liu et al., 2012), ProSight Lite (Fellers et al., 2015), TopPIC (Kou et al., 2016), MASH Suite (Cai et al., 2016) and Informed-Proteomics (Park et al., 2017).

Assessment of Error Rates in Identification Results and Protein Inference

Since only a fraction of all annotations generated by search engines are correct, proper assessment and control of error rates in identification results is a critical step in the informatics pipeline to distinguish correct from incorrect identifications. In proteomics, the most commonly used and accepted statistical confidence measure is the false discovery rate (FDR) (Benjamini and Hochberg, 1995) initially adapted by Elias and Gygi (2007) and also known as the "target-decoy approach" (TDA). In this strategy, a decoy database is first generated by reversing or shuffling the amino acids in the sequences of the reference or target database. All MS/MS spectra are searched against a composite target plus decoy protein sequence database, the best peptide match for each spectrum is selected for further analysis and the number of decoy peptide matches are used to estimate the fraction of incorrect annotations at given thresholds (**Fig. 2**, assessment and control of error rates).

The FDR analysis can be performed by the identification software or externally using stand-alone tools, represented in **Table 2**.

Table 2 Selected list of open source tools freely available for error rate assessment in identification results and protein inference

Software	Programming language	Features	Base modelling	Reference and website
PeptideProphet[a]	C++	Peptide FDR analysis	Target-decoy, mixture model, expectation-maximization	(Choi and Nesvizhskii, 2007; Keller et al., 2002) http://tools.proteomecenter.org
ProteinProphet[a]	C++	Protein inference and FDR analysis	Mixture model, expectation-maximization	(Nesvizhskii et al., 2003) http://tools.proteomecenter.org
iProphet[a]	C++	Peptide FDR analysis. Combination of results from multiple search engines	Target-decoy, mixture model, expectation-maximization	(Shteynberg et al., 2011) http://tools.proteomecenter.org
Percolator	C++	Peptide and protein FDR analysis. Protein inference	Target-decoy, support vector machine	(Käll et al., 2007; The et al., 2016) http://percolator.ms
MAYU	Perl	Peptide and protein FDR analysis. Protein inference	Target-decoy, hypergeometric model	(Reiter et al., 2009) http://proteomics.ethz.ch
PeptideShaker	Java	Peptide and protein FDR analysis. Combination of results from multiple search engines. Protein inference	Target-decoy	(Vaudel et al., 2015) http://compomics.github.io/projects/peptide-shaker.html
mProphet	Perl and R	Peptide FDR for SRM, adapted later for DIA	Target-decoy, linear discriminant analysis	(Reiter et al., 2011) http://www.mprophet.org/

[a]Available through TPP.

The vast majority of the peptide FDR analysis tools utilize the target-decoy strategy. In the case of PeptideProphet, the initial parametric and unsupervised mixture model approach for posterior probability calculation was later combined with the target-decoy strategy within a single semisupervised framework (Choi and Nesvizhskii, 2007; Keller et al., 2002).

Tools such as Percolator, MAYU and PepideShaker, also include the protein inference step to roll up identified peptide sequences into the corresponding proteins, and then calculate a statistical confidence score for each protein identification.

Inferring protein identities given a set of identified peptides is a difficult task due to several factors such as the presence of distinct proteins having a high degree of sequence homology and alternative splice forms of the same gene (Nesvizhskii and Aebersold, 2005).

It has been reported that the combination of several search engines further improves the results, by taking advantage of the specific strengths of each search engine to generate maximal results (Audain et al., 2017; Kwon et al., 2011; Shteynberg et al., 2013). Tools like iProphet and PeptideShaker are of special considerations in the FDR analysis and necessary to properly combine multiple results while avoiding the accumulation of false positives from the different search engines.

Moreover, the FDR is a summary statistics for the entire identification results either at the PSM, peptide or protein level, however it does not indicate the identification confidence for a single hit at each level. There is a distinction between a global and local FDR. The global FDR is the fraction of incorrect hits among all hits that pass the threshold. The local FDR is the fraction of incorrect hits within a subset of hits that share the same score. In the data interpretation guidelines recently published by the Human proteome project (Deutsch et al., 2016), detail reports of FDR values at each of the three levels are required for publication. The PSM-level FDR should be lower than the peptide-level FDR, which should be lower than the protein-level FDR. The larger the dataset, the more extreme these differences become. Extraordinary and novel detection claims require the evidence of two distinct peptides with length of nine amino acids or more.

Despite the FDR is a well-accepted concept, it has been reported that more advance statistical methods are needed in proteomics (Serang and Käll, 2015). An alternative to the target-decoy database approach was proposed and implemented in MSGF+, which uses E-values to evaluate statistical significance of individual PSMs through the generating function approach (Kim and Pevzner, 2014; Kim et al., 2008).

The mProphet algorithms was originally implemented to automate data processing and statistical validation of large-scale selected reaction monitoring (SRM) experiments by scoring decoy transitions. Due to the similarity of the validation procedures, it has been extended and re-implemented in several tools for DIA.

Software to generate decoy spectral libraries are discussed in (Lisacek).

Quantification

Bioinformatics has played an essential role on the large-scale characterization of complex proteomes, traditionally by cataloguing proteins, and more recently, towards quantifying their expression profiles at different cellular states.

Due to differences in ionization efficiency, the signal detected by MS is not directly proportional to the protein concentration. Several combinations of analytical and computational methods have been developed and applied to quantify proteins by MS. Examples of studies evaluating quantification strategies include (Ahrné et al., 2013; Blein-Nicolas and Zivy, 2016; Dowle et al., 2016; Nahnsen et al., 2013; Neilson et al., 2011). **Table 3** shows a list of selected quantification software currently available.

Quantification methods can be label-free based or include stable isotope labels. Labeling methods are discussed in (Lisacek).

In label-free methods different samples are analyzed by LC-MS individually (**Fig. 2**, quantification), requiring reproducible chromatography and advanced informatics algorithms for retention time alignment across multiple samples and intensity normalization to minimize systematic biases. The so-called spectral counting technique has been conventionally used as a rapid and semi-quantitative measure of protein abundance based on the number of MS/MS spectra identified per peptide (Liu et al., 2004). However, biological samples usually comprise a great variety of proteins across a wide dynamic range of abundances, where spectral counting has been found to often give irreproducible results, more notably for low abundance proteins (Cappadona et al., 2012; Ahrné et al., 2013; Blein-Nicolas and Zivy, 2016). In contrast, label-free methods based on the area under the peak or ion chromatogram provide a level of accuracy comparable to labeling approaches (Neilson et al., 2011).

Multiple software packages are available to analyze the integrated peak areas and determine candidate proteins differentially expressed (**Fig. 2**, differential expression analysis). For instance, the R statistical package MSstats (Choi et al., 2014) is a popular choice for analyzing DDA, DIA and SRM quantification results.

Data-Dependent Acquisition and MS1 Quantification

In shotgun proteomics, confidently identified peptides can be quantified using the continuous precursor ion signals from MS1 data or survey scans. MS1 quantification can be performed in two ways, differing in the approach used for chromatogram processing. One strategy, referred here as LC-MS feature-based, consists in first detecting and characterizing every LC-MS peak (often called LC-MS feature) in the MS1 data and then annotating them using the peptide identifications from the MS2 data. Among the first open source proteomics software implementing this approach are SuperHirn (Mueller et al., 2007) and VIPER (Monroe et al., 2007), and more recently MzMine 2 (Pluskal et al., 2010). In contrast to attempt finding every MS1 peak, the second strategy is guided or targeted, i.e., extracted ion chromatograms (XIC or EIC) are constructed given a list of

Table 3   Selected software frameworks freely available for protein quantification

Software	Programming language	Open source	Identification	Chromatogram processing	Labelling	DIA	Reference and website
OpenMS	C++, Python	Yes	X!Tandem, MSGF+, OMSSA, MyriMatch, PepNovo	LC-MS feature-based, XIC-based	SILAC, iTRAC, TMT	Yes[a]	(Röst *et al.*, 2016; Sturm *et al.*, 2008; Pfeuffer *et al.*, 2017) http://www.openms.de/
TPP	Perl, C++, C, Java	Yes	X!Tandem, SpectraST, Comet	–	ICAT, SILAC, iTRAC, TMT	No	(Deutsch *et al.*, 2010, 2015) http://tools. proteomecenter.org
MaxQuant	C#	No	Andromeda	LC-MS feature-based	ICAT, SILAC, iTRAC, TMT	No	(Cox and Mann, 2008; Tyanova *et al.*, 2016) http://www.maxquant.org
Skyline	C#	Yes	BiblioSpec. Results import from a variety of search engines	XIC-based Advanced processing for SRM and PRM	Customizable heavy labeled stable-isotope modifications	Yes	(MacLean *et al.*, 2010; Pino *et al.*, 2017) https://skyline.ms

[a]OpenSWATH.

mass-to-charge targets. Referred here as XIC-based, information from confidently identified peptides are used to extracts the precursor ion signals of each peptide from the MS1 data. This is the case of Skyline, which compute the area under the peak from the XIC corresponding to each targeted peptide.

Data-Independent Acquisition and MS2 Quantification

DDA-based methods have been widely used to identify and quantify proteins. However, the semi-stochastic sampling in DDA leads to irreproducible results, particularly for highly complex proteomic samples. With recent developments in MS instrumentation and data analysis software, it has become practical the systematic and parallel fragmentation of all ions from peptides eluting during LC-MS analysis by DIA. These methods avoid the selection of individual peptide ions (Bilbao *et al.*, 2015a; Chapman *et al.*, 2014; Hu *et al.*, 2016). The label-free quantification performance of DIA methods has seen a growing popularity and adoption, providing a valid alternative to isotope-labeling-based methods (Navarro *et al.*, 2016).

DIA requires more sophisticated post-acquisition data analysis compared to DDA and several informatics tools have been developed to effectively process these complex datasets. The initial SWATH strategy (Gillet *et al.*, 2012) describing a targeted data extraction method using information from spectral libraries has been automated and improved in several tools: Skyline, DIANA (Teleman *et al.*, 2014), OpenSWATH (Röst *et al.*, 2014), SWATHProphet (Keller *et al.*, 2015). More recently, alternative workflows that circumvent the need of DDA for creating a spectral library have been developed: DIA-Umpire (Tsou *et al.*, 2015, 2016), GroupDIA (Li *et al.*, 2015), MSPLIT-DIA (Wang *et al.*, 2015) and Pecan (Ting *et al.*, 2017). A discussion about the challenges associated with error-rate control in the analysis of DIA data has been recently reported (Rosenberger *et al.*, 2017a).

Finally, a drawback of DIA is the increased likelihood of interference due to the overlap of fragment ions from co-fragmented precursors and several computational strategies can tackle this issue to further expand the benefits of DIA (Bilbao *et al.*, 2015b, 2016; Zhang *et al.*, 2015).

Selected Reaction Monitoring

Also referred to as multiple reaction monitoring (MRM), SRM is the MS method of choice to accurately and consistently quantify proteins across hundreds of complex samples. However, it is applicable when only a limited number of predefined proteins are compared. Therefore, it is typically used as the validation method, to confirm a list of target proteins obtained in a previous global or discovery study. For SRM data acquisition, the two mass filters of a triple quadrupole MS instrument select and monitor, over chromatographic elution, the signal of a series of transitions (precursor/fragment ion pairs and expected times) in a predefined list of target peptides (Lange *et al.*, 2008; Picotti and Aebersold, 2012).

Several informatics tools are available to assist the time-consuming process of defining the acquisition method. Examples of software to select proteins, proteotypic peptides, transitions and best acquisition settings include SRMcolider (Röst *et al.*, 2012), MRMOptimizer (Alghanem *et al.*, 2017), PREGO (Searle *et al.*, 2015) and the PNNL Biodiversity Plugin as one of the external tools available in Skyline (Degan *et al.*, 2016).

Since SRM data consist of a set of chromatographic traces for each monitored peptide, data analysis is simpler, compared to shotgun methods. The Skyline software tool is widely used for the integration of the chromatographic peaks for each transition, allowing the relative or, if heavy isotope–labeled reference standards are used, absolute quantification of the targeted peptides, which are used as a surrogate measure of the proteins of interest.

Alternatively, targeted MS methods such as parallel reaction monitoring (PRM) (Peterson *et al.*, 2012), use hybrid instruments where the third quadrupole is substituted with a high resolution and accurate mass analyzer, acquiring high-resolution full MS/MS spectra for each target peptide. Consequently, in PRM, the best fragment ions can be computationally selected post-acquisition. A web-based application for assay development was recently reported for targeted and label-free phosphoproteome analysis by PRM (Lawrence *et al.*, 2016).

Discussions regarding the best practices for targeted MS measurements in biology and medicine, including the computational and statistical tools, can be found in Carr *et al.* (2014).

How it all Fits Together

MS-based proteomics studies are supported by publicly accessible bioinformatics tools for processing and interpreting the large amounts of data that are generated in complex projects (Aebersold and Mann, 2016). To illustrate how various software tools are connected into processing pipelines, two recent representative applications are described.

The first study was based on MS1 quantification. Rieckmann and colleagues characterized 28 primary human hematopoietic cell populations in either steady or activated state (Rieckmann *et al.*, 2017). Measurements from three to four donors generated a total of 175 immune cell proteomes. Samples were analyzed by LC-MS with DDA and selection of the top-10 most intense precursor ions for fragmentation. Raw MS files were analyzed with MaxQuant. Identification was performed by the Andromeda search engine. The Label Free Quantification algorithm (MaxLFQ) was applied. Data analysis and visualization was performed using Perseus and R. More than 10,000 proteins were identified in total.

The second study was based on the SWATH strategy. Williams *et al.* (2016) analyzed phenotypes associated with liver mitochondrial metabolism. The authors profiled 386 individuals from 80 cohorts of the BXD mouse genetic reference population across two environmental states. Measurements in the livers of the entire population involved: transcriptome, proteome and metabolome. In particular for proteomics, a spectral library was built from a pooled sample analyzed by LC-MS with DDA and selection of the top-20 most intense precursor ions for fragmentation. Raw files were converted to mzML using AB Sciex Data Converter, applying the algorithm for centroid mode. Files were processed through TPP for identification. Search results were combined using iProphet and consensus spectra for a total of 22,208 peptides were constructed using SpectraST. Retention time was transformed to indexed retention time (iRT) (Escher *et al.*, 2012). The top-5 most abundant *b* and *y* fragment ions of each peptide were selected to generate the assays for the targeted data extraction. The DIA LC-MS analysis with SWATH was performed using a set of 32 overlapping isolation windows (width of 26 Da), covering the 400 to 1200 *m/z* precursor range. Raw files were converted to profile mzXML files using ProteoWizard. Targeted data extraction was performed using the OpenSWATH workflow. Identified peptides were analyzed with ProteinProphet. Peptide quantities and network graphs were calculated in R using the custom package imsbInfer (see Relevant Websites Section). A total of 2622 proteins were quantified in all 80 cohorts.

As it can be perceived, the processing and interpretation of hundreds of MS files and more than a million of spectra generated from these projects would have been impossible without bioinformatics tools properly assembled into workflows.

Related to network biology for MS-based proteomics (**Fig. 2**, protein network analysis), more information can be found in Bensimon *et al.* (2012). Examples of other available software include STRING (see Relevant Websites Section) as an online resource for querying and visualizing interaction networks and Cytoscape (Saito *et al.*, 2012; Shannon *et al.*, 2003) as one of the most popular open-source software for integration, visualization and analysis.

Closing Remarks

MS-based proteomics methodologies have evolved rapidly over recent years along with a remarkable progress on specialized software. These approaches are nowadays used in combination with other omics disciplines such as genomics, transcriptomics and metabolomics (Nesvizhskii, 2014; White *et al.*, 2017a,b). With the increasingly high throughput large scale studies in the big data era, robust and efficient MS software is as important as advanced MS hardware for comprehensively exploiting the information and generating knowledge. In this sense, progress in instrumentation and analytical methods are expected, and importantly, development of new computational methods to cope with these advances will continue to be an active area of research. Therefore, an increasing number of available tools including proprietary and open source can be expected.

The data processing workflow in MS-based proteomics is a set of consecutive complex tasks. It has been previously stated that this stepwise approach should be replaced by a joint model in which all relevant aspects of the experiment are taken into account (Noble and MacCoss, 2012). However, it is not possible tackling the complexity and variety of applications with a single joint model. Software modularity is thus required, yet, with high transparency and automation to support efficient integration of all available information. Open source software plays a key role towards this goal. Software modules with well-defined interfaces and input/output formats facilitate integration at different levels, avoiding information loss at every processing step and opens the potential to integrate state-of-the-art machine learning to model dependencies among variables at the spectral, peptide, and protein level that are currently decoupled, as well as other relevant information such as clinical or environmental data. This is enhancing our understanding of the state of the proteome and its response to perturbations.

The diversity of questions possibly addressed with MS and the evolving technologies require software customization and/or extension. Robust open source software accelerates adaptations for technological innovations. For instance, the additional ion mobility (IM) dimension currently available in commercial instruments provides reproducible separation and structural information with numerous benefits in proteomic analyses (Baker et al., 2015). However, this brings challenges for data processing because existing software tools for LC-MS data do not work. Skyline was updated to support LC-IM-MS and optimization of IM informatics is in progress for analysis of these large and multi-dimensional datasets (Pino et al., 2017).

In this regard, several useful comments can be summarized about scientific software development. First, bioinformaticians are strongly encouraged to embrace practices that ensure software quality, sustainability and reproducible research (Jiménez et al., 2017; Perez-Riverol et al., 2016). Development of new algorithms and data analysis tools is challenging, it is a common practice developing prototypes first, with at least some basic design and testing. However, as soon as a software tool proves its usefulness, and before it is "too late" because it is too extensive and complex, time and effort must be dedicated for more testing, re-factoring and documentation. And this implies financial support.

Similarly, bioinformaticians are encouraged to join community efforts in order to avoid reinventing the wheel. Improving upon something is typically much easier than starting from scratch and allows for incremental progress, as one can see further by standing on the shoulders of giants. Crucially, any previous related work should be properly cited and acknowledged to avoid the term "research parasites". This term has been recently and unjustifiably used for people reusing data (Martens and Vizcaíno, 2017), and should be also avoided for people reusing code.

Second, a software tool should have both command line functionality and GUI, either as stand-alone tool or integrated into current pipelines. Command line interfaces facilitate integration with other tools and are necessary for automation in large scale studies. On the other hand, a GUI expedites a quick execution in a few clicks, potentially increasing dissemination and number of users. In addition, the GUI facilitates testing, quality control, optimization of parameters and re-execution from any step to the end. Related to this point, visualization, at least as static figures in output files, is critical to support quick evaluation of results.

Third, a person other than the developer should try to follow the documentation (e.g., user guide, tutorial) and test the tool. The provided feedback will increase the robustness and usability. For instance, in a recent study of benchmarking DIA label-free quantification tools (Navarro et al., 2016), feedback enabled developers to improve their software tools and after optimization, all tools provided highly convergent identification and reliable quantification performance. Documentation should include installation steps, software dependencies with versions, file formats and test data. Testing should be performed also with real data. For example, a simple approach producing acceptable results for a simple protein mixture, may produce poor results when processing a complex one such as a cell lysate.

Fourth, to ensure reproducibility, scientific articles related to both benchmarking and application studies should properly document the usage of any software, including details about the parameters and software version used, and data description with experimental and technical metadata related with the type of experiment. The metadata requirements for proteomics are in general much less comprehensive than those from more mature omics fields (Martens and Vizcaíno, 2017).

Finally, multidisciplinary collaborations will facilitate the design, implementation and validation of more advanced computational methods for proteomics and integration with other omics. Development of more powerful and user-friendly MS tools that can be used by researchers non-specialized in mass spectrometry or informatics is encouraged.

Acknowledgements

The author would like to acknowledge the following researchers for various insightful discussions: Frédérique Lisacek (software tools, text editions and article organization), Samuel H. Payne (scoring functions and MSGF+), John Fjeldsted (software workflows and automation) and Richard Allen White III (open source software).

See also: Clinical Proteomics. Experimental Platforms for Extracting Biological Data: Mass Spectrometry, Microarray, Next Generation Sequencing. Identification of Proteins from Proteomic Analysis. Natural Language Processing Approaches in Bioinformatics. Protein Post-Translational Modification Prediction. Protein Properties. Proteomics Data Representation and Databases. Quantification of Proteins from Proteomic Analysis. Transmembrane Domain Prediction

References

Aebersold, R., 2003. Mass spectrometry-based proteomics. Nature 422, 198–207.
Aebersold, R., 2009. A stress test for mass spectrometry-based proteomics. Nature Methods 6 (6), 411–412.
Aebersold, R., Mann, M., 2016. Mass-spectrometric exploration of proteome structure and function. Nature 537 (7620), 347–355.

Afgan, E., Baker, D., Van den Beek, M., et al., 2016. The Galaxy platform for accessible, reproducible and collaborative biomedical analyses: 2016 update. Nucleic Acids Research 44 (W1), W3–W10.

Ahrné, E., Molzahn, L., Glatter, T., Schmidt, A., 2013. Critical assessment of proteome-wide label-free absolute abundance estimation strategies. Proteomics 13 (17), 2567–2578.

Ahrné, E., Müller, M., Lisacek, F., 2010. Unrestricted identification of modified proteins using MS/MS. Proteomics 10 (4), 671–686.

Aiche, S., Sachsenberg, T., Kenar, E., et al., 2015. Workflows for automated downstream data analysis and visualization in large-scale computational mass spectrometry. Proteomics 15 (8), 1443–1447.

Alghanem, B., Nikitin, F., Stricker, T., et al., 2017. Optimization by infusion of multiple reaction monitoring transitions for sensitive peptides LC-MS quantification. Rapid Communications in Mass Spectrometry.

Audain, E., Uszkoreit, J., Sachsenberg, T., et al., 2017. In-depth analysis of protein inference algorithms using multiple search engines and well-defined metrics. Journal of Proteomics 150 (Suppl. C), 170–182.

Baker, E.S., Burnum-Johnson, K.E., Ibrahim, Y.M., et al., 2015. Enhancing bottom-up and top-down proteomic measurements with ion mobility separations. Proteomics 15 (16), 2766–2776.

Benjamini, Y., Hochberg, Y., 1995. Controlling the false discovery rate: A practical and powerful approach to multiple testing. Journal of the Royal Statistical Society Series B (Methodological) 57 (1), 289–300.

Bensimon, A., Heck, A.J., Aebersold, R., 2012. Mass spectrometry-based proteomics and network biology. Annual Review of Biochemistry 81, 379–405.

Berthold, M.R., Cebron, N., Dill, F., et al., 2009. KNIME-the Konstanz information miner: Version 2.0 and beyond. ACM SIGKDD Explorations Newsletter 11 (1), 26–31.

Bilbao, A., Lisacek, F., Hopfgartner, G., 2016. Dedicated software enhancing data-independent acquisition methods in mass spectrometry. Chimia International Journal for Chemistry 70 (4), 293.

Bilbao, A., Varesio, E., Luban, J., et al., 2015a. Processing strategies and software solutions for data-independent acquisition in mass spectrometry. Proteomics 15 (5-6), 964–980.

Bilbao, A., Zhang, Y., Varesio, E., et al., 2015b. Ranking fragment ions based on outlier detection for improved label-free quantification in data-independent acquisition LC-MS/MS. Journal of Proteome Research 14, 4581–4593.

Blein-Nicolas, M., Zivy, M., 2016. Thousand and one ways to quantify and compare protein abundances in label-free bottom-up proteomics. Biochimica et Biophysica Acta (BBA) – Proteins and Proteomics 1864 (8), 883–895.

Cai, W., Guner, H., Gregorich, Z.R., et al., 2016. MASH Suite Pro: A comprehensive software tool for top-down proteomics. Molecular & Cellular Proteomics 15 (2), 703–714.

Cappadona, S., Baker, P.R., Cutillas, P.R., Heck, A.J., Breukelen, B.van., 2012. Current challenges in software solutions for mass spectrometry-based quantitative proteomics. Amino Acids 43 (3), 1087–1108.

Carr, S.A., Abbatiello, S.E., Ackermann, B.L., et al., 2014. Targeted peptide measurements in biology and medicine: Best practices for mass spectrometry-based assay development using a fit-for-purpose approach. Molecular & Cellular Proteomics 13 (3), 907–917.

Chambers, M.C., Maclean, B., Burke, R., et al., 2012. A cross-platform toolkit for mass spectrometry and proteomics. Nature Biotechnology 30 (10), 918–920.

Chapman, J.D., Goodlett, D.R., Masselon, C.D., 2014. Multiplexed and data-independent tandem mass spectrometry for global proteome profiling. Mass Spectrometry Reviews 33, 452–470.

Choi, H., Nesvizhskii, A.I., 2007. Semisupervised model-based validation of peptide identifications in mass spectrometry-based proteomics. Journal of Proteome Research 7 (01), 254–265.

Choi, M., Chang, C.-Y., Clough, T., et al., 2014. MSstats: An R package for statistical analysis of quantitative mass spectrometry-based proteomic experiments. Bioinformatics 30 (17), 2524–2526.

Cox, J., Mann, M., 2008. MaxQuant enables high peptide identification rates, individualized ppb-range mass accuracies and proteome-wide protein quantification. Nature Biotechnology 26 (12), 1367–1372.

Cox, J., Neuhauser, N., Michalski, A., et al., 2011. Andromeda: A peptide search engine integrated into the MaxQuant environment. Journal of Proteome Research 10 (4), 1794–1805.

Craig, R., Beavis, R.C., 2004. TANDEM: Matching proteins with tandem mass spectra. Bioinformatics 20 (9), 1466–1467.

da Veiga Leprevost, F., Grüning, B.A., Alves Aflitos, S., et al., 2017. BioContainers: An open-source and community-driven framework for software standardization. Bioinformatics. btx192.

Degan, M.G., Ryadinskiy, L., Fujimoto, G.M., et al., 2016. A skyline plugin for pathway-centric data browsing. Journal of The American Society for Mass Spectrometry 27 (11), 1752–1757.

Deutsch, E.W., 2012. File formats commonly used in mass spectrometry proteomics. Molecular & Cellular Proteomics 11 (12), 1612–1621.

Deutsch, E.W., Csordas, A., Sun, Z., et al., 2017. The ProteomeXchange consortium in 2017: Supporting the cultural change in proteomics public data deposition. Nucleic Acids Research 45 (D1), D1100–D1106.

Deutsch, E.W., Mendoza, L., Shteynberg, D., et al., 2010. A guided tour of the trans-proteomic pipeline. Proteomics 10 (6), 1150–1159.

Deutsch, E.W., Mendoza, L., Shteynberg, D., et al., 2015. Trans-proteomic pipeline, a standardized data processing pipeline for large-scale reproducible proteomics informatics. Proteomics – Clinical Applications 9 (7-8), 745–754.

Deutsch, E.W., Overall, C.M., Van Eyk, J.E., et al., 2016. Human proteome project mass spectrometry data interpretation guidelines 2.1. Journal of Proteome Research 15 (11), 3961–3970.

Dowle, A.A., Wilson, J., Thomas, J.R., 2016. Comparing the diagnostic classification accuracy of iTRAQ, peak-area, spectral-counting, and emPAI methods for relative quantification in expression proteomics. Journal of Proteome Research 15 (10), 3550–3562.

Elias, J.E., Gygi, S.P., 2007. Target-decoy search strategy for increased confidence in large-scale protein identifications by mass spectrometry. Nature Methods 4 (3), 207–214.

Eng, J.K., Hoopmann, M.R., Jahan, T.A., et al., 2015. A deeper look into Comet–implementation and features. Journal of the American Society for Mass Spectrometry 26 (11), 1865–1874.

Eng, J.K., Jahan, T.A., Hoopmann, M.R., 2013. Comet: An open-source MS/MS sequence database search tool. Proteomics 13 (1), 22–24.

Eng, J.K., McCormack, A.L., Yates, J.R., 1994. An approach to correlate tandem mass spectral data of peptides with amino acid sequences in a protein database. Journal of the American Society for Mass Spectrometry 5 (11), 976–989.

Escher, C., Reiter, L., MacLean, B., et al., 2012. Using iRT, a normalized retention time for more targeted measurement of peptides. Proteomics 12 (8), 1111–1121.

Fellers, R.T., Greer, J.B., Early, B.P., et al., 2015. ProSight lite: Graphical software to analyze top-down mass spectrometry data. Proteomics 15 (7), 1235–1238.

Frank, A.M., Savitski, M.M., Nielsen, M.L., Zubarev, R.A., Pevzner, P.A., 2007. De novo peptide sequencing and identification with precision mass spectrometry. Journal of Proteome Research 6 (1), 114–123.

Frewen, B.E., Merrihew, G.E., Wu, C.C., Noble, W.S., MacCoss, M.J., 2006. Analysis of peptide MS/MS spectra from large-scale proteomics experiments using spectrum libraries. Analytical Chemistry 78 (16), 5678–5684.

Gillet, L.C., Navarro, P., Tate, S., et al., 2012. Targeted data extraction of the MS/MS spectra generated by data-independent acquisition: A new concept for consistent and accurate proteome analysis. Molecular & Cellular Proteomics. 11.

Horlacher, O., Lisacek, F., Müller, M., 2015a. Mining large scale tandem mass spectrometry data for protein modifications using spectral libraries. Journal of Proteome Research 15 (3), 721–731.

Horlacher, O., Nikitin, F., Alocci, D., et al., 2015b. MzJava: An open source library for mass spectrometry data processing. Journal of Proteomics 129, 63–70.

Hu, A., Noble, W.S., Wolf-Yadlin, A., 2016. Technical advances in proteomics: New developments in data-independent acquisition. F1000Research. 5.

Ison, J., Rapacki, K., Ménager, H., *et al.*, 2016. Tools and data services registry: A community effort to document bioinformatics resources. Nucleic Acids Research 44 (D1), D38–D47.

Jiménez, R., Kuzak, M., Alhamdoosh, M., *et al.*, 2017. Four simple recommendations to encourage best practices in research software. F1000Research 6, 876.

Käll, L., Canterbury, J.D., Weston, J., Noble, W.S., MacCoss, M.J., 2007. Semi-supervised learning for peptide identification from shotgun proteomics datasets. Nature Methods 4 (11), 923–925.

Keller, A., Bader, S.L., Shteynberg, D., Hood, L., Moritz, R.L., 2015. Automated validation of results and removal of fragment ion interferences in targeted analysis of data independent acquisition MS using SWATHProphet. Molecular & Cellular Proteomics. (mcp-O114).

Keller, A., Nesvizhskii, A.I., Kolker, E., Aebersold, R., 2002. Empirical statistical model to estimate the accuracy of peptide identifications made by MS/MS and database search. Analytical Chemistry 74 (20), 5383–5392.

Kim, S., Gupta, N., Pevzner, P.A., 2008. Spectral probabilities and generating functions of tandem mass spectra: A strike against decoy databases. Journal of Proteome Research 7 (8), 3354–3363.

Kim, S., Pevzner, P.A., 2014. MS-GFmathplus makes progress towards a universal database search tool for proteomics. Nature Communications 5, 5277.

Kong, A.T., Leprevost, F.V., Avtonomov, D.M., Mellacheruvu, D., Nesvizhskii, A.I., 2017. MSFragger: Ultrafast and comprehensive peptide identification in mass spectrometry-based proteomics. Nature Methods 14 (5), 513–520.

Kou, Q., Xun, L., Liu, X., 2016. TopPIC: A software tool for top-down mass spectrometry-based proteoform identification and characterization. Bioinformatics 32 (22), 3495–3497.

Kwon, T., Choi, H., Vogel, C., Nesvizhskii, A.I., Marcotte, E.M., 2011. MSblender: A probabilistic approach for integrating peptide identifications from multiple database search engines. Journal of Proteome Research 10 (7), 2949–2958.

Lam, H., Deutsch, E.W., Eddes, J.S., *et al.*, 2007. Development and validation of a spectral library searching method for peptide identification from MS/MS. Proteomics 7 (5), 655–667.

Lange, V., Picotti, P., Domon, B., Aebersold, R., 2008. Selected reaction monitoring for quantitative proteomics: A tutorial. Molecular Systems Biology 4 (1), 1–14.

Lawrence, R.T., Searle, B.C., Llovet, A., Villén, J., 2016. Plug-and-play analysis of the human phosphoproteome by targeted high-resolution mass spectrometry. Nature Methods.

Liu, H., Sadygov, R.G., Yates III, J.R., 2004. A model for random sampling and estimation of relative protein abundance in shotgun proteomics. Analytical Chemistry 76 (14), 4193–4201.

Liu, X., Sirotkin, Y., Shen, Y., *et al.*, 2012. Protein identification using top-down spectra. Molecular & Cellular Proteomics 11 (6), 111–8524.

Li, Y., Zhong, C.-Q., Xu, X., *et al.*, 2015. Group-DIA: Analyzing multiple data-independent acquisition mass spectrometry data files. Nature Methods.

MacLean, B., Tomazela, D.M., Shulman, N., *et al.*, 2010. Skyline: An open source document editor for creating and analyzing targeted proteomics experiments. Bioinformatics 26 (7), 966–968.

Martens, L., Chambers, M., Sturm, M., *et al.*, 2011. mzML – A community standard for mass spectrometry data. Molecular & Cellular Proteomics 10 (1), 110–133.

Martens, L., Vizcaíno, J.A., 2017. A golden age for working with public proteomics data. Trends in Biochemical Sciences.

Monroe, M.E., Tolić, N., Jaitly, N., *et al.*, 2007. VIPER: An advanced software package to support high-throughput LC-MS peptide identification. Bioinformatics 23 (15), 2021.

Mueller, L.N., Rinner, O., Schmidt, A., *et al.*, 2007. SuperHirn – A novel tool for high resolution LC-MS-based peptide/protein profiling. Proteomics 7 (19), 3470–3480.

Nahnsen, S., Bielow, C., Reinert, K., Kohlbacher, O., 2013. Tools for label-free peptide quantification. Molecular & Cellular Proteomics 12 (3), 549–556.

Na, S., Bandeira, N., Paek, E., 2012. Fast multi-blind modification search through tandem mass spectrometry. Molecular & Cellular Proteomics 11 (4), 111–10199.

Navarro, P., Kuharev, J., Gillet, L.C., *et al.*, 2016. A multicenter study benchmarks software tools for label-free proteome quantification. Nature Biotechnology.

Neilson, K.A., Ali, N.A., Muralidharan, S., *et al.*, 2011. Less label, more free: Approaches in label-free quantitative mass spectrometry. Proteomics 11 (4), 535–553.

Nesvizhskii, A.I., 2010. A survey of computational methods and error rate estimation procedures for peptide and protein identification in shotgun proteomics. Journal of Proteomics 73 (11), 2092–2123.

Nesvizhskii, A.I., 2014. Proteogenomics: Concepts, applications and computational strategies. Nature Methods 11 (11), 1114–1125.

Nesvizhskii, A.I., Aebersold, R., 2005. Interpretation of shotgun proteomic data: The protein inference problem. Molecular & Cellular Proteomics 4 (10), 1419–1440.

Nesvizhskii, A.I., Keller, A., Kolker, E., Aebersold, R., 2003. A statistical model for identifying proteins by tandem mass spectrometry. Analytical Chemistry 75 (17), 4646–4658.

Noble, W.S., MacCoss, M.J., 2012. Computational and statistical analysis of protein mass spectrometry data. PLOS Computational Biology 8 (1), 1–6.

Park, J., Piehowski, P.D., Wilkins, C., *et al.*, 2017. Informed-Proteomics: Open-source software package for top-down proteomics. Nature Methods.

Pedrioli, P.G., Eng, J.K., Hubley, R., *et al.*, 2004. A common open representation of mass spectrometry data and its application to proteomics research. Nature Biotechnology 22 (11), 1459–1466.

Perez-Riverol, Y., Gatto, L., Wang, R., *et al.*, 2016. Ten simple rules for taking advantage of Git and GitHub. PLOS Computational Biology: Public Library of Science 12 (7), e1004947.

Perkins, D.N., Pappin, D.J., Creasy, D.M., Cottrell, J.S., 1999. Probability-based protein identification by searching sequence databases using mass spectrometry data. Electrophoresis 20 (18), 3551–3567.

Peterson, A.C., Russell, J.D., Bailey, D.J., Westphall, M.S., Coon, J.J., 2012. Parallel reaction monitoring for high resolution and high mass accuracy quantitative, targeted proteomics. Molecular & Cellular Proteomics 11, 1475–1478.

Pfeuffer, J., Sachsenberg, T., Alka, O., *et al.*, 2017. OpenMS – A platform for reproducible analysis of mass spectrometry data. Journal of Biotechnology 261 (Suppl. C), 142–148.

Picotti, P., Aebersold, R., 2012. Selected reaction monitoring-based proteomics: Workflows, potential, pitfalls and future directions. Nature Methods 9 (6), 555–566.

Pino, L.K., Searle, B.C., Bollinger, J.G., *et al.*, 2017. The skyline ecosystem: Informatics for quantitative mass spectrometry proteomics. Mass Spectrometry Reviews.

Pluskal, T., Castillo, S., Villar-Briones, A., Orešic, M., 2010. MZmine 2: Modular framework for processing, visualizing, and analyzing mass spectrometry-based molecular profile data. BMC Bioinformatics 11 (1), 395.

Reiter, L., Claassen, M., Schrimpf, S.P., *et al.*, 2009. Protein identification false discovery rates for very large proteomics data sets generated by tandem mass spectrometry. Molecular & Cellular Proteomics 8 (11), 2405–2417.

Reiter, L., Rinner, O., Picotti, P., *et al.*, 2011. mProphet: Automated data processing and statistical validation for large-scale SRM experiments. Nature Methods 8 (5), 430–435.

Rieckmann, J.C., Geiger, R., Hornburg, D., *et al.*, 2017. Social network architecture of human immune cells unveiled by quantitative proteomics. Nature Immunology 18 (5), 583–593.

Rosenberger, G., Bludau, I., Schmitt, U., *et al.*, 2017a. Statistical control of peptide and protein error rates in large-scale targeted data-independent acquisition analyses. Nature Methods 14 (9), 921–927.

Rosenberger, G., Liu, Y., Rost, H.L., *et al.*, 2017b. Inference and quantification of peptidoforms in large sample cohorts by SWATH-MS. Nature Biotechnology 35 (8), 781–788.

Röst, H., Malmström, L., Aebersold, R., 2012. A computational tool to detect and avoid redundancy in selected reaction monitoring. Molecular & Cellular Proteomics 11 (8), 540–549.

Röst, H.L., Rosenberger, G., Navarro, P., *et al.*, 2014. OpenSWATH enables automated, targeted analysis of data-independent acquisition MS data. Nature Biotechnology 32 (3), 219–223.

Röst, H.L., Sachsenberg, T., Aiche, S., *et al.*, 2016. OpenMS: A flexible open-source software platform for mass spectrometry data analysis. Nature Methods 13 (9), 741–748.

Saito, R., Smoot, M.E., Ono, K., *et al.*, 2012. A travel guide to cytoscape plugins. Nature Methods 9 (11), 1069–1076.

Searle, B.C., Egertson, J.D., Bollinger, J.G., Stergachis, A.B., MacCoss, M.J., 2015. Using data independent acquisition (DIA) to model high-responding peptides for targeted proteomics experiments. Molecular & Cellular Proteomics 14 (9), 2331–2340.

Serang, O., Käll, L., 2015. Solution to statistical challenges in proteomics is more statistics, not less. Journal of Proteome Research.

Shannon, P., Markiel, A., Ozier, O., et al., 2003. Cytoscape: A software environment for integrated models of biomolecular interaction networks. Genome Research 13 (11), 2498–2504.

Shilov, I.V., Seymour, S.L., Patel, A.A., et al., 2007. The paragon algorithm, a next generation search engine that uses sequence temperature values and feature probabilities to identify peptides from tandem mass spectra. Molecular & Cellular Proteomics 6, 1638.

Shteynberg, D., Deutsch, E.W., Lam, H., et al., 2011. iProphet: Multi-level integrative analysis of shotgun proteomic data improves peptide and protein identification rates and error estimates. Molecular & Cellular Proteomics 10 (12), 111–7690.

Shteynberg, D., Nesvizhskii, A.I., Moritz, R.L., Deutsch, E.W., 2013. Combining results of multiple search engines in proteomics. Molecular & Cellular Proteomics 12 (9), 2383–2393.

Sidoli, S., Schwämmle, V., Ruminowicz, C., et al., 2014. Middle-down hybrid chromatography/tandem mass spectrometry workflow for characterization of combinatorial post-translational modifications in histones. Proteomics 14 (19), 2200–2211.

Smith, L.M., Kelleher, N.L., 2013. Proteoform: A single term describing protein complexity. Nature Methods 10 (3), 186–187.

Sturm, M., Bertsch, A., Gröpl, C., et al., 2008. OpenMS – An open-source software framework for mass spectrometry. BMC Bioinformatics 9 (1), 163.

Szabo, Z., Janaky, T., 2015. Challenges and developments in protein identification using mass spectrometry. TrAC Trends in Analytical Chemistry. 76–87.

Teleman, J., Röst, H., Rosenberger, G., et al., 2014. DIANA-algorithmic improvements for analysis of data-independent acquisition MS data. Bioinformatics. btu686.

The, M., MacCoss, M.J., Noble, W.S., Käll, L., 2016. Fast and accurate protein false discovery rates on large-scale proteomics data sets with percolator 3.0. Journal of the American Society for Mass Spectrometry 27 (11), 1719–1727.

Ting, Y.S., Egertson, J.D., Bollinger, J.G., et al., 2017. PECAN: Library-free peptide detection for data-independent acquisition tandem mass spectrometry data. Nature Methods 14 (9), 903–908.

Tran, N.H., Zhang, X., Xin, L., Shan, B., Li, M., 2017. De novo peptide sequencing by deep learning. Proceedings of the National Academy of Sciences 114 (31), 8247–8252.

Tsou, C.-C., Avtonomov, D., Larsen, B., et al., 2015. DIA-Umpire: Comprehensive computational framework for data independent acquisition proteomics. Nature Methods 12 (3), 258–264.

Tsou, C.-C., Tsai, C.-F., Teo, G., Chen, Y.-J., Nesvizhskii, A.I., 2016. Untargeted, spectral library-free analysis of data independent acquisition proteomics data generated using Orbitrap mass spectrometers. Proteomics 16, 2257–2271.

Tyanova, S., Temu, T., Cox, J., 2016. The MaxQuant computational platform for mass spectrometry-based shotgun proteomics. Nature Protocols 11 (12), 2301–2319.

Vaudel, M., Barsnes, H., Berven, F.S., Sickmann, A., Martens, L., 2011. SearchGUI: An open-source graphical user interface for simultaneous OMSSA and X! Tandem searches. Proteomics 11 (5), 996–999.

Vaudel, M., Burkhart, J.M., Zahedi, R.P., et al., 2015. PeptideShaker enables reanalysis of MS-derived proteomics data sets. Nature Biotechnology 33 (1), 22–24.

Wang, J., Tucholska, M., Knight, J.D., et al., 2015. MSPLIT-DIA: Sensitive peptide identification for data-independent acquisition. Nature Methods.

White, R.A., Borkum, M.I., Rivas-Ubach, A., et al., 2017a. From data to knowledge: The future of multi-omics data analysis for the rhizosphere. Rhizosphere 3 (Part 2), 222–229.

White, R.A., Rivas-Ubach, A., Borkum, M.I., et al., 2017b. The state of rhizospheric science in the era of multi-omics: A practical guide to omics technologies. Rhizosphere 3 (Part 2), 212–221.

Wilkinson, M.D., Dumontier, M., Aalbersberg, I.J., et al., 2016. The FAIR guiding principles for scientific data management and stewardship. Scientific Data 3, 160018.

Williams, E.G., Wu, Y., Jha, P., et al., 2016. Systems proteomics of liver mitochondria function. Science 352 (6291), aad0189.

Yates III, J.R., 2015. Pivotal role of computers and software in mass spectrometry – SEQUEST and 20 years of tandem MS database searching. Journal of The American Society for Mass Spectrometry 26 (11), 1804–1813.

Zhang, H., Ge, Y., 2011. Comprehensive analysis of protein modifications by top-down mass spectrometry. Circulation: Cardiovascular Genetics 4 (6), 711–724.

Zhang, Y., Bilbao, A., Bruderer, T., et al., 2015. The use of variable Q1 isolation windows improves selectivity in LC–SWATH–MS acquisition. Journal of Proteome Research 14, 4359–4371.

Relevant Websites

http://biocontainers.pro/
  BioContainers.
https://bio.tools/
  BioTools.
http://compomics.github.io/projects/searchgui.html
  Compomics Docs.
https://www.knime.com/community/bioinf/openms
  OpenMS Nodes for KNIME.
https://galaxyproject.org/proteomics/use-cases
  Protoeomics Use Cases.
http://www.proteomexchange.org
  ProteomeXchange.
http://string-db.org
  STRING.
https://github.com/wolski/imsbInfer
  Wolski/imsbInfer.

Biological Databases

Sarinder K Dhillon, University of Malaya, Kuala Lumpur, Malaysia

Introduction

Database is one of the most important innovation in computer science and information technology. Since its birth in 1960s, it has evolved many folds and is still evolving until today. Previously file based systems were used but were limited in its use due to issues such as data redundancy and inconsistency, difficulty in accessing data, data integrity and security, data isolation and many more. In order to resolve these concerns and limitations, the database approach was introduced and it was received well by the community. This is because database approach is more advanced, as the data is structured in a more formalised manner while the relationships between the data items are meaningful and consistent. Additionally, the database redundancy problem is very much minimised in databases compared to file based systems. In this way, the data can be used more efficiently either by automated or even manual systems. Database system applications are widely used in enterprises, banking and finance, universities, airlines, telecommunication and science due to the need for a computerised management of data which includes manipulation of data. Database systems promotes the use of advanced techniques such as data visualisation, data mining and data integration.

Currently in the era of data explosion, particularly in biology, the use and application of database technology is seen to be rising. Many areas such as bioinformatics, biodiversity and medical are applying the database technology to store, retrieve, manipulate and analyse data. The trend of big data has taken over conventional approaches in managing experimental data in biology. It is envisaged that a marriage between biology and database will soon churn out new trends, applications and discoveries.

This article presents the database fundamentals which includes different types of databases, the evolvement of biological databases, the approaches used in developing biological databases, some case studies using the approaches described, the results and discussion and finally the future direction of biological databases.

Database Fundamentals

Databases

In the simplest form, a database can be defined as an organised and formal collection of information stored in a computer readable format. In this section, some of the common approaches used in developing databases are presented.

Relational databases

The relational database has evolved over the years since its first introduction in 1970 and was largely adopted by industries across the world. Data in the relational model is usually represented by a structured schema (**Fig. 1**) in order to provide the data dictionary or metadata concerning the actual data (Elmasri and Navathe, 2011). Data which are represented by attributes are clustered based on the similarity, type and format in the form of columns of a table. The row on the other hand, represents the records in a database. A primary key is used to link data between multiple tables, whereas a foreign key is a reference to a primary key in another table. The data in a database is organised based on the relationships among the entities (and their corresponding attributes) while redundancy is minimized by performing data normalization.

Relational database comes with a default declarative language named Structured Query Language (SQL) developed at IBM by Donald D. Chamberlin and Raymond F. Boyce in the early 1970s. It provides interaction with structure data via CRUD (Create, Read, Update and Delete) statements and allows combination with multiple statement to perform more complex actions (Cancelo, 2014) (**Fig. 2**).

One of the main drawback of the relational model is that searching is burdened whereby the value being searched for is compared against every row in the table, from the first row to the last. This process is time consuming and has higher computational burden especially for large databases unless it uses other methods for searching such as indexes (Segaran and Hammerbacher, 2009a).

Multidimensional databases (OLAP)

In principle, a multidimensional database (MDD) is created using inputs from a relational database. In a MDD, data is stored in a cube, or a multidimensional array. When talking about MDDs, the term OLAP is prominent. The On-Line Analytical Processing (OLAP) is a term introduced by Dr. E.F. Codd (Mallach, 2000) which encompasses relational databases. The rule based OLAP system described by Dr. Codd are explained by Pendse (2008) formulating them into five simple definitions; Fast, Analysis, Shared, Multidimensional and Information. The OLAP is typically used in the business domain where it meets the needs of business analysts in business management, generating reports in addition to data mining. OLAP systems has also been used in

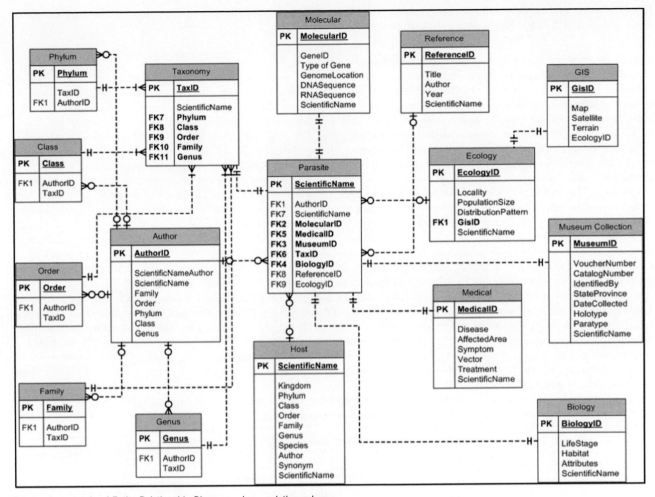

Fig. 1 An example of Entity Relationship Diagram using a relation schema.

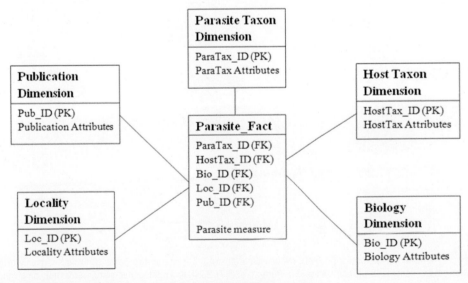

Fig. 2 Example of an OLAP Model where data is represented in a cube dimensional model.

Biology, but are vaguely reported. Information stored in a MDD database are often used for effective decision making that usually deals with complex relationships with a short response time.

Graph and ontology based databases (NoSQL databases)

Ontology based databases are in principle made of the object oriented approach, where the datasets are treated as objects, with values and properties, similar to objects represented by object oriented programming like C++ and Java. Ontologies can be defined as formal representation of the types, properties, and interrelationships of the entities on a specific domain and form the basic component of the semantic web architecture. These entities, being linked with other entities by the means of relationships, lay the underlying foundation of the semantic web (Guarino, 1995). The ontologies, when linked with other ontologies, generate a network of nodes and relationships, laying the foundations of linked data (Heath and Bizer, 2011).

The ontology development in a particular domain is facilitated by standardization of vocabulary that is accepted and used by domain experts. These vocabularies are shared for reuse, monitored by a body of experts to oversee the development of such vocabularies. An ontology can enhance information providing capability for a database by the use of metadata to discover and gather new information from other databases, or by linking them to create a better information network. An ontology can also be used as an underlying framework for building a database (Abu *et al.*, 2013a,b; Ali *et al.*, 2017).

NoSQL is a term used to refer to schemaless databases, which do not fit within the traditional relational paradigm. It describes a new category of non-relational databases, which are also known as distributed data stores that are capable of scaling to large datasets spanning to multiple data centres (Ercan, 2014). Recently, NoSQL database systems started to receive much attention as the demand increases for high speed data access over large volume of data without much effort in scaling and tuning (Ferreira *et al.*, 2013). NoSQL architecture is designed to overcome the limited scalability, flexibility, performance, availability, and infrastructure cost issues associated with relational databases (Stonebraker, 2009). They depend on horizontal scalability which scales performance and capacity by increasing the number of nodes, instead of increasing the computer power of a single node as seen in relational databases (Abramova and Bernardino, 2013). An example of a no SQL database is a graph database **Fig. 3**. Data are stored as graph in the form of nodes (objects) and edges (relationships). The primary goal is to visualize the relationships between nodes. There are now many tools available to design and develop moSQL databases, such as Neo4j (2018) and MongoDB (2018).

Graph databases, which forms the base for noSQL databases use a graph model for semantic queries with nodes, edges and properties to represent and store data. Unlike the relational structures, in graph dabases, millions of connections can traversed

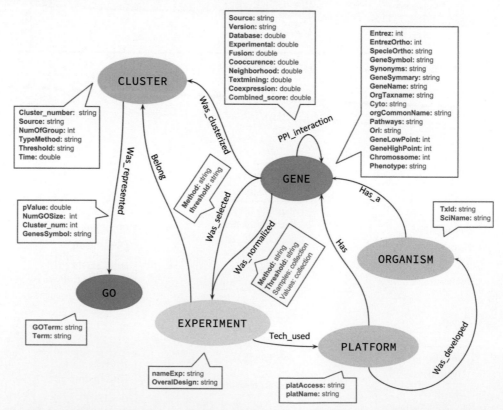

Fig. 3 Graph data model representing the nodes (in oval), relationships (arrows). The descriptive boxes showing the mainly properties in nodes and edges. *Source*: Costa, R.L., Gadelha, L., Ribeiro-Alves, M., Porto, F., 2017. GeNNet: An integrated platform for unifying scientific workflows and graph databases for transcriptome data analysis. PeerJ, 5, p. e3509.

per second per core (Neo4j, 2018). Having a conceptual design, similar to network-model databases in 1970s, graph databases excel at managing linked data and complex queries, independent of the total size of the dataset (Angles and Gutierrez, 2008). Graph databases explore the larger neighborhood around the initial starting points with a pattern and a set of starting points by collecting and aggregating information from thousands or millions of nodes and relationships (Neo4j, 2018). Contrary to relational databases that operate with SQL language, in graph databases, there is no single query language that operates in same fashion as SQL. Query languages such as Gremlin, SPARQL, and Cypher can be used to access graph databases using application programming interfaces (APIs) (Wood, 2012).

Spatial databases

Data in a geometric space contains objects such as coordinates, points, lines, polygons and regions are stored in a spatial database as in objects, rather than pictures or images of a space. It is important to note that spatial databases are different from image databases. The data model encompasses the spatial data types (SDTs) which is used as underlying framework to construct Geographic Information Systems (GIS). Typically, spatial databases uses the relational schema discussed in Section "Relational databases".

Database System Architectures

Distributed databases

In a typical distributed databases environment, the databases are stored on several computers, which are interconnected via a communication media such as a network connection. These databases are loosely coupled in terms of physical specifications and hardware however in terms of the design schema they might have both a local schema and a global schema. A local schema is unique to a particular database whereas a global schema is used to connect all the databases to achieve a specific goal. Distributed databases are also used in situations where there is an attempt to link or integrate a few existing databases, which are built upon different formats or using different architectures. The reason behind a potential integration could be to share the data stored in these distributed databases.

A general structure of distributed database is shown in **Fig. 4**.

Federated databases

In a federated database environment, several homogeneous or heterogeneous databases are harmonised to function as a single entity in a unified view, through a common structured metadata. Each of the constituent database serves as a distinct, self-sustainable and autonomous component in the federated system. These components acts as nodes in a computer network and may be geographically decentralized. Federated database is one of most common mainstream model in life sciences data integration. An example of federated database architecture is shown in **Fig. 5**.

Data warehouse

A data warehouse is a repository for storing data which may have been gathered from a source or multiple sources, manually or automatically, via an integration layer that transforms data to meet the criteria of the warehouse. Data warehouse can be

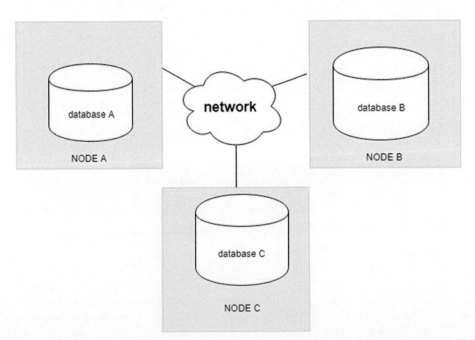

Fig. 4 Distributed Database Structure. The nodes (computers) storing database A, B, and C are distributed and connected via the network.

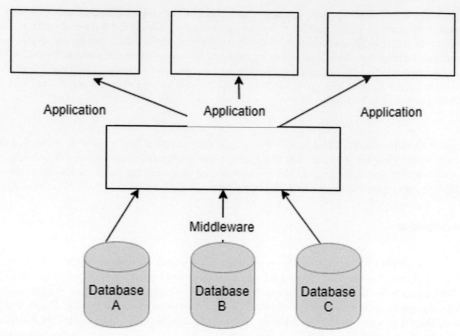

Fig. 5 Federated Database Architecture.

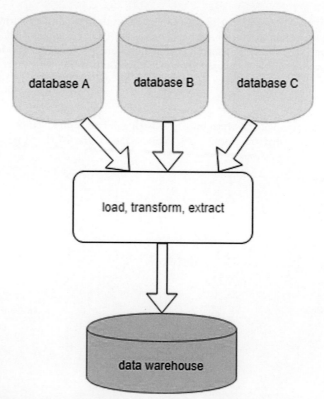

Fig. 6 Data Warehouse Architecture.

conceptualised as a one stop information center large volume of data which is designed under a common framework. In a centralized data warehouse, since data is unified and in homogeneous format, queries are easier to write and gives user better performance than accessing multiple distributed data warehouses. Integration of data in bioinformatics data warehouse and the issues affecting the integration has been presented by Kormeier (2014). An example of data warehouse architecture is shown in **Fig. 6**.

Uses of Databases

The role of database has evolved from a stand alone specialised system into an integral component of an advanced and sophisticated computing environment. Conventionally, databases were constructed for storage, retrieval and manipulation of data. Today, the role has gone far beyond to knowledge extraction, data mining and visualisation.

Knowledge extraction

Databases are built to support front end applications which ranges from simple query based systems to sophisticated artificial intelligence enabled systems. With the concept of big data that has taken Information Technology into a new era altogether, databases are extensively used for knowledge extraction using algorithms that are run through large volumes of data. Thus, databases now have bigger role to play in many data centric domains to function as platforms for machine readable and machine interpretable format of information. Thiele et al. (2010) presented an interesting tool, the Protein Information and Knowledge Extractor (PIKE) which allows researchers to extract information from genomics and proteomics databases. Another good example is Lynx (see "Relevant Websites section"), a database and knowledge extraction engine for integrative medicine (Sulakhe et al., 2016). Thessen and Parr (2014) presented a method on knowledge extraction and semantic annotation of text from the Encyclopedia of Life (Parr et al., 2014).

Data mining

Databases are very useful for storing massive and ever growing amount of data, which can be utilised for discovering patterns and rules, automatically or even semi-automatically. This method is widely known as data mining and is conceived based on logics in database systems (Paramasivam et al., 2014). A clean set of integrated data is usually stored in a data warehouse before the data mining techniques are applied. Some of the key techniques commonly used in data mining include classification, clustering, prediction, association, text mining, link analysis and regression. Data mining is seen as an important concept in databases as the process involves collecting data, creating database and management, analyzing data and finally interpreting data (Han et al., 2000).

Visualisation

Visualizations can be used to interpret meaning from data through comprehension and exploration as large part of human nervous system has evolved to process visual information (Segaran and Hammerbacher, 2009b). An effective visual design aids decision making and interpretation to generate knowledge. One of the main component of data visualisation is the primary data, which conceived in a data warehouse. A data visualisation software or a script can be used to find pattern, correlations or trends to generate knowledge, which could not be seen in textual or numerical content in a database. One good example of data visualisation is using graph based databases whereby the data is retrieved in a visual context.

Biological Databases

Since its birth, database has become a core element of any discipline that produces large amounts of data. In Biology (particularly in genomics), massive amounts of data are being accumulated on a daily basis via experimental work, field work, software analysis, illustrations, microscopic photographs and image acquisition. Stephens et al. (2015) compared other Big Data Domains such as, astronomy, youtube and twitter in their recent study on Genomics perceiving to be Big Data science (**Fig. 7**). They estimated that genomics is leading or at par with the other big domains in terms of terms of data acquisition, storage, distribution, and analysis. Hence, the need to accommodate the data surge in biology is a challenge. This scenario has lead to the construction of not only single databases but also the consortiums managing large amounts of biological data.

Data Phase	Astronomy	Twitter	YouTube	Genomics
Acquisition	25 zetta-bytes/year	0.5–15 billion tweets/year	500–900 million hours/year	1 zetta-bases/year
Storage	1 EB/year	1–17 PB/year	1–2 EB/year	2–40 EB/year
Analysis	In situ data reduction	Topic and sentiment mining	Limited requirements	Heterogeneous data and analysis
	Real-time processing	Metadata analysis		Variant calling, ~2 trillion central processing unit (CPU) hours
	Massive volumes			All-pairs genome alignments, ~10,000 trillion CPU hours
Distribution	Dedicated lines from antennae to server (600 TB/s)	Small units of distribution	Major component of modern user's bandwidth (10 MB/s)	Many small (10 MB/s) and fewer massive (10 TB/s) data movement

doi:10.1371/journal.pbio.1002195.t001

Fig. 7 Four domains of Big Data in 2025. *Source*: Stephens, Z.D., Lee, S.Y., Faghri, F., *et al.*, 2015. Big data: Astronomical or Genomical? PLOS Biology 13(7), e1002195. Available at: https://doi.org/10.1371/journal.pbio.1002195.

The most current online biological databases can be found in the yearly issue of the journal Nucleic Acids Research (NAR: see "Relevant Websites section") and The Journal of Biological Database and Curation (DATABASE: see "Relevant Websites section"). Both these journals are freely published by the Oxford Academic Journals (see "Relevant Websites section"). Another important source of biological databases is the BMC Bioinformatics (see "Relevant Websites section").

In this section, a range of biological databases is presented. Biological databases are defined under three broad categories which are primary databases, secondary databases and specialised databases. Types of biological data includes (but not limited to): nucleotide and protein sequences (b) protein structure (c) microarray and gene expression (d) metabolic pathways (e) species profile (taxonomy) (f) animal model (g) human disease (h) clinical database and others.

Primary Databases

Nucleotide sequence databases were propelled by the necessity to store primary experimental data for further analysis in Bioinformatics. Major sequence databases include the European Nucleotide Archive (ENA) (see "Relevant Websites section") (Rasko et al., 2011), GenBank (see "Relevant Websites section") (Benson et al., 2008) and the DNA Data Bank of Japan (DDBJ) (see "Relevant Websites section") (Jun et al., 2017) and EMBL Nucleotide Sequence Database (see "Relevant Websites section") (Kanz et al., 2005). These databases are matured sequence databases with search functionality and basic Bioinformatics tools, which are useful to the Bioinformatics community.

A variety of primary sequence databases exist and are widely used in bioinformatics. A significant contribution was made by UniProt (see "Relevant Websites section") to combine Swiss-Prot, TrEMBL and PIR-PSD protein sequence databases into a unified view. The widespread protein aid (uniprot) is a complete resource for protein sequence and annotation statistics. The uniprot databases consists of uniprot knowledgebase (uniprotkb), the uniprot reference clusters (uniref), and the uniprot archive (uniparc) (Uniport, 2017). Protein Data Bank (PDB: see "Relevant Websites section") is one of the oldest database that contains database containing experimentally determined three-dimensional structures of proteins, nucleic acids and other biological macromolecules (Burley et al., 2017).

Microarray database contain microarray gene expression data, which can be further analysed, and interpreted. Recently microarray databases have become very popular with the emergence of high-throughput technology, enabling the expression for thousands of genes simultaneously. Example of very common microarray databases are ArrayExpress (see "Relevant Websites section") from EBI (Brazma et al., 2003) and GEO (Gene Expression Omnibus) (see "Relevant Websites section") from NCBI (Barrett et al., 2006) and the Stanford Microarray Database (see "Relevant Websites section") (Sherlock et al., 2001).

As an important and one of the most studied branch of Biology, biochemistry related databases have emerged, such as the metabolic pathway databases. A few good examples are BRaunschweig ENzyme Database, BRENDA (see "Relevant Websites section") (Schomburg et al., 2002), KEGG PATHWAY Database (see "Relevant Websites section") (Kanehisa and Goto, 2000) and Metacyc (Caspi et al., 2007).

Phylogenetics is the study the evolutionary history and relationships among individuals or groups of organisms. Despite being one of the important aspect of Biology, there are not many databases available online for phylogenetics. There are two databases worth mentioning which are Phylomedb (see "Relevant Websites section") (Huerta-Cepas et al., 2010), a public database for complete catalogs of gene phylogenies (phylomes) and treebase (see "Relevant Websites section") (Sanderson et al., 1994).

Secondary Databases

Secondary databases are more focused as these databases contain derived data (by analyzing primary data) to address specific requirements. This section presents significant and popular secondary biological databases which are well represented in their domain.

The Eukaryotic Promoter Database (EPD: see "Relevant Websites section") comprehensive organisms-specific transcription start site (TSS) collections automatically derived from next generation sequencing (NGS) data (Dreos et al., 2016).

NCBI Reference Sequence Database (RefSeq (see "Relevant Websites section")) is a comprehensive, integrated, non-redundant, well-annotated set of reference sequences including genomic, transcript, and protein (Pruitt, 2006).

Pfam (see "Relevant Websites section") is a database of protein families that consists of their annotations and a couple of series alignments generated using fashions. The most current version, Pfam 31.0, was launched in March 2017 and incorporates 16,712 families (Finn et al., 2017a,b). Pfam has moreover been used inside the advent of different assets which includes iPfam, which catalogs region-area interactions inside and between proteins, based totally on information in structure databases and mapping of Pfam domain names onto those structures (Xu and Dunbrack, 2012).

Prosite (see "Relevant Websites section") is a protein database, which includes entries describing the protein families, domains and practical websites in addition to amino acid styles and profiles in them, manually curated via a team of the Swiss Institute of Bioinformatics and tightly included into Swiss-Prot protein annotation. Prosite was developed in 1988 by Amos Bairoch, who also directed the institution for more than two decades (Prosite introduction, 2017).

The PRIDE PRoteomics IDEntifications (PRIDE) (see "Relevant Websites section") is a data warehouse on proteomics data, including protein and peptide identifications, post-translational modifications and supporting spectral evidence.

SCOP2 (see "Relevant Websites section") is a successor of Structural classification of proteins (SCOP) focusing on proteins that are structurally characterized and deposited in the PDB. These proteins are organised according to their structural and evolutionary relationships in a complex graph network. Each node represents a relationship of a particular type and is exemplified by a region of protein structure and sequence (Andreeva et al., 2013).

CATH (Class, Architecture, Topology, Homology) (see "Relevant Websites section") is a protein structure classification database containing annotations for over 235,000 protein domain structures and includes 25 million domain predictions (Sillitoe *et al.*, 2014).

Protein-protein interaction database are useful source of secondary databases. Two major works in this area include BioGRID and STRING. The Biological General Repository for Interaction Datasets (BioGRID: see "Relevant Websites section") is a public database of genetic and protein interactions curated from the primary biomedical literature for model organism species and humans (Chatr-Aryamontri *et al.*, 2014) whereas the Search Tool for the Retrieval of Interacting Genes (STRING: see "Relevant Websites section") contains the Predicted functional associations between proteins (Szklarczyk *et al.*, 2014).

Specialised Databases

Other than the above mentioned categories, there are specialised biological databases available for the scientific community. A few examples are Phylogenomic Database for Plant Comparative Genomics (GreenPhylDB: see "Relevant Websites section"), the Plant Secretome and Subcullular Proteome Knowledgebase (PlantSecKB: see "Relevant Websites section"), TreeBASE (see "Relevant Websites section"), an open-access database of phylogenetic trees and associated data, the Plant Alternative Splicing database (PASD: see "Relevant Websites section"), Comparative Toxicogenomics Database (CTD: see "Relevant Websites section"), Drosophila Genes & Genomes (FlyBase: see "Relevant Websites section"), WormBase (see "Relevant Websites section"), AceDB (see "Relevant Websites section"), the Arabidopsis Information Resource (TAIR: see "Relevant Websites section"), Online Mendelian Inheritance in Man (OMIM: see "Relevant Websites section") which is an online catalogue of human genes and genetic disorders and the *Saccharomyces* Genome Database (SGD: see "Relevant Websites section").

Integrated Databases

Integrated databases are databases that are unified under a platform, with the aim to offer a suite of applications for secondary uses of the data. These days integrated databases are preferred by researchers due to their usability as one stop centres for knowledge extraction. A large number of biological databases are produced or are managed by the National Center for Biotechnology Information (Coordinators, 2016), which contains a myriad of biological content. Other initiatives are the Kyoto Encyclopedia of Genes and Genomes (KEGG: see "Relevant Websites section"), GeneCards (see "Relevant Websites section"), the InterPro Consortium (Finn *et al.*, 2017a,b), the European Molecular Biology Laboratory (EMBL) Nucleotide Sequence Database (see "Relevant Websites section") and the Global Biodiversity Information Facility (GBIF: see "Relevant Websites section"). These databases are more like consortiums managing and integrating sources of information to provide a unified access to users.

Other Biological Databases

Biodiversity (taxonomy) databases
Taxonomy databases, also referred as biodiversity databases have been developed much before sequence databases are introduced. Taxonomy, being a very old and matured branch of Biology, has produced abundant of data, particularly on species profiling. Almost every country listed as a megabiodiversity country has produced a database of their very own indigenous organisms. Some of the most important biodiversity initiatives are started by GBIF (see "Relevant Website section"), Taxonomic Data Working Group (TDWG: see "Relevant Websites section") and Encyclopedia of Life (EoL: see "Relevant Website section"). These global initiatives encourage sharing of biodiversity data which was previously stored in data silos managed by smaller groups. The unprecedented volume of biodiversity data has caused a momentous increase in the number of online biodiversity databases hosted in separate infrastructure which are managed by research groups. Example of some very popular databases include the Catalogue of Life (see "Relevant Websites section"), Integrated Taxonomic Information System (ITIS: see "Relevant Websites section") and FishBase (see "Relevant Websites section").

Animal model database
Animal models are mostly used in biomedical research which is an important research area to demonstrate biological significance in experiments. Due to this, animal model, particularly mouse model databases are adevent these days. A few examples are mentioned here to demonstrate the kind of databases produced in this domain.

The MUGEN mouse database (MMdb: see "Relevant Websites section") is a repository of murine models of immune processes and immunological diseases.

Mouse Genome Informatics (MGI: see "Relevant Websites section") MGI is the international database resource for the laboratory mouse, providing integrated genetic, genomic, and biological data to facilitate the study of human health and disease.

The Mouse Phenome Database (MPD: see "Relevant Websites section") is one of the most widely used resource for primary experimental trait data and genotypic variation.

International Mouse Strains Resource (IMSR: see "Relevant Websites section") is a database on mouse strains, stocks, and mutant ES cell lines available worldwide, including inbred, mutant, and genetically engineered strains.

Other mouse model databases are The European Mouse Mutant Archive (see "Relevant Websites section") and the Rat Resource and Research Center (see "Relevant Websites section").

Besides mouse, other animals database on Zebra, Zebrafish and Drosophila have been published. One good example is the Zebrafish models (ZF-Models: see "Relevant Websites section"), which is an information system on zebrafish that can be referred for producing and identifying disease models, drug targets and insight into pathways of gene regulation applicable to human development and disease. Other examples are Zebra International Resource Center (see "Relevant Websites section") and Bloomington Drosophila Stock Center (BDSC: see "Relevant Websites section").

Clinical/health database

Clinical and Health databases must not be sidelined in the discussion about biological databases. While we will not be able to mention all the works done in this area, it is crucial to highlight a few significant examples to demonstrate the kind of databases produced for this purpose. The National Center of Biotechnology Information (NCBI), as part of their database initiative, has produced two important clinical and health databases which are ClinVar (see "Relevant Websites section"), a database on the reports of the relationships among human variations and phenotypes; and MedGen (see "Relevant Websites section") a database containing information related to human medical genetics, such as attributes of conditions with a genetic contribution. Database of Genotypes and Phenotypes (dbGaP: see "Relevant Websites section") is an archive of data and results from studies that have investigated the interaction of genotype and phenotype in humans. PubMed Health (see "Relevant Websites section") is the world's largest digital medical library. MEDLINE (see "Relevant Websites section") contains journal citations and abstracts for biomedical literature from around the world while PubMed (see "Relevant Websites section") is an archive of more than 28 million citations (links to full text content from PubMed and publisher websites) for biomedical literature from MEDLINE, life science journals, and online books.

Clinical and Health databases focusing on diseases are also useful resource for biologists and medical scientists. There are some prominent disease databases published online such as MalaCrads (see "Relevant Websites section"), the Autoimmune Disease Database (see "Relevant Websites section"), Inflammatory Bowel Diseases (IBD: see "Relevant Websites section"), KEGG Disease Database (see "Relevant Websites section") and LiverWiki (see "Relevant Websites section"), a wiki-based database for human liver. The Diseases Database ver 2.0 contain a whole range of database in their portal, which is available at the website provided in "Relevant Websites section". Finally, a comprehensive collection of human related biological databases is presented by Zou *et al.* (2015).

Biological Databases Case Studies: Genecards and Cancer Genome Atlas

Genecards (see "Relevant Websites section")

In the early 80s, storing sequence data was not easy due to cost and infrastructure limitations. Nevertheless, many institutions have started to store their data in primary databases which are available in isolated platforms and are focused on specific subject of study. In order to gather these scattered data, The Weizmann Institute of Science Crown Human Genome Center (http://www.weizmann.ac.il) developed a database called *GeneCards* in 1997. In the early stage, this database was dealing mostly with human genome information, human genes, the encoded protein's function and related diseases.

Later, it incorporated automatically-mined genomic, proteomic and transcriptomic data, as well as orthologies, disorder relationships, snps, gene expression, gene characteristic, and hyperlinks for ordering assays and antibodies (Genecards, 2018). Currently it serves as a complete, authoritative compendium of annotative information about human genes that has been broadly used for almost 15 years (Safran *et al.*, 2010). Genecards has blossomed into a collection of equipment (along with Genedecks, Genealacart, Geneloc, Genenote and Geneannot) for a spread of analyses of each single human genes and units due to its viewing information and deep links associated with unique genes and constant improvements, over the years. (Stelzer *et al.*, 2011) (**Figs. 8–11**).

GeneCards versions

GeneCards Version 1.xx- contained gene-centric information, automatically mined and integrated from a myriad of databases resulting in a web-based *card* for each of human gene entry (GeneCards, 2018).

GeneCards Version 2.xx- the information is stored in flat files, one file per gene, all indexed to enable full-text searches (GeneCards, 2018).

GeneCards Version 3.xx- introduced a persistent object/relational model of all the data entities and relationships displaying diverse functions of single genes by extracting various slices of attributes of the genes (GeneCards, 2018). It also hosts the new seek facility by offering hyperlinks to a sample gene and its diverse sections (Safran *et al.*, 2010).

GeneCards Version 4.xx- the hybrid system is eliminated, moving to a robust infrastructure based entirely on a new and well defined relational database (GeneCards, 2018).

GeneCards applications

GeneDecks exploits Genecards unique wealth of combinatorial annotations to discover similar (associate) genes, and to perform quantitative descriptor enrichment analyses for figuring out set-shared annotations. Some of these functions are available online, through the Genecards website, while others independent studies. Some of the Genecards applications are described below.

SYNLET – Regulatory control networks of synthetic lethality via Genedecks mainly addresses robustness of phenotypic characteristic on the basis of artificial lethality, as proposed for novel most cancers remedy regimes. It derives novel standards, methodologies and algorithms for annotation and evaluation of regulatory networks, with attention on tumorigenesis and drug resistance. It aims

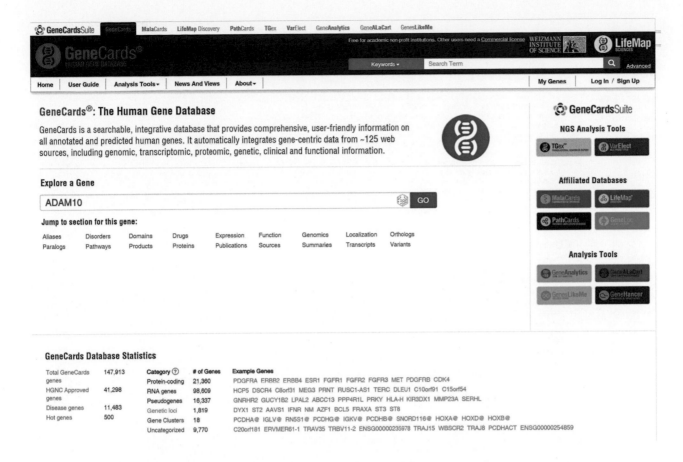

Fig. 8 The screenshot image of the domain GeneCards, human genome database. Retrieved from: https://www.genecards.org/.

to explore key proteins of cellular escape mechanisms that conquer lethality of medicine and find approaches to block them, utilizing siRNA. Genedecks became one of the major equipment which served the consortium to choose candidates for the siRNA experiments. The set distiller mode of Genedecks permits the identification of annotation descriptors shared via experimentally-received gene sets, permitting the evaluation in their belonging to precise practical lessons, for this reason a really appropriate choice of inactivation targets (Safran *et al.*, 2010).

SysKid – Systems biology towards novel chronic kidney disease diagnosis and treatment explores chronic kidney ailment, with diabetes mellitus and high blood pressure being the maximum widely wide-spread causative conditions. It assumes that despite it being numerous in etiology, the underlying molecular pathophysiology of different manifestations of this condition can be similar. The venture goals at acquiring a unified view on the disease within the scope of systems biology, primarily based on high-throughput omics analysis. The Genecards version 3.xx database and superior seek engine are instrumental for the combination procedure and in figuring out the maximum promising sickness markers (Stelzer *et al.*, 2011). For instance, Genecards might be used to perceive applicable genes for metabolomics hints. In this framework, a 'Genecards 100k' attempt is presently beneath way, aiming at increasing the range of gene entries closer to one hundred 000. This will be finished mainly through a big growth of Genecards' scope within the realm of non-protein-coding RNA genes, and resolving a massive wide variety of genes currently entitled 'uncategorized'. In parallel, a challenge-particular database (Genekid) could be created, to residence incoming omics facts in Genecards-well suited tables, for that reason facilitating the systems biology analyses (Safran *et al.*, 2010).

Future of genecards

The success of applications described above will further upgrade GeneCards core features. An exciting destiny undertaking is to plot a set of rules for unifying disorder names/descriptions and permit those tables to be merged. An increase in pathway repertoire and inclusion of public area (e.g., reactome) and/or additional industrial pathways are expected. In the future, the 'feature' phase will also consist of animal models from species other than mouse (Safran *et al.*, 2010).

The Cancer Genome Atlas (TCGA) (see "Relevant Websites section")

The National Institutes of Fitness (NIH) mounted The cancer Genome Atlas (TCGA) to generate complete, multi-dimensional maps of the important thing genomic changes in important types and subtypes of cancer (National Institutes of Health, 2017).

SYNLET home

consortium

disseminations

meetings

internal

SYNLET Project

Project Summary

This project addresses robustness of **phenotypic function** on the basis of highly interlinked **network topologies** on a very practical level: **Synthetic Lethality** – as proposed for novel cancer treatment regimes.
Function in this context is the property of transformed cells to continuously divide, and robustness is encoded by regulatory elements on the genomic and proteomic level triggering drug treatment escape as an emergent property: chemotherapy resistance. This project will derive novel concepts, methodologies, and their algorithmic implementation for annotation and analysis of general regulatory networks - with particular focus on network robustness and escape routes toward maintaining function. In parallel, computational genomics will be applied on a gene expression data set derived from a unique, systematic collection of chemotherapy resistant cancer cell lines. These real world data encode a set of regulatory escape mechanisms to overcome the lethality imposed by a specific drug. The general concepts and approaches – focusing on dynamical levels - will subsequently be calibrated and merged with results from computational genomics – focusing on a static system representation – to gain insight into the mechanisms of cancer cell robustness, but in particular to identify key proteins of cellular escape mechanisms to overcome lethality of drugs. Blocking these proteins may result in synthetic lethality including re-sensitization of chemoresistant cancer cells to cytotoxic drugs. To prove the validity of our Complex Systems Analysis approach we will test generated hypotheses on synthetically lethal hubs in the cellular control network via experimental knock down experiments utilizing siRNA.

EU foundation

Proposal Full Title	Regulatory Control Networks of Synthetic Lethality
Proposal Acronym	SYNLET
Date of Preparation	February 15, 2006
Version No.	2.0
Type of Instrument	Specific Targeted Research Project
Submission Stage	Full Proposal
Activity Code Addresses	NEST-2005-Path-COM

Coordinator:
Dr. Michaelis Martin
e-mail: martin.michaelis@blue-drugs.com
Fax: +49 69 67 86 65-9172

Frankfurter Stiftung für Krebskranke Kinder

http://www.kinderkrebsforschung-frankfurt.de

Hilfe für krebskranke Kinder e.V.

http://www.hilfe-fuer-krebskranke-kinder.de

Fig. 9 The screenshot image of the domains SYNLET. Retrieved from: http://synlet.izbi.uni-leipzig.de/.

The Cancer Genome Atlas (TCGA) is a venture began in 2005, to catalogue genetic mutations answerable for cancer, the usage of genome sequencing and bioinformatics. TCGA applies excessive-throughput genome analysis strategies to enhance the ability to diagnose and treat cancer via a detailed investigation of the genetic basis of this ailment (Tomczak *et al.*, 2015; The National Genome Atlas, 2017).

In 2006, TCGA targeted on characterization of 3 varieties of human cancers: Glioblastoma Multiforme, lung, and ovarian cancer. This attempt initiated by NCI and NHGRI in 2006 developed the rules, production pipeline, collaborative studies network, databases and analytical equipment important for TCGA's big-scale analysis of cancer genomics. The pilot's characterizations of brain tumours and ovarian cancer have proven that this systematic, excessive-quantity approach can generate excellent novel discoveries, useful for scientific researchers around the globe (The National Genome Atlas, 2017).

In 2009, TCGA accelerated into Phase II, which planned to complete the genomic characterization and collection evaluation of 20–25 special tumour types through 2014. TCGA passed that aim, characterizing 33 cancer types together with 10 rare cancers. The project was split between genome characterization centres (GCCs), which carry out the sequencing, and genome records analysis facilities (GDACs), which carry out the bioinformatics analyses (The National Genome Atlas, 2017). With its pipeline in area, TCGA correctly acquired samples of tumours and coupled ordinary tissue, rapidly sequenced the DNA and RNA of the samples, and comprehensively characterized the genomes the usage of seven genomic platforms (The National Genome Atlas, 2017).

In December 2013, while TCGA finalized its collection of matched tumour and everyday samples, sufficient tissue have been accumulated to map the genomic traits of extra than 30 cancer types. TCGA evaluation operating agencies, groups of scientists from across the United States of America, have now comprehensively studied 33 cancer types and subtypes, along with 10 rare cancers (The National Genome Atlas, 2017).

This pioneering effort to map and analyse most cancers genomes in a huge-scale, systematic way has modified the way most cancers are studied, and could ultimately rework the way most cancers are dealt with because TCGA information is loose to use, it significantly reduces the costs and expedites the manner of drug development, allowing both public and private quarter researchers

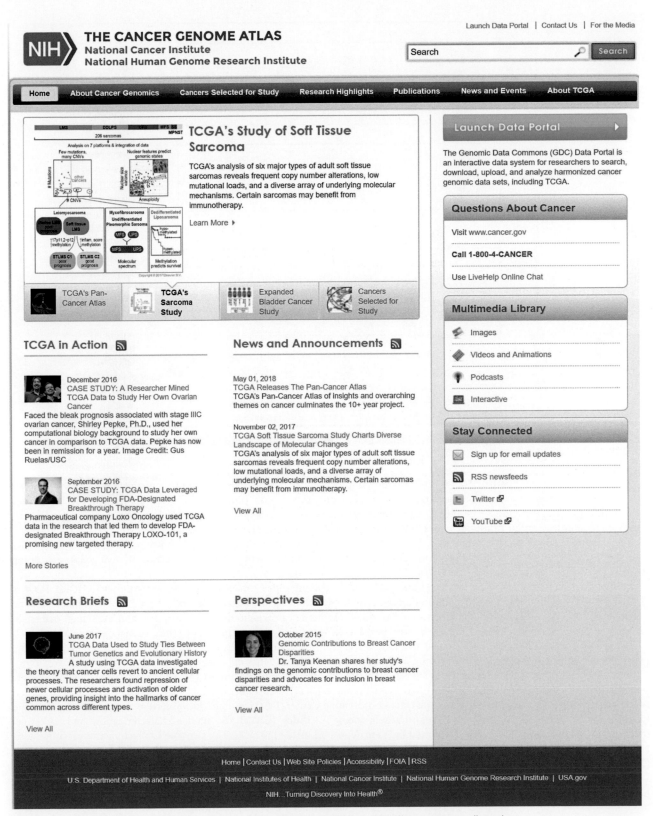

Fig. 10 The screenshot image of The Cancer Genome Atlas (TCGA). Available from: https://cancergenome.nih.gov/.

to pursue centred cures geared toward the precise pathways in a positive most cancers kind or subtype. This new level of insight, supplied by means of TCGA will chart a brand new path for most cancers research and pioneer the manner to extra effective, individualized techniques for helping every affected person with most cancers (The National Genome Atlas, 2017).

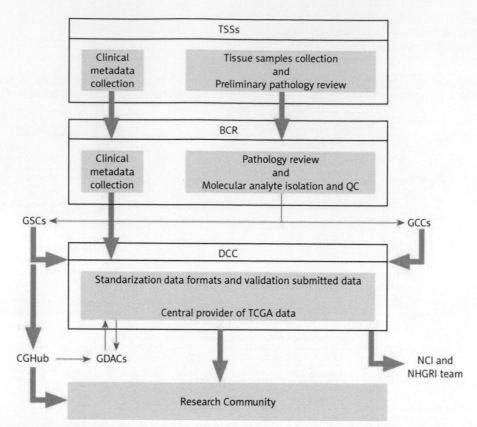

Fig. 11 The Cancer Genome Atlas (TCGA) Research Network Centres flowchart. Available from: https://www.ncbi.nlm.nih.gov/pmc/articles/ PMC4322527/figure/F0001/.

The TCGA initiative scheduled 500 affected person samples and used distinctive strategies to investigate the patient samples. Techniques include gene expression profiling, reproduction quantity version profiling, SNP genotyping, genome huge DNA methylation profiling, microRNA profiling, and exon sequencing. TCGA is sequencing the whole genomes of some tumours, which include at least 6000 candidate genes and microRNA sequences. (The National Genome Atlas, 2017).

This catalogue serves as an effective useful resource for a brand new generation of research aimed towards developing higher strategies for diagnosing, treating and preventing most cancers. TCGA also presents a version for plenty different genome mapping projects (National Cancer Institute, 2017).

Mutually led via the countrywide cancer Institute (NCI) and the country wide Human Genome Research Institute (NHGRI), TCGA is changing the manner genetic most cancers studies are completed. Following the instance set with the aid of the Human Genome undertaking, TCGA's collaborative studies network has added together researchers from diverse disciplines and more than one establishment to provide precious statistics units to be used by using the worldwide studies community (The National Genome Atlas, 2017).

The shape of TCGA is properly organised and includes several cooperating centres liable for collection and sampleprocessing, observed through high-throughput sequencing and sophisticated bioinformatics data analyses.

First, special tissue supply websites (TSSs) accumulate the desired bio specimens (blood, tissue) from eligible cancers patients and deliver them to the Biospecimen Core Resource (BCR). Subsequently, the BCR catalogue validates and then submit the scientific data and metadata to the Data Coordinating Center (DCC) and provide molecular analyses for the Genome Characterization Centres (GCCs) and Genome Sequencing Facilities (GSCs) for similar characterization and excessive-throughput sequencing. Then, series-associated facts are deposited inside the DCC. The Genome Characterization facilities also post hint documents, sequences, and alignment mappings to NCI's cancer Genomics Hub (CGHub) secure repository. The generated genomic records is made available to the research community and Genome Data Analysis Centers (GDACs). The GDACs provide new facts-processing, evaluation, and visualisation gear to the whole studies community to facilitate broader use of TCGA data. Furthermore, the statistics generated by means of the TCGA studies network is centrally managed on the DCC and deposited into public databases (TCGA Portal, NCBI's hint Archive, CGHub), allowing scientists to obtain cancer datasets.

TCGA applications

The Cancers Imaging Archive, TCIA (see "Relevant Websites section") TCIA is a service created through the NCI to host a huge range of medical images (radiological imaging statistics), from TCGA instances, therefore it supports imaging phenotype-genotype research. The data is de-identified and available for public download.

Berkeley Morphometric Visualisation and Quantification from H&E sections (see "Relevant Websites section") is a data repository of computed histology-primarily based snap shots of various tumour samples from TCGA cases, and is backed via the Lawrence Berkeley National Laboratory.

The Cancer Digital Slide Archive (CDSA: see "Relevant Websites section"), is an online interactive device for viewing and annotating diagnostic and tissue slide photographs of different tumour kinds from TCGA assignment.

The Broad GDAC Firehose (see "Relevant Websites section") is an analytical infrastructure created to coordinate the flow of cancers datasets, using various quantitative algorithms along with GISTIC, MutSig, Clustering, and Correlation.

The MD Anderson GDAC's MBatch (see "Relevant Websites section") is a website that permits scientists to discover and quantify the batch effects accompanying TCGA facts set, currently in line with hierarchical clustering and more desirable PCA plots.

Cancers Genome Workbench, (CGWB: see "Relevant Websites section"), is a utility evolved with the aid of the National Cancer Institute (NCI) to combine and display pattern-level genomic and transcription changes in numerous cancers, by obtaining information from several cancers projects, including TCGA. The major visitors in CGWB are incorporated tracks view, heat map view, and an alignment viewer known as Bambino.

Future perspectives of TCGA

Systematic advances in cancer genomics provided via TCGA have revealed a brand new approach of molecular biology in cancer. The application of high-throughput technology, boosted with bioinformatics analytics has been successful in highlighting the similarities and differences in the genomic structure of every cancer and across more than one kind. The TCGA has furnished a large amount of publicly available data on most cancers genetic and epigenetic profiles, highlighting candidate cancer biomarkers and drug objectives. Moreover, translation of cancer genomics into therapeutics and diagnostics will offer potential to increase personalized cancer medicinal drug. The ultimate goal for scientists is to develop very high end bioinformatics tools to get rid of noise in data and enhance the decision of the analysis, for cutting edge discoveries. In the future, all novel findings will facilitate diagnosis, treatment, and cancer prevention. Lately, researchers have new discoveries in cancer research by trying to "teach" a device – an artificially clever laptop, called Watson – to aid doctors in diagnosing patients. In time to come, these rapid advances can hopefully be incorporated into clinics (Tomczak *et al.*, 2015).

Results and Discussion

Results are presented in terms of database models, tools and architecture employed in developing biological databases.

Database Models

Predominantly, traditional relational model has been used in the design and development of biological databases. Some prominent example of relational databases are GenBank, EMBL, SWISS-PROT and Protein Information Resource(PIR) and CBS Genome Atlas Database.

However, with the recent development in the semantic web technologies, many biological ontologies have been created that uses object based database models using ontologies and nosql database tools. One good example is BioPortal (see "Relevant Websites section") which is a repository for biomedical ontologies. It currently contains 704 ontologies. Other prominent examples are BioPAX (see "Relevant Websites section"), Cell Cycle Ontology (CCO: see "Relevant Websites section"), Gene Expression Knowledgebase (GexKB: see "Relevant Websites section"), Disease Ontology (see "Relevant Websites section"), Gene Ontology Consortium (see "Relevant Websites section"), Sequence Ontology (see "Relevant Websites section") and SNOMED CT (see "Relevant Websites section"). A good example of ontology based database is the Fish Ontology (FISHO: see "Relevant Websites section") which is shown in **Fig. 12**.

The Human Protein Reference Database (HPRD: see "Relevant Websitessection") (Prasad *et al.*, 2009), on the contrary, was developed using the traditional object based data model.

A summary of database models and development tools employed in biological databases are presented in **Table 1**.

Database Architecture

In terms of database architecture, all three approaches: Distributed, Federated and Data Warehouse have been utilised in biological databases. Example of distributed databases include Ensembl, WormBase, and the Berkeley Drosophila Genome Project whereas examples of federated databases are TwinNET (Muilu *et al.*, 2007), ENCODE (Blankenberg *et al.*, 2007), EBI search (Park *et al.*, 2017), SPINE2 (Goh *et al.*, 2003), Cancer Biomedical Informatics Grid (Saltz *et al.*, 2006), NIF (Gardner *et al.*, 2008), Biomedical Informatics Research Network (Ashish *et al.*, 2010), Biomedical Investigations (Taylor *et al.*, 2008), EdgeExpressDB (Severin *et al*, 2009) and Minimal Information About Neural Electromagnetic Ontologies (Frishkoff *et al.*, 2011). Examples of biological databases that are built on data warehouse concept are Pathway Commons (Cerami *et al.*, 2010), String (Szklarczyk *et al.*, 2010), CBS Genome Atlas Database (Hallin and Ussery, 2004), BioMart (Haider *et al.*, 2009), BrainMap (see "Relevant Websites section"), PubBrain (see "Relevant Websites section").

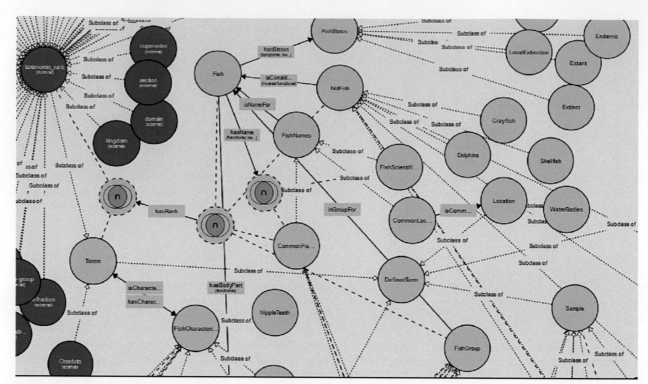

Fig. 12 The Fish Ontology (FO) model. A portion of the FO is shown here on how the classes are related to each other and to other ontology classes. The dark blue circles represent terms from other ontologies while light blue circles represent terms from the FO. *Source*: Ali, N.M., Khan HA, Then AY, Ving Ching C, Gaur M, Dhillon SK. (2017) Fish Ontology framework for taxonomy-based fish recognition. PeerJ 5:e3811 https://doi.org/10.7717/peerj.3811.

Issues Pertaining Biological Databases

The revolutions caused by advanced computing power, advanced informatics and the Internet have changed the way biological data are stored, located and disseminated. Today almost every scientist realizes the need of storing their data sets using relevant technology and models.The architecture of a database system is greatly influenced by the underlying computer system on which the database system runs.In the past 20 years, scientists have engaged themselves in the informatics discipline of querying multiple remote or local heterogeneous data sources, integrating manually received data and manipulating it with advanced data analyzing and visualizing tools. The access to relevant data, combining a myriad data sources and coping with distinct and heterogeneous systems is a tremendously difficult task. In recent times, however, information retrieval has become easier through networking of related datasets. Searching integrated multiple biological databases simultaneously have significantly alleviated the process of information retrieval in this field. The underlying mechanism which integrates different data sources has greatly helped researchers in information gathering and mining.

Undoubtedly, the problem faced in biological information is technically due to the reason that the data is still in various forms, locations and in diverse formats. One of the most intriguing phase in promoting the growth of biological information centres is to correlate, synthesize, disseminate, share and retrieve information in the form of databases. Development of advanced systems to link the scattered data so as to facilitate exchange of data amongst the different sources is indeed quite challenging. The fundamental problem affecting data integration is the adoption of data standards. In Biology, data standards in terms of vocabularies and ontologies have facilitated data integration tremendously. However an undesirable situation could arise if developers are from the computing background and do not really understand the available data standards in Biology, while, biologists are not very keen to undertake technical tasks such as developing databases. Having said this, on the whole there are some very established database initiatives by the National Center of Biotechnology Information, European Bioinformatics Institute, Swiss Institute of Bioinformatics, Center for Bioinformatics, International Nucleotide Sequence Database Collaboration and Computational Biology at University of Delaware to integrate biological resources. With the exponential growth in biological data, many more such initiatives must be established especially in the Asia Pacific Region, where some very good efforts have been initiated by Asia Pacific Bioinformatics Network (APBioNet) which is Asia's oldest bioinformatics organisation to champion the advancement of bioinformatics in the Asia Pacific (Ranganathan *et al.*, 2006).

There is also another concern affecting the establishment of biological databases. Publishing databases online is still not favored by scientists especially if they are the data owners, as it may give rise to copyright issues as well as misuse of the data by

Table 1   Summary of data modeling techniques in biological databases

Database	Data modeling	Function	References (Harvard Style)
European Nucleotide Archive (ENA) (https://www.ebi.ac.uk/ena)	noSQL	Database of raw sequencing data, sequence assembly information and functional annotation	Nicole Silvester, Blaise Alako, Clara Amid, Ana Cerdeño-Tarrága, Laura Clarke, Iain Cleland, Peter W Harrison, Suran Jayathilaka, Simon Kay, Thomas Keane, Rasko Leinonen, Xin Liu, Josué Martínez-Villacorta, Manuela Menchi, Kethi Reddy, Nima Pakseresht, Jeena Rajan, Marc Rossello, Dmitriy Smirnov, Ana L Toribio, Daniel Vaughan, Vadim Zalunin, Guy Cochrane; The European Nucleotide Archive in 2017, Nucleic Acids Research, Volume 46, Issue D1, 4 January 2018, Pages D36–D40. https://doi.org/10.1093/nar/gkx1125
GenBank (https://www.ncbi.nlm.nih.gov/genbank/)	Flat file	An annotated collection of all publicly available DNA sequences	NCBI GenBank. https://www.ncbi.nlm.nih.gov/genbank/
Protein Data Bank (PDB) (https://www.wwpdb.org/)	Flat file (XML)	Expertise in the techniques of X-ray crystal structure determination, NMR, cryoelectron microscopy and theoretical modeling.	Berman, H. M., Westbrook, J., Feng, Z., Gilliland, G., Bhat, T. N., Weissig, H., Shindyalov I. N. & Bourne, P. E. (2000). The Protein Data Bank. Nucleic Acids Research, 28(1), 235–242.
MEDLINE (https://www.nlm.nih.gov/bsd/pmresources.html)	Flat file (XML)	Database on journal citations and abstracts for biomedical literature from around the world	
ArrayExpress (https://www.ebi.ac.uk/arrayexpress/)	Flat file (XML)	Database of microarray gene expression data	Brazma, A., Parkinson, H., Sarkans, U., Shojatalab, M., Vilo, J., Abeygunawardena, N., Holloway, E., Kapushesky, M., Kemmeren, P., Lara, G. G., Oezcimen, A., Rocca-Serra, P. & Sansone, S.-A. (2003). ArrayExpress – A public repository for microarray gene expression data at the EBI. Nucleic Acids Research, 31(1), 68–71.
GEO (https://www.ncbi.nlm.nih.gov/geo/)	Flat file and relational	Database of high-throughput gene expression data	Barrett, T., Troup, D. B., Wilhite, S. E., Ledoux, P., Rudnev, D., Evangelista, C., Edgar, R. (2009). NCBI GEO: archive for high-throughput functional genomic data. Nucleic Acids Res., 37 (Database issue), D885-D890. https://doi.org/10.1093/nar/gkn764
EdgeExpressDB (http://fantom.gsc.riken.jp/4/edgeexpress/about/)	Relational (snow flake schema design)	A database for interpreting biological networks and comparing large high-throughput expression datasets	Severin, J., Waterhouse, A.M., Kawaji, H., Lassmann, T., van Nimwegen, E., Balwierz, P.J., Hoon, M.J., Hume, D.A., Carninci, P., Hayashizaki, Y., Suzuki, H., Daub, C.O. & Forrest, A.R. (2009). FANTOM4 EdgeExpressDB: An integrated database of promoters, genes, microRNAs, expression dynamics and regulatory interactions. Genome Biology, 10(4), R39. https://doi.org/10.1186/gb-2009-10-4-r39
BRENDA (https://www.brenda-enzymes.org/)	Relational	Database on identified enzymes	Barthelmes, J., Ebeling, C., Chang, A., Schomburg, I., & Schomburg, D. (2007). BRENDA, AMENDA and FRENDA: The enzyme information system in 2007. Nucleic Acids Research, 35(Database issue), D511–D514. https://doi.org/10.1093/nar/gkl972
KEGG PATHWAY (http://www.genome.jp/kegg/pathway.html)	Flat file (XML)	A database resource for network of protein products, including functional RNAs.	Kanehisa and Goto (2000) KEGG: kyoto encyclopedia of genes and genomes. Nucleic acids research, 28(1), pp.27-30.
Metacyc (https://metacyc.org/)	Flat file	Database describing metabolic pathways and enzymes from all domains of life.	Caspi et al., 2016, "The MetaCyc database of metabolic pathways and enzymes and the BioCyc collection of Pathway/Genome Databases," Nucleic Acids Research 44:D471-D480
Human metabolic atlas (http://www.metabolicatlas.org)	Relational model	Resource for human metabolism	Natapol Pornputtapong, Intawat Nookaew, Jens Nielsen; Human metabolic atlas: An online resource for human metabolism, Database, Volume 2015, 1 January 2015, bav068. https://doi.org/10.1093/database/bav068
Encyclopedia of Life (eol.org)	Flat File (JSON)	Database containing information about life on Earth	Parr, C.S., Wilson, N., Leary, P., Schulz, K.S., Lans, K., Walley, L., ... Corrigan, Jr., R.J. (2014). The Encyclopedia of Life v2: Providing Global Access to Knowledge About Life on Earth. Biodiversity Data Journal, (2), e1079. Advance online publication. https://doi.org/10.3897/BDJ.2.e1079
Pfam (https://pfam.xfam.org/)	Relational	A collection of protein multiple sequence alignments	The Pfam protein families database: towards a more sustainable future: R.D. Finn, P. Coggill, R.Y. Eberhardt, S.R. Eddy, J. Mistry, A. L. Mitchell, S.C. Potter, M. Punta, M. Qureshi, A. Sangrador-Vegas, G.A. Salazar, J. Tate, A. Bateman Nucleic Acids Research (2016) Database Issue 44:D279-D285

(Continued)

Table 1   Continued

Database	Data modeling	Function	References (Harvard Style)
SCOP2 (http://scop2.mrc-lmb.cam.ac.uk/)	Relational model Graph Data Model	Structural classification of proteins	Andreeva, A., Howorth, D., Chothia, C., Kulesha, E., & Murzin, A.G. (2014). SCOP2 prototype: A new approach to protein structure mining. *Nucleic Acids Res., 42*(Database issue), D310-D314. https://doi.org/10.1093/nar/gkt1242
BioGRID (https://thebiogrid.org/)	Relational model	An interaction repository with data compiled through comprehensive curation efforts.	Stark, C., Breitkreutz, B.J., Reguly, T., Boucher, L., Breitkreutz, A. and Tyers, M., 2006. BioGRID: a general repository for interaction datasets. *Nucleic acids research, 34*(suppl_1), pp. D535–D539.
STRING (https://string-db.org/)	Relational Model	A database of known and predicted protein-protein interactions.	Christian von Mering, Martijn Huynen, Daniel Jaeggi, Steffen Schmidt, Peer Bork, Berend Snel; STRING: a database of predicted functional associations between proteins, *Nucleic Acids Research*, Volume 31, Issue 1, 1 January 2003, Pages 258–261. https://doi.org/10.1093/nar/gkg034
GreenPhylDB (http://greenphyl.cirad.fr)	Relational model	Platform for comparative functional genomics in *Oryza sativa* and *Arabidopsis thaliana* genomes	Conte, M. G., Gaillard, S., Lanau, N., Rouard, M., & Périn, C. (2008). GreenPhylDB: a database for plant comparative genomics. *Nucleic Acids Research, 36*(Database issue), D991–D998. https://doi.org/10.1093/nar/gkm934
TreeBASE (treebase.org)	Relational model	Database on phylogenetic relationships	Piel, W.H., Chan, L., Dominus, M.J., Ruan, J., Vos, R.A. and Tannen, V., 2009. TreeBASE v. 2: a database of phylogenetic knowledge.
FlyBase (Flybase.org)	Relational	Contain fundamental genomic and genetic data on the major genetic model Drosophila melanogaster and related species	The FlyBase database of the Drosophila genome projects and community literature. (2003). *Nucleic Acids Res., 31*(1), 172–175.
WormBase (Wormbase.org)	Flat file (XML)	Database for Caenorhabditis elegans genome and its biology.	Stein, L., Sternberg, P., Durbin, R., Thierry-Mieg, J., & Spieth, J. (2001). WormBase: network access to the genome and biology of Caenorhabditis elegans. *Nucleic Acids Research, 29*(1), 82–86. https://doi.org/10.1093/nar/29.1.82
RefSeq (https://www.ncbi.nlm.nih.gov/refseq/)	Relational model	Contains sequences, including genomic DNA, transcripts, and proteins	O'Leary, N.A., Wright, M.W., Brister, J.R., Ciufo, S., Haddad, D., McVeigh, R., Rajput, B., Robbertse, B., Smith-White, B., Ako-Adjei, D. and Astashyn, A., 2015. Reference sequence (RefSeq) database at NCBI: current status, taxonomic expansion, and functional annotation. *Nucleic acids research, 44*(D1), pp. D733–D745.
OMIM (omim.org)	Flat file	Database of human genes and genetic disorders	Amberger, J.S., Bocchini, C.A., Schiettecatte, F., Scott, A.F., & Hamosh, A. (2015). OMIM.org: Online Mendelian Inheritance in Man (OMIM®), an online catalogue of human genes and genetic disorders. *Nucleic Acids Research, 43*(Database issue), D789–D798. https://doi.org/10.1093/nar/gku1205
AceDB (http://www.sanger.ac.uk/science/tools/acedb)	Hierarchical object	A database for handling genome and bioinformatics data	Srinivasan, J., Otto, G.W., Kahlow, U., Geisler, R., & Sommer, R.J. (2004). AppaDB: An AcedB database for the nematode satellite organism Pristionchus pacificus. *Nucleic Acids Research, 32*(Database issue), D421–D422. https://doi.org/10.1093/nar/gkh057
PRoteomics IDEntifications database	Relational model	Database on mass spectrometry (MS) – based proteomics data	Vizcaíno, J. A., Csordas, A., del-Toro, N., Dianes, J. A., Griss, J., Lavidas, I., Hermjakob, H. (2016). 2016 update of the PRIDE database and its related tools. *Nucleic Acids Research, 44*(Database issue), D447–D456. https://doi.org/10.1093/nar/gkv1145

some fraudulent parties. Scientists who may not want to share unpublished data have their valuable data stored in personal computers which are most of the time untapped. These kind of speckled data, both textual and image date, will not allow discovery of new knowledge and its evident that correlated data can give rise to new discoveries in science.

Finally, the development of databases require data to be in the digital form and digitizing biological data is a daunting task. It requires extensive manpower, enough equipments and all of these requires funds. However, this can be overcomed if world class biological data centres makes databasing lenient enough for underprivileged scientists.

Future Directions

In line with the advancement of the digital world and high performance computing technology, biological databases have been mushrooming and the number is expected grow consistently. These biological databases have become an integral part of research in many scientific areas, from data extraction, knowledge discovery and biological simulations to advanced analysis.

In line with the exponential increase in size and heterogeneity of biological data, new database models have been proposed, which are more scalable compared to traditional relational databases deployed on single computer system. Recent studies by Valduriez (2011), Han *et al.* (2011) and Konishetty *et al.* (2012) have explored the practical implementations of NoSQL databases to help in developing a viable and usable practice in information management. While new biological databases are being introduced every year, it is now important to relook at models and infrastructure deployed in construction of these databases. Older database models such as relational, which have been the core technology behind biological databases, are losing its relevance with regards to handling exponential increase of data. Therefore, traditional database concepts are evolving in order to meet and map the data management requirements of today's environment.

The growing data in genomics, metabolomics, proteomics and metagenomics can be regarded as Big Data if these data sources are harmonised. It is critical to explore back end data integration that are streamlined into front end resilient computer systems for performing seamless transactions, whether for simple search algorithms or high level data science. In order to achieve this vision, these data sources need to be mediated via ontologies and noSQL databases by adopting parallel computing technology and artificial intelligence. In order to facilitate the new paradigm, automated methods need to map relational models (SQL based) into noSQL models which are the essence of big data applications.

NoSQL graph model has recently become very popular as it presents a solution to many of today's challenges such as visualisation of data with complex connections and rich relationships.

Closing Remarks

This article presents a holistic overview of biological databases, with regards to modeling and the architecture of the databases. A comprehensive literature on biological databases is presented with a focus on two case studies which are GeneCards and Cancer Genome Atlas. This literature, however, by no means covers the whole range of biological databases that are currently available. The Nucleic Acids Research Journal features the latest biological databases with a very good coverage on the most recent research in this area (see "Relevant Websites section"). Finally, the future direction of biological databases is discussed focusing on the new concepts in the digital world such as Big Data, Data Science and NoSQL databases which holds the future of biological databases.

Acknowledgement

I would like to thank all my postgraduate students for assisting in my biological database research over the years. The research contribution made by these students ranging from relational models, semantic web, ontologies, image databases, noSQL and graph models as well as medical databases has been overwhelming.

See also: Biological Database Searching. Information Retrieval in Life Sciences. Integrative Bioinformatics. Integrative Bioinformatics of Transcriptome: Databases, Tools and Pipelines. Natural Language Processing Approaches in Bioinformatics

References

About uniport, 2017. Uniport (online). Available at: http://www.uniprot.org/ (14 February 2018).

Abramova, V., Bernardino, J., 2013. NoSQL databases. In: Proceedings of the International C* Conference on Computer Science and Software Engineering - C3S2E '13. ACM Press New York, New York, USA, pp. 14 – 22. Available at: http://doi.org/10.1145/2494444.2494447.

Abu, A., Lim, S.L.H., Sidhu, A.S., Dhillon, S.K., 2013a. Biodiversity image retrieval framework for monogeneans. Systematics and Biodiversity 11 (1), 19–33.

Abu, A., Susan, L.L.H., Sidhu, A.S., Dhillon, S.K., 2013b. Semantic representation of monogenean haptoral Bar image annotation. BMC Bioinformatics 14 (1), 48.

Ali, N.M., Khan, H.A., Amy, Y., *et al.*, 2017. Fish Ontology framework for taxonomy-based fish recognition. PeerJ 5, e3811.

Andreeva, A., Howorth, D., Chothia, C., Kulesha, E., Murzin, A.G., 2013. SCOP2 prototype: A new approach to protein structure mining. Nucleic Acids Research 42 (D1), D310–D314.

Angles, R., Gutierrez, C., 2008. Survey of graph database models. ACM Computing Surveys 40 (1), 1–39. Available at: http://doi.org/10.1145/1322432.1322433.

Ashish, N., Ambite, J.L., Muslea, M., Turner, J.A., 2010. Neuroscience data integration through mediation: An (F) BIRN case study. Frontiers in Neuroinformatics 4, 118.

Barrett, T., Troup, D.B., Wilhite, S.E., *et al.*, 2006. NCBI GEO: Mining tens of millions of expression profiles – database and tools update. Nucleic Acids Research 35 (suppl_1), D760–D765.

Benson, D.A., Karsch-Mizrachi, I., Lipman, D.J., Ostell, J., Wheeler, D.L., 2008. GenBank. Nucleic Acids Research 36 (Database issue), D25.

Blankenberg, D., Taylor, J., Schenck, I., *et al.*, 2007. A framework for collaborative analysis of ENCODE data: Making large-scale analyses biologist-friendly. Genome Research 17 (6), 960–964. doi:10.1101/gr.5578007.

Brazma, A., Parkinson, H., Sarkans, U., *et al.*, 2003. ArrayExpress – A public repository for microarray gene expression data at the EBI. Nucleic Acids Research 31 (1), 68–71.

Burley, S.K., Berman, H.M., Kleywegt, G.J., *et al.*, 2017. Protein Data Bank (PDB): The single global macromolecular structure archive. Protein Crystallography: Methods and Protocols. 627–641.

Cancelo, N., 2014. Not Only SQL as a Alternative to Relational Database Systems. 2.

Caspi, R., Foerster, H., Fulcher, C.A., *et al.*, 2007. The MetaCyc database of metabolic pathways and enzymes and the BioCyc collection of pathway/genome databases. Nucleic Acids Research 36 (Suppl_1), D623–D631.

Cerami, E.G., Gross, B.E., Demir, E., *et al.*, 2010. Pathway commons, a web resource for biological pathway data. Nucleic acids research 39 (Suppl_1), D685–D690.

Chatr-Aryamontri, A., Breitkreutz, B.J., Oughtred, R., *et al.*, 2014. The BioGRID interaction database: 2015 update. Nucleic Acids Research 43 (D1), D470–D478.

Coordinators, N.R., 2016. Database resources of the national center for biotechnology information. Nucleic Acids Research 44 (Database issue), D7.

Dreos, R., Ambrosini, G., Groux, R., Cavin Périer, R., Bucher, P., 2016. The eukaryotic promoter database in its 30th year: Focus on non-vertebrate organisms. Nucleic Acids Research 45 (D1), D51–D55.

Elmasri, R., Navathe, S., 2011. Fundamentals of Database Systems. Addison-Wesley.

Ercan, M.Z., 2014. Evaluation of NoSQL databases for EHR systems. In: Proceedings of the *25th* Australasian Conference on Information Systems, pp. 1–10. Auckland, New Zealand.

Ferreira, G.D.S., Calil, A., Mello, R.D.S., 2013. On Providing Ddl Support for a Relational Layer over a Document Nosql Database, in: Proceedings of International Conference on Information Integration and Web-based Applications & Services. Vienna, Austria: ACM, 125-132.

Finn, R.D., Attwood, T.K., Babbitt, P.C., 2017a. InterPro in 2017—beyond protein family and domain annotations. Nucleic Acids Research 45, D190–D199. Available at: https://www.ncbi.nlm.nih.gov/pmc/articles/PMC5210578/ (accessed 14.02.18).

Finn, R.D., Coggill, P., Eberhardt, R.Y., 2017b. The Pfam protein families database: Towards a more sustainable future. Nucleic Acids Research. Available at: http://pfam.xfam.org/ (accessed 14.02.18).

Gardner, D., Akil, H., Ascoli, G.A., *et al.*, 2008. The neuroscience information framework: A data and knowledge environment or neuroscience. Neuroinformatics 6 (3), 49–160.

Frishkoff, G., Sydes, J., Mueller, K., *et al.*, 2011. Minimal Information for Neural Electromagnetic Ontologies (MINEMO): A standards-compliant method for analysis and integration of event-related potentials (ERP) data. Standards in Genomic Sciences 5 (2), 211.

Goh, C.S., Lan, N., Echols, N., *et al.*, 2003. SPINE 2: A system for collaborative structural proteomics within a federated database framework. Nucleic Acids Research 31 (11), 2833–2838.

Guarino, N., 1995. Formal ontology, conceptual analysis and knowledge representation. International Journal of Human-Computer Studies 43 (5-6), 625–640.

Hallin, Peter F., Ussery, David W., 2004. CBS Genome Atlas Database: A dynamic storage for bioinformatic results and sequence data. Bioinformatics 20 (18), 3682–3686. Available at: https://doi.org/10.1093/bioinformatics/bth423.

Haider, S., Ballester, B., Smedley, D., *et al.*, 2009. BioMart Central Portal—unified access to biological data. Nucleic Acids Research 37 (suppl_2), W23–W27.

Han, J., Kamber, M., Pei, J., 2000. Data mining: Concepts and Techniques (the Morgan Kaufmann Series in data management systems). Morgan Kaufmann.

Han, J., Haihong, E., Le, G., Du, J., 2011. Survey on NoSQL database. In Pervasive computing and applications (ICPCA), 2011 6th international conference on, IEEE, pp. 363-366.

Heath, T., Bizer, C., 2011. Linked Data: Evolving the Web Into a Global Data Space. Morgan & Claypool Publishers.

Huerta-Cepas, J., Capella-Gutierrez, S., Pryszcz, L.P., *et al.*, 2010. PhylomeDB v3. 0: An expanding repository of genome-wide collections of trees, alignments and phylogeny-based orthology and paralogy predictions. Nucleic Acids Research 39 (suppl_1), D556–D560.

Mashima, J., Kodama, Y., Fujisawa, T., *et al.*, 2017. DNA Data Bank of Japan. Nucleic Acids Research 45 (D1), D25–D31 https://doi.org/10.1093/nar/gkw1001.

Kanehisa, M., Goto, S., 2000. KEGG: Kyoto encyclopedia of genes and genomes. Nucleic Acids Research 28 (1), 27–30.

Kanz, C., Aldebert, P., Althorpe, N., *et al.*, 2005. The EMBL nucleotide sequence database. Nucleic Acids Research 33 (suppl_1), D29–D33.

Kormeier, B., 2014. Data Warehouses in Bioinformatics. In: Chen, M., Hofestädt, R. (Eds.), Approaches in Integrative Bioinformatics. Berlin, Heidelberg: Springer.

Konishetty, V.K., Kumar, K.A., Voruganti, K., Rao, G.V., 2012. August. Implementation and evaluation of scalable data structure over HBase. In Proceedings of the International Conference on Advances in Computing, Communications and Informatics, ACM, 1010-1018.

Mallach, E.G., 2000. Decision Support and Data Warehouse Systems. Irwin/McGraw-Hill: Pennsylvania State University.

MongoDB, 2018. MongoDB. Retrieved from NoSQL Databases Explained: https://www.mongodb.com/nosql-explained.

Muilu, J., Peltonen, L., Litton, J.-E., 2007. The federated database – A basis for biobankbasedpost-Genome studies, integrating phenome and genome data from 600 000 twin pairs in Europe. European Journal of Human Genetics. 1–6.

Neo4j, 2018. Neo4j. Retrieved from Neo4j: https://neo4j.com/developer/graph-database/.

Paramasivam, V., Yee, T.S., Dhillon, S.K., Sidhu, A.S., 2014. A methodological review of data mining techniques in predictive medicine: An application in hemodynamic prediction for abdominal aortic aneurysm disease. Biocybernetics and Biomedical Engineering 34 (3), 139–145.

Park, Y.M., Squizzato, S., Buso, N., Gur, T., Lopez, R., 2017. The EBI search engine: EBI search as a service – making biological data accessible for all. Nucleic Acids Research 2, 2017 https://doi.org/10.1093/nar/gkx359.

Pendse, N., 2008. What is OLAP? An Analysis of What the Often Misused OLAP Term is Supposed to Mean. Retrieved February 8, 2013, from www.olapreport.com/fasmi.htm.

Prasad, T.S.K., 2009. Human Protein Reference Database - 2009 Update. Nucleic Acids Research 37, D767–D772. [PubMed].

Pruitt, K.D., Tatusova, T., Maglott, D.R., 2006. NCBI reference sequences (RefSeq): A curated non-redundant sequence database of genomes, transcripts and proteins. Nucleic Acids Research 35 (suppl_1), D61–D65.

Prosite introduction, 2017. PROSITE (online). Available at: https://prosite.expasy.org/ (accessed 14.02.18).

Ranganathan, S., Tammi, M., Gribskov, M., Tan, T.W., 2006. Establishing bioinformatics research in the Asia Pacific. BMC Bioinformatics 7 (Suppl 5), S1. Available at: http://doi.org/10.1186/1471-2105-7-S5-S1.

Saltz, J., Oster, S., Hastings, S., *et al.*, 2006. caGrid: Design and implementation of the core architecture of the cancer biomedical informatics grid. Bioinformatics 22 (15), 1910–1916.

Safran, M., Dalah, I., Alexander, J., *et al.*, 2010. GeneCards Version 3: The human gene integrator. Database 2010.

Sanderson, M.J., Donoghue, M.J., Piel, W.H., Eriksson, T., 1994. TreeBASE: A prototype database of phylogenetic analyses and an interactive tool for browsing the phylogeny of life. American Journal of Botany 81 (6), 183.

Schomburg, I., Chang, A., Hofmann, O., *et al.*, 2002. BRENDA: A resource for enzyme data and metabolic information. Trends Biochemical Sciences 27 (1), 54–56.

Segaran, T., Hammerbacher, J., 2009a. Beautiful data, first ed. Canada: O'Reilly, p. 112.

Segaran, T., Hammerbacher, J., 2009b. Beautiful data, first ed. Canada: O'Reilly, p. 184.

Severin, J., Waterhouse, A.M., Kawaji, H., *et al.*, 2009. FANTOM4 EdgeExpressDB: An integrated database of promoters, genes, microRNAs, expression dynamics and regulatory interactions. Genome Biology 10 (4), R39.

Sherlock, G., Hernandez-Boussard, T., Kasarskis, A., *et al.*, 2001. The stanford microarray database. Nucleic Acids Research 29 (1), 152–155.

Sillitoe, I., Lewis, T.E., Cuff, A., *et al.*, 2014. CATH: Comprehensive structural and functional annotations for genome sequences. Nucleic Acids Research 43 (D1), D376–D381.

Stelzer, G., Dalah, I., Stein, T.I., *et al.*, 2011. In-silico human genomics with GeneCards. Human Genomics 5 (6), 709.

Stephens, Z.D., Lee, S.Y., Faghri, F., *et al.*, 2015. Big data: Astronomical or genomical? PLOS Biology 13 (7), e1002195. Available at: https://doi.org/10.1371/journal.pbio.1002195.

Stonebraker, M., 2009. SQL Databases vs NoSQL Databases. Retrieved November 15, 2016, from: http://cacm.acm.org/blogs/blog-cacm/50678-the-nosql-discussion-has-nothing-to-do-with-sql/fulltext.

Sulakhe, D., Xie, B., Taylor, A., et al., 2016. Lynx: A knowledge base and an analytical workbench for integrative medicine. Nucleic Acids Research 44 (Database issue), D882–D887. Available at: http://doi.org/10.1093/nar/gkv1257.

Szklarczyk, D., Franceschini, A., Kuhn, M., et al., 2010. The STRING database in 2011: Functional interaction networks of proteins, globally integrated and scored. Nucleic Acids Research 39 (suppl_1), D561–D568.

Szklarczyk, D., Franceschini, A., Wyder, S., et al., 2014. STRING v10: Protein–protein interaction networks, integrated over the tree of life. Nucleic Acids Research 43 (D1), D447–D452.

The National Genome Atlas, 2017. National Cancer Institute, Available at: https://cancergenome.nih.gov/abouttcga (accessed 26.02.18).

National Cancer Institute, 2017. (online). Available at https://www.cancer.gov/ [14 February 2018].

Thessen, A.E., Parr, C.S., 2014. Knowledge extraction and semantic annotation of text from the encyclopedia of life. PLOS ONE 9 (3), e89550. Available at: https://doi.org/10.1371/journal.pone.0089550.

Thiele, H., Glandorf, J., Hufnagel, P., 2010. Bioinformatics strategies in life sciences: From data processing and data warehousing to biological knowledge extraction. Journal of Integrative Bioinformatics (JIB) 7 (1), 32–40.

Tomczak, K., Czerwińska, P., Wiznerowicz, M., 2015. The Cancer Genome Atlas (TCGA): An immeasurable source of knowledge. Contemporary Oncology 19 (1A), A68–A77. Available at: https://www.ncbi.nlm.nih.gov/pmc/articles/PMC4322527/#CIT0044 (accessed 26.02.18).

Valduriez, P., 2011. Principles of Distributed Data Management in 2020? In: Hameurlain, A., Liddle, S.W., Schewe, K.D., Zhou, X. (Eds.), Database and Expert Systems Applications. DEXA 2011. Lecture Notes in Computer Science, vol 6860. Berlin, Heidelberg: Springer.

Weizmann Institute of Science, 2018. GeneCards (online). Available at https://www.genecards.org/ 14 February 2018.

Wood, P.T., 2012. Query languages for graph databases. ACM SIGMOD Record 41 (1), 50. Available at: http://doi.org/10.1145/2206869.2206879.

Xu, Q., Dunbrack, R.L., 2012. Assignment of protein sequences to existing domain and family classification systems: Pfam and the PDB. Bioinformatics 28 (21), 2763–2772. Available at: https://www.ncbi.nlm.nih.gov/pmc/articles/PMC3476341/ (accessed 14.02.18).

Zou, D., Ma, L., Yu, J., Zhang, Z., 2015. Biological databases for human research. Genomics, Proteomics & Bioinformatics 13 (1), 55–63.

Further Reading

Bioinformatics: Converting Data to Knowledge: Workshop Summary, 2000. Washington (DC): National Academies Press (US), Available from: https://www.ncbi.nlm.nih.gov/books/NBK44936/ (accessed 20.02. 2018).

Relevant Websites

https://www.infrafrontier.eu/resources-and-services/access-emma-mouse-resources
    Access to EMMA mouse resources - Infrafrontier.
https://www.acedb.org/
    AceDB.
https://www.ebi.ac.uk/arrayexpress
    ArrayExpress < EMBL-EBI.
http://tcga.lbl.gov/biosig/tcgadownload.do
    Berkeley Cancer Morphometric Data - TCGA - Lawrence Berkeley.
http://thebiogrid.org
    BioGRID.
http://www.biopax.org/
    BioPAX (on Github).
https://bdsc.indiana.edu/
    Bloomington Drosophila Stock Center.
https://bmcbioinformatics.biomedcentral.com/
    BMC Bioinformatics.
www.brainmap.org
    brainmap.org.
http://cancer.digitalslidearchive.net/
    Cancer Digital Slide Archive.
https://cgwb.nci.nih.gov
    Cancer Genome Workbench.
http://www.catalogueoflife.org/
    Catalogue of Life.
http://www.cathdb.info
    CATH: Protein Structure Classification Database at UCL.
https://bioportal.bioontology.org/ontologies/CCO
    Cell Cycle Ontology - Summary.
https://www.clinicalgenome.org/data-sharing/clinvar/
    ClinVar - ClinGen.
https://academic.oup.com/database
    Database - Oxford Academic.
http://www.ddbj.nig.ac.jp/
    DDBJ Center.
http://www.diseasesdatabase.com/
    Diseases Database Ver 2.0.
http://disease-ontology.org/
    Disease Ontology.
http://www.dnachip.org
    DNAchip.

https://www.brenda-enzymes.org/index.php
    Enzyme Database - BRENDA.
https://www.eol.org
    EOL.org.
http://epd.vital-it.ch
    EPD The Eukaryotic Promoter Database.
http://www.ebi.ac.uk/embl/index.html
    European Nucleotide Archive.
https://prosite.expasy.org/
    ExPASy - PROSITE.
http://flybase.org/
    FlyBase Homepage.
https://confluence.broadinstitute.org/display/GDAC/home
    FIREHOSE.
http://www.fishbase.org/
    FishBase.
.https://bioportal.bioontology.org/ontologies/FISHO
    Fish Ontology - Summary.
https://www.gbif.org
    GBIF.org.
https://www.ncbi.nlm.nih.gov/genbank/
    GenBank Overview.
www.genecards.org
    GeneCards - Human Genes.
http://www.greenphyl.org/cgi-bin/index.cgi
    GreenPhyl v4.
http://www.geneontology.org/
    Gene Ontology Consortium.
https://sourceforge.net/projects/gexkb/
    Gene Expression Knowledge Base.
https://www.ncbi.nlm.nih.gov/gap
    Home - dbGaP - NCBI - NIH.
https://www.ncbi.nlm.nih.gov/geo
    Home - GEO - NCBI - NIH.
https://www.ncbi.nlm.nih.gov/medgen/
    Home - MedGen - NCBI - NIH.
https://www.ncbi.nlm.nih.gov/pubmed/
    Home - PubMed - NCBI - NIH.
https://cancergenome.nih.gov/
    Home - The Cancer Genome Atlas - Cancer Genome - TCGA - NIH.
https://www.hprd.org
    HPRD.org.
https://www.itb.cnr.it/ibd/
    IBDsite - CNR-ITB.
http://www.findmice.org/
    International Mouse Strain Resource.
https://www.itis.gov/
    Integrated Taxonomic Information System.
https://academic.oup.com/journals
    Journals - Oxford Academic.
http://www.genome.jp/kegg/
    KEGG.
http://www.genome.jp/kegg/disease/
    KEGG DISEASE Database - GenomeNet.
http://www.kegg.jp/kegg/kegg2.html
    KEGG - Table of Contents.
http://liverwiki.hupo.org.cn/
    LiverWiki.
http://lynx.ci.uchicago.edu
    Lynx.
https://www.malacards.org
    MalaCards - human disease database.
https://www.nlm.nih.gov/bsd/pmresources.html
    MEDLINE/PubMed Resources Guide.
http://www.informatics.jax.org/
    MGI-Mouse Genome Informatics.
https://phenome.jax.org
    MPD: Mouse Phenome Database.
http://bioit.fleming.gr/mugen/mde.jsp
    mugen database environment.
https://bioportal.bioontology.org/
    NCBO BioPortal.

https://www.ncbi.nlm.nih.gov/pubmedhealth
 NCBI - National Library of Medicine - PubMed Health - NIH.
https://academic.oup.com/nar
 Nucleic Acids Research - Oxford Academic.
https://www.omim.org/
 OMIM.
www.pubbrain.org
 OpenfMRI.
https://pfam.xfam.org/
 Pfam: Home page.
http://phylomedb.org/
 PhylomeDB v4.
http://proteomics.ysu.edu/altsplice/
 Plant Alternative Splicing Database.
http://bioinformatics.ysu.edu/secretomes/plant/index.php
 PlantSecKB.
https://www.ebi.ac.uk/pride/archive/
 PRIDE Archive.
https://www.ncbi.nlm.nih.gov/refseq/
 RefSeq: NCBI Reference Sequence Database - NIH.
http://www.rrrc.us/
 RRRC.
https://www.yeastgenome.org/
 Saccharomyces Genome Database.
http://scop2.mrc-lmb.cam.ac.uk/
 SCOP2.
http://www.sequenceontology.org/
 Sequence Ontology.
https://www.snomed.org/snomed-ct
 Snomed CT - SNOMED International.
http://string.embl.de/
 STRING: functional protein association networks.
https://www.arabidopsis.org/
 TAIR.
http://bioinformatics.mdanderson.org/tcgabatcheffects
 TCGA Batch Effects Tool.
www.tdwg.org
 TDWG.
http://www.autoimmuneregistry.org/
 The Autoimmune Registry.
http://www.cancerimagingarchive.internet
 The Cancer Imaging Archive (TCIA) - A growing archive of medical.
http://ctdbase.org/
 The Comparative Toxicogenomics Database.
https://treebase.org/
 TreeBASE Web.
www.uniprot.org
 UniProt.Org.
http://www.wormbase.org
 WormBase Home.
https://www.wwpdb.org
 wwPDB: Worldwide Protein Data Bank.
http://zebrafish.org
 Zebrafish International Resource Center.
http://www.zf-health.org/zf-models/
 zf-models - zf-health.

Functional Genomics

Shalini Kaushik, Indian Institute of Technology Roorkee, India
Sandeep Kaushik, European Institute of Excellence on Tissue Engineering and Regenerative Medicine, Guimaraes, Portugal and University of Minho, Braga, Portugal
Deepak Sharma, Indian Institute of Technology Roorkee, India

Introduction

Functional genomics is a branch that integrates molecular biology and cell biology studies, and deals with the whole structure, function and regulation of a gene in contrast to the gene-by-gene approach of classical molecular biology technique. It aims to relate the phenotype and genotype on genome level and includes processes such as transcription, translation, protein-protein interaction and epigenetic regulation. This involves comprehensive analysis to understand genes, their functional roles and variable levels of protein expression. The Human Genome Project (HGP) (Collins *et al.*, 2003; Green *et al.*, 2015) is an integral part of functional genomics. It elucidated that the human genome contains 3164.7 million nucleotide bases with the total number of ~20,000 genes. The size of chromosomes is in between 47 and 250 Mb. Chromosome 1 has a maximum number of genes (5078) and the MT has fewest (37). Functional genomics is characterized by the following distinct research areas:

Genetic Interaction Mapping

Gene Interaction

Gene interactions occur when two or more allelic or non-allelic genes of same genotype influence the outcome of particular phenotypic characters. To understand the molecular basis of this complex biological phenomena, there is a need of genetic interaction mapping where the effects on one gene are modified by one or several other genes. Furthermore, systematic pairwise deletion of genes or inhibition of gene expression can be used to identify genes with related function, even if they do not interact physically. Genetic interaction is the set of functional association between genes. One such relationship is epistasis, which is the interaction of non-allelic genes where the effect of one gene is masked by another gene to result either in the suppression of the effect or they both combine to produce a new trait (character). Mapping genetic interactions by simultaneously perturbing pairs of genes, which report on how genes work together, provides a powerful tool for systematically defining gene function and pathways.

The complex etiology of most human diseases that cannot be explained by single nucleotide polymorphism (SNPs) is thought to be caused by combination of multiple genetic variations. By finding the (i) similarities in sequences, gene order and fusion events, (ii) common structural motifs, or (iii) resemblance in gene expression, the role of uncharacterized proteins can be deduced computationally. Furthermore in gene networks, based on the comparison of neighbors, functionality of related genes can be deciphered (Schlitt *et al.*, 2003). The statistical epistasis (deviation from additivity in a mathematical model) enabled the identification of an array of different gene interactions that include aggravating or synergistic interactions (Boucher and Jenna, 2013).

Computational Tools for Gene-Gene Interactions

Till date, almost 100 computational tools are available for epistasis detection (see "Relevant Websites section"). Some of the prominent ones are: (i) BEAM (Bayesian Epistasis Association Mapping), which treats the disease-associated markers and their interactions via a bayesian partitioning model and computes the posterior probability that each marker set is related with the disease by employing Markov chain Monte Carlo (Zhang and Liu, 2007) (ii) TEAM (Tree-Based Epistasis Association Mapping), for fast detection of gene-gene interactions without ignoring any epistatic interaction in genome-wide association study (GWAS) of humans (Zhang *et al.*, 2010a), (iii) BOOST (BOolean Operation-based Screening and Testing), a computationally and statistically powerful tool for the detection of unrevealed gene-gene interactions, which is the basis of many complex diseases (Wan *et al.*, 2010), and (iv) TS-GSIS (Two Stage-Grouped Sure Independence Screening), for studying SNP-SNP interactions with/without marginal effects as well as making direct inference and easy interpretation on the biologically important genes (**Table 1**; Fang *et al.*, 2017).

Single Nucleotide Polymorphism

A single-nucleotide polymorphism (SNP) is a variation in a single nucleotide that occurs at a specific position in the genome. These genetic variations in different individuals are the cause of differences in human genomes and are at the forefront of many disease-gene association studies. Based on SNPs, there are many techniques [for instance, Restriction Fragment Length Polymorphism (RFLP)] used for determining suspected individuals of crime such as DNA fingerprinting. Microsatellites or simple sequence repeats (SSRs) and minisatellites are abundant in whole genome and show polymorphism at a high level. SSRs are

Table 1   Tools implicated in gene interaction mapping, SNP and gene regulation

Research areas	Website	Key feature(s)	CI[a]	Access[b]	URL
Gene interaction mapping	BEAM	Investigates disease-associated markers and their interactions	38.1	D	https://sites.fas.harvard.edu/~junliu/BEAM/
	BOOST	Discovers unknown gene-gene interactions	42.1	D	http://snpboost.sourceforge.net
	TEAM	Identifies genetic markers underlying phenotypic studies	17.1	W,D	http://csbio.unc.edu/epistasis/client-team2.php
	TS-GSIS	Detects SNP-SNP interactions and their whole genome effects	–[c]	D	https://cran.r-project.org/web/packages/TSGSIS/index.html
SNP	kSNP3.0	Identifies SNPs and phylogenetic analysis of genomes	26.4	D	https://sourceforge.net/projects/ksnp/files/
	SP2TFBS	Collects specific annotation of human SNPs	6.0	W,D	https://ccg.vital-it.ch/snp2tfbs/
Gene regulation	AlignACE	Finds conserved sequence elements and identifies motifs in a set of DNA sequences	66.9	W,D	http://arep.med.harvard.edu/mrnadata/mrnasoft.html
	Contra	Identifies TFBSs	6.3	W	http://bioit.dmbr.ugent.be/ConTra/index.php
	CpGProD	Predicts promoters associated with CpG islands in mammalian genomic sequences	12.2	W,D	http://pbil.univ-lyon1.fr/software/cpgprod.html
	DBTSS	Database of TSSs, majorly of human adult and embryonic tissues	15.9	W,D	https://dbtss.hgc.jp/
	JASPAR	Repository of transcription binding factor profiles in eukaryotes	102.0	W	http://jaspar.genereg.net/
	MATCH	Weight matrix-based program used to predict TFBSs	73.9	W	http://www.gene-regulation.com/pub/programs.html#match
	MEME	Portal for discovery and analysis of motifs	380.0	W,D	http://meme-suite.org/tools/meme
	microTSS	Identifies TSSs	9.8	D	https://github.com/geo2mandos/microTSS
	MoPP	Motif discovery tool	1.9	W	http://compbio.iitr.ac.in/reganalyst/mopp/
	MyPatternFinder	Finds matches to a user-defined query motif	1.9	W	http://compbio.iitr.ac.in/reganalyst/mypatternfinder/
	oPOSSUM	Detects over-represented TFBSs in co-expressed genes	28.2	W	http://www.cisreg.ca/oPOSSUM/
	PRODORIC	Organizes information on prokaryotic gene expression and integrate it into regulatory networks	24.4	W,D	http://prodoric.tu-bs.de/
	SCPD	Catalogs TFBSs and TSSs in *S. cerevisiae*	30.0	W	http://rulai.cshl.edu/SCPD/
	TRANSCompel	Provides data on experimenta-proven binding-sites, consensus binding sites and regulated genes	152.5	D	http://gene-regulation.com/pub/databases/transcompel/compelSM.html
	TRANSFAC	Repository of DNA binding sites with their transcription factors	44.1	W	http://genexplain.com/transfac/

[a]CI, citation index (number of citations per year).
[b]W, Web based; D, downloadable; W,D, both Web based and downloadable.
[c]–, no citations.

distributed both in coding and non-coding DNA and are implicated in diverse biological functions. Protein coding SNPs can be further divided into synonymous and non-synonymous (nsSNPs); synonymous SNPs do not change the amino acid sequence while nsSNPs do.

Tools Implicated in Analysis of SNPs

Several computational tools for the analysis of SNPs have been developed (**Table 1**; for a more exhaustive list of tools see Karchin, 2009). The algorithm CHOISS (Choosing Optimal-Interval SNP Set) finds a set of SNP markers based on interval regularity (Lee and Kang, 2004). Subsequently, an informative SNP selection method for unphased genotype data based on multiple linear regression (MLR) was implemented in the software package MLR-tagging (He and Zelikovsky, 2006). A neural network-derived method, SNAP (screening for non-acceptable polymorphisms), has also been established that predicts functional effects of nsSNPs by incorporating evolutionary information (residue conservation within sequence families), aspects of protein structure (secondary structure, solvent accessibility) and other relevant information (Bromberg and Rost, 2007). kSNP3.0 is a program for SNP detection and phylogenetic analysis of genomes without genome alignment or reference genome (Gardner et al., 2015). Recently, a database SP2TFBS was developed that catalogs regulatory SNPs affecting predicted transcription factor binding site (TFBS) affinity (Kumar et al., 2017).

Gene Regulation

Promoter Analysis

There are two main promoters, one TATA-box enriched promoter, a well-defined binding site, and another CpG island promoter. Using the Cap Analysis of Gene Expression (CAGE) approach, it is possible to re-design promoter for differential gene expression. One can tag several thousand transcription start sites and quantitatively analyze promoter usage of different tissues with their differential expression in such sites. The identification and analysis of promoters is essential to understand the transcriptional regulatory networks. Several transcriptional start sites databases have been developed (**Table 1**), amongst which many are dedicated to eukaryotes and mammals for instance, DBTSS (Database of Transcription Start Sites), SCPD (*S. cerevisiae* Promoter Database) and microTSS (hosts miRNA transcription start sites). DBTSS covers the data of the majority of human adult and embryonic tissues, and contains 418 million TSS tag sequences from 28 different tissues/cell cultures (Wakaguri *et al.*, 2008). On the other hand, SCPD catalogs experimentally mapped TFBSs and TSSs in yeast *S. cerevisiae* (Zhu and Zhang, 1999). These databases not only identify and retrieve promoters but also provide information on genes with mapped regulatory regions.

Various Regulatory Regions

Different types of regulatory regions are present in our human genome sequence which is a huge source of data of functional genes (Collado-Vides *et al.*, 2009). Different expression of promoter depends on various factors like transcription factor, enhancer, activator, repressor and they define the regulatory region. In addition, there are conserved sequences in human genome that can be employed in phylogenetic footprinting; it is an area of bioinformatics used to identify TFBSs within a non-coding region of DNA by comparing it to the orthologous sequences in different species (Impey *et al.*, 2004; Lenhard *et al.*, 2003). JASPAR is a database for transcription binding factor profiles and it catalogs matrix-based nucleotide profiles describing the binding preference of transcription factors from multiple species (**Table 1**; Sandelin *et al.*, 2004). In a similar vein, TRANSFAC database maintains data on eukaryotic transcription factors, their experimentally proven binding sites, consensus binding sequences and regulated genes (Wingender *et al.*, 1996). Likewise, TRANSCompel database also contains data on eukaryotic transcription factors experimentally proven to act together in a synergistic or antagonistic manner (Matys *et al.*, 2006). On the other hand, PRODORIC hosts data on environmental stimuli with *trans*-acting transcription factors, *cis*-acting promoter elements and regulon definition (Munch *et al.*, 2003). It includes graphical representations of operon, gene and promoter structures including regulator-binding sites, transactivational start sites and also provides supplemented data on regulatory proteins.

In the vast area of functional genomics, finding the TFBSs on a whole-genome level, is a formidable challenge for gene-regulation studies including, annotation of regulatory sites and interpreting gene regulatory networks. MATCH (Kel *et al.*, 2003), oPOSSUM (Ho Sui *et al.*, 2005) and ConTra (Broos *et al.*, 2011) are some of the notable tools used to find TFBSs.

Pattern Discovery

Pattern identification is another area of bioinformatics that analyzes expression pattern of genes as well as sequence patterns in DNA, RNA or proteins. Additionally, in bioinformatics analysis, feature selection technique (Saeys *et al.*, 2007), also plays an important role in extracting huge data from genomic sequences (Sandelin and Wasserman, 2004).

Bioinformatics Tools Used for Regulation of Genes

The availability of numerous genomic sequences and the huge genome-scale experimental data allows the use of computational techniques to investigate *cis*-acting sequences controlling transcriptional regulation. Various tools have been developed to find new sites for a given transcription factor based on a set of known sites (for instance, MyPatternFinder Sharma *et al.*, 2009) and Fuzznuc (emboss.sourceforge.net/apps/cvs/emboss/apps/fuzznuc.html) as well as to find unknown DNA binding motifs for unspecified transcription factors by searching the regions upstream of the translational start sites of a set of potentially coregulated genes (for instance, MEME Bailey *et al.*, 2009; Tavan *et al.*, 1990) and MoPP (Sharma *et al.*, 2009; **Table 1**). AlignACE is a computational tool that implements Gibbs sampling algorithm for identifying motifs that are present in a set of DNA sequences in *S. cerevisiae* (Hughes *et al.*, 2000).

Microarray Data Analysis

Implications of Microarrays in Functional Genomics

Rapid and systematic study of the expression of many genes can be analyzed by microarray technology. On the solid surface (silicon chip, nylon membrane or glass slide), thousands of tiny spots (oligonucleotides) are immobilized at defined positions. These known oligonucleotides act as probes and permit similarity analysis by binding with corresponding unknown samples via hybridization for the purpose of gene discovery and their function. Thus, microarrays facilitate the finding of absolutely novel and unexpected functional roles of genes (Slonim and Yanai, 2009). Furthermore, it compares the expression pattern of genes in normal and diseased tissues. Identification of SNP in the related genes can be observed by using guide DNA. Such kind of detection helps in mutation analysis as genes may differ from each other by single nucleotide base. In addition, microarrays are also being employed in comparative genome hybridization that permits the detection of altered chromosomal copy number and these alterations can be measured by hybridizing fluorescently labeled DNA from a diseased specimen to the microarray plate.

Applications of Microarrays

DNA microarrays are useful in the identification of various diseases like cancer, in the field of pharmacogenomics, in the discovery of drugs or development of more effective drugs for specific treatment strategies. Comparative study of biochemical composition of the gene product that is responsible for the disease with the proteins formed by the normal genes, helps researchers to synthesize drugs that can minimize the effect of such defective proteins. DNA microarrays help in gene discovery, their expression profiling, their regulation, genetic disease screening, SNP discrimination, microbial detection and typing, and observe variation between transcript levels of genes under different conditions. It also helps in toxicological research by tracking the modifications in the genetic profiles of the cells when exposed to certain toxicants. Additionally, microarray technology is useful in clinical microbiology as it is a potential tool for bacterial recognition and for diagnosis of bloodstream infections caused by certain pathogens. Another application of microarray technique is the determination of antimicrobial drug resistance by detecting the mutations due to drug resistance in the genome of microbes (Miller and Tang, 2009).

Computational Prospective for Microarray Technology

Various bioinformatics tools and programs are available for microarray data analysis and mining (Selvaraj and Natarajan, 2011). Computational analysis of microarray data becomes more important to address the problems of data quality, to handle the large amount of microarray data and standardization of this technology. For this, data mining needs to be carried out which includes clustering and classification. Clustering, as name suggests, is to categorize data into groups of related genes (genes with similar characteristics falls into single group). It is an unsupervised learning approach of classifying data into clusters of related observations and is based on the algorithm of K-means clustering and Self-Organizing Map (SOM) (Tavan et al., 1990). On the other hand, classification is a supervised learning approach which predicts the class of the unknown gene by observing the pre-classified examples and is based on the algorithm such as Artificial Neural Network (ANN), k-Nearest Neighbors (kNN) and Support Vector Machines (SVM). One of the main application of classification is the prediction of disease type. Prominent tools available for the clustering analysis include MAGIC Tools (Heyer et al., 2005) and dChip (Amin et al., 2011) and for classification purpose include LIBSVM (see "Relevant Websites section") (**Table 2**).

High-Throughput GoMiner provides outcomes across multiple microarrays one at a time, for the broad scale of applications such as various drug treatments, time-courses evaluation, collective gene comparison (Zeeberg et al., 2005). It organizes the list of under- and over-expressed genes, particularly those genes that have some biological importance in Gene Ontology. iArray (Simon et al., 2007), EDGE (Leek et al., 2006) and Cyber-T (Kayala and Baldi, 2012) are some of the programs available for gene expression analysis. Such studies are important for finding potential drug targets and biomarkers.

SAGE (Serial Analysis of Gene Expression)

Unlike microarray technology, where hybridization is the principle for analysing expression of genes, SAGE depends on RNA sequencing for global analysis of gene expression in a cell. The main benefit of using this technique over others is that it does not require any prior understanding of transcripts and it provides a clear illustration of both qualitative study by comparing gene expression between samples and quantitative study by identifying novel transcripts. The method involves the isolation of short (9–11 base pairs) oligonucleotides sequence tag from discrete mRNAs. These short distinctive sequence tags (SAGE tags) are unique to each gene, allowed to concatenate for efficient sequencing and analysis of transcripts in a serial fashion. The resulting sequence data are analyzed to identify each gene expressed in cell and the level of gene expression. This information can further be used to test the differences in the level of expression of gene between cells.

SAGE Bioinformatics Packages

The role of SAGE tools is very clear as we need them for the isolation of the unique sequence tags from raw sequence files and their tabulation as well as comparison of SAGE tag abundance. Furthermore, SAGE software helps in matching tags to reference sequences in other databases.

Various bioinformatics methods have been developed for compiling and analysing the SAGE data (**Table 2**; for a detailed list of tools see Tuteja and Tuteja, 2004). In order to identify the SAGE tags, SAGE300 (see "Relevant Websites section") compiles the database of tags extracted from human sequences in Genbank. In contrast, POWER_SAGE software is a useful tool for planning SAGE experiments and it has the capability of handling large number of transcripts/cDNAs and different sample size combinations

Table 2   Tools implicated in microarray, SAGE and mutation analysis as well as other prominent tools used in functional genomics

Research areas	Website	Key feature(s)	CI^a	$Access^b$	URL
Microarray	Cyber-T	Analysis of high-throughput data particularly from microarrays, NGS and mass spectroscopy	19.0	D	http://cybert.microarray.ics.uci.edu/
	dChip	For unsupervised sample clustering and identification of novel cluster and corresponding genes	$-^c$	D	https://sites.google.com/site/dchipsoft/
	EDGE	For differential gene expression analysis	19.6	D	http://www.genomine.org/edge/
	High-Throughput GoMiner	Organizes lists of under- and over-expressed genes for biological interpretation in context to Gene Ontology	22.1	W,D	https://discover.nci.nih.gov/gominer/htgm.jsp
	iArray	Use meta-analysis of multiple microarray datasets	–	D	http://zhoulab.usc.edu/iArrayAnalyzer.htm
	LIBSVM	For classifying groups based on Support Vector Machine	567.4	D	http://www.csie.ntu.edu.tw/~cjlin/libsvmtools/
	MAGIC	Clustering tool to analyze all gene-expression data	3.6	D	http://www.bio.davidson.edu/MAGIC/
SAGE	SAGEmap	Public gene expression data repository	27.0	W	https://www.ncbi.nlm.nih.gov/projects/SAGE/
	SAGEnet	For analysing gene expression patterns	–	W	http://www.sagenet.org/
Mutation	Cas-OFFinder	Algorithm for searching potential off- target sites of cas9 RNA-guided endonuclease	69.3	D	http://www.rgenome.net/cas-offinder/
	COSMIC	Catalogue of somatic mutations in cancer	46.7	W	https://cancer.sanger.ac.uk/cosmic
	HGMD	Collates all known gene lesions responsible for human inherited diseases	364.9	W	http://www.hgmd.cf.ac.uk/ac/index.php
	Mudi	Mutation detecting tool in different organisms from whole genome sequence data	1.3	D	http://naoii.nig.ac.jp/mudi_top.html
	MUFFINN	For cancer gene discovery via network analysis of somatic mutation data	11.4	W,D	http://www.inetbio.org/muffinn/search.php
	PANTHER	Detects amino acid substitutions using Hidden Markov Model algorithm	205.8	W	http://pantherdb.org/
	Polyphen-2	Predicts the adverse effects of a missense mutation on protein stability	164.7	W	http://genetics.bwh.harvard.edu/pph2
	PROVEAN	A tool that predicts non-synonymous variations	91.5	W	http://provean.jcvi.org/index.php
	SIFT	Predicts the occurrence of non-synonymous polymorphisms or laboratory-induced missense mutations	207.9	W	http://sift.jcvi.org/
	SNAP2	For identifying functional effects of mutation using neural network algorithm	49.6	W	https://www.rostlab.org/services/snap/
General tools	COMPLEAT	Analysing high throughput data, particularly for analysis of network dynamics	42.1	D	https://fgr.hms.harvard.edu/compleat
	DAVID	For functional classification and annotation of genes	450.3	W,D	https://david.ncifcrf.gov/
	GSEA	For identifying over-expressed genes/proteins among large set of data	22.0	D	http://software.broadinstitute.org/gsea/index.jsp

[a]CI, citation index (number of citations per year).
[b]W, Web based; D, downloadable; W,D, both Web based and downloadable.
[c]–, no citations.

(Man *et al.*, 2000). Expression Profile Viewer (ExProView) is a tool for the analysis of gene expression profiles derived from expressed sequence tags and SAGE (Larsson *et al.*, 2000). It visualizes the transcript data in a two-dimensional array of dots, in which each dot represents a known gene as specified in the transcript database. In addition, SAGEnet (see "Relevant Websites section") catalogs tags for colon cancer, pancreatic cancer and corresponding normal tissues while SAGEmap is a public database for gene expression (Lash *et al.*, 2000).

Mutations

A mutation is an alteration in the primary sequence of a gene. Mutations may lead to an alteration in the structural and/or functional properties of proteins and result in some of the deadly diseases. Mutations can be categorized into two types (i) spontaneous and (ii) induced mutation. The spontaneous mutations involve deamination, depurination, tautomerism and slipped strand mispairing (Drake *et al.*, 1998). On the other hand, induced mutations occur due to the effects induced by chemicals such as hydroxylamine, oxidative radicals, DNA intercalating agents (e.g., ethidium bromide), alkylating agents, base analogs and nitrous acid. Some other physical factors that have been known to induce mutations include ultraviolet light, ionizing radiation and DNA crosslinkers. The scale of mutations may be small or large, and may alter function of some of the essential proteins. Mutations may be classified as (i) Insertions (addition of DNA bases); (ii) Deletions (deletion of DNA bases); (iii) Duplications (a DNA segment is copied one or more times); (iv) Frameshift mutations (addition/loss of bases that shifts the reading frame of the codons); (v) Nonsense mutations (a sense codon changes to a premature stop codon that results in shorter, unfinished proteins); and (vi) Missense mutations (nucleotide substitutions that result in a different codon which codes for different amino acid). Even small mutations could lead to a defective phenotype. However, such unexpected consequences due to mutations are normally kept in check by an error-prone repair system of our highly developed replication system.

Mutation studies have been applied in pharmacological investigations for carrying out translational work in functional genomics. Mutations in a specific gene are detected using TILLING and AMES tests. With the advent of advanced bioinformatics techniques, it has been possible to analyze the mutations in various diseases like colon cancer and thus indicating a positive role of microsatellites in mutagenesis (Brinkmann et al., 1998). Hereditary nonpolyposis colorectal cancer (HNPCC) is thought to increase the risk of developing colon cancer. Bioinformatics approaches have been employed to investigate the role of nsSNPs in HNPCC genes, by identifying functional SNPs and by exploring phenotypic variation due to genetic mutations (Doss and Sethumadhavan, 2009). These bioinformatic investigations have led to the study of various HNPCC related genes, pathways, diseases and post-translational modifications.

Bioinformatics in Mutational Studies

In recent years, bioinformatics studies have provided a new direction in mutational studies. Monte Carlo simulation methods deal with stochastic models in biology that are interactable mathematically. These methods have been used in the incorporation of mutation and natural selection into Wright-Fisher gametic sampling model (Mode and Gallop, 2008). It helps to predict an infinite model of various mutations, gene conversions, migration among subpopulations and allows recombination. In addition, Human Gene Mutation Database (HGMD) is a collection of germline mutations associated with human inherited diseases that contains 141,000 different lesions detected in over 5700 different genes, with new mutation entries (**Table 2**; Stenson et al., 2014).

In vitro mutational study is a difficult and time taking process. However, the structural and functional study of mutational genes are possible through computational tools. For instance, there are various software tools available for driver mutation prioritization (see "Relevant Websites section"). The most notable tools among them are mentioned in this section (**Table 2**). The effects of missense mutations on the structure of a protein product can be analyzed even if its 3D (3 dimensional) structure is not available. The 3D structure of a protein can be deduced by using the primary sequence through homology modeling or other approaches like threading. The analysis of the effects of mutations on the structural stability of a protein can also be carried out. Investigations of steric, anomeric carbon, chiral center and stereochemical consequences of substitutions, provide insights on the molecular fit using structural studies. ProDrg (van Aalten et al., 1996) and PyMOL (Lill and Danielson, 2011) are tools used to observe the effects of mutations on the 3D structure and also assist in carrying out computer-aided drug designing. The effects of missense mutations have been analyzed on the structure and function of p53, a tumor suppressor gene (Kato et al., 2003). CRISPR (Type II Clustered Regularly Interspaced Short Palindromic Repeats)/Cas system helps in adaptive immune response by protecting the host cells against pathogens. Cas-OFFinder algorithm has been developed to search for potential off-target sites in a given genome or user-defined sequences (Bae et al., 2014).

Over 4 million coding mutations have been identified across all human cancer disease types. COSMIC (Catalogue Of Somatic Mutations In Cancer database) describes 10 million non-coding mutations, 1 million copy-number aberrations, 9 million gene-expression variants, and almost 8 million differentially methylated CpGs (Bamford et al., 2004). Mutated protein sequence analysis can also lead to evolutionary relationships. Evolutionary trees are constructed for prediction of orthologs, xenologs and paralogs using Hidden Markov Model (HMM) algorithm. PANTHER (Protein ANalysis THrough Evolutionary Relationships) provides a single nucleotide polymorphism (SNP) scoring tool that uses multiple sequence alignments to identify amino acid substitutions (Mi et al., 2016). Similarly, SIFT (Sorting Intolerant From Tolerant) predicts the occurrence of non-synonymous polymorphisms or laboratory-induced missense mutations based on PSI-BLAST (Ng and Henikoff, 2003). On the other hand, Mudi identifies the causative mutations in the whole-genome sequence data and performs mapping, annotation, prioritization and detection of variant alleles (Iida et al., 2014).

Cancer is caused by DNA sequence abnormalities. A major challenge in cancer is to detect a defected gene via somatic mutation profiling. The Cancer Genome ATLAS [TCGA] (Cancer Genome Atlas Research et al., 2013) and the International Cancer Corsortium [ICGC] (Zhang et al., 2011) are two comprehensive and coordinated projects that deal with systematic profiling of genome alterations in many cancer types. MUFFINN (Mutations for Functional Impact on Network Neighbors) introduced an intuitive, frequency-based approach to quantify the significance of the mutation frequency of each gene or a genomic region compared with a background mutation rate [BMR] (Cho et al., 2016). MUFFINN prioritizes genes based on four ways: (i) the mutation frequency of each gene as well as that of its neighbors in a functional network; (ii) mutation frequency of direct neighbor genes of maximum mutation frequency; (iii) mutation frequency of all direct neighbors with normalization by the number of their network neighbors; and (iv) mutational frequency of all genes of the entire network with diffused weight through the network. On the other hand, Polyphen-2 is a tool implemented in R language to predict the adverse effects of a missense mutation on the protein stability. It uses eight sequence-based and three structure-based predictive features, which are selected automatically by an iterative greedy algorithm (Adzhubei *et al.*, 2013). The prediction pipeline is characterized by comparing two properties like wild-type allele (normal) corresponding to mutant allele (disease). Likewise, PROVEAN webserver also predicts functional effects of amino acid substitution using alignment-based approach (Choi and Chan, 2015).

miRNAs/Transposons

Genomics, Biogenesis and Mechanism of Regulatory RNA

MicroRNAs (miRNAs) are approximately 22 nucleotides long double stranded endogenous non-coding RNAs. miRNAs play an important role in the suppression of target mRNAs (Krol *et al.*, 2010). For instance, Lin-4 is the predominant regulatory RNA that controls the larval development in *C. elegans*. It is now recognized as the founding member of a large class of miRNAs (Bracht *et al.*, 2010). Two RNase III enzymes, named as DORSHA and DICER, cleave primary mRNA and hence, release a hairpin loop in the nucleus (Han *et al.*, 2004). This mRNA is then exported to cytoplasm and cleaved by DICER that results in mature miRNA duplex. The resulted mature miRNA duplex again gets processed by DICER protein and finally two strands act as guide RNA and passenger RNA. The function of guide RNA strand is to cleave the target mRNA specifically, in order to repress its function. On the other hand, passenger mRNA strand acts as an helper for guide RNA to specifically bind to the target mRNA. The pathway of short-interfering RNA (siRNA) or silencing RNA is similar to miRNA, except that the maturation of siRNA is in the cytosol rather than in the nucleus (Bartel, 2004; He and Hannon, 2004).

Computational Approaches for Studying miRNAs

Experimentally, the functional study of miRNA is a time-consuming process. Hence, various computational approaches have been developed for the study of orthologous and paralogous gene identification of miRNAs. Furthermore, several bioinformatics tools have been developed to manage the exponential growth in the generation of miRNA-related data. The main goals of currently available tools are (i) miRNA expression analysis, (ii) miRNA target prediction, (iii) miRNA detection, (iv) studying miRNA in the

Table 3 Various resources and tools for miRNAs/Transposons

Website	Key feature(s)	CI[a]	Access[b]	URL
BUFET	Boosting the unbiased miRNA functional enrichment analysis using bitsets	–[c]	D	https://github.com/diwis/BUFET/
DASHR	Database of small non-coding RNAs in humans	8.0	W,D	http://lisanwanglab.org/DASHR/smdb.php
deepBase	Identification, expression, evolution and function of small non-coding RNAs	25.7	D	http://rna.sysu.edu.cn/deepBase/
MatureBayes	Algorithm for identifying the mature miRNA within novel precursors	8.2	D	http://mirna.imbb.forth.gr/MatureBayes.html
microRNA.org	For miRNA target prediction as well as hosts expression profiles	187.7	W	www.microrna.org/
miRBase	Repository of miRNA sequences, targets and gene nomenclature	323.2	W	http://www.mirbase.org/
mirDB	For miRNA target prediction and functional annotation	112.7	W,D	http://mirdb.wustl.edu/
miRmine	Database of human miRNA expression profiles	19.2	W	http://guanlab.ccmb.med.umich.edu/mirmine/
miR-PREFeR	Prediction of plant miRNAs using small RNA-Seq data	9.0	D	https://github.com/hangelwen/miR-PREFeR
miRsig	For identification of pan-cancer miRNA-miRNA interaction signatures	3.9	W	http://bnet.egr.vcu.edu/miRsig/
Repbase	Database for repetitive DNA elements	85.0	W,D	http://www.girinst.org/repbase/
RetroSeq	For transposable element discovery from NGS data	16.3	W	https://github.com/tk2/RetroSeq
siDirect	Target-specific siRNA design software that avoids off-target silencing	11.1	W,D	http://sidirect2.rnai.jp/
SRF	Identification of all types repetitive sequences, including transposons	10.2	W	http://compbio.iitr.ac.in/srf/
Target-align	Tool for miRNA target identification	10.1	W	http://www.leonxie.com/targetAlign.php
tasiRNAdb	Resource for known ta-siRNA regulatory pathways in plants	5.2	W	http://bioinfo.jit.edu.cn/tasiRNADatabase/
YM500	RNA sequencing database for small RNAs in human and mice	8.2	W	http://driverdb.tms.cmu.edu.tw/ym500v3/index.php

[a]CI, citation index (number of citations per year).
[b]W, Web based; D, downloadable; W,D, both Web based and downloadable.
[c]–, no citations.

context of metabolic and signaling pathways, (v) analysis of interplay between miRNA and transcription factors, (vi) study miRNA in therapeutics, and (vii) identification of miRNA regulatory networks (Akhtar *et al.*, 2016). Here, we discuss some of the major bioinformatics resources and tools that deal with these aspects of miRNA research. High throughput sequencing of mRNA is possible through RNA sequencing. On the basis of small-RNAseq data and expression values, several databases such as miRbase, miRDB, microRNA.org, YM500, DASHR and deepBase have been developed (**Table 3**). These databases help to predict the functional annotations, target predictions and expression profiles of various miRNAs.

miRBase is an online repository of miRNA sequences and annotations (Griffiths-Jones *et al.*, 2008). It has more than 28,645 entries representing hairpin miRNAs that expressed 35,828 mature miRNA products in 223 different species (Moore *et al.*, 2015). The sequence data, nomenclature, annotations and target predictions have been mentioned systematically. In a similar vein, miRDB is a resource for miRNA target predictions and functional annotations, which are known to play a crucial role in understanding various physiological and disease processes (Wong and Wang, 2015). Till date, it has 2588 and 1912 functional miRNAs from humans and mice, respectively. For better prediction of targets, Support Vector Machine (SVM) has been incorporated as MirTarget algorithm in this resource. Likewise, microRNA.org predicts candidate targets using miRanda algorithm and hosts expression profiles to provide functional information. The current version of the database has target sites for 1100 human, 717 mouse, 387 rat, 186 *D. melanogaster*, and 233 *C. elegans* miRNAs. In comparison, YM500 database has about 11,000 cancer-related smRNA and 10,000 RNA-seq datasets (Cheng *et al.*, 2013). It allows visualization of expression data, enables search for existing evidence of the novel miRNAs, analyzes arm switching phenomenon, performs survival analysis of a specific small non-coding RNA (sncRNA) in different cancer types and finds the isomiRs of the known miRNAs in miRBase. DASHR (Database of small human non-coding RNAs) is another repository for human sncRNA genes, precursor and mature sncRNA annotations, sequence, expression levels (Leung *et al.*, 2016). Currently, it stores 187 small-RNA deep sequencing datasets from 42 tissues and cell types for more than 48,000 sncRNA loci, genes and mature sncRNA products. DASHR allows searches using sncRNA gene name or RNA sequence, location of sncRNA loci in a given interval of genomic coordinates, and visualization of expression of sncRNA genes and mature products.

miRmine is a human miRNA expression database developed using 304 high-quality microRNA sequencing (miRNA-seq) datasets from NCBI-SRA (Panwar *et al.*, 2017). The expression profiles for one or more miRNAs from a specific tissue or cell-line, either normal or with disease information can be retrieved easily. In addition, miRsig (miRNA signature) is a web tool developed for prediction and visualization of the common signatures or core sets of miRNA-miRNA interactions for different diseases (Nalluri *et al.*, 2017). It uses six network inference algorithms, generates scores using each of the algorithms and generates disease-specific miRNA-interaction networks using a consensus-based approach. Finally, it makes the graph intersection analysis on multiple diseases to find core interactions. Amongst siRNA (small interfering RNA) based research studies, siDirect has been developed for computing functional and off-target minimized siRNA design for mammalian RNA interference (RNAi). The server accepts an input sequence and returns siRNA candidates that can be used for systematic functional genomics and therapeutic gene silencing (Ui-Tei *et al.*, 2004). Another distinct tool MatureBayes has been established that uses a machine learning approach for the prediction of miRNA genes products (Gkirtzou *et al.*, 2010). It employs Naive Bayes classifier and uses sequence and secondary structural information of miRNA precursors to determine mature miRNA candidates. It can predict the start position of mature miRNA using information from miRNA precursors in humans and mouse. BUFET (Bitset-based Unbiased miRNA Functional Enrichment Tool) is a recent and novel computational approach that helps to discover potential biological functions that would be affected by the differential expression of a group of miRNAs (Zagganas *et al.*, 2017). It implements efficient data structures that lead to a significant reduction in the execution time of the unbiased enrichment analysis. In addition, this tool exploits parallel architecture of multi-core systems to increase its performance.

Plant miRNAs

The plant miRNAs are non-coding regulatory RNAs that perform important roles in the regulation of plant development and physiological processes like growth, development and stress responses. Plant miRNAs have also played a significant role in the phenotypic evolution, for instance, adaptation of plants to life on land (Taylor *et al.*, 2014). Till date, 20 unique *Arabidopsis* miRNAs have been reported; a few are closely related to each other representing 15 distinct miRNA families. Since some could be derived from multiple genomic loci, the 20 miRNAs could represent more than 40 *Arabidopsis* genes (Bartel, 2004).

In both plants and animals, miRNAs regulate the expression of the target mRNA, often in the context of regulating developmental events. However, in spite of similarities between them, the repression mechanism of target mRNAs varies between plants and animals. Repression of gene expression in animals is mediated by translational attenuation of 3' UTRs (untranslated regions) of target mRNAs that contains multiple miRNA binding sites. Whereas, in almost all plants, miRNAs regulate the gene expression of their target mRNAs by cleaving at a single site within the coding region (Millar and Waterhouse, 2005). Besides their roles in growth, development and maintenance of genome integrity, sncRNAs are also important components in plant stress responses (Khraiwesh *et al.*, 2012). The expression profiles of most miRNAs involved in plant growth and development are significantly altered during stress (Sunkar *et al.*, 2012). miRNAs in different plant species are modulated differently in abiotic and biotic stresses. Furthermore, miRNA398 (miR398) has been proposed to regulate plant stress responses due to oxidative stress, water deficit, salt stress, abscisic acid stress and ultraviolet stress (Zhu *et al.*, 2011).

A critical step in elucidating miRNA function is identifying potential miRNA target. Target-align is one such prediction tool which uses dynamic programming alignment techniques (**Table 3**; Xie and Zhang, 2010). Trans-acting siRNAs (ta-siRNAs) play a significant role in regulating gene networks involved in development, metabolism, DNA methylation and stress responses. TasiRNAdb is a database developed for a comprehensive understanding of ta-siRNA regulatory pathways in plants (Zhang *et al.*, 2014). This database also provides a tool, TasExpAnalysis, for mapping user-submitted RNA and degradome libraries to a ta-siRNAs-producing locus (TAS). In addition, the users can perform TAS cleavage analysis and sRNA phasing analysis within tasiRNAdb. Concurrently, a distinct tool miR-PREFeR (miRNA PREdiction From small RNA-Seq data) has also been developed that uses the expression profiles of miRNAs from small RNA-Seq data samples, for the precise prediction of plant miRNAs (Lei and Sun, 2014).

miRNA in Therapeutics

By regulating gene expression, miRNAs play an important role in various biological processes like cell proliferation, apoptosis, biogenesis, transcription and signal transduction. It is not surprising that the deregulation of miRNA results in many human diseases. This can be illustrated by some examples: (i) miRNA expression pattern in heart disease shows that miR-1, miR-29, miR-30, miR-133, and miR-150 have often been found to be down-regulated while miR-21, miR-125, miR-195, miR-199, and miR-214 are up-regulated with hypertrophy (Shah *et al.*, 2009); (ii) miR-17-92 is overexpressed in lung cancer (Hayashita *et al.*, 2005), and (iii) let-7 has been found to be a suppressor miRNA in tumor cells (Johnson *et al.*, 2005).

The recent advances in application of antisense RNAi (RNA interference) and mRNAs has made it possible to manipulate miRNA regulations and hence show considerable promise as therapeutic agents (Vidal *et al.*, 2005). Decreased level of miR-122 in hepatitis B virus (Jopling *et al.*, 2008), miR-21 upregulation in hepatocellular carcinoma (Xu *et al.*, 2011) and high expression of miR-10b in glioblastoma (Guessous *et al.*, 2013) are few signatures associated with their respective diseases. Dysregulated miRNAs have been observed in cancer therefore miRNAs have also been used as biomarkers for cancer diagnosis. miRNA studies have great potential in terms of drug discoveries due to their involvement in different biological pathways. Recently, clinical trials have been carried out for miRNAs- and siRNAs-targeted therapeutics (Rupaimoole and Slack, 2017). The *in silico* approaches and potential use of bioinformatics to predict miRNAs and their targets has seen a lot of development especially in the context of diseases (Tombol *et al.*, 2009; Banwait and Bastola, 2015).

Transposons in Functional Genomics

Transposable elements or jumping genes are moderately repeated mobile DNA sequences, interspersed throughout the genomes. These elements are of two types, retrotransposons (type I elements) and transposons (type II elements). They regulate the functioning of a gene and the evolution of genome. Transposase enzyme helps in the mobilization of transposons. Generally eukaryotes contain retrotransposons, but plants have transposons in their genome. Retrotransposons follow a copy and paste mechanism *i.e.*, first DNA is copied to a RNA molecule that is reverse transcribed again into a DNA using the reverse transcriptase enzyme. This reverse transcribed DNA sequence is then inserted at new position into the genome. In contrast, transposons follow cut and paste mechanism that do not involve any RNA intermediate. They can bind to target site in both specific or non-specific manner. Retrotransposons are of two types LTRs (long terminal repeats) and Non-LTRs (non-long terminal repeats). The most abundant Non-LTRs present in mammals are of two types, LINEs (long interspersed nuclear elements) and SINEs (short interspersed nuclear elements). Multiple copies of transgenes and transposons are often integrated within the genome and they are considered to be identical as endogenous sequences. Repetitive sequences and their transposition is very important in the context of therapeutic studies of diseases, like cancer. Several bioinformatics approaches have been developed to understand various aspects of these elements (**Table 3**). Repbase is a database for transposable elements (TEs) and other types of repeats from eukaryotic genomes (Bao *et al.*, 2015). The current version of the database contains 38,000 sequences of different families or subfamilies of repeats. In addition, RetroSeq is a tool that can be used for the prediction and genotyping of non-reference transposable element (TE) insertions from whole genome sequencing data (Keane *et al.*, 2013). On the other hand, SRF (Spectral Repeat Finder) is a widely used tool for the detection of all the types/classes of repeats, including transposons, within genomic DNA (Sharma *et al.*, 2004).

Epigenomics

Epigenomics (i.e., genome science of epigenetics) provides the functional context of genome sequences and proposes that organization of genes in a particular order on a chromosome and arrangement of these chromosomes in a cell play an important role in deciphering the function of genes. Epigenetics encompasses heritable changes in the DNA structure without changing the genetic code, i.e., change in phenotype without any change in genotype, resulting in chromatin remodeling and consequently affecting transcriptional potential and expression of genes. This is a natural process that takes place regularly in the cell/s of an organism. Epigenetics plays a crucial role in different cellular events, for instance, cell development and differentiation. Several factors, like environment, age, and disease states, may have disruptive effects on a gene's expression pattern. It has been shown that the epigenetic states such as gene silencing can be induced by processes like DNA methylation or by expression of non-coding RNA (Egger *et al.*, 2004; Schubeler, 2015).

Functional anatomy of genome, provided by epigenomics has disclosed various connections among genes and with the nearby non-genic DNA. One such example is beta-globin cluster of human hemoglobin, in which genes are distinctively expressed throughout the development (Proudfoot *et al.*, 1980). During the development stages of cluster, the genes are arranged in order of their expressions. Another astonishing interaction is found between intra- and inter-chromosomes, mediated by chromatin proteins (van Steensel and Dekker, 2010). In fact, DNA loop structures were also found to be associated with functions that can be illustrated by multiple interleukin genes in Th2 cytokine locus of mouse (Cai *et al.*, 2006). An example of large-scale genomic organization mediated by chromatin is the link between long RNAs, heterochromatin modification and gene activity (Feinberg, 2010). Furthermore, the organization of genome is affected by Large Organized Chromatin K9-modifications (LOCKs) and these large regions provide a vital mechanism for functional genomic organization (Wen *et al.*, 2009).

Epigenetic diseases such as cancer, play an important role in epigenetic pathology. The combination of mutations, structural variations and epigenetic alterations differ between each tumor, making patient-specific diagnosis and treatment strategies necessary (Schweiger *et al.*, 2013). Epigenetic modifications play a critical role in the regulation of transcription, DNA repair and replication. The cause of a disease may be due to the environmentally driven variation in epigenetics, particulary as a substitute for mutational change (Jiang *et al.*, 2004). The risk of a disease might also be due to (i) chromatin remodeling or post transcriptional

Table 4   Tools and databases used in epigenetics

Website	Key feature(s)	CI[a]	Access[b]	URL
4D Genome	Resources of chromatin interaction for structure and function relationship of genome	24.0	W	https://4dgenome.research.chop.edu
Catalogue of Imprinted Genes	Catalogue of imprinted genes and parent-of-origin effects in humans and animals	7.31	W	http://igc.otago.ac.nz/home.html
ChromatinDB	Genome wide histone modifications in *S. cerevisiae*	3.4	W	http://www.chromdb.org/
CpGCluster	A distance-based algorithm for CpG islands detection	15.8	D	https://github.com/bioinfoUGR/cpgcluster
CpGProD	For identifying CpG islands associated with transcription start sites in mammalian genomes	12.2	W	http://pbil.univ-lyon1.fr/software/cpgprod.html
CR Cistrome	A chip-seq database for histone modification and chromatin regulation in human as well as mouse	20.0	W	http://cistrome.org/cr
DBCAT	Database of CpG islands and various analytical tools for comprehensive methylation profiles in cancer	2.7	W	http://dbcat.cgm.ntu.edu.tw/
DeepCpG	Accurate prediction of single-cell DNA methylation states using deep learning	10.0	W,D	http://deepcpg.readthedocs.io/en/latest/
DiseaseMeth	Repository for DNA methylation in human diseases	10.5	W	http://www.biobigdata.com/diseasemeth
EMDN algorithm	For integrative study of DNA methylation and gene expression	1.3	D	https://github.com/william0701/EMDN
EpiFactors	Provides knowledge about various epigenetic regulators, their complex, targets and products	19.0	W	http://epifactors.autosome.ru/
Gene Imprint	Portal for imprinted genomes of various species	–[c]	W	http://www.geneimprint.com/
HHMD	Comprehensive database for human histone modifications	6.8	W	http://bioinfo.hrbmu.edu.cn/hhmd
Histome	Database of human histone variants, different sites of post-translational modifications and histone modifying enzymes	14.1	W	http://www.actrec.gov.in/histome/
Histone DB	Repository of histone protein sequences, classified by histone types and variants	9.0	W	https://www.ncbi.nlm.nih.gov/projects/HistoneDB2.0
MethBank	Whole-genome single-base-resolution methylomes of gametes and early embryos	3.7	W	http://www.dnamethylome.org/
MethBase	Database of annotated methylomes from the public domain	18.3	W	http://smithlabresearch.org/software/methbase/
MethDB	Resource for DNA methylation data	5.4	W	http://www.methdb.de
MethSMRT	Database hosting 6 mA and 4 mC methylations	2.0	W	http://sysbio.sysu.edu.cn/methsmrt/
MethylomeDB	Genome-wide DNA methylation profiles of brain	9.3	W	http://www.neuroepigenomics.org/methylomedb/
NAB	Program to generate three-dimensional molecular structures	0.1	D	http://casegroup.rutgers.edu/casegr-sh-2.2.html
NGSmethDB	Repository for single-base whole-genome methylome maps for the best-assembled eukaryotic genomes	2.0	W	http://bioinfo2.ugr.es/NGSmethDB/gbrowse/
PathoYeastract	Provides information about genomic transcriptional regulation in yeasts	6.8	W	http://pathoyeastract.org/
QSEA	Modeling of genome-wide DNA methylation from sequencing enrichment experiments	2.0	W,D	http://www.bioconductor.org/packages/qsea
WAMIDEX	Database of murine genomic imprinting and differential expression	4.8	W	http://atlas.genetics.kcl.ac.uk/

[a]CI, citation index (number of citations per year).
[b]W, Web based; D, downloadable; W,D, both Web based and downloadable.
[c]–, no citations.

modification of histones (e.g., acetylation, ubiquitinylation or methylation), (ii) binding factors, and (iii) DNA replication, repair and transcription based processes (Dawson and Kouzarides, 2012). The database PathoYeastract (PATHOgenic YEAst Search for Translational Regulators and Consensus Tracking) helps in the prediction of the regulatory associations in pathogenic yeasts *Candida albicans and C. glabrata* (**Table 4**; Monteiro *et al.*, 2017). High throughput sequencing technologies provide the comprehensive maps of nucleosome positioning and chromatin conformation (de Wit and de Laat, 2012). In addition, epigenomics is transforming the search for genetic causes of common human diseases on the basis of epigenome anatomy with relevant possible disease list and a new approach to their search (Feinberg, 2010).

During the cell division, DNA exists in highly condensed thread-like structures called chromosomes. The DNA in a eukaryotic chromosome is traditionally bound by two classes of proteins called histones and non-histone proteins. Histones (namely H1, H2A, H2B, H3, H4) are responsible for the most basic level of chromosome organization termed as nucleosomes. Nucleosome exists in a fully super condensed form with the help of histone proteins. However, the histone proteins can be chemically modified during the process of replication and transcription by acetylation, methylation, phosphorylation, sumoylation, ubiquitylation and ADP ribosylation process. These chemical modifications of histones lead to chromosomal rearrangement and consequently help in the regulation of gene expression. Such alterations in DNA organization do not change DNA sequences but still change the expression pattern of a gene and possibly the phenotype of an organism.

DNA Methylation, Genome Imprinting and Paramutation

In mammals, 70%–80% of the cytosines have to be methylated to form 5-methylcytosine within a gene, thereby altering the gene expression. Occasionally, some genes remain unmethylated, for instance, Toll-like receptors in dendritic cells, monocytes and natural killer cells. Various studies have been carried out to decipher how CpG islands remain methylation-free in a globally methylated genome (Deaton and Bird, 2011). Interestingly, in cancer cells, CpG islands are known to contribute to gene silencing. The method BsRADseq (bisulfite-converted restriction site associated DNA sequencing) has the ability to quantify the level of DNA methylation differentiation across multiple individuals (Trucchi *et al.*, 2016). The epigenetic phenomenon that causes genes to be expressed in a parent-of-origin specific manner in daughter cells is termed as 'genome imprinting'. The expression or suppression of genes in progeny is determined by an epigenetic factor of the parental genome. Indeed, the epigenetic mechanisms that are involved in genomic imprinting and X chromosome inactivation (Barr body) share marked similarities. The imprinted genome of various species are listed in Gene Imprint database (See Relevant Websites Section) (**Table 4**). In comparison, the Catalogue of Imprinted Genes is a collation of genes and phenotypes for which parent-of-origin effects have been reported (Morison *et al.*, 2001). WAMIDEX (Web Atlas of Murine genomic Imprinting and Differential EXpression) is another useful tool that integrates murine imprinted genes with a genome browser that makes microarray data easily accessible in annotation rich genomic context (Schulz *et al.*, 2008).

Sometimes changes in one allele induce the heritable changes in other allele, for example, the color of corn kernels in maize. This phenomenon is called paramutation. Gene expression is regulated by two ways - *cis* or *trans*. In *cis*-interactions, the binding site is on the same molecule of DNA as the gene(s) to be transcribed, while for *trans*-interactions it is on the other DNA molecule or chromosome. In *trans* interactions, paramutation occurs between homologous DNA sequences on different chromosomes that lead to changes in gene expression that are transmitted to next generation through mitosis and meiosis. Paramutation has been observed mainly between homologous DNA sequences present in both allelic and non-allelic positions.

Computational Tools in Epigenetics

The Nucleic acid builder (NAB) is a molecular mechanics program which is used for generation of three-dimensional molecular structures, for instance, in the library in DNA wrapping study (**Table 4**; Cheatham *et al.*, 2001). Flexibility is one of the factors involved in DNA binding to nucleosomes, and the salt concentration and DNA sequence affects this flexibility. The NAB package provides an alternative to molecular dynamics in implicit solvent and can be used to produce a picture of the vibrational motions for small nucleic acids. DNA duplexes are usually long due to which they have different open and closed curves. The NAB library can be used to produce simple closed circles with or without supercoiled form, simple nucleosome core fragment model and duplex winding around any open curve. Furthermore, various tools and databases have been associated with methylation, CpG islands, histones and chromatin structures (**Table 4**).

DNA methylation databases

As discussed above, DNA methylation has an important role in epigenetics. The first database developed was MethDB which provides information about DNA methylation profiles and associated gene expression (Amoreira *et al.*, 2003). The database contains experimentally compiled data of a DNA, where cytosine is converted into 5-methylcytosine through methylation process and then regulates epigenetic process in different genes, cells and tissues. It mainly describes methylation profile, pattern and content. The methylation profile gives positional values of cytosine in a sequence of interests. Subsequently, NGSmethDB and MethBase that are based on DNA methylation in bisulfite sequencing data were developed. NGSmethDB retrieves methylation data derived from next-generation sequencing. Two cytosine methylation contexts (CpG and CAG/CTG) are considered. Through a browser interface, the user can search for methylation states and retrieve methylation values for a set of tissues in a given chromosomal region, or display the methylation of promoters among different tissues (Hackenberg *et al.*, 2011). Whereas, MethBase (a database of annotated methylomes from the public domain) along with MethPipe (a pipeline for both low and high-level methylome analysis) enable researchers to extract interesting features from methylomes and compare them with those identified in public methylomes (Song et al., 2013). MethylomeDB is the first comprehensive source methylation profile of brain that facilitates the comparative epigenomic investigation, as well as analysis of schizophrenia, Alzheimer disease and depression methylomes (Xin et al., 2012). On the other hand, MethBank is another DNA methylome programming database dedicated to storing, browsing and visualizing the genome-wide single-base nucleotide methylomes of gametes and early embryos at different developmental stages in mouse and zebrafish (Zou et al., 2015). In contrast to other methylation databases, MethSMRT is the first database hosting 6-mA and 4-mC methylations (Ye et al., 2017). Additionally, Epigenetic Module based on Differential Networks (EMDN) algorithm was developed for simultaneous analysis of DNA methylation and gene expression data (Ma et al., 2017). Quantitative Sequence Enrichment Analysis (QSEA) is a comprehensive workflow for predicting exact levels of DNA methylation from sequence enrichment experiments (Lienhard et al., 2017). Since aberrant changes in DNA methylation leads to diseases, DiseaseMeth, a repository for DNA methylation in human diseases, was established (Xiong et al., 2017).

CpG island related databases

DBCAT (database of CpG islands and analytical tools) is a database for investigating comprehensive methylation profiles of DNA in human cancers. The database contains information regarding the human genes and CpG islands while the analytical tools comprise of a CpG island finder, a genome query browser and a methylation microarray analytical tool. The analytical tool identifies genes with methylated regions, provides comparison across microarrays and also displays the information about probe sites, intensities, and transcription binding sites. It is the first methylation analytical tool investigating epigenetic regulation related to a human disease (Kuo et al., 2011). CpGProD (CpG island Promoter Detection) is another tool for identifying CpG island associated with transcription start sites in mammalian genome sequences (Ponger and Mouchiroud, 2002). In contrast, CpGcluster is based on the physical distance between neighboring CpGs on the chromosome and is able to predict directly clusters of CpGs. By assigning a p-value to each of these clusters, the most statistically significant ones are predicted as CpG islands (Hackenberg et al., 2006). Recently, machine learning methods have also been incorporated in this field. The neural networks based DeepCpG method has been formulated to predict methylation states in single cells (Angermueller et al., 2017).

Histone related databases

HistoneDB is a resource for the interactive comparative analysis of histone protein sequences and their implications for chromatin function (Draizen et al., 2016). Over the past decade, several enzymes that catalyze and modify histones have been discovered, enabling investigations of their roles in normal cellular processes and adverse pathological conditions. HIstome (The Histone Infobase) is a browsable, manually curated and relational database of 55 human histone proteins, 106 distinct sites of their post-translational modifications and 152 histone-modifying enzymes (Khare et al., 2012). In addition, Human Histone Modification Database (HHMD) focuses on storage of histone modification datasets that can be searched for functional annotations, gene ID, chromosome location and cancer name (Zhang et al., 2010b). EpiFactors, is a curated information-based database to facilitate epigenetic regulators, their complexes, targets and products. EpiFactors contains information on 815 proteins, including 95 histones and protamines, as well as 69 protein complexes that are involved in epigenetic regulation (Medvedeva et al., 2015).

Chromatin related databases

DNA is wrapped around histone proteins to form highly condensed chromatin structure in order to fit into the nucleus of a cell. Post-translational modifications of histone proteins play an important role in gene expression. ChromatinDB is a database for histone modification of Saccharomyces cerevisiae (O'Connor and Wyrick, 2007). On the other hand, 4DGenome is a distinct database that hosts chromatin interaction data from five organisms and covers all experimental/computational technologies for detecting chromatin interactions including 3C, 4C-Seq, 5C, ChIA-PET, Hi-C, Capture-C and IM-PET (Teng et al., 2015). In addition, CR Cistrome is an online resource for the linkage of chromatin regulators with histone modifications and datasets collected from ChIP-sequencing (Wang et al., 2014).

General Tools Implicated in Functional Genomics

The Database for Annotation, Visualization and Integrated Discovery (DAVID) is a widely used tool that allows functional classification of genes, discovers functionally related genes, enables conversion of gene/protein identifiers from one type to another as well as explores gene names in a batch (Huang et al., 2007; Dennis et al., 2003; **Table 2**). On the other hand, GSEA (Gene Set Enrichment Analysis) is Gene Ontology based functional enrichment analysis method that recognizes the over-expressed and statistically significant set of genes or proteins among large set of data (Hung et al., 2012; Subramanian et al., 2005). In addition, a protein Complex Enrichment Analysis Tool (COMPLEAT) has also been developed for analysing high-throughput data, particularly for analysis of network dynamics (Vinayagam et al., 2013).

Conclusion

This article will not only enable researchers to get a comprehensive overview of tools available in the field of functional genomics but will also enable them to select the best tool(s) to carry out their task. Additionally, it will assist the development of new and better resources by highlighting the areas where no (or a few) resources exist.

See also: Cell Modeling and Simulation. Comparative Genomics Analysis. Computational Systems Biology. Exome Sequencing Data Analysis. Functional Enrichment Analysis. Functional Enrichment Analysis Methods. Gene Prioritization Tools. Gene Prioritization Using Semantic Similarity. Gene Regulatory Network Review. Gene-Gene Interactions: An Essential Component to Modeling Complexity for Precision Medicine. Genome Analysis – Identification of Genes Involved in Host-Pathogen Protein-Protein Interaction Networks. Genome Annotation: Perspective From Bacterial Genomes. Genome-Wide Haplotype Association Study. Hidden Markov Models. Integrative Analysis of Multi-Omics Data. Metagenomic Analysis and its Applications. Natural Language Processing Approaches in Bioinformatics. Next Generation Sequence Analysis. Next Generation Sequencing Data Analysis. Ontology in Bioinformatics. Pathway Informatics. Predicting Non-Synonymous Single Nucleotide Variants Pathogenic Effects in Human Diseases. Protein Functional Annotation. Regulation of Gene Expression. Sequence Composition. Single Nucleotide Polymorphism Typing. Standards and Models for Biological Data: FGED and HUPO. Whole Genome Sequencing Analysis

References

Adzhubei, I., Jordan, D.M., Sunyaev, S.R., 2013. Predicting functional effect of human missense mutations using PolyPhen-2. Curr. Protoc. Hum. Genet. 7, 7–20.

Akhtar, M.M., Micolucci, L., Islam, M.S., Olivieri, F., Procopio, A.D., 2016. Bioinformatic tools for microRNA dissection. Nucleic Acids Res. 44, 24–44.

Amin, S.B., Shah, P.K., Yan, A., *et al.*, 2011. The dChip survival analysis module for microarray data. BMC Bioinform. 12, 72.

Amoreira, C., Hindermann, W., Grunau, C., 2003. An improved version of the DNA Methylation database (MethDB). Nucleic Acids Res. 31, 75–77.

Angermueller, C., Lee, H.J., Reik, W., Stegle, O., 2017. DeepCpG: Accurate prediction of single-cell DNA methylation states using deep learning. Genome Biol. 18, 67.

Bae, S., Park, J., Kim, J.S., 2014. Cas-OFFinder: A fast and versatile algorithm that searches for potential off-target sites of Cas9 RNA-guided endonucleases. Bioinformatics 30, 1473–1475.

Bailey, T.L., Boden, M., Buske, F.A., *et al.*, 2009. MEME SUITE: Tools for motif discovery and searching. Nucleic Acids Res. 37, W202–W208.

Bamford, S., Dawson, E., Forbes, S., *et al.*, 2004. The COSMIC (Catalogue of Somatic Mutations in Cancer) database and website. Br. J. Cancer 91, 355–358.

Banwait, J.K., Bastola, D.R., 2015. Contribution of bioinformatics prediction in microRNA-based cancer therapeutics. Adv. Drug Deliv. Rev. 81, 94–103.

Bao, W., Kojima, K.K., Kohany, O., 2015. Repbase Update, a database of repetitive elements in eukaryotic genomes. Mob. DNA 6, 11.

Bartel, D.P., 2004. MicroRNAs: Genomics, biogenesis, mechanism, and function. Cell 116, 281–297.

Boucher, B., Jenna, S., 2013. Genetic interaction networks: Better understand to better predict. Front. Genet. 4, 290.

Bracht, J.R., V.A.N. Wynsberghe, P.M., Mondol, V., Pasquinelli, A.E., 2010. Regulation of lin-4 miRNA expression, organismal growth and development by a conserved RNA binding protein in C. elegans. Dev. Biol. 348, 210–221.

Brinkmann, B., Klintschar, M., Neuhuber, F., Huhne, J., Rolf, B., 1998. Mutation rate in human microsatellites: Influence of the structure and length of the tandem repeat. Am. J. Hum. Genet. 62, 1408–1415.

Bromberg, Y., Rost, B., 2007. SNAP: Predict effect of non-synonymous polymorphisms on function. Nucleic Acids Res. 35, 3823–3835.

Broos, S., Hulpiau, P., Galle, J., *et al.*, 2011. ConTra v2: A tool to identify transcription factor binding sites across species, update 2011. Nucleic Acids Res. 39, W74–W78.

Cai, S., Lee, C.C., Kohwi-Shigematsu, T., 2006. SATB1 packages densely looped, transcriptionally active chromatin for coordinated expression of cytokine genes. Nat. Genet. 38, 1278–1288.

Cancer Genome Atlas Research, N., Weinstein, J.N., Collisson, E.A., *et al.*, 2013. The cancer genome atlas pan-cancer analysis project. Nat. Genet. 45, 1113–1120.

Cheatham 3rd, T.E., Brooks, B.R., Kollman, P.A., 2001. Molecular modeling of nucleic acid structure. Curr. Protoc. Nucleic Acid Chem. 7, 7. 5.

Cheng, W.C., Chung, I.F., Huang, T.S., *et al.*, 2013. YM500: A small RNA sequencing (smRNA-seq) database for microRNA research. Nucleic Acids Res. 41, D285–D294.

Cho, A., Shim, J.E., Kim, E., *et al.*, 2016. MUFFINN: Cancer gene discovery via network analysis of somatic mutation data. Genome Biol. 17, 129.

Choi, Y., Chan, A.P., 2015. PROVEAN web server: A tool to predict the functional effect of amino acid substitutions and indels. Bioinformatics 31, 2745–2747.

Collado-Vides, J., Salgado, H., Morett, E., *et al.*, 2009. Bioinformatics resources for the study of gene regulation in bacteria. J. Bacteriol. 191, 23–31.

Collins, F.S., Morgan, M., Patrinos, A., 2003. The Human Genome Project: Lessons from large-scale biology. Science 300, 286–290.

Dawson, M.A., Kouzarides, T., 2012. Cancer epigenetics: From mechanism to therapy. Cell 150, 12–27.

Deaton, A.M., Bird, A., 2011. CpG islands and the regulation of transcription. Genes Dev. 25, 1010–1022.

Dennis Jr., G., Sherman, B.T., Hosack, D.A., *et al.*, 2003. DAVID: Database for annotation, visualization, and integrated discovery. Genome Biol. 4, P3.

Doss, C.G., Sethumadhavan, R., 2009. Investigation on the role of nsSNPs in HNPCC genes – A bioinformatics approach. J. Biomed. Sci. 16, 42.

Draizen, E.J., Shaytan, A.K., Marino-Ramirez, L., *et al.*, 2016. HistoneDB 2.0: A histone database with variants – An integrated resource to explore histones and their variants. Database (Oxford) 2016.

Drake, J.W., Charlesworth, B., Charlesworth, D., Crow, J.F., 1998. Rates of spontaneous mutation. Genetics 148, 1667–1686.

Egger, G., Liang, G., Aparicio, A., Jones, P.A., 2004. Epigenetics in human disease and prospects for epigenetic therapy. Nature 429, 457–463.

Fang, Y.H., Wang, J.H., Hsiung, C.A., 2017. TSGSIS: A high-dimensional grouped variable selection approach for detection of whole-genome SNP-SNP interactions. Bioinformatics 33, 3595–3602.

Feinberg, A.P., 2010. Epigenomics reveals a functional genome anatomy and a new approach to common disease. Nat. Biotechnol. 28, 1049–1052.

Gardner, S.N., Slezak, T., Hall, B.G., 2015. kSNP3.0: SNP detection and phylogenetic analysis of genomes without genome alignment or reference genome. Bioinformatics 31, 2877–2878.

Gkirtzou, K., Tsamardinos, I., Tsakalides, P., Poirazi, P., 2010. MatureBayes: A probabilistic algorithm for identifying the mature miRNA within novel precursors. PLOS ONE 5, e11843.

Green, E.D., Watson, J.D., Collins, F.S., 2015. Human Genome Project: Twenty-five years of big biology. Nature 526, 29–31.

Griffiths-Jones, S., Saini, H.K., Van Dongen, S., Enright, A.J., 2008. miRBase: Tools for microRNA genomics. Nucleic Acids Res. 36, D154–D158.

Guessous, F., Alvarado-Velez, M., Marcinkiewicz, L., *et al.*, 2013. Oncogenic effects of miR-10b in glioblastoma stem cells. J. Neurooncol. 112, 153–163.

Hackenberg, M., Previti, C., Luque-Escamilla, P.L., *et al.*, 2006. CpGcluster: A distance-based algorithm for CpG-island detection. BMC Bioinform. 7, 446.

Hackenberg, M., Barturen, G., Oliver, J.L., 2011. NGSmethDB: A database for next-generation sequencing single-cytosine-resolution DNA methylation data. Nucleic Acids Res. 39, D75–D79.

Han, J., Lee, Y., Yeom, K.H., et al., 2004. The Drosha-DGCR8 complex in primary microRNA processing. Genes Dev. 18, 3016–3027.

Hayashita, Y., Osada, H., Tatematsu, Y., et al., 2005. A polycistronic microRNA cluster, miR-17-92, is overexpressed in human lung cancers and enhances cell proliferation. Cancer Res. 65, 9628–9632.

He, J., Zelikovsky, A., 2006. MLR-tagging: Informative SNP selection for unphased genotypes based on multiple linear regression. Bioinformatics 22, 2558–2561.

He, L., Hannon, G.J., 2004. MicroRNAs: Small RNAs with a big role in gene regulation. Nat. Rev. Genet. 5, 522–531.

Heyer, L.J., Moskowitz, D.Z., Abele, J.A., et al., 2005. MAGIC tool: Integrated microarray data analysis. Bioinformatics 21, 2114–2115.

Ho Sui, S.J., Mortimer, J.R., Arenillas, D.J., et al., 2005. oPOSSUM: Identification of over-represented transcription factor binding sites in co-expressed genes. Nucleic Acids Res. 33, 3154–3164.

Huang, D.W., Sherman, B.T., Tan, Q., et al., 2007. The DAVID Gene Functional Classification Tool: A novel biological module-centric algorithm to functionally analyze large gene lists. Genome Biol. 8, R183.

Hughes, J.D., Estep, P.W., Tavazoie, S., Church, G.M., 2000. Computational identification of cis-regulatory elements associated with groups of functionally related genes in Saccharomyces cerevisiae. J. Mol. Biol. 296, 1205–1214.

Hung, J.H., Yang, T.H., Hu, Z., Weng, Z., Delisi, C., 2012. Gene set enrichment analysis: Performance evaluation and usage guidelines. Brief. Bioinform. 13, 281–291.

Iida, N., Yamao, F., Nakamura, Y., Iida, T., 2014. Mudi, a web tool for identifying mutations by bioinformatics analysis of whole-genome sequence. Genes Cells 19, 517–527.

Impey, S., Mccorkle, S.R., Cha-Molstad, H., et al., 2004. Defining the CREB regulon: A genome-wide analysis of transcription factor regulatory regions. Cell 119, 1041–1054.

Jiang, Y.H., Bressler, J., Beaudet, A.L., 2004. Epigenetics and human disease. Annu. Rev. Genom. Hum. Genet. 5, 479–510.

Johnson, S.M., Grosshans, H., Shingara, J., et al., 2005. RAS is regulated by the let-7 microRNA family. Cell 120, 635–647.

Jopling, C.L., Schutz, S., Sarnow, P., 2008. Position-dependent function for a tandem microRNA miR-122-binding site located in the hepatitis C virus RNA genome. Cell Host Microbe 4, 77–85.

Karchin, R., 2009. Next generation tools for the annotation of human SNPs. Brief. Bioinform. 10, 35–52.

Kato, S., Han, S.Y., Liu, W., et al., 2003. Understanding the function-structure and function-mutation relationships of p53 tumor suppressor protein by high-resolution missense mutation analysis. Proc. Natl. Acad. Sci. USA 100, 8424–8429.

Kayala, M.A., Baldi, P., 2012. Cyber-T web server: Differential analysis of high-throughput data. Nucleic Acids Res. 40, W553–W559.

Keane, T.M., Wong, K., Adams, D.J., 2013. RetroSeq: Transposable element discovery from next-generation sequencing data. Bioinformatics 29, 389–390.

Kel, A.E., Gossling, E., Reuter, I., et al., 2003. MATCH: A tool for searching transcription factor binding sites in DNA sequences. Nucleic Acids Res. 31, 3576–3579.

Khare, S.P., Habib, F., Sharma, R., et al., 2012. HIstome – A relational knowledgebase of human histone proteins and histone modifying enzymes. Nucleic Acids Res. 40, D337–D342.

Khraiwesh, B., Zhu, J.K., Zhu, J., 2012. Role of miRNAs and siRNAs in biotic and abiotic stress responses of plants. Biochim. Biophys. Acta 1819, 137–148.

Krol, J., Loedige, I., Filipowicz, W., 2010. The widespread regulation of microRNA biogenesis, function and decay. Nat. Rev. Genet. 11, 597–610.

Kumar, S., Ambrosini, G., Bucher, P., 2017. SNP2TFBS – A database of regulatory SNPs affecting predicted transcription factor binding site affinity. Nucleic Acids Res. 45, D139–D144.

Kuo, H.C., Lin, P.Y., Chung, T.C., et al., 2011. DBCAT: Database of CpG islands and analytical tools for identifying comprehensive methylation profiles in cancer cells. J. Comput. Biol. 18, 1013–1017.

Larsson, M., Stahl, S., Uhlen, M., Wennborg, A., 2000. Expression profile viewer (ExProView): A software tool for transcriptome analysis. Genomics 63, 341–353.

Lash, A.E., Tolstoshev, C.M., Wagner, L., et al., 2000. SAGEmap: A public gene expression resource. Genome Res. 10, 1051–1060.

Lee, S., Kang, C., 2004. CHOISS for selection of single nucleotide polymorphism markers on interval regularity. Bioinformatics 20, 581–582.

Leek, J.T., Monsen, E., Dabney, A.R., Storey, J.D., 2006. EDGE: Extraction and analysis of differential gene expression. Bioinformatics 22, 507–508.

Lei, J., Sun, Y., 2014. miR-PREFeR: An accurate, fast and easy-to-use plant miRNA prediction tool using small RNA-Seq data. Bioinformatics 30, 2837–2839.

Lenhard, B., Sandelin, A., Mendoza, L., et al., 2003. Identification of conserved regulatory elements by comparative genome analysis. J. Biol. 2, 13.

Leung, Y.Y., Kuksa, P.P., Amlie-Wolf, A., et al., 2016. DASHR: Database of small human noncoding RNAs. Nucleic Acids Res. 44, D216–D222.

Lienhard, M., Grasse, S., Rolff, J., et al., 2017. QSEA-modelling of genome-wide DNA methylation from sequencing enrichment experiments. Nucleic Acids Res. 45, e44.

Lill, M.A., Danielson, M.L., 2011. Computer-aided drug design platform using PyMOL. J. Comput. Aided Mol. Des. 25, 13–19.

Ma, X., Liu, Z., Zhang, Z., Huang, X., Tang, W., 2017. Multiple network algorithm for epigenetic modules via the integration of genome-wide DNA methylation and gene expression data. BMC Bioinform. 18, 72.

Man, M.Z., Wang, X., Wang, Y., 2000. POWER_SAGE: Comparing statistical tests for SAGE experiments. Bioinformatics 16, 953–959.

Matys, V., Kel-Margoulis, O.V., Fricke, E., et al., 2006. TRANSFAC and its module TRANSCompel: Transcriptional gene regulation in eukaryotes. Nucleic Acids Res. 34, D108–D110.

Medvedeva, Y.A., Lennartsson, A., Ehsani, R., et al., 2015. EpiFactors: A comprehensive database of human epigenetic factors and complexes. Database (Oxford) 2015, bav067.

Mi, H., Poudel, S., Muruganujan, A., Casagrande, J.T., Thomas, P.D., 2016. PANTHER version 10: Expanded protein families and functions, and analysis tools. Nucleic Acids Res. 44, D336–D342.

Millar, A.A., Waterhouse, P.M., 2005. Plant and animal microRNAs: Similarities and differences. Funct. Integr. Genom. 5, 129–135.

Miller, M.B., Tang, Y.W., 2009. Basic concepts of microarrays and potential applications in clinical microbiology. Clin. Microbiol. Rev. 22, 611–633.

Mode, C.J., Gallop, R.J., 2008. A review on Monte Carlo simulation methods as they apply to mutation and selection as formulated in Wright-Fisher models of evolutionary genetics. Math. Biosci. 211, 205–225.

Monteiro, P.T., Pais, P., Costa, C., et al., 2017. The PathoYeastract database: An information system for the analysis of gene and genomic transcription regulation in pathogenic yeasts. Nucleic Acids Res. 45, D597–D603.

Moore, A.C., Winkjer, J.S., Tseng, T.T., 2015. Bioinformatics resources for microRNA discovery. Biomark. Insights 10, 53–58.

Morison, I.M., Paton, C.J., Cleverley, S.D., 2001. The imprinted gene and parent-of-origin effect database. Nucleic Acids Res. 29, 275–276.

Munch, R., Hiller, K., Barg, H., et al., 2003. PRODORIC: Prokaryotic database of gene regulation. Nucleic Acids Res. 31, 266–269.

Nalluri, J.J., Barh, D., Azevedo, V., Ghosh, P., 2017. miRsig: A consensus-based network inference methodology to identify pan-cancer miRNA-miRNA interaction signatures. Sci. Rep. 7, 39684.

Ng, P.C., Henikoff, S., 2003. SIFT: Predicting amino acid changes that affect protein function. Nucleic Acids Res. 31, 3812–3814.

O'Connor, T.R., Wyrick, J.J., 2007. ChromatinDB: A database of genome-wide histone modification patterns for Saccharomyces cerevisiae. Bioinformatics 23, 1828–1830.

Panwar, B., Omenn, G.S., Guan, Y., 2017. miRmine: A database of human miRNA expression profiles. Bioinformatics 33, 1554–1560.

Ponger, L., Mouchiroud, D., 2002. CpGProD: Identifying CpG islands associated with transcription start sites in large genomic mammalian sequences. Bioinformatics 18, 631–633.

Proudfoot, N.J., Shander, M.H., Manley, J.L., Gefter, M.L., Maniatis, T., 1980. Structure and in vitro transcription of human globin genes. Science 209, 1329–1336.

Rupaimoole, R., Slack, F.J., 2017. MicroRNA therapeutics: Towards a new era for the management of cancer and other diseases. Nat. Rev. Drug Discov. 16, 203–222.

Saeys, Y., Inza, I., Larranaga, P., 2007. A review of feature selection techniques in bioinformatics. Bioinformatics 23, 2507–2517.

Sandelin, A., Wasserman, W.W., 2004. Constrained binding site diversity within families of transcription factors enhances pattern discovery bioinformatics. J. Mol. Biol. 338, 207–215.

Sandelin, A., Alkema, W., Engstrom, P., Wasserman, W.W., Lenhard, B., 2004. JASPAR: An open-access database for eukaryotic transcription factor binding profiles. Nucleic Acids Res. 32, D91–D94.

Schlitt, T., Palin, K., Rung, J., et al., 2003. From gene networks to gene function. Genome Res. 13, 2568–2576.

Schubeler, D., 2015. Function and information content of DNA methylation. Nature 517, 321–326.

Schulz, R., Woodfine, K., Menheniott, T.R., et al., 2008. WAMIDEX: A web atlas of murine genomic imprinting and differential expression. Epigenetics 3, 89–96.

Schweiger, M.R., Barmeyer, C., Timmermann, B., 2013. Genomics and epigenomics: New promises of personalized medicine for cancer patients. Brief. Funct. Genom. 12, 411–421.

Selvaraj, S., Natarajan, J., 2011. Microarray data analysis and mining tools. Bioinformation 6, 95–99.

Shah, P.P., Hutchinson, L.E., Kakar, S.S., 2009. Emerging role of microRNAs in diagnosis and treatment of various diseases including ovarian cancer. J. Ovarian Res. 2, 11.

Sharma, D., Mohanty, D., Surolia, A., 2009. RegAnalyst: A web interface for the analysis of regulatory motifs, networks and pathways. Nucleic Acids Res. 37, W193–W201.

Sharma, D., Issac, B., Raghava, G.P., Ramaswamy, R., 2004. Spectral Repeat Finder (SRF): Identification of repetitive sequences using Fourier transformation. Bioinformatics 20, 1405–1412.

Simon, R., Lam, A., Li, M.C., et al., 2007. Analysis of gene expression data using BRB-ArrayTools. Cancer Inform. 3, 11–17.

Slonim, D.K., Yanai, I., 2009. Getting started in gene expression microarray analysis. PLOS Comput. Biol. 5, e1000543.

Song, Q., Decato, B., Hong, E.E., et al., 2013. A reference methylome database and analysis pipeline to facilitate integrative and comparative epigenomics. PLOS ONE 8, e81148.

Stenson, P.D., Mort, M., Ball, E.V., et al., 2014. The Human Gene Mutation Database: Building a comprehensive mutation repository for clinical and molecular genetics, diagnostic testing and personalized genomic medicine. Hum. Genet. 133, 1–9.

Subramanian, A., Tamayo, P., Mootha, V.K., et al., 2005. Gene set enrichment analysis: A knowledge-based approach for interpreting genome-wide expression profiles. Proc. Natl. Acad. Sci. USA 102, 15545–15550.

Sunkar, R., Li, Y.F., Jagadeeswaran, G., 2012. Functions of microRNAs in plant stress responses. Trends Plant Sci. 17, 196–203.

Tavan, P., Grubmuller, H., Kuhnel, H., 1990. Self-organization of associative memory and pattern classification: Recurrent signal processing on topological feature maps. Biol. Cybern. 64, 95–105.

Taylor, R.S., Tarver, J.E., Hiscock, S.J., Donoghue, P.C., 2014. Evolutionary history of plant microRNAs. Trends Plant Sci. 19, 175–182.

Teng, L., He, B., Wang, J., Tan, K., 2015. 4DGenome: A comprehensive database of chromatin interactions. Bioinformatics 31, 2560–2564.

Tombol, Z., Szabo, P.M., Molnar, V., et al., 2009. Integrative molecular bioinformatics study of human adrenocortical tumors: MicroRNA, tissue-specific target prediction, and pathway analysis. Endocr. Relat. Cancer 16, 895–906.

Trucchi, E., Mazzarella, A.B., Gilfillan, G.D., et al., 2016. BsRADseq: Screening DNA methylation in natural populations of non-model species. Mol. Ecol. 25, 1697–1713.

Tuteja, R., Tuteja, N., 2004. Serial analysis of gene expression (SAGE): Unraveling the bioinformatics tools. Bioessays 26, 916–922.

Ui-Tei, K., Naito, Y., Takahashi, F., et al., 2004. Guidelines for the selection of highly effective siRNA sequences for mammalian and chick RNA interference. Nucleic Acids Res. 32, 936–948.

van Aalten, D.M., Bywater, R., Findlay, J.B., et al., 1996. PRODRG, a program for generating molecular topologies and unique molecular descriptors from coordinates of small molecules. J. Comput. Aided Mol. Des. 10, 255–262.

van Steensel, B., Dekker, J., 2010. Genomics tools for unraveling chromosome architecture. Nat. Biotechnol. 28, 1089–1095.

Vidal, L., Blagden, S., Attard, G., De Bono, J., 2005. Making sense of antisense. Eur. J. Cancer 41, 2812–2818.

Vinayagam, A., Hu, Y., Kulkarni, M., et al., 2013. Protein complex-based analysis framework for high-throughput data sets. Sci. Signal. 6, rs5.

Wakaguri, H., Yamashita, R., Suzuki, Y., Sugano, S., Nakai, K., 2008. DBTSS: Database of transcription start sites, progress report 2008. Nucleic Acids Res. 36, D97–D101.

Wang, Q., Huang, J., Sun, H., et al., 2014. CR Cistrome: A ChIP-Seq database for chromatin regulators and histone modification linkages in human and mouse. Nucleic Acids Res 42, D450–8.

Wan, X., Yang, C., Yang, Q., et al., 2010. BOOST: A fast approach to detecting gene-gene interactions in genome-wide case-control studies. Am. J. Hum. Genet. 87, 325–340.

Wen, B., Wu, H., Shinkai, Y., Irizarry, R.A., Feinberg, A.P., 2009. Large histone H3 lysine 9 dimethylated chromatin blocks distinguish differentiated from embryonic stem cells. Nat. Genet. 41, 246–250.

Wingender, E., Dietze, P., Karas, H., Knuppel, R., 1996. TRANSFAC: A database on transcription factors and their DNA binding sites. Nucleic Acids Res. 24, 238–241.

de Wit, E., de Laat, W., 2012. A decade of 3C technologies: Insights into nuclear organization. Genes Dev. 26, 11–24.

Wong, N., Wang, X., 2015. miRDB: An online resource for microRNA target prediction and functional annotations. Nucleic Acids Res. 43, D146–D152.

Xie, F., Zhang, B., 2010. Target-align: A tool for plant microRNA target identification. Bioinformatics 26, 3002–3003.

Xin, Y., Chanrion, B., O'donnell, A.H., et al., 2012. MethylomeDB: A database of DNA methylation profiles of the brain. Nucleic Acids Res. 40, D1245–D1249.

Xiong, Y., Wei, Y., Gu, Y., et al., 2017. DiseaseMeth version 2.0: A major expansion and update of the human disease methylation database. Nucleic Acids Res. 45, D888–D895.

Xu, J., Wu, C., Che, X., et al., 2011. Circulating microRNAs, miR-21, miR-122, and miR-223, in patients with hepatocellular carcinoma or chronic hepatitis. Mol. Carcinog. 50, 136–142.

Ye, P., Luan, Y., Chen, K., et al., 2017. MethSMRT: An integrative database for DNA N6-methyladenine and N4-methylcytosine generated by single-molecular real-time sequencing. Nucleic Acids Res. 45, D85–D89.

Zagganas, K., Vergoulis, T., Paraskevopoulou, M.D., et al., 2017. BUFET: Boosting the unbiased miRNA functional enrichment analysis using bitsets. BMC Bioinform. 18, 399.

Zeeberg, B.R., Qin, H., Narasimhan, S., et al., 2005. High-Throughput GoMiner, an 'industrial-strength' integrative gene ontology tool for interpretation of multiple-microarray experiments, with application to studies of Common Variable Immune Deficiency (CVID). BMC Bioinform. 6, 168.

Zhang, C., Li, G., Zhu, S., Zhang, S., Fang, J., 2014. tasiRNAdb: A database of ta-siRNA regulatory pathways. Bioinformatics 30, 1045–1046.

Zhang, J., Baran, J., Cros, A., et al., 2011. International Cancer Genome Consortium Data Portal – A one-stop shop for cancer genomics data. Database (Oxford) 2011, bar026.

Zhang, X., Huang, S., Zou, F., Wang, W., 2010a. TEAM: Efficient two-locus epistasis tests in human genome-wide association study. Bioinformatics 26, i217–i227.

Zhang, Y., Liu, J.S., 2007. Bayesian inference of epistatic interactions in case-control studies. Nat. Genet. 39, 1167–1173.

Zhang, Y., Lv, J., Liu, H., et al., 2010b. HHMD: The human histone modification database. Nucleic Acids Res. 38, D149–D154.

Zhu, C., Ding, Y., Liu, H., 2011. MiR398 and plant stress responses. Physiol. Plant 143, 1–9.

Zhu, J., Zhang, M.Q., 1999. SCPD: A promoter database of the yeast Saccharomyces cerevisiae. Bioinformatics 15, 607–611.

Zou, D., Sun, S., Li, R., et al., 2015. MethBank: A database integrating next-generation sequencing single-base-resolution DNA methylation programming data. Nucleic Acids Res. 43, D54–D58.

Relevant Websites

http://omictools.com/epistasis-detection-category
Epistasis detection software tools I GWAS analysis
OMICtools.
http://www.geneimprint.com
Geneimprint.
https://www.csie.ntu.edu.tw/ ~ cjlin/libsvm/
LIBSVM.
https://omictools.com/driver-mutations-category
OMICtools.
http://www.sagenet.org
SageNet.

Transcription Factor Databases

Edgar Wingender, geneXplain GmbH, Wolfenbüttel, Germany and University Medical Center Göttingen, Göttingen, Germany
Alexander Kel and Mathias Krull, geneXplain GmbH, Wolfenbüttel, Germany

Background

Transcription, i.e., the synthesis of an RNA strand that is complementary to a piece of genomic DNA of an organism, is an essential step of gene expression. To our present knowledge, it is the most tightly and subtly regulated mechanism of the whole process of making a gene product according to genomic instructions. Transcribing DNA into RNA is done by a small class of enzymes, the DNA-dependent RNA polymerases. To direct polymerase molecules to a specific place in the genome, the transcription start site (TSS), the assistance of additional factors is required, which are called transcription factors (TFs). In the beginning of the research on these factors, the focus was on a set of co-factors of the polymerase that we call today general transcription factors (GTFs). Best studied for bacteria are the sigma factors, for eukaryotic a set of GTFs is known such as TFIIA, TFIIB, TFIID-F, TFIIH. Some of these GTFs bind directly to the DNA with a more or less relaxed sequence-specificity, others are included through protein-protein interactions. They are involved in the assembly of the transcription complex at the core promoter, the region around the TATA-box at ~ -30 bp of the TSS and the TSS itself. A much larger set of DNA-binding TFs that bind to specific regulatory sequences, or cis-regulating elements, have been discovered later. In addition, there are factors involved in transcriptional regulation that do not bind directly to DNA by themselves, but are incorporated into DNA-bound complexes through protein-protein interactions. Depending on the TF definition applied, they may be included or some of them may be split off as separate class of, e.g., co-activators etc. In bacteria, we may have up to 300 TFs (Gama-Castro et al., 2016), while the human genome may encode up to 2000 TFs (Wingender et al., 2013), the exact number depending on the definition.

Aims of TF Databases

A TF database (TFDB) may first of all store information about TFs. In doing so, it should provide additional information rather than just serving as a specialized subsection of a comprehensive protein resource such as UniProt. Therefore, already the first compilation of TFs from 1988 that developed into the database TRANSFAC combined information about (eukaryotic) TFs with specifications of their DNA binding sites (Wingender, 1988). As an increasing amount of transcription factor binding sites (TFBSs) was gathered from the scientific literature, it was possible to complement the TFDB contents by binding sites models, for instance as consensus sequences, as positional weight matrices (PWMs) or as Hidden Markov Models (HMMs). Libraries of such models were compiled and are used now to computationally predict potential TFBSs in yet uncharacterized DNA sequences. Also on the side of the binding proteins, the increase of the number of known TFs and their characterization, in particular the mapping of their DNA-binding domains (DBDs), allowed to come up with corresponding domain models that became part of general protein motif collections like InterPro or of specialized TFBDs (see **Fig. 1** for a simplified scheme). They are used to predict potential TFs in not yet annotated genomes, or to classify TFs according to their DBDs, e.g., in TFClass (Wingender et al., 2013).

For the purpose of predicting corresponding entities (TFBSs or TFs), several of the TFDBs available today are accompanied with pieces of software that support these tasks.

Among the TFDBs that have been developed so far, some have a more comprehensive scope and try to cover the whole field of prokaryotic or eukaryotic TFs, while others are more focussed on certain taxa such as insects or plants. Some will be described in the following, others can be found in **Table 1**. Moreover, the individual TFDBs differ in the way their content has been curated, whether manually or by text mining approaches, or a mix of both. The exact method by which the contents have been gathered is not always clear.

TRANSFAC

The oldest still actively maintained database in the field is the TRANSFAC database. Its contents are basically structured in tables that provide information about the TFs (FACTOR table) or their genomic binding sites (SITE). All information in these tables has been manually extracted from the original publications. Semi-automatically retrieved are, however, data from high-throughput experiments such as ChIPseq data. Large numbers of genomic TF binding fragments can arise from these approaches, millions of them are stored in a separate table of the database (FRAGMENT).

On top of these entities, abstract models of DNA binding specificities as PWMs (MATRIX) or of DBDs (CLASS) have been implemented (**Fig. 1**), the contents of which are semi-manually curated. The entries of the MATRIX table are assigned to TFs of one of four taxonomic groups (vertebrates, insects, plants, and fungi). Many entries have been generated by TRANSFAC curators from information compiled in the SITE table. Other MATRIX entries were published in scientific reports and were included for documentary purposes. This dual purpose made it necessary to set up filters discriminating between those matrices that are suitable for predicting TFBSs and others, which were incorporated for archiving.

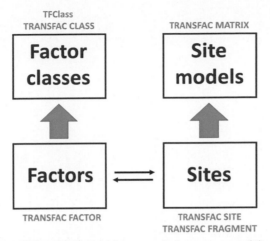

Fig. 1 Simplified scheme of content types of TFDBs. Individual databases usually focus on some of these content areas. As examples, the corresponding tables of the TRANSFAC database and associated resources have been indicated. "Factors" stands for information about transcription factors (TFs), "Sites" for TF binding sites (TFBS), either conventionally identified or obtained from high-throughput experiments such as ChIPseq, both pointing to each other. On top of these entities, abstract models of DNA-binding domains of TFs and of TFBSs are frequently provided for both descriptive and predictive purposes.

An accompanying resource is the database TRANSCompel, the concept of which has emerged from the COMPEL database on composite elements (CEs) (Kel *et al.*, 1995). These are binary or higher-order combinations of individual TFBSs, which together convey a qualitatively and/or quantitatively new regulatory feature to the associated gene. They serve as integrative modules, where two regulatory activities exert a synergistic effect; for instance, one TFBS of a CE may interact with a tissue-specific TF, while the partner site may convey the response to a certain signal transduction pathway. Another accompanying database is TRANSPro, a comprehensive compilation of the promoters of several genomes (human, mouse, rat, and the plant *Arabidopsis thaliana*). TRANSFAC is connected with TFClass, a classification system for TFs according to their DBDs, so far applied to mammalian TFs (Wingender *et al.*, 2018). Since 1999, another complementary information resource has been established that represents signaling pathways aiming at TFs (Heinemeyer *et al.*, 1999). This database, TRANSPATH, has been focussing on pathways that lead to the activation or inactivation of TFs, although its scope was expanded later on.

TRANSFAC strives to cover the whole area of eukaryotic TFs, although the coverage of different taxa may be significantly different.

First attempts to use TRANSFAC contents for the prediction of TFBSs made use of the information about the experimentally validated sites compiled in the SITE table. Soon after, partly based on TARNSFAC's matrix library, several algorithms were developed that utilize PWMs for TF search, e.g., ConsInspector (Frech *et al.*, 1993), MATRIX SEARCH (Chen *et al.*, 1995), MatInspector (Quandt *et al.*, 1995), MATCH (Kel *et al.*, 2003a,b) and several others (see Relevant Websites Section). TRANSFAC itself is accompanied by MATCH (Kel *et al.*, 2003a,b). The program identifies TF binding sites at three different levels of stringency either minimizing the false positive or the false negative rate, or the rate of the sum of both error rates. The different algorithms mainly differ in the way how site scores are computed. In some of the earlier approaches some algorithms use matrix regularization (pseudo-counts) and forbidden nucleotides in order to account for specific constraints of the nucleotide distribution in the TF motifs. Similar to the MatInspector approach, MATCH applies specific weighting of PWM positions by information content of the nucleotide composition in the given position of the matrix. It was shown that this approach often increases precision of TF site recognition (Kondrakhin *et al.*, 2016). On top of MATCH results, the program F-Match identifies TF binding sites that are enriched in a set of regulatory sequences (e.g. a set of promoter sequences of co-regulated genes or a set of ChIP-seq peaks of active chromatin region). Combinations of predicted TF sites can be identified with a specific genetic algorithm implemented in the program CMA (Composite Module Analyst) (Kel *et al.*, 2006). CMA and other algorithms with similar scope were independently reviewed by Klepper *et al.* (2008).

The TRANSFAC database started in 1988 first as a private initiative and continued later as a public project. To secure long-term financial support, a company (BIOBASE) was established in 1999 to maintain and distribute the database commercially. An older version was kept in the public domain. Today, TRANSFAC is maintained, further developed and distributed by the company geneXplain. The FACTOR table of TRANSFAC has recently been made freely accessible.

Since its publication, the aims and ideas of the TRANSFAC database were adopted by a number of other databases, many of them also adding their own concepts.

JASPAR

JASPAR started as a carefully selected, non-redundant compilation of high-quality profiles, or position-specific scoring matrices that were derived from experimentally validated TF binding sites (Sandelin *et al.*, 2004). The database expanded over the years and more recently included third-generation binding models, which in contrast to PWMs or PSSMs consider position

Table 1 Overview of transcription factor databases, their scope and associated tools

Database name	URL	Species focus	Project start	Last updated	Brief summary	Tools	Country
3D-footprint	http://floresta.eead.csic.es/3dfootprint/	Eukaryotes		2017	Dedicated to structural data of DNA-binding proteins. Comprises clustering and classification of protein-DNA complexes.	Structure analysis, consensus literature miner, BLAST	Spain
AnimalTFDB2.0	http://bioinfo.life.hust.edu.cn/AnimalTFDB/	Vertebrates		2015?	General transcription factor (TF) database that focuses on curation and classification of genome-wide TFs, chromatin modifying proteins and transcription co-factors. The classification is based on DNA-binding domain characteristics.	TF prediction; BLAST	China
ArchaeaTF	http://bioinformatics.zj.cn/archaeatf/	Archaea	2006	2008	Database on (putative) transcription factors in Archaeal genomes. Due to annotated sequence characteristics, it can be useful for comparative genomics analysis of TFs in Archaea	Phylogenetic tree; BLAST	China
AthaMap	http://www.athamap.de/	Plants; Arabidopsis	2004	2016	Species-specific database on genome-wide putative transcription factor and small RNA binding sites in Arabidopsis thaliana. The sites are either experimentally verified or predicted by positional weight matrices representing TF binding specificities	TFBS and small/miRNA RNA target site search	Germany
ATRM (Arabidopsis Transcriptional Regulatory Map)	http://atrm.cbi.pku.edu.cn/	Plants; Arabidopsis	2014?	2017	Depicts the transcriptional regulatory network in Arabidopsis thaliana based on literature text mining and manual annotation		China
AtTFDB	http://arabidopsis.med.ohio-state.edu/AtTFDB/	Plants; Arabidopsis	2003?	2012?	Contains information on Arabidopsis thaliana transcription factors. The TF are classified as members of 50 families.		US
CIS-BP	http://cisbp.ccbr.utoronto.ca/	Eukaryotes	2014	2015	The Catalog of Inferred Sequence Binding Preferences (CIS-BP). It documents a large number of TF binding motifs/matrices, also from a variety of other sources.	TF binding site prediction tools	Canada
CollecTF	http://www.collectf.org/browse/home/	Prokaryotes	2013?	2015	Database for TFBS in bacteria. All sites have been experimentally verified and manually curated.		US
CoryneRegNet	https://coryneregnet.compbio.sdu.dk/v6/index.html	Prokaryotes; Corynebacteria		2012?	Reference database for TF in Corynebacteria and their regulatory networks. Comprises experimentally-verified as well as automatically predicted regulation data.	TFBS prediction tool	Denmark/Germany
CTCFBSDB	http://insulatordb.uthsc.edu/	Eukaryotes		2012	Dedicated to CTCF binding sites and genome organization. Contains experimentally-verified as well as automatically predicted sites.	Site prediction tool	US
cTFbase	http://bioinformatics.zj.cn/cTFbase/Home.php	Cyanobacteria		2007	Database focused on putative transcription factors in the genomes of Cyanobacteria. Classification of Tfs based on DBD characteristics.	BLAST, sequence alignment, phylogenetic tree	China
DBD	http://www.transcriptionfactor.org/index.cgi?Home	Eukaryotes	2008	2011	Genome-wide collection of putative TF with assignments to sequence specific DNA-binding domain families. The DBD database covers 930 sequenced genomes.		UK
DBTBS	http://dbtbs.hgc.jp/	Prokaryotes; Bacillus subtilis	1999	2008	Resource dedicated to experimentally validated TFBS in gene regulatory regions of Bacillus subtilis. All cis-elements (TFBS) in a promoter are depicted together with their positions.		Japan
iDMMPMM	http://autosome.ru/iDMMPMM/	Drosophila	2009	2009	Drosophila melanogaster collection of transcription factor DNA-binding motifs from several experimental sources		Russia
DOOR[2]	http://csbl.bmb.uga.edu/DOOR/index.php#tabs-1	Prokaryotes	2009	2013	Comprehensive prokaryotic operon database with data from 2072 bacterial genomes. The operons are predicted based on essential genomic features. Contains transcription units from experiments and RNA-seq predictions.	Genome Browser	US

Name	URL	Created	Updated	Species	Description	Tools/Features	Country
ENCODE	https://www.encodeproject.org/	2017		Human, mouse, worm, fly	Collection of high-throughput data about functional DNA elements, including TF binding regions from ChIP-seq experiments conducted in different cell types.		US
Factorbook	http://www.factorbook.org/human/chipseq/tf/	2017?	2012?	Human, mouse	Factorbook summarizes the results of ENCODE TF ChIP-seq data, together with ChIP-seq of histone modification and nucleosome occupancy experiments		US
Fly Factor Survey	http://mccb.umassmed.edu/ffs/	2013?	2008?	Drosophila	Focuses on DNA-binding specificities of TF in Drosophila melanogaster based on results obtained from systematic application of the bacterial one-hybrid method.	Motif search, MEME	USA
footprintDB	http://floresta.eead.csic.es/footprintdb/	2017	2016	Eukaryotes	Resource that collects TFs, TFBS and their DNA-binding specificity representing matrices from various sources.		Spain
Grassius	http://grassius.org/	2017	2008	Plants; Rice, Maize, Sorghum etc.	Collection of databases focused on gene regulation control in grass plants. Captures interactions of transcription factors with DNA-binding sites in promoter regions by literature curation and community contribution.	Network visualization, BLAST	USA
GTRD	http://gtrd.biouml.org/	2017	2017	Human, mouse	The Gene Transcription Regulation Database (GTRD) has a comprehensive collection of uniformly processed ChIP-seq data sets, allowing to deduce TFBS in human and mouse.	Genome Browser	Russia
hmChIP	http://jilab.biostat.jhsph.edu/database/cgi-bin/hmChIP.pl	2011?	2011?	Human, mouse	Resource to collect ChIP-chip and ChIP-seq data sets from experiments conducted in human and mouse cell lines or tissues.		USA
HOCOMOCO	http://hocomoco.autosome.ru/	2017	2017	Human, mouse	The Homo sapiens COmprehensive MOdel COllection (HOCOMOCO) contains mono- and dinucleotide TF DNA-binding models for more than 1100 human and mouse factors.	TF binding site prediction, model comparison, rSNP functional annotation	Russia
HOMER	http://homer.ucsd.edu/homer/index.html	2017	2010	Eukaryotes	The HOMER website combines TF motif databases with a software for binding site prediction and next-generation sequencing analysis.	TF binding site prediction tools	USA
HTPSELEX	http://ccg.vital-it.ch/htpselex/	2010	2010		This database comprises sets of in vitro selected TFBS sequences obtained by SELEX and high-throughput SELEX experiments.		Switzerland
HTRIdb	http://www.lbbc.ibb.unesp.br/htri/	2012?	2009	Human	Database on regulatory interactions between human transcription factors and their target genes. The interactions are experimentally verified and manually curated from the biomedical literature.		Brazil
Islet Regulome Browser	http://gattaca.imppc.org/isletregulome/info	2010	2010		Visualization tool for maps of ChIP-seq derived TFBS in pancreatic islet and progenitor cells	Network visualization	Spain
iTAK database	http://itak.feilab.net/cgi-bin/itak/db_home.cgi	2016	2010	Plants	Database for putative plant transcription factors and protein kinases predicted by the classifier software iTAK.	TF prediction	USA
JASPAR	http://jaspar.genereg.net/	2017	2004	Eukaryotes	Collection of non-redundant transcription factor DNA-binding models for multiple species. They are stored as Position Frequency Matrices (PFMs) and TF flexible models (TFFMs).	Profile inference, matrix comparison, API	UK
MtbRegList	http://mtbreglist.genap.ca/MtbRegList/www/index.php	2005	2005	Prokaryotes; M. tuberculosis	Resource focused on the analysis of gene regulation mechanisms in Mycobacterium tuberculosis. Contains transcription start sites and TFBS as well as the respective transcription factors.	Genome Browser	Canada
newPLACE	https://sogo.dna.affrc.go.jp/cgi-bin/sogo.cgi?lang=en&pj=64C&action=page&page=newplace	2007?	1999	Plants	Database of motifs detected in plant cis-acting regulatory DNA elements, extracted from published literature.		Japan
NURSA	https://www.nursa.org/nursa/index.jsf	2017	2003	Human, mouse, rat	A website dedicated to accrue information on nuclear receptors and co-regulators and their physiological roles. The data includes NR ligands and transcriptional targets.		USA
P2TF	http://www.p2tf.org/	2012?		Prokaryotes		BLAST, Genome Browser	France

(Continued)

Table 1   Continued

Database name	URL	Species focus	Project start	Last updated	Brief summary	Tools	Country
PAZAR	http://www.pazar.info/	Eukaryotes		will retire end of 2018	The Predicted Prokaryotic Transcription Factors (P2TF) database includes putative TFs of a large number of bacterial and archaeal genomes as well as a TF classification.		Canada
Plant Cistrome Database	http://neomorph.salk.edu/dap_web/pages/index.php	Plants; Arabidopsis; Maize		2016	Website hosting TF and regulatory sequence collections. It provides a software framework for regulatory sequence annotation tasks.		USA
PlantPAN 2.0	http://plantpan2.itps.ncku.edu.tw/index.html	Plants	2008	2015	Resource that provides TF binding profiles generated out of DAP-seq experiments in plants.	Genome Browser, Matrix comparison and clustering	Taiwan
PlantRegulome	http://plantregulome.org/	Plants; Arabidopsis	2013	2013	The Plant Promoter Analysis Navigator (PlantPAN) contains TFs, binding sites, and motifs for 76 plant species and allows to analyze promoters and to construct regulatory networks.	Network visualization, promoter analysis	USA
PlantTFDB	http://planttfdb.cbi.pku.edu.cn/	Plants	2010	2017	Focuses on creating maps of gene regulatory information in various developmental stages, tissues and cell types of Arabidopsis thaliana.	Regulatory network visualization	China
PlnTFDB	http://plntfdb.bio.uni-potsdam.de/v3.0/	Plants	2009		Database storing TF from 165 plant species as a result from an automated prediction pipeline. The TF are classified by DBD family assignments.	TF prediction server, tools for regulation prediction and functional enrichment analyses	Germany
PRODORIC	http://www.prodoric.de/	Prokaryotes	2001	2008	The resource aims to identify and classify all plant genes involved in transcriptional control. The TF family classification is based on characteristic domains described in the literature.	BLAST	Germany
RegulonDB	http://regulondb.ccg.unam.mx/	Prokaryotes; E. coli	1998	2017	A database containing comprehensive gene regulation and expression information in prokaryotes. It includes experimentally validated and manually curated transcription factor binding sites. It's accompanied by bioinformatics tools for the prediction, analysis and visualization of gene regulatory networks.	Genome Browser, network analysis and visualization	Mexico
ReMap 2018	http://tagc.univ-mrs.fr/remap/index.php	Human	2015	2017	Database dedicated to comprehensively model the transcriptional regulatory network in E. Coli K-12. It includes information on transcription units, operons and regulons. The data is manually curated from peer-reviewed literature.	Regulatory network visualization, genome browser, text mining, PWM clustering, classification browser	France
Riken Tfdb	http://genome.gsc.riken.jp/TFdb/	Mouse		2004?	A platform that collects TFBS ChIP-seq data sets from ENCODE, GEO and ArrayExpress. Peaks have been uniformly processed and are available for either in total or as non-redundant merged peaks.	Statistical search of TF binding peaks	Japan
ScerTF	http://stormo.wustl.edu/ScerTF	Yeast		2012?	A database containing a set of mouse non-redundant transcription factors and their related genes.	TFBS prediction	USA
STIFDB	http://caps.ncbs.res.in/stifdb2/	Plants; Arabidopsis, rice		2012	A resource that collects non-redundant position weight matrices (PWMs) for yeast transcription factors from literature and public database sources.	Blast	India
TcoF	http://www.cbrc.kaust.edu.sa/tcof/	Human	2010	2017	The Stress Responsive Transcription Factor Database (STIFDB) is a comprehensive set of biotic and abiotic stress responsive genes and their potential TFBS in respective promoter regions in Arabidopsis thaliana and Oryza sativa.		Saudi-Arabia
TEC	https://shigen.nig.ac.jp/ecoli/tec/top/	Prokaryotes; E. coli	2010?	2016?	A database focusing on human transcription co-factors and TF interacting proteins.		Japan

Name	URL	Organism	Year	Year	Description	Tools	Country
					TEC contains regulatory targets for TFs and sigma factors in E. coli K-12 determined by SELEX-chip screening.		
TFBSshape	http://rohslab.cmb.usc.edu/TFBSshape/	Eukaryotes		2014?	TFBSshape includes predicted DNA shape features for TFBS, characterizing DNA binding specificities of TFs from 23 species.		USA
TFCat	http://www.tfcat.ca/	Human, mouse		2009?	Catalog of mouse and human transcription factors (TF) based manual curation from the biomedical literature. In addition, TF encoding genes have been predicted based on sequence analysis. DNA-binding TF are assigned to a structural classification system.		Canada
TFcheckpoint	http://www.tfcheckpoint.org/	Human, mouse, rat		2014?	Manually curated database for human, mouse and rat TFs based on public database and literature sources.		Norway
Tfe	http://cisreg.cmmt.ubc.ca/cgi-bin/tfe/home.pl	Human, mouse, rat	2012	2017	The Transcription factor encyclopedia (Tfe) aims to create encyclopedic articles for each well-studied TF.		Canada
TFinDit	http://bioinfozen.uncc.edu/tfindit/	Eukaryotes	2012	2015	Database and web service for structural information of transcription factor (TF)-DNA interactions. It contains pairs of bound and unbound transcription factor structures and homologous TF-DNA complex structures.		USA
TRANSFAC	http://genexplain.com/transfac/ Public: http://gene-regulation.com/pub/databases.html Factor search: http://factor.genexplain.com/cgi-bin/transfac_factor/search.cgi	Eukaryotes	1988	2017	TRANSFAC® is the database of eukaryotic transcription factors, their genomic binding sites and DNA-binding profiles. Its TFBS data is experimentally verified and manually curated from peer-reviewed literature.	TFBS prediction, composite element search	Germany
TRRD	http://wwwmgs.bionet.nsc.ru/mgs/gnw/trrd/	Eukaryotes	1995	2005	The Transcription Regulatory Regions Database (TRRD) collects information on structural and functional organization of transcriptional regulatory regions of eukaryotic genes. All data in TRRD has been experimentally verified.	TFBS prediction	Russia
TRUST	http://www.grnpedia.org/trust/	Human, mouse	2014	2017	Manually curated database of human and mouse transcriptional regulatory networks. For a large part of the stored TF-target gene relations, the mode of regulation (activation or repression) is provided.	Key regulator search	South Korea
UniProbe	http://the_brain.bwh.harvard.edu/uniprobe/	Eukaryotes	2008	2017	The UniPROBE (Universal PBM Resource for Oligonucleotide Binding Evaluation) database contains data on in vitro DNA-binding specificities of proteins created by the universal protein binding microarray (PBM) experimental method.	TFBS prediction; motif similarity comparison	USA
Yeastract	http://www.yeastract.com/index.php	Yeast	2006	2017	YEASTRACT (Yeast Search for Transcriptional Regulators And Consensus Tracking) is a curated database of regulatory interactions between TF and target genes in Saccharomyces cerevisiae.	Motif identification in promoters	Portugal
YeTFaSCo	http://yetfasco.ccbr.utoronto.ca/	Yeast		2012	The site provides a collection of TF specificities for Saccharomyces cerevisiae as position frequency matrices (PFM) or position weight matrices (PWM).	TFBS prediction, genome browser, motif comparison	Canada

interdependencies within and variable lengths of the TF a binding motifs. These Transcription Factor Flexible Models (TFFMs) are built with the help of hidden Markov models (HMMs) (Mathelier and Wasserman, 2013). The TFFMs of the provided library have been created from ChIP-seq data coming from the ENCODE project. Software tools for the application of these advanced binding models accompany the library. They are complemented by tools that support the creation of new TFFMs as well as to scan sequences with them (Mathelier *et al.*, 2016). – As for the factor side, JASPAR is complemented by TFe, a transcription factor encyclopedia, which so far as compiled more than 800 expert-written articles about individual TFs.

Plant TFs

Two actively maintained resources for plant TFs have to be highlighted. One is **PlantTFDB** from Peking University, which covers properties of TFs from so far 165 plant species. Using an automated TF prediction pipeline, more than 320,000 TFs were identified to be encoded in plant genomes and can be browsed by (DBD) family assignments. Making use of a related resource of the same group, PlantRegMap, binding site and binding motif information has been made available as well (Jin *et al.*, 2017). TFBSs have been partly taken from literature, where experimental evidence was given, or predicted by using the binding motif information. – Another program to identify TFs encoded in plant genomes is iTAK. It was trained with contents of PlantTFDB and others, and the results are provided online as database (Zheng *et al.*, 2016). In addition, there are several databases that focus on individual species, mostly the best-studied plant organism *Arabidopsis thaliana*, such has AthaMap (see also **Table 1**) (Hehl *et al.*, 2016).

Other Model Organisms

There are several other TFDBs that provide corresponding information with different emphasis on individual binding sites, high-throughput data on binding regions, or binding motifs for the factors of, e.g., *Drosophila* (Fly Factor Survey; Zhu *et al.*, 2011) or yeast (Yeastract; Teixeira *et al.*, 2014).

TFDBs on Prokaryotic Transcription Factors

One of the most popular databases about entities of prokaryotic transcription is **RegulonDB** (Huerta *et al.*, 1998). Structurally similar to TRANSFAC, the early database version provided in-depth information about *E. coli* TFs and their binding sites. As a unique feature, RegulonDB put the co-regulation of genes or operons in the focus. By time, the database broadened its scope even further by integrating information about regulatory interactions at genes encoding components of defined metabolic pathways with the signals triggering them into so-called Genetic Sensory-Response units (GENSOR units). – A similar resource focussing on *Bacillus subtilis*, **DBTBS**, has been released in 2000 by a team from Tokyo University (Ishii *et al.*, 2001).

With regard to the range of organisms covered, the database **PRODORIC** had a broader approach since it aimed at the whole plethora of prokaryotic organisms. Otherwise, it has similar aims as RegulonDB and a structure resembling TRANSFAC (Münch *et al.*, 2003). Later, the database was complemented with a rich set of software tools (Grote *et al.*, 2009). Among them are a site scanning software, Virtual Footprint, supporting sensitive searches for single or composite TF sites in bacterial genomes. It also visualizes search results in the context of the whole bacterial genome.

Additional Collections of Binding Sites and Motifs

ENCODE (Encyclopedia of DNA Elements) is a consortial project that collects high-throughput data about functional DNA elements, among them TF binding regions from, e.g., ChIPseq studies (Davis *et al.*, 2017). Its main focus is on the human genome, but several model organisms are covered as well. Binding motifs have been extracted from ENCODE datasets that refer to defined TFs, usually with the MEME-ChIP software suite (Machanick and Bailey, 2011), and have been compiled in Factorbook (Wang *et al.*, 2012). At present, it shows binding motifs of 167 TFs derived from 837 experiments.

A principally similar approach, but using pre-processed high-throughput data provided by GTRD (Gene Transcription Regulation Database; Yevshin *et al.*, 2017) and a novel motif discovery tool (ChIPMunk; Kulakovskiy *et al.*, 2010), led to the generation of HOCOMOCO, a collection of mono- and dinucleotide matrices for human and mouse TFs (Kulakovskiy *et al.*, 2018).

Based on a novel experimental approach, the UniPROBE (Universal PBM Resource for Oligonucleotide Binding Evaluation) database has been generated (Hume *et al.*, 2015). The underlying data were obtained by using protein binding microarrays (PBMs) of synthetic oligonucleotides (decamers) to determine those sequences that are suitable to interact with a test protein or its DNA-binding domain in vitro (Mukherjee *et al.*, 2004).

Acknowledgement

TRANSFAC and TRANSPATH are registered trademarks of QIAGEN.

See also: Bioinformatics Data Models, Representation and Storage. Biological Database Searching. Data Storage and Representation. Genome-Wide Scanning of Gene Expression. Natural Language Processing Approaches in Bioinformatics. Prediction of Protein-Binding Sites in DNA Sequences. Protein-DNA Interactions. Regulation of Gene Expression

References

Chen, Q.K., Hertz, G.Z., Stormo, G.D., 1995. MATRIX SEARCH 1.0: A computer program that scans DNA sequences for transcriptional elements using a database of weight matrices. Comput. Appl. Biosci. 11, 563–566.

Davis, C.A., Hitz, B.C., Sloan, C.A., et al., 2017. The encyclopedia of DNA elements (ENCODE): Data portal update. Nucleic Acids Res. 2017 (26), gkx1081 (ahead of print).

Frech, K., Herrmann, G., Werner, T., 1993. Computer-assisted prediction, classification, and delimitation of protein binding sites in nucleic acids. Nucleic Acids Res. 21, 1655–1664.

Gama-Castro, S., Salgado, H., Santos-Zavaleta, A., et al., 2016. RegulonDB version 9.0: High-level integration of gene regulation, coexpression, motif clustering and beyond. Nucleic Acids Res. D133–D143.

Grote, A., Klein, J., Retter, I., et al., 2009. PRODORIC (release 2009): A database and tool platform for the analysis of gene regulation in prokaryotes. Nucleic Acids Res. 37, D61–D65.

Hehl, R., Norval, L., Romanov, A., Bülow, L., 2016. Boosting AthaMap database content with data from protein binding microarrays. Plant Cell Physiol. 57, e4.

Heinemeyer, T., Chen, X., Karas, H., et al., 1999. Expanding the TRANSFAC database towards an expert system of regulatory molecular mechanisms. Nucleic Acids Res. 27, 318–322.

Huerta, A.M., Salgado, H., Thieffry, D., Collado-Vides, J., 1998. RegulonDB: A database on transcriptional regulation in Escherichia coli. Nucleic Acids Res. 26, 55–59.

Hume, M.A., Barrera, L.A., Gisselbrecht, S.S., Bulyk, M.L., 2015. UniPROBE, update 2015: New tools and content for the online database of protein-binding microarray data on protein-DNA interactions. Nucleic Acids Res. 43, D117–D122.

Ishii, T., Yoshida, K., Terai, G., Fujita, Y., Nakai, K., 2001. DBTBS: A database of Bacillus subtilis promoters and transcription factors. Nucleic Acids Res. 29, 278–280.

Jin, J.P., Tian, F., Yang, D.C., et al., 2017. PlantTFDB 4.0: Toward a central hub for transcription factors and regulatory interactions in plants. Nucleic Acids Res. 45, D1040–D1045.

Kel, A.E., Gößling, E., Reuter, I., et al., 2003a. MATCH™: A tool for searching transcription factor binding sites in DNA sequences. Nucleic Acids Res. 31, 3576–3579.

Kel, A.E., Gossling, E., Reuter, I., et al., 2003b. MATCH: A tool for searching transcription factor binding sites in DNA sequences. Nucleic Acids Res. 31, 3576–3579.

Kel, A., Konovalova, T., Waleev, T., et al., 2006. Composite module analyst: A fitness-based tool for identification of transcription factor binding site combinations. Bioinformatics 22, 1190–1197.

Kel, O.V., Romaschenko, A.G., Kel, A.E., Wingender, E., Kolchanov, N.A., 1995. A compilation of composite regulatory elements affecting gene transcription in vertebrates. Nucleic Acids Res. 23, 4097–4103.

Klepper, K., Sandve, G.K., Abul, O., Johansen, J., Drablos, F., 2008. Assessment of composite motif discovery methods. BMC Bioinform. 9, 123.

Kondrakhin, Y., Valeev, T., Sharipov, R., et al., 2016. Prediction of protein-DNA interactions of transcription factors linking proteomics and transcriptomics data. EuPA Open Proteom. 13, 14–23.

Kulakovskiy, I.V., Boeva, V.A., Favorov, A.V., Makeev, V.J., 2010. Deep and wide digging for binding motifs in ChIP-Seq data. Bioinformatics 26, 2622–2623.

Kulakovskiy, I.V., Vorontsov, I.E., Yevshin, I.S., et al., 2018. HOCOMOCO: Towards a complete collection of transcription factor binding models for human and mouse via large-scale ChIP-Seq analysis. Nucleic Acids Res 8 (46), D252–D259.

Machanick, P., Bailey, T.L., 2011. MEME-ChIP: Motif analysis of large DNA datasets. Bioinformatics 27, 1696–1697.

Mathelier, A., Fornes, O., Arenillas, D.J., et al., 2016. JASPAR 2016: A major expansion and update of the open-access database of transcription factor binding profiles. Nucleic Acids Res. 44, D110–D115.

Mathelier, A., Wasserman, W.W., 2013. The next generation of transcription factor binding site prediction. PLOS Comput. Biol. 9, e1003214.

Mukherjee, S., Berger, M.F., Jona, G., et al., 2004. Rapid analysis of the DNA-binding specificities of transcription factors with DNA microarrays. Nat. Genet. 36, 1331–1339.

Münch, R., Hiller, K., Barg, H., et al., 2003. PRODORIC: Prokaryotic database of gene regulation. Nucleic Acids Res. 31, 266–269.

Quandt, K., Frech, K., Karas, H., Wingender, E., Werner, T., 1995. MatInd and MatInspector: New fast and versatile tools for detection of consensus matches in nucleotide sequence data. Nucleic Acids Res. 23, 4878–4884.

Sandelin, A., Alkema, W., Engström, P., Wasserman, W.W., Lenhard, B., 2004. JASPAR: An open-access database for eukaryotic transcription factor binding profiles. Nucleic Acids Res. 32, D91–D94.

Teixeira, M.C., Monteiro, P.T., Guerreiro, J.F., et al., 2014. The YEASTRACT database: An upgraded information system for the analysis of gene and genomic transcription regulation in Saccharomyces cerevisiae. Nucleic Acids Res. 42, D161–D166.

Wang, J., Zhuang, J., Iyer, S., et al., 2012. Sequence features and chromatin structure around the genomic regions bound by 119 human transcription factors. Genome Res. 22, 1798–1812.

Wingender, E., 1988. Compilation of transcription regulating proteins. Nucleic Acids Res. 16, 1879–1902.

Wingender, E., Schoeps, T., Dönitz, J., 2013. TFClass: An expandable hierarchical classification of human transcription factors. Nucleic Acids Res. 41, D165–D170.

Wingender, E., Schoeps, T., Haubrock, M., Krull, M., Dönitz, J., 2018. TFClass: Expanding the classification of human transcription factors to their mammalian orthologs. Nucleic Acids Res. 46, D343–D347.

Yevshin, I.S., Sharipov, R.N., Valeev, T.F., Kel, A.E., Kolpakov, F.A., 2017. GTRD: A database of transcription factor binding sites identified by ChIP-seq experiments. Nucleic Acids Res. 45, D61–D67.

Zheng, Y., Jiao, C., Sun, H., et al., 2016. iTAK: A program for genome-wide prediction and classification of plant transcription factors, transcriptional regulators, and protein kinases. Mol. Plant 9, 1667–1670.

Zhu, L.J., Christensen, R.G., Kazemian, M., et al., 2011. FlyFactorSurvey: A database of Drosophila transcription factor binding specificities determined using the bacterial one-hybrid system. Nucleic Acids Res. 39, D111–D117.

Relevant Websites

http://gene-regulation.com/
   Gene regulation.
https://bip.weizmann.ac.il/toolbox/seq_analysis/promoters.html
   Promoters.
https://omictools.com/transcription-factor-binding-site-prediction-category
   Transcription Factor Staining.

Protein-DNA Interactions

Preeti Pandey, Sabeeha Hasnain, and Shandar Ahmad, Jawaharlal Nehru University, New Delhi, India

Introduction

The central dogma of molecular biology involves transcription and translation of genes leading to formation of proteins to carry out specific biological functions. Activation and suppression of this so-called "gene expression" by DNA binding proteins is a fundamental regulatory mechanism involving chromatin modification and transcription complexes to initiate RNA synthesis (Announcement, 1996). In human and other higher organisms, transcription involves binding of a specific group of proteins (known as transcription factors) to and unzipping of local regions in the double helical genomic DNA into two separate strands and synthesis of messenger RNA (mRNA) from one of them. Apart from transcription, DNA recognition by proteins is crucial in host response to certain pathogens (some pathogen DNAs can be recognized by "DNA sensors"), transfer of genetic information, cell differentiation, packaging, rearrangement, replication and repair (Wold, 1997; Dalrymple *et al.*, 2001; Wang *et al.*, 2000; Lieb *et al.*, 2001; Ofran *et al.*, 2007; Luscombe *et al.*, 2000; Walter *et al.*, 2009; Hashimoto *et al.*, 2003; Ptashne, 2005; Dillon and Dorman, 2010; Thanbichler *et al.*, 2005; Kow *et al.*, 2007; Kamashev, 2000). All such interactions proceed by the binding of proteins to (usually specific) DNA sequences and the formation of a protein-DNA complex (Luscombe *et al.*, 2000). The understanding of DNA-protein interactions is therefore extremely important to understand the fundamental processes of life.

The interaction between DNA-binding proteins and their target DNA is often accompanied by large conformational changes in either or both of them (Thompson and Landy, 1988; Paull *et al.*, 1993; Masse *et al.*, 2002; Andrabi *et al.*, 2014). These conformational changes facilitate and stabilize the formation of a protein-DNA complex, enabling the desired molecular events to take place. Different types of conformational changes have been confirmed by experimental data, e.g., an increase in superhelical density of DNA and bending and twisting of DNA are well documented (Beloin *et al.*, 2003; Tapias *et al.*, 2000; Noy *et al.*, 2016).

This review focuses on the understanding of the mechanism of protein-DNA interactions and their role in various biological processes, gives a broad overview of the nature of conformational changes in both protein and DNA in the formation of protein-DNA complex and the factors affecting its stability. We also aim to review the experimental and computational techniques to understand the structural and conformational changes in protein and DNA to form a stable complex. This review also investigates the current understanding of the role of DNA structure and dynamics in protein DNA interactions and highlights various unanswered questions on the subject.

Representative Systems of Protein-DNA Interactions

As stated above, the interactions between protein and DNA play a central role in a variety of biological processes such as transcription factors binding to DNA, DNA modifications, viral infection and chromosomal packing. Some of these interactions, such as transcription factor binding occur while the interacting sequence is still a part of a genomic DNA, which is millions of times larger, whereas other interactions may potentially be with a free and small DNA fragment in the cytoplasm (Beloin *et al.*, 2003). A protein on the other hand mostly acts as a complete unit, as a monomer or a complex formed by oligomerization on its own copies, with other co-factors or as a small number of subunits (e.g., histones) (Rouvière-Yaniv *et al.*, 1979; Luger *et al.*, 1997; Dillon and Dorman, 2010; Thanbichler *et al.*, 2005; Lia *et al.*, 2003). DNA-recognition in a genome towards gene expression requires rapid binding of modifying enzymes, polymerases, repressors and activators to their site of action on a large, packaged DNA, not always easily "visible" to the acting proteins (Gasser and Laemmli, 1987). The interaction between a protein and its "target" DNA in transcription factor binding promotes or suppresses specific genes and the corresponding proteins are called *activators* and *repressors* respectively. One of the most widely studied systems is the repressor protein responsible for switching on and off the *lac* and *lambda* phage genes of *Escherichia coli* (Matthews *et al.*, 1982; Kamashev *et al.*, 1995). The inactivation of *lac* and *lambda* repressor proteins leads to the expression of *lac* and *lambda* gene respectively. The process of binding of transcription factor is crucial, in general, for any such processes. The inappropriate binding of transcription factors results in alterations in gene expression, which has been found to be a cause of disorders (Jimenez-Sanchez *et al.*, 2001) and medical conditions in humans (Lee and Young, 2013; Farnham, 2009).

Another widely studied model system of protein-DNA interactions is related to the packing of chromosomes (Dillon and Dorman, 2010; Thanbichler *et al.*, 2005). The chromosomes consist of incredibly long DNA chains, which fold and compactify to protein-DNA complexes called chromatids (Fischle *et al.*, 2003). The DNA binding proteins involved in this process can be classified into histones and nonhistones chromosomal proteins. Histone proteins form nucleosomes which are attached by DNA string, forming a *bead-spring* kind of structure at their first level of organization.

Role of protein-DNA recognition in host defense against invading pathogens is also of great significance so much so that in some cases pathogen DNA itself can act as a vaccine (Rice *et al.*, 1999; Modlin, 2000). Examples of protein-DNA interactions in host defense are CpG-rich viral DNA like simplex virus (HSV-1), HSV-2, and murine cytomegalovirus (MCMV), which are recognized by TLR9 resulting in activation of inflammatory cytokinesis and type 1 IFN secretion (Hochrein *et al.*, 2004; Krug *et al.*, 2004a,b; Lund *et al.*, 2003; Tabeta *et al.*, 2004; Akira *et al.*, 2006; Herzner *et al.*, 2015; Manders and Thomas, 2000; Dell'Oste *et al.*, 2015).

It may be noted that, although more common and widely investigated, it is not always the protein, which acts on DNA and hunts for its targets. As an example of the opposite, DNA fragments, binding to protein are known to stabilize proteins protecting them against misfolding and aggregation and thus leading to formation of a stable structure (Jolly and Morimoto, 2000; Morimoto, 1998; Wu, 1995).

DNA-Binding Protein Structures, Dynamics and Conformational Changes

Experimental techniques like X-ray crystallography and NMR have been employed to give high resolution structures of DNA-protein complexes in crystalline state and in solution respectively (Spolar and Record, 1994; Garvie and Wolberger, 2001). Many studies have shown large conformational changes in the structures of both the DNA and proteins during sequence-specific binding (Kalodimos, 2004; Kalodimos *et al.*, 2002; Spolar and Record, 1994). The conformational changes in DNA vary from smooth change in bending deformations to drastic changes resulting in disruption of base pair stacking, base flipping, sharp bends and kinks in the helical axis (Schultz *et al.*, 1991; McClarin *et al.*, 1986; Kim *et al.*, 1990; Honnappa *et al.*, 2005). Two key questions that have been investigated in details are:

"How a protein's binding nonspecifically to DNA, preceding the specific interaction, brings out a change in the structural and dynamic properties of DNA?" and *"How does a complex undergo transitions from nonspecific to specific interactions".*

To address these questions, Kalodimos (2004) studied *lac* repressor and categorized its DNA-binding domain on the basis of structure and dynamics in the free state and non-specifically bound state to DNA. In the study, the authors have reported the high-resolution structures of the dimeric lac repressor DNA-binding domain bound to an 18-base pair non-specific DNA fragment by imposing 2412 experimental restraints on multidimensional NMR spectroscopy. The authors did a comparison of the nonspecific and specific binding modes to determine the changes that occurred due to specific binding. The calculations for protein-DNA docking were performed (Brünger *et al.*, 1998) using HADDOCK setup and protocols, which are a handy set of tools for studying protein-DNA interactions *de novo* (Brünger *et al.*, 1998; Dominguez *et al.*, 2003). The protein in a non-specific complex is rotated by an additional ∼25 degrees relative to DNA, when compared with the specific complex. Some of the residues (Tyr7, Gln18, and Arg22 in this complex), which play a key role in specific interaction, are found to undergo shifting and twisting to participate in hydrogen bonding and electrostatic interactions in the case of their corresponding non-specific complex. The interface of protein-DNA in case of non-specific complex in this system is found to be highly flexible in contrast to the rigid interface in the case of specific complex, thus implicating a role of conformational dynamics in specificity. In this system and in general, the atomic arrangements of protein-DNA complex are significantly altered in their nonspecific to specific transitions of interactions, which may be crucial in target search followed by complex formation with the genomic target DNA. In a recent study by Corona and Guo (2016), authors performed a comparative analysis of this protein-DNA complex to investigate the structural features that contributes to the binding specificity. Based on the binding specificity DNA-protein complex has been categorized as (1) Highly specific (HS) (2) multi-specific (MS) and (3) nonspecific (NS). From the analysis of conformational changes between bound and unbound states, authors found HS and MS to have larger conformational changes and flexibility in both bound and unbound states.

In another study by Velmurugu *et al.* (2016), dynamic conformational changes were studied for a DNA-repair protein radiation-sensitive 4 (Rad4; yeast XPC ortholog) recognition of the damaged site on DNA by using temperature-jump spectroscopy. One of the key finding of the study was the role of DNA deformability which is found to be the key factor that helps in a rapid search of the damaged site by DNA-repair protein. Specifically, the RAD4 recognizes the damaged site through twist-open mechanism in which twisting assists in inter-converting RAD4 from search mode to interrogation mode. The role of β-hairpin here is essential for the nucleotide flipping. The conformational changes compatible with rapid diffusion on DNA makes RAD4 stall preferentially at these sites offering time for DNA to open.

Insights into the conformational dynamics of protein-DNA interactions, not accessible to crystal structures determined by X-ray can be gleaned by NMR relaxation of backbone amides (Akke, 2002). Such studies have resulted in the understanding that most of the functional processes, such as docking, protein folding, allosteric transitions are associated with slow motion of *motor* domains (on the time scale of *s* to *ms*). Most of the residues that perform the motion at this time scale are found to be involved in both specific and nonspecific binding. Motion on this time scale and the flexibility of the complex facilitates the conformational transitions and the search of specific sites respectively.

To understand the effect of conformational dynamics in protein-DNA recognition Chu *et al.*, (2014) developed a two-basin structure based model to understand the dynamics of *Sulfolobus solfataricus* DNA Y-family polymerase 1V (DPO4) when it binds to DNA. A two-basin structure based model (SBM) was developed with electrostatic interaction given by *Debye-Huckel* model. Thermodynamic and kinetic simulations were performed to investigate the conformational transitions of DPO4 during the process of binding to the target site on DNA. In the study, authors have found that DPO4 maintains multiple conformational equilibrium stages during the process of binding and the distribution of conformations vary at different stages of binding. By varying the strength of electrostatic interactions and the flexibility of the linkers, the authors showed the way DPO4 dynamically regulates the DNA recognition.

In yet another recent study using NMR spectroscopy, Desjardins *et al.* (2016) investigated the mechanism of binding of the Ets-1 transcription factor to DNA. Ets-1 belonging to ETS family of eukaryotic transcription factors is auto-inhibited by serine-rich region (SRR) and helical inhibitory module (IM). The structural and dynamic features of the Ets-1 were characterized for both specific and nonspecific binding of ETS-1 to 12 bp of oligonucleotides. On binding to both nonspecific and specific DNA, the N-terminal sequences are found to be predominantly unfolded, but still, sample ordered conformations. With specific DNA, the backbone undergoes more structural and dynamic changes and arginine/lysine side chains of the core ETS domains are much larger in comparison to non-specific binding. The study supports the general model that the Ets-1 form non-specific binding *via* electrostatic interactions and hydrogen bonding assist in the formation of well-ordered specific complexes with DNA.

Conformational changes in protein-DNA complexes are not always associated with target search as the physical cellular environments such as varying pH, temperature as well as site-specific chemical modification (phosphorylation and methylation) often lead to the formation and alteration of the oligomeric states of protein-DNA complexes (Andrabi *et al.*, 2014). Irrespective of the mechanisms involved in the conformational changes, two or more conformations of proteins in DNA-bound and unbound states may be compared to investigate the differences caused by complex formation. Based on this approach, we recently surveyed and analyzed the conformational changes in a large number of protein-DNA complex (Andrabi *et al.*, 2014). We addressed three issues of conformational changes in DNA-binding proteins: (a) types of conformational changes and their distribution among different proteins with respect to charge and dipole moments, (b) Intrinsic flexibility of unbound DNA and its role in conformational change (c) contributions of conformational change to the stability and specificity of DNA-protein complex. Six groups of conformational changes were identified and related to their biological implications. Proteins in one of these groups undergoing little conformational change "the rigid group" were found to form smaller number of contacts with DNA in comparison to proteins, which undergo large conformational change. We also reported that the degree of conformational change in protein-DNA complex increases the stability and specificity of the complex. Normal mode analysis was used to analyze the extent and direction of conformational changes in protein in some classes of DNA-binding proteins. The amount of conformational changes occurring in a protein was found to depend on the polarity of proteins. The proteins with positive charge interact mostly with the negatively charged backbone of DNA and thereby undergo least conformational change. However, for hydrophobic residues like Cys, Ala and Gly, proteins undergo large conformational changes at the interface to adjust for the binding.

Role of DNA Structure and Dynamics

Binding of proteins to its specific site of action on DNA is usually accompanied by diffusion and nonspecific binding to DNA (Berg and von Hippel, 1985). It has been proposed that the proteins undergoing three-dimensional diffusive motion in cytoplasm bind nonspecifically to DNA and then undergo one-dimensional sliding motion to reach its specific binding site on DNA (Shimamoto, 1999; Halford and Marko, 2004; Kalodimos, 2004; Esadze *et al.*, 2014). This mechanism is termed as facilitated diffusion mechanism (Berg *et al.*, 1981; Richter and Eigen, 1974) and has been proposed to justify the fast rate of binding of protein to genomic DNA (see **Fig. 1** for the general process of TFs finding their targets in genomic DNA). This non-specific binding of protein and DNA is a result of electrostatic interactions between the negatively charged phosphate backbone and protein.

The specific binding of proteins to DNA involves two types of mechanisms: one is the hydrogen bond formation with specific sequence along the major groove and the other is sequence dependent deformation of DNA (Fuxreiter *et al.*, 2011; Rohs *et al.*, 2009b; Ahmad *et al.*, 2006). They are respectively termed as *sequence* and *shape* readout (earlier known as *direct* and *indirect* readout). In some protein-DNA complexes, the bending of DNA ends up with huge conformational changes in the structure of

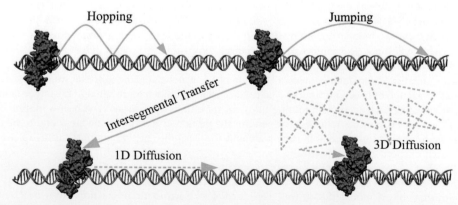

Fig. 1 Diffusion mechanisms of transcription factors trying to find their targets in genomic DNA (for simplicity genomic DNA is shown as a straight helix instead of chromatin folding): The TFs are reported to undergo electrostatically dominated non-specific interactions with DNA, which undergoes a non-specific to specific hydrogen-bond dominated interaction transition upon reaching the targets. Diffusions processes themselves during non-specific association between protein and DNA occurs through multiple events such as hopping, jumping, inter-segmental transfer and three-dimensional diffusion.

DNA and is found to deviate from B-form of double helix. The role of dynamics is not just limited to non-specific interactions but it also plays an important role in specific interactions between protein and DNA. As an example, Sox2 proteins cause a large DNA bend when they bind to their genomic targets requiring substantial interaction energy, which is often provided by partner protein-protein interactions with cofactors (Hou *et al.*, 2017; Kamachi *et al.*, 1999; Scaffidi and Bianchi, 2001). The flexible and positively charged tails of proteins assist in fine-tuning the specificity in case of transcription factors (Fuxreiter *et al.*, 2011; Rohs *et al.*, 2010).

Molecular dynamics (MD) is the principal technique to investigate the detailed molecular mechanism of protein-DNA interactions, partly because actual experimental procedures to observe the process of protein-DNA interactions are beyond the technological capabilities of researchers. There have been a number of successful efforts in applying MD to understand and interpret observed biological knowledge of protein-DNA interactions in the cellular contexts. For example, in a recent MD study of telomere repeat binding factors (TRF1 and TRF2), the dynamics of individual amino acids chains suggested that they could contribute to the recognition of more than one base pair (Etheve *et al.*, 2016; Garton and Laughton, 2013; Jolma *et al.* 2010). The study helped to resolve conflicting experimental data (Luscombe *et al.*, 2000; Garton and Laughton, 2013). In a study by Tan *et al.* (2016) authors emphasized the dynamic aspects of nonspecific protein-DNA interactions. The dynamic coupling between protein binding to DNA and the change in the structure of DNA was investigated. Protein binding to DNA changes the DNA structure, which in turn modifies the mechanics of proteins along the DNA scaffold and thereby its interaction dynamics (Bhattacherjee and Levy, 2014a,b; Rohs *et al.*, 2009a,b; Stella *et al.*, 2010). Molecular dynamics simulations have been performed on Coarse-grained model of DNA-binding proteins to understand the dynamic couplings among protein-DNA binding, sliding of protein along DNA and bending of DNA. Simulations have been performed on bacterial architectural protein HU (Rouvière-Yaniv *et al.*, 1979) and 14 other DNA-binding proteins. At long time-scales, it has been found that the sliding motion of HU is associated with DNA bending different from *induced-fit* and *population-shift*. At shorter time scales, the HU is found to pause when the DNA is highly bent and starts to move when DNA returns to less bent structure. Thus the sliding motion is found to largely depend on the DNA bending dynamics.

Issues in Protein-DNA Interactions Predictions

The interplay between protein and DNA are ubiquitous in nature and governs several cellular processes (replication, transcription, translation, chromosomal packing, *etc.*) crucial for growth and survival of any living organism. Given its importance, the field of protein DNA interaction has been studied by many researchers with a focus on the development of tools and techniques for identification of DNA binding site in proteins, DNA binding proteins from sequence or structure and transcription factor binding site on genomic DNA. In the following sections, we review the progress made on these issues.

Prediction of DNA Binding Residues and DNA Binding Proteins

The binding of DNA to the protein is very specific where DNA binds to a particular region of the protein usually termed as the binding site, defined by the group of residues that are essential for the biological function and whose mutations can result in attenuation/loss of function. An accurate estimation of DNA-binding site and DNA-binding proteins thus is of utmost significance and can provide an insight into the various functions of the protein and also in designing novel therapeutics. Many experimental and computational methods have been developed for prediction of DNA binding site and DNA-binding proteins. Experimental methods for direct assessment of DNA-binding specificities include chromatin immunoprecipitation (ChIP) (Bardet *et al.*, 2013; Johnson *et al.*, 2007), MicroChIP (Luscombe *et al.*, 2001), Fast ChIP (Mandel-Gutfreund and Margalit, 1998), electrophoretic mobility shift assays (EMSAs) (Jones *et al.*, 1999, 2003), peptide nucleic acid (PNA)-assisted identification of RNA binding proteins (RBPs) (PAIR) (Olson *et al.*, 1998), X-ray crystallography (Orengo *et al.*, 1997) and nuclear magnetic resonance (NMR) spectroscopy (Ponting *et al.*, 1999). With the growing advancement in associated computational technologies and machine learning methods, several bioinformatics methods have also been developed for the prediction of DNA binding site complementing experimental procedures. The computational approaches for prediction of DNA binding site and DNA-binding proteins can be broadly classified into three categories: sequenced based approaches, structure-based approaches and homology modeling and threading (discussed in Section "Computational Methods for Investigating Protein-DNA Interactions") (Si *et al.*, 2015).

Many sequence- and structure-based methods for prediction of DNA-binding site and DNA-binding proteins rely on the identification of homologous sequences/structures with known DNA-binding site information. In sequence-based methods, based on sequence alignment of the homologous sequence and query protein, DNA binding site is inferred from the homologous sequence. Structure-based methods are typically utilized when the crystal structure of the protein of interest is known. In this approach, the probable DNA binding site is inferred by comparing the binding site of the known DNA binding proteins and the query protein **(Fig. 2)**. Homology based methods are effective when similar proteins with known binding sites are available, which is often not the case for many DNA-binding proteins. Several studies have utilized non-comparative sequence and structure-based methods for identification of DNA-binding site and DNA binding proteins. These methods try to develop an association between the *descriptors* or *features* of subsequences or substructures in a protein and implicitly model their propensity individually or cumulatively to binding DNA (Ahmad and Sarai, 2005; Yan *et al.*, 2006; Hwang *et al.*, 2007; Ofran *et al.*, 2007; Wu *et al.*, 2009; Carson *et al.*, 2010; Alibés *et al.*, 2010b; Xiong *et al.*, 2011; Li *et al.*, 2013, 2014a,b; Zhao *et al.*, 2014; Peled *et al.*, 2016). Various sequence-based features include amino acid sequence/residue type, sequence conservation, global composition of amino acids.

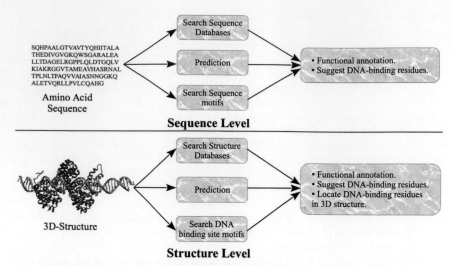

Fig. 2 Sequence and structure based predictions of protein-DNA interactions. Two groups of widely investigated problems are (a) identification of DNA-binding residues for known DBPs and (b) identifying novel candidate DBPs from a list of proteins.

Structure-based features include structural motifs, structural neighborhood, structural flexibility, secondary structure, accessible surface area (ASA), hydrophobicity, electrostatic potentials, net charge, and dipole and quadrupole moments (Si *et al.*, 2015, Ahmad and Sarai, 2004). Once the features have been identified, the next step is to develop a computational model that can find a quantitative predictability of DNA binding from these features. Among the earliest efforts, Ahmad *et al.* (2004) examined the protein DNA binding residues in terms of local amino acid composition, solvent accessible surface area, and secondary structure. Employing this knowledge a neural network model was developed to identify the DNA binding site and DNA binding proteins directly from sequence. Later on, the position-specific scoring matrix (PSSM) profiles were also incorporated into the model to improve the accuracy of the DNA-binding site prediction (Ahmad and Sarai, 2005). Utilizing the same dataset as by Ahmad *et al.*, Kuznetsov *et al.* (2006) used sequence and structure features to identify DNA binding site employing Support Vector Machine, a supervised pattern recognition method. Yan *et al.* (2006) utilized the Naive Bayes classifier and sequence based features (identity of central and neighboring amino acid residues) to predict the DNA binding residues in DNA binding proteins. Using amino acid residue properties *i.e.* molecular mass, hydrophobicity index and side chain *pka* values, Wang and Brown (2006) developed SVM classifier for identification of DNA binding site from primary sequence data. Hwang *et al.* (2007) employed machine learning techniques (kernel logistic regression, support vector machine and penalized logistic regression) for prediction of DNA binding sites from primary sequences with/without the use of PSSM profiles. Employing features derived from sequence and structure, Bhardwaj and Lu (2007) used the SVM classifier to predict DNA binding residues. Chu *et al.* (2009) used protein secondary structures for identification of DNA binding sites in transcription factors. More recently, Zhou *et al.* (2017) have utilized a novel residue encoding method termed as Position Specific Score Matrix (PSSM) Relation Transformation (PSSM-RT), to encode amino acid residues using the evolutionary information between residues and an ensemble learning classifier (EL_PSSM-RT) for prediction of DNA binding site. In a study by Yan and Kurgan (2017), authors have developed a novel method that accurately predicts DNA- and RNA-binding residues as well as DNA- and RNA-binding proteins using sequence-derived features (physicochemical and biochemical properties) and logistic regression model.

Several machine-learning approaches have also been developed for the related problem of identifying the DNA binding sites in proteins using evolutionary information. Similar to prediction of DNA binding residues, both sequence based (Cai and Lin, 2003; Yu *et al.*, 2006; Kumar *et al.*, 2007, 2009; Langlois and Lu, 2010; Huang *et al.*, 2011; Lin *et al.*, 2011) and structure-based methods (Stawiski *et al.*, 2003; Ahmad and Sarai, 2004; Bhardwaj and Lu, 2007; Szilágyi and Skolnick, 2006; Nimrod *et al.*, 2009, 2010; Zhou and Yan, 2011; Zhao *et al.*, 2014) have been developed for the identification of DNA binding proteins. Cai and Lin (2003) utilized pseudo amino acid composition and a group of nonlinear features derived from primary protein sequence to develop SVM for prediction of DNA binding proteins. Yu *et al.* (2006) utilizing sequence derived physicochemical properties and SVM proposed the binary classifications for rRNA-, RNA-, and DNA-binding proteins. Shao *et al.* (2009) developed a classifier based on SVM to differentiate between DNA/RNA binding proteins from non-nucleic acid binding proteins. Kumar *et al.* (2009) proposed a method, DNA-Prot, to determine DNA-binding proteins from protein sequence based on random forest. Later, Lin *et al.* (2011) developed a new method, iDNA-Prot, for annotating uncharacterized proteins as DNA-binding proteins or non-DNA-binding proteins. Later, in a study by Ma and Wu (2013), authors developed a new approach for identifying DNA-binding proteins using sequence-derived knowledge and support vector machine-sequential minimal optimization (SVM-SMO) algorithm. In a study by Peled *et al.* (2016), authors have utilized biophysical features and random forest to discover DNA and RNA binding proteins. In a recent study by Paz *et al.* (2016) authors have developed a non-homology based approach for prediction of DNA- and RNA-binding proteins. The method utilizes the structure-based features (electrostatic features and general properties of the protein), given the three-dimensional structure of the protein. Wei *et al.* (2017) have developed a novel approach, Local-DPP, combining local Pse-PSSM (Pseudo Position-Specific Scoring Matrix) with random forest classifier to predict DNA-binding proteins.

Prediction of TF Binding Sites

In the preceding section, we have discussed the identification of DNA-binding site on protein; however, in protein-DNA interaction, DNA also contains specific sites, which are recognized by the protein. These transcription factors (TF), the proteins which bind at specific sites on genomic DNA regulate the process of transcription. The presence of transcription factor on the DNA will attract or impede RNA polymerase, thereby promoting or repressing gene expression, respectively. To gain insight into the overall picture of the functioning genome, it is crucial to determine the genomic positions to which transcription factors bind. Transcription Factor Binding Sites (TFBSs) prediction is one of the well-studied domains in the area of protein-DNA interaction and continues to progress rapidly, as the subject moves from the knowledge of only a few binding events to genome-wide data sets. Various experimental and computational methods have been developed to characterize the binding site of TFs (reviewed in Xie *et al.*, 2011; Bulyk, 2003). One of the simplest computational approaches, used for the identification of TFBSs is searching for the over-represented nucleotide sequences using motif-detection algorithms. But, since the binding sites of transcription factors are short nucleotide sequences and are tolerant to sequence variations, plus this approach (motif identification) requires a large number of sequences for pattern identification, this approach is severely limited by low accuracy. Sequence-based computational methods for prediction of transcription factor binding sites (TFBSs) are often based on Position Weight Matrices (PWMs), which reflects the degeneracy observed TFBSs among the binding motifs corresponding to a particular transcription factor. PWMs have been utilized by various computational methods to identify the binding site of TFs (reviewed in Bulyk, 2003). In a recent study Andrabi *et al.* (2017) have shown that the information about the TF binding lies not only at specific binding positions but also extends as far as 200 nt away from the TFBSs.

Another successful approach employed for the prediction of TFBSs is structure-based prediction of TFBSs, which makes use of the available three-dimensional structure of protein-DNA complexes. Structure-based prediction of TFBSs has numerous advantages over sequence-based methods. For example, they can also describe the biophysics of interaction, not possible from a simple sequence-level analysis. However, there are several issues in structure-based prediction of TFBSs. One of the crucial issues is the choice of the scoring function used for estimating the binding energy or affinity of the protein-DNA complex. Two types of scoring functions are usually utilized, physics-based molecular mechanics force fields and the knowledge-based statistical potentials. Molecular mechanics energy function comprises of physicochemical interactions which include van der Waals (VDW) interaction, electrostatic interactions, solvation energy, *etc.* (Liu and Bradley, 2012). Molecular mechanics energy functions are based on approximations and assume fixed charges. They have been successfully utilized in investigating protein–DNA interactions (Havranek *et al.*, 2004; Morozov *et al.*, 2005; Siggers and Honig, 2007; Alibés *et al.*, 2010a). Besides the electrostatic and VDW interactions, the physics-based energy function also takes into account the hydrogen bonds, π-cation, and π–π interactions, on the thought that these interactions play a significant role in the stability of the protein-ligand complex. Although the molecular mechanics energy functions can comprehensively illustrate the protein–DNA interactions, they are typically time-consuming. Structure-based methods only consider few snapshots for the prediction of TFBSs, which represents a very small fraction of possible conformations of protein–DNA complexes and because molecular mechanics energy functions are sensitive to conformational alterations, these may result in incorrect identification of TFBSs.

Knowledge-based statistical potentials are deduced from the statistical study of known protein–DNA complexes. They are frequently utilized because they are computationally inexpensive and are comparatively less susceptible to conformational transitions than physics-based molecular mechanics energy functions. Knowledge-based statistical potentials ranges in resolution from atom-based potentials (Zhang *et al.*, 2005; Donald *et al.*, 2007; Robertson and Varani, 2007) to residue-based potentials (Mandel-Gutfreund and Margalit, 1998; Aloy *et al.*, 1998; Liu, 2005a,b; Takeda *et al.*, 2013). Recent studies have shown that residue-based statistical potentials also work well in the case of protein-DNA interaction (Liu, 2005a,b; Takeda *et al.*, 2013). Knowledge-based statistical potentials also differ in terms of their applicability to distance ranges e.g., distance-independent (Mandel-Gutfreund and Margalit, 1998; Aloy *et al.*, 1998) and distance dependent functions (Liu, 2005a,b; Zhang *et al.*, 2005; Robertson and Varani, 2007; Takeda *et al.*, 2013). Knowledge-based statistical potentials are limited by two factors. First being the mean force makeup of the statistical potentials. For example, lysine and arginine make specific and nonspecific interactions with DNA in the form of hydrogen bonds and electrostatic interactions. Though knowledge-based statistical potentials could capture the hydrogen bonds implicitly, they are averaged out with the nonspecific interactions. The other originates due to low frequency problem. Recent studies have shown that the interaction between the DNA and aromatic amino acids are abundant; however, very limited knowledge is available about these specific interactions (Wilson *et al.*, 2014; Wilson and Wetmore, 2015). Corona and Guo (2016), through comparative analysis have shown that the interaction between the DNA bases and amino acid residues Tyr and His are enriched in highly specific DNA-binding proteins and might contribute directly to the TF binding specificity (Farrel *et al.*, 2016). Some studies have also shown that combining knowledge-based potential with other interactions like hydrogen bond and π interaction improves the accuracy of the TFBSs prediction when compared with atomic-level and residue-level knowledge-based potentials.

Experimental Methods to Investigate Protein-DNA Interactions

In view of various essential roles played by protein-DNA interactions, various experimental methods have evolved over time to elucidate them. Apart from constantly improving technologies to analyze interactions in greater detail and accuracy, some method

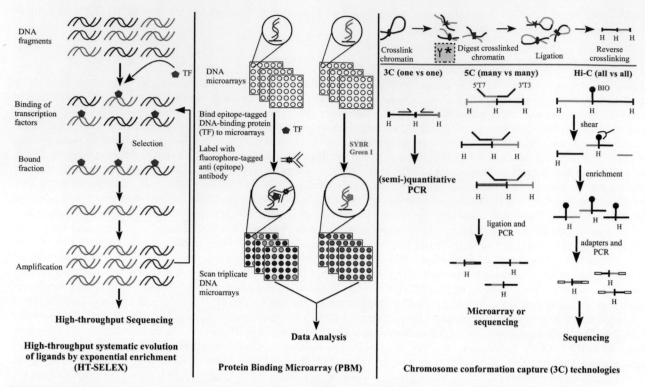

Fig. 3 Representative experimental techniques to study protein-DNA interactions: (a) HT-SELEX in which TF is allowed to bind a diverse set of DNA sequences, which are picked up using an antibody and then the sequencing of attached DNA is carried out to infer binding patterns (b) Protein-binding Microarray use the technique of hybridization with pre-selected DNA oligomers. A relatively long sequence can be used to interrogate interactions with many sub-sequences and Bioinformatics approaches help in designing an optimum chip and (c) Chromatin conformation capture (3C) and their variants which can reveal the larger-scale contact maps between different parts of genomic DNA as a result of nucleosome formation.

serve specific purposes and have their own trade-offs. In this section, we will discuss popularly used experimental methods to interrogate protein-DNA interactions. Overall, experimental techniques can be grouped into (a) *In vitro* binding experiments (b) NGS-based methods and (c) other methods. **Fig. 3** gives the schematic representation of the experimental techniques used in investigating protein-DNA interactions. They include protein-binding microarrays, high throughput systematic evolution of ligands by exponential enrichment (HT-SELEX) and Chromosome Conformational Capture (3C). They are briefly reviewed in the following:

In Vitro Binding Experiments

Various *in vitro* techniques popularly used to assay protein-DNA interactions are Protein binding microarray (PBM), High-throughput systematic evolution of ligands by exponential enrichment (HT-SELEX) (Jolma *et al.*, 2010), High-throughput sequencing–fluorescent ligand interaction profiling (HiTS–FLIP), Mechanically induced trapping of molecular interactions (MITOMI), and DIP–chip and DIP–seq (DNA immunoprecipitation followed by microarray (DIP–chip) and DIP followed by high-throughput sequencing (DIP–seq)). All these methods allow high-throughput characterization of DNA-binding proteins. A brief description of these methods is provided below.

Protein binding microarray (PBM)

PBM is a DNA microarray-based technology to ascertain the binding preferences of a transcription factor towards DNA. In this technique, first, the DNA-binding protein is expressed, purified and then is allowed to bind directly to a DNA microarray. The array is then tagged with a fluorescent antibody, which is further utilized to measure the relative amount of protein bound to each probe. The probes in the DNA microarray often consist of large number of possible k-mers. Unique k-mers needed to represent much larger set of subsequences can be designed using *de Bruijn* algorithm (Vassallo and Ralston, 1992).

High-throughput systematic evolution of ligands by exponential enrichment (HT-SELEX)

In this technique, a double-stranded DNA mixture is incubated with a DNA-binding protein immobilized onto a well-plate. The mixture is then washed, and the bound oligonucleotides are retrieved, amplified by PCR and sequenced to identify the TF binding specificities (Ogawa and Biggin, 2012).

High-throughput sequencing–fluorescent ligand interaction profiling (HiTS–FLIP)

It is a high throughput technology to perform DNA sequencing and systematic protein-DNA binding experiment on the flowcell. This allows determining the DNA-binding specificities of a transcription factor (Nutiu *et al.*, 2011).

Mechanically induced trapping of molecular interactions (MITOMI)

It is a high-throughput microfluidic platform widely used to study protein-DNA interactions. In this method, a microfluidic equipment is aligned to a microarray containing programmed DNA, such that the TFs (located onto the surface) can bind to the DNA (Maerkl and Quake, 2007). The equipment allows mechanical trapping of the protein-DNA complexes, thus protecting them from being washed out. The device is then examined to quantify binding of TFs.

DNA immunoprecipitation followed by microarray (DIP–chip) and DIP followed by high-throughput sequencing (DIP–seq)

It is a high throughput method for identifying DNA regions that bind to DNA-binding proteins or transcription factors on a genome-wide scale. In this technique, DNA-binding protein is allowed to bind with genomic DNA, and the bound complexes are then segregated using immunoprecipitation or affinity purification (Gossett and Lieb, 2008; Liu, 2005a,b). The DNA fragments are then purified, amplified, fluorescently labeled and characterized either by hybridization to a DNA microarray or high-throughput sequencing.

NGS-Based Methods

ChIP-Seq and DNase-Seq

With the advancement in the next generation sequencing technologies, ChIP-Seq and DNase-Seq have become the preferred choice of *in vivo* techniques to elucidate the protein-DNA interactions (Meyer and Liu, 2014). Both these technologies estimate the binding occupancy of a DNA-binding protein to the nucleotide sequences.

In ChIP-Seq, the cells are first treated with chemical crosslinks which allows cross linking (covalent binding) of the proteins to each other and then to the DNA. Once the protein-DNA are crosslinked, the chromatin is extracted and DNA is sheared using sonication and the bound complex is isolated utilizing antibodies specific to the protein of interest. The cross-links are then reversed and purified DNA segments are characterized using sequencing.

DNase-seq is a more recent technology developed to obtain the genome-wide protein-DNA binding map. In DNase-seq, protein-DNA complexes are digested with DNase I which preferentially digests the DNA not bound to proteins. The DNA fragments are then extracted, purified and sequenced.

DNAse-Seq can be used to interrogate binding sites of large number of TFs in a single experiment. However, the challenge is to assign the identified site to their corresponding TF and powerful Bioinformatics techniques have been developed to address this issue exclusively (Rajagopal *et al.*, 2016).

Conformation capture by 3C, 5C and HiC

The topological structures of the genomic DNA play a significant role in coordinating transcription and other DNA-dependent metabolic processes and thus are of great importance. Chromosome conformation capture (3C) technology developed by Dekker *et al.* (2002) are one of the recent technologies which allow study of chromatin structure in their native cellular environment with high resolution. All 3C and 3C-derived technologies (4C (chromosome conformation capture-on-chip), 5C (chromosome conformation capture carbon copy), HiC, ChIP-loop and ChIA-PET) follow a similar set of protocols to study the 3D organization of the genomic DNA. In 3C methods, first, the genomic DNA is fixed, utilizing fixative agents like formaldehyde (Dekker *et al.*, 2002). Then the chromatin is cut using restriction enzymes and the topologically proximal regions are re-ligated to allow intramolecular crosslinking. The fragments close to each other ligate, thus quantifying the proximity of fragments. The primary difference between 3C methods lies in the fragments of interest. In 3C, also referred as "one-vs-one", the two specific regions are probed for interaction, which is usually known *a priori*. In 5C, also known as "many-vs-many", all fragments in a given region are probed for interaction between them. Hi-C (all-vs-all) is an unbiased technique to identify interaction between all potential fragments.

Other Methods

Apart from the techniques discussed above, a plethora of other *in-vitro* and *in-vivo* techniques are also available to interrogate protein-DNA interactions. They include Electrophoretic mobility shift assay (EMSA) (Hellman and Fried, 2007), Yeast one-hybrid assay (Y1H) (Reece-Hoyes and Marian Walhout, 2012), Proximity ligation assay (PLA) (Gustafsdottir *et al.*, 2007), X-ChIP, Fast ChIP, Carrier ChIP, Native-ChIP (N-ChIP), Matrix ChIP, ChIP-Chip (Tong and Falk, 2009), ChIP display (Barski, 2004), other variations of ChIP such as Quick and Quantitative ChIP (Q^2 ChIP) and MicroChIP, ChiP-loop and ChIA-PET (Li *et al.*, 2014a,b; de Wit and de Laat, 2012), for which the readers are directed to relevant literature.

Computational Methods for Investigating Protein-DNA Interactions

As discussed above, protein-DNA interactions occur in complex and diverse cellular conditions, which are difficult to observe in their native form. Crystal structures do give us a closer look at the atomic level determinants of molecular interactions. However, on the one hand, many protein structures remain unsolved and on the other, it remains unknown if the crystal structures solved *in vitro* represent real cellular context. Some of our data suggests that the three dimensional structures of protein-DNA complexes are cell-type dependent (Andrabi *et al.* Bioarxive 2015). To address the first of these two issues efforts are made to model protein structures using comparative modeling and predictive models and for the second first principle computations have been attempted. Some of the specific cases have been discussed above. From a technical perspective, the techniques used for modeling DNA-binding proteins are often based on homology modeling and threading (Morozov *et al.*, 2005). A high quality template is an essential requirement for homology modeling and in the absence of a similar protein structure available, data driven approach such as machine learning are a handy choice. For the second issue, *ab initio* physics based techniques can try to emulate the cellular contexts and efforts are going on to bring such techniques as close to the experimental conditions as possible. Homology modeling, Molecular dynamics (MD), Monte Carlo (MC) and machine learning (ML) principles applied to model protein-DNA interactions are not unique to them but are essentially an adaptation from other applications. The technical considerations behind the application of MD, MC and ML techniques have been frequently reviewed and the same can be referred to elsewhere (Tuckerman and Martyna, 2000; Ding *et al.*, 2010; Paquet and Viktor, 2015).

Summary and Conclusions

Protein-DNA interactions are essential components of all biological systems, fundamental to almost all biological processes. Tremendous efforts have been made to understand them employing (a) first principle science such as MD and MC (b) experimental techniques such as ChIP-Seq and HiC and (c) data driven approaches such as ML and statistics. It is not possible to predict DNA-binding sites from sequence with high confidence build homology based models for many novel proteins as well as direct query the nature of genome-wide interactions between TFs and their target DNAs. The studies have so far revealed the role of sequence and structure/shape of the protein and DNA sequences in the recognition process. The current trends are to look at the DNA topology in genome-wide recognition, cell-type specific structures and role for non-binding flanking regions of DNA playing an allosteric role in protein-DNA interactions. Data driven techniques are also going to be a crucial tool to interrogate such large-scale interactions, due to the availability of unprecedented amounts of data emerging from NGS-based techniques under many phenotypic contexts. Even though, no universal recognition code of DNA is yet described, more and more insights are emerging and controlling protein-DNA interactions will be become easier with the growing knowhow.

See also: Chromatin: A Semi-Structured Polymer. Computational Prediction of Nucleic Acid Binding Residues from Sequence. Computational Systems Biology. Genome-Wide Scanning of Gene Expression. Natural Language Processing Approaches in Bioinformatics. Nucleic-Acid Structure Database. Prediction of Protein-Binding Sites in DNA Sequences. Protein–Protein Interaction Databases. Regulation of Gene Expression. Sequence Analysis. Transcription Factor Databases

References

Ahmad, S., Gromiha, M.M., Sarai, A., 2004. Analysis and prediction of DNA-binding proteins and their binding residues based on composition, sequence and structural information. Bioinformatics 20, 477–486. Available at: http://www.ncbi.nlm.nih.gov/pubmed/14990443.

Ahmad, S., Kono, H., Araúzo-Bravo, M.J., Sarai, A., 2006. ReadOut: Structure-based calculation of direct and indirect readout energies and specificities for protein-DNA recognition. Nucleic Acids Res. 34, W124–W127. Available at: http://www.ncbi.nlm.nih.gov/pubmed/16844974.

Ahmad, S., Sarai, A., 2004. Moment-based prediction of DNA-binding proteins. J. Mol. Biol. 341, 65–71. Available at: http://www.sciencedirect.com/science/article/pii/S0022283604006382.

Ahmad, S., Sarai, A., 2005. PSSM-based prediction of DNA binding sites in proteins. BMC Bioinformatics 6, 33. Available at: http://www.pubmedcentral.nih.gov/articlerender.fcgi?artid=550660&tool=pmcentrez&rendertype=abstract.

Akira, S., Uematsu, S., Takeuchi, O., 2006. Pathogen recognition and innate immunity. Cell 124, 783–801. Available at: http://www.ncbi.nlm.nih.gov/pubmed/16497588.

Akke, M., 2002. NMR methods for characterizing microsecond to millisecond dynamics in recognition and catalysis. Curr. Opin. Struct. Biol. 12, 642–647. Available at: http://www.ncbi.nlm.nih.gov/pubmed/12464317.

Alibés, A., Nadra, A.D., De Masi, F., *et al.*, 2010a. Using protein design algorithms to understand the molecular basis of disease caused by protein-DNA interactions: The Pax6 example. Nucleic Acids Res. 38, 7422–7431. Available at: http://www.ncbi.nlm.nih.gov/pubmed/20685816.

Alibés, A., Serrano, L., Nadra, A.D., 2010b. Structure-based DNA-binding prediction and design. In: Mackay, J.P., Segal, D.J. (Eds.), Engineered zinc finger proteins: Methods and protocols. Totowa, NJ: Humana Press, pp. 77–88. Available at: https://doi.org/10.1007/978-1-60761-753-2_4.

Aloy, P., Moont, G., Gabb, H.A., *et al.*, 1998. Modelling repressor proteins docking to DNA. Proteins 33, 535–549.

Andrabi, M., Hutchins, A.P., Miranda-Saavedra, D., *et al.*, 2017. Predicting conformational ensembles and genome-wide transcription factor binding sites from DNA sequences. Sci. Rep. 7, 4071. Available at: http://www.nature.com/articles/s41598-017-03199-6.

Andrabi, M., Mizuguchi, K., Ahmad, S., 2014. Conformational changes in DNA-binding proteins: Relationships with precomplex features and contributions to specificity and stability. Proteins Struct. Funct. Bioinform. 82, 841–857. Available at: http://www.ncbi.nlm.nih.gov/pubmed/24265157.

Announcement, 1996. Mol. Cell. Biochem. 159, 170.

Bardet, A.F., Steinmann, J., Bafna, S., et al., 2013. Identification of transcription factor binding sites from ChIP-seq data at high resolution. Bioinformatics 29, 2705–2713.

Barski, A., 2004. ChIP Display: Novel method for identification of genomic targets of transcription factors. Nucleic Acids Res., 32. e104–e104. Available at: https://academic.oup.com/nar/article-lookup/doi/10.1093/nar/gnh097.

Beloin, C., Jeusset, J., Révet, B., et al., 2003. Contribution of DNA conformation and topology in right-handed DNA wrapping by the bacillus subtilis LrpC protein. J. Biol. Chem. 278, 5333–5342. Available at: http://www.jbc.org/lookup/doi/10.1074/jbc.M207489200.

Berg, O.G., von Hippel, P.H., 1985. Diffusion-controlled macromolecular interactions. Annu. Rev. Biophys. Biophys. Chem. 14, 131–160. Available at: http://www.ncbi.nlm.nih.gov/pubmed/3890878.

Berg, O.G., Winter, R.B., von Hippel, P.H., 1981. Diffusion-driven mechanisms of protein translocation on nucleic acids. 1. Models and theory. Biochemistry 20, 6929–6948. Available at: http://www.ncbi.nlm.nih.gov/pubmed/7317363.

Bhardwaj, N., Lu, H., 2007. Residue-level prediction of DNA-binding sites and its application on DNA-binding protein predictions. FEBS Lett. 581, 1058–1066. Available at: http://www.sciencedirect.com/science/article/pii/S0014579307001263.

Bhattacherjee, A., Levy, Y., 2014a. Search by proteins for their DNA target site: 1. The effect of DNA conformation on protein sliding. Nucleic Acids Res. 42, 12404–12414. Available at: http://www.ncbi.nlm.nih.gov/pubmed/25324308.

Bhattacherjee, A., Levy, Y., 2014b. Search by proteins for their DNA target site: 2. The effect of DNA conformation on the dynamics of multidomain proteins. Nucleic Acids Res. 42, 12415–12424. Available at: https://academic.oup.com/nar/article-lookup/doi/10.1093/nar/gku933.

Brünger, A.T., Adams, P.D., Clore, G.M., et al., 1998. Crystallography & NMR system: A new software suite for macromolecular structure determination. Acta Crystallogr. D: Biol. Crystallogr. 54, 905–921. Available at: http://www.ncbi.nlm.nih.gov/pubmed/9757107.

Bulyk, M.L., 2003. Computational prediction of transcription-factor binding site locations. Genome Biol. 5, 201. Available at: http://www.ncbi.nlm.nih.gov/pubmed/14709165%5Cnhttp://www.pubmedcentral.nih.gov/articlerender.fcgi?artid=PMC395725.

Cai, Y., Lin, S.L., 2003. Support vector machines for predicting rRNA-, RNA-, and DNA-binding proteins from amino acid sequence. Biochim. Biophys. Acta 1648, 127–133. Available at: http://www.sciencedirect.com/science/article/pii/S1570963903001122.

Carson, M.B., Langlois, R., Lu, H., 2010. NAPS: A residue-level nucleic acid-binding prediction server. Nucleic Acids Res. 38, W431–W435. Available at: http://www.ncbi.nlm.nih.gov/pubmed/20478832.

Chu, W.-Y., Huang, Y.-F., Huang, C.-C., et al., 2009. ProteDNA: A sequence-based predictor of sequence-specific DNA-binding residues in transcription factors. Nucleic Acids Res. 37, W396–W401. Available at: http://www.ncbi.nlm.nih.gov/pubmed/19483101.

Chu, X., Liu, F., Maxwell, B.A., et al., 2014. Dynamic conformational change regulates the protein-DNA recognition: An investigation on binding of a Y-family polymerase to its target DNA. PLOS Comput. Biol. 10, e1003804. Available at: http://www.ncbi.nlm.nih.gov/pubmed/25188490.

Corona, R.I., Guo, J.-T., 2016. Statistical analysis of structural determinants for protein-DNA-binding specificity. Proteins 84, 1147–1161. Available at: http://www.ncbi.nlm.nih.gov/pubmed/24655651.

Dalrymple, B.P., Kongsuwan, K., Wijffels, G., Dixon, N.E., Jennings, P.A., 2001. A universal protein-protein interaction motif in the eubacterial DNA replication and repair systems. Proc. Natl Acad. Sci. 98, 11627–11632. Available at: http://www.pnas.org/cgi/doi/10.1073/pnas.191384398.

Dekker, J., Rippe, K., Dekker, M., Kleckner, N., 2002. Capturing chromosome conformation. Science 295, 1306–1311. Available at: http://www.sciencemag.org/cgi/doi/10.1126/science.1067799.

Dell'Oste, V., Gatti, D., Giorgio, A.G., et al., 2015. The interferon-inducible DNA-sensor protein IFI16: A key player in the antiviral response. New Microbiol. 38, 5–20. Available at: http://www.ncbi.nlm.nih.gov/pubmed/25742143.

Desjardins, G., Okon, M., Graves, B.J., McIntosh, L.P., 2016. Conformational dynamics and the binding of specific and nonspecific DNA by the autoinhibited transcription factor Ets-1. Biochemistry 55, 4105–4118. Available at: http://www.ncbi.nlm.nih.gov/pubmed/27362745.

de Wit, E., de Laat, W., 2012. A decade of 3C technologies: Insights into nuclear organization. Genes Dev. 26, 11–24. Available at: http://genesdev.cshlp.org/cgi/doi/10.1101/gad.179804.111.

Dillon, S.C., Dorman, C.J., 2010. Bacterial nucleoid-associated proteins, nucleoid structure and gene expression. Nat. Rev. Microbiol. 8, 185–195. Available at: http://www.nature.com/doifinder/10.1038/nrmicro2261.

Ding, X.-M., Pan, X.-Y., Xu, C., Shen, H.-B., 2010. Computational prediction of DNA-protein interactions: A review. Curr. Comput. Aided-Drug Des. 6, 197–206. Available at: http://www.eurekaselect.com/openurl/content.php?genre=article&issn=1573-4099&volume=6&issue=3&spage=197

Dominguez, C., Boelens, R., Bonvin, A.M.J.J., 2003. HADDOCK: A protein — protein docking approach based on biochemical or biophysical information. J. Am. Chem. Soc. 125, 1731–1737. Available at: http://pubs.acs.org/doi/abs/10.1021/ja026939x.

Donald, J.E., Chen, W.W., Shakhnovich, E.I., 2007. Energetics of protein–DNA interactions. Nucleic Acids Res. 35, 1039–1047. Available at: https://academic.oup.com/nar/article-lookup/doi/10.1093/nar/gkl1103.

Esadze, A., Kemme, C.A., Kolomeisky, A.B., Iwahara, J., 2014. Positive and negative impacts of nonspecific sites during target location by a sequence-specific DNA-binding protein: Origin of the optimal search at physiological ionic strength. Nucleic Acids Res. 42, 7039–7046. Available at: http://www.ncbi.nlm.nih.gov/pubmed/24838572.

Etheve, L., Martin, J., Lavery, R., 2016. Protein–DNA interfaces: A molecular dynamics analysis of time-dependent recognition processes for three transcription factors. Nucleic Acids Res. gkw841. Available at: https://academic.oup.com/nar/article-lookup/doi/10.1093/nar/gkw841.

Farnham, P.J., 2009. Insights from genomic profiling of transcription factors. Nat. Rev. Genet. 10, 605–616. Available at: http://www.nature.com/doifinder/10.1038/nrg2636.

Farrel, A., Murphy, J., Guo, J., 2016. Structure-based prediction of transcription factor binding specificity using an integrative energy function. Bioinformatics 32, i306–i313. Available at: https://academic.oup.com/bioinformatics/article-lookup/doi/10.1093/bioinformatics/btw264.

Fischle, W., Wang, Y., Allis, C.D., 2003. Histone and chromatin cross-talk. Curr. Opin. Cell Biol. 15, 172–183. Available at: http://linkinghub.elsevier.com/retrieve/pii/S0955067403000139.

Fuxreiter, M., Simon, I., Bondos, S., 2011. Dynamic protein-DNA recognition: Beyond what can be seen. Trends Biochem. Sci. 36, 415–423. Available at: http://www.ncbi.nlm.nih.gov/pubmed/21620710.

Garton, M., Laughton, C., 2013. A comprehensive model for the recognition of human telomeres by TRF1. J. Mol. Biol. 425, 2910–2921. Available at: http://www.ncbi.nlm.nih.gov/pubmed/23702294.

Garvie, C.W., Wolberger, C., 2001. Recognition of specific DNA sequences. Mol. Cell 8, 937–946. Available at: http://www.ncbi.nlm.nih.gov/pubmed/8303294.

Gasser, S.M., Laemmli, U.K., 1987. A glimpse at chromosomal order. Trends Genet. 3, 16–22. Available at: https://doi.org/10.1016/0168-9525(87)90156-9.

Gossett, A.J., Lieb, J.D., 2008. DNA Immunoprecipitation (DIP) for the determination of DNA-binding specificity. Cold Spring Harb. Protoc. 2008.pdb.prot4972-prot4972 Available at: http://www.cshprotocols.org/cgi/doi/10.1101/pdb.prot4972.

Gustafsdottir, S.M., Schlingemann, J., Rada-Iglesias, A., et al., 2007. In vitro analysis of DNA-protein interactions by proximity ligation. Proc. Natl. Acad. Sci. 104, 3067–3072. Available at: http://www.pnas.org/cgi/doi/10.1073/pnas.0611229104.

Halford, S.E., Marko, J.F., 2004. How do site-specific DNA-binding proteins find their targets? Nucleic Acids Res. 32, 3040–3052. Available at: http://www.ncbi.nlm.nih.gov/pubmed/10336412.

Hashimoto, M., Imhoff, B., Ali, M.M., Kow, Y.W., 2003. HU protein of Escherichia coli has a role in the repair of closely opposed lesions in DNA. J. Biol. Chem. 278, 28501–28507. Available at: http://www.ncbi.nlm.nih.gov/pubmed/12748168.

Havranek, J.J., Duarte, C.M., Baker, D., 2004. A simple physical model for the prediction and design of protein-DNA interactions. J. Mol. Biol. 344, 59–70. Available at: http://www.ncbi.nlm.nih.gov/pubmed/15504402.

Hellman, L.M., Fried, M.G., 2007. Electrophoretic mobility shift assay (EMSA) for detecting protein–nucleic acid interactions. Nat. Protoc. 2, 1849–1861. Available at: http://www.nature.com/nprot/journal/v2/n8/abs/nprot.2007.249.html.

Herzner, A.-M., Hagmann, C.A., Goldeck, M., et al., 2015. Sequence-specific activation of the DNA sensor cGAS by Y-form DNA structures as found in primary HIV-1 cDNA. Nat. Immunol. 16, 1025–1033. Available at: http://www.ncbi.nlm.nih.gov/pubmed/16497588.

Hochrein, H., Schlatter, B., O'Keeffe, M., et al., 2004. Herpes simplex virus type-1 induces IFN-alpha production via Toll-like receptor 9-dependent and -independent pathways. Proc. Natl. Acad. Sci. USA. 101, 11416–11421. Available at: http://www.ncbi.nlm.nih.gov/pubmed/15272082.

Honnappa, S., John, C.M., Kostrewa, D., Winkler, F.K., Steinmetz, M.O., 2005. Structural insights into the EB1-APC interaction. EMBO J. 24, 261–269. Available at: http://emboj.embopress.org/cgi/doi/10.1038/sj.emboj.7600529.

Hou, L., Srivastava, Y., Jauch, R., 2017. Molecular basis for the genome engagement by Sox proteins. Semin. Cell Dev. Biol. 63, 2–12. Available at: https://doi.org/10.1016/j.semcdb.2016.08.005.

Huang, H.-L., Lin, I.-C., Liou, Y.-F., et al., 2011. Predicting and analyzing DNA-binding domains using a systematic approach to identifying a set of informative physicochemical and biochemical properties. BMC Bioinformatics 12 (Suppl 1), S47. Available at: http://www.biomedcentral.com/1471-2105/12/S1/S47.

Hwang, S., Gou, Z., Kuznetsov, I.B., 2007. DP-Bind: A web server for sequence-based prediction of DNA-binding residues in DNA-binding proteins. Bioinformatics 23, 634–636. Available at: http://www.ncbi.nlm.nih.gov/pubmed/17237068.

Jimenez-Sanchez, G., Childs, B., Valle, D., 2001. Human disease genes. Nature 409, 853–855. Available at: http://www.nature.com/doifinder/10.1038/35057050.

Johnson, D.S., Mortazavi, A., Myers, R.M., Wold, B., 2007. Genome-wide mapping of in vivo protein-DNA interactions. Science 316 (80-), 1497–1502. Available at: http://www.ncbi.nlm.nih.gov/pubmed/17540862.

Jolly, C., Morimoto, R.I., 2000. Role of the heat shock response and molecular chaperones in oncogenesis and cell death. J. Natl. Cancer Inst. 92, 1564–1572. Available at: http://www.ncbi.nlm.nih.gov/pubmed/11018092.

Jolma, A., Kivioja, T., Toivonen, J., et al., 2010. Multiplexed massively parallel SELEX for characterization of human transcription factor binding specificities. Genome Res. 20, 861–873.

Jones, S., Barker, J.A., Nobeli, I., Thornton, J.M., 2003. Using structural motif templates to identify proteins with DNA binding function. Nucleic Acids Res. 31, 2811–2823. Available at: http://www.ncbi.nlm.nih.gov/pubmed/12771208.

Jones, S., van Heyningen, P., Berman, H.M., Thornton, J.M., 1999. Protein-DNA interactions: A structural analysis. J. Mol. Biol. 287, 877–896. Available at: http://www.ncbi.nlm.nih.gov/pubmed/10222198.

Kalodimos, C.G., 2004. Structure and flexibility adaptation in nonspecific and specific protein-DNA complexes. Science 305 (80-), 386–389. Available at: http://www.sciencemag.org/cgi/doi/10.1126/science.1097064.

Kalodimos, C.G., Bonvin, A.M.J.J., Salinas, R.K., et al., 2002. Plasticity in protein-DNA recognition: LAC repressor interacts with its natural operator 01 through alternative conformations of its DNA-binding domain. EMBO J. 21, 2866–2876. Available at: http://www.ncbi.nlm.nih.gov/pubmed/12065400.

Kamachi, Y., Cheah, K.S., Kondoh, H., 1999. Mechanism of regulatory target selection by the SOX high-mobility-group domain proteins as revealed by comparison of SOX1/2/3 and SOX9. Mol. Cell. Biol. 19, 107–120. Available at: http://www.ncbi.nlm.nih.gov/pubmed/9858536.

Kamashev, D., 2000. The histone-like protein HU binds specifically to DNA recombination and repair intermediates. EMBO J. 19, 6527–6535. Available at: http://www.pubmedcentral.nih.gov/articlerender.fcgi?artid=305869&tool=pmcentrez&rendertype=abstract%5Cnhttp://emboj.embopress.org/content/19/23/6527.abstract.

Kamashev, D.E., Esipova, N.G., Ebralidse, K.K., Mirzabekov, A.D., 1995. Mechanism of Lac repressor switch-off: Orientation of the Lac repressor DNA-binding domain is reversed upon inducer binding. FEBS Lett. 375, 27–30. Available at: http://www.ncbi.nlm.nih.gov/pubmed/7498473.

Kim, Y.C., Grable, J.C., Love, R., Greene, P.J., Rosenberg, J.M., 1990. Refinement of Eco RI endonuclease crystal structure: A revised protein chain tracing. Science 249, 1307–1309. Available at: http://www.sciencemag.org/cgi/doi/10.1126/science.2399465.

Kow, Y.W., Imhoff, B., Weiss, B., et al., 2007. Escherichia coli HU protein has a role in the repair of abasic sites in DNA. Nucleic Acids Res 35, 6672–6680. Available at: http://www.ncbi.nlm.nih.gov/pubmed/17916578.

Krug, A., French, A.R., Barchet, W., et al., 2004a. TLR9-dependent recognition of MCMV by IPC and DC generates coordinated cytokine responses that activate antiviral NK cell function. Immunity 21, 107–119. Available at: http://www.ncbi.nlm.nih.gov/pubmed/15345224.

Krug, A., Luker, G.D., Barchet, W., et al., 2004b. Herpes simplex virus type 1 activates murine natural interferon-producing cells through toll-like receptor 9. Blood 103, 1433–1437. Available at: http://www.ncbi.nlm.nih.gov/pubmed/14563635.

Kumar, K.K., Pugalenthi, G., Suganthan, P.N., 2009. DNA-Prot: Identification of DNA binding proteins from protein sequence information using random forest. J. Biomol. Struct. Dyn. 26, 679–686. Available at: http://www.ncbi.nlm.nih.gov/pubmed/19385697.

Kumar, M., Gromiha, M.M., Raghava, G.P.S., 2007. Identification of DNA-binding proteins using support vector machines and evolutionary profiles. BMC Bioinformatics 8, 463. Available at: http://www.ncbi.nlm.nih.gov/pubmed/18042272.

Kuznetsov, I.B., Gou, Z., Li, R., Hwang, S., 2006. Using evolutionary and structural information to predict DNA-binding sites on DNA-binding proteins. Proteins 64, 19–27. Available at: http://www.ncbi.nlm.nih.gov/pubmed/17705269.

Langlois, R.E., Lu, H., 2010. Boosting the prediction and understanding of DNA-binding domains from sequence. Nucleic Acids Res. 38, 3149–3158. Available at: http://www.ncbi.nlm.nih.gov/pubmed/20156993.

Lee, T.I., Young, R.A., 2013. Transcriptional regulation and its misregulation in disease. Cell 152, 1237–1251. Available at: https://doi.org/10.1016/j.cell.2013.02.014.

Lia, G., Bensimon, D., Croquette, V., et al., 2003. Supercoiling and denaturation in Gal repressor/heat unstable nucleoid protein (HU)-mediated DNA looping. Proc. Natl. Acad. Sci. 100, 11373–11377. Available at: http://www.pnas.org/cgi/doi/10.1073/pnas.2034851100.

Li, B.-Q., Feng, K.-Y., Ding, J., Cai, Y.-D., 2014a. Predicting DNA-binding sites of proteins based on sequential and 3D structural information. Mol. Genet. Genomics 289, 489–499. Available at: http://link.springer.com/10.1007/s00438-014-0812-x.

Lieb, J.D., Liu, X., Botstein, D., Brown, P.O., 2001. Promoter-specific binding of Rap1 revealed by genome-wide maps of protein-DNA association. Nat. Genet. 28, 327–334. Available at: http://www.ncbi.nlm.nih.gov/pubmed/11455386.

Li, G., Cai, L., Chang, H., et al., 2014b. Chromatin interaction analysis with paired-end tag (ChIA-PET) sequencing technology and application. BMC Genomics 15.S11 Available at: http://bmcgenomics.biomedcentral.com/articles/10.1186/1471-2164-15-S12-S11.

Lin, W.-Z., Fang, J.-A., Xiao, X., Chou, K.-C., 2011. iDNA-Prot: Identification of DNA binding proteins using random forest with grey model. PLoS One 6, e24756. Available at: http://dx.plos.org/10.1371/journal.pone.0024756.

Li, T., Li, Q.-Z., Liu, S., et al., 2013. PreDNA: Accurate prediction of DNA-binding sites in proteins by integrating sequence and geometric structure information. Bioinformatics 29, 678–685. Available at: http://www.ncbi.nlm.nih.gov/pubmed/23335013.

Liu, L.A., Bradley, P., 2012. Atomistic modeling of protein–DNA interaction specificity: Progress and applications. Curr. Opin. Struct. Biol. 22, 397–405. Available at: http://linkinghub.elsevier.com/retrieve/pii/S0959440X12000991.

Liu, X., 2005a. DIP-chip: Rapid and accurate determination of DNA-binding specificity. Genome Res. 15, 421–427. Available at: http://www.genome.org/cgi/doi/10.1101/gr.3256505.

Liu, Z., 2005b. Quantitative evaluation of protein-DNA interactions using an optimized knowledge-based potential. Nucleic Acids Res, 33. . pp. 546–558. Available at: https://academic.oup.com/nar/article-lookup/doi/10.1093/nar/gki204.

Luger, K., Mäder, A.W., Richmond, R.K., Sargent, D.F., Richmond, T.J., 1997. Crystal structure of the nucleosome core particle at 2.8 A resolution. Nature 389, 251–260. Available at: http://www.nature.com/doifinder/10.1038/38444.

Lund, J., Sato, A., Akira, S., Medzhitov, R., Iwasaki, A., 2003. Toll-like receptor 9-mediated recognition of Herpes simplex virus-2 by plasmacytoid dendritic cells. J. Exp. Med. 198, 513–520. Available at: http://www.ncbi.nlm.nih.gov/pubmed/12900525.

Luscombe, N.M., Austin, S.E., Berman, H.M., Thornton, J.M., 2000. An overview of the structures of protein-DNA complexes. Genome Biol. 1.REVIEWS001 Available at: http://genomebiology.com/2000/1/1/reviews/001.1%5Cnhttp://genomebiology.com/2000/1/1/reviews/001.

Luscombe, N.M., Laskowski, R.A., Thornton, J.M., 2001. Amino acid-base interactions: A three-dimensional analysis of protein-DNA interactions at an atomic level. Nucleic Acids Res, 29. . pp. 2860–2874. Available at: https://academic.oup.com/nar/article-lookup/doi/10.1093/nar/29.13.2860.

Maerkl, S.J., Quake, S.R., 2007. A systems approach to measuring the binding energy landscapes of transcription factors. Science 315 (80-), 233–237. Available at: http://www.sciencemag.org/cgi/doi/10.1126/science.1131007.

Mandel-Gutfreund, Y., Margalit, H., 1998. Quantitative parameters for amino acid-base interaction: Implications for prediction of protein-DNA binding sites. Nucleic Acids Res. 26, 2306–2312. Available at: http://www.ncbi.nlm.nih.gov/pubmed/9580679.

Manders, P., Thomas, R., 2000. Immunology of DNA vaccines: CPG motifs and antigen presentation. Inflamm. Res. 49, 199–205. Available at: http://www.ncbi.nlm.nih.gov/pubmed/10893042.

Masse, J.E., Wong, B., Yen, Y.-M., et al., 2002. The S.cerevisiae architectural HMGB Protein NHP6A Complexed with DNA: DNA and protein conformational changes upon binding. J. Mol. Biol. 323, 263–284. Available at: http://linkinghub.elsevier.com/retrieve/pii/S0022283602009385.

Matthews, B.W., Ohlendorf, D.H., Anderson, W.F., Takeda, Y., 1982. Structure of the DNA-binding region of lac repressor inferred from its homology with CRO repressor. Proc. Natl. Acad. Sci. 79, 1428–1432. Available at: http://www.pnas.org/content/79/5/1428.short%5Cnhttp://www.pnas.org/cgi/doi/10.1073/pnas.79.5.1428.

Ma, X., Wu, J., 2013. Identification of DNA-binding proteins using support vector machine with sequence information. Comput. Math. Methods Med. 2013.Available at: http://downloads.hindawi.com/journals/cmmm/aip/524502.pdf.

McClarin, J.A., Frederick, C.A., Wang, B.C., et al., 1986. Structure of the DNA-Eco RI endonuclease recognition complex at 3 A resolution. Science 234, 1526–1541. Available at: http://www.ncbi.nlm.nih.gov/pubmed/3024321.

Meyer, C.A., Liu, X.S., 2014. Identifying and mitigating bias in next-generation sequencing methods for chromatin biology. Nat. Rev. Genet. 15, 709–721. Available at: http://www.nature.com/doifinder/10.1038/nrg3788.

Modlin, R.L., 2000. Immunology. A Toll for DNA vaccines. Nature 408, 659–660. Available at: http://www.ncbi.nlm.nih.gov/pubmed/11130055.

Morimoto, R.I., 1998. Regulation of the heat shock transcriptional response: Cross talk between a family of heat shock factors, molecular chaperones, and negative regulators. Genes Dev. 12, 3788–3796. Available at: http://www.ncbi.nlm.nih.gov/pubmed/9869631.

Morozov, A.V., Havranek, J.J., Baker, D., Siggia, E.D., 2005. Protein-DNA binding specificity predictions with structural models. Nucleic Acids Res. 33, 5781–5798.

Nimrod, G., Schushan, M., Szilágyi, A., Leslie, C., Ben-Tal, N., 2010. iDBPs: A web server for the identification of DNA binding proteins. Bioinformatics 26, 692–693. Available at: http://www.ncbi.nlm.nih.gov/pubmed/20089514.

Nimrod, G., Szilágyi, A., Leslie, C., Ben-Tal, N., 2009. Identification of DNA-binding proteins using structural, electrostatic and evolutionary features. J. Mol. Biol. 387, 1040–1053. Available at: http://www.ncbi.nlm.nih.gov/pubmed/19233205.

Noy, A., Sutthibutpong, T., A. Harris, S., 2016. Protein/DNA interactions in complex DNA topologies: Expect the unexpected. Biophys. Rev. 8, 233–243. Available at: http://www.ncbi.nlm.nih.gov/pubmed/27738452.

Nutiu, R., Friedman, R.C., Luo, S., et al., 2011. Direct measurement of DNA affinity landscapes on a high-throughput sequencing instrument. Nature biotechnology 29, 659–664.

Ofran, Y., Mysore, V., Rost, B., 2007. Prediction of DNA-binding residues from sequence. Bioinformatics 23, i347–i353. Available at: http://www.ncbi.nlm.nih.gov/pubmed/1764.

Ogawa, N., Biggin, M.D., 2012. High-throughput SELEX determination of DNA sequences bound by transcription factors in vitro. Methods Mol. Biol. 786, 51–63.

Olson, W.K., Gorin, A.A., Lu, X.J., Hock, L.M., Zhurkin, V.B., 1998. DNA sequence-dependent deformability deduced from protein-DNA crystal complexes. Proc. Natl Acad. Sci. USA. 95, 11163–11168. Available at: http://www.pnas.org/cgi/doi/10.1073/pnas.95.19.11163.

Orengo, C.A., Michie, A.D., Jones, S., et al., 1997. CATH – A hierarchic classification of protein domain structures. Structure 5, 1093–1108. Available at: http://www.ncbi.nlm.nih.gov/pubmed/9309224.

Paquet, E., Viktor, H.L., 2015. Molecular dynamics, monte carlo simulations, and Langevin dynamics: A computational review. Biomed Res. Int. 2015, 1–18. Available at: http://www.hindawi.com/journals/bmri/2015/183918/.

Paull, T.T., Haykinson, M.J., Johnson, R.C., 1993. The nonspecific DNA-binding and -bending proteins HMG1 and HMG2 promote the assembly of complex nucleoprotein structures. Genes Dev. 7, 1521–1534. Available at: http://www.genesdev.org/cgi/doi/10.1101/gad.7.8.1521.

Paz, I., Kligun, E., Bengad, B., Mandel-Gutfreund, Y., 2016. BindUP: A web server for non-homology-based prediction of DNA and RNA binding proteins. Nucleic Acids Res. 44, W568–W574. Available at: http://www.ncbi.nlm.nih.gov/pubmed/27198220%5Cnhttp://www.pubmedcentral.nih.gov/articlerender.fcgi?artid=PMC4987955.

Peled, S., Leiderman, O., Charar, R., et al., 2016. De-novo protein function prediction using DNA binding and RNA binding proteins as a test case. Nat. Commun. 7, 13424. Available at: http://www.nature.com/doifinder/10.1038/ncomms13424.

Ponting, C.P., Schultz, J., Milpetz, F., Bork, P., 1999. SMART: Identification and annotation of domains from signalling and extracellular protein sequences. Nucleic Acids Res. 27, 229–232. Available at: http://www.ncbi.nlm.nih.gov/pubmed/9847187.

Ptashne, M., 2005. Regulation of transcription: From lambda to eukaryotes. Trends Biochem. Sci. 30, 275–279. Available at: http://www.ncbi.nlm.nih.gov/pubmed/15950866.

Rajagopal, N., Srinivasan, S., Kooshesh, K., et al., 2016. High-throughput mapping of regulatory DNA. Nat. Biotechnol. 34, 167–174. Available at: http://www.nature.com/doifinder/10.1038/nbt.3468.

Reece-Hoyes, J.S., Marian Walhout, A.J., 2012. Yeast one-hybrid assays: A historical and technical perspective. Methods, 57. . pp. 441–447. Available at: http://link.springer.com/10.1007/978-1-62703-673-3.

Rice, J., King, C.A., Spellerberg, M.B., Fairweather, N., Stevenson, F.K., 1999. Manipulation of pathogen-derived genes to influence antigen presentation via DNA vaccines. Vaccine 17, 3030–3038. Available at: http://www.ncbi.nlm.nih.gov/pubmed/10462238.

Richter, P.H., Eigen, M., 1974. Diffusion controlled reaction rates in spheroidal geometry. Application to repressor–operator association and membrane bound enzymes. Biophys. Chem 2, 255–263. Available at: http://www.ncbi.nlm.nih.gov/pubmed/4474030.

Robertson, T.A., Varani, G., 2007. An all-atom, distance-dependent scoring function for the prediction of protein-DNA interactions from structure. Proteins 66, 359–374. Available at: http://www.ncbi.nlm.nih.gov/pubmed/17078093.

Rohs, R., Jin, X., West, S.M., et al., 2010. Origins of specificity in protein-DNA recognition. Annu. Rev. Biochem. 79, 233–269. Available at: http://www.ncbi.nlm.nih.gov/pubmed/20334529.

Rohs, R., West, S.M., Liu, P., Honig, B., 2009a. Nuance in the double-helix and its role in protein–DNA recognition. Curr. Opin. Struct. Biol. 19, 171–177. Available at: http://linkinghub.elsevier.com/retrieve/pii/S0959440X09000347.

Rohs, R., West, S.M., Sosinsky, A., et al., 2009b. The role of DNA shape in protein–DNA recognition. Nature 461, 1248–1253. Available at: http://www.nature.com/doifinder/10.1038/nature08473.

Rouvière-Yaniv, J., Yaniv, M., Germond, J.E., 1979. E. coli DNA binding protein HU forms nucleosomelike structure with circular double-stranded DNA. Cell 17, 265–274. Available at: http://www.ncbi.nlm.nih.gov/pubmed/222478.

Scaffidi, P., Bianchi, M.E., 2001. Spatially precise DNA bending is an essential activity of the sox2 transcription factor. J. Biol. Chem. 276, 47296–47302. Available at: http://www.ncbi.nlm.nih.gov/pubmed/11584012.

Schultz, S.C., Shields, G.C., Steitz, T.A., 1991. Crystal structure of a CAP-DNA complex: The DNA is bent by 90 degrees. Science 253, 1001–1007. Available at: http://www.ncbi.nlm.nih.gov/pubmed/1653449.

Shao, X., Tian, Y., Wu, L., et al., 2009. Predicting DNA- and RNA-binding proteins from sequences with kernel methods. J. Theor. Biol. 258, 289–293. Available at: http://www.sciencedirect.com/science/article/pii/S0022519309000289.

Shimamoto, N., 1999. One-dimensional diffusion of proteins along DNA. Its biological and chemical significance revealed by single-molecule measurements. J. Biol. Chem. 274, 15293–15296. Available at: http://www.ncbi.nlm.nih.gov/pubmed/10336412.

Siggers, T.W., Honig, B., 2007. Structure-based prediction of C_2H_2 zinc-finger binding specificity: Sensitivity to docking geometry. Nucleic Acids Res. 35, 1085–1097. Available at: http://www.ncbi.nlm.nih.gov/pubmed/17264128.

Si, J., Zhao, R., Wu, R., 2015. An overview of the prediction of protein DNA-binding sites. Int. J. Mol. Sci. 16, 5194–5215.

Spolar, R.S., Record, M.T., 1994. Coupling of local folding to site-specific binding of proteins to DNA. Science 263, 777–784. Available at: http://www.ncbi.nlm.nih.gov/pubmed/8303294.

Stawiski, E.W., Gregoret, L.M., Mandel-Gutfreund, Y., 2003. Annotating nucleic acid-binding function based on protein structure. J. Mol. Biol. 326, 1065–1079. Available at: http://www.sciencedirect.com/science/article/pii/S0022283603000317.

Stella, S., Cascio, D., Johnson, R.C., 2010. The shape of the DNA minor groove directs binding by the DNA-bending protein Fis. Genes Dev. 24, 814–826. Available at: http://www.ncbi.nlm.nih.gov/pubmed/20395367.

Szilágyi, A., Skolnick, J., 2006. Efficient prediction of nucleic acid binding function from low-resolution protein structures. J. Mol. Biol. 358, 922–933. Available at: http://www.sciencedirect.com/science/article/pii/S002228360600252X.

Tabeta, K., Georgel, P., Janssen, E., et al., 2004. Toll-like receptors 9 and 3 as essential components of innate immune defense against mouse cytomegalovirus infection. Proc. Natl. Acad. Sci. USA. 101, 3516–3521. Available at: http://www.ncbi.nlm.nih.gov/pubmed/14993594.

Takeda, T., Corona, R.I., Guo, J.T., 2013. A knowledge-based orientation potential for transcription factor-DNA docking. Bioinformatics 29, 322–330.

Tan, C., Terakawa, T., Takada, S., 2016. Dynamic coupling among protein binding, sliding, and DNA bending revealed by molecular dynamics. J. Am. Chem. Soc. 138, 8512–8522. Available at: http://www.ncbi.nlm.nih.gov/pubmed/27309278.

Tapias, A., López, G., Ayora, S., 2000. Bacillus subtilis LrpC is a sequence-independent DNA-binding and DNA-bending protein which bridges DNA. Nucleic Acids Res. 28, 552–559. Available at: http://www.ncbi.nlm.nih.gov/pubmed/10606655.

Thanbichler, M., Wang, S.C., Shapiro, L., 2005. The bacterial nucleoid: A highly organized and dynamic structure. J. Cell. Biochem., 96. . pp. 506–521. Available at: http://doi.wiley.com/10.1002/jcb.20519.

Thompson, J.F., Landy, A., 1988. Empirical estimation of protein-induced DNA bending angles: Applications to lambda site-specific recombination complexes. Nucleic Acids Res. 16, 9687–9705. Available at: http://www.ncbi.nlm.nih.gov/pubmed/2972993.

Tong, Y., Falk, J., 2009. Genome-wide analysis for protein — DNA interaction: Chip-chip. Methods in Mol. Biol. (Clifton, N. J.). 235–251. United States Available at: http://link.springer.com/10.1007/978-1-60327-378-7_15.

Tuckerman, M.E., Martyna, G.J., 2000. Understanding modern molecular dynamics: Techniques and applications. J. Phys. Chem. B, 104. . pp. 159–178. Available at: http://pubs.acs.org/doi/abs/10.1021/jp992433y.

Vassallo, M., Ralston, A., 1992. Algorithms for De Bruijn sequences – a case study in the empirical analysis of algorithms. Comput. J, 35. . pp. 88–90. Available at: https://academic.oup.com/comjnl/article-lookup/doi/10.1093/comjnl/35.1.88.

Velmurugu, Y., Chen, X., Slogoff Sevilla, P., Min, J.-H., Ansari, A., 2016. Twist-open mechanism of DNA damage recognition by the Rad4/XPC nucleotide excision repair complex. Proc. Natl. Acad. Sci. USA 113, E2296–E2305. Available at: http://www.pnas.org/content/pnas/early/2016/03/30/1514666113.abstract.html?collection.

Walter, M.C., Rattei, T., Arnold, R., et al., 2009. PEDANT covers all complete RefSeq genomes. Nucleic Acids Res., 37. pp. D408–D411. Available at: https://academic.oup.com/nar/article-lookup/doi/10.1093/nar/gkn749.

Wang, L., Brown, S.J., 2006. BindN: a web-based tool for efficient prediction of DNA and RNA binding sites in amino acid sequences. Nucleic Acids Res. 34, W243–W248. Available at: http://www.ncbi.nlm.nih.gov/pubmed/16845003.

Wang, Y., Cortez, D., Yazdi, P., et al., 2000. BASC, a super complex of BRCA1-associated proteins involved in the recognition and repair of aberrant DNA structures. Genes Dev. 14, 927–939. Available at: http://www.ncbi.nlm.nih.gov/pubmed/10783165.

Wei, L., Tang, J., Zou, Q., 2017. Local-DPP: An improved DNA-binding protein prediction method by exploring local evolutionary information. Inf. Sci. (NY). 384, 135–144. Available at: http://www.sciencedirect.com/science/article/pii/S0020025516304509.

Wilson, K.A., Kellie, J.L., Wetmore, S.D., 2014. DNA-protein π-interactions in nature: Abundance, structure, composition and strength of contacts between aromatic amino acids and DNA nucleobases or deoxyribose sugar. Nucleic Acids Res. 42, 6726–6741. Available at: http://www.ncbi.nlm.nih.gov/pubmed/24744240.

Wilson, K.A., Wetmore, S.D., 2015. A survey of DNA-protein π-interactions: A comparison of natural occurrences and structures, and computationally predicted structures and strengths. In: Scheiner, S. (Ed.), Noncovalent Forces. Cham: Springer International Publishing, pp. 501–532. Available at: https://doi.org/10.1007/978-3-319-14163-3_17.

Wold, M.S., 1997. REPLICATION PROTEIN A: A heterotrimeric, single-stranded dna-binding protein required for eukaryotic dna metabolism. Annu. Rev. Biochem. 66, 61–92. Available at: http://www.ncbi.nlm.nih.gov/pubmed/9242902.

Wu, C., 1995. Heat shock transcription factors: Structure and regulation. Annu. Rev. Cell Dev. Biol. 11, 441–469. Available at: http://www.ncbi.nlm.nih.gov/pubmed/8689565.

Wu, J., Liu, H., Duan, X., et al., 2009. Prediction of DNA-binding residues in proteins from amino acid sequences using a random forest model with a hybrid feature. Bioinformatics 25, 30–35.

Xie, Z., Hu, S., Qian, J., Blackshaw, S., Zhu, H., 2011. Systematic characterization of protein-DNA interactions. Cell. Mol. Life Sci. 68, 1657–1668. Available at: http://www.ncbi.nlm.nih.gov/pubmed/21207099.

Xiong, Y., Xia, J., Zhang, W., Liu, J., 2011. Exploiting a reduced set of weighted average features to improve prediction of DNA-binding residues from 3D structures. PLOS One 6, e28440. Available at: http://www.ncbi.nlm.nih.gov/pubmed/22174808.

Yan, C., Terribilini, M., Wu, F., et al., 2006. Predicting DNA-binding sites of proteins from amino acid sequence. BMC Bioinformatics 7, 262. Available at: http://bmcbioinformatics.biomedcentral.com/articles/10.1186/1471-2105-7-262.

Yan, J., Kurgan, L., 2017. DRNApred, fast sequence-based method that accurately predicts and discriminates DNA-and RNA-binding residues. Nucleic Acids Res. 45.

Yu, Y., Cao, J., Cai, Y., Shi, T., Li, Y., 2006. Predicting rRNA-, RNA-, and DNA-binding proteins from primary structure with support vector machines. J. Theor. Biol. 240, 175–184. Available at: http://www.sciencedirect.com/science/article/pii/S0022519305004212.

Zhang, C., Liu, S., Zhu, Q., Zhou, Y., 2005. A knowledge-based energy function for protein-ligand, protein-protein, and protein-DNA complexes. J. Med. Chem. 48, 2325–2335. Available at: http://www.ncbi.nlm.nih.gov/pubmed/15801826.

Zhao, H., Wang, J., Zhou, Y., Yang, Y., 2014. Predicting DNA-binding proteins and binding residues by complex structure prediction and application to human proteome. PLOS One 9, 26–28.

Zhou, J., Lu, Q., Xu, R., He, Y., Wang, H., 2017. EL _ PSSM-RT: DNA-binding residue prediction by integrating ensemble learning with PSSM. Relation Transformation. 1–16.

Zhou, W., Yan, H., 2011. Prediction of DNA-binding protein based on statistical and geometric features and support vector machines. Proteome Sci. 9 (Suppl 1), S1. Available at: http://www.pubmedcentral.nih.gov/articlerender.fcgi?artid=3289070&tool=pmcentrez&rendertype=abstract.

Gene Regulatory Network Review

Enze Liu, Indiana University School of Medicine, Indianapolis, IN, United States and Indiana University, Indianapolis, IN, United States
Lang Li, The Ohio State University, Columbus, OH, United States
Lijun Cheng, The Ohio State University, Columbus, OH, United States

Nomenclature

BN	Bayesian network	MI	Mutual information
CMI	Conditional mutual information	PPI	Protein–protein interactions
DBN	Dynamic Bayesian Network	ODE	Ordinary differential equation
EST	Expressed sequence tags	PTM	Posttranslational modification
GRN	Gene regulatory networks	ROC	Receiver-operating characteristic
KNN	k-nearest neighbors	TF	Transcription factors

Introduction

Genes act as a secret key in governing the entire biological system's function. Genes' cooperation and interactions form a dynamic network that brings about our vivid biodiversity (Cheng *et al.*, 2013). If one gene is regulated, then it affects its targets, which further affect their next targets, causing a domino effect towards the entire cellular environment. A gene regulatory network (GRN) is the collection of molecular regulators (DNA segments, miRNAs, or histones) in a cell that interact with each other directly or indirectly (through their RNA and protein expression products) and with other substances in the cell (Vijesh *et al.*, 2013). By a comprehensive network perspective, we can mine the direct regulations among genes and gain deep insights into various biological processes, which could further facilitate medical related fields such as drug design or drug target discovery.

A GRN is composed of functional linkages between regulators and targets to which their products bind. *Cis*-regulatory element and *trans*-regulatory element are two main regulatory types (Peters, 2008). *Cis*-regulatory elements are present near the structural portion of the gene/protein as the gene they regulate, such as, the photosynthetic protein family, are expressed at the same time in development. Whereas *trans*-regulatory elements can distantly regulate genes from which they were transcribed. Enhancers and multiple *trans*-acting factors are essential for *trans*-control transcription initiations. Regulators contain DNA epigenetic modifications by methylation, miRNA, transcription factors (TFs), and posttranslational modification (PTM) that include histone proteins and other proteins, which are involved in methylation, phosphorylation, acetylation, ubiquitylation, and sumoylation (**Fig. 1**). Motif is a basic unit of GRN subgraph for these regulators and gene sets regulated with similar functions in networks. Normally, TFs coded by a gene can regulate gene expression by binding to specific motifs. MicroRNAs (miRNAs) regulate gene expression via RNA silencing or posttranscription regulations. Histones can alter the chromatin structure, which further controls the access of TFs and polymerases to genes thus resulting in an expression regulation (Dong and Weng, 2013). Methylation plays a crucial role in regulating gene expression by blocking the promoters that can activate TFs (Phillips, 2008). Large experimental evidence has been gathered to verify biology gene interactions, such as KEGG (see "Relevant Websites section"), Pathway Commons (see "Relevant Websites section"), MetaCyc Metabolic Pathway Database (see "Relevant Websites section"), JASPAR CORE (see "Relevant Websites section"), HistoneDB 2.0 (see "Relevant Websites section") and miRGator v3.0 (see "Relevant Websites section"), GeneHancer(see "Relevant Websites section"), etc. However, this is only the tip of the iceberg in terms of genome knowledge. Huge knowledge mining from genome is still in its infancy. Comprehensive understanding of GRNs is a major challenge in the field of systems biology. Recent advances in high-throughput techniques provide an opportunity to reconstruct regulatory networks, by offering a huge amount of binding data, like ChIP-Seq, miRNA-Seq, ATAC-Seq with expression data RNA-Seq (Cheng *et al.*, 2011). By performing various topological analyses, including its hierarchical organization and motif enrichment, a potential molecular mechanism would be detected. Gene expression microarrays and RNA-Seq provide us another chance to evaluate gene expression dependencies between TFs and its target genes systematically for understanding systematical biological activities (Madan Babu and Teichmann, 2003). Accurate reconstruction of the gene regulatory interactions, forming GRN, is one of the key tasks in systems biology.

The inference of a GRN is often accomplished through the use of gene expression data. So far, there are numerous computational methods and models developed for restoring GRNs in a real cellular environment. However, each of them have their own assumptions and methods, drawing different blueprints that the GRN described. There is still much confusion about the basic meaning of GRN, ways of assessment, and possible biomedical application. Typically, the relationships between genes are directional in nature and they can change over time or in response to external stimulus. Researchers are facing the choice of whether to include extra features such as causality and temporal behaviors when modeling gene networks or not. In this paper, we aim to clarify the meaning and usage conditions for GRNs and also provide a discussion on next steps to bring GRNs closer to clinical and medical application. In the section "Expression Data for Reconstructing GRN," we briefly introduced the expression data composition and two major platforms that generate expression data. In the section "Static Network Models and Inference

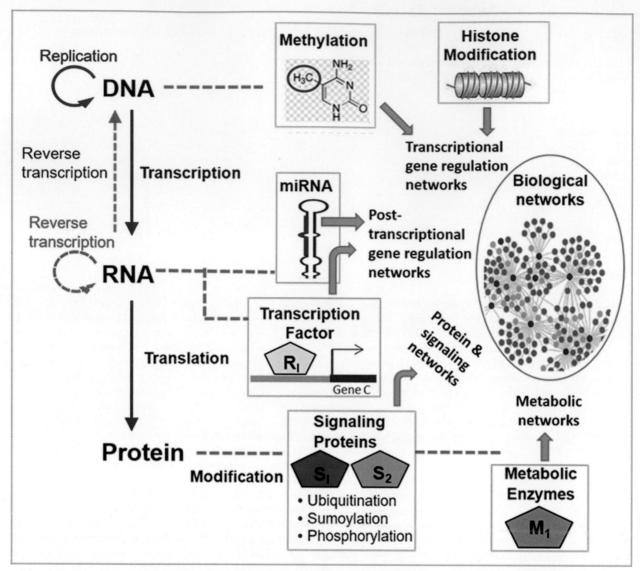

Fig. 1 Central dogma and main regulating elements for biological networks at each molecular level.

Methods," we focus on network inference methods for reconstructing a static GRN. We highlighted methods themselves, the underlying assumptions and advantages/drawbacks by indicating their features and comparing them. For some representative methods we illustrated examples of how they had been applied in real study. In the section "Dynamic Network Models and Inference Methods," we discussed inferring methods that simulate temporal changes of gene regulations over time in a dynamic GRN. In the section "Network Validation and Assessment of the Network Inference Methods," we expanded to how to validate the inferred network and brought up several measurements for assessing the performance of different inference methods. In the section "Gene Regulatory Network Inference Development Trend," we forecast the trend of further development of GRN inference methods, including building comprehensive GRNs by integrating other omics data besides expression data and methods updating trends for resolving current issues during GRN reconstruction.

Expression Data for Reconstructing Gene Regulatory Network

Nowadays, GRNs are mainly reconstructed from gene expression data, which is generated by high-throughput platforms such as microarray and RNA-seq. Microarray technology uses a set of short expressed sequence tags (ESTs) generated from the cDNA library to hybridize target RNAs and then reflect fluorescent intensity, which is used to measure the expression level of certain RNAs. RNA-seq utilizes next generation sequencing (NGS) technology to sequence the cDNA library generated from sample RNAs and then directly generates read (RNA) counts to represent expression level of RNAs. Although their principles and experiments differ, data preprocessing and normalization methods for microarray and RNA-seq vary drastically. Normalized gene expression data generated from either platform have the same form associated with a matrix with M rows of genes under N columns of conditions (samples). Static network modeling and dynamic network modeling with time series are two typical reconstruction types for GRN. Five representative models related to the two types will be discussed in detail (**Fig. 2**).

Static Network Models and Inference Methods

Static models reflect interactions among genes at any given time point. These models are often used to describe network topologies and qualitative features of gene relationships (Wang and Huang, 2014).

CoExpression Networks and Relevance Networks

A GRN built with a correlation based method is called a coexpression network (Muhammad *et al.*, 2017). A coexpression network is constructed by calculating pairwise correlations for each gene pairs to form a fully connected graph, and then by applying a threshold scheme to remove edges between genes that are not significantly correlated (**Fig. 2(a)**). Pearson correlation, Spearman correlation and Euclidean distance are widely used for measuring linear correlations while mutual information (MI) and kernel correlations between two genes (Cheng *et al.*, 2013) are the representative measures for detecting nonlinear correlations. These measures are used to calculate association on a pair of genes to seek if they are coexpression. P-value, Z-score, clustering coefficient (Elo *et al.*, 2007), and random matrix theory (Luo *et al.*, 2007) are typical threshold schemes for filtering out low content coexpression.

   Pearson correlation ρ is the most basic correlation that measures pairwise correlation between two continuous variables.

$$\rho_{X,Y} = \frac{cov(X,Y)}{\sigma_X \sigma_Y} \tag{1}$$

Here X and Y denote expression levels of two genes. *Cov* is the covariance between gene X, Y and sigma σ indicates the individual standard deviation. Pearson correlation measures the collinearity between two genes with the assumption that both genes' expression level follow an approximately normal distribution, making it unsuitable for measuring genes with nonlinear relationships.

   Rank based correlation describes correlations by comparing the rank of the variables (genes) instead of covariance. Take Spearman correlation as an example, which compares the monotonic relationship between two variables (gene). If two genes follow a normal expression pattern and have a clearly linear relationship, then Pearson correlation and Spearman correlation would be very similar. However, if two genes are monotonically correlated, not linearly correlated, only the Spearman correlation could measure the nonlinear relationships. In other words, the Spearman correlation is more general and robust compared to the Pearson correlation, even though information gets lost during the ranking transformation.

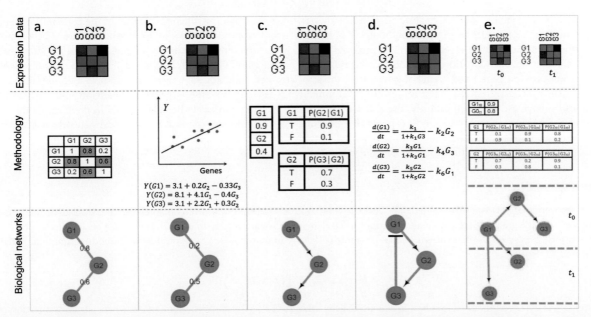

Fig. 2 Representative models and steps for reconstructing gene regulatory network. (a) Coexpression network (b) regression based regulatory network (c) Bayesian network (d) differential equation based regulatory network (e) dynamic Bayesian network.

Distance based correlation mainly refers to Euclidean distance, which is the geometric distance between two genes if two genes are assumed to be in the same p-dimensional Euclidean space. Euclidean distance is sensitive to transformations, whereas the other two are normally invariant to linear transformations.

Information theoretic entropy and mutual information concepts provide a new perspective to scale the correlation between variables. Entropy measures the quantity of information within a given system and two variables that contain more entropy tend to be more related. MI measures all the entropy within a system, including marginal entropy from variables and joint entropy between/among variables. A MI $I(X, Y)$ measures a system of two variables (genes) and can be defined as:

$$I(X, Y) = H(X) + H(Y) - H(X, Y) \tag{2}$$

where

$$H(X) = -\sum_{x=1}^{m} f(x)\log(f(x)) \tag{3}$$

is the marginal entropy and

$$H(X, Y) = -\sum_{x=1, y=1}^{m.n} f(x, y)\log(f(x, y)) \tag{4}$$

is the joint entropy for two continuous variables of gene X and Y.

A gene association network built with MI is called a relevance network (Butte and Kohane, 1999). ARACNe (Margolin *et al.*, 2006) is the most famous and representative algorithm for building a relevance network. ARACNe assumes genes following a normal distribution and performed a Gaussian Kernel estimator (Scott and Sheather, 1985) to estimate marginal distribution $f(x)$ and joint distribution $f(x,y)$ in formula 3 and 4. ARACNe defined its edge-pruning thresholds named data processing inequality (DPI) as: For all available triplets g_i, g_j, and g_k, if g_i interacts with g_k via g_j, then the edges (pairwise MI) among them should follows the following inequality: $I(E_i, E_j) \leq min(I(E_i, E_k), I(E_k, E_j))$, meaning that the indirect interaction in the system should contain the smallest MI value. Any violated indirect interactions will be pruned.

A GRN based on correlation provides coexpression or cofunctionality from a network scale with no directionality, meaning that if no prior knowledge is provided (e.g., a gene is confirmed to be a transcription factor in a two-gene interaction system), we cannot identify the causations and results of genes in a GRN. MI detects underlying correlations that linear Pearson correlation cannot discover, especially for variables with nonlinear correlation.

Regression Based Approaches

Regression based approaches are a conventional and straightforward method that is widely used in inferring GRNs (**Fig. 2(b)**). In a regression model, the problem of constructing a GRN is converted to the problem of selecting regulators (e.g., TFs) that have a significant impact towards particular genes. Given an expression matrix with m genes and k conditions. A multiregression model for each gene can be formulated as:

$$y_i = \beta_0 + \beta_1 X_{i1} + \beta_2 X_{i2} + \ldots + \beta_k X_{ik} + \varepsilon_i$$

Where: y_i is ith gene of whose regulating relationship towards other genes is to be inferred. β_0 denotes the constant/intercept. X_{ik} indicates the expression values of gene i in condition k. ε_i is the noise/error terms within gene i. For each gene i, a series of beta indicating the contributing towards gene i expression will be predicted. Various feature selection methods can be applied during the inference such as forward, backward, stepwise, ridge, and LASSO. After this step, a p-value cutoff (< 0.05 or less) is usually used to determine whether there is a link between gene i and predictor gene j. If so, the associated beta indicates the weight and the correlation directions.

Regression based approaches not only estimate the underlying associations among genes in a network, but also infer the association intensities. More importantly, the correlations among predictors can easily be incorporated in the formula by adding the correlated predictors' term, making it capable of mimicking real coregulating scenarios in biological networks. However, the model assumes that all the predictors follow a multivariate normal distribution similar to a coexpression network, therefore regression does not infer the directionality of associations.

Bayesian Networks

A Bayesian network (BN) is a probabilistic network model that takes a group of random variables and measures their relationships as a directed acyclic graph (DAG) with all the conditional dependencies among them (**Fig. 2(c)**). Given a certain network structure G, a BN can be expressed as the following joint probability density function (PDF) p_B (a product of individual marginal PDFs and conditional PDFs):

$$p_B(x_1, \ldots, x_n) = \prod_{i=1}^{n} p\left(x_i | P_{a_i}^g, \theta_i\right)$$

Where $P_{a_i}^g$ is the set of parents of node x_i in a graph G, gene node $g \in G$, θ_i is the ith distribution mean, where all gene expression obeys a normal distribution $N(\theta_i, \delta_i)$. All conditional probabilities denote all the edges within the DAG. The product of them clearly indicate the local Markov property of BN: Every node (gene) is conditionally independent of its nondescendants given its parents.

Reconstructing a BN from the given expression data requires two types of inferences: (1) Parameter learning, which infers marginal and conditional distributions of all nodes, and (2) structure learning, which infers the optimal topology that has the biggest overall probability. In parameter learning, the marginal distributions can be either given by prior knowledge or inferred by various methods such as principle of maximum entropy (Clarke, 2007) while conditional distributions are often inferred with approaches such as maximum likelihood estimator or expectation maximization (Friedman *et al.*, 2000). Learning the optimal structure of a BN has been a great challenge for decades. One can always use a brute force approach to go through every possible structure and finally identify the best, which would cost superexponential time complexity to achieve. Thus, most of the existing algorithms for searching the optimal structures are heuristic and progressive. For instance, Monte Carlo Markov chain (MCMC) (Mukherjee and Speed, 2008) methods use approximate sampling approaches. Greedy equivalence search (GES) (Chickering, 2002) proposed a local optimal constraint and performed a stepwise searching scheme. As for the scoring scheme, usually a Bayesian information criterion (BIC) or Akaike's information criterion (AIC) is chosen (Liu, Malone and Yuan, 2012).

Compared to coexpression networks, BNs can clearly point out the directionality of each edge thus revealing a causation relationship among all genes, which is significantly advantageous. Large number experiments have shown that BN can offer a better accuracy and tolerance with noise expression input on small number of genes (Hill *et al.*, 2016). As more variables (genes) are incorporated into the model, the computational time grows superexponentially, which brings up the great challenge of building a whole genomic BN for higher organisms like humans (Hill *et al.*, 2016). Moreover, for one dataset, there might be more than one optimal BN structure that has the equivalent overall probabilities. These models (structures) are called equivalence class (Chickering, 2002), in which each model is probabilistically indistinguishable. Consider two BNs A->B->C and C->B->A are probabilistically equivalent since their overall probabilities are P(A)P(B|A)P(C|B) and P(C)P(B|C)P(A|B), which is the same. Causal network (CN) refers to a more strict form of BN, meaning not only conditional dependencies, but also Markov conditions among variables need to be followed in the DAG, which provide a clear causation of one variable over another. Hence, the above BN are not causally equivalent due to different Markov conditions: A_||_C|B (A is independent of C given B) versus C_||_A|B (C is independent of A given B), where in the first network, A causes C, in the second, C causes A. GES algorithm that is mentioned above is a searching algorithm designed for CN models.

Another significant drawback of the BN model lies in its assumption: A DAG doesn't allow the existence of feedback loops, which have been widely proven to exist in biological networks. This can be resolved by dynamic Bayesian network(DBN) (Yu and Liu, 2004; Murphy and Russell, 2002). Additionally, there is no way to observe the type of regulations (activation/inhibition) in a GRN generated with a BN model. Hence, prior knowledge or differential expression analysis is usually incorporated.

Applications

The applications of static models can be found in various fields such as cancer biology and biotechnology (Marbach *et al.*, 2012). Moreover, Chuang *et al.* (2007) constructed a relevance network that uses mutual information to measure protein–protein interactions with protein interaction information from databases and gene expression profiles from breast cancer patients. They were able to identify several important biomarkers that highly correlated to breast cancer metastasis. Zhang *et al.* (2017) utilized linear regression and Cox regression and created a semisupervised learning approach and built a comprehensive drug–disease--gene network for drugs repositioning on ovarian cancer. Yavari *et al.* (2008) clustered genes based on gene ontology (GO) and used a BN to build a regulatory network containing 84 yeast genes based on coclustered information.

Dynamic Network Models and Inference Methods

Dynamic network models capture the dynamic nature of real networks thus yielding quantitative predictions of gene behaviors (Wang and Huang, 2014). If you don't like static models, self-regulations of genes can be reflected in dynamic models.

Dynamic Bayesian Networks

At a given time slice, DBNsshow exactly the forms and properties as BNs. However, as an extension of BN, DBNs incorporate time dimensions into the network reconstruction to detect features of dynamic GRNs with time series, including feedback loops and self-regulation that commonly occur in nature and are not able to be simulated by BNs (**Fig. 2(e)**). A DBN with N genes can be described as:

$$P(X_t|X_{t-1}) = \prod_{i=1}^{N} P\big(X_{i,t}|Pa_{Bt}\big(X_{i,t}\big)\big)$$

Where $X_{i,t}$ is a given gene i at time slice t and Pa is its parent's set. In a DBN, each node (gene) has duplicates corresponding to the number of time slice t in order to model the cyclic interactions and dynamic behaviors of GRN (Aluru, 2005). Temporal relationships among genes are represented by interconnections between time slices (Chai *et al.*, 2014), which means even though a DBN generates DAGs similar to BN, a DBN could have both interslice connections and intraslice connections. The structural learning and parameter learning of a DBN is similar to a BN. However, to infer the future state, a DBN needs to be converted to hidden Markov models first and then applied with forward or backward algorithms (Murphy and Russell, 2002).

Boolean Network Models

In system biology, a Boolean network refers to a set of genes whose states are binary (activate or inactivate) (D'haeseleer *et al.*, 2000). At a given time slice t, each gene x will only be in one state $S(t)$ where $S(t) = [x_1(t), x_2(t),...,x_n(t)]$. States of each gene are determined by one Boolean logic function B, which takes a subset of genes as input and generates the current state for associated genes. Boolean networks use a series of discrete states of the network to represent the complete dynamics of the whole GRN. Current state of the whole network at time slice $t+1$ is inferred from the network state at time slice t. Hence, to generate a GRN using a Boolean network model, the expression value of genes that are continuous needs to be discretized. The objective of the searching process is to identify a series of logic functions that optimally explain the expression data.

A Boolean network is a simple way to simulate the sophisticated dynamics of GRN. There are several studies that take advantage of Boolean network model and successfully capture some common patterns of regulator genes and the suppressed/activated genes (Shmulevich *et al.*, 2002). However, limitations of the Boolean network model are still obvious: a GRN is a real-time dynamic system and states of genes are more complicated than binary. Hence a Boolean network cannot accurately mimic the real dynamic interactions among genes since gene expression data normally contains a small number of time slices. Moreover, a Boolean network also masks important information during discretization. As gene expression patterns are generally more complex than having only two states, data discretization often leads to information loss.

Constructing GRN Using Ordinary Differential Equations

Unlike Boolean networks or DBNs, GRN modeling by ordinary differential equation uses a series of discrete states to simulate the complete dynamics of GRNs (**Fig. 2(d)**). Each gene expression variation uses continuous variables to describe the complete concentration change of RNA or protein with generation, degradation, interaction, regulations, and other events. Hence the concentration change of each gene expression is simulated by one or more independent gene variables and their derivatives with an ordinary differential equation (ODE). The following formula is a standard ODE, which denotes the dynamic concentration change of gene x_i expression under the influence from N other genes in the system (Cao *et al.*, 2012).

$$\frac{dx_i}{dt} = f_i(x_1, x_2, ..., x_n, p, u)$$

Where x_i is the target gene, $x_1, x_2... x_n$ are n regulator genes that could affect the expression level of x_i. p is the parameter set and u is the external perturbation, all of which are included in a node function fi.

A differential equation is said to be *linear* if f can be written as a linear combination of the derivatives of x in time t:

$$x^{(t)} = \sum_{i=1}^{n} pi(x)x_i^{(t)} + u^{(t)}(x)$$

where $p_i(x)$ and $u(x)$ are continuous functions in x. For a linear system, regression methods can be used to identify the solutions. However, an optimal evolutionary algorithm is normally used to estimate the nonlinear system to obtain the steady-state parameter equation (Kimura *et al.*, 2004).

GRNs based on differential equations have several limitations: using differential equations assumes the concentration of RNA or protein depends solely on the concentration of other genes within the system. The influence from other molecules or mechanisms (e.g., TFs) is not directly taken into account (Vijesh *et al.*, 2013). Moreover, in reality, parameters in differential equations are usually difficult to estimate in an experiment.

Applications

Van Gerven *et al.* (2008) illustrated a DBN based model that is developed to provide prognostics of carcinoid patients. Lähdesmäki *et al.* (2003) presented an algorithm for searching optimal Boolean network models that best mimic yeast regulatory network from 15 time-series expression data. Geva-Zatorsky *et al.* (2006) proposed a differential equation based model for reconstructing the P53 dynamic system, which plays a key role in tumor suppression.

Network Validation and Assessment of the Network Inference Methods

To prove the correctness and fidelity of the inferred GRN, a network validation is essential after the inference. Normally, various microarray experiments in silico are conducted to record the expression change after gene perturbations (Yeang *et al.*, 2005), such as knockout, drug treatment, or overexpression. The regulation that exists between two genes can be directly validated by using variation analysis of gene expression and investigate associated biological functions. Moreover, the direction of regulation can be identified (Chen *et al.*, 2015).

Assessing the performance of inference methods is also an important aspect in GRN inference. Generally, for an undirected network, the inferred network model generated from validation data needs to be compared with the associated gold standard to see the consistency of the presence or absence of edges in gold standard. For a directed network, not only consistency of edge presence and absence, but also the consistency of edge directionality needs to be evaluated. True positive (TP), true negative (TN), false positive (FP), and false negative (FN) edge counts should be calculated. And precision (specificity) and recall (sensitivity), defined as $P=TP/(TP+FP)$ and $R=TP/(TP+FN)$ respectively, can be calculated and used as performance assessments. The higher the P and R values are, the better gold standard network has been restored by the method. To measure the performance of methods under a series of changes, for instance, a parameter change from 0 to 1 with a 0.1 interval, or different inputs with 10 to 100 genes within. A receiver-operating characteristic (ROC) curve (Balakrishnan, 1992) denotes the performance consistency of inference method under different conditions. In a ROC graph, y and x axis indicate true positive rate (TPR) and false positive rate (FPR), which can be further calculated with the respective formula: $TPR=TP/(TP+FN)$, $FPR=FP/(FP+TN)$.

Gold standards that are based on real world data have been systematically built up in both machine learning and system biology fields. ALARM network (Beinlich *et al.*, 1989) is a popular tool that is firstly applied for comparison of reconstructed network in machine learning. DREAM project (Marbach *et al.*, 2009; Prill *et al.*, 2010; Marbach *et al.*, 2010) consists of a series of well-studied regulatory network of prokaryotes, eukaryotes, including *Escherichia coli* (Simmons *et al.*, 2008) and in silico microarray datasets of knock-down, knock-out and their associated transcriptome data are collected and published systematically. Causal molecular networks inference focus specifically on reconstructing the signaling downstream of receptor tyrosine kinases based on phosphoprotein data from cancer cell lines as well as in silico data. Linear regression and pair-wise correlation obtained the best-scoring model for the experimental data and in silico data respectively compared with prior network, Ensemble method, nonlinear regression, ODEs, and BNs. A similar result showed linear correlation outperforms over 30 network inference methods on *E. coli*, *Staphylococcus aureus*, *Saccharomyces cerevisiae*, and in silico microarray data (Marbach *et al.*, 2012). A recent large scale study (Saelens *et al.*, 2018) evaluated 42 methods for detecting coexpression networks on nine genome-wide gene expression datasets, including *E. coli*, *Saccharomyces cerevisae*, human datasets, and two synthetic datasets. Benchmark results suggested that decomposition methods can detect local coexpression networks while other methods cannot. Topology inference of GRNs from high-throughput omics data is a long-standing open challenge in computational systems biology. Accuracy and speed become a bottleneck for optimum network reconstruction in such big data.

Existing gold standards of biological networks suffer from the following issues: most of the gold standards describe certain biological processes most likely in primitive organisms such as *E. coli* and yeast, which has a significant regulation mechanism compared with humans. Hence they cannot fully reflect the performance of certain methods that are specifically designed for humans. Secondly, many of these gold standards are the product of curated and repetitive studies among different labs, across different time periods, hence under different experiment conditions, which means they represent more general patterns of certain biological processes, which may not necessarily cover some extreme cases. Thus, using these gold standards to benchmark methods developed for extreme cases could result in bias. For instance, a method for inferring GRNs especially for cancer cells will show a biased benchmark result when using gold standards that mimic the GRN in a normal human cell.

Gene Regulatory Network Inference Development Trend

Multi-Omics Data Integration and Hybrid Model

Due the development of genomic, transcriptomic, and proteomic research, numerous novel experiments and computational technologies lead to the explosion of multiomics data curation. For instance, genomic database NCBI (Coordinators, 2016), transcriptomic database ExPASy (Gasteiger *et al.*, 2003), protein database Pfam (Bateman *et al.*, 2002), pathway database KEGG (Ogata *et al.*, 1999), expression data hub GEO (Edgar *et al.*, 2002), and ArrayExpress (Brazma *et al.*, 2003) store thousands of high-throughput datasets across all kinds of diseases. These databases have a considerable overlaps in terms of describing same biological processes or functions. These overlapped data actually measure the same thing however from various different angles. Hence, when building GRNs from expression data, if associated data (e.g., mutation or binding information) can be utilized, then the inference could be more solid, accurate, and comprehensive (Coordinators, 2016). Take Gong *et al.* (2015) as an example; in the article, they integrated data from RNA-seq, histone modification ChIP-seq, transcription factor, and gene perturbation experiments into a DBN model that resembles a temporal control of gene expression patterns during human cardiac differentiation. Moreover, Hore *et al.* (2016) created a comprehensive GRN by incorporating multiomics data into a Bayesian sparse tensor decomposition model. The common features across multiple tissues are discovered from one GRN. Lu *et al.* (2017) combined ChIP-seq data with expression data from RNA-seq to identify transcription factor binding sites and regulatory modules, which could be considered as another novel method of reconstructing a comprehensive GRN. Due to the rapid advancing single-cell techniques, mapping GRNs from single-cell multiomics data could provide opportunities to systematically understand cellular heterogeneity, which can greatly facilitate precision medicine studies (Fiers *et al.*, 2018).

As more types of omics data were incorporated into the model, the number of variables significantly increased, which could cause convergence difficulties of time consuming models such as BN and regression (De Smet and Marchal, 2010). To overcome this issue, novel methods have been developed under two ideas. (1) New models and learning algorithms that contain a higher learning efficiency. Take deep learning as an example; it has been extensively developed and applied during recent years and proven to be successful in many machine learning tasks mainly due to its multiple layers and hierarchical representation. "D-GEX" (Chen *et al.*, 2016), a deep learning method, has been developed for training GRNs from expression data and predicting expression of target genes in cancer. The new model has been proven to have a better outcome compared to linear regression and k-nearest neighbors (KNN) regression (Bentley, 1975). (2) Hybrid models and learning algorithms that incorporate existing approaches together to get a better performance. For instance, Liu *et al.* (2016) developed a hybrid method called a local , which can be summarized as follows: use pair-wise conditional mutual information (CMI) to build a skeleton; then apply a BN model for each of the genes and their first-order neighbors to determine the direction of regulations; integrate subnetworks into candidate network; and perform CMI again to trim necessary edges so as to obtain comprehensive optimization. Integrating different models to solve the problem might be a feasible way to model expression data with high dimensions. Taken together, these approaches give us new thoughts on using newly developed methods to solve GRN related problems.

GRN Application for Precision Medicine

GRNs can be used to reveal numerous biological functions, which also include cancer initiation and progression. Through a GRN, one may be able to better understand how genes and proteins interact, which further leads to a better understanding of functions of oncogenes and related pathways, offering a better discovery towards drug target identification. A typical application is a GRN approach for precision medicine. Mounika Inavolu *et al.* (2017) developed a novel algorithm that integrates expression profiles with protein-–protein interaction (PPI) networks to identify key deregulated subnetworks, which facilitated drug target discovery especially for breast cancer. Precision medicine target-drug selection (PMTDS) is another novel method for detecting individual interaction networks from multiomics data and recommending the optimum targets and associated drugs for precision medicine style (Vasudevaraja *et al.*, 2017).

See also: Algorithms for Graph and Network Analysis: Clustering and Search of Motifs in Graphs. Cell Modeling and Simulation. Community Detection in Biological Networks. Community Detection in Biological Networks. Computational Systems Biology. Data Mining: Mining Frequent Patterns, Associations Rules, and Correlations. Gene-Gene Interactions: An Essential Component to Modeling Complexity for Precision Medicine. Gene-Gene Interactions: An Essential Component to Modeling Complexity for Precision Medicine. Graphlets and Motifs in Biological Networks. Natural Language Processing Approaches in Bioinformatics. Network Inference and Reconstruction in Bioinformatics. Network Models. Network Properties. Network-Based Analysis for Biological Discovery. Networks in Biology. Pathway Informatics. Protein-DNA Interactions. Regulation of Gene Expression

References

Aluru, S., 2005. Handbook of Computational Molecular Biology. CRC Press.
Balakrishnan, N., 1992. Handbook of the Logistic Distribution. New York, Basel: Marcel Dekker. Inc.
Bateman, A., Birney, E., Cerruti, L., *et al.*, 2002. The Pfam protein families database. Nucleic Acids Research 30 (1), 276–280.
Beinlich, I.A., Suermondt, H.J., Chavez, R.M., Cooper, G.F., 1989. The ALARM monitoring system: A case study with two probabilistic inference techniques for belief networks. In: Proceedings of the Conference on Artificial Intelligence in Medical Care 89, pp. 247–256. Berlin, Heidelberg: Springer.
Bentley, J.L., 1975. Multidimensional binary search trees used for associative searching. Communications of the ACM 18 (9), 509–517.
Brazma, A., Parkinson, H., Sarkans, U., *et al.*, 2003. ArrayExpress – A public repository for microarray gene expression data at the EBI. Nucleic Acids Research 31 (1), 68–71.
Butte, A.J., Kohane, I.S., 1999. Mutual information relevance networks: Functional genomic clustering using pairwise entropy measurements. Pacific Symposium on Biocomputing 2000, 418–429.
Cao, J., Qi, X., Zhao, H., 2012. Modeling gene regulation networks using ordinary differential equations. In: Proceedings of the Next Generation Microarray Bioinformatics, pp. 185-197. Humana Press.
Chai, L.E., Loh, S.K., Low, S.T., *et al.*, 2014. A review on the computational approaches for gene regulatory network construction. Computers in Biology and Medicine 48, 55–65.
Cheng, L., Khorasani, K., Ding, Y., Guo, X., 2013. Gene interaction networks based on kernel correlation metrics. International Journal of Computational Biology and Drug Design 6 (1-2), 72–92.
Cheng, C., Yan, K.K., Hwang, W., *et al.*, 2011. Construction and analysis of an integrated regulatory network derived from high-throughput sequencing data. PLOS Computational Biology 7 (11), e1002190.
Chen, Y., Li, Y., Narayan, R., Subramanian, A., Xie, X., 2016. Gene expression inference with deep learning. Bioinformatics 32 (12), 1832–1839.
Chen, Y.H., Yang, C.D., Tseng, C.P., Huang, H.D., Ho, S.Y., 2015. GeNOSA: Inferring and experimentally supporting quantitative gene regulatory networks in prokaryotes. Bioinformatics 31 (13), 2151–2158.
Chickering, D.M., 2002. Optimal structure identification with greedy search. Journal of Machine Learning Research 3 (2002), 507–554.
Chuang, H.Y., Lee, E., Liu, Y.T., Lee, D., Ideker, T., 2007. Network-based classification of breast cancer metastasis. Molecular Systems Biology 3 (1), 140.
Clarke, B., 2007. Information optimality and Bayesian modelling. Journal of Econometrics 138 (2), 405–429.
Coordinators, N.R., 2016. Database resources of the national center for biotechnology information. Nucleic Acids Research 44 (Database issue), D7.
De Smet, R., Marchal, K., 2010. Advantages and limitations of current network inference methods. Nature Reviews Microbiology 8 (10), 717.
Dong, X., Weng, Z., 2013. The correlation between histone modifications and gene expression. Epigenomics 5 (2), 113–116.
D'haeseleer, P., Liang, S., Somogyi, R., 2000. Genetic network inference: From co-expression clustering to reverse engineering. Bioinformatics 16 (8), 707–726.
Edgar, R., Domrachev, M., Lash, A.E., 2002. Gene expression omnibus: NCBI gene expression and hybridization array data repository. Nucleic Acids Research 30 (1), 207–210.
Elo, L.L., Järvenpää, H., Orešič, M., Lahesmaa, R., Aittokallio, T., 2007. Systematic construction of gene coexpression networks with applications to human T helper cell differentiation process. Bioinformatics 23 (16), 2096–2103.
Fiers, M.W., Minnoye, L., Aibar, S., *et al.*, 2018. Mapping gene regulatory networks from single-cell omics data. Briefings in Functional Genomiscs.
Friedman, N., Linial, M., Nachman, I., Pe'er, D., 2000. Using Bayesian networks to analyze expression data. Journal of Computational Biology 7 (3-4), 601–620.
Gasteiger, E., Gattiker, A., Hoogland, C., *et al.*, 2003. ExPASy: The proteomics server for in-depth protein knowledge and analysis. Nucleic Acids Research 31 (13), 3784–3788.
Geva-Zatorsky, N., Rosenfeld, N., Itzkovitz, S., *et al.*, 2006. Oscillations and variability in the p53 system. Molecular Systems Biology 2 (1).
Gong, J., Liu, C., Liu, W., *et al.*, 2015. An update of miRNASNP database for better SNP selection by GWAS data, miRNA expression and online tools. Database 2015.
Hill, S.M., Heiser, L.M., Cokelaer, T., *et al.*, 2016. Inferring causal molecular networks: Empirical assessment through a community-based effort. Nature Methods 13 (4), 310.
Hore, V., Viñuela, A., Buil, A., *et al.*, 2016. Tensor decomposition for multiple-tissue gene expression experiments. Nature Genetics 48 (9), 1094.
Kimura, S., Ide, K., Kashihara, A., *et al.*, 2004. Inference of S-system models of genetic networks using a cooperative coevolutionary algorithm. Bioinformatics 21 (7), 1154–1163.
Lähdesmäki, H., Shmulevich, I., Yli-Harja, O., 2003. On learning gene regulatory networks under the Boolean network model. Machine Learning 52 (1-2), 147–167.
Liu, Z., Malone, B., Yuan, C., 2012. . Empirical evaluation of scoring functions for Bayesian network model selection. BMC Bioinformatics 13 (Suppl. 15), S14.
Liu, F., Zhang, S.W., Guo, W.F., *et al.*, 2016. Inference of gene regulatory network based on local bayesian networks. PLOS Computational Biology 12 (8), e1005024.

Lu, X.M., Batugedara, G., Lee, M., et al., 2017. Nascent RNA sequencing reveals mechanisms of gene regulation in the human malaria parasite Plasmodium falciparum. Nucleic Acids Research 45 (13), 7825–7840.

Luo, F., Yang, Y., Zhong, J., et al., 2007. Constructing gene co-expression networks and predicting functions of unknown genes by random matrix theory. BMC Bioinformatics 8 (1), 299.

Madan Babu, M., Teichmann, S.A., 2003. Evolution of transcription factors and the gene regulatory network in Escherichia coli. Nucleic Acids Research 31 (4), 1234–1244.

Marbach, D., Costello, J.C., Küffner, R., et al., 2012. Wisdom of crowds for robust gene network inference. Nature Methods 9 (8), 796.

Marbach, D., Prill, R.J., Schaffter, T., et al., 2010. Revealing strengths and weaknesses of methods for gene network inference. Proceedings of the National Academy of Sciences 107 (14), 6286–6291.

Marbach, D., Schaffter, T., Mattiussi, C., Floreano, D., 2009. Generating realistic in silico gene networks for performance assessment of reverse engineering methods. Journal of Computational Biology 16 (2), 229–239.

Margolin, A.A., Nemenman, I., Basso, K., et al., 2006. ARACNE: An algorithm for the reconstruction of gene regulatory networks in a mammalian cellular context. BMC Bioinformatics 7 (Suppl. 1), S7.

Mounika Inavolu, S., Renbarger, J., Radovich, M., et al., 2017. IODNE: An integrated optimization method for identifying the deregulated subnetwork for precision medicine in cancer. CPT: Pharmacometrics & Systems Pharmacology 6 (3), 168–176.

Muhammad, D., Schmittling, S., Williams, C., Long, T.A., 2017. More than meets the eye: Emergent properties of transcription factors networks in Arabidopsis. Biochimica et Biophysica Acta (BBA)-Gene Regulatory Mechanisms 1860 (1), 64–74.

Mukherjee, S., Speed, T.P., 2008. Network inference using informative priors. Proceedings of the National Academy of Sciences 105 (38), 14313–14318.

Murphy, K.P., Russell, S., 2002. Dynamic bayesian networks: Representation, inference and learning.

Ogata, H., Goto, S., Sato, K., et al., 1999. KEGG: Kyoto encyclopedia of genes and genomes. Nucleic Acids Research 27 (1), 29–34.

Peters, A.D., 2008. A combination of cis and trans control can solve the hotspot conversion paradox. Genetics 178 (3), 1579–1593.

Phillips, T., 2008. The role of methylation in gene expression. Nature Education 1 (1), 116.

Prill, R.J., Marbach, D., Saez-Rodriguez, J., et al., 2010. Correction: Towards a rigorous assessment of systems biology models: The DREAM3 challenges. PLOS ONE 5 (2), e9202.

Saelens, W., Cannoodt, R., Saeys, Y., 2018. A comprehensive evaluation of module detection methods for gene expression data. Nature Communications 9 (1), 1090.

Scott, D.W., Sheather, S.J., 1985. Kernel density estimation with binned data. Communications in Statistics-Theory and Methods 14 (6), 1353–1359.

Shmulevich, I., Dougherty, E.R., Kim, S., Zhang, W., 2002. Probabilistic Boolean networks: A rule-based uncertainty model for gene regulatory networks. Bioinformatics 18 (2), 261–274.

Simmons, L.A., Foti, J.J., Cohen, S.E., Walker, G.C., 2008. The SOS regulatory network. EcoSal Plus 2008.

Van Gerven, M.A., Taal, B.G., Lucas, P.J., 2008. Dynamic Bayesian networks as prognostic models for clinical patient management. Journal of Biomedical Informatics 41 (4), 515–529.

Vasudevaraja, V., Renbarger, J., Shah, R.G., et al., 2017. PMTDS: A computational method based on genetic interaction networks for Precision Medicine Target-Drug Selection in cancer. Quantitative Biology 5 (4), 380–394.

Vijesh, N., Chakrabarti, S.K., Sreekumar, J., 2013. Modeling of gene regulatory networks: A review. Journal of Biomedical Science and Engineering 6 (02), 223.

Wang, Y.R., Huang, H., 2014. Review on statistical methods for gene network reconstruction using expression data. Journal of Theoretical Biology 362, 53–61.

Yavari, F., Towhidkhah, F., Gharibzadeh, S., 2008. Gene regulatory network modeling using Bayesian networks and cross correlation. In: Proceedings of the Cairo International Biomedical Engineering Conference, CIBEC 2008, pp. 1–5. IEEE.

Yeang, C.H., Mak, H.C., McCuine, S., et al., 2005. Validation and refinement of gene-regulatory pathways on a network of physical interactions. Genome Biology 6 (7), R62.

Yu, L., Liu, H., 2004. Efficient feature selection via analysis of relevance and redundancy. Journal of Machine Learning Research 5 (2004), 1205–1224.

Zhang, W., Chien, J., Yong, J., Kuang, R., 2017. Network-based machine learning and graph theory algorithms for precision oncology. npj Precision Oncology 1 (1), 25.

Relevant Websites

https://metacyc.org/
    BIOCYC.
http://www.genecards.org/
    GeneCards.
http://www.ncbi.nlm.nih.gov/projects/HistoneDB2.0
    HistoneDB 2.0.
http://jaspar.genereg.net/
    JASPAR.
www.genome.jp/kegg/
    KEGG.
http://mirgator.kobic.re.kr/index.html
    miRGator v3.0.
http://www.pathwaycommons.org/
    Pathway Commons.

Biographical Sketch

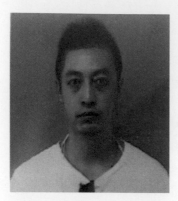

Enze Liu, Ms. Enze Liu is currently enrolled as a PhD student at the Indiana University School of Informatics and Computing. Before that, he graduated from Royal Institute of Technology, Stockholm, Sweden in 2013 with a master's degree in computational and system biology. He then worked at the Science for Life Laboratory in Stockholm, Sweden and Biomarker Bio-tech corporations in Beijing, China. His research interests mainly lie in the computational biology domain, in cancer biology and personalized medicine.

Lang Li, PhD Dr. Li gained his PhD of Biostatistics from University of Michigan in 2000. He then served Indiana University School of Medicine for 11 years and became the chair of Computational Biology and Bioinformatics and associate director of the Indiana Institute of Personalized Medicine. He recently joined The Ohio State University College of Medicine as the new chair of Department of Biomedical Informatics. Dr. Li's research interests include numerous areas across system biology, drug interaction, and personalized medicine.

Lijun Cheng, PhD Dr. Cheng gained her PhD from Donghua University in China. She started her research career as a postdoctoral fellow at the Indiana University School of Medicine in 2013. Dr. Cheng currently works in the Department of Biomedical Informatics, College of Medicine, Ohio State University as an assistant professor. Her research interests include computational and systems biology mainly in the cancer biology and personalized medicine domains.

MicroRNA and lncRNA Databases and Analysis

Xiangying Sun and Michael Gribskov, Purdue University, West Lafayette, IN, United States

Nomenclature

elncRNA	Enhancer lncRNA	ncRNA	Non-coding RNA
lincRNA	Long intergenic non-coding RNA	PIN	Partially intronic RNA, a kind on lncRNA
lncRNA	Long non-coding RNA	piRNA	Piwi-interacting RNA
miR	miRNA, microRNA	plncRNA	Promoter lncRNAs
miRNA	microRNA	pri-miRNA	miRNA primary transcript
NAT	Natural antisense transcript	TIN	Totally intronic RNA, a kind of lncRNA

Introduction

Over half of the transcription in mammals originates from non-coding regions of the genome, producing noncoding RNA (ncRNA). The function of some non-coding RNA is known, for instance, the ribosomal and tRNA components of the translation system (rRNA and tRNA), small nuclear RNA (snRNA) components of the splicosome, small nucleolar RNAs (snoRNA) which are mostly involved in modification of RNA bases, and others such as the telomerase guide RNA, tmRNA, which releases stalled ribosomes, and RNaseP RNA, the catalytic component of the tRNA modification enzyme RNaseP. But genome-wide analysis of transcription has revealed that a large fraction of transcripts are from neither known structural classes nor from mRNA. Because RNA synthesis is an energetically expensive process, it is unlikely that miRNAs and lncRNAs, which together make up as much as 65% of transcription (Kapranov *et al.*, 2007), are accidental noise produced by random read-through from coding gene transcription. Indeed, in the last twenty years, a great deal has been learned about microRNAs (miRNA) and their important role in cellular regulation, but the function of the remaining ncRNAs, loosely referred to as long noncoding RNAs (lncRNA) is, except for a few examples, still obscure. Computational approaches have played an important role in identifying both miRNAs and lncRNAs, and methods are still undergoing rapid development.

MicroRNA

Biological Background

MicroRNAs are typically 22 base long RNAs (lengths can vary from 16–24 bases in different species) that play an important role in gene regulation in eukaryotic organisms, usually acting by targeting the mRNA for degradation or acting as a translation repressor (see (Catalanotto *et al.*, 2016) for a review of nuclear functions). MicroRNA genes are common in eukaryotic genomes; usually there are thousands of miRNA genes, for instance in human, the ENCODE project reported 11,000 small RNA genes (The Encode Project Consortium, 2012), and about 2600 mature miRNAs are listed in miRBase (Griffiths-Jones, 2006). MiRNA genes are located in the introns of protein coding genes (sometimes called miRtrons), in UTRs of coding transcripts, or as completely separate transcripts. The primary transcript (pri-miRNA) is processed to produce a 60–100 base precursor RNA by the splicing process, in the case of intron encoded miRNAs, or by Drosha (DCL1 in plants) in the case of non-intron miRNAs. In either case, the result is a precursor RNA (pre-miRNA), with an extended base-paired hairpin structure. After export from the nucleus, the pre-miRNA is asymmetrically cleaved, near the loop of the hairpin structure, by the Dicer endonuclease to produce a mature miRNA. One strand of the mature miRNA, usually with a strong preference for the strand originating from the 5′ end of the pre-miRNA, referred to as the 5p strand, is loaded into the RNA-induced silencing complex, RISC, and the other, the 3p or * strand is degraded. Within the RISC, the miRNA is bound by an Argonaute (Ago) protein, and is used to locate its complementary target mRNA, usually binding in the 3′ UTR. The first 2–7 bases of the miRNA, the seed sequence, are particularly important in binding to the target, although extended complementarity or the miRNA and target mRNA is common. The seed region is also of interest because miRNAs with the same seed sequence are generally assigned to the same family. The mRNA is ultimately degraded by one of several pathways once it is bound to RISC. There are many exceptions and differences from the canonical process described above, but the general aspects are highly conserved. In addition to classical miRNA, there are additional classes of regulatory small RNAs including piwi-interacting RNA (piRNA), which interact with piwi proteins, a subtype of Argonaute proteins, and appear to act primarily to repress transcription of transposable elements (for a review see (Tang, 2010)), and small interfering RNAs (siRNA) which are also produced from double stranded precursors by a Dicer-like system. One of the difficulties in predicting miRNAs is that exceptions to almost every aspect of the canonical process, described above, have been found, and while the process is very similar, there are significant differences between plants, animals, and fungi.

From a computational viewpoint, the focus is usually on 1) miR discovery, identifying the miRNA genes or pri-miRNA transcripts from genome or transcriptome sequence and predicting the mature miRNA sequence from the gene/precursor sequence, and 2) predicting the mRNA targets.

miR discovery

Mature miRNAs can be identified experimentally by isolating and sequencing small RNAs, typically 17–28 bases long. The sequences can then be mapped to the reference genome using standard short-read mapping programs such as BWA or Bowtie2 (Friedländer et al., 2012; Hendrix et al., 2010). Mismatches must be allowed since miRNAs may be the target of adenosine to inosine editing (Cai et al., 2009). Pre-miRNAs, which are typically capped and polyadenylated can also be identified in typical RNA-Seq experiments. In this case, the two arms of the miRNA hairpin stem are often detected as separate reads (Kozomara and Griffiths-Jones, 2013). Crosslinking-immunoprecipitation has been used to identify miRNAs bound in vivo to Argonaute. This should, in principle, provide much better experimental datasets, but Agarwal et al. (2015) suggest that many of these putative sites may be non-functional.

Computational identification of novel miRNA genes in genomic sequence relies on a combination of sequence similarity to known mature miRNAs, typically based on Blast (Altschul et al., 1997) searches, and secondary structure prediction using UnaFold (Markham and Zuker, 2008), RNAFold (Mathews et al., 2004), or the Vienna RNA package (Lorenz et al., 2011). MiRNAs are often highly conserved between species, and the conservation of the mature miRNA should be nearly perfect, however due to "arm switching" (Griffiths-Jones et al., 2011), the shift of the mature miRNA from the 5p to 3p side of the precursor hairpin, matching to just mature miRNAs can be problematic, and matching to the pre-miRNA is likely to be more reliable. The second approach to identification of miRNAs lies in detection of the long base-paired hairpin stem of the miRNA. Minimum free energy RNA secondary structure prediction methods are typically used to detect potential hairpin structures (see references, above), usually after pre-screening to restrict the analysis to only 3'UTRs, to identify sequences similar to known miRNA, or to remove protein coding sequences, structural RNAs and transposable elements. Because there are many sequences that are predicted to fold as an acceptable hairpin stem, for instance Bentwich et al. (2005) identified about 11 million in the human genome, this basic approach is typically augmented by additional ad hoc criteria. These so-called context criteria examine features such as the presence of a base paired region adjacent to the precursor hairpin with a typical 2 base overhang on the 3p arm, require fewer than 4 mismatches between the 5p and 3p arms in the mature miRNA region, absence of loops in the mature miRNA region, constraints on GC-content, minimum predicted folding free energy (Zhang et al., 2006) or alignment score, structural "exposure" of the seed binding region in the mRNA, exclusion of perfect inverted repeats forming the putative pre-miRNA hairpin (possible transposable element) (see, for instance, (Lucas and Budak, 2012; Meyers et al., 2008)), and continuous pairing in the precursor stem. RNA sequencing data is frequently included, requiring that sequences for both the 5p and 3p arms be detected with a minimum number of reads. Most of these features have been determined by inspection from specific sets of predicted or validated miRNAs and their power and generality are often unclear.

miRNA target prediction

There is general consensus that complementarity of the mRNA with the miRNA seed sequence, conservation of the miRNA and mRNA target sequence across species, the predicted stability of the miRNA-mRNA duplexes, presence of multiple sites (abundance), and accessibility of the mRNA target site are among the most import features in predicting mRNA target sites (Peterson, 2014). However many of the same ad hoc features listed above may be incorporated. There are literally dozens of predictive methods (for a recent comparison see (Fan and Kurgan, 2015)). The recall (fraction of known miRNA targets identified in known data) and precision (fraction of correctly predicted targets) vary widely, and offer the classic trade-off between high recall-low precision and low recall-low precision.

Many machine learning methods have been applied to miRNA discovery (see (Demirci et al., 2017) for a review), most often trained using data from miRBase as training data. Negative datasets are usually obtained from coding regions (or equivalently, exons), or by random sampling from whole genomes. It has been suggested that an average decision tree is best, and generalizes across species. Recently, a number of groups have attempted to determine which of the many proposed features are most discriminative. Agarwal et al. (2015) found that 3'-UTR site abundance, predicted seed-and downstream pairing stability, the base at position 1 and 8 of the miRNA seed sequence, the base at position 8 of the target site, local UA content, predicted structural accessibility, distance from the miRNA site to the stop codon of polyA site, site conservation, ORF length, 3'UTR length, and the number of offset sites in the UTR are the most important features in identifying miRNA targets. Lopes et al. (2014) found that minimum predicted free energy index, ensemble free energy, normalized number of sequence variants, normalized Shannon entropy, and normalized base-pair distance were the most important features in a random forest approach. Tran et al. (2015) found, using a boosted support vector machine approach, that the most important features are folding free energy of the longest nonexact stem, maximum number of consecutive G's in the longest nonexact stem, percentage of CC and GA dinucleotides, maximum number of consecutive C's, maximum number of consecutive G's in the hairpin, percentage of G–U pairs, folding free energy normalized by hairpin size, percentage of paired U, average folding free energy, percentage of unpaired–unpaired A–paired triplets, and size of bulges. As just these three examples show, there is considerable disagreement, even today, over what features are most relevant and powerful for miRNA target identification.

Databases/Resources

There are many online resources related to miRNA (for a recent review see (Singh, 2017). A large fraction of these have been created for a particular organism or purpose, and then not updated. Below, we give our recommendations for the most useful and reliable resources (in our opinion).

- MiRBase (miRBase, 2016; Kozomara and Griffiths-Jones, 2013) is the original miRNA resource and still hosts the miRBase registry which provides unique names for novel miRNA genes prior to publication. Release 21 of miRBase contains 28,645 entries from 223 species, including extensive annotation of functions, experimental evidence, and links to other databases.

- DIANA-TarBase (Paraskevopoulou, 2016; DIANA-TOOLS, 2016; Vlachos et al., 2015) focuses on experimentally validated miRNAs and includes more than half a million experimentally supported miRNA-mRNA interactions. In addition TarBase includes computational predictions made with the MicroT-CDS method.
- Plant Non-coding RNA Database (PNRD, 2016; Yi et al., 2014) focuses on all types of non-coding RNAs in plants, not just miRNAs. Since miRNAs are structurally somewhat different in plants, a plant specific resource is sometimes useful. The earlier Plant MicroRNA Database (PMRD) appeared to be inactive at the time this article was written.
- Rfam (Rfam, 2017) contains a large amount of information about miR families, including sequences, species of occurrence, secondary structure (usually predicted), and matching motifs.

lncRNA

Biological Background

By definition, long noncoding RNA (lncRNA) collectively refers to the group of transcribed RNAs, with a length longer than 200 nucleotides and low coding potential. However, the 200 nucleotide threshold is an arbitrary threshold, which was selected based on a convenient biochemical cutoff in RNA isolation protocols. BC1 and snaR, for example, are examples of ncRNAs that are shorter than 200 nucleotides but still classified as lncRNAs. Therefore, in 2011, Amaral et al. refined this definition: lncRNAs are noncoding RNAs that may have a function as either primary or spliced transcripts, that do not encode proteins, and are neither members of structural RNA families (tRNAs, snoRNAs, spliceosomal RNAs, etc.), nor processed into known classes of small RNAs, such as microRNAs (miRNAs), piwi-interacting (piRNAs) and small nucleolar RNA (snoRNAs)[1]. Clearly this is still somewhat unsatisfying as lncRNAs are primarily defined as those that do not belong to known classes.

Because of the generous definition, lncRNAs are diverse in their biogenesis, stability, sub-cellular localization, evolutionary conservation, structures and functions (Ayupe et al., 2015; Johnsson et al., 2014). LncRNAs are typically capped, spliced, and poly-adenylated. Compared with mRNAs, lncRNAs have relatively a lower expression level and lower stability. They are more tissue specific and cell-type specific, and are often expressed in a narrower time window. LncRNAs are mostly located in the nucleus, presumably to regulate gene expression, at the epigenetic level, but a minority of lncRNAs are present in the cytoplasm where they regulate translation. For example, Xist is a well-studied lncRNA that is involved in X inactivation in placental mammals. Xist is localized in the nucleus, and is highly expressed from the inactivated X chromosome at the onset of X chromosome inactivation. Xist binds at many locations in the inactivated X chromosome (by an, as yet, not well understood process) and recruits silencing factors such as the Polycomb repressive complex 2 (PRC2) to silence X chromosome genes (Brown et al., 1992; Clemson et al., 1996).

Beyond primates, little sequence conservation is typically observed in lncRNAs. Unlike mRNAs, the lack of conservation in the sequences of lncRNAs does not indicate a lack of common functions; An increasing number of examples have shown that lncRNAs are conserved in structure rather than sequence, and the secondary (or higher) structure of lncRNAs constitutes the main functional unit. For example, HOTAIR is a trans-acting lncRNA whose sequence is poorly conserved in mammals beyond primates (Bhan and Mandal, 2016). Covariance analysis across 33 mammalian sequences of HOTAIR revealed a significant number of covarying positions and half-flips localized in all four domains of HOTAIR, which maintain a similar structure (Somarowthu et al., 2015).

According to NONCODE (v 5.0, a database of lncRNAs documented in the literature), 354,855 lncRNAs have been identified in 17 species. However, the functional roles of these lncRNAs remain mostly unknown. According to lncRNAdb (a database of eukaryotic lncRNA annotations), fewer than 300 have annotated functions confirmed by overexpression or knockdown experiments. LncRNAs, in general, can either repress or activate gene expression, which is associated with cell-fate programming (Flynn and Chang, 2014) and numerous human diseases (Esteller, 2011). The number of lncRNAs whose mechanisms are known in detail are even more limited, less than 20. But these examples have already shown that lncRNAs are involved in important biological processes, such as genomic imprinting, chromatin remodeling, post-transcriptional RNA processing, and regulation of translation. Based on our current knowledge, lncRNAs consummate their regulatory roles in 3 major ways: 1). Decoys: that is they bind to regulatory proteins and preclude their access to DNA; 2). Scaffolds: they recruit epigenetic complexes to regulate chromatin states; and 3). Guides: the lncRNA binds proteins and guides the ribonucleoprotein complex to a target (Rinn and Chang, 2012). PANDA is an example of a lncRNA decoy. It sequesters a transcription factor called NF-YA, and keeps NF-YA from binding to its target genes, and prevents p53-mediated apoptosis (Hung et al., 2011). HOTAIR, which is located in the HOXC cluster, is an example of a lncRNA scaffold. It can simultaneously bind PRC2 in its 5′ domain and LSD1 in its 3′ domain. PRC2 has the function of histone H3 lysine-27 trimethylation, and LSD1 is involved in demethylation of histone H3 at lysine 4. This combination of interactions ensures epigenetic silencing of multiple cancer related genes (Hajjari and Salavaty, 2015). As mentioned above, Xist is an example of a lncRNA guide. Xist recruits Polycomb 1 and 2 complexes and guides them to the inactivated X chromosome to establish and maintain its silencing. Because the mechanisms of so few lncRNAs are known in detail, it is likely that many other mechanisms will be uncovered, ultimately revealing a more complex picture of the role of lncRNAs in regulatory networks.

Classification of lncRNAs

In GENCODE (GENCODE, 2017), lncRNA is classified based on its genomic location with respect to nearby protein-coding genes. This is also one of the most commonly used methods to classify lncRNAs. Initially, lncRNAs are classified as either intergenic lncRNAs or intragenic lncRNAs. The transcripts of Intergenic lncRNAs (lincRNAs) do not overlap protein coding transcripts, while intragenic lncRNAs are transcribed from regions that overlap protein coding genes and can be further classified into sense and antisense lncRNAs. Sense lncRNAs are transcribed from regions of protein-coding genes on the same strand as the mRNA. They can overlap with both introns and exons of protein-coding genes. Totally Intronic RNAs (TINs), are lncRNAs that are located entirely within intronic regions of protein-coding genes. Partially Intronic RNAs (PINs), (Nakaya et al., 2007) are lncRNAs that partially or entirely cover the introns of protein-coding gene. Antisense lncRNAs, or Natural Antisense Transcripts (NATs), are lncRNAs transcribed from the opposite strand of protein-coding genes.

Another way to classify lncRNAs is to distinguish their roles in the regulation of gene expression, distinguishing cis-acting and trans-acting RNAs. Cis-acting lncRNAs regulate the expression of genes that are positioned at the same, or a nearby, genomic locus. They may function through transcriptional interference or chromatin modification. Promoters and enhancers are two natural targets of cis-regulatory lncRNAs, which can recruit transcription factors, or chromatin modification complexes which remodel the structure of adjacent protein coding genes, to increase transcription. Promoter lncRNAs (sometimes called bidirectional promoter lncRNAs), plncRNAs, are transcribed from regions near the transcription start site (usually within 1500 bp of the transcription start site) of protein-coding genes, whereas enhancer lncRNAs (elncRNAs) are located up to 1 Mbp upstream or downstream of the regulated gene. Trans-acting lncRNAs can control the expression of a gene at an independent loci, for example, genes on a different chromosomes.

Discovery and Exploration of lncRNAs

Determining the nature and possible biological functions of lncRNAs has become a focus of intense research. Expression profiling is often a first step in uncovering the function of a lncRNA. Identifying differentially expressed lncRNAs in developmental stages or conditions can imply their potential functions. Alternatively, an informatic method termed "Guilt by Association" identifies functions of lncRNAs by looking for protein-coding genes whose expression is significantly correlated with a lncRNA (Guttman et al., 2009).

Even though some researchers have successfully identified lncRNAs using polyadenylated RNA sequencing (mRNA-Seq), total cellular RNA sequencing (total RNA-Seq) is the usually the method of choice for comprehensive expression profiling of lncRNAs. This is because some lncRNAs, particularly elncRNAs, may not be spliced or polyadenylated. By using total cellular RNA-Seq, both mRNAs and lncRNAs can be identified, regardless of whether they are polyadenylated.

In the following section, we describe a computational pipeline for the identification of lncRNAs using total RNA-Seq data.

1. First the RNA-Seq reads must be mapped, separately, to a reference.
   a. If using a reference genome, reads are first mapped to the genome using an intron aware mapper (e.g., using Tophat2 (Kim et al., 2013). Transcripts from different samples are merged (e.g., using Cufflinks (Trapnell et al., 2012)).
   b. If not using a reference genome, reads from all samples are combined to construct a de novo transcript assembly (e.g., using Trinity (Grabherr et al., 2011)).
2. Reads from the individual samples are separately mapped to the reference.
3. Possible protein coding transcripts are excluded by multiple filtering steps, for example, removing
   a. annotated protein coding transcripts,
   b. transcripts with high coding potential (e.g., using the Coding Potential Calculator (Kong et al., 2007)),
   c. transcripts that match conserved known proteins or motifs (e.g., using Blastx (Altschul et al., 1997)),
   d. transcripts that have a high rate of synonymous versus nonsynonymous substitutions (e.g., using PhyloCSF (Lin et al., 2011)).
4. ChIP-Seq (Chromatin Immunoprecipitation Sequencing) can be used to identify lncRNAs involved in gene activation involving transcription factors, or histone modification.

lncRNA Databases/Resources

Many online resources are available for lncRNAs. As with miRNAs, the spectrum of resources rapidly changes as many database are created for particular purposes, but not maintained over time. In **Table 1** we list a few of the currently active resources. Online searches for "lncRNA database" or similar terms will typically provide an updated list of resources, and a list is also maintained on Wikipedia (see the source citation in **Table 1**).

Closing Remarks and Future Directions

MicroRNA and lncRNA play important roles in biological regulation. In the case of miRNA, a good deal is known about the identity of miRNAs and their targets. In spite of this depth of knowledge, there are a dismaying number of online resources that have been created and, apparently, never updated. This makes it particularly hard to make recommendations about electronic resources; miRBase, Diana-TarBase, and Rfam seem to have the best long-term availability.

Table 1   lncRNA databases and resources

Database	Species	Last update	URL http:// Description
lncRNAdb	69 species	23-Nov-15	www.lncrnadb.org Includes lnc RNAs shown to be functional by overexpression or knockdown experiments
RNAcentral	37 species	1-Apr-17	rnacentral.org Combines 25 well maintained ncRNA databases. Provides integrated text search, sequence similarity search, and programmatic data access (API)
NONCODE	17 species	6-Sep-17	www.noncode.org Includes lncRNAs from published literature, GenBank, and specialized Databases such as Ensembl, RefSeq, lncRNAdb and LNCipedia. Functions of lncRNA are predicted by lnc-GFP
LNCipedia	Human	4-May-17	lncipedia.org Includes 146,742 annotated human lncRNAs. Provides basic transcript information, predicted 2° structure, calculated protein coding potential, and predicted microRNA binding sites
GreeNC	45 species	19-Sep-16	greenc.sciencedesigners.com Includes lncRNAs annotated in plants and algae that are identified by using self-developed pipelines. Provides information about sequence, genomic coordinates, coding potential, and predicted folding energy
PLAR2	17 vertebrates	Unknown	www.weizmann.ac.il/Biological_Regulation/IgorUlitsky/pipeline-lncrna-annotation-rna-seq-data-plar Includes lncRNAs identified using self-developed pipelines. 3P-seq information are included
LncRNADisease	human	26-Jul-17	www.cuilab.cn/lncrnadisease Includes experimentally supported lncRNA-disease association data and lncRNA interactions in various levels, including protein, RNA, miRNA, and DNA
Lnc2Cancer	human	4-Jul-16	www.bio-bigdata.com/lnc2cancer A manually curated database that include 1488 entries of associations between 666 human lncRNAs and 97 human cancers

Source: Wikipedia (http://en.wikipedia.org/wiki//List_of_long_non-coding_RNA_databases).

The situation with lncRNAs is very different. Very few lncRNAs have experimentally determined functions, and many have been predicted only computationally. We expect that methods for identifying lncRNA and for assigning functions will develop rapidly over the next few years as more biological information about their function becomes available. In the meantime, it is likely that there will considerable flux in the availability of available websites and tools.

See also: Applications of Ribosomal RNA Sequence and Structure Analysis for Extracting Evolutionary and Functional Insights. Bioinformatics Data Models, Representation and Storage. Biological Database Searching. Cell Modeling and Simulation. Cell Modeling and Simulation. Characterizing and Functional Assignment of Noncoding RNAs. Computational Approaches for Modeling Signal Transduction Networks. Computational Systems Biology. Data Storage and Representation. Natural Language Processing Approaches in Bioinformatics. Nucleic-Acid Structure Database. Prediction of Coding and Non-Coding RNA. RNA Structure Prediction. Sequence Analysis

---

References

Agarwal, V., Bell, G.W., Nam, J.-W., Bartel, D.P., 2015. Predicting effective microRNA target sites in mammalian mRNAs. eLIFE 4, e05005.

Altschul, S.F., Madden, T.L., Schäffer, A.A., *et al.*, 1997. Gapped BLAST and PSI-BLAST: A new generation of protein database search programs. Nucleic Acids Research 25, 3389–3402.

Ayupe, A.C., Tahira, A.C., Camargo, L., *et al.*, 2015. Global analysis of biogenesis, stability and sub-cellular localization of lncRNAs mapping to intragenic regions of the human genome. RNA Biology 12, 877–892.

Bentwich, I., Avniel, A., Karov, Y., *et al.*, 2005. Identification of hundreds of conserved and nonconserved human microRNAs. Nature Genetics 37, 766–770.

Bhan, A., Mandal, S.S., 2016. Estradiol-induced transcriptional regulation of long non-coding RNA, HOTAIR. Methods in Molecular Biology 1366, 395–412.

Brown, C.J., Hendrich, B.D., Rupert, J.L, *et al.*, 1992. The human XIST gene: Analysis of a 17 kb inactive X-specific RNA that contains conserved repeats and is highly localized within the nucleus. Cell 71, 527–542.

Cai, Y., Yu, X., Hu, S., Yu, J., 2009. A brief review on the mechanisms of miRNA regulation. Genomics Proteomics Bioinformatics 7, 147–154.

Catalanotto, C., Cogoni, C., Zardo, G., 2016. MicroRNA in control of gene expression: MicroRNA in control of gene expression. International Journal Molecular Sciences 17, 1712.

Clemson, C.M., McNeil, J.A., Willard, H.F., Lawrence, J.B, 1996. XIST RNA paints the inactive X chromosome at interphase: Evidence for a novel RNA involved in nuclear/chromosome structure. Journal of Cell Biology 132, 259–275.

Demirci, M.D.S., Baumbach, J., Allmer, J., 2017. On the performance of pre-microRNA detection algorithms. Nature Communications 8, 330.

DIANA-TOOLS, 2016. DIANA-TOOLS. Available at: diana.imis.athena-innovation.gr (accessed 21.10.17).

Esteller, M., 2011. Non-coding RNAs in human disease. Nature Reviews Genetics 12, 861–874.

Fan, X., Kurgan, L., 2015. Comprehensive overview and assessment of computational prediction of microRNA targets in animals. Breifings in Bioinformatics 16, 780–794.

Flynn, R.A., Chang, H.Y., 2014. Long noncoding RNAs in cell-fate programming and reprogramming. Cell Stem Cell 14, 752–761.

Friedländer, M.R., Mackowiak, S.D., Li, N., Chen, W.I., Rajewsky, N., 2012. miRDeep2 accurately identifies known and hundreds of novel microRNA genes in seven animal clades. Nucleic Acids Research 40, 37–52.

GENCODE, 2017. GENCODE. Available at: https://www.gencodegenes.org/ (accessed 21.10.17).

Grabherr, M.G., Haas, B.J., Yassour, M., et al., 2011. Full-length transcriptome assembly from RNA-Seq data without a reference genome. Nature Biotechnology 29, 644–652.

Griffiths-Jones, S., Hui, J.H., Marco, A., Ronshaugen, M., 2011. MicroRNA evolution by arm switching. EMBO Reports 12, 172–177.

Griffiths-Jones, S., Grocock, R.J., van Dongen, S., Bateman, A., Enright, A.J., 2006. miRBase: MicroRNA sequences, targets and gene nomenclature. Nucleic Acids Research 34, D140–D144.

Guttman, M., Amit, I., Garber, M., et al., 2009. Cromatin signature reveals over a thousand highly conserved large non-coding RNAs in mammals. Nature 458, 233–237.

Hajjari, M., Salavaty, A., 2015. HOTAIR: An oncogenic long non-coding RNA in different cancers. Cancer Biology & Medicine 12, 1–9.

Hendrix, D., Levine, M., Shi, W., 2010. miRTRAP, a computational method for the systematic identification of miRNAs from high throughput sequencing data. Genome Biology 11, R39.

Hung, T., Wang, Y., Lin, M.F., et al., 2011. Extensive and coordinated transcription of noncoding RNAs within cell-cycle promoters. Nature Genetics 43, 621–629.

Johnsson, P., Lipovich, L., Grandér, D., Morris, K.V., 2014. Evolutionary conservation of long non-coding RNAs; sequence, structure, function. Biochimica Biophysica Acta 1840, 1063–1071.

Kapranov, P., Cheng, J., Dike, S., et al., 2007. RNA maps reveal new RNA classes and a possible function for pervasive transcription. Science 316, 1484–1488.

Kim, D., Pertea, G., Trapnell, C., et al., 2013. TopHat2: Accurate alignment of transcriptomes in the presence of insertions, deletions and gene fusions. Genome Biology 14, R36.

Kong, L., Zhang, Y., Ye, Z.Q., et al., 2007. CPC: Assess the protein-coding potential of transcripts using sequence features and support vector machine. Nucleic Acids Research 35, W345–W349.

Kozomara, A., Griffiths-Jones, S., 2013. miRBase: Annotating high confidence microRNAs using deep sequencing data. Nucleic Acids Research 42, D68–D73.

Lin, M.F., Jungreis, I., Kellis, M., 2011. PhyloCSF: A comparative genomics method to distinguish protein coding and non-coding regions. Bioinformatics 27, i275–i282.

Lopes, I.O.N., Schliep, A., de Carvalho, A.C.P., 2014. The discriminant power of RNA features for pre-miRNA recognition. BMC Bioinformatics 15, 124.

Lorenz, R., Bernhart, S.H., Höner, Z., et al., 2011. ViennaRNA Package 2.0. Algorithms for Molecular Biology 6, 26.

Lucas, S.J., Budak, H., 2012. Sorting the wheat from the chaff: Identifying miRNAs in genomic survey sequences of Triticum aestivum chromosome 1AL. PLOS One 7, e40859.

Markham, N.R., Zuker, M., 2008. UNAFold: Software for nucleic acid folding and hybridization. Methods in Molecular Biology 453, 3–31.

Mathews, D.H., Disney, M.D., Childs, J.L., et al., 2004. Incorporating chemical modification constraints into a dynamic programming algorithm for prediction of RNA secondary structure. Proceedings of the National Academy of Sciences, USA 101, 7287–7292.

Meyers, B.C., Axtell, M.J., Barte, I.B., et al., 2008. Criteria for annotation of plant MicroRNAs. Plant Cell 20, 3186–3190.

miRBase, 2016. miRBase. Available at: http://www.mirbase.org/ (accessed 21.10.17).

Nakaya, H.I., Amaral, P.P., Louro, R., et al., 2007. Genome mapping and expression analyses of human intronic noncoding RNAs reveal tissue-specific patterns and enrichment in genes related to regulation of transcription. Genome Biology 8, R43.

Paraskevopoulou, M.D., Vlachos, I.S., Hatzigeorgiou, A.G., 2016. DIANA-TarBase and DIANA suite tools: Studying experimentally supported microRNA targets. Current Protocols in Bioinformatics 55, 12.14.1–12.14.18.

Peterson, S.M., Thompson, J.A., Ufkin, M.L., et al., 2014. Common features of microRNA target prediction tools. Fronteirs in Genetics 18, 5–23.

PNRD, 2016. PNRD. Available at: http://structuralbiology.cau.edu.cn/PNRD/.

Rfam, 2017. Rfam. Available at: http://rfam.xfam.org/ (accessed 21.10.17).

Rinn, J.L., Chang, H.Y., 2012. Genome regulation by long noncoding RNAs. Annual Review of Biochemistry 81, 145–166.

Singh, N.K., 2017. MicroRNAs databases: Developmental methodologies, structural and functional annotations. Interdisciplinary Science and Computational Life Science 9, 357–377.

Somarowthu, S., Legiewicz, M., Chillón, I., et al., 2015. HOTAIR forms an intricate and modular secondary structure. Molecular Cell 58, 353–361.

Tang, F., 2010. Small RNAs in mammalian germline: Tiny for immortal. Differentiation 79, 141–146.

The Encode Project Consortium, 2012. An integrated encyclopedia of DNA elements in the human genome. Nature 489, 91–100.

Tran, V.T., Tempel, S., Zerath, B., Zehraoul, F., Tahi, F., 2015. miRBoost: Boosting support vector machines for microRNA precursor classification. RNA 21, 775–785.

Trapnell, C., Roberts, A., Goff, L., et al., 2012. Differential gene and transcript expression analysis of RNA-seq experiments with TopHat and Cufflinks. Nature Protocols 7, 562–578.

Vlachos, I.S., Paraskevopoulou, M.D., Karagkouni, D., et al., 2015. DIANA-TarBase v7.0: Indexing more than half a million experimentally supported miRNA:mRNA interactions. Nucleic Acids Research 43, D153–D159.

Yi, X., Zhang, Z., Ling, Y., Xu, W., Su, Z., 2014. PNRD: A plant non-coding RNA database. Nucleic Acids Research 43, D982–D989.

Zhang, B.H., Pan, X.H., Cox, S.B., Anderson, T.A., 2006. Evidence that miRNAs are different from other RNAs. Cellular and Molecular Life Sciences 63, 246–254.

Relevant Website

http://en.wikipedia.org/wiki//List_of_long_non-coding_RNA_databases
    Wikipedia.

Gene-Gene Interactions: An Essential Component to Modeling Complexity for Precision Medicine

Molly A Hall, The Pennsylvania State University, University Park, PA, United States
Brian S Cole and Jason H Moore, University of Pennsylvania, Philadelphia, PA, United States

Genetic Interaction Modeling is Needed to Better Understand and Predict Complex Diseases

Imagine you are prescribed a medication in your doctor's office and a warning pops up on the computer screen as your doctor enters the prescription into your electronic health record. This warning tells your doctor that you have a genetic variant that predisposes you to serious side effects from the medicine being prescribed. This scenario is the goal of precision medicine: that medications, therapies, disease risk assessment, and interventions are informed by each individual patient's genetic background as well as their environmental history. Presently, however, clinicians typically do not have access to patient genetic data or a comprehensive list of past and current environmental exposures. More importantly, even if warning systems like these were available in every clinic, this scenario would not be possible for every drug therapy and disease because, to date, there is simply not enough known about the complex mechanisms that lead to common diseases and adverse drug reactions.

For many rare, Mendelian disorders like cystic fibrosis and Huntington's disease, the genetic component has been largely explained. These diseases involve one or a small set of rare but highly penetrant mutations. For example, early linkage studies found mutations in *cystic fibrosis transmembrane conductance regulator (CFTR)* that lead to cystic fibrosis (Kerem *et al.*, 1989; Riordan *et al.*, 1989; Rommens *et al.*, 1989). Conversely, due to their complex etiologies, the genetic basis of common traits like type 2 diabetes, height, body mass index (BMI), and many cancers have remained elusive.

Genetic association studies are often used to study complex traits like these. In a genetic association test, allele frequencies for a locus are compared between individuals who have a disease of interest (e.g., type 2 diabetes cases) and those who do not (e.g., type 2 diabetes controls). Before the advent of low-cost and high-throughput genotyping technologies, genetic association tests were performed for genetic loci that were predicted to be involved with a disease (candidate genetic association studies). Alternatively, genome-wide association studies (GWAS) allow agnostic testing of a set of single nucleotide polymorphisms (SNPs) across the genome to identify novel genetic regions predictive of disease.

GWAS have been successful at identifying numerous loci across the genome that are associated with a diverse and wide range of phenotypes. However, as predicted by Wang *et al.* (2005), if the odds ratios of all SNP-phenotype associations curated in the NHGRI GWAS Catalog (Welter *et al.*, 2014) are considered, the vast majority of SNPs have a low effect on the traits with which they are associated, and relatively few SNPs have a high effect (Wang *et al.*, 2005; Hall *et al.*, 2016a,b). Manolio *et al.* (2009) suggested a 2-fold risk increase (odds ratio of 2) as indicative of a variant exhibiting a large effect (Manolio *et al.*, 2009). Yet, the majority of GWAS-identified associations demonstrate an odds ratio less than 1.5, which has led many in the field to look beyond GWAS (Manolio *et al.*, 2009; Maher, 2008; Hall *et al.*, 2016a,b; Moore *et al.*, 2010).

Incorporating gene-gene interactions is an essential component to modeling the complexity that exists in biology and the processes that lead to common diseases. Gene products interact physically with one another or within one or multiple pathways. Further, physical interactions between and within DNA, RNA, and proteins are a major part of gene regulation (Martinez, 2002) and translation (Gallie, 2002). Considering loci individually does not allow for the complexity known to exist in molecular and cell biology and in physiology. Embracing complexity by exploring gene-gene interactions is likely to explain much of the yet unknown etiology of common diseases, as these complex traits arise from vastly dynamic and interconnected systems in biology. Such analyses will also identify the genotype combinations that confer the greatest risk to adverse health outcomes. Discussed in this article is the history of epistasis, examples of genetic interactions in model systems and humans, major challenges to identifying gene-gene interactions and common methods for overcoming these challenges, as well as consideration of additional types of complexity that must be explored in combination with epistasis to further reveal the etiology of common diseases.

Historical Perspectives on Analysis and Interpretation of Gene–Gene Interactions

The analysis of gene-gene interactions long predated the elucidation of the structure of DNA. After the rediscovery of Mendel's pioneering experiments of pea crosses that spawned the field of genetics, an explosive period of genetic discovery, driven by experiments in model systems and mathematical analysis at the population level, dominated the first two decades of the twentieth century (Sturtevant, 2001). It was during this gilded age of genetics that pioneering analysis of model systems extended and refined Mendel's laws into a cohesive theory of genetics that formed the basis of our modern understanding. During this period, epistasis was discovered by William Bateson, the biologist who coined the term "genetics" to name the nascent field of the study of heritable variation (Bateson, 1909).

Bateson used the term "epistasis" to describe a cross between two strains in which the phenotypic distribution of the resulting offspring departs from the ratios expected by Mendel's laws (Cordell, 2002). Specifically, Bateson used the term epistasis to describe one mutation blocking or masking the effects of another, hence the use of term "epistasis" which may be translated as "resting upon." Bateson's usage of the term epistasis described an interaction between two genetic variants in which one variant negates the effects of another (Phillips, 2008). This type of genetic interaction, sometimes called modification, was the first form of gene-gene interactions to be observed in experimental crosses. Working together with Reginald Punnett, Bateson developed a two-locus Punnett square to describe the phenotypic ratios of F_2 progeny from crosses of two strains of the flowering sweet pea *Lathyrus odoratus* which displayed a flower coloration trait only when two separate dominant alleles were present at separate loci (Sturtevant, 2001). This two-locus epistasis model extended Mendel's original postulations of a two-locus model to incorporate an interaction between genetic variants, without which the phenotypic ratios of Bateson's and Punnett's sweet peas did not conform to Mendel's laws.

Bateson and Punnett's description of epistasis is the result of crosses between self-fertilizing strains of plants, which are essentially controlled for genetic background, allowing the analysis of the effects of one or a small number of genetic variants on visible phenotypes such as morphological traits. Natural populations of organisms, including humans and wild populations of other organisms which are used as model systems in experimental genetics, contain genetic variation across the genome, eliminating the ability to analyze the effects of one or a small number of genetic variants against a controlled background (Moore and Williams, 2005). The statistical geneticist R. A. Fisher extended the description of epistasis to populations which are not controlled for genomic background by defining "epistasy" as deviations from additivity in a linear model (Fisher, 1919). This definition of gene-gene interactions allows for the statistical detection of epistasis in a population which contains a large number of polymorphic sites in the genome by defining epistasis as a statistical deviation from additivity, a definition which incorporates the mean effect of two or more genetic variants in a given population of organisms (Doust et al., 2014.). Importantly, Bateson's definition of epistasis involved organisms which share almost all of their genome (inbred strains) and Fisher's definition involved organisms which contain polymorphisms across the genome (wild populations). Modern scientists have synthesized these concepts into biological and statistical epistasis (Moore and Williams, 2005). Biological epistasis refers to experimental crosses in which the distribution of phenotypes in offspring deviate from Mendelian ratios (as described by Bateson and Punnett), and statistical epistasis indicates genetic effects which deviate significantly from additivity in highly polymorphic populations (as described by Fisher). As a hypothetical example to demonstrate statistical epistasis consider two loci: LocusA and LocusB. If the relationship between the two loci is additive, we would expect the combined effect of the two on a phenotype to be the addition of the main effect of LocusA and the main effect of LocusB. For example, if there is a 2-fold and 3-fold risk associated with the risk alleles for LocusA and LocusB, respectively, the additive result from both loci is a 5-fold increase risk. If the relationship is epistatic, however, the effect of the two loci together will significantly differ from the combined main effects of the two loci. In the scenario described, the presence of both risk alleles under an epistatic relationship could be a 15-fold risk increase; alternately, risk could decrease to 1.1-fold. In other words, when statistical epistasis occurs, there is a non-linear relationship between the effects of two or more loci when combined. While these two forms of epistasis are experimentally distinct, the underlying theory is identical: epistasis, defined broadly, is the interaction between distinct genetic variants (Phillips, 1998, 2008). This definition encompasses both statistical epistasis which might be detectable in population-scale studies and biological epistasis which might be observable in controlled crosses.

Epistasis research has continued to play a central role in genetics since the early work of Bateson, Punnett, Fisher, and others at the dawn of the twentieth century. An important application of epistasis to biological discovery came in the form of pathway ordering, in which multiple strains of a model organism are crossed together and phenotypes observed such that the ordering of a biological pathway becomes evident (Avery and Wasserman, 1992). This important genetic tool can be used to discover which gene products are upstream or downstream of other gene products, providing evidence of gene product function without molecular or biochemical analysis. This can be achieved by crossing together separate mutant strains of a model organism which display different phenotypes (Beadle, 1945). If a double mutant displays the same phenotype as one of the mutants does individually, one mutation likely occurs in a gene whose product functions downstream in a biochemical pathway. While this is certainly not always the case, epistasis as a tool for pathway ordering elucidated the ordering of mutations (and thereby their gene products, even if the molecular functions were only later established) in the biological pathways which control cell cycle in yeast, sex determination in *C. elegans*, embryonic development in *D. melanogaster*, and other pathways (Phillips, 2008).

As described by Moore (2003), epistasis is thought to be ubiquitous in biology (Moore, 2003; Templeton, 2000). Examples of epistasis have been observed in many model organisms (Mackay, 2014), including yeast (Wagner, 2000; Boone et al., 2007; Tong et al., 2004; Szappanos et al., 2011; Moore, 2005; Baryshnikova et al., 2013), *C. elegans* (Lehner et al., 2006; Gaertner et al., 2012; Byrne et al., 2007), *D. melanogaster* (Horn et al., 2011; Huang et al., 2012; Lloyd et al., 1998), *M. musculus* (Cheng et al., 2011; Hanlon et al., 2006; Gale et al., 2009), and *A. thaliana* (Rowe et al., 2008; Kroymann and Mitchell-Olds, 2005). While examples of epistasis have accrued in model organisms over the years, epistasis is not confined to Mendelian traits, for which a small set of highly penetrant mutations explain much of the variance in observed phenotypes. Epistatic interactions between genetic loci have been discovered in human traits ranging from blood types and eye coloration to complex, polygenic, and multifactorial traits such as disease susceptibility (Moore, 2005). Nelson et al. (2001) identified interactions between ApoB and ApoE in females as well as between the low-density lipoprotein receptor and the ApoAI/CIII/AIV complex in males for triglyceride levels (Nelson et al., 2001). Interactions between SNPs in three estrogen metabolism genes, *COMT*, *CYP1B1*, and *CYP1A1*, were identified by Ritchie et al. (2001) that were predictive of sporadic breast cancer (Ritchie et al., 2001). Further, a study by Hemani et al. (2014) identified and replicated a large number of genetic interactions involved in gene expression regulation (Hemani et al., 2014).

Epistasis research was spawned shortly after the rediscovery of Mendel's foundational work that gave rise to the field of genetics and has found application in the understanding of how genetic variants interact to determine phenotypic outcomes. While genetic interactions have provided insight into traits which cannot be adequately explained by additive models or Mendelian ratios, epistasis research remains an active application of genetics in the modern scientific research enterprise. Particularly, detection of epistasis in studies of complex traits in humans presents methodological and computational challenges that remain an active area of development.

Major Challenges to Exploring Genetic Interactions and Common Solutions

Genome-wide association studies most commonly utilize regression to identify statistically significant genotype-phenotype associations and to quantify the certainty and effect size of the association. These regression analyses typically utilize multivariable linear or generalized linear models in which genotypes and covariates of interest such as age, gender, and Body Mass Index are represented as additive terms in a linear equation for which the parameters are estimated by regression. As described above, epistasis is defined as significant deviation from additivity. It is necessary to consider this definition when testing for epistasis using regression. Testing for independent locus main effects in regression, even at the genome-wide level with GWAS, typically involves a model such as $Phenotype \sim \beta_0 + \beta_1 SNP + covariates$. In this linear model formula, the main effect of a SNP (as represented by the parameter estimate of β_1 and its associated significance) on a phenotype (or outcome) of interest is tested. In GWAS, a model is tested for each SNP included in the data set. A likelihood ratio test is recommended when performing interaction tests using regression. The likelihood ratio test compares the difference between a full and reduced model, where the full model in this case is $Phenotype \sim \beta_0 + \beta_1 SNP + \beta_2 SNP2 + \beta_3 SNP1 \times SNP2 + covariates$ and the reduced model is $Phenotype \sim \beta_0 + \beta_1 SNP + \beta_2 SNP2 + covariates$. The difference between these two models is the interaction term, and thus, a significant likelihood ratio p-value indicates a significant effect of the interaction (β_3) above and beyond the additive effects from SNP1 (β_1) and SNP2 (β_2). If the result of interest is the effect of the interaction of the two SNPs when adjusting for the main effects of each SNP, it would be sufficient to consider the effect of the interaction term in the full model; however, when attempting to identify statistical epistasis, it is recommended to use the likelihood ratio p-value to identify significant SNP-SNP interactions (Hall et al., 2017).

The interaction models above involve two SNPs, but how are the SNPs included in the model chosen? One option is to consider every pairwise combination of SNPs within a data set. This approach is not ideal, however, when dealing with large sets of genetic variants. The post-genome era has led to a consistent decline in sequencing cost and a commensurate explosion in high-throughput genotype data. Technologies such as exome sequencing and whole genome sequencing can identify millions to tens of millions of variants within a population of humans, lending the ability for researchers to study the associations of traits with genotypes at historically unprecedented depth and breadth. With this increase in genotyping power comes an increase in the number of potential interactions between genotypes, creating a steep multiple hypothesis testing burden (Eichler et al., 2010; Maher, 2008; Manolio et al., 2009). Because of this, exhaustive searches for significant interactions between multiple genetic loci within the context of trait association become incredibly difficult because associations of ever greater significance are needed to pass a multiple hypothesis testing correction with increasing numbers of genotyped loci. For instance, an exhaustive test of every two-way SNP-SNP combination from 500,000 SNPs will amount to 1×10^{11} pairwise SNP-SNP tests. If the Bonferroni correction at an alpha level of 0.05 is utilized as is common statistical practice, a model would need a p-value below 5×10^{-13} to be considered statistically significant when adjusting for the number of tests. With such a stringent significance threshold, it is likely that many true positives will go undiscovered with this approach. This problem is exponentially exacerbated when higher order genetic interactions are taken into account, for example the interaction between three loci instead of two (Greene et al., 2010).

The rapid expanse of genotyping depth and breadth, accompanied by the exponentially increasing burden of multiple hypothesis testing, combine to create a situation in which exhaustive searches for significant interactions between genetic variants become untenable. Because of this, new techniques for searching for significant interactions only where they are likely to occur are motivated. These techniques fall into broad categories, including main effect filtering, knowledge-based searches, and machine learning approaches.

Logistically speaking, main effect filtering is a straightforward and simple method used to reduce the number of SNP-SNP models to test. It involves first testing SNPs for association with a trait of interest and choosing those exhibiting a main effect at a chosen significance threshold for subsequent SNP × SNP interaction. However, the major limitation of this method is that the SNPs chosen are only those that have a significant effect on the phenotype on their own, which precludes the possibility of identifying SNPs that only have an effect when combined with other variants (Ritchie, 2011).

Knowledge-based filtering offers an appealing alternative to main effect filtering, as the SNP-SNP pairs are chosen for interaction testing based on demonstrated biological relationships (Ma et al., 2015; Ritchie, 2011, 2015; Sun et al., 2014; Thomas et al., 2009). Previous studies have shown success in applying prior knowledge to genetic interaction analyses for ovarian cancer (Kim et al., 2014), multiple sclerosis (Bush et al., 2011; Ritchie, 2009), HIV pharmacogenetics (Grady et al., 2010), age-related cataract (Hall et al., 2015; Pendergrass et al., 2015), HDL cholesterol (Turner et al., 2011; Ma et al., 2015), and other lipid traits (De et al., 2015). Knowledge-based searches for epistasis utilize external knowledge such as the organization of genetic variants into genes, biological pathways, and biophysical interaction networks to identify sets of genetic variants which are likely candidates for epistatic interactions without the need to evaluate or investigate all sets of genetic variants (Greene et al., 2015; Moore et al., 2012;

Table 1   Traditional encodings for genotypes in linear and multivariable linear regression analyses, including generalized linear models

Genotype	AA	Aa	aa
Additive encoding	0	1	2
Dominant encoding	0	1	1
Recessive encoding	0	0	1
Codominant encoding	0, 0	0, 1	1, 0

A biallelic locus with two possible alleles, here denoted as A and a, may be encoded in multiple manners intended to reflect the putative mechanism of action according to Mendel's laws. In this example, the non-reference allele (a) is considered dominant to the reference allele (A) for the purposes of defining dominance and recessivity. In the codominant encoding, two "dummy" or "one-hot" variables are used to denote the three genotype states.

Moore and White, 2006). As biological knowledge of the organization of genetic variants into genes, pathways, and networks increases, knowledge-guided design of epistasis search strategies is likely to improve. Importantly, advances in high-throughput biological interaction screens and functional genomics analyses are currently unlocking knowledge which can be expected to inform targeted searches for epistasis by discarding unlikely interactions, for example between genetic variants within genes that are not co-expressed. Conversely, interactions which are empirically likely based on biological knowledge may be prioritized for search, such as at protein-protein interaction domains between proteins which biophysically interact to transduce an intracellular signal.

In contrast to knowledge-based approaches, advances in machine learning techniques have led to development of algorithmic approaches to prioritize genetic variants which are likely to interact. Feature selection and feature construction methods aim to reduce the complexity of the search space, in this case representing combinations of genetic variants under study (Kira and Rendell, 1992a). The Multifactor Dimensionality Reduction (MDR) algorithm developed by Ritchie and Moore and modified by many others reduces the complexity of an epistasis search by constructing a new two-level feature from the distributions of the trait under study among the 9 possible two-locus genotype combinations between two biallelic variants (Ritchie et al., 2001). Importantly, this new feature can be tested in linear and generalized linear models, as well as being appropriate for input into classifiers such as decision trees (Lou et al., 2007; Moore and Andrews, 2015). MDR has been generalized to handle missing data, imbalanced classes, quantitative traits, permutation analysis, adjustment for covariates, and family pedigree information. Originally written in Java, MDR has recently been implemented in the scikit-learn framework for Python, a popular machine learning distribution (Pedregosa et al., 2011).

In contrast to MDR, which is a feature construction method, feature selection methods can reduce the search space by identifying subsets of features themselves which are likely to interact without creating new features like that created by MDR (Kira and Rendell, 1992a). The Relief family of algorithms has seen application in epistasis discovery (Kira and Rendell, 1992b). Modifications of the Relief algorithm, including SURF, SURF*, multiSURF, and TURF can identify subsets of genetic variants which may be tested directly for epistasis (Greene et al., 2009, 2010; Moore and White, 2007).

Beyond the issues relevant to reducing the multiple testing burden, a further challenge to identifying SNP models with epistatic action is the way SNPs are genetically encoded. Traditional encoding methods include additive, dominant, and recessive, and each make critical assumptions about a given SNP's biological action (**Table 1**). It is common for investigators to choose one encoding, most often additive. However, every SNP across the genome does not demonstrate the same genetic action. If a SNP is encoded under a different assumption than its true genetic action, this can lead to a reduction in power or inflation of false positive results. The traditional approach is inflexible and does not reflect the diversity known to exist in genetics. One approach to circumvent the bias in choosing a traditional encoding method is to use the codominant encoding when testing for interactions, which is a dummy encoding for the different genotype states. Still, further method development is needed to create encoding alternatives that are flexible to the diverse genetic action of SNPs.

Feature construction and feature selection methods are examples of machine learning methods which can provide information about likely interactions without incorporating expert knowledge, instead using the data itself. In contrast, knowledge-based methods leverage a growing base of orthogonal biological knowledge gleaned from increasingly high-throughput experimentation, including interaction screens. In the future, both machine learning methods and biological knowledge are likely to expand. Knowledge of pathways and biophysical interactions is steadily increasing, and software which can organize genetic variants under study into higher-order units such as genes, pathways, and networks will likely find increasing application in the search for epistasis. Commensurately, statistical advances in the form of novel association tests which can control type I and type II error and methods to assign appropriate genetic encodings offer to improve the quality of epistatic hypotheses generated in genome-wide studies, offering the power to quantify uncertainty in the search for gene-gene interactions (Wang et al., 2016; Pattin et al., 2009; Kooperberg and LeBlanc, 2008).

In sum, epistasis research faces a considerable challenge from the multiple hypothesis testing problem which has motivated the development and application of new approaches that aim to reduce the complexity of an exhaustive search through external knowledge of the search space or machine learning methods. It is reasonable to expect further developments in both areas in the future as biological knowledge accumulates. Simultaneously, advances in statistical and machine learning methodologies could lead to breakthroughs in epistasis discovery by utilizing patterns within data, including the most powerful ways in which to genetically encode SNPs. Equipped with both machine learning tools and external knowledge, the epistasis researcher of the future could circumvent the incredible complexity of all possible interactions between genotypes within a population and focus directly on discovery of biologically relevant interactions.

Beyond Epistasis: Investigating Additional Complexity in Precision Medicine

In this article, we have described the importance of exploring genetic interactions as well as the challenges surrounding these analyses and common methods to overcome those hurdles. For precision medicine to be a reality, though, investigating genetic interaction in combination with other types of complexity is a necessity. Here we will discuss the ways in which the environment, rare and structural variation, and multi-omics data are important features to consider for disease prediction in the context of epistasis.

The exposome, herein defined as all environmental exposures in an individual's life, has a major impact on health outcomes (Rappaport and Smith, 2010). A large portion of the complexities leading to common diseases are missed if the environment is not considered. Gene-gene-environment interaction models are especially difficult to test for the same multiple-testing reasons outlined previously. Currently, much of the work done in gene-environment interactions include candidate environmental exposures, but this prevents identification of gene interactions with novel exposures. Environment-wide association studies (EWAS) have successfully identified novel exposures associated with disease which can subsequently be used for gene-exposure interaction analysis (Patel et al., 2010, 2013, 2014; Patel and Ioannidis, 2014; Tzoulaki et al., 2012; Hall et al., 2014, 2017). This approach decreases the search space to only include exposures identified through EWAS. Just as knowledge-based methods have been successful at identifying gene-gene interactions predictive of disease, database knowledge may be employed for integrating the exposome with epistatic models. The Human Metabolome Database (HMDB) (Wishart et al., 2007), the Toxic Exposome Database (Wishart et al., 2015), and the Comparative Toxicogenomics Database (CTD) (Davis et al., 2017) are examples of databases with knowledge of gene-gene and gene-exposure relationships, which may be leveraged to build putative gene-gene-exposure models for testing. Understanding the connection between gene networks and the environment is an essential component to uncovering the etiology of disease.

Whether it be gene-gene or gene-environment interactions, the vast majority of studies have focused on common SNPs (Hall et al., 2016a,b). However, rare variants and copy number variation (CNV) have demonstrated influence on many diseases, and methods that consider these types of data are therefore essential in discovering complex genetic and gene-environment interactions. Successful examples have been found for epistasis involving rare variation and between CNV and the environment (Kim et al., 2017; Hall et al., 2017; Albers et al., 2012). Further, other -omics data such as from the transcriptome, proteome, and metabolome are important features in elucidating the biological impact of interactions at the gene level. There is need for development of methods to handle these types of data that have unique challenges such as limited power for detection and heterogeneity.

Finally, to investigate epistasis and gene-environment interactions with common and rare variants, CNVs, and multi-omics data, studies that generate multiple types of data (e.g., exposure, genotype, sequence, CNV, gene expression, metabolomic, proteomic) are critical. Costs associated with generating multiple data types on large sample sizes are limiting, however. Examples of studies with multiple data types available include the electronic MEdical Record and GEnomics Network (eMERGE Network), the National Health and Nutrition Examination Surveys (NHANES), the Marshfield Personalized Medicine Research Project (PMRP), the Women's Health Initiative (WHI), and the Million Veterans Project (MVP).

As more data types become available for an increasing number of samples, the opportunity arises for improved understanding of biology and the complex processes at play in development of disease. Yet, further unmet methodological development needs will also emerge in finding the most appropriate ways to aggregate, quality control, integrate, and analyze multiple disparate data types. As such, there is immense opportunity to uncover the hidden etiology of disease as further data and methods become available.

Conclusions

To realize the goal of precision medicine, genetic interactions are an important part of modeling complexity that leads to common disease, and their consideration is essential. There are many examples of successful searches for epistasis in humans and non-human model systems. However, as data availability increases, so do the challenges of finding true genetic interactions amidst the noise. Many approaches offer promise in circumventing these hurdles. Still, to understand the mechanisms that lead to diseases, and leverage that knowledge in order make informed and individualized decisions about health care, epistasis must be considered in the context of the environment, rare variation, CNVs, and other types of inter-individual variation. For these reasons, multi-omics data must be incorporated in the search for epistasis. This is an exciting time for the field of biomedical informatics, as there is an abundance of available data and methods, and the further unmet needs for complex and integrative analyses using multiple data types leave ample room for exploration, innovation, and discovery.

See also: Computational Systems Biology. Data Mining: Mining Frequent Patterns, Associations Rules, and Correlations. Gene Regulatory Network Review. MicroRNA and lncRNA Databases and Analysis. Natural Language Processing Approaches in Bioinformatics. Network Inference and Reconstruction in Bioinformatics. Networks in Biology. Pathway Informatics

References

Albers, C.A., *et al.*, 2012. Compound inheritance of a low-frequency regulatory SNP and a rare null mutation in exon-junction complex subunit RBM8A causes TAR syndrome. Nature Genetics 44 (4), 435–439. Available at: http://www.ncbi.nlm.nih.gov/pubmed/22366785 (accessed 31.10.17).

Avery, L., Wasserman, S., 1992. Ordering gene function: The interpretation of epistasis in regulatory hierarchies. Trends in Genetics 8 (9), 312–316.

Baryshnikova, A., *et al.*, 2013. Genetic interaction networks: Toward an understanding of heritability. Annual Review of Genomics and Human Genetics 14, 111–133. Available at: http://www.ncbi.nlm.nih.gov/pubmed/23808365.

Bateson, W., 1909. Mendel's Principles of Heredity. Cambridge University Press.

Beadle, G.W., 1945. Genetics and metabolism in neurospora. Physiological Reviews 25 (4).

Boone, C., Bussey, H., Andrews, B.J., 2007. Exploring genetic interactions and networks with yeast. Nature Reviews. Genetics 8 (6), 437–449.

Bush, W.S., *et al.*, 2011. A knowledge-driven interaction analysis reveals potential neurodegenerative mechanism of multiple sclerosis susceptibility. Genes and Immunity 12 (5), 335–340. Available at: http://www.pubmedcentral.nih.gov/articlerender.fcgi?artid=3136581&tool=pmcentrez&rendertype=abstract (accessed 17.08.15).

Byrne, A.B., *et al.*, 2007. A global analysis of genetic interactions in *Caenorhabditis elegans*. Journal of Biology 6 (3), 8.

Cheng, Y., *et al.*, 2011. Mapping genetic loci that interact with myostatin to affect growth traits. Heredity 107 (6), 565–573.

Cordell, H.J., 2002. Epistasis: What it means, what it doesn't mean, and statistical methods to detect it in humans. Human Molecular Genetics 11 (20), 2463–2468.

Davis, A.P., *et al.*, 2017. The Comparative Toxicogenomics Database: Update 2017. Nucleic Acids Research 45 (D1), D972–D978. Available at: http://www.ncbi.nlm.nih.gov/pubmed/27651457 (accessed 31.10.17).

De, R., *et al.*, 2015. Identifying gene-gene interactions that are highly associated with Body Mass Index using Quantitative Multifactor Dimensionality Reduction (QMDR). BioData Mining 8 (1), 41. Available at: http://biodatamining.biomedcentral.com/articles/10.1186/s13040-015-0074-0 (accessed 09.04.16).

Doust, A.N., *et al.*, 2014. Beyond the single gene: How epistasis and gene-by-environment effects influence crop domestication. Proceedings of the National Academy of Sciences of the United States of Anetica 111, 6178–6183.

Eichler, E.E., *et al.*, 2010. Missing heritability and strategies for finding the underlying causes of complex disease. Nature Reviews Genetics 11 (6), 446–450.

Fisher, R.A., 1919. XV. The Correlation between relatives on the supposition of Mendelian inheritance. Transactions of the Royal Society of Edinburgh 52 (2), 399–433. Available at: http://www.journals.cambridge.org/abstract_S0080456800012163 (accessed 31.10.17).

Gaertner, B.E., *et al.*, 2012. More than the sum of its parts: A complex epistatic network underlies natural variation in thermal preference behavior in Caenorhabditis elegans. Genetics 192 (4), 1533–1542.

Gale, G.D., *et al.*, 2009. A genome-wide panel of congenic mice reveals widespread epistasis of behavior quantitative trait loci. Molecular Psychiatry 14 (6), 631–645.

Gallie, D.R., 2002. Protein-protein interactions required during translation. Plant Molecular Biology 50 (6), 949–970.

Grady, B.J., *et al.*, 2010. Finding unique filter sets in PLATO: A precursor to efficient interaction analysis in GWAS data. Pacific Symposium on Biocomputing. Pacific Symposium on Biocomputing, pp. 315–26. Available at: http://www.pubmedcentral.nih.gov/articlerender.fcgi?artid=2903053&tool=pmcentrez&rendertype=abstract (accessed 17.08.15).

Greene, C.S., *et al.*, 2009. Spatially Uniform ReliefF (SURF) for computationally-efficient filtering of gene-gene interactions. BioData Mining 2, 5.

Greene, C.S., *et al.*, 2015. Understanding multicellular function and disease with human tissue-specific networks. Nature Publishing Group 47 (6).

Greene, C.S., *et al.*, 2010. Enabling personal genomics with an explicit test of epistasis. *Pacific Symposium on Biocomputing. Pacific Symposium on Biocomputing*, pp.327–36.

Hall, M.A. *et al.*, 2014. Environment-wide association study (EWAS) for type 2 diabetes in the Marshfield Personalized Medicine Research Project Biobank. *Pacific Symposium on Biocomputing. Pacific Symposium on Biocomputing*, pp. 200–11. Available at: http://www.pubmedcentral.nih.gov/articlerender.fcgi?artid=4037237&tool=pmcentrez&rendertype=abstract (accessed 08.07.15).

Hall, M.A., *et al.*, 2015. Biology-driven gene–gene interaction analysis of age-related cataract in the eMERGE network. Genetic Epidemiology 39 (5), 376–384. Available at: http://www.ncbi.nlm.nih.gov/pubmed/25982363 (accessed 23.07.15).

Hall, M.A., *et al.*, 2017. PLATO software provides analytic framework for investigating complexity beyond genome-wide association studies. Nature Communications 8 (1), 1167. Available at: http://www.nature.com/articles/s41467-017-00802-2 (accessed 31.10.17).

Hall, M.A., Moore, J.H., Ritchie, M.D., 2016a. Embracing complex associations in common traits: Critical considerations for precision medicine. Trends in Genetics 32 (8), 470–484. Available at: http://www.ncbi.nlm.nih.gov/pubmed/27392675 (accessed 12.05.17).

Hall, M.A., Moore, J.H., Ritchie, M.D., *et al.*, 2016b. Embracing complex associations in common traits: Critical considerations for precision medicine. Trends in Genetics 32 (8), 470–484. Available at: http://linkinghub.elsevier.com/retrieve/pii/S0168952516300506 (accessed 05.09.16).

Hanlon, P., *et al.*, 2006. Three-locus and four-locus QTL interactions influence mouse insulin-like growth factor-I. Physiological Genomics 26 (1), 46–54.

Hemani, G., *et al.*, 2014. Detection and replication of epistasis influencing transcription in humans. Nature 508 (7495), 249–253. Available at: http://www.ncbi.nlm.nih.gov/pubmed/24572353.

Horn, T., *et al.*, 2011. Mapping of signaling networks through synthetic genetic interaction analysis by RNAi. Nature Methods 8 (4), 341–346.

Huang, W., *et al.*, 2012. Inaugural article: Epistasis dominates the genetic architecture of Drosophila quantitative traits. Proceedings of the National Academy of Sciences 109 (39), 15553–15559.

Kerem, B., *et al.*, 1989. Identification of the cystic fibrosis gene: Genetic analysis. Science (New York, NY) 245 (4922), 1073–1080.

Kim, D., *et al.*, 2014. Knowledge-driven genomic interactions: An application in ovarian cancer. BioData Mining 7, 20. Available at: http://www.pubmedcentral.nih.gov/articlerender.fcgi?artid=4161273&tool=pmcentrez&rendertype=abstract (accessed 17.08.15).

Kim, D., *et al.*, 2017. The joint effect of air pollution exposure and copy number variation on risk for autism. Autism Research 10 (9), 1470–1480. Available at: http://doi.wiley.com/10.1002/aur.1799 (accessed 26.10.17).

Kira, K., Rendell, L.A., 1992a. A practical approach to feature selection. In: Proceedings of the Ninth International Workshop on Machine Learning. pp. 249–256.

Kira, K., Rendell, L.A., 1992b. The feature selection problem: Traditional methods and a new algorithm. AAAI. pp. 129–134.

Kooperberg, C., LeBlanc, M., 2008. Increasing the power of identifying gene × gene interactions in genome-wide association studies. Genetic Epidemiology.

Kroymann, J., Mitchell-Olds, T., 2005. Epistasis and balanced polymorphism influencing complex trait variation. Nature 435 (7038), 95–98.

Lehner, B., *et al.*, 2006. Systematic mapping of genetic interactions in *Caenorhabditis elegans* identifies common modifiers of diverse signaling pathways. Nature Genetics 38 (8), 896–903.

Lloyd, V., Ramaswami, M., Krämer, H., 1998. Not just pretty eyes: Drosophila eye-colour mutations and lysosomal delivery. Trends in Cell Biology 8 (7), 257–259.

Lou, X.-Y., *et al.*, 2007. A generalized combinatorial approach for detecting gene-by-gene and gene-by-environment interactions with application to nicotine dependence. The American Journal of Human Genetics 80 (6), 1125–1137.

Mackay, T.F.C., 2014. Epistasis and quantitative traits: Using model organisms to study gene–gene interactions. Nature Reviews. Genetics 15 (1), 22–33. Available at: http://www.pubmedcentral.nih.gov/articlerender.fcgi?artid=3918431&tool=pmcentrez&rendertype=abstract (accessed 21.05.15).

Maher, B., 2008. Personal genomes: The case of the missing heritability. Nature 456 (7218), 18–21. Available at: http://www.ncbi.nlm.nih.gov/pubmed/18987709 (accessed 09.01.15).

Ma, L., Keinan, A., Clark, A.G., 2015. Biological knowledge-driven analysis of epistasis in human GWAS with application to lipid traits. Methods in Molecular Biology (Clifton, NJ) 1253, 35–45. Available at: http://www.ncbi.nlm.nih.gov/pubmed/25403526.

Manolio, T.A., et al., 2009. Finding the missing heritability of complex diseases. Nature 461 (7265), 747–753.

Martinez, E., 2002. Multi-protein complexes in eukaryotic gene transcription. Plant Molecular Biology 50 (6), 925–947.

Moore, C.B., et al., 2012. BioBin: A bioinformatics tool for automating the binning of rare variants using publicly available biological knowledge. From Second Annual Translational Bioinformatics Conference BMC Medical Genomics 6 (2), 13–16.

Moore, J.H., 2003. The ubiquitous nature of epistasis in determining susceptibility to common human diseases. Human Heredity. 73–82.

Moore, J.H., 2005. A global view of epistasis. Nature Genetics 37 (1), 13–14.

Moore, J.H., Andrews, P.C., 2015. Epistasis analysis using multifactor dimensionality reduction. Methods in Molecular Biology (Clifton, NJ). 301–314.

Moore, J.H., Asselbergs, F.W., Williams, S.M., 2010. Bioinformatics challenges for genome-wide association studies. Bioinformatics (Oxford, England) 26 (4), 445–455. Available at: http://www.pubmedcentral.nih.gov/articlerender.fcgi?artid=2820680&tool=pmcentrez&rendertype=abstract (accessed 15.07.15).

Moore, J.H., White, B.C., 2007. LNCS 4447 – Tuning ReliefF for Genome-Wide Genetic Analysis.

Moore, J.H., White, B.C., 2006. Exploiting Expert Knowledge in Genetic Programming for Genome-Wide Genetic Analysis. Berlin, Heidelberg: Springer, pp. 969–977.

Moore, J.H., Williams, S.M., 2005. Traversing the conceptual divide between biological and statistical epistasis: Systems biology and a more modern synthesis. BioEssays 27 (6), 637–646.

Nelson, M.R., et al., 2001. A combinatorial partitioning method to identify multilocus genotypic partitions that predict quantitative trait variation. Genome Research 11 (3), 458–470.

Patel, C.J., et al., 2013. Systematic evaluation of environmental and behavioural factors associated with all-cause mortality in the United States national health and nutrition examination survey. International Journal of Epidemiology 42 (6), 1795–1810. Available at: http://www.pubmedcentral.nih.gov/articlerender.fcgi?artid=3887569&tool=pmcentrez&rendertype=abstract (accessed 16.08.15).

Patel, C.J., et al., 2014. Investigation of maternal environmental exposures in association with self-reported preterm birth. Reproductive Toxicology (Elmsford, NY) 45, 1–7. Available at: http://www.pubmedcentral.nih.gov/articlerender.fcgi?artid=4316205&tool=pmcentrez&rendertype=abstract (accessed 16.08.15).

Patel, C.J., Bhattacharya, J., Butte, A.J., 2010. An Environment-Wide Association Study (EWAS) on type 2 diabetes mellitus. PlOS One 5 (5), e10746. Available at: http://www.pubmedcentral.nih.gov/articlerender.fcgi?artid=2873978&tool=pmcentrez&rendertype=abstract (accessed 29.06.15).

Patel, C.J., Ioannidis, J.P.A., 2014. Studying the elusive environment in large scale. Journal of American Medical Association 311 (21), 2173–2174. Available at: http://www.pubmedcentral.nih.gov/articlerender.fcgi?artid=4110965&tool=pmcentrez&rendertype=abstract (accessed 16.08.15).

Pattin, K.A., et al., 2009. A computationally efficient hypothesis testing method for epistasis analysis using multifactor dimensionality reduction. Genetic Epidemiology 33 (1), 87–94.

Pedregosa, Fabianpedregosa, F., et al., 2011, Scikit-learn: Machine learning in Python Gaël Varoquaux. Journal of Machine Learning Research 12, 2825–2830. Available at: http://delivery.acm.org/10.1145/2080000/2078195/p2825-pedregosa.pdf?

Pendergrass, S.A. et al., 2015. Next-generation analysis of cataracts: Determining knowledge driven gene-gene interactions using biofilter, and gene-environment interactions using the Phenx Toolkit*. Pacific Symposium on Biocomputing. Pacific Symposium on Biocomputing, pp. 495–505. Available at: http://www.ncbi.nlm.nih.gov/pubmed/25741542 (accessed 17.08.15).

Phillips, P.C., 1998. The language of gene interaction. Genetics 149 (3).

Phillips, P.C., 2008. Epistasis the essential role of gene interactions in the structure and evolution of genetic systems. Nature Reviews Genetics 9 (11), 855–867.

Rappaport, S.M., Smith, M.T., 2010. Epidemiology. Environment and disease risks. Science (New York, NY) 330 (6003), 460–461. Available at: http://www.ncbi.nlm.nih.gov/pubmed/20966241 (accessed 16.08.15).

Riordan, J.R., et al., 1989. Identification of the cystic fibrosis gene: Cloning and characterization of complementary DNA. Science (New York, NY) 245 (4922), 1066–1073.

Ritchie, M.D., et al., 2001. Multifactor-dimensionality reduction reveals high-order interactions among estrogen-metabolism genes in sporadic breast cancer. The American Journal of Human Genetics 69 (1), 138–147.

Ritchie, M.D., 2009. Using prior knowledge and genome-wide association to identify pathways involved in multiple sclerosis. Genome Medicine 1 (6), 65. Available at: http://www.pubmedcentral.nih.gov/articlerender.fcgi?artid=2703874&tool=pmcentrez&rendertype=abstract (accessed 30.06.15).

Ritchie, M.D., 2011. Using biological knowledge to uncover the mystery in the search for epistasis in genome-wide association studies. Annals of Human Genetics 75 (1), 172–182.

Ritchie, M.D., 2015. Finding the epistasis needles in the genome-wide haystack. Methods in Molecular Biology (Clifton, NJ) 1253, 19–33. Available at: http://www.ncbi.nlm.nih.gov/pubmed/25403525.

Rommens, J.M., et al., 1989. Identification of the cystic fibrosis gene: Chromosome walking and jumping. Science (New York, NY) 245 (4922), 1059–1065.

Rowe, H.C., et al., 2008. Biochemical networks and epistasis shape the *Arabidopsis thaliana* metabolome. The Plant Cell 20 (5), 1199–1216.

Sturtevant, A.H., 2001. A History of Genetics. CSHL Press.

Sun, X., et al., 2014. Analysis pipeline for the epistasis search – Statistical versus biological filtering. Frontiers in Genetics 5 (APR).

Szappanos, B., et al., 2011. An integrated approach to characterize genetic interaction networks in yeast metabolism. Nature Genetics 43 (7), 656–662.

Templeton, 2000. Epistasis and Complex Traits. New York: Oxford University Press.

Thomas, D.C., et al., 2009. Use of pathway information in molecular epidemiology. Human Genomics 4 (1), 21–42.

Tong, A.H.Y., et al., 2004. Global mapping of the yeast genetic interaction network. Science (New York, NY) 303 (5659), 808–813.

Turner, S.D., et al., 2011. Knowledge-driven multi-locus analysis reveals gene-gene interactions influencing HDL cholesterol level in two independent EMR-linked biobanks. PlOS one 6 (5), e19586. Available at: http://journals.plos.org/plosone/article?id=10.1371/journal.pone.0019586 (accessed 17.08.15).

Tzoulaki, I., et al., 2012. A nutrient-wide association study on blood pressure. Circulation 126 (21), 2456–2464. Available at: http://www.pubmedcentral.nih.gov/articlerender.fcgi?artid=4105584&tool=pmcentrez&rendertype=abstract (accessed 16.08.15).

Wagner, A., 2000. Robustness against mutations in genetic networks of yeast. Nature Genetics 24 (4), 355–361.

Wang, M.H., et al., 2016. A fast and powerful W-test for pairwise epistasis testing. Nucleic Acids Research 44 (12), e115.

Wang, W.Y.S., et al., 2005. Genome-wide association studies: Theoretical and practical concerns. Nature Reviews Genetics 6 (2), 109–118. Available at: http://www.ncbi.nlm.nih.gov/pubmed/15716907 (accessed 12.05.15).

Welter, D., et al., 2014. The NHGRI GWAS Catalog, a curated resource of SNP-trait associations. Nucleic Acids Research 42 (D1),

Wishart, D., et al., 2015. T3DB: The toxic exposome database. Nucleic Acids Research 43 (D1), D928–D934. Available at: http://www.ncbi.nlm.nih.gov/pubmed/25378312 (accessed 31.10.17).

Wishart, D.S., et al., 2007. HMDB: The Human Metabolome Database. Nucleic Acids Research 35 (Database), D521–D526. Available at: http://www.ncbi.nlm.nih.gov/pubmed/17202168 (accessed October 31.10.17).

Genome Informatics

Anil K Kesarwani, Ankit Malhotra, Anuj Srivastava, Guruprasad Ananda, Haitham Ashoor, Parveen Kumar, Rupesh K Kesharwani, Vishal K Sarsani, Yi Li, Joshy George, and R Krishna Murty Karuturi, The Jackson Laboratory, Farmington, CT, United States

Introduction

Inherited characteristics of an organism are encoded in a molecule called deoxyribonucleic acid (DNA), represented as a string of four-element code (ACGT). This molecule encodes the information required for the synthesis of all the proteins required for the functioning of cell that ultimately determines the phenotype of the organism. Variations introduced in this code introduces the changes in the organisms for natural selection to work on. The variable fitness level of organisms results in certain variations to be selected over time resulting in the incredibly different variety of viable organisms we see around. The genetic information in the DNA is transcribed into RNA by an enzyme called RNA polymerase. The RNA is then decoded into a ribosome, outside the nucleus, to produce a specific amino acid chain, or polypeptide. The polypeptide later folds into an active protein and performs its functions in the cell which then determines the phenotype of the cell and the organism (Felix, 2016).

The particular genetic encoding for an organism is called its *genotype* and the resulting physical characteristics is called its *phenotype*. In sexually reproducing organisms including humans, sexual recombination as well as mutations introduce variation in the *genotype* resulting in the variations in its *phenotype* enabling natural selection to select organisms with better fitness in their current environment. Small changes in the genotype space can have significant differences in the phenotypes resulting in the enormous diversity of organisms we see around. Our understanding of the genotype-phenotype relationship has helped us to understand the genetic cause of several diseases and have helped us to treat them with better efficacy (Lehner, 2007).

The DNA of an organism is typically packaged into several chromosomes in eukaryotic cells, exist in discrete chromosome territories. There is compelling evidence that suggest that each chromosome is comprised of many distinct chromatin domains, referred to variably as topological domains or topologically associating domains (TADs), that are hundreds of kilobases to several million bases in length. These chromatin domains are stable for many cell divisions, cell-type specific and demonstrated to play important roles in transcriptional regulation, DNA replication, and recombination. Additionally, regulatory DNA elements such as enhancers, silencers and insulators are embedded in chromosomes and they control gene expression during development and maintenance (Pope *et al.*, 2014).

A significant step in our understanding of our genetic makeup happened when the first release of the human genome was announced by the Human Genome Project (HGP) in 2001. The Genome Reference Consortium (GRC) – an international, collaborative research program whose goal was the complete mapping and understanding of all the genes of a genome – facilitate the curation of genome assemblies for human, mouse or zebrafish (Hudson, 2002).

Once the genome sequence was established, the next task is to identify the functional elements in the genome. The ENCODE (Encyclopedia of DNA Elements) Consortium is an international collaboration of research groups funded by the National Human Genome Research Institute (NHGRI). The goal of ENCODE is to build a comprehensive parts list of functional elements in the human genome, including elements that act at the protein and RNA levels, and regulatory elements that control cells and circumstances in which a gene is active. The ENCODE consortium not only produces high-quality data, but also analyzes the data in an integrative fashion and provides tools to search and visualize them (Consortium *et al.*, 2007; Consortium *et al.*, 2012).

The gene expression levels in a cell is also controlled using epigenetic mechanisms, defined as heritable alterations that are not encoded in DNA sequence. These modifications include mechanisms such as DNA methylation, histone modification, DNA accessibility and chromatin structure. The NIH Roadmap Epigenomics Mapping Consortium was launched with the goal of producing a public resource of human epigenomic data to catalyze basic biology and disease-oriented research. The Consortium leverages experimental pipelines built around next-generation sequencing technologies to map the epigenetic landscapes in primary ex vivo tissues selected to represent the normal counterparts of tissues and organ systems frequently involved in human disease (Roadmap Epigenomics *et al.*, 2015).

The basic mechanisms involved in the translation of the molecular codes encoded in the genetic material to the proteins that do the functions are very similar in almost all organisms. Therefore, biologists can use other organisms to study the basic mechanisms to understand gene regulation. Organisms like mouse that are closer to humans in the evolutionary tree are used to model diseases in humans and have proved to be of significant advantage to our understanding of the genetic mechanisms of several diseases. The goal of the modENCODE consortium project is to provide the biological research community with a comprehensive encyclopedia of genomic functional elements in model organisms such as *C. elegans* and *D. melanogaster*, including the domains of gene structure, mRNA and ncRNA expression profiling, transcription factor binding sites, histone modifications and replacement, chromatin structure, DNA replication initiation and timing, and copy number variation (Boley *et al.*, 2014).

The DNA sequence of any two individuals will be almost identical in the sense that the sequence will be matching in more than 99% of the genome. The 1% variations, however, may greatly affect an individual's disease risk. One of the significant DNA variation commonly studied is single nucleotide polymorphisms (SNPs). Sets of nearby SNPs on the same chromosome are inherited in blocks and is called a haplotype. The HapMap is a map of these haplotype blocks and the specific SNPs that identify the haplotypes are called tag SNPs. The International HapMap Project is a partnership of scientists and funding agencies from Canada, China, Japan, Nigeria, the United Kingdom and the United States. This will make genome scan approaches to finding regions with genes that affect diseases much more efficient and comprehensive (International HapMap, 2003).

The high quality curated genomes made available by the GRC and the hapmap data from the consortium coupled with advanced sequencing technologies enable us to study the genomic variants associated with diseases or predisposition to certain diseases, develop or identify appropriate therapeutic strategies. Examples of such initiatives are the Alzheimer's Disease Sequencing Project (ADSP) to identify new genomic variants contributing to increased or reduced risk of developing Late-Onset Alzheimer's Disease (LOAD), the GoT2D consortia to understand the genetic architecture of type 2 diabetes, and The Cancer Genome Atlas (TCGA) – a collaboration between the National Cancer Institute (NCI) and the National Human Genome Research Institute (NHGRI) – has generated comprehensive, multi-dimensional maps of the key genomic changes in 33 types of cancer. Likewise, the International Cancer Genome consortium (ICGC) project generated a comprehensive description of genomic, transcriptomic and epigenomic changes in 50 different tumor types and/or subtypes which are of clinical and societal importance across the globe (Devarakonda et al., 2013; Cancer Genome Atlas Research et al., 2013; Zhang et al., 2011).

The Trans-Omics for Precision Medicine (TOPMed), is a program to generate scientific resources to enhance our understanding of fundamental biological processes that underlie heart, lung, blood and sleep disorders. The projects aim to integrate whole-genome sequencing data, metabolic profiles, protein levels and RNA expression data with molecular, behavioral, imaging, environmental, and clinical data to uncover factors that increase or decrease the risk of disease and develop more targeted and personalized treatments (Auer et al., 2016).

The identification of SNPs and other differences in an individual is complicated because of the difficulty in identifying the disease-causing alterations from the large number of SNPs observed in any individual. The 1000 Genomes Project was initiated to find most genetic variants with frequencies of at least 1% in the populations studied. Though the project started with the aim of profiling 1000 individuals from three populations, the final data set contains data for 2504 individuals from 26 populations. Low coverage and exome sequence data are present for all of these individuals, 24 individuals were also sequenced to high coverage for validation purposes. The variations present in apparently normal individuals helps to remove germline events and help to identify disease causing mutations (Kuehn, 2008).

Putting together, studying genome and variations within is crucial to understanding the evolution, diversity and diseases. Plethora of tools and techniques have been developed to elicit genomic structure, variations and their functional characterization. Hence, this article hi-lights the informatics of all important aspects of genomic study. However, before diving into genome informatics, we briefly discuss the evolution of genomic technologies in Section "Genomic Technologies in Genomic Studies".

Genomic Technologies in Genomic Studies

The completion of the human genome project (Lander et al., 2001) (HGP) in 2004 heralded a new era in biomedical sciences and brought the world of genomics to the mainstream. The HGP costed about 500 million dollars and took about a decade to finish a human genome sequence. Over the next decade, the NIH championed new initiatives to foster technology development at a massive scale to be able to sequence the human genome at a fraction of the cost ($1000) in a few hours. This provided a massive impetus that gave rise to whole slew of genomic technologies to not only generate the sequence of the genome but to also interrogate it for functional characterization.

Before the HGP, genomics was limited to site based analysis. No single technology existed that could provide a complete view of the genome. Light based microscopy techniques to detect signals from bio-markers that have been added to the genome have been around since the middle of the last century. However, these technologies either provided a very coarse view of the genome, or required the investigator to know precisely the location of the genome to query. In addition, these methods were tedious and expensive requiring hours of a highly skilled technician's time. These were followed by the advent of polymerase chain reaction (PCR) based techniques to interrogate the presence or absence of genomic segments. However once again these technologies were low throughput, time consuming and required prior knowledge of the genome of interest to design appropriate primers. To circumvent this problem, sequencing of the genome was adopted as a viable strategy to get de novo information on genomic regions of interest. The sequencing technologies evolved in 3 generations of advancements.

First generation sequencing technology: These methods were the first of the sequencing based methods – developed by Fred Sanger and colleagues in 1977 (Sanger et al., 1977; Prober et al., 1987) that relied on incorporation of chain-terminating dideoxynucleotides during the DNA replication process. 'Sanger sequencing' was the mainstay of genomic interrogation technologies for close to 30 years and is even used currently for small projects and for validation experiments. The mean sequencing reads are ~ 700 bp in length. However, the throughput still remains severely inadequate compared to the more recent next generation sequencing methods described below.

With the successful completion of the HGP, most of which was achieved using Sanger sequencing methods, the genomic community focused on developing new platforms that would allow for cheaper, faster and less error prone DNA interrogation. Over the past decade, rapid research and development of genomic technologies, both microarray based and more critically sequencing based ("Next Generation Sequencing") technologies have allowed us to get close to that aim, leading to the two generations of advanced hi-throughput sequencing technologies.

Second generation sequencing technologies: Next generation sequencing technologies were inspired by a new development called pyrosequencing (Nyren et al., 1993), pioneered by Pal Nyren and colleagues, and relied on luminescence produced and measured during new DNA strand synthesis. Pyrosequencing is a sequencing by synthesis (SBS) based method that uses the DNA polymerase to construct nascent DNA strand using a template strand. Pyrosequencing was licensed by 454 Life Sciences (later acquired by Roche). They produced the first commercially available massively parallel sequencing technology (Ronaghi et al., 1998). The 454 machines produced long contiguous reads in the range of 4–500 bps in millions of wells at the same time. However, the data from the 454 systems suffered from errors arising in nucleotide runs. In long nucleotide runs, the phospho-luminiscent based method had difficulties determining the precise number of nucleotides that were present (Ronaghi et al., 1998).

At around the same time, Solexa and the SOLiD method of sequencing also gained momentum (Voelkerding et al., 2009). Solexa (later acquired by Illumina) relied on the incorporation of dNTPs that are end terminated i.e., they can only incorporate one nucleotide at a time (Bentley et al., 2008). Although this prevented the system from generating errors in the long nucleotide runs, it also limited the system to only generate about 35 basepairs at a time. Sequencing chemistry improvements and other developments allowed for read sizes of up to 250 basepairs to be sequenced (using the MiSeq system). In the most current state-of-the art, the Illumina HiSeq method can produce about a 100 gigabases of nucleotide sequences in about 2 days for about a thousand dollars in price – there by realizing the goal of bringing low cost sequencing to the disease genomic studies and the clinic (Balasubramanian, 2011; Quail et al., 2012).

Third generation sequencing technologies: There is considerable debate in the genomic community on how to set the various milestones that define a generational change between the continuous developments of sequencing technologies. We choose to define these milestones whenever there is a significant improvement in technology resulting in much better sequencing data.

One example of such a technology is the single molecule real time (SMRT) sequencing platform from Pacific Biosciences or PacBio (Quail et al., 2012). The platform relies on polymerase based sequencing of single molecules of DNA in specially made wells known as zero-mode waveguides (Harris et al., 2008). SMRT sequencing and its newest iterations (Sequel) produce read lengths in the range of 10–20 kb range (Rhoads and Au, 2015). The long reads (although plagued by a slightly higher error rate) provide much more sequence context than the short-read sequencing platforms and have enabled an unprecedented view of the genome (Nakano et al., 2017). It allowed development of pipelines that can deliver de-novo assembled genomes of an individual in a few hours.

More recently nanopore based DNA sequencing technologies have come to the fore. While the idea of using channels (nanopores) to linearize DNA molecules and drive them through is not new, the Oxford Nanopore Minion machines have successfully commercialized the concept (Clarke et al., 2009). The nanopore channel is capable of detecting the change in electric current as different nucleotides of the DNA molecule pass through the channel and thus can read the DNA sequence. The platform allows for potentially generating majority of DNA sequences in the same size range as the SMRT system, however the largest read lengths could top 100–200 kbp (Branton et al., 2008).

There are several other genomic technologies that could be added to the gamut of third generation sequencing technologies, such as the BioNano Genomics' Irys platform and the Fluidigm C1 single cell system. Irys systems produced by Bio Nano Genomics are optical mapping based solution for interrogating large DNA molecules. It relies on DNA nicking enzymes that allow for a large DNA molecule to be tagged at certain known locations across the genome (Schwartz et al., 1993). DNA molecule is then linearized and run through a sensitive system that can identify the nick sites. Identification of the genomic location requires prior knowledge of genome sequence and nicking sites.

The data and technologies greatly enhances our ability to understand the underlying genetics in the development and disease. However, we need to cross the hurdle of plethora of informatics problems. In the remainder of the article, we review the fundamental genome informatics problems and the respective methods based on Illumina sequencing data, long-read sequencing data and microarray data.

Genomic Structural Informatics

Eliciting genomic structure, identifying functional elements and variants that affect their function is critical to eliciting the genomic factors associated with cellular function, cellular state, cellular state transition, diversity, disease and ultimately devising appropriate treatment options. Plethora of informatics methods and procedures have been devised in conjunction with the technologies discussed in Section "Genomic Technologies in Genomic Studies" to elicit genomic structure and variation. This section outlines the informatics procedures to elicit linear as well as 3D organization of genome, identifying functional elements, and genomic variants that lead to diversity and disease.

Eliciting Linear Structure of Genome

A typical eukaryotic chromosome is at least millions of base-pair long while currently available sequencing technology can decode only hundreds to tens of thousands of base-pair in one stretch. Thus, in shotgun (short-read) sequencing approach, chromosomes are broken down in fragments and sequenced. The process of rejoining these fragments to build the linear structure of chromosomes is called assembly and the program that performs the merging of smaller fragments into larger contigs/scaffolds is called *assembler*. The assembly procedures are classified as (1) reference or template based assembly, and (2) *de novo assembly*. The reference based assembly uses genome assembly of a nearest species as starting point for assembly. It involves mapping the reads to the reference genome, resolving conflicts, restructuring reference genome based on the structural patterns observed in the genome to be assembled. On the other hand, *de novo* methods assemble the genome from scratch, discussed in detail below.

de novo assembly starts assembling the genome by connecting short sequence reads based on their similarity which gives rise to longer reads and eventually contigs and scaffolds. Long insert libraries and long reads, which can be generated using Illumina mate-pair and Pacific BioSciences (PacBio) technology (Shi *et al.*, 2016) respectively, have been used due to their capability to span larger repeats and provide better continuity to the assembly. *de novo* assembly algorithms can be divided primarily into 3 types: Greedy, Overlap-layout-consensus (OLC) and de Bruijn graph based algorithms (Miller *et al.*, 2010).

Greedy algorithms, which perform local optimization of an objective function, involve iteratively merging the reads based on sequence similarity until no more merging is feasible. The criterion for merging typically involves length and percentage identity of overlapping regions among the reads (Miller *et al.*, 2010). Greedy algorithms have been used effectively to assemble smaller genomes that are of repeat free. Popular greedy assemblers are Phrap (de la Bastide and McCombie, 2007), TIGR Assembler (Pop and Kosack, 2004), and CAP3 (Huang and Madan, 1999).

The OLC algorithms use a different overlap approach, which works in 3 steps: (1) The Overlap step involves all vs all pairwise overlap comparison among the reads and use it to create the overlap graph where nodes corresponds to reads and edges represents overlap between the reads; (2) The layout step involves analysis of graph to identify paths in which each node is being traversed just once (hamiltonian path); (3) The final step involves the consensus determination where reads overlapping the same base vote for the identity of that base (Miller *et al.*, 2010). Compared to the greedy algorithms, OLC approach can use long distance information and can be used for assembly of larger genomes with repeats. Some of the popular OLC assembler are ARACHNE (Batzoglou *et al.*, 2002), Celera Assembler (Myers *et al.*, 2000), Newbler (Margulies *et al.*, 2005) and Minimus (Sommer *et al.*, 2007).

Most popular category of *de novo* algorithms are *de Bruijn* based (k-mer graph) methods. They work in 2 steps: 1) reads are broken into k-mers (typically k=25) and de Bruijn graph is being constructed where each k-mer is a node and edges correspond to overlap of k–1 nucleotides between connected nodes. A simple de Bruijn graph of k=3 with 5 nodes is shown in the **Fig. 1**. For typical genomes, de Bruijn graph for carefully chosen k-mer size, most of neighboring nodes will be connected and distal connection (loop formation) will occur in the repeating region of the genomes only. The second step of the de Bruijn graph algorithm is to find an Eulerian path (a path in a graph where each edge is visited once) (Miller *et al.*, 2010). de Bruijn graph algorithms are fast and can be efficiently used for eukaryotic genomes' assemblies. Limitation includes sensitivity to parameter k and vulnerability to sequencing errors. Popular de Bruijn graph assemblers are Velvet (Zerbino and Birney, 2008), SOAPdenovo2 (Luo *et al.*, 2012), ABySS (Simpson *et al.*, 2009) and ALLPATHS-LG (Gnerre *et al.*, 2011).

The quality of de novo assembled genomes is tested using various metrics (Haiminen *et al.*, 2011): Contig N50/Scaffold N50, total number of known genes/transcripts found in the genome, the rate of discordant alignments (which quantifies mis-assembly) determined by mapping the paired end data back to the genome, completeness of the assembly detected by using CEGMA (Core Eukaryotic Genes Mapping Approach) (Parra *et al.*, 2007) which uses a set of 248 ultra-conserved CEGs conserved among *A. thaliana*, *C. elegans*, *D. melanogaster*, *H. sapiens*, *S. cerevisiae*, and *S. pombe*. One of the popular software to automatically detect the quality of the assembly is reapr (Recognition of Errors in Assemblies using Paired Reads) (Hunt *et al.*, 2013). It primarily measures the mis-assemblies and break incorrect scaffolds at mis-assembly points. More software packages for evaluation of quality of assembly can be found at "Relevant Websites section".

Putting together, a typical complete whole genome de novo assembly pipeline is shown in **Fig. 2**. It is based on sequencing of Illumina paired end, Illumina mate pair and low-coverage PacBio data. The steps of the assembly include QC of the raw data by NGS QC Toolkit (Patel and Jain, 2012), k-mer correction by corrector module of SOAPdenovo2, k-mer size estimation by KmerGenie (Chikhi and Medvedev, 2014), Assembly using SOAPdenovo2 (Luo *et al.*, 2012), gap closer using GapCloser module of SOAPdenovo2, evaluation & correction of assembly using reapr (Hunt *et al.*, 2013), and finally extending the contigs using low coverage (∼5X) data from PacBio using PBJelly (English *et al.*, 2012).

Inferring 3D Structure of Genome

DNA is tightly packaged in 3D space of the nucleus. Understanding the 3D genome architecture helps explain events such as promoter-enhancer interaction and disease development (Javierre *et al.*, 2016; Au-Yeung *et al.*, 2016; Martin *et al.*, 2015).

Identifying 3D genome interactions has been a popular approach to infer 3D structure of genome. Imaging technologies, such as microscopy, have enabled the identification of 3D genome interactions for specific predefined regions. On the other hand, chromatin conformation capture methods enabled a flexible platform to identify 3D genome interactions. For instance, 3C technology (Dekker *et al.*, 2002) enabled identification of the interactions between two predefined genomic loci. 4C (Zhao *et al.*, 2006) technology identifies 3D interactions between one locus and all other genomic loci. 5C (Dostie *et al.*, 2006) technology

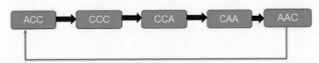

Fig. 1   An example of a de Bruijn graph with k=3. It has one repeating k-mer thus instead of six it has only 5 nodes; blue line is showing the formation of loop pointing to the repeating k-mer (http://www.homolog.us/Tutorials/index.php?p=2.1&s=1).

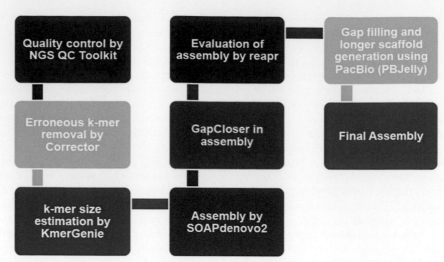

Fig. 2   A schematic of de novo assembly pipeline using Illumina paired end, Illumina mate pair and low coverage PacBio data.

identifies interaction between two sets of predefined loci. Finally, the development of HiC (Lieberman-Aiden *et al.*, 2009) technology enabled the identification of 3D genome interaction at genome scale.

HiC technology generates a massive amount of data which introduced new computational challenges including speed and scalability for data processing, data normalization, and data visualization. To address speed and scalability challenges in processing HiC data, many specialized high-performance tools were developed, e.g., HiCPro (Servant *et al.*, 2015), HiFive (Sauria *et al.*, 2015), HiCBench (Lazaris *et al.*, 2017) and Juicer (Durand *et al.*, 2016a,b). In general, the output of these tools is a contact matrix quantifying the strength of interactions for all pairs of genomic bins. The resolution of the contact matrix increases with increasing sequencing depth.

HiC data normalization is a crucial step before proceeding into downstream data analysis. Sequencing process may introduce several biases innate in the HiC data. HiC data normalization methods fall into two main categories: explicit and implicit normalization. For explicit normalization, HiC data is normalized for each possible covariate (e.g., GC-content bias and sequencing depth bias) independently (Yaffe and Tanay, 2011). HiCNorm (Hu *et al.*, 2012) is an example of explicit HiC data normalization method. For implicit normalization, the main assumption is that any given genomic bin has an equal chance to interact with another bin. Most implicit normalization methods implement normalization using matrix balancing concept. Several implicit approaches have been developed including KR normalization (Rao *et al.*, 2014) and iterative correction and eigenvector decomposition (ICE) (Imakaev *et al.*, 2012) normalization.

As a typical genome browsers are not suitable to visualize 3D genome data, several groups developed specialized tools to visualize 3D genome data. These tools provide convenient ways to visualize 3D genome data including heatmaps, circular plots, and arc tracks. Examples of tools to visualize 3D genome are HiCBrowse (Paulsen *et al.*, 2014a), Juicebox (Durand *et al.*, 2016b), my5C (Lajoie *et al.*, 2009), 3D genome browser (see "Relevant Websites section"), and Epigenome browser (Zhou *et al.*, 2013).

Although HiC data is providing useful information, its coarse resolution limits its ability to map HiC interactions between functional elements (enhancer, promoter, insulators, etc) and the genome. To overcome HiC limitations, many technologies have been developed that focus on interactions that are related to genomic functional elements. These technologies include: chromatin interaction analysis by paired-end tag sequencing (ChIA-PET) (Fullwood *et al.*, 2009), and captureHiC (Martin *et al.*, 2015). ChIA-PET technology reports chromatin interactions that are mediated by specific DNA binding proteins such as CTCF or Pol II. For instance, interactions mediated by Pol II are more likely to contain enhancer-promoter interactions. On the other hand, captureHiC enriches interactions that involve active promoter regions to generate interaction maps between active promoters and the rest of the genome.

The informatics methods developed for HiC data analysis are no longer applicable to process ChIA-PET and captureHiC data. Thus, several methods were developed to process ChIA-PET data: ChIA-PET tool (Li *et al.*, 2010), ChIASig (Paulsen *et al.*, 2014b), and Mango (Phanstiel *et al.*, 2015) are a few examples. CHiCAGO (Cairns *et al.*, 2016) pipeline was developed to process captureHiC data.

Identifying Functional Elements in Genome

Immediate follow-up to genome assembly and 3D architecture prediction is to identify the functional elements such as genes, promoters and distant regulatory elements (e.g., protein binding elements, enhancers and insulators) to make sense of the genome.

Gene prediction approaches are mainly classified into two categories: (a) evidence-based, and (b) *ab initio* based (Picardi and Pesole, 2010). In the Evidence-based (Similarity or Homology-based) algorithms, the sequence of interest is searched against extrinsic homologous (shared evolutionary history) expression-sequence tags (ESTs), mRNA and protein sequences for complete or partial matches using exact algorithms like Smith-Watermann (Smith and Waterman, 1981) or approximate procedures like BLAST (Altschul *et al.*, 1990) and BLAT (Kent, 2002). A high sequence-similarity, for example, with a protein-product suggests that the target sequence constitutes a protein-coding gene. Latest technologies like RNA-sequencing and Chip-sequencing data can also be exploited as external evidence for accurate gene prediction and validation (Pombo et al., 2017; Hoff et al., 2016; Sikora-Wohlfeld et al., 2013). The extensive dependence on mRNA or protein sequences makes this approach an expensive and limited to known genes. In addition, gene prediction may be hampered by low quality sequences, frameshift mutations or horizontal gene transfers (in prokaryotes). In contrast, *Ab initio* methods rely on the intrinsic sequence content and pre-defined signatures in protein-coding genes i.e., the protein-coding genes are comprised of open reading frames (ORFs) which starts with an initiation codon (ATG) and ends with a triplet stop codon (TAA, TAG or TGA). In a double stranded DNA sequence, there can be six ORFs, three from forward direction and three from reverse (complimentary strand) direction that are always read from 5′ to 3′ direction. The average gene length in E. *coli*, S. *cerevisiae* and H. *Sapiens* is 317, 483 and 450 codons respectively (Brown, 2002). In general, prokaryotes' genomes contain only 6%–14% of non-coding DNA (Rogozin et al., 2002) and their genes mainly contain continuous ORFs with well characterized promoter sequences, making ORF finding an effective way to identify genes. In contrast to prokaryotes' genomes, eukaryotes' genomes such as humans may contain up to 98% non-coding DNA (Elgar and Vavouri, 2008) and the gene structure is complex due to the presence of introns. They also contain complex promoter or regulatory regions comprised of variable length elements like CpG islands, making ORF finding challenging in eukaryotes. Moreover, eukaryotes may encounter selection of different exons from the same gene, producing multiple gene isoforms, further hampering the gene prediction. However, exons and introns in a gene can be traced using their specific sequence. The upstream sequence (5′ end) at an exon-intron boundary is evolutionarily conserved and is mainly 5′-AG↓GTAAGT-3′ where arrow indicates the exact breakpoint. The downstream sequence (3′ end) at an exon-intron boundary is less evolutionary conserved comparatively and is 5′-PPPPPPNCAG↓-3′ where P stands for Pyrimidine nucleotide i.e., Thymidine or Cytosine and N can be any nucleotide.

Gene regulation is combinatorial in nature and is achieved by the interaction of proteins binding to gene regulatory elements on the genome such as promoters, enhancers, insulators and transcription factor binding sites. Among these gene regulatory elements, promoters are the easiest to locate as they are mainly present upstream (5′ end) of transcription start sites (TSS) and can be classified into core and proximal promoters. Core promoter region lies 100 base pairs around the TSS and acts as a binding site for general transcription factors (e.g., Pol II). On the other hand, proximal promoter is located few hundred bases upstream of TSS region. Prokaryotes generally contain two short consensus sequences in promoter, TATAAT (also called as Pribnow Box) and TTGACA at 10 and 35 upstream of TSS. On the contrary, eukaryotes promoter regions are more complex and are not well characterized. Many eukaryotic promoters contain a highly conserved TATA box (TATAAA) sequence within first 50 bases of TSS upstream region. Alternatively, some eukaryotic genes may also contain another promoter element called initiator (Kugel and Goodrich, 2017). Unlike TATA box, an initiator is comprised of a highly degenerate sequence 5′-PPA^{+1}N[T/A]$^{+3}$PPP-3′ where P is Pyrimidine (C or T) nucleotide, N can be any nucleotide, A^{+1} is the transcription starting point and T/A is T or A nucleotide downstream ($+3$) of TSS. In contrast, the gene regulatory elements such as enhancers are located far away from the promoter or TSS region, up to few kilo bases or even mega bases in either direction (upstream or downstream) of the gene. The genomic approaches to identify these elements use chromatin immunoprecipitation (ChIP) based techniques like ChIP-Chip or ChIP-seq (Chen et al., 2012; Heintzman et al., 2007). Here, target protein is first cross-linked to cell's chromatin and then protein-DNA complexes are precipitated using protein-specific antibodies. The identity of DNA sequences, present in the protein-DNA complexes, are revealed by mapping them back to the genome and identifying enriched signals by peak finding algorithms such as MACS (Zhang et al., 2008) and QUEST (Valouev et al., 2008).

In eukaryotes, DNA is wrapped around a structure called nucleosome. Nucleosome's length is about 185 bp (Zhou et al., 2011) and it is built from a family of proteins called histones. The combination of DNA and nucleosome is called chromatin. Histone protein tails may go through epigenetic post-transcriptional chemical modifications (Zhou et al., 2011). These modifications include methylation, acetylation, and phosphorylation. Similar to the gene regulatory elements, ChIp-chip and ChIP-seq technologies are used in conjunction with bioinformatics methods (Ashoor et al., 2013; Xu et al., 2010) to identify chromatin modifications.

Chemical modifications of histone tails may correlate with gene expression; certain modifications (e.g., H3K27ac) correlate with increase in gene expression, while others (e.g., H3K27me3) correlate with decrease in gene expression. Many studies (Ernst and Kellis, 2012; Hoffman et al., 2012, 2013; Hon et al., 2008) used combinations of these modifications to demarcate functional regions in human and drosophila genomes. These methods use semi-supervised approaches to identify different patterns of several chromatin modifications and then annotate those patterns based on expert knowledge. Patterns help demarcate the genome into euchromatin and hetero-chromatin regions, identify gene regulatory elements such as enhancers and promoters. Moreover, specific chromatin modifications like H3k27ac, are used to identify super-enhancers in the genome (Pott and Lieb, 2015). Super-enhancers are defined as a set of closely localized enhancers, which are usually located near cell identity genes (Pott and Lieb, 2015). Super-enhancer detection programs work on constructing longer stitches of enhancer marks like H3K27ac. It then scores such longer stitches based on the abundance of mark reads to separate normal enhancers from super-enhancers. Examples of these programs are ROSE (Whyte et al., 2013) and LILY (Boeva et al., 2017) which is specialized to detect super-enhancers for cancer data.

Identifying Genomic Variants

The population diversity and disease are driven by the genomic variations present in the population and disease tissues. These variations include single nucleotide variants (SNVs) and structural variants (SVs). Identifying these variants help understand a variety of population traits as well as disease. Hi-throughput genomic technologies in combination with a variety of informatics methods made their identification feasible at genome scale. However, a good quality control, alignment of the sequencing reads to the reference genome, and downstream processing are important steps for effective use of these technologies. A typical SNV calling workflows is given in **Fig. 3** and informatics of these steps is described below.

Quality Control: Sequencing technologies are not perfect and reads generated from every technology suffers from technical errors. Sequencing by synthesis method from Illumina have problem of deteriorating quality toward the 3′ end of reads due to phasing (incomplete removal of 3′ terminators and sequences in the cluster) (Kircher *et al.*, 2011). With increasing read length, more error accumulates which is problematic in signal detection and base calling. Moreover, if fragment length is shorter than the read length, added adaptors during the library-prep step occasionally get sequenced with the reads (Bolger *et al.*, 2014). This is problematic in the alignment the sequenced adaptors will appear as mismatches (due to non-reference bases). Thus removal of poor quality bases/adaptors is essential. The basic diagnostic involves generating the quality report using fastqc (See Relevant Websites Section), performing trimming/filtering using FASTQ/A Trimmer (see "Relevant Websites section") & trimmomatic (Bolger *et al.*, 2014), and adaptor removal using cut-adapt (Martin, 2011) & trimmomatic. High quality reads passing the trimming & filtering criterion will be used for downstream analysis.

Alignment: Alignment of the reads to the genome of interest follows quality control. Read alignment is a "pattern matching" problem. Aligners account for sequencing errors and align millions of reads to the genome at reasonable speed. There are various aligners available and can be broadly classified into two categories (Lindner and Friedel, 2012):

A) Hash table based aligners use the seed and extend approach. To quickly perform the seeding, these algorithms store reads or reference genome as a hash table (Lindner and Friedel, 2012). They attempt to reduce the sequence search space without discarding the correct alignment candidates. Popular aligners in this category are Mosaik (Lee *et al.*, 2014), BFAST (Homer *et al.*, 2009), Novoalign (see "Relevant Websites section"), PASS (Campagna *et al.*, 2009) and SHRiMP2 (David *et al.*, 2011).

B) FM-index based aligners work in two steps: first identify the exact matches and then build in-exact alignment from the exact matches. Various representations of suffix/prefix trie namely suffix tree, FM-index and enhanced suffix array are used to find exact matches (Li and Homer, 2010). In contrast to hash table based approach, alignment is needed only once for identical copies of substring in reference genome in FM-index based aligners. Popular alignment algorithms in this category are BWA (Li and Durbin, 2010), BOWTIE 2 (Langmead and Salzberg, 2012) and SOAP2 (Li *et al.*, 2009a).

Post Alignment Processing is an important step to increase accuracy of alignment and variant calling:

A) Duplicate removal: Reads aligning to the same position could represent true alignments or PCR artifacts. It is recommended to remove the PCR duplicates prior to variant calling as duplicates from PCR could skew variant allele frequency estimates (Bao *et al.*, 2014). Popular methods for removing PCR duplicates include Picard, MarkDuplicates (see "Relevant Websites section") and samtools (rmdup) (Li *et al.*, 2009b). These tools identify duplicates based on their 5′ alignment position and orientation with

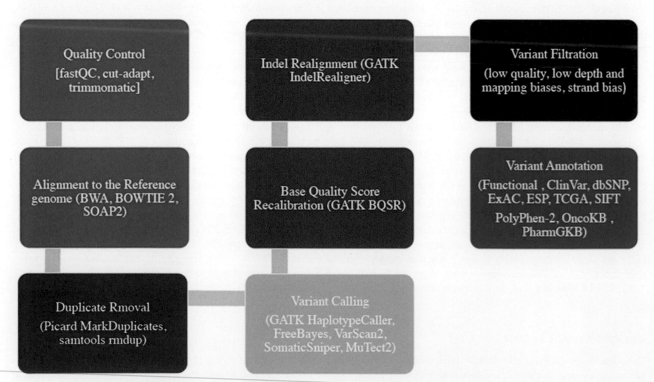

Fig. 3 A schematic depicting typical variant calling workflow.

respect to the genome. The 3′ position is not taken into account as trimming of 3′ end of reads may result in reads of varying lengths.

B) Realignment around indels: Various loci in a genome have indels (insertions and deletions) in the sample's genome with respect to reference genome. Alignment of reads with indels is problematic and often leads to inaccurate placement of reads with respect to reference as Indels cause nearby regions shifted. To circumvent this problem, GATK IndelRealigner (McKenna et al., 2010) creates the list of problematic regions and perform the local realignment across these regions to create a consensus indel and thus minimizing the false SNP calls.

C) Base Quality Score Recalibration (BQSR): Per base quality score emitted by the sequencing machine play important role in variant calling. However, quality values emitted by the sequencer are affected by systematic technical error which leads to over/under assessment of base quality in the data. BQSR step is designed to apply machine learning to model these errors and adjust the quality score for increased accuracy in variant calling. It works in 2 steps: 1) builds a model of covariation based on the data and set of known variation, covariation is analyzed for various features of the base such as machine cycle, quality score and dinucleotide context; and, 2) it adjusts the quality score in the data based on this recalibration model (McKenna et al., 2010).

Variant calling: The primary challenge in variant calling is identifying/separating the real mutation signature from the noise embedded in the sequence data. The current popular algorithm from GATK suite for variant calling is GATK HaplotypeCaller (McKenna et al., 2010) which works in four steps: 1) Active Region search, identifies the regions of genome containing the evidence of significant variation; 2) Identification of haplotypes by performing the assembly of Active Regions; 3) Perform pairwise alignment of reads with haplotype using PairHMM algorithm to determine the haplotype likelihoods based on read data; and, finally, 4) assign genotype to sample based on Bayes' rule for each candidate variant site. Other popular variant callers are FreeBayes (Garrison and Marth, 2012), VarScan2 (Koboldt et al., 2012), SomaticSniper (Larson et al., 2012) and MuTect2 (Cibulskis et al., 2013).

Variant filtration: Errors are introduced during sequencing and bioinformatics steps, which can lead to misleading results. It is therefore necessary to assess various quality metrics to identify real variants. A few common metrics output by most variant callers are coverage and allele frequency. Filters based on these metrics would help get rid of low coverage and low allele frequency variants which are typically false positives. Similarly, variants in low-complexity regions (such as homopolymers and tandem repeats) or in regions with known assembly issues are likely to be errors and would need to be filtered out. Additionally, tool-specific filters would be useful to deduce the authenticity of a variant – for instance, the GATK variant callers (McKenna et al., 2010; De Summa et al., 2017) output information on the mapping quality, genotype quality, evidence of strand bias, position of variant allele along reads – which could be effective in identification and removal of erroneous variant calls.

Structural Variant (SV) calling: In addition to SNVs and microIndels, the structural variants (SVs) such as copy number variations (CNVs), Deletions, Insertions, Duplications, Inversions and Translocations are significant determinants of human genetic diversity (Genomes Project et al., 2010; Mills et al., 2011; Sudmant et al., 2015). SVs can span in size from a few base pairs to chromosome scale, can cause diverse phenotypes and genomic diseases such as cancer. Over the next decade, we expect accurate detection of SVs would become regular practice at the clinic.

Traditional approach to the detection of SVs relied on karyotyping, PCR, FISH and microarray based approaches. However, these legacy approaches permit only a very coarse view of the SVs in the genome (Conrad et al., 2010; Zhao et al., 2013). With the advent of next generation sequencing technologies there has been an unprecedented growth in technological and analytical approaches for the identification of SVs. The massive amounts of sequence data generated from the first, second and third generation sequencing technologies has enabled us to detect SVs and resolve them down to the base pair level.

Currently there are three basic approaches to SV detection: (1) read-depth (RD); (2) split-read (SR) and (3) paired end (PR) (**Fig. 4**) (Medvedev et al., 2009). RD aligns sequences against a reference genome and counts the depth to estimate the copy number of that particular genomic segment. Using alignment information, SR uses reads that have multiple alignments from different parts of the same read – as these could potentially be overlapping SV breakpoints. PR relies on the paired end nature of the sequencing data from the Illumina sequencing platform to find read-pairs that align abnormally to the reference genome and could have spanned of SV breakpoints.

However, each individual method has its own biases that prevent it from detecting the full spectrum of structural variations in the human genome. Therefore, there is a need for new and improved methods of detecting structural variations. One such method is fusorSV (Published at: https://genomebiology.biomedcentral.com/articles/10.1186/s13059-018-1404-6T, see "Relevant Websites section") which takes an ensemble approach to the problem. fusorSV is an open source framework that includes a data fusion model built using the 1000 genomes project as the truth set. The model is then used to integrate the output from eight most popular SV calling methods to come up with unified and comprehensive SV call set that maximizes discovery while reducing false positives. Plewczynski et al. (2016) provide a comprehensive overview of recent approaches to SV detection using sequence data from whole genome and whole exome sequencing platforms.

Genomic Functional Informatics

Annotating the function of genomics elements and variants is a critical step for meaningful interpretation of data. In this section, we review the *in silico* approaches to functional annotation of genomic elements and variants.

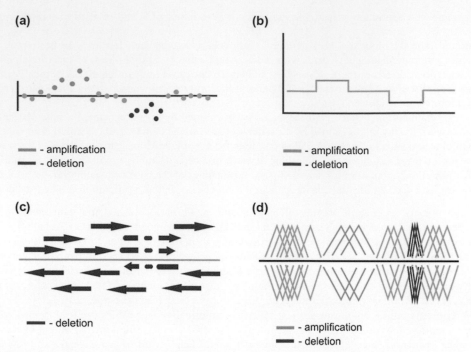

Fig. 4 Typical methods for SV detection (a) aCGH, (b) RD, (c) split read, and (d) paired-end. Adapted from Iskow, R.C., Gokcumen, O., Lee, C., 2012. Exploring the role of copy number variants in human adaptation. Trends Genet. 28 (6), 245–257.

Functional Analysis of Variants

The first step towards meaningful interpretation of variants and their functional impact on the genome is variant annotation. SNVs in coding regions can be characterized as transitions if they change a purine base to a purine base or a pyrimidine base to a pyrimidine base; or as transversions if they change a purine base to a pyrimidine base or vice-versa. Further, based on the type of amino-acid change, coding SNVs can be synonymous (no change in amino acid), missense (change in amino acid), or nonsense (introduction of a stop codon). Similarly, coding insertions and deletions (indels) can be classified as inframe if the size of the indel event is a multiple of three bases or frameshift if the size of the indel event is not a multiple of three bases and therefore shifts the reading frame.

The advent of projects such as 1000 Genomes Project (Kuehn, 2008), the Exome Aggregation Consortium (EXAC) (Lek *et al.*, 2016), Exome Sequencing Project (ESP) (See Relevant Websites Section), The Cancer Genome Atlas (TCGA) (Cancer Genome Atlas Research *et al.*, 2013) and others have provided rich catalogs of variants in a number of healthy and disease populations. Population level annotation helps in characterizing a variant as common or rare based on its frequency in different populations, which in turn is interpretation of the variant's impact on disease. In cancer studies, normal population allele frequencies are often used to characterize a variant as germline or somatic depending on whether its population allele frequency is above or below a certain threshold.

Variants can be further functionally characterized based on how conserved they are across different species, with the idea that functionally important variants are likely to be highly conserved. Conservation based scores provided by PhyloP (Pollard *et al.*, 2010), phastCons (Siepel *et al.*, 2005), GERP (Cooper *et al.*, 2005), SiPhy (Garber *et al.*, 2009), and others can be used to assess the evolutionary conservation of variants. Furthermore, prediction based scores such as SIFT (Ng and Henikoff, 2003), PolyPhen (Adzhubei *et al.*, 2013), FATHMM (Shihab *et al.*, 2013), MutationTaster (Schwarz *et al.*, 2010), etc can help evaluate the pathogenic nature of variants based on their the impact on the structure and/or function of the protein.

To understand the impact of variants on drug response and disease, they are annotated with information from clinical knowledge bases such as the JAX Clinical Knowledge Base (JAX CKB) (Patterson *et al.*, 2016), Clinical Interpretation of Variants in Cancer (CIViC) (Griffith *et al.*, 2017), OncoKB (Chakravarty *et al.*, 2017), Pharmacogenomics Knowledgebase (PharmGKB) (Whirl-Carrillo *et al.*, 2012) amongst others. The clinical annotation of variants associates variants with therapies, diagnostic/prognostic information, and clinical trials, and thus helps assess the clinical relevance of variants.

Functional Analysis of Non-Coding Regulatory Elements

Functional analysis of non-coding regulatory elements is carried out by mapping their target genes and conducting functional analysis of the target genes using experimental or enrichment based *in silico* methods. As promoters were defined using a fixed short window around transcription start sites (TSSs) (− 1 kb, + 200 bp for example), mapping promoters to gene targets is straightforward. In addition, recently, functional annotation of mammalian (FANTOM) consortium generated genome-wide and cell specific maps for promoters (Consortium et al., 2014), transcribed enhancers (Andersson et al., 2014) and other functional elements to genes including lncRNAs (Hon et al., 2017) and miRNAs (de Rie et al., 2017).

In the in silico approach, protein binding elements/sites and other distant functional elements are mapped to a nearest gene or all genes within a linear genomic window. The mapped genes are used to conduct enrichment analysis of Gene Ontologies, pathways, gene signature such as MSigDB. However, such approach has inherent bias due to non-uniform distribution of genes across the genome and widely varying gene lengths which leads to so called varying lengths of 'assignment domains' (Jair Zhou, 2012). The regulatory elements within an assignment domain of a gene are mapped to the gene. Assignment domain depends on the criterion used for mapping regulatory elements to genes. The bias is minimal if short windows (<5 kb) are used with reference to TSS of genes. However, the bias was shown to be unacceptably high for other choice of reference (e.g., whole gene) or longer window sizes of the order of tens of kilobases. To circumvent this problem, GREAT (McLean et al., 2010) and reFABS (Jair Zhou, 2012) were proposed for unbiased functional analysis. GREAT contrasts total length of assignment domains of the mapped genes to that of the genes in the functional category. Whereas, reFABS performs resampling based test to correct for the bias.

The target mapping can be more precise using experimental approaches that map interaction of the distant functional elements with promoters (e.g., enhancer-promoter interaction and TFBS-promoter interaction). 3D genome technologies such as HiC and ChIA-PET enable such precise mapping. Given the high resolution and targeted nature of ChIA-PET data, it is more suitable to discover promoter interactions compared to HiC data. Recently, more sophisticated methods were developed to identify enhancer-promoter interaction. For instance, promoter enhancer predictor (PEP) (Yang et al., 2017) uses deep learning from sequence based features only to predict enhancer promoter interactions. Moreover, TargetFinder method (Whalen et al., 2016), integrates information from 3D genome data, in particular HiC data, gene expression, epigenomic data (ChIP-seq and DNase-seq) to predict enhancer targets which can be used for the functional analysis of enhancers as described above.

Functional Analysis of Genes and Their Isoforms

Functional analysis of genes can involve determination of biochemical function, molecular regulation, interaction and expression of genes. Prediction of gene function mostly relies on sequence homology, domain annotation and interaction to functionally known proteins (Clark and Radivojac, 2011; Mostafavi et al., 2008). The homology-based predictions rely on the hypothesis that significant sequence similarity of genes suggests the predicted functional similarity. Homology search for coding region of a gene against the non-redundant protein database, such as NCBI (National Center for Biotechnology Information), SwissProt, allows the initial determination of gene function. Existing methods such as, PSI-BLAST and HMMer (Bork et al., 1998; Bork and Koonin, 1998; Koonin et al., 1998) can determine subtle conservation of protein sequence, which can be used to assign function to an uncharacterized protein. Identifying common evolutionary origin of genes based on homology method can provide further insights into gene function. However, homology approach may have caveats in functional characterization of gene, as fewer amino acid differences can alter the protein conformation and so function, and mutations in genes although may retain the structure, but may confer distinct function (Sjolander, 2004). Gene ontology (Ashburner et al., 2000) and UniProt Gene Ontology Annotation (UniProt-GOA) database (Barrell et al., 2009) are the widely used resources to determine cellular component, molecular function and biological process.

Function prediction of RNA-isoforms: Alternative splicing (AS) is known to be the primary source of transcriptome and proteome diversity in eukaryotes. Functional characterization of specific isoforms and their relevance to cell growth, development and differentiation as well in disease has been reported. However, for most isoforms, the function annotations remain missing. Thus, the extent of functional diversity of alternate mRNA isoforms generated from AS is largely unknown.

Not all mRNA isoforms derived from AS are protein-coding. In addition, AS produces unstable mRNAs which are degraded by nonsense-mediated decay (NMD) machinery (Lykke-Andersen and Jensen, 2015). Transcriptome-wide analysis predicted that ~ 10% of mRNAs are NMD targets in human (Mendell et al., 2004). Furthermore, some studies suggest that a large number of low abundant isoforms are the consequence of splicing noise, and therefore can be considered as non-functional isoforms (Melamud and Moult, 2009; Pickrell et al., 2010). While, splicing error or noisy splicing is possible, the low abundance of the alternate isoforms can also be due to their high-turnover rate, for instance their recognition by NMD pathway and consequent degradation. Alternate isoforms can have tissue-specific expression and distinct functions (Flouriot et al., 2000; Himeji et al., 2002). Such variations in isoforms are difficult to capture based on sequence-based analysis.

Unlike annotating gene function, the isoform-specific functional annotation is limited by several constraints: (1) AS can lead to skipping of exons that encode protein domains (Liu and Altman, 2003; Resch et al., 2004); (2) AS can cause disorder in amino acid sequence and structure, and thus functional change (Buljan et al., 2012; Romero et al., 2006); (3) AS resulting in a few amino acid changes in the open-reading frame can modulate the protein structure and function (Vogan et al., 1996; Yan et al., 2000); and, (4) mRNA isoforms can have opposite functions, for example, BCLX gene generate two isoforms, where one being pro-apoptotic and the other one being anti-apoptotic (Revil et al., 2007). Most functional annotations are documented at gene levels (Barrell et al., 2009) and genome-wide catalogue for isoform-specific function is still missing. However, methods are available that infer isoform functions based on domains and binding sites (Murvai et al., 2001; Thibert et al., 2005; Vacic et al., 2010; Verspoor et al., 2012). Several computational approaches have been applied to determine isoform-specific function (Eksi et al., 2013; Jia et al., 2010; Li et al., 2014a,b; Tseng et al., 2015). Another method modelled gene-isoform relationship as multiple-instance data and developed a multiple instance-based label propagation method to predict functions. In addition, assessing isoform ortho-logues has been found to provide functional information (Jia *et al.*, 2010; Zambelli *et al.*, 2010). Further improvements in isoform function prediction method include integration of heterogeneous data: RNA-Seq, protein-sequence and structure, amino acid composition, protein isoform docking and post-translational modification (Li *et al.*, 2014a).

Functional Analysis of ncRNAs – miRNAs and lncRNAs

Computational functional analyses of microRNAs (miRNAs): Short RNAs are part of the broad family of non-coding RNAs, which perform many regulatory functions. In particular, miRNAs have major interest because of their activity in gene silencing through post-transcriptional repression (Bartel, 2009). Mature miRNAs are about 21–23 nucleotides (nt) in length (Winter *et al.*, 2009), single stranded molecules that bind to 3' untranslated regions (3' UTR) of their target mRNAs and regulate many biological processes, including cell differentiation (Bray *et al.*, 2009), organ development (Ambros, 2011). It has also been shown that miRNAs are involved in various diseases, such as cancer (Esquela-Kerscher and Slack, 2006; Bailey *et al.*, 2009) and cardiovascular disorders (Urbich *et al.*, 2008). The mRNA:miRNA interaction depends on a functional motif (2–7 nt) present in the 5' end of the miRNA which is known as "seed" region (Lewis *et al.*, 2005). In most cases, these short sequences negatively influence the expression of miRNA target genes via translation repression or degradation.

Similar to distant functional elements, the function of miRNAs is also elicited by mapping miRNAs to their target genes, followed by functional analysis of target genes in an *in silico* approach. Numerous online and stand-alone bioinformatics tools have been developed to find miRNA target genes, see **Table 1**. These algorithms depend on so called *sequence to sequence base pairing*, i.e., complementarity of the base pairing (aka Watson-Crick base pairing) between the seed sequences of the miRNA and the 3' UTR of the genes, to map miRNA target genes. Further improvement of the method is possible by investigating the nature of the evolutionary conservation of the miRNA and target sequences (Lewis *et al.*, 2005). Some methods predict secondary structure of the duplex to calculate thermodynamic stability of the mRNA:miRNA interaction (Rehmsmeier *et al.*, 2004; John *et al.*, 2004). Next, these predicted targets are used to determine biological processes (gene ontology (The Gene Ontology, 2017), disease ontology (Hillen *et al.*, 1991)) and pathways (Reactome (Fabregat *et al.*, 2016) and KEGG (Kanehisa *et al.*, 2017)) affected by the miRNAs and hence the function of miRNAs.

Computational functional analysis of long non-coding RNAs (lncRNAs): lncRNAs are RNA molecules with a length of more than 200 nucleotides (Perkel, 2013). lncRNAs typically do not encode proteins, certain lncRNAs were shown to have protein coding potential (Guo *et al.*, 2013) though. They are located throughout the genome (Derrien *et al.*, 2012) and exclusively localized in the nucleus (Vance and Ponting, 2014). However, a recent study also reported that it can be found in cytoplasm of the cell (Sun and Kraus, 2015). The lncRNAs are important epigenetic, post transcriptional and transcriptional regulators, and have a variety of functions in developmental, cellular, molecular and disease processes such as cancer (Sun and Kraus, 2015). Several bioinfor-matics resources are given in **Table 2** to access the annotated and predicted lncRNAs with their function, disease association and expression analysis. Along with these resources, multiple tools have been developed to determine the functional properties of the lncRNAs. The tools use homology-based sequence similarity search such as BLASTn (Zhang *et al.*, 2008) and BLAT (Kent, 2002), alignment-based coding potential search (Kong *et al.*, 2007), alignment-free coding region search with CPAT (Wang *et al.*, 2013), motif-based analyses using HOMER (Heinz *et al.*, 2010) and MEME (Bailey *et al.*, 2009). Several R/ Bioconductor packages, for instance, ClusterProfiler (Yu *et al.*, 2012), GDCRNATools (Ruidong *et al.*, 2017) and WGCNA (Langfelder and Horvath, 2008) are also available to explore the gene enrichment, pathways and clustering-based network analyses for functional analysis of lncRNAs.

Population Genome Informatics

Over the last two decades, the availability of entire genomes of multiple species (Mouse Genome Sequencing *et al.*, 2002; International Chicken Genome Sequencing, 2004; International Human Genome Sequencing, 2004; Chimpanzee and Analysis, 2005; Rhesus Macaque Genome *et al.*, 2007) has enabled an in-depth investigation of different species at the molecular level.

Table 1   Online and standalone bioinformatics resources for miRNA target predictions

Tool	Website	Organisms	Features	Reference
TargetScan	www.targetscan.org/	Mammals, worm, fly, fish	Seed match, conservation	Lewis *et al.* (2005)
DIANA-microT-CDS	www.microrna.gr/microT-CDS	Human, mouse, fly, worm	Seed match, conservation, free energy, target site accessibility	Paraskevopoulou *et al.* (2013)
miRanda	www.microrna.org	Human, mouse, rat, fly, worm	Seed match, conservation, free energy	John *et al.* (2004)
RNAhybrid	www.bibiserv2.cebitec.uni-bielefeld.de/rnahybrid	Mouse, rat, fly, worm	Seed match, conservation, free energy	Rehmsmeier *et al.* (2004)

Note: Min, H., Yoon, S., 2010. Got target?: Computational methods for microRNA target prediction and their extension. Exp. Mol. Med. 42 (4), 233–244.

Table 2   Databases for lncRNAs

Database	Website (Reference)
RNAcentral	http://rnacentral.org (Bateman *et al.*, 2011)
lncRNAdb	http://www.lncrnadb.org (Quek *et al.*, 2015)
NONCODE	http://www.noncode.org (Zhao *et al.*, 2016)
Rfam	http://rfam.org (Kalvari *et al.*, 2017)
LncRNAWiki	http://lncrna.big.ac.cn/index.php (Ma *et al.*, 2015)
LNCipedia	https://lncipedia.org (Volders *et al.*, 2013)
fRNAdb	http://www.ncrna.org/frnadb (Kin *et al.*, 2007)
Human Body Map lincRNAs	http://www.broadinstitute.org/genome_bio/human_lincrnas (Khalil *et al.*, 2009)
LNCediting	http://bioinfo.life.hust.edu.cn/LNCediting (Gong *et al.*, 2017)
Co-LncRNA	http://www.bio-bigdata.com/Co-LncRNA (Zhao *et al.*, 2015)
Lnc2Cancer	http://www.bio-bigdata.net/lnc2cancer (Ning *et al.*, 2016)
LncRNADisease	http://cmbi.bjmu.edu.cn/lncrnadisease (Chen *et al.*, 2013)

Note: Signal, B., Gloss, B.S., Dinger, M.E., 2016. Computational approaches for functional prediction and characterisation of long noncoding RNAs. Trends Genet. 32 (10), 620–637.

Along with this, the advent of sophisticated sequence alignment algorithms (Schwartz *et al.*, 2003; Blanchette *et al.*, 2004) and the field of comparative genomics (Miller *et al.*, 2004) have contributed tremendously to a better understanding of evolution and function of genomes. Comparative genomics involves comparison of genomic DNA among different species to understand various genomic differences and similarities as well as the evolutionary relationship between them. This approach has been used by many scientists to study the pattern and rate variation of mutations across multiple species and it has been clearly demonstrated that mutation rates not only vary between species and organisms but also between different regions of a genome (Chiaromonte *et al.*, 2001; Hardison *et al.*, 2003; Ananda *et al.*, 2011). In fact, regions with high and low neutral substitution rates in the human genome have been shown to contain different functional classes of genes – neutral substitution hot spots are generally biased towards surface receptor and immune response genes while cold spots are biased towards regulatory genes (Chuang and Li, 2004). The knowledge of neutral mutation rate variation can also be valuable to the discovery of non-coding functional elements – specifically, to distinguish between functional conservation and low background neutral mutation rates. Functional prediction algorithms incorporating background rates demonstrate improved separation of functional and neutral sites (Siepel *et al.*, 2005; Taylor *et al.*, 2006; Tyekucheva *et al.*, 2008). Understanding mutation rate variation is also critical for elucidating how biochemical processes (e.g., replication, repair, recombination etc.) and errors in them drive mutagenesis.

The next generation sequencing (NGS) technologies, which have facilitated cost-effective sequencing of entire genomes, are being efficiently employed by projects such as the 1000 Genomes Project (Kuehn, 2008) the Exome Aggregation Consortium (EXAC) (Lek *et al.*, 2016), Exome Sequencing Project (ESP) (See Relevant Websites Section) and others to generate large population-scale resources. These large-scale population genomics projects provide rich comprehensive catalogs of genetic variation of various types (SNVs, indels, structural variants etc.) across human populations of different ancestries (African, European, East Asian, South Asian, South American etc.). They have helped understand the nature and extent of intra- and inter-population differences in genetic variation and have facilitated population-based interpretation of clinical findings, which is important to curtail misdiagnosis (Husten, 2016). Additionally, population genomics also helped identify candidate genes for different diseases based on the distribution of functional variants in them. Scores such as Residual Variance Intolerance Score (RVIS) (Petrovski *et al.*, 2013) and probability of loss of function intolerance (pLI) (Lek *et al.*, 2016) compare the observed genetic variation in a population based on the expected variation to identify genes that are more or less tolerant to functional genetic variation. Thus, they help identify genes under different selective constraints – for instance, genes with the highest tolerance for genetic variation are likely under positive selection, and those with the lowest tolerance under purifying selection. This information is pivotal to identification of candidate genes in a disease population.

Statistical genetics, based on the understanding laid above, makes use of statistical methods to study evolutionary processes that shape genetic variations, and to infer the effect of genetic variants on normal phenotypes and diseases. Besides analyzing the evolutionary constraints and patterns that help understand the genomic function, statistical genetics has proven to be a highly successful approach to identify genetic markers associated with phenotypes.

The early efforts to map disease genes were through linkage analysis in familial data (Laird and Lange, 2011). A marker gene and a disease susceptibility gene are in linkage if they are inherited together during the meiosis phase of sexual reproduction, namely the recombination fraction between them is smaller than one half. Parametric linkage analysis (Abecasis *et al.*, 2002) need to specify appropriate disease inheritance mode, disease allele frequency and penetrance, which may not be always obtainable. Non-parametric linkage analysis (Kruglyak *et al.*, 1996), which is based on identity-by-descent (IBD) sharing (Keith *et al.*, 2008), does not have such requirement, hence is more general. However, linkage analysis is most successful for Mendelian diseases (monogene for a disease), and much less powerful for complex diseases where polygenes dictate a disease. Since family data are hard to collect, association studies in unrelated subjects prevail for complex diseases.

Association analysis is based on linkage disequilibrium (LD) (Pritchard and Przeworski, 2001) among genetic variants. Two alleles of two variants are in linkage disequilibrium if their co-occurrence probability in a haplotype is larger than that when they are independent. r^2 is a LD coefficient for measuring such allelic associations (Pritchard and Przeworski, 2001). Disease variants, which may not be typed in a study, can be tagged by its LD proxy variants. To achieve the same detection power, the sample size required in an association study which analyzes marker variants is proportion to the reciprocal of r^2 of that when disease causing variant is directly genotyped.

In order to tag most of the common SNPs, the number of typed SNPs in genome-wide association studies (GWAS) has increased from several hundred thousand to five million. Majority of the typed SNPs are common (minor allele frequency > 5%), and non-coding in GWAS. To increase SNP coverage, LD-based imputation methods (Marchini *et al.*, 2007; Marchini and Howie, 2010; Browning and Browning, 2009) for untyped SNPs were developed, which also made the meta-analysis among GWASs on different typing platforms feasible.

Exome array (Huyghe *et al.*, 2013; Liu *et al.*, 2017) for low frequency coding SNPs and exome sequencing (Kiezun *et al.*, 2012; Liu *et al.*, 2014) for rare coding SNPs are alternative ways to investigate roles of coding genetic variants. Two main approaches have been developed for rare variant association analysis which typically has low power. One is burden test (AL *et al.*, 2010) which sums up the counts of alleles of all rare variants in a gene region. Burden test is powerful when the majority of rare variants are functional and have the same effect direction. The other is SKAT-O (L. *et al.*, 2012), which calculates the weighted sum of test statistics for all variants in a gene region. SKAT-O is powerful when a large portion of the variants are neutral and/or have different effect directions.

In GWAS, the most common confounding factor is population structure, where population subgroups are associated with both disease and variant allele frequency, inducing the crude odds ratio different from the stratified odds ratio. The confounding effect of population structure can be adjusted by several ways. For example λ_{GC} correction (Devlin and Roeder, 1999), principal component correction (Price *et al.*, 2006), and linear or generalized linear mixed model (Chen *et al.*, 2016; Zhou and Stephens, 2012) correction.

Discussion

Genome informatics is critical to eliciting genomic structure and its relationship to individual and population phenotypes, including disease. Multiple aspects of genome informatics play important role in a modern genomic project. Though plethora of informatics methods have been developed for each aspect of genome analysis, none of them serve as single go-to tool. Hence, a combination of methods may need to be used for accurate analysis. In addition, while the emerging technologies offer opportunities to better understand the role of genome by generating more precise data, they pose brand new informatics challenges including developing new tools and integrative analysis.

Author Contributions

Karuturi and George conceptualized and led the writing of the article. All authors contributed equally to the article.

See also: Bioinformatics Approaches for Studying Alternative Splicing. Chromatin: A Semi-Structured Polymer. Comparative and Evolutionary Genomics. Comparative Genomics Analysis. Computational Immunogenetics. Detecting and Annotating Rare Variants. Exome Sequencing Data Analysis. Experimental Platforms for Extracting Biological Data: Mass Spectrometry, Microarray, Next Generation Sequencing. Functional Enrichment Analysis. Functional Genomics. Gene Mapping. Genome Alignment. Genome Analysis – Identification of Genes Involved in Host-Pathogen Protein-Protein Interaction Networks. Genome Annotation. Genome Annotation: Perspective From Bacterial Genomes. Genome Databases and Browsers. Genome-Wide Association Studies. Genome-Wide Haplotype Association Study. Hidden Markov Models. Integrative Analysis of Multi-Omics Data. Integrative Bioinformatics of Transcriptome: Databases, Tools and Pipelines. Learning Chromatin Interaction Using Hi-C Datasets. Linkage Disequilibrium. Metagenomic Analysis and its Applications. Natural Language Processing Approaches in Bioinformatics. Nucleosome Positioning. Phylogenetic Footprinting. Pipeline of High Throughput Sequencing. Predicting Non-Synonymous Single Nucleotide Variants Pathogenic Effects in Human Diseases. Repeat in Genomes: How and Why You Should Consider Them in Genome Analyses?. Sequence Analysis. Single Nucleotide Polymorphism Typing. Transcription Factor Databases. Transcriptomic Databases. Whole Genome Sequencing Analysis

References

Abecasis, G.R., *et al.*, 2002. Merlin – Rapid analysis of dense genetic maps using sparse gene flow trees. Nat. Genet. 30, 97–101.
Adzhubei, I., Jordan, D.M., Sunyaev, S.R., 2013. Predicting functional effect of human missense mutations using PolyPhen-2. Curr. Protoc. Hum. Genet. (Chapter 7: p. Unit7 20).
Altschul, S.F., *et al.*, 1990. Basic local alignment search tool. J. Mol. Biol. 215 (3), 403–410.
Price, A.L., *et al.*, 2010. Pooled association tests for rare variants in exon-resequencing studies. Am. J. Hum. Genet. 86, 832–838.
Ambros, V., 2011. MicroRNAs and developmental timing. Curr. Opin. Genet. Dev. 21 (4), 511–517.
Ananda, G., Chiaromonte, F., Makova, K.D., 2011. A genome-wide view of mutation rate co-variation using multivariate analyses. Genome Biol. 12 (3), R27.
Andersson, R., *et al.*, 2014. An atlas of active enhancers across human cell types and tissues. Nature 507 (7493), 455–461.
Ashburner, M., *et al.*, 2000. The Gene Ontology Consortium, Gene ontology: Tool for the unification of biology. Nat. Genet. 25 (1), 25–29.

Ashoor, H., *et al.*, 2013. HMCan: A method for detecting chromatin modifications in cancer samples using ChIP-seq data. Bioinformatics 29 (23), 2979–2986.

Au-Yeung, G., *et al.*, 2016. Selective targeting of Cyclin E1 amplified high grade serous ovarian cancer by cyclin-dependent kinase 2 and AKT inhibition. Clin. Cancer Res.

Auer, P.L., *et al.*, 2016. Guidelines for large-scale sequence-based complex trait association studies: Lessons learned from the NHLBI exome sequencing project. Am. J. Hum. Genet. 99 (4), 791–801.

Bailey, T.L., *et al.*, 2009. MEME SUITE: Tools for motif discovery and searching. Nucleic Acids Res. 37 (Web Server issue), W202–W208.

Balasubramanian, S., 2011. Sequencing nucleic acids: From chemistry to medicine. Chem. Commun. (Camb.) 47 (26), 7281–7286.

Bao, R., *et al.*, 2014. Review of current methods, applications, and data management for the bioinformatics analysis of whole exome sequencing. Cancer Inform. 13 (Suppl. 2), 67–82.

Barrell, D., *et al.*, 2009. The GOA database in 2009 – An integrated Gene Ontology Annotation resource. Nucleic Acids Res. 37 (Database issue), D396–D403.

Bartel, D.P., 2009. MicroRNAs: Target recognition and regulatory functions. Cell 136 (2), 215–233.

Bateman, A., *et al.*, 2011. RNAcentral: A vision for an international database of RNA sequences. RNA 17 (11), 1941–1946.

Batzoglou, S., *et al.*, 2002. ARACHNE: A whole-genome shotgun assembler. Genome Res. 12 (1), 177–189.

Bentley, D.R., *et al.*, 2008. Accurate whole human genome sequencing using reversible terminator chemistry. Nature 456 (7218), 53–59.

Blanchette, M., *et al.*, 2004. Aligning multiple genomic sequences with the threaded blockset aligner. Genome Res. 14 (4), 708–715.

Boeva, V., *et al.*, 2017. Heterogeneity of neuroblastoma cell identity defined by transcriptional circuitries. Nat. Genet. 49 (9), 1408–1413.

Boley, N., *et al.*, 2014. Navigating and mining modENCODE data. Methods 68 (1), 38–47.

Bolger, A.M., Lohse, M., Usadel, B., 2014. Trimmomatic: A flexible trimmer for Illumina sequence data. Bioinformatics 30 (15), 2114–2120.

Bork, P., *et al.*, 1998. Predicting function: From genes to genomes and back. J. Mol. Biol. 283 (4), 707–725.

Bork, P., Koonin, E.V., 1998. Predicting functions from protein sequences—where are the bottlenecks? Nat. Genet. 18 (4), 313–318.

Branton, D., *et al.*, 2008. The potential and challenges of nanopore sequencing. Nat. Biotechnol. 26 (10), 1146–1153.

Bray, I., *et al.*, 2009. Widespread dysregulation of MiRNAs by MYCN amplification and chromosomal imbalances in neuroblastoma: Association of miRNA expression with survival. PLOS ONE 4 (11), e7850.

Brown, T.A., 2002. Genomes, second ed. Oxford.

Browning, B.L., Browning, S.R., 2009. A unified approach to genotype imputation and haplotype-phase inference for large data sets of trios and unrelated individuals. Am. J. Hum. Genet. 84, 210–223.

Buljan, M., *et al.*, 2012. Tissue-specific splicing of disordered segments that embed binding motifs rewires protein interaction networks. Mol. Cell 46 (6), 871–883.

Cairns, J., *et al.*, 2016. CHiCAGO: Robust detection of DNA looping interactions in Capture Hi-C data. Genome Biol. 17 (1), 127.

Campagna, D., *et al.*, 2009. PASS: A program to align short sequences. Bioinformatics 25 (7), 967–968.

Cancer Genome Atlas Research, N., *et al.*, 2013. The cancer genome atlas pan-cancer analysis project. Nat. Genet. 45 (10), 1113–1120.

Chakravarty, D., *et al.*, 2017. OncoKB: A precision oncology knowledge base. JCO Precis. Oncol. 2017.

Chen, C.Y., Morris, Q., Mitchell, J.A., 2012. Enhancer identification in mouse embryonic stem cells using integrative modeling of chromatin and genomic features. BMC Genom. 13, 152.

Chen, G., *et al.*, 2013. LncRNADisease: A database for long-non-coding RNA-associated diseases. Nucleic Acids Res. 41 (Database issue), D983–D986.

Chen, H., *et al.*, 2016. Control for population structure and relatedness for binary traits in genetic association studies via logistic mixed models. Am. J. Hum. Genet. 98 (4), 653–666.

Chiaromonte, F., *et al.*, 2001. Association between divergence and interspersed repeats in mammalian noncoding genomic DNA. Proc. Natl. Acad. Sci. USA 98 (25), 14503–14508.

Chikhi, R., Medvedev, P., 2014. Informed and automated k-mer size selection for genome assembly. Bioinformatics 30 (1), 31–37.

Chimpanzee, S., Analysis, C., 2005. Initial sequence of the chimpanzee genome and comparison with the human genome. Nature 437 (7055), 69–87.

Chuang, J.H., Li, H., 2004. Functional bias and spatial organization of genes in mutational hot and cold regions in the human genome. PLOS Biol. 2 (2), E29.

Cibulskis, K., *et al.*, 2013. Sensitive detection of somatic point mutations in impure and heterogeneous cancer samples. Nat. Biotechnol. 31 (3), 213–219.

Clarke, J., *et al.*, 2009. Continuous base identification for single-molecule nanopore DNA sequencing. Nat. Nanotechnol. 4 (4), 265–270.

Clark, W.T., Radivojac, P., 2011. Analysis of protein function and its prediction from amino acid sequence. Proteins 79 (7), 2086–2096.

Conrad, D.F., *et al.*, 2010. Origins and functional impact of copy number variation in the human genome. Nature 464 (7289), 704–712.

Consortium, E.P., *et al.*, 2007. Identification and analysis of functional elements in 1% of the human genome by the ENCODE pilot project. Nature 447 (7146), 799–816.

Consortium, E.P., 2012. An Integrated encyclopedia of DNA elements in the human genome. Nature 489 (7414), 57–74.

Consortium, F., *et al.*, 2014. A promoter-level mammalian expression atlas. Nature 507 (7493), 462–470.

Cooper, G.M., *et al.*, 2005. Distribution and intensity of constraint in mammalian genomic sequence. Genome Res. 15 (7), 901–913.

David, M., *et al.*, 2011. SHRiMP2: Sensitive yet practical SHort Read Mapping. Bioinformatics 27 (7), 1011–1012.

Dekker, J., *et al.*, 2002. Capturing chromosome conformation. Science 295 (5558), 1306–1311.

de la Bastide, M., McCombie, W.R., 2007. Assembling genomic DNA sequences with PHRAP. Curr. Protoc. Bioinform. (Chapter 11: p. Unit11 4).

de Rie, D., *et al.*, 2017. An integrated expression atlas of miRNAs and their promoters in human and mouse. Nat. Biotechnol. 35 (9), 872–878.

De Summa, S., *et al.*, 2017. GATK hard filtering: Tunable parameters to improve variant calling for next generation sequencing targeted gene panel data. BMC Bioinform. 18 (Suppl. 5), 119.

Derrien, T., *et al.*, 2012. The GENCODE v7 catalog of human long noncoding RNAs: Analysis of their gene structure, evolution, and expression. Genome Res. 22 (9), 1775–1789.

Devarakonda, S., Morgensztern, D., Govindan, R., 2013. Clinical applications of The Cancer Genome Atlas project (TCGA) for squamous cell lung carcinoma. Oncology (Williston Park) 27 (9), 899–906.

Devlin, B., Roeder, K., 1999. Genomic control for association studies. Biometrics 55, 997–1004.

Dostie, J., *et al.*, 2006. Chromosome Conformation Capture Carbon Copy (5C): A massively parallel solution for mapping interactions between genomic elements. Genome Res. 16 (10), 1299–1309.

Durand, N.C., *et al.*, 2016a. Juicer provides a one-click system for analyzing loop-resolution Hi-C experiments. Cell Syst. 3 (1), 95–98.

Durand, N.C., *et al.*, 2016b. Juicebox provides a visualization system for Hi-C contact maps with unlimited zoom. Cell Syst. 3 (1), 99–101.

Eksi, R., *et al.*, 2013. Systematically differentiating functions for alternatively spliced isoforms through integrating RNA-seq data. PLOS Comput. Biol. 9 (11), e1003314.

Elgar, G., Vavouri, T., 2008. Tuning in to the signals: Noncoding sequence conservation in vertebrate genomes. Trends Genet. 24 (7), 344–352.

English, A.C., *et al.*, 2012. Mind the gap: Upgrading genomes with Pacific Biosciences RS long-read sequencing technology. PLOS ONE 7 (11), e47768.

Ernst, J., Kellis, M., 2012. ChromHMM: Automating chromatin-state discovery and characterization. Nat. Methods 9 (3), 215–216.

Esquela-Kerscher, A., Slack, F.J., 2006. Oncomirs – MicroRNAs with a role in cancer. Nat. Rev. Cancer 6 (4), 259–269.

Fabregat, A., *et al.*, 2016. The Reactome pathway Knowledgebase. Nucleic Acids Res. 44 (D1), D481–D487.

Felix, M.A., 2016. Phenotypic evolution with and beyond genome evolution. Curr. Top. Dev. Biol. 119, 291–347.

Flouriot, G., *et al.*, 2000. Identification of a new isoform of the human estrogen receptor-alpha (hER-alpha) that is encoded by distinct transcripts and that is able to repress hER-alpha activation function 1. EMBO J. 19 (17), 4688–4700.

Fullwood, M.J., *et al.*, 2009. An oestrogen-receptor-alpha-bound human chromatin interactome. Nature 462 (7269), 58–64.

Garber, M., *et al.*, 2009. Identifying novel constrained elements by exploiting biased substitution patterns. Bioinformatics 25 (12), i54–i62.

Garrison, E., Marth, G., 2012. Haplotype-based variant detection from short-read sequencing. arXiv. (1207.3907).

Genomes Project, C., *et al.*, 2010. A map of human genome variation from population-scale sequencing. Nature 467 (7319), 1061–1073.

Gnerre, S., *et al.*, 2011. High-quality draft assemblies of mammalian genomes from massively parallel sequence data. Proc. Natl. Acad. Sci. USA 108 (4), 1513–1518.

Gong, J., *et al.*, 2017. LNCediting: A database for functional effects of RNA editing in lncRNAs. Nucleic Acids Res. 45 (D1), D79–D84.

Griffith, M., *et al.*, 2017. CIViC is a community knowledgebase for expert crowdsourcing the clinical interpretation of variants in cancer. Nat. Genet. 49 (2), 170–174.

Guo, X., *et al.*, 2013. Long non-coding RNAs function annotation: A global prediction method based on bi-colored networks. Nucleic Acids Res. 41 (2), e35.

Haiminen, N., *et al.*, 2011. Evaluation of methods for de novo genome assembly from high-throughput sequencing reads reveals dependencies that affect the quality of the results. PLOS ONE 6 (9), e24182.

Hardison, R.C., *et al.*, 2003. Covariation in frequencies of substitution, deletion, transposition, and recombination during eutherian evolution. Genome Res. 13 (1), 13–26.

Harris, T.D., *et al.*, 2008. Single-molecule DNA sequencing of a viral genome. Science 320 (5872), 106–109.

Heintzman, N.D., *et al.*, 2007. Distinct and predictive chromatin signatures of transcriptional promoters and enhancers in the human genome. Nat. Genet. 39 (3), 311–318.

Heinz, S., *et al.*, 2010. Simple combinations of lineage-determining transcription factors prime cis-regulatory elements required for macrophage and B cell identities. Mol. Cell 38 (4), 576–589.

Hillen, U., Weidenmeier, W., Haug, C., 1991. Ruptured aneurysm of an aberrant subclavian artery. Dtsch. Med. Wochenschr. 116 (48), 1832–1836.

Himeji, D., *et al.*, 2002. Characterization of caspase-8L: A novel isoform of caspase-8 that behaves as an inhibitor of the caspase cascade. Blood 99 (11), 4070–4078.

Hoff, K.J., *et al.*, 2016. BRAKER1: Unsupervised RNA-seq-based genome annotation with GeneMark-ET and AUGUSTUS. Bioinformatics 32 (5), 767–769.

Hoffman, M.M., *et al.*, 2012. Unsupervised pattern discovery in human chromatin structure through genomic segmentation. Nat. Methods 9 (5), 473–476.

Hoffman, M.M., *et al.*, 2013. Integrative annotation of chromatin elements from ENCODE data. Nucleic Acids Res. 41 (2), 827–841.

Homer, N., Merriman, B., Nelson, S.F., 2009. BFAST: An alignment tool for large scale genome resequencing. PLOS ONE 4 (11), e7767.

Hon, C.C., *et al.*, 2017. An atlas of human long non-coding RNAs with accurate 5′ ends. Nature 543 (7644), 199–204.

Hon, G., Ren, B., Wang, W., 2008. ChromaSig: A probabilistic approach to finding common chromatin signatures in the human genome. PLOS Comput. Biol. 4 (10), e1000201.

Hu, M., *et al.*, 2012. HiCNorm: Removing biases in Hi-C data via Poisson regression. Bioinformatics 28 (23), 3131–3133.

Huang, X., Madan, A., 1999. CAP3: A DNA sequence assembly program. Genome Res. 9 (9), 868–877.

Hudson, K., 2002. The Human Genome Project: A public good. Health Matrix Clevel 12 (2), 367–375.

Hunt, M., *et al.*, 2013. REAPR: A universal tool for genome assembly evaluation. Genome Biol. 14 (5), R47.

Husten, L., 2016. Imprecise medicine: Genetic tests lead to misdiagnosis.

Huyghe, J.R., *et al.*, 2013. Exome array analysis identifies novel loci and low-frequency variants for insulin processing and secretion. Nat. Genet. 45 (2), 197–201.

Imakaev, M., *et al.*, 2012. Iterative correction of Hi-C data reveals hallmarks of chromosome organization. Nat. Methods 9 (10), 999–1003.

International Chicken Genome Sequencing, C., C., 2004. Sequence and comparative analysis of the chicken genome provide unique perspectives on vertebrate evolution. Nature 432 (7018), 695–716.

International HapMap, C., C., 2003. The International HapMap Project. Nature 426 (6968), 789–796.

International Human Genome Sequencing, C., C., 2004. Finishing the euchromatic sequence of the human genome. Nature 431 (7011), 931–945.

Iskow, R.C., Gokcumen, O., Lee, C., 2012. Exploring the role of copy number variants in human adaptation. Trends Genet. 28 (6), 245–257.

Jair Zhou, H.R.L.A.R.K.M.K., 2012. Bias in genome scale functional analysis of transcription factors using binding site data. J. Phys. Chem. Biophys.

Javierre, B.M., *et al.*, 2016. Lineage-specific genome architecture links enhancers and non-coding disease variants to target gene promoters. Cell 167 (5), 1369–1384. e19.

Jia, Y., *et al.*, 2010. Refining orthologue groups at the transcript level. BMC Genom. 11 (Suppl. 4), S11.

John, B., *et al.*, 2004. Human MicroRNA targets. PLOS Biol. 2 (11), e363.

Kalvari, I., *et al.*, 2017. Rfam 13.0: Shifting to a genome-centric resource for non-coding RNA families. Nucleic Acids Res.

Kanehisa, M., *et al.*, 2017. KEGG: New perspectives on genomes, pathways, diseases and drugs. Nucleic Acids Res. 45 (D1), D353–D361.

Keith, J.M., *et al.*, 2008. Calculation of IBD probabilities with dense SNP or sequence data. Genet. Epidemiol. 32, 513–519.

Kent, W.J., 2002. BLAT – The BLAST-like alignment tool. Genome Res. 12 (4), 656–664.

Khalil, A.M., *et al.*, 2009. Many human large intergenic noncoding RNAs associate with chromatin-modifying complexes and affect gene expression. Proc. Natl. Acad. Sci. USA 106 (28), 11667–11672.

Kiezun, A., *et al.*, 2012. Exome sequencing and the genetic basis of complex traits. Nat. Genet. 44 (6), 623–630.

Kin, T., *et al.*, 2007. fRNAdb: A platform for mining/annotating functional RNA candidates from non-coding RNA sequences. Nucleic Acids Res. 35 (Database issue), D145–D148.

Kircher, M., Heyn, P., Kelso, J., 2011. Addressing challenges in the production and analysis of illumina sequencing data. BMC Genom. 12, 382.

Koboldt, D.C., *et al.*, 2012. VarScan 2: Somatic mutation and copy number alteration discovery in cancer by exome sequencing. Genome Res. 22 (3), 568–576.

Kong, L., *et al.*, 2007. CPC: Assess the protein-coding potential of transcripts using sequence features and support vector machine. Nucleic Acids Res. 35 (Web Server issue), W345–W349.

Koonin, E.V., Tatusov, R.L., Galperin, M.Y., 1998. Beyond complete genomes: From sequence to structure and function. Curr. Opin. Struct. Biol. 8 (3), 355–363.

Kruglyak, L., *et al.*, 1996. Parametric and nonparametric linkage analysis: A unified multipoint approach. Am. J. Hum. Genet. 58, 1347–1363.

Kuehn, B.M., 2008. 1000 Genomes Project promises closer look at variation in human genome. JAMA 300 (23), 2715.

Kugel, J.F., Goodrich, J.A., 2017. Finding the start site: Redefining the human initiator element. Genes Dev. 31 (1), 1–2.

L., S., W., M.C., L., X., 2012. Optimal tests for rare variant effects in sequencing association studies. Biostatistics 13, 762–775.

Laird, N., Lange, C., 2011. The Fundamentals of Modern Statistical Genetics. New York: Springer Science.

Lajoie, B.R., *et al.*, 2009. My5C: Web tools for chromosome conformation capture studies. Nat. Methods 6 (10), 690–691.

Lander, E.S., *et al.*, 2001. Initial sequencing and analysis of the human genome. Nature 409 (6822), 860–921.

Langfelder, P., Horvath, S., 2008. WGCNA: An R package for weighted correlation network analysis. BMC Bioinform. 9, 559.

Langmead, B., Salzberg, S.L., 2012. Fast gapped-read alignment with Bowtie 2. Nat. Methods 9 (4), 357–359.

Larson, D.E., *et al.*, 2012. SomaticSniper: Identification of somatic point mutations in whole genome sequencing data. Bioinformatics 28 (3), 311–317.

Lazaris, C., *et al.*, 2017. HiC-bench: Comprehensive and reproducible Hi-C data analysis designed for parameter exploration and benchmarking. BMC Genom. 18 (1), 22.

Lee, W.P., *et al.*, 2014. MOSAIK: A hash-based algorithm for accurate next-generation sequencing short-read mapping. PLOS ONE 9 (3), e90581.

Lehner, B., 2007. Modelling genotype-phenotype relationships and human disease with genetic interaction networks. J. Exp. Biol. 210 (Pt 9), 1559–1566.

Lek, M., *et al.*, 2016. Analysis of protein-coding genetic variation in 60,706 humans. Nature 536 (7616), 285–291.

Lewis, B.P., Burge, C.B., Bartel, D.P., 2005. Conserved seed pairing, often flanked by adenosines, indicates that thousands of human genes are microRNA targets. Cell 120 (1), 15–20.

Li, G., *et al.*, 2010. ChIA-PET tool for comprehensive chromatin interaction analysis with paired-end tag sequencing. Genome Biol. 11 (2), R22.

Li, H., Durbin, R., 2010. Fast and accurate long-read alignment with Burrows-Wheeler transform. Bioinformatics 26 (5), 589–595.

Li, H., Homer, N., 2010. A survey of sequence alignment algorithms for next-generation sequencing. Brief. Bioinform. 11 (5), 473–483.

Li, H., *et al.*, 2009a. The Sequence Alignment/Map format and SAMtools. Bioinformatics 25 (16), 2078–2079.

Li, H.D., *et al.*, 2014a. The emerging era of genomic data integration for analyzing splice isoform function. Trends Genet. 30 (8), 340–347.

Li, R., *et al.*, 2009b. SOAP2: An improved ultrafast tool for short read alignment. Bioinformatics 25 (15), 1966–1967.

Li, W., *et al.*, 2014b. High-resolution functional annotation of human transcriptome: Predicting isoform functions by a novel multiple instance-based label propagation method. Nucleic Acids Res. 42 (6), e39.

Lieberman-Aiden, E., *et al.*, 2009. Comprehensive mapping of long-range interactions reveals folding principles of the human genome. Science 326 (5950), 289–293.

Lindner, R., Friedel, C.C., 2012. A comprehensive evaluation of alignment algorithms in the context of RNA-seq. PLOS ONE 7 (12), e52403.

Liu, S., Altman, R.B., 2003. Large scale study of protein domain distribution in the context of alternative splicing. Nucleic Acids Res. 31 (16), 4828–4835.

Liu, H., *et al.*, 2017. Genome-wide analysis of protein-coding variants in leprosy. J. Investig. Dermatol. 137, 2544–2551.

Liu, H., *et al.*, 2014. Genome-wide linkage, exome sequencing and functional analyses identify ABCB6 as the pathogenic gene of dyschromatosis universalis hereditaria. PLOS ONE 9, e87250. doi:10.1371/journal.pone.0087250.

Luo, R., *et al.*, 2012. SOAPdenovo2: An empirically improved memory-efficient short-read de novo assembler. Gigascience 1 (1), 18.

Lykke-Andersen, S., Jensen, T.H., 2015. Nonsense-mediated mRNA decay: An intricate machinery that shapes transcriptomes. Nat. Rev. Mol. Cell Biol. 16 (11), 665–677.

Ma, L., *et al.*, 2015. LncRNAWiki: Harnessing community knowledge in collaborative curation of human long non-coding RNAs. Nucleic Acids Res. 43 (Database issue), D187–D192.

Marchini, J., *et al.*, 2007. A new multipoint method for genome-wide association studies by imputation of genotypes. Nat. Genet. 39, 906–913.

Marchini, J., Howie, B., 2010. Genotype imputation for genome-wide association studies. Nat. Rev. Genet. 11, 499–511.

Margulies, M., *et al.*, 2005. Genome sequencing in microfabricated high-density picolitre reactors. Nature 437 (7057), 376–380.

Martin, M., 2011. Cutadapt removes adapter sequences from high-throughput sequencing reads. EMBnet.journal 17 (1), 10–12.

Martin, P., *et al.*, 2015. Capture Hi-C reveals novel candidate genes and complex long-range interactions with related autoimmune risk loci. Nat. Commun. 6, 10069.

McKenna, A., *et al.*, 2010. The Genome Analysis Toolkit: A MapReduce framework for analyzing next-generation DNA sequencing data. Genome Res. 20 (9), 1297–1303.

McLean, C.Y., *et al.*, 2010. GREAT improves functional interpretation of cis-regulatory regions. Nat. Biotechnol. 28 (5), 495–501.

Medvedev, P., Stanciu, M., Brudno, M., 2009. Computational methods for discovering structural variation with next-generation sequencing. Nat. Methods 6 (Suppl. 11), S13–S20.

Melamud, E., Moult, J., 2009. Stochastic noise in splicing machinery. Nucleic Acids Res. 37 (14), 4873–4886.

Mendell, J.T., *et al.*, 2004. Nonsense surveillance regulates expression of diverse classes of mammalian transcripts and mutes genomic noise. Nat. Genet. 36 (10), 1073–1078.

Miller, J.R., Koren, S., Sutton, G., 2010. Assembly algorithms for next-generation sequencing data. Genomics 95 (6), 315–327.

Miller, W., *et al.*, 2004. Comparative genomics. Annu. Rev. Genom. Hum. Genet. 5, 15–56.

Mills, R.E., *et al.*, 2011. Mapping copy number variation by population-scale genome sequencing. Nature 470 (7332), 59–65.

Mostafavi, S., *et al.*, 2008. GeneMANIA: A real-time multiple association network integration algorithm for predicting gene function. Genome Biol. 9 (Suppl. 1), S4.

Mouse Genome Sequencing, C, *et al.*, 2002. Initial sequencing and comparative analysis of the mouse genome. Nature 420 (6915), 520–562.

Murvai, J., *et al.*, 2001. Prediction of protein functional domains from sequences using artificial neural networks. Genome Res. 11 (8), 1410–1417.

Myers, E.W., *et al.*, 2000. A whole-genome assembly of Drosophila. Science 287 (5461), 2196–2204.

Nakano, K., *et al.*, 2017. Advantages of genome sequencing by long-read sequencer using SMRT technology in medical area. Hum. Cell 30 (3), 149–161.

Ng, P.C., Henikoff, S., 2003. SIFT: Predicting amino acid changes that affect protein function. Nucleic Acids Res. 31 (13), 3812–3814.

Ning, S., *et al.*, 2016. Lnc2Cancer: A manually curated database of experimentally supported lncRNAs associated with various human cancers. Nucleic Acids Res. 44 (D1), D980–D985.

Nyren, P., Pettersson, B., Uhlen, M., 1993. Solid phase DNA minisequencing by an enzymatic luminometric inorganic pyrophosphate detection assay. Anal. Biochem. 208 (1), 171–175.

Paraskevopoulou, M.D., *et al.*, 2013. DIANA-microT web server v5.0: Service integration into miRNA functional analysis workflows. Nucleic Acids Res. 41 (Web Server issue), W169–W173.

Parra, G., Bradnam, K., Korf, I., 2007. CEGMA: A pipeline to accurately annotate core genes in eukaryotic genomes. Bioinformatics 23 (9), 1061–1067.

Patel, R.K., Jain, M., 2012. NGS QC Toolkit: A toolkit for quality control of next generation sequencing data. PLOS ONE 7 (2), e30619.

Patterson, S.E., *et al.*, 2016. The clinical trial landscape in oncology and connectivity of somatic mutational profiles to targeted therapies. Hum. Genom. 10, 4.

Paulsen, J., *et al.*, 2014a. HiBrowse: Multi-purpose statistical analysis of genome-wide chromatin 3D organization. Bioinformatics 30 (11), 1620–1622.

Paulsen, J., *et al.*, 2014b. A statistical model of ChIA-PET data for accurate detection of chromatin 3D interactions. Nucleic Acids Res. 42 (18), e143.

Perkel, J.M., 2013. Visiting "noncodarnia". Biotechniques 54 (6), 303–304.

Petrovski, S., *et al.*, 2013. Genic intolerance to functional variation and the interpretation of personal genomes. PLOS Genet. 9 (8), e1003709.

Phanstiel, D.H., *et al.*, 2015. Mango: A bias-correcting ChIA-PET analysis pipeline. Bioinformatics 31 (19), 3092–3098.

Picardi, E., Pesole, G., 2010. Computational methods for ab initio and comparative gene finding. Methods Mol. Biol. 609, 269–284.

Pickrell, J.K., *et al.*, 2010. Noisy splicing drives mRNA isoform diversity in human cells. PLOS Genet. 6 (12), e1001236.

Plewczynski, D., *et al.*, 2016. Chapter 3 – Analysis of Structural Chromosome Variants by Next Generation Sequencing Methods, in Clinical Applications for Next-Generation Sequencing. Boston: Academic Press, pp. 39–61.

Pollard, K.S., *et al.*, 2010. Detection of nonneutral substitution rates on mammalian phylogenies. Genome Res. 20 (1), 110–121.

Pombo, M.A., *et al.*, 2017. Use of RNA-seq data to identify and validate RT-qPCR reference genes for studying the tomato-Pseudomonas pathosystem. Sci. Rep. 7, 44905.

Pope, B.D., *et al.*, 2014. Topologically associating domains are stable units of replication-timing regulation. Nature 515 (7527), 402–405.

Pop, M., Kosack, D., 2004. Using the TIGR assembler in shotgun sequencing projects. Methods Mol. Biol. 255, 279–294.

Pott, S., Lieb, J.D., 2015. What are super-enhancers? Nat. Genet. 47 (1), 8–12.

Price, A.L., *et al.*, 2006. Principal components analysis corrects for stratification in genome-wide association studies. Nat. Genet. 38, 904–909.

Pritchard, J.K., Przeworski, M., 2001. Linkage disequilibrium in humans: Models and data. Am. J. Hum. Genet. 69, 1–14.

Prober, J.M., *et al.*, 1987. A system for rapid DNA sequencing with fluorescent chain-terminating dideoxynucleotides. Science 238 (4825), 336–341.

Quail, M.A., *et al.*, 2012. A tale of three next generation sequencing platforms: Comparison of Ion Torrent, Pacific Biosciences and Illumina MiSeq sequencers. BMC Genom. 13, 341.

Quek, X.C., *et al.*, 2015. lncRNAdb v2.0: Expanding the reference database for functional long noncoding RNAs. Nucleic Acids Res. 43 (Database issue), D168–D173.

Rao, S.S., *et al.*, 2014. A 3D map of the human genome at kilobase resolution reveals principles of chromatin looping. Cell 159 (7), 1665–1680.

Rehmsmeier, M., *et al.*, 2004. Fast and effective prediction of microRNA/target duplexes. RNA 10 (10), 1507–1517.

Resch, A., *et al.*, 2004. Assessing the impact of alternative splicing on domain interactions in the human proteome. J. Proteome Res. 3 (1), 76–83.

Revil, T., *et al.*, 2007. Protein kinase C-dependent control of Bcl-x alternative splicing. Mol. Cell Biol. 27 (24), 8431–8441.

Rhesus Macaque Genome, S., *et al.*, 2007. Evolutionary and biomedical insights from the rhesus macaque genome. Science 316 (5822), 222–234.

Rhoads, A., Au, K.F., 2015. PacBio sequencing and its applications. Genom. Proteom. Bioinform. 13 (5), 278–289.

Roadmap Epigenomics, C., *et al.*, 2015. Integrative analysis of 111 reference human epigenomes. Nature 518 (7539), 317–330.

Rogozin, I.B., *et al.*, 2002. Congruent evolution of different classes of non-coding DNA in prokaryotic genomes. Nucleic Acids Res. 30 (19), 4264–4271.

Romero, P.R., *et al.*, 2006. Alternative splicing in concert with protein intrinsic disorder enables increased functional diversity in multicellular organisms. Proc. Natl. Acad. Sci. USA 103 (22), 8390–8395.

Ronaghi, M., Uhlen, M., Nyren, P., 1998. A sequencing method based on real-time pyrophosphate. Science 281 (5375), 363–365.

Li, R., Qu, H., Wang, S., *et al.*, 2017. GDCRNATools: An R/Bioconductor package for integrative analysis of lncRNA, miRNA, and mRNA data in GDC. bioRxiv.

Sanger, F., Nicklen, S., Coulson, A.R., 1977. DNA sequencing with chain-terminating inhibitors. Proc. Natl. Acad. Sci. USA 74 (12), 5463–5467.

Sauria, M.E., *et al.*, 2015. HiFive: A tool suite for easy and efficient HiC and 5C data analysis. Genome Biol. 16, 237.

Schwartz, D.C., *et al.*, 1993. Ordered restriction maps of Saccharomyces cerevisiae chromosomes constructed by optical mapping. Science 262 (5130), 110–114.

Schwartz, S., *et al.*, 2003. Human-mouse alignments with BLASTZ. Genome Res. 13 (1), 103–107.

Schwarz, J.M., *et al.*, 2010. MutationTaster evaluates disease-causing potential of sequence alterations. Nat. Methods 7 (8), 575–576.

Servant, N., *et al.*, 2015. HiC-Pro: An optimized and flexible pipeline for Hi-C data processing. Genome Biol. 16, 259.

Shi, L., *et al.*, 2016. Long-read sequencing and de novo assembly of a Chinese genome. Nat. Commun. 7, 12065.

Shihab, H.A., *et al.*, 2013. Predicting the functional, molecular, and phenotypic consequences of amino acid substitutions using hidden Markov models. Hum. Mutat. 34 (1), 57–65.

Siepel, A., *et al.*, 2005. Evolutionarily conserved elements in vertebrate, insect, worm, and yeast genomes. Genome Res. 15 (8), 1034–1050.

Sikora-Wohlfeld, W., *et al.*, 2013. Assessing computational methods for transcription factor target gene identification based on ChIP-seq data. PLOS Comput. Biol. 9 (11), e1003342.

Simpson, J.T., *et al.*, 2009. ABySS: A parallel assembler for short read sequence data. Genome Res. 19 (6), 1117–1123.

Sjolander, K., 2004. Phylogenomic inference of protein molecular function: Advances and challenges. Bioinformatics 20 (2), 170–179.

Smith, T.F., Waterman, M.S., 1981. Identification of common molecular subsequences. J. Mol. Biol. 147 (1), 195–197.

Sommer, D.D., *et al.*, 2007. Minimus: A fast, lightweight genome assembler. BMC Bioinform. 8, 64.

Sudmant, P.H., *et al.*, 2015. An integrated map of structural variation in 2,504 human genomes. Nature 526 (7571), 75–81.

Sun, M., Kraus, W.L., 2015. From discovery to function: The expanding roles of long noncoding RNAs in physiology and disease. Endocr. Rev. 36 (1), 25–64.

Taylor, J., *et al.*, 2006. ESPERR: Learning strong and weak signals in genomic sequence alignments to identify functional elements. Genome Res. 16 (12), 1596–1604.

The Gene Ontology, C., C., 2017. Expansion of the Gene Ontology knowledgebase and resources. Nucleic Acids Res. 45 (D1), D331–D338.

Thibert, B., Bredesen, D.E., del Rio, G., 2005. Improved prediction of critical residues for protein function based on network and phylogenetic analyses. BMC Bioinform. 6, 213.

Tseng, Y.T., *et al.*, 2015. IIIDB: A database for isoform-isoform interactions and isoform network modules. BMC Genom. 16 (Suppl. 2), S10.

Tyekucheva, S., *et al.*, 2008. Human-macaque comparisons illuminate variation in neutral substitution rates. Genome Biol. 9 (4), R76.

Urbich, C., Kuehbacher, A., Dimmeler, S., 2008. Role of microRNAs in vascular diseases, inflammation, and angiogenesis. Cardiovasc. Res. 79 (4), 581–588.

Vacic, V., *et al.*, 2010. Graphlet kernels for prediction of functional residues in protein structures. J. Comput. Biol. 17 (1), 55–72.

Valouev, A., *et al.*, 2008. Genome-wide analysis of transcription factor binding sites based on ChIP-Seq data. Nat. Methods 5 (9), 829–834.

Vance, K.W., Ponting, C.P., 2014. Transcriptional regulatory functions of nuclear long noncoding RNAs. Trends Genet. 30 (8), 348–355.

Verspoor, K.M., *et al.*, 2012. Text mining improves prediction of protein functional sites. PLOS ONE 7 (2), e32171.

Voelkerding, K.V., Dames, S.A., Durtschi, J.D., 2009. Next-generation sequencing: From basic research to diagnostics. Clin. Chem. 55 (4), 641–658.

Vogan, K.J., Underhill, D.A., Gros, P., 1996. An alternative splicing event in the Pax-3 paired domain identifies the linker region as a key determinant of paired domain DNA-binding activity. Mol. Cell Biol. 16 (12), 6677–6686.

Volders, P.J., *et al.*, 2013. LNCipedia: A database for annotated human lncRNA transcript sequences and structures. Nucleic Acids Res. 41 (Database issue), D246–D251.

Wang, L., *et al.*, 2013. CPAT: Coding-Potential Assessment Tool using an alignment-free logistic regression model. Nucleic Acids Res. 41 (6), e74.

Whalen, S., Truty, R.M., Pollard, K.S., 2016. Enhancer-promoter interactions are encoded by complex genomic signatures on looping chromatin. Nat. Genet. 48 (5), 488–496.

Whirl-Carrillo, M., *et al.*, 2012. Pharmacogenomics knowledge for personalized medicine. Clin. Pharmacol. Ther. 92 (4), 414–417.

Whyte, W.A., *et al.*, 2013. Master transcription factors and mediator establish super-enhancers at key cell identity genes. Cell 153 (2), 307–319.

Winter, J., *et al.*, 2009. Many roads to maturity: MicroRNA biogenesis pathways and their regulation. Nat. Cell Biol. 11 (3), 228–234.

Xu, H., *et al.*, 2010. A signal-noise model for significance analysis of ChIP-seq with negative control. Bioinformatics 26 (9), 1199–1204.

Yaffe, E., Tanay, A., 2011. Probabilistic modeling of Hi-C contact maps eliminates systematic biases to characterize global chromosomal architecture. Nat. Genet. 43 (11), 1059–1065.

Yang, Y., *et al.*, 2017. Exploiting sequence-based features for predicting enhancer-promoter interactions. Bioinformatics 33 (14), i252–i260.

Yan, M., *et al.*, 2000. Two-amino acid molecular switch in an epithelial morphogen that regulates binding to two distinct receptors. Science 290 (5491), 523–527.

Yu, G., *et al.*, 2012. clusterProfiler: An R package for comparing biological themes among gene clusters. OMICS 16 (5), 284–287.

Zambelli, F., *et al.*, 2010. Assessment of orthologous splicing isoforms in human and mouse orthologous genes. BMC Genom. 11, 534.

Zerbino, D.R., Birney, E., 2008. Velvet: Algorithms for de novo short read assembly using de Bruijn graphs. Genome Res. 18 (5), 821–829.

Zhang, J., *et al.*, 2011. International cancer genome consortium data portal – A one-stop shop for cancer genomics data. Database (Oxford) 2011, bar026.

Zhang, Y., *et al.*, 2008. Model-based analysis of ChIP-Seq (MACS). Genome Biol. 9 (9), R137.

Zhao, Z., *et al.*, 2006. Circular chromosome conformation capture (4C) uncovers extensive networks of epigenetically regulated intra- and interchromosomal interactions. Nat. Genet. 38 (11), 1341–1347.

Zhao, M., *et al.*, 2013. Computational tools for copy number variation (CNV) detection using next-generation sequencing data: Features and perspectives. BMC Bioinform. 14 (Suppl. 11), S1.

Zhao, Y., *et al.*, 2016. NONCODE 2016: An informative and valuable data source of long non-coding RNAs. Nucleic Acids Res. 44 (D1), D203–D208.

Zhao, Z., *et al.*, 2015. Co-LncRNA: Investigating the lncRNA combinatorial effects in GO annotations and KEGG pathways based on human RNA-Seq data. Database (Oxford) 2015.

Zhou, X., *et al.*, 2013. Exploring long-range genome interactions using the WashU Epigenome Browser. Nat. Methods 10 (5), 375–376.

Zhou, V.W., Goren, A., Bernstein, B.E., 2011. Charting histone modifications and the functional organization of mammalian genomes. Nat. Rev. Genet. 12 (1), 7–18.

Zhou, X., Stephens, M., 2012. Genome-wide efficient mixed-model analysis for association studies. Nat. Genet. 44, 821–824.

Relevant Websites

https://www.bioinformatics.babraham.ac.uk/projects/fastqc/
   Babraham Bioinformatics.
http://evs.gs.washington.edu/EVS/
   Exome Variant Server.
http://hannonlab.cshl.edu/fastx_toolkit/
   FASTX-Toolkit
   Hannon Lab.
https://www.jax.org/research-and-faculty/tools/fusorsv
   FusorSV
   The Jackson Laboratory.
http://www.3dgenome.org
   Genome3D.
http://broadinstitute.github.io/picard/
   GitHub Pages.
http://www.novocraft.com
   Novocraft.
https://omictools.com/assembly-evaluation-category
   OMICtools.

Genome Annotation

Josep F Abril, University of Barcelona (UB), Barcelona, Spain and Institute of Biomedicine of the University of Barcelona (IBUB), Barcelona, Spain

Sergi Castellano, Great Ormond Street Institute of Child Health (ICH), University College London (UCL), London, United Kingdom and UCL Genomics, University College London (UCL), London, United Kingdom

"The genome sequence of an organism is an information resource unlike any that biologists have previously had access to. But the value of the genome is only as good as its annotation. It is the annotation that bridges the gap from the sequence to the biology of the organism." Lincoln Stein, 2001. Nat. Rev. Genet. 2(7), 493–503.

Introduction

The precise ordering of nucleotides and amino acids confer specific functional properties to biological sequences. Functional and structural features can then be understood as regions contained within those sequences, defined as segments delimited by an interval of positions, initial and terminal, on the coordinates axis provided by the sequence itself. In a wide sense, sequence annotation is defined by the procedures leading to determine the location and properties of these sequence features. Genomes are the organisms' blueprints, encoding for all the functional elements that lead to a fully developed life form. Therefore, genomes are complex and contain a mixture of features with different roles, which are interspersed with non-functional sequences under no evolutionary constraint. Genome annotation is the process of identifying any functional element along the DNA sequence of a genome, yet at initial stages often focuses on genes. The annotation gives meaning to the genome by providing the location and function of genes (protein coding or otherwise) and regulatory regions, which underlie the biology of living organisms. Completing the catalog of functional regions of the genome can facilitate further downstream analyses, as for instance the assessment of the effect of mutations caused by single nucleotide variants in individuals or populations. As it has been remarked since the start of the genomics era, raw genomes are thus of limited use to scientists and their annotation has become central to genome research.

Genome annotation itself has evolved over the years and it is possible to distinguish three, possibly four, major stages based on the techniques and data used. Each succeeding, yet overlapping, stage has refined previous annotations of the genome, characterized some of its previously unannotated parts, and provided novel insights into genome biology. The development in the mid 1990s of computational techniques to predict protein-coding genes, which span thousands of nucleotides in eukaryotes, marks the start of the efforts to provide genome-wide annotations. This first stage centered on the annotation of individual reference genomes (one per species), using a combination of the statistical properties of the sequences of the protein-coding genes themselves (to predict genes *ab initio*) and the similarity of mRNA and protein sequences to their genome of origin (to align known transcripts and proteins). Both approaches though rely on the existence of a set of already characterized genes, *a priori* knowledge based on experimental evidences, which will be used either as training set to build the models describing the genes species-specific signals and sequence content biases for subsequent computational predictions, or to be compared against other anonymous sequences to map their location based on similarity, thus annotating the genes in them (**Fig. 1**).

When the reference genomes of more species became available in the early and mid 2000s, they ushered in a comparative and second stage of genome annotation, where mRNA, protein and genome sequences from one species could be homologously aligned to annotate the genome of another. A third and regulatory stage of genome annotation was initiated in the mid and late 2000s, after genome-wide methods to profile DNA binding, conformation and alteration (*e.g.*, in their copy number) were introduced. Such genome-wide and multi-omics regulatory annotations have become standard in the last few years and short (in the tens or hundreds of nucleotides) regulatory elements are commonly annotated. Finally, genome annotation definitely moved away from one or few genomes per species in the late 2000s and early 2010s. This fourth and population stage of genome annotation, which uses hundreds if not thousands of genomes from the same species to annotate individual nucleotides, is where we find ourselves in the late 2010s. We discuss in some detail these different stages of genome annotation and how their increased resolution and inclusiveness of multi-omics approaches leads to more precise genome biology.

The *ab Initio* and Similarity Years

Ab initio gene prediction (also known as *de novo*) refers to the identification of protein-coding genes using the signals that define and characterize their gene structures along the genome. This is done without the aid of their encoded mRNAs and proteins. When the sequences of these mRNAs and proteins are sequenced they can be aligned to the genome to locate the genes that encode them. Thus, allowing the prediction of protein-coding genes by similarity. We discuss both approaches and the increased accuracy of similarity approaches over the *ab initio* ones.

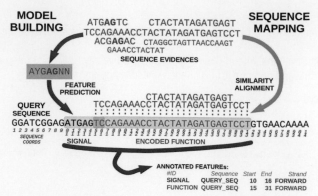

Fig. 1 In order to annotate an anonymous sequence (query) we have two main approaches. The first approach, depicted on the left, it is to build models that, based on known sequences, summarize some property or bias characteristic of the functional elements being annotated, and then apply an algorithm to score these elements along the sequence, taking the optimal prediction as the one to annotate. The second approach uses the alignment against the query of the set of known sequences, retrieved either from the same (similarity search) or other species (homology search). Again, the optimal alignment is taken as the one to annotate the functional element. The coordinates – relative to the original sequence – define the location for each of the functional elements found in the anonymous genome, also known as the annotation set.

Ab Initio Gene Prediction

The structure of prokaryotic and eukaryotic genes is such that start (AUG) and stop (UAA/UAG/UGA) codons signal the initiation and termination steps of protein translation. These codons, together with some nucleotide context around them, define the coding DNA sequence (CDS) of the exon or exons in a gene. In eukaryote genomes, in particular, codons in exons along the CDS are not necessarily adjacent but often interrupted by introns of hundreds or thousands of nucleotides in length (Sharp, 2005). These breaks are defined, in most cases, by the canonical donor [AG｜GTRAGT, where R=A or G] and acceptor [YYTTYYYYYYNCAG｜G, where Y=C or T and N=A, C, T or G] splice sites in mammalian genomes (Burset *et al.*, 2000; Abril *et al.*, 2005). Albeit GT-AG and the surrounding context defines a highly conserved motif, splicing is a complex biological process that tolerates small nuclear RNA variants in the spliceosome (Kyriakopoulou *et al.*, 2006), the RNA-protein complex that mediates the splicing process, and even non-canonical splice sites, such as those defined by the donor and acceptor pair AT-AC (Burset *et al.*, 2000; Patel and Steitz, 2003). Together with transcription initiation and termination sites, these signals have allowed the computational definition of prokaryotic and eukaryotic genes.

In addition, the genetic code is degenerate and the same amino acid can be encoded by different codons. These codons are synonymous and the frequency of their occurrence is known as codon usage or bias. Codon usage varies between species and between genes in a genome due to mutational biases and differences in the strength of natural selection on translational optimization (Plotkin and Kudla, 2011). Predating the sequencing of genomes there were tools that used codon bias to annotate expressed sequence tags (ESTs), fragments of mRNAs, which were even capable to correct frameshifts from sequencing errors like ESTSCAN (Iseli *et al.*, 1999). Thus, codon usage biases within the signals that define prokaryotic and eukaryotic genes can be used to computationally identify the CDS of genes along a particular species genome. Some of the first *ab initio* programs to predict genes this way did so in the short genomes of prokaryotes. For example, the seminal GENEMARK (Lukashin and Borodovsky, 1998) and GLIMMER (Delcher *et al.*, 1999) programs made use of Markov chains – a stochastic model describing a sequence of nucleotide-emitting states in which the probability of the next state only depends on the previous one – to capture the usage and dependence between successive nucleotides and codon positions, while defining the start and end of genes with models of the nucleotides around these signals (their sequence context). Interestingly, Markov chains of order-five, in which the probability of the next nucleotide depends on the five previous ones, work best in gene prediction as they also capture dependencies between consecutive amino acids in proteins (Durbin *et al.*, 1998). These models are used to score potential genes along the genome. The final prediction is then obtained using algorithms that find the optimal gene structure, that is, the combination of predicted exons that maximizes the score of the predicted genes. GENEMARK, for example, uses the Viterbi algorithm (Durbin *et al.*, 1998). This is a dynamic programming algorithm (Eddy, 2004a) – an optimization technique based on breaking down a problem into smaller ones and using them recursively to find the optimal solution – which, in a Hidden Markov Model (HMM) (Eddy, 2004b) – a statistical model following a Markov Chain process with hidden states that either emit nucleotides for coding or noncoding sequences – finds the best scoring gene structures. More recent developments include Prodigal (Hyatt *et al.*, 2010), which extends dynamic programming with a special set of rules to cope with overlapping genes while looking for the optimal gene structure prediction along the genome.

Hyatt *et al.* also established a reference annotation set to evaluate the accuracy of available prokaryotic gene-finders; a difficult task due to the lack of an experimentally verified translation start in many genes that changes the average length of genes, as well as the huge variability in GC content across prokaryota. The accuracy of gene prediction in prokaryotic genomes is generally high, with more than 95% of true protein coding genes being detected in most species (Hyatt *et al.*, 2010). Though the accuracy predicting the start of the gene is lower, usually around 80%. As a result, the 5′ end (upstream) of prokaryotic genes is less likely to be correct. In addition, prokaryotic gene prediction may still slightly overpredict the number of genes in a genome, with the resulting false positive genes being often short. It is however possible that some of these genes are indeed real. In any case, prokaryotic gene prediction has become remarkably accurate in the last two decades.

The identification of eukaryotic genes is, compared to their prokaryotic counterparts, more demanding due to their length (often in the hundreds of thousands of bases) and their split structure. Still, similar approaches to those in prokaryotes can be used. Early eukaryotic gene predictors included GENEID (Guigo *et al.*, 1992) and GENSCAN (Burge and Karlin, 1997). Both of these programs use Markov Chains of order five to assess codon usage and score coding exons as a combination of their coding bias and their start, end and splice sites signals; either in a rule-based (the GENEID Gene Model; (Guigo *et al.*, 1992)) or a generalized HMM (GHMM) framework for GENSCAN (Burge and Karlin, 1997). To cope with the combinatorial explosion of putative exons and predict the optimal gene structure along the genome dynamic programming was used (Burge and Karlin, 1997; Guigo, 1998).

The accuracy of such predictions is lower than in prokaryotic genomes (**Table 1**). An influential work on the assessment of gene prediction accuracy in eukaryotes (Burset and Guigo, 1996) established the ground metrics at the nucleotide, exon and gene levels, and provided one of the first reference gene test sets on this field. To begin with, there is in gene prediction a compromise between sensitivity (*Sn* in **Table 1**), the proportion of true coding nucleotides correctly predicted as such, and specificity (*Sp* in **Table 1**), the proportion of predicted coding nucleotides actually coding, so increasing one often decreases the other. At the nucleotide level this is not much of an issue, as fragments of most genes are predicted; yet missing the right start and end of an exon reduces the sensitivity and specificity of gene prediction at the exon level (**Table 1**). The accuracy for the overall gene is even lower due the difficulty in determining the right combination of exons along the genome. On the other hand, short genes encoding small proteins (Su *et al.*, 2013) are usually missed (missing exons, ME, in **Table 1**) as most genome annotation protocols apply a minimum length threshold, *i.e.*, of 100 amino acids, to reduce the number of falsely predicted genes (wrong exons, WE, in **Table 1**). These measures show that classic *ab initio* gene prediction programs identified no more than 80% of the true coding nucleotides while predicting as coding a large fraction of noncoding ones (**Table 1**). Also, a substantial fraction of coding exons had their boundaries mispredicted. Furthermore, around 40% of the predicted exons are completely wrong, a rather unreasonable figure. Still, *ab initio* gene prediction provided at the time a first set of annotations that could be later refined using comparative and/or experimental approaches (see below). To cope with annotation changes and multiple transcripts in a gene further accuracy measures have been proposed such as the Annotation Edit Distance, which quantifies changes to a gene annotation, and the Splice Complexity, which quantifies the complexity of alternative splicing in a gene (Eilbeck *et al.*, 2009). The terms recall and precision were later adopted to measure gene prediction accuracy. Recall is equivalent to the sensitivity measure described above, whereas precision is equivalent to specificity. Importantly, specificity or precision in the gene prediction field avoid the estimation of true negatives from incomplete genomes. Nowadays gene predictions are also compared with respect to the area under the curve in a receiver operating characteristics plot (known as AUC-ROC plot) (Powers, 2011).

The first large-scale assessment on gene-finding accuracy was performed over 2.4 megabases of the *Drosophila melanogaster* genome, which contained the *alcohol dehydrogenase* gene (*Adh*), in a collaborative experiment known as the Genome Annotation Assessment Project (GASP) (Reese *et al.*, 2000). This region was first annotated by human curators, with the raw sequence and a subset of the reference fly annotations later provided to the participants performing computational gene predictions. Similar assessments have subsequently been done in the worm (nGASP) (Coghlan *et al.*, 2008) and human genomes (EGASP and RGASP, see **Figs. 2** and **3**) (ENCODE Consortium, 2012; Guigo *et al.*, 2006).

Human curators have been invaluable in producing reference annotations for the above assessment experiments and in integrating the experimental evidence that validates (or refutes) individual gene predictions. For example, *D. melanogaster* was the first eukaryotic genome assembled from shotgun sequences (where DNA is broken up randomly) (Adams *et al.*, 2000) and Celera, the company sequencing it, organized an annotation jamboree with bioinformatics experts, protein specialists, and fruit fly biologists to functionally annotate it (Pennisi, 2000). Similar community annotation projects have since then become a standard for other genomes, among those it is worth citing the VERtebrate Genome Annotation database (VEGA, http://vega.sanger.ac.uk) maintained by the human and vertebrate analysis and annotation (HAVANA) group at the Wellcome Trust Sanger Institute (Harrow *et al.*, 2012), which also has defined guidelines to integrate gene annotations (Madupu *et al.*, 2010; Loveland *et al.*, 2012). To assist curators in the visualization and editing of gene annotations different tools have been created. First, to produce static maps of annotations along a genome sequence with GFF2PS (Abril and Guigo, 2000), GFF2APLOT (Abril *et al.*, 2003), and more recently CIRCOS (Krzywinski *et al.*, 2009). Later, tools to manually review gene predictions and integrate experimental evidence, such as APOLLO (Lewis *et al.*, 2002), the Integrative Genomics

Table 1   Accuracy of gene prediction programs on human chromosome 22

Program	Nucleotide			Exon				
	Sn	Sp	CC	Sn	Sp	$\frac{Sn+Sp}{2}$	ME	WE
"*ab initio*" gene finding								
GENEID	0.73	0.67	0.70	0.65	0.55	0.60	0.21	0.33
GENSCAN	0.79	0.53	0.64	0.68	0.41	0.55	0.15	0.48
Comparative genomics approach								
SGP2	0.75	0.73	0.73	0.66	0.58	0.62	0.19	0.28
TWINSCAN	0.72	0.67	0.69	0.69	0.59	0.64	0.18	0.29

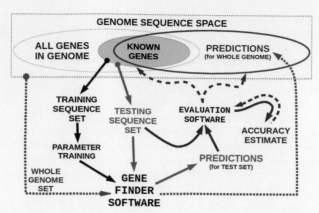

Fig. 2 The assessment of the accuracy of gene predictions (or other functional features) is usually done in three steps. First, a set of well characterized genes is split into the training and the testing sequence sets. Second, the training set is used to estimate the species-specific parameters, such as the frequencies of the different codons in coding exons. Finally, the test set is used to predict genes with the computational approach under assessment. The sequence coordinates of the predicted genes are then compared against those of the known genes. Accuracy metrics, calculated at the nucleotide, exon and gene levels, are used to assess accuracy of the method and the reliability of the predictions. This assessment can be used later on either to estimate overall genome predictions reliability or to iterate through the train/test sets again in order to improve the software parameters (dashed lines).

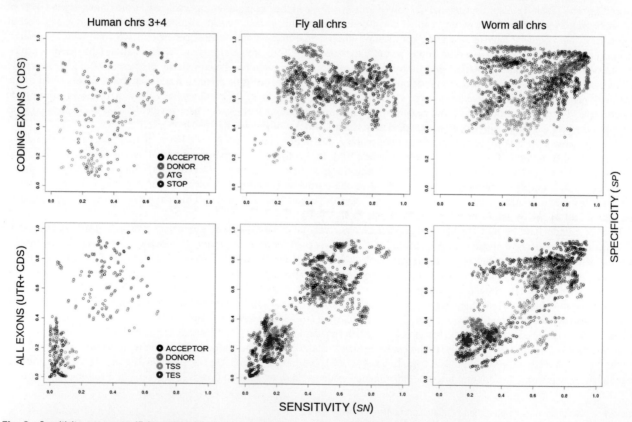

Fig. 3 Sensitivity versus specificity scatterplots for the prediction of a number of defining signals in coding exons (CDS), on top row, and full-length exons, on bottom row, for the submitted predictions to RGASP. The accuracy metrics were computed over the genome sequences from three species: the human chromosomes three and four, the *Drosophila melanogaster* (fly) genome and the *Caenorhabditis elegans* (worm) genome. Each dot corresponds to a set of predictions made by one of each RGASP participant over one of the species sequences. Note the lower accuracy of predictions in the larger human genome as well as the difficulty to predict the untranslated part of exons due to missing the correct transcription start (TSS) and end (TES) sites in the three species.

Viewer (IGV) (Thorvaldsdottir *et al.*, 2013) and the GALAXY server (Giardine *et al.*, 2005) grew in importance. Over time, genome browsers providing access to databases storing the analyses from genome annotation pipelines have become the standard. These include GBROWSE (Stein *et al.*, 2002), the UCSC Genome Browser (Kent *et al.*, 2002) and ENSEMBL (Hubbard *et al.*, 2002; Aken *et al.*, 2017), which are widely used today (**Fig. 4**).

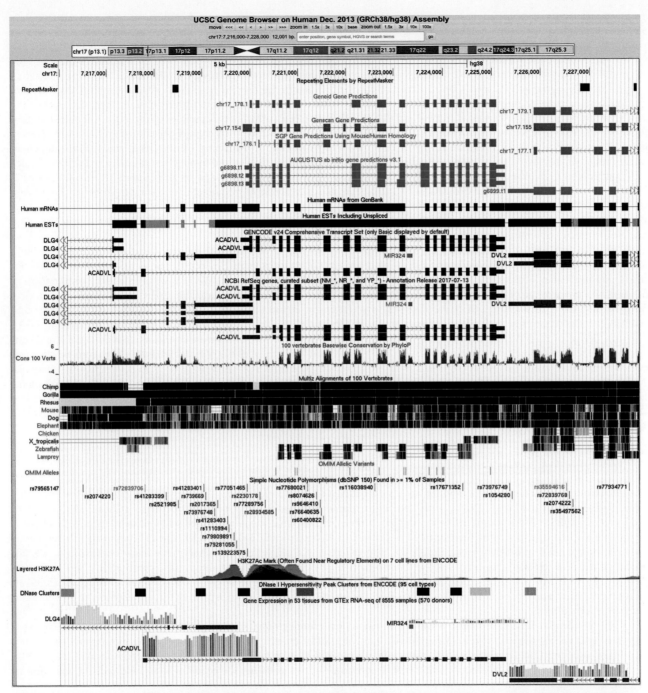

Fig. 4 UCSC Genome Browser view of the human chromosome 17, focused on the sequence annotation around the *ACADVL* gene. Topmost annotation tracks show several gene/transcript structures obtained by different methods; mid tracks summarize the conservation score and the human sequence aligned to different vertebrate genomes; bottom tracks include sequence variants (OMIM and common variants), histone marks from the ENCODE project and transcript gene expression. The gene annotation tracks on top displays predictions from GENEID, GENSCAN, SGP2, and AUGUSTUS, which can be compared with the current curated annotations from two reference sets, GENCODE and NCBI-RefSeq. From the aforementioned gene-finding programs, only AUGUSTUS predicts the three alternative spliced transcripts as it integrates experimental evidence, although it fails to predict some conserved exons detected by the other programs. Note that downstream *DVL2* gene was also detected by the gene-finders, but the upstream *DLG4* gene was missed.

Using mRNAs and Proteins

The sequencing of fragments (ESTs) or full-length mRNAs and proteins from the species whose genome is being annotated has been the main approach to increase the accuracy of *ab initio* gene annotation. Instead of relying on the statistical properties of the CDS and the signals that define them, the direct alignment of transcribed or translated sequences from a genome provides experimental evidence to a gene structure. The alignment of such sequences to a naked genome has been computationally addressed yet again as a dynamic programming algorithm that takes into account the splice sites and exon/intron structure of eukaryotic genes. In particular, ESTs or full-length transcripts can be properly aligned to the genome using variations of the Smith-Waterman (Smith and Waterman, 1981; Gotoh, 1982) and Needleman-Wunsch (Needleman and Wunsch, 1970) algorithms, the classic local and global sequence alignment algorithms, respectively. This is the approach used by the pioneering EST_GENOME program (Mott, 1997). For the sake of speed, other approaches use the heuristic BLAST algorithm (Altschul *et al.*, 1990) to produce the starting alignments of coding exons in the genome, which are later recursively chained together to obtain the best gene structure. SIM4 (Florea *et al.*, 1998) and SPIDEY (Wheelan *et al.*, 2001) were examples of the latter. A more recent tool to align transcripts to a genomic sequence is EXONERATE (Slater and Birney, 2005), which provides (slower) exhaustive or (faster) heuristic approaches to align transcripts using different dynamic programming algorithms. On the other hand, spliced-alignment algorithms with proteins require the conceptual translation of the genomic sequence and finding the optimal alignment of a multi-exon structure to a related protein. PROCRUSTES (Gelfand *et al.*, 1996) and GENEWISE (Birney and Durbin, 1997, 2000) are well-known examples, with the HMM-based GENEWISE at the core of the ENSEMBL gene annotation system (Curwen *et al.*, 2004; Aken *et al.*, 2016). The alignment of proteins to predict genes in large scale genomes can become extremely time- and space-demanding. This investment, however, is usually at the benefit of prediction accuracy. In this regard, the faster EXONERATE can also align proteins to DNA sequences and is also heavily used in the ENSEMBL pipeline.

In the last few years, the widespread use of next-generation sequencing technologies for genomes and transcriptomes (RNA sequencing, RNA-seq) and high-throughput mass spectrometry approaches for proteins have put the approaches above at the forefront of genome annotation. For example, the alignment of transcripts from RNA-seq experiments has allowed the comprehensive annotation of alternative splice forms of protein-coding genes. Humans, for example, produce on average four different protein-coding sequences per gene (ENCODE Consortium, 2012). Pseudogenes, which derived from functional genes through retrotransposition or duplication but have lost the original functions of their parental genes, are sometimes (about 10%) also transcribed (ENCODE Consortium, 2012). The high degree of conservation and similarity to functional genes of recent pseudogenes are significant enough to confound conventional gene prediction approaches (Zheng *et al.*, 2007). Other non-standard types of genes that have benefited from experimental evidence are fused genes, which combine part or all of the exons from transcripts of two collinear genes. In this way, they may contribute to the increase the protein repertoire in eukaryotes (Parra *et al.*, 2006). Similarly, trans-splicing, where the starting exons from one transcript are merged with exons from another transcript, far downstream or even located in another chromosome sequence, can be annotated using sequenced transcripts. Remarkably, up to 70% of the worm *Caenorhabditis elegans* mRNAs begin with the spliced leader sequence (SL, 22bp), which is not associated with the gene (Hastings, 2005).

Annotation of Non-Standard Protein Coding Genes

The identification of eukaryotic selenoprotein genes has challenged most computational gene prediction methods. The reason for this is the alternative use of the UGA codon, typically a termination signal, to code for selenocysteine (Chambers *et al.*, 1986; Zinoni *et al.*, 1986). The recoding of UGA confounds the computational identification of selenoprotein genes, whose exons containing a selenocysteine codon are truncated or skipped in their prediction. Selenocysteine, the 21st amino acid in the genetic code, is incorporated into selenoprotein with the help of an mRNA element, the selenocysteine insertion sequence (SECIS), located in the 3′ untranslated region of selenoprotein genes (48,49). This RNA structure directs the incorporation of selenocysteine through the interaction with several trans-acting factors. Selenocysteine mediates the biological functions of selenium, an essential micronutrient whose deficiency is associated to infertility, preterm birth and serious children's disorders of the bone and heart (Rayman, 2000).

Efforts to identify selenoproteins were first directed to the computational prediction of the SECIS element in Expressed Sequence Tags (ESTs, sequenced mRNA fragments). Known SECIS elements were used to derive deterministic models of their secondary structures (**Fig. 5**) which, in turn, were used to scan and identify two novel selenoprotein genes (Kryukov *et al.*, 1999; Lescure *et al.*, 1999). Efforts to identify additional selenoprotein genes in eukaryotic genomes soon followed. The inclusion of UGA as a selenocysteine codon in the prediction of optimal gene structures along the genome, however, led to the prediction of many erroneous selenoprotein genes (false positives). The low specificity of selenoprotein gene prediction was due to the difficulty of using the typical codon usage characteristic of proteins to extend them beyond an in-frame termination codon (Castellano *et al.*, 2001). A solution to this problem was to use the prediction of SECIS elements along the genome to limit the number of exons with a TGA in-frame that are incorporated into the dynamic programming recursion used to predict optimal genes. Several new selenoproteins in different species were identified this way (Castellano *et al.*, 2001; Kryukov *et al.*, 2003; Castellano *et al.*, 2005). Another example of stop codon recoding is pyrrolysine, which is encoded by UAG but may not require a specific stem loop structure in the 3′ untranslated region for translation (Zhang *et al.*, 2005).

The Comparative and Homology Years

The Zoo Blot is an experimental technique which uses Southern blot analysis to test the ability of a nucleic acid probe from one species to hybridize with DNA of a range of species. Zoo Blots were already used in the 1980s to characterize the protein coding

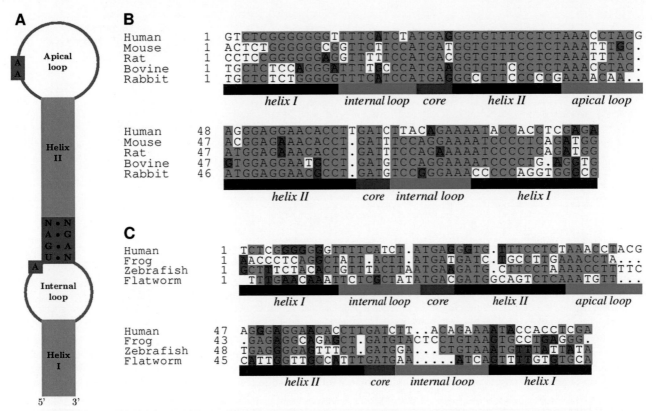

Fig. 5 (a) Schematic SECIS element (form 1) divided into structural units. Mostly invariant nucleotides are indicated. (b) Alignment of human and other mammalian glutathione peroxidase 1 SECIS sequences. Note the strong conservation among orthologous sequences. (c) Alignment of human and other non-mammalian glutathione peroxidase 1 SECIS sequences. Compared to the previous alignment, sequence conservation drops significantly.

regions of newly discovered genes (Monaco *et al.*, 1986; Katzav *et al.*, 1989). The rationale behind this approach was that protein coding regions will be conserved across species and would thus retain the capability to hybridize. Testing for hybridization of genomic probes to genomic libraries from many species will reveal the fragments likely to correspond to coding exons. As often happens, bioinformatics techniques based on sequence similarity replicate in the computer those experimental techniques based on the hybridization properties of nucleic acid molecules. In this regard, comparative gene prediction methods that rely on the fact that protein coding regions are generally more conserved during evolution that non-functional ones (**Fig. 6**), can be though as a sort of electronic zoo blots. While phylogenetic footprinting – the phenomenon of the conservation of functional regions across species – has been used to discover, identify and characterize functional regions other than protein coding genes (Margulies *et al.*, 2003).

Essentially, there are two main classes of comparative approaches for gene identification: the comparison of the DNA query sequence with a target protein or mRNA sequence, or a database of such sequences (the computational equivalent of a Northern zoo blot), and the comparison of two or more genomic sequences. In both approaches query and target sequences may be from the same (as discussed above) or different species. If the latter, the selection of the target organism is not a negligible step; the organism should be chosen, when possible, at the phylogenetic distance maximizing the correlation between sequence conservation and coding function. Thus, allowing the inference of homology.

Using mRNAs and Proteins

The backbone of similarity-aided or similarity-based gene structure determination is constituted by those methods that rely on a comparison of the query sequence with protein, or mRNA sequences. Although mostly known as a database search program, BLASTX (Altschul *et al.*, 1990; Gish and States, 1993) illustrates the rationale behind this approach. With BLASTX a genomic query is translated into a set of amino acid sequences in the six possible frames and compared against a database of known protein sequences. The assumption is that those segments in the genomic query similar to database proteins are likely to correspond to homologous coding exons. A similar assumption is behind the comparison of the genomic query against a database of mRNA sequences (such as ESTs), using BLASTN (Altschul *et al.*, 1990), FASTA (Pearson and Lipman, 1988; Pearson, 2000), or similar programs. The National Center for Biotechnology Information GENBANK (Benson *et al.*, 2013) and UNIPROT (Bateman *et al.*, 2017), and their European and Asian counterparts, have been the main sources of nucleotide and protein sequences, respectively, for homology-based searches.

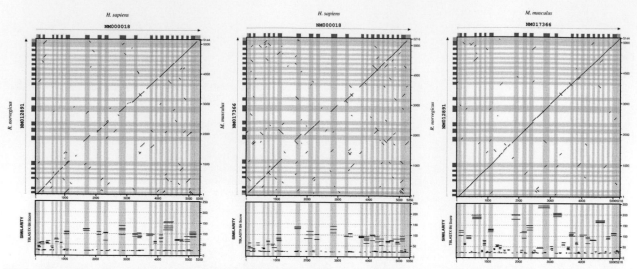

Fig. 6 Comparison of the orthologous genomic region of the acyl-Coenzyme A dehydrogenase very long chain (*ACADVL*) in human, mouse, and rat. The comparison was done at the amino acid level using the ungapped TBLASTX. Coding exons are shown as red boxes and non-coding ones (untranslated regions) are depicted in grey. Exons are projected to highlight their conservation (orange/grey ribbons). Note that introns are also conserved, albeit to a lesser extent than coding exons between human and any of the rodent genomes. The plot was obtained with GFF2APLOT.

Database search programs, however, are not dedicated gene prediction tools; they are not capable of automatically identifying start and stop codons or splice sites, the signals defining the exonic structure of the genes. Thus, after a database search and the identification of potential targets of homology, additional tools are required to define these structures. One approach is to use the top database match as target sequence and obtain a so-called spliced alignment between this and the genomic query. In such an alignment, large gaps – likely to correspond to introns – are only allowed at legal splice junctions. To calculate such spliced alignments between transcript and genomic sequences several programs we have already mentioned have been used: SIM4 (Florea *et al.*, 1998), EST_GENOME (Mott, 1997), SPIDEY (Wheelan *et al.*, 2001), and EXONERATE (Slater and Birney, 2005). Regarding the alignment of proteins of one species to another species genome, GENEWISE (Birney and Durbin, 1997, 2000) and EXONERATE (Slater and Birney, 2005) are used by ENSEMBL. In this direction, Meyer and Durbin (Meyer and Durbin, 2004) have developed the program PROJECTOR, which makes explicit use of the conservation of the exon-intron structure between two related genes. Because human and mouse orthologous genes show a remarkable conservation of their exonic structure, (Mouse Genome Sequencing Consortium, 2002), the PROJECTOR program outperforms GENEWISE predictions when human targets are used in mouse genomic sequences, and vice versa, in particular, when the conservation at the amino level is weak.

In an alternative approach, the results of a database search can be integrated *ad-hoc* into the framework of a typical *ab initio* gene prediction program. In essence, these methods promote candidate exons in the query sequence for which similar known coding sequences exist. Indeed, the score of the candidate exon – initially a function of the score of the splice (start, stop) sites and of the coding potential of the exon sequence – is increased as a function of the similarity between the candidate exon and the known coding sequences. In this way, candidate exons showing similarity to known coding sequences and thus likely orthologous, are promoted into the final gene prediction. In theory, this approach should produce predictions as accurate as pure *ab initio* programs when no similar target sequences exist, but more accurate ones (ideally, as accurate as those from splicing alignment tools) when such target sequences do exist. One example of this approach is the program GENOMESCAN (Yeh *et al.*, 2001), an extension of GENSCAN (Burge and Karlin, 1997) over which it reports increased accuracy. GRAILEXP (Xu *et al.*, 1997), CRASA (Chuang *et al.*, 2003) and AUGUSTUS (Stanke *et al.*, 2006a) are examples of methods using ESTs or mRNAs instead.

Using Whole-Genomes

With the increasing availability of the genome sequences for a wide range of eukaryotic organisms, whole genome sequence comparisons gained popularity as a mean of identifying protein coding genes. Under the assumption that regions conserved in genome sequence alignments will tend to correspond to coding exons from homologous genes, a number of programs were developed. The program EXOFISH (Crollius *et al.*, 2000) was one of the first such programs. Basically, it predicted human exons aided by the comparison of the human genome with sequences from *Tetraodon nigroviridis*, a puffer fish species (within Actinopterygii) that diverged about 400 Myr ago from the lineage later leading to humans. Latter developments followed notably different approaches.

In one such approach (Pedersen and Scharling, 2002; Blayo *et al.*, 2003), the problem was stated as a generalization of pairwise sequence alignment: given two genomic sequences coding for homologous genes, the goal was to obtain the predicted exonic structure in each sequence maximizing the score of the alignment of the resulting amino acid sequences. Both, (Blayo *et al.*, 2003) and Pedersen and Scharling (Pedersen and Scharling, 2002) solved this problem through a complex extension of the classical dynamic programming algorithm for sequence alignment.

In a different approach, the programs SLAM (Alexandersson *et al.*, 2003) and DOUBLESCAN (Meyer and Durbin, 2002) combined sequence alignment pair HMMs (Durbin *et al.*, 1998), with gene prediction generalized HMMs (Burge and Karlin, 1997) into the so-called generalized pair HMMs. In these, gene prediction is not the result of the sequence alignment, as in the programs above, but both gene prediction and sequence alignment are obtained simultaneously.

A third class of programs adopted a more heuristic approach, and clearly separated gene prediction from sequence alignment. The programs ROSETTA (Batzoglou *et al.*, 2000), SGP1 (from Syntenic Gene Prediction, (Wiehe *et al.*, 2001)), and CEM (from the Conserved Exon Method, (Bafna and Huson, 2000)) are representative of this approach. All these programs start by aligning two syntenic sequences – understood here as sequences from different species, but containing homologous genes in a similar order–, and then predict gene structures in which the exons are compatible with the alignment.

Although similarity-based gene prediction with homologous genomic sequences produced high quality results (Miller, 2001), an obvious shortcoming was the need for two homologous sequences. Also, genes without a homologue in the partner sequence will escape detection. This is particularly problematic if species are compared at genomic regions where synteny is not preserved. Given only a single query sequence, it is therefore desirable to automatically search for homologous sequences or syntenic chromosome stretches in other species that are suited to similarity-based approaches. The programs TWINSCAN (Korf *et al.*, 2001) and SGP2 (Parra *et al.*, 2003) attempted to address this limitation. The approach in these programs was reminiscent of that used in GENOMESCAN (Yeh *et al.*, 2001). Essentially, the sequence from the query genome is compared against a collection of sequences from the target or informant genome – which can be a single homologous sequence to the query sequence, a whole assembled genome, or a collection of shotgun reads –, and the results of the comparison are used to modify accordingly the scores of the exons produced by *ab initio* gene predictors. In TWINSCAN, the query and target genome sequences are compared at the nucleotide level with BLASTN and the results serve to modify the underlying probability of the potential exons predicted by GENSCAN. In SGP2, the genome sequences are compared using TBLASTX (Altschul *et al.*, 1990), and the results used to modify the scores of the potential exons predicted by GENEID.

TWINSCAN, SGP2, and SLAM were successfully applied to the annotation of the mouse genome (Mouse Genome Sequencing, C, 2002), and helped to identify previously unconfirmed genes (Guigo *et al.*, 2003). Importantly, a number of studies consistently indicate that comparative gene finders improve over their *ab initio* counterparts (Parra *et al.*, 2003; Wu *et al.*, 2004). Indeed, in **Table 1**, we show that, when evaluated in human chromosome 22, TWINSCAN and SGP2 outperformed GENSCAN and GENEID, respectively. The exhaustive scrutiny to which the sequence of human chromosome 22 (Dunham *et al.*, 1999) was subjected through the Vertebrate Genome Annotation database project at the Wellcome Sanger Institute offers an excellent platform to obtain estimates of the accuracy of current gene finders.

Annotating a genome has required a complex craftsmanship from its inception, given the many different layers of annotations that have to be considered and therefore put together. Common to the tools and approaches described above is the first task of masking the repetitive sequences, often taxa-specific, that populate the genome; this step facilitates and speeds up the downstream analyses just by replacing the segments where the repeats are found by N's, a common nucleotide wildcard – meaning A or C or G or T–. RepeatMasker (Smit *et al.*, 2013-2015), has been the classical tool for this purpose, yet it has been described as sensitive but computationally costly when applied to large eukaryotic genomes (Bedell *et al.*, 2000). After repeat-masking, genome sequences are scanned for protein-coding and non-coding genes using different prediction tools while, simultaneously, homology-based searches and whole-genome alignments to other species are performed. The result are sets of annotations with varying degrees of supporting evidence that need to be integrated into one set of annotated genes, also known as the gene-build of a genome. Additional genome annotation layers result from high-throughput experimental approaches, such as RNA-seq and the transcriptional evidence it produces (which may be used on the gene-build itself), or the typing and resequencing experiments that sample sequence variation across populations (human or otherwise), and so on. Genes from the gene-build, and the proteins they encode, require these other annotation layers to assign them a molecular function, a process known as functional annotation. These functions rely on controlled vocabularies to describe them, for instance the Gene Ontology (GO) (Ashburner *et al.*, 2000; The Gene Ontology Consortium, 2017).

As the complexity of the annotation process increased due to the need to integrate new layers of information, guidelines and/ or automatic or semiautomatic workflows to annotate novel genomes or to re-annotate existing ones (every time the assemblies are improved) were developed (Mudge and Harrow, 2016; Elsik *et al.*, 2014). For example, the NCBI prokaryotic (Tatusova *et al.*, 2016) and eukaryotic (Thibaud-Nissen *et al.*, 2016) pipelines, or EnsEMBL (Potter *et al.*, 2004), which rely on in-house protocols for large-scale genome annotation. In addition, gene-finding tools were wrapped into specialized genome annotation workflows. For instance, the Fgenesh (Salamov and Solovyev, 2000), SNAP+Exonerate, and GeneMark+AUGUSTUS were integrated into the Fgenesh++ (Solovyev *et al.*, 2006), MAKER (Cantarel *et al.*, 2008), and BRAKER (Hoff *et al.*, 2016) annotation pipelines, respectively.

Comparative genomics can be also useful to assess the completeness of genome annotations. REFSEQ (Pruitt *et al.*, 2012) has been used as a reference annotation set for the annotation of novel gene-builds, but also to validate the completeness of genome assemblies. A set of 458 highly reliable core proteins, conserved across plants, fungi and mammals, was collected and integrated into a gene-finding validation pipeline known as CEGMA (Parra *et al.*, 2007). More recently, it has been superseded by a set of single-copy orthologs in the BUSCO pipeline (Simao *et al.*, 2015), which can be used to benchmark the completeness of both the annotation of genes and the genome assembly.

Annotation on Non-Standard Protein-Coding Genes

Comparative methods for gene prediction have been also useful to annotate selenoprotein genes. The fact that downstream of a recoded stop codon lies a real protein implies this region to have the typical pattern of sequence conservation among its protein homologues. At the least, to the extent of the conservation in the upstream protein sequence. Therefore, sequence conservation beyond a TGA codon in the genome strongly suggests selenocysteine coding function. It is then feasible to predict selenoprotein genes in two different species genomes, without information from their SECIS elements, and compare them to identify a pattern of symmetrical protein sequence conservation around a predicted selenocysteine codon. Application of this approach to the human and fugu genomes identified a new selenoprotein in this fish (Castellano *et al.*, 2004). More recently, selenoproteins have been annotated across eukaryotes, in a SECIS-independent manner, using alignment profiles derived from the known selenoprotein families (Mariotti and Guigo, 2010).

The RNA and Regulatory Years

The functional annotation of the regulatory and non-coding part of the genome has lagged behind the annotation of protein-coding genes. Systematic analysis of RNA genes and regulatory regions was made possible in the mid and late 2000s by a number of experimental approaches for the characterization of transcribed regions, transcription factor binding sites, DNA methylation sites and chromatin structure. The ENCODE project was an early adopter of these methods to systematically characterize the human genome, (2012ENCODE Project Consortium). Among these, tiling arrays and RNA-seq have been important to understand the transcriptional landscape of the human genome, which is pervasively transcribed (Djebali *et al.*, 2012; Bertone *et al.*, 2004). This was perhaps not unexpected if one acknowledged the high initiation promiscuity of *Pol II*, the eukaryotic RNA polymerase complex (Struhl, 2007). In any case, transcription from regions where protein-coding genes or known RNA genes – such as ribosomal and transfer RNAs, small nuclear RNAs involved in splicing, microRNAs, short interfering RNAs, and a few others – have so far not been annotated, produced thousands of long noncoding RNAs with no obvious function. These unknown RNAs (Derrien *et al.*, 2012), which standard gene prediction programs cannot detect using codon bias measures, constitute a vast array of potential genes that do not encode for proteins (Kowalczyk *et al.*, 2012), albeit some may do (Ji *et al.*, 2015). While a few of these new RNAs have been shown to have functions (Nesterova *et al.*, 2001; Sleutels *et al.*, 2002), there is little functional evidence for the majority of them. Their transcription is often extremely low and they tend to have limited sequence or RNA structure conservation (Rivas *et al.*, 2017), raising the possibility that they are transcriptional noise. Indeed, between and within species analyses indicate that most of these RNA genes are not under strong evolutionary constraint (purifying selection) (Wiberg *et al.*, 2015). Alternatively, it could be their transcription itself rather than their sequence that is under selection. The functional annotation of RNA genes thus remains open and additional work is needed before this controversy is solved.

The high-throughput sequencing of transcriptomes and other experimental approaches has required the development of fast aligners like BWA (Li and Durbin, 2009) and Bowtie (Langmead and Salzberg, 2012), which can align millions of short sequences to a reference genome sequence. Specialized tools can integrate those alignments into predicted gene structures, such as AUGUSTUS (Stanke *et al.*, 2006b), mGENE (Schweikert *et al.*, 2009), or the Transomics pipeline (Solovyev *et al.*, 2006). Other tools, like SpliceGrapher (Rogers *et al.*, 2012), have been developed to refine the alignments of mapped sequences to predict "splicing graphs", a representation of the splicing variants of a gene in which exons are shown as nodes that are connected by the intervening introns represented as edges. As discussed, obtaining simultaneously the genome and transcriptome of a species facilitates the task of gene annotation, by aligning the transcriptome sequences along the newly assembled genome, and provides an estimation of gene expression levels when translating the alignment coverage of read sequences into normalized counts (Mortazavi *et al.*, 2008). An extensive assessment of the sequence mappers (Engstrom *et al.*, 2013) and the gene-finders (Steijger *et al.*, 2013) was undertaken by the GENCODE consortium (Harrow *et al.*, 2012). Among the results obtained it is worth mentioning that 5′- and 3′-UTR exons were still difficult to predict (see **Fig. 3**) as well as the different exon combinations for splicing transcript isoforms, and that prediction accuracy correlated with the gene expression level, as highly expressed genes are often more abundant in the training sets. However, this is changing as single molecule sequencing methods, such as PacBio or ONT nanopore, produce long sequences recovering the full length splicing isoforms of a gene (Sharon *et al.*, 2013). Thus easing the task of protein-coding and non-coding mRNA genes annotation, even at single cell resolution (Byrne *et al.*, 2017).

Regulatory regions where proteins (*e.g.*, transcription factors) bind have been studied using chromatin immunoprecipitation followed by sequencing (ChIP-seq). Around 8% of the genome across different cell types is typically protein-bound with sequence-specific signals known to bind proteins being common in them. In addition, *DNase I* hypersensitivity has been used to identify those regions of the genome open to transcription. These regions can be bound by transcription factors and were found to be upstream of the predicted transcription start sites, coinciding with the regions defined by the ChIP-seq experiments. Methylation of cytosine, usually at CpG dinucleotides, is involved in the epigenetic regulation of gene expression, with promoter and gene methylation leading to repression and promotion of transcription, respectively. Levels of DNA methylation correlate with chromatin accessibility as defined by the *DNase I* hypersensitivity experiments. Interestingly, most variable regions in their methylation correspond to genes rather than their promoters.

One surprising result from the ENCODE project was the finding that the majority (about 80%) of the human genome participated in one or more of the experiments above (ENCODE Project Consortium, 2012), mostly in the RNA transcription experiments discussed above. This was initially interpreted as providing evidence for a functional role to most of the human DNA sequence but low specificity of the assays used is a simpler explanation. These assays often do not distinguish between meaningful function – some molecular function that impacts fitness in the individual – and one that does not. As a result, having the majority of the genome performing important roles contrasts with the small fraction of it – probably 5%–10% – that has been shown to be under evolutionary constraint (Asthana *et al.*, 2007; Davydov *et al.*, 2010). It follows that some fraction of non-coding DNA is functional in the typical meaning of this word but that the majority of it – mostly transposable elements that make up about half of the human genome (Eddy, 2012) – is not (albeit being biochemically active in different experiments). Indeed, one of the common tasks in gene prediction is masking about half of the human genome for repeats and transposons using the venerable REPEATMASKER program.

The Population Years

The functional annotation of a species genome provides a reference to which additional genomes from the same species can be compared. Thousands of human exomes (protein-coding fraction of the genome) or genomes have now been sequenced (1k Genomes Project Consortium, 2012, 2015) and, in doing so, many differences have been found between them and their reference – *e.g.*, between four and five million nucleotide differences in each human genome –. These differences, which include single nucleotide variants, insertions and deletions, copy number variants and others, are often novel and unannotated, which each ensuing genome having more of them. These rare variants in humans, which are found in specific populations and at low frequency, are to a large the result of the recent growth of human populations (Keinan and Clark, 2012). Indeed, humans have accumulated in the last 5000–10,000 years an enormous number of rare variants, particularly in non-African populations. Protein-coding variants at low frequency are often slightly deleterious as measured by approaches that infer the probability of amino acid changes impacting the structure or function of proteins (Fu *et al.*, 2013). Programs like POLYPHEN (Adzhubei *et al.*, 2010) and SIFT (Kumar *et al.*, 2009), using amino acid conservation (within and between species) and different amino acid properties, annotate their deleteriousness across the genome. In addition, a few hundred variants disrupt the translation of proteins in a typical human genome. Non-coding variants in regulatory regions have also been annotated with regard to their deleteriousness using, for example, their conservation among species with PHASTCONS (Siepel *et al.*, 2005). It follows that the functional annotation of single nucleotide variants, particularly those that change (amino acid substitutions) or disrupt proteins (e.g., changing splice sites or creating stop codons), are important to understand interindividual differences linked to diseases that are both rare and severe. Indeed, rare diseases, understood as those that occur in one out of 2000 individuals, have often a genetic basis and the annotation of the deleteriousness of the causal mutation(s) in single-gene disorders is the first step in their diagnosis and treatment.

Still, most of the genetic variation between individuals genomes of a species is shared among some (or most) of them. This common variation was described in a landmark study that provided not only a catalog of single nucleotide variants in humans but the genetic association among them (International HapMap Consortium, 2005). This association, known as linkage disequilibrium, reflects the coinheritance of sets of single nucleotide variants (haplotypes) and their functional annotation is needed to investigate, using genome-wide association studies, the hereditary factors involved in disease. This is important as each human genome has around 10,000 rare and common amino acid changes, a few hundred protein-disrupting ones and about 500,000 non-coding changes in regulatory regions (1k Genomes Project Consortium, 2015). Of these, about 2000 are linked through genome-wide studies with common disease whereas a few dozen are implicated with rare disease per genome. Today, thousands of mutations have been associated to common or rare disease in humans and their genome and functional annotation can be obtained from different databases, for example OMIM (http://omim.org), CLINVAR (Landrum *et al.*, 2016) and the NHGRI-EBI GWAS catalog (MacArthur *et al.*, 2017).

Conclusion

Genome annotation has come a long way since the first protein-coding genes were annotated using *ab initio* and later comparative approaches. Experimental evidence is now routinely integrated in the annotation of these genes and their multiple alternative splice forms. At the same time, promoters and smaller regulatory elements, as well as RNA genes (including controversial ones), are being annotated with a variety of experimental approaches genome-wide. Knowing the location of these coding and noncoding functional features has provided the framework to understand the role of sequence variants in health and disease, which remains largely undetermined. Still, since the early nineties, when the Human Genome Project was launched, our knowledge of genome biology has greatly improved. This is not only due to the ever increasing computational capabilities but also to novel discoveries in molecular biology from experimental methods with increased resolution and throughput, which has been harnessed in annotation consortiums like the ENCODE project. As a result, well annotated genomes like that of human, fly or mice, the result of years of intense human curation, as well as gold standards such REFSEQ or VEGA genes, provide reference annotations to other species.

This is important as the pace of sequencing new species genomes, transcriptomes, and environmental (metagenomics) samples is only accelerating given the ever decreasing cost of sequencing technologies. Thus, a shift towards more automated annotation pipelines is warranted to cope with the scale of the analyses in genome annotation. Automated annotation pipelines have room to improve though, with gene-builds still having a significant impact in the downstream analyses. For example, on the variants found in a genome and their predicted functional consequences (Frankish *et al.*, 2015). This is due to the fact that most gene-finding tools return a single and often different set of optimal gene structures. This also poses a problem for overlapping genes, including those on different strands, nested genes, genes located within introns of other genes, non-standard gene structures, and genes with translation recoding like selenoprotein genes, as they are often not included in the optimal prediction of genes. Furthermore, most gene prediction tools do not consider frame shifts, RNA editing, and the impact of sequencing errors in the identification of gene structures. Still, some tools are capable to infer more than one alternative splicing isoform provided that there is some sort of experimental evidence, for example, from RNA-seq experiments. However, problems may appear when transcripts contain start or stop codons split by introns. Together with non-cannonical splice-sites, these are some of the special cases that human curators solve in the final refinement steps of genome annotation when all sort of evidences are integrated. Many of aforementioned issues will be overcome by the use of single molecule sequencing technologies, which make possible to map the full length of individual protein-coding and non-coding transcript isoforms. In combination with single cell gene expression experiments, it will be finally feasible to produce gene-builds and their corresponding functional annotation at the single molecule and cell resolution, an astounding promise just two decades ago.

Acknowledgements

This study was partially funded (Sergi Castellano) by the NIHR HS&DR Programme (14/21/45) and supported by the NIHR GOSH BRC. The views expressed are those of the author(s) and not necessarily those of the NHS, the NIHR or the Department of Health.

See also: Algorithms for Strings and Sequences: Multiple Alignment. Algorithms for Strings and Sequences: Pairwise Alignment. Bioinformatics Approaches for Studying Alternative Splicing. Comparative and Evolutionary Genomics. Comparative Genomics Analysis. Data Mining: Classification and Prediction. Data Mining: Prediction Methods. Detecting and Annotating Rare Variants. Exome Sequencing Data Analysis. Functional Enrichment Analysis. Functional Genomics. Gene Mapping. Genome Annotation: Perspective From Bacterial Genomes. Genome Databases and Browsers. Genome Informatics. Genome-Wide Haplotype Association Study. Hidden Markov Models. Integrative Analysis of Multi-Omics Data. Linkage Disequilibrium. Metagenomic Analysis and its Applications. Natural Language Processing Approaches in Bioinformatics. Next Generation Sequencing Data Analysis. Ontology in Bioinformatics. Ontology in Bioinformatics. Ontology-Based Annotation Methods. Phylogenetic Footprinting. Prediction of Coding and Non-Coding RNA. Protein Functional Annotation. Quantitative Immunology by Data Analysis Using Mathematical Models. Sequence Analysis. Sequence Composition. Single Nucleotide Polymorphism Typing. Whole Genome Sequencing Analysis

References

Abril, J.F., Castelo, R., Guigo, R., 2005. Comparison of splice sites in mammals and chicken. Genome Res. 15 (1), 111–119.
Abril, J.F., Guigo, R., 2000. gff2ps: Visualizing genomic annotations. Bioinformatics 16 (8), 743–744.
Abril, J.F., Guigo, R., Wiehe, T., 2003. gff2aplot: Plotting sequence comparisons. Bioinformatics 19 (18), 2477–2479.
Adams, M.D., *et al.*, 2000. The genome sequence of *Drosophila melanogaster*. Science 287 (5461), 2185–2195.
Adzhubei, I.A., *et al.*, 2010. A method and server for predicting damaging missense mutations. Nat. Methods 7 (4), 248–249.
Aken, B.L., *et al.*, 2016. The Ensembl gene annotation system. Database (Oxford) 2016.
Aken, B.L., *et al.*, 2017. Ensembl 2017. Nucleic Acids Res. 45 (D1), D635–D642.
Alexandersson, M., Cawley, S., Pachter, L., 2003. SLAM: Cross-species gene finding and alignment with a generalized pair hidden Markov model. Genome Res. 13 (3), 496–502.
Altschul, S.F., *et al.*, 1990. Basic local alignment search tool. J. Mol. Biol. 215 (3), 403–410.
Ashburner, M., *et al.*, 2000. Gene ontology: Tool for the unification of biology. The Gene Ontology Consortium. Nat. Genet. 25 (1), 25–29.
Asthana, S., *et al.*, 2007. Widely distributed noncoding purifying selection in the human genome. Proc. Natl. Acad. Sci. USA 104 (30), 12410–12415.
Bafna, V., Huson, D.H., 2000. The conserved exon method for gene finding. Proc. Int. Conf. Intell. Syst. Mol. Biol. 8, 3–12.
Bateman, A., *et al.*, 2017. UniProt: The universal protein knowledgebase. Nucleic Acids Res. 45 (D1), D158–D169.
Batzoglou, S., *et al.*, 2000. Human and mouse gene structure: Comparative analysis and application to exon prediction. Genome Res. 10 (7), 950–958.
Bedell, J.A., Korf, I., Gish, W., 2000. MaskerAid: A performance enhancement to RepeatMasker. Bioinformatics 16 (11), 1040–1041.
Benson, D.A., *et al.*, 2013. GenBank. Nucleic Acids Res. 41 (D1), D36–D42.
Bertone, P., *et al.*, 2004. Global identification of human transcribed sequences with genome tiling arrays. Science 306 (5705), 2242–2246.
Birney, E., Durbin, R., 1997. Dynamite: A flexible code generating language for dynamic programming methods used in sequence comparison. In: Proceedings of Ismb-97 – Fifth International Conference on Intelligent Systems for Molecular Biology, pp. 56–64.
Birney, E., Durbin, R., 2000. Using GeneWise in the *Drosophila* annotation experiment. Genome Res. 10 (4), 547–548.
Blayo, P., Rouze, P., Sagot, M.F., 2003. Orphan gene finding – An exon assembly approach. Theor. Comput. Sci. 290 (3), 1407–1431.
Burge, C., Karlin, S., 1997. Prediction of complete gene structures in human genomic DNA. J. Mol. Biol. 268 (1), 78–94.
Burset, M., Guigo, R., 1996. Evaluation of gene structure prediction programs. Genomics 34 (3), 353–367.
Burset, M., Seledtsov, I.A., Solovyev, V.V., 2000. Analysis of canonical and non-canonical splice sites in mammalian genomes. Nucleic Acids Res. 28 (21), 4364–4375.
Byrne, A., *et al.*, 2017. Nanopore long-read RNAseq reveals widespread transcriptional variation among the surface receptors of individual B cells. Nat. Commun. 8, 16027.
Cantarel, B.L., *et al.*, 2008. MAKER: An easy-to-use annotation pipeline designed for emerging model organism genomes. Genome Res. 18 (1), 188–196.

Castellano, S., *et al.*, 2001. *In silico* identification of novel selenoproteins in the *Drosophila melanogaster* genome. EMBO Rep. 2 (8), 697–702.

Castellano, S., *et al.*, 2004. Reconsidering the evolution of eukaryotic selenoproteins: A novel nonmammalian family with scattered phylogenetic distribution. EMBO Rep. 5 (1), 71–77.

Castellano, S., *et al.*, 2005. Diversity and functional plasticity of eukaryotic selenoproteins: Identification and characterization of the SelJ family. Proc. Natl. Acad. Sci. USA 102 (45), 16188–16193.

Chambers, I., *et al.*, 1986. The structure of the mouse glutathione peroxidase gene: The selenocysteine in the active site is encoded by the 'termination' codon, TGA. EMBO J. 5 (6), 1221–1227.

Chuang, T.J., *et al.*, 2003. A complexity reduction algorithm for analysis and annotation of large genomic sequences. Genome Res. 13 (2), 313–322.

Coghlan, A., *et al.*, 2008. nGASP – The nematode genome annotation assessment project. BMC Bioinform. 9, 549.

E.PENCODE, Project Consortium, 2012. An integrated encyclopedia of DNA elements in the human genome. Nature 489 (7414), 57–74.

Crollius, H.R., *et al.*, 2000. Estimate of human gene number provided by genome-wide analysis using *Tetraodon nigroviridis* DNA sequence. Nat. Genet. 25 (2), 235–238.

Curwen, V., *et al.*, 2004. The Ensembl automatic gene annotation system. Genome Res. 14 (5), 942–950.

Davydov, E.V., *et al.*, 2010. Identifying a high fraction of the human genome to be under selective constraint using GERP plus. PLOS Comput. Biol. 6 (12),

Delcher, A.L., *et al.*, 1999. Improved microbial gene identification with GLIMMER. Nucleic Acids Res. 27 (23), 4636–4641.

Derrien, T., *et al.*, 2012. The GENCODE v7 catalog of human long noncoding RNAs: Analysis of their gene structure, evolution, and expression. Genome Res. 22 (9), 1775–1789.

Djebali, S., *et al.*, 2012. Landscape of transcription in human cells. Nature 489 (7414), 101–108.

Dunham, I., *et al.*, 1999. The DNA sequence of human chromosome 22. Nature 402 (6761), 489–495.

Durbin, R., *et al.*, 1998. Biological Sequence Analysis: Probabilistic Models of Proteins and Nucleic Acids. Cambridge, United Kingdom: Cambridge University Press, (xi, 356 pages: Illustrations; 26 cm).

Eddy, S.R., 2004a. What is dynamic programming? Nat. Biotechnol. 22 (7), 909–910.

Eddy, S.R., 2004b. What is a hidden Markov model? Nat. Biotechnol. 22 (10), 1315–1316.

Eddy, S.R., 2012. The C-value paradox, junk DNA and ENCODE. Curr. Biol. 22 (21), R898–R899.

Eilbeck, K., *et al.*, 2009. Quantitative measures for the management and comparison of annotated genomes. BMC Bioinform. 10, 67.

Elsik, C.G., *et al.*, 2014. Finding the missing honey bee genes: Lessons learned from a genome upgrade. BMC Genomics 15, 86.

Engstrom, P.G., *et al.*, 2013. Systematic evaluation of spliced alignment programs for RNA-seq data. Nat. Methods 10 (12), 1185–1191.

Florea, L., *et al.*, 1998. A computer program for aligning a cDNA sequence with a genomic DNA sequence. Genome Res. 8 (9), 967–974.

Frankish, A., *et al.*, 2015. Comparison of GENCODE and RefSeq gene annotation and the impact of reference geneset on variant effect prediction. BMC Genom. 16 (Suppl. 8), S2.

Fu, W.Q., *et al.*, 2013. Analysis of 6515 exomes reveals the recent origin of most human protein-coding variants. Nature 493 (7431), 216–220.

Gelfand, M.S., Mironov, A.A., Pevzner, P.A., 1996. Gene recognition via spliced sequence alignment. Proc. Natl. Acad. Sci. USA 93 (17), 9061–9066.

1k Genomes Project Consortium, *et al.*, 2012. An integrated map of genetic variation from 1092 human genomes. Nature 491 (7422), 56–65.

1k Genomes Project Consortium, *et al.*, 2015. A global reference for human genetic variation. Nature 526 (7571), 68–74.

Giardine, B., *et al.*, 2005. Galaxy: A platform for interactive large-scale genome analysis. Genome Res. 15 (10), 1451–1455.

Gish, W., States, D.J., 1993. Identification of protein coding regions by database similarity search. Nat. Genet. 3 (3), 266–272.

Gotoh, O., 1982. An improved algorithm for matching biological sequences. J. Mol. Biol. 162 (3), 705–708.

Guigo, R., *et al.*, 1992. Prediction of gene structure. J. Mol. Biol. 226 (1), 141–157.

Guigo, R., 1998. Assembling genes from predicted exons in linear time with dynamic programming. J. Comput. Biol. 5 (4), 681–702.

Guigo, R., *et al.*, 2003. Comparison of mouse and human genomes followed by experimental verification yields an estimated 1019 additional genes. Proc. Natl. Acad. Sci. USA 100 (3), 1140–1145.

Guigo, R., *et al.*, 2006. EGASP: The human ENCODE genome annotation assessment project. Genome Biol. 7 (Suppl. 1), S2 1–31.

Harrow, J., *et al.*, 2012. GENCODE: The reference human genome annotation for The ENCODE project. Genome Res. 22 (9), 1760–1774.

Hastings, K.E.M., 2005. SL trans-splicing: Easy come or easy go? Trends Genet. 21 (4), 240–247.

Hoff, K.J., *et al.*, 2016. BRAKER1: Unsupervised RNA-seq-based genome annotation with GeneMark-ET and AUGUSTUS. Bioinformatics 32 (5), 767–769.

Hubbard, T., *et al.*, 2002. The Ensembl genome database project. Nucleic Acids Res. 30 (1), 38–41.

Hyatt, D., *et al.*, 2010. Prodigal: Prokaryotic gene recognition and translation initiation site identification. BMC Bioinform. 11, 119.

International HapMap Consortium, C., 2005. A haplotype map of the human genome. Nature 437 (7063), 1299–1320.

Iseli, C., Jongeneel, C.V., Bucher, P., 1999. ESTScan: A program for detecting, evaluating, and reconstructing potential coding regions in EST sequences. Proc. Int. Conf. Intell. Syst. Mol. Biol. 138–148.

Ji, Z., *et al.*, 2015. Many lncRNAs, 5′ UTRs, and pseudogenes are translated and some are likely to express functional proteins. eLife 4.

Katzav, S., Martin-Zanca, D., Barbacid, M., 1989. vav, a novel human oncogene derived from a locus ubiquitously expressed in hematopoietic cells. EMBO J. 8 (8), 2283–2290.

Keinan, A., Clark, A.G., 2012. Recent explosive human population growth has resulted in an excess of rare genetic variants. Science 336 (6082), 740–743.

Kent, W.J., *et al.*, 2002. The human genome browser at UCSC. Genome Res. 12 (6), 996–1006.

Korf, I., *et al.*, 2001. Integrating genomic homology into gene structure prediction. Bioinformatics 17 (Suppl. 1), S140–S148.

Kowalczyk, M.S., Higgs, D.R., Gingeras, T.R., 2012. Molecular biology: RNA discrimination. Nature 482 (7385), 310–311.

Kryukov, G.V., *et al.*, 2003. Characterization of mammalian selenoproteomes. Science 300 (5624), 1439–1443.

Kryukov, G.V., Kryukov, V.M., Gladyshev, V.N., 1999. New mammalian selenocysteine-containing proteins identified with an algorithm that searches for selenocysteine insertion sequence elements. J. Biol. Chem. 274 (48), 33888–33897.

Kumar, P., Henikoff, S., Ng, P.C., 2009. Predicting the effects of coding non-synonymous variants on protein function using the SIFT algorithm. Nat. Protoc. 4 (7), 1073–1082.

Kyriakopoulou, C., *et al.*, 2006. U1-like snRNAs lacking complementarity to canonical 5′ splice sites. RNA 12 (9), 1603–1611.

Krzywinski, M., *et al.*, 2009. Circos: an Information Aesthetic for Comparative Genomics. Genome Res. 19, 1639–1645.

Landrum, M.J., *et al.*, 2016. ClinVar: Public archive of interpretations of clinically relevant variants. Nucleic Acids Res. 44 (D1), D862–D868.

Langmead, B., Salzberg, S.L., 2012. Fast gapped-read alignment with Bowtie 2. Nat. Methods 9 (4), 357–359.

Lescure, A., *et al.*, 1999. Novel selenoproteins identified *in silico* and *in vivo* by using a conserved RNA structural motif. J. Biol. Chem. 274 (53), 38147–38154.

Lewis, S.E., *et al.*, 2002. Apollo: A sequence annotation editor. Genome Biol. 3 (12), (RESEARCH0082).

Li, H., Durbin, R., 2009. Fast and accurate short read alignment with Burrows-Wheeler transform. Bioinformatics 25 (14), 1754–1760.

Loveland, J.E., *et al.*, 2012. Community gene annotation in practice. Database – J. Biol. Databases Curation.

Lukashin, A.V., Borodovsky, M., 1998. GeneMark.hmm: New solutions for gene finding. Nucleic Acids Res. 26 (4), 1107–1115.

MacArthur, J., *et al.*, 2017. The new NHGRI-EBI Catalog of published genome-wide association studies (GWAS Catalog). Nucleic Acids Res. 45 (D1), D896–D901.

Madupu, R., *et al.*, 2010. Meeting report: A workshop on best practices in genome annotation. Database (Oxford) 2010, baq001.

Margulies, E.H., *et al.*, 2003. Identification and characterization of multi-species conserved sequences. Genome Res. 13 (12), 2507–2518.

Mariotti, M., Guigo, R., 2010. Selenoprofiles: Profile-based scanning of eukaryotic genome sequences for selenoprotein genes. Bioinformatics 26 (21), 2656–2663.

Meyer, I.M., Durbin, R., 2002. Comparative *ab initio* prediction of gene structures using pair HMMs. Bioinformatics 18 (10), 1309–1318.

Meyer, I.M., Durbin, R., 2004. Gene structure conservation aids similarity based gene prediction. Nucleic Acids Res. 32 (2), 776–783.

Miller, W., 2001. Comparison of genomic DNA sequences: Solved and unsolved problems. Bioinformatics 17 (5), 391–397.

Monaco, A.P., et al., 1986. Isolation of candidate cDNAs for portions of the Duchenne muscular dystrophy gene. Nature 323 (6089), 646–650.

Mortazavi, A., et al., 2008. Mapping and quantifying mammalian transcriptomes by RNA-Seq. Nat. Methods 5 (7), 621–628.

Mott, R., 1997. EST_GENOME: A program to align spliced DNA sequences to unspliced genomic DNA. Comput. Appl. Biosci. 13 (4), 477–478.

Mouse Genome Sequencing Consortium, 2002. Initial sequencing and comparative analysis of the mouse genome. Nature 420 (6915), 520–562.

Mudge, J.M., Harrow, J., 2016. The state of play in higher eukaryote gene annotation. Nat. Rev. Genet. 17 (12), 758–772.

Needleman, S.B., Wunsch, C.D., 1970. A general method applicable to the search for similarities in the amino acid sequence of two proteins. J. Mol. Biol. 48 (3), 443–453.

Nesterova, T.B., et al., 2001. Characterization of the genomic Xist locus in rodents reveals conservation of overall gene structure and tandem repeats but rapid evolution of unique sequence. Genome Res. 11 (5), 833–849.

Parra, G., et al., 2003. Comparative gene prediction in human and mouse. Genome Res. 13 (1), 108–117.

Parra, G., et al., 2006. Tandem chimerism as a means to increase protein complexity in the human genome. Genome Res. 16 (1), 37–44.

Parra, G., Bradnam, K., Korf, I., 2007. CEGMA: A pipeline to accurately annotate core genes in eukaryotic genomes. Bioinformatics 23 (9), 1061–1067.

Patel, A.A., Steitz, J.A., 2003. Splicing double: Insights from the second spliceosome. Nat. Rev. Mol. Cell Biol. 4 (12), 960–970.

Pearson, W.R., 2000. Flexible sequence similarity searching with the FASTA3 program package. Methods Mol. Biol. 132, 185–219.

Pearson, W.R., Lipman, D.J., 1988. Improved tools for biological sequence comparison. Proc. Natl. Acad. Sci. USA 85 (8), 2444–2448.

Pedersen, C.N.S., Scharling, T., 2002. Comparative methods for gene structure prediction in homologous sequences. Proc. Algorithms Bioinform. 2452, 220–234.

Pennisi, E., 2000. Ideas fly at gene-finding jamboree. Science 287 (5461), 2182. +.

Plotkin, J.B., Kudla, G., 2011. Synonymous but not the same: The causes and consequences of codon bias. Nat. Rev. Genet. 12 (1), 32–42.

Potter, S.C., et al., 2004. The Ensembl analysis pipeline. Genome Res. 14 (5), 934–941.

Powers, D., 2011. Evaluation: From precision, recall and F-measure to ROC, informedness, markedness and correlation. Int. J. Mach. Learn. Technol. 2, 37–63.

Pruitt, K.D., et al., 2012. NCBI Reference Sequences (RefSeq): Current status, new features and genome annotation policy. Nucleic Acids Res. 40 (Database issue), D130–D135.

Rayman, M.P., 2000. The importance of selenium to human health. Lancet 356 (9225), 233–241.

Reese, M.G., et al., 2000. Genome annotation assessment in *Drosophila melanogaster*. Genome Res. 10 (4), 483–501.

Rivas, E., Clements, J., Eddy, S.R., 2017. A statistical test for conserved RNA structure shows lack of evidence for structure in lncRNAs. Nat. Methods 14 (1), 45–48.

Rogers, M.F., et al., 2012. SpliceGrapher: Detecting patterns of alternative splicing from RNA-Seq data in the context of gene models and EST data. Genome Biol. 13 (1), R4.

Salamov, A.A., Solovyev, V.V., 2000. *Ab initio* gene finding in *Drosophila* genomic DNA. Genome Res. 10 (4), 516–522.

Schweikert, G., et al., 2009. mGene: Accurate SVM-based gene finding with an application to nematode genomes. Genome Res. 19 (11), 2133–2143.

Sharon, D., et al., 2013. A single-molecule long-read survey of the human transcriptome. Nat. Biotechnol. 31 (11), 1009–1014.

Sharp, P.A., 2005. The discovery of split genes and RNA splicing. Trends Biochem. Sci. 30 (6), 279–281.

Siepel, A., et al., 2005. Evolutionarily conserved elements in vertebrate, insect, worm, and yeast genomes. Genome Res. 15 (8), 1034–1050.

Simao, F.A., et al., 2015. BUSCO: Assessing genome assembly and annotation completeness with single-copy orthologs. Bioinformatics 31 (19), 3210–3212.

Slater, G.S., Birney, E., 2005. Automated generation of heuristics for biological sequence comparison. BMC Bioinform. 6, 31.

Sleutels, F., Zwart, R., Barlow, D.P., 2002. The non-coding Air RNA is required for silencing autosomal imprinted genes. Nature 415 (6873), 810–813.

Smith, T.F., Waterman, M.S., 1981. Identification of common molecular subsequences. J. Mol. Biol. 147 (1), 195–197.

Smit, A.F.A., Hubley, R., Green, P., *RepeatMasker Open-4.0*, 2013-2015, http://www.repeatmasker.org.

Solovyev, V., et al., 2006. Automatic annotation of eukaryotic genes, pseudogenes and promoters. Genome Biol. 7 (Suppl. 1), S10 1–12.

Stanke, M., et al., 2006a. Gene prediction in eukaryotes with a generalized hidden Markov model that uses hints from external sources. BMC Bioinform. 7.

Stanke, M., Tzvetkova, A., Morgenstern, B., 2006b. AUGUSTUS at EGASP: Using EST, protein and genomic alignments for improved gene prediction in the human genome. Genome Biol. 7.

Steijger, T., et al., 2013. Assessment of transcript reconstruction methods for RNA-seq. Nat. Methods 10 (12), 1177–1184.

Stein, L.D., et al., 2002. The generic genome browser: A building block for a model organism system database. Genome Res. 12 (10), 1599–1610.

Struhl, K., 2007. Transcriptional noise and the fidelity of initiation by RNA polymerase II. Nat. Struct. Mol. Biol. 14 (2), 103–105.

Su, M., et al., 2013. Small proteins: Untapped area of potential biological importance. Front. Genet. 4, 286.

Tatusova, T., et al., 2016. NCBI prokaryotic genome annotation pipeline. Nucleic Acids Res. 44 (14), 6614–6624.

The Gene Ontology Consortium,, 2017. Expansion of the Gene Ontology knowledgebase and resources. Nucleic Acids Res. 45 (D1), D331–D338.

Thibaud-Nissen, F., et al., 2016. The NCBI eukaryotic genome annotation pipeline. J. Anim. Sci. 94, 184.

Thorvaldsdottir, H., Robinson, J.T., Mesirov, J.P., 2013. Integrative Genomics Viewer (IGV): High-performance genomics data visualization and exploration. Brief. Bioinform. 14 (2), 178–192.

Wheelan, S.J., Church, D.M., Ostell, J.M., 2001. Spidey: A tool for mRNA-to-genomic alignments. Genome Res. 11 (11), 1952–1957.

Wiberg, R.A.W., et al., 2015. Assessing recent selection and functionality at long noncoding RNA loci in the mouse genome. Genome Biol. Evol. 7 (8), 2432–2444.

Wiehe, T., et al., 2001. SGP-1: Prediction and validation of homologous genes based on sequence alignments. Genome Res. 11 (9), 1574–1583.

Wu, J.Q., et al., 2004. Identification of rat genes by TWINSCAN gene prediction, RT-PCR, and direct sequencing. Genome Res. 14 (4), 665–671.

Xu, Y., Mural, R.J., Uberbacher, E.C., 1997. Inferring gene structures in genomic sequences using pattern recognition and expressed sequence tags. In: Proceedings Ismb-97 – Fifth International Conference on Intelligent Systems for Molecular Biology, pp. 344–353.

Yeh, R.F., Lim, L.P., Burge, C.B., 2001. Computational inference of homologous gene structures in the human genome. Genome Res. 11 (5), 803–816.

Zhang, Y., et al., 2005. Pyrrolysine and selenocysteine use dissimilar decoding strategies. J. Biol. Chem. 280 (21), 20740–20751.

Zheng, D., et al., 2007. Pseudogenes in the ENCODE regions: Consensus annotation, analysis of transcription, and evolution. Genome Res. 17 (6), 839–851.

Zinoni, F., et al., 1986. Nucleotide sequence and expression of the selenocysteine-containing polypeptide of formate dehydrogenase (formate-hydrogen-lyase-linked) from *Escherichia coli*. Proc. Natl. Acad. Sci. USA 83 (13), 4650–4654.

Biographical Sketch

Josep F. Abril, PhD Associate Professor at the Department of Genetics, Microbiology and Statistics of the Universitat de Barcelona. He earned his Bachelor's degree in Biology from the Universitat de Barcelona, and his PhD in Bioinformatics from the Universitat Pompeu Fabra in Barcelona. His research focuses on the computational analysis of sequences and their annotated features. He has worked on a variety of genomic and transcriptomic projects, but also on the functional characterization of proteins, and on different areas in computational biology. These include the development of protocols for the assembly and annotation of genomes, the use of comparative genomics methods, the analysis of gene expression, the modeling of splice sites and exonic structure of eukaryotic genes, the visualization of whole-genome annotations – this includes the human genome map published in *Science* in 2001 – the integration of expression and variation data into interaction networks, as well as the characterization of viral samples from metagenomic experiments. His organisms of interest are planarians, flies, and humans, not necessarily in that order, although he contributed to the annotation of other species too. Finally, he has been involved in the organization and analysis of three of the most relevant gene-prediction accuracy assessments in computational gene prediction, namely GASP in the *Drosophila* genome and EGASP and RGASP in the human one.

Sergi Castellano, PhD Associate Professor at the Genetics and Genomics Medicine Programme at University College London. He received his Bachelor's degree in Biology from the Universitat de Barcelona, and his PhD in Bioinformatics from the Universitat Pompeu Fabra, also in Barcelona. His group works on understanding the role of essential micronutrients, with particular emphasis on selenium, in the adaptation of human metabolism to the different environments encountered by archaic and modern humans. Much of his early work focused on the identification and characterization of functional elements in eukaryotic genomes, primarily selenium-containing genes, which are missannotated in standard databases. At the University of Hawaii, he developed the first database, SELENODB, with correct annotations for selenium-containing genes across animal genomes. Later, while at Cornell University and Howard Hughes Medical Institute, he used population genetics and molecular evolution approaches to study the evolution of the use of selenocysteine, the selenium-containing amino acid, compared to cysteine in proteins. More recently, at Max Planck in Germany, his group contributed to the population genetics analysis of Neandertals as it relates to their coding variation and their first encounters – as early as 100,000 years ago – with modern humans. His group is currently working on the role of micronutrient deficiencies in common and rare disease.

Repeat in Genomes: How and Why You Should Consider Them in Genome Analyses?

Emmanuelle Lerat, Université Lyon, CNRS, Villeurbanne, France

Nomenclature

Bp	Base pairs	NGS	Next Generation Sequencing
ChIP	Chromatin Immuno Precipitation	SINE	Short Interspersed Nuclear Elements
Kb	Kilo base pairs	SSA	Sequence self-alignment
LINE	Long Interspersed Nuclear Elements	SSR	Simple Sequence Repeat
MITEs	Miniature Inverted-repeat Transposable Elements	STR	Short Tandem Repeats
		TEs	Transposable Elements
		TIR	Terminal inverted repeat

Introduction

Repeats: A Large Bestiary Grouping Very Different Sequences

Although the genome size has been shown in prokaryotes to largely correlate to its complexity, i.e. the number of genes, it is not the case in eukaryotes. This observation, often referred to as the C-value paradox, can be partly explained by the fact that coding genes in eukaryotic genomes may represent only a tiny fraction of the total genome (Elliott and Gregory, 2015). In fact, it has been shown that the proportion of non-coding sequences in genomes may be particularly huge. For example, the coding genes in the human genome only represent 2% (Lander et al., 2001). A large proportion of the non-genic sequences are represented by repeats. In a genome, repeats correspond to non-coding sequences present in several occurrences. Two main categories of repeats are described, the tandem repeats and the interspersed repeats (**Fig. 1**). Another type of repeats exists in a genome which correspond to segmental duplications, which are very large nearly identical sequences (from 1 to 400 kb) often referred to as "low-copy repeats" (Sharp et al., 2005).

The tandem repeats correspond to sequences having a size from few to hundred of base pairs (bp) occurring in tandem and over several hundred of kilo base pairs (kb), which characterized them as "highly repeated sequences". We can distinguish among this category the Simple Sequence Repeats (SSR) also called Short Tandem Repeats (STR) or microsatellites, which are a short tract of adjacent DNA motifs (around 1 to 13 bp) and the minisatellites which are longer (around 14 to 500 bp), both types being repeated several times (from 5 to 50 times for example). This type of repeats usually occurs at particular regions of the chromosomes corresponding to the telomeres and centromeres, but they also can be found throughout the genome, especially in regulatory and coding regions of genes (Richard et al., 2008).

The interspersed repeats stand for sequences that are more often known under the name transposable elements (TEs). These particular sequences were discovered during the 1950s by Barbara McClintock (1950). TEs have the particularity to be able to move from one position to another along the chromosomes since the majority of them encode all the proteins necessary for their

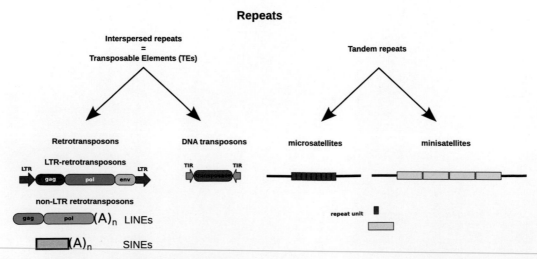

Fig. 1 Schematic representation of the different types of repeats.

transposition. Various types exist, according to structural features, the transposition intermediate (RNA or DNA), and their evolutionary origin (Wicker *et al.*, 2007; Kapitonov and Jurka, 2008). Retrotransposons use an RNA intermediate and form the class I, in which are found the Long Terminal Repeat (LTR)-retrotransposons (endogenous retrovirus-like mobile elements) which possess from two to three open reading frames (gag, pol, and env) and the non-LTR retrotransposons grouping the LINE and the SINE elements (standing for Long and Short Interspersed Nuclear Elements respectively). DNA transposons use a DNA intermediate and form the class II. They possess short terminal inverted repeats (TIRs) at their extremities surrounding one open reading frame coding for a transposase. Different types of DNA transposons have been classified mainly on the basis of the presence or absence of a catalytic site in the protein responsible for their transposition. A specific category of non-autonomous elements among the DNA transposons exist, the Miniature Inverted-repeat Transposable Elements (MITEs) that may be particularly numerous in some genomes. Depending on the organism, the proportion of TEs can be highly variable and at times very large. For example, their proportion in genomes represent 3% in yeast (Kim *et al.*, 1998), 15% in *Drosophila* (Dowsett and Young, 1982), 45% in human and in the mouse (Lander *et al.*, 2001; Waterston *et al.*, 2002), and more than 80% in maize (Schnable *et al.*, 2009).

"From Junk to Funk": The Importance of Repeats in the Genome Functioning and Its Evolution

Repeats, and more particularly TEs, have long been considered as selfish and unnecessary components of the genomes. However, since the work of Britten and Davidson (1969) where TEs were for the first time considered as potentially playing a role in the gene regulation, numerous examples have flourished allowing to show the genuine importance of such sequences in genomes (Biemont, 2010). By their ability to move and because they are repeated, TEs can promote various types of mutations, which are expected to be mostly deleterious when affecting functional regions. When TE insertions occur in or near protein-coding genes, they can result in coding sequence modification or alteration of their splicing or polyadenylation patterns, therefore disrupting the protein coding capacity of the gene. Moreover, because TEs possess their own regulatory sequences, they can alter the normal expression pattern of neighboring genes while inserted in an intergenic region (Kidwell and Lisch, 2000; Biémont and Vieira, 2006). The possibility of homologous recombination between copies can also promote illegitimate recombinations, chromosome breakages, deletions and genome rearrangements (Kidwell and Lisch, 2000; Biémont and Vieira, 2006). In human, 0.3% of TE insertions have been suggested for causing disease (Belancio *et al.*, 2008) and approximately 96 new transposition events were directly linked to single-gene diseases (Hancks and Kazazian, 2012). For example, the *Alport* syndrome has been shown to be due to a TE mediated rearrangement resulting in the partial deletion and fusion of two genes (Segal *et al.*, 1999). More specific to cancer, the disruption of the APC gene caused by the TE insertion is involved in a colon cancer (Miki *et al.*, 1992). Despite the deleterious effects they may have, TEs have also been associated with useful adaptation for their host genome. For example, the antigen receptor gene assembly by V(D)J recombination in vertebrates is performed by genes that originated from a DNA transposon (Agrawal *et al.*, 1998). TE insertions near specific genes confer resistance to insecticide for some insects (Rostant *et al.*, 2012). These examples make TEs to be now considered as major players in genome evolution due to the genetic and epigenetic diversity they can promote (Biémont and Vieira, 2006). Similarly, tandem repeats have been found to play a fundamental role in the organization of the genome (Dumbovic *et al.*, 2017). For example, changes in mini- and micro-satellites correlate with various diseases but are also associated with gene regulation (Dumbovic *et al.*, 2017).

In addition to the fundamental biological role repeats have in genomes, in the current big sequencing era, they also represent a technical challenge that may complicate the task of genome assembly and sequence alignment. For example, the presence of repeats in a genome is the major source of genome mis-assemblies via rearrangement assembly errors and collapsed repeats but also in the assignment of splicing events and gene expression estimate in transcriptome analyses (Treangen and Salzberg, 2011). Since simply ignoring these sequences is not an issue in genomics, it is important to be able to identify them.

Bioinformatic Approaches to Identify, Annotate and Analyze Repeats in Genomes

Since several decades, numerous bioinformatics tools have been developed to allow a better identifications of repeats in genome assemblies (Lerat, 2010; Modolo and Lerat, 2013; Saha *et al.*, 2008; Bergman and Quesneville, 2007). New tools continue to arise regularly to follow the progress in sequencing technology in particular, but also to response to specific biological questions. According to the type of data on which they can work and the biological question, the different tools can be separated into various categories (**Table 1**).

Detection of Repeats in Assembled Genomes

Diverse methods exist to allow the detection and annotation of repeats in assembled genomes. These methods depend on the type of repeats and on the knowledge we have concerning their content inside the organism under investigation. Repeat annotation is a particularly complex computing problem due to the nature of the sequences, especially in the case of the TEs. Indeed, the TEs are not always exact repeats since a large divergence among copies can occur and TEs also can be found in the genome inserted inside each others (nested insertions). The detection of TEs has led to the development of a large number of different tools that fall in different categories according to the approach they use. Two main types of approaches exist: the library- or signature-based methods and the ab initio methods.

Table 1   Non-exhaustive list of tools to identify and analyze repeats

Program name	Type of repeats	Input data	Approach	References	Websites
Tandem Repeat Finder	Tandem repeat	Assembled genome	Identification	Benson (1999)	https://tandem.bu.edu/trf/trf.html
XSTREAM	Tandem repeat	Assembled genome	Identification	Newman and Cooper (2007)	http://jimcooperlab.mcdb.ucsb.edu/xstream/
MREPS	Tandem repeat	Assembled genome	Identification	Kolpakov et al. (2003)	http://mreps.univ-mlv.fr/
STAR	Tandem repeat	Assembled genome	Identification	Delgrange and Rivals (2004)	http://www.atgc-montpellier.fr/star/
TRED	Tandem repeat	Assembled genome	Identification	Sokol et al. (2007)	http://tandem.sci.brooklyn.cuny.edu/tandem/
ReD Tandem	Tandem repeat	Assembled genome	Identification	Audemard et al. (2012)	NA
RepeatMasker	TEs, tandem repeat	Assembled genome	Identification	Smit et al. (1996–2010)	http://www.repeatmasker.org/
LTR_STRUC	LTR-retrotransposons	Assembled genome	Identification	McCarthy and McDonald (2003)	http://www.mcdonaldlab.biology.gatech.edu/ltr_struc.htm
LTR_FINDER	LTR-retrotransposons	Assembled genome	Identification	Xu and Wang (2007)	http://tlife.fudan.edu.cn/ltr_finder/
find_LTR	LTR-retrotransposons	Assembled genome	Identification	Rho et al. (2007)	http://darwin.informatics.indiana.edu/cgi-bin/evolution/ltr.pl
LTR_harvest	LTR-retrotransposons	Assembled genome	Identification	Ellinghaus et al. (2008)	http://www.zbh.uni-hamburg.de/forschung/arbeitsgruppe-genominformatik/software/ltrharvest.html
TSDfinder	Non-LTR retrotransposons	Assembled genome	Identification	Szak et al. (2002)	https://www.ncbi.nlm.nih.gov/CBBresearch/Landsman/TSDfinder/
SINEDR	Non-LTR retrotransposons	Assembled genome	Identification	Tu et al. (2004)	NA
TRANSPO	MITEs	Assembled genome	Identification	Santiago et al. (2002)	NA
MUST / MUSTv2	MITEs	Assembled genome	Identification	Chen et al. (2009); Ge et al. (2017)	http://www.healthinformaticslab.org/supp/resources.php
MITE-hunter	MITEs	Assembled genome	Identification	Han and Wessler (2010)	http://target.iplantcollaborative.org/
One code to find them all	TEs, tandem repeat	Output files from RepeatMasker	Analysis	Bailly-Bechet et al. (2014)	http://doua.prabi.fr/software/one-code-to-find-them-all
LTRdigest	LTR-retrotransposons	GFF3 format	Analysis	Steinbiss et al. (2009)	http://www.zbh.uni-hamburg.de/forschung/arbeitsgruppe-genominformatik/software/ltrdigest.html
RECON	TEs, tandem repeat	Assembled genome	Identification	Bao and Eddy (2003)	http://eddylab.org/software/recon/
PILER	TEs, tandem repeat	Assembled genome	Identification	Edgar and Myers (2005)	https://www.drive5.com/piler/
ReAS	TEs, tandem repeat	Assembled genome	Identification	Li et al. (2005)	Freely available via ReAS@genomics.org.cn
RepeatScout	TEs, tandem repeat	Assembled genome	Identification	Price et al. (2005)	https://bix.ucsd.edu/repeatscout/
TEclass	TEs	Individual sequences	Classification	Abrusán et al. (2009)	http://www.mybiosoftware.com/teclass-2-1-classification-te-consensus-sequences.html
REPCLASS	TEs	Individual sequences	Classification	Feschotte et al. (2010)	https://github.com/feschottelab/REPCLASS
PASTEC	TEs	Individual sequences	Classification	Hoede et al. (2014)	https://urgi.versailles.inra.fr/Tools/PASTEClassifier
REPET	TEs, tandem repeat	Assembled genome	Pipeline (Identification and classification)	Flutre et al. (2011)	https://urgi.versailles.inra.fr/Tools/REPET
RepeatModeler	TEs, tandem repeat	Assembled genome	Pipeline (Identification and classification)	Smit and Hubley, unpublished	http://www.repeatmasker.org/RepeatModeler/
AAARF	TEs, tandem repeat	Short reads	Identification	DeBarry et al. (2008)	https://sourceforge.net/projects/aaarf
Tedna	TEs, tandem repeat	Short reads	Identification	Zytnicki et al. (2014)	https://urgi.versailles.inra.fr/Tools/Tedna

Tool	Type	Data	Purpose	Reference	URL
RepeatExplorer (SeqGraphR)	TEs, tandem repeat	Short reads	Identification	Novák et al. (2010); Novák et al. (2013)	http://www.repeatexplorer.org/
DNApipeTE	TEs, tandem repeat	Short reads	Identification	Goubert et al. (2015)	http://w3lamc.univ-lyon1.fr/-dnaPipeTE-.html
RepARK	TEs, tandem repeat	Short reads	Identification	Koch et al. (2014)	https://github.com/PhKoch/RepARK
REPdenovo	TEs, tandem repeat	Short reads	Identification	Chu et al. (2016)	https://github.com/Reedwarbler/REPdenovo
Transposome	TEs, tandem repeat	Short reads	Identification	Staton and Burke (2015)	https://github.com/sestaton/Transposome
VNTRseek	Tandem repeat	Short reads	Identification	Gelfand et al. (2014)	https://github.com/yzhernand/VNTRseek
TAREAN	Tandem repeat	Short reads	Identification	Novák et al. (2017)	http://w3lamc.umbr.cas.cz/lamc/?page_id=312
TE-locate	TEs	Short reads	TE insertion variants	Platzer et al. (2012)	https://sourceforge.net/projects/te-locate/
TraFiC	TEs	Short reads	TE insertion variants	Tubio et al. (2014)	https://gitlab.com/mobilegenomes/TraFiC
TE-Tracker	TEs	Short reads	TE insertion variants	Gilly et al. (2014)	http://www.genoscope.cns.fr/externe/tetracker/
RelocaTE	TEs	Short reads	TE insertion variants	Robb et al. (2013)	https://github.com/srobb1/RelocaTE
Ngs-TE-mapper	TEs	Short reads	TE insertion variants	Linheiro and Bergman (2012)	https://github.com/bergmanlab/ngs_te_mapper
TIDAL	TEs	Short reads	TE insertion variants	Rahman et al. (2015)	https://github.com/laulabbrandeis/TIDAL
ITIS	TEs	Short reads	TE insertion variants	Jiang et al. (2015)	http://bioinformatics.psc.ac.cn/software/ITIS/
T-Lex2	TEs	Short reads	TE insertion variants	Fiston-Lavier et al. (2015)	http://petrov.stanford.edu/cgi-bin/Tlex.html
Tea	TEs	Short reads	TE insertion variants	Lee et al. (2012)	http://compbio.med.harvard.edu/Tea/
RetroSeq	TEs	Short reads	TE insertion variants	Keane et al. (2013)	https://github.com/wtsi-svi/RetroSeq
TEMP	TEs	Short reads	TE insertion variants	Zhuang et al. (2014)	https://github.com/JialiUMassWengLab/TEMP
Mobster	TEs	Short reads	TE insertion variants	Thung and et al. (2014)	https://sourceforge.net/projects/mobster/
Tangram	TEs	Short reads	TE insertion variants	Wu et al. (2014)	https://github.com/jiantao/Tangram/issues
TranspoSeq	TEs	Short reads	TE insertion variants	Helman et al. (2014)	http://archive.broadinstitute.org/cancer/cga/transposeq
Jitterbug	TEs	Short reads	TE insertion variants	Hénaff et al. (2015)	https://github.com/elzbth/jitterbug
DD-Detection	TEs	Short reads	TE insertion variants	Kroon et al. (2016)	https://bitbucket.org/mkroon/dd_detection
popoolation_TE2	TEs	Short reads	TE insertion variants	Kofler et al. (2016)	https://sourceforge.net/p/popoolation-te2/wiki/Home/
MELT	TEs	Short reads	TE insertion variants	Gardner et al. (2017)	http://melt.igs.umaryland.edu/manual.php
LorTE	TEs	Long reads	TE insertion variants	Disdero and Filée (2017)	http://www.egce.cnrs-gif.fr/?p=6422
Repeat Histone enrichment	TEs	ChipSeq data	Histone enrichment	Day et al. (2010)	http://compbio.med.harvard.edu/repeats/
piPipes	TEs	RNAseq data	Differential expression analyses	Han et al. (2015)	https://github.com/bowhan/piPipes
TEtools	TEs	RNAseq data	Differential expression analyses	Lerat et al. (2017)	https://github.com/l-modolo/TEtools
EpiTEome	TEs	Short reads	TE insertion variants and DNA methylation analysis	Daron and Slotkin (2017)	https://github.com/jdaron/epiTEome

The library- or signature-based methods all require a certain amount of knowledge concerning the searched TEs. Library-based methods compare the genome sequence to a set of reference sequences corresponding to TE consensus (i.e. library) to search for their occurrences in the genome. It is thus an approach by sequence homology. The most known and used program from this category is *RepeatMasker* (Smit *et al.*, 1996–2010) whose search engines include *nhmmer, cross_match, ABBlast/WUBlast, RMBlast* and *Decypher*. It relies on a library of consensus sequences of TEs called Repbase (Bao *et al.*, 2015). It is also able to identify low complexity DNA sequences, which can correspond to tandem repeats, by using sequence homology and the *Tandem Repeat Finder* program (Benson, 1999). The main outputs of the program are a global annotation of the repeats that are present in the query sequence as well as a modified version of the query sequence in which all the annotated repeats have been masked. Recently, a program has been developed to allow a more detailed analyses of one output as well as an easy way to retrieve the identified sequences (Bailly-Bechet *et al.*, 2014). Although *RepeatMasker* is fast and quite efficient, the main drawback consists in the need to already know the sequences of the TEs that are present in the genome under investigation or at least in not too far closely related ones. It is thus not possible by this approach to detect new TE families. The signature-based methods allow to have less knowledge concerning the searched TEs since they seek the genome sequence for nucleotidic or proteic motifs, or particular structural features from a specific class of TEs. LTR-retrotransposons can be identified based on their structure (presence of two direct repeats (LTR) at their extremities, size, presence of protein motifs inside the coding parts etc.) by different programs like *LTR_STRUC* (McCarthy and McDonald, 2003), *find_LTR* (Rho *et al.*, 2007), *LTR-finder* (Xu and Wang, 2007) and *LTRharvest* (Ellinghaus *et al.*, 2008). Outputs from *LTRharvest* can further be analyzed by *LTRdigest* (Steinbiss *et al.*, 2009) to annotate internal features of LTR-retrotransposons. These programs have various amount of success since they can find a lot of false positives, meaning that their output files need to be manually curated. They also are designed to find only full-length elements with two LTRs at both extremities of the element and sufficiently conserved. The identification of non-LTR retro-transposons has been the goal of some programs like *TSDfinder* (Szak *et al.*, 2002) and *SINEDR* (Tu *et al.*, 2004). DNA transposons can also be searched for using structural and sequence characteristics of such elements using programs like *TRANSPO* (Santiago *et al.*, 2002), *MUST* (Chen *et al.*, 2009), recently updated into a new version *MUSTv2* (Ge *et al.*, 2017), and *MITE-hunter* (Han and Wessler, 2010). All signature-based programs are thus mainly designed to help the researcher finding very specific types of TEs and thus concerning very specific biological questions since with these approaches, a large proportion of repeats remains ignored.

Alternatively to the precedent approaches, *ab initio* methods have been developed to detect virtually all kind of repeats in a genome without any *a priori* knowledge. These approaches can be separated into two distinct categories. In the first category, the methods first use self-comparison approaches of sequences to identify repeats and then use clustering methods to group them into families, before generating a consensus sequence for each detected family (method implemented in *RECON* (Bao and Eddy, 2003) or *PILER* (Edgar and Myers, 2005)). In the second category are grouped programs using k-mer and spaced seed approaches, which count "words" (method implemented in *ReAS* (Li *et al.*, 2005) or *RepeatScout* (Price *et al.*, 2005)). In that case, a repeat is defined as a sub-sequence that appears more than once in a longer sequence. All *ab-initio* programs have very varying rates of success in the identification of repeats depending on the data (quality of the genome assembly, proportion and age of the repeats in the genome). Moreover, the results produced by these methods are generally raw implying further analyses to identify the different type of repeats. To help with this step, some classification programs have been proposed like *TEclass* (Abrusán *et al.*, 2009), *REPCLASS* (Feschotte *et al.*, 2010), and *PASTEC* (Hoede *et al.*, 2014), which use structural features and sequence similarities to determine to which type of TE a sequence can be associated. The level of precision is variable according to the tool, *PASTEC* having currently the finer level in the assignation of TE type according to the classification proposed by Wicker and colleagues (Wicker *et al.*, 2007). Globally, no stand alone *ab initio* program can discover all repeated sequences present in a genome (Platt *et al.*, 2016). This is why approaches using several different programs to optimize repeat finding have been developed like, for example, the pipelines *REPET* (Flutre *et al.*, 2011) and *RepeatModeler* (Smit and Hubley; see Relevant Website section). They tend to give better results and have been largely used in the scientific community. For example, *RepeatModeler* has recently been used to identify TEs in the genome of a fungi (Castanera *et al.*, 2016) while this program was used in addition to *REPET* in the genome of the Atlantic salmon (Lien *et al.*, 2016).

A large number of tools also exist to detect specifically tandem repeats (see for a review Lim *et al.*, 2013). One of the most used tools to identify these sequences is the *Tandem Repeat Finder* program, which has been developed almost 20 years ago and that is still maintained by his developer (Benson, 1999). The algorithm of this program uses the approach of matching k-tuples, i.e. two windows of k consecutive characters from a nucleotide sequence that have identical content. This requires no *a priori* knowledge concerning the pattern of the repeat, its size or the number of copies. A similar approach is used in programs like *XSTREAM* (Newman and Cooper, 2007) and *MREPS* (Kolpakov *et al.*, 2003). Other programs use improved dynamic programming algorithms like *STAR* (Delgrange and Rivals, 2004) and *TRED* (Sokol *et al.*, 2007). More recently, the program *ReD tandem* (Audemard *et al.*, 2012) was developed using a flow based chaining algorithm. Some programs, like any type of dot-plot programs, are based solely on sequence self-alignment (SSA) algorithms, which are particularly efficient for the detection of long repeats. Their drawbacks are that these programs are relatively slow due to their time complexity and usually fail to identify short repeats.

Detection of Repeats in Unassembled Genomes

The development of the next generation sequencing (NGS) technologies has overturned our approach to genomics with a huge amount of data being produced everyday (Margulies *et al.*, 2005; Mardis, 2017). We thus now have access to more data on very various organisms at low cost and with fewer bias than the previous sequencing technologies (Wicker *et al.*, 2006). Currently however, the data produced by regular NGS technology like Illumina, correspond to rather short sequences due to the small size of the reads (maximum of 300 bp length) (Mardis, 2017). This short size implies that the assembly of the original DNA sequences is the most challenging and time-consuming step especially when the considered organism is rich in repeats. To assemble a genome in this condition often leads to unfinished drafts of very numerous scaffolds, a large number of them corresponding to unplaced repeats on chromosomes.

By their repetitive nature, repeats represent portions of the genome with the best coverage, especially in the case of genome survey sequencing, where a sample of a complete genome is actually sequenced. Indeed, for a genomic coverage of 0.01X, each repeat having 1000 occurrences will theoretically have a coverage of 10X (Macas *et al.*, 2007). This situation has allowed the development of new approaches to detect repeats directly from the raw data, without the need for any homology search nor assembly.

The first method to have been developed based on this assumption and that is able to work with short reads is the *AAARF* algorithm (DeBarry *et al.*, 2008). This approach uses *BLAST* (Altschul *et al.*, 1997) to compare a read against all the others and obtain its nucleotide coverage. This value is used to determine overlapping reads that will be aligned to reconstruct a new sequence. Iteratively, the program will elongate each new sequence to assemble a set of TE contigs. Another type of approach that is also working on genomic sample is based on the construction of sequence clusters like the *SeqGraphR* program implemented in the *RepeatExplorer* pipeline (Novák *et al.*, 2010, 2013). In this method, reads are clustered using a hierarchical agglomeration algorithm. Various graph metrics are computed to discriminate between different types of repeats and the assembly of TE sequences gives consensus sequences. Based on a similar approach, the *Transposome* program has been proposed more recently that also uses a graph-based analysis of similarity between reads (Staton and Burke, 2015). As an alternative to read clustering, another approach, *DNApipeTE* (Goubert *et al.*, 2015), proposes to use the RNAseq assembler *Trinity* (Grabherr *et al.*, 2011) to build repeat contigs from genomic samples of less than 1X of coverage. The use of *Trinity* allows to recover alternative repeat consensus of a given family by producing distinct contigs for each structural variant. Alternatively, other tools like *Tedna* (Zytnicki *et al.*, 2014), *RepARK* (Koch *et al.*, 2014) and *REPdenovo* (Chu *et al.*, 2016) use directly a de Bruijn graph on the most represented k-mers to perform TE assembly.

Although the previous methods are supposed to be able to find any kind of repeats, they usually are best suited to discover TEs. They may be less powerful concerning tandem repeats. This is why some specific tools have been designed to specifically uncover tandem repeats from raw reads. The first tool, *VNTRseek* (Gelfand *et al.*, 2014), is a variant detection tool that compares reads in which tandem repeats have been detected using *Tandem Repeat Finder* (Benson, 1999) to a set of tandem repeats present in a reference genome. By this comparison, the variable number of tandem repeats is determined. A recent pipeline called *TAREAN* (Novák *et al.*, 2017) uses the principles of graph-based repeat clustering, as implemented in the *RepeatExplorer* pipeline, as well as tools facilitating the unsupervised identification and characterization of satellite repeats from raw reads. The reconstruction of the satellite sequences is based on k-mer decomposition and counting.

Globally, all these programs allow to determine the global proportion of repeats inside a genome, with sometime the possibility to estimate the copy number, as well as a catalog of the different types of repeats with the production of consensus sequences. However, since these programs work on raw reads, the information concerning the exact positions of these repeats is missing. Other approaches have thus been developed to specifically answer this question by comparing the copy number variation of repeats between two genomes.

Identification of the Repeat Copy Number Variation

When the first genomes were sequenced, it was considered that only one would be enough to understand the functioning and evolution of a given species. However, having only one genome is not enough to uncover the polymorphism existing among individuals. Particularly for repeats, some of which being able to move and replicate themselves in the genome, it is known that variations exist in term of copy number and insertion sites among natural populations (Petrov *et al.*, 2011; Boulesteix *et al.*, 2006). With the decrease in cost of sequencing, it is now possible to obtain data from several individuals of a given population and of several populations of a given species. This has open the door to perform high throughput population genomics when using pooled sequencing data. Since it is not always possible to obtain good assembly with these data, new tools have been developed to specifically search for differences when compared to a reference genome. Thus, the goal of the bioinformatic tools developed for this purpose is to determine either one or the three types of TE insertions: insertions shared between the reference and the analyzed data (fixed insertions), and polymorphic insertions corresponding to insertions either absent from the analyzed data or absent from the reference genome (new insertions) (**Fig. 2**).

Several tools have been developed to determine the structural variation due to TEs in genomes and some attempts have been made to evaluate and review them (Ewing, 2015; Rishishwar *et al.*, 2016). The various methods have all in common in their process to first map the reads on a reference genome and/or on a set of annotated TE sequences before applying various filters and metrics to retain the informative ones. Then, two approaches exist that may be combined to analyze the results and to detect the presence/absence of a TE insertion. In the first approach, the program considers discordant read pairs, which are read pairs with one member matching uniquely on the reference genome sequence and the other matching on different copies from a TE family (**Fig. 3(A)**). This type of approach is used in the programs *TE-locate* (Platzer *et al.*, 2012), *TraFiC* (Tubio *et al.*, 2014) and *TE-Tracker* (Gilly *et al.*, 2014).

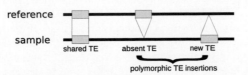

Fig. 2 The different types of TE insertions when comparing a reference genome and a newly sequenced one.

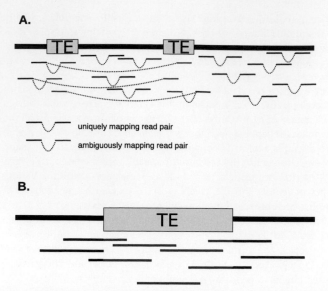

Fig. 3 Schematic representation of discordant read pairs (A) and split reads (B) that are used to identify TE insertion in raw data by comparison to a reference genome. Lines in black correspond to reads mapping on unique genomic regions and lines in red correspond to reads mapping on TE sequences.

The other approach consists in considering split reads, i.e., reads which overlap a junction between the genome and an inserted TE copies, with one part of the read mapping uniquely on the genome while the other part maps on TE sequences (**Fig. 3(B)**). The programs *RelocaTE* (Robb *et al.*, 2013), *ngs-TE-mapper* (Linheiro and Bergman, 2012), *TIDAL* (Rahman *et al.*, 2015), *ITIS* (Jiang *et al.*, 2015), and *T-Lex2* (Fiston-Lavier *et al.*, 2015) use this approach. Other programs use both approaches like *Tea* (Lee *et al.*, 2012), *RetroSeq* (Keane *et al.*, 2013), *TEMP* (Zhuang *et al.*, 2014), *Mobster* (Thung *et al.*, 2014), *Tangram* (Wu *et al.*, 2014), *TransposSeq* (Helman *et al.*, 2014), *Jitterbug* (Hénaff *et al.*, 2015), *DD_Detection* (Kroon *et al.*, 2016), *popoolation_TE2* (Kofler *et al.*, 2016) and *MELT* (Gardner *et al.*, 2017). These different programs have been developed to answer specific biological questions, which make them either consider individual or population data to estimate the insertion frequency, to consider in majority polymorphic insertions (especially new insertions in the case of cancer research) but sometimes also shared insertions, and in some cases to take into account the genotype status of the detected insertions, i.e. if the insertion is present on only one (heterozygous) or two (homozygous) homologous chromosomes. For example, the program *T-Lex2* has been developed to compute the frequency insertions of TE copies that are present in the reference genome of *Drosophila melanogaster* in natural populations, whereas the program *Tea* has been developed to identify only new TE insertions in human cancers corresponding to somatic insertions.

Discussion

Repeats, and more particularly TEs, are important component of the eukaryote genomes that cannot be simply ignored when performing sequencing and assembly tasks. A lot of various bioinformatics tools have been developed during the last 20 years to allow a better handling of these particular sequences. Such tools need to evolve jointly with the evolution of sequencing technologies. Currently, one of the major problem with the current whole genome sequencing data produced to identify repeat insertions is the size of the available sequence reads (< 300 bp). Full-length TE insertions ranged between 500 bp to 10 kb, which implies that several reads are needed to cover an entire TE copy. This step may be particularly difficult since several insertions may present a very high degree of nucleotide identity between themselves. In that case, it is often impossible to recover some insertions and the only way to distinguish them is to take into account the genomic environment, i.e. the flanking regions around the insertion. However, sequence reads being shorter than the majority of TE insertions and reads overlapping the junction of the genomic region and the TE insertions being not numerous and difficult to map, this task is particularly challenging and lead to a loss of information. A way to tackle this difficulty is to produce longer reads. The third generation of DNA sequencing is currently under development and some already used techniques may be helpful, although still expensive or not optimized. For example, PacBio sequencing allows to produce sequences up to 20 kb but the rate of errors is still quite high (Mardis, 2017). However, this technique may be used in addition to short-read sequencing techniques to enhance the quality of a genome assembly (Pendleton *et al.*, 2015). The Illumina synthetic long-read technique has been successfully tested to perform the *de novo* assembly and resolve TE sequences in the genome of *Drosophila melanogaster* (McCoy *et al.*, 2014). However this technique is still expensive since the coverage that is needed is very high (Mardis, 2017). The MinION technology, which performs a single molecule sequencing, has been recently successfully used to detect new TE insertions in the plant genome *Arabidopsis thaliana* allowing to show that the high error rate of the technique could be compensated by the read length in this particular question (Debladis *et al.*, 2017). Of course, as soon as these technologies will become the new standard, the current tools for TE analyses will become obsolete and new methodological procedures will need to be developed. In this way, a bioinformatic tool has recently been proposed to determine the presence/absence of TE insertions using reads produced by the PacBio technology (*LorTE*, Disdero and Filée, 2017).

Future Directions

The identification of TE insertions is also becoming very important in the field of epigenetics research. Indeed, TEs are known to be associated to particular epigenetic modifications that may impact the neighboring genes (Eichten *et al.*, 2012; Estécio *et al.*, 2012). Very few tools have been developed to allow the direct association between TEs and epigenetic modifications. For example, a web interface has been developed to specifically study histone modification enrichment of repeats taking advantage of their increased sequence coverage in ChIP-seq data (Day *et al.*, 2010). An advantage of this method is that it incorporates both ambiguously and uniquely mapped reads to avoid bias due to read mapping on consensus TE sequences. However, the results are not given at the insertion level and thus the information concerning the genomic environment of each insertion is lost. Several efforts have been made to develop methods allowing the association of small RNA data to TE sequences either at a global (*piPipes*, Han *et al.*, 2015) or at a individual copy level (*TEtools*, Lerat *et al.*, 2017). More recently, a new program has been developed allowing both the identification of new TE insertions and their associated DNA methylation in MethylC-seq data (*EpiTEome*, Daron and Slotkin, 2017). In summary, methodological efforts are still needed to study the epigenetic modifications directly associated to TE sequences. It is particularly important to consider the global sequence diversity of a TE family that may be very variable but also the genomic environment of a given insertion that may have consequences on the associated epigenetic modifications. These issues are currently difficult to handle due to the short size of sequence reads. However, as long read technologies will continue to be developed, it should open the door to new developments in a close future.

Closing Remarks

The domain of TE annotation in genome sequences is constantly producing new tools, which are supposed to outperform the previous ones and to offer new ways to handle the ever growing sequence data. However, the question of the impartial evaluation of these tools still remains. Indeed, the different tools have usually different competencies and may be at some point complementary. A clear problem underlying this situation is the lack of common standard data that would allow an unbiased estimation of any new tools (Hoen *et al.*, 2015). There is still room in that domain to propose various benchmarks according to the biological questions asked behind each tool.

See also: Functional Enrichment Analysis. Gene Mapping. Genome Annotation. Genome Annotation: Perspective From Bacterial Genomes. Genome Databases and Browsers. Genome Informatics. Integrative Analysis of Multi-Omics Data. Natural Language Processing Approaches in Bioinformatics. Next Generation Sequencing Data Analysis. Quantitative Immunology by Data Analysis Using Mathematical Models. Sequence Analysis. Sequence Composition. Whole Genome Sequencing Analysis

References

Abrusán, G., Grundmann, N., DeMester, L., Makalowski, W., 2009. TEclass – A tool for automated classification of unknown eukaryotic transposable elements. Bioinformatics 25, 1329–1330.

Agrawal, A., Eastman, Q.M., Schatz, D.G., 1998. Transposition mediated by RAG1 and RAG2 and its implications for the evolution of the immune system. Nature 394, 744–751.

Altschul, S.F., *et al.*, 1997. Gapped BLAST and PSI-BLAST: A new generation of protein database search programs. Nucleic Acids Research 25, 3389–3402.

Audemard, E., Schiex, T., Faraut, T., 2012. Detecting long tandem duplications in genomic sequences. BMC Bioinformatics 13, 83.

Bailly-Bechet, M., Haudry, A., Lerat, E., 2014. "One code to find them all": A perl tool to conveniently parse RepeatMasker output files. Mobile DNA 5, 13.

Bao, W., Kojima, K.K., Kohany, O., 2015. Repbase update, a database of repetitive elements in eukaryotic genomes. Mobile DNA 6, 11.

Bao, Z., Eddy, S.R., 2003. Automated *de novo* identification of repeat sequence families in sequenced genomes. Genome Research 13, 1269–1276.

Belancio, V., Hedges, D.J., Deininger, P., 2008. Mammalian non-LTR retrotransposons: For better or worse, in sickness and in health. Genome Research 18, 343–358.

Benson, G., 1999. Tandem repeats finder: A program to analyze DNA sequences. Nucleic Acids Research 27, 573–580.

Bergman, C.M., Quesneville, H., 2007. Discovering and detecting transposable elements in genome sequences. Briefings in Bioinformatics 8, 382–392.

Biemont, C., 2010. A brief history of the status of transposable elements: From junk DNA to major players in evolution. Genetics 186, 1085–1093.

Biémont, C., Vieira, C., 2006. Genetics: Junk DNA as an evolutionary force. Nature 443, 521–524.

Boulesteix, M., Simard, F., Antonio-Nkondjio, C., *et al.*, 2006. Insertion polymorphism of transposable elements and population structure of *Anopheles gambiae* M and S molecular forms in Cameroon. Molecular Ecology 16, 441–452.

Britten, R.J., Davidson, E.H., 1969. Gene regulation for higher cells: A theory. Science 165, 349–357.

Castanera, R., López-Varas, L., Borgognone, A., *et al.*, 2016. Transposable elements versus the fungal Genome: Impact on whole-genome architecture and transcriptional profiles. PLOS Genetics 12, e1006108.

Chen, Y., Zhou, F., Li, G., Xu, Y., 2009. MUST: A system for identification of miniature inverted-repeat transposable elements and applications to *Anabaena variabilis* and *Haloquadratum walsbyi*. Gene 436, 1–7.

Chu, C., Nielsen, R., Wu, Y., 2016. REPdenovo: Inferring *de novo* repeat motifs from short sequence reads. PLOS ONE 11, e0150719.

Daron, J., Slotkin, R.K., 2017. EpiTEome: Simultaneous detection of transposable element insertion sites and their DNA methylation levels. Genome Biology 18, 91.

Day, D.S., Luquette, L.J., Park, P.J., Kharchenko, P.V., 2010. Estimating enrichment of repetitive elements from high-throughput sequence data. Genome Biology 11, R69.

DeBarry, J.D., Liu, R., Bennetzen, J.L., 2008. Discovery and assembly of repeat family pseudomolecules from sparse genomic sequence data using the Assisted Automated Assembler of Repeat Families (AAARF) algorithm. BMC Bioinformatics 9, 235.

Debladis, E., Llauro, C., Carpentier, M.C., Mirouze, M., Panaud, O., 2017. Detection of active transposable elements in *Arabidopsis thaliana* using Oxford Nanopore Sequencing technology. BMC Genomics 18, 537.

Delgrange, O., Rivals, E., 2004. STAR: An algorithm to search for tandem approximate repeats. Bioinformatics 20, 2812–2820.

Disdero, E., Filée, J., 2017. LoRTE: Detecting transposon-induced genomic variants using low coverage PacBio long read sequences. Mobile DNA 8, 5.

Dowsett, A., Young, M.W., 1982. Differing levels of dispersed repetitive DNA among closely related species of *Drosophila*. Proceedings of the National Academy of Sciences of the United States of America 79, 4570–4574.

Dumbovic, G., Forcales, S.-V., Perucho, M., 2017. Emerging roles of macrosatellite repeats in genome organization and disease development. Epigenetics 12, 515–526.

Edgar, R.C., Myers, E.W., 2005. PILER: Identification and classification of genomic repeats. Bioinformatics 21 (Suppl 1), 152–158.

Eichten, S.R., Ellis, N.A., Makarevitch, I., *et al.*, 2012. Spreading of heterochromatin is limited to specific families of maize retrotransposons. PLOS Genetics 8, e1003127.

Ellinghaus, D., Kurtz, S., Willhoeft, U., 2008. LTRharvest, an efficient and flexible software for de novo detection of LTR retrotransposons. BMC Bioinformatics 9, 18.

Elliott, T.A., Gregory, T.R., 2015. What's in a genome? The C-value enigma and the evolution of eukaryotic genome content. Philosophical transactions of the Royal Society of London. Series B: Biological Sciences 370, 20140331.

Estécio, M.R., Gallegos, J., Dekmezian, M., *et al.*, 2012. SINE retrotransposons cause epigenetic reprogramming of adjacent gene promoters. Molecular Cancer Research 10, 1332–1342.

Ewing, A.D., 2015. Transposable element detection from whole genome sequence data. Mobile DNA 6, 24.

Feschotte, C., Keswani, U., Ranganathan, N., Guibotsy, M.L., Levine, D., 2010. Exploring repetitive DNA landscapes using REPCLASS, a tool that automates the classification of transposable elements in eukaryotic genomes. Genome Biology and Evolution 1, 205–220.

Fiston-Lavier, A.S., Barrón, M.G., Petrov, D.A., González, J., 2015. T-lex2: Genotyping, frequency estimation and re-annotation of transposable elements using single or pooled next-generation sequencing data. Nucleic Acids Research 43, e22.

Flutre, T., Duprat, E., Feuillet, C., Quesneville, H., 2011. Considering transposable element diversification in *de novo* annotation approaches. PLOS ONE 6, e16526.

Gardner, E.J., Lam, V.K., Harris, D.N., *et al.*, 2017. The Mobile Element Locator Tool (MELT): Population-scale mobile element discovery and biology. Genome Research 218032, 116.

Gelfand, Y., Hernandez, Y.2., Loving, J., Benson, G., 2014. VNTRseek – A computational tool to detect tandem repeat variants in high-throughput sequencing data. Nucleic Acids Research 42, 8884–8894.

Ge, R., Mai, G., Zhang, R., *et al.*, 2017. MUSTv2: An improved de novo detection program for recently active Miniature Inverted repeat Transposable Elements (MITEs). Journal of Integrative Bioinformatics 14.

Gilly, A., Etcheverry, M., Madoui, M.A., *et al.*, 2014. TE-Tracker: Systematic identification of transposition events through whole-genome resequencing. BMC Bioinformatics 15, 377.

Goubert, C., Modolo, L., Vieira, C., *et al.*, 2015. *De novo* assembly and annotation of the Asian tiger mosquito (*Aedes albopictus*) repeatome with dnaPipeTE from raw genomic reads and comparative analysis with the yellow fever mosquito (*Aedes aegypti*). Genome Biology and Evolution 7, 1192–1205.

Grabherr, M.G., Haas, B.J., Yassour, M., *et al.*, 2011. Full-length transcriptome assembly from RNA-Seq data without a reference genome. Nature Biotechnology 29, 644–652.

Han, B.W., Wang, W., Zamore, P.D., Weng, Z., 2015. piPipes: A set of pipelines for piRNA and transposon analysis via small RNA-seq, RNA-seq, degradome- and CAGE-seq, ChIP-seq and genomic DNA sequencing. Bioinformatics 31, 593–595.

Han, Y., Wessler, S.R., 2010. MITE-Hunter: A program for discovering miniature inverted-repeat transposable elements from genomic sequences. Nucleic Acids Research 38, 1–8.

Hancks, D.C., Kazazian, H.H., 2012. Active human retrotransposons: Variation and disease. Current Opinion in Genetics and Development 22, 191–203.

Helman, E., Lawrence, M.S., Stewart, C., *et al.*, 2014. Somatic retrotransposition in human cancer revealed by whole-genome and exome sequencing. Genome Research 24, 1053–1063.

Hénaff, E., Zapata, L., Casacuberta, J.M., Ossowski, S., 2015. Jitterbug: Somatic and germline transposon insertion detection at single-nucleotide resolution. BMC Genomics 16, 768.

Hoede, C., Arnoux, S., Moisset, M., *et al.*, 2014. PASTEC: An automatic transposable element classification tool. PLOS ONE 9, e91929.

Hoen, D.R., Hickey, G., Bourque, G., *et al.*, 2015. A call for benchmarking transposable element annotation methods. Mobile DNA 6, 13.

Jiang, C., Chen, C., Huang, Z., Liu, R., Verdier, J., 2015. ITIS, a bioinformatics tool for accurate identification of transposon insertion sites using next-generation sequencing data. BMC Bioinformatics 16, 72.

Kapitonov, V.V., Jurka, J., 2008. A universal classification of eukaryotic transposable elements implemented in Repbase. Nature reviews Genetics 9, 411–412.

Keane, T.M., Wong, K., Adams, D.J., 2013. RetroSeq: Transposable element discovery from next-generation sequencing data. Bioinformatics 29, 389–390.

Kidwell, M.G., Lisch, D.R., 2000. Transposable elements and host genome evolution. Trends in Ecology and Evolution 15, 95–99.

Kim, J.M., Vanguri, S., Boeke, J.D., Gabriel, A., Voytas, D.F., 1998. Transposable elements and genome organization: A comprehensive survey of retrotransposons revealed by the complete *Saccharomyces cerevisiae* genome sequence. Genome Research 8, 464–478.

Koch, P., Platzer, M., Downie, B.R., 2014. RepARK – *De novo* creation of repeat libraries from whole-genome NGS reads. Nucleic Acids Research 42, 1–12.

Kofler, R., Gómez-Sánchez, D., Schlötterer, C., 2016. PoPoolationTE2: Comparative population genomics of transposable elements using Pool-Seq. Molecular Biology and Evolution 33, 2759–2764.

Kolpakov, R., Bana, G., Kucherov, G., 2003. mreps: Efficient and flexible detection of tandem repeats in DNA. Nucleic Acids Research 31, 3672–3678.

Kroon, M., Lameijer, E.W., Lakenberg, N., *et al.*, 2016. Detecting dispersed duplications in high-throughput sequencing data using a database-free approach. Bioinformatics 32, 505–510.

Lander, E.S., *et al.*, 2001. Initial sequencing and analysis of the human genome. Nature 409, 860–921.

Lee, E., Iskow, R., Yang, L., *et al.*, 2012. Landscape of somatic retrotransposition in human cancers. Science 337, 967–971.

Lerat, E., 2010. Identifying repeats and transposable elements in sequenced genomes: How to find your way through the dense forest of programs. Heredity 104, 520–533.

Lerat, E., Fablet, M., Modolo, L., Lopez-Maestre, H., Vieira, C., 2017. TEtools facilitates big data expression analysis of transposable elements and reveals an antagonism between their activity and that of piRNA genes. Nucleic Acids Research 45, e17.

Lien, S., Koop, B.F., Sandve, S.R., *et al.*, 2016. The Atlantic salmon genome provides insights into rediploidization. Nature 533, 200–205.

Lim, K.G., Kwoh, C.K., Hsu, L.Y., Wirawan, A., 2013. Review of tandem repeat search tools: A systematic approach to evaluating algorithmic performance. Briefings in Bioinformatics 14, 67–81.

Linheiro, R.S., Bergman, C.M., 2012. Whole-genome resequencing reveals natural target site preferences of transposable elements in Drosophila melanogaster. PLOS ONE 7, e30008.

Li, R., Ye, J., Li, S., et al., 2005. ReAS: Recovery of ancestral sequences for transposable elements from the unassembled reads of a whole genome shotgun. PLOS Computational Biology 1, 313–321.

Macas, J., Neumann, P., Navrátilová, A., 2007. Repetitive DNA in the pea (Pisum sativum L.) genome: Comprehensive characterization using 454 sequencing and comparison to soybean and Medicago truncatula. BMC Genomics 8, 427.

Mardis, E.R., 2017. DNA sequencing technologies: 2006–2016. Nature Protocols 12, 213–218.

Margulies, M., Egholm, M., Altman, W.E., et al., 2005. Genome sequencing in microfabricated high-density picolitre reactors. Nature 437, 376–380.

McCarthy, E.M., McDonald, J.F., 2003. LTR_STRUC: A novel search and identification program for LTR retrotransposons. Bioinformatics 19, 362–367.

McClintock, B., 1950. The origin and behavior of mutable loci in maize. Proceedings of the National Academy of Sciences of the United States of America 36, 344–355.

McCoy, R.C., Taylor, R.W., Blauwkamp, T.A., et al., 2014. Illumina TruSeq Synthetic Long-Reads empower de novo assembly and resolve complex, highly-repetitive transposable elements. PLOS ONE 9, e106689.

Miki, Y., Nishisho, I., Horii, A., et al., 1992. Disruption of the APC gene by a retrotransposal insertion of L1 sequence in a colon cancer. Cancer Research 52, 643–645.

Modolo, L., Lerat, E., 2013. Identification and analysis of transposable elements in genomic sequences. In: Poptsova, M. (Ed.), Genome Analysis: Current Procedures and Applications. Poole: Caister Academic Press, pp. 165–181.

Newman, A.M., Cooper, J.B., 2007. XSTREAM: A practical algorithm for identification and architecture modeling of tandem repeats in protein sequences. BMC Bioinformatics 8, 382.

Novák, P., Ávila Robledillo, L., Koblížková, A., et al., 2017. TAREAN: A computational tool for identification and characterization of satellite DNA from unassembled short reads. Nucleic Acids Research 45, e111.

Novák, P., Neumann, P., Macas, J., 2010. Graph-based clustering and characterization of repetitive sequences in next-generation sequencing data. BMC Bioinformatics 11, 378.

Novák, P., Neumann, P., Pech, J., Steinhais, J., Macas, J., 2013. RepeatExplorer: A galaxy-based web server for genome-wide characterization of eukaryotic repetitive elements from next-generation sequence reads. Bioinformatics 29, 792–793.

Pendleton, M., Sebra, R., Pang, A.W., et al., 2015. Assembly and diploid architecture of an individual human genome via single-molecule technologies. Nature Methods 12, 780–786.

Petrov, D.A., Fiston-Lavier, A.S., Lipatov, M., Lenkov, K., González, J., 2011. Population genomics of transposable elements in Drosophila melanogaster. Molecular Biology and Evolution 28, 1633–1644.

Platt, R.N., Blanco-Berdugo, L., Ray, D.A., 2016. Accurate transposable element annotation is vital when analyzing new genome assemblies. Genome Biology and Evolution 8, 403–410.

Platzer, A., Nizhynska, V., Long, Q., 2012. TE-Locate: A tool to locate and group transposable element occurrences using paired-end next-generation sequencing data. Biology 1, 395–410.

Price, A.L., Jones, N.C., Pevzner, P.A., 2005. De novo identification of repeat families in large genomes. Bioinformatics 21 (Suppl. 1), 351–358.

Rahman, R., Chirn, G.W., Kanodia, A., et al., 2015. Unique transposon landscapes are pervasive across Drosophila melanogaster genomes. Nucleic Acids Research 43, 10655–10672.

Rho, M., Choi, J.H., Kim, S., Lynch, M., Tang, H., 2007. De novo identification of LTR retrotransposons in eukaryotic genomes. BMC Genomics 8, 90.

Richard, G.-F., Kerrest, A., Dujon, B., 2008. Comparative genomics and molecular dynamics of DNA repeats in eukaryotes. Microbiology and Molecular Biology Reviews 72, 686–727.

Rishishwar, L., Mariño-Ramírez, L., Jordan, I.K., 2016. Benchmarking computational tools for polymorphic transposable element detection. Briefings in Bioinformatics. bbw072.

Robb, S.M.C., Lu, L., Valencia, E., et al., 2013. The use of RelocaTE and unassembled short reads to produce high-resolution snapshots of transposable element generated diversity in rice. G3 3, 949–957.

Rostant, W.G., Wedell, N., Hosken, D.J., 2012. Transposable elements and insecticide resistance. Advances in Genetics 78, 169–201.

Saha, S., Bridges, S., Magbanua, Z.V., Peterson, D.G., 2008. Computational approaches and tools used in identification of dispersed repetitive DNA sequences. Tropical Plant Biology 1, 85–96.

Santiago, N., Herráiz, C., Goñi, J.R., Messeguer, X., Casacuberta, J.M., 2002. Genome-wide analysis of the Emigrant family of MITEs of Arabidopsis thaliana. Molecular Biology and Evolution 19, 2285–2293.

Schnable, S., Ware, D., Fulton, R.S., et al., 2009. The B73 maize genome: Complexity, diversity, and dynamics. Science 326, 1112–1115.

Segal, Y., Peissel, B., Renieri, A., et al., 1999. LINE-1 elements at the sites of molecular rearrangements in Alport syndrome-diffuse leiomyomatosis. American Journal of Human Genetics 64, 62–69.

Sharp, A.J., Locke, D.P., McGrath, S.D., et al., 2005. Segmental duplications and copy-number variation in the human genome. American Journal of Human Genetics 77, 78–88.

Smit, A.F.A., Hubley, R., Green, P., 1996–2010. RepeatMasker Open-3.0. Available at: http://www.repeatmasker.org.

Sokol, D., Benson, G., Tojeira, J., 2007. Tandem repeats over the edit distance. Bioinformatics 23, e30–e35.

Staton, S.E., Burke, J.M., 2015. Transposome: A toolkit for annotation of transposable element families from unassembled sequence reads. Bioinformatics 31, 1827–1829.

Steinbiss, S., Willhoeft, U., Gremme, G., Kurtz, S., 2009. Fine-grained annotation and classification of de novo predicted LTR retrotransposons. Nucleic Acids Research 37, 7002–7013.

Szak, S.T., Pickeral, O.K., Makalowski, W., et al., 2002. Molecular archeology of L1 insertions in the human genome. Genome Biology 3.(research0052).

Thung, D.T., de Ligt, J., Vissers, L.E., et al., 2014. Mobster: Accurate detection of mobile element insertions in next generation sequencing data. Genome Biology 15, 488.

Treangen, T.J., Salzberg, S.L., 2011. Repetitive DNA and next-generation sequencing: Computational challenges and solutions. Nature Reviews Genetics 13, 36–46.

Tu, Z., Li, S., Mao, C., 2004. The changing tails of a novel short interspersed element in Aedes aegypti: Genomic evidence for slippage retrotransposition and the relationship between 3′ tandem repeats and the poly(dA) tail. Genetics 168, 2037–2047.

Tubio, J.M.C., Li, Y., Ju, Y.S., et al., 2014. Mobile DNA in cancer. Extensive transduction of nonrepetitive DNA mediated by L1 retrotransposition in cancer genomes. Science 345, 1251343.

Waterston, R.H., Lindblad-Toh, K., Birney, E., et al., 2002. Initial sequencing and comparative analysis of the mouse genome. Nature 420, 520–562.

Wicker, T., Schlagenhauf, E., Graner, A., et al., 2006. 454 Sequencing put to the test using the complex genome of barley. BMC Genomics 7, 275.

Wicker, T., Sabot, F., Hua-Van, A., et al., 2007. A unified classification system for eukaryotic transposable elements. Nature Reviews Genetics 8, 973–982.

Wu, J., Lee, W.P., Ward, A., et al., 2014. Tangram: A comprehensive toolbox for mobile element insertion detection. BMC Genomics 15, 795.

Xu, Z., Wang, H., 2007. LTR_FINDER: An efficient tool for the prediction of full-length LTR retrotransposons. Nucleic Acids Research 35 (Web Server), W265–W268.

Zhuang, J., Wang, J., Theurkauf, W., Weng, Z., 2014. TEMP: A computational method for analyzing transposable element polymorphism in populations. Nucleic Acids Research 42, 6826–6838.

Zytnicki, M., Akhunov, E., Quesneville, H., 2014. Tedna: A transposable element de novo assembler. Bioinformatics 30, 2656–2658.

Relevant Website

http://www.repeatmasker.org/RepeatModeler.html
Institute for Systems Biology – RepeatModeler Download.

Biographical Sketch

Emmanuelle Lerat is a CNRS researcher since 2005 working in the Laboratory "Biométrie et Biologie Evolutive" at the University Lyon 1, France. Her major research interest concerns the evolution of genomes in the light of their repeat content and more particularly the transposable elements (TEs). She has a strong background in molecular biology, in evolution, and in bioinformatics, with a long history in the annotation and analysis of TEs in various eukaryotic genomes, especially in Drosophila. She currently coordinates different subjects allowing to link informatic to biological questions concerning the functional impact of TEs on genome and gene evolution. She is a member of the Editorial Board of the international SMBE journal "Genome Biology and Evolution" since 2008.

Bioinformatics Approaches for Studying Alternative Splicing

Prasoon K Thakur, Institute of Molecular Genetics of the ASCR, Prague Czech Republic
Hukam C Rawal, ICAR – National Research Centre on Plant Biotechnology, New Delhi, India
Mina Obuca, Institute of Molecular Genetics of the ASCR, Prague, Czech Republic
Sandeep Kaushik, European Institute of Excellence on Tissue Engineering and Regenerative Medicine, Guimaraes, Portugal and University of Minho, Braga, Portugal

Introduction

Genes

Genes are genomic regions that contain necessary information for the expression of a protein or a molecule that ultimately helps in the survival, reproduction and function of the organism. The information stored in the DNA sequence of a gene is first transcribed into an RNA molecule, by RNA polymerase. For the protein coding genes, this transcript is called a messenger RNA (mRNA) which is then translated to synthesize proteins. The nascent RNA (pre-mRNA) transcripts contain intervening sequences, known as the *introns*, that do not become part of the final mRNA (Gilbert, 1978). The regions of pre-mRNA that are retained and ligated for translation are known as *exons*. The *introns* are removed from nascent RNA by a mechanism known as *splicing*. The spliced mRNA molecule forms a continuous protein-coding region ready to be translated into a protein molecule. Nearly all eukaryotes share the presence of *introns* and the mechanism of RNA *splicing*. The presence of introns plays a major role in facilitating recombination of RNA sequences and helps in developing protein diversity (Rogozin *et al.*, 2012). For a better understanding of the bioinformatics tools and approaches used to study alternative splicing, it is necessary to have a brief overview of the elements and mechanisms involved in splicing.

Eukaryotic vs Prokaryotic Splicing

Eukaryotes have devised a complex mechanism of modifying their primary RNA transcripts. Using this mechanism, called *"splicing"*, the nascent mRNA is processed so that the *introns* are selectively excised out and *exons* are ligated together. Splicing is catalyzed by a mega-dalton multi-ribonucleoprotein (RNP) complex known as the *"spliceosome"* (Will and Lührmann, 2011). The spliceosome is composed of five small nuclear RNPs (snRNPs) and numerous proteins. *Splicing* is more prevalent in higher eukaryotes and leads to the expression of a higher number of unique proteins from a single gene (Nilsen and Graveley, 2010). On the other hand, protein coding genes in prokaryotes lack introns which is complemented by the lack of a spliceosomal machinery (Rogozin *et al.*, 2012). This is why bacterial genes have a one-to-one correspondence of bases between a gene and its mRNA. However, some bacteria and viruses do contain some rare introns in non-coding RNAs that have been observed to self-splice *in vitro* (Woodson, 1998; Belfort, 1990; Schmidt, 1985; Berget *et al.*, 1977; Chow *et al.*, 1977). Presence of introns in early life forms has been discovered recently and this still remains a confounding fact (Roy and Gilbert, 2006).

Spliceosome and the Mechanism of Splicing

The post-transcriptional modification of the nascent mRNA (pre-mRNA) for the removal of introns is catalyzed by the spliceosome. The spliceosome is a large RNP complex composed of five snRNPs (U1, U2, U4, U5 and U6) and other accessory proteins (Staley and Guthrie, 1998; Jurica and Moore, 2003). **Fig. 1** depicts various elements of the spliceosome. Each snRNP contains the corresponding uridine rich non-coding small RNA molecule (snRNA) and a variable number of complex-specific proteins (Hoskins *et al.*, 2011). Another variant of the spliceosome is known to exist; it is composed of other functionally analogous snRNPs and splices a rare class of introns (Patel and Steitz, 2003). Below, we describe the mechanism that relates to the well-explored and major spliceosome.

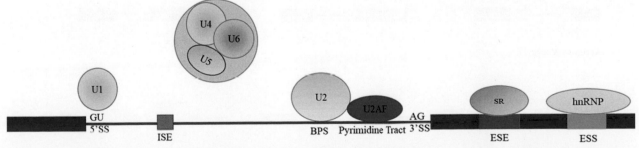

Fig. 1 Alternative splicing mechanism and events. Splicing factors recognize different sequences along the pre-mRNA. Conserved recognition sequences include intron boundaries named 5 ss and 3 ss. These elements interact with the spliceosome, recruiting snRNPs (U1, U2, U4, U5 and U6). Other regulatory elements are embedded in the exon (ESE, ESS) or in the intron (ISE) and could be recognized by proteins of the SR and/or hnRNP families.

The splicing reaction is a coordinated series of RNA–RNA, RNA–protein and protein–protein interactions (Trowitzsch *et al.*, 2009; Hoskins and Moore, 2012). First, U1 binds to the 5′ splice site of an exon and U2 binds near branch point sequence (BPS) just upstream of the 3′ splice site of the adjacent exon (Peled-Zehavi *et al.*, 2001). The U1 and U2 binding is important for the intron definition, thus for the accuracy of the splicing reaction. Later, a tri-snRNP complex, composed of U4/U6 and U5, joins in and leads to the formation of an active complex that catalyzes splicing. Once the splicing is over, the spliceosome disassembles and all components are recycled for future splicing reactions (Hnilicova and Stanek, 2011). Each splicing alternative is regulated through an interplay between constitutive splicing motifs, such as 5′ splice sites, branch points, poly-pyrimidine tracts and 3′ splice sites and components of the core splicing machinery. Exons and introns often contain *cis*-regulatory sequences, intronic enhancers or silencers that affect the splicing in a positive and negative manner. Splicing regulatory sequences (enhancer or silencer) are usually 6–10 base pair (bp) in length and induce binding of specific regulatory proteins such as serine/arginine rich (SR) proteins or heterogeneous nuclear ribonucleoproteins (hnRNPs) (Black, 2003; Wang and Burge, 2008; Chasin, 2007; Han *et al.*, 2010).

Types of Splicing

Splicing of a multi-exon pre-mRNA transcripts leads to two or more distinct mRNA products (Black, 2003; Wang and Burge, 2008; Nilsen and Graveley, 2010; Calarco *et al.*, 2011). Based on the complexity of the events, two different types of splicing have been described: *constitutive* splicing and *alternative* splicing (van den Hoogenhof *et al.*, 2016). *Constitutive splicing* (CS) is the process where the introns are removed from the pre-mRNA and the exons are joined together to form a mature mRNA. On the other hand, *alternative splicing* (AS), is the process where exons can be included or excluded in different combinations to create a diverse range of mRNA transcripts from a single pre-mRNA (Nilsen and Graveley, 2010). Computational and experimental studies have concurred that splicing signals at AS sites are weaker, the length of AS exons are shorter than that of CS exons, skipped exons tend to preserve the reading frame at a greater frequency than CS, and orthologous AS are more conserved than orthologous CS (Zheng *et al.*, 2005; Clark and Thanaraj, 2002; Garg and Green, 2007). Because of a rising interest towards AS in past years, this article is focused on it.

Alternative Splicing and its Mechanism

Alternative splicing (AS) is a mechanism by which one gene gives rise to multiple protein products with different functions (Roy *et al.*, 2013; Brett *et al.*, 2002). It was discovered when two different mRNA species were observed to be encoded by the same gene (Early *et al.*, 1980; Rosenfeld *et al.*, 1982). Recent high-throughput transcriptome analysis has suggested that about 95% of the mammalian multi-exonic genes are subjected to AS (Pan *et al.*, 2008). Four main AS events have been reported in literature namely, exon skipping, intron retention, alternative donor and alternative acceptor splice sites. AS events can be classified into different splicing patterns that include cassette exons, mutually exclusive exons, retained intron, alternative 5′ splice sites, alternative 3′ splice sites, alternative promoters, and alternative poly-A sites (**Fig. 2**; Keren *et al.*, 2010; Chen, 2011). Among them, the most common type of alternative splicing is including or skipping a cassette exon in the mature mRNA. In cassette exons, internal

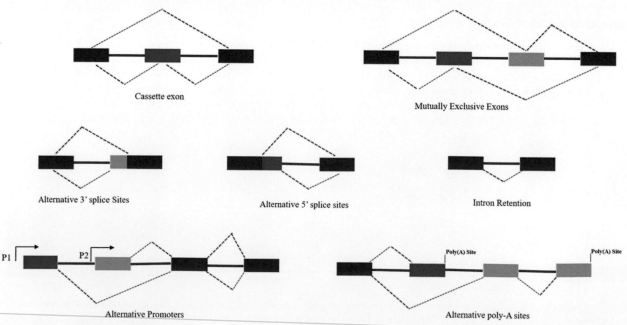

Fig. 2 Types of alternative splicing.

exons can either be included or excluded from the mature mRNA (Zhang et al., 2016). In case of intron retention, the excision of an intron can be suppressed, which results in the retention of the entire intron and exons can be extended or shortened through the use of alternative 5′ or 3′ splice sites. In contrast, alternative promoters and alternative poly-A sites are alternative selection of transcription start sites or poly-A sites and are not due to alternative splicing. Incompletely spliced transcripts may contain intron fragments that can be inaccurately identified as intron retention, so it is hard to distinguish them from experimental artifacts. In general, intron retention is the most difficult AS event to detect. Many genes have multiple alternative splicing events with complex combinations of exons and produce diverse transcript isoforms. Most interesting example is, Drosophila melanogaster's Down syndrome cell adhesion molecule (Dscam) gene contains 20 constitutive and 95 cassette exons. These 95 cassette exons can express 38,016 different mature mRNAs (Graveley et al., 2004; Missler and Sudhof, 1998).

Alternative Splicing Regulation

As already mentioned, AS increases the proteomic diversity in cells to adapt to various conditions. Therefore, it is important that AS be precisely regulated. Except the minimal core elements 5′ splice site, 3′ splice site and BP that are crucial for defining splicing sites, there are others instructions that help in discriminating between correct and incorrect splice sites and play important roles in AS regulation. These instructions are called as (exonic or intronic) splicing enhancers and/or silencers which are represented as 4–10 nucleotides. Two major classes of widely expressed trans-acting factors that are involved in binding these sequences and thus in controlling of splice site recognition are the SR proteins and heterogeneous ribonucleoproteins (hnRNPs). SR proteins got their name according to the domain enriched in dipeptides arginine/serine (RS). They bind to the pre-mRNA by their N terminal domains which is an RNA recognition motif (RRM) and mediates protein-protein interaction by SR domain facilitating assembly of spliceosome proteins (Long and Javier, 2009). Bradley and colleagues found that all eight SR proteins in Drosophila affect approximately 560 endogenous simple ASS events proving that they are indeed the key regulators of AS (Bradley et al., 2015). The splicing repressor family of proteins are hnRNPs that were first identified as proteins that interact with heterogeneous nuclear RNA. They have their roles in telomerase maintenance, mRNA stability, regulation of transcription and translation (Geuens et al., 2016). Several of the hnRNPs are identified as splicing repressors that suppress splicing by different mechanisms such as directly displacing snRNP binding, blocking interaction between U1 and U2 snRNP or looping out exons (Martinez-Contreras et al., 2007). It is worth to mention that nuclear concentration of regulators help proper splicing patterns and that SR as well as hnRNP proteins control their own splicing patterns. At the end, these proteins can compete for binding to an overlapping ESE/ESS element (Zahler et al., 2004) and this binding competition between them provides a combinatorial control in final decision of splice site choice.

Tissue-Specific AS in Humans

AS is considered an important mechanism for tissue-specific expression of transcript isoforms in humans. Recent EST based analyses have shown that the human brain contains highest fraction of alternatively spliced genes, followed by the liver and testis whereas human muscle, uterus, breast, stomach and pancreas were observed to have the lowest levels of genes undergoing AS (Lee et al., 2003; Xu et al., 2002; Stamm et al., 2000). Moreover, human AS events have different usage pattern of alternative 5′ splice site exons (A5Es) or alternative 3′ splice site exons (A3Es). These studies suggest that the fraction of genes containing A3Es and A5Es have significantly high in the liver as in any other human tissue (Yeo et al., 2004). Brain was observed to be the tissue with the second highest level of alternative usage for both 5′ splice sites and 3′ splice sites whereas muscle, uterus, breast, pancreas and stomach have shown the lowest level of A5Es and A3Es (Yeo et al., 2004). AS events variation in human tissues are primarily regulated by trans-acting factors that bind on exonic and intronic cis-acting RNA elements (CAEs). A computational study using publicly available Affymetrix Genechip Human Exon Array dataset identified 652 cis-acting RNA elements (CAEs) across 11 human tissues (Wang et al., 2009). Approximately, one third of all predicted CAEs matched with exonic splicing regulator databases and the vast majority of predicted CAEs were observed in the intronic regulatory regions. Most of these CAEs contribute to the AS between two tissues, while some are important in multiple tissues. Overall, the analysis suggests that genome-wide AS patterns are regulated by a combination of tissue-specific cis-acting elements and "general elements" whose functional activities are important but differ across multiple tissues (Wang et al., 2009).

Bioinformatics Approaches for AS Detection

Until recently, systematic analysis of AS was done using expressed sequence tags (EST) (Gupta et al., 2004; Sorek et al., 2004; Xie et al., 2002) or splicing-specific microarrays (Castle et al., 2008; Clark et al., 2002; Pan et al., 2008). However, the detection of alternative splicing events has been significantly improved by next-generation sequencing (NGS). From this data, AS is identified through bioinformatics approaches by aligning the EST sequences with mRNA and genomic sequences using sequence alignment tools (Kim and Lee, 2008). Genomic sequence acts as the main reference sequence for detection and validation of AS events. With the availability of high-throughput techniques, multiple genomes have been sequenced and it has become feasible to study AS on a genomic scale (Pan et al., 2008; Wang et al., 2008). Therefore, a large number of alternative transcripts have been discovered along with extraction of distinctive features of alternatively spliced exons using bioinformatics (Roy et al., 2013). With the advances in tools used to study AS, it has also become easier to uncover the splicing dysregulations that lead to diseases. It is observed that dysregulation of alternative splicing affects various human conditions, including cancers (Ghigna *et al.*, 2010; Yap and Makeyev, 2013; Mills and Janitz, 2012; Poulos *et al.*, 2011; Mittendorf *et al.*, 2012; Manetti *et al.*, 2011; Endo-Umeda *et al.*, 2012; Medina and Krauss, 2013; Bogdanov, 2006; Lara-Pezzi *et al.*, 2012; Cooper, 2005; Miura *et al.*, 2011; Sampath and Pelus, 2007; Hagen and Ladomery, 2012; Yi and Tang, 2011; Omenn *et al.*, 2010). Databases like dbSNP (see "Relevant Websites section") and ssSNPTarget (see "Relevant Websites section") help to search for the splice site SNPs located in the genes of interest for the identification of any association to diseases (Tang *et al.*, 2013). These tools prove helpful in developing new splicing-targeted drugs or therapeutics.

AS detection usually involves three distinct stages. During the first step, transcript and genomic sequence data are processed to eliminate repetitive or ambiguous sequences. Poly-A tails are also removed during this pre-processing even though there may be some genuine poly-A genomics sequences. Second, the transcript sequences are aligned to the genomic sequence and "gene models" are deduced. The alignment to genomic sequence can be performed using tools like BLAT, GMAP or SPA (Kent, 2002; Wu and Watanabe, 2005; van Nimwegen *et al.*, 2006) with or independent of SIM4 (Florea *et al.*, 1998). Next, several possible alignments for a given sequence are screened out by choosing only the best hit as measured by percent identity and alignment coverage. Rest of the unspliced alignments can be removed if strictly focused on AS detection. Third, each of the alternative splicing events are identified and various alternative "gene models" are constructed. In order to exclude artifacts, individual AS events only consist of a pair of genomically non-overlapping splicing events. From here, liberal methods generate all possible combinations of the splicing events (splicing graphs) whereas more conservative methods seek only a minimal set of isoforms. Below, we present an overview of the role of bioinformatics approaches and tools in detecting alternative splicing along with its relevance in present era and challenges involved.

AS detection using sequence alignment: We can identify alternative splicing events by aligning ESTs with genomic and mRNA sequences using alignment tools like BLAST, BLAT (BLAST-Like Alignment Tool) or ClustalW. Using the alignment between ESTs and genomic sequences, one can detect the locations of exons and introns, and then by comparing their structures we can identify the alternative splicing events. For identifying alternative splicing by sequence alignment, we can further use alignment tools like GMAP or SPA to correct the genome alignments and generate valid alignments. Once we have identified the alternative splicing events, we can construct full-length alternatively spliced transcripts using graphical display tools including "Splice graph". Although these alignments based methods are rapid, possible limitations that could affect the analysis include high sequencing errors, contamination, misalignments, low sequence coverage of ESTs.

AS detection using sequence conservation: AS events are conserved among different organisms *e.g.*, human and mouse, and therefore are of biological importance. Comparative genomics can provide us different means to predict the conserved exons that were spliced alternatively featuring the evolutionarily conserved alternative splicing events. As alternative exons have comparatively higher conservation level than constitutive exons in flanking intronic regions, hence can be used as good identifier for the alternative splicing (Chen, 2011).

AS detection using microarray data analysis: The above mentioned two approaches of sequence alignment and comparative genomics for identification of alternative splicing events, provides us only with the existence of an alternative splicing event but neither the degree nor the regulation of these alternative splicing events. The microarray study can provide us quantity of an alternative splicing event for a particular stage, condition or tissue of a cell. Previously, Affymetrix exon arrays were used by researchers to identify tissue-specific exons (Clark *et al.*, 2007) and differentially expressed AS between different human-cells (Yeo *et al.*, 2007). Some of the tools or methods that have been used widely for the alternative splicing microarray data analysis includes the splicing index calculation, Analysis of splice variation (ANOSVA), Finding isoforms using robust multichip analysis (FIRMA), a gene structure-based splice variant deconvolution method (DECONV), splicing prediction and concentration estimation (SPACE), and Generative model for the alternative splicing array platform (GenASAP). The details along with pros and cons of these methods is well described (Chen, 2011).

AS detection using RNA-seq data analysis: Alternative splicing is considered as main mechanism governing protein diversity and gene regulation. With the advancement of RNA-seq technology. It is possible to analyze the global impact and regulation of this biological process. There are increasing number of studies illustrates that the selection of wrong splice sites causes human disease. Therefore, the identification and quantification of differentially spliced transcripts are crucial for RNA-seq analysis.

Splicing Efficiency

Splicing efficiency (SE) (or splicing score; or splicing index) is used for the quantification of alternative splicing. SE is calculated as the ratio between the amounts of (spliced) mRNA and the (nascent) pre-mRNA. The conventional approach of using real-time quantitative PCR (RT-qPCR) with primers spanning exon-intron and exon-exon junctions (Hao and Baltimore, 2013) to determine SE is feasible for only a limited number of genes. However, with the availability of strong computational hardware and bioinformatics tools that can analyze the large amounts of RNA-seq data, it is now possible to determine genome-wide SE. Using RNA-seq data, several approaches have been used for calculating SE based on RNA-seq read counts from intronic, exonic or exon-exon junctions (Herzel and Neugebauer, 2015; Volanakis *et al.*, 2013). Below, we provide the main approaches for calculating SE (or scores):

1) *Exon-centered splicing score (ECSS)*: ECSS is calculated using the splicing frequency around a given exon, by subtracting the read coverage over 2 kb of the upstream intron from the read coverage over 2 kb of the 5′ end of the downstream intron (Pandya-Jones, 2011; Tilgner *et al.*, 2012).

2) *Intron-centered splicing score (ICSS)*: ICSS for each intron is calculated as the ratio of reads around the 3′ splice site. The read coverage over the last 25 bp of a given intron is divided by the read coverage over the first 25 bp of the downstream exon (Carrillo Oesterreich *et al.*, 2010).

3) *Gene-based splicing score (GBSS)*: It is the splicing frequency for each gene, by dividing the read coverage over exons by the read coverage over the whole locus (Bhatt *et al.*, 2012).

ECSS is optimal for alternative cassette exon usage. However, it is not suitable for the first and terminal exons which have to be analyzed differently. Using ICSS, first and terminal exons can be included in the analysis but short exons are a disadvantage. In GBSS, the noise is reduced by the involvement of many reads in the analysis. However, this approach neither provides SE for individual introns nor any information about AS events per gene. The user may refer to Brugiolo et al for further discussion regarding calculation of splicing efficiency using bioinformatics analysis (Brugiolo *et al.*, 2013).

Application of RNA-seq for Analyzing AS

RNA-sequencing (RNA-seq) uses next generation sequencing technology to sequence DNA molecules reversely transcribed from mRNAs. It has allowed the identification of RNA populations like miRNAs, lncRNAs, lincRNAs, and several other species of RNA that in fact have opened new areas of biology (Lynch, 2015). Using RNA-seq data, it is indeed easier to quantify gene expression, previously unknown coding or non-coding transcripts, the composition of various isoforms, and study events like RNA editing, gene fusion and nucleotide variations. RNA-seq also provides an unprecedented opportunity to study AS quantitatively in a systematic way (Feng *et al.*, 2013). In recent years, many researchers used RNA-seq analysis to study AS. The bioinformatics tools and steps involved in the quantitative study of AS with RNA-seq are now well established, including visualization approaches (Feng *et al.*, 2013; Liu *et al.*, 2012; Song *et al.*, 2016; Mezlini *et al.*, 2013; Li *et al.*, 2011; Rogers *et al.*, 2012; Barann *et al.*, 2017). Therefore, several groups have studied AS using RNA-seq data. Pan *et al.* (2008) used RNA-seq for the first time to analyze the complexity of AS in 5 different human tissues. They estimated that about 95% of multi-exon genes undergo AS and discovered thousands of new splice junctions, as demonstrated by earlier studies as well (Sultan *et al.*, 2008). This observation was later corroborated by Wang *et al.* (2008) who analyzed the RNA-seq data from 15 different human tissue samples and cell line transcriptomes and observed that 92%–94% of human genes undergo AS. The same study also analyzed tissue-specific AS regulation, individual specific AS isoform variation, and alternative cleavage and polyadenylation (APA) events using RNA-seq data. Several important diseases that precipitate due to erroneous splicing have been studied using RNA-seq data (Garcia-Blanco *et al.*, 2004). Many studies have studied AS in the context of highly complicated diseases like cancer (Kalari *et al.*, 2012; Ren *et al.*, 2012; Shapiro *et al.*, 2011). In order to study genome-wide splicing regulation and mechanisms involved in transcriptome-splicing coupling, it is also possible to integrate many human RNA-seq datasets to identify splicing modules – a unit in the splicing regulatory network (Dai *et al.*, 2012; Li *et al.*, 2012). Schreiber *et al.* (2015) used the RNA-seq data from *Saccharomyces cerevisiae* assess the impact of AS on its transcriptome. The detection of AS events has rapidly advanced with NGS in spite of the limitation due to the short sequencing read length. The upcoming 3rd generation sequencing technology like PacBio and Nanopore meant for long reads of RNA-seq will hopefully resolve these limitations in future.

Bioinformatics Tools for AS Detection

Several software tools have been developed for the identification of alternatively spliced exons and isoforms during the last decade. Herein, we are highlighting basic characteristics of some the commonly used tools:

AStalavista (2007): AStalavista (Alternative Splicing transcriptional landscape visualization tool) is the first tool (a web server) for the dynamically and exhaustive extraction of complex AS events from annotated genes in order to compare different types of AS and distributions (Foissac and Sammeth, 2007). It is a JAVA-based tool available as a web server provided in "Relevant Websites section" and the latest version can be downloaded from the website provided in "Relevant Websites section". For any given set of transcripts annotated with known exon–intron structure, AStalavista first performs an exhaustive pairwise comparison between all transcripts in a locus. To ensure an exhaustive detection of AS events, it pools all transcripts from a single transcriptional locus that overlap on the same strand of the genome sequence. It then dynamically assigns an *AS code* to the splicing events observed based on the relative position of splice sites. This code helps in automatic identification and exhaustive extraction of variations in the exon–intron structure (Sammeth *et al.*, 2008). AS events of the same type are given an identical code and classified in the same structural group whereas a new, concise and univocal AS code is assigned to each of the variant splicing structures. This generic protocol is applicable to any genome with or without annotation. Hence, AStalavista has an advantage over other methods that otherwise require a predefined splice form (as a reference transcript) for comparison. It detected more than 24,000 AS mechanistically un-elucidated events involving more than two alternatives in humans (Sammeth, 2009). Using this analysis, AStalavista generates a ranked list for each detected event type with its unique code (relative-position notation) and builds a AS landscape (in the form of a pie chart) for the given set of transcripts. This landscape of AS events can be used to investigate the transcriptome diversity across genes, chromosomes, and species.

AltAnalyze (2010): AltAnalyze (see "Relevant Websites section") is an comprehensive tool the for performing the analysis of alternative splicing data from Affymetrix Exon and Gene Arrays and their functional prediction at proteins and domains level (Emig *et al.*, 2010). It requires neither any programming knowledge nor any exon-array analysis expertise. It was designed for computing alternative exon statistics based on the widely used rigorous statistical methods like FIRMA and MiDAS using 'detection above background' (DABG) *P*-value thresholds with other alternative exon analysis parameters for any number of raw Affymetrix CEL files. AltAnalyze provides outputs in tab-delimited text file format which be either opened with Microsoft Excel like spreadsheet programs or can be visualized with DomainGraph (see "Relevant Websites section") for further analysis.

GPSeq (2010): GPSeq is a tool to analyze RNA-seq data to estimate gene and exon expression, identify differentially expressed genes (DEGs), and differentially spliced exons (DSEs) through log-likelihood ratio approaches (Srivastava and Chen, 2010). It is based on a two-parameter (i.e., θ and λ) generalized Poisson (GP) model to fit to the position-level read counts across all of the positions of a gene/exon. Here the estimated parameter θ represents the transcript amount for the gene and λ represents the average bias during the sample preparation and sequencing process. After normalization of mapped reads a likelihood ratio test is used to identify DEGs or DSEs by treating the λ estimates as true values. Moreover, a simulation strategy can be adopted to better estimate the *P*-values. It deals with the fundamental problem of the distribution of the position-level read counts by proposing a two-parameter generalized Poisson (GP) model. This GP model fits the data much better than the traditional Poisson model by separating true signals from sequencing bias and aids in the better estimation of gene or exon expression, in performing a more reasonable normalization across different samples, and hence improve the identification of DEGs as well as DSEs. More importantly, it can deal with multiple RNA-seq data sets. The codes for the GP model were written in C and 'GPseq' is an R-package to implement all these methods available for download at the website provided in "see Relevant Websites section". This tool is not available anymore.

MISO (2010): Mixture of Isoforms (MISO) is a statistical model that quantitates the expression level of alternatively spliced genes from RNA-seq data, and identifies differentially regulated isoforms or exons across samples (Katz *et al.*, 2010). MISO model uses Bayesian inference by modeling the generative process to reproduce the reads from isoforms in RNA-seq for calculating the probability that a read originated from a particular isoform. It treats the expression level of a set of isoforms as a random variable and estimates a distribution over the values of this variable using a sampling based algorithm known as Markov Chain Monte Carlo ("MCMC"). These estimates can be quantified by the confidence intervals. It not just estimates the expression level of a single alternatively spliced exon ("exon-centric"), or of each transcript belonging to a gene ("isoform-centric"), but can quantify the levels of multiple isoforms produced by several nearby alternative splicing events. ISO is available as a Python package at the website provided in "Relevant Websites section". It outputs the results as the exon/isoform expression levels in each sample along with confidence intervals which can be visualize alongside the RNA-seq data with sashimi-plot.

JuncBASE (2011): JuncBASE is used to identify and classify alternative splicing events from RNA-seq data (Brooks *et al.*, 2010). JuncBASE is available from the GitHub repository located at the website provided in "Relevant Websites section". Alternative splicing events are identified from splice junction reads from RNA-seq read alignments and annotated exon coordinates. JuncBASE also uses read counts to quantify the relative expression of each isoform and identifies splice events that are significantly differentially expressed across two or more samples.

SpliceTrap (2011): SpliceTrap is a method to quantify local exon inclusion levels by estimating the expression-levels of each exon using paired-end RNA-seq data (Wu *et al.*, 2011). SpliceTrap generates alternative splicing profiles for different splicing patterns, such as exon skipping, alternative 5′ or 3′ splice sites, and intron retention. It can also identify major classes of alternative splicing events under a single cellular condition, without requiring a background set of reads to estimate relative splicing changes. It utilizes a comprehensive human exon database called TXdb (see section "Bioinformatics Approaches for AS Detection") to estimate the expression level of every exon as an independent Bayesian inference problem. Unlike microarray-based methods, SpliceTrap relies on RNA-seq, and therefore it can determine the inclusion level of every exon within a single cellular condition, without requiring a background set of reads. Compared to Cufflinks and Scripture, it was shown to improve the accuracy, robustness and reliability in quantifying a large fraction of AS activity. SpliceTrap is useful in studying changes at the single-exon levels and can be helpful in the discovery of nearby *cis*-regulatory elements in diverse applications. It can also be implemented online through the CSH Galaxy server the website provided in "Relevant Websites section" and is also available for download and installation at the website provided in "Relevant Websites section".

MATS (2012): MATS is a computational tool to detect differential AS events from RNA-seq data (Shen *et al.*, 2012). The statistical model of MATS calculates the P-value and false discovery rate that the difference in the isoform ratio of a gene between two conditions exceeds a given user-defined threshold. From the RNA-seq data, MATS can automatically detect and analyze AS events corresponding to all major types of AS patterns. MATS handles replicate RNA-seq data from both paired and unpaired study design and has been designed to handle two RNA-seq samples.

SpliceSeq (2012): SpliceSeq is a Java based application, freely available at the website provided in "Relevant Websites Section", for visualization and quantitation of RNA-seq reads for alternative splicing and to identify potential functional changes resulted from splice variation (Ryan *et al.*, 2012). It aligns RNA-seq reads to gene splice graphs for accurate analysis of large and complex transcript variants. Firstly, an alignment database is generated from the imported RNA-seq data by using Bowtie and then SpliceSeq aligns reads to the splice graphs or gene models. Further, it evaluates the patterns of predicted transcript splicing for AS events and classifies these events in different types including exon skip, cassette exons, alternate promoter *etc*. The resulted alternative protein sequences are predicted with the weighted traversal of each gene's splice graph.

SplicingViewer (2012): SplicingViewer is an integrated tool, freely accessed at the website provided in "Relevant Websites section" for detecting the splice junctions from known gene models or RNA-seq data, annotating the alternative splicing patterns using the splice junctions and visualizing these patterns (Liu *et al.*, 2012). Firstly, the RNA-seq short reads are mapped to the provided or selected reference genome using aligners like BWA, Bowtie and SOAP. The mapped data are then converted to SAM/

BAM format by SAMtools to be used in SplicingViewer as input for further analysis and display alternative splicing patterns and the RNA-seq mapping result with a user-friendly interface, in a memory efficient and quick manner.

ASprofile (2013): ASprofile is computational framework for the identification of AS events in different RNA-seq samples (Florea et al., 2013). It comprises of three main programs for extracting (*extract-as*), quantifying (*extract-as-fpkm*) and comparing (*collect-fpkm*) AS events from transcripts assembled from RNA-seq data in multiple conditions. First, *extract-as* program takes as input a GTF transcript file and compares all pairs of transcripts within a gene to determine exon-intron structure differences that indicate an AS event. Second, *extract-as-fpkm* calculates the FPKM of each AS event from those of transcripts. Finally, *collect-fpkm* collects the FPKM event values for all RNA-seq samples, calculates and compares splicing ratios across samples.

ASprofile are providing following types of AS event such as exon skipping (SKIP), cassette exons (MSKIP), alternative transcript start and termination (TSS, TTS), retention of single or multiple introns (IR, MIR), and alternative exon (AE). ASprofile has been implemented using RNA-seq data from Illumina's Human Body Map project and authors have provided global view of alternative splicing events in 16 different human tissues. The list of all AS events catalog and the ASprofile software are freely available and can be access through their web site the website provided in "Relevant Websites section".

DiffSplice (2013): DiffSplice is tools for the identification and quantification of alternative splicing events (Hu et al., 2013). This software is available at the website provided in "Relevant Websites section". This approach starts with the identification of alternative splicing module (ASMs) from the splice graph that created directly from the exons and introns predicted from RNA-seq read alignments. The abundance of alternative splicing isoforms residing in each ASM is estimated for each sample and is compared across sample groups. DiffSplice takes as input the SAM files and results are summarized as a decomposition of the genome. It can be visualized using the UCSC genome browser. The approach does not depend on transcript or gene annotations. It evades the need for full transcript inference and quantification, which is a usually difficult because of short read lengths, as well as various sampling biases.

DSGseq (2013): DSGseq is method to identify differentially spliced genes between two groups of samples by comparing read counts on all exons (Wang et al., 2013). DSGseq software is available at the website provided in "Relevant Websites section". DSGseq tools is use negative binomial (NB) distribution to model sequencing reads on exons, and propose a NB-statistic to detect differentially spliced genes between two groups of RNA-seq samples by comparing read counts on all exons. It produces the tabular output the differences in the relative abundance of the isoforms of each gene in two groups of samples and relative abundance of the isoforms of each exon in each gene. It also ranks the exon that the most significant difference. This is a novel exon-based approach and it does not need isoform composition information and isoform expression estimation. Experiment on simulated and real RNA-seq data shows that this method has good performance and applicability. DSGseq method can identify the exons that contribute the best to the differential splicing. It can also identify previously unknown alternative splicing events.

GLiMMPS (2013): GLiMMPS (Generalized Linear Mixed Model Prediction of sQTL) is a robust statistical method for detecting splicing quantitative trait loci (sQTLs) from RNA-seq data (Zhao et al., 2013). It works at a low false positive rate and characterizes the genetic variation of alternative splicing. It takes into account the individual variation in sequencing coverage and the noise prevalent in RNA-seq data. To begin with, it needs a set of identified AS events and RNA-seq reads mapped to splice junctions detected for the estimation of the exon inclusion levels. GLiMMPS uses the reads information from both exon inclusion and skipping isoforms to model the estimation uncertainty of exon inclusion level. The source code used to be available from the website provided in "Relevant Websites section". However, currently, it appears to be discontinued.

SpliceR (2014): SpliceR is an R package for classification of alternative splicing and prediction of coding potential (Vitting-Seerup et al., 2014). SpliceR is implemented as an R package based on standard Bioconductor classes and is freely available from the Bioconductor repository (see "Relevant Websites section"). spliceR uses the full-length transcript output from RNA-seq assemblers (like *Cufflinks*) to detect single or multiple exon skipping, alternative donor and acceptor sites, intron retention, alternative first or last exon usage, and mutually exclusive exon events. For each gene, spliceR constructs the hypothetical pre-RNA based on the exon information from all transcripts originating from that gene. Subsequently, all transcripts are compared to this hypothetical pre-RNA in a pairwise manner. AS events are classified and annotated. It is flexible and can be easily integrated in existing pipelines or workflows. It predicts coding potential of transcripts, calculates untranslated region (UTR) and open reading frame (ORF) lengths. Moreover, it also predicts whether transcripts are nonsense mediated decay (NMD)-sensitive based on compatible annotated start codon positions and their downstream ORF.

GESS (2014): Graph-based exon-skipping scanner (GESS) is computational method to detect *de novo* exon-skipping events directly from raw RNA-seq data without the prior knowledge of gene annotation information (Wang et al., 2017). It can also detect the dominant isoform among the skipping- or inclusion- isoforms, generated for these skipping event sites and presents a more accurate and comprehensive data of skipping events associated with a particular physiological condition within a cell. GESS tools is available at the website provided in "Relevant Websites section". First, a splice-site-link graph is created from the splicing-aware aligned reads using a greedy algorithm. The sub-graphs are navigated iteratively by implementing a walking strategy to reveal a pattern corresponding to an exon-skipping event. Finally, the MISO model is implemented to obtain the ratio of skipping isoform vs. inclusion isoform and the dominated isoform is determined. In this method, the input must be sorted bam file and exon-skipping sites output from GESS can be use by MISO for Ψ –value (Percent Spliced Isoform) calculation. GESS method is capable of capturing *de novo* exon-skipping events directly from using raw RNA-seq reads.

rMATS (2014): As described previously, MATS works for detecting differential AS between two RNA-seq samples. Similarly, rMATS can be used to detect differential AS from replicate RNA-seq data using a hierarchical model to highlight the sampling uncertainty in individual replicates and the possible variability among them (Shen *et al.*, 2014). rMATS is flexible in testing the splicing difference above any user-defined threshold. Being a general method for analyzing mRNA isoform ratios from read-counts, rMATS method can also be used for sequencing-based analyses of other types of mRNA isoform variations including alternative polyadenylation and RNA editing. The rMATS source code is freely available at the website provided in "Relevant Websites section". It takes the raw RNA-seq reads, a genome sequence file, and a transcript annotation file as the input to identify the alternative splicing events corresponding to all major types of alternative splicing patterns and calculates the *P* value and FDR for differential splicing.

FineSplice (2014): FineSplice is a Python wrapped splice junction detection algorithm combined with TopHat2 for a reliable identification of expressed exon junctions from RNA-seq data (Gatto *et al.*, 2014). FineSplice software is freely available at the website provided in "Relevant Websites section". During the first step, it performs the transcriptome alignment with *de novo* splice junction discovery using TopHat2 with available annotations for known transcript isoforms. The resulting binary alignment map (BAM) file is used as input by FineSplice. At next step, it computes the set of split-read overhangs across each junction for each of the uniquely mapping read. It labels the splice junctions with no matching overhang and defines a potential false positives subset. Then for each junction it constructs a feature vector based on the log$_2$ deviation. With a defined class label and feature vector, it fits a L1-regularized logistic regression model over the whole set of junctions. FineSplice outputs a confident set of expressed splice junctions with the corresponding read counts. Potential false positives arising from spurious alignments are filtered out via a semi-supervised anomaly detection strategy based on logistic regression. Multiple mapping reads with a unique location after filtering are rescued and reallocated to the most reliable candidate location. This is conjugate approach for an efficient mapping solution with a semi-supervised anomaly detection scheme to filter out false positives. It allows reliable estimation of expressed junctions from the alignment output.

Splicing Code Prediction

Mutations in the splicing signals (5′ splice site, 3′ splice site and branch point), splicing silencer and enhancer motifs can affect the proper binding of spliceosome and other RNA binding proteins that changes AS patterns. These changes would lead to altered gene products and several diseases as a consequence. Therefore, it is important to be able to predict splicing codes that potentially affect AS. There are several online bioinformatics tools available for the prediction of splicing code that could be useful to understand human variation on splicing and its consequences in human diseases. These predictions are taking advantage of experimentally binding data of RNA binding proteins such as SR proteins as well as experimentally validated splicing signals sequences. These tools take primary sequence as input and predict the splicing regulatory binding sites, properties of exon/intron such as splice site strength, branch point and other regulatory binding motifs.

Tools for Splice Code Prediction: **Table 1** summarizes some of such online bioinformatics tools, like Human Splicing Finder (HSF), RegRNA, ESEfinder, Alternative Splice Site Predictor (ASSP), SplicePort, EX-SKIP, RESCUE-ESE, Maxentscan and SROOGLE (Desmet *et al.*, 2009; Huang *et al.*, 2006; Cartegni *et al.*, 2003; Wang and Marin, 2006; Dogan *et al.*, 2007; Raponi *et al.*, 2011; Fairbrother *et al.*, 2004; Eng *et al.*, 2004; Schwartz *et al.*, 2009). Maxentscan (Yeo and Burge, 2004), SplicePort (Dogan *et al.*, 2007), HSF (Desmet *et al.*, 2009) and SFmap (Paz *et al.*, 2010) can be used for the prediction of the basic *cis*-acting elements (5′ splice site, 3′ splice site and branch point). All these programs provide a useful indication of whether donor, acceptor and branch-point are well defined compared with ideal consensus sequences. ESEfinder and RESCUE-ESE are dedicated to disruption or creation of splicing regulatory elements (SREs) (Cartegni *et al.*, 2003; Fairbrother *et al.*, 2004). In addition, SFmap server is useful to search for the splicing factor binding motifs in the sequence of interest (Paz *et al.*, 2010). RegRNA and SROOGLE are servers can be useful in searching for both basic splicing signals and SREs (Huang *et al.*, 2006; Schwartz *et al.*, 2009). SROOGLE, is one of the most used tools, which provides a graphic output and displays the four core splicing signals with scores based on nine different algorithms. It also highlights the sequences belonging to 13 different groups of SREs. EX-SKIP application compares the ESE/ESS profile of a wild-type and a mutated allele to quickly determine which exonic variant has the highest chance to skip this exon (Raponi *et al.*, 2011).

Performance of Splice Code Predictors: One of the key questions that naturally arises is regarding the performance of these programs for the identification of possible splicing mutations. In humans, SREs tend to be reasonably conserved (Burset *et al.*, 2001) and the programs that evaluate their relative strengths seem to be more successful than those that aim to target the much more loosely conserved SRE elements. MaxEntScan takes the nucleotide dependencies within donor site sequences into account and has been shown to be the best predictors of cryptic splice site activation in disease-causing mutations (Houdayer *et al.*, 2008). Currently, SROOGLE is the most used tool that provides information on splice-sites or enhancer/silencer disruption in a single interface (Schwartz *et al.*, 2009). Some programs such as ESEfinder have worked well in some contexts (Zatkova *et al.*, 2004; Kralovicova and Vorechovsky, 2007; Wang *et al.*, 2005) and have led to a scientific debate in others (Cartegni and Krainer, 2002; Cartegni *et al.*, 2006; Fairbrother *et al.*, 2004; Pfarr *et al.*, 2005; Deburgrave *et al.*, 2007). Combination of all these resources still represents the best chance of "predicting" putative splicing mutations whether in conserved or less conserved regions (Houdayer *et al.*, 2008; Soukarieh *et al.*, 2016). Therefore, it appears that the future trend in developing such prediction algorithms or software will be to integrate all analyses on a single platform. This way, it will be easy to obtain most of the in depth details on global splicing signals on the same platter/platform. Due the progress in this field and improvement in bioinformatics predictions, methodologies related to splicing diagnostics are enticing. In such methodologies, the information obtained from *in silico* bioinformatics approaches are translated to wet lab (*in vitro* and *in vivo*) systems to evaluate splicing efficiencies (Baralle and Buratti, 2017).

Table 1   Online tools for splicing code prediction

Tools	Input	Output	Weblink
Human splicing finder	mRNA sequence	Consensus values of potential splice sites and search for branch points	http://www.umd.be/HSF3/
RegRNA	mRNA sequence	Motifs in mRNA 5'-UTR and 3'-UTR, motifs involved in mRNA splicing, motifs involved in transcriptional regulation, other motifs in mRNA, such as riboswitches, prediction of the splice sites, such as splicing donor/acceptor sites, RNA structural features, such as inverted repeat, and miRNA target sites	http://regrna.mbc.nctu.edu.tw/html/prediction.html
ESEfinder	Exonic sequence	ESE finder use to find the presence of exonic splicing enhancer elements	http://exon.cshl.edu/ESE/
Alternative Splice Site Predictor (ASSP)	Raw sequence, DNA/RNA	Putative alternative exon isoform, cryptic, and constitutive splice sites of internal (coding) exons	http://wangcomputing.com/assp/index.html
SplicePort	Multiple or single sequence	Splice-site predictions and user can also browse feature associated with the prediction	http://spliceport.cbcb.umd.edu
EX-SKIP	Two exonic sequence (up to 4000bp)	It calculates the total number of ESSs, ESEs and their ratio	http://ex-skip.img.cas.cz/
RESCUE-ESE	RNA or DNA Sequence (4 k)	Exonic splicing enhancers (ESEs) prediction	http://genes.mit.edu/burgelab/rescue-ese/
Maxentscan	Each sequence must be the same length	Building distributions over short sequence motifs	http://genes.mit.edu/burgelab/maxent/Xmaxent.html
	5' splice site sequence [3 bases in exon + 6 bases in intron]	5' splice site score	http://genes.mit.edu/burgelab/maxent/Xmaxentscan_scoreseq.html
	3' splice site sequence. [20 bases in the intron + 3 base in the exon]	3' splice site score	http://genes.mit.edu/burgelab/maxent/Xmaxentscan_scoreseq_acc.html
SROOGLE	Exon along with the introns	Graphic display of splicing related data on DNA segments	http://sroogle.tau.ac.il/
SFmap	Human genomic sequence or a list of sequences in FASTA format	Splicing factor binding motifs	http://sfmap.technion.ac.il/

Concluding Remarks

Field of alternative splicing has become very active and attractive in recent years, especially due to the development of high throughput technologies. Consequently, a large number of bioinformatics tools are available for various analyses regarding AS. In fact, it is hard to list every tool developed as of yet, nonetheless, we have reviewed bioinformatics software and algorithms that are used for the detection of splicing events using RNA-seq or other data. Drawbacks or advantages of each approach has been put forth. Some popular online tools and software that can be used to study intronic and exonic mutations leading to splicing defects, have also been discussed.

Acknowledgment

PKT was supported by the Czech Science Foundation (P305/12/G034) and the institutional support (RVO68378050).

See also: Exome Sequencing Data Analysis. Functional Enrichment Analysis. Genome Annotation. Genome Annotation: Perspective From Bacterial Genomes. Genome Databases and Browsers. Genome Informatics. Integrative Analysis of Multi-Omics Data. Metabolome Analysis. Natural Language Processing Approaches in Bioinformatics. Next Generation Sequencing Data Analysis. Prediction of Coding and Non-Coding RNA. Quantitative Immunology by Data Analysis Using Mathematical Models. Sequence Analysis. Whole Genome Sequencing Analysis

References

Baralle, D., Buratti, E., 2017. RNA splicing in human disease and in the clinic. Clin. Sci. (Lond.) 131 (5), 355–368.

Barann, M., Zimmer, R., Birzele, F., 2017. Manananggal – A novel viewer for alternative splicing events. BMC Bioinform. 18 (1), 120.

Belfort, M., 1990. Phage T4 introns: Self-splicing and mobility. Annu. Rev. Genet. 24 (1), 363–385.

Berget, S.M., Moore, C., Sharp, P.A., 1977. Spliced segments at the 5′ terminus of adenovirus 2 late mRNA. Proc. Natl. Acad. Sci. 74 (8), 3171–3175.

Bhatt, D.M., et al., 2012. Transcript dynamics of proinflammatory genes revealed by sequence analysis of subcellular RNA fractions. Cell 150 (2), 279–290.

Black, D.L., 2003. Mechanisms of alternative pre-messenger RNA splicing. Annu. Rev. Biochem. 72, 291–336.

Bogdanov, V.Y., 2006. Blood coagulation and alternative pre-mRNA splicing: An overview. Curr. Mol. Med. 6 (8), 859–869.

Bradley, T., Cook, M.E., Blanchette, M., 2015. SR proteins control a complex network of RNA-processing events. RNA 21 (1), 75–92.

Brett, D., et al., 2002. Alternative splicing and genome complexity. Nat. Genet. 30 (1), 29–30.

Brooks, A.N., et al., 2010. Conservation of an RNA regulatory map between Drosophila and mammals. Genome Res.

Brugiolo, M., Herzel, L., Neugebauer, K.M., 2013. Counting on co-transcriptional splicing. F1000Prime Rep. 5, 9.

Burset, M., Seledtsov, I.A., Solovyev, V.V., 2001. SpliceDB: Database of canonical and non-canonical mammalian splice sites. Nucleic Acids Res. 29 (1), 255–259.

Calarco, J.A., Zhen, M., Blencowe, B.J., 2011. Networking in a global world: Establishing functional connections between neural splicing regulators and their target transcripts. RNA 17 (5), 775–791.

Carrillo Oesterreich, F., Preibisch, S., Neugebauer, K.M., 2010. Global analysis of nascent RNA reveals transcriptional pausing in terminal exons. Mol. Cell 40 (4), 571–581.

Cartegni, L., et al., 2003. ESEfinder: A web resource to identify exonic splicing enhancers. Nucleic Acids Res. 31 (13), 3568–3571.

Cartegni, L., et al., 2006. Determinants of exon 7 splicing in the spinal muscular atrophy genes, SMN1 and SMN2. Am. J. Hum. Genet. 78 (1), 63–77.

Cartegni, L., Krainer, A.R., 2002. Disruption of an SF2/ASF-dependent exonic splicing enhancer in SMN2 causes spinal muscular atrophy in the absence of SMN1. Nat. Genet. 30 (4), 377–384.

Castle, J.C., et al., 2008. Expression of 24,426 human alternative splicing events and predicted cis regulation in 48 tissues and cell lines. Nat. Genet. 40 (12), 1416–1425.

Chasin, L.A., 2007. Searching for splicing motifs. Adv. Exp. Med. Biol. 623, 85–106.

Chen, L., 2011. Statistical and computational studies on alternative splicing. In: Lu, H.H.-S., Schölkopf, B., Zhao, H. (Eds.), Handbook of Statistical Bioinformatics. Berlin, Heidelberg: Springer Berlin Heidelberg, pp. 31–53.

Chow, L.T., et al., 1977. An amazing sequence arrangement at the 5′ ends of adenovirus 2 messenger RNA. Cell 12 (1), 1–8.

Clark, F., Thanaraj, T.A., 2002. Categorization and characterization of transcript-confirmed constitutively and alternatively spliced introns and exons from human. Hum. Mol. Genet. 11 (4), 451–464.

Clark, T.A., et al., 2007. Discovery of tissue-specific exons using comprehensive human exon microarrays. Genome Biol. 8 (4), R64.

Clark, T.A., Sugnet, C.W., Ares Jr., M., 2002. Genomewide analysis of mRNA processing in yeast using splicing-specific microarrays. Science 296 (5569), 907–910.

Cooper, T.A., 2005. Alternative splicing regulation impacts heart development. Cell 120 (1), 1–2.

Dai, C., et al., 2012. Integrating many co-splicing networks to reconstruct splicing regulatory modules. BMC Syst. Biol. 6 (1), S17.

Deburgrave, N., et al., 2007. Protein- and mRNA-based phenotype-genotype correlations in DMD/BMD with point mutations and molecular basis for BMD with nonsense and frameshift mutations in the DMD gene. Hum. Mutat. 28 (2), 183–195.

Desmet, F.O., et al., 2009. Human Splicing Finder: An online bioinformatics tool to predict splicing signals. Nucleic Acids Res. 37 (9), e67.

Dogan, R.I., et al., 2007. SplicePort – An interactive splice-site analysis tool. Nucleic Acids Res. 35 (Web Server issue), W285–W291.

Early, P., et al., 1980. Two mRNAs can be produced from a single immunoglobulin mu gene by alternative RNA processing pathways. Cell 20 (2), 313–319.

Emig, D., et al., 2010. AltAnalyze and DomainGraph: Analyzing and visualizing exon expression data. Nucleic Acids Res. 38 (Web Server issue), W755–W762.

Endo-Umeda, K., et al., 2012. Differential expression and function of alternative splicing variants of human liver X receptor alpha. Mol. Pharmacol. 81 (6), 800–810.

Eng, L., et al., 2004. Nonclassical splicing mutations in the coding and noncoding regions of the ATM Gene: Maximum entropy estimates of splice junction strengths. Hum. Mutat. 23 (1), 67–76.

Fairbrother, W.G., et al., 2004. RESCUE-ESE identifies candidate exonic splicing enhancers in vertebrate exons. Nucleic Acids Res. 32 (Web Server issue), W187–W190.

Fairbrother, W.G., et al., 2004. Single nucleotide polymorphism-based validation of exonic splicing enhancers. PLOS Biol. 2 (9), E268.

Feng, H., Qin, Z., Zhang, X., 2013. Opportunities and methods for studying alternative splicing in cancer with RNA-Seq. Cancer Lett. 340 (2), 179–191.

Florea, L., et al., 1998. A computer program for aligning a cDNA sequence with a genomic DNA sequence. Genome Res. 8 (9), 967–974.

Florea, L., Song, L., Salzberg, S.L., 2013. Thousands of exon skipping events differentiate among splicing patterns in sixteen human tissues. F1000Res 2, 188.

Foissac, S., Sammeth, M., 2007. ASTALAVISTA: Dynamic and flexible analysis of alternative splicing events in custom gene datasets. Nucleic Acids Res. 35 (Web Server issue), W297–W299.

Garcia-Blanco, M.A., Baraniak, A.P., Lasda, E.L., 2004. Alternative splicing in disease and therapy. Nat. Biotechnol. 22, 535.

Garg, K., Green, P., 2007. Differing patterns of selection in alternative and constitutive splice sites. Genome Res. 17 (7), 1015–1022.

Gatto, A., et al., 2014. FineSplice, enhanced splice junction detection and quantification: A novel pipeline based on the assessment of diverse RNA-Seq alignment solutions. Nucleic Acids Res. 42 (8), e71.

Geuens, T., Bouhy, D., Timmerman, V., 2016. The hnRNP family: Insights into their role in health and disease. Hum. Genet. 135, 851–867.

Ghigna, C., et al., 2010. Pro-metastatic splicing of Ron proto-oncogene mRNA can be reversed: Therapeutic potential of bifunctional oligonucleotides and indole derivatives. RNA Biol. 7 (4), 495–503.

Gilbert, W., 1978. Why genes in pieces? Nature 271 (5645), 501.

Graveley, B.R., et al., 2004. The organization and evolution of the dipteran and hymenopteran Down syndrome cell adhesion molecule (Dscam) genes. RNA 10 (10), 1499–1506.

Gupta, S., et al., 2004. Strengths and weaknesses of EST-based prediction of tissue-specific alternative splicing. BMC Genom. 5, 72.

Hagen, R.M., Ladomery, M.R., 2012. Role of splice variants in the metastatic progression of prostate cancer. Biochem. Soc. Trans. 40 (4), 870–874.

Han, S.P., et al., 2010. Functional implications of the emergence of alternative splicing in hnRNP A/B transcripts. RNA 16 (9), 1760–1768.

Hao, S., Baltimore, D., 2013. RNA splicing regulates the temporal order of TNF-induced gene expression. Proc. Natl. Acad. Sci. USA 110 (29), 11934–11939.

Herzel, L., Neugebauer, K.M., 2015. Quantification of co-transcriptional splicing from RNA-Seq data. Methods 85, 36–43.

Hnilicova, J., Stanek, D., 2011. Where splicing joins chromatin. Nucleus 2 (3), 182–188.

Hoskins, A.A., et al., 2011. Ordered and dynamic assembly of single spliceosomes. Science 331 (6022), 1289–1295.

Hoskins, A.A., Moore, M.J., 2012. The spliceosome: A flexible, reversible macromolecular machine. Trends Biochem. Sci. 37 (5), 179–188.

Houdayer, C., et al., 2008. Evaluation of in silico splice tools for decision-making in molecular diagnosis. Hum. Mutat. 29 (7), 975–982.

Hu, Y., et al., 2013. DiffSplice: The genome-wide detection of differential splicing events with RNA-seq. Nucleic Acids Res. 41 (2), e39.

Huang, H.Y., et al., 2006. RegRNA: An integrated web server for identifying regulatory RNA motifs and elements. Nucleic Acids Res. 34 (Web Server issue), W429–W434.

Jurica, M.S., Moore, M.J., 2003. Pre-mRNA splicing: Awash in a sea of proteins. Mol. Cell 12 (1), 5–14.

Kalari, K.R., et al., 2012. Deep sequence analysis of non-small cell lung cancer: Integrated analysis of gene expression, alternative splicing, and single nucleotide variations in lung adenocarcinomas with and without oncogenic KRAS mutations. Front. Oncol. 2, 12.

Katz, Y., et al., 2010. Analysis and design of RNA sequencing experiments for identifying isoform regulation. Nat. Methods 7 (12), 1009–1015.

Kent, W.J., 2002. BLAT – The BLAST-like alignment tool. Genome Res. 12 (4), 656–664.

Keren, H., Lev-Maor, G., Ast, G., 2010. Alternative splicing and evolution: Diversification, exon definition and function. Nat. Rev. Genet. 11 (5), 345–355.

Kim, N., Lee, C., 2008. Bioinformatics detection of alternative splicing. In: Keith, J.M. (Ed.), Bioinformatics: Data, Sequence Analysis and Evolution. Totowa, NJ: Humana Press, pp. 179–197.

Kralovicova, J., Vorechovsky, I., 2007. Global control of aberrant splice-site activation by auxiliary splicing sequences: Evidence for a gradient in exon and intron definition. Nucleic Acids Res. 35 (19), 6399–6413.

Lara-Pezzi, E., Dopazo, A., Manzanares, M., 2012. Understanding cardiovascular disease: A journey through the genome (and what we found there). Dis. Model Mech. 5 (4), 434–443.

Lee, C., et al., 2003. ASAP: The alternative splicing annotation project. Nucleic Acids Res. 31 (1), 101–105.

Li, J.J., et al., 2011. Sparse linear modeling of next-generation mRNA sequencing (RNA-Seq) data for isoform discovery and abundance estimation. Proc. Natl. Acad. Sci. USA 108 (50), 19867–19872.

Liu, Q., et al., 2012. Detection, annotation and visualization of alternative splicing from RNA-Seq data with SplicingViewer. Genomics 99 (3), 178–182.

Li, W., et al., 2012. Algorithm to identify frequent coupled modules from two-layered network series: Application to study transcription and splicing coupling. J. Comput. Biol. 19 (6), 710–730.

Long, J.C., Javier, F.C., 2009. The SR protein family of splicing factors: Master regulators of gene expression. Biochem. J. 417 (1), 15–27.

Lynch, K.W., 2015. Thoughts on NGS, alternative splicing and what we still need to know. RNA 21 (4), 683–684.

Manetti, M., et al., 2011. Impaired angiogenesis in systemic sclerosis: The emerging role of the antiangiogenic VEGF(165)b splice variant. Trends Cardiovasc. Med. 21 (7), 204–210.

Martinez-Contreras, R., et al., 2007. hnRNP proteins and splicing control. Adv. Exp. Med. Biol. 623, 123–147.

Medina, M.W., Krauss, R.M., 2013. Alternative splicing in the regulation of cholesterol homeostasis. Curr. Opin. Lipidol. 24 (2), 147–152.

Mezlini, A.M., et al., 2013. iReckon: Simultaneous isoform discovery and abundance estimation from RNA-seq data. Genome Res. 23 (3), 519–529.

Mills, J.D., Janitz, M., 2012. Alternative splicing of mRNA in the molecular pathology of neurodegenerative diseases. Neurobiol. Aging 33 (5), 1012. e11-24.

Missler, M., Sudhof, T.C., 1998. Neurexins: Three genes and 1001 products. Trends Genet. 14 (1), 20–26.

Mittendorf, K.F., et al., 2012. Tailoring of membrane proteins by alternative splicing of pre-mRNA. Biochemistry 51 (28), 5541–5556.

Miura, K., Fujibuchi, W., Sasaki, I., 2011. Alternative pre-mRNA splicing in digestive tract malignancy. Cancer Sci. 102 (2), 309–316.

Nilsen, T.W., Graveley, B.R., 2010. Expansion of the eukaryotic proteome by alternative splicing. Nature 463, 457–463.

Omenn, G.S., Yocum, A.K., Menon, R., 2010. Alternative splice variants, a new class of protein cancer biomarker candidates: Findings in pancreatic cancer and breast cancer with systems biology implications. Dis. Markers 28 (4), 241–251.

Pan, Q., et al., 2008. Deep surveying of alternative splicing complexity in the human transcriptome by high-throughput sequencing. Nat. Genet. 40, 1413–1415.

Pandya-Jones, A., 2011. Pre-mRNA splicing during transcription in the mammalian system. Wiley Interdiscip. Rev. RNA 2 (5), 700–717.

Patel, A.A., Steitz, J.A., 2003. Splicing double: Insights from the second spliceosome. Nat. Rev. Mol. Cell Biol. 4, 960.

Paz, I., et al., 2010. SFmap: A web server for motif analysis and prediction of splicing factor binding sites. Nucleic Acids Res. 38 (Web Server issue), W281–W285.

Peled-Zehavi, H., et al., 2001. Recognition of RNA branch point sequences by the KH domain of splicing factor 1 (mammalian branch point binding protein) in a splicing factor complex. Mol. Cell. Biol. 21 (15), 5232–5241.

Pfarr, N., et al., 2005. Linking C5 deficiency to an exonic splicing enhancer mutation. J. Immunol. 174 (7), 4172–4177.

Poulos, M.G., et al., 2011. Developments in RNA splicing and disease. Cold Spring Harb. Perspect. Biol. 3 (1), a000778.

Raponi, M., et al., 2011. Prediction of single-nucleotide substitutions that result in exon skipping: Identification of a splicing silencer in BRCA1 exon 6. Hum. Mutat. 32 (4), 436–444.

Ren, S., et al., 2012. RNA-seq analysis of prostate cancer in the Chinese population identifies recurrent gene fusions, cancer-associated long noncoding RNAs and aberrant alternative splicings. Cell Res. 22 (5), 806–821.

Rogers, M.F., et al., 2012. SpliceGrapher: Detecting patterns of alternative splicing from RNA-Seq data in the context of gene models and EST data. Genome Biol. 13 (1), R4.

Rogozin, I.B., et al., 2012. Origin and evolution of spliceosomal introns. Biol. Direct 7, 11-11.

Rosenfeld, M.G., et al., 1982. Calcitonin mRNA polymorphism: Peptide switching associated with alternative RNA splicing events. Proc. Natl. Acad. Sci. 79 (6), 1717–1721.

Roy, S.W., Gilbert, W., 2006. The evolution of spliceosomal introns: Patterns, puzzles and progress. Nat. Rev. Genet. 7 (3), 211–221.

Roy, B., Haupt, L.M., Griffiths, L.R., 2013. Review: Alternative splicing (AS) of genes as an approach for generating protein complexity. Curr. Genom. 14 (3), 182–194.

Ryan, M.C., et al., 2012. SpliceSeq: A resource for analysis and visualization of RNA-Seq data on alternative splicing and its functional impacts. Bioinformatics 28 (18), 2385–2387.

Sammeth, M., 2009. Complete alternative splicing events are bubbles in splicing graphs. J. Comput. Biol. 16 (8), 1117–1140.

Sammeth, M., Foissac, S., Guigó, R., 2008. A general definition and nomenclature for alternative splicing events. PLOS Comput. Biol. 4 (8), e1000147.

Sampath, J., Pelus, L.M., 2007. Alternative splice variants of survivin as potential targets in cancer. Curr. Drug Discov. Technol. 4 (3), 174–191.

Schmidt, F.J., 1985. RNA splicing in prokaryotes: Bacteriophage T4 leads the way. Cell 41 (2), 339–340.

Schreiber, K., et al., 2015. Alternative splicing in next generation sequencing data of saccharomyces cerevisiae. PLOS ONE 10 (10), e0140487.

Schwartz, S., Hall, E., Ast, G., 2009. SROOGLE: Webserver for integrative, user-friendly visualization of splicing signals. Nucleic Acids Res. 37 (Web Server issue), W189–W192.

Shapiro, I.M., et al., 2011. An EMT – Driven alternative splicing program occurs in human breast cancer and modulates cellular phenotype. PLOS Genet. 7 (8), e1002218.

Shen, S., et al., 2012. MATS: A Bayesian framework for flexible detection of differential alternative splicing from RNA-Seq data. Nucleic Acids Res. 40 (8), e61.

Shen, S., et al., 2014. rMATS: Robust and flexible detection of differential alternative splicing from replicate RNA-Seq data. Proc. Natl. Acad. Sci. USA 111 (51), E5593–E5601.

Song, L., Sabunciyan, S., Florea, L., 2016. CLASS2: Accurate and efficient splice variant annotation from RNA-seq reads. Nucleic Acids Res. 44 (10), e98.

Sorek, R., Shamir, R., Ast, G., 2004. How prevalent is functional alternative splicing in the human genome? Trends Genet. 20 (2), 68–71.

Soukarieh, O., et al., 2016. Exonic splicing mutations are more prevalent than currently estimated and can be predicted by using in silico tools. PLOS Genet. 12 (1), e1005756.

Srivastava, S., Chen, L., 2010. A two-parameter generalized Poisson model to improve the analysis of RNA-seq data. Nucleic Acids Res. 38 (17), e170.

Staley, J.P., Guthrie, C., 1998. Mechanical devices of the spliceosome: Motors, clocks, springs, and things. Cell 92 (3), 315–326.

Stamm, S., et al., 2000. An alternative-exon database and its statistical analysis. DNA Cell Biol. 19 (12), 739–756.

Sultan, M., et al., 2008. A global view of gene activity and alternative splicing by deep sequencing of the human transcriptome. Science 321 (5891), 956–960.

Tang, J.Y., et al., 2013. Alternative splicing for diseases, cancers, drugs, and databases. Sci. World J. 2013, 703568.

Tilgner, H., et al., 2012. Deep sequencing of subcellular RNA fractions shows splicing to be predominantly co-transcriptional in the human genome but inefficient for lncRNAs. Genome Res. 22 (9), 1616–1625.

Trowitzsch, S., et al., 2009. Crystal structure of the Pml1p subunit of the yeast precursor mRNA retention and splicing complex. J. Mol. Biol. 385 (2), 531–541.

van den Hoogenhof, M.M., Pinto, Y.M., Creemers, E.E., 2016. RNA splicing: Regulation and dysregulation in the heart. Circ. Res. 118 (3), 454–468.

van Nimwegen, E., et al., 2006. SPA: A probabilistic algorithm for spliced alignment. PLOS Genet. 2 (4), e24.

Vitting-Seerup, K., et al., 2014. spliceR: An R package for classification of alternative splicing and prediction of coding potential from RNA-seq data. BMC Bioinform. 15, 81.

Volanakis, A., et al., 2013. Spliceosome-mediated decay (SMD) regulates expression of nonintronic genes in budding yeast. Genes Dev. 27 (18), 2025–2038.

Wang, E.T., et al., 2008. Alternative isoform regulation in human tissue transcriptomes. Nature 456, 470–476.

Wang, J., et al., 2017. Computational methods and correlation of exon-skipping events with splicing, transcription, and epigenetic factors. Methods Mol. Biol. (Clifton, N.J.) 1513, 163–170.

Wang, J., et al., 2005. Distribution of SR protein exonic splicing enhancer motifs in human protein-coding genes. Nucleic Acids Res. 33 (16), 5053–5062.

Wang, M., Marin, A., 2006. Characterization and prediction of alternative splice sites. Gene 366 (2), 219–227.

Wang, W., et al., 2013. Identifying differentially spliced genes from two groups of RNA-seq samples. Gene 518 (1), 164–170.

Wang, X., et al., 2009. Genome-wide prediction of cis-acting RNA elements regulating tissue-specific pre-mRNA alternative splicing. BMC Genom. 10 (1), S4.

Wang, Z., Burge, C.B., 2008. Splicing regulation: From a parts list of regulatory elements to an integrated splicing code. RNA 14 (5), 802–813.

Will, C.L., Lührmann, R., 2011. Spliceosome structure and function. Cold Spring Harb. Perspect. Biol. 3 (7).

Woodson, S.A., 1998. Ironing out the kinks: Splicing and translation in bacteria. Genes Dev. 12 (9), 1243–1247.

Wu, J., et al., 2011. SpliceTrap: A method to quantify alternative splicing under single cellular conditions. Bioinformatics 27 (21), 3010–3016.

Wu, T.D., Watanabe, C.K., 2005. GMAP: A genomic mapping and alignment program for mRNA and EST sequences. Bioinformatics 21 (9), 1859–1875.

Xie, H., et al., 2002. Computational analysis of alternative splicing using EST tissue information. Genomics 80 (3), 326–330.

Xu, Q., Modrek, B., Lee, C., 2002. Genome-wide detection of tissue-specific alternative splicing in the human transcriptome. Nucleic Acids Res. 30 (17), 3754–3766.

Yap, K., Makeyev, E.V., 2013. Regulation of gene expression in mammalian nervous system through alternative pre-mRNA splicing coupled with RNA quality control mechanisms. Mol. Cell. Neurosci. 56, 420–428.

Yeo, G., et al., 2004. Variation in alternative splicing across human tissues. Genome Biol. 5 (10), R74.

Yeo, G.W., et al., 2007. Alternative splicing events identified in human embryonic stem cells and neural progenitors. PLOS Comput. Biol. 3 (10), 1951–1967.

Yeo, G., Burge, C.B., 2004. Maximum entropy modeling of short sequence motifs with applications to RNA splicing signals. J. Comput. Biol. 11 (2–3), 377–394.

Yi, Q., Tang, L., 2011. Alternative spliced variants as biomarkers of colorectal cancer. Curr. Drug Metab. 12 (10), 966–974.

Zahler, A.M., Tuttle, J.D., Chisholm, A.D., 2004. Genetic suppression of intronic +1G mutations by compensatory U1 snRNA changes in Caenorhabditis elegans. Genetics 167 (4), 1689–1696.

Zatkova, A., et al., 2004. Disruption of exonic splicing enhancer elements is the principal cause of exon skipping associated with seven nonsense or missense alleles of NF1. Hum. Mutat. 24 (6), 491–501.

Zhang, X., et al., 2016. Recognition of alternatively spliced cassette exons based on a hybrid model. Biochem. Biophys. Res. Commun. 471 (3), 368–372.

Zhao, K., et al., 2013. GLiMMPS: Robust statistical model for regulatory variation of alternative splicing using RNA-seq data. Genome Biol. 14 (7), R74.

Zheng, C.L., Fu, X.-D., Gribskov, M., 2005. Characteristics and regulatory elements defining constitutive splicing and different modes of alternative splicing in human and mouse. RNA 11 (12), 1777–1787.

Relevant Websites

http://www.altanalyze.org
  AltAnalyze.
http://ccb.jhu.edu/software/ASprofile
  ASprofile
  CCB.
http://genome.imim.es/astalavista
  AStalavista.
http://www.domaingraph.de
  DomainGraph.
http://sammeth.net/confluence/display/ASTA/2 + - + Download
  2
  Download
  AStalavista
  Confluence.
http://www.netlab.uky.edu/p/bioinfo/DiffSplice
  DiffSplice.
http://bioinfo.au.tsinghua.edu.cn/software/DSGseq
  DSGseq.
https://sourceforge.net/p/finesplice/
  FineSplice.
https://github.com/anbrooks/juncBASE
  GitHub
  anbrooks/juncBASE.
http://compbio.uthscsa.edu/GESS_Web/
  GESS-RNA.
rnaseq-mats.sourceforge.net/
  Multivariate Analysis of Transcript Splicing.
http://www.ncbi.nlm.nih.gov/snp
  NCBI
  NIH.
http://www.rcf.usc.edu/~liangche/software.html
  GPSeq.
https://pypi.python.org/pypi/misopy/
  Python Package Index.
http://variome.kobic.re.kr/ssSNPTarget/
  ssSNPTarget.
http://cancan.cshl.edu/splicetrap
  SpliceTrap.
http://rulai.cshl.edu/splicetrap/
  SpliceTrap.

http://bioinformatics.mdanderson.org/main/SpliceSeq:Overview
    SpliceSeq:Overview.
http://bioinformatics.zj.cn/splicingviewer
    SpliceSeq:Overview.
https://codeload.github.com/Xinglab/GLiMMPS/zip/master
    GLiMMPS.
http://www.bioconductor.org/packages/2.13/bioc/html/spliceR.html
    SpliceR
    Bioconductor.

Biographical Sketch

Prasoon Kumar Thakur is presently working as a bioinformatician at the laboratory of RNA Biology at the Institute of Molecular genetics, Prague. He is pursuing a PhD degree in Charles University, Prague, Czech Republic. He has a bachelor's degree in Bioinformatics from the Institute of Advance Studies in Education, Sardarshahr (IASES), India and M.Sc (Bioinformatics) from the Jamia Millia Islamia, New Delhi, India. The focus of his current research is RNA splicing and high-throughput data analysis.

Hukam C. Rawal is presently working as a Research Associate at NRCPB, New Delhi, India. He has over 10 years' experience in different aspects of computational biology. He has good exposure to a wide range of computational approaches for the analysis related to human disease, pathogen infection and crop sciences. He has expertise on advance and popular genomics studies and has worked on different aspects of comparative genomics. He is quite familiar with high throughput sequencing data. He has several publications in the field of comparative genomics, transcriptome assembly, genome level analysis, chloroplast and mitochondrial genome in reputed journals like Genome Biology & Evolution, Genes, Scientific Reports, etc.

Mina Obuca finished her bachelor and master studies in the Faculty of Sciences at the University of Novi Sad, Serbia. Currently, she is a PhD student at the Charles University in Prague, Czech Republic where she is doing her PhD at the Institute of Molecular Genetics in the Laboratory of RNA Biology. She is interested in answering why mutations in splicing proteins are causing retinitis pigmentosa.

Sandeep Kaushik is a passionate computational biologist with a wide exposure of computational approaches. Presently, he is presently carrying out his research as an Assistant Researcher (equivalent to Assistant Professor) at 3B's Research Group, University of Minho, Portugal. He has a PhD in Bioinformatics from National Institute of Immunology, New Delhi, India. He has gained a wide exposure on analysis of scientific data ranging from transcriptomic data on mycobacteria and human samples to genomic data from wheat. His research experience and expertise entails molecular dynamics simulations, RNA-sequencing data analysis, *de novo* genome assembly, protein prediction and annotation, database mining, agent-based modeling and simulations. He has a cumulative experience of more than 10 years of programming (using PERL, R and NetLogo languages). He has published his research in reputed international journals like Molecular Cell, Biomaterials, Biophysical Journal and others.

Genome-Wide Association Studies

William S Bush, Case Western Reserve University, Cleveland, OH, United States

Glossary

Common disease/common variant hypothesis The concept that genetic susceptibility to diseases common in a population is likely due to genetic variation that is also common in that population.

Direct association An association between a trait and SNP that is directly genotyped in a study and itself induces an influential biological change.

False positive A result that is statistically significant when the null hypothesis is true.

Genome-wide significance A significance threshold used to account for multiple hypothesis tests conducted in a GWAS which corresponds to the number of effective independent regions in the human genome.

Genotype imputation The use of population-based genome reference panels to infer the state of ungenotyped variants from GWAS data.

Indirect association An association between a trait and a SNP that is in linkage disequilibrium with another nearby SNP that induces an influential biological change.

Linkage disequilibrium The genetic correlation among alleles of co-located SNPs within the genome that is due to limited recombination events between those SNPs within a population.

Manhattan plot A plot illustrating the magnitude and genomic location of GWAS associations.

Meta-analysis An approach for combining the results of multiple independent studies by analyzing summary statistics.

Population stratification An effect where many SNPs within a study are systematically associated with a trait due to differences in both allele frequency and trait distribution across genetic ancestries.

P-value The probability of observing a value as extreme or more extreme than the test statistic when the null hypothesis is true.

QQ-plot A diagnostic plot used in GWAS to identify systematic inflation of test statistics across all SNPs.

Single nucleotide polymorphism A single base-pair change to the DNA sequence that is common within a population.

TagSNPs A SNP that is used to predict the state of other nearby SNPs via linkage disequilibrium.

Introduction: The Common Disease-Common Variant Hypothesis

Early successes in human genetics led to the discovery of disease genes for a variety of inherited disorders such as cystic fibrosis (Rommens *et al.*, 1989) and Huntington disease (MacDonald *et al.*, 1992). These Mendelian diseases were studied by genotyping families affected by the disease using a collection of genetic markers across the genome, and examining how those genetic markers segregate with the disease across multiple families (Bush and Haines, 2010). This approach, called linkage analysis, was largely unsuccessful when applied to more common disorders, like heart disease, type 2 diabetes, or stroke, implying that the genetic architecture of commonly occurring diseases is likely different from that of rare disorders (Hirschhorn and Daly, 2005). This observation, along with the discovery of some relatively frequent risk alleles for common diseases disease such as *APOE* alleles for Alzheimer's Disease (Corder *et al.*, 1993) and *PPARG* alleles for type II diabetes (Lander *et al.*, 2000), led to the formalization of a *common disease/common variant* (CD/CV) hypothesis (Reich and Lander, 2001), which simply states that disorders which are common in a population are likely influenced by genetic variation which is also common in that population. To test the CD/CV hypothesis, a new type of population-based study was developed, Genome-Wide Association Studies (GWAS). GWAS was designed to broadly capture common genetic variation, generally defined as DNA base-pair changes with a minor allele frequency (MAF) greater than 5% for the population being studied. Before GWAS could be effectively implemented, however, a catalog of common variation assessed across multiple populations was needed as a reference.

Genetic Variation and Linkage Disequilibrium

Through a large-scale coordinated effort, the International HapMap Project (The International HapMap Project, 2003) captured much of the common variation among 11 human subpopulations, and established the groundwork for GWAS (Manolio *et al.*, 2008). The HapMap Project focused primarily on characterizing *single nucleotide polymorphisms* (SNPs), a single base-pair change to the DNA sequence. The project also produced extensive maps of *linkage disequilibrium* (LD), a property of SNPs that lie on a single chromosome and are co-inherited through the transmission of the chromosome within a population. Humans are diploid, having two copies of each chromosome (with exceptions for X and Y), and recombination events break apart contiguous stretches of SNPs on a chromosome during meiosis when germ cells are created. Through successive generations within a population, repeated recombination events distributed across each chromosome produce stretches of alleles that are correlated to one another. The degree to which LD exists for a given chromosomal region depends on many factors, including the population size, the number of generations for which the population has existed, the recombination rate, and the number of potential recombination points along the chromosome. The influence of these factors is clear from examining different global populations; African-descent populations on average have smaller regions of LD compared to European-descent or East Asian-descent populations, which were the result of a series of founder events and as a result have larger regions of LD.

Maps of LD were instrumental in the initial design of many fluorescence-based genotyping arrays, which relied on the use of *tagSNPs*, specific variants that were highly correlated (and predictive of) other nearby sites. The use of tagSNPs prevented genotyping SNPs that essentially provide redundant information, and allowed the capture of greater than 80% of all common SNPs in Europeans using 500,000 to 1 Million SNPs from across the genome (Li *et al.*, 2008). Notably, as LD patterns are population-dependent, early genotyping arrays were potentially biased toward European-descent populations as different tagSNPs were needed for optimal capture of variation in other populations.

While tagSNPs greatly improved the capture of genetic information, they also limit any causal inference that can be made from GWAS. When a tagSNP is associated to a disease, this association can either be a *direct association*, where the genotyped tagSNP is the true variant inducing a functional effect, or it may be an *indirect association*, where the true disease-associated variant is in high LD with the genotyped tagSNP. As a result, significant SNP associations from GWAS are not always the true functional/causal variant, and additional fine-mapping studies may be necessary to elucidate the true disease variant.

GWAS Experimental Designs

A critical first step for any genetic study is to precisely define the trait or outcome thought to be influenced by genetic variation. Though there are some exceptions, GWAS generally examine either case/control or continuous outcomes. Continuous, quantitative traits are often preferred as they increase statistical power to detect a genetic effect, and provide a unit change per risk allele. Quantitative traits such as body mass index (BMI) and low-density lipoprotein (LDL) levels are routinely collected through standardized procedures and represent intermediate outcomes for more severe conditions such as heart disease. However, quantitative traits are not always easy to define or collect for certain disorders. For these conditions, dichotomous or case/control categorical variables are used as the primary outcome. Categorical variables however are more susceptible to problematic definitions of "cases" and "controls", and often require adjudication by clinical experts. For many conditions, consensus criteria have been established for defining case/control status, such as the McDonald criteria for multiple sclerosis (Polman *et al.*, 2011), or the NINCDS-ADRDA criteria for Alzheimer's disease (McKhann *et al.*, 1984). These standardized, evidence-based approaches can be uniformly applied by multiple diagnosing clinicians to ensure that consistent phenotype definitions are used for a genetic study.

Standardized criteria for study phenotypes are especially critical for multi-center studies. If different case/control definitions are used at different study sites, a site-based effect may be introduced into the study. Even when standardized criteria are used across all study sites, there may still be variability among how clinicians interpret them to assign case/control status. While generally thought to be more accurate, quantitative traits are also susceptible to bias in measurement. For example, in studies of cataract severity, lens (eye) photography is used to assign cataract cases to one of three grades of lens opacity. These photographs are adjudicated by clinicians whose characterizations of cataract stage may not be consistent. Interrater agreement is often evaluated across multiple grading clinicians to measure phenotype consistency (Chew *et al.*, 2010). Significant disagreement between raters typically indicates that disease criteria are not being uniformly applied or interpreted, and a narrower set of phenotype rules are needed.

Once a phenotype has been selected and standardized for a particular study population, genetic information must be collected on study participants. While DNA can be extracted from any tissue, the most common source of DNA for GWAS is whole blood. An individual blood sample is separated into fractions via centrifugation to isolate the buffy coat, a layer between erythrocytes (which do not contain nuclei or DNA) and plasma. The buffy coat contains white blood cells which are subsequently lysed to capture participant DNA. DNA is then processed via vendor specifications for large-scale genotyping using microarray-based technologies, generally from either Illumina (San Diego, CA) or Affymetrix (Santa Clara, CA). These two technologies have relative strengths and weaknesses (DiStefano and Taverna, 2011), and other approaches exist to measure genetic variation, but are not generally used for primary GWAS data collection.

Statistical Analysis

The primary results for a typical GWAS are generated through systematic statistical evaluation of each genetic variant captured by the genotyping process. As part of conducting the statistical analysis for GWAS, genotype data must be encoded, and different encoding schemes reflect different underlying models of allelic effect, and have implications on the degrees of freedom and statistical power of a test. An allelic encoding assumes that each person in the dataset represents two separate chromosomes, and examines the association of the allele to the phenotype. A genotypic encoding assumes that each person is a distinct observation of a combination of two alleles, and genotypes can further be grouped into specific modes of action such as dominant, recessive, multiplicative, or additive models (Lewis, 2002). While these various encoding schemes exist and are used to different degrees in GWAS, it is common practice to examine additive genotypic models only, as this has the best statistical power to detect a variety of modes of action (Lettre *et al.*, 2007).

After choosing an encoding, each SNP is examined independently for association the study phenotype, and the statistical test employed depends multiple factors, most critically whether the phenotype is a quantitative trait or a case/control outcome. In either case, generalized linear models (GLM) are generally used. In the case of quantitative traits, Analysis of Variance (ANOVA) is used to test the deviation from the null hypothesis that there is no difference between the mean trait value across individuals by genotype. Assumptions of the ANOVA apply, most critically that trait follows a normal distribution and that the variance of the trait is the same across groups (the groups are homoscedastic). Case/control traits are typically analyzed using a variant of GLM, logistic regression, which transforms a linear outcome to a probability using a logit curve. While some early GWAS used contingency table analysis methods, such as a chi-square test, logistic regression is now the preferred approach as it allows for adjustment for other relevant factors and can provide adjusted effect estimates for each allele of a SNP. Logistic regression is a mature approach with well-established interpretations and diagnostic procedures to assess the goodness of model fit and the influence of potential outliers (Hosmer *et al.*, n.d.).

Each SNP-based statistical model should also be adjusted for other factors with established effects on the study trait. This often includes demographic factors (i.e., age, sex), clinical factors (i.e., body mass index, comorbidity status), and design factors (i.e., study site, experimental batch). Adjustment for covariates like these reduces artifacts in the statistical analysis that may differentially inflate the frequency of alleles/genotypes between cases and controls. By far the most common confounding factor in GWAS is population substructure. For nearly any phenotype, there are often known differences in the occurrence rate or prevalence across different racial, ethnic, and geographic backgrounds. Similarly, due to human migration patterns, the founding of new populations, and dramatic changes to population size, allele frequencies are highly variable across human subpopulations. Therefore, if a study collects individuals from cases and controls from different ancestral populations, any systematic differences in allele frequency between the ancestral populations will appear as differences between cases and controls – a phenomenon known as *population stratification*. In GWAS, population stratification is generally assessed by estimating the ancestry of each sample in the dataset using either sample clustering within the STRUCTURE software (Falush *et al.*, 2003), or using principal components implemented in the EIGENSTRAT (Price *et al.*, 2006), GCTA (Yang *et al.*, 2011), and PLINK 1.9 (Chang *et al.*, 2015) software packages. Principal component values can then be used as a covariate in statistical models to adjust for minute degrees of ancestry effects present in the study. Linear mixed models (also known as mixed linear models) are also now commonly used to adjust for stratification due to either population differences or familial relationships, including GCTA (Yang *et al.*, 2011), REACTA (Cebamanos *et al.*, 2014), EMMAX (Kang *et al.*, 2010), FaST-LMM (Lippert *et al.*, 2011), GEMMA (Zhou and Stephens, 2012), and GRAMMAR-Gamma (Svishcheva *et al.*, 2012).

For each SNP, a statistical model including the encoded effect of the SNP alleles along with any covariate adjustments is evaluated. The effect (also referred to as the β or coefficient) for the encoded alleles is evaluated (by T-test in the case of linear regression or Wald test in the case of logistic regression) to produce a *p-value* corresponding to the significance of the SNP. Assuming a null hypothesis that there is no effect of the SNP on the phenotype, a *p-value* is the probability of observing a test statistic equal to or greater than the statistic reported from the analysis of the SNP in the GWAS data. In effect therefore, very small p-values indicate that if there is no real association between the SNP and the phenotype, observing the reported test statistic is extremely unlikely. Statistical tests like the T-test or Wald test generally conducted in GWAS are often labeled "significant", meaning the null hypothesis of no association is rejected, if the p-value falls below a pre-defined significance threshold (often referred to as α). For most statistical tests, the α is generally pre-set to 0.05, meaning that for 5% of all statistical tests, the null hypothesis is rejected when it is in fact true – a phenomenon called a *false positive*. This 5% probability of a false positive is relative to a single statistical test, and in the case of GWAS, many thousands or even millions of individual hypothesis tests are conducted with each one having a probability of false positive. As a result, assuming α of 0.05, the cumulative probability of finding one or more false positives over a GWAS is much higher. Put a different way, assume you paint one side of a twenty-sided die red; a single roll of this die is unlikely to show the red side, but if you roll this die 500,000 times, it is extremely likely that you roll the red side at least once. The easiest way to adjust for the multiple statistical comparisons conducted in GWAS is to use a Bonferroni correction. This correction adjusts the alpha value for each statistical test from $\alpha = 0.05$ to $\alpha = (0.05/k)$ where k is the number of statistical tests conducted. Assuming that there is some degree of linkage disequilibrium and thus correlation among statistical tests of a GWAS, this correction is rather conservative as it assumes completely independent tests. Perhaps the most commonly used approach is to assume a level of *genome-wide significance*, which is an estimate of the effective number of independent genomic regions and thus statistical tests for which the analysis should be corrected. This assessment is population specific, owing to differences in LD patterns among different ancestral populations, but in the case of European-descent studies, the genome-wide significance threshold has been estimated at $7.2e - 8$ (Dudbridge and Gusnanto, 2008).

There are two commonly used data visualizations shown in GWAS publications, a quantile-quantile (QQ) plot, and a "Manhattan" plot, both implemented in the *qqman* R package (Turner, 2014). A QQ plot (**Fig. 1**) is a comparison of the p-value distribution across all evaluated SNPs to an expected theoretical uniform distribution. Values are typically plotted as the negative log (base 10) of the p-values, which provides a unit increase per order of magnitude decrease in the p-value (a value of 7 is equivalent to 1×10^{-7}). A *QQ plot* of log p-values is an easy way to visually identify systematic deviations from the null hypothesis – an indicator that other factors like population stratification produce systematic biases across all test statistics. In some cases, the deviation of observed p-values from the expected uniform distribution is quantified as a *genomic inflation factor*. A Manhattan plot (**Fig. 2**) is a similar visualization, showing the negative log p-value plotted by chromosome position. These plots often color-code chromosomes on the X-axis to help identify where significant SNPs lie in the genome. A Manhattan plot illustrates two properties of GWAS results, first the physical location of SNPs with extreme p-values, and secondly the degree to which a SNP association is corroborated by other nearby SNPs in linkage disequilibrium. Often these plots are labeled with additional information, such as a line demarking the statistical significance criteria used for the analysis, and previously established true positive SNP associations.

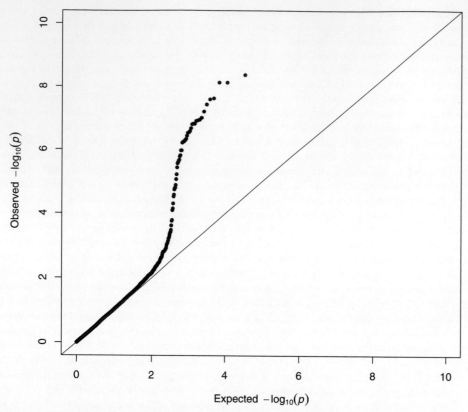

Fig. 1 A quantile-quantile plot of GWAS p-values. The negative log10 p-values for each analyzed SNP are plotted on the y-axis versus the expected p-values under the null hypothesis of no associated SNPs on the x-axis. Points deviating from the diagonal line are more significant than expected under the null hypothesis.

The vast majority of published GWAS were conducted using PLINK (Chang *et al.*, 2015; Purcell *et al.*, 2007) software for statistical analysis. PLINK has numerous functions for conducting genotype and sample quality control, and for performing basic analyses of continuous and binary traits. As the number of genetic variants analyzed in GWAS studies have grown, genetic datasets are often split (by chromosome or other metrics) to facilitate more rapid processing. Analyzing data subsets, however, requires more detailed file and process management, and can introduce errors. To address these issues, software such as HAIL (Hail, n.d.) has been developed to enable distributed genomic data processing. HAIL provides much of the functionality of PLINK, but performs its operations over a HADOOP/SPARK distributed computing environment. By leveraging this memory and process rich environment, an entire GWAS analysis can be completed in minutes.

Meta-Analysis, Replication, and GWAS Follow-Up

Shortly after initial publications of GWAS for many traits, there was a growing realization within the genomics community that ever larger sample sizes are needed to improve statistical power for the detection of new risk variants. To reduce the need to share individual-level genomic data, research groups began to combine the statistical results from multiple independent GWAS through *meta-analysis*. Techniques for meta-analysis were originally developed to summarize statistical results from multiple independent published studies which examined the same hypothesis. When applied to GWAS, these approaches allow for multiple sets of genome-wide statistical results to be combined using sample size based or inverse variance based approaches as implemented in the software METAL (Sanna *et al.*, 2008; Willer *et al.*, 2008), or the *metafor* R package, among others. GWAS meta-analysis can be challenging in practice. The multiple studies participating in a meta-analysis must take care to conduct their quality control procedures in a standardized way, include consistent covariate adjustments, and ensure that all statistical results are reported relative to a pre-defined reference allele for each SNP (Zeggini and Ioannidis, 2009). Even with standardization across studies, there is often some degree to which studies differ, including some planned differences in design and analysis. Heterogeneity across multiple studies participating in a GWAS meta-analysis is often assessed using either the Q statistic or the I^2 index (Huedo-Medina *et al.*, 2006), with I^2 typically favored as it represents the approximate proportion of variability in a SNP coefficient attributed to heterogeneity between studies (Higgins, 2008). High degrees of heterogeneity may point to specific outlier studies that may be excluded for legitimate reasons, but may also simply point to unstable associations for a given SNP.

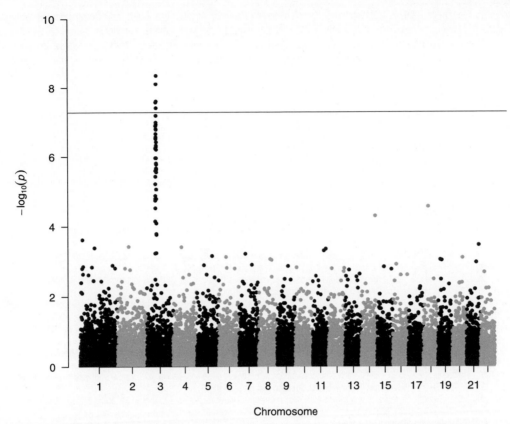

Fig. 2 GWAS Manhattan Plot. The y-axis represents the negative log10 of the p-value for each analyzed SNP. The x-axis illustrates the chromosome (note color coding) and position of each SNP. Horizontal lines mark different levels of statistical significance.

Meta-analyses of GWAS require estimates of consistent alleles from a consistent set of SNPs. In reality, different GWAS often use different genotyping platforms and technologies, leading to different collections of genotyped SNPs for each study. To resolve this issue, studies perform *genotype imputation*, which exploits known patterns of linkage disequilibrium from a selected reference panel to estimate the alleles of ungenotyped SNPs (Li *et al.*, 2009). Popular algorithms for genotype imputation include BimBam (Guan and Stephens, 2008), IMPUTE (Howie *et al.*, 2009), MaCH (Biernacka *et al.*, 2009), and Beagle (Browning and Browning, 2009). Genotype imputation produces an estimate for the allele state at each ungenotyped SNP for each individual, often reported as the probability of each possible allele. Allele probabilities for imputed SNPs should be used in the GWAS analysis to take into account the uncertainty of the genotype imputation process (Marchini *et al.*, 2007). Also, as with other GWAS design factors, population ancestry is of critical importance; the reference panel used in the imputation process must contain individuals drawn from the same population as the GWAS dataset.

Once a key set of SNP-trait associations are discovered from a GWAS, the next critical step is to replicate these associations in an additional independent sample (Chanock *et al.*, 2007). A well-designed replication study should have a sample size to detect the predicted effect of any discovered SNP – ideally large enough to detect an effect slightly smaller than the original (Zollner and Pritchard, 2007). Replication studies should be drawn from the same population as the original GWAS to confirm the effect within the target population. Studies in additional populations are sometimes conducted to determine if the effect of a SNP "generalizes" to other populations (Carlson *et al.*, 2013). Replication studies should use identical phenotype criteria to ensure a clear interpretation of the genetic results. Finally, analyses of replication datasets should show a similar magnitude and direction of effect across both studies.

Conclusions

Genome-wide association studies have played a critical role in identifying thousands of new SNP-trait associations. GWAS as a study design is sometimes considered as both a success and a failure. They are a failure in that even though tens or hundreds of SNPs associate to any given common disease, the collective effect of these SNPs explain very little of the disease risk that is known to be genetic (Maher, 2008). They are a success in that for many diseases, entirely new biological mechanisms of disease have been elucidated for common diseases which have long had little genetic understanding (Hirschhorn, 2009). While many large-scale GWAS have been conducted on many thousands of samples, ever larger studies, coupled with decreasing costs of genotyping and sequencing will likely yield an even better understanding of the genomics of human disease.

See also: Comparative Genomics Analysis. Data Mining: Mining Frequent Patterns, Associations Rules, and Correlations. Detecting and Annotating Rare Variants. Functional Enrichment Analysis. Functional Enrichment Analysis Methods. Functional Genomics. Genetics and Population Analysis. Genome Databases and Browsers. Genome Informatics. Genome-Wide Haplotype Association Study. Integrative Analysis of Multi-Omics Data. Introduction to Biostatistics. Natural Language Processing Approaches in Bioinformatics. Next Generation Sequencing Data Analysis. Population Analysis of Pharmacogenetic Polymorphisms. Quantitative Immunology by Data Analysis Using Mathematical Models. Single Nucleotide Polymorphism Typing. Whole Genome Sequencing Analysis

References

Biernacka, J.M., Tang, R., Li, J., *et al.*, 2009. Assessment of genotype imputation methods. BMC Proceedings 3 (Suppl. 7), S5. Retrieved from: http://www.ncbi.nlm.nih.gov/pubmed/20018042.

Browning, B.L., Browning, S.R., 2009. A unified approach to genotype imputation and haplotype-phase inference for large data sets of trios and unrelated individuals. American Journal of Human Genetics 84 (2), 210–223. Available at: https://doi.org/10.1016/j.ajhg.2009.01.005.

Bush, W.S., Haines, J., 2010. Overview of linkage analysis in complex traits. Current Protocols in Human Genetics 64 (1), Chapter 1, Unit 1.9.1–18. Available at: https://doi.org/10.1002/0471142905.hg0109s64.

Carlson, C.S., Matise, T.C., North, K.E., *et al.* PAGE Consortium, 2013. Generalization and dilution of association results from European GWAS in populations of non-European ancestry: The PAGE study. PLOS Biology 11 (9), e1001661. Available at: https://doi.org/10.1371/journal.pbio.1001661.

Cebamanos, L., Gray, A., Stewart, I., Tenesa, A., 2014. Regional heritability advanced complex trait analysis for GPU and traditional parallel architectures. Bioinformatics (Oxford, England) 30 (8), 1177–1179. Available at: https://doi.org/10.1093/bioinformatics/btt754.

Chang, C.C., Chow, C.C., Tellier, L.C., *et al.*, 2015. Second-generation PLINK: Rising to the challenge of larger and richer datasets. GigaScience 4 (1), 7. Available at: https://doi.org/10.1186/s13742-015-0047-8.

Chanock, S.J., Manolio, T., Boehnke, M., *et al.*, 2007. Replicating genotype–phenotype associations. Nature 447 (7145), 655–660. Available at: https://doi.org/10.1038/447655a.

Chew, E.Y., Kim, J., Sperduto, R.D., *et al.*, 2010. Evaluation of the age-related eye disease study clinical lens grading system AREDS report No. 31. Ophthalmology 117 (11), 2112–2119. e3. Available at: https://doi.org/10.1016/j.ophtha.2010.02.033.

Corder, E.H., Saunders, A.M., Strittmatter, W.J., *et al.*, 1993. Gene dose of apolipoprotein E type 4 allele and the risk of Alzheimer's disease in late onset families. Science (New York, N.Y.) 261 (5123), 921–923. Retrieved from: http://www.ncbi.nlm.nih.gov/pubmed/8346443.

DiStefano, J.K., Taverna, D.M., 2011. Technological issues and experimental design of gene association studies. Methods in Molecular Biology (Clifton, N.J.) 700, 3–16. Available at: https://doi.org/10.1007/978-1-61737-954-3_1.

Dudbridge, F., Gusnanto, A., 2008. Estimation of significance thresholds for genomewide association scans. Genetic Epidemiology 32 (3), 227–234. Available at: https://doi.org/10.1002/gepi.20297.

Falush, D., Stephens, M., Pritchard, J.K., 2003. Inference of population structure using multilocus genotype data: Linked loci and correlated allele frequencies. Genetics 164 (4), 1567–1587. Retrieved from: http://www.ncbi.nlm.nih.gov/pubmed/12930761.

Guan, Y., Stephens, M., 2008. Practical issues in imputation-based association mapping. PLOS Genetics 4 (12), e1000279. Available at: https://doi.org/10.1371/journal.pgen.1000279.

Hail (n.d.). Retrieved from: https://github.com/hail-is/hail.

Higgins, J.P.T., 2008. Commentary: Heterogeneity in meta-analysis should be expected and appropriately quantified. International Journal of Epidemiology 37 (5), 1158–1160. Available at: https://doi.org/10.1093/ije/dyn204.

Hirschhorn, J.N., 2009. Genomewide association studies – Illuminating biologic pathways. New England Journal of Medicine 360 (17), 1699–1701. Available at: https://doi.org/10.1056/NEJMp0808934.

Hirschhorn, J.N., Daly, M.J., 2005. Genome-wide association studies for common diseases and complex traits. Nature Reviews Genetics 6 (2), 95–108. Available at: https://doi.org/10.1038/nrg1521.

Hosmer, D.W., Lemeshow, S., Sturdivant, R.X. (n.d.). Applied logistic regression. Retrieved from: https://books.google.co.in/books?hl=en&lr=&id=64JYAwAAQBAJ&oi=fnd&pg=PA313&dq=HOSMER+LEMESHOW+logistic+regression&ots=DscL3YathK&sig=52_KJ7diR7s39CH4MD3idTdLzd8&redir_esc=y#v=onepage&q=HOSMER LEMESHOW logistic regression&f=false.

Howie, B.N., Donnelly, P., Marchini, J., 2009. A flexible and accurate genotype imputation method for the next generation of genome-wide association studies. PLOS Genetics 5 (6), e1000529. Available at: https://doi.org/10.1371/journal.pgen.1000529.

Huedo-Medina, T.B., Sánchez-Meca, J., Marín-Martínez, F., Botella, J., 2006. Assessing heterogeneity in meta-analysis: Q statistic or I2 index? Psychological Methods 11 (2), 193–206. Available at: https://doi.org/10.1037/1082-989X.11.2.193.

Kang, H.M., Sul, J.H., Service, S.K., *et al.*, 2010. Variance component model to account for sample structure in genome-wide association studies. Nature Genetics 42 (4), 348–354. Available at: https://doi.org/10.1038/ng.548.

Lander, E.S., Altshuler, D., Hirschhorn, J.N., *et al.*, 2000. The common PPARgamma Pro12Ala polymorphism is associated with decreased risk of type 2 diabetes. Nature Genetics 26 (1), 76–80. Available at: https://doi.org/10.1038/79216.

Lettre, G., Lange, C., Hirschhorn, J.N., 2007. Genetic model testing and statistical power in population-based association studies of quantitative traits. Genetic Epidemiology 31 (4), 358–362. Available at: https://doi.org/10.1002/gepi.20217.

Lewis, C.M., 2002. Genetic association studies: Design, analysis and interpretation. Briefings in Bioinformatics 3 (2), 146–153. Retrieved from: http://www.ncbi.nlm.nih.gov/pubmed/12139434.

Li, M., Li, C., Guan, W., 2008. Evaluation of coverage variation of SNP chips for genome-wide association studies. European Journal of Human Genetics 16 (5), 635–643. Available at: https://doi.org/10.1038/sj.ejhg.5202007.

Li, Y., Willer, C., Sanna, S., Abecasis, G., 2009. Genotype imputation. Annual Review of Genomics and Human Genetics 10 (1), 387–406. Available at: https://doi.org/10.1146/annurev.genom.9.081307.164242.

Lippert, C., Listgarten, J., Liu, Y., *et al.*, 2011. FaST linear mixed models for genome-wide association studies. Nature Methods 8 (10), 833–835. Available at: https://doi.org/10.1038/nmeth.1681.

MacDonald, M.E., Novelletto, A., Lin, C., *et al.*, 1992. The Huntington's disease candidate region exhibits many different haplotypes. Nature Genetics 1 (2), 99–103. Available at: https://doi.org/10.1038/ng0592-99.

Maher, B., 2008. Personal genomes: The case of the missing heritability. Nature 456 (7218), 18–21. Available at: https://doi.org/10.1038/456018a.

Manolio, T.A., Brooks, L.D., Collins, F.S., 2008. A HapMap harvest of insights into the genetics of common disease. The Journal of Clinical Investigation 118 (5), 1590–1605. Availble at: https://doi.org/10.1172/JCI34772.

Marchini, J., Howie, B., Myers, S., McVean, G., Donnelly, P., 2007. A new multipoint method for genome-wide association studies by imputation of genotypes. Nature Genetics 39 (7), 906–913. Available at: https://doi.org/10.1038/ng2088.

McKhann, G., Drachman, D., Folstein, M., *et al.*, 1984. Clinical diagnosis of Alzheimer's disease: Report of the NINCDS-ADRDA Work Group under the auspices of Department of Health and Human Services Task Force on Alzheimer's Disease. Neurology 34 (7), 939–944. Retrieved from: http://www.ncbi.nlm.nih.gov/pubmed/6610841.

Polman, C.H., Reingold, S.C., Banwell, B., *et al.*, 2011. Diagnostic criteria for multiple sclerosis: 2010 Revisions to the McDonald criteria. Annals of Neurology 69 (2), 292–302. Available at: https://doi.org/10.1002/ana.22366.

Price, A.L., Patterson, N.J., Plenge, R.M., *et al.*, 2006. Principal components analysis corrects for stratification in genome-wide association studies. Nature Genetics 38 (8), 904–909. Available at: https://doi.org/10.1038/ng1847.

Purcell, S., Neale, B., Todd-Brown, K., *et al.*, 2007. PLINK: A tool set for whole-genome association and population-based linkage analyses. American Journal of Human Genetics 81 (3), 559–575. Available at: https://doi.org/10.1086/519795.

Reich, D.E., Lander, E.S., 2001. On the allelic spectrum of human disease. Trends in Genetics 17 (9), 502–510. Retrieved from: http://www.ncbi.nlm.nih.gov/pubmed/11525833.

Rommens, J., Iannuzzi, M., Kerem, B., *et al.*, 1989. Identification of the cystic fibrosis gene: Chromosome walking and jumping. Science 245 (4922), 1059–1065. Available at: https://doi.org/10.1126/science.2772657.

Sanna, S., Jackson, A.U., Nagaraja, R., *et al.*, 2008. Common variants in the GDF5-UQCC region are associated with variation in human height. Nature Genetics 40 (2), 198–203. Available at: https://doi.org/10.1038/ng.74.

Svishcheva, G.R., Axenovich, T.I., Belonogova, N.M., van Duijn, C.M., Aulchenko, Y.S., 2012. Rapid variance components-based method for whole-genome association analysis. Nature Genetics 44 (10), 1166–1170. Available at: https://doi.org/10.1038/ng.2410.

The International HapMap Project, 2003. Nature 426 (6968), 789–796. Available at: https://doi.org/10.1038/nature02168.

Turner, S.D., 2014. qqman: An R package for visualizing GWAS results using Q-Q and manhattan plots. bioRxiv. 5165. Available at: https://doi.org/10.1101/005165.

Willer, C.J., Sanna, S., Jackson, A.U., *et al.*, 2008. Newly identified loci that influence lipid concentrations and risk of coronary artery disease. Nature Genetics 40 (2), 161–169. Available at: https://doi.org/10.1038/ng.76.

Yang, J., Lee, S.H., Goddard, M.E., Visscher, P.M., 2011. GCTA: A tool for genome-wide complex trait analysis. American Journal of Human Genetics 88 (1), 76–82. Available at: https://doi.org/10.1016/j.ajhg.2010.11.011.

Zeggini, E., Ioannidis, J.P.A., 2009. Meta-analysis in genome-wide association studies. Pharmacogenomics 10 (2), 191–201. Available at: https://doi.org/10.2217/14622416.10.2.191.

Zhou, X., Stephens, M., 2012. Genome-wide efficient mixed-model analysis for association studies. Nature Genetics 44 (7), 821–824. Available at: https://doi.org/10.1038/ng.2310.

Zollner, S., Pritchard, J.K., 2007. Overcoming the winner's curse: Estimating penetrance parameters from case-control data. American Journal of Human Genetics 80 (4), 605–615. Available at: https://doi.org/10.1086/512821.

Gene Mapping

Kelly L Williams, Macquarie University, Sydney, NSW, Australia

Introduction

Gene mapping is the sequential allocation of loci to a relative position on a chromosome. Genetic maps are species-specific and comprised of genomic markers and/or genes and the genetic distance between each marker. These distances are calculated based on the frequency of chromosome crossovers occurring during meiosis, and not on their physical location on the chromosome. There are existing dense genetic marker maps available for humans, and the introduction of next-generation sequencing technologies is facilitating increased construction of genetic maps for other species. Genetic maps are a necessary tool for mapping of disease genes or trait loci, a method also commonly known as linkage mapping. Integrating genetic mapping and disease gene mapping with next-generation sequencing has proven to be a powerful strategy in genetic research.

From Meiosis to centiMorgans

Observation of the frequent co-inheritance of traits in *Drosophila* led Morgan, one of the pioneers of genetic research, to first describe the concept of linkage in 1911 as *"associations of factors" that are located near together in the chromosomes* (Morgan, 1911). Morgan's student Sturtevant, another pioneer in genetic research, extended these observations and determined that the proportion of recombinant gametes resulting from crossovers during meiosis allows for the calculation of genetic distance between two loci in *Drosophila* (Sturtevant, 1913). Chromosomal crossing-over is the biological phenomenon occurring during meiosis that underpins genetic variation in populations. No two gametes produced during meiosis are the same. In the process of meiosis, the paternal and maternal inherited chromosomes replicate to form sister chromatids. During prophase I, homologous chromosomes are paired in the cell. Then, two non-sister chromatids undergo recombination (or 'crossing-over') where homologous portions of the chromatids exchange places. This results in recombinant chromatids (which are a combination of the maternal and paternal chromosomes) and non-recombinant chromatids (identical to the parental chromosomes) per chromosome (**Fig. 1**). The recombination fraction (θ) is the probability of a crossing over event occurring between two loci. Crossovers occur semi-randomly throughout the genome and thus allow genetic maps to be constructed.

Genetic distance between two loci is measured in Morgans (M) and is the expected number of crossovers to occur between those two loci on a chromosome in a gamete. A centiMorgan (cM) is a more commonly used measure of distance and corresponds to $\frac{1}{100}$ of a Morgan. Based on Sturtevant's seminal work, if two loci are unlinked (for example on difference chromosomes), we would expect to see 50% recombinant gametes, and 50% non-recombinant gametes. However if we see lower percentage of recombinant gametes that differs to the expected, such as 13% recombinants, the two loci are considered linked at a genetic distance of 13 cM. One cM roughly corresponds to 1 Mb (1 megabase, or 1,000,000 base pairs), but differs substantially throughout the genome. Recombination rate can vary from 0.1 cM Mb^{-1} to 3 cM Mb^{-1} across a chromosome (Kong *et al.*, 2002). Females display a higher crossover frequency than males, therefore the human female genetic map is approximately 44 Morgans, compared to 27 Morgans for males (Broman *et al.*, 1998; **Table 1**). Dense genetic maps are usually released in three versions: male, female and sex-averaged. **Table 1** outlines the map distances and physical distance of each chromosome.

Evolution of Genetic Markers and Human Genetic Maps

The Human Genome Project represented a time of enormous progress in human genetics, and resulted in the construction of dense genetic maps that are still in use by geneticists today. Prior to the publication of the first draft human genome sequence completed under the Human Genome Project, all genetic maps were constructed without having a reference genome sequence. Simply determining the order of genetic markers was a considerable and time-consuming task. The first genetic map of the human genome comprised 403 polymorphic markers, the majority of which were low-informative bi-allelic RFLPs (restriction fragment

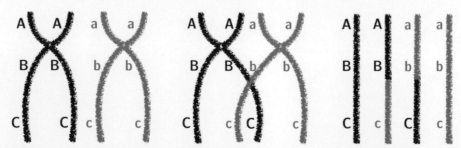

Fig. 1 Crossing over between sister chromatids of the same chromosome. Alleles at three loci (A, B and C) are shown. There is a single recombination event between B and C. The parental *BC* and *bc* haplotypes will become *Bc* or *bC* in the recombinant gametes.

Table 1   Chromosome lengths in terms of Sex-Averaged, Female and Male genetic distances and physical distances

Chromosome	Marshfield map length (cM)			Physical length (Mb)
	Sex-averaged	Female	Male	GRCh38
1	290	365	216	249
2	269	331	209	242
3	228	270	190	198
4	212	264	160	190
5	198	245	151	182
6	193	254	131	171
7	182	231	134	159
8	168	224	113	145
9	169	193	143	138
10	173	211	138	134
11	148	180	115	135
12	171	214	128	133
13	115	130	95	114
14	138	155	117	107
15	122	136	110	102
16	134	169	101	90
17	126	149	105	83
18	126	156	97	80
19	105	114	96	59
20	101	121	81	64
21	58	65	51	47
22	62	74	49	51
X		184		156
Total	3488	4435	2730	3031

Source: Genetic distances obtained from Broman, K.W., Murray, J.C., Sheffield, V.C., White, R.L., Weber, J.L., 1998. Comprehensive human genetic maps: Individual and sex-specific variation in recombination. American Journal of Human Genetics 63, 861–869. Physical distances obtained from The Genome Reference Consortium. Available from: https://www.ncbi.nlm.nih.gov/grc/human/data.

length polymorphisms), with an average resolution of 10 cM (Donis-Keller *et al.*, 1987). When microsatellites were identified in the genome, these became the genotyping tool of choice by geneticists, as they were multi-allelic and therefore highly heterozygous in the population (Litt and Luty, 1989; Weber and May, 1989). These polymorphic short tandem repeat sequences, usually di-, tri- or tetra-nucleotide repeats, are able to be amplified by PCR and therefore less cumbersome to genotype when compared to RFLPs. Their prevalence throughout the genome, and informativeness for recombination, led to the construction of the first dense human genetic maps, some of which are still in use today. The Marshfield genetic map (see "Relevant Websites section") provided the first complete genome-wide comprehensive genetic map compromising more than 8000 microsatellite markers, including hetero-zygosity values for each marker (Broman *et al.*, 1998). Once the first draft of the human genome sequence was released, geneticists began to construct denser microsatellite marker genetic maps in a comparatively shorter period of time (Kong *et al.*, 2002). Comprehensive sequencing of the human genome identified an increasing number of bi-allelic single nucleotide polymorphisms (SNPs), which would ultimately become the next genetic marker of choice. SNPs are less informative than microsatellites (they only have a maximum heterozygosity of 0.5), but their abundance throughout the human genome has allowed for construction of very dense genetic maps. Due to their bi-allelic nature, hundreds of thousands of SNPs can be genotyped on a microarray in parallel with a small input volume of DNA. The next major human genetic map to be published was the deCODE genetic map in 2010 (Kong *et al.*, 2010). In contrast to previous microsatellite marker genetic maps, the deCODE genetic map comprises 300,000 less informative bi-allelic SNPs, but at a higher density to get resolution to 10 kb.

With the publication of the first reference human genome (Lander *et al.*, 2001; Venter *et al.*2001) and the public repository of genetic mapping data, combined genetic-physical maps were able to be constructed (Kong *et al.*, 2004; Matise *et al.*, 2007). These maps are able to be used for interpolation of genetic map positions of unmapped markers or genomic variants. Web-based interpolation tools exist (such as that at Rutgers: see "Relevant Websites section"), or scripts from Github Repository for inter-polation on a larger scale (an example python script: see "Relevant Websites section"). Interpolation of genetic maps (see Section Using Interpolated Genetic Maps for Shared IBD Segment Analysis below) is now particularly relevant with whole-genome sequencing datasets available with genotypes at each nucleotide position in the genome.

Generating a Genetic Map

Construction of genetic marker maps begins with a small set of loci, with the gradual addition of more loci. Somatic cell genetics and cytogenetics was used in the 1970s to first assign genetic markers to specific regions on single chromosomes. New genetic markers, whether RFLPs, microsatellites or SNPs, are then genotyped in large families and recombination fractions are determined. As part of the Human Genome Project, DNA samples from the CEPH Reference Family Panel was used collaboratively worldwide for genotyping for the construction of genetic maps (Dausset *et al.*, 1990). CEPH Reference Family Panel DNA samples that were made available to the genetics community, allowed standardised genotyping by using a common set of families. Using observed recombination fractions from genotyping data, there are established software tools to construct a genetic map including LINKMAP (Lathrop *et al.*, 1984), MAPMAKER (Lander and Green, 1987; Lander *et al.*, 1987) and CRI-MAP (P. Green, K. Falls, and S. Crooks, documentation for CRI-MAP, version 2.4). These tools utilise map functions, which are mathematical relationships that transform non-additive recombination fractions (θ) into additive map distances (x). The most commonly used transformations are Haldane's and Kosambi's mapping functions (Haldane, 1919; Kosambi, 1944), both of which are still used by geneticists for mapping genomes. To calculate genetic distances (x) from recombination fractions (θ):

$$x = -1/2 ln(1 - 2\theta) \text{ Haldane's map function}$$

$$x = 1/4 ln\frac{1 + 2\theta}{1 - 2\theta} \text{ Kosambi's map function}$$

The mapping function that Kosambi derived assumes genetic interference, whereas the Haldane-derived mapping function does not allow for genetic interference. Genetic interference is the reduction in randomness of crossovers on a chromosome due to a decreased likelihood of a new crossover in the immediate vicinity of an existing crossover. Conversions between θ and x using several map functions can be performed interactively using the MAPFUN linkage utility program (see "Relevant Websites section"). Nonetheless, with today's dense human genetic maps, there is little need for mapping functions.

However in plant and animal genetics, construction of genetic maps of rare species is still an active field of research, and will be outlined below in Section Gene Mapping in Plants and Animals.

Linkage Analysis for Mapping Mendelian Disease Genes

When elucidating the cause of a Mendelian disease, identifying the causal gene would be considered the most important aspect of the research. Pinpointing the approximate location of the disease locus on a genetic map is the first step in identifying the causal gene. Subsequent fine mapping using additional genetic markers, and then sequencing of gene exons contained within the linkage region to find a mutation, is performed. Once a causal gene is identified, the mechanisms by which the mutation, and therefore mutated gene, causes disease is able to be further investigated. Identifying disease genes through linkage mapping offers the clinical potential for diagnostic tests and treatment.

Linkage Simulation and Genotyping Prior to Analysis

Disease gene mapping differs to constructing a genetic map of markers, as disease loci are mapped based on genotypes from families specifically collected for the study. Historically, to successfully identify a locus significantly linked to disease, a large multigenerational family with many DNA samples from both affected and unaffected individuals was required. Prior to generating genotype data, geneticists commonly perform a computer simulation on the family of interest. SLINK (Ott, 1989; Weeks *et al.*, 1990) and SIMLINK (Boehnke, 1986; Ploughman and Boehnke, 1989) are examples of such simulation software and require input data including pedigree, number of genetic markers used in the study, allele frequencies, recombination fractions between loci, mode of inheritance and disease penetrance. This prospective analysis can determine the statistical power of the proposed linkage study, and therefore the suitability of the family to conduct a linkage study.

DNA samples are then collected from available family members, including both individuals affected with the disease of interest and those unaffected. Extensive genotyping is then performed on the family members using either microsatellite markers or SNP microarrays, or a combination of both. Current SNP microarrays can genotype 500,000 – 2 million SNPs on a single chip. Although the density of markers is significantly greater when using SNP microarrays, they are less informative, frequently occur in LD (linkage disequilibrium) and have a maximum heterozygosity of 0.5. Since multiallelic microsatellite markers are highly informative due to their heterozygosity (which can be as high as 0.95), less markers are required for a comparable amount of linkage information (Dunn *et al.*, 2005). Genotype data and pedigree information is then converted into input files for linkage programs.

Linkage Programs and Lod Scores

The statistical measure of linkage for a disease locus is known as a lod score (or logarithm of the odds) and was first defined by Morton (1955). A lod score $z(x)$ is the logarithm (base 10) of the likelihood ratio: The likelihood of the data if the disease locus is linked to the genetic marker and the likelihood of the data if the disease locus is unlinked to the marker, where x represents a particular value of recombination fraction $0 \le \theta \le 1/2$:

$$z(x) = log_{10}\left[\frac{L(\text{pedigree if linked, } \theta = x)}{L(\text{pedigree if unlinked, } \theta = 0.5)}\right]$$

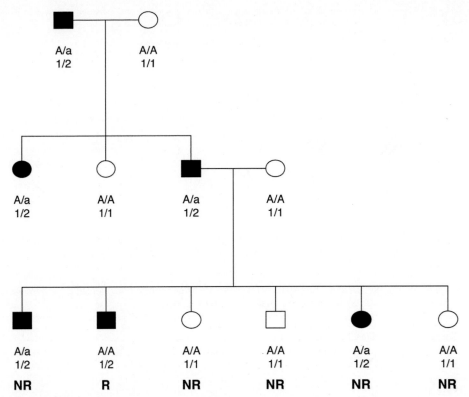

Fig. 2 Pedigree showing segregation of an autosomal dominant trait and a phase-known genetic marker. Recombinants (R) and non-recombinants (NR) between the trait locus and the marker locus are shown.

A lod score is a function of θ in the range of $0 \le \theta \le 1/2$ that always has a value of 0 at $\theta = 1/2$. The maximum of the lod score function occurs at the maximum likelihood estimate to θ. In the example pedigree in **Fig. 2**, the likelihood of observing linkage data is:

$$L = \theta^R (1 - \theta^{NR})$$

where θ is the recombination fraction, R is the number of recombinants and NR is the number of non-recombinants at a specific genetic marker. The likelihood of not observing linkage is

$$L = (0.5)^{(R+NR)}$$

Since grandparent genotypes are available this pedigree is considered phase-known, and therefore there is 1 recombinant and 5 non-recombinants in this pedigree. To calculate a two-point lod score at a range of recombination fractions is as follows:

$$z(\theta) = log_{10} \left[\frac{(\theta)^1 (1 - \theta)^5}{(0.5)^6} \right]$$

$$z(x) = log_{10} \left[\frac{0.5(\theta^1(1-\theta)^5) + 0.5(\theta^5(1-\theta)^1)}{(0.5)^6} \right]$$

At $\theta = 0.01$, 0.05, 0.1, 0.2, 0.3, and 0.4, the calculated lod scores are -0.22, 0.40, 0.58, 0.62, 0.51, and 0.30. However, if the pedigree only has genotype information in one or two generations, the pedigree is considered phase-unknown, and this must be taken into consideration in lod score calculations. Consider the example pedigree in **Fig. 3** where there are two possible phases. Phase 1 has 1 recombinant and 5 non-recombinants, whereas Phase 2 has 5 recombinants and 1 non-recombinant. To calculate lod scores for this example, both phases need to be considered:

At $\theta = 0.01$, 0.05, 0.1, 0.2, 0.3, and 0.4, the calculated lod scores are -0.51, 0.09, 0.28, 0.32, 0.22, and 0.08. Calculating lod scores by hand is a very laborious task. A ground-breaking development in human genetics studies was the development of LIPED, the first generally available computer program to perform linkage analysis, thus computational calculation of lod scores (Ott, 1974, 1976). LIPED is based on the Elston-Stewart algorithm (Elston and Stewart, 1971) and is restricted to two-point linkage analysis, which is pairwise comparisons between a disease (or trait) locus and each marker locus. Linkage programs output lod scores for each marker at different recombination fractions (such as the example for Arts Syndrome in **Table 2**), and can also include the exact recombination fraction where the lod score is at a maximum (Z_{max}).

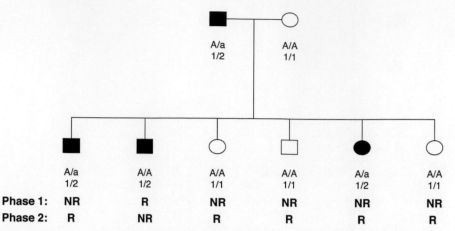

Fig. 3 Pedigree showing segregation of an autosomal dominant trait and a phase-unknown genetic marker. Recombinants (R) and non-recombinants (NR) between the trait locus and the marker locus are shown for the two different haplotype phases.

Table 2 Pairwise lod scores of Arts Disease and markers at Xq21.33 – Xq24. The descending order of markers is proximal to distal. The underlying disease gene was found to be *PRPS1* and maps to proximal of marker DXS1105.

Marker	Locus	$\theta = 0.00$	$\theta = 0.10$	$\theta = 0.20$	$\theta = 0.30$	$\theta = 0.40$
DXS1231	Xq21.33–q22.1	$-\infty$	-1.00	-0.33	-0.06	0.03
DXS178	Xq22.1	3.69	2.98	2.26	1.53	0.78
DXS1106	Xq22.2	6.97	5.57	4.08	2.53	1.06
DXS1105	Xq22.3	6.84	5.45	3.94	2.35	0.85
COL4A5	Xq22.3	5.38	4.18	2.95	1.78	0.76
DXS456	Xq22.3–q23	5.77	4.36	2.96	1.73	0.70
DXS424	Xq23	4.23	3.57	4.24	1.99	1.05
DXS1001	Xq24	$-\infty$	2.01	1.97	1.47	0.76

Note: Modified from Kremer, H., Hamel, B.C., van den Helm, B., *et al.*, 1996. Localization of the gene (or genes) for a syndrome with X-linked mental retardation, ataxia, weakness, hearing impairment, loss of vision and a fatal course in early childhood. Human genetics 98, 513–517 and de Brouwer, A.P.M., Williams, K.L., Duley, J.A., *et al.*, 2007. 'Arts syndrome is caused by loss-of-function mutations in PRPS1'. American Journal of Human Genetics 81, 507–518.

Since lod scores can be additive across families with an identical disease, two-point lod scores were historically reported at the following fixed set of θ values: 0, 0.01, 0.05, 0.1, 0.2, 0.3 and 0.4 (Conneally *et al.*, 1985). Lod scores (or sums of lod scores) of 3.0 or more are indicative of genome-wide significant evidence for linkage corrected for multiple testing (or greater than 2.0 on the X chromosome), whereas lod scores of less than -2.0 indicate non-linkage and can be used for exclusion purposes (Morton, 1955). Lod scores between -2.0 and 3.0 are inconclusive and require additional families, and thus more linkage analysis to be performed. Lander and Kruglyak (Lander and Kruglyak, 1995) built on Morton's work to determine that a more statistically accurate genome-wide significant lod score would be 3.3, and that a lod score of 1.86 is evidence of suggestive linkage.

The LIPED program was subsequently extended to provide multipoint analysis, resulting in the suite of LINKAGE programs for Linux and Windows ("see Relevant Websites section"), which are still in use today (Lathrop *et al.*, 1984). Dense genetic marker maps have made multipoint analysis possible, where simultaneous analysis of several linked loci is performed. Other multipoint linkage analysis programs are outlined in **Table 3**. In multipoint linkage analysis studies, geneticists will evaluate whether to apply parametric or non-parametric multipoint linkage analysis. Parametric multipoint analysis is usually applied to a few large pedigrees that have a known (or assumed) mode of inheritance of a disease or trait. A 'parameter' is set during linkage analysis: An explicit model of inheritance specifying the relationship between genotype and disease/trait. Parametric multipoint analysis is to identify linkage between the disease/trait gene and markers on the genetic map. In comparison, non-parametric multipoint analysis does not consider inheritance of the disease/trait. This analysis is usually considered an 'allele sharing' method that can look at many small families with the same disease.

Limitations of Current Linkage Analysis Programs

Although the introduction of computerised calculation of lod scores provided a platform for geneticists to regularly perform disease gene mapping studies, linkage programs have computational limitations. Improvement and enhancement in the algorithms over recent decades has improved efficiency of the programs, but computational burden is the major limitation, and is an ongoing problem with the current linkage programs in **Table 3**. Computational complexity increases with an increasing number of genetic markers or pedigree individuals. The Elston–Stewart algorithm (Elston and Stewart, 1971), which is used by LINKAGE and

Table 3 Multipoint linkage analysis software

Program	Multipoint analysis	Algorithm	Reference
LINKAGE	Parametric	Elston-Stewart	Ott (1974)
FASTLINK	Parametric	Elston-Stewart	Cottingham et al. (1993)
GENHUNTER	Parametric and Non-parametric	Lander-Green	Kruglyak et al. (1996)
ALLEGRO[a]	Parametric and Non-parametric	Lander-Green	Gudbjartsson et al. (2000)
MERLIN	Parametric and Non-parametric	Lander-Green	Abecasis et al. (2002), Abecasis and Wigginton (2005)

[a]Deprecated.

FASTLINK programs, scales exponentially with the number of marker loci considered, and generally cannot handle more than 5–10 markers at a time. This is a serious limitation in the current environment where hundreds of thousands of genetic markers can be used per person in linkage analysis. Conversely, GENEHUNTER, ALLEGRO AND MERLIN use the Lander-Green algorithm which uses a hidden Markov model (HMM) to calculate probability distributions of inheritance vectors (Lander and Green, 1987). Computational time and memory requirements only increase linearly with additional markers, but increase exponentially as the number of meioses in a pedigree increases. In addition, there is an upper limit of 32 meioses in a pedigree because of the representation of inheritance vectors as 32-bit integers (Markianos et al., 2001).

To date, there is no widely used linkage analysis program that can handle both large pedigrees and a large number of markers. There is a need for more powerful methods and enhancements to gene-mapping algorithms.

Human Linkage Analysis in the Age of Next-Generation Sequencing

Although linkage mapping for disease gene discovery was largely surpassed by large-scale GWAS studies due to the introduction of SNP microarrays, the era of next-generation sequencing has enabled linkage mapping to be revisited by geneticists. Combining linkage analysis results, with either whole-exome or whole-genome sequencing results has become an extremely powerful tool for identifying disease genes. **Fig. 4** outlines standard procedures geneticists use when combining data from these technologies.

In the current genetics/genomics environment, usually smaller families undergo linkage analysis using either microsatellites or SNP microarrays. The main reason for this being that many large families with clear Mendelian inheritance have already undergone linkage analysis in the previous decade, and thus have data already available. Prospective simulation analysis will frequently determine that these smaller families cannot meet genome-wide significance (i.e., a lod score >3.3), however lod scores of < − 2.0 can exclude large regions of the genome that will not harbour the causal mutation. DNA from all available family members will undergo linkage analysis, however due to higher costs, only a few informative individuals from the pedigree will undergo next-generation sequencing.

Standard bioinformatic workflows applied to raw next-generation sequencing data results in a VCF file, comprised of all identified variants in the entire genome. These bioinformatic workflows are covered in detail in the Next generation sequence analysis Reference Module. With genomic variant data, instead of geneticists spending years of research performing laborious Sanger sequencing for every exon in every gene in each linked region, next-generation sequencing provides a list of all the variants present in a comparably short period of time (often weeks). For example, a recent gene discovery publication for a large autosomal dominant amyotrophic lateral sclerosis/frontotemporal dementia family used genome-wide microsatellite linkage analysis from all available family members combined with exome sequencing data from four affected individuals within the family. Linkage analysis pinpointed the disease gene to a 7.5 Mb linked region on chromosome 16, containing more than 100 genes comprised of over 2000 exons. A considerable effort, and cost, would be required to Sanger sequence each of these exons. In a demonstration of the power of this combined technique, whole-exome sequencing and standardised filtering of the VCF (unshared variants, MAF >0.001 in public databases) identified only a single non-synonymous mutation in the linkage region, in the *CCNF* gene. Subsequent Sanger sequencing of this gene in all family members demonstrated strong segregation of the mutation with disease (Williams et al., 2016).

Using Interpolated Genetic Maps for Shared IBD Segment Analysis

Linkage maps are not only used for disease gene mapping, but can be integrated with other software packages to identify cryptic relatedness within a cohort of interest. Currently programs exist to work with large GWAS cohorts. These programs include PLINK (Purcell et al., 2007) and KING (Manichaikul et al., 2010) which can determine first and second degree relatives. Often, geneticists want to identify these relations to remove them from a study to prevent confounding, biasing or inflating GWAS results. However, these tools are not accurate past 2nd degree relatives. More complex programs exist to identify more distantly related individuals (up to 7th degree) using IBD segment analysis, and can utilise either SNP microarray data or whole-genome sequencing variant files (VCF). First the genotype data needs to be phased using Beagle (Browning and Browning, 2007), usually in comparison to a phased references population. The phased data is compared alongside interpolated genetic maps to discover long shared segments of Identity by Descent (IBD) between pairs of individuals (GERLINE program; Gusev et al., 2009). To then identify cryptic relatedness through shared ancestry, a maximum likelihood method is applied to the IBD segment data, using the number and genetic lengths of IBD segments, for the estimation of recent shared ancestry (ERSA) (Huff et al., 2011).

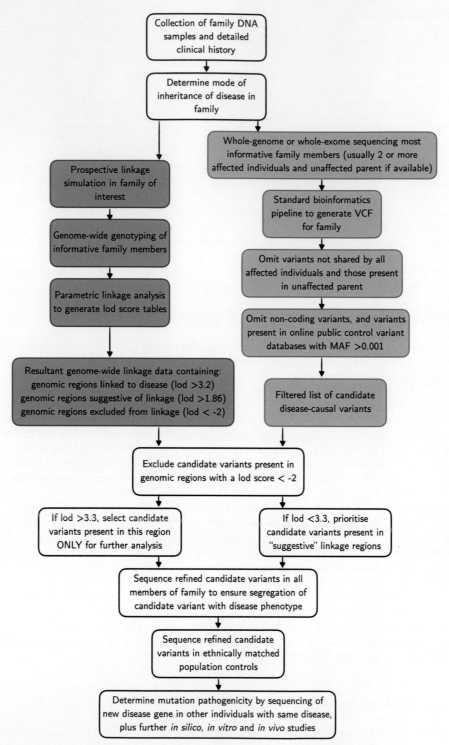

Fig. 4 Workflow for disease gene mapping using a combination of linkage analysis and next-generation sequencing methods. Linkage analysis workflow is outlined in blue, next-generation sequencing workflow is outlined in green, combined analysis is in black.

Gene Mapping in Plants and Animals

Existing human genetic maps are dense, well constructed, highly informative and amenable to interpolation, thus there is a limited need to develop new human genetic maps. Yet in plants and other animals, there is still a strong desire by geneticists to develop genetic maps for their species of interest. In contrast to human gene mapping studies, gene mapping in plants and animals, particularly those with agricultural interests, can be for identification of beneficial traits (quantitative trait loci, QTL). Genetic markers linked to QTLs are then used for marker-assisted plant selection to expedite genetic improvement in the species. Next-generation sequencing technologies have opened a new article in gene mapping for rare species. Plant and animal geneticists can use whole-genome sequencing to construct draft reference genomes, identify species-specific genetic markers (such as SNPs), and then to use these markers in species-specific genetic maps. Existing genetic maps of similar species can also be used to refine and complete genetic maps of other species. For example, a recently completed genetic map of the Atlantic Salmon was used to improve the genome sequence of the Rainbow Trout (Lien *et al.*, 2016).

One of the major challenges for rare species is that draft reference genomes constructed by *de novo* assemblies are often incomplete and with alignment mistakes in the assembly. *De novo* assembly has good fine-scale accuracy, but poor large-scale accuracy for constructing chromosomes (Fierst, 2015). Geneticists are able to use genetic maps and recombination to resolve these errors by ordering, correctly orientating and concatenating the contigs into chromosomes. New tools coupling *de novo* assembly with linkage mapping provide a powerful method to generate a high-quality reference sequence (Rastas *et al.*, 2013; Liu *et al.*, 2014).

Conclusions

Gene mapping has gone through peaks and troughs over the past several decades. In the 1980s and 1990s, increasing number of publications were reported linking human Mendelian disease genes to loci. The Human Genome Project represented the peak of gene mapping efforts to develop dense gene maps of the human genome containing thousands of informative markers. In the 2000s, these genetic maps were highly used to pinpoint disease loci for more Mendelian disorders. However, for complex diseases, genome-wide association studies (GWAS) using SNP microarrays became increasingly popular. In the last decade, the introduction of next-generation sequencing dramatically decreased the use of linkage mapping in human disease. Instead, geneticists sequence the entire exome or genome of multiple affected individuals and search for shared variants as putative causal mutations. This does not require large families, usually only a few affected individuals in a family. This has proven more fruitful in recessive diseases compared to dominant diseases, and geneticists are beginning to return to use linkage mapping coupled with next-generation sequencing as a powerful approach to identify disease genes. Conversely, animal and plant genetics are thriving by using next-generation sequencing technologies to facilitate generating genetic maps of their chosen species. Prior to next generation sequencing, the cost, time and difficulty in generating a genetic map of a rare species was prohibitive. Now using variant identification following *de novo* assembly, coupled with linkage maps of similar species, an increased number of publications are coming from these fields of research. This has the potential for major impact on agricultural genetics, as high-density linkage maps facilitate mapping quantitative trait loci (QTL) for important traits. The main challenge facing geneticists now is overcoming, or reducing, sequencing errors encountered in next-generation sequencing.

See also: Exome Sequencing Data Analysis. Functional Enrichment Analysis. Genome Annotation. Genome Databases and Browsers. Genome Informatics. Genome-Wide Association Studies. Integrative Analysis of Multi-Omics Data. Linkage Disequilibrium. Natural Language Processing Approaches in Bioinformatics. Next Generation Sequencing Data Analysis. Quantitative Immunology by Data Analysis Using Mathematical Models. Repeat in Genomes: How and Why You Should Consider Them in Genome Analyses?. Sequence Analysis. Sequence Composition

References

Abecasis, G.R., Cherny, S.S., Cookson, W.O., Cardon, L.R., 2002. Merlin-rapid analysis of dense genetic maps using sparse gene flow trees. Nature Genetics 30, 97–101.

Abecasis, G.R., Wigginton, J.E., 2005. Handling marker-marker linkage disequilibrium: Pedigree analysis with clustered markers. American Journal of Human Genetics 77, 754–767.

Boehnke, M., 1986. Estimating the power of a proposed linkage study: A practical computer simulation approach. American Journal of Human Genetics 39, 513–527.

Broman, K.W., Murray, J.C., Sheffield, V.C., White, R.L., Weber, J.L., 1998. Comprehensive human genetic maps: Individual and sex-specific variation in recombination. American Journal of Human Genetics 63 (3), 861–869.

Browning, S.R., Browning, B.L., 2007. Rapid and accurate haplotype phasing and missing-data inference for whole-genome association studies by use of localized haplotype clustering. American Journal of Human Genetics 81, 1084–1097.

Conneally, P.M., Edwards, J.H., Kidd, K.K., et al., 1985. Report of the committee on methods of linkage analysis and reporting. Cytogenetics and Cell Genetics 40, 356–359.

Cottingham, R.W., Idury, R.M., Schaffer, A.A., 1993. Faster sequential genetic linkage computations. American Journal of Human Genetics 53, 252–263.

Dausset, J., Cann, H., Cohen, D., et al., 1990. Centre d'etude du polymorphisme humain (CEPH): Collaborative genetic mapping of the human genome. Genomics 6 (3), 575–577.

de Brouwer, A.P.M., Williams, K.L., Duley, J.A., et al., 2007. Arts syndrome is caused by loss-of-function mutations in PRPS1. American Journal of Human Genetics 81, 507–518.

Donis-Keller, H., Green, P., Helms, C., et al., 1987. A genetic linkage map of the human genome. Cell 51, 319–337.

Dunn, G., Hinrichs, A.L., Bertelsen, S., et al., 2005. Microsatellites versus single-nucleotide polymorphisms in linkage analysis for quantitative and qualitative measures. BMC Genetics 6 (Suppl 1), S122.

Elston, R.C., Stewart, J., 1971. A general model for the genetic analysis of pedigree data. Human Heredity 21, 523–542.

Fierst, J.L., 2015. Using linkage maps to correct and scaffold de novo genome assemblies: Methods, challenges, and computational tools. Frontiers in Genetics 6, 220.

Gudbjartsson, D.F., Jonasson, K., Frigge, M.L., Kong, A., 2000. Allegro, a new computer program for multipoint linkage analysis. Nature Genetics 25, 12–13.

Gusev, A., Lowe, J.K., Stoffel, M., et al., 2009. Whole population, genome-wide mapping of hidden relatedness. Genome Research 19, 318–326.

Haldane, J.B.S., 1919. The combination of linkage values and the calculation of distances between the loci of linked factors. Journal of Genetics 8, 299–309.

Huff, C.D., Witherspoon, D.J., Simonson, T.S., et al., 2011. Maximum-likelihood estimation of recent shared ancestry (ERSA). Genome Research 21, 768–774.

Kong, A., Gudbjartsson, D.F., Sainz, J., *et al.*, 2002. A high-resolution recombination map of the human genome. Nature Genetics 31, 241–247.

Kong, A., Thorleifsson, G., Gudbjartsson, D.F., *et al.*, 2010. Fine-scale recombination rate differences between sexes, populations and individuals. Nature 467 (7319), 1099–1103.

Kong, X., Murphy, K., Raj, T., *et al.*, 2004. A combined linkage-physical map of the human genome. American Journal of Human Genetics 75, 1143–1148.

Kosambi, D.D., 1944. The estimation of map distances from recombination values. Annals of Eugenics 12, 172–175.

Kremer, H., Hamel, B.C., van den Helm, B., *et al.*, 1996. Localization of the gene (or genes) for a syndrome with X-linked mental retardation, ataxia, weakness, hearing impairment, loss of vision and a fatal course in early childhood. Human Genetics 98, 513–517.

Kruglyak, L., Daly, M.J., Reeve-Daly, M.P., Lander, E.S., 1996. Parametric and nonparametric linkage analysis: A unified multipoint approach. American Journal of Human Genetics 58, 1347–1363.

Lander, E., Kruglyak, L., 1995. Genetic dissection of complex traits: Guidelines for interpreting and reporting linkage results. Nature Genetics 11, 241–247.

Lander, E.S., Green, P., 1987. Construction of multilocus genetic linkage maps in humans. Proceedings of the National Academy of Sciences of the United States of America 84, 2363–2367.

Lander, E.S., Green, P., Abrahamson, J., *et al.*, 1987. Mapmaker: An interactive computer package for constructing primary genetic linkage maps of experimental and natural populations. Genomics 1, 174–181.

Lander, E.S., Linton, L.M., Birren, B., *et al.*, 2001. Initial sequencing and analysis of the human genome. Nature 409, 860–921.

Lathrop, G.M., Lalouel, J.M., Julier, C., Ott, J., 1984. Strategies for multilocus linkage analysis in humans. Proceedings of the National Academy of Sciences of the United States of America 81 (11), 3443–3446.

Lien, S., Koop, B.F., Sandve, S.R., *et al.*, 2016. The Atlantic salmon genome provides insights into rediploidization. Nature 533, 200–205.

Litt, M., Luty, J.A., 1989. A hypervariable microsatellite revealed by in vitro amplification of a dinucleotide repeat within the cardiac muscle actin gene. American Journal of Human Genetics 44, 397–401.

Liu, D., Ma, C., Hong, W., *et al.*, 2014. Construction and analysis of high-density linkage map using high-throughput sequencing data. PLOS ONE 9 (6), e98855.

Manichaikul, A., Mychaleckyj, J.C., Rich, S.S., *et al.*, 2010. Robust relationship inference in genome-wide association studies. Bioinformatics 26, 2867–2873.

Markianos, K., Daly, M.J., Kruglyak, L., 2001. Efficient multipoint linkage analysis through reduction of inheritance space. American Journal of Human Genetics 68, 963–977.

Matise, T.C., Chen, F., Chen, W., *et al.*, 2007. A second-generation combined linkage physical map of the human genome. Genome Research 17, 17831786.

Morgan, T.H., 1911. Random segregation versus coupling in Mendelian inheritance. Science 34, 384.

Morton, N.E., 1955. Sequential tests for the detection of linkage. American Journal of Human Genetics 7 (3), 277–318.

Ott, J., 1974. Estimation of the recombination fraction in human pedigrees: Efficient computation of the likelihood for human linkage studies. American Journal of Human Genetics 26 (5), 588–597.

Ott, J., 1976. A computer program for linkage analysis of general human pedigrees. American Journal of Human Genetics 28, 528–529.

Ott, J., 1989. Computer-simulation methods in human linkage analysis. Proceedings of the National Academy of Sciences of the United States of America 86, 4175–4178.

Ploughman, L.M., Boehnke, M., 1989. Estimating the power of a proposed linkage study for a complex genetic trait. American Journal of Human Genetics 44, 543–551.

Purcell, S., Neale, B., Todd-Brown, K., *et al.*, 2007. PLINK: A tool set for whole-genome association and population-based linkage analyses. American Journal of Human Genetics 81, 559–575.

Rastas, P., Paulin, L., Hanski, I., Lehtonen, R., Auvinen, P., 2013. Lep-MAP: Fast and accurate linkage map construction for large SNP datasets. Bioinformatics 29 (24), 3128–3134.

Sturtevant, A.H., 1913. The linear arrangement of six sex-linked factors in Drosophila, as shown by their mode of association. Journal of Experimental Zoology 14, 43–59.

Venter, J.C., Adams, M.D., Myers, E.W., *et al.*, 2001. The sequence of the human genome. Science (New York, NY) 291, 1304–1351.

Weber, J.L., May, P.E., . Abundant class of human DNA polymorphisms which can be typed using the polymerase chain reaction. American Journal of Human Genetics 44, 388–396.

Weeks, D., Ott, J., Lathrop, G.M., 1990. SLINK: A general simulation program for linkage analysis. American Journal of Human Genetics 47, A204.

Williams, K.L., Topp, S., Yang, S., *et al.*, 2016. CCNF mutations in amyotrophic lateral sclerosis and frontotemporal dementia. Nature Communications 7, 11253.

Further Reading

Ott, J., 1999. Analysis of Human Genetic Linkage. Johns Hopkins University Press. Available at: https://books.google.com.au/books?id=BKJqAAAAMAAJ.

Ott, J., Wang, J., Leal, S.M., 2015. Genetic linkage analysis in the age of whole-genome sequencing. Nature Reviews Genetics 16 (5), 275–284.

Dudbridge, F., 2003. A survey of current software for linkage analysis. Human Genomics 1, 63–65.

Fierst, J.L., 2015. Using linkage maps to correct and scaffold de novo genome assemblies: Methods, challenges, and computational tools. Frontiers in Genetics 6, 220.

Relevant Websites

http://compgen.rutgers.edu/map_interpolator.shtml
Computational Genetics.

http://www.jurgott.org/linkage/LinkagePC.html
Genetic LINKAGE programs for Windows and Linux.

https://github.com/joepickrell/1000-genomes-genetic-maps/blob/master/scripts/interpolate_maps.py
GitHub.

http://www.bli.uzh.ch/BLI/Projects/genetics/maps/marsh.html
Marshfield Genetic Maps.

http://www.jurgott.org/linkage/util.htm
Statistical Genetics Utility programs.

Genome Databases and Browsers

Dean Southwood and Shoba Ranganathan, Macquarie University, Sydney, NSW, Australia

Introduction

"In God we trust. All others must bring data." – W. Edwards Deming

"Data matures like wine, applications like fish." – James Governor

The effective use of genomic data is vital for many different fields, from disease research, to drug development, to biosecurity, to agriculture. Analysing this data lets us determine how organisms function, what causes certain traits and features, and what variation there is between different organisms, both within the same species as well as across species. As ideas and desires such as personalised medicine gain traction, the importance of careful and considered genomic analysis will only increase. However, the field of genomics is growing at an unprecedented rate, spurred on in recent times by the development and improvement of various genomic sequencing techniques. In the 15 years since the completion of the Human Genome Project in 2003, the time, manpower and financial costs of sequencing the genome of an organism have all dramatically decreased. It is therefore no surprise that there has been an explosion of sequencing data on both familiar and novel organisms in the years since. However, while it is of great benefit to science to have such a wealth of data available, it is difficult use effectively without the right tools. Appropriate methods of storing, organising, curating, and accessing the data (in the form of databases), as well as viewing, editing, and analysing the data (in the form of browsers) are vital in progressing the field.

As whole-genome sequencing (WGS) becomes more routine, smaller sequencing projects of specific genes or groups of genes have become even more common. The amount of data that these projects produce needs to be managed effectively, so that it can be stored, accessed and shared among researchers. However, the sheer volume of the data, as well as the number of organisms, data sources, and study types mean that there are thousands of different databases available to access genomic data, depending on what is being sought. Individual research communities will have specific databases that are more useful than others, but there are still a number of common features across them, and a small number of repositories that span multiple organisms.

Being able to access data is an important first step in doing high-quality research (or research at all), but if the data cannot be visualised and analysed effectively, there is little point in accessing it in the first place. This is especially true for genomic data, where its sheer volume coupled with its repetitive nature creates a situation in which annotation and visualisation are vital to making the data have any meaning. Genome browsers have been developed alongside databases to deliver on this requirement. However, like databases, there are many different browsers available with which to analyse data. These largely vary on how robust they are, whether they are server-based or local, and what specialised tools they have embedded in them. In general however, there are a small number of key browsers that tend to be suitable for broad, generic purposes, which have then served as the basis for the development of more specialised browsers and tools.

The current article is divided into four sections. The first section discusses currently available databases for accessing genomic data, including important multi-organism databases, as well as more specialised ones. The second section discusses the state-of-the-art genome browsers available to visualise and analyse genomic data, including their range of different capabilities and specialised tools. The third section uses WormBase as an example database, and visualises data from WormBase using GBrowse as an example of a broad-purpose genome browser. The fourth section discusses the future directions and issues involved in developing genome databases and browsers, and provides further reading to the interested reader.

Fundamentals

Genome Databases

Genomic databases allow for the storing, sharing and comparison of data across research studies, across data types, across individuals and across organisms. These are not a new invention – even before the popularisation of the modern internet, 'online' databases have been available in order to share data on key organisms, such as *Escherichia coli* (Blattner *et al.*, 1997) and *Saccharomyces cerevisiae* (Cherry *et al.*, 2012). Recent advances in both data sharing technology and genome sequencing technology have created an explosion of databases, based around particular organisms, as has been historically the case, as well as around particular data types, such as transcriptional data or short-read sequencing data. This section discusses a number of key classes of genome databases, namely broad cross-organism databases such as NCBI Genome and Ensembl Genome, specific organism databases and the GMOD project family, and the broad variety of databases available for human genome research. The IDs provided in brackets correspond to those in the ID column of **Table 1**.

Broad Databases

From the early development of genome databases several decades ago, there have been ongoing efforts to combine and index multiple databases in central repositories for ease of access. With the avalanche of genomic data in the last decade, these efforts have become more difficult. However, there are still two key databases which serve as first-point-of-call references for genomes, especially if the desired genomic information is likely to be accessed by many researchers. These are the NCBI Genome database (1), co-ordinated by the National Centre for Biotechnology Information in the United States, and the Ensembl Genomes database (2), jointly co-ordinated by EMBL-EBI and the Wellcome Trust Sanger Institute in the United Kingdom.

At time of writing, NCBI Genome contains genomic information on 35,211 distinct organisms, which are further index as eukaryotes (5413 entries), prokaryotes (134,749 entries), viruses (14,054 entries), plasmids (11,946 entries), and organelles (11,504 entries). These can contain multiple assemblies, and can be downloaded or searched on the database itself through a number of tools.

Ensembl Genomes also acts as a repository of genomic data across multiple kingdoms of organism, and is complemented by five sub-sites: Bacteria-Ensembl, Protist-Ensembl, Fungi-Ensembl, Plants-Ensembl, and Metazoa-Ensembl, containing the genomes of bacteria (50,364 entries), protists (200 entries), fungi (882 entries), plants (55 entries), and invertebrate metazoa (71 entries) respectively. This is in addition to the 12 central Ensembl organisms, including zebrafish (*Danio renio*), chicken (*Gallus gallus*), human (*Homo sapiens*), and mouse (*Mus musculus*). Ensembl Genomes also allows for data to be downloaded or analysed using online tools. There is also the option to analyse the genomic data in Ensembl, the genome browser associated with the database itself (discussed further below).

Specific Organism Databases and the GMOD Project

It is possibly unsurprising that with the evolution of sequencing technology and the power to sequence the genome of most any organism, given a reasonable amount of time and a reasonable amount of research effort, individual databases have developed around the genomes of specific organisms. In the past, this was mostly focussed around so-called 'model' organisms, or ones with large research bases, such as the mouse (*Mus musculus*) (Smith *et al.*, 2018) and nematode (*Caenorhabditis elegans*) (Lee *et al.*, 2018). In many cases, these were created by their own research communities, to suit their own needs, both in terms of how the data could be accessed, as well as what tools were provided to dissect the data. As efforts continued, there have been moves to create some consistency between databases and the tools they offer, meaning new organism databases are not required to 're-invent the wheel' so to speak. In this regard, the Generic Model Organism Database (GMOD) project has served to provide a framework of tools and database methods to allow new databases to be created. The 'users' of the GMOD project are no longer limited to 'model' organisms, and now consist of a variety of different species and databases. The GMOD project also has its own genome browser associated with it, GBrowse (as discussed further below), which can be integrated into the participating databases as a web-based genome browser.

Human Genome Databases

The breadth and depth of human genome databases is vast, as is to be expected when an organism attempts to study itself, and analyse its own biological problems. These databases are often structured around various data sources, such as transcriptional data, as is the case for the H-Invitational database (H-InvDB) (4). Particular study types have also given rise to specific databases: genome-wide association study databases such as GWASCentral (5), and structural variant study databases such as dbVar (6) and DGV (7). As is the case with other organisms, there are also some databases which seek to be more comprehensive in scope: DNA

Table 1   Prominent databases of genomic data; links current as of April 2018

Name	ID	URL	References
NCBI Genome	1	https://www.ncbi.nlm.nih.gov/genome	NCBI Resource Coordinators (2018)
Ensembl Genome	2	http://ensemblgenomes.org/	Kersey *et al.* (2018)
GMOD Project	3	http://gmod.org/wiki/Main_Page	N/A
H-InvDB	4	http://www.h-invitational.jp/	Yamasaki *et al.* (2009)
GWASCentral	5	https://www.gwascentral.org/	Beck *et al.* (2014)
dbVar	6	https://www.ncbi.nlm.nih.gov/dbvar	Lappalainen *et al.* (2013)
DGV	7	http://dgv.tcag.ca/dgv/app/home	MacDonald *et al.* (2014)
ENCODE	8	https://www.encodeproject.org/	Sloan *et al.* (2016)
IGSR (1000 Genomes)	9	http://www.internationalgenome.org/	The Genomes Project Consortium (2015)

element databases such as ENCODE (8), and the 1000 Genomes project database, now hosted as the International Genome Sample Resource (IGSR) (9). Databases for even more specific purposes exist, such as a wealth of databases on cancer genomic data, and will need to be searched for on a case-by-case basis depending on need.

Genome Browsers

Data access and quality means very little if no meaning can be gained from it. In a field with as complex and abstract data as genomics, methods for data visualisation and analysis are of even greater importance. These must be able to cope with vast amounts of data, in the order of gigabytes or terabytes, as well as be able to connect these to tangible, biological meaning in the form of genes and products. Genome browsers seek to fill this need by providing a pre-existing software basis to visualise and analyse genomic data. Due to the sheer variety of researchers, purposes, expectations, and goals involved in the field, a number of genome browsers are available. For the new user, there are three broad-class, easy-to-pick-up databases for generic uses that stand out at present: the UCSC Genome Browser, managed by the University of California, Santa Cruz (Casper *et al.*, 2018); GBrowse, managed by the GMOD project (see "Relevant Website section"); and Ensembl, managed by EMBL-EBI and the Wellcome Trust Sanger Institute (Zerbino *et al.*, 2018). This section will consider each of these browsers in turn, and then give an overview of the more specific browsers which have been created based on these three forerunners.

UCSC Genome Browser

UCSC Genome Browser is the one of the most widely regarded broad-class browser, and has been integrated into a number of major databases. Its initial conception in 2000 was to visualise the first working draft of the Human Genome Project, but has been adapted in the following years to include a broad variety of organisms, and a vast suite of tools for visualising and analysing data.

Gbrowse

Due to the 'generic' nature of the GMOD project (discussed above at 2.2), there was a need for a generic browser to accompany the suite of tools provided for new databases. GBrowse developed from this idea, and is therefore one of the more flexible genome browsers available. It has had a number of spin-off browsers created since its conception, tailored for particular purposes. As it is a part of the GMOD project, it is also available across many different databases.

Ensembl

The Ensembl genome browser created by EMBL-EBI and the Wellcome Trust Sanger Institute is the native genome browser for the Ensembl Genomes databases. Due to the broad nature of the databases it is used for, it contains a broad variety of tools for visualisation and analysis across a variety of kingdoms of organisms.

Specialised Browsers

Browsers for more specialised purposes have been developed by particular groups, largely based on one of the primary three browsers. Due to its generic nature, the majority of 'subsidiary' browsers are based on GBrowse in particular. Lighter implementations have been created, such as JBrowse, as well as browsers more suited to collaboration and annotation, such as Apollo.

Applications

Wormbase and Gbrowse

An illustrative example of a user-friendly genome database, as well as a common all-purpose genome browser, is the WormBase-GBrowse pairing. WormBase (Lee *et al.*, 2018) is a member of the GMOD family, meaning it has the GBrowse genome browser embedded in the site. This makes quick searches extremely easy, and allows for visualisation of data before download.

The basic functionality of WormBase is illustrated in **Fig. 1**, through the homepage, search, and GBrowse features. The homepage is very user-friendly with logical menus and news items. The search function allows the user to access gene information across a variety of species, including *C. elegans*, *C. briggsae*, and *C. brenneri*. It also allows for search by class of information, including gene, strain, protein, or sequence feature. Searching for a particular gene gives the option to open the gene in GBrowse, as shown in **Fig. 1(c)** for the *set-23* gene in *C. elegans*. The *set-23* gene has been shown to be an embryonic lethal gene, and as such could be a possible CRISPR target in the future – knowing the location (chromosome IV), length (2625 bp), and protein coding regions (as shown in pink) of the gene are important, and easily accessible from the browser view. GBrowse is also equipped with an extensive functionality and toolbox – further information can be found through the GMOD wiki (see "Relevant Website section"), including tutorials on more specialised features of GBrowse.

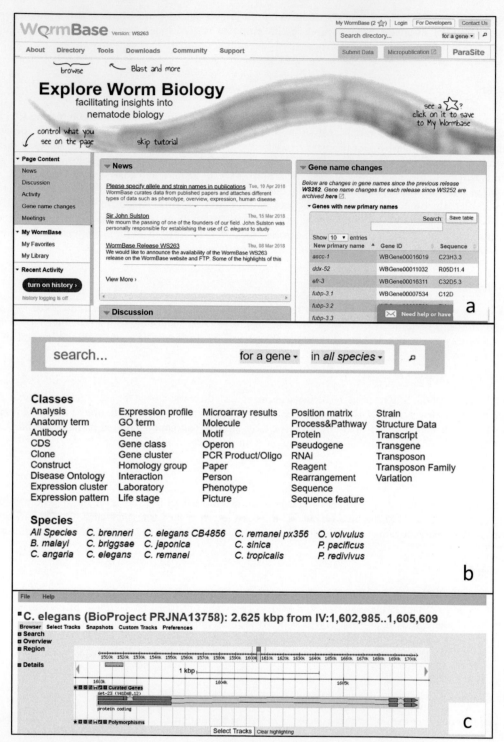

Fig. 1 Example of a genome database and browser pairing using WormBase and GBrowse, showing (a) the homepage of WormBase, (b) the search tool by search class and by organism, and (c) GBrowse, with the Set-23 gene of *C. elegans* selected as the track.

Discussion and Future Directions

With the wealth of databases and browsers available today for genomic research, researchers are in some ways spoilt for choice. However, as new pursuits such as personalised medicine progress, databases and browsers will need to developed or adapted to meet these new demands. In a field with such fast-paced progress as genomics, not all future issues will be able to be predicted, and many will have to be dealt with as they arise. However, a few key directions are evident. The continued development of high-throughput sequencing technology will likely continue to bring sequencing time and money costs down, increasing the availability of genomic data. Therefore, current databases will need to be optimised and adapted, and new databases created, to allow for such an influx of data. The curation and annotation of genomic data is already an issue, significant efforts put into generating the data, but less plentiful efforts put into assigning it with accurate annotations before adding it to databases; as data creation increases, this problem will only continue to get more and more relevant. More effective strategies around who is charged with curating and annotating the data in order to give it meaning, while keeping it reliable and reputable, need to devised.

Closing Remarks

There are a multitude of databases and browsers available for genomic purposes, and with future increases in genomic data, there will only be more and more new databases formed, and new browsers created. No static article can keep up with the furious developments in the field; the hope is that this article can provide an introduction to the researcher looking into genomic databases and browsers for the first time, or looking for a refresher. This article is also not comprehensive, nor does it purport to be; with the already vast sea of resources available, no article could hope to offer a view of the whole current offering. If the reader requires more specialised databases and tools, hopefully the notions discussed herein provide a guide for what to look for.

See also: Bioinformatics Approaches for Studying Alternative Splicing. Bioinformatics Data Models, Representation and Storage. Biological Database Searching. Comparative and Evolutionary Genomics. Comparative Genomics Analysis. Data Storage and Representation. Exome Sequencing Data Analysis. Functional Enrichment Analysis. Functional Enrichment Analysis Methods. Functional Genomics. Gene Mapping. Genome Alignment. Genome Annotation. Genome Annotation: Perspective From Bacterial Genomes. Genome Informatics. Integrative Analysis of Multi-Omics Data. Integrative Bioinformatics of Transcriptome: Databases, Tools and Pipelines. Integrative Bioinformatics of Transcriptome: Databases, Tools and Pipelines. Linkage Disequilibrium. Natural Language Processing Approaches in Bioinformatics. Next Generation Sequencing Data Analysis. Ontology-Based Annotation Methods. Phylogenetic Footprinting. Quantitative Immunology by Data Analysis Using Mathematical Models. Repeat in Genomes: How and Why You Should Consider Them in Genome Analyses?. Transcriptomic Databases. Whole Genome Sequencing Analysis

References

Beck, T., Hastings, R.K., Gollapudi, S., et al., 2014. GWAS Central: A comprehensive resource for the comparison and interrogation of genome-wide association studies. Eur. J. Human Genet. 22 (7), 949–952.

Blattner, F.R., Plunkett 3rd, G., Bloch, C.A., et al., 1997. The complete genome sequence of Escherichia coli K-12. Science 277 (5331), 1453–1462.

Casper, J., Zweig, A.S., Villarreal, C., et al., 2018. The UCSC genome browser database: 2018 update. Nucleic Acids Res. 46, D762–D769.

Cherry, J.M., Hong, E.L., Amundsen, C., et al., 2012. Saccharomyces genome database: The genomics resource of budding yeast. Nucleic Acids Res. 40, D700–D705.

Kersey, P.J., Allen, J.E., Allot, A., Barba, M., et al., 2018. Ensembl genomes 2018: An integrated omics infrastructure for non-vertebrate species. Nucleic Acids Res. 46, D802–D808.

Lappalainen, I., Lopez, J., Skipper, L., et al., 2013. DbVar and DGVa: Public archives for genomic structural variation. Nucleic Acids Res. 41, D936–D941.

Lee, R.Y.N., Howe, K.L., Harris, T.W., et al., 2018. WormBase 2017: Molting into a new stage. Nucleic Acids Res. 46, D869–D874.

MacDonald, J.R., Ziman, R., Yuen, R.K., et al., 2014. The Database of genomic variants: A curated collection of structural variation in the human genome. Nucleic Acids Res. 42, D986–D992.

NCBI Resource Coordinators, 2018. Database resources of the National Center for Biotechnology Information. Nucleic Acids Res. 46, D8–D13.

Sloan, C.A., Chan, E.T., Davidson, J.M., et al., 2016. ENCODE data at the ENCODE portal. Nucleic Acids Res. 44, D726–D732.

Smith, C.L., Blake, J.A., Kadin, J.A., et al., 2018. Mouse Genome Database (MGD)-2018: Knowledgebase for the laboratory mouse. Nucleic Acids Res. 46, D836–D842.

The Genomes Project Consortium, 2015. A global reference for human genetic variation. Nature 526, 68–74.

Yamasaki, C., Murakami, K., Takeda, J.I., et al., 2009. H-InvDB in 2009: Extended database and data mining resources for human genes and transcripts. Nucleic Acids Res. 38, D626–D632.

Zerbino, D.R., Achuthan, P., Akanni, W., et al., 2018. Ensembl 2018. Nucleic Acids Res. 46, D754–D761.

Relevant Website

http://gmod.org/wiki/Main_Page
GMOD.

Biographical Sketch

Dean Southwood is a PhD student in the Department of Molecular Sciences at Macquarie University. His background is in theoretical quantum physics, quantum computation, and bioinformatics. His current research interests involve genome analysis and improving basecalling algorithms using machine learning.

Shoba Ranganathan holds a Chair in Bioinformatics at Macquarie University since 2004. She has held research and academic positions in India, USA, Singapore and Australia as well as a consultancy in industry. Shoba's research addresses several key areas of bioinformatics to understand biological systems using computational approaches. Her group has achieved both experience and expertise in different aspects of computational biology, ranging from metabolites and small molecules to biochemical networks, pathway analysis and computational systems biology. She has authored as well as edited several books in as well as contributed several articles to Springer's Encyclopedia of Systems Biology. She is currently the Editor-in-Chief of Elsevier's Encyclopedia of Bioinformatics and Computational Biology as well as the Bioinformatics Section Editor for Elsevier's Reference Module in Life Sciences.

Comparative and Evolutionary Genomics

Takeshi Kawashima, National Institute of Genetics, Mishima, Japan

Introduction

Comparative genomics is a fundamental tool of genome analysis. Typically, DNA sequences from whole genomes and whole gene sets are compared to elucidate the common and different genomic features among two or more target organisms. Features common to two species are considered properties originating from the genome of their common ancestry, whereas different features reflect the distinct histories of each genome traced since the common ancestor. This rationale can be applied to any two species in a comparative analysis because all organisms on the Earth can be connected to a single phylogenetic tree (**Fig. 1**). Seen in this sense, comparative genomics is equivalent to evolutionary genomics.

However, caution is required in comparative analyses because exceptional events, such as horizontal gene transfer and convergent evolution, can often be detected at levels more than expected. In such cases, it is necessary to carefully confirm the accuracy of the data and the correctness of the analytical process. Whether the result can be verified by other independent methods should also be considered.

Previous reports on comparative genomics are rich examples of the types of analyses and modifications performed for each case. In this report, the achievements of comparative genome analysis are discussed in order to unravel the evolution of metazoans, and the analytical methods used in these studies and the knowledge obtained from application of these methods are outlined.

Aspects of Comparative Genomics: Structure, Type of Change, and Evolutionary Distances

When performing comparative genomic analyses, the following three biological aspects should be considered: genome structure, types of structural change, and evolutionary distance.

Genome Structure and Types of Change

For convenience in genome analyses, a genomic DNA sequence can be divided into genomic "parts", such as protein-coding genes, noncoding genes, and intergenic regions that include repeat and interspersed sequences, and other functional regions represented by cis-/trans-regulatory elements.

Each of these parts undergoes changes throughout a generation through the combination of several evolutionary mechanisms. In the evolution of protein-coding genes, Koonin classified the major mechanisms into the following five types of change: (i) direct descendants with some nucleotide changes; (ii) gene duplication; (iii) gene deletion; (iv) gene gain by horizontal gene transfer;

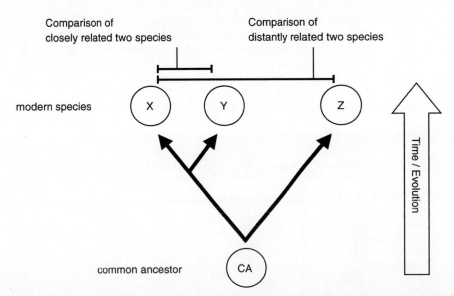

Fig. 1 Conceptual Diagram of Relationship of Taxonomic Distances in Comparative Genomics. Genome data can be obtained for X, Y and Z, each of which is an extant organism. X and Y are closer to each other than to Z. X and Z have evolved since they diverged from the common ancestral organism CA. When comparing the three types of X, Y, Z, the commonly observed features may be the characteristics of CA.

and (v) fusion, fission, and other genetic restructuring, including loss or gain of new domains on the gene (Koonin, 2005). In addition to these, (vi) emergence of a de novo gene is also mentioned often.

"Emergence of a de novo gene (have been frequently occurred)" is a relatively new concept. In this mechanism, a protein-coding gene emerges from noncoding DNA with some mutations. There are many unresolved problems in gene evolution, including the contribution rate. The mechanism was mainly discovered from the comparative genome analysis of closely related fly species (Begun et al., 2006). The fixation rate of a de novo gene is still unknown. However, it seems certain that de novo genes are constantly created from noncoding regions at a certain evolutionary rate (Zhao et al., 2014).

The above six mechanisms (i) to (vi) were originally described for the evolution of protein-coding genes. However, the same classification may also be applied to the other genomic parts by replacing them more general concepts, such as: (i) direct descendants, (ii) duplication, (iii) deletion, (iv) gain by horizontal transfer, (v) changes in partial structure, and (vi) de novo emergence.

The proportion of each genomic part differs greatly among organisms (Lindblad-Toh et al., 2005). This suggests that each evolutionary mechanism works independently on each genomic part, resulting in the creation of differences in the characteristics of various genomes as a basis for forming biodiversity from common ancestry (Simakov and Kawashima, 2017).

For example, the number of gene losses since the common mammalian ancestor has been estimated as about 1000 for humans and about 2000 for dogs, which is almost twice the rate (Demuth et al., 2006). As other examples from an comparative analysis of interspersed sequences, the proportion of repetitive sequences in the genome of limpet (*Lottia gigantea*) is reported to be only around 21%, whereas those in the genomes of Hydra (*Hydra magnipapillata*) and humans are 57% and 43%, respectively (Simakov et al., 2013; Chapman et al., 2010). In addition, the evolutionary rate of the relative position of each part also differs in genome evolution. Preservation of the gene position on a chromosome is represented by the concepts of synteny and linkage, each of which forms a genomic feature. Many microsyntenies derived from the common ancestry are detected between the genomes of amphioxi and chickens, but are not detected between the genomes of amphioxi and tunicates. This suggests that the evolutionary rate of the chromosomal structure has considerable differences among animals.

Although there are major differences in the genomic features of each organism, we can infer those features of the common ancestor. **Table 1** shows an example of the estimation value of genomic features of the metazoan ancestor (Simakov and Kawashima, 2017).

In any case, the genome has evolved by changing all genomic parts based on the above six mechanisms, making up the genomic features of extant organisms.

Evolutionary Distances and Taxonomic Data Bias

Evolutionary distances for comparative analysis

Evolutionary distances between the target species of comparative analysis are largely related to choosing an appropriate data type and the method to be used.

By way of one of the most basic methods of comparative genomics, finding a common genetic set is relatively simple in most cases, for any pair of organisms (e.g., even for human and bacteria, in The C. elegans Sequencing Consortium, 1998). In such analyses, it is necessary to correctly understand the concepts of "orthologs," "paralogs," and other related concepts. These concepts are detailed in the next Section "Homologous Genes: Orthologs, Paralogs, and Others".

In contrast, biological discovery becomes difficult if the distance between the compared species is too far or too close in some cases. This can be easily understood by considering the sequence comparison of intergenic regions as an example. Generally, in comparing intergenic regions between two genetically distant species, few conserved sequences are found. On another front, comparison between very closely related or among the same species reveals nearly identical DNA sequences, such that almost no differences may be detected between two species unless a considerable number of sequences from multiple individuals are compared.

Conservation and differences among relative positions of genes on the DNA between two species provides a history of chromosomal evolution, also called as "conserved-synteny" or "conserved linkage." Although more than hundreds of orthologous genes are conserved between fungi and animals (Li et al., 2003), almost no conservation is found in most of the gene-order between these

Table 1   An example of the estimation value of genomic features of the metazoan ancestor

Genomic feature	Inferred ancestral values
Presumptive genome size	~300 Mb
Gene family number	7,000–8,000
Total gene number	>20,000
Average exon count	3–4
Repeat content	~30%
Macro-syntenic linkage groups	~10–17
Micro-syntenic linkage groups	~400

Note: Reproduced from Simakov O., Kawashima, T., 2017. Independent evolution of genomic characters during major metazoan transitions. Dev. Biol. 427, 179–192.

two possibly because they are highly distant evolutionarily. Only exceptional example is reported in which the order of three genes, psmb 1, Tbp, and Pdcd 2, is conserved among yeast, humans, and some other animals (Trachtulec and Forejt, 2001).

As described above, it is necessary to carefully ensure appropriate evolutionary distance among subject data before initiating comparative studies.

Biased genomic data

The result of comparative genome analyses is susceptible to taxon sampling. Unfortunately, considerable taxonomic-bias is commonly found in any usable sequence data in practice. To perform a comparative analysis correctly, researchers need to be recognize the extent of bias in the genomic data which they used.

In metazoans for example, the genome sequences of chordate animals are overwhelmingly numerous by total published data, followed by those of Arthropoda (the column of "Number of decoded species" in **Fig. 2**). In contrast, for other animal taxonomic groups at the phylum level, it is often observed that no genomic data have been decoded or only one or two species have decoded genomes (e.g., gray vertical bar A and B in **Fig. 2**). Of the genome data for the phylum Chordata, most of the data are overly concentrated in sub-phylum Vertebrata compared to the other two subphyla, Urochordata and Cephalochordata.

The main reason for the sequence data bias towards vertebrates is presumably because human-related research accounts for a major portion of biological studies. Especially from a medical perspective, the amount of sequence data for humans and their closely related species tends to be increased. Genome analysis of domestic animals is also actively pursued as it is probably considered valuable from an agricultural point of view, regardless mammals have a larger genome size being more than 1 Gb (Anderson and Georges, 2004). 16 metazoan species are known as major domestic animals including 15 of mammals (dog, goats, sheep, cattle, pig, horse, cat, camel, ferret, chicken, turkey, rabbit, guinea pig, llama, and alpaca) and an insect, silkworm (Morey, 1994; Banno *et al.*, 2010). Of the 16 major domestic animals, genomes of 12 mammals and one insect have been decoded

Phylum Level Taxonomy *	Number of Decoded Spec.**	Common Name	Species Name	Size (Mb)	Year	Authors	Remarkable Genomic Findings
Deuterostomia							
Chordata	291	Round Worm	C. elegans	97	1998	CSC.,	The first sequenced metazoan genome. Olfactory receptor expansion.
Hemichordata	2						
Echinodermata	7	Fly	D. melanogaster	180	2000	Adams et al.,	The first metazoan WGS genome. Hetero-euchromatin transition zone.
		Human	H. sapiens	3,289	2001	IHGSC.,, Venter et al.,	Total human genes fewer than thought.
Ecdysozoa							
Nematoda	81						
Priapulida	1						
Nematomorpha	0	Purple sea urchin	S. purpuratus	814	2006	SUGSC et al.,	Genes associated with vision, balance, and chemosensation in urchin.
Loricifera	0						
Kinorhyncha	0	Starlet sea anemone	N. vectensis	357	2007	Putnam et al.,	Ancient metazoan genome complement and organization.
Onychophora	0						
Tardigrada	1	Placozoa	T. adhaerens	98	2008	Srivastava et al.,	Diversity of developmental genes more than cell types of this organism.
Arthropoda	239	Schistosoma	S. mansoni	363	2009	Berriman et al.,	Micro exon and unusual intron size distribution.
Lophotrochozoa							
Mollusca	10	Demosponge	A. queenslandica	167	2010	Srivastava et al.,	Six hallmarks of animal multicellularity and the evolution of relevant genes.
Annelida	2						
Brachiopoda	1	Owl limpet	L. gigantea	348	2012	Simakov et al.,	Multiple duplications of Hox complements in the leech.
Cycliophora	0	Polychaeta	C. teleta	324			
Chaetognatha	0	Freshwater leech	H. robusta	228			
Phoronida	1	Bdelloid rotifer	A. vaga	224	2013	Flot et al.,	Intragenomic synteny and ameiotic evolution.
Bryozoa (Ectoprocta)	0	Comb jellies	M. leidyi	150	2013	Ryan et al.,	Mesodermal genes absent from Ctenophore genome.
Entoprocta	0		P. bachei	156	2014	Moroz et al.,	Independent evolution of the nervous system.
Acanthocephala	0						
Gastrotricha	0	Lingula	L. anatina	425	2015	Luo et al.,	Expansion of chitin synthase corresponding to the shell formation.
Gnathostomulida	0	Water bear	H. dujardini	212	2015	Boothby et al.,	
Nemertea	1	Acornworm (Atlantic)	S. kowalevskii	758	2015	Simakov et al.,	Deuterostome novelties and the pharyngeal cluster.
Rhombozoa	0	Acornworm (Pacific)	P. flava	1229			
Orthonectida	1	Orthonectid	I. linei	43	2016	Mikhailov et al.,	Elemental genes of the metazoan sensory systems.
Rotifera	2	Ribbon worm	N. geniculatus	859	2017	Luo et al.,	Lophotrochozoans share an ancestral bilaterian gene repertoire with deuterostomes.
Platyhelminthes	29	Horseshoe worm	P. australis	498			
Basal-Bilateria							
Xenacoelomorpha	0						
Non-Bilateria							
Cnidaria	10						
Placozoa	1						
Ctenophora	2						
Porifera	1						

Fig. 2 History of animal genomics. On the left side of the figure, 32 phyla of animals (metazoa) are described. There are various opinions as to how many are appropriate number for phylum level classification. Recently it classified often as fewer than 32 phylum. In the recent phylogenetic interpretation, all animal phyla are divided into two large groups, non-bilaterian and bilaterian. And the latter is classified further into three groups, Ecdysozoa, Lophotrochozoa and Deuterostomia. In the column of "the number of decoded species", how many species of its genome was decoded from each animal phylum was described based on the data of NCBI Taxonomy Database at the end of 2017. On the left side of the figure, among the respective animal phyla, the animal species whole genome was first decoded are shown in chronological form. Looking at the gray bar (A and B), we can see that there are considerable bias in the number of genomes decoded for each animal phylum.

(Kirkness *et al.*, 2003; Lindblad-Toh *et al.*, 2005; Dong *et al.*, 2013; Jiang *et al.*, 2014; The Bovine Genome Sequencing and Analysis Consortium *et al.*, 2009; Rubin *et al.*, 2012; Wade *et al.*, 2009; Pontius *et al.*, 2007; The Bactrian Camels Genome Sequencing and Analysis Consortium, 2012; Peng *et al.*, 2014; Rubin *et al.*, 2010; Dalloul *et al.*, 2010; Carneiro *et al.*, 2014; Mita *et al.*, 2004; Xia *et al.*, 2004). For the remaining three other animals (guinea pig, llama, and alpaca), genome sequencing projects are still ongoing.

In several cases for the genome projects of the domestic animals, multiple strains and individuals have been re-sequenced for applying various methods of population genetics to identify the genes responsible for genetic features of domestic animals that have been acquired through the history of domestication. Some examples of such genome projects of domestic animals will be described later.

Early studies decoded the insect genome presumably because it includes the fruit-fly, which is a major model organism of basic experimental biology, and also because many of them have relatively small genome sizes (Adams *et al.*, 2000; Gregory, 2011). Another reason is that several insects are deeply involved in the human society; some of them are useful, such as bees and silkworms whereas others are pests including aphids and flour beetles (Grimmelikhuijzen *et al.*, 2007).

Thus, analysis of genome sequence data cannot avoid bias for various reasons and it may simply be correlated with the researcher's population in most cases. Thus, it is important to determine the existence of bias in published data when performing comparative genome analysis.

Comparison of Homologous Genes

Homologous Genes: Orthologs, Paralogs, and Others

During comparative analysis of genes, it is necessary to understand the types of homologous genes (**Fig. 3**). "Homologous genes" are defined as genes with a common evolutionary ancestry (Brown, 2007). "Homologous genes" are sometimes called homologs (or homologues), but homolog(ue) is originally a histological term introduced by Owen, and defined as "the same organ in different animals under every variety of form and function (Owen, 1843)". Thus, this term is often inappropriate for use in genomics. Instead "orthologs" and "paralogs" are used frequently, both of which are simply basic concepts for genes that have a common ancestor with respect to molecular evolution (**Fig. 3(A)**). Orthologs are determined as homologous genes that diverged as a result of a speciation event. Paralogs are defined as homologous genes that diverged as a result of a gene duplication event. Another related term that is not used frequently, is "xenologs," defined as "homologous genes that diverged as a result of lateral gene transfer" (Kristensen *et al.*, 2011).

"Outparalogs (Alloparalogs)," "Inparalogs (Symparalogs)," and "Co-orthologs" are also defined as terms for further fine classification (Sonnhammer and Koonin, 2002). Outparalogs are defined as paralogs that were duplicated before the speciation event. Inparalogs are defined as paralogs that were duplicated after the speciation event. If lineage-specific gene-duplications (or expansion) has occurred in one or both of a pair of orthologs after speciation, sister genes of such duplicated orthologs constitute one-to-many or many-to-many relationships, respectively. Such gene relationships are called co-orthologs. Co-orthologs are thus defined as paralogs produced by duplication of orthologs after a given speciation event.

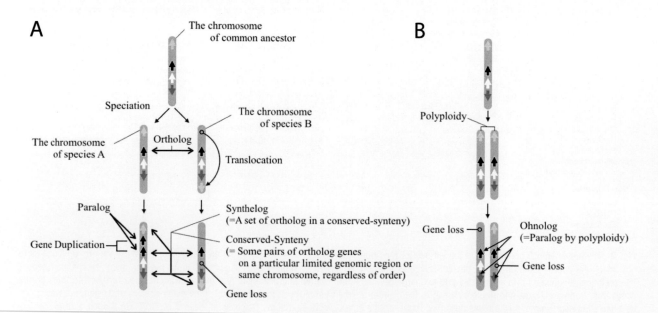

Fig. 3   Homolog Related Concepts. Illustrated various homolog related concepts shown in the main text.

Homology Search and Functional Prediction

One of the main purposes of searching homologous genes is prediction of gene function. This is based on the knowledge that homologous genes often have similar functions (Toh et al., 1983). Since orthologs and paralogs are descendants of a common ancestral gene, it is estimated that both their functions and sequences are likely to be preserved. For this reason, determining orthologs and paralogs is the most basic step in prediction of gene function in bioinformatics (Gabaldón and Koonin, 2013).

Since the number of genes for which function has been elucidated experimentally is overwhelmingly smaller than the number of sequenced genes, it is frequently noted that gene prediction based on homology search may have problems with accuracy (Benner, 1999). Meaningless annotations like "similar to a gene of unknown function," are also accumulated in genome data-bases, and these are likely to be reused in other databases. Despite this, the search for homologous genes has not lost its significance as the first step in annotation of genes (Joshi and Xu, 2007).

Various methods are still being developed for ortholog determination in comparative genomics because it is difficult to accurately decide true homologous genes pairs for the entire gene set between two species. Therefore, in practice, "bidirectional best-hits" (BBH, also called "mutual best-hits" or "reciprocal best-hits") from all gene-sets are treated as conventional candidates of homologous genes (Tatusov et al., 1997; Kristensen et al., 2011). Although BBH are commonly used, a little study has been performed to ascertain their reliability. BBH are considered to be appropriate for use in prokaryotes (Wolf and Koonin, 2012). In contrast, a marked increase in mistakes is reported when BBH are used for comparisons among metazoans or plants, probably because these have many duplicated genes (Dalquen and Dessimoz, 2013). Ortholog determination in the comparison of whole gene sets is thus a product of some compromise. The fact that various ortholog-databases continue to be generated based on various criteria also implies the difficulty of ortholog determination (Dessimoz et al., 2012).

Besides predicting gene function, as will be described later, determination of homologous genes is required to find conserved-synteny and linkage in the entire genome to reconstruct chromosomal evolution.

Whole Genome Duplication

Whole genome duplication (WGD) is an evolutionary event in which the whole gene set of an organism has been duplicated, probably by chromosomal polyploidy and its fixation. WGD is postulated as one of the major sources of evolutionary innovations.

Although polyploidy is a common phenomenon in eukaryotes, evolutionary fixed WGDs are exceedingly rare. In metazoa, known WGD events are almost limited to vertebrate lineages includes famous two round WGD (2R) which is occurred on the common ancestor of vertebrates (Ohno, 1970; Van de Peer et al., 2009, 2010). In invertebrates, it is also suggested that some lineages have experienced WGD although those traces are not so clearly like vertebrates (Yoshida et al., 2011; Kenny et al., 2016). In plants and fungi, WGD occurs relatively more frequently than in animals (Van de Peer et al., 2009) but there is still some bias.

Each type of duplicated genes created by WGD has named based on their forming mechanisms, respectively **(Fig. 3(B))**. Gene pairs produced by WGD are a special subset of paralogs (Wolfe, 2000). Because of their evolutionary significance as described above, they are sometimes distinguished by terms such as "ohnologs," or its synonym "syntenic paralogs" (Wolfe, 2004; Schnable et al., 2012). A similar word is "homoeologs" (or "homeologs" as another spelling) defined as homologous genes resulting from allopolyploidy (Glover et al., 2016). The term "ohnolog" is derived from the name of Susumu Ohno who first proposed the hypothesis that gene duplication plays a key role in evolution (Ohno, 1970; Wolfe, 2000). Allopolyploid is a polyploid individual which composed of a chromosome set composed of two (or more) chromosome set. It formed from the hybridization of two (or more) separate but closely related species. The difference between the definitions of ohnolog and homoeolog is that the former refers to paralog made by autopolyploidy and the latter refers to paralog made by allopolyploidy. However, if polyploidization arose very long ago, it can not distinguish which mechanism truly occurred for creating the paralogus gene-sets. In such cases where autopolyploidy or allopolyploidy can not be distinguished, the term "paleolog" is sometimes used conveniently. Another similar term is "syntelogs" representing orthologous genes shared across species (or isolates for viruses) with a common syntenic genome location (Hatcher et al., 2014).

The evolutionary process of WGD is one of the key themes in evolutionary genomics. In general, duplicated genes tend to be fixed only when those are to take on different functions but otherwise to be lost immediately after the duplication event. However, WGD is an event that a large number of duplication has occurred at once, it was quite difficult to investigate the process of gene losses until the era of genomics.

Because the function of an ohnolog pairs is presumably redundant each other just after WGD, it is theoretically expected that one of the duplicated genes to be lost gradually. Interestingly, the rate of gene losses may differ depending on the types of genes. For example in metazoa, it is known that ohnolog genes which relevant to embryonic development tend to be conserved as paralog gene-set (Holland et al., 1994).

To distinguish a true pair of ohnolog or homoeolog from other common paralogs is technically difficult. Various methods have been devised to extract the set of ohnolog that were thought to be made by WGD. Several models have been proposed why the conservation rate of developmental genes after WGD is high (for example, Holland et al., 1994). A future explanation will be awaited as to why the same trend is shown for cell cycle regulators.

To extract the ohnolog created by 2R in vertebrate evolution, Dehal and Boore utilized a genome of an invertebrate which is branched before the 2R. Using a graph-based method for this purpose, they have developed a way to extract homologous genes of which number encoded in the genome of invertebrates and vertebrates at a ratio of approximately 1:2 to 1:4, respectively (Dehal and Boore, 2005).

Nishida and his colleagues developed a method to determine ohnolog gene-pair and examined the rate of gene loss after the WGD which occurred the common ancestor of teleost fish (teleost genome duplication: TGD) (Inoue *et al.*, 2015). As a result of the analysis for the obtained ohnolog gene set, they found that the rate of loss of ohnolog is roughly approximated by a double-exponential curve (2-phase model). According to their "2 phase model" of gene loss after TGD, duplicated genes are rapidly lost immediately after the WGD event, and then slowly lose at a roughly constant rate.

It is estimated that the rapid rate of gene loss in the first phase probably depends on the deletion of contiguous clusters of genes or large chromosomal segments ("block loss"), but the details have not been clarified yet.

It may be even more difficult to accurately identify the set of paralogs in an allopolyploid (homoeolog). The african clawed frog Xenopus laevis is proposed to be an allotetraploid animal that arose via the interspecific hybridization of diploid progenitors followed by subsequent genome doubling. In the sequence-decoding and reconstruction of the whole-genome of this frog, researchers noticed that the possibility that the subgenome derived from the two progenitor frogs each retained unique transposable elements in their genome and that those elements had been fossilized in each genome (Session *et al.*, 2016). They have actually found three of such transposable element and succeeded in separating reconstructed DNA sequences into two homoeologous subgenomes using those fossilized transposable elements as markers.

Homoeologs in the *X. laevis* were extracted from those separated subgenomes thus obtained and the tendency of the gene-loss after WGD was investigated. Interestingly, it was found that not only the developmental genes but also cell cycle regulators are tend to be conserved as paralogs after WGD.

Comparative Domain Structure of Genes

Many proteins retain several functional domains in their amino acid sequences. Particularly in the evolution of eukaryotic protein coding regions, the complexity of the domain structure may have played a significant role in gene evolution (Patthy, 1999; Chothia *et al.*, 2003). As mentioned in Section "Genome Structure and Types of Change", the evolution of the domain architecture of a gene is a known mechanism for the changes in the partial structures of genes. The presence of genes with a lineage-specific or clade-specific domain structure makes it difficult to find true orthologous relationships (Uchiyama, 2006).

In eukaryotes, because genes are divided into several segmented structures of exons, duplication, shuffling, or deletion of functional domains can occur relatively easily by DNA recombination, potentially producing genes with a novel domain structure.

Choanoflagellates are unicellular eukaryotes and the closest known relatives of animals. King *et al.* (2008) decoded the genome of a choanoflagellate, *Monosiga brevicollis*, and compared the protein structure against those of metazoans. They found that choanoflagellates and metazoans shared many domains in their genomes, but the multidomain structure in their genes differed. These findings suggested that abundant domain shuffling followed the separation of the choanoflagellate and metazoan lineages. Such shuffling may have created animal-specific proteins during evolution.

Acquisition of a new domain structure may also be involved in the evolution of metazoan traits. In vertebrates, the domain structure of aggrecan protein, a major component of cartilage, was created by a domain-shuffling event that occurred in the common ancestry of vertebrates (Kawashima *et al.*, 2009).

Comparative Genomics of Intergenic Regions

Studies of intergenic regions have exploited comparative genomics methodologies. Such strategies are based on the known tendency that not only the coding-region but also other functional elements are likely to be well conserved in DNA among closely related species; thus, functional genomic regions can be predicted to some degree from sequence comparisons. Unlike the coding region, which can be predicted using transcriptome data, prediction of functional motifs in intergenic regions has been difficult. One of the established strategies is collating with known elements in databases such as TRANSFAC, JSAPAR, and PAZAR. However, it is difficult to obtain strong candidates of the true element using this method alone because usually many false positives are detected. Additionally, this method is often not suitable for taxonomically separated species. Thus, comparative genome sequences provide a simple and widely applicable method for narrowing the candidates.

Nowadays, more direct and powerful methods for experimentally targeting such regions have been developed. Typical examples of these experimental methods include chromatin immunoprecipitation-seq (ChIP-seq), DNase-seq, formaldehyde-assisted isolation of regulatory elements-seq (FAIRE-seq), and assay for transposase-accessible chromatin-seq (ATAC-seq); of these, the former is a method for determining target positions on DNA of particular DNA-binding proteins, and the latter three are methods for detecting the open-chromatin regions on DNA. However, comparative sequences are still useful for surveying functional elements through intergenic regions because they can be combined with a variety of other methods.

Boffelli reviewed some examples of early studies of surveying functional elements in intergenic regions by comparative genomics among vertebrates (Boffelli *et al.*, 2004). Two representative researches introduced in the review are referred here. Nobrega and Rubin compared the 2.6-Mb region of the human genome with the homologous region of the mouse, frog, and puffer fish genomes and found 20 short conserved sequences in the vicinity of the *DACH* gene (Nobrega *et al.*, 2003). At least six of the sequences have been confirmed to function as enhancers in mouse embryos. As another example, Lettice and colleagues compared the sequences in the neighboring area of *SHH* genes among humans, chickens, and puffer fish and found a sequence preserved in the introns of another gene separated by 1 Mb or more. The sequence was experimentally confirmed as a strong candidate enhancer of the *SHH* gene (Lattice *et al.*, 2003).

Some tools (e.g., VISTA and PipMaker) can be used to detect the conserved element in noncoding regions by comparative sequences (Mayor *et al.*, 2000; Schwartz *et al.*, 2000). Using these tools, comparative analyses of intergenic regions has become easy for two or more organisms if the taxonomic distances are selected appropriately. However, if the distance between two species is far, detecting functional motifs is still difficult. Future studies in comparative genomics are needed to overcome this unsolved problem.

Comparison of Gene Order

Linkage and Synteny

During evolution, chromosomal rearrangements change the gene order little by little as generations recapetulate due to deletions, insertions, duplications, inversions, and translocations of some chromosomal parts. Nevertheless, the signatures of the gene order in the genome of the common ancestor have been preserved for a long period in their offspring. Such traces of gene order are called "synteny", which is defined as the presence at least two pairs of homologous genes on a particular limited genomic region or the same chromosome, regardless of order (Nadeau and Sankoff, 1998; Trachtulec and Forejt, 2001). Another similar concept is conserved linkage, which is defined as the synteny of two or more pairs of orthologous genes in the same order and orientation. However, the nomenclature for the evolutionarily conserved gene order is not unified. Although the original definition is as described above, it is often confusing (Passarge *et al.*, 1999). Therefore, researchers should consider the intentions of authors when consulting the literature. Some reports have referred to synteny as the conservation of blocks of gene orders within two sets of chromosomes or DNA fragments (scaffolds) that are being compared with each other. Since draft genome analysis has become popular, comparative genomic analysis of fragmental DNAs with only several kb or Mb can be easily performed. In such analyses, analysts cannot find the conservation of genes at the chromosomal level, but can find groups of orthologs in the relatively short fragmented DNAs of two organisms. Such synteny is sometimes called conserved synteny-block.

Long-Term Conservation of Synteny

Particularly in eukaryotes, synteny and linkage are often conserved for a surprisingly extended period. For example, many synteny blocks have been found in humans and mice, although the two species diverged from a common ancestor about 75 million years ago. The synteny blocks between these two species are approximately 65 Mb in length. The total length of these synteny blocks has been reported to cover over 90% of the entire genome of humans and mice.

Synteny is found even among the more evolutionarily distant species of eukaryotes, in contrast to bacteria, in which traces of gene orders disappeared almost immediately in some cases (Koonin *et al.*, 2000). However, as the evolutionary distance increased, the conservation degree of gene order, i.e., both the number and size of detectable blocks, tends to decrease gradually. When analyzing such slight traces of synteny, we can distinguish between two types of conserved gene orders that span different genomic scales, i.e., microsynteny and macrosynteny (Putanam *et al.*, 2007). Microsynteny is a local genomic region in which a very limited number of gene orders are conserved. Compared with microsynteny, macrosynteny describes much larger ranges, those on the scale of (partial) chromosomes and including several (or sometimes hundreds of) genes.

In a comparison of two markedly distant species, it is necessary to determine whether synteny and linkage found in modern organisms are indeed derived from the gene order in the common ancestor. If traces of synteny are very faint, we should focus on short stretches of conserved gene orders (microsynteny), composed of only two or three genes with some inserted genes, in order to consider the possibility of false positives due to coincidence. By validating the conservation in more than three species, false positives for surveying synteny blocks under such soft filtering can be eliminated.

In the following, three examples of micro- and macrosynteny are shown. In all cases, the human genome was compared with the genome of another organism. The first was a comparison of humans with lancelets, the invertebrate that is closest to vertebrates; the second was a comparison of humans with sea anemones, the non-bilaterial animal closest to bilaterial animals; and the third was a comparison of humans and fungi.

Lancelets, or amphioxi, belong to the subphylum cephalochordata and are classified as the phylum chordata together with the subphyla Tunicata and Vertebrata. Molecular phylogenetic analysis has revealed that cephalochordata were the first branched organisms among these three about 550 million years ago. Putnam and colleagues reported that more than 1000 microsyntenies were found between lancelets and humans under conditions allowing up to 20 genes to intervene between consecutive pairs of orthologs on the respective genomes. Although this may include some false positives, the fact that the number was over six times the discovery rate compared with the randomly reordered control strongly suggested that most of the gene orders were conserved since the common ancestor (Putnam *et al.*, 2008).

It is unclear why microsyntenies preserved for so long-period. In an attempt to explain this problem, Irimia *et al.* (2011) found that genes encoding transcription factors were more likely to preserve the order with neighboring genes than other types of genes.

Based on this trend, they proposed a model in which regulatory elements of the intergenic region moved into the intron of the adjacent gene and in which the order of two genes was often fixed under the evolutionary timescale.

Domestication and Comparative Animals

Domestic Organisms are Invaluable Resources In Comparative Genomics

Domestic animals and agricultural plants are good research subjects for comparative genomic analysis between closely related species (Andersson and Georges, 2004). These strains have well-known traits that have been chosen over the years, and multiple DNA samples of closely related strains are available. To utilize the advantages of domestic animals, we typically attempt two types of analyses for comparing genomes to elucidate the domestication process of target organisms. In the first approach, the main subject is the elucidation of the origin of domestication and the history of subsequent propagation. In contrast, in the second approach, the aim is to find genes associated with traits relevant to domestication. In both of the approaches, data for sequence variations (primarily single nucleotide polymorphisms [SNPs]) are collected from several strains.

Origin and History of Domestication

Molecular phylogenetic analyses are used for the first approach to elucidate the domestication process. For this purpose, in addition to the genomes of multiple domestic strains, the genomes of the wild species closest to those domesticated strains are decoded, as are genomes of as many extant domestic strains as possible. When creating a species tree from amino acid or RNA sequences, using a long sequence created by concatenating a set of orthologs, rather than using only one gene, is the most common method (Delsuc *et al.*, 2005). For comparing closely related species, such as domestic organisms of different lines, SNP data gathered from comparative whole genomes are also useful for constructing a species tree (Guo *et al.*, 2013). In order to correct information on branching times, it is useful to calibrate the molecular phylogenetic tree based on paleontological data, such as fossil records and geological information. Some projects have succeeded in decoding even the DNA sequence from the fossils of ancestral animals to trace the process of domestication. In cases in which the fossil DNA of ancestral-related organisms can be obtained, it is possible to deduce the evolutionary rate up to the branching time with the extant taxa.

As an example of research on horse domestication, one research group succeeded in sequencing a draft genome from a horse bone recovered from permafrost dated to approximately 560–780 thousand years ago (Orland *et al.*, 2013). The group compared the obtained genome with the genomes of several modern horses and concluded that the common ancestor of all contemporary horses, zebras, and donkeys originated 4.0–4.5 million years ago.

Dogs are the first domesticated animal for which the genome was decoded. In the dog genome, many genes are missing compared with other mammals. Indeed, approximately 2000 genes may have been substantially mutated or lost from the dog genome after branching of dogs from the common ancestor of other mammals (Kirkness *et al.*, 2003; The Bovine Genome Sequencing and Analysis Consortium *et al.*, 2009). However, in the decoding of other domesticated animals, e.g., cows and pigs, such substantial gene losses have not been observed. Although it is difficult to determine the specific number of genes, and no reports have clearly described the number of genes lost during dog evolution, the results of comparative gene numbers among mammals in some papers are all similar; all results show that approximately 2000 genes have been lost in dogs. These findings related to gene losses may involve the evolution of dogs with various unique morphologies; further studies are needed to clarify these findings.

Notably, dogs were the first domesticated animals, separated from a group of wolves living with humans. The first branching time was estimated to be about 14,000 years ago, after which, two major changes occurred. A wide range of multiplications have also been made, particularly in the last 200 years.

Genome-Wide Association Study (GWAS) and Selective Sweep

The second approach is called GWAS, which uses sequence variations in the whole genome together with the phenotype and pedigree information to perform association analysis and to identify the loci of genes or regulatory elements that are important for the traits of interest (Zhang *et al.*, 2012). Each domestic animal has evolved specific traits, such as meat, egg, or fur productivity. Those traits are expected to be acquired by strong selection pressures. Therefore, by comparing the genomes among strains, the loci dominantly selected in each strain can be identified.

When analyzing the distribution of SNPs from a GWAS, "selective sweep" is frequently noted (Boffelli *et al.*, 2004). Selective sweep refers to a process by which a new advantageous mutation eliminates or reduces variation in linked neutral sites as it increases in frequency in the population (Nielsen *et al.*, 2005). This phenomenon is also called "genetic hitchhiking". Since specific traits of domestic animals work predominantly by artificial selection, selective sweeps are expected to be observed in the neighboring region of the causative gene(s) of the trait. For example, according to the report of the first whole genome SNP analysis in cattle, the *MSTN* gene, which is related to muscle formation, was present in one of the regions where a strong signal of a selective sweep was detected in the genome of two beef cattle strains (The Bovine HapMap Consortium, 2009).

Another similar example is the genomic analyses of chickens. In chicken genomics, SNPs have been collected from eight different populations of domestic chickens, including broiler (meat-producing) and layer (egg-producing) lines as well as a red jungle fowl, the major wild ancestor. A selective sweep at the locus for thyroid-stimulating hormone receptor (TSHR) was performed for all domestic chickens. The *TSHR* gene is known to be involved in the regulation of photoperiod control of reproduction in birds and mammals. The authors demonstrated that mutations in this gene and its neighboring region were related to early domestication of chickens, as shown by the absence of strict regulation of seasonal reproduction in natural populations.

Future Directions

In the 2000s, whole genome analysis became possible in both prokaryotes and eukaryotes, and a comparative genome analysis was undertaken. Initially, there was a focus on the presence or absence of orthologs as well as the detection of conserved intergenic regions in major comparative genomics. Subsequently, the performance of the high-throughput DNA sequencing has improved, and it is now feasible to compare the genome sequences of multiple individuals for the same species or closely related strains. For this reason, the theories advocated in classical population genetics are now being actively tested with various actual data. This trend is expected to continue.

With the development of several "finishing" technologies, many projects aiming to reconstruct genomic sequences that had been stopped at the "draft" level may be able to be completed to the number of chromosomal arms in the near future (Eisenstein, 2015). If more genomes are completely sequenced, researchers in the field of evolutionary genomics will be able to elucidate the history of structural changes in the chromosome, such as chromosomal rearrangement or segmental duplication.

Closing Remarks

Both the terms "comparative genomics" and "evolutionary genomics" tend to be used when referring to a wide range of fields. In this report, the following definition was applied: "comparison of all gene sets in two or more species of organisms". However, this definition has expanded to include comparisons of intergenic regions, SNPs, and chromosomal structures. Examples in this report focuses on comparative and evolutionary genomics of eukaryotes. Because comparative genomic analysis of bacteria also differs greatly from that of eukaryotes, it has not been described in this paper. Additionally, the examples provided herein are biased toward the study of animals; however, it is expected that the information presented in this report will still be helpful because the basic trends of the described studies were not different from those in fungi and plants. Notably, the genomic structure of unicellular eukaryotes has not yet been elucidated. It is possible that the basic structures of genomes in these organisms may deviate considerably from those of previously decoded genomes. Accordingly, further genome decoding and comparative analyses are needed.

Acknowledgements

I thank Yi-jun Luo, Oleg Simakov, Jun Inoue, and Hirokazu Chiba who kindly commented on the very early version of this document. I'm grateful to anonymous reviewer K.K. who reads the final manuscript intensively and gave useful comments. I greatly appreciate that Masanori Arita, Naoko Murakata and Takako Ohnuki supported the office procedure related to writing work. Part of this work was supported by the Kaken-hi Grant (no.: 17K19248).

See also: Algorithms for Strings and Sequences: Multiple Alignment. Bioinformatics Approaches for Studying Alternative Splicing. Comparative Genomics Analysis. Exome Sequencing Data Analysis. Functional Enrichment Analysis. Gene Duplication and Speciation. Genome Alignment. Genome Annotation. Genome Databases and Browsers. Genome Informatics. Genome-Wide Association Studies. Homologous Protein Detection. Identification of Homologs. Inference of Horizontal Gene Transfer: Gaining Insights Into Evolution via Lateral Acquisition of Genetic Material. Integrative Analysis of Multi-Omics Data. Metagenomic Analysis and its Applications. Natural Language Processing Approaches in Bioinformatics. Next Generation Sequencing Data Analysis. Phylogenetic analysis: Early evolution of life. Phylogenetic Footprinting. Quantitative Immunology by Data Analysis Using Mathematical Models. Repeat in Genomes: How and Why You Should Consider Them in Genome Analyses?. Sequence Analysis. Whole Genome Sequencing Analysis

References

Adams, M.K., *et al.*, 2000. The genome sequence of Drosophila melanogaster. Science 287, 2185–2195.

Anderson, L., Georges, M., 2004. Domestic-animal genomics: Deciphering the genetics of complex traits. Nat. Rev. Genet. 5, 202–212.

Banno, Y., Shimada, T., Kajiura, Z., Sezutsu, H., 2010. The silkworm — An attractive bioresource supplied by Japan. Exp. Anim. 59, 139–146.

Begun, D.J., Lindfors, H.A., Thompson, M.E., Holloway, A.K., 2006. Recently evolved genes identified from Drosophila yakuba and D. erecta accessory gland expressed sequence tags. Genetics 172, 1675–1681.

Benner, S.E., 1999. Errors in genome annotation. Trends Genet. 15, 132–133.

Boffelli, D., Nobrega, M.A., Rubin, E.M., 2004. Comparative genomics at the vertebrate extremes. Nat. Rev. Genet. 5, 456–465.

Brown, T.A., 2007. Genomes 3. Garland Science Publishing.

Carneiro, M., Rubin, C., Di Palma, F., Albert, F.W., et al., 2014. Rabbit genome analysis reveals a polygenic basis for phenotypic change during domestication. Science 345, 1074–1079.

Chapman, J.A., Kirkness, E.F., Simakov, O., Hampson, S.E., et al., 2010. The dynamic genome of Hydra. Nature 464, 592–596.

Chothia, C., et al., 2003. Evolution of the protein repertoire. Science 300, 1701–1703.

The C. elegans Sequencing Consortium, 1998. Genome sequence of the nematode C. elegans: A platform for investigating biology. Science 282, 2012–2018.

Dalloul, R.A., Long, J.A., Zimin, A.V., Aslam, L., et al., 2010. Multi-platform next-generation sequencing of the domestic turkey (Meleagris gallopavo): Genome assembly and analysis. PlOS Biol. 8, e1000475.

Dalquen, D.A., Dessimoz, C., 2013. Bidirectional best hits miss many orthologs in duplication-rich clades such as plants and animals. Genome Biol. Evol. 5, 1800–1806.

The Bactrian Camels Genome Sequencing and Analysis Consortium, 2012. Genome sequences of wild and domestic bactrian camels. Nat. Commun. 3, 1202.

Dehal, P., Boore, J.L., 2005. Two rounds of whole genome duplication in the ancestral vertebrate. PLOS Biol. 3, e314.

Delsuc, F., Brinkmann, H., Philippe, H., 2005. Phylogenomics and the reconstruction of the tree of life. Nat. Rev. Genet. 6, 361–375.

Demuth, J.P., et al., 2006. The evolution of mammalian gene families. PLOS ONE 1, e85.

Dessimoz, C., Gabaldón, T., Roos, D.S., Sonnhammer, E.L.L., et al., 2012. Toward community standards in the quest for orthologs. Bioinformatics 28, 900–904.

Dong, Y., Xie, M., Jiang, Y., et al., 2013. Sequencing and automated whole-genome optical mapping of the genome of a domestic goat (Capra hircus). Nat. Biotechnol. 31, 135–141.

Eisenstein, M., 2015. Startups use short-read data to expand long-read sequencing market. Nat. Biotechnol. 33, 433–435.

Gabaldón, T., Koonin, E.V., 2013. Functional and evolutionary implications of gene orthology. Nat. Rev. Genet. 14, 360–366.

Glover, N.M., Redestig, H., Dessimoz, C., 2016. Homoeologs: What are they and how do we infer them? Trends Plant Sci. 21, 609–621.

Gregory, R.R., 2011. The Evolution of the Genome. Academic Press.

Grimmelikhuijzen, C.J.P., et al., 2007. The promise of insect genomics. Pest Manag. Sci. 63, 413–416.

Guo, S., et al., 2013. The draft genome of watermelon (Citrullus lanatus) and resequencing of 20 diverse accessions. Nat. Genet. 45, 51–58.

Hatcher, E.L., Hendrickson, R.C., Lefkowitz, E.J., 2014. Identification of nucleotide-level changes impacting gene content and genome evolution in orthopoxviruses. J. Virol. 88, 13651–13668.

Holland, P.W.D., Garcia-Fernandez, J., Williams, N.A., Sidow, A., 1994. Gene duplications and the origins of vertebrate development. Dev. Suppl. 125–133.

Inoue, J., Sato, Y., Sinclair, R., Tsukamoto, K., Nishida, M., 2015. Rapid genome reshaping by multiple-gene loss after whole-genome duplication in teleost fish suggested by mathematical modeling. Proc. Natl. Acad. Sci. USA 112, 14918–14923.

Jiang, Y., Xie, M., Chen, W., Talbot, R., et al., 2014. The sheep genome illuminates biology of the rumen and lipid metabolism. Science 344, 1168–1173.

Joshi, T., Xu, D., 2007. Quantitative assessment of relationship between sequence similarity and function similarity. BMC Genom. 8, 222.

Kawashima, T., et al., 2009. Domain shuffling and the evolution of vertebrates. Genome Res. 19, 1393–1403.

Kenny, N.J., Chan, K.W., Nong, W., Qu, Z., et al., 2016. Ancestral whole-genome duplication in the marine chelicerate horseshoe crabs. Heredity 116, 190–199.

King, N., et al., 2008. The genome of the choanoflagellate Monosiga brevicollis and the origin of metazoans. Nature 451, 783–788.

Kirkness, E.F., Bafna, V., Halpern, A.L., Levy, S., et al., 2003. The dog genome: Survey sequencing and comparative analysis. Science 301, 1898–1903.

Koonin, E.V., 2005. Orthologs, paralogs, and evolutionary genomics. Annu. Rev. Genet. 39, 308–309.

Kristensen, D.M., Wolf, Y.I., Mushegian, A.R., Koonin, E.V., 2011. Computational methods for gene orthology inference. Brief. Bioinform. 12, 379–391.

Lattice, L.A., et al., 2003. A long-range Shh enhancer regulates expression in the developing limb and fin and is associated with preaxial polydactyly. Hum. Mol. Genet. 12, 1725–1735.

Li, L., Stoeckert, C.J., Roos, D., 2003. OrthoMCL: Identification of ortholog groups for eukaryotic genomes. Genome Res. 13, 2178–2189.

Lindblad-Toh, K., Wade, C.M., Mikkelsen, T.S., Karlsson, E.K., et al., 2005. Genome sequence, comparative analysis and haplotype structure of the domestic dog. Nature 438, 803–819.

Mayor, C., et al., 2000. VISTA: Visualizing global DNA sequence alignments of arbitrary length. Bioinformatics 16, 1046–1047.

Mita, K., Kasahara, M., Sasaki, S., Nagayasu, Y., et al., 2004. The genome sequence of silkworm, Bombyx mori. DNA Res. 11, 27–34.

Morey, D.F., 1994. The early evolution of the domestic dog. Am. Sci. 82, 336–347.

Nielsen, R., Williamson, S., Kim, Y., Hubisz, M.J., et al., 2005. Genomic scans for selective sweeps using SNP data. Genome Res. 15, 1566–1575.

Ohno, S., 1970. Evolution by Gene Duplication. Springer-Verlag New York Inc..

Orland, L., et al., 2013. Recalibrating Equus evolution using the genome sequence of an early Middle Pleistocene horse. Nature 499, 74–78.

Owen, R., 1843. Lectures on comparative anatomy and physiology of the invertebrate animals, delivered at the royal college of surgeons in 1843. Longman, Brown, Green and Longman, London.

Passarge, E., Horsthemke, B., Farber, R.A., 1999. Incorrect use of the term synteny. Nat. Genet. 23, 387.

Patthy, L., 1999. Genome evolution and the evolution of exon-shuffling – A review. Gene 238, 103–114.

Peng, X., Alföldi, J., Gori, K., Eisfeld, A.J., et al., 2014. The draft genome sequence of the ferret (Mustela putorius furo) facilitates study of human respiratory disease. Nat. Biotechnol. 32, 1250–1255.

Pontius, J.U., Mullikin, J.C., Smith, D.R., et al., 2007. Agencourt Sequencing Team, Initial sequence and comparative analysis of the cat genome. Genome Res. 17, 1675–1689.

Putnam, N.H., et al., 2008. The amphioxus genome and the evolution of the chordate karyotype. Nature 453, 1064–1071.

Rubin, C., Zody, M.C., Eriksson, J., Meadows, J.R.S., et al., 2010. Whole-genome resequencing reveals loci under selection during chicken domestication. Nature 464, 587–591.

Rubin, C., Megens, H., Barrio, A.M., Maqbool, K., et al., 2012. Strong signatures of selection in the domestic pig genome. Proc. Nat. Acad. Sci. USA 109, 19529–19536.

Schnable, J.C., Freeling, M., Lyons, E., 2012. Genome-wide analysis of syntenic gene deletion in the grasses. Genome Biol. Evol. 4, 265–277.

Schwartz, S., et al., 2000. PipMaker – A web server for aligning two genomic DNA sequences. Genome Res. 10, 577–586.

Session, A.M., Uno, Y., Kwon, T., Chapman, J.A., et al., 2016. Genome evolution in the allotetraploid Frog Xenopus Laevis. Nature 538, 336–343.

Simakov, O., Kawashima, T., 2017. Independent evolution of genomic characters during major metazoan transitions. Dev. Biol. 427, 179–192.

Simakov, O., Marletaz, F., Cho, S., et al., 2013. Insights into bilaterian evolution from three spiralian genomes. Nature 493, 526–531.

Sonnhammer, E.L., Koonin, E.V., 2002. Orthology, paralogy and proposed classification for paralog subtypes. Trends Genet. 18, 619–620.

Tatusov, R.L., Koonin, E.V., Lipman, D.J., 1997. A genomic perspective on protein families. Science 278, 631–637.

The Bovine Genome Sequencing and Analysis Consortium, Elsik, C.G., Tellam, R.L., et al., 2009. The genome sequence of taurine cattle: A window to ruminant band evolution. Science 324, 522–528.

The Bovine HapMap Consortium, 2009. Genome-wide survey of SNP variation uncovers the genetic structure of cattle breeds. Science 324, 528–532.

Toh, H., Hayashida, H., Miyata, T., 1983. Sequence homology between retroviral reverse transcriptase and putative polymerases of hepatitis B virus and cauliflower mosaic virus. Nature 305, 827–829.

Trachtulec, Z., Forejt, J., 2001. Synteny of orthologous genes conserved in mammals, snake, fly, nematode, and fission yeast. Mammalian Genome. 12, 227–231.

Uchiyama, I., 2006. Hierarchical clustering algorithm for comprehensive orthologous-domain classification in multiple genomes. Nucleic Acids Res. 34, 647–658.

Van de Peer, Y., Maere, S., Meyer, A., 2009. The evolutionary significance of ancient genome duplications. Nat. Rev. Genet. 10, 725–732.

Van de Peer, Y., Maere, S., Meyer, A., 2010. 2R or not 2R is not the question anymore. Nat. Rev. Genet. 11, 166.

Wade, C.M., Giulotto, E., Sigurdsson, S., Zoli, M., et al., 2009. Genome sequence, comparative analysis, and population genetics of the domestic horse. Science 326, 865–867.

Wolf, Y.I., Koonin, E.V., 2012. A tight link between orthologs and bidirectional best hits in bacterial and archaeal genomes. Genome Biol. Evol. 4, 1286–1294.

Wolfe, K., 2000. Robustness – It's not where you think it is. Nat. Genet. 25, 3–4.

Wolfe, K., 2004. Evolutionary genomics: Yeasts accelerate beyond BLAST. Curr. Biol. 14, R392–R394.

Xia, Q., Zhou, Z., Lu, C., Cheng, D., et al., 2004. A draft sequence for the genome of the domesticated silkworm (Bombyx mori). Science 306, 1937–1940.

Yoshida, M., Ishikura, Y., Moritaki, T., Shoguchi, E., et al., 2011. Genome structure analysis of molluscs revealed whole genome duplication and lineage specific repeat variation. Bioinformatics, 483, 63–71.

Zhang, H., Wang, Z., Wang, S., Li, H., 2012. Progress of genome wide association study in domestic animals. J. Anim. Sci. Biotechnol. 3, 26.

Zhao, L., Saelao, P., Jones, C.D., Begun, D.J., 2014. Origin and spread of de novo genes in Drosophila melanogaster populations. Science 343, 769–772.

Genome Alignment

Tetsushi Yada, Kyushu Institute of Technology, Fukuoka, Japan

Introduction

Genome alignment is the alignment of genome sequences; it can be considered as a way to reformat a genome sequence dataset in order to facilitate the comparison of evolutionarily related genomes. We can also consider a genome alignment as a prediction of homology (i.e., shared origin, see the next section) at the nucleotide level between two or more genomes (Dewey and Pachter, 2006; Dewey, 2012).

Regarding the question of why we (re)construct genome alignments, the following sentence would neatly summarize its answers that have been given thus far (e.g., Dubchak and Pachter, 2002; Frazer et al., 2003; Brudno and Dubchak, 2005; Dewey and Pachter, 2006; Margulies and Birney, 2008; Dewey, 2012; Thompson, 2016a; Bleidorn, 2017): "we (re)construct genome alignments *in order to exploit the evolutionary information inscribed in the genome sequences.*" (Note: This sentence itself is our homage to Dr. Hitoshi Kihara, a renowned plant geneticist who once wrote: "The history of the earth is recorded in the layers of its crust; the history of all organisms is *inscribed* in the chromosomes" (originally in Japanese) (Kihara, 1947)).

Here, let us elaborate on this sentence. The genome sequences of all extant living organisms are, after all, the products of billions of years of evolution, starting from the genome(s) of the most recent common ancestor(s) of life.

During the course of evolution, the genome sequences have gradually (or suddenly) changed, and have accumulated mutations such as base substitutions, insertions, deletions (e.g., Graur and Li, 2000), and large-scale rearrangements such as inversions, translocations, duplications, etc. (e.g., Lupski, 2007; Lynch, 2007; Gu et al., 2008; Lee et al., 2017). In other words, information on the evolution is 'inscribed' in the genome sequences in the form of accumulated mutations. Broadly speaking, the amount of difference between two genome sequences should be bigger as they diverge earlier. Therefore, by comparing the genome sequences of two species (or individuals) and quantifying the differences, we can estimate the *evolutionary distance* between them, that is, how closely (or distantly) the two species (or individuals) are related to each other (e.g., Waterston et al., 2002; Gibbs et al., 2004). If we extend the comparison to that of three or more genomes, we can infer the phylogenetic relationships among them, that is, the order in which each genome (or lineage) diverged from others (e.g., Clark et al., 2007; Sheng et al., 2011; Sahl et al., 2012; Prabha et al., 2014). Moreover, by meticulously examining the patterns of differences between the genomes, we can identify what types of mutations (like substitutions, insertions, deletions, inversions, translocations and duplications) occurred, and how and when (or where in the phylogenetic tree of the genomes) they occurred (e.g., Kent et al., 2003; Watanabe et al., 2004; Ma et al., 2006; Clark et al., 2007; Wang et al., 2015). Thus, genome alignments provide integral materials for the study of molecular evolution and phylogenetics (e.g., Gascuel, 2005; Saitou, 2013).

But this is not all; genome alignments also enable us to identify functionally important regions of the genomes. According to the neutral theory of molecular evolution (Kimura, 1968, 1983), non-functional parts of a genome by default evolve neutrally, i.e., without any selective pressure, and mutations on them are occasionally fixed in the population due to genetic drift. On the other hand, if the parts have some biologically important functions, mutations on them are likely to disrupt the functions and thus less likely to be fixed than neutrally (Kimura and Ohta, 1974). Therefore, the alignment of genomes that are appropriately evolutionarily far apart will highlight potentially functional parts of the genomic sequences as those showing significantly less frequent changes than the neutral background (e.g., Kellis et al., 2003; Birney et al., 2007; Stark et al., 2007; Tseng and Tompa, 2009; Hupalo and Kern, 2013). In contrast, if the parts show significantly more frequent changes than the neutral background, the parts may have been under (continuous) positive selection pressure, maybe because they are involved in the responses to incessantly changing environments, including interactions between hosts and pathogens or parasites and between males and females (e.g., Hughes and Nei, 1988; Swanson and Vacquier, 2002; Sackton et al., 2007; Wong, 2011).

For that matter, the conservation of synteny (i.e., colocalization within a single chromosome, see Section "Conserved Synteny and Collinearity") of genomic elements (including genes) during a long evolutionary period may indicate either that their synteny itself is functionally significant and/or that the region in between the elements is also functionally essential (e.g., Ferrier and Holland, 2001; Hurst et al., 2004; Kikuta et al., 2007; Engström et al., 2007; Dewey, 2012; Schnable, 2015; Bleidorn, 2017), because otherwise the synteny would eventually be disrupted by some genomic rearrangements. Also in this sense, genome alignments are useful because they enable us to reveal the regions of conserved synteny and possibly to infer the large-scale structures of ancestral genomes (e.g., Tesler, 2002; Kent et al., 2003; Ma et al., 2006, 2008).

Because of such multi-faceted utility, genome alignment has been a corner stone of comparative genomics, whose main purposes are to identify conserved (and thus functionally important) elements and regions in the genomes and to study underlying evolutionary mechanisms and forces operating on the genomes (e.g., Frazer et al., 2003; Brudno and Dubchak, 2005; Margulies and Birney, 2008; Dewey, 2012).

During the past two decades, genome alignment has been a subject of intensive study and vigorous efforts, and numerous algorithms, methods, and programs have been published thus far. Therefore, it is virtually impossible to exhaust the past studies. Hence we will overview the past developments in this area of research while discussing a number of methods that particularly caught our attention. If the readers want more extensive, or nearly comprehensive, information on this subject, we recommend them to read not only this article but also some other reviews, such as (Dubchak and Pachter, 2002; Frazer et al., 2003; Brudno and Dubchak, 2005; Jayaraj, 2006; Dewey and Pachter, 2006; Margulies and Birney, 2008) for early developments, and (Dewey, 2012;

Earl *et al.*, 2014; Kehr *et al.*, 2014; Thompson, 2016a) for relatively recent developments. Moreover, because of the space limitation, we will focus on the methods to align two or more long sequences (at least on the order of 100 kb), and will not discuss methods to align shorter sequences, such as the standard aligners for, e.g., protein-coding or RNA genes (reviewed, *e.g.*, in Kumar and Filipski, 2007; Notredame, 2007; Aniba *et al.*, 2010; Löytynoja, 2012; Iantorno *et al.*, 2014) or methods to map the reads from Next-Generation Sequencers to a reference genome (reviewed, *e.g.*, in Cordero *et al.*, 2012; Lee and Tang, 2012; Bahassi and Stambrook, 2014; Mielczarek and Szyda, 2016; Smith and Yun, 2017). Those who are interested in these topics should refer to the reviews cited just above and/or the relevant articles in this Encyclopedia.

Background/Fundamentals

This section provides some basic information that is necessary for discussing genome alignment.

Homology

In molecular evolution (including evolutionary genomics), the term "homology" means the sharing of evolutionary origin between two genomic elements (or among three or more elements). That is, two genomic elements are "homologous" to each other if they are descended from a common ancestral element (e.g., Fitch, 1970, 2000). The concept of homology can be extended (or reduced) to the nucleotide level (Fitch, 2000; Dewey and Pachter, 2006; Dewey, 2012).

In the traditional matrix (row-column) format of a sequence alignment, each column consists of homologous nucleotides from the aligned sequences, and each row represents an aligned sequence interspersed with gaps that indicate the lack of the corresponding nucleotides. Usually, each nucleotide is represented by a one-letter abbreviation code of its base ('A' for adenine, 'C' for cytosine, 'G' for guanine, and 'T' for thymine), and each gap is represented by a hyphen ('−'). Actually, this traditional matrix representation is not enough to represent a genome alignment, because it is not flexible enough to accommodate genome rearrangements (such as inversions) and duplications. Nevertheless, the matrix representation still remains a crucial building block of almost all formats of genome alignments.

In molecular evolution, the homology relationship can be broadly classified into three categories: "orthology", "paralogy", and "xenology" (Fitch, 1970, 2000; Gray and Fitch, 1983). The two homologous elements are "orthologous" if they diverged from the common ancestor as a result of speciation; they are "paralogous" if their divergence is due to duplication; they are "xenologous" if at least one horizontal transfer contributed to the evolutionary path between them.

When discussing different types of genome alignments, it is sometimes convenient to further sub-classify the categories of orthology and paralogy, because such sub-classes could help clarify particular goals of different methods. Especially, it is important to note that, in general, orthology is not necessarily a one-to-one correspondence, because a genome element may have duplicated after speciation. "Topoorthology" (or "positional orthology") is a special subclass of orthology (Dewey and Pachter, 2006; Dewey, 2012); a pair of orthologous elements are topoorthologous if the evolutionary path between them does not go through any duplication event that changed the genomic position of the element. (Such duplication events include retrotranspositions and interspersed segmental duplications, but do not include tandem duplications.) In other words, a pair of topoorthologous elements have never changed their genomic positions due to duplications (though they may have changed positions via rearrangements,) since their divergence due to speciation. This notion of "topoorthology" may be closer to what some readers might have considered as "orthology"; because topoorthologous genes usually share the same ancestral genomic context, they are much more likely to conserve functions than non-topoorthologous orthologous genes are (Dewey, 2012).

Conserved Synteny and Collinearity

During the course of evolution, a genome has undergone rearrangements, thus some long-range organizations of the genomic sequence have been disrupted. When aligning whole genomes, we are supposed to align segments that have escaped (especially large-scale) rearrangements since the divergence of the genomes. Thus, some terms have been defined to represent such situations (e.g., Gregory *et al.*, 2002; Frazer *et al.*, 2003; Margulies and Birney, 2008). When two or more elements in a genome reside on the same chromosome, they are referred to as "syntenic". When two or more elements reside on a single chromosome in one species and their orthologous counterparts also reside on a single chromosome in the other species, the situation is referred to as "conserved synteny". It should be noted that conserved syntenic blocks (or regions) may have undergone small-scale rearrangements confined within them, thus the order among the elements may not be conserved. When the order is also conserved within the pair of conserved syntenic regions, Gregory *et al.* (2002) (and also Frazer *et al.*, 2003) referred to them as "conserved segments". However, they did not mention the relative orientations of the elements, which may change via (small-scale) inversions. When both the order and the orientations are also conserved within a pair of conserved syntenic regions, we usually refer to the regions as having "conserved collinearity" or, for short, being "collinear" (to each other). (Conserved) collinearity implies that the regions have not undergone any rearrangements since their divergence. Considering these original definitions, the term "synteny mapping" should rigorously mean pairing up conserved syntenic blocks in two genomes (or its extension to more genomes), though the term has come to imply finding collinear regions (as pointed out by Margulies and Birney, 2008). Moreover, the notions of conserved synteny and collinearity can be extended to any homologous segments (including paralogous ones), not limited to orthologous ones. Thus, Margulies and Birney (2008) proposed to use the term "reconstructing homologous collinearity" instead of "synteny mapping". It should be noted, however, that some programs are *literally* synteny mappers, because they do not care much about the order or orientations of the elements within each block. Therefore, we will refer to this kind of programs collectively as "synteny mappers", while keeping in mind that some of them are actually programs to reconstruct homologous collinearity. (When necessary, we will clearly distinguish whether a program reconstructs conserved synteny or collinearity).

Local, Global, and "*Glocal*" Alignment

Traditionally, there have been two types of sequence alignment methods, local and global. Local alignment methods align only those regions of the sequences that are suspected to be homologous (based on their similarities). In contrast, global alignment methods "force" to align the sequences from the beginning to the end. Merits of local alignment are: (i) it could avoid false alignment of non-homologous regions, especially those neighbouring homologous regions; and (ii) it can even align regions that are no longer collinear as a result of rearrangements. Demerits of local alignment are: (i) it could miss weakly homologous regions located near highly homologous regions; and (ii) it could mistake by-chance similarities as homologies, even if they are isolated from other homologous regions. Pros and cons of global alignment are opposite from those of local alignment.

In 2003, an intermediate concept, "*glocal* alignment" was proposed (Brudno *et al.*, 2003c). Glocal alignment is intended to be like local alignment in that it can align those regions that have undergone rearrangements and duplications (and, in addition, suggests the presence of these events), and like global alignment in that it can potentially detect weak homologies if they are neighbouring strongly homologous regions.

Originally, Brudno *et al.* (2003c) defined a (pairwise) glocal alignment as "a series of operations that transform one sequence into the other". They considered that the operations include insertions, deletions, point mutations, inversions, translocations and duplications, and that the total edit distance between the two sequences is the sum of penalties incurred by individual operations.

Topic

Major Challenges on Genome Alignment

When aligning genome sequences, some major challenges need to be overcome (*e.g.*, Dubchak and Pachter, 2002; Margulies and Birney, 2008; Thompson, 2016a); especially pressing among them are: (1) time (and space) requirement, and (2) rearrangements and duplications.

(1) When optimising simple objective functions for pair-wise sequence alignment (PWA), dynamic programming (DP) algorithms have long been used, such as Needleman-Wunsch (NW) (Needleman and Wunsch, 1970) for global alignment, Smith-Waterman (SW) (Smith and Waterman, 1981) for local alignment, and their various improvements (e.g., Hirschberg, 1975; Gotoh, 1982; Myers and Miller, 1988). These algorithms have the time complexity quadratic in the sequence length and space complexity quadratic or linear in the sequence length. When aligning sequences such as protein-coding genes and RNA genes, which are not so long, they do their jobs in a reasonable amount of time. However, when aligning genome sequences, each of which consists of millions to billions of bases, these DP algorithms could consume an enormous amount of time, making such computation impractical.

One heuristics to avoid this problem is to first find short regions showing high similarity (commonly called "seeds") via a fast heuristic method and to restrict the (2-dimensional) search space of DP to those around, and/or in between, the seeds. This could drastically reduce the computational time. There are two major approaches based on this "seeding" heuristics (*e.g.*, Dubchak and Pachter, 2002; Jayaraj, 2006; Dewey and Pachter, 2006; Margulies and Birney, 2008; Dewey, 2012; Thompson, 2016a): one is to extend each seed on both sides via some modified version of the SW algorithm ("seed-and-extend"), somewhat similarly to gapped BLAST (Altschul *et al.*, 1997), and the other is to chain mutually consistent seeds first and then to use the chained seeds as "anchors" and to align the regions in between the neighbouring anchors ("seed-and-chain", or "anchoring"). (For the chaining algorithms in early days, see, e.g., Ohlebusch and Abouelhoda, 2005). Usually, the regions in between the neighbouring anchors are globally aligned, but some aligners locally align the regions. (The "seed-and-extend" strategy usually produces a set of local alignments, which may sometimes be chained afterwards (e.g., Kent *et al.*, 2003)). When aligning three or more sequences, even if the sequences are not so long, it could take an enormous amount of time even for a DP to exhaustively search the alignment space (e.g., Gusfield, 1997); in fact, it has been shown that such a problem with some-of-pairs scoring is NP hard (Wang and Jiang, 1994). Thus, to reconstruct a multiple sequence alignment (MSA), we must resort to some heuristics (e.g., Gusfield, 1997; Notredame, 2007; Kemena and Notredame, 2009; Löytynoja, 2012). Among the most popular heuristics is the progressive alignment (Hogeweg and Hesper, 1984; Feng and Doolittle, 1987), which iteratively performs pairwise alignment of sequences or sub-alignments at each internal node of a guide tree, from leaves upwards, and finally constructs an MSA of all input sequences at the root. Multiple genome aligners that have been developed thus far mostly combine some heuristics for MSA with either "seed-and-extend" or "anchoring" heuristics.

(2) Traditionally, sequence alignment algorithms have been built under simple models that only consider base substitutions, insertions, and deletions (e.g., Gusfield, 1997; Kumar and Filipski, 2007; Notredame, 2007; Löytynoja, 2012), thus they have only been able to produce collinear alignments, where the order and orientations of the homologous elements are conserved. However, it is undeniable that genome sequences have undergone other types of mutations such as rearrangements (inversions, translocations, etc.) and duplications (*e.g.*, Kent *et al.*, 2003; Pevzner and Tesler, 2003; Ma *et al.*, 2006; Lupski, 2007; Lynch, 2007; Clark *et al.*, 2007; Lee *et al.*, 2017). Therefore, when aligning genome sequences, it becomes inevitable to handle these rearrangements and duplications, in addition to substitutions, insertions, and deletions that have been traditionally dealt with. Currently, a common strategy to address this challenge is splitting the genome alignment problem into two sub-problems (e.g., Dewey and Pachter, 2006; Margulies and Birney, 2008; Dewey, 2012): (1) finding the regions of conserved collinearity (or conserved synteny), and (2) aligning each collinear (or syntenic) region at the nucleotide level. In early days, these two subproblems were addressed by separate programs (or separate components of a pipeline) (e.g., Couronne *et al.*, 2003; Brudno *et al.*, 2004), namely, a homologous collinearity reconstruction method (or a "synteny mapper") and a collinear aligner. Around since two important concepts, "glocal alignment" (Brudno *et al.*, 2003c) (see also Section "Local, Global, and *"glocal"* Alignment" above) and a "threaded blockset" (Blanchette *et al.*, 2004), were proposed, however, many research groups have developed "non-collinear aligners", each of which addresses the two subproblems in a single run (see also the "Approaches" Section below). Such programs may be advantageous, because they may potentially become able to refine the result of the first component taking account of the result of the second component (e.g., Paten *et al.*, 2011). Nevertheless, the development of collinear genome aligners is still an area of active research, because they are important components of the non-collinear aligners.

Blanchette *et al.* (2004) defined the "threaded blockset" as a set of "blocks" that are "threaded" by each of the original input sequences. (Here, a "block" is a set of mutually homologous collinear segments from the original input sequences (or their reverse complements), and a sequence is said to "thread" the blockset if the blockset contains completely and non-redundantly the sequence as a whole. It should be noted that a block is permitted to consist only of a segment of a single sequence.) In our view, this threaded blockset could be regarded as a representation of a glocal alignment, if the threading information is explicitly presented. In fact, a threaded blockset is highly similar to a graph-based alignment (reviewed *e.g.* in, Kehr *et al.*, 2014), which could be considered as a sort of glocal alignment (as argued by Darling *et al.*, 2010). And, indeed, the developers of a graph-based aligner (the A-Bruijn alignment method) even stated that their aligner "is able to automatically generate threaded blockset for genomic sequence alignment" (Raphael *et al.*, 2004).

A yet another similar, but distinct, class of non-collinear aligners is the positional homology genome alignment method (Darling *et al.*, 2004, 2010), which may be traced back to MUMmer (version 2) (Delcher *et al.*, 2002). Like glocal aligners, they can deal with rearrangements, such as inversions and translocations. Unlike glocal aligners, however, they will not attempt to align paralogous regions within the same genome, because their primary goal is to align positional orthologs (topoorthologs) of different genomes. This type of aligner will be useful if the focus of the downstream study is on identifying and analyzing topoorthologs, although it may not be suitable for detailed analyses of duplication-related phenomena (Darling *et al.*, 2010).

Major Steps in Genome Alignment

As somewhat discussed above, alignment of genome sequences poses several challenges. Therefore, traditionally, genome alignments have been constructed in several steps. The major steps involved in pairwise alignment are: (1) reconstructing homologous collinearity (e.g., Margulies and Birney, 2008) (also known as "synteny mapping" (e.g., Thompson, 2016a) or "homology mapping" (e.g., Dewey and Pachter, 2006; Dewey, 2012)); then, for each collinear region, (2) finding "seeds" (i.e., short, highly similar pairs (or sets) of subsequences), (3) selecting promising seeds, (4) completing the alignments around the selected seeds (for seed-and-extend) or in between them (for anchoring); and (5) displaying and viewing the resulting genome alignment.

Many programs perform steps (2)–(3) or (2)–(4) recursively, and gradually narrows down the search space, aiming to enhance the sensitivity and/or accuracy of the resulting alignment, and/or computational speed. When a pipeline consists mainly of a synteny mapper and a collinear aligner, all the steps from (1) to (4) are usually performed separately. Many graph-based aligners also seem to follow this practice; in this case, step (1) is usually performed by a component of the program separate from the component dealing with steps (2)–(4). Some non-collinear aligners, however, functionally merges step (1) with steps (2)–(3). Furthermore, some seed-and-extend type local aligners perform steps (2)–(4) first and, when necessary, performs step (1) later, using the output of step (4).

As briefly mentioned above, multiple genome alignment involves a heuristic MSA step (such as progressive alignment) in addition to the aforementioned steps (1)–(4). (Note: Here, we consider step (5) as a separate, absolutely final step.) Therefore, another dimension of variety arises regarding when to perform the MSA-dedicated heuristics; it could be either before step (1), before step (2) or (3), before step (4), or after all steps, (1)–(4).

In early days, step (1) mostly focused on finding orthologous collinear (or syntenic) regions. However, some later methods can also find paralogous collinear (or syntenic) regions.

Step (2) can vary widely depending on the types of seeds and the types of data structure to store the seeds. Seeds can be un-spaced (i.e., consecutive) or spaced (i.e., containing some positions that will be ignored), and also can be exact matches or inexact matches, or even with some gaps. Particular types of spaced seed patterns were shown to enhance the sensitivity of the alignment as well as computational speed, because they substantially reduce the correlation between overlapping seeds

(Ma *et al.*, 2002); these seed patterns have been used by some aligners. Regarding the seed-storing data structure, suffix-trees (or suffix-arrays) and hash-tables seem to be most widely used. Suffix-trees are good at finding exact or nearly exact matches (e.g., Delcher *et al.*, 1999), whereas hash-tables can be used for finding considerably in-exact matches (e.g., Schwartz *et al.*, 2003). A notable seed-finding method is CHAOS, which can find even gapped matches and use them as seeds (Brudno *et al.*, 2003a); it is thus expected to work also on genomes that are somewhat distantly related to each other. More recently, a new type of seeds, "adaptive seeds", have been proposed (Kiełbasa *et al.*, 2011). An adaptive seed is a short match (of any length) that occurs in a target genome less frequently than a specified threshold. They are robust against regional heterogeneity of base composition and repetitive elements.

Step (3) is performed either by imposing some thresholds on each seed (mainly for seed-and-extend) and/or by chaining mutually consistent seeds (for "anchoring").

Methods for step (4) could be classified in terms of the broad category (local, global, or probabilistic), the scoring scheme (a traditional score with match rewards, mismatch penalties and affine gap penalties, a consistency score (e.g., Paten *et al.* 2008a), an expected accuracy (e.g., Bradley *et al.* 2009), etc.), and so forth. Some methods even apply external collinear aligners (e.g., Thompson *et al.*, 1994; Edgar, 2004; Katoh and Toh, 2008) to this step.

Regarding step (5), there are some programs (viewers and browsers) dedicated particularly to displaying the alignment (e.g., Mayor *et al.*, 2000; Schwartz *et al.*, 2000; Frazer *et al.*, 2004; Elnitski *et al.*, 2010; Nguyen *et al.*, 2014; Poliakov *et al.*, 2014). But many aligners come with their own internal (or accompanying) viewers (e.g., Darling *et al.*, 2010). In addition, some web-sites, such as Ensembl (Herrero *et al.*, 2016), Galaxy (Afgan *et al.*, 2016), UCSC Genome Browser (Miller *et al.*, 2007; Tyner *et al.*, 2017), and VISTA browser (e.g., Poliakov *et al.*, 2014), store a number of genome alignments that are too large to be reconstructed by a single researcher (or laboratory), and their own dedicated browsers enable us to view the alignments in interactive and convenient manners, possibly with some annotations.

Approaches

To the best of our knowledge, the development of genome aligners started with MUMmer (Delcher *et al.*, 1999), a pairwise collinear local genome aligner employing "seed-and-chain" strategy based on a suffix-tree. (Note: MUMmer evolved into a non-collinear local aligner afterwards (Delcher *et al.*, 2002; Kurtz *et al.*, 2004)). After that, some (either local or collinear global) pairwise genome aligners were developed (e.g., Batzoglou *et al.*, 2000; Kent and Zahler, 2000; Ma *et al.*, 2002; Bray *et al.*, 2003; Schwartz *et al.*, 2003). Then, in 2002, "multiple genome aligner" (MGA), the first collinear multiple global genome aligner, was published (Höhl *et al.*, 2002), and some (either global or local) collinear multiple aligners followed (e.g., Brudno *et al.*, 2003a, b; Blanchette *et al.*, 2004; Bray and Pachter, 2004). (Note: Later, MLAGAN and TBA evolved into non-collinear multiple aligners; the former is glocal (Dubchak *et al.*, 2009), and the latter is local (discussed, *e.g.*, in: Raphael *et al.*, 2004; Ovcharenko *et al.*, 2005; Margulies *et al.*, 2007)). It was in 2003 that the first (pairwise) non-collinear "glocal" genome aligner, Shuffle-LAGAN, was published (Brudno *et al.*, 2003c). And, in 2004, some non-collinear multiple genome aligners of different types were published (e.g., Darling *et al.*, 2004; Raphael *et al.*, 2004). Since then, non-collinear multiple aligners have been in the mainstream of genome aligner developments (e.g., Darling *et al.*, 2010; Angiuoli and Salzberg, 2011; Paten *et al.*, 2011), while pairwise aligners and collinear multiple aligners have also been vigorously developed with a variety of advanced features (e.g., Harris, 2007; Rausch *et al.*, 2008; Bradley *et al.*, 2009; Paten *et al.*, 2009; Kryukov and Saitou, 2010; Nakato and Gotoh, 2010; Kiełbasa *et al.*, 2011).

Before non-collinear genome aligners became popularized, "synteny mappers" played a crucial role in whole-genome alignment pipelines (e.g., Couronne *et al.*, 2003; Kent *et al.*, 2003; Brudno *et al.*, 2004; Dewey and Pachter, 2006; Margulies and Birney, 2008; Paten *et al.*, 2008a). Even after the advent of non-collinear aligners, though, synteny mappers have been advanced steadily (Hachiya *et al.*, 2009; Pham and Pevzner, 2010; Drillon *et al.*, 2014; Tang *et al.*, 2015), maybe partly because gene-based analyses of synteny (and collinearity) remain handy and important (both by themselves and as preliminaries of functional and evolutionary analyses). Now, some state-of-the-art synteny mappers can handle quite a large number of genomes (e.g., Rödelsperger and Dieterich, 2010; Proost *et al.*, 2012; Wang *et al.*, 2012; Minkin *et al.*, 2013).

In Supplementary Table S1, we have listed the approaches (such as methods, programs, and algorithms) discussed above and in the previous "Topic" section, as well as some additional ones that also caught our attention. In the table, the approaches are classified from several points of view, and each of them is also accompanied by some comments (on algorithms, scope, output formats, etc.) and notes, if at all. (The relevant references are listed in a separate sheet).

Supplementary data associated with this article can be found in the online version at https://doi.org/10.1016/B978-0-12-809633-8.20237-9.

Here are some caveats. It should be kept in mind that Supplementary Table S1 is far from exhaustive. Information that somewhat complements the table can be obtained, e.g., from the following sources: regarding aligners, the references cited in Introduction of this article; regarding synteny mappers, table 1 of Rödelsperger and Dieterich (2010), supplementary table S1 of Proost *et al.* (2012), table 3 of Wang *et al.* (2012), table 1 of Tang *et al.* (2015), and a review (Ghiurcuta and Moret, 2014); regarding browsers and viewers, Introduction of Nguyen *et al.* (2014), and a review (Wang *et al.*, 2013).

As far as we know, among the numerous (stand-alone) synteny mappers that have been developed thus far, only Mercator, Enredo, Shuffle-LAGAN and its successor, SuperMap (Dubchak *et al.*, 2009), and some methods based on BLAT (Kent, 2002) (e.g., Brudno *et al.*, 2004) were combined with collinear aligners to produce multiple genome alignments. Other synteny mappers have been used mainly for the evolutionary analyses of genome architectures (including whole-genome duplications, segmental duplications, and tandem arrays of repeats) or genomic contexts of genes (including orthology, regulatory elements and gene families). It may be worth exploring the combinations of such unused synteny mappers and (especially collinear) multiple aligners.

Regarding visualization, many aligners (e.g., Darling *et al.*, 2010) are internally equipped with, or packaged with, their own viewers, which may be sufficient for usual purposes.

Illustrative Examples

In genome alignment programs, multi-FASTA (MFA) format and multiple alignment format (MAF) files are frequently adopted as input and output, respectively (e.g., Schwartz *et al.*, 2003; Blanchette *et al.*, 2004; Angiuoli and Salzberg, 2011; Paten *et al.*, 2011). A MFA file contains multiple FASTA-formatted sequence data, each of which consists of a sequence identifier line followed by one or more sequence lines. The FASTA format is used in a wide range of sequence analysis programs, such as sequence alignment, motif discovery and gene finding. See the NCBI web site (see "Relevant Websites section") for more details on the FASTA format. The MAF, which was also adopted as the submission format in Alignathon (see Section "Alignathon" below) (Earl *et al.*, 2014), stores a series of MSAs which covers some (or most) parts of entire genomes. Each MSA is expected to consist of genome sequences which are descended from a common ancestral sequence. Depending on what types of genome rearrangements occurred during the course of evolution, MSAs in MAF files show characteristic composition. For instance, duplications lead to MSAs which include more than one sequence from a single species, while deletions lead to MSAs which do not include sequences from a specific species. Moreover, inversions lead to MSAs which include sequences from the reverse strand. See the UCSC Genome Browser web site (see "Relevant Websites section") for more details on the MAF.

A traditional way to view (the large-scale structure of) a (typically non-collinear) genome alignment is to draw dot-plots (Gibbs and McIntyre, 1970). A dot-plot graphically visualizes similarities between two nucleotide or amino acid sequences. Pairs of similar regions, which are identified by sequence alignments, are represented as diagonal segments in a dot-plot. **Fig. 1** shows the dot-plot of a pairwise alignment between simulated mouse and rat genomes. To create the figure, we downloaded a MAF file, which stores the "correct" alignment of artificial mammalian genomes, from the Alignathon web site (see "Relevant Websites section") and extracted the pairwise alignment between simulated mouse and rat genomes. Then, we drew its dot-plot by using mafTools (Earl *et al.*, 2014), the HAL package (Hickey *et al.*, 2013) and a few of our perl and R scripts (available on request to TY). The artificial mammalian genomes were generated by a forward-time simulation of genome evolution. MafTools is a toolkit for MAF file (pre)processing, and the HAL package is a toolkit for comparative genome analyses.

A dot-plot is useful to visualize the overall picture of a pairwise genome alignment and infer what types of genome rearrangements occurred during the course of evolution. **Fig. 1** also shows typical patterns of diagonal segments in dot-plots and **Fig. 2** shows the corresponding parsimonious genome rearrangements. (And **Fig. 3** shows some typical patterns that are obscure in **Fig. 1**.) The more complicated patterns can be understood as combinations of the typical patterns.

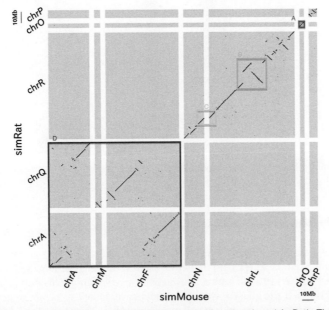

Fig. 1　Dot-plot of pairwise alignment between genomes of simulated mouse (simMouse) and rat (simRat). The colored rectangles enclose typical patterns indicative of genome rearrangements. See **Fig. 2** for their interpretations. This dot-plot is based on a simulated alignment used as input for Alignathon. Adapted from Earl, D., Nguyen, N., Hickey, G., *et al.*, 2014. Alignathon: A competitive assessment of whole-genome alignment methods. Genome Res. 24, 2077–2089.

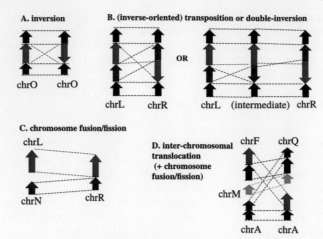

Fig. 2 Interpretation of diagnostic patterns in dot-plot (**Fig. 1**). Panels A-D in this figure correspond to the colored rectangles labeled A-D, respectively, in **Fig. 1**. In panel D, we ignored small-scale rearrangements such as inversions, in order to focus only on the inter-chromosomal translocation and the chromosome fusion/fission. In each panel, an arrow represents a chromosomal region. The arrows are oriented according to the genome on the left, which corresponds to the horizontal axis of the dot-plot (**Fig. 1**). A black arrow indicates that the region remain unrearranged. A colored arrow indicates that the region was rearranged. A dashed line indicates the homology of the region ends that it ties.

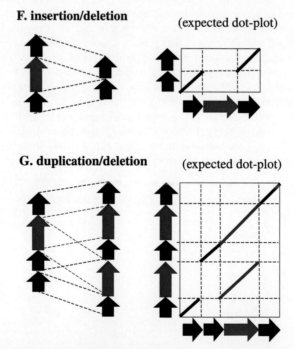

Fig. 3 Other diagnostic patterns common in dot-plots. These patterns are generally common in dot-plots, but are missing or obscure in **Fig. 1**.

Results and Discussion: Evaluation of Alignment Methods

It is utterly crucial to evaluate the accuracy of alignments output by each aligner, because an aligner would be useless unless its resulting alignments are reasonably accurate at least, no matter how fast it can align genome sequences. It is desirable that a third party should evaluate a number of aligners in a fair setting. However, an overwhelming majority of assessments thus far have been conducted by the groups that developed the aligners, thus the results have tended to favor the developers' own aligners. Nevertheless there have been a few studies where independent groups evaluated some aligners (e.g., Margulies *et al.*, 2007; Chen and Tompa, 2010), and a competitive assessment called the "Alignathon", in which a number of aligner-developing groups participated under a considerably fair setting (Earl *et al.*, 2014). In this section, we discuss these evaluations, after briefly discussing major evaluation methods.

Major Evaluation Methods

It is notoriously difficult to evaluate the accuracy of a genome alignment, because we do not know the "correct" alignment that resulted from the true evolution of the genome sequences, and that can be used as the "gold standard" to be compared against alignments reconstructed by aligners. Therefore, it is inevitable to settle for some surrogate methods for evaluating the alignment accuracy. According to Earl *et al.* (2014), such evaluation methods conducted thus far can be classified into four broad categories: (i) those using simulation, (ii) those using expert information, (iii) those using direct statistical assessment, and (iv) those that assess how well an alignment functions for a downstream analysis. Each of these methods has its own pros and cons. Here, we briefly summarize the discussions on them by Earl *et al.* (2014). (See Iantorno *et al.*, 2014 for a more comprehensive review on the topic).

In category (i), researchers create some "correct" alignments via simulations of genome evolution, and compare them with alignments reconstructed by aligners. A merit of this category is that we know the "correct" alignment that can be compared to the reconstructed ones. Its major demerit is that it is not certain whether the simulation adequately reflects the evolutionary forces operating on the true genomes.

In category (ii), researchers evaluate reconstructed alignments using some expert biological information, such as annotations of functional elements, structural alignments, and expertly curated alignments (e.g., Blackshields *et al.*, 2006; Wilm *et al.*, 2006; Kemena *et al.*, 2013). A merit of this category is that the methods provide objective evolutionary contexts derived from the real genomes, giving us some relief. Its demerit is that each method may address only a small fraction of the alignment and that each method may itself rely on other forms of inference, thus making it uncertain whether the assessment is indeed correct or not.

In category (iii), evaluations are made using statistical measures that indicate the reliability (or uncertainty) of each portion of reconstructed alignments (e.g., Prakash and Tompa, 2007; Landan and Graur, 2008; Penn *et al.*, 2010; Kim and Ma, 2011; Chang *et al.*, 2014). A merit of this category is that we can assess the complete alignments of real sequences. Its demerit is that, unless carefully calibrated using gold standard alignments, they are only a proxy to a true assessment of accuracy.

In category (iv), aligners are assessed, typically by biologists, according to intuition or downstream analyses. Because of their ad-hoc nature, these assessments are difficult if not impossible to generalize from. For whole-genome alignments (WGAs), their effects were assessed, e.g., on de novo ncRNA predictions (Gorodkin *et al.*, 2010) and on the prediction of conserved elements (Margulies *et al.*, 2007).

Before Alignathon

Before the Alignathon (Earl *et al.*, 2014) took place, there have been relatively few independent or community organized assessments of WGA pipelines. Notable among such evaluations are the assessment as a part of the ENCODE pilot project (Margulies *et al.*, 2007) and the evaluation by Chen and Tompa (2010).

As a part of the ENCODE pilot project (Birney *et al.*, 2007), Margulies *et al.* (2007) aligned sequences orthologous to the ENCODE-regions obtained from 23 mammalian species (including human and two non-eutherian mammals). They used four pipelines (each represented by its major components): TBA/BLASTZ (abbreviated as TBA) (Blanchette *et al.*, 2004; Schwartz *et al.*, 2003), MLAGAN/Shuffle-LAGAN (abbr. as MLA-GAN) (Brudno *et al.*, 2003b,c; Dubchak *et al.*, 2009), MAVID/Mercator (abbr. as MAVID) (Bray and Pachter, 2004; Dewey, 2007), and Pecan (Paten *et al.*, 2008a, 2009). (NOTE: The Pecan pipeline actually used Shuffle-LAGAN as a pre-processor, because Pecan is a collinear aligner.) Their comparisons of the four pipelines revealed large-scale consistency but substantial differences in terms of small genomic rearrangements, sensitivity (sequence coverage) and specificity (alignment accuracy).

In terms of the above classification, their alignment assessments belong to category (ii). Because their alignments are of real genomic sequences, they cannot be compared with the true alignments, which are inherently unknown. Therefore Margulies *et al.* (2007) used some surrogates for sensitivity and specificity. For sensitivity, they used two surrogates. One is the coverage of annotated protein-coding sequences in each alignment. And the other is the coverage of ancestral repeats (ARs) in each alignment. Overall, these coverage measures varied considerably among different alignments: MAVID showed the smallest coverage, MLA-GAN showed the largest coverage for protein-coding sequences, and Pecan showed the largest coverage for ARs. For specificity as well, they used two surrogates. One was the "*Alu exclusion*", which is how well each pipeline can avoid the false-positive alignments of human *Alu* elements to any non-primate mammalian sequence. (It is well known that *Alu* elements are specific to primates.) And the other was the substitution periodicity in coding regions, which takes advantage of the fact that the third codon positions are enriched in synonymous sites and thus prone to substitutions. TBA was the most specific in terms of both measures, and Pecan almost tied with TBA in terms of the substitution periodicity. MLAGAN was the least specific in terms of the *Alu* exclusion, and MAVID was the least specific in terms of the substitution periodicity.

Chen and Tompa (2010) also evaluated the alignments of the ENCODE-regions generated by the aforementioned four pipelines. One difference from the study of Margulies *et al.* (2007) is that Chen and Tompa (2010) included non-mammalian vertebrates as well. (Like Margulies *et al.*, 2007, they analyzed the pairwise alignments of human and each non-human species extracted from the reconstructed multiple genome alignments.) Another difference is that the evaluation of Chen and Tompa (2010) was more comprehensive, involving all sites of the aligned sequences. The evaluation results were also broken down into four location categories, namely, coding, untranslated (UTR), intronic, and intergenic. In the comparison of the alignments generated by TBA, MLAGAN, and MAVID, the degree of agreement substantially decreased from coding to non-coding regions, and as the evolutionary distance between species increases. Their alignment evaluation belongs to category (iii). As a surrogate for the sensitivity, they examined the coverage of each alignment, and to evaluate the alignment accuracy, they used a statistical measure computed via StatSigMa-w (Prakash and Tompa, 2007). Regarding the coverage, the results were consistent with those of Margulies et al. (2007), that is, MLAGAN showed the highest coverage and MAVID showed the lowest coverage. For every pipeline, the coverage decreased as the divergence between species increases, especially for non-coding regions. Regarding the alignment accuracy, Pecan was the most accurate, and MLAGAN was the least accurate. TBA was nearly as accurate as Pecan for eutherian mammals, but was substantially less accurate than Pecan for more distant species. For every pipeline, the accuracy suddenly dropped from eutherian mammals to more divergent species, and from coding to non-coding regions. These results indicated that reconstructing whole-genome multiple sequence alignments remains a challenge, especially for non-coding regions and distant species (like non-eutherian mammals and non-mammalian vertebrates). One caveat given by Chen and Tompa (2010) is that their accuracy measure based on StatSigMa-w may somewhat overestimate the true accuracy; thus, the reconstructed alignments should have more misalignments than their accuracy measure indicates. Chen and Tompa (2010) attributed this to StatSigMa-w's conservative calls of suspicious regions in the alignment.

Alignathon

The Alignathon (Earl et al., 2014) is a competitive evaluation of whole-genome alignment tools in which teams submitted their alignments and then assessments were performed collectively after all the submissions were received. As the input, they used three data sets. Two were simulated DNA data: one based on the phylogeny of four primates (human, chimpanzee, gorilla, and orangutan), and the other based on the phylogeny of five eutherian mammals (human, mouse, rat, dog, and cow). And the remaining one was the real DNA sequence data, comprised of 20 fly genomes. In total, the Alignathon received 35 submissions, which were generated by 12 different alignment pipelines. The 12 pipelines they used were: three pipelines used by genome browsers (VISTA-LAGAN for the VISTA Browser (Frazer et al., 2004; Dubchak et al., 2009), MULTIZ for the UCSC Genome Browser (Miller et al., 2007; Meyer et al., 2013), and Pecan and EPO for the Ensembl Browser (Paten et al., 2008a, 2009; Flicek et al., 2013); seven standalone WGA tools (progressiveMauve (Darling et al., 2010), TBA (Blanchette et al., 2004), Cactus (Paten et al., 2011), Mugsy (Angiuoli and Salzberg, 2011), Robusta (Notredame, 2012), which combines results from multiple stand-alone tools, PSAR-Align (Kim and Ma, 2014), which was used to re-align MULTIZ-based alignments and the pipelines of AutoMZ and GenomeMatch teams. (Specific settings of the pipelines are described in the Supplemental material of Earl et al., 2014).

For each reconstructed alignment of the simulated data, they computed the "precision" (the proportion of aligned pairs of nucleotides in the reconstructed alignment that are also true) and the "recall" (the proportion of aligned pairs of nucleotides in the true alignment that are also reconstructed). And they also defined the F-score as the harmonic mean of precision and recall. This evaluation belongs to category (i) in Section "Major Evaluation Methods". They found that many of the submissions were able to align primate data set with both relatively high recall and precision, consistently across annotation types. For the mammalian simulations, they found a much wider variation of results, both between alignments and between different annotation classes. Especially, recall varied much more widely than precision, which was relatively high for most aligners. Among submissions, Cactus showed the highest F-score. Generally, submissions performed best in genic regions and most poorly in repetitive regions. And precision and recall values got lower as the distance between species increased.

For each reconstructed alignment of the real gnomic data, the aforementioned recall or precision cannot be computed. Thus, they resorted to a statistical evaluation (belonging to category (iii) in Section "Major Evaluation Methods"). As a proxy to the precision, they computed the *PSAR-precision* based on the output of PSAR statistical alignment tool (Kim and Ma, 2011). As a proxy to the recall, they used the *coverage*. Then, as a proxy to the F-Score, they also defined the *pseudo F-Score* as the harmonic mean of PSAR-precision and coverage. Before applying these statistical measures to alignments of the real fly genomes, they examined their consistency with the measures for simulations, after calculating both for the simulated alignments. In the simulated mammals, they found a very good, linear correlation between recall and coverage ($r^2 = 0.984$), but no linear correlation between PSAR-precision and precision. Regarding the F-Score and the pseudo F-Score, they linearly correlated strongly ($r^2 = 0.975$), likely because recall varied much more widely among the alignments than precision. Then, using the statistical measures, they evaluated the reconstructed alignments of the real 20 fly genomes. In general, they saw good concordance between the fly and simulated results.

These results are more or less consistent with those of the aforementioned two evaluations (Margulies et al., 2007; Chen and Tompa, 2010). In addition, the Alignathon produced some interesting results. For example, direct comparisons between the submitted alignments suggested that sharing the same synteny mapper influenced the results more strongly than sharing the same (collinear) synteny block aligner. Moreover, they found that no pipelines except Cactus can capture significantly nonzero duplications into the reconstructed alignment. In their conclusions, they cautioned against overinterpretation of the results because of the limitations of their assessment methods, and they encouraged more independent lines of assessment: more simulations, more statistical assessments, and assessments at different scales, etc.

Future Directions

In the past two decades, vigorous efforts and dedication of many researchers have greatly improved genome alignment methods in terms of their speed, accuracy, and scope (such as the number of input sequences, evolutionary distance, the range of acceptable mutations, etc.). However, as the field of comparative genomics gets more advanced and more and more genome sequences become available, genome aligners will need to become faster, more accurate, more robust and more flexible. Indeed, researchers eagerly attempt to improve the aligners' performance even today (e.g., Frith and Kawaguchi, 2015; Uricaru *et al.*, 2015; Mai *et al.*, 2017; Suarez *et al.*, 2017). Here, we discuss a few representatives among possible future directions that the study of genome alignment should head for.

Pursuit of Further Space- and Time-Efficiency

Genome alignment methods have become substantially efficient in computational time and space (memory) requirement (e.g., Paten *et al.*, 2009; Nakato and Gotoh, 2010; Angiuoli and Salzberg, 2011). However, as the number of available genome sequences has still been growing exponentially (e.g., Margulies and Birney, 2008; Dewey, 2012; Earl *et al.*, 2014; Thompson, 2016a), the demand also keeps growing for a method that can align a large number of genomes in a reasonable amount of time (e.g., Dewey, 2012).

Regarding the multiple alignment of small to medium size sequences, some new heuristic methods have recently been proposed that enable the alignment of thousands to even millions of sequences (e.g., Katoh and Toh, 2007; Sievers *et al.*, 2011; Mirarab *et al.*, 2015; Nguyen *et al.*, 2015). (For a review of such aligners, see, e.g., Thompson, 2016b and Warnow (2017)). By combining in some way these heuristics with the heuristics that have been developed for genome alignment thus far (such as "anchoring" (e.g., Delcher *et al.*, 1999; Höhl *et al.*, 2002; Brudno *et al.*, 2003a), also known as the "divide-and-conquer" strategy for a long alignment (e.g., Kryukov and Saitou, 2010)), it may be possible to develop aligners that can align, e.g., hundreds to thousands of genomes in a reasonable amount of time. A framework using the meta-alignment methods, Crumble and Prune (Roskin *et al.*, 2011), would probably provide a good starting point for such efforts. And it would also be important to extend the framework to non-collinear aligners.

Pursuit of Further Alignment Accuracy

Although the accuracy of genome aligners have improved considerably (e.g., Chen and Tompa, 2010; Earl *et al.*, 2014), there is still big room for improvement. The problem of alignment accuracy may be broadly divided into two subproblems: (a) it seems that the specificity (or precision) have reached a plateau that is still considerably below (near) perfection; (b) the sensitivity (or recall) is far below perfection even with the current best aligner (e.g., Cactus). The subproblem (a) may be overcome, *e.g.*, by (1) incorporating information other than the primary structures (i.e., the strings of bases) of DNA sequences (reviewed, *e.g.*, in Kemena and Notredame, 2009), and by (2) taking account of the intrinsic stochasticity of genome evolution more seriously. The subproblem (b) may be addressed by (3) more accurate detection of genome rearrangements and duplications. And both subproblems may be partly solved by (4) properly dealing with errors in sequencing and assembly as well as false-positive matches. Let us discuss them in more details.

(1) One possible reason for the imperfect specificity of current aligners is that the scoring schemes for them (including the quite sophisticated probabilistic scoring (*e.g.*, Paten *et al.*, 2008b; Bradley *et al.*, 2009; Paten *et al.*, 2009)), are mostly based on simple "space-homogeneous" models, which provide homogeneous mutation rates along the sequence. In reality, it is well-known that functionally important regions evolve more slowly than neutral regions, and that evolution tend to conserve features related to functions (such as the secondary and tertiary structures of proteins and RNA genes) than the primary structures. Therefore, if these functional features are known prior to the alignment (via some experiments), such information could help reconstruct more accurate genome alignments (*e.g.*, Kemena and Notredame, 2009). One implementation of this idea is via template-based aligners, which align sequences guided by templates that represent such functional information (e.g., Pei *et al.*, 2008; Wilm *et al.*, 2008; Katoh and Toh, 2008).

A possible challenge would be how to make template-based aligners scale to genome alignments, because they tend to be slower than simple aligners. Besides, this method would work only for functional regions with experimental information, which generally account for only a small fraction of genome sequences.

(2) Another possible reason for the imperfect specificity is that the evolutionary processes are inherently stochastic; in consequence, a true alignment resulting from an evolutionary process is very frequently *not* optimal even under the ideal objective score function, i.e., (the logarithm of) the probability that natural evolutionary processes give rise to the alignment. To support this idea, it was shown that taking account of the stochasticity of evolution substantially improves the accuracy and consistency of the pairwise sequence comparisons (*e.g.*, Lunter *et al.*, 2008; Cartwright, 2009). Moreover, a simulation study demonstrated that, in a (near) majority of errors in multiple sequence alignments, the true alignments indeed do *not* optimize the alignment probabilities calculated with the particular evolutionary model used for the simulations. These results strongly suggest that, in order to further improve the alignment accuracy, it is essential to present the stochastic nature of evolutionary processes as well, rather than merely attempting to reconstruct a single optimal alignment.

Attempts to cast the sequence alignment problems into a statistical framework with some probabilistic modelling of gaps (often referred to as "statistical alignment" (Hein *et al.*, 2000)) started with two groundbreaking studies (Bishop and Thompson, 1986; Thorne *et al.*, 1991). Then, the efforts extended to various problems (including multiple sequence alignments) using Hidden Markov Models (HMMs) (including transducers) (e.g., Holmes and Bruno, 2001; Holmes, 2003; Bradley and Holmes, 2007; Miklós *et al.*, 2009; Herman *et al.*, 2015; Rivas and Eddy, 2015). Currently, HMMs are the dominating models in the study of statistical alignment, partly because they are convenient enough to enable various dynamic programming algorithms (for, e.g., finding an optimum alignment and calculating the posterior probabilities of matches) (Durbin *et al.*, 1998), and partly because they are flexible enough to accommodate site-specific mutation rates (e.g., Durbin *et al.*, 1998; Rivas and Eddy, 2015) and mixed geometric length distributions of insertions/deletions (indels) (e.g., Miklós *et al.*, 2004; Do *et al.*, 2005; Lunter *et al.*, 2008; Bradley *et al.*, 2009; Paten *et al.*, 2009), which can approximate reasonably (up to dozens of residues) the power-law indel length distributions that have been repeatedly observed in molecular evolutionary analyses (e.g., Gonnet *et al.*, 1992; Benner *et al.*, 1993; Gu and Li, 1995; Zhang and Gerstein, 2003; Chang and Benner, 2004; Yamane *et al.*, 2006; Fan *et al.*, 2007). (For a recent review on these developments, see, e.g., Holmes, 2017).

Unfortunately, however, some recent studies revealed the limitations of alignment methods based on HMMs (e.g., Darling *et al.* 2010; Rivas and Eddy, 2015).

These results imply that, in order to pursue near perfect accuracy, we will ultimately need a probabilistic aligner that is based on a genuine sequence evolutionary model with realistic instantaneous insertion/deletion rates, such as the substitution/insertion/deletion models (Miklós *et al.* 2004). (NOTE: Roughly speaking, in a genuine sequence evolutionary model, the probability of an evolutionary process of a sequence is calculated as a multiplicative accumulation of infinitesimal-time transition probabilities of the states of an entire sequence over a given time interval). However, there still remain some technical challenges, such as the development of adequately fast algorithms, to achieve this crucial goal of developing a probabilistic aligner based on a genuine evolutionary model of a sequence.

(3) In most of the current genome alignment methods, detecting genome rearrangements and duplications is the job of a separate synteny mapping program or the synteny mapping component of a non-collinear aligner. These methods have been based on heuristics, including the graph construction and modification procedures, and a number of involved parameters had to be adjusted via an extensive trial and error. Therefore, more principled methods based on probabilistic or, ideally, evolutionary modelling of synteny mapping may improve the accuracy (both sensitivity and specificity) of the detection of rearrangements and duplications (as suggested, e.g., by Dewey, 2012; Ghiurcuta and Moret, 2014). (Although some maximum likelihood methods have already been proposed (e.g., Hachiya *et al.*, 2009), the underlying probabilistic models seem somewhat heuristic). Besides, because synteny blocks (or regions of homologous collinearity) are products of inherently stochastic evolutionary processes, the true synteny maps may not optimize even the true occurrence probability, as in the case of collinear alignments. Therefore, also in synteny mapping, it will become essential to take adequate account of inherent stochastic nature of evolution, for example by presenting multiple near-optimal solutions and their probabilities (as suggested, *e.g.*, by Dewey, 2012).

Probably, the best way would be to construct a genuine stochastic evolutionary model of a genome that incorporate rearrangements and duplications in addition to substitutions, insertions, and deletions, and to establish a probabilistic framework for synteny mapping based on the model. Thus far, some evolutionary models of genome rearrangements (and duplications) seem to have been proposed (e.g., Tesler, 2002; Ma *et al.*, 2008; Lin and Moret, 2011; Da Silva *et al.*, 2012; Shao *et al.*, 2013; Ghiurcuta and Moret, 2014; Braga and Stoye, 2015). To the best of our knowledge, however, there have been no genuine stochastic evolutionary models describing genome rearrangements and duplications applicable to the genome alignment problem. If such genuine stochastic evolutionary models are developed in the future, they may be applied to the problem of "statistical whole-genome alignment", including "statistical synteny mapping". We do admit that lots of challenges lie ahead in order to achieve this formidable goal. Especially, in addition to some methodological breakthroughs to enable the efficient computation of the likelihoods of possible synteny mappings, it should become indispensable to determine feature-dependent rates of rearrangements and duplications via data analyses that are more extensive and more meticulous than the past ones (e.g., Kent *et al.*, 2003; Church *et al.*, 2009; Mills *et al.*, 2011).

(4) Because errors in the sequencing and assembly processes are likely to cause serious errors in the final alignment, dealing with these errors prior to or during the alignment process could significantly improve the alignment accuracy (as discussed, *e.g.*, in Pevzner and Tesler, 2003; Margulies and Birney, 2008). An almost equally important issue is the handling of false-positive matches (or seeds) that are erroneously chosen as anchors. Traditionally, chaining (or clustering) algorithms (e.g., Bourque *et al.*, 2004; Ohlebusch and Abouelhoda, 2005) have served to filter potentially false-positive matches, and more recently, some graph-based non-collinear aligners and synteny mappers seem to deal with these problems by "cleaning" or "smoothening" the graphs via some intuition-based heuristics (e.g., Paten *et al.*, 2008a; Pham and Pevzner, 2010). Theoretically, these problems could also be addressed in a more principled way, if we model these errors as some stochastic processes (e.g., Verzotto *et al.*, 2016). And the error-handling and non-collinear alignment (or synteny mapping) might be performed in a unified manner if we merge the stochastic model of the errors with the genuine stochastic evolutionary model of genomes, as in some population genetic studies.

Closing Remarks

Genome sequence alignment is a cornerstone of comparative genomic study. Unlike traditional alignment of e.g., protein coding sequences or RNA sequences, genome alignment has been extremely challenging because of the sheer sizes of genome sequences and of the fact that genome sequences have undergone rearrangements such as inversions, translocations, and duplications during their evolution, in addition to substitutions, insertions, and deletions that have been dealt with by traditional alignment methods. Over the past two decades, a number of methodological innovations have greatly advanced the genome alignment methods, to cope with these challenges and to improve the accuracy of the resulting alignments. However, as the number of sequenced genomes has been growing exponentially, and as further accurate comparative genomic analyses have been getting required, the genome alignment methods must still keep improving, e.g., via new strategies discussed in the "Future Directions" section.

Acknowledgments

We appreciate the sincere efforts of reviewers, editors, and other dedicated members of Elsevier and its contractors. We are also grateful to Dr. Glenn Hickey at UC Santa Cruz for his technical support for the use of the HAL package.

Appendix A Supplementary Information

Supplementary data associated with this article can be found in the online version at https://doi.org/10.1016/B978-0-12-809633-8.20237-9.

> *See also*: Algorithms for Strings and Sequences: Pairwise Alignment. Comparative and Evolutionary Genomics. Comparative Genomics Analysis. Exome Sequencing Data Analysis. Functional Enrichment Analysis. Genome Informatics. Identification of Homologs. Integrative Analysis of Multi-Omics Data. Metagenomic Analysis and its Applications. Natural Language Processing Approaches in Bioinformatics. Next Generation Sequencing Data Analysis. Phylogenetic Footprinting. Quantitative Immunology by Data Analysis Using Mathematical Models. Repeat in Genomes: How and Why You Should Consider Them in Genome Analyses?. Sequence Analysis. Sequence Composition. Whole Genome Sequencing Analysis

References

Afgan, E., Baker, D., Van Den Beek, M., *et al.*, 2016. The Galaxy platform for accessible, reproducible and collaborative biomedical analyses: 2016 update. Nucleic Acids Res. 44, W3–W10.

Altschul, S., Madden, T., Schäffer, A., *et al.*, 1997. Gapped BLAST and PSI-BLAST: A new generation of protein database search programs. Nucleic Acids Res. 25, 3389–3402.

Angiuoli, S., Salzberg, S., 2011. Mugsy: Fast multiple alignment of closely related whole genomes. Bioinformatics 27, 334–342.

Aniba, M., Poch, O., Thompson, J., 2010. Issues in bioinformatics benchmarking: The case study of multiple sequence alignment. Nucleic Acids Res. 38, 7353–7363.

Bahassi, E.M., Stambrook, P., 2014. Next-generation sequencing technologies: Breaking the sound barrier of human genetics. Mutagenesis 29, 303–310.

Batzoglou, S., Pachter, L., Mesirov, J., *et al.*, 2000. Human and mouse gene structure: Comparative analysis and application to exon prediction. Genome Res. 10, 950–958.

Benner, S., Cohen, M., Gonnet, G., 1993. Empirical and structural models for insertions and deletions in the divergent evolution of proteins. J. Mol. Biol. 229, 1065–1082.

Birney, E., Stamatoyannopoulos, J., Dutta, A., *et al.*, 2007. Identification and analysis of functional elements in 1% of the human genome by the ENCODE pilot project. Nature 447, 799–816.

Bishop, M., Thompson, E., 1986. Maximum likelihood alignment of DNA sequences. J. Mol. Biol. 190, 159–165.

Blackshields, G., Wallace, I., Larkin, M., *et al.*, 2006. Analysis and comparison of benchmarks for multiple sequence alignment. In Silico Biol. 6, 321–339.

Blanchette, M., Kent, W., Riemer, C., *et al.*, 2004. Aligning multiple genomic sequences with the threaded blockset aligner. Genome Res. 14, 708–715.

Bleidorn, C., 2017. Phylogenomics: An introduction..Cham: Springer, (Chapter 6).

Bourque, G., Pevzner, P., Tesler, G., 2004. Reconstructing the genomic architecture of ancestral mammals: Lessons from human, mouse, and rat genomes. Genome Res. 14, 507–516.

Bradley, R., Roberts, A., Smoot, M., *et al.*, 2009. Fast statistical alignment. PLOS Comput Biol. 5, e1000392.

Bradley, R., Holmes, I., 2007. Transducers: An emerging probabilistic framework for modeling indels on trees. Bioinformatics 23, 3258–3262.

Braga, M., Stoye, J., 2015. Sorting linear genomes with rearrangements and indels. IEEE/ACM Trans. Comput. Biol. Bioinform. 12, 500–506.

Bray, N., Dubchak, I., Pachter, L., 2003. AVID: A global alignment program. Genome Res. 13, 97–102.

Bray, N., Pachter, L., 2004. MAVID: Constrained ancestral alignment of multiple sequences. Genome Res. 14, 693–699.

Brudno, M., Chapman, M., Göttgens, B., *et al.*, 2003a. Fast and sensitive multiple alignment of large genomic sequences. BMC Bioinform. 4, 66.

Brudno, M., Do, C., Cooper, G., *et al.*, 2003b. LAGAN and Multi-LAGAN: Efficient tools for large-scale multiple alignment of genomic DNA. Genome Res. 13, 721–731.

Brudno, M., Dubchak, I., 2005. Comparisons of long genomic sequences: Algorithms and applications. In: ALURU, S. (Ed.), Handbook of Computational Molecular Biology. Chapman and Hall/CRC.

Brudno, M., Poliakov, A., Salamov, A., *et al.*, 2004. Automated whole-genome multiple alignment of rat, mouse, and human. Genome Res. 14, 685–692.

Brudno, M., Malde, S., Poliakov, A., *et al.*, 2003c. Glocal alignment: Finding rearrangements during alignment. Bioinformatics 19, i54–i62.

Cartwright, R., 2009. Problems and solutions for estimating indel rates and length distributions. Mol. Biol. Evol. 26, 473–480.

Chang, J., Di Tommaso, P., Notredame, C., 2014. TCS: A new multiple sequence alignment reliability measure to estimate alignment accuracy and improve phylogenetic tree reconstruction. Mol. Biol. Evol. 31, 1625–1637.

Chang, M., Benner, S., 2004. Empirical analysis of protein insertions and deletions determining parameters for the correct placement of gaps in protein sequence alignments. J. Mol. Biol. 341, 617–631.

Chen, X., Tompa, M., 2010. Comparative assessment of methods for aligning multiple genome sequences. Nat. Biotechnol. 28, 567–572.

Church, D., Goodstadt, L., Hillier, L., *et al.*, 2009. Lineage-specific biology revealed by a finished genome assembly of the mouse. PLOS Biol. 7, e1000112.

Clark, A., Eisen, M., Smith, D., *et al.*, 2007. Evolution of genes and genomes on the Drosophila phylogeny. Nature 450, 203–218.

Cordero, F., Beccuti, M., Donatelli, S., *et al.*, 2012. Large disclosing the nature of computational tools for the analysis of next generation sequencing data. Curr. Top. Med. Chem. 12, 1320–1330.

Couronne, O., Poliakov, A., Bray, N., *et al.*, 2003. Strategies and tools for whole-genome alignments. Genome Res. 13, 73–80.

Darling, A., Mau, B., Blattner, F., *et al.*, 2004. Mauve: Multiple alignment of conserved genomic sequence with rearrangements. Genome Res. 14, 1394–1403.

Darling, A., Mau, B., Perna, N., 2010. progressiveMauve: Multiple genome alignment with gene gain, loss and rearrangement. PLOS ONE 5, e11147.

Da Silva, P., Machado, R., Dantas, S., *et al.*, 2012. Restricted DCJ-indel model: Sorting linear genomes with DCJ and indels. BMC Bioinform. 13 (Suppl. 19), S14.

Delcher, A., Kasif, S., Fleischmann, R., *et al.*, 1999. Alignment of whole genomes. Nucleic Acids Res. 27, 2369–2376.

Delcher, A., Phillippy, A., Carlton, J., *et al.*, 2002. Fast algorithms for large-scale genome alignment and comparison. Nucleic Acids Res. 30, 2478–2483.

Dewey, C., 2012. Whole-genome alignment. In: Anisimova, M. (Ed.), Evolutionary Genomics. Methods in Molecular Biology (Methods and Protocols), vol. 855. Totowa, NJ: Humana Press, pp. 237–257.

Dewey, C., 2007. Aligning multiple whole genomes with Mercator and MAVID. Methods Mol Biol. 395, 221–236.

Dewey, C., Pachter, L., 2006. Evolution at the nucleotide level: The problem of multiple whole-genome alignment. Hum. Mol. Genet. 15, R51–R56.

Do, C., Mahabhashyam, M., Brudno, M., *et al.*, 2005. ProbCons: Probabilistic consistency-based multiple sequence alignment. Genome Res. 15, 330–340.

Drillon, G., Carbone, A., Fischer, G., 2014. SynChro: A fast and easy tool to reconstruct and visualize synteny blocks along eukaryotic chromosomes. PLOS ONE 9, e92621.

Dubchak, I., Poliakov, A., Kislyuk, A., *et al.*, 2009. Multiple whole-genome alignments without a reference organism. Genome Res. 19, 682–689.

Dubchak, I., Pachter, L., 2002. The computational challenges of applying comparative-based computational methods to whole genomes. Brief. Bioinform. 3, 18–22.

Durbin, R., Eddy, S., Krogh, A., *et al.*, 1998. Biological Sequence Analysis: Probabilistic Models of Proteins and Nucleic Acids. Cambridge, UK: Cambridge University Press.

Earl, D., Nguyen, N., Hickey, G., *et al.*, 2014. Alignathon: A competitive assessment of whole-genome alignment methods. Genome Res. 24, 2077–2089.

Edgar, R., 2004. MUSCLE: Multiple sequence alignment with high accuracy and high throughput. Nucleic Acids Res. 32, 1792–1797.

Elnitski, L., Burhans, R., Riemer, C., *et al.*, 2010. MultiPipMaker: A comparative alignment server for multiple DNA sequences. Curr. Protoc. Bioinform. 30, 10.4.1–10.4.14. https://doi.org/10.1002/0471250953.bi1004s30.

Engström, P., Ho Sui, S., Drivenes, O., *et al.*, 2007. Genomic regulatory blocks underlie extensive microsynteny conservation in insects. Genome Res. 17, 1898–1908.

Fan, Y., Wang, W., Ma, G., *et al.*, 2007. Patterns of insertion and deletion in mammalian genomes. Curr. Genom. 8, 370–378.

Feng, D., Doolittle, R., 1987. Progressive sequence alignment as a prerequisite to correct phylogenetic trees. J. Mol. Evol. 25, 351–360.

Ferrier, D., Holland, P., 2001. Ancient origin of the Hox cluster. Nat. Rev. Genet. 2, 33–38.

Fitch, W., 1970. Distinguishing homologous from analogous proteins. Syst. Zool. 19, 99–113.

Fitch, W., 2000. Homology a personal view on some of the problems. Trends Genet. 16, 227–231.

Flicek, P., Ahmed, I., Amode, M.R., *et al.*, 2013. Ensembl 2013. Nucleic Acids Res. 41, D48–D55.

Frazer, K., Elnitski, L., Chrch, D., *et al.*, 2003. Cross-species sequence comparisons: A review of methods and available resources. Genome Res. 13, 1–12.

Frazer, K., Pachter, L., Poliakov, A., *et al.*, 2004. VISTA: Computational tools for comparative genomics. Nucleic Acids Res. 32, W273–W279.

Frith, M., Kawaguchi, R., 2015. Split-alignment of genomes finds orthologies more accurately. Genome Biol. 16, 106. https://doi.org/10.1186/s13059-015-0670-9.

Gascuel, O. (Ed.), 2005. Mathematics of Evolution and Phylogeny. New York, NY: Oxford University Press.

Ghiurcuta, C., Moret, B., 2014. Evaluating synteny for improved comparative studies. Bioinformatics 30, i9–i18.

Gibbs, R., McIntyre, G., 1970. The diagram, a method for comparing sequences. Its use with amino acid and nucleotide sequences. Eur. J. Biochem. 16, 1–11. https://doi.org/10.1111/j.1432-1033.1970.tb01046.x.

Gibbs, R., Weinstock, G., Metzker, M., *et al.*, 2004. Genome sequence of the brown Norway rat yields insights into mammalian evolution. Nature 428, 493–521. https://doi.org/10.1038/nature02426.

Gonnet, G., Cohen, M., Benner, S., 1992. Exhaustive matching of the entire protein sequence database. Science 256, 1443–1445.

Gorodkin, J., Hofacker, I., Torarinsson, E., *et al.*, 2010. De novo prediction of structured RNAs from genomic sequences. Trends Biotechnol. 28, 9–19.

Gotoh, O., 1982. An improved algorithm for matching biological sequences. J. Mol. Biol. 162, 705–708.

Graur, D., Li, W., 2000. Fundamentals of Molecular Evolution, second ed. Sunderland, MA: Sinauer Associates.

Gray, G., Fitch, W., 1983. Evolution of antibiotic resistance genes: The DNA sequence of a kanamycin resistance gene from Staphylococcus aureus. Mol. Biol. Evol. 1, 57–66.

Gregory, S., Sekhon, M., Schein, J., *et al.*, 2002. A physical map of the mouse genome. Nature 418, 743–750.

Gusfield, D., 1997. Algorithms on Strings, Trees, and Sequences: Computer Science and Computational Biology. Cambridge, UK: Cambridge University Press.

Gu, X., Li, W., 1995. The size distribution of insertions and deletions in human and rodent pseudogenes suggests the logarithmic gap penalty for sequence alignment. J. Mol. Evol. 40, 464–473.

Gu, W., Zhang, F., Lupski, J., 2008. Mechanisms for human genomic rearrangements. Pathogenetics 1, 4.

Hachiya, T., Osana, Y., Popendorf, K., *et al.*, 2009. Accurate identification of orthologous segments among multiple genomes. Bioinformatics 25, 853–860.

Harris, R., 2007. Improved Pairwise Alignment of Genomic DNA. The Pennsylvania State University. (PhD thesis).

Hein, J., Wiuf, C., Knudsen, B., *et al.*, 2000. Statistical alignment: Computational properties, homology testing and goodness-of-fit. J. Mol. Biol. 302, 265–279.

Herman, J., Novák, A., Lyngsø, R., *et al.*, 2015. Efficient representation of uncertainty in multiple sequence alignments using direct acyclic graphs. BMC Bioinform. 16, 108. https://doi.org/10.1186/s12859-015-0516-1.

Herrero, J., Muffato, M., Beal, K., *et al.*, 2016. Ensembl comparative genomics resources. Database (Oxford) 2016, bav096.

Hickey, G., Paten, B., Earl, D., *et al.*, 2013. HAL: A hierarchical format for storing and analyzing multiple genome alignments. Bioinformatics 29, 1341–1342.

Hirschberg, D., 1975. A linear space algorithm for computing maximal common subsequences. Commun. ACM 18, 341–343.

Hogeweg, P., Hesper, B., 1984. The alignment of sets of sequences and the construction of phyletic trees: An integrated method. J. Mol. Evol. 20, 175–186.

Höhl, M., Kurtz, S., Ohlebusch, E., 2002. Efficient multiple genome alignment. Bioinformatics 18, S312–S320.

Holmes, I., 2003. Using guide trees to construct multiple-sequence evolutionary HMMs. Bioinformatics 19, i147–i157.

Holmes, I., 2017. Solving the master equation for indels. BMC Bioinform. 18, 255. https://doi.org/10.1186/s12859-017-1665-1.

Holmes, I., Bruno, W., 2001. Evolutionary HMMs: A Bayesian approach to multiple alignment. Bioinformatics 17, 803–820.

Hughes, A., Nei, M., 1988. Pattern of nucleotide substitution at major histocompatibility complex class I loci reveals overdominant selection. Nature 335, 167–170.

Hupalo, D., Kern, A., 2013. Conservation and functional element discovery in 20 angiosperm plant genomes. Mol. Biol. Evol. 30, 1729–1744.

Hurst, L., Pál, C., Lercher, M., 2004. The evolutionary dynamics of eukaryotic gene order. Nat. Rev. Genet. 5, 299–310.

Iantorno, S., Gori, K., Goldman, N., *et al.*, 2014. Who watches the watchmen? an appraisal of benchmarks for multiple sequence alignment. In: Russell, D. (Ed.), Multiple Sequence Alignment Methods. Methods in Molecular Biology (Methods and Protocols), vol. 1079. Totowa, NJ: Humana Press, pp. 59–73.

Jayaraj, J., 2006. Computational methods for multiple genome alignment and synteny detection. Available at: https://www.semanticscholar.org.

Katoh, K., Toh, H., 2007. PartTree: An algorithm to build an approximate tree from a large number of unaligned sequences. Bioinformatics 23, 372–374.

Katoh, K., Toh, H., 2008. Recent developments in the MAFFT multiple sequence alignment program. Brief Bioinform. 9, 286–298.

Kehr, B., Trappe, K., Holtgrewe, M., *et al.*, 2014. Genome alignment with graph data structures: A comparison. BMC Bioinform. 15, 99.

Kellis, M., Patterson, N., Endrizzi, M., *et al.*, 2003. Sequencing and comparison of yeast species to identify genes and regulatory elements. Nature 423, 241–254.

Kemena, C., Bussotti, G., Capriotti, E., *et al.*, 2013. Using tertiary structure for the computation of highly accurate multiple RNA alignments with the SARA-Coffee package. Bioinformatics 29, 1112–1119.

Kemena, C., Notredame, C., 2009. Upcoming challenges for multiple sequence alignment methods in the high-throughput era. Bioinformatics 25, 2455–2465.

Kent, W., 2002. BLAT-the BLAST-like alignment tool. Genome Res. 12, 656–664.

Kent, W., Zahler, A., 2000. Conservation, regulation, synteny, and introns in a large-scale C. briggsae-C. elegans genomic alignment. Genome Res. 10, 1115–1125.

Kent, W., Baertsch, R., Hinrichs, A., *et al.*, 2003. Evolution's cauldron: Duplication, deletion, and rearrangement in the mouse and human genomes. Proc. Natl. Acad. Sci. USA 100, 11484–11489.

Kiełbasa, S., Wan, R., Sato, K., *et al.*, 2011. Adaptive seeds tame genomic sequence comparison. Genome Res. 21, 487–493.

Kikuta, H., Laplante, M., Navratilova, P., *et al.*, 2007. Genomic regulatory blocks encompass multiple neighbouring genes and maintain conserved synteny in vertebrates. Genome Res. 17, 545–555.

Kimura, M., 1968. Evolutionary rate at the molecular level. Nature 217, 624–626.

Kimura, M., 1983. The Neutral Theory of Molecular Evolution. Cambridge, UK: Cambridge University Press.

Kimura, M., Ohta, T., 1974. On some principles governing molecular evolution. Proc. Natl. Acad. Sci. USA 71, 2848–2852.

Kim, J., Ma, J., 2014. PSAR-align: Improving multiple sequence alignment using probabilistic sampling. Bioinformatics. 30, 1010–1012.

Kim, J., Ma, J., 2011. PSAR: Measuring multiple sequence alignment reliability by probabilistic sampling. Nucleic Acids Res. 39, 6359–6368.

Kihara, H., 1947. Ancestors of Common Wheat. Tokyo: Sogensha, (in Japanese).

Kryukov, K., Saitou, N., 2010. MISHIMA – A new method for high speed multiple alignment of nucleotide sequences of bacterial genome scale data. BMC Bioinform. 11, 142.

Kumar, S., Filipski, A., 2007. Multiple sequence alignment: In pursuit of homologous DNA positions. Genome Res. 17, 127–135.

Kurtz, S., Phillippy, A., Delcher, A., *et al.*, 2004. Versatile and open software for comparing large genomes. Genome Biol. 5, R12.

Landan, G., Graur, D., 2008. Local reliability measures from sets of co-optimal multiple sequence alignments. Pac. Symp. Biocomput. 13, 15–24.

Lee, T., Kim, J., Robertson, J., *et al.*, 2017. Plant genome duplication database. In: Van Dijk, A. (Ed.), Plant Genomic Databases. Methods in Molecular Biology, vol. 1533. New York, NY: Humana Press, pp. 267–277.

Lee, H., Tang, H., 2012. Next-generation sequencing technologies and fragment assembly algorithms. Methods Mol. Biol. 855, 155–174.

Lin, Y., Moret, B., 2011. A new genomic evolutionary model for rearrangements, duplications, and losses that applies both eukaryotes and prokaryotes. J. Comput Biol. 18, 1055–1064.

Löytynoja, A., 2012. Alignment methods: Strategies, challenges, benchmarking, and comparative overview. In: Anisimova, M. (Ed.), Evolutionary Genomics. Methods in Molecular Biology (Methods and Protocols), vol. 855. Totowa, NJ: Humana Press, pp. 203–235.

Lunter, G., Rocco, A., Mimouni, N., *et al.*, 2008. Uncertainty in homology inferences: Assessing and improving genomic sequence alignment. Genome Res. 18, 298–309.

Lupski, J., 2007. Genomic rearrangements and sporadic disease. Nat. Genet. 39, S43–S47.

Lynch, M., 2007. The Origins of Genome Architecture. Sunderland, MA: Sinauer Associates.

Mai, H., Lam, T., Ting, H., 2017. A simple and economical method for improving whole genome alignment. BMC Genom. 18 (Suppl 4), 362.

Margulies, E., Birney, E., 2008. Approaches to comparative sequence analysis: Towards a functional view of vertebrate genomes. Nat. Rev. Genet. 9, 303–313.

Margulies, E., Cooper, G., Asimenos, G., *et al.*, 2007. Analyses of deep mammalian sequence alignments and constraint predictions for 1% of the human genome. Genome Res. 17, 760–774.

Mayor, C., Brudno, M., Schwartz, J., *et al.*, 2000. VISTA: Visualizing global DNA sequence alignments of arbitrary length. Bioinformatics 16, 1046–1047.

Ma, J., Ratan, A., Raney, B., *et al.*, 2008. The infinite sites model of genome evolution. Proc. Natl. Acad. Sci. USA 105, 14254–14261.

Ma, B., Tromp, J., Li, M., 2002. PatternHunter: Faster and more sensitive homology search. Bioinformatics 18, 440–445.

Ma, J., Zhang, L., Suh, B., *et al.*, 2006. Reconstructing contiguous regions of an ancestral genome. Genome Res. 16, 1557–1565.

Meyer, L.R., Zweig, A.S., Hinrichs, A.S., *et al.*, 2013. The UCSC Genome Browser database: Extensions and updates 2013. Nucleic Acids Res. 41, D64–D69.

Mielczarek, M., Szyda, J., 2016. Review of alignment and SNP calling algorithms for next-generation sequencing data. J. Appl. Genet. 57, 71–79.

Miklós, I., Lunter, G., Holmes, I., 2004. A "long indel" model for evolutionary sequence alignment. Mol. Biol. Evol. 21, 529–540.

Miklós, I., Novák, A., Satija, R., *et al.*, 2009. Stochastic models of sequence evolution including insertion-deletion events. Stat. Methods Med. Res. 18, 453–485.

Miller, W., Rosenbloom, K., Hardison, R., *et al.*, 2007. 28-way vertebrate alignment and conservation track in the UCSC Genome Browser. Genome Res. 17, 1797–1808.

Mills, R., Walter, K., Steward, C., *et al.*, 2011. Mapping copy number variation by population-scale genome sequencing. Nature 470, 59–65.

Minkin, I., Patel, A., Kolmogorov, M., *et al.*, 2013. Sibelia: A scalable and comprehensive synteny block generation tool for closely related microbial genomes. In: DARLING, A., STOYE, J. (Eds.), Algorithms in Bioinformatics. WABI 2013. Lecture Notes in Computer Science, vol. 8126. Berlin, Heidelberg: Springer.

Mirarab, S., Nguyen, N., Guo, S., *et al.*, 2015. PASTA: Ultra-large multiple sequence alignment for nucleotide and amino-acid sequences. J. Comput. Biol. 22, 377–386.

Myers, E., Miller, W., 1988. Optimal alignments in linear space. Comput. Appl. Biosci. 4, 11–17.

Nakato, R., Gotoh, O., 2010. Cgaln: Fast and space-efficient whole-genome alignment. BMC Bioinform. 11, 224.

Needleman, S., Wunsch, C., 1970. A general method applicable to the search for similarities in the amino acid sequences of two proteins. J. Mol. Biol. 48, 443–453.

Nguyen, N., Hickey, G., Raney, B., *et al.*, 2014. Comparative assembly hubs: Web-accessible browsers for comparative genomics. Bioinformatics 30, 3293–3301.

Nguyen, N., Mirarab, S., Kumar, K., *et al.*, 2015. Ultra-large alignments using phylogeny-aware profiles. Genome Biol. 16, 124. https://doi.org/10.1186/s13059-015-0688-z.

Notredame C., 2012. Robusta: A meta-multiple genome alignment tool. http://www.tcoffee.org/Projects/robusta/.

Notredame C., 2007. Recent evolutions of multiple sequence alignment algorithms. PLOS Comput Biol. 3, e123.

Ohlebusch, E., Abouelhoda, M., 2005. Chaining algorithms and applications in comparative genomics. In: ALURU, S. (Ed.), Handbook of Computational Molecular Biology. Chapman and Hall/CRC.

Ovcharenko, I., Loots, G., Giardine, B., *et al.*, 2005. Mulan: Multiple-sequence local alignment and visualisation for studying function and evolution. Genome Res. 15, 184–194.

Paten, B., Earl, D., Nguyen, N., *et al.*, 2011. Cactus: Algorithms for genome multiple sequence alignment. Genome Res. 21, 1512–1528.

Paten, B., Herrero, J., Beal, K., *et al.*, 2009. Sequence progressive alignment, a framework for practical large-scale probabilistic consistency alignment. Bioinformatics 25, 295–301.

Paten, B., Herrero, J., Beal, K., *et al.* 2008a. Enredo and Pecan: Genome-wide mammalian consistency-based multiple alignment with paralogs. Genome Res. 18, 1814–1828.

Paten, B., Herrero, J., Fitzgerald, S., *et al.*, 2008b. Genome-wide nucleotide-level mammalian ancestor reconstruction. Genome Res. 18, 1829–1843.

Pei, J., Kim, B., Grishin, N., 2008. PROMALS3D: A tool for multiple protein sequence and structure alignments. Nucleic Acids Res. 36, 2295–2300.

Penn, O., Privman, E., Ashkenazy, H., *et al.*, 2010. GUIDANCE: A web server for assessing alignment confidence scores. Nucleic Acids Res. 38, W23–W28.

Pevzner, P., Tesler, G., 2003. Genome rearrangements in mammalian evolution: Lessons from human and mouse genomes. Genome Res. 13, 37–45.

Pham, S., Pevzner, P., 2010. DRIMM-Synteny: Decomposing genomes into evolutionary conserved segments. Bioinformatics 26, 2509–2516.

Poliakov, A., Foong, J., Brudno, M., *et al.*, 2014. Genome VISTA – An integrated software package for whole-genome alignment and visualization. Bioinformatics 30, 2654–2655.

Prabha, R., Singh, D., Gupta, S., *et al.*, 2014. Whole genome phylogeny of Prochlorococcus marinus group of cyanobacteria: Genome alignment and overlapping gene approach. Interdiscip. Sci. Comput. Life Sci. 6, 149–157.

Prakash, A., Tompa, M., 2007. Measuring the accuracy of genome-size multiple alignments. Genome Biol. 8, R124.

Proost, S., Fostier, J., De Witte, D., *et al.*, 2012. i-ADHoRe 3.0-fast and sensitive detection of genomic homology in extremely large data sets. Nucleic Acids Res. 40, e11.

Raphael, B., Zhi, D., Tang, H., *et al.*, 2004. A novel method for multiple alignment of sequences with repeated and shuffled elements. Genome Res. 14, 2336–2346.

Rausch, T., Emde, A., Weese, D., *et al.*, 2008. Segment-based multiple sequence alignment. Bioinformatics 24, i187–i192.

Rivas, E., Eddy, S., 2015. Parameterizing sequence alignment with an explicit evolutionary model. BMC Bioinformatics 16, 406. https://doi.org/10.1186/s12859-015-0832-5.

Rödelsperger, C., Dieterich, C., 2010. CYNTENATOR: Progressive gene order alignment of 17 vertebrate genomes. PLOS ONE 5, e8861.

Roskin, K., Paten, B., Haussler, D., 2011. Meta-alignment with Crumble and Prune: Partitioning very large alignment problems for performance and parallelization. BMC Bioinform. 12, 144.

Sahl, J., Matalka, M., Rasko, D., 2012. Phylomark, a tool to identify conserved phylogenetic markers from whole-genome alignments. Appl. Environ. Microbiol. 78, 4884–4892.

Saitou, N., 2013. Introduction to Evolutionary Genomics. Computational Biology. vol. 17. London, UK: Springer.

Sackton, T., Lazzaro, B., Schlenke, T., *et al.*, 2007. Dynamic evolution of the innate immune system in Drosophila. Nat. Genet. 39, 1461–1468.

Schwartz, S., Kent, W., Smit, A., *et al.*, 2003. Human-mouse alignments with BLASTZ. Genome Res. 13, 103–107.

Schwartz, S., Zhang, Z., Frazer, K., *et al.*, 2000. PipMaker – A web server for aligning two genomic DNA sequences. Genome Res. 10, 577–586.

Shao, M., Lin, Y., Moret, B., 2013. Sorting genomes with rearrangements and segmental duplications through trajectory graphs. BMC Bioinform. 14 (Suppl. 15), S9.

Sheng, X., Liang, D., Wen, J., et al., 2011. Multiple genome alignments facilitate development of NPCL markers: A case study of tetrapod phylogeny focusing on the position of turtles. Mol. Biol. Evol. 28, 3237–3252. https://doi.org/10.1093/molbev/msr148.

Sievers, F., Wilm, A., Dineen, D., et al., 2011. Fast, scalable generation of high-quality protein multiple sequence alignments using Clustal Omega. Mol. Syst. Biol. 7, 539. https://doi.org/10.1038/msb.2011.75.

Smith, T., Waterman, M., 1981. Identification of common molecular subsequences. J. Mol. Biol. 147, 195–197.

Smith, H., Yun, S., 2017. Evaluating alignment and variant-calling software for mutation identification in C. elegans by whole-genome sequencing. PLOS ONE 12, e0174446.

Schnable, J., 2015. Genome evolution in maize: From genomes back to genes. Annu. Rev. Plant Biol. 66, 329–343.

Stark, A., Lin, M., Kheradpour, P., et al., 2007. Discovery of functional elements in 12 Drosophila genomes using evolutionary signatures. Nature 450, 219–232.

Suarez, H., Langer, B., Ladde, P., et al., 2017. chainCleaner improves genome alignment specificity and sensitivity. Bioinformatics 33, 1596–1603.

Swanson, W., Vacquier, V., 2002. The rapid evolution of reproductive proteins. Nat. Rev. Genet. 3, 137–144.

Tang, H., Bomhoff, M., Briones, E., et al., 2015. SynFind: Compiling syntenic regions across any set of genomes on demand. Genome Biol. Evol. 7, 3286–3298.

Tesler, G., 2002. GRIMM: Genome rearrangements web server. Bioinformatics 18, 492–493.

Thompson, J., 2016a. Statistics for Bioinformatics: Methods for Multiple Sequence Alignment, first ed. ISTE Press - Elsevier (Chapter 6).

Thompson, J., 2016b. Statistics for Bioinformatics: Methods for Multiple Sequence Alignment, first ed. ISTE Press - Elsevier (Chapter 7).

Thompson, J., Higgins, D., Gibson, T., 1994. CLUSTAL W: Improving the sensitivity of progressive multiple sequence alignment through sequence weighting, position-specific gap penalties and weight matrix choice. Nucleic Acids Res. 22, 4673–4680.

Thorne, J., Kishino, H., Felsenstein, J., 1991. An evolutionary model for maximum likelihood alignment of DNA sequences. J. Mol. Biol. 33, 114–124.

Tseng, H., Tompa, M., 2009. Algorithms for locating extremely conserved elements in multiple sequence alignments. BMC Bioinform. 10, 432.

Tyner, C., Barber, G., Casper, J., et al., 2017. The UCSC Genome Browser database: 2017 update. Nucleic Acids Res. 45, D626–D634.

Uricaru, R., Michotey, C., Chiapello, H., et al., 2015. YOC, a new strategy for pairwise alignment of collinear genomes. BMC Bioinform. 16, 111. https://doi.org/10.1186/s12859-015-0530-3.

Verzotto, D., Teo, A., Hillmer, A., et al., 2016. OPTIMA: Sensitive and accurate whole-genome alignment of error-prone genomic maps by combinatorial indexing and technology-agnostic statistical analysis. Gigascience 5, 2. https://doi.org/10.1186/s13742-016-0110-0.

Wang, J., Kong, L., Gao, G., et al., 2013. A brief introduction to web-based genome browsers. Brief. Bioinform. 14, 131–143.

Wang, L., Jiang, T., 1994. On the complexity of multiple sequence alignment. J. Comput. Biol. 1, 337–348.

Wang, X., Wang, J., Jin, D., et al., 2015. Genome alignment spanning major Poaceae lineages reveals heterogeneous evolutionary rates and alters inferred dates for key evolutionary events. Mol. Plant 8, 885–898.

Wang, Y., Tang, H., Debarry, J., et al., 2012. MCScanX: A toolkit for detection and evolutionary analysis of gene synteny and collinearity. Nucleic Acids Res. 40, e49.

Warnow, T., 2017. Computational Phylogenetics: An Introduction to Designing Methods for Phylogeny Estimation. Cambridge University Press (Chapter 9).

Watanabe, H., Fujiyama, A., Hattori, M., et al., 2004. DNA sequence and comparative analysis of chimpanzee chromosome 22. Nature 429, 382–388.

Waterston, R., Lindblad-Toh, K., Birney, E., et al., 2002. Initial sequencing and comparative analysis of the mouse genome. Nature 420, 520–562.

Wilm, A., Higgins, D., Notredame, C., 2008. R-Coffee: A method for multiple alignment of non-coding RNA. Nucleic Acids Res. 36, e52.

Wilm, A., Mainz, I., Steger, G., 2006. An enhanced RNA alignment benchmark for sequence alignment programs. Algorithms Mol Biol. 1, 19.

Wong, A., 2011. The molecular evolution of animal reproductive tract proteins: What have we learned from mating-system comparisons? Int. J. Evol. Biol. 2011, 908735. https://doi.org/10.4061/2011/908735.

Yamane, K., Yano, K., Kawahara, T., 2006. Pattern and rate of indel evolution inferred from whole chloroplast intergenic regions in sugarcane, maize and rice. DNA Res. 13, 197–204.

Zhang, Z., Gerstein, M., 2003. Patterns of nucleotide substitution, insertion and deletion in the human genome inferred from pseudogenes. Nucleic Acids Res. 31, 5338–5348.

Further Reading

Brudno, Dubchak, 2005. A fairly comprehensive and systematic review of genome alignment methods in the early days (until 2004), as well as of visualization and applications of the alignments. In: Aluru, S. (Ed.), Handbook of Computational Molecular Biology. Chapman and Hall/CRC. (ISBN: 1420036270, 9781420036275).

Dewey, 2012. A quite comprehensive and well-organized review of whole-genome alignment methods (including quite recent ones) and related issues, based on the concept of 'topoorthology' and on the broad classification of alignment strategies into the "hierarchical" and "local" approaches. https://doi.org/10.1007/978-1-61779-582-4_8.

Dubchak, Pachter, 2002. This review discusses challenges that computational biologists would face when they address genome alignment, gene finding and regulatory element discovery. https://doi.org/10.1093/bib/3.1.18.

Earl et al., 2014. This paper describes "Alignathon", the biggest-to-date competitive evaluation of genome alignment methods, in which 10 different teams (with 12 different alignment pipelines) participated. https://doi.org/10.1101/gr.174920.114.

Frazer et al., 2003. It discusses resources and tools that were available around 2003 for comparative genomic study, including genome alignment. https://doi.org/10.1101/gr.222003.

Jayaraj, 2006. Though being unpublished, it reviews quite a few genome alignment methods developed in the early days (until 2005). (Available at: https://www.semanticscholar.org.)

Kehr et al., 2014. This article compares different types of graphs, which are essential for some whole-genome aligners (and synteny mappers), and discusses relationships between these graphs. https://doi.org/10.1186/1471-2105-15-99.

Kim, J., Sinha, S., 2007. Indelign: A probabilistic framework for annotation of insertions and deletions in a multiple alignment. Bioinformatics 23, 289–297.

Margulies, Birney, 2008. It critically overviews the sequence data and computational methods available (around 2008) for comparative genomic analyses, especially genome sequence alignment and functional element detection. https://doi.org/10.1038/nrg2185.

Thompson, 2016a. A fairly comprehensive and well-organized review of whole-genome alignment methods (including recent ones) and related issues. In: Thompson, J., Statistics for Bioinformatics: Methods for Multiple Sequence Alignment, first ed. ISTE Press-Elsevier. (ISBN: 0081019610, 9780081019610).

Relevant Websites

https://compbio.soe.ucsc.edu/alignathon/
    Alignathon.
http://www.ncbi.nlm.nih.gov/BLAST/fasta.shtml
    BLAST TOPICS (FASTA format) – NCBI.
http://www.ensembl.org
    Ensembl Browser.

https://usegalaxy.org/
  Galaxy.
http://genome.ucsc.edu/FAQ/FAQformat.html
  Genome Browser FAQ – UCSC Genome Browser.
https://genome.ucsc.edu
  UCSC Genome Browser.
http://genome.lbl.gov/vista/index.shtml
  VISTA Browser.

Phylogenetic Footprinting

Hiroyuki Toh, Kwansei Gakuin University, Sanda, Japan

Introduction

Identification of regulatory motifs such as transcription factor binding sites in genomic non-coding regions is one of the approaches to understand the mechanisms of gene expression regulation. Recent progress of the sequencing technologies such as 'ChIP-Seq' has made it possible to identify the genome-wide identification of regulatory regions, although such 'wet' studies are not enough to reveal overall regulatory motifs in a genome. The development of the sequencing technologies has also contributed to the accumulation of genomic sequences from various organisms. Thereby, a wide variety of computational approaches for the prediction of regulatory motifs with the sequence data have been developed to complement or accelerate the experimental studies on the gene expression regulation. The computational approaches are roughly classified into two types, which are referred to as 'single species, multiple genes' approach and 'singe gene, multiple species' approach, respectively (Wang and Stormo, 2003). The former examines the regulatory motifs of multiple genes in a single genome, which are indicated to be regulated under the same regulatory mechanism. The latter analyzes the regulatory motifs of an orthologous set of a single gene from multiple organisms, in which the conservation of the regulatory mechanism for the orthologous genes from different organisms is implicitly assumed. 'Phylogenetic footprinting' belongs to the latter, the 'single gene, multiple species' approach.

Phylogenetic Footprinting and Related Approaches

Phylogenetic footprinting was first developed by Tagle *et al.* (1988) to predict *cis*-regulatory motifs of embryonic ε and γ globin genes of primates. The idea behind the method is quite simple. The motifs involved in gene expression regulation are assumed to be subject to strong functional constraints. Suppose that a residue in such a region is replaced with other residue. If the mutation has ill effects on the regulatory mechanism, the progeny with the mutation would be rapidly excluded from a population by negative selection or purifying selection. By that means, genomic regions responsible for regulatory functions are more conserved than non-coding regions without function. So, it is the first step of phylogenetic footprinting to collect nucleotide sequences of the orthologous non-coding regions from multiple species. For example, the 5′ upstream region of a gene is often used for the analysis, although *cis*-regulatory elements are not always restricted in the 5′ regions. Next, a multiple alignment of the nucleotide sequences thus collected is made. Standard tools for global multiple alignment can be used for this step. Finally, the degree of conservation was evaluated at each site or with a sliding window. Highly conserved non-coding regions detected from the alignment are predicted as putative regulatory motifs (see **Fig. 1**). Several methods to improve the performance of phylogentic footprinting have been reported. For example, an approach to use statistical alignment instead of a fixed alignment was proposed for phylogenetic footprinting (Satija *et al.*, 2008). A Bayesian method named BigFoot (Satija *et al.*, 2009), which was developed as an extension of the method with the statistical alignment, used Markov chain Monte Carlo sampling for both alignment and conserved regions.

There are several limitations in phylogenetic footprinting. One of them is related to sequence divergence. Moderate sequence divergence can highlight the conserved regions in a multiple sequence alignment. When sequences under consideration are highly diverged, however, an inaccurate alignment may be generated. Regulatory motifs are often short, comparing to the lengths of the

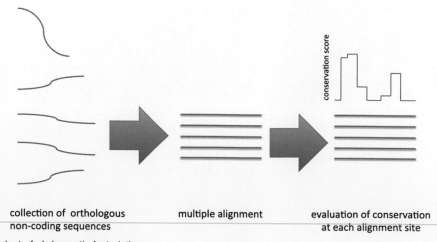

Fig. 1 Standard flowchart of phylogenetic footprinting.

entire sequences under examination. In that case, even if each sequence has the regulatory motif, they may not be correctly matched each other in the alignment, which leads to failure in detection of the regulatory motifs. Inversely, when sequences under consideration are closely related, the footprinting procedure described above would not work well to detect regulatory motifs, since the number of substituted sites in the closely related sequences is too small to reveal conserved sites. The other limitation is that the implicit assumption behind the analysis does not always hold true. The regulatory motifs have been sometimes deleted or added by the lineage-specific manner. Therefore, complete conservation of regulatory motifs among all the collected orthologous sequences is not always expected. Various methods have been developed to overcome the limitations of the original phylogenetic footprinting.

The problem related to high sequence divergence has been addressed by searching for the short segments shared by the collected orthologous non-coding sequences, instead of making a global multiple alignment. The idea is the same as those of the motif discovery tools for the 'single species, multiple genes' approach. For example, McCue et al. (2001) used Gibbs sampler (Lawrence et al., 1993) as a tool for phylogentic footprinting to detect motifs from bacterial genomes. However, the subject of the motif discovery tools for the 'single species, multiple genes' approach is different from that of the phylogenetic footprinting. The set of non-coding regions of the genes under the same regulatory mechanism, which are indicated by co-expression or immunoprecipitation, is the subject of the approach (Wang and Stormo, 2003). That is, the genes used for the analysis are not always homologous and basically independent. Blanchette and Tompa (2002) pointed out a problem to apply the motif discovery tools for the 'single species, multiple genes' approach to phylogenetic footprinting. Most of the motif discovery tools are designed to find motifs from a set of independent sequences as described above. However, the subject of phylogenetic footprinting consists of orthologous sequences. Consider a set of orthologous sequences, in which the majority of the sequences are closely related to each other and only a few sequences are distantly related to the majority. Then, the patterns present in the closely related sequences would be highly weighted, even if the patterns were not conserved in the distantly related sequences. Therefore, such taxonomic bias should be removed for the motif discovery from the orthologous sequences. FootPrinter (Blanchette et al., 2002; Blanchette and Tompa, 2003), PhyME (Sinha et al., 2004), PhyloGibbs (Siddharthan et al., 2005), and a phylogenetic Gibbs sampler (Newberg et al., 2007) are the tools for phylogenetic footprinting developed by introducing the idea.

A different approach is required when only closely related sequences are available. Boffelli et al. (2003) developed a method referred to as 'phylogenetic shadowing' to identify primate specific functional regions in human genome by comparing closely related primate sequences. In the analysis, the log likelihood ratio between the fast and slow mutation models is calculated for each column of the alignment with the phylogenetic tree and mutation rates for 'non-conserved' and 'conserved' regions estimated from a training set. A low log likelihood ratio indicates that the corresponding site is subject to strong constraints. Later, the group extended their method to develop an online shadowing tool, eShadow (Ovcharenko et al., 2004). Ray et al. (2008) devised a new shadowing method named CSMET by combining a probabilistic graphical model with a hidden Markov model.

The problem of addition or deletion of regulatory elements is an important problem. Actually, it is considered that such events may have occurred by lineage specific manners. The problem is explicitly treated in an algorithmic method developed by Blanchette et al. (2002).

There are studies to integrate phylogenyetic footprinting with the motif discovery methods for the 'single species, multiple genes' approach. Wang and Stormo (2003) first identified conserved regions from the alignments of non-coding regions for the co-regulated genes. The conserved regions are expressed as profiles and are compared the profiles from different co-regulated genes to identify the shared motifs. ConSite is also a tool to predict transcription factor binding sites by integrating phylogentic footprinting with a motif discovery method (Lenhard et al., 2003). PFR-Sampler can integrate the phylogenyetic footprints with the information of co-regulation of genes with the Markov chain discrimination algorithm to identify the similar phylogenyetic footprinted regions (PFRs) near the co-regulated genes (Grad et al., 2004). Bigelow et al. (2004) developed a pipeline named CisOrtho, in which non-coding sequences from two species under consideration are first scanned by a weight matrix respectively. Then, the hit regions thus obtained are compared to identify the conserved motifs between the two species. The latter step was regarded as phylogenetic footprinting. TargetOrtho is an expanded version of CisOrtho (Glenwinkel et al., 2014). One of the interesting expansions is the introduction of relaxed definition of motif conservation based on the alignment-free approach (Elemento and Tavazoie, 2005; Gordân et al., 2010). Probabilistic frameworks to integrate phylogenyetic footprinting and motif discovery are described in a review by van Nimwegen (2007).

Some of the online tools currently available are listed in **Table 1**.

Table 1   List of the online tools for phylogenyetic footprinting and shadowing

Tool	Ref.	URL
MicroFootprinter	Neph and Tompa (2006)	http://bio.cs.washington.edu/microfootprinter
PhyloGibbs	Siddharthan et al. (2005)	http://www.phylogibbs.unibas.ch/cgi-bin/phylogibbs.pl
eShadow	Ovcharenko et al. (2004)	https://eshadow.dcode.org/
CSMET	Ray et al. (2008)	http://www.sailing.cs.cmu.edu/main/?page_id=360
TargetOrtho	Glenwinkel et al. (2014)	http://hobertlab.org/targetortho/

Examples of Application of Phylogenetic Footprinting

As described above, phylogenetic footprinting was first applied to the prediction of the *cis*-regulatory motifs of embryonic ε and γ globin genes of primates in 1988 (Tagle *et al.*, 1988). The same group continued the studies of the *cis*-regulatory regions of globin genes with phylogenetic footprinting. In 1992, they identified several conserved sequences in the 5′-flanking regions of the γ and ε globin genes by phylogenyetic footprinting, some of which were suggested to function as repressors by experiments (Gumucio *et al.*, 1992). They also applied the method to the 5′ region of the ε globin gene to identify conserved sequences in 1993. The combination of the computational analysis with the experimental study revealed the complex binding profile of trans factors in the 5′ region of the ε globin gene (Gumucio *et al.*, 1993).

After the pioneering works, the method has been applied to various organisms. Sauer *et al.* (2006) evaluated the validity and usability of phylogenetic footprinting through the comparison of human and rodent based on the known 2678 transcription factor binding sites. They discussed the dependency of phylogenetic footprinting on the criterion of conservation, alignment algorithm and the types of transcription factors. von Bubnoff *et al.* (2005) compared the upstream regions of *Xenopus* and human *Vent2* genes to identify a 21 bp conserved sequence. The conserved sequence overlaps the known BMP response element. Their experimental study suggested the necessity and sufficiency of the conserved sequence for the BMP responsiveness. Matsunami *et al.* (2010) applied the phylogenetic footprinting to the identification of the conserved non-coding regions among the paralogues of Hox clusters generated by the 2-round whole genome duplications. Cliften *et al.* (2003) compared six *Saccharomyces* genomes. They found that some of the conserved sequences identified by the analysis are present in the upstream of the genes with similar annotations, similar expression patterns, or are in the regions bound by the same transcription factor. Buchanan *et al.* (2004) compared the 5′-non-coding regions of *rab*16/17 genes from sorghum, maize, and rice to identify the *cis*-regulatory elements. Lutz *et al.* (2017) aligned the genomic sequence of *Arabidopsis FLM* (*FLOWERING LOCUS M*) with those of *FLM* homologues from five other *Brassicaceae* species, together with that of its closest homologue *MAF3*, for the phylogenetic footprinting analysis. They found conserved non-coding sequences in the promotor region and the first intron. They revealed that the conserved regions thus detected are involved in the basal expression of the gene by experimental studies. Aceto *et al.* (2010) combined the motif search with the phylogenetic footprinting to identify putative transcription factor binding sites corresponding to the conserved 5′-non-coding sequences of the *OrcPI* gene from the orchids *Orchis italic*. The homologous non-coding sequences from two other monocots, *Lilium regale* and *Oryza sativa*, and one dicots, *Arabidopsis thaliana* were used for the footprinting analysis. Phylogenetic footprinting is applied not only to eukaryotes, but also to prokaryotes (Neph and Tompa, 2006; Katara *et al.*, 2012; Liu *et al.*, 2016).

Phylogenetic shadowing is also widely utilized for the detection of conserved non-coding regions. The method was first introduced by Boffelli *et al.* (2003) They compared the sequences of primates consisting of Old World monkeys, New World monkeys and hominoids, to identify primate specific regulatory elements and exon boundaries. Clark *et al.* (2005) applied the phylogenetic shadowing to the identification of the conserved non-coding regions of *CAPN10* (calpain 10), a genetic factor of type 2 diabetes, through the comparison of 10 primates sequences. They discussed the possibility that some of the detected regions may be associated with the expression of *CAPN10*.

What phylogenetic footprinting does is to identify the conserved or slowly evolving regions. So, the application of the method is not restricted to the detection of regulatory motif. Actually, both phylogenetic footprinting and shadowing are applied not only to regulatory motif discovery but also to the detection of microRNA genes (Berezikov *et al.*, 2005; Gao *et al.*, 2013).

The examples described above are quite limited, and there are more literatures on the applications of phylogenetic footprinting and shadowing. However, the term, "phylogenetic footprinting" or "phylogenetic shadowing", is often missing from abstracts of the literatures. The search with combination of keywords such as "conservation" and "regulatory elements" could lead to the literatures of the studies with phylogenetic footprinting or shadowing.

Concluding Remarks

The idea behind the phylogenetic footprinting and the development of the approach were described in this section. All the methods described above utilize only sequence data to predict regulatory elements as conserved non-coding regions. However, an insightful review by Maeso *et al.* (2013) suggests novel directions to predict regulatory elements beyond the sequence-based techniques of current phylogenetic footprinting. Metazoans share basic developmental process such as genetic toolkits. So, it was expected that the *cis*-regulatory elements related to the process are conserved across animal phyla. However, it turned out that the prediction of *cis*-regulatory elements based on sequence conservation such as phylogenetic footprinting is difficult when genomes of different phyla are compared. They stressed that the approach with the next generation sequencer (NGS) such as ChIP-seq and DNase I footprint is efficient for the genome-wide identification of *cis*-regulatory elements. They also suggested the possibility that the functionally equivalent *cis*-regulatory elements occupy similar syntenic positions, and that the inclusion of such positional information is useful to avoid the false positives of sequence-based predictions. The introduction of NGS data and the syntenic information could shed new light on the conservation and evolution of *cis*-regulatory elements across phyla.

See also: Algorithms for Strings and Sequences: Pairwise Alignment. Comparative and Evolutionary Genomics. Comparative Genomics Analysis. Genome Alignment. Genome Annotation. Genome Databases and Browsers. Genome Informatics. Natural Language Processing Approaches in Bioinformatics. Prediction of Protein-Binding Sites in DNA Sequences. Quantitative Immunology by Data Analysis Using Mathematical Models. Sequence Analysis. Sequence Composition

References

Aceto, S., Cantone, C., Chiaiese, P., *et al.*, 2010. Isolation and phylogenetic footprinting analysis of the 5′-regulatory region of the floral homeotic gene *OrcPI* from *Orchis italica* (Orchidaceae). J. Heredity 101, 124–131.

Berezikov, E., Guryev, V., van de Belt, J., *et al.*, 2005. Phylogenetic shadowing and computational identification of human microRNA genes. Cell 120, 21–24.

Bigelow, H.R., Wenick, A.S., Wong, A., Hobert, O., 2004. CisOrtho: A program pipeline for genome-wide identification of transcription factor target genes using phylogenetic footprinting. BMC Bioinform. 5, 27.

Blanchette, M., Schwikowski, B., Tompa, M., 2002. Algorithms for phylogenetic footprinting. J. Comput. Biol. 9, 211–223.

Blanchette, M., Tompa, M., 2002. Discovery of regulatory elements by a computational method for phylogenyetic footprinting. Genome Res. 12, 739–748.

Blanchette, M., Tompa, M., 2003. FootPrinter: A program designed for phylogenyetic footprinting. Nucleic Acids Res. 31, 3840–3842.

Boffelli, D., McAuliffe, J., Ovcharenko, D., *et al.*, 2003. Phylogenetic shadowing of primate sequences to find functional regions of the human genome. Science 299, 1391–1394.

Buchanan, C.D., Klein, P.E., Mullet, J.E., 2004. Phylogenetic analysis of 5′-noncoding regions from the ABA-responsive *rab*16/17 gene family of sorghum, maize and rice provides insight into the composition, organization and function of *cis*-regulatory modules. Genetics 168, 1639–1654.

Clark, V.J., Cox, N.J., Hammond, M., Hanis, C.L., Rienzo, A.D., 2005. Haplotype structure and phylogenetic shadowing of a hypervariable region in the *CAPN10* gene. Hum. Genet. 117, 258–266.

Cliften, P., Sudarsanam, P., Desikan, A., *et al.*, 2003. Finding functional features in *Saccharomyces* genomes by phylogenetic footprinting. Science 301, 71–76.

Elemento, O., Tavazoie, S., 2005. Fast and systematic genome-wide discovery of conserved regulatory elements using a non-alignment based approach. Genome Biol. 6, R18.

Gao, D., Middleton, R., Rasko, J.E.J., Ritchie, W., 2013. miREval 2.0: A web tool for simple microRNA prediction in genome sequences. Bioinformatics 29, 3225–3226.

Glenwinkel, L., Wu, D., Minevich, G., Hobert, O., 2014. TargetOrtho: A phylogenetic footprinting tool to identify transcription factor targets. Genetics 197, 61–76.

Gordân, R., Narlikar, L., Hartemink, A.J., 2010. Finding regulatory DNA motifs using alignment-free evolutionary conservation information. Nucleic Acids Res. 38, e90.

Grad, Y.H., Roth, F.P., Halfon, M.S., Church, G.M., 2004. Prediction of similarly acting *cis*-regulatory modules by subsequence profiling and comparative genomics in *Drosophila melanogaster* and *D. pseudoobscura*. Bioinformatics 20, 2738–2750.

Gumucio, D.L., Heilstedt-Williamson, H., Gray, T.A., *et al.*, 1992. Phylogenetic footprinting reveals a nuclear protein which binds to silencer sequences in the human γ and ε globin genes. Mol. Cell. Biol. 12, 4919–4929.

Gumucio, D.L., Shelton, D.A., Bailey, W.J., Slightom, J.L., Goodman, M., 1993. Phylogenetic footprinting reveals unexpected complexity in trans factor binding upstream from the ε-globin gene. Proc. Natl. Acad. Sci. USA 90, 6018–6022.

Katara, P., Grover, A., Sharma, V., 2012. Phylogenetic footprinting: A boost for microbial regulatory genomics. Protoplasma 249, 901–907.

Lawrence, C.E., Altschul, S.F., Boguski, M.S., *et al.*, 1993. Detecting subtle sequence signals: A Gibbs sampling strategy for multiple alignment. Science 262, 208–214.

Lenhard, B., Sandelin, A., Mendoza, L., *et al.*, 2003. Identification of conserved regulatory elements by comparative genome analysis. J. Biol. 2, 13.

Liu, B., Zhang, H., Zhou, C., *et al.*, 2016. An integrative and applicable phylogenetic footprinting framework for *cis*-regulatory motifs identification in prokaryotic genomes. BMC Genom. 17, 578.

Lutz, U., Nussbaumer, T., Spannagl, M., *et al.*, 2017. Natural haplotypes of FLM non-coding sequences fine-tune flowering time in ambient spring temperatures in Arabidopsis. eLife 6, e22114.

Maeso, I., Irimia, M., Tena, J.J., Casares, F., Gómez-Skarmeta, J.L., 2013. Deep conservation of *cis*-regulatory elements in metazoans. Philos. Trans. R. Soc. *Lond.* B *Biol. Sci.* 368, 20130020.

Matsunami, M., Sumiyama, K., Saitou, N., 2010. Evolution of conserved non-coding sequences within the vertebrate Hox clusters through the two-round whole genome duplications revealed by phylogenetic footprinting analysis. J. Mol. Evol. 71, 427–436.

McCue, L., Thompson, W., Carmack, C.S., *et al.*, 2001. Phylogenetic footprinting of transcription factor binding sites in proteobacterial genomes. Nucleic Acids Res. 29, 774–782.

Neph, S., Tompa, M., 2006. MicroFootPrinter: A tool for phylogenetic footprinting in prokaryotic genomes. Nucleic Acids Res. 34, W366–W368.

Newberg, L.A., Thompson, W.A., Conlan, S., *et al.*, 2007. A phylogenetic Gibbs sampler that yields centroid solutions for *cis*-regulatory site prediction. Bioinformatics 23, 1718–1727.

Ovcharenko, I., Boffelli, D., Loots, G.G., 2004. eShadow: A tool for comparing closely related sequences. Genome Res. 14, 1191–1198.

Ray, P., Shringarpure, S., Kolar, M., Xing, E.P., 2008. CSMET: Comparative genomic motif detection via multi-resolution phylogenetic shadowing. PLOS Comput. Biol. 4, e1000090.

Satija, R., Novák, Á., Miklós, I., Lyngsø, R., Hein, J., 2009. BigFoot: Bayesian alignment and phylogenetic footprinting with MCMC. BMC Evol. Biol. 9, 217.

Satija, R., Pachter, L., Hein, J., 2008. Combining statistical alignment and phylogenetic footprinting to detect regulatory elements. Bioinformatics 24, 1236–1242.

Sauer, T., Shelest, E., Wingender, E., 2006. Evaluating phylogenetic footprinting for human–rodent comparisons. Bioinformatics 22, 430–437.

Siddharthan, R., Siggia, E.D., van Nimwegen, E., 2005. PhyloGibbs: A gibbs sampling motif finder that incorporates phylogeny. PLOS Comput. Biol. 1, e67.

Sinha, S., Blanchette, M., Tompa, M., 2004. PhyME: A probabilistic algorithm for finding motifs in sets of orthologous sequences. BMC Bioinform. 5, 170.

Tagle, D.A., Koop, B.F., Goodman, M., *et al.*, 1988. Embryonic ε and γ globin genes of a prosimian primate (*Galago crassicaudatus*): Nucleotide and amino acid sequences, developmental regulation and phylogenetic footprints. J. Mol. Biol. 203, 439–455.

van Nimwegen, E., 2007. Finding regulatory elements and regulatory motifs: a general probabilistic framework. BMC Bioinform. 8, S4.

von Bubnoff, A., Peiffer, D.A., Blitz, I.L., *et al.*, 2005. Phylogenetic footprinting and genome scanning identify vertebrate BMP response elements and new target genes. Dev. Biol. 281, 210–226.

Wang, T., Stormo, G.D., 2003. Combining phylogenetic data with co-regulated genes to identify regulatory motifs. Bioinformatics 19, 2369–2380.

Chromatin: A Semi-Structured Polymer

Wayne K Dawson, CeNT, University of Warsaw, Warsaw, Poland and BiLab, The University of Tokyo, Tokyo, Japan
Michal Lazniewski, CeNT, University of Warsaw, Warsaw, Poland and Medical University of Warsaw, Warsaw, Poland
Dariusz Plewczynski, CeNT, University of Warsaw, Warsaw, Poland and MINI, Warsaw University of Technology, Warsaw, Poland

Introduction

It has long been a question whether the genome has spatial organization and to what extent it influences its function. A critical part of modern medicine might depend on our understanding how the genome organization works in the context of replication and gene expression and its regalia of promoters, enhancers, the DNA associated proteins, including RNA polymerase II (Rpol2), and the spliceosome apparatus (SA). The most crucial question is does gene expression depend on the spatial proximity of the promotors, inhibitors, and enhancers? What role, if any, does proximity play? How do Rpol2 and SA operate in conjunction with unraveling the nucleosome and how does this factor into exon/intron splicing? How does a developing organism determine what genes to express and when and how is that controlled? What happens when proper functioning of the cell fails? There are many unanswered questions in the field of genomics; nevertheless, recent research has opened the door to at least a partial understanding of some of these important issues.

Progress in our understanding of nuclear architecture has advanced due to the ongoing development of various imaging techniques like FISH (fluorescent in situ hybridization), confocal microscopy (O'connor, 2008; Wurtele and Chartrand, 2006; Wang et al., 2015; Lomvardas et al., 2006; Rosa et al., 2010; Fraser et al., 2015; Giorgetti et al., 2014; Muller et al., 2010) and electron microscopy (Ou et al., 2017) as well as in contact mapping techniques especially chromatin conformation capture (3C) method (Wurtele and Chartrand, 2006; Dekker et al., 2002; Lomvardas et al., 2006) and various techniques derived from 3C; e.g., Hi-C (Rao et al., 2014; Lieberman-Aiden et al., 2009; Jin et al., 2013; Fraser et al., 2015; Dixon et al., 2012), ChiA-PET (Ji et al., 2016; Tang et al., 2015; Li et al., 2017; Fullwood et al., 2009a) or HiChIP (Mumbach et al., 2016). The details of these methods are available elsewhere (Fraser et al., 2015) and in this review we provide only a brief summary regarding some of these techniques. There is a growing consensus that there is considerable (quasi) organization in the arrangement of eukaryotic chromatin in the nucleus and that the chromatin fiber exhibits a considerable degree of compartmentalization.

It should seem puzzling that a 6 billion base pair (bp, 6 Gbp) somatic chromatin fiber arranges itself in such a way that it does not become tied up in knots, and organizes itself efficiently into 46 chromosomes in a repeatable way during its mitosis phase. As a thought experiment, consider what happens to a pair of earphone when dumped into a carry bag with other stuff and rattled around for a while. The earphones quickly become tangled into at least one knot. Moreover, the cell itself is a highly dynamic yet in many ways disordered system too. How then does a 6 Gbp genome keep from getting thoroughly tangled up like the earphones yet at the same time remain only partially ordered?

In this review, we will look briefly through the levels of the DNA organization starting from the double stranded DNA (dsDNA) up to the interactions between different parts of the chromatin. We will briefly discuss some of the current experimental and computational methods aiming to predict chromatin conformations.

Chromatin 3D Interactions: Short Range to Long Range

Here we take a progressive picture of chromatin from the components at the coarse-grained scale of histones ($\sim$nm) to the long range binding between distant parts of the chromatin, which extends to about 1 Mbp and perhaps even as large as 3 Mbp.

The Local Structure of the Nucleosomes

Chromatin is a complex heteropolymer comprised of many different components. Double-stranded DNA (dsDNA) in eukaryotic organisms is bundled into packages called nucleosomes. The dsDNA wraps about an octamer of core histone proteins (roughly 7/4 turns; about 147 bp) like on a solenoid (**Fig. 1**). The octamer contains a tetrameric core of H3 and H4 proteins bound together and forming a dimer structure $(H3-H4)_2$ and this dimer is capped on both sides of the solenoid like structure by a heterodimer of H2A–H2B (Nacheva et al., 1989). In contrast to core histones, the linker histones, like H1, and its isoforms are not components of the nucleosome core, but instead bind to the linker DNA at the nucleosome entry and exit sites, stabilizing the entire complex (Bustin et al., 2005; Kalashnikova et al., 2016). Some examples of structural fragments of dsDNA wrapped around a core of histone proteins – obtained via X-ray diffraction – can be found in the Protein Data Bank (PDB); these include 5DNM, 5VA6, 5B31 (**Fig. 1**) or 5KGF.

Unless specified otherwise, the word "genome" represents a sequence corresponding to one complete set of autosomes (chromosomes not associated with the sex of the organism) and one of each type of sex chromosome; the result representing both possible sexes. In addition, the "genome sequence" is often a composite read from various individuals. Therefore, the length of a "genome" roughly represents to the length of the haploid cell or gamete (sperm/ovum). The more typical somatic cells of an

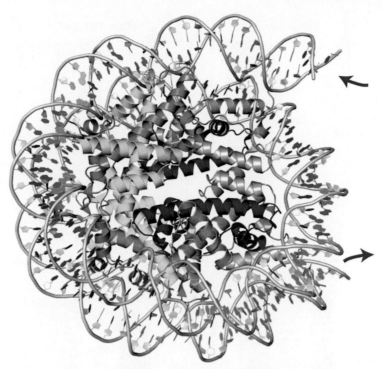

Fig. 1 Example of the histone and double stranded DNA (dsDNA) complex (PDB id: 5B31) showing the 1 and 3/4th turn of the dsDNA around the histone core. The coloring scheme for the histone subunits and DNA are as follows: silver blue – H3 (chains A + E), blue – H4 (chains B + F), orange – H2A (chains C + G), yellow – H2B (chains D + H), silver – dsDNA (chain I, J). The Figure is created using Pymol (Delano, 2004; pymol available from http://www.pymol.org).

organism contain a fused copy of both the male and female haploid cell DNA. Haploid cells are formed through meiosis (splitting in half the diploid). Somatic cells are replicated through the process of mitosis.

The human genome consists of three billion base pairs (3 Gbp) of dsDNA. The haploid cells have 23 chromosomes and the dsDNA sequence is a monoploid roughly 3 Gbp long. The somatic cells are diploids with 46 chromosomes and a sequence length of approximately 6 Gbp. Naked dsDNA (the dsDNA without the histones) from somatic cells would reach two meters $(2 \text{ [haploid]})(3 \times 10^9 \text{ [bp/haploid]})(0.34 \text{ [nm/bp]})$ if it were stretched out over its full contour length; overstretched (Olson and Zhurkin, 2000), this length would even double. When packaged in the form of chromatin fiber, the overall length is reduced by about one third (Rosa and Zimmer, 2014) where each solenoid shaped nucleosome core is about 6 nm by 11 nm. The distance between each nucleosome is called the *nucleosome repeat length*, (NRL (Routh *et al.*, 2008)). The histone core is extended by H1 linkers. There is nearly always at least one H1 between adjoining nucleosome cores leading to a *n* + 1 rule (Routh *et al.*, 2008). Each H1 linker tends to increase the NRL in increments of 10 bp; coincidentally *almost* a full 360° turn of the dsDNA helix (Voong *et al.*, 2016; Brogaard *et al.*, 2012). As a result, the NRL ranges from 167 bp (the histone core plus H1 plus one H1 linker, 10 bp; e. g., yeast) to 240 bp (core plus H1 plus an 80 bp linker, echinoderm sperm) (Bascom and Schlick, 2017). The nucleosomes appear to be connected as in an irregular zig-zag (Woodcock *et al.*, 1993; Bednar *et al.*, 1998), which resembles a loose coil formed when "flaking" a rope (Bascom and Schlick, 2017); i.e., rolling out a mountain climbing rope. The globules of dsDNA and histones (the nucleosomes), and the linker(s) form the basic chromatin fiber of somatic cells and gametes.

The H1 linker and other related factors can affect the compactness of the chromatin. For example, the length of the H1 linker appears to influence the organization of the chromatin into an ordered structure. In a study by Routh *et al.* (2008), chromatin compaction was analyzed when the NRL was either 167 or 197 bp and the linker histone H5 was present or not. The H5 histone is found in chicken erythrocytes and is a variant of the linker histone H1. H5 has a greater binding affinity to DNA than H1, but the influence of H5 on compaction is rather similar. In the absence of H5 (**Fig. 2**, top two panels), the dsDNA appear to be disordered, especially in the case when NRL was 197 bp. When H5 is present, the structure appears to be ordered. In **Fig. 2** (bottom left panel, " + H5"), when H5 is present and the NRL is equal to 167 bp, the nucleosome appear to form a structure similar to a 20 nm fiber. On the other hand, the 197 bp NRL yields something closer to the 30 nm fiber (**Fig. 2**, lower right panel).

Depending on the concentration of salt in the buffer, at the length-scale of several nucleosomes and corresponding linkers (NRL approximately 200 bp), there is a tendency for the nucleosomes to become partially ordered. It was long thought that the chromatin beaded up into 30 nm globules (Finch and Klug, 1976); however, such structural regularity has also been challenged (Davie, 1997; Woodcock *et al.*, 1993). Based on observations of small angle x-ray scattering (SAXS) studies on chromatin from HeLa cells, the structure of the chromatin fiber appears to take on a more complex local structure or even possibly random

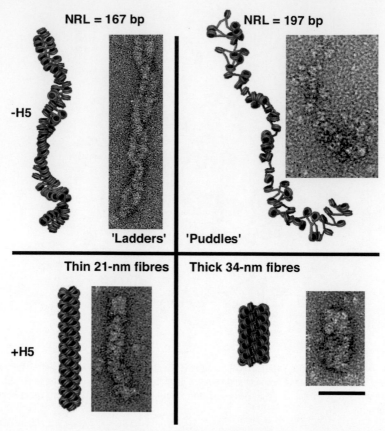

Fig. 2   The role of the linker histone in determining higher-order chromatin structure for different nucleosome repeat lengths (NRL); 167 bp and 197 bp. (*Upper*) Without the linker histone. (*Lower*) With the linker-histone. To the left of each example, a sample of a simulated structure is shown to provide an impression of the overall arrangement of the nucleosomes. (*Upper*) Here, it can be seen that both the 167 and 197 bp NRL are largely disordered or certainly rather loose in the absence of H5; the 197 case is more so than the 167 case, which is at least wrapped. (*Lower*) Here, it can be seen that both the 167 and 197 bp fibers are quite compact in the presence of H5. (Scale bar: 50 nm.). Reproduced from Routh, A., Sandin, S., Rhodes, D., 2008. Nucleosome repeat length and linker histone stoichiometry determine chromatin fiber structure. Proc. Natl. Acad. Sci. USA 105, 8872–8877.

structure (Maeshima *et al.*, 2010; Joti *et al.*, 2012; Nishino *et al.*, 2012; Maeshima *et al.*, 2016). The studies using SAXS to analyze the rough structure of the nuclear material of the HeLa cells revealed a clear 30 nm signal (Joti *et al.*, 2012; Nishino *et al.*, 2012) in the presence of Rpol2. However, after washing away the Rpol2, the peak disappeared. The SAXS data also shows clear 6 nm and 11 nm peaks and a broadband peak for longer length scales (Maeshima *et al.*, 2014) indicating a local lattice-like clustering of 3 or 4 nucleosomes; consistent with a local stacking and tandem arrangement nucleosome units.

The H1 histone protein is particularly sensitive to environmental factors such as ionic strength (Davie, 1997). For example, the existence of 30 nm fibers appears to be partially dependent on the concentration of Mg^{2+}. Later work from Maeshima and coworkers focused more on the salt buffer in the absence of Rpol2 (Maeshima *et al.*, 2016). In **Fig. 3(a)**, the diameter of the HeLa cell nucleus is found to be significantly inflated in the absence of Mg^{2+} (16.4 μm) and becomes increasingly compact with addition of Mg^{2+} (around 8.6 μm, where normal HeLa cell diameters appear to be 7.36 μm (Milo *et al.*, 2010) BNID 104990). In **Fig. 3(b)**, SAXS measurements of nucleosome oligomers (blue) and nuclear material (red) show similar profiles; both show no obvious structure other than a broad peak in the absence of Mg^{2+} and sharper images of 6 and 11 nm peaks at increasing concentrations of Mg^{2+}. In the same study by Maeshima *et al.*, the authors also examined isolated native chicken chromatin fragments that showed similar profiles as the human HeLa cells in **Fig. 3(b)**. Hence, using the chicken linker histone (H5) in **Fig. 2** (Routh *et al.*, 2008) instead of human H1 does not change the behavior of chromatin. Finally, **Fig. 3(c)** shows a proposed progression from a disorder local structure of the nucleosomes in chromatin in the absence of Mg^{2+}, to very compact and possibly disordered structure at higher concentrations of Mg^{2+}. The ion concentrations within the cell of most animals is roughly 140 mM K^+, 12 mM Na^+ and about 1 mM Mg^{2+} whereas outside the cell it is about 150 mM Na^+ (Auffinger *et al.*, 2011). Therefore, the middle panel in **Fig. 3(c)** represents the chromatin state closest to the physiological conditions typically found in human cells.

It might be tempting to assume that a regular double helix of dsDNA has almost no influence on the positioning of the nucleosomes on the dsDNA; however, this is not the case. In Todolli *et al.* (2017), it was observed that the bases in the dsDNA sequence can influence the positioning of the histones and consequently the arrangement of the nucleosomes. Presently, there are no first principles prediction methods at the bp level that can explain the arrangement of the nucleosomes (Todolli *et al.*, 2017). If

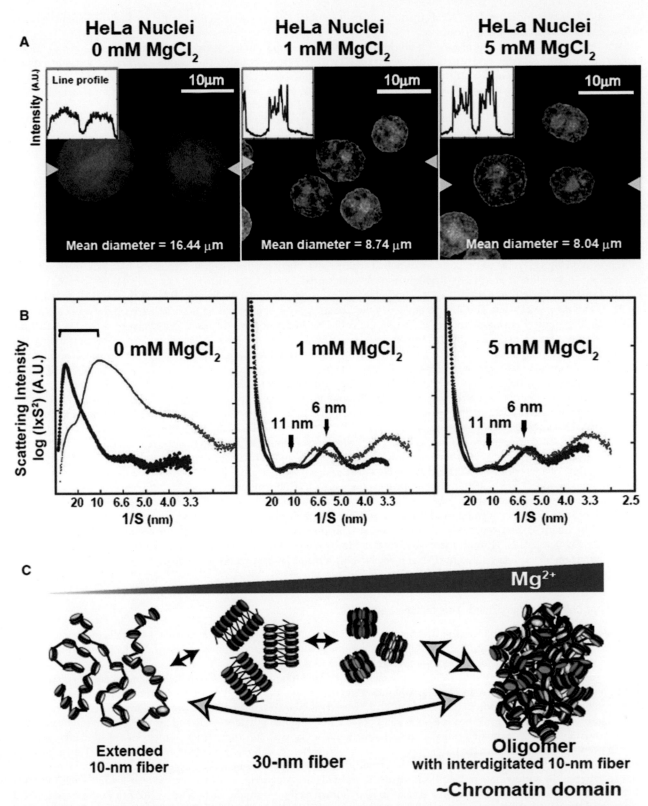

A

HeLa Nuclei
0 mM MgCl₂

HeLa Nuclei
1 mM MgCl₂

HeLa Nuclei
5 mM MgCl₂

Line profile

10μm

10μm

10μm

Intensity (A.U.)

Mean diameter = 16.44 μm

Mean diameter = 8.74 μm

Mean diameter = 8.04 μm

B

Scattering Intensity
log (I×S²) (A.U.)

0 mM MgCl₂

1 mM MgCl₂

5 mM MgCl₂

11 nm   6 nm

11 nm   6 nm

20 10 6.6 5.0 4.0 3.3
1/S (nm)

20 10 6.6 5.0 4.0 3.3
1/S (nm)

20 10 6.6 5.0 4.0 3.3 2.5
1/S (nm)

C

Mg²⁺

Extended
10-nm fiber

30-nm fiber

Oligomer
with interdigitated 10-nm fiber

~Chromatin domain

Fig. 3 Isolated HeLa nuclei showing the effects of Mg²⁺ concentration on the overall chromatin structure (nuclear material). (A) FM images of chromatin structure in the nuclei with 0 (left), 1 (center), and 5 mM Mg²⁺ (right). The insets show the intensity line profiles as measured by scanning (horizontally) between the two yellow arrow heads in each of the images. (B) Sax profiles of nucleosome arrays formed by an oligomer of H1-linked nucleosomes (red) and by HeLa nuclei (red); 0 mM (left), 1 mM (center) and 5 mM Mg²⁺ (right). (C) A schematic of the effects of Mg²⁺ concentration on the characteristic form of chromatin fibers for a 12-mer nucleosomal array. In 0 mM Mg²⁺, the nucleosomes are highly disordered and separated, in 1–2 mM Mg²⁺, there one sees some evidence or regular arrangements of nucleosomes that to some extent generated a 30 nm chromatin fiber structure, though the 30 nm peak is not so visible in the spectrum. With additional increases in Mg²⁺, the structure collapses into a structure which is largely disordered but compact. Reproduced from Maeshima, K., Rogge, R., Tamura, S., *et al.*, 2016. Nucleosomal arrays self-assemble into supramolecular globular structures lacking 30-nm fibers. EMBO J. 35, 1115–1132.

we were able to better understand this local structure, it would be possible to understand the short-range nucleosome-nucleosome (histone coated dsDNA and linker) structure.

It is a well-established fact that histone modifications impact gene expression by altering the overall nucleosome-nucleosome interactions in the chromatin structure through acylation (Davie, 1997), methylation (Kouzarides, 2002; Cheung et al., 2000), phosphorylation (Cheung et al., 2000) and sometimes sumoylation (Shiio and Eisenman, 2003) and ubiquitylation (Nathan et al., 2003). Acylation and deacetylation of the exposed parts of the histone assists in turning transcription on and off (Davie, 1997). It also promotes the recruitment of proteins that can bind diverse parts of the chromatin fiber into loops, which ultimately form transcription factories that become the loci of highly expressed regions through Rpol2 activity (Dunn et al., 2003; He et al., 2008; Jahan et al., 2016; Warns et al., 2016).

A recent molecular dynamics simulation (Zhang et al., 2017) shows that lysine acetylation of the H4 tail with K16Ac yields an increased diversity of conformations and, consequently, the interactions between nearest neighboring nucleosome units. The positively charged side chains of lysine and arginine tend to repel each other in an isolated tail of H4. K16Ac relieves some of this tension and contributes to the folding and stabilization for a variety of conformations of the chromatin. At the same time, since the dsDNA has negatively charged phosphates, there are extensive electrostatic interactions, particularly salt bridges. When the lysine is acetylated, these electrostatic interactions are disrupted yielding a more unstable interaction at the H4 tail and resulting in a more diverse conformation space.

The role of these various histone modifications is quite complex, but the effects appear to strongly influence gene expression and suppression by exposing some parts in unbound regions and burying other parts rendering them largely inaccessible to transcription. In Rychkov et al. (2017), Rychkov and coworkers used atomic force microscopy to study various combination of partially assembled nucleosome states: (dsDNA bound to partial histone subunits); hexasome $(H2A \cdot H2B) \cdot (H3 \cdot H4)_2$, tetrasome $(H3 \cdot H4)_2$ and disome $H3 \cdot H4$. The authors argue that the free energy cost of unwrapping the histones is on the order of 40 kcal/mol, which is extremely strong for proteins. Hence, the collective interaction would result in regions of the dsDNA that are protected by the histone octamer subunits and likely to strongly impede the progress of Rpol2 and SA from unwrapping and transcribing and splicing, respectively. Such a large free energy barrier would result in kinetics where unwrapping the nucleosome could consume as much as 10 min of time, yet the unimpeded rate appears to be around 1.4 kbp/s (Shermoen and O'farrell, 1991) and certainly at least a single nucleosome (~ 200 bp/s) per second (Tims et al., 2011).

To overcome this issue, it has been argued that the histone binding may very well be rather dynamic (Bustin et al., 2005; Bascom and Schlick, 2017) because the histone octamer appear to exist in dynamic equilibrium, regularly binding and unbinding to the dsDNA. This may also result from the partial separation of a much smaller structural unit like a disome $(H3 \cdot H4)$ separating with only a 40 bp sequence being in contact with the positively charged core. In such a case, the structure would only involve some 10 kcal per mole (because the 40 kcal/mol binding is spread over 8 histone proteins). Such a free energy would only require an unwrapping time of a few milliseconds (Rychkov et al., 2017), which is more consistent with the observed speed of transcription of Rpol2 (Shermoen and O'farrell, 1991), though perhaps there are other possibilities.

The details of nucleosome, the interplay of the histones and dsDNA, dynamics of histone modifications such as above mentioned acetylation, other histone modifications, and how these factors influence transcription remain largely unanswered questions. The interplay between these modified histones and the overall structure of chromatin is currently largely unknown (Bascom and Schlick, 2017). How all these variables combine together to determine the distance between the nucleosomes and the local arrangement of nucleosome-nucleosome interactions remains an area extensive research.

Flexibility of Chromatin: Persistence Length/Kuhn Length

In polymer chemistry (Flory, 1953, 1969; Grosberg and Khokhlov, 1994), the stiffness of a polymer is typically expressed in terms of the persistence length (l_p) or the Kuhn length (ξ). The Kuhn length can be thought of as the average length of segments of a chain and persistence length as describing the average curvature of a wire.

In the case of Kuhn length (Kuhn, 1936), one imagines a polymer that is generated as though one were taking equal-distant steps in random directions. For example, **Fig. 4(A)** shows the mer-to-mer distance (distance between monomers, the red beads) laid on top of a green cord. On top of the green cord, purple beads are shown at a fixed link distance (yellow), which represents the Kuhn length. The overlay of purple beads is sometimes called a freely-jointed polymer chain, and the overall path can be modeled as a Markov chain and approximated using a Gaussian function. The Gaussian function has the property of self-similarity; a feature where at every scale of resolution, the random walk looks the same. Hence, Kuhn length reflects a selected uniform length scale of the random walk; it is often described as beads on a chain, where the "beads" represent the purple dots and the Kuhn length is the distance between the yellow bars in **Fig. 4(A)**. In general, real polymers only reflect this self-similarity at discrete length scales.

The persistence length (l_p) is based on a similar concept known as the worm-like-chain (Flory, 1969) or Kratky-Porod model ((Kratky and Porod, 1949), op cit. (Flory, 1969)). In this model, one imagines a long flexible cord like a rope or a rubber hose of length l. Consider two consecutive points along this chain (k and k', where $k \neq k'$), with each point along the cord represented by its unit vectors $\mathbf{u}_k$ and $\mathbf{u}_{k'}$ and the angle between $\mathbf{u}_k$ and $\mathbf{u}_{k'}$: $\mathbf{u}_k \cdot \mathbf{u}_{k'} = \cos(\theta_{k,k'})$. It can be shown that the integration with respect to all the angles $\theta_{k,k'}$ would result in $\cos(\theta) = \exp(-l/l_p)$. The l_p (persistence length) expresses the minimum length of a cord that will flex to a curvature of one radian (about 57.3°) without kinking and remembering its original straight position (bending back). In other words, imagine there are two points on the straight polymer with distance d between them. If we can bend the polymer so that the

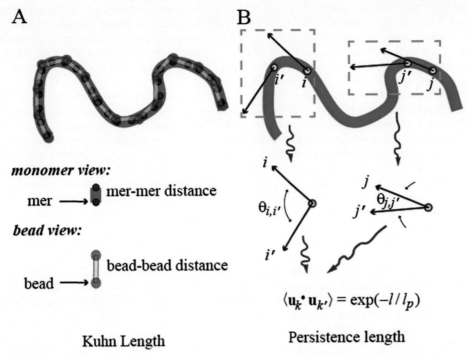

A

B

monomer view:

mer ⟶ ▮ mer-mer distance

bead view:

bead ⟶ ▮ bead-bead distance

$$\langle \mathbf{u}_k \cdot \mathbf{u}_{k'} \rangle = \exp(-l / l_p)$$

Kuhn Length   Persistence length

Fig. 4 Comparison of flexibility in terms of Kuhn length and persistence length. (A) Kuhn length where blue beads represent monomers and purple beads represent the beads. (B) Persistence length where vectors i, i' involve a large bend of angle $\theta_{i,i'}$ and j, j' involve only a small bend of angle $\theta_{j,j'}$. All the angles $\theta_{k,k'}$ are integrated yielding $\langle u_k \cdot u_{k'} \rangle = \exp(-n/l_p)$, where $n = l_p$ is the persistence length of the polymer chain.

angle between the unit vectors from these points (e.g., $\theta_{i,i'}$ or $\theta_{j,j'}$ in **Fig 4B**) is 1 radian but after we remove the "bending force" the polymer can return to the straight position, then d is effectively l_p. For example, compare l_p for a rubber hose and a thick steel cable; the thick steel cable cannot bend to $57.3°$ over the same distance as the rubber hose. How about an iron bar? Given a long enough bar, one will see some visible bending but that flexing may only be obvious on length scales on the order of a football field.

For most real polymers, it is found that the distance between the monomers is shorter than the Kuhn length or persistence length. Hence, both models reflect the tendency of the polymer to behave as though the fundamental structural unit is larger than the size of a single monomer (in most cases).

Moreover, real systems are not completely self-similar at all scales; therefore, there is typically a defined Kuhn length for every polymer. The minimum Kuhn length is typically greater than the distance between the monomers. In the case of *naked* dsDNA (without the histones), this can exceed more than 100 bp, depending on salt concentration (Hagerman, 1988). It can be thought of as a kind of coarse-grained resolution length. In typical polymers, it is treated as a statistical parameter and averaged over the entire structure; however, the value should depend on a variety of factors such as chemical binding interactions, local structural interactions, and other considerations. In RNA, the variability depends strongly on the length of the duplex stem and to some extent single-stranded interactions (Dawson *et al.*, 2013).

In the case of chromatin, the polymer comprises a composite of histone proteins and dsDNA. Because the dsDNA is wrapped around the histone core (**Fig. 1**), there is a local bending of the dsDNA on the scale of 5–10 bp. Because the wrapping of the dsDNA to the histone core has a binding energy of about 40 kcal/mol (Rychkov *et al.*, 2017), a second Kuhn length exists at the scale of the nucleosome (around 200 bp). **Figs. 2** and **3** suggest yet a further hierarchical level; The nucleosomes, (about 10 nm in diameter) appear to aggregate into larger collective structures as a result of their mutual attraction. Therefore, the bead size of the cluster of nucleosomes may not necessarily be precisely 30 nm units (Hansen, 2002; Maeshima *et al.*, 2016); however, at physiological conditions, there is little dispute that there is structure encompassing at least several nucleosomes (Maeshima *et al.*, 2016). The collective interaction of the nucleosomes appears to range from 90 nm (Davie, 1997) (roughly the size of 9 nucleosomes) to possibly 40 nm (about 4 nucleosomes).

Although the length scale is probably better understood in terms of clusters of nucleosomes, the NRL is a variable in the cell; therefore, a single base pair of dsDNA is the most regular and invariant unit to define as the "monomer" and five nucleosomes comprise roughly 1 kbp.

Since the differences in the NRL generate very different patterns in **Fig. 1**, one should probably surmise that the Kuhn length of chromatin is a variable. Bystricky *et al.* (2004) found an overall persistence length of about 170–220 nm, which corresponds to a Kuhn length of 340–440 nm (comprising 34 to 44 nucleosomes). This latter number seems too excessive but heterochromatin is a highly compact material that may have an effective Kuhn length in this range. The variability might also be explained by various histone modifications including (I) the H1 linker (Bocharova *et al.*, 2012; Bascom and Schlick, 2017), (II) H3 phosphorylation,

(III) H4 K16 acylation (H4K16Ac, (Zhang *et al.*, 2017)), (IV) various expression related CpG islands with their corresponding histone modifications (H4K20me1, H2BK5me1 and H3K79me1/2/3) near the transcription start site, and (V) dynamic histone mobility (Bustin *et al.*, 2005). A reasonable estimate to derive from these considerations is a Kuhn length ranging between 3 and 9 nucleosomes (and possibly up to as many as 22). Further, the difference is partly reflected in whether the region is euchromatin (typically loose) or heterochromatin (typically packed). Thus it must be considered that a truer representation of the chromatin chain is that it has a variable bead-to-bead length (Todolli *et al.*, 2017).

Presently, due to both computational issues and the lack of sufficient resolution of experimental methods, computational models of chromatin typically assume that the chromatin chain is composed of beads of a uniform size ranging between 1 kbp to several Mbp. The distance between beads is, therefore, usually much longer than any Kuhn length associated with either het-erochromatin or euchromatin. For the shortest lengths of 1 kbp, there may be consequences to assuming complete plasticity of the chromatin beads if we assume the maximum range of 44 nucleosomes (roughly 8 kbp). However, the SAXS data in **Fig. 3** Maeshima *et al.* (2016) suggests that 44 is probably rather extreme. The evidence suggests something akin to 30 nm fibers at physiological salt, possibly ranging between 3 and 6 nucleosomes. Hence, certainly at 5 kbp resolution, chromatin can be treated as a set of uncoupled bead on a chain, and, it would be better to be thinking of corrections for the fact that the true bead size is actually *shorter* than the 5 kbp segments that a computational program would bin the data into. Hence, there is likely to be some loss of information in coarse-graining to such length scales.

It seems arguable that using 5 kbp "beads" represents a range where it is possible to describe chromatin as a Gaussian polymer chain (with a self-avoiding walk correction) and possibly some non-ideal polymer behavior associated with the solvent and buffering characteristics (Flory, 1953, 1969; Grosberg and Khokhlov, 1994; Dawson and Kawai, 2015) of the environment (the nucleus) where the chromatin resides. To this polymer model, one would compute the long-range chromatin interactions mediated by various protein factors (such as CTCF, or Rpol2). Nevertheless, we should be mindful that this uniform length scale homo-polymer Kuhn length is a very rough approximation.

The Role of CTCF Protein and Cohesin in Genome Organization

In addition to the obvious structure of chromosomes during cell division, the chromatin in the interphase state appears to organize into compartments (Munkel *et al.*, 1999; Bystricky *et al.*, 2004). Moreover, on the length scale of a 20 kbp to about 3 Mbp, there appears to be considerable long-range pairwise organization of the chromatin according to both FISH (Guo *et al.*, 2015; Xu and Corces, 2016; Munkel *et al.*, 1999; Bystricky *et al.*, 2004) and any of the chromosome conformation capture (3C) methods particularly ChIA-PET (Guo *et al.*, 2015; Xu and Corces, 2016; Tang *et al.*, 2015) and Hi-C (Dixon *et al.*, 2012; Lieberman-Aiden *et al.*, 2009; Rao *et al.*, 2014). These methods will be discussed in the Subsection titled "Experimental Methods".

At the level of genetics, there is a plethora of interactions within the chromatin (Ji *et al.*, 2016; Beagan *et al.*, 2016). Several kinds of genetic features that are of special interest here are insulators, enhancers and promotors. The genetic boundary elements, known as insulators, have the ability to effectively block the interaction between enhancers and promotors. The promoters mark the region of the dsDNA sequence that attracts proteins that initiate transcription and are typically located near the transcription start sites (upstream on the sense strand). Enhancers are relatively short fragments of dsDNA where transcription factors can bind to increase the likelihood of transcription of a particular gene. Enhancers amplify the transcription rate of a specified gene (Sapoj-nikova *et al.*, 2009). Unlike promotors, enhancers can be located far away from the transcription start site on the dsDNA chain (Fraser *et al.*, 2015).

The CTCF protein factor has long been associated with insulators (Lewis and Murrell, 2004; Dunn *et al.*, 2003), tending to mark the boundaries between salt soluble regions (often expressed) and insoluble regions (rarely expressed) (Jahan *et al.*, 2016). The CTCF protein is an 11 zinc-fingers protein that binds CCCTC motifs (Ji *et al.*, 2016; Tang *et al.*, 2015; West *et al.*, 2002; Rao *et al.*, 2014). The binding of CTCF involves at least 4 of the 11 zinc-fingers and is to specific sequences (Renda *et al.*, 2007). It is generally thought that two CTCFs form a homodimer that can cause two distal regions of DNA to form into a loop (Ji *et al.*, 2016); though a recently published article describing the structure of CTCF bound to dsDNA noted that no dimerization was observed (Yin et al., 2017). By creating a loop mediated by CTCF proteins, the promoter(s) and enhancers of a given gene, and distant associated parts of the gene on the DNA sequence, are brought in close spatial proximity (Doyle *et al.*, 2014). The CTCF binding sites also correlate with alternative splicing (Warns *et al.*, 2016) and differentially methylated domains (Lewis and Murrell, 2004).

Cohesin is also a key factor in the formation of many CTCF contacts. Cohesin is a large complex of proteins and is closely associated with the formation of the chromosomes during mitosis and meiosis and plays a general role in chromatin stability (Hirano, 2015; Nasmyth and Haering, 2009; Kudo *et al.*, 2009; Ohta *et al.*, 2011). However, cohesin also appears to play an important role in the binding of some CTCF homodimers (Ji *et al.*, 2016; Nasmyth and Haering, 2009; De Wit *et al.*, 2015; Tang *et al.*, 2015; Tark-Dame *et al.*, 2014; Dowen *et al.*, 2014). The chromatin loops are often closed by a pair of CTCF proteins that preferentially adopt a parallel orientation where the CTCF binding partners point into the loop (De Wit *et al.*, 2015; Tang *et al.*, 2015; Szalaj *et al.*, 2016; Guo *et al.*, 2015) as shown schematically in **Fig. 5(A)**. There are some indications that the CTCF tends to bend the chromatin at its binding position (Macpherson and Sadowski, 2010). Since the dimer was not observed (Yin *et al.*, 2017), dimer formation may require the presence of cohesin and possibly other factors. When the CTCF partners are pointing away from the loop in divergent loops (**Fig. 5(B)**), the binding is far weaker, and likewise, tandem loops (**Fig. 5(C)** and **(D)**) appear to have an intermediate frequency of occurrence (Tang *et al.*, 2015). It is currently not clear whether divergent and tandem loops also involve

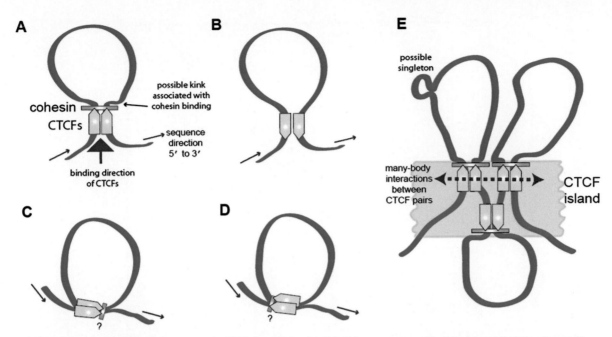

Fig. 5 Examples of.various types of loops formed by CTCF dimerization (combined in some cases with cohesin) and its interactions with chromatin. (A) A convergent loop, (B) a divergent loop, (C) a tandem right loop, (D) a tandem left loop, and (E) CTCF islands composed of loops from A and, in the top left hand region of the loop, a singleton is also indicated. The 5′ to 3′ direction of the sense strand of the DNA sequence is indicated by the red arrows, cohesin is indicated by the blue bar that closes one end of the CTCF, and the implied sequence direction for the CTCF motifs are indicated by the pointed-end of the CTCF objects.

cohesin. Models for the interplay of CTCF and cohesin will be discussed in the last section of this review (particularly in the subsection on loop extrusion).

Multiple loops can co-localize within the same three dimensional genomic loci; the process is mediated by many CTCF dimers that form together into CTCF islands, or rafts (see **Fig. 5(E)**). Such proximate collections of loops are defined here as chromatin contact domains (CCDs), similar to the term topologically associating domains (TADs), which is used widely in the field of 3D genomics. Such CCDs lengths often range between several kbp to several Mbp and show considerable similarity between cells and/ or stages of cell development (Dekker and Mirny, 2016; Beagan *et al.*, 2016; Ji *et al.*, 2016; Tang *et al.*, 2015; Rao *et al.*, 2014). To a lesser extent, Rpol2 also influences large regions of chromatin structure (Dunn *et al.*, 2003; He *et al.*, 2008; Jahan *et al.*, 2016; Warns *et al.*, 2016).

On the global scale of chromatin associated with this histone-dsDNA complex, the overall topology of the chromatin appears to regulate gene expression by delineating regions of transcription where active and repressed chromatin can be found because distant parts of the chromatin chain are brought into proximity of each other in loops (Ji *et al.*, 2016; Ulianov *et al.*, 2016; Tang *et al.*, 2015; Beagan *et al.*, 2016; Dekker and Mirny, 2016; Buenrostro *et al.*, 2015; Rao *et al.*, 2014). Chromosomal rearrangements and changes in the shape of the cell are also strongly correlated with malignant cells (He *et al.*, 2008; Davie, 1997; Schuster-Bockler and Lehner, 2012) suggesting that some cancers occur when normal patterns of CTCF binding begin to break down. The interplay of cohesin with enhancers, repressors, insulators and promotors and various histone modifications is very intricate (Merkens-chlager and Odom, 2013) and beyond the scope of this work. Beside CTCF and cohesin, several other proteins are postulated to be involved in forming chromatin loops. They include Rpol2 mediator, Yin Yang 1 (YY1) and BORIS (Pugacheva *et al.*, 2015; Weintraub *et al.*, 2017; Kagey *et al.*, 2010).

It remains unclear how the nucleosomes are unwound and transcribed into pre-mRNA and processed by the spliceosome to produce mRNA. Moreover, although the nucleosome loci affect the type of exon/intron splicing that occurs (Warns *et al.*, 2016; Voong *et al.*, 2016; Shukla *et al.*, 2011), the details of the mechanism are still unclear. There appear to be promoter-proximal nucleosomes, where some of the positively charged nucleosome contributes to the pausing of Rpol2 (Voong *et al.*, 2016; Shukla *et al.*, 2011). How these processes are carried out remains an area of intense research.

Methods for Evaluating Chromatin Interactions

In this section, we consider both the essential experimental and computational techniques that are used to study chromatin interactions and structure. We start by reminding the reader that the field of biology is very complex and demands considerable openness to multidisciplinary approaches to interpret and understand what we observe. For example, many research papers in the field have been written by people unaware of many decades of research in polymer physics. Likewise, as we forge ahead in new

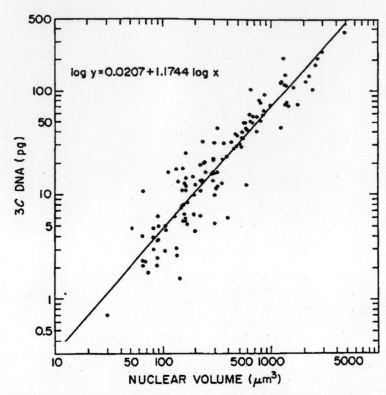

Fig. 6 Comparison of the mass of the DNA (directly proportional to the length of the sequence) with respect to the volume of the cell nucleus for plant species. (Taken from government document Nauman, A.F. 1979. Plant Nuclear Data Handbook: Radiosensitivity in Plants. Available at: https://www.osti.gov/scitech/servlets/purl/5369196, public domain).

techniques and are faced with the explosive sea of literature, we often think that work 50 years ago is irrelevant today. For example, in the early '60s in a somewhat obscure study (Baetcke *et al.*, 1967), it was discovered that the volume of the cell depends linearly on the length of the genome or the number of base pairs (bp) of the chromatin. Moreover, a correlation between cell volume and nuclear volume has been known since the late 19th century (explained in Gregory (2005)). Hence, to some extent, once again, we have reinvented the wheel – we should have expected this result, not been surprised by it. The important thing, however, is not just "what" but "why" we observe certain features of chromatin and recent developments in the field have helped us understand more of the "why" question and are an exciting area of new discoveries.

Experimental Methods

This section summarizes some of the most commonly used experimental techniques in studying chromatin and partly reviews some of the history. The optical techniques of fluorescence in situ hybridization (FISH) have forged ahead on discovering 3D compartment in the nucleus (Fraser *et al.*, 2015). Recently developed techniques like chromatin conformation capture (3C) and derivative technologies such as Hi-C (Rao *et al.*, 2014) and ChIA-PET (Li *et al.*, 2017) allow us to propose 2D maps of the interacting parts of chromatin. These method allow us not only to see which parts of the genome are in spatial proximity but also, in case of ChIA-PET, which proteins are responsible for maintaining the genome stability. This invaluable knowledge relies heavily on being able to propose a 3D structure of the chromatin using computational techniques. Beside the short introduction to the abovementioned methods we briefly discuss of their strengths and weaknesses.

Size of the Cell Nucleus Correlates with Chromatin Chain Length

Since the early '50s, numerous studies have shown that there is a strong correlation between the quantity of DNA in a given cell and the size of its nucleus in both the animal and plant kingdoms (Baetcke *et al.*, 1967; Price *et al.*, 1973; Gregory, 2017, 2005; Jovtchev *et al.*, 2006; Beaulieu *et al.*, 2008; Nauman, 1979). To obtain the size of the genome, most of these studies measured the mass (C-value) of the DNA (in pg). Very similar conclusions were even reached in the 19th century when the volume of a given cell was compared with the volume of its nucleus (*op. cit.*, (Gregory, 2005)). Likewise, DNA content and chromosome volume (Baetcke *et al.*, 1967) correlate. This phenomenon is also observed throughout the plant and animal kingdoms (Hodgson *et al.*, 2010; Mirsky and Ris, 1951; Beaulieu *et al.*, 2008). Furthermore, when the size of the nucleus deviates from this tendency, it is usually an indication of diseases like cancer (Webster *et al.*, 2009). In **Fig. 6** we show the correlation between nuclear volume and C-value (the mass of the DNA) from Nauman (1979) for various plant species. **Fig. 7(A)** shows a similar fit for animal species reproduced from

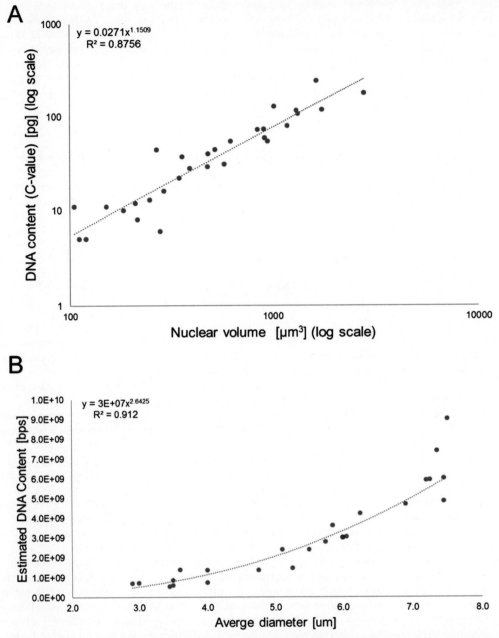

Fig. 7 Similar examples to **Fig. 6** (A) A reproduction of the data from Baetcke *et al.* (1967). (B) A comparison of the quantity of nuclear DNA (measured in base pairs) with respect to average diameter of the nucleus for measured species of fish (the red dot in the figure indicates the position of the human genome and is included for reference and comparison). From data obtained from the Animal Genome Size Database. Reproduced from Gregory, T.R., 2017. Animal genome size database: Cell size dataBase. In: Gregory, T.R. (Ed.). Available at: http://www. genomesize.com/cellsize/. **Fig. 7(B)** is plotted with linear axes for easier readability, plotted log-log, it looks similar to 5A.

Baetcke *et al.* (1967) and **Fig. 7(B)** shows a fit of C-values with respect to the average nuclear diameter for available information on ichthyic genomes (from Gregory (2017)) and the human genome for reference and comparison (the red dot). Whereas there is some scatter, the conclusions from the abovementioned articles are quite reproducible.

To understand and appreciate the significance of this behavior, we need to delve into some of the essential mathematical formulas. In the traditional model of an ideal polymer (Flory, 1953; Grosberg and Khokhlov, 1994), the distance between the ends of the polymer chain (*r*, also known as the end-to-end distance) can be shown to be a function of the square root of the number of monomers (*N*); $r \propto N^{1/2}$. This is the general property of the random walk (Flory, 1953; Grosberg and Khokhlov, 1994) and is known as the Gaussian polymer chain (gpc) model. The formal definition is given with Eq. (1):

$$r = b\sqrt{\xi N} \tag{1}$$

where b is the distance between adjacent monomers on the polymer chain, ξ is the Kuhn length (Kuhn, 1936) – a measure of the stiffness of the polymer – and r is measured in units monomer-monomer separation distance b. It must be noted that, in the literature, the parameter b often incorporates the Kuhn length. Thus in some cases, b would be equivalent to ξb in Eq. (1); $b \Rightarrow \xi b$.

This *ideal polymer* relationship often requires additional modification to Eq. (1) because real polymers can end up in good solvent (e.g., polystyrene in hexane) where the polymer tends to soak up the surrounding solvent and consequently swell in size. Likewise, a polymer can end up in poor solvent (e.g., polyvinylchloride in water) and become very compact because the monomers in the polymer tend to exclude solvent in favor of its own residues. This behavior is known as the excluded volume and it affects the value of the exponent in Eq. (1), which is often labeled ν;

$$r = \xi^{1-(\nu)} N^{(\nu)} b \tag{2}$$

Thus for the ideal polymer where $\nu = 0.5$, we obtain a formula identical to Eq. (1); $r = (\xi N)^{0.5} b$. For non-ideal polymer ν can range from $1/3 < \nu < 2/3$.

Examining **Figs. 6** and **7** in terms of volume relationships it can be seen that the typical volume of the nucleus (V_{nuc}) grows essentially linearly with genomic content: $V_{nuc} \approx \rho N$, where ρ is a proportionality constant. In **Fig. 7(A)**, the exponent is close to 1 (when fitted to volume) and in **Fig. 7(B)** (when fitted to diameter) the exponent is close to 3. Although there is some scatter, the weight of the measured data suggests a strong tendency toward $V_{nuc} \propto r_{nuc}^3 \propto N$. If we use the formula for radius from Eq. (1) (i.e., $r_{nuc} = b\sqrt{\xi N}/2$) to calculate the volume of an ideal sphere we get $V_p = (4\pi/3)(r/2)^3 = (\pi b^3/6)(\xi N)^{3/2}$. We can see that the volume of an ideal polymer (V_p) should thus grow as $V_p \sim N^{3/2}$. The ratio of V_p and V_{nuc} is given in Eq. (3).

$$\frac{V_p}{V_{nuc}} \approx \frac{(\pi/6)(\xi N b^2)^{3/2}}{\rho N} = \frac{\pi \xi^{3/2} b^3 N^{1/2}}{6\rho} \propto N^{1/2} \tag{3}$$

Thus it can be clearly seen that chromatin in the nucleus must be far more compact than what can be expected for an ideal polymer. Since the size of a genome can range from 10^4 (for viruses) all the way to more than 10^{10} (for some species of plants and some animals, particularly amphibians), this compacted structure is probably essential for optimal functioning of any cell and some specific mechanism must exist that allows chromatin to remain so condensed.

In eukaryotic cells, the nucleus provides confinement for the chromatin chain. When a polymer is inserted into a capillary where the inner radius of the capillary (r_{cap}) is smaller than the size of the ideal polymer (r_p from Eq. (1); i.e., $r_{cap} < r_p$), the maximum radius of the polymer vertical to the axis of the capillary approaches the inner radius of the capillary and the polymer balls up into as many globules as required to pack the polymer into the capillary; a process known as reptation (Grosberg and Khokhlov, 1994; Thirumalai and Shi, 2017).

One might therefore imagine the chromatin fiber to behave as an ideal polymer chain embedded in a capillary, to some extent. To preserve $V_{nuc} \propto N$, one might assume that $\nu \sim 1/3$. However, the environment of the cell is not particularly ideal for DNA with its somewhat hydrophobic bases; however, its sugars are quite hydrophilic and the cellular environment is a vast heterogeneous mixture of many proteins, small molecules, water, ions and DNA/RNA. Hence, it would be better to imagine that the heterogeneous properties of the nucleus (without other considerations) should correspond to V_p, an ideal polymer, in the "polymer melt" consisting of all the different components of the cell's nucleus. Nevertheless, $\nu \sim 1/3$ would explain how a nuclear membrane could compress a range of genome sizes from 10 Mbp to 10 Gbp; i.e., it doesn't have to.

More recently, independent of the above findings, a linear relationship between nucleus volume and the amount of DNA ($V \propto N$) was observed with the first high throughput sequencing experiment using the Hi-C strategy (Lieberman-Aiden *et al.*, 2009). In that study, it was noticed that the likelihood of observing contact positions was inversely proportional to the genomic distance s between the contact points (wherein the number of contacts vanished as $1/s$ rather than $1/s^{3/2}$ (what would be expected if the chromatin behaved like an ideal polymer). The measuring of contact frequency is obviously different from the volume but whether the contacts had been random or specific; their frequency should correlate with the end-to-end distance, for a given genomic distance s. The frequency of observed contacts in any polymer should drop off as $1/s^\alpha$, where α is a constant. In the work of Lieberman-Aiden and co-workers, they proposed a fractal globule to explain this discrepancy from the behavior expected for an ideal polymer.

This again prompts the proposal that the chromatin is essentially a polymer in poor solvent and, therefore, $\nu \approx 1/3$; even though it is a heterogeneous polymer melt of diverse materials. However, studies of various models of chromatin tend to disagree with the simplistic collapsed polymer model (Rosa and Zimmer, 2014; Jerabek and Heermann, 2014). What has been discovered over the years, particularly from fluorescence in situ hybridization (FISH) experiments (Wang *et al.*, 2015; Brackley *et al.*, 2016; Giorgetti *et al.*, 2014; Bickmore and Van Steensel, 2013) and more recently experiments using Hi-C (Thurman *et al.*, 2012; Lieberman-Aiden *et al.*, 2009; Rao *et al.*, 2014; Ulianov *et al.*, 2016; Jin *et al.*, 2013; Dixon *et al.*, 2012) and ChIA-PET (Fullwood and Ruan, 2009; Tang *et al.*, 2015; Szalaj *et al.*, 2016; Li *et al.*, 2017; Ji *et al.*, 2016; Dowen *et al.*, 2014) strongly suggest that the genome in the cell is highly organized and different segments of the genome are compartmentalized for protein expression and maintenance. The main counter argument about organization of the genome has been that the diffusion rates for various factors could also be roughly 10^{-6} s; hence, the promotor is just generated in large enough quantities that it can diffuse to the desired gene in reasonable time. The rates are arguably fast enough. However, it suggests that a lot of factors are generated (indeed, wasted) and the cell machinery must remove these factors. With the introduction of CTCF binding and cohesin, this argument has become increasingly moot in recent years.

In terms of structure, there is nothing in the theory of a collapsed polymer (Grosberg and Khokhlov, 1994; Mckenzie, 1976) stating that the agents associated with this collapse simply cannot be the result of protein factors located at specific locations on the dsDNA chain (such as the CTCF and cohesin, discussed in the previous section). What the theory assumes is that there are mutual weak binding interactions between the residues of the polymer. In a simple collapsed polymer model, only random contacts are observed and the dynamics of the polymer chain would quickly lead to knots. (As pointed out in the Introduction, an earphone dropped in a bag quickly knots.) This is because there are just so many degrees of freedom in a flexible cord (Mckenzie, 1976). Because the cause of the globular structure is uniform attractive self-interaction forces between the polymer itself (Mckenzie, 1976), the enormous number of degrees of freedom mean that even if the contacts have a definite organization, natural selection pressures will have forced that organization to adopt a structure around the minimum free energy that resembles one of the many random configurations that could be generated with the same free energy for a set of heterogeneous contacts. Hence, whereas in principle ν represents a globular structure with uniform attractive contacts, the overall size of the organized structure with heterogeneous contacts can also follow an $r^3 \propto N$ rule. The nuclear envelope is reflecting just exactly the size of the nuclear material, the chromatin fiber. Hence, although more organized than a traditional globular structure (Grosberg and Khokhlov, 1994; Flory, 1953), the overall distribution of the contacts still resembles a collapsed polymer.

Optical Approach: FISH

Before the advent of high-throughput sequencing techniques that can directly quantify proximity ligation products in contact libraries, the commonly used method for determining nuclear organization and chromatin conformations was fluorescence in situ hybridization (FISH).

This optical technique was first developed in the '80s (Langer-Safer et al., 1982) and has been used from the beginning for various cytogenetic studies. In the earliest stages, it was used to identify the presence or absence of certain DNA sequences and therefore effectively to map the location of various genes on the chromosomes. This was done by attaching a probe that carried a fluorophore to a sequence of DNA using in situ hybridization and next observing the specific part of DNA to which probed attached using fluorescence (O'connor, 2008). Fluorescence microscopy can be used to track gene expression with probes binding to mRNA and be used to identify the relative distance between two positions on the chromosome, producing a 2D map. It can also be used to identify the overall structure of a region of the nucleus and it can be used to model dynamics.

FISH is a cytogenetic approach; i.e., a study of genetics in terms of the structure and function of the chromosomes, the structure, architecture and function of the genome. Since the observed information comes in the form of visible light (about 900 to 400 nm), this limits the direct resolution to the scale of tenths of micrometers in the best case (less than 200 nm in the xy-plane and less than 500 nm along z-axis (optical axis)). However, with knowledge of the DNA sequence and targeted placement of the fluorescence probes in the genome sequence the method permits to infer that distant regions on the linear sequence are in proximity of each other. We will not discuss super-resolution microscopy or EM microscopy, which are also rapidly advancing.

Specific examples of the interaction between the cohesin complex and the CTCFs at the level of x-ray crystallography have yet to be obtained. Therefore, we can only infer the interactions between these components by their regular association. What is more typically done is to use FISH to tag different loci of the chromatin and in this way the dynamic association of diverse parts of the chromatin chain in live cells can be visualized (Chuang et al., 2006; Tumbar and Belmont, 2001; Chubb et al., 2002; Muller et al., 2010). In such studies, it has been shown that the arrangement of the chromosomes can change from a largely peripheral location to a more internal region of the nucleus and that there is considerable compartmentalization in the architectural arrangement of the chromosomes within the nucleus; at least that there is a clear tendency for various sectors of the chromatin to appear in a similar distribution from cell to cell, though dependent upon the particular phase of the cell. It would seem reasonable that an efficient way to use the transcription production machinery is to place it in various central locations and bring the diverse parts of the chromatin to those loci rather than move the very complex spliceosomal machinery with a plethora of moving parts to specific locations in the cell. Efficient, shared usage of common but complex factory resources is used in designing assembly lines.

FISH Methods: Experimental Considerations

Whereas the broad 3D features of chromatin can be deduced by FISH experiments, to set up an experiments that identify the interactions within the chromatin on the length scales that might be considered to influence expression and transcription requires an ingenious choice of the right markers on the right genes. Only then FISH can be used to directly infer the loci between two locations in the genome and other information. One can imagine that this is very much a trial and error approach. It should be clear that this process of identifying these interactions requires considerable labor, and one is still largely limited to a resolution of maybe 0.5 μm. In the end, one will need to employ a wealth of computational tools to interpret the observed data. A different way to approach the interactions of chromatin is to map what areas of the chromatin are in contact with each other; i.e., showing the interactions between diverse parts of the chromatin chain. The number of instances of a given contact in a very large collection of the same type of cell in a similar phase of the cells is a measure of how likely a given interaction occurs. Once a 2D map is constructed, it is possible to infer the local 3D structure of the chromatin – or more correctly, the ensemble of structures that comprise the chromatin in the given region of the cell. At some point, the larger picture of the whole cell has too many degrees of freedom to be modeled solely with a mere contact map, but in terms of the many compartments within the cell nucleus, these length scales border closer to the minimum resolution of the 3D FISH experiments.

3C Methods

There are several experimental contact mapping techniques (Hakim and Misteli, 2012) based largely on the "chromosome conformation capture" (3C), which was developed by Dekker *et al.* (2002). The essence of the concept is that a population of cells is chemically fixed with formaldehyde. The formaldehyde forms cross-links between neighboring parts of the chromatin in the form of covalent bonds (Dekker *et al.*, 2002; Jackson, 1999; Orlando *et al.*, 1997). Restriction enzymes are used in the next step to digest the chromatin leaving small fragments that represent the parts of the chromatin that were in proximity of each other. After diluting the DNA, ligase is employed to link pairwise fragments (also sometimes called anchors) that are then amplified with PCR (Dekker *et al.*, 2002; Abou El Hassan and Bremner, 2009; Hagege *et al.*, 2007) and finally sequenced. During analysis, the sequences at similar locations of the genome are clustered according to the resolution; roughly on the order of the square of the number of cells analyzed. This basic protocol was often modified to improve the signal-to-noise ratio or to limit the contacts to ones resulting from mutual interactions of specific protein factors (like CTCF or Rpol2).

The term "anchor" is used to refer both to a uniquely identifiable sequence of DNA (Mckernan *et al.*, 2009) and to a pair of uniquely identifiable sequences that interact with each other in the form of loops (Fullwood and Ruan, 2009). The term "intra-ligated" indicates contacts (cross-links) that result from the coincidental interactions of the chromatin or local bending and compaction of the chromatin. Such interactions are not likely to serve any significant physiological purpose in terms of gene expression or regulation. Inter-ligated contacts are ones that result from long-distance interactions mediated by, for example, CTCFs and cohesin.

The chromosome conformation capture-on-chip (4C) technique was designed to help speed up the throughput and as a consequence to increase the resolution of the 3C (Lomvardas *et al.*, 2006; Simonis *et al.*, 2006; Wurtele and Chartrand, 2006; Zhao *et al.*, 2006). As with 3C, the interactions are cross-linked and selected restriction enzymes are used to digest the chromatin away from the contacts of interest. The method relies on generating circular ligation products between a selected fragment and the rest of the genome. The circular products are quantified using microarrays or sequencing. The advantage is that one can deduce all the interactions between a given, specified location and the rest of the genome. The chromosome conformation capture carbon copy (5C) is an extension of 3C in that, rather than the ligation products from specific restriction fragments (like 4C), all the restriction fragments are used and amplified in a microarray. The "carbon copy" is because it essentially amplifies the 3C data. The advantage is that rather than detecting the interactions of a specific fragment, it measures the relative proportions of the amplified product.

Hi-C Methods

The Hi-C method is a high resolution genomic approach that can capture chromatin loops, domains (in this case called topologically associated domains TADs), and many local features related to unspecific long-range protein-mediated chromatin interactions. The TADs have borders strongly enriched with the CTCF binding sites (Nichols and Corces, 2015; Alipour and Marko, 2012) and their sizes are typically found over length scales of 1 to 3 Mbps.

The Hi-C method essentially adds high-throughput sequencing to help build libraries that quantify the proximity ligation products. The method is able to cover the entire genome at a at a certain size resolution. Like 3C, the population of cells are fixed with formaldehyde and digested with different types of restriction enzymes. At this stage, the overhangs in the restriction fragments

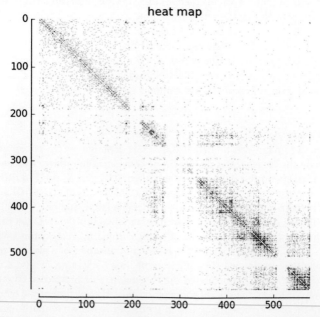

Fig. 8   Example of ChIA-PET data of the X chromosome from 150882227 to 153764804 in human lymphoblastoid cells (GM12878).

are filled with Klenow enzyme and various deoxynucleoside triphosphates (dNTPs) including biotin-14-dCTP and the products ligated and purified. The libraries are sonicated and subjected to pulldown with streptavidin-coated beads.

ChIA-PET Methods

The "chromatin interaction analysis by paired-end tag sequencing" (ChIA-PET) is an extension of Hi-C in the sense that ChIA-PET selects for ligation products of interactions of specific proteins; typically Rpol2 or CTCF (Fullwood et al., 2009a; Handoko et al., 2011; Tang et al., 2015; Sandhu et al., 2012). The paired-end tag sequence (PET) refers to a short DNA sequence that is unique enough to identify a particular segment of the genome, a sequence at least 13 bp long (Fullwood et al., 2009b). Rather than applying restriction enzymes, chromatin fragments are created by sonicating the cells and selection for specific proteins is achieved by applying chromatin immunopreciptation (ChIP); specifically targeting proteins like CTCF that are bound to the chromatin and retained in the cross-linked fragments. To the ends of the co-immunopreciptated DNA segments, biotinylated DNA linkers are added and the DNA fragments are ligated together. This is the paired end tag (PET). Purification is carried out using streptavidin beads and paired-end sequencing. The ChIA-PET experiments yield pair-end-tag (PET) data; ligation products of different parts of the chromatin chain. Each half of the ChIA-PET products is next mapped to the reference genome. An example of this method for a part of the X chromosome from 150,882,227 to 153,764,804 from human lymphoblastoid cells (GM12878) is shown in **Fig. 8**.

Results of ChIA-PET/Hi-C Methods

There are several important observations to consider that result from these genome mapping experiments. First, we compare the general outcomes of 3DFISH and the more recent ligation products. Second, we examine some of the larger differences found in the ligation techniques.

3DFISH vs Ligation Products of ChIA-PET and Hi-C Methods

The ligation products of both Hi-C and ChIA-PET experiments yield both intra-ligated and inter-ligated products. With Hi-C data, any contact can be either intra-ligature or inter-ligated, irrespective of whether the interactions involve CTCF or other target proteins. On the other hand, for ChIA-PET data, the intra-ligation is limited to a range of about 8 kbp and the remaining data can all be treated as inter-ligation pairs of sequence fragments (Szalaj et al., 2016). This distinction can be made because ChIA-PET data involves the targeting of ChIP products: the CTCF proteins located at chromatin binding sites or other well defined markers such as proteins associated with Rpol2. Since there is a significant hyper-exponential shift in the number of counts for data less than 8 kbp and the highly flexible chromatin chain is likely to cross within this range, the majority of contributions within 8 kbp clearly must be coming from self-ligated products. On the other hand, beyond this 8 kbp range, the behavior is quite different. The paired end tag (PET) counts or anchors can show a drastic increase in counts at specific locations on the chromatin chain. This is certainly characteristic of the expected behavior of CTCF protein-protein binding sites and similar observations are obtained from Rpol2 sites (a collection of protein subunits) using ChIA-PET. Therefore, whereas we cannot completely rule out the possibility of *some* small fraction of self-ligated products within the data beyond 8 kbp, such contributions are certainly small and cannot contribute significantly to the statistical weight even if given they are present.

After filtering, both Hi-C data and the singleton ChIA-PET data appear to be rather similar overall, characterized by high Pearson correlation coefficient between both heatmaps (Tang et al., 2015; Szalaj et al., 2016). With Hi-C, there is a mixture of ligation between diverse parts of the chromatin chain and self-ligation interactions. Therefore it is more difficult to untangle self-ligation for Hi-C data compared to ChIA-PET data.

Although the positions of the CTCF dimers is accurate to roughly the length scale of the nucleosome (*roughly* between 150 and 200 bps), practical considerations currently set limits on the amount of statistics that can be acquired; e.g., limits on data acquisition, storage, and processing, and limits due to time and budget constraints. Currently, these limits are perhaps on the order of 1 kbp (Tang et al., 2015; Rao et al., 2014) – at least five times larger than the approximate size of a nucleosome.

There remain some questions about the differences of FISH and the various 3C methods. One important issue is coverage. It remains unclear if the cross-links created during fixing the cells are uniform throughout the chromatin sample or if they favor the boundaries of the nucleus; for example (Gavrilov et al., 2013; Ulianov et al., 2016). So specificity and stability may be issues.

Important Considerations of Ligation 2D Mapping

A major limiting factor in ChIA-PET and Hi-C measurements is statistics. As a practical example, there are some 30 million nucleosomes (nuc) in the human genome ($2 \times 3 \times 10^9$ [bp]/200 [bp/nuc]). In a recent ChIA-PET measurement (Tang et al., 2015), roughly 30.8 million reads were obtained from roughly 100 million human lymphoblastoid cells (GM12878). Hence, there would be roughly 1 count per nucleosome (on average). However, if the data is binned into 5 kbp segments (about 25 nucleosomes per bead: 5000 [bp]/200 [bp/nuc]), then this leads to an average coverage of about 25 counts per bin. For ChIA-PET data exceeding the 8 kbp boundary, the CTCF or Rpol2 only yield inter-ligated products. In other words, the short range intra-ligated contacts that generally result from local chromatin looping, would be discarded, and the random long-distance spurious intra-ligated contacts are eliminated because ChIA-PET, as already mentioned, selects only specific interactions such as CTCF and Rpol2. Clearly, single counts in this type of measurement can be spurious. This is the advantage of a thermodynamic model involving statistical weights; the most significant interaction show up constantly and the weak interactions contribute minimally to the overall structure.

With these considerations, for typical sampling of some 100 M reads, it is probably more instructive to focus on 5 kbp resolution data (5 kbp beads) because it represented the best compromise between obtaining sufficient details of the CTCF binding sites without rarefying the intensity of individual pair contacts in the heat maps to such an extent as to render the noise from potential spurious contacts excessive. In terms of the polymer behavior of the chromatin, a 5 kbp resolution means that each 5 kbp genomic segment is represented as single bead. At this length scale, the detailed conformations of the individual nucleosomes ($\sim$200 bp) and mutual interactions within each segment are blurred out rendering a segment that can be treated as though it had the flexibility of a single bead on a string in the polymer chain model. Finally, a 5 kbp resolution permits detection of interligation products just at the boundary of the 8 kbp window. Unlike the data in a range less than 8 kbp, the number of PET counts (from 31 M reads) appears to be less than four for nearest interactions at 5 kbp resolution but anchors describing very distant contact points can possess hundreds of PET counts.

Measured in terms of the genomic distance (s) between bead i and bead j (for $j > i$), where $s = (j-i+1)$, the likelihood of finding contact points on the chromatin chain between bead i and bead j (the contact probability) for Hi-C and ChIA-PET data tends to decrease as a function of the inverse of the genomic distance

$$p_c(s) \propto s^{-\alpha} \tag{4}$$

where α is a dimensionless scaling parameter. This is also known as the pair interaction frequency (PIF).

Presently, it appears that estimates of α have varied drastically (Barbieri et al., 2013a,b; Chiariello et al., 2014) depending on the organism (e.g., drosophila, 0.7 for active regions and 0.85 for inactive; yeast, 1.5), cell type (e.g., human embryonic stem cells, 1.6, interphase lymphoblast cell, 1.1, and metaphase HeLa cells, 0.5), and theoretical models (e.g., (Meluzzi and Arya, 2013) indicates that unrestrained fits yielded 2.23, and restrained fits 3–8). These values for the α-parameter are estimated using Hi-C data. In the latter examples, the large values for α may reflect the harmonic potential that was used. Nevertheless, predictions of $p(s)$ are not simple in general. If we presume the polymer characteristics of the chromatin as discussed in the previous section, then α should be assumed to be around 1/3 if measured in one dimension, and 1 if the probability is measured in volume. Since s (Lieberman-Aiden et al) is specifically genomic distance (the number of base pairs or the linear distance covered by said base pairs), this is probably a source of confusion in the literature.

Computational Models of the Mechanism

There are a variety of computational models that have emerged in recent years based on polymer science methods that try to explain the behavior of chromatin and satisfy the volume probability measure of $\alpha =1$. A comprehensive review of all the approaches is beyond the scope of this work. Here, we will focus on two such models that have become most popular in recent years: the loop extrusion model and the strings and binders model. We will not discuss molecular dynamics and Monte Carlo simulation techniques – such as the random walk giant-loop model, Multi-loop subcompartment (MLS) model (Munkel et al., 1999), Entropy-driven chromosome organization (Rosa et al., 2010) and many, many others – because these topics are covered in detail in a recent review (Rosa and Zimmer, 2014) and follow the general concepts of polymers outlined above.

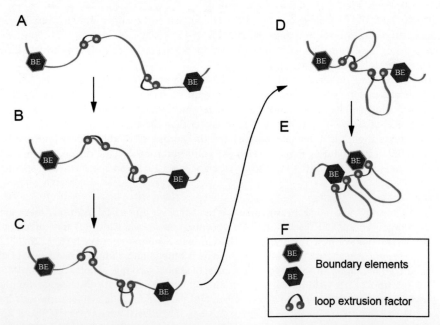

Fig. 9 The loop extrusion (LE) model. In this model, the loop extrusion factor (F) binds to the polymer chain in A, and progressively ratchets along the polymer chain one unit at a time (perhaps nucleosome by nucleosome) gradually creating a larger loop (B, C, D and E). Eventually, the loop cannot grow any larger because the loop extrusion factor collides with a boundary element (labeled BE) or it runs into an additional loop extrusion factor as shown in E.

Loop Extrusion Model

The loop extrusion (LE) model is an example how to understand the process of loop formation in chromatin features. The model was introduced with the notion that it behaves conceptually like the long-distance intracellular delivery systems like kinesin and dynein motor proteins that transport their cargoes along microtubule tracks (Hirokawa and Takemura, 2005; Vallee *et al.*, 2004).

The LE model was originally proposed in Marko and coworkers (Alipour and Marko, 2012) and further developed by Mirny and coworkers (Dekker and Mirny, 2016; Fudenberg *et al.*, 2016) and is depicted in **Fig. 9**. The chromatin is represented as a polymer consisting of beads separated by some specific distance; e.g., 1 kbp, 5 kbp, etc. Initially, loop extruding factors (LEF) – probably cohesin – binds to a region on the chromatin comprising neighboring beads or, possibly, a small loop formed by a few beads. Although the role of LEF was not assigned to any particular protein, it is now widely accepted that this role is realized by cohesion, a complex of several proteins (Ji *et al.*, 2016; Hirano, 2015; Nasmyth and Haering, 2009; De Wit *et al.*, 2015). Recent work also suggests that the cohesin loading complex localizes near the transcription initiation sites, thus it is plausible that the extrusion process begins in the vicinity of actively transcribed genes (Busslinger *et al.*, 2017). The LEF then begins to traverse along the chromatin with a biased rate of hopping between adjacent beads, where (for illustration) the forward rate $r+$ on the right side hops faster than the reverse rate ($r-$) on the left hand side. (Here we presume the beginning of the chain to be on the left). This motor behavior is known to occur with transport of cargo on microtubules (Vallee *et al.*, 2004; Hirokawa and Takemura, 2005). The extrusion process stops when cohesin reaches the insulator (also called boundary element) or another cohesion ring traversing in opposite direction (Fudenberg *et al.*, 2016). How cohesion slides along DNA is still an open question; however, it was recently postulated that this role can be fulfilled by RPol2 (Davidson *et al.*, 2016).

Strings and Binders Switch Model

The strings and binders switch (SBS) model (Barbieri *et al.*, 2012, 2013a, 2013b; Chiariello *et al.*, 2016; Bianco *et al.*, 2016) is worked out from the concepts of chemical equilibrium and the polymer physics of many body interactions containing various components. In its simplest form shown in **Fig. 10**, one imagines a polymer chain composed of beads (green and maroon circles)

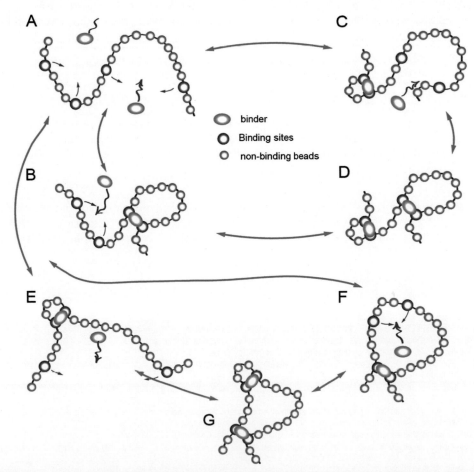

Fig. 10 The strings and binder switch (SBS) model. The green beads are effectively non-interacting, the maroon beads mark the binding sites (a CTCF-like binding sites), and the blue beads indicate the binder (cohesin-like unit). The unfolded chain is shown in A. There are two general pathways; ABCD and AEFG. These pathways are relatively mutually exclusive; however, B, C, E and F are all potential initial directions for the SBS to go and they exist in competition with one another in thermodynamic equilibrium.

where the green beads are effectively non-interacting elements and the maroon beads tend to bind with the mobile factor (represented as blue beads). The analogy is that the maroon beads would resemble something like the CTCF binding sites and cohesin, and the blue beads would represent the CTCF/cohesin binder. The system is in equilibrium, meaning that there is dynamic exchange between all the possible configurations shown in the Figure. In **Fig. 10**, the pathways of ABCD form one way that the strings and binders can combine; however, additional states might be also present like in pathway AEFG. Note that the arrows indicate that everything is in equilibrium. Each binding configuration resembles a switch that can be turned on and off for a given pattern.

One clear difference between the strings and binder model and the loop extrusion model is that the loop extrusion model presumes that the cohesin-like system forms an ATP-driven molecular machine that begins with binding between adjacent beads and mechanically extends in both directions until it encounters a CTCF-like obstruction. Hence, in its most rigid form, the structure in **Fig. 10(G)** would not form. However, in the earliest version of this model (Alipour and Marko, 2012), it was assumed that somehow the machine could skip positions. Since the precise hopping mechanism is not established, there is no reason presently to assume any restrictions per se. On the other hand, because there are many more possibilities for the dynamics of the chromatin contacts in the SBS model, the model often requires that the binding sites and binders are distinguishable and relatively mutually exclusive.

Summary

In this review, we have briefly indicated the critical features of chromatin structure, the interactions and relationships between enhancers and inhibiters, the role of CTCF proteins in forming the overall architecture of the chromatin, the volume of the nucleus using various experimental techniques, the essential methods of measuring chromatin structure and finally some examples of computational models. In the study of genomics, our understanding of how expression is regulated and how the regalia of spliceosome proteins, transcription factors, promotors and RNA polymerase II interactions will be greatly enhanced as our knowledge of chromatin binding structures and their control and regulation grows. Ultimately, it is increasingly clear that the genome is not an amorphous blob of DNA and histones randomly squeezed into a nucleus; rather, it is a well-organized system that, while possibly lacking poster-child-like publishable shapes, still exhibits a deep and highly organized meta-structural landscape.

Acknowledgments

This work was supported by grants from the Polish National Science Center (2014/15/B/ST6/05082, 2013/09/B/NZ2/00121); Foundation for Polish Science (TEAM to DP); and by grant 1U54DK107967-01 "Nucleome Positioning System for Spatiotemporal Genome Organization and Regulation" within 4DNucleome NIH program. We appreciate the assistance of Michal Kadlof in reviewing the text.

See also: Biomolecular Structures: Prediction, Identification and Analyses. Genome Informatics. Learning Chromatin Interaction Using Hi-C Datasets. Natural Language Processing Approaches in Bioinformatics. Nucleosome Positioning. Protein Post-Translational Modification Prediction. Protein Structural Bioinformatics: An Overview. Protein-DNA Interactions. Quantitative Immunology by Data Analysis Using Mathematical Models

References

Abou El Hassan, M., Bremner, R., 2009. A rapid simple approach to quantify chromosome conformation capture. Nucleic Acids Res. 37, e35.

Alipour, E., Marko, J.F., 2012. Self-organization of domain structures by DNA-loop-extruding enzymes. Nucleic Acids Res. 40, 11202–11212.

Auffinger, P., Grover, N., Westhof, E., 2011. Metal ion binding to RNA. Met. Ions Life Sci. 9, 1–35.

Baetcke, K.P., Sparrow, A.H., Nauman, C.H., Schwemmer, S.S., 1967. The relationship of DNA content to nuclear and chromosome volumes and to radiosensitivity (LD50). Proc. Natl. Acad. Sci. USA 58, 533–540.

Barbieri, M., Chotalia, M., Fraser, J., et al., 2012. Complexity of chromatin folding is captured by the strings and binders switch model. Proc. Natl. Acad. Sci. USA 109, 16173–16178.

Barbieri, M., Fraser, J., Lavitas, L.M., et al., 2013a. A polymer model explains the complexity of large-scale chromatin folding. Nucleus 4, 267–273.

Barbieri, M., Scialdone, A., Piccolo, A., et al., 2013b. Polymer models of chromatin organization. Front. Genet. 4, 113.

Bascom, G., Schlick, T., 2017. Linking chromatin fibers to gene folding by hierarchical looping. Biophys. J. 112, 434–445.

Beagan, J.A., Gilgenast, T.G., Kim, J., et al., 2016. Local genome topology can exhibit an incompletely rewired 3D-folding state during somatic cell reprogramming. Cell Stem Cell 18, 611–624.

Beaulieu, J.M., Leitch, I.J., Patel, S., Pendharkar, A., Knight, C.A., 2008. Genome size is a strong predictor of cell size and stomatal density in angiosperms. New Phytol. 179, 975–986.

Bednar, J., Horowitz, R.A., Grigoryev, S.A., et al., 1998. Nucleosomes, linker DNA, and linker histone form a unique structural motif that directs the higher-order folding and compaction of chromatin. Proc. Natl. Acad. Sci. USA 95, 14173–14178.

Bianco, S., Chiariello, A.M., Annunziatella, C., Esposito, A., Nicodemi, M., 2016. Polymer physics of the large-scale structure of chromatin. Methods Mol. Biol. 1480, 201–206.

Bickmore, W.A., Van Steensel, B., 2013. Genome architecture: Domain organization of interphase chromosomes. Cell 152, 1270–1284.

Bocharova, T.N., Smirnova, E.A., Volodin, A.A., 2012. Linker histone H1 stimulates DNA strand exchange between short oligonucleotides retaining high sensitivity to heterology. Biopolymers 97, 229–239.

Brackley, C.A., Brown, J.M., Waithe, D., et al., 2016. Predicting the three-dimensional folding of cis-regulatory regions in mammalian genomes using bioinformatic data and polymer models. Genome Biol. 17, 59.

Brogaard, K., Xi, L., Wang, J.P., Widom, J., 2012. A map of nucleosome positions in yeast at base-pair resolution. Nature 486, 496–501.

Buenrostro, J.D., Wu, B., Litzenburger, U.M., et al., 2015. Single-cell chromatin accessibility reveals principles of regulatory variation. Nature 523, 486–490.

Busslinger, G.A., Stocsits, R.R., Van Der Lelij, P., et al., 2017. Cohesin is positioned in mammalian genomes by transcription, CTCF and Wapl. Nature 544, 503–507.

Bustin, M., Catez, F., Lim, J.H., 2005. The dynamics of histone H1 function in chromatin. Mol. Cell 17, 617–620.

Bystricky, K., Heun, P., Gehlen, L., Langowski, J., Gasser, S.M., 2004. Long-range compaction and flexibility of interphase chromatin in budding yeast analyzed by high-resolution imaging techniques. Proc. Natl. Acad. Sci. USA 101, 16495–16500.

Cheung, P., Tanner, K.G., Cheung, W.L., et al., 2000. Synergistic coupling of histone H3 phosphorylation and acetylation in response to epidermal growth factor stimulation. Mol. Cell 5, 905–915.

Chiariello, A., Bianco, S., Piccolo, A., et al., 2014. Polymer Models of the Organization of Chromosomes in the Nucleus of Cells. Napoli, Italy: Complesso Universitario di Monte S. Angelo.

Chiariello, A.M., Annunziatella, C., Bianco, S., Esposito, A., Nicodemi, M., 2016. Polymer physics of chromosome large-scale 3D organisation. Sci. Rep. 6, 29775.

Chuang, C.H., Carpenter, A.E., Fuchsova, B., et al., 2006. Long-range directional movement of an interphase chromosome site. Curr. Biol. 16, 825–831.

Chubb, J.R., Boyle, S., Perry, P., Bickmore, W.A., 2002. Chromatin motion is constrained by association with nuclear compartments in human cells. Curr. Biol. 12, 439–445.

Davidson, I.F., Goetz, D., Zaczek, M.P., et al., 2016. Rapid movement and transcriptional re-localization of human cohesin on DNA. EMBO J. 35, 2671–2685. doi:10.15252/embj.201695402.

Davie, J.R., 1997. Nuclear matrix, dynamic histone acetylation and transcriptionally active chromatin. Mol. Biol. Rep. 24, 197–207.

Dawson, W., Kawai, G., 2015. A new entropy model for RNA: Part V, Incorporating the Flory-Huggins model in structure prediction and folding. J. Nucl. Acids Investig. 6, 1. DOI: 10.4081/jnai.2015.2657.

Dawson, W., Yamamoto, K., Shimizu, K., Kawai, G., 2013. A new entropy model for RNA: Part II. Persistence-related entropic contributions to RNA secondary structure free energy calculations. J. Nucl. Acids Investig. 4, e2.

Dekker, J., Mirny, L., 2016. The 3D genome as moderator of chromosomal communication. Cell 164, 1110–1121.

Dekker, J., Rippe, K., Dekker, M., Kleckner, N., 2002. Capturing chromosome conformation. Science 295, 1306–1311.

Delano, W.L., 2004. Pymol. San Carlos, California: Now Distributed by Schrodinger, 101 SW Main Street, Suite 1300, Portland, OR 97204. Available at: http://www.pymol.org.

De Wit, E., Vos, E.S., Holwerda, S.J., et al., 2015. CTCF binding polarity determines chromatin looping. Mol. Cell 60, 676–684.

Dixon, J.R., Selvaraj, S., Yue, F., et al., 2012. Topological domains in mammalian genomes identified by analysis of chromatin interactions. Nature 485, 376–380.

Dowen, J.M., Fan, Z.P., Hnisz, D., et al., 2014. Control of cell identity genes occurs in insulated neighborhoods in mammalian chromosomes. Cell 159, 374–387.

Doyle, B., Fudenberg, G., Imakaev, M., Mirny, L.A., 2014. Chromatin loops as allosteric modulators of enhancer-promoter interactions. PLOS Comput. Biol. 10, e1003867.

Dunn, K.L., Zhao, H., Davie, J.R., 2003. The insulator binding protein CTCF associates with the nuclear matrix. Exp. Cell Res. 288, 218–223.

Finch, J.T., Klug, A., 1976. Solenoidal model for superstructure in chromatin. Proc. Natl. Acad. Sci. USA 73, 1897–1901.

Flory, P.J., 1953. Principles of Polymer Chemistry. Ithaca: Cornell University Press.

Flory, P.J., 1969. Statistical Mechanics of Chain Molecules. New York: Wiley.

Fraser, J., Williamson, I., Bickmore, W.A., Dostie, J., 2015. An overview of genome organization and how we got there: From FISH to Hi-C. Microbiol. Mol. Biol. Rev. 79, 347–372.

Fudenberg, G., Imakaev, M., Lu, C., et al., 2016. Formation of chromosomal domains by loop extrusion. Cell Rep. 15, 2038–2049.

Fullwood, M.J., Liu, M.H., Pan, Y.F., et al., 2009a. An oestrogen-receptor-alpha-bound human chromatin interactome. Nature 462, 58–64.

Fullwood, M.J., Wei, C.L., Liu, E.T., Ruan, Y., 2009b. Next-generation DNA sequencing of paired-end tags (PET) for transcriptome and genome analyses. Genome Res. 19, 521–532.

Fullwood, M.J., Ruan, Y.J., 2009. ChIP-based methods for the identification of long-range chromatin interactions. J. Cell Biochem. 107, 30–39.

Gavrilov, A.A., Gushchanskaya, E.S., Strelkova, O., et al., 2013. Disclosure of a structural milieu for the proximity ligation reveals the elusive nature of an active chromatin hub. Nucleic Acids Res. 41, 3563–3575.

Giorgetti, L., Galupa, R., Nora, E.P., et al., 2014. Predictive polymer modeling reveals coupled fluctuations in chromosome conformation and transcription. Cell 157, 950–963.

Gregory, T.R., 2005. The C-value enigma in plants and animals: A review of parallels and an appeal for partnership. Ann. Bot. 95, 133–146.

Gregory, T.R., 2017. Animal genome size database: Cell size database. In: Gregory, T.R. (Ed.), Available at: http://www.genomesize.com/cellsize/.

Grosberg, A.Y., Khokhlov, A.R., 1994. Statistical Physics of Macromolecules. New York: AIP Press.

Guo, Y., Xu, Q., Canzio, D., et al., 2015. CRISPR inversion of CTCF sites alters genome topology and enhancer/promoter function. Cell 162, 900–910.

Hagege, H., Klous, P., Braem, C., et al., 2007. Quantitative analysis of chromosome conformation capture assays (3C-qPCR). Nat. Protoc. 2, 1722–1733.

Hagerman, P.J., 1988. Flexibility of DNA. Annu. Rev. Biophys. Biophys. Chem. 17, 265–286.

Hakim, O., Misteli, T., 2012. SnapShot: Chromosome confirmation capture. Cell 148 (1068), e1–e2.

Handoko, L., Xu, H., Li, G., et al., 2011. CTCF-mediated functional chromatin interactome in pluripotent cells. Nat. Genet. 43, 630–638.

Hansen, J.C., 2002. Conformational dynamics of the chromatin fiber in solution: Determinants, mechanisms, and functions. Annu. Rev. Biophys. Biomol. Struct. 31, 361–392.

He, S., Dunn, K.L., Espino, P.S., et al., 2008. Chromatin organization and nuclear microenvironments in cancer cells. J. Cell Biochem. 104, 2004–2015.

Hirano, T., 2015. Chromosome dynamics during mitosis. Cold Spring Harb. Perspect. Biol. 7.

Hirokawa, N., Takemura, R., 2005. Molecular motors and mechanisms of directional transport in neurons. Nat. Rev. Neurosci. 6, 201–214.

Hodgson, J.G., Sharafi, M., Jalili, A., et al., 2010. Stomatal vs. genome size in angiosperms: The somatic tail wagging the genomic dog? Ann. Bot. 105, 573–584.

Jackson, V., 1999. Formaldehyde cross-linking for studying nucleosomal dynamics. Methods 17, 125–139.

Jahan, S., Xu, W., He, S., et al., 2016. The chicken erythrocyte epigenome. Epigenet. Chromatin 9, 19.

Jerabek, H., Heermann, D.W., 2014. How chromatin looping and nuclear envelope attachment affect genome organization in eukaryotic cell nuclei. Int. Rev. Cell Mol. Biol. 307, 351–381.

Jin, F., Li, Y., Dixon, J.R., et al., 2013. A high-resolution map of the three-dimensional chromatin interactome in human cells. Nature 503, 290–294.

Ji, X., Dadon, D.B., Powell, B.E., et al., 2016. 3D chromosome regulatory landscape of human pluripotent cells. Cell Stem Cell 18, 262–275.

Joti, Y., Hikima, T., Nishino, Y., et al., 2012. Chromosomes without a 30-nm chromatin fiber. Nucleus 3, 404–410.

Jovtchev, G., Schubert, V., Meister, A., Barow, M., Schubert, I., 2006. Nuclear DNA content and nuclear and cell volume are positively correlated in angiosperms. Cytogenet. Genome Res. 114, 77–82.

Kagey, M.H., Newman, J.J., Bilodeau, S., et al., 2010. Mediator and cohesin connect gene expression and chromatin architecture. Nature, 467, pp. 430–435. doi:10.1038/nature09380.

Kalashnikova, A.A., Rogge, R.A., Hansen, J.C., 2016. Linker histone H1 and protein-protein interactions. Biochim. Biophys. Acta 1859, 455–461.

Kouzarides, T., 2002. Histone methylation in transcriptional control. Curr. Opin. Genet. Dev. 12, 198–209.

Kratky, O., Porod, G., 1949. Röntgenuntersuchung gelöster Fadenmoleküle. Rec. Trav. Chim. Pays-Bas. 68, 1106–1123.

Kudo, N.R., Anger, M., Peters, A.H., *et al.*, 2009. Role of cleavage by separase of the Rec8 kleisin subunit of cohesin during mammalian meiosis I. J. Cell Sci. 122, 2686–2698.

Kuhn, W., 1936. Beziehungen zwischen Molekulgroshe, statistischer Molekulgestalt und elastischen Eigenschaften hochpolymerer Stoffe (Relations between molecular size, statistical molecular shape and elastic properties of high polymers). Kolloidzeitschrift 76, 258.

Langer-Safer, P.R., Levine, M., Ward, D.C., 1982. Immunological method for mapping genes on Drosophila polytene chromosomes. Proc. Natl. Acad. Sci. USA 79, 4381–4385.

Lewis, A., Murrell, A., 2004. Genomic imprinting: CTCF protects the boundaries. Curr. Biol. 14, R284–R286.

Lieberman-Aiden, E., Van Berkum, N.L., Williams, L., *et al.*, 2009. Comprehensive mapping of long-range interactions reveals folding principles of the human genome. Science 326, 289–293.

Li, X., Luo, O.J., Wang, P., *et al.*, 2017. Long-read ChIA-PET for base-pair-resolution mapping of haplotype-specific chromatin interactions. Nat. Protoc. 12, 899–915.

Lomvardas, S., Barnea, G., Pisapia, D.J., *et al.*, 2006. Interchromosomal interactions and olfactory receptor choice. Cell 126, 403–413.

Macpherson, M.J., Sadowski, P.D., 2010. The CTCF insulator protein forms an unusual DNA structure. BMC Mol. Biol. 11, 101.

Maeshima, K., Hihara, S., Eltsov, M., 2010. Chromatin structure: Does the 30-nm fibre exist in vivo? Curr. Opin. Cell Biol. 22, 291–297.

Maeshima, K., Imai, R., Hikima, T., Joti, Y., 2014. Chromatin structure revealed by X-ray scattering analysis and computational modeling. Methods 70, 154–161.

Maeshima, K., Rogge, R., Tamura, S., *et al.*, 2016. Nucleosomal arrays self-assemble into supramolecular globular structures lacking 30-nm fibers. EMBO J. 35, 1115–1132.

Mckenzie, D.S., 1976. Polymers and scaling. Phys. Rep. 27C, 35–88.

Mckernan, K.J., Peckham, H.E., Costa, G.L., *et al.*, 2009. Sequence and structural variation in a human genome uncovered by short-read, massively parallel ligation sequencing using two-base encoding. Genome Res. 19, 1527–1541.

Meluzzi, D., Arya, G., 2013. Recovering ensembles of chromatin conformations from contact probabilities. Nucleic Acids Res. 41, 63–75.

Merkenschlager, M., Odom, D.T., 2013. CTCF and cohesin: Linking gene regulatory elements with their targets. Cell 152, 1285–1297.

Milo, R., Jorgensen, P., Moran, U., Weber, G., Springer, M., 2010. BioNumbers – The database of key numbers in molecular and cell biology. Nucleic Acids Res. 38, D750–D753.

Mirsky, A.E., Ris, H., 1951. The desoxyribonucleic acid content of animal cells and its evolutionary significance. J. Gen. Physiol. 34, 451–462.

Muller, I., Boyle, S., Singer, R.H., Bickmore, W.A., Chubb, J.R., 2010. Stable morphology, but dynamic internal reorganisation, of interphase human chromosomes in living cells. PLOS ONE 5, e11560.

Mumbach, M.R., Rubin, A.J., Flynn, R.A., *et al.*, 2016. HiChIP: Efficient and sensitive analysis of protein-directed genome architecture. Nat. Methods 13, 919–922.

Munkel, C., Eils, R., Dietzel, S., *et al.*, 1999. Compartmentalization of interphase chromosomes observed in simulation and experiment. J. Mol. Biol. 285, 1053–1065.

Nacheva, G.A., Guschin, D.Y., Preobrazhenskaya, O.V., *et al.*, 1989. Change in the pattern of histone binding to DNA upon transcriptional activation. Cell 58, 27–36.

Nasmyth, K., Haering, C.H., 2009. Cohesin: Its roles and mechanisms. Annu. Rev. Genet. 43, 525–558.

Nathan, D., Sterner, D.E., Berger, S.L., 2003. Histone modifications: Now summoning sumoylation. Proc. Natl. Acad. Sci. USA 100, 13118–13120.

Nauman, A.F., 1979. Plant Nuclear Data Handbook: Radiosensitivity in Plants. Available at: https://www.osti.gov/scitech/servlets/purl/5369196.

Nichols, M.H., Corces, V.G., 2015. A CTCF code for 3D genome architecture. Cell 162, 703–705.

Nishino, Y., Eltsov, M., Joti, Y., *et al.*, 2012. Human mitotic chromosomes consist predominantly of irregularly folded nucleosome fibres without a 30-nm chromatin structure. EMBO J. 31, 1644–1653.

O'connor, C., 2008. Fluorescence in situ hybridization (FISH). Nat. Educ. 1, 171.

Ohta, S., Wood, L., Bukowski-Wills, J.C., Rappsilber, J., Earnshaw, W.C., 2011. Building mitotic chromosomes. Curr. Opin. Cell Biol. 23, 114–121.

Olson, W.K., Zhurkin, V.B., 2000. Modeling DNA deformations. Curr. Opin. Struct. Biol. 10, 286–297.

Orlando, V., Strutt, H., Paro, R., 1997. Analysis of chromatin structure by in vivo formaldehyde cross-linking. Methods 11, 205–214.

Ou, H.D., Phan, S., Deerinck, T.J., *et al.*, 2017. ChromEMT: Visualizing 3D chromatin structure and compaction in interphase and mitotic cells. Science. 357.

Price, H.J., Sparrow, A.H., Nauman, A.F., 1973. Correlations between nuclear volume, cell volume and DNA content in meristematic cells of herbaceous angiosperms. Experientia 29, 1028–1029.

Pugacheva, E.M., Rivero-Hinojosa, S., Espinoza, C.A., *et al.*, 2015. Comparative analyses of CTCF and BORIS occupancies uncover two distinct classes of CTCF binding genomic regions. Genome Biology, 16, p. 161. doi:10.1186/s13059-015-0736-8.

Rao, S.S., Huntley, M.H., Durand, N.C., *et al.*, 2014. A 3D map of the human genome at kilobase resolution reveals principles of chromatin looping. Cell 159, 1665–1680.

Renda, M., Baglivo, I., Burgess-Beusse, B., *et al.*, 2007. Critical DNA binding interactions of the insulator protein CTCF: A small number of zinc fingers mediate strong binding, and a single finger-DNA interaction controls binding at imprinted loci. J. Biol. Chem. 282, 33336–33345.

Rosa, A., Becker, N.B., Everaers, R., 2010. Looping probabilities in model interphase chromosomes. Biophys. J. 98, 2410–2419.

Rosa, A., Zimmer, C., 2014. Computational models of large-scale genome architecture. Int. Rev. Cell Mol. Biol. 307, 275–349.

Routh, A., Sandin, S., Rhodes, D., 2008. Nucleosome repeat length and linker histone stoichiometry determine chromatin fiber structure. Proc. Natl. Acad. Sci. USA 105, 8872–8877.

Rychkov, G.N., Ilatovskiy, A.V., Nazarov, I.B., *et al.*, 2017. Partially assembled nucleosome structures at atomic detail. Biophys. J. 112, 460–472.

Sandhu, K.S., Li, G., Poh, H.M., *et al.*, 2012. Large-scale functional organization of long-range chromatin interaction networks. Cell Rep. 2, 1207–1219.

Sapojnikova, N., Thorne, A., Myers, F., Staynov, D., Crane-Robinson, C., 2009. The chromatin of active genes is not in a permanently open conformation. J. Mol. Biol. 386, 290–299.

Schuster-Bockler, B., Lehner, B., 2012. Chromatin organization is a major influence on regional mutation rates in human cancer cells. Nature 488, 504–507.

Shermoen, A.W., O'farrell, P.H., 1991. Progression of the cell cycle through mitosis leads to abortion of nascent transcripts. Cell 67, 303–310.

Shiio, Y., Eisenman, R.N., 2003. Histone sumoylation is associated with transcriptional repression. Proc. Natl. Acad. Sci. USA 100, 13225–13230.

Shukla, S., Kavak, E., Gregory, M., *et al.*, 2011. CTCF-promoted RNA polymerase II pausing links DNA methylation to splicing. Nature 479, 74–79.

Simonis, M., Klous, P., Splinter, E., *et al.*, 2006. Nuclear organization of active and inactive chromatin domains uncovered by chromosome conformation capture-on-chip (4C). Nat. Genet. 38, 1348–1354.

Szalaj, P., Tang, Z., Michalski, P., *et al.*, 2016. An integrated 3-dimensional genome modeling engine for data-driven simulation of spatial genome organization. Genome Res. 26, 1697–1709.

Tang, Z., Luo, O.J., Li, X., *et al.*, 2015. CTCF-mediated human 3D genome architecture reveals chromatin topology for transcription. Cell 163, 1611–1627.

Tark-Dame, M., Jerabek, H., Manders, E.M., *et al.*, 2014. Depletion of the chromatin looping proteins CTCF and cohesin causes chromatin compaction: Insight into chromatin folding by polymer modelling. PLOS Comput. Biol. 10, e1003877.

Thirumalai, D., Shi, G., 2017. Chromatin is stretched but intact when the nucleus is squeezed through constrictions. Biophys. J. 112, 411–412.

Thurman, R.E., Rynes, E., Humbert, R., *et al.*, 2012. The accessible chromatin landscape of the human genome. Nature 489, 75–82.

Tims, H.S., Gurunathan, K., Levitus, M., Widom, J., 2011. Dynamics of nucleosome invasion by DNA binding proteins. J. Mol. Biol. 411, 430–448.

Todolli, S., Perez, P.J., Clauvelin, N., Olson, W.K., 2017. Contributions of sequence to the higher-order structures of DNA. Biophys. J. 112, 416–426.

Tumbar, T., Belmont, A.S., 2001. Interphase movements of a DNA chromosome region modulated by VP16 transcriptional activator. Nat. Cell Biol. 3, 134–139.

Ulianov, S.V., Khrameeva, E.E., Gavrilov, A.A., *et al.*, 2016. Active chromatin and transcription play a key role in chromosome partitioning into topologically associating domains. Genome Res. 26, 70–84.

Vallee, R.B., Williams, J.C., Varma, D., Barnhart, L.E., 2004. Dynein: An ancient motor protein involved in multiple modes of transport. J. Neurobiol. 58, 189–200.

Voong, L.N., Xi, L., Sebeson, A.C., *et al.*, 2016. Insights into nucleosome organization in mouse embryonic stem cells through chemical mapping. Cell 167, 1555–1570. (e15).

Wang, S., Xu, J., Zeng, J., 2015. Inferential modeling of 3D chromatin structure. Nucleic Acids Res. 43, e54.

Warns, J.A., Davie, J.R., Dhasarathy, A., 2016. Connecting the dots: Chromatin and alternative splicing in EMT. Biochem. Cell Biol. 94, 12–25.

Webster, M., Witkin, K.L., Cohen-Fix, O., 2009. Sizing up the nucleus: Nuclear shape, size and nuclear-envelope assembly. J. Cell Sci. 122, 1477–1486.

Weintraub, A.S., Li, C.H., Zamudio, A.V., et al., 2017. YY1 Is a Structural Regulator of Enhancer-Promoter Loops. Cell, 172, pp. 1573–1588. doi:10.1016/j.stem.2015.11.007.

West, A.G., Gaszner, M., Felsenfeld, G., 2002. Insulators: Many functions, many mechanisms. Genes Dev. 16, 271–288.

Woodcock, C.L., Grigoryev, S.A., Horowitz, R.A., Whitaker, N., 1993. A chromatin folding model that incorporates linker variability generates fibers resembling the native structures. Proc. Natl. Acad. Sci. USA 90, 9021–9025.

Wurtele, H., Chartrand, P., 2006. Genome-wide scanning of HoxB1-associated loci in mouse ES cells using an open-ended chromosome conformation capture methodology. Chromosome Res. 14, 477–495.

Xu, C., Corces, V.G., 2016. Towards a predictive model of chromatin 3D organization. Semin. Cell Dev. Biol. 57, 24–30Yin, M., Wang, J., Wang, M. et al., 2017. Molecular mechanism of directional CTCF recognition of a diverse range of genomic sites. Cell Research 27, 1365–1377doi:10.1038/cr.2017.131.

Yin, M., Wang, J., Wang, M., et al., 2017. Molecular mechanism of directional CTCF recognition of a diverse range of genomic sites. Cell Research 27, 1365–1377. doi:10.1038/cr.2017.131.

Zhang, R., Erler, J., Langowski, J., 2017. Histone acetylation regulates chromatin accessibility: Role of H4K16 in inter-nucleosome interaction. Biophys. J. 112, 450–459.

Zhao, Z., Tavoosidana, G., Sjolinder, M., et al., 2006. Circular chromosome conformation capture (4C) uncovers extensive networks of epigenetically regulated intra- and interchromosomal interactions. Nat. Genet. 38, 1341–1347.

Nucleosome Positioning

Vladimir B Teif and Christopher T Clarkson, University of Essex, Colchester, United Kingdom

Introduction

The genome of a eukaryotic cell is stored inside the nucleus in the form of the nucleoprotein complex called chromatin. If one would gently remove chromatin from the human cell nucleus, the $\sim$2 m long DNA would appear as a string with beads (nucleosomes). Each nucleosome consists of two copies of each of histones H2A, H2B, H3 and H4 (so called histone octamer) and about 147 DNA base pairs (bp) wrapped around these histones (Luger *et al.*, 1997). The distances between neighbouring nucleosomes can vary from zero to tens of base pairs (van Holde, 1989). Nucleosomes can be formed at any location along the DNA, but their positioning in the genome is not random – it is affected by the DNA sequence and other factors described below, which allow the same genome to be packed differently depending on the cell state. Nucleosome positioning determines the accessibility of DNA to transcription factors and other proteins, and is thus an important regulator of gene expression. Therefore, nucleosome positioning has been an active area of research since the discovery of the nucleosome (Kornberg, 1974; Olins and Olins, 1974). These investigations have further intensified in the 2000s with the developments of new methods allowing direct measurements of genome-wide nucleosome locations using high-throughput sequencing (Ioshikhes *et al.*, 2006b; Segal *et al.*, 2006b; Yuan *et al.*, 2005). Nowadays nucleosome positioning studies have advanced up to the level of human patients, down to single cells and cell-free DNA in blood (Snyder *et al.*, 2016).

Parameters Characterizing Nucleosome Positioning

The regulation of DNA accessibility by nucleosome positioning is performed through different mechanisms that can be characterised by the following four major parameters:

Nucleosome Dyad Positions

This term refers to the position of the center of the DNA segment symmetrically wrapped around the histone octamer (so called dyad), that can be directly measured in vitro or inferred from averaging in high-throughput sequencing experiments. Thus, the term "nucleosome positioning" can be understood either in its general sense (including all the parameters listed below), or in a narrow sense, referring to the location of the nucleosome dyads (see **Fig. 1(A)**).

Nucleosome Occupancy

This term refers to the probability of a given DNA site to be occupied by a histone octamer. It can be inferred from averaging over an ensemble of cells studied via Next Generation Sequencing (NGS) experiments. In such experiments the nucleosome occupancy at a given locus is reflected by the density of sequencing reads corresponding to the nucleosomal DNA, which determines the shape of the continuous nucleosome occupancy landscape (**Fig. 1(B)**).

Nucleosome Stability, Accessibility and Fuzziness

In the context of genomic nucleosome positioning, nucleosome stability is usually determined by comparing the nucleosome occupancy landscape of the same genomic region obtained at different levels of chromatin digestion. The better the correlations between nucleosome landscapes in different experiments the higher the nucleosome stability. Physically, the nucleosome stability is characterised by the energetic cost to partially unwrap the nucleosome (which depends on the DNA sequence, covalent modifications of DNA and histones, and the length of the unwrapped DNA part). This thermodynamic value can be measured directly in single-molecule experiments (Poirier *et al.*, 2008) or inferred indirectly from genome-wide experiments in the living cells. In the latter case the nucleosome stability is usually defined as a measure of its sensitivity to different levels of chromatin digestion (Chereji *et al.*, 2017; Mueller *et al.*, 2017; Teif *et al.*, 2014). One can similarly define the opposite parameter called accessibility, which refers to the probability to make the nucleosomal DNA accessible for protein binding (Mueller *et al.*, 2017). Another related parameter is called nucleosome fuzziness, which is proportional to the standard error of determining the nucleosome occupancy at a given genomic location based on the averaging of several replicate experiments (Vainshtein *et al.*, 2017) (**Fig. 1(C)**).

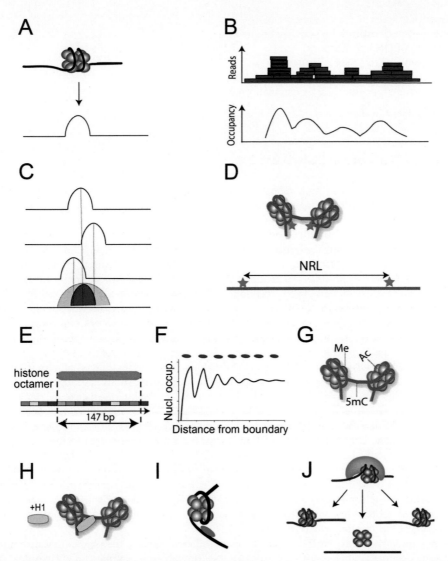

Fig. 1 Schematic representations of the main characteristics of nucleosome positioning and main factors affecting it. (A-D): Main characteristics of nucleosome positioning include the dyad position (A), nucleosome occupancy landscape (B), nucleosome stability/fuzziness (C), nucleosome repeat length (NRL) (D). (E-J): Main factors affecting nucleosome positioning include the DNA sequence affinity of the histone octamer (E), statistical positioning by the boundaries (F), covalent modifications of DNA and histones (G), interaction with linker histones (H), competition with transcription factors (I), and action of chromatin remodellers (J).

Nucleosome Repeat Length (NRL)

NRL is an integrative parameter equal to the average distance between the centres of neighbouring nucleosomes (**Fig. 1(D)**). NRL can be calculated either for large enough genomic regions or the whole genome. Genome wide average NRL depends on the cell type and state. For example, it can vary from as small as 160 bp in Yeast to up to 210 bp in human or mouse (van Holde, 1989). NRL also can change during cells differentiation (Teif *et al.*, 2012; van Holde, 1989) and can be different within the same cells in specific genomic regions, e.g., a 10-bp NRL decrease with respect to genome-average NRL has been reported near bound transcription factors CTCF or SP1 (Teif *et al.*, 2017), and NRLs associated with transcriptionally active and inactive genomic regions are also known to be different (Valouev *et al.*, 2011).

Genetic and Epigenetic Factors Affecting Nucleosome Positioning

Genomic nucleosome positioning is non-random, and represents a unique characteristic of a given cell state and type. Several counteracting processes affect nucleosome positioning both at the level of the genome (DNA sequence) and epigenome (beyond the DNA sequence). One can distinguish six major determinants of genomic nucleosome positioning:

- Intrinsic DNA sequence affinity of the histone octamer
- Statistical positioning of nucleosomes by genomic boundaries
- Chemical modifications of DNA or histones
- Interaction of nucleosomes with linker histones
- Binding of transcription factors and other chromatin proteins
- ATP-dependent nucleosome repositioning by chromatin remodellers

These factors will be considered in detail in the next sections.

Intrinsic DNA Sequence Affinity of the Histone Octamer

DNA has a natural affinity for histones as its backbone is constituted by negatively charged phosphate residues and can be neutralised by histones which bear positive charges. Thus, nucleosomes can form at any location, but some locations are more likely to be selected for nucleosome formation for a number of reasons. The geometry of the DNA double helix is characterised by six parameters: twist, shift, slide, roll, rise and tilt (van Holde, 1989). In the first approximation most of these parameters can be determined from the dinucleotide content neglecting the effect of longer than nearest-neighbour nucleotides either based on available crystal structures (Olson *et al.*, 1998) or molecular dynamics simulations (Lankas *et al.*, 2003; Lavery *et al.*, 2010). Therefore, studies of dinucleotide compositions of genomic DNA have contributed significantly to the establishment of the nucleosome positioning field. Almost 40 years ago it was noticed that genomic DNA is characterised by ~10 bp periodicity of dinucleotide distribution, which theoretically could play a role in histone recruitment to specific sites in the DNA (Trifonov and Sussman, 1980). Later such periodicities were indeed observed in the nucleosome core DNA sequences (Satchwell *et al.*, 1986).

In order to find the strongest nucleosome positioning DNA sequence, in vitro nucleosome reconstitution via salt-dialysis was performed for a large pool of randomly generated DNA sequences (Lowary and Widom, 1998). As a result of this study the sequence named "601" (later referred to as the "Widom 601" was found to have the highest affinity for the histone octamer). Furthermore, it has been demonstrated by genome-wide sequencing that G/C rich regions have higher nucleosome density, while A/T rich regions are more nucleosome depleted (Segal and Widom, 2009). At the intermediate scale, one can distinguish longer, nucleosome-size motifs, which are either attractive for the histone octamer, such as e.g., the Widom 601 sequence (Lowary and Widom, 1998) and so called Trifonov's "strong nucleosomes" (Trifonov and Nibhani, 2015a), or present nucleosome excluding barriers (Drillon *et al.*, 2016). The cumulative evidence from many experiments conducted in the pre-NGS era confirmed the hypothesis that the DNA sequence at least partially affects the positions of nucleosomes. This hypothesis was further supported when first genome-wide locations of nucleosomes have been mapped (Ioshikhes *et al.*, 2006b; Lee *et al.*, 2007; Segal *et al.*, 2006b; Yuan *et al.*, 2005). Later genome-wide studies have revealed that nucleosome positions change with cell differentiation and during the cell cycle, which has led to the refinement of the concept of DNA-sequence dependent nucleosome organization (Schones *et al.*, 2008; Whitehouse *et al.*, 2007). It is currently believed that ~9% of nucleosomes are arranged consistently by the DNA sequence across different types of human cells (Gaffney *et al.*, 2012).

Since the core histone proteins are extremely conserved among different species, the physics that determines DNA sequence preferences for the histone octamer is universal from yeast to human. However, the ensemble of nucleosomal DNA sequences can differ between species, and so are the uses of nucleosomes in gene regulation. For example, nucleosome positioning at functional genomic elements can be either favoured by default (the genomic element is always closed by the nucleosome unless active nucleosome displacement happens leading to the activation of this element), or it may be unfavourable (nucleosome depletion by default allowing TF binding to a given site unless a nucleosome is actively repositioned there). A recent study showed that the former mechanism is probably preferred in higher eukaryotes while the latter is implemented in unicellular organisms (Tompitak *et al.*, 2017). This study reported that promoters of multicellular organisms are in general characterised by nucleosome-favouring sequences, while unicellular organisms have nucleosome-disfavouring promoters.

Statistical Positioning of Nucleosomes by Genomic Boundaries

In 1988 Kornberg and Stryer proposed that nucleosome positioning, at least to some extent, is governed by the boundary effect: nucleosomes will form ordered arrays next to any physical boundary in the genome (Kornberg and Stryer, 1988). In the post-NGS era genome-wide maps of nucleosome binding have proved this hypothesis. The average nucleosome occupancy shows regular oscillations near genomic locations that act as a boundary, such as the nucleosome depleted transcription start sites (Mavrich *et al.*, 2008), replication origins (Eaton *et al.*, 2010) or large DNA-bound transcription factors such as CTCF (Cuddapah *et al.*, 2009; Fu *et al.*, 2008). Even large genomic intervals such as genes experience boundary effects by their nucleosome-depleted starts and ends (Chevereau *et al.*, 2009). Nucleosomes are different from the analogy of "beads on the string" in that nucleosomes are not freely moving along the DNA. Indeed, nucleosome reconstitution experiments show that the nucleosome occupancy oscillations around TSS can be recapitulated in vitro only in the presence of ATP and energy-dependent chromatin remodellers which facilitate nucleosome movements (Zhang *et al.*, 2011). Nevertheless, even with the non-equilibrium component introduced by chromatin remodellers, nucleosome density oscillation near genomic barriers can be quantitatively predicted from equilibrium thermodynamics consistently with the experiments (Beshnova *et al.*, 2014; Mobius and Gerland, 2010; Riposo and Mozziconacci, 2012; Rube and Song, 2014). Interestingly, recent studies show that up to 37.5% of human nucleosome positions can be accounted for by considering boundary effects created by nucleosome-excluding sequences (Drillon *et al.*, 2016).

Chemical Modifications of DNA or Histones

Covalent modifications of DNA and histones represent a large layer of epigenetic regulation (Portela and Esteller, 2010). Most common DNA modifications are cytosine methylation and hydroxymethylation in the context of CpG dinucleotides ("CpG" means cytosine followed by phosphate followed by guanine). Histone modifications are numerous and may have combinatorial nature (several different amino acids of the same histone can be covalently modified, and different histones in the same nucleosome can carry different modifications). Epigenetic modifications are established by the "writer" enzymes, and recognized by specific "reader" proteins, which modulate the genetic program based on these modifications (Strahl and Allis, 2000). The molecular action of covalent modifications of DNA and histones is often through nucleosome repositioning. A number of statistically defined relations between nucleosome positioning and covalent chromatin modification have been reported, but there seems to be no simple code connecting DNA/histone modifications with the presence/absence of a nucleosome at a given location. For example, the action of DNA methylation is context-dependent – it is rarely seen in CpG islands and when this is observed, it is associated with recruited nucleosomes. Outside of CpG islands DNA methylation is most likely between nucleosomes (Teif *et al.*, 2014). On the other hand, DNA hydroxymethylation can decrease the nucleosome stability. A methylation/hydroxymethylation switch model has been proposed to explain some of the changes in nucleosome occupancy and stability during embryonic stem cell differentiation when an almost 10% fraction of genomic CpGs change their state from hydroxymethylation to methylation, increasing the nucleosome density of the corresponding chromatin regions (Teif *et al.*, 2014). Histone modifications also have characteristic relations with nucleosome positioning. For example, regions with "active" histone modification marks such as acetylation of histone H3 lysine 4 or 9 are characterised by lower nucleosome density, while "inactive" chromatin marks such as methylation of histone H3 lysine 9 or 27 are characterised by higher nucleosome density (Teif *et al.*, 2014).

Interaction of Nucleosomes With Linker Histones

Proteins involved in nucleosome positioning are not limited to the core histones composing the nucleosome. Nucleosome packing also critically depends on the linker histones, which belong to a class of basic nuclear proteins binding the DNA entry/exit from the nucleosome and providing a separation and interaction medium between nucleosomes in chromatin (Bednar *et al.*, 2017). It is well established that the abundance of linker histones affects the NRL (Woodcock *et al.*, 2006), which may in turn determine different types of chromatin packing (Bascom *et al.*, 2017; Routh *et al.*, 2008). Usually NRL increases with the increase of linker histone concentration in the nucleus, but it can be also affected by the competition of linker histones with other abundant chromatin proteins (Beshnova *et al.*, 2014).

Binding of Transcription Factors and Other Chromatin Proteins

Transcription factors (TFs), which bind DNA sequence-specifically, can affect nucleosome positioning in a number of ways considered below (see **Fig. 1(E)–(J)**).

TF Competition With Histone Octamer

The energy of complete nucleosome unwrapping or partial dissociation of the histone core from the DNA is much larger than typical energies of TF-DNA binding. Therefore, for most TFs nucleosomes can be viewed as almost immobile obstacles that do not really compete with TFs on a timescale of usual TF binding to the DNA. If TF/nucleosome arrangement would have been determined by equilibrium thermodynamics, nucleosome positioning would be determining TF binding and not the other way around (Teif and Rippe, 2010). Only few chromatin proteins such as CTCF have DNA binding energy comparable to that of the histone octamer, allowing them to act as a boundary affecting up to several nucleosomes in its vicinity (Cuddapah *et al.*, 2009; Fu *et al.*, 2008). It is important to keep in mind that the chromatin is never in a true thermodynamic equilibrium, and thus active energy-dependent processes, as well as the kinetics of binding determine the order of events in terms of the nucleosome/TF competition (**Table 1**).

TF Binding to DNA Partially Unwrapped From the Histone Octamer

Since the nucleosome is not a single entity, it can "breath" by partially uncoiling the nucleosomal DNA. Transcription factors then can bind the nucleosomal DNA, depending on how far their binding site is hidden inside the nucleosome (North *et al.*, 2012). An

Table 1   Computational tools to predict nucleosome positioning (sorted alphabetically)

Description	Online/local installation	Di(tri)nucleotide/periodicity/specific motifs/empirical feature
FineStr: Single-base-resolution nucleosome mapping server (Trifonov, 2010; Gabdank et al., 2009, 2010). The analysis is performed using the probe based on the 117-bp DNA bendability matrix derived from C. elegans. The authors suggested the universality of this pattern for other species. http://www.cs.bgu.ac.il/~nucleom/	+/−	+/+/+/−
ICM Web: ICM Web allows users to assess nucleosome stability and fold any sequence of DNA into a 3D model of chromatin (Stolz and Bishop, 2010b; Sereda and Bishop, 2010). The model is displayed in the visual browser JSmol or can be downloaded. ICM takes a DNA sequence and generates (i) a nucleosome energy level diagram, (ii) coarse-grained representations of free DNA and chromatin and (iii) plots of the helical parameters (Tilt, Roll, Twist, Shift, Slide and Rise) as a function of position. http://dna.engr.latech.edu/icm-du	+/−	−/−/−/+
iNuc-PhysChem: Identifying nucleosomal or linker sequences from physicochemical properties (Chen et al., 2012). The algorithm identifies nucleosomal sequences by incorporating twelve physicochemical properties defined elsewhere, such as A-philicity, base stacking, B-DNA twist, bendability, bending stiffness, DNA denaturation energy, Z-DNA potential. The model was trained on data from H. sapiens, C. elegans and D. melanogaster. http://lin.uestc.edu.cn/server/iNuc-PhysChem	+/+	+/−/+/+
iNuc-PseKNC: A sequence-based predictor for nucleosome positioning in genomes with pseudo k-tuple nucleotide composition (Guo et al., 2014). This is another software package from the developers of **iNuc-PhysChem**. Here, the samples of DNA sequences were formulated using six basic DNA local structural properties trained on datasets from H. sapiens, C. elegans and D. melanogaster. http://lin.uestc.edu.cn/server/iNuc-PseKNC	+/−	+/−/+/+
Mapping_CC: Displays the nucleosome predictions based on the DNA dinucleotide correlation pattern. This algorithm was initially associated with one of the first high-throughput genome-wide nucleosome maps in Yeast (Ioshikhes et al., 2006a). An updated version is available at http://nuclbrowser.ucoz.com/load/	−/+	+/+/−/−
MOSAICS: Methodologies for Optimization and Sampling in Computational Studies (Minary and Levitt, 2014). Perl scripts and a precompiled package to perform training-free atomistic prediction of nucleosome occupancy based on all-atom force field calculations. The effect of DNA methylation can be taken into account. http://www.cs.ox.ac.uk/mosaics/nucleosome/nucleosome.html	−/+	+/−/−/+
NucEnerGen: Nucleosome energetics predictions based on high throughput sequencing (Locke et al., 2010). It utilizes dynamic programming to calculate allowed nucleosome configurations and the Percus equation to infer sequence-dependent energies from the experimental occupancy profiles. http://nucleosome.rutgers.edu/nucenergen/	−/+	+/+/+/−
nuMap: A web application implementing the YR and W/S schemes to predict nucleosome positioning (Alharbi et al., 2014; Cui and Zhurkin, 2010, 2014). The methodology is based on the sequence-dependent anisotropic bending, which dictates how DNA is wrapped around a histone octamer. This application allows users to specify a number of options such as schemes and parameters for threading calculation and provides multiple layout formats. http://numap.rit.edu/app/dna/index.xhtml	+/−	+/+/+/+
NuPoP: Nucleosome Positioning Prediction Engine (Xi et al., 2010; Wang et al., 2008). NuPoP is built upon a duration hidden Markov model, in which the linker DNA length is explicitly modeled. NuPoP outputs the Viterbi prediction, nucleosome occupancy score (from backward and forward algorithms) and nucleosome affinity score. NuPoP has three formats including a web server prediction engine, a stand-alone Fortran program, and an R package. The latter two can predict nucleosome positioning for a DNA sequence of any length. http://nucleosome.stats.northwestern.edu	+/+	+/+/−/−
Nu-OSCAR: Nucleosome-Occupancy Study for Cis-elements Accurate Recognition. It is devoted to identifying binding sites of known transcription factors, which further incorporates nucleosome occupancy around sites on promoter regions. The derivation of the algorithm is based on a biophysical view of interactions between protein factors and nucleosome DNA. http://bioinfo.au.tsinghua.edu.cn/nu_oscar/oscar.html	+/−	+/+/−/−
nuScore: A nucleosome-positioning score calculator based on the DNA curvature properties (Tolstorukov et al., 2008). This software allows an important type of analysis, where a user enters many sequences to calculate the average nucleosome energy profile. http://compbio.med.harvard.edu/nuScore	+/−	+/+/−/+
N-score: MATLAB and Python codes using a wavelet analysis based model for predicting nucleosome positions from DNA sequence (Yuan and Liu, 2008). http://bcb.dfci.harvard.edu/~gcyuan/software.html	−/+	+/+/−/−

NXSensor: Prediction of nucleosome-excluding sequences based on DNA bending properties (Luykx *et al.*, 2006). It takes as input DNA sequences in FASTA format, and outputs nucleosome-excluding or nucleosome favouring segments. http://www.sfu.ca/~ibajic/NXSensor/	+/-	+/+/-/+
Online nucleosomes position prediction by genomic sequence (Segal Lab) (Field *et al.*, 2008; Kaplan *et al.*, 2009a; Segal *et al.*, 2006a). Although it has no specific name, this is one of the most popular tools in this class, realized as a web server (allows analyzing a limited number of DNA sequences), and a stand-alone application which can be installed on a local cluster. It allows calculating nucleosome occupancy or nucleosome start site probability profiles of non-overlapping nucleosomes; alternatively, it is possible to calculate the net nucleosome formation energy profile. It uses machine learning for energy assignment based on the training datasets and dynamic programming to sample nucleosome configurations (similar to **NucEnerGen, NuPoP** and the algorithm of van Noort and co-authors). http://genie.weizmann.ac.il/software/nucleo_prediction.html	+/+	+/+/+/-
Online nucleosome position prediction (van Noort and co-authors) (van der Heijden *et al.*, 2012). This algorithm is based on dinucleotide distributions, but unlike other methods based on dinucleotide distributions it does not use machine learning and accounts only for the dinucleotide periodicity. In addition, this method uses dynamic programming to account for size exclusion and the Percus equation to assign nucleosome affinities (similar to **NucEnerGen, NuPoP** and the algorithm of Segal and co-authors mentioned above). http://bio.physics.leidenuniv.nl/~noort/cgi-bin/nup3_st.py	+/-	+/+/+/-
Phase: A web server for prediction of the nucleosome formation probability based on (i) the 10–11 bp periodicities of dinucleotides and (ii) the typical pattern "linker–nucleosome - linker" defined by the authors (Levitsky *et al.*, 2014). http://wwwmgs.bionet.nsc.ru/mgs/programs/phase/	+/-	+/+/-/-
RECON: A web server for prediction of the nucleosome formation potential learned from dinucleotide frequencies distribution for nucleosome positioning sequences (Levitsky *et al.*, 1999; Levitsky, 2004). http://wwwmgs.bionet.nsc.ru/mgs/programs/recon/	+/-	+/+/-/-
SymCurv: A program for nucleosome positioning prediction (Nikolaou *et al.*, 2010). It calculates the curvature of the DNA sequence and uses a greedy algorithm to parse the sequence in nucleosome-bound and nucleosome-free segments. http://genome.crg.es/SymCurv/documentation.html	-/+	-/-/-/+
Strong nucleosomes: Based on a recent discovery of strong nucleosome positioning sequences which are visually seen as regular arrays in genomic sequence (Salih *et al.*, 2015; Nibhani and Trifonov, 2015; Trifonov and Nibhani, 2015b), the program from Trifonov's lab is finding a specific class of strongly positioned nucleosomes of the RR/YY and TA periodic types (Nibhani and Trifonov, 2015). http://strn-nuc.haifa.ac.il:8080/mapping/home.jsf	+/-	+/+/-/-

effect called "collaborative competition" allows two TFs help each other to bind the nucleosomal DNA, because it is easier to the second TF to bind if the first TF is already bound and the DNA is already partially unwrapped from the nucleosome (Polach and Widom, 1996). This can lead to nonlinear effects of nucleosome positioning at regulatory sites such as enhancers (Mirny, 2010; Teif and Rippe, 2011).

TF-Mediated Recruitment of ATP-Dependent Enzymes

Sequence-specific binding of transcription factors can be used as a strategy to recruit to a given site an enzyme that chemically modifies DNA or histones, or actively translocates the nucleosome (Agalioti *et al.*, 2000). Mode details on this effect can be found in the next section.

ATP-Dependent Nucleosome Repositioning by Chromatin Remodellers

Random sliding of nucleosomes along the DNA is very slow at physiological conditions, and this effect is mostly limited to in vitro studies (Meersseman *et al.*, 1992), while nucleosome repositioning in live systems is usually an active process requiring ATP-dependent molecular motors, so called chromatin remodellers (Clapier *et al.*, 2017). For example, it has been shown that the oscillatory nucleosome density pattern near Yeast promoters can be recovered by in vitro nucleosome reconstitution on the same DNA sequences only in the presence of the cell extract and added ATP (Zhang *et al.*, 2011). In the absence of ATP the remodeller activity is abolished and nucleosome positioning tends to reflect the thermodynamically favoured pattern. Remodellers can act both in favour and against the thermodynamically preferred pattern depending on the context. A simplistic mathematical representation of remodeller rules is that a remodeller randomly binds a nucleosome, and then depending on the remodeller type, DNA sequence and chromatin context it either moves the nucleosome left/right, or evicts it completely (Teif and Rippe, 2009). Remodellers usually move nucleosomes in discrete steps such as 5 or 10 base pairs along the DNA, which is explained by the mechanics of the double helix repositioning along the nucleosome through formation of small loops relocated along the histone octamer (Clapier *et al.*, 2017). In many instances remodellers bind their target nucleosome non-randomly, and their recruitment is achieved by multiple remodeller subunits that recognize certain TFs or histone modifications (Ho and Crabtree, 2010).

Theoretical Approaches to Predict Nucleosome Positioning

Computational algorithms for predicting nucleosome positions can be roughly categorized into biophysical (taking into account physical properties of DNA and histones), bioinformatical (learning the rules of preferred nucleosome distributions without knowing details of molecular interactions), and hybrid models representing the mixture of these two approaches. A list of more than 20 web servers that offer nucleosome positioning prediction has been compiled elsewhere (Teif, 2016).

In typical bioinformatics-inspired models no assumptions are made about the underlying forces that determine nucleosome positions. Instead, features of the DNA content are analysed for available experimental nucleosome positioning datasets and it is assumed that nucleosome positioning in other systems follows the same rules. For example, the nucleosomal and linker DNA sequences can be collected from experiments in order to be able to find some common themes in these genomic features, and then analysed using e.g., wavelet analysis to detect changes in di-nucleotide frequency allowing to distinguish between linker and nucleosomal sequences (Yuan and Liu, 2008). The problem with training a model based on nucleotide content is that there are 4^{147} possible arrangements within a 147 bp DNA window, which is larger than the length of any known eukaryotic genome. Even if the model is trained on many datasets from different genomes, these still do not cover all possible nucleotide combinations that can be theoretically encountered. For this reason, models trained on one species are in general expected to underperform on other species (Kaplan *et al.*, 2009b), although few universal nucleosome sequence combinations are also expected (Trifonov and Nibhani, 2015a).

In typical biophysical models the prediction of nucleosome positioning is based on available nucleosome crystal structures to infer bending energies corresponding to all dinucleotides and then calculate the nucleosome score for each sliding window of 147 bp with a given dinucleotide distribution. In contrast to bioinformatics models which are trained on data that may or may not be representative of all genomic sequences, biophysical models are based on the assumptions about the underlying molecular dynamics of the process of nucleosome formation (Chereji and Morozov, 2015; Chevereau *et al.*, 2009; Stolz and Bishop, 2010a; Tompitak *et al.*, 2017). Most biophysical models are quite demanding in terms of the computational power if they perform realistic simulations. The power of these models is that they allow connecting intrinsic DNA sequence affinities with active repositioning scenarios such as TF/nucleosome competition and remodeller action (Teif and Rippe, 2009). Furthermore, biophysical models allow treating such basic phenomena as statistical nucleosome positioning by the boundary (Chevereau *et al.*, 2009), the exclusion of nucleosomes by their neighbours (Segal *et al.*, 2006b) and partial nucleosome unwrapping (Teif and Rippe, 2011). In principle, biophysical models allow treating any level of complexity in nucleosome positioning, provided there is a way to parameterise the models based on some experimental data. The latter is frequently becoming the bottleneck.

Future Directions

The number of computational algorithms and experimental datasets devoted to nucleosome positioning continues to increase, but the fundamental questions of how to predict nucleosome cell type/state-specific nucleosome re-positioning and the corresponding changes in gene expression are still not solved. One major complication is that the rules of chromatin remodelling are difficult to derive from experiments. Another complication is that even when the remodeller rules will be determined experimentally, it will be still a challenge to integrate them in the computational models where both equilibrium and non-equilibrium processes coexist. A third complication is the 3D genome structure – nucleosome positioning happening in the 1D is influenced by and influences the 3D chromatin packing. In order to address these challenges one can expect three major directions of the nucleosome positioning field. Firstly, new high-throughput sequencing methods will continue to emerge, with a special emphasis on single-molecule and single-patient studies. Secondly, the gap in theoretical descriptions will require developing new approaches integrating the dynamics of nucleosome positioning in the description of in vivo processes. Thirdly, new biophysical models will have to be developed to account for the interplay of the 1D and 3D nucleosome arrangements based on the experimental data on 3D chromatin packing. The latter is already a very active area of research supported by large funding initiatives (Dekker *et al.*, 2017).

Acknowledgement

This work was supported by the Wellcome Trust grant 200733/Z/16/Z.

See also: Chromatin: A Semi-Structured Polymer. Data Mining: Classification and Prediction. Genome Informatics. Learning Chromatin Interaction Using Hi-C Datasets. Natural Language Processing Approaches in Bioinformatics. Protein-DNA Interactions. Quantitative Immunology by Data Analysis Using Mathematical Models

---

References

Agalioti, T., Lomvardas, S., Parekh, B., *et al.*, 2000. Ordered recruitment of chromatin modifying and general transcription factors to the IFN-beta promoter. Cell 103 (4), 667–678.

Alharbi, B.A., Alshammari, T.H., Felton, N.L., *et al.*, 2014. nuMap: A web platform for accurate prediction of nucleosome positioning. Genom. Proteom. Bioinform. 12, 249–253.

Bascom, G.D., Kim, T., Schlick, T., 2017. Kilobase pair chromatin fiber contacts promoted by living-system-Like DNA linker length distributions and nucleosome depletion. J. Phys. Chem. B 121 (15), 3882–3894.

Bednar, J., Garcia-Saez, I., Boopathi, R., *et al.*, 2017. Structure and dynamics of a 197 bp nucleosome in complex with linker histone H1. Mol. Cell 66 (3), 384–397. e8.

Beshnova, D.A., Cherstvy, A.G., Vainshtein, Y., Teif, V.B., 2014. Regulation of the nucleosome repeat length in vivo by the DNA sequence, protein concentrations and long-range interactions. PLOS Comput. Biol. 10 (7), e1003698.

Chen, W., Lin, H., Feng, P.M., *et al.*, 2012. iNuc-PhysChem: A sequence-based predictor for identifying nucleosomes via physicochemical properties. PLOS ONE 7, e47843.

Chereji, R.V., Morozov, A.V., 2015. Functional roles of nucleosome stability and dynamics. Brief Funct. Genom. 14 (1), 50–60.

Chereji, R.V., Ocampo, J., Clark, D.J., 2017. MNase-sensitive complexes in yeast: Nucleosomes and non-histone barriers. Mol. Cell 65 (3), 565–577. e3.

Chevereau, G., Palmeira, L., Thermes, C., Arneodo, A., Vaillant, C., 2009. Thermodynamics of intragenic nucleosome ordering. Phys. Rev. Lett. 103 (18), 188103.

Clapier, C.R., Iwasa, J., Cairns, B.R., Peterson, C.L., 2017. Mechanisms of action and regulation of ATP-dependent chromatin-remodelling complexes. Nat. Rev. Mol. Cell Biol. 18, 407–422.

Cuddapah, S., Jothi, R., Schones, D.E., *et al.*, 2009. Global analysis of the insulator binding protein CTCF in chromatin barrier regions reveals demarcation of active and repressive domains. Genome Res. 19 (1), 24–32.

Cui, F., Zhurkin, V.B., 2010. Structure-based analysis of DNA sequence patterns guiding nucleosome positioning in vitro. J. Biomol. Struct. Dyn. 27, 821–841.

Cui, F., Zhurkin, V.B., 2014. Rotational positioning of nucleosomes facilitates selective binding of p53 to response elements associated with cell cycle arrest. Nucleic Acids Res. 42, 836–847.

Dekker, J., Belmont, A.S., Guttman, M., *et al.*, 2017. The 4D nucleome project. Nature 549 (7671), 219–226.

Drillon, G., Audit, B., Argoul, F., Arneodo, A., 2016. Evidence of selection for an accessible nucleosomal array in human. BMC Genom. 17, 526.

Eaton, M.L., Galani, K., Kang, S., Bell, S.P., MacAlpine, D.M., 2010. Conserved nucleosome positioning defines replication origins. Genes Dev. 24 (8), 748–753.

Field, Y., Kaplan, N., Fondufe-Mittendorf, Y., *et al.*, 2008. Distinct modes of regulation by chromatin encoded through nucleosome positioning signals. PLOS Comput. Biol. 4, e1000216.

Fu, Y., Sinha, M., Peterson, C.L., Weng, Z., 2008. The insulator binding protein CTCF positions 20 nucleosomes around its binding sites across the human genome. PLOS Genet. 4 (7), e1000138.

Gabdank, I., Barash, D., Trifonov, E.N., 2009. Nucleosome DNA bendability matrix (C. elegans). J. Biomol. Struct. Dyn. 26, 403–411.

Gabdank, I., Barash, D., Trifonov, E.N., 2010. FineStr: A web server for single-base-resolution nucleosome positioning. Bioinformatics 26, 845–846.

Gaffney, D.J., McVicker, G., Pai, A.A., *et al.*, 2012. Controls of nucleosome positioning in the human genome. PLOS Genet. 8 (11), e1003036.

Guo, S.H., Deng, E.Z., Xu, L.Q., *et al.*, 2014. iNuc-PseKNC: A sequence-based predictor for predicting nucleosome positioning in genomes with pseudo k-tuple nucleotide composition. Bioinformatics 30, 1522–1529.

Ho, L., Crabtree, G.R., 2010. Chromatin remodelling during development. Nature 463 (7280), 474–484.

Ioshikhes, I.P., Albert, I., Zanton, S.J., *et al.*, 2006a. Nucleosome positions predicted through comparative genomics. Nat. Genet. 38, 1210–1215.

Ioshikhes, I.P., Albert, I., Zanton, S.J., Pugh, B.F., 2006b. Nucleosome positions predicted through comparative genomics. Nat. Genet. 38 (10), 1210–1215.

Kaplan, N., Moore, I.K., Fondufe-Mittendorf, Y., *et al.*, 2009a. The DNA-encoded nucleosome organization of a eukaryotic genome. Nature 458, 362–366.

Kaplan, N., Moore, I.K., Fondufe-Mittendorf, Y., *et al.*, 2009b. The DNA-encoded nucleosome organization of a eukaryotic genome. Nature 458 (7236), 362–366.

Kornberg, R.D., 1974. Chromatin structure: A repeating unit of histones and DNA. Science 184 (4139), 868–871.

Kornberg, R.D., Stryer, L., 1988. Statistical distributions of nucleosomes: Nonrandom locations by a stochastic mechanism. Nucleic Acids Res. 16, 6677–6690.

Lankas, F., Sponer, J., Langowski, J., Cheatham 3rd, T.E., 2003. DNA basepair step deformability inferred from molecular dynamics simulations. Biophys. J. 85 (5), 2872–2883.

Lavery, R., Zakrzewska, K., Beveridge, D., et al., 2010. A systematic molecular dynamics study of nearest-neighbor effects on base pair and base pair step conformations and fluctuations in B-DNA. Nucleic Acids Res. 38 (1), 299–313.

Lee, W., Tillo, D., Bray, N., et al., 2007. A high-resolution atlas of nucleosome occupancy in yeast. Nat. Genet. 39 (10), 1235–1244.

Levitsky, V.G., 2004. RECON: A program for prediction of nucleosome formation potential. Nucleic Acids Res. 32, W346–W349.

Levitsky, .V.G., Babenko, V.N., Vershinin, A.V., 2014. The roles of the monomer length and nucleotide context of plant tandem repeats in nucleosome positioning. J. Biomol. Struct. Dyn. 32, 115–126.

Levitsky, V.G., Ponomarenko, M.P., Ponomarenko, J.V., et al., 1999. Nucleosomal DNA property database. Bioinformatics 15, 582–592.

Locke, G., Tolkunov, D., Moqtaderi, Z., et al., 2010. High-throughput sequencing reveals a simple model of nucleosome energetics. Proc. Natl. Acad. Sci. USA 107, 20998–21003.

Lowary, P.T., Widom, J., 1998. New DNA sequence rules for high affinity binding to histone octamer and sequence-directed nucleosome positioning. J. Mol. Biol. 276 (1), 19–42.

Luger, K., Mader, A.W., Richmond, R.K., Sargent, D.F., Richmond, T.J., 1997. Crystal structure of the nucleosome core particle at 2.8 A resolution. Nature 389 (6648), 251–260.

Luykx, P., Bajic, I.V., Khuri, S., 2006. NXSensor web tool for evaluating DNA for nucleosome exclusion sequences and accessibility to binding factors. Nucleic Acids Res. 34, W560–W565.

Mavrich, T.N., Ioshikhes, I.P., Venters, B.J., et al., 2008. A barrier nucleosome model for statistical positioning of nucleosomes throughout the yeast genome. Genome Res. 18 (7), 1073–1083.

Meersseman, G., Pennings, S., Bradbury, E.M., 1992. Mobile nucleosomes – A general behavior. EMBO J. 11 (8), 2951–2959.

Minary, P., Levitt, M., 2014. Training-free atomistic prediction of nucleosome occupancy. Proc. Natl. Acad. Sci. USA 111, 6293–6298.

Mirny, L.A., 2010. Nucleosome-mediated cooperativity between transcription factors. Proc. Natl. Acad. Sci. USA 107 (52), 22534–22539.

Mobius, W., Gerland, U., 2010. Quantitative test of the barrier nucleosome model for statistical positioning of nucleosomes up- and downstream of transcription start sites. PLOS Comput. Biol. 6 (8), e1000891.

Mueller, B., Mieczkowski, J., Kundu, S., et al., 2017. Widespread changes in nucleosome accessibility without changes in nucleosome occupancy during a rapid transcriptional induction. Genes Dev. 31 (5), 451–462.

Nibhani, R., Trifonov, E.N., 2015. TA-periodic ("601"-like) centromeric nucleosomes of A. thaliana. J. Biomol. Struct. Dyn. 1–4.

Nikolaou, C., Althammer, S., Beato, M., et al., 2010. Structural constraints revealed in consistent nucleosome positions in the genome of S. cerevisiae. Epigenetics Chromatin 3, 20.

North, J.A., Shimko, J.C., Javaid, S., et al., 2012. Regulation of the nucleosome unwrapping rate controls DNA accessibility. Nucleic Acids Res. 40 (20), 10215–10227.

Olins, A.L., Olins, D.E., 1974. Spheroid chromatin units (v bodies). Science 183 (4122), 330–332.

Olson, W.K., Gorin, A.A., Lu, X.J., Hock, L.M., Zhurkin, V.B., 1998. DNA sequence-dependent deformability deduced from protein-DNA crystal complexes. Proc. Natl. Acad. Sci. USA 95 (19), 11163–11168.

Poirier, M.G., Bussiek, M., Langowski, J., Widom, J., 2008. Spontaneous access to DNA target sites in folded chromatin fibers. J. Mol. Biol. 379 (4), 772–786.

Polach, K.J., Widom, J., 1996. A model for the cooperative binding of eukaryotic regulatory proteins to nucleosomal target sites. J. Mol. Biol. 258 (5), 800–812.

Portela, A., Esteller, M., 2010. Epigenetic modifications and human disease. Nat. Biotechnol. 28 (10), 1057–1068.

Riposo, J., Mozziconacci, J., 2012. Nucleosome positioning and nucleosome stacking: Two faces of the same coin. Mol. Biosyst. 8, 1172–1178.

Routh, A., Sandin, S., Rhodes, D., 2008. Nucleosome repeat length and linker histone stoichiometry determine chromatin fiber structure. Proc. Natl. Acad. Sci. USA 105 (26), 8872–8877.

Rube, T.H., Song, J.S., 2014. Quantifying the role of steric constraints in nucleosome positioning. Nucleic Acids Res. 42 (4), 2147–2158.

Salih, B., Tripathi, V., Trifonov, E.N., 2015. Visible periodicity of strong nucleosome DNA sequences. J. Biomol. Struct. Dyn. 33, 1–9.

Satchwell, S.C., Drew, H.R., Travers, A.A., 1986. Sequence periodicities in chicken nucleosome core DNA. J. Mol. Biol. 191, 659–675.

Schones, D.E., Cui, K., Cuddapah, S., et al., 2008. Dynamic regulation of nucleosome positioning in the human genome. Cell 132 (5), 887–898.

Segal, E., Fondufe-Mittendorf, Y., Chen, L., et al., 2006a. A genomic code for nucleosome positioning. Nature 442, 772–778.

Segal, E., Fondufe-Mittendorf, Y., Chen, L., et al., 2006b. A genomic code for nucleosome positioning. Nature 442 (7104), 772–778.

Segal, E., Widom, J., 2009. Poly(dA:dt) tracts: Major determinants of nucleosome organization. Curr. Opin. Struct. Biol. 19 (1), 65–71.

Sereda, Y.V., Bishop, T.C., 2010. Evaluation of elastic rod models with long range interactions for predicting nucleosome stability. J. Biomol. Struct. Dyn. 27, 867–887.

Snyder, M.W., Kircher, M., Hill, A.J., Daza, R.M., Shendure, J., 2016. Cell-free DNA comprises an in vivo nucleosome footprint that informs its tissues-of-origin. Cell 164 (1–2), 57–68.

Stolz, R.C., Bishop, T.C., 2010a. ICM Web: The interactive chromatin modeling web server. Nucleic Acids Res. 38 (Web Server issue), W254–W261.

Stolz, RC, Bishop, TC., 2010b. ICM Web: The interactive chromatin modeling web server. Nucleic Acids Res. 38, W254–W261.

Strahl, B.D., Allis, C.D., 2000. The language of covalent histone modifications. Nature 403 (6765), 41–45.

Teif, V.B., 2016. Nucleosome positioning: Resources and tools online. Brief Bioinform. 17 (5), 745–757.

Teif, V.B., Beshnova, D.A., Vainshtein, Y., et al., 2014. Nucleosome repositioning links DNA (de)methylation and differential CTCF binding during stem cell development. Genome Res. 24 (8), 1285–1295.

Teif, V.B., Mallm, J.-P., Sharma, T., et al., 2017. Nucleosome repositioning during differentiation of a human myeloid leukemia cell line. Nucleus 8 (2), 188–204.

Teif, V.B., Rippe, K., 2009. Predicting nucleosome positions on the DNA: Combining intrinsic sequence preferences and remodeler activities. Nucleic Acids Res. 37 (17), 5641–5655.

Teif, V.B., Rippe, K., 2010. Statistical-mechanical lattice models for protein-DNA binding in chromatin. J. Phys. Condens. Matter 22 (41), 414105.

Teif, V.B., Rippe, K., 2011. Nucleosome mediated crosstalk between transcription factors at eukaryotic enhancers. Phys. Biol. 8, 044001.

Teif, V.B., Vainshtein, Y., Caudron-Herger, M., et al., 2012. Genome-wide nucleosome positioning during embryonic stem cell development. Nat. Struct. Mol. Biol. 19 (11), 1185–1192.

Tolstorukov, M.Y., Choudhary, V., Olson, W.K., et al., 2008. nuScore: A web-interface for nucleosome positioning predictions. Bioinformatics 24, 1456–1458.

Tompitak, M., Vaillant, C., Schiessel, H., 2017. Genomes of multicellular organisms have evolved to attract nucleosomes to promoter regions. Biophys. J. 112 (3), 505–511.

Trifonov, E.N., 2010. Base pair stacking in nucleosome DNA and bendability sequence pattern. J. Theor. Biol. 263, 337–339.

Trifonov, E.N., Nibhani, R., 2015a. Review fifteen years of search for strong nucleosomes. Biopolymers 103 (8), 432–437.

Trifonov, E.N., Nibhani, R., 2015b. Review fifteen years of search for strong nucleosomes. Biopolymers 103, 432–437.

Trifonov, E.N., Sussman, J.L., 1980. The pitch of chromatin DNA is reflected in its nucleotide sequence. Proc. Natl. Acad. Sci. USA 77 (7), 3816–3820.

Vainshtein, Y., Rippe, K., Teif, V.B., 2017. NucTools: Analysis of chromatin feature occupancy profiles from high-throughput sequencing data. BMC Genom. 18 (1), 158.

Valouev, A., Johnson, S.M., Boyd, S.D., et al., 2011. Determinants of nucleosome organization in primary human cells. Nature 474 (7352), 516–520.

van der Heijden, T., van Vugt, J., Logie, C., et al., 2012. Sequence-based prediction of single nucleosome positioning and genome-wide nucleosome occupancy. Proc. Natl. Acad. Sci. USA 109, E2514–E2522.

van Holde, K.E., 1989. Chromatin. New York: Springer-Verlag.

Wang, J.P., Fondufe-Mittendorf, Y., Xi, L., et al., 2008. Preferentially quantized linker DNA lengths in Saccharomyces cerevisiae. PLOS Comput. Biol. 4, e1000175.

Whitehouse, I., Rando, O.J., Delrow, J., Tsukiyama, T., 2007. Chromatin remodelling at promoters suppresses antisense transcription. Nature 450 (7172), 1031–1035.

Woodcock, C.L., Skoultchi, A.I., Fan, Y., 2006. Role of linker histone in chromatin structure and function: H1 stoichiometry and nucleosome repeat length. Chromosome Res. 14 (1), 17–25.

Xi, L., Fondufe-Mittendorf, Y., Xia, L., et al., 2010. Predicting nucleosome positioning using a duration Hidden Markov Model. BMC Bioinform. 11, 346.

Yuan, G.-C., Liu, J.S., 2008. Genomic sequence is highly predictive of local nucleosome depletion. PLOS Comp. Biol. 4, e13.

Yuan, G.C., Liu, Y.J., Dion, M.F., *et al.*, 2005. Genome-scale identification of nucleosome positions in S. cerevisiae. Science 309 (5734), 626–630.
Zhang, Z., Wippo, C.J., Wal, M., *et al.*, 2011. A packing mechanism for nucleosome organization reconstituted across a eukaryotic genome. Science 332 (6032), 977–980.

Further Reading

Arneodo, A., Drillon, G., Argoul, F., Audit, B., 2017. The role of nucleosome positioning in genome function and evolution. In: Lavelle, C., Victor, J.M. (Eds.), Nuclear Architecture and Dynamics. Academic Press, Elsevier.
Blossey, R., 2017. Chromatin: Structure, Dynamics, Regulation. Chapman & Hall/CRC.
Clapier, C.R., Iwasa, J., Cairns, B.R., Peterson, C.L., 2017. Mechanisms of action and regulation of ATP-dependent chromatin-remodelling complexes. Nat. Rev. Mol. Cell Biol. 18, 407–422.
Hughes, A.L., Rando, O.J., 2014. Mechanisms underlying nucleosome positioning in vivo. Annu. Rev. Biophys. 43, 41–63.
Jiang, C., Pugh, B.F., 2009. Nucleosome positioning and gene regulation: Advances through genomics. Nat. Rev. Genet. 10 (3), 161–172.
Lai, W.K.M., Pugh, B.F., 2017. Understanding nucleosome dynamics and their links to gene expression and DNA replication. Nat. Rev. Mol. Cell Biol. (advance online publication).
Längst, G., Teif, V.B., Rippe, K., 2011. Chromatin remodeling and nucleosome positioning. In: Rippe, K. (Ed.), Genome Organization and Function in the Cell Nucleus. Weinheim: Wiley-VCH Verlag GmbH & Co. KGaA, pp. 111–138.
Struhl, K., Segal, E., 2013. Determinants of nucleosome positioning. Nat. Struct. Mol. Biol. 20 (3), 267–273.
Teif, V.B., 2016. Nucleosome positioning: Resources and tools online. Brief Bioinform. 17 (5), 745–757.
Trifonov, E.N., Nibhani, R., 2015. Review fifteen years of search for strong nucleosomes. Biopolymers 103 (8), 432–437.

Relevant Website

http://generegulation.info/index.php/nucleosome-positioning
   Catalog of nucleosome positioning resources.

Biographical Sketch

Vladimir Teif is a Lecturer in the University of Essex. Previously he has been working in the German Cancer Research Center, University of California in San Diego, Hebrew University in Jerusalem, CEA/Saclay and Belarus National Academy of Sciences. His current interests include modelling of epigenetic regulation in chromatin, broadly defined.

Christopher Clarkson is a PhD student in the University of Essex. Previously he did MSc in Bioinformatics in the Imperial College London. Current interests include chromatin topology and prediction of genetic pathways.

Learning Chromatin Interaction Using Hi-C Datasets

Wing-Kin Sung, National University of Singapore, Singapore and Genome Institute of Singapore, Singapore

Introduction

Our genome consists of a set of chromosomes where each chromosome is a linear sequence of four nucleotides A, C, G and T. This linear model organizes the genes in functional units that help to understand many biological processes and diseases. For example, the histone genes are mostly clustered in Albig and Doenecke (1997). We also know that genes in the same pathways are more likely to be close proximity in our genome (Lee and Sonnhammer, 2003). For disease, Down syndrome is caused by the amplification of chromosome 21. Charcot-Marie-Tooth disease type I and Hereditary neuropathy with pressure palsies are caused by duplication and deletion of Roa et al. (1993). Although the linear model is useful for studying our genome, the linear model is insufficient to explain every biological processes. For instance, we cannot explain how enhances control the expressions of their target genes which can be millions base pairs away or even in different chromosomes.

One possible explanation is that our genome folds into a 3 dimensional structure. The folding brings the enhancers close to the target genes to control their expressions. It also can explain why cells of different tissue types have different expression profiles. Although the genome in different tissue cells are the same, the 3 dimensional structures of our genome are different in different tissues (Parada et al., 2004). The tissue-specific spatial organization of our genome may change the spatial distance between the enhancers and their target genes, that finally create the tissue-specific expression profile (Krijger and de Laat, 2016). Hence, decoding the 3D structure of our genome is important to learn the mystery of our life and understand diseases.

To understand the 3D organization of our genome, a few biotechnologies are developed. They can be classified into two approaches: Fluorescence in situ hybridization (FISH) and Chromosome Conformation Capture (3C).

FISH experiments use fluorescence probes to examine the co-localization of different DNA fragments with microscopies. It allows us to measure the 3D distance between two genomic loci. By the distance, we determine if these two genomic loci are interacting. However, since FISH is based on fluorescence probes, we can only measure the 3D distance of the genomic loci for a few loci. It is difficult to expand this approach to the genome-wide scale.

Chromosome Conformation Capture (3C) was proposed by Dekker et al. (2002). Instead of observing the 3D distance of the interacting loci, the 3C method digests the genomic DNA with enzyme, like HindIII, followed by ligation. If two different loci are spatially close, the corresponding DNA fragments have higher chance to be ligated. Using specific PCR primer pair, the interaction between the two loci can be captured. Unlike FISH, 3C cannot tell the interaction of two loci in the single cell level and the 3D distance between them. It only tells the average contact frequency.

Moreover, it is possible to enhance the 3C method to detect the genomic interactions of multiple loci. Many 3C-based methods are developed, they include: 4C (Circular Chromosome Conformation Capture (Hakim and Misteli, 2012) or Chromosome Conformation Capture–on-Chip (Simonis et al., 2006)), 5C (Chromosome Conformation Capture Carbon Copy method (Dostie et al., 2006)), Hi-C (Lieberman-Aiden et al., 2009) and ChIA-PET (Fullwood et al., 2009). 4C methods are used to identify the long chromatin interactions from one locus to all other loci as "one-to-all" approach. 5C method is used to identify the chromatin interactions between many loci specified by primers as "many-to-many" approach. By taking advantage of high-throughput sequencing, Hi-C and ChIA-PET were introduced in 2009 to call genome-wide chromatin interaction.

However, the raw sequencing datasets from Hi-C and ChIA-PET do not provide direct evidences of chromatin interactions. Computational methods are required to extract the interactions among loci and determine the three dimensional structure of our genome. This review aims to review the computational methods for analyzing Hi-C datasets.

The manuscript is organized as follows. We first review the experimental method for Hi-C. Then, we give an overview of the analysis pipeline for Hi-C, which consists of 7 parts. Finally, we detail the currently available tools for these 7 steps.

Review of Hi-C Experiment

Hi-C is proposed by Lieberman-Aiden et al. (2009). It aims to recover the genome-wide chromatin interactions. Precisely, for every two loci in the genome that are spatially close, HiC experiment will generate a chimeric DNA fragment formed by the ligation of the two DNA fragments in these two loci. The steps of Hi-C experiment are as follows. First, cells are cross-linked with formaldehyde. This ensures the interacting DNA fragment pairs are covalently linked. Then, the chromatin is digested with restriction enzyme (say, HindIII). The sticky ends are filled in with biotinylated nucleotides. Ligation is performed under extremely dilute conditions to create chimeric molecules that link the two interacting DNA fragments. These chimeric molecules represent chromatin interactions. Then, these resulting DNA fragments are purified and sheared. By pulling down the biotinylated junctions, we obtain a set of DNA fragments that are formed by connecting two interacting DNA fragments. These fragments are called Hi-C ligation products. Through high-throughput sequencing, Hi-C paired-end reads are obtained from the two ends of the Hi-C ligation products.

Overview of Hi-C Analysis Pipeline

The Hi-C experiment generates paired-end reads which provide information on chromatin interaction. The aim of Hi-C analysis is to decrypt such interactions. Hi-C analysis involves the following steps:

(1) *Mapping*: Given a Hi-C library, this step aligns the Hi-C reads on the reference genome.
(2) *Read-level filtering*: Some aligned read pairs may not be coming from Hi-C ligation products. This step aims to filter the noise and retain only valid Hi-C read pairs.
(3) *Normalization*: Due to experimental noises and mapping biases, systematic errors may occur. This step aims to correct the Hi-C signal.
(4) *Finding local structures*: Our genome is folded into multiple local structures like compartments and TADs. This step aims to call these local structures.
(5) *Calling mid-range or long-range chromatin interactions*: Locally, the genome folds into local structure. Globally, the loci are interacted through mid-range or long-range chromatin interactions. This step aims to call these global interactions.
(6) *Visualization*: A large number of local structures, mid-range and long-range chromatin interactions are predicted in the previous steps. To help users to study these results, a number of tools are developed to visualize those results.
(7) *3D structure prediction*: The Hi-C experiment provides the pairwise interaction information. However, it does not explicitly report the actual chromatin structure. This step aims to predict the 3D structure based on the information provided by Hi-C.

Below, we will detail tools for these 7 steps.

Hi-C Paired Read Mapping

Hi-C paired reads are extracted from the two ends of Hi-C ligation products, where each Hi-C ligation product is formed by ligating two genomic fragments crossing a ligation junction. Depending on the location of the ligation junction, there are 5 cases as shown in **Fig. 1**. Mapping of such Hi-C read pair is complicated by the location of the ligation junction.

Early methods assume that, for most cases, the ligation junctions do not occur near to the ends of the Hi-C ligation products (like the form of **Fig. 1(a)**). Then, simple approaches (like (Yaffe and Tanay, 2011)) use the first 50bp of each read for mapping. They assume the first 50bp does not cross the ligation junction.

HiCLib (also called ICE) (Imakaev *et al.*, 2012) is the first ligation junction-aware method. It first aligns the 25bp of each read in the HiC read pair. If it is aligned uniquely, we accept the alignment. Otherwise, extends by 5bp and realign again until the read is uniquely aligned.

HiCLib is slow since it requires to extend the alignment 5bp by 5bp. Later, chimeric alignment algorithms are proposed. They include: HIPPIE (Hwang *et al.*, 2015), HiCCUPS (Rao *et al.*, 2014), diffHic and HiC-Pro (Servant *et al.*, 2015). These methods run STAR, BWA or Bowtie2 to align the read pair first. If both reads align uniquely, the alignment is done. Otherwise, one of the reads is soft-clipped. These methods trim the reads and realign the clipped portion with the potential to identify the ligation junction.

Read-Level Filtering

After read mapping, the uniquely mapped paired-end reads are subjected to read-level filtering. The read-level filtering removes paired-end reads that are not coming from Hi-C ligation products. These noisy paired-end reads include:

(1) Self-ligation read pairs: Self-ligation product is formed by a DNA fragment that is circularized and self-ligated. The corresponding paired-end read forms an outward alignment and has short insert size.
(2) Dangling read pairs: Dangling product is a normal DNA fragment with no ligation. The corresponding paired-end read forms an inward alignment and has short insert size.
(3) Redundant read pairs: When multiple paired-end reads map to exactly the same genome location, these are redundant read pairs generated by PCR amplification.
(4) Random breaking read pairs: A random breaking read pair Paired alignments where both ends are far from restriction enzyme cut sites:

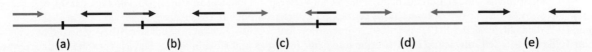

(a)   (b)   (c)   (d)   (e)

Fig. 1 Possible forms of Hi-C ligation product formed by ligating the blue and red DNA fragments. Depending on the location of the ligation junction, there are 5 forms. (a) is the common form where the ligation junction is in the middle, and is not covered by the sequenced reads, (b) is the form where the ligation junction is near to the left end and covered by one of the reads in a pair, (c) is the form where the ligation junction is near to the right end and covered by one of the reads in a pair, (d) and (e) are the form where there is no ligation.

In this step, the alignment of every read pair is examined. If the read pair is classified as self-ligation read pair, dangling read pair, redundant read pair, or random breaking read pair, it is filtered.

The read-level filtering is implemented in a number of software. They include: HICUP (Wingett *et al.*, 2015), HiC-inspector, HiC-Box, HiCdat, HiCLib (Imakaev *et al.*, 2012), HiC-Pro (Servant *et al.*, 2015), HIPPIE (Hwang *et al.*, 2015).

Hi-C Data Normalization

Due to biases in the wet-lab steps and mapping, the Hi-C paired reads are also biasedly mapped on the genome. One (arguably the most) important step is to normalize the Hi-C signal. There are two approaches: (1) explicitly · model the bases and (2) matrix balancing approach.

Yaffe and Tanay (2011) proposed Hicpipe to explicitly model the biases. For any two fragment ends a and b, they observed three types of biases:

(1)   Mappability: For any fragment end x, the mappability M(x) is defined to be the proportion of loci in [x-500, x + 500] that can be uniquely aligned. Yaffe and Tanay observed that the number of interaction reads between two fragment ends a and b is proportional to M(a)M(b).
(2)   GC content of the fragment ends: Yaffe and Tanay observed that GC content around the fragment ends affects the ligation efficiency.
(3)   Fragment length: Restriction fragment length is the length of the fragment after cutting the genome with the restriction enzyme. Yaffe and Tanay observed that restriction fragment length affects the ligation efficiency.

By analyzing these three biases, Yaffe and Tanay (2011) proposed a model to compute the by-chance probability of the interaction of two fragment ends. Normalized contact map is computed by dividing the observed number of contacts between 1-Mb chromosomal bins by the expected number of contacts predicted by the model.

HiCNorm (Hu *et al.*, 2012) is another method for normalizing the contact map. It also aims to remove the above three biases. HiCNorm uses a generalized linear regression-based method. HiCNorm directly models the contact frequency between two bins using Poisson or negative binomial distribution, which simplify the set of parameters. Hence, HiCNorm is computationally more efficient. A few pipelines use HICNorm for normalization. They include: HiCdat (Schmid *et al.*, 2015), HIPPIE (Hwang *et al.*, 2015).

Apart from explicitly models the biases, another approach is called matrix-balancing approach. The matrix-balancing approach does not assume the knowledge of the sources of biases. It assumes that, in the "true" model, every locus has the same number of interactions with other genomic regions. This is known as the equal visibility assumption.

Under the equal visibility assumption, the simplest solution is to use vanilla coverage (Lieberman-Aiden *et al.*, 2009). Consider any two bins i and j, suppose the observed number of paired end reads between two bins i and j is $O(i,j)$. Then, the number of paired end reads with one end in bin i is $O(i) = \sum_j O(i,j)$. The normalized count between two bins i and j is $N(i,j) = \frac{O(i,j)}{O(i)O(j)}$.

Vanilla coverage is not good enough, since the noise in the original interaction frequency will affect the normalized count significantly. Later, ICE (Imakaev *et al.*, 2012) is proposed. ICE is an iterated version of Vanilla coverage. It repeatedly runs Vanilla coverage until the signal is converged. A few methods use ICE for normalization. They include: HiCdat (Schmid *et al.*, 2015), Fit-Hi-C (Ay *et al.*, 2014), HiC-Pro (Servant *et al.*, 2015), HOMER (Heinz *et al.*, 2010).

Local Structure Calling

The 3 dimensional structure of our genome has two levels of structures: local and global structures. Locally, our genome is folded into multiple hierarchical of local structures, from compartments (Lieberman-Aiden *et al.*, 2009) to TADs (Dixon *et al.*, 2012) and nested sub-TAD (Rao *et al.*, 2014). Then, those local structures interact and form the global structure. Here, we detail the local structure calling methods.

Compartments partition the genome into regions with high frequency of interactions and regions with low frequency of interactions. The initial approach (Lieberman-Aiden *et al.*, 2009; Imakaev *et al.*, 2012) uses principal component analysis (PCA) to identify them. Given the normalized contact probability T_{ij} for every pair of bins i and j, $T_{ij} = \sum_{k=1}^{N} \lambda_k \cdot E_i^k \cdot E_j^k + \overline{T}$, where $\overline{T}$ is the mean value of T, λ_k is the Eigen value and E^k is the Eigen vector. It is observed that the first Eigen vector divides the genome into compartments with high and low frequency of interactions (Sometimes, the first Eigen vector represents the short and long chromosome arms. In this case, the compartments are represented by the second Eigen vector.) Compartments typically are of size several megabases. Compartments with high frequencies of interactions correspond to Euchromatin region with high density of genes while compartments with low frequency of interactions correspond to heterochromatin and is largely made up of gene deserts.

Using Hi-C data, Dixon *et al.* (2012) visualized 2D-interaction matrices using a variety of bin sizes. They noticed that at bin sizes less than 100kb, highly self-interacting regions begin to emerge (triangles in the heatmap, see **Fig. 2**). These regions are termed as "Topologically-Associated Domain" (TAD). Many methods are proposed to call TADs. They include: DomainCaller(DI/HMM) (Dixon *et al.*, 2012), Insulation score (Crane *et al.*, 2015), Armatus (Filippova *et al.*, 2014), HiCseg (Levy-Leduc *et al.*, 2014), DomainCaller (Dixon *et al.*, 2012), TopDom (Shin *et al.*, 2016), IC-Finder (Haddad *et al.*, 2017), TADbit (Serra *et al.*, 2017) and HiFive (Sauria *et al.*, 2015).

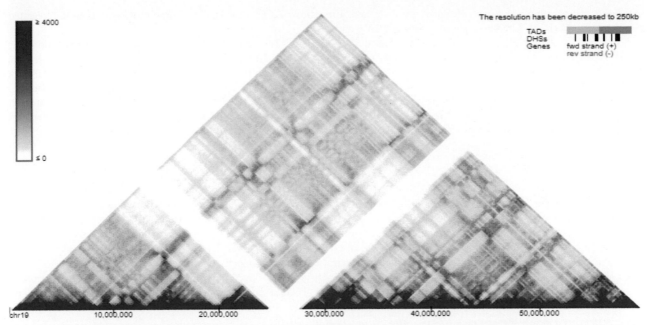

Fig. 2 This is the visualization of the 2D-interaction matrices of chr19 using bins of size 25 k. A number of self-interacting regions (as visualized as triangles) are observed. (The figure is generated using 3D Genome Browser. Reproduced from Wang, Y., Zhang, B., Zhang, L., *et al.*, 2017b. The 3d genome browser: A web-based browser for visualizing 3d genome organization and long-range chromatin interactions. bioRxiv).

Rao *et al.* (2014) observed that within a TAD, it contains some smaller structures called subTADs. Juicer (Arrowhead) (Rao *et al.*, 2014), TADtree (Weinreb and Raphael, 2016), CaTCH (Zhan *et al.*, 2017), HBM (Shavit *et al.*, 2016), and HiTAD (Wang *et al.*, 2017a) are developed to predict these subTADs.

Recently, Forcato *et al.* (2017) and Dali and Blanchette (2017) have assessed the performance of these methods. Since different TAD prediction tools are based on different assumptions, they report different sets of TADs. In the future, more works are needed to improve the predictions of TADs and subTADs.

The above methods predict TADs using Hi-C data only. There are methods that predict TADs using other omics datasets. One example is HubPredictor (Huang *et al.*, 2015), which predicts TADs using both Hi-C and histone mark ChIP-seq data.

Calling Mid-Range or Long-Range Chromatin Interactions

Our genome locally forms structures like compartments, TADs and subTADs. Globally, the loci are interacting through mid-range or long-range chromatin interactions. A number of methods have been proposed to call these global interactions.

Duan *et al.* (2010). gives the early method to call global interactions. A background expected contact frequency is learnt for each genomic distance. When the actual contact frequency between two loci is higher than expected, these two loci are deemed to be interacting.

After that, a number of papers are proposed for calling global interactions. The major improvement is on the modeling of the background contact frequency. AFC (Jin *et al.*, 2013) models the background contact frequencies as a negative binomial distribution and a global background model was devised that consists of both systematic bias factors and the linear genomic distance factor. Fit-Hi-C (Ay *et al.*, 2014) uses a non-parametric spline approach to model the background-chromatin contact frequency. The above two methods have many false positives. GOTHiC (Mifsud *et al.*, 2017) stated that the background contact frequency also depends on the read coverage. It improves the global background interaction frequency of two loci to depend also on the relative genome-wide coverage.

All above methods assume the chromatin interaction frequencies for different loci pairs are independent. Instead of assuming a global background, HiCCUPS (Rao *et al.*, 2014) uses a combination of local and global background to model the contact frequency. HMRF (Xu *et al.*, 2016a) proposed to use a hidden Markov random field based Bayesian method to explicitly model the spatial dependency among adjacent loci. HMRF is slow. The same authors proposed FastHiC (Xu *et al.*, 2016b), which uses a simulated field approximation to approximate the joint distribution of the hidden peak status by a set of independent random variables, leading to more tractable computation.

Visualization

After the Hi-C data is processed, local structures like TADs and subTADs, mid-range and long-range chromatin interactions are predicted. To help the users, it is good to visualize them. A number of software is available for Hi-C data visualization. First, we can use traditional genome browser to visualize the Hi-C data. These browsers include: USCS genome browser, Ensembl, Integrated genome browser (IGB), Integrated genome viewer (IGV), etc.

Some specialized visualization tools for HiC data are also available. They include: WashU Epigenome Browser (Zhou *et al.*, 2013), CytoHiC (Shavit and Lio, 2013), HiCPlotter (Akdemir and Chin, 2015), QuIN (Thibodeau *et al.*, 2016), 3D Genome Browser (Wang et al., 2017b). WashU Epigenome Browser is modified that allows us to view long-range contact data. CytoHiC is a Cytsocape plugin that allow users to view and compare spatial maps of genomic landmarks, based on normalized Hi-C datasets. HiCPlotter and 3D Genome Browser are simple tools that visualize Hi-C data as a Hi-C matrix. QuIN (Thibodeau *et al.*, 2016) represents the Hi-C interactions as a network. Precisely, the genomic loci are represented as vertices while the interactions between two genomic loci are represented as edges.

3D Modeling

The above steps extract interactions among pairs of loci. They did not recover the actual three dimensional structure of the genome. This step aims to reconstruct the three dimensional chromatin structure from the pairwise interactions predicted from Hi-C data. Given the contact frequencies among the genomic loci, most existing methods run in two steps. The first step is to convert the contact frequency between any two loci into their spatial distance. Then, the second step learns the 3D structure from the spatial distance matrix. There are two approaches for the second step: Consensus approach and ensemble approach.

Consensus approach uses the spatial distance (or the contact frequency directly) as the constraint, then the 3D structure is reconstructed. This approach is first used by Duan *et al.* (2010). Then, a number of methods are developed using the same approach. They include: (Tanizawa *et al.*, 2010), ChromSDE (Zhang *et al.*, 2013), ShRec3D (Lesne *et al.*, 2014), 3D-GNOME (Szalaj *et al.*, 2016), PASTIS (Varoquaux *et al.*, 2014), and HSA (Zou *et al.*, 2016).

Another approach is ensemble approach. Observe that the chromatin interactions extracted from the Hi-C experiment are actually the interactions from a pool of millions of cells. The conformations in different cells are different. The ensemble approach aims to produce a set of 3D models which can fit the observed contact frequencies. There are two strategies. The first strategy identifies multiple independent models that can fit the Hi-C data. Hence, this strategy can identify multiple models instead of just one local optimal model. Trieu and Cheng (Trieu and Cheng, 2014), MOGEN (Trieu and Cheng, 2016), LorDG (Trieu and Cheng, 2017), AutoChrom3D (Peng *et al.*, 2013), (Bau and Marti-Renom, 2012) (and TADbit (Serra *et al.*, 2017)), MCMC5C (Rousseau *et al.*, 2011) use this strategy.

Another strategy finds multiple models that can fit the Hi-C data in a coordinated manner. BACH-MIX (Hu *et al.*, 2013) uses MCMC to find a mixture of models that can fit the Hi-C data. InfMod3DGen (Wang *et al.*, 2015) uses EM algorithm to infer an ensemble of models that best fit the data. Other methods include Kalhor *et al.* (2011)

Summary

Understanding the 3 dimensional organization of the genome is an important topic. We started to understand that our genome is not simply organized linearly. Instead, our genome is folded into local structures called TADs; then, the local structures are interacting through mid-range and long-range interactions. Researchers found that the 3 dimensional structure of our genome can help to explain different biological processes like the enhancer and promoter interaction and the tissue-specific expression profiles.

In the last decades, many tools have been developed to recover the three dimensional structure of our genome. Although those tools provide useful information, the predictions are still far from accurate. In the future, more works need to be done.

See also: Chromatin: A Semi-Structured Polymer. Experimental Platforms for Extracting Biological Data: Mass Spectrometry, Microarray, Next Generation Sequencing. Genome Informatics. Natural Language Processing Approaches in Bioinformatics. Nucleosome Positioning. Protein-DNA Interactions. Quantitative Immunology by Data Analysis Using Mathematical Models

References

Akdemir, K.C., Chin, L., 2015. Hicplotter integrates genomic data with interaction matrices. Genome Biology 16, 198.
Albig, W., Doenecke, D., 1997. The human histone gene cluster at the D6S105 locus. Hum. Genet. 101, 284–294.
Ay, F., Bailey, T.L., Noble, W.S., 2014. Statistical confidence estimation for hi-c data reveals regulatory chromatin contacts. Genome Research 24, 999–1011.
Bau, D., Marti-Renom, M.A., 2012. Genome structure determination via 3c-based data integration by the integrative modeling platform. Methods (San Diego, CA) 58, 300–306.
Crane, E., Bian, Q., McCord, R.P., *et al.*, 2015. Condensin-driven remodelling of X chromosome topology during dosage compensation. Nature 523, 240–244.
Dali, R., Blanchette, M., 2017. A critical assessment of topologically associating domain prediction tools. Nucleic Acids Research 45, 2994–3005.
Dekker, J., Rippe, K., Dekker, M., Kleckner, N., 2002. Capturing chromosome conformation. Science (New York, N.Y.) 295, 1306–1311.
Dixon, J.R., Selvaraj, S., Yue, F., *et al.*, 2012. Topological domains in mammalian genomes identified by analysis of chromatin interactions. Nature 485, 376–380.
Dostie, J., Richmond, T.A., Arnaout, R.A., *et al.*, 2006. Chromosome conformation capture carbon copy (5c): A massively parallel solution for mapping interactions between genomic elements. Genome Research 16, 1299–1309.
Duan, Z., Andronescu, M., Schutz, K., *et al.*, 2010. A three-dimensional model of the yeast genome. Nature 465, 363–367.
Filippova, D., Patro, R., Duggal, G., Kingsford, C., 2014. Identification of alternative topological domains in chromatin. Algorithms for Molecular Biology 9, 14.

Forcato, M., Nicoletti, C., Pal, K., 2017. Carmen Maria Livi, Francesco Ferrari, and Silvio Bicciato. Comparison of computational methods for hi-c data analysis. Nature Methods 14, 679–685.

Fullwood, M.J., Liu, M.H., Pan, Y.F., et al., 2009. An oestrogen-receptor-alpha-bound human chromatin interactome. Nature 462, 58–64.

Haddad, N., Vaillant, C., Jost, D., 2017. Ic-finder: Inferring robustly the hierarchical organization of chromatin folding. Nucleic Acids Research 45, e81.

Hakim, O., Misteli, T., 2012. Snapshot: Chromosome confirmation capture. Cell 148, 1068.e1–1068.e2.

Heinz, S., Benner, C., Spann, N., et al., 2010. Simple combinations of lineage-determining transcription factors prime cis-regulatory elements required for macrophage and b cell identities. Molecular Cell 38, 576–589.

Huang, J., Marco, E., Pinello, L., Yuan, G.-C., 2015. Predicting chromatin organization using histone marks. Genome Biology 16, 162.

Hu, M., Deng, K., Qin, Z., et al., 2013. Bayesian inference of spatial organizations of chromosomes. PLOS Computational Biology 9, e1002893.

Hu, M., Deng, K., Selvaraj, S., et al., 2012. Hicnorm: Removing biases in hi-c data via poisson regression. Bioinformatics (Oxford) 28, 3131–3133.

Hwang, Y.-C., Lin, C.-F., Valladares, O., et al., 2015. Hippie: A high-throughput identification pipeline for promoter interacting enhancer elements. Bioinformatics (Oxford) 31, 1290–1292.

Imakaev, M., Fudenberg, G., McCord, R.P., et al., 2012. Iterative correction of hi-c data reveals hallmarks of chromosome organization. Nature Methods 9, 999–1003.

Jin, F., Li, Y., Dixon, J.R., et al., 2013. A high-resolution map of the three-dimensional chromatin interactome in human cells. Nature 503, 290–294.

Kalhor, R., Tjong, H., Jayathilaka, N., Alber, F., Chen, L., 2011. Genome architectures revealed by tethered chromosome conformation capture and population-based modeling. Nature Biotechnology 30, 90–98.

Krijger, P.H.L., de Laat, W., 2016. Regulation of disease-associated gene expression in the 3d genome. Nature Reviews. Molecular Cell Biology 17, 771–782.

Lee, J.M., Sonnhammer, E.L.L., 2003. Genomic gene clustering analysis of pathways in eukaryotes. Genome Research 13, 875–882.

Lesne, A., Riposo, J., Roger, P., Cournac, A., Mozziconacci, J., 2014. 3d genome reconstruction from chromosomal contacts. Nature Methods 11, 1141–1143.

Levy-Leduc, C., Delattre, M., Mary-Huard, T., Robin, S., 2014. Two-dimensional segmentation for analyzing hi-c data. Bioinformatics (Oxford) 30, i386–i392.

Lieberman-Aiden, E., van Berkum, N.L., Williams, L., et al., 2009. Comprehensive mapping of long-range interactions reveals folding principles of the human genome. Science (New York, N.Y.) 326, 289–293.

Mifsud, B., Martincorena, I., Darbo, E., et al., 2017. Gothic, a probabilistic model to resolve complex biases and to identify real interactions in hi-c data. PLOS ONE 12, e0174744.

Parada, L.A., McQueen, P.G., Misteli, T., 2004. Tissue-specific spatial organization of genomes. Genome Biology 5, R44.

Peng, C., Fu, L.-Y., Dong, P.-F., et al., 2013. The sequencing bias relaxed characteristics of hi-c derived data and implications for chromatin 3d modeling. Nucleic Acids Research 41, e183.

Rao, B.B., Garcia, C.A., Suter, U., et al., 1993. Charcot-Marie-Tooth Disease Type 1A – Association with a Spontaneous Point Mutation in the PMP22 Gene. N Engl J Med. 329, 96–101.

Rao, S.S.P., Huntley, M.H., Durand, N.C., et al., 2014. A 3d map of the human genome at kilobase resolution reveals principles of chromatin looping. Cell 159, 1665–1680.

Rousseau, M., Fraser, J., Ferraiuolo, M.A., Dostie, J., Blanchette, M., 2011. Three-dimensional modeling of chromatin structure from interaction frequency data using markov chain monte carlo sampling. BMC Bioinformatics 12, 414.

Sauria, M.E., Phillips-Cremins, J.E., Corces, V.G., Taylor, J., 2015. Hifive: A tool suite for easy and efficient hic and 5c data analysis. Genome Biology 16, 237.

Schmid, M.W., Grob, S., Grossniklaus, U., 2015. Hicdat: A fast and easy-to-use hi-c data analysis tool. BMC Bioinformatics 16, 277.

Serra, F., Bau, D., Goodstadt, M., et al., 2017. Automatic analysis and 3d-modelling of hi-c data using tadbit reveals structural features of the fly chromatin colors. PLOS Computational Biology 13, e1005665.

Servant, N., Varoquaux, N., Lajoie, B.R., et al., 2015. Hic-pro: An optimized and flexible pipeline for hi-c data processing. Genome Biology 16, 259.

Shavit, Y., Lio, P., 2013. Cytohic: A cytoscape plugin for visual comparison of hi-c networks. Bioinformatics (Oxford) 29, 1206–1207.

Shavit, Y., Walker, B.J., Lio, P., 2016. Hierarchical block matrices as efficient representations of chromosome topologies and their application for 3c data integration. Bioinformatics (Oxford) 32, 1121–1129.

Shin, H., Shi, Y., Dai, C., et al., 2016. Topdom: An efficient and deterministic method for identifying topological domains in genomes. Nucleic Acids Research 44, e70.

Simonis, M., Klous, P., Splinter, E., et al., 2006. Nuclear organization of active and inactive chromatin domains uncovered by chromosome conformation capture-on-chip (4c). Nature Genetics 38, 1348–1354.

Szalaj, P., Michalski, P.J., Wroblewski, P., et al., 2016. 3d-gnome: An integrated web service for structural modeling of the 3d genome. Nucleic Acids Research 44, W288–W293.

Tanizawa, H., Iwasaki, O., Tanaka, A., et al., 2010. Mapping of long-range associations throughout the fission yeast genome reveals global genome organization linked to transcriptional regulation. Nucleic Acids Research 38, 8164–8177.

Thibodeau, A., Marquez, E.J., Luo, O., et al., 2016. Quin: A web server for querying and visualizing chromatin interaction networks. PLOS Computational Biology 12, e1004809.

Trieu, T., Cheng, J., 2014. Large-scale reconstruction of 3d structures of human chromosomes from chromosomal contact data. Nucleic Acids Research 42, e52.

Trieu, T., Cheng, J., 2016. Mogen: A tool for reconstructing 3d models of genomes from chromosomal conformation capturing data. Bioinformatics (Oxford) 32, 1286–1292.

Trieu, T., Cheng, J., 2017. 3d genome structure modeling by lorentzian objective function. Nucleic Acids Research 45, 1049–1058.

Varoquaux, N., Ay, F., Noble, W.S., Vert, J.-P., 2014. A statistical approach for inferring the 3d structure of the genome. Bioinformatics (Oxford) 30, i26–i33.

Wang, X.-T., Cui, W., Peng, C., 2017a. Hitad: Detecting the structural and functional hierarchies of topologically associating domains from chromatin interactions. Nucleic Acids Research 45, e163.

Wang, S., Xu, J., Zeng, J., 2015. Inferential modeling of 3d chromatin structure. Nucleic Acids Research 43, e54.

Wang, Y., Zhang, B., Zhang, L., et al., 2017b. The 3d genome browser: A web-based browser for visualizing 3d genome organization and long-range chromatin interactions. bioRxiv.

Weinreb, C., Raphael, B.J., 2016. Identification of hierarchical chromatin domains. Bioinformatics (Oxford) 32, 1601–1609.

Wingett, S., Ewels, P., Furlan-Magaril, M., et al., 2015. Hicup: Pipeline for mapping and processing hi-c data. F1000Research 4, 1310.

Xu, Z., Zhang, G., Jin, F., et al., 2016a. A hidden markov random field-based bayesian method for the detection of long-range chromosomal interactions in hi-c data. Bioinformatics (Oxford) 32, 650–656.

Xu, Z., Zhang, G., Wu, C., Li, Y., Hu, M., 2016b. Fasthic: A fast and accurate algorithm to detect long-range chromosomal interactions from hi-c data. Bioinformatics (Oxford) 32, 2692–2695.

Yaffe, E., Tanay, A., 2011. Probabilistic modeling of hi-c contact maps eliminates systematic biases to characterize global chromosomal architecture. Nature Genetics 43, 1059–1065.

Zhang, Z.Z., Li, G., Toh, K.-C., Sung, W.-K., 2013. Inference of spatial organizations of chromosomes using semi-definite embedding approach and hi-c data. In: RECOMB, pp. 317–332.

Zhan, Y., Mariani, L., Barozzi, I., et al., 2017. Reciprocal insulation analysis of hi-c data shows that tads represent a functionally but not structurally privileged scale in the hierarchical folding of chromosomes. Genome Research 27, 479–490.

Zhou, X., Lowdon, R.F., Li, D., et al., 2013. Exploring long-range genome interactions using the washu epigenome browser. Nature Methods 10, 375–376.

Zou, C., Zhang, Y., Ouyang, Z., 2016. Hsa: Integrating multi-track hi-c data for genome-scale reconstruction of 3d chromatin structure. Genome Biology 17, 40.

Transcriptome Informatics

Liang Chen and Garry Wong, University of Macau, Macau, China

Introduction

Since the elucidation of the double-helix as the structure of DNA in biological systems, and the advent of genetic engineering techniques in laboratory science, DNA has been at the forefront of the mind of many scientists. However, RNA has for long served as a central character in the process of life. This can be easily seen in the central dogma which states that DNA is transcribed into RNA and RNA is translated into protein (Crick, 1970). Whereas DNA maintains and replicates the information, it requires an intermediary, RNA, in order to produce protein. Indeed, RNA can perform the functions of DNA as a repository of information and template for replication, as well as catalyze chemical reactions in the form of a ribozyme (Serganov and Patel, 2007). Moreover, RNA is capable of regulating itself from various feedback and feed forward loops. Furthermore, RNA's ability to form secondary structure, which allows it to fold into tertiary structure, makes it a good structural molecule in large protein complexes ranging from those needed in transcription to capping the ends of telomeres (Azzalin et al., 2007; Lodish et al., 1995). Because RNA can perform so flexibly and has been so well conserved, it has been speculated that evolution began with RNA in a RNA world and only later evolved to include DNA and protein (Neveu et al., 2013). It should be no surprise that even the founding group of molecular biologists in the 1950s (James D. Watson, George Gamow, Sydney Brenner and others) formed their club with the name "RNA tie club" rather than DNA (Kay, 2000). Basic fundamental discoveries concerning the structure, function, and roles of RNA in cellular systems continue to amaze and surprise.

Theoretical Background

Transcription

The process of transcription describes the steps required to copy the genetic information encoded in DNA to RNA. Each step involves specific chemical reactions and actions at the subcellular level to bring molecular components together to form these reactions. RNA (ribonucleic acid) differs from DNA (deoxyribonucleic acid) by having a hydrogen atom in the 2′ position of the ribose sugar. However, both form polymers based on the ribose sugar backbone and connecting phosphodiester bonds with orientations at 5′ and 3′ ends. In transcription, the thymidine residue of DNA is replaced by uracil in RNA. Transcription occurs in 4 discrete steps: initiation, promoter escape, elongation, and termination (Kaiser et al., 2007). Initiation describes the step where double stranded DNA, which is wrapped around histones becomes unwound and accessible to the transcription initiation complex (Murakami et al., 2002a,b; Sainsbury et al., 2015). This complex consists of a large number of proteins and ribosomal RNA factors that separate the double stranded DNA and begin synthesizing a RNA polymer complementary to the DNA sequence. The initiation site on DNA is defined by a promoter sequence, CpG islands, and a transcription factor binding site on the double stranded DNA that is proximal, and enhancer DNA sequences that are distal to the initiation site. Activators, co-activators, repressors, and co-repressors are proteins that form part of the transcription initiation complex and provide an additional level of control (Juven-Gershon and Kadonaga, 2010; Ptashne and Gann, 1997). Promoter escape describes the step where the RNA polymer has begun to be synthesized (Dvir, 2002). A RNA polymerase recruited by the transcription initiation complex synthesizes a RNA strand in reverse complement (A→U, C→G, G→C, T→A) using one DNA strand as the template. RNA is synthesized as antiparallel: 5′ to 3′ orientation. The DNA template used is termed the antisense DNA strand and the opposite DNA strand the sense strand. Different classes of RNA molecules are synthesized depending upon the RNA polymerase used: RNA polymerase I synthesizes ribosomal RNA (rRNA), RNA polymerase II synthesizes messenger RNA (mRNA), and RNA polymerase III synthesizes tRNA as well as other small RNAs. After a ~15-mer RNA nucleotide is synthesized, the RNA polymerase may continue. At this step, promoter escape is completed and elongation begins (Fouqueau and Werner, 2017). Transcription termination following elongation of the RNA strand is signalled by DNA sequences downstream of the promoter and results in RNA polymerase stopping its synthesis (Richard and Manley, 2009). The DNA sequences recognized by the transcription initiation complex to the termination signal define a transcription unit and are also referred to as a genes (Alberts, 2007). Post-transcriptional processing of newly synthesized RNAs occurs following termination. These include chemical modifications to RNA such as 5′ capping, polyadenylation, and splicing of mRNAs, RNA editing, and degradation. Regulatory processes that influence the synthesis and abundance of RNAs that occur before and during transcription are termed pre-transcriptional and transcriptional regulation, respectively, while those that occur after the RNA has been synthesized are termed post-transcriptional (Jacob and Monod, 1961). Prokaryotes differ from eukaryotes in transcription by less complexity, but basic principles are maintained (Eick et al., 1994). Some RNA based viruses use RNA as the template in transcription (Baltimore, 1971). Reverse transcription describes the process where RNA serves as the template for the synthesis of DNA (Baltimore, 1970; Temin and Mizutani, 1970). It has been estimated that greater than 85% of an eukaryotic genome is transcribed into RNA at some time during the life of the organism (Hangauer et al., 2013).

Major RNA Classes

While most RNA classes have been historically defined by their chemical and functional properties, the power of next-generation sequencing has blurred some of these distinctions. A recent example shows how a single molecule can be classified as either an eRNA or lncRNA (Espinosa, 2016; Paralkar *et al.*, 2016). RNA classes may also be defined via the proteins family members they bind such as Argonauts to distinguish between miRNAs and piRNAs (Meister, 2013). Below, we list major RNA classes as defined in the current literature.

mRNA – messenger RNAs represent protein coding and noncoding RNAs transcribed by RNA polymerase II. They are capped by addition of a methyl-7 guanine moiety and modification of 2′-O-methyl ribose (Fabrega *et al.*, 2004; Kiss, 2001). After synthesis, adenine ribonucleic acid residues (poly A +) are added by polyadenylate polymerase via a complex that recognizes a poly A signal sequence (AAUAAA) in the last 10–30 nucleotides of the mRNA (Colgan and Manley, 1997). The poly A + mRNA is exported from the nucleus to the cytoplasm where it is edited to remove introns and splices together the exons. These edited mRNA molecules are translated into proteins by ribosomes located either in rough endoplasmic reticulum or in the cytosol (Stephens *et al.*, 2008). Noncoding RNAs may undergo further processing. It is estimated that humans have around 20,000 protein coding genes which may be sliced and edited to form approximately 10 different forms for each gene on average.

tRNA – also known as transfer RNA, it is the adaptor molecule that translates triplet codon information from mRNA to amino acid. It is part of the translation machinery called the ribosome that also contains ribosomal RNAs and proteins. The size of a tRNA is ~75 to 90 nucleotides. tRNAs are transcribed from clusters in the genome by RNA polymerase III and is highly conserved across species (Sharp *et al.*, 1985).

rRNA – also known as ribosomal RNA, it is a structural molecule that forms the ribosomal complex responsible for translating mRNA into protein (Palade, 1955). The ribosomal complex consists of multiple rRNAs and over a dozen proteins depending upon the species, along with tRNAs (Wilson and Doudna Cate, 2012). rRNAs are also highly conserved and present in all taxa and therefore are often used in evolutionary studies (Woese, 1987). rRNA is the most abundant class of RNAs and constitutes ~50% of all RNA species in a cell. rRNAs range in size from 1.5 to 5 kb (Noller, 1991).

Ribozymes – these are RNA molecules that have catalytic ability, typically to generate RNA polymers by ligating or cleaving RNA fragments together or in the ribosome to catalyze protein peptide production (Guerrier-Takada *et al.*, 1983; Kruger *et al.*, 1982). Ribozymes can be very small: around 50 nucleotides in total. Ribozymes were originally naturally occurring RNA molecules, however, they have been actively pursued and engineered as biotechnology products especially to cleave unwanted RNA molecules such as from viruses or those generated in diseases (Pyle, 1993).

eRNA – enhancer RNAs are recently discovered non coding RNAs transcribed from enhancer regions (Kim *et al.*, 2010). They are short sequences and form multiple overlapping transcripts that are not typically spliced or polyadenylated. They commonly occur at distal enhancers several kb away from the promoter and are believed to act by aiding in the recruitment of co-activators to the transcription initiation complex (Kowalczyk *et al.*, 2012). The size of eRNA varies but can be as short as 50 nucleotides and as large as several kb (Natoli and Andrau, 2012).

lncRNA – long non-coding RNAs define a class of RNAs that are >200 nucleotides long but do not code for proteins. It is now estimated that the human genome may contain >50,000 lncRNAs (Iyer *et al.*, 2015). They may be transcribed antisense to a coding RNA transcript or in intergenic spaces. They are polyadenylated and spliced and therefore share many of the same biosynthesis properties as mRNAs. While they are abundant, the precise functions of only a few are known, such as Xist which acts in dosage compensation of the X chromosome via recruitment of a polycomb repressor complex that lays down repressive epigenetic marks (Ballabio and Willard, 1992; Calabrese *et al.*, 2012). Another example is the telomerase RNA component, TERC that acts as scaffold and template to add DNA ends to telomeres (Greider and Blackburn, 1989; Kapranov *et al.*, 2007; Kung *et al.*, 2013).

miRNA – microRNAs are small noncoding RNAs transcribed by RNA polymerase II mostly but sometimes by RNA polymerase III as well (Borchert *et al.*, 2006; Lee *et al.*, 1993; Lee *et al.*, 2004). They may also be spliced from introns. These are termed mirtrons. They may also be transcribed in intergenic spaces. The initial transcript is a primary miRNA (pri-miRNA) and after its cleavage to a 75–90 nucleotide hairpin secondary structure, it is termed a pre-miRNA (precursor miRNA) (Ha and Kim, 2014). The finally processed 21 or 22 nucleotide form is the mature miRNA. miRNAs function as down-regulators of mRNA either pre-transcriptionally or post-transcriptionally. Human genomes may express >6000 different miRNAs with a very wide range of expression levels (Londin *et al.*, 2015). Isomeric forms of miRNAs termed isomiRs are also known to exist and are derived from differential processing of pre-miRNAs by Dicer enzyme (He and Hannon, 2004).

moRNA – microRNA off-set RNAs are small RNA fragments 15–22 nucleotides in length that are derived from the pre-miRNA but are "offset" on the hairpin either located upstream of the 5p mature miRNA sequence or downstream of the 3p mature miRNA sequence (Langenberger *et al.*, 2009; Shi *et al.*, 2009). The abundance of moRNAs compared to miRNAs is very low: they typically represent <1% of miRNAs derived from the same pre-miRNA. moRNAs have been found most abundantly in stem cells, embryonic bodies, and germ cells (Asikainen *et al.*, 2015). moRNAs may act to regulate mRNA expression in molecular pathways similar or shared with miRNAs (Langenberger *et al.*, 2009; Shi *et al.*, 2009).

snRNA – small nuclear RNAs are 100–300 nucleotides in length and highly abundant in the nucleus. Their sequences are Uridine rich and they function as a scaffold in splicing introns within the spliceosome complex. snRNAs exist in riboprotein complexes termed snrps (small nuclear riboproteins). They are a family of molecules and their high abundance in cells often leads to them being used as a loading control in experiments involving quantitation of small non coding RNAs (Matera *et al.*, 2007).

snoRNA – small nucleolar RNAs are 70 nucleotides in length and processed from intron fragments. Vertebrates express several hundred snoRNAs. snoRNAs act as a guide for rRNA processing, rRNA maturation, or chemical modification of complementary sequences in rRNA (Matera *et al.*, 2007).

piRNA – piwi interacting RNAs are 26–31 nucleotide RNAs transcribed from specific loci (Saito *et al.*, 2006; Vagin *et al.*, 2006). Their sequences are highly diverse and thousands of different piRNAs may be transcribed from different regions of a chromosome. Simple model organisms like fruit flies and nematodes may express tens of thousands of piRNAs while more complex organisms like mice and humans may express even more (Das *et al.*, 2008). piRNAs typically begin with Uridine, however remaining sequences may be highly variable and no consistent secondary structure has so far been found. Their initial function has been described as inhibiting transcription of transposon sequences in the genome, but more recently they have been shown as essential for controlling transcription in sperm cells (Iwasaki *et al.*, 2015).

circRNA – circular RNAs are recently discovered covalently closed RNA molecules derived from splicing of exonic sequences of mRNA or lncRNAs (Hansen *et al.*, 2013; Memczak *et al.*, 2013). Some circRNAs also contain intronic sequences. The covalent linkage to form a circular molecule occurs via a 3–5′ or 2–5′ phosphate. Although initially found in higher organisms, they now appear to be found in a wide variety of organisms including most eukaryotes studied (Ebbesen *et al.*, 2016). While their function remains under active investigation, their high abundance in brain tissues and during development suggests involvement in specific molecular functions in particular biological processes. It is hypothesized that they may act as molecular sponges by sequestering active mature miRNAs, although other potential mechanisms of action have been suggested. Sizes of circRNAs range from less than 200 nucleotides to greater than 4 kb (Lasda and Parker, 2014; Salzman, 2016).

endo-siRNA – endogenous silencing RNAs are short ~20 to 27 nucleotide sequences that are similar to miRNAs as short RNA species that attenuate gene expression (Asikainen *et al.*, 2008; Okamura and Lai, 2008). They can be derived from dsRNA structures, but unlike miRNAs do not form hairpin structures. They may however share components of the RNA silencing machinery with miRNAs including amplification pathways and binding to Argonautes. They are highly abundant in many different species and can be found in germline cells such as sperm and thus may be important in developmental processes (Nilsen, 2008; Yuan *et al.*, 2016).

rasiRNA - Repeat associated silencing RNAs are short 24–29 nucleotide RNAs that form a related but distinct family of small noncoding RNAs. Similar to piRNAs, they act in transposon and retrotransposon silencing in the genome, but also control transcription of other repeat sequences in the genome (Aravin *et al.*, 2001). Similar to piRNAs, they bind to piwi family Argonaute proteins, but are expressed in organisms such as yeast and plants where piRNAs have not been found, and are therefore distinct species from piRNAs (Aravin *et al.*, 2001).

ceRNA – Competing endogenous RNAs comprise a class including all RNA molecules that may compete for miRNAs via complementary pairing and thus attenuate miRNA actions (Salmena *et al.*, 2011). The concept has been considered as a type of molecular sponge that absorbs miRNAs and prevents their function. The growing list of ceRNAs includes molecules from other RNA classes including mRNAs, lncRNAs, circRNAs, and pseudogenes (Hansen *et al.*, 2013). ceRNAs may also act in coordination to regulate a specific gene (Tay *et al.*, 2011) (**Fig. 1**).

Biological Sources of RNA

Early sources of RNA for biological study came from dissected tissues that were easily accessible, abundant, soft, non-fibrous, and were easy to homogenize in order to perform biochemical fractionations (Chomczynski and Sacchi, 1987). These tissues included liver, kidney, brain, spleen, and heart from mammals. Skin, although the most abundant organ by weight from humans was typically not used in isolations because of the presence of RNAase that degraded RNA upon contact (Probst *et al.*, 2006). As different biological questions arose and more sensitive methods for biochemical analysis developed, specific tissue regions were dissected and RNA isolated. These included different brain regions such as hippocampus from brain, lymphocytes from blood, and veins from liver. As even more tissue specific resolution was sought, laser capture microdissection methods became available and isolation of ~50 cell samples from tissues were possible (Mikulowska-Mennis *et al.*, 2002). More recently, RNA isolation from single cells has come to the forefront of experimental research (Saliba *et al.*, 2014). However, even higher resolution is still being pursued and while not routine, subcellular RNA fractions from nucleus, cytoplasm, or mitochondria to name just a few structures, have been available for many years (Derrien *et al.*, 2012).

In contrast to intracellular organelles with RNA components, an extracellular structure termed the exosome, has become an exciting and unusual place to isolate RNA (Johnstone *et al.*, 1987). Exosomes are small < 150 nm lipid vesicles that contain RNA, usually miRNAs but other species as well (Valadi *et al.*, 2007). Exosomes are excreted into the extracellular space and can eventually end up in body fluids or the circulation. The body fluids from where exosomes have been isolated include blood, urine, and amniotic fluid (Keller *et al.*, 2007). The amount of exosomal RNA secreted into these tissues is small and estimates of their abundance include 1–10 ng/mL fluid such as plasma (Enderle *et al.*, 2015). Exosomal sources of RNA are different than cfRNA which are cellular free RNAs which have nonspecific cellular origins. cfRNAs can be isolated like DNA from a large list of body fluids including amniotic fluid, spinal fluid, saliva, tears, and breast milk (Li *et al.*, 2014a). Like DNA used for diagnostics, it is possible to isolate RNA from a fetal amniotic fluid and this species is termed cffRNA, cell free fetal RNA (Vora *et al.*, 2017). The principle advantage of isolating RNAs from body fluids is due to its noninvasive nature, thus sparing the organism from losing tissues. Moreover many of these fluids such as urine and blood are highly abundant (Gahan, 2015; Huang *et al.*, 2013).

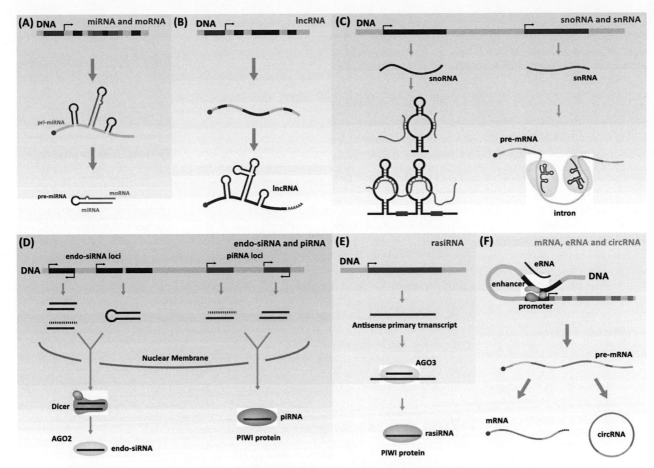

Fig. 1 Basic biogenesis of different RNA classes. (A) Biogenesis of miRNA and moRNA. A brief cartoon shows the origin of miRNA and moRNA. miRNA is marked in orange and moRNA is marked in green. (B) Biogenesis of lncRNA. lncRNA undergoes a splicing processing event. (C) Biogenesis of snoRNA and snRNA. The left side shows snoRNAs base pairing with ribosomal RNA to guide folding and cleavage of precursor rRNA. The right side shows snRNAs removing introns from pre-mRNA. (D) Biogenesis of endo-siRNA and piRNA. endo-siRNA and piRNA are exported from the nucleus to cytoplasm and different RNA binding proteins (RBPs)(Argonaute 2, AGO2 and PIWI protein are shown as examples) are required to process them. endo-siRNA and piRNA are marked by blue and green, respectively. (E) Biogenesis of rasiRNA. The process requires the Argonaute protein AGO3, and PIWI protein. (F) Biogenesis of mRNA, eRNA and circRNA. eRNAs are transcribed from enhancer regions. Canonical splicing of mRNAs produces linear transcripts while back-splicing produces circRNAs.

Meta-samples, derived from the metagenomics field, describe samples with mixtures of organisms (Hugenholtz *et al.*, 1998). These are typically obtained from the environment such as soil or sea water, but can also be complex cellular mixtures from discrete sources such as fecal matter (David *et al.*, 2014). Both RNA and DNA can be sequenced from these samples. A classical application of sequencing RNA from these samples is to identify ribosomal RNA as these are unique identifications for species in the sample and thus make it possible to identify individual species in a population living in a complex biological mixture.

Laboratory cell culture remains one of the most common sources of RNA. These can include primary cell cultures, transformed cell cultures, and stem cell derived cultures including induced pluripotent stem cells (iPSC), embryonic stem cells (eSC), or embryoid bodies (EB) to name a few. The laboratory success in growing and maintaining these different types of cells including adherent and suspended in liquid type cultures has made these cells a very flexible and dynamic source of RNA for studies.

Fractionation and enrichment or purification of specific classes of RNAs occurs independent of the sources. A total RNA purification from a biological sample contains > 80% rRNAs, while mRNA and lncRNAs together account for 1%–2% of the total RNA (Kaiser *et al.*, 2007). The amount of small RNAs by weight are less than this. mRNAs can be separated from total RNAs by isolation methods using oligodT binding to take advantage of the polyadenylation of this RNA class (Morrison *et al.*, 1979). Small noncoding RNAs can be enriched by size fractionation on columns or more cleanly by size fractionation followed by elution on polyacrylamide gels followed by elution. Any carryover DNA from these fractionation procedures are digested with an RNAse free DNAse. As detection methods become more sensitive, a greater amount of attention has been paid to increasing the purity of each fraction.

RNA Modifications

Chemical modifications to RNA molecules has been known for a long time, however, interest in their study has been hampered by lack of experimental tools to detect their presence and low abundance of some modifications (Song and Yi, 2017). This is now changing rapidly and with highly sensitive mass spectrometry techniques as well as other analytical chemistry methods, more than 140 chemical modifications of RNA are known of which >100 are known to occur naturally (Helm and Motorin, 2017). More than half of these modifications have been characterized on rRNA and tRNA with at least 13 described for mRNA. The most common modification for mRNA is addition of a m^{7}G (methylation of guanine nucleoside moiety) cap followed on the same molecule by 2'-O-methyl (methylation of ribose 2') secondary cap (Shatkin, 1976). The purpose of this modification is to increase stability of the RNA. Similarly, tRNA has a m^{5}C modification that functions to prevent degradation. The most prevalent internal mRNA modification is N(6)-methyladenosine, m^{6}A (Dominissini et al., 2012). The most common mRNA modification in vertebrates are adenosine to inosine (6-deaminated adenosine) which is catalyzed enzymatically via a family of adenosine deaminase acting on RNA (ADAR) proteins. This RNA modification is not randomly distributed along the transcriptome but appears to occur at double stranded RNA regions such as in 3' UTRs (untranslated regions) and miRNAs. These modifications have been shown to have effects on levels of transcripts and thus can be considered a means for post-transcriptional modification. Another abundant modification in mRNAs and lncRNAs is m6A (methyl-6-Adenosine) which has been shown to affect transcript stability and translation as regions in the 5' UTR and stop codons are enriched for this modification. RNA modifications at the transcriptomic level have led to the advent of a new field termed "epitranscriptomics". This field shares many similar concepts with the epigenetics field including the identification of RNA epitranscriptomic readers (Li et al., 2016), writers, and erasers (Frye et al., 2016). The identity of specific RNA metabolism pathways combined with transcriptome scale experiments has made this young field move very quickly.

Driving the field of epitranscriptomics is next generation sequencing. RNA modifications at a transcriptomic scale are possible by immune precipitating RNA fragments with the modification via an antibody and subsequent sequencing, commonly referred to as RIP-seq (Zhao et al., 2010). Using this and modifications of this method enable single nucleotide resolution of modified RNAs. A challenge for this approach are contaminants since RNA modification events, with a few exceptions, are not very abundant in the cell. Another challenge is in mapping RNA modification events to DNA sequences, so called RNA-DNA difference (RDD) comparisons (Chi, 2017). For example, adenosine→inosine modifications can be identified by comparing RNA to DNA since adenosine base pairs with thymine and inosine pairs with adenine, cytosine, or uracil. After synthesis of cDNA from RNA for sequencing, these differences between the original DNA sequence transcribed and the RNA sequence can identify specific locations of differences. However, mapping algorithms may not correctly assign mismatched sequences.

RNA Transcriptomics Nomenclature

Gene nomenclature committees representing various scientific communities have the responsibility to produce, maintain, and update gene names. These committees are necessary to maintain the stability of gene names and to ensure their accuracy. An example of such a group is the nomenclature committee of the Human Genome Organization (HUGO) that maintains *Homo sapiens* genes (Povey et al., 2001). Similar committees exist to maintain gene names for mouse, rat, and *Drosophila* scientific research communities.

In general, genes names are italicized while protein names are not. INS refers to the human insulin protein while *INS* refers to the human insulin gene. Gene names are one type of identifier for a gene which may have many depending upon the database. The National Center of Biotechnology Information GENE ID for human insulin is 3630, while the HUGO nomenclature committee GENE ID is 6081. Gene names provided by a nomenclature committee are useful as they provide a common identifier that may be used as a standard among many users.

Gene names are typically 3–4 letters. For example, the human insulin gene is *INS* and the human insulin receptor gene is *INSR*. Model organisms such as worms (*Caenorhabditis elegans*), and fruit flies (*Drosophila melanogaster*) have gene names provided by the phenotypes of mutants recovered from forward genetic screens and maintained by their scientific communities (Gelbart et al., 1997; Stein et al., 2001). Examples include *Arg⁻* for a strain requiring arginine as a phenotype in yeast, *dpy-3* for dumpy appearance in worms, and *w* or *white* for white eye in fruit flies. Names are also assigned based on orthology to other organisms such as *cyp35a* for cytochrome P450 family 35a gene in *C. elegans* (Aarnio et al., 2011). In bacteria, a 3 letter designation of the molecular pathway or biological process the gene product is involved in plus a capital letter denotes a gene, for example, *hisA* for gene A of the histidine biosynthesis pathway.

RNA transcripts can always be mapped to a chromosome and nucleotide position on a sequenced genome (Hubbard et al., 2002; Kent et al., 2002). The units are essentially coordinates on the genome. For example, the human insulin gene is located at Chromosome 11, nucleotides 2,159,779–2,161,341. Likewise, all transcript elements such as exons, enhancers, miRNAs etc. can also be identified by their coordinates. Another coordinate system identifies the continuous sequence (contig) and the order of genes. A contig may originate artificially from a yeast artificial chromosome library, phage library, or naturally from a chromosome. For example, the *daf-2* gene of *C. elegans* can also be named Y55D5A.5 as it is 5th gene on contig Y55D5A which is derived from a yeast artificial chromosome library.

miRNA names begin with the species (hsa- for *Homo sapiens*), followed by miR-, and then a number (e.g., hsa-miR-128). The mature miRNA has the miR (uppercase R) designation while the precursor is denoted with mir (lower case r, e.g., hsa-mir-128)

(Griffiths-Jones *et al.*, 2006). The numbers are provided sequentially for new miRNAs, but if an ortholog exists, they will be given the same number as the other orthologs (e.g., mus-mir-128). New mir ortholog names simply require the species prefix to be changed. Numbered suffixes indicate distinct precursors from unique genomic loci but identical mature miRNA sequences (e.g., mus-mir-128-1). Lettered suffixes indicate mature miRNAs closely related but from distinct precursors (e.g., mir-128b). The final suffix is given for the side of the hairpin from which the miRNA is located: either the 5′ side designated 5p or 3′ side designated 3p (e.g., miR-128-1-3p) (Budak *et al.*, 2016).

Other RNA classes also have their own nomenclature and naming conventions such as those for rRNAs, lncRNAs, and piRNAs provided by the HUGO Gene Nomenclature Committee (Gray *et al.*, 2015).

Transcriptome Measurements

Three essential measurements can be made on the transcriptome level of an organism: sequence, abundance, and structure (Conesa *et al.*, 2016). From a gene's sequence, specific information regarding the transcription start site, intron-exon usage, transcript termination, sequence variation, can be determined, and gene identity and orthology can be inferred. Gene abundance measurements can be determined by the number of transcripts representing a gene sequenced while secondary structure can be determined by various programs that fold the RNA sequence based on minimal free energy constraints as well as other structural factors (Miao and Westhof, 2017). Orthology is inferred by best alignment for similarity with other nucleotide sequences from other organisms while paralogy is inferred by alignment within the same organisms to detect genes belonging to the same family (Doyle and Gaut, 2000).

Whole transcriptome sequencing is performed on a sequencing platform that integrates sample preparation protocols, sequencing chemistry and detection, and data analysis as a single activity (Korpelainen *et al.*, 2014). Each transcriptome level sequencing has its own properties, history, biases, advocates, and advantages and disadvantages. The productivity and accuracy of sequencing platforms are constantly and rapidly improving (van Dijk *et al.*, 2014). Since the chemistry and technology for sequencing DNA is more mature and advanced than RNA, transcriptomic platforms typically reverse transcribe RNA into DNA and then sequence the complementary DNA. Among the commonly used platforms are microarrays, Illumina, Ion Torrent, Pacific Biosystems, and Oxford Nanopore.

Transcriptome Sequencing Platforms

Microarrays use the chemical principle of nucleic acid hybridization (DeRisi *et al.*, 1997). For a given sequence A, C, G, and T, hybridize with their complementary nucleotides via hydrogen bonding T, G, C, and A, respectively. Short DNA fragments of known sequence or identity are bound or directly synthesized to a solid support and a fluorescently labelled cDNA reverse transcribed from an RNA sample is hybridized to the fragments on the support. Since the abundance of a specific cDNA is directly correlated with the original amount of RNA in the sample, the amount hybridized is an indicator for the abundance of an RNA sequence. Since DNA sequences bound to the solid support are known and are fixed to a specific location, the identification of the gene can be determined by the location of the hybridization signal on the support. This platform has several advantages including relatively simple workflow and chemistry for detection and analysis. It is reasonably cost effective as well and entire transcriptomes can be interrogated in single replicates for a couple of hundred dollars. A disadvantage is that sequences to be interrogated need to be determined beforehand for synthesis of the microarray, cross hybridization of cDNAs can occur, and sensitivity of detection is limited to fluorescent labelling. In the early use of microarrays, many academic labs produced and printed their own microarrays, however, the field has lately been dominated by Affymetrix due to their ability to synthesize oligonucleotides directly onto a substrate.

Next-generation sequencing technologies (NGS) exploit massively parallel sequencing by synthesis (MPSS) as a chemical principle for transcriptome level sequencing (Marioni *et al.*, 2008). In the Illumina platform, RNA isolated from a sample is fragmented, and then indexes/bar codes, sequencing primer, and amplification primers are added as an adaptor molecule to the RNA. The RNA is reverse transcribed to cDNA and a second DNA strand is synthesized to yield a double stranded DNA. This DNA molecule is amplified and then sequenced by synthesis one nucleotide at a time. After each addition a specific fluorescent signal is produced for each nucleotide.

On the Illumina platform, the number of nucleotides on a DNA fragment can be up to 200 nucleotides, however 100 is more common. The number of templates sequenced, equivalent to the number of reactions on a single chip, can reach 2×10^9. Each sequence obtained from a template is called a "read". The number of reads can doubled to 4×10^9 using a dual flow cell, and further can be doubled to 8×10^9 by sequencing from both ends of the DNA called paired end sequencing. This can then generate up to 1.6×10^{12} of sequence information. Since the chemical reactions occur on the chip and reactants need to be removed for each chemical reaction, the solid substrate is termed a flow cell, and each flow cell is divided into 8 lanes. Indexing allows a user to add a tag, also known as a bar code, to each RNA molecule that will be sequenced. All RNAs from a single library have a specific bar code, and this allows mixing of libraries to be sequenced on a flow cell. The result is that the 2×10^9 reaction capacity can be divided into an unlimited number of libraries to be sequenced. This allows multiplexing of reactions, but more importantly drives down the cost of sequencing on a per reaction basis. The cost per reaction which equates to a read sequence can be as low as 0.001 US cents. While this might seem very cheap per reaction, there are other viable and competitive MPSS platforms.

Among these are Ion Torrent's personal genome sequencers (Liu *et al.*, 2012; Merriman and Rothberg, 2012; Quail *et al.*, 2012). While also based on a sequencing by synthesis principle, the detection of each added nucleotide is based on the release of a hydrogen ion which occurs after each addition. A pH change is measured in real time using semiconductors that are multiplexed in the platform. The advantage of this chemistry is that no modified nucleotides are needed. If homopolymer regions are encountered, for example, TTTTT regions, then multiple reactions will occur leading to bigger signals. This platform, while not generating as many reads as Illumina, has lower instrumentation cost and the total run time for generating sequence is substantially lower, for example, 2 h compared to approximately 24 h. Also, for infrequently run instruments or low through-put, it provides an advantage since it is easier to fill and run at full capacity. Length of reads is also important and Ion Torrent routinely uses 200 nucleotide reads with 400 read chemistry available. The number of reads produced on the Ion Torrent platform depends upon the chip used but can be as low as 0.5×10^6 to as high as 5×10^6 per run. At 200 nucleotide chemistry, a run generating 5×10^6 reads would produce 1×10^9 of sequence information.

Another sequencing platform is Pacific Biosystems which also utilizes a MPSS principle but differs from Illumina in utilizing the detection of newly added nucleotides at single molecule resolution enabled by detection of fluorescent phospholinked nucleotides (Rhoads and Au, 2015). In this platform, a single well contains a synthesis reaction of an enzyme, the DNA template, and labelled nucleotides, of which the addition of a specific nucleotide results in a specific signal. The signal is detected at the bottom of the well using a zero mode wave guide that directs one molecule of template and one molecule of DNA polymerase per well. Signals are detected as the synthesis of each nucleotide occurs, thus is giving this technology the name single molecule real-time (SMRT) sequencing. The addition of nucleotides produces a natural DNA molecule and thus can produce very long reads on average of above 10,000 nucleotides (Rhoads and Au, 2015) with up to 75,000 reads giving 0.75×10^9 of sequencing data. While the super long reads are advantageous especially in genome sequencing applications, the Pac Biosystems platform has also been of use in de novo transcriptome assembly of novel transcriptomes in RNA sequence informatics. The instrumentation cost is quite high compared to other platforms so this has been another limitation as well as a higher error rate. However, the ability to produce large reads has made it indispensable for applications such as genome assembly, transcriptome assembly, and sequencing complex genetic samples such as those from metagenomics.

Oxford Nanopore is another single molecule sequencing platform (Jain *et al.*, 2016). The technology depends upon small changes in electrical current when characteristic nucleotides pass through the pores. The small size of the platform, similar to a USB memory stick makes this an attractive technology. A recent version contains 500 nanopores from which single DNA molecules pass at 450 nucleotides per second with a generation of more than 5×10^9 of sequence according to a technical update from the company. While accuracy has been a problem for this platform in the past, technology improvements have increased this parameter dramatically in recent months.

Table 1 shows the comparison of sequencing platforms (Fox *et al.*, 2014; Li *et al.*, 2014b; Mardis, 2017).

Transcriptome Mapping

Microarrays contain spatially addressed genes already identified. NGS platforms require the reads obtained from sequencing to be mapped onto a transcriptome if one exists, and to produce a *de novo* transcriptome if it does not. In the former case mapping is typically performed by alignment of the read to a sequence from the transcriptome and then assignment of the read to a particular gene (Mortazavi *et al.*, 2008). In the latter, reads are aligned with each other to obtain continuous sequences (contigs) to generate an assembly, and then assigned to that particular contig (Martin and Wang, 2011). While mapping is intuitively straightforward, the large number of reads and large size of the transcriptome require efficient algorithms to improve upon the computational task of mapping. One algorithm, the Burrows-Wheeler transform (BWT), converts the transcriptome sequences into blocks, sorts these, and because of the many repeats of sequences in the transcriptome, makes it more efficient to search through rather than going nucleotide to nucleotide through the transcriptome. Another algorithm to align reads is to use a hash table, which is simply a table of possible short nucleotide sequences that uses a hash function to compute their locations. Finally, de Brujin graphs are used since a transcript might have many alternative exons and introns and these need to be considered in the alignment.

In **Table 2**, different tools which align reads as un-spliced or spliced are shown.

Table 1   Comparison of sequencing platforms

	Microarrays	*Illumina*	*Ion Torrent*	*Pacific biosystems*	*Oxford nanopore*
Chemistry	Hybridization	MPSS	Hydrogen ion	SMRT	SMRT
Read lengths	N/A	50–300 bp	200–600 bp	250 bp to 40 kb	– 150 kb
Read modalities	N/A	Single, paired end	Single, paired end	Single	1D or 2D reads
Throughput	10–100 MB	7.5–300 Gb	1.5–4.5 Gb	500 Mb to 1 Gb	10–20 Gb
Accuracy		100.00%	99.99%	99.99%	99.96%

Note: Columns indicate different sequencing platforms.

Rows indicate common features of platforms. Chemistry provides technical differences of each platform. Read Lengths measures the input read size. Read Modalities provide the type of read each platform can generate. Throughput estimates the amount of data that is processed by each platform. Accuracy indicates the accuracy of reads generated by the platform.

Abbreviations: bp, basepair; Gb, gigabyte; kb, kilobase; MPSS, massively parallel sequencing by synthesis; MB, megabyte; SMRT, single molecule real-time.

Table 2   Aligner and *de novo* assembler

Tools	Alignment type	Read length	Algorithm	System	Link
SOAP	Un-spliced	Short	BWT-based	Linux/Unix/MAC OS	http://soap.genomics.org.cn/
Bowtie	Un-spliced	Short	BWT-based	Linux/Unix/MAC OS	http://bowtie-bio.sourceforge.net/
Bowtie2	Un-spliced	Long	BWT-based	Linux/Unix/MAC OS	http://bowtie-bio.sourceforge.net/bowtie2
BWA	Un-spliced	Short/Long	BWT-based	Linux/Unix/MAC OS	http://bio-bwa.sourceforge.net/
TopHat	Spliced	Short/Long	BWT-based	Linux/Unix/MAC OS	https://ccb.jhu.edu/software/tophat
HISAT	Spliced	Short/Long	BWT-based	Linux/Unix/MAC OS	http://ccb.jhu.edu/software/hisat2
MAQ	Un-spliced	Short	Hash tables	Linux/Unix/MAC OS/Windows	http://maq.sourceforge.net/
PerM	Un-spliced	Short/Long	Hash tables	Linux/Unix/MAC OS	https://code.google.com/archive/p/perm/
SHRiMP	Un-spliced	Short/Long	Hash tables	Linux/Unix/MAC OS	http://compbio.cs.toronto.edu/shrimp/
Trinity	de novo	Long	de Bruijn graph	Linux/Unix/MAC OS	http://trinityrnaseq.github.io
Oases	de novo	Short	de Bruijn graph	Linux/Unix/MAC OS	https://www.ebi.ac.uk/~zerbino/oases/

Note: Rows indicate aligner examples. Columns list the common features of the aligners. The algorithm gives the background algorithm used in the aligner. The System lists the operating system where the aligner works. Link shows the home web page for each aligner.

Mapping values for a data set are provided as a percentage of reads mapped to the transcriptome or genome. More precise values can be given as percentage of read nucleotides mapped to a transcriptome calculated by taking all nucleotides aligned to the transcriptome divided by the number of nucleotides sequenced. These are usually termed % on-target.

Mapping data from NGS platforms generates an alignment file that contains read information and location on transcriptome where it maps. These are sometimes provided as coordinates in the genome where the transcriptome is located that also provides exon and intron data. These files are known as BAM, SAM, WIGGLE files, and have .bam, .sam, .wig extensions (Zhang, 2016). This data is very dense, and an index file can provide a means of searching through the alignment file. Index files have names such as BAI and have .bai extensions.

Transcriptome Output and Normalization

Microarray or NGS sequence data can typically be output as a tab-delimited text file with gene names in rows and samples names in columns (Korpelainen *et al.*, 2014; Laine *et al.*, 2014). Each field contains an abundance value that corresponds to a specific gene and specific sample. For microarrays the value is a fluorescence measurement and for NGS platforms the value is for number of reads. These values need to be "normalized" since fluorescence signals may vary from experiment to experiment in microarrays. Per-chip normalization of microarrays divides all values on the microarray by a number while per-gene normalization divides only a specific gene across different microarrays by a number. Other microarray normalization procedures may convert values to Z-scores, indicating a fluorescence measurement at Z-value, equivalent to one standard deviation from the mean. Thus, a value might be converted to -3.5 meaning its intensity is -3.5 standard deviations from the mean intensity of values on the microarray. Another microarray normalization method converts values to ordinals, so a gene might be the 503rd most highly expressed from 20,000 genes.

For data from NGS platforms, normalization also needs to be performed because while reads are an absolute measurement of transcript abundance, the values for each gene are also a function of the number of overall reads obtained from a sample (Korpelainen *et al.*, 2014). One normalization method is to divide the number of reads for each gene per million reads that map to the transcriptome. This value is called RPM. Values can also take into account the size of a transcript because a transcript 4 kb in size will have 8 times more reads mapping to it than a gene 0.5 kb in size if all other factors are equal. To normalize for transcript size, values are divided by the size of the gene over 1 kb. These values are termed RPKM for reads per kilo bp of the transcript per million mapped reads. Because Illumina can produced paired-end reads by sequencing from both sides of a transcript, the number of reads can be converted into fragments, so that reads from 2 ends of a transcript produce a fragment and these values are normalized as FPKM for fragments per 1 kbp gene size per million mapped reads.

Analysis of Transcriptome Data

Differential gene expression, commonly abbreviated as DG or DGE analysis refers to the analysis and interpretation of differences in abundance of gene transcripts within a transcriptome (Conesa *et al.*, 2016). Lists of genes that differ between 2 sample sets are often provided by RNA-seq data analysis tools, or can be generated manually by statistical testing of data sets. Due to the large number of genes to be tested, (e.g., >20,000 in the human genome), multiple testing correction such as Bonferroni correction is usually applied. Because the number of gene that are differentially expressed between samples may still be high (e.g., >1000), a method to understand and interpret the meaning of so many gene expression changes is needed. One method to solve this problem is termed "gene set enrichment" or GSE (Hung *et al.*, 2012; Subramanian *et al.*, 2005). In this method, a gene or group of genes that belong to a particular category that are enriched in one sample is compared to another sample. For example, if a breast cancer sample has more genes regulated that are annotated to the "cell cycle genes group" than a control sample. Grouping genes can be performed based on their annotation to a number of sources including the Kyoto Encyclopedia Gene Group (KEGG) pathway. Genes can also be annotated to a group based on a particular encoded protein motif such as F-box binding domain containing genes. Gene ontology (GO) annotations provide another method to annotate and group genes (Ashburner et al., 2000; Schulze-Kremer, 1997). Ontologies are a structured vocabulary with a hierarchy so that genes that belong in a lower level of the vocabulary also belong to higher levels. For example, genes that are annotated with lower level mitochondrion cristae (GO:0030061), are also annotated with the higher level GO term mitochondrion (GO:0005739). Genes can also belong to multiple groups at the same level. Thus, a differentially expressed gene list may be statistically over-represented in a GO category such as rough ER. There are actually 3 GO categories including molecular function, biological process, and cell component.

A gene co-expression network is a group of genes whose level of expression across different samples and conditions for each sample are similar (Gardner et al., 2003). This network identifies similarly behaving genes from the perspective of abundance and infers a common function that can then be hypothesized to work on the same biological process. A good example of a co-expression network are genes that are coregulated during diauxic shift when changing the energy consumption of yeast from glucose to galactose. During this shift, a new set of enzymes are needed and so genes in the galactose metabolism pathway are activated together (DeRisi et al., 1997). Inverse co-expression can also be useful as a network of co-expressed genes that are inversely related include miRNAs and their targets, where in conditions miRNA levels increase, the mRNA targets should be decreased (Ambros, 2004). DG expression has two measures, magnitude and direction. Co-expressed genes can also be identified by using correlation measures (e.g., Pearson Correlation coefficient) between 2 or more genes.

Transcriptomes can be assembled *de novo*, without other information, or based on a reference sequence. The reference sequence can be a genome sequence from the same organism as the source RNA or a closely related species. Reference sequences can also be closely related transcriptome sequences. The number of reads needed to sequence a transcriptome can be determined by the concept of depth. Given that the human transcriptome accounts for 3% of the nucleotides of the human genome consisting of 3 billion nucleotides, sequencing $0.03 \times 3,000,000,000 = 90$ M nucleotides. Thus, sequencing 90 M would be $1 \times$ depth and on average cover each nucleotide once. However, some genes are more highly expressed than others and some genes are rarely expressed, so even $1000 \times$ depth at 90×10^9 would only provide an even chance of sequencing a transcript that is 1 in a thousand in a cell. Therefore, it is likely that many transcriptome assemblies are not complete. Once assembled, several statistical measures reveal the quality and completeness of the assembly. The number and distribution of contigs indicate whether sufficient sequencing has been performed. Since the human transcriptome contains approximately 20,000 genes, a primate transcriptome might be expected to yield the same number. A smaller number might indicate insufficient coverage while a larger number might indicate contigs that need to be joined. Another measure is N50 size, which indicates the median size of a contig, another is the average size of a contig. Since the average size of a gene in humans is approximately 1.5 kb, values in this range suggest a good transcriptome assembly. Following assembly, novel transcriptomes are annotated usually by alignment with known genes from a closely related species. This is performed either by comparison of gene sequences, or translated protein sequences. In both cases, the annotation provided by the alignment can then be grouped to known functional categories of proteins known to be present in all metazoan species. Examples of these complete protein sets include CEGMA or core genes expected to be present in most metazoan species and BUSCO for benchmarking universal single copy orthologs (Parra et al., 2007; Simao et al., 2015). The inclusion and percentage measurement of genes present in these core gene data sets can give a rough idea of the transcriptome completeness.

RNA-seq data can also identify novel transcripts such as fusion genes (Levin et al., 2009). Fusion genes can be produced based on chromosome rearrangements including translocations, inversions, deletions, and duplications. The junction site for each of the rearrangements can potentially give rise to a Fusion gene. These genes are important as an indicator for cytogenetic events, but also some fusion genes can produce protein products that are functional and their activity may have consequences on the organism where it is expressed (Mitelman et al., 2007). Detecting fusion genes is a complex problem since it may be difficult to tell the difference between an authentic fusion gene and a sequencing artifact. More than 20 software are available for fusion gene detection (Kumar et al., 2016). Generally, these software tools look for reads that begin alignment at one gene and continue at another gene in the genome. Filters are used to decrease the number of false positives, however, it is typical to validate identified fusion genes using experimental laboratory tools such as PCR.

Single nucleotide polymorphisms (SNPs) can be detected from RNA-seq data as well by comparison of the obtained read data to a reference genome. This requires a reference genome with identified SNPs such as those from the 1000 genomes project (Quinn et al., 2013). To differentiate true SNPs from false positives, high coverage of the transcript ($>10 \times$) is usually required. Another strategy is termed "exome sequencing" where only RNA representing exons are sequenced (Ng et al., 2009). The purpose of this approach is to identify any nucleotide changes, both SNPs and other polymorphisms, in protein coding genes that might account for a specific phenotype (Biesecker, 2010). Sequencing the exome is more efficient by about 100 fold than sequencing the entire genome. In addition, some target enrichment strategies are used to sequence only specific exomes but perhaps in a larger number of samples. This type of sample design fits well sequencing platforms built for speed and simplicity rather than throughput.

RNA-Seq/Immunoprecipitation Methods

RNA sequencing (RNA-seq) has become arguably the most powerful approach for characterizing the abundance and structure of the transcriptome. However, variations in how the RNA is isolated, enriched, and purified have made this approach even more useful. Coupled with the modest cost of sequencing RNA via conversion to a cDNA library, RNA-seq will likely yield huge amounts of important biological data for many years to come (Wang et al., 2009). Below are some common variations of RNA-seq.

Biochemically, the approaches can be broken down into those based on size selection (miRNA-seq), selection based upon immunoprecipitation of RNA-proteins or RNA-protein complexes (e.g., RIP-seq, CLIP-seq, GRO-seq), or selection based on modification of RNA.

Cel-seq: Cell sequencing collects RNAs that are sourced from a single cell and is also known as single cell RNA sequencing or scRNA-seq (Hashimshony et al., 2012; Picelli, 2017; Ziegenhain et al., 2017). Single cells may be isolated in situ or manually from cultures. In some experiments, single cells may also be pooled to gather sufficient RNA for sequencing. Single cells can also be separated from whole animal embryos or complex tissues. Once separated, an emerging technique termed drop-seq joins each individual cell with a unique DNA barcoded bead into a nanoliter droplet using microfluidics. Once in a droplet, a library can be synthesized that contains a unique barcode identifier for each individual cell (Macosko et al., 2015).

CLIP-seq, PAR-CLIP-seq, HITS-CLIP-seq: Coss-linking immunoprecipitation sequencing refers to techniques that physically cross-links RNA to protein, usually via UV light, and subsequently immunoprecipitates the product (Licatalosi et al., 2008). In the PAR-CLIP-seq variation, photo-activatable RNA incorporation into RNA provides a more efficient cross-linking solution (Hafner et al., 2010). HITS-CLIP-seq simply refers to High-throughput sequencing of cross-linked immunoprecipitated RNA products.

GRO-seq: Global Run-On sequencing captures nascently transcribed RNA molecules by selecting for BrUTP nucleotides that are incorporated into the RNA strand during transcription (Danko et al., 2015). This method is generally used to identify active transcriptional regulatory elements.

i-CLIP-seq: Individual nucleotide resolution cross-linking immunoprecipitation sequencing refers to an approach to sequence specific nucleotides (Rossbach et al., 2014; Yao et al., 2014). Access to single nucleotides are provided by first selecting via immunoprecipation and subsequently by selecting RNAs using sequence specific DNA primers to construct the library for sequencing.

miRNA-seq: microRNA sequencing uses size selection of small RNA molecules to enrich for small noncoding RNAs typically mature miRNAs (Luo, 2012).

Ribo-seq: Ribosome sequencing captures RNAs as they are being translated in the ribosome complex (Calviello and Ohler, 2017). The large size of the ribosome allows this approach to be performed using classical biochemical techniques such as ultracentrifuge gradients. This approach has the same goal as TRAP-seq which is to sequence RNAs actively translated.

RIP-seq: RNA Immunoprecipitation sequencing refers to any technique that uses antibody reactivity to a protein-RNA complex in order select RNAs (Zhao et al., 2010).

SHAPE-seq: This is one of several approaches to identify RNA structure motifs via chemical modifications to the RNA at these motifs. The general principle is to chemically probe RNA structural motifs, perform reverse transcription of the RNA which will terminate at the modified site, and then generate cDNA libraries for RNA-sequencing to uncover the specific RNA sites where the motif was located. SHAPE-seq stands for selective-2′-hydroxyl acylation primer extension and follows these principles (Lucks et al., 2011; Mortimer et al., 2012; Watters and Lucks, 2016).

TN-seq: Transposon insertion sequencing refers to an approach to sequence regions where transposons have inserted into the genome (van Opijnen et al., 2009). Transposon insertion is a molecular biology tool to mutate the genome by insertion of DNA sequences into random genomic loci and then uncover phenotypes resulting from insertional mutagenesis. The method relies on massive parallel sequencing of transposon adjacent areas in order to map insertion sites. It is the most commonly used approach of IN-seq or insertion sequencing. This approach has been used with much success in microorganisms.

TRAP-seq: Translating ribosome affinity purification sequencing is an approach to purify RNAs as they are being translated (Reynoso et al., 2015). The method relies on affinity tagging of a ribosome component (e.g., 80S rRNA) and then utilizing immunoprecipitation to pull down associated RNAs. If successful, this provides transcripts that are part of the translatome.

Fig. 2 shows different RNA transcriptome sequencing methods mapped onto the process of transcription.

Data Integration

Data integration in the context of transcriptional informatics refers to the process of combining RNA structure, abundance, or sequence information with other types of biological information. Other types of information can include DNA sequence, epigenetic marks, protein abundance, protein-protein interaction, or metabolomics data. For example, RNA-seq and ChIP-seq data can be integrated (Angelini and Costa, 2014; Klein and Schafer, 2016). A very common data integration combines transcriptomic and proteomic data (Haider and Pal, 2013). Data integration can also occur between different RNA classes. The scale at which integration is performed is typically at the genome, transcriptome, or proteome level, although sub-genome scale integrations are common especially in pathway or biological process studies.

The purpose of data integration is to gain both a broader and a more precise view of RNA dynamics as it relates to other biologically active molecules and the molecular level (Hawkins et al., 2010). Early data integration efforts were aimed at correlating RNA levels with protein levels to confirm aspects of the central dogma. These basic correlations are still performed and have been extended to relate RNA abundance within RNA classes. For example, negative correlations would be expected of miRNAs with their target mRNAs (Camps et al., 2014). Positive correlations would be expected between mRNAs of transcription factors and their targets. More complex correlations can be found by integrating RNA abundance of genes within a biological pathway with products of that pathway. Data integration has been performed for a large group of physiologic processes such during the cell cycle, cell stress, cell death, and energy metabolism.

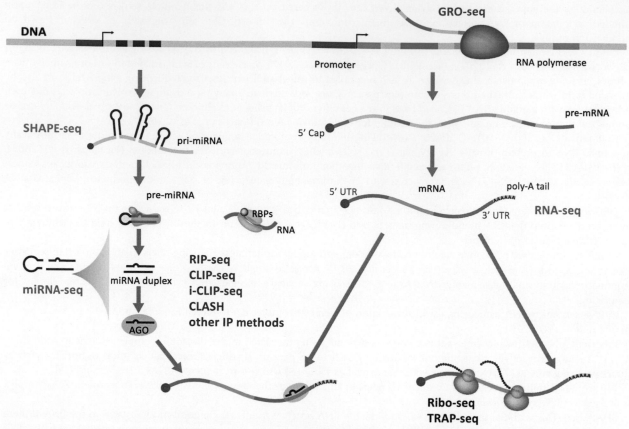

Fig. 2 Targets of RNA transcriptome sequencing methods. The image shows biogenesis of miRNAs and mRNAs. Each sequencing method is marked near the target sequences they focus on. GRO-seq (purple) captures nascent transcripts from actively engaged RNA polymerase. RNA-seq (gray) captures mRNAs (in any status, with or without poly-A tail) as input. Ribo-seq and TRAP-seq (blue) captures the mRNAs which are being translated. SHAPE-seq (green) target RNA with a specific structure motif. miRNA-seq aims to capture the small non coding RNA. RIP-seq, CLIP-seq, i-CLIP-seq, CLASH, and other IP methods (red) aim to sequence RNA bound to RNA-binding proteins (RBPs).

At the whole transcriptome level the integration of abundance data, so called expression RNA or eRNA data with other data has proven to be a powerful tool to identify both important genes in disease and physiological processes. For example, integrating eRNA (RNA abundance data) with SNP data from a single tissue of an individual, using a large population, has made it possible to identify specific DNA sequence variations that influence the expression levels of specific genes or sets of genes including both mRNAs and miRNAs (Lappalainen *et al.*, 2013). These SNPs that influence RNA expression levels have been termed eQTLs for expression quantitative trait loci. Such a strategy has also been used to integrate transcriptomic data with metabolomics data in order to identify genes whose expression levels correlate with the abundance of biologically active small molecules in human blood (Bartel *et al.*, 2015).

Although there have been many exceptionally performed integration studies, data integration in the transcriptomic field remains challenging. One bottleneck is that the biological sources of material should be the same and this is not always possible. DNA from cells is relatively stable and can be isolated thousands of years after the organism has died, however, RNA and protein are extremely labile. Another problem is that for genome scale statistics to be performed, the number of samples needed for sufficient power is large. Finally, for reasons still yet to be understood, data do not always correlate as expected on the genome, transcriptome, and protein level. While this complexity is frustrating, it also presents an opportunity to better understand the relationship between all of these molecules in living systems.

Visualization

The RNA research community is currently experiencing the challenge of rapid growth in volume, variety, and velocity of big data. A key step in understanding and learning from RNA structure, transcriptome expression, RNA-RNA interactions, and other RNA properties is visualization (Shabash and Wiese, 2017). An advanced visualization figure is worth a thousand words.

Visualization of RNA is a critical element in interpreting the structure and function of RNA molecules. At its best, visualization provides a means to analyze and interpret RNA data. As the function of RNA depends on its primary sequence, the structure in 2D and 3D provides insight into its putative function as well as possible mechanism. Most visualization programs provide 2D views of RNA based on thermodynamic considerations although other factors such as base pairing probabilities are also considered. More recently, chemical probing has been used to determine structures (Siegfried *et al.*, 2014). RNA visualization tools are also becoming more sophisticated as required by the discovery of long noncoding RNAs which also function via their secondary structures but may be several kb in length compared to classical tRNAs which are at least 10 times smaller.

Visualization of transcriptome assemblies, alignments, expression, variations and annotations is an efficient way to reveal the unknown. Visualization of transcriptome data can be obtained in various viewers and portals which plot the abundances of transcripts versus different conditions or samples. The most common and simple way to visualize transcriptome expression data is as a heat map with rows representing genes and columns representing conditions. Integrated viewers can provide sequence, structure, function, and abundance visualization within a single tool (Thorvaldsdottir *et al.*, 2013). These views can also integrate genomic information with transcriptomic data.

Visualization of interactions as a network is one way of coping with such data complexity. Since the RNA information for a single gene is related to the information about others, for example, by being co-expressed in cells, or co-regulated by inducers, RNA networks provide an efficient way to visualize the relationships between different RNA molecules. One example is a ceRNA network where miRNAs are connected to mRNAs, lncRNAs, and circRNAs (Salmena *et al.*, 2011). These networks permit discovery of new knowledge and allows inference to form conclusions. Hubs in these networks can help to identity central RNA molecules in a biological process or specific disease.

Artificial Intelligence Approaches

Artificial intelligence (A.I.) is generally defined as the property of machines that mimic human intelligence as characterized by behaviours such as cognitive ability, memory, learning, and decision making. It is an exploding field that encompasses many applied fields including language translation, self-driving cars, and game playing. The success of A.I. techniques in industry fields, has aroused a great deal of interest among biologists. Modern A.I., such as deep learning, promises to take advantage of large data sets for finding hidden structure within them, and for making accurate predictions (LeCun *et al.*, 2015). In the context of transcriptional informatics, a branch of A.I. that has found abundant applications is machine learning. Given a training set of data, A.I. algorithms can determine characteristics or properties of the data and based on this knowledge can act on new input data in various intelligent ways such as classification, prediction, and performing calculations (Swan *et al.*, 2015). Such an approach has been used, for example, to identify biomarkers during disease processes. Machine learning applied to transcription informatics includes neural network algorithms to cluster gene expression data from microarrays, predict mRNA targets from miRNA sequences, fold RNA molecules to secondary structures, and identify and correct biases in RNA-seq data (Heikkinen *et al.*, 2011; Toronen *et al.*, 1999). The common characteristic of these problems in RNA biology is that the precise answer is not clear, although there may be some basic principles to be followed. A large training set exists that can be used to identify and weigh specific parameters that will provide the best possible way to solve the problem given that set of information. Cross validation via external sample data sets or independent experimental validation have been used to provide confidence and to verify answers provided by A.I.

Tools and Resources: Databases, Browsers, and Portals

Recent advances in NGS technologies have created unprecedented amounts of big RNA-seq data sets for study, and many software and database resources are being developed for assisting biology researchers to deal with RNA-seq datasets. Many analysis tools have been developed for different study purposes, by diverse implementation techniques and in distinct interfaces. Many bioinformatics tools have been implemented to analyze RNA-seq data, including tools for preprocessing, read alignment, and other downstream analysis tools such as transcriptome quantification, differential expression analysis, peak finding etc. **Fig. 3** shows the basic steps of RNA-seq analysis and example tools involved in each step.

Tools for RNA-seq analysis include standalone software, databases, and web services. The Sequence Read Archive (SRA) is a database portal which collects many RNA-seq raw data sets and makes them publically available for new discoveries, meta-analysis, and reanalysis (O'Sullivan *et al.*, 2017). The most basic and essential part of RNA-seq is mapping reads to the reference genome. MAQ, BWA, Bowtie, SOAP, STAR, etc. are the commonly used aligners and many of them are command line tools running on the Linux system (Bao *et al.*, 2011). Visualization tools help view the alignment, such as the stand-alone software Integrative Genomics Viewer (IGV) or the web-based tools Genome Browser (Pavlopoulos *et al.*, 2015). While the basic tools are limited to a single function like a piece from a LEGO building block, workflow tools (including integrated analysis tools, workflow management systems) aim to combine and efficiently use separate tools and resources. Galaxy and Chipster are two outstanding representatives of a workflow management system where the user can choose certain tools from a large tools collection to make their own custom workflow through a web-based user friendly interface (Poplawski *et al.*, 2016). Workflow tools are typically hard to configure, but easy to use. Take Chipster as an example. It starts from a java web start desktop client. The user then chooses the tools from a large catalogue and selects the input, sets parameters, uploads raw data files and clicks a button to submit jobs. The whole process is program-free and combined with advanced visualization can more easily become available to the broader community. Also workflow tools can be deployed in a private cloud or local high performance cluster (HPC) for advanced users.

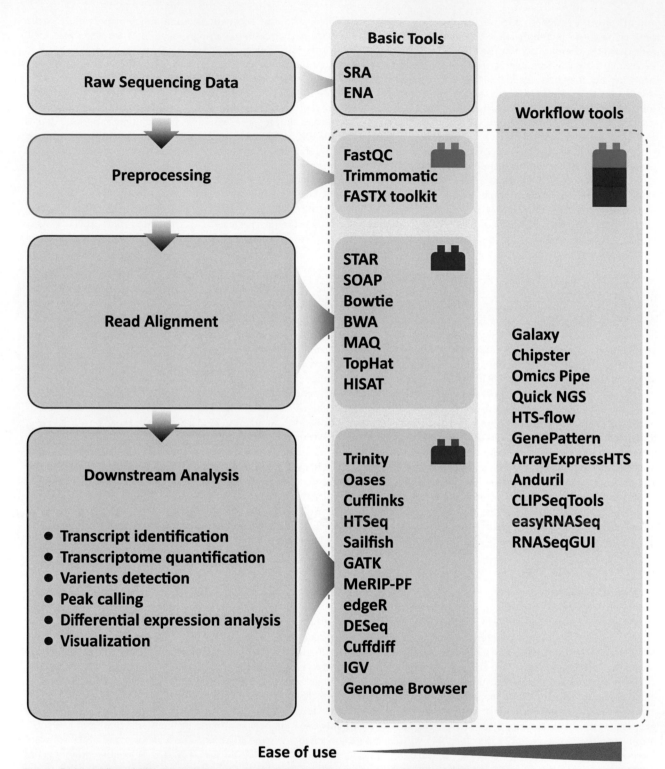

Fig. 3 Basic workflow of RNA-seq analysis and available tools. The left is the basic workflow for NGS data analysis, including raw sequence data (orange), preprocessing (green), read alignment (red), and other downstream analyses (blue). Names of tools are provided that correspond to specific sections in the workflow. In the figure, we only list a small portion of existing tools for each analysis step. All-in-one integrated analysis workflow tools are also being developed to utilize a variety tools and these are indicated with multiple colored blocks.

Many tools have a unique feature for a particular application. Workflows may report different results although they have the same analysis goals. These differences may be caused by a different combination of tools in the workflow, or different parameter settings. Large scale tools are available, but no gold standard for how to integrate them exists. Due to bandwidth, network speed, storage space and other hardware limitations, it is still difficult to analyze an RNA-seq dataset through internet based workflow tools only. Even Galaxy, a popular workflow tool, can only handle a small sized data set. To perform batch analysis, a large and private data set still needs to run on the workflow locally and rely on large computational resources. Although tools are diverse and extremely fragmented, biologists or bioinformaticians can nearly always find handy tools for their own study.

Summary

The RNA field is broad. From structure to function to sequence, the vastness from which biological systems have modified, edited, extended, and synthesized RNA is truly astounding. Scientists are only now beginning to grasp the depths and variety of RNAs in living systems. Major advancements in engineering, computer science, and biological understanding have greatly aided RNA transcriptomics as a field of study and research. Indeed, the tailwind created by the decreased cost of sequencing RNA, storing, and processing the information has expanded the field enormously. Yet there is still a sense that hardware and software is just keeping up with the demands of the field. Finally, while many important discoveries in RNA biology have only been reported recently (e.g., RNA interference, long noncoding RNAs, circular RNAs) we wonder what new discoveries await us. As one would think of an encyclopedia as a reference work of factual information, we look forward to when we can add more details at greater accuracy and in abundance. Indeed, it is likely that next iterations of this encyclopedia will change over the years in large and unexpected ways. Such is life in an RNA world.

Acknowledgments

We thank members of the Wong Laboratory Dr. Merja Lakso, Yang Yang, Menglei Zhang, and Changliang Wang for critical reading and discussion. This manuscript would not be possible without the guidance and enthusiasm of Dr. Eija Korpelainen. This work was supported by a grant from the University of Macau MYRG-2016-00101-FHS.

See also: Algorithms for Graph and Network Analysis: Clustering and Search of Motifs in Graphs. Analyzing Transcriptome-Phenotype Correlations. Bioinformatics Approaches for Studying Alternative Splicing. Characterization of Transcriptional Activities. Characterizing and Functional Assignment of Noncoding RNAs. Codon Usage. Comparative Transcriptomics Analysis. Computational Systems Biology. Computing Languages for Bioinformatics: R. Data Cleaning. Data Mining: Clustering. Data Mining: Mining Frequent Patterns, Associations Rules, and Correlations. Data Reduction. Differential Expression from Microarray and RNA-seq Experiments. Exome Sequencing Data Analysis. Experimental Platforms for Extracting Biological Data: Mass Spectrometry, Microarray, Next Generation Sequencing. Expression Clustering. Functional Enrichment Analysis. Functional Enrichment Analysis Methods. Gene-Gene Interactions: An Essential Component to Modeling Complexity for Precision Medicine. Genome Informatics. Genome-Wide Scanning of Gene Expression. Integrative Analysis of Multi-Omics Data. Integrative Bioinformatics of Transcriptome: Databases, Tools and Pipelines. Introduction to Biostatistics. MicroRNA and lncRNA Databases and Analysis. Natural Language Processing Approaches in Bioinformatics. Next Generation Sequence Analysis. Next Generation Sequencing Data Analysis. Pre-Processing: A Data Preparation Step. Regulation of Gene Expression. Sequence Analysis. Survey of Antisense Transcription. Transcriptome Analysis. Transcriptome During Cell Differentiation. Transcriptomic Data Normalization. Transcriptomic Databases. Utilising IPG-IEF to Identify Differentially-Expressed Proteins. Whole Genome Sequencing Analysis

References

Aarnio, V., Lehtonen, M., Storvik, M., *et al.*, 2011. *Caenorhabditis elegans* mutants predict regulation of fatty acids and endocannabinoids by the CYP-35A gene family. Frontiers in Pharmacology 2, 12.

Alberts, B., 2007. Molecular biology of the cell. In: Artificial Life, fourth ed., vol. 10. New York, NY: Garland Publishing, pp. 82–95.

Ambros, V., 2004. The functions of animal micrornas. Nature 431, 350–355.

Angelini, C., Costa, V., 2014. Understanding gene regulatory mechanisms by integrating ChIP-seq and RNA-seq data: Statistical solutions to biological problems. Frontiers in Cell and Developmental Biology 2, 51.

Aravin, A.A., Naumova, N.M., Tulin, A.V., *et al.*, 2001. Double-stranded RNA-mediated silencing of genomic tandem repeats and transposable elements in the *D. melanogaster* germline. Current Biology 11, 1017–1027.

Ashburner, M., Ball, C.A., Blake, J.A., *et al.*, 2000. gene ontology: Tool for the unification of biology. The gene Ontology Consortium. Nature Genetics 25, 25–29.

Asikainen, S., Heikkinen, L., Juhila, J., *et al.*, 2015. Selective microRNA-offset RNA expression in human embryonic stem cells. PLOS ONE 10, e0116668.

Asikainen, S., Heikkinen, L., Wong, G., Storvik, M., 2008. Functional characterization of endogenous siRNA target genes in *Caenorhabditis elegans*. BMC Genomics 9, 270.

Azzalin, C.M., Reichenbach, P., Khoriauli, L., *et al.*, 2007. Telomeric repeat containing RNA and RNA surveillance factors at mammalian chromosome ends. Science 318, 798–801.

Ballabio, A., Willard, H.F., 1992. Mammalian X-chromosome inactivation and the XIST gene. Current Opinion in Genetics & Development 2, 439–447.

Baltimore, D., 1970. RNA-dependent DNA polymerase in virions of RNA tumour viruses. Nature 226, 1209–1211.

Baltimore, D., 1971. Expression of animal virus genomes. Bacteriological Reviews 35, 235–241.

Bao, S., Jiang, R., Kwan, W., *et al.*, 2011. Evaluation of next-generation sequencing software in mapping and assembly. Journal of Human Genetics 56, 406–414.

Bartel, J., Krumsiek, J., Schramm, K., *et al.*, 2015. The human blood metabolome-transcriptome interface. PLOS Genetics 11, e1005274.

Biesecker, L.G., 2010. Exome sequencing makes medical genomics a reality. Nature Genetics 42, 13–14.

Borchert, G.M., Lanier, W., Davidson, B.L., 2006. RNA polymerase III transcribes human microRNAs. Nature Structural and Molecular Biology 13, 1097–1101.

Budak, H., Bulut, R., Kantar, M., Alptekin, B., 2016. MicroRNA nomenclature and the need for a revised naming prescription. Briefings in Functional Genomics 15, 65–71.

Calabrese, J.M., Sun, W., Song, L., et al., 2012. Site-specific silencing of regulatory elements as a mechanism of X inactivation. Cell 151, 951–963.

Calviello, L., Ohler, U., 2017. Beyond read-counts: Ribo-seq data analysis to understand the functions of the transcriptome. Trends in Genetics 33, 728–744.

Camps, C., Saini, H.K., Mole, D.R., et al., 2014. Integrated analysis of microRNA and mRNA expression and association with HIF binding reveals the complexity of microRNA expression regulation under hypoxia. Molecular Cancer 13, 28.

Chi, K.R., 2017. The RNA code comes into focus. Nature 542, 503–506.

Chomczynski, P., Sacchi, N., 1987. Single-step method of RNA isolation by acid guanidinium thiocyanate–phenol–chloroform extraction. Analytical Biochemistry 162, 156–159.

Colgan, D.F., Manley, J.L., 1997. Mechanism and regulation of mRNA polyadenylation. Genes & Development 11, 2755–2766.

Conesa, A., Madrigal, P., Tarazona, S., et al., 2016. A survey of best practices for RNA-seq data analysis. Genome Biology 17, 13.

Crick, F., 1970. Central dogma of molecular biology. Nature 227, 561–563.

Danko, C.G., Hyland, S.L., Core, L.J., et al., 2015. Identification of active transcriptional regulatory elements from GRO-seq data. Nature Methods 12, 433–438.

Das, P.P., Bagijn, M.P., Goldstein, L.D., et al., 2008. Piwi and piRNAs act upstream of an endogenous siRNA pathway to suppress Tc3 transposon mobility in the Caenorhabditis elegans germline. Molecular Cell 31, 79–90.

David, L.A., Maurice, C.F., Carmody, R.N., et al., 2014. Diet rapidly and reproducibly alters the human gut microbiome. Nature 505, 559–563.

DeRisi, J.L., Iyer, V.R., Brown, P.O., 1997. Exploring the metabolic and genetic control of gene expression on a genomic scale. Science 278, 680–686.

Derrien, T., Johnson, R., Bussotti, G., et al., 2012. The GENCODE v7 catalog of human long noncoding RNAs: Analysis of their gene structure, evolution, and expression. Genome Research 22, 1775–1789.

Dominissini, D., Moshitch-Moshkovitz, S., Schwartz, S., et al., 2012. Topology of the human and mouse m6A RNA methylomes revealed by m6A-seq. Nature 485, 201–206.

Doyle, J.J., Gaut, B.S., 2000. Evolution of genes and taxa: A primer. Plant Molecular Biology 42, 1–23.

Dvir, A., 2002. Promoter escape by RNA polymerase II. Biochimica et Biophysica Acta 1577, 208–223.

Ebbesen, K.K., Kjems, J., Hansen, T.B., 2016. Circular RNAs: Identification, biogenesis and function. Biochimica et Biophysica Acta 1859, 163–168.

Eick, D., Wedel, A., Heumann, H., 1994. From initiation to elongation: Comparison of transcription by prokaryotic and eukaryotic RNA polymerases. Trends in Genetics 10, 292–296.

Enderle, D., Spiel, A., Coticchia, C.M., et al., 2015. Characterization of RNA from exosomes and other extracellular vesicles isolated by a novel spin column-based method. PLOS ONE 10, e0136133.

Espinosa, J.M., 2016. Revisiting lncRNAs: How do you know yours is not an eRNA? Molecular Cell 62, 1–2.

Fabrega, C., Hausmann, S., Shen, V., et al., 2004. Structure and mechanism of mRNA cap (guanine-N7) methyltransferase. Molecular Cell 13, 77–89.

Fouqueau, T., Werner, F., 2017. The architecture of transcription elongation. Science 357, 871–872.

Fox, E.J., Reid-Bayliss, K.S., Emond, M.J., Loeb, L.A., 2014. Accuracy of next generation sequencing platforms. Next Generation, Sequencing & Applications 1.

Frye, M., Jaffrey, S.R., Pan, T., et al., 2016. RNA modifications: What have we learned and where are we headed? Nature Reviews. Genetics 17, 365–372.

Gahan, P.B., 2015. Circulating Nucleic Acids in Early Diagnosis, Prognosis and Treatment Monitoring. Springer.

Gardner, T.S., di Bernardo, D., Lorenz, D., Collins, J.J., 2003. Inferring genetic networks and identifying compound mode of action via expression profiling. Science 301, 102–105.

Gelbart, W.M., Crosby, M., Matthews, B., et al., 1997. FlyBase: A Drosophila database. The FlyBase consortium. Nucleic Acids Research 25, 63–66.

Gray, K.A., Yates, B., Seal, R.L., et al., 2015. Genenames.org: The HGNC resources in 2015. Nucleic Acids Research 43, D1079–D1085.

Greider, C.W., Blackburn, E.H., 1989. A telomeric sequence in the RNA of Tetrahymena telomerase required for telomere repeat synthesis. Nature 337, 331–337.

Griffiths-Jones, S., Grocock, R.J., van Dongen, S., et al., 2006. MiRBase: Microrna sequences, targets and gene nomenclature. Nucleic Acids Research 34, D140–D144.

Guerrier-Takada, C., Gardiner, K., Marsh, T., et al., 1983. The RNA moiety of ribonuclease P is the catalytic subunit of the enzyme. Cell 35, 849–857.

Ha, M., Kim, V.N., 2014. Regulation of microRNA biogenesis. Nature Reviews Molecular Cell Biology 15, 509–524.

Hafner, M., Landthaler, M., Burger, L., et al., 2010. Transcriptome-wide identification of RNA-binding protein and microRNA target sites by PAR-CLIP. Cell 141, 129–141.

Haider, S., Pal, R., 2013. Integrated analysis of transcriptomic and proteomic data. Current Genomics 14, 91–110.

Hangauer, M.J., Vaughn, I.W., McManus, M.T., 2013. Pervasive transcription of the human genome produces thousands of previously unidentified long intergenic noncoding RNAs. PLOS Genetics 9, e1003569.

Hansen, T.B., Jensen, T.I., Clausen, B.H., et al., 2013. Natural RNA circles function as efficient microRNA sponges. Nature 495, 384–388.

Hashimshony, T., Wagner, F., Sher, N., Yanai, I., 2012. CEL-Seq: Single-cell RNA-Seq by multiplexed linear amplification. Cell Reports 2, 666–673.

Hawkins, R.D., Hon, G.C., Ren, B., 2010. Next-generation genomics: An integrative approach. Nature Reviews Genetics 11, 476–486.

He, L., Hannon, G.J., 2004. MicroRNAs: Small RNAs with a big role in gene regulation. Nature Reviews Genetics 5, 522–531.

Heikkinen, L., Kolehmainen, M., Wong, G., 2011. Prediction of microRNA targets in Caenorhabditis elegans using a self-organizing map. Bioinformatics 27, 1247–1254.

Helm, M., Motorin, Y., 2017. Detecting RNA modifications in the epitranscriptome: Predict and validate. Nature Reviews. Genetics 18, 275–291.

Huang, X., Yuan, T., Tschannen, M., et al., 2013. Characterization of human plasma-derived exosomal RNAs by deep sequencing. BMC Genomics 14, 319.

Hubbard, T., Barker, D., Birney, E., et al., 2002. The Ensembl genome database project. Nucleic Acids Research 30, 38–41.

Hugenholtz, P., Goebel, B.M., Pace, N.R., 1998. Impact of culture-independent studies on the emerging phylogenetic view of bacterial diversity. Journal of Bacteriology 180, 4765–4774.

Hung, J.H., Yang, T.H., Hu, Z., et al., 2012. Gene set enrichment analysis: Performance evaluation and usage guidelines. Briefings in Bioinformatics 13, 281–291.

Iwasaki, Y.W., Siomi, M.C., Siomi, H., 2015. PIWI-interacting RNA: Its biogenesis and functions. Annual Review of Biochemistry 84, 405–433.

Iyer, M.K., Niknafs, Y.S., Malik, R., et al., 2015. The landscape of long noncoding RNAs in the human transcriptome. Nature Genetics 47, 199–208.

Jacob, F., Monod, J., 1961. Genetic regulatory mechanisms in the synthesis of proteins. Journal of Molecular Biology 3, 318–356.

Jain, M., Olsen, H.E., Paten, B., Akeson, M., 2016. The Oxford nanopore MinION: Delivery of nanopore sequencing to the genomics community. Genome Biology 17, 239.

Johnstone, R.M., Adam, M., Hammond, J.R., et al., 1987. Vesicle formation during reticulocyte maturation. Association of plasma membrane activities with released vesicles (exosomes). Journal of Biological Chemistry 262, 9412–9420.

Juven-Gershon, T., Kadonaga, J.T., 2010. Regulation of gene expression via the core promoter and the basal transcriptional machinery. Developmental Biology 339, 225–229.

Kaiser, C.A., Krieger, M., Lodish, H., Berk, A., 2007. Molecular Cell Biology. WH Freeman.

Kapranov, P., Cheng, J., Dike, S., et al., 2007. RNA maps reveal new RNA classes and a possible function for pervasive transcription. Science 316, 1484–1488.

Kay, L.E., 2000. Who Wrote the Book of Life? A History of the Genetic Code. Stanford University Press.

Keller, S., Rupp, C., Stoeck, A., et al., 2007. CD24 is a marker of exosomes secreted into urine and amniotic fluid. Kidney International 72, 1095–1102.

Kent, W.J., Sugnet, C.W., Furey, T.S., et al., 2002. The human genome browser at UCSC. Genome Research 12, 996–1006.

Kim, T.K., Hemberg, M., Gray, J.M., et al., 2010. Widespread transcription at neuronal activity-regulated enhancers. Nature 465, 182–187.

Kiss, T., 2001. Small nucleolar RNA-guided post-transcriptional modification of cellular RNAs. EMBO Journal 20, 3617–3622.

Klein, H.U., Schafer, M., 2016. Integrative analysis of histone ChIP-seq and RNA-seq data. Current Protocols in Human Genetics, 90. pp. 20 23 21–20 23 16.

Korpelainen, E., Tuimala, J., Somervuo, P., et al., 2014. RNA-seq Data Analysis: A Practical Approach. CRC Press.

Kowalczyk, M.S., Hughes, J.R., Garrick, D., et al., 2012. Intragenic enhancers act as alternative promoters. Molecular Cell 45, 447–458.

Kruger, K., Grabowski, P.J., Zaug, A.J., et al., 1982. Self-splicing RNA: Autoexcision and autocyclization of the ribosomal RNA intervening sequence of tetrahymena. Cell 31, 147–157.

Kumar, S., Vo, A.D., Qin, F., Li, H., 2016. Comparative assessment of methods for the fusion transcripts detection from RNA-Seq data. Scientific Reports 6, 21597.

Kung, J.T., Colognori, D., Lee, J.T., 2013. Long noncoding RNAs: Past, present, and future. Genetics, 193. pp. 651–669.

Laine, M.M., Pasanen, T., Saarela, J. et al., 2014. DNA Microarray Data Analysis, second edition.

Langenberger, D., Bermudez-Santana, C., Hertel, J., et al., 2009. Evidence for human microRNA-offset RNAs in small RNA sequencing data. Bioinformatics 25, 2298–2301.

Lappalainen, T., Sammeth, M., Friedlander, M.R., et al., 2013. Transcriptome and genome sequencing uncovers functional variation in humans. Nature 501, 506–511.

Lasda, E., Parker, R., 2014. Circular RNAs: Diversity of form and function. RNA (New York, NY) 20, 1829–1842.

LeCun, Y., Bengio, Y., Hinton, G., 2015. Deep learning. Nature 521, 436–444.

Lee, R.C., Feinbaum, R.L., Ambros, V., 1993. The C. elegans heterochronic gene lin-4 encodes small RNAs with antisense complementarity to lin-14. Cell 75, 843–854.

Lee, Y., Kim, M., Han, J., et al., 2004. MicroRNA genes are transcribed by RNA polymerase II. EMBO Journal 23, 4051–4060.

Levin, J.Z., Berger, M.F., Adiconis, X., et al., 2009. Targeted next-generation sequencing of a cancer transcriptome enhances detection of sequence variants and novel fusion transcripts. Genome Biology 10, R115.

Li, M., Zeringer, E., Barta, T., et al., 2014a. Analysis of the RNA content of the exosomes derived from blood serum and urine and its potential as biomarkers. Philosophical Transactions of the Royal Society of London. Series B, Biological Sciences 369.

Li, S., Tighe, S.W., Nicolet, C.M., et al., 2014b. Multi-platform assessment of transcriptome profiling using RNA-seq in the ABRF next-generation sequencing study. Nature Biotechnology 32, 915–925.

Li, X., Xiong, X., Yi, C., 2016. Epitranscriptome sequencing technologies: Decoding RNA modifications. Nature Methods 14, 23–31.

Licatalosi, D.D., Mele, A., Fak, J.J., et al., 2008. HITS-CLIP yields genome-wide insights into brain alternative RNA processing. Nature 456, 464–469.

Liu, L., Li, Y., Li, S., et al., 2012. Comparison of next-generation sequencing systems. Journal of Biomedicine & Biotechnology 2012, 251364.

Lodish, H., Berk, A., Zipursky, S.L., et al., 1995. Molecular Cell Biology. New York, NY: Scientific American Books.

Londin, E., Loher, P., Telonis, A.G., et al., 2015. Analysis of 13 cell types reveals evidence for the expression of numerous novel primate- and tissue-specific microRNAs. Proceedings of the National Academy of Sciences of the United States of America 112, E1106–E1115.

Lucks, J.B., Mortimer, S.A., Trapnell, C., et al., 2011. Multiplexed RNA structure characterization with selective 2′-hydroxyl acylation analyzed by primer extension sequencing (SHAPE-Seq). Proceedings of the National Academy of Sciences of the United States of America 108, 11063–11068.

Luo, S., 2012. MicroRNA expression analysis using the Illumina microRNA-seq platform. Methods in Molecular Biology 822, 183–188.

Macosko, E.Z., Basu, A., Satija, R., et al., 2015. Highly parallel genome-wide expression profiling of individual cells using nanoliter droplets. Cell 161, 1202–1214.

Mardis, E.R., 2017. DNA sequencing technologies: 2006–2016. Nature Protocols 12, 213–218.

Marioni, J.C., Mason, C.E., Mane, S.M., et al., 2008. RNA-seq: An assessment of technical reproducibility and comparison with gene expression arrays. Genome Research 18, 1509–1517.

Martin, J.A., Wang, Z., 2011. Next-generation transcriptome assembly. Nature Reviews. Genetics 12, 671–682.

Matera, A.G., Terns, R.M., Terns, M.P., 2007. Non-coding RNAs: Lessons from the small nuclear and small nucleolar RNAs. Nature Reviews Molecular Cell Biology 8, 209–220.

Meister, G., 2013. Argonaute proteins: Functional insights and emerging roles. Nature Reviews Genetics 14, 447–459.

Memczak, S., Jens, M., Elefsinioti, A., et al., 2013. Circular RNAs are a large class of animal RNAs with regulatory potency. Nature 495, 333–338.

Merriman, B., Rothberg, J.M., 2012. Progress in Ion Torrent semiconductor chip based sequencing. Electrophoresis 33, 3397–3417.

Miao, Z., Westhof, E., 2017. RNA structure: Advances and assessment of 3D structure prediction. Annual Review of Biophysics 46, 483–503.

Mikulowska-Mennis, A., Taylor, T.B., Vishnu, P., et al., 2002. High-quality RNA from cells isolated by laser capture microdissection. BioTechniques 33, 176–179.

Mitelman, F., Johansson, B., Mertens, F., 2007. The impact of translocations and gene fusions on cancer causation. Nature Reviews Cancer 7, 233–245.

Morrison, M.R., Brodeur, R., Pardue, S., et al., 1979. Differences in the distribution of poly(A) size classes in individual messenger RNAs from neuroblastoma cells. Journal of Biological Chemistry 254, 7675–7683.

Mortazavi, A., Williams, B.A., McCue, K., et al., 2008. Mapping and quantifying mammalian transcriptomes by RNA-Seq. Nature Methods 5, 621–628.

Mortimer, S.A., Trapnell, C., Aviran, S., et al., 2012. SHAPE-Seq: High-throughput RNA structure analysis. Current Protocols in Chemical Biology 4, 275–297.

Murakami, K.S., Masuda, S., Campbell, E.A., et al., 2002a. Structural basis of transcription initiation: An RNA polymerase holoenzyme–DNA complex. Science 296, 1285–1290.

Murakami, K.S., Masuda, S., Darst, S.A., 2002b. Structural basis of transcription initiation: RNA polymerase holoenzyme at 4 A resolution. Science 296, 1280–1284.

Natoli, G., Andrau, J.C., 2012. Noncoding transcription at enhancers: General principles and functional models. Annual Review of Genetics 46, 1–19.

Neveu, M., Kim, H.J., Benner, S.A., 2013. The "strong" RNA world hypothesis: Fifty years old. Astrobiology 13, 391–403.

Ng, S.B., Turner, E.H., Robertson, P.D., et al., 2009. Targeted capture and massively parallel sequencing of 12 human exomes. Nature 461, 272–276.

Nilsen, T.W., 2008. Endo-siRNAs: Yet another layer of complexity in RNA silencing. Nature Structural and Molecular Biology 15, 546–548.

Noller, H.F., 1991. Ribosomal RNA and translation. Annual Review of Biochemistry 60, 191–227.

Okamura, K., Lai, E.C., 2008. Endogenous small interfering RNAs in animals. Nature Reviews Molecular Cell Biology 9, 673–678.

O'Sullivan, C., Busby, B., Mizrachi, I.K., 2017. Managing sequence data. Methods in Molecular Biology (Clifton, NJ) 1525, 79–106.

Palade, G.E., 1955. A small particulate component of the cytoplasm. Journal of Biophysical and Biochemical Cytology 1, 59–68.

Paralkar, V.R., Taborda, C.C., Huang, P., et al., 2016. Unlinking an lncRNA from its associated cis element. Molecular Cell 62, 104–110.

Parra, G., Bradnam, K., Korf, I., 2007. CEGMA: A pipeline to accurately annotate core genes in eukaryotic genomes. Bioinformatics 23, 1061–1067.

Pavlopoulos, G.A., Malliarakis, D., Papanikolaou, N., et al., 2015. Visualizing genome and systems biology: Technologies, tools, implementation techniques and trends, past, present and future. GigaScience 4, 38.

Picelli, S., 2017. Single-cell RNA-sequencing: The future of genome biology is now. RNA Biology 14, 637–650.

Poplawski, A., Marini, F., Hess, M., et al., 2016. Systematically evaluating interfaces for RNA-seq analysis from a life scientist perspective. Briefings in Bioinformatics 17, 213–223.

Povey, S., Lovering, R., Bruford, E., et al., 2001. The HUGO gene nomenclature Committee (HGNC). Human Genetics 109, 678–680.

Probst, J., Brechtel, S., Scheel, B., et al., 2006. Characterization of the ribonuclease activity on the skin surface. Genetic Vaccines and Therapy 4, 4.

Ptashne, M., Gann, A., 1997. Transcriptional activation by recruitment. Nature 386, 569–577.

Pyle, A.M., 1993. Ribozymes: A distinct class of metalloenzymes. Science 261, 709–714.

Quail, M.A., Smith, M., Coupland, P., et al., 2012. A tale of three next generation sequencing platforms: Comparison of ion torrent, Pacific biosciences and Illumina MiSeq sequencers. BMC Genomics 13, 341.

Quinn, E.M., Cormican, P., Kenny, E.M., et al., 2013. Development of strategies for SNP detection in RNA-seq data: Application to lymphoblastoid cell lines and evaluation using 1000 Genomes data. PLOS One 8, e58815.

Reynoso, M.A., Juntawong, P., Lancia, M., et al., 2015. Translating Ribosome Affinity Purification (TRAP) followed by RNA sequencing technology (TRAP-SEQ) for quantitative assessment of plant translatomes. Methods in Molecular Biology 1284, 185–207.

Rhoads, A., Au, K.F., 2015. PacBio sequencing and its applications. Genomics, Proteomics & Bioinformatics 13, 278–289.

Richard, P., Manley, J.L., 2009. Transcription termination by nuclear RNA polymerases. Genes Dev 23, 1247–1269.

Rossbach, O., Hung, L.H., Khrameeva, E., et al., 2014. Crosslinking-immunoprecipitation (iCLIP) analysis reveals global regulatory roles of hnRNP L. RNA Biology 11, 146–155.

Sainsbury, S., Bernecky, C., Cramer, P., 2015. Structural basis of transcription initiation by RNA polymerase II. Nature Reviews. Molecular Cell Biology 16, 129–143.

Saito, K., Nishida, K.M., Mori, T., et al., 2006. Specific association of Piwi with rasiRNAs derived from retrotransposon and heterochromatic regions in the Drosophila genome. Genes and Development 20, 2214–2222.

Saliba, A.E., Westermann, A.J., Gorski, S.A., Vogel, J., 2014. Single-cell RNA-seq: Advances and future challenges. Nucleic Acids Research 42, 8845–8860.

Salmena, L., Poliseno, L., Tay, Y., et al., 2011. A ceRNA hypothesis: The Rosetta stone of a hidden RNA language? Cell 146, 353–358.

Salzman, J., 2016. Circular RNA expression: Its potential regulation and function. Trends in Genetics 32, 309–316.

Schulze-Kremer, S., 1997. Adding semantics to genome databases: Towards an ontology for molecular biology. Proceedings. International Conference on Intelligent Systems for Molecular Biology 5, 272–275.

Serganov, A., Patel, D.J., 2007. Ribozymes, riboswitches and beyond: Regulation of gene expression without proteins. Nature Reviews Genetics 8, 776–790.

Shabash, B., Wiese, K.C., 2017. RNA Visualization: Relevance and the current state-of-the-art focusing on pseudoknots. IEEE/ACM Transactions on Computational Biology and Bioinformatics, 14. pp. 696–712.

Sharp, S.J., Schaack, J., Cooley, L., et al., 1985. Structure and transcription of eukaryotic tRNA genes. CRC Critical Reviews in Biochemistry 19, 107–144.

Shatkin, A.J., 1976. Capping of eucaryotic mRNAs. Cell 9, 645–653.

Shi, W., Hendrix, D., Levine, M., Haley, B., 2009. A distinct class of small RNAs arises from pre-miRNA-proximal regions in a simple chordate. Nature Structural and Molecular Biology 16, 183–189.

Siegfried, N.A., Busan, S., Rice, G.M., et al., 2014. RNA motif discovery by SHAPE and mutational profiling (SHAPE-MaP). Nature Methods 11, 959–965.

Simao, F.A., Waterhouse, R.M., Ioannidis, P., et al., 2015. BUSCO: Assessing genome assembly and annotation completeness with single-copy orthologs. Bioinformatics 31, 3210–3212.

Song, J., Yi, C., 2017. Chemical modifications to RNA: A new layer of gene expression regulation. ACS Chemical Biology 12, 316–325.

Stein, L., Sternberg, P., Durbin, R., et al., 2001. WormBase: Network access to the genome and biology of Caenorhabditis elegans. Nucleic Acids Research 29, 82–86.

Stephens, S.B., Dodd, R.D., Lerner, R.S., et al., 2008. Analysis of mRNA partitioning between the cytosol and endoplasmic reticulum compartments of mammalian cells. Methods in Molecular Biology 419, 197–214.

Subramanian, A., Tamayo, P., Mootha, V.K., et al., 2005. Gene set enrichment analysis: A knowledge-based approach for interpreting genome-wide expression profiles. Proceedings of the National Academy of Sciences of the United States of America 102, 15545–15550.

Swan, A.L., Stekel, D.J., Hodgman, C., et al., 2015. A machine learning heuristic to identify biologically relevant and minimal biomarker panels from omics data. BMC Genomics 16 (Suppl. 1), S2.

Tay, Y., Kats, L., Salmena, L., et al., 2011. Coding-independent regulation of the tumor suppressor PTEN by competing endogenous mRNAs. Cell 147, 344–357.

Temin, H.M., Mizutani, S., 1970. RNA-dependent DNA polymerase in virions of Rous sarcoma virus. Nature 226, 1211–1213.

Thorvaldsdottir, H., Robinson, J.T., Mesirov, J.P., 2013. Integrative Genomics Viewer (IGV): High-performance genomics data visualization and exploration. Briefings in Bioinformatics 14, 178–192.

Toronen, P., Kolehmainen, M., Wong, G., Castren, E., 1999. Analysis of gene expression data using self-organizing maps. FEBS Letters 451, 142–146.

Vagin, V.V., Sigova, A., Li, C., et al., 2006. A distinct small RNA pathway silences selfish genetic elements in the germline. Science 313, 320–324.

Valadi, H., Ekstrom, K., Bossios, A., et al., 2007. Exosome-mediated transfer of mRNAs and microRNAs is a novel mechanism of genetic exchange between cells. Nature Cell Biology 9, 654–659.

van Dijk, E.L., Auger, H., Jaszczyszyn, Y., Thermes, C., 2014. Ten years of next-generation sequencing technology. Trends in Genetics 30, 418–426.

van Opijnen, T., Bodi, K.L., Camilli, A., 2009. Tn-seq: High-throughput parallel sequencing for fitness and genetic interaction studies in microorganisms. Nature Methods 6, 767–772.

Vora, N.L., Smeester, L., Boggess, K., Fry, R.C., 2017. Investigating the role of fetal gene expression in preterm birth. Reproductive Sciences (Thousand Oaks, CA) 24, 824–828.

Wang, Z., Gerstein, M., Snyder, M., 2009. RNA-Seq: A revolutionary tool for transcriptomics. Nature Reviews Genetics 10, 57–63.

Watters, K.E., Lucks, J.B., 2016. Mapping RNA structure in vitro with SHAPE chemistry and next-generation sequencing (SHAPE-Seq). Methods in Molecular Biology (Clifton, NJ) 1490, 135–162.

Wilson, D.N., Doudna Cate, J.H., 2012. The structure and function of the eukaryotic ribosome. Cold Spring Harbor Perspectives in Biology 4.

Woese, C.R., 1987. Bacterial evolution. Microbiological Reviews 51, 221–271.

Yao, C., Weng, L., Shi, Y., 2014. Global protein-RNA interaction mapping at single nucleotide resolution by iCLIP-seq. Methods in Molecular Biology (Clifton, NJ) 1126, 399–410.

Yuan, S., Schuster, A., Tang, C., et al., 2016. Sperm-borne miRNAs and endo-siRNAs are important for fertilization and preimplantation embryonic development. Development (Cambridge, England) 143, 635–647.

Zhang, H., 2016. Overview of sequence data formats. Methods in Molecular Biology (Clifton, NJ) 1418, 3–17.

Zhao, J., Ohsumi, T.K., Kung, J.T., et al., 2010. Genome-wide identification of polycomb-associated RNAs by RIP-seq. Molecular Cell 40, 939–953.

Ziegenhain, C., Vieth, B., Parekh, S., et al., 2017. Comparative analysis of single-cell RNA sequencing methods. Molecular Cell 65 (631–643), e634.

Further Reading

Asikainen, S., Heikkinen, L., Juhila, J., et al., 2015. Selective microRNA-Offset RNA expression in human embryonic stem cells. PLOS One 10, e0116668.

Korpelainen, E., Tuimala, J., Somervuo, P., Huss, M., Wong, G., 2014. RNA-seq Data Analysis: A Practical Approach. New York: CRC Press.

Quail, M.A., Smith, M., Coupland, P., et al., 2012. A tale of three next generation sequencing platforms: Comparison of ion torrent, Pacific biosciences and Illumina MiSeq sequencers. BMC Genomics 13, 341.

Relevant Websites

https://github.com/samtools/hts-specs
    GitHub.
https://nanoporetech.com/about-us/news/highlights-clive-g-browns-technical-update
    Oxford Nanopore Technologies.
http://www.pacb.com/wp-content/uploads/2015/09/PacBio_RS_II_Brochure.pdf
    Pacific Biosciences.
http://simpsonlab.github.io/2016/08/23/R9/
    Simpson Lab.

Transcriptomic Databases

Askhat Molkenov, Altyn Zhelambayeva, Akbar Yermekov, Saule Mussurova, Aliya Sarkytbayeva, Yerbol Kalykhbergenov, Zhaxybay Zhumadilov, and Ulykbek Kairov, Nazarbayev University, Astana, Kazakhstan

Introduction

Since the late 90s, the rapid development of the high-throughput omics technologies has led to a generation of the available transcriptomic data that demands effective solutions for organization, storage and further progress of new analysis tools. Database platforms have therefore been extensively developed to address these needs. Data repositories containing both curated and user-submitted entries are vital sources of knowledge for both researchers across various fields such as bioinformaticians, biologists, and clinicians for data analysis and for elucidating numerous processes within a disease or its development stages. As the development of omics technologies and bioinformatics continues, possibilities for data integration would allow inferring many more conclusions that could be crucial for a variety of problems arising from fundamental researches or practical applications in biomedicine and agriculture. The development and appearance of transcriptomic databases, in particular, can be useful to prevent pharmacological disease and even increase yield by determination of biomarkers or potential effector genes in cancer and plant pathogens, as well as identification of favorable alleles from the crop landraces.

We have conducted a Pubmed publications analysis for the occurrences of the terms "database" and "transcriptome" in both the title and the abstract of a publication to infer the "chronology" of this development. Thus, the first occurrence of the term "database" in a publication's title is dated as early as 1974; "transcriptome" made the first appearance much later in 1998 and both terms were used together in both the title and the abstract of a publication for the first time in 1999.

As **Fig. 1** illustrates, a log10 scale increase trend in the terms' use showed a corresponding growth of interest in the omics research, however, even nowadays, the occurrences of both "transcriptome" and "database" terms in the titles and abstract reflect, to the best of our knowledge, an absence of publications that could provide comprehensive overviews of the currently available transcriptomic resources. Thus, the aim of this review is to provide an overview of the current state of transcriptomic databases as broadly divided into the categories of human, animal and plant sources. A short summary of the database types will be provided for each category.

Human Transcriptome Databases

The human databases are mostly disease-specific sources with a particular focus on cancer data in addition to tissue-, interaction- and development-specific data platforms. Other human data sources include circRNA/ncRNA repositories, platforms containing raw full transcriptome data with the SNP information, the sources containing transcriptional signatures from selected experiments in the GEO databases, as well as functional databases.

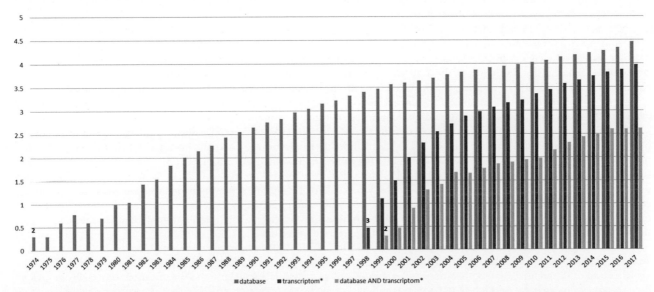

Fig. 1 Appearances of the terms "database" and "transcriptome" in the titles and abstracts of the available PubMed publications (as of March 2018); the number of publications is presented on a log10 scale.

Multiple Cancer Sources, Single Cancer-Specific Sources and an RNA-Specific Cancer Database

The largest source of cancer data estimated at 2.5 petabytes is represented by TGCA (Chang *et al.*, 2013) (The Cancer Genome Atlas). TGCA is a result of a concentrated effort between the National Cancer Institute (NCI) and the National Human Genome Research Institute (NHGRI) with 20 collaborating institutions to present the range of genetic changes and comparison between the samples and with the normal tissue profiles for 23 most common types of cancer in addition to 10 rare cancers. TGCA provides information on molecular, DNA, protein and epigenetic aberrations resulting in a better understanding of cancer subtypes and higher precision of treatment options as the types of cancer chosen for study have a poor prognosis and result in a major detrimental public health impact (e.g., pancreatic ductal adenocarcinoma, acute myeloid leukemia, cutaneous melanoma). As of now, the data is compiled from 11,000 patients and includes whole genome and exome profiles, mRNA/microRNA sequences, copy number and methylation profiles as well as protein activation status information. Clinical data and data regarding the sample (e.g., weight) are available for some of the sample entries.

Another major data source which contains cancer data in addition GEO-NCBI (the Gene Expression Omnibus) (Barrett *et al.*, 2012) is a public submission data repository containing high-throughput sequencing and microarray data that was created and is under maintenance of the National Center for Biotechnology Information (NCBI). The submitted data comes from the scientific communities worldwide resulting in MIAME (Minimum Information About a Microarray Experiment)-compliant data entries representing more than 1 min samples and 1600 organisms (humans, animals, plants, bacteria and fungi) with links to 20,000 related publications, 90% of the data coming from gene expression studies characterizing such processes as development, metabolism and disease. The database is a user-friendly archival resource that has a number of useful options such as "similar studies" and "find pathways" (i.e., gene mapping to a list of frequency weighted pathways in NCBI BioSystems database) search options, repository browser and GEO2R web app which allows to perform differential gene expression analysis on more than 90% of GEO entries. The data is divided into 3 categories of Platform (array/sequencer description), Sample (biological materials/ experimental protocol descriptions) and Series (a link between a group of related Sample and the point of the whole study). GEO DataSet compiles the comparable Samples into a consistently processed format that is divided into categories based on experimental variables and has a description of the study. Thus, a search for term "cancer" in GEO Datasets (all fields) results in 646,300 entries with 1272 DataSets and 616,885 Samples (as of March 2018). The cancer information is available for both humans, animals (e.g., mouse and zebrafish) and constructs (e.g., studies on protein expression data from using viral constructs and cancer xenografts); a number of DataSets contain expression information on metastatic cancers.

Another database which includes transcriptome information on multiple types of cancer is BioXpress (Wan *et al.*, 2015). BioXpress is a database of gene expression and cancer association. It includes gene expression data from 64 cancer types from 6361 patients. It also provides 17,469 genes with 9513 of them showing differential gene expression between normal and tumor samples. To facilitate pan-cancer analysis, all cancer types are mapped to Disease Ontology terms. BioXpress database includes mappings of expression levels to genes using RNA-seq data extracted from The Cancer Genome Atlas, International Cancer Genome Consortium, Expression Atlas. Expression levels are also obtained from publications of large-scale studies.

CoReCG, HNdb and curatedOvarianData databases are cancer-specific and discussed below. CoRECG database contains a comprehensive review of previous studies on colon-rectal cancer (Agarwal *et al.*, 2016). They collected the information about the genes associated with colon-rectal cancer from high-quality published literature and organized this data into the CoReCG database. CoReCG contains 2056 cancer genes annotation involved in distinct stages of colon-rectal cancer. CoReCG database contains comprehensive information about genes such as locations in chromosomes, homology, ontology, SNP information, genomic coordinates, mRNA, information about proteins, etc.

The Head and Neck (HNdb) database (Henrique *et al.*, 2016) was created to provide a comprehensive resource of the Head and Neck squamous cell carcinoma genes (HNSCC) and associated proteins. The retrieved previously performed studies from PubMed database and developed a methodology for literature mining to identify genes associated with HNSCC. The information about the gene can be searched by the name of the gene or its associated protein, by gene symbol, aliases. Also, the information about chromosome regions and gene associated with HNSCC can also be obtained. The database also includes the information about gene ontologies, metabolic pathways and links for comparison of the expression features in normal tissues. The protein page includes the proteins' 3D structures, posttranslational modifications, metabolite and protein-protein interaction as well as drug targets.

Transcriptomic database of ovarian cancer (Ganzfried *et al.*, 2013) was implemented as the *curatedOvarianData* Bioconductor package in R. It contains microarray data of 2970 patients from 23 studies with curated and documented clinical metadata. They pre-processed available Affymetrix microarray data by the same normalization algorithm. For the Affymetrix U133A and U133 Plus 2.0, they used fRMA normalizing algorithm. Then, the analysis of expression levels of *CXCL12* was performed in all primary tumor samples included in the database. The samples were centered by their means and divided by their standard deviation to ensure that expression levels are compared on the same scale. They confirmed the hypothesis that poor outcome in 2108 patients from 13 independent studies with different stages and histologies was associated with upregulation of *CXCL12*. They emphasized the importance of biomarker validation in sufficiently large collections of data.

Cancer RNA-seq Nexus (Li *et al.*, 2016) is the first publicly available RNA-seq database that contains comprehensive data analyses and archives for transcript profiles of coding sequences, long non-coding RNA (lncRNA), coexpression networks, visualization of differential gene expression profiles and phenotype-based data searching and organization. It contains in total 54 human cancer RNA-seq datasets for 11,030 samples and 326 phenotype-specific subsets where each subset represents the group of samples under specific cancer type and stage conditions. mRNA-lncRNA coexpression networks can be visualized, compared and analyzed for different pairs of phenotype-specific subsets. Coding transcript/lnc expression profiles can also be visualized. Its functionality and user-friendly interface makes Cancer RNA-seq Nexus serve as a handful tool for personalized medicine research.

Other Disease-Specific Sources: Kidney Diseases, Parkinson's, Inner Ear Diseases, Prion Disease, Complex Diseases

Renal Gene Expression Database (RGED) was developed in Zhang et al. (2014) to provide researchers with an access to gene expression profiles of various kidney diseases. RGED includes profiles obtained from renal cell lines, human kidney tissues and murine model kidneys. They searched for gene expression profiling experiments from NCBI and selected a total of 88 studies that published between 2004 and 2014. The platforms are varying across different studies but most of them used Affymetrix, Inc., Illumina, Inc. and Agilent Technologies, Inc. RGED contains profiles of kidney carcinoma, acute kidney injury, autosomal dominant polycystic kidney disease, IgA nephropathy, chronic kidney disease. RGED release 1.0 has a collection of 5354 samples which include cell lines, human tissues and model animals. Previously created databases for kidney research contains only genes or genomic loci in a human which may not be sufficient for a comprehensive study of kidney diseases. RGED allows exploration of gene profiles, relationships between genes of interests, identification of biomarkers and drug targets in different kidney diseases.

The first queryable database ParkDB for gene expression in Parkinson disease was developed in Taccioli et al. (2011). ParkDB has three main functions. The first is to collect all microarray data related to Parkinson disease from publicly available sources such as Array express, Gene Expression Omnibus (GEO) and Stanford microarray database. The second is to provide information about differentially expressed genes across different tissues, treatments and species. The third is to provide information about statistical analysis of microarray data to provide users with the comparison between different experiments.

Shared Harvard Inner Ear Laboratory Database (SHIELD) was developed by Shen et al. (2015). SHIELD is an integrated resource of genomic, transcriptomic and proteomic information of the inner ear. It contains five datasets that are combined into a relational database and provides experimental data and annotations. They derived annotations from public databases and publications. Annotations include gene symbols, gene name, synonyms, human and mouse chromosome cytogenetic banding, genomic coordinations, RefSeq RNA, Ensembl, ontology, etc. Five datasets represent different developmental stages of the variety of model organism species and different cell types. Most commonly used platforms are RNAseq (3*DGE and full-length mRNA), Microarray (GeneChip Mouse Gene 2.0 ST and MOE430) and mass spectrometry.

Prion Disease Database (PDDB) (Gehlenborg et al., 2009) developed by is a comprehensive transcriptomic database of mouse prion disease, which provides primary experimentally obtained data. The PDDB contains mRNA measurements across the time spanning the interval from prion inoculation through the onset of clinical signs in eight mouse strain-prion strain combinations.

dbPTB database was created by Uzun et al. (2012) to facilitate the identification of genetic variants associated with complex diseases. The database combines information about genes, their SNPs, genetic variations and various pathways involved in the preterm birth. They searched publicly available databases and publications that describe genome- and transcriptome-wide studies. Genes, which expression levels were increased or decreased in the association with preterm delivery were included in the database.

Tissue-, Process- and Interaction-Specific Sources: Human Oral Microbiome, Human Brain Transcriptome, Human Ageing, Neurological Ageing, Genotype-Tissue and Genotype-Phenotype Interactions

Human Oral Microbiome Database (HOMD) (Chen et al., 2010) provides datasets for more than 600 prokaryote species of a human oral cavity. HOMD provides both taxonomic and genomic types of information. Each species was given a unique Human Oral Taxon (HOT) number based on the 16S rRNA gene sequence analysis. Each HOT contains genomic, clinical, phenotypic, phylogenetic and bibliographic information for each taxon. Gene sequences can be matched to full-length 16s rRNA reference dataset by using BLAST tool. HOMD can serve as a model for microbiome database of gut, skin and human body cavities.

The HBT (Human Brain Transcriptome) database (Johnson et al., 2009) is a public database for transcriptome data and the associated metadata for the developing and adult brain, from data on males and females of different ethnicities, from different brain regions and neocortical areas. The database is built upon the results of whole-genome, exon-level expression analysis of 16 brain regions, generated using Affymetrix GeneChip Human Exon 1.0 ST Arrays. Over 1340 tissue samples from both hemispheres of postmortem human brains were analyzed. The authors applied ANOVA for differential expression analysis and unsupervised hierarchical co-clustering for analyzing brain regions with respect to their transcriptional profiles. The online database enables searching for individual genes and assessing their temporal activity in an age-intensity graph, or searching for the most active genes with regard to their spatial (by brain structure) and temporal (by start and end period) activity.

GenAge, a database of genes potentially associated with human ageing (Tacutu et al., 2013), was developed as a part of The Human Ageing Genomic Resources (HAGR) framework. GenAge is a publicly available, continuously curated database containing a comprehensive list of genes potentially associated with ageing in humans and model organisms, as well as a list of differentially expressed genes in mammals, including humans, from a meta-analysis of microarray studies. The manually-curated list of human genes in GenAge is 288, annotated with respect to their role in the human ageing process as a whole, as a result of analysis of more than 2200 references. In addition, it includes over 1700 genes associated with longevity in model organisms, which are based on experimentally validated results from the multitude of studies, such as overexpression of certain transcripts or RNA interference with age. Each entry in the database a description of its main functions and its relation to the ageing process, as well as the reason for inclusion into the database.

Brain transcriptome studies on gene dysregulation in Alzheimer's disease (AD) and closely related processes such as ageing and neurological disorders were collected and curated in AlzBase database (Bai *et al.*, 2016). Aside from gene dysregulation in AD, it includes information on correlation of gene dysregulation with severity of AD, functional and regulatory annotations, and a network of the corresponding gene-gene interactions, built via WGCNA. It includes a total of 14,145 dysregulated genes. For gene annotations, eQTL experiments (11,055), in situ hybridization experiments (1167), and GWAS experiments (12,287) were used. The authors constructed a gene coexpression network using several large datasets on brain transcriptome taken from GEO, which covered AD-associated data on several brain regions and peripheral blood, and four stages of progression of the disease. For the selection of differentially expressed genes, the RankProd algorithm (p < 0.05) was used. To further group the most significant genes (277), the authors used hierarchical clustering and grouped them into 11 clusters based on the temporal pattern of gene expression.

The Genotype-Tissue Expression (GTEx) database was created as part of GTEx project which aimed at establishing a data resource for studying the relationship between genetic variation and gene expression in human tissues (GTEx Consortium, 2017). The GTEx represents the results of high-throughput sequencing and microarray analysis used for eQTL analysis from samples (7051) of 44 different tissue types from 449 donors. The authors compared Spearman's similarity between various tissues for cis- and trans- eQTLs and used agglomerative hierarchical clustering to group the tissues that share any given eQTL. The GTEx's web portal enables searching for genes by Gene or SNP Id and outputs a visualization of its median expression level across all samples for each tissue.

dbGaP, the database of Genotypes and Phenotypes, was developed for storing and distributing the data from studies of genotype-phenotype interaction in humans (Mailman *et al.*, 2007). It is a large NCBI-sponsored public repository that assigns unique identifiers to studies and datasets. The main objective of the project is to store, collect and distribute phenotype data that can be associated with other data types, such as sequence-derived genotypes and haplotypes, gene expression data, proteomic, metabolomic, and epigenetic data. Currently, the main focus of the GWAS studies covered in dbGaP is microarray data. dbGaP has integration with Sequence Read Archive (SRA), and thus the sequences for the queried genes can be downloaded from the NCBI web portal. Aside from that, users can view the association results in Genome Browser's chromosomal viewer, where loci with the smallest p-value are color-coded, and in the sequencing view, users can see the corresponding gene/ transcript/protein annotations.

Other Tools and Sources: Circular RNA/Small Non-Coding RNA Databases, Raw Read Archives, a Transcriptome Browser and a Human Transcriptome Functional Annotation Source

CircNet is a database for circular RNAs which is the type of regulatory noncoding RNA (Liu *et al.*, 2016). The expression of circRNAs was identified by using transcriptome sequencing datasets in 464 RNA-seq samples. CircNet provides an access novel circRNAs, integrated miRNA-target networks, expression profiles, genomic annotations and sequences of circRNA isoforms. It also displays the regulation between circRNAs, miRNAs, and genes by generating the integrated regulatory network. By using the search box, users can find the gene-miRNA-circRNA regulatory network of particular genes. CircNet allows studying circRNA tissue-specific functions and correlation to diseases due to its expression profile and accessibility.

Human genome data generated from next-generation sequencing and high-resolution comparative genomic hybridization array were collected into the Total Integrated Archive of short-Read and Array (TIARA) database (Hong *et al.*, 2013). TIARA provides an access for variant information and raw data for 16 sequenced transcriptomes, 13 sequenced whole genomes and 33 high-resolution array assays (for September 2012). Genomic variants data includes a total of approximately 9.56 million SNPs, where 23,025 constitute non-synonymous SNPs and approximately 1.19 million indels. The abundance of genomic variants data makes it a useful resource for personalized genomics studies. Genome browser allows the matching of whole-genome sequence with transcriptome sequencing data. Also, the levels of gene expression can be observed with allele-specific quantification.

For storing next-generation sequence data (Leinonen *et al.*, 2011), the Sequence Read Archive (SRA), which serves as NIH's primary archive of high-throughput sequencing data, was created. The SRA stores raw sequencing data and alignment information through such high-throughput sequencing platforms as Roche 454, Illumina Genome Analyzer, SOLiD™, Helicos HeliScope™, Complete Genomics, and SMRT™. The SRA archives raw oversampling NGS data with EMBL and DDBJ, as well as the data from human clinical samples. The database enables users searching for genes of interest via metadata in the query request, as well as performing a sequence-based search using BLAST.

TranscriptomeBrowser (TBrowser) (Lopez *et al.*, 2008) was created for hosting a large integrated database of transcriptional signatures extracted from more than 1400 (currently ~4000) experiments stored in the GEO database. The authors used a modified version of the Markov Clustering algorithm – DBF-MCL, to systematically extract clusters of co-regulated genes, which allowed to define over 18,000 transcriptional signatures (currently around ~40,000). The authors tested established transcriptional signatures for functional enrichment using DAVID knowledgebase, and over-representation of functional terms was found in a large proportion (84%) of these transcriptional signatures. The TBrowser enables searching by GeneID and annotation, and supports Boolean operators.

The GENCODE is an online database developed as a part of the ENCODE (Encyclopedia of DNA Elements) project, which contains functional annotations for 20,687 protein-coding genes, 9640 long non-coding RNA loci for 15,512 transcripts, and 33,977 coding transcripts not represented in RefSeq and UCSC (Harrow *et al.*, 2012). The gene annotations are derived from a combination of manual curation and automatic gene annotation from Ensembl, different computational analysis techniques (PhyloCSF, APRIS annotation pipeline), and targeted experiments. GENCODE contains a comprehensive and manually annotated set of pseudogenes, in addition to the sets of protein-coding and non-coding genes. The authors also used transcriptome sequencing data facilitated by high-depth RNA sequencing for a very detailed overview of alternative splicing. The web portal of GENCODE enables searching by ENS id, ENS gene id, and gene name (for a corresponding BioDalliance genome visualization), as well as downloading all the relevant GTF/GFF3 datasets, FASTA datasets, and the metadata files.

Database of small human noncoding RNA (DASHR) is the most comprehensive database on human sncRNA genes and mature sncRNA products (Leung *et al.*, 2016). DASHR integrates over 180 smRNA smRNA-seq datasets over 2.5 billion reads and annotation data from various publicly available sources. DASHR contains 7641 sncRNA gene records, and 9703 mature sncRNA product records, corresponding to ~48000 genomic loci. Among the sncRNA genes, 82% are expressed in one or more tissues. As such, 6301 annotated sncRNA genes can be queried in DASHR. DASHR has an intuitive web interface that allows users to search by full name, partial name or ID of sncRNA, compare expression levels in various human tissues, locate RNA loci overlapping with the given genomic coordinates using h19 transcript annotation as a reference, or query by raw RNA sequence. It also provides a summary of annotation and expression information, with an expression heatmap, an expression profile across tissues, and a genome browser for mapped sequences.

Animal Transcriptome Databases

Animal transcriptome databases are mainly represented by the data sources containing the genome and transcriptome sequencing/RNASeq/expression/functional annotation information on the model and mammalian organisms such as *Xenopus*, *Mus musculus*, *Rattus norvigicus* and *Sus scrofa*, combining sequencing and expression data. *M.musculus*-focused databases are represented by data repositories on embryo and brain development as well ageing, reflecting the fact that mouse is a model organism for most of the crucial mammalian processes as important to human health and development research. Ageing-related databases also include entries from *Caenorhabditis elegans*, *Drosophila melanogaster*, *Saccharomyces cerevisiae* and *Schizosaccharomyces pombe* as the organisms are widely used for understanding ageing processes in *vivo*. Arthropods and parasites are also represented by the relevant database source reflecting the importance of the omics data for studies in comparative genomics, evolutionary and developmental biology as well as ecology and plant/human disease management (e.g., rice pest *Chilo suppressalis* and Leishmania databases). Another important animal transcriptome database source is a *Ctenopharyngodon idellus* (grass carp) GCGD database; the grass carp is an important farmed fish for the global freshwater aquaculture business. In addition, as mentioned in the previous section, GEO-NCBI also contains a wide range of data from animal sources.

Model and Mammalian Organism Sources

Xenbase is an online database that provides an access to the sequencing data for *Xenopus laevis* and *Xenopus tropicalis* (Karimi *et al.*, 2018). The data were obtained from different resources such as publications and bioinformatics analyses. The database includes genomes for both *X. laevis* and *X. tropicalis*, chromosomal assemblies, annotations, genome segmentation, visualization of RNA-Seq data, protein interaction, gene ontologies, updated ChIP-Seq mapping, etc. Users can use BLAST pages from the main navigator bar and select the genes of interest. They also provide the graphical representation of the genomic and chromosomal structure by using the gene coordinates data.

cSynechococcus sp. PCC 7002 omics studies were combined into the CyanOmics database (Yang *et al.*, 2015). It comprises one genomic dataset, one proteomic dataset and 29 transcriptomic datasets. CyanOmics provides functional annotation for complete genome sequence, transcriptomes under different experimental conditions and proteomic analyses. They also integrated tools for browsing, navigation, sequence alignment and data visualization. The database provides links.

The Mammalian Transcriptomic Database (MTD) was developed by Sheng *et al.* (2017) for characterization and comparative analysis of transcriptome data, and it comprises RNA-seq data from most available tissues of certain mammalian species. Currently, MTD contains RNA-seq data from *Homo sapiens* (83), *Sus scrofa* (27), *Rattus norvegicus* (36) and *Mus musculus* (108). Among the core features of the database is that it allows finding genes based on their neighboring genomic coordinates or joint KEGG pathway, and displays detailed expression information on exons, transcripts, and genes in a genome browser (GBrowse). The MTD enables comparative transcriptomic analysis in both intraspecies and interspecies manner. The database is implemented as a web interface, which provides the following functionalities: Browse (by chromosome, region, or pathway), Search (by gene feature or isoform feature), Analysis (intraspecies and interspecies), and Download (for raw SRA data or gene expression data by tissue/cell line).

Database of Transcriptome in Mouse Embryos (DBTMEE) (Park *et al.*, 2015) was developed for systematic analysis of *mus musculus* fertilization dynamics. DBTMEE is based on ultra-large-scale RNA-seq analysis, with more than 1.5×10^5 oocytes sequenced. The authors also included data from other publicly available resources, such as microarray gene expression data, RNA-seq data prepared by different sequencing protocols and at various cell types and embryonic stages, as well as mass-spectrometry proteomic data, DNA-methylation data and ChIP-seq of histone variants in spermatozoa. The database has a comprehensive web interface, through which the users can identify fertilization-specific genes, assess the impact of histone modifications on early embryo development, and use genetic and epigenetic characteristics to investigate molecular characteristics among totipotent, pluripotent, and differentiated cells. The authors applied hierarchical clustering to categorize the gene expression patterns with respect to mouse embryonic development, and the results of the clustering (with further details and links for each gene within each cluster) are accessible from DBTMEE's front page.

The Brain Transcriptome Database (BrainTx) was developed by Sato *et al.* (2008) as a successor to Cerebellar Development Transcriptome Database (CD-DBT, alternative/former name), an online database centered around transcriptomes related to a cerebellar development of a mouse. The current version of BrainTx represents an integrated online platform for visualization and analysis of transcriptomes related to various stages and states of the mammalian brain (primarily brain of *mus musculus*, but also includes vast amounts of data on other species). Aside from search by NCBI Gene ID and CD ID, it enables searching for genes and transcription factors via associated keywords, e.g., name of the related gene, name of the associated protein structure or function, or cell function. The online database enables searching for differentially expressed genes during the different stages of brain development, analyze expression patterns with respect to the region (Brain distribution) and tissue distribution pattern (Brain specificity). BrainTx's Gene Ontology Tree Browser includes 6603 genes annotated for their associated biological processes, 5782 genes annotated for their associated cellular components, and 6968 genes annotated for their associated molecular functions.

AGEMAP (Atlas of Gene Expression in Mouse Aging Project) (Zahn *et al.*, 2007) is a gene expression database that catalogs changes in gene expression profiles as a function of age in mice. The authors of the database generated mRNA transcript profiles in mice with respect to changes in age and included 8932 genes in 16 tissues at 4 times during the aging. The authors used hierarchical clustering to categorize the data on age-related transcriptional changes into 3 groups (3 modes of aging for different mouse tissues): neural tissues, vascular tissues, and steroid-responsive group (thymus, gonads), which suggests that different types of tissues have different rates of aging. Certain mice tissues exhibit the strongest transcriptional differences as a function of age, while other mice tissues show little to no difference with age. The database can be downloaded as Z-normalized expression data for each tissue.

GenDR is a database also developed under The Human Ageing Genomic Resources (HAGR) framework (Wuttke *et al.*, 2012). GenDR is the first database that contains consistent gene expression changes induced by dietary restriction, in addition to gene mutations interfering with dietary restriction-mediated lifespan extension. The authors established transcriptional signatures induced by dietary restriction (DR), analyzing differentially expressed genes from microarray data in model organisms (*Caenorhabditis elegans, Drosophila melanogaster, Mus musculus, Saccharomyces cerevisiae, Schizosaccharomyces pombe*). For all the model organisms, they assessed the influence of gene expression changes upon DR on the multiple interactomes within an interaction network (using ExprEssence algorithm, analysis of suppressed and stimulated interactions, and DAVID functional enrichment analysis). They identified more than 100 genes that are essential for an effect of dietary restriction to take place (DR-essential genes) in model organisms. The web interface of GenDR enables matching genes with the list of DR-essential genes by Entrez ID, gene symbol or name, or browsing the list of DR-essential genes by species. Finally, it allows comparing the DR-essential genes with their homologs across other species, including *Homo sapiens*.

Arthropod and Parasite Sources

Genomic and transcriptomic database ChiloDB for rice insect *Chilo suppressalis* was developed by Yin *et al.* (2014). The aim of ChiloDB was to provide researchers with access to recent genomic and transcriptomic data to allow comparative genomic studies. They integrated the data with protein-coding genes, microRNAs, piwi-interacting RNAs (piRNAs) and RNA-Seq data. They also provide gene annotations for *Chilo suppressalis*. ChiloDB contains the first version of the *Chilo suppressalis* gene set that includes 10,221 annotated protein-coding genes and 80,479 scaffolds. They identified 82,639 piRNAS with piRNApredictor software and predicted 262 microRNA genes from small RNA library. They also integrated 37,040 transcripts from midgut transcriptome and 69,977 transcripts from a mixed sample.

Assembled Searchable Giant Arthropod Read Database (ASGARD) database (Zeng and Extavour, 2012) is a resource for transcriptomic data of arthropod species. Transcriptomes were assembled *de novo* from milkweed bug *Oncopeltus fasciatus*, the cricket *Gryllus bimaculatus* and amphipod crustacean *Parhyale hawaiensis*. Putative ontology, coding region determination, protein domain identification was annotated. Assemblies can be searched by orthology annotation, gene ontology term annotation or Basic Local Alignment Search Tool. Users can download FASTA format assembly product sequences, find links to NCBI predicted orthologs data. Also, the locations of protein domains can be graphically visualized and matched to similar sequences from the NCBI. The database is useful for studies in different fields such as comparative genomics, evolutionary biology, physiology, developmental biology, phylogenomics, and ecology.

The database LeishDB of *Leishmania braziliensis* was developed by Torres *et al.* (2017) to update and improve publicly available data for non-coding RNAs (ncRNAs) and coding gene annotations of *L. braziliensis*. They used updated databases and five prediction algorithms: GENAN, GLIMMER, SNAP, RATT, and AUGUSTUS. A total of 11491 Open reading frames (ORF), including 5263 (45.80%) of them that are associated with proteins were identified. Also, they found 11,243 ncRNAs that belong to different classes distributed along the whole genome. The accuracy of their predictions was verified by the comparison of predicted sequences with those obtained from RNA-seq. The automatic predictions and annotations of non-coding sequences were performed by covariance models comparisons and sequence similarity searches.

GCGD: A Farmed Fish Database

Grass carp genome database (GCGD) was developed by Chen *et al.* (2017) to provide researchers with the centralized database of genomic data for *Ctenopharyngodon idellus*. Genome, amino-acid sequences of predicted protein-coding genes, annotations and their visualizations are included into the database. GCGD also provides an access to transcriptomic datasets, genetic linkage maps, Short Sequence Repeats. They integrated different useful tools which enable efficient data retrieval, analysis, and data visualization. For example, JBrowse provides genomic annotation navigation, BLAST enables to perform sequence alignment, EC2KEGG facilitates the comparison of metabolic pathways, IDConvert allows the conversion of terms across databases and ReadContigs facilitates the extraction of sequences from the genome of grass carp.

Plant Transcriptome Databases

Most transcriptome plant databases are dedicated to agronomically important species such as crop, tree and vegetable plants. For the crop plants, the main data repositories were created for bread wheat, rice, the burclover (*Medicago truncatula*) legume and sorghum. Vegetable and tree species such as carrot and mulberry tree (important for the mulberry-silkworm interaction), eucalyptus (commercially important for the production of paper and cellulosic ethanol) have corresponding transcriptome database resources. Other important sources include transcriptome databases of medicinal plants and metabolic pathway information.

Crop Plants

As wheat provides an estimated 20% of all the consumed calories worldwide, considerable interest in the crop drives efforts into developing more commercially advantageous varieties. Omics resources are the first port of knowledge to identify potential traits of interest which could further help to narrow down to alleles for the use in breeding programs. With this goal, the functional annotation database called dbWFA for bread wheat *Triticum aestivum* was developed in Vincent *et al.* (2013). dbWFA is based on the reference 56,954 transcripts NCBI wheat UniGene set which is an expressed gene catalogue created based on full-length 17,541 TriFLBD database coding sequences by using expressed sequence tag clustering. Annotation information from the model plant *Arabidopsis thaliana* and *Oryza sativa* L. was linked back to the *T.aestivum* sequences using BLAST-based homology searches. Putative functions were assigned to 45% of the UniGene transcripts and 81% of the TriFLDB full-length coding sequences. The database contains information on functional annotations, Gene Ontologies, gene families and metabolic pathways.

Burclover is also a forage crop of commercial interest. MtDB is a relational database (Lamblin *et al.*, 2003) contains transcriptomic data for *Medicago truncatula* legume *M. truncatula* samples that represent various developmental stages and pathogen-challenged tissues. Unigene sets generated by different assemblers such as Phrap, Cap3 and Cap4 can be selected and compared. MtDB provides users with quick access and identification of sequences for particular research interests. To facilitate comparison between different sites, the authors performed cross-reference of sequence identifiers from all public *M. Truncatula* sites such as GenBank ID, CCGB, TIGR, NCGR, INRA. Also, the database provides hypertext links for databases for all queries' results.

Sorghum genomic data and functional annotations were collected by Tian *et al.* (2016) to create SorghumFDB database. The database provides detailed gene annotations, orthologous pairs in Arabidopsis, rice and maize, miRNA and target gene information, genome browser and gene loci conversations. It also includes the information about sorghum gene family classification such as transcription factors and regulators, carbohydrate-active enzymes, ubiquitins, protein kinases, etc. Co-expression data, protein-protein interactions, and miRNA-target pairs were used to construct a dynamic network of biological relationships.

TENOR (Transcriptome Encyclopedia Of Rice) is a database for comprehensive mRNA-seq experiments in rice that was developed by Kawahara *et al.* (2016). The database includes large-scale mRNA-seq data from rice in a wide variety of conditions. TENOR was developed to enable the scientists better understand the regulatory networks of genes responsible for adaptation to different environmental conditions. As such, the authors used mRNA-seq and applied time-course transcriptome analysis of rice (*Oryza sativa* L.), under 2 plant hormone treatment and 10 abiotic stress conditions. Differential expression analysis detected a large number of genes responsive to abiotic stress and plant hormones treatment. The online database enables users to download rice transcriptome data, co-expression data of the responsive genes, search relevant genes by keyword, genomic coordinates, and by responsive expression patterns. TENOR also includes functionality for visualization of the expression levels under various conditions for each transcript.

Tree and Vegetable Species

The CarrotDB database of *Daucus carota* L. was assembled and analyzed in Xu *et al.* (2014). They downloaded 14 carrot RNA-seq raw data from Sequence Read Archive (SRA) and National Center for Biotechnology Information (NCBI) and mapped them to the whole genome sequence. They classified a total of 2826 transcription factors (TF) into 57 families and identified them in the entire genomic sequence. Carrot-DB integrates carrot genome sequence, coding sequences, putative genes sequences, transcript sequences, etc. The part of CarrotDB called 'GERMPLASM' provides 45 carrot genotypes taproot photos and information about names, accession numbers, colors of the cortex, countries of origin, phloem and xylem.

Morus Notabilis is an important plant for sericulture and plant-insect interaction studies (Li *et al.*, 2014). The genome and transcriptome of *M. Notabilis* were sequenced and analyzed. They developed open-access Morus Genome Database (MorusDB). The database provides access to large-scale genomic sequencing and assembly, genes and functional annotation transposable elements (TEs), gene ontology (GO) and expressed sequence tags (ESTs). To provide an access for identification of orthologous and paralogous groups of *M. Notabilis*, the most recent versions of protein sequences of related plants such as *Arabidopsis thaliana*, *Populus trichocarpa*, *Malus domestica* and *Fragaria vesca* were downloaded. Gene ontologies of *M. Notabilis* can be analyzed using Mulberry GO tool package. Gene ontologies can be searched using Search GO or obtained using gene ID and Fetch GO.

EUCANEXT (Nascimento, 2017) database is an integrated gene expression (RNAseq) and ESTs data platform with a set of the web-based tool allowing sequence similarity/keyword/transcript ID searches as well as allowing to find differentially expressed sets of genes under different custom-chosen conditions. The database is based on 14 Illumina RNA-Seq libraries and 165,268 ESTs; genome annotation information is also available with GO, orthoMCL, NR, Uniref90, Swiss-Prot, Tair and CDD terms. A genome

Table 1   Transcriptomic databases

#	Database name (short name)	Organism	Type of data	URL	Reference (PMID/DOI)
1	TCGA	human	Whole genome and exome sequencing profiles, mRNA/microRNA sequences, copy number and methylation profiles	https://cancergenome.nih.gov	10.1038/ng.2764
2	GEO	human/animal/plant/fungi	RNA-seq, microarrays	https://www.ncbi.nlm.nih.gov/geo/	10.1093/nar/gks1193
3	BioXpress	human	RNA-seq	http://hive.biochemistry.gwu.edu/tools/bioxpress	https://doi.org/10.1093/database/bav019
4	CoReCG	human	SNP, gene expression	lms.snu.edu.in/corecg	10.1093/database/baw059
5	HNdb	human	Gene expression, microarray	http://www.gencapo.famerp.br/hndb	10.1093/database/baw026
6	curatedOvarianData	human	Gene expression, microarray, RT-PCR	http://bcb.dfci.harvard.edu/ovariancancer	10.1093/database/bat013
7	Cancer RNA-Seq Nexus	human	Gene expression, RNA-seq	http://syslab4.nchu.edu.tw/CRN	10.1093/nar/gkv1282
8	RGED	human	Gene expression, microarray, RNA-seq	http://rged.wall-eva.net	10.1093/database/bau092
9	ParkDB	human	Microarrays, gene expression	http://www2.cancer.ucl.ac.uk/Parkinson_Db2/	10.1093/database/bar007
10	SHIELD	human	Gene expression, microarray, ChIP-seq	https://shield.hms.harvard.edu	10.1093/database/bav071
11	PDDB	human	Gene expression, microarray	http://prion.systemsbiology.net	10.1093/database/bap011
12	dbPTB	human	SNPs, gene expression	http://ptbdb.cs.brown.edu/dbPTBv1.php	10.1093/database/bar069
13	HOMD	human	Microarray	http://www.homd.orghttp://www.homd.org http://www.homd.org	10.1093/database/baq013
14	HBT	human	Microarrays	http://hbatlas.org/	10.1016/j.neuron.2009.03.027
15	GenAge	human	Microarray data	http://genomics.senescence.info/genes/	10.1093/nar/gks1155
16	AlzBase	human	SNP and expression microarray data	http://alz.big.ac.cn/alzBase	10.1007/s12035-014-9011-3
17	GTEx	human	RNA-seq, microarray data	https://www.gtexportal.org/	10.1038/nature24277
18	dbGaP	human	SNP and expression microarray data, RNA-seq	https://www.ncbi.nlm.nih.gov/gap/	10.1038/ng1007-1181
19	CircNet	human	Gene expression, RNA-seq	http://circnet.mbc.nctu.edu.tw/	10.1093/nar/gkv940
20	TIARA	human	Gene expression, RNA-seq	http://tiara.gmi.ac.kr	10.1093/database/bat003
21	SRA	human	RNA-seq, ChIP-seq	http://www.ncbi.nlm.nih.gov/Traces/sra	10.1093/nar/gkq1019
22	TranscriptomeBrowser	human	Microarray data	http://tagc.univ-mrs.fr/tbrowser/	10.1371/journal.pone.0004001
23	GENCODE	human	RNA-seq, RT-PCR-seq, RACE-seq	https://www.gencodegenes.org/	10.1101/gr.135350.111
24	DASHR	human	RNA-seq	http://lisanwanglab.org/DASHR/	10.1093/nar/gkv1188
25	Xenbase	Xenopus laevis and Xenopus tropicalis	Gene expression, RNA-seq, ChIP-seq	http://www.xenbase.org/	https://doi.org/10.1093/nar/gkx936
26	CyanOmics	cyanobacterium Synechococcus	Gene expression, RNA-seq	http://lag.ihb.ac.cn/cyanomics	10.1093/database/bau127
27	MTD	Sus scrofa, Rattus norvegicus, Mus musculus, Homo sapiens	RNA-seq	http://mtd.cbi.ac.cn/	10.1093/bib/bbv117

No.	Database	Species	Data type	URL	DOI/Reference
28	DBTMEE	Mus musculus	RNA-seq, ChIP-seq, DNA methylation data, MS proteomic data	http://dbtmee.hgc.jp/	10.1093/nar/gku1001
29	BrainTx	Mus musculus	Microarray data, RT-PCR-seq	http://www.cdtdb.neuroinf.jp/CDT/Top.jsp	10.1016/j.neunet.2008.05.004
30	AGEMAP	Mus musculus	Microarray data	http://cmgm.stanford.edu/~kimlab/aging_mouse/	10.1371/journal.pgen.0030201
31	GenDR	Caenorhabditis elegans, Drosophila melanogaster, Mus musculus, Saccharomyces cerevisiae, Schizosaccharomyces pombe	Microarray data, RNA-seq	http://genomics.senescence.info/diet/	10.1371/journal.pgen.1002834
32	ChiloDB	Chilo suppressalis	RNA-seq, RT-PCR	http://ento.njau.edu.cn/ChiloDB	10.1093/database/bau065
33	ASGARD	arthropods	Gene expression, RNA-seq	http://asgard.rc.fas.harvard.edu/	doi:10.1093/database/bas048
34	LeishDB	Leishmania braziliensis	RNA-seq	www.leishdb.com	10.1093/database/bax047
35	GCGD	Ctenopharyngodon idellus	Genomic data	http://bioinfo.ihb.ac.cn/gcgd	10.1093/database/bax051
36	dbWFA	Triticum aestivum	Microarray	urgi.versailles.inra.fr/dbWFA/	10.1093/database/bat014
37	MtDB	Medicago truncatula	Transcriptomic data	http://www.medicago.org/MtDB	PMID: 12519981
38	SorghumFDB	Sorghum bicolor	Gene expression, RNA-seq, microarray	http://structuralbiology.cau.edu.cn/sorghum/index.html	10.1093/database/baw099
39	TENOR	Oryza sativa L.	mRNA-seq data	http://tenor.dna.affrc.go.jp/	10.1093/pcp/pcv179
40	CarrotDB	Daucus carota L	RNA-seq data	http://apiaceae.njau.edu.cn:8080/carrotdb/	10.1093/database/bau096
41	MorusDB	Morus notabilis	Gene expression	http://morus.swu.edu.cn/morusdb	10.1093/database/bau054
42	EUCANEXT	Eucalyptus genus	RNAseq and ESTs	http://bioinfo03.ibi.unicamp.br/eucalyptusdb/	10.1093/database/bax079
43	EGENES	93 plant species (monocots, eudicots, ferns, mosses, green/red algae, basal magnoliophyta)	ESTs	http://www.genome.jp/kegg-bin/create_kegg_menu?category=plants_egenes	10.1104/pp.106.095059

browse tool is also available as well as the statistical testing for GO terms' enrichment. Overall, the resources integrated the omics data with a user-friendly platform with a potential to incorporate more data such as microRNAs or SNPs.

Other Sources

EGENES (Masoudi-Nejad *et al.*, 2007) is an extended plant transcriptome-based database of metabolic pathway genes integrated into the KEGG (Kyoto Encyclopedia of Genes and Genomes) database. EGENES is different from the set of other transcriptome data platforms because it does not concentrate mostly on the molecular properties but covers the information on metabolism, signal transduction and the cellular cycle steps. The data comes from assembled EST contigs (base on the public ESTs) with the information on the sources as well as functional annotation and a gene/EST index available for each genome. The main distinguishing feature of EGENS is the integration of genomic and pathway information. Functional assignments are made by connecting gene sets with a list of molecules that interact with it in the cell so that a higher order biological function pathway is created. EGENES pathways are created according to the same protocols which KEGG uses for functional assignment in other species.

At the start of the database in 2006, the pathway information was compiled for 25 species with 178 manually curated unique diagrams and divided into 4 main categories: (1) metabolism (2) genetic information processing (3) environmental information processing (4) cellular processes. As of now, the current EGENES version consists of 79 plant species.

Conclusion

According to the recent of 2018 NAR Molecular Biology Database Collection report the current number of databases is 1737 (Rigden and Fernández, 2018). However, this comprehensive list covers the broad spectrum of different primary and secondary biological databases containing multiple types of "omics" data. Considering the increasing number of biological databases year by year it becomes time inefficient to find a useful resource that would fit exactly to a researcher's specific task. This review is no means comprehensive but aimed to present a current overview of the major available transcriptome sources for human, animal and plant data. This review covers 43 main transcriptomic data sources. The following **Table 1** is the list of the sources discussed and provides a link to the databases as well as the publications. At a glance, most of the databases are built on human data with a particular focus on cancer and other major diseases such as Parkinson's, Alzheimer's and kidney diseases. Animal databases also reflect this focus on understanding the transcriptomic profile of major diseases with most of the transcriptome databases dedicated to model organism data crucial for understanding the processes of ageing, embryonic development (in mammals), brain transcriptome and oxygenic photosynthesis (in blue-green algae). Other animal database sources are also focused on arthropod, parasite transcriptome information with a focus on a rice pest and a leishmaniasis-causing parasite. Another database is dedicated to an important farmed fish. Plant transcriptome databases are also created and maintained with an emphasis on agriculturally-relevant species such as wheat, burclover legume, rice, sorghum and carrot. Other databases are dedicated to the economically important mulberry tree and eucalyptus as well as 79 other plant species with extended pathway information, which integrates genomic and biological function pathways. Overall, it is easy to see that technological advances and an overabundance of available data allow rapid integration with other findings and possibilities of performing advanced bioinformatics analysis and in-silico queries (e.g., gene enrichment, pathway construction, ID searches) in a user-friendly interface.

Competing Interests

The authors declared that there are no competing interests.

Acknowledgements

This work was supported by the grant research project "Pan-cancer deconvolution of omics data using Independent Component Analysis" (IRN: AP05135430) and research project "Analysis of cancer transcriptome data using Independent Component Analysis" (budget program "Creation and development of genomic medicine in Kazakhstan" – 0115RKO1931) of the Ministry of Education and Science of the Republic of Kazakhstan.

See also: Analyzing Transcriptome-Phenotype Correlations. Bioinformatics Data Models, Representation and Storage. Biological Database Searching. Characterization of Transcriptional Activities. Characterizing and Functional Assignment of Noncoding RNAs. Comparative Transcriptomics Analysis. Data Storage and Representation. Exome Sequencing Data Analysis. Functional Enrichment Analysis. Information Retrieval in Life Sciences. Integrative Analysis of Multi-Omics Data. Integrative Bioinformatics of Transcriptome: Databases, Tools and Pipelines. Natural Language Processing Approaches in Bioinformatics. Next Generation Sequencing Data Analysis. Quantitative Immunology by Data Analysis Using Mathematical Models. Regulation of Gene Expression. Standards and Models for Biological Data: FGED and HUPO. Statistical and Probabilistic Approaches to Predict Protein Abundance. Survey of Antisense Transcription. Transcriptome Analysis. Transcriptome During Cell Differentiation. Transcriptome Informatics. Utilising IPG-IEF to Identify Differentially-Expressed Proteins. Whole Genome Sequencing Analysis

References

Agarwal, R., *et al.*, 2016. CoReCG: A comprehensive database of genes associated with colon-rectal cancer. Database: The Journal of Biological Databases and Curation 2016.
Bai, Z., *et al.*, 2016. AlzBase: An integrative database for gene dysregulation in alzheimer's disease. Molecular Neurobiology 53 (1), 310–319.
Barrett, T., *et al.*, 2012. NCBI GEO: Archive for functional genomics data sets – Update. Nucleic Acids Research 41 (D1), D991–D995.
Chang, K., *et al.*, 2013. The Cancer Genome Atlas Pan-Cancer analysis project. Nature Genetics 45 (10), 1113–1120.
Chen, T., *et al.*, 2010. The Human Oral Microbiome Database: A web accessible resource for investigating oral microbe taxonomic and genomic information. Database: The Journal of Biological Databases and Curation 2010, baq013.
Chen, Y.X., *et al.*, 2017. The Grass Carp Genome Database (GCGD): An online platform for genome features and annotations. Database-The Journal of Biological Databases and Curation 2017, bax051.
Ganzfried, B.F., *et al.*, 2013. curatedOvarianData: Clinically annotated data for the ovarian cancer transcriptome. Database: The Journal of Biological Databases and Curation 2013, bat013.
Gehlenborg, N., *et al.*, 2009. The Prion Disease Database: A comprehensive transcriptome resource for systems biology research in prion diseases. Database: The Journal of Biological Databases and Curation 2009, bap011.
GTEx Consortium, *et al.*, 2017. Genetic effects on gene expression across human tissues. Nature 550 (7675), 204–213.
Harrow, J., *et al.*, 2012. GENCODE: The reference human genome annotation for The ENCODE Project. Genome Research 22 (9), 1760–1774.
Henrique, T., *et al.*, 2016. HNdb: An integrated database of gene and protein information on head and neck squamous cell carcinoma. Database: The Journal of Biological Databases and Curation 2016, baw026.
Hong, D., *et al.*, 2013. TIARA genome database: Update 2013. Database: The Journal of Biological Databases and Curation 2013, bat003.
Johnson, M.B., *et al.*, 2009. Functional and evolutionary insights into human brain development through global transcriptome analysis. Neuron 62 (4), 494–509.
Karimi, K., *et al.*, 2018. Xenbase: A genomic, epigenomic and transcriptomic model organism database. Nucleic Acids Research 46 (D1), D861–D868.
Kawahara, Y., *et al.*, 2016. TENOR: Database for comprehensive mRNA-Seq experiments in rice. Plant & Cell Physiology 57 (1), doi:10.1093/pcp/pcv179.
Lamblin, A.F., *et al.*, 2003. MtDB: A database for personalized data mining of the model legume Medicago truncatula transcriptome. Nucleic Acids Research 31 (1), 196–201.
Leinonen, R., *et al.*, 2011. The sequence read archive. Nucleic Acids Research 39, D19–D21. doi:10.1093/nar/gkq1019.
Leung, Y.Y., *et al.*, 2016. DASHR: Database of small human noncoding RNAs. Nucleic Acids Research 44 (D1), D216–D222. doi:10.1093/nar/gkv1188.
Li, J.R., *et al.*, 2016. Cancer RNA-Seq Nexus: A database of phenotype-specific transcriptome profiling in cancer cells. Nucleic Acids Research 44 (D1), D944–D951.
Liu, Y.C., *et al.*, 2016. CircNet: A database of circular RNAs derived from transcriptome sequencing data. Nucleic Acids Research 44 (D1), D209–D215.
Li, T., *et al.*, 2014. MorusDB: A resource for mulberry genomics and genome biology. Database: The Journal of Biological Databases and Curation 2014.
Lopez, F., *et al.*, 2008. TranscriptomeBrowser: A powerful and flexible toolbox to explore productively the transcriptional landscape of the Gene Expression Omnibus database. PLOS ONE 3 (12).
Mailman, M.D., *et al.*, 2007. The NCBI dbGaP database of genotypes and phenotypes. Nature Genetics 39 (10), 1181–1186.
Masoudi-Nejad, A., *et al.*, 2007. EGENES: Transcriptome-based plant database of genes with metabolic pathway information and expressed sequence tag indices in KEGG. Plant Physiology 144 (2), 857–866.
Nascimento, L.C., 2017. EUCANEXT: An integrated database for the exploration of genomic and transcriptomic data from Eucalyptus species. Database: The Journal of Biological Databases and Curation 2017, bax079.
Park, S.J., *et al.*, 2015. DBTMEE: A database of transcriptome in mouse early embryos. Nucleic Acids Research. 43), D771–D776. doi:10.1093/nar/gku1001.
Rigden, D.J., Fernández, X.M., 2018. The 2018 Nucleic Acids Research database issue and the online molecular biology database collection. Nucleic Acids Research 46 (D1), D1–D7.
Sato, A., *et al.*, 2008. Cerebellar development transcriptome database (CDT-DB): Profiling of spatio-temporal gene expression during the postnatal development of mouse cerebellum. Neural Networks 21 (8), 1056–1069.
Shen, J., *et al.*, 2015. SHIELD: An integrative gene expression database for inner ear research. Database: The Journal of Biological Databases and Curation 2015, bav071.
Sheng, X., *et al.*, 2017. MTD: A mammalian transcriptomic database to explore gene expression and regulation. Briefings in Bioinformatics 18 (1), 28–36. doi:10.1093/bib/bbv117.
Taccioli, C., *et al.*, 2011. ParkDB: A Parkinson's disease gene expression database. Database: The Journal of Biological Databases and Curation 2011, bar007.
Tacutu, R., *et al.*, 2013. Human Ageing Genomic Resources: Integrated databases and tools for the biology and genetics of ageing. Nucleic Acids Research 41, D1027–D1033. doi:10.1093/nar/gks1155.
Tian, T., *et al.*, 2016. SorghumFDB: Sorghum functional genomics database with multidimensional network analysis. Database: The Journal of Biological Databases and Curation 2016.
Torres, F., *et al.*, 2017. LeishDB: A database of coding gene annotation and non-coding RNAs in Leishmania braziliensis. Database-The Journal of Biological Databases and Curation 2017, bax047.
Uzun, A., *et al.*, 2012. dbPTB: A database for preterm birth. Database: The Journal of Biological Databases and Curation 2012, bar069.
Vincent, J., *et al.*, 2013. dbWFA: A web-based database for functional annotation of Triticum aestivum transcripts. Database: The Journal of Biological Databases and Curation 2013, bat014.
Wan, Q., *et al.*, 2015. BioXpress: An integrated RNA-seq-derived gene expression database for pan-cancer analysis. Database: The Journal of Biological Databases and Curation 2015.
Wuttke, D., *et al.*, 2012. Dissecting the gene network of dietary restriction to identify evolutionarily conserved pathways and new functional genes. PLOS Genetics 2012.
Xu, Z.S., *et al.*, 2014. CarrotDB: A genomic and transcriptomic database for carrot. Database: The Journal of Biological Databases and Curation 2014.
Yang, Y., *et al.*, 2015. CyanOmics: An integrated database of omics for the model cyanobacterium Synechococcus sp. PCC 7002. Database: The Journal of Biological Databases and Curation 2015.
Yin, C., *et al.*, 2014. ChiloDB: A genomic and transcriptome database for an important rice insect pest Chilo suppressalis. Database: The Journal of Biological Databases and Curation 2014.
Zahn, J.M., *et al.*, 2007. AGEMAP: A gene expression database for aging in mice. PLOS Genetics 3 (11), e201.
Zeng, V., Extavour, C.G., 2012. ASGARD: An open-access database of annotated transcriptomes for emerging model arthropod species. Database: The Journal of Biological Databases and Curation 2012, bas048.
Zhang, Q., *et al.*, 2014. Renal Gene Expression Database (RGED): A relational database of gene expression profiles in kidney disease. Database: The Journal of Biological Databases and Curation 2014.

Next Generation Sequence Analysis

Christian Rockmann, Christoph Endrullat, Marcus Frohme, and Heike Pospisil, University of Applied Sciences, Wildau, Germany

Introduction

The Human Genome Project (HGP), launched in 1990, was a 13-year long, collaborative research project to unravel the DNA sequence of the complete human genome. It involved numerous scientific groups from universities and research centers of the United States, the United Kingdom, France, Germany, Japan and China and costs $2.7 billion. The original sequencing technology, which enabled the HGP, was developed by Sanger *et al.* (1977) The advent of Next Generation Sequencing (NGS) technologies enabled individual research groups to realize genome sequencing projects for any organism of choice. Nowadays a genome can be sequenced within weeks for a few thousand dollars. The advances in terms of speed, throughput, accuracy and cost-effectiveness expanded the field of applications to targeted resequencing, transcriptome sequencing, methylation sequencing, chromatin immunoprecipitation sequencing, metagenomics and a lot more. The variety in these applications requires different approaches in data analysis and led to a vast amount of available algorithms and software.

This article aims to give an overview of a basic NGS analysis (There is no claim for completeness and further reading is suggested for more detailed insights). Typical steps (**Fig. 1**) are introduced in the following chapters and shall help to get an overview. Preprocessing, including quality control, is mandatory for every analysis. Alignments or assemblies are necessary for nearly all downstream analyses. The following procedures are depended from the aim of the respective research. The rapid development of technologies and methods leads to a constantly evolving environment of analysis tools and workflows.

Preprocessing and Quality Assessment

File Formats: FASTQ

Various Next Generation Sequencing (NGS) technologies with distinct approaches arose in the last decade. Each platform measures signals (e.g., light intensity, electrical pulse) in different ways and produces different types of raw signal data formats. The process of assigning nucleobases to those signals is called base calling. This is done by software, which is typically integrated into the sequencer. Often this data is stored in device specific formats like per-cycle .bcl files, flowgrams or ionograms, which are not suitable for downstream analysis applications.

FASTQ

The most common file format is FASTQ (Cock *et al.*, 2009b), which represent raw reads as input for alignments and other secondary analyses. It extends the FASTA format (Pearson and Lipman, 1988) and contains a quality score string besides the nucleotide sequence for each read. An example of FASTQ file format is as follows:

```
@SIM:1:FCX:1:15:6329:1045 1:N:0:2
TCGCACTCAACGCCCTGCATATGACAAGACAGAATC
+
<>;##=><9=AAAAAAAAAA9#:<#<;<<<????#=
```

Each entry consists of four line types. (1) The first line is the sequence identifier, beginning with the @. Additional annotations or comments, like read length or sequencing specifications (e.g., Illuminas systematic identifier), could be included here. (2) The second line contains the sequence in upper case IUPAC nucleotide codes. (3) The following quality score identifier line begins with a + and signals the end of the sequence. Additional information is optional here but using just the + character reduces the filesize significantly and is recommended by the MAQ convention (see "Relevant Websites section"). (4) The fourth line encodes quality values in ASCII characters for the sequences. Every base has a corresponding quality value Q, so the sequence length equals the Q-score string. Parsing FASTQ files requires caution because the @ and + symbol can also occure as quality score in the fourth line.

Quality Values

There are two different equations for the quality value Q which describes the reliability of a base call. The first one is called Phred score and is defined as:

$$Q_{Phred} = -10 \log_{10} p \tag{1}$$

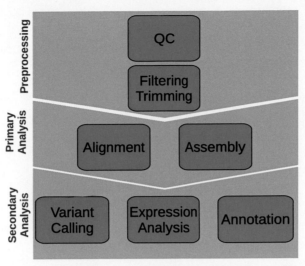

Fig. 1 Example workflow of basic NGS analysis on the levels of preprocessing, primary analysis and subsequent steps.

Herein, p is the probability for an incorrect base call. A second definition was introduced in 2004 by Bennett (2004):

$$\text{QSolexa} = -10\log_{10}\frac{p}{1-p} \tag{2}$$

Different variants of FASTQ evolved with those two variations of quality values and are described in **Table 1**. The Q scores in Sanger FASTQ format are encoded in printable ASCII characters 33–126, which allows assigning quality values between 0 and 93. A Q_{30} Phred score means a base call accuracy of 99.9% and a error probability of 1 in 1.000 ($10^{-\frac{30}{10}}$).

The underlying FASTQ variant should be aware. There are also several tools to interconvert FASTQ formats such as Biopython (Cock *et al.*, 2009a), EMBOSS (Rice *et al.*, 2000) or MAQ (Li *et al.*, 2008a).

Quality Control of Raw Data

Quality control is a crucial aspect in nearly every step of the NGS analysis pipeline. The results of sequencing data analysis are strongly dependent on the quality of raw data. Checking their quality should be the initial step of every data analysis. Errors can origin from several factors, including the library protocol, the sequencing and the wet-lab preparation. Commercial sequencing platforms provide their own quality control but erroneous reads often remain in raw data. Therefore, various metrics should be checked in the first step of a data processing pipeline. Many tools for this purpose are publicly available. One of the most common programs is FastQC. It allows a direct import of raw reads in different file formats and offers an initial quality control validation with multiple metrics to get a quick overview of the data. FastQC reports provide graphical summaries, tables and highlighted problems. But it is important to evaluate those results as they are impacted by their sequencing platforms or library types. The most important parameters to check with FastQC are listed below:

Basic Statistics (Total sequences, sequence length, %GC): The basic statistics section provide the total number of sequence, their length (that depend on the sequencer and the library protocol) and the GC content. The latter varies across species but there should not be a significant deviation (>10%) from the expected value. All values should conform with the experiment design.

Table 1   FASTQ Variants. Quality types are explained in formula (1) and (2). The ASCII offset is needed to produce printable characters

Description	Quality type	ASCII offset	Quality range
Sanger	Phred	33	0–93
Solexa	Solexa	64	−5–62
Illumina 1.3+	Phred	64	0–62
Illumina 1.5+	Phred	64	3–62
Illumina 1.8+	Phred	33	0–93

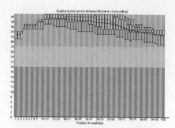

Per Base Sequence Quality: The distribution of quality scores across all bases at each read position is shown. The Phred-scaled scores are assigned by the base calling algorithms of the sequencer. Due to differences in chemistry, hardware and image processing, the Q-scores are not comparable among platforms. There is a general degradation of quality with increasing read length and it is advised to truncate bases with poor quality scores. Also mid-read drops in quality can effect the mapping efficiency.

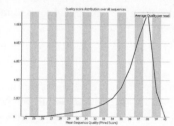

Per Sequence Quality Scores: This reports the number of reads (y-axis) versus their mean sequence quality (x-axis). The most frequently observed mean Phred-score should be higher than 27 with an error rate below 0.2%. A significant amount of reads with an overall low quality could indicate problems during sequencing. This can commonly occur to a subset of sequences. Long-read sequencing runs tend to lower read scores. This could by rectified by quality trimming.

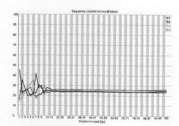

Per Base Sequence Content: The plot shows the proportion of each base across all read positions. Four parallel lines would be expected in a random library. Some types of libraries will induce biases in sequence composition (e.g., hexamer priming during cDNA generation produces biases at the beginning of nearly all RNA-Seq reads) (Hansen *et al.*, 2010). Furthermore bisulfite treatment converts most cytosines to thymines and results in a high difference of the relative amount of C and T.

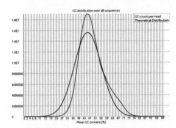

Per Sequence GC Content: The distribution of the mean GC content of each read should follow the shape of a normal distribution. Sharper shapes indicate a contamination with a specific organism or adaptor dimers. Biologically overrepresented sequences result in small spikes. A broader shape hints at different contaminants. The peak is expected at the overall %GC of the genome. Peak shifts are caused by systematic bias.

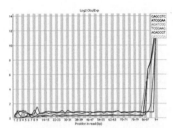

k-mer Content: The coverage of 7-mers (subsequences of 7 nt) over each read position in the whole library should be evenly distributed. The comparison between expected and observed frequency with binomial test reports possible positionally biased enrichments of k-mers. Overrepresented sequences and random hexamer primers in RNA-Seq libraries lead to highly enriched k-mers.

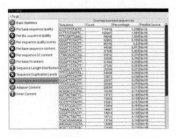

Overrepresented Sequences: A high diversity of sequences is expected in a random fragmented library. If a sequence is represented by a significant proportion of the total library, it is considered as overrepresented. This could be caused by contamination, a low diverse library (e.g., small RNA libraries) or artifacts of adaptor sequences. These highly duplicated reads can be searched against databases of common contaminants or primer sequences and should be filtered out.

Trimming and Filtering

There are various causes for low quality bases/reads in sequencing data, as already mentioned in Section "Quality Control of Raw Data". Some NGS applications (e.g., *de novo* assembly or detection of low frequency mutations) require high accuracy reads, because poor quality input data would result in erroneous outputs. Therefore it is desired to discard ambiguous sequence parts and keep as much as possible. A quality control of trimmed/corrected reads is recommended to exclude remaining low quality reads.

Trimming

There are two approaches to handle this problem. The first removes low quality regions from reads, which is called "trimming". Different implementations, for instance with sliding window approaches, try to achieve the aim of keeping the longest possible high quality subsequence while cutting off poor quality bases. Trimming is advised before most NGS applications.

Available trimming tools are Cutadapt (Martin, 2011), Trimmomatic (Bolger *et al.*, 2014), PRINSEQ (Schmieder and Edwards, 2011), FASTX (Gordon, 2008) or SolexaQA (Cox *et al.*, 2010).

Filtering

The second method of correcting reads is superimpose them to each other. The essential assumption behind this approach is the observation, that erroneous bases occur infrequently and randomly and can be modified by the most frequent reads with this pattern. These approaches base on using multiple sequence alignments, hidden Markov models, suffix trees/arrays or k-mers, whereby the last one seems to be the most suitable (Akogwu *et al.*, 2016). Corrected reads can improve alignments or assemblies significantly. However, it is ineligible for applications with an non-uniform read distribution (e.g., RNA-Seq, see chapter 4.3) or if a low coverage is expected (e.g., ctDNA analysis or discovering of rare variants). Common filtering tools are QUAKE (Kelley *et al.*, 2010), Reptile (Yang *et al.*, 2010), Musket (Liu *et al.*, 2012), SEECER (Le *et al.*, 2013) or Rcorrector (Song and Florea, 2015).

Primary Analysis

File Formats: SAM/BAM

The *de factor* standard for storing alignments is the Sequence Alignment/Map (SAM) format (Li *et al.*, 2009a). Alignment information is organized in a tabular structure in plain text. The human readable SAM format can be easily converted into Binary Alignment/Map (BAM) format, which contains the same information in a compressed, indexable, binary form. It is suitable for many alignment programs because of its flexibility. This also allows the conversion from other alignment formats. The ability of streaming records directly reduces the RAM requirements. Indexing allows a fast retrieving of alignments at a specific genomic position.

SAM & BAM (see "Relevant Websites section") files consist of an optional header section followed by the alignments. Each line of the header begins with the @ character followed by a two letter code, specified tags and their value. Supplemental information about SAM version, reference sequence, library protocol and various comments can be placed here. Each alignment line contains 11 mandatory fields, shown in **Table 2**.

The FLAG value of the second column can be converted in a binary number, where each bit holds the value (0 or 1) of a specific property (e.g., pairing, mapping and stranding informations). The CIGAR string (6th column) describes how the read is aligned against the reference with the possible operations (M: match; I: insert; D: deletion; N: skipped region; S: soft clipped; H: hard clipped; P: padding; =: read match; X: read mismatch) and how often they occur(As an example, the CIGAR string 34M3D specifies a sequence with 34 nucleotide bases matching to the reference and 3 deleted bases). The associated software package and library is called samtools (Li *et al.*, 2009a). It provides numerous options to:

- Convert between SAM and BAM and other formats,
- Sort alignments in BAM file according position on reference sequence,
- Index sorted BAM file for faster access,
- Extract header information or specific alignments (based on FLAG or region),
- Visualize interactively,
- Mark/remove (PCR) duplicates.

It should be mentioned here that the BAM format is also used to store raw sequence reads (called unaligned BAM), since it requires less disk space than FASTQ and the conversion between FASTQ and SAM/BAM can be performed easily. But compressing Deorowicz and Grabowski (2011) the FASTQ format also reduces its size significantly and gziped data can be directly processed by a lot of tools.

Table 2   Description of the 11 mandatory fields of the SAM/BAM file format

Field	Name	Description
1	QNAME	Query name of read (pair)
2	FLAG	Bitwise flag (combination for various properties)
3	RNAME	Reference sequence name
4	POS	Leftmost mapping position of alignment
5	MAPQ	Mapping quality (Phred-scale)
6	CIGAR	CIGAR string (notation for variants)
7	RNEXT	Reference name of mate
8	PNEXT	Position of mate
9	TLEN	Template length (insert size)
10	SEQ	Query sequence
11	QUAL	Query quality (Phred score)

Note: Modified from Li H., *et al.*, 2009. The sequence alignment/map format and SAMtools. Bioinformatics 25 (16), 2078–2079.

Sequence Alignment

Sequence alignment describes the arrangement of two or more DNA/RNA/protein sequences based on their similarity. In the field of NGS, this method is used to align short reads to a reference sequence, also called read mapping. Alignments are crucial for whole genome sequencing, exome sequencing, transcriptome sequencing, ChIP-Seq and much more applications. Reference sequences can be obtained from public databases and are usually stored in FASTA format. Current next-generation sequencers produce hundreds of million reads with a typically length >200 bp per run and represent them in FASTQ files.

Sequence alignment is not as CPU-intensive as assembly but yet not trivial. Depending on the size of the input files (sequencers with one terabases throughput are available), powerful hardware is still required. Due the short lengths of reads, there may be multiple matching positions on the reference sequence. Genomes of many organisms contain highly repetitive elements (half of the human genome belongs to known families of transposable elements Schmid and Deininger, 1975, Treangen and Salzberg, 2013), therefore it is commonly encountered that reads can not span the repeats and they map at multiple positions. The resequenced DNA always differs from the reference sequence. Small structural variations such as InDels (insertions or deletions of a single base) and substitutions must be considered. Ambiguous reads can not be aligned with confidence if there occure several multiple alignments with the same similarity scores. Furthermore, these problems are exacerbated by sequencing errors of reads and reference sequence.

To process the ever increasing number of reads in an acceptable amount of time, traditional dynamic programming algorithms, Needleman Wunsch alike, are not applicable anymore. Efficiency can be gained by preprocessing the reads, the reference sequence, or sometimes both. Two methods stand out here: hashing and Burrows-Wheeler transformation. All algorithms relying on hash tables follow the seed-and-extend strategy, which was already used by BLAST (Altschul *et al.*, 1990). Notable aligners basing on this methods are MAQ (Li *et al.*, 2008a; Homer *et al.*, 2009) and SSAHA2 (Ning *et al.*, 2001).

Another clever approach is using data structures such as prefix/suffix trees/arrays. Several tools employ those with Burrows-Wheeler transform (BWT), an algorithm origins from text data compression. The authors of BWA (Li and Durbin, 2009) explain this method in detail in their paper. Further aligners using this strategy are Bowtie2 (Langmead *et al.*, 2009), and SOAP (Li *et al.*, 2008b). HISAT2 (Kim *et al.*, 2015), TopHat2 (Kim *et al.*, 2013) and STAR (Dobin *et al.*, 2013) provide extended functionality, such as spliced alignments and should be considered for mapping RNA-Seq data.

Both methods have their strengths and weaknesses to be aware of. Hashing methods show good results with erroneous reads. A drawback is the increased runtime with seeds on highly repeated sequences, because the extend step is performed for every seed. This is the advantage of BWT aligners, as all matching positions are joined in one path. On the downside, they have difficulties to resolve sequencing errors.

Last but not least, the correct implementation is important regarding feasibility and runtime. Hard disk I/O operation are extremely slow. Keeping required data in CPU cache or RAM speeds up analysis. Furthermore, mapping is highly parallelizable and many aligners provide the option to use multiple cores/threads. To increase sensitivity and specificity of the alignment, paired end (PE) and mate-pair (MP) mapping should be regarded. Although the size of FASTA is significantly smaller than that of FASTQ, the additional quality values improve alignment accuracy remarkably.

Alignment quality control is essential, even if used reads passed their raw QC. Distinct NGS applications have different control parameters to be regarded. Overall mapping rate is a general metric to be observed. In case of exome sequencing the proportion of reads mapped to targeted exon regions is of importance to estimate the capture efficiency. The mean/median read coverage and percentage of the covered genome are relevant, especially for whole genome sequencing. Moreover, SAM/BAM files provide the mapping quality and insert size for each read.

Programs to be considered for QC are BAMStats (Ashelford, 2011), Picard Tools (Picard-Team, 2017), QPLOT (Li *et al.*, 2013) and the Genome Analysis Tool Kit (GATK) (McKenna *et al.*, 2010).

Genome/Transcriptome Assembly

Assembling in bioinformatics describes the reconstruction of a longer sequence from smaller fragments. This is a computational challenging task, especially for large and complex genomes (e.g., from humans and plants), because each read has to be compared with every other read. Repetitive DNA sequences, heterozygosity, high ploidy, and transposons makes it arduous to assemble the genome from short (paired end) reads. Using very long reads (e.g., PacBio Systems) and mate pair libraries with various insert sizes will facilitate and improve the assembly. The general procedure is assembling NGS reads into larger contiguous sequences, called contigs. This is the first step of assembling followed be the arranging of contigs in scaffolds in their right order, orientation and distance. Gaps (caused by repeats or no coverage) between contigs are filled by N's instead of nucleotides. Next, all scaffolds that belong to the same chromosome are arranged in one sequence. Closing all gaps completes the assembly.

It is hard, labor-intensive and expensive to achieve a complete assembly at the chromosome-level. Many genome assemblies are already available in public databases. It is recommended to always align to accessible reference genomes, because it improves and accelerates the assembly. Reconstructing genome or transcriptome sequences without using a reference is called *de novo* assembly. Numerous software tools are available, but they vary strongly with respect to their performance, hardware requirements and output. Therefore no general recommendation is possible. Various teams compete periodically in the Assemblathon (see "Relevant Websites section") to compare different assemblers and methods. Assemblathon 1 (Earl *et al.*, 2011) and 2 (Bradnam *et al.*, 2013) revealed that one assembler may perform well in one situation (e.g., one species) but will not work as well in other situations. Two methods are utilized by most assemblers and will be described in the next sections.

Overlap-layout-consensus method

As the name implies, Overlap-Layout-Consensus (OLC) works in three stages:

O pairwise alignments of all reads find overlaps and construct a graph of reads (nodes) and overlaps (edges),

L overlapping reads are ordered to identify redundant reads which are removed during the layout step,

C finally, the most likely nucleotides are chosen to build a consensus sequence.

CELERA (Myers, 2000), ARACHNE (Batzoglou et al., 2002), SGA (Simpson et al., 2012) and NEWBLER (Margulies et al., 2006) are exemplary mentioned assemblers using the OLC method.

de Bruijn graph method

The de Bruijn Graph (DBG) approach chops all reads into short strings of a fixed size *k*. A directed graph is built by connecting k-mers, where adjacent nodes overlap by k-1 nucleotides. The number of matches from short reads to each node is tracked. The genome sequence can be constructed following the paths in the graph and those node coverages. The layout step is a Hamiltonian path problem (NP hard) and building the consensus sequence requires multiple sequence alignments, which is very CPU consuming. Constructing the contig sequence in DBG approach infers the Euler path problem and does not require read alignments, which is algorithmically easier to solve and more CPU/RAM efficiently. DBG approaches seem to perform better with large NGS datasets.

Common DBG-tools are SOAPDENOVO2 (Luo et al., 2012) ALLPATHS-LG (Gnerre et al., 2011), ABYSS (Simpson et al., 2009) and VELVET (Zerbino and Birney, 2008).

Transcriptome reconstruction

Assemblers, as e.g., TRINITY (Grabherr et al., 2013), SOAPDENVO-TRANS (Xie et al., 2014), OASES (Schulz et al., 2012) and TRANS-ABYSS (Robertson et al., 2010) are mentioned here for the transcriptome reconstruction using RNA-Seq data instead of DNA. The challenge of RNA-based assembly is the high coverage caused by gene expression. These tools are also capable to detect splicing events and can reconstruct different isoforms. It is suggested to combine several assemblers and methods for better results.

The assembly quality can be evaluated by different metrics, such as N50, multiplicity or coverage. Especially the contig or scaffold lengths are used to describe the completeness of draft assemblies. The N50 statistic characterizes the length distribution of contigs and can be seen as a weighted median. In an set of contigs, ordered by their lengths, N50 is the length of the contig at 50% of total assembly length. That means at least 50% of the nucleotides of the entire assembly are contained in contigs with the N50 length or longer. Linked to this value is the L50 metric, which represents the number of contigs whose lengths form the N50.

The authors of the Assemblathon (Bradnam et al., 2013) proposed ten different metrics, which should be weighted to the own demands.

Secondary Analysis

Formats: VCF

The Variant Call Format (VCF) (Danecek et al., 2011) is a text file format for storing DNA variations, such as single nucleotide variant (SNV), InDels and structural variations (SVs) together with annotations. It was developed for the *1000 Genomes Project* with the aim of representing a wide variety of genomic variations regarding one reference. Therefore it stores only variations and no redundant genetic data are stored e.g., in the General Feature Format (GFF). Compressing saves disk space and indexing allows fast data access. A collection of tools to filter, compare, summarize and convert VCF files is combined in the VCFTOOLS package.

A VCF file is segmented in a header and a data section. Lines starting with ## define the header and provide space for meta information. Special keywords allow an easy parsing. The line starting with a single # character marks the field definition and beginning of the data section. Each data line corresponds to a variant at one genomic position or region. They are tab separated in 8 mandatory fields, which are described in **Table 3**.

Table 3   Description of the fields of the VCF format

Field	Name	Description
1	Chrom	name of reference sequence (typically chromosome)
2	POS	position of variation on reference
3	ID	unique identifier, e.g., dbSNP ID
4	REF	Reference allele
5	ALT	comma-separated list of alternate alleles
6	QUAL	associated quality score (Phred-scale)
7	FILTER	FLAG for site filtering information
8	INFO	user extensible annotation
9	FORMAT	(optional) genotype fields
10	SAMPLE	(optional) sample identifiers

Variant Discovery

Single nucleotide variants

Variant discovery is an important procedure for a lot of whole genome sequencing (WGS) and whole exome sequencing (WES) experiments due the fact of differences between the reference and the sequenced individual. The most abundant genomic variation is the substitution of a single nucleotide base, called single nucleotide variant (SNV). Synonymous SNVs are called single nucleotide polymorphism (SNP). SNVs are often used as genetic markers in genome-wide association studies (GWAS), especially if they are related to diseases.

Another form of genomic variations are insertions or deletions of a single base, also called InDels (Note that in this context an InDel is only *a* single base variation and that an insertion/deletion of multiple bases is considered multiple consecutive InDels). There is a variety of algorithms for SNV detection based on probabilistic or heuristic methods. One has to distinguish SNVs from sequencing errors or other systematic biases. Another common source of error are misalignments. GATK offers a local realignment function around SNVs and InDels to minimize errors due misalignments.

Common tools for SNP calling are Freebayes (Garrison and Marth, 2012), samtools mpileup, SOAPsnp (Li *et al.*, 2009b), VarScan (Koboldt *et al.*, 2009) and GATKs Haplotypecaller.

Structural variations

Structural variations (SV) affect larger sequence fragments (> 50 nucleotides (Mills *et al.*, 2011)). The include insertions, deletions, duplications, translocations and inversions. SVs can be assessed by four approaches, which are shortly described below:

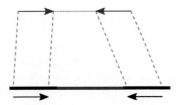

Assembly In general the assembly method is capable to detect all possible SV. However, assembling is error-prone, as described in chapter 3.3. It is recommended to use a reference-assisted local assembly instead a *de novo* whole genome assembly. It is also adviced to reduce the set of reads to ones, that are missing a mate or had other mapping difficulties in this region.

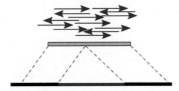

Read pair (RP) SVs are detected by odd mapping patterns. Differences between expected library insert size and distances of mapped read pairs point as insertions, if the mapping distance shorter than the insert size or as deletions (vice versa). Translocations and inversions can cause different relative mapping orientation of a read pair.

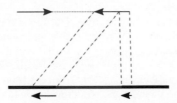

Split read (SR) Split read analysis show similarities to the RP method but the breakpoint of the variant is located directly at the reads. It is necessary to split the read in two or more fragments and realign them independently. Breakpoints can be detected on single base level, however short fragments may map at many genomic positions. After aligning the fragments successful, SVs can be determined by distance and orientation.

─────── Reference Sequence	─────── Deletion	▭▭▭▭ Assembled contig	───────▶ Read	············· PE-Insert

Read depth (RD) The read coverage of unique genomic regions should be equally distributed. Deletions and duplications result in significantly lower or higher read depth and copy number variants of large regions can be discovered easily. Insertions of (novel) sequences or inversions are not detectable through this method.

Numerous tools utilize a combination of these methods to reach a higher specificity and sensitivity for SV detection. BREAKDANCER (Chen *et al.*, 2009), CNVNATOR (Abyzov *et al.*, 2011), DELLY (Rausch *et al.*, 2012), HYDRA (Quinlan *et al.*, 2010) and PINDEL (Ye *et al.*, 2009) are be mentioned here as examples.

Expression Analysis

Sequencing mRNA, an intermediate between genome and proteome has a broad variety of applications in life sciences. RNA-seq enables a combined identification and quantification of gene activity and associated mechanisms. The diversity of scientific questions of interests makes it impossible to provide an universal analysis approach. One primary objective is gene expression profiling between samples.

A typical workflow begins with the mapping of quality controlled RNA-Seq reads to a reference genome/transcriptome or *de novo* assembled contigs. This step is followed up by an identification of transcripts, exons or genes. Tools, such as CUFFLINKS (Trapnell *et al.*, 2010) and STRINGTIE (Pertea *et al.*, 2015), assemble transcripts of each data set and estimate their expression levels. Using a spliced alignment as input (e.g., from TOPHAT2 or HISAT2) allows to summarize novel transcripts and spliced isoforms along with annotated genes.

Normalization

First, all transcript assemblies from different RNA-seq libraries must be merged to one master transcriptome. The next essential step is the normalization of expression values to enable comparisons within and between samples. Intralibrary normalization allows the quantification of genes with different lengths. A common method is calculating RPKM/FPKM (reads/fragments per kilobase exon per million mapped reads) values. Comparing expression levels of individual genes between samples requires interlibrary normalization. A simple approach is the adjustment with respect to the library size but there are further methods, such as different scaling factors or quantile normalization. Multiple methods are described and compared by Risso *et al.* (2014), Li *et al.* (2015), Bullard *et al.* (2010).

Differential expression

Differential expression analysis aims to detect genes, which significantly differ in their expression level across samples/conditions. RNA-Seq offers a wide dynamic range and a low background noise but inherits other difficulties, such as non-uniform read coverage distribution and sequencing biases through nucleotide compositions. Different statistical models and methods to counter these are implemented in several tools and packages. Many of them are part of Bioconductor (Huber *et al.*, 2015) and based on the statistical programming language R (R Core Team, 2017).

DESEQ2 (Love *et al.*, 2014), EDGER (Robinson *et al.*, 2010), LIMMA (Ritchie *et al.*, 2015), EBSEQ (Leng and Kendziorski, 2015), BALLGOWN (Frazee *et al.*, 2015) and BAYSEQ (Hardcastle, 2012) are notable representatives. The Basic Tuxedo Suite (Trapnell *et al.*, 2012) (including BOWTIE TOPHAT CUFFLINKS CUMMERBUND (Goff *et al.*, 2013)) and the "new Tuxedo" protocol (Pertea *et al.*, 2016) (HISAT, STRINGTIE, BALLGOWN) offer a complete RNA-Seq pipeline. There are also fast methods to estimate abundances without aligning reads. SALMON (Patro *et al.*, 2015), SAILFISH (Patro *et al.*, 2014), KALLISTO (Bray *et al.*, 2015) and RSEM (Li and Dewey, 2011) should be noted. However, it is important to choose tools and workflow fitting to the experiment settings. All tools offer detailed documentations and examples, which should be regarded.

Genome Annotation

Even perfect assembled genome sequences are mostly useless if they are not complemented by biological relevant information. This information includes location, structure and function of genes and gene models. The process of assigning biological information to pure sequence data is called annotation. A high quality annotation is expensive and labor-intensive, so most genome projects rely on automated gene annotation, but still requires editing by human experts. The basic steps are described below.

First annotation step

First of all, it should be noted that annotation strongly depends on the quality of the genome assembly. The automated annotation process can roughly be divided in two phases: A computational prediction phase and a synthesizing annotation phase.

Repeat identification is the first step to mask repetitive sequences, low complexity regions and transposable elements. Excluding those repeats with tools like REPEATMASKER (Smit et al., 2013) prevents numerous false annotations.

The next step is the AB INITIO gene prediction to identify genes using mathematical models. This is done by scanning for open reading frames (ORF) and training for species-specific traits like codon frequencies. Programs like AUGUSTUS (Stanke et al., 2006) or GENEMARK (Lomsadze et al., 2014) implemented this approach but are usually limited to find coding sequences (CDS). Improvements can be gained by aligning external evidence, such as transcript or protein sequences from related organisms.

Mapping ESTs and RNA-Seq data extends evidence for UTRs and provides additional splice sites. GNOMON (Souvorov et al., 2010) and the PASA (Haas et al., 2011) pipeline support this feature.Protein sequences from related organisms can be obtained from UniProt/SwissProt (Wasmuth and Lima, 2016) and are aligned by BLAST (Altschul et al., 1990).

Second annotation step

The second annotation phase synthesizes the results of the *ab initio* prediction into a consensus set of gene annotations. Evidence from external sources are weighted and the most representative gene model is chosen. This is incorporated in automated pipelines, such as MAKER (Cantarel et al., 2008), PASA and GLEAN (Elsik et al., 2007). However, manual curation is advised to detect systematic problems.

Conclusion

The field of NGS technologies is constantly evolving and novel methods become a routine part in biological and medical applications. e.g., ultra deep sequencing or read lengths of several hundred kb were considered as impossible a few years ago. There are also approaches which seem to be very promising today. Beside their advantages, challenges which cannot be solved satisfactorily, may cause a future disappearance of this technology. Nevertheless, every advancement or shift in technology requires adjustments of analyzing software. Hardware- and time requirements must be considered, too. Best results can be achieved by combining different methods and tools which fulfill the desired demands.

See also: Bioinformatics Approaches for Studying Alternative Splicing. Exome Sequencing Data Analysis. Experimental Platforms for Extracting Biological Data: Mass Spectrometry, Microarray, Next Generation Sequencing. Experimental Platforms for Extracting Biological Data: Mass Spectrometry, Microarray, Next Generation Sequencing. Functional Enrichment Analysis. Integrative Analysis of Multi-Omics Data. MicroRNA and lncRNA Databases and Analysis. Natural Language Processing Approaches in Bioinformatics. Next Generation Sequencing Data Analysis. Pipeline of High Throughput Sequencing. Pre-Processing: A Data Preparation Step. Sequence Analysis. Transcriptome Informatics. Whole Genome Sequencing Analysis

References

Abyzov, A., Urban, A.E., Snyder, M., Gerstein, M., 2011. CNVnator: An approach to discover, genotype, and characterize typical and atypical CNVs from family and population genome sequencing. Genome Research 21 (6), 974–984.

Akogwu, I., Wang, N., Zhang, C., et al., 2016. A comparative study of k-spectrum-based error correction methods for next-generation sequencing data analysis. Human Genomics 10 (S2), 20.

Altschul, S.F., Gish, W., Miller, W., Myers, E.W., Lipman, D.J., 1990. Basic local alignment search tool. Journal of Molecular Biology 215 (3), 403–410.

Ashelford, K., 2011. BAMStats: An interactive desktop GUI tool for summarising next generation sequencing alignments. Available at: http://bamstats.sourceforge.net/.

Batzoglou, S., Jaffe, D.B., Stanley, K., et al., 2002. ARACHNE: A whole-genome shotgun assembler. Genome Research 12, 177–189.

Bennett, S., 2004. Solexa Ltd. Pharmacogenomics 5, 433–438.

Bolger, A.M., Lohse, M., Usadel, B., 2014. Trimmomatic: A flexible trimmer for Illumina sequence data. Bioinformatics 30 (15), 2114–2120.

Bradnam, K.R., Fass, J.N., Alexandrov, A., et al., 2013. Assemblathon 2: Evaluating de novo methods of genome assembly in three vertebrate species. GigaScience 2 (1), 10.

Bray, N., Pimentel, H., Melsted, P., Pachter, L., 2015. Near-optimal RNA- Seq quantification. arXiv. 1505.02710.

Bullard, J.H., Purdom, E., Hansen, K.D., Dudoit, S., 2010. Evaluation of statistical methods for normalization and differential expression in mRNA-Seq experiments. BMC Bioinformatics 11 (1), 94.

Cantarel, B.L., Korf, I., Robb, S.M., et al., 2008. MAKER: An easy-to-use annotation pipeline designed for emerging model organism genomes. Genome Research 18 (1), 188–196.

Chen, K., Wallis, J.W., McLellan, M.D., et al., 2009. BreakDancer: An algorithm for high-resolution mapping of genomic structural variation. Nature Methods 6 (9), 677–681.

Cock, P.J.A., Antao, T., Chang, J.T., et al., 2009a. Biopython: Freely available Python tools for computational molecular biology and bioinformatics. Bioinformatics 25 (11), 1422–1423.

Cock, P.J.A., Fields, C.J., Goto, N., Heuer, M.L., Rice, P.M., 2009b. The Sanger FASTQ file format for sequences with quality scores, and the Solexa/Illumina FASTQ variants. Nucleic Acids Research 38 (6), 1767–1771.

Cox, M.P., Peterson, D.A., Biggs, P.J., 2010. SolexaQA: At-a-glance quality assessment of Illumina second-generation sequencing data. BMC Bioinformatics 11 (1), 485.

Danecek, P., Auton, A., Abecasis, G., et al., 2011. The variant call format and VCFtools. Bioinformatics 27 (15), 2156–2158.

Deorowicz, S., Grabowski, S., 2011. Compression of DNA sequence reads in FASTQ format. Bioinformatics 27 (6), 860–862.

Dobin, A., Davis, C.A., Schlesinger, F., et al., 2013. STAR: Ultrafast universal RNA-seq aligner. Bioinformatics 29 (1), 15–21.

Earl, D., Bradnam, K., St. John, J., et al., 2011. Assemblathon 1: A competitive assessment of de novo short read assembly methods. Genome Research 21 (12), 2224–2241.

Elsik, C.G., Mackey, A.J., Reese, J.T., et al., 2007. Creating a honey bee consensus gene set. Genome Biology 8 (1), R13.

Frazee, A.C., Collado-Torres, L., Jaffe, A.E., Leek, J.T., 2015. Ballgown: Flexible, isoform-level differential expression analysis. R package version 2.8.4.

Garrison, E., Marth, G., 2012. Haplotype-based variant detection from short-read sequencing. arXiv preprint arXiv:1207.3907. 9. (q-bio.GN).

Gnerre, S., Maccallum, I., Przybylski, D., et al., 2011. High-quality draft assemblies of mammalian genomes from massively parallel sequence data. Proceedings of the National Academy of Sciences of the United States of America 108 (4), 1513–1518.

Goff, L., Trapnell, C., Kelley, D., 2013. cummeRbund: Analysis, exploration, manipulation, and visualization of Cufflinks high-throughput sequencing data. R package version 2.12.1.

Gordon, A., 2008. FASTX-Toolkit: FASTQ/A short-reads pre-processing tools. Available at: http://hannonlab.cshl.edu/fastx_toolkit/.

Grabherr, M.G., Haas, B.J., Yassour, M., et al., 2013. Trinity: Reconstructing a full-length transcriptome without a genome from RNA-Seq data. Nature Biotechnology 29 (7), 644–652.

Haas, B.J., Zeng, Q., Pearson, M.D., Cuomo, C.A., Wortman, J.R., 2011. Approaches to fungal genome annotation. Mycology 2 (3), 118141.

Hansen, K.D., Brenner, S.E., Dudoit, S., 2010. Biases in Illumina tran-scriptome sequencing caused by random hexamer priming. Nucleic Acids Research 38 (12), 1–7.

Hardcastle, T.J., 2012. baySeq: Empirical Bayesian analysis of patterns of differential expression in count data. R package version 2.4.1.

Homer, N., Merriman, B., Nelson, S.F., 2009. BFAST: An alignment tool for large scale genome resequencing. PLOS ONE 4, 11.

Huber, W., Carey, V.J., Gentleman, R., et al., 2015. Orchestrating high-throughput genomic analysis with Bioconductor. Nature Methods 12 (2), 115–121.

Kelley, D.R., Schatz, M.C., Salzberg, S.L., 2010. Quake: Quality-aware detection and correction of sequencing errors. Genome Biology 11 (11), R116.

Kim, D., Langmead, B., Salzberg, S.L., 2015. HISAT: A fast spliced aligner with low memory requirements. Nature Methods 12 (4), 357–360.

Kim, D., Pertea, G., Trapnell, C., et al., 2013. TopHat2: Accurate alignment of transcriptomes in the presence of insertions, deletions and gene fusions. Genome Biology 14 (4), R36.

Koboldt, D.C., Chen, K., Wylie, T., et al., 2009. VarScan: Variant detection in massively parallel sequencing of individual and pooled samples. Bioinformatics 25 (17), 2283–2285.

Langmead, B., Trapnell, C., Pop, M., Salzberg, S.L., 2009. Ultrafast and memory-efficient alignment of short DNA sequences to the human genome. Genome Biology 10 (3), R25.

Le, H.S., Schulz, M.H., Mccauley, B.M., Hinman, V.F., Bar-Joseph, Z., 2013. Probabilistic error correction for RNA sequencing. Nucleic Acids Research 41 (10), 1–11.

Leng, N., Kendziorski, C., 2015. EBSeq: An R package for gene and isoform differential expression analysis of RNA-seq data. R package version 1.10.0.

Li, B., Dewey, C.N., 2011. RSEM: Accurate transcript quantification from RNA-Seq data with or without a reference genome. BMC Bioinformatics 12 (1), 323. arXiv: NIHMS150003.

Li, B., Zhan, X., Wing, M.-K., et al., 2013. QPLOT: A quality assessment tool for next generation sequencing data. BioMed Research International 2013, 1–4.

Li, H., Handsaker, B., Wysoker, A., et al., 2009a. The Sequence Alignment/Map format and SAMtools. Bioinformatics 25 (16), 2078–2079.

Li, H., Ruan, J., Durbin, R., 2008a. Mapping short DNA sequencing reads and calling variants using mapping quality scores. Genome Research 18, 1851–1858.

Li, H., Durbin, R., 2009. Fast and accurate short read alignment with Burrows-Wheeler transform. Bioinformatics 25 (14), 1754–1760.

Li, P., Piao, Y., Shon, H.S., Ryu, K.H., 2015. Comparing the normalization methods for the differential analysis of Illumina high-throughput RNA-Seq data. BMC Bioinformatics 16 (1), 347.

Li, R., Li, Y., Kristiansen, K., Wang, J., 2008b. SOAP: Short oligonucleotide alignment program. Bioinformatics 24 (5), 713–714.

Li, R., Li, Y., Fang, X., et al., 2009b. SNP detection for massively parallel whole-genome resequencing SNP detection for massively parallel whole-genome resequencing. Genome Research 19 (6), 1124–1132.

Liu, Y., Schröder, J., Schmidt, B., 2012. Musket: A multistage k-mer spectrum based error corrector for Illumina sequence data. Bioinformatics 29 (3), 308–315.

Lomsadze, A., Burns, P.D., Borodovsky, M., 2014. Integration of mapped RNA-Seq reads into automatic training of eukaryotic gene finding algorithm. Nucleic Acids Research 42 (15), 1–8.

Love, M.I., Huber, W., Anders, S., 2014. Moderated estimation of fold change and dispersion for RNA-seq data with DESeq2. Genome Biology 15 (12), 550.

Luo, R., Liu, B., Xie, Y., et al., 2012. SOAPdenovo2: An empirically improved memory-efficient short-read de novo assembler. GigaScience 1 (1), 18.

Margulies, M., Egholm, M., Altman, W.E., et al., 2006. Genome sequencing in open microfabricated high density picoliter reactors. Nature Biotechnology 437 (7057), 376–380.

Martin, M., 2011. Cutadapt removes adapter sequences from high-throughput sequencing reads. EMBnet.Journal 17 (1), 10.

McKenna, A., Hanna, M., Banks, E., et al., 2010. The Genome Analysis Toolkit: A MapReduce framework for analyzing next-generation DNA sequencing data. Genome Research 20 (9), 1297–1303.

Mills, R., Walter, K., Stewart, C., et al., 2011. Mapping copy number variation by population-scale genome sequencing. Nature 470 (2010), 59–65.

Myers, E.W., 2000. A whole-genome assembly of drosophila. Science 287 (5461), 2196–2204.

Ning, Z., Ning, Z., Cox, A.J., et al., 2001. SSAHA: A fast search method for large DNA databases. Genome Research 2, 1725–1729.

Patro, R., Duggal, G., Love, M.I., Irizarry, R.A., Kingsford, C., 2015. Salmon provides accurate, fast, and bias-aware transcript expression estimates using dual-phase inference. bioRxiv 14 (4), 417–419. arXiv: 1308.3700.

Patro, R., Mount, S.M., Kingsford, C., 2014. Sailfish enables alignment-free isoform quantification from RNA-seq reads using lightweight algorithms. Nature Biotechnology 32 (5), 462–464. arXiv: 1505.02710.

Pearson, W.R., Lipman, D.J., 1988. Improved tools for biological sequence comparison. Proceedings of the National Academy of Sciences of the United States of America 85 (8), 2444–2448.

Pertea, M., Kim, D., Pertea, G.M., Leek, J.T., Salzberg, S.L., 2016. Transcript-level expression analysis of RNA-seq experiments with HISAT, StringTie and Ballgown. Nature Protocols 11 (9), 1650–1667.

Pertea, M., Pertea, G.M., Antonescu, C.M., et al., 2015. StringTie enables improved reconstruction of a tran-scriptome from RNA-seq reads. Nature Biotechnology 33 (3), 290–295.

Picard-Team, 2017. A set of command line tools (in Java) for manipulating high-throughput sequencing (HTS) data and formats such as SAM/BAM/CRAM and VCF. Availble at: http://broadinstitute.github.io/picard.

Quinlan, A.R., Clark, R.A., Sokolova, S., et al., 2010. Genome-wide mapping and assembly of structural variant breakpoints in the mouse genome Genome-wide mapping and assembly of structural variant breakpoints in the mouse genome. Genome Research 20, 623–635.

Rausch, T., Zichner, T., Schlattl, A., et al., 2012. DELLY: Structural variant discovery by integrated paired-end and split-read analysis. Bioinformatics 28 (18), 333–339.

R Core Team, 2017. R: A Language and Environment for Statistical Computing. Vienna, Austria: R Foundation for Statistical Computing.

Rice, P., Longden, I., Bleasby, A., 2000. EMBOSS: The European molecular biology open software suite. Trends in Genetics 16 (1), 276–277.

Risso, D., Ngai, J., Speed, T.P., Dudoit, S., 2014. Normalization of RNA-seq data using factor analysis of control genes or samples. Nature Biotechnology 32 (9), 896–902.

Ritchie, M.E., Phipson, B., Wu, D., *et al.*, 2015. limma powers differential expression analyses for RNA-sequencing and microarray studies. Nucleic Acids Research 43 (7), e47.

Robertson, G., Schein, J., Chiu, R., *et al.*, 2010. *De novo* assembly and analysis of RNA-seq data. Nature Methods 7 (11), 909–912.

Robinson, M.D., McCarthy, D.J., Smyth, G.K., 2010. edgeR: A Bioconductor package for differential expression analysis of digital gene expression data. Bioinformatics 26, 1.

Sanger, F., Nicklen, S., Coulson, A.R., 1977. DNA sequencing with chain-terminating inhibitors. Proceedings of the National Academy of Sciences 74 (12), 5463–5467.

Schmid, C.W., Deininger, P.L., 1975. Sequence organization of the human genome. Cell 6 (3), 345–358.

Schmieder, R., Edwards, R., 2011. Quality control and preprocessing of metagenomic datasets. Bioinformatics 27 (6), 863–864.

Schulz, M.H., Zerbino, D.R., Vingron, M., Birney, E., 2012. Oases: Robust *de novo* RNA-seq assembly across the dynamic range of expression levels. Bioinformatics 28 (8), 1086–1092.

Simpson, J.T., Durbin, R., Zerbino, D.R., *et al.*, 2012. Efficient *de novo* assembly of large genomes using compressed data structures sequence data. Genome Research. 549–556.

Simpson, J.T., Wong, K., Jackman, S.D., *et al.*, 2009. ABySS: A parallel assembler for short read sequence data. Genome Research 19, 1117–1123.

Smit, A., Hubley R., Green P., 2013. RepeatMasker Open-4.0. Available at: http://www.repeatmasker.org.

Song, L., Florea, L., 2015. Rcorrector: Efficient and accurate error correction for Illumina RNA-seq reads. GigaScience 4 (1), 48.

Souvorov, A., Kapustin, Y., Kiryutin, B., *et al.*, 2010. GnomonNCBI eukaryotic gene prediction tool. National Center for Biotechnology Information. 1–24.

Stanke, M., Schoffmann, O., Morgenstern, B., Waack, S., 2006. Gene prediction in eukaryotes with a generalized hidden Markov model that uses hints from external sources. BMC Bioinformatics 7 (1), 62.

Trapnell, C., Roberts, A., Goff, L., *et al.*, 2012. Differential gene and transcript expression analysis of RNA-seq experiments with TopHat and Cufflinks. Nature Protocols 7 (3), 562–578.

Trapnell, C., Williams, B.A., Pertea, G., *et al.*, 2010. Transcript assembly and quantification by RNA-Seq reveals unannotated transcripts and isoform switching during cell differentiation. Nature Biotechnology 28 (5), 511–515.

Treangen, T.J., Salzberg, S.L., 2013. Repetitive DNA and next-generation sequencing: Computational challenges and solutions. Nature reviews. Genetics 13 (1), 36–46.

Wasmuth, E.V., Lima, C.D., 2016. UniProt: The universal protein knowl-edgebase. Nucleic Acids Research 45 (2016), 1–12.

Xie, Y., Wu, G., Tang, J., *et al.*, 2014. SOAPdenovo-Trans: *De novo* transcriptome assembly with short RNA-Seq reads. Bioinformatics 30 (12), 1660–1666.

Yang, X., Dorman, K.S., Aluru, S., 2010. Reptile: Representative tiling for short read error correction. Bioinformatics 26 (20), 2526–2533.

Ye, K., Schulz, M.H., Long, Q., Apweiler, R., Ning, Z., 2009. Pindel: A pattern growth approach to detect break points of large deletions and medium sized insertions from paired-end short reads. Bioinformatics 25 (21), 2865–2871.

Zerbino, D.R., Birney, E., 2008. Velvet: Algorithms for *de novo* short read assembly using de Bruijn graphs. Genome Research 18 (5), 821–829.

Relevant Websites

http://maq.sourceforge.net/fastq.shtml
  FASTQ Format
  Maq
  SourceForge.
https://github.com/samtools/hts-specs
  SamTools · GitHub.
http://assemblathon.org/
  The Assemblathon.

Transcriptomic Data Normalization

Fatemeh Vafaee, University of New South Wales, Sydney, NSW, Australia
Hamed Dashti, Sharif University of Technology, Tehran, Iran
Hamid Alinejad-Rokny, University of New South Wales, Sydney, NSW, Australia

Introduction

Cells are built, developed, and reproduced by genetic instructions stored in the form of DNA, a long double-stranded molecule of nucleic acids. DNA is *transcribed* into RNA, which is then *translated* into proteins to carry out the duties specified by the instructions encoded in the genes. The complete set of RNA transcripts produced by the genome, under a specific condition or in a specific cell, is called transcriptome (Adams, 2008). Transcriptomics is the study of the transcriptome through which researchers gain insight into gene activities and better understand how cells normally function or how changes in RNA activities can contribute to disease (Kalia and Gupta, 2005). Transcriptomes can basically enable researchers to depict a comprehensive, genome-wide picture of gene activities across different cells (Adams, 2008). In addition, comparative transcriptomics facilitates the study of the evolution of gene regulation and provides deeper understanding of the species conservation at the molecular level (Breschi *et al.*, 2017).

Transcriptomics has been characterized by technological innovations that continuously transform the field. Key contemporary transcriptomic technologies include DNA microarrays and NGS technologies called RNA-seq. Microarrays measure the expressions of a predetermined set of sequences while RNA-seq uses high-throughput sequencing to capture all sequences (Lowe *et al.*, 2017a). Transcription can also be studied at the level of individual cells using single-cell transcriptomic techniques, for example, single-cell RNA-seq (Kanter and Kalisky, 2015).

Analysis of microarray or RNA-seq data relies on the hypothesis that the measured values (e.g., florescent intensities or read counts) for each RNA represent its relative expression level in a specific sample under study. To compare RNAs across multiple samples and to reveal biologically relevant patterns of gene expression, transcriptomic data shall undergo a preprocessing transformation step referred to as normalization. Normalization adjusts for technical biases and other systematic variations to balance intensities or read counts appropriately so that meaningful biological comparisons can be made (Quackenbush, 2002; Mortazavi *et al.*, 2008). Although microarray and RNA-seq experiments carry different sources of systematic variations, it has been shown that normalization is an essential step in the analysis of expression data derived from both technologies (e.g., Park *et al.*, 2003a; Bullard *et al.*, 2010).

Several normalization methods developed during the past decades. They mainly focus on the mean and the variance, which are quantifying the center and the spread of the data, respectively. These methods try to adjust mean and variance between different samples. **Fig. 1**, for instance, shows how normalization would adjust the distribution of RNA expressions across different samples.

Microarray Data Normalization

A microarray is a laboratory tool used to quantify the expression of tens of thousands of genes simultaneously. DNA microarrays (a.k.a. biochips or DNA chips) are glass, plastic, or silicon-based slides printed with thousands of microscopic spots in defined positions, with each spot containing picomoles (10^{-12} mol) of a specific DNA sequence. The DNA molecules attached to a microarray slide act as probes to capture messenger RNA (mRNA) transcripts expressed by genes.

A microarray experiment typically starts with purification of mRNAs from the biological sample (i.e., RNA isolation). RNAs are converted into complementary DNA (cDNA) using reverse transcriptase enzyme and labeled with fluorescent dyes (e.g., cyanine dyes, a synthetic dye family belonging to polymethine group). After labeling, cDNA molecules are allowed to bind or *hybridize* to the DNA probes on the microarray slide. The microarray is then scanned to visualize fluorescence intensities that measure the expression of each gene printed on the slide – that is, brighter spots indicates higher level of expressions.

DNA microarrays come in *two-channel* versus *one-channel* designs. Two-channel or two-color microarrays are hybridized with cDNA prepared from two samples to be compared (e.g., diseased versus healthy tissues) and labeled with two different fluorophores (i.e., Cy3 and Cy5 for green and red florescence emission, respectively). Relative intensities of each fluorophore through a ratio-based analysis can then identify downregulated and upregulated genes. A single-channel or one-color microarray is hybridized with one sample and the array provides intensity data for each probe as the relative level of hybridization with the labeled cDNA or the expression level of the corresponding gene. **Fig. 2** illustrates key steps of a typical DNA microarray experiment and analysis of resulting expression values.

Several technical errors may occur during the microarray experimental procedure such as irregular spot printing, nonuniform intensity of the fluorescent compound, dusty arrays, purification errors, difference in efficiency of labeling via fluorescent dyes, hybridization efficiencies, and systematic biases in quantified expression levels (Quackenbush, 2002; Park *et al.*, 2003b; Yang *et al.*, 2002). These artifacts have bearings on capturing data leading to different measurements of same expression values. Hence, this is important to eliminate technical noises prior to any downstream analysis and normalization plays an important role in reducing potential systematic noises.

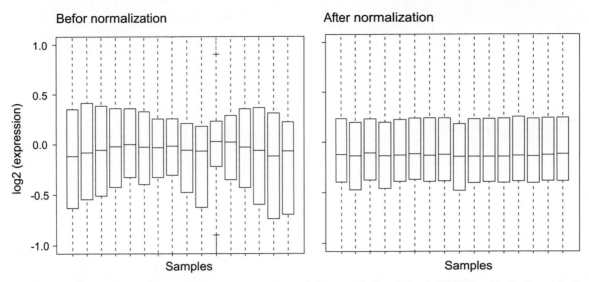

Fig. 1 Sample boxplots. Box plots of log2(expression) array data before and after normalization. Dataset: GSE4158, normalization method: Quantile normalization implemented using "Rnits" R package (Sangurdekar, 2014).

Key steps of microarray experiment & analysis pipeline

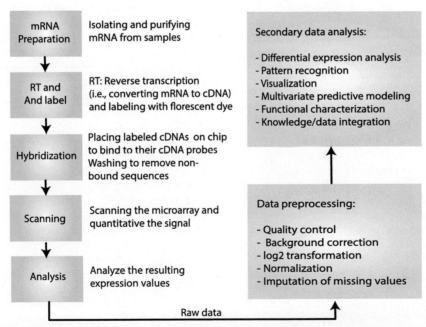

Fig. 2 Microarray workflow. Key steps of performing a microarray experiment for gene expression studies after sample preparation. Raw expression data will be preprocessed to assure the data quality and then undergo various downstream analyses.

There are several normalization methods. Some have been developed for particular platforms and the analysis may be proprietary for commercial platforms. Yet, several freely accessible resources are available to normalize microarray data including various Bioconductor R packages (see "Relevant Websites section"). Selecting the best normalization method is not an obvious decision and suitable choices may vary depending on the data under study. Complex methods do not guarantee better performance and background assumptions may add bias and eliminate important data (Park *et al.*, 2003a). Nonetheless, there are several scientific works comparing the performance of different normalization methods with respect to different performance criteria (Smyth and Speed, 2003; Bolstad *et al.*, 2003; Park *et al.*, 2003b) which can help to opt for an appropriate normalization method based on the experimental platform and the quality of the data under study.

MA plot (Dudoit *et al.*, 2002) is a visualization tool traditionally used to assess the performance of normalization methods by revealing systematic intensity-dependent effects in the measurements taken from two samples. We denote two, that is, query and

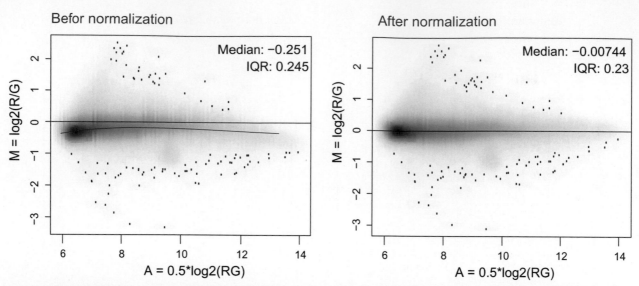

Fig. 3 Sample MA plots. MA plots before and after normalization using smoothScatter method which produces a smoothed color density representation of a scatterplot. Red line is the lowess smoothed line to visually show the bias. Dataset: Dilution (array 20B v 10A) from "affydata" R package. Reproduced from Gautier, L. 2011. Affydata: Affymetrix Data for Demonstration Purpose. R package version, 1, 18. Normalization method: Quantile normalization implemented using "affy" R package. Reproduced from Gautier, L., Cope, L., Bolstad, B.M., Irizarry, R.A., 2004. Affy – Analysis of Affymetrix GeneChip data at the probe level. Bioinformatics 20, 307–315.

reference, samples as R and G (for the red and green colors commonly used to represent Cy5 and Cy3 intensities in two-channel microarrays). MA plot is a scatter plot whose y-axis and x-axis respectively display $M = \log_2(R_i/G_i)$ and $A = \log_2(R_i * G_i)$ where R_i and G_i represent the intensity of the ith gene in R and G samples. An underlying assumption in many microarray gene expression experiments is that most of the genes do not change in their expressions, and thus the majority of the points would have 0 y-coordinate and lie on the x-axis, that is, $M = \log_2(1) = 0$. If this is not the case, a proper normalization method should be applied to the data to adjust imbalanced intensities. **Fig. 3** shows MA plots for a sample before and after normalization.

Lowess Normalization

Lowess/loess normalization is one of the earliest developments in microarray normalization proposed by Terry Speed's group (Dudoit *et al.*, 2002; Yang *et al.*, 2002). They observed nonlinear bias in two-color log ratios changing as a function of average intensities in two-channel microarrays (as seen in MA plots). Similar bias has been also observed in single-channel arrays when plotting intensities in one chip against intensities averaged over all chips (Reimers, 2010).

The purpose of lowess normalization is to estimate this bias using a nonparametric curve known as *local weighted regression* (lowess). At each point on MA plot, M value is adjusted by subtracting the estimated bias (the height of the loess curve) at the same A value, that is, $\log_2(R/G) = \log_2(R/G) - c(A)$, where $c(A)$ is the lowess fit to MA plot. Lowess normalization operates on individual chips (within-chip normalization), but was intended to make measures comparable across chips as well (between-chip normalization). It can perform robust locally linear fits not affected by outliers in the MA plot (Yang *et al.*, 2002).

Lowess normalization was further improved by Speed's group to account for spatial dependence in dye biases (Yang *et al.*, 2002) which was based on the observation that there were pronounced differences in the lowess curves fit to log ratios in different regions of the same chip. Accordingly, they proposed *within print-tip group normalization*, which aims at fitting separate lowess curves to each "print tip," that is, $\log_2(R/G) = \log_2(R/G) - c_i(A)$, where $c_i(A)$ is the lowess fit to MA plot for the ith print tip only. A print tip is a set of probes on a common grid of a robotically printed cDNA array.

Researchers continued to further complicate lowess normalization adjusting for other possible biases. Examples include rescaling variation within each print-tip group or estimating a plate order effect (see, e.g., Smyth, 2002; Smyth and Speed, 2003).

Quantile Normalization

Quantile normalization is a simple nonparametric normalization method initially proposed for single-channel arrays (Bolstad *et al.*, 2003). It is a between-array normalization method that makes the distribution of all arrays identical in statistical properties. The algorithm maps every expression value on each chip to the corresponding quantile of a reference distribution that is determined by pooling across distributions of all individual chips. It is motivated by the idea that a quantile–quantile plot shows that the distribution of two data vectors is the same only if the plot is a straight diagonal line.

While being simple and fast, quantile normalization could outperform most of the more complex contemporary methods including loess. It has been widely used for normalization of single-channel high intensity oligonucleotide arrays, such as

Affymetrix arrays (one of the most frequently used platforms for microarray experiments), and is also applicable to dual-channel chips (Reimers, 2010). Quantile normalization was made available as the default normalization method in the widely used "affy" package of R Bioconductor, which contains functions for exploratory oligonucleotide array analysis (Gautier *et al.*, 2004).

Despite its clear advantages, quantile normalization explicitly assumes that the distribution of gene expression measures is identical across the samples. This can be violated when there are biologically meaningful differences between samples' transcriptomic distributions (e.g., when studying treatments with severe effects on the transcriptome, or cancer samples with severe genomic aberrations). Also, in its canonical form, quantile normalization forces genes with high intensities into the reference distribution shape which may reduce biological differences (Reimers, 2010). The latter has been addressed in a later version of the "affy" package.

Overall, due to its simplicity, efficiency, generality, and availability, quantile normalization has become a comparatively popular choice of normalization for microarray data.

Variance Stabilization

Various widely used standard statistical methods (e.g., regression, ANOVE) assume that data is normally or symmetrically distributed with a constant variance not depending on the mean of the data. However, microarray data fail to comply with this canonical assumption underlying standard methodologies (Rocke and Durbin, 2001). It is therefore a common practice to stabilize the variance of microarray expression values by transforming the expression values to a logarithmic base 2 scale. While logarithmic transformation makes variations among measurements that span several orders of magnitudes comparable, it can inflate variances of low-level, near-background observations (Durbin *et al.*, 2002), which may necessitate removal of low-intensity probes a priori. To address this issue, (Durbin *et al.*, 2002) introduced a transformation that stabilizes the variance of microarray data across the full range of expression values without needing to remove low-level observations. For more details and other options, interested readers may wish to consult (Durbin *et al.*, 2002; Huber *et al.*, 2002; Durbin and Rocke, 2004; Lin *et al.*, 2008).

Nonetheless, while more sophisticated methods may provide more coherent variance stabilization, the ultimate gain is possibly small compared to the simplicity of log transformation (Speed, 2001) and thus, researchers usually stick by log2 transformation.

RNA-seq Data Normalization

RNA-seq uses next-generation sequencing (NGS) to quantify the presence and abundance of RNAs in a biological sample at a specific moment of time. It is a revolutionary tool that provides a comprehensive view of the transcriptome, and enables the detection of novel transcripts and characterization of alternative splicing (Wang *et al.*, 2009). It is rapidly replacing gene expression microarrays for whole-genome transcriptome profiling (Mortazavi *et al.*, 2008; Nagalakshmi *et al.*, 2010). That is basically due to limitations involved in microarray-based experiments that have been resolved using RNA-seq technologies. For example, RNA-seq intrinsically avoids errors introduced during hybridization of microarrays as it does not depend on genome annotation for prior probe selection. RNA-seq also captures a broader dynamic range that ensures the detection of more differentially expressed genes with higher fold-change compared to microarrays (Zhao *et al.*, 2014).

A typical RNA-seq experiment consists of RNA purification and isolation, library preparation (i.e., converting RNA to cDNA and adding sequencing adapters), sequencing using a NGS platform, and analysis of resulting sequencing reads (**Fig. 4(a)**). RNA-seq analysis pipeline usually involves demultiplexing and trimming sequencing reads, mapping sequencing reads to a reference transcriptome, annotating transcripts to which reads are mapped, counting mapped reads to estimate the abundance of transcripts, normalizing reads, and performing downstream statistical analyses – for example, identifying differentially expressed genes, visualization, etc. (**Fig. 4(b)**).

Although RNA-seq experiments do not suffer from technical noises inherent to microarray technologies, they are still subject to some systematic variations such as library size (i.e., sequencing depth) differences between samples, as well as transcript length bias and GC content within a specific sample (Oshlack and Wakefield, 2009b). It is therefore essential to normalize data in order to adjust for such biases. Several normalization methods have been developed for RNA-seq data. Some of the most frequently used ones have been reviewed in here.

Total Count

Differences in the library size (i.e., sequencing depth) account for the most apparent source of intersample variations in RNA-seq experimentations. Therefore, the simplest form of between-sample normalization has been realized by scaling raw read counts in each sample (or sequencing "lane") by a sample-specific factor (Dillies *et al.*, 2013).

In total count (TC) normalization, for each sample, gene counts are divided by the sample's library size (i.e., total number of mapped reads associated with that sample) and multiplied by the mean TC across all the samples of the dataset.

Upper Quartile

Upper quartile (UQ) normalization is very similar to TC and only replaces the TCs (i.e., library size) by the UQ (75th percentile) of read counts, after excluding genes with zero reads across all lanes (Bullard *et al.*, 2010).

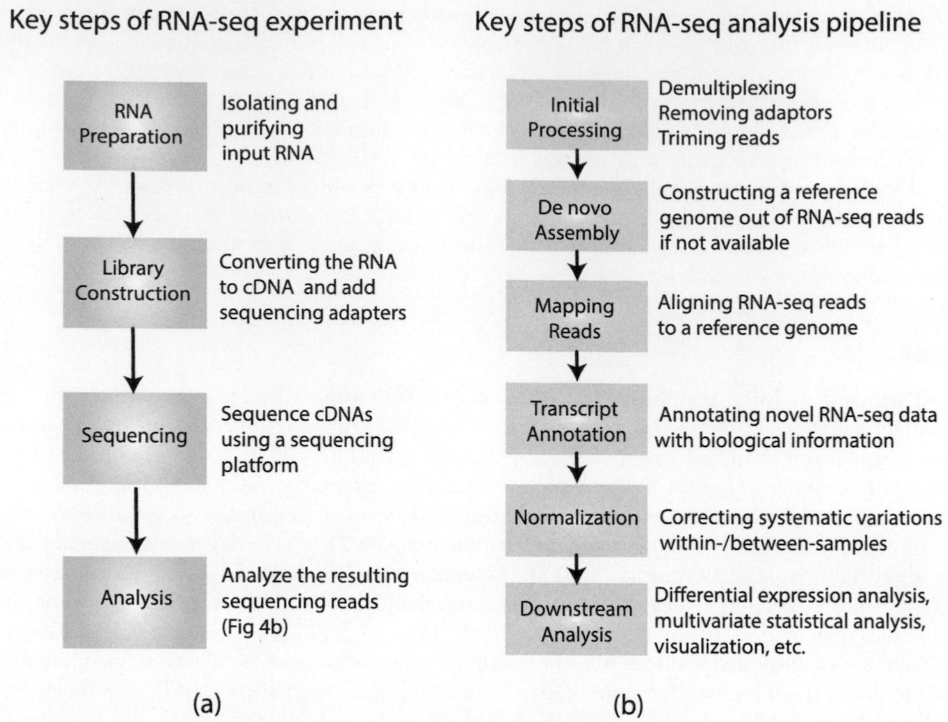

Key steps of RNA-seq experiment

Key steps of RNA-seq analysis pipeline

RNA Preparation	Isolating and purifying input RNA
Library Construction	Converting the RNA to cDNA and add sequencing adapters
Sequencing	Sequence cDNAs using a sequencing platform
Analysis	Analyze the resulting sequencing reads (Fig 4b)

Initial Processing	Demultiplexing Removing adaptors Triming reads
De novo Assembly	Constructing a reference genome out of RNA-seq reads if not available
Mapping Reads	Aligning RNA-seq reads to a reference genome
Transcript Annotation	Annotating novel RNA-seq data with biological information
Normalization	Correcting systematic variations within-/between-samples
Downstream Analysis	Differential expression analysis, multivariate statistical analysis, visualization, etc.

(a)   (b)

Fig. 4   RNA-seq workflow. (a) Key steps of a typical RNA-seq workflow after designing an RNA-seq experiment. (b) Main steps of RNA-seq analysis pipeline. For more details refer to RNA-seqlopedia (https://rnaseq.uoregon.edu).

Reads Per Kilo Base Per Million Mapped Reads

This approach was aimed to facilitate comparison of transcript levels both within and between samples as it rescales gene counts to correct for differences in both library sizes and gene length (Mortazavi *et al.*, 2008). Two replicate samples with different total library size would have proportionally different sequenced reads mapped to the same gene. Also, longer transcripts can potentially produce more fragments than shorter ones. To adjust for these variations, reads per kilo base per million mapped reads (RPKM) divides gene counts by the library size as well as the length of the transcript in kilobase. It has been shown that adjusting differences in gene length can potentially introduce bias in per-gene variances, in particular for lowly expressed genes (Oshlack and Wakefield, 2009a). Nonetheless, RPKM is a widely held choice in many practical applications. ERANGE tool (website available at "Relevant Websites section") implements RPKM normalization method (Mortazavi *et al.*, 2008).

Trimmed Mean of M-Values

This normalization method assumes that most genes are not differentially expressed. Trimmed Mean of M-values (TMM) considers one sample as a reference sample and the others as test samples. For each test sample, it then computes the TMM normalization factor as the weighted mean of log ratios (i.e., M values) between this test and the reference, after trimming genes based on a predefined percentage of M values and absolute intensities. This TMM factor should be close to 1 according to the underlying assumption (i.e., low proportion of differentially expressed genes). Otherwise, the library sizes shall be corrected by TMM normalization factor to satisfy the assumption. These normalization factors are rescaled by the mean of the normalized library sizes. Raw read counts are then divided by rescaled normalization factors to obtain normalized read counts. TMM normalization has been implemented in 'edgeR' Bioconductor R package, version 2.5 (Robinson *et al.*, 2010).

DESeq

Similar to TMM, DESeq normalization method assumes that most genes are not differentially expressed – that is, have the same read counts across all samples (Anders and Huber, 2010). For each gene in a given sample, a DESeq scaling factor is computed as the median of the ratio of the gene's read count over its geometric mean across all samples. The underlying hypothesis (i.e., low number of differentially expressed genes) assumes that this ratio is close to 1. Accordingly, the median of this ratio for the sample provides an estimate of the normalization factor to be applied to all read counts of this sample to fulfill the assumption. DESeq normalization is included in Bioconductor R package DESeq, vesion1.30.0 (Anders and Huber, 2010). DESeq normalization factor adjusts for differences in sequencing depth between samples. An improved version of this method, implemented in DESeq2

Bioconductor package version 1.18.1 (Love *et al.*, 2014), calculates *gene-specific* normalization factors to account for further sources of technical biases such as differing dependence on GC content, gene length, etc. Interested readers can consult (Love *et al.*, 2014) for more details.

Single-Cell RNA-seq

Single-cell RNA sequencing (scRNA-seq) is becoming a powerful tool for high-throughput, high-resolution transcriptomic analysis of individual cells yielding new insights into the cellular diversity underlying homogeneous populations (Tang *et al.*, 2009). It has enabled scientists to investigate fundamental research questions not previously addressable by bulk-level experiments – for example, identifying novel cell types, unraveling tumor heterogeneity, and finding novel microbiome identities (Shapiro *et al.*, 2013).

Currently, scRNA-seq experimental pipeline typically includes isolation of single cell and RNA, reverse transcription (i.e., creation of cDNA from RNA templates), amplification, library generation, sequencing, data preprocessing and downstream analyses. Normalization is a critical data preprocessing step to adjust for unwanted biological effects and technical biases specific to the technology of interest. A distinctive feature of RNA-seq data is "zero inflation," that is, the high proportion of zero read counts (Pierson and Yau, 2015). Single-cell RNA-seq data are also highly heterogeneous as individual cells are captured from potentially very different cell types (Finak *et al.*, 2015).

Nonetheless, the majority of currently available scRNA-seq analysis tools rely on normalization methods developed for bulk RNA-seq (Vallejos *et al.*, 2017). Given that the assumptions underlying most bulk RNA-seq normalization techniques do not necessarily apply to the single-cell setting, using existing normalization methods for scRNA-seq data may introduce artifacts that bias downstream analyses (Bacher *et al.*, 2017).

Recent methods have been proposed for scRNA-seq normalization specifically. Some of these methods rely on RNA *spike-in* controls (i.e., RNA transcripts of known sequence and quantity used to calibrate expression measurements) such as GRM (Ding *et al.*, 2015) and SAMstrt (Katayama *et al.*, 2013). However, it has been recently observed that the performance of spike-ins in scRNA-seq is often compromised, and have not been used for normalization in many laboratories (Bacher *et al.*, 2017). Here, we therefore only elaborate on two recently developed, state-of-the-art scRNA-seq normalization methods that do not require spike-ins – that is, Scnorm (Bacher *et al.*, 2017) and Scran (Lun *et al.*, 2016). Nevertheless, bulk-based approaches as well as methods particularly tailored to scRNA-seq data have been compared in a recent review (Vallejos *et al.*, 2017) to which interested readers are referred for further details.

Scran

This method estimates cell-specific size factors from pools of cells to enhance normalization accuracy in the presence of zero inflation and unbalanced differential expression of genes across groups of cells. Scran normalization consists of the following key steps (Lun *et al.*, 2016):
   (1) Defining pools of cells: to alleviate the problem of data heterogeneity, scran preclusters cells into smaller, more homogeneous sets and perform normalization for each cluster before between-cluster normalization.
   (2) Summing expression values across all cells in each pool: summation across cells results in fewer zeroes in each pool, which implies that the subsequent normalization is less susceptible to zero inflation.
   (3) Normalizing pool-based size factors (i.e., summed expression values) against an average reference.
   (4) Deconvolving the pool-based size factors into the size factors for its constituent cells to adjust for cell-specific biases.
The scran normalization method is publicly available using the computeSumFactors function in the "scran" package, version 1.6.6 on Bioconductor (see "Relevant Websites section").

SCnorm

This method is based on the observation that scRNA-seq data shows systematic biases in the relationship between transcript-specific expression and sequencing depth – referred to as count-depth relationship – that cannot be adjusted by a single global scale factor common to all genes in a cell. Accordingly, normalization methods based upon global scale factors can contribute to overcorrection of weakly/moderately expressed genes and undernormalization of highly expressed genes (Bacher *et al.*, 2017). To address this, SCnorm performs the following key steps:
   (1) Estimating the dependence of transcript expression on sequencing depth for every gene using quantile regression.
   (2) Grouping genes with similar dependence.
   (3) Estimating scale factors within each group through a second quantile regression.
   (4) Providing normalized estimates of expressions by performing within-group adjustment for sequencing depth using the estimated scale factors.
The SCnorm normalization is publicly available as an R package "SCnorm" version 1.0.0 on Bioconductor (see "Relevant Websites section").

Conclusions

Transcriptomics has paved the way for a comprehensive understanding of how genes are expressed and interconnected. Over the last three decades, methodological breakthroughs have repeatedly revolutionized transcriptome profiling and redefined what is possible to investigate. Integration of transcriptomic data with other omics is giving an increasingly integrated view of cellular complexities facilitating holistic approaches to biomedical research (Lowe *et al.*, 2017b).

Normalization of transcriptomic data is an essential preprocessing step aimed at correcting unwanted biological effects and technical noises prior to any downstream analysis. Normalization methods shall be chosen according to the undertaken technology and can be platform-specific. While there is no consensus on the best normalization methods across different transcriptomic technologies, several efforts have been taken to develop additional robust and effective normalization techniques and to systematically assess their performance on individual data sets.

Author Contributions

FV designed and supervised the study. FV, HD, and HAR wrote and revised the article. FV prepared the final version of the article. HD and HAR generated **Fig. 1** and FV generated **Figs. 2–4**. All authors have read and approved the final version of the article.

> *See also*: Analyzing Transcriptome-Phenotype Correlations. Characterization of Transcriptional Activities. Comparative Transcriptomics Analysis. Data Mining: Outlier Detection. Introduction to Biostatistics. Natural Language Processing Approaches in Bioinformatics. Next Generation Sequence Analysis. Pre-Processing: A Data Preparation Step. Regulation of Gene Expression. Statistical and Probabilistic Approaches to Predict Protein Abundance. Survey of Antisense Transcription. Transcriptome Analysis. Transcriptome During Cell Differentiation. Transcriptome Informatics. Utilising IPG-IEF to Identify Differentially-Expressed Proteins

References

Adams, J., 2008. Transcriptome: Connecting the genome to gene function. Nature Education 1 (1), 195.
Anders, S., Huber, W., 2010. Differential expression analysis for sequence count data. Genome Biology 11, R106.
Bacher, R., Chu, L.-F., Leng, N., et al., 2017. SCnorm: Robust normalization of single-cell RNA-seq data. Nature Methods 14, 584–586.
Bolstad, B.M., Irizarry, R.A., Åstrand, M., Speed, T.P., 2003. A comparison of normalization methods for high density oligonucleotide array data based on variance and bias. Bioinformatics 19, 185–193.
Breschi, A., Gingeras, T.R., Guigó, R., 2017. Comparative transcriptomics in human and mouse. Nature Reviews Genetics 18, 425–440.
Bullard, J.H., Purdom, E., Hansen, K.D., Dudoit, S., 2010. Evaluation of statistical methods for normalization and differential expression in mRNA-Seq experiments. BMC Bioinformatics 11.
Dillies, M., Rau, A., Aubert, J., et al., 2013. A comprehensive evaluation of normalization methods for Illumina high-throughput RNA sequencing data analysis. Briefings in Bioinformatics 14, 671–683.
Ding, B., Zheng, L., Zhu, Y., et al., 2015. Normalization and noise reduction for single cell RNA-seq experiments. Bioinformatics 31, 2225–2227.
Dudoit, S., Yang, Y.H., Callow, M.J., Speed, T.P., 2002. Statistical methods for identifying differentially expressed genes in replicated cDNA microarray experiments. Statistica Sinica. 111–139.
Durbin, B.P., Hardin, J.S., Hawkins, D.M., Rocke, D.M., 2002. A variance-stabilizing transformation for gene-expression microarray data. Bioinformatics 18, S105–S110.
Durbin, B.P., Rocke, D.M., 2004. Variance-stabilizing transformations for two-color microarrays. Bioinformatics 20, 660–667.
Finak, G., Mcdavid, A., Yajima, M., et al., 2015. MAST: A flexible statistical framework for assessing transcriptional changes and characterizing heterogeneity in single-cell RNA sequencing data. Genome Biology 16, 278.
Gautier, L., Cope, L., Bolstad, B.M., Irizarry, R.A., 2004. Affy – Analysis of Affymetrix GeneChip data at the probe level. Bioinformatics 20, 307–315.
Huber, W., von Heydebreck, A., Sültmann, H., Poustka, A., Vingron, M., 2002. Variance stabilization applied to microarray data calibration and to the quantification of differential expression. Bioinformatics 18, S96–S104.
Kalia, A., Gupta, R.P., 2005. Proteomics: A paradigm shift. Critical Reviews in Biotechnology. 173–198.
Kanter, I., Kalisky, T., 2015. Single cell transcriptomics: Methods and applications. Frontiers in Oncology 5.
Katayama, S., Töhönen, V., Linnarsson, S., Kere, J., 2013. SAMstrt: Statistical test for differential expression in single-cell transcriptome with spike-in normalization. Bioinformatics 29, 2943–2945.
Lin, S.M., Du, P., Huber, W., Kibbe, W.A., 2008. Model-based variance-stabilizing transformation for Illumina microarray data. Nucleic Acids Research 36, e11.
Love, M.I., Huber, W., Anders, S., 2014. Moderated estimation of fold change and dispersion for RNA-seq data with DESeq2. Genome Biology 15, 550.
Lowe, R., Shirley, N., Bleackley, M., Dolan, S., Shafe, T., 2017a. Transcriptomics technologies. PLOS Computational Biology 13.
Lowe, R., Shirley, N., Bleackley, M., Dolan, S., Shafee, T., 2017b. Transcriptomics technologies. PLOS Computational Biology 13, e1005457.
Lun, A.T., Bach, K., Marioni, J.C., 2016. Pooling across cells to normalize single-cell RNA sequencing data with many zero counts. Genome Biology 17, 75.
Mortazavi, A., Williams, B.A., Mccue, K., Schaeffer, L., Wold, B., 2008. Mapping and quantifying mammalian transcriptomes by RNA-Seq. Nature Methods 5, 621–628.
Nagalakshmi, U., Waern, K., Snyder, M., 2010. RNA-Seq: A method for comprehensive transcriptome analysis. Current Protocols in Molecular Biology. 4.11. 1–4.11. 13.
Oshlack, A., Wakefield, M., 2009a. Transcript length bias in RNA-seq data confounds systems biology. Biology Direct 4.
Oshlack, A., Wakefield, M.J., 2009b. Transcript length bias in RNA-seq data confounds systems biology. Biology Direct 4, 14.
Park, T., Yi, S.-G., Kang, S.-H., et al., 2003a. Evaluation of normalization methods for microarray data. BMC Bioinformatics 4.
Park, T., Yi, S.-G., Kang, S.-H., et al., 2003b. Evaluation of normalization methods for microarray data. BMC Bioinformatics 4, 33.
Pierson, E., Yau, C., 2015. ZIFA: Dimensionality reduction for zero-inflated single-cell gene expression analysis. Genome Biology 16, 241.
Quackenbush, J., 2002. Microarray data normalization and transformation. Nature Genetics 32, 496–501.
Reimers, M., 2010. Making informed choices about microarray data analysis. PLOS Computational Biology 6, e1000786.

Robinson, M., Mccarthy, D., Smyth, G., 2010. EdgeR: A Bioconductor package for differential expression analysis of digital gene expression data. Bioinformatics 26, 139–140.

Rocke, D.M., Durbin, B., 2001. A model for measurement error for gene expression arrays. Journal of Computational Biology 8, 557–569.

Sangurdekar, D.P., (2014). *Rnits: R Normalization and Inference of Time Series data.* R package version 1.12.0.

Shapiro, E., Biezuner, T., Linnarsson, S., 2013. Single-cell sequencing-based technologies will revolutionize whole-organism science. Nature Reviews Genetics 14, 618–630.

Smyth, G.K., 2002. Print-order normalization of cDNA microarray. Available at: http://www.statsci.org/smyth/pubs/porder/porder.html.

Smyth, G.K., Speed, T., 2003. Normalization of cDNA microarray data. Methods 31, 265–273.

Speed, T. 2001. Always log spot intensities and ratios. Speed Group Microarray Page: Available at: https://www.stat.berkeley.edu/~terry/zarray/Html/log.html.

Tang, F., Barbacioru, C., Wang, Y., *et al.*, 2009. mRNA-Seq whole-transcriptome analysis of a single cell. Nature Methods 6, 377–382.

Vallejos, C.A., Risso, D., Scialdone, A., Dudoit, S., Marioni, J.C., 2017. Normalizing single-cell RNA sequencing data: Challenges and opportunities. Nature Methods 14 (6), 565–571.

Wang, Z., Gerstein, M., Snyder, M., 2009. RNA-Seq: A revolutionary tool for transcriptomics. Nature Reviews Genetics 10, 57–63.

Yang, Y.H., Dudoit, S., Luu, P., *et al.*, 2002. Normalization for cDNA microarray data: A robust composite method addressing single and multiple slide systematic variation. Nucleic Acids Research 30, e15.

Zhao, S., Fung-Leung, W.-P., Bittner, A., Ngo, K., Liu, X., 2014. Comparison of RNA-Seq and microarray in transcriptome profiling of activated T cells. PLOS ONE 9.

Relevant Websites

https://bioconductor.org/packages
  Bioconductor all packages.
https://bioconductor.org/packages/3.6/bioc/html/SCnorm.html
  Bioconductor
  SCnorm.
http://bioconductor.org/packages/scran
  Bioconductor scran.
http://woldlab.caltech.edu/rnaseq
  RNASeq
  Wold Lab
  Caltech.

Differential Expression From Microarray and RNA-seq Experiments

Marc Delord, Université Paris Diderot-Paris 7, Sorbone Paris Cité, Paris, France

Introduction

Gene transcription is the process by which genes are copied into different types of RNAs, such as messenger RNA (mRNA) leading to the synthesis of proteins through translation, or noncoding RNA such as micro RNA (mRNA), transfer RNA (tRNA), or ribosomal RNA (rRNA) (**Fig. 3**) (Bolsover *et al.*, 1997). In the past decades, various methods have been developed for quantifying mRNA from biological samples at different scale, including quantitative reverse transcription-polymerase chain reaction (qRT-PCR) (Heid *et al.*, 1996), serial analyses of gene expression (SAGE) (Velculescu *et al.*, 1995), hybridization of cDNA (Schena *et al.*, 1995) or oligonucleotide chips (Lockhart *et al.*, 1996), and high-throughput RNA sequencing (RNA-seq) (Nagalakshmi *et al.*, 2008; Holt and Jones, 2008; Wang *et al.*, 2009) (**Fig. 1**).

Differential expression refers to the supervised analysis of gene expression data. This definition supposes therefore that expression of one or more genes has been quantified in various experimental conditions, or with respect to a quantitative trait or a censored time to event outcome. In large scale gene profiling experiments, the expression of thousands of genes or transcribed regions is quantified from relatively few biological samples. This situation where the number of measured items far exceeds the sample size is known as the *large p small n paradigm* (West, 2003; Kosorok and Ma, 2007) as it reverses the standard frequentist paradigm in which experiments are designed and powered in order to highlight a meaningful difference of one or few response variables observed in various experimental conditions.

Gene profiling experiments are therefore relatively underpowered and can be considered numerically as ill-posed problems, the uniqueness and stability of results being not guaranteed (Kabanikhin, 2008; Philippe, 2008). Besides, allowances that have to be made for the amount of multiple testing by the control of the type I error rate (familywise error rate or false discovery rate (FDR) (Benjamini and Hochberg, 1995; Dudoit *et al.*, 2003)) also reduce the statistical power available for detecting differential expression at the gene level. Differential analysis of gene expression is therefore a challenging task from a statistical point of view.

The challenge of the analysis of gene expression data and more generally high-throughput genomic data has favored the development and the spreading of statistical methods and procedures that are able to take advantage of the massively parallel structure of genomic data. By *borrowing strength* across features (Morris, 1983; Kerr and Churchill, 2001; Newton *et al.*, 2001; Smyth, 2004), the so-called empirical Bayes approaches (Efron and Morris, 1973; Efron and Morris, 1977; Carlin and Louis, 1996) outperform the naive feature-by-feature estimations. The idea behind these methods stems from the pioneering work of Stein Charles, (1955) who shows that the simultaneous estimate of multiple parameters is more accurate on average than sequential optimal estimations.

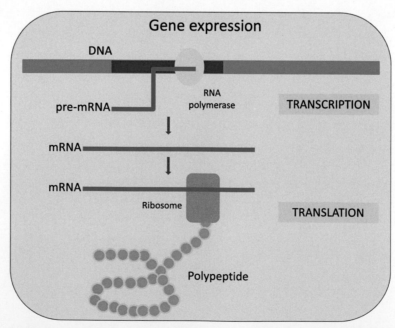

Fig. 1 Gene expression: Transcription and translation are the two steps of gene expression. During transcription, DNA is used as a template by RNA polymerase (green) to produce a pre-mRNA transcript. The pre-mRNA is then processed into a mature messenger RNA molecule (mRNA) finally translated into a protein (polypeptide).

However, having a better estimator on average does not guarantee that a particular single estimate is more accurate. Therefore, analysts and practitioners have to keep in mind that the aim of gene profiling experiments is less to test for a prespecified hypothesis on the expression of single genes than to set expression signatures associated to particular experimental conditions, cell types, tumors, or clinical outcomes. It represents therefore a shift from an hypothesis testing paradigm to an hypothesis generating paradigm (Biesecker, 2013) where a relevant task would be to prioritize genes with respect to some statistical summaries that are not directly used to assess for departure from the null hypothesis (that is no association between expression of a given gene and a covariate) in favor of the alternative hypothesis of differential expression (Noma *et al.*, 2010).

Since the emergence of high-throughput gene profiling technologies such as the serial analysis of gene expression (SAGE) (Velculescu *et al.*, 1995), cDNA microarrays (Schena *et al.*, 1995; Lennon and Lehrach, 1991) and oligonucleotide microarrays (Lockhart *et al.*, 1996), or RNA sequencing (NRA-seq) (Holt and Jones, 2008; Wang *et al.*, 2009; Wilhelm *et al.*, 2008; Mortazavi *et al.*, 2008; Cloonan *et al.*, 2008; Sultan *et al.*, 2008), the consensus that gene expression data should be analyzed at once in the framework of empirical Bayes procedures has emerged (Cui and Churchill, 2003; Soneson and Delorenzi, 2013). This idea has been translated into numerous statistical libraries allowing to perform differential analysis of gene expression data in computational frameworks such as the *R Language and Environment for Statistical Computing* (R. Core Team, 2017) and the *Bioconductor* project (Gentleman *et al.*, 2004; Huber *et al.*, 2015).

Though differential expression analysis refers preferentially to methods and tools developed in the context of gene expression data arising from microarray or RNA-seq, the paradigm of differential expression analysis extends to the analysis of data arising from any massively parallelized technologies producing intensity measurements such as array-based proteomics methods (MacBeath and Schreiber, 2000) or overdispersed count data such as ChIP-Seq data (Robertson *et al.*, 2007) or 4C (Simonis *et al.*, 2006).

This article is organized as follows: Section "Background" gives an overview of gene profiling experiments and describes principles of hybridization- (microarrays) and sequencing-based (RNA-seq) approaches. Section "Differential Expression" introduces differential expression of gene profiling experiments in the framework of microarrays (first subsection) and in the framework of RNA-seq experiments (second subsection). The Section "Conclusion" proposes a summary and some concluding remarks.

Background

Transcriptomics refers to the mapping and quantification of the sum of all transcripts present in particular cell type or tissue (de la Grange *et al.*, 2010; Aguet *et al.*, 2017), in determined conditions (Hughes *et al.*, 2000), development stages (Graveley *et al.*, 2011), cell cycle (Cho *et al.*, 1998), treatments or states of health (Lenz *et al.*, 2008). Gene expression microarrays (Schena *et al.*, 1995; Lockhart *et al.*, 1996) and deep sequencing of RNA (RNA-seq) (Nagalakshmi *et al.*, 2008; Holt and Jones, 2008; Wang *et al.*, 2009) are the two dominant high-throughput technologies used to quantify transcripts at the genome-wide scale. These techniques allow to capture a snapshot of the transcriptional activity of a sample of cells or a tissue at reasonable costs and labor (Holt and Jones, 2008; Wang *et al.*, 2009; Aguet *et al.*, 2017; Heller, 2002; Lowe *et al.*, 2017) (**Figs. 2** and **4**).

Differential expression is arguably one of the most common uses of gene expression profiling experiments (Soneson and Delorenzi, 2013; Pan, 2002). Its application includes the molecular characterization of tissues (Mortazavi *et al.*, 2008; Aguet *et al.*, 2017), the classification and prediction of diseases subtypes (Golub, 1999; Perou *et al.*, 2000), the development of prognostic

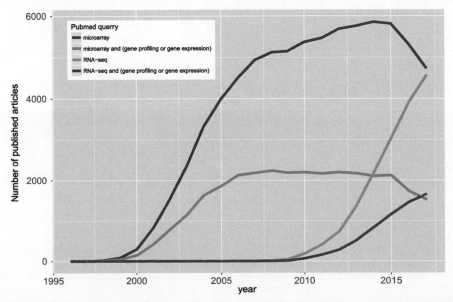

Fig. 2 Reference of gene expression profiling methods over time: Number of published papers referenced in pubmed referring to "microarray," "microarray + gene expression or gene profiling," "RNA-seq," "RNA-seq + gene expression or gene profiling" over time in their title or abstract.

tools (van't Veer *et al.*, 2002; Paik *et al.*, 2004) and diagnostic tools (Spira *et al.*, 2007; Griffith *et al.*, 2006), or the functional annotation of gene through the analysis of clusters of genes associated to developmental stages (Graveley *et al.*, 2011; Bassett *et al.*, 1999) or diseases subtypes (Hughes *et al.*, 2000; Figueroa *et al.*, 2013). In drug discovery, differential expression analysis allows to validate candidate targets and get further insight on the functional connections among diseases and drug action at the transcriptome level (Hughes *et al.*, 2000; Lamb *et al.*, 2006). Gene expression profiling and differential expression are also powerful tools to characterize at the molecular level biological systems perturbed by genes knock-out or mutations (Hughes *et al.*, 2000).

Of note, although the molecular classification and characterization of oncogenic subtypes have represented an early success of gene profiling experiments (Golub, 1999; van't Veer *et al.*, 2002), their clinical utility remain unclear (Edén *et al.*, 2004; Bonastre *et al.*, 2014; Blok *et al.*, 2018). Indeed, assessment of molecular signatures for personalizing patient care is pursued preferentially using dedicated prospective clinical trials (Simon and Wang, 2006; Delord, 2015; Tajik *et al.*, 2013) in which the marginal contribution of molecular signatures is assessed over gold standards of clinical care. In practice, however, it is frequent that the information provided by molecular profiling of tumors overlaps with standard phenotypic or anatomopathological biomarkers (Edén *et al.*, 2004; Simon *et al.*, 2009), making it therefore unlikely to provide an improvement of the clinical practice. Also, when gene expression signatures are found to be clinically relevant, cost-effectiveness analyses are needed to evaluate their utility in the daily clinical practice (Bonastre *et al.*, 2014; Blok *et al.*, 2018).

Gene expression profiling technologies are based on the following principles: in the microarrays approach (Schena *et al.*, 1995; Lockhart *et al.*, 1996; Lennon and Lehrach, 1991; Heller, 2002), gene expression is estimated via hybridization of transcript cDNA to an array of predefined complementary probes spotted or synthesized at determined locations. One the other hand, in the sequencing approach, transcripts cDNA are sequences using high-throughput sequencing technologies (Holt and Jones, 2008; Wang *et al.*, 2009) and quantification is derived from counts of sequences associated (in silico) to a given transcript (**Fig. 3**) (Wang *et al.*, 2009).

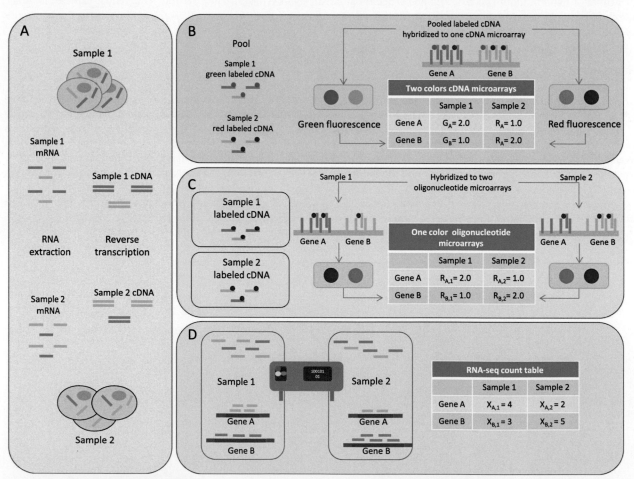

Fig. 3   Differential gene expression profiling using two-color microarrays (B), one-color microarray (C), and RNA-seq (D): In the proposed experiment, differential expression is assessed between biological samples 1 and 2. mRNA is extracted from biological samples and reverse transcribed to double strand cDNA (A). In the two-color microarray approach (B), single strand cDNA from both samples are labeled using different dyes, pooled, and incubated for hybridization into a single microarray. In the one-color microarray approach (C), single strand cDNA is labeled using the same dye and incubated for hybridization into separate oligonucleotide microarrays. Once hybridization is completed, microarrays are scanned and row expression values are derived from intensity of fluorescence for each color and each probe. In the RNA sequencing procedure (D), mRNA is reverse transcribed and amplified to cDNA followed by deep sequencing. Resulting short reads are aligned in silico and quantified.

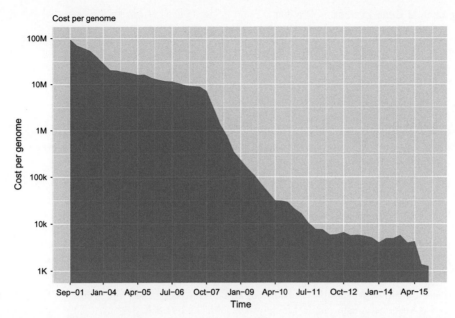

Fig. 4 Estimated cost per genome: Estimation includes global production costs such as labor, utilities, reagents and consumables, sequencing instruments, informatics activities, and indirect costs such as general laboratory supplies, etc. Data from the National Human Genome Research Institute (NHGRI) Genome Sequencing Program (GSP) (https://www.genome.gov/27541954/dna-sequencing-costs-data/NationalHumanGenomeResearchInstitute).

Since the publications of the first RNA sequencing studies using high-throughput sequencing platforms in 2008 (Nagalakshmi *et al.*, 2008; Wilhelm *et al.*, 2008; Mortazavi *et al.*, 2008; Cloonan *et al.*, 2008; Sultan *et al.*, 2008; Denoeud *et al.*, 2008), this approach has become widespread (**Fig. 2**). This evolution, which is due to the effectiveness of this technique (higher sensitivity and dynamic range and lower technical variations (Griffith *et al.*, 2010)), has been also favored by the dramatic decrease in sequencing costs (**Fig. 4**) (Holt and Jones, 2008; Wang *et al.*, 2009), and by the development of bioinformatic tools and dedicated statistical software (Soneson and Delorenzi, 2013; Huber *et al.*, 2015; Oshlack *et al.*, 2010). Nevertheless, the use of microarrays remains a powerful tool to assess the expression level of large sets of genes in the biological and clinical setting (Oshlack *et al.*, 2010). The use of microarrays is also favored by the availability of numerous preprocessing and normalizing procedures as well as an established theoretical and computational framework (Smyth, 2004; Cui and Churchill, 2003; R. Core Team, 2017; Huber *et al.*, 2015; Li and Wong, 2001; Lönnstedt and Speed, 2002; Efron and Tibshirani, 2002; Irizarry *et al.*, 2003; Storey *et al.*, 2007; Ritchie *et al.*, 2015).

The Microarray Approach to Gene Expression Profiling

In the late 1990s and into the 2000s, advances in genome sequencing of multiple species have provided the necessary information to produce microarrays allowing therefore to quantify the expression of large sets of genes (Brown and Botstein, 1999). Two microarray technologies have been widely used: the two-color approach (Schena *et al.*, 1995), and the one-color (or single channel) approach (Lockhart *et al.*, 1996). In two-color microarrays, two samples of transcript cDNA are labeled with different fluorophores and mixed. The resulting solution is incubated for competitive hybridization into a single microarray (**Fig. 2(A)** and **(B)**). In one-color microarrays, transcript cDNA from a single sample is labeled and incubated for hybridization into a single microarray (**Fig. 2(A)** and **(C)**).

While the two-color approach produces ratios of fluorescence intensities that reflect the relative abundance of labeled transcripts from the samples and the reference, the one-color approach yields absolute fluorescence intensities. In both cases, intensity of fluorescence is assumed to be related to the abundance of transcripts extracted from a biological sample hybridized to a microarray.

Dedicated studies have investigated the validity and equivalence of both methods (Patterson *et al.*, 2006; Larkin *et al.*, 2005; Oberthuer *et al.*, 2010) and have concluded in an overall good concordance between them. Of note, due to advantages in terms of logistics, simpler experimental designs, and data interpretation, the one-color oligonucleotide microarray has progressively supplanted the two-color spotted microarray in the 2000s (Bumgarner, 2013).

However, despite the successful use of microarrays in gene expression profiling experiments, this technology presents a number of inherent limitations such as dye-based detection issues, cross-hybridization artifacts, and a range of detection limited by a high background level and saturation of signals (Okoniewski and Miller, 2006). The hybridization-based approach relies also on prior knowledge on organisms under study including the coverage of all possible genes and associate isoforms, noncoding RNAs, and splicing events. Microarray may also produce poor results in profiling genes from highly variables genomes such as those from bacteria (Bumgarner, 2013). Advances of high-throughput DNA sequencing technologies in the mid-2000s have allowed to overcome previous constraints associated to the sequencing-based approach and provided an effective alternative to the hybridization-based approach to gene profiling experiments.

The Sequencing Approach to Gene Expression Profiling

The sequencing approach has emerged as an indirect application of DNA sequencing technologies (Velculescu *et al.*, 1995; Holt and Jones, 2008; Wang *et al.*, 2009). The SAGE (Velculescu *et al.*, 1995) proposed in the 1990s is one of the first successful attempts for quantifying expression of multiple genes in a single experiment. This approach uses Sanger sequencing and provides *digital* gene expression through the following steps:

- Mature mRNA is reverse transcribed into cDNA,
- cDNA is fragmented into 11 base-pair (bp) fragments (*tags*) using restriction enzymes,
- Tags are concatenated into long strands of cDNA and sequenced using Sanger technology,
- The resulting sequence tags are aligned along the reference genome and identify corresponding genes,
- Gene expression can be derived from counts of associated tags.

Differential gene expression can then be performed by using the SAGE approach on biological samples from various experimental conditions.

However, SAGE and other tag-based approaches that are based on traditional Sanger sequencing technologies are expensive and lack specificity as a significant proportion of tags cannot be unambiguously mapped to the reference genome. Following the advent of high-throughput sequencing technologies, such as 454 technology allowing to sequence transcripts of random length (from 25 to 300 base pair (bp)) the principle of digital gene expression has evolved and allowed the quantification of transcripts by sequencing (RNA-seq) through the following steps:

- mRNA is extracted from a biological sample, fragmented, and reverse transcribed into a library of double strand cDNA,
- Double strand cDNA is sequenced using high-throughput sequencing technologies,
- Resulting reads are aligned to the reference genome and identify corresponding genes,
- Gene expression is derived from counts of associated reads.

Different options are available at each step, for instance the cDNA library can be amplified by PCR to allow sequencing of low amounts of input RNA (Shanker *et al.*, 2015), high-throughput sequencing of cDNA can be performed using one of the different technologies mentioned above, from one end (single-end) or both ends (pair-end). However, the single-end sequencing is faster, cheaper, and sufficient for simple gene profiling experiments and differential gene expression analysis and is therefore recommended for these applications. Each biological sample produces up to 10^9 short sequences, which are associated to quality scores and listed in the fastq format in so-called fastq files. A human transcriptome can be adequately captured with approximately 30 million 100 bp reads (Hart *et al.*, 2013; Conesa *et al.*, 2016), corresponding to 1.8 gigabytes compressed fastq files.

Preprocessing of RNA-seq data

Prior to differential expression analysis using RNA-seq experiments the following three steps are needed: (1) quality control of raw reads, (2) reads alignment, and (3) read counts or transcript quantification (**Fig. 5**).

Quality control: Fastq files contain the list of raw reads sequence data and associated quality scores. Tools such as FastQC (Babraham Bioinformatics - FastQC A, 2017) (Illumina platform) or NGSQC (Dai *et al.*, 2010) (any platform) allow dealing with sequencing errors such as duplicated reads, GC and k-mers content, or presence of adapters. A common trait to all sequencing platforms is to produce reads whose quality decreases toward their 3′ end. Software such as FASTX-Toolkit (Hannon lab, 2010) or Trimmomatic (Bolger *et al.*, 2014) can be used to discard low-quality reads and trim low-quality bases. Prior to the mapping step, the filtering out of poor quality reads allows to prevent for mismatches and downstream errors and bias.

Alignment: Different options are available for the alignment step: if no reference genome is available or is incomplete, the *de novo* transcript reconstitution is required using tools such as SOAPdenovo-Trans (Xie *et al.*, 2014), Oases (Schulz *et al.*, 2012), trans-AByss (Grabherr *et al.*, 2011), or Trinity (Haas *et al.*, 2013) (for a comprehensive evaluation of *de novo* transcriptome programs in differential gene expression analysis see Wang and Gribskov (2016)). When a reference genome is available, the possible options are mapping to the genome or mapping to the transcriptome. Mapping to the genome requires the use of mapper allowing for gaps to account for (alternative) splice junctions. This approach allows the discovery of novel transcripts, gene

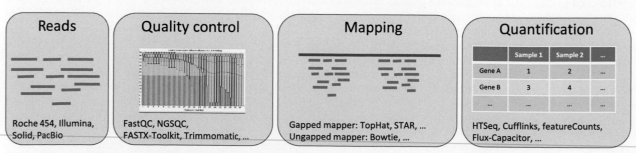

Fig. 5 RNA-seq workflow prior to differential expression analysis.

isoforms, or fusion transcripts using tools such as Tophat (Trapnell *et al.*, 2009; Kim *et al.*, 2013), MapSplice (Wang *et al.*, 2010), STAR (Dobin *et al.*, 2013), or OSA (Hu *et al.*, 2012) (for a review of genome-based sequence reads mappers, see Xie *et al.* (2014)). Finally, if a comprehensively annotated transcriptome is available, sequence reads alignment can be performed using an unspliced mapper such as Bowtie (Langmead *et al.*, 2009) allowing for gene identification and quantification simultaneously.

Transcript quantification: Once a genomic position is allocated to as many sequence reads as possible the next task is the quantification of transcripts. Depending on a chosen biologically meaningful unit, i.e., exon, transcript, or gene, this task is performed following specific sets of rules and results in a table of read counts. Quantification tools include HTSeq-count (Anders *et al.*, 2015), featureCounts (Liao *et al.*, 2014), or Cufflinks (Trapnell *et al.*, 2010). For instance, the basic strategy proposed by HTSeq is to account for a given transcript all read mapping unambiguously to the corresponding location while discarding multimapping reads. On the other hand, the Cufflinks strategy is to allocate ambiguous reads to the most probable transcribed or gene.

From fastq files containing reads for each sample to gene quantification, many options are therefore available at each step. The choice of an analysis pipeline will depend on subjective preferences of practitioners and can result in quite different expression level estimated in subsets of genes (Fonseca *et al.*, 2014). It can be noted also that other factors, such as overall sequencing coverage or gene characteristics (length or GC-content), may also affect gene profiling results (Fonseca *et al.*, 2014). When prior transcriptome knowledge is required, a paired strand-specific sequencing strategy is recommended as it provides more information to compute a *de novo* assembly of the transcriptome (Haas *et al.*, 2013). However, when a reference genome or transcriptome is available, this strategy does not improve significantly gene expression estimates (Fonseca *et al.*, 2014). Different studies have reported overall good performance of gapped mappers including Tophat, STAR, and OSA (Hu *et al.*, 2012; Fonseca *et al.*, 2014) and suggest that any of these tools could be used in the context of gene profiling experiments. On the other hand, Fonseca *et al.* (2014) have reported the critical role played by quantification methods in numerous pipelines and the good performance of the simple counting strategy of HTSeq over more sophisticated tools such as Cufflinks or Flux-Capacitor.

Differential Expression

Microarrays

Microarray experiments provide indirect estimates of gene expression through the measurement of fluorescent *spots* supposed to be related to the abundance of labeled transcripts hybridized at given positions on an array.

On a two-color microarray, where mRNA samples are hybridized against a common reference using red and green dyes respectively, relative expression values are computed as a log-ratio of fluorescence intensities:

$$M_g = \log_2(R_g) - \log_2(G_g), \ g = 1,...,N \tag{1}$$

where R_g and G_g are respectively the red and green intensities for gene g, and N is the total number of genes. Most of the time, the scatter plot of $M_g = \log_2(R_g) - \log_2(G_g)$ versus $A = (\log_2(R_g) + \log_2(G_g))/2$, $g = 1,...,N$ (MA plots) reveals a systematic relation between log-ratios versus mean intensities. Making the assumption that most of the genes share the same expression between samples, this effect can therefore be considered as a dyes bias and corrected for using a method such as the locally weighted scatterplot smoothing (LOWESS) regression of Cleveland (1979). Of note, the assumption of equal expression of the majority of genes between biological samples can be verified by conducting a dye-swap experiment. In this experiment, the two-color procedure is duplicated from the same biological samples except for dyes, which are inverted. Dyes swap experiments allow therefore to verify that the relation between log-ratios and the mean intensity is a dye artifact that has to be corrected for.

The assumption of equal global expression level between biological samples allows also to consider that variations in global intensity between microarrays are technical artifacts that should be corrected for by scaling expression level between microarrays using a method such as quantile normalization. Once log ratios have been properly corrected for within-sample (dyes bias) and between-sample artifacts, differential expression related to treatment, conditions, or phenotype of interest can be conducted.

For one-color microarrays, a single vector of log-intensities is derived from each microarray. Summarization and normalization of intensities between arrays are usually computed following methods such as model based expression index (MBEI) implemented in dChip (Li and Wong, 2001), robust multichip average (RMA) (Irizarry *et al.*, 2003) or RMA using sequence information (GCRMA) (Wu *et al.*, 2004). These routines produce expression values suitable for further analysis. These procedures are also based on the assumption of equal global expression level between biological samples.

We introduce now differential expression analysis using data arising from a microarray experiment in a simple two-condition experimental design. In this experiment the expression of N genes was estimated from n_1 wild type samples and n_2 muted samples, using single channel microarrays. We assume that expression values were summarized and normalized between arrays and represented in a $N \times (n = n_1 + n_2)$ matrix (Y). The response vector for gene g (i.e., the estimated expression of gene g measured in the n biological samples) is $y_g = \left(y_{g,1}^{wt}, ..., y_{g,n_1}^{wt}, y_{g,1}^{m}, ..., y_{g,n_2}^{m} \right)$. We compute also for gene g, $\bar{y}_{wt,g}$ and $\bar{y}_{m,g}$, the sample mean of expression, and $s_{wt,g}$ and $s_{m,g}$ the sample standard deviation in wild type and mutates samples respectively. From the above we obtain the difference in mean log-expression or log fold change for gene g:

$$D_g = \bar{y}_{m,g} - \bar{y}_{wt,g}$$

and the sample variance:

$$s_g^2 = \frac{(n_1 - 1)s_{wt,g}^2 + (n_2 - 1)s_{m,g}^2}{n_1 - n_2 - 2}$$

Using these simple quantities, different procedures have been used for identifying differentially expressed genes in microarray experiments. Their basic principle is to rank genes according to some relevant statistic and to consider genes whose statistics stand above an arbitrary chosen cut-off as differentially expressed (Noma et al., 2010; Cui and Churchill, 2003; Draghici, 2002). The log fold change has been used as a simple criterion to rank genes in early gene expression profiling studies (Schena et al., 1995), however as its computation does not integrate a measure of the variability observed in each group, this statistic lacks confidence and therefore is not recommended in practice (Cui and Churchill, 2003).

The *signal to noise ratio* integrates a measure of the observed variance:

$$\frac{D_g}{s_{wt,g} + s_{m,g}}$$

Though this statistic allows to rank genes according the their differential expression, it is not related to a standard probability distribution and therefore does not allow to compute a type I error rate when testing the hypothesis of differential expression: $D_g \neq 0$.

Despite small sample size, the need for controlling the type I error rate has led to use of statistics following tabulated distributions such as the *t*-statistic. Assuming normality of log-expression values and equal variance between groups, under H_0 (i. e., $D_g = 0$), the t_g statistic:

$$\frac{D_g}{s_g \sqrt{\frac{1}{n_1} + \frac{1}{n_2}}}$$

follows a Student distribution with $(n_1 + n_2 - 2)$ degrees of freedom and can be approximated by a standard normal distribution if samples size are large enough. This probabilist framework allows to compute a *p*-value:

$$2 \times P(t_{(n_1+n_2-2)} > |t_g|)$$

which can be interpreted as a measure of the evidence against H_0 given the data. If the assumption of equal variance between groups does not stand, the Welch t-test can be used by computing the statistic:

$$\frac{D_g}{\sqrt{\frac{s_{wt,g}}{n_1} + \frac{s_{m,g}}{n_2}}}$$

which follows a Student distribution with modified degrees of freedom if $D_g = 0$ (Welch, 1947).

The *t*-statistic presents appealing properties as it allows to rank genes in order of evidence against H_0 (Smyth, 2004; Noma et al., 2010) and to choose a *p*-value cutoff associated to a desired overall type I error rate. However, the ranking obtain by the *t*-statistic would not be optimal if the sample variance is not reliable due to the few degrees of freedom available. Underestimated sample variances would lead to large values of the *t*-statistics and high rate of false positives, while overestimated sample variances would lead to low values of the *t*-statistics and therefore high rate of false negative, i.e., a loss of power.

To overcome this difficulty, the assumption of equal variance across genes has been proposed (Arfin et al., 2000). This assumption leads to use the empirical mean of the sample variance estimated from all genes:

$$s^2 = \frac{1}{N} \sum_{i=1}^{N} s_i^2$$

This average pooled variance has therefore $N \times (n_1 + n_2 - 2)$ degrees of freedom, and the denominator of all t statistics becomes $s\sqrt{n_1^{-1} + n_2^{-1}}$. The resulting test statistics can be compared to a standard normal variable to assess the probability of observed expression values under H_0. This approach suffers however from the following two major restrictions: (1) it relies on highly unlikely assumptions and (2) it is equivalent to a log fold change ranking procedure and shares the same limitations, it is therefore not recommended to asses differential expression.

To prevent erratic values of the *t*-statistic that may be due to unreliable estimates of the sample variance Tusher et al. (2001) have proposed to add a small positive constant to the denominator of the *t*-statistic. In practice the constant can be derived as the 90th percentile of the standard deviation of all genes (Efron et al., 2000). The statistical significance is then assessed using a permutation procedure.

Alternative approaches have been proposed in which advantage is taken from the parallel structure of the data while gene-specific variance is also considered (Smyth, 2004; Lönnstedt and Speed, 2002; Baldi and Long, 2001; Wright and Simon, 2003).

The regularized *t*-statistic proposed by Baldi and Long (2001) can be presented as follows: In a Bayesian framework, prior information on the gene-specific variance is obtained from d_0 genes sharing similar intensity (σ_0^2). The *Bayesian adjusted denominator* of the regularized *t*-statistic is computed as a weighted average of the gene-specific sample standard error (s_g) and σ_0. The regularize *t*-statistic is:

$$\tilde{t}_g = \frac{D_g}{\tilde{s} \sqrt{\frac{1}{n_1} + \frac{1}{n_2}}}$$

with

$$\tilde{s}^2 = \frac{(n-1)s_g^2 + d_0\sigma_0^2}{(n-2) + d_0}$$

The *moderated* t-statistic of Smyth (2004) is similar to the Baldi and Long statistic but with a prior information on the gene-specific variance, which is derived from the proportion (p_0) of genes assumed to be differentially expressed between conditions where p_0 is estimated from the data.

In both cases, a background pooled variance σ_0^2 on d_0 degrees of freedom is estimated from the data and used as prior information on the gene-specific sample variance to moderate the gene-specific sample variance. The resulting modified t-statistic follows a Student distribution with $(n-2) + d_0$ degrees of freedom under H_0. That is, an extra d_0 degrees of freedom are considered for $\tilde{t}_g$ as compared to the $n-2$ degrees of freedom used for the classical gene-wise t-statistic t_g.

These methods are part of empirical Bayes approaches as prior information is derived from the whole dataset. The added degrees of freedom used to infer from individual genes is borrowed from all genes involved in the experiment. Individual sample variances are therefore shrunk toward a common variance and result in a far more stable inference, especially when the sample size is small, and with overall gain in statistical power (Smyth, 2004). Finally, even in cases where the normality assumption does not hold, empirical Bayes test statistics lead to more accurate ranking of genes as compared to gene-wise ranking procedures.

It is worth mentioning also the B statistic of Lönnstedt and Speed (2002), which is a empirical Bayes log posterior odds-ratio of differential expression. The numerator of the B statistic is the probability that a given gene is differentially expressed, while the denominator is the probability that the gene is not differentially expressed. Both quantities are posterior probabilities combining gene-specific information, and information derived from all other genes, producing therefore more stable inference and overall more powerful procedure than the naive gene-wise approach.

Review of tests and procedures used for differential expression using microarrays can be found in Pan (2002) and Cui and Churchill (2003). In practice, the approach proposed by Smyth (2004) is the most widely used. It is implemented in the R library *limma* (Ritchie *et al.*, 2015) which provides also facilities for reading and normalizing data from many microarray technologies including two-color microarrays and single-channel oligonucleotide platforms such as Affymetrix microarrays. It allows also to handle complex experimental designs and multiple comparison experiments in the framework of linear models.

RNA-seq

Counts resulting from RNA-seq experiments are supposed to be linked proportionally to the number of mRNA copies extracted from a biological sample and yield therefore a direct estimation of the transcriptional activity of cells from biological samples under study. However, as read counts not only depend on gene expression levels but also on length of transcripts (Oshlack and Wakefield, 2009) and overall coverage level, (Robinson and Oshlack, 2010) an appropriate normalization procedure is needed before differential expression analysis. This situation is illustrated by the RNA-seq experiment of **Fig. 6(A)** where gene A is assumed to be expressed twice as much as gene B in both samples. Gene A is also assumed to have the same expression between samples. Genes A and B have length $l_A = 2$ and $l_B = 3$ respectively, and total read counts in sample 1 and 2 are $N_1 = 7$ and $N_2 = 14$ respectively. $Y_{i,j}$ denotes read count of gene i in sample j. Computing fold-change of gene expression using raw read counts gives the following results:

- Fold change of gene expression between genes A and B in sample 1:

$$\frac{Y_{A,1}}{Y_{B,1}} = \frac{3}{7} = 0.75$$

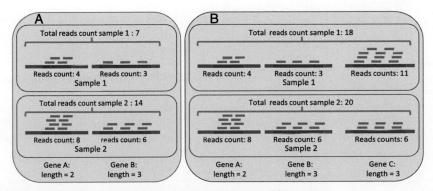

Fig. 6 The need for normalization procedure in differential expression analysis using RNA-seq data.

- Fold change of gene expression between conditions 2 and 1 for gene B:

$$\frac{Y_{B,2}}{Y_{B,1}} = \frac{6}{3} = 2$$

which does not reflect the *true* relative expression level of gene A versus gene B. This discrepancy is due to differences in gene lengths within samples and overall coverage between samples. Therefore, alike for the hybridization approach, the need arises for normalization procedures to adjust for within-sample and/or between-sample characteristics prior to differential analysis in RNA-seq experiments (Bullard *et al.*, 2010).

Within-Sample Normalization

Taking into account the gene length as a normalization factor for the within-sample comparison of gene expression gives the expected results:

$$\frac{\frac{Y_{A,1}}{l_A}}{\frac{Y_{B,1}}{l_B}} = \frac{4/2}{3/3} = 2$$

In practice, for within-sample normalization of RNA-seq experiments, different metrics of gene expression have been proposed among which are RPKM (reads per kilobase of exon per million mapped reads) (Mortazavi *et al.*, 2008), FPKM (replace reads by fragments) (Grabherr *et al.*, 2011; Trapnell *et al.*, 2010), and TPM (transcript per million) (Li *et al.*, 2010), all of which are essentially identical.

To compute these quantities, we can obtain first the count per million (CPM) for gene i (Law *et al.*, 2014):

$$\text{CPM}_i = \frac{Y_i}{N} \cdot 10^6 \tag{2}$$

where Y_i is the read count for gene i and N the total number of reads. This quantity represents simply the number of reads from gene i per million reads.

RPKM and FPKM are analogous quantities. FPKM is a more generic quantity as it explicitly accommodates data from single- or paired-end RNA-seq experiments. It is equivalent to RPKM in single-end RNA-seq experiments.

FPKM$_i$ is proportional to CPM$_i$ divided by the length of gene i (l_i):

$$\text{FPKM}_i = \frac{\text{CPM}_i}{l_i} \cdot 10^3 = \frac{Y_i}{l_i N} \cdot 10^9 \tag{3}$$

Finally, to obtain TPM$_i$, transcript i reads count needs to be normalized by length of gene i to get the rate of reads of type i, then to adjust for the total number of reads, divide rate i by the sum of all rates and scale to get a *per million reads* quantity:

$$\text{TPM}_i = \frac{Y_i}{l_i} \cdot \left(\frac{1}{\sum_j \frac{Y_j}{l_j}} \right) \cdot 10^6 \tag{4}$$

Remarks

- As shown in (3) the FPKM metric can be obtained by dividing the rate of transcript i by the total number of reads N and multiplying the result by 10^9. Using this formulation in (4) the TPM metric can be expressed as (Pachter, 2011):

$$\text{TPM}_i = \frac{Y_i}{l_i N} \cdot 10^9 \cdot \left(\frac{1}{\sum_j \frac{Y_j}{l_j N} \cdot 10^9} \right) \cdot 10^6$$

$$= \frac{\text{FPKM}_i}{\sum_j \text{FPKM}_j} \cdot 10^6$$

Basically, both FPKM and TPM give the number of reads per base of feature i, per $N/10^6$ million reads in the FPKM metric and per million reads in the TPM metric. TPM is therefore more generic than FPKM (or RPKM) and its use is therefore recommended (Wagner *et al.*, 2012).

- The number of reads for a given gene depends also on the overall coverage and therefore the number of reads in a given sample. Though TPM or FPKM take into account the total number of reads, the use of these metrics would allow for comparison between samples, only if samples show similar expression profiles such as in the experiment shown in **Fig. 6(A)**:

$$\frac{\frac{Y_{B,2}}{l_B N_2}}{\frac{Y_{B,1}}{l_B N_1}} = \frac{\frac{6}{3 \times 14}}{\frac{3}{3 \times 7}} = 1$$

However, this assumption is highly unlikely in any RNA-seq experiments. **Fig. 6(B)** shows a more realistic RNA-seq experiment where expression profiles differ between samples. Total read counts in samples 1 and 2 are $N_1^* = 18$ and $N_2^* = 20$ respectively. Differential expression of gene B between samples using FPKM gives a misleading result:

$$\frac{\frac{Y_{B,2}}{L_B N_2^*}}{\frac{Y_{B,1}}{L_B N_1^*}} = \frac{\frac{6}{3 \times 20}}{\frac{3}{3 \times 18}} = 1.8$$

This simple example illustrates the need for normalizing procedures allowing to estimate the *true* differences in overall coverage between samples. It suggests also that a good normalizing procedure between samples should reduce as much as possible the influence of differentially expressed genes as reads from these genes may represent a significant proportion of the total number of reads.

Between Sample Normalization

Recall that the goal of differential expression is to estimate ratios of gene expression between two or more experimental conditions. Using the sampling framework of Robinson and Oshlack (2010), in RNA-seq experiments, the expected count of reads associated to transcript i in condition k is assumed to be related to the expected number of transcript i per cell ($\mu_{i,k}$), the length of gene i ($l_{i,k}$), the total number of reads (N_k), and the length of transcriptome under study (S_k) in condition k:

$$E[C_{i,k}] = \frac{\mu_i l_i}{S_k}\, N_k \tag{5}$$

with

$$S_k = \sum_i \mu_{i,k} l_i$$

Using the delta method, it can be shown from expression (5) (Robinson and Oshlack, 2010) that the expected value of the ratio of read counts between conditions 1 and 2 for gene i is:

$$E\left[\frac{Y_{i,2}}{Y_{i,1}}\right] = \frac{\mu_{i,2}}{\mu_{i,1}} \frac{S_1}{S_2} \frac{N_2}{N_1} \tag{6}$$

Expression (6) shows clearly that for gene i, the expected fold change of read counts between conditions 1 and 2 depends on the ratio of the expected number of transcripts ($\mu_{i,2}/\mu_{i,1}$), the relative overall coverage (N_2/N_1), and the relative size of transcriptomes under study (S_1/S_2), which can vary drastically between conditions (Robinson and Oshlack, 2010). An appropriate normalization procedure would therefore aim at estimating Y_1^* and Y_2^* such that $E\left[Y_{i,1}^*/Y_{i,2}^*\right] = \mu_{i,1}/\mu_{i,2}$.

Among the many approaches available for between-sample normalization (Dillies *et al.*, 2013), three can be used to deal with both bias due to transcriptome size and sequencing depth: the trimmed mean of M-values procedure (TMM) (Robinson and Oshlack, 2010) implemented in the edgeR R library (Robinson *et al.*, 2010), the relative log-expression procedure (RLE) implemented in the DESeq2 R library (Anders and Huber, 2010; Anders *et al.*, 2013; Love *et al.*, 2014), and the mean ratio normalization procedure (MRN) (Maza *et al.*, 2013).

These schemes rely on the hypothesis that technical bias affects uniformly all genes in a given sample. Therefore, non-differentially expressed genes should have the same level of normalized counts between conditions, on average. A normalization factor can therefore be derived for each sample by equalizing read counts of nondifferentially expressed genes to a reference sample or a pseudoreference. In the TMM procedure, the reference sample is the first replicate of the first condition, whereas in the RLE and MRN procedures, a pseudoreference sample is defined as the geometric mean of all samples and the simple mean of the replicate of a given condition respectively.

For each sample, computation of the relative size of transcriptome with respect to the (pseudo) reference sample is based on gene-wise count ratio. As the bias is assumed to affect identically all genes, the same normalization factor can therefore be used for all genes within a given sample.

These normalization schemes are widely used and give roughly similar results (for a detailed description and comparison of these procedures see Maza (2016)).

It remains possible nevertheless that assumptions on which these normalization schemes rely are not verified in practice. This is especially true if all or most of the genes are overexpressed in one condition only (Athanasiadou *et al.*, 2016; Zheng *et al.*, 2014; Lin *et al.*, 2012; Zuqin *et al.*, 2012). In this special case, negative controls are needed. The negative control can be a gene for which relative expression between experimental conditions is known or a spike-in control. In the resulting normalizing procedure, scaling factors will be determined such that relative expression of the control gene (or spike-in control) matches its expected value between experimental conditions.

Of note, from (6), it appears that the expected fold change of read counts for a given feature between two conditions does not depend on length of that feature. Therefore, normalization within sample is not needed when differential expression is computed between samples (Soneson and Delorenzi, 2013; Oshlack *et al.*, 2010). Moreover, in this situation, any approximation resulting from a within-sample normalization procedure is likely to introduce bias in further differential expression analysis, therefore within-sample normalization is not recommended when between-sample differential expression analysis is pursued (Dillies *et al.*, 2013).

Differential Expression From RNA-seq

At this stage, we assume an RNA-seq experiment where read counts of N genes were derived from biological samples from two conditions, say n_1 wild type samples and n_2 muted samples, which is arguably the most encountered experimental design for differential expression experiments (Soneson and Delorenzi, 2013). We assume also that a procedure such as TMM, RLE, or MRN has been used for read count normalization. Recall that the use of these normalization procedures relies on the assumption that most genes are not differentially expressed, the remaining genes being overexpressed and underexpressed in a balanced proportion (Dillies *et al.*, 2013; Bolstad *et al.*, 2003; Calza and Pawitan, 2010). Finally and despite an appropriate experimental design (Auer and Doerge, 2010), principal component analysis may reveal presence of batch effects, which can be removed using methods such as COMBAT (Johnson *et al.*, 2007) or ARSyN (Nueda *et al.*, 2012).

Discrete distribution of read counts

An RNA-seq experiment can be considered as a sampling procedure in which the probability of drawing a read mapping gene i is equal to the proportion of cDNA fragments arising from that gene. Under this sampling design, Y_i, the number of reads mapping gene i follows a binomial distribution with parameters N, the total number of reads for that sample and p_i, the unknown proportion of cDNA fragment arising from gene i:

$$Y_i \sim \mathrm{B}(N, p_i) \tag{7}$$

As N far exceeds Y_i, the binomial distribution can be approximated by a Poisson distribution with parameter Np_i. This approximation is supported empirically in RNA-seq experiments even with technical replicates (Marioni *et al.*, 2008). In such experiments, as technical replicates can be seen as drawn from the same large pool of cDNA fragments, the resulting sum of counts for gene i will also follow a Poisson distribution with expected mean $p_i(N_1 + \cdots + N_k)$ for k technical replicates.

On the other hand, in presence of biological replicates, counts mapping to a given gene are *overdispersed* as compared to counts arising from technical replicates for the same gene. In this situation, the Poisson approximation still holds for individual samples with expected counts for gene i in sample j being:

$$\mathrm{E}\left[Y_{i,j}\right] = N_j p_{i,j} \tag{8}$$

However, in a given experimental condition the sampling design becomes a two-step scheme:

1. draw the parameter of the Poisson distribution for gene i and sample j from a distribution with support on the nonnegative real numbers, and
2. draw the count of reads for gene i in sample j ($Y_{i,j}$) from a Poisson distribution with parameter obtained in 1.

That is, the parameter of the Poisson distribution is also a random parameter and the resulting distribution is a compound or a mixture of Poisson distributions. In practice the gamma distribution satisfies properties required for the parameter of the Poisson distribution (Anders and Huber, 2010; Robinson and Smyth, 2007a). The Poisson–gamma mixture distribution is commonly referred to as the negative binomial (NB) distribution with parameters μ and v ($\mathrm{NB}(\mu,v)$), with mean and variance:

$$\mathrm{E}(Y) = \mu, \text{ and } \mathrm{Var}(Y) = \left(\mu + (v)\mu^2\right) \tag{9}$$

Other possible discrete distributions for modelizing counts in RNA-seq experiments are the Poisson log-normal distribution (Bulmer, 1974; Lu *et al.*, 2005) or the quasi-Poisson distribution (Lund *et al.*, 2012).

Counts from a Poisson log-normal distribution are supposed to be drawn from a Poisson distribution with parameter e^z, Z being drawn from the normal distribution with parameters μ and σ^2. The NB and Poisson log-normal distributions look similar for low values of μ and σ but diverge for larger values of σ, the NB distribution giving more weight to $P(y_{i,j}=0)$, whereas the Poisson log-normal distribution puts more weight on its right tail (Love, 2013). Like the NB distribution, the quasi-Poisson distribution has two parameters. However, the variance of the quasi-Poisson distribution is a linear function of its mean, whereas in the NB distribution the variance is a quadratic function of the mean. The choice of an appropriate distribution can therefore be based on the variance-to-mean relationship (Ver Hoef and Boveng, 2007). As RNA-seq experiments involves few replicates, the diagnostic plot may include all read counts of an experiment.

Once a distribution is chosen, the hypothesis of equal mean in read counts between the two conditions can be tested using the generalized linear model (GLM) framework. Using the NB distribution, we have:

$$Y_{i,j} = \mathrm{NB}(\mu_{i,j}, (v)_i) \tag{10}$$

and

$$\log_2(\mu_{i,j}) = x_j^t \beta_i \tag{11}$$

where $\mu_{i,j}$ is the mean parameter of the NB distribution, that is the expected value of read count for gene i in sample j, v_i is the overdispersion parameter for gene i, x_j is a column vector of covariates for sample j, and β_i a column vector of coefficients associated to gene i. The link function here is the $\log_2$ function for interpretation convenience. In a simple RNA-seq experiment with two conditions, β_i can be written as:

$$\beta_i = \begin{pmatrix} \beta_{i,0} \\ \beta_{i,1} \end{pmatrix} \tag{12}$$

In this formalism, $\beta_{i,0}$ represents the $\log_2$ of the mean of normalized counts in the first condition and $\beta_{i,1}$ represents $\log_2$ fold change between conditions 1 and 2. For instance, with $\beta_a^t = (10, 1)$, and sample 1 and 2 being in the first and the second condition respectively, we have:

$$x_1 = \begin{pmatrix} 1 \\ 0 \end{pmatrix} \quad \text{and} \quad x_2 = \begin{pmatrix} 1 \\ 1 \end{pmatrix}$$

then for sample 1:

$$\log_2(\mu_{a,1}) = x_1^t \beta_a, = (1 \quad 0)\begin{pmatrix} 10 \\ 1 \end{pmatrix} = 10$$

$$\Rightarrow \mu_{a,1} = 2^{10} = 1024$$

and for sample 2:

$$\log_2(\mu_{a,2}) = x_1^t \beta_a, = (1 \quad 1)\begin{pmatrix} 10 \\ 1 \end{pmatrix} = 11$$

$$\Rightarrow \mu_{a,2} = 2^{11} = 2048$$

Then, the hypothesis of nondifferential expression between both conditions (H_0) can be tested with the Wald test. Using the standard iteratively reweighted least squares (IRLS), the maximum likelihood estimate is derived for $\beta_{a,1}$ as well as its standard deviation $\left(\text{se}\left(\hat{\beta}_{a,1}\right)\right)$. Under H_0, $\hat{\beta}_{a,1}/\text{se}\left(\hat{\beta}_{a,1}\right)$ follows approximately a standard normal distribution. If $|\beta_{a,1}|/\text{se}\left(\hat{\beta}_{a,1}\right) > U_{1-\alpha/2}$, where α is a prespecified type I error rate, gene a is called differentially expressed between conditions 1 and 2.

Borrowing strength across genes

From a practical point of view however, as RNA-seq experiments involve often small number of biological replicates, estimation of both parameters of the NB model remain challenging. Especially, a poor estimation of the overdispersion parameter can lead to unreliable estimates of the standard error of the mean parameter and therefore leads to false positive or false negative calls. To overcome this difficulty and take advantage of the massively parallel structure of the data, various methods were developed. Robinson and Smyth (2007) proposed to reduce (Robinson and Smyth, 2007b) or shrink (Robinson and Smyth, 2007a; Wu et al., 2013) the gene-wise estimate of v to a common value estimated from the whole dataset. Love et al. (2014) moderate the gene-wise dispersions by assuming a log normal prior on this parameter conditional on mean counts for a given gene. These approaches are implemented in the edgeR (Robinson et al., 2010) and the DESeq2 (Love et al., 2014) R libraries respectively.

Parametric approach on transformed counts

It has been argued that after extensive normalization and/or batch effect removal the read counts data have eventually lost their discrete nature, allowing practitioners to use parametric model and statistical workflows developed for microarray data. This approach has been used since the emergence of RNA-seq experiments after appropriate counts transformation (Law et al., 2014). However, the main challenge here is to transform RNA-seq data in order to meet hypotheses allowing the use of parametric linear regression models also used for microarray data, assumed to follow log-linear distributions.

In the linear regression model framework, the response variable k is assumed to be a linear combination of the predictors X plus an error term and the hypothesis of homoscedasticity implies that the error term is independent of X and k. However, one of the characteristics that have led to the use of Poisson distributions or compound Poisson distributions for modelizing RNA-seq data is heterogeneous dispersion, i.e., heteroscedasticity. When a linear mean–variance relationship is suspected (relation (9)), the square root transformation of counts can stabilize the variance of mean counts ('t Hoen et al., 2008), but more sophisticated procedures such as variance stabilizing transformations developed for microarray data (Huber et al., 2002; Durbin et al., 2002), or the $\log_2$ transformation can be used if counts are overdispersed (Zwiener et al., 2014).

It has been shown nevertheless that variance stabilizing transformations such as the log-transformation would result in a reduced variance for large counts as compared to small counts (Law et al., 2014). To overcome this difficulty two approaches developed to stabilize variance of RNA-seq count data have been proposed. In the first approach denoted as *vst* (variance stabilizing transformations), the moderate gene-wise dispersions of a NB model (Love et al., 2014) are used as a variance stabilizing parameter (Anders and Huber, 2010). In the second approach, denoted as the *voom* method (variance modeling at the observation-level), a mean-variance trend is estimated nonparametrically (Cleveland, 1979) from the dataset of normalized log-counts. Fitted values associated to each individual normalized log-count are then incorporated into *precision weights*, which are used in the downstream statistical inference step of dedicated statistical pipelines such as the *limma* suite (Smyth, 2004; Ritchie *et al.*, 2015). Of note it has been reported that this approach performs well as compared to methods based on discrete distribution of read counts (Soneson and Delorenzi, 2013; Law *et al.*, 2014).

Conclusion

Since the 1990s, progresses in genetic engineering and information technologies have allowed the sequencing of the human genome (Human Genome Sequencing Consortium International, 2004) as well as many other model organisms. At the same time, these advances have provided prior knowledge needed to perform gene expression profiling experiments using cDNA and oligonucleotide microarrays (Schena *et al.*, 1995; Lockhart *et al.*, 1996; Brown and Botstein, 1999). Since the pioneering work of Schena *et al.* (1995) tens of thousands of scientific publications have referred to gene profiling experiments using microarrays (**Fig. 2**). More than a decade latter, and owing for the advent of high-throughput sequencing platforms, the sequencing-base approach or RNA-seq has emerged as a promising tool for gene expression profiling, as it allows both the mapping and the quantification of the sum of transcripts extracted from a biological sample. RNA-seq presents many advantages over array-based approaches, such as an higher sensitivity and dynamic range as well as a lower background noise (Wang *et al.*, 2009; Griffith *et al.*, 2010; 't Hoen *et al.*, 2008). This approach benefits also from the decrease of sequencing costs and is now the most widespread method used in gene profiling experiments (**Fig. 2**).

Results of gene profiling experiments using high-throughput technologies such as microarrays or RNA-seq consist of large numerical matrices in which expression level of many genes are recorded from few biological samples. Intensity records from microarrays are assumed to follow a log-normal distribution, whereas RNA-seq experiments produce counts assumed to follow Poisson related distributions. Though nonparametric approaches have also been proposed such as in Li and Tibshirani (2013) or Tarazona *et al.* (2011) most of the methods used in practice assume probabilistic distributions to expression values in order to fit parametric models in the framework of linear models or GLM theory (McCullagh and Nelder, 1989). Despite allowances that have to be made for a specific distributional hypothesis, and the choice of a model, the analysis of large scale expression data remains a high-dimensional inference problem.

In this context, empirical Bayes methods (Efron and Morris, 1973; Efron and Morris, 1977; Carlin and Louis, 1996) have provided a general theoretical framework to analyze large scale gene expression data (Newton *et al.*, 2001; Smyth, 2004; Lönnstedt and Speed, 2002; Baldi and Long, 2001; Robinson *et al.*, 2010; Anders and Huber, 2010; Love *et al.*, 2014; Robinson and Smyth, 2007a; Hardcastle and Kelly, 2010; Leng *et al.*, 2013; Van De Wiel *et al.*, 2013; Yanming *et al.*, 2011). The key point of empirical Bayes methods, and related methods, states that optimal statistical procedures to infer from a single feature became suboptimal for high-dimensional inference problems (Stein Charles, 1955). This result implies that more efficient procedures involving sharing information across features exist. In practice, estimation of a given statistic, say the standard deviation of the expression of a gene through biological samples, is shrunk toward a common value, or distribution, estimated from other genes. This principle has been applied to microarray data to moderate the denominator of test statistics and therefore prevent erratic values of the test statistic observed by chance. By preventing false positive and false negative calls, these procedures result in more stable inferences and provide therefore overall higher statistical power.

This principle has been also acknowledged for RNA-seq statistical procedures in which estimation of the dispersion parameter for each gene remains crucial while small sample size is generally involved. This situation has therefore motivated the development of empirical Bayes approaches involving information sharing across genes to obtain more reliable estimates of single dispersion parameters and improved overall inference procedure. Empirical Bayes approaches are also *de facto* convened when RNA-seq read counts are transformed using variance stabilizing procedures to enter subsequent microarray analysis pipelines.

Many tools exist to compute differential expression, none of which perform optimally in all circumstances (Soneson and Delorenzi, 2013; Conesa *et al.*, 2016). The practitioner challenge remains therefore to design a strategy that suits its particular needs, possibly in view of performances of the different tools as reported in similar experimental conditions, for instance in terms of their sensitivity/specificity and control of the FDR (Conesa *et al.*, 2016). Of note, analysis requirements such as the need to adjust results for confounding covariates and to perform multiple comparisons may also limit the list of available options to software allowing for complex statistical designs, generally in the theoretical background of GLMs and empirical Bayes procedures such as limma (Smyth, 2004) svt/voom-limma (Smyth, 2004; Law *et al.*, 2014; Love *et al.*, 2014), edgeR (Robinson *et al.*, 2010), or DESeq (Love *et al.*, 2014). The choice of an analysis strategy may also depend on individual preferences and habits, the quality of the documentation of software, and the existence of an active community of users.

See also: Analyzing Transcriptome-Phenotype Correlations. Characterization of Transcriptional Activities. Comparative Transcriptomics Analysis. Data Mining: Mining Frequent Patterns, Associations Rules, and Correlations. Expression Clustering. Functional Enrichment Analysis. Functional Enrichment Analysis Methods. Genome-Wide Scanning of Gene Expression. Introduction to Biostatistics. Natural Language Processing Approaches in Bioinformatics. Regulation of Gene Expression. Statistical and Probabilistic Approaches to Predict Protein Abundance. Survey of Antisense Transcription. Transcriptome Analysis. Transcriptome During Cell Differentiation. Transcriptome Informatics. Utilising IPG-IEF to Identify Differentially-Expressed Proteins

References

Aguet, F., Ardlie, K.G., Cummings, B.B., et al., 2017. Genetic effects on gene expression across human tissues. Nature 550, 204–213.
Anders, S., Huber, W., 2010. Differential expression analysis for sequence count data. Genome Biology 11, R106.
Anders, S., McCarthy, D.J., Chen, Y., et al., 2013. Count-based differential expression analysis of RNA sequencing data using R and Bioconductor. Nature Protocols 8, 1765–1786.
Anders, S., Pyl, P.T., Huber, W., 2015. HTSeq a Python framework to work with high-throughput sequencing data. Bioinformatics 31, 166–169.
Arfin, S.M., Long, A.D., Ito, E.T., Tolleri, L., et al., 2000. Global gene expression profiling in *Escherichia coli* K12. The effects of integration host factor. The Journal of Biological Chemistry 275, 29672–29684.
Athanasiadou, N., Neymotin, B., Brandt, N., et al., 2016. Growth rate-dependent global amplification of gene expression. bioRxiv. 044735.
Auer, P.L., Doerge, R.W., 2010. Statistical design and analysis of RNA sequencing data. Genetics 185, 405–416.
Babraham Bioinformatics – FastQC A, 2017. Quality Control tool for High Throughput Sequence Data. Available at: https://www.bioinformatics.babraham.ac.uk/projects/fastqc/ (accessed 24.11.17).
Baldi, P., Long, A.D., 2001. A Bayesian framework for the analysis of microarray expression data: Regularized t-test and statistical inferences of gene changes. Bioinformatics 17, 509–519.
Bassett, D.E., Eisen, M.B., Boguski, M.S., 1999. Gene expression informatics – It's all in your mine. Nature Genetics 21, 51–55.
Benjamini, Y., Hochberg, Y., 1995. Controlling the false discovery rate: A practical and powerful approach to multiple testing. Journal of the Royal Statistical Society. Series B (Methodological). 289–300.
Biesecker, L.G., 2013. Hypothesis-generating research and predictive medicine. Genome Research 23, 1051–1053.
Blok, E.J., Bastiaannet, E., Hout, W.B., et al., 2018. Systematic review of the clinical and economic value of gene expression profiles for invasive early breast cancer available in Europe. Cancer Treatment Reviews 62, 74–90.
Bolger, A.M., Lohse, M., Usadel, B., 2014. Trimmomatic: A flexible trimmer for Illumina sequence data. Bioinformatics (Oxford, England) 30, 2114–2120.
Bolsover, S., Hyams, J., Jones, S., et al., 1997. From Genes to Cells. New York: John Wiley & Sons, Inc.
Bolstad, B.M., Irizarry, R.A., Astrand, M., Speed, T.P., 2003. A comparison of normalization methods for high density oligonucleotide array data based on variance and bias. Bioinformatics 19, 185–193.
Bonastre, J., Marguet, S., Lueza, B., et al., 2014. Cost effectiveness of molecular profiling for adjuvant decision making in patients with node-negative breast cancer. Journal of Clinical Oncology: Official Journal of the American Society of Clinical Oncology 32, 3513–3519.
Brown, P.O., Botstein, D., 1999. Exploring the new world of the genome with DNA microarrays. Nature Genetics 21 (33), 37.
Bullard, J.H., Purdom, E., Hansen, K.D., Dudoit, S, 2010. Evaluation of statistical methods for normalization and differential expression in mRNA-Seq experiments. BMC Bioinformatics 11, 94.
Bulmer, M.G., 1974. On fitting the Poisson lognormal distribution to species-abundance data. Biometrics 30, 101–110.
Bumgarner, R, 2013. Overview of DNA microarrays: Types, applications, and their future. Current Protocols in Molecular Biology. (Chapter 22:Unit 22.1).
Calza S., Pawitan Y., 2010. Normalization of gene-expression microarray data. In: Computational Biology. Totowa, NJ: Humana Press, pp. 37–52.
Carlin, B.P., Louis, T.A., 1996. Bayes and Empirical Bayes Methods For Data Analysis. Chapman & Hall.
Cho, R.J., Campbell, M.J., Winzeler, E.A., et al., 1998. A Genome-wide transcriptional analysis of the mitotic cell cycle. Molecular Cell 2, 65–73.
Cleveland, W.S, 1979. Robust locally weighted regression and smoothing scatterplots. Journal of the American Statistical Association 74, 829–836.
Cloonan, N., Forrest Alistair, R.R., Kolle, G., et al., 2008. Stem cell transcriptome profiling via massive-scale mRNA sequencing. Nature Methods 5, 613–619.
Conesa, A., Madrigal, P., Tarazona, S., et al., 2016. A survey of best practices for RNA-seq data analysis. Genome Biology 17, 13.
Cui, X., Churchill, G.A., 2003. Statistical tests for differential expression in cDNA microarray experiments. Genome Biology 4, 210.
Dai, M., Thompson, R.C., Maher, C., et al., 2010. NGSQC: Cross-platform quality analysis pipeline for deep sequencing data. BMC Genomics 11, S7.
de la Grange, P., Gratadou, L., Delord, M., et al., 2010. Splicing factor and exon profiling across human tissues. Nucleic Acids Research 38, 2825–2838.
Delord M., 2015. Pharmacogénétique de l'Imatinib dans la Leucémie Myéloïde Chronique et Données Censurées par Intervalles en présence de Compétition. PhD Thesis, Université Paris-Saclay.
Denoeud, F., Aury, J.-M., Da Silva, C., et al., 2008. Annotating genomes with massive-scale RNA sequencing. Genome Biology 9, R175.
Dillies, M.-A., Rau, A., Aubert, J., et al., 2013. A comprehensive evaluation of normalization methods for Illumina high-throughput RNA sequencing data analysis. Briefings in Bioinformatics 14, 671–683.
Di, Y., Schafer, D.W., Cumbie, J.S., Chang, J.H., 2011. The NBP negative binomial model for assessing differential gene expression from RNA-Seq. Statistical Applications in Genetics and Molecular Biology 10.
Dobin, A., Davis, C.A., Schlesinger, F., et al., 2013. STAR: Ultrafast universal RNA-seq aligner. Bioinformatics 29, 15–21.
Draghici, S, 2002. Statistical intelligence: Effective analysis of high-density microarray data. Drug Discovery Today 7, S55–S63.
Dudoit, S., Shaffer, J.P., Boldrick, J.C, 2003. Multiple hypothesis testing in microarray experiments. Statistical Science Statistical Science 18, 71–103.
Durbin, B.P., Hardin, J.S., Hawkins, D.M., Rocke, D.M., 2002. A variance-stabilizing transformation for gene-expression microarray data. Bioinformatics (Oxford, England) 18 (Suppl 1), S105–S110.
Edén, P., Ritz, C., Rose, C., et al., 2004. "Good Old" clinical markers have similar power in breast cancer prognosis as microarray gene expression profilers. European Journal of Cancer 40, 1837–1841.
Efron, B., Morris, C, 1973. Stein's estimation rule and its competitors – An empirical Bayes approach. Journal of the American Statistical Association Journal of the American Statistical Association 68, 117–130.
Efron, B., Morris, C, 1977. Stein's paradox in statistics. Scientific American 236, 119–127.
Efron, B., Tibshirani, R, 2002. Empirical bayes methods and false discovery rates for microarrays. Genetic Epidemiology 23, 70–86.
Efron, B., Tibshirani, R., Goss, V., Chu, G., 2000. Microarrays and their use in a comparative experiment. Technical Report, Division of Biostatistics, Stanford University.
Figueroa, M.E., Chen, S.-C., Andersson, A.K., et al., 2013. Integrated genetic and epigenetic analysis of childhood acute lymphoblastic leukemia. The Journal of Clinical Investigation 123, 3099–3111.
Fonseca, N.A., Marioni, J., Brazma, A, 2014. RNA-Seq gene profiling – A systematic empirical comparison. PLOS ONE 9, e107026.
Gentleman, R.C., Carey, V.J., Bates, D.M., et al., 2004. Bioconductor: Open software development for computational biology and bioinformatics. Genome Biology 5, R80.
Golub, T.R., 1999. Molecular classification of cancer: Class discovery and class prediction by gene expression monitoring. Science 286, 531–537.
Grabherr, M.G., Haas, B.J., Yassour, M., et al., 2011. Full-length transcriptome assembly from RNA-Seq data without a reference genome. Nature Biotechnology 29, 644–652.
Graveley, B.R., Brooks, A.N., Carlson, J.W., et al., 2011. The developmental transcriptome of *Drosophila melanogaster*. Nature 471, 473–479.
Griffith, M., Griffith, O.L., Mwenifumbo, J., et al., 2010. Alternative expression analysis by RNA sequencing. Nature Methods 7, 843–847.
Griffith, O.L., Melck, A., Jones, S.J., Wiseman, S.M., 2006. Meta-analysis and meta-review of thyroid cancer gene expression profiling studies identifies important diagnostic biomarkers. Journal of Clinical Oncology: Official Journal of the American Society of Clinical Oncology 24, 5043–5051.
Haas, B.J., Papanicolaou, A., Yassour, M., et al., 2013. De novo transcript sequence reconstruction from RNA-seq using the Trinity platform for reference generation and analysis. Nature Protocols 8, 1494–1512.

Hannon Lab, 2010. FASTX-Toolkit: FASTQ/A short-reads pre-processing tools (accessed 24.11.17).

Hardcastle, T.J., Kelly, K.A., 2010. baySeq: Empirical Bayesian methods for identifying differential expression in sequence count data. BMC Bioinformatics 11, 422.

Hart, S.N., Therneau, T.M., Zhang, Y., et al., 2013. Calculating sample size estimates for rna sequencing data. Journal of Computational Biology: A Journal of Computational Molecular Cell Biology 20, 970–978.

Heid, C.A., Stevens, J., Livak, K.J., Williams, P.M., 1996. Real time quantitative PCR. Genome Research 6, 986–994.

Heller, M.J., 2002. DNA microarray technology: Devices, systems, and applications. Annual Review of Biomedical Engineering 4, 129–153.

Holt, R.A., Jones, S.J., 2008. The new paradigm of flow cell sequencing. Genome Research 18, 839–846.

Huber, W., Carey, J.V., Gentleman, R., et al., 2015. Orchestrating high-throughput genomic analysis with Bioconductor. Nature Methods 12, 115–121.

Huber, W., Heydebreck, A., Sultmann, H., et al., 2002. Variance stabilization applied to microarray data calibration and to the quantification of differential expression. Bioinformatics 18, S96–S104.

Hughes, T.R., Marton, M.J., Jones, A.R., et al., 2000. Functional discovery via a compendium of expression profiles. Cell 102, 109–126.

Human Genome Sequencing Consortium International, 2004. Finishing the euchromatic sequence of the human genome. Nature 431, 931–945.

Hu, Z., Chen, K., Xia, Z., et al., 2014. Nucleosome loss leads to global transcriptional up-regulation and genomic instability during yeast aging. Genes & Development 28, 396–408.

Hu, J., Ge, H., Newman, M., Liu, K, 2012. OSA: A fast and accurate alignment tool for RNA-Seq. Bioinformatics 28, 1933–1934.

Irizarry, R.A., Hobbs, B., Collin, F., et al., 2003. Exploration, normalization, and summaries of high density oligonucleotide array probe level data. Biostatistics 4, 249–264.

Johnson, W.E., Li, C., Rabinovic, A., 2007. Adjusting batch effects in microarray expression data using empirical Bayes methods. Biostatistics 8, 118–127.

Kabanikhin, S.I., 2008. Definitions and examples of inverse and ill-posed problems. Journal of Inverse and Ill-Posed Problems 16, 317–357.

Kerr, M.K., Churchill, G.A., 2001. Experimental design for gene expression microarrays. Biostatistics 2, 183–201.

Kim, D., Pertea, G., Trapnell, C., et al., 2013. TopHat2: Accurate alignment of transcriptomes in the presence of insertions, deletions and gene fusions. Genome Biology 14, R36.

Kosorok, M.R., Ma, S, 2007. Marginal asymptotics for the "large p , small n" paradigm: With applications to microarray data. The Annals of Statistics 35, 1456–1486.

Lamb, J., Crawford, E.D., Peck, D., et al., 2006. The connectivity map: Using gene-expression signatures to connect small molecules, genes, and disease. Science 313, 1929–1935.

Langmead, B., Trapnell, C., Pop, M., Salzberg, S.L., 2009. Ultrafast and memory-efficient alignment of short DNA sequences to the human genome. Genome Biology 10, R25.

Larkin, J,E., Frank, B.C., Gavras, H., et al., 2005. Independence and reproducibility across microarray platforms. Nature Methods 2, 337–344.

Law, C.W., Chen, Y., Shi, W., Smyth, G.K. 2014. Voom: Precision weights unlock linear model analysis tools for RNA-seq read counts. Genome Biology 15, R29.

Leng, N., Dawson, J.A., Thomson, J.A., et al., 2013. EBSeq: An empirical Bayes hierarchical model for inference in RNA-seq experiments. Bioinformatics 29, 1035–1043.

Lennon, G.G., Lehrach, H., 1991. Hybridization analyses of arrayed cDNA libraries. Trends in Genetics: TIG 7, 314–317.

Lenz, G., Wright, G., Dave, S.S., et al., 2008. Stromal gene signatures in large-B-cell lymphomas. The New England Journal of Medicine 35922359, 2313–2323.

Liao, Y., Smyth, G.K., Shi, W., 2014. FeatureCounts: An efficient general purpose program for assigning sequence reads to genomic features. Bioinformatics 30, 923–930.

Lin, C.Y., Lovén, J., Rahl, P.B., et al., 2012. Transcriptional amplification in tumor cells with elevated c-Myc. Cell 151, 56–67.

Li, B., Ruotti, V., Stewart, R.M., et al., 2010. RNA-Seq gene expression estimation with read mapping uncertainty. Bioinformatics 26, 493–500.

Li, J., Tibshirani, R, 2013. Finding consistent patterns: A nonparametric approach for identifying differential expression in RNA-Seq data. Statistical Methods in Medical Research 22, 519–536.

Li, C., Wong, W.H., 2001. Model-based analysis of oligonucleotide arrays: Expression index computation and outlier detection. Proceedings of the National Academy of Sciences 98, 31–36.

Lockhart, D.J., Dong, H., Byrne, M.C., et al., 1996. Expression monitoring by hybridization to high-density oligonucleotide arrays. Nature Biotechnology 14, 1675–1680.

Lönnstedt, I., Speed, T, 2002. Replicated microarray data. Statistica Sinica 12, 31–46.

Love M.I., 2013. Statistical analysis of high-throughput sequencing count data. PhD Thesis, Freien Universität Berlin.

Love, M.I., Huber, W., Anders, S, 2014. Moderated estimation of fold change and dispersion for RNA-seq data with DESeq2. Genome Biology 15, 550.

Lowe, R., Shirley, N., Bleackley, M., et al., 2017. Transcriptomics technologies. PLOS Computational Biology 13, e1005457.

Lund, S.P., Nettleton, D., McCarthy, D.J., Smyth, G.K., 2012. Detecting differential expression in RNA-sequence data using quasi-likelihood with shrunken dispersion estimates. Statistical Applications in Genetics and Molecular Biology 11.

Lu, J., Tomfohr, J.K., Kepler, T.B., 2005. Identifying differential expression in multiple SAGE libraries: An overdispersed log-linear model approach. BMC Bioinformatics 6, 165.

MacBeath, G., Schreiber, S.L., 2000. Printing proteins as microarrays for high-throughput function determination. Science (New York, N.Y.) 289, 1760–1763.

Marioni, J.C., Mason, C.E., Mane, S.M., et al., 2008. RNA-seq: An assessment of technical reproducibility and comparison with gene expression arrays. Genome Research 18, 1509–1517.

Maza, E, 2016. In Papyro Comparison of TMM (edgeR), RLE (DESeq2), and MRN normalization methods for a simple two-conditions-without-replicates RNA-Seq experimental design. Frontiers in Genetics 7, 164.

Maza, E., Frasse, P., Senin, P., et al., 2013. Comparison of normalization methods for differential gene expression analysis in RNA-Seq experiments: A matter of relative size of studied transcriptomes. Communicative & Integrative Biology 6, e25849.

McCullagh, P., Nelder, J.A., 1989. Generalized linear models, Monograph on Statistics and Applied Probability, no. 37.

Morris, C.N., 1983. Parametric empirical Bayes inference: Theory and applications. Journal of the American Statistical Association 78, 47–55.

Mortazavi, A., Williams, B.A., McCue, K., et al., 2008. Mapping and quantifying mammalian transcriptomes by RNA-Seq. Nature Methods.

Nagalakshmi, U., Wang, Z., Waern, K., et al., 2008. The transcriptional landscape of the yeast genome defined by RNA sequencing. Science 320, 1344–1349.

Newton, M.A., Kendziorski, C.M., Richmond, C.S., et al., 2001. On differential variability of expression ratios: Improving statistical inference about gene expression changes from microarray data. Journal of Computational Biology 8, 37–52.

Nie, Z., Hu, G., Wei, G., et al., 2012. c-Myc is a universal amplifier of expressed genes in lymphocytes and embryonic stem cells. Cell 151, 68–79.

Noma, H., Matsui, S., Omori, T., Sato, T, 2010. Bayesian ranking and selection methods using hierarchical mixture models in microarray studies. Biostatistics 11, 281–289.

Nueda, M.J., Ferrer, A., Conesa, A., 2012. ARSyN: A method for the identification and removal of systematic noise in multifactorial time course microarray experiments. X_i Biostatistics 13, 553–566.

Oberthuer, A., Juraeva, D., Li, L., et al., 2010. Comparison of performance of one-color and two-color gene-expression analyses in predicting clinical endpoints of neuroblastoma patients. The Pharmacogenomics Journal 10, 258–266.

Okoniewski, M.J., Miller, C.J., 2006. Hybridization interactions between probesets in short oligo microarrays lead to spurious correlations. BMC Bioinformatics 7, 276.

Oshlack, A., Robinson, M.D., Young, M.D., 2010. From RNA-seq reads to differential expression results. Genome Biology 11, 220.

Oshlack, A., Wakefield, M.J., 2009. Transcript length bias in RNA-seq data confounds systems biology. Biology Direct 4, 14.

Pachter, L., 2011. Models for transcript quantification from RNA-Seq, arXiv:1104.3889.

Paik, S., Shak, S., Kim, C., et al., 2004. A multigene assay to predict recurrence of tamoxifen-treated, node-negative breast cancer. The New England Journal of Medicine 35127351, 2817–2826.

Pan, W, 2002. A comparative review of statistical methods for discovering differentially expressed genes in replicated microarray experiments. Bioinformatics 18, 546–554.

Patterson, T.A., Lobenhofer, E.K., Fulmer-Smentek, S.B., et al., 2006. Performance comparison of one-color and two-color platforms within the Microarray Quality Control (MAQC) project. Nature Biotechnology 24, 1140–1150.

Perou, C.M., Sørlie, T., Eisen, M.B., et al., 2000. Molecular portraits of human breast tumours. Nature 406, 747–752.

Philippe, B., 2008. Regression pls et donnees censurees. PhD Thesis, Conservatoire National des Arts et Métiers.

R. Core Team, 2017. R: A Language and Environment for Statistical Computing. Vienna, Austria: R Foundation for Statistical Computing.

Ritchie, M.E., Phipson, B., Wu, D., et al., 2015. Limma powers differential expression analyses for RNA-sequencing and microarray studies. Nucleic Acids Research 43, e47.

Robertson, G., Hirst, M., Bainbridge, M., et al., 2007. Genome-wide profiles of STAT1 DNA association using chromatin immunoprecipitation and massively parallel sequencing. Nature Methods 4, 651–657.

Robinson, M.D., McCarthy, D.J., Smyth, G.K., 2010. edgeR: A Bioconductor package for differential expression analysis of digital gene expression data. Bioinformatics (Oxford, England) 26, 139–140.

Robinson, M.D., Oshlack, A, 2010. A scaling normalization method for differential expression analysis of RNA-seq data. Genome Biology 11, R25.

Robinson, M.D., Smyth, G.K., 2007a. Moderated statistical tests for assessing differences in tag abundance. Bioinformatics 23, 2881–2887.

Robinson, M.D., Smyth, G.K., 2007b. Small-sample estimation of negative binomial dispersion, with applications to SAGE data. Biostatistics 9, 321–332.

Schena, M., Shalon, D., Davis, R.W., Brown, P.O., 1995. Quantitative monitoring of gene expression patterns with a complementary DNA microarray. Science (New York, N.Y.) 270, 467–470.

Schulz, M.H., Zerbino, D.R., Vingron, M., Birney, E, 2012. Oases: Robust de novo RNA-seq assembly across the dynamic range of expression levels. Bioinformatics 28, 1086–1092.

Shanker, S., Paulson, A., Edenberg, H.J., et al., 2015. Evaluation of commercially available RNA amplification kits for RNA sequencing using very low input amounts of total RNA. Journal of Biomolecular Techniques: JBT 26, 4–18.

Simon, R.M., Paik, S., Hayes, D.F., 2009. Use of archived specimens in evaluation of prognostic and predictive biomarkers. Journal of the National Cancer Institute 101, 1446–1452.

Simonis, M., Klous, P., Splinter, E., et al., 2006. Nuclear organization of active and inactive chromatin domains uncovered by chromosome conformation capture-on-chip (4C). Nature Genetics 38, 1348–1354.

Simon, R., Wang, S.-J., 2006. Use of genomic signatures in therapeutics development in oncology and other diseases. The Pharmacogenomics Journal 6, 166–173.

Smyth, G.K., 2004. Linear models and empirical Bayes methods for assessing differential expression in microarray experiments. Statistical Applications in Genetics and Molecular Biology 3, 1–25.

Soneson, C., Delorenzi, M, 2013. A comparison of methods for differential expression analysis of RNA-seq data. BMC Bioinformatics 14, 91.

Spira, A., Beane, J.E., Shah, V., et al., 2007. Airway epithelial gene expression in the diagnostic evaluation of smokers with suspect lung cancer. Nature Medicine 13, 361–366.

Stein Charles, 1955. Inadmissibility of usual estimator for the mean of multivariate normal distributions. In: Proceedings of the Third Berkeley Symposium on Mathematical Statistics and Probability, 4, 197-206.

Storey, J.D., Dai, J.Y., Leek, J.T., 2007. The optimal discovery procedure for large-scale significance testing, with applications to comparative microarray experiments. Biostatistics 8, 414–432.

Sultan, M., Schulz, M.H., Richard, H., et al., 2008. A global view of gene activity and alternative splicing by deep sequencing of the human transcriptome. Science (New York, N.Y.) 321, 956–960.

Tajik, P., Zwinderman, A.H., Mol, B.W., Bossuyt, P.M., 2013. Trial designs for personalizing cancer care: A systematic review and classification. Clinical Cancer Research. 19, 4578–4588.

Tarazona, S., Garca-Alcalde, F., Dopazo, J., et al., 2011. Differential expression in RNA-seq: A matter of depth. Genome Research 21, 2213–2223.

't Hoen, P.A.C., Ariyurek, Y., Thygesen, H.H., et al., 2008. Deep sequencing-based expression analysis shows major advances in robustness, resolution and inter-lab portability over five microarray platforms. Nucleic Acids Research 36, e141.

Trapnell, C., Pachter, L., Salzberg, S.L., 2009. TopHat: Discovering splice junctions with RNA-Seq. Bioinformatics 25, 1105–1111.

Trapnell, C., Williams, B.A., Pertea, G., et al., 2010. Transcript assembly and quantification by RNA-Seq reveals unannotated transcripts and isoform switching during cell differentiation. Nature Biotechnology 28, 511–515.

Tusher, V.G., Tibshirani, R., Chu, G., 2001. Significance analysis of microarrays applied to the ionizing radiation response. Proceedings of the National Academy of Sciences of the United States of America 98, 5116–5121.

van't Veer, L.J., Dai, H., Vijver, M.J., et al., 2002. Gene expression profiling predicts clinical outcome of breast cancer. Nature 415, 530–536.

Velculescu, V.E., Zhang, L., Vogelstein, B., Kinzler, K.W., 1995. Serial analysis of gene expression. Science (New York, N.Y.) 270, 484–487.

Ver Hoef, J.M., Boveng, P.L., 2007. Quasi-poisson vs. negative binomial regression: How should we model overdispersed count data? Ecology 88, 2766–2772.

Wagner, G.P., Kin, K., Lynch, V.J., 2012. Measurement of mRNA abundance using RNA-seq data: RPKM measure is inconsistent among samples. Theory in Biosciences 131, 281–285.

Wang, Z., Gerstein, M., Snyder, M, 2009. RNA-Seq: A revolutionary tool for transcriptomics. Nature Reviews Genetics 10, 57–63.

Wang, S., Gribskov, M, 2016. Comprehensive evaluation of de novo transcriptome assembly programs and their effects on differential gene expression analysis. Bioinformatics 33, btw625.

Wang, K., Singh, D., Zeng, Z., et al., 2010. MapSplice: Accurate mapping of RNA-seq reads for splice junction discovery. Nucleic Acids Research 38, e178.

Welch, B.L., 1947. The generalization of student's problem when several different population variances are involved. Biometrika 34, 28–35.

West, M., 2003. Bayesian factor regression models in the "Large p, Small n" paradigm. Bayesian Statistics 7, 723–732.

Van De Wiel, M.A., Leday, G.G.R., Pardo, L., et al., 2013. Bayesian analysis of RNA sequencing data by estimating multiple shrinkage priors. Biostatistics 14, 113–128.

Wilhelm, B.T., Marguerat, S., Watt, S., et al., 2008. Dynamic repertoire of a eukaryotic transcriptome surveyed at single-nucleotide resolution. Nature 453, 1239–1243.

Wright, G.W., Simon, R.M., 2003. A random variance model for detection of differential gene expression in small microarray experiments. Bioinformatics 19, 2448–2455.

Wu, Z., Irizarry, R.A., Gentleman, R., et al., 2004. A model-based background adjustment for oligonucleotide expression arrays. Journal of the American Statistical Association 99, 909–917.

Wu, H., Wang, C., Wu, Z., 2013. A new shrinkage estimator for dispersion improves differential expression detection in RNA-seq data. Biostatistics 14, 232–243.

Xie, Y., Wu, G., Tang, J., et al., 2014. SOAPdenovo-Trans: De novo transcriptome assembly with short RNA-Seq reads. Bioinformatics 30, 1660–1666.

Zwiener, I., Frisch, B., Binder, H, 2014. Transforming RNA-Seq data to improve the performance of prognostic gene signatures. PLOS ONE 9, e85150.

Expression Clustering

Xiaoxin Ye and Joshua WK Ho, Victor Chang Cardiac Research Institute, Darlinghurst, NSW, Australia and University of New South Wales, Sydney, NSW, Australia

Introduction

Quantification of gene expression is very important in biological and biomedical research. A major breakthrough in gene expression analysis is the development of DNA microarrays in the 1990s (DeRisi *et al.*, 1997; Schena *et al.*, 1995). Prior to the development of DNA microarrays, expression of genes is mainly measured by relatively low-throughput experimental techniques such as quantitative polymerase chain reaction (qPCR), Southern and Northern blotting. All these techniques measure the absolute or relative amount of a small number of target complementary DNA (cDNA) or RNA molecules in a sample. One key innovation of microarray technology is its ability to simultaneously measure the expression of thousands of genes in a single assay. It enables researchers to ask new questions about the entire gene expression system as a whole, instead of focusing on individual molecules one at a time.

In late 2000s, the advent of next-generation sequencing (NGS) brought about the next major breakthrough in gene expression analysis, namely RNA sequencing (RNA-seq) (Cloonan *et al.*, 2008; Mortazavi *et al.*, 2008; Nagalakshmi *et al.*, 2008; Wilhelm *et al.*, 2008). The idea is very simple. Expressed RNA transcripts are reversed transcribed into cDNA, which can be amplified, fragmented, and sequenced using NGS technologies. Expression of a gene can be quantified by the number of sequence reads aligned to the reference transcriptome. The raw readout of RNA-seq is therefore read counts of each gene, which can then be further normalised by the total number of reads in a library and sometimes by gene length. RNA-seq is found to have better detection sensitivity and specificity than microarrays (Bottomly *et al.*, 2011; Fu *et al.*, 2009; Marioni *et al.*, 2008). It is also much more flexible since we are no longer restricted by the collection of probes that are available on a microarray. RNA-seq is currently the preferred method for gene expression profiling (Wang *et al.*, 2009).

Regardless of the technology used to obtain the gene expression profiles, the aim of gene expression profiling is to identify similarities and differences in gene expression patterns among a group of samples from different biological conditions (i.e., phenotypic, genotypic, temporal, or environmental differences). Each sample can be derived from a cell line or tissue from an individual or an experimental animal model. A microarray data set is commonly represented by an nxm matrix,

$$\begin{bmatrix} x_{1,1} & x_{1,2} & \cdots & x_{1,m} \\ x_{2,1} & x_{2,2} & \cdots & x_{2,m} \\ \cdots & \cdots & \ddots & \vdots \\ x_{n,1} & x_{n,1} & \cdots & x_{n,m} \end{bmatrix}$$

where n is the number of genes and m is the number of samples. Each element $x_{i,j}$ in the matrix represents the expression value of gene i and sample j. For example, in a study that profiles the genome-wide mRNA expression levels of 50 heart disease patients and 50 matched healthy individuals, n is approximately 20,000 (the number of protein coding genes in human), and m is 100. Gene expression analysis refers to the manipulation of this matrix.

The use of genome-wide gene expression profiling in biological or biomedical research has grown rapidly in the last decade. One of the most widely used public repositories of gene expression data is the National Centre for Biotechnology Information (NCBI) Gene Expression Omnibus (GEO) (Barrett *et al.*, 2005, 2013). To date (October 2017), GEO contains gene expression profiles from close to 2 million samples. These gene expression profiles cover many species, cell types, and biological conditions. Proper analysis of these data is crucial in bioinformatics.

One of the most basic analyses that one can perform on a gene expression matrix is clustering (Eisen *et al.*, 1998). This includes the clustering of genes and clustering of samples. Roughly speaking, the goal of clustering is to group objects into clusters such that similar objects are grouped in a cluster, whereas dissimilar objects are grouped in different clusters. In the context of gene expression analysis, clustering can be performed to cluster genes (where each gene is represented by its expression across a number of samples) or samples (where each sample is represented by the expression of the genes).

A small data set is shown in **Fig. 1** to illustrate the power of gene expression clustering. This toy data set consists of 10 genes and 8 samples. A popular way to visualise a gene expression matrix is to use a heat map. In the heat map in **Fig. 1**, red indicates high gene expression and blue indicates low expression. Without clustering, it is quite difficult to discern any patterns in this gene expression matrix. After clustering is performed on the genes and samples (hierarchical clustering is used in this case), we could reorder the genes and samples such that genes/samples that are within the same cluster are closer together. Clear patterns emerge from the clustering analysis. We can readily identify three groups of samples: the brown cluster (sample 2, 6, and 7), blue cluster (sample 1 and 3), and the red cluster (sample 4, 5, and 8). We could also identify three distinct groups of genes: the green cluster (gene 6, 7, and 10), the yellow cluster (gene 4 and 9), and the purple cluster (gene 2, 3, and 8). There are also 2 genes that have expression patterns that are quite distinct from other gene clusters. They can be viewed as outliers.

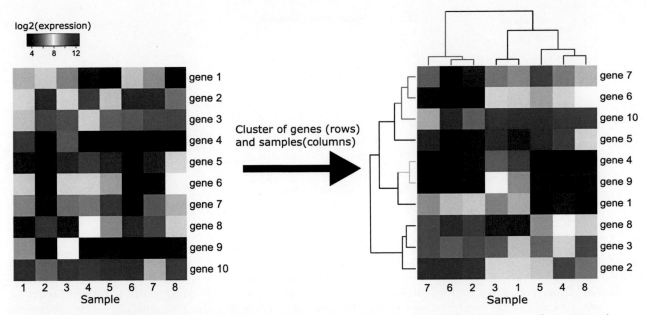

Fig. 1 A toy example illustrating the power of gene expression clustering in discovering groupings of samples and genes. A gene expression matrix consists of 10 genes (rows) and 8 samples (columns) are subjected to clustering analysis. After hierarchical clustering, it is much easier to identify three distinct groups of samples, and three sets of co-expressed genes, along with two 'outlier' genes that have very distinct expression patterns compared to other gene clusters.

Michael Eisen and colleagues were one of the first groups that applied clustering to analyse and visualise gene expression data (Eisen *et al.*, 1998). Through clustering of genes, they were able to show that genes of similar functions tend group to together in a cluster. This is a major observation because it suggests that genes that have similar expression patterns across multiple samples may be co-regulated at the cellular level. This motivated the use of clustering to identify co-expression modules as a way to discover cellular gene regulation programs (Basso *et al.*, 2005; Friedman *et al.*, 2000; Stuart *et al.*, 2003). Furthermore, clustering of samples can lead to discovery of patient subtypes and rare cell types (Golub *et al.*, 1999; Ramaswamy *et al.*, 2001; Su *et al.*, 2001). Nowadays, clustering is one of the most routinely performed analyses on gene expression data.

Clustering Methods

Clustering is an unsupervised method to discover patterns in the data. It is unsupervised because it depends on the data alone (i.e., gene expression values only), without the need to include additional information about how the data should be grouped. Clustering can be framed as an optimisation problem in which the objective is to maximize the similarity between every item within each subset (i.e., cluster), and maximize the dissimilarity between every subset (i.e., cluster). Solving this optimisation problem exactly is computationally difficult because in general one needs to enumerate all k^n possible clustering configurations to identify the optimal clustering for n points and k clusters. Instead, various algorithms have been developed to find a good local optimal solution (Jain *et al.*, 1999). In this section, we will review three of the most popular clustering methods used in gene expression analysis: k-means clustering, hierarchical clustering, and DBSCAN (**Fig. 2**). All these methods are easily accessible via the scikit-learn package in python (Pedregosa *et al.*, 2011) and the cluster (Maechler *et al.*, 2017) and dbscan (Hahsler *et al.*, 2017) packages in R.

K-Means Clustering

K-means clustering is a partitional clustering method, in which a set of items is partitioned into k disjoint subsets such that each item belongs to exactly one subset. The basic idea of k-means is that each cluster is assigned a cluster centre (i.e., cluster mean), and the optimisation objective is to minimise the distance between each data point and its cloest cluster centre. The classical k-means clustering algorithm starts with selecting k random data points as the initial cluster centres, and iteratively perform the following two steps:

1) Assigns each point to its cloest cluster centre
2) Based on the points in each cluster after step 1, calculate the cluster mean and make it the new cluster centre

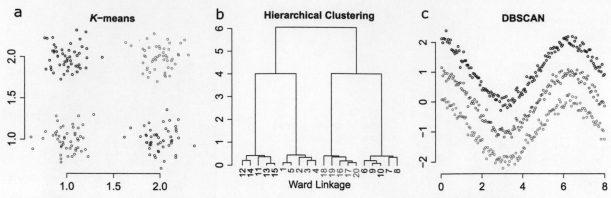

Fig. 2   Illustration of three popular clustering methods: (a) k-means clustering, (b) hierarchical clustering, and (c) DBSCAN.

These two steps are repeated until the maximum number of iteration is reached or the cluster assignment no longer change (i.e., no points are re-assigned in step 1). The k-means algorithm is fast and simple. The algorithm works very well for ball-shape and cubic-shape clusters, especially when k (the number of cluster) is somehow known. K-means is memory efficient, and it is easy to parallelise (Zhang et al., 2011). Nonetheless, k-means is sensitive to the initialisation of cluster centres. To improve the quality of clustering, k-means++ has been developed (Arthur and Vassilvitskii, 2007). The idea is to perform a simple randomised seeding step to obtain good initial cluster centres. This simple step has been shown to significantly improve both clustering accuracy and speed (by reducing the number of iterations required).

Hierarchical Clustering

The key difference between hierarchical clustering and k-means clustering is the structure of clusters. While k-means clustering groups all points into distinct non-overlapping clusters, hierarchical clustering groups all the points in a tree-like hierarchy, also known as a dendrogram. The root of the tree represent the whole data set and every branch node represents a subset of the data. Hierarchical clustering methods are categorised into agglomerative (bottom-up) and divisive(top-down) algorithms. In bioinformatics, the agglomerative algorithm is more popular, and therefore will be the focus of this section.

The basic idea of agglomerative hierarchical clustering is recursive agglomeration of two nearest clusters or data points. Agglomeration criteria include single-linkage, complete-linkage, averaged-linkage clustering, and the Ward's method. Different linkage criteria can lead to different results, with the Ward's method generally considered to produce consistently good results for expression data (Meunier et al., 2007). The Ward method (Ward, 1963) merges two clusters with minimum weighted squared distance between the new cluster centres.

In the initial step of the agglomerative hierarchical clustering algorithm, each of the n points is labelled as a distinct cluster. Then an nxn dissimilarity matrix is calculated to reflect the dissimilarity (or distance) between every pair of clusters. Clusters are then recursively merged to form larger clusters based on the following steps:

1) Merge two clusters with minimum cluster distance (based on the an agglomeration criterion)
2) Update the dissimilarity matrix

The two steps are repeated until only one cluster is left.

The main benefit of hierarchical clustering is that there is no need to determine the number of clusters prior to the analysis. This allows information about the relationships between different clusters (or data points) to be preserved. If it is desirable to obtain discrete clusters, one can always 'cut' the dendrogram at a certain level to obtain them. In practice, one drawback of hierarchical clustering is that the pairwise dissimilarity matrix can become prohibitively large when the number of data point is large, which makes clustering impractical for data sets with more than several thousand points using standard computers.

DBSCAN

Density based clustering (Kriegel et al., 2011) is a category of algorithm that could detect clusters of any shape by dividing a data set into a set of connected 'high density' components. It includes DBSCAN (Ester et al., 1996), OPTICS (Ankerst et al., 1999) and DENCLUE (Hinneburg and Keim, 1998). The key idea is to define a cluster as a region in the data space that has high density, separated from other clusters by low density regions. The definition is intuitive, does not require the number of clusters to be predetermined, and has the potential to recover complex non-spherical clusters. DBSCAN is a frequently used density-based clustering method in bioinformatics. The basic idea of DBSCAN is outlined as follows (Schubert et al., 2017):

1) Go through every point in the data set to check if each point p has at least m neighbours within a distance of ε. If p satisfies this criterion, it is said to be a core point. Otherwise, it is a non-core point.

2) Join neighbouring core points into clusters
3) Go through each non-core point, and try to joint it to a reachable cluster. If this point can be joint, it becomes part of that cluster. Otherwise this point is said to be an outlier (or noise).

With suitable user-specified parameters (m and ε), DBSCAN can work very well in detecting clusters with complex shape, and in the presence of outliers. DBSCAN is fast and has low memory requirement. However, its performance is sensitive to the choice of parameters and data dimensionality. A large ε may merge too many clusters into one, while a small ε may incorrectly create too many outliers. In high dimension, finding suitable parameters becomes very difficult because of the increased sparsity. Various improvements have been made to the original DBSCAN algorithm (Rehman *et al.*, 2014). For example, FDBSCAN dramatically reduces the query time of neighbours and the effect of user-specified parameters (Liu, 2006).

Evaluation and Interpretation of Clustering Results

Of the three clustering algorithms discussed in the previous sections, k-means algoritm generally scales very well with increasing number of data points because its algorithmic complexity is linear with respect to the number of data points. K-means is especially helpful if the user has prior knowledge about the expected number of clusters. Nonetheless, k-means clustering is ineffective in discovery clusters that have a non-spherical shape. Also, the results are highly sensitive to the presence of outliers (Gan and Ng, 2017). Therefore, it is important to visualise and carefully inspect the quality of the resulting clusters to ensure they are not subjected to significant biases.

Hierarchical clustering provides more information than flat partitional clustering. It is particularly useful in the case where we have no idea what the correct number of cluster should be. Nonetheless, the algorithm requires quadratic runtime and space. This means in practice it is often not possible to cluster more than several thousand points on a standard desktop machine. If the number of points is small, hierarchical clustering can often provide a highly informative view of the clustering structure of the data.

Density-based clustering algorithms such as DBSCAN is much more flexible in terms of detecting complex cluster structures, and they can deal with larger data sets as these algorithms generally scale quasi-linearly with the number of points. One major drawback with DBSCAN is that the determination of local density is very sensitive to increased dimensionality . Also, choosing the right parameters can often be quite difficult.

It should be noted that most clustering methods find clusters even when there is no natural clustering structure in the data. Therefore it is important to evaluate the quality of the clustering results and intepret its significance cautiously (Altman and Krzywinski, 2017; Ronan *et al.*, 2016). Data preprocessing and dimensionality reduction can dramatically affect the quality of clustering results (see discussion in the following section).

In terms of evaluation of the quality of the clustering results, if there is external information (such as sample labels or even partial labels), such information can be used to determine whether a clustering is reasonable or not. For example, if all the labels are given, one can calculate an adjusted rand index to quantify the accuracy of the clustering. Often only partial information is given, such as which samples are replicates. In this case, one might consider the quality of the overall clustering based on whether replicates are correctly grouped together.

Practical Considerations

In gene expression analysis, one of the most important decisions is to identify the appropriate feature set. A genome-wide gene expression data set typically contains the expression levels of tens of thousands of genes. Prior to clustering, it is often useful to perform dimensionality reduction to reduce the number of features used in the clustering analysis. This may be necessary because of both biological reasons and technical reasons.

In terms of biological reasons, sometimes scientists want to focus on a specific set of key genes, such as transcription factors, receptors, or signalling molecules. In this case, it is useful to only include these key genes in clustering analysis to discover clusters based on these key genes.

If there is no prior knowledge about which genes should be included in the analysis, there are still a number of technical reasons why dimensionality reduction is necessary in practice. In microarray data analysis, often a gene is represented by several probes. One might wish to first combine the expression levels of multiple probes to generate a single expression value for each gene. Several methods are possible, such as only including the probe with the highest median expression level, or simply taking the average of the probe expression values.

Furthermore, it is often observed that many genes have no detectable expression across all the samples profiled in the entire data set. In this case, including those non-expressed genes in the clustering might artificially inflat the dimensionality of the data without providing any real signal for clustering. It is often advisable to remove these 'undetectable' genes, using criteria such as removing all genes which have detectable expression in less than 2 samples.

After all these filtering steps, a genome-wide gene expression data set may still have a large number of genes (i.e., in the order of thousands). Clustering high dimensional data can often be problematic, because as the number of dimensions increases, the space become more sparse, and the difference between within- and between-cluster distances become smaller – this phenomena is known as the curse of dimensionality. Many methods have been designed to deal with clustering of high dimensional data (Assent, 2012), but as a general rule, doing sensible feature selection and dimensionality reduction is often critical in gene expression analysis.

A simple method for feature selection involves only selecting features that have the high variance (or some measure of variability) across all the samples in the data set. For example, one might consider selecting only those genes in which the maximum and minimum expression levels are at least 2 fold different. Besides feature selection, another popular dimensionality reduction method is Principal Component Analysis (PCA) (Ma and Dai, 2011). PCA identifies a series of orthogonal principal components (PCs), in which each PC corresponds to a linear combination of the original features. The PCs are constructed such that the first PC (PC1) contains the largest proportion of variance of the data, followed by PC2, then by PC3, and so on. Dimensionality reduction can be achieved by selecting only the first k PCs that cumulatively explain sufficient proportion of variance. In other words, PCA is used to project a data set of n dimensions into k projected PCs.

Another important consideration in gene expression clustering analysis is the definition of distance between two points (two genes or two samples). In practice, the most common distance measures are Euclidean distance and Pearson correlation coefficient (when used as a distance, it is typically represented as 1-|correlation| or (1+ correlation)/2). A more comprehensive comparison of distance measures for gene expression clustering can be found elsewhere (Jaskowiak *et al.*, 2014).

Correlation coefficient is less sensitive to the scale of data than Euclidean distance. Consider the case of a pair of co-expressed genes in which one gene is highly expressed while other one has significantly lower expression. These two genes would have a large Euclidean distance but strong correlation. In this case the choice of Euclidean distance or correlation as a measure of dissimilarity is largely driven by the underlying biological question. Sometimes it might be important to distinguish genes that have large difference in expression levels, whereas sometimes it is more important to determine correlation without worrying too much about th absolute expression levels. Another possible analysis strategy is to first normalise the expression of both genes to have a mean of zero and a standard deviation of one (i.e., conversion to standardised scores). In this case, both Eucledian distance and correlation would yield similar results.

Euclidean distance and correlation coefficient are both sensitive to outliers or missing data. Missing data is a major issue in the analysis of single-cell RNA-seq (scRNA-seq) data because many transcripts are not detectable due to dropout events in the experimental process. In the case of scRNA-seq data, it is often not possible to distinguish whether a lack of gene expression (zero read count per gene) is an artefact caused by dropout or it reflects true non-expression of that gene. One simple approach to alleviate the impact of dropouts in scRNA-seq data set is to use 'implicit imputation' in the calculation of Eucledian distance (Lin *et al.*, 2017). This approach is implemented in an R package CIDR for dimensionality reduction and clustering of scRNA-seq data.

Case Study

Let's put everything together. In this case study, we analysed a time series gene expression data set of mouse embryonic dental epithelium (Epi) and dental mesenchym (Mes) between embryonic day 11.5 and 13.5 (E11.5-E13.5) (O'Connell *et al.*, 2012), with three replicates per time point per tissue. This data set consists of 30 gene expression profiles in total (5 time points x 2 tissues x 3 replicates). The gene expression profiles were obtained by an Illumina microarray platform. The data can be downloaded from GEO using accession number GSE32321. The goal of the analysis is to identify the gene expression dynamics in these two adjacent developing dental tissues, and to identify genes that are driving those dynamic patterns.

After downloading the data, we first checked that the quality of the data. **Fig. 3(a)** shows the histogram of expression levels (in logarithmic scale) of one representative sample in the data set. This plot illustrates the characteristics of a microarray-based gene expression profile: a sharp peak on the left which consists of mostly unexpressed or lowly expressed genes, and a long tail on the right which consists of expressed genes. Based on this plot, approximately half of the genes are unexpressed or have low expression (the left peak), and the other half has a wide range of expression levels (the long right tail). The plot also shows that there is an absence of outliers (values that are exceptionally small or large). These are all characteristics of a good quality gene expression profile. In RNA-seq data, the distinction between the left peak and the right tail is often much clearer.

We next checked whether the gene expression distributions were consistent across all samples in a data set. **Fig. 3(b)** is a boxplot showing the distribution of all 30 samples in this data set. The interquartile range and median of all the samples are highly consistent, suggesting that further normalisation is not necessary.

We then wanted to visualise the gene expression data using a heat map. Nonetheless, rendering a heat map of the entire data set (consisting of 45,281 gene probes) is computationally difficult on standard computer screens and difficult to visually interpret by humans. We therefore decided to analyse and visualise the 100 most variably expressed genes, as defined as the 100 genes with the highest variance across the 30 samples. Analysis of these high-variance genes often allows us to gain insight into the major significant patterns in the full data set. We performed hierarchical clustering on the genes and samples. The hierarchical clustering was performed using Eucledian distance and the Ward method for aggregation. The results are displayed using a heat map in which a gene represent as a row and a sample is represented as a column (**Fig. 3(c)**). One striking feature is that samples from epithelial and mesenchymal tissues are grouped into two distinct clusters. The replicates generally group together, with only one minor exception (E11.5 epithelium), suggesting that the replicates are generally highly consistent. Within each tissue, the samples appear to be grouped based on temporal ordering of samples. In particular, samples from E11.5 and E12 tend to group together, and samples from E12.5-E13.5 tend to group together.

We further visualised the relationships among the samples by performing PCA, and displaying the sample based on the first two principal components (PC1 and PC2; **Fig. 3(d)**). The first two components collectively explains about 90% of the variance in the data set, with most of the variance explained by the PC1 (76%). We can see that PC1 clearly separates epithelial and mesenchymal tissues. PC2

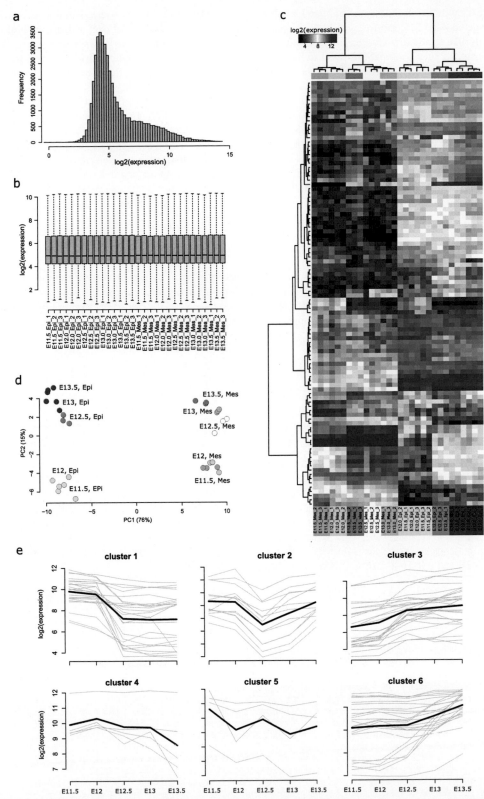

Fig. 3 A case study on expression clustering of an embryonic dental tissue development data set. (a) The gene expression distribution of a typical sample in this data set. (b) A box plot showing the distribution of gene expression values is largely consistent across all the samples in the data set. (c) A heat map showing the gene expression levels of the 100 most variably expressed genes in this data set. Hierarchical clustering has been performed on the samples and genes. The clustering results are displayed as dendrograms around the heat map. (d) The data set is subjected to the principal component analysis (PCA), and the samples are displayed based on first two principal components (PC1 and PC2). (e) The top 100 variably expressed genes are subjected to k-means clustering to identify 6 clusters. The dark line in each plot represents the mean expression pattern of all the genes in a cluster.

generally separates early time points (E11.5 and E12) and later time points (E12.5-E13.5) in both tissues. This implies that there is a major transcriptomic change between E12 and E12.5 in both tissues. This plot clearly shows the spatial temporal clustering of samples.

Finally we would like to identify clusters of genes with similar gene expression patterns. Here we only analysed the epithelial samples for simplicity (the mesenchymal samples can be analysed in a similar manner). We first normalised the data such that each gene has a mean value of zero and a standard deviation of one. This process ensures that all the genes are comparable in scale. After normalisation, we performed k-means clustering to identify clusters of genes. In this example, we set $k=6$, but in general one should attempt a reasonable range of k. The resulting 6 clusters are displayed in **Fig. 3(e)**. One salient pattern is that all clusters have a major transition point at E12.5, corroborating the pattern observed in the PCA plot (**Fig. 3(d)**). Through looking at the plots, some clusters can clearly be further divided into smaller clusters. For example, cluster 6 clearly consists of two major clusters – one characterised by a smooth and gradual increase in expression of mostly highly expressed genes, and the other characterised by a massive up-regulation of lowly expressed genes starting at E12. Further downstream bioinformatics analyses, such as gene set analysis (Djordjevic *et al.*, 2016; de Leeuw *et al.*, 2016), can be performed to identify what are genes that are enriched in each of these clusters.

This case study illustrates how clustering of genes and samples of a gene expression data set can lead to discovery of biologically relevant patterns, and highlight the overall workflow of clustering in gene expression bioinformatics analysis.

Summary

Clustering is one of the most fundamental bioinformatics tools, especially in terms of gene expression analysis. As discussed in this article, care must be taken to ensure that proper feature filtering, dimensionality reduction and quality controls are undertaken prior to clustering. Also, clustering results must be evaluated and interpreted carefully, especially mindful of the characteristics of different clustering algorithms. When performed properly, clustering is a powerful tool to discover co-regulated genes, patient groupings, and sub-population of cells.

Acknowledgement

The authors are supported by the Victor Chang Cardiac Research Institute. JWKH is also supported by a Career Development Fellowship by the National Health and Medical Research Council of Australia and a Future Leader Fellowship from the National Heart Foundation of Australia.

See also: Algorithms for Graph and Network Analysis: Clustering and Search of Motifs in Graphs. Analyzing Transcriptome-Phenotype Correlations. Characterization of Transcriptional Activities. Comparative Transcriptomics Analysis. Computation Cluster Validation in the Big Data Era. Data Mining: Clustering. Data Mining: Mining Frequent Patterns, Associations Rules, and Correlations. Differential Expression from Microarray and RNA-seq Experiments. Functional Enrichment Analysis Methods. Genome-Wide Scanning of Gene Expression. Natural Language Processing Approaches in Bioinformatics. Regulation of Gene Expression. Statistical and Probabilistic Approaches to Predict Protein Abundance. Survey of Antisense Transcription. Transcriptome Analysis. Transcriptome During Cell Differentiation. Transcriptome Informatics. Unsupervised Learning: Clustering. Utilising IPG-IEF to Identify Differentially-Expressed Proteins

References

Altman, N., Krzywinski, M., 2017. Points of significance: Clustering. Nat. Methods 14, 545–546.

Ankerst, M., Breunig, M.M., Kriegel, H.-P., Sander, J., 1999. OPTICS: Ordering points to identify the clustering structure. In: Proceedings of the 1999 ACM SIGMOD International Conference on Management of Data, (New York, NY, USA: ACM), pp. 49–60.

Arthur, D., Vassilvitskii, S., 2007. K-means ++ : The advantages of careful seeding. In: Proceedings of the 18th Annual ACM-SIAM Symposium on Discrete Algorithms.

Assent, I., 2012. Clustering high dimensional data. Wiley Interdiscip. Rev. Data Min. Knowl. Discov. 2, 340–350.

Barrett, T., Suzek, T.O., Troup, D.B., *et al.*, 2005. NCBI GEO: Mining millions of expression profiles – Database and tools. Nucleic Acids Res. 33, D562–D566.

Barrett, T., Wilhite, S.E., Ledoux, P., *et al.*, 2013. NCBI GEO: Archive for functional genomics data sets – Update. Nucleic Acids Res. 41, D991–D995.

Basso, K., Margolin, A.A., Stolovitzky, G., *et al.*, 2005. Reverse engineering of regulatory networks in human B cells. Nat. Genet. 37, 382–390.

Bottomly, D., Walter, N.A.R., Hunter, J.E., *et al.*, 2011. Evaluating gene expression in C57BL/6J and DBA/2J mouse striatum using RNA-Seq and microarrays. PLOS ONE 6, e17820.

Cloonan, N., Forrest, A.R.R., Kolle, G., *et al.*, 2008. Stem cell transcriptome profiling via massive-scale mRNA sequencing. Nat. Methods 5, 613–619.

de Leeuw, C.A., Neale, B.M., Heskes, T., Posthuma, D., 2016. The statistical properties of gene-set analysis. Nat. Rev. Genet. 17, 353–364.

DeRisi, J.L., Iyer, V.R., Brown, P.O., 1997. Exploring the metabolic and genetic control of gene expression on a genomic scale. Science 278, 680–686.

Djordjevic, D., Kusumi, K., Ho, J.W.K., 2016. XGSA: A statistical method for cross-species gene set analysis. Bioinformatics 32, i620–i628.

Eisen, M.B., Spellman, P.T., Brown, P.O., Botstein, D., 1998. Cluster analysis and display of genome-wide expression patterns. Proc. Natl. Acad. Sci. USA 95, 14863–14868.

Ester, M., Kriegel, H.-P., Sander, J., Xu, X., 1996. A density-based algorithm for discovering clusters a density-based algorithm for discovering clusters in large spatial databases with noise. In: Proceedings of the Second International Conference on Knowledge Discovery and Data Mining, (Portland, Oregon: AAAI Press), pp. 226–231.

Friedman, N., Linial, M., Nachman, I., Pe'er, D., 2000. Using Bayesian networks to analyze expression data. J. Comput. Biol. 7, 601–620.

Fu, X., Fu, N., Guo, S., *et al.*, 2009. Estimating accuracy of RNA-Seq and microarrays with proteomics. BMC Genomics 10, 161.

Gan, G., Ng, M.K.-P., 2017. k-means clustering with outlier removal. Pattern Recognit. Lett. 90, 8–14.

Golub, T.R., Slonim, D.K., Tamayo, P., *et al.*, 1999. Molecular classification of cancer: Class discovery and class prediction by gene expression monitoring. Science 286, 531–537.

Hahsler, M., Piekenbrock, M., Doran, D., 2017. dbscan: Fast density-based Clustering with R.

Hinneburg, A., Keim, D.A., 1998. An efficient approach to clustering in large multimedia databases with noise. In: Proceedings of the Fourth International Conference on Knowledge Discovery and Data Mining, (New York, NY: AAAI Press), pp. 58–65.

Jain, A.K., Murty, M.N., Flynn, P.J., 1999. Data clustering: A review. ACM Comput. Surv. CSUR 31, 264–323.

Jaskowiak, P.A., Campello, R.J., Costa, I.G., 2014. On the selection of appropriate distances for gene expression data clustering. BMC Bioinform. 15, S2.

Kriegel, H.-P., Kröger, P., Sander, J., Zimek, A., 2011. Density-based clustering. Wiley Interdiscip. Rev. Data Min. Knowl. Discov. 1, 231–240.

Lin, P., Troup, M., Ho, J.W.K., 2017. CIDR: Ultrafast and accurate clustering through imputation for single-cell RNA-seq data. Genome Biol. 18, 59.

Liu, B., 2006. A fast density-based clustering algorithm for large databases. In: 2006 International Conference on Machine Learning and Cybernetics, pp. 996–1000.

Ma, S., Dai, Y., 2011. Principal component analysis based methods in bioinformatics studies. Brief. Bioinform. 12, 714–722.

Maechler, M., Rousseeuw, P., Struyf, A., Hubert, M., Hornik, K., 2017. Cluster: Cluster analysis basics and extensions.

Marioni, J.C., Mason, C.E., Mane, S.M., Stephens, M., Gilad, Y., 2008. RNA-seq: An assessment of technical reproducibility and comparison with gene expression arrays. Genome Res. 18, 1509–1517.

Meunier, B., Dumas, E., Piec, I., et al., 2007. Assessment of hierarchical clustering methodologies for proteomic data mining. J. Proteome Res. 6, 358–366.

Mortazavi, A., Williams, B.A., McCue, K., Schaeffer, L., Wold, B., 2008. Mapping and quantifying mammalian transcriptomes by RNA-Seq. Nat. Methods 5, 621–628.

Nagalakshmi, U., Wang, Z., Waern, K., et al., 2008. The transcriptional landscape of the yeast genome defined by RNA sequencing. Science 320, 1344–1349.

O'Connell, D.J., Ho, J.W.K., Mammoto, T., et al., 2012. A Wnt-Bmp feedback circuit controls Intertissue signaling dynamics in tooth organogenesis. Sci. Signal 5, ra4.

Pedregosa, F., Varoquaux, G., Gramfort, A., et al., 2011. Scikit-learn: Machine learning in python. J. Mach. Learn. Res. 12, 2825–2830.

Ramaswamy, S., Tamayo, P., Rifkin, R., et al., 2001. Multiclass cancer diagnosis using tumor gene expression signatures. Proc. Natl. Acad. Sci. 98, 15149–15154.

Rehman, S.U., Asghar, S., Fong, S., Sarasvady, S., 2014. DBSCAN: Past, present and future. In: Proceedings of the Fifth International Conference on the Applications of Digital Information and Web Technologies (ICADIWT 2014), pp. 232–238.

Ronan, T., Qi, Z., Naegle, K.M., 2016. Avoiding common pitfalls when clustering biological data. Sci. Signal 9, re6.

Schena, M., Shalon, D., Davis, R.W., Brown, P.O., 1995. Quantitative monitoring of gene expression patterns with a complementary DNA microarray. Science 270, 467–470.

Schubert, E., Sander, J., Ester, M., Kriegel, H.P., Xu, X., 2017. DBSCAN revisited, revisited: Why and how you should (still) use DBSCAN. ACM Trans. Database Syst. 42, 19:1–19:21.

Stuart, J.M., Segal, E., Koller, D., Kim, S.K., 2003. A gene-coexpression network for global discovery of conserved genetic modules. Science 302, 249–255.

Su, A.I., Welsh, J.B., Sapinoso, L.M., et al., 2001. Molecular classification of human carcinomas by use of gene expression signatures. Cancer Res. 61, 7388–7393.

Wang, Z., Gerstein, M., Snyder, M., 2009. RNA-Seq: A revolutionary tool for transcriptomics. Nat. Rev. Genet. 10, 57–63.

Ward, J.H., 1963. Hierarchical grouping to optimize an objective function. J. Am. Stat. Assoc. 58, 236–244.

Wilhelm, B.T., Marguerat, S., Watt, S., et al., 2008. Dynamic repertoire of a eukaryotic transcriptome surveyed at single-nucleotide resolution. Nature 453, 1239–1243.

Zhang, J., Wu, G., Hu, X., Li, S., Hao, S., 2011. A parallel K-means clustering algorithm with MPI. In: 2011 Proceedings of the Fourth International Symposium on Parallel Architectures, Algorithms and Programming, pp. 60–64.

Biographical Sketch

Xiaoxin Ye is a Masters student at the University of New South Wales (UNSW), and an intern at the Bioinformatics and Systems Medicine Laboratory at the Victor Chang Cardiac Research Institute. He received his BSc in Statistics from Guangzhou Medical University and a Graduate Certificate in Mathematics and Statistics from UNSW. His research focuses on developing fast and scalable algorithms for clustering single cell RNA-seq (scRNA-seq) data. His interests include scalable clustering algorithms, shared neighbours clustering algorithms, as well as statistical and parallelised machine learning.

Joshua Ho is the Head of Bioinformatics and Systems Medicine Laboratory at the Victor Chang Cardiac Research Institute in Sydney Australia. He is also a conjoint Senior Lecturer at UNSW Sydney. Dr Ho completed his BSc (Hon 1, Medal) and PhD in Bioinformatics at the University of Sydney, and undertook postdoctoral research at the Harvard Medical School in the United States. His current research focuses on developing fast and reliable bioinformatics methods to identify the genetic cause and mechanism of heart diseases, using a range of approaches such as whole genome sequencing, single-cell RNA-seq, ChIP-seq, machine learning, systems biology, cloud computing, and software testing and quality assurance. His research excellence has been recognised by the 2015 NSW Ministerial Award for Rising Star in Cardiovascular Research, the 2015 Australian Epigenetics Alliance' Illumina Early Career Research Award, and the 2016 Young Tall Poppy Science Award.

Metabolome Analysis

Saravanan Dayalan, The University of Melbourne, Parkville, VIC, Australia
Jianguo Xia, McGill University, Sainte Anne de Bellevue, QC, Canada and McGill University, Montreal, QC, Canada
Rachel A Spicer and Reza Salek, European Bioinformatics Institute (EMBL-EBI), Hinxton, United Kingdom
Ute Roessner, The University of Melbourne, Parkville, VIC, Australia

Introduction

Metabolites, including lipids, are biological active compounds important as building blocks for macromolecules and carriers of energy, both required in all living cells to participate in all biological processes. They are members of thousands of biosynthetic pathways providing the chemical diversity to fulfil specific functions. The chemical diversity of metabolites is enormous and has challenged biochemists and physiologists to provide knowledge on many biological processes.

This challenge has led to the development of metabolomics utilizing technological advances in analytical chemistry; it is today a scientific field used in many sectors, including biomedicine, plant and agricultural sciences, food and beverage sciences, microbial and environmental sciences, forensic, and many more. The analytical challenge is to tackle mostly the chemical diversity of metabolites which are characterized by thousands of unique chemical structures with diverse physical and chemical properties, including molecular weight, molecular size and shape, polarity, stability, volatility, solubility, and so on. These chemical and physical features determine the requirements for extraction, separation, detection and quantification. Therefore, a range of complementary methodologies for extraction and analysis must be used to cover the range of metabolites present in any living cell, making metabolomics an exciting playground for analytical chemists and biochemists.

A range of analytical instrumentation has been successfully utilized to analyze metabolites in many different organisms, tissue types or biofluids. These include Mass Spectrometry (MS) and Nuclear Magnetic Resonance (NMR) for the detection and quantification of metabolites, the former often coupled to chromatographic separation techniques to separate the highly complex metabolite extracts obtained from any biological sample. One of the well-established "work horses" in metabolomics is gas chromatography (GC) coupled to mass spectrometry. It is characterized by its great separation power and reproducibility, which allowed for the development of comprehensive mass spectral libraries for compound identification which are easily adaptable for routine applications on any GC–MS system available (Hill and Roessner, 2013). GC–MS in general enables the analysis of around 400 compounds in a single run depending on tissue type, including amino, organic and fatty acids, sugars, amines and sterols. Important to note here is that GC–MS also plays an important role in metabolic flux analysis using stable isotope labelling techniques. This is because of its great ability to detect isotopic patterns in each metabolite detected therefore allowing quantification of stable isotope enrichments which can then be refereed back to actual carbon or nitrogen flux within a pathway of interest. One major drawback of GC–MS is, however, that only compounds which are either volatile or can be made volatile using chemical derivatization are amenable for analysis, which are mainly low molecular weight compounds, therefore complementary techniques such as liquid chromatography (LC) coupled to MS need to be used to cover the diversity of the metabolome.

LC-MS in metabolomics has attracted enormous attention in the past few years, mainly driven by great technological step-changes making LC-MS much for reproducible, sensitive and easy to use (Hill and Roessner, 2014). Today, common LC-MS methodologies allow analysis of thousands of metabolites simultaneously, therefore covering a large portion of the metabolome, making LC-MS an ideal tool for biomarker discovery and for generating novel hypotheses. An important aspect for LC-MS-based metabolomics is the choice of separation chemistry and suitable elution procedures addressing the chemical diversity of the metabolome. Unfortunately, no single methodology is able to separate all compound classes found in any biological system. The main used separation chemistries are based on reverse phase and hydrophilic interactions, when considering different ionization techniques available, e.g. electrospray ionization or atmospheric pressure ionization, which will again ionize different compounds, and positive and negative ionization to cover differentially charged ions produced in the ionization process, the number of potential methods increase enormously. However, if one aims to measure "all" metabolites within a sample, all these different options need to be explored and utilized. Each individual approach may detect between 500 and 2000 compounds (depending on tissue type and extraction procedure), represented as mass features with an accurate mass, isotopic pattern and chromatographic retention time. Accurate mass and isotopic pattern in addition to secondary and tertiary MS-based fragmentation (MS^2 or MS^3) can aid structural elucidation of compounds. A difficulty in this approach are identifications of metabolites which are similar in structure, mass and their fragmentation patterns, such as sugars. This makes unambiguous identification almost impossible and therefore requires good chromatographic separation and confirmation of identification with authentic standards. Commercial and open source mass spectral libraries such as MassBank (see Relevant Websites section), Human Metabolome Database (HMD; see Relevant Websites section), Metlin (see Relevant Websites section) or Chemspider (see Relevant Websites section) can aid identification, however often these libraries only record accurate mass of compounds but not their structure-specific fragmentation patterns or chromatographic retention times. Therefore unambiguous identification of compounds detected using LC-MS approaches remain limited leading to limited ability of biochemical and biological interpretation (Sumner et al., 2014). There have been many initiatives globally attempting to develop better solutions for metabolite identification such as Creek et al. (2014), Bingol et al. (2016), Lynn et al. (2015), Pahler and Brink (2013), Shen et al. (2014).

As described above, methodologies in metabolomics are as diverse as metabolites themselves, but taken together allow detection and quantification of thousands of compounds in any given sample. This leads to the almost greater challenge of extracting information from the analytical instrumentation, handling complex and multi-dimensional data sets and ultimately interpreting the vast amount of information in a biological context. This offers an exciting opportunity for collaboration between biologists, analytical chemists, bioinformaticians and statisticians to develop and apply sophisticated computational and statistical tools for metabolomics data analysis and interpretation. Below, we discuss in detail the various software and statistical tools and methods that are used for the analysis, visualization and interpretation of metabolomics data including metabolomics workflows, metadata, standards in metabolomics, data processing and workflow softwares, metabolomics experiments repositories, metabolomics databases and statistical analysis methods for analyzing metabolomics data. A collation of these software tools, repositories and databases are shown in **Table 1**.

Metabolomics Workflow and Lims Solutions

The typical workflow of a metabolomics experiment starts with establishing the experimental design followed by sample collection, sample preparation, running of samples in analytical instruments, data processing, statistical analysis and pathway analysis (**Fig. 1**).

Of these steps, the first step of establishing the experimental design could easily be termed as one of the most important steps in the experimental workflow. Experiments that are setup with careful consideration of how different variables may affect the experimental outcome would produce the best results. Some of the basic aspects that needs careful consideration are establishing the correct groups, sampling the population appropriately such that the samples best describe the underlying population, establishing the number of sample (biological replicates) and reducing variability of samples within groups. The adage used so frequently in Computer Science 'Garbage In, Garbage Out' is well-suited for experimental designs and the eventual results they would produce. In order to establish the most suited experimental design that would best tease out the answer to the scientific query, consulting with a biostatistician before the beginning of the experiment is imperative. RA Fisher, the father of modern statistics and experimental design rightly said, "To call in the statistician after the experiment is done may be no more than asking him to perform a post-mortem examination: he may be able to say what the experiment died of." ("RA Fisher" 1938). Therefore, statistical considerations of metabolomics experiments should not come into play after the data has been processed, but even before the experiment is designed.

Bioinformatics analyses of metabolomics data starts once the samples are run in analytical instruments and raw data generated. There are various algorithms and softwares that process the instrument generated raw data thereby converting them into a two-dimensional data matrix consisting of a measure of the different metabolites found in the samples. A detailed discussion of the data processing softwares of different platforms is presented below. After the raw data is processed and the data matrix obtained, various types of statistical analyses are performed in order to find metabolites that are significantly different between the compared groups. A detailed discussion on the different types of statistical tests is presented below. More often than not, statistical significance is determined using the p-value. In recent years, there have been substantial discussion in the scientific community about the dangers of relying only on measures such as the p-value to draw conclusions on the importance of a measured variable (Baker, 2016; Huber, 2016). While p-values give a measure of statistical significance of the measured metabolites, this may not necessarily translate to biological significance. Other measures such as the 95% confidence intervals needs to be taken into consideration before deciding on the biological significance of metabolites. In metabolomics experiments, hundreds if not thousands of metabolites are measured and while investigating the significance of these metabolites, care must be taken to correct for multiple tests. When multiple tests of significance are performed as part of a single experiment, the resulting statistical significance needs to be controlled for false positives (Curran-Everett, 2000).

There are several softwares and software systems that focus on different parts of the metabolomics experimental workflow. Software systems known as Laboratory Information Management Systems (LIMS) focus on capturing all meta data and any resulting data of an experiment. Generally, LIMS solutions capture information about the project, the several experiments under the project, information on the type of analytical platforms the experiments are run, details of the runs themselves such as the analytical methods and any SOPs used and detailed information about the samples that are run. This include details such as the biological nature of the samples and any metadata associated with the samples such as age, sex, treatment, quantity, weight etc. There are several LIMS solutions currently present for metabolomics.

MASTR-MS (Hunter et al., 2017) is a web-based LIMS solution that can be deployed either within a single laboratory or in a collaborative multi-site network. It comprises a Node Management System that can be used to link and manage projects across one or multiple collaborating laboratories; a User Management System which defines different user groups and privileges of users; a Quote Management System where client quotes are managed; a Project Management System in which metadata is stored and all aspects of project management, including experimental setup, sample tracking and instrument analysis, are defined, and a Data Management System that allows the automatic capture and storage of raw and processed data from the analytical instruments to the LIMS. SetupX (Scholz and Fiehn, 2007) is a web based metabolomics LIMS solution that is XML compatible and built around a relational database management core. It is particularly oriented towards the capture and display of GC–MS metabolomics data through its metabolic annotation database, BinBase (Skogerson et al., 2011). SetupX is able to handle a wide variety of BioSources (spatial, historical, environmental and genotypic descriptions of biological objects undergoing metabolomic investigations) and

Table 1   Metabolomics software tools, repositories and databases

LIMS solutions

Software tool	Platform	Reference
MASTR-MS	Web-based	Hunter *et al.* (2017)
SetupX	Stand-alone	Scholz and Fiehn (2007)
Sesame	Web-based	"Sesame LIMS" (2017)

Data preprocessing tools

Software tool	Instrument data type	Reference
XCMS	LC-MS, GC–MS	Smith *et al.* (2006)
MS-DIAL	LC-MS	Tsugawa *et al.* (2015)
mzMatch	LC-MS	Scheltema *et al.* (2011)
MetAlign	LC-MS	Lommen and Kools (2012)
AMDIS	GC–MS	Meyer *et al.* (2010)
MetaboliteDetector	GC–MS	Hiller *et al.* (2009)
MET-IDEA	GC–MS	Broeckling *et al.* (2006)
MeltDB	LC-MS, GC–MS	Kessler *et al.* (2013)
MSeasy	GC–MS	Nicolè *et al.* (2012)

Metabolite annotation tools

Software tool	Identification	Reference
CAMERA	Level 4	Kuhl *et al.* (2012)
MZedDB	Level 4	Draper *et al.* (2009)
SIRIUS	Level 4	Bocker *et al.* (2009)
MI-PACK	Level 3	Weber and Viant (2010)
PUTMEDID-LCMS	Level 3	Brown *et al.* (2011)
CFM-ID	Level 2a	Allen *et al.* (2014)
MAGMa	Level 2a	Ridder *et al.* (2013)
MetFrag	Level 2a	Ruttkies *et al.* (2016)
BATMAN	NMR	Hao *et al.* (2012)
Bayesil	NMR	Ravanbakhsh *et al.* (2015)

Data analysis softwares	*Software system*	*Instrument data type*	*Reference*
Workflow Management Systems (WMS)	Workflow4metabolomics	LC-MS, GC–MS NMR	Giacomoni *et al.* (2015)
	Galaxy-M	LC-MS, DIMS	Davidson *et al.* (2016)
	PhenoMeNal	LC-MS NMR, Isotope labelled data	http://phenomenal-h2020.eu/
Non-WMS tools	XCMS Online	LC-MS, GC–MS	Tautenhahn *et al.* (2012)
	MZmine2	LC-MS, GC–MS	Pluskal *et al.* (2010)
	FOCUS	NMR	Alonso *et al.* (2014)
	MatNMR	NMR	Van Beek (2007)

Metabolomics repositories

Repository	Reference
MetaboLights	Haug *et al.* (2013)
Metabolomics Workbench	Sud *et al.* (2016)
GNPS	Wang *et al.* (2016)
MeRY-B	Ferry-Dumazet *et al.* (2011)
MetaPhen	Carroll *et al.* (2010)
MetabolomeXchange	http://www.metabolomexchange.org/
OmicsDI	Perez-Riverol *et al.* (2016)

Metabolomics databases

Database type	Database	Reference
Mass Spectral Databases	NIST 17	"NIST Standard Reference Database 1A v17" (2017)
	Wiley Registry	"Wiley Registry of Mass Spectral Data, 11th Edition" (2017)
	Massbank	Horai *et al.* (2010)
	METLIN	Smith *et al.* (2005)

(Continued)

	Golm	Kopka *et al.* (2005)
	MMCD	Cui *et al.* (2008)
	SDBS	"SDBSWeb" (2017)
Species-Specific Databases	HMDB	Wishart *et al.* (2013)
	KNApSAcK	Nakamura *et al.* (2014)
	ReSpect	Sawada *et al.* (2012)
Pathway Databases	KEGG	Arakawa *et al.* (2005)
	MetaCyc	Caspi *et al.* (2016)
	BioCyc	Caspi *et al.* (2016)
	Reactome	Fabregat *et al.* (2016)
	PMN	Kruger and Ratcliffe (2012)

Treatments (experimental alterations that influence the metabolic states of BioSources). Sesame ("Sesame LIMS", 2017) is also a web-based, platform-independent LIMS. It was originally developed to facilitate NMR-based structural genomics studies. The Sesame module for metabolomics is called 'Lamp'. The Lamp module was originally designed to process NMR metabolomic analyses of Arabidopsis, although it is flexible enough to be easily adapted to other biological systems and other analytical methods.

Data Processing and Workflow Management Solutions

In metabolomics, a number of analytical techniques are typically used for analysis, with the most popular being liquid chromatography-mass spectrometry (LC-MS), gas chromatography-mass spectrometry (GC–MS) and nuclear magnetic resonance (NMR) spectroscopy. The data produced by these analytical techniques is distinct and requires different data handling methods. There is therefore the need for analysis tools and workflows designed for handling each data format.

The analysis of metabolomics data typically includes: pre processing, annotation, post-processing (pre-treatment) and statistical analysis as stages of data processing. However, the order of processing varies based on the analytical technique used. Untargeted LC-MS data analysis is generally performed in the above order, whilst with targeted LC-MS data only metabolites of interest are measured and there is no further annotation stage. Typically, with GC–MS pre-processing and annotation are performed concurrently due to the wide availability of reference libraries. NMR is an inherently more quantitative technique than MS and when metabolites are identified they are also typically relatively quantified.

Typically for mass spectrometry data, prior to preprocessing, data must be converted into an open format e.g. mzML, mzXML and netCDF, as the majority of tools do not accept proprietary formats. Most proprietary formats can be converted to mzML or mzXML using msconvert, a proteowizard (Chambers *et al.*, 2012) project tool. For NMR, there is also an open data exchange format known as JCAMP-DX version 6.0 11, with many different vendor-dependent variants. Recently, as part of the COSMOS (COordination of Standards in MetabOlomicS) EU FP7 consortium initiative (Salek *et al.*, 2015), the XML open standard nmrML, has been introduced (nrmML, 2017).

The most popular MS metabolomics software for preprocessing is XCMS (Weber *et al.*, 2016). XCMS (Smith *et al.*, 2006) is a well established software, having first been released over a decade ago. It is a bioconductor R package that can be used for the preprocessing of LC-MS or GC–MS data. Other popular software for LC-MS data preprocessing includes mzMatch (Scheltema *et al.*, 2011), Mass Spectrometry-Data Independent AnaLysis (MS-DIAL) (Tsugawa *et al.*, 2015) and MetAlign (Lommen and Kools, 2012). Like XCMS, mzMatch is also an R package. As well as preprocessing, it additionally provides isotopic labelling analysis (Chokkathukalam *et al.*, 2013) and probabilistic metabolite annotation (Daly *et al.*, 2014). Both MS-DIAL and MetAlign are exclusively available for windows operating systems. MS-DIAL can preprocess LC- and GC- MS and tandem MS (MS/MS) data, whereas MetAlign is only able to process MS1 LC-MS and GC–MS data. For GC–MS data, AMDIS (Meyer *et al.*, 2010) (Automated Mass Spectral Deconvolution and Identification System) is the most widely used processing tool. Pure compound peaks, free of overlapping signals, are extracted by deconvoluting spectra. A user defined target library can then be used for metabolite identification. However, it is worth noting that AMDIS does not include spectral alignment.

There are also many alternative software specifically for GC–MS preprocessing, including: MetaboliteDetector (Hiller *et al.*, 2009), MET-IDEA (Broeckling *et al.*, 2006), MeltDB (Kessler *et al.*, 2013), metaMS (Wehrens *et al.*, 2014) and MSeasy (Nicolè *et al.*, 2012). Baseline correction, smoothing, peak detection and deconvolution are provided by MetaboliteDetector. MET-IDEA quantifies AMDIS outputs, also generating a list of mass spectral tags. MeltDB is a suite of molecular tools, including peak picking, retention indices calculation and sum formula annotation. metaMS is based on XCMS, but performs pseudospectra analysis to avoid the alignment stage that can be hard to execute with GC–MS.

Compared to for MS data preprocessing, there are far less tools for NMR data. Examples include Dolphin (Gómez *et al.*, 2014) and rNMR (Lewis *et al.*, 2009). Both 1D and 2D NMR spectra can be analysed by Dolphin and rNMR. Dolphin automatically quantifies a set of target metabolites. rNMR uses a regions of interest (ROIs) based approach for analysis.

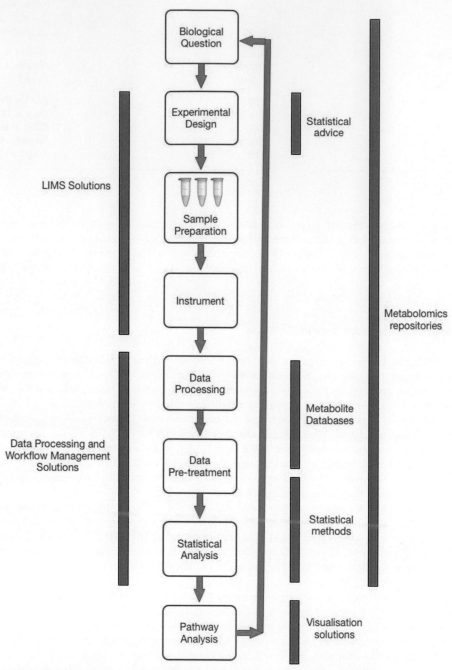

Fig. 1 Metabolomics experimental workflow.

Various software tools for the annotation of preprocessed MS data provide different levels of identification. Some annotate adducts, isotopes and fragments (e.g. CAMERA Kuhl *et al.*, 2012, MZedDB Draper *et al.*, 2009) or empirical formula (e.g. SIRIUS Bocker *et al.*, 2009). Others automatically search metabolite databases to provide tentative metabolite candidates (e.g. MI-PACK Weber and Viant, 2010, PUTMEDID-LCMS Brown *et al.*, 2011). MS/MS or MS^n data can be assigned a candidate metabolite at a higher level of confidence by searching against spectral libraries acquired using authenticated chemical standards. Examples of software that perform automatic database matching includes: CFM-ID (Allen *et al.*, 2014), CSI: FingerID (Dührkop *et al.*, 2015) and MCID (Li *et al.*, 2013a). MAGMa (Ridder *et al.*, 2013) and MetFrag (Ruttkies *et al.*, 2016) are both *in silico* fragmentation tools. Metabolite identification and quantification are often performed simultaneously for NMR data. BATMAN (Hao *et al.*, 2012; Ravanbakhsh *et al.*, 2015) deconvolutes and quantifies ^{1}H NMR spectra and Bayesil (Ravanbakhsh *et al.*, 2015) performs automatic spectral processing and identification of serum, plasma and cerebrospinal fluid (CSF) 1D ^{1}H NMR spectra. These tools provide different levels of identification (levels 1 to 4) as described by the Metabolomics Standards Initiative (Sumner *et al.*, 2007).

For a more in depth look of metabolomics data handling, data processing and data analysis see reviews (Spicer *et al.*, 2017; O'sullivan *et al.*, 2011; Hoffmann and Stoye, 2012; Beisken *et al.*, 2015).

Workflows in Metabolomics

Data analysis workflows or pipelines function to provide all stages of data analysis in a single space, including sequential data processing steps provided by multiple interconnected tools. This benefits the researcher by increasing computational reproducibility and removing problems caused by lack of interoperability between tools. In metabolomics, the majority of workflows have been designed to handle LC-MS data, as it is the most widely used instrumental technique.

Workflow management systems (WMS) are platforms that provide an infrastructure where workflows can be set-up, executed and monitored. Examples of WMS that have been utilized for metabolomics data analysis include Galaxy (Afgan *et al.*, 2016), Taverna (Wolstencroft *et al.*, 2013) and Konstanz Information Miner (KNIME) (Berthold *et al.*, 2009).

Galaxy was initially designed for the analysis of genomics data, but it's use has since widely expanded across many bioinformatics disciplines. The Galaxy project has three stated project goals: accessibility, reproducibility and transparency. It has been designed for ease-of-use, so that researchers without programming experience are able to run workflows without difficulty. Galaxy enables the exact replication of analysis by capturing information on all modules and parameters used. Users are able to share Galaxy workflows, histories, datasets and pages (interactive, web-based documents describing analysis), either publicly or with specific individuals, facilitating transparent research. Examples of tools that utilise Galaxy for metabolomics data analysis include Workflow4metabolomics (Giacomoni *et al.*, 2015), Galaxy-M (Davidson *et al.*, 2016) and PhenoMeNal (see Relevant Websites section).

Workflow4metabolomics 3.0 is a web-based tool that includes workflows for the analysis of LC-MS, GC–MS and NMR data. Both MS workflows consist of three analysis stages: preprocessing, statistics and annotation. In the LC-MS workflow, data are preprocessing using XCMS (Smith *et al.*, 2006) (XCMS can additionally be used for preprocessing GC–MS data). Adducts and fragments are then annotated using CAMERA (Kuhl *et al.*, 2012). Alternatively users can upload a peak list of data that has already undergone preprocessing for further analysis. Modules for filtering and batch correction can then optionally be performed. The available statistical analysis modules include both univariate (Student's *t*-test, Wilcoxon *t*-test, ANOVA, Kruskal-Wallis test, etc.) and multivariate analysis (PCA, PLS, OPLS and PLS-DA). These modules are applicable to all data types. Metabolite databases can be searched for the annotation of LC-MS data, including the Human Metabolome Database (HMDB) (Wishart *et al.*, 2013), LipidMaps Structure database (LMSD) (Sud *et al.*, 2007) and MassBank (Horai *et al.*, 2010). For GC–MS data the Golm database (Kopka *et al.*, 2005) is available for annotation, or users can upload a local database. Bruker bucketing and integration, and normalization are modules supplied specifically for NMR data.

Galaxy-M is designed for the analysis of LC-MS and DIMS metabolomics data. It provides Galaxy installations of MI-PACK (Weber and Viant, 2010), XCMS and CAMERA. Preprocessing is performed by XCMS for raw LC-MS data and SIM-stitch for raw DIMS data. Following peak picking and alignment, data can be filtered by appearance in blank samples or by removing peaks that are not present in x-out-of-n samples. In preparation for statistical analysis missing values can be imputed, and data can be normalized and transformed. For statistical analysis PCA and univariate analysis are provided. If users have the commercial MATLAB toolbox, PLS-Toolbox available, this can be installed for chemometric multivariate analysis. Tools from MI-PACK are implemented for metabolite annotation.

The PhenoMeNal App library consists of ~50 tools or apps, not including individual algorithms as some tools such as OpenMS includes multiple submodules. These tools cover analysis of a wide variety of experimental data types including LC-MS, NMR and isotopically labelled data. As well as users being able to design their own workflows with these tools, there are pre-existing workflows also available. All the tools are available via Galaxy workflows and Jupyter libraries through the Cloud Research Environment (see Relevan Websites section).

Like Galaxy, KNIME is available as a graphical user interface (GUI) with a modular environment, constructed for usability. This allows for new tools or algorithms to be easily integrated into to workflow. For metabolomics, the workflows OpenMS (Aiche *et al.*, 2015) and MassCascade (Beisken *et al.*, 2014) are available as KNIME implementations. MassCascade is specifically designed for the analysis of LC-MS^n data. OpenMS was originally designed for proteomics data, but has been expanded to include metabolomics data analysis (Kenar *et al.*, 2014). Similarly to MassCascade, OpenMS focuses on high resolution LC-MS data. It provides an infrastructure for the rapid development of tools. There are over 185 different tools and ready-made workflows for common MS data processing tasks (Röst *et al.*, 2016; Sturm *et al.*, 2008).

The Taverna Workflow Management System provides a suite of tools enabling users to design, edit, execute and share workflows. Publicly available workflows can be found at http://www.myexperiment.org/. As of 5th October 2017, searching for 'metabolomics' returns 19 workflows.

There are also some widely used tools for metabolomics data analysis that do not use WMS. These include XCMS Online (Tautenhahn *et al.*, 2012), MetaboAnalyst (Xia *et al.*, 2015) and MZmine 2 (Pluskal *et al.*, 2010). XCMS Online is an online implementation of XCMS, that includes the entire data processing pipeline for MS data, from preprocessing to metabolite identification, statistical and pathway analysis. Users can also share their data and workflows either with collaborators or publicly. MetaboAnalyst 3.0 is a web server, that contains eight independent modules for analysis, including statistical, pathway, biomarker and enrichment analysis. The majority of modules require data to have been already preprocessed using other software. MZmine 2 has a GUI and provides a toolkit of modules for MS data analysis. It is able to profile, centroid and MS^n data. Modules include raw data processing, peak detection and visualization.

The majority of workflows for NMR processing are written in MATLAB, including FOCUS (Alonso *et al.*, 2014) and MatNMR (Van Beek, 2007). FOCUS is specifically designed for ^{1}H NMR data, whilst MatNMR is more general and can process both 1D and 2D NMR data.

Metabolomics Experiments Repositories and Metabolite Databases

Metabolomics data repositories can store both raw and annotated data. Raw data can be deposited as either property data formats (e.g. Thermo . RAW, Agilent .d) or open formats, such as .mzML, .mzXML, .mzData or .netCDF. Studies in repositories can also contain lists of identified metabolites, along with their relative abundance, concentration, what spices and type of samples they are found Salek *et al.* (2017), Haug *et al.* (2017).

Currently there are two global repositories for metabolomics data: the European Bioinformatics Institute's MetaboLights (Haug *et al.*, 2013) and and United States Government National Institute of Health (NIH) Metabolomics Workbench (Sud *et al.*, 2016). These repositories accept submissions of all types of metabolomics data. Both repositories include multiple layers, storing data and metadata from metabolomics studies along with reference libraries of metabolites.

The two layers of the MetaboLights repository are: (i) the repository, where raw data, along with annotated metadata, are stored, and (ii) the reference layer that contains a library of reference compounds (Kale *et al.*, 2016). The repository and compound library can be queried by species, instrumental analysis technology, organism part (e.g. serum, leaf) or study validation status. As of 31st August 2017 there are 279 publicly available studies on MetaboLights and 24957 compounds, with 3171 including MS spectra and 407 NMR spectra.

There is a focus of the curation of data in MetaboLights (Salek *et al.*, 2013), and it therefore has more stringent submission guidelines that Metabolomics Workbench. Experimental metadata must be submitted to MetaboLights in the ISA-TAB format (Rocca-Serra *et al.*, 2010). In order to submit a study, users must first register. There are four stages of curation a submitted study must pass through until it becomes publicly available: submitted, in curation, in review and public. During the submitted stage the submission is not yet complete and mandatory metadata is missing. When all mandatory fields are complete curators will appraise the study, looking for any missing optional fields or errors not detected by automatic validation (Kale *et al.*, 2016). When curation is finished, the submitter is given a read-only reviewer link that can be shared with a journal or collaborators. The study will not become public until a submitter specified publication date has been reached. The analysis tools section of MetaboLights is currently under development. A new tool that plots study factor distribution using parallel coordinates is in the beta development stage.

Compared to MetaboLights, Metabolomics Workbench provides greater analysis capability. Data in publicly available studies can be normalized, clustered, and univariate and multivariate statistical analysis (PCA, linear discriminate analysis and PLS-DA) can be performed. For MS data additional analyses are provided: classification and feature analysis, pathway mapping, MS search for metabolite identification and comparison of metabolites between studies (meta-analysis). Users also have the option of performing statistical analysis on their own uploaded dataset.

There are 440 publicly available studies in Metabolomics Workbench as of October 2017. Bubble plots classifying studies by disease, sample source, species, pathway or metabolite class allow users to select a specific subset of studies. Metabolomics Workbench also provides a REST service where HTTP requests can be used to access data and metadata, including metabolite structures and experimental results. The Metabolomics Workbench Metabolite Database stores metabolite structures and annotations. It contains >61,000 entries collected from public databases including LIPID MAPS (Sud *et al.*, 2007), ChEBI (Hastings *et al.*, 2013), HMDB (Wishart *et al.*, 2013), BMRB (Ulrich *et al.*, 2008), PubChem (Kim *et al.*, 2016) and KEGG (Kanehisa *et al.*, 2012). Users can employ either text based search or *m/z* value from untargeted MS search to search for metabolites.

To submit a study to Metabolomics Workbench, users must register and obtain authorization. Once this has been completed (i) the study is registered, (ii) users must submit metadata in the specified format, either via an online form or a supplied excel template and (iii) raw data and/or supplementary material must be uploaded. It is encouraged that common names from the RefMet (A Reference list of Metabolite names) database are used for annotation.

The Global Natural Products Social Molecular Networking (GNPS) (Wang *et al.*, 2016) platform provides analysis and storage of natural products mass spectrometry data. Data are stored in the Mass Spectrometry Interactive Virtual Environment (MassIVE) repository, originally developed for proteomics studies (Perez-Riverol *et al.*, 2015). Compared to other repositories, GNPS is unique in providing continuous metabolite identification for MS/MS spectra, with data sets being reanalysed for new identifications once a month. Unlike other repositories GNPS has no requirements for reporting experimental metadata.

As well as the global metabolomics repositories, there are also a number of smaller, more specific repositories. These include Metabolomics Repository Bordeaux (MeRy-B) (Ferry-Dumazet *et al.*, 2011) and Metabolic Phenotype Database (MetaPhen) (Carroll *et al.*, 2010, 2015). MeRy-B is a repository specifically designed for plant ^{1}H NMR data, but also contains GC–MS data.

MetaPhen is a data repository that is available as part of MetabolomeExpress, a GC–MS metabolomics data analysis platform. It contains a variety of tools for the analysis of the publicly available data, including ResponseFinder, MetaAnalyser, PhenoMeter, and MSRI Library manager. Users can directly search for metabolites of interest using ResponseFinder, which includes search by species and organ/tissue/fluid type. MetaAnalyser performs meta-analysis by comparing the results of multiple experiments, plotting heatmaps of aligned and clustered data. PhenoMeter allows users to search the reference library via metabolite response patterns.

The MetabolomeXchange (see Relevant Websites section) consortium is based on the successful ProteomeXchange (Vizcaíno *et al.*, 2014). It serves as a data aggregator, collecting studies from the MetaboLights, Metabolomics Workbench and MeRy-B repositories. Users have the option of subscribing in order to be notified about new datasets or updates to existing datasets.

OmicsDI is an open-source platform based at EMBL-EBI, that captures and integrates proteomics, genomics, metabolomics and transcriptomics datasets (Perez-Riverol *et al.*, 2016). For metabolomics the data repositories: MetaboLights, GNPS, MetaPhen and Metabolomics Workbench, are included. Since different repositories use different data models, ontologies, metadata representation and identifiers, OmicsDI uses metadata normalization and annotation expansion to harmonise searching across the various resources. Users are therefore able to discover multi-omics datasets that are linked to publications, as well as similar experimental datasets, suggested by metadata matching.

Most omics, including metabolomics, aim to follow FAIR principles by making data Findable, Accessible, Interoperable and Reusable (Van Rijswijk *et al.*, 2017; Wilkinson *et al.*, 2016). The data handling, analysis tools, data standards and resources discussed here should help users to adheres to FAIR principles.

Metabolomics Databases

The substantial growth of the field of Metabolomics during the past decade has instilled a need for sharable resources. Amongst other resources such as publicly available data processing softwares and workflow pipelines, data repositories, libraries and databases that store experiment and metabolite related information have been on the rise.

As detailed above, public repositories such as MetaboLights (Salek *et al.*, 2013) and the Metabolomics Workbench (Sud *et al.*, 2016) attempt to capture the data output of the entire workflow of a metabolomics experiment. These repositories store information on the experimental design, details on projects, experiments and samples as well as all the instrument generated raw data and the processed data of experiments. Commercial and public libraries and databases store exhaustive details about metabolites such as their physicochemical properties, related structures and links to other databases. Metabolite databases can be divided into three categories, a) metabolite mass spectral databases, b) Species-specific metabolite databases and c) metabolic pathway databases.

Metabolite mass spectral databases

Once the raw data is acquired from the instruments, data processing softwares are used to convert the raw data into a data matrix of molecular features with measures of their abundance in the samples. One of the last step in processing of raw data is to identify these molecular features as specific metabolites and this identification is performed using metabolite spectral libraries or databases. Metabolite spectral libraries store information on the physicochemical properties of metabolites such as molecular name, molecular weight, chemical formula, structure information, retention time, high-quality spectral data and links to external databases. Currently, there are several mass spectral libraries available for metabolomics.

NIST 17 ("NIST Standard Reference Database 1A v17", 2017) is the latest mass spectral database release of the National Institute of Standards and Technology, USA. NIST 17 contains an Electron Ionization (EI) mass spectral library of over 306,000 spectral belonging to over 267,000 unique compounds, a Gas Chromatography (GC) data library with over 404,000 retention index values for over 99,000 compounds covering both polar and nonpolar columns and an MS/MS library of over 650,000 spectra of over 14,000 compounds. It also includes the NIST MS search software for searching and identifying compounds and the automated mass spectral deconvolution and identification system (AMDIS) software used for deconvoluting chromatograms. The Wiley Registry of Mass Spectral Data is one of the largest collection of EI mass spectra. The 11th edition of Wiley ("Wiley Registry of Mass Spectral Data, 11th Edition", 2017) stores about 775,500 mass spectra for over 599,000 unique compounds and over 740,000 searchable structures. Massbank (Horai *et al.*, 2010) is a public repository that stores mass spectral data of high-resolution MS^n spectra. It is a distributed database in the sense where participating research groups are able to submit mass spectral information from local MassBank data servers. Users are then able to access spectral information from multiple groups through Massbank's web interface. Currently Massbank stores over 41,000 spectra. The METLIN Metabolite Database (Smith *et al.*, 2005) is a repository of MS and MS/MS data that stores details of more than 961,000 molecules. In addition, it also stores over 200,000 high-resolution in silico MS/MS spectra. The Golm Metabolome Database (Kopka *et al.*, 2005) stores information of EI mass spectra and associated retention indices of peaks quantified by GC–MS. The Madison-Qingdao Metabolomics Consortium Database (MMCD) (Cui *et al.*, 2008) is a library that stores details of metabolites analysed through NMR and MS. MMCD currently contains details of over 20,000 compounds. The Spectral Database for Organic Compounds (SDBS) ("SDBSWeb", 2017) stores mass spectrum information for around 34,000 organic compounds. The different types of spectra it stores are electron impact Mass spectrum (EI-MS), Fourier transform infrared spectrum (FT-IR), 1H nuclear magnetic resonance (NMR) spectrum, 13C NMR spectrum, laser Raman spectrum, and electron spin resonance (ESR) spectrum.

Species-specific metabolite databases

The Human Metabolome Database (HMDB) (Wishart *et al.*, 2013) is a public repository containing detailed information about metabolites found in the human body. HMDB currently stores more than 330,000 spectra, stores details of more than 114,000 metabolites and has information derived from 17 different biofluids. HMDB is designed to contain or link externally to chemical data, clinical data and biochemistry data. The KNApSAcK family databases (Nakamura *et al.*, 2014) are an integrated metabolite-plant species databases for multifaceted plant research. It currently has over 111,000 species-metabolite relationships encompassing 22,350 species and over 50,000 metabolites. The RIKEN MSn spectral database for phytochemicals (ReSpect) (Sawada *et al.*, 2012) stores plant-specific MS/MS spectral data. ReSpect currently stores around 9000 spectral records.

Metabolic pathway databases

Metabolic pathway databases store information on biochemical pathways and the various enzyme reactions along the pathways. In addition, they store data on genes, gene products and compounds that are part of these reactions along with links to external databases on further information on the participants.

The KEGG Pathway Database (Arakawa *et al.*, 2005) is a collection of manually drawn pathway maps detailing metabolic reactions and the resulting molecular interactions. KEGG is a collection of fifteen databases that are divided into four categories: Systems Information, Genomic Information, Chemical Information and Health Information. The KEGG PATHWAY database that stores metabolic maps currently contains 519 pathway maps totalling to more than 530,000 different pathways and are detailed under the Systems Information category.

The MetaCyc database (Caspi *et al.*, 2016) stores information on metabolic pathways and enzymes that are experimentally determined and are curated from scientific literature. The database stores pathways for a wide range of organisms and currently contains 2572 pathways from 2844 different organisms. MetaCyc is widely used as the online encyclopaedia of metabolism, used to predict metabolic pathways in sequenced genomes and also used in metabolic engineering.

BioCyc (Caspi *et al.*, 2016) is a collection of pathway and genome databases for organisms whose genomes have been sequenced. Each of these databases contain the full genome, predicted metabolic network and additional information on enzymes, reactions and predicted operons. The BioCyc database also includes the Pathway Tools Software, a symbolic systems biology software system. The tools allow the creation of new pathway and genome database, has options to query visualize and analyze the databases, supports the development of metabolic flux models and allows the interactive editing of the underlying genome databases.

The Reactome pathway knowledgebase (Fabregat *et al.*, 2016) is a manually curated database of pathways. Reactome's pathways are peer reviewed where experts and the Reactome editorial staff work together in finalising the contents and are cross-referenced to other bioinformatics resources such as NCBI Gene, Ensembl, UniProt, KEGG Compound, PubMed and Gene Ontology. Reactome also provides pathway analysis tools that performs ID mapping, pathway assignment and enrichment analysis.

The Plant Metabolic Network (Kruger and Ratcliffe, 2012) is a collection of one multi-species reference database called PlantCyc and 76 species/taxon-specific databases. PlantCyc currently contains more than 1000 pathways and compounds from over 350 plant species. The pathways have been manually curated from literature and also includes hypothetical pathways and computationally predicted pathways that are published in peer-reviewed journals.

Statistical Analysis and Functional Interpretation

The widespread adoption of metabolomics in basic research and large-scale clinical investigations have resulted in the generation of increasingly complex data sets. Making sense of such high-dimensional data requires a good understanding of modern statistical and functional analysis approaches. In the previous sections, we have described experimental design, raw data processing, databases and data repositories. In this section, we will introduce the key concepts and main approaches commonly used in statistical analysis and functional interpretation of metabolomics data.

Data Input

The standard input for statistical data analysis is a high-dimensional data matrix with variables in rows and samples in columns (or its transposed format). Such data can be produced from most metabolomics data processing tools (Smith *et al.*, 2006; Pluskal *et al.*, 2010; Hao *et al.*, 2012; Ravanbakhsh *et al.*, 2015; Tautenhahn *et al.*, 2012). For data from targeted metabolomics, variables are compound IDs; and for untargeted metabolomics data, variables are typically peak locations or spectral bins. Data dimension refers to the number of variables in the matrix, which can vary from 100s to 1000s, with untargeted metabolomics data typically containing more variables. Samples are biological replicates associated with class labels or other metadata reflecting experimental designs. Quality control (QC) samples are often included for quality assurance during analysis.

Data Cleaning

The first, but often neglected step in metabolomics data analysis, is data cleaning or data filtering. Data from comprehensive metabolomics experiments usually contains a large amount of noise that can negatively impact downstream data analysis. Proper data cleaning will not only improve data quality, but also enhance statistical power and reduce false positives (Bourgon *et al.*, 2010). To identify low-quality data for cleaning, several empirical rules are frequently used, including a) hard-to-measure variables such as those close to baseline noises or showing large variations in QC samples or technical replicates; b) uninformative variables that are relatively constant throughout the experimental conditions based on their inter-quantile ranges, standard deviations, etc.; and c) outlier samples with extreme deviations from the main clusters. As outliers could be generated by unexpected but biologically important processes, their exclusion should be better justified with technical or experimental reasons, in addition to visual inspections.

Data Normalization

This step aims to reduce overall undesirable variations in the data to facilitate biologically and statistically meaningful comparisons. As different analytical platforms are currently used in metabolomics, a wide variety of normalization methods have been developed over the past decade (Van Den Berg *et al.*, 2006). Some of them are generic statistical methods such as log transformation and auto-scaling; while others are biologically motivated approaches such as the probabilistic quotient normalization that was initially developed to address the dilution effects in NMR-based metabolomics on urine samples (Dieterle *et al.*, 2006). The MetaboAnalyst web server currently supports a comprehensive selection of 13 different normalization methods (Xia and Wishart, 2016; Xia *et al.*, 2015). A new web-based tool, NOREVA allows users to evaluate the effects of 24 normalization methods for MS-based metabolomics data (Li *et al.*, 2017). No consensus has been reached on the best normalization methods for metabolomics data. In general, we should start with simple methods with minimal transformations of the original data. More complicated approaches may be necessary for complex data from large-scale studies.

Unsupervised Methods

After cleaning and normalization, the data is now suitable for statistical analysis. The first step is to examine if there are any interesting patterns or trends in the data, and whether these patterns are likely to be technical artifacts or biological effects. Unsupervised methods are designed for such tasks. As the name suggests, these type of methods use only the data (X) itself without considering the class labels (Y) during the analysis. Here we will introduce two main approaches - principal component analysis (PCA) and cluster analysis. PCA is the most widely used method in metabolomics for data overview and pattern discovery. PCA is a dimensionality reduction method which aims to capture the most variation of the data in the top few principal components (PCs) created by linear combinations of the original variables. These PCs can be visualized as a 2D or 3D scatter plot (also known as scores plot) to see if there are major grouping patterns among samples. The corresponding loading plots can be used to identify the variables that contribute to such separation patterns. In addition to PCA, cluster analysis (or clustering) is also widely used to help reveal inherent patterns in data. These type of methods work by first computing similarities or distances among samples or variables, and then assigning them to different clusters so that items within the same cluster will be more similar to each other than they are to those outside the cluster. The result of the cluster analysis can often be visualized in heatmaps, a technique that uses color gradients to represent data values. In clustered heatmaps, main patterns or clusters will stand out as distinct "patches" displaying more homogenous colors as compared to the surrounding background. Several clustering methods are often used in metabolomics data analysis, including hierarchical clustering, k-means clustering, self-organizing maps (SOM), etc. More detailed introductions on these clustering algorithms can be find in this excellent review paper (Ren *et al.*, 2015).

Supervised Approaches

These methods are used to formally evaluate the associations between the variations in data (X) and class labels (Y). Metabolomics has a long history of using various multivariate statistical methods and machine learning algorithms for classification and regression analysis (Broadhurst and Kell, 2006). Here we will introduce a widely used multivariate statistical method - partial least squares discriminant analysis (PLS-DA), and then briefly comment on a powerful machine learning methods - support vector machine (SVM). Similar to PCA, PLS-DA is a dimensionality reduction method by creating latent variables (LV) through linear combinations of original variables in the data (X). These LVs are computed by referring to the class labels (Y) to maximize the explained variance in Y in the top few LVs. The PLS-DA results can also be visualized in scores and loadings plots for interpretation. Due to the supervised nature of the method, the performance (as seen from the separation patterns in the scores plot) is usually much better compared to PCA. However, this may be misleading. It has been shown that PLS-DA can produce good separation patterns even with randomly generated group labels (Westerhuis *et al.*, 2008). Therefore, PLS-DA is very susceptible to "overfitting" – a well-known issue associated with using supervised methods. To address this issue, results from PLS-DA needs to be evaluated using both cross-validations (or double cross-validation if sample size is sufficient) and permutation tests. In cross-validation, samples are divided into trainings and testing groups, and the training group is then used to build PLS-DA models whose performances are evaluated using the test group. The permutation tests evaluate whether similar performances can be obtained using models created based on randomly labelled data, compared to the model created using the original group labels. SVM is a powerful machine learning methods commonly employed in metabolomics data analysis (Mahadevan *et al.*, 2008; Heinemann *et al.*, 2014). The algorithm uses kernel functions to map the original data to a higher dimensional feature space and separating it by means of a maximum margin hyperplane (Noble, 2006). Compared to PLS-DA, SVM generally gives better performance and is less susceptible to over-fitting issues. However, the interpretation of the underlying models is very complicated and much less intuitive (i.e. black box).

Functional Analysis

Direct functional analysis is generally limited to data from targeted metabolomics. For untargeted metabolomics, researchers need to first identify significant peaks or spectral bins, and then annotate the underlying metabolites before performing functional analysis. A key concept in functional analysis is that the unit of analysis has been shifted from individual metabolites to a set of metabolites (termed a metabolite set). Metabolite sets are groups of metabolites that share a particular property, for instance, a group of metabolites involved in the same pathways, altered in the same diseases, or associated with the same genetic or epigenetic variations. If particular metabolite sets are significantly enriched in the data, the corresponding shared properties (i.e. pathways) are considered to be relevant to the experimental conditions of interest. Many different algorithms have been developed to perform enrichment tests. For instance, the over-representation analysis (ORA) methods accept a list of significant metabolites and evaluate whether certain metabolite sets appear more often than expected by chance, using Fisher's Exact tests or hyper-geometrics tests. For instance, the Metabolites Biological Role (MBROLE) web application uses hypergeometric tests to perform comprehensive functional enrichment analysis (Lopez-Ibanez et al., 2016). The quantitative enrichment analysis (QEA, also known as functional class scoring) is a relatively new approach for enrichment tests. Compared to ORA, QEA utilizes the complete data to calculate enrichment scores for individual metabolite sets, and is more sensitive for detecting subtle but consistent changes. A wide range of QEA tools are available, showing many statistical and practice differences (Tarca et al., 2013). For instance, the QEA module in MSEA (now part of MetaboAnalyst) is based on the *globaltest* R package (Goeman et al., 2004; Xia and Wishart, 2010). Detailed discussions on the performance characteristics and statistical properties for these different approaches can be found in a recent excellent review paper (De Leeuw et al., 2016). Due to the inherent analytical bias in metabolomics platforms and the limited metabolome coverage, results from current enrichment analysis should be interpreted with caution.

From Peaks to Biological Activities

The increased application of sensitive and high-resolution MS-based metabolomics makes it possible to perform functional analysis directly from raw spectral peaks. The key idea is to redefine the metabolite sets using the spectral features of the corresponding metabolites, and then evaluate the enrichment of these chemical signals directly from spectral peaks without formal compound identification. The *mummichog* method is the first bioinformatics approach in this direction (Li et al., 2013b). The algorithm accepts two lists of spectral peaks - a significant peak list obtained through statistical analysis (i.e. t-tests) and a reference peak list (i.e. all peaks detected in the experiment). It computes matching scores for different pathways from the significant peak lists, and enrichment p-values are calculated by comparing the scores to those obtained using peak lists of the same size, but randomly drawn from the reference peaks. Further development in this direction holds the potential to greatly facilitate metabolomics functional analysis.

Due to space limitations, this section covers only those most common methods used in statistical and functional analysis of metabolomics data, while ignoring very basic methods such as univariate statistics (i.e. t-tests and ANOVA) for the detection of significant variables. For the same reason, we have also intentionally omitted some advanced topics including biomarker analysis and time-series data analysis. Readers are referred to several comprehensive reviews and tutorials for more detailed discussions on these topics (Broadhurst and Kell, 2006; Smilde et al., 2010; Xia et al., 2013; Worley and Powers, 2013; Ren et al., 2015). Most of these concepts and procedures have been implemented and are available thorough easy to use and freely accessible XCMS online and MetaboAnalyst web applications.

Acknowledgements

SD and UR thank for support from Metabolomics Australia Pty Ltd which is funded through the National Collaborative Research Infrastructure Strategy (NCRIS), 5.1 Biomolecular Platforms and Informatics with co-investments from the University of Melbourne. U.R is funded through an Australian Research Council Future Fellowship program. JX would like to acknowledge the financial support from the Natural Sciences and Engineering Research Council of Canada (NSERC) and the Canada Research Chairs (CRC) program.

See also: Experimental Platforms for Extracting Biological Data: Mass Spectrometry, Microarray, Next Generation Sequencing. Investigating Metabolic Pathways and Networks. Mass Spectrometry-Based Metabolomic Analysis. Metabolic Models. Metabolic Profiling. Metabolome Analysis. Natural Language Processing Approaches in Bioinformatics. Standards and Models for Biological Data: FGED and HUPO. Two Decades of Biological Pathway Databases: Results and Challenges

References

Afgan, E., Baker, D., Van Den Beek, M., et al., 2016. The Galaxy platform for accessible, reproducible and collaborative biomedical analyses: 2016 update. Nucleic Acids Res. 44, W3–W10.

Aiche, S., Sachsenberg, T., Kenar, E., et al., 2015. Workflows for automated downstream data analysis and visualization in large-scale computational mass spectrometry. Proteomics 15, 1443–1447.

Allen, F., Pon, A., Wilson, M., Greiner, R., Wishart, D., 2014. CFM-ID: A web server for annotation, spectrum prediction and metabolite identification from tandem mass spectra. Nucleic Acids Res. 42, 94–99.

Alonso, A., Rodríguez, M.A., Vinaixa, M., et al., 2014. Focus: A robust workflow for one-dimensional NMR spectral analysis. Anal. Chem. 86, 1160–1169.

Arakawa, K., Kono, N., Yamada, Y., Mori, H., Tomita, M., 2005. KEGG-based pathway visualization tool for complex omics data. In Silico Biol. 5, 419–423.

Baker, M., 2016. Statisticians issue warning over misuse of P values. Nature 531, 151.

Beisken, S., Earll, M., Portwood, D., Seymour, M., Steinbeck, C., 2014. MassCascade: Visual programming for LC-MS data processing in metabolomics. Mol. Inform. 33, 307–310.

Beisken, S., Eiden, M., Salek, R.M., 2015. Getting the right answers: Understanding metabolomics challenges. Expert Rev. Mol. Diagn. 15, 97–109.

Berthold, M.R., Cebron, N., Dill, F., et al., 2009. KNIME – The Konstanz Information Miner: Version 2.0 and Beyond. SIGKDD Explor. Newsl. 11, 26–31.

Bingol, K., Bruschweiler-Li, L., Li, D., et al., 2016. Emerging new strategies for successful metabolite identification in metabolomics. Bioanalysis 8, 557–573.

Bocker, S., Letzel, M.C., Liptak, Z., Pervukhin, A., 2009. SIRIUS: Decomposing isotope patterns for metabolite identification. Bioinformatics 25, 218–224.

Bourgon, R., Gentleman, R., Huber, W., 2010. Independent filtering increases detection power for high-throughput experiments. Proc. Natl Acad. Sci. USA 107, 9546–9551.

Broadhurst, D.I., Kell, D.B., 2006. Statistical strategies for avoiding false discoveries in metabolomics and related experiments. Metabolomics 2, 171–196.

Broeckling, C.D., Reddy, I.R., Duran, A.L., et al., 2006. MET-IDEA: Data Extraction Tool for Mass Spectrometry-Based Metabolomics. Anal. Chem. 78, 4334–4341.

Brown, M., Wedge, D.C., Goodacre, R., et al., 2011. Automated workflows for accurate mass-based putative metabolite identification in LC/MS-derived metabolomic datasets. Bioinformatics 27, 1108–1112.

Carroll, A.J., Badger, M.R., Harvey Millar, A., 2010. The MetabolomeExpress Project: Enabling web-based processing, analysis and transparent dissemination of GC/MS metabolomics datasets. BMC Bioinformatics 11, 376.

Carroll, A.J., Zhang, P., Whitehead, L., et al., 2015. PhenoMeter: A metabolome database search tool using statistical similarity matching of metabolic phenotypes for high-confidence detection of functional links. Frontiers Bioeng. Biotechnol. 3, 106.

Caspi, R., Billington, R., Ferrer, L., et al., 2016. The MetaCyc database of metabolic pathways and enzymes and the BioCyc collection of pathway/genome databases. Nucleic Acids Res 44, D471–D480.

Chambers, M.C., Maclean, B., Burke, R., et al., 2012. A cross-platform toolkit for mass spectrometry and proteomics. Nat. Biotechnol. 30, 918–920.

Chokkathukalam, A., Jankevics, A., Creek, D.J., et al., 2013. mzMatch–ISO: An R tool for the annotation and relative quantification of isotope-labelled mass spectrometry data. Bioinformatics 29, 281–283.

Creek, D.J., Dunn, W.B., Fiehn, O., et al., 2014. Metabolite identification: Are you sure? And how do your peers gauge your confidence? Metabolomics 10, 350–353.

Cui, Q., Lewis, I.A., Hegeman, A.D., et al., 2008. Metabolite identification via the Madison Metabolomics Consortium Database. Nat. Biotechnol. 26, 162–164.

Curran-Everett, D., 2000. Multiple comparisons: Philosophies and illustrations. Am. J. Physiol. Regul. Integr. Comp. Physiol. 279, R1–R8.

Daly, R., Rogers, S., Wandy, J., et al., 2014. MetAssign: Probabilistic annotation of metabolites from LC–MS data using a Bayesian clustering approach. Bioinformatics 30, 2764–2771.

Davidson, R.L., Weber, R.J.M., Liu, H., Sharma-Oates, A., Viant, M.R., 2016. Galaxy-M: A Galaxy workflow for processing and analyzing direct infusion and liquid chromatography mass spectrometry-based metabolomics data. Gigascience 5, 10.

De Leeuw, C.A., Neale, B.M., Heskes, T., Posthuma, D., 2016. The statistical properties of gene-set analysis. Nat. Rev. Genet. 17, 353–364.

Dieterle, F., Ross, A., Schlotterbeck, G., Senn, H., 2006. Probabilistic quotient normalization as robust method to account for dilution of complex biological mixtures. Application in ^{1}H NMR metabonomics. Anal. Chem. 78, 4281–4290.

Draper, J., Enot, D.P., Parker, D., et al., 2009. Metabolite signal identification in accurate mass metabolomics data with MZedDB, an interactive m/z annotation tool utilising predicted ionisation behaviour 'rules'. BMC Bioinformatics 10, 227.

Dührkop, K., Shen, H., Meusel, M., Rousu, J., Böcker, S., 2015. Searching molecular structure databases with tandem mass spectra using CSI: Fingerid. Proc. Natl. Acad. Sci. USA 112, 12580–12585.

Fabregat, A., Sidiropoulos, K., Garapati, P., et al., 2016. The Reactome pathway Knowledgebase. Nucleic Acids Res. 44, D481–D487.

Ferry-Dumazet, H., Gil, L., Deborde, C., et al., 2011. MeRy-B: A web knowledgebase for the storage, visualization, analysis and annotation of plant NMR metabolomic profiles. BMC Plant Biol. 11.

Giacomoni, F., Le Corguillé, G., Monsoor, M., et al., 2015. Workflow4Metabolomics: A collaborative research infrastructure for computational metabolomics. Bioinformatics 31, 1493–1495.

Goeman, J.J., Van De Geer, S.A., De Kort, F., Van Houwelingen, H.C., 2004. A global test for groups of genes: Testing association with a clinical outcome. Bioinformatics 20, 93–99.

Gómez, J., Brezmes, J., Mallol, R., et al., 2014. Dolphin: A tool for automatic targeted metabolite profiling using 1D and 2D ^{1}H NMR data. Anal. Bioanal. Chem. 406, 7967–7976.

Hao, J., Astle, W., De Iorio, M., Ebbels, T.M.D., 2012. Batman-an R package for the automated quantification of metabolites from nuclear magnetic resonance spectra using a bayesian model. Bioinformatics 28, 2088–2090.

Hastings, J., De Matos, P., Dekker, A., et al., 2013. The ChEBI reference database and ontology for biologically relevant chemistry: Enhancements for 2013. Nucleic Acids Res. 41, D456–D463.

Haug, K., Salek, R.M., Conesa, P., et al., 2013. MetaboLights – an open-access general-purpose repository for metabolomics studies and associated meta-data. Nucleic Acids Res. 41, D781–D786.

Haug, K., Salek, R.M., Steinbeck, C., 2017. Global open data management in metabolomics. Curr. Opin. Chem. Biol. 36, 58–63.

Heinemann, J., Mazurie, A., Tokmina-Lukaszewska, M., Beilman, G.J., Bothner, B., 2014. Application of support vector machines to metabolomics experiments with limited replicates. Metabolomics 10, 1121–1128.

Hill, C., Roessner, U., 2013. Plant metabolomics using GC–MS. In: Weckwerth, W. (Ed.), GC–MS Metabolomics Handbook. Springer.

Hill, C., Roessner, U., 2014. Advances in high-throughput untargeted LC-MS analysis for plant metabolomics. Advanced LC-MS Applications for Metabolomics. UK: Future Science Group.

Hiller, K., Hangebrauk, J., Jäger, C., et al., 2009. Metabolite detector: Comprehensive analysis tool for targeted and nontargeted GC/MS based metabolome analysis. Anal. Chem. 81, 3429–3439.

Hoffmann, N., Stoye, J., 2012. Generic software frameworks for GC–MS based metabolomics. Metabolomics. InTech.

Horai, H., Arita, M., Kanaya, S., et al., 2010. MassBank: A public repository for sharing mass spectral data for life sciences. J. Mass Spectrom. 45, 703–714.

Huber, W., 2016. A clash of cultures in discussions of the P value. Nat. Methods 13, 607.

Hunter, A., Dayalan, S., De Souza, D., et al., 2017. MASTR-MS: A web-based collaborative laboratory information management system (LIMS) for metabolomics. Metabolomics 13, 14.

Kale, N.S., Haug, K., Conesa, P., et al., 2016. MetaboLights: An Open-Access Database Repository for Metabolomics Data. Curr. Protoc. Bioinformatics, 53. . pp. 14.13.1–14.13.18.

Kanehisa, M., Goto, S., Sato, Y., Furumichi, M., Tanabe, M., 2012. KEGG for integration and interpretation of large-scale molecular data sets. Nucleic Acids Res. 40, D109–D114.

Kenar, E., Franken, H., Forcisi, S., et al., 2014. Automated label-free quantification of metabolites from liquid chromatography–mass spectrometry data. Mol. Cell. Proteomics 13, 348–359.

Kessler, N., Neuweger, H., Bonte, A., et al., 2013. MeltDB 2.0-advances of the metabolomics software system. Bioinformatics 29, 2452–2459.

Kim, S., Thiessen, P.A., Bolton, E.E., et al., 2016. PubChem Substance and Compound databases. Nucleic Acids Res. 44, D1202–D1213.

Kopka, J., Schauer, N., Krueger, S., et al., 2005. GMD@CSB.DB: The Golm metabolome database. Bioinformatics 21, 1635–1638.

Kruger, N.J., Ratcliffe, R.G., 2012. Pathways and fluxes: Exploring the plant metabolic network. J. Exp. Bot. 63, 2243–2246.

Kuhl, C., Tautenhahn, R., Böttcher, C., Larson, T.R., Neumann, S., 2012. CAMERA: An integrated strategy for compound spectra extraction and annotation of liquid chromatography/mass spectrometry data sets. Anal. Chem. 84, 283–289.

Lewis, I.A., Schommer, S.C., Markley, J.L., 2009. rNMR: Open source software for identifying and quantifying metabolites in NMR spectra. Magn. Reson. Chem. 47.

Li, B., Tang, J., Yang, Q., et al., 2017. NOREVA: Normalization and evaluation of MS-based metabolomics data. Nucleic Acids Res.

Li, L., Li, R., Zhou, J., et al., 2013a. MyCompoundID: using an evidence-based metabolome library for metabolite identification. Anal. Chem. 85, 3401–3408.

Li, S., Park, Y., Duraisingham, S., et al., 2013b. Predicting network activity from high throughput metabolomics. PLOS Comput Biol. 9, e1003123.

Lommen, A., Kools, H.J., 2012. MetAlign 3.0: Performance enhancement by efficient use of advances in computer hardware. Metabolomics 8, 719–726.

Lopez-Ibanez, J., Pazos, F., Chagoyen, M., 2016. MBROLE 2.0-functional enrichment of chemical compounds. Nucleic Acids Res. 44, W201–W204.

Lynn, K.S., Cheng, M.L., Chen, Y.R., et al., 2015. Metabolite identification for mass spectrometry-based metabolomics using multiple types of correlated ion information. Anal. Chem. 87, 2143–2151.

Mahadevan, S., Shah, S.L., Marrie, T.J., Slupsky, C.M., 2008. Analysis of metabolomic data using support vector machines. Anal. Chem. 80, 7562–7570.

Meyer, M.R., Peters, F.T., Maurer, H.H., 2010. Automated mass spectral deconvolution and identification system for GC–MS screening for drugs, poisons, and metabolites in urine. Clin. Chem. 56, 575–584.

Nakamura, Y., Afendi, F.M., Parvin, A.K., et al., 2014. KNApSAcK Metabolite Activity Database for retrieving the relationships between metabolites and biological activities. Plant Cell Physiol. 55, e7.

Nicolè, F., Guitton, Y., Courtois, E.A., et al., 2012. MSeasy: Unsupervised and Untargeted GC–MS data processing. Bioinformatics 28, 2278–2280.

NIST Standard Reference Database 1A v17, 2017. Available at: https://www.nist.gov/srd/nist-standard-reference-database-1a-v17. [accessed 20.10.17].

nmrML, 2017. Available at: http://nmrml.org/. [accessed 20.10.2017].

Noble, W.S., 2006. What is a support vector machine? Nat. Biotechnology 24, 1565–1567.

O'sullivan, A., Avizonis, D., Bruce German, J., Slupsky, C.M., 2011. Software Tools for NMR Metabolomics. Encyclopedia of Magnetic Resonance.

Pahler, A., Brink, A., 2013. Software aided approaches to structure-based metabolite identification in drug discovery and development. Drug Discov. Today Technol. 10, e207–e217.

Perez-Riverol, Y., Alpi, E., Wang, R., Hermjakob, H., Vizcaíno, J.A., 2015. Making proteomics data accessible and reusable: Current state of proteomics databases and repositories. Proteomics 15, 930–949.

Perez-Riverol, Y., Bai, M., Leprevost, F., et al., 2016. Omics Discovery Index – Discovering and Linking Public Omics Datasets. bioRxiv.

Pluskal, T., Castillo, S., Villar-Briones, A., Oresic, M., 2010. MZmine 2: Modular framework for processing, visualizing, and analyzing mass spectrometry-based molecular profile data. BMC Bioinformatics 11, 395.

Ravanbakhsh, S., Liu, P., Bjordahl, T.C., et al., 2015. Accurate, fully-automated NMR spectral profiling for metabolomics. PLOS One 10, e0124219.

Ren, S., Hinzman, A.A., Kang, E.L., Szczesniak, R.D., Lu, L.J., 2015. Computational and statistical analysis of metabolomics data. Metabolomics 11, 1492–1513.

Ridder, L., Van Der Hooft, J.J.J., Verhoeven, S., et al., 2013. Automatic chemical structure annotation of an LC–MSn based metabolic profile from green tea. Anal. Chem. 85, 6033–6040.

Rocca-Serra, P., Brandizi, M., Maguire, E., et al., 2010. ISA software suite: Supporting standards-compliant experimental annotation and enabling curation at the community level. Bioinformatics 26, 2354–2356.

Röst, H.L., Sachsenberg, T., Aiche, S., et al., 2016. OpenMS: A flexible open-source software platform for mass spectrometry data analysis. Nat. Methods 13, 741–748.

Ruttkies, C., Schymanski, E.L., Wolf, S., Hollender, J., Neumann, S., 2016. MetFrag relaunched: Incorporating strategies beyond in silico fragmentation. J. Cheminform. 8, 3.

Salek, R.M., Conesa, P., Cochrane, K., et al., 2017. Automated assembly of species metabolomes through data submission into a public repository. Gigascience 6, 1–4.

Salek, R.M., Haug, K., Conesa, P., et al., 2013. The MetaboLights repository: Curation challenges in metabolomics. Database 2013, bat029.

Salek, R.M., Neumann, S., Schober, D., et al., 2015. COordination of Standards in MetabOlomicS (COSMOS): Facilitating integrated metabolomics data access. Metabolomics, 11. . pp. 1587–1597.

Sawada, Y., Nakabayashi, R., Yamada, Y., et al., 2012. RIKEN tandem mass spectral database (ReSpect) for phytochemicals: A plant-specific MS/MS-based data resource and database. Phytochemistry 82, 38–45.

Scheltema, R.A., Jankevics, A., Jansen, R.C., Swertz, M.A., Breitling, R., 2011. PeakML/mzMatch: A file format, Java library, R library, and tool-chain for mass spectrometry data analysis. Anal. Chem. 83, 2786–2793.

Scholz, M., Fiehn, O., 2007. SetupX – A public study design database for metabolomic projects. Pac. Symp. Biocomput. 169–180.

SDBSWeb, 2017. Spectral Database for Organic Compounds (SDBS). Available at: http://sdbs.db.aist.go.jp/ [accessed 20.10.17].

Sesame LIMS, 2017. Available at: http://www.sesame.wisc.edu/sesame_home.html. [accessed 20.10.17].

Shen, H., Duhrkop, K., Bocker, S., Rousu, J., 2014. Metabolite identification through multiple kernel learning on fragmentation trees. Bioinformatics 30, i157–i164.

Skogerson, K., Wohlgemuth, G., Barupal, D.K., Fiehn, O., 2011. The volatile compound BinBase mass spectral database. BMC Bioinformatics 12, 321.

Smilde, A.K., Westerhuis, J.A., Hoefsloot, H.C., et al., 2010. Dynamic metabolomic data analysis: A tutorial review. Metabolomics 6, 3–17.

Smith, C.A., O'maille, G., Want, E.J., et al., 2005. METLIN: Ametabolite mass spectral database. Ther. Drug Monit. 27, 747–751.

Smith, C.A., Want, E.J., O'maille, G., Abagyan, R., Siuzdak, G., 2006. XCMS: Processing mass spectrometry data for metabolite profiling using nonlinear peak alignment, matching, and identification. Anal. Chem. 78, 779–787.

Spicer, R., Salek, R.M., Moreno, P., Cañueto, D., Steinbeck, C., 2017. Navigating freely-available software tools for metabolomics analysis. Metabolomics 13, 106.

Sturm, M., Bertsch, A., Gröpl, C., et al., 2008. OpenMS – an open-source software framework for mass spectrometry. BMC Bioinformatics 9, 163.

Sud, M., Fahy, E., Cotter, D., et al., 2016. Metabolomics Workbench: An international repository for metabolomics data and metadata, metabolite standards, protocols, tutorials and training, and analysis tools. Nucleic Acids Res. 44, D463–D470.

Sud, M., Fahy, E., Cotter, D., et al., 2007. LMSD: Lipid MAPS structure database. Nucleic Acids Res., 35. pp. D527–D532.

Sumner, L.W., Amberg, A., Barrett, D., et al., 2007. Proposed minimum reporting standards for chemical analysis Chemical Analysis Working Group (CAWG) Metabolomics Standards Initiative (MSI). Metabolomics 3, 211–221.

Sumner, L.W., Lei, Z., Nikolau, B.J., et al., 2014. Proposed quantitative and alphanumeric metabolite identification metrics. Metabolomics 10, 1047–1049.

Tarca, A.L., Bhatti, G., Romero, R., 2013. A comparison of gene set analysis methods in terms of sensitivity, prioritization and specificity. PLOS One 8, e79217.

Tautenhahn, R., Patti, G.J., Rinehart, D., Siuzdak, G., 2012. XCMS online: A web-based platform to process untargeted metabolomic data. Anal. Chem. 84, 5035–5039.

Tsugawa, H., Cajka, T., Kind, T., et al., 2015. MS-DIAL: Data-independent MS/MS deconvolution for comprehensive metabolome analysis. Nat. Methods. 12.

Ulrich, E.L., Akutsu, H., Doreleijers, J.F., et al., 2008. BioMagResBank. Nucleic Acids Res. 36, 402–408.

Van Beek, J.D., 2007. matNMR: A flexible toolbox for processing, analyzing and visualizing magnetic resonance data in Matlab. J. Magn. Reson. 187, 19–26.

Van Den Berg, R.A., Hoefsloot, H.C., Westerhuis, J.A., Smilde, A.K., Van Der Werf, M.J., 2006. Centering, scaling, and transformations: Improving the biological information content of metabolomics data. BMC Genomics 7, 142.

Van Rijswijk, M., Beirnaert, C., Caron, C., et al., 2017. The future of metabolomics in ELIXIR. F1000Res. 6.

Vizcaíno, J.A., Deutsch, E.W., Wang, R., et al., 2014. ProteomeXchange provides globally coordinated proteomics data submission and dissemination. Nat. Biotechnol. 32, 223–226.

Wang, M., Carver, J.J., Phelan, V.V., et al., 2016. Sharing and community curation of mass spectrometry data with Global Natural Products Social Molecular Networking. Nat. Biotechnol. 34, 828–837.

Weber, R.J.M., Lawson, T.N., Salek, R.M., *et al.*, 2016. Computational tools and workflows in metabolomics: An international survey highlights the opportunity for harmonisation through Galaxy. Metabolomics 13, 12.

Weber, R.J.M., Viant, M.R., 2010. MI-Pack: Increased confidence of metabolite identification in mass spectra by integrating accurate masses and metabolic pathways. Chemometrics Intellig. Lab. Syst. 104, 75–82.

Wehrens, R., Weingart, G., Mattivi, F., 2014. metaMS: An open-source pipeline for GC–MS-based untargeted metabolomics. J. Chromatogr. B 966, 109–116.

Westerhuis, J.A., Hoefsloot, H.C.J., Smit, S., *et al.*, 2008. Assessment of PLSDA cross validation. Metabolomics 4, 81–89.

Wiley Registry of Mass Spectral Data, 2017. eleventh ed. Available at: http://au.wiley.com/WileyCDA/WileyTitle/productCd-1119171016.html. [accessed 20.10.2017].

Wilkinson, M.D., Dumontier, M., Aalbersberg, I.J.J., *et al.*, 2016. The FAIR Guiding Principles for scientific data management and stewardship. Sci. Data, 3. . p. 160018.

Wishart, D.S., Jewison, T., Guo, A.C., *et al.*, 2013. HMDB 3.0 – The Human Metabolome Database in 2013. Nucleic Acids Res. 41, 801–807.

Wolstencroft, K., Haines, R., Fellows, D., *et al.*, 2013. The Taverna workflow suite: Designing and executing workflows of Web Services on the desktop, web or in the cloud. Nucleic Acids Res. 41, W557–W561.

Worley, B., Powers, R., 2013. Multivariate analysis in metabolomics. Curr. Metabolomics 1, 92–107.

Xia, J., Broadhurst, D.I., Wilson, M., Wishart, D.S., 2013. Translational biomarker discovery in clinical metabolomics: An introductory tutorial. Metabolomics 9, 280–299.

Xia, J., Sinelnikov, I.V., Han, B., Wishart, D.S., 2015. MetaboAnalyst 3.0 – making metabolomics more meaningful. Nucleic Acids Res. 43, W251–W257.

Xia, J., Wishart, D.S., 2010. MSEA: A web-based tool to identify biologically meaningful patterns in quantitative metabolomic data. Nucleic Acids Res. 38, W71–W77.

Xia, J., Wishart, D.S., 2016. Using MetaboAnalyst 3.0 for Comprehensive Metabolomics Data Analysis. Curr. Protoc. Bioinformatics 55, 14 10 1–14 10 91.

Relevant Websites

http://www.chemspider.com/
   ChemSpider.
http://www.hmdb.ca/
   HMDB The Human Metabolome Database.
http://www.massban.jp
   MassBank.
http://www.metabolomexchange.org/
   MetabolomeXchange.
http://metlin.scripps.edu/
   METLIN.
http://phenomenal-h2020.eu/
   PhenoMeNal.
http://www.myexperiment.org/
   WorkFlows.

Mass Spectrometry-Based Metabolomic Analysis

Russell Pickford, University of New South Wales, Kensington, NSW, Australia

Introduction

Metabolites are the intermediates and end products of biological metabolism, the definition typically being limited to small molecules such as amino acids, sugars and lipids.

The field of metabolomics aims to measure metabolites within biological samples to obtain profiles or 'metabolomes'. These can be compared across sample types – different disease states or gene knockouts for example – to learn more about the underlying biology or to find metabolites which can be used to recognise or define the disease state – biomarkers.

The term 'metabolome' was defined as "the complete set of metabolites/low-molecular-weight intermediates, which are context dependent, varying according to the physiology, developmental or pathological state of the cell, tissue, organ or organism" (Oliver, 2002).

Metabolomics is somewhat of an ancient science – ants were used by the ancient Chinese (2000–1500 BC) to detect high levels of glucose in urine – and hence diabetes. Currently, Doctors often test a panel of metabolites within patient's blood – for example glucose or cholesterol – to ensure that they are within a normal range. These are examples of targeted metabolomics.

Science is now aiming to study organisms as integrated systems to obtain as complete a picture as possible. This systems biology approach combines experimental data from a range of 'omics experiments':

- Genomics measure the DNA, a 'blueprint' for the organism.
- Transcriptomics measure the RNA transcripts produced by the genome under specific circumstances. Proteomics measures the proteins produced under these circumstances, including metabolic enzymes.
- Metabolomics measures the small molecules or metabolites of the system.

As the final downstream product, the metabolome can be considered the closest reflection of the phenotype (Kell et al., 2005).

The analytical chemistry measurements and subsequent data analysis required to examine a metabolome are complex. Trying to simultaneously measure many different metabolites of widely-varying properties – such as concentration, molecular weight and polarity – is very difficult. No current analytical technique can measure an entire metabolome within a single experiment, or perhaps even with a combination of experiments.

The two techniques most frequently used for metabolomic analysis are mass spectrometry (MS) and Nuclear Magnetic Resonance (NMR). Both have advantages and disadvantages.

NMR has a much poorer detection limit and so can measure fewer metabolites, but is fast, simple and non-destructive to the sample. Also, metabolites observed can almost always be identified from the rich data obtained.

Mass spectrometry is more complex experimentally, destructive to the sample and identification of detected species can be a major challenge.

MS does however offer some key strengths:

- wide dynamic range – the ability to simultaneously measure analytes of very different concentrations, important as the range in biological samples can be many orders of magnitude.
- high sensitivity – the ability to measure analytes at very low concentration.
- reproducible, quantitative analysis – the signal from a metabolite is proportional to concentration.

Data analysis of metabolomics data acquired using mass spectrometry relies on a good understanding of how the data is generated and what information it can and cannot provide. Data analysis ranges from simple univariate statistics to complex manipulations and clustering algorithms.

Instrumentation Employed for Mass Spectrometry-Based Metabolomics

Mass Spectrometry

A mass spectrometer is an instrument that measures the molecular weight and abundance of molecular species within a sample.

The instrument consists of several functional stages (**Fig. 1**):

- Ionisation
- Mass Analysis
- Detection

Manipulating and detecting molecules with a mass spectrometer requires them to be charged. Neutral species cannot be observed. Thus, the first part of mass spectrometry analysis is ionisation. This stage converts neutral species into gaseous charged ions which can then be manipulated, measured and detected.

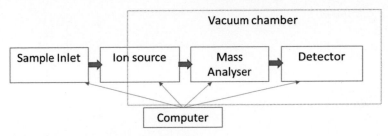

Fig. 1 Schematic detailing the major components of a mass spectrometer.

There is no universal ionisation procedure – different choices here will allow observation of different panels of analytes, but hide others.

Choices of ionisation are typically electrospray (ES) (Fenn *et al.*, 1989; Gaskell, 1997) and atmospheric pressure chemical ionisation (APCI) (Carroll *et al.*, 1974; Byrdwell, 2001) for solution phase analysis and electron ionisation (EI) (Bleakney, 1929) or chemical ionisation (CI) (Munson and Field, 1966) for gaseous analysis. ES, APCI and CI are 'gentle' ionisation methods which generally send an intact cationised molecule into the mass analyser. This is important as fragmentation during ionisation can greatly complicate the data produced. EI however produces a radical cation which then usually fragments, sometimes so completely that the intact precursor molecule is not observed.

Except for EI, which only produces positively-charged species, each can be employed in positive or negative polarity. The choice of polarity will again allow observation of different analytes. For example, basic species generally are easier to observe in positive mode and acidic species in the negative mode as they will easily become charged by gaining or losing a proton respectively.

The generated ions are directed into a mass analyser where they are sorted by molecular weight (really mass to charge ratio, m/z). All mass analysers require a vacuum to be functional and operate using a combination of electric fields and (less frequently) magnetic fields to manipulate the ions.

Mass analysers can be divided into two main categories – low resolution and high resolution. Resolution refers to the ability to discriminate analytes which are close in m/z.

Low resolution mass analysers include linear quadrupoles (Douglas, 2009) and quadrupole ion traps (March, 2009), typically operating at unit resolution – they will discriminate mass 100 from mass 99 and mass 101.

High resolution mass analysers include Time-of-Flight (Boesl, 2017), Fourier Transform Ion Cyclotron Resonance (Marshall *et al.*, 1998) and Orbitrap (Zubarev and Makarov, 2013). These are a finer toothed comb – they will separate mass 100.0 from 99.9 and 100.1, and potentially much finer than this – allowing measurement of molecular weight much more accurately and discrimination of species of very similar molecular weight.

Mass analysers should however not be judged by resolution alone. Each has specific advantages and disadvantages and are employed for different experiments (Brunee, 1987). For example, advantages of quadrupoles include high transmission, robustness, tolerance of higher pressures and low cost.

When looking for specific targets and requiring high sensitivity then a quadrupole is an excellent choice. When screening samples for as many metabolites as possible then resolution and mass accuracy become important, therefore Time-of-Flight or FT-ICR or Orbitrap are ideal for these profiling experiments.

Different mass analysers and combinations thereof allow different experiments to be performed.

The final component is detection – the abundance of each ion is measured, and a mass spectrum produced. The mass spectrum is a 2D plot with m/z on the x-axis and abundance on the y-axis. The units of the abundance axis vary with instrument, with detector and with manufacturer. It is frequently scaled to 'relative abundance' where the most intense signal in the spectrum is assigned 100% and the other signals shown relative to this (**Fig. 2**).

Separation Techniques

Whilst direct analysis of complex samples is possible, it is generally preferable to 'hyphenate' the mass spectrometer with a separation technique. This separates the components in time and space according to some chosen physical property, therefore simplifying the mass spectra obtained at any individual time point. Addition of a separation technique also facilitates automated analysis as well as providing potential enhancements such as potential discrimination of isomeric species (which have the same molecular weight but different structure) and enhancement of signal due to decreased competition during ionisation.

Gas chromatography (GC) is typically employed where the analytes are heat stable and volatile (for example gas, flavour, fragrance analysis) or can be made volatile using simple chemical derivatisations.

Samples are injected into a hot liner where they are vaporised and then directed through a fused-silica column using a carrier gas (mobile phase). Interactions between the analytes and column surface (stationary phase) will differ based on their chemistry, so that some are retained longer and some shorter as they pass through to the detector – providing separation in time. GC separations can be tuned and optimised using column chemistry and column temperature.

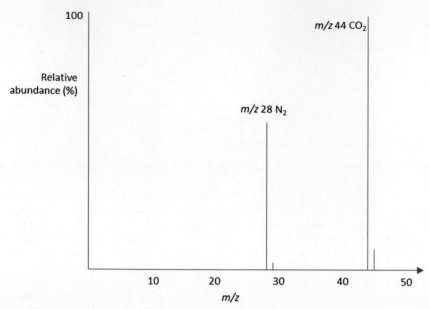

Fig. 2   Mass Spectrum obtained from electron ionisation of a mixture of nitrogen and carbon dioxide gases.

Liquid chromatography (LC) is typically employed for involatile analytes (for example amino acids, sugars, lipids) and does not generally require any prior chemical derivatisation. This makes it a popular choice for untargeted metabolomics as the sample preparation is minimal, reducing experimental bias.

Samples are injected into a stream of liquid – mobile phase – and pass through a stainless-steel or plastic column packed with silica beads derivatised to some specific functionality.

For example, C18 columns are widely used and have an 18-carbon hydrocarbon chain attached to the beads. This provides a hydrophobic surface (stationary phase) for the analytes to interact with as they pass through, therefore non-polar analytes will have a strong interaction with this surface and be retained longer than polar analytes. LC separations can be optimised using a wide range of parameters such as column chemistry, choice of mobile phase solvents and temperature.

Less frequently used, but no less useful, is capillary electrophoresis where analytes are separated by charge and mobility within a buffer solution as they move down a fused silica capillary under the influence of high voltage.

GC–MS and LC–MS

During these analysis methods the output of the separation device is connected to the ionisation source of the mass spectrometer. The output is continuously ionised, and the ions produced directed into the mass spectrometer. Mass spectra are recorded, typically every second or so, throughout the whole run time. The data therefore consists of several hundred to thousand consecutive mass spectra – it is now a 3D plot with the axes of time, m/z and abundance. This can be visualised as an 'ion map' with time and m/z as the axes and intensity of colour as the abundance (**Fig. 3**).

The use of mass spectrometry as a detector places some limits on what can be used for the separation part of the experiment – for example LC–MS mobile phases and sample components need to be volatile – salts cannot be used.

Tandem Mass Spectrometry – MS/MS

Mass spectra generated by ES, APCI and CI generally only contain intact cationised analytes, yielding the molecular weights of the compounds within the sample. An extra dimension of information can be obtained using tandem mass spectrometry. This process is performed within the mass spectrometer and involves a specific precursor ion (m/z) being isolated and fragmented, normally via collisions with a neutral gas such as nitrogen or argon. These product ions are then detected and can yield a 'fingerprint' helping to identify the analyte. Two analytes can have identical molecular weight – isobaric species – but very different MS/MS fingerprints and can therefore be discriminated by MS/MS but not by MS (Yost and Fetterolf, 1983; de Hoffmann, 1996).

The tandem mass spectrum produced from MS/MS analysis of the precursor ion at m/z 784 is shown (**Fig. 4**). Some remaining (unfragmented) precursor can be observed together with a large product ion at m/z 184. This ion is structurally diagnostic – it is formed from lipid molecules containing a choline group and therefore yields information that the precursor at m/z 784 is a phosphatidylcholine.

The use of MS/MS can add an additional dimension to the data if employed together with MS scans, such as in the case of profiling experiments. The data now contain a time component and both MS and MS/MS scans.

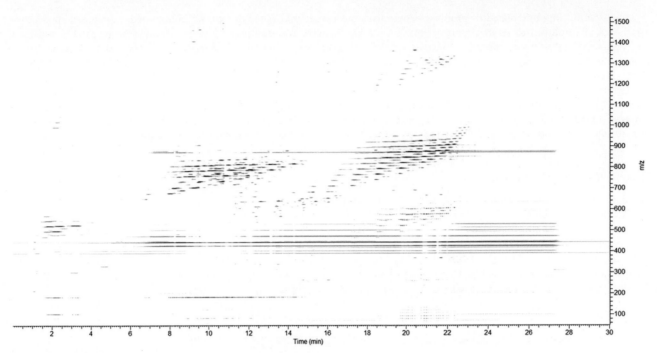

Fig. 3 Ion map obtained from LC–MS analysis of a plasma lipid extract.

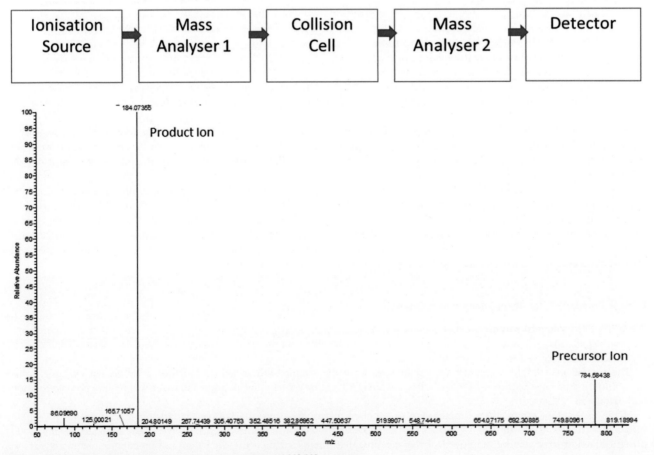

Fig. 4 Tandem mass spectrometry instrument components and MS/MS spectrum.

When performing untargeted or profiling experiments the analyte masses are not known. Automated, rules-based approaches are therefore employed to select precursor ions for MS/MS analysis. For example, a full MS 'survey' scan performed and then consecutive MS/MS scans on the 6 most intense species in this survey scan, and the cycle repeats throughout the run. Rules can be added – to include or exclude selected masses or charge states for example. This is generally termed data-dependent analysis – DDA (Mann *et al.*, 2001).

Performing MS/MS with a very wide mass selection window is becoming popular, for example simultaneously fragmenting everything over the mass range 150–250 Th. These have less specificity and the data are harder to interpret since the connectivity between precursor and product ions is no longer certain. They do however provide much greater MS/MS coverage. This approach is termed data independent analysis – DIA (Doerr, 2014), MS^E (Plumb *et al.*, 2006) and SWATH (Gillet *et al.*, 2012)

Mass Spectrometer Data Files

The data structure remains similar, but each instrument manufacturer produces their own data format – .d from Agilent, .wiff from Applied Biosystems and .raw from ThermoFisher Scientific for example.

These formats are proprietary and need to be analysed within the vendor's own software or be converted to a community format such as NetCDF (Unidata, 2016), mzDATA (Orchard *et al.*, 2007) or MzXML (Pedrioli *et al.*, 2004; Lin *et al.*, 2005). Since 2008 the community has largely deprecated prior formats and adopted mzML (Deutsch, 2008) as the unified standard. These formats are readable and accessible to freeware or web-based data processing tools. Open-source tools such as ProteoWizard (Kessner *et al.*, 2008) can be employed to batch convert files from proprietary to open source.

General Considerations for Metabolomics Experiments using Mass Spectrometry

Experimental Design

The experimental design is important to consider prior to any practical work being performed. The design will depend upon the statistical rigour required (power analysis) as well as the experimental goals and practical limitations. If we are comparing changes across two sample types, then large changes in metabolite concentration are easier to see and have confidence in. Small changes require much more rigour to be able to observe confidently.

Replication is important – technical replicates are repeat analyses of the same sample which allow the variability of the measurement to be calculated. Biological replicates allow the natural variability of the target metabolites within a sample set to be examined. These are both very important – if the target analyte varies 10% between test and control but varies 20% within control then it is very hard to draw any confident conclusions about the result. Acceptable minimum standards for replication vary between scientists and experiments. The data is generally assessed using at least mean and standard deviation (coefficient of variation). These require a normal distribution of data and therefore at least 6 replicates to describe this. Generally, more is better!

It is also important to randomise the samples during instrumental acquisition. The instrument's performance can change over time and the effect of this on the data needs to be minimised, so the comparison is really between test vs control and not data from clean instrument vs data from dirty instrument. For the same reason all samples should be run in a single batch and not on different days.

There are many confounding variables when dealing with biological samples, which must be controlled as much as possible. The experiment may be testing the blood of cancer patients vs healthy patients, but not all differences observed will be due to the cancer. Metabolite profiles can vary for many reasons, for example, age, sex, race, obesity, exercise…. Even diet is an influence as what is eaten makes its way into the blood, this is one reason that pathology labs prefer to take patient blood in a fasted state.

It is therefore important to know about the variability in the sample population, and control as best possible. It is no good looking for a cure for cancer and instead finding a biomarker that differentiates males from females……

Sample Preparation for Mass Spectrometry Analysis

Preparing the biological sample so that it may be analysed by mass spectrometry is a key part of metabolomics experiments (Raterink *et al.*, 2014). The sample preparation will have a large effect on the result – changes to sample preparation will change the metabolites observed in the experiment.

The general principle is to extract the metabolites from the sample with as light a touch as possible. For example, urine metabolomics may just require a simple dilution, plasma will require protein precipitation and tissue will require homogenisation and extraction.

If the experiment is targeting a metabolite or panel of metabolites then the sample preparation can be optimised experimentally – choices of solvents, pH, volumes etc can all be optimised for maximum signal. A preconcentration step such as solid phase extraction may also be employed. If the experiment is untargeted then the extraction and preparation normally seek to extract as wide a range of metabolites as possible with as little perturbation to the sample as possible.

Targeted Metabolomics

Targeted metabolomic experiments seek to accurately quantify specific metabolites within a sample (Roberts *et al.*, 2012; Griffiths *et al.*, 2010; Koal and Deigner, 2010). They target a limited number of analytes. For example, to observe what happens to the concentration of a drug in a patient's body over time only detection of this drug is required. To study heart disease a panel of key analytes involved in cardiovascular metabolism may be chosen (Griffin *et al.*, 2011).

Only the targeted analytes are observed in the data, the rest of what is happening with the sample is hidden. This is therefore a trade off, achieving excellent data concerning one panel of metabolites but no data about the rest of what may be happening with the sample.

Absolute quantification obtains the amount of analyte within a sample, normally measured in moles or by weight, for example nanograms analyte per gram of tissue.

Quantification with mass spectrometry is performed by regression analysis of a calibration curve plotting signal against concentration, therefore an authentic standard is required for every individual compound quantified. Internal standards are employed to correct for experimental error, these are generally isotopically-labelled (C13, N15) analogues of the analyte, or close structural analogues which are known not to be contained in the sample, again with each analyte ideally having its own internal standard (Green, 1996).

It is possible to perform relative quantification across samples and sample sets when authentic standards are not available, however the results are reported only as a fold-change difference in peak areas obtained, which is difficult to draw biologically-relevant conclusions from. Typically, untargeted metabolomics will produce relative quantification results and then the more interesting changes can be confirmed and more accurately characterised using targeted experiments.

Data Characteristics of Targeted Metabolomics Experiments

The data output of the mass spectrometer during targeted quantification experiments is a plot of signal (relative abundance) against time for each analyte (**Fig. 5**). The signal can be from the intact analyte – normally termed Selected Ion Monitoring – or from a selected fragment ion produced by MS/MS of the analyte – normally termed Selected Reaction Monitoring. The fragment ion is selected to optimise for sensitivity and specificity.

The plot is then integrated, and the area of the peak produced by the sample signal recorded. Integration is performed in an automated manner with each mass spectrometry vendor providing their own set of algorithms. The shape of the peaks produced are somewhat variable with analyte and with separation method, therefore the integration process requires some manual checking and tuning to do the best job possible.

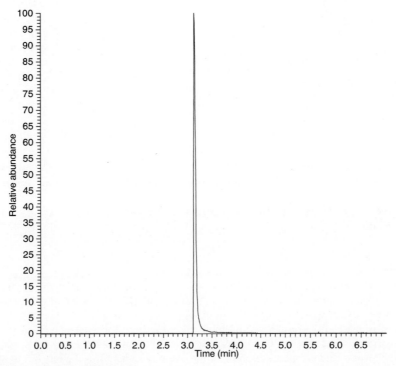

Fig. 5 LC-SRM trace obtained from analysis of Verapamil.

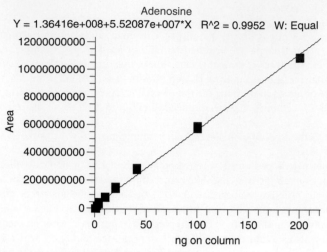

Adenosine
Y = 1.36416e+008+5.52087e+007*X   R^2 = 0.9952   W: Equal

Fig. 6   Calibration Curve obtained from LC-SRM analysis of Adenosine Standard.

Filename	Sample Type	Inj Vol	Retention	Calculated Amount	Units
4ngc	Standard	30.00	1.2	3.579	ng on column
4ngd	Standard	30.00	1.2	3.816	ng on column
27	Unknown Sample	20.00	1.2	0.000	ng on column
28	Unknown Sample	20.00	1.2	3.155	ng on column
29	Unknown Sample	20.00	1.2	3.979	ng on column
3	Unknown Sample	20.00	1.2	0.021	ng on column
30	Unknown Sample	20.00	1.2	3.744	ng on column
31	Unknown Sample	20.00	1.2	4.586	ng on column
4	Unknown Sample	20.00	1.2	0.055	ng on column
5	Unknown Sample	20.00	1.2	0.029	ng on column
6	Unknown Sample	20.00	1.2	0.029	ng on column
7	Unknown Sample	20.00	1.2	0.284	ng on column
8	Unknown Sample	20.00	1.2	0.122	ng on column
9	Unknown Sample	20.00	1.2	0.107	ng on column

Fig. 7   Part of an output table after LC-SRM quantitative analysis.

Metabolite Quantification

A calibration curve is obtained by regression analysis after plotting the peak area against concentration for a known set of standard concentrations (**Fig. 6**). If an internal standard is used for normalisation, then the plot becomes (peak area standard/peak area internal standard) vs concentration.

The concentration of the analyte is calculated using regression analysis of the standard curve – the peak area obtained is transformed into concentration.

Data Analysis

Targeted metabolomic data (**Fig. 7**) is typically analysed in a spreadsheet using univariate analysis – the comparison of a single variable (analyte amount) between samples and sample sets (Vinaixa *et al.*, 2012). For example, the amount of glucose in the blood of diabetics vs healthy individuals.

Analysis is normally performed with replicates – both biological and technical. Therefore, a mean and a standard deviation or coefficient of variation are the required outputs. Outliers can be removed to improve the data quality after appropriate analysis, although this requires much care – the sample measurement may be very different to the rest of the set, but it may be due to genuine biological variation not experimental error.

Typical statistical tests employed are T-tests (to compare the mean value across two groups) and Analysis of Variance (ANOVA). to compare multiple groups.

Fluxomics

A subset of targeted metabolomics is fluxomics (Winter and Krömer, 2013; Munger *et al.*, 2008), which seeks to determine the rates of metabolic reactions within the samples. This differs from targeted metabolomics which provide a snapshot of metabolite concentrations.

Fluxomics can be performed using ^{13}C or 15N labelled precursors introduced to the biological system – these have the same chemical properties as the unlabelled counterpart but an increased mass. They can therefore be discriminated using a mass spectrometer. The progress of the label down metabolic pathways can be quantified – the appearance of 'heavy' downstream metabolites – and the metabolic flux estimated.

Untargeted Metabolomics

Untargeted metabolomics experiments seek to measure as many compounds/metabolites as possible within a sample – it is an untargeted approach which 'casts the fishing net' as widely as possible. Samples are most frequently analysed using LC–MS (although GC–MS can be used) with a high-resolution mass analyser scanning a wide mass range, e.g., 50–2000 Th.

Detected components are compared across sample sets to find differences and yield biologically-relevant information.

Experimental Considerations

The components detected will depend upon the LC(GC)-MS experiment performed. Electrospray ionisation will allow observation of a different set of ions than APCI. Positive polarity will show different compounds than negative polarity. Even the choice of separation method, and derivatisation method for GC, will change the panel of metabolites observed. It is important to remember there is no global 'see everything' metabolomics assay.

Biological replicates are very important here – the mean peak area obtained from compound X is compared between 'test' and 'control' sample sets to search for (significant) differences which may provide information about the biological state of the samples – an upregulation in a certain metabolic pathway for example. Sufficient replication to accurately describe the within-sample-set variation is therefore required.

It is also a good idea to include process blanks. Mass spectrometry is a very sensitive detection technique and will detect many background and contaminant ions from the solvents, solutions, plasticware etc used in the experiment – interfering signals not due to the sample. By including blanks in experiments and subsequent data analysis these 'background components' can be recognised and dealt with appropriately.

Workflow

Untargeted metabolomics experiments generally follow the same workflow, shown schematically in **Fig. 8**.

There are many software packages available to facilitate the analysis of untargeted metabolomics data.

Instrument manufacturers generally produce a commercially-available software to work with data from their own instruments. SIEVE and Compound Discoverer (ThermoFisher), ProfileAnalysis (Bruker), Progenesis QI (Waters) and MassHunter (Agilent) for example. Progenesis QI also will analyse data from other instrument manufacturers.

There are many freely-available tools available on the web, which are more universally-applicable. The most popular being XCMS(2) (Smith *et al.*, 2006; Gowda *et al.*, 2014), MetaboAnalyst (Xia *et al.*, 2009, 2015) and MzMine2 (Pluskal (2010)).

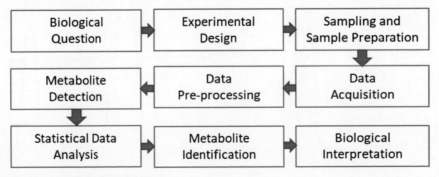

Fig. 8 Workflow of an untargeted metabolomics experiment.

MetabolomeExpress (Carroll *et al.*, 2010) is also used for GC–MS data analysis. Data generated from Data Independent Analysis (DIA) experiments can be analysed using MS-DIAL (Tsugawa *et al.*, 2015).

At the time of writing there is no universally recognised 'best' software or approach, usage depending on the experiment, user preference and software availability.

Data Pre-Processing

Data conversion

The output from the mass spectrometer is encoded in a format specific to the instrument manufacturer. These generally need to be converted into open source formats prior to software interrogation, particularly when using vendor-independent or freely available software both on- and off-line.

Widely usable data formats are NetCDF, mzDATA, mzXML and mzML. Instrument manufacturers frequently include an option to export data to this format from within their own data analysis software.

A useful data conversion tool is the open source ProteoWizard (Kessner *et al.*, 2008), this tool accepts data from a wide variety of instrument manufacturers and will batch convert into open source formats.

Data import

The first step in data analysis is to import the data into the metabolomics analysis software, this is normally performed in an automated manner. The 3-dimensional data sets (time vs *m/z* vs intensity) are then prepared for analysis. The data may contain an extra dimension – MS/MS – which is useful to aid in compound identification but is ignored during the early processing stages.

The three-dimensional data is generally converted into stacks of two-dimensional chromatograms by plotting the signal intensity of narrow *m/z* windows against time to produce extracted ion chromatograms. The size of these bins will depend on the mass analyser used – the data obtained from high resolution instruments allows many small bins to be used and therefore have high discriminating power between compounds of very similar molecular weight. The data from low resolution instruments can only be put into large windows or bins.

Data alignment

The data is then aligned – the instrument time required to analyse large numbers of samples can be significant, and variations during the data acquisition produce small shifts or drifts in chromatographic retention time. Signal from *m/z* X at time Y cannot therefore simply be compared across samples. There needs to be some alignment (shifting/warping) in the time dimension to line up the data and ensure correct comparison across samples and sample groups (**Fig. 9**). The shift is not a linear one and so warping algorithms are preferred. These are numerous (Smith *et al.*, 2013) and include Time Correlation Optimised Warping, Dynamic Time Warping and Parametric Time Warping (Tomasi *et al.*, 2004). There is no clear 'best' algorithm, with results being context and data dependent.

Missing values can be a significant problem for metabolomics statistical analysis – what is done when a signal is present in one sample set or one sample within a sample set but not the other(s)? Various approaches have been employed to solve this including missing value imputation (Armitage *et al.*, 2015; Hrydziuszko and Viant, 2012; Di Guida *et al.*, 2016) or producing an aggregate run from all samples in the dataset and then using this for peak picking. This ensures the same ion is detected in every individual run and produces a complete data set which is required for valid multivariate analysis. Backfilling with zeros for missing values does not produce a valid dataset for multivariate analysis and should be avoided. However, there is no clear 'best' solution to the missing values problem, with results being context and data dependent.

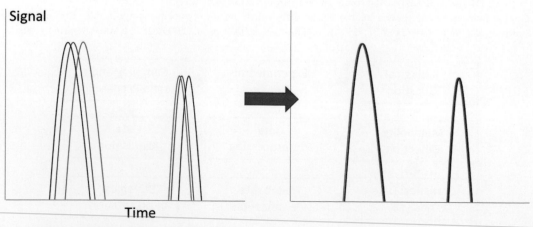

Fig. 9   Cartoon example of unaligned and aligned LC–MS data.

Data normalisation

After alignment the data is then normalised (scaled) to adjust for experimental noise and variation. The goal is to remove experimental bias whilst keeping the signals relating to genuine biological differences (De Livera *et al.*, 2015). Examples of sample systematic bias include differences in sample concentration, particularly with biofluids – for example urine can be more dilute or more concentrated depending on the individual's hydration status. Experimental variations in signal due to sample preparation and instrumental analysis can also arise. Targeted metabolomic analysis can correct to a 'housekeeping' endogenous metabolite such as creatinine in urine, although this can be problematic due to the assumptions made. The ideal solution is the use of internal standards to correct for experimental bias, however a single internal standard will not correct properly for all the compounds detected in an untargeted experiment. The use of multiple internal standards improves the situation although this is still not ideal. Normalisation methods based on statistical methods (unit norm, median and quantile), scaling methods (auto scaling, range scaling, Pareto scaling, vast scaling and level scaling) and data transformation (log and power) are all employed (Sugimoto *et al.*, 2012). There is no consensus best approach, and several should be tested (Di Guida *et al.*, 2016).

Metabolite Detection

Peak picking and deconvolution

Peak picking algorithms are then applied to the aligned and normalised data to identify peaks over a user specified threshold, for example peak height or peak area. Wavelet transformation and Gaussian-curve fitting is commonly used to distinguish signal to noise (Smith *et al.*, 2006; Tautenhahn *et al.*, 2008), although manual tuning and checking is required due to the compound-specific variation in chromatographic peak shape, particularly with LC–MS and CE–MS. The output is generally a list of mass-retention time pairs e.g., *mz* 100-4.5 min refers to a compound of *m/z* 100 which elutes with a peak intensity maximum at 4.5 min. Each of these will have an associated integrated peak area.

This data can then be deconvoluted or reduced – each compound may ionise in different ways or fall victim to fragmentation in the source region of the mass spectrometer. For example, a compound may be protonated, sodiated and be observed as protonated with loss of water $[M + H\text{-}H_2O]+$. These can be recognised as they will have matching chromatographic profiles (signal vs time) and can then be deconvoluted to a single data entry of mass and retention time. Reducing the size of the dataset at this point simplifies the downstream data analysis. Caution must be applied however – these could be separate species which happen to have the same chromatographic profile.

Feature list

The final output of this process is a feature list or data matrix. The matrix contains the features detected and their abundance (integrated peak area) within the different samples.

Data can then be filtered – for example only those features showing a fold change greater than 2 × between test and control or only those features observed with high significance or with a low coefficient of variation.

Exploratory data analysis methods are then used to find patterns in the filtered or unfiltered data

Statistical Data Analysis

Data analysis normally employs multivariate analysis (Worley and Powers, 2013). This is used to identify metabolite patterns that correlate in a statistically significant manner with a sample set/phenotype so that they can be used for diagnostics.

The main reason for using multivariate as compared to univariate analysis for metabolomics is simply since there are many variables in each sample, and these usually correlate to the phenotype in a concerted fashion rather than correlating independently. For example, your fingerprint contains many data points which combine to a unique whole. Consequently, it is important to identify metabolic profiles or 'fingerprints' that relate to the phenotype rather than individual, independent biomarkers.

It is important to apply appropriate statistical tools for the experiment. For example, comparing 'test' and 'control' samples can be done using one-factor ANOVA to search for statistically significant differences between groups. However, if the samples are a time course study with several samples being taken from the same source (for example a cell culture being sampled at 0, 4, 8, 24 h) then one-factor ANOVA is not appropriate as the groups being compared are not truly independent. In this case different tools such as repeated measures ANOVA must be employed.

Multivariate analysis tools can be supervised or unsupervised. Unsupervised methods are exploratory in nature, showing previously unknown patterns in the data (**Fig. 10**) or helping to confirm suspected patterns. Supervised methods use experimental knowledge (for example sample groups) to discover associated patterns.

Principal Component Analysis (PCA) (Jolliffe, 2002) is frequently used for untargeted metabolomics data analysis. This is an unsupervised data analysis tool – the algorithm is unaware of which are test, and which are control samples. PCA allows clustering of sample sets and to determine which variables (compounds/features) have the highest ability to discriminate between the sample sets.

PCA is an unbiased process and may reveal previously unknown clusters in the data – for example when comparing healthy vs diseased patients, we may see further splitting into groups male and female or fed and fasted. This can help to recognise limitations or errors in the experimental design.

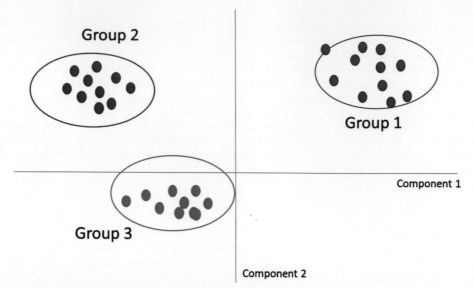

Fig. 10 Cartoon Principal Component Analysis (PCA) Plot demonstrating clustering of 3 sample groups.

The clusters/groupings also allow recognition of samples which may be outliers – they do not cluster with the rest of their biological group.

However, PCA only reveals group structure when within-group variation is less than between-group variation and so may not provide separation between very similar samples.

Alternative tools employ Partial Lease Squares algorithms to discriminate groups (Barker and Rayens, 2003). Partial Least Squares Discriminant Analysis (PLS-DA) and Orthogonal Partial Least Squares Discriminant Analysis (OPLS-DA) are the most well-known tools for multivariate analysis in metabolomics (Brereton and Lloyd, 2014). These are supervised analysis tools – information about which samples belong to which group is given to the algorithm. The algorithm then seeks to maximise the differences between the sample groups and show the variables responsible for this separation.

A danger of employing these algorithms is that they can aggressively over-fit models to the data, almost always managing to separate the defined groups or classes. Consequently, PLS-DA and OPLS-DA can generate excellent class separation from random data! This can be controlled for somewhat by creating training sets and testing them with other data or by performing permutation tests.

An alternative statistical tool is to use clustering methods on the data. These are unsupervised machine learning techniques which seek to group samples together using similarities in their measurements. Frequently used clustering techniques in metabolomics are K-Means (Hartigan and Wong, 1979), Hierarchical Clustering (Johnson, 1967), and Self-Organising Maps (Kohonen, 1990).

Further data analysis

Most metabolomics software packages will allow export of the data matrix (feature list) in .csv format. This file can then be further analysed in statistical packages such as the commercial SIMCA-P or the freely available R software using scripts.

Metabolite Identification

The final stage of data analysis is to identify those compounds (Kind *et al.*, 2017; Vaniya and Fiehn, 2015) found to be of likely biological interest – statistically different between groups and/or discriminating between groups for example.

Identification is not a trivial task, hence efforts are generally directed towards 'interesting' metabolites rather than trying to identify every feature identified in the experiment.

The information obtained about a compound will be molecular weight, retention time on the LC or GC and perhaps an MS/MS fingerprint. These properties can be targeted in database searches.

Several online and accessible chemical and metabolomic databases are available

- ChemSpider (Pence and, Williams, 2010) is a large chemical database, containing over 60 million structures from hundreds of data sources. It is the most complete and therefore the biggest chance of a 'hit'. It is also the biggest chance of a false positive.
- PubChem (Wang *et al.*, 2009) is another large general chemical database provided by the NIH.
- Metlin (Smith *et al.*, 2005) is curated specifically for metabolomics experiments, containing MS and MS/MS spectra.
- Human Metabolome Database – HMDB (Wishart *et al.*, 2017), contains detailed information about metabolites found in the human body.
- Small Molecule Pathway Database – SMPDB (Jewison *et al.*, 2014) contains more than 700 small molecule pathways found in humans and can assist with pathway elucidation and discovery.

- Kyoto Encyclopaedia of Genes and Genomes – KEGG (Kanehisa *et al.*, 2017) contains pathway information including genomes, proteomes and metabolomes from a variety of species.
- MassBank (Horai *et al.*, 2010) is a large shared mass spectral data repository.

The largest database is not always the best – it will contain many species not related to the samples under analysis, therefore increasing the search space and complicating the analysis. A reasonable strategy would be to attempt identification using the smaller and better curated databases first, moving on to larger databases as required.

Identifying a compound by molecular weight alone is very difficult – imagine trying to identify a single person in a stadium full of people using only their weight. It is obvious that there will be many people/compounds of the same or similar (molecular) weight, and that discriminating these will be difficult. High resolution and accurate mass data is helpful here as it significantly narrows the search space. However, even measuring with sufficient accuracy to yield an elemental composition does not usually provide sufficient discrimination to narrow down to a single compound – isomeric or isobaric alternatives are possible.

Chromatographic retention time information can be helpful to narrow the search space, but is not frequently employed due to the wide variation in separation methods employed in different laboratories and consequent wide variation in experimental retention times obtained. An exception to this is the Golm Metabolome Database which contains GC–MS data and uses retention times scaled to a set of standards.

Further information within the mass spectrum can be useful – the relative intensities of the lower abundance isotopes of the compound to confirm isotopic composition for example – but are not often utilised.

Tandem mass spectra or MS/MS fingerprints are useful discriminatory tools but currently only a limited number exist in the available databases for LC–MS based metabolomics (Vinaixa *et al.*, 2016). This is due to experimental variation in spectra obtained from different instrument configurations and different instrument manufacturers, which makes production of a library difficult. However, in the case of Electron Ionisation GC–MS the spectra obtained are very similar across different instruments and different manufacturers and hence a large generic GC–MS database is commercially-viable. The most commonly used of these are from Wiley and NIST. Employing these large spectral libraries makes confident identification of species observed in GC–MS currently much easier than LC–MS.

An increasingly popular alternative is to compare the experimental spectra with those obtained from rules based in silico fragmentation of a large library of molecules using software such as the commercial Mass Frontier, Progenesis QI, or the freely available MetFrag (Ruttkies *et al.*, 2016).

Once a tentative identification has been made using database searching then experimental confirmation should be performed using a purchased or synthesised authentic standard. The standard should have the same behaviour – observed *m/z*, retention time and MS/MS fingerprint – as the unknown. It is preferable to perform large scale purification and NMR as a 'gold standard' confirmation. Standards are not always available, making this the only option in these cases.

Pathway Analysis

Metabolomics data can be employed as part of a systems biology approach looking to understand the interactions between genes, proteins and metabolites (Krumsiek *et al.*, 2016; Johnson *et al.*, 2016).

This can be employed using various 'pathway analysis' tools. Some are incorporated into online metabolomics software such as XCMSOnline and MetPa within MetaboAnalyst. Others are commercially-available packages such as Ingenuity Pathway Analysis from Qiagen Bioinformatics.

The tools will input metabolite information – identity and concentration/peak area – and combine it with protein and gene data from different experiments. The combination allows a bigger overview and provides mechanistic insight – upregulated enzymes resulting in an increase of their metabolic output for example.

This valuable information comes at significant time and money cost – the input must be heavily curated, only including truly significant differences between sample groups. Often further experiments must be performed to confirm the identities of the metabolites observed.

Biological Interpretation

The final stage in metabolomics analysis is of course biological interpretation – making sense of the how the observed changes in metabolites map to changes in the phenotype. Having well-curated data at this stage is paramount. Interpretation can be somewhat of an iterative process – if unexpected changes are observed are we confident about the statistics and identity of this metabolite? If expected changes are not observed have we missed something by over filtering the data or misidentifying the metabolites?

Major Challenges in Mass Spectrometry-Based Metabolomics

Mass spectrometry powered metabolomics is a powerful tool, providing rich, reproducible and high-quality data from the samples analysed. It is not without shortcomings however, which can be broadly categorised into experimental and data analysis.

Experimental

- Control of samples – requirements for large numbers of well-controlled biological replicates.
- High cost of equipment.
- High level of expertise required to perform sample preparation reliably and reproducibly.
- High level of expertise required to operate mass spectrometers and obtain reliable, reproducible data.
- Importance of experimental design – the data obtained must be able to answer the questions asked.

Data analysis

- Variety of data formats produced from different instruments – often require conversion before analysis.
- Data quantity and large file sizes (up to several GB per sample) make data handling difficult and time consuming.
- Appropriate handling of outliers can be difficult to decide.
- Requires appropriate selection of data analysis tools, PCA vs PLS-DA for example.
- Difficult to identify metabolites from mass spectrometric data alone.
- Observation of a metabolite change between phenotypes does not mean that this metabolite is responsible for the phenotype – causality or coincidence cannot be established directly from the data.

Lack of standardisation of rigour, experimental approach, outputs and reporting between different labs and instrument manufacturers is currently a large problem in metabolomics (Rocca-Serra *et al.*, 2016). This makes comparisons and reproducing results between different laboratories very difficult. Also, for example, the data output format of the metabolomics experiments may not match the requirements for input for pathway analysis, depending on the software used. Consortia such as the Metabolomics Standards Initiative (Fiehn *et al.*, 2007) and COSMOS (Salek *et al.*, 2015) have been set up to tackle some of these issues. There is movement in the right direction, for example the adoption of International Chemical Identifiers (InChiKeys) to describe metabolite structure. The Metabolomics Workbench (Sud *et al.*, 2016) also offer a public repository for metabolomics data, helping standardisation and research efforts.

Applications

Targeted and untargeted metabolomics have an incredibly wide variety of applications. These include toxicology, study of disease, nutrition and functional genomics.

Future Directions

Metabolomics has a bright future and is an area of intensive research around the globe. Future areas for improvement include

- Advances in instrumentation such as more accurate and more sensitive mass spectrometers and more powerful separation technologies.
- Advances in community analysis such as widely-available cloud-based databases for compound identification.
- Increasing numbers of MS/MS libraries suitable for searching for identities within LC–MS/MS data.
- Community databases of 'known unknowns' where the metabolite's identity could not be ascertained, but is observed by other laboratories in their samples.

See also: Computational Lipidomics. Experimental Platforms for Extracting Biological Data: Mass Spectrometry, Microarray, Next Generation Sequencing. Investigating Metabolic Pathways and Networks. Metabolic Profiling. Metabolome Analysis. Metabolome Analysis. Natural Language Processing Approaches in Bioinformatics

---

References

Armitage, E.G., Godzien, J., Alonso-Herranz, V., López-Gonzálvez, Á., Barbas, C., 2015. Missing value imputation strategies for metabolomics data. Electrophoresis 36 (24), 3050–3060. Available at: https://doi.org/10.1002/elps.201500352.

Barker, M., Rayens, W., 2003. Partial least squares for discrimination. Journal of Chemometrics 17, 166–173. Available at: https://doi.org/10.1002/cem.785.

Bleakney, W., 1929. A new method of positive ray analysis and its application to the measurement of ionization potentials in mercury vapor. Physical Review 34 (1), 157–160. Available at: https://doi.org/10.1103/PhysRev.34.157.

Boesl, U., 2017. Time-of-flight mass spectrometry: Introduction to the basics. Mass Spectrometry Reviews. Available at: https://doi.org/10.1002/mas.21520.

Brereton, R.G., Lloyd, G.R., 2014. Partial least squares discriminant analysis: Taking the magic away. Journal of Chemometrics 28 (4), Available at: https://doi.org/10.1002/cem.2609.

Brunee, C., 1987. The ideal mass analyser: Fact or fiction? International Journal of Mass Spectrometry and Ion Processes 76, 125–237.

Byrdwell, W.C., 2001. Atmospheric pressure chemical ionization mass spectrometry for analysis of lipids. Lipids 36 (4), 327–346. Available at: https://doi.org/10.1007/s11745-001-0725-5.

Carroll, A.J., Badger, M.R., Millar, A.H., 2010. The metabolomeexpress project: Enabling web-based processing, analysis and transparent dissemination of GC/MS metabolomics datasets. BMC Bioinformatics 11, 376–388. Available at: https://doi.org/10.1016/j.febslet.2005.01.029.

Carroll, D.I., Dzidic, I., Stillwell, R.N., Horning, M.G., Horning, E.C., 1974. Subpicogram detection system for gas phase analysis based upon atmospheric pressure ionization (API) mass spectrometry. Analytical Chemistry 46 (6), 706–710. Available at: https://doi.org/10.1021/ac60342a009.

de Hoffmann, E., 1996. Tandem mass spectrometry: A primer. Journal of Mass Spectrometry 31 (2), 129–137. Available at: https://doi.org/10.1002/(SICI)1096-9888(199602)31:2 129::AID-JMS305 3.0.CO;2-T.

De Livera, A.M., Sysi-Aho, M., Jacob, L., et al., 2015. Statistical methods for handling unwanted variation in metabolomics data. Analytical Chemistry 87 (7), Available at: https://doi.org/10.1021/ac502439y.

Deutsch, E., 2008. mzML: A single, unifying data format for mass spectrometer output. Proteomics 8 (14), 2776–2777. Available at: https://doi.org/10.1002/pmic.200890049.

Di Guida, R., Engel, J., Allwood, J.W., et al., 2016. Non-targeted UHPLC-MS metabolomic data processing methods: A comparative investigation of normalisation, missing value imputation, transformation and scaling. Metabolomics 12 (5), 1–14. Available at: https://doi.org/10.1007/s11306-016-1030-9.

Doerr, A., 2014. DIA mass spectrometry. Nature Methods 12 (1), 35. Available at: https://doi.org/10.1038/nmeth.3234.

Douglas, D.J., 2009. Linear quadrupoles in mass spectrometry. Mass Spectrometry Reviews. 937–960. Available at: https://doi.org/10.1002/mas.

Fenn, J.B., Mann, M., Meng, C.K.A.I., Wong, S.F., Whitehouse, C.M., 1989. Electrospray ionization for mass spectrometry of large biomolecules. Science 246 (4926), 64–71.

Fiehn, O., Robertson, Æ.D., Griffin, Æ.J., et al., 2007. The metabolomics standards initiative (MSI). Metabolomics 3, 175–178. Available at: https://doi.org/10.1007/s11306-007-0070-6.

Gaskell, S.J., 1997. Special feature: Electrospray: Principles and practice. Journal of Mass Spectrometry 32, 677–688.

Gillet, L.C., Navarro, P., Tate, S., et al., 2012. Targeted data extraction of the MS/MS spectra generated by data-independent acquisition: A new concept for consistent and accurate proteome analysis. Molecular & Cellular Proteomics 11 (6), (O111.016717) Available at: https://doi.org/10.1074/mcp.O111.016717.

Gowda, H., Ivanisevic, J., Johnson, C.H., et al., 2014. Interactive XCMS online: Simplifying advanced metabolomic data processing and subsequent statistical analyses. Analytical Chemistry 86 (14), Available at: https://doi.org/10.1021/ac500734c.

Green, J.M., 1996. Analytical method validation. Analytical Chemistry 305A.

Griffin, J.L., Atherton, H., Shockcor, J., Atzori, L., 2011. Metabolomics as a tool for cardiac research. Nature Reviews Cardiology 8 (11), 630–643. Available at: https://doi.org/10.1038/nrcardio.2011.138.

Griffiths, W.J., Koal, T., Wang, Y., et al., 2010. Targeted metabolomics for biomarker discovery. Angewandte Chemie International Edition. Available at: https://doi.org/10.1002/anie.200905579.

Hartigan, J., Wong, M., 1979. A K-means clustering algorithm. Journal of the Royal Statistical Society Series C (Applied Statistics) 28 (1), 100–108. Available at: https://doi.org/10.2307/2346830.

Horai, H., Arita, M., Kanaya, S., et al., 2010. MassBank: A public repository for sharing mass spectral data for life sciences. Journal of Mass Spectrometry 45 (7), 703–714. Available at: https://doi.org/10.1002/jms.1777.

Hrydziuszko, O., Viant, M.R., 2012. Missing values in mass spectrometry based metabolomics: An undervalued step in the data processing pipeline. Metabolomics 8, 161–174. Available at: https://doi.org/10.1007/s11306-011-0366-4.

Jewison, T., Su, Y., Disfany, F.M., et al., 2014. SMPDB 2.0: Big improvements to the small molecule pathway database. Nucleic Acids Research 42 (D1), 478–484. Available at: https://doi.org/10.1093/nar/gkt1067.

Johnson, C.H., Ivanisevic, J., Siuzdak, G., 2016. Metabolomics: Beyond biomarkers and towards mechanisms. Nature Reviews Molecular Cell Biology 17 (7), Available at: https://doi.org/10.1038/nrm.2016.25.

Johnson, S.C., 1967. Hierarchical clustering schemes. Psychometrika 32, 241–254.

Jolliffe, I.T., 2002. Principal Component Analysis, second ed. Springer.

Kanehisa, M., Furumichi, M., Tanabe, M., Sato, Y., Morishima, K., 2017. KEGG: New perspectives on genomes, pathways, diseases and drugs. Nucleic Acids Research 45 (D1), D353–D361. Available at: https://doi.org/10.1093/nar/gkw1092.

Kell, D.B., Brown, M., Davey, H.M., et al., 2005. Metabolic footprinting and systems biology: The medium is the message. Nature Reviews 3, 557. Available at: https://doi.org/10.1038/nrmicro1177.

Kessner, D., Chambers, M., Burke, R., Agus, D., Mallick, P., 2008. ProteoWizard: Open source software for rapid proteomics tools development. Bioinformatics 24 (21), 2534–2536. Available at: https://doi.org/10.1093/bioinformatics/btn323.

Kind, T., Tsugawa, H., Cajka, T., et al., 2017. Identification of small molecules using accurate mass MS/MS search. Mass Spectrometry Reviews. Available at: https://doi.org/10.1002/mas.21535.

Koal, T., Deigner, H.-P., 2010. Challenges in mass spectrometry based targeted metabolomics. Current Molecular Medicine 10, 216–226. Available at: https://doi.org/10.2174/156652410790963312.

Kohonen, T., 1990. The self-organizing map. Proceedings of the IEEE 78 (9), 1464–1480. Available at: https://doi.org/10.1109/5.58325.

Krumsiek, J., Bartel, J., Theis, F.J., 2016. Computational approaches for systems metabolomics. Current Opinion in Biotechnology. Available at: https://doi.org/10.1016/j.copbio.2016.04.009.

Lin, S.M., Zhu, L., Winter, A.Q., Sasinowski, M., Kibbe, W.A., 2005. What is mzXML good for? Expert Review of Proteomics 2 (6), 839–845. Available at: https://doi.org/10.1586/14789450.2.6.839.

Mann, M., Hendrickson, R.C., Pandey, A., 2001. Analysis of proteins and proteomes by mass spectrometry. Annual Reviews in Biochemistry 70, 437–473. Available at: https://doi.org/10.1146/annurev.biochem.70.1.437.

March, R.E., . Quadrupole ion traps. Mass Spectrometry Reviews. 961–989. Available at: https://doi.org/10.1002/mas.

Marshall, A.G., Hendrickson, C.L., Jackson, G.S., 1998. Fourier transform ion cyclotron resonance mass spectrometry: A primer. Mass Spectrometry Reviews 17, 1–35. Available at: https://doi.org/10.1002/(SICI)1098-2787(1998)17:1 1::AID-MAS1 3.0.CO;2-K.

Munger, J., Bennett, B.D., Parikh, A., et al., 2008. Systems-level metabolic flux profiling identifies fatty acid synthesis as a target for antiviral therapy. Nature Biotechnology 26 (10), 1179–1186. Available at: https://doi.org/10.1038/nbt.1500.

Munson, M.S.B., Field, F.H., 1966. Chemical ionization mass spectrometry. I. General introduction. Journal of the American Chemical Society 88 (12), 2621–2630. Available at: https://doi.org/10.1021/ja00964a001.

Oliver, S.G., 2002. Functional genomics: Lessons from yeast. Philosophical Transactions of the Royal Society B: Biological Sciences 357 (1417), 17–23. Available at: https://doi.org/10.1098/rstb.2001.1049.

Orchard, S., Montechi-Palazzi, L., Deutsch, E.W., et al., 2007. Five years of progress in the standardization of proteomics data 4th annual spring workshop of the HUPO-proteomics standards initiative – April 23–25, 2007. Ecole Nationale Supérieure (ENS), Lyon, France. Proteomics 7 (19), 3436–3440. Available at: https://doi.org/10.1002/pmic.200700658.

Pedrioli, P.G.A., Eng, J.K., Hubley, R., et al., 2004. A common open representation of mass spectrometry data and its application to proteomics research. Nature Biotechnology 22 (11), 1459–1466. Available at: https://doi.org/10.1038/nbt1031.

Pence, H.E., Williams, A., 2010. Chemspider: An online chemical information resource. Journal of Chemical Education 87 (11), 1123–1124. Available at: https://doi.org/10.1021/ed100697w.

Plumb, R.S., Johnson, K.A., Rainville, P., et al., 2006. UPLC/MSE; A new approach for generating molecular fragment information for biomarker structure elucidation. Rapid Communications in Mass Spectrometry 20 (13), 1989–1994. Available at: https://doi.org/10.1002/rcm.2550.

Raterink, R.J., Lindenburg, P.W., Vreeken, R.J., Ramautar, R., Hankemeier, T., 2014. Recent developments in sample-pretreatment techniques for mass spectrometry-based metabolomics. Trends in Analytical Chemistry. Available at: https://doi.org/10.1016/j.trac.2014.06.003.

Roberts, L.D., Souza, A.L., Gerszten, R.E., Clish, C.B., 2012. Targeted metabolomics. Current Protocols in Molecular Biology 1 (Suppl. 98), Available at: https://doi.org/10.1002/0471142727.mb3002s98.

Rocca-Serra, P., Salek, R.M., Arita, M., et al., 2016. Data standards can boost metabolomics research, and if there is a will, there is a way. Metabolomics 12 (1), 1–13. Available at: https://doi.org/10.1007/s11306-015-0879-3.

Ruttkies, C., Schymanski, E.L., Wolf, S., Hollender, J., Neumann, S., 2016. MetFrag relaunched: Incorporating strategies beyond in silico fragmentation. Journal of Cheminformatics 8 (1), Available at: https://doi.org/10.1186/s13321-016-0115-9.

Salek, R.M., Neumann, S., Schober, D., et al., 2015. Coordination of Standards in MetabOlomicS (COSMOS): Facilitating integrated metabolomics data access. Metabolomics 11 (6), Available at: https://doi.org/10.1007/s11306-015-0810-y.

Smith, C.A., Want, E.J., O'Maille, G., Abagyan, R., Siuzdak, G., 2006. XCMS: Processing mass spectrometry data for metabolite profiling using nonlinear peak alignment, matching, and identification. Analytical Chemistry 78 (3), 779–787. Available at: https://doi.org/10.1021/ac051437y.

Smith, Colin A., O'Maille, Grace, Want, Elizabeth J., et al., 2005. A metabolite mass spectral database. Therapeutic Drug Monitoring 27 (6), 747–751.

Smith, R., Ventura, D., Prince, J.T., 2013. LC-MS alignment in theory and practice: A comprehensive algorithmic review. Briefings in Bioinformatics 16 (1), 104–117. Available at: https://doi.org/10.1093/bib/bbt080.

Sud, M., Fahy, E., Cotter, D., et al., 2016. Metabolomics Workbench: An international repository for metabolomics data and metadata, metabolite standards, protocols, tutorials and training, and analysis tools. Nucleic Acids Research 44 (D1), Available at: https://doi.org/10.1093/nar/gkv1042.

Sugimoto, M., Kawakami, M., Robert, M., Soga, T., 2012. Bioinformatics tools for mass spectroscopy-based metabolomic data processing and analysis. Current Bioinformatics 7, 96–108.

Tautenhahn, R., Bottcher, C., Neumann, S., 2008. Highly sensitive feature detection for high resolution LC/MS. BMC Bioinformatics 9, 1–16. Available at: https://doi.org/10.1186/1471-2105-9-504.

Tomasi, G., Van Den Berg, F., Andersson, C., 2004. Correlation optimized warping and dynamic time warping as preprocessing methods for chromatographic data. Journal of Chemometrics 18 (5), 231–241. Available at: https://doi.org/10.1002/cem.859.

Tsugawa, H., Cajka, T., Kind, T., et al., 2015. MS-DIAL: Data-independent MS/MS deconvolution for comprehensive metabolome analysis. Nature Methods 12 (6), Available at: https://doi.org/10.1038/nmeth.3393.

Unidata, Available at: https://ww.unidata.ucar.edu, https://doi.org/10.5065/D6H70CW6.

Vaniya, A., Fiehn, O., 2015. Using fragmentation trees and mass spectral trees for identifying unknown compounds in metabolomics. Trends in Analytical Chemistry. Available at: https://doi.org/10.1016/j.trac.2015.04.002.

Vinaixa, M., Samino, S., Saez, I., et al., 2012. A guideline to univariate statistical analysis for LC/MS-based untargeted metabolomics-derived data. Metabolites 2 (4), 775–795. Available at: https://doi.org/10.3390/metabo2040775.

Vinaixa, M., Schymanski, E.L., Neumann, S., et al., 2016. Mass spectral databases for LC/MS- and GC/MS-based metabolomics: State of the field and prospects. Trends in Analytical Chemistry. Available at: https://doi.org/10.1016/j.trac.2015.09.005.

Wang, Y., Xiao, J., Suzek, T.O., et al., 2009. PubChem: A public information system for analyzing bioactivities of small molecules. Nucleic Acids Research 37 (Suppl. 2), S623–S633. Available at: https://doi.org/10.1093/nar/gkp456.

Winter, G., Krömer, J.O., 2013. Fluxomics – Connecting 'omics analysis and phenotypes. Environmental Microbiology 15 (7), 1901–1916. Available at: https://doi.org/10.1111/1462-2920.12064.

Wishart, D.S., Feunang, Y.D., Marcu, A., et al., 2017. HMDB 4.0: The human metabolome database for 2018. Nucleic Acids Research. 1–10. Available at: https://doi.org/10.1093/nar/gkx1089.

Worley, B., Powers, R., 2013. Multivariate analysis in metabolomics. Current Metabolomics 1, 92–107.

Xia, J., Psychogios, N., Young, N., Wishart, D.S., 2009. MetaboAnalyst: A web server for metabolomic data analysis and interpretation. Nucleic Acids Research. Available at: https://doi.org/10.1093/nar/gkp356.

Xia, J., Sinelnikov, I.V., Han, B., Wishart, D.S., 2015. MetaboAnalyst 3.0-making metabolomics more meaningful. Nucleic Acids Research 43 (W1), Available at: https://doi.org/10.1093/nar/gkv380.

Yost, R.A., Fetterolf, D.D., 1983. Tandem mass spectrometry (MS/MS) instrumentation. Mass Spectrometry Reviews 2 (1), 1–45. Available at: https://doi.org/10.1002/mas.1280020102.

Zubarev, R.A., Makarov, A., 2013. Orbitrap mass spectrometry. Analytical Chemistry 85 (11), 5288–5296. Available at: https://doi.org/10.1021/ac4001223.

Further Reading

Barnes, S., Benton, H.P., Casazza, K., et al., 2016. Training in metabolomics research. II. Processing and statistical analysis of metabolomics data, metabolite identification, pathway analysis, applications of metabolomics and its future. Journal of Mass Spectrometry. Available at: https://doi.org/10.1002/jms.3780.

Beger, R.D., Dunn, W., Schmidt, M.A., et al., 2016. Metabolomics enables precision medicine: "A white paper, community perspective." Metabolomics 12 (10), Available at: https://doi.org/10.1007/s11306-016-1094-6.

Bouatra, S., Aziat, F., Mandal, R., et al., 2013. The human urine metabolome. PLOS ONE 8 (9), Available at: https://doi.org/10.1371/journal.pone.0073076.

Coble, J.B., Fraga, C.G., 2014. Comparative evaluation of preprocessing freeware on chromatography/mass spectrometry data for signature discovery. Journal of Chromatography A. 1358. Available at: https://doi.org/10.1016/j.chroma.2014.06.100.

Dunn, W.B., Lin, W., Broadhurst, D., et al., 2014. Molecular phenotyping of a UK population: Defining the human serum metabolome. Metabolomics 11 (1), 9–26. Available at: https://doi.org/10.1007/s11306-014-0707-1.

Engskog, M.K.R., Haglöf, J., Arvidsson, T., Pettersson, C., 2016. LC–MS based global metabolite profiling: The necessity of high data quality. Metabolomics. Available at: https://doi.org/10.1007/s11306-016-1058-x.

Gates, S.C., Sweeley, C.C., 1978. Quantitative metabolic profiling based on gas chromatography. Clinical Chemistry 24 (10), 1663–1673.

Gorrochategui, E., Jaumot, J., Lacorte, S., Tauler, R., 2016. Data analysis strategies for targeted and untargeted LC-MS metabolomic studies: Overview and workflow. Trends in Analytical Chemistry 82, 425–442. Available at: https://doi.org/10.1016/j.trac.2016.07.004.

Griffin, J.L., Shockcor, J.P., 2004. Metabolic profiles of cancer cells. Nature Reviews Cancer 4 (7), 551–561. Available at: https://doi.org/10.1038/nrc1390.

Huan, T., Forsberg, E.M., Rinehart, D., et al., 2017. Systems biology guided by XCMS Online metabolomics. Nature Methods 14 (5), 461–462. Available at: https://doi.org/10.1038/nmeth.4260.

Johnson, C.H., Ivanisevic, J., Benton, H.P., Siuzdak, G., 2015. Bioinformatics: The next frontier of metabolomics. Analytical Chemistry. Available at: https://doi.org/10.1021/ac5040693.

Johnson, C.H., Ivanisevic, J., Siuzdak, G., 2016. Metabolomics: Beyond biomarkers and towards mechanisms. Nature Reviews Molecular Cell Biology 17 (7), Available at: https://doi.org/10.1038/nrm.2016.25.

Krumsiek, J., Theis, F.J., 2013. Statistical methods for the analysis of high-throughput metabolomics data. Computational and Structural Biotechnology Journal 4 (5).

Mahieu, N.G., Genenbacher, J.L., Patti, G.J., 2016. A roadmap for the XCMS family of software solutions in metabolomics. Current Opinion in Chemical Biology. Available at: https://doi.org/10.1016/j.cbpa.2015.11.009.

Martin, J.C., Maillot, M., Mazerolles, G., Verdu, A., et al., 2015. Can we trust untargeted metabolomics? Results of the metabo-ring initiative, a large-scale, multi-instrument inter-laboratory study. Metabolomics 11 (4), 807–821. Available at: https://doi.org/10.1007/s11306-014-0740-0.

Misra, B.B., van der Hooft, J.J.J., 2016. Updates in metabolomics tools and resources: 2014–2015. Electrophoresis. Available at: https://doi.org/10.1002/elps.201500417.

Pluskal, T., Castillo, S., Villar-briones, A., Ore, M., 2010. MZmine 2: Modular framework for processing, visualizing, and analyzing mass spectrometry-based molecular profile data. BMC Bioinformatics 11 (395),

Ren, S., Hinzman, A.A., Kang, E.L., Szczesniak, R.D., Lu, L.J., 2015. Computational and statistical analysis of metabolomics data. Metabolomics 11. Available at: https://doi.org/10.1007/s11306-015-0823-6.

Rocca-Serra, P., Salek, R.M., Arita, M., et al., 2016. Data standards can boost metabolomics research, and if there is a will, there is a way. Metabolomics 12 (1), 1–13. Available at: https://doi.org/10.1007/s11306-015-0879-3.

Spicer, R., Salek, R.M., Moreno, P., Cañueto, D., Steinbeck, C., 2017. Navigating freely-available software tools for metabolomics analysis. Metabolomics. Available at: https://doi.org/10.1007/s11306-017-1242-7.

Steuer, R., Morgenthal, K., Weckwerth, W., Selbig J., 2007. A gentle guide to the analysis of metabolomic data. Methods in Molecular Biology 358, 105–128.

Sugimoto, M., Kawakami, M., Robert, M., Soga, T., 2012. Bioinformatics tools for mass spectroscopy-based metabolomic data processing and analysis. Current Bioinformatics, 96–108.

Viant, M.R., Kurland, I.J., Jones, M.R., Dunn, W.B., 2017. How close are we to complete annotation of metabolomes? Current Opinion in Chemical Biology 36, 64–69. Available at: https://doi.org/10.1016/j.cbpa.2017.01.001.

Vinaixa, M., Schymanski, E.L., Neumann, S., et al., 2016. Mass spectral databases for LC/MS- and GC/MS-based metabolomics: State of the field and prospects. Trends in Analytical Chemistry. Available at: https://doi.org/10.1016/j.trac.2015.09.005.

Wu, Y., Li, L., 2016. Sample normalization methods in quantitative metabolomics. Journal of Chromatography A. Available at: https://doi.org/10.1016/j.chroma.2015.12.007.

Xia, J., Broadhurst, D.I., Wilson, M., Wishart, D.S., 2013. Translational biomarker discovery in clinical metabolomics: An introductory tutorial. Metabolomics 9 (2), Available at: https://doi.org/10.1007/s11306-012-0482-9.

Xia, J., Wishart, D.S., 2016. Using MetaboAnalyst 3.0 for comprehensive metabolomics data analysis. Current Protocols in Bioinformatics. Available at: https://doi.org/10.1002/cpbi.11.

Yi, L., Dong, N., Yun, Y., et al., 2016. Chemometric methods in data processing of mass spectrometry-based metabolomics: A review. Analytica Chimica Acta 914, 17–34. Available at: https://doi.org/10.1016/j.aca.2016.02.001.

Relevant Websites

http://metabolomicssociety.org/resources/metabolomics-software
   Metabolomics Software
   Metabolomics Society.
http://metabolomicssociety.org/resources/metabolomics-standards
   Metabolomics Standards
   Metabolomics Society.
http://metabolomicssociety.org/resources/metabolomics-databases
   Metabolomics Society: Databases.
http://www.metaboanalyst.ca/
   MetaboAnalyst.
https://mzmine.github.io/
   MZmine 2.
https://xcmsonline.scripps.edu/
   XCMS Online
   The Scripps Research Institute.

Biographical Sketch

Doctor Russell Pickford was first introduced to metabolite mass spectrometry during his undergraduate degree in 1999 (Sheffield University). He then went on to study an MSc in Analytical Chemistry (Warwick University) and a PhD in biological mass spectrometry (University of Manchester and University of York).Following this he moved to the Bioanalytical Mass Spectrometry Facility at the University of New South Wales in 2003. He is responsible for the 'small molecule' mass spectrometric analysis team. He enjoys work on a wide range of projects including small molecule quantification and characterisation, metabolomics and lipidomics.

Metabolic Profiling

Joram M Posma, Imperial College London, London, United Kingdom

Nomenclature

^{13}C Carbon-13
^{1}H Hydrogen, proton
ANOVA Analysis of variance
BMRB Bio-magnetic resonance bank
COSY Correlation spectroscopy
δ Chemical shift
DIMS Direct-infusion mass spectrometry
FDR False discovery rate
FID Free induction decay
FT-ICR-MS Fourier-transform ion cyclotron resonance mass spectrometry
GC–MS Gas chromatography coupled mass spectrometry
GWAS Genome-wide association study
HMBC Hetero-nuclear multiple bond correlation
HMDB Human metabolome database
HSQC Hetero-nuclear single quantum coherence
Hz Hertz
ICA Independent component analysis
KEGG Kyoto encyclopaedia of genes and genomes
LC-MS Liquid chromatography coupled mass spectrometry
m/z Mass-to-charge
MHz Megahertz
MS Mass spectrometry
MS/MS Tandem mass spectrometry
MWAS Metabolome-wide association study
n Number of samples
NMR Nuclear magnetic resonance
OLS Ordinary least squares
OPLS Orthogonal projections to latent structures
OSC Orthogonal signal correction
p Number of variables (predictors)
PC Principal component
PCA Principal component analysis
PLS Partial least squares
QqQ Triple quadrupole mass analyzer
QTOF Quadrupole time-of-flight
RF Radio frequency
STOCSY Statistical total correlation spectroscopy
TOCSY Total correlation spectroscopy
TOF Time-of-flight
TSP Trimethylsilyl-[^{2}H$_4$]-propionate
X Data matrix
Y Outcome variable

Introduction: Metabolic Profiling, Phenotyping and Fingerprinting

Metabolic profiling offers a powerful manner for providing a top-down systems level overview that captures the information from both genetic (host and microbiome – metagenome) and environmental (diet, environmental factors, exposures and lifestyle) origins. It involves measuring metabolites, small molecules that are typically under 1 kDa in molecular weight, in biofluids and other body samples (e.g., tissues, cells, breath) using spectroscopic or spectrometric methods to provide knowledge of the metabolic phenotype of an individual. The metabolic phenotype is the combined contribution of genetic and environmental impacts on the metabolic state under a particular set of conditions (Nicholson, 2006). The metabolic profile associated with a particular phenotype captures information that cannot be directed obtained from a genotype, gene expressions, epigenetic changes or the individual proteome, and in addition all these operate on different time scales adding to the complexity of drawing causal inferences from observed patterns (Nicholson and Lindon, 2008) – metabolic profiling aims to combat these problems by monitoring the global outcome and phenotypic traits of all these factors (environmental exposure, physiological status, system biochemistry) as a whole for utilization in personalized medicine (Holmes *et al.*, 2008b).

There are two main starting points of modern metabolic profiling using high-throughput spectroscopic (Nuclear Magnetic Resonance (NMR)) and spectrometric (mass spectrometry (MS)) techniques. The first is 'metabonomics' (Nicholson *et al.*, 1999) which is the analysis of the time-related, global, dynamic metabolic response in living systems to biological stimuli – it was exemplified using NMR spectroscopy and pattern recognition methods and focussed on finding the metabolites that are different between for instance disease states using multivariate statistical analysis for the first time in 1999. The second is 'metabolomics' (Fiehn, 2002) which was described in 2002 as the quantification, characterization and identification of the global metabolic profile in living systems – this initially focussed on using MS to identify as many metabolites in a sample. Over the years, these initial terms have been used interchangeably for studies using MS and/or NMR, with/without multivariate statistical methods and targeted or untargeted approaches. In the same way, as time passed, many different terms were introduced for these types of analyses such as metabolic fingerprinting, metabolic foot-printing, metabolic profiling and metabolite target analysis, for example, see review in Ellis *et al.* (2007), each aimed at a specific type of analysis of the metabolites with overlap in use of these terms by researchers. The first mentions of metabolic profiling in literature date back to the 1970s, way before the current time of high-throughput analysis that characterizes the current use of the term.

A metabolic profile is the result from a chain of conditional probabilistic interactions between metabolites themselves and with other components of the biochemical network, including gene-environment interactions amongst which bacteria, fungi, yeasts, archaea, viruses and parasites (Hooper et al., 2012; Nicholson et al., 2012a). Specific changes in metabolic flow that are out of the systemic homeostatic control can be mapped to pathway perturbations to connect with disparate biological events that operate on different time scales and in different compartments to provide a deeper understanding of the multi-layered perturbations and disruptions in disease development and therapeutic interventions. Metabolites that are shown to be associated with a type of physiological status, both reproducibly and quantitatively, can be used as potential biomarkers for disease diagnosis, therapeutic targets or for monitoring of treatment efficacy.

Backgrounds and Fundamentals of Metabolic Profiling

Analytical Workhorses

Metabolic profiling relies on two main platforms that combine the gathering of systemic information from non- or minimally-invasive sampling of biological samples, for instance biological fluids such as blood/plasma/serum and urine, with high-throughput technologies capable to analyse thousands of samples. Both techniques have their advantages and disadvantages for the detection of metabolites, but because the biologically interesting molecules have different physicochemical properties the methods are complementary as no single platform is able measure all metabolites in a biological sample (Bouatra et al., 2013).

Nuclear magnetic resonance (NMR) spectroscopy

Metabonomics was exemplified using ^{1}H NMR spectroscopy in the first instance (Nicholson et al., 1999). NMR spectroscopy measures specific quantum mechanical properties of the nuclei from atoms with an odd number of (protons + neutrons). The simplest example of an atom that has an intrinsic angular momentum, and thus non-zero spin, is hydrogen (^{1}H) which only has a single proton in the nucleus. The quantum mechanical spin can either be up or down and it is this property that is used to analyse molecules with NMR spectroscopy.

A very strong magnetic field (in MHz) forces all quantum mechanical spins of nuclei to align either parallel to the field ("spin up," lower energy state) or anti-parallel ("spin down," higher energy state). After which a broad range of radio frequency (RF) pulses are used to excite all protons and bring them in a higher energy state perpendicular to the magnetic field (90° pulse), the difference in energy between states is dependent on the magnetic field strength and on the molecule the proton is attached to, as electrons shield protons slightly from applied magnetic fields. Immediately after the pulse stops, all spins are perpendicular to the magnetic field in a higher energy state and they go back to the lower energy, or relaxation, states (aligned with the field) and they emit the absorbed radiation which induces a current in a coil that is wrapped around the sample tube. The signal, known as the free induction decay (FID), contains all the different observed frequencies in the time domain with the difference in frequencies (in Hz) known as the 'chemical shift.' In order to obtain a frequency domain NMR spectrum the FID is Fourier transformed and this results in a spectrum of peaks, the average over multiple FIDs is used to improve the signal-to-noise ratio.

Comparison of spectra in metabolic profiling is made possible by optimized pulse sequences and by the addition of an internal standard which usually is the sodium salt of trimethylsilyl-$[^2H_4]$-propionate (TSP) in aqueous (biofluid) media (Beckonert et al., 2007). The nine protons of TSP are magnetically equivalent and are maximally shielded because the carbon atoms they are attached to are attached to the most electronegative atom, silicon. This means the emitted energy of TSP is highest and therefore it is used as reference at a chemical shift (δ) of 0. To standardize over different magnetic field strengths, the δ of each signal is expressed as parts per million (ppm) – Hz per MHz.

A number of other properties can be used to differentiate different molecules from each other. First, the integral of different peaks from the same molecule are directly proportional to the number of protons. Second, nuclei have a small magnetic moment which influences the signals observed from other nuclei in the vicinity. This is called spin-spin coupling and causes the splitting of NMR peaks in specific patterns, also known as the multiplicity of a peak, and the relative intensities of the peaks in the multiplet follow Pascal's triangle. The most important of the coupling types is scalar coupling, which is the interaction of two nuclei through at most 3 chemical bonds. In a sample there are many molecules of the same compound of which the protons emit different energies when excited due to the different arrangements (up/down) of the spins. The fine structure of peaks from each metabolite is different and this is used to structurally identify metabolites. Most biological molecules have a carbon backbone, thus it is the CH bonds that are typically observed. NH and OH signals can still be observed; however they give weaker signals and are less shielded.

The advantage of NMR spectroscopy is that it requires relatively little sample preparation (Beckonert et al., 2007; Dona et al., 2014), the sample is not destroyed in the process of analysing it, analysis takes only a few minutes and a spectrum contains information about the structure of molecules. This makes NMR spectroscopy a good method for the high-throughput analysis of biofluid samples (Dona et al., 2014) and a very reproducible method. The downside is that because many CH signals will have a similar 'molecular environment' the observed peaks/peak patterns can overlap which makes it difficult to distinguish between peaks from the same metabolite, in addition NMR spectroscopy is typically less sensitive compared to mass spectrometry.

Two-dimensional NMR spectroscopy techniques can be used for metabolite identification to uncover molecular interactions between nuclei. 2D J-resolved NMR spectroscopy separates homonuclear J-couplings along an orthogonal dimension from the standard axis where both the chemical shifts and J-couplings appear. This has the benefit of reducing the overlap between peaks in an NMR spectrum. Other 2D experiments include the COrrelation SpectroscopY (COSY) experiment, which measures correlations between nuclei separated by at most 3 chemical bonds, to study neighbouring atoms and TOtal Correlation SpectroscopY (TOCSY) which does the same but over longer distances to create 2D $^1H - ^1H$ spectra. Hetero-nuclear 2D experiments do the same but for $^1H - ^{13}C$ in the Hetero-nuclear Single Quantum Coherence (HSQC) spectroscopy (one bond) and Hetero-nuclear Multiple Bond Correlation (HMBC) spectroscopy (more than one bond). With the exception of J-resolved spectroscopy, ^2D NMR experiments currently take too much time to be used for profiling studies and are therefore solely used for metabolite identification purposes.

Mass Spectrometry (MS)

The description of metabolomics in 2002 (Fiehn, 2002) focussed heavily on MS approaches, including gas chromatography prefaced MS (GC–MS), liquid chromatography prefaced MS (LC-MS) and Fourier-transform ion cyclotron resonance MS (FT-ICR-MS), but also considered NMR spectroscopy. The concept of mass spectrometry is to measure the mass of a compound that is ionized, expressed as a mass-to-charge (m/z) ratio, and can determine the elemental composition and isotopic signature of a molecule. For detecting the m/z ratios there are two main types of detectors used in metabolic profiling, most discovery based research is done using time-of-flight (TOF) detectors and for analyses requiring higher mass accuracies different types of 'ion traps' are used.

For TOF detectors ions are accelerated by an electric field of a specified strength so that each identically charged ion has the same kinetic energy. Ions with large m/z ratios have lower velocities than lower m/z do, therefore the time it takes for an ion to reach the detector (typically an electron multiplier tube) is directly proportional to the m/z ratio. Quadrupole mass analyzers use four parallel metal rods of which the opposing rods are connected electrically. An RF voltage is applied between the pairs resulting in ions with small m/z ranges being able to pass through the quadrupole, whereas others will collide with the rods. This allows specific m/z to pass through and reach the detector based on the variable voltage. Quadrupoles can be followed by an electron multiplier tube detector alone or precede a TOF detector (QTOF).

The second type of detectors are ion traps and they come in many different forms; the main ones used in metabolic profiling are triple quadrupole mass analyzers (QqQ) where the second quadrupole serves as a collision cell for fragmentation of ions, FT-ICR ion traps and Orbitraps. QqQ's work the same as a normal quadrupole, however now the third quadrupole now selects the fragmented ions (from the second quadrupole) to reach the detector – this often used as method for identification of specific molecules based on their fragmentation patterns and is called tandem-MS or MS/MS. FT-ICR and Orbitraps work based on the concept of having ions go in orbit around a central point and measuring the image current generated by the orbiting ions. In FT-ICR this is achieved by using a magnetic field to force the ions to spiral and in an orbitrap ions are orbiting around a central electrode in a spiral due to electrostatic forces. The different frequencies of the image currents depend on the oscillations of ions of different m/z ratios. Fourier transforming the image current results in a mass spectrum. Different methods exist to perform tandem-MS experiments in FT-ICR and Orbitraps.

It is important to realize is that all these detectors require the molecules to be in the gas phase. In metabolic profiling aqueous samples are usually ionized using electrospray ionization, which disperses the sample to create an aerosol and applies high voltage to create either positively charged ions ("positive mode"), usually by the addition of a cation (H or Na), or negative ions ("negative mode") by loss of a hydrogen.

MS has the advantage over ^1H NMR in that it can provide information of the mass of a compound that can be used to identify a compound. Measuring all masses simultaneously in mass spectrometry can be done using direct-infusion MS (DIMS) and a spectrum can be obtained within minutes (Draper et al., 2013) which makes it a good method for high-throughput analysis of biofluid samples.

However, it has some disadvantages such as reduced ionization efficiency due to the simultaneous ionization of all compounds in the sample, which may contain less volatile compounds, and the corresponding co-elution of metabolites into the mass detector (Draper et al., 2013), the inability to distinguish between molecules with molecular masses that fall into the same mass-to-charge (m/z) bin and batch differences due to changed experimental conditions (Kirwan et al., 2013). The former two of the disadvantages can be solved using a separation technique before ionization, such as LC, GC or capillary electrophoresis to separate compounds based on physicochemical properties. The separation step improves the identification of compounds not only because less metabolites co-elute, but also because additional information is available from the separation set (retention time) that can be used to identify metabolites based on physicochemical properties in addition to the m/z-ratio (see **Fig. 1**).

However, these 'hyphenated' techniques also have disadvantages, for example, the increased run-time makes high-throughput experiments less feasible, the conditions of the chromatographic separation step can change when analysing many samples which then makes the identification of important metabolites more difficult and also metabolites may ionize favourably in either positive or negative mode, therefore experiments would have to be done using both modes of ionization. Dedicated data processing packages, such as XCMS (Smith et al., 2006), focus on retention time alignment, and peak detection, filtering and matching, between spectra. Recent advancements in sample preparation, automated measurements and methodologies for MS-based metabolic profiling are improving its reproducibility for the analysis of large data sets (Want et al., 2010; Dunn et al., 2011). Specific assays and libraries exist for different applications, such as those specifically tailored to the analysis of lipids (Fahy et al., 2009; Kind et al., 2013), and online platforms for data processing (Tautenhahn et al., 2012b).

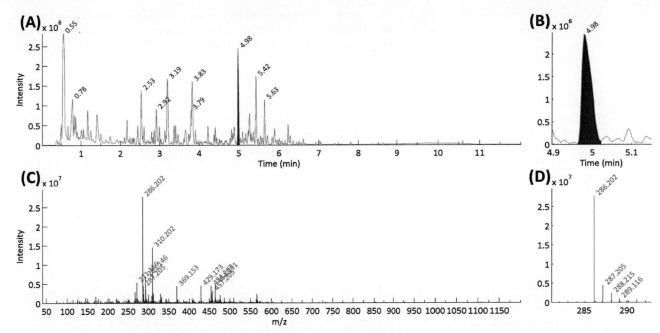

Fig. 1 Example of LC-MS global metabolic profiling analysis of a human urine sample in positive ionization mode. (A) Baseline corrected chromatogram with the ten most intense peaks labelled by their retention time and one selected peak highlighted in blue. (B) Zoom of the selected peak. (C) Mass spectrum extracted from the selected peak, the ten most intense masses are labelled by their m/z ratio. (D) Zoom of the most intense peak in the mass spectrum shows the isotope pattern for a compound with likely elemental composition $C_{15}H_{28}NO_4$.

Statistical Methods

The high-throughput analysis of samples using NMR spectroscopy and MS yields data (X) with many thousands of variables, thus for data analysis this results in data sets where the number of variables far exceeds the number of samples ($n<p$). This is problematic in standard regression methods, such as ordinary least squares (OLS), because variables will no longer be linearly independent and thus X^TX might become (close to) singular and non-invertible. In metabolic profiling different methods are used to deal with the multivariate nature of the data and deal with collinearity of variables to find models that are able to predict samples well and simultaneously provide a metric of assessing the relative contribution of each variable in the predictive model.

Dimension reduction techniques

One approach used in metabolic profiling is to reduce the dimension of the data by decomposing the data into multiple orthogonal linear combinations of the original variables. Principal Component Analysis (PCA) (Hotelling, 1933), also known as Singular Value Decomposition (SVD), is used to decompose X into three parts: left singular vectors (U), a diagonal matrix with singular values ((Σ)) and right singular vectors (V), where the original data can be reconstructed using these three matrices using $X=U(\Sigma)V^T$–PCA is a form of a linear-mixture model. The PC scores ($T=U(\Sigma)$) are used to look for differences between groups of samples in specific PCs by using the scores as a new coordinate system. The sum of the squared elements on the diagonal of (Σ) is equal to the total variance of X and indicates the relative importance of each component. V is used to pinpoint the variables responsible for separation by looking at the magnitude of a variable in a PC that gives information about the weight of a variable ('loading') on the transformation of a spectrum into a score. PCA is often used to 'get a feel' for the data, to see if unsupervised decomposition into latent variables based on the variance can show some signs of clustering according to the experimental design. The benefit of PCA is that components are ranked based on how much variance of X they describe, which can be used as a proxy for their importance.

However, in modern metabolic profiling PCA is rarely sufficient on its own to show differences between groups. Many different processes might impact the concentration of metabolites that may not be easily captured by linear methods. Decomposition of data in specific components according to experimental design can be performed in a supervised manner using ANOVA-based decomposition of the data into effect (data that co-varies with experimental design) and residual (data whose variation is unique and unrelated to experimental design) matrices before analysis of both of these matrices, or only the effect matrices, by PCA (Zwanenburg *et al.*, 2011).

A related approach to PCA is Independent Component Analysis (ICA) which can be used to capture non-linear, and non-Gaussian, metabolic processes from the data. It does so by decomposing X into a mixing (A) and a separating matrix (W), where A gives information about the contribution of samples in each independent component and W indicates the relative importance of variables in each component (Hyvarinen and Oja, 2000). The benefit of ICA is that it aims to find non-Gaussian, thus potentially informative, latent variables and that these are orthogonal to each other as with PCA, a related technique to ICA called Projection

Pursuit does not have the orthogonality constraint. One potential disadvantage of ICA is that it is not straightforward to assess *a priori* how many independent components are required, however the variance explained by k ICs is the same as for the first k PCs. Therefore similar methods can be used as with PCA to determine this for instance to select a cut-off for the eigenvalues or the percentage of variance of X explained by the components.

Classification algorithms

High variance is not necessarily equivalent to high information content. Therefore most metabolic profiling studies make use of supervised methods that aim to identify linear combinations of variables that co-vary with an outcome vector or response variable (Y). The most commonly used methods in metabolic profiling for multivariate regression and discriminant analysis are Partial Least Squares (PLS) (Wold, 1973) and its extension Orthogonal Projections to Latent Structures (OPLS) (Trygg and Wold, 2002).

PLS can be seen as an as a supervised extension of PCA, in that it seeks direction in the data with highest (co-)variance with the response variable, like PCA it calculates multiple orthogonal components. However, unlike PCA – or SVD specifically – these components cannot be obtained in a single step, but are found using a sequential procedure. The weights of the first PLS component are found based on the decomposition of X^TY, and the first score vector is found by multiplying X with the weights, $t=Xw=X(X^TY)$ – for a univariate Y, decomposition of X^TY is X^TY itself. The corresponding loading vector is found by multiplying X^T with normalized scores, $p=\frac{X^Tt}{t^Tt}$. The residual, or 'deflated,' X is found by subtracting tp^T from X. The same process, finding the corresponding loading and finding the residual, is performed for Y. Sequential components are obtained from the residual X and Y using the same process.

The popularity of PLS in metabolic profiling can be explained by the simple linear algebraic nature of PLS, and the low computational costs associated with it. Optimizing the number of components is performed by minimizing the cross-validation error. Including more components will improve the prediction of the model (training) set, but comes at a cost of over-fitting and possibly sub-optimal prediction of the validation (test) set. In order to assess a model's predictive ability a second validation set must be set aside prior to data analysis to avoid biased results, this can be incorporated into a double cross-validation framework (Filzmoser *et al.*, 2009). In addition, to avoid reliance on a single model, some recent metabolic profiling studies have used Monte Carlo double cross-validation which has an important benefit in terms of providing a means of assessing variable importance (Garcia-Perez *et al.*, 2017).

Traditionally, the assessment of important variables in PLS models in metabolic profiling are based on either (a visual inspection of) the magnitude of variables in the loadings, the magnitude of the regression coefficients (by combining multiple loadings into a single regression coefficient vector), calculating the correlation between the predicted scores and the original data, or to assess the importance of a variable in terms of magnitude of each component weighted by the contribution to the model of that component (variable importance on projection) (Mehmood *et al.*, 2012). The difficulty lies in the fact that all variables contribute in a PLS model, i.e., the coefficients are all non-zero – unlike coefficients in penalized regression techniques such as the lasso and elastic net. Some approaches can borrow from univariate statistics and multiple testing corrections. For example, calculating the correlation between the predicted scores and the original data will give information about which variables are likely to be associated with the predicted scores. Despite the fact that the scores come from a multivariate model, calculating correlations of it with all original variables has the potential to give false positive associations with the predicted outcome and as well with the true outcome. A simple form of multiple testing correction on the P-values associated with each of the correlations can limit the number of false positives for correlations with predicted scores. However, outliers in the predicted scores, or even a bad predictive model, can still give rise to 'significant' correlations. Therefore these correlations may not explain the true outcome, as the predicted scores are used as proxy for the true outcome. In addition, reliance on a single model may not generalize well. From all individual models from a Monte Carlo model the mean regression coefficient can be used as a more robust estimate of the true coefficients. Dividing the regression coefficient by its variance can give a t-score which can be converted into a P-value and adjusted for multiple testing. Estimating the mean and variance from the same models can give biased results. Garcia-Perez *et al.* (2017) therefore opted to include a number of bootstrap resamplings of each individual model to obtain an unbiased estimate of the variance of regression coefficients in a Monte Carlo model.

The methods to determine variable importance above are all filtering methods, in that they are calculated *a posteriori*. If a subset of variables is deemed important enough, a new sparser model using only those variables can be calculated, however, it will need to account for the fact that all data that has been used are now biased. Therefore, this requires a third independent validation set that has been completely set aside from the get go to inform on the predictive ability of the new model. Most metabolic profiling studies are not set up for this to be possible.

OPLS is an extension of PLS where data is separated into variation orthogonal to Y and non-orthogonal (informative for outcome) to Y. Two main methods are used that use Orthogonal Signal Correction (OSC) (Eriksson *et al.*, 2000) to extract orthogonal variation. The first is to adjust the data matrix by removing variation from X that is orthogonal to Y. The first score from PCA is used and orthogonalized to Y, once stable the corresponding loading of the orthogonal score is calculated as the loading in PLS is calculated, and this variation is removed from X. This process can be repeated as much as deemed necessary (based on cross-validation). This process of removing non-orthogonal variation from X allows more variation non-orthogonal to Y to be captured in the first few PLS components. Analysis is carried out using regular PLS on the OSC corrected X matrix, X_{OSC}. The second method is to find the first component using PLS and adjust this *a posteriori* for non-orthogonal variation. Often the first predictive and first orthogonal components are used to visualize the class separation in a two dimensional score plot for this method.

Molecular epidemiology

Molecular epidemiology data is characterized by a wealth of metadata in addition to metabolic profiling data. Data analysis focusses on correcting for possible confounding effects that can be identified or gathered from the metadata. Typically this is done using univariate regression including possible confounding factors into a model with one variable at a time. The regression coefficients from all variables are converted to P-values and these are adjusted for multiple testing. Recently, these corrections have moved away from controlling the Family-Wise Error Rate, by for instance a Bonferroni correction, toward False Discovery Rates (FDRs) as some FDR methods are able to account for correlation between variables, whereas the Bonferroni correction assumes independent variables which for $n < p$ does not hold and in addition metabolites are intrinsically correlated. The Benjamini-Hochberg (1995) FDR and the q-value method (Storey and Tibshirani, 2003) are most commonly used, but owing to dedicated software packages (Strimmer, 2008) other methods able to deal with correlated data are now commonly used.

Bottlenecks in Metabolic Profiling

Following data analysis, important variables need to be identified in order to study them in relation to perturbations in the biological system under study. Metabolite identification is simultaneously one of the key aspects as well as a major bottleneck of metabolic profiling for both NMR spectroscopy and MS, as without it the interpretation of results is not-feasible (Wishart, 2011). Both statistical and bioinformatic, and analytical approaches are used complimentarily for NMR spectroscopy and MS (Posma et al., 2017).

Due to the accurate determination of the molecular mass of a metabolite, a chemical formula can be deduced from an MS experiment and both the mass and formula can be used to search in databases for chemical compounds with matching masses or formulae. A number of databases are dedicated to MS data specifically, such as METLIN (Tautenhahn et al., 2012a) and MassBank (Horai et al., 2010), to specific classes of compounds such as lipids, such as LIPID MAPS (Fahy et al., 2009) and LipidBlast (Kind et al., 2013), or tailored to specific platforms such as GC–MS and the FiehnLib (Kind et al., 2009). Some of these databases include species that have not been formally identified or measured, for instance by including all possible combinations of di- and tripeptides in METLIN or inclusion of putative lipids in LIPID MAPS. MS/MS fragmentation patterns of compounds are often also included and searchable in the databases. Other databases focus on data from multiple sources other than MS, with the most widely used resource being the Human Metabolome DataBase (HMDB) (Wishart et al., 2013). HMDB includes background information gathered from different sources and MS and NMR spectra of authentic chemical standards. PubChem (Kim et al., 2016) and ChEBI (Hastings et al., 2016) are the most widely used chemically annotated and curated chemical databases that can be used to find chemical structures and cross-reference with other databases, the Kyoto Encyclopaedia of Genes and Genomes (KEGG) (Kanehisa and Goto, 2000) also contains information on the mass and structure of metabolites and links metabolites via chemical reactions and maps them in biochemical pathways. HMDB, as mentioned, contains a huge collection of NMR spectra, including 2D spectra, of authentic chemical standards. The Bio-Magnetic Resonance Bank (BMRB) (Markley et al., 2008) is another online resource that contains many 1D and 2D NMR spectra of biologically relevant compounds, as well as MS spectra.

The benefit of NMR spectroscopy is that multiple peaks in a spectrum can belong to one and the same metabolite, their chemical shifts and multiplicities can give more information about the chemical structure. The fact that peaks from the same metabolite always appear in the same ratio to one another can be exploited in NMR metabolic profiling by calculating the correlation between one variable and all other variables in the data set, this has been dubbed statistical TOCSY (STOCSY) (Cloarec et al., 2005) and has been one of major workhorses in metabolite identification for NMR metabolic profiling to identify the peaks with high correlations to a 'driver' peak as these are likely from the same metabolite. Different extensions have been proposed that improve on certain aspects, such as to use subset selection and multiple testing to improve the information recovery (Posma et al., 2012), as well as methods to extract metabolite–metabolite interactions from the data and calculate cross-platform correlations (Robinette et al., 2013). Once multiple peaks have been found to be associated with each other, these combinations of chemical shifts and multiplicities can be searched in databases such as HMDB and BMRB to find a match with known compounds and spectra (see **Fig. 2**).

However, not always do databases give an exact match, they might give no (exact) matches or a whole ranges of possible compounds. For the former, the only option is to perform additional experiments using 2D NMR and/or MS/MS to uncover more and complimentary information about an unknown metabolite (Posma et al., 2017). For the latter, experiments can be done by spiking in chemical standards to find a match, or more recently a new approach has emerged that uses metabolic network information, for example, from KEGG, to predict which match has the highest probability of being true based on a joint statistical model with other compounds that have already been identified in the sample (Cai et al., 2017). This uses both the data and prior biological knowledge to narrow down the number of compounds that can be matched to a specific peak. Once a match has been found, it is confirmed using spike-ins with increased concentrations of an authentic chemical standard in a sample which are re-run using the same experiment and experimental conditions to confirm the identity.

Databases rely on the quality of the data provided to them, and in recent years a number of consortia have focussed on harmonising research in metabolic profiling, including analysis protocols, data processing, data analysis, annotation and data storage. This allows for research and data to be more effectively shared within the research community when specific reporting

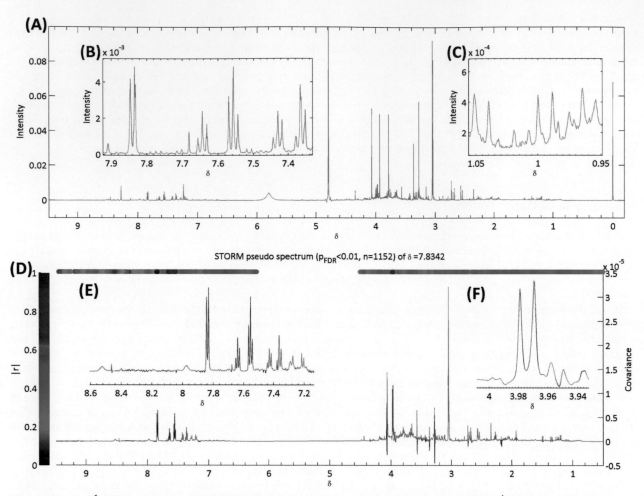

Fig. 2   Example of ^{1}H NMR spectroscopy metabolic profiling of a human urine sample. (A) Unprocessed 600 MHz ^{1}H NMR spectrum, including the internal standard TSP at δ 0 and water resonance at δ 4.74. (B) and (C) Zooms of two regions showing the splitting patterns (multiplicity) of peaks along the chemical shift axis. (D) Statistical spectroscopy analysis using STORM for metabolite identification (of δ 7.83) on a data set with 1208 urine samples shows correlation and covariance between variables, adjusted for multiple testing (pFDR at 1%) and after subset selection (n = 1152). Top bar is a 1D projection of the correlations and highly correlated variables belong to the same metabolite. (E) and (F) Zooms of two regions containing peaks from hippuric acid.

guidelines are followed. One example is the 'Coordination Of Standards in MetabOlomicS' (COSMOS) (Salek *et al.*, 2015) which proposes to use specific data formats that allow simple linkage between data and metadata to facilitate improved data curation and uses the already existing and dedicated repository for metabolic profiling data MetaboLights (Haug *et al.*, 2013). More research focus has to go into automated annotation and (statistical) identification of metabolites in large data sets as these are cost-effective methods when compared to wet-lab chemistry methods that can narrow down the actual experiments that need to be performed for metabolite identification and biomarker profile validation.

Another challenge in metabolic profiling pertains to the analysis of datasets with $n < p$ and pinpointing important variables, as well as the often overlooked aspect of data processing. For instance in MS it is important to filter out noise peaks from the data prior to data analysis. One approach is to filter variables based on the number of non-zero values, i.e., measurements higher than the limit of detection. This procedure is prone to introducing bias into a model when class information is used for filtering, for instance to only select variables that are present in all samples from at least one class. Ideally the filtering step is included in the data analysis, where any noise variables do not contribute to the model. However, when methods are used for which all variables have non-zero coefficients this does not hold. Random Forests and other recursive partitioning methods are able to deal with huge numbers of variables, at relatively low computational cost, and simultaneously perform variable selection and are ideally suited for the analysis of high-dimensional MS data (Enot *et al.*, 2008). However, for NMR these methods are not ideal because of the correlated nature of NMR data, each metabolite can have multiple peaks which each have multiple data points (variables). Likewise, penalized regression methods such as the Lasso will not pinpoint all important variables, whereas the Elastic Net (Zou and Hastie, 2005) could do this, however optimizing the mixing and penalty parameters over a whole range of possible values comes at a heavy computational cost. This opens up the gap for methods, and specifically dedicated software packages, that simultaneously are able to deal with multi-collinearity, perform variable selection including filtering out noise and are able to deal with large numbers of variables that can be used for MS- and NMR-based metabolic profiling alike.

Different Approaches in Experimental and Statistical Design

Some approaches focus on 'global metabolic profiling' in that they assay as many metabolites as possible, whereas others perform targeting of specific metabolites. While targeting can be more accurate experimentally, as the assay can be tailored to the measurement of a number of compounds, the resulting statistical model may be limited in the information it provides of the biological system for exactly this reason. Performing a global profiling analysis first and uncovering important metabolites which are then as a second step targeted for inclusion in a more sparse model, is potentially less biased than starting out using targeted methodology from the get go, and will provide a full(er) systems view.

Another major difference between studies is the statistical analysis of the data, specifically whether a univariate or multivariate approach is chosen. From a statistical point of view, there is no single right (or wrong) way to go about the analysis. From a more conceptual point of view, multivariate analysis make more sense considering the metabolites interact in a metabolic network themselves and can thus influence changes in disparate metabolic compartments. The most important aspect is that the results are properly validated, tested and that models are free of any bias.

Illustrative Examples of Metabolic Profiling

Metabolome-Wide Association Studies (MWAS) (Holmes *et al.*, 2008a; Elliott *et al.*, 2015) are used to find links between epidemiological and metabolic data, analogous to Genome-Wide Associations Studies (GWAS), that are collected at the same time to make inferences about the connections between disease risk factors and phenotype variation. The first study of its kind (Holmes *et al.*, 2008a) showed specific changes in the metabolic profile related to dietary, genetic, gut microbial co-metabolic and xenometabolic factors in relation to blood pressure in four different populations which provided testable hypotheses and important starting points for further research, such as investigating some blood pressure biomarker metabolites in relation to renal ion balance and possible kidney stone formation (Garcia-Perez *et al.*, 2012). The second study (Elliott *et al.*, 2015) focussed on the global profiling of urine and the relation to adiposity as a first stage using univariate statistics and multiple testing correction, and as a second stage specific additional metabolites were targeted, based on other targeted studies performed in other biofluids (Newgard *et al.*, 2009), and modelled in a penalized regression model (Elastic Net) to predict Body Mass Index. This was done on a reduced list of variables to alleviate the computational cost. The model was validated using an additional set of data of the same individuals obtained at a later time and tested in another population. The benefit of using a global profiling approach opposed to a targeted one was demonstrated by the fact that multiple areas known to be associated with obesity were all found simultaneously, such as muscle turnover, diet, amino acids and gut microbial metabolism, whereas targeted studies only identified bits and pieces.

Metabolic profiling can be used to provide information about disease aetiology, diagnosis or progression, for instance for prostate cancer progression (Sreekumar *et al.*, 2009), development of diabetes (Wang *et al.*, 2011) and prediction of survival from sepsis (Langley *et al.*, 2013). For prostate cancer it is important to have non-invasive measures of tracking progression, Sreekumar *et al.* found that one metabolite detectable in urine could be used to indicate progression and further experiments showed that knock-out of enzymes involved with sarcosine synthesis and degradation showed attenuation and increases of cancer invasion, respectively. Likewise, for diabetes and sepsis it was shown that a panel of metabolites in blood could be used to predict future diabetes and survival in sepsis patients.

Then there are the examples that highlight the capabilities of metabolic profiling to be used for personalized medicine, nutrition and care, for instance to predict drug effects (Clayton *et al.*, 2009), detect tumour margin status during surgery *in vestigio* (Balog *et al.*, 2013) and to assess adherence to dietary advice without relying on biased patient reports/records (Garcia-Perez *et al.*, 2017). Pharmacogenomics aims to stratify patients in specific groups that benefit from specific (drug) treatments, however the downside is that it does not capture the environmental variation, pharmacometabonomics aims to predict post-dose response to a drug or treatment based on pre-dose metabolic profile (Everett, 2015). For instance, paracetamol (acetaminophen) was shown to be metabolized differently, in the glucuronide rather than the sulfate form, by individuals based on the excretion of 4-cresylsulfate in urine (Clayton *et al.*, 2009) and, in addition, this observation highlighted the contribution of the gut microbiota that cannot be captured by other approaches such as pharmacogenomics. The second example details the potential of metabolic profiling to be used for the instantaneous detection of cancer tissue based on the metabolic profile of lipids in tissue. In cancer surgery it is imperative that no cancer tissue is left behind after a surgery and at the same time that as little healthy tissue is removed as possible. It is well known that cancer and healthy tissue, as well as different cancer types are characterized by different lipid profiles. The demonstration of the intelligent knife, mass spectrometry coupled to electrosurgery tools to analyse the biochemical composition of smoke after cutting tissue, has shown that on-line metabolic profiling using statistical modelling can match histopathology diagnosis of tissues perfectly and at the same time provide information on tumour biochemistry (Balog *et al.*, 2013). Finally there is public health nutrition which suffers from misreporting that confounds associations between diet and health. Metabolic profiling of urine was used to show that adherence to different diets could be assessed from the urine profile and that the metabolic profile from free-living individuals could be matched to different diets from a controlled clinical trial (Garcia-Perez *et al.*, 2017).

Discussion

Global metabolic profiling allows for an unbiased assessment of metabolites in a sample and can reveal novel and possibly unexpected perturbations in a biological system. The bottleneck point is the identification of metabolites that are deemed most important in a particular model. Specifically those that are considered 'unknown unknowns,' for instance compounds for which no match is found in any database or for which multiple matches remain, even after further structure elucidation using MS/MS and/or 2D NMR experiments. Identifying a compound that does not appear in any database and for which no authentic chemical standard is available is a tedious and cumbersome process requiring many different experiments (Posma et al., 2017). Therefore most metabolic profiling studies focus on identifying unknowns that can be found using information in databases and validated using authentic chemical standards. Similarly, targeted profiling focuses on measurement of predefined groups of metabolites and thus may only provide a limited view. Combining a global profiling as a first stage and targeted analysis as second stage is a more powerful approach in which biomarker metabolites can be validated.

Most NMR-based metabolic profiling studies focus on global profiling, as all metabolites are measured in the same sample simultaneously. Hyphenated MS-based approaches have the benefit that they can be used both to target specific molecules based on retention-time and m/z, as well as to do global profiling. The benefit of NMR is that the sample is not destroyed in the process, which is beneficial when additional experiments are performed on the same sample for identification. Whereas MS is more sensitive than NMR which allows for more metabolites, specifically those present in low concentrations, to be profiled in a single analysis. In addition NMR can be used to analyse tissue samples in a similar way as biofluid samples are, whereas for MS this requires different techniques such as evaporative ionization MS techniques (Balog et al., 2013).

In order to uncover multidimensional metabolic trajectories underlying a particular phenotype, different statistical methods can be used to establish a 'biomarker profile'. Unlike other omics approaches, metabolites interact and are more likely to influence the concentrations of others, even those in disparate metabolic compartments. Intuitively this makes multivariate statistical methods more suited to analysing metabolic profiling data, however whether the choice is made for multi- or univariate statistical methods, the key aspect to take under consideration is to avoid introducing bias into a model, for instance by careful selection of training, test and validation sets, and utilization of robust methods for the identification of important metabolites in a model.

Future Directions for Metabolic Profiling

The future prospects of metabolic profiling do not solely relate to the information it provides, but more in its integration into global systems biology. Specifically, what it can provide for patient stratification in personalized medicine (Nicholson et al., 2012b) and how it can be used in conjunction with other omic data for diagnosis, prognosis and treatment monitoring, as well as for prevention in precision medicine (Collins and Varmus, 2015). Specific attention will be focussed on understanding the environmental variability and trans-genomic interactions that together with the host genome determine the molecular phenotype (Li et al., 2008; Nicholson et al., 2011). This includes understanding how gut microbes are part of host metabolism and integrate these data for modelling metabolic reactions (Posma et al., 2014) and metabolic signalling information (Rodriguez-Martinez et al., 2017), and to better understand metabolism by studying the metabolic fluxes and regulation in a biological system (Zamboni et al., 2015).

Owing to the data tsunami that the omic sciences create, the effective storage of biological samples in biobanks (Elliott et al., 2008) as well as the storage of omic, clinical and other metadata (Haug et al., 2013) has a major impact on future science. In the same light, reproducibility efforts (Fiehn et al., 2016) of previous studies (Ward et al., 2010) aim to address concerns about possible reproducibility issues in scientific research (Ioannidis, 2005). These approaches, either the effective storage of biological specimens for analysis at a later date by analytical techniques, publically available data sets that can be re-analysed using more sophisticated statistical methods and reproduction efforts of high profile studies, contribute to reducing effects of reporting bias.

Development and improvement of bioinformatic and statistical methods to make suitable for analysis of metabolic profiling data, for instance using linear-mixed effects models to correct for confounders as in GWAS (Listgarten et al., 2012), or those tackling specific tasks such as expanding automated and robust data processing pipelines and tools to facilitate metabolite identification, are essential to understand the systems-level effects of metabolites and to deal with the high-dimensionality and collinearity of metabolic profiling data. In addition, the integration of metabolic profiling with other omic data such as metabolic Quantitative Trait Loci mapping and metabolic GWAS (Robinette et al., 2012) or investigating the interplay between the microbiome and metabolism will play an important role in advancing health and understanding of the aetiology truly at a systems level (Johnson et al., 2016).

Closing Remarks

Metabolic profiling has established itself as a field that can provide information about a biological system that are not possible using other systems biology methods. More focus on automated and robust processing pipelines, data analysis algorithms to reduce the level of bias and confounding in the model, development of new approaches to aid in the computational aspect of metabolite identification and efforts that aim to improve the reproducibility of research will increase the impact that metabolic profiling has in human health and disease.

Acknowledgement

The author has no funding information relevant to this work to report.

See also: Epidemiology: A Review. Investigating Metabolic Pathways and Networks. Mass Spectrometry-Based Metabolomic Analysis. Metabolome Analysis. Metabolome Analysis. Natural Language Processing Approaches in Bioinformatics

References

Balog, J., Sasi-Szabo, L., Kinross, J., *et al.*, 2013. Intraoperative tissue identification using rapid evaporative ionization mass spectrometry. Science Translational Medicine 5, 194ra93.

Beckonert, O., Keun, H.C., Ebbels, T.M.D., *et al.*, 2007. Metabolic profiling, metabolomic and metabonomic procedures for NMR spectroscopy of urine, plasma, serum and tissue extracts. Nature Protocols 2, 2692–2703.

Benjamini, Y., Hochberg, Y., 1995. Controlling the false discovery rate – A practical and powerful approach to multiple testing. Journal of the Royal Statistical Society Series B-Methodological 57, 289–300.

Bouatra, S., Aziat, F., Mandal, R., *et al.*, 2013. The human urine metabolome. Plos One 8, e73076.

Cai, Q.P., Alvarez, J.A., Kang, J., Yu, T.W., 2017. Network marker selection for untargeted LC-MS metabolomics data. Journal of Proteome Research 16, 1261–1269.

Clayton, T.A., Baker, D., Lindon, J.C., Everett, J.R., Nicholson, J.K., 2009. Pharmacometabonomic identification of a significant host-microbiome metabolic interaction affecting human drug metabolism. Proceedings of the National Academy of Sciences of the United States of America 106, 14728–14733.

Cloarec, O., Dumas, M.E., Craig, A., *et al.*, 2005. Statistical total correlation spectroscopy: An exploratory approach for latent biomarker identification from metabolic H-1 NMR data sets. Analytical Chemistry 77, 1282–1289.

Collins, F.S., Varmus, H., 2015. A new initiative on precision medicine. New England Journal of Medicine 372, 793–795.

Dona, A.C., Jimenez, B., Schafer, H., *et al.*, 2014. Precision high-throughput proton NMR spectroscopy of human urine, serum, and plasma for large-scale metabolic phenotyping. Anal Chem 86, 9887–9894.

Draper, J., Lloyd, A.J., Goodacre, R., Beckmann, M., 2013. Flow infusion electrospray ionisation mass spectrometry for high throughput, non-targeted metabolite fingerprinting: A review. Metabolomics 9, S4–S29.

Dunn, W.B., Broadhurst, D., Begley, P., *et al.*, 2011. Procedures for large-scale metabolic profiling of serum and plasma using gas chromatography and liquid chromatography coupled to mass spectrometry. Nature Protocols 6, 1060–1083.

Elliott, P., Peakman, T.C., UK Biobank Consortium, 2008. The UK Biobank sample handling and storage protocol for the collection, processing and archiving of human blood and urine. International Journal of Epidemiology 37, 234–244.

Elliott, P., Posma, J.M., Chan, Q., *et al.*, 2015. Urinary metabolic signatures of human adiposity. Science Translational Medicine 7, 285ra62.

Ellis, D.I., Dunn, W.B., Griffin, J.L., Allwood, J.W., Goodacre, R., 2007. Metabolic fingerprinting as a diagnostic tool. Pharmacogenomics 8, 1243–1266.

Enot, D.P., Lin, W., Beckmann, M., *et al.*, 2008. Preprocessing, classification modeling and feature selection using flow injection electrospray mass spectrometry metabolite fingerprint data. Nature Protocols 3, 446–470.

Eriksson, L., Trygg, J., Johansson, E., Bro, R., Wold, S., 2000. Orthogonal signal correction, wavelet analysis, and multivariate calibration of complicated process fluorescence data. Analytica Chimica Acta 420, 181–195.

Everett, J.R., 2015. Pharmacometabonomics in humans: A new tool for personalized medicine. Pharmacogenomics 16, 737–754.

Fahy, E., Subramaniam, S., Murphy, R.C., *et al.*, 2009. Update of the LIPID MAPS comprehensive classification system for lipids. J Lipid Res 50, S9–S14.

Fiehn, O., 2002. Metabolomics – The link between genotypes and phenotypes. Plant Mol Biol 48, 155–171.

Fiehn, O., Showalter, M.R., Schaner-Tooley, C.E., 2016. Registered Report: The common Feature of Leukemia-Associated IDH1 and IDH2 Mutations is a Neomorphic Enzyme Activity Converting Alpha-Ketoglutarate to 2-Hydroxyglutarate, 5. Cambridge: Elife.

Filzmoser, P., Liebmann, B., Varmuza, K., 2009. Repeated double cross validation. Journal of Chemometrics 23, 160–171.

Garcia-Perez, I., Posma, J.M., Gibson, R., *et al.*, 2017. Objective assessment of dietary patterns by use of metabolic phenotyping: A randomised, controlled, crossover trial. The Lancet Diabetes and Endocrinology 5, 184–195.

Garcia-Perez, I., Villasenor, A., Wijeyesekera, A., *et al.*, 2012. Urinary metabolic phenotyping the slc26a6 (chloride-oxalate exchanger) null mouse model. Journal of Proteome Research 11, 4425–4435.

Hastings, J., Owen, G., Dekker, A., *et al.*, 2016. ChEBI in 2016: Improved services and an expanding collection of metabolites. Nucleic Acids Res 44, D1214–D1219.

Haug, K., Salek, R.M., Conesa, P., *et al.*, 2013. MetaboLights – An open-access general-purpose repository for metabolomics studies and associated meta-data. Nucleic Acids Res 41, D781–D786.

Holmes, E., Loo, R.L., Stamler, J., *et al.*, 2008a. Human metabolic phenotype diversity and its association with diet and blood pressure. Nature 453, 396–400.

Holmes, E., Wilson, I.D., Nicholson, J.K., 2008b. Metabolic phenotyping in health and disease. Cell 134, 714–717.

Hooper, L.V., Littman, D.R., Macpherson, A.J., 2012. Interactions between the microbiota and the immune system. Science 336, 1268–1273.

Horai, H., Arita, M., Kanaya, S., *et al.*, 2010. MassBank: A public repository for sharing mass spectral data for life sciences. Journal of Mass Spectrometry 45, 703–714.

Hotelling, H., 1933. Analysis of a complex of statistical variables into principal components. Journal of Educational Psychology 24, 417–441.

Hyvarinen, A., Oja, E., 2000. Independent component analysis: Algorithms and applications. Neural Networks 13, 411–430.

Ioannidis, J.P.A., 2005. Why most published research findings are false. PLOS Medicine 2, 696–701.

Johnson, C.H., Ivanisevic, J., Siuzdak, G., 2016. Metabolomics: Beyond biomarkers and towards mechanisms. Nature Reviews Molecular Cell Biology 17, 451–459.

Kanehisa, M., Goto, S., 2000. KEGG: Kyoto encyclopedia of genes and genomes. Nucleic Acids Res 28, 27–30.

Kim, S., Thiessen, P.A., Bolton, E.E., *et al.*, 2016. PubChem substance and compound databases. Nucleic Acids Res 44, D1202–D1213.

Kind, T., Liu, K.H., Lee, D.Y., *et al.*, 2013. LipidBlast in silico tandem mass spectrometry database for lipid identification. Nat Methods 10, 755–758.

Kind, T., Wohlgemuth, G., Lee, D.Y., *et al.*, 2009. FiehnLib: Mass spectral and retention index libraries for metabolomics based on quadrupole and time-of-flight gas chromatography/mass spectrometry. Anal Chem 81, 10038–10048.

Kirwan, J.A., Broadhurst, D.I., Davidson, R.L., Viant, M.R., 2013. Characterising and correcting batch variation in an automated direct infusion mass spectrometry (DIMS) metabolomics workflow. Analytical and Bioanalytical Chemistry 405, 5147–5157.

Langley, R.J., Tsalik, E.L., Van Velkinburgh, J.C., *et al.*, 2013. An integrated clinico-metabolomic model improves prediction of death in sepsis. Science Translational Medicine 5, 195ra95.

Li, M., Wang, B.H., Zhang, M.H., *et al.*, 2008. Symbiotic gut microbes modulate human metabolic phenotypes. Proceedings of the National Academy of Sciences of the United States of America 105, 2117–2122.

Listgarten, J., Lippert, C., Kadie, C.M., *et al.*, 2012. Improved linear mixed models for genome-wide association studies. Nat Methods 9, 525–526.

Markley, J.L., Ulrich, E.L., Berman, H.M., *et al.*, 2008. BioMagResBank (BMRB) as a partner in the Worldwide Protein Data Bank (wwPDB): New policies affecting biomolecular NMR depositions. Journal of Biomolecular Nmr 40, 153–155.

Mehmood, T., Liland, K.H., Snipen, L., Saebo, S., 2012. A review of variable selection methods in partial least squares regression. Chemometrics and Intelligent Laboratory Systems 118, 62–69.

Newgard, C.B., An, J., Bain, J.R., *et al.*, 2009. A branched-chain amino acid-related metabolic signature that differentiates obese and lean humans and contributes to insulin resistance. Cell Metabolism 9, 311–326.

Nicholson, G., Rantalainen, M., Maher, A.D., *et al.*, 2011. Human metabolic profiles are stably controlled by genetic and environmental variation. Molecular Systems Biology 7.

Nicholson, J.K., 2006. Global systems biology, personalized medicine and molecular epidemiology. Molecular Systems Biology 2, 52.

Nicholson, J.K., Holmes, E., Kinross, J., *et al.*, 2012a. Host-gut microbiota metabolic interactions. Science 336, 1262–1267.

Nicholson, J.K., Holmes, E., Kinross, J.M., *et al.*, 2012b. Metabolic phenotyping in clinical and surgical environments. Nature 491, 384–392.

Nicholson, J.K., Lindon, J.C., 2008. Systems biology: Metabonomics. Nature 455, 1054–1056.

Nicholson, J.K., Lindon, J.C., Holmes, E., 1999. 'Metabonomics': Understanding the metabolic responses of living systems to pathophysiological stimuli via multivariate statistical analysis of biological NMR spectroscopic data. Xenobiotica 29, 1181–1189.

Posma, J.M., Garcia-Perez, I., De Iorio, M., *et al.*, 2012. Subset optimization by reference matching (STORM): An optimized statistical approach for recovery of metabolic biomarker structural information from (1)H NMR spectra of biofluids. Analytical Chemistry 84, 10694–10701.

Posma, J.M., Garcia-Perez, I., Heaton, J.C., *et al.*, 2017. Integrated analytical and statistical two-dimensional spectroscopy strategy for metabolite identification: Application to dietary biomarkers. Anal Chem 89, 3300–3309.

Posma, J.M., Robinette, S.L., Holmes, E., Nicholson, J.K., 2014. MetaboNetworks, an interactive Matlab-based toolbox for creating, customizing and exploring sub-networks from KEGG. Bioinformatics 30, 893–895.

Robinette, S.L., Holmes, E., Nicholson, J.K., Dumas, M.E., 2012. Genetic determinants of metabolism in health and disease: From biochemical genetics to genome-wide associations. Genome Med 4, 30.

Robinette, S.L., Lindon, J.C., Nicholson, J.K., 2013. Statistical spectroscopic tools for biomarker discovery and systems medicine. Anal Chem 85, 5297–5303.

Rodriguez-Martinez, A., Ayala, R., Posma, J.M., *et al.*, 2017. MetaboSignal: A network-based approach for topological analysis of metabotype regulation via metabolic and signaling pathways. Bioinformatics 33, 773–775.

Salek, R.M., Neumann, S., Schober, D., *et al.*, 2015. COordination of Standards in MetabOlomicS (COSMOS): Facilitating integrated metabolomics data access. Metabolomics 11, 1587–1597.

Smith, C.A., Want, E.J., O'Maille, G., Abagyan, R., Siuzdak, G., 2006. XCMS: Processing mass spectrometry data for metabolite profiling using nonlinear peak alignment, matching, and identification. Anal Chem 78, 779–787.

Sreekumar, A., Poisson, L.M., Rajendiran, T.M., *et al.*, 2009. Metabolomic profiles delineate potential role for sarcosine in prostate cancer progression. Nature 457, 910–914.

Storey, J.D., Tibshirani, R., 2003. Statistical significance for genomewide studies. Proceedings of the National Academy of Sciences of the United States of America 100, 9440–9445.

Strimmer, K., 2008. fdrtool: A versatile R package for estimating local and tail area-based false discovery rates. Bioinformatics 24, 1461–1462.

Tautenhahn, R., Cho, K., Uritboonthai, W., *et al.*, 2012a. An accelerated workflow for untargeted metabolomics using the METLIN database. Nature Biotechnology 30, 826–828.

Tautenhahn, R., Patti, G.J., Rinehart, D., Siuzdak, G., 2012b. XCMS Online: A web-based platform to process untargeted metabolomic data. Anal Chem 84, 5035–5039.

Trygg, J., Wold, S., 2002. Orthogonal projections to latent structures (O-PLS). Journal of Chemometrics 16, 119–128.

Wang, T.J., Larson, M.G., Vasan, R.S., *et al.*, 2011. Metabolite profiles and the risk of developing diabetes. Nature Medicine 17, 448–453.

Want, E.J., Wilson, I.D., Gika, H., *et al.*, 2010. Global metabolic profiling procedures for urine using UPLC-MS. Nature Protocols 5, 1005–1018.

Ward, P.S., Patel, J., Wise, D.R., *et al.*, 2010. The common feature of leukemia-associated IDH1 and IDH2 mutations is a neomorphic enzyme activity converting alpha-ketoglutarate to 2-hydroxyglutarate. Cancer Cell 17, 225–234.

Wishart, D.S., 2011. Advances in metabolite identification. Bioanalysis 3, 1769–1782.

Wishart, D.S., Jewison, T., Guo, A.C., *et al.*, 2013. HMDB 3.0 – The Human Metabolome Database in 2013. Nucleic Acids Res 41, D801-7.

Wold, H.O., 1973. Operative aspects of econometric and sociological models current developments of Fp (Fix-Point) estimation and nipals (nonlinear iterative partial least squares) modelling. Economie Appliquee 26, 385–421.

Zamboni, N., Saghatelian, A., Patti, G.J., 2015. Defining the metabolome: Size, flux, and regulation. Mol Cell 58, 699–706.

Zou, H., Hastie, T., 2005. Regularization and variable selection via the elastic net. Journal of the Royal Statistical Society Series B-Statistical Methodology 67, 301–320.

Zwanenburg, G., Hoefsloot, H.C.J., Westerhuis, J.A., Jansen, J.J., Smilde, A.K., 2011. ANOVA-Principal component analysis and ANOVA-simultaneous component analysis: A comparison. Journal of Chemometrics 25, 561–567.

Further Reading

Bouatra, S., Aziat, F., Mandal, R., *et al.*, 2013. The human urine metabolome. PLOS ONE 8, e73076.

De Iorio, M., Ebbels, T.M.D., Stephens, D.A., 2008. Statistical techniques in metabolic profiling. Handbook of Statistical Genetics. John Wiley & Sons, Ltd.. [Chapter 11].

Hastie, T., Tibshirani, R., Friedman, J.H., 2009. The Elements of Statistical Learning: Data Mining, Inference, and Prediction. New York, NY: Springer.

Holmes, E., Loo, R.L., Stamler, J., *et al.*, 2008a. Human metabolic phenotype diversity and its association with diet and blood pressure. Nature 453, 396–400.

Holmes, E., Wilson, I.D., Nicholson, J.K., 2008b. Metabolic phenotyping in health and disease. Cell 134, 714–717.

Nicholson, J.K., Lindon, J.C., 2008. Systems biology: Metabonomics. Nature 455, 1054–1056.

Nicholson, J.K., Lindon, J.C., Holmes, E., 2007. The Handbook of Metabonomics and Metabolomics. Amsterdam; Boston: Elsevier.

Nicholson, J.K., Lindon, J.C., Holmes, E., 2018. The Handbook of Metabolic Phenotyping. Amsterdam; Boston: Elsevier.

Salek, R.M., Neumann, S., Schober, D., *et al.*, 2015. COordination of Standards in MetabOlomicS (COSMOS): Facilitating integrated metabolomics data access. Metabolomics 11, 1587–1597.

Sreekumar, A., Poisson, L.M., Rajendiran, T.M., *et al.*, 2009. Metabolomic profiles delineate potential role for sarcosine in prostate cancer progression. Nature 457, 910–914.

Trygg, J., Holmes, E., Lundstedt, T., 2007. Chemometrics in metabonomics. Journal of Proteome Research 6, 469–479.

Relevant Websites

Databases.
http://www.bmrb.wisc.edu
    Bio-Magnetic Resonance Bank.

http://www.hmdb.ca
   Human Metabolome DataBase.
http://www.lipidmaps.org
   LIPID MAPS.
http://metlin.scripps.edu
   METLIN.
Online software.
http://www.metaboanalyst.ca
   MetaboAnalyst.
http://xcmsonline.scripps.edu
   XCMS-online.

Biographical Sketch

Joram Matthias Posma's work involves the development of new algorithms and methods to maximize information recovery from omic data including multivariate regression and classification algorithms, software for the interactive and immersive visualization of metabolic reaction network data, statistical spectroscopy methods for metabolite identification, and bioinformatic and statistical workflows for the analysis of metabolic profiling data. He is first author of publications that appeared in *Science Translational Medicine, The Lancet Diabetes & Endocrinology, Bioinformatics, Analytical Chemistry* and *Analytical and Bioanalytical Chemistry,* and is co-author of publications in *Nature Communications, Hypertension, Bioinformatics, Analytical Chemistry* and *Journal of Proteome Research.* He obtained his MSc in Chemistry *cum laude* from Radboud University Nijmegen in 2011 and his PhD in Bioinformatics and Clinical Medicine Research from Imperial College London in 2014. He is a Member of the Royal Society of Chemistry.

Metabolic Models

Jean-Marc Schwartz, University of Manchester, Manchester, United Kingdom
Zita Soons, Maastricht University, Maastricht, The Netherlands

Introduction

Metabolism is an essential part of living organisms, constituted by the set of biochemical reactions that process nutrients and produce the molecular components needed by cells to maintain themselves and grow. Metabolic reactions are biochemical transformations but most of them are not spontaneous, they are catalysed by enzyme proteins which are specifically adapted to process a precise transformation. Metabolic reactions have been studied since the 19th century using both *in vitro* and *in vivo* analyses, and are well characterised for many species. Therefore metabolism is well suited for the development of mathematical models, which are aimed at calculating the levels of flux going through metabolic reactions and the amounts of substrates and products used by these reactions.

Different types of metabolic models can be produced, which differ in their level of complexity and in the precision of their outcomes. In the following we explain the principle of the most important approaches for metabolic modelling, with a focus on those that have shown the greatest range of uses and applications.

Kinetic Models

Principle

Kinetic models are the most flexible type of metabolic models. They generally use differential equations to represent detailed mechanisms of enzymatic reactions and enable a detailed computation of reaction fluxes and metabolite concentrations over time.

The formulation of a kinetic model of a metabolic system uses two main types of relations. The first type are mass conservation equations, which reflect the fact that matter is conserved in any type of chemical transformation (**Fig. 1**). For each metabolite, the mass conservation equation is written by expressing that the change in the metabolite concentration is equal to the sum of all incoming reaction fluxes minus the sum of all outgoing reaction fluxes.

The second type of relations are kinetic equations, which are generally expressed in the form of ordinary differential equations. The kinetic equation expresses the dependence of the reaction rate (flux) on the concentration of metabolites participating in the reaction and any regulators. It is worth noting that if the reaction is reversible, the flux can take positive or negative values and the positive direction can be chosen arbitrarily; if the reaction is irreversible only positive values are allowed.

Types of Kinetic Equations

There are many types of possible kinetic equations, depending on the complexity of the reaction mechanism and the desired level of precision (Cornish-Bowden, 2004). The simplest type is mass action kinetics, in which the reaction rate is directly proportional to the concentration of substrate(s). While this relation is sometimes used due to its simplicity, it is only strictly valid for elementary chemical reactions and is generally not an accurate depiction of enzymatic reaction kinetics. In practice, when all enzyme active sites are occupied the reaction rate reaches a plateau determined by the amount of available enzyme, but this property is not captured by the mass action relation.

The Michaelis-Menten relation offers a more accurate depiction of enzymatic reaction kinetics, which captures the saturation property. It is based on the assumption that the enzyme forms a complex with the substrate in a reversible process, followed by an irreversible decomposition of the complex into enzyme and product; moreover the concentration of enzyme is assumed to be much lower than the concentration of substrate. More complex forms of the Michael-Menten relation can be derived depending on

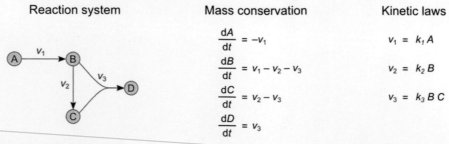

Reaction system	Mass conservation	Kinetic laws
	$\dfrac{dA}{dt} = -v_1$	$v_1 = k_1 A$
	$\dfrac{dB}{dt} = v_1 - v_2 - v_3$	$v_2 = k_2 B$
	$\dfrac{dC}{dt} = v_2 - v_3$	$v_3 = k_3 B C$
	$\dfrac{dD}{dt} = v_3$	

Fig. 1 From metabolic network to kinetic model.

the number of substrates and products, the order of binding of different substrates to the enzyme-substrate complex, the presence of activators and inhibitors, and whether the reaction is reversible or not.

In the case of reactions with multiple substrates and products, it can be difficult to know the exact reaction mechanism and to derive the most accurate form of Michaelis-Menten equation. For this reason, generic forms of enzyme reaction kinetics have been proposed which are applicable to any type of reaction at the expense of simplifying assumptions. One of the most versatile forms is the convenience kinetics equation, which assumes random-order and reversible binding and dissociation of substrates with the enzyme (Liebermeister and Klipp, 2006).

Simulation and Applications

When the kinetic model is built, the time course of flux and concentration values is obtained by resolving the system of differential equations. This is generally not manually feasible, except for very simple systems, because of the mathematical complexity of equations. A number of computational tools are available that are more specifically targeted at simulating biochemical systems, in particular Copasi, CellDesigner, VCell (see "Relevant Website sections").

The construction and use of kinetic metabolic models is challenging, not only because complex equations and numerous parameters need to be determined for their construction, but also because large amounts of quantitative experimental data need to be measured for their validation. For this reason, relatively few practical applications have been reported compared to other methods, despite the fact that such models have been made for a long time. For example, several kinetic models of yeast glycolysis were proposed which were validated against measured concentrations and fluxes, and were further used to explain glycolytic oscillations in yeast (Hynne et al., 2001; Williamson et al., 2012). Kinetic models were developed of the red blood cell and used to predict effects of enzymopathies (Jamshidi et al., 2002), and of iron metabolism in order to simulate haemochromatosis (Mitchell and Mendes, 2013). The Grape software has been developed more specifically to facilitate the construction of large metabolic models using generic rate equations and to enable parameter estimation using 'omics' datasets (Adiamah et al., 2010).

Petri Nets

Although they are generally not classified as a kinetic modelling method, Petri nets are another technique that can be used to simulate dynamic metabolic networks. Petri nets are a concept derived from informatics which is well suited for the modelling of biochemical reaction networks (Chaouiya, 2007). They are based on a bipartite graph representation consisting of *places* and *transitions*, connected by directed arcs. In a model of metabolic pathway, places correspond to metabolites and transitions correspond to enzymatic reactions. Each place holds a certain number of *tokens*, which represents the amount or concentration of the metabolite. When the required number of tokens is present, a transition is *fired* meaning that tokens of substrates are consumed and those of products are produced, in quantities defined by the weights of the arcs connecting them to the transition. This process represents a good analogy of what occurs in a metabolic reaction, therefore many concepts derived from Petri net theory have a real-world interpretation in metabolic networks (Baldan et al., 2010). For example, steady-state flux distributions are the equivalent of *T-invariants* in Petri net theory, and elementary flux modes correspond to *minimal T-invariants*. Several models of metabolic pathways using Petri nets were developed, including both qualitative and quantitative models, for example of sucrose breakdown in potato, human iron metabolism, glycolysis in liver, etc.

Stoichiometric Models

Stoichiometric Matrix

Stoichiometric models are a class of metabolic models that do not use kinetic information but only rely on the mass conservation principle. Mass conservation is embedded in the stoichiometric coefficients of metabolic reactions, which indicate the number of molecules of each substrate and product needed to complete the transformation. For example in the adenylate kinase reaction:

$$2\,ADP \rightarrow ATP + AMP$$

two molecules of adenosine diphosphate (ADP) are transformed into one molecule of adenosine triphosphate (ATP) and one molecule of adenosine monophosphate (AMP).

For computational modelling, it is generally convenient to represent stoichiometric coefficients in the form of a matrix, where each column represents a reaction and each row a metabolic compound. For example the stoichiometric matrix of the adenylate kinase reaction becomes:

$$\begin{array}{c} ADP \\ ATP \\ AMP \end{array} \begin{bmatrix} -2 \\ 1 \\ 1 \end{bmatrix}$$

It is worth noting that the stoichiometric coefficient of the substrate (ADP) is negative, indicating that it is consumed by the reaction, and that of products (ATP and AMP) is positive, indicating that they are produced by the reaction. In the case of reversible reactions, an arbitrary positive orientation needs to be chosen.

Quasi Steady State Assumption

In a metabolic network, the stoichiometric matrix can be used to express mass conservation relations in a condensed form:

$$\mathbf{S} \cdot \mathbf{v} = \frac{d\mathbf{C}}{dt}$$

where $\mathbf{S}$ is the stoichiometric matrix, $\mathbf{v}$ is the vector of reaction fluxes and $\mathbf{C}$ is the vector of metabolite concentrations.

The quasi steady state assumption consists in considering that metabolite concentrations do not change over time, then the previous equation becomes:

$$\mathbf{S} \cdot \mathbf{v} = 0$$

The advantage of this approach is to reduce the system to a set of linear equations, which can be solved according to standard linear algebraic techniques. In practice, this assumption is generally valid in biological systems because metabolism adjusts fast to changes in gene expression, therefore at any time metabolic fluxes can be considered balanced (Fell, 1997). Changes in the flux distribution over time can still be modelled as a succession of steady states resulting from changing boundary conditions. Moreover, it is generally admitted that internal metabolites cannot accumulate or deplete over a long period of time in living systems (Reimers and Reimers, 2016).

Flux Balance Analysis

The linear system of equations obtained from the quasi steady state assumption is generally underdetermined, which means that there is an infinite number of possible solutions. It is possible to reduce the range of possible solutions by adding an assumption of optimality: living organisms do not behave randomly but are expected to have evolved to maximise the production of certain compounds that are essential for their survival and growth. The production of these compounds can be represented by an artificial biomass reaction whose composition reflects the overall molecular composition of the cell. In some cases a different objective is more relevant, for example ATP production may be maximised in order to optimise energetic availability, or in a metabolic engineering context one may seek to maximise the production of certain compounds of biotechnological interest. In all cases, the fluxes leading to the production of objective compounds can be represented by an *objective function Z*. The optimisation of Z is added as a constraint to the linear system of equations defined by $\mathbf{S} \cdot \mathbf{v} = 0$, resulting in a considerable reduction of the solution space. Additional constraints apply to metabolic fluxes, such as the maximum capacity of enzymes which sets an upper limit on the allowable flux, and the irreversibility of some reactions which forbids the flux from running in the reverse direction.

Taken together with the conservation of mass in steady state $\mathbf{S} \cdot \mathbf{v} = 0$, the optimisation of the objective function Z, and information about the cellular environment such as uptake rates of certain nutrients, this set of constraints define the method of Flux Balance Analysis (FBA) (Orth *et al.*, 2010). These equations can be solved by *linear programming* algorithms using computational tools such as the COBRA toolbox (Schellenberger *et al.*, 2011) (see "Relevant Websites" section).

A major advantage of FBA is that this method is fast enough to calculate flux distribution in genome-scale metabolic models containing several thousands of reactions. For this reason, FBA has been one of the most widely used methods for metabolic modelling and has led to numerous applications. Among others it has been used for determining optimal growth conditions of organisms, effects of environmental changes and mutations, understanding how certain pathogens rely on their host's metabolism, determining drug targets, designing optimised strains to produce compounds of interest for biotechnology, etc. (Bordbar *et al.*, 2014; Chapman *et al.*, 2015).

Range of Optimal Solutions

Although the FBA algorithm results in a unique value of the objective function Z, there may still be a range of flux distributions satisfying the optimal constraints of FBA. For example, flux may be distributed along several internal branches in the metabolic network that are neutral in terms of the objective function. The method of Flux Variability Analysis was developed to calculate the upper and lower flux values of all reactions that can lead to the optimal value of Z (Gudmundsson and Thiele, 2010). Besides making it possible to identify degeneracies leading to multiple solutions, this method can also be used to identify blocked reactions, whose flux is constrained to zero due to inaccuracies in the model or unavailable nutrients.

However, any combination of fluxes comprised between the bounds of FVA does not necessarily make a feasible flux distribution. Hence, other methods were developed to sample feasible flux distributions that optimise the objective function, making it possible to get a comprehensive view of the range of possible solutions (Schellenberger and Palsson, 2009).

A different possibility to reduce the multiplicity of solutions is offered by the method of parsimonious FBA (pFBA). pFBA adds to standard FBA an assumption of minimising the overall flux in the metabolic network, which amounts to minimise the amount of enzyme required. This assumption may not hold true in all cases but was shown to produce good agreement with experimental data, in particular for evolved strains where there is competition for fastest growth (Lewis *et al.*, 2010).

Genome-Scale Metabolic Models

Advances in genome sequencing have made it possible to get a good knowledge of the enzyme content of many species. This in turn has made it possible to reconstruct the full set of metabolic reactions occurring in several of these species, which are then called genome-scale metabolic models (GEMs). A GEM contains not only all known metabolic reactions with their substrates and stoichiometries, but also all the genes and proteins associated to these reactions. Some methods were developed to automate in part the reconstruction process of GEMs (Swainston *et al.*, 2011), but a high level of manual input and curation is generally needed to achieve the best quality (Feist *et al.*, 2009). A functional model must not only integrate as many as possible of the known metabolic reactions of the species, but also be able to produce all the essential cellular compounds under known experimental growth conditions.

GEMs are currently available for a range of species, including bacteria, archaea, fungi, plants and human cell types. Among the most complete and curated are the models of *E. coli* (Orth *et al.*, 2011), yeast (Heavner *et al.*, 2013) and human (Swainston *et al.*, 2016). These models were developed through several iterations over years, and the latter two involved a collaborative approach associating several research groups around the world. More recently, methods were developed that use transcriptomics or proteomics data to construct specific models for different types of human cells or tissues (Uhlén *et al.*, 2015).

GEMs have been used for numerous types of applications. They greatly facilitate the interpretation of high-throughput data such as transcriptomics, proteomics and metabolomics data, since overlaying them on a GEM can show what type of metabolic functions are active (Oberhardt *et al.*, 2009). They can be used to understand metabolic functions and their regulation in an organism, effects of environmental or nutritional changes on an organism's growth or adaptation, interactions between multiple species in an ecosystem, and define genetic manipulations in order to improve some functionalities of an organism in metabolic engineering.

Elementary Mode Analysis

Principle

Constrained based methods such as FBA are biased towards a certain biological objective. Pathway analysis has the advantage of identifying all pathways inherent to a metabolic network. A popular method for pathway analysis, Elementary Modes (EMs), identifies all admissible minimal pathways connecting substrates with biomass and products inherent to a metabolic network (Schuster *et al.*, 1999). Here, admissible means that the pathway satisfies internal mass balances and the direction constraints on reactions arising from thermodynamic conditions. For example, in the toy network shown in **Fig. 2**, four EMs exist, which cannot be further decomposed. EM_1 presents a cycle between metabolites B and C with no net production or consumption. EM_2 presents the formation of extracellular metabolite D from substrate Aext. EM_3 and EM_4 produce biomass from A through either reaction F6 or through reactions F7 and F9.

In addition to EMs, several other concepts such as Extreme Pathways (EPs) (Schilling *et al.*, 2000) and Generating Vectors (GVs) (Wagner and Urbanczik, 2005) are promising. EP analysis identifies the minimal set of independent pathways through the network, which is a subset of the EMs. GVs are in turn a subset of the EPs. These methods for pathway analysis have been evaluated by several authors, amongst others (Klamt and Stelling, 2003). Several tools have been developed to enumerate the pathways, amongst others the widely used METATOOL (Pfeiffer *et al.*, 1999) and efmtool (Terzer and Stelling, 2008).

Large-Scale Enumeration

Despite improvements in the algorithms for computation of EMs, EPs and GVs, their application to larger metabolic networks has been hampered by the combinatorial explosion in the number of modes as the network size increases. The enumeration of the complete set of EMs for genome-scale networks has been infeasible, and perhaps even undesirable due to the hardly manageable number of modes that would be generated. An alternative approach is the enumeration of a subset of pathways representing the

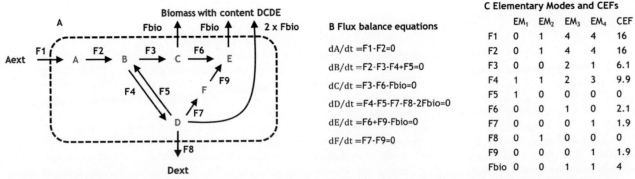

Fig. 2 Control effective flux analysis on an example network. **(A)** Toy network modified from (Rabinowitz and Vastlag, 2012). **(B)** Flux balance equations **(C)** Elementary modes and CEFs.

complete system. Examples are enumeration of the shortest pathways (De Figueiredo *et al.*, 2009), enumeration of pathways including a specific target reaction (Kaleta *et al.*, 2009a,b), enumeration based on available measurements (Jungers *et al.*, 2011), elementary flux patterns (Kaleta *et al.*, 2009a,b), decomposition of the network in modules (Schuster *et al.*, 2002; Schwartz *et al.*, 2007) and random sampling (Machado *et al.*, 2012).

Applications

EMs have been used to understand the cellular metabolism through analysis of the network structure, regulations and characterization of all possible phenotypes (Çakir *et al.*, 2007; Schuster *et al.*, 1999, 2002; Stelling *et al.*, 2002; Soons *et al.*, 2013; Schwartz *et al.*, 2015). Examples of other applications of pathway analysis are the determination of minimum medium requirements (Schilling and Palsson, 2000) and the derivation of reduced kinetic models (Provost and Bastin, 2004). They also play an essential role in the development of model-based metabolic engineering strategies for strain optimisation by identification of suitable intervention targets (Hädicke and Klamt, 2010; Trinh *et al.*, 2008; Boghigian *et al.*, 2010). Of particular interest, Hädicke and Klamt (2011) proposed a gene deletion strategy based on minimal cut sets to identify a minimal set of knockouts disabling the operation of a specified set of target elementary modes, while keeping a set of desired modes.

Control Effective Fluxes and Structural Fluxes

Structural fluxes combine the advantages of both objective function-centred and pathway enumeration-centred approaches. They account for a biological objective function and are derived from the enumeration of all pathways in a given metabolic network. Structural fluxes are inspired from the concept of control effective flux (CEF) to understand changes in transcriptional regulation (Stelling *et al.*, 2002). In the original formulation of CEF, the efficiency of each elementary mode *i* is defined as the ratio of EM's output (the cellular objective) to the investment required to establish the EM (the sum of the absolute flux values in the EM) for a specific carbon source:

$$\varepsilon_i = \frac{Y_i^{objective}}{\sum_k |e_i^k|} \tag{1}$$

The control effective flux of each reaction *k* is then obtained by the weighted average of the product of mode-specific efficiencies and reaction-specific fluxes over the sum of all mode efficiencies:

$$CEF_k = \frac{1}{Y_{X/S}^{max}} \cdot \frac{\sum_i \varepsilon_i \cdot |e_i^k|}{\sum_i \varepsilon_i} \tag{2}$$

where $Y_{X/S}^{max}$ denotes the maximum yield obtained for the specified cellular objective.

In order to find a measure that can predict fluxes across mutants and for growth on multiple substrates, Soons *et al.* (2013) and Schwartz *et al.* (2015) proposed three modifications in Eqs. (1)–(2) and thereby introduced Structural Fluxes. In summary, the yield is computed towards total substrate uptake (in terms of C-moles or molecular weight) and an additional normalization was added to account for networks of different sizes.

As an example, we determined the yields, efficiencies and CEFs for the toy network in **Fig. 2(C)** with biomass as the cellular objective (reaction Fbio) and uptake of external metabolite A as the input (reaction F1). Both EM₃ and EM₄ produce biomass with the same yield towards the substrate uptake reaction F1 (1/4). EM4 involves one additional enzyme (through F7 and F9) compared to EM3 (through F6) and is hence less efficient in terms of investment of enzymes (Eq. (1)). This is reflected by a slightly higher CEF for reaction F6 compared with F7 and F9. In real life networks, these trade-offs may be for instance presented in cancer cells by a high ATP yield per substrate for the TCA cycle, but also high investment costs in the number of enzymes in comparison with aerobic glycolysis.

Future Directions

As procedures for building genome-scale metabolic models are getting well established and the number of available models is growing, new research directions aim at enriching metabolic models with additional information. Thermodynamics plays an important role for metabolic reactions, since the energy levels of substrates and products determine if a reaction is feasible or not. Several studies have attempted to integrate thermodynamic data into metabolic models in order to estimate metabolite concentrations and determine reaction directionality (Ataman and Hatzimanikatis, 2015), determine which elementary modes are feasible or not (Gerstl *et al.*, 2016) and discover reactions that favour growth at higher or lower temperature (Paget *et al.*, 2014).

Another direction to enrich metabolic models is to incorporate transcriptional regulation mechanisms. The metabolic flux is controlled by enzymes which are themselves regulated by transcription factors and other regulatory mechanisms, therefore these interactions must be taken into account to achieve better quantitative predictions. The construction of integrated transcriptional and metabolic models has started, using both kinetic and constraint-based approaches, and constitutes an important objective for future research (Imam *et al.*, 2015).

Eventually, metabolic models from different species need to be connected. In biological environments ranging from human microbiota to global ecosystems, different organisms exchange compounds with each other or compete for resources. The behaviour of each species cannot be fully determined without considering its interactions with other species, which requires moving from species-level to community-level metabolic models (Biggs *et al.*, 2015).

Closing Remarks

Metabolic models have shown broad utility for both biomedical and biotechnological applications. Different approaches can be used to model metabolic systems and each of them has specific advantages and limitations, which is why it is important to choose the best suited technique depending on the expected outcomes of the model. Metabolic reactions are not isolated but are interconnected with other cellular processes, therefore a challenge in any model is often to define suitable boundaries; this choice should always be directed by biological knowledge to include enough but no superfluous components to explain the expected biological functions and reach the desired level of precision.

See also: Biological Pathways. Cell Modeling and Simulation. Investigating Metabolic Pathways and Networks. Metabolome Analysis. Metabolome Analysis. Natural Language Processing Approaches in Bioinformatics. Network Models. Pathway Informatics. Two Decades of Biological Pathway Databases: Results and Challenges

References

Adiamah, D.A., Handl, J., Schwartz, J.M., 2010. Streamlining the construction of large-scale dynamic models using generic kinetic equations. Bioinformatics 26, 1324–1331.

Ataman, M., Hatzimanikatis, V., 2015. Heading in the right direction: Thermodynamics-based network analysis and pathway engineering. Current Opinion in Biotechnology 36, 176–182.

Baldan, P., Cocco, N., Marin, A., Simeoni, M., 2010. Petri nets for modelling metabolic pathways: A survey. Natural Computing 9, 955–989.

Biggs, M.B., Medlock, G.L., Kolling, G.L., Papin, J.A., 2015. Metabolic network modeling of microbial communities. WIREs Systems Biology and Medicine 7, 317–334.

Boghigian, B.A., Shi, H., Lee, K., Pfeifer, B.A., 2010. Utilizing elementary mode analysis, pathway thermodynamics, and a genetic algorithm for metabolic flux determination and optimal metabolic network design. BMC Systems Biology 4.

Bordbar, A., Monk, J.M., King, Z.A., Palsson, B.Ø., 2014. Constraint-based models predict metabolic and associated cellular functions. Nature Reviews Genetics 15, 107–120.

Çakir, T., Kirdar, B., Onsan, Z.I., Ulgen, K.O., Nielsen, J., 2007. Effect of carbon source perturbations on transcriptional regulation of metabolic fluxes in Saccharomyces cerevisiae. BMC Systems Biology 1.

Chaouiya, C., 2007. Petri net modelling of biological networks. Briefings in Bioinformatics 8, 210–219.

Chapman, S.P., Paget, C.M., Johnson, G.N., Schwartz, J.M., 2015. Flux balance analysis reveals acetate metabolism modulates cyclic electron flow and alternative glycolytic pathways in *Chlamydomonas reinhardtii*. Frontiers in Plant Science 6, 474.

Cornish-Bowden, A., 2004. Fundamentals of Enzyme Kinetics, third ed. London: Portland Press.

Feist, A.M., Herrgård, M.J., Thiele, I., Reed, J.L., Palsson, B.Ø., 2009. Reconstruction of biochemical networks in microorganisms. Nature Reviews Microbiology 7, 129–143.

Fell, D., 1997. Understanding the Control of Metabolism. London: Portland Press.

De Figueiredo, L.F., Podhorski, A., Riblo, A., *et al.*, 2009. Computing the shortest elementary flux modes in genome-scale metabolic networks. Bioinformatics 25, 3158–3165.

Gerstl, M.P., Jungreuthmayer, C., Müller, S., Zanghellini, J., 2016. Which sets of elementary flux modes form thermodynamically feasible flux distributions? FEBS Journal 283, 1782–1794.

Gudmundsson, S., Thiele, I., 2010. Computationally efficient flux variability analysis. BMC Bioinformatics 11, 489.

Hädicke, O., Klamt, 2010. CASOP: A computational approach for strain optimization aiming at high productivity. Journal of Biotechnology 147, 88–101.

Hädicke, O., Klamt, S., 2011. Computing complex metabolic intervention strategies using constrained minimal cut sets. Metabolic Engineering 13, 204–213.

Heavner, B.D., Smallbone, K., Price, N.D., Walker, L.P., 2013. Version 6 of the consensus yeast metabolic network refines biochemical coverage and improves model performance. Database 2013, bat059.

Hynne, F., Danø, S., Sørensen, P.G., 2001. Full-scale model of glycolysis in Saccharomyces cerevisiae. Biophysical Chemistry 94, 121–163.

Imam, S., Schäuble, S., Brooks, A.N., Baliga, N.S., Price, N.D., 2015. Data-driven integration of genome-scale regulatory and metabolic network models. Frontiers in Microbiology 6, 409.

Jamshidi, N., Wiback, S.J., Palsson, B.Ø., 2002. In silico model-driven assessment of the effects of single nucleotide polymorphisms (SNPs) on human red blood cell metabolism. Genome Research 12, 1687–1692.

Jungers, R.M., Zamorano, F., Blondel, V.D., Wouwer, A.V., 2011. Fast computation of minimal elementary decompositions of metabolic flux vectors. Automatica 47, 1255–1259.

Kaleta, C., De Figueiredo, L.F., Behre, J., Schuster, S., 2009a. In: Gross, I., Neumann, S., Posch, S. (Eds.), EFMEvovler: Computing Elementary Flux Modes in Genome-Scale Metabolic Networks. Bonn, pp. 179–189.

Kaleta, C., De Figueiredo, L.F., Schuster, S., 2009b. Can the whole be less than the sum of its parts? Pathway analysis in genome-scale metabolic networks using elementary flux patterns. Genome Research 19, 1872–1883.

Klamt, S., Stelling, J., 2003. Two approaches for metabolic pathway analysis? Trends in Biotechnology 21, 64–69.

Lewis, N.E., Hixson, K.K., Conrad, T.M., *et al.*, 2010. Omic data from evolved E. coli are consistent with computed optimal growth from genome-scale models. Molecular Systems Biology 6, 390.

Liebermeister, W., Klipp, E., 2006. Bringing metabolic networks to life: Convenience rate law and thermodynamic constraints. Theoretical Biology and Medical Modelling 3, 41.

Machado, D., Soons, Z.I.T.A., Patil, K.R., Ferreira, E.C., Rocha, I., 2012. Random sampling of elementary flux modes in large-scale metabolic networks. Bioinformatics 28, i515–i521.

Mitchell, S., Mendes, P., 2013. A computational model of liver iron metabolism. PLOS Computational Biology 9, e1003299.

Oberhardt, M.A., Palsson, B.Ø., Papin, J.A., 2009. Applications of genome-scale metabolic reconstructions. Molecular Systems Biology 5, 320.

Orth, J.D., Conrad, T.M., Na, J., *et al.*, 2011. A comprehensive genome-scale reconstruction of Escherichia coli metabolism–2011. Molecular Systems Biology 7, 535.

Orth, J.D., Thiele, I., Palsson, B.Ø., 2010. What is flux balance analysis? Nature Biotechnology 28, 245–248.

Paget, C.M., Schwartz, J.M., Delneri, D., 2014. Environmental systems biology of cold-tolerant phenotype in *Saccharomyces* species adapted to grow at different temperatures. Molecular Ecology 23, 5241–5257.

Pfeiffer, T., Sanchez-Valdenebro, I., Nuno, J.C., Montero, F., Schuster, S., 1999. METATOOL: For studying metabolic networks. Bioinformatics 15, 251–257.

Provost, A., Bastin, G., 2004. Dynamic metabolic modelling under the balanced growth condition. Journal of Process Control 14, 717–728.

Rabinowitz, J.D., Vastlag, L., 2012. Teaching the design principles of metabolism. Nature Chemical Biology 8, 497–501.

Reimers, A.M., Reimers, A.C., 2016. The steady-state assumption in oscillating and growing systems. Journal of Theoretical Biology 406, 176–186.

Schellenberger, J., Palsson, B.Ø., 2009. Use of randomized sampling for analysis of metabolic networks. Journal of Biological Chemistry 284, 5457–5461.

Schellenberger, J., Que, R., Fleming, R.M., *et al.*, 2011. Quantitative prediction of cellular metabolism with constraint-based models: The COBRA Toolbox v2.0. Nature Protocols 6, 1290–1307.

Schilling, C.H., Letscher, D., Palsson, B.O., 2000. Theory for the systemic definition of metabolic pathways and their use in interpreting metabolic function from a pathway-oriented perspective. Journal of Theoretical Biology 203, 229–248.

Schilling, C.H., Palsson, B.O., 2000. Assessment of the metabolic capabilities of Haemophilus influenzae Rd through a genome-scale pathway analysis. Journal of Theoretical Biology 203, 249–283.

Schuster, S., Dandekar, T., Fell, D.A., 1999. Detection of elementary flux modes in biochemical networks: A promising tool for pathway analysis and metabolic engineering. Trends in Biotechnology 17, 53–60.

Schuster, S., Pfeiffer, T., Moldenhauer, F., Koch, I., Dandekar, T., 2002. Exploring the pathway structure of metabolism: Decomposition into subnetworks and application to Mycoplasma pneumoniae. Bioinformatics 18, 351–361.

Schwartz, J.M., Barber, M., Soons, Z., 2015. Metabolic flux prediction in cancer cells with altered substrate uptake. Biochemical Society Transactions 43, 1177–1181.

Schwartz, J.M., Gaugain, C., Nacher, J.C., De, D.A., Kanehisa, M., 2007. Observing metabolic functions at the genome scale. Genome Biology 8, R123.

Soons, Z.I., Ferreira, E.C., Patil, K.R., Rocha, I., 2013. Identification of metabolic engineering targets through analysis of optimal and sub-optimal routes. PLOS ONE 8, e61648.

Stelling, J., Klamt, S., Bettenbrock, K., Schuster, S., Gilles, E.D., 2002. Metabolic network structure determines key aspects of functionality and regulation. Nature 420, 190–193.

Swainston, N., Smallbone, K., Hefzi, H., *et al.*, 2016. Recon 2.2: From reconstruction to model of human metabolism. Metabolomics 12, 109.

Swainston, N., Smallbone, K., Mendes, P., Kell, D., Paton, N., 2011. The SuBliMinaL toolbox: Automating steps in the reconstruction of metabolic networks. Journal of Integrative Bioinformatics 8, 186.

Terzer, M., Stelling, J., 2008. Large-scale computation of elementary flux modes with bit pattern trees. Bioinformatics 24, 2229–2235.

Trinh, C.T., Unrean, P., Srienc, F., 2008. Minimal Escherichia coli cell for the most efficient production of ethanol from hexoses and pentoses. Applied and Environmental Microbiology 74, 3634–3643.

Uhlén, M., Fagerberg, L., Hallström, B.M., *et al.*, 2015. Tissue-based map of the human proteome. Science 347, 1260419.

Wagner, C., Urbanczik, R., 2005. The geometry of the flux cone of a metabolic network. Biophysical Journal 89, 3837–3845.

Williamson, T., Adiamah, D., Schwartz, J.M., Stateva, L., 2012. Exploring the genetic control of glycolytic oscillations in *Saccharomyces cerevisiae*. BMC Systems Biology 6, 108.

Relevant Websites

http://bigg.ucsd.edu/
  BiGG database.
http://www.ebi.ac.uk/biomodels-main/
  BioModels database.
http://celldesigner.org/
  CellDesigner modelling software.
http://opencobra.github.io/
  COBRA toolbox for constraint-based modelling.
http://copasi.org/
  COPASI biochemical system simulator.
http://metabolicatlas.com/
  Human Metabolic Atlas.
https://omictools.com/metatool-tool
  Metatool for elementary modes analysis.
http://vcell.org/
  Virtual Cell.

Biographical Sketch

Jean-Marc Schwartz is a senior lecturer in the Faculty of Biology, Medicine and Health at the University of Manchester. He has been working for 20 years at the interface between physical and biological sciences, focusing on computational techniques for modelling biological systems. He holds a PhD in computer engineering from Laval University and has worked as a postdoctoral researcher at the Kyoto University Bioinformatics Center, which develops the KEGG genomic and metabolic database, before joining the university of Manchester. His research applies a range of computational techniques, including kinetic, logical and constraint-based modelling, to understand the functioning of biological pathways and cellular systems. Of particular interest are the development of metabolic models of human cells to understand diseases such as cancer, and of microorganisms to discover how they are affected by environmental change and develop biotechnological applications.

Zita Soons works as an assistant professor at the Department of Surgery of Maastricht University, The Netherlands. Her main topic is the integration of stable isotope labelling and MS-Imaging to understand cancer metabolism. She is a specialist in computational systems biology of human and microbial metabolism. In previous jobs, she integrated empirical data on metabolism with metabolic models ranging from biotechnological production to cancer. She received a personal grant to work as a postdoctoral researcher at the University of Minho (Portugal) in 2009–2012, which also involved a stay at the European Molecular Biology Laboratory in Heidelberg (Germany). She received a PhD degree on systems and control theory at Wageningen University/Netherlands Vaccine Institute (The Netherlands). In her current job, she also enjoys incorporating the latest advances in her field of research into teaching.

Protein Structural Bioinformatics: An Overview

M Michael Gromiha, Tokyo Institute of Technology, Tokyo, Japan
Raju Nagarajan, Indian Institute of Technology Madras, Chennai, India and Vanderbilt University Medical Center, Nashville, TN, United States
Samuel Selvaraj, Bharathidasan University, Tiruchirappalli, India

Introduction

Proteins

Proteins are one of the important biological macromolecules built from 20 amino acids. They perform several functions in living organisms such as enzymatic catalysis, transportation of ions and molecules, antibodies and regulation of cellular and physiological activities. The functions of proteins mainly depend on their three-dimensional (3D) structures. Currently, more than 130,000 structures of proteins and their complexes are deposited in Protein Data Bank, PDB (Burley *et al.*, 2017) whereas the amino acid sequences are known for more than 80 million proteins (The UniProt Consortium, 2017). Anfinsen (1973) stated that the amino acid sequence of a protein contains all the information necessary to specify its 3D structure. Deciphering the 3D structure of a protein from its amino acid sequence is a longstanding goal in molecular and computational biology. Further, the structures of proteins and their complexes provide ample insights to understand inter-residue interactions, protein folding mechanisms, folding and unfolding rates, stability of protein structures, stability upon mutations, recognition mechanism of protein-protein, protein-nucleic acid and protein-ligand complexes, and structure-based drug design (Gromiha, 2010). **Fig. 1** illustrates the major themes covered in this review.

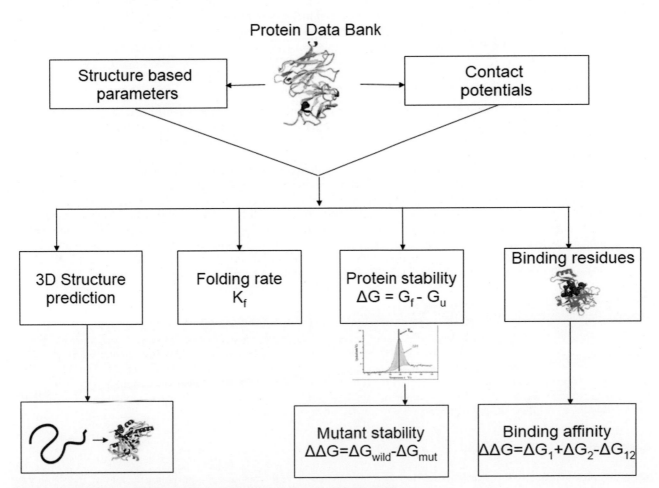

Fig. 1 Flowchart showing various aspects of computational studies on protein structures.

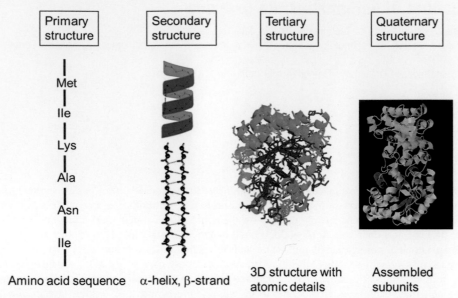

Fig. 2 For major levels of protein structures. Adopted from Gromiha, M.M., 2010. Protein Bioinformatics. Academic Press, New Delhi.

Structural Organization of Proteins

Protein structures are usually described in terms of four levels: primary, secondary, tertiary and quaternary (**Fig. 2**). Primary structure describes the linear sequence of amino acid residues in a protein. Secondary structure is the regular, recurring arrangements of adjacent amino acid residues in a polypeptide chain and major secondary structures are α-helices and β-strands.

They are formed by hydrogen bonds between amide hydrogens and carbonyl oxygens of the peptide backbone. Tertiary structure provides atomic and spatial level information of all residues in a protein. Quaternary structure refers to the association of two or more polypeptide chains into a multisubunit or an oligomeric protein.

Classification of Proteins Based on Structure and Function

Proteins fall into three major groups: fibrous, globular and membrane proteins. Fibrous proteins are insoluble in solvents and the polypeptide chains are arranged in long strands or sheets. They play important structural roles, providing external protection, support and shape. Examples of fibrous proteins include α-keratin, the major component of hair and nails, and collagen, the major protein component of tendons, skin, bones and teeth.

Globular proteins are important for their functional roles and the amino acid residues in a globular protein fold to a proper, unique three-dimentional structure. Globular proteins are classified into four structural classes, namely, all-α, all-β, $\alpha + \beta$ and α/β (**Fig. 3**). The all-α and all-β classes are dominated by α-helices and β-strands, respectively. The $\alpha + \beta$ and α/β classes contain both α-helices and β-strands. In $\alpha + \beta$ class, helices and strands usually segregate while they are mixed together in the α/β class.

Proteins that are embedded in biological membranes are called membrane proteins and they serve as receptors, transporters and signaling molecules. Membrane proteins are of two kinds: transmembrane helical proteins that span the cellular membrane with α-helices and transmembrane β-sheet proteins that traverse the outer membrane of gram negative bacteria with β-strands. In most of genomes (bacteria, archaea and eukaryota), 20–30% of the proteins are estimated to be membrane proteins (Almeida *et al.*, 2017) and are identified as targets for drug design.

Structure Databases for Proteins and Their Complexes

The Protein Data Bank (PDB) is a unique resource for experimentally determined structures of proteins and their complexes (Burley *et al.*, 2017). Utilizing the information available in PDB, several secondary databases have been developed for structural classes and architectures of proteins such as SCOP (Structural classification of proteins) (Andreeva *et al.*, 2008), CATH (Class, Architecture, Topology and Homologous superfamily) (Sillitoe *et al.*, 2015), PDBTM (Protein Data Bank of Trams Membrane proteins) (Kozma *et al.*, 2013), and protein-protein, protein-nucleic acid, protein-carbohydate and protein-ligand complexes. **Table 1** lists the available structural databases for proteins and their complexes.

Analysis of Protein 3D Structures

The availability of protein 3D structures paves the way for researchers to derive the principles governing the folding and stability of proteins, binding sites in protein complexes and development of several structural parameters. Further, various structure-based

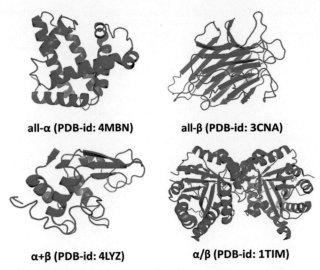

all-α (PDB-id: 4MBN)   all-β (PDB-id: 3CNA)

α+β (PDB-id: 4LYZ)   α/β (PDB-id: 1TIM)

Fig. 3 Major structural classes of globular proteins.

parameters for proteins and physicochemical, energetic and conformational parameters of amino acids have been developed, using 3D structures of proteins deposited in PDB.

Assignment of Secondary Structure and Solvent Accessibility

Kabsch and Sander (1983) utilized the hydrogen bonding pattern in protein structures and assigned secondary structures to each residue in a protein. Lee and Richards (1971) proposed the concept of accessible surface area (ASA) to locate the interior seeking and exposed residues in a protein. ASA is defined as the locus of the centre of the solvent molecule as it rolls over the van der Waals surface of the protein. Generally, water is assumed to be the solvent, represented by a sphere of radius 1.4 Å. The solute molecules are represented by a set of interlocking spheres of appropriate van der Waals radii assigned to each atom and the solvent molecule is rolled along the envelope of the van der Waals surface at planes conveniently sectioned. Several methods have been proposed to compute the solvent accessibility of amino acid residues from protein structures (Gromiha and Ahmad, 2005). These programs provide numerical data for accessible surface area and hence Ahmad *et al.* (2004) developed a tool, ASAview to represent them pictorially. The ASA values for the Zif268 zinc finger protein are shown in **Fig. 4** as spiral circles with different colors to visualize the exposed and interior seeking residues, based on their chemical behavior and ASA values (see Relevant Websites section).

Residue-Residue Contacts

The contacts between amino acid residues in a protein as well as with the surrounding medium mainly dictate the folding and stability of protein structures. These contacts influence all major non-covalent interactions such as hydrophobic, electrostatic, hydrogen bonding and van der Waals interactions as well as disulfide bonds (**Fig. 5**). Inter-residue contacts are classified as short, medium and long-range, based on the location of the contacting residues in the 3D structure as well as in the sequence (Gromiha and Selvaraj, 2004). The information about preferred contacts in protein structures has been used to understand the residue-residue co-operativity in protein folding as well as to develop contact potentials.

The most widely used approach for deriving contact potentials is computing frequencies of sequence and structure features, and converting them into free energies using the following relation

$$\Delta G = - kT\log(f) \tag{1}$$

where k is the Boltzmann's constant, f denotes frequencies and T is the temperature (Sippl, 1990). Further, contacts between amino acid residues in proteins structures reveal the importance of specific contact pairs in different structural classes and folding types of globular proteins, α-helical and β-barrel membrane proteins, mesophilic and thermophilic proteins (Gromiha, 2010).

Non-Covalent Interactions in Protein Structures

Hydrophobic interactions play a dominant role in the folding and stability of protein structures. Hydrophobicity is inversely proportional to ASA and hence, hydrophobic free energy is estimated indirectly using the computable parameter, ASA. It is computed by (Gromiha, 2010):

$$G_{hy} = \Sigma\Delta\sigma_i[A_i(\text{folded})-A_i(\text{unfolded})] \tag{2}$$

where, i stands for different atom types, and $A_i(\text{folded})$ and $A_i(\text{unfolded})$ are the ASA of each atom type in the folded and

Table 1   List of databases and servers for protein structures, structure prediction, stability and interactions

Name of the database/web server	URL
Structure and classification	
PDB	http://www.rcsb.org/pdb/home/home.do
SCOP	http://scop.mrc-lmb.cam.ac.uk/scop/
CATH	http://www.cathdb.info/
PDBTM	http://pdbtm.enzim.hu/
Structure-based computations and parameters	
ASAview	http://www.abren.net/asaview/
PDB param	http://www.iitm.ac.in/bioinfo/pdbparam/
PIC	http://crick.mbu.iisc.ernet.in/~PIC
CMWeb	http://cmweb.enzim.hu/
INTAA	http://bioinfo.uochb.cas.cz/INTAA/
Structure prediction	
Homology modelling	
MODELLER	https://salilab.org/modeller/
SWISS-MODEL	https://swissmodel.expasy.org/
PRIMO	https://primo.rubi.ru.ac.za/
PyMod	http://schubert.bio.uniroma1.it/pymod/index.html
MaxMod	http://www.immt.res.in/maxmod/
Fold recognition	
GenTHREADER	http://bioinf.cs.ucl.ac.uk/psipred/
Phyre2	http://www.sbg.bio.ic.ac.uk/phyre2/html/page.cgi?id=index
MUSTER	http://zhanglab.ccmb.med.umich.edu/MUSTER/
ORION	http://www.dsimb.inserm.fr/orion/
DN-Fold	http://iris.rnet.missouri.edu/dnfold/
Ab initio structure prediction	
Robetta	http://www.robetta.org/submit.jsp
I-TASSER	http://zhanglab.ccmb.med.umich.edu/I-TASSER/
QUARK	http://zhanglab.ccmb.med.umich.edu/QUARK/
BHAGEERATH	http://www.scfbio-iitd.res.in/bhageerath/index.jsp
EVfold	http://evfold.org/evfold-web/evfold.do
CABS-fold	http://biocomp.chem.uw.edu.pl/CABSfold/
PEP-FOLD	http://bioserv.rpbs.univ-paris-diderot.fr/services/PEP-FOLD/
Thermodynamic data and effect of mutations on stability	
ProTherm	http://www.abren.net/protherm/
PoPMuSiC	https://soft.dezyme.com/
CUPSAT	http://cupsat.tu-bs.de/
SDM	http://mordred.bioc.cam.ac.uk/~sdm/sdm.php
STRUM	http://zhanglab.ccmb.med.umich.edu/STRUM/
Interaction databases	
DIP	http://dip.doe-mbi.ucla.edu/dip/Main.cgi
IntAct	http://www.ebi.ac.uk/intact/
MINT	http://mint.bio.uniroma2.it/mint/
BioGRID	https://thebiogrid.org
SKEMPI	https://life.bsc.es/pid/mutation_database/
PROXiMATE	http://www.iitm.ac.in/bioinfo/PROXiMATE/
NDB	http://ndbserver.rutgers.edu/
ProNIT	http://www.abren.net/pronit/
PDIdb	http://melolab.org/pdidb
PROCARB	http://www.procarb.org/
Druggable binding sites	
Sc-PDB	http://bioinfo-pharma.u-strasbg.fr/scPDB/

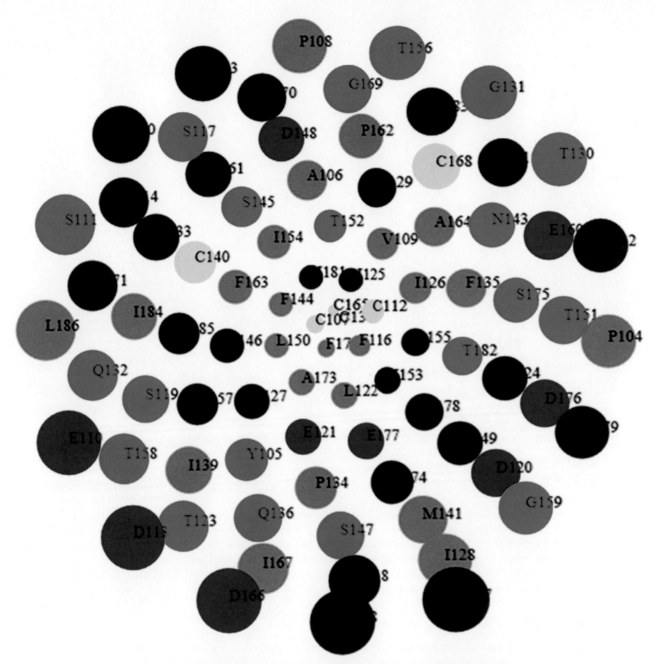

Fig. 4 Graphical representation of ASA values of amino acid residues in Zif268 zinc finger protein (PDB-id: 1AAY).

unfolded (extended) states of the protein, respectively. $\Delta\sigma_i$ refers to atomic solvation parameters obtained by fitting ΔG and ASA in extended conformation of each atom type in the 20 amino acid residues. It is calculated as:

$$\Delta G = \Sigma \Delta\sigma_i A_i \tag{3}$$

Electrostatic free energy is generally computed by Coulomb's law using the charge of interacting atoms and the distance between them.

$$E_{es} \propto q_i q_j / r_{ij} \tag{4}$$

where q_i and q_j are the effective partial charges for atoms i and j, respectively and r_{ij} is the distance between them.

Hydrogen bonds are responsible for the formation and stability of protein secondary structures, α-helices and β-strands. Hydrogen bonds are formed between two electronegative atoms mediating with a hydrogen atom. The contribution of hydrogen bonds is computed using the distance between the heavy atoms and the angle formed by the three atoms involved.

van der Waals interaction energy is generally computed using Lennard-Jones 6–12 potential:

$$G_{vw} = \left(A_{ij}/r_{ij}^{12} - B_{ij}/r_{ij}^6 \right) \tag{5}$$

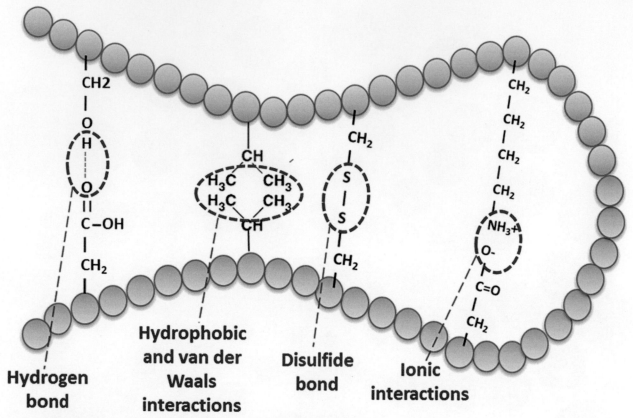

Fig. 5 Various non-covalent interactions and disulfide bonds in protein structures.

where $A_{ij} = \varepsilon_{ij}^*(R_{ij}^*)^{12}$ and $B_{ij} = 2\,\varepsilon_{ij}^*(R_{ij}^*)^6$; $R_{ij}^* = (R_i^* + R_j^*)$ and $\varepsilon_{ij}^* = (\varepsilon_i^*\,\varepsilon_j^*)^{1/2}$, R^* and ε^* are, respectively, the van der Waals radius and well depth.

In addition, cation-π interactions (Dougherty, 1996) are recognized to play an important role in globular, membrane, DNA binding, mesophilic and thermophilic proteins as well as in the recognition of protein-protein complexes. These are formed by the interaction between the positive charged residues Lys and Arg and aromatic residues, Phe, Trp and Tyr. Cation-π interactions in protein structures have been identified using distance and energy-based criteria: (i) distance between the cationic group and the centers of all aromatic rings should be less than 10Å and (ii) the interaction energy between the positive charged and aromatic residues (sum of electrostatic and van der Waals) should be less than -2 kcal/mol.

Amino Acid Properties Derived from Protein Structural Data

Protein structures have been used to derive several properties for amino acid residues as well as the whole protein.

(i) Medium and long-range contacts: Medium-range contacts refer to the contacts between amino acid residues within the distance of 8Å and a sequence separation of three or four residues while the sequence separation is more than four residues in long-range contacts.

(ii) Number of neighbouring residues: For each central residue, number of contacting residues within a specific cut-off distance (8Å).

(iii) Surrounding hydrophobicity: For a specific residue, this is the sum of the experimental hydrophobicity indices of residues, which occur within a sphere of radius 8 Å (Ponnuswamy, 1993). It is estimated using

$$H_p(i) = \sum_{j=1}^{20} n_{ij} h_j \tag{6}$$

where, n_{ij} is the total number of surrounding residues of type j around ith residue of the protein and h_j is the experimental hydrophobic index of residue type j in kcal/mol (Jones, 1975).

(iv) Transfer free energy: The ASA of each residue provides information about the occurrence of that residue either in the interior of the protein or the surface and the ratio between them is used to compute the partition coefficient (f), which has been used to derive the transfer free energy:

$$\Delta G_t = -RT \ln f \tag{7}$$

(v) Buriedness and reduction in accessibility: Buriedness is defined as the ratio between the number of residues (of type i) in the interior of the protein to the total number of residues (of type i) in a protein. Similarly, reduction in accessibility is given by:

$$\langle R_A \rangle = A^0 - \langle A \rangle / A^0, \tag{8}$$

where, A^0 and $\langle A \rangle$ represent the accessible area in the unfolded (extended) and folded states of the protein, respectively.

(vi) Flexibility: This parameter deals with the root mean square displacement of amino acid residues and is also related to the temperature factor (B factor) of amino acid residues in protein structures.

(vii) Numerical values for amino acid properties: Kawashima *et al.* (2008) developed a database, AAindex, which contains the numerical indices for 516 physicochemical and biochemical properties of amino acids and pairs of amino acids. Gromiha *et al.* (1999a) compiled a set of 49 physicochemical, energetic and conformational properties of amino acid residues for understanding the folding and stability of proteins.

On the other hand, several parameters have been derived for a whole protein or protein chains.

(viii) Contact Order: This measure reflects the relative importance of local and non-local contacts to the native structure of a protein (Plaxco *et al.*, 1998). It is defined as

$$CO = \Sigma \Delta S_{ij} / L.N, \tag{9}$$

where ΔS_{ij}, is the sequence separation between contacting residues i and j within a distance of 6Å between any heavy atoms, L is the total number of residues in the protein and N is the total number of contacts.

(ix) Long-range order: A knowledge of long-range contacts (contacts between two residues that are close in space and far in the sequence) in protein structure (Gromiha and Selvaraj, 2001) defines long-range order:

$$LRO = \Sigma n_{ij} / N; \; n_{ij} = 1 \text{ if } |i - j| > 12; \; = 0 \text{ otherwise}, \tag{10}$$

where i and j are two contacting residues in space within a distance of 8Å and N is the total number of residues in a protein.

(x) Multiple contact index: When a residue has multiple contacts, the multiple contact index is used for computing long-range order (Gromiha, 2009) and is defined as

$$n_{ci} = \Sigma n_{ij}; \; n_{ij} = 1 \text{ if } r_{ij} < 7.5 \text{ Å}; \; |i - j| > 12 \text{ residues}; \; 0 \text{ otherwise}$$

$$MCI = \Sigma n_{mi} / N; \; n_{mi} = 1 \text{ if } n_{ci} \geq 4; \; 0 \text{ otherwise} \tag{11}$$

where n_c is the number of contacts for each residue, and r_{ij} is the distance between the residues i and j.

Nagarajan *et al.* (2016) developed a web-based tool, PDBparam, to derive a set of more than 50 features on four different themes, viz., (i) identification of binding sites, (ii) inter-residue interactions, (iii) secondary structure propensities and (iv) physicochemical properties. Similar webservers have also been developed for various weak and strong interactions, PIC (Tina *et al.*, 2007), contact between amino acid residues, CMWeb (Kozma *et al.*, 2012) and binding site residues in protein-DNA complexes, INTAA (Galgonek *et al.*, 2017).

Prediction of Protein Tertiary Structures

Deciphering the native conformation of a protein from its amino acid sequence, termed as the "protein folding problem" is a longstanding challenge in molecular and computational biology. Several methods have been proposed to achieve the goal and the performances of these methods have been critically evaluated by the forum, Critical Assessment of Techniques for Protein Structure Prediction (CASP), using blind prediction experiments. The developments of reliable methods are outlined in this section. Three major approaches namely, homology modeling, fold recognition and *ab initio* have been developed for the prediction of protein tertiary structures. In recent years, the above three approaches have been grouped into template-based modeling and template-free modeling (Li *et al.*, 2015; Källberg *et al.*, 2012; Xu *et al.*, 2009).

Homology Modeling

Homology modeling, also known as comparative modeling, is based on the biological fact that when two sequences share high similarity/identity, their respective structures are also similar. In this method, the 3D structure of a protein is obtained with the following steps: (i) for a given target sequence, identifying the proper template using BLAST search, (ii) sequence alignment (iii) alignment corrections to ensure the conserved or functionally important residues are aligned, (iv) backbone generation (v) loop modeling, (vi) side chain modeling using rotamer libraries, (vii) optimizing the model using energy minimization and (viii) validating the model by stereochemical evaluation using the residues in the allowed regions of Ramachandran plot as well as favourable energies. MODELLER (Webb and Sali, 2014) is one of the widely used computational tools for predicting protein 3D structures using homology modeling. MaxMod, PyMod, and PRIMO are other recent methods/servers for homology modeling and their URLs are included in **Table 1**.

Fold Recognition

If the sequence identity between the target sequence and template structures fall below 30%, homology modeling methods may not be able to produce a reliable model and under such situations fold recognition methods are used for prediction (Buchan and Jones, 2017; Lhota and Xie, 2016; Jo et al., 2015). These methods follow different steps to predict the 3D structure of a protein, which include (i) advanced sequence comparison methods, (ii) comparison of predicted secondary structure strings with those of known folds, (iii) tests of the compatibility of sequences with 3D folds and (iv) the use of human expert knowledge. Consequently, several methods and webservers have been developed for fold recognition such as, 3D-1D profiles (David et al., 2004), GenTHREADER (Jones, 1999), MUSTER (Wu and Zhang, 2008), ORION (Ghouzam et al., 2016), Phyre2 (Kelley et al., 2015), TMFoldRec (Kozma and Tusnády, 2015), DescFold (Yan et al., 2009) and DN-Fold (Jo et al., 2015). **Table 1** includes the web addresses of these servers.

Ab Initio Protein Structure Prediction

Ab initio prediction methods consider all bonded and non-bonded interactions involved in the process of folding, and choose the structure with lowest free energy. This approach is based on the 'thermodynamic hypothesis', which states that the native structure of a protein is the one for which the free energy achieves the global minimum. Two major components to *ab initio* prediction are (i) devising a scoring function that can distinguish between correct (native or native-like) structures from incorrect ones and (ii) a search method to explore the conformational space. One of the best performing methods for *ab initio* structure prediction is the Rosetta algorithm (Simons et al., 1997). Several other recent methods include I-TASSER (Roy et al., 2010; Yang et al., 2015), Robetta (Kim et al., 2004), EVfold (Marks et al., 2011), CABS-fold (Blaszczyk et al., 2013), QUARK (Xu and Zhang, 2012), PEP-FOLD (Thévenet et al., 2012; Shen et al., 2014) and Bhageerath (Jayaram et al., 2014) **(Table 1)**.

Modeling on the Web

Few 3D structure prediction methods are available online with user friendly interface. SWISS-MODEL is one of the widely used comparative protein modeling servers to predict the structure for a given amino acid sequence directly (Bienert et al., 2017). It has different options such as (i) 'first approach mode', which requires only an amino acid sequence of a protein; the server automates subsequent steps including template selection, alignment and model building, (ii) 'alignment mode', which is the modeling process based on a user-defined target-template alignment and (iii) 'project mode' in which complex modeling tasks can be handled.

Protein Folding Kinetics

Many small proteins are known to fold rapidly by simple two-state kinetics and the mechanism of protein folding has been elaborately addressed with folding rates and Φ-value analysis. Φ-value is given by the equation:

$$\Phi_\# = \Delta\Delta G_{\#-U}/\Delta\Delta G_{F-U} \tag{12}$$

where $\Delta\Delta G_{F-U}$ is the change in stability between the folded (F) and unfolded states (U), and $\Delta\Delta G_{\#-U}$ is the energy difference between mutant and wild type in the transition state, #. $\Phi=0$ corresponds to no structure formation in the transition state as in the unfolded state and a Φ value of 1 is interpreted that the residue has a native like structure in transition state (Itzhaki et al., 1995). The analysis of experimental Φ values of 296 mutants in seven single domain proteins showed that more than 82% of them exhibit a Φ value below 0.6 (Raleigh and Plaxco, 2005). In addition, negative Φ values are interpreted as mutations which stabilize the folding transition state while it destabilizes the native state (De los Rios et al., 2005).

Folding Nuclei

The residues with high Φ values form folding nuclei in protein structures, which mediate important interactions for folding and assembly of their native states (Dobson, 2003). Folding nuclei could be computationally identified using different techniques: (i) searching free-energy saddle points on a network of protein unfolding pathways, (ii) elucidating the correspondence between conserved hydrophobic positions in amino acid sequences and folding nuclei, (iii) computing the conservation scores of amino acid residues based on physicochemical properties and (iv) coupling ensembles of pathways with the lowest effective contact order and identifying contacts that are crucial for folding.

Protein Folding Rates

Protein folding rate is a measure of the slow/fast folding of a protein from its unfolded state to the native 3D structure. The data for folding rates for proteins and their mutants are available in literature (Gromiha, 2010; Chaudhary et al., 2015).

Relating Protein Folding Rates Using Structural Parameters

Several structure-based parameters have been proposed to understand and predict protein folding rates. Plaxco *et al.* (1998) proposed the concept of contact order (CO) by estimating the average sequence separation of all contacting residues in the native state within a distance of 6Å normalized with total number of contacts and total number of residues. Gromiha and Selvaraj (2001) considered only the long range contacts with a sequence separation of 12 residues, which are within 8Å limit and proposed the concept of LRO. Istomin *et al.* (2007) showed that LRO is the only parameter that shows good correlation in all structural classes of proteins. Further, protein folding rates are well studied with total contact distance, first principles of protein folding, combination of CO and stability, topomer search model, topological properties of protein conformations, cliquishness, multiple contact index, helix parameter, n-order contact distance and chain connectivity due to crosslinks (Gromiha, 2010).

Prediction of Protein Folding Rates From Amino Acid Sequence

The analysis of the relationship between amino acid properties and protein folding rates showed a poor correlation between them. On the other hand, the secondary structure content of a protein and chain length showed a linear relationship with protein folding rates. Although chain length showed a good inverse correlation with the folding rates of three-state proteins no correlation between them was observed in two-state proteins (Galzitskaya *et al.*, 2003).

Punta and Rost (2005) predicted amino acid contacts using amino acid sequence and utilize the information to predict protein folding rates using the concept of long-range order. We have developed multiple regression models for predicting the folding rates of protein using amino acid properties (Gromiha *et al.*, 2006). Furthermore, protein folding rates are predicted with amino acid composition, rigidity and quadratic response surface models, n-order contact distance, secondary structure, solvent accessibility, compression capability of protein sequence, reactability of proteins in solvents, contact surface of the unfolded chain and solvent, Gibbs free energy of hydration in denatured proteins, the average range of amino acid contact, the polarity value of amino acids, number of torsion angles of the side chain and residue-level coevolutionary information. The different algorithms for predicting the folding rates of proteins have been detailed in recent review articles (Gromiha, 2010; Gromiha and Huang, 2011; Chang *et al.*, 2015).

Prediction of Protein Folding Rates Upon Mutation

Amino acid substitutions may accelerate or decelerate the folding rate of a protein and it mainly depends on the location and type of mutations. Huang and Gromiha (2010) proposed the first approach to distinguish between accelerating and decelerating mutants using amino acid properties. The method has been extended further to predict the real value change in folding rates (Huang and Gromiha, 2012). The integrated approach using secondary structure information, location of mutants in a protein as well as the distribution along the sequence is reported to substantially improve the prediction accuracy of protein folding rate change upon mutation (Chaudhary *et al.*, 2015; see Relevant Websites section). The analysis also showed that folding rate is primarily influenced with hydrophobicity, emphasizing the dominant influence of hydrophobic effect during the folding process. Mallik *et al.* (2016) utilized residue-level coevolutionary information for predicting protein folding rates upon mutations.

Protein Stability

Protein stability (ΔG) is the net balance between free energies in the folded and unfolded states. While the folded state is stabilized through disulphide bonds and non-covalent interactions, hydrophobic, electrostatic, hydrogen bonding and van der Waals interactions, the unfolded state is influenced by entropy. However, the difference between the folded and unfolded states is marginal and ΔG lies in the range of 5–25 kcal/mol.

Thermodynamic Database for Proteins and Mutants

Gromiha *et al.* (2016) compiled the data on stability of proteins and mutants from the literature and developed a database, ProTherm, which contains the thermodynamic parameters for more than 25,000 normal (or wild-type) and mutant proteins along with sequence, structure, function and literature information. The analysis of ΔG values of protein structures along with the contribution of free energies showed that the stability is mainly achieved by hydrophobic interactions by overcoming the entropic factors whereas other interactions help to maintain stability, provide the shape and keep the folded state from falling apart.

Stability of Thermophilic Proteins

Thermophilic proteins maintain their stability at high temperatures (80–100°C) and there is a direct relationship between environmental growth temperature and melting temperature (Gromiha *et al.*, 1999b; Gaucher *et al.*, 2008). Analysis of amino acid sequences in thermophilic proteins as well as their mesophilic homologues showed a preference of Ala, Thr, Arg and Glu over Gly, Ser, Lys and Asp, respectively (Gromiha and Suresh, 2008). Additionally, the role of specific amino acid properties such as shape,

Gibbs free energy change of hydration in native proteins, dipeptide composition, contacts between amino acid residues, number of ion pairs, hydrogen bonds, packing and aromatic clusters are reported to be important to enhance the stability of thermophilic proteins (Gromiha *et al.*, 1999b; Pica and Graziano, 2016). On the other hand, available pairs of mesophilic and thermophilic protein structures have been used effectively to detect interactions important for the stability of thermophilic proteins. Gromiha *et al.* (2013) carried out a comprehensive analysis on the influence of all interactions and showed that hydrophobicity is the major factor in the stability of thermophilic proteins followed by ion pairs and hydrogen bonds. The systematic elimination of mesophilic proteins based on surrounding hydrophobicity, interaction energy and ion pairs/hydrogen bonds can be used to identify 95% of the thermophilic proteins correctly.

Stability of Proteins Upon Mutations

The availability of thermodynamic data for proteins and mutants in the ProTherm database prompted researchers to understand the factors influencing the stability of mutants as well as predicting the free energy change upon mutation. It has been shown that buried mutations are influenced mainly by hydrophobicity whereas partially buried and exposed mutants are affected by neighboring and surrounding residues (Gromiha *et al.*, 1999b).

The effect of mutation on free energy change of a protein has been systematically analyzed with the development of torsion and distance potentials and environment dependent substitution tables, amino acid properties and energy based approach. Further, the mutational information as well as amino acid properties have been trained with machine learning techniques for predicting the change in free energy upon mutation. Saraboji *et al.* (2006) proposed an "average assignment method" based on the average free energy change of all possible 380 mutants, which provides first-hand information on the capability of a method for predicting the free energy change upon mutation. Further, several methods were developed to estimate the protein stability changes upon pouint mutations using sequence/structure information such as PoPMuSiC, CUPSAT, SDM, STRUM etc (**Table 1**). Recent reviews comprehensively describe the methods and applications for predicting the stability of mutants and their applications (Gromiha and Huang, 2011; Gromiha *et al.*, 2016).

Protein Interactions

The interactions of proteins with other molecules govern various cellular processes. The functions of protein complexes have been studied with a view to understand the recognition mechanism, predicting the binding partners and binding site residues as well as the binding affinity of protein-protein, protein-nucleic acid and protein-ligand complexes.

Protein-Protein Interactions

Protein-protein interactions have been studied with two major aspects, (i) within protein-protein complexes and (ii) large scale analysis on protein-protein interaction networks. Protein-protein complex level approach includes the analysis and identification of binding sites, development of structural, functional, thermodynamic and kinetic databases for protein-protein complexes, understanding the binding affinity and recognition mechanism of protein-protein complexes. Interaction network analysis deals with the interaction of each protein with other proteins within the same organism or between different organisms (E.g. host-pathogen interactions) and the complexity of interactions among a set of proteins.

Several databases have been developed to relate various aspects of protein-protein interactions such as interacting pairs of proteins (DIP; Xenarios *et al.*, 2002, IntAct; Orchard *et al.*, 2014), functional interactions between proteins (MINT; Zanzoni *et al.*, 2002), interactions between proteins in human proteome and other organisms (BioGRID; Chatr-Aryamontri *et al.*, 2017), kinetic data on binding affinity (SKEMPI; Moal and Fernández-Recio, 2012) and thermodynamic data for protein-protein interactions and mutants (PROXiMATE; Jemimah *et al.*, 2017) (**Table 1**). The 3D structures of protein-protein complexes are deposited in Protein Data Bank (Burley *et al.*, 2017).

The availability of experimental data on protein-protein complexes in several databases as well literature prompted researchers to understand the recognition mechanism using statistical analysis of binding sites, identification of binding site residues, prediction of protein-protein complex structures as demonstrated with Critical Assessment of Protein-protein Interactions (CAPRI; Lensink *et al.*, 2017), distinguishing high and low affinity complexes, predicting the binding affinity of protein-protein complexes using sequence and structure based parameters and energy based approach for understanding the recognition mechanism. Recently, Kulandaisamy *et al.* (2017) dissected the important residues, which are involved in both folding and binding, and revealed the characteristic features of these residues. The developments on binding site prediction, binding affinity, complex structure prediction and scoring schemes have been reviewed in detail (Gromiha and Yugandhar, 2017; Gromiha *et al.*, 2017).

Protein-Nucleic Acid Interactions

Owing to the importance of protein-DNA interactions in gene regulation, DNA replication and repair, several computational studies have been performed to understand the recognition mechanism (Gromiha and Nagarajan, 2013). Databases have been developed to accumulate the experimental structures of protein-nucleic acid complexes (NDB) (Narayanan *et al.*, 2013), pairs of contacting amino acid residues and DNA bases (PDIdb) (Norambuena and Melo, 2010), and thermodynamic data of protein-nucleic acid interactions (ProNIT) (Kumar *et al.*, 2006) **(Table 1)**. Utilizing these information, computational analyses have been carried out to extract important interactions, such as hydrogen bonds, physicochemical properties at the interface, interface surface area, structure based potentials, cation-π interactions, moments of electric charge distribution, DNA stiffness, clustering of protein-DNA complexes and thermodynamics for binding. Further, the role of direct and indirect readout mechanisms (Gromiha *et al.*, 2004) and the importance of scoring functions have been explored for understanding the recognition mechanism of protein-DNA complexes (Gromiha and Fukui, 2011). On the other hand, several methods have been proposed to discriminate DNA binding proteins and predicting the binding sites (Gromiha and Nagarajan, 2013). Nagarajan *et al.* (2013) showed that the performance of prediction methods depends on the structure and function of the specific protein and DNA.

Similar to protein-DNA interactions, the structures of protein-RNA complexes have been effectively used to understand the importance of specific interactions in the recognition mechanism and to predict binding sites (Nagarajan and Gromiha, 2014). In addition, Nagarajan *et al.* (2015) utilized statistical analysis and molecular dynamics simulation of the same protein-RNA complex from different organisms and showed that the recognition mechanism is specific to the target organism.

Protein-Carbohydrate and Protein-Ligand Interactions

The availability of data for protein-carbohydrates is relatively limited, compared to protein-protein and protein-nucleic acid complexes. Malik *et al.* (2010) developed the only database for protein-carbohydrate complex structures (PROCARB) **(Table 1)** and annotated the binding sites. The available 3D structures have been utilized to reveal the important factors and to understand the recognition mechanism of protein-carbohydrate complexes (Gromiha *et al.*, 2014).

Several databases are available for protein-ligand complexes, specifically for binding free energies, structural information, and druggable binding sites in protein structures (sc-PDB) (Kellenberger *et al.*, 2006). Accordingly, several methods have been proposed to predict the ligand binding sites and binding affinity of complexes upon binding (Tsujikawa *et al.*, 2016; Singh *et al.*, 2016; Pai *et al.*, 2017).

Proteins and Drug Design

Proteins, the key functional molecules in all organisms, are one of the major drug targets to combat diseases. Small organic molecules obtained from natural sources or synthesized in the laboratory are used to inhibit the biological activity of enzymes, receptors etc., which are involved in disease processes. When the 3D structure of a target receptor is known, computational methods (*in silico*) such as virtual screening or docking are used to screen a large number of compounds available in chemical databases to identify potential inhibitors. Protein-ligand docking involves different steps such as identifying the active sites, ligand flexibility and interaction energy between ligand and protein. In the absence of 3D structures of target receptors, homology modeling is used to construct a 3D model of the receptor to be used for virtual screening/docking. In recent years, several attempts have been made to identify hit compounds as potential inhibitors to target proteins for addressing various diseases such as dengue, chikungunya, cancer etc. (Anusuya *et al.*, 2016; Akhtar and Jabeen, 2017; Leal *et al.*, 2017; Zou *et al.*, 2017; Ramakrishnan *et al.*, 2017). These compounds are taken further for experimental verifications and drug development. Chiba *et al.* (2015) proposed a contest based approach for identifying potential inhibitors for cYes kinase using *in silico* screening and experimental validations. In protein-ligand docking, two important bottlenecks are conformational sampling and scoring functions. AutoDock and Glide are widely used freely available and commercial docking programs to identify probable drug compounds. Recently, Anusuya *et al.* (2017) reviewed the methods and applications of drug-target interactions.

On the other hand, quantitative structure-activity relationship (QSAR) model(s) are developed for relating the biological activity of compounds such as IC50 (Half maximal inhibitory concentration) and/or EC50 (Half maximal effective concentration) with structure based features of ligands such as hydrophobic, electronic, steric and topological parameters. The structural features of the ligands are combined using multiple regression to relate with experimental activity values. This procedure is widely used to identify hit compounds for several diseases (Kanakaveti *et al.*, 2017). Alternatively, a set of molecules could be analyzed to identify key pharmacophore features, such as hydrogen bond donors, hydrogen bond acceptors, aromatic rings/hydrophobic groups and related with biological activity.

Conclusions and Future Perspectives

The 3D structures of proteins and their complexes have been effectively utilized to derive several structure based parameters and contact potentials. These features have been successfully used to predict protein folding rates, 3D structures of proteins, protein stability, stability upon mutations, binding sites and binding affinity of protein complexes. The refinement of these parameters would improve the prediction performance as well as successful prediction of diverse systems. We have reviewed the methods and applications as well as latest developments on protein folding, stability and interactions. Most of the prediction methods could be refined with larger datasets and effective learning methodologies such as machine learning algorithms and deep learning. In addition, high throughput in silico screening of small molecules and next generation sequence analysis would aid the identification of lead compounds for drug discovery.

Acknowledgements

MMG thanks Prof. Shoba Ranganathan for her invitation to contribute the review article. The work is partially supported by the Department of Science and Technology, Government of India to MMG (EMR/2016/001476).

See also: *Ab initio* Protein Structure Prediction. Algorithms for Structure Comparison and Analysis: Docking. Algorithms for Structure Comparison and Analysis: Homology Modelling of Proteins. Algorithms for Structure Comparison and Analysis: Prediction of Tertiary Structures of Proteins. Assessment of Structure Quality (RNA and Protein). Biomolecular Structures: Prediction, Identification and Analyses. Cloud-Based Molecular Modeling Systems. Computational Protein Engineering Approaches for Effective Design of New Molecules. Computational Tools for Structural Analysis of Proteins. Drug Repurposing and Multi-Target Therapies. Epitope Predictions. Identifying Functional Relationships Via the Annotation and Comparison of Three-Dimensional Amino Acid Arrangements in Protein Structures. In Silico Identification of Novel Inhibitors. Molecular Dynamics and Simulation. Mutation Effects on 3D-Structural Reorganization Using HIV-1 Protease as A Case Study. Natural Language Processing Approaches in Bioinformatics. Pharmacophore Development. Population-Based Sampling and Fragment-Based De Novo Protein Structure Prediction. Prediction of Protein-Protein Interactions: Looking Through the Kaleidoscope. Protein Design. Protein Properties. Protein Structure Analysis and Validation. Protein Structure Classification. Protein Structure Databases. Protein Structure Visualization. Protein Three-Dimensional Structure Prediction. Protocol for Protein Structure Modelling. Rational Structure-Based Drug Design. Secondary Structure Prediction. Small Molecule Drug Design. Structural Genomics. Structure-Based Design of Peptide Inhibitors for Protein Arginine Deiminase Type IV (PAD4). Structure-Based Drug Design Workflow. Study of The Variability of The Native Protein Structure. Transmembrane Domain Prediction

References

Ahmad, S., Gromiha, M.M., Fawareh, H., Sarai, A., 2004. ASAView: Database and tool for solvent accessibility representation in proteins. BMC Bioinform 5, 51.

Akhtar, N., Jabeen, I., 2017. Pharmacoinformatic approaches to design novel inhibitors of protein kinase B pathways in cancer. Curr Cancer Drug Targets. in press.

Almeida, J.G., Preto, A.J., Koukos, P.I., Bonvin, A.M.J.J., Moreira, I.S., 2017. Membrane proteins structures: A review on computational modeling tools. Biochim Biophys Acta 1859, 2021–2039.

Andreeva, A., Howorth, D., Chandonia, J.-M., et al., 2008. Data growth and its impact on the SCOP database: New developments. Nucleic Acids Res 36 (Database Issue), D419–D425.

Anfinsen, C.B., 1973. Principles that govern the folding of protein chains. Science 181 (96), 223–230.

Anusuya, S., Kesherwani, M., Priya, K.V., et al., 2017. Drug–target interactions: Prediction methods and applications. Curr Protein Pept Sci. in press.

Anusuya, S., Velmurugan, D., Gromiha, M.M., 2016. Identification of dengue viral RNA-dependent RNA polymerase inhibitor using computational fragment-based approaches and molecular dynamics study. J Biomol Struct Dyn 34 (7), 1512–1532.

Bienert, S., Waterhouse, A., de Beer, T.A., et al., 2017. The SWISS-MODEL repository – New features and functionality. Nucleic Acids Res 45 (D1), D313–D319.

Blaszczyk, M., Jamroz, M., Kmiecik, S., Kolinski, A., 2013. CABS-fold: Server for the de novo and consensus-based prediction of protein structure. Nucleic Acids Res 41 (Web Server Issue), W406–W411.

Buchan, D.W., Jones, D.T., 2017. EigenTHREADER: Analogous protein fold recognition by efficient contact map threading. Bioinformatics 33 (17), 2684–2690.

Burley, S.K., Berman, H.M., Kleywegt, G.J., et al., 2017. Protein Data Bank (PDB): The single global macromolecular structure archive. Methods Mol Biol 1607, 627–641.

Chang, C.C., Tey, B.T., Song, J., Ramanan, R.N., 2015. Towards more accurate prediction of protein folding rates: A review of the existing Web-based bioinformatics approaches. Brief Bioinform 16 (2), 314–324.

Chatr-Aryamontri, A., Oughtred, R., Boucher, L., et al., 2017. The BioGRID interaction database: 2017 Update. Nucleic Acids Res 45 (Database Issue), D369–D379.

Chaudhary, P., Naganathan, A., Gromiha, M.M., 2015. Folding RaCe: A robust method for predicting changes in protein folding rates upon point mutations. Bioinformatics 31 (13), 2091–2097.

Chiba, S., Ikeda, K., Ishida, T., et al., 2015. Identification of potential inhibitors based on compound proposal contest: Tyrosine-protein kinase Yes as a target. Sci Rep 5, 17209.

David, R., Korenberg, M.J., Hunter, I.W., 2000. 3D–1D threading methods for protein fold recognition. Pharmacogenomics 1 (4), 445–455.

De los Rios, M.A., Daneshi, M., Plaxco, K.W., 2005. Experimental investigation of the frequency and substitution dependence of negative phi-values in two-state proteins. Biochemistry 44 (36), 12160–12167.

Dobson, C.M., 2003. Protein folding and misfolding. Nature 426, 884–890.

Dougherty, D.A., 1996. Cation-pi interactions in chemistry and biology: A new view of benzene, Phe, Tyr, and Trp. Science 271 (5246), 163–168.

Galgonek, J., Vymetal, J., Jakubec, D., Vondrášek, J., 2017. Amino Acid Interaction (INTAA) web server. Nucleic Acids Res 45, W388–W392.

Galzitskaya, O.V., Garbuzynskiy, S.O., Ivankov, D.N., Finkelstein, A.V., 2003. Chain length is the main determinant of the folding rate for proteins with three-state folding kinetics. Proteins 51, 162–166.

Gaucher, E.A., Govindarajan, S., Ganesh, O.K., 2008. Palaeotemperature trend for Precambrian life inferred from resurrected proteins. Nature 451 (7179), 704–707.

Ghouzam, Y., Postic, G., Guerin, P.E., de Brevern, A.G., Gelly, J.C., 2016. ORION: A web server for protein fold recognition and structure prediction using evolutionary hybrid profiles. Sci Rep, 6. . p. 28268.

Gromiha, M.M., 2009. Multiple contact network is a key determinant to protein folding rates. J Chem Inf Model 49 (4), 1130–1135.

Gromiha, M.M., 2010. Protein Bioinformatics. New Delhi: Academic Press.

Gromiha, M.M., Ahmad, S., 2005. Role of solvent accessibility in structure based drug design. Curr Comp Aided Drug Des 1, 65–72.

Gromiha, M.M., Anoosha, P., Huang, L.T., 2016. Applications of protein thermodynamic database for understanding protein mutant stability and designing stable mutants. Methods Mol Biol 1415, 71–89.

Gromiha, M.M., Fukui, K., 2011. Scoring function based approach for locating binding sites and understanding recognition mechanism of protein–DNA complexes. J Chem Inf Model 51 (3), 721–729.

Gromiha, M.M., Huang, L.T., 2011. Machine learning algorithms for predicting protein folding rates and stability of mutant proteins: Comparison with statistical methods. Curr Protein Pept Sci 12 (6), 490–502.

Gromiha, M.M., Nagarajan, R., 2013. Computational approaches for predicting the binding sites and understanding the recognition mechanism of protein–DNA complexes. Adv Protein Chem Struct Biol 91, 65–99.

Gromiha, M.M., Oobatake, M., Kono, H., Uedaira, H., Sarai, A., 1999a. Relationship between amino acid properties and protein stability: Buried mutations. J Protein Chem 18, 565–578.

Gromiha, M.M., Oobatake, M., Sarai, A., 1999b. Important amino acid properties for enhanced thermostability from mesophilic to thermophilic proteins. Biophys Chem 82 (1), 51–67.

Gromiha, M.M., Pathak, M.C., Saraboji, K., Ortlund, E.A., Gaucher, E.A., 2013. Hydrophobic environment is a key factor for the stability of thermophilic proteins. Proteins 81 (4), 715–721.

Gromiha, M.M., Selvaraj, S., 2001. Comparison between long-range interactions and contact order in determining the folding rates of two-state proteins: Application of long-range order to folding rate prediction. J Mol Biol 310, 27–32.

Gromiha, M.M., Selvaraj, S., 2004. Inter-residue interactions in protein folding and stability. Prog Biophys Mol Biol 86 (2), 235–277.

Gromiha, M.M., Siebers, J.G., Selvaraj, S., Kono, H., Sarai, A., 2004. Intermolecular and intramolecular readout mechanisms in protein–DNA recognition. J Mol Biol 337, 285–294.

Gromiha, M.M., Suresh, M.X., 2008. Discrimination of mesophilic and thermophilic proteins using machine learning algorithms. Proteins 70 (4), 1274–1279.

Gromiha, M.M., Thangakani, A.M., Selvaraj, S., 2006. FOLD-RATE: Prediction of protein folding rates from amino acid sequence. Nucleic Acids Res 34, W70–W74.

Gromiha, M.M., Veluraja, K., Fukui, K., 2014. Identification and analysis of binding site residues in protein–carbohydrate complexes using energy based approach. Protein Pept Lett 21 (8), 799–807.

Gromiha, M.M., Yugandhar, K., 2017. Integrating computational methods and experimental data for understanding the recognition mechanism and binding affinity of protein–protein complexes. Prog Biophys Mol Biol 128, 18–33.

Gromiha, M.M., Yugandhar, K., Jemimah, S., 2017. Protein-protein interactions: Scoring schemes and binding affinity. Curr Opin Struct Biol 44, 31–38.

Huang, L.T., Gromiha, M.M., 2010. First insight into the prediction of protein folding rate change upon point mutation. Bioinformatics 26 (17), 2121–2127.

Huang, L.T., Gromiha, M.M., 2012. Real value prediction of protein folding rate change upon point mutation. J Comput Aided Mol Des 26 (3), 339–347.

Istomin, A.Y., Jacobs, D.J., Livesay, D.R., 2007. On the role of structural class of a protein with two-state folding kinetics in determining correlations between its size, topology, and folding rate. Protein Sci 16 (11), 2564–2569.

Itzhaki, L.S., Otzen, D.E., Fersht, A.R., 1995. The structure of the transition state for folding of chymotrypsin inhibitor 2 analysed by protein engineering methods: Evidence for a nucleation-condensation mechanism for protein folding. J Mol Biol 254, 260–288.

Jayaram, B., Dhingra, P., Mishra, A., et al., 2014. Bhageerath-H: A homology/ab initio hybrid server for predicting tertiary structures of monomeric soluble proteins. BMC Bioinform 15. Suppl 16:S7.

Jemimah, S., Yugandhar, K., Gromiha, M.M., 2017. PROXiMATE: A database of mutant protein-protein complex thermodynamics and kinetics. Bioinformatics. in press.

Jones, D.D., 1975. Amino acid properties and side-chain orientation in proteins: A cross correlation approach. J Theor Biol 50, 167–183.

Jones, D.T., 1999. GenTHREADER: An efficient and reliable protein fold recognition method for genomic sequences. J Mol Biol 287 (4), 797–815.

Jo, T., Hou, J., Eickholt, J., Cheng, J., 2015. Improving protein fold recognition by deep learning networks. Sci Rep 5, 17573.

Kabsch, W., Sander, C., 1983. Dictionary of protein secondary structure: Pattern recognition of hydrogen-bonded and geometrical features. Biopolymers 22, 2577–2637.

Källberg, M., Wang, M., Wang, S., et al., 2012. Template-based protein structure modeling using the RaptorX web server. Nat Protoc 7, 1511–1522.

Kanakaveti, V., Sakthivel, R., Rayala, S.K., Gromiha, M.M., 2017. Importance of functional groups in predicting the activity of small molecule inhibitors for Bcl-2 and Bcl-xL. Chem Biol Drug Des 90 (2), 308–316.

Kawashima, S., Pokarowski, P., Pokarowska, M., et al., 2008. AAindex: Amino acid index database, progress report 2008. Nucleic Acids Res 36 (Database Issue), D202–D205.

Kellenberger, E., Muller, P., Schalon, C., et al., 2006. sc-PDB: An annotated database of druggable binding sites from the Protein Data Bank. J Chem Inf Model 46 (2), 717–727.

Kelley, L.A., Mezulis, S., Yates, C.M., Wass, M.N., Sternberg, M.J., 2015. The Phyre2 web portal for protein modeling, prediction and analysis. Nat Protoc 10 (6), 845–858.

Kim, D.E., Chivian, D., Baker, D., 2004. Protein structure prediction and analysis using the Robetta server. Nucleic Acids Res 32 (Web Server Issue), W526–W531.

Kozma, D., Simon, I., Tusnády, G.E., 2012. CMWeb: An interactive on-line tool for analysing residue–residue contacts and contact prediction methods. Nucleic Acids Res 40 (Web Server Issue), W329–W333.

Kozma, D., Simon, I., Tusnády, G.E., 2013. PDBTM: Protein Data Bank of transmembrane proteins after 8 years. Nucleic Acids Res 41 (Database issue), D524–D529.

Kozma, D., Tusnády, G.E., 2015. TMFoldRec: A statistical potential-based transmembrane protein fold recognition tool. BMC Bioinformatics 16, 201.

Kulandaisamy, A., Lathi, V., ViswaPoorani, K., Yugandhar, K., Gromiha, M.M., 2017. Important amino acid residues involved in folding and binding of protein-protein complexes. Int J Biol Macromol 94 (Pt A), 438–444.

Kumar, M.D., Bava, K.A., Gromiha, M.M., et al., 2006. ProTherm and ProNIT: Thermodynamic databases for proteins and protein–nucleic acid interactions. Nucleic Acids Res 34 (Database Issue), D204–D206.

Leal, E.S., Aucar, M.G., Gebhard, L.G., et al., 2017. Discovery of novel dengue virus entry inhibitors via a structure-based approach. Bioorg Med Chem Lett 27 (16), 3851–3855.

Lee, B., Richards, F.M., 1971. The interpretation of protein structures: Estimation of static accessibility. J Mol Biol 55, 379–400.

Lensink, M.F., Velankar, S., Wodak, S.J., 2017. Modeling protein–protein and protein–peptide complexes: CAPRI 6th edition. Proteins 85 (3), 359–377.

Lhota, J., Xie, L., 2016. Protein-fold recognition using an improved single-source K diverse shortest paths algorithm. Proteins 84 (4), 467–472.

Li, J., Adhikari, B., Cheng, J., 2015. An improved integration of template-based and template-free protein structure modeling methods and its assessment in CASP11. Protein Pept Lett 22 (7), 586–593.

Malik, A., Firoz, A., Jha, V., Ahmad, S., 2010. PROCARB: A database of known and modelled carbohydrate-binding protein structures with sequence-based prediction tools. Adv Bioinform. 436036.

Mallik, S., Das, S., Kundu, S., 2016. Predicting protein folding rate change upon point mutation using residue-level coevolutionary information. Proteins 84 (1), 3–8.

Marks, D.S., Colwell, L.J., Sheridan, R., et al., 2011. Protein 3D structure computed from evolutionary sequence variation. PLOS One 6 (12), e28766.

Moal, I.H., Fernández-Recio, J., 2012. SKEMPI: A structural kinetic and energetic database of mutant protein interactions and its use in empirical models. Bioinformatics 28, 2600–2607.

Nagarajan, R., Gromiha, M.M., 2014. Prediction of RNA binding residues: An extensive analysis based on structure and function to select the best predictor. PLOS One 9 (3), e91140.

Nagarajan, R., Ahmad, S., Gromiha, M.M., 2013. Novel approach for selecting the best predictor for identifying the binding sites in DNA binding proteins. Nucleic Acids Res 41 (16), 7606–7614.

Nagarajan, R., Archana, A., Thangakani, A.M., et al., 2016. PDBparam: Online Resource for Computing Structural Parameters of Proteins. Bioinform Biol Insights 10, 73–80.

Nagarajan, R., Chothani, S.P., Ramakrishnan, C., Sekijima, M., Gromiha, M.M., 2015. Structure based approach for understanding organism specific recognition of protein–RNA complexes. Biol Direct 10, 8.

Narayanan, B.C., Westbrook, J., Ghosh, S., et al., 2013. The Nucleic Acid Database: New features and capabilities. Nucleic Acids Res. 42, D114–D122.

Norambuena, T., Melo, F., 2010. The Protein-DNA Interface database. BMC Bioinformatics 11, 262.

Orchard, S., Ammari, M., Aranda, B., et al., 2014. The MIntAct project – IntAct as a common curation platform for 11 molecular interaction databases. Nucleic Acids Res 42 (Database Issue), D358–D363.

Pai, P.P., Dattatreya, R.K., Mondal, S., 2017. Ensemble architecture for prediction of enzyme-ligand binding residues using evolutionary information. Mol Inform. in press.

Pica, A., Graziano, G., 2016. Shedding light on the extra thermal stability of thermophilic proteins. Biopolymers 105 (12), 856–863.

Plaxco, K.W., Simons, K.T., Baker, D., 1998. Contact order, transition state placement and the refolding rates of single domain proteins. J Mol Biol 277, 985–994.
Ponnuswamy, P.K., 1993. Hydrophobic characteristics of folded proteins. Prog Biophys Mol Biol 59 (1), 57–103.
Punta, M., Rost, B., 2005. Protein folding rates estimated from contact predictions. J Mol Biol 348, 507–512.
Raleigh, D.P., Plaxco, K.W., 2005. The protein folding transition state: What are Phi-values really telling us? Protein Pept Lett 12 (2), 117–122.
Ramakrishnan, C., Thangakani, A.M., Velmurugan, D., et al., 2017. Identification of type I and type II inhibitors of c-Yes kinase using in silico and experimental techniques. J Biomol Struct Dyn. in press.
Roy, A., Kucukural, A., Zhang, Y., 2010. I-TASSER: A unified platform for automated protein structure and function prediction. Nat Protoc 5, 725–738.
Saraboji, K., Gromiha, M.M., Ponnuswamy, M.N., 2006. Average assignment method for predicting the stability of protein mutants. Biopolymers 82 (1), 80–92.
Shen, Y., Maupetit, J., Derreumaux, P., Tufféry, P., 2014. Improved PEP-FOLD approach for peptide and miniprotein structure prediction. J Chem Theor Comput 10, 4745–4758.
Sillitoe, I., Lewis, T.E., Cuff, A.L., et al., 2015. CATH: Comprehensive structural and functional annotations for genome sequences. Nucleic Acids Res 43 (Database Issue), D376–D381.
Simons, K.T., Kooperberg, C., Huang, E., Baker, D., 1997. Assembly of protein tertiary structures from fragments with similar local sequences using simulated annealing and Bayesian scoring functions. J Mol Biol 268, 209–225.
Singh, H., Srivastava, H.K., Raghava, G.P., 2016. A web server for analysis, comparison and prediction of protein ligand binding sites. Biol Direct 11 (1), 14.
Sippl, M.J., 1990. Calculation of conformational ensembles from potentials of mean force. An approach to the knowledge-based prediction of local structures in globular proteins. J Mol Biol 213, 859–883.
The UniProt Consortium, 2017. UniProt: The universal protein knowledgebase. Nucleic Acids Res 45, D158–D169.
Thévenet, P., Shen, Y., Maupetit, J., et al., 2012. PEP-FOLD: An updated de novo structure prediction server for both linear and disulfide bonded cyclic peptides. Nucleic Acids Res 40, W288–W293.
Tina, K.G., Bhadra, R., Srinivasan, N., 2007. PIC: Protein interactions calculator. Nucleic Acids Res 35 (suppl_2), W473–W476.
Tsujikawa, H., Sato, K., Wei, C., et al., 2016. Development of a protein–ligand-binding site prediction method based on interaction energy and sequence conservation. J Struct Funct Genomics 17 (2–3), 39–49.
Webb, B., Sali, A., 2014. Comparative protein structure modeling using modeller. Current Protocols in Bioinformatics. John Wiley & Sons, Inc. pp. 5.6.1–5.6.32.
Wu, S., Zhang, Y., 2008. MUSTER: Improving protein sequence profile-profile alignments by using multiple sources of structure information. Proteins 72 (2), 547–556.
Xenarios, I., Salwínski, L., Duan, X.J., et al., 2002. DIP, the Database of Interacting Proteins: A research tool for studying cellular networks of protein interactions. Nucleic Acids Res 30 (1), 303–305.
Xu, D., Zhang, Y., 2012. Ab initio protein structure assembly using continuous structure fragments and optimized knowledge-based force field. Proteins 80, 1715–1735.
Xu, J., Peng, J., Zhao, F., 2009. Template-based and free modeling by RAPTOR++ in CASP8. Proteins 77 (Suppl. 9), 133–137.
Yang, J., Roy, A., Zhang, Y., 2015. The I-TASSER Suite: Protein structure and function prediction. Nature Methods 12, 7–8.
Yan, R., Si, J., Wang, C., Zhang, Z., 2009. DescFold: A web server for protein fold recognition. BMC Bioinform 10, 416.
Zanzoni, A., Montecchi-Palazzi, L., Quondam, M., et al., 2002. MINT: A Molecular INTeraction database. FEBS Lett 513 (1), 135–140.
Zou, Y., Wang, Y., Wang, F., et al., 2017. Discovery of potent IDO1 inhibitors derived from tryptophan using scaffold-hopping and structure-based design approaches. Eur J Med Chem 138, 199–211.

Relevant Websites

http://www.abren.net/asaview/
    ASAView: Solvent Accessibility Graphics for proteins.
http://www.iitm.ac.in/bioinfo/proteinfolding/foldingrace.html
    Indian Institute of Technology Madras.

Biographical Sketch

M. Michael Gromiha received his PhD in Physics from Bharathidasan University, India and served as STA fellow, RIKEN Researcher, Research Scientist and Senior Scientist at Computational Biology Research Center, AIST, Japan till 2010. Currently, he is working as an Associate Professor at Indian Institute of Technology (IIT) Madras, India. His main research interests are structural analysis, prediction, folding and stability of globular and membrane proteins, protein interactions and development of bioinformatics databases and tools. He has published over 200 research articles, 40 reviews, 5 editorials and a book on Protein Bioinformatics: From Sequence to Function by Elsevier/Academic Press. His papers received more than 9000 citations and h-index is 51. He is an Associate Editor of BMC Bioinformatics as well as Editorial Board Member of Scientific Reports, Biology Direct, Journal of Bioinformatics and Computational Biology and Current Computer Aided Drug design. He has received several awards including Oxford University Press Bioinformatics prize, Okawa Science Foundation Research Grant, Young Scientist Travel awards from ISMB, JSPS, AMBO, ICTP etc., Best paper award at ICIC2011, ICTP Associateship award, ICMR International fellowship for Senior Biomedical Scientists, INSA senior scientist award, Best paper award in Bioinformatics by Department of Biotechnology, India, Institute Research and Development Award at IIT Madras and Outstanding Performance award from Initiative for Parallel Bioinformatics (IPAB), Tokyo Institute of Technology, Japan. He is a member of the National Academy of Sciences, India.

R. Nagarajan completed his PhD degree in Computational Biology from Department of Biotechnology, Indian Institute of Technology Madras, India. He is currently working as a Postdoctoral Research Fellow in Vanderbilt Vaccine Center, Vanderbilt University Medical Center, USA. His main research interests are computational analysis of protein–nucleic acid interactions, recognition mechanism, protein aggregation, computational immunology and development of prediction algorithms, web servers and databases. He has 14 publications (including two book chapters) in reputed journals and received more than 100 citations with h-index 6. He has received Research award from IIT Madras, AU-CBT Excellence award from Biotech Research Society of India, Incentive award from DBT, Govt of India and travel awards from International Society of Computational Biology and Asia Pacific Bioinformatics network.

S. Selvaraj received his PhD degree in Biophysics from the University of Madras, Chennai, India. He has worked as a Collaborative Scientist in RIKEN Life Science Centre, Tsukuba, Japan, and later as Assistant Professor in the Department of Bioinformatics, School of Life Sciences, Bharathidasan University, Tiruchirappalli, India. Post-retirement he has been awarded the Emeritus Fellowship by the University Grants Commission, New Delhi, India. He has published about 60 research articles and 6 reviews. His publications have received more than 2000 citations with h-index of 20. He has been awarded the National Associateship by the Department of Biotechnology, Government of India.

Protein Structure Databases

David R Armstrong, John M Berrisford, Matthew J Conroy, Alice R Clark, Deepti Gupta, and Abhik Mukhopadhyay, Protein Data Bank in Europe, EMBL-EBI, Hinxton, United Kingdom

Introduction

The Value of Understanding Protein Structure

Our scientific understanding of the molecular basis of life has expanded dramatically since the very first macromolecular structures were determined in late 1950s. These huge achievements in building atomic models of myoglobin (Kendrew *et al.*, 1958) and hemoglobin (Perutz *et al.*, 1960) paved the way for resolving structures of more complex proteins. More recently the structures of enormous complexes such as the ribosome and drug targets such as G protein-coupled receptors have been determined. These highly informative structures have provided answers to the basic understanding of the molecular machinery in many biological processes and have contributed enormously to human health. An early example of drug design informed by knowledge of the target protein structure is that of the carbonic anhydrase inhibitor dorzolamide which was approved in 1995 (Kubinyi, 1999). Several anti-virals directed against HIV protease were designed using mainly structure-based methods (Greer *et al.*, 1994). The 22 Nobel prizes awarded in the field of structural biology between 1946 and 2016 highlight its importance.

While technological advances and talented scientists have been key to these discoveries, no small part has been played by the structural biology community's willingness to make their data, in the form of atomic models, freely available. To enable this, the community has instigated and supported resources which store and distribute these data. Some of these resources archive experimental data and make it available, others take this archive data and classify it according to their own analyses and annotations. This chapter reviews some of these resources, all of which are available online. URLs of those mentioned are given in **Table 1**.

The Protein Data Bank

The oldest and largest database of protein structure is the Protein Data Bank (PDB) which archives, curates and distributes experimentally determined structural information. Analysis of citations of the PDB archive, and the data it contains indicate how vital this resource is to the biomedical community (Bousfield *et al.*, 2016). In maintaining a centralised, consistent archive of protein structures, the PDB has played a crucial role in the field of structural bioinformatics (Lesk and Chothia, 1980; Chothia and Lesk, 1986).

The PDB is a freely accessible repository for experimentally determined 3D structures of proteins, nucleic acids, and complex assemblies. Many journals now require scientists to have deposited their data with the PDB before a manuscript is accepted for publication. When the PDB was formed as a collaboration between the Brookhaven National Laboratory (BNL) in the US, and the Cambridge Crystallographic Data Center in the UK (Protein Data Bank, 1971), it contained just seven structures. All of these were of proteins and solved by X-ray crystallography. 46 years later, the archive contains more than 130,000 structures of over 40,000 unique macromolecules (see **Fig. 1**). Methodology too has expanded and the PDB archives structures determined by several experimental techniques. Around 90% of structures in the archive are solved by diffraction techniques with structures solved by nuclear magnetic resonance (NMR) and electron microscopy (EM) making up most of the remaining 10%. Though historically this was not the case, experimental data from which the models are derived must also be deposited to support the atomic coordinates. In the case of diffraction studies, structure factors are deposited to the PDB and for NMR derived models, the PDB archives chemical shifts and geometric restraints. Further NMR data can be archived with the Biological Magnetic Resonance Data Bank (Ulrich *et al.*, 2008). In the case of electron microscopy studies, data are deposited to the Electron Microscopy Data Bank (Tagari *et al.*, 2002).

By 1977 the PDB had added a further distribution center in Japan (Bernstein *et al.*, 1977), and over time multiple sites mirrored PDB data- a common phenomenon in the days of low speed web connections. PDB format data could be deposited either to BNL or the Macromolecular Structure Database (MSD) at the European Bioinformatics Institute (Keller *et al.*, 1998; Sussman *et al.*, 1999). In 1998 the Research Collaboratory for Structural Bioinformatics (RCSB) took over from BNL as the US PDB center. By the turn of the millennium, the Protein Data Bank Japan (PDBj) based at the Institute for Protein Research, Osaka University was also handling deposition of PDB data. In 2002, to recognise the growing international nature of structural biology and its data archiving activities, the Worldwide Protein Data Bank (wwPDB) was formed (Berman *et al.*, 2003). Founder members were RCSB, MSD (now Protein Data Bank in Europe (PDBe)) and PDBj. In 2008, Biological Magnetic resonance Bank (BMRB (Markley *et al.*, 2008)) which deals specifically with NMR data, became a wwPDB partner. The history of the PDB and wwPDB is well documented elsewhere (Berman *et al.*, 2016; Berman *et al.*, 2014; Berman *et al.*, 2012; Meyer, 1997). The wwPDB manages the PDB archive and ensures that the PDB is freely and publicly available to the global community, conforming to the FAIR (Findable, Accessible, Interoperable, and Re-usable) guiding principles (Wilkinson *et al.*, 2016).

Depositing data

Since 2014, data has been deposited to the wwPDB via a unified deposition interface across all wwPDB sites. This system, OneDep (Young *et al.*, 2017), directs data depositions to the relevant wwPDB partner site for processing based on geography. Thus RCSB

Table 1   URLs of resources described in the text.

Resource	url
wwPDB resources	
Worldwide PDB	wwpdb.org
Protein Data Bank in Europe	PDBe.org
Protein Data Bank Japan	PDBj.org
RCSB PDB	rcsb.org
BioMagResBank	BMRB.org
EMDB Resources	
EBI	emdb-empiar.org
RCSB	emsearch.rutgers.edu
PDBj	pdbj.org/emnavi
EmDataBank	emdatabank.org
Other Experimental Archives	
SASBDB	sasbdb.org
BIOISIS	bioisis.net
PRIDE	www.ebi.ac.uk/pride/
Model repositories	
SWISS-MODEL Repository	swissmodel.expasy.org/repository
Protein Model Portal	proteinmodelportal.org
Secondary protein structure atlases	
PDBsum	www.ebi.ac.uk/pdbsum/
OCA	oca.weizmann.ac.il/oca-bin/ocamain
Proteopedia	proteopedia.org/
JenaLib	jenalib.leibniz-fli.de/
MMDB	ncbi.nlm.nih.gov/structure/
Fold Classification Databases	
SCOP	scop.mrc-lmb.cam.ac.uk/scop/
SCOPe	scop.berkeley.edu
CATH	cathdb.info/
ECOD	prodata.swmed.edu/ecod/
ArchDB	sbi.imim.es/archdb/
Subject-specific resources	
VIPERdb	viperdb.scripps.edu/
GPCRdb	gpcrdb.org/
PyIgClassify	dunbrack2.fccc.edu/PyIgClassify/
SAbDb	opig.stats.ox.ac.uk/webapps/sabdab-sabpred/
SpliProt3D	iimcb.genesilico.pl/SpliProt3D/
PDBTM	pdbtm.enzim.hu/
Membrane Protein Data Bank	www.lipidat.tcd.ie/
MemProt	memprot.bicpu.edu.in/
Membrane Proteins of Known Structure	blanco.biomol.uci.edu/mpstruc
Indian PDB	iris.physics.iisc.ernet.in/ipdb/
Non-structural resources which integrate structural data	
COSMIC 3D	cancer.sanger.ac.uk/cosmic3d
EBI search	www.ebi.ac.uk/
NCBI	ncbi.nlm.nih.gov/
Open Targets	opentargets.org/
Genome3D	genome3d.eu/

PDB handle data from the Americas and Oceania, PDBj process data from Asia and PDBe annotate all data from Europe and Africa. NMR data not linked to a PDB structure deposition or additional experimental data (e.g., free induction decay) can be deposited directly to BMRB.

Once deposited, data can be held privately for up to one year and are usually released on publication of the accompanying paper. the PDB archive is updated weekly at 00.00 UTC each Wednesday and the complete updated archive is available from the master FTP site (see Relevant Websites section) or from any of the wwPDB partners.

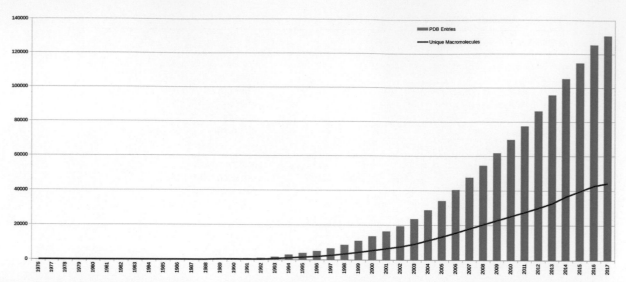

Fig. 1 Growth of the PDB archive over time. The green bars show total number of PDB entries and the red line unique macromolecules (as determined by standardised molecule names).

Curation and validation

Atomic coordinates merely define the shape of a molecule. To make the structure data truly useful, they must be related to other information (Gerstein, 2000). Thus structures are almost always accompanied by a host of metadata. In the PDB archive, these are either entered by the scientists solving the structure, or added by the curation procedures of wwPDB (Dutta *et al.*, 2008; Sen *et al.*, 2014). These annotations include details of the experiment performed to produce the model, cross references to polymer sequence databases, and any discrepancies between the polymer sequences in the model and in a reference database. Where a structure is described in a scientific paper, the citation is included.

Curation also includes adding information about the quaternary structure of the entry. The convention in crystallography is to deposit only the smallest repeating unit of the crystal to the PDB (the asymmetric unit, ASU) thus the true quaternary structure may contain only a part of the ASU or consist of multiple ASUs.

There is frequently a misconception that the atomic coordinates define the absolute positions of all the atoms in a given macromolecule. It should be remembered however that the coordinates are a model which is an interpretation of experimental data. Some data may enable a more precise model to be built than other data. Users of structural databases need to be aware of the limitations of the model when drawing conclusions from it. To address this issue, wwPDB convened expert task forces representing the structural biology community which published their recommendations on how the wwPDB should assess the quality of structures in the PDB (Read *et al.*, 2011; Henderson *et al.*, 2012; Montelione *et al.*, 2013; Adams *et al.*, 2016). Following these recommendations, extensive validation reports now accompany the structures in the PDB archive.

File formats

Early in the history of structural biology, as the possibility of viewing atomic coordinates on a computer rather than as a physical model became reality, the community realised that standards were needed if structure data were to be read computationally between different research groups. Thus the PDB file format was introduced using the 80-column Hollerith card format.

The PDB format defined the standard for representation of atomic coordinates in structural biology for decades and as a human readable text file, is still used by many today. The historical fixed width of the PDB format fields imposes severe limitations for modern structural biology and in 2014 the wwPDB officially moved to use PDBx/mmCIF (Bourne *et al.*, 1997; Westbrook and Bourne, 2000) as the master format for the PDB archive. This format is free of the limitations inherent in the PDB format and is much more extensible to record the diverse metadata which accompanies modern coordinate data.

PDB data is also distributed by the wwPDB in PDBML format (Westbrook *et al.*, 2005), which is an XML-like format, and as RDF format. Both of these are direct translations of the PDBx/mmCIF formatted files.

Access to PDB Data

Websites of wwPDB Members

"The PDB" refers to the FTP archive of files distributed by the wwPDB partners. While the wwPDB website provides details of archive policies and procedures, it is the wwPDB partners who provide tools to search, download, visualize, and analyse protein structures and annotations from the archive. Data from each entry are presented on a series of pages which have been termed 'atlases' detailing the structure and its annotations. Whilst hosting identical data, each site develops its own website, providing unique views of the primary data, and a variety of tools and resources to analyse that data. All sites also have RESTful API calls which grant programmatic access to the archive. A brief summary of the wwPDB partner sites is given below.

Protein Data Bank in Europe

The Protein Data Bank in Europe (PDBe) (Velankar *et al.*, 2016) is an interactive and image-rich site providing access to all data in the PDB archive via a faceted search system. Search results can be ranked based on a series of metrics including the quality of structures. Each entry is detailed on its own series of pages on which macromolecules can viewed in three dimensions using the in-browser viewer LiteMol. Visualisation of electron density maps for X-ray structures and electric potential maps for those derived by electron microscopy is possible via this viewer. Annotations such as sequence and structure domains, and validation data can be displayed on the model in this viewer and also on sequence and topology views, which are all interlinked (see **Fig. 2**).

PDBe works with Europe PMC to list publications which cite the paper describing a particular structure, as well as those which mention the PDB code but do not cite the paper.

Advanced tools to analyse the data include PDBePISA (Krissinel and Henrick, 2007), which assesses stability of interactions within a crystal structure, and PDBeFold (Krissinel and Henrick, 2004), which is used to search for structures of a similar three dimensional topology.

Structure integration with function, taxonomy and sequence (SIFTS)

PDBe, along with UniProt (The UniProt Consortium, 2017) produces the SIFTS resource (Velankar *et al.*, 2013) which maintains up-to-date cross-references between the UniProt sequence database and the sample sequence in the PDB entry. Through mapping

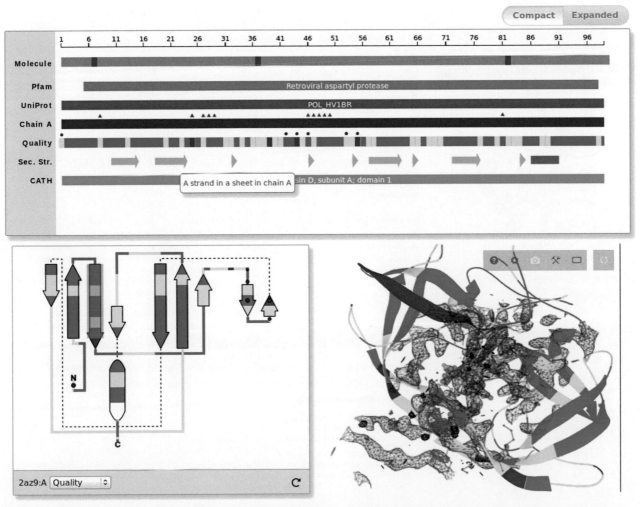

Fig. 2 1D, 2D and 3D interactive views of a HIV protease (PDB entry 2az9) on the PDBe website. The topology diagram and 3D structure are shown coloured by model geometry quality according to the wwPDB validation report and electron density is displayed.

to UniProt the SIFTS service adds cross references to Gene Ontology (GO) (Jones *et al.*, 2014; Ashburner *et al.*, 2000) terms, to sequence motifs from InterPro (Jones *et al.*, 2014), and Pfam sequence domains (Finn *et al.*, 2016). SIFTS also collates cross references to structural domains (SCOP, CATH). This service is used by the other wwPDB partners to maintain up-to date cross-references.

The Protein Data Bank Japan

The Protein Data Bank Japan (PDBj) (Kinjo *et al.*, 2017) provides search, advanced search and visualisation facilities for all entries in the PDB, not only in English but also in Japanese, Chinese and Korean. On each entry page, structures can be viewed in 3D in a variety of molecular viewers. The layout of the PDBj site is user-configurable and all PDB entries are provided in Resource Description Framework (RDF) format.

A variety of advanced services are supplied by PDBj including calculations of electrostatic surfaces and eF-seek, which searches protein structures for similar binding sites. GIRAF (Kinjo and Nakamura, 2007) finds similarities in small molecule binding sites between proteins which might lack detectable sequence similarity. Omokage is a PDBj service which finds structures of similar shapes across PDB and EMDB archives (Suzuki *et al.*, 2016).

Research Collaboratory for Structural Bioinformatics PDB

The Research Collaboratory for Structural Bioinformatics (RCSB) PDB website (Berman *et al.*, 2002) also has a search, advanced search, and pages for each structure in the PDB archive. Search results can be saved in customisable reports. Each PDB entry has a summary page (see **Fig. 3**) with data described across a series of subpages which tabulate data and annotations, including sequence and structure similarity with other molecules in the PDB archive. Structures can be viewed in 3D using the online NGL viewer (Rose and Hildebrand, 2015) and various other visualisation programs. Recent updates have included linking to other databases

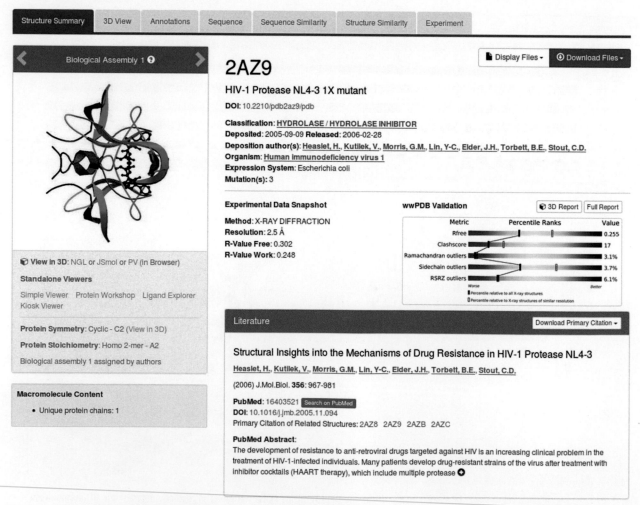

Fig. 3   Summary page for HIV-1 protease PDB entry 2az9 at the RCSB PDB website.

and mapping human proteins to their locations on the genome (Rose *et al.*, 2017). Mentions of the PDB code in pubmed are plotted per year.

RCSB enables pairwise structure and sequence alignments using a variety of different algorithms.

Statistics of the PDB archive are readily available at RCSB displayed as interactive graphs across a wide range of data criteria.

In common with its partner sites, RCSB has the facility to search the PDB archive for small molecules which are bound to macromolecules.

Biological Magnetic Resonance Data Bank (BMRB)

The Biological Magnetic Resonance Data Bank (Ulrich *et al.*, 2008) collects and distributes data from nuclear magnetic resonance experiments on proteins and other biological molecules. In addition to the atomic model and data directly used to derive it, BMRB serves as a repository for other kinds of NMR experimental data, such as kinetic and NMR relaxation parameters, and free induction decay (FID) data BMRB webpages for NMR derived protein structures display validation data and links to the experimental data in addition to the model. BMRB has mirror sites in Osaka and Florence in addition to the master site in Wisconsin.

Protein Structure Databases outside of the wwPDB

In addition to members of the wwPDB consortium, several other databases archive protein structural models and/or the experimental data used to derive them. Some of these are listed below.

Electron Microscopy Data Bank

The Electron Microscopy Data Bank (EMDB) is a repository for three dimensional volumes of macromolecular complexes and subcellular structures derived by electron microscopy. It was founded by the MSD team (now PDBe) at EBI in 2002 (Tagari *et al.*, 2002). EMDB archives data from a range of electron microscopic techniques including single-particle reconstruction, tomography, and electron crystallography. The scale of data in the EMDB archive ranges from tomography of whole cells, to electron crystallographic studies of peptides at atomic resolution. In the last few years, improvements in instrumentation and software have been enormously beneficial to the field. These advances, termed the 'resolution revolution' (Kühlbrandt, 2014) have resulted in a huge increase in both the quantity of data in EMDB and its quality, certainly in terms of resolution (Patwardhan, 2017).

EMDataBank, a collaboration between PDBe at EMBL-EBI, the RCSB PDB and the National Center for Macromolecular Imaging (NCMI) (Lawson *et al.*, 2011), has pushed forward data standards in the EM community. Outcomes have included EMDB files being archived jointly in the wwPDB FTP archive, electron microscopy validation reports and the integration of EM deposition with the wwPDB deposition system allowing simultaneous deposition of EM maps to EMDB and the fitted model to PDB. EMDB has a weekly release cycle, synchronized with that of the wwPDB.

Accessing EMDB data

Whilst not formally a wwPDB partner, the data in EMDB archive is available in the wwPDB FTP tree and it can be further accessed through four websites: the EMDB pages at EMBL-EBI, at the RCSB PDB website, PDBj and via the emdatabank.org website. Each website presents the data in different ways. A summary is provided below.

EMDB at EBI provides an advanced search, with EM maps available for download or viewable online using a number of tools and viewers. Many of these visualisation tools also serve as validation tools. In the visual analysis page (Lagerstedt *et al.*, 2013), the map data is presented to the user as a surface, projection, slices and overlaid with the fitted model. The volume slicer tool (Salavert-Torres *et al.*, 2016) is of particular use when visualising tomogram volumes and allows navigation through the slices of a tomogram. Statistics for the EMDB archive are provided in live graphs.

PDBj also provides a search functionality (including shape recognition) and access to EMDB data through the EM Navigator which also provides visual analysis and validation tools. Maps can be viewed as images and movies from UCSF Chimera (Pettersen *et al.*, 2004), the session files for which are available for download. The Yorodumi viewer allows interactive viewing of the map and the PDB model. To analyse overall trends of the data in the EMDB archive an interactive statistics table is provided. The RCSB PDB provide a basic search tool, links to download and view the data in the EMDB.

The EMDataBank site provides an access point for both deposition to and retrieval of data from the EMDB archive in addition to the derived models in the PDB. It also provides data for various 'challenges' which are helping drive developments in the electron microscopy field.

Solution Scattering Databases

X-ray and neutron scattering techniques can provide data on the size and molecular envelope (i.e., shape) of a macromolecule. Small Angle Scattering Biological Databank (SASBDB) (Valentini *et al.*, 2015) and BIOISIS (Hura *et al.*, 2009) are repositories of structural information from these experimental techniques. They allow researchers to deposit and access experimental scattering data and derived models from the scattering experiment. Both provide experimental and derived information and other relevant manually curated data such as experimental conditions, sample details and structural models that have been provided by the depositor during the time of data deposition.

Hybrid and Integrative Techniques

With the advancement of scientific methods, researchers working on challenging biological systems are now able to use a whole suite of different experimental techniques, sometimes supplemented by theoretical data, to determine three dimensional structures of proteins and large complexes. Such approaches are referred to as integrative or hybrid methods (Sali *et al.*, 2015). For instance, mass spectrometry and cross linking data can give insights into a protein's fold or the composition of a protein-protein complex. PRIDE at EBI (Vizcaíno *et al.*, 2016) archives such mass spectrometry data. More generally, where such data should be archived, and vitally, how they should be validated, remains a challenge for the scientific community.

Modelling Archives

On the advice of the community, from 2006 the wwPDB ceased accepting structures determined *in silico* (i.e., by homology or *ab initio* methods) (Berman *et al.*, 2007) and such models which were in the PDB archive were removed. Nevertheless, homology modelling and protein structure prediction is a critical area of research in structural biology with competitions such as CASP (Kryshtafovych *et al.*, 2014) and CAPRI (Lensink *et al.*, 2016) assessing structure prediction and docking methodologies. There is therefore a community demand to store the information collected from theoretical protein modelling.

SWISS-MODEL Repository

The SWISS-MODEL repository (Bienert *et al.*, 2017) is a database of annotated 3D protein structure models generated by the SWISS-MODEL homology-modelling pipeline. Currently the repository contains models for protein sequences in the UniProt database as well as experimentally determined structures from PDB with mapping to UniProt. If a particular protein is absent from the repository, users have the option to attempt building the model. Model quality is indicated over a variety of validation criteria.

Protein Model Portal

Protein Model Portal (Haas *et al.*, 2013) provides access to experimentally determined protein structures and also theoretical models based on sequences available in UniProt. Currently, over 5 million sequences are covered by a model. Crucially, the service also gives an indication of the quality of the model.

Other services exist which will generate a protein model based on homology, for example I-TASSER (Yang *et al.*, 2015) and Phyre2 (Yang *et al.*, 2015; Kelley *et al.*, 2015), but unlike the SWISS-MODEL repository these lack archives of previously calculated structures.

'Derived' Protein Structure Databases

A plethora of specialist databases exist which do not archive structures, but take data from the PDB archive and provide further manual or computational analysis of the data. Some use the whole PDB archive, whilst others might focus on specific protein families. Approximately 25% of resources described in the Nucleic Acids Research database issue each year make use of PDB data in some way. An extensive list of protein structure databases is available online at the Nucleic Acids Research website (see Relevant Websites section).

Here we will detail just a few of these as space allows. For the sake of clarity, we have divided these databases into different types.

(1) Secondary protein structure atlases
(2) Fold classification databases
(3) Subject-specific resources

Secondary Protein Structure Atlases

Several resources integrate information from other resources with three dimensional protein structures and provide a wealth of information to the research community. Some could be considered 'atlases' providing individual webpages for each structure in the PDB archive. Often these were created to provide functionality which wwPDB partner sites lacked, but in many cases, this functionality is now present at one or more of the wwPDB partners. While the members of the wwPDB consortium release new data each Wednesday, other sites necessarily do not have access to these data prior to release and will clearly lag behind the wwPDB partners' sites.

PDBsum

PDBsum (Laskowski *et al.*, 1997) is a pictorial database that provides an overview of the contents of each entry in the PDB archive. It provides several tools to analyse three dimensional macromolecular structures. PDBSum's Ligplot tool gives a pictorial description of a ligand binding site while Nucplot displays interactions between proteins and nucleic acids, and the Prot-prot tool generates a schematic diagram of protein-protein interfaces.

OCA

OCA displays information from the PDB archive and integrates data from several other sources. Its front page is an advanced search form for users to generate specific search queries. Unusually for protein structure databases, the pages for PDB entries in OCA are image free.

Proteopedia

Proteopedia (Hodis *et al.*, 2010) is a wiki-based encyclopedia of macromolecular structures in which annotations are community sourced. Thus the content of each page is extremely variable. Each page has a 3D structure visualiser adjacent to user-editable text describing the structure. The 3D views are scriptable such that users can illustrate a text description with specific views in 3D. Proteopedia has pages that describe generic molecules and complexes in addition to pages for specific PDB entries.

Jenalib

Jenalib (Reichert and Sühnel, 2002) has an emphasis on visualisation and analysis of all entries from the PDB. It provides tools such as SiteDB, GO2PDB and HeteroDB, which are useful to find information on ligand binding sites, Gene ontology and bound small molecules, respectively.

Molecular modelling database

Molecular modelling database, MMDB (Madej *et al.*, 2014) contains experimental structures derived from PDB. MMDB links proteins of similar sequence and structure and displays information about bound small molecules. It allows users to analyse the macromolecular structure in an interactive 3D manner.

Fold Classification Databases

It was recognised relatively early in the history of structural biology that structure is better conserved than sequence (Rao and Rossmann, 1973; Chothia and Lesk, 1986). Several resources classify proteins by their three dimensional fold and, through fold similarity, help infer evolutionary and functional relationships.

SCOP

Structural Classification of Proteins (SCOP) (Murzin *et al.*, 1995) is a database which classifies protein domains according to regions of structural similarity, utilizing both manual and automatic curation. This database uses known structures from the PDB to classify protein folds hierarchically, based on structural and evolutionary relationships. Users can identify known protein folds within their structure of interest and use this information to suggest potential functions of the protein.

More recently, SCOP2 (Andreeva *et al.*, 2014) was launched which presents structural relationships as a network, rather than a tree-like hierarchy. While it is currently a prototype version, SCOP2 enables more complex representations of the relationships between different folds than was possible in the original SCOP. The SCOPextended (SCOPE) database (Chandonia *et al.*, 2017) adds many more structures from the PDB archive to the original SCOP classification.

CATH

The CATH database (Sillitoe *et al.*, 2015) also classifies protein folds. It is named according to the hierarchy used in classification: Class, Architecture, Topography and Homology. Philosophically, CATH and SCOP are very similar, differing in precisely what folds and domains of proteins are identified. CATH also links with Gene3D (Lam *et al.*, 2016) to extend this classification to protein sequences for which a 3D structure is not yet available but can be predicted. In addition to browsing the CATH database, CATH's web pages allow searches based on an uploaded sequence or coordinate file.

ECOD

Both SCOP and CATH databases require manual curation and neither of the two databases has kept up to date with releases in the PDB (though recent developments are addressing this). A further fold classification methodology, Evolutionary Classification of Protein Domains (ECOD) (Cheng *et al.*, 2014; Schaeffer *et al.*, 2017) aims to overcome this problem by offering only automatic annotation of structural domains. ECOD also allows users to upload their own coordinate file for fold classification.

ArchDB

CATH and SCOP classify folds based on the secondary structure elements within a protein, but ArchDB (Bonet *et al.*, 2014) curates instead those non-regular portions of a protein, termed loops, which still have defined structures. Loops are classified on their geometry into 10 separate types. The interface allows users to find structures of loops based on geometry and classification and to search the database using a variety of other criteria.

Subject-Specific Resources

There are many resources that provide functional annotation for specific types of proteins, dependant on the interests of the creators. The following descriptions provide just a taste of what is available, many are the work of individual labs and their longevity is not guaranteed.

VIPERdb (Carrillo-Tripp *et al.*, 2009) focusses on capsid structures from icosahedral viruses. VIPERdb also provides data on computational analyses on icosahedral virus systems. GPCRdb (Isberg *et al.*, 2017) provides annotations and tools focussing on G-protein coupled receptors.

PyIgClassify (Adolf-Bryfogle *et al.*, 2015) and SAbDab (Dunbar *et al.*, 2014) provide extra annotation on antibodies, Spli-Prot3D (Korneta *et al.*, 2012) brings together structures and models of the human spliceosome components.

The structures of membrane proteins are also notoriously difficult to solve experimentally but of great biological relevance. Several resources focus on membrane proteins. Among these resources are PDBTM (Kozma *et al.*, 2013) which takes proteins from the PDB with likely membrane spanning regions, classifies them and indicates the probable membrane embedded region. The Membrane Protein Data Bank, MPDB (Raman *et al.*, 2006), MemPROT and "Membrane Proteins of known structure" also list transmembrane structures from the PDB.

Rather than subject specificity, Indian PDB provides a database of structures solved by Indian structural biologists, sorted by institute.

Integrating Structure into the Bigger Picture

Protein structures do not exist without a context, and many databases are integrating structures into their resources or ensuring that they link to relevant structural resources. Thus biologists searching for data relevant to their field of research may encounter protein structure outside of the specialist protein structure databases. A few are mentioned below.

COSMIC 3D

COSMIC 3D maps cancer mutation data from COSMIC (Forbes *et al.*, 2017), a database of somatic mutations in cancer, to three dimensional protein structures, allowing users to visualize the location of these mutations in the three dimensional structure.

EBI search

Whilst all of the data resources at the European Bioinformatics Institute have their own search systems, the EBI-wide search engine provides a facility to search all of these resources, highlighting the interconnectivity of biological data resources. In a similar way, searches at NCBI (NCBI Resource Coordinators, 2016) integrate MMDB, which is a secondary atlas for the PDB (Park et al., 2017).

Open Targets

Open Targets (Koscielny *et al.*, 2017) is a platform which brings together data from a variety of different biomedical data resources, including structure data, with the aim of aiding its user community in identification of potential drug targets.

Genome3D

Genome3D (Lewis *et al.*, 2015) integrates a variety of resources of fold classification and structure prediction to extend structural annotation across the genomes of several model organisms.

Challenges for the Future

Technological advances in crystallography, not least robotics enabling high throughput, microfocus beamlines, free electron lasers and the amazing progress in electron microscopy are resulting in a huge leap in the volume of protein structure data. As mentioned above, integrative and hybrid techniques combine very diverse types of data.

Advances in speed and reduction in cost of genome sequence have resulted in sequences of many related proteins being easily available. New modelling techniques based on the sequence comparisons can predict even novel protein folds (Ovchinnikov *et al.*, 2017).

With such an avalanche of data, protein structure databases have an exciting challenge going forward in archiving these data. They need to be made available in a meaningful way such that specialists and non-specialists alike can locate the data they need and draw relevant conclusions from it.

Acknowledgements

We thank EMBL and the Wellcome Trust for funding and Drs Aleksandras Gutmanas and Sameer Velankar for critical feedback on the manuscript.

See also: *Ab initio* Protein Structure Prediction. Algorithms for Structure Comparison and Analysis: Docking. Algorithms for Structure Comparison and Analysis: Homology Modelling of Proteins. Algorithms for Structure Comparison and Analysis: Prediction of Tertiary Structures of Proteins. Bioinformatics Data Models, Representation and Storage. Biological Database Searching. Biomolecular Structures: Prediction, Identification and Analyses. Computational Tools for Structural Analysis of Proteins. Data Storage and Representation. Drug Repurposing and Multi-Target Therapies. Identifying Functional Relationships Via the Annotation and Comparison of Three-Dimensional Amino Acid Arrangements in Protein Structures. In Silico Identification of Novel Inhibitors. Information Retrieval in Life Sciences. Information Retrieval in Life Sciences. Natural Language Processing Approaches in Bioinformatics. Pharmacophore Development. Protein Structural Bioinformatics: An Overview. Protein Structure Classification. Protein Three-Dimensional Structure Prediction. Rational Structure-Based Drug Design. Secondary Structure Prediction. Small Molecule Drug Design. Structure-Based Design of Peptide Inhibitors for Protein Arginine Deiminase Type IV (PAD4). Study of The Variability of The Native Protein Structure. The Evolution of Protein Family Databases

References

Adams, P.D., Aertgeerts, K., Bauer, C., *et al.*, 2016. Outcome of the first wwPDB/CCDC/D3R ligand validation workshop. Structure 24 (4), 502–508.

Adolf-Bryfogle, J., Xu, Q., North, B., Lehmann, A., Dunbrack, R.L. Jr., 2015. PyIgClassify: a database of antibody CDR structural classifications. Nucleic Acids Research 43 (Database issue), D432–D438.

Andreeva, A., Howorth, D., Chothia, C., Kulesha, E., Murzin, A.G., 2014. SCOP2 prototype: a new approach to protein structure mining. Nucleic Acids Research 42 (Database issue), D310–D314.

Ashburner, M., Ball, C.A., Blake, J.A., *et al.*, 2000. Gene ontology: tool for the unification of biology. The Gene Ontology Consortium. Nature genetics 25 (1), 25–29.

Berman, H., Henrick, K., Nakamura, H., 2003. Announcing the worldwide Protein Data Bank. Nature Structural Biology 10 (12), 980.

Berman, H., Henrick, K., Nakamura, H., Markley, J.L., 2007. The worldwide Protein Data Bank (wwPDB): ensuring a single, uniform archive of PDB data. Nucleic acids research 35 (Database issue), D301–D303.

Berman, H.M., Battistuz, T., Bhat, T.N., *et al.*, 2002. The Protein Data Bank. Acta Crystallographica D: Biological Crystallography 58 (Pt 61), 899–907.

Berman, H.M., Kleywegt, G.J., Nakamura, H., Markley, J.L., 2012. The Protein Data Bank at 40: reflecting on the past to prepare for the future. Structure 20 (3), 391–396.

Berman, H.M., Kleywegt, G.J., Nakamura, H., Markley, J.L., 2014. The Protein Data Bank archive as an open data resource. Journal of Computer-Aided Molecular Design 28 (10), 1009–1014.

Berman, H.M., Burley, S.K., Kleywegt, G.J., *et al.*, 2016. The archiving and dissemination of biological structure data. Current Opinion in Structural Biology 40, 17–22.

Bernstein, F.C., Koetzle, T.F., Williams, G.J., *et al.*, 1977. The Protein Data Bank: a computer-based archival file for macromolecular structures. Journal of Molecular Biology 112 (3), 535–542.

Bienert, S., Waterhouse, A., de Beer, T.A.P., *et al.*, 2017. The SWISS-MODEL Repository-new features and functionality. Nucleic Acids Research 45 (D1), D313–D319.

Bonet, J., Planas-Iglesias, J., Garcia-Garcia, J., *et al.*, 2014. ArchDB 2014: structural classification of loops in proteins. Nucleic Acids Research 42 (Database issue), D315–D319.

Bourne, P.E., Berman, H.M., McMahon, B., *et al.*, 1997. Macromolecular crystallographic information file. Methods in Enzymology 277, 571–590.

Bousfield, D., McEntyre, J., Velankar, S., *et al.*, 2016. Patterns of database citation in articles and patents indicate long-term scientific and industry value of biological data resources. F1000Research 5.

Carrillo-Tripp, M., Shepherd, C.M., Borelli, I.A., *et al.*, 2009. VIPERdb2: an enhanced and web API enabled relational database for structural virology. Nucleic Acids Research 37 (Database issue), D436–D442.

Chandonia, J.-M., Fox, N.K., Brenner, S.E., 2017. SCOPe: manual curation and artifact removal in the structural classification of proteins - extended Database. Journal of Molecular Biology 429 (3), 348–355.

Cheng, H., Schaeffer, R.D., Liao, Y., *et al.*, 2014. ECOD: an evolutionary classification of protein domains. PLoS Computational Biology 10 (12), e1003926.

Chothia, C., Lesk, A.M., 1986. The relation between the divergence of sequence and structure in proteins. The EMBO Journal 5 (4), 823–826.

Dunbar, J., Krawczyk, K., Leem, J., *et al.*, 2014. SAbDab: the structural antibody database. Nucleic Acids Research 42 (Database issue), D1140–D1146.

Dutta, S., Burkhardt, K., Swaminathan, G.J., *et al.*, 2008. Data deposition and annotation at the worldwide protein data bank. Methods in Molecular Biology 426, 81–101.

Finn, R.D., Coggill, P., Eberhardt, R.Y., *et al.*, 2016. The Pfam protein families database: towards a more sustainable future. Nucleic Acids Research 44 (D1), D279–D285.

Forbes, S.A., Beare, D., Boutselakis, H., *et al.*, 2017. COSMIC: somatic cancer genetics at high-resolution. Nucleic Acids Research 45 (D1), D777–D783.

Gerstein, M., 2000. Integrative database analysis in structural genomics. Nature Structural Biology 7, 960–963.

Greer, J., Erickson, J.W., Baldwin, J.J., Varney, M.D., 1994. Application of the three-dimensional structures of protein target molecules in structure-based drug design. Journal of Medicinal Chemistry 37 (8), 1035–1054.

Haas, J., Roth, S., Arnold, K., *et al.*, 2013. The Protein Model Portal–a comprehensive resource for protein structure and model information. Database: The Journal of Biological Databases and Curation 2013, bat031.

Henderson, R., Sali, A., Baker, M.L., *et al.*, 2012. Outcome of the first electron microscopy validation task force meeting. Structure 20 (2), 205–214.

Hodis, E., Prilusky, J., Sussman, J.L., 2010. Proteopedia: a collaborative, virtual 3D web-resource for protein and biomolecule structure and function. Biochemistry and Molecular Biology Education: A Bimonthly Publication of the International Union of Biochemistry and Molecular Biology 38 (5), 341–342.

Hura, G.L., Menon, A.L., Hammel, M., *et al.*, 2009. Robust, high-throughput solution structural analyses by small angle X-ray scattering (SAXS). Nature Methods 6 (8), 606–612.

Isberg, V., Mordalski, S., Munk, C., *et al.*, 2017. GPCRdb: an information system for G protein-coupled receptors. Nucleic Acids Research 45 (5), 2936.

Jones, P., Binns, D., Chang, H.-Y., *et al.*, 2014. InterProScan 5: genome-scale protein function classification. Bioinformatics 30 (9), 1236–1240.

Keller, P.A., Henrick, K., McNeil, P., Moodie, S., Barton, G.J., 1998. Deposition of macromolecular structures. Acta Crystallographica. Section D, Biological Crystallography 54 (Pt 6 Pt 1), 1105–1108.

Kelley, L.A., Mezulis, S., Yates, C.M., Wass, M.N., Sternberg, M.J.E., 2015. The Phyre2 web portal for protein modeling, prediction and analysis. Nature Protocols 10 (6), 845–858.

Kendrew, J.C., Bodo, G., Dintzis, H.M., et al., A three-dimensional model of the myoglobin molecule obtained by x-ray analysis. Nature 181 (4610), 662–666.

Kinjo, A.R., Bekker, G.-J., Suzuki, H., et al., 2017. Protein Data Bank Japan (PDBj): updated user interfaces, resource description framework, analysis tools for large structures. Nucleic Acids Research 45 (D1), D282–D288.

Kinjo, A.R., Nakamura, H., 2007. Similarity search for local protein structures at atomic resolution by exploiting a database management system. Biophysics 3, 75–84.

Korneta, I., Magnus, M., Bujnicki, J.M., 2012. Structural bioinformatics of the human spliceosomal proteome. Nucleic Acids Research 40 (15), 7046–7065.

Koscielny, G., An, P., Carvalho-Silva, D., 2017. Open Targets: a platform for therapeutic target identification and validation. Nucleic Acids Research 45 (D1), D985–D994.

Kozma, D., Simon, I., Tusnády, G.E., 2013. PDBTM: protein Data Bank of transmembrane proteins after 8 years. Nucleic Acids Research 41 (Database issue), D524–D529.

Krissinel, E., Henrick, K., 2004. Secondary-structure matching (SSM), a new tool for fast protein structure alignment in three dimensions. Acta Crystallographica D: Biological Crystallography 60 (Pt 12 Pt 1), 2256–2268.

Krissinel, E., Henrick, K., 2007. Inference of macromolecular assemblies from crystalline state. Journal of Molecular Biology 372 (3), 774–797.

Kryshtafovych, A., Monastyrskyy, B., Fidelis, K., 2014. CASP prediction center infrastructure and evaluation measures in CASP10 and CASP ROLL. Proteins 82 (Suppl 2), 7–13.

Kubinyi, H., 1999. Chance favors the prepared mind – from serendipity to rational drug design. Journal of Receptor and Signal Transduction Research 19 (1–4), 15–39.

Kühlbrandt, W., 2014. Biochemistry. The resolution revolution. Science 343 (6178), 1443–1444.

Lagerstedt, I., Moore, W.J., Patwardhan, A., et al., 2013. Web-based visualisation and analysis of 3D electron-microscopy data from EMDB and PDB. Journal of Structural Biology 184 (2), 173–181.

Lam, S.D., Dawson, N.L., Das, S., et al., 2016. Gene3D: expanding the utility of domain assignments. Nucleic Acids Research 44 (D1), D404–D409.

Laskowski, R.A., Hutchinson, E.G., Michie, A.D., et al., 1997. PDBsum: a Web-based database of summaries and analyses of all PDB structures. Trends in Biochemical Sciences 22 (12), 488–490.

Lawson, C.L., Baker, M.L., Best, C., et al., 2011. EMDataBank.org: unified data resource for CryoEM. Nucleic Acids Research 39 (Database issue), D456–D464.

Lensink, M.F., Velankar, S., Kryshtafovych, A., et al., 2016. Prediction of homoprotein and heteroprotein complexes by protein docking and template-based modeling: a CASP-CAPRI experiment. Proteins 84 (Suppl 1), 323–348.

Lesk, A.M., Chothia, C., 1980. How different amino acid sequences determine similar protein structures: the structure and evolutionary dynamics of the globins. Journal of Molecular Biology 136 (3), 225–270.

Lewis, T.E., Sillitoe, I., Andreeva, A., et al., 2015. Genome3D: exploiting structure to help users understand their sequences. Nucleic Acids Research 43 (Database issue), D382–D386.

Madej, T., Lanczycki, C.J., Zhang, D., et al., 2014. MMDB and VAST+: tracking structural similarities between macromolecular complexes. Nucleic Acids Research 42 (Database issue), D297–D303.

Markley, J.L., Ulrich, E.L., Berman, H.M., et al., 2008. BioMagResBank (BMRB) as a partner in the Worldwide Protein Data Bank (wwPDB): new policies affecting biomolecular NMR depositions. Journal of Biomolecular NMR 40 (3), 153–155.

Meyer, E.F., 1997. The first years of the Protein Data Bank. Protein Science: A Publication of the Protein Society 6 (7), 1591–1597.

Montelione, G.T., Nilges, M., Bax, A., et al., 2013. Recommendations of the wwPDB NMR validation task force. Structure 21 (9), 1563–1570.

Murzin, A.G., Brenner, S.E., Hubbard, T., Chothia, C., 1995. SCOP: a structural classification of proteins database for the investigation of sequences and structures. Journal of Molecular Biology 247 (4), 536–540.

NCBI Resource Coordinators, 2016. Database resources of the national center for biotechnology information. Nucleic Acids Research 44 (D1), D7–D19.

Ovchinnikov, S., Park, H., Varghese, N., et al., 2017. Protein structure determination using metagenome sequence data. Science 355 (6322), 294–298.

Park, Y.M., Squizzato, S., Buso, N., Gur, T., Lopez, R., et al., 2017. The EBI search engine: EBI search as a service-making biological data accessible for all. Nucleic Acids Research. Available at: http://dx.doi.org/10.1093/nar/gkx359.

Patwardhan, A., 2017. Trends in the Electron Microscopy Data Bank (EMDB). Acta Crystallographica D: Structural Biology 73 (Pt 6), 503–508.

Perutz, M.F., Rossmann, M.G., Cullis, A.F., et al., 1960. Structure of haemoglobin: a three-dimensional Fourier synthesis at 5.5-A. resolution, obtained by X-ray analysis. Nature 185 (4711), 416–422.

Pettersen, E.F., Goddard, T.D., Huang, C.C., et al., 2004. UCSF Chimera –a visualization system for exploratory research and analysis. Journal of Computational Chemistry 25 (13), 1605–1612.

Protein Data Bank, 1971. Protein Data Bank. Nature New Biology 233, p223.

Raman, P., Cherezov, V., Caffrey, M., 2006. The Membrane Protein Data Bank. Cellular and Molecular Life Sciences: CMLS 63 (1), 36–51.

Rao, S.T., Rossmann, M.G., 1973. Comparison of super-secondary structures in proteins. Journal of Molecular Biology 76 (2), 241–256.

Read, R.J., Adams, P.D., Arendall, W.B., 3rd, et al., 2011. A new generation of crystallographic validation tools for the protein data bank. Structure 19 (10), 1395–1412.

Reichert, J., Sühnel, J., 2002. The IMB Jena Image Library of Biological Macromolecules: 2002 update. Nucleic Acids Research 30 (1), 253–254.

Rose, A.S., Hildebrand, P.W., 2015. NGL Viewer: a web application for molecular visualization. Nucleic Acids Research 43 (W1), W576–W579.

Rose, P.W., Prlić, A., Altunkaya, A., et al., 2017. The RCSB protein data bank: integrative view of protein, gene and 3D structural information. Nucleic Acids Research 45 (D1), D271–D281.

Salavert-Torres, J., Iudin, A., Lagerstedt, I., et al., 2016. Web-based volume slicer for 3D electron-microscopy data from EMDB. Journal of Structural Biology 194 (2), 164–170.

Sali, A., Berman, H.M., Schwede, T., et al., 2015. Outcome of the First wwPDB Hybrid/Integrative Methods Task Force Workshop. Structure 23 (7), 1156–1167.

Schaeffer, R.D., Liao, Y., Cheng, H., Grishin, N.V., et al., 2017. ECOD: new developments in the evolutionary classification of domains. Nucleic Acids Research 45 (D1), D296–D302.

Sen, S., Young, J., Berrisford, J.M., et al., 2014. Small molecule annotation for the Protein Data Bank. Database: the Journal of Biological Databases and Curation 2014, bau116.

Sillitoe, I., Lewis, T.E., Cuff, A., et al., 2015. CATH: comprehensive structural and functional annotations for genome sequences. Nucleic Acids Research 43 (Database issue), D376–D381.

Sussman, J.L., Abola, E.E., Lin, D., et al., 1999. The protein data bank. Bridging the gap between the sequence and 3D structure world. Genetica 106 (1–2), 149–158.

Suzuki, H., Kawabata, T., Nakamura, H., 2016. Omokage search: shape similarity search service for biomolecular structures in both the PDB and EMDB. Bioinformatics 32 (4), 619–620.

Tagari, M., Newman, R., Chagoyen, M., Carazo, J.M., Henrick, K., et al., 2002. New electron microscopy database and deposition system. Trends in Biochemical Sciences 27 (11), 589.

The UniProt Consortium, 2017. UniProt: the universal protein knowledgebase. Nucleic Acids Research 45 (D1), D158–D169.

Ulrich, E.L., Akutsu, H., Doreleijers, J.F., et al., 2008. BioMagResBank. Nucleic Acids Research 36 (Database issue), D402–D408.

Valentini, E., Kikhney, A.G., Previtali, G., Jeffries, C.M., Svergun, D.I., et al., 2015. SASBDB, a repository for biological small-angle scattering data. Nucleic acids research 43 (Database issue), D357–D363.

Velankar, S., Dana, J.M., Jacobsen, J., et al., 2013. SIFTS: structure integration with function, taxonomy and sequences resource. Nucleic Acids Research 41 (Database issue), D483–D489.

Velankar, S., van Ginkel, G., Alhroub, Y., et al., 2016. PDBe: improved accessibility of macromolecular structure data from PDB and EMDB. Nucleic Acids Research 44 (D1), D385–D395.

Vizcaíno, J.A., Csordas, A., del-Toro, N., et al., 2016. update of the PRIDE database and its related tools. Nucleic Acids Research 44 (D1), D447–D456.

Westbrook, J., Ito, N., Nakamura, H., Henrick, K., Berman, H.M., 2005. PDBML: the representation of archival macromolecular structure data in XML. Bioinformatics 21 (7), 988–992.

Westbrook, J.D., Bourne, P.E., 2000. STAR/mmCIF: an ontology for macromolecular structure. Bioinformatics 16 (2), 159–168.

Wilkinson, M.D., Dumontier, M., Aalbersberg, I.J.J., *et al.*, 2016. The FAIR guiding principles for scientific data management and stewardship. Scientific Data 3, 160018.

Yang, J., Yan, R., Roy, A., *et al.*, 2015. The I-TASSER Suite: protein structure and function prediction. Nature Methods 12 (1), 7–8.

Young, J.Y., Westbrook, J.D., Feng, Z., *et al.*, 2017. OneDep: unified wwPDB system for deposition, biocuration, and validation of macromolecular structures in the PDB archive. Structure 25 (3), 536–545.

Further Reading

Branden, C., Tooze, J., 1991. Introduction to Protein Structure. New York: Garland publishing.

Burley, S.K., Berman, H.M., Kleywegt, G.J., *et al.*, 2017. Protein Data Bank (PDB): the single global macromolecular structure archive. Methods in Molecular Biology 1607, 627–641.

Lamb, A.L., Kappock, T.J., Silvaggi, N.R., 2015. You are lost without a map: navigating the sea of protein structures. Biochimica et Biophysica Acta 1854 (4), 258–268.

Mackay, J.P., Landsberg, M.J., Whitten, A.E., Bond, C.S. Whaddaya know: a guide to uncertainty and subjectivity in structural biology. Trends in Biochemical Sciences 42 (2), 155–167.

Mackenzie, C.O., Grigoryan, G., 2017. Protein structural motifs in prediction and design. Current Opinion in Structural Biology 44, 161–167.

Patwardhan, A., Lawson, C.L., 2016. Databases and archiving for CryoEM. Methods in Enzymology 579, 393–412.

Paxman, J.J., Heras, B., 2017. Bioinformatics tools and resources for analyzing protein structures. Methods in molecular biology 1549, 209–220.

Sillitoe, I., Dawson, N., Thornton, J., Orengo, C., 2015. The history of the CATH structural classification of protein domains. Biochimie 119, 209–217.

Westbrook, J.D., Fitzgerald, P.M.D., 2003. The PDB format, mmCIF, and other data formats.

Relevant Websites

http://www.oxfordjournals.org/nar/database/subcat/4/14
  Oxford University Press.
ftp.wwpdb.org
  wwPDP.

Protein Structure Classification

Natalie L Dawson, Sayoni Das, Jonathan G Lees, and Christine Orengo, Institute of Structural and Molecular Biology, University College London, London, United Kingdom

Introduction to Protein Structure Classifications

The first three-dimensional protein structure, myoglobin, was solved by X-ray crystallography in the late 1950s. The number of solved protein structures grew at a steady rate and a computational archive was built in 1971 to collect, standardise and distribute this data. This archive, known as the Protein Data Bank (PDB) has been growing at an exponential rate since the mid 1990s, increasing from 3000 structures to 135,000 entries in November 2017. The dramatic increase in the data made many manual analyses impractical. Organising the data into structural and evolutionary classifications, gave biologists insights into common structural units and evolutionary relationships.

Several publicly available, structural classifications have been developed providing information on fold groups, evolutionary relationships and sequence families (see **Table 1**). These include: SCOP (Hubbard *et al.*, 1997), SCOP2 (Andreeva *et al.*, 2014), CATH (Orengo *et al.*, 1997; Dawson *et al.*, 2017), ECOD (Cheng *et al.*, 2014), and HOMSTRAD (Mizuguchi *et al.*, 1998). These classifications are built on a combination of sequence and structure comparison methods (e.g., SSAP (Taylor and Orengo, 1989), Dali (Holm and Sander, 1995), SSM (Krissinel and Henrick, 2004), and CE (Shindyalov and Bourne, 2001)).

There are also a number of structural comparison web servers that provide lists of structural neighbours, including: the protein structure comparison service, PDBeFold at the European Bioinformatics Institute (Krissinel and Henrick, 2004); VAST at the NCBI (Madej *et al.*, 1995; Gibrat *et al.*, 1996); Combinatorial Extension (CE) (Shindyalov and Bourne, 2001) at the RCSB Protein Data Bank; Dali (Holm and Sander, 1995) at the University of Helsinki.

Table 1 Protein structure classifications and nearest-neighbour resources

Acronym/Name	Description	URL
RCSB PDB Protein Comparison Tool	The Combinatorial Extension method is integrated into the RCSB PDB to compare pairs of protein structures. It compares a structure with the PDB and identifies its nearest neighbours. The RCSB PDB protein comparison tool provides access to CE to perform pairwise structural alignments.	http://www.rcsb.org/pdb/workbench/workbench.do
CATH	CATH assigns protein domain structures into a structural classification hierarchy: Class, Architecture, Topology (fold), Homologous superfamily. A mixture of automated and manual curated steps are used in its continual development.	http://cathdb.info/
Dali server	The Dali server allows users to perform structural comparisons for: one structure against the whole PDB; one structure against user-defined list of PDBs; a list of structures to get an all-against-all comparison.	http://ekhidna2.biocenter.helsinki.fi/dali/
ECOD	Evolutionary Classification of Domains. The classification consists of proteins with experimentally-determined 3D structures and focuses on establishing remote evolutionary relationships.	http://prodata.swmed.edu/ecod/
HOMSTRAD	HOMologous STRucture Alignment Database. Comprises protein structural alignments for homologous families.	http://mizuguchilab.org/homstrad/
PDBeFOLD	Uses the Secondary Structure Matching (SSM) algorithm to compare a given structure with the whole PDB archive, and identifies the nearest structural neighbours.	http://www.ebi.ac.uk/msd-srv/ssm/
SCOP	Structural Classification of Proteins. Classifies protein domains into a hierarchy consisting of: class, fold, superfamily and family.	http://scop.mrc-lmb.cam.ac.uk/scop/
SCOP2	Structural Classification of Proteins 2. A prototype successor of SCOP, which uses a directed acyclic graph to describe complex structural and evolutionary relationships. Each graph node represents a particular type of relationship.	http://scop2.mrc-lmb.cam.ac.uk/
VAST	Vector Alignment Search Tool. Identifies 3D domains for a given query and then finds a list of nearest structural neighbours within the Molecular Modelling Database (MMDB).	https://www.ncbi.nlm.nih.gov/Structure/VAST/vast.shtml
VAST +	Vector Alignment Search Tool Plus. Uses all the domains identified by VAST for a query structure, then finds macromolecular structures that have a structurally similar biological unit. This differs from VAST, which focuses on individual protein molecules (e.g., domains).	https://www.ncbi.nlm.nih.gov/Structure/vastplus/vastplus.cgi

Modified from Pearl, F.M.G., Sillitoe, I., Orengo, C.A., 2015. Protein Structure Classification. eLS, pp. 1–10.

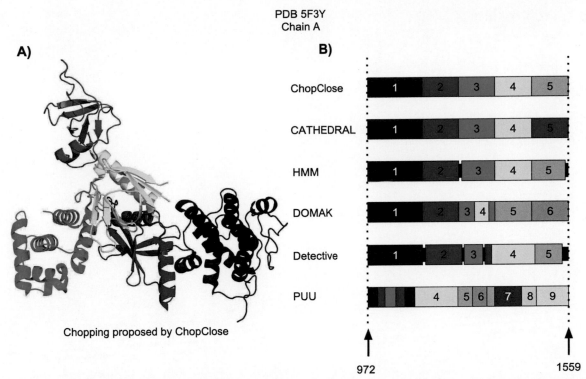

Fig. 1 Figure showing how automatic detection of domains can be non-trivial for multi-domain proteins. (A) The 3D structure of the PDB 5F3Y, chain A. The domain chopping proposed by the method ChopClose is shown, where each colour represents a different domain. (B) The proposed domain choppings by the different algorithms used to identify domains for classification in CATH. The different colours represent different domains proposed by each method, as in (A). Se descriptions of the methods below. The chain start (972) and stop (1559) PDB residue labels are shown at the bottom of the figure. The domains have been numbered where space permits.

Protein Domains and Methods Used to Identify Them

Early analysis of protein structures revealed that proteins can have one or more distinct structural units, or domains. A structural domain is created by the protein polypeptide chain folding into a compact unit with its own hydrophobic core. Domains consist of secondary structures (e.g., alpha helices and beta strands) and supersecondary structures (i.e., any structural motif, e.g., helix-turn-helix, beta hairpin, alpha-beta motif), and have an average size of 150 ± 50 residues (see Pearl *et al.*, 2015 for a review). They are typically formed of contiguous parts of the polypeptide chain but may be formed of discontiguous sections. Domains can be defined as semi-independently evolving units, and sometimes have their own independent functions (Hadley and Jones, 1999).

The multi-domain architecture (MDA) of a protein describes the sequential order of domains within the protein polypeptide chain. Studies have found that over two-thirds of known prokaryotic proteins, and almost 80% of eukaryotic proteins, contain at least two domains (Teichmann *et al.*, 1998; Gerstein, 1998). Domains within multi-domain proteins can be difficult to identify if they have extensive contacts between them, making it hard to recognise boundaries between the domains. **Fig. 1** illustrates the difficulties of identifying domain boundaries in a multi-domain protein when the methods aiding manual curation disagree with each other (see subsequent sections for details on these algorithms). The curator uses these results, and the literature, to decide on the correct domain boundaries.

Although the protein domain structure data is still increasing at an exponential rate, the number of new folds discovered has plateaued in recent years, indicating that existing fold libraries cover the majority of structural space. The comprehensive fold libraries now available can be searched with newly solved structures, using structure based algorithms, to find related domain folds and inherit the domain boundaries (see Section Sequence-based methods for recognising domains in proteins). Protein sequence data, i.e., for domains of known and predicted domain structure, is also very comprehensive and can be searched with powerful HMM-based algorithms to find domain matches and inherit boundaries (see Section Structure-based methods for recognising domains in proteins). CATH and ECOD use their own automated protocols for domain assignment, based on structure- and sequence-based algorithms. In addition, CATH uses a selection of ab initio algorithms, which are used to detect domains with novel, uncharacterised folds (see Section *Ab-initio* methods for recognising domains in proteins). These different types of algorithms are described below.

Algorithms to identify protein structural domains
Structure-based methods for recognising domains in proteins

To identify protein structural domains for the CATH resource, the ChopClose automated protocol (Greene *et al.*, 2007) is first used to scan newly determined protein structures from the PDB against the structures of multidomain proteins in CATH, that have

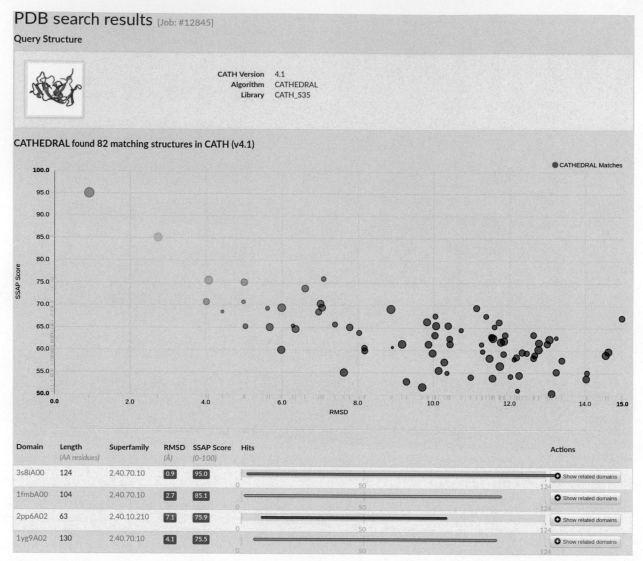

Fig. 2 Results from running a structure-based search with the CATHEDRAL web server using an example case PDB 4RGH, downloaded from the PDBe. A total of 82 matching structures were found in CATH with the most confident domain matches shown in green, through to the least confident matches in red.

already had their domain boundaries identified. For query structures that have high structure similarity to multi-domain structures already classified in CATH (a score ≥80 by the SSAP algorithm (range 0–100) (Taylor and Orengo, 1989), RMSD <=6A) the domain boundaries from the match are inherited to the query.

If there is no significant match, the chain is submitted to a number of additional algorithms (structure-based: CATHEDRAL, sequence-based: CATH-HMMscan) to propose domain boundaries (see Sections "Sequence-based methods for recognising domains in proteins", "Ab-initio methods for recognising domains in proteins", "Hierarchical Structural Classification", "Class", "Architecture", "Topology or fold similarity", "Homologous superfamily", "Homologous family", "Nearest-Neighbour Clustering", "Integration of the SCOP And CATH Resources", "Current Status of the Protein Structural Universe", "Structural Families in the Genomic Era"). The CATHEDRAL algorithm (Redfern et al., 2007) is used to perform a structure-based search of a query chain against a library of previously classified protein domains in CATH. First, an initial fast secondary structure comparison is run between structures using graph theory to find putative fold matches. These putative-fold-matching domains are realigned to the query with the slower, more accurate SSAP to produce a structural alignment, from which the domain boundaries can be inferred. High scoring matches (i.e., SSAP score >=80 and RMSD <=6 Å) provide confidence to infer the domain boundaries of the match (see **Fig. 2**).

In ECOD, query protein structures from the PDB are first scanned against protein chains and domains in ECOD fast sequence based approaches (see below). If the sequence-based methods do not return any significant match, the query chain is searched against a library of representative domain structures in ECOD using a DaliLite search (Holm and Park, 2000). The domain boundaries are inherited if there is significant structural similarity and an acceptable BLOSUM-based alignment score (Cheng et al., 2014). If this does not return any results, the chain is marked for manual curation.

Table 2   Algorithms used in the CATH Update protocol to automatically identify protein domains. The best result from each algorithm is also provided to the CATH curators for manual curation purposes

Acronym/Name	Description	Reference
ChopClose	Compares query chains against previously chopped chains in CATH to find the best match.	Greene *et al.* (2007)
CATHEDRAL	CATH's Existing Domain Recognition ALgorithm. Uses GRATH to find the best fold match and then SSAP to produce an accurate structural alignment of these fold matches. Publically available to use through: http://cathdb.info/search/by_structure	Redfern *et al.* (2007)
HMMs	Hidden Markov Models. Probabilistic models are created from multiple sequence alignments to capture information on the amino acid probabilities at each alignment position. Protein sequences are searched against these models to identify homology and to aid the inference of domain boundaries.	Karplus *et al.* (1998)
DETECTIVE	*Ab initio* algorithm that identifies protein domains by searching for the hydrophobic core and residues in contact with the core.	Swindells (1995)
DOMAK	DOmain MAKer. *Ab initio* algorithm that splits a given protein into arbitrary pieces (i.e. domains) until it has found the pieces that maximise intra-domain contacts and minimise inter-domain contacts.	Siddiqui and Barton (1995); Holm and Sander (1994)
PUU	*Ab initio* algorithm that studies the motion of residue clusters to analyse how different parts of the protein move in relation to each other. Each domain in a multi-domain protein has its own continuous, relative motion, which is used to identify domain boundaries.	Holm and Sander (1994)

Modified from Pearl, F.M.G., Sillitoe, I., Orengo, C.A., 2015. Protein Structure Classification. eLS, pp. 1–10.

Sequence-based methods for recognising domains in proteins

In CATH, all new query structures from the PDB are also scanned against sequence profiles or hidden Markov models (HMMs) built with SAM (Sequence Alignment and Modelling) (Karplus *et al.*, 1998) from representative domain structures in CATH. The best-matching HMM to the query is that with the lowest E-value, which is significant if it is less than 0.001.

BLAST (Altschul *et al.*, 1990) and HHsearch (Söding, 2005) are used in the ECOD pipeline to search PDB chain queries against existing domain data in the ECOD database. The protein chain is first searched with BLAST against a library of full-length chains from ECOD that have had domains identified. The domains in the most significantly-matching chain (i.e., a match with a sequence similarity E-value $<2 \times 10^{-03}$) are used to infer domain boundary positions onto the query chain. Next, the protein chain is search with BLAST against a library of ECOD domain sequences. Domains are assigned individually to the best match (i.e., sequence similarity E-value $<2 \times 10^{-03}$ and hit coverage $>80\%$). Finally, to detect more remotely related domains, HHblits (Söding, 2005) is used to generate a sequence profile of the query, which is then searched against the ECOD representative domain profiles using HHsearch (Cheng *et al.*, 2014).

Ab-initio methods for recognising domains in proteins

When there are no existing fold and domain matches to assist in defining domain boundaries, it is often useful to apply ab initio methods. Three ab-initio methods are used in CATH to identify protein domains: DETECTIVE (Swindells, 1995), DOMAK (Siddiqui and Barton, 1995), and PUU (Holm and Sander, 1994). Each algorithm uses a different method to identify domain boundaries, but they all use information on contacts between residues in the structure. They are all able to predict boundaries for both contiguous and discontiguous domains. DETECTIVE searches for the hydrophobic core of each domain. DOMAK randomly divides up a protein to maximise intra-domain contacts and minimise inter-domain contacts. PUU analyses the motion of predicted domains in relation to each other.

The majority of the chains that come into CATH are automatically chopped into domains using the ChopClose, CATHEDRAL and HMM protocols. However, due to the strict criteria imposed for quality control purposes, around 5% of chains need manual curation which is guided by the ab-initio methods.

Table 2 summarises the structure, sequence and ab-initio methods used to assign domain boundaries for domains classified in CATH.

Hierarchical Structural Classification

The three major domain structure classifications: SCOP, CATH and ECOD, use similar hierarchical classifications, with a few differences between them (**Fig. 3**). In the CATH classification the main levels from top to bottom are: class, architecture, topology or fold, homologous superfamily, and family. In the SCOP classification the different levels are: class, topology or fold, super-family, and family. In the ECOD classification, which specialises in detecting remote homologies, there are five main levels: architecture, possible homology, homology, topology or fold, and family. These different levels will be discussed below.

Class

The protein class (C-level) considers the proportion of residues within a structure that fall within α-helix and/or β-strand secondary structure elements. There are three major classes: mainly α, mainly β, and a mixture of α and β. SCOP separates the latter into $\alpha + \beta$

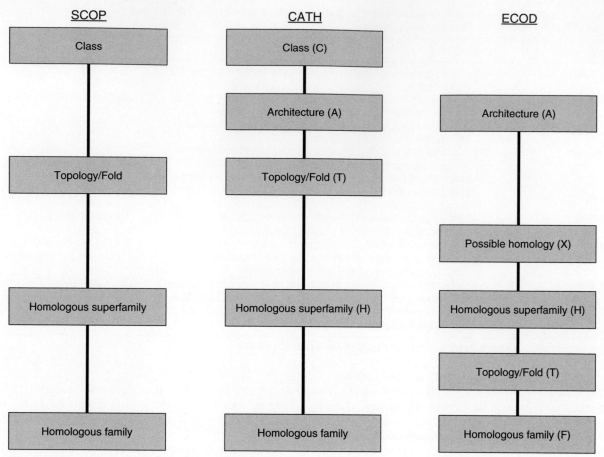

Fig. 3 Comparison of the structural classification hierarchies used by SCOP, CATH and ECOD. The single letter used to represent a particular level has been included. While there are many similarities between the resources, there are clear differences such as the addition of the possible homology (X) level in ECOD, the lack of an architecture level in SCOP, and no class level in ECOD.

and α/β, giving this resource four main classes. CATH considers these as one class after finding considerable overlap between them since their original description in Levitt and Chothia (1976). CATH has a fourth class that represents domains with few secondary structures. Class is not featured in ECOD.

Architecture

Protein architecture (the A-level) describes the 3D spatial arrangement of secondary structure elements, independent of any connectivity information. Currently in CATH there are 41 different architectures: 5 mainly α, 21 mainly β, 14 α and β, and 1 in few secondary structures. Each architecture can represent 3D arrangements of secondary structures with diverse connectivities, which is represented by the topology level. There are more β-based architectures as constraints on the packing of β-sheets lead to very distinct and easily recognisable architectures such as β-barrels, β-prisms, and β-propellers.

Topology or fold similarity

The topology of a protein domain is given by its 3D spatial arrangement together with connectivity information, i.e., two domains having the same fold should have the same sequence of secondary structure elements packed together in similar orientations in 3D. In CATH, protein domains that have significant structural similarity to each other (i.e., a SSAP score of $> = 70$), but no sequence or functional similarity, are grouped into the same T-level. ECOD sub-classifies its A-levels into possible homology (X-levels), to represent groups of domains that have incomplete evidence for being homologous (Cheng *et al.*, 2014). The T-level in ECOD comes after the homologous superfamily, H-level, rather than the A-level, so homologous domains are subdivided into topology-based groups. This is because in some superfamilies, relatives can diverge structurally to such an extent that they could be considered to have different folds. In CATH this structural variation is captured by grouping homologues into different structurally-similar clusters (SC5s), where relatives superpose within 5 Å RMSD or less.

Homologous superfamily

Domains are grouped at the homologous superfamily level (H-level) where there is clear evidence of homology, i.e., the domains have evolved from a common ancestor. There can be considerable divergence across some superfamilies in terms of structure, sequence and function. **Fig. 4** shows that for representative relatives (i.e., representatives of relatives clustered at 95% sequence identity) within the same superfamily there can be a wide range of sequence identity scores.

For these very remote homologues, structure and function information can be used to detect an evolutionary relationship. Protein structure is generally much more conserved than sequence, making it easier to use in homology detection. As with sequence, there can also be considerable variation of structure across some superfamilies (**Fig. 5**).

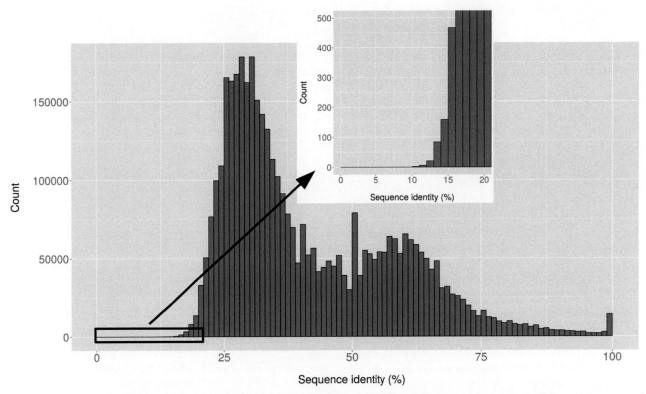

Fig. 4 The distribution of sequence identity scores between pairs of representative homologous domains (non-redundant at 95% sequence identity) in CATH. All-against-all comparisons were performed using BLASTp with the default maximum E-value (i.e., 10) and default maximum number of target sequences (i.e., 500). If we relax both of these thresholds then we can see results from comparing more relatives which have even lower sequence identities, however there are too many false positives (data not shown).

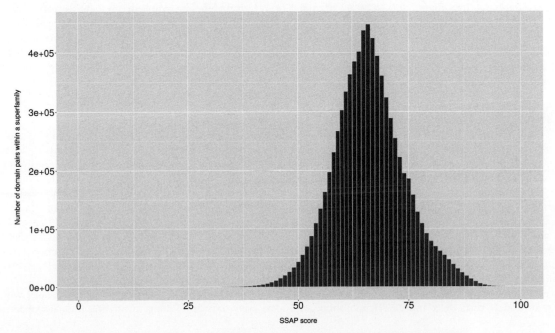

Fig. 5 The distribution of structural similarity scores (using SSAP) between pairs of homologous domains in CATH.

Homologous family

Protein families consist of close relatives which generally have at least 30% pairwise sequence identity to each other, or evidence of close functional similarity, or taxonomic similarity. SCOP places domains in the same family if they have at least 30% sequence identity, or if they have some level of sequence identity and a similar function (Hubbard *et al.*, 1997). These families generally consist of taxonomically close relatives. CATH recognises functionally-related families of homologues through similarities in specificity-determining residues (see FunFHMMer in Section "Capturing Functional Diversity by Sub-Classification of Super-families"). ECOD domains are scanned against Pfam family HMMs to define family memberships and those domains that cannot be assigned to a Pfam family are grouped together by HHsearch (Cheng *et al.*, 2014).

Nearest-Neighbour Clustering

Nearest-neighbour clustering is an alternative to hierarchical classification. A structural comparison program such as Dali or VAST+ is used to compare one or more structures against each other or the whole PDB archive, for example. These resources provide lists of the closest structural neighbours, where the significance of the matches is determined by taking the distribution of scores and calculating a Z-score for each match. In Dali (Holm and Sander, 1995; Holm and Laakso, 2016), the structures from the Z-score-ranked list can then be clustered to visualise the structural relationships as a dendrogram. These resources simply provide information on structural similarity and do not take into account any other information (e.g., similarity in functional features or rare sequence motifs). Hence they do not distinguish homologous and analogous relationships.

The resources providing nearest-neighbour clustering and hierarchical classification are listed in **Table 1**.

Integration of the SCOP and CATH Resources

While SCOP and CATH are both hierarchical classifications, there are differences in their methodologies. For example, while both resources use both manual and automated methods to identify domains and assign them to superfamilies, SCOP relies more on manual curation. There are also differences in the philosophies behind what defines a structural domain: SCOP recognises domains when they are found to occur across different multi-domain contexts, whereas CATH also uses physical information such as the domain size, globularity, and compactness (Hadley and Jones, 1999).

The Genome3D collaborative project (Lewis *et al.*, 2013, 2015) was developed to integrate the important information in these two resources and use it to annotate genomic sequence data. Well-known structure prediction resources from six structural bioinformatics groups across the UK were incorporated: DomSerf (Buchan *et al.*, 2010), FUGUE (Shi *et al.*, 2001), Gene3D (Yeats *et al.*, 2011), pDomTHREADER (Lobley *et al.*, 2009), Phyre2 (Kelley and Sternberg, 2009), SUPERFAMILY (Gough and Chothia, 2002), and VIVACE. UniProt sequence data from ten genomes were annotated by these resources, which used SCOP- and CATH-based structural data to predict structural domains within the sequences and build structural models.

Consensus superfamilies were identified between SCOP and CATH to also facilitate mapping between the structure predictions. This was done by comparing pairs of SCOP and CATH domains, and calculating the residue overlap between them. SCOP and CATH superfamilies sharing a high proportion of significantly overlapping domains were considered equivalent. The level of equivalence was graded into Bronze, Silver and Gold Standard categories, with gold being the best. A pair of superfamilies (SCOP-CATH) meeting the Bronze Standard are "more similar to each other than to any other superfamily" (Lewis *et al.*, 2013). A Silver Standard pair meets the Bronze Standard and, in addition, at least 80% of domains from each superfamily map between them with an average residue overlap of at least 80%. A Gold Standard pair, in addition to the other two Standards, has domains that map between both superfamilies with a minimum of 80% residue overlap. Mapping SCOP v1.75 to CATH v4.0 domains: 938 Gold Standard, 213 Silver Standard and 388 Bronze Standard superfamily consensus pairs were found. The Gold Standard superfamily pairs are currently being integrated into the InterPro resource (Finn *et al.*, 2017), where they have been added in as homologous superfamily entries. This work has greatly increased the coverage of both Gene3D and SUPERFAMILY domain and superfamily entries in InterPro.

Current Status of the Protein Structural Universe

As previously mentioned, as of November 2017, there are almost 135,000 entries in the PDB. **Table 3** describes the statistics for SCOP, CATH and ECOD. The SCOP resource was last updated in 2009, which accounts for the large difference in the number of PDBs processed and domains identified between the three resources.

Of the 110,800 domains in SCOP v1.75, 15% of domains are mainly α-helical in structure, 24% are mainly β-structure, and 49% have mixed α/β or $\alpha + \beta$ structural content. The CATH classification has processed 90% of PDBs, and contains almost 445,000 structural domains. Currently, 22% of the domains in CATH superfamilies are mainly α-helical in structure, 25% are mainly β structure, and over half (52%) have mixed $\alpha\beta$ secondary structure content. The ECOD classification has processed slightly more PDBs than CATH ($\sim$3500) and has detected more remote homologies, leading to a smaller number of superfamilies. It is expected that the domains in ECOD follow a similar distribution, but as the ECOD hierarchy starts at the architecture level rather than class, it was not possible to separate the domains by class information.

All three of these resources provide free, publically available data. CATH and ECOD provide frequent updates, with ECOD publishing an update each week and CATH publishing an update every day. In addition to offering the daily-updated CATH

Table 3   Statistics for the SCOP, CATH and ECOD resources

	SCOP v1.75	CATH v4.2	ECOD v199
PDBs processed	38,221	131,091	134,556
Domains identified	110,800	434,857	577,454
Number of folds	1195	1391	3738[a]
Number of superfamilies	1962	6119	3537
Number of families	3902	68,069	13,426

[a]All SCOP and ECOD data were obtained from their websites, apart from the number of folds, which was calculated by finding the number of unique X.H.T strings in their download file.

structural domain and superfamily assignment data, there are also CATH-Plus releases with a wealth of extra data. The latest CATH-Plus (version 4.2) release was recently published in October 2017. This release comprises 434,857 protein structural domains, grouped into 6119 homologous superfamilies. Other data included in CATH-Plus includes: functional family data, function annotations from Gene Ontology (GO) and Enzyme Commission (EC) data; enzyme function evolution information with FunTree; species diversity information from taxonomic information, structural diversity information through structurally similar groups (SSGs) and superposition images; domain organisation information with ArchSchema (see **Fig. 6**). Additional sequence domain data is also provided by Gene3D (see Section "Structural Families in the Genomic Era"), which contributes ∼95 million sequence domains to CATH. Briefly, Gene3D uses the curated structural information in CATH to predict domain boundaries for UniProt sequences, which do not have a known 3D structure.

There is an uneven distribution of domains across the protein fold universe. For example, of the 1391 folds in CATH, the top five most populated folds account for 30% for all structural domain sequences in CATH. The top five most populated folds, including the: Rossmann fold, TIM barrel, and Immunoglobulin fold, are shown in **Fig. 7**. This distribution bias is also clearly seen in the homologous superfamilies as the most populated 100 CATH superfamilies represent almost half (49%) of all domain sequences.

Structural Families in the Genomic Era

Structural databases have increased massively over recent years. However, the sequence databases continue to grow even faster, with ∼100 million sequences now in UniProt-KB. These numbers will continue to grow rapidly over the coming years. New sequence read technologies that provide longer reads will make sequencing animal genomes more easy and affordable, and will contribute to an explosion in genome sequences over the coming years. Despite this, the various resources for predicting domain families from sequences (see **Table 4**) have continued to cope with the current increase in datasets (Lewis et al., 2017; Finn et al., 2017).

Many of the family annotation resources such as Pfam and Gene3D use sequence profiles in the form of HMMs, built from structural representatives of a family, to recognise relatives in genome sequences. The sensitivity of HMMs comes from their ability to capture information of conserved residue positions across a given family, and positions where residue insertions and deletions are more tolerated.

One reason for this ability to scale arises from improvements in HMM-based software with large speedups produced recently by the HMMER package (Eddy, 2011). The Gene3D v16 resource assigns domains for over 52 million protein sequences, producing over 95 million domain assignments (Lewis et al., 2017). For each CATH S95 sequence, a HMM is built using 1–2 rounds of Jackhmmer (for details see Lewis et al., 2017). These HMMs are then scanned against the protein sequences using the hmmsearch program as part of the HMMER package. Sometimes different HMM models overlap the same region and it is necessary to resolve a final set of non-overlapping domain annotations. Gene3D makes use of the cath-resolve-hits (CRH) software (available from the cath-tools resource pages (see Relevant Websites Section)), which uses dynamic programming to efficiently select the optimal set of non-overlapping domains, using the bit score of each hit as a measure of the hits overall quality. Importantly CRH is both extremely fast, memory efficient and can deal with discontinuous domain assignments whilst still finding an optimal solution.

Evolution of Protein Structure, Sequence and Function

Evolution of Protein Folds

As more and more structural data and functional annotations accumulated in protein structure databases, such as CATH and SCOP, it appeared likely that only a limited number of folding arrangements exist in nature. For example, there now are 131,091 protein structures in the CATH database (version 4.2) that have been classified into 434,857 structural domains, which is a nearly 50% increase in the number of domain structures over the last five years. However, the number of folds identified has increased very slowly with still only 1391 different folds identified (Dawson et al., 2017).

Protein fold similarity can often provide important clues regarding the functional role of a protein. However, studies on different folds have shown that protein domains that share the same fold can show functional divergence (Martin et al., 1998). As mentioned already above, some folds, also known as 'superfolds', are highly populated and occur more frequently than others.

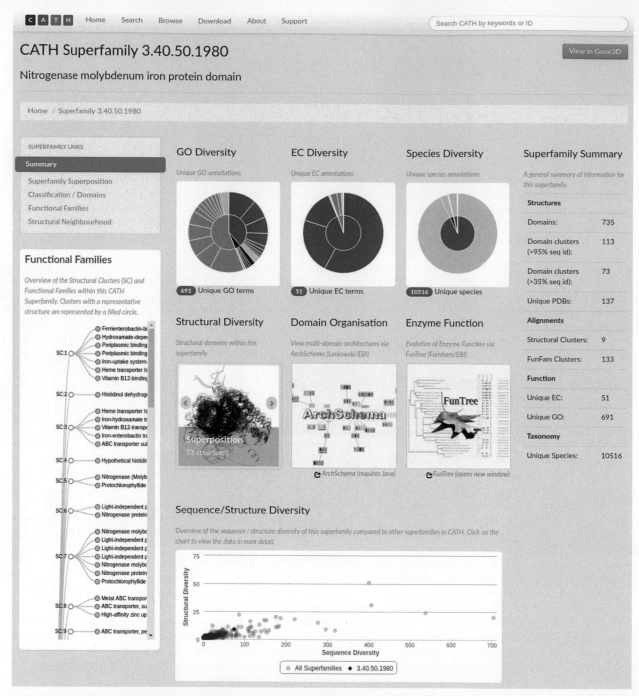

Fig. 6   Example of a superfamily summary web page, which is available for each superfamily in CATH. These pages contain information on: function annotation (GO and EC terms), species diversity, structural diversity, sequence diversity, and functional families.

These generally adopt Rossmann and TIM barrel folds or other folds which possess very regular architectures such as layers of beta-sheets and/or alpha-helices (Orengo *et al.*, 1994). It has been suggested that the regularity of these folding arrangements may provide a stable core to support diverse sequences. Hence, it is not surprising that they comprise multiple superfamilies. For example, the Rossmann fold comprises 130 superfamilies and the TIM barrel constitutes 30 superfamilies in the CATH database.

Exploiting CATH Superfamilies to Examine the Evolution of Protein Functions

The CATH-Gene3D resource (Dawson *et al.*, 2017; Lewis *et al.*, 2017), which brings together evolutionarily-related structural and sequence domains into superfamilies along with their functional annotations, provides an ideal platform to systematically analyse and explore how function is modulated by sequence and structural changes.

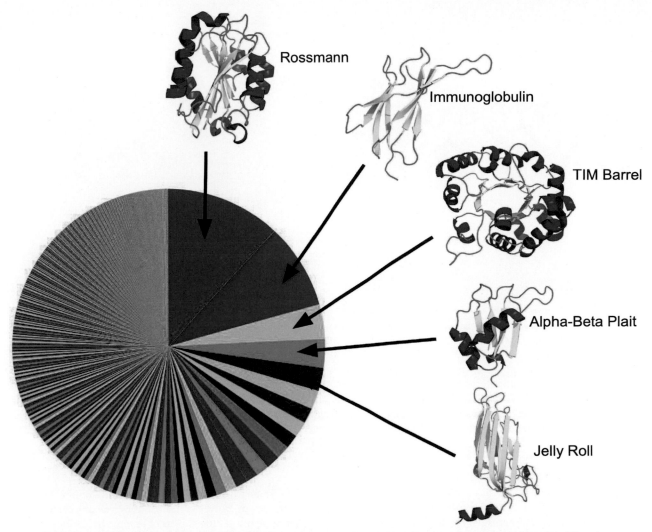

Fig. 7 Fold counts based on the number of CATH structural domain sequences. The top five folds account for 30% of all structural domain sequences in CATH. A structural representative from each of the top five folds is illustrated, coloured according to their secondary structure content.

Table 4   Example resources with genome annotations of structural assignments along with details of some of their unique features

Resource	Description	Useful Features
PFAM http://pfam.xfam.org/	Comprehensive prediction of protein families using HMMs.	User friendly tools to scan datasets. Extensive annotations of the Pfam families that helps to functionally interpret results. Combines families into larger Clans based on shared evolutionary origin.
InterPro https://www.ebi.ac.uk/interpro/	Integrates multiple resources assigning protein families domains and functional sites.	Scope and annotations has proven it to be a cornerstone of function prediction. Annotates entries with extensive text, GO terms etc.
Gene3D http://gene3d.biochem.ucl.ac.uk/	Predicts CATH structural domains.	Powered by CATH and so is largely up to date relative to the PDB. Functional Family level provides state of the art function and structure prediction.
SUPERFAMILY http://supfam.org/SUPERFAMILY	Predicts SCOP structural domains.	Established resource that is widely used in research projects e.g., provides tools for comparative genomics and genome annotations.
Genome3D http://genome3d.eu/	Provides consensus structural annotations and 3D models for model organisms.	Integrates SCOP, CATH, SUPERFAMILY, Gene3D, FUGUE, THREADER, PHYRE.

Modified from Pearl, F.M.G., Sillitoe, I., Orengo, C.A., 2015. Protein Structure Classification. eLS, pp. 1–10.

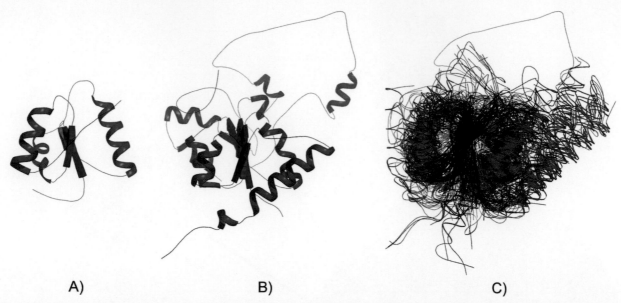

A)   B)   C)

Fig. 8 Structural diversity in the "Nitrogenase molybdenum iron protein domain" superfamily (CATH ID: 3.40.50.1980). This superfamily is highly structurally diverse with nine structurally similar clusters. The figure shows structures of: (A) the smallest, (B) largest domain and (C) superposition of representative non-redundant superfamily members to highlight the conserved structural core.

Some CATH superfamilies are very highly populated and universal to all kingdoms of life (Reid *et al.*, 2010). As previously mentioned, the 100 most highly populated superfamilies in CATH account for half of all known protein domains. The domain relatives in these superfamilies can incorporate large amounts of structural and functional diversities even though they share a conserved structural core. In some superfamilies relatives can vary in size by as much as 5-fold (**Fig. 8(A)** and **(B)**). This has been described as the 'Russian doll' effect (Swindells *et al.*, 1998). However, most relatives across a superfamily share the same common structural core comprising at least 50% of their structure (**Fig. 8(C)**).

Mechanisms of functional divergence

Various studies (Das *et al.*, 2015; Todd *et al.*, 2001; Reeves *et al.*, 2006; Brown and Babbitt, 2014; Galperin and Koonin, 2012) on the evolution of function in diverse superfamilies have illustrated that proteins can evolve new functions by one or a combination of different molecular mechanisms (**Fig. 9**). Some of these mechanisms include use of different sets of residues in binding or active site, addition of secondary structure embellishments to the core protein structure which alters the geometry of the active site or an interface on the protein or due to domain-shuffling in multi-domain proteins which can alter the context of the domain. Examples of some these mechanisms of functional divergence have been briefly discussed below.

Molecular tinkering

Relatively small changes in a binding or active site can result in changes in shape, physicochemical or other characteristics of the site which can alter the function of a protein significantly. Although, the active site may be highly conserved in some diverse enzyme superfamilies, for example, functionally diverse enzymes of the α/β hydrolase superfamily utilize the same catalytic triad (Todd *et al.*, 2001), a recent study indicated that relatives of a large number of diverse enzyme superfamilies show changes in catalytic machinery (Furnham *et al.*, 2016). One such superfamily is the Aldolase Class I superfamily, the members of which can carry out reactions with 31 different enzyme chemistry by using different sets of catalytic residues.

Structural mechanisms

Superfamily relatives can also vary in size to a great extent. For more than 150 CATH superfamilies, at least a 2-fold variation in the size has been observed between the most diverse domains and up to 5-fold in some superfamilies (Reeves *et al.*, 2006). The structural diversity among the domains can range from small changes within the active site to extensive embellishments which can alter their function, as seen in the HUP (High-signature proteins, UspA, and PP-ATPase) superfamily (Dessailly *et al.*, 2013).

Multi-domain context

Changes in the multi-domain architecture (MDA) of a protein can significantly alter its context, thereby modifying its function. For example, enzymes of the Thiamine pyrophosphate (TPP)-dependent superfamily, catalyse a large number of different reactions using TPP as the co-factor, by changing the domain partnership which alters the size and physicochemical properties of the active site pocket (Vogel and Pleiss, 2014).

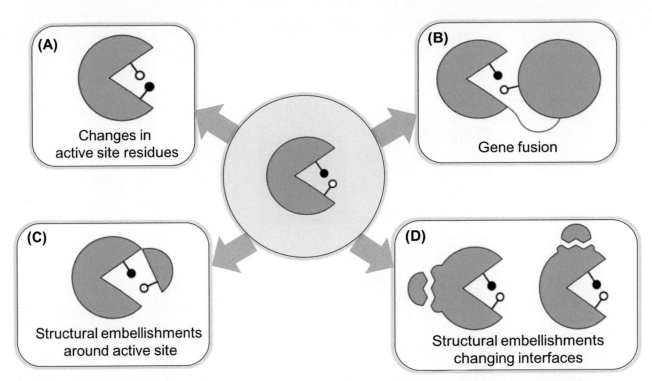

Fig. 9 Functional divergence in protein domains can arise due to one or combination of some of the following molecular mechanisms: (A) changes in active site residue, (B) gene fusion, (C) structural embellishments around active site and (D) structural embellishments changing interfaces. Modified from Das, S., Dawson, N.L., Orengo, C.A., 2015. Diversity in protein domain superfamilies. *Current Opinion* in *Genetics* and *Development* 35, 40–49.

In the majority of the large functionally diverse superfamilies, functional diversity generally results from a combination of different molecular mechanisms.

Capturing Functional Diversity by Sub-Classification of Superfamilies

One way to capture and understand this functional diversity among protein homologues is to further classify the superfamilies into functional groups. This functional classification also helps in inheriting annotations between homologous proteins that are most likely to perform the same function. Additionally, it allows identification of residues that are highly conserved among family members and which are most likely to have a functional role.

The CATH superfamilies have been functionally classified into Functional Families (FunFams) that group together relatives likely to have highly similar functions. The functional classification is derived from sequence similarities of superfamily relatives by optimal partitioning of a hierarchical sequence clustering tree for the superfamily which is generated by an agglomerative clustering algorithm (Lee *et al.*, 2010). This optimal partition is done by the FunFHMMer algorithm (Das *et al.*, 2016) which is based on maximising coherence of conserved residues within clusters and separation of differentially conserved residue positions (also known as functional determinants or FDs) between different clusters. The resulting clusters from the partition make up the FunFams for the superfamily. The FunFams have been shown to be functionally coherent based on different benchmarks using known functional annotations from the EC, GO, and annotations from the gold-standard SFLD superfamilies (Das *et al.*, 2016).

Hidden Markov Models (HMMs) are generated for all FunFams in CATH superfamilies and are used for predicting functions for uncharacterized sequences. When an uncharacterised query sequence is scanned against the library of CATH FunFam HMMs, it provides FunFam assignments. This provides both structural and functional insights for identified sequence domains. Functions for the query sequence domains are predicted by inheriting the known functional annotations of the sequence relatives of the assigned FunFam. This function prediction pipeline has been consistently ranked among the top 10 performing methods by the Critical Assessment of Function Annotation (CAFA) challenges (Radivojac *et al.*, 2013; Jiang *et al.*, 2016), an independent large-scale assessment of function prediction methods using different benchmarks and evaluation metrics. A web server (See Relevant Websites Section), **Fig. 10(A)** has been set up to allow users to submit query sequences in FASTA format for assignment to CATH FunFams where possible for the constituent domains of the query sequence (Das *et al.*, 2015). For each of the constituent FunFams of the query sequence, the server (**Fig. 10(B)**) reports functional annotations of the FunFam domain relatives and highly conserved residues for alignments with sufficient information content.

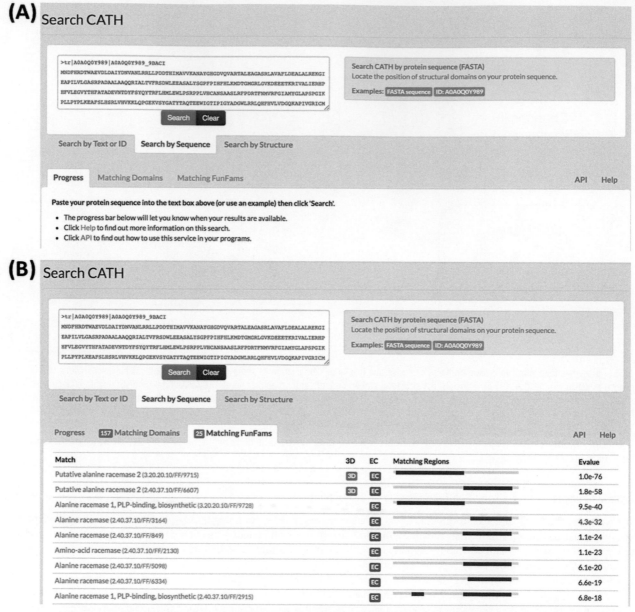

Fig. 10 CATH sequence search: (A) The server (http://www.cathdb.info/search/by_sequence) takes a query protein sequence (in FASTA format) as input and scans it against the library of CATH-FunFam HMMs. (B) The CATH FunFam matches for the constituent domains of the query sequence are reported along with their corresponding E-values.

The functional classification also provides a platform to detect subtle changes in conservation patterns that can suggest shifts in binding specificities or catalytic machineries of domain relatives in a superfamily. This kind of information would be helpful to guide experiments for target selection which would help to characterise unusual domain relatives of all protein superfamilies.

Conclusions

Protein structural classifications categorise protein structures deposited in the PDB according to their 3D structural characteristics and evolutionary relationships. There are three major classifications publically available to use: SCOP, CATH, and ECOD. A range of algorithms have been developed to recognise domain boundaries within protein structures using sequence-, structure-, and ab-initio-based methods. Algorithms have also been developed to detect homologous relationships, which are used to assign these domains into homologous superfamily groups and thereby classifying them within a classification hierarchy.

Nearest-neighbour clustering methods are an alternative way of searching for structural homologues. Structural comparison methods (e.g., Dali and VAST +) are used to compare one or more query structures against each other or the whole PDB archive, for example. Only information on structural similarity is provided, offering a fast way to generate lists of potential structural relatives and analogues.

SCOP, CATH and ECOD have many similarities in their protein structural classification hierarchies, which are discussed in Section "Hierarchical Structural Classification". Consensus SCOP and CATH superfamily data have been identified through the Genome3D project and are currently being integrated into the InterPro resource to increase the coverage of data for the SUPERFAMILY and Gene3D resources (see Section "Integration of the SCOP and CATH Resources").

A look into the current status of the protein structural universe highlights the large amounts of structural data that has been analysed and made publically available. As seen over recent years, there is an uneven distribution of domains assigned to protein folds and superfamilies. For example, the 100 most-populated CATH superfamilies have consistently accounted for around half of all structural domain sequences.

Taking a look at the vast amounts of genomic sequence data available, Section "Structural Families in the Genomic Era" describes how structural family data is used to annotate these data, which do not have any known structures. HMM libraries are built from structural domain data in CATH, which are then used by the sister resource Gene3D to scan genome sequences against to find significant structural matches for domain and homology detection. These scanning methods and libraries provide a very powerful way to annotate genomic data.

CATH superfamilies and functional families can be used to examine the evolution of protein sequence, structure and function. The mechanisms behind these divergences are discussed in Section "Evolution of Protein Structure, Sequence and Function". Superfamily superpositions illustrate any structural diversity within a superfamily, yet clearly show that even diverse superfamily have a highly conserved structural core. Functional family data are an important part in exploring functional diversity within a superfamily, determining functional relatives and in annotating protein function for unknown sequences.

Acknowledgements

N.L.D. acknowledges funding from the Wellcome Trust (Award number: 104960/Z/14/Z). J.G.L. acknowledges funding from the BBSRC [BB/L002817/1]. S.D. acknowledges funding from the BBSRC.

See also: *Ab initio* Protein Structure Prediction. Algorithms for Structure Comparison and Analysis: Docking. Algorithms for Structure Comparison and Analysis: Homology Modelling of Proteins. Algorithms for Structure Comparison and Analysis: Prediction of Tertiary Structures of Proteins. Biomolecular Structures: Prediction, Identification and Analyses. Computational Tools for Structural Analysis of Proteins. Drug Repurposing and Multi-Target Therapies. Identifying Functional Relationships Via the Annotation and Comparison of Three-Dimensional Amino Acid Arrangements in Protein Structures. In Silico Identification of Novel Inhibitors. Natural Language Processing Approaches in Bioinformatics. Protein Structural Bioinformatics: An Overview. Protein Structure Databases. Protein Three-Dimensional Structure Prediction. Secondary Structure Prediction. Small Molecule Drug Design. Structure-Based Design of Peptide Inhibitors for Protein Arginine Deiminase Type IV (PAD4). Study of The Variability of The Native Protein Structure. Supervised Learning: Classification. Supervised Learning: Classification. The Evolution of Protein Family Databases. Transmembrane Domain Prediction

References

Altschul, S.F., *et al.*, 1990. Basic local alignment search tool. Journal of Molecular Biology 215 (3), 403–410.

Andreeva, A., *et al.*, 2014. SCOP2 prototype: A new approach to protein structure mining. Nucleic Acids Research 42 (Database issue), D310–D314.

Brown, S.D., Babbitt, P.C., 2014. New insights about enzyme evolution from large scale studies of sequence and structure relationships. The Journal of Biological Chemistry 289 (44), 30221–30228.

Buchan, D.W.A., *et al.*, 2010. Protein annotation and modelling servers at University College London. Nucleic Acids Research 38 (Web Server issue), W563–W568.

Cheng, H., *et al.*, 2014. ECOD: An evolutionary classification of protein domains. PLOS Computational Biology 10 (12), e1003926.

Das, S., *et al.*, 2016. Functional classification of CATH superfamilies: A domain-based approach for protein function annotation. Bioinformatics 32 (18), 2889.

Das, S., Dawson, N.L., Orengo, C.A., 2015. Diversity in protein domain superfamilies. Current Opinion in Genetics & Development 35, 40–49.

Das, S., Sillitoe, I., *et al.*, 2015. CATH FunFHMMer web server: Protein functional annotations using functional family assignments. Nucleic Acids Research 43 (W1), W148–W153.

Dawson, N.L., *et al.*, 2017. CATH: An expanded resource to predict protein function through structure and sequence. Nucleic Acids Research 45 (D1), D289–D295.

Dessailly, B.H., *et al.*, 2013. Functional site plasticity in domain superfamilies. Biochimica Et Biophysica Acta 1834 (5), 874–889.

Eddy, S.R., 2011. Accelerated profile HMM searches. PLOS Computational Biology 7 (10), e1002195.

Finn, R.D., *et al.*, 2017. InterPro in 2017-beyond protein family and domain annotations. Nucleic Acids Research 45 (D1), D190–D199.

Furnham, N., *et al.*, 2016. Large-scale analysis exploring evolution of catalytic machineries and mechanisms in enzyme superfamilies. Journal of Molecular Biology 428 (2 Pt A), 253–267.

Galperin, M.Y., Koonin, E.V., 2012. Divergence and convergence in enzyme evolution. The Journal of Biological Chemistry 287 (1), 21–28.

Gerstein, M., 1998. How representative are the known structures of the proteins in a complete genome? A comprehensive structural census. Folding and Design 3 (6), 497–512.

Gibrat, J.-F., Madej, T., Bryant, S.H., 1996. Surprising similarities in structure comparison. Current Opinion in Structural Biology 6 (3), 377–385.

Gough, J., Chothia, C., 2002. SUPERFAMILY: HMMs representing all proteins of known structure. SCOP sequence searches, alignments and genome assignments. Nucleic Acids Research 30 (1), 268–272.

Greene, L.H., *et al.*, 2007. The CATH domain structure database: New protocols and classification levels give a more comprehensive resource for exploring evolution. Nucleic Acids Research 35 (Database issue), D291–D297.

Hadley, C., Jones, D.T., 1999. A systematic comparison of protein structure classifications: Scop, CATH and FSSP. Structure 7 (9), 1099–1112.

Holm, L., Laakso, L.M., 2016. Dali server update. Nucleic Acids Research 44 (W1), W351–W355.

Holm, L., Park, J., 2000. DaliLite workbench for protein structure comparison. Bioinformatics 16 (6), 566–567.

Holm, L., Sander, C., 1995. Dali: A network tool for protein structure comparison. Trends in Biochemical Sciences 20 (11), 478–480.

Holm, L., Sander, C., 1994. Parser for protein folding units. Proteins 19 (3), 256–268.

Hubbard, T.J.P., et al., 1997. SCOP: A Structural Classification of Proteins database. Nucleic Acids Research 25 (1), 236–239.

Jiang, Y., et al., 2016. An expanded evaluation of protein function prediction methods shows an improvement in accuracy. Genome Biology 17 (1), 184.

Karplus, K., Barrett, C., Hughey, R., 1998. Hidden Markov models for detecting remote protein homologies. Bioinformatics 14 (10), 846–856.

Kelley, L.A., Sternberg, M.J.E., 2009. Protein structure prediction on the Web: A case study using the Phyre server. Nature Protocols 4 (3), 363–371.

Krissinel, E., Henrick, K., 2004. Secondary-structure matching (SSM), a new tool for fast protein structure alignment in three dimensions. Acta Crystallographica Section D, Biological Crystallography 60 (Pt 12 Pt 1), 2256–2268.

Lee, D.A., Rentzsch, R., Orengo, C., 2010. GeMMA: Functional subfamily classification within superfamilies of predicted protein structural domains. Nucleic Acids Research 38 (3), 720–737.

Levitt, M., Chothia, C., 1976. Structural patterns in globular proteins. Nature 261 (5561), 552–558.

Lewis, T., et al., 2017. Gene3D: Extensive prediction of globular domains in proteins. Nucleic Acids Research. Available at: https://doi.org/10.1093/nar/gkx1069.

Lewis, T.E., et al., 2013. Genome3D: A UK collaborative project to annotate genomic sequences with predicted 3D structures based on SCOP and CATH domains. Nucleic Acids Research 41 (Database issue), D499–D507.

Lewis, T.E., et al., 2015. Genome3D: Exploiting structure to help users understand their sequences. Nucleic Acids Research 43 (Database issue), D382–D386.

Lobley, A., Sadowski, M.I., Jones, D.T., 2009. pGenTHREADER and pDomTHREADER: New methods for improved protein fold recognition and superfamily discrimination. Bioinformatics 25 (14), 1761–1767.

Madej, T., G.ibrat, J.F., Bryant, S.H., 1995. Threading a database of protein cores. Proteins 23 (3), 356–369.

Martin, A.C., et al., 1998. Protein folds and functions. Structure 6 (7), 875–884.

Mizuguchi, K., et al., 1998. HOMSTRAD: A database of protein structure alignments for homologous families. Protein Science: A Publication of the Protein Society 7 (11), 2469–2471.

Orengo, C.A., et al., 1997. CATH – A hierarchic classification of protein domain structures. Structure 5 (8), 1093–1109.

Orengo, C.A., Jones, D.T., Thornton, J.M., 1994. Protein superfamilies and domain superfolds. Nature 372 (6507), 631–634.

Pearl, F.M.G., Sillitoe, I., Orengo, C.A., 2015. Protein Structure Classification. eLS. pp. 1–10.

Radivojac, P., et al., 2013. A large-scale evaluation of computational protein function prediction. Nature Methods 10, 221.

Redfern, O.C., et al., 2007. CATHEDRAL: A fast and effective algorithm to predict folds and domain boundaries from multidomain protein structures. PLOS Computational Biology 3 (11), e232.

Reeves, G.A., et al., 2006. Structural diversity of domain superfamilies in the CATH database. Journal of Molecular Biology 360 (3), 725–741.

Reid, A.J., Ranea, J.A., Orengo, C.A., 2010. Comparative evolutionary analysis of protein complexes in E. coli and yeast. BMC Genomics 11, 79.

Shi, J., Blundell, T.L., Mizuguchi, K., 2001. FUGUE: Sequence-structure homology recognition using environment-specific substitution tables and structure-dependent gap penalties. Journal of Molecular Biology 310 (1), 243–257.

Shindyalov, I.N., Bourne, P.E., 2001. A database and tools for 3-D protein structure comparison and alignment using the Combinatorial Extension (CE) algorithm. Nucleic Acids Research 29 (1), 228–229.

Siddiqui, A.S., Barton, G.J., 1995. Continuous and discontinuous domains: An algorithm for the automatic generation of reliable protein domain definitions. Protein Science: A Publication of the Protein Society 4 (5), 872–884.

Söding, J., 2005. Protein homology detection by HMM-HMM comparison. Bioinformatics 21 (7), 951–960.

Swindells, M.B., 1995. A procedure for detecting structural domains in proteins. Protein Science: A Publication of the Protein Society 4 (1), 103–112.

Swindells, M.B., et al., 1998. Contemporary approaches to protein structure classification. BioEssays: News and Reviews in Molecular, Cellular and Developmental Biology 20 (11), 884–891.

Taylor, W.R., Orengo, C.A., 1989. Protein structure alignment. Journal of Molecular Biology 208 (1), 1–22.

Teichmann, S.A., Park, J., Chothia, C., 1998. Structural assignments to the Mycoplasma genitalium proteins show extensive gene duplications and domain rearrangements. Proceedings of the National Academy of Sciences of the United States of America 95 (25), 14658–14663.

Todd, A.E., Orengo, C.A., Thornton, J.M., 2001. Evolution of function in protein superfamilies, from a structural perspective. Journal of Molecular Biology 307 (4), 1113–1143.

Vogel, C., Pleiss, J., 2014. The modular structure of ThDP-dependent enzymes. Proteins: Structure, Function, and Bioinformatics. Available at: http://onlinelibrary.wiley.com/doi/10.1002/prot.24615/full.

Yeats, C., et al., 2011. The Gene3D Web Services: A platform for identifying, annotating and comparing structural domains in protein sequences. Nucleic Acids Research 39 (suppl), W546–W550.

Further Reading

Branden, C., Tooze, J., 1999. Introduction to Protein Structure, second ed. New York: Garland Science.

Dawson, N.L., Sillitoe, I., Lees, J.G., Lam, S.D., Orengo, C.A., 2017. CATH-Gene3D: Generation of the Resource and its use in Obtaining Structural and Functional Annotations for Protein Sequences. Protein Bioinformatics: From Protein Modifications and Networks to Proteomics. New York, NY: Humana Press, pp. 79–110.

Gu, J., Bourne, P.E. (Eds.), 2009. Structural Bioinformatics. Hoboken, NJ: Wiley-Blackwell.

Lesk, A.M., 2010. Introduction to Protein Science: Architecture, Function and Genomics, second ed. Oxford: Oxford University Press.

Mount, S., 2004. Bioinformatics: Sequence and Genome Analysis, second ed. Cold Spring Harbor Press.

Mukherjee, S., Seshadri, R., Varghese, N.J., et al., 2017. 1003 reference genomes of bacterial and archaeal isolates expand coverage of the tree of life. Nature Biotechnology 35, 676–683.

Orengo, C.A., Bateman, A., Uversky, V. (Eds.), 2013. Protein Families: Relating Protein Sequence, Structure, and Function. Wiley-Blackwell.

Orengo, C.A., Thornton, J.M., Jones, D.T. (Eds.), 2003. Bioinformatics: Genes, Proteins and Computers. Oxford: BIOS Scientific.

Shendure, J., Balasubramanian, S., Church, G.M., et al., 2017. DNA sequencing at 40: Past, present and future. Nature 550, 345–353.

Williamson, M., 2012. How Proteins Work. New York: Garland Science, Taylor & Francis Group, LLC.

Relevant Websites

http://cath-tools.readthedocs.io/
    cath-tools.
http://www.cathdb.info/search/by_sequence
    CATH.
http://www.cathdb.info/search/by_structure
    CATH.

Secondary Structure Prediction

Jonas Reeb and Burkhard Rost, Technical University of Munich (TUM), Garching/Munich, Germany

Introduction

The study of proteins is fuelled by the desire to understand their function. The molecular understanding of function requires the understanding of structure. However, determining the three-dimensional (3D) structure of proteins is a highly non-trivial process. Despite substantial advances in technologies, the experimental determination of a single protein is still costly and might take years. Conversely, the advent of next generation sequencing techniques has tremendously increased the amount of raw sequence data about which there is very little structural knowledge (Goodwin *et al.*, 2016; Martinez and Nelson, 2010). Due to ever decreasing costs, the speed at which this sequencing data is being generated far outpaces the annotation of the data with structural (or functional) knowledge. This creates a divide that has been referred to as the sequence-structure gap. For instance, the UniProtKB database currently contains almost 60 million proteins which are at most 90% identical (The Uniprot Consortium, 2016; Suzek *et al.*, 2015). In contrast, the Protein Data Bank (PDB), the major database of protein 3D structures, contains just 48,000 protein structures at that similarity cutoff (Berman *et al.*, 2000).

The prediction of the protein secondary structure is one step toward bridging this gap. Due to its simplicity and elementary importance, secondary structure is also one of the earliest features that has been predicted from sequence and has grown over many decades into a mature field where predictions can be performed at relatively little computational cost, yet at a high performance.

If the 3D structure could be predicted directly, then one would also have an implicit prediction of secondary structure. However, despite crucial breakthroughs in the last 5–10 years, the number of proteins for which 3D structure can be predicted from sequence remains limited (Kinch *et al.*, 2016; Hopf *et al.*, 2012; Moult *et al.*, 2016; Marks *et al.*, 2012). Even for most proteins for which comparative modelling can infer 3D structures (Biasini *et al.*, 2014; Yang *et al.*, 2011), dedicated secondary structure prediction methods still tend to perform better (Eyrich *et al.*, 2001). Additionally, the computational costs of more advanced methods are substantial which is in contrast to the decrease in resources needed for generating novel sequence (Muir *et al.*, 2016). Hence, both effective and efficient annotation of this sequencing data is necessary.

On top of the knowledge about the structural elements, secondary structure predictions also serve as input to predict more complex protein features such as disordered regions, interaction sites or aspects of protein function. Solvent accessibility is another feature of protein structure and often predicted by many of the tools for secondary structure prediction.

This contribution succinctly summarizes the underlying biological concepts of protein secondary structure and continues discussing some of the main concepts in the field of secondary structure prediction before presenting a selection of such methods in more detail. Finally, the level at which state-of-the-art methods perform is examined.

Background

Secondary Structure Elements in Protein Folding

Protein 3D structure is encoded by its amino acid sequence which is sometimes misleadingly referred to as the "primary structure". The order in which the 20 biogenic amino acids are arranged uniquely defines the structure and determines its folding in 3D (Anfinsen, 1973). Amino acids are linked by a covalent, so-called peptide bond which forms the backbone of the protein. Given the chemical properties of the amino acids and the physical constraints of this bond, the number of possible protein conformations is immense. The forces that drive the folding process that strives for a stable energetic minimum, are non-covalent interactions such as electrostatic interactions, much weaker but more common van der Waals interactions, or the hydrophobic effect, i.e., non-polar components aggregating in water (Kessel and Ben-Tal, 2011).

For the formation of secondary structure elements, a specific type of electrostatic interactions, the hydrogen bonds, are often claimed as the main stabilizing force. While this is most likely not the case, they still give structure to the elements in question and are important to overall protein stability (Kessel and Ben-Tal, 2011). Hydrogen bonds form between two dipoles which act as donor and acceptor, such as two atoms of amino acid side chains. However, in the case of secondary structure elements, the relevant atoms are part of the peptide bonds. Here, a hydrogen atom that is covalently bound to a nitrogen, gets in proximity of, and is therefore attracted by, an electronegative oxygen (**Fig. 1**). The stable secondary structure elements then further develop interactions between themselves and form a tertiary structure.

Various types of secondary structure are observed repeatedly out of which helices and strands are the most prominent ones. Both are characterized by specific patterns of hydrogen bonds. Due to these patterns, those two types are also often referred to as "regular secondary structure". Most prediction methods target three "states" using those two regular types and merging everything else into the state "other".

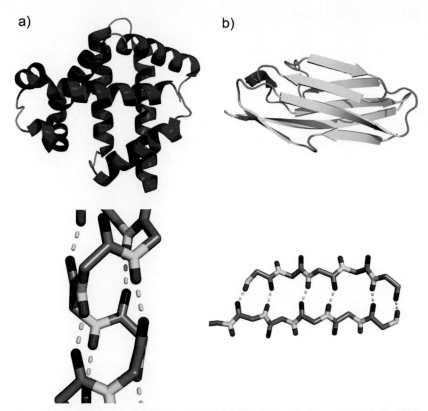

Fig. 1 α-helical and β-sheet secondary structure elements. The two most common secondary structure elements are visualized in PyMol (Schrodinger, 2015). Above, the protein is visualized in a cartoon view that highlights the secondary structure elements and hides details such as the amino acid side chains. Below, the protein backbone, without amino acid sidechains, of parts of the same secondary structures are shown in a sticks view. Here, oxygen atoms are colored blue, nitrogen yellow, hydrogen red and all others grey. Hydrogen bonds between atoms of the protein backbone are visualized as yellow dashed lines. (a) The human protein myoglobin is shown with α-helices highlighted in red and the rest of the protein in green (PDB identifier 3rgk). This is an "all-alpha" protein fold which consists of only α-helices and loops. (b) A part of the human vascular cell adhesion molecule is shown with β-sheets highlighted in yellow and loops in green. A small, single-turn α-helix can be seen in red at the left side.

Helices (α, 3_{10}, π)

As mentioned above, helices are defined by a repeating pattern of hydrogen bonds between residues. In an ideal α-helix, these bonds from between residues i and i + 4. This type of helix allows a relatively compact formation where four residues form a single "turn" of a helix. 3_{10} helices are more compact with just three residues (or ten atoms) forming a single "turn". π-helices on the other hand are wider and complete one "turn" every five residues. Among these three types, α-helices are energetically most favorable and thus most often observed. In literature and the output of prediction methods, helices are typically abbreviated by the letter H.

Strands and sheets (β-strands, parallel and antiparallel β-sheets)

β-sheets are a more spacious type of secondary structure formed from β-strands. Strands consist of the protein backbone "zig-zagging", typically for four to ten residues. Single β-strands are not energetically favorable. However, they can form β-sheets which are characterized by a pattern of hydrogen bonds between the residues on two different β-strands. Residue i in strand 1 forms a hydrogen bond with residue j in strand 2. The next bond can be formed in two different ways: either residue i + 2 in strand 1 binds to j + 2 in strand 2 (referred to as a parallel β-sheet), or residue i + 2 in strand 1 binds to j − 2 in strand 2 (referred to as anti-parallel β-sheet). In contrast to the situation for helices, the hydrogen bonding residues in β-sheets might be far away in sequence. Most β-sheets are formed from more than two strands. Typically, β-strands and -sheets are abbreviated by the letter E.

Other/loops

All residues that do not participate in helices or strands are often joined in one category as "other". Historically, this is also referred to as "loop" or, very misleadingly, as "(random) coil". Although the residues in question do not form hydrogen-bonded regular secondary structure, they are often constrained by bonds to other residues and are certainly not all random. Some methods distinguish different types within the class "other" such as "(reverse) turns" or "bends" where the residues in question change the direction of the amino acid chain. This non-regular class is typically abbreviated by the letter L.

Inference of Secondary Structure Elements From 3D Protein Structures

DSSP

Define Secondary Structure of Proteins (DSSP) is the standard tool for the annotation of secondary structure elements from protein structures (Kabsch and Sander, 1983; Touw *et al.*, 2015). Based primarily on hydrogen bonding patterns and some geometric constraints, it assigns every residue to one of eight possible states. States corresponding to helical structures are α-, 3_{10}- and π-helices. β-bridges are short fragments that show β-sheet like binding patterns. Multiple consecutive β-bridges form the state referred to as β-ladder. Multiple ladders can then form the above-mentioned β-sheets. However, DSSP does not have a separate state for β-sheet residues and the one for β-ladders is used. The remaining three states, called turn, bend and other, describe loop structures. Typically, secondary structure prediction methods simplify these eight states into just three: α-helix, β-sheet and loop.

STRIDE

Given a high-resolution 3D structure, annotation of secondary structure elements remains a matter of definition to some degree. STRIDE represents an alternative approach to DSSP. It aims to provide secondary structure assignments that are more consistent with the assignments performed by experimentalists who determined the protein structure (Frishman and Argos, 1995; Heinig and Frishman, 2004). Next to hydrogen bonds it includes other aspects such as the backbone geometry in the form of dihedral angle propensities. Another difference to DSSP is that some of its decision thresholds have been optimized on data from the PDB which makes it a knowledge-based approach. Nonetheless, the agreement between DSSP and STRIDE annotations is very high at 95% (Martin *et al.*, 2005).

Assessing Secondary Structure Predictions

Assessing secondary structure predictions requires all the care generally exercised when evaluating the performance of machine learning models. One aspect is the careful choice of a meaningful scoring model.

The simplest and most commonly used performance measure is the three-state per-residue accuracy referred to as Q_3. It represents the fraction of all residues that have been predicted accurately in the three states helix, strand, other. Despite its popularity, this score has several shortcomings. Q_3 ignores the uneven distribution in the three states: most residues are in the state other, i.e., methods that poorly predict the more important regular states (in particular the least common state of strand) might still reach high performance. Additional scores to measure these aspects include Q_H and Q_E, the percentages of correctly predicted helix and strand residues. Many other scores have been proposed and are useful in different contexts (Rost and Sander, 1993b). All these per-residue ignore the fact that secondary structure segments span over several residues. This shortcoming is corrected by per-segment scores (Rost *et al.*, 1994). The simplest such score compares the average length of predicted and observed segments. More advanced is a composite score referred to as segment overlap (SOV) (Fidelis *et al.*, 1999). Instead of measuring single-residue accuracy, SOV scores the correct placement of secondary structure segments. Such scores tend to give leeway with respect to predicting the precise begin and end positions of segments. Predictions that roughly predict the location of most segments score high.

Given performance measures, one also needs a test dataset to evaluate predictions. Authors exercise varying degrees of care in trying to report unbiased performance estimates. Best would be to assess all relevant methods based on proteins that differ substantially in sequence from those used for development of any of the methods.

One effort for such an assessment was EVA, a server that collected newly released protein structures as an unknown test set and then automatically queried secondary structure prediction methods for their predictions of those proteins' secondary structure elements (Eyrich *et al.*, 2001). This process was performed on a weekly basis and the results published online. A similar concept was followed by LiveBench (Rychlewski and Fischer, 2005). Unfortunately, both servers have now been offline for more than 10 years. CAMEO is a current effort in automated protein structure prediction evaluation and its operators have stated that it may include an assessment of secondary structure prediction in the future if there is a community interest (Haas *et al.*, 2013).

Approaches

Amino Acid Propensities

Since the very beginning, the prediction of secondary structure elements was based on the fact that amino acid occurrences are not evenly distributed between those elements but show preferences (Kessel and Ben-Tal, 2011). For example, due to their compactness helices often contain amino acids with small, or more precisely linear, side chains such as alanine, glutamic acid or methionine. On the other hand, proline is rarely found in helices. Its pyrrolidine ring and bound backbone amide group disturb the geometry and hydrogen bonding pattern of the helix. This leads to a kink in the helix that is energetically unfavorable and gives proline the nickname "helix breaker". In the same way β-sheets have preferences for certain amino acids, in particular at their termini. Although weak, these propensities suffice to predict secondary structure better than random. In one way or another even the most recent methods developed today still exploit such features. However, finding the right relationships is far from trivial as the formation of secondary structure depends on more than just the local amino acid content. The most extreme evidence for this are so-called chameleon sequences. These sequence fragments can fold into both helix and strand conformations depending on context (Jacoboni *et al.*, 2000; Guo *et al.*, 2007; Li *et al.*, 2015).

Basic Concepts of Secondary Structure Prediction

Even before high-resolution protein structures became available, the earliest methods used exactly these amino acid propensities mentioned in the previous section together with empirically defined rules to gain insights into the secondary structure content of proteins of interest. For example, Szent-Györgyi and Cohen described the relationship between helices and proline as outlined above (Szent-Györgyi and Cohen, 1957). Chou and Fasman determined a simple rule as to which a clustering of certain helix or strand "former" residues leads to formation of the respective secondary structure element (Chou and Fasman, 1974). One could refer to those early approaches as "first generation" methods because they ultimately scanned for features of single residues.

Stepping up, the "second generation" methods improved performance by including more information. The basic idea was to consider the sequence neighborhood of a residue, i.e., for a residue at position i those before it $(i-1, i-2, \ldots)$ and those directly after it $(i+1, i+2, \ldots)$. At the end of the 80s such statistical approaches were complemented by the first machine learning solutions. While any of the usual machine learning models is potentially applicable, (artificial) neural networks (NNs) have been particularly popular in the field. Early machine learning-based methods used a limited set of input features which may be as simple as the amino acid sequence itself. Typically, the sequence would be parsed in a sliding window of uneven size X, i.e., to predict the secondary structure of the central amino acid residue, all $\lfloor X/2 \rfloor$ neighboring residues are considered. With increasingly complex models and more computational power becoming available, a multitude of other features was added to increase the chances of the model finding the intricate correlations between amino acids which lead to a respective secondary structure. This allowed to predict strands and helices at equal levels of accuracy (Rost and Sander, 1993b). However, all those tools seemingly hit a performance limit since the information available within the window was too small and increasing the window size introduced more noise than signal.

This led to the leap into the "third generation" methods that combined machine learning with evolutionary information (Rost and Sander, 1993a). To use evolutionary information, one first needs to create a multiple sequence alignment (MSA). Toward this end, one needs to find all proteins with sufficiently high sequence similarity to a query which guarantees that those proteins have similar secondary structure. Most important here is not the number of related proteins, but the amount of diversity, i.e., the sequences more distantly related are more important to improve the prediction than those that are very similar. The MSA is then converted into a position-specific scoring matrix (PSSM), often also referred to as "profile". This profile can be used by the prediction method to replace a much simpler 20-dimensional binary vector with 1 for the amino acid at that position and 0 for all 19 amino acids not at that position. An example of a simple neural network with homology information as input is shown in **Fig. 2**.

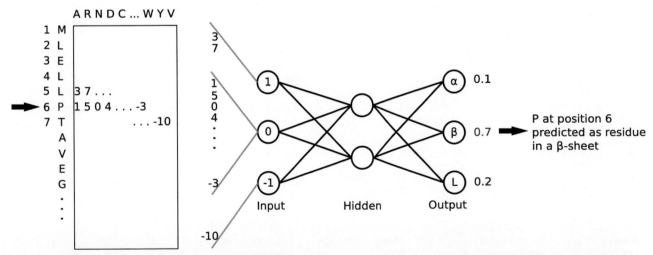

Fig. 2 A simple feed-forward neural network using homology information for the prediction of secondary structure elements. On the left, the input sequence for which the secondary structure is to be predicted runs from top to bottom. Right of it, the position specific scoring matrix (PSSM) for this input sequence is shown. Such a PSSM is created for example by performing a PSI-BLAST query with the input sequence against a large sequence database. The matrix contains information about which amino acids are commonly found in homologous sequences and which substitutions are rare. Typically, these values are used as log-odds such that negative values determine an unlikely substitutions and positives ones a likely one. These values can then be used as input to the neural network which leads to 20 inputs for every residue. To allow for a better visualization, the network shown here uses a very small sliding window of just three residues. Therefore, when predicting the secondary structure state at position 6, the PSSM values of positions 5, 6 and 7 are used as input to the network. The three input nodes labelled − 1, 0 and 1 are in fact 20 nodes each, leading to a total of 60 inputs. These values are then fed through a hidden layer to the output layer which predicts one of the three secondary structure states. Weights on the connections between these layers are optimized during the training of the network. A typical neural network for the prediction of secondary elements may use a sliding window of size 13 and have hundreds of nodes in the hidden layer. In deep learning, more than one hidden layer is used and in recurrent neural networks nodes from one layer may be connected to those of a previous layer.

Over the last few years, increasingly cheap, yet powerful hardware capable of highly parallelized calculations has given a new rise to deep learning approaches that has also reached the field of secondary structure prediction. These approaches are also based on NNs. However, they have more than one hidden layer and their architecture often differs significantly from that of traditional feed-forward NNs.

Results and Discussions

Secondary Structure Prediction Methods

In the following, a few secondary structure prediction methods are described in more detail starting with some of the oldest approaches to the current state-of-the-art (**Table 1**). The methods were chosen to represent some of the historically most valuable publications, to give an overview of how the field developed and highlight some of the currently most promising methods.

Prediction based on amino acid composition statistics

GOR, named by the initials of its authors is an information theoretical approach first published in 1978 (Garnier *et al.*, 1978, 1996). Using a sliding window of 17 residues the method calculates the most likely secondary structure element based on amino acid propensities which were extracted from a database of proteins with known secondary structure. The method was later extended by updating the underlying database with more recent data in GOR II. In addition to the previous single-residue statistics, GOR III included correlations between pairs of residues, i.e., between the central amino acid and amino acids at all other positions in the window. GOR IV further increased the number of residue pairs considered. Instead of only pairs with the central residue, all possible residue pairs in the window were now included in the calculation. In 2002 the final version of the method, GOR V, further extended the approach by adding statistics of amino acid triplets, as well as by including homology information, choosing the windows size depending on sequence length and again updating the underlying database (Kloczkowski *et al.*, 2002).

Prediction based on two levels of neural networks

PHDsec is one of the earliest methods to apply machine learning to the task of secondary structure prediction and the first to harness homology information to further improve it (Rost and Sander, 1993a). The method is based on NNs and uses only sequence information as input. For a given input sequence an MSA with homologous sequences is constructed from the HSSP database (Sander and Schneider, 1991). Then the amino acid occurrences at every position in the sequence are calculated. The method uses a sliding window of 13 residues and every position translates to 21 input units which are fed with the amino acid occurrences from the MSA at that position. 21 units are required (as opposed to just 20) to model cases near the N- and C-terminus of the input sequence where parts of the sliding window are "empty". The total 273 input units are connected to a hidden layer and an output layer which contains three nodes corresponding to an α-helix, β-sheet or loop prediction for the central residue in the sliding window. The output of this first NN is used as input to another NN with a 17-residue sliding window. This second layer aims to smooth the output such that segments are predicted more continuously. Later extensions to the method added more input features derived from the MSA as well as global amino acid composition information (Rost and Sander, 1994). The method has also been renamed several times, with the most recent version, ReProf, being available as part of the PredictProtein webserver (Yachdav *et al.*, 2014). PSIPRED is another common NN based prediction method with an approach similar to that of the PHDsec family (Jones, 1999). It uses sequence profiles

Table 1   A selection of secondary structure prediction methods discussed in this article

Method	Learning model	Other predicted features	Primary reference	Webserver address
GOR I-V	Bayes rules on amino acid propensities	None	Kloczkowski *et al.* (2002)	
PhDsec/ReProf	Neural Network	Solvent accessibility	Rost and Sander (1994)	https://predictprotein.org/
PSIPRED	Neural network	None	Jones (1999)	http://bioinf.cs.ucl.ac.uk/psipred/
JPred4	Neural network	None	Drozdetskiy *et al.* (2015)	http://www.compbio.dundee.ac.uk/jpred/
Frag1D	Database Lookup (Fragment matching)	None	Zhou *et al.* (2010)	http://frag1d.bioshu.se/
SSpro5	Database lookup + Deep Neural Network	None	Magnan and Baldi (2014)	http://scratch.proteomics.ics.uci.edu/
S2D	Neural Network	Disorder	Sormanni *et al.* (2015)	http://www-mvsoftware.ch.cam.ac.uk/
SPIDER2	Deep Neural Network	Solvent accessibility, torsion angles	Heffernan *et al.* (2015)	http://sparks-lab.org/server/SPIDER2/
Raptor X Property	Deep Neural Network	Solvent accessibility, disorder	Wang *et al.* (2016b)	http://raptorx.uchicago.edu/StructurePropertyPred/predict/

generated by PSI-BLAST (Altschul, 1997) as input to a NN with a sliding window of size 15, followed by a second smoothing NN with the same window size. The method has been continually updated since its first publication and the network architecture was recently improved (Buchan *et al.*, 2013). Finally, JPred4, a webserver built on the underlying Jnet also employs the two-level neural network methodology and has been updated many times since its first publication in 2000 (Drozdetskiy *et al.*, 2015; Cole *et al.*, 2008; Cuff and Barton, 2000). The method builds sequence profiles using PSI-BLAST as well as HMMER3, a hidden Markov model-based sequence search tool (Eddy, 2011), and trains two separate NN-pairs on the respective data. If these two models do not agree on a residue's prediction a third "jury" NN is used on the output of the previous two.

Prediction based on deep learning

SSpro5 is the most recent iteration of a secondary structure predictor which first performs a lookup of already known structures (Magnan and Baldi, 2014). The input sequence is compared to structures in the PDB using BLAST (Camacho *et al.*, 2009). If the resulting (local) alignments satisfy a set of similarity criteria the most common secondary structure class annotated in DSSP is used as the prediction for the respective residue. If no similar positions exist or the DSSP annotations do not form a majority, a neural network predictor is invoked to fill in predictions for the missing residues. This model consists of 100 bi-directional recurrent NNs. In contrast to the window-based feed-forward NNs covered so far, these networks can parse information of an arbitrarily large input sequence at once. In addition to the middle segment with the residue of interest and its neighbors, two other components of the network parse the rest of the sequence (Pollastri *et al.*, 2002). This could allow the recognition of global contacts which go beyond the correlations captured in the local window. The number of 100 networks result from the double cross-validation procedure which first splits the training set into 10 folds and then performs 10-fold cross-validation on each of them.

SPIDER2 is a deep-learning approach that combines the prediction of secondary structure elements with the prediction of solvent accessibility, backbone torsion angles and two more angles around the backbone C_α atoms (Heffernan *et al.*, 2015). The underlying idea is that these properties all correlate with each other to some degree. This makes the machine learning model aware of all classes at the same time which might improve prediction performance. SPIDER2 achieves this with an iterative approach: Given a PSI-BLAST PSSM of the input sequence and seven physicochemical indices, three NNs are separately trained for the prediction of secondary structure, solvent accessibility and the backbone angles. In the two following iterations, the input to each respective network is extended by the input from the other two networks. For example, predicted angles and solvent accessibility are used as input to the secondary structure predictor. In sum, this leads to three deep NN predictors with three hidden layers each. An improvement of the method using long short-term memory neural networks is currently in development.

Similar to SPIDER2, RaptorX Property is a webserver predicting several sequence features, including secondary structure and solvent accessibility (Wang *et al.*, 2016a,b). However, the prediction models are treated separately and do not interfere with each other. The authors use a deep convolutional neural field, which is the combination of a deep convolutional NN (DCNN) with seven hidden layers and a final output layer which together with the last hidden layer forms a conditional neural field (CNF). The rationale behind this design is that the DCNN can capture global relationships that go beyond the local window size of 11 residues while the final CNF learns the correlations between the secondary structure of adjacent residues. The input to the network is either just the sequence or a PSI-BLAST profile of that sequence.

Prediction based on other approaches

Frag1D performs secondary structure prediction based on a fragment matching approach (Zhou *et al.*, 2010). Given a query sequence, a sliding window of nine residues is used to find the 100 best-scoring fragments in a database of proteins with known structure. These 100 fragments are selected by comparing PSI-BLAST profiles and structural profiles between the input and database sequences. The structural profiles were created from so-called ShapeStrings which are defined based on the torsion angle pairs of the peptide bond. Same as for the direct sequence comparison, the database was searched for similar nine-residue stretches of ShapeStrings to build the structural profile which represents the conservation pattern in similar local structures. Further weighting the 100 fragments results in a preliminary secondary structure prediction. Given this prediction, the structural profile of the query sequence can now be computed and the whole process is repeated once to yield the final prediction.

s2D combines secondary structure prediction with the identification of disordered regions in proteins (Sormanni *et al.*, 2015). Such regions do not adapt a well-defined structure but instead fluctuate between different conformations (Habchi *et al.*, 2014). Typically, other methods extract secondary structure elements mostly from X-Ray crystallography structures through tools such as DSSP. In contrast, the s2D training data consists of annotations extracted from the chemical shifts of structures determined by nuclear magnetic resonance. In this way, all residues that show neither α-helical nor β-sheet properties can be considered as some form of disordered segments. Given these annotations, two NNs are trained with window sizes 11 and 15 but otherwise equal architecture. PSI-BLAST profiles are used as input to these networks. Next, a global network is trained with the same profiles and the output of the previous two networks resulting in a prediction of the mean secondary structure content in the whole sequence. Finally, the output of that network is combined with the initial inputs in a NN that smooths the prediction with a sliding window of five residues.

Estimates of Prediction Performance

As mentioned above in Section Assessing Secondary Structure Predictions two previous efforts for an automated independent assessment of secondary structure prediction are not maintained any longer. A related effort is CASP, a biannual challenge in which organizers collect target proteins with known but unpublished structures (Moult *et al.*, 1995). The target sequences are given to the community who can submit their predictions for the proteins' structures. After the structures are publicly disclosed, the assessors evaluate the performance of the prediction methods on these previously unseen proteins. This enables an independent evaluation where neither the competition organizers nor the participating groups know the answer at the time of submission. CASP has a focus on protein 3D structure prediction but in the past also included an evaluation of secondary structure prediction methods. In CASP5 assessors found that helices are predicted better than sheets and that sequences with no homology to a solved structure were harder to predict (Aloy *et al.*, 2003).

Overall, methods seem to have approached a saturation point with only small improvements over the previous years. Therefore, public assessments were stopped. This means that current estimates of prediction performance are only regularly given by the authors of newly developed methods. Although these may have the best intentions in providing a fair comparison to previous approaches they typically focus only on the flawed Q_3 score and their reported numbers may still be biased to favor their own method – for example, by the selection of a specific dataset. Traditionally, reported performance values have often proven to be overestimated. Apart from over-training, over-estimates can also be caused by discovering completely new types of structures. Ultimately, one therefore has two ways to evaluate performance by either using old results from the "last public" assessments, thereby excluding all new methods, or using method-skewed values from publications, likely over-estimating performance for the methods reported. In this chapter, the second solution was favored. However, readers must be aware that the published evaluation numbers need to be viewed with substantial skepticism.

Keeping that in mind, the most recent methods such as s2D, JPred4, RaptorX Property and SSpro5, all report Q_3 values around 80–85%. These values seem realistic, given the last independently determined performance estimates along with the increase in database size. However, choosing the best method among them is impossible given their heterogeneous evaluation schemes.

Future Directions

Without any independent evaluations, it is hard to say whether the top has already been reached in secondary structure prediction. Reinstating secondary structure prediction evaluation through a CAMEO revival of approaches such as EVA and LiveBench may help greatly. Certainly, creating such an independent evaluation entity is far from trivial and great care will be necessary when choosing the underlying datasets and scoring procedure. Clearly the limit is not given by a Q_3 of 100%. Due to errors in experimental structure determination and since proteins are in motion, secondary structure assignments will always contain some amount of error (Kihara, 2005). For the time being, the main problem with the assessment of secondary structure prediction is that although the tools are essential as input to many subsequent "higher level" prediction methods, there are very few incentives to invest substantial efforts to doing evaluations right. At the same time, it has become so difficult to publish new prediction methods in this field that almost nothing less than the claim that the ultimate has been reached will even appear on the screen of experts. There are many good reasons for these two opposing realities but they clearly make it difficult to assess today's state-of-the-art. It also remains debatable whether there is a large value in another method that shows increasingly smaller improvements over the current state-of-the-art. With the recent advances in the prediction of 3D protein structures from sequence, the interest in dedicated secondary structure prediction methods may continue to fade. However, at this point they cannot be substituted yet, since the difference in runtime between the two fields reaches several orders of magnitude and secondary structure prediction methods currently still perform better than inferring helices and sheets from predicted 3D structures (Faraggi *et al.*, 2012).

Closing Remarks

Predicting protein secondary structure elements from sequence is among the oldest prediction approaches in the field. Benefitting from decades of work, predictions are now both fast and highly accurate although determining just how accurate exactly is proving hard.

Recently, interest in these predictions has decreased, partly owing to a focus shift towards higher goals such as the prediction of protein 3D structures. Nonetheless, the respective methods are still relevant and a crucial help in making sense of the current sequence data deluge.

See also: *Ab initio* Protein Structure Prediction. Algorithms for Structure Comparison and Analysis: Docking. Algorithms for Structure Comparison and Analysis: Homology Modelling of Proteins. Algorithms for Structure Comparison and Analysis: Prediction of Tertiary Structures of Proteins. Artificial Intelligence and Machine Learning in Bioinformatics. Biomolecular Structures: Prediction, Identification and Analyses. Data Mining: Classification and Prediction. Data Mining: Prediction Methods. Drug Repurposing and Multi-Target Therapies. In Silico Identification of Novel Inhibitors. Machine Learning in Bioinformatics. Multilayer Perceptrons. Natural Language Processing Approaches in Bioinformatics. Protein Structural Bioinformatics: An Overview. Protein Three-Dimensional Structure Prediction. Protocol for Protein Structure Modelling. Sequence Analysis. Sequence Composition. Small Molecule Drug Design. Structure-Based Design of Peptide Inhibitors for Protein Arginine Deiminase Type IV (PAD4). Supervised Learning: Classification. Transmembrane Domain Prediction

References

Aloy, P., Stark, A., Hadley, C., Russell, R.B., 2003. Predictions without templates: New folds, secondary structure, and contacts in CASP5. Proteins: Structure, Function and Genetics 53, 436–456.

Altschul, S., 1997. Gapped BLAST and PSI-BLAST: A new generation of protein database search programs. Nucleic Acids Research 25, 3389–3402.

Anfinsen, C.B., 1973. Principles that govern the folding of protein chains. Science 181, 223–230.

Berman, H.M., Westbrook, J., Feng, Z., et al., 2000. The protein data bank. Nucleic Acids Research 28, 235–242.

Biasini, M., Bienert, S., Waterhouse, A., et al., 2014. SWISS-MODEL: Modelling protein tertiary and quaternary structure using evolutionary information. Nucleic Acids Research 42, W252–W258.

Buchan, D.W.A., Minneci, F., Nugent, T.C.O., Bryson, K., Jones, D.T., 2013. Scalable web services for the PSIPRED protein analysis workbench. Nucleic Acids Research 41, 349–357.

Camacho, C., Coulouris, G., Avagyan, V., et al., 2009. BLAST + : Architecture and applications. BMC Bioinformatics 10, 421.

Chou, P.Y., Fasman, G.D., 1974. Prediction of protein conformation. Biochemistry 13 (2), 222–245.

Cole, C., Barber, J.D., Barton, G.J., 2008. The Jpred 3 secondary structure prediction server. Nucleic Acids Research 36, 197–201.

Cuff, J.A., Barton, G.J., 2000. Application of multiple sequence alignment profiles to improve protein secondary structure prediction. Proteins 40, 502–511.

Drozdetskiy, A., Cole, C., Procter, J., Barton, G.J., 2015. JPred4: A protein secondary structure prediction server. Nucleic Acids Research 43, W389–W394.

Eddy, S.R., 2011. Accelerated profile HMM searches. PLOS Computational Biology 7. doi:10.1371/journal.pcbi.1002195.

Eyrich, V., Martí-Renom, M.A., Przybylski, D., et al., 2001. EVA: Continuous automatic evaluation of protein structure prediction servers. Bioinformatics 17, 1242–1243.

Faraggi, E., Zhang, T., Yang, Y., Kurgan, L., Zhou, Y., 2012. SPINE X: Improving protein secondary structure prediction by multistep learning coupled with prediction of solvent accessible surface area and backbone torsion angles. Journal of Computational Chemistry 33, 259–267.

Fidelis, K., Rost, B., Zemla, A., 1999. A modified definition of sov, a segment-based measure for protein secondary structure prediction assessment. Proteins 223, 220–223.

Frishman, D., Argos, P., 1995. Knowledge-based protein secondary structure assignment. Proteins-Structure Function and Genetics 23, 566–579.

Garnier, J., Gibrat, J.-F., Robson, B., 1996. GOR method for predicting protein secondary structure from amino acid sequence. Methods in Enzymology 266, 540–553.

Garnier, J., Osguthorpe, D.J., Robson, B., 1978. Analysis of the accuracy and implications of simple methods for predicting the secondary structure of globular proteins. Journal of Molecular Biology 120, 97–120.

Goodwin, S., Mcpherson, J.D., Mccombie, W.R., 2016. Coming of age: Ten years of next-generation sequencing technologies. Nature Reviews Genetics 17, 333–351.

Guo, J.-T., Jaromczyk, J.W., Xu, Y., 2007. Analysis of chameleon sequences and their implications in biological processes. Proteins 67, 548–558.

Haas, J., Roth, S., Arnold, K., et al., 2013. The protein model portal – A comprehensive resource for protein structure and model information. Database 2013, 1–8.

Habchi, J., Tompa, P., Longhi, S., Uversky, V.N., 2014. Introducing protein intrinsic disorder. Chemical Reviews 114, 6561–6588.

Heffernan, R., Paliwal, K., Lyons, J., et al., 2015. Improving prediction of secondary structure, local backbone angles, and solvent accessible surface area of proteins by iterative deep learning. Scientific Reports 5, 11476. doi:10.1038/srep11476.

Heinig, M., Frishman, D., 2004. STRIDE: A web server for secondary structure assignment from known atomic coordinates of proteins. Nucleic Acids Research 32, 500–502.

Hopf, T.A., Colwell, L.J., Sheridan, R., et al., 2012. Three-dimensional structures of membrane proteins from genomic sequencing. Cell 149, 1607–1621.

Jacoboni, I., Martelli, P.L., Fariselli, P., Compiani, M., Casadio, R., 2000. Predictions of protein segments with the same aminoacid sequence and different secondary structure: A benchmark for predictive methods. Proteins 41, 535–544.

Jones, D.T., 1999. Protein secondary structure prediction based on position-specific scoring matrices. Journal of Molecular Biology 292, 195–202.

Kabsch, W., Sander, C., 1983. Dictionary of protein secondary structure: Pattern recognition of hydrogen-bonded and geometrical features. Biopolymers 22, 2577–2637.

Kessel, A., Ben-Tal, N., 2011. Protein structure. In: Introduction to Proteins. Boca Raton, FL: CRC Press.

Kihara, D., 2005. The effect of long-range interactions on the secondary structure formation of proteins. Protein Science: A Publication of the Protein Society 14, 1955–1963.

Kinch, L.N., Li, W., Monastyrskyy, B., Kryshtafovych, A., Grishin, N.V., 2016. Evaluation of free modeling targets in CASP11 and ROLL. Proteins: Structure, Function and Bioinformatics. doi:10.1002/prot.24973.

Kloczkowski, A., Ting, K.L., Jernigan, R.L., Garnier, J., 2002. Combining the GOR V algorithm with evolutionary information for protein secondary structure prediction from amino acid sequence. Proteins: Structure, Function and Genetics 49, 154–166.

Li, W., Kinch, L.N., Karplus, P.A., Grishin, N.V., 2015. ChSeq: A database of chameleon sequences. Protein Science 24, 1075–1086.

Magnan, C.N., Baldi, P., 2014. SSpro/ACCpro 5: Almost perfect prediction of protein secondary structure and relative solvent accessibility using profiles, machine learning and structural similarity. Bioinformatics (Oxford, England) 30, 2592–2597.

Marks, D.S., Hopf, T.A., Chris, S., Sander, C., 2012. Protein structure prediction from sequence variation. Nature Biotechnology 30, 1072–1080.

Martinez, D.A., Nelson, M.A., 2010. The next generation becomes the now generation. PLOS Genetics 6, e1000906.

Martin, J., Letellier, G., Marin, A., et al., 2005. Protein secondary structure assignment revisited: A detailed analysis of different assignment methods. BMC Structural Biology 5, 17.

Moult, J., Fidelis, K., Kryshtafovych, A., Schwede, T., Tramontano, A., 2016. Critical assessment of methods of protein structure prediction: Progress and new directions in round XI. Proteins 84 (Suppl. 1), 4–14.

Moult, J., Pedersen, J.T., Judson, R., Fidelis, K., 1995. A large-scale experiment to assess protein structure prediction methods. Proteins: Structure, Function, and Genetics 23, ii–iv.

Muir, P., Li, S., Lou, S., et al., 2016. The real cost of sequencing: Scaling computation to keep pace with data generation. Genome Biology 17, 53.

Pollastri, G., Przybylski, D., Rost, B., Baldi, P., 2002. Improving the prediction of protein secondary structure in three and eight classes using recurrent neural networks and profiles. Proteins: Structure, Function, and Bioinformatics 47, 228–235.

Rost, B., Sander, C., 1993a. Improved prediction of protein secondary structure by use of sequence profiles and neural networks. Proceedings of the National Academy of Sciences 90, 7558–7562.

Rost, B., Sander, C., 1993b. Prediciton of protein secondary structure at better than 70% accuracy. Journal of Molecular Biology 232 (2), 584 599.

Rost, B., Sander, C., 1994. Combining evolutionary information and neural networks to predict protein secondary structure. Proteins: Structure, Function, and Bioinformatics 19, 55–72.

Rost, B., Sander, C., Schneider, R., 1994. Redefining the goals of protein secondary structure prediction. Journal of Molecular Biology 235, 13–26.

Rychlewski, L., Fischer, D., 2005. LiveBench-8: The large-scale, continuous assessment of automated protein structure prediction. Protein Science 14, 240–245.

Sander, C., Schneider, R., 1991. Database of homology-derived protein structures and the structural meaning of sequence alignment. Proteins: Structure, Function, and Bioinformatics 9, 56–68.

Schrodinger, L.L.C., 2015. The PyMOL molecular graphics system, Version 1.8.

Sormanni, P., Camilloni, C., Fariselli, P., Vendruscolo, M., 2015. The s2D method: Simultaneous sequence-based prediction of the statistical populations of ordered and disordered regions in proteins. Journal of Molecular Biology 427, 982–996.

Suzek, B.E., Wang, Y., Huang, H., Mcgarvey, P.B., Wu, C.H., 2015. UniRef clusters: A comprehensive and scalable alternative for improving sequence similarity searches. Bioinformatics 31, 926–932.

Szent-Györgyi, A.G., Cohen, C., 1957. Role of proline in polypeptide chain configuration of proteins. Science 126, 697.

The Uniprot Consortium, 2016. UniProt: The universal protein knowledgebase. Nucleic Acids Research 45, 1–12.

Touw, W.G., Baakman, C., Black, J., *et al.*, 2015. A series of PDB-related databanks for everyday needs. Nucleic Acids Research 43, D364–D368.

Wang, S., Li, W., Liu, S., Xu, J., 2016a. RaptorX-Property: A web server for protein structure property prediction. Nucleic Acids Res 44, W430–W435.

Wang, S., Peng, J., Ma, J., Xu, J., 2016b. Protein secondary structure prediction using deep convolutional neural fields. Scientific Reports 6, 18962.

Yachdav, G., Kloppmann, E., Kajan, L., *et al.*, 2014. PredictProtein – An open resource for online prediction of protein structural and functional features. Nucleic Acids Research 42, W337–W343.

Yang, Z., Lasker, K., Schneidman-Duhovny, D., *et al.*, 2011. UCSF Chimera, MODELLER, and IMP: An integrated modeling system. Journal of Structural Biology 179 (3), 269–278.

Zhou, T., Shu, N., Hovmöller, S., 2010. A novel method for accurate one-dimensional protein structure prediction based on fragment matching. Bioinformatics (Oxford, England) 26, 470–477.

Relevant Websites

https://www.cameo3d.org
    CAMEO.
https://www.wwpdb.org/
    The Protein Data Bank.

Biographical Sketch

Jonas Reeb is a PhD student in the Rostlab at the Technical University of Munich, Germany (TUM). During his studies at TUM he has worked on evaluating methods for the prediction of transmembrane helices from sequence and on developing a crystallization propensity predictor for the NYCOMPS structural genomics pipeline. His doctoral thesis focusses on predicting the effects of sequence variants.

Burkhard Rost is a professor and Alexander von Humboldt award recipient at the Technical University of Munich, Germany (TUM). Rost was the first to combine machine learning with evolutionary information, and realized this combination to predict secondary structure. Since, his group has repeated this success in many other tools predicting aspects of protein structure and protein function. All tools are available through the first internet server in the field of protein structure prediction (PredictProtein), which has been online for over 25 years. Over the last years, the Rostlab has been shifting focus on the development of methods that predict and annotate the effect of sequence variation and their implications for precision medicine and personalized health.

Protein Three-Dimensional Structure Prediction

Sanne Abeln and Klaas Anton Feenstra, Center for Integrative Bioinformatics VU (IBIVU), The Netherlands
Jaap Heringa, Vrije Universiteit Amsterdam, Amsterdam, The Netherlands

What is the Protein Structure Prediction Problem?

Predicting the Structure for a Protein Sequence

This article revolves around a simple question: "given an amino acid sequence, what is the folded structure of the protein?" (**Fig. 1**) Even though this seems like a simple question, the answer is far from straightforward. In fact, whether we can give an answer at all depends heavily on the sequence in question and available protein structures that can be used as modelling templates. While the number of structures deposited in the Protein Data Bank (PDB) (Berman *et al.*, 2000) continues to rise rapidly (see "Relevant Websites section"), the number of sequenced genes rises much faster. The large and widening gap between protein structures and sequences makes structure prediction an important problem to solve. Fortunately, recently developed methods can use these large resources of sequence data to increase the quality of some predictions. Here, we will give an overview of current structure prediction methods, and describe some tools that provide insight into how reliable the structure predicted will be.

The typical problem is that we want to generate a structural model for a protein with a sequence, but without an experimentally determined structure. In this article, we will build up a workflow for tackling this problem, starting from the easy options that, if applicable, are likely to generate a good structural model, and gradually working up to the more hypothetical options whose results are much more uncertain.

Another very important remark is in place here: the modelling strategy should depend heavily on what we want to do with the structure. Do we want to predict where the functional site of the protein is, whether a specific substrate binds, or if a certain residue may be exposed to the surface? These different questions imply a different degree of accuracy in the answer, and may lead to choices regarding technology and methods to carry out these predictions. It is important to keep in mind that one of the most important aspects of any scientific model is whether a research question may be answered with the model produced or not. Even if we do have an experimental structure available, some of these questions may not be straightforward to answer; we will come back to this issue later in the article.

Structure is More Conserved Than Sequence

Almost all structure prediction relies on the fact that, for two homologous proteins, structure is more conserved than (Bajaj and Blundell, 1984; see **Fig. 2**). The real power of this observation manifests itself when we turn this statement around: if two protein sequences are similar, these two proteins are likely to have a very similar structure. The latter statement has very important consequences. It means that if our sequence of interest is similar to a protein sequence with a known structure, we have a good starting point for a structural model. In such a scenario we use sequence similarity, suggesting an homologous relation between the proteins, to predict the structure. The vast majority of accurate structure prediction methods use structure conservation as an underlying principle; while methods that have been developed to deal with the more difficult modelling questions, exploit the sequence-structure-conservation relation in an advanced manner, as discussed towards the end of this article.

Terminology in Structure Prediction

Firstly, we should take care to lay down a good problem definition. Here we will generously borrow the nomenclature from the Critical Assessment of Protein Structures (CASP). CASP is a scientific competition, in which structure prediction groups and

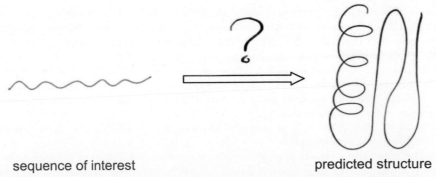

sequence of interest   predicted structure

Fig. 1 Structure prediction methods try to answer the question: Given an amino acid sequence, what is the folded protein structure?

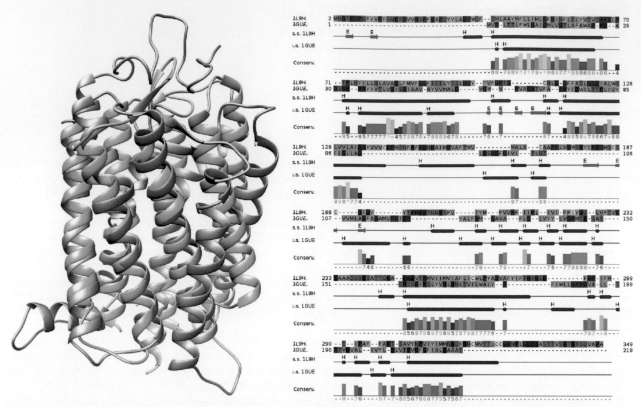

Fig. 2 Protein structure is more conserved than sequence. Here the output of a structural alignment is shown on the left, created using *Chimera* (Molecular graphics and analyses were performed with the UCSF Chimera package. Chimera is developed by the Resource for Biocomputing, Visualisation, and Informatics at the University of California, San Francisco (supported by NIGMS P41-GM103311)). Reproduced from Pettersen, E. F., Goddard, T.D., Huang, C.C., *et al.*, 2004. UCSF Chimera – A visualization system for exploratory research and analysis. Journal of Computational Chemistry, 25 (13), 1605–1612. The structural alignment shows both proteins are highly similar; the RMSD is 2.3 over 144 aligned residues. Furthermore, the function of the two proteins, one from cattle (PDB:1L9H, light brown) and one from a archaeon (PDB:1GUE, light blue), is similar: both are light sensitive rhodopsins, used for vision and phototaxis, respectively. However, as can be seen in the sequence alignment on the right, the sequence identity is only 7%. This is lower than would be expected for any two random sequences. The alignment shown is based on the structural alignment on the left, and visualised using *JalView*. Reproduced from Waterhouse, A.M., Procter, J.B., Martin, D.M.A., Clamp, M., Barton, G.J., 2009. Jalview Version 2 – A multiple sequence alignment editor and analysis workbench. Bioinformatics 25 (9), 1189–1191.

structure prediction servers compete to predict the structure for an unknown sequence, that has been running since 1995 (Moult *et al.*, 1995). The sequence for which we will predict a structure is called the *target* sequence. If there is a suitable structure to build a model for our query sequence we call this structure a *template*, see also **Fig. 3**. Using the structure of the template and using the *sequence alignment* between the template and the target sequence, we can create one or more structural *models*: the predicted structure, for a target sequence. In CASP structural models from different prediction methods, are compared to the experimentally determined solution or target structure.

Different Classes of Structure Prediction Methods

We can classify structure prediction strategies into two categories of difficulty: template-based modelling, and template-free modelling (see **Fig. 4**). In the first case, it is possible to find a suitable template for the target sequence in the PDB, as a basis for the model, whereas for template-free modelling no such experimental structure is available. Note that it may not be trivial at all to find out in which of these two categories a structure prediction problem falls. Only if we can find a close homolog – based on sequence similarity – in the PDB we can be sure that a template based modelling strategy will suffice; this is also referred to as homology modelling. With a template, the constraints from the alignment between the model and the template sequence, in addition to the template structure, will give sufficient constraints to build a structural model for the target sequence. Even in this case, small missing substructures in the alignment, e.g., loops, may require a template-free modelling strategy.

If no close homologs are available in the PDB, we may need to use more advanced template finding strategies, such as remote homology detection or fold recognition methods.

If no suitable template is available, we will need to resort to a template-free modelling strategy. In the *ab initio* approach knowledge-based energy terms are used to generate structural models based on the sequence of the template alone. Small, suitable fragments, from various PDB structures are assembled to generate possible structural models. In some cases, we can find

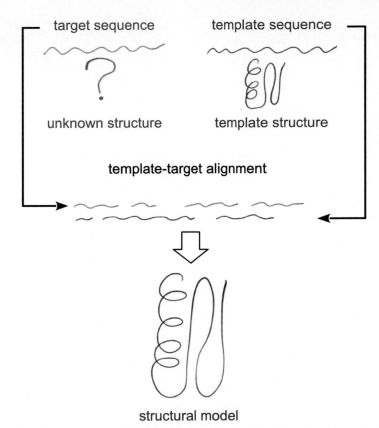

Fig. 3   Terminology used in protein structure prediction. We start from our protein of interest (with no known structure): the target sequence. First step is find a matching protein: a template sequence with known structure; the template structure. We then create a template-target sequence alignment, and from this alignment create the structural model which is the solution structure for our target protein.

additional constraints, for example from experiments, such as NMR or cryoEM, or from contact prediction methods; in that case we have a much better chance of building a suitable model (Moult *et al.*, 2016). In fact, we could consider such constraints an alternative for the constraints provided by homology.

Lastly, several steps may be taken to refine the model, and to select the most likely model, from several model building attempts. Note that some structure prediction methods, may also include variations of model refinement and model selection steps higher up in the modelling workflow.

Domains

So far we have implied that we may follow the above strategy for an entire protein, however, this generally is not the case. In fact many proteins consist of multiple domains. If this is the case, it is wise to also run one or multiple disorder prediction methods on the target sequence; Any large regions (>25 residues) predicted to be disordered should be left out for further structure prediction and template finding.

Most structure prediction methods only work well at the domain level. This means that a sequence first needs to be split in multiple domains, before we can start to make models. However, domain splitting is often ambiguous both given the sequence and the structure, while combining models built from various domains is far from trivial. In practice, this means that multiple templates might be necessary for a single target sequence and that it is difficult to resolve the orientation of the modelled domains with respect to each other.

Predicting the orientation for several domains is currently an unsolved problem, unless there is a suitable, homologous, template available – with the domains in the same orientation. In some cases, coarse constraints on the domain orientations such as data from small-angle scattering experiments, or distance restraints from NMR, chemical cross-links or co-evolution may help to put different homology models in the correct orientation.

Typically, it only makes sense to generate a model, be it template-based or template-free, for a single domain. In fact, in CASP model predictions are assessed per structural domain, separately. Therefore, it is essential to split the target sequence into its constituent domains – which is a non-trivial task, particularly if no homologous templates are available for each of the domains.

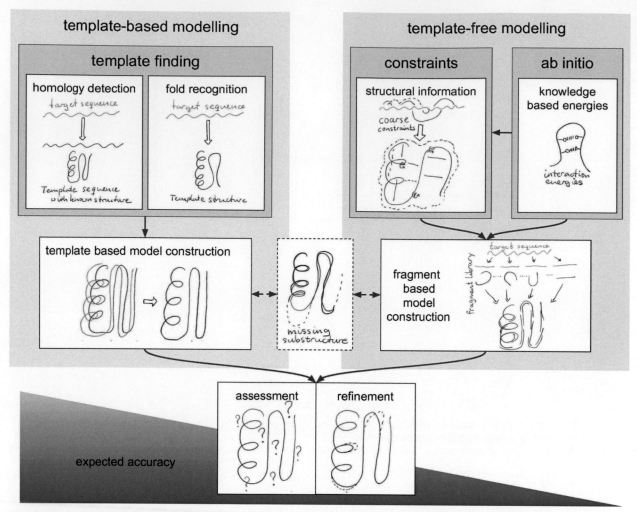

Fig. 4 Overview of Structure Prediction. Template-based modelling: a template is found on the basis of homology between the template and the target. Fold recognition: no obvious homologous structure can be found in the PDB, we need fold recognition methods to find a suitable template. Template-free modelling: no suitable template for protein domains can be found. Without template, we need to use a combination of coarse constraints from experiment or co-evolution analysis, and *ab initio* prediction. *Ab initio* methods typically work with taking fragment templates from various proteins, and assemble these into a model or decoy structure. Expected model accuracy declines from left to right: good accuracy is expected if based on homology; in contrast, *ab initio* modelling should only be considered if no other options remain.

Assessing the Quality of Structure Prediction Methods

Generally, as with any prediction problem, we can assess the quality of a prediction if we have a true answer to the question. Here, truth will be represented by an experimentally determined protein structure (of high quality). Fortunately, there are now (November 2017) over 120,000 deposited protein structures in the PDB (structures in the PDB: see "Relevant Websites section"). However, simply assessing how well a method performs over this set is problematic. The methods have been trained on structures contained in this data set, which implies there may be a strong bias in these methods to predict good models for sequences in their training dataset, and homologs of these sequences. In order to truly assess a method, a completely independent data set is required.

Critical Assessment of Protein Structure Prediction

Every other year CASP, a Critical Assessment of Protein Structure Prediction, provides such an independent validation benchmark. CASP is a blind test or competition: experimentalists provide sequences for which they know the structures will be solved imminently; modelling groups and servers try to predict the structure (Moult *et al.*, 1995). Once the structure is solved, the models can be evaluated using the solution structure of the target (see also **Fig. 5**).

CASP was started because the protein structure prediction problem was claimed to have been solved several times. The problem was, that algorithms were trained on databases that contained the structures that were evaluated in benchmarking tests. CASP overcomes this problem.

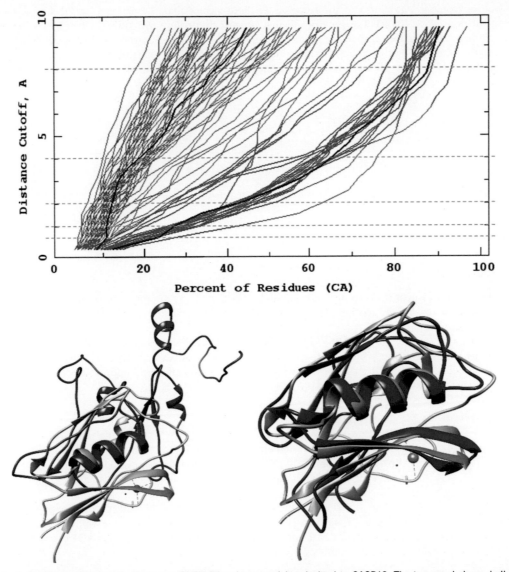

Fig. 5 Example of structural comparison for the target *T0886-D2* and two models submitted to CASP12. The top panel shows individual traces for all models generated for this target; the distance cutoff (vertical axis, in Å) is plotted against the fraction of residues (horizontal axis, in %) that can be aligned within this cutoff. The traces were obtained from *predictioncenter.org/casp12*. The dotted lines indicate the thresholds used in the GDT_TS (1, 2, 4, 8 Å) and GDT_HA (0.5, 1, 2, 4 Å) scores. Two models are highlighted in blue: a bad model (TS236, GDT_TS = 18.90) on the left, and a good model (TS173; GDT_TS = 51.97) on the right. Both model structures are also shown in the panels below in red, superposed onto the solution crystal structure in blue (PDB:5FHY). Structural superposition created using LGA at proteinmodel.org/AS2TS/LGA/. Reproduced from Zemla, A., 2003. LGA: A method for finding 3D similarities in protein structures. Nucleic Acids Research, 31 (13), 3370–4. 3D visualisation using *Chimera 1.11.2*. Reproduced from Pettersen, E.F., Goddard, T.D., Huang, C.C., *et al.*, 2004. UCSF Chimera – A visualization system for exploratory research and analysis. Journal of Computational Chemistry, 25 (13), 1605–1612.

Note that the very first step in any practical structure prediction approach, should be to inspect the results from the latest CASP round (Moult *et al.*, 2016) via the CASP website (CASP website: see "Relevant Websites section") to see what the state of the art methods are, and what their expected performance is.

Root-Mean-Square Deviation (RMSD)

If we want to asses the quality of a method, we need to measure the quality of the predictions made by the method. Hence, one would like to structurally compare atomic coordinates of the model and of the solution structure, and quantify the (dis)similarity.

The problem of comparing a model to a solution structure, is less difficult than the comparison between two homologous protein structures. This is because the alignment is trivial: the model has the same sequence as the solution structure; we know which residues, and atoms should correspond in the two structures.

The easiest way to compare structures, is the calculate the Root-Mean-Square Deviation (RMSD) after a structural super-positioning (Marti-Renom *et al.*, 2009). The superpositioning is required, because two arbitrary structures will typically not be positioned at coordinates suitable for comparison; first a translation and rotation needs to be applied to one of the two structures, to minimise the RMSD; the resulting RMSD after superpositioning can be used as a dissimilarity measure.

The Root-Mean-Square Deviation (RMSD) calculates the squared difference between two sets of atoms, and can be defined as follows:

$$RMSD(v, w) = \sqrt{\frac{1}{n}\sum_{i=1}^{n} \|v_i - w_i\|^2}$$

$$= \sqrt{\frac{1}{n}\sum_{i=1}^{n} (v_{ix} - w_{ix})^2 + (v_{iz} - w_{iz})^2 + (v_{iz} - w_{iz})^2}$$

Here, v_i is the position vector of ith atom of structure v; w_i is the position vector of ith atom of structure w; and n is the total number of aligned atoms.

The RMSD takes the average over all aligned pairs. In protein structures typically one representative atom per residue is chosen, such as Cα or Cβ.

GDT – Global Distance Test

If a model gets a loop very wrong, it tends to stick out and can be positioned very distant from the true structure, even though the remaining structure may be reasonably accurate. This partial outlier weighs heavily on the average distance calculated. Hence the RMSD is oversensitive to such outliers.

The global distance test total score (GDT_TS) is a more robust structural similarity measure that is well defined given an alignment between two structures. The key idea is to count the number of residues that can maximally be fitted within a certain distance cutoff, see also **Fig. 5**. The GDT score will therefore produce a percentage. In the formula below, the final score is the average over four different distance cutoffs (1, 2, 4, 8 Å).

$$GDT\_TS = \frac{1}{4}\sum_{v=1,2,4,8\text{Å}} \frac{G(v)}{t} \tag{1}$$

Here, $G(v)$ is the number of aligned residues within given RMSD cutoff v (in Ångstrom $- 10^{-10}m$) and t is the total number of aligned residues. A related score called GDT_HA was introduced in CASP some time ago (Read and Chavali, 2007) using stricter distance cutoffs (0.5, 1, 2, 4 Å), to cater for targets in the template-based modelling category where very high accuracies can be realized:

For a typical "difficult" CASP target no model even comes close to the experimentally solved structure; typical results would be similar to the left-most model in **Fig. 5**. If we have a look at the latest CASP results one will see that a performance of GDT_TS < 20% is not an exception. In other words, the protein structure prediction problem has NOT yet been solved, especially not if one considers targets without a good template structure.

How Difficult is it to Predict?

Overall, if one can find a good template, the quality of the predicted model will be relatively good. CASP results show that for homology modelling based on close homologs, it is possible to obtain models similar to the experimentally determined structure (Moult *et al.*, 2016). The modelled structure will typically have a good accuracy for the regions that can be well aligned between the target and template (using the sequences). The top two bars in **Fig. 6** shows that one may expect the majority of such models to be accurate for > 50% of their residues. Gaps in an alignment will typically lie in loop regions of a structure and are more difficult to model. So, if we are interested in a large loop region that is not present in our template, we still may not be able to answer our scientific question with the resulting model structure (Moult *et al.*, 2016).

If no acceptable template can be found, the chances of successfully answering our scientific question will become very low. As a last resort, *ab initio* modelling can provide us with structural models. Typically, *ab initio* methods use very small templates from various proteins (see **Fig. 4**). The state of the art is that on average one may expect to find one structure that looks somewhat like the solution structure for the target among the top five or ten models (Moult *et al.*, 2016). However, be aware that the best model is typically not recognised as being the best through the scores of the prediction program. In **Fig. 6** one sees that very clearly in the bottom few bars: without template, even with predicted contacts, one may have less than 20% of the structure correct in the majority of models; even in the best cases at most 40% of the residues are modelled accurately.

For Which Gene Sequences can we Predict a Three-Dimensional Structure?

If and only if there is a structure of a homologous protein present in the PDB, it is possible to generate a structural model of reasonable accuracy. Based on this notion, we can estimate for which (fraction of) gene sequences it is possible to predict a structure. This way it

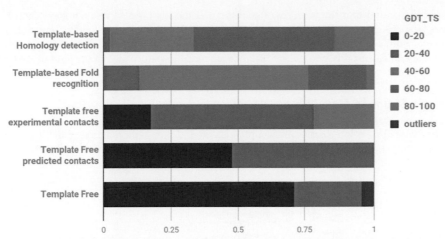

Fig. 6 Distribution of GDT_TS scores for the different model categories in CASP11 for template-based, template-free with contact information and template-free. The legend coloring corresponds to the GDT_TS scores, the bars indicate the fraction of models in each GDT_TS range for the six categories (GDT_TS scores for Modi *et al.* were estimated from the reported GDT_HA scores using their **Fig. 4(A)**). "Outliers" targets have unusually high GDT_TS due to being very short (~50 residue) with extended structures. Targets selected for server prediction (top bar) were considered easier than those for human prediction (second from top), average sequence identity was 26% vs. 20%, respectively. It is clear that overall prediction accuracy sharply declines going down this list of categories. For template-free modelling, the quality of contact information used is crucial. Experimental information (from chemical cross linking or simulated NMR) can give reasonable models. Predicted contacts do not guarantee that an acceptable model can be obtained, but without even predicted contacts, more than two-thirds of models are at most 20% correct. Reproduced from Modi, V., Xu, Q., Adhikari, S., Dunbrack, R.L., 2016. Assessment of template-based modeling of protein structure in CASP11. Proteins: Structure, Function, and Bioinformatics 84 (S1), S200–S220. Kinch, L.N., Li, W., Monastyrskyy, B., Kryshtafovych, A., Grishin, N.V., 2016a. Assessment of CASP11 contact-assisted predictions. Proteins: Structure, Function, and Bioinformatics 84 (S1), S164–S180. Kinch, L. N., Li, W., Monastyrskyy, B., Kryshtafovych, A., Grishin, N.V., 2016b. Evaluation of free modeling targets in CASP11 and ROLL. Proteins: Structure, Function, and Bioinformatics 84 (S1), S51–S66.

has been estimated that for a about 44% of residues in Eukaryotic gene sequences, we cannot yet make a homology model, and 15% of these residues lie within a gene for which we can not make a homology model for a single domain (Perdigão *et al.*, 2015). Especially membrane proteins are underrepresented in the PDB, due to the experimental difficulty of determining these structures. Note that these residues, may also lie in natively disordered regions (see also Section "Is there such a Concept as a Single Native Fold?").

Similarly, it is possible to predict the range of protein structures present in an organism, based on the gene sequences identified in their completed genome. Gene prediction over a large number of organisms has revealed that there is a subset of protein structures that is present in nearly all organisms, for example TIM-barrels or Rossmann-folds (Abeln and Deane, 2005; Edwards *et al.*, 2013). Nevertheless, there is also a group of structures that is extremely lineage specific. It is to be expected that for this group of proteins many new structures remain to be discovered. This also implies that it will remain difficult to find suitable templates for homology modelling for these lineage-specific protein families.

How Accurate do we Need to be?

We already mentioned that we may approach the modelling of a protein structure of interest differently, depending on the biological question we want to ask, e.g., which residues are likely to be crucial for the functioning of the protein. Sometimes an answer to the research question may be possible in a simpler way, without full-scale prediction of the protein structure, e.g., by direct prediction of the impact of certain mutations or of protein-protein interaction sites. Examples of fully-automated webservers that do just that, are HOPE- (Venselaar *et al.*, 2010) and SeRenDIP (Hou *et al.*, 2017). In some cases, a rough homology model inspires the understanding of experimental results, spurring forward the project and eventually ending with crystal structures highlighting the protein function (in this case, protein-protein interactions) of interest (e.g., De Vries-van Leeuwen *et al.*, 2013). Also, specifically for enzymes, such as for example cytochromes P450, modelling of the protein structure should be done in combination with that of the ligand (de Graaf *et al.*, 2005).

In CASP11, three functional aspects, selected on being amenable to qualitative evaluation, were explicitly scored: multimeric state, (small) ligand binding, and mutation impact. Targets were selected that in solved crystal structure were dimeric, or had a ligand bound, or where from the crystallographers or in literature interest was expressed for evaluating mutants (Huwe *et al.*, 2016).

For prediction of dimer structures, only in two out of ten cases a dimer model with reasonable accuracy could be generated for the majority of monomer model structures (Huwe *et al.*, 2016). In the critical assessment of prediction of protein interaction (CAPRI), between 30%–80% of models were of 'acceptable' or 'medium' quality for easy dimer targets, while for harder targets (difficult dimers, multimers and heteromers), this fraction dropped to below 10% (Lensink *et al.*, 2016). Encouragingly, it was seen that also structure models of lower quality could sometimes lead to acceptable or even medium quality models of the bound proteins (Lensink *et al.*, 2016).

For ligand binding, it was found that the accuracy of even the best models (~ 2 Å) was not good enough for accurate ligand docking; the best ligands were around 5 Å RMSD (Huwe *et al.*, 2016). Something similar was found for mutation impact prediction; for most targets, model accuracy did not correlate with accuracy of impact prediction (Huwe *et al.*, 2016). Apparently, either homology models are not yet accurate enough for this purpose, or methods are tuned to particular characteristics of crystal structures.

Template Based Protein Structure Modelling

In the next two sections, we will in detail discuss first template-based and then template-free modelling. An overview of protein structure modelling, including both template-based and template-free modelling is given in **Fig. 7**; see also **Fig. 3** for the terminology used.

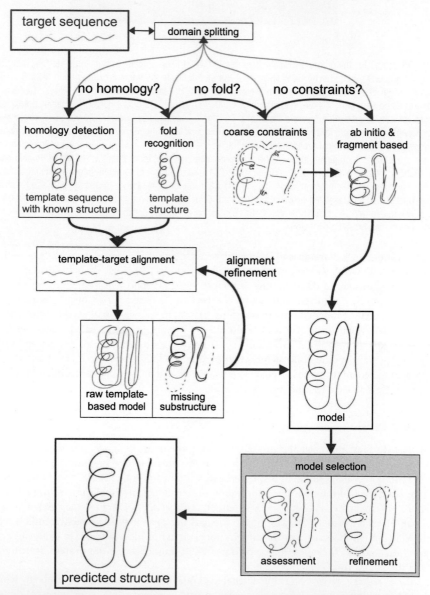

Fig. 7 Flowchart of protein three-dimensional structure prediction. It starts at the top left with the target protein sequence of interest, and ends with a predicted 3D structure at the bottom. Depending on the availability of a homologous template, a suitable fold, or coarse/experimental constraints, different options are available, with sharply decreasing expected model accuracy for each step. Homology-based models (far left) are most accurate, while ab-initio modelling (far right) is notoriously unreliable. Template-based (by homology or fold recognition) models require an alignment with the target sequence, from which the initial model will be built. Course constraints may sometimes be incorporated in this stage. The raw model may need to be completed by separate modelling of the missing substructures. Template-free modelling can benefit greatly if such constraints are available, if not the only sources left are fragment libraries and knowledge-based energy functions. The model, whether template-based or -free, is usually refined until a desired level of (estimated) quality is reached to produce the final predicted structure for our target protein sequence of interest.

Homology Based Template Finding

Homology modelling is a type of template based modelling with a template that is homologous to the target protein. As mentioned before, homology modelling works well, because structure is more conserved than sequence (Bajaj and Blundell, 1984; see **Fig. 2**).Typically, we will use a sequence-based homology detection method, such as BLAST, to search for homologous protein sequences in the full PDB dataset. If we find a sequence that has significant sequence similarity to the full length of our target sequence we have found a template. Of course it is possible that a template only covers part of the target sequence, see also Section "Domains" on domains above.

If a simple BLAST search against the PDB gives no good results we may need to start using alignment methods that can detect more distant homologs based on sequence comparison, such a PSI-BLAST or HMMs. PSI-BLAST uses hits from a previous iteration of BLAST to create a profile for our query sequence (Altschul et al., 1997). This allows successive iterations to give more weight to very conserved residues, allowing to find more distant homologs. Even more sensitive homology detection tools typically consider sequence profiles of both the query and the template sequence, and try to align or score these profiles against each other (Pietrokovski, 1996), with more recent implementations including Compass (Sadreyev and Grishin, 2003), HMMer (Finn et al., 2011) and HHpred (Söding et al., 2005). Profile-profile matching is known to increase detection of evolutionary sequence signals, leading to optimised alignment or sensitive sequence searching (Wang and Dunbrack, 2004). Note that to make full use of such methods it is important that the searching profiles are generated using extended sequence databases (so not just on the PDB). The evolutionary sequence profiles used in these methods can ensure that the most conserved, and therefore structurally most important, residues are more likely to be aligned accurately.

Once we have a good template, and an alignment of the target sequence with the template structure, we can build a structural model; this is called homology modelling in cases of clear homology between the target and the template.

Fold Recognition

If no obvious homologs with a known structure can be found in the PDB, it becomes substantially more difficult to predict the structure for a sequence. We may then use another trick to find a template: remember that the template does not only have a sequence, but also a structure. Therefore, we may be able to use structural information in this search. Fold recognition methods look explicitly for a plausible match between our query sequence and a *structure* from a database. Typically, such methods use structural information of the putative template to determine a match between the target sequence and the putative template sequence and structure.

Note that, in contrast to homology-based template search, it is not strictly necessary for a target sequence and a template sequence to be homologous – they may have obtained similar structures through convergent evolution. Therefore some fold recognition methods are trained and optimised for detection of structural similarity, rather than homology. In fact, for very remote homologs we may not be able to differentiate between convergent and divergent evolution. For the purpose of structure prediction this may not be an important distinction, nevertheless divergent evolution, i.e., homology through a shared common ancestor, invokes the principle of structure being more conserved, giving more confidence in the final model.

Threading is an example of a fold recognition method (Jones et al., 1992). The query sequence is threaded through the proposed template structure. This means structural information of the template can be taken into account when score if the (threaded) alignment between the target and the potential template is a good fit. Typically threading works by scoring the pairs in the structure that make a contact, given the sequence composition. A sequence should also be aligned (threaded) onto the structure giving the best score. Scores are typically based on knowledge based potentials. For example, if two hydrophobic residues are in contact this would give a better score than a contact between a hydrophobic and polar amino acid. Note that threading does not necessarily search for homology. Threading remains a popular fold recognition methods, with several implementations available (Jones et al., 1992; Zhang, 2008; Song et al., 2013).

Another conceptually different fold recognition approach is to consider that amino-acid conservation rates may strongly differ between different structural environments. For example, one would expect residues on the surface to be less conserved, compared to those buried in the core. Similarly, residues in a α-helix or β-strand or typically more conserved than those in loop regions. In fact, the chance to form an insertion or deletion in a loop regions is seven times more likely than within (other) secondary structure elements. Since we know the structural environment of the residues for the potential template, we can use this to score an alignment between the target sequence and potential template sequence. FUGUE is a method that scores alignments using structural environment-specific substitution matrices and structure-dependent gap penalties (Shi et al., 2001).

Generating the Target-Template Alignment

Once a suitable template has been found, one can start building a structural model. Typical model building methods will need the following inputs: (1) the target sequence, (2) the template structure, (3) sequence alignment between target and template and (4) any additional known constraints. The output will be a structural model, based on the constraints defined by the template structure and the sequence alignment.

Here, it is important to note that the methods that recognize good potential templates, are not necessarily the methods that will produce the most accurate alignments between the target and template sequence. As the final model will heavily depend upon the alignment used, it is important to consider different methods, including for example structure and profile based methods, multiple sequence alignment programs or methods that can include structural information of the template in constructing the alignment. Examples of good sequence-based alignment methods, which can also exploit evolutionary signals and profiles, are *Praline* (Simossis *et al.*, 2005) and *T-coffee* (Notredame *et al.*, 2000). The T-coffee suite also includes the structure-aware alignment method *3d-coffee* (O'Sullivan *et al.*, 2004). Some aligners are context-aware, taking into account for example secondary structure (Simossis and Heringa, 2005) or trans-membrane (TM) regions explicitly (Pirovano *et al.*, 2008, 2010; Floden *et al.*, 2016), which are useful extension as TM proteins are severely underrepresented in the PDB. Of particular interest to the general user are automated template detection tools such as HHpred (Söding *et al.*, 2005). Bawono *et al.* (2017) give a good and up-to-date overview of multiple sequence alignment methodology, including profile-based and hidden-Markov-based methods.

Moreover, once a first model is created, it may be wise to interactively adapt the alignment, based on the resulting model, which might lead to an updated model. This procedure may also be carried out in an interactive fashion.

Generating a Model

Here we consider the *MODELLER* software to generate template based alignments (Sali and Blundell, 1993), which is one of several alternative approaches to construct homology models (Schwede *et al.*, 2003; Zhang, 2008; Song *et al.*, 2013). Firstly, the known template 3D structures should be aligned with the target sequence. Secondly, spatial features, such as Cα–Cα distances, hydrogen bonds, and mainchain and sidechain dihedral angles, are transferred from the template(s) to the target. *MODELLER* uses "knowledge based" constraints. The constraints are based on the template distances, the alignment, but also on knowledge- Based energy functions (probability distribution). The constraints are optimised using molecular dynamics with simulated annealing. Finally, a 3D model can be generated by satisfying all the restraints as well as possible.

Loop or Missing Substructure Modelling

We now have a model for all the residues that were aligned well between the template and the target. The remaining substructure(s), that are not covered by the template, will show as gaps in the alignment between target sequence and template. 'Loop modelling' is used to determine the structure for these missing parts. Loop models are typically based on fragment libraries, knowledge based potentials and constraints from the aligned structure. This problem is in fact closely related to the template-free modelling procedure, as we need to generate a structure without a readily available template (template-free approaches are discussed in more detail in the next section). In CASP11, consistent refinement overall as well as for loop regions was achieved; the limiting factor for effective refinement was concluded to be the energie functions used, in particular missing physicochemical effects and balance of energy terms (Lee *et al.*, 2016).

Template-Free Protein Structure Modelling

What if no Suitable Template Exists?

If on the other hand, no suitable template is available for our target protein of interest, we will need to follow a 'template-free' modelling strategy. Without a direct suitable template, we need an *ab initio* strategy that can suggest possible structural models based on the sequence of the template alone. In this case, we need to resort to fragment based approaches. Here small, suitable fragments, from various PDB structure, are assembled to generate possible structural models. As the fragments are typically matched using sequence similarity, one may even consider this as template based modelling at a smaller scale. However, since the sequence match is based on a limited number of residues, this would not generally imply a homologous relation between the fragment template and the target sequence. It is also important to note that this type of *ab initio* modelling is still "knowledge based": the structural models are generated from small substructures present in the PDB, assessed by energy scoring functions generated by mining the PDB. In other words, these model are not based on physical principles and physico-chemical properties alone. This also means, that such models are likely to share any of the biases that are present in the PDB, such as lack of trans-membrane proteins and absence of disordered regions.

Generating Models From Structural Fragments

Here we follow the keys ideas in the *Rosetta* based *ab initio* modelling suite (Simons *et al.*, 1999). Quark, another fragment-based approach provides a similar performing alternative (Xu and Zhang, 2012). The overall approach is to split the target sequence into 3 and 9-residue overlapping sequence fragments, i.e., sliding windows, and find matching structural fragments from PDB.

A fragment library is generated by taking 3 and 9-residue fragments from the PDB and clustering these together into groups of similar structure. For each fragment in the database sequence profiles are created. These profiles subsequently are used to search for suitable fragments for our query/target sequence, as we only have sequence information (see **Fig. 8** on the left-hand side).

The target sequence also will be split into 3 and 9-residue overlapping sequence fragments. These target sequence fragments are then matched with the structural PDB based fragment library using profile-profile matching. Note that this procedure will generate multiple fragments for each fragment window in the sequence. The fragment windows on the sequence typically also overlap (see **Fig. 8** middle panel).

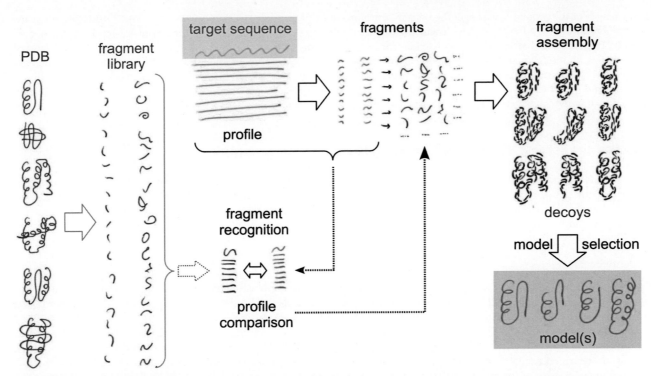

Fig. 8 Overview of the fragment-based modelling strategy. A library of structure fragments was created once from the PDB; all small 3-residues and larger 9-residue fragments are collected and clustered. A target sequence of interest is also separated into 3- and 9-residue sequence fragments. For each of these, a profile-profile search is performed to find matching fragments from the fragment library; typically for each target fragment, multiple hits with different structure are retrieved. This collection of fragments of alternate structure are then assembled through a Monte Carlo algorithm into a large set of possible structures, called 'decoys'. Using knowledge-based potentials and overall statistics, from the decoy set, a final selection of model structures is made.

Fragment Assembly Into Decoys

The above scenario leads to many possible structural fragments to cover a sequence position. To find the best structural model is a combinatorial problem in terms of fragment combinations, see **Fig. 8** top right. A Monte Carlo algorithm is used to search through different fragment combinations. Good combinations are those that give a low energy. Each MC run will produce a different model, since it is a stochastic algorithm. Note that to be able to optimise a model, we need a scoring function. A knowledge based energy function is used, including the number of neighbours, given amino acid type, residue pair interactions, backbone hydrogen bonding, strand arrangement, helix packing, radius of gyration, Van der Waals repulsion. These are all terms that are relatively cheap to compute when a new combination of fragments is tried. The structural model is optimised by slowly lowering the MC temperature – this is also called simulated annealing. The generated models are called 'decoys'.

Thereafter, decoys are refined using additional Monte Carlo cycles, and a more fine-grained energy function including: back-bone torsion angles, Lennard-Jones interactions, main chain and side-chain hydrogen bonding, solvation energy, rotamers and a comparison to unfolded state.

Finally, the most difficult task is to select, from all the refined decoys, a structure that is a suitable model for the target sequence, see **Fig. 8** bottom right. Again, using a more detailed, knowledge-based energy function, decoys can be scored to assess how 'protein-like' they are. Such a selection procedure may get rid of very wrong models. However, selecting the best model, without any additional information (from for example experiments or co-evolution-based contact-prediction), is likely to lead to poor results, as is shown in **Fig. 6**.

Constraints From Co-Evolution Based Contact Prediction or Experiments

As already mentioned, valuable additions to the modelling process are coarse constraints from experimental data or contact prediction. Experimental data from NMR and chemical cross-linking can yield distance restraints that are particularly useful in the template-free modelling to narrow down the conformational space to be searched; still average accuracy of models produced remains extremely limited, (see **Fig. 6**; Kinch *et al.*, 2016a). Other sources of information are contours or surfaces that can be obtained from cryo-EM, or small angle scattering experiments, either with electrons, neutrons, or x-ray radiation. However, since these techniques are employed mostly for elucidating larger macromolecular complexes, they are considered out of scope for the current article.

Of more general applicability may be methods for predicting intra-protein residue contacts; the main approach currently is based on some form of co-evolution information obtained by direct-coupling methods from 'deep' alignments (Marks *et al.*, 2011; Jones *et al.*, 2012; Morcos *et al.*, 2011). Depth here signifies the amount of sequence variation present in the alignment in relation to the length of the protein (the longer the protein, the more variation is needed). Ovchinnikov *et al.* (2017) expresses this as the *effective protein length*: $Nf = N80\%ID/\sqrt{l}$ where l is the protein length, and $N80\%ID$ the number of cluster at 80% sequence identity. They showed that Nf can be greatly enhanced by the use of metagenomic sequencing data, and that this leads to a marked improvement in model quality, and estimate that this would triple the number of protein families for which the correct fold might be predicted (Ovchinnikov *et al.*, 2017). Wuyun *et al.* (2016) investigated 'consensus'-based methods, which combine both direct-coupling and machine-learning approaches, and find that the machine-learning methods are less sensitive to alignment depth and target difficulty, which are crucial factors for success for the direct-coupling methods.

Selecting and Refining Models From Structure Prediction

Once we have created (several) models, we need to assess which model is the best one. Typically this can be done by scoring models on several properties using model quality assessment programs and visual inspection with respect to "protein like" features. Moreover, if any additional knowledge about the structure or function of our target protein is available, this may also help to assess the quality of the model(s). In addition, one may in some cases want to improve a model, or parts of it; this is called model refinement.

Model Refinement

For many years in CASP, model refinement was a no-go area; the rule of thumb was: build our homology model and do not touch it! An impressive example of the failure of refinement methods was shown by the David Shaw group, who concluded that "simulations initiated from homology models drift away from the native structure" (Raval *et al.*, 2012). Since CASP10 in 2014 (Nugent *et al.*, 2014) and continuing in the latest CASP11, there is reason for moderate optimism. General refinement strategies report small but significant improvements of 3%–5% over 70% of models (Modi and Dunbrack, 2016). Interestingly, and in stark contrast to the earlier results by Raval *et al.* (2012), the average improvement of GDT_HA using simulation-based refinement now also is about 3.8, with an improvement (more than 0.5) for 26 models (Feig and Mirjalili, 2016). For five models, the scores became worse (by more than 0.5), and another five showed no significant change. Particularly, for very good initial models (GDT_HA > 65), models were made worse. Moreover, they also convincingly showed that both more and longer simulations consistently improved these results; note however that protocol details such as using $C\alpha$ restraints, are thought to be the limiting factor (Feig and Mirjalili, 2016), as already used previously (e.g., Keizers *et al.*, 2005; Feenstra *et al.*, 2006), and replicated by others (e.g., Cheng *et al.*, 2017). Most successful refinement appears to come from correctly placing //-sheet or coil regions at the termini (Modi and Dunbrack, 2016).

Model Quality Assessment Strategies

It may be generally helpful to compare models generated by different prediction methods; if models from different methods look alike (more precisely if the pair has a low RMSD and/or high GDT_TS) they are more likely to be correct. Similarly, templates found by different template finding strategies, found for example both by homology sequence searches and fold recognition methods are more likely to yield good modelling results (Moult *et al.*, 2016; Kryshtafovych *et al.*, 2016). Such a consensus template is generally more reliable than the predictions from individual methods – especially if the individual scores are barely significant. Lastly, one can consider biological context to select good models.

Whether a model is built using homology modelling, fold recognition and modelling or *ab initio* prediction, all models can be given to a Model Quality Assessment Program (MQAP) for model validation. A validation program provides a score predicting how reliable the model is. These scores typically take into account to what extent a model resembles a "true" protein structure. The best performing validation programs take a large set of predicted models, and indicate which out of these is expected to be the most reliable.

Validation scoring may be based on similar ideas as validation for experimental structures or may be specific to structure prediction. For example, it can be checked if the amount of secondary structure, e.g., helix and strand vs. loop, has a similar ratio as in known protein structures; if a model for a sequence of 200 amino acids does not contain a single helix or β-strand, the model does not resemble true protein structures, and is therefore very unlikely to be the true structural solution for the sequence. A similar type of check may be done for the amount of buried hydrophobic groups and globularity of the protein.

Different models may also be compared to each other. One trick that is commonly used, is that if multiple prediction methods create structurally similar models, these models are more likely to be correct. Hence, a good prediction strategy is to use several prediction methods, and pick out the most consistent solution. A pitfall here is that if all models are based on the same, or very similar, templates, they will look similar but this may not indicate the likelihood of them being correct.

Secondary Structure Prediction

Secondary structure prediction is relatively accurate, and is in fact much easier to solve than three-dimensional structure prediction, see, e.g., the review by Pirovano and Heringa (2010). The accuracy of assigning strand, helix or loops to a certain residue can go up to 80% with the most reliable methods. Typically such methods use (hydrophobic) periodicity in the sequence combined with phi and psi angle preferences of certain amino acid types to come to accurate predictions. The real challenge lies in assembling the secondary structure element in a correct topology. Nevertheless, secondary structure prediction may be used to assess the quality of a model built with a (tertiary) structure prediction method. Many (automated) methods also incorporate secondary structure information during (Simossis and Heringa, 2005), homology detection (Söding *et al.*, 2005; Shi *et al.*, 2001) and contact prediction (Terashi and Kihara, 2017; Wang *et al.*, 2017).

Is There Such a Concept as a Single Native Fold?

Before we conclude, we should consider a more physical description of protein structure. In fact, protein folding from a physical point of view is a very interesting process: given a sequence, a protein tends to fold always, and exactly into the same functional structure. In material design, it is extremely difficult to mimic such high specificity. The apparent observation of folding specificity also leads to the question, is there such a concept as a single native fold? Or, more pragmatically, is sequence-to-structure truly a one-to-one relation?

In fact, if one wants to start making quantitative predictions, such as the stability of a protein fold, or the binding strength between two proteins in terms of free energy, it is much more helpful to think in ensembles of structural configurations for a protein sequence (e.g., May *et al.*, 2014; Pucci *et al.*, 2017). The probability to find a protein in a specific ensemble of structural configurations will depend on conditions such as the presence or absence of binding partners, the pressure, the pH or the temperature (e.g., van Dijk *et al.*, 2015, 2016). There are a few specific cases, common cases, for which even the functional or biologically relevant structural ensembles do not resemble a single globular folded structure.

Disordered Proteins

Not all proteins fold into single configurations, some proteins stay natively unfolded, i.e., they can take up a large variety of more extended, and very different configurations (Uversky *et al.*, 2000; Mészáros *et al.*, 2007). Some disordered regions contain elements that do form stable structures upon binding. The regions that remain disordered are thought to be important to prevent aggregation within the cell (Abeln and Frenkel, 2008). Missing residues in X-ray structures are typically removed for crystallization; for this reason disorder prediction methods have been developed. Disordered regions are relatively easy to predict in protein sequences just like secondary structures; broadly speaking, prediction can be based on the large amount of charged/polar (hydrophilic) amino acids in combination with the presence of amino acids that disrupt the secondary structure (proline and glycine) in these regions (Oldfield *et al.*, 2005; Wang *et al.*, 2016). We know sequences of many proteins contain large disordered segments (33% of eukaryotic, 2% archaeal, and 4% bacterial proteins; Ward *et al.* (2004)).

Allostery and Functional Structural Ensembles

It is important to realize that one protein, typically, does not correspond to one defined three-dimensional structure. Disordered regions or proteins are one particularly salient case, but also proteins which fold into specific three-dimensional configurations, may exist in multiple functional states each with a specific structure. The biological question of interest dictates which state is relevant. Most proteins have only been crystallized in one particular state, and often it is not known to which biological condition this crystal structure may correspond. One may have cases where a homology model of the relevant state may be preferred over a crystal structure of a different or unknown state (e.g., de Graaf *et al.*, 2005).

Amyloid Fibrils

Lastly, we should consider a competing state of folded proteins: the aggregated state, where multiple peptide chains clog together in fibrillar structures or amorphous aggregates. Amyloid fibres are formed by β-strands formed between different protein or peptide (small protein) chains. Fibril formation is associated with various neurodegenerative diseases, such as Alzheimer's, Creutzfeldt-Jakob and Parkinson's (Chiti and Dobson, 2006). In fact, the fibrillar state is more favorable than the state of separately folded structures for several protein types. The general cellular toxicity of such aggregates, puts evolutionary pressure on avoiding structural characteristics on the surface of proteins; hence it is extremely rare to observe solvent accessible β-strand edges or large hydrophobic surface patches (Richardson and Richardson, 2002; Abeln and Frenkel, 2011). The propensity proteins have to form Amyloid fibrils is relatively easy to predict (Graña-Montes *et al.*, 2017). However, reference databases are still small so it is difficult to verify such methods.

Acknowledgements

We thank Nicola Bonzanni, Kamil K. Belau, Ashley Gallagher and Jochem Bijlard for insightful discussions and critical proof-reading of early versions.

See also: *Ab initio* Protein Structure Prediction. Algorithms for Structure Comparison and Analysis: Homology Modelling of Proteins. Algorithms for Structure Comparison and Analysis: Prediction of Tertiary Structures of Proteins. Assessment of Structure Quality (RNA and Protein). Biomolecular Structures: Prediction, Identification and Analyses. Cloud-Based Molecular Modeling Systems. Drug Repurposing and Multi-Target Therapies. In Silico Identification of Novel Inhibitors. Natural Language Processing Approaches in Bioinformatics. Pharmacophore Development. Population-Based Sampling and Fragment-Based De Novo Protein Structure Prediction. Protein Structural Bioinformatics: An Overview. Protein Structure Analysis and Validation. Protein Structure Classification. Protein Structure Databases. Protocol for Protein Structure Modelling. Rational Structure-Based Drug Design. Secondary Structure Prediction. Small Molecule Drug Design. Structural Genomics. Structure-Based Design of Peptide Inhibitors for Protein Arginine Deiminase Type IV (PAD4). Structure-Based Drug Design Workflow. Transmembrane Domain Prediction

References

Abeln, S., Deane, C.M., 2005. Fold usage on genomes and protein fold evolution. Proteins 60 (4), 690–700.

Abeln, S., Frenkel, D., 2008. Disordered flanks prevent peptide aggregation. PLOS Computational Biology 4 (12), e1000241.

Abeln, S., Frenkel, D., 2011. Accounting for protein-solvent contacts facilitates design of nonaggregating lattice proteins. Biophysical Journal 100 (3), 693–700.

Altschul, S.F., Madden, T.L., Schäffer, A.A., *et al.*, 1997. Gapped BLAST and PSI-BLAST: A new generation of protein database search programs. Nucleic Acids Research 25 (17), 3389–3402.

Bajaj, M., Blundell, T., 1984. Evolution and the tertiary structure of proteins. Annual Review of Biophysics and Bioengineering 13 (1), 453–492.

Bawono, P., Dijkstra, M., Pirovano, W., *et al.* 2017. Multiple Sequence Alignment. In: Methods in Molecular Biology – Bioinformatics – Volume I: Data, Sequence Analysis, and Evolution. New York: Humana Press, pp. 167–189.

Berman, H.M., Westbrook, J., Feng, Z., *et al.*, 2000. The protein data bank. Nucleic Acids Research 28 (1), 235–242.

Cheng, Q., Joung, I., Lee, J., 2017. A simple and efficient protein structure refinement method. Journal of Chemical Theory and Computation 13 (10), 5146–5162.

Chiti, F., Dobson, C.M., 2006. Protein misfolding, functional amyloid, and human disease. Annual Review of Biochemistry 75, 333–366.

de Graaf, C., Vermeulen, N.P.E., Feenstra, K.A., 2005. Cytochrome P450 in silico: An integrative modeling approach. Journal of Medicinal Chemistry 48 (8), 2725–2755.

De Vries-van Leeuwen, I.J., da Costa Pereira, D., Flach, K.D., *et al.*, 2013. Interaction of 14-3-3 proteins with the estrogen receptor alpha F domain provides a drug target interface. Proceedings of the National Academy of Sciences of the United States of America 110 (22), 8894–8899.

Edwards, H., Abeln, S., Deane, C.M., 2013. Exploring fold space preferences of new-born and ancient protein superfamilies. PLOS Computational Biology 9 (11), e1003325.

Feenstra, K.A., Hofstetter, K., Bosch, R., *et al.*, 2006. Enantioselective substrate binding in a monooxygenase protein model by molecular dynamics and docking. Biophysical Journal 91 (9), 3206–3216.

Feig, M., Mirjalili, V., 2016. Protein structure refinement via molecular-dynamics simulations: What works and what does not? Proteins: Structure, Function, and Bioinformatics 84 (S1), S282–S292.

Finn, R.D., Clements, J., Eddy, S.R., 2011. HMMER web server: Interactive sequence similarity searching. Nucleic Acids Researh 39 (Suppl. 2), W29–W37.

Floden, E.W., Tommaso, P.D., Chatzou, M., *et al.*, 2016. PSI/TM-Coffee: A web server for fast and accurate multiple sequence alignments of regular and transmembrane proteins using homology extension on reduced databases. Nucleic Acids Research 44 (W1), W339–W343.

Graña-Montes, R., Pujols-Pujol, J., Gómez-Picanyol, C., Ventura, S. 2017. Prediction of protein aggregation and amyloid formation. In: From Protein Structure to Function with Bioinformatics. Dordrecht: Springer, pp. 205–263.

Hou, Q., De Geest, P., Vranken, W., Heringa, J., Feenstra, K., 2017. Seeing the trees through the forest: Sequencebased homo- and heteromeric protein-protein interaction sites prediction using random forest. Bioinformatics 33 (10),

Huwe, P.J., Xu, Q., Shapovalov, M.V., *et al.*, 2016. Biological function derived from predicted structures in CASP11. Proteins: Structure, Function, and Bioinformatics 84 (S1), 370–391.

Jones, D.T., Buchan, D.W.A., Cozzetto, D., Pontil, M., 2012. PSICOV: Precise structural contact prediction using sparse inverse covariance estimation on large multiple sequence alignments. Bioinformatics 28 (2), 184–190.

Jones, D.T., Taylor, W.R., Thornton, J.M., 1992. A new approach to protein fold recognition. Nature 358 (6381), 86–89.

Keizers, P.H.J., de Graaf, C., de Kanter, F.J.J., *et al.*, 2005. Metabolic Regio- and Stereoselectivity of Cytochrome P450 2D6 towards 3,4-Methylenedioxy-N-alkylamphetamines: In silico predictions and experimental validation. Journal of Medicinal Chemistry 48 (19), 6117–6127.

Kinch, L.N., Li, W., Monastyrskyy, B., Kryshtafovych, A., Grishin, N.V., 2016a. Assessment of CASP11 contact-assisted predictions. Proteins: Structure, Function, and Bioinformatics 84 (S1), S164–S180.

Kryshtafovych, A., Barbato, A., Monastyrskyy, B., *et al.*, 2016. Methods of model accuracy estimation can help selecting the best models from decoy sets: Assessment of model accuracy estimations in CASP11. Proteins: Structure, Function, and Bioinformatics 84 (S1), S349–S369.

Lee, G.R., Heo, L., Seok, C., 2016. Effective protein model structure refinement by loop modeling and overall relaxation. Proteins: Structure, Function, and Bioinformatics 84 (S1), S293–S301.

Lensink, M.F., Velankar, S., Kryshtafovych, A., *et al.*, 2016. Prediction of homoprotein and heteroprotein complexes by protein docking and template-based modeling: A CASP-CAPRI experiment. Proteins: Structure, Function, and Bioinformatics 84 (S1), S323–S348.

Marks, D.S., Colwell, L.J., Sheridan, R., *et al.*, 2011. Protein 3D structure computed from evolutionary sequence variation. PLOS ONE 6 (12), e28766.

Marti-Renom, M.A., Capriotti, E., Shindyalov, I.N., Bourne, P.E., 2009. Structure comparison and alignment. In: Gu, J., Bourne, P.E. (Eds.), Structural Bioinformatics, second ed. John Wiley & Sons, Inc., pp. 397–418.

May, A., Pool, R., van Dijk, E., *et al.*, 2014. Coarse-grained versus atomistic simulations: Realistic interaction free energies for real proteins. Bioinformatics 30 (3), 326–334.

Mészáros, B., Tompa, P., Simon, I., Dosztányi, Z., 2007. Molecular principles of the interactions of disordered proteins. Journal of Molecular Biology 372 (2), 549–561.

Modi, V., Dunbrack, R.L., 2016. Assessment of refinement of template-based models in CASP11. Proteins: Structure, Function, and Bioinformatics 84 (S1), S260–S281.

Morcos, F., Pagnani, A., Lunt, B., *et al.*, 2011. Direct-coupling analysis of residue coevolution captures native contacts across many protein families. Proceedings of the National Academy of Sciences of the United States of America 108 (49), E1293–E1301.

Moult, J., Fidelis, K., Kryshtafovych, A., Schwede, T., Tramontano, A., 2016. Critical assessment of methods of protein structure prediction: Progress and new directions in round XI. Proteins: Structure, Function and Bioinformatics 84 (S1), S4–S14.

Moult, J., Pedersen, J.T., Judson, R., Fidelis, K., 1995. A large-scale experiment to assess protein structure prediction methods. Proteins: Structure, Function, and Genetics 23 (3), ii–iv.

Notredame, C., Higgins, D.G., Heringa, J., 2000. T-coffee: A novel method for fast and accurate multiple sequence alignment. Journal of Molecular Biology 302 (1), 205–217.

Nugent, T., Cozzetto, D., Jones, D.T., 2014. Evaluation of predictions in the CASP10 model refinement category. Proteins: Structure, Function, and Bioinformatics 82 (S2), S98–S111.

O'Sullivan, O., Suhre, K., Abergel, C., Higgins, D.G., Notredame, C., 2004. 3DCoffee: Combining protein sequences and structures within multiple sequence alignments. Journal of Molecular Biology 340 (2), 385–395.

Oldfield, C.J., Cheng, Y., Cortese, M.S., *et al.*, 2005. Comparing and combining predictors of mostly disordered proteins. Biochemistry 44 (6), 1989–2000.

Ovchinnikov, S., Park, H., Varghese, N., *et al.*, 2017. Protein structure determination using metagenome sequence data. Science 355 (6322), 294–298.

Perdigão, N., Heinrich, J., Stolte, C., *et al.*, 2015. Unexpected features of the dark proteome. Proceedings of the National Academy of Sciences 112 (52), 15898–15903.

Pietrokovski, S., 1996. Searching databases of conserved sequence regions by aligning protein multiple-alignments. Nucleic Acids Research 24 (19), 3836–3845.

Pirovano, W., Abeln, S., Feenstra, K. A., Heringa, J., 2010. Multiple alignment of transmembrane protein sequences. In: Structural Bioinformatics of Membrane Proteins, Vienna: Springer, pp. 103–122

Pirovano, W., Feenstra, K.A., Heringa, J., 2008. PRALINETM: A strategy for improved multiple alignment of transmembrane proteins. Bioinformatics 24 (4), 492–497.

Pirovano, W., Heringa, J., 2010. Protein secondary structure prediction. In: Carugo, O., Eisenhaber, F. (Eds.), Data Mining Techniques for the Life Sciences. Humana Press, pp. 327–348.

Pucci, F., Kwasigroch, J.M., Rooman, M., 2017. SCooP: An accurate and fast predictor of protein stability curves as a function of temperature. Bioinformatics.

Raval, A., Piana, S., Eastwood, M.P., Dror, R.O., Shaw, D.E., 2012. Refinement of protein structure homology models via long, all-atom molecular dynamics simulations. Proteins: Structure, Function, and Bioinformatics 80 (8), 2071–2079.

Read, R.J., Chavali, G., 2007. Assessment of CASP7 predictions in the high accuracy template-based modeling category. Proteins: Structure, Function, and Bioinformatics 69 (S8), S27–S37.

Richardson, J.S., Richardson, D.C., 2002. Natural beta-sheet proteins use negative design to avoid edge-to-edge aggregation. Proceedings of the National Academy of Sciences of the United States of America 99 (5), 2754–2759.

Sadreyev, R., Grishin, N., 2003. COMPASS: A tool for comparison of multiple protein alignments with assessment of statistical significance. Journal of Molecular Biology 326 (1), 317–336.

Sali, A., Blundell, T.L., 1993. Comparative protein modelling by satisfaction of spatial restraints. Journal of Molecular Biology 234 (3), 779–815.

Schwede, T., Kopp, J., Guex, N., Peitsch, M.C., 2003. SWISS-MODEL: An automated protein homology-modeling server. Nucleic Acids Research 31 (13), 3381–3385.

Shi, J., Blundell, T.L., Mizuguchi, K., 2001. FUGUE: Sequence-structure homology recognition using environment-specific substitution tables and structure-dependent gap penalties. Journal Molecular Biology 310 (1), 243–257.

Simons, K.T., Bonneau, R., Ruczinski, I., Baker, D., 1999. Ab initio protein structure prediction of CASP III targets using ROSETTA. Proteins. Suppl. 3), S171–S176.

Simossis, V.A., Heringa, J., 2005. PRALINE: A multiple sequence alignment toolbox that integrates homology-extended and secondary structure information. Nucleic Acids Research 33 (Web Server), W289–W294.

Simossis, V.A., Kleinjung, J., Heringa, J., 2005. Homology-extended sequence alignment. Nucleic Acids Research 33 (3), 816–824.

Söding, J., Biegert, A., Lupas, A.N., 2005. The HHpred interactive server for protein homology detection and structure prediction. Nucleic Acids Research 33 (Web Server issue), W244–W248.

Song, Y., Dimaio, F., Wang, R.Y.R., *et al.*, 2013. High-resolution comparative modeling with RosettaCM. Structure 21 (10), 1735–1742.

Terashi, G., Kihara, D., 2017. Protein structure model refinement in CASP12 using short and long molecular dynamics simulations in implicit solvent. Proteins: Structure, Function and Bioinformatics 86 (Suppl. 1), S189–S201.

Uversky, V.N., Gillespie, J.R., Fink, A.L., 2000. Why are "natively unfolded" proteins unstructured under physiologic conditions? Proteins 41 (3), 415–427.

van Dijk, E., Hoogeveen, A., Abeln, S., 2015. The hydrophobic temperature dependence of amino acids directly calculated from protein structures. PLOS Computational Biology 11 (5), e1004277.

van Dijk, E., Varilly, P., Knowles, T.P.J., Frenkel, D., Abeln, S., 2016. Consistent treatment of hydrophobicity in protein lattice models accounts for cold denaturation. Physical Review Letters 116 (7), 078101.

Venselaar, H., te Beek, T.A., Kuipers, R.K., Hekkelman, M.L., Vriend, G., 2010. Protein structure analysis of mutations causing inheritable diseases. An e-Science approach with life scientist friendly interfaces. BMC Bioinformatics 11 (1), 548.

Wang, G., Dunbrack, R.L., 2004. Scoring profile-to-profile sequence alignments. Protein Sciences 13 (6), 1612–1626.

Wang, S., Ma, J., Xu, J., 2016. AUCpreD: Proteome-level protein disorder prediction by AUC-maximized deep convolutional neural fields. Bioinformatics 32 (17), i672–i679.

Wang, S., Sun, S., Xu, J., 2017. Analysis of deep learning methods for blind protein contact prediction in CASP12. Proteins 86 (Suppl.1), S67–S77.

Ward, J.J., Sodhi, J.S., McGuffin, L.J., *et al.*, 2004. Prediction and functional analysis of native disorder in proteins from the three kingdoms of life. Journal of Molecular Biology 337, 635645.

Wuyun, Q., Zheng, W., Peng, Z., Yang, J., 2016. A large-scale comparative assessment of methods for residue-residue contact prediction. Briefings in Bioinformatics. bbw106.

Xu, D., Zhang, Y., 2012. Ab initio protein structure assembly using continuous structure fragments and optimized knowledge-based force field. Proteins: Structure, Function and Bioinformatics 80 (7), 1715–1735.

Zhang, Y., 2008. I-TASSER server for protein 3D structure prediction. BMC Bioinformatics 9 (1), 40.

Relevant Websites

http://predictioncenter.org/
Prediction Center.
https://www.rcsb.org/pdb/statistics/holdings.do
RCSB PDB - Holdings Report.
https://www.rcsb.org/pdb/statistics/contentGrowthChart.do?content=total&seqid=100
Yearly Growth of Total Structures.

Protein Structure Analysis and Validation

Tsuyoshi Shirai, Nagahama Institute of Bio-Science and Technology, Nagahama, Japan

Glossary

Clashscore Representative score for validating the quality of a molecular structural model, that is, the number of too-close atom–atom contacts per 1000 atoms in the model.

Density map Electron density maps and EM maps are maps of a material's density; they are direct outputs of X-ray crystallography and electron microscopy, respectively, and models of molecular structures are fitted into them.

Distance constraint Nuclear magnetic resonance spectroscopy generates a list of atom–atom distances in a molecule, which are used to determine the molecule's conformation.

Electron microscopy Method to determine a structure by observing molecular images with an electron microscope and projecting a density map of the molecules.

Nuclear magnetic resonance spectroscopy Method to determine a structure by irradiating radioisotope-labeled samples with microwaves and evaluating interatom distances and bond angles constraints.

Ramachandran plot Two-dimensional plot for validating a protein main-chain conformation in which main-chain torsion ϕ angles are plotted against ψ angles, and the plot area is divided into allowed and disallowed regions.

RMSZ score Representative score for validating the quality of a molecular structural model by the Z-score deviations of atom bond lengths and angles of models from canonical values of reference structures.

X-ray crystallography Method to determine a structure by measuring X-ray diffractions from crystals and synthesizing an electron density map of molecules.

Introduction

Proteins are the major working biological molecules in the cell. Proteins fold into three-dimensional (3D or tertiary) structures under natural conditions. Knowing the 3D structures is required to understand the chemical and physical bases of their functions. It is also important that most biological molecules work in collaboration by forming complexes (supramolecules) of subunits including proteins, nucleic acids, carbohydrates, lipids, and small molecules. Recently, analysis of supramolecules is increasingly important in structural biology, and many innovative improvements have been made in the methods of structural analyses and validations for application to larger supramolecules.

Experimental Methods to Determine Structure

There are three major experimental methods to determine the structure of biological molecules: X-ray crystallography (XRC or often referred to as XP, which stands for protein crystallography) (Drenth and Mesters, 2007), nuclear magnetic resonance spectroscopy (NMR) (Wuthrich, 1986), and electron microscopy (EM) (Frank, 2006). Each method has both advantages and disadvantages (**Table 1**). XP and NMR have been used so far mainly to determine structures at atomic resolution. XP has a much higher resolution and limit of molecular weight than NMR, and has been the most utilized method. However, XP first requires preparation of crystals of molecules, which is often difficult. In contrast, NMR does not need crystals, and structures in solution, which are probably closer to their native structures, can be determined. EM is, to date, lower in resolution in comparison to the other methods, but it has a very high molecular weight limit and can be used to determine supramolecular structures. Thus, these methods are often applied collaboratively, and such approaches are called integrative/correlative structure analyses.

X-ray Crystallography

XP has been the most used method to date, and more than 90% of structures in the Protein Data Bank (PDB) were determined by this method (Berman *et al.*, 2000). Biological molecules like proteins in 1–30 mg/mL concentrations produce an aggregate (precipitate) of regularly assembled molecules, that is, crystals, under appropriate solution conditions. A unit cell, which contains defined number of molecules in defined structures, is a virtual building block of crystals (Drenth and Mesters, 2007). The form of a crystal is defined by the axis dimensions (a, b, c), angles between axes (α, β, γ), and symmetry (space groups) of the unit cell. The unit cell is further divided into asymmetric units (ASU), which is the minimum set of atoms that must be determined independently. The unit cell can be reproduced by applying symmetry operations defined by the space group to the ASU, and a whole crystal is constructed by simply compiling unit cells.

Crystals are exposed to an X-ray beam of uniform wavelength and high coherency from a synchrotron or an X-ray generator for diffraction experiments (**Fig. 1(a)**). An X-ray is light with a wavelength 0.1–100 Å (0.01–10 nm). The electrons in a crystal scatter X-rays, and the scattered X-rays, which have delayed phases depending on the locations of the scatter, interfere with each other.

Encyclopedia of Bioinformatics and Computational Biology, Volume 2   doi:10.1016/B978-0-12-809633-8.20282-3

Table 1   Brief summary of protein structure analysis and validation methods

Method	X-ray crystallography (XP/XRC)	Nuclear magnetic resonance spectroscopy (NMR)	Electron microscopy (EM)
Sample form	Crystal	Solution/powder	Stained/frozen solution on mesh
Probe	X-ray	Microwave	Electron
Direct data	Diffraction intensities	Spectrograms	Particle images
Resolution	<1.0–4.0 Å	~2.5 Å equivalent (to XP)	10–20 Å (2–4 Å in cryo-EM)
Model constraints	Electron density map	Distance/angle constraints	EM density map
Model characteristic (see also **Fig. 2**)	Atomic model without hydrogens	Multiple atomic models with hydrogens	Subunits fit in map
Specific validation metrics of data	Resolution, completeness, $<I/\sigma I>$, R_{sym}	Completeness of signal assignment, number of constraints	Resolution, number of particle images
Specific validation metrics of model	R-factor (R), free R-factor (R_{free})	Average numbers of distance/angle constraint violations, largest distance/angle violations	Map-model cross correlation
Advantage	Atomic resolution	Dynamic structure	Large supramolecule
Disadvantage	Crystallization required (partly fixed in XFEL)	Low MW limit	Low resolution (partly fixed in cryo-EM)

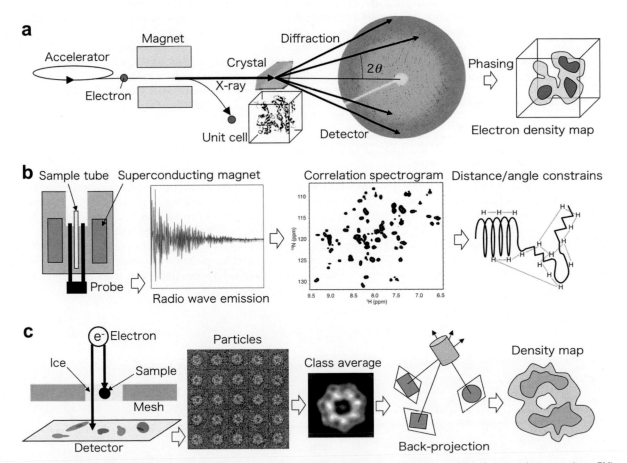

Fig. 1   Schematic processes of (a) X-ray crystallography, (b) nuclear magnetic resonance spectroscopy, and (c) electron microscopy (cryo-EM).

Therefore, scattered X-rays yield information about the locations of electrons/atoms. Scattered X-rays are very weak and usually cancel each other out. Due to the regular structure of crystals, however, scattered X-rays are not muted in the directions where the phase delay between crystal unit cells is equal to an integer multiple of the wavelength, and such outputs, observed as X-ray spots from the crystal, are called diffractions. Thus, crystals are necessary for XP as "amplifiers" of scattered X-rays.

Resolution is the smallest distance between imaginary mirror planes defined by the unit cell lattice, which diffract (reflect) the incident X-ray, and therefore is the important parameter in judging how finely the structures are measured. The smaller the value, the higher the resolution, and the resolution is typically from 1.0 (very high) to 3.0 Å (low) for protein crystals. Each diffraction spot is indexed with a Miller index (h, k, l), which defines the mirror plane reflecting the corresponding diffraction (mirror plane is defined by three points a/h, b/k, and c/l, where a, b, and c are unit cell axes).

Diffractions are the Fourier transform of electron density in the unit cell. Their intensities are recorded by diffractometers, such as an imaging plate or digital camera. The intensities of diffraction are integrated from the diffraction images. The programs HKL2000 and iMosflm are often used for this process (Battye *et al.*, 2011; Otwinowski and Minor, 1997). The electron density map in a unit cell is retrievable as the inverse Fourier transform of the diffractions. The phase information, however, cannot be recorded by diffractometers, and must be recovered by other means. This is called the phasing problem of XP. Multiple isomorphous replacement (MIR) and anomalous scattering (AS) are experimental methods for recovering phases, in which scatterings from a few heavy atoms introduced into the crystals are overlaid in diffractions in order to deduce phases from the amplitude differences. Molecular replacement (MR) is an alternative computational method, in which calculated phases from homologous protein structures (placed appropriately in the unit cell) are used instead of experimental phases.

Once the electron density is successfully synthesized, an atomic model of the molecule is built so that it fits the density as far as possible, usually half-manually using a computer graphics program (**Fig. 2(a)**). This process of adjusting the model (set of atomic coordinates) in order to explain diffractions is called refinement, and CCP4, X-PLOR-NIH, SHELEX, and PHENIX program suites are frequently used for this purpose (Adams *et al.*, 2010; Schwieters *et al.*, 2003; Sheldrick, 2008; Winn *et al.*, 2011). The modeling process can be automated if a very fine electron density is obtained.

The small angle X-ray scattering (SAXS) method also utilizes X-rays to determine the overall shape of molecules. When a protein solution (not crystallized) is irradiated with X-rays, the protein molecules scatter X-rays in a spherically averaged manner. The observed X-ray intensity over the scattering angle is converted into the frequency distribution of pairs of points within the molecule, and this pair distance distribution can be used to define the shape of a molecule. Although SAXS does not provide structures in atomic detail, this method is often employed to determine the size, molecular association, and interaction of proteins (Tuukkanen *et al.*, 2017).

Recently, XFEL (X-ray free electron laser) facilities are provided in major synchrotron X-ray sources (Spence, 2017). XFEL is a laser oscillated from accelerated free electrons. Because of its very high intensity and coherency, XFEL can be used for diffraction of microcrystals or even a single particle. Each single microcrystal or particle is exposed to a pulse X-ray of fs (10^{-12} s) duration only once. Therefore, the XFEL diffraction data are theoretically unaffected by X-ray radiation damage, and it enables time-resolved tracing of chemical reactions on 3D structures.

Nuclear Magnetic Resonance Spectroscopy

Atoms behave like magnets (due to the spin magnetic moment of nuclei), and these magnets can be manipulated by radio waves externally. The properties of these atomic magnets depend on the local molecular environment, and their measurement provides information about how the atoms are linked chemically, and how close they are in space. This information can be used to determine the distance between atoms, and then the overall structure of biological molecules (Wuthrich, 2001).

NMR samples are either in solution or solid (typically powder) phases. Usually the protein molecules are labeled with isotopes, such as ^{15}N, ^{13}C, and/or ^{2}H, to facilitate the measurements. A highly purified protein solution at a concentration of 0.1–3 mM is placed in the superconductance magnet, the sample is irradiated with a series of radio waves (pulse sequence) to activate particular nuclei (spin magnetization), and the radio wave emissions from the sample are measured (**Fig. 1(b)**).

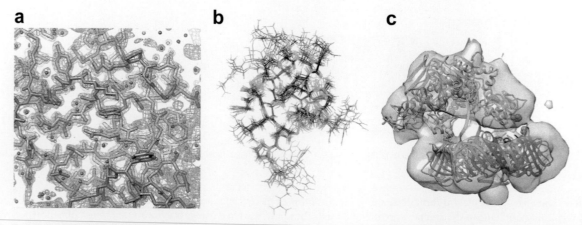

Fig. 2   Typical model images of (a) X-ray crystallography (atomic models fit into an electron density map), (b) nuclear magnetic resonance spectroscopy (multiple atomic models satisfying distance/angle constraints), and (c) electron microscopy (models fit into a density map).

A series of observed radio wavelengths, expressed as per-million difference (units in ppm) from reference wavelengths are the "fingerprints" of nuclei, and are called chemical shifts. Spin magnetization is transferred through chemical bonds among covalently connected atoms and also through space among atoms in close proximity. The chemical shifts of nuclei are perturbed by the status of adjacent nuclei. Correlation spectrograms, which depict the spin coupling (correlations among nuclei), are measured by magnetizing specific nuclei and observing whether the other nuclei are affected by them. The introduction of multiple isotopes is required to resolve the spectra, which tend to overlap each other (multidimensional NMR spectroscopy).

Various types of multidimensional NMR spectroscopic techniques have been devised to efficiently correct spin coupling, and they are usually combined step by step to construct molecular structural models. Typically, the first step is assigning the chemical shifts to specific residues and atoms using the heteronuclear single-quantum correlation (HSQC) method, in which one amino acid residue (except for proline) gives one signal. The $^1H^{15}N$-HSQC spectra are often referred to as the fingerprint of a protein because each protein has a unique pattern of signal positions. These experiments allow each ^1H –^{15}N peak to be linked to the preceding carbonyl carbon by tracing the coupling between nuclei through chemical bonds with the HNCA (measuring coupling through Cα–NH–Cα atoms of a peptide) and the HN(CO)CA (measuring coupling through NH–CO–Cα atoms of a peptide) methods (Bax and Ikura, 1991). The peaks of sidechain atoms are assigned through COSY (correlation spectroscopy) and TOCSY (total correlation spectroscopy). In this way, chemical shifts are assigned to an amino acid sequence.

Nuclear Overhauser effect (NOE) spectroscopy (NOESY) is additionally required in modeling protein conformation. Because magnetizations are transferred through space in this experiment, NOESY will show cross peaks for all ^1H atoms that are close in space regardless of whether they belong to same residue or not. A cross-peak in a NOESY experiment signifies spatial proximity (~ 6.0 Å) between the two nuclei in question, and thus it is used for distance constraints. NOESY is also used to obtain torsion angle constraints. The experimentally determined constraints of distance and angle are used as input for the structure calculation process. Usually, multiple models, which equally satisfy the constraints, are generated in this process. Computer programs such as XPLOR-NIH, CYANA, or ARIA/CNS attempt to construct atomic models that satisfy as many of the constraints as possible (**Fig. 2(b)**) (Linge et al., 2001; Schwieters et al., 2017; Wurz et al., 2017).

Important parameters, such as the interactions and dynamics of molecules, can be additionally obtained with NMR experiments. NMR spectroscopy is nucleus specific; thus, ^1H (hydrogen) and ^2H (deuterium) are distinguished. The amide protons (labeled with ^1H for example) in a peptide exchange readily with the hydrogen of a solvent containing a different isotope (^2H_2O). This H–D (hydrogen–deuterium) exchange can be used to investigate the folding process of proteins, because the parts of a protein folded early in the folding process are protected from hydrogen–deuterium exchanges (Englander and Kallenbach, 1983).

NOEs are also observed between atoms of a protein and interacting molecules. If the NOE spectrum of a residue is different in the presence and absence of ligands (chemical shift perturbation), it suggests that the residue takes part in the molecular interactions (Meyer and Peters, 2003). NMR can also yield information on the dynamics of a protein by measuring relaxation times, that is, times before magnetizations of nuclei degenerate. NMR relaxation is a consequence of overall or local molecular motions and can be measured by various types of HSQC methods. The motions occur on a time-scale of about 10^{-12}–10^{-9} s, and usually the relaxation time becomes longer when the molecular motion is faster (Sapienza and Lee, 2010).

Electron Microscopy

EM is a rather intuitive method for determining molecular structures because it directly observes images of the molecule (Adrian et al., 1984; Crowther and Klug, 1975). An electron microscope uses electrons, instead of the light in traditional microscopes, for imaging. High-energy electrons are generated and condensed, and samples are irradiated and focused on detectors by using electromagnets as lenses (**Fig. 1(c)**). The protein solution samples are adsorbed on a carbon grid and stained by heavy atoms, such as uranium acetate, for contrasting images. In a frequently employed method, the interprotein regions rather than the proteins are stained with heavy atoms, and the method is called negative staining. Recently, a method of rapidly freezing the sample solution in mesh pores, which is called cryo-electron microscopy (cryo-EM), is becoming popular (Cheng et al., 2015). Damage of samples by irradiation and the heavy atom stain is reduced in cryo-EM, although image contrast is generally lower than with the negative staining method.

Transmission electron microscopy (TEM), which observes the electrons transmitted through a sample, is generally used for determining molecular structures. First, images of the biological molecules (particles) spread on a mesh are taken. The images of each particle are generally obscure, and further image processing is required. In the canonical procedure, the particle images are picked up separately, and categorized into clusters of those viewing the same molecule in the same direction. The images in each cluster are aligned and averaged into class averages. The class averages are then assigned Euler angles (of viewing angles), and a 3D molecular density map is constructed with the back-projection method (Orlov et al., 2006).

Usually, back-projection is done in an iterative manner. First, a spherical density map is prepared, and the density map and the assigned Euler angles of the class average are refined step-by-step to be more consistent with each other. Back-projection might be done in either real or reciprocal spaces. The latter method is similar to the synthesis of an electron density map in XP: the 2D class averages are Fourier transformed and assigned angles, and the density map is synthesized as a reverse Fourier transformation. Although these processes might be executed automatically, human intervention is usually required because appropriately selecting particles and/or class averages is often critical for the quality of the final density map. The most used program suites for this process are EMAN2, SPIDER, and Relion (Fernandez-Leiro and Scheres, 2017; Shaikh et al., 2008; Tang et al., 2007).

Molecular structural models are constructed to fit the density map. For standard EM density maps of typical resolution, 10–20 Å, detailed atomic models cannot be constructed based on the maps (**Fig. 2(c)**). Usually, therefore, the atomic models that have been determined by XP or NMR are used to build pieces of the model. The pieces (partial models) are placed in the density map so that the experimental and model-derived densities correlate with each other. This process is often executed manually by inspecting the density map on computer graphic programs.

Recently, mainly owing to the development of the direct electron detector (DED), advanced cryogenic technique, and sophisticated image processing software such as Relion, atomic resolution (2–4 Å resolution) analyses have become possible with cryo-EM (Callaway, 2015; Subramaniam et al., 2016). The density map in this resolution range can be interpreted like the electron density maps from XP, and refinement procedures/programs for XP might be used for constructing and refining models (Murshudov, 2016).

Structure Validation Methods

Since protein structures are "theoretical" models that fit experimental data as far as possible, it is very important to detect modeling errors and validate how accurate the constructed models are determined. A structure validation process consists of three major parts: evaluations of (1) experimental data quality/quantity; (2) consistency between experimental data and structural models, that is, to what extent the experimental data is realized in the models; and (3) quality of model atomic coordinates, that is, how much the model parameters deviate from the canonical (dictionary) values derived from certified examples. The former two criteria are mainly specific to each structure-determination method, and only the last one is largely common among all methods (**Table 1**). The last one is based on a set of rules that any molecules must observe as a real entity, and thus also applicable to evaluation of the models from various computational predictions.

X-ray Crystallography

The quality of XP experimental data can be defined by a large variety of parameters. Among them, resolution, completeness, $<I/\sigma I>$, and R_{sym} (also called R_{merg}) are the most fundamental set of parameters (Drenth and Mesters, 2007). Resolution (d) is inversely proportional to the sine of the scattering angle of diffraction (θ) as $2d\sin\theta = n\lambda$ (Bragg's law; λ is X-ray wavelength). Thus, the higher (smaller in figure) the resolution, the larger the number of diffraction data points, and the more finely a structure might be defined. Completeness is the fraction of diffraction data actually observed of that theoretically observable for the resolution range (usually more than 90% is required). The average of diffraction intensity (I) over the standard deviation of the intensity (σI) in the proximal background is $<I/\sigma I>$, and the higher the values, the greater the significance of the signals. There are sets of independent diffractions, the intensities of which are theoretically equal due to the nature of diffraction and crystal symmetry. The equivalent diffraction intensities are scaled and averaged into a single datum during the X-ray data reduction process. R_{sym} is the residual of the observed intensities (I) from the averaged one ($<I>$) over total intensities, which is defined as $R_{sym} = \sum_{hkl}(I - <I>)/\sum_{hkl}<I>$. R_{sym} below 0.1 (10%) is preferable.

The consistency of models and experimental data in XP is also evaluated by various criteria, and R-factor and free R-factor are the basic components of such values. R(-factor) represents the residual of observed structure factors (F_o; square root of I) from those evaluated from the model (F_c), as $R = \sum_{hkl}(|F_o| - |F_c|)/\sum_{hkl}|F_o|$. F_c values are calculated by simulating X-ray diffraction from the model atomic coordinates. R is used as a target function to be minimized during model refinement. Usually, a fraction (5%–10%) of observed diffractions are selected randomly and uniformly over resolution ranges, and are not referenced during refinement for cross-validation purposes. The R-factor calculated on those set-aside diffractions alone is called the free R (R_{free} or fR). Usually, an R below 0.2 (20%) and a difference between R and R_{free} less than 0.1 are acceptable for XP models.

Nuclear Magnetic Resonance Spectroscopy

Due to the difference in methodology, resolution is not definable for NMR data. The quality of data in NMR is defined by the completeness of resonance assignments, that is, to what extent the observed chemical shifts are assigned to specific atoms, and the number of conformational constraints derived from correlation spectroscopy, where a higher number of constraints, if consistent, is preferable (Rosato et al., 2013).

The consistency of models and experimental data in NMR is assessed by the numbers of experimental constraints that were not satisfied by the models, that is, average numbers of distance constraint violations (typically separately evaluated as short 0.1–0.2 Å, middle 0.2–0.5 Å, and long >0.5 Å ranges, for example) and dihedral angle constraints (also separately evaluated in small 1–10° and large $>10°$ angle ranges. for example) per structure. The largest distance and dihedral angle violations should also be indicated. Because NMR generates multiple models, these metrics are usually presented as an average and a standard deviation over models. An NOE completeness score, which is defined as the ratio of the number of experimentally observed NOEs to the number of expected NOEs from the model, is also used (Doreleijers et al., 1999).

Electron Microscopy

The major parameters characterizing EM data are the number of particles and map resolution. The number of particles is the number of individual molecular images cut out from the original microscopic images, which often contain hundreds of molecules in the field of vision. More than 10,000 to 100,000 particle images are usually collected for a structure determination. Because a 3D density map is constructed by back-projection of 2D class averages, a nonbiased distribution of views (Euler angles assigned to class averages) is also required to justify the reconstruction.

Resolution in EM analysis is defined differently from that of XP. To estimate EM map resolution, particle images are randomly divided into two sets of equal numbers, and a map is constructed for each set independently. Then, the density maps are Fourier-transformed, and correlation coefficients (FCS) between sets are evaluated in resolution (wavelength) shells. Generally, the FCS decreases as resolution increases, and the resolution is defined as the point where the FCS falls at 0.5 (low resolution analyses) or 0.143 (high resolution analyses) (Baker *et al.*, 2010).

Consistency of models and experimental data in EM is mainly evaluated as cross correlation between the density of the experimental map and that calculated from the model (Monroe *et al.*, 2017; Xu and Volkmann, 2015).

Consistency With Dictionary Values

The stereochemical parameters of model atomic coordinates, such as bond length, bond angle, dihedral angle, and chiral volume, should be statistically consistent with the authentic reference values **(Fig. 3(a))** (Evans, 2007). Methods/programs such as PRO-CHECK or Molprobity are often used for evaluating stereochemical parameters and detecting errors in models (Chen *et al.*, 2010; Hintze *et al.*, 2016; Laskowski *et al.*, 1993). Deposition of structural models in the PDB before publication is recommended/ required, and the model atomic coordinates are checked through a program pipeline of the PDB validation server (Gore *et al.*, 2017). The set of scores in the PDB validation report are representative metrics of the stereochemical quality of the models.

The deviations of chemical bond lengths and angles of models from high-quality reference structures (including crystal structures of small molecules) are some of the most fundamental metrics (Bruno *et al.*, 2004). These values are presented as RMSZ, which are the root-mean-square value of the Z-scores (normalized values with average and standard deviation of reference structures) (Tickle, 2007). RMSZ scores of bond length and bond angles are around 0.5 for well-defined models. The root-mean-square deviations in an unnormalized scale are typically ~0.01 Å and ~1.2° for bond lengths and angles, respectively.

A Ramachandran plot is a traditional yet still-in-use method to characterize the main-chain conformation of proteins (Rama-chandran *et al.*, 1963). When main-chain torsion angles ϕ (horizontal axis) are plotted against ψ (vertical axis) **(Fig. 3(b))**, the plot field is divided into allowed (originally further divided into favored, allowed, and generously allowed regions) and disallowed regions **(Fig. 3(c))**. Empirically, nearly 100% of residues fall into the allowed region, and the residues outside of this region are thought to be outliers (probable modeling defects). High-quality reference structures contain less than 0.5% of residues as outliers.

The clashscore depicts atomic steric hindrance as the number of too-close contacts per 1000 atoms. An all-atom model is constructed by providing hydrogen atoms on the deposited atomic coordinates, and when van der Waals surfaces overlap more than ~0.4 Å between nonbonded atoms, the pair of atoms is identified as a clash **(Fig. 3(d))** (Word *et al.*, 1999). An average clashscore in reference structures is 5–20.

Sidechain conformation of amino acid residues can be defined as a set of χ torsion angles. These angles adopt a certain preferred set of values depending on the amino acid. A frequently observed sidechain conformation is called a rotamer or a rotameric conformer. The sidechain conformations observed in less than 0.3% of the residues in reference structures are defined as outliers. Models are scored by the percent of outlier sidechains in total (Williams *et al.*, 2017).

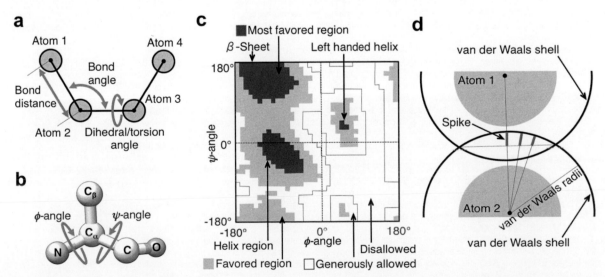

Fig. 3 (a) Schema of basic stereochemical parameters. (b) Definition of main-chain torsion angles ϕ and ψ. (c) Schema of the Ramachandran plot and area participations. (d) An overlap between van der Waals shells of two nonbonded atoms is defined by the lengths of spikes.

Other knowledge-based validation methods are also proposed. For example, a protein structural model with perfect stereo-chemical parameters may be still "unnatural" because amino acids show preferred distributions in 3D structures depending on their chemical properties, such as hydrophilicity or hydrophobicity. Programs such as PROSA2 or Verify3D are used for evaluating the molecular environments of each residue in a model, and numerically assess the fitness of the residues to the model structure by using an empirically derived function, which is used in threading methods (method to detect fitting of amino acid sequences to protein folds) (Luthy et al., 1992; Wiederstein and Sippl, 2007). These knowledge-based validation methods are often useful to find local mistakes in modeling.

Conclusion and Prospects

Since 1950, when the structure of myoglobin was first determined using XP (Kendrew et al., 1958), many improvements have been made in structure-determination methods, and several revolutionary changes, represented by cryo-EM and XFEL, are currently ongoing. These efforts are largely dedicated to determining large supramolecular structures, and elucidating their functions in atomic detail. Although current validation methods mainly concern the structure of each molecule (subunit), methods suitable to evaluate the correctness of complex structures as a whole will be required in the near future.

See also: *Ab initio* Protein Structure Prediction. Algorithms for Structure Comparison and Analysis: Docking. Algorithms for Structure Comparison and Analysis: Homology Modelling of Proteins. Algorithms for Structure Comparison and Analysis: Prediction of Tertiary Structures of Proteins. Assessment of Structure Quality (RNA and Protein). Biomolecular Structures: Prediction, Identification and Analyses. Drug Repurposing and Multi-Target Therapies. Identifying Functional Relationships Via the Annotation and Comparison of Three-Dimensional Amino Acid Arrangements in Protein Structures. In Silico Identification of Novel Inhibitors. Natural Language Processing Approaches in Bioinformatics. Protein Structural Bioinformatics: An Overview. Protein Structure Databases. Protein Three-Dimensional Structure Prediction. Protocol for Protein Structure Modelling. Rational Structure-Based Drug Design. Secondary Structure Prediction. Small Molecule Drug Design. Structure-Based Design of Peptide Inhibitors for Protein Arginine Deiminase Type IV (PAD4). Study of The Variability of The Native Protein Structure

References

Adams, P.D., Afonine, P.V., Bunkoczi, G., *et al.*, 2010. PHENIX: A comprehensive Python-based system for macromolecular structure solution. Acta Crystallogr. D Biol. Crystallogr. 66, 213–221.

Adrian, M., Dubochet, J., Lepault, J., McDowall, A.W., 1984. Cryo-electron microscopy of viruses. Nature 308, 32–36.

Baker, M.L., Zhang, J., Ludtke, S.J., Chiu, W., 2010. Cryo-EM of macromolecular assemblies at near-atomic resolution. Nat. Protoc. 5, 1697–1708.

Battye, T.G., Kontogiannis, L., Johnson, O., Powell, H.R., Leslie, A.G., 2011. iMOSFLM: A new graphical interface for diffraction-image processing with MOSFLM. Acta Crystallogr. D Biol. Crystallogr. 67, 271–281.

Bax, A., Ikura, M., 1991. An efficient 3D NMR technique for correlating the proton and 15N backbone amide resonances with the alpha-carbon of the preceding residue in uniformly 15N/13C enriched proteins. J. Biomol. NMR 1, 99–104.

Berman, H.M., Westbrook, J., Feng, Z., *et al.*, 2000. The Protein Data Bank. Nucleic Acids Res. 28, 235–242.

Bruno, I.J., Cole, J.C., Kessler, M., *et al.*, 2004. Retrieval of crystallographically-derived molecular geometry information. J. Chem. Inf. Comput. Sci. 44, 2133–2144.

Callaway, E., 2015. The revolution will not be crystallized: A new method sweeps through structural biology. Nature 525, 172–174.

Cheng, Y., Grigorieff, N., Penczek, P.A., Walz, T., 2015. A primer to single-particle cryo-electron microscopy. Cell 161, 438–449.

Chen, V.B., Arendall 3rd, W.B., Headd, J.J., *et al.*, 2010. MolProbity: All-atom structure validation for macromolecular crystallography. Acta Crystallogr. D Biol. Crystallogr. 66, 12–21.

Crowther, R.A., Klug, A., 1975. Structural analysis of macromolecular assemblies by image reconstruction from electron micrographs. Annu. Rev. Biochem. 44, 161–182.

Doreleijers, J.F., Raves, M.L., Rullmann, T., Kaptein, R., 1999. Completeness of NOEs in protein structure: A statistical analysis of NMR. J. Biomol. NMR 14, 123–132.

Drenth, J., Mesters, J., 2007. Principles of Protein X-Ray Crystallography, third ed. New York: Springer.

Englander, S.W., Kallenbach, N.R., 1983. Hydrogen exchange and structural dynamics of proteins and nucleic acids. Q. Rev. Biophys. 16, 521–655.

Evans, P.R., 2007. An introduction to stereochemical restraints. Acta Crystallogr. D Biol. Crystallogr. 63, 58–61.

Fernandez-Leiro, R., Scheres, S.H.W., 2017. A pipeline approach to single-particle processing in RELION. Acta Crystallogr. D Struct. Biol. 73, 496–502.

Frank, J., 2006. Three-Dimensional Electron Microscopy of Macromolecular Assemblies: Visualization of Biological Molecules in Their Native State. New York: Oxford University Press.

Gore, S., Sanz Garcia, E., Hendrickx, P.M.S., *et al.*, 2017. Validation of structures in the Protein Data Bank. Structure 25 (12), 1916–1927.

Hintze, B.J., Lewis, S.M., Richardson, J.S., Richardson, D.C., 2016. Molprobity's ultimate rotamer-library distributions for model validation. Proteins 84, 1177–1189.

Kendrew, J.C., Bodo, G., Dintzis, H.M., *et al.*, 1958. A three-dimensional model of the myoglobin molecule obtained by X-ray analysis. Nature 181, 662–666.

Laskowski, R.A., MacArthur, M.W., Moss, D.S., Thornton, J.M., 1993. PROCHECK: A program to check the stereochemical quality of protein structures. J. Appl. Crystallogr. 26, 283–291.

Linge, J.P., O'Donoghue, S.I., Nilges, M., 2001. Automated assignment of ambiguous nuclear overhauser effects with ARIA. Methods Enzymol. 339, 71–90.

Luthy, R., Bowie, J.U., Eisenberg, D., 1992. Assessment of protein models with three-dimensional profiles. Nature 356, 83–85.

Meyer, B., Peters, T., 2003. NMR spectroscopy techniques for screening and identifying ligand binding to protein receptors. Angew. Chem. Int. Ed. Engl. 42, 864–890.

Monroe, L., Terashi, G., Kihara, D., 2017. Variability of protein structure models from electron microscopy. Structure 25, 592–602. e592.

Murshudov, G.N., 2016. Refinement of atomic structures against cryo-EM maps. Methods Enzymol. 579, 277–305.

Orlov, I.M., Morgan, D.G., Cheng, R.H., 2006. Efficient implementation of a filtered back-projection algorithm using a voxel-by-voxel approach. J. Struct. Biol. 154, 287–296.

Otwinowski, Z., Minor, W., 1997. Processing of X-ray diffraction data collected in oscillation mode. Methods Enzymol. 276, 307–326.

Ramachandran, G.N., Ramakrishnan, C., Sasisekharan, V., 1963. Stereochemistry of polypeptide chain configurations. J. Mol. Biol. 7, 95–99.

Rosato, A., Tejero, R., Montelione, G.T., 2013. Quality assessment of protein NMR structures. Curr. Opin. Struct. Biol. 23, 715–724.

Sapienza, P.J., Lee, A.L., 2010. Using NMR to study fast dynamics in proteins: Methods and applications. Curr. Opin. Pharmacol. 10, 723–730.

Schwieters, C.D., Bermejo, G.A., Clore, G.M., 2017. Xplor-NIH for molecular structure determination from NMR and other data sources. Protein Sci 27 (1), 26–40.

Schwieters, C.D., Kuszewski, J.J., Tjandra, N., Clore, G.M., 2003. The Xplor-NIH NMR molecular structure determination package. J. Magn. Reson. 160, 65–73.

Shaikh, T.R., Gao, H., Baxter, W.T., et al., 2008. SPIDER image processing for single-particle reconstruction of biological macromolecules from electron micrographs. Nat. Protoc. 3, 1941–1974.

Sheldrick, G.M., 2008. A short history of SHELX. Acta Crystallogr. A 64, 112–122.

Spence, J.C.H., 2017. XFELs for structure and dynamics in biology. IUCrJ 4, 322–339.

Subramaniam, S., Kuhlbrandt, W., Henderson, R., 2016. CryoEM at IUCrJ: A new era. IUCrJ 3, 3–7.

Tang, G., Peng, L., Baldwin, P.R., et al., 2007. EMAN2: An extensible image processing suite for electron microscopy. J. Struct. Biol. 157, 38–46.

Tickle, I.J., 2007. Experimental determination of optimal root-mean-square deviations of macromolecular bond lengths and angles from their restrained ideal values. Acta Crystallogr. D Biol. Crystallogr. 63, 1273–1282.

Tuukkanen, A.T., Spilotros, A., Svergun, D.I., 2017. Progress in small-angle scattering from biological solutions at high-brilliance synchrotrons. IUCrJ 4, 518–528.

Wiederstein, M., Sippl, M.J., 2007. ProSA-web: Interactive web service for the recognition of errors in three-dimensional structures of proteins. Nucleic Acids Res. 35, W407–W410.

Williams, C.J., Headd, J.J., Moriarty, N.W., et al., 2017. MolProbity: More and better reference data for improved all-atom structure validation. Protein Sci. 27 (1), 293–315.

Winn, M.D., Ballard, C.C., Cowtan, K.D., et al., 2011. Overview of the CCP4 suite and current developments. Acta Crystallogr. D Biol. Crystallogr. 67, 235–242.

Word, J.M., Lovell, S.C., LaBean, T.H., et al., 1999. Visualizing and quantifying molecular goodness-of-fit: Small-probe contact dots with explicit hydrogen atoms. J. Mol. Biol. 285, 1711–1733.

Wurz, J.M., Kazemi, S., Schmidt, E., Bagaria, A., Guntert, P., 2017. NMR-based automated protein structure determination. Arch. Biochem. Biophys. 628, 24–32.

Wuthrich, K., 1986. NMR of Proteins and Nucleic Acids. John Wiley & Sons Inc.

Wuthrich, K., 2001. The way to NMR structures of proteins. Nat. Struct. Biol. 8, 923–925.

Xu, X.P., Volkmann, N., 2015. Validation methods for low-resolution fitting of atomic structures to electron microscopy data. Arch. Biochem. Biophys. 581, 49–53.

Protein Structure Visualization

Sandeep Kaushik, European Institute of Excellence on Tissue Engineering and Regenerative Medicine, Guimaraes, Portugal and University of Minho, Braga, Portugal
Soumya Lipsa Rath, Nagoya University, Nagoya, Japan

Introduction

The usage of graphical presentations greatly helps to successfully describe, communicate or understand structural concepts related to various biological phenomena (Sánchez-Ferrer et al., 1995). A new era of structural biology dawned when the first 3D structure of a protein (myoglobin) was solved by Kendrew et al. (1958). They built a physical "ball and spoke" model made of brass at a scale of 5 cm/Å and supported by 2500 vertical rods and colored clips that signified the electron density. The size of the model made it cumbersome and problematic to move whereas the forest of rods obscured the view of the model and made it hard to adjust. Nonetheless, this step was so important that for their studies on the structures of globular proteins, Max Perutz and John C Kendrew won the 1962 Nobel Prize in chemistry. In the meanwhile, Byron Rubin invented a machine for bending wire to follow the backbone trace of a protein and build small backbone wire models. These "wireframe models" were the most manipulable and portable models available at the time. Later during the 1960s and 1970s, the earliest of computer representations for molecules surfaced. Cyrus Levinthal and his colleagues at MIT developed a system that displayed rotating "wireframe" representations of macromolecular structures on an monochrome oscilloscope (Levinthal, 1966). Later, a program known as ORTEP was released by Carroll K. Johnson that could generate stereoscopic drawings of molecular and crystal structures with a pen-plotter. David and Jane Richardson and colleagues (Beem et al., 1977) solved the structure of superoxide dismutase, modeled the metal sites of the enzyme using an interactive density-fitting computer system called "GRIP" and visualized entirely with computers for the first time. Building models for newly solved protein crystals with computers rather than with physical Kendrew-style models became more and more popular.

Since then, scientists all over the world have been studying proteins and other macromolecules at atomic level using structure determination techniques like X-ray, NMR, etc. The availability of protein structures enabled the study of their surfaces and surface characteristics, based on atomic contribution. Consequently, a huge amount of protein structures (134436 on 23 Oct 2017) have been determined and deposited in the Protein Data Bank (PDB; see "Relevant Websites section") (Berman et al., 2000). PDB stores each structure in a specified format (see "Relevant Websites section") that describes each molecule as a list of its constituent atoms and their three dimensional (3D) coordinates. This format allows researchers to exchange protein coordinates through a database system. The availability of structural data and advances in computer graphics technologies together have pushed development of computational tools for molecular visualization and analysis (Lesk and Hardman, 1982). Biologists, computer scientists, and application developers usually get together for the development of methods or software for molecular visualization. Some examples of molecular visualization software include VMD (Humphrey et al., 1996), Chimera (Pettersen et al., 2004), and PyMOL (Schrodinger, 2015). In this article, first we describe molecular visualization within the required level of details.

Protein Topology (2D) Visualization

Structural Topology

After determining and analyzing many protein structures, it became clear that protein structures for a particular biological function remained conserved even if the sequences diverged. Wherever 3D structure information is absent, working with secondary structure information can provide a lot of relevant information. In fact, secondary structure information is as important as the final 3D structure of a protein. Therefore, another way of describing the structure of a protein is by describing its 3D fold as a sequence of secondary structure elements (SSEs). These SSEs can be easily assigned using DSSP algorithm (Kabsch and Sander, 1983) and later represented graphically in a diagram. Topology, in terms of protein secondary structure, can be defined as the structural elements which remain unchanged as a protein undergoes continuous deformation, or simply the sequence of SSEs (Martin, 2000). Using these SSEs, several authors have tried to develop a system for the topological classification of proteins (Flower, 1994; Koch et al., 1992; Richardson, 1977). Below we provide the most accepted tools regarding the algorithms and methods used for protein topology depiction and searching/matching.

HERA

HERA is an application for generating hydrogen bonding diagrams for proteins (Hutchinson and Thornton, 1990). The program uses SSEs determined by DSSP algorithm and plots a diagram in which helical residues are represented as helical wheels or helical nets. It also provides an option to automatically search for simple β-sheet motifs in a database. PDBsum (Laskowski et al., 2018), a database which provides summary information about each experimentally determined structure in PDB, uses HERA for generating hydrogen bonding plots (Laskowski, 2009). An example hydrogen bonding diagram for rabbit ʟ-gulonate 3-dehydrogenase [3ADO] is given in **Fig. 1(a)**.

TOPS

Another such method for plotting protein topology is known as topology cartoons or diagrams (Westhead *et al.*, 1998; Levitt and Chothia, 1976). It has been popular among several others approaches (Richardson, 1977; Tsukamoto *et al.*, 1997; Bansal *et al.*, 2000; Nagano, 1977). The TOPS algorithm was developed to generate topology cartoons for proteins depicting α-helices as circles and β strands with triangles, both of which are used as nodes in a mathematical graph connected by five edge sets (Flores *et al.*, 1994; Westhead *et al.*, 1999). The five edge sets include the directed N- to C-terminal sequential edges, and the four other undirected pairwise relationships between SSEs. The undirected relationships include hydrogen bonding, helix packing, chirality and neighbor relationships. Direction information for strands is duplicated, upward pointing triangles indicating 'up' strands and *vice versa* (Gilbert *et al.*, 1999). It ignores the details like the lengths of SSEs and loops however describes their spatial adjacency within the fold and approximate orientation. One example database that stores this topological descriptions of protein structures is TOPS database (Michalopoulos *et al.*, 2004). TOPS representations and database has been used for visualizing, searching and comparing (especially within a given structural family) fold topologies (Gilbert *et al.*, 2001; Víksna and Gilbert, 2001). Computer scientists and bioinformaticians used it for automated extraction of patterns relating protein sequence to topology and function. Unfortunately, the TOPS servers located at the website provided in "Relevant Websites section" are no more functional. The TOPS graph model was enhanced by replacing the graph model with a new string model (TOPS +) which included previously omitted loops and other structural and biochemical features, like ligand interaction and length of the SSEs (Veeramalai and Gilbert, 2008).

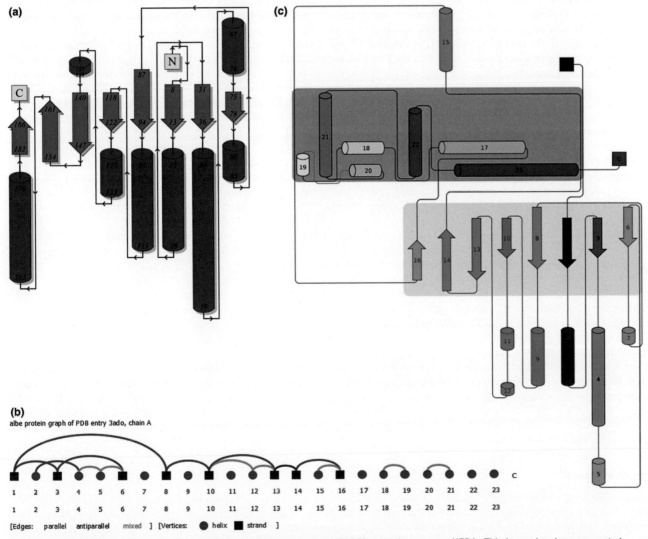

Fig. 1 (a) A hydrogen bonding diagram for rabbit gulonate dehydrogenase [3ADO] using the program HERA. This image has been generated using PDBsum website. (b) Protein topology graph generated for same 3ADO using PTGL. (c) Pro-Origami diagram for 3ADO topology.

Other Approaches for Protein Topology Analysis

Bostick *et al.* developed a topological representation of proteins by identifying natural nearest neighboring residues and the distance between residues in sequence space (Bostick *et al.*, 2004; Bostick and Vaisman, 2003). This representation allowed a quantitative and computationally inexpensive comparison of protein structural topologies. TOPSCAN was developed for a rapid comparison of protein structures (Martin, 2000). It uses a string notation which reduces the protein structure to a topology string where the letters encode the primary or secondary topology. The topology strings describe about 52 vectors of characters and incorporates information (like direction) from the three-dimensional structure. It can be accessed at the website provided in "Relevant Websites section". On the other hand, Protein Topology Graph Library (PTGL; see "Relevant Websites section") relies simply on the graph definitions without explicitly considering the order of SSEs (May *et al.*, 2010). Helices and strands are drawn as red circles (or cylinders) and black squares (or arrows), respectively (**Fig. 1(b)**). It also serves as a database on protein topologies with facilities for search, visualization, and getting additional information. Circuit topology (CT) is the latest method proposed for describing topology in proteins and nucleic acids (Mashaghi *et al.*, 2014). The idea behind CT is that the arrangement of a linear polymer's intra-molecular contacts (like disulfide bonds) determines its topology. The fundamental rules that intra-chain contacts must obey have been identified and CT can now be used to determine structural equivalence, study the topological structure of genomes and even protein folding. The latest approach for automatically generating protein structure cartoons is Pro-origami (Stivala *et al.*, 2011). Arrows and cylinders are used to depict β-strands and α-helices, respectively, and are drawn proportional in length to the number of residues (**Fig. 1(c)**). The Pro-origami web server (see "Relevant Websites section") allows diagrams to be generated from any PDB file.

3D Structure Visualization

Different molecular visualization (MV) software have provided different levels of flexibility to visualize or perform analytical computations. However, one common feature of these programs is that the structural information contained in a PDB file is translated into the position of atoms in 3D space. This positional information when combined with other chemical information like bond connectivity, atom types, atomic radius, electric charges, electron density maps, molecular surfaces, hydrophobicity scales and so on, helps a user observe molecular structure and various other molecular features (Andrei *et al.*, 2012). When not available, the bond connectivity for each atom is usually determined through a nearest-neighbor distance search (Humphrey *et al.*, 1996). Other dimensions of molecular data such as scalar (e.g., charge, hydrophobicity), vector (e.g., force fields, electric field) or tensor (e.g., anisotropic temperature factors) may be spatially located at isolated points in two dimensional (2D) or 3D spaces and may change with time (Duncan *et al.*, 1995). The translation of all this structural data into a geometric representation is a key element of the visualization process. These geometric representations usually are depicted using spheres, lines, polygons, tubes, cones and ellipsoids. Atoms in a molecule are drawn using (solid or dotted) spheres under space-filling (CPK) and variable ball-and-stick representations. These atomic spheres have a radius equal to the van der Waals radius of the depicted atom. Bonds can be drawn either as lines or as cylinders with variable, user selectable radii. Other display attributes such as color, shininess, and opacity can be set independently for each item. The secondary structure elements of a molecule are usually represented using ribbons, licorice, tube, or cartoon representations. Molecular surfaces can also be drawn for the molecules.

Advanced molecular visualization: Besides the basic MV application, most of the MV software can also calculate and represent advanced surface features such as electrostatic potential and hydropathy using field lines and/or as ranges of colors. The electrostatic potential is usually calculated with algorithms like APBS or DelPhi (Baker *et al.*, 2001; Rocchia *et al.*, 2001) whereas the hydropathy is calculated using Kyte-Doolittle scale (Kyte and Doolittle, 1982). These basic and advanced components are drawn and displayed as a scene by the MV software. This generated scene can then be output as an image file for storage or publication. The visual quality of a scene can be further enhanced by a technique known as ray-tracing. In ray-tracing, a path is traced from an imaginary eye through each pixel in a virtual screen and the effects of its encounters with virtual objects are simulated. Some examples of algorithms used to ray-trace are POV-Ray and Rayshade. A very high degree of visual realism is achieved using these techniques but at a greater computational cost. The visual experience is further enhanced with additional features like transparency. Transparency is introduced using a variable known as the alpha (α) channel. A concept closely tied to the transparency of an object is the notion that the rendered appearance of that object contributes only partially to the net color assigned to pixels it occludes. Another technique used to enhance the visual appearance of a scene is known as "antialiasing". The term antialiasing is applied to techniques that reduce the jaggedness of lines and edges in a raster image. Combined together, these algorithms and approaches generate a highly appealing and detailed image of a molecular structure.

Representations

Various ways of representing the constituent atoms of a molecule have been used. Each of them have different advantages and disadvantages. Here we describe some details about the most common representations used by various molecular visualizations software. And, to create a visual distinction between various representations, we depict the structure of insulin, 2HIU (Hua *et al.*, 1995), using VMD (Humphrey *et al.*, 1996) in **Fig. 2**.

Points: In this representation, each atom is drawn as a point. The bonds are not drawn at all.

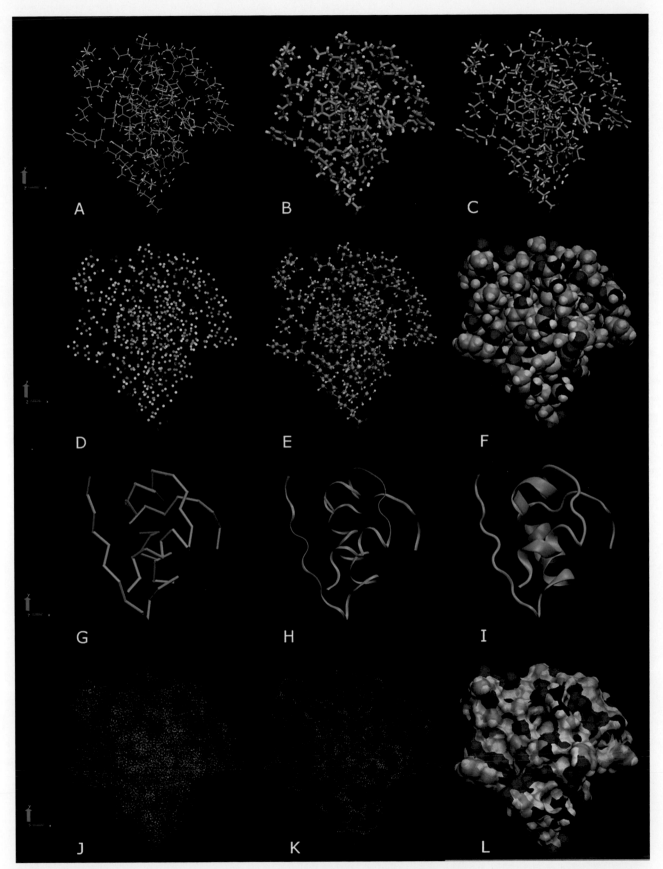

Fig. 2 Various molecular representations used to display molecules especially proteins and nucleic acids. Structure of insulin (2HIU) has been used to depict various representations using VMD. Twelve representations have been shown namely (a) lines (b) bonds (c) licorice or Drieden's structure (d) points (e) CPK (f) van der Waals (g) backbone or trace (h) ribbons (i) cartoon (j) dotted (k) solvent and (l) surface.

Lines: In the 'Lines' (or 'wireframe') representation, a line is drawn between each of the bonded atoms. The first half of each bond is colored according to the color of the first atom, while the second half is colored according to the second atom. The user can adjust the line thickness.

Bonds: In this representation, a cylinder is drawn between two atoms instead of a line. This cylinder is drawn as an *n*-sided prism, where the number of sides and the radius can be specified.

Backbone (or Trace): In this representation, successive residues of unbroken chains are connected by cylinders with adjustable width, in a fashion similar to *"Tube"*. Only the main backbone atoms, Cα atoms in case of proteins and C4'/C3' atoms for RNA/DNA, respectively, are considered.

Ball and stick: In this representation, the atoms are displayed as spheres (or balls) whereas the bonds are shown as cylinders (sticks). It is the most fundamental and common representation used where the atoms and bonds are colored according to atom type. The original function of physical ball and stick models were the support of measurements of structure angles and bonds lengths, leaving the real structure representation to space filling models.

VDW: In this representation, the atoms are drawn as spheres with a radius equal to the *van der Waals* radius. The spheres are built using many polygons.

Dotted: This representation is same as *VDW* except that the spheres are drawn as dotted instead of solid. Like mentioned above, spheres are built using many polygons. Here, a dot is placed at each of the vertices of the triangle making up each sphere. It can be used to speculate about the surface of the molecule.

CPK: This scheme is named after chemists Robert Corey, Linus Pauling and Walter Koltun. This representation draws the atoms as spheres and the bonds as cylinders. The radius of the sphere drawn in CPK mode is usually smaller than VDW mode. Therefore, it can be called as a combination of both *Bonds* and *VDW*. It is also called as the space filling model.

Licorice or Dreiding model: In this representation, the atoms and bonds are drawn as spheres and cylinders, respectively. The sphere radius is not controllable and is same size as the bond. Therefore, it contains less distractors compared to the normal ball-and-stick.

Tube: The tube representation is a smooth spline curve that passes through backbone atom of proteins (Cα) or nucleic acids (P atoms of phosphates). The spline radius and resolution can be set by the user. The curve connecting the two Cα atoms is broken into six line segments by determining five evenly spaced interpolation points along the spline curve. The first three segments are colored by the color assigned to first Cα and the last three segments are colored by the color of the second Cα.

Ribbon: This representation also follows the same spline curve for both the protein and nucleic acids as in Tube. Additionally, it uses the O atom of the protein backbone or some of the phosphate oxygens for nucleic acids to find a normal for drawing the oriented ribbon. Given the coordinates of each atom and the offset vector for the ribbon vector, the drawing code finds the spline curves for the top and bottom of the ribbon. The two splines are connected by triangles and both splines are drawn as small tubes.

Cartoon: This representation is based on the secondary structure of the protein. Helices are drawn as cylinders, beta sheets as solid ribbons, and all other structures (coils and turns) as a tube. A least squares linear fit along at least three Cα atoms of the helix is used to construct a helix cylinder. The solid beta ribbon is constructed by building a spline along the center points between each beta sheet residue.

Solvent: This representation gives a quick estimate of the molecular surface with a collection of dots. It is similar to the *Dotted* representation and provides a more uniform coverage of the surface. The "probe radius" and "density of the dots" can be set or adjusted by the user.

Surface: This representation uses a molecular surface renderer for example MSMS which allows to compute very efficiently triangulations of solvent excluded surfaces. Representation of molecular surfaces needs a special detailed description. Therefore, below we have provided some details on the rendering of molecular surfaces.

Molecular Surface

Molecular surface (MS) is one of the most important geometric feature of a protein as the size and shape of indentations in the surface play an important role in various functions like protein folding, docking and interactions between proteins (Connolly, 1983). The computation of MS area is important for the understanding of interactions of protein surfaces and internal cavities with ligands and hence crucial in structure-based drug design. It also allows one to incorporate the effects of solvent in the potential energy calculations (Juba and Varshney, 2008). Molecular surfaces are most usually displayed as dot-surfaces or solid renderings. Various algorithms have been developed for molecular surface area computation for example Connolly's (Connolly, 1983), MSMS (Sanner *et al.*, 1996), GETAREA (Lesk and Hardman, 1982), LSMS (Humphrey *et al.*, 1996), 3V (Pettersen *et al.*, 2004), and an adaptive grid-based algorithm (Schrodinger, 2015) included in TexMol (Kabsch and Sander, 1983). Some software that have been used for surface depiction include Grasp, AVS, VOIDOO, etc.

Broadly, three types of molecular surface models have been described which are van der Waals (vdW) surface, solvent-accessible-surface (SAS), and solvent excluded surface (SES) (Deanda and Pearlman, 2002). Ligand excluded surface is an extension of the SES concept. Below these surfaces have been described.

Van der Waals surface: This surface is a reasonable approximation of the molecular surface. Here the atomic radius of each atom is its van der Waals radius and each atom is represented as a hard sphere. Therefore, the van der Waals surface of the molecule can be defined as the union of all portions of all atomic sphere surfaces not occluded by neighboring atomic spheres.

Solvent accessible surface (SAS): Atoms located in narrow crevices may not be accessible to the solvent if the solvent molecule is bigger than the crevice. Therefore, SAS is calculated by rolling a probe sphere of a defined radius (e.g., 1.50 Å for a water molecule) over the entire van der Waals surface of a molecule of interest. The molecular surface created by the center of the probe is called as SAS (Lee and Richards, 1971).

Solvent excluded surface (SES): SES is also calculated by rolling a probe sphere of a defined radius over the entire van der Waals surface of a molecule. However unlike SAS, SES is defined as the surface traced out by the inward-facing (towards the macromolecule) surface of a probe sphere (Richards, 1977; Greer and Bush, 1978). In other words, SES resembles the van der Waals surface of the molecule except that crevices too small for the probe sphere to enter are eliminated whereas the clefts between atoms are smoothed over. SES area provides a better model for describing hydration effects and a better choice to visualize and study molecular properties. Surface computation is easier for the vdW and SAS but not for the SES.

Ligand excluded surface (LES): LES represents a more accurate approximation of the regions accessible to the ligand under consideration than SES because here the probe sphere is replaced with the full vdWs geometry of the ligand.

Other advanced approaches for surface representation: Voronoi diagram of atoms, is an approach that is independent of the probe size therefore invariable for a given molecule, have also been used to calculate MS of a molecule (Ryu *et al.*, 2007). With advancements in GPU hardware, several alternative approaches have been proposed for faster rendering of molecular surfaces like ambient occlusion, halos, GPU ray-casting, improved ball-and-stick representation called "*HyperBalls*", theory of implicit surfaces, depth peeling, tessellation shaders, and dynamic view-dependent level-of-detail (LOD) representation have been proposed (Guo *et al.*, 2015).

Ray Tracing

Shaded images for displaying molecules have been a popular approach for conveying 3D information especially in the absence of animation. Therefore, algorithms have been developed that can handle reflections, transparency with refraction, shadows, intensity depth cuing, texturing and multiple local light sources. Ray-tracing algorithms provide the best solution to these aspects and are capable of generating realistic visual effects that are difficult with other rendering techniques (Palmer *et al.*, 1989). During ray tracing, a ray is fired from the eye point through each pixel in the image where it can either get reflected or transmitted through the surface (refraction). This reflected or transmitted ray can further interact with other surfaces, until it gets terminated after either leaving the scene or being absorbed. Ray tracing is capable of simulating motion blur, soft shadows, depth of field, diffuse interreflections and caustics by using various extensions.

Ray tracing in a scene requires a substantial additional computational cost compared to a non-ray-traced scene which is a significant problem. Nonetheless, ray tracing amplifies the clarity of a structure enormously. **Fig. 3** compares a non-ray-traced image of insulin (2HIU (Hua *et al.*, 1995)) with a ray-traced version using the *spheres* (van der Waals) representation using PyMOL (Brunger and Wells, 2009). Ray tracing is an effective and flexible technique for producing high-quality visualizations of molecular structures especially for scientific publications. Persistence-of-vision (POV) ray (see "Relevant Websites section"), RADIANCE (see "Relevant Websites section") and OSPRay (see "Relevant Websites section") are some of the many ray tracing software available online.

Depth Perception

Depth perception in 2D is conveyed by motion cues during rotation and (directional) lighting. Surfaces where light bounces off the surface and is reflected toward the viewer appear bright and give a strong sense of depth for smooth, slowly varying surfaces. Lighting

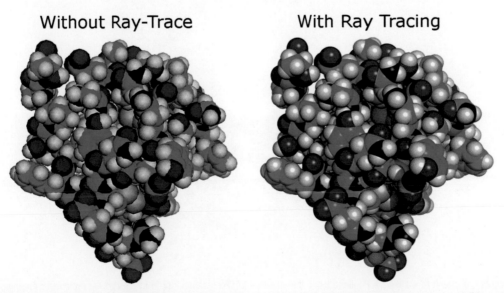

Fig. 3 The advantage of ray tracing in amplifying the clarity of a structure has been depicted using (a) non ray-traced and (b) ray-traced images of insulin (2HIU) in *spheres* (van der Waals) representation using PyMOL.

techniques like ambient occlusion are computationally demanding but appear highly effective for such space-filling sphere rendering as surface regions in recesses or concavities appear darker. **Fig. 4** depicts how PyMol overlays fog on objects to assist in emphasizing what is in the foreground and backgound of the image with respect to the camera. Another technique to enhance 3D appearance to highlight protrusions of a surface overlaying an identically colored surface is to add black outlines at edges where a jump in depth occurs in an image. An improvement to this algorithm was introduced to make the line thickness depend on the magnitude of the depth change so that larger depth changes are highlighted with thicker lines while small depth jumps may not be highlighted to simplify appearance (Goddard and Ferrin, 2007). Similarly, eliminating depth is another alternative approach that is getting attention. For larger assemblies, the individual residues appear like countries on a map divided by boundary lines with text labels. Recent graphics processing units (GPUs) frequently have an order of magnitude more floating-point computation speed than the main "computer processing unit". Some GPU-based techniques called textural and procedural impostors enable rendering of atomic models faster by a factor of 10 or more. Therefore, new effects such as the depth-dependent edge highlighting are possible.

Visualizing Molecular Dynamics Trajectories

Molecular dynamics (MD) studies of the function of the structurally known biopolymers are being widely applied. The results of such studies are typically large molecular trajectory files consisting of hundreds or thousands of frames of atomic coordinates. The trajectories contain substantial amounts of dynamical data that require suitable visualization tools to display sequences of structures (Humphrey *et al.*, 1996). Therefore, programs were developed that could display the molecular structure in motion. The investigator may relate various thermodynamic quantities better to molecular motions by the complementarity provided by molecular visualization. Indeed, by visualizing a computational model it becomes easier to have a subjective assessment or make a non-quantitative judgment.

Visualization of Molecular Assemblies

The hierarchy of organization of atoms in a biomolecule may start from atoms and go to secondary, tertiary and quaternary structures. For certain systems this hierarchy may even be followed by further organization into molecular complexes such as virus

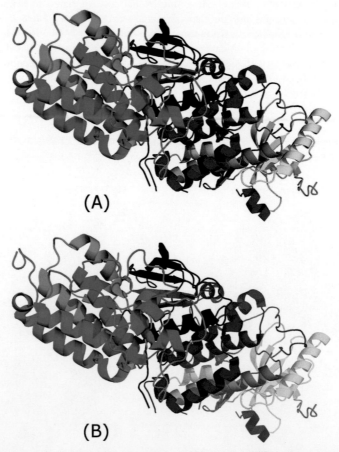

(A)

(B)

Fig. 4   The perception of depth in a scene is depicted using the structure of EFG receptor (2JIT) in the *cartoon* representation using PyMOL. (a) Without depth cueing and (b) with depth cueing. The obscurity and haziness of the distant domain of EFGR clearly depicts the impact of depth cueing.

capsids. The visualization software would be required to render millions to billions of atoms in real time. Therefore, visualization of larger biological structures needs a humongous computational power and is still a challenging task. To circumvent this problem, lesser details for each component of the larger assemblies are used to decrease the computational burden (Goddard and Ferrin, 2007). The approximate shape of domains, molecules, or complexes are drawn as smooth surfaces at any desired resolution using methods like spherical harmonics, density contours, and surface decimation. Each method has its own advantage and disadvantage but discussing that is beyond the scope of this article.

Evolution of Molecular Visualization Software

A number of molecular visualization programs have been developed since the days of first molecular visualization using computers. These programs use component modules each specifically written to create the structural components like spheres, tubes, ribbons, etc. Not all the programs can be described in this article but we try to describe the best known molecular visualization programs. The programs are arranged according to their year of release to generate a sense of evolution in terms of molecular visualization. Each program is listed and the capabilities and mode of action of the underlying algorithm(s) have been described.

RIBBONS (1987)

Mike Carson constructed an algorithm that generated aesthetically pleasing and smooth 3D representations of a protein using a set of nearly parallel B-spline curves fitted to the peptide plane (Carson, 1987). The peptide planes were used to generate a series of 'guide coordinates' equally spaced along the desired ribbon width. Smooth and regular cubic B-spline curves fitted to these guide coordinates formed the roughly parallel threads of the ribbon. Ribbon models clearly displayed the secondary structure of the backbone. It was also possible to code by residue to depict information such as residue type or temperature factors.

MAGE (1992)

A "kinemage" or kinetic image was a term designed for a structure presented on a computer display (Richardson and Richardson, 1992). A utility called PREKIN was used to prepare kinemage files from input PDB files which were then viewed using a graphics program called MAGE. The kinemages were plain text files with commented display lists and accompanying explanations that could be edited by user. The interactive image could be rotated in real time, parts of the display could be turned on or off, cursor selection of atoms or points was possible, and different forms could be changed. The depth-cueing, rotation, rocking, distance measurement and animation of two or more conformations was possible.

RasMol (1993)

RasMol (stands for *raster molecules*) is an easy to use molecular visualization tool for proteins, nucleic acids and small molecules (Sayle and Milner-White, 1995). It was developed by Roger A. Sayle (*also forms R.A.S. of RasMol*) for Unix, Windows and Macintoshes and is available freely from the websites provided in " Relevant Websites section". Rasmol can read a PDB file both locally or directly from PDB database, which can then be displayed interactively with a variety of color schemes and molecule representations including depth-cued wireframes, 'Dreiding' sticks, spacefilling (CPK) spheres, ball and stick, solid and strand biomolecular ribbons, hydrogen bonding and dot surfaces (Sayle and Milner-White, 1995). The users could interact with a molecule by rotating, translating, zooming and z-clipping (slabbed) using either the mouse, scroll bars, command line or an attached dial box. Ligands, active sites, multiple subunits, hydrogen bonds and various parts of the molecule could be displayed selectively or in a combination of display modes. It could measure interactive distances, angles and torsion angles. RasMol allowed depiction of stereo images, multiple NMR models and labeling of atoms. The structural features like β-sheets, protein backbones, double and triple bonds were represented as arrows, smoothed-tube, multiple lines and, respectively. RasMol can also take a list of commands using a 'script' file to (re)generate a given image or viewpoint that may be written out in a variety of formats, e.g., PostScript, GIF, BMP, etc. No doubt, RasMol quickly became the most popular molecular graphics software with over 15,000 sites using it.

AVS (1995)

AVS or Advanced Visual Systems, was designed to be a general-purpose visualization environment with no elaborate data structures for specific applications areas (Duncan *et al.*, 1995). Several programs, e.g., AutoDock, Harmony and SURFDOCK use AVS as an important visualization tool as the rendering capability of AVS is of very high quality. AVS uses several modules to translate molecular coordinate data to a geometric representation, e.g., SPHERES (for spheres), DISJLINES (for bonds), POLYLINE, POLYGON (to define triangles, and other convex polygons), POLYTUBE (a cylindrical object from a polyline), ELBOW (for smooth joints), COLORTUBE (produces a tube from disjoint line), MSMS (for molecular surfaces), MV102 (to control the display style of selected subset of atoms). Intermolecular interfaces can be calculated using the INTERFACE module. The PLAY BINPOS module displays MD simulation trajectories with facilities for selecting individual frames and controlling the speed of the display. Both, a graphical user interface (GUI) and a command line interpreter (CLI) can be used to control most interactions with AVS. AVStk, a Tcl/Tk interface to AVS implements the facilities for expressions, variable assignment, loop control, or functions. The NAB (nucleic acid builder) language is used for building chemistry modules as it has data structures and operations designed specifically for molecular data. AVS is suitable for biomolecular visualization as it provided general-purpose and chemistry-specific modules, and AVS-tool and NAB to facilitate network control and module development.

Raster3D (1996)

Raster3D is a hardware independent suite of programs used to generate photorealistic molecular graphics (Merritt and Bacon, 1997). It is available for various Windows, various linux distributions and macintosh operating systems. It offered several advantages compared to other available programs. For example, it provided platform-independent tools for composition and rendering. It was much faster than general-purpose ray-tracing programs. Raster3D suite has four different types of programs namely, composition tools (*balls, ribbon,* and *rods*), input conversion utilities, the central rendering program, and output conversion filters. The composition tools read the atomic coordinates found in a PDB formatted file and generate van der Waals surface (*balls*), a peptide backbone trace (*ribbons*) and a ball-and-stick model (*rods*) of the bonded atoms. This information is stored or converted into a series of header lines specifying global rendering options, followed by a series of individual object descriptors. The input stream can have further input from one or more files. Also, the size of the rendered image in pixels could be controlled using header records input to the render program. The main program of the Raster3D suite, *render,* takes this input and produces a raster image in one of three standard formats (AVS, TIFF, or SGI libimage). Rendered images could be processed further to add labels, edit colors, form composite images, apply *"gamma correction"* or be converted to other formats. Raster3D could also import objects from other molecular visualization tools using two conversion utilities *ungrasp* and *normal3d.*

   Raster3D generated *shadows* to convey an impression of depth, obviate the need for stereo pairs and to convey information that otherwise would not be apparent in the rendered image (Merritt and Bacon, 1997). Raster3D implemented *"Z buffer"* algorithms rather than true ray tracing to produce a remarkably photorealistic appearance for the rendered objects. The rendering algorithm can be optimized to output images that include effects like shading, transparency and effective transparency (or α channel). It required much less computation than full ray tracing of the same scene. One can generate various kinds of pictures in Raster3d such as black and white, colored, and animated images. We can also add labels, shadows, transparency and stereo images. Ribbons, ring, surface and rods are the common kinds of representations available. It has been implemented in other protein visualization packages such as VMD, Xtalview, Molscript, GRASP, MSMS, ORTEX, Conscript, etc.

MD Display (1996)

MD Display was developed initially as a means of visualizing molecular dynamic trajectories generated by Amber (Callahan *et al.,* 1996). The program runs on Silicon Graphics workstations, and features a simple user interface, and convenient display and analysis options. The program can accept input from several other molecular dynamics programs. A preprocessor handles output files from different computational programs (like Amber, CHARMM, Discover, and GROMOS) and to generate a consistent input for the display program. It can load PDB files and allows construction of a pseudo-trajectory using multiple PDB files. It provides animation control in terms of speed (and *hyperspeed*) and direction of animation so that freezing one frame and single step in either direction is possible in the frame sequence. Periodic boundary conditions are handled and multiple frames can also be superimposed to create a "smear" image. The user can monitor the interatomic distances, angles, dihedrals, hydrogen bonds (based on donor-acceptor distances), and Ramachandran plot (ϕ-ψ values) in the trajectory.

MOLMOL (1996)

MOLMOL is developed specifically for the display, analysis, and manipulation of sets of conformers in nuclear magnetic resonance (NMR) structures of proteins and nucleic acids (Koradi *et al.,* 1996). It is downloadable from the website provided in "Relevant Websites section". It also implemented special functions to represent the structural uncertainty by the spatial spread among groups of NMR conformers. It could be run on Silicon Graphics workstations, unix machines using either OpenGL or X11 library. The selection of sets of atoms, bonds, distances, or primitives was possible by expression syntax or mouse. This selected set could be represented as ribbons, ellipsoids, or other simple geometric shapes. The secondary structures were either identified using the DSSP algorithm or read directly from the PDB file. MOLMOL can draw dots or shaded molecular surfaces for three different kinds of surfaces namely van der Waals surfaces, solvent accessible surfaces and contact (SES) surfaces.

   This program was unique as it allowed modification of chemical structures by adding and deleting atoms and bonds, and generation of new 3D structures by variation of dihedral angles about individual covalent bonds. It could also analyze the trajectories from MD simulations. For quantitative comparisons, the program calculated root mean square distances (RMSDs), short interatomic contacts, dihedral angles, hydrogen bonds, radius of gyration and solvent-accessible surfaces. The program could generate high-quality plots as PostScript, or FrameMaker (MIF), TIFF format or as an input file for ray-tracing programs.

Cn3D (1996)

Cn3D (pronounced 'see in threedee') is a visualization tool developed by NCBI to simultaneously display biomolecular structures, sequences, and sequence alignments for all platforms (Hogue, 1997; Wang *et al.*, 2000). It provided all the functionality of RASMOL along with other enhanced rendering and labeling options. It could also play animations for multiple conformations of NMR structures or multiple superposed structures. To provide a better understanding of the 3D structure, by default the protein's α-carbon trace is colored by secondary structure and α-helices are rendered as hollow cylinders whereas a DNA molecule is colored by CPK color scheme and metal ions are shown as spheres. It is also possible to 'hide' chains, domains or residues at one's discretion. It is also easy to set custom rendering or labeling for a set of atoms or residues. It offers a programming layer as MMDB-API along with a GUI called Viewer3D. The co-ordinates data from a PDB file is translated into MMDB-API representation and converted to graphics primitives (like spheres, cylinders, lines, etc.) to be displayed in Viewer3D.

Cn3D includes algorithms to import and align multiple sequences or structures. Homologous sequences to a protein with known 3D structure can be imported and aligned into the sequence viewer. After importing multiple homologous sequences, the structure window displays the 3D superposition while the sequence window shows a sequence view of the structural alignment. A seamless combination of structure visualization and comparative analysis makes Cn3D a powerful tool to map functional sites between structure and sequence. This proves helpful in finding conserved regions among aligned proteins, locating the residues corresponding to known disease mutations in a crystal structure or mapping the locations of conserved binding site or active site residues from a family of homologous sequences. Because of these capabilities, Cn3D is the default structure viewer for NCBI's Entrez Structure database and the primary alignment curation tool for the CDD project (Wang *et al.*, 2000). Cn3D also has powerful annotation and alignment editing features.

VMD (1996)

VMD, stands for Visual Molecular Dynamics, is a tool written in C++ and designed to visualize structural features of various biopolymers like proteins and nucleic acids using a wide variety of rendering styles and coloring methods (Humphrey *et al.*, 1996). The source code and documentation for VMD is available at the website provided in "Relevant Websites section". VMD can display standard PDB files and animate trajectories obtained from molecular dynamics (MD) simulations. The trajectories can be loaded both from stored files and from a direct connection to a MD simulation running on a remote supercomputer or high-performance workstation. For remote trajectory sources, VMD uses a set of daemons and library routines known as the MDCOMM that buffer data transfer from the remote connection. A portable MD simulation program called NAMD, available with VMD, has been designed for performance, scalability and modularity to take advantage of the parallel architecture of processors. VMD together with NAMD constituted a larger set of computational tools for structural biology known as MDScope (Nelson *et al.*, 1995).

VMD requires Silicon Graphics GL library or the OpenGL library for 3D graphics rendering. The GUI can be used to perform tasks such as changing the current molecular display characteristics or animating selected molecules using molecular dynamics trajectories. All actions in VMD are available *via* text commands so that users can write and execute their own *tcl* scripts. To read data from file formats other than PDB, VMD has an interface to Babel (Walters and Stahl, 1996) to read data from other formats. Molecules are drawn as one or more *"representations"* for all or a subset of constituent atoms. A particular set of atoms can be selected using Boolean operators and regular expressions and assigned one of the various rendering styles and coloring schemes. The solid objects are illuminated using up to four hardware-accelerated, independent, infinitely distant light sources. VMD provides an extensive program control, using both graphical user interface (GUI) and a text interface using the *Tcl* embeddable parser, to generate high-resolution raster images for photorealistic image-rendering applications.

Trajectory display and analysis: As the name suggests, VMD was primarily developed with the ability to animate and study molecular dynamics (MD) simulations. A molecular trajectory may be read (from PDB files or from direct connection) and edited (using a trajectory editor) that provides options to delete or write (*to a new file*) specific frames or set of frames. VMD can also perform complex analysis on a molecular dynamics trajectory such as computing the RMS deviation or correlation functions. With the advent of petascale computers (e.g., Blue Waters supercomputer) and GPU-accelerated ray tracing engine, atomic-detail simulation and visualization of really large cellular apparatus and processes like the 100 M-atom photosynthetic membrane complex in purple bacteria, is feasible with VMD (Stone *et al.*, 2016).

WebMol (1997)

With the advent of Java and web-browsers with in-built support for Java applications, programs could be run virtually and remotely using regular web-browsers. The client computer system doesn't need to install the programs code ('program once - run anywhere') because Java programs are directly transferred and interpreted to the client upon request. One such application for molecular visualization, called WebMol, was developed at EMBL.

WebMol is a Java based interactive graphical program to display and analyze molecular structures stored locally or remotely in the PDB database using Java-supporting web-browsers only (Walther, 1997). It provided a similar functionality online as RasMol provided offline but suffered slow rendering and java-dependent security problems. The main advantage was that novice users didn't require to install programs or their source codes. They could directly go to WebMol and start visualizing or analyzing molecular structures. A new version of WebMol (see "Relevant Websites section") is under development in HTML5 and Javascript with the support of PhiloGL. The application is designed to be accessible from Internet and compatible with the JSON format of protein provided by the project WebPdb (see "Relevant Websites section").

PyMOL (1999)

PyMOL is an elegant open source and cross-platform program for visualization of complex macromolecular systems based on OpenGL and Python (Schrodinger, 2015). PyMOL can be downloaded freely for common platforms like Windows, Linux, and Mac OSX from the website provided in "Relevant Websites section" (v1.8) or from the website provided in "Relevant Websites section" (v2.0). Along with a graphical user interface (GUI), PyMOL offers a command line mode that resembles the Python interpreter. In fact, 'Py' part in PyMOL refers to the programming language Python. PyMOL commands are a series of Python function calls that function as a superset of the Python language. It also supports reusable scripts written in the PyMOL command language or in Python. PyMOL supports most of the common representations like wire, cylinders, backbone and cartoon ribbons, spheres, ball-and-stick, dot surfaces, solid surfaces, wire mesh surfaces. It supports labels, dashed bonds (for hydrogen bonding interactions and distances), transparent surfaces, reads and displays CCP4 and XPLOR electron density map files, generate symmetry-related molecules and load multiple common file formats. PyMOL was the first intuitive molecular graphics program that provided the click-and-drag functionality that we now see in almost all the molecular visualization software, no doubt it became really popular (Brunger and Wells, 2009). PyMOL was originally designed to dynamically visualize single or multiple conformations (trajectories or conformational ensembles) of a single structure with professional strength graphics. It has an integrated ray-tracing engine (in addition to PovRay engine) for generating publication quality figures that are complete with lighting, specular reflections, and shadows. **Fig. 5** depicts some of the representations available in PyMOL for the same insulin structure (2HIU (Hua *et al.*, 1995)). Indeed, figures generated with PyMol appear superior than many other software. This obviated the need for command-line operations which were otherwise required by other programs such as Molscript, Raster3D, and ImageMagick. Different states of molecule can be assembled into QuickTime or AVI movie by rendering simple or Ray-traced frames. Multiple atom selection was made possible using arbitrary extended algebraic expressions. Molecular editing allowed users to create new objects out of atom selections (across any number of other objects) and edit (delete, replace, or grow) bonds, angles, torsions, and positions on an atomic basis using just click-and-drag operation. Another important aspect of PyMOL is that other applications can utilize it as a molecular display window while providing their own external menus, windows, dialog boxes, and controls or even add additional geometries. PyMOL can also display the results of the macromolecular electrostatics calculations by APBS (Baker *et al.*, 2001) as an electrostatic potential molecular surface using the APBS Plugin. PyMOL also provides an automated qualitative electrostatic representation for a protein's contact potential by generating a charge-smoothed surface. It performs a "charge smoothing" using a quasi-Coulombic-shaped convolution function to average the charges over a small region of space. **Fig. 6** displays such an electrostatic contact potential for insulin generated using PyMOL.

UCSF Chimera (2004)

The developments in electron microscopy (EM) led to the determination of the atomic resolution structure of large-scale subcellular systems. Compared to other crystallographic and NMR structures, which are usually not very large, the structural complexity for large-scale molecular assemblies like viral particles, chaperonins and chromosomes, increases rapidly. Such large assemblies or molecules may contain several million atoms and therefore normal visualization software may not be able display it properly. The reason is that normal representations are so detailed that it requires impractically large computer memory and leads to insufficient graphics rendering speed (Goddard *et al.*, 2005). Therefore, to display such large molecules, better algorithms are required that can perform higher order calculations efficiently on desktop computing resources. UCSF Chimera was launched as a solution to this problem (Goddard *et al.*, 2007). Chimera is Python based freely available to academic and nonprofit users from the website provided in "Relevant Websites section". Its GUI and command-line interfaces provide rich and overlapping sets of functionality under Windows, Mac OSX and Linux or other unix based systems. It generates a mixture of low-resolution depictions of assembly components with high resolution depiction of regions of interest to avoid the above-mentioned drawbacks. It offers interactive visualization and analysis of atomic-resolution molecular models as well as specialized capabilities for low resolution depictions, contact calculations and quaternary structure navigation to study large multimeric assemblies such as virus capsids, ribosomes, microtubules, etc. To interactively explore such large complexes, it is important to have the abilities to build multimeric forms, display low-resolution representations and define levels of structure. The widely used molecular visualization programs can also display such density maps but do not offer a broad selection of tools for studying macromolecules.

Chimera was designed from its predecessor MIDAS (Ferrin *et al.*, 1988) with extensibility as a primary goal. Therefore, the architecture of Chimera is composed of a core, for basic services and state-of-the-art visualization, and extensions that provide all extra higher level functionality (Pettersen *et al.*, 2004). The C++ based functions in "core" can perform molecular file input/output, graphical display as representations (e.g., wire-frame, ball-and-stick, ribbon, and sphere), generation of molecular surface using MSMS algorithm, select parts of structures, control transparency, near and far clipping planes, and lenses. On the other hand, the extensions include *Multiscale*, to visualize large-scale molecular assemblies; *Collaboratory*, to share Chimera sessions remotely in real time; *Multalign Viewer*, for reading, writing and displaying multiple sequence alignments together with associated structures; *ViewDock*, facilitates interactive screening of docked ligand orientations from DOCK; *Movie*, for displaying and saving MD

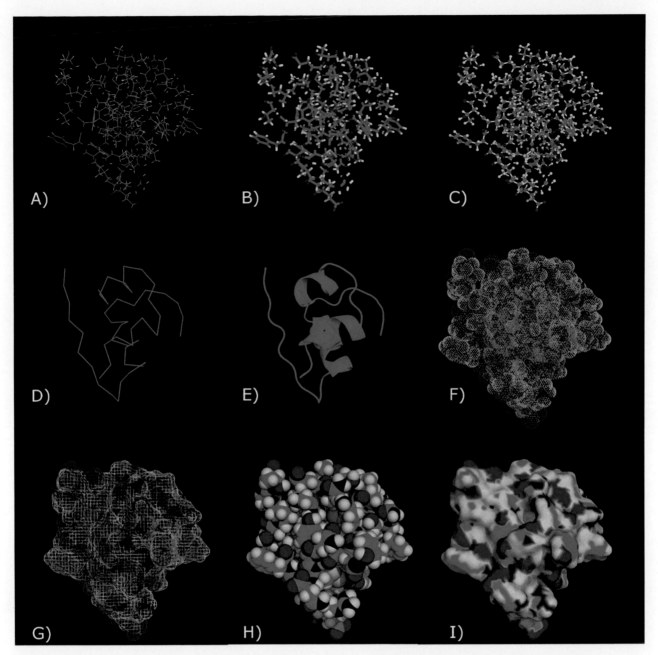

Fig. 5 Molecular representations used in PyMOL to display proteins and other types molecules. Structure of insulin (2HIU) for all these representations namely (a) lines (b) sticks (c) ball and stick (d) ribbons (e) cartoon (f) dots (g) mesh (h) spheres (i) and (j) surface.

trajectories as a video; and *Volume Viewer*, for displaying volumetric data as isosurfaces, meshes, and translucent solids. It now also integrates MODELLER and other modeling tools for structural modeling of multiprotein complexes from sequence to 3D structure (Yang *et al.*, 2012). High-quality images and animations can be generated for publication using Chimera.

Jmol (2004)

Jmol has been developed as an open source Java-based molecular visualization software by an active user community of professional crystallographers, educators and students over many years (Herraez, 2006). Initially, it was made available at the website provided in "Relevant Websites section" however currently, it is available from the website provided in "Relevant Websites section". Jmol creates every pixel of the model on the fly, using exceptionally efficient rastering techniques and without any external 3D graphics packages. That is why Jmol runs independent of the operating system (like Linux, Windows or Macintosh OS) and the browsers. It was designed as a web-based replacement for RasMol or Chime (Rasmol as a web browser plug-in) for crystallographic visualization and analysis. Being web-based, Jmol offered possibilities of revolutionizing the way we learned

Fig. 6 A qualitative representation for insulin's (2HIU) electrostatic contact potential generated by PyMOL using a charge-smoothed surface. The "charge smoothing" is performed by averaging the charges over a small region of space.

about molecular structure (Hanson, 2010; Cass *et al.*, 2005). Jmol can read, display, manipulate, analyze, and output data from molecular structures stored in various file formats. It allowed diverse molecular representations like wireframe, sticks, balls and sticks, spheres, ribbons and cartoons. Its extensive mathematical scripting capability allows extensive analysis of molecular structure by writing functions/scripts using the popular command flow syntaxes. It can produce high-quality images in popular formats like JPG and PNG with or without point-of-view ray (POV-Ray) tracing. Jmol can also depict electron density maps and isotropic B-value ('temperature') data. Finally, it can also be used to read, load and display MD simulations trajectories.

Bioblender (2012)

Traditionally, the electrostatic and hydrophobic characteristics of proteins are visualized as range of colors that vary according to the tool used. The simultaneous visualization of both these properties has been almost impossible, except when visualized in different images. A novel and intuitive software, BioBlender, attempts to describe protein motion with simultaneous visualization of their chemical and physical features (Andrei *et al.*, 2012). It can be downloaded freely from the website provided in "Relevant Websites section" for Windows, Linux and Macintosh. It has been developed as an add-on to an open-source, free, and cross-platform application called as Blender that can access several scientific programs. BioBlender uses Python based scripts for building the interface, various calculations and managing other component programs like PyMOL (for loading PDB files, molecular surface calculations), PDB2PQR (for continuum electrostatics calculations), APBS (for calculating the electrostatic potential), and Blender (for 3D graphics). Blender itself has many benefits, one of them is that physics-based animations can be achieved by simulating forces such as gravity, magnetic, vortex, wind, etc.

BioBlender has introduced simultaneous representation of surface physico-chemical properties of proteins, even in motion. Using features different from color permits their simultaneous delivery in photo-realistic images leaving the utilization of color space for the description of other biochemical information. The new visual code introduced in Bioblender entails the depiction of molecular lipophilic potential (MLP) as a range of optical features going from smooth-shiny for hydrophobic regions to rough-dull for hydrophilic ones. Similarly, animated line particles are used for electrostatic potential (EP) that flow along field lines, proportional to the total charge of the protein (Andrei *et al.*, 2012). A python script named "*pyMLP.py*" is used to calculate the MLP for the protein space. Finally, a custom software "*scivis.exe*" is used to calculate field lines. Therefore, the outcome is the ability to visualize the molecular surface, the EP grid, the gradient grid and the field lines. A continuous perception and visualization of molecular features can be generated by the program using various conformations of macromolecules during their motion. For a protein in motion or with multiple conformations, BioBlender uses Blender Game Engine (BGE). BGE is equipped with special rules to simulate atomic behavior, to interpolate between known conformations and obtain a physically plausible sequence of intermediate conformations. The program gives as output both the intermediate PDB files and the rendered images as a movie. This approach makes nanoscale "un-seen" phenomena such as hydropathy or charges of proteins become more understandable and familiar to our everyday life by drawing viewer's attention to the most active regions of the protein.

JSmol (2013)

JSmol is a JavaScript-only version of Jmol (Hanson, 2010) with the complete set of Jmol functionalities (Hanson *et al.*, 2013). Being Java-based has its own demerits in terms of security. Therefore, Java does not run on some handheld devices but JavaScript

(JS) does. Therefore, Jmol has been completely rewritten in JavaScript to produce identical graphical results. Thus, JSmol is the first full-featured molecular viewer based on JS, and the first ever JS-based viewer for proteins. For small molecules it performs at par with Jmol, but for larger molecules it doesn't scale as Jmol. Nonetheless, JSmol has been successfully implemented in Proteopedia (Hanson *et al.*, 2013; Prilusky *et al.*, 2011).

YASARA View (2015)

The spherical representation of an atom usually requires about 320 triangles, therefore visualizing larger proteins with atomic details requires tens of millions of triangles, far too many for smooth interactive frame rates. A new algorithm called YASARA View was designed to solve this problem (Krieger and Vriend, 2014). It does not depend on high-end shader tricks and therefore works on smartphones too. Too complex objects are replaced with 'impostors', i.e., entities that have pre-calculated textures attached, which make them look like the original object. It allowed sharing the GPU and multiple CPU cores for producing high-quality scenes with perfectly round spheres, shadows and ambient lighting. For changes in the position of the light source, the texture is updated from a collection of 200 different pre-rendered views. Using this approach, YASARA View can display larger complexes on smartphones as well.

NGL Viewer (2015)

With HTML5 feature set and tremendous performance gains of JavaScript, web-browsers have now become more powerful in integrating GPU accelerated image processing, physics and other graphical effects as part of the web page canvas (Yuan *et al.*, 2017). This is due to the development and integration of a JavaScript based Web Graphics Library (WebGL) into modern compatible browsers. WebGL renders interactive 2D and 3D graphics without the use of plug-ins. NGL Viewer makes use of this capability of web browsers to interactively display molecular structures (Rose and Hildebrand, 2015). It can create rich visualization for common structural file formats like PDB, GRO or mmCIF using various molecular representations like *line, point, cartoon, space fill, tube, ribbon, licorice*, etc. In order to greatly reduce the geometric complexity in displaying large macromolecules, the image rendering includes ray-casted impostors. NGL viewer can be embedded into structural database websites like RCSB PDB (see "Relevant Websites section") to allows loading, quick depiction and subsequent manipulation of molecular structures online (Rose *et al.*, 2017). In fact, only NGL Viewer is used by PDB database for structure larger than 10,000 residues (Rose *et al.*, 2016).

3Dmol.js (2015)

3Dmol.js is a pure JavaScript, hardware-accelerated, object-oriented molecular visualization library (Rego and Koes, 2015). It has been developed as an alternative to the software-based rendering of Jmol and JSmol. The major focus is online interactive visualization for molecular structures with near native performance. It uses WebGL (*as described above for NGL viewer*) which comes preinstalled in modern browsers. Therefore, 3Dmol.js does not require any additional plugin. The users can either embed the viewer or use it through a hosted viewer using a URL. It can read various file formats like pdb, sdf, mol2, xyz and cube. Its performance is at par with Jmol, JSmol.

Web3DMol (2017)

Like NGL Viewer, besides allowing interactive displays and manipulation of 3D structures, Web3DMol provides some extra functions such as sequence plot, fragment segmentation, measure tool and meta-information display for a better understanding (Shi *et al.*, 2017). Additionally, NGL Viewer does not allow rendering different parts of a protein under different modes. Web3DMol, on the other hand, can display both the primary, secondary structures and related meta-information (e.g., molecular classification, structural resolution and experimental details) for a structure. It can render different parts of a molecule as different modes using the fragment segmentation and interactively measure distances, angles and area between a group of atoms. Another interesting feature is Web3DMol's graphical display sharing using an automatically generated *sharing-URL*. Anyone can share the complete interactive behaviors such as rotating, zooming and translating, can be easily shared with others using this URL.

Additional Insights From Structure

The growth of tools that help in online rendering of structures have boosted development of several web resources that provide more insights related to protein structures. Such resources are more helpful for undergraduate students in understanding the relationship between 3D protein structure and function (Terrell and Listenberger, 2017; Hayes, 2014). With the same goal in mind, several journals now provide interactive 3D structure visualization while others have started using interactive 3D figures. However, there is nothing better than a dedicated online server or tool or database for this purpose. Below are details on some excellent tools that provide additional insights along with structural visualization.

Aquaria

Aquaria (see "Relevant Websites section") generates a concise visual summary of all related PDB structures for any sequence or PDB structure of interest (O'Donoghue *et al.*, 2015). It is capable of loading information regarding domains, post-translational modifications or single nucleotide polymorphisms (**Fig. 7**). The server generates an all-against-all sequence comparison monthly from PDB and SWISS-Prot. Using this pre-calculated data or comparison, Aquaria ranks and serves the list of structures related to the sequence loaded. Using such information, Aquaria contains a median of 35 structures per protein and no other resources can match this depth of sequence-to-structure information.

PDBsum

PDBsum is a web server capable of performing various analyses that include protein secondary structure, protein-ligand and protein-DNA interactions, structural quality, among others. It is freely available at the website provided in "Relevant Websites section". Previously, PDBsum used static images generated using PyMOL however now it uses 3Dmol.js, RasMol, Jmol, PyMOL, and Strap to display the 3D structures interactively (Schrodinger, 2015; Laskowski *et al.*, 2018; Sayle and Milner-White, 1995; Rego and Koes, 2015; Gille and Frommel, 2001). It uses the SURFNET program (Laskowski, 1995) to calculate cleft regions which can then be visualized. PDBsum provides links to homologous proteins using links to the Sequences Annotated by Structure (SAS) server (Milburn *et al.*, 1998). Secondary structure assignments and topology diagram are computed using PROMOTIF and HERA (Hutchinson and Thornton, 1990; Hutchinson and Thornton, 1996), respectively. Interactions of a protein with other proteins, ligands, or metal ions are generated using HBPLUS and LIGPLOT (Wallace *et al.*, 1995; McDonald and Thornton, 1994). PDBsum shows information related to domain architecture from Pfam and CATH (Finn *et al.*, 2016; Sillitoe *et al.*, 2015). Pores and tunnels in a protein are calculated using Mole (Sehnal *et al.*, 2013). For enzymes, PDBsum shows a reaction diagram obtained from Kyoto Encyclopedia of Genes and Genomes (KEGG) database (Kanehisa *et al.*, 2017). A subset server of PDBsum, DrugPort, identifies all "drug targets" in the PDB using the DrugBank database (Law *et al.*, 2014) and any drug molecules that occur as ligands in PDB structures.

Proteopedia

Proteopedia (available at the website provided in "Relevant Websites section") is a collaborative effort to generate an interactive wiki-based encyclopedia for visualizing 3D structures of protein, nucleic acid and other biomolecules (Hodis *et al.*, 2008). The structural annotation here is intuitive, interactive, and the knowledgeable users can even contribute to the annotations pages in the website (like Wikipedia; see "Relevant Websites section"). It can also be used as a pedagogic tool for teaching the relationships between biomolecular structure and function in the classroom. Initially, Jmol (Hanson, 2010; Cass *et al.*, 2005) was used to display the structures but it required users to have Java installed on their computers. Therefore, Proteopedia has now shifted to JSmol, which does not require Java to be installed. There are now more than 130,000 pages in Proteopedia which have been contributed by more than 3300 users worldwide. Users can also generate content for their scientific publications using the interactive 3D visualizations while keeping these articles absolutely private so that only they may view and edit (Prilusky *et al.*, 2011).

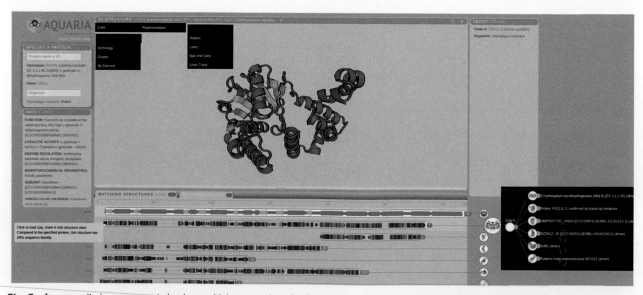

Fig. 7 A composite image generated using multiple screenshots for Aquaria tool. Different options available in the tool have been brought together to provide a general idea of the interface. We loaded the 3ADO, rabbit gulonate dehydrogenase, crystal structure from PDB as an example.

MolviZ.Org

MolviZ.org is another similar web resource that brings together molecular visualization resources. It is available freely at the website provided in "Relevant Websites section". It has a lot of content related to molecular visualization, for example, interactive tutorials, software related to molecular exploration, an atlas that contains images and descriptions of selected macromolecules, etc.

POLYVIEW (-2D/3D/MM)

The POLYVIEW is a 2D and 3D molecular visualization server that offers a flexible annotation tool for generating protein sequence annotations, including secondary structures, relative solvent accessibilities (RSA), functional motifs and polymorphic sites (Porollo and Meller, 2007). 2D graphical representations in a customizable format may be generated for both known protein structures and predictions obtained using protein structure prediction servers like CASP. For example, one can identify putative globular soluble as well as membrane domains to make preliminary conclusions about the domain structure of a protein (Porollo et al., 2004). POLYVIEW may be used for automated generation of pictures with structural and functional annotations like secondary structure (SS) states (H, E, C), RSA, hydrophobicity, polarity and charge profiles for publications. POLYVIEW-3D, is a major update of POLYVIEW. It uses PyMol (Schrodinger, 2015) for high quality rendering of models and structures (Porollo and Meller, 2007). It can also perform advanced structure and function analysis, like mapping interaction interfaces, binding pockets, and comparison and scoring of protein docking models. POLYVIEW also provides extensive cross-linking with several rigorously validated annotation and prediction servers, such as ConSurf, CASTp, ClusPro, and SPPIDER. POLYVIEW-MM (for molecular motion) enables integration of high-quality animation, e.g., trajectories generated by molecular dynamics and related simulation techniques, with structural annotation (Porollo and Meller, 2010).

Future Applications in Molecular Visualization

The molecular visualization software for viewing 3D models is improving in terms of many aspects for example, level of details, perception of depth using lighting, etc. People from various fields like experimental biologists, database developers, computer scientists, and package developers join hands to bring about these advances. The latest and most exotic computer hardware like 3D printers that create three-dimensional objects or displays that produce stereoscopic depth without special glasses and force feedback (haptic) devices that provide a sense of touch are finding applications to molecular assemblies (Nagata et al., 2002; Stocks et al., 2009). With force feedback technology, the user can 'touch' and sense the electrostatic potential field of a protein molecule and aid in various molecular modeling approaches (Haptic-driven, 2012). A globular probe is used to scan the surface of a protein, the electrostatic forces between the protein and the probe are calculated in real-time and fed into the force feedback device (Wollacott and Merz, 2007; Bolopion et al., 2010; Hou et al., 2014; Subasi and Basdogan, 2006). Multi-modality enhancements of such tangible models are being created by superimposing graphical information on physical models, by adding voice commands and by providing haptic feedback (Sankaranarayanan et al., 2003). However, it would require another huge space to discuss these advances in molecular visualization.

Concluding Remarks

By describing the state-of-the-art features used for structural visualization of protein and other macromolecules, this article has provided a brief overview of topic. We hope that its content successfully explains most of the concepts related to molecular visualization, especially related to biomolecules. In fact, there is a continuous need for improvement in the field as there is always new details made available by the advances in other technologies related to structure determination. This keeps creating a void which needs to be filled by more advances and improvements in the existing visualization approaches. The development of better and newer software for molecular visualization continues and keeps advancing the field.

See also: *Ab initio* Protein Structure Prediction. Algorithms for Structure Comparison and Analysis: Docking. Algorithms for Structure Comparison and Analysis: Homology Modelling of Proteins. Algorithms for Structure Comparison and Analysis: Prediction of Tertiary Structures of Proteins. Biomolecular Structures: Prediction, Identification and Analyses. Computational Tools for Structural Analysis of Proteins. Drug Repurposing and Multi-Target Therapies. Identifying Functional Relationships Via the Annotation and Comparison of Three-Dimensional Amino Acid Arrangements in Protein Structures. In Silico Identification of Novel Inhibitors. Natural Language Processing Approaches in Bioinformatics. Pharmacophore Development. Protein Structural Bioinformatics: An Overview. Protein Structure Classification. Protein Structure Databases. Protein Three-Dimensional Structure Prediction. Protocol for Protein Structure Modelling. Secondary Structure Prediction. Small Molecule Drug Design. Structure-Based Design of Peptide Inhibitors for Protein Arginine Deiminase Type IV (PAD4). Study of The Variability of The Native Protein Structure

References

Andrei, R.M., Callieri, M., Zini, M.F., Loni, T., Maraziti, G., et al., 2012. Intuitive representation of surface properties of biomolecules using BioBlender. BMC Bioinformatics 13, S16.

Baker, N.A., Sept, D., Joseph, S., Holst, M.J., McCammon, J.A., 2001. Electrostatics of nanosystems: Application to microtubules and the ribosome. Proceedings of the National Academy of Sciences of the United States of America 98, 10037–10041.

Bansal, M., Kumart, S., Velavan, R., 2000. HELANAL: A program to characterize helix geometry in proteins. Journal of Biomolecular Structure and Dynamics 17, 811–819.

Beem, K.M., Richardson, D.C., Rajagopalan, K.V., 1977. Metal sites of copper-zinc superoxide dismutase. Biochemistry 16, 1930–1936.

Berman, H.M., Westbrook, J., Feng, Z., Gilliland, G., Bhat, T.N., et al., 2000. The protein data bank. Nucleic Acids Research 28, 235–242.

Bolopion, A., Cagneau, B., Redon, S., Régnier, S., 2010. Comparing position and force control for interactive molecular simulators with haptic feedback. Journal of Molecular Graphics and Modelling 29, 280–289.

Bostick, D., Vaisman, I.I., 2003. A new topological method to measure protein structure similarity. Biochemical and Biophysical Research Communications 304, 320–325.

Bostick, D.L., Shen, M., Vaisman, I.I., 2004. A simple topological representation of protein structure: Implications for new, fast, and robust structural classification. Proteins: Structure Function and Genetics 56, 487–501.

Brunger, A.T., Wells, J.A., 2009. Warren L. DeLano 21 June 1972–3 November 2009. Nature Structural & Molecular Biology 16, 1202–1203.

Callahan, T.J., Swanson, E., Lybrand, T.P., 1996. MD Display: An interactive graphics program for visualization of molecular dynamics trajectories. Journal of Molecular Graphics 14 (39–41), 32.

Carson, M., 1987. Ribbon models of macromolecules. Journal of Molecular Graphics 5, 103–106.

Cass, M.E., Rzepa, H.S., Rzepa, D.R., Williams, C.K., 2005. The use of the free, open-source program Jmol to generate an interactive web site to teach molecular symmetry. Journal of Chemical Education 82, 1736.

Connolly, M., 1983. Analytical molecular surface calculation. Journal of Applied Crystallography 16, 548–558.

Deanda, F., Pearlman, R.S., 2002. A novel approach for identifying the surface atoms of macromolecules. Journal of Molecular Graphics and Modelling 20, 415–425.

Duncan, B.S., Macke, T.J., Olson, A.J., 1995. Biomolecular visualization using AVS. Journal of Molecular Graphics 13, 271–282.

Ferrin, T.E., Huang, C.C., Jarvis, L.E., Langridge, R., 1988. The MIDAS display system. Journal of Molecular Graphics 6, 13–27.

Finn, R.D., Coggill, P., Eberhardt, R.Y., Eddy, S.R., Mistry, J., et al., 2016. The Pfam protein families database: Towards a more sustainable future. Nucleic Acids Research 44, D279–D285.

Flores, T.P., Moss, D.S., Thornton, J.M., 1994. An algorithm for automatically generating protein topology cartoons. Protein Engineering 7, 31–37.

Flower, D.R., 1994. β-Sheet topology. A new system of nomenclature. FEBS Letters 344, 247–250.

Gilbert, D., Westhead, D., Nagano, N., Thornton, J., 1999. Motif-based searching in TOPS protein topology databases. Bioinformatics 15, 317–326.

Gilbert, D., Westhead, D., Viksna, J., Thornton, J., 2001. A computer system to perform structure comparison using TOPS representations of protein structure. Computers & Chemistry 26, 23–30.

Gille, C., Frommel, C., 2001. STRAP: Editor for STRuctural Alignments of Proteins. Bioinformatics 17, 377–378.

Goddard, T.D., Ferrin, T.E., 2007. Visualization software for molecular assemblies. Current Opinion in Structural Biology 17, 587–595.

Goddard, T.D., Huang, C.C., Ferrin, T.E., 2005. Software extensions to UCSF chimera for interactive visualization of large molecular assemblies. Structure 13, 473–482.

Goddard, T.D., Huang, C.C., Ferrin, T.E., 2007. Visualizing density maps with UCSF Chimera. Journal of Structural Biology 157, 281–287.

Greer, J., Bush, B.L., 1978. Macromolecular shape and surface maps by solvent exclusion. Proceedings of the National Academy of Sciences of the United States of America 75, 303–307.

Guo, D., Nie, J., Liang, M., et al., 2015. View-dependent level-of-detail abstraction for interactive atomistic visualization of biological structures. Computers & Graphics 52, 62–71.

Hanson, R., 2010. Jmol – A paradigm shift in crystallographic visualization. Journal of Applied Crystallography 43, 1250–1260.

Hanson, R.M., Prilusky, J., Renjian, Z., Nakane, T., Sussman, J.L., 2013. JSmol and the next-generation web-based representation of 3D molecular structure as applied to proteopedia. Israel Journal of Chemistry 53, 207–216.

Hayes, J.M., 2014. An integrated visualization and basic molecular modeling laboratory for first-year undergraduate medicinal chemistry. Journal of Chemical Education 91, 919–923.

Herraez, A., 2006. Biomolecules in the computer: Jmol to the rescue. Biochemistry and Molecular Biology Education 34, 255–261.

Hodis, E., Prilusky, J., Martz, E., et al., 2008. Proteopedia – A scientific 'wiki' bridging the rift between three-dimensional structure and function of biomacromolecules. Genome Biology 9, R121.

Hogue, C.W.V., 1997. Cn3D: A new generation of three-dimensional molecular structure viewer. Trends in Biochemical Sciences 22, 314–316.

Hou, X., Sourina, O., Klimenko, S., 2014. Visual haptic-based collaborative molecular docking. In: Proceedings of the 15th International Conference on Biomedical Engineering.

Hua, Q.X., Gozani, S.N., Chance, R.E., et al., 1995. Structure of a protein in a kinetic trap. Nature Structural Biology 2, 129–138.

Humphrey, W., Dalke, A., Schulten, K., 1996. VMD: Visual molecular dynamics. Journal of Molecular Graphics 14 (33–38), 27–38.

Hutchinson, E.G., Thornton, J.M., 1990. HERA – A program to draw schematic diagrams of protein secondary structures. Proteins 8, 203–212.

Hutchinson, E.G., Thornton, J.M., 1996. PROMOTIF – A program to identify and analyze structural motifs in proteins. Protein Science 5, 212–220.

Juba, D., Varshney, A., 2008. Parallel, stochastic measurement of molecular surface area. Journal of Molecular Graphics and Modelling 27, 82–87.

Kabsch, W., Sander, C., 1983. Dictionary of protein secondary structure: Pattern recognition of hydrogen-bonded and geometrical features. Biopolymers 22, 2577–2637.

Kanehisa, M., Furumichi, M., Tanabe, M., Sato, Y., Morishima, K., 2017. KEGG: New perspectives on genomes, pathways, diseases and drugs. Nucleic Acids Research 45, D353–D361.

Kendrew, J.C., Bodo, G., Dintzis, H.M., et al., 1958. A three-dimensional model of the myoglobin molecule obtained by X-ray analysis. Nature 181, 662–666.

Koch, I., Kaden, F., Selbig, J., 1992. Analysis of protein sheet topologies by graph theoretical methods. Proteins: Structure Function and Bioinformatics 12, 314–323.

Koradi, R., Billeter, M., Wuthrich, K., 1996. MOLMOL: A program for display and analysis of macromolecular structures. Journal of Molecular Graphics 14 (51–55), 29–32.

Krieger, E., Vriend, G., 2014. YASARA view – Molecular graphics for all devices – From smartphones to workstations. Bioinformatics 30, 2981–2982.

Kyte, J., Doolittle, R.F., 1982. A simple method for displaying the hydropathic character of a protein. Journal of Molecular Biology 157, 105–132.

Laskowski, R.A., 1995. SURFNET: A program for visualizing molecular surfaces, cavities, and intermolecular interactions. Journal of Molecular Graphics 13 (323–330), 307–328.

Laskowski, R.A., 2009. PDBsum new things. Nucleic Acids Research 37, D355–D359.

Laskowski, R.A., Jabłońska, J., Pravda, L., Vařeková, R.S., Thornton, J.M., 2018. PDBsum: Structural summaries of PDB entries. Protein Science 27, 129–134.

Law, V., Knox, C., Djoumbou, Y., Jewison, T., Guo, A.C., et al., 2014. DrugBank 4.0: Shedding new light on drug metabolism. Nucleic Acids Research 42, D1091–D1097.

Lee, B., Richards, F.M., 1971. The interpretation of protein structures: Estimation of static accessibility. Journal of Molecular Biology 55, 379. (IN374).

Lesk, A.M., Hardman, K.D., 1982. Computer-generated schematic diagrams of protein structures. Science 216, 539–540.

Levinthal, C., 1966. Molecular model-building by computer. Scientific American 214, 42–52.

Levitt, M., Chothia, C., 1976. Structural patterns in globular proteins. Nature 261, 552.

Martin, A.C.R., 2000. The ups and downs of protein topology; Rapid comparison of protein structure. Protein Engineering Design and Selection 13, 829–837.

Mashaghi, A., van Wijk Roeland, J., Tans Sander, J., 2014. Circuit topology of proteins and nucleic acids. Structure 22, 1227–1237.

May, P., Kreuchwig, A., Steinke, T., Koch, I., 2010. PTGL: A database for secondary structure-based protein topologies. Nucleic Acids Research 38, D326–D330.

McDonald, I.K., Thornton, J.M., 1994. Satisfying hydrogen bonding potential in proteins. Journal of Molecular Biology 238, 777–793.

Merritt, E.A., Bacon, D.J., 1997. Raster3D: Photorealistic Molecular Graphics. Methods in Enzymology. Academic Press. pp. 505–524.

Michalopoulos, I., Torrance, G.M., Gilbert, D.R., Westhead, D.R., 2004. TOPS: An enhanced database of protein structural topology. Nucleic Acids Research 32, D251–D254.

Milburn, D., Laskowski, R.A., Thornton, J.M., 1998. Sequences annotated by structure: A tool to facilitate the use of structural information in sequence analysis. Protein Engineering 11, 855–859.

Nagano, K., 1977. Triplet information in helix prediction applied to the analysis of super-secondary structures. Journal of Molecular Biology 109, 251–274.

Nagata, H., Mizushima, H., Tanaka, H., 2002. Concept and prototype of protein–ligand docking simulator with force feedback technology. Bioinformatics 18, 140–146.

Nelson, M., Humphrey, W., Kufrin, R., et al., 1995. MDScope – A visual computing environment for structural biology. In: Atluri, S.N., Yagawa, G., Cruse, T. (Eds.), Computational Mechanics '95: Theory and Applications. Berlin, Heidelberg: Springer, pp. 476–481.

O'Donoghue, S.I., Sabir, K.S., Kalemanov, M., Stolte, C., 2015. Aquaria: Simplifying discovery and insight from protein structures. Nature Methods 12, 98–99.

Palmer, T.C., Hausheer, F.H., Saxe, J.D., 1989. Applications of ray tracing in molecular graphics. Journal of Molecular Graphics 7, 160–164.

Pettersen, E.F., Goddard, T.D., Huang, C.C., et al., 2004. UCSF Chimera – A visualization system for exploratory research and analysis. Journal of Computational Chemistry 25, 1605–1612.

Porollo, A., Meller, J., 2007. Versatile annotation and publication quality visualization of protein complexes using POLYVIEW-3D. BMC Bioinformatics 8, 316.

Porollo, A., Meller, J., 2010. POLYVIEW-MM: Web-based platform for animation and analysis of molecular simulations. Nucleic Acids Research 38, W662–W666.

Porollo, A., Adamczak, R., Meller, J., 2004. POLYVIEW: A flexible visualization tool for structural and functional annotations of proteins. Bioinformatics 20, 2460–2462.

Prilusky, J., Hodis, E., Canner, D., et al., 2011. Proteopedia: A status report on the collaborative, 3D web-encyclopedia of proteins and other biomolecules. Journal of Structural Biology 175, 244–252.

Rego, N., Koes, D., 2015. 3Dmol.js: Molecular visualization with WebGL. Bioinformatics 31, 1322–1324.

Ricci, A., Anthopoulos, A., Massarotti, A., Grimstead, I., Brancale, A., 2012. Haptic-driven applications to molecular modeling: State-of-the-art and perspectives. Future Medicinal Chemistry 4, 1219–1228.

Richards, F.M., 1977. Areas, volumes, packing and protein structure. Annual Review of Biophysics and Bioengineering 6, 151–176.

Richardson, D.C., Richardson, J.S., 1992. The kinemage: A tool for scientific communication. Protein Science 1, 3–9.

Richardson, J.S., 1977. β-Sheet topology and the relatedness of proteins. Nature 268, 495.

Rocchia, W., Alexov, E., Honig, B., 2001. Extending the applicability of the nonlinear Poisson–Boltzmann equation: Multiple dielectric constants and multivalent ions. The Journal of Physical Chemistry B 105, 6507–6514.

Rose, A.S., Hildebrand, P.W., 2015. NGL Viewer: A web application for molecular visualization. Nucleic Acids Research 43, W576–W579.

Rose, P.W., Prlić, A., Altunkaya, A., et al., 2017. The RCSB protein data bank: Integrative view of protein, gene and 3D structural information. Nucleic Acids Research 45, D271–D281.

Rose, A.S., Bradley, A.R., Valasatava, Y., et al., 2016. Web-based molecular graphics for large complexes. In: Proceedings of the 21st International Conference on Web3D Technology, pp. 185–186. Anaheim, California: ACM.

Ryu, J., Park, R., Kim, D.-S., 2007. Molecular surfaces on proteins via beta shapes. Computer-Aided Design 39, 1042–1057.

Sánchez-Ferrer, A., Núñez-Delicado, E., Bru, R., 1995. Software for viewing biomolecules in three dimensions on the internet. Trends in Biochemical Sciences 20, 286–288.

Sankaranarayanan, G., Weghorst, S., Sanner, M., Gillet, A., Olson, A., 2003. Role of haptics in teaching structural molecular biology. In: Proceedings of the 11th Symposium on Haptic Interfaces for Virtual Environment and Teleoperator Systems (HAPTICS'03), pp. 363–366, March 22–23, 2003.

Sanner, M.F., Olson, A.J., Spehner, J.C., 1996. Reduced surface: An efficient way to compute molecular surfaces. Biopolymers 38, 305–320.

Sayle, R.A., Milner-White, E.J., 1995. RASMOL: Biomolecular graphics for all. Trends in Biochemical Sciences 20, 374.

Schrodinger, L.L.C., 2015. The PyMOL Molecular Graphics System, Version 1.8.

Sehnal, D., Svobodová Vařeková, R., Berka, K., et al., 2013. MOLE 2.0: Advanced approach for analysis of biomacromolecular channels. Journal of Cheminformatics 5, 39.

Shi, M., Gao, J., Zhang, M.Q., 2017. Web3DMol: Interactive protein structure visualization based on WebGL. Nucleic Acids Research.

Sillitoe, I., Lewis, T.E., Cuff, A., et al., 2015. CATH: Comprehensive structural and functional annotations for genome sequences. Nucleic Acids Research 43, D376–D381.

Stivala, A., Wybrow, M., Wirth, A., Whisstock, J.C., Stuckey, P.J., 2011. Automatic generation of protein structure cartoons with Pro-origami. Bioinformatics 27, 3315–3316.

Stocks, M.B., Hayward, S., Laycock, S.D., 2009. Interacting with the biomolecular solvent accessible surface via a haptic feedback device. BMC Structural Biology 9, 69.

Stone, J.E., Sener, M., Vandivort, K.L., et al., 2016. Atomic detail visualization of photosynthetic membranes with GPU-accelerated ray tracing. Parallel Computing 55, 17–27.

Subasi, E., Basdogan, C., 2006. A New Approach to Molecular Docking in Virtual Environments with Haptic Feedback, pp. 141–145.

Terrell, C.R., Listenberger, L.L., 2017. Using molecular visualization to explore protein structure and function and enhance student facility with computational tools. Biochemistry and Molecular Biology Education 45, 318–328.

Tsukamoto, Y., Takiguchi, K., Satou, K., et al., 1997. Application of a deductive database system to search for topological and similar three-dimensional structures in protein. Computer Applications in the Biosciences. 183–190.

Veeramalai, M., Gilbert, D., 2008. A novel method for comparing topological models of protein structures enhanced with ligand information. Bioinformatics 24, 2698–2705.

Víksna, J., Gilbert, D., 2001. Pattern matching and pattern discovery algorithms for protein topologies. In: Gascuel, O., Moret, B.M.E. (Eds.), Proceedings of the First International Workshop on Algorithms in Bioinformatics, WABI 2001, pp. 98–111. Århus Denmark, Berlin, Heidelberg: Springer, August 28–31, 2001.

Wallace, A.C., Laskowski, R.A., Thornton, J.M., 1995. LIGPLOT: A program to generate schematic diagrams of protein-ligand interactions. Protein Engineering 8, 127–134.

Walters, P., Stahl, M., 1996. Babel.

Walther, D., 1997. WebMol – A Java-based PDB viewer. Trends in Biochemical Sciences 22, 274–275.

Wang, Y., Geer, L.Y., Chappey, C., Kans, J.A., Bryant, S.H., 2000. Cn3D: Sequence and structure views for Entrez. Trends in Biochemical Sciences 25, 300–302.

Westhead, D.R., Hatton, D.C., Thornton, J.M., 1998. An atlas of protein topology cartoons available on the world-wide web. Trends in Biochemical Sciences 23, 35–36.

Westhead, D.R., Slidel, T.W.F., Flores, T.P.J., Thornton, J.M., 1999. Protein structural topology: Automated analysis and diagrammatic representation. Protein Science 8, 897–904.

Wollacott, A.M., Merz, K.M., 2007. Haptic applications for molecular structure manipulation. Journal of Molecular Graphics and Modelling 25, 801–805.

Yang, Z., Lasker, K., Schneidman-Duhovny, D., et al., 2012. UCSF Chimera, MODELLER, and IMP: An integrated modeling system. Journal of Structural Biology 179, 269–278.

Yuan, S., Chan, H.C.S., Hu, Z., 2017. Implementing WebGL and HTML5 in macromolecular visualization and modern computer-aided drug design. Trends in Biotechnology 35, 559–571.

Relevant Websites

http://aquaria.ws
  Aquaria.
http://www.bioblender.eu
  Bioblender.

www.jmol.org
   Jmol.
https://sourceforge.net/projects/jmol/
   Jmol download.
https://sourceforge.net/projects/molmol/
   Molmol.
https://www.umass.edu/microbio/chime/index.html
   MolviZ.Org.
https://sourceforge.net/projects/-openrasmol/
   OpenRasMol.
http://www.ospray.org/
   OSPRay.
https://www.rcsb.org/pdb/home/home.do
   PDB.
http://www.wwpdb.org/documentation/file-format
   PDB File Format Documentation.
http://www.ebi.ac.uk/pdbsum
   PDBsum - EMBL-EBI.
http://munk.csse.unimelb.edu.au/pro-origami/
   Protein Structure Cartoons.
http://www.proteopedia.org
   Proteopedia, life in 3D.
http://ptgl.uni-frankfurt.de/
   PTGL - The Protein Topology Graph Library.
https://pymol.org/
   PyMOL.
https://sourceforge.net/projects/pymol/
   PyMOL Molecular Graphics System download.
http://radsite.lbl.gov/radiance/framew.html
   RADIANCE.
http://rasmol.org/
   Rasmol.
http://www.rcsb.org
   RCSB PDB: Homepage.
http://www.povray.org/
   The Persistence of Vision Raytracer.
http://www.bioinf.org.uk/topscan/
   TOPSCAN - Rapid protein structure comparison.
http://tops.ebi.ac.uk/tops
   TOPS - Protein topology atlas - EMBL-EBI.
http://www.cgl.ucsf.edu/chimera/
   UCSF Chimera Home Page - RBVI.
http://www.ks.uiuc.edu/Research/vmd/
   VMD - Visual Molecular Dynamics.
https://github.com/cvdlab-projects/webmol
   webmol: Web Protein Viewer.
https://github.com/cvdlab-projects/webpdb
   webpdb: Web Protein Data Bank.
https://www.wikipedia.org/
   Wikipedia.

Computational Tools for Structural Analysis of Proteins

Luciano A Abriata, Swiss Federal Institute of Technology in Lausanne, Lausanne, Switzerland and Swiss Institute of Bioinformatics, Lausanne, Switzerland

Glossary

Coevolution coupling Extent of correlated amino acid substitutions between pairs of positions in a protein sequence.

Force field (in molecular dynamics simulations) Set of equations and parameters that describe the potential energy of a system, required to compute and propagate forces during molecular dynamics simulations.

HTML The code standard for writing web pages.

NOE restraint Restraint about upper distances between pairs of – typically – hydrogen atoms, as retrieved from the nuclear Overhauser effect in special nuclear magnetic resonance (NMR) spectra.

Site (in the context of proteins) Position in a protein's sequence, residue.

SMILES (small molecules) A computer-readable text code to encode the chemical constitution of molecules.

Introduction

Work on any protein begins by defining the exact amino acid sequence of the relevant protein form and by retrieving annotations from resources like UniProt in **Table 1** (Pundir *et al.*, 2016). One can then make extensive analysis of the protein's sequence in structural and functional terms; but at the deepest residue and atomic levels, protein-specific details are best available from analyses of experimental structures or models of reasonable confidence. Such structural analyses help to readily explain biochemical observations and suggest experiments, toward the final aim of dissecting structure–function relationships for the system and a full physicochemical description of the protein's role in organismal physiology. This article deals with a variety of tools to carry out such structural analyses of proteins.

Table 1 Sources of protein sequences with structural annotations, databases of three dimensional structures, and methods to model biomacromolecules

Retrieval of protein sequences	
UniProt	http://www.uniprot.org/
P-FAM	http://pfam.xfam.org/
Basic local alignment search tool (BLAST) (and BLAST Protein Data Bank (PDB))	http://blast.ncbi.nlm.nih.gov/Blast.cgi
PDBFINDER2	ftp://ftp.cmbi.ru.nl/pub/molbio/data/pdbfinder2/
Experimental structures	
WorldWide PDB	http://wwpdb.org/
RCSB PDB	http://www.rcsb.org/
PDB Europe	http://www.ebi.ac.uk/pdbe/node/1
PDB Japan	http://pdbj.org/
PDB-derived databases	(see Abriata *Briefings in Bioinformatics* 2016)
Primary data for experimental structures	
Nuclear magnetic resonance	http://bmrb.wisc.edu/
X-ray diffraction	http://eds.bmc.uu.se/eds/
Electron microscopy	http://www.rcsb.org/
Resources for structural modeling excluding homology modeling	
Rosetta Suite	https://www.rosettacommons.org/software
QUARK	http://zhanglab.ccmb.med.umich.edu/QUARK/
EVFold	http://evfold.org/evfold-web/evfold.do
RBO Aleph	http://compbio.robotics.tu-berlin.de/rbo_aleph/
RaptorX contact predictor	http://raptorx.uchicago.edu/ContactMap/
Tools for modeling complexes and assemblies	
PowER	http://lbm.epfl.ch/resources
HADDOCK	http://haddock.science.uu.nl/
Packmol	http://www.ime.unicamp.br/~martinez/packmol/home.shtml
LipidBuilder	http://lipidbuilder.epfl.ch/home
CHARMM-GUI	http://charmm-gui.org/

Retrieving Experimental Protein Structures and Modeling Unknown Structures for Analysis

All experimentally determined structures are deposited in a validated form at the servers of the worldwide Protein Data Bank (PDB) partnership (**Table 1**; Berman *et al.*, 2003), containing at the moment over 130,000 structures (Berman *et al.*, 2013). One can reach a given structure either from a PDB ID indicated in a publication, by performing a BLAST or HHpred search against the PDB, or through detailed searches in the PDB data centers themselves, which typically allow extensive filtering based on sequence similarity, the presence of specific molecules, the experimental technique used to solve the structure, structure quality parameters, etc. Note that PDB searches can yield the structure of the protein of interest itself or can yield template structures for homology modeling of the subject protein. Also, PDB searches often return structures of mutants, complexes with antibodies or small molecules, and other forms that do not correspond exactly to the isolated, native protein.

Importantly, many PDB-derived webservers and databases specialize on specific kinds of molecules and even on specific protein families, or in PDB entries containing specific interactions, or in sequence-structure or structure–dynamics relationships, etc. They are extremely useful for the structural biologist and bioinformatician, as they facilitate many kinds of analyses, as recently reviewed in Abriata (2016a).

When no experimental structures are available for the protein or assembly of interest, there are still several avenues that can be exploited to obtain at least a structural model of it, or parts of it, from various sources of information. Such modeling strategies might be exclusively computer-based or might integrate sparse experimental data with computational modeling tools (Tamo *et al.*, 2015). An excellent, up-to-date briefing on protein modeling strategies is available (Kc, 2016), while a very important resource is the collection of articles produced after each biannual critical assessment of structure prediction (CASP) competition (Molt *et al.*, 2016). Here only a brief overview of modeling methods based on protein homology, residue coevolution and ab initio folding is given.

Homology modeling consists in building a model of the coordinates of a protein's atoms using the experimental structure of another protein as a template, both assumed structurally similar given their high sequence similarity (Khan *et al.*, 2016). Since high sequence similarity with a good coverage of the subject sequence is needed, modeling efforts focus mainly on well-defined domains. A new emerging strategy, far less established than homology modeling but applicable when no structures of homologs are available, exploits the strong coevolution couplings undergone by pairs of residues that make contacts in folded proteins. After several works reporting the proof of concept and extended testing (Hopf *et al.*, 2012, 2014; Kamisetty *et al.*, 2013; Marks *et al.*, 2011; Morcos *et al.*, 2011; Ovchinnikov *et al.*, 2014), actual use to model structures began in the last ∼2 years (Abriata, 2016b; Kassem *et al.*, 2016; Ovchinnikov *et al.*, 2015, 2017; Tian *et al.*, 2015). Some online tools like EVFold and RaptorX take a sequence as input, build an alignment from it, compute coevolution couplings and finally build 3D models. Others like Gremlin input a sequence, build an alignment and provide a list of couplings and structural restraints that can be used in external modeling programs.

Last, a series of methods attempt to fold proteins "*ab initio*" or "*de novo*" (Blaszczyk *et al.*, 2013; Bradley *et al.*, 2005; Mabrouk *et al.*, 2015; Xu and Zhang, 2012). The success rate of these methods is relatively low, so extensive external validation is important. (Plus, note that even in successful cases, the folding protocols and pathways do not necessarily mimic the true events that lead to the folded protein.) Among ab initio methods, there are examples based on classical molecular dynamics simulations started from extended conformations; these approaches are computationally very expensive, even if enhanced sampling methods are used, and are currently limited to peptides, although some small proteins have been folded using specialized supercomputers (Lindorff-Larsen *et al.*, 2011; Piana *et al.*, 2013).

Analyzing Protein Structures in Their Biological Contexts: Protein–Protein, Protein–Nucleic Acid and Protein–Membrane Assemblies

Very often, protein structures are determined in nonnative conditions, for example, a membrane protein is solved in micelle lipids instead of a lipid bilayer, or a protein is solved isolated but physiologically exists only within the framework of a larger assembly. One can use a number of tools to build up models of their complexes, and even high-order assemblies, using contact information from experiments or computation, and experimental data about shape and volume of the complex, even accounting for subunit dynamics (Tamo *et al.*, 2015). It is also possible to explore protein oligomerization through – for example, coarse-grain molecular dynamics simulations or docking, typically aided by or tested through mutagenesis experiments. Atomistic molecular dynamics simulations are also potentially useful but are still far from becoming routine due to the extensive sampling required, the typically large systems involved, and many shortcomings of current force fields (Abriata and Dal Peraro, 2015; Bandaranayake *et al.*, 2012; Petrov and Zagrovic, 2014); however, they have potential for scoring predicted complexes (Sarti *et al.*, 2016).

When insertion of proteins into membranes is important, there are tools that help assemble proteins into lipid bilayers, disks, micelles and other lipid-based structures (Bovigny *et al.*, 2015; Jo *et al.*, 2009; Lomize *et al.*, 2006; Stansfeld *et al.*, 2015). Some of these tools further provide starting points for simulations with popular programs. For combining even more elements, in principle with no limitations, programs like Packmol and Cellpack (Johnson *et al.*, 2015; Martinez *et al.*, 2009) become very handy. These programs facilitate the setup of atomistic or coarse-grained systems for visualization, comparison of geometries, serving as starting points for molecular dynamics simulations of complex multimolecular systems as in Spiga *et al.* (2014).

A Note on Computational Models of Proteins and Assemblies

As an important note, each computational model is useful for tasks and analysis of varying complexity depending on the approach and the quality of the underlying experimental and computational information utilized to derive it. For example, a model of a transmembrane protein based on packing of predicted transmembrane helices will be enough to approximate how the protein embeds in biological membranes and to estimate the relative location of specific residues, especially along the membrane normal, possibly even leading to propose mutations that can be tested experimentally. However, it will very likely be unstable in atomistic molecular dynamics simulations and certainly will not serve for docking or virtual screening campaigns.

Alignments in a Structural Context

It is standard practice in biology to think about a subject protein in the context of its family, as this helps to identify residues that are ultimately important for the protein to achieve its function. Moreover, protein 3D structures are conserved much more than their sequences, such that highly similar structures often imply a similar function, which is very important for function prediction of new protein sequences.

These observations have practical parallels in structural bioinformatics, with the advantage that (1) typically much larger numbers of related sequences are analyzed, and – as a consequence of this – (2) that specific reasons for why different amino acids are more or less tolerated at each position can be explored deeply. The second advantage stems from an often overlooked aspect of amino acid variation in proteins, namely that the probability of observing a given substitution in a given condition arises as the combination of a variety of factors that can be affected on top of effects on function itself. The list includes trivial deleterious effects of mutations at active site residues and well-known effects on the stability of the protein, but also effects on protein internal dynamics (González et al., 2016), on the response to unexpected interactions forced at high concentrations (Abriata et al., 2016a), and even effects not directly related to the protein itself, for example, at the transcription level (Weatheritt and Babu, 2013). Moreover, substitutions at multiple sites are often coupled through structural constraints or might display nonadditive effects on some protein traits, facts that often contain structural and functional information. Keeping all these points in mind while analyzing alignments is hard, but as shown in this section many computational tools help to understand some of these effects and thus gain more insights into the subject protein.

On one hand, the new methods for calculating *de novo* protein structures and protein complexes from coevolution patterns as described above, have also proved powerful to find out relevant alternative protein conformations (Morcos et al., 2013). Therefore, given a protein structure or model, mapping of contacts predicted through coevolution methods can reveal alternative conformations of functional significance (**Table 2**). On the other hand, alignments that are consistent with structure according to residue coevolution analysis have also proved informative about the physical and biological chemistry that underlie sequence variation in the family, in such a way that helps to derive structural and functional information (Abriata et al., 2016b). The PsychoProt webserver, originally developed to dissect amino acid variability from deep-sequence-based experiments on saturated mutational libraries (Abriata et al., 2015), helps to easily quantify conservation and variability throughout a protein's sequence and structure and to extract structural and functional information about the protein. For example, PsychoProt can easily unveil polarity/hydrophobicity requirements in soluble and transmembrane proteins, or residues that evolve under functionally relevant constraints for steric hindrance or conformational dynamics. Such kind of analysis is useful for augmenting the amount of information of both experimental structures and models, as well as to justify the position of hydrophobic and polar amino acids in models and to hypothesize about a protein's dynamic features and functional elements.

Tools for Analysis of Protein Structures and Models

This section focuses on two main kinds of analysis one can perform on structures and models. On one hand, treated first, one can (and should, for many applications) check the quality of the structural information; on the other hand, there are those analyses that deal with extracting information from the models.

Checking the Qualities of Experimental and Modeled Structures

Quality checks are important to estimate how well the protein structure or model satisfies the known geometries and the underlying data. These checks are critical when atomic-level details affect the interpretation or further studies that one wants to achieve from the structure or model. This is the case when, for example, setting up systems for molecular docking, virtual screening and molecular dynamics simulations; but on the other hand is less of a problem, for example, for mapping nuclear magnetic resonance (NMR) data on a model or structure, or if a model/structure is used for solving the X-ray structure of a related protein through molecular replacement.

A large array of tools (RPF, PROCHECK, MolProbity, Verify3D, Prosa, WHAT_CHECK, ERRAT, PROVE (Chen et al., 2010; Colovos and Yeates, 1993; Eisenberg et al., 1997; Huang et al., 2012; Wiederstein and Sippl, 2007; Willard et al., 2003) exist that check for geometries and chemistry, such as reasonable backbone and side chain dihedrals, correct stereochemistry, correct packing without clashes, fulfilment of hydrogen-bonding networks, swapped side chains, etc. Other tools are more specialized for checking the quality of the structures in relation to the underlying X-ray or NMR data. Many of these geometry-, chemistry- and data-related tests are actually compulsory upon submission of new structures to the PDB, which is required to ensure the good quality of PDB

Table 2   Online resources for structural analysis of protein sequences, and for structural interpretation of sequence alignments

Disordered regions, residue exposure and (super)secondary structures from sequence	
Disordered regions	http://dis.embl.de/
	http://prdos.hgc.jp/cgi-bin/top.cgi
	http://iupred.enzim.hu/
	http://bioinf.cs.ucl.ac.uk/psipred/?disopred=1
Disordered regions prone to interaction	http://morf.chibi.ubc.ca/
	http://webapp.yama.info.waseda.ac.jp/fang/MoRFs.php
	http://bioinf.cs.ucl.ac.uk/psipred/?disopred=1
	http://anchor.enzim.hu/
Solvent accessibility	http://sable.cchmc.org/
	http://www.cbs.dtu.dk/services/NetSurfP/
Secondary structure	http://bioinf.cs.ucl.ac.uk/psipred/
	http://www.compbio.dundee.ac.uk/jpred/
	https://npsa-prabi.ibcp.fr/cgi-bin/npsa_automat.pl?page=/NPSA/npsa_phd.html
	http://www.ibi.vu.nl/programs/yaspinwww/
	http://split.pmfst.hr/split/4/
Transmembrane helices,	http://www.cbs.dtu.dk/services/TMHMM/
Transmembrane topology and signal peptide	http://www.ch.embnet.org/software/TMPRED_form.html
	http://bioinf.cs.ucl.ac.uk/software_downloads/memsat/
	http://harrier.nagahama-i-bio.ac.jp/sosui/
	http://octopus.cbr.su.se/
	http://phobius.binf.ku.dk/
Coiled-coil prediction	http://www.ch.embnet.org/software/COILS_form.html
	http://paircoil2.csail.mit.edu/
	http://multicoil2.csail.mit.edu/cgi-bin/multicoil2.cgi
	https://npsa-prabi.ibcp.fr/cgi-bin/npsa_automat.pl?page=/NPSA/npsa_lupas.html
Coiled-coil analysis	Fit to a structure: http://arteni.cs.dartmouth.edu/cccp/index.fit.html
	Generate: http://www.grigoryanlab.org/cccp/index.gen.html
	Generate: http://coiledcoils.chm.bris.ac.uk/app/cc_builder/
	Draw wheel diagram: http://www.grigoryanlab.org/drawcoil/
Repeat detection	http://www.ebi.ac.uk/Tools/pfa/radar/
Coevolution from alignments	
Intragenic residue–residue coevolution (for single proteins)	http://evfold.org/evfold-web/
	http://gremlin.bakerlab.org/submit.php
	http://dca.rice.edu/portal/dca/
	http://mistic.leloir.org.ar/index.php
Intergenic residue–residue coevolution (for protein complexes and interactions)	http://gremlin.bakerlab.org/cplx_submit.php
	https://evcomplex.hms.harvard.edu/
	http://csbg.cnb.csic.es/mtserver/
	http://i-coms.leloir.org.ar/index.php
Contact map against coevolution patterns	http://lucianoabriata.altervista.org/evocoupdisplay/gremlin.html
	http://lucianoabriata.altervista.org/evocoupdisplay/evcouplings.html
Structural and physicochemical features from alignments	
PsychoProt	http://psychoprot.epfl.ch/
Variability from alignment	http://psychoprot.epfl.ch/aln2data.html
ProtParam, ProtScale	http://web.expasy.org/protparam/, http://web.expasy.org/protscale/
MultiProtScale	http://lucianoabriata.altervista.org/multiprotscale/multiprotscale.html
Alignment filter	http://lucianoabriata.altervista.org/multiprotscale/alignmentfilterer.html

entries, derived experimentally. It must be said on the other hand, though, that as shown in recent CASP experiments, computational models of better geometry/chemistry scores do not warrant agreement to "true" structures.

Most checking tools are available as servers, and moreover there are servers that integrate multiple tools: PSVS, SAVES, PROSESS, VADAR, CING, ResProx, Vivaldi (Berjanskii *et al.*, 2010, 2012; Bhattacharya *et al.*, 2007; Doreleijers *et al.*, 2012; Hendrickx *et al.*, 2013). Some servers, notably MolProbity, return an overall score that puts together different quality statistics, in this case such that lower numbers overall indicate "better" structures.

For initial checks on specific PDB entries, there are databases of precomputed quality checks (Hooft *et al.*, 1996) and of optimized PDB entries (Joosten *et al.*, 2014). Yet other servers focus on specific problems structures might suffer from, for example, on the refinement of metal sites (Zheng *et al.*, 2014), on trans-cis, amide side chain and peptide plane flips (Touw *et al.*, 2015; Weichenberger and Sippl, 2007), etc.

Visualization of Biomolecular Structures and Models

Moving on to the second kind of structural analyses, more directed to extract information out of the experimental or modeled structure, there are several questions that computational approaches can solve or at least shed light into. These include analysis of structural features such as surface properties, electrostatics, dynamics; identification of active sites, allosteric sites and interaction sites, mapping of conserved and variable regions, testing for the effects of mutations, measuring the degree of exposure of given residues, etc.

Visualization of molecular structures is key to work in structural biology, and a key step of structural analysis. Moreover, careful inspection of structures and models is critical before running automated analysis or molecular dynamics simulations on them, especially to highlight problems that are invisible to the analyses and could actually introduce errors on their results.

Among the most popular visualization tools that are free for academics, there are COOT (especially suited for working with X-ray data), MOLMOL (specialized for NMR structures), VMD (mainly tailored to setup and analyze MD simulations), PyMOL and Chimera (both with very simple ways to model secondary structures and nonnatural groups and to get stunning images). Many of these visualization programs actually do much more than displaying structures, containing tens of commands for molecular building and analysis. With some advanced knowledge of these tools one can also create detailed movies; on the other hand, simple movies about rotations, morphing and vibrations among other few possibilities, can be rendered online with the MovieMaker webserver (Maiti et al., 2005).

Nowadays, integration of biomolecular structures directly inside HTML for online visualization without plug-ins is possible thanks to tools like JSmol (Hanson et al., 2013), PV, GLmol (Virag et al., 2016), 3dmol.js (Rego and Koes, 2015), Molmil (Bekker et al., 2016), Litemol, and the NGLviewer (Rose and Hildebrand, 2015) soon to be updated to handle large macromolecular complexes online (Rose et al., 2016). Such tools make it extremely easy to openly share customized, interactive 3D views of molecules (see Relevant Websites).

Properties of Protein Surfaces and Cores

The physicochemical properties of a protein's core and surface are important to its stability, particularly regarding solubility, and to its interactions with specific substrates and proteins or with other solutes present at high concentrations in crowded media. An easy first qualitative look at a protein's surface properties is to simply inspect its surface colored by residue type in any visualization program, or more quantitatively, by mapping the amino acid hydrophobic moments on the protein surface (Eisenberg et al., 1984). You expect mostly polar and charged amino acids at the surface of a globular soluble protein, and mostly hydrophobic amino acids in membrane-embedded regions. Patches of inconsistent polarity point at sites of potential functional importance (e.g., for protein–protein interactions) or that can be mutated to improve solubility (e.g., in an enzyme) or simply to errors in the case of models. Note that a given protein might contain an isolated amino acid that does not match polarity requirements while the whole family does contain information about the correct polarity preferred at the position; here is where the joint analysis of alignments in the context of protein structures with tools like PsychoProt as mentioned above (Abriata et al., 2016b) becomes useful. Last, there are also tools to simplify visualization and comparison of the shape and physicochemical features of protein surfaces through mapping in 2D (Yang et al., 2012).

Closely related to surface polarity is the global electrostatics of a protein or complex (Dong et al., 2008). The APBS program and online server (Baker et al., 2001) computes electrostatic fields and electrostatic surface maps from the coordinates of a molecular system, which can be visualized easily in programs like PyMOL and VMD. This requires assignment of charges to all atoms, which can be carried out with PDB2PQR (Dolinsky et al., 2007) through a handful of charge models and which might in turn require assessment of the protonation states of several groups as automatically carried out by PROPKA (Olsson et al., 2011). The main PDB2PQR server (Unni et al., 2011) performs all these calculations, in its simplest mode taking a PDB file and returning an APBS file for visualization (as electrostatic surfaces or fields) in external programs like PyMOL and VMD. VMD can also itself calculate the electric moment of a selection of atoms, which can then be used to track its evolution over time.

Regarding protein cores, structure-based analyses of alignments with PsychoProt revealed that not only hydrophobicity is required in a protein's interior but also precise combinations of volume and steric hindrance, adorned with regions of certain flexibility or conformational preferences closer to the protein surface (Abriata et al., 2015, 2016b). Stability of the core of globular proteins upon mutation can be tested with tools like FoldX, I-Mutant, Dmutant, PopMusic and Rosetta, among others. These programs tend to be better predictors of destabilizing than stabilizing mutations (for an assessment of different tools see (Khan and Vihinen, 2010)). Last, another informative way to test packing in a protein structure or model is to compare its contact map to the coevolutionary couplings computed from a related alignment (tools for this are available in **Table 2**).

As a final point, for precise quantification of solvent accessibility, programs like VMD have built-in functions that can easily be executed on multiple structures or on simulation trajectories, while servers like POPS (Cavallo et al., 2003) provide easier ways to run the calculation but on single structures.

Small-Molecule Pockets and Protein-Small Molecule Interactions

Testing the interactions between small molecules and proteins requires special treatments, and the procedure is usually split into finding pockets where small molecules can fit, and finding how small molecules can bind to it.

Many tools allow users to search for potential small molecule-binding sites in protein structures. Among them, fpocket (Schmidtke *et al.*, 2010) implements a high-performance algorithm which allows detection and tracking of pockets in large ensembles of structures or molecular dynamics trajectories. Another very recent and complete tool, PockDrug (Hussein *et al.*, 2015), is a server specialized in finding druggable pockets, which inputs a PDB file and returns the detected pockets with complete physicochemical descriptions plus scores that estimate how druggable they are.

Once a druggable pocket has been identified in a protein, or if it is already known, one can choose among a large list of academic and paid programs for docking small molecules to the target pockets (full searches are also possible but much more time consuming and leading to less significant results). These programs can be used to test how a given molecule would dock in a pocket, often called "virtual docking," and can be run iteratively on a library to predict which molecules are expected to bind stronger, often called "virtual screening." It is important that such campaigns usually seek to enrich libraries in drug leads and potentially active molecules, but are currently not capable of delivering definitive answers on the perfect drug for a given pocket. For a recent review of potentials and limitations for realistic uses in biology, see Forli (2015), Irwin and Shoichet (2016). For the special case of docking peptides as small molecules onto proteins, there are specialized tools from the Rosetta and HADDOCK families (London *et al.*, 2011; Spiliotopoulos *et al.*, 2016).

Note that there are many challenges in the field of docking and virtual screening, related to poor sampling of flexibility especially in the target protein and to overlooked or only poorly described effects of water (Spyrakis and Cavasotto, 2015). Such problems could in principle be solved by using atomistic molecular dynamics simulations; however, they are orders of magnitude more expensive and hence currently outperformed by docking-based methods, although they are increasingly becoming useful to score poses obtained through docking-based methods (for a thorough discussion on the role of molecular dynamics simulations in drug discovery, see De Vivo *et al.* (2016).

Regarding the structural universe of small molecules, among the most important servers for structural research there are some (with Chemicalize.org (Southan and Stracz, 2013) and CORINA standing out) that facilitate conversion between compound names, SMILES codes, 2D schemes and 3D structures of small molecules, some of them linking directly to existing annotations and retrieving or computing on-the-fly physicochemical descriptors for the small molecules. Other resources for small molecules act as mere databases of small molecules, useful along the drug discovery process, such as ZINC and PubChem (Irwin and Shoichet, 2005; Kim *et al.*, 2016).

Dynamics From Structures

Experimentally, protein dynamics are accessible at the residue level from crystallographic B-factors if properly analyzed (see the B-factor DataBase (Touw and Vriend, 2014)) and from NMR ^{15}N relaxation (note that simply variability in NMR structures is not a good indication of dynamics). Computationally, the ultimate tool is provided by atomistic molecular dynamics simulations, but such methods are costly and often unnecessary if only identification of flexible regions is required.

It is possible to predict flexibility as "disorder" from sequences with variable confidence, or to inspect a sequence manually by mapping coil propensity and other amino acid descriptors with programs like ProtScale (**Table 2**). Meanwhile, analysis of alignments with tools like PsychoProt can point at residues where amino acids of small volumes, low steric hindrance, large flexibility or low conformational preference are better tolerated, again indicating dynamics.

Adding a structural level to improve confidence, there are programs that predict flexible regions from structures. Some of them exploit the fact that regions of low atomic packing or high solvent accessibility tend to be more flexible (Marsh, 2013), whereas others like PDBFlex (Hrabe *et al.*, 2016) are based on the comparison of PDB entries for related proteins or even same proteins crystallized in different conditions. Another tool especially useful to identify domain motions mediated by loops acting as hinges is Normal Mode Analysis, implemented together with other related tools in the program ProDy (Bakan *et al.*, 2011). Last and probably highest in detail but only reporting on loops, a tool developed by the Brüschweiler group from a dataset of sub-microsecond MD simulations validated with NMR data, looks directly at protein loops in a structure returning their predicted flexibility and motion timescales (Gu *et al.*, 2015).

Predicting Active Sites and Functions

Prediction of function for a novel protein, or of even subtle functional changes upon mutation or upon interactions, are particularly challenging even if experimental structures are available. While annotation databases or deep searching on Pubmed, for example, with text-mining tools like ChiliBot (Chen and Sharp, 2004) can lead to functional hints directly from written information, there are also tools that can propose approximate functions from sequences and structures. The critical assessment of functional annotation (CAFA) regularly evaluates the state of function evaluation methods from information at the sequence level (Wass *et al.*, 2014), which includes methods based on sequence comparison, on co-occurrence in operons, regulons and genomes, on analysis of gene fusions, on protein colocation, and on structure. Other efforts and groups focus specifically on predicting function from structures, as briefly covered here.

Methods for predicting protein function from sequence and structural homology have been around for some time, as reviewed by Petrey *et al.* (2015). They perform reasonably well at least at the level of broadly defining function, but cannot give precisions on, for example, the exact substrates of enzymes, which is often very sensitive to point mutations (Abriata *et al.*, 2012). As a proxy for function assignment, many programs actually predict ligand-binding sites through comparison with known data, as recently evaluated in a subcategory of CASP10 (Gallo Cassarino *et al.*, 2014). Some modeling servers like I-TASSER actually provide such analysis on their top models.

When a large alignment is available, there is evidence that gradients of amino acid substitution rates mapped on an enzyme's structure can approximate the location of its active site (Jack *et al.*, 2016), which one might better define through experimental evaluation of mutants of the most conserved residues. One can also analyze what physicochemical properties shape the preference for each of the 20 amino acids at each position of the alignment, with tools like the PsychoProt server, under the premise that preferences for flexibility, steric hindrance, volumes and backbone conformational descriptors could hint at functionally relevant regions of a protein as exemplified in Abriata *et al.* (2016b).

Data Mining

The far end within the field of computational analysis of protein structures consists in (semi)automated mining of PDB entries, either on the full PDB or on PDB-derived databases. Mining the PDB with the help of clever connections to other sources of information, powerful analytical methods, and often using specialized sub-databases, has brought to light several fine details about protein structure, dynamics and hydration, features of transmembrane proteins, metal centers, cofactor and ligand binding, and noncovalent interactions between biomolecules (a few examples among a few hundred such works in the last decade: (Gallivan and Dougherty, 1999; Jackson *et al.*, 2007; Kumar and Balaji, 2014; Lanzarotti *et al.*, 2011; Ringer *et al.*, 2007; Shi *et al.*, 2002; Valley *et al.*, 2012)). But also, mining studies can disclose effects of different experimental methods, diffraction resolution, processing software, etc. (Abriata, 2012, 2013; Djinovic-Carugo and Carugo, 2015; Harding, 2002).

Mining protocols usually involve writing specific programs and scripts for the analysis of interest, but tools aimed at simplifying mining exist (Sehnal *et al.*, 2015; Valasatava *et al.*, 2014).

Comparing Structures in 3D

Comparison of biomolecular structures is at the heart of structural biology, as this science attempts to explain function in relation to structure, for example, when classifying domains, inferring function from structure, modeling proteins from homologs, and at higher level of detail when explaining functional differences as consequences of structural differences.

The problem of comparing two structures has two main steps, namely the detection of similar elements in both structures, which are to be superimposed upon alignment, and the subsequent computation that aligns the coordinates of the proteins based on such similarities. Multiple structure alignments usually start by computing all possible pairwise alignments, and then running a global optimization.

At the computer, most visualization tools allow the user to align and compare structures in 3D with simple methods that work fairly well when there is obvious similarity. However, more complex protocols are available from *ad hoc* programs and servers. Among the most widely used, currently up-to-date tools, there are CE, DALI, FATCAT and FATCAT POSA, MAMMOTH, SALIGN, SSP, TopMatch, TM-align and VAST/VAST+ (Holm and Laakso, 2016; Madej *et al.*, 2014; Ortiz *et al.*, 2002; Shindyalov and Bourne, 2001; Sippl and Wiederstein, 2008; Ye and Godzik, 2004a, 2004b; Zhang and Skolnick, 2005). It is important to highlight that different programs usually incur in inconsistencies among each other, even for relatively similar proteins (Sadowski and Taylor, 2012). Some of these tools allow comparisons between multiple provided structures, others (most notably DALI) allow quering a structure against all PDB entries; some facilitate instant online visualization; and in particular, FATCAT includes internal flexibility in its alignments, with a version (POSA) for multiple protein comparisons and another for scanning structural databases. The RCSB PDB currently facilitates online pairwise comparisons of its entries through some of these programs.

Other More Specific Analyses

A multitude of computational tools are continuously developed tailored to more specific questions, tasks and molecules. Examples are the Protein Knot Server to detect knots in structures (Kolesov *et al.*, 2007), the CheckMyMetal server to validate metal ions in protein structures (Zheng *et al.*, 2014), servers for advanced superposition and 3d-searching of proteins structures (Holm and Rosenström, 2010; Maiti *et al.*, 2004), structural databases containing precomputed descriptions for specific molecules (Abriata, 2016a), servers to predict spectroscopic observables like circular dichroism and NMR spectra from protein structures (Bulheller and Hirst, 2009; Han *et al.*, 2011), servers to validate protein structures and quantify their qualities (Willard *et al.*, 2003), and certainly much more.

A Glance at the Power of Molecular Simulation Methods

This article has focused on modeling and analyzing static structures; however, macromolecules undergo dynamics over orders-of-magnitude timescales ranging from the fastest pico-to-nanosecond vibrations to the slowest loop motions relevant for catalysis and allosterism, conformational transitions, breathing motions and folding events in the microsecond to second timescales. Understanding protein folding, function, regulation, interactions, assembly processes, etc. requires consideration of these dynamic features of molecules, for which computational molecular simulations are as important as experiments. Moreover, simulations have the potential to embrace most other kinds of analysis, for example, predicting secondary and even tertiary structures through folding simulations, predicting small-molecule binding even after free diffusion, exploring reactivity, etc., although far much less efficiently and not necessarily more accurately due to problems in the force fields.

The most important tools to study protein dynamics can be split into those based on quantum methods (which allow probing bond breaking and formation, i.e., reactions, as well as energy-exchange processes related to spectroscopic observables), and those based on classical molecular mechanics either at all-atom level (where bonds cannot be broken or formed, but all interatomic dynamics are accessible allowing the sampling of conformational dynamics) or at coarse-grain level (where atoms are grouped into beads of variable size according to the coarse-graining scheme, loosing atomic detail but allowing for extended diffusional sampling). The three levels of detail can in turn be mixed, being particularly popular in structural biology the combination of atomistic molecular mechanics with quantum-level calculations (QM/MM) to study, for example, enzyme catalysis.

Note that one way or another, all simulations methods require parametrizations (most importantly in the form of force fields for classical molecular mechanics simulations, or orbital descriptions in quantum calculations) and assumptions on how to calculate and propagate forces. Therefore, just like results from most other computational tools, results from simulations need to be tested against experimental data, or eventually taken as mere predictions to guide experiments or explain existing results. Classical dynamics can, for example, be compared against NMR dynamic data, quantum calculations can be tested by predicting and comparing spectroscopic observables.

For classical dynamics, several force fields and functional forms are available to treat proteins, nucleic acids, and some ions and lipids. If small molecules are to be used in molecular dynamics simulations, there are a few online databases of ready-to-use parameter files and also online tools that assist in parametrization (Kirschner et al., 2008; Malloci et al., 2015; Zoete et al., 2011) and building complex systems (Bovigny et al., 2015; Jo et al., 2009; Martinez et al., 2009).

See also: Biomolecular Structures: Prediction, Identification and Analyses. Computational Protein Engineering Approaches for Effective Design of New Molecules. Drug Repurposing and Multi-Target Therapies. Identification of Homologs. Identifying Functional Relationships Via the Annotation and Comparison of Three-Dimensional Amino Acid Arrangements in Protein Structures. In Silico Identification of Novel Inhibitors. Natural Language Processing Approaches in Bioinformatics. Protein Design. Protein Structural Bioinformatics: An Overview. Protein Structure Analysis and Validation. Protein Structure Classification. Protein Structure Visualization. Protein Three-Dimensional Structure Prediction. Small Molecule Drug Design. Structure-Based Design of Peptide Inhibitors for Protein Arginine Deiminase Type IV (PAD4). The Evolution of Protein Family Databases

---

References

Abriata, L.A., 2012. Analysis of copper-ligand bond lengths in X-ray structures of different types of copper sites in proteins. Acta Crystallographica Section D, Biological Crystallography 68, 1223–1231.

Abriata, L.A., 2013. Investigation of non-corrin cobalt(II)-containing sites in protein structures of the Protein Data Bank. Acta Crystallographica Section D, Biological Crystallography 69, 176–183.

Abriata, L.A., 2016a. Structural database resources for biological macromolecules. Briefings in Bioinformatics. pii: bbw049 (Epub ahead of print).

Abriata, L.A., 2016b. Homology- and coevolution-consistent structural models of bacterial copper-tolerance protein CopM support a "metal sponge" function and suggest regions for metal-dependent protein-protein interactions. bioRxiv.

Abriata, L.A., Bovigny, C., Dal Peraro, M., 2016b. Detection and sequence/structure mapping of biophysical constraints to protein variation in saturated mutational libraries and protein sequence alignments with a dedicated server. BMC Bioinformatics 17, 242.

Abriata, L.A., Dal Peraro, M., 2015. Assessing the potential of atomistic molecular dynamics simulations to probe reversible protein–protein recognition and binding. Scientific Reports 5, 10549.

Abriata, L.A., Palzkill, T., Dal Peraro, M., 2015. How structural and physicochemical determinants shape sequence constraints in a functional enzyme. PLOS ONE 10, e0118684.

Abriata, L.A., Salverda, M.L.M., Tomatis, P.E., 2012. Sequence-function-stability relationships in proteins from datasets of functionally annotated variants: The case of TEM β-lactamases. FEBS Letters 586, 3330–3335.

Abriata, L.A., Spiga, E., Peraro, M.D., 2016a. Molecular effects of concentrated solutes on protein hydration, dynamics, and electrostatics. Biophysical Journal 111, 743–755.

Bakan, A., Meireles, L.M., Bahar, I., 2011. ProDy: Protein dynamics inferred from theory and experiments. Bioinformatics (Oxford, England) 27, 1575–1577.

Baker, N.A., Sept, D., Joseph, S., Holst, M.J., McCammon, J.A., 2001. Electrostatics of nanosystems: Application to microtubules and the ribosome. Proceedings of the National Academy of Sciences of the United States of America 98, 10037–10041.

Bandaranayake, R.M., Ungureanu, D., Shan, Y., et al., 2012. Crystal structures of the JAK2 pseudokinase domain and the pathogenic mutant V617F. Nature Structural & Molecular Biology 19, 754–759.

Bekker, G.-J., Nakamura, H., Kinjo, A.R., 2016. Molmil: A molecular viewer for the PDB and beyond. Journal of Cheminformatics 8, 42.

Berjanskii, M., Liang, Y., Zhou, J., et al., 2010. PROSESS: A protein structure evaluation suite and server. Nucleic Acids Research 38, W633–W640.

Berjanskii, M., Zhou, J., Liang, Y., Lin, G., Wishart, D.S., 2012. Resolution-by-proxy: A simple measure for assessing and comparing the overall quality of NMR protein structures. Journal of Biomolecular NMR 53, 167–180.

Berman, H., Henrick, K., Nakamura, H., 2003. Announcing the worldwide Protein Data Bank. Nature Structural & Molecular Biology 10, 980.

Berman, H.M., Coimbatore Narayanan, B., Di Costanzo, L., et al., 2013. Trendspotting in the Protein Data Bank. FEBS Letters 587, 1036–1045.

Bhattacharya, A., Tejero, R., Montelione, G.T., 2007. Evaluating protein structures determined by structural genomics consortia. Proteins 66, 778–795.

Blaszczyk, M., Jamroz, M., Kmiecik, S., Kolinski, A., 2013. CABS-fold: Server for the de novo and consensus-based prediction of protein structure. Nucleic Acids Research 41, W406–W411.

Bovigny, C., Tamò, G., Lemmin, T., Maïno, N., Dal Peraro, M., 2015. LipidBuilder: A framework to build realistic models for biological membranes. Journal of Chemical Information and Modeling 55, 2491–2499.

Bradley, P., Malmström, L., Qian, B., et al., 2005. Free modeling with Rosetta in CASP6. Proteins 61 (Suppl 7), 128–134.

Bulheller, B.M., Hirst, J.D., 2009. DichroCalc – Circular and linear dichroism online. Bioinformatics (Oxford, England) 25, 539–540.

Cavallo, L., Kleinjung, J., Fraternali, F., 2003. POPS: A fast algorithm for solvent accessible surface areas at atomic and residue level. Nucleic Acids Research 31, 3364–3366.

Chen, H., Sharp, B.M., 2004. Content-rich biological network constructed by mining PubMed abstracts. BMC Bioinformatics 5, 147.

Chen, V.B., Arendall, W.B., Headd, J.J., et al., 2010. MolProbity: All-atom structure validation for macromolecular crystallography. Acta Crystallographica Section D, Biological Crystallography 66, 12–21.

Colovos, C., Yeates, T.O., 1993. Verification of protein structures: Patterns of nonbonded atomic interactions. Protein Science 2, 1511–1519.

De Vivo, M., Masetti, M., Bottegoni, G., Cavalli, A., 2016. Role of molecular dynamics and related methods in drug discovery. Journal of Medicinal Chemistry 59, 4035–4061.

Djinovic-Carugo, K., Carugo, O., 2015. Structural biology of the lanthanides-mining rare earths in the Protein Data Bank. Journal of Inorganic Biochemistry 143, 69–76.

Dolinsky, T.J., Czodrowski, P., Li, H., et al., 2007. PDB2PQR: Expanding and upgrading automated preparation of biomolecular structures for molecular simulations. Nucleic Acids Research 35, W522–W525.

Dong, F., Olsen, B., Baker, N.A., 2008. Computational methods for biomolecular electrostatics. Methods in Cell Biology 84, 843–870.

Doreleijers, J.F., Sousa da Silva, A.W., Krieger, E., et al., 2012. CING: An integrated residue-based structure validation program suite. Journal of Biomolecular NMR 54, 267–283.

Eisenberg, D., Lüthy, R., Bowie, J.U., 1997. VERIFY3D: Assessment of protein models with three-dimensional profiles. Methods in Enzymology 277, 396–404.

Eisenberg, D., Schwarz, E., Komaromy, M., Wall, R., 1984. Analysis of membrane and surface protein sequences with the hydrophobic moment plot. Journal of Molecular Biology 179, 125–142.

Forli, S., 2015. Charting a path to success in virtual screening. Molecules 20, 18732–18758.

Gallivan, J.P., Dougherty, D.A., 1999. Cation-pi interactions in structural biology. Proceedings of the National Academy of Sciences of the United States of America 96, 9459–9464.

Gallo Cassarino, T., Bordoli, L., Schwede, T., 2014. Assessment of ligand binding site predictions in CASP10. Proteins 82 (Suppl 2), 154–163.

González, M.M., Abriata, L.A., Tomatis, P.E., Vila, A.J., 2016. Optimization of conformational dynamics in an epistatic evolutionary trajectory. Molecular Biology and Evolution 33 (7), 1768–1776.

Gu, Y., Li, D.-W., Brüschweiler, R., 2015. Decoding the mobility and time scales of protein loops. Journal of Chemical Theory and Computation 11, 1308–1314.

Han, B., Liu, Y., Ginzinger, S.W., Wishart, D.S., 2011. SHIFTX2: Significantly improved protein chemical shift prediction. Journal of Biomolecular NMR 50, 43–57.

Hanson, R.M., Prilusky, J., Renjian, Z., Nakane, T., Sussman, J.L., 2013. JSmol and the next-generation web-based representation of 3D molecular structure as applied to proteopedia. Israel Journal of Chemistry 53, 207–216.

Harding, M.M., 2002. Metal-ligand geometry relevant to proteins and in proteins: Sodium and potassium. Acta Crystallographica Section D, Biological Crystallography 58, 872–874.

Hendrickx, P.M.S., Gutmanas, A., Kleywegt, G.J., 2013. Vivaldi: Visualization and validation of biomacromolecular NMR structures from the PDB. Proteins 81, 583–591.

Holm, L., Laakso, L.M., 2016. Dali server update. Nucleic Acids Research 44, W351–W355.

Holm, L., Rosenström, P., 2010. Dali server: Conservation mapping in 3D. Nucleic Acids Research 38, W545–W549.

Hooft, R.W., Vriend, G., Sander, C., Abola, E.E., 1996. Errors in protein structures. Nature 381, 272.

Hopf, T.A., Colwell, L.J., Sheridan, R., et al., 2012. Three-dimensional structures of membrane proteins from genomic sequencing. Cell 149, 1607–1621.

Hopf, T.A., Schärfe, C.P.I., Rodrigues, J.P.G.L.M., et al., 2014. Sequence co-evolution gives 3D contacts and structures of protein complexes. eLife 3, e03430.

Hrabe, T., Li, Z., Sedova, M., et al., 2016. PDBFlex: Exploring flexibility in protein structures. Nucleic Acids Research 44, D423–D428.

Huang, Y.J., Rosato, A., Singh, G., Montelione, G.T., 2012. RPF: A quality assessment tool for protein NMR structures. Nucleic Acids Research 40, W542–W546.

Hussein, H.A., Borrel, A., Geneix, C., et al., 2015. PockDrug-server: A new web server for predicting pocket druggability on holo and apo proteins. Nucleic Acids Research 43, W436–W442.

Irwin, J.J., Shoichet, B.K., 2005. ZINC – A free database of commercially available compounds for virtual screening. Journal of Chemical Information and Modeling 45, 177–182.

Irwin, J.J., Shoichet, B.K., 2016. Docking screens for novel ligands conferring new biology. Journal of Medicinal Chemistry 59, 4103–4120.

Jack, B.R., Meyer, A.G., Echave, J., Wilke, C.O., 2016. Functional sites induce long-range evolutionary constraints in enzymes. PLOS Biology 14, e1002452.

Jackson, M.R., Beahm, R., Duvvuru, S., et al., 2007. A preference for edgewise interactions between aromatic rings and carboxylate anions: The biological relevance of anion-quadrupole interactions. The Journal of Physical Chemistry B 111, 8242–8249.

Jo, S., Lim, J.B., Klauda, J.B., Im, W., 2009. CHARMM-GUI membrane builder for mixed bilayers and its application to yeast membranes. Biophysical Journal 97, 50–58.

Johnson, G.T., Autin, L., Al-Alusi, M., et al., 2015. cellPACK: A virtual mesoscope to model and visualize structural systems biology. Nature Methods 12, 85–91.

Joosten, R.P., Long, F., Murshudov, G.N., Perrakis, A., 2014. The PDB_REDO server for macromolecular structure model optimization. IUCrJ 1, 213–220.

Kamisetty, H., Ovchinnikov, S., Baker, D., 2013. Assessing the utility of coevolution-based residue-residue contact predictions in a sequence- and structure-rich era. Proceedings of the National Academy of Sciences of the United States of America 110, 15674–15679.

Kassem, M.M., Wang, Y., Boomsma, W., Lindorff-Larsen, K., 2016. Structure of the bacterial cytoskeleton protein bactofilin by NMR chemical shifts and sequence variation. Biophysical Journal 110, 2342–2348.

Kc, D.B., 2016. Recent advances in sequence-based protein structure prediction. Briefings in Bioinformatics. doi:10.1093/bib/bbw070.

Khan, F.I., Wei, D.-Q., Gu, K.-R., Hassan, M.I., Tabrez, S., 2016. Current updates on computer aided protein modeling and designing. International Journal of Biological Macromolecules 85, 48–62.

Khan, S., Vihinen, M., 2010. Performance of protein stability predictors. Human Mutation 31, 675–684.

Kim, S., Thiessen, P.A., Bolton, E.E., et al., 2016. PubChem substance and compound databases. Nucleic Acids Research 44, D1202–D1213.

Kirschner, K.N., Yongye, A.B., Tschampel, S.M., et al., 2008. GLYCAM06: A generalizable biomolecular force field. Carbohydrates. Journal of Computational Chemistry 29, 622–655.

Kolesov, G., Virnau, P., Kardar, M., Mirny, L.A., 2007. Protein knot server: Detection of knots in protein structures. Nucleic Acids Research 35, W425–W428.

Kumar, M., Balaji, P.V., 2014. C–H...pi interactions in proteins: Prevalence, pattern of occurrence, residue propensities, location, and contribution to protein stability. Journal of Molecular Modeling 20, 2136.

Lanzarotti, E., Biekofsky, R.R., Estrin, D.A., Marti, M.A., Turjanski, A.G., 2011. Aromatic-aromatic interactions in proteins: Beyond the dimer. Journal of Chemical Information and Modeling 51, 1623–1633.

Lindorff-Larsen, K., Piana, S., Dror, R.O., Shaw, D.E., 2011. How fast-folding proteins fold. Science 334, 517–520.

Lomize, M.A., Lomize, A.L., Pogozheva, I.D., Mosberg, H.I., 2006. OPM: Orientations of proteins in membranes database. Bioinformatics (Oxford, England) 22, 623–625.

London, N., Raveh, B., Cohen, E., Fathi, G., Schueler-Furman, O., 2011. Rosetta FlexPepDock web server – High resolution modeling of peptide-protein interactions. Nucleic Acids Research 39, W249–W253.

Mabrouk, M., Putz, I., Werner, T., et al., 2015. RBO Aleph: Leveraging novel information sources for protein structure prediction. Nucleic Acids Research 43, W343–W348.

Madej, T., Lanczycki, C.J., Zhang, D., et al., 2014. MMDB and VAST +: Tracking structural similarities between macromolecular complexes. Nucleic Acids Research 42, D297–D303.

Maiti, R., Van Domselaar, G.H., Wishart, D.S., 2005. MovieMaker: A web server for rapid rendering of protein motions and interactions. Nucleic Acids Research 33, W358–W362.

Maiti, R., Van Domselaar, G.H., Zhang, H., Wishart, D.S., 2004. SuperPose: A simple server for sophisticated structural superposition. Nucleic Acids Research 32, W590–W594.

Malloci, G., Vargiu, A.V., Serra, G., et al., 2015. A database of force-field parameters, dynamics, and properties of antimicrobial compounds. Molecules Basel Switzerland 20, 13997–14021.

Marks, D.S., Colwell, L.J., Sheridan, R., et al., 2011. Protein 3D structure computed from evolutionary sequence variation. PLOS ONE 6, e28766.

Marsh, J.A., 2013. Buried and accessible surface area control intrinsic protein flexibility. Journal of Molecular Biology 425, 3250–3263.

Martinez, L., Andrade, R., Birgin, E.G., Martinez, J.M., 2009. Packmol: A package for building initial configurations for molecular dynamics simulations. Journal of Computational Chemistry 30, 2157–2164.

Morcos, F., Jana, B., Hwa, T., Onuchic, J.N., 2013. Coevolutionary signals across protein lineages help capture multiple protein conformations. Proceedings of the National Academy of Sciences of the United States of America 110, 20533–20538.

Morcos, F., Pagnani, A., Lunt, B., et al., 2011. Direct-coupling analysis of residue coevolution captures native contacts across many protein families. Proceedings of the National Academy of Sciences of the United States of America 108, E1293–E1301.

Moult, J., Fidelis, K., Kryshtafovych, A., Schwede, T., Tramontano, A., 2016. Critical assessment of methods of protein structure prediction (CASP) – Progress and new directions in Round XI. Proteins 84 (S1), 4–14.

Olsson, M.H.M., Søndergaard, C.R., Rostkowski, M., Jensen, J.H., 2011. PROPKA3: Consistent treatment of internal and surface residues in empirical pKa predictions. Journal of Chemical Theory and Computation 7, 525–537.

Ortiz, A.R., Strauss, C.E.M., Olmea, O., 2002. MAMMOTH (matching molecular models obtained from theory): An automated method for model comparison. Protein Science 11, 2606–2621.

Ovchinnikov, S., Kamisetty, H., Baker, D., 2014. Robust and accurate prediction of residue-residue interactions across protein interfaces using evolutionary information. eLife 3, e02030.

Ovchinnikov, S., Kinch, L., Park, H., et al., 2015. Large-scale determination of previously unsolved protein structures using evolutionary information. eLife 4. doi:10.7554/eLife.09248.

Ovchinnikov, S., Park, H., Varghese, N., et al., 2017. Protein structure determination using metagenome sequence data. Science 355, 294–298.

Petrey, D., Chen, T.S., Deng, L., et al., 2015. Template-based prediction of protein function. Current Opinion in Structural Biology 32, 33–38.

Petrov, D., Zagrovic, B., 2014. Are current atomistic force fields accurate enough to study proteins in crowded environments? PLOS Computational Biology 10, e1003638.

Piana, S., Lindorff-Larsen, K., Shaw, D.E., 2013. Atomic-level description of ubiquitin folding. Proceedings of the National Academy of Sciences of the United States of America 110, 5915–5920.

Pundir, S., Martin, M.J., O'Donovan, C., UniProt Consortium, 2016. UniProt Tools. Current Protocols in Bioinformatics 53, 1.29.1–1.29.15.

Rego, N., Koes, D., 2015. 3Dmol.js: Molecular visualization with WebGL. Bioinformatics (Oxford, England) 31, 1322–1324.

Ringer, A.L., Senenko, A., Sherrill, C.D., 2007. Models of S/pi interactions in protein structures: Comparison of the H2S benzene complex with PDB data. Protein Science 16, 2216–2223.

Rose, A.S., Bradley, A.R., Valasatava, Y., et al., 2016. Web-based molecular graphics for large complexes. In: Proceedings of the 21st International Conference on Web3D Technology, New York, NY: ACM, pp. 185–186.

Rose, A.S., Hildebrand, P.W., 2015. NGL Viewer: A web application for molecular visualization. Nucleic Acids Research 43, W576–W579.

Sadowski, M.I., Taylor, W.R., 2012. Evolutionary inaccuracy of pairwise structural alignments. Bioinformatics (Oxford, England) 28, 1209–1215.

Sarti, E., Gladich, I., Zamuner, S., Correia, B.E., Laio, A., 2016. Protein–protein structure prediction by scoring molecular dynamics trajectories of putative poses. Proteins 84, 1312–1320.

Schmidtke, P., Le Guilloux, V., Maupetit, J., Tufféry, P., 2010. fpocket: Online tools for protein ensemble pocket detection and tracking. Nucleic Acids Research 38, W582–W589.

Sehnal, D., Pravda, L., Svobodová Vařeková, R., Ionescu, C.-M., Koča, J., 2015. PatternQuery: Web application for fast detection of biomacromolecular structural patterns in the entire Protein Data Bank. Nucleic Acids Research 43, W383–W388.

Shi, Z., Olson, C.A., Bell, A.J., Kallenbach, N.R., 2002. Non-classical helix-stabilizing interactions: C–H...O H-bonding between Phe and Glu side chains in alpha-helical peptides. Biophysical Chemistry 101–102, 267–279.

Shindyalov, I.N., Bourne, P.E., 2001. A database and tools for 3-D protein structure comparison and alignment using the combinatorial extension (CE) algorithm. Nucleic Acids Research 29, 228–229.

Sippl, M.J., Wiederstein, M., 2008. A note on difficult structure alignment problems. Bioinformatics (Oxford, England) 24, 426–427.

Southan, C., Stracz, A., 2013. Extracting and connecting chemical structures from text sources using chemicalize.org. Journal of Cheminformatics 5, 20.

Spiga, E., Abriata, L.A., Piazza, F., Dal Peraro, M., 2014. Dissecting the effects of concentrated carbohydrate solutions on protein diffusion, hydration, and internal dynamics. The Journal of Physical Chemistry B 118, 5310–5321.

Spiliotopoulos, D., Kastritis, P.L., Melquiond, A.S.J., et al., 2016. dMM-PBSA: A new HADDOCK scoring function for protein-peptide docking. Frontiers in Molecular Biosciences 3, 46.

Spyrakis, F., Cavasotto, C.N., 2015. Open challenges in structure-based virtual screening: Receptor modeling, target flexibility consideration and active site water molecules description. Archives of Biochemistry and Biophysics 583, 105–119.

Stansfeld, P.J., Goose, J.E., Caffrey, M., et al., 2015. MemProtMD: Automated insertion of membrane protein structures into explicit lipid membranes. Structure 23, 1350–1361.

Tamo, G., Abriata, L., Dal Peraro, M., 2015. The importance of dynamics in integrative modeling of supramolecular assemblies. Current Opinion in Structural Biology 31, 28–34.

Tian, P., Boomsma, W., Wang, Y., et al., 2015. Structure of a functional amyloid protein subunit computed using sequence variation. Journal of the American Chemical Society 137, 22–25.

Touw, W.G., Vriend, G., 2014. BDB: Databank of PDB files with consistent B-factors. Protein Engineering, Design and Selection 27, 457–462.

Touw, W.G., Joosten, R.P., Vriend, G., 2015. Detection of trans-cis flips and peptide-plane flips in protein structures. Acta Crystallographica Section D, Biological Crystallography 71, 1604–1614.

Unni, S., Huang, Y., Hanson, R.M., et al., 2011. Web servers and services for electrostatics calculations with APBS and PDB2PQR. Journal of Computational Chemistry 32, 1488–1491.

Valasatava, Y., Rosato, A., Cavallaro, G., Andreini, C., 2014. MetalS(3), a database-mining tool for the identification of structurally similar metal sites. Journal of Biological Inorganic Chemistry 19, 937–945.

Valley, C.C., Cembran, A., Perlmutter, J.D., et al., 2012. The methionine-aromatic motif plays a unique role in stabilizing protein structure. Journal of Biological Chemistry 287, 34979–34991.

Virag, I., Stoicu-Tivadar, L., Crişan-Vida, M., 2016. Gesture interaction browser-based 3D molecular viewer. Studies in Health Technology and Informatics 226, 17–20.

Wass, M.N., Mooney, S.D., Linial, M., Radivojac, P., Friedberg, I., 2014. The automated function prediction SIG looks back at 2013 and prepares for 2014. Bioinformatics (Oxford, England) 30, 2091–2092.

Weatheritt, R.J., Babu, M.M., 2013. Evolution. The hidden codes that shape protein evolution. Science 342, 1325–1326.

Weichenberger, C.X., Sippl, M.J., 2007. NQ-Flipper: Recognition and correction of erroneous asparagine and glutamine side-chain rotamers in protein structures. Nucleic Acids Research 35, W403–W406.

Wiederstein, M., Sippl, M.J., 2007. ProSA-web: Interactive web service for the recognition of errors in three-dimensional structures of proteins. Nucleic Acids Research 35, W407–W410.

Willard, L., Ranjan, A., Zhang, H., *et al.*, 2003. VADAR: A web server for quantitative evaluation of protein structure quality. Nucleic Acids Research 31, 3316–3319.

Xu, D., Zhang, Y., 2012. Ab initio protein structure assembly using continuous structure fragments and optimized knowledge-based force field. Proteins 80, 1715–1735.

Yang, H., Qureshi, R., Sacan, A., 2012. Protein surface representation and analysis by dimension reduction. Proteome Science 10 (Suppl 1), S1.

Ye, Y., Godzik, A., 2004a. FATCAT: A web server for flexible structure comparison and structure similarity searching. Nucleic Acids Research 32, W582–W585.

Ye, Y., Godzik, A., 2004b. Database searching by flexible protein structure alignment. Protein Science 13, 1841–1850.

Zhang, Y., Skolnick, J., 2005. TM-align: A protein structure alignment algorithm based on the TM-score. Nucleic Acids Research 33, 2302–2309.

Zheng, H., Chordia, M.D., Cooper, D.R., *et al.*, 2014. Validation of metal-binding sites in macromolecular structures with the CheckMyMetal web server. Nature Protocols 9, 156–170.

Zoete, V., Cuendet, M.A., Grosdidier, A., Michielin, O., 2011. SwissParam: A fast force field generation tool for small organic molecules. Journal of Computational Chemistry 32, 2359–2368.

Relevant Websites

http://webchemdev.ncbr.muni.cz/Litemol/
   LiteMol.
https://lucianoabriata.altervista.org/modelshome.html
   Luciano A. Abriata.
https://www.mn-am.com/online_demos/corina_demo
   MNAM.
http://psychoprot.epfl.ch/
   PsychoProt.
https://biasmv.github.io/pv/
   PV - JavaScript Protein Viewer.
http://research.bmh.manchester.ac.uk/bryce/amber
   University of Manchester.

Biographical Sketch

Luciano Abriata is a postdoctoral researcher at the Laboratory for Biomolecular Modeling at École Polytechnique Fédérale de Lausanne and the Swiss Institute of Bioinformatics, Switzerland. His research focuses on the use of experiments and computation to achieve integrative models of biomolecular structure and function. Website: http://lucianoabriata.altervista.org/

Molecular Dynamics and Simulation

Liangzhen Zheng, Amr A Alhossary, Chee-Keong Kwoh, and Yuguang Mu, Nanyang Technological University, Singapore

Introduction

History

Molecular dynamics (MD) simulations have been evolved as powerful tools to explore atomic level of mechanisms, which are usually out the scopes of current experimental tools. Molecular dynamics are numeric tools based on the Newton's equations of motions for N-body system simulations by iteratively updating particle forces and system potential energies composed by inter-atomic/intra-atomic potentials or mechanic force fields. During MD simulations, coordinates, velocities, forces and potential energies are calculated every iterative step, whose temporal gap between each other is called "time step". The smaller the time step, the more accurate but more computational intensive the simulation is.

The original idea of MD simulations came up in 1950s in theoretical physics, but later it was also introduced to chemical, material science, and biological molecular simulations. Later in 1960s, Lennard-Jones potentials were introduced for lipid simulations (Rahman, 1964). However, at that time, computational power limited the simulation system size and time scale.

McCammon *et al.* (1977) reported the first MD simulation study of the dynamics of a protein bovine pancreatic trypsin inhibitor. Ever since then, more and more researches were carried out, and MD simulation has developed as a powerful tool in studying macromolecules and their interactions. Ogata *et al.* (2013) reported a work using MD simulation to study the photo-synthesis system II, which is a signal that we are embracing a period where simulations of complicated systems become possible.

Anton, as well as Anton II, created by D.E. Shaw's group, as a specialized simulation machine, outperforms any other same scale computational devices in speed and efficiency. It was successfully used to perform long time scale all atomic simulations (Lindorff-Larsen *et al.*, 2011; Piana *et al.*, 2011; Shaw *et al.*, 2010).

As the knowledge of chemistry and physics increase, we would know much more about the micro-world behaviors. Along with the development of hardware of computers and the algorithms, we are now towards a theoretic biology era, when every experiment could be performed in silico rather than by bench works. This goal, using computers to simulate every aspect and every level of life sciences, is difficult to imagine but would be finally achieved.

Background/Fundamentals

Molecular dynamics (MD) simulation is the computer based N-Body simulation of systems containing atoms and molecules. In MD simulation, the system components are left to interact together for a predetermined period, while the system physical properties are monitored. Those properties are either macroscopic system properties, {**V** (volume), **P** (pressure), **T** (temperature), **N** (number of particles)}, or microscopic system properties, {$\mathbf{v_i}$ (velocities), $\mathbf{r_i}$ (positions)}.

Ensemble and trajectory

The output of the simulation comes in the form of ensemble of frames. All frames share the same macroscopic/thermodynamic state but may differ in the microscopic states. Each frame represents the system at a specific point of time (a specific microscopic state). If the ensemble is sequence (time) dependent, it is called a trajectory. In this case, the trajectory represents the time-dependent evolution of the system.

Canonical ensemble (NVT)

The canonical ensemble contains all possible states in thermal equilibrium with a heat bath. The system remains in the absolute temperature T but may exchange energy with the heat bath. Three parameters of the system are fixed throughout the simulation: the absolute temperature (T), the number of atoms (N), and the volume (V). T is the most influential parameter on the system states among them.

Micro canonical ensemble (NVE)

The micro canonical ensemble represents an isolated system. No change in mass/number of atoms (N), Volume (V), nor exchange of Energy (E) is allowed.

Isothermal–isobaric ensemble (NPT)

The system has a fixed temperature (T), hence it is isothermal, and fixed pressure (P), hence it is isobaric.

Force field

Nearly all the simulation methods rely on force field for calculation. Force field, a potential energy (Eq. (1)) field in atomic and molecular level, describes the topology and motion behavior of atoms in molecules. When we use force field to describe the properties of molecules, spectrum constant force field and empirical potential function force field are often introduced. There are also Dreiding force field and universal force field (Mayo *et al.*, 1990; Setubal and Meidanis, 1997) which are not discussed here in detail.

$$E_{total} = E_{Kinetic} + U_{potential} \tag{1}$$

In spectrum constant field, the energy is simply linked with distance between two atoms. (See Eq. (2), K_i is a constant measured by spectrum data, ΔR_i is the distance between atom i and atom j.) This only applies for simple molecules such as water molecules.

$$E_{vibrant} = \frac{1}{2}\sum_i K_i(\Delta R_i)^2 \tag{2}$$

For multi-atom molecules, internal coordinates are introduced. Four internal coordinates are commonly used as depicted in **Fig. 1**. These four coordinates (bond stretching, bond angle, torsion angle and out-of-plane angle) do not affect the overall movements in the space, but they are enough to describe the internal motion of the molecules.

In other simulation methods, potential energy is composed by several terms: non-binding potential, bonding stretching term potential, angle bending term potential, torsion (dihedral) angle term potential, out-of-plane bending term potential, columbic interaction term potential (Frenkel and Smit, 2002). For many force fields, the calculations are based on the terms mentioned above. The most frequently applied non-bonding potential is described by Lennard-Jones (LJ) potential force. (See Eq. (3), r is distance between two atoms, ε and σ are atomic specific constants).

$$U_r = 4\varepsilon\left[\left(\frac{\sigma}{r}\right)^{12} - \left(\frac{\sigma}{r}\right)^6\right] \tag{3}$$

In Eq. (3), for the U_r of two atoms A and B, the parameter σ could be approximated by the following equation. (See Eq. (4)).

$$\sigma_{AB} = \frac{1}{2}(\sigma_A + \sigma_B), \varepsilon_{AB} = \sqrt{\varepsilon_A\varepsilon_B}; \tag{4}$$

$$U_b = \frac{1}{2}\sum_i k_b\left(r_i - r_i^0\right)^2 \tag{5}$$

Eq. (5) is a common expression of bonding stretching term potential function (a simple harmonic vibration model). r_i is the bond length of the no. ith bond, while r_i^0 is the average bond length of the bond i:

$$U_\theta = \frac{1}{2}\sum_i k_\theta\left(\theta_i - \theta_i^0\right)^2 \tag{6}$$

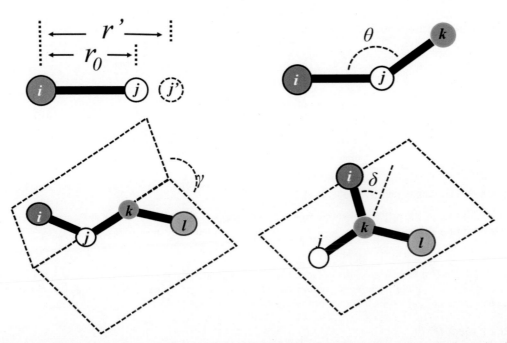

Fig. 1 Four types of internal coordinates used to potential energies in MD force fields. The 4 types of internal coordinates are bond, angle, dihedral angle, and out-of-plane angle.

$$U_\tau = \frac{1}{2}\sum_i [V_1(1+\cos\tau) + V_2(1-\cos2\tau) + V_3(1+\cos3\tau)] \tag{7}$$

Common bending angle term potential is defined by Eq. (6). θ_i is the angle between the two bonds on the atom i. The Eq. (7) concerns the torsion angle term potential energy. While τ is torsion angle using the no. i atom as the center. V_1, V_2 and V_3 are force constants.

$$U_\alpha = \frac{1}{2}\sum_i k_\alpha.\alpha^2 \tag{8}$$

Eq. (8) is out-of-plane term potential energy function. k_α is a constant and α is the out-of-plane angle.

$$U_{columbic} = \sum_{ij} \frac{q_i.q_j}{D.r_{ij}} \tag{9}$$

This columbic term potential energy is the electric energy between the no. i and no. j atom. D is dielectric constant, and r_{ij} is the distance between two atoms. q is the electric charge of atoms.

Most famous classical force fields used for modeling of *Macromolecules* are GROMOS (Groningen Molecular Simulation) (Christen et al., 2005; van Gunsteren et al., 1996), AMBER (Assisted Model Building with Energy Refinement) (Cornell et al., 1995), and CHARMM (Chemistry at HARvard Macromolecular Mechanics) (Brooks et al., 1983). CHARMM can also be used for *small molecules*. OPLS (Optimized Potentials for Liquid Simulations) (Jorgensen et al., 1996; Jorgensen and Tirado-Rives, 1988) is used to model *liquids* and it can as well be used in *Energy minimization*.

Another group of force fields called second generation force fields, integrates more complicated quantum mechanics and experimental results to pursuit more precise simulation. Some examples are CFF91, PCFF, CFF95, MMFF94 (Halgren, 1996), etc. There are also some special force fields (such as COMPASS and Gay-Berne) "costumed" for specific molecules, and thus are not explained in detail here.

Methodology

The total energy of a system is the sum of both kinetic and potential energies. The kinetic energy can be calculated from the atoms/particles velocities, while the potential energy is related to the positions of all atoms/particles in the 3 directions (that is $3N$, where N is number of atoms). This is too big number of parameters to have any analytical solution. Therefore, numerical methods only can be used to calculate.

Integrators

The MD simulation is based on Newton's second law of motion, ($F=ma$), where F is the force affecting a particle, m is the particle mass and a is its acceleration. Most integrators assume that the velocity and acceleration can be approximated using taylor's expansion. From particles' relative position (r), velocity (v), and acceleration (a), the time series time-dependent evolution of the system can be predicted step wise.

$$r(t+\delta t) = r(t) + v(t)\delta t + \frac{1}{2}a(t)\delta t^2 + \ldots$$

$$v(t+\delta t) = v(t) + a(t)\delta t + \frac{1}{2}b(t)\delta t^2 + \ldots$$

$$a(t+\delta t) = a(t) + b(t)\delta t + \ldots$$

Most famous integrators are *Verlet*, *Leapfrog*, and *Velocity Verlet*.

Methods to Overcome Limitations and Decrease Load

The calculation complexity of the MD simulation problem is a $O(n^2)$ problem. This limits the ability to simulate large systems. Methods exist to minimize the computational load. Some of them are mentioned next sections.

Neighborhood list
Having a neighborhood list for each atom, reduces the calculation complexity from $O(n^2)$ to $O(n)$. The neighborhood list needs to be updated periodically, in a suitable frequency: Neither too infrequent so that important updates may be missed, nor so frequent that the calculation efficiency gained in reduced complexity calculations may be lost.

Periodic boundary check (PBC)
To neglect the solvent surface tension effect, there are two options: having an infinite amount of solvent, which is impractical, or to have the periodic boundary check (PBC). In the PBC, every atom is considered replicated to image atoms in all directions.

PME, PPPM

Another advantage of the PBC is the ability to model long interactions as periodic interactions to be solved in the frequency domain. Particle Mesh Ewald summation (PME) is an efficient way to achieve this goal which depends on sorting and calculating image atoms in a certain order that simplifies the calculation. Another advancement of PME method is "Particle-Particle, Particle-Mesh" (PPPM) method. In this method, the short range interactions are calculated as particle-particle interactions, while for the long range interactions, it is calculated as follows: first, all particles effect is approximated to a set of charges spread on a mesh. The long range interactions with the mesh are calculated efficiently in the frequency domain.

Advanced Techniques

The large-scale conformation differences between meta-stable states of macromolecules are often separated by high energy barriers which are unlikely accessible through conventional MD simulations. Based on the transition state theory, there is an exponential relationship between the time scale of state transition and the height of an energy barrier (Huang, 1987). Therefore, the rare dynamics events of interest, e.g., the folding process of a polypeptide, are often happening in a rather large time scale (Kubelka et al., 2004). For example, according to Kubelka et al. (2004) the folding time limit of a generic single-domain protein from unfolded to folded state, could be expressed as following: $N/100$ μs, where N is the number of residues in the protein. In order to obtain a statistical meaningful transition free energies, multiple times trans-passing of the energy local minima are required based on the assumption of the ergodicity (Tolman, 1938) for conventional MD simulations, which probably requires extremely long simulation time. Therefore, for large proteins, using conventional MD simulations to sample the large energy barrier crossing events could be an impossible mission. However, a number of enhanced sampling methods, such as the replica exchange MD simulation (REMD) method (Sugita and Okamoto, 1999a), solute tempering (Liu et al., 2005), Hamiltonian replica exchange MD (HREMD) (Liu et al., 2005), Wang-Landau method and simulated tempering (ST) (Marinari and Parisi, 1992; Zhang et al., 2015), have been developed to address the difficulties. Besides, other enhanced sampling methods, including umbrella sampling (Torrie and Valleau, 1977) and metadynamics (Laio and Parrinello, 2002), based on introduced bias potentials, could reconstruct the probability distribution along one or a few CVs (Sutto et al., 2012). What's more, the combinations of these methods are also proved to be efficient tools to explore the free energy surface (FES). For example, the hybrid Hamiltonian method by introducing the Generalized Born energy for polar solvation energy calculations, is a useful tool to quickly reconstruct the FES which is similar to the standard REMD simulation (Mu et al., 2007). Also the combination of parallel tempering with metadynamics by Bussi et al. is also proved to be powerful (Bussi et al., 2006b).

Umbrella sampling

Umbrella sampling (Torrie and Valleau, 1974, 1977), as an enhanced sampling method, could sample the conformational dynamics along reaction coordinates thus we could estimate the relative free energy of different states along the reaction coordinates. A reaction coordinate (ξ) could be a kind of continuous parameter which could describe the system from a higher dimensional space. If the reaction coordinate is good enough to differentiate distinct states, by biasing the system along the reaction coordinate, thus we would calculate the free energy differences between the states.

In a general umbrella sampling scheme, multiple windows were set with initial structures with different reaction coordinate values. A bias potential is applied to each window (Eq. (10)) according to a harmonic bias function (Eq. (11)) or an adaptive bias function (Kästner, 2011). Therefore, the system in each window would be constrained to sample a narrow phase space along the reaction coordinate to ensure potential energy distribution overlap between adjacent windows. After the simulations, a post-processing method, weighted histogram analysis method (WHAM) (Rosenbergl, 1992), could recover the unbiased free energy by the umbrella integration. Multiple integration methods and biasing potential styles are existed, among them, WHAM as the integration method, harmonic bias potential as the biasing potential style are relatively widely used. The following part is a brief introduction to the general procedures of umbrella sampling using harmonic bias potential and WHAM post-processing method.

When we add a reaction coordinate dependent bias potential $\omega_i(\xi)$ to the system in a window, the total biased energy could be expressed as following:

$$E_{bias} = E_{unbias} + \omega_i(\xi) \tag{10}$$

where i represents the ith window of the umbrella sampling. The harmonic bias potential adding to the system is a simple bias potential expressed like this:

$$\omega_i(\xi) = \frac{1}{2}K(\xi - \xi_i)^2 \tag{11}$$

Here ξ_i works as a reference coordinate point. During the simulations, if the system is escaping the reaction coordinate, then a bias potential is added to push the system back.

To calculate the unbiased free energy, we need to obtain the unbiased distribution of the reaction coordinate, according the following equation:

$$P_i^u(\xi) = \frac{\int exp[-\beta E(r)]\delta[\xi^r(r) - \xi]d^{N_r}}{\int exp[-\beta E(r)]d^{N_r}} \tag{12}$$

Furthermore, the unbiased probability $P_i^u(\xi)$ could be determined by:

$$P_i^u(\xi) = P_i^b(\xi)exp[\beta\omega_i(\xi)]exp[-\beta\omega_i(\xi)] \tag{13}$$

From the simulation in each window, the biased probability $P_i^b(\xi)$ is known, and the free energy of the window thus could be induced by:

$$E_{unbias} = -\left(\frac{1}{\beta}\right)\ln P_i^b(\xi) - \omega_i(\xi) + F_i \tag{14}$$

where F_i is a constant which could be solved by self-iteration until a convergence reached (Banavali and Roux, 2005). Besides, the unbiased free energy could be combined by all the windows to produce a whole free energy along the reaction coordinate.

Metadynamics simulation

Different from the conventional MD simulations, metadynamics (Laio and Parrinello, 2002) adopts gaussian-like bias to bypass high energy barriers in FES. By adding bias potentials, the predefined collective variables (CVs), metadynamics sampling could reconstruct the FES based on the biased CVs (**Fig. 2**) and be able to accelerate the sampling process to capture rare dynamics events, such as the large-scale conformation changes.

Ever since the first algorithm of the metadynamics (Laio and Parrinello, 2002), many modified versions of metadynamics have been proposed, such as well-tempered (WT) metadynamics (Barducci *et al.*, 2008), the bias-exchange approach (Piana and Laio, 2007) and the parallel tempering (PT) metadynamics (Bussi *et al.*, 2006a). Besides, new CVs as reviewed in the reference (Sutto *et al.*, 2012), such as widely used path variables, principle component analysis (PCA) variables, alpha RMSD, beta RMSD, and contact map variables, have accelerating the sampling efficiency vastly.

For a general understand of the metadynamics, we take the well-tempered metadynamics (Barducci *et al.*, 2008) as an example, and briefly introduce the rationales behind. Supposing that the microscopic coordinate **R** at temperature T has a potential $U(\mathbf{R},t)$, the ultimate goal of metadynamics is to obtain the FES from the unbiased probability distribution $P(s(\mathbf{R}))$ along the CV $s(\mathbf{R})$ by solving the following equation:

$$F(s) = -\frac{1}{\beta}logP(s) \tag{15}$$

where the $\beta = \frac{1}{k_B T}$, k_B is the Boltzmann constant. For the unbiased, $P(s)$ we could have:

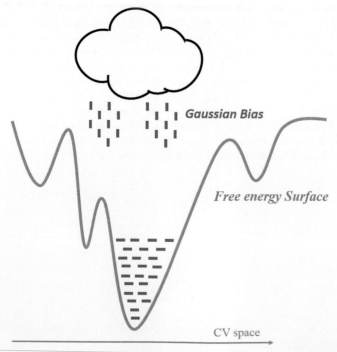

Fig. 2 A cartoon representation of metadynamics. During metadynamics simulations, gaussian shape biases are deposited along specific coordinates (or CVs).

$$P(R,t) = \frac{e^{-\beta U(R)}}{\int dR e^{-\beta U(R)}} \tag{16}$$

By adding a history-dependent bias potential as the following

$$V(s,t) = \Delta T \ln\left(1 + \frac{\omega N(s,t)}{\Delta T}\right) \tag{17}$$

where $N(s,t) = \int_0^t \delta_{s,s(t')}dt'$ is the histogram of the CV $s(\mathbf{R})$, ω is energy related term, ΔT is a temperature which we would explain in detail later. For this equation, $V(s,t)$ disfavors the frequently visited space in CV $s(\mathbf{R})$. The first order derivate of the $V(s,t)$ could be expressed as following:

$$\dot{V}(s,t) = \frac{\omega \Delta T \delta_{s,s(t)}}{\Delta T + \omega N(s,t)} = \omega e^{-\frac{V(s,t)}{\Delta T}} \delta_{s,s(t)} \tag{18}$$

Then we use τ_G to replace $\delta_{s,s(t)}$, and let $w = \omega e^{-\frac{V(s,t)}{\Delta T}} \tau_G$, where τ_G is the Gaussian height deposit time step. By incorporating the Eq. (18) into the standard metadynamics, where the height of Gaussian deposited is a constant, we could modulate the bias potential as a history dependent style.

As the time t flows, $\dot{V}(s,t)$ could be determined by $\frac{V(s,t)}{\Delta T}$, which indicates at position where $V(s,t)$ is large, the $\dot{V}(s,t)$ is close to zero to resemble a thermodynamic equilibrium. At this situation, one might get $P(s,t)ds \propto e^{-\frac{F(s)-V(s,t)}{T}}ds$, based on that, then one could have the following relationship between $\dot{V}(s,t)$ and $F(s)$:

$$\dot{V}(s,t) = \omega e^{-\frac{V(s,t)}{\Delta T}} P(s,t) = \omega e^{-\frac{V(s,t)}{\Delta T}} \frac{e^{\frac{-F(s)-V(s,t)}{T}}}{\int ds e^{\frac{-F(s)-V(s,t)}{T}}} \tag{19}$$

While the time $t \to \infty$,

$$V(s) = -\frac{\Delta T}{T + \Delta T} F(s) \tag{20}$$

If we go a step further, we could have $F(s) + V(s) = \frac{T}{T+\Delta T}F(s)$. Until now, we have established the relationship between $V(s,t)$ with $F(s)$ in a simpler way. Combining Eq. (19) with Eq. (20), we then can deduce the estimation of $F(s)$:

$$\tilde{F}(s,t) = -\frac{T + \Delta T}{\Delta T}V(s,t) = -(T + \Delta T)\ln\left(1 + \frac{\omega N(s,t)}{\Delta T}\right) \tag{21}$$

From Eq. (21), we could estimate the final FES by calculating the histogram $N(s,t)$ of CV $s(\mathbf{R})$. The modulation of the constant ΔT could define different behavior of the Gaussian deposit. For $\Delta T = 0$, the bias potential added in the system is always 0 along the simulation time t. Another limiting case, if $T \to \infty$, then $-\frac{T+\Delta T}{\Delta T} \to -1$, thus we could give the conclusion that $\tilde{F}(s,t) \approx -V(s,t)$, which is the case in the standard metadynamics where the bias potential energy, added in the system along the CVs space, has the same absolute value as the FES along the same CVs space. What's more, a finite value of ΔT, therefore restricts the system from exploring all the physical space and ensures a thermodynamic equilibrium at local minima in FES. Fine tuning of this ΔT could facilitate us to better explore the CVs space. By calculating the $V(s,t)$ in the following equation:

$$V(s,t) = w \sum_{t' = \tau_G, 2\tau_G, 3\tau_G, \cdots} e^{\left(\frac{-s(\mathbf{R})+s(R_G(t'))}{2\delta s^2}\right)^2} \tag{22}$$

where w and δ have been defined already. Finally, we could recover the $\tilde{F}(s,t)$ from the biased probability distribution of $s(\mathbf{R})$.

In practice, we often use the following bias factor $\gamma = \frac{T+\Delta T}{T}$ to set up the ΔT. τ_G is the Gaussian deposition stride, and another input value ω in $w = \omega e^{-\frac{V(s,t)}{\Delta T}}\tau_G$, is the initial Gaussian height. And the δ is the width of the Gaussian for the CV $s(\mathbf{R})$.

Plumed (Bonomi et al., 2009) and METAGUI (Biarnés et al., 2012) (a VMD plugin), are example softwares which make the simulation and the post-process of the data much more convenient. The usage and tutorials of the tool could be found here: see "Relevant Websites section".

Replica exchange MD (REMD)

Conventional MD simulation takes long time simulation to sample large conformational changes, or observe biological relevant behaviors. In most cases, simulation system would be trapped in local potential energy minima, which limit the sampling efficiency. To overcome the potential energy barriers and search for the global energy basins, REMD (sometimes also called generalized ensemble method), emerged as an efficient sampling technique (Sugita and Okamoto, 1999b; La Penna et al., 2004; Mitsutake et al., 2001; Tai, 2004). A lot of researches have demonstrated that the REMD method is much more efficient than the conventional MD (García and Sanbonmatsu, 2002; Sugita and Okamoto, 1999b).

In practice, M replicas in canonical ensemble are simulated with fixed distinct temperatures independently and spontaneously for some MD (or MC) steps, though the optimal number of steps between exchanges are controversial (Sindhikara et al., 2008; Cecchini et al., 2004; Zhang et al., 2005b). Then, the adjacent replicas (say m and $m + 1$) could exchange their configurations based

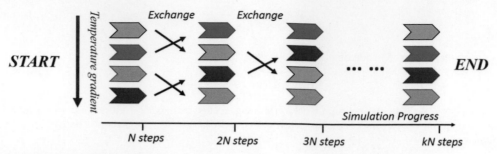

Fig. 3   A cartoon representation of REMD methods.

on the Metropolis criterion (Swendsen and Wang, 1987). Once the exchange occurs, replica m swamps the configuration with replica $m+1$ and continues at the fixed temperature. **Fig. 3** shows a sketch representation of the idea.

The idea was based on the hypothesis that at higher temperature, the potential energy of the simulation system is much higher and the system is much easier to escape the local energy minima to explore global energy minima. The potential energies of the parallel systems (replicas) in different simulating temperatures may overlap, thus the low energy conformations from high temperature replicas could be exchanged to lower temperature replicas and stabilized and sampled in lower temperature replicas. This way, in normal temperature range, the replicas could sample the structures with low potential energy, and the sampling efficiency therefore suppresses the conventional single trajectory MD simulation.

Here, we briefly illustrate the exchange method, the Metropolis scheme, used in REMD. The Metropolis criterion specifies that the exchange between neighboring replicas m and $m+1$ at temperature T_m and T_m+1 could take place with an acceptance probability P to ensure the detailed balance, $P = e^{(\beta_m - \beta_{m+1})(F_m - F_{m+1})}$, where F_m and F_m+1 are the potential energies of replica m and replica $m+1$. And $\beta_m = 1/(k_B T_m)$, where k_B is the Boltzmann's constant. To achieve acceptable exchange ratio, certain temperature intervals between replicas should be selected to ensure enough overlapping of the potential energies of neighboring replicas (Kone and Kofke, 2005; Sugita and Okamoto, 1999b).

The convergence problem is one of the major issue to achieve a statistical meaningful REMD simulation (Lin and Shell, 2009; Denschlag et al., 2008; Okur et al., 2007; Zhang et al., 2005a). It has been widely discussed in many researches and reviews, therefore we are not going to elucidate it in detail here.

Hamiltonian REMD

The conventional REMD tends to sample low energy structure and then deposits it in the low temperature replicas, thus it is blind sampling and requires no prior knowledge of the simulation system. During REMD simulation, the number of required replicas to ensure enough potential energy overlapping between adjacent replicas is exponentially increased along with the system size, therefore relies on tremendous amount of computational power, and increases the difficulty of convergence, thus those limit the usage of REMD for huge simulation systems.

Instead of exchanging between different temperatures, Hamiltonian could also be exchanged between replicas if there are possible potential energies overlapping (Affentranger et al., 2006; Fukunishi et al., 2002; Curuksu and Zacharias, 2009). Hamiltonian REMD is quite similar to REMD, where each replica samples independently in different environments (temperature for REMD, Hamiltonian for Hamiltonian REMD respectively). The exchanges every N step in Hamiltonian REMD also obey the metropolitan rule.

For Hamiltonian REMD, the Hamiltonian of each replica is scaled by a factor lambda. It has to be ensured that the first replica has a normal Hamiltonian, while other lambda values increase or decrease sequentially. In the same time, the lambda values should be set appropriately thus enough overlapping of potential energies between adjacent replicas must be retained. There are several ways to modulate the Hamiltonian of simulation systems, for example, reducing the atomic charges of a molecule, increasing intra-molecular repulsions (Mu, 2009), decreasing interatomic attractions, increasing molecular hydrophobic interactions, or modifying the atomic charges of atoms.

The combination of Hamiltonian simulation with REMD is proved to be a good practice. Using different Hamiltonian, we could actually bias the systems towards different directions. Hamiltonian REMD thus is not blindly sampling, nor the CV based sampling, but something in between.

Monte Carlo Simulation

The essence of Monte Carlo (MC) simulation lies on the "randomness". Multiple repeated random sampling using MC simulation could obtain statistical data for systems with multiple degrees of freedom. However, in general, in order to sample the configurational space of the system sufficiently, brute force MD method requires tons of millions of repeats to complete, therefore, analytic techniques must be used (Frenkel and Smit, 2001; Heermann, 1990)

Metropolis method was introduced to facilitate random sampling in MC simulation. In each step, a small random displacement Δ of a particle is tried. The displacement Δ is generated from a relative probability distribution proportional to the Boltzmann factor $e^{-\beta U(o)}$, where β is $\frac{1}{k_B T}$, U(o) is the potential of the system in previous step. This trail displacement is then assessed to accept or reject based on many rules, one of which is the Metropolis scheme, simple but applicable (Kalos and Whitlock, 2008).

To make it simpler, the system in previous state is called o (old), the system state after the displacement is donated as n (new). The transition probability from o to n, therefore is determined here:

$$P(o \rightarrow n) = e^{-\beta[U(n)-U(o)]} \tag{23}$$

U(n) and U(o) are the potential energies of the old state and the new state. After the trial displacement, a random number r from the uniform distribution between 0 and 1 is generated and compared to $P(o \rightarrow n)$. The displacement would be accepted if $r < P(o \rightarrow n)$, otherwise, it would be rejected. This way, it is ensured that the acceptance probability from state o to n is equal to $P(o \rightarrow n)$.

The Metropolis scheme is efficient, but only thought to be ergodic for simple systems, therefore, the mixed schemes with efficient Metropolis scheme with other in principle ergodic scheme would achieve an statistical reliable simulation (Frenkel and Smit, 2001). In contrast to MD simulation, in general, MC simulation focuses on sampling, rather than time-dependent events, therefore it is not advisable to extract time dependent phenomena from the MC simulations.

Systems/Applications

Molecular dynamics have been widely adopted in molecular modeling of organic chemicals, short peptides (Woutersen et al., 2002; Snow et al., 2002; Marinelli et al., 2009; Gnanakaran et al., 2003; Blanco et al., 1994; Ma and Nussinov, 2002), protein-protein interactions (Arkin et al., 2014; Baaden and Marrink, 2013; Rajamani et al., 2004), lipid-protein complexes (Shrivastava and Sansom, 2000; Im and Roux, 2002; Khalili-Araghi et al., 2009; Gumbart et al., 2005; Woutersen et al., 2002; Snow et al., 2002; Marinelli et al., 2009; Gnanakaran et al., 2003; Blanco et al., 1994; Ma and Nussinov, 2002), even virus capsid (Freddolino et al., 2006; Miao et al., 2010). By harnessing the computation power from the supercomputing facilities, the physical properties of sub-micrometer particles could be revealed. Using MD simulations, as well as other simulation methods, stable conformations, the folding process, energy landscape, transitions between macro-states, enzyme catalytic mechanisms, ligand binding/association and aggregation states could be examined. Besides, MD simulations are also commonly used in computational drug discovery and design (Borhani and Shaw, 2012; De Vivo et al., 2016; Durrant and McCammon, 2011).

For example, the enhanced sampling methods, umbrella sampling and metadynamics simulation could be applied to recover the free energy landscape of a macromolecule system along some specific collective variables, such as RMSD, PCA components, helical contents, angles, and other high dimensional coordinates. Bias potentials are deposited in simulations and the post-processing reweight techniques would then recover the true energy landscape of the simulation system along selected dimensions. The combination of metadynamics with REMD, or parallel exchanges between replicas biasing difference collective variables are proved to be more efficient and converges faster than the single-trajectory metadynamics simulations (Abrams and Bussi, 2013; Piana and Laio, 2007; Barducci et al., 2011).

By incorporating multiple short MD simulations, Markov state model (HMM) could be applied to analyze the trajectories and compute the transition dynamics of proteins or protein-ligand systems between macro-states of the systems (Beauchamp et al., 2012; McGibbon and Pande, 2013; Suárez et al., 2016). Starting from different initial structures, it is expected that multiple short simulations starting in different directions are more powerful in conformational sampling than a single long trajectory. Combining the short simulations, several macro-states of the system could be identified using different methods, such as clustering and independent component analysis. The transition probabilities between the microstate then would be derived from the trajectories, and More importantly, the transition time scales could be estimated.

What's more, combining MD simulation with quantum calculation, more accurate enzyme activation mechanisms (van der Kamp and Mulholland, 2013; Hu et al., 2011; Riccardi et al., 2010; Senn and Thiel, 2007; Zhang et al., 2000). The statistic properties could also be calculated and compared to experiments.

MD Simulation With Molecular Docking

Molecular docking is a simplified form of MD simulation. It can be used on intervals to replace lengthy segments of MD simulation trajectories, especially in cases where certain domains undergo large translations, rotations, and conformation changes. A typical example is biological interactions that include large protein folding, like capsid or vesicle formation. To be able to replace lengthy MD simulation intervals with docking, it is necessary to be able to perform "blind docking" of a protein (or a domain) to the whole surface of another protein. Some tools claim the ability to perform blind docking, for example QuickVina-W (111), PatchDock (222), BSP-SLIM (333) (Nafisa et al., 2017; Schneidman-Duhovny et al., 2005; Lee and Zhang, 2012).

Markov State Model (MSM)

The dynamics and kinetics properties are extremely valuable for a deep understanding of the structure-function relationships. MD simulation method has its innate advantage over current experimental techniques in exploring macromolecules kinetics and dynamics. By using MSM, or Hidden Markov State Model (HMM), the dynamics properties of proteins, as well as other macromolecules, could be estimated with near to experiment accuracy (Beauchamp *et al.*, 2012; Suárez *et al.*, 2016; McGibbon and Pande, 2013; Noé *et al.*, 2009; McGibbon *et al.*, 2014; Husic and Pande, 2018).

Here we briefly explain the four general steps applied in most MSM studies for macromolecules dynamics and kinetics.

1. Extensive short MD simulations. The idea of constructing transitions between macro-states is based on such a hypothesis that we could adequately sample all the microstates, as well as their transitions. Therefore, in practice, other than a single long MD trajectory, multiple short MD simulations could be performed to sample as many intermediate states as possible (Noé *et al.*, 2009). Certain intermediate structures are selected as "seeding" initial structures for further high throughput short MD simulations, or adaptive sampling (Singhal and Pande, 2005; Pande *et al.*, 2010; Doerr and De Fabritiis, 2014; Doerr *et al.*, 2016).
2. Features dimension reduction. The features harnessed in previous step are then transformed into low dimensional vectors and clustering of the vectors could be applied to capture the microstates of the conformations. The coordinates of the trajectories are transformed into vector features. The features could be RMSD of a structure, dihedral angles of backbone atoms, distance between alpha-carbon atoms, or contacts between alpha-carbon atoms (Naritomi and Fuchigami, 2011; Pérez-Hernández *et al.*, 2013). Radius of gyrate, coordination number, PCA and time lagged independent component analysis (tICA) (Schwantes *et al.*, 2015; Noé and Clementi, 2015) are commonly applied in this dimension reduction step.
3. Microstates clustering. The distance matrices with clustering strategy is proved to be reliable for partitioning the phase space (Pande *et al.*, 2010; Chodera and Noé, 2014; McGibbon and Pande, 2013). The conformations from trajectories would be clustered, based on the low dimensional vectors calculated in step 2, to form microstates which could be rapidly interconverted between each other. Several commonly used clustering methods include K-mean-like clustering and hierarchical clustering. For an easy visualization of the MSM network, fewer macrostate are chosen, using the so-called "lumping" scheme using principal canonical *correlation* analysis (PCCA) or PCCA+ methods.
4. MSM construction. Then using maximum likelihood method to estimate the transition rates between the states, and then construct the MSM, and estimate the kinetics of the MSM network (Weber, 2011; Röblitz and Weber, 2013).

Currently, there are majorly two tools, Pyemma (Scherer *et al.*, 2015) and MSMBuilder (Harrigan *et al.*, 2017), available for MSM or HMM construction. Their tutorials and manuals could be found here: see "Relevant Websites section" and see "Relevant Websites section".

MD Simulation With Machine Learning

The combination of machine learning and chemical informatics in drug discovery has been proved to be very successful for predicting ligand binding affinity (Ballester and Mitchell, 2010; Cheng *et al.*, 2012), drug toxicity (Walters and Murcko, 2002; Judson *et al.*, 2008), membrane permeability (Doniger *et al.*, 2002; Hou *et al.*, 2006; Walters and Murcko, 2002) and so on. In structural biology, machine learning models could advance the protein-protein binding interface prediction (Lise *et al.*, 2009), protein folding prediction, and protein structure prediction (Hua and Sun, 2001; Cai *et al.*, 2002; Heffernan *et al.*, 2015; Wang *et al.*, 2016).

Recently, machine learning algorithms have also been introduced into MD simulations for faster, more accurate sampling and predictions (Behler, 2016), for example, in metadynamics simulations. Proper CVs should be carefully chosen to achieve successful sampling. A recent work illustrates the probability of using neural network to reduce dimensions in tICA calculation and aiding the enhanced sampling (Sultan *et al.*, 2018). Or using machine learning algorithms to perform or instruct the way of bias potential deposit to achieve high dimensional enhanced sampling (Galvelis and Sugita, 2017).

Quantum calculation based first-principles molecular dynamics (FPMD) are useful in understanding chemical processes, however, FPMD could only simulate up to hundreds of atoms within a very short simulation time to the limitations of computation power. There have been a lot of studies trying to adopt machine learning methods, such as neural network and Gaussian Processes, to predict the quantum level potential energy surface, thus further to obtain the atomic forces imposed in atoms (Behler and Parrinello, 2007; Bartók *et al.*, 2010; Botu and Ramprasad, 2015; Li *et al.*, 2015).

Illustrative Example(s) or Case Studies

Here, we use two examples to illustrate the applications of molecular dynamics and simulations in biophysics field.

Conformations and Energy Landscapes of *apo*-form PR LBD

Progesterone receptor (PR) is a member of the nuclear receptor (NR) family. PR is the natural receptor of the important female hormone progesterone, which affects every aspect in female development, maintenance and reproductive. The ligand binding domain (LBD) of PR, shares a similar sandwich like folding as other LBDs in NR. Though, several crystal structures of ligand bound (and co-peptides bound) PR LBD have been deposited in PBD databank, the *apo*-form (non-ligand bound state) PR LBD conformation, which is valuable for structural based virtual screening to identify druggable ligands, is still void. Combining conventional MD, umbrella sampling (Wang *et al.*, 2014), and metadynamics (Laio and Parrinello, 2002; Laio and Gervasio, 2008), the *apo*-form PR LBD has been thoroughly examined using the popular amber99SB-ildn force in explicit water environment (Zheng *et al.*, 2016). Multiple analysis methods, such as dihedral PCA, correlation analysis, reweighting, were applied. The free energy landscape of the *apo*-form PR LBD was constructed, indicating the agonistic conformation of *apo*-form PR LBD is a meta-stable state, though not the most stable one. Several stable conformations were identified and are useful for further virtual screening study. E723, N719 and R899 are identified as key residues for conformation dynamics for helix 3 and helix 12., and metadynamics (Laio and Parrinello, 2002; Laio and Gervasio, 2008), the *apo*-form PR LBD has been thoroughly examined using the popular amber99SB-ildn force in explicit water environment (Zheng *et al.*, 2016). Multiple analysis methods, such as dihedral PCA (Mu *et al.*, 2005), correlation analysis (Göbel *et al.*, 1994), reweighting (Tiwary and Parrinello, 2014), were applied. The free energy landscape of the *apo*-form PR LBD was constructed, indicating the agonistic conformation of *apo*-form PR LBD is a meta-stable state, though not the most stable one. Several stable conformations were identified and are useful for further virtual screening study. E723, N719 and R899 are identified as key residues for conformation dynamics for helix 3 and helix 12.

Understanding SAC6 Inter-Domain Peptides Flexibility

Another example is the dynamics study of a short peptide from yeast SAC6 inter-domain loop region (Miao *et al.*, 2016). SAC6 composed by two domains, is actin binding protein upon phosphorylation.T103 is a high potent phosphorylation site. However, why phosphorylated SAC6 has higher actin binding activity is poorly understood. We firstly hypothesized that the inter-domain peptides where T103 resides, behave differently with or without phosphorylation.

By comparing the dynamics property of two simulation systems (T103 phosphorylated and non-phosphorylated respectively), it turns out that phosphorylated T103 could restrict the peptide flexibility through electrostatic interactions with lysine and arginine residues, meanwhile phosphorylated T103 could form a hydrogen bond with G109. This hydrogen bond maintains the turn structure around G109. Free energy landscape and clustering analysis both suggest that phosphorylation at T103 stabilizes the peptides, thus facilitates the actin binding with the actin binding domains in SAC6.

Future Directions

The future of molecular simulations moves towards a more accurate, large system scale, longer simulation time scale. The polarized force field, though computational intensive, is supposed to be more precise than conventional single change force field. The combination of quantum simulation, MMQM simulations are powerful for enzyme catalytic center accurate enough comparing to experimental results.

The fast advancement of supercomputer facilities, as well as the adaption from CPU simulations towards GPU simulations, vastly increases our confidences of future of molecular simulation. In another hand, less accurate models, such as coarse graining methods, have drawn great attention due to their ability of decreasing computational power and increasing simulation time scale.

GPU Accelerated MD Simulations

GPUs have been gradually adopted in many analytic fields, such as machine learning and simulations. Recent years, many popular simulation packages, such as Amber, Gromacs, and NAMD, have all embraced the powerful GPUs based using CUDA library. Other

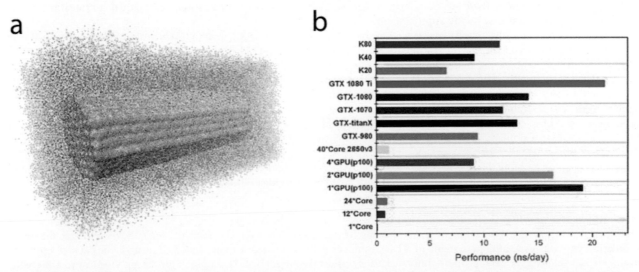

Fig. 4 GPU benchmark for cellulose system MD simulations with Amber 16. a) the simulation box of cellulose in explicit waters, with 408,609 atoms in total. b), performance of different GPUs and different CPU cores. The CPUs used in this benchmark are Intel Xeon E5-2683 V4 and E2650 V3.

new generation of simulation packages, such as openmm, htmd, and acemd all support GPU calculations. Gromacs, starting from version 4.5, implemented GPU calculations and combined them with CPUs to accelerate MD simulations (Abraham *et al.*, 2015; van der Spoel and Hess, 2011). Later in version 4.6 and higher, native GPU support has been developed to handle the most computational intensive non-bonded interactions based on CUDA library, while the PME electrostatics, bonded interactions and input-output are subjected to CPUs. Whereas, in Amber, GPUs handle nearly all calculations, such as energy evaluation, bonded and non-bonded forces.

Taking cellulose-water system ($\sim$41,000 atoms) as a benchmark using Amber 16 for *NPT* MD simulations, mid-end GPU cards (GTX 1080 and 1080Ti) show rather surprising performance comparing to high-end GPUs, such as Tesla P100, Tesla K40 and K80, as shown in **Fig. 4**. Given that the prices of the mid-end Nvidia GPU cards are relatively cheap (see "Relevant Websites section"), these data thus indicate that GPUs outperform CPUs, and could be adopted for efficient MD simulations with an affordable budget.

Coarse Graining

Current all-atom molecular modeling techniques enable biological systems description with details limited to simulation times and system sizes less than 1000 ns and 100 nm respectively. Coarse Graining is developed for bridging the "time-scale" and "length-scale" gaps between computational and experimental methods. The general contract of all CG methods is representing a system by a reduced number of degrees of freedom (DOFs) and fine interaction details, by mapping from atomically detailed configuration (fine grains) into a CG configuration called super atom (coarse grain) (Zhang *et al.*, 2008). That reduces the system size about *ten* folds and allows time steps up to 25–50 fs (Shih *et al.*, 2006). This way, the simulation would go faster and require less computational resources than the all-atom representation while achieving an increase of orders of magnitude in the simulated time and length scales.

Defining a coarse grained system needs three components: 1) a method to map a set of fine grains to a coarse grain, 2) a method to define the location of the center of the newly introduced coarse grain, and 3) a method to derive a suitable force field for the system. The mapping process depends on the size and complexity of the biomolecules. Small and low details molecules mapping is straightforward while the mapping process of large and complex molecules is a challenging one.

Although coarse graining is not a new concept, it is expected to take bigger rule in the near future, as the direction would be towards simulating systems on larger time scale and size scale.

The *Implicit hydrogen* model (AKA *united atom*) is an old and simple form of coarse graining, where nonpolar hydrogens are omitted and implicitly included in their connected carbon atoms with reduced number of interaction sites to around half. The resulting interaction sites for CH_2, CH_3, and CH_4 are usually parametrized Lennard-Jones potentials.

There are two general CG approaches: the residue-based CG (Shih *et al.*, 2007), and the shape-based CG (Arkhipov *et al.*, 2008). Zhang *et al.* (2008) designed a method called "essential dynamics coarse-graining (ED-CG)" to define CG sites from protein primary sequence that reflect the essential dynamics characterized by PCA of an atomistic trajectory.

Multiscale coarse-graining (MC-CG)

The multiscale coarse-graining method (MS-CG, aka "force matching") is a method of decreasing the computational load of simulating a system by "optimizing a CG potential to reproduce many-body potentials of mean force (PMF) calculated from atomistic configurations" initially for non-bonded potentials and then for bonded interactions (Lu *et al.*, 2013).

Instead of the commonly used two-body approximation, Larini *et al.* (2010) in 2010 proved that using explicit three-body potentials is superior to the two-body approximation.

LY Lu and G. Voth in 2009 implemented B-spline basis function in the MS-CG method. Which shows dramatic reduction of memory requirements and increase in computational efficiency of the MS-CG calculation. Their work showed that the MS-CG force field can approximate the many-body PMF in the coarse-grained coordinates (Lu and Voth, 2009). Then in 2011, they showed that MS-CG method can also be used to analyze the CG interactions from atomistic MD trajectories via PMF calculations and its performance is comparable to various free energy computation methods (Lu and Voth, 2011). The distribution function defined in the coarse-grained potential could be reproduced by iteratively applying the MS-CG algorithm (Lu *et al.*, 2013). It has been proven that both of the two common methods "postprocessing method" and "fixed part method" (Das and Andersen, 2009; Noid *et al.*, 2008) can similarly make MS-CG potentials which are limited in accuracy only by the incompleteness of the basis set usable (Das *et al.*, 2012).

Finally, it is worth noting that it is necessary to intervene in the variational calculation in some way, in case we need to produce accurate MS-CG potential in both high and average potential energy regions (Das *et al.*, 2012).

Multilevel coarse graining

Multilevel coarse graining is the future of coarse graining. This direction is not yet well defined. Generally, it is expected that future simulation packages can simulate systems in a heterogeneous way. Fine details can be simulated using a fine detailed model on minor steps, while coarser details can be simulated using less detailed models and over bigger time steps. One way to achieve this goal can be including the defined coarse grains into *further coarser* grains in a hierarchal way. Examples of coarse details may include diffusion and movement in the solvent, which can be simulated using models like Brownian dynamics, ignoring the detailed atom-atom interactions, but rather modeling their overall effect as a continuum.

Force fields used in CG MD simulation

Pair-potentials vs. many-body potentials

One essential requirement for any simulation package is the definition of a potential function, which describes how the simulated particles in the system would interact. Such function is known as the Force Field. Force Fields are usually empirical, and they usually employ preset bonding arrangements, which makes them capable of modeling structural and conformational changes but not chemical reactions (bond breaking and reformation) explicitly.

Force field potential function can be either *pair-potential* function, where the force is calculated as a sum of pairwise interactions between particles (e.g., Lennard-Jones potential of Van der Waals force and Born (ionic) model of the ionic lattice), or *many-body potentials*, where three or more particles interact with each other. Examples of the later are Tersoff potential (Tersoff, 1989), embedded-atom method (EAM) (Daw *et al.*, 1993), and Tight-Binding Second Moment Approximation (TBSMA) (Cleri and Rosato, 1993) potentials.

Static partial charges vs. polarizable potentials

Most force fields (especially experimental Force fields) model the effect of polarizability as static partial charges. While new techniques model polarizability as fluctuating charges, electronic structural theory, induced dipole, distributed multipoles, point charges, Bond Polarization Theory (BPT). Increased accuracy was achieved using fluctuating charges in water molecules (Lamoureux *et al.*, 2006) and with some promising results for proteins (Patel *et al.*, 2004).

Coarse-grained force fields

Coarse-Grained Force Fields can be based on generic atom types (e.g., MARTINI Force Field), or on secondary structure (e.g., VAMM).

MARTINI (Marrink *et al.*, 2004, 2007) is a very common Coarse Grained Force Field, based on having four categories of particles/beads (Q (charged), P (polar), N (nonpolar) and C (apolar)). They are then split in 4 or 5 levels, summing up to a total of 20 bead types. It has parameters for lipids, proteins, carbohydrates, DNA, RNA, and molecule types available for download on their website.

On the other hand, VAMM (Virtual Atom Molecular Mechanics) (Korkut and Hendrickson, 2009) is a "knowledge based" force field developed to model large scale conformational transitions based on the virtual interactions of C-alpha atoms and on secondary structure and residue specific contact features.

Finite Element Method

Finite element method is a common method of numerical solution of complex problems including force and heat acting on irregular structures. It is mostly used for structural mechanics applications. In this method, the system is modelled as a set of finite elements of appropriate properties (e.g., stiffness, toughness, etc.), interconnected at points called nodes.

Although the method is famous for use in civil engineering applications to solve displacement versus boundary and load conditions, it can be applied on smaller scale objects with high efficiency in terms of speed and accuracy. For instance, in 1999 O'Brien *et al.* simulated different fractured objects of different toughness property after collisions, using continuously remeshing technique and it was comparative to experimental results (O'Brien and Hodgins, 1999).

Even on the molecular levels there have been promising trails to combine FEM and MD simulation to model silicon (Izumi *et al.*, 1999) and LASER induced pressure wave propagation (Smirnova *et al.*, 1999). In 2013, Lee at al. published a multiscale modeling technique for bridging molecular dynamics with finite element method, based on weighted average momentum principle. Their work was applied on 2-D problems, but to the best of our knowledge, no 3D applications are available so far (Lee and Basaran, 2013).

Closing Remarks

Along with the vast increasing in supercomputing power, MD simulations, as well as other simulation variants, are gradually adopted to explore more complex biological systems, from electron level to molecule level, and evolve to achieve free energy surface construction and states-transition with affordable time and resources. In this chapter, we went through the basics of MD simulation, and then explained the advanced sampling methods. Besides, we also illustrated some applications in alignment with MD simulation in biophysical and drug design field, and examples of applying MD simulation to address biological systems have been presented. Lastly, we discussed possible future developments and directions of MD and simulations. This review thus serves as a general introduction to MD simulation and its applications.

Acknowledgement

This work was partially supported by MOE Tier 1 Grant [RG 138/15], [2015-T1-001-169]; as well as Nanyang Technological University School of Computer Science and Engineering RSS [M060020005-7070680]. We'd like to thank Dr Ning Lulu for her contribution of the GPU benchmark data we used in **Fig. 4**.

See also: *Ab initio* Protein Structure Prediction. Algorithms for Structure Comparison and Analysis: Docking. Algorithms for Structure Comparison and Analysis: Docking. Biomolecular Structures: Prediction, Identification and Analyses. Cloud-Based Molecular Modeling Systems. Drug Repurposing and Multi-Target Therapies. Identifying Functional Relationships Via the Annotation and Comparison of Three-Dimensional Amino Acid Arrangements in Protein Structures. In Silico Identification of Novel Inhibitors. Molecular Dynamics Simulations in Drug Discovery. Multi-Scale Modelling in Biology. Natural Language Processing Approaches in Bioinformatics. Prediction of Protein-Protein Interactions: Looking Through the Kaleidoscope. Protein Structural Bioinformatics: An Overview. Protein Structure Visualization. Protein Three-Dimensional Structure Prediction. Rational Structure-Based Drug Design. Small Molecule Drug Design. Structure-Based Design of Peptide Inhibitors for Protein Arginine Deiminase Type IV (PAD4). Structure-Based Drug Design Workflow. Study of The Variability of The Native Protein Structure

References

Abrams, C., Bussi, G., 2013. Enhanced sampling in molecular dynamics using metadynamics, replica-exchange, and temperature-acceleration. Entropy 16 (1), 163–199.

Abraham, M.J., Murtola, T., Schulz, R., et al., 2015. GROMACS: High performance molecular simulations through multi-level parallelism from laptops to supercomputers. SoftwareX 1, 19–25.

Affentranger, R., Tavernelli, I., Di Iorio., E.E., 2006. A novel Hamiltonian replica exchange MD protocol to enhance protein conformational space sampling. J. Chem. Theory Computation 2 (2), 217–228.

Arkhipov, A., Yin, Y., Schulten, K., 2008. Four-scale description of membrane sculpting by BAR domains. Biophysical Journal 95, 2806–2821.

Arkin, M.R., Tang, Y., Wells, J.A., 2014. Small-molecule inhibitors of protein-protein interactions: Progressing toward the reality. Chemistry & Biology 21, 1102–1114.

Baaden, M., Marrink, S.J., 2013. Coarse-grain modelling of protein–protein interactions. Current Opinion in Structural Biology 23, 878–886.

Ballester, P.J., Mitchell, J.B., 2010. A machine learning approach to predicting protein–ligand binding affinity with applications to molecular docking. Bioinformatics 26, 1169–1175.

Banavali, N.K., Roux, B., 2005. Free energy landscape of A-DNA to B-DNA conversion in aqueous solution. Journal of the American Chemical Society 127, 6866–6876.

Barducci, A., Bussi, G., Parrinello, M., 2008. Well-tempered metadynamics: A smoothly converging and tunable free-energy method. Physical Review Letters 100, 020603.

Barducci, A., Bonomi, M., Parrinello, M., 2011. Metadynamics. Wiley Interdiscip. Rev. Comput. Mol. Sci. 1 (5), 826–843.

Bartók, A.P., Payne, M.C., Kondor, R., Csányi, G., 2010. Gaussian approximation potentials: The accuracy of quantum mechanics, without the electrons. Physical Review Letters 104, 136403.

Beauchamp, K.A., McGibbon, R., Lin, Y.S., Pande, V.S., 2012. Simple few-state models reveal hidden complexity in protein folding. Proceedings of the National Academy of Sciences of the United States of America 109, 17807–17813.

Behler, J., 2016. Perspective: Machine learning potentials for atomistic simulations. The Journal of Chemical Physics 145, 170901.

Behler, J., Parrinello, M., 2007. Generalized neural-network representation of high-dimensional potential-energy surfaces. Physical Review Letters 98, 146401.

Biarnés, X., Pietrucci, F., Marinelli, F., Laio, A., 2012. METAGUI. A VMD interface for analyzing metadynamics and molecular dynamics simulations. Computer Physics Communications 183, 203–211.

Blanco, F.J., Rivas, G., Serrano, L., 1994. A short linear peptide that folds into a native stable β-hairpin in aqueous solution. Nature Structural & Molecular Biology 1, 584–590.

Bonomi, M., Branduardi, D., Bussi, G., et al., 2009. PLUMED: A portable plugin for free-energy calculations with molecular dynamics. Computer Physics Communications 180, 1961–1972.

Borhani, D.W., Shaw, D.E., 2012. The future of molecular dynamics simulations in drug discovery. J. Comput.-Aided Mol. Des. 26 (1), 15–26.

Botu, V., Ramprasad, R., 2015. Adaptive machine learning framework to accelerate ab initio molecular dynamics. International Journal of Quantum Chemistry 115, 1074–1083.

Brooks, B.R., Bruccoleri, R.E., Olafson, B.D., et al., 1983. CHARMM A program for macromolecular energy, minimization, and dynamics calculations. Journal of Computational Chemistry 4, 187–217.

Bussi, G., Gervasio, F.L., Laio, A., Parrinello, M., 2006a. Free-energy landscape for beta hairpin folding from combined parallel tempering and metadynamics. Journal of the American Chemical Society A 128, 13435–13441.

Bussi, G., Gervasio, F.L., Laio, A., Parrinello, M., 2006b. Free-energy landscape for β hairpin folding from combined parallel tempering and metadynamics. Journal of the American Chemical Society 128, 13435–13441.

Cai, Y.-D., Liu, X.-J., Xu, X.-B., Chou, K.-C., 2002. Prediction of protein structural classes by support vector machines. Computers & Chemistry 26, 293–296.

Cecchini, M., Rao, F., Seeber, M., Caflisch, A., 2004. Replica exchange molecular dynamics simulations of amyloid peptide aggregation. Journal of Chemical Physics 121, 10748–10756.

Cheng, T., Li, Q., Zhou, Z., Wang, Y., Bryant, S.H., 2012. Structure-based virtual screening for drug discovery: A problem-centric review. The AAPS journal 14, 133–141.

Chodera, J.D., Noé, F., 2014. Markov state models of biomolecular conformational dynamics. Curr. Opin. Struct. Biol. 25, 135–144.

Christen, M., Hünenberger, P.H., Bakowies, D., et al., 2005. The GROMOS software for biomolecular simulation GROMOS05. Journal of Computational Chemistry 26, 1719–1751.

Cleri, F., Rosato, V., 1993. Tight-binding potentials for transition-metals and alloys. Physical Review B 48, 22–33.

Cornell, W.D., Cieplak, P., Bayly, C.I., et al., 1995. A second generation force field for the simulation of proteins, nucleic acids, and organic molecules. Journal of the American Chemical Society 117, 5179–5197.

Curuksu, J., Zacharias, M., 2009. Enhanced conformational sampling of nucleic acids by a new Hamiltonian replica exchange molecular dynamics approach. J. Chem. Phys. 130 (10), 03B610.

Das, A., Andersen, H.C., 2009. The multiscale coarse-graining method. III. A test of pairwise additivity of the coarse-grained potential and of new basis functions for the variational calculation. Journal of Chemical Physics 131, 034102.

Das, A., Lu, L., Andersen, H.C., Voth, G.A., 2012. The multiscale coarse-graining method. X. Improved algorithms for constructing coarse-grained potentials for molecular systems. Journal of Chemical Physics 136, 194115.

Daw, M.S., Foiles, S.M., Baskes, M.I., 1993. The embedded-atom method – A review of theory and applications. Materials Science Reports 9, 251–310.

De Vivo, M., Masetti, M., Bottegoni, G., Cavalli, A., 2016. Role of molecular dynamics and related methods in drug discovery. J. Med. Chem. 59 (9), 4035–4061.

Denschlag, R., Lingenheil, M., Tavan, P., 2008. Efficiency reduction and pseudo-convergence in replica exchange sampling of peptide folding-unfolding equilibria. Chemical Physics Letters 458, 244–248.

Doniger, S., Hofmann, T., Yeh, J., 2002. Predicting CNS permeability of drug molecules: Comparison of neural network and support vector machine algorithms. Journal of Computational Biology 9, 849–864.

Doerr, S., De Fabritiis, G., 2014. On-the-fly learning and sampling of ligand binding by high-through put molecular simulations. J. Chem. Theory Comput. 10 (5), 2064–2069.

Doerr, S., Harvey, M.J., Noe, F., De Fabritiis, G., 2016. HTMD: high-throughput molecular dynamics for molecular discovery. J. Chem. Theory Comput. 12 (4), 1845–1852.

Durrant, J.D., McCammon, J.A., 2011. Molecular dynamics simulations and drug discovery. BMC biology 9 (1), 71.

Freddolino, P.L., Arkhipov, A.S., Larson, S.B., Mcpherson, A., Schulten, K., 2006. Molecular dynamics simulations of the complete satellite tobacco mosaic virus. Structure 14, 437–449.

Frenkel, D., Smit, B., 2002. Understanding molecular simulation: From algorithms to applications. Comput. Sci. Ser 1, 1–638.

Frenkel, D., Smit, B., 2001. Understanding Molecular Simulation: From Algorithms to Applications, vol. 1. Elsevier.

Fukunishi, H., Watanabe, O., Takada, S., 2002. On the Hamiltonian replica exchange method for efficient sampling of biomolecular systems: Application to protein structure prediction. The Journal of chemical physics 116 (20), 9058–9067.

Galvelis, R., Sugita, Y., 2017. Neural network and nearest neighbor algorithms for enhancing sampling of molecular dynamics. Journal of Chemical Theory and Computation 13, 2489–2500.

García, A.E., Sanbonmatsu, K.Y., 2002. α-Helical stabilization by side chain shielding of backbone hydrogen bonds. Proceedings of the National Academy of Sciences 99, 2782–2787.

Gnanakaran, S., Nymeyer, H., Portman, J., Sanbonmatsu, K.Y., Garcia, A.E., 2003. Peptide folding simulations. Current Opinion in Structural Biology 13, 168–174.

Göbel, U., Sander, C., Schneider, R., Valencia, A., 1994. Correlated mutations and residue contacts in proteins. Proteins: Structure, Function, and Bioinformatics 18, 309–317.

Gumbart, J., Wang, Y., Aksimentiev, A., Tajkhorshid, E., Schulten, K., 2005. Molecular dynamics simulations of proteins in lipid bilayers. Current Opinion in Structural Biology 15, 423–431.

Halgren, T.A., 1996. Merck molecular force field. I. Basis, form, scope, parameterization, and performance of MMFF94. Journal of Computational Chemistry 17, 490–519.

Harrigan, M.P., Sultan, M.M., Hernández, C.X., et al., 2017. MSMBuilder: Statistical models for biomolecular dynamics. Biophysical Journal 112, 10–15.

Heermann, D.W., 1990. Computer-simulation methods. In: Computer Simulation Methods in Theoretical Physics. Springer.

Heffernan, R., Paliwal, K., Lyons, J., et al., 2015. Improving prediction of secondary structure, local backbone angles, and solvent accessible surface area of proteins by iterative deep learning. Scientific Reports 5, 11476.

Hou, T., Wang, J., Zhang, W., Wang, W., Xu, X., 2006. Recent advances in computational prediction of drug absorption and permeability in drug discovery. Current Medicinal Chemistry 13, 2653–2667.

Huang, K., 1987. Statistical Mechanics. 18. Wiley. 3.

Hua, S., Sun, Z., 2001. A novel method of protein secondary structure prediction with high segment overlap measure: Support vector machine approach. Journal of Molecular Biology 308, 397–407.

Hu, L., Soderhjelm, P., Ryde, U., 2011. On the convergence of QM/MM energies. Journal of Chemical Theory and Computation 7, 761–777.

Husic, B.E., Pande, V.S., 2018. Markov state models: from an art to a science. J. Am. Chem. Soc. 140 (7), 2386–2396.

Im, W., Roux, B.T., 2002. Ions and counterions in a biological channel: A molecular dynamics simulation of OmpF porin from Escherichia coli in an explicit membrane with 1M KCl aqueous salt solution. Journal of Molecular Biology 319, 1177–1197.

Izumi, S., Kawakami, T., Sakai, S., 1999. A FEM-MD combination method for silicon. In: Proceeding of the 1999 International Conference on Simulation of Semiconductor Processes and Devices, SISPAD '99, pp. 143–146.

Jorgensen, W.L., Maxwell, D.S., Tiradorives, J., 1996. Development and testing of the OPLS all-atom force field on conformational energetics and properties of organic liquids. Journal of the American Chemical Society 118, 11225–11236.

Jorgensen, W.L., Tirado-Rives, J., 1988. The OPLS [optimized potentials for liquid simulations] potential functions for proteins, energy minimizations for crystals of cyclic peptides and crambin. Journal of the American Chemical Society 110, 1657–1666.

Judson, R., Elloumi, F., Setzer, R.W., Li, Z., Shah, I., 2008. A comparison of machine learning algorithms for chemical toxicity classification using a simulated multi-scale data model. BMC Bioinformatics 9, 241.

Kalos, M.H., Whitlock, P.A., 2008. Monte Carlo Methods. John Wiley & Sons.

Kästner, J., 2011. Umbrella sampling. Wiley Interdisciplinary Reviews: Computational Molecular Science 1, 932–942.

Khalili-Araghi, F., Gumbart, J., Wen, P.-C., et al., 2009. Molecular dynamics simulations of membrane channels and transporters. Current Opinion in Structural Biology 19, 128–137.

Kone, A., Kofke, D.A., 2005. Selection of temperature intervals for parallel-tempering simulations. The Journal of Chemical Physics 122, 206101.

Korkut, A., Hendrickson, W.A., 2009. A force field for virtual atom molecular mechanics of proteins. Proceedings of the National Academy of Sciences of the United States of America 106, 15667–15672.

Kubelka, J., Hofrichter, J., Eaton, W.A., 2004. The protein folding 'speed limit'. Current Opinion in Structural Biology 14, 76–88.

Laio, A., Gervasio, F.L., 2008. Metadynamics: A method to simulate rare events and reconstruct the free energy in biophysics, chemistry and material science. Reports on Progress in Physics 71, 126601.

Laio, A., Parrinello, M., 2002. Escaping free-energy minima. Proceedings of the National Academy of Sciences 99, 12562–12566.

Lamoureux, G., Harder, E., Vorobyov, I.V., Roux, B., Mackerell, A.D., 2006. A polarizable model of water for molecular dynamics simulations of biomolecules. Chemical Physics Letters 418, 245–249.

Larini, L., Lu, L., Voth, G.A., 2010. The multiscale coarse-graining method. VI. Implementation of three-body coarse-grained potentials. The Journal of Chemical Physics 132, 164107.

Lee, H.S., Zhang, Y., 2012. (333) BSP-SLIM: A blind low-resolution ligand-protein docking approach using theoretically predicted protein structures. Proteins 80, 93–110.

Lee, Y.C., Basaran, C., 2013. A multiscale modeling technique for bridging molecular dynamics with finite element method. Journal of Computational Physics 253, 64–85.

Lindorff-Larsen, K., Piana, S., Dror, R.O., Shaw, D.E., 2011. How fast-folding proteins fold. Science 334, 517–520.

Lin, E., Shell, M.S., 2009. Convergence and heterogeneity in peptide folding with replica exchange molecular dynamics. Journal of Chemical Theory and Computation 5, 2062–2073.

Lise, S., Archambeau, C., Pontil, M., Jones, D.T., 2009. Prediction of hot spot residues at protein-protein interfaces by combining machine learning and energy-based methods. BMC Bioinformatics 10, 365.

Liu, P., Kim, B., Friesner, R.A., Berne, B., 2005. Replica exchange with solute tempering: A method for sampling biological systems in explicit water. Proceedings of the National Academy of Sciences of the United States of America 102, 13749–13754.

Li, Z., Kermode, J.R., De Vita, A., 2015. Molecular dynamics with on-the-fly machine learning of quantum-mechanical forces. Physical Review Letters 114, 096405.

Lu, L., Dama, J.F., Voth, G.A., 2013. Fitting coarse-grained distribution functions through an iterative force-matching method. Journal of Chemical Physics 139, 121906.

Lu, L., Voth, G.A., 2011. The multiscale coarse-graining method. VII. Free energy decomposition of coarse-grained effective potentials. The Journal of Chemical Physics 134, 224107.

Lu, L.Y., Voth, G.A., 2009. Systematic coarse-graining of a multicomponent lipid bilayer. Journal of Physical Chemistry B 113, 1501–1510.

Ma, B., Nussinov, R., 2002. Stabilities and conformations of Alzheimer's β-amyloid peptide oligomers (Aβ16–22, Aβ16–35, and Aβ10–35): Sequence effects. Proceedings of the National Academy of Sciences 99, 14126–14131.

Mayo, S.L., Olafson, B.D., Goddard, W.A., 1990. DREIDING: A generic force field for molecular simulations. J. Phys. Chem. 94 (26), 8897–8909.

Marinari, E., Parisi, G., 1992. Simulated tempering: A new Monte Carlo scheme. EPL (Europhysics Letters) 19, 451.

Marinelli, F., Pietrucci, F., Laio, A., Piana, S., 2009. A kinetic model of trp-cage folding from multiple biased molecular dynamics simulations. PLOS Computational Biology 5, e1000452.

Marrink, S.J., De Vries, A.H., Mark, A.E., 2004. Coarse grained model for semiquantitative lipid simulations. Journal of Physical Chemistry B 108, 750–760.

Marrink, S.J., Risselada, H.J., Yefimov, S., Tieleman, D.P., De Vries, A.H., 2007. The MARTINI force field: Coarse grained model for biomolecular simulations. Journal of Physical Chemistry B 111, 7812–7824.

McCammon, J.A., Gelin, B.R., Karplus, M., 1977. Dynamics of folded proteins. Nature 267, 585–590.

McGibbon, R., Ramsundar, B., Sultan, M., Kiss, G., Pande, V., 2014. Understanding protein dynamics with L1-regularized reversible hidden Markov models. In: Proceeding of the International Conference on Machine Learning, pp. 1197–1205.

McGibbon, R.T., Pande, V.S., 2013. Learning kinetic distance metrics for Markov state models of protein conformational dynamics. J. Chem. Theory Comput. 9 (7), 2900–2906.

Miao, Y., Johnson, J.E., Ortoleva, P.J., 2010. All-atom multiscale simulation of cowpea chlorotic mottle virus capsid swelling. The Journal of Physical Chemistry B 114, 11181–11195.

Miao, Y.S., Han, X.M., Zheng, L.Z., et al., 2016. Fimbrin phosphorylation by metaphase Cdk1 regulates actin cable dynamics in budding yeast. Nature Communications 7.

Mitsutake, A., Sugita, Y., Okamoto, Y., 2001. Generalized-ensemble algorithms for molecular simulations of biopolymers. Peptide Science 60, 96–123.

Mu, Y., 2009. Dissociation aided and side chain sampling enhanced Hamiltonian replica exchange. Journal of Chemical Physics 130, 164107.

Mu, Y., Nguyen, P.H., Stock, G., 2005. Energy landscape of a small peptide revealed by dihedral angle principal component analysis. Proteins: Structure, Function, and Bioinformatics 58, 45–52.

Mu, Y., Yang, Y., Xu, W., 2007. Hybrid Hamiltonian replica exchange molecular dynamics simulation method employing the Poisson–Boltzmann model. The Journal of Chemical Physics 127, 084119.

Naritomi, Y., Fuchigami, S., 2011. Slow dynamics in protein fluctuations revealed by time-structure based independent component analysis: The case of domain motions. The Journal of Chemical Physics 134, 02B617.

Nafisa, M. Hassan, 2017. (111) Protein-Ligand Blind Docking Using QuickVina-W With Inter-Process Spatio-Temporal Integration. Scientific Reports 7 (1), doi:10.1038/s41598-017-15571-7.

Noé, F., Schütte, C., Vanden-Eijnden, E., Reich, L., Weikl, T.R., 2009. Constructing the equilibrium ensemble of folding pathways from short off-equilibrium simulations. Proceedings of the National Academy of Sciences 106, 19011–19016.

Noé, F., Clementi, C., 2015. Kinetic distance and kinetic maps from molecular dynamics simulation. J. Chem. Theory Comput. 11 (10), 5002–5011.

Noid, W.G., Liu, P., Wang, Y., et al., 2008. The multiscale coarse-graining method. II. Numerical implementation for coarse-grained molecular models. Journal of Chemical Physics 128, 244115.

O'Brien, J.F., Hodgins, J.K., 1999. Graphical modeling and animation of brittle fracture. In: Siggraph 99 Conference Proceedings, pp. 137–146.

Ogata, K., Yuki, T., Hatakeyama, M., Uchida, W., Nakamura, S., 2013. All-atom molecular dynamics simulation of photosystem II embedded in thylakoid membrane. Journal of the American Chemical Society 135, 15670–15673.

Okur, A., Roe, D.R., Cui, G., Hornak, V., Simmerling, C., 2007. Improving convergence of replica-exchange simulations through coupling to a high-temperature structure reservoir. Journal of Chemical Theory and Computation 3, 557–568.

Patel, S., Mackerell Jr., A.D., Brooks 3rd, C.L., 2004. CHARMM fluctuating charge force field for proteins: II protein/solvent properties from molecular dynamics simulations using a nonadditive electrostatic model. Journal of Computational Chemistry 25, 1504–1514.

La Penna, G., Morante, S., Perico, A., Rossi, G.C., 2004. Designing generalized statistical ensembles for numerical simulations of biopolymers. Journal of Chemical Physics 121, 10725–10741.

Pande, V.S., Beauchamp, K., Bowman, G.R., 2010. Everything you wanted to know about Markov State Models but were afraid to ask. Methods 52 (1), 99–105.

Pérez-Hernández, G., Paul, F., Giorgino, T., De Fabritiis, G., Noé, F., 2013. Identification of slow molecular order parameters for Markov model construction. The Journal of Chemical Physics 139, 07B604_1.

Piana, S., Laio, A., 2007. A bias-exchange approach to protein folding. Journal of Physical Chemistry B 111 (17), 4553–4559.

Piana, S., Lindorff-Larsen, K., Shaw, D.E., 2011. How robust are protein folding simulations with respect to force field parameterization? Biophysical Journal 100, L47–L49.

Rahman, A., 1964. Correlations in the motion of atoms in liquid argon. Physical Review 136, A405.

Rajamani, D., Thiel, S., Vajda, S., Camacho, C.J., 2004. Anchor residues in protein–protein interactions. Proceedings of the National Academy of Sciences of the United States of America 101, 11287–11292.

Riccardi, D., Yang, S., Cui, Q., 2010. Proton transfer function of carbonic anhydrase: Insights from QM/MM simulations. Biochimica et Biophysica Acta 1804, 342–351.

Röblitz, S., Weber, M., 2013. Fuzzy spectral clustering by PCCA+: Application to Markov state models and data classification. Adv. Data Anal. Classif. 7 (2), 147–179.

Rosenbergl, J.M., 1992. The weighted histogram analysis method for free-energy calculations on biomolecules. I. The method. Journal of Computational Chemistry 13, 1011–1021.

Scherer, M.K., Trendelkamp-Schroer, B., Paul, F., et al., 2015. PyEMMA 2: A software package for estimation, validation, and analysis of Markov models. Journal of Chemical Theory and Computation 11, 5525–5542.

Schwantes, C.R., Pande, V.S., 2015. Modeling molecular kinetics with tICA and the kernel trick. J. Chem. Theory Comput. 11 (2), 600–608.

Schneidman-Duhovny, D., Inbar, Y., Nussinov, R., Wolfson, H.J., 2005. (222) PatchDock and SymmDock: servers for rigid and symmetric docking. Nucl. Acids. Res 33, W363–367.

Senn, H.M., Thiel, W., 2007. QM/MM studies of enzymes. Current Opinion in Chemical Biology 11, 182–187.

Setubal, Joao Carlos, Meidanis, Joao, Setubal-Meidanis, 1997. Introduction to computational molecular biology 1997 (No. 04), QH506, S4. PWS Pub.

Shaw, D.E., Maragakis, P., Lindorff-Larsen, K., et al., 2010. Atomic-level characterization of the structural dynamics of proteins. Science 330, 341–346.

Shih, A.Y., Arkhipov, A., Freddolino, P.L., Schulten, K., 2006. Coarse grained protein – lipid model with application to lipoprotein particles. The Journal of Physical Chemistry B 110, 3674–3684.

Shih, A.Y., Freddolino, P.L., Arkhipov, A., Schulten, K., 2007. Assembly of lipoprotein particles revealed by coarse-grained molecular dynamics simulations. Journal of Structural Biology 157, 579–592.

Shrivastava, I.H., Sansom, M.S., 2000. Simulations of ion permeation through a potassium channel: Molecular dynamics of KcsA in a phospholipid bilayer. Biophysical Journal 78, 557–570.

Sindhikara, D., Meng, Y.L., Roitberg, A.E., 2008. Exchange frequency in replica exchange molecular dynamics. Journal of Chemical Physics 128.

Singhal, N., Pande, V.S., 2005. Error analysis and efficient sampling in Markovian state models for molecular dynamics. J. Chem. Phys. 123 (20), 204909.

Smirnova, J.A., Zhigilei, L.V., Garrison, B.J., 1999. A combined molecular dynamics and finite element method technique applied to laser induced pressure wave propagation. Computer Physics Communications 118, 11–16.

Snow, C.D., Zagrovic, B., Pande, V.S., 2002. The Trp cage: Folding kinetics and unfolded state topology via molecular dynamics simulations. Journal of the American Chemical Society 124, 14548–14549.

Suárez, E., Adelman, J.L., Zuckerman, D.M., 2016. Accurate estimation of protein folding and unfolding times: Beyond Markov state models. Journal of Chemical Theory and Computation 12, 3473–3481.

Sugita, Y., Okamoto, Y., 1999a. Replica-exchange molecular dynamics method for protein folding. Chemical Physics Letters 314, 141–151.

Sugita, Y., Okamoto, Y., 1999b. Replica-exchange molecular dynamics method for protein folding. Chemical Physics Letters 314, 141–151.

Sultan, M.M., Wayment-Steele, H.K., Pande, V.S., 2018. Transferable neural networks for enhanced sampling of protein dynamics. arXiv preprint arXiv:1801.00636.

Sutto, L., Marsili, S., Gervasio, F.L., 2012. New advances in metadynamics. Wiley Interdisciplinary Reviews: Computational Molecular Science 2, 771–779.

Swendsen, R.H., Wang, J.-S., 1987. Nonuniversal critical dynamics in Monte Carlo simulations. Physical Review Letters 58, 86.

Tai, K., 2004. Conformational sampling for the impatient. Biophysical Chemistry 107, 213–220.

Tersoff, J., 1989. Modeling solid-state chemistry – Interatomic potentials for multicomponent systems. Physical Review B 39, 5566–5568.

Tiwary, P., Parrinello, M., 2014. A time-independent free energy estimator for metadynamics. The Journal of Physical Chemistry B 119, 736–742.

Tolman, R.C., 1938. The principles of statistical mechanics. Courier Corporation.

Torrie, G.M., Valleau, J.P., 1974. Monte Carlo free energy estimates using non-Boltzmann sampling: Application to the sub-critical Lennard-Jones fluid. Chemical Physics Letters 28, 578–581.

Torrie, G.M., Valleau, J.P., 1977. Nonphysical sampling distributions in Monte Carlo free-energy estimation: Umbrella sampling. Journal of Computational Physics 23, 187–199.

van der Kamp, M.W., Mulholland, A.J., 2013. Combined quantum mechanics/molecular mechanics (QM/MM) methods in computational enzymology. Biochemistry 52, 2708–2728.

van Gunsteren, W.F., Billeter, S.R., Eising, A.A., et al., 1996. Biomolecular simulation: The {GROMOS96} manual and userguide. Hochschuleverlag AG an der ETH Zürich.

van der Spoel, D., Hess, B., 2011. GROMACS—the road ahead. Wiley Interdiscip. Rev.: Comput. Molecular Science 1 (5), 710–715.

Walters, W.P., Murcko, M.A., 2002. Prediction of 'drug-likeness'. Advanced Drug Delivery Reviews 54, 255–271.

Wang, J., Shao, Q., Xu, Z., *et al.*, 2014. Exploring transition pathway and free-energy profile of large-scale protein conformational change by combining normal mode analysis and umbrella sampling molecular dynamics. Journal of Physical Chemistry B 118, 134–143.

Wang, S., Peng, J., Ma, J., Xu, J., 2016. Protein secondary structure prediction using deep convolutional neural fields. Scientific Reports 6.

Weber, M., 2011. A subspace approach to molecular Markov state models via a new infinitesimal generator.

Woutersen, S., Pfister, R., Hamm, P., *et al.*, 2002. Peptide conformational heterogeneity revealed from nonlinear vibrational spectroscopy and molecular-dynamics simulations. The Journal of Chemical Physics 117, 6833–6840.

Zhang, T., Nguyen, P.H., Nasica-Labouze, J., Mu, Y., Derreumaux, P., 2015. Folding atomistic proteins in explicit solvent using simulated tempering. The Journal of Physical Chemistry B.

Zhang, W., Wu, C., Duan, Y., 2005a. Convergence of replica exchange molecular dynamics. Journal of Chemical Physics 123, 154105.

Zhang, W., Wu, C., Duan, Y., 2005b. Convergence of replica exchange molecular dynamics. Journal of Chemical Physics 123.

Zhang, Y., Liu, H., Yang, W., 2000. Free energy calculation on enzyme reactions with an efficient iterative procedure to determine minimum energy paths on a combined ab initio QM/MM potential energy surface. The Journal of Chemical Physics 112, 3483–3492.

Zhang, Z.Y., Lu, L.Y., Noid, W.G., *et al.*, 2008. A systematic methodology for defining coarse-grained sites in large biomolecules. Biophysical Journal 95, 5073–5083.

Zheng, L., Lin, V.C., Mu, Y., 2016. Exploring flexibility of progesterone receptor ligand binding domain using molecular dynamics. PLOS ONE 11, e0165824.

Further Reading

Anderson, J.A., Lorenz, C.D., Travesset, A., 2008. General purpose molecular dynamics simulations fully implemented on graphics processing units. Journal of Computational Physics 227 (10), 5342–5359.

Bonomi, M., Parrinello., M., 2010. Enhanced sampling in the well-tempered ensemble. Physical Review Letters 104 (19), 190601.

Borodin, O., 2009. Polarizable force field development and molecular dynamics simulations of ionic liquids. The Journal of Physical Chemistry B 113 (33), 11463–11478.

Braun, G.H., *et al.*, 2008. Molecular dynamics, flexible docking, virtual screening, ADMET predictions, and molecular interaction field studies to design novel potential MAO-B inhibitors. Journal of Biomolecular Structure and Dynamics 25 (4), 347–355.

Chmiela, S., *et al.*, 2018. Towards exact molecular dynamics simulations with machine-learned force fields. Bulletin of the American Physical Society.

Relevant Websites

https://www.geforce.com/hardware/compare-buy-gpus
 GeForce
  GeForce.com.
http://www.emma-project.org
 MD with MSM (or HMM) Pyemma.
http://msmbuilder.org
 MSMBuilder.
https://plumed.github.io/doc.html
 Plumed tutorials and manual.
http://docs.markovmodel.org/
 Principle of MSM.

Biographical Sketch

Zheng Liangzhen is a PhD candidate in School of Biological Sciences, Nanyang Technological University, Singapore since 2014. He obtained his bachelor's degree in biological sciences from College of Life Science, Wuhan University, in 2010. His current interests are protein dynamics, protein-nucleic acid complex dynamics, and machine learning aided drug discovery.

Amr Alhossary received his M.B.B.S. from School of Medicine, Cairo University in 2003, then his Graduate diploma and M.Sc. in Bioinformatics from Faculty of Computers and Information, Helwan University in 2008 and 2012 respectively. He is currently a PhD candidate in School of Computer Science and Engineering, Nanyang Technological University, Singapore. His research interests include Structural Computational Biology, Molecular Docking, and Drug Design.

Kwoh Chee Keong is currently in the School of Computer Science and Engineering since 1993. He is currently the Assistant Chair (Graduate Studies). He received his Bachelor degree in Electrical Engineering (1st Class) and Master in Industrial System Engineering from the National University of Singapore in 1987 and 1991 respectively. He received his Ph.D. degrees from the Imperial College, University of London in 1995. He is often invited for conferences and journals, including GIW, IEEE BIBM, RECOMB, PRIB etc. He is a member of The Institution of Engineers Singapore, Association for Medical and Bio-Informatics, Imperial College Alumni Association of Singapore (ICAAS). He was conferred the Public Service Medal, the President of Singapore in 2008. His research interests include Data Mining and Soft Computing and Graph-Based inference; applications areas include Bioinformatics and Biomedical Engineering.

Mu Yuguang received his B.Sc. in Physics, M.S. in Quantum Chemistry, and Ph.D. Physics from Shandong University, China in 1991, 1994, and 1997 respectively. He worked as a Lecturer in Department of photoelectronics, Shandong University before moving to Germany in 2009 to work as a research fellow in Fraunhofer Institute of Integrated Systems and Device Technology (IISB) then Institute of Theoretical Chemistry Frankfurt. He has been in School of Biological Sciences, NTU, since 2003, as a Lee Kuan Yew Fellowship Scholar then Assistant and associate professor until now. His research interests include MD simulation method and data-analysis method development; protein folding and unfolding; RNA dynamics; DNA dynamics; DNA-protein; and DNA-counterions interaction study.

Nucleic-Acid Structure Database

Airy Sanjeev and Venkata SK Mattaparthi, Tezpur University, Tezpur, India
Sandeep Kaushik, European Institute of Excellence on Tissue Engineering and Regenerative Medicine, Barco, Guimaraes, Portugal and University of Minho, Braga, Portugal

Introduction

Nucleic acids have a significant role in the development of living systems. They contain wholesome interesting information related to the molecular biologists. They are extensively studied both for biological applications and for various databases. The nucleic acids that consist of RNAs are considered as the main target for several specialized databases. The sequence of nucleic acids consists of numerous control regions such as promoters, splice junctions, origins of replication, etc. The 3D structure of nucleic acid can be well understood by investigating the different types of nucleic acid functions such as nucleic acid–protein interactions, functional RNAs, etc. Several databases have been developed till date based on the nucleic acid sequence apart from some that have been developed from the structures storing information related to the secondary and tertiary structures.

Background

The molecules of nucleic acids allow to transfer the genetic information from one generation to the next generation. The nucleic acids are abundantly found in the nucleus of the cell. These nucleic acids are long polymers that are built up of several units of nucleotide monomers such as C (cytosine), A (adenine), G (guanine), T (thymine), and U (uracil). There are two types of nucleic acids, that is, deoxyribonucleic acid (DNA) and ribonucleic acid (RNA). The structure of DNA is coiled together by both the strands of nucleotide polymer chains (C, A, G, T) to form a double-helical pattern. The double-helical pattern of DNA is further stabilized by the formation of hydrogen bonds between the bases of nucleotides characterized on the chemical affinity. These chemical affinities defined by the nucleotides signify adenine with thymine and cytosine with guanine. On the other hand, RNA consists of a single strand of polymer with nucleotides C, A, G, and U. The structure of RNA has the ability to fold itself and obtain a secondary structure that is important for carrying out the various biological processes. Among both the nucleic acids, the DNA molecule stores the genetic information of the cell and carries that information from parent to offspring. In case of some viruses, the RNA stores the genetic information that is needed for the cell. However the information that is related to the life cycle of the cell and also its activities are stored in the DNA. The molecule of RNA is being built up when the complex biochemical decodification machinery present in the cell reacts on the DNA molecule to acquire information required for a specific function. The RNA molecule is the primary constituent for the process of protein synthesis. It is also accountable for transmitting the genetic information that is encoded within the DNA to construct a specific protein that is required for a specific function in a specific procedure. There exist three main types of RNA, which are *mRNA, tRNA, and rRNA*. mRNA is messenger RNA, which initiates the information required for building the protein molecule from the nucleus to the cytoplasm. tRNA, also known as transfer RNA, on the other hand transports the amino acids to the ribosomes for making the proteins. The rRNA, or ribosomal RNA, is connected with a variety of proteins that form the ribosomes. The complex structures thus formed catalyze the association of amino acids to form protein chains by moving along with the mRNA molecule. Also they are involved in the process of protein synthesis by binding to the tRNAs and other accessory molecules. Apart from these, there also exists regulatory RNA, which has the ability to regulate the gene expression from the various mechanisms such as blocking or interference (Alberts *et al.*, 2002; Bailey, 2017; Chen *et al.*, 1995; Darnell, 1985; Dickerson *et al.*, 1982; Felsenfeld, 1985; Katsuyuki *et al.*, 2016; Krieger *et al.*, 2004; Mirkin, 2001; Travers and Muskhelishvili, 2015; Watson and Crick, 1953; Weinberg, 1985).

The proteins interact with the nucleic acids, mainly DNA and RNA, by means of similar physical forces that consist of dipolar interactions (hydrogen bonding, H-bonds), electrostatic interactions (salt-bridges), effect of entropy (hydrophobic interactions), and also forces of dispersion (base stacking). These physical forces, which are acted upon by the interactions, help in varying degrees for binding the proteins in a sequence-specific (*tight*) or non-sequence-specific (*loose*) orientation. Apart from that, the affinity and specificity of a particular protein–nucleic acid interaction is increased via the protein oligomerization or multiprotein complex formation. The secondary and tertiary structures that are formed from the nucleic acid sequences give an additional mechanism from which the proteins recognize themselves and then bind to the specific nucleic acid sequences. The DNA is a linear macromolecule that is found in most of the cells of living beings. Unlike proteins, it is made up of four diverse building blocks known as *nucleotides*. These nucleotides are made up of a base (purine or pyrimidine group) and a phosphate group (*2'-deoxyribosyl-tri-phosphate*). The bases present in the nucleotides are of four types. They are adenine and guanine present in *purine* and thymine and cytosine present in *pyrimidine*. The sugar present in the DNA is the 2'-deoxy ribose that is phosphorylated at position 5' of the hydroxyl group. The free nucleotide radical consists of any of one, two, or three phosphates that represent mono-, di-, or triphosphate, which are better known as *dATP, dGTP, dTTP, and dCTP*. These nucleotides are however linked covalently to the DNA via *5'-phosphate to 3'-hydroxyl* of the subsequent nucleotide. The single strands of DNA are not stable, although they remain connected with another strand in order to form a double-helical structure such that both strands interlink around each other. The four bases situated at the center of the helix are H-bonded to one another in a very precise manner. The specific way that the bases point towards the center of the helix and H-bond is as follows: *G with C (3 H-bonds)* and *A with T (2 H-bonds)*. The linear geometry and rigidity associated with the H-bonds restricts the formation of *base pairs* (bp). The helical axis and the plane of the base pair are perpendicular to one another.

The B-DNA is the right-handed physiological form of double-helical DNA structure which was described by Watson and Crick in the year 1953. It is an antiparallel double-helical structure with a diameter of 2 nm. The B-DNA consists of *10 bp per turn, 36° per bp twist, 3.4 Å per base, 34 Å pitch, 2 nm diameter*. These characteristics describe B-DNA as an idealized helical structure. The actual DNA conformation deviates from the B-form DNA conformation in both a sequence dependent manner and DNA-binding interaction protein.

The A-form of the DNA is a wider and flatter helical conformer with the base pair plane tilted to the helical axis. The helix of A-DNA possess the major and minor groove that consists of the respective parameters such as *11 bp per turn, 33° per turn twist, 0.26 nm per base, 2.8 nm pitch, 2.6 nm diameter, the base plane tilting to the helical axis with 20°*.

Another form of DNA is the Z-DNA, which is a left-handed conformation consisting of the subsequent structural parameters, which are *12 bp per turn, 0.37 nm per base, 1.8 nm diameter, a tilt base of 7°, 4.5 nm pitch*.

RNA is another nucleic acid, which is similar to DNA and is built up from four different building blocks that are the *ribonucleotides*. In case of RNA, thymine is modified such that it lacks a methyl group and results in the formation of uracil in the base-pairing mechanism. The ribose sugar present in the RNA comes along with a hydroxylated form. The presence of uracil instead of thymine and 2'-OH in the ribose sugar are the main chemical differences between the two nucleic acids, DNA and RNA. RNA is basically a single stranded helical structure. But both RNA and DNA form bp that result in the formation of a heteromeric double-helical secondary structure consisting of a single DNA and RNA strand. This process of annealing of RNA strand to the complementary DNA strand is known as the *hybridization process*. And this process however plays a significant role in the transcription and translation mechanism of the genetic sequences into protein sequences. As the RNA forms a short double strand structure on itself, hence it is known as the stem and loop structure. The double-helical structures DNA/RNA and RNA/RNA form an A-DNA-like conformation, better known as A-RNA or RNA-II. This helix consists of the following parameters: *11 bp per turn, 3.0 nm pitch (30 Å)*. In the RNA conformation, three major types exist, which are *messenger RNA (mRNA), transfer RNA (tRNA), and ribosomal RNA (rRNA)* (Alberts et al., 2002; Bailey, 2017; Bansal, 2003; Chen et al., 1995; Churchill, 1996; Darnell, 1985; Dickerson et al., 1982; Felsenfeld, 1985; Katsuyuki et al., 2016; Krieger et al., 2004; Mirkin, 2001; Travers and Muskhelishvili, 2015; Watson and Crick, 1953; Weinberg, 1985).

The binding function of DNA or RNA is restricted in the distinct conserved domains that are within the tertiary structure. The affinity and specificity can be increased by binding the single protein with multiple nucleic acid binding domains for a specific target nucleic acid sequence. This can however lead to the development of a hierarchy of interactions as the proteins bind directly or indirectly to the nucleic acids via the other bound proteins. However we can gain a significant binding affinity for the protein and nucleic acids complex by understanding the strength of interactions influencing the approaches for the complex assembly. Among these interactions, some are recognized as transient, which need stabilization through chemical cross-linking prior to isolation of the complexes. Hence by studying the protein–nucleic acid interactions, the identification of protein in the complexes can be known and also the structure that is required to assemble these complexes is vital to understanding the role these complexes play in regulating cellular processes (Govil et al., 1985; Luger and Phillips, 2010; Park et al., 2014; Sathyapriya et al., 2008; Siggers and Gordan, 2014; Valuchova et al., 2016).

It has been observed that small, cationic, and planar molecules interpolate between the various bp whereas the large nonpolar molecules interact by the mechanism of groove binding. Various molecules for in vitro and in vivo activities have been tested and also some molecules from them have entered clinical trials. Hence various parameters that contain the experimental dataset required for drug-nucleic acid interactions have been established. These datasets contain the various types of ligands and binding parameters for their interaction with the associated nucleic acid. Although it becomes a challenging task to gather the whole information specifically for the ligand and its targets. However, by understanding the various interactions improvement in the projects of drug discovery can be achieved. Hence it is essential to combine those experimental data and information at a particular platform.

Nucleic Acid Database

NDB History, Sources and Number of NDB Data

The Nucleic Acid Database (NDB) was the first database for nucleic acid structures (Coimbatore Narayanan et al., 2014), which was designed in the early 1990s and is considered as the main reference in the field of nucleic acids. The newly designed database (see "Relevant Websites section"), is a web portal that can provide access to various information related to 3D nucleic acid structures and their complexes. Apart from the contents of primary data, NDB also contains some derived geometric data, classification of structure and motifs, standards for describing nucleic acid features, and also tools and software for analysis of nucleic acids. In this review, we describe the recently redesigned NDB website with some special features related to new RNA-derived data and annotations and their implementation and integration into the search capabilities. This NDB database was founded in 1991 to assemble and distribute structural information about nucleic acids (Berman et al., 1992). It contains annotations that are specific to the structure of nucleic acid and function and tools that allow users to search, download, analyze, and to learn more about nucleic acids apart from the contents available in PDB (Berman et al., 2003). Thus, this database is a value-added one that provides services that are specific to the nucleic acid community. The main importance of NDB on its discovery was to focus on the DNA structural biology. However various tools and annotations have been developed that can address the various characteristics of these molecules as more RNA structures have been determined. It serves as a central medium for the nucleic acid structural information and annotations. NDB at present consists of more than 8000 experimental structures for DNA and RNA molecules that are acquired and curated from the Protein Data Bank (PDB). The structures available in the NDB are annotated with various information that are unique to the nucleic acids but are almost not available from the original entry of PDB. The structures are further analyzed by means of computation of structural geometries and thereby classification was done by searching the sequence, structure levels, and secondary structure by incorporating the PDB entries to the NDB. By using the various tools that are available at the NDB server linked to the other resources, the specific analyses and online visualization in 2D and 3D can be carried out by the users. Apart from that, various new tools were developed in 2014 that are required to analyze the RNA sequences, DNA molecules, aligning RNA and DNA molecules, calculating and visualizing RNA structural geometries and base-pairing patterns that are comparatively more complex and diverse than DNA. This server also includes various statistics related to ideal geometries that are well suited for sugars, bases, and also for various hydrogen bonding pairing patterns that consist of a new catalog of base-pairing in RNA, an educational section, and software for further offline analyses.

NDB Entry Format

The main content of NDB is the primary structure information that is related to the nucleic acids that are acquired from the PDB and also from classification and derived data. The primary data present in the NDB that is obtained from PDB consists of experimental files, identifiers, structural descriptions, citations, crystal data, coordinate information, and various details related to the crystallization, data collection, and structure refinement. NDB also contains ERFs (*external reference files*), which mainly consist of nucleic acid classifications, calculated derived data, and additional derived data on RNA structural features. The web interface provides search, reporting, and also download functionalities. The results of these searches are recurred as either group of structures or individual structures based on the query. These search results are related to a number of reporting features that include predefined feature reports, navigation to individual structure summary reports that further permit primary data files, derived data, and molecular images download. There are about three search options that are present from the secondary persistent header, that is, *ID search*, *search*, and *advanced search*. These search options can however be accessed in the *Search Structures*, a section available on the homepage.

The asymmetric unit's atomic coordinates and the biological assemblies are accessible from the *Downloads* section obtained from the corresponding NDB structure summary report, which is available both in *PDB* as well as *mmCIF* formats. This *mmCIF* format contains the structure factor data whereas the NMR restraint information is available in deposited program format. During the database update, the files are updated weekly and are downloadable from the NDB ftp server at website provided in "Relevant Websites section".

From the advanced search query, the featured reports for any result set are available. The predefined set of reports is such that it contains *NDB status, cell dimensions, citation, refinement data, backbone torsion, base pair and base step parameters, descriptor, sequence, and RNA motifs*. The report containing NDB status consists of parameters such as *NDB and PDB IDs, structure title, authors, deposition, and release dates* is the default report for any advanced search query.

The structures available in the NDB database possess their own personal summary page that is mostly appropriate to that particular structure. There are about four sections of data in the structure summary page, including (a) *primary structure information*, (b) *downloads*, (c) *derived structural data*, and (d) *images*. The primary structure information, which is the main division, encompasses the entry title, sequence, citation, experimental details, refinement information, and various structural descriptions. In the *Download Data* section, the details of the atomic coordinates, structure factors, and NMR restraint information are present. An RNA View Image is provided for the entries of RNA that represents 2D base pairing. Under the *More Images* link, various images are available including options such as *biological assemblies, crystal packing*, and *ensemble images*. Also in the *Structural Features* option, more advanced RNA-containing structures are now available. The motif for the base pair signature is provided that is available as the common name of the motif.

The various annotations used in nucleic acids are nucleic acid type and conformation, secondary structure, and description. Also information pertaining to the drug binding mode and bound proteins are also available. A schematic representation is shown that involves the various steps involved in the extraction and monitoring of data from the NDB (**Fig. 1**). A schematic representation of the NDB is depicted in **Fig. 1**.

The resources available for NDB can be retrieved via two constant headers that are available on top page of the website. There are six tabs that are contained in the first constant header. They are: *About NDB, Standards, Education, Tools, Software*, and *Download*. Introduction to nucleic acids; definitions of terms used in the website; nucleic acid-related features from PDB-101, an educational component of RCSB PDB (Rose *et al.*, 2013), and links to other educational activities and sites. The software is downloadable includes geometric and symbolic 3D motif search (*FR3D*) (Sarver *et al.*, 2008), visualization of secondary structure (*RNA View*) (Yang *et al.*, 2003), and visualization of 3D structures (*3DNA*) (Lu and Olson, 2003). There are various links to the software packages that include various software for statistical folding of nucleic acids and studies of regulatory RNAs (Sfold; see "Relevant Websites section") (Ding and Lawrence, 2003; Ding *et al.*, 2005), a visualization applet for RNA (*VARNA*) (Darty *et al.*, 2009) and the *UNAFold* web server (Markham and Zuker, 2005, 2008).

A representation of the structural summary report for the NDB is clearly shown that represents the various parameters and information obtained from the protein (**Fig. 2**). From the schematic representation, a snapshot for the reported data summary of the information related to ribozymes from the NDB can be obtained.

The information illustrated in the NDB has a good connection with the information related to the PDB. The summary report obtained from the PDB that is retained in the NDB is summarized in **Table 1**. From the table, we can observe the association between the NDB and PDB in terms of their contents and other specifiers.

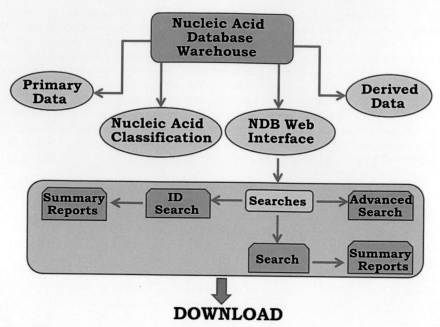

Fig. 1 A database curator workflow for the various steps involved in the extraction and monitoring of data obtained from the NAD. Understanding bioinformatics, Marketa Zvelebil & Jeremy O. Baum.

NDB User Interface

Both the primary and derived data classification is obtained in the NDB. The derived and calculated data based on the structural features of RNA from the manually annotated nucleic acid classification are separately managed from the entries of primary structure. The data thus recorded and stored are kept as *external reference files* (ERFs). The respective PDB structures contain the primary data entries including the experimental files, structural descriptions, citations, crystal data, identifiers, coordinate information, details regarding crystallization, data collection, and structure refinement. The coordinate data that are stored in the searchable database store the various derived structural features such as bond distances, angles, torsions, and base morphology (Lu and Olson, 2003, 2008). The information related to the RNA structural features including pair-wise nucleotide interactions for each RNA structure, equivalence classes, and NR sets of RNA structure files and RNA 3D motifs extracted from structures were recently reported.

NDB Applications

For RNA base-pairing and base stacking interactions the *FR3D* server (Sarver *et al.*, 2008) is used to annotate the pair-wise interactions between RNA nucleotides, and is described in (Zirbel *et al.*, 2009) for base phosphate interactions. The basis for RNA Base Triple Atlas and RNA 3D Motif Atlas are formed from these annotations. Also statistics about pair-wise interaction frequency are provided. For modelers and computational scientists who are keen to determine the characteristics of structured RNAs, these data can be a useful target. Reports suggest that the entries of RNA are clumped into *'equivalence classes'* of structures that share the same, or nearly the same, sequence and geometry (Leontis and Zirbel, 2012). For every week the equivalence classes are computed such that the 3D structure databases are reflected with new additions. The same equivalence class contains different structures of the same RNA from the same organism whereas difference class contains the homologous RNAs structures from different organisms.

Other Databases

Apart from NDB, there also exist other databases that are unique for RNA and that combine data from the other sources including NDB and also from directed PDB sites. These however supply an elaborated annotation that is combined with the sequence based predictions in most of the cases (Appasamy *et al.*, 2016; Chojnowski *et al.*, 2014; Firdaus-Raih *et al.*, 2014; Popenda *et al.*, 2010; Zirbel *et al.*, 2015). The majority of the servers however permit secondary structures and RNA sequences searches. They also help in analyzing and searching from the uploaded PDB file the likely coordinates of RNA. These findings depend on the geometries but they are not sustained presently by NDB. RNA 3D hub is another useful database related to the structure of RNA that gives details about the various information associated with the precomputed structural analyses part of RNA molecules available at NDB (Zirbel *et al.*, 2015).

Databases such as NDB, G4LDB, SMMRNA, etc., have been provided by various groups earlier. These databases are dedicated particular type of nucleic acid or its special structure. Databases such as NDB consist of the various information related to three-dimensional structures of nucleic acids and their complexes.

A Portal for Three-dimensional Structural Information about Nucleic Acids
As of 8-May-2018 number of released structures: 9479

Search DNA   Search RNA   Advanced Search

Enter an NDB ID or PDB ID 🔍
Search for released structures

NDB ID: 4P95   PDB ID: 4P95 🔗

Title:

SPECIATION OF A GROUP I INTRON INTO A LARIAT CAPPING RIBOZYME (CIRCULARLY PERMUTATED RIBOZYME)

Molecular Description:

RNA (189-MER)

Structural Keywords:

Nucleic Acid Sequence:

Click to show/hide 1 nucleic acid sequences

Protein Sequence:

No Protein Sequence Found

Primary Citation:

Meyer, M., Nielsen, H., Olieric, V., Roblin, P., Johansen, S.D., Westhof, E., Masquida, B. Speciation of a group I intron into a lariat capping ribozyme. 🔗 *Proc.Natl.Acad.Sci.USA*, **111**, pp. 7659 - 7664, 2014.

Experimental Information:

X-RAY DIFFRACTION

Space Group:

P 21 21 21

Cell Constants:

a = 57.63   b = 85.71   c = 108.53 (Ångstroms)

α = 90.0   β = 90.0   γ = 90.0 (degrees)

Refinement:

The structure was refined using the PHENIX program. The R value is 0.1906 for 34663 reflections in the resolution range 47.824 to 2.5 Ångstroms with Fobs > 1.59 sigma(Fobs) and with I > 0.0 sigma(I)

Download Data:

Asymmetric Unit coordinates (pdb format, Unix compressed(.gz))

Asymmetric Unit coordinates (cif format, Unix compressed(.gz))

Biological Assembly coordinates (pdb format) 1

Structure Factors (cif format)

XML | Complete with coordinates (xml format, GNU compressed(.gz))

XML | Coordinates only (xml format, GNU compressed(.gz))

XML | Header only (xml format, GNU compressed(.gz))

Structural Features

RNA Base-Pairs, Stacking, etc.

Similar Structures

Interactive basepair map with 3D fragment visualization for: 4P95 🔗

Symbolic WebFR3D search for:4P95 🔗

Geometric WebFR3D search for:4P95 🔗

RNAML

Base Pair Hydrogen Bonding Classification

Nucleic Acid Backbone Torsions

Base Pair Morphology Parameters

Base Pair Morphology Step Parameters

Biological Assembly 1

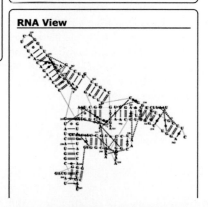

RNA View

Fig. 2 A reported summary of the particular ribozyme that depicts the headers of gray and red color with the individual sections: (a) Primary structural information, (b) atomic coordinate and experimental file download, (c) derived structural data, (d) images, and (e) for RNA structures an additional RNA view image. Understanding bioinformatics, Marketa Zvelebil & Jeremy O. Baum.

The G4LDB database (G4LDB, see "Relevant Websites section") is related to the G-quadruplex ligands with the docking tool. This database gives information about the unique collection of the reported G-quadruplex ligands associated with the streamline ligand/drug discovery targeting G-quadruplexes. The G-quadruplexes are those nucleic acid sequences that are guanine-rich found in the human telomeres and the promoter region of the gene. This database at present consists of more than 800 G-quadruplex ligands with 4000 activity records, which is the most extensive collection to our knowledge. It also provides a user-friendly interface that necessitates various data inquired from researchers (Li *et al.*, 2012).

The SMMRNA database relates to the RNA structure and its complexes with small molecules (Li, *et al.*, 2013; Mehta *et al.*, 2014). This is an interactive database that is available at website provided in " Relevant Websites section" and focuses on the various small ligand

Table 1   NDB primary content acquired from PDB and its description

Main content of Nucleic Acids Database (NDB)	Explanation about the contents
Coordinate information	Description about the atomic coordinates for the asymmetrical and biological unit
Experimental files	The details about the *NMR* restraints and the structure factor files
Identifiers	The *IDs* for *NDB* and *PDB*
Structural description	The description about the asymmetric and biological unit, sequence, mismatches, base pairing, and modification
Citation	The details about the authors, title, journals, year, pages, volume, *DOI*, and *PUBMED ID*
Crystal data	A description about the space group information and cell parameters
Crystallization details	The conditions for temperature, *pH*, method, and crystallization
Data collection information	The information related to wavelength, temperature, radiation source, detector, resolution, *R*-merge, and number of reflections
Refinement details	Information regarding programs and method used, reflection number, resolution, *R*-factor, refinement of occupancies, and temperature factors

Table 2   Summary of the various nucleic acid databases with vivid description

Examples of Nucleic Acid Database (NDB)	Description of database
ProNIT	Provides experimentally determined thermodynamic interaction data between proteins and nucleic acids. It contains the properties of the interacting protein and nucleic acid, bibliographic information, and several thermodynamic parameters such as the binding constants, changes in free energy, enthalpy, and heat capacity
RNA FRABASE	An engine with database to search the 3D fragments within 3D RNA structures using as an input the sequences and/or secondary structures given in the dot-bracket notation
RNA Bricks	Database of RNA 3D structure motifs and their contacts, both with themselves and with proteins. The database provides structure-quality score annotations and tools for the RNA 3D structure search and comparison
RNA 3D Motif Atlas	RNA 3D motifs are recurrent structural modules that are essential for many biological functions and RNA folding. Usually drawn as unstructured hairpin and internal loops, these motifs are organized by noncanonical base pairs, supplemented by characteristic stacking and base–backbone interactions
RNAJunction	Provides a user-friendly way of searching structural elements by PDB code, structural classification, sequence, and keyword or interhelix angles. Contains structure and sequence information for RNA structural elements such as helical junctions, internal loops, bulges, and loop–loop interactions

molecules that target the RNA. At present this database contains approximately 770 unique ligands with the structural images pertaining to the RNA molecules. Also it gives information related to comprehensive resources that are required for further design, development, and refinement of small molecule modulators necessary for selective target of the RNA molecules (Mehta *et al.*, 2014).

Apart from G4LDB, there is also an important database consisting of detailed information related to the proteins' interaction with the nucleic acids forming a G-quadruplex structure, *Nucleic acid G-quadruplex structure (G4) Interacting Proteins DataBase (G4IPDB)*. This first database gives accurate knowledge related to the interaction at a single platform. It also consists of more than 200 entries with various interaction details such as interacting protein names and their synonyms, their UniProt-ID, source organism, target name and its sequences, ΔTm, binding/dissociation constants, protein gene name, protein FASTA sequence, interacting residue in protein, related PDB entries, interaction ID, graphical view, PMID, author's name, and techniques that were used to detect their interactions. A vivid summary of the various examples of NDB with their detailed explanation is illustrated in **Table 2**. From the table, examples of various NDB have been described that give a clear picture about the interconnection with the nucleic acids.

Future Directions and Closing Remarks

After the redesign of NDB, new improvements can be noticed highlighting the data content that consists of derived data, annotations, and presentation. However, NDB will further continue to develop and expand its scope. The original NDB contained fewer than 100 crystal structures of nucleic acids but now has more than 700. As the technology is growing faster day by day, it will help us to observe a wide variety of structures in the NDB as it has permitted to determine number of structures. Also this database will facilitate the addition of more search options, annotations, visualizations, and various reports related to the nucleic acid-containing structures in the near future. Hence to our knowledge and information, no such database exists containing detailed experimental information related to drug-nucleic acid binding (Kd, Tm, IC50, etc.,) for all the majorly available types and forms of nucleic acid. A clear view of the NDB can provide us much information related to the structure and functionality and also about the various sequences. Meanwhile the website has refurbished the information related to the query and its reporting capabilities. And this can help us to observe a wide variety of structures in the NDB as it has permitted to determine number of structures.

Acknowledgment

We thank the Central Library of Tezpur University for providing us resources to complete the article.

See also: Applications of Ribosomal RNA Sequence and Structure Analysis for Extracting Evolutionary and Functional Insights. Assessment of Structure Quality (RNA and Protein). Bioinformatics Data Models, Representation and Storage. Biological Database Searching. Biomolecular Structures: Prediction, Identification and Analyses. Computational Design and Experimental Implementation of Synthetic Riboswitches and Riboregulators. Data Storage and Representation. Engineering of Supramolecular RNA Structures. Genome-Wide Probing of RNA Structure. Information Retrieval in Life Sciences. Natural Language Processing Approaches in Bioinformatics. Predicting RNA-RNA Interactions in Three-Dimensional Structures. Protein Structure Databases

References

Alberts, B., Johnson, A., Lewis, J., *et al.*, 2002. Molecular Biology of the Cell, fourth ed. New York: Garland Science the Structure and Function of DNA.

Appasamy, S.D., Hamdani, H.Y., Ramlan, E.I., *et al.*, 2016. InterRNA: A database of base interactions in RNA structures. Nucleic Acids Research 44, D266–D271.

Bailey, Regina. (2017). Learn about nucleic acids. ThoughtCo, thoughtco.com/nucleic-acids-373552.

Bansal, M., 2003. DNA structure: Revisiting the Watson – Crick double helix. Current Science 85 (11), 1556–1563.

Berman, H.M., Henrick, K., Nakamura, H., 2003. Announcing the worldwide protein data bank. Nature Structural & Molecular Biology 10, 980.

Berman, H.M., Olson, W.K., Beveridge, D.L., *et al.*, 1992. The Nucleic Acid Database – A comprehensive relational database of three-dimensional structures of nucleic acids. Biophysical Journal 3, 751–759.

Chen, X., Ramakrishnan, B., Sundaralingam, M., 1995. Crystal structures of B-form DNA-RNA chimers complexed with distamycin. Nature Structural Biology 2 (9), 733–735. doi:10.1038/nsb0995-733.

Chojnowski, G., Walen, T., Bujnicki, J.M., 2014. RNA Bricks – A database of RNA 3D motifs and their interactions. Nucleic Acids Research 42, D123–D131.

Churchill, M.E.A., 1996. The latest on DNA form and function. In: Eckstein, Fritz, Lilley, David MJ (Eds.), Nucleic Acids and Molecular Biology,, vol. 9. Berlin, Heidelberg: SpringerVerlag, p. 376.

Coimbatore Narayanan, B., Westbrook, J., Ghosh, S., *et al.*, 2014. The nucleic acid database: New features and capabilities. Nucleic Acids Research 42, D114–D122.

Darnell Jr., J.E., 1985. RNA. Scientific American 253, 54–72.

Darty, K., Denise, A., Ponty, Y., 2009. VARNA: Interactive drawing and editing of the RNA secondary structure. Bioinformatics, 25. . pp. 1974–1975.

Dickerson, R.E., Drew, H.R., Conner, B.N., *et al.*, 1982. The anatomy of A-, B-, and Z-DNA. Science 216 (4545), 475–485. doi:10.1126/science.7071593. PMID 7071593.

Ding, Y., Chan, C.Y., Lawrence, C.E., 2005. RNA secondary structure prediction by centroids in a Boltzmann weighted ensemble. RNA 11, 1157–1166.

Ding, Y., Lawrence, C.E., 2003. A statistical sampling algorithm for RNA secondary structure prediction. Nucleic Acids Research 31, 7280–7301.

Felsenfeld, G., 1985. DNA. Scientific American 253, 54–72.

Firdaus-Raih, M., Hamdani, H.Y., Nadzirin, N., *et al.*, 2014. COGNAC: A web server for searching and annotating hydrogen-bonded base interactions in RNA three-dimensional structures. Nucleic Acids Research, 42. pp. W382–W388.

Govil, G., Kumar, N.Y., Ravi Kumar, M., *et al.*, 1985. Recognition schemes for protein–nucleic acid interactions. Journal of Bioscience 8, 645. doi:10.1007/BF02702763.

Katsuyuki, A., Kazutaka, M., Hu, N., 2016. Chapter 3, section 3. Nucleic Acid Constituent complexes. In: Sigel, A., Sigel, H., Sigel, R.K.O. (Eds.), The Alkali Metal Ions: Their Role in Life. Metal Ions in Life Sciences 16. Springer, pp. 43–66. doi:10.1007/978-3-319-21756-7_3.

Krieger, M., Scott, M.P., Matsudaira, P.T., *et al.*, 2004. Section 4.1: Structure of nucleic acids. In: Molecular Cell Biology. New York: W.H. Freeman and CO, ISBN 0-7167-4366-3.

Leontis, N.B., Zirbel, C.L., 2012. Nonredundant 3D structure datasets for RNA knowledge extraction and benchmarking. In: Leontis, N.B., Westhof, E. (Eds.), RNA 3D Structure Analysis and Prediction 27. Berlin, Heidelberg: Springer, pp. 281–298.

Li, Q., Xiang, J.F., Zhang, H., *et al.*, 2012. Searching drug-like anti-cancer compound(s) based on G-quadruplex ligands. Curr. Pharm. Des. 18, 1973–1983.

Li, Q., Xiang, J.F., Yang, Q.F., *et al.*, 2013. G4LDB: A database for discovering and studying G-quadruplex ligands. Nucleic Acids Research 41, D1115–D1123.

Luger, K., Phillips, S.E.V., 2010. Protein-nucleic acid interactions. Current Opinion in Structural Biology 20 (1), 70–72. doi:10.1016/j.sbi.2010.01.006.

Lu, X.J., Olson, W.K., 2003. 3DNA: A software package for the analysis, rebuilding and and visualization of three-dimensional nucleic acid structures. Nucleic Acids Research 31 (17), 5108–5121.

Lu, X.J., Olson, W.K., 2008. 3DNA: A versatile, integrated software system for the analysis, rebuilding and visualization of three-dimensional nucleic-acid structures. Nature Protocols 3, 1213–1227.

Markham, N.R., Zuker, M., 2005. DINAMelt web server for nucleic acid melting prediction. Nucleic Acids Research 33, W577–W581.

Markham, N.R., Zuker, M., 2008. UNAFold: Software for nucleic acid folding and hybridization. Methods in Molecular Biology 453, 3–31.

Mehta, A., Sonam, S., Gouri, I., *et al.*, 2014. SMMRNA: A database of small molecule modulators of RNA. Nucleic Acids Research 42, D132–D141.

Mirkin, S.M., 2001. DNA Topology: Fundamentals. Encyclopedia of Life Sciences.. doi:10.1038/npg.els.0001038.

Park, B., Kim, H., Han, K., 2014. DBBP: Database of binding pairs in protein-nucleic acid interactions. BMC Bioinformatics 15 (Suppl. 15), S5. doi:10.1186/1471-2105-15-S15-S5.

Popenda, M., Szachniuk, M., Blazewicz, M., *et al.*, 2010. RNA FRABASE 2.0: An advanced web-accessible database with the capacity to search the three-dimensional fragments within RNA structures. BMC Bioinformatics 11, 231.

Rose, P.W., Bi, C., Bluhm, W.F., *et al.*, 2013. The RCSB Protein Data Bank: New resources for research and education. Nucleic Acids Research 41, D475–D482.

Sarver, M., Zirbel, C.L., Stombaugh, J., Mokdad, A., Leontis, N.B., 2008. FR3D: Finding local and composite recurrent structural motifs in RNA 3D structures. Journal of Mathematical Biology 56, 215–252.

Sathyapriya, R., Vijayabaskar, M.S., Vishveshwara, S., 2008. Insights into Protein – DNA interactions through structure network analysis. PLOS Computational Biology 4 (9), e1000170. doi:10.1371/journal.pcbi.1000170.

Siggers, T., Gordan, R., 2014. Protein – DNA binding: Complexities and multi-protein codes. Nucleic Acids Research 42 (4), 2099–2111. doi:10.1093/nar/gkt1112.

Travers, A., Muskhelishvili, G., 2015. DNA Structure and Function. FEBS Journal 282, 2279–2295.

Valuchova, S., *et al.*, 2016. A rapid method for detecting protein-nucleic acid interactions by protein induced fluorescence enhancement. Scientific Reports 6, 39653. doi:10.1038/srep39653.

Watson, J.D., Crick, F.H.C., 1953. Molecular structure of nucleic acids: A Structure for deoxyribose nucleic acid. Nature 4356, 737.

Weinberg, R.A., 1985. The Molecules of Life. Scientific American 253, 34–43.

Yang, H., Jossinet, F., Leontis, N., *et al.*, 2003. Tools for the automatic identification and classification of RNA base pairs. Nucleic Acids Research 31, 3450–3460.

Zirbel, C.L., Sponer, J.E., Sponer, J., Stombaugh, J., Leontis, N.B., 2009. Classification and energetics of the base-phosphate interactions in RNA. Nucleic Acids Research 37, 4898–4918.

Zirbel, C.L., Roll, J., Sweeney, B.A., *et al.*, 2015. Identifying novel sequence variants of RNA 3D motifs. Nucleic Acids Research 43, 7504–7520. doi:10.1093/nar/gkv651.

Relevant Websites

http://www.g4ldb.org
    G-Quadruplex Ligands Database.
http://ndbserver.rutgers.edu
    Nucleic Acid Database.
ftp://ndbserver.rutgers.edu/NDB/
    Nucleic Acid Database.
http://sfold.wadsworth.org
    Sfold.
http://www.smmrna.org
    SMMRNA.

Biographical Sketch

Airy Sanjeev is presently working as a research scholar in the Department of Molecular Biology and Biotechnology in Tezpur University under the supervision of Dr. Venkata Satish Kumar Mattaparthi. Her research work is based on understanding the aggregation pathway of α-synuclein which is responsible for Parkinson's Disease. She has expertise in various softwares such as AutoDock, PyRex, Schrodinger, AMBER, GROMACS, and also programming languages such as Perl, Java, C, C++. She has also skills in operating systems, protein modeling, graphing and data analysis suites, etc. She has published almost 10 papers in the peer-reviewed journals. She has completed her Masters from Pondicherry University in 2015 in Centre for Bioinformatics. She did her undergraduate work at Cotton College, Guwahati, Assam with Botany as her honors subject.

Dr. Venkata Satish Kumar Mattaparthi received his PhD in Biotechnology from Indian Institute of Technology Guwahati, Assam, India in 2010. From 2010 to 2012, he was with the Centre for Condensed Matter Theory (CCMT), Department of Physics, Indian Institute of Science Bangalore, India. He is currently working as Assistant Professor in the Department of Molecular Biology and Biotechnology, School of Sciences, Tezpur University, Assam, India. His research interests include understanding the functioning of intrinsically disordered proteins (IDPs), pathways of protein aggregation mechanism in neurodegenerative diseases like Alzheimer, Parkinson, etc., and also the features of protein–RNA interactions using computational techniques. He has published his research in reputed international journals like *Journal of Biomolecular Structure and Dynamics*, *PLoS ONE*, *Journal of Physical Chemistry B*, *Biophysical Journal*, *Journal of Bioinformatics and Computational Biology*, and others.

Dr. Sandeep Kaushik is a passionate computational biologist with a wide exposure of computational approaches. Presently, he is carrying out his research as an Assistant Researcher (equivalent to Assistant Professor) at 3B's Research Group, University of Minho, Portugal. He has a PhD in Bioinformatics from National Institute of Immunology, New Delhi, India. He has a wide exposure on analysis of scientific data ranging from transcriptomic data on mycobacterial and human samples to genomic data from wheat. His research experience and expertise entails molecular dynamics simulations, RNA-sequencing data analysis using Bioconductor (R language based package), de novo genome assembly using various software like *AbySS*, *automated protein prediction* and *annotation*, *database mining*, *agent-based modeling and simulations*. He has a cumulative experience of more than 10 years of programming using PERL, R language and NetLogo. He has published his research in reputed international journals like *Molecular Cell*, *Biomaterials*, *Biophysical Journal*, and others. Currently, his research interest involves in silico modeling of disease conditions like breast cancer, using agent-based modeling and simulations approach.

RNA Structure Prediction

Junichi Iwakiri and Kiyoshi Asai, University of Tokyo, Kashiwa, Japan and Artificial Intelligence, Research Center, Koto-ku, Japan

Introduction

Three major types of RNAs, namely messenger RNA (mRNA) as coding RNA, as well as transfer RNA (tRNA) and ribosomal RNA (rRNA) as non-coding RNAs, are involved in the translation of genetic code. Among them, tRNAs fold into the common clover leaf-like secondary structure, having anti-codons that interact with the codons of the mRNAs, to perform their function as adaptors during the translation of mRNAs into proteins. Such common secondary structures are often observed in other non-coding RNAs such as snoRNAs and various snRNAs involved in splicing. It is also known that the secondary structural stability of mRNAs, especially around the start codon, affects the efficiency of translation. Micro RNAs (miRNAs) repress the translation of target mRNAs, are first transcribed as pri-miRNAs to form the stem-loop secondary structures for processing to double-stranded pre-miRNAs and then to mature miRNAs. Many non-coding RNAs other than miRNAs can be transcribed, but few of their functions have been identified. The functions of RNA molecules often depend on their high sequence specificity for interacting with other RNAs. These interactions are based on the base pairing of complementary bases, which are also observed in the secondary structures. Thus, it is important to identify the secondary structures of RNAs to elucidate and understand their functions. In this article, we introduce currently available knowledge on the structure prediction of RNAs with a particular focus on secondary structures.

A number of software tools related to RNA secondary structures are available. Researchers should select the appropriate tools for their specific purposes. It should be noted that Vienna RNA package (Lorenz *et al.*, 2011) is the most popular and convenient suite of the tools and the software libraries.

Models of RNA Secondary Structures

Secondary Structure

Single-stranded RNA molecules form hydrogen bonds between the Watson-Crick (WC) edges of the complementary pairs of bases, namely the canonical WC pairs *A-U* and *G-C*, and the wobble pair *G-U*. The set of base-pairs of an RNA molecule is called the secondary structure of the RNA. RNA molecules fold into tertiary structures via complicated interactions inside of the molecules, but the secondary structures are often used as abstract forms of the RNA structure. It is known that the secondary structures are often well conserved between RNAs of the same functional family.

The secondary structure of an RNA sequence $x = x_1 \dots x_L$ of the length L is represented by an upper triangle binary matrix, specifically the secondary structure matrix (2DSM), $\{\sigma_{i,j}\}$ ($1 \leq i < j \leq L$) where in the secondary structure $\sigma_{i,j} = 1$ shows that x_i and x_j form a base pair, $\sigma_{i,j} = 0$ shows that there is no base pair between them, and $\sigma_{i,j}$ satisfies $\sum_i \sigma_{i,j} \leq 1$ and $\sum_j \sigma_{i,j} \leq 1$. The interactions of the base pairs are illustrated in **Fig. 1**.

Pseudoknots

Fig. 2 presents examples of non-crossing and crossing interactions in secondary structures. Secondary structures with such crossing interactions are called *pseudo-knots*. The 2DSM of a pseudoknot-free secondary structure $\sigma_{i,j}$ satisfies: $\sigma_{i,j}\sigma_{k,\ell} = 0$ if $i < k < j < \ell$. The majority of software tools for RNA secondary structure prediction only produce pseudoknot-free structures.

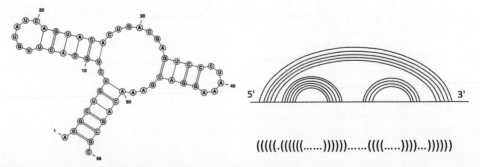

Fig. 1 Representations of a secondary structure (Hammerhead ribozyme). Two-dimensional (2D) representation (left), linear representation (right top) and sequence representation using parentheses (right bottom).

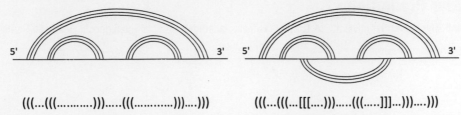

$$(((...(((..........)))....(((..........)))....)))\qquad(((...(((...[[[....)))....(((.....]]]...)))....)))$$

Fig. 2 Interactions in RNA secondary structures. The interactions of base pairs without pseudoknots (left) and with pseudoknots (right) are shown.

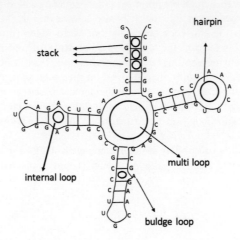

Fig. 3 Loops in secondary structures.

Extended secondary structures

Three-dimensional (3D) RNA structures consist of WC base pairs and non-WC base pairs and various hydrogen bonds. Such base-pairing interactions are categorized into 12 groups based on the combination of interacting edges (WC, Hoogsteen and sugar) in nucleotide bases and the relative orientations of glycosidic bonds (cis or trans) (Leontis *et al.*, 2002), which totally form the *extended secondary structures*. Among these groups, cis WC-WC base pairs are the standard WC base pairs, and the remaining 11 groups are non-WC base pairs.

Energy Models

An energy model of RNA secondary structures and their parameters define the free energy of a given RNA sequence. The most widely used energy model for pseudoknot-free 2D structures is Turner nearest neighbor model (Mathews *et al.*, 1999; Mathews, 2004). The Turner model assumes that the free energy of a secondary structure is the sum of all the *loops* surrounded by the backbone and the base-pairing hydrogen bonds, such as stacks, internal loops, bulge loops, hairpin loops, and multi-loops (**Fig. 3**). The free energy of the whole RNA structure is given by

$$E(x,\sigma) = \sum_{\sigma_\xi \in \sigma} E(\sigma_\xi) \tag{1}$$

The parameters of the Turner model represent the *nearest neighbors* of those loops and the length of the included nucleotides, as determined experimentally (Mathews *et al.*, 1999; Mathews, 2004).

The constraint generation (CG) method (Andronescu *et al.*, 2007) estimates Turner parameters from structural and experimental data. It applies the constraint that the reference structure has lower energy than that of alternative structures and achieved better accuracy in secondary structure prediction. Boltzmann Likelihood (BL) method for the Turner model further improved the predictive accuracy of secondary structures. The energy parameters have also been calculated by combination of experiments and molecular dynamics calculations (Sakuraba *et al.*, 2015).

The number of model parameters is the key issue in secondary structure prediction. Approximately 70,000 free parameters were used in ContextFold (Zakov *et al.*, 2011), and 50% reduction in error rate was reported, but there is serious danger for overfitting.

Boltzmann distribution

Given an RNA sequence x, the probability distribution of the secondary structure σ with free energy $E(x,\sigma)$ is written by the Boltzmann distribution,

$$P(\sigma|x) = \frac{1}{Z(x)} e^{\frac{-E(x,\sigma)}{RT}} \tag{2}$$

where $Z(x)$, the partition function of this distribution, is the sum of the Boltzmann factors for the set of all possible secondary structures S, as follows:

$$Z(x) = \sum_{\sigma \in S} e^{\frac{-E(\sigma,x)}{RT}} \tag{3}$$

This partition function $Z(x)$ can be calculated efficiently, costing $O(L^3)$ in time and $O(L^2)$ in memory for an RNA sequence of the length L, via the McCaskill algorithm (McCaskill, 1990), which is a dynamic programming (DP) method that will be explained in a subsequent section.

Stochastic Models

The Turner energy model and its parameters represent the thermodynamic character of the secondary structure and consequently define the probability via the Boltzmann distribution. Meanwhile, stochastic models can directly define the probabilities of the secondary structures. Stochastic context free grammar (SCFG) was first used to model the secondary structure of tRNA (Sakakibara et al., 1994), and it has been widely used for pseudoknot-free secondary structure prediction.

For example, the following set of stochastic production rules gives a type of stochastic model of RNA secondary structure:

$$S \rightarrow x S \bar{x} \quad \text{with probability} \quad P(S \rightarrow x S \bar{x}) \tag{4}$$

$$S \rightarrow x \quad \text{with probability} \quad P(S \rightarrow x) \tag{5}$$

where $\bar{x}$ represents the complementary base of x. Using more complicated production rules, SCFG can represent the nearest neighbor parameters (Knudsen and Hein, 2003; Rivas et al., 2012). In the stochastic models of RNA secondary structures, the set of parameters are learned from the set of RNA sequences for which the secondary structures are determined.

Do et al. maximized the conditional likelihood of the stochastic log-linear model for a structural dataset by using a gradient descent for optimization in their secondary structure prediction software CONTRAfold (Do et al., 2006). The CONTRAfold model does not reliably determine free-energy changes, but it is important for estimating bindings.

To treat pseudoknotted secondary structures using stochastic models, formal grammars beyond context-free grammar are necessary (Rivas and Eddy, 2000). The complexity of the formal grammar depends on the complexity of pseudoknots.

Secondary Structure Prediction

We mainly treat the simplest problem in RNA secondary structure prediction to predict the secondary structure of a single RNA sequence for pseudoknot-free structures without any additional information. To solve this problems, several software tools are available, including Mfold (Zuker, 2003), RNAfold (Lorenz et al., 2011), Pfold (Knudsen and Hein, 2003), RNAstructure (Mathews, 2014), CONTRAfold (Do et al., 2006), CentroidFold (Sato et al., 2009; Hamada et al., 2009a), TurboFold (Harmanci et al., 2011), and ContextFold (Zakov et al., 2011).

Meanwhile, tools for pseudoknotted secondary structures including Pknots (Rivas and Eddy, 1999), IPknot (Sato et al., 2011), tend to incorrectly predict pseudoknot-free structures. They are useful for finding pseudoknotted local structures, but often show lower total accuracy even for RNAs including pseudoknotted structures.

RNA secondary structure prediction is reasonably accurate for RNAs shorter than 1000, but the accuracy is often unsatisfactory for longer RNAs. The accuracy is improved using homologous sequences (Hamada et al., 2009b) because the local secondary structures are often evolutionarily conserved.

There are several variations of secondary structure prediction problems, such as the prediction of suboptimal structures, the common secondary structure and alignment of a group of sequences, the common secondary structure from aligned sequences, the secondary structure of an interacting pair of sequences, and the secondary structure combining experimental data.

Criteria of Secondary Structure Prediction

Prediction of minimum free energy structures
The Boltzmann distribution Eq. (2) of the secondary structure indicates that the maximum likelihood estimator (MLE) $\hat{\sigma}^{MLE}$ of the secondary structure is the MFE structure $\hat{\sigma}^{MFE}$ as follows.

$$\hat{\sigma}^{MLE} = \underset{\sigma}{\mathrm{argmax}}\, P(\sigma|x) = \underset{\sigma}{\mathrm{argmin}}\, E(x,\sigma) = \hat{\sigma}^{MFE} \tag{6}$$

The dynamic programming algorithm proposed by Zuker and Stiegler (1981) computes the MFE/MLE structure efficiently costing $O(L^3)$ in time and (L^2) in memory. Mfold and RNAfold of the Vienna RNA package are popular tools for finding MFE structures via dynamic programming.

It is natural to believe that the MFE/MLE structures are the *best* structures we can predict, but the probability of a specific secondary structure is extremely small. For example, the probability of MFE structure of a tRNA is often less than 1%, and for longer RNAs it is easily less than 10^{-8}. Whereas the MFE/MLE is an exremely important criterion, the MFE/MLE structures are not as reliable as they appear considering their thermodynamic probability.

Predictions Based on the Maximum Expected Accuracy

The MLE structure $\hat{\sigma}^{MLE}$ in Eq. (6) can be re-written as follows:

$$\hat{\sigma}^{MLE} = \underset{\sigma}{\mathrm{argmax}}\, P(\sigma|x) = \underset{\hat{\sigma}}{\mathrm{argmax}} \sum_{\sigma} \delta(\sigma, \hat{\sigma}) P(\sigma|x) \tag{7}$$

where $\delta(a,b)$ is the δ function, which is 1 only if the two parameters are equal and 0 otherwise. According to the aforementioned equation, the MLE structure, or equivalently the MFE structure, maximizes the probability that the predicted structure is exactly the same as the *true* structure. This means that no neighboring structure different from the *true* structure, is considered in the evaluation.

In CONTRAfold, which applies the criterion of MEA, the gain function, which reflects an accuracy measure of the secondary structure prediction, was used instead of using the δ function in Eq. (7) as follows:

$$G^{MEA} = TN_{position} + \gamma \times TP_{position} \tag{8}$$

where $TN_{position}$ and $TP_{position}$ are, the numbers of positions correctly predicted as unpaired (*true negative*), and paired (*true positive*), respectively. The γ is the sensitivity/specificity tradeoff parameter. Whereas this gain function uses the true negatives and positives with respect to the positions, the accuracy of secondary structure prediction has been usually measured using the number of correctly paired/unpaired *base pairs* which CentroidFold adopted in its gain function

$$G^{\gamma} = TN_{base-pair} + \gamma \times TP_{base-pair} \tag{9}$$

Marginal Probabilities and Reliability of Predictions

Marginal Probabilities of Secondary Structures

Although the probability that an RNA sequence will form a specific secondary structure is extremely small, the marginal probabilities of the structure are often reasonably large. The most popular marginal probability is the base-pairing probability (BPP), or the probability that the specific pair of bases form a base pair in the secondary structure. BPPs include rich information regarding the probability distribution of the RNA secondary structure. They are often illustrated in an upper triangular matrix (Lorenz *et al.*, 2011) suggesting the existence of hidden alternative structures, or presented on the predicted secondary structure with colors that illustrate the reliability of the stem structures.

The marginal probabilities of various structural contexts, such as stems, loops, and bulges, also carry rich information about the secondary structure. Although it is difficult to detect distinct structural motifs from a set of functional RNAs, local pattern of marginal probabilities has been observed (Fukunaga *et al.*, 2014).

Reliability of Predicted Structures

The accuracies of predictions for secondary structures are often evaluated regarding the accuracy of base pair prediction using sensitivity (SEN, or recall), the positive predicted value (PPV, or precision) and balanced measures such as Matthew's correlation coefficient (MCC) and F-measures, defined as follows.

$$\mathrm{SEN} = \frac{TP}{TP + FN} = \frac{\#\ of\ correctly\ predicted\ base-pairs}{\#\ of\ true\ base-pairs} \tag{10}$$

$$\mathrm{PPV} = \frac{TP}{TP + FP} = \frac{\#\ of\ correctly\ predicted\ base-pairs}{\#\ of\ predicted\ base-pairs} \tag{11}$$

$$\mathrm{MCC} = \frac{TP \cdot TN + FT \cdot FN}{\sqrt{(TP + FP)(TP + FN)(TN + FP)(TN + FN)}} \tag{12}$$

$$\text{F-measure} = 2\,\frac{\mathrm{PPV} \cdot \mathrm{SEN}}{\mathrm{PPV} + \mathrm{SEN}} \tag{13}$$

Whereas base-pairing probabilities are convenient for partially representing the probability distribution, they do not clearly reveal the entire picture of the distribution. RNAbor (Freyhult *et al.*, 2007) computes the marginal probability of secondary structures on the Hamming distance from the reference structure. Mori *et al.* (2014) has formalized the calculation of Distribution of the Hamming distance based on the general scheme proposed by Newberg and Lawrence (2009). When the sum of the probabilities within a rage of Hamming distance exceeds c%, the smallest such Hamming distance is called the *c% credibility limit*. A lower credibility limit with a higher percent indicates that the distribution is more concentrated, and that the prediction is more reliable.

Algorithms on RNA Secondary Structures

Dynamic Programming Algorithms

As described previously, pseudoknot-free RNA secondary structure can be modeled using SCFGs corresponding to pseudoknot-free energy models. The DP algorithms for pseudoknot-free RNA secondary structures therefore corresponds to the standard SCFG algorithms.

Dynamic programming for partition function

The partition function Eq. (3) of the Boltzmann distribution Eq. (2) can be calculated by the McCaskill algorithm (McCaskill, 1990), a DP algorithm corresponding to the inside algorithm of SCFG.

The McCaskill algorithm
Initialization: for $1 \leq i \leq N$

$$Z_{i,i} = 1.0, \quad Z_{i,i}^1 = Z_{i,i}^b = Z_{i,i}^m = Z_{i,i-1}^m = Z_{i,i}^{m1} = 0.0$$

Recursion: for $1 \leq i < j \leq N$ (from short intervals to long intervals)

$$Z_{i,j} = 1.0 + \sum_{h=i-1}^{j-1} Z_{i,h} Z_{h+1,j}^1 \tag{14}$$

$$Z_{i,j}^1 = \sum_{h=i+1}^{j} Z_{i,h}^b$$

$$Z_{i,j}^b = e^{-f_1(i,j)/kT} + \sum_{h=i+1}^{j-2} \sum_{h=\ell+1}^{j-1} Z_{h,\ell}^b e^{-f_2(i,j,h,\ell)/kT}$$

$$+ \sum_{h=i+2}^{j-1} Z_{i+1,h-1}^m Z_{h,j-1}^{m1} e^{-f_3(i,j)/kT} \tag{15}$$

$$Z_{i,j}^m = \sum_{h=i+1}^{j-1} (e^{-f_4(i,h-1)/kT} + Z_{i,h-1}^m) Z_{h,j}^{m1} e^{-f_5/hT} \tag{16}$$

$$Z_{i,j}^{m1} = \sum_{h=i+1}^{j} Z_{i,h}^b e^{-f_4(h+1,j)/kT}$$

$Z_{i,j}$ is the partition function of subsequence $[i, j]$, $Z_{i,j}^b$ is that of i-j base pairing, and $Z_{i,j}^m$ is that of multi-loop. $Z_{i,j}^1$ and Z^{m1} are the partitions functions that include exactly one outmost base pair. The functions f_1 to f_5 reflect the free energy of unpaired bases, depending on the structural contexts.

Dynamic programming for MFE

The first reported DP algorithm for secondary structure prediction was the Nussinov algorithm (Nussinov *et al.*, 1978). It maximizes the number of base pairs in the secondary structure, which we can regard as a rough approximation of MFE, by iterating the following equations from short intervals to long intervals:

$$M(i,j) = \max \begin{cases} M(i+1, j) \\ M(i, j-1) \\ M(i+1, j-1) + 1 \text{ if } (i,j) \text{ can form a base pair} \\ \max_h M(i, h) M(h+1, j) \end{cases}$$

The Nussinov algorithm correspond to the CYK algorithm (Kasami *et al.*, 1965; Younger, 1967) for the parsing of SCFGs. with complexities $O(L^3)$ in time and $O(L^2)$ in memory for the length L of the sequence. Assuming a Turner type energy model for pseudoknot-free structures, MFE structures can be calculated via Zuker algorithm (Zuker and Stiegler, 1981), a CYK-like DP algorithm, using more complicated iterations, but with complexities $O(L^3)$ in time and $O(L^2)$ in memory.

DP for Base-Pairing Probability

The BPP of RNA is the marginal probability that a specific pair of bases will form a base pair in the secondary structure. BPPs are calculated using inside/outside partition functions as follows:

$$P_{i,j}^b = \frac{Z_{ij}^b W_{ij}^b}{Z_{1L}}$$

where Z_{ij}^b is the *inside* partition function of subsequence $[i, j]$ when the i-th and j-th bases form a base pair, which appeared in Eq. (15) of the McCaskill algorithm, and W_{ij}^b is the corresponding *outside* partition function, which is the sum of all the Boltzmann factors outside of $[i, j]$ when the i-th and j-th bases form a base pair. The W^b (i, j) is calculated by a DP algorithm, which is the outside algorithm of the McCaskill (inside) algorithm for partition function. A DP algorithm that directly calculate BPPs similar to outside algorithm for W^b (i, j) was reported in the same study reporting the McCaskill algorithm (McCaskill, 1990).

The computational complexity of those DP algorithms is also $O(L^3)$ in time and $O(L^2)$ in memory. For long RNA sequences, $O(L^3)$ in time and $O(L^2)$ are too expensive for practical analysis. By limiting the maximal span of base pairs to W, Rfold (Kiryu *et al.*, 2008) reduces them to $O(W^2L)$ in time and $O(L + W^2)$ in memory. The Rfold algorithm has been extended to parallel computations in ParasoR (Kawaguchi and Kiryu, 2016), including calculations of marginal probabilities implemented in CapR (Fukunaga *et al.*, 2014). Because $O(W^2L)$ is linear with respect to sequence length L, these tools are applicable for analyzing long non-coding RNAs.

Posterior Decoding for Maximum Expected Gain Prediction

As described in Predictions Based on the Maximum Expected Accuracy (MEA), predictions based on MEA are maximum expected gain estimators with the associated gain functions. The MEA estimator of CONTRAfold and the γ-centroid estimator of CentroidFold are calculated by the using dynamic programming as follows (Do *et al.*, 2006):

$$M(i,j) = \max \begin{bmatrix} M(i+1,j) \\ M(i,j-1) \\ M(i+1,j-1) + g(\gamma, P_{ij}^b) \\ \max_h M(i,h)M(h+1,j) \end{bmatrix}$$

where $g(\gamma, P_{ij}^b) = 2\gamma P_{ij}^b - (1 - \sum_i P_{ij}^b) - (1 - \sum_j P_{ij}^b)$ for the MEA estimator, and $g(\gamma, P_{ij}^b) = (\gamma + 1)P_{ij}^b - 1$ for the γ-centroid estimator. P_{ij}^b is the BPP.

Such DP is called *posterior decoding* of the marginal probability. Note that the DP in this case is exremely simple and similar to the Nussinov algorithm.

Additional Information on Secondary Structure Prediction

RNA Secondary Structure Prediction Incorporating Experimental Data

In recent decades, experimental investigations have been common approaches for understanding RNA structures. In the experimental approaches, recently emerging high-throughput sequencing technology changed the scale of target RNAs from a single transcript to the whole transcriptome. Currently, experimental investigations of RNA structures are categorized into the following two approaches: (1) *in vitro/in vivo* probing of RNA base pairs, (2) *in vivo* probing of RNA duplexes.

In the first approach, the pairing state (paired or unpaired) of individual bases in RNA molecules is analyzed using chemical and enzymatic probing. Selective 2'-hydroxyl acylation analyzed via primer extension and sequencing (SHAPE-seq) (Loughrey *et al.*, SHAPE; Lucks *et al.*, 2011), SHAPE and mutational profiling (SHAPE-MaP) (Siegfried *et al.*, 2014; Smola *et al.*, 2015), *in vivo* click selective 2'-hydroxyl acylation and profiling experiment (icSHAPE) (Flynn *et al.*, 2016; Spitale *et al.*, 2015), Structure-seq (Ding *et al.*, 2015), dimethyl sulfate (DMS) treatment and sequencing (DMS-seq) (Rouskin *et al.*, 2014), DMS mutational profiling with sequencing (DMS-MaPseq) (Zubradt *et al.*, 2017), Mod-seq (Talkish *et al.*, 2014), and chemical inference of RNA structures followed by massive parallel sequencing (CIRS-seq) (Incarnato *et al.*, 2014) are categorized as chemical probing methods employing various chemical compounds reacting with unpaired nucleotides (single-stranded regions) in RNA molecules. Fragmentation sequencing (Fragseq) (Underwood *et al.*, 2010), parallel analysis of RNA structure (PARS) (Kertesz *et al.*, 2010; Wan *et al.*, 2014) and parallel analysis of RNA structures with temperature elevation (PARTE) (Wan *et al.*, 2012) are enzymatic probing methods using structure-specific RNase, including RNase V1 and S1, for identifying both paired and unpaired nucleotides. For all of the aforementioned methods, raw sequencing reads must be processed to obtain the base-pairing or single-stranded signals of RNA bases at single-nucleotide resolution. The signal data derived from the limited number of structure probing experiments, such as icSHAPE and DMS-seq, is pre-computed and provided by Structure Surfer database (Berkowitz *et al.*, 2016) and FoldAtlas (Norris *et al.*, 2017).

As the second approach, RNA proximity ligation (RPL) (Ramani *et al.*, 2015), psoralen analysis of RNA interactions and structures (PARIS) (Lu *et al.*, 2016), ligation of interacting RNA and high-throughput sequencing (LIGR-seq) (Sharma *et al.*, 2016), sequencing of psoralen crosslinked, ligated and selected hybrids (SPLASH) (Aw *et al.*, 2016), and mapping RNA interactome *in vivo* (MARIO) (Nguyen *et al.*, 2016) employ proximity ligation of RNA duplexes to directly identify two RNA regions involved in base-pairing interactions.

The base-pairing or single-stranded signals derived from the probing experiments can be combined with computational prediction of RNA secondary structures to improve their prediction accuracy. In MC-fold prediction (Parisien and Major, 2008), three grades of energetic penalties are applied according to the SHAPE signal of each nucleotide. RNAstructure (Deigan et al., 2009), RNAsc (Zarringhalam et al., 2012), and RNApbfold (Washietl et al., 2012) incorporate the signal data into their energy calculation as pseudo-energy contribution. Currently, these three approaches are also implemented in the RNAfold Vienna RNA package (Lorenz et al., 2016). As a sampling-based approach, Seqfold (Ouyang et al., 2013) was developed for incorporating various type of probing data, such SHAPE-seq, PARS, and FragSeq data.

Non-DP Algorithms

Stochastic sampling
In order to estimate marginal probabilities of the secondary structures of RNA, stochastic sampling can be used (Ding and Lawrence, 2003). In stochastic sampling, the structures are sampled from Boltzmann distribution of the secondary structures.

Integer programing
It is computationally expensive to predict pseudoknotted secondary structures using DP algorithms. Integer programming has been used to predict pseudoknotted secondary structures in IPknot (Sato et al., 2011). Predicting the joint secondary structure of two interacting RNA sequences has the same computational complexity as pseudoknotted secondary structure prediction, and integer programming has been applied in RactIP (Kato et al., 2010).

Common Secondary Structure Prediction

RNA secondary structures often conserved during evolution. Therefore, it is important to detect the common secondary structure of the related sequences and find the functions correlated to the secondary structures. Several tools exist for locating the consensus secondary structures from multiple alignments of RNA sequences, such as CentroidAlifold (Hamada et al., 2011) and RNAalifold (Bernhart et al., 2008). PPFold (Sukosd et al., 2012) incorporates SHAPE data in predicting the consensus secondary structures. The aforementioned tools require multiple alignments of RNA sequences to detect common secondary structures. As inputs of those tools, we can use RNA structural multiple alignment tools, which find the multiple alignments considering secondary structures, such as MAFFT (Katoh and Toh, 2008), LocARNA (Will et al., 2012), LARA (Bauer et al., 2007), MXSCARNA (Tabei et al., 2008), and CentroidAlign (Hamada et al., 2009c).

Joint Secondary Structures Prediction

Functional RNAs interact with other molecules to perform their functions. Predictions of the joint secondary structures of two RNA molecules require predictions of inter-molecular and intra-molecular base pairs. RactIP (Kato et al., 2010) and IntaRNA (Busch et al., 2008) efficiently predicts joint secondary structures. PETcofold (Seemann et al., 2011) predicts the conserved joint secondary structures.

Prediction of RNA Three Dimensional Structure

Accurate prediction of the 3D structures of RNA molecules is one of the most challenging problems in the RNA bioinformatics field. In the RNA-Puzzles (Cruz et al., 2012; Miao et al., 2015, 2017), a community-wide blind test for RNA 3D structure prediction, most participant groups could not predict near-native structures for several problems (target RNA molecules). RNA 3D structures consist of not only WC base-pairs but also non-WC base-pairs and various hydrogen bonds, in contrast to RNA 2D structures, which are simply constructed from WC base-pairs. Currently, base-pairing interactions are categorized into 12 groups based on the combination of interacting edges (Watson-Crick, Hoogsteen and sugar) in nucleotide bases and the relative orientations of glycosidic bonds (cis or trans) (Leontis et al., 2002). Among these groups, cis WC-WC base-pair are known as standard WC base pairs, and the remaining 11 groups are non-WC base-pairs. Currently, there are two types of 3D structure prediction methods: template-based and physical chemistry-based methods.

In template-based methods, small structural elements derived from several known structures, referred to modules/motifs or templates, are assembled to construct a predicted structure as building blocks. These structural modules or motifs contain many non-WC base pairs. An appropriate selection of these structural modules leads to accurate prediction of the RNA 3D structures. Template sizes and template selection methods depend on the prediction software. As the largest template-based methods, ModeRNA (Rother et al., 2011) and RNABuilder (Flores et al., 2010) use the whole structures as the templates for their predictions. MC-fold/MC-sym (Parisien and Major, 2008), RNAComposer (Popenda et al., 2012) and 3dRNA (Zhao et al., 2012) employ medium-sized templates derived from small elements in RNA secondary structures that are computationally predicted or provided by users. FARNA (Das and Baker, 2007) is a fragment assembly method, which is also known as ROSETTA framework in protein structure prediction fields, which uses a few nucleotides as fragments. The FARNA-predicted structures could be refined at atomic resolution using FARFAR (Das *et al.*, 2010).

In physical chemistry-based methods, each nucleotide of an input RNA sequence is represented with coarse-grained beads instead of the full atoms in 3D space. Monte Carlo simulation or molecular dynamics simulation is used for conformational sampling of the coarse-grained RNA structure. There are two types of energy function for scoring the simulated (predicted) structures. The first type is knowledge-based statistical potential derived from known 3D structures as employed in SimRNA (Boniecki *et al.*, 2016). The second type of energy function is based on physicochemical interactions, such as hydrogen bonds of base pairing and electrostatic repulsion of the phosphates, implemented in HiRE-RNA (Cragnolini *et al.*, 2015). These methods could provide accurate predictions for small (10–20 nucleotides) to medium (50–80 nucleotides) of RNA molecules, whereas an application of these methods to moderately large (>100 nucleotides) RNA molecules remains challenging. The physical chemistry-based methods require high computational cost for efficient conformational sampling of RNA structures.

Conclusion

The structures of RNAs are important for understanding their functions. Among them, RNA secondary structures are important and well studied. Most functional RNAs interact with other RNAs via complementary base paring, which complete the intra-molecule interactions that form the secondary structures. The minimum free energy secondary structures are usually selected as the predicted structures, but the MEA estimator or gamma-centroid estimator have higher expected accuracy with respect to the number of correctly predicted base-pairs. Recent progress in experiments has allowed us to incorporate the experimentally predicted base pairs into computational predictions of the secondary structures. Even if using experimental data, the probabilities of the predicted secondary structures remain low because the number of possible secondary structures are large. The marginal probabilities, such as base-pairing probabilities, represent structurally important information, and they are not necessarily low. Predicting pseudoknotted structures, the interactions of two RNA molecules, the common secondary structures of two RNA sequences, are all computationally expensive if the strict DP algorithms are applied, but stochastic sampling or integer programming is often useful.

Computational methods have been developed for RNA secondary structures, and they are becoming useful for the structural analysis of RNA. Although the point estimations of RNA secondary structures have limited reliability, we can understand the total picture of the secondary structures via combinations of bioinformatics tools. The probability distribution of Hamming distance and the credibility limits are relevant examples. The prediction of the 3D structures remains challenging, but its accuracy is expected to improve due to the increased availability of experimentally validated 3D structures.

Acknowledgements

The authors also thank to the members of Artificial Intelligence Research Center, Hisanory Kiryu's laboratory, and Kiyoshi Asai's laboratory for useful discussions. This work was supported in part by JSPS KAKENHI Grant Numbers JP16H02484 and JP16H06279.

See also: Algorithms for Strings and Sequences: Pairwise Alignment. Algorithms for Strings and Sequences: Pairwise Alignment. Applications of Ribosomal RNA Sequence and Structure Analysis for Extracting Evolutionary and Functional Insights. Assessment of Structure Quality (RNA and Protein). Biomolecular Structures: Prediction, Identification and Analyses. Characterizing and Functional Assignment of Noncoding RNAs. Computational Design and Experimental Implementation of Synthetic Riboswitches and Riboregulators. Engineering of Supramolecular RNA Structures. Genome-Wide Probing of RNA Structure. MicroRNA and lncRNA Databases and Analysis. Natural Language Processing Approaches in Bioinformatics. Nucleic-Acid Structure Database. Predicting RNA-RNA Interactions in Three-Dimensional Structures. Prediction of Coding and Non-Coding RNA

References

Andronescu, M., Condon, A., Hoos, H.H., Mathews, D.H., Murphy, K.P., 2007. Efficient parameter estimation for RNA secondary structure prediction. Bioinformatics 23 (13), 19–28.

Aw, J.G., Shen, Y., Wilm, A., *et al.*, 2016. In vivo mapping of eukaryotic RNA interactomes reveals principles of higher-order organization and regulation. Mol. Cell 62 (4), 603–617.

Bauer, M., Klau, G.W., Reinert, K., 2007. Accurate multiple sequence-structure alignment of RNA sequences using combinatorial optimization. BMC Bioinform. 8, 271.

Berkowitz, N.D., Silverman, I.M., Childress, D.M., *et al.*, 2016. A comprehensive database of high-throughput sequencing-based RNA secondary structure probing data (Structure Surfer). BMC Bioinform. 17 (1), 215.

Bernhart, S.H., Hofacker, I.L., Will, S., Gruber, A.R., Stadler, P.F., 2008. RNAalifold: Improved consensus structure prediction for RNA alignments. BMC Bioinform. 9, 474.

Boniecki, M.J., Lach, G., Dawson, W.K., *et al.*, 2016. SimRNA: A coarse-grained method for RNA folding simulations and 3D structure prediction. Nucleic Acids Res. 44 (7), e63.

Busch, A., Richter, A.S., Backofen, R., 2008. IntaRNA: Efficient prediction of bacterial sRNA targets incorporating target site accessibility and seed regions. Bioinformatics 24 (24), 2849–2856.

Cragnolini, T., Laurin, Y., Derreumaux, P., Pasquali, S., 2015. Coarse-grained HiRE-RNA model for ab Initio RNA folding beyond simple molecules, including noncanonical and multiple base pairings. J. Chem. Theory Comput. 11 (7), 3510–3522.

Cruz, J.A., Blanchet, M.F., Boniecki, M., et al., 2012. RNA-Puzzles: A CASP-like evaluation of RNA three-dimensional structure prediction. RNA 18 (4), 610–625.

Das, R., Baker, D., 2007. Automated de novo prediction of native-like RNA tertiary structures. Proc. Natl. Acad. Sci. USA 104 (37), 14664–14669.

Das, R., Karanicolas, J., Baker, D., 2010. Atomic accuracy in predicting and designing noncanonical RNA structure. Nat. Methods 7 (4), 291–294.

Deigan, K.E., Li, T.W., Mathews, D.H., Weeks, K.M., 2009. Accurate SHAPE-directed RNA structure determination. Proc. Natl. Acad. Sci. USA 106 (1), 97–102.

Ding, Y., Kwok, C.K., Tang, Y., Bevilacqua, P.C., Assmann, S.M., 2015. Genome-wide profiling of in vivo RNA structure at single-nucleotide resolution using structure-seq. Nat. Protoc. 10 (7), 1050–1066.

Ding, Y., Lawrence, C.E., 2003. A statistical sampling algorithm for RNA secondary structure prediction. Nucleic Acids Res. 31 (24), 7280–7301.

Do, C.B., Woods, D.A., Batzoglou, S., 2006. CONTRAfold: RNA secondary structure prediction without physics-based models. Bioinformatics 22 (14), e90–98.

Flores, S.C., Wan, Y., Russell, R., Altman, R.B., 2010. Predicting RNA structure by multiple template homology modeling. Pac. Symp. Biocomput. 216–227.

Flynn, R.A., Zhang, Q.C., Spitale, R.C., et al., 2016. Transcriptome-wide interrogation of RNA secondary structure in living cells with icSHAPE. Nat. Protoc. 11 (2), 273–290.

Freyhult, E., Moulton, V., Clote, P., 2007. Boltzmann probability of RNA structural neighbors and riboswitch detection. Bioinformatics 23 (16), 2054–2062.

Fukunaga, T., Ozaki, H., Terai, G., et al., 2014. CapR: Revealing structural specificities of RNA-binding protein target recognition using CLIP-seq data. Genome Biol. 15 (1), R16.

Hamada, M., Kiryu, H., Sato, K., Mituyama, T., Asai, K., 2009a. Prediction of RNA secondary structure using generalized centroid estimators. Bioinformatics 25 (4), 465–473.

Hamada, M., Sato, K., Asai, K., 2011. Improving the accuracy of predicting secondary structure for aligned RNA sequences. Nucleic Acids Res. 39 (2), 393–402.

Hamada, M., Sato, K., Kiryu, H., Mituyama, T., Asai, K., 2009b. Predictions of RNA secondary structure by combining homologous sequence information. Bioinformatics 25 (12), i330–i338.

Hamada, M., Sato, K., Kiryu, H., Mituyama, T., Asai, K., 2009c. CentroidAlign: Fast and accurate aligner for structured RNAs by maximizing expected sum-of-pairs score. Bioinformatics 25 (24), 3236–3243.

Harmanci, A.O., Sharma, G., Mathews, D.H., 2011. TurboFold: Iterative probabilistic estimation of secondary structures for multiple RNA sequences. BMC Bioinform. 12, 108.

Incarnato, D., Neri, F., Anselmi, F., Oliviero, S., 2014. Genome-wide profiling of mouse RNA secondary structures reveals key features of the mammalian transcriptome. Genome Biol. 15 (10), 491.

Kasami, T., 1965. An efficient recognition and syntax-analysis algorithm for context-free languages, Coordinated Science Laboratory Report R257.

Katoh, K., Toh, H., 2008. Improved accuracy of multiple ncRNA alignment by incorporating structural information into a MAFFT-based framework. BMC Bioinform. 9, 212.

Kato, Y., Sato, K., Hamada, M., et al., 2010. RactIP: Fast and accurate prediction of RNA-RNA interaction using integer programming. Bioinformatics 26 (18), i460–466.

Kawaguchi, R., Kiryu, H., 2016. Parallel computation of genome-scale RNA secondary structure to detect structural constraints on human genome. BMC Bioinform. 17 (1), 203.

Kertesz, M., Wan, Y., Mazor, E., et al., 2010. Genome-wide measurement of RNA secondary structure in yeast. Nature 467 (7311), 103–107.

Kiryu, H., Kin, T., Asai, K., 2008. Rfold: An exact algorithm for computing local base pairing probabilities. Bioinformatics 24 (3), 367–373.

Knudsen, B., Hein, J., 2003. Pfold: RNA secondary structure prediction using stochastic context-free grammars. Nucleic Acids Res. 31 (13), 3423–3428.

Leontis, N.B., Stombaugh, J., Westhof, E., 2002. The non-Watson-Crick base pairs and their associated isostericity matrices. Nucleic Acids Res. 30 (16), 3497–3531.

Lorenz, R., Bernhart, S.H., Honer Zu Siederdissen, C., et al., 2011. , ViennaRNA Package 2.0. Algorithms Mol. Biol. 6, 26.

Lorenz, R., Luntzer, D., Hofacker, I.L., Stadler, P.F., Wolfinger, M.T., 2016. SHAPE directed RNA folding. Bioinformatics 32 (1), 145–147.

Loughrey, D., Watters, K.E., Settle, A.H., Lucks, J.B., 2014. SHAPE-Seq 2.0: Systematic optimization and extension of high-throughput chemical probing of RNA secondary structure with next generation sequencing. Nucleic Acids Res. 42 (21), doi: 10.1093/nar/gku909.

Lucks, J.B., Mortimer, S.A., Trapnell, C., et al., 2011. Multiplexed RNA structure characterization with selective 2'-hydroxyl acylation analyzed by primer extension sequencing (SHAPE-Seq). Proc. Natl. Acad. Sci. USA 108 (27), 11063–11068.

Lu, Z., Zhang, Q.C., Lee, B., et al., 2016. RNA duplex map in living cells reveals higher-order transcriptome structure. Cell 165 (5), 1267–1279.

Mathews, D.H., Sabina, J., Zuker, M., Turner, D.H., 1999. Expanded sequence dependence of thermodynamic parameters improves prediction of RNA secondary structure. J. Mol. Biol. 288 (5), 911–940.

Mathews, D.H., 2004. Using an RNA secondary structure partition function to determine confidence in base pairs predicted by free energy minimization. RNA 10 (8), 1178–1190.

Mathews, D.H., 2014. Using the RNAstructure software package to predict conserved RNA structures. Curr. Protoc. Bioinform. 46, 1–22.

McCaskill, J.S., 1990. The equilibrium partition function and base pair binding probabilities for RNA secondary structure. Biopolymers 29 (6-7), 1105–1119.

Miao, Z., Adamiak, R.W., Antczak, M., et al., 2017. RNA-Puzzles Round III: 3D RNA structure prediction of five riboswitches and one ribozyme. RNA 23 (5), 655–672.

Miao, Z., Adamiak, R.W., Blanchet, M.F., et al., 2015. RNA-Puzzles Round II: Assessment of RNA structure prediction programs applied to three large RNA structures. RNA 21 (6), 1066–1084.

Mori, R., Hamada, M., Asai, K., 2014. Efficient calculation of exact probability distributions of integer features on RNA secondary structures. BMC Genomics 15 (Suppl 10), S6.

Newberg, L.A., Lawrence, C.E., 2009. Exact calculation of distributions on integers, with application to sequence alignment. J. Comput. Biol. 16 (1), 1–18.

Nguyen, T.C., Cao, X., Yu, P., et al., 2016. Mapping RNA-RNA interactome and RNA structure in vivo by MARIO. Nat. Commun. 7, 12023.

Norris, M., Kwok, C.K., Cheema, J., et al., 2017. FoldAtlas: A repository for genome-wide RNA structure probing data. Bioinformatics 33 (2), 306–308.

Nussinov, R., Pieczenik, G., Griggs, J., Kleitman, D., 1978. Algorithms for Loop Matchings. SIAM J. Appl. Math. 35 (1), 68–82.

Ouyang, Z., Snyder, M.P., Chang, H.Y., 2013. SeqFold: Genome-scale reconstruction of RNA secondary structure integrating high-throughput sequencing data. Genome Res. 23 (2), 377–387.

Parisien, M., Major, F., 2008. The MC-Fold and MC-Sym pipeline infers RNA structure from sequence data. Nature 452 (7183), 51–55.

Popenda, M., Szachniuk, M., Antczak, M., et al., 2012. Automated 3D structure composition for large RNAs. Nucleic Acids Res. 40 (14), e112.

Ramani, V., Qiu, R., Shendure, J., 2015. High-throughput determination of RNA structure by proximity ligation. Nat. Biotechnol. 33 (9), 980–984.

Rivas, E., Eddy, S.R., 2000. The language of RNA: A formal grammar that includes pseudoknots. Bioinformatics 16 (4), 334–340.

Rivas, E., Eddy, S.R., 1999. A dynamic programming algorithm for RNA structure prediction including pseudoknots. J. Mol. Biol. 285 (5), 2053–2068.

Rivas, E., Lang, R., Eddy, S.R., 2012. A range of complex probabilistic models for RNA secondary structure prediction that includes the nearest-neighbor model and more. RNA 18 (2), 193–212.

Rother, M., Rother, K., Puton, T., Bujnicki, J.M., 2011. ModeRNA: A tool for comparative modeling of RNA 3D structure. Nucleic Acids Res. 39 (10), 4007–4022.

Rouskin, S., Zubradt, M., Washietl, S., Kellis, M., Weissman, J.S., 2014. Genome-wide probing of RNA structure reveals active unfolding of mRNA structures in vivo. Nature 505 (7485), 701–705.

Sakakibara, Y., Brown, M., Hughey, R., et al., 1994. Stochastic context-free grammars for tRNA modeling. Nucleic Acids Res. 22 (23), 5112–5120.

Sakuraba, S., Asai, K., Kameda, T., 2015. Predicting RNA duplex dimerization free-energy changes upon mutations using molecular dynamics simulations. J. Phys. Chem. Lett. 6 (21), 4348–4351.

Sato, K., Hamada, M., Asai, K., Mituyama, T., 2009. CENTROIDFOLD: A web server for RNA secondary structure prediction. Nucleic Acids Res. 37 (Web Server issue), W277–W280.

Sato, K., Kato, Y., Hamada, M., Akutsu, T., Asai, K., 2011. IPknot: Fast and accurate prediction of RNA secondary structures with pseudoknots using integer programming. Bioinformatics 27 (13), 85–93.

Seemann, S.E., Menzel, P., Backofen, R., Gorodkin, J., 2011. The PETfold and PETcofold web servers for intra- and intermolecular structures of multiple RNA sequences. Nucleic Acids Res. 39 (Web Server issue), W107–111.

Sharma, E., Sterne-Weiler, T., O'Hanlon, D., Blencowe, B.J., 2016. Global mapping of human RNA-RNA interactions. Mol. Cell 62 (4), 618–626.

Siegfried, N.A., Busan, S., Rice, G.M., Nelson, J.A., Weeks, K.M., 2014. RNA motif discovery by SHAPE and mutational profiling (SHAPE-MaP). Nat. Methods 11 (9), 959–965.

Smola, M.J., Rice, G.M., Busan, S., Siegfried, N.A., Weeks, K.M., 2015. Selective 2′-hydroxyl acylation analyzed by primer extension and mutational profiling (SHAPE-MaP) for direct, versatile and accurate RNA structure analysis. Nat. Protoc. 10 (11), 1643–1669.

Spitale, R.C., Flynn, R.A., Zhang, Q.C., et al., 2015. Structural imprints in vivo decode RNA regulatory mechanisms. Nature 519 (7544), 486–490.

Sukosd, Z., Knudsen, B., Kjems, J., Pedersen, C.N., 2012. PPfold 3.0: Fast RNA secondary structure prediction using phylogeny and auxiliary data. Bioinformatics 28 (20), 2691–2692.

Tabei, Y., Kiryu, H., Kin, T., Asai, K., 2008. A fast structural multiple alignment method for long RNA sequences. BMC Bioinform. 9, 33.

Talkish, J., May, G., Lin, Y., Woolford, J.L., McManus, C.J., 2014. Mod-seq: High-throughput sequencing for chemical probing of RNA structure. RNA 20 (5), 713–720.

Underwood, J.G., Uzilov, A.V., Katzman, S., et al., 2010. FragSeq: Transcriptome-wide RNA structure probing using high-throughput sequencing. Nat. Methods 7 (12), 995–1001.

Wan, Y., Qu, K., Ouyang, Z., et al., 2012. Genome-wide measurement of RNA folding energies. Mol. Cell 48 (2), 169–181.

Wan, Y., Qu, K., Zhang, Q.C., et al., 2014. Landscape and variation of RNA secondary structure across the human transcriptome. Nature 505 (7485), 706–709.

Washietl, S., Hofacker, I.L., Stadler, P.F., Kellis, M., 2012. RNA folding with soft constraints: Reconciliation of probing data and thermodynamic secondary structure prediction. Nucleic Acids Res. 40 (10), 4261–4272.

Will, S., Joshi, T., Hofacker, I.L., Stadler, P.F., Backofen, R., 2012. LocARNA-P: Accurate boundary prediction and improved detection of structural RNAs. RNA 18 (5), 900–914.

Younger, D., 1967. Recognition and parsing of context-free languages in time n^3. Inf. Control 10 (2), 189–208.

Zakov, S., Goldberg, Y., Elhadad, M., Ziv-Ukelson, M., 2011. Rich parameterization improves RNA structure prediction. J. Comput. Biol. 18 (11), 1525–1542.

Zarringhalam, K., Meyer, M.M., Dotu, I., Chuang, J.H., Clote, P., 2012. Integrating chemical footprinting data into RNA secondary structure prediction. PLOS ONE 7 (10), e45160.

Zhao, Y., Huang, Y., Gong, Z., et al., 2012. Automated and fast building of three-dimensional RNA structures. Sci. Rep. 2, 734.

Zubradt, M., Gupta, P., Persad, S., et al., 2017. DMS-MaPseq for genome-wide or targeted RNA structure probing in vivo. Nat. Methods 14 (1), 75–82.

Zuker, M., 2003. Mfold web server for nucleic acid folding and hybridization prediction. Nucleic Acids Res. 31 (13), 3406–3415.

Zuker, M., Stiegler, P., 1981. Optimal computer folding of large RNA sequences using thermodynamics and auxiliary information. Nucleic Acids Res. 9 (1), 133–148.

Rational Structure-Based Drug Design

Varun Khanna, Flinders University, Adelaide, SA, Australia and Vaxine Pty Ltd., Adelaide, SA, Australia
Shoba Ranganathan, Macquarie University, Sydney, NSW, Australia
Nikolai Petrovsky, Flinders University, Adelaide, SA, Australia and Vaxine Pty Ltd., Adelaide, SA, Australia

Introduction

The discovery, optimization and evaluation of small molecules which interact with a biological target of pharmaceutical relevance is the core of the drug discovery process. So far, this has been mostly dominated by the conventional brute-force approach of lead discovery via high throughput screening (HTS) of the natural product, synthetic or combinatorial compound libraries to identify potential leads (Macarron et al., 2011). While effective in many cases, HTS is expensive, requires a readily available diverse set of compounds to screen, often yields hits with poor efficacy and provides little information to guide further optimization. This has driven the development of more rational and cost-efficient screening methods. The advancement in molecular biology, crystallographic techniques and computational power over the last 30 years have provided researchers access to high-quality structural information on a wide variety of biological targets. Structural information is the ultimate rational drug design tool which can streamline all aspects of drug discovery from target selection to lead optimization and can significantly reduce the development cost. Indeed, structure-guided drug design efforts have led to the discovery of many high profile drugs including amprenavir (Agenerase) and nelfinavir (Viracept) developed against the HIV protease (Kaldor et al., 1997), zanamivir (Relenza) against neuraminidase (Varghese, 1999), imatinib mesylate (Gleevec) which inhibits Abl tyrosine kinase (Schindler et al., 2000), erlotinib (Tarceva) for the treatment of metastatic lung cancer (Pollack et al., 1999) and the thrombin inhibitor ximelagatran (Exanta) as an oral anticoagulant (Gustafsson et al., 2001). More recently, lapatinib, an ErbB2 inhibitor for cancer, sitagliptin (Januvia) for type-2 diabetes, vorinostat (Zolinza) to inhibit histone deacetylase, and rivaroxaban, as an oral Factor Xa inhibitor are examples of successful application of rational structure-based methods in drug development.

Structure-based drug design (SBDD) is the process in which novel drug-like molecules are designed based on structural knowledge of the relevant macromolecular target (Wang et al., 2016). In SBDD, virtual chemical libraries containing millions of compounds and structural data of macromolecules are used in tandem to assess the ability of compounds to interact with the target of interest. Many virtual libraries are publicly available, a few notable examples being PubChem (Kim et al., 2016), ChEMBL (Bento et al., 2014), BindingDB (Gilson et al., 2016), DrugBank (Law et al., 2014) and ZINC (Irwin et al., 2012). In addition to public sources, commercial vendors (Maybridge, Enamine) maintain virtual libraries. Structural information on biological targets can be obtained from Protein Data Bank (PDB) (Parasuraman, 2012). If the structure of the desired target is not available, it may be possible to create a homology model using related structures available in PDB. In either case, the compounds in the virtual library are subsequently docked into the binding cavity of the target to determine the relative rank for the entire dataset. Automatic and manual data analysis of the docking result is then carried out to predict the compounds with highest binding affinity to the target. This results in a smaller subset of compounds as a starting point for downstream analysis.

This article summarizes the process of structure-based drug design and discusses the choice of target, lead identification and the various docking-based methods. Key concepts in SBDD will be illustrated through a case study that explores the identification of potential novel anti-malarial agents targeting the hemoglobin degrading enzyme, Plasmepsin II, of *Plasmodium falciparum*. The aim of this article is to guide novice computational users by elucidating the general steps involved in such a drug discovery exercise.

Background and Overview of the Process

The process of drug discovery and development generally requires extensive examination of target proteins and potential drug-like compounds before a lead is ready for phase 1 clinical trials. The first stage is identification, purification and structure determination of the target (usually a protein). The selection of the target is the most crucial step as it sets the course for all future aspects of research in the drug discovery pipeline. Once the biological target has been selected, its structure is determined by one of three methods: *X-ray diffraction* generally known as *X-ray crystallography*, *nuclear magnetic resonance spectroscopy* generally known as *NMR* and *comparative modeling* generally known as *in silico homology modeling*. In the second stage, computational techniques are used to screen and rank compound or fragment libraries based on their electrostatic and steric interaction with the target. Promising hits are then tested in biological assays to identify the compounds with desired activity (lead discovery). Once a candidate series has been identified, the lead optimization stage begins. In this stage, key lead compound properties like adsorption, metabolism, distribution and excretion are optimized while avoiding risk of toxicity. Further steps include synthesis of the optimized leads, testing, determination of the target lead complex and additional optimization. After several rounds of the process, optimized compounds usually show marked improvement in these pharmacokinetic properties and specificity for the target. The lead optimization stage usually concludes with the successful demonstration of *in vivo* efficacy of the compound in an appropriate animal model.

Rational Structure-Based Drug Design

Choice of the Drug Target and Structure Determination

The choice of the target protein is primarily made based on therapeutic and biological relevance. The 'druggable' target is a protein, peptide or nucleic acid whose activity can be modulated by a small molecule (Owens, 2007). Proteins are often used as drug targets due to their significant role in metabolic or signaling pathways specific to a disease condition and can either be activated or inhibited by small molecule drugs. Once the target has been identified it is essential to determine its three-dimensional (3D) structure ideally with a well-defined drug binding pocket to enable high throughput virtual screening in the SBDD approach. The structure of the target can be determined by one of the following methods:

X-ray crystallography and NMR

Both X-ray crystallography and NMR produce data on the relative position of atoms of a molecule. The basis of X-ray crystallography is the scattering of X-rays from electron clouds of atoms whereas NMR measures the interaction of atomic nuclei. The end-product of crystallographic structure determination is an electron density map which is essentially a contour plot indicating positions in the crystal structure where electrons are most likely to be found. This data must be interpreted in terms of a 3D model using semi-automatic computational methodologies. On the other hand, the end-product of an NMR experiment is usually a set of distances between atomic nuclei that define both bonded and non-bonded close contacts in a molecule. These must be interpreted manually to produce a 3D molecular structure using computational tools. The structure determination in each case requires assumptions and approximations, hence the resulting molecular structures obtained may have errors. The choice of technique depends on many factors including molecular weight, ease of solubility and crystallization of the macromolecule under study. X-ray crystallography remains the main workhorse of structure determination for SBDD. Currently, the RCSB Protein Data bank (Parasuraman, 2012) database contains over 130,000 structures of which 90% were solved through X-ray crystallography (**Fig. 1**).

Homology modeling

In the absence of an experimentally determined 3D structure of a protein, *in silico* homology modeling can provide structural models that are comparable to the best results achieved experimentally. In general, 30% target-template sequence identity is required to generate a useful structural model (Forrest *et al.*, 2006). This allows researchers to use the generated *in silico* models for

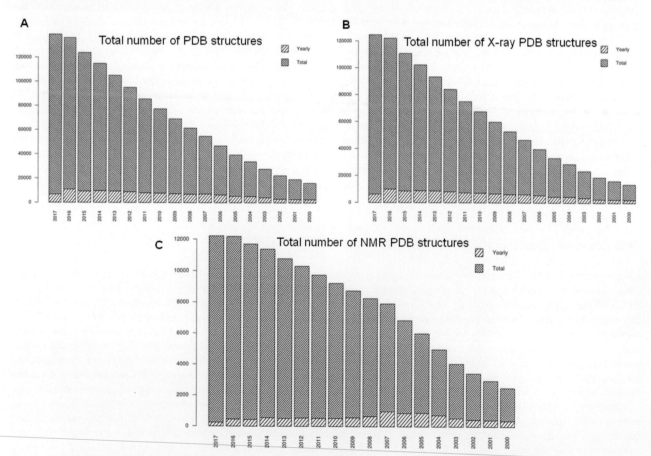

Fig. 1   Incremental increase in the structures of macromolecules in PDB database since year 2000. (Data obtained from PDB).

functional analysis and to predict interactions with other molecules. Homology modeling predicts the structure of the target protein primarily by aligning the target sequence called a query with the sequence of one or more known structures called templates and is based on two major assumptions:

1. The structure of a protein is encoded in its sequence thus knowing the sequence should at least, in theory, suffice to obtain the structure. This observation was elegantly demonstrated by Anfinsen when he showed that bovine pancreatic ribonuclease, following exposure to a denaturant, could spontaneously regain its native folded structure (Anfinsen, 1973).
2. The structure of a protein is more stable and conserved than its sequence. Therefore, closely related similar sequences will essentially adopt similar structures and more distantly related sequences will at least have similar folding, a relationship first identified by Chothia and Lesk (1986).

In practice homology modeling is a multistep process which can be summarized in the following seven steps: a). Template identification, b). Alignment correction, c). Model building d). Loop modeling, e). Side-chain modeling f). Model optimization and g). Model validation. A comprehensive review of homology modeling is beyond the scope of this article. However, interested readers are referred to a series of publications and reviews on homology modeling (Eswar *et al.*, 2006; Fiser and Sali, 2003; Martí-Renom *et al.*, 2000). A list of tools for homology modeling are given in **Table 1**. Repositories such as Protein model portal (Arnold *et al.*, 2009), Modbase (Pieper *et al.*, 2011) and SWISS-MODEL (Kiefer *et al.*, 2009) contain proteins models generated using various methods.

Identification of Binding Site

Once the 3D structure of the target protein is generated, its potential ligand binding sites are determined to facilitate docking computation and virtual screening (VS). Typically, the ligand binding site is a pocket with a variety of potential hydrogen bond donors, acceptors, hydrophobic characteristics and a well-defined stereochemical arrangement. The ligand binding site can be the active site where the substrate binds or can be an assembly site where another macromolecule binds or a communication site necessary to relay the information. Although empirical data suggest that actual ligand binding sites tend to coincide with the largest and deepest pockets on the target surface (Weisel *et al.*, 2007), there exist cases where ligands are known to bind exposed and shallow clefts (Nisius *et al.*, 2012). Due to the importance of binding sites in molecular recognition and interaction, a large number of computational methods for binding site recognition and analysis have been developed that scan the surface of the target for cavities or pockets which might be potential binding sites. All these methods take a target structure as input and then output an ordered list of putative binding sites. Generally, not all reported sites correspond to true binding sites, however, it is expected that entries at the top will correspond to regions with a higher probability of being true binding sites. The methods can be broadly divided into sequence-, structure- and energy-based methods.

Table 1   Tools for homology modeling

Name	URL	Summary	Reference
RaptorX	http://raptorx.uchicago.edu/	Predict secondary, tertiary, solvent accessibility, disordered regions and binding sites. Remote homology detection	Källberg *et al.* (2012)
Biskit	http://biskit.pasteur.fr/	Python library for typical tasks of structural bioinformatics including homology modeling, docking and dynamics	Grünberg *et al.* (2007)
Phyre2	http://www.sbg.bio.ic.ac.uk/~phyre2	Uses HMM to predict, analyze and build protein 3D structure and binding site. Analyzes the effect of SNPs.	Kelley *et al.* (2015)
EsyPred3D	http://www.unamur.be/sciences/biologie/urbm/bioinfo/esypred/	Automated webserver based on MODELLER. Uses neural networks to improve the alignment	Lambert *et al.* (2002)
Modeller	https://salilab.org/modeller/	Predicts three-dimensional structure by satisfying spatial restraints also performs multiple sequence alignments, clustering, de novo modeling of loops and optimization of the models	Eswar *et al.* (2006)
Robetta	http://robetta.bakerlab.org/	Robetta parses proteins chains into putative domains and model these domains either by homology modeling or by ab inito modeling	Kim *et al.* (2004)
I-TASSER	https://zhanglab.ccmb.med.umich.edu/I-TASSER/	I-TASSER generated 3D models from multiple threading alignments. The function of the protein is inferred by structural matching the model to protein function database	Roy *et al.* (2010)
Bhageerath-H	http://www.scfbio-iitd.res.in/bhageerath/bhageerath_h.jsp	Hybrid of ab initio folding and homology methods	Jayaram *et al.* (2014)

Sequence-based methods

Sequence-based methods are based on evolutionary conservation and exploit the propensity of conserved residues in the binding site. In the LIGSISITEcsc algorithm (Huang and Schroeder, 2006) a sequence conservation measure of neighboring residues is used to re-rank the top three sites predicted by LIGSITE (explained under structure-based methods), which leads to an improved success rate in binding site prediction. Unlike LIGSITEcsc, in ConCavity (Capra *et al.*, 2009) the conservation information is not only used to re-rank the predictions but is also incorporated into the binding site detection procedure.

Structure-based methods

In contrast to sequence-based methods, structure-based method predict ligand binding site by analyzing geometrical features like clefts or cavities. Some methods are solely based on geometrical features (LIGSITE (Hendlich *et al.*, 1997), PocketPicker (Weisel *et al.*, 2007)) while others take into account additional features like physicochemical information, polarity or charge (Fpocket (Le Guilloux *et al.*, 2009), SiteFinder by MOE (Locating Binding Sites in Protein Structures, 2017)).

Energy-based methods

Energy-based methods depend on the calculation of interaction energy between various probes placed on the grid points around the target surface and subsequent clustering of the probes with low interaction energy to identify the most energetically favorable binding pockets. Notable examples of energy-based methods are Q-SiteFinder (Laurie and Jackson, 2005) and SiteHound (Ghersi and Sanchez, 2009).

Consensus method

A consensus method is essentially a meta approach that combines results from several algorithms mentioned above. For example, Metapocket 2.0 (Huang, 2009) collects results of eight different methods by taking the top three sites from each method with the authors demonstrating that Metapocket 2.0 performs better than any one individual method alone.

For a detailed review of binding site prediction methods, readers can refer to Henrich *et al.* (2010), Nisius *et al.* (2012), Xie and Hwang (2015).

Molecular Docking Based Virtual Screening

Once the structure and the target site are determined, molecular docking based virtual screening of compound and fragment libraries can be carried out to identify potential leads. VS is regarded as a computational counterpart of the experimental high-throughput screening method and is one of the most widely used strategies for SBDD (Stahura and Bajorath, 2004). The major advantage of docking approaches is their speed and the guidance they provide to follow-on wet lab experiments (López-Vallejo *et al.*, 2011). Molecular docking is also useful in the study of ligand-target interactions as the docking programs can analyze the microscopic atomic interactions between a ligand and a target.

In VS, typically a database of small molecules is docked onto the region of interest and is scored based on the predicted interactions with the site. However, in the fragment based approach, small drug-like fragments such as benzene rings, amino, hydroxy, carbonyl or other functional groups can be positioned, scored, optimized in the binding site and finally linked *in silico* to generate lead candidates. The compounds obtained *in silico* by linking fragments can then be synthesized and tested. In general, molecular docking based VS consists of several steps; (i) target preparation, (ii) compound or fragment database selection, (iii) molecular docking, (iv) post-docking evaluation analysis, and (v) molecular dynamics and binding free energy calculation. Careful and thorough literature survey regarding the target, binding site and known ligands (if available) is often necessary to select the docking algorithms best suited for the given target.

Target preparation

The preparation of the target protein requires great care because experimentally determined structures frequently have many problems like incomplete side chains, missing loops and result in inaccurate models due to incorrect interpretation of the experimental data. Structural characterization of the targeted protein includes a choice of tautomeric forms of histidine residues, correct assignment of protonation states of amino acids and conformation of residues, especially in the binding site. It is also important to add missing hydrogen atoms, build missing residues or loops, identify overlapping atoms to reduce clash and optimize hydrogen bonding network. Water molecules and cofactors (metal) in the protein active site may need to be removed or retained (if they are known to be critical for ligand binding). Metals and cofactors which form an integral part of the binding interaction with the ligand are considered part of the docking site and hence are retained. If the software allows for flexible docking in order to account for conformational changes during ligand binding, the number, the identity of flexible residues and degrees of flexibility needs to be defined.

Compound library preparation

Prior to generating the 3D structure of the library of ligands, it is important to clean up the 2D structures by removing the ions, salts, incomplete structures, inorganic molecules, duplicates and water molecules. All reactive, promiscuous and otherwise undesirable compounds should also be removed. Additionally, compounds may be filtered based on several physicochemical

Table 2   List of publically available databases of chemical compounds

Database	No. of compounds	Description	URL	Reference
PubChem	~120 million	Contains compounds and associated bioassays	https://pubchem.ncbi.nlm.nih.gov/	Kim *et al.* (2016)
ZINC	~35 million	Ready to dock, purchasable compounds	http://zinc.docking.org/	Irwin *et al.* (2012)
ChemSpider	~25 million	Data linked to 300 diverse sources	http://www.chemspider.com/	Williams (2010)
ChEMBL	~2 million	Bioactive compounds with drug target information	https://www.ebi.ac.uk/chembl/	Bento *et al.* (2014)
CCDC	~900,000	Crystal structures of small molecules	https://www.ccdc.cam.ac.uk/	Groom *et al.* (2016)
BindingDB	621,060	Binding affinities of drug-targets	www.bindingdb.org/bind/index.jsp	Gilson *et al.* (2016)
NCI	265,242	Anticancer and anti-HIV screening data	https://cactus.nci.nih.gov/download/nci/index.html	Ihlenfeldt *et al.* (2002)
HMDB	74,507	Metabolites in human body	http://www.hmdb.ca/	Wishart *et al.* (2013)
DrugBank	9,591	Drugs and drug like compounds	https://www.drugbank.ca/	Law *et al.* (2014)

parameters such as molecular weight, number of hydrogen bond acceptors, number of hydrogen bond donors, number of rotatable bonds and log P to increase the bioavailability of the remaining compounds (Doak *et al.*, 2014; Lipinski, 2000). Ligands should also be checked for proper valence states and geometry, including reasonable bond distances and angles. Energy minimization and protonation of ligands should also be performed to achieve correct ligand-protein binding (ten Brink and Exner, 2009). **Table 2** list the major organizations which maintain the public repositories of chemical compounds from where 2D and 3D structures can be obtained for molecular docking.

Molecular docking

Molecular docking is a method which predicts the preferred relative orientation of one molecule (key) when bound in an active site of another molecule (lock) to form a stable complex such that free energy of the overall system is minimized. It exploits the concept of molecular shape and physicochemical complementarity. The structures interact like a hand in a glove, where both shape and physicochemical properties contribute to the fit. Molecular docking process can be separated into two major steps i) searching and ii) scoring.

Search algorithm

The search algorithm implemented in any docking tool should enumerate an optimum number of ways two molecules can be put together. The size of the search space grows exponentially with the increase in the size of the molecules. For example, the number of possible conformations for a small molecule with 10 rotatable bonds with 30 degrees of increments are 10^{12}. If the target protein is also allowed to be flexible, this quickly becomes an intractable problem. Docking applications frequently employ one or more of the following search algorithms; Simulated Annealing, Fast shape matching, Incremental construction, Particle Swarm optimization and Evolutionary algorithms (Heberlé and de Azevedo, 2011).

Scoring functions

The scoring function measures and ranks the binding of a ligand-receptor complex. It is therefore not only important to have an efficient, accurate scoring function that gives the best rank to true bound structures, but it must give the correct relative rank of each ligand in the database. Ideally, the score should directly correspond to the binding affinity of the ligand for the protein so that the top scoring compounds are also the best binders. There are three main types of scoring functions that are used by various docking programs; 1) Force field based – GLODscore, DOCK, AutoDOCK derived from AMBER and CHARMm; 2) Empirical scoring – PLP, CHemscore, Glide SP/XP and PLANTSchemplp and PLANTSplp and 3) Knowledge-based – PMF, DrugScore and its derivates DrugScoreCSD, Astex statistical potential.

For a detailed review of molecular docking, the reader can refer to the following publications (Halperin *et al.*, 2002; Meng *et al.*, 2011; Yuriev and Ramsland, 2013). A substantial number of docking tools are now available which have been applied for discovery of novel bioactive molecules. **Table 3** summarizes basic features, license terms, algorithms and scoring functions of currently available docking software. The list is not exhaustive as many docking tools have become available in recent years. Selecting a particular docking tool is always a challenge. In one study it was reported that AutoDock offered a combination of accuracy and speed as opposed to other methods for prediction of protein kinase inhibitors (Buzko *et al.*, 2002). Kellenberger *et al.* (2004) when compared the performance of eight docking programs to recover the X-ray poses of 100 ligands, they reported that GLIDE, GOLD and SURFLEX had the best docking accuracy. Therefore, choosing a docking program largely depend upon the protein family under consideration.

Table 3   Currently available docking tools

Docking tool	License terms	Short description	URL	Reference
Autodock	Freeware	Genetic algorithm, Lamarckian genetic algorithm and Simulated annealing	http://autodock.scripps.edu/	Forli *et al.* (2016)
PatchDock	Freeware	Rigid docking	https://bioinfo3d.cs.tau.ac.il/PatchDock/	Schneidman-Duhovny *et al.* (2005)
GEMDOCK	Freeware	Generic evolutionary method for docking	http://gemdock.life.nctu.edu.tw/dock/	Yang and Chen (2004)
Autodock VINA	Open source	New generation of Autodock	http://vina.scripps.edu/manual.html	Trott and Olson (2010)
rDOCK	Open source	For HTVS of small molecules against proteins and nucleic acids	http://rdock.sourceforge.net/	Ruiz-Carmona *et al.* (2014)
PLANTS	Free for academic use	Stochastic optimization algorithms	http://www.uni-tuebingen.de/	Korb *et al.* (2009)
DOCK	Free for academic use	Geometric matching algorithm, docks either small molecules or fragments and include solvent effect	http://dock.compbio.ucsf.edu/	Allen *et al.* (2015)
FRED	Free for academic	Exhaustive examination of all possible poses of small molecules	https://www.eyesopen.com/oedocking	McGann (2011)
HADDOCK	Free for academic	Protein-protein docking	http://www.bonvinlab.org/software/haddock2.2/	Dominguez *et al.* (2003)
ICM	Commercial	Pseudo brownian sampling and Monte Carlo minimization for protein ligand docking	https://www.molsoft.com/docking.html	Neves *et al.* (2012)
GLIDE	Commercial	Exhaustive search and two different scoring function to rank-order compounds	https://www.schrodinger.com/glide	Repasky *et al.* (2007)
GOLD	Commercial	Genetic algorithm, flexible ligand, partial flexibility of proteins	https://www.ccdc.cam.ac.uk/solutions/csd-discovery/components/gold/	Verdonk *et al.* (2003)
FlexX	Commercial	Incremental construction algorithm for protein ligand docking	https://www.biosolveit.de/FlexX/	Kramer *et al.* (1999)

The individual docking procedure associated with different docking tools may differ, for example, formats of the ligands and target files, scoring functions and algorithms for ligand placement or the ligand may be docked in entirety or in fragments. However, the general principles of docking remain similar; compounds are first placed inside the binding pocket using the algorithm implemented by the docking software and then evaluated for non-covalent interactions using a scoring algorithm available in the same package. If the position of the ligand is known *a priori*, this information can be exploited during and after the docking process. Some programs such as DOCK allow an anchor fragment, ligand placement can thus be guided during the docking process.

Post-docking evaluation and analysis

The output of docking and scoring is a ranked list of predicted bound ligands. Using computer graphics software, these can be visually evaluated for goodness of fit, the formation of key hydrogen bonds, electrostatic interactions, surface complementarities and stability of the bound conformation compared to free conformation. Compounds that get selected in the initial run are then subjected to further *in silico* optimization using focused libraries of potential ligands. Furthermore, the 3D structure of the complex comprising the target protein and the compound is often determined experimentally in order to validate the binding mode predicted by the docking software. Eventually, several high scoring ligands are purchased or synthesized and evaluated for binding and biological activity in the wet lab.

Molecular dynamics and binding free energy calculation

Molecular dynamics (MD) is a technique where the time evolution of a set of interacting atoms is followed by numerical integration of Newton's equation of motion (Adcock and McCammon, 2006). The field of MD simulations is rapidly progressing with improvement in simulation methodology and increasing accuracy of bio-molecular force fields (Hansson *et al.*, 2002). Experiments that are difficult or even impossible to perform in the wet lab can be simulated using MD. From a drug discovery perspective, MD simulation can be used to study the stability of docked complexes and to ensure that molecular docking has yielded accurate results. The stability of the docked complex is often estimated using the root mean square deviation (RMSD) and root mean square fluctuation (RMSF) during the period of the simulations (Kuzmanic and Zagrovic, 2010). RMSF values are calculated to study the thermal stability and structural flexibility (Kuzmanic and Zagrovic, 2010) while the RMSD value measures the changes in the structure during simulation. Changes in the order of 1–3 Å are considered acceptable.

Limitations of molecular docking

Docking protocols are the combination of search algorithms and scoring functions. The scoring function is a mathematical construct that is used to calculate the strength of non-covalent interactions or binding affinity (Halperin et al., 2002). Some docking experiments fail due to the inability of docking methods to account for conformational changes that occur during the binding process of protein and ligand while searching the potential binding pose of the ligand in the target cavity. Predicting target receptor structural rearrangements during ligand binding is a complex problem. Unfortunately, docking tools have limited ability to follow the exact modeling of flexibility available to the protein during the binding process (Teodoro and Kavraki, 2003). This problem can be solved by MD simulations although it needs to be remembered that MD simulations are computationally expensive.

Studies comparing different docking tools on a large test set, reported 30% to 80% success rate (Cross et al., 2009; Lape et al., 2010). Modifying basic parameters in the docking software can drastically affect the docking and virtual screening results, indicating that expert knowledge is critical for optimizing the accuracy of docking predictions. Particular docking tool works best for specific targets, it is possible to use multiple docking tools and scoring functions in order to enhance accuracy (Charifson et al., 1999). A universal docking tool (algorithm and scoring function) is not available at this time.

Illustrative Example or Case Study

Plasmodium Falciparum Plasmepsin-II Case Study

Malaria is a widespread problem estimated to kill about two million people each year (Fidock, 2010). Malaria is caused by a protozoan parasite which belongs to the genus *Plasmodium*. Despite considerable progress, researchers are yet to develop a vaccine to prevent malaria. Meanwhile, the emergence of anti-malarial drug resistance has become a major problem, particularly in Africa and south-east Asia (Murray et al., 2012). Development of resistance to artemisinin derivatives and the majority of the potential partner drugs has contributed to the failures of Artemisinin-based combination treatment (ACT), recommended by World Health Organization as the current cure for malaria (Ariey et al., 2014). Apart from drug resistance, other current drug issues such as poor efficacy, toxicity and high production cost also hamper malaria control.

Several targets play a vital role in the survival of *P. falciparum* in human blood stages. Hemoglobin (Hb) metabolism is a critical process during the parasite's life cycle as it provides the main source of amino acids for the growth and maturation of the parasite (Francis et al., 1997). Plasmepsins (PM) are a class of enzymes that play a key role in the degradation of host hemoglobin and have been validated as potential malaria drug targets (Ersmark et al., 2006). The active site cleft of PM II is different from its human ortholog, therefore, selective inhibition is possible. Moreover, PM II appears to be indispensable for parasite survival. It has been shown that PM inhibitors induce malaria parasite death both in culture and in animal models (Ersmark et al., 2006). To date, several peptidic and non-peptidic inhibitors of PM II have been identified including molecules with the statin-based core (**Fig. 2**). However, most of these exhibits low bioavailability and are difficult to synthesize. The availability of the X-ray crystal structure of PM II makes these enzymes a useful choice for illustrating the process of structure-based drug design to discover novel anti-malarial compounds.

Three different ligand binding modes of *P. falciparum* PM II viz. 'close', 'partially open' and 'open' were described by Luksch et al. in 2008 based on the extent of coverage of the flap region on the binding pocket (Luksch et al., 2008) (**Fig. 3**).

The flap region in the first group folds on the inhibitors and puts the binding pocket in the closed state (e.g., 1XE5, 1M43, 1SME). In the second group, the binding pocket is in partially open conformation (e.g., 1LEE, 1LF2, 1LF3) whereas in the third group the binding pocket is wide open due to the movement of flap lid away from the pocket (e.g., 2BJU, 2IGX). However, since then several new structures have become available in PDB which prompted us to re-look at the data. To classify the new structures, RMSD values from the structural alignment of all the available PM II inhibitor were calculated, which was later used to build a dendrogram for clustering new structures (**Fig. 4**).

The RMSD dendrogram confirms the three ligand-binding modes of PM II and classifies all the available PM II structures into the close, open or partially open groups. Out of all the structures available in open conformation, a high-resolution crystal structure of 2BJU was selected to demonstrate the principle of SBDD.

Materials and Methods

UCSF DOCK v6.7 (also referred to as DOCK) (Allen et al., 2015) and AutoDock Vina (also referred to as VINA) (Trott and Olson, 2010) docking programs were used in this study. UCSF Chimera (Pettersen et al., 2004) was used to visualize and prepare the receptor structure. A molecular format conversion program called OpenBabel (O'Boyle et al., 2011) and ChemAxon tools

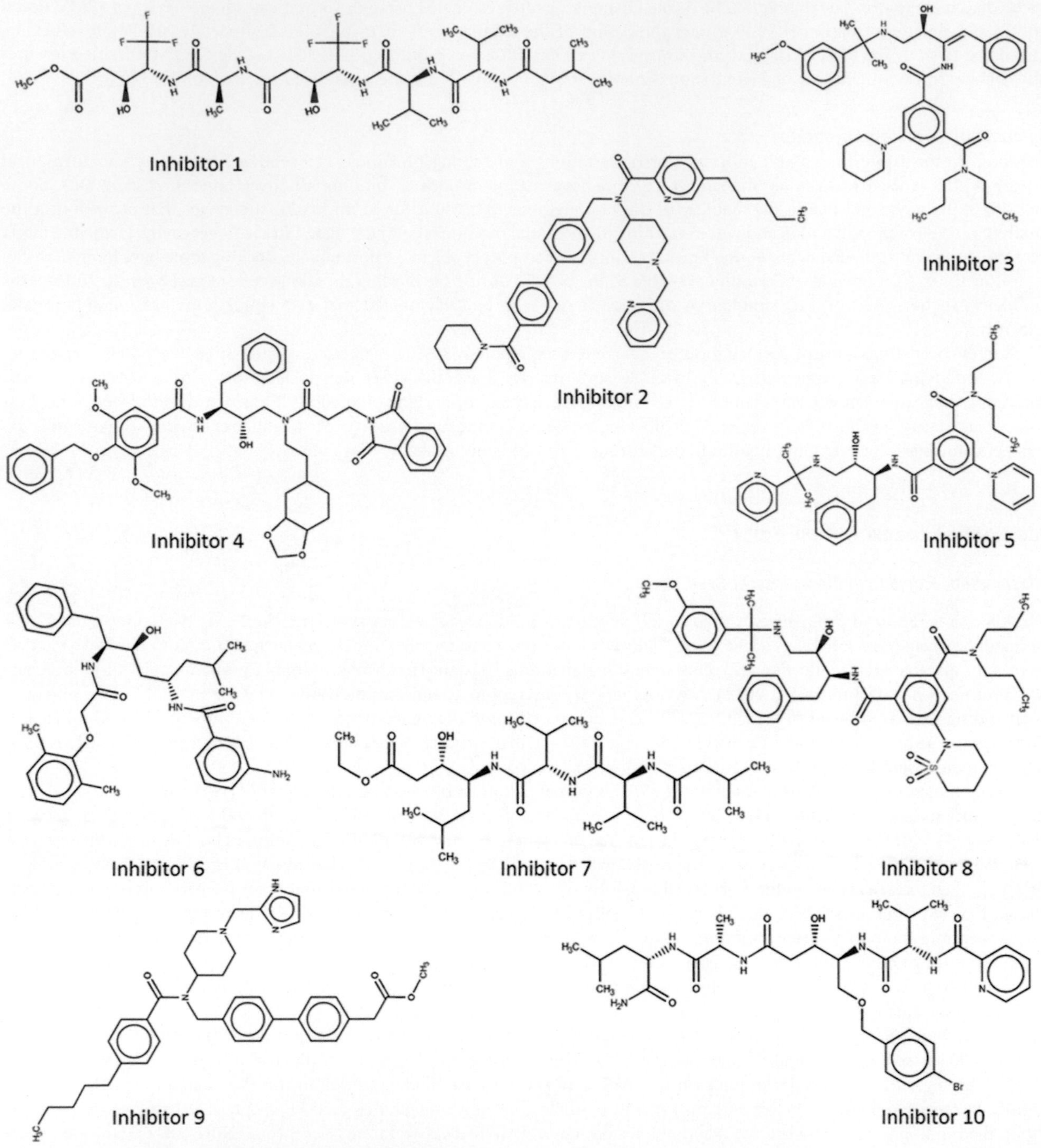

Fig. 2 Known inhibitors of *Plasmodium falciparum* Plasmepsin II.

(ChemAxon – Software for Chemistry and Biology, 2017) were used to prepare ligands for virtual screening and docking. The preparation of receptor files and the determination of the grid box were carried out using AutoDock Tools version 1.5.4.

Receptor preparation

The crystal structure of *P. falciparum* Plasmepsin, 2BJU, in complex with a potent inhibitor IH4 (IC50: 34 nM) was retrieved from PDB. The bound water molecules were removed. The incomplete side chains were replaced using Dunbrack rotamer library (Dunbrack and Karplus, 1993) and the charges on the ionizable residues were made consistent with the acidic conditions found in food vacuoles (pH 5). Subsequently, Dock Prep tool in UCSF Chimera was used to complete the initial receptor preparation.

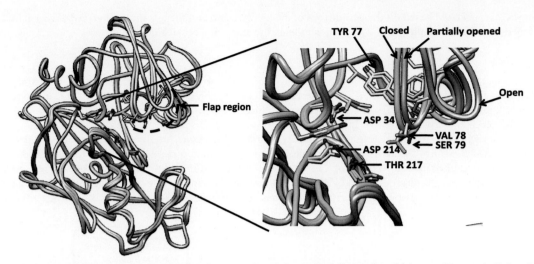

Fig. 3 Three different ligand binding conformations (closed, open and partially open) of *Plasmodium falciparum* Plasmepsin II described in the literature.

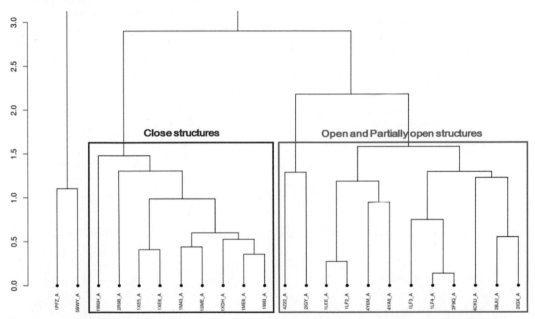

Fig. 4 Dendrogram based on RMSD calculated by structurally aligning all available *Plasmodium falciparum* Plasmepsin II structures in PDB. Structures belonging to closed and open or partially open conformations of Plasmepsin II clearly separate in distinct groups.

UCSF DOCK

After the initial preparation of the receptor, DOCK specific steps were implemented prior to virtual screening as described in DOCK manual. The first step was to calculate the solvent accessible surface area of the receptor without hydrogen atoms using a probe radius of 1.4 Å. The negative image of the surface was created by overlapping spheres using SPHGEN program. The final cluster used for docking had 44 spheres. The binding site was defined comprising all the atoms within the area of 8 Å of the co-crystallized ligand. The definition of the active site covered the catalytic dyad ASP 34 and ASP 214, along with other residues of known relevance. Subsequently, the box was constructed around the active site using SHOWBOX program and the grids were calculated using an accessory program called GRID using default parameters to complete receptor preparation.

AutoDock VINA

For VINA, the receptor PDB files were converted to PDBQT format using prepare_receptor4.py in MGL Tools (version 1.5.4). The default box size was calculated following the protocol mentioned by the authors of VINA (Trott and Olson, 2010). Briefly, an initial docking box is generated based on the coordinates of the native ligand and the box dimensions are increased by 10 Å. If the box size in any dimension is smaller than 22.5 Å, it is extended to this value on all sides.

Library preparation

Around 4.5 million compounds were downloaded from the "lead-like" subset of the ZINC database (Irwin *et al.*, 2012). The dataset was cleaned by applying a number of filters to remove duplicate compounds, incorrect valence compounds and compounds violating Rule of five (Ro5). This step removed over a million compounds. The overall methodology followed for library preparation and molecular docking is shown in **Fig. 5**.

Next, compounds with sub-structural features commonly found in Pan Assay Interference Compounds (PAINS) were also filtered out using SMARTS strings inspired from the original Sybyl Line Notation patterns provided in the reference (Baell and Holloway, 2010). This resulted in the removal of 111,305 compounds. Additionally, the dataset was screened for potentially reactive and promiscuous compounds described by a set of 275 medicinal chemistry rules, developed at Lilly over a period of 18 years (Bruns and Watson, 2012). Reasons for rejection include reactivity, interference with assay measurements, instability, lack of

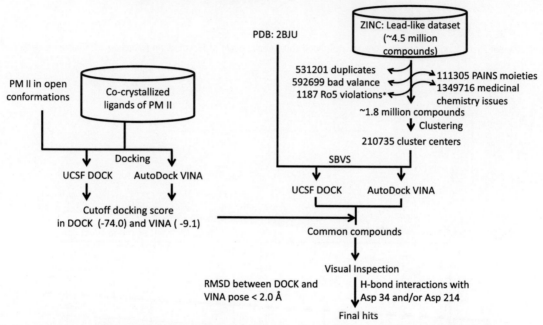

Fig. 5   The overall virtual screening methodology adopted to find novel drug leads against *Plasmodium falciparum* Plasmepsin II.

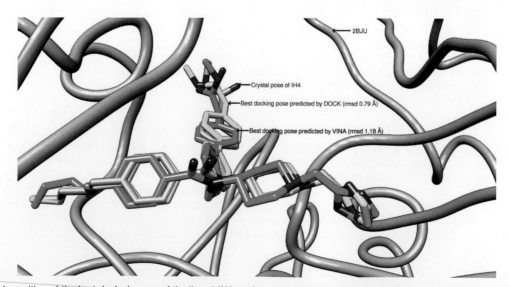

Fig. 6   Superimposition of the best docked poses of the ligand IH4 on the co-crystallized pose in the binding pocket of 2BJU. The DOCK pose (in yellow color) and the VINA pose (in cyan color) has the positional RMSD of 0.79 Å and 1.18 Å, respectively.

drug-ability and activities that damage proteins, among others. A total of 1,349,716 compounds were removed and leaving 1,864,623 filtered compounds. To ensure chemical diversity, the approximately 1.8 million remaining molecules were clustered using PubChem fingerprints as implemented in the ChemmineR package (Cao *et al.*, 2008). The maximum Tanimoto distance between a compound and the cluster center was set at 0.3. This resulted in the multi-molecule file containing 210,735 molecules which were first protonated for the acidic pH 5 and then minimized using the Ghemical force field. Subsequently, the minimized structures were used as the input to VINA and DOCK for molecular docking analysis.

Results and Discussions

Validation of the Docking Methodology

To access the suitability of the prepared target structure for structure-based drug discovery, a preliminary validation of the docking protocol was carried out by re-docking the IH4 ligand to the prepared 2BJU structure with VINA and DOCK programs. The 3D structure of the crystal ligand was obtained by removing it from protein crystallographic complex. The crystal ligand was docked to the receptor protein and from the list of conformations the pose in which the ligand bound with the maximum number of vander Waal and electrostatic interactions and lowest binding energy was chosen as the top ranked binding pose. The experimental pose of IH4 was reproduced. **Fig. 6** shows the superimposition of the docked ligand (IH4) with the crystal structure position. The structure is seen to overlap well with the positional RMSD of 1.18 Å in VINA and 0.79 Å in DOCK.

As both VINA and DOCK softwares were able to reproduce the bound conformation of the co-crystallized ligand, therefore, we considered them suitable for further analysis. Overall, VINA produced better results than DOCK in ranking co-crystallized ligands in open and partially open conformations of PM II (**Table 4**). However, DOCK was superior in ranking 2BJU and 2IGX ligands (both in open conformation) with positional RMSD of 0.7 Å and 0.8 Å respectively, as compared to VINA with positional RMSD of 1.3 Å and 1.1 Å.

Table 4   Comparison of docking scores and rmsd values calculated by redocking co-crystallized ligands of Plasmepsin II in open and partially open conformations

PDB ID	Ligand ID	IC50	DOCK score	DOCK rmsd	VINA score	VINA rmsd	No. of rotatable bonds
Open conformation							
2BJU	1H4	34 nM	− 87.23	0.7 Å	− 10.8	1.3 Å	14
2IGX	A1T	54 nM	− 90.28	0.8 Å	− 11.5	1.1 Å	14
2IGY	A2T	113 nM	− 74.63	2.4 Å	− 9.1	1.3 Å	14
4Z22	4KG	340 nM	− 42.13	10.4 Å	− 8.2	5.9 Å	5
Partially open conformation							
1LEE	R36	18 nM	− 60.35	3.8 Å	− 9.9	1.4 Å	15
1LF2	R37	30 nM	− 58.52	2.4 Å	− 9.3	1.4 Å	17
4Y6M	48Q	70 nM	− 71.89	1.7 Å	− 8.2	1.3 Å	17
4YA8	49W	80 nM	− 76.67	1.0 Å	− 8.5	1.7 Å	16
1LF3	EH5	100 nM	− 74.54	2.6 Å	− 8.5	9.9 Å	19
4CKU	P2F	150 nM	− 82.39	1.6 Å	− 8.5	0.6 Å	17

Table 5   Novel predicted PM II lead compounds with DOCK score, VINA score, hydrogen bond interaction residues and vendor information

S. no.	ID	DOCK score	VINA score	Hbond residues	Vendors
1	ZINC09467542	− 74.57	− 10.2	SER 37, ASP 214, THR 217	eMolecules: 357481 Enamine: Z734242026
2	ZINC89512612	− 75.11	− 9.5	ASP 34, ASP 214, THR 217, GLY 36	NA
3	ZINC97236464	− 94.72	− 9.3	ASP 34, ASP 214, GLY 36	Molport: 030–015–561 Enamine: Z1631599836
4	ZINC72478028	− 75.84	− 9.3	ASP 214	eMolecules: 43171874 ChemBridge: 89912227
5	ZINC7189459	− 79.91	− 9.2	ASP 34, ASP 214	NA
6	ZINC67868828	− 84.67	− 9.1	TYR 77, ASP 214, THR 217	eMolecule: 36456659 ChemBridge: 78484059
7	ZINC25344579	− 79.04	− 9.1	ASP 34, ASP 214	Ambinter: Amb14002467 Enamine: Z339889742
8	ZINC54837115	− 81.18	− 9.1	ASP 34, ASP 214	eMolecules: 32048500 Enamine: Z734242026

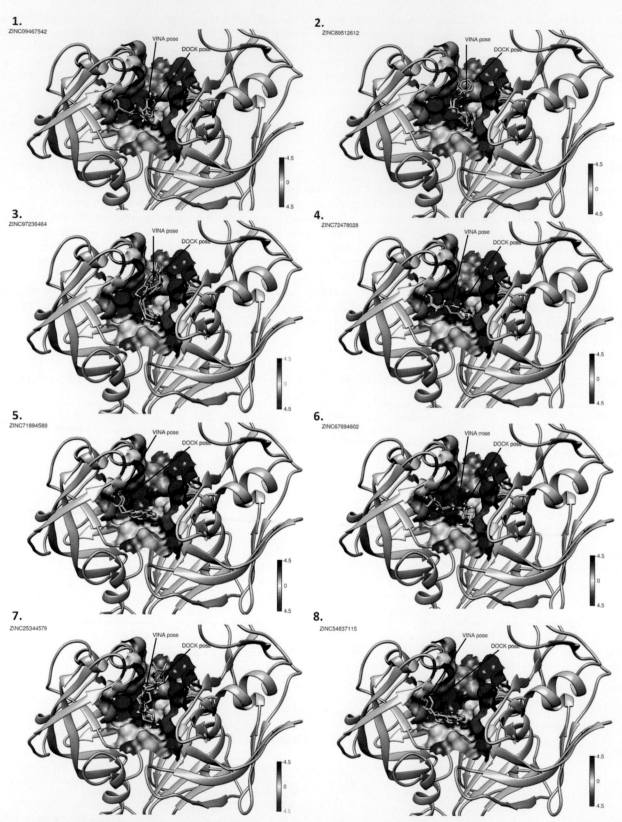

Fig. 7 Structure of eight novel potential Plasmepsin II inhibitors lead molecules. The predicted binding modes of the molecules by DOCK (cyan) and VINA (pink) in the cavity of Plasmepsin II are depicted.

Structure-Based Virtual Screening

In this study, two different docking programs were used to predict novel ligands with high binding affinity to Plasmepsin II. The final library of 210,735 compounds was screened and compounds were ranked based on the docking score to select molecules that had a score better than the cut-off score of -74.0 for DOCK and -9.1 for VINA. This resulted in a set of 823 and 1525 docked compounds from DOCK and VINA, respectively. Compounds that did not form hydrogen bonds with catalytic residues Asp34 and Asp214 were filtered out form further analysis. Subsequently, 25 common compounds were retained and visualized for binding conformations. Finally, compounds with greater than 2.0 Å RMSD between binding conformations predicted by DOCK and VINA were removed resulting in the final set of 8 compounds which are listed in **Table 5** and shown in **Fig. 7**.

Since binding of a ligand produces a conformational change in the protein, we proposed that compounds binding in a similar fashion to the co-crystallized complex (ligand IH4 bound to 2BJU) will have a high probability to inhibit PM II *in vivo*.

Future Directions

Structure-based drug design is a powerful method, especially when combined with combinatorial chemistry and high throughput screening. Where access to large compounds libraries is limited or where HTS fails to produce a viable ligand, structure-based virtual screening continues to gain credibility as an alternative source of initial lead identification. The major developments that have facilitated the rise of SBDD to a place of prominence in drug development are dramatic increase in computational speed and efficiency of docking algorithms, ability to incorporate receptor flexibility information during docking, easily available structural data of proteins and compounds, better understanding of interaction of small molecules in the binding site and more accurate methods for estimating protein-ligand binding free energy. The application of computer-aided drug discovery approaches has been extended in both directions of the drug development pipeline, upstream for target identification and validation and downstream for ADME/T predictions.

In the modern genomic era, there has been a dramatic increase in biological data from gene sequences to structural data of proteins and small molecules. The number of potential new drug targets identified has increased more rapidly than the discovery of new drugs. One of the major challenges is to bring the wealth of structural and bioassay data together in an integrated way to design novel, improved compounds that can be readily synthesized.

Currently, there are over 60 docking tools available, evaluation and comparison of different docking programs and scoring functions is an active area of research. Further, protein flexibility during docking has only been addressed recently and the development of computational methods for protein flexibility is still in its infancy and is thereby one of the major future directions (Lill, 2011).

Closing Remarks

This article should provide the reader with an overall perspective of the field of structure-based drug discovery. We have listed and summarized several cheminformatics tools and databases with a special emphasis on software packages that are freely available. We have also outlined the major steps involved in the process of rational structure-based drug design, namely target identification, structure determination, binding site prediction, receptor and compound library preparation, molecular docking, virtual screening and binding free energy calculation. Subsequently, we have applied different SDBB approaches to predict novel leads for hemoglobin degrading enzyme of *Plasmodium falciparum viz.* Plasmepsin II. In addition, we 1) compared the performance of two different docking tools on the Plasmepsin II inhibitor dataset and 2) demonstrated the use of consensus docking methodology for virtual screening of novel PM II inhibitors.

In conclusion, computer-aided drug discovery approaches like ligand- or structure-based pharmacophore modeling and VS can successfully complement their wet-lab counterparts to expedite the drug discovery process. Although computer-aided approaches hold a great promise, it is an evolving technology and still has some shortcomings that should be remembered. The major challenges in the field are 1) to determine the structure of the target, 2) identify suitable hits, 3) progress rapidly to a molecule with the desired activity that does not fail during development and 4) ability to design novel patentable chemical compounds that are easy to synthesize. Finally, *in vitro* testing of the discovered lead molecules is necessary for further optimization of the leads.

See also: Algorithms for Structure Comparison and Analysis: Docking. Biomolecular Structures: Prediction, Identification and Analyses. Chemical Similarity and Substructure Searches. Chemoinformatics: From Chemical Art to Chemistry in Silico. Comparative Epigenomics. Computational Protein Engineering Approaches for Effective Design of New Molecules. Drug Repurposing and Multi-Target Therapies. Fingerprints and Pharmacophores. In Silico Identification of Novel Inhibitors. Molecular Dynamics Simulations in Drug Discovery. Natural Language Processing Approaches in Bioinformatics. Pharmacophore Development. Protein Structural Bioinformatics: An Overview. Protein Three-Dimensional Structure Prediction. Public Chemical Databases. Quantitative Structure-Activity Relationship (QSAR): Modeling Approaches to Biological Applications. Small Molecule Drug Design. Structure-Based Design of Peptide Inhibitors for Protein Arginine Deiminase Type IV (PAD4). Structure-Based Drug Design Workflow

References

Adcock, S.A., McCammon, J.A., 2006. Molecular dynamics: Survey of methods for simulating the activity of proteins. Chem. Rev. 106, 1589–1615. Available at: https://doi.org/10.1021/cr040426m.

Allen, W.J., Balius, T.E., Mukherjee, S., et al., 2015. DOCK 6: Impact of new features and current docking performance. J. Comput. Chem. 36, 1132–1156. Available at: https://doi.org/10.1002/jcc.23905.

Anfinsen, C.B., 1973. Principles that govern the folding of protein chains. Science 181, 223–230.

Ariey, F., Witkowski, B., Amaratunga, C., et al., 2014. A molecular marker of artemisinin-resistant Plasmodium falciparum malaria. Nature 505, 50–55. Available at: https://doi.org/10.1038/nature12876.

Arnold, K., Kiefer, F., Kopp, J., et al., 2009. The protein model portal. J. Struct. Funct. Genomics 10, 1–8. Available at: https://doi.org/10.1007/s10969-008-9048-5.

Baell, J.B., Holloway, G.A., 2010. New substructure filters for removal of pan assay interference compounds (PAINS) from screening libraries and for their exclusion in bioassays. J. Med. Chem. 53, 2719–2740. Available at: https://doi.org/10.1021/jm901137j.

Bento, A.P., Gaulton, A., Hersey, A., et al., 2014. The ChEMBL bioactivity database: An update. Nucleic Acids Res. 42, D1083–D1090. Available at: https://doi.org/10.1093/nar/gkt1031.

Bruns, R.F., Watson, I.A., 2012. Rules for identifying potentially reactive or promiscuous compounds. J. Med. Chem. 55, 9763–9772. Available at: https://doi.org/10.1021/jm301008n.

Buzko, O.V., Bishop, A.C., Shokat, K.M., 2002. Modified AutoDock for accurate docking of protein kinase inhibitors. J. Comput. Aided Mol. Des. 16, 113–127.

Cao, Y., Charisi, A., Cheng, L.-C., Jiang, T., Girke, T., 2008. ChemmineR: A compound mining framework for R. Bioinformatics 24, 1733–1734. Available at: https://doi.org/10.1093/bioinformatics/btn307.

Capra, J.A., Laskowski, R.A., Thornton, J.M., Singh, M., Funkhouser, T.A., 2009. Predicting protein ligand binding sites by combining evolutionary sequence conservation and 3D structure. PLOS Comput. Biol. 5, e1000585. Available at: https://doi.org/10.1371/journal.pcbi.1000585.

Charifson, P.S., Corkery, J.J., Murcko, M.A., Walters, W.P., 1999. Consensus scoring: A method for obtaining improved hit rates from docking databases of three-dimensional structures into proteins. J. Med. Chem. 42, 5100–5109. Available at: https://doi.org/10.1021/jm990352k.

ChemAxon – Software for Chemistry and Biology, 2017. (WWW Document). ChemAxon – Softw. Chem. Biol. RD. Available at: https://www.chemaxon.com/ (accessed 09.11.17).

Chothia, C., Lesk, A.M., 1986. The relation between the divergence of sequence and structure in proteins. EMBO J. 5, 823–826.

Cross, J.B., Thompson, D.C., Rai, B.K., et al., 2009. Comparison of several molecular docking programs: Pose prediction and virtual screening accuracy. J. Chem. Inf. Model. 49, 1455–1474. Available at: https://doi.org/10.1021/ci900056c.

Doak, B.C., Over, B., Giordanetto, F., Kihlberg, J., 2014. Oral druggable space beyond the rule of 5: Insights from drugs and clinical candidates. Chem. Biol. 21, 1115–1142. Available at: https://doi.org/10.1016/j.chembiol.2014.08.013.

Dominguez, C., Boelens, R., Bonvin, A.M.J.J., 2003. HADDOCK: A protein-protein docking approach based on biochemical or biophysical information. J. Am. Chem. Soc. 125, 1731–1737. doi:10.1021/ja026939x.

Dunbrack, R.L., Karplus, M., 1993. Backbone-dependent rotamer library for proteins. Application to side-chain prediction. J. Mol. Biol. 230, 543–574. Available at: https://doi.org/10.1006/jmbi.1993.1170.

Ersmark, K., Samuelsson, B., Hallberg, A., 2006. Plasmepsins as potential targets for new antimalarial therapy. Med. Res. Rev. 26, 626–666. Available at: https://doi.org/10.1002/med.20082.

Eswar, N., Webb, B., Marti-Renom, M.A., et al., 2006. Comparative Protein Structure Modeling Using Modeller. Curr. Protoc. Bioinforma. Ed. Board Andreas Baxevanis Al 0 5, Unit-5.6. Available at: https://doi.org/10.1002/0471250953.bi0506s15.

Fidock, D.A., 2010. Drug discovery: Priming the antimalarial pipeline. Nature 465, 297–298. Available at: https://doi.org/10.1038/465297a.

Fiser, A., Sali, A., 2003. Modeller: Generation and refinement of homology-based protein structure models. Methods Enzymol. 374, 461–491. Available at: https://doi.org/10.1016/S0076-6879(03)74020-8.

Forli, S., Huey, R., Pique, M.E., Sanner, M.F., Goodsell, D.S., Olson, A.J., 2016. Computational protein-ligand docking and virtual drug screening with the AutoDock suite. Nat. Protoc. 11, 905–919. doi:10.1038/nprot.2016.051.

Forrest, L.R., Tang, C.L., Honig, B., 2006. On the accuracy of homology modeling and sequence alignment methods applied to membrane proteins. Biophys. J. 91, 508–517. Available at: https://doi.org/10.1529/biophysj.106.082313.

Francis, S.E., Sullivan, D.J., Goldberg, D.E., 1997. Hemoglobin metabolism in the malaria parasite Plasmodium falciparum. Annu. Rev. Microbiol. 51, 97–123. Available at: https://doi.org/10.1146/annurev.micro.51.1.97.

Ghersi, D., Sanchez, R., 2009. EasyMIFs and SiteHound: A toolkit for the identification of ligand-binding sites in protein structures. Bioinformatics 25, 3185–3186. Available at: https://doi.org/10.1093/bioinformatics/btp562.

Gilson, M.K., Liu, T., Baitaluk, M., et al., 2016. BindingDB in 2015: A public database for medicinal chemistry, computational chemistry and systems pharmacology. Nucleic Acids Res. 44, D1045–D1053. Available at: https://doi.org/10.1093/nar/gkv1072.

Grünberg, R., Nilges, M., Leckner, J., 2007. Biskit—a software platform for structural bioinformatics. Bioinforma. Oxf. Engl. 23, 769–770. doi:10.1093/bioinformatics/btl655.

Groom, C.R., Bruno, I.J., Lightfoot, M.P., Ward, S.C., 2016. The Cambridge Structural Database. Acta Crystallogr. Sect. B Struct. Sci. Cryst. Eng. Mater. 72, 171–179. doi:10.1107/S2052520616003954.

Gustafsson, D., Nyström, J., Carlsson, S., et al., 2001. The direct thrombin inhibitor melagatran and its oral prodrug H 376/95: Intestinal absorption properties, biochemical and pharmacodynamic effects. Thromb. Res. 101, 171–181.

Halperin, I., Ma, B., Wolfson, H., Nussinov, R., 2002. Principles of docking: An overview of search algorithms and a guide to scoring functions. Proteins 47, 409–443. Available at: https://doi.org/10.1002/prot.10115.

Hansson, T., Oostenbrink, C., van Gunsteren, W., 2002. Molecular dynamics simulations. Curr. Opin. Struct. Biol. 12, 190–196.

Heberlé, G., de Azevedo, W.F., 2011. Bio-inspired algorithms applied to molecular docking simulations. Curr. Med. Chem. 18, 1339–1352.

Hendlich, M., Rippmann, F., Barnickel, G., 1997. LIGSITE: Automatic and efficient detection of potential small molecule-binding sites in proteins. J. Mol. Graph. Model. 15, 359–363. (389).

Henrich, S., Salo-Ahen, O.M.H., Huang, B., et al., 2010. Computational approaches to identifying and characterizing protein binding sites for ligand design. J. Mol. Recognit. JMR 23, 209–219. Available at: https://doi.org/10.1002/jmr.984.

Huang, B., 2009. MetaPocket: A meta approach to improve protein ligand binding site prediction. OMICS J. Integr. Biol. 13, 325–330. Available at: https://doi.org/10.1089/omi.2009.0045.

Ihlenfeldt, W.-D., Voigt, J.H., Bienfait, B., Oellien, F., Nicklaus, M.C., 2002. Enhanced CACTVS Browser of the Open NCI Database. J. Chem. Inf. Comput. Sci. 42, 46–57. doi:10.1021/ci010056s.

Huang, B., Schroeder, M., 2006. LIGSITEcsc: Predicting ligand binding sites using the Connolly surface and degree of conservation. BMC Struct. Biol. 6, 19. Available at: https://doi.org/10.1186/1472-6807-6-19.

Irwin, J.J., Sterling, T., Mysinger, M.M., Bolstad, E.S., Coleman, R.G., 2012. ZINC: A free tool to discover chemistry for biology. J. Chem. Inf. Model. 52, 1757–1768.

Jayaram, B., Dhingra, P., Mishra, A., Kaushik, R., Mukherjee, G., Singh, A., Shekhar, S., 2014. Bhageerath-H: A homology/ab initio hybrid server for predicting tertiary structures of monomeric soluble proteins. BMC Bioinformatics 15, S7. doi:10.1186/1471-2105-15-S16-S7.

Källberg, M., Wang, H., Wang, S., Peng, J., Wang, Z., Lu, H., Xu, J., 2012. Template-based protein structure modeling using the RaptorX web server. Nat. Protoc. 7, 1511–1522. doi:10.1038/nprot.2012.085.

Kaldor, S.W., Kalish, V.J., Davies, J.F., *et al.*, 1997. Viracept (nelfinavir mesylate, AG1343): A potent, orally bioavailable inhibitor of HIV-1 protease. J. Med. Chem. 40, 3979–3985. Available at: https://doi.org/10.1021/jm9704098.

Kellenberger, E., Rodrigo, J., Muller, P., Rognan, D., 2004. Comparative evaluation of eight docking tools for docking and virtual screening accuracy. Proteins 57, 225–242. Available at: https://doi.org/10.1002/prot.20149.

Kelley, L.A., Mezulis, S., Yates, C.M., Wass, M.N., Sternberg, M.J.E., 2015. The Phyre2 web portal for protein modeling, prediction and analysis. Nat. Protoc. 10, 845–858. doi:10.1038/nprot.2015.053.

Kiefer, F., Arnold, K., Kunzli, M., Bordoli, L., Schwede, T., 2009. The SWISS-MODEL repository and associated resources. Nucleic Acids Res. 37, D387–D392. Available at: https://doi.org/10.1093/nar/gkn750.

Kim, D.E., Chivian, D., Baker, D., 2004. Protein structure prediction and analysis using the Robetta server. Nucleic Acids Res. 32, W526–W531. doi:10.1093/nar/gkh468.

Kim, S., Thiessen, P.A., Bolton, E.E., *et al.*, 2016. PubChem substance and compound databases. Nucleic Acids Res. 44, D1202–D1213. Available at: https://doi.org/10.1093/nar/gkv951.

Korb, O., Stützle, T., Exner, T.E., 2009. Empirical scoring functions for advanced protein-ligand docking with PLANTS. J. Chem. Inf. Model. 49, 84–96. doi:10.1021/ci800298z.

Kramer, B., Rarey, M., Lengauer, T., 1999. Evaluation of the FLEXX incremental construction algorithm for protein-ligand docking. Proteins 37, 228–241.

Kuzmanic, A., Zagrovic, B., 2010. Determination of ensemble-average pairwise root mean-square deviation from experimental B-factors. Biophys. J. 98, 861–871. Available at: https://doi.org/10.1016/j.bpj.2009.11.011.

Lambert, C., Léonard, N., De Bolle, X., Depiereux, E., 2002. ESyPred3D: Prediction of proteins 3D structures. Bioinforma. Oxf. Engl. 18, 1250–1256.

Lape, M., Elam, C., Paula, S., 2010. Comparison of current docking tools for the simulation of inhibitor binding by the transmembrane domain of the sarco/endoplasmic reticulum calcium ATPase. Biophys. Chem. 150, 88–97. Available at: https://doi.org/10.1016/j.bpc.2010.01.011.

Laurie, A.T.R., Jackson, R.M., 2005. Q-SiteFinder: An energy-based method for the prediction of protein-ligand binding sites. Bioinformatics Oxford 21, 1908–1916. Available at: https://doi.org/10.1093/bioinformatics/bti315.

Law, V., Knox, C., Djoumbou, Y., *et al.*, 2014. DrugBank 4.0: Shedding new light on drug metabolism. Nucleic Acids Res. 42, D1091–D1097. Available at: https://doi.org/10.1093/nar/gkt1068.

Le Guilloux, V., Schmidtke, P., Tuffery, P., 2009. Fpocket: An open source platform for ligand pocket detection. BMC Bioinform. 10, 168. Available at: https://doi.org/10.1186/1471-2105-10-168.

Lill, M.A., 2011. Efficient incorporation of protein flexibility and dynamics into molecular docking simulations. Biochemistry (Moscow) 50, 6157–6169. Available at: https://doi.org/10.1021/bi2004558.

Lipinski, C.A., 2000. Drug-like properties and the causes of poor solubility and poor permeability. J. Pharmacol. Toxicol. Methods 44, 235–249. Available at: https://doi.org/10.1016/S1056-8719(00)00107-6.

Locating Binding Sites in Protein Structures, 2017. (WWW Document). Available at: https://www.chemcomp.com/journal/sitefind.htm (accessed 09.09.17).

López-Vallejo, F., Caulfield, T., Martínez-Mayorga, K., *et al.*, 2011. Integrating virtual screening and combinatorial chemistry for accelerated drug discovery. Comb. Chem. High Throughput Screen. 14, 475–487.

Luksch, T., Chan, N.-S., Brass, S., *et al.*, 2008. Computer-aided design and synthesis of nonpeptidic plasmepsin II and IV inhibitors. Chem. Med. Chem. 3, 1323–1336. Available at: https://doi.org/10.1002/cmdc.200700270.

Macarron, R., Banks, M.N., Bojanic, D., *et al.*, 2011. Impact of high-throughput screening in biomedical research. Nat. Rev. Drug Discov. 10, 188–195. Available at: https://doi.org/10.1038/nrd3368.

Martí-Renom, M.A., Stuart, A.C., Fiser, A., *et al.*, 2000. Comparative protein structure modeling of genes and genomes. Annu. Rev. Biophys. Biomol. Struct. 29, 291–325. Available at: https://doi.org/10.1146/annurev.biophys.29.1.291.

McGann, M., 2011. FRED Pose Prediction and Virtual Screening Accuracy. J. Chem. Inf. Model. 51, 578–596. doi:10.1021/ci100436p.

Meng, X.-Y., Zhang, H.-X., Mezei, M., Cui, M., 2011. Molecular docking: A powerful approach for structure-based drug discovery. Curr. Comput. Aided Drug Des. 7, 146–157.

Murray, C.J.L., Rosenfeld, L.C., Lim, S.S., *et al.*, 2012. Global malaria mortality between 1980 and 2010: A systematic analysis. Lancet London 379, 413–431. Available at: https://doi.org/10.1016/S0140-6736(12)60034-8.

Neves, M.A.C., Totrov, M., Abagyan, R., 2012. Docking and scoring with ICM: The benchmarking results and strategies for improvement. J. Comput. Aided Mol. Des. 26, 675–686. doi:10.1007/s10822-012-9547-0.

Nisius, B., Sha, F., Gohlke, H., 2012. Structure-based computational analysis of protein binding sites for function and druggability prediction. J. Biotechnol. 159, 123–134. Available at: https://doi.org/10.1016/j.jbiotec.2011.12.005.

O'Boyle, N.M., Banck, M., James, C.A., *et al.*, 2011. Open Babel: An open chemical toolbox. J. Cheminform. 3, 33. Available at: https://doi.org/10.1186/1758-2946-3-33.

Owens, J., 2007. Target validation: Determining druggability. Nat. Rev. Drug Discov. 6, nrd2275. Available at: https://doi.org/10.1038/nrd2275.

Parasuraman, S., 2012. Protein data bank. J. Pharmacol. Pharmacother. 3, 351–352. Available at: https://doi.org/10.4103/0976-500X.103704.

Pettersen, E.F., Goddard, T.D., Huang, C.C., *et al.*, 2004. UCSF Chimera – A visualization system for exploratory research and analysis. J. Comput. Chem. 25, 1605–1612. Available at: https://doi.org/10.1002/jcc.20084.

Pieper, U., Webb, B.M., Barkan, D.T., *et al.*, 2011. ModBase, a database of annotated comparative protein structure models, and associated resources. Nucleic Acids Res. 39, D465–D474. Available at: https://doi.org/10.1093/nar/gkq1091.

Pollack, V.A., Savage, D.M., Baker, D.A., *et al.*, 1999. Inhibition of epidermal growth factor receptor-associated tyrosine phosphorylation in human carcinomas with CP-358,774: Dynamics of receptor inhibition in situ and antitumor effects in athymic mice. J. Pharmacol. Exp. Ther. 291, 739–748.

Repasky, M.P., Shelley, M., Friesner, R.A., 2007. Flexible ligand docking with Glide. Chapter 8, Unit 8.12.Curr. Protoc. Bioinforma. doi:10.1002/0471250953.bi0812s18.

Roy, A., Kucukural, A., Zhang, Y., 2010. I-TASSER: A unified platform for automated protein structure and function prediction. Nat. Protoc. 5, 725–738. doi:10.1038/nprot.2010.5.

Ruiz-Carmona, S., Alvarez-Garcia, D., Foloppe, N., *et al.*, 2014. rDock: A Fast, Versatile and Open Source Program for Docking Ligands to Proteins and Nucleic Acids. PLoS Comput. Biol. 10. doi:10.1371/journal.pcbi.1003571.

Schindler, T., Bornmann, W., Pellicena, P., *et al.*, 2000. Structural mechanism for STI-571 inhibition of abelson tyrosine kinase. Science 289, 1938–1942.

Schneidman-Duhovny, D., Inbar, Y., Nussinov, R., Wolfson, H.J., 2005. PatchDock and SymmDock: Servers for rigid and symmetric docking. Nucleic Acids Res. 33, W363–W367. doi:10.1093/nar/gki481.

Stahura, F.L., Bajorath, J., 2004. Virtual screening methods that complement HTS. Comb. Chem. High Throughput Screen. 7, 259–269.

ten Brink, T., Exner, T.E., 2009. Influence of protonation, tautomeric, and stereoisomeric states on protein-ligand docking results. J. Chem. Inf. Model. 49, 1535–1546. Available at: https://doi.org/10.1021/ci800420z.

Teodoro, M.L., Kavraki, L.E., 2003. Conformational flexibility models for the receptor in structure based drug design. Curr. Pharm. Des. 9, 1635–1648.

Trott, O., Olson, A.J., 2010. AutoDock Vina: Improving the speed and accuracy of docking with a new scoring function, efficient optimization and multithreading. J. Comput. Chem. 31, 455–461. Available at: https://doi.org/10.1002/jcc.21334.

Varghese, J.N., 1999. Development of neuraminidase inhibitors as anti-influenza virus drugs. Drug Dev. Res. 46, 176–196. Available at: https://doi.org/10.1002/(SICI)1098-2299(199903/04)46:3/4<176:AID-DDR4>3.0.CO;2-6.

Verdonk, M.L., Cole, J.C., Hartshorn, M.J., Murray, C.W., Taylor, R.D., 2003. Improved protein-ligand docking using GOLD. Proteins 52, 609–623. doi:10.1002/prot.10465.

Wang, T., Wu, M.-B., Zhang, R.-H., *et al.*, 2016. Advances in computational structure-based drug design and application in drug discovery. Curr. Top. Med. Chem. 16, 901–916.

Weisel, M., Proschak, E., Schneider, G., 2007. PocketPicker: Analysis of ligand binding-sites with shape descriptors. Chem. Cent. J. 1, 7. Available at: https://doi.org/10.1186/1752-153X-1-7.

Williams, A.J., 2010. ChemSpider: Integrating Structure-Based Resources Distributed across the Internet, in: Enhancing Learning with Online Resources, Social Networking, and Digital Libraries, ACS Symposium Series. American Chemical Society, pp. 23–39. https://doi.org/10.1021/bk-2010-1060.ch002

Wishart, D.S., Jewison, T., Guo, A.C., *et al.*, 2013. HMDB 3.0–The Human Metabolome Database in 2013. Nucleic Acids Res. 41, D801–807. doi:10.1093/nar/gks1065.

Xie, Z.-R., Hwang, M.-J., 2015. Methods for predicting protein-ligand binding sites. Methods Mol. Biol. Clifton 1215, 383–398. Available at: https://doi.org/10.1007/978-1-4939-1465-4_17.

Yang, J.-M., Chen, C.-C., 2004. GEMDOCK: A generic evolutionary method for molecular docking. Proteins 55, 288–304. doi:10.1002/prot.20035.

Yuriev, E., Ramsland, P.A., 2013. Latest developments in molecular docking: 2010–2011 in review. J. Mol. Recognit. JMR 26, 215–239. Available at: https://doi.org/10.1002/jmr.2266.

Biographical Sketch

Varun Khanna is a research scientist at Vaxine Pty Ltd, a biotechnology company involved in vaccine design. He completed his PhD in Cheminformatics from Macquarie University, Sydney, Australia in 2011. During his PhD he analyzed physicochemical properties, occurrence and co-occurrence patterns of molecular fragments in pharmaceutically important chemical compounds for advancing the knowledge of computational library design in drug discovery. He then moved to University of California, Riverside, USA as a post-doc where he was involved in the chemogenomics, data mining project aimed at carrying out a comprehensive selectivity-centric analysis of ligand-target interactions in public activity profile databases such as PubChem Bioassay. His current research interest includes developing novel techniques for computational drug design, immuno-informatics, fragment-based virtual screening, machine learning and reverse vaccinology. He has authored several research papers in international peer-reviewed journals and two book chapters.

Shoba Ranganathan holds a Chair in Bioinformatics at Macquarie University since 2004. She has held research and academic positions in India, USA, Singapore and Australia as well as a consultancy in industry. Shoba's research addresses several key areas of bioinformatics to understand biological systems using computational approaches. Her group has achieved both experience and expertise in different aspects of computational biology, ranging from metabolites and small molecules to biochemical networks, pathway analysis and computational systems biology. She has authored as well as edited several books in as well as contributed several articles to Springer's Encyclopedia of Systems Biology. She is currently the Editor-in-Chief of Elsevier's Encyclopedia of Bioinformatics and Computational Biology as well as the Bioinformatics Section Editor for Elsevier's Reference Module in Life Sciences.

Nikolai Petrovsky is a Professor in the School of Medicine at Flinders University, Director of Endocrinology at Flinders Medical Center and the founder of Vaxine Pty Ltd, a biotechnology company involved in vaccine design. His research interests including vaccine design, adjuvants, immuno-informatics, and autoimmunity. He has been a principal investigator on multiple grants from National Institutes of Health and has won prestigious awards including the Ernst & Young Entrepreneur of the Year Award and a Biopharma Asian Executive of the Year Award. He has taken vaccines for seasonal and pandemic influenza, hepatitis B and insect sting allergy from the bench to human clinical trials and authored over 140 research papers

Chemoinformatics: From Chemical Art to Chemistry in Silico

Jaroslaw Polanski, University of Silesia Institute of Chemistry, Katowice, Poland

Introduction

Chemoinformatics (cheminformatics) a term associating chemistry and informatics (computer science) on the lexical level describes a discipline organizing and coordinating the application of computers in chemistry. However, some branches of computational chemistry usually are not included in the field, although they depend on computers. The example here is quantum chemistry which is more frequently insisted to be a part of theoretical physics than chemoinformatics because, in principle, it is just the physics of atoms and mathematics that allow the correct modeling of molecular objects, and pure mathematics, hypothetically, can be done without computers (Polanski, 2009b). However, the larger the molecules, the more remote and inaccessible the precise mathematical model. The history of fundamental concepts or theories in chemistry illustrates the complexity of chemical researches. Consider chemical bonding. The molecular orbitals or valence bonding theories describe atomic scale aggregation into molecules. Both models have been competing with each other and chemists still discuss which is correct and which is better. In physics, it is possible to develop a relatively simple model to explain certain facts of nature. In contrast, in chemistry, a theory often partially interprets some data, also offering partial solutions. Thus, we need several high-quality models for the correct theory (Brock, 1993). Theoretical chemistry is an empirical science based on the Schrödinger equation; however, we are aware that its general solution will not be available. In science we use mathematics for modeling and developing theories. In the language of informatics this means mathematics is a compression tool unifying the facts of nature. Now we do not need individual facts any longer. Details disappear and information is coded into a form of a model (occasionally a single equation) that explains the reality. Compressing the complexity of natural information to such models is used typically for an approach called reductionism (Cohen and Stewart, 1994). What if, the Nature appears too complex for an accurate mathematical description or the developed model does not have a precise solution. The answers could be still important but unavailable by precise mathematics. In such situations, we need further substantial simplifications, even in cost of reliability. This brings into the game speculation or educated guess which are still better than blind guess or no answer. Chemists discovered that "educated guess is being supported by the computer" (Kowalski and Bender, 1975). This describes the first application of computers in chemistry, which is to assist a chemist in a calculation or computation that requires the calculation or simulation by computers. Why can computations still be possible, flexible, and efficient in data processing when human calculations fail? The efficiency of in silico mathematics is first of all, not by computer intuition or flexibility but by a brute force under relatively simple mathematical formalism (Audibert, 2013; Bock and Goode, 2002). The methods played in computers are evidently different than those involving human brain. Coupling human intelligence with the speed and competence in low-level manipulations observed in computer appeared efficient enough to solve "formerly intractable problems, and explore areas beyond the reach of human calculation" (Audibert, 2013; Jager and Krebs, 2003). Accordingly, the domain of chemoinformatics can be preferentially outlined by calculations that cannot do without mathematics in silico, or in other words, those chemistry branches that process massive or ill-defined and fuzzy data for which standard mathematical modeling is unavailable. It was suggested that, chemistry branches that "generate more numbers than information, e.g., physical and chemical property calculation" are not in chemoinformatics (Oprea, 2003). It seems however that even though the calculations in the latter case can be performed efficiently in silico, hypothetically, they can do completely without computer assistance basing on the relatively simple mathematics.

The Scope of Chemoinformatics

Imitation or simulation of natural systems by building physical or virtual models is a core of scientific method. Basically, chemoinformatics is a tool for chemistry simulation in computer. Recent definitions: *using informatics method to solve chemical problems* (Gasteiger and Engel, 2004) or just in silico chemistry (Polanski, 2009b) well define the objectives and the scope of the discipline. A term chemoinformatics has been coined in the late 90′, when computer applications in chemistry had been already controlled by chemometrics, another discipline which associates chemistry and informatics. At the same time chemists were already aware of the potential of computer applications in advanced molecular problems at the same time however also realized that such problems as drug (molecular) design will not proceed as smooth as previously expected. The experience that chemists gained here quenched the early enthusiasm. We understood far more advanced methods would be needed for the efficient in silico chemistry. At the same time chemometrics which developed from analytical chemistry only partially cover molecular problems. Therefore, chemoinformatics originated as a result of early disappointment of computer assisted drug design as a discipline conceived for "the combination of all the information resources that a scientist needs to optimize the properties of a ligand to become a drug" (Brown, 1998). In **Fig. 1** we briefly illustrated the scope of chemical investigations performed in computer in the association with other disciplines. Eventually, a new drug is not only biophysics, but also economic market with its unpredictable behavior. Therefore, the scope of chemoinformatics is broad, extending from physics to economy. The large scope decides the complexity of problems is enormous high. In turn, high complexity implies low precision. This fact explains the

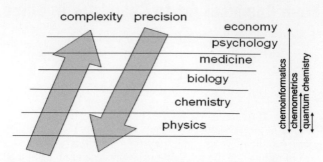

Fig. 1   A precision and complexity in chemoinformatics which scope extends from physics to economy.

autonomous character of chemoinformatics which had to develop their own methodology to trigger the computer to zoom on chemical problems recognizing specific chemical patterns and replying to them accordingly. The history of the computer chemistry can be found in Polanski (2009b), Noordik (2004), Polanski and Gasteiger (2016).

Teaching Computers Chemistry: From Molecular Data to Chemical Information and Chemical Ontology

The development of technology, in particular, scientific equipment capable of describing human environment by any type of records provided us with data, where data can be interpreted in the broadest sense as anything that is recorded, including metadata, i.e., data referring to other data. This can be both ordered and unordered collections of both nominal and numerical values, whereas the latter can be discrete numbers, intervals or ratios. Another type of data (binary large objects, BLOBs) is used to describe audio, video and graphic files (Maheshwari, 2014). With the enormous increase of data volume the so-called big data appeared recently. It is generally believed that big data brings new value and innovation. For example, Szlezak et al. cited the recent McKinsey research that suggests that the potential use of big data in US health care could reduce costs by $300 billion a year (Szlezak et al., 2014). Current impact of big data for drug design is however far less noticeable that could be expected. The reason is that, first, this kind of information is much less clearly defined and messy. Accordingly, its analysis causes serious problems. Second, 120 million compounds were synthesized and registered in databases; however, this should be compared to 7 billion individuals in human society. Data gets bigger while going from chemistry to biology or economy (where humans are interacting). A single genome for a single individual can be interpreted as special type "big data", e.g., already a structure of the genome is a large record. A phonotype yielded by the genome is a specific property. In comparison, a population of drugs or bioactive molecules is represented by much smaller data, in particular measured properties. Therefore, the generation of big data here must go by the increase of the number of properties registered for a single molecule, e.g., polypharmacology. Third, the availability of big data for drugs is limited because of the need to keep the secret during drug development. Therefore, in order to replace such data a library of building blocks is analyzed to probe big data behavior of the large molecular populations (Polanski et al., 2016). Formal classifications of big data and their analyses were discussed in Polanski (2017).

Science, in particular, chemistry is knowledge organized from data, that is, facts and numbers making up information. In turn, information develops chemical knowledge. Atoms, molecules, substances and their properties and transformations are the main objectives of chemistry. An unbelievably large number of molecules can be arranged from a matter available in the universe. The efficient data storage and management is needed for the control of this molecular representation, which indicates the next field of computer application in chemistry. Generally, a structure of chemical information is often chaotic which prevents its obvious transformation to in silico form. This can be illustrated by a fact that chemists often underlines chemistry is an art. This means that a human expert is needed for the efficient playing with uncertainty in the field. In turn, computer is an unsophisticated *device* for which uncertainty is a problem. Therefore, computer-understandable chemistry needed to be developed to store and process chemical data and information. Translating molecular data into a machine-readable and -processable structure is far from triviality. A problem is also the interaction between chemist and computer.

Chemical compounds (CC) are the main objects of chemical investigations. Surprisingly, although commonly used in chemistry a term chemical compound has never been defined by IUPAC. Historically, the ambiguity of the term CC can be related to the early term of *mixts*, a result of *mixing different bodies* (not necessarily in the contemporary chemical sense). Mixts was replaced in 18th century by composition or compounds (Bensaude-Vincent and Simon, 2012). What originally related to substances can be however also associated with molecules which are the combinations of chemical elements. Therefore, CC refers both to molecules and substances (chemical species). Actually, the meaning of the term CC is intuitively interpreted by a chemist and can indicate (Polanski and Gasteiger, 2016):

- A *synonym of a molecule* (composed of at least two different atoms) representing a single chemical body for virtual needs (coding, in silico processing, visualization, etc.)
- A *molecule* representing a single chemical body under measurement (e.g., a single DNA strand that can be observed, e.g., under microscopy).

- A *substance* composed of a replicated but single molecular entity under measurement.
- A *chemical species* under measurement when replicated is more than one molecule type.

However, a simplified meaning of a molecule is also a model-like representation that can refer both to a chemical compound that has been yet synthesized and described, or to a virtual structure (hypothetical compound) that is under design or speculation. Historical reasons decide that molecules belong to the domain of organic or inorganic chemistry. Although the organic vs. inorganic typology is more and more fuzzy the proportions of chemical compounds are approximately 1:200 in favor of organic chemistry, if we follow standard rules. The management of chemical compounds needs the efficient machine-searchable databases registering all virtual structures and real compounds that have been synthesized and described in chemical literature from the very early days to today. This problem of substantial importance for the organization of chemistry in silico is known under the term *structure representation and searching* (Willet, 2003) which basic concepts has been formulated as early as 1960s. Atomic composition given by molecular formulae is not sufficient to unambiguously identify a molecule. Therefore, in the broadest sense the structure of the chemical entity is defined by constitution and stereochemistry. What we mean by constitution is "the description of the identity and connectivity (and corresponding bond multiplicities) of the atoms in a molecular entity (omitting any distinction arising from their spatial arrangement, i.e., molecular stereochemistry"). Isomers are chemical entities of the same atomic composition but different constitution and/or stereochemistry or precisely isomers are "one of several species (or molecular entities) that have the same atomic composition (molecular formulae) but different line formulae or different stereochemical formulae and hence different physical and/ or chemical properties." In particular, a line formula is an example of molecular representation showing atoms that are "joined by lines representing single or multiple bonds, without any indication or implication concerning the spatial direction of bonds."

We more and more understand in chemistry the importance of the precise shape of chemical formalism, typical for mathematics and physics. However, at the same time we would like to preserve the completeness of chemical information with its uncertainty. Therefore, we need an intelligent in silico system fitted to receive the *soft* chemical data without information reduction or deterioration. In this context *chemical ontology* is a recent solution of this problem (Hastings *et al.*, 2011). It is a flexible dictionary like chemistry representation, a system capable of organizing the complete chemical information in silico. In other words, the strategy here is that the chaotic nature of chemical information is tamed not by the significant information reduction but by building a structure capable of accepting all variety of data and facts. The inspiration for the chemical ontology seems to be object-oriented programming where a processing of information attempts to imitate reality. We are defining here not only the data type of a data structure, but also the types of operations or functions that can be applied to the data structure. Therefore, data are structured by objects that includes both data and functions. Furthermore, relationships between the objects are defined. In particular a category of class is an extensible program-code-template for creating objects, providing initial values for state (member variables) and implementations of behavior (member functions or methods). Accordingly, let focus on chemical compound which can be defined as a class having a fixed ratio of defined atoms and a chemical structure that can be expressed by one connection table of non'hydrogen atoms and one or more connection tables that include hydrogen atoms and bond orders as well, connected by OR logic. This allows, for example, to clearly classified all tautomeric forms of vitamin C as a single chemical compound (Bobach *et al.*, 2012).

Between Descriptors and Properties

Basically, chemical compounds can be represented by descriptors or properties, whereas a property is to be measured in experiment and a descriptor is calculated or more precisely is a final result of a logic and mathematical procedure transforming chemical information encoded within a symbolic representation of a molecule into a useful number (Todeschini and Consonni, 2000). As in real world we usually have a contact with substances the properties will mainly refer to the substances, while descriptors to the molecules. However, this is not always true because we can also measure the property of a single molecule, e.g., MW in the MS spectrometer like conditions. A fully consistent differentiation between properties and descriptors could not always be easy. Interestingly, while IUPAC identifies only properties (descriptors are also classified as properties); in chemoinformatics all parameters are often referred to as descriptors. Properties are recorded here as physicochemical descriptors, e.g., molecular descriptors are divided into two main classes: experimental measurements, such as log P, molar refractivity, dipole moment, polarizability, and, in general, physicochemical properties, and theoretical molecular descriptors" (Consonni and Todeschini, 2010). In turn Polanski and Gasteiger argued for a clear and consistent descriptors versus properties differentiation (Polanski and Gasteiger, 2016). It has been shown recently how important and informative this taxonomy can appear (Polanski *et al.*, 2016; Polanski and Tkocz, 2017). Accordingly it has been demonstrated that *ligand efficiency* (LE), a molecular descriptor designed for an early evaluation of drug candidates (Hopkins *et al.*, 2014); if interpreted not as a descriptor but as statistical property allows to explain an evident preference of small ligands. Actually, LE does not characterize a real binding potency but is related to the statistics involved in the binding of the 1 g sample of the ligands (Polanski and Tkocz, 2017). This indicated the next difference between descriptor which can be clearly related to a single molecule and a property of the substance which can also be a statistical parameter not so obviously connected to a single molecule.

From Problems to Concepts

In the previous articles we illustrate the difficulties of the adaptation of chemistry to an in silico method. A variety of terms that are interpreted by human chemists intuitively are not well suited for computers. This should have been first understood well enough to be transferred to a form understandable for computers. Structure representation is an example of the early problem that should have been solved to teach computers chemistry. We briefly presented the examples of the molecular problems and in silico concepts to deal with them after (Polanski and Gasteiger, 2016).

Problems

- *Chemical data documentation and searches*: database searches and management
- *Calculating molecular descriptors*: 2D (chemical graphs) and 3D (molecular modeling) representations are mapped into single numbers (0D); vectors, fingerprints(1D); matrixes, surface maps (2D); surface, atomic representations, force fields, virtual or real receptor data (3D); etc.
- *Molecular modeling*: molecular structure 3D data generation in silico from their 2D or 1D representations. This includes both predictions of molecular topography and molecular descriptors i.e., atomic annotations data by calculated (simulated) atomic properties.
- *Structure elucidation*: (synonym: structure-spectra correlations) property to molecular structure is mapped in factual chemical space (FCS) when we are attempting to find a structure having a certain spectra or in structure to property version we are simulating a spectra for a given molecule (substance).
- *Mapping structure to activity* (SAR): a series of structures (FCS substances) is needed to study SAR, which are usually synthesized. The real goal of SAR is usually to predict the training series for designing new structures and substances by property predictions in a qualitative or quantitative procedures.
- *Property prediction*: compounds' series (FCS) are mapped from FCS into virtual chemical space (VCS); two basic versions are available, i.e., property vs. property or structure vs. property. However, in the design step for novel compounds in VCS (we can design both compounds VCS or FCS where some properties can be registered in databases and/or literature) we are always to use a structure version (no property is available in VCS).

More Important Concepts

- chemical space (CS), factual CS (FCS) or virtual CS (VCS) SMILES coding
- connectivity (graph theory) approaches, additivity concept, molecular modeling, force fields
- molecular mechanics, molecular dynamics, force field, molecular interaction field, partial atomic charges, lipophilic potential
- RDF structure coding (2D), 2D structure–2D spectra correlation
- SAR, QSAR, QSAR domain, similarity measures, privileged structures, fragonomics
- QSAR, QPAR, logP versus partition coefficient, fragonomics, virtual screening
- Synthons, retrosynthetic analysis, synthesis tree

Some of them are adopted directly from chemistry where they originated not necessarily with a though of computer applications. For some others in silico methods were an obvious inspiration.

Chemical Space: From Storing Information to Advanced Ontology Formalism

The population of chemical data has significantly increased in recent years. In particular, Chemical Abstracts Service (CAS) has registered almost 130 million unique organic and inorganic chemical substances, 67 million sequences. The daily update records 15,000 substances. Substance information, includes both molecular descriptors, property and predicted property data includes more than 7.6 billion property values, data tags and spectra. CASREACT contains 83.6 million single and multistep reaction data entries. The Pubchem BioAssay Database records count to millions (Wang *et al.*, 2017). The number of potential compounds, i.e., the population of chemical space is however much higher. This number is estimated between 10^{18} and 10^{200}, whereas 10^{60} is probably the most representative value found in the literature. We can compare this to the factual CS of the order of $10^7/10^8$. Alternatively, a number of stars in the universe is estimated to 10^{22} (Polanski and Gasteiger, 2016; Baldi, 2005; Bohacek *et al.*, 1996). The enormous population of CS can be realized better if we evaluate the potential expansion of *n*-hexane into a family of analogues decorated with 150 different substituents. Accordingly, if we bring together a full collection of mono- to 14-substituted analogues count to a population of 10^{29} (Lipinski and Hopkins, 2004). Therefore, chemical synthesis or the construction of new chemical molecules cannot do without a well-organized data mining system that allows us for a verification of physical and/or chemical data by screening a large population of the variety of compounds described. This problem is fundamental for chemistry, where the access to information is a common need. Moreover we would like to have proper data delivered at the proper time to the chemist's desk. An efficient chemical information system on chemical compounds needed to be developed from the very beginning of chemical investigations. Actually, *Chemisches Zentralblatt* appeared as early as 1830; while the first edition of Beilstein's *Handbuch der Organischen Chemie* (BH) was published in 1881. Even in that time the latter contained two volumes, registering 1500 compounds, with more than 2000 pages. In chemistry we realized quite early that the improvement in data storage could be of the crucial importance for the development of chemical sciences. However, searching a compound within thousands of pages of the printed book was tedious and impractical. Instead, the computer desktop could provide an efficient environment for the management of chemical data and chemistry could evidently profited from computers. Accordingly, chemical information is a core component of chemoinformatics.

This means that we need new tools for storing and manipulating chemical data. *Chemical space* (CS) is an example of the idea that contributes not only into the computations but also to the chemical documentation. The original CS concept was inspired by the cosmological universe populated by stars, where chemical compounds replace cosmological objects to navigate chemical space (Lipinski and Hopkins, 2004). The vague cosmological analogy can be replaced by more precise parallels. Accordingly, mathematical model was designed. Practically, clear illustration of the chemical operations by in vitro and in silico operators appeared to be the most important functionality in this method (Polanski, 2009b; Polanski and Gasteiger, 2016). Summing up CS is a concept which appeared in the hope of organizing a whole population of chemical compounds. Essentially, CS is a structure for the mapping of chemical compounds by descriptors and properties. Furthermore, potential operations schemes accompany crude data. If we realize these functionalities we can easily understand that in the current chemoinformatics *chemical ontology* is an advanced formal structure that represents a CS system for in silico chemistry that allows for a flexible multipurpose property and descriptor registering and manipulation in silico. Ontologies encode human knowledge in computationally accessible forms. They are designed to narrow the gap between the knowledge of human experts and the functionality available in computer systems, by expressing expert knowledge in a manner computers can manipulate and reason over (Hastings *et al.*, 2011).

Basic In Silico Functionalities

Coding Chemical Information Into Molecular Descriptors

Molecular graphs encode chemical information that can be transformed to a variety of useful numbers, or in other words molecular descriptors (Todeschini and Consonni, 2000). A variety of molecular descriptors, their taxonomy and detailed algorithms or software for their calculations from molecular representation are available nowadays. This decides that usually a large number of calculated molecular descriptors are accompanied by few measured property data. This effect described as a property deficit (Polanski and Gasteiger, 2016; Polanski, 2017) results from a fact that the measurements are much more expensive than in silico calculations. On the other side this illustrates the success of in silico chemistry.

Computer-Processable Molecular Codes

Two-dimensional atom arrangement of chemical entities can be coded into molecular graphs which are easy-to-read for chemists; however; are not computer-friendly. Accordingly, we need a method for a proper translation of molecular graphs to a form that would be understandable for a computer system. Coding molecular graphs by linear notation or connection table forms a typical computer representation. In the linear notation a molecule is represented by a string of the line formulae type. An example of SMILES (Simplified Molecular Input Line Entry) notation can be an up-to-date illustration of such a system (Weininger, 2003). The detailed tutorial can be found online in the Daylight Chemical Information Systems. Matrix representation is an alternative for coding chemical graphs. The examples here are adjacency matrix, atom connectivity matrix, incidence matrix, bond electron matrix. Molecular structures can also be coded by connection tables which record atoms and bonds within a molecule. The need of the individual application decides a form of the computer representation that we use. An in-depth description of the matrix and connection table codes can be found in the Handbook of Chemoinformatics edited by Gasteiger (2003). A precise definition of chemical compound could need a combination of connection tables, which can be achieved by the chemical ontology method.

A connection table or linear notation can be formed arbitrarily. This means that numbers can be assigned to the atoms differently and there is none standard molecular representation. Canonical labeling is a solution for this problem. This provides a unique representation for a certain molecular graph. Unique SMILES are example of such canonical labeling system (Daylight Chemical Information System). Chirality is an important chemical structure property and ISOMERIC SMILES is a system allowing for various chiral and isotopic specifications.

Molecular Editors

Molecular graphs are unambiguous, chemist friendly and illustrative way for the presentation of molecules constitution and stereochemistry. Molecular editor is an interface that allows a user not only to draw professionally presented molecular structures but it is also a tool for the translation of such a structure into the computer processable molecular codes. A number of systems have been developed that are capable of the translation of molecular formulas introduced into computer by its user in a form of direct drawing, e.g., using a mouse, to the machine readable code. ISIS, ChemSketch ACDLAB, JME, RasMol are molecular editors available free of charge at the web sites, respectively (Polanski, 2009b).

Coding Chemical Reactions

Chemical reaction refers to a dynamic behavior involving both a breaking and formation of chemical bonds, in which chemical compounds are transformed from reagents to products. Accordingly, chemical reaction changes the atom bonding system in molecules. The problems encountered while naming chemical reactions resemble those that are typical in the nomenclature of chemical

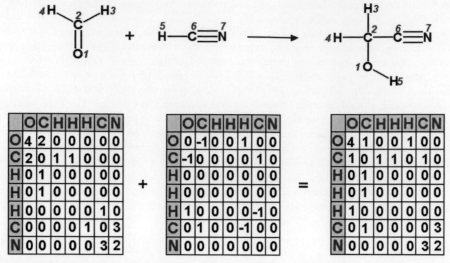

Fig. 2 Reaction coded by the B + R = E matrix. Reproduced from Gasteiger, J., Engel, T., 2004. Chemoinformatics: A Textbook. Weinheim: Wiley-VCH, p. 186.

compounds. For example common reaction nomenclature often honors distinguished chemists, the discoverers of the certain reactions. This remind us the trivial chemical compounds nomenclature, because trivial name encodes no information on the reaction itself. Merck Index can serve as a dictionary guiding us through the name reaction chemistry (Smith *et al.*, 2001). An online alternative is the name reaction database of the Organic Chemistry Portal (Organic Chemistry Portal). Similarly to the nomenclature of chemical compounds IUPAC developed a systematic reaction classification capable of coding a variety of useful information on the molecular transformation; however, this is used only very seldom (Polanski, 2009b). Molecular transformations in silico can be coded by an algebraic model based on logical connectivity and matrix equation B + R = E, where B (beginning) represents an initial reaction stage, E (end) codes a final state and R is a reaction matrix. **Fig. 2** illustrates an example of the reaction noted in such an approach (Gasteiger and Engel, 2004).

Other representations and classifications of chemical reaction, including SMARTS, has been developed but will not be discussed here further and a reader is referred to the references available in Gasteiger and Engel (2004), Chen (2003) or at the internet site of the Daylight Chemical Information System.

Organizing Chemical Facts Into Databases

Finally, what we need to enable chemists using computers to perform efficient chemistry is the access to chemical information, i.e., chemical data represented by chemical facts and numbers. Thus, for example, coding chemical transformation as described in section "Coding Chemical Reactions" does not bring information on this specific reaction gathered by experimental chemistry which are often fundamental for the chemists. Therefore, the additional data on the reaction conditions, solvents, temperatures, catalysts or byproducts have to be shaped to a form compatible with the computer platform. Accordingly, organizing chemical data in searchable databases is a focal point. A variety of chemical databases are available on-line having user-friendly interfaces. This includes chemical reaction or substances databases as Reaxys (formerly Beilstein) or Chemical Abstracts, patent databases as esp@cenet, chemical substance catalogues, e.g., Aldrich, or on-line chemical journals.

Managing interactions between a chemist and computer needs chemical information and software which puts together to a chemoinformatic platform capable of understanding chemical data efficiently assisting computer-aided chemistry or advising chemists in more traditional chemistry branches. Below we discussed some of typical problems of data processing and data output.

Computer Generated Chemical Names

Chemical structures can be nowadays easily transformed to chemical names. Chemical name generators typically are supported by a molecular editor. Historically, Beilstein pioneered in this field with the AutoNom program, which succeeded in 86.3%, if tested for more than 63,000 structures. Currently, AutoNom can generate both Beilstein or ACS nomenclature forms. Alternative software has been developed by Advanced Chemistry Development Inc. The ChemSketch freeware is a part of the extensive platform available at the ACD/Labs internet site (ACD Lab). Coding name generator was discussed in the Wisniewski (2003).

Molecular Modeling

Basically, what we mean by molecular modeling is simulating molecular structures in silico. This involves molecular manipulations such as visualizing, merging, superimposing or rotating individual molecules in space but also rotating bonds within individual molecules, etc.

More generally, molecular modeling involves the predictions of various molecular behaviors and properties. In particular, this includes predicting molecular shape, e.g., by 3D structure generation, simulation of chemical or biological effects. Modeling virtual molecular structures is not a trivial problem and includes a variety of computational schemes on the different level of approximation (Holtje et al., 2003). The importance of molecular modeling can be understood better if we realized that chemistry often needs to investigate virtual or hypothetical structures, for example, active complexes that would be extremely unstable or even could have never been synthesized. In such cases a single possibility offering 3D molecular representation is computer modeling.

2D and 3D Structure Generators

Current methods often demand to analyze thousands or even millions of structures. Obviously, this cannot be realized by any system operated manually. Instead, we need the automated algorithms. Actually, nowadays this can be easily programmed in a variety of environments. Alternatively, a ready to use databases of virtual compounds are available (Reymond et al., 2010). It is important to realize that chemists usually simplify a real structure of a chemical molecule to its molecular configuration. For example, the shape and stability of E and Z isomers which are two different configuration series are commonly estimated solely on their configuration type, irrespective of the actual 3D structure of the individual molecules. Such simplification is often sufficient, in particular, in organic chemistry where more detailed molecular representation will be even too complex and impractical. However, molecules are three dimensional and we can indicate the exact space location of individual atoms with high resolution. In some areas as medicinal chemistry such high resolution molecular representation can be of substantial importance. Traditionally, we use X-ray diffraction on crystals. However, as this effect is limited to the condensed matter (crystals), many further approaches appeared that allow to disclose various structural data defining 3D structure; however; general technology is still unavailable here. Despite an impressive progress X-ray crystallography is still a tedious process demanding producing crystals. Moreover, we can doubt a full correspondence between atom configuration in condensed matter vs. this in other environments. Accordingly, even nowadays only a small fraction of factual chemical space is described by the respective 3D structures despite a fact that such a data are sought after in particular for peptides or drug–ligand complexes (Motherwell, 2004). In other words 3D structures as measured properties are rare.

The importance of molecular modeling can be illustrated by a variety of descriptors and properties influenced by 3D molecular structure. Shape is an example of molecular descriptor that directly depend upon 3D structure. This is, however, a fuzzy category because molecules are capable of adopting different configurations and/or conformations (representing different shapes), depending upon the requirements of the environment. Accordingly, a population of the available shapes depends on the energy of the molecule. An uncertainty of this category explains a fuzzy nature of the properties and effects of various chemical or biological effects that are controlled by shape. This includes drug-receptor recognition and interactions. All shape effects can be currently modeled in silico (Polanski, 2003).

The generators of 3D structures are designed for a high-speed conversion of 2D into 3D molecular representations. Formally, this will be molecular descriptors, obtained by property prediction. How such descriptors can be calculated? Usually, X-ray structures are a kind of standard for experimental atomic 3D coordinates. 3D structure generators are programmed to use a model atom types for certain hybridizations to calculate standard bond length and bond angles. Typically, the strategy includes additional rules as preferred conformations of ring structures. As the speed and program performance is preferred, non-typical atom input should not crash the software. Instead, reasonable value would be expected.

LHASA is an early example of 3D generator designed to evaluate influence of steric hindrance on the reactivity of cyclohexane conformations. Model builders are another example of the program block capable of generating 3D atomic coordinates in an interactive mode. This allows to convert a molecular graph drew by a user at the program interface into a 3D molecular structure. Typically, the programs use built-in rules for the generation of the approximate structures having standard bond lengths, bond angles, torsion angles and stereochemistry. Usually generated structures require further refinement by geometry optimization. High-speed conversion of 2D to 3D representations of the large molecular populations are the main task of such programs. The examples of the software for automated structure data conversion includes CORINA, Cobra, Alcogen, Chem-X, Molgeo. Obvious priority here is a speed performance which can achieve more than 10^5 small to medium-sized molecules in ca. 2000 s on a 1.0 GHz workstation with a performance of 23 ms/cpd yielding a 99.99% conversion rate for the Corina program (Sadowski and Gasteiger, 1993). Optionally we can generate multiple ring conformers, rotamers or tautomers (ROTATE and STERGEN).

Modeling 3D-structures

Virtual molecular models in silico provides a useful platform widely used in nowadays chemistry. At the same time, theoretical chemistry developed a variety of methods aimed at modeling structure and bonding systems using a various molecular or quantum mechanics approaches which owe their origins to the Born-Oppenheimer (BO) approximation. In particular Schrödinger equation:

$$H\Psi = E\Psi$$

relating the wave function Ψ, Hamiltonian operator H, and energy E terms can be given in a simplified form. Formally, H includes components responsible for nuclear kinetic energy, electron kinetic energy, nuclear repulsion energy, electron repulsion energy, electron-nuclear attraction. In the BO approximation we assume that electron distribution depends on the fixed nuclear position only, which means that nuclear kinetic energy term can be neglected in the H operator. Introductory tutorials to molecular modeling with representative references are widely available (Holtje *et al.*, 2003). A term *molecular mechanics* (MM) has been coined in 1970'. This refers to a so-called force field method (Hinchliffe, 2003) based on the assumption that a population of atoms in the molecule can be described by the potential energy depending upon the space locations of the atoms. This allows us for the optimization of the molecular geometry. The mechanics needed for the calculations of potential energy have been developed using classical physics. A suggestive model where atoms are balls and bonding are springs connecting the balls are commonly associated with this method in the chemical audience. Analytical functions are used to describe atomic interactions defining the so called force field. A variety of force fields have been developed to adopt the method for the different chemical functionalities and compounds' classes. Illustrative examples are MM +, AMBER, BIO +, OPLS. Via parameterization a force field concept can incorporate electronic energy in MM calculations. The MM method give us an insight into the molecular geometries and energies which minimization provides us the simulated low energy molecular model within the accessible molecular geometry space. Currently, a variety of MM software packages are available, e.g., HYPERCHEM, Sybyl including freeware options, e.g., the Tinker package. Informative tutorial and review of the methods and applications can be found in Goodman (1999), Keseru and Kolossvary (1999). Alternative online materials are offered by Rzepa.

Semiempirical quantum mechanical methods (SM) can be used for further approximation of the molecular models. This provides us with the deeper insight into the electron distribution within a molecule. Various levels of approximation are available in SM for the simulation of molecular orbitals. The calculations involve only valence electrons and are based on a set of experimental parameters forming a database referring to a certain set of chemical compounds. Such a strategy makes the calculations much easier. The examples are the AM1, MNDO, PM3, CNDO or INDO methods (Holtje *et al.*, 2003).

Molecular dynamics (MD) is an approach in which a single point atom location is modified to cover the dynamic atom behavior. In this model a nuclear system goes into a motion under a certain temperature forces. The motion is simulated by the Newtonian dynamic equations which solution indicates the set of possible atom locations. In particular, this allows to determine a so-called conformational ensemble profile for a molecule. Thermodynamic and dynamic properties of the molecules can be calculated using the MD method. The MD appeared especially useful for the simulations or refinement of protein shapes including X-ray structures. For further discussion of MD methods and applications compare Rapaport (2004). Many recent advances involve molecular dynamics simulations of proteins in explicit mixed solvents applied to various problems in protein biophysics and drug discovery including protein folding, protein surface characterization, fragment screening, allostery or druggability assessment (Kimura *et al.*, 2017). AQUA-DUCT: is an example of the up-to-date tool for the identification and tracking of molecules which enter active site cavity among thousands of single molecules along several thousand molecular dynamic steps (Magdziarz *et al.*, 2017).

Structure and Substructure Searches

Comparison is an important method in chemistry. In particular, this includes molecular structures. Therefore, structure or substructure searches is a substantial problem in a variety of the issues discussed in this article. An illustrative example is a need for the registration of novel chemical compounds and reaction within databases. A structure query is the essential operation for which molecular graph has to be translated to a canonical molecular representation. An example of a function that enable data structure organization is the so-called hash key. In turn, hash codes can be obtained from canonical SMILES strings or Augmented Connectivity Molecular Formula, e.g., in ACS Registry System. Additional information and references are available in Leach and Gillet (2003).

Structure query is the most simple database operation. Nowadays substructure search is also a routine. This provides us a method for the identification of the defined molecular fragment included in a series of molecules (Noordik, 2004). Graph theory, in particular the so-called subgraph isomorphism, can provide theoretical framework while practically a binary string molecular representation enables rapid structure screening (Leach and Gillet, 2003). The procedures developed include structural key (Boolean array) representation which can be a bitmap where each element coding a true or false refers to the presence or the absence of a certain structural feature or pattern. Alternatively, high-speed structure queries can involve fingerprint representations. In turn, a concept of molecular similarity measures enables the similarity based searches. For the essentials compare (Kochev *et al.*, 2004). Occasionally, for example, in drug discovery finding a common 3D pattern by mining 3D structure databases is an important problem. Database mining and identifying such a pattern known as a pharmacophore can assist a search for new ligands in the ligand based drug discovery paradigm (Nicklaus, 2003). Software available for the procedures discussed in this article can be found in the Kochev *et al.* (2004). Recently a concept of the so-called chemotype appeared. This represents molecules, chemical substructures and patterns, reaction rules, and reactions. This method is capable of integrating types of information beyond what is possible using current representation methods (e.g., SMARTS patterns) or reaction transformations (e.g., SMIRKS, reaction SMILES). For example new publicly available chemical query language, CSRML, to support chemotype representations for application to data mining and modeling has been developed recently by Yang *et al.* (2015).

Molecular Graphics

Formally, molecular graphics refers to a visualization of molecular objects, but this term is also used as a synonym of molecular modeling. Because molecular objects are complicated solving a chemical problem needs high speed and quality in virtual reality. Therefore, molecular graphics has significantly contributed to the currently available computer visualization technology.

Computer screen has been used for scientific visualization of physical models as early as in 1960s in the MIT during the Mathematics and Computation project. Developing technology for molecular visualization engages both chemistry and computer sciences. Currently, virtual chemistry on screen is a routine and highly interactive user-friendly systems are expected to be a part of each molecular platform. Molecular representations here can range from the atoms to the surfaces including a variety of others symbolic forms (Keil *et al.*, 2003). For a brief discussion on the differences between physical and virtual model compare Morris (2002).

Advanced Functionalities: In Silico Chemistry

Chemical Syntheses and Retro-Syntheses (Disconnections)

Chemical objects can be constructed via chemical synthesis. We can understand the importance of this problem in chemistry if we realize that even today chemical synthesis can be a bottleneck in research and a variety of applications. Chemists still believe chemical synthesis is an art. For the better illustration of this issue let us take a textbook example. Atropine, a natural product available from nightshade or belladonna was isolated and used already in ancient Rome and India. Of course the structure of this compound cannot be identified at that time. A core atropine moiety was proved to be tropinone in 1901 when Willstätter synthesized this compound. This needed however a complex more than 20 step synthesis which resulted in a total yield of no more than 1%. Low yield suggests inefficiency and makes easy to overlook the stunning success of this masterpiece of a synthetic work performed by a nobel prize winner. Even today this project would pose a complex organizational challenge, despite an availability of precise and rapid NMR or MS spectrometers easily and rapidly identifying the structure of compounds synthesized. True excitement comes however with the Robinson approach, **Fig. 3**. One step condensation and decarboxylation can result in more than 90% yield of tropinone (42% yield was originally reported).

It is still true, that the efficiency of synthesis critically depends on the skills and creativity of chemists. Can computers play a role in synthesis design providing us useful hints and supporting our skills?

The Development of Product to Reagents Strategy in Synthesis Design

Practical applications needs novel substances which are to be designed and produced for the industrial, pharmaceutical, agricultural use. This indicates the importance of a chemical product. Products are formed in chemical reactions; however; products appears at the very end of the process which starts from reagents. Therefore, in order to know how to design the products in a rational way we should have first explored a conversion of a variety of chemical compounds and functionalities. This explains how historical development of the chemical knowledge has imprinted into the organic chemistry textbooks where chemical reactivity of a molecule is a core issue. This decides also chemists are trained in a reagents to product chemistry where products can be designed basically only by a nondirect screening of possible reactions. However, an efficient general solution in such a strategy would need for example a database for the direct searches of products. This however, cannot be realized because a database registering all possible virtual products would be impracticable due to a fact that chemical space is too highly populated. Therefore, for years

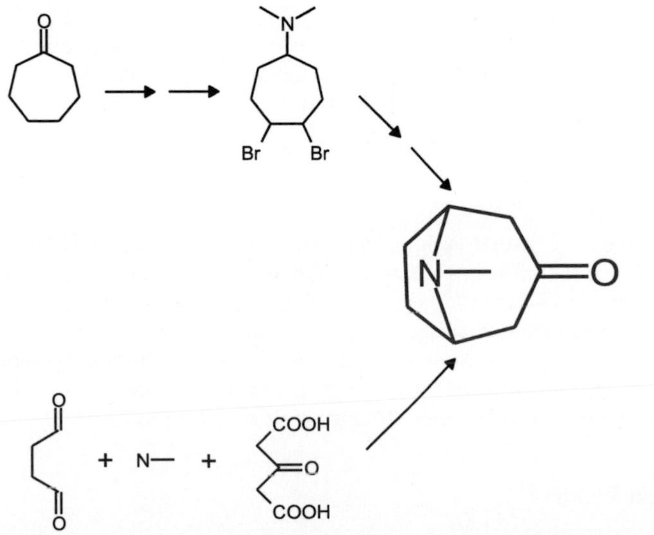

Fig. 3 Willstätter (upper) and Robinson (lower) approaches to the tropinone synthesis. Reproduced from Polanski, J., 2009. Chemoinformatics. In: Walczak, B., Tauler, R., Brown, S. (Eds.), Comprehensive Chemometrics vol. 4. Amsterdam: Elsevier, pp. 459–505.

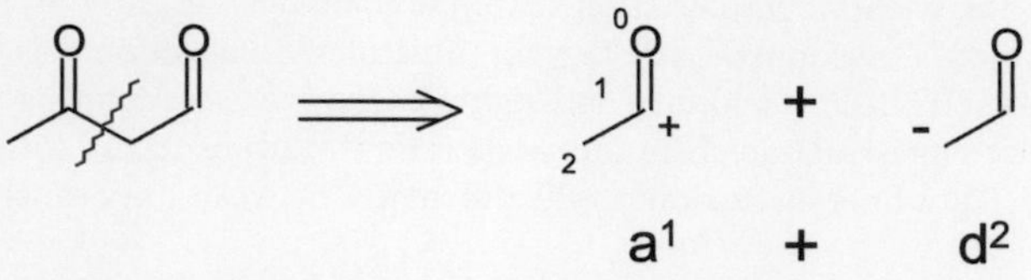

Fig. 4 A simple example of the disconnection to an acceptor a1 and donor d2 synthon. Reproduced from Polanski, J., 2009. Chemoinformatics. In: Walczak, B., Tauler, R., Brown, S. (Eds.), Comprehensive Chemometrics vol. 4. Amsterdam: Elsevier, pp. 459–505.

synthesis design were focused on a reagent to product method using a variety of nondirect approaches which can involve (Barone and Chanon, 2003):

- exploring synthetic availability of a substructure of the target molecule (TM), which can provide the major synthetic step opening the way to target molecule,
- an identification of chemical compound (reagent) similar to the target molecule. This allows to transform synthesis design to a problem of the conversion of the structure of this reagent to the structure of the target molecule,
- an identification of a substructure of the target molecule available as a natural product, if present that can be included as a ready-to-use fragment or so-called building block, e.g., an example can be chiral pool synthesis where building blocks are enantiopure substances.

Eventually, we know a bunch of serendipitous discoveries that provided us with the efficient reagents to product conversions.

A direct product to reagent strategy for designing synthesis has been developed by Corey and Cheng (1989). This allows to start synthesis design from a product, commonly referred as a target molecule (TM) or synthetic target. Operations called retro-synthesis, transform or disconnection identify retron or synthon representation of reagents. The meaning of retron and synthon is very close. Retron means a subunit structural pattern identified in the molecule of product designed as TM, provided that we can discover also a transformation representing (retro-reaction). Synthon represents any group of atom(s) indicated by disconnecting chemical bonds in TM in such a way that we can find real reagents representing these synthons and chemical reactions yielding the product called TM. Synthons are virtual chemical entities. Heterolytic bond disconnection results in the moiety having a potential for the electrophilic or nucleophilic behavior, coded by plus (+) or minus (−) signs indicating the acceptor or donor synthon types, respectively. In **Fig. 4** we illustrated a simple example of the disconnection. Instead of clearly indicated virtual synthons in many advanced analyses we can find directly reagents, for example Corey did not use this term in his recent monograph (Corey and Cheng, 1989). Some essentials of synthon chemistry is discussed below. For further details a reader should compare Smit et al. (1998).

Synthon Nomenclature

Carbon chains formed of C-C and C-H bonds are the most common motifs in organic compounds. Although some reactions may be possible in such systems, in order to observe useful reactions we usually need to differentiate individual atoms by bonding to chemical elements other than C or H. The resulted reactivity differentiation is called chemoselectivity. For example oxygen or nitrogen if bonded to the carbon atom significantly modifies the reactivity of their chemical neighborhood. Therefore, molecular modules other than C-C and C-H are called functional groups or functionalities (FG). FG are of key importance in the contemporary synthetic chemistry approaches. In particular, the mode in which FG influences the reactivity of the neighboring hydrocarbon plays a role in designing potential chemical reactions of the systems. Accordingly, synthon atoms are numbered in relation to the location of the FG and a disconnected (usually carbon) atom (Fuhrhop and Penzlin, 1983) **Fig. 4** illustrates a simple disconnection yielding a common carbonyl acceptor synthon a^1 and carbonyl donor synthon d^2.

Operations on Synthons

Synthons formed in disconnections are virtual entities that need additional operations in order to indicate a corresponding reagents. Synthon to reagent transformation needs at least an addition of an ending at the terminal atom, but can also be much more complex. The basic assumption of organic chemistry is that a certain FG determines a certain reactivity for a series of chemical compounds, e.g., aldehydes react with alcohols to form acetals. A concept of synthons allows us further generalization. Synthon chemistry indicates the reactivity type, comprising various FGs. Some common operations beyond synthetic procedures includes modifications to control synthon reactivity (in blocking, activating or umpolung, i.e., activity reversal mode). A brief introduction to these topics can be found in Polanski (2009b), Smit et al. (1998).

Computer-Assisted Synthesis Design

Total synthesis is a term that describes a synthesis yielding natural products. Mimicking Nature and designing synthesis of natural products demands experience, talent and art. Longifolene is an illustrative example of a first disconnection noted from product to

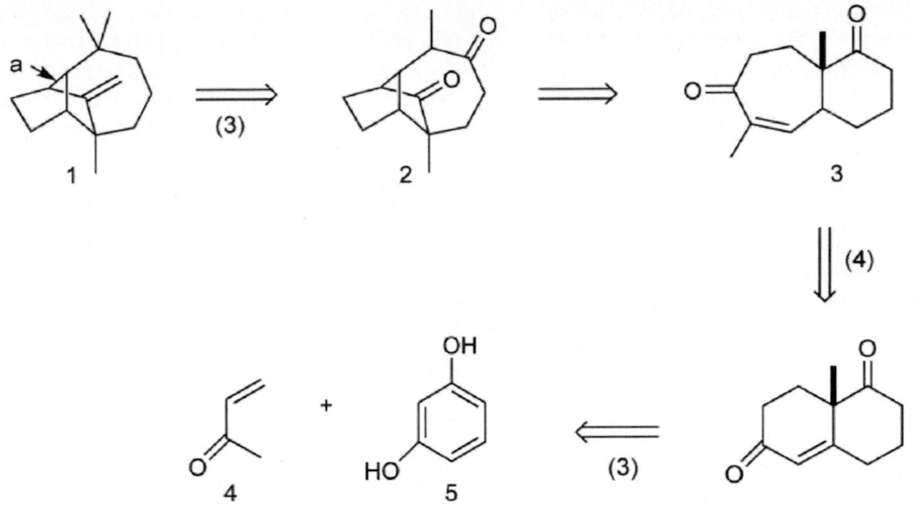

Fig. 5 First disconnections by Corey. Arrows indicate the direction from product to reagent. Modified from Polanski, J., 2009. Chemoinformatics. In: Walczak, B., Tauler, R., Brown, S. (Eds.), Comprehensive Chemometrics vol. 4. Amsterdam: Elsevier, pp. 459–505.

reagent scheme (**Fig. 5**) (Corey and Cheng, 1989; Corey, 1993). Precisely the scheme designed by Corey in 1957 refers not to synthons but directly to reagents. The application of computers in the CASD method was also pioneered by Corey. As usually a single disconnection of TM does not solve the problem but the first level reactants must be disconnected further and further to the next level regents, forming the so-called synthesis tree. Eventually, the availability of the reactants indicates a moment for the last level disconnection. Practically a full disconnection architecture appeared to be an extremely complex computational problem due to an expansion of a number of possible routes analyzed. For example, medium complexity synthesis can involve 10 level synthesis tree in which a number of precursors can exceed a thousand. The expansion of synthetic tree needs to be pruned (Barone and Chanon, 2003).

The strategy of CASD can be compared to the chess strategy where relatively simple laws result in a complex architecture of decision-making where human intuition and experience sometime can find optimal solutions at a first glance (Todd, 2005). Various successful paths are possible; however, the strategy and further paths are strongly limited by the early choices. The expansion of the decision-making tree usually needs pruning, usually by rules of thumb. Both games often rewards a gambit strategy. In chess we can open a game sacrificing a chess piece for the strategic advantages gained. In turn, an unexpected disconnection that breaks common chemical rules may appear a winning strategy. Basically, the CASD software is programmed to find synthons. An important part is also the generation of precursors and sketching synthetic route that corresponds to the decided disconnection tree. A first CASD software, Logic and Heuristics Applied to Synthetic Analysis (LHASA), was programmed in the late 1960′ by Corey group (LHASA). Searching for disconnections was coded here by CHeMistry TRaNslator (CHMTRN), chemical language constructed exceptionally for this purpose (Ott, 2004). In turn, *Chematica* is a recent retrosynthetic system based on the complete chemical information available (Szymkuc et al., 2016). Accordingly, seven million chemicals connected by the number of reactions and catalogues 86,000 chemical rules. Cost-cutting measure is an important indication controlling synthesis paths, shorter and more economical paths are preferred, which nicely illustrates the scope of chemoinformatics (**Fig. 1**). A variety of other retrosynthetic software is currently available.

Computer Assisted Structure Elucidation

Chemical syntheses are designed to provide chemical compounds which structures usually are to be proved by analytical procedures and/or spectroscopic methods (MS, IR NMR) which use nowadays is a routine in modern chemistry. Chemistry as a soft science relatively early took advantage from the artificial intelligence methods which were adopted by chemometrics. This allowed us to develop an efficient Computer Assisted Structure Elucidation which engages both the domains of organic and analytical chemistry (Steinbeck, 2003).

Computer Assisted Knowledge Discovery

Currently, a database is an efficient replacement for a printed version of documentation that enables registration and identifications of chemical compounds. The database system can also be used for the extensive investigations and analyses of chemical data working in a teacher-like mode, modifying, falsifying, or improving existing models, or in other words extending our chemical knowledge. Such data extraction and analysis by database mining is nontrivial and can be an attractive tool giving us a novel and fresh insight into the architecture of chemistry in the approach that is referred to as the discovery of knowledge (Frawley et al., 1991).

Typically, chemical databases equipped with molecular editor and the efficient user friendly searching machine allow us the advanced data queries of the complex syntax. The management of traditional printed information system cannot be any longer a competition for such a technology. Therefore, the last update of the printed BH version appeared in 1998. The printed handbook that has been developed as a breaking innovation in the 19th century science, has been finally overpowered by in silico technologies. Data mining enabled Computer Assisted Knowledge Discovery bringing a substantial improvement into chemical investigations. chemical data can be efficiently accessed and processed all together in a real time. A question appears if this has significantly influenced chemistry. The increased performance is the first issue. But can we really discover chemical laws on the basis of database mining? The answer is positive. To show several examples. The analysis of molecular weight MW of chemical compounds registered in the Beilstein database indicated for example the mean MW of the most commonly used substrates are those near 150 g/mol and the most common products near 250 gmol/1. Interestingly this values did not changed between 1850 and 2004. In turn, the distribution of MW for druggable substances and this for all registered compounds are virtually identical (Fialkowski *et al.*, 2005). An *architecture of organic chemistry* was further outlined by Rucker and Meringer who described the difference between theoretical graphs and the graphs mapped by real chemical compounds registered in Beilstein (Rucker and Meringer, 2002). Technically, self-organizing neural network was used for knowledge discovery in the reaction databases (Chen and Gasteiger, 1997). For review in this area a reader can compare Ester and Sander (2000).

Biological data at unprecedented tera- and petabyte scales were generated recently and the integrated computational platforms (D'Souza *et al.*, 2017) or new visualization tools (Tsiolaki *et al.*, 2017) have been developed for the efficient data mining, accordingly.

Chemometrics – Adopting Mathematics to Chemistry and Chemistry to Mathematics

The origins of chemometrics are in analytical chemistry where the first applications of computers basically engaged the problems which were more of the mathematical than strictly chemical background. Clearly, this application is available on the relatively early stage of the development of computer science. The experience we got by chemometrics can be then applied to fully molecular problems which originated a domain of chemoinformatics. Chemometrics helped to increase the potential of informatics for in silico chemistry expending at the same time the scope of interest of chemical computer simulations. Historically, chemometrics focuses on mathematics, including statistics, programming and so on (Wold, 1995). This explains why we use a term chemometrical analysis as a synonym of the application of the advanced statistical methods for the extraction of chemical knowledge from chemical data (Myshkin and Wang, 2003; Pierce *et al.*, 2005; Pytela *et al.*, 1994) or even in the more narrow sense as PCA or PLS analyses (Rodriguez-Barrios and Gago, 2004). Mathematics forms a common language generally understood by science. However, each science branch has its specific language. Accordingly, an efficient translator support can significantly enhance the quality and understanding of mathematics in chemistry. At the same time chemical problems can be formulated in a form suitable for mathematical applications.

Clearly, chemoinformatics is a broader term than chemometrics and sometime chemometrics is included here as a sub-discipline. Usually, however, chemometricians prefer to form an autonomous science.

Computer Assisted Molecular Design

With the increasing number of chemical compounds the kind of inflation of new compounds have appeared. Therefore, we need to indicate and design chemical compounds that would be interesting objects of investigations. The architecture of organic chemistry indicates chemical space interesting for chemists (Fialkowski *et al.*, 2005). Actually, a term molecular design focuses on two different rationale categories. First is a design of any compound or in other words synthesis design discussed in article Chemical Syntheses and Retro-Syntheses. Second design rationale relates more to property than structural design. Chemistry contributes to industry constructing a variety of chemicals of the crucial importance for a nowdays civilization such as drugs, preservatives, flavors, etc. However, individual chemical molecules having a certain property can be integrated with different chemical structures. Accordingly, we can arranged atoms in the molecules into a variety of combinations of the different compounds hopefully possessing individual properties as desired. In practice, the available molecular configurations even today depend upon synthetic capability of individual laboratories and current chemical technology. This mean molecular design is to some extent limited by synthetic capability. Are there any simple rules indicating a formation of useful chemical compounds? Surprisingly, some chemical frameworks are suggested to be preferred patterns in the drug molecules (Fattori, 2004; Kubinyi, 2006; Schneider and Schneider, 2017).

From Data to Drug Candidates and Drugs

Basically, a molecule of drug is a man-made moiety designed or discovered in other way to fit the biological counterpart and produce the required action. Accordingly, drugs are to manipulate biological effects generated by macromolecular proteins designated as targets or receptors (a precise definition of the receptor can be found in Cohen, 1996). Thus, the drug-receptor interactions define a potential activity. Currently we attempt to simulate these interactions in silico. Drug design or molecular design are two terms used as synonyms to refer to in silico technology in this area. This duality nicely illustrates the difference between a molecule and a substance (drug). A term drug design scores high on optimism because currently we do not have an efficient technology to virtually construct a real drug in a single step engineering like process that would resemble for example the design of a bridge or a building. Therefore, precisely we can define molecular design as a method attempting to find *drug candidates* on the basis of computational techniques. Drug discovery or drug development are broader categories referring to the investigations (both in silico and in vitro) of drug candidates as potential drugs. Basically drugs can be developed by target based or phenotypic strategies (Swinney and Anthony, 2011). The analysis of the economic efficiency of current drug design R&D efficiency is pessimistic. The decreasing financial success is revealed and described in the so-called backward Moors (Erooms) law (Scannell *et al.*, 2012). Big pharma is driven by innovations. This means new drugs are needed for new patents capable of bringing new incomes. Therefore, companies are forced to fierce R&D search. At the same time regulations decided that development strategy was turned to target oriented approaches which, apparently, appeared too optimistic. Stiff quality and safety requirements and organizational factors could also be a success barrier here (Leeson and St-Gallay, 2011). On the other hand there are good examples of the successful drug design (Borman, 2005) and drugs get better and better imitating natural biological effectors.

Structure Based Design

What we mean by structure based design is a method that is based on the known receptor structure. Basically, two basic approaches of docking and de-novo design are available to fill of a receptor with a ligand structure (Schneider and Fechner, 2005). By docking we understand fitting the ligand structure into the receptor cavity (Kitchen *et al.*, 2004). Available programs are specified in Warren *et al.* (2006). In turn, in de-novo design a potential receptor ligand is constructed directly in the receptor cavity from the molecular fragments or even individual atoms. Available programs and illustrative examples are given in Warren *et al.* (2006), Kirkpatrick (2005). Docking protocols were used for High Throughput Screening (HTS) mode of in silico searching for the receptor ligands. The Intel-United Devices Cancer Research Project is here an example of internet distributed computing in which 3.5 billion molecules were investigated as potential anticancer drugs in the simulation performed by downloadable screensaver (Richards, 2002).

Recent advances in technology and novel approaches to drug discovery are now available in the context of the description of the target structure. This includes *structure, dynamics, mutational activation and inactivation, and signaling mechanisms.* An illustrative example can be small-molecule inhibitors of KRAS (Ostrem and Shokat, 2016). *Mechanism based design* is a term used as a formal term referring to such methods.

Ligand Based Design

Often not enough data for target receptor is available. In such a case the ligands that are known forms the basis for the comparison of the new molecular representations. This approach is known as a *ligand based design*. Accordingly, this method is based on the nondirect investigation of the ligand-target interactions or in other words *receptor* or *pharmacophore mapping* where a pharmacophore refers to the receptor or receptor sector model *reconstructed* from a series of its ligands. Depending upon individual method, pharmacophores or pseudoreceptors are usually assembled from virtual elements which can be represented by various functionalities, interaction types, or even fragments of molecules, e.g., aminoacids, atom or atom types. Actually, generally only very rarely any relation between a real receptor and a pharmacophore exists. Extensive review of the methods is available for example in Cohen (1996), Horvath *et al.* (2005).

Mapping Structure to Property in QSAR Approach

Quantitative structure activity relationship (QSAR) is a strategy of the essential importance for chemistry and pharmacy, based on the idea that when we change a structure of a molecule then also the activity or property of the substance will be modified. Structure modification can result from the virtual in silico operations, intentional in vitro projects (usually synthetic research) or we can just investigate available substances. The importance of the QSAR concept can be recognized by a volume of the literature and variants of the methods described. A brief introduction to this area can be found in Polanski (2009a,b), while current state of QSAR is thoroughly reviewed by Cherkasov *et al.* (2014). Basically, in QSAR we map chemical space to property space modeling a function relating property to chemical structure or more accurately to the molecular descriptors that are related to this chemical structure. This function should work like a dictionary between two spaces hopefully predicting the activity of novel compounds of the desired properties. Theoretically, these molecules can be then synthesized and tested as drug candidates while iterative QSAR modeling will further optimize second level synthetic targets and the property itself. Because a QSAR function relates property to structure, this is an indirect design of the molecular representations having this property (Polanski *et al.*, 2006).

Multidimensionality of QSAR (0-7D) relates to a complexity of the ligand or ligand-receptor data coded by molecular descriptors used during modeling. In the majority of examples QSARs are modeled without receptor data (receptor independent mode) but receptor dependent QSARs are also possible (Polanski, 2009a). In the majority of described applications QSAR realizes a strategy from molecules to property, however, an inverse strategy from property to molecules can also be hypothesized

(De Julian-Ortiz, 2000). According to the scope of chemoinformatics (**Fig. 1**) analyzing and/or modeling economic properties are hot issues that could significantly contribute to molecular design. Accordingly, the analysis of big data from molecular market was performed to obtain the first quantitative structure economy relationship (Polanski *et al.*, 2016).

Drug-Likeness and Druggability Concept

Data dependency and uncertainty is an immanent part of QSAR modeling (Polanski, 2009a; Polanski *et al.*, 2006) and we still do not have a technology that would be capable of designing in a single act a molecule having properties desired. Instead, molecular design is a method that improve the success ratio in a search for new drug candidates. Therefore, this kind of design prefers rather general than rigorous rules. The example can be the so-called drug-likeness concept insisting that some common features or even privileged substructures can be indicated in drugs. The Lipinski rule of five lists here molecular weight < 500, $\log P < 5$, a number of hydrogen bond acceptors < 10, and a number of hydrogen bond donors < 5 that rule advantageous drug-likeness for the orally available drugs (Lipinski *et al.*, 1997). Precisely, the Lipinski rule focuses on molecular descriptors. In turn, the ADMET concept shifted our interest to the properties, in particular, these deciding compounds bioavailability and/or biocompatibility, i.e., Adsorption, Distribution, Metabolism, Elimination and Toxicology (Davis and Riley, 2004; Hodgson, 2001; Van de Waterbeemd, Gifford, 2003). Finally, a concept of privileged structures appeared. For example, it has been discovered that a "group of 32 common shapes or frameworks accounted for 50% of the 5120 molecules considered. Whether these fragments had intrinsic characteristics that gave them drug-like properties or their presence was a result of chemists' habits, familiarities or synthetic versatility was an issue that was recognized but not addressed" (Fattori, 2004). NCI DTP AIDS Antiviral Screen database (40,000 compounds) was searched for the common frameworks and molecular fragment distribution typical for anti-HIV activity (Helma *et al.*, 2002). The related approach allowed Oprea to cluster more than 12,000 compounds into different activity levels (Oprea, 2004). Some privileged drug-like moieties can be found in Kubinyi (2006). Eventually, the comparative or combined QSAR databases appeared in a hope for identifying more molecular drug-likeness rules (Hansch *et al.*, 2002; Oprea, 2002). Shen *et al.* developed a concept of "the application of predictive QSAR as a virtual screening tool for database mining" (Shen *et al.*, 2004). Protein and gene expression data can be processed similarly to molecular representations (Helma *et al.*, 2002). Accordingly, druggability relates to the selected receptors that provide the privileged fit to drug-like molecules while we can look for druggable genomes expressing druggable proteoms (Lipinski and Hopkins, 2004).

Bioinformatics in Drug Design

In bioinformatics we focuses of the biological space rather than on the chemical one. Formally, we are mapping here the biological to chemical space. Pharmacogenomics is a concept that by the analysis of genomic data we can obtain novel drugs. Novel data measured form DNA microarray gene expression experiments pose new computational and processing challenges that formed bioinformatics to an important interdisciplinary research.

The relation of bioinformatics and chemoinformatics resembles this of chemoinformatics and chemometrics. Accordingly, chemoinformatics can be interpreted as a part of bioinformatics.

Internet Resources for Chemistry and Chemoinformatics

A first cable computer connection and networking took place in 1960' in the United States. Nobody could have expected the expansion of social interaction by Internet. Nowadays, the impact of Internet on the economy and science is clear. Web technologies are more and more common in computer science and chemoinformatics. A number of chemical resources available in the web steadily increases. First of all, we can indicate a number of sites with chemical data. A variety of other web resources include educational materials, software (free or commercial), on-line chemistry journals and many others. E-commerce is also a common way of the distribution of chemicals. The enumeration of all available web sites that are of the interest for chemoinformatics is not possible any-longer in such a brief review. The addiction to Wikipedia and Google and the enormous wide SMILES presence in web technologies are examples of the success of chemoinformatics.

Current Practices and Capabilities

Basically, the methods of chemoinformatics have been formed in 1990' and early 2000'. Accordingly, a four volume Handbook of Chemoinformatics edited by Gasteiger (2003) is still the most comprehensive up-to-date introduction to the topic. Generally, what we observe recently is a move from method development to the efficient applications and vigorous software development. On one hand chemical informatics was moved into public domain. On the other hand still a lot of payable services are under development. In this context current practices follow general directions of computer related technology development. A lot of the services are available. A mobile CAS service can be an illustrative example here. Chemoinformatics has answered the expectations and needs of pharma for new methods in computer-aided molecular design. Drug discovery and development is still the most important research field. Organization and utilization of drug discovery data involving small molecules, analysis of structure-activity relationships, and prediction of active compounds have remained hallmarks (Bajorath, 2015). Much higher computation capability, significantly higher and less expensive storage and memory provided significantly better opportunities for the information storage and processing. Docking is routinely used to identify potential hits from large compound libraries and Zanamivir and oseltamivir (Tamiflu) are examples of anti-influenza drugs developed in structure-based strategies which combined computer-based approaches, such as docking and pharmacophore-based virtual screening with X-ray crystallographic structural analyses (Mallipeddi *et al.*, 2014). Big data is a good example illustrating a level of the development of chemo and info counterparts of chemoinformatics. We need a proper structure of the data, in particular, the measured properties should be reliable, precise and extend enough to describe the complexity of the problem. For example, we usually analyze drug candidate action by the activity value versus a single target. In fact, however, a xenobiotic in an organism can induce a significant changes in signaling of various targets. This concept is known as polypharmacology. In this context, the polypharmacology and lipidomics inspired scheme of HTS data screening has been developed recently by Gabriele Cruciani group (Goracci *et al.*, 2017). The method is limited by an inexpensive and rapid procedure for the extraction and MS based determination of a medium size population of lipid species that can characterize the complex answer of any organism for a drug (drug candidate) administration. In 20 min this provides us a data on the serum concentration of 1000 different lipids in serum. Such a data can be arranged in a fingerprint-like lipid profile which if tested as a function of time, can give us a real information on the activity and toxicity of the xenobiotics used. This illustrated that more efficient big data methods are limited not necessarily by the statistical issues but by new technology in inexpensive property measurements (Polanski and Gasteiger, 2016). New HTS methods for property measurements should undoubtedly bring new quality in big data analyses in drug design. Another recent big data study is an attempt to model the Alzheimer disease. The extend experimental property measurement has been referred here as *smart data* approach (Geerts *et al.*, 2016). Femto or atosecond serial X-ray protein nanocrystallography which generates as much as 3 million. X-ray patterns (measured 3D structure data) in seconds (Chapman *et al.*, 2011) is another illustrative example.

Closing Remarks and Future Directions

We can interpret a formation of chemoinformatics as an answer to a fascination of the potential of in silico chemistry but also to the disappointment of computer assisted drug design in early 90′. A need to teach computers the essentials of chemistry not only provided us with new tools but also allowed us for better understanding of chemistry. Currently chemoinformatics is intelligent and flexible enough to adopt the uncertainty of fuzzy data produced by in vitro chemistry. In this context chemical ontology is an example of the trend which will probably be followed in silico. Molecular design and drug design are two terms used for the description of the main objective of chemoinformatics. This nicely illustrates the dichotomy between molecules and substances forming a class of chemical compounds. While molecules are represented mainly by molecular descriptors, properties represent substances. A flood of molecular descriptors is accompanied by property deficit. Properties are rare because their measurements are expensive. This means property predictions are often required. Property deficit explains low success ratio of drug design. In fact, the basic objective of chemoinformatics is a design of drug candidates by the search of advantageous molecular descriptors on the basis of few properties available. Then, we hope some of candidates will appear successful. Hopefully, with the physicalization of chemistry, properties will appear more available making drug candidates better. The example are *omics* methods which provided us with new technologies providing more and more property data. Big pharma is driven by innovation. To develop innovation capable of wining on the market these data should be cheap and capable for rapid measurements. Basically, we know how to efficiently transform the data into chemical knowledge and further into drug candidates. With the increasing availability of properties better quality of drug candidates will increase the success ratio in developing new drugs. New in silico methods are appearing and still are to appear with wider availability of the property data to efficiently treat this kind of data.

Nowdays Internet is a philosophy of the informatics, in particular, chemoinformatics. Chemistry depended on data and the web has conserved this dependency making the data broadly available. Our addiction to the Internet will definitely increase, stimulating also further cooperative development of the discipline.

Eventually, this text is a significantly shortened but at the same time revised and upgraded version of the article in the Elsevier Encyclopedia of Comprehensive Chemometrics.

See also: Binding Site Comparison – Software and Applications. Chemical Similarity and Substructure Searches. Drug Repurposing and Multi-Target Therapies. Fingerprints and Pharmacophores. In Silico Identification of Novel Inhibitors. Natural Language Processing Approaches in Bioinformatics. Pharmacophore Development. Public Chemical Databases. Quantitative Structure-Activity Relationship (QSAR): Modeling Approaches to Biological Applications. Rational Structure-Based Drug Design. Small Molecule Drug Design

References

Audibert, P., 2013. Mathematics for Informatics and Computer Science. Hoboken, London: Wiley-ISTE.

Bajorath, J., 2015. Entering new publication territory in chemoinformatics and chemical information science. F1000Research 4, 35.

Baldi, P., 2005. Chemoinformatics, drug design, and systems biology. Genome Inform. 16, 281–285.

Barone, R., Chanon, M., 2003. Computer-assisted synthesis design. In: Gasteiger, J. (Ed.), Handbook of Chemoinformatics From Data to Knowledge. Weinheim: Wiley-VCH, pp. 1428–1456.

Bensaude-Vincent, B., Simon, J., 2012. Chemistry – The Impure Science. Rosewood Drive: Imperial College Press.

Bobach, C., Boehme, T., Laube, U., Pueschel, A., Weber, L., 2012. Automated compound classification using a chemical ontology. J. Cheminf. 4 (40), 1–12.

Bock, G., Goode, J.A., 2002. Silico' Simulation of Biological Processes. Chichester: John Wiley and Sons.

Bohacek, R.S., McMartin, C., Guida, W.C., 1996. The art and practice of structure-based drug design: A molecular modelling perspective. Med. Res. Rev. 16, 3–50.

Borman, S., 2005. Drugs by design. Chem. Eng. News 83, 28–30.

Brock, W.H., 1993. The Fontana History of Chemistry. London: Fontana Press.

Brown, F., 1998. Chemoinformatics: What is it and how does it impact drug discovery. Annu. Rep. Med. Chem. 33, 375–384.

Chapman, H.N., Fromme, P., Barty, A., et al., 2011. Femtosecond X-ray protein nanocrystallography. Nature 470, 73–77.

Chen, L., 2003. Reaction classification and knowledge acquisition. In: Gasteiger, J. (Ed.), Handbook of Chemoinformatics From Data to Knowledge. Weinheim: Wiley-VCH, pp. 348–388.

Chen, L.R., Gasteiger, J., 1997. Knowledge discovery in reaction databases: Landscaping organic reactions by a self-organizing neural network. J. Am. Chem. Soc. 119, 4033–4042.

Cherkasov, A., Muratov, E.N., Fourches, D., et al., 2014. QSAR modeling: Where have you been? Where are you going to? J. Med. Chem. 57, 4977–5010.

Cohen, J., Stewart, J., 1994. The Collapse of Chaos: Discovering Simplicity in a Complex World. New York: Viking.

Cohen, N.C., 1996. Guidebook on Molecular Modelling in Drug Design. San Diego: Academic Press.

Consonni, V., Todeschini, R., 2010. Molecular descriptors. In: Puzyn, T., et al. (Eds.), Recent Advances in QSAR Studies. Amsterdam: Springer, pp. 29–102.

Corey, E.J., 1993. The logic of chemical synthesis: Multistep synthesis of complex Carbogenic molecules. In: Malmstrom, B.G. (Ed.), Nobel Lectures in Chemistry 1981–1990. Singapore: World Scientific, pp. 686–708.

Corey, E.J., Cheng, X.-M., 1989. The Logic of Chemical Synthesis. New York: Wiley.

De Julian-Ortiz, J., 2000. Virtual Darwinian drug design: Qsar inverse problem. Comb. Chem. High Throughput Screening 4, 295–310.

D'Souza, M., Sulakhe, D., Wang, S., et al., 2017. Strategic integration of multiple bioinformatics resources for system level analysis of biological networks. Methods Mol. Biol. 1613, 85–99.

Davis, A.M., Riley, R.J., 2004. Predictive ADMET studies, the challenges and the opportunities. Curr. Opin. Chem. Biol. 8, 378–386.

Ester, M., Sander, J., 2000. Knowledge discovery in databases Techniken und Anwendungen. Berlin: Springer.

Fattori, D.D., 2004. Molecular recognition: The fragment approach in lead generation. Drug Discovery Today 9, 229–238.

Fialkowski, M., Bishop, K.J., Chubukov, V.A., Campbell, C.J., Grzybowski, B.A., 2005. Architecture and evolution of organic chemistry. Angew. Chem. Int. Ed. Engl. 44, 7263–7269.

Frawley, W.J., Piatetsky-Shapiro, G., Matheus, C., 1991. Knowledge discovery in databases: An overview. In: Piatetsky-Shapiro, G., Frawley, W.J. (Eds.), Knowledge Discovery in Databases. Cambridge, MA: AAAI Press/MIT Press, pp. 1–30.

Fuhrhop, J., Penzlin, G., 1983. Organic Synthesis Concepts, Methods, Starting Materials. Weinheim: VCH.

Gasteiger, J., Engel, T., 2004. Chemoinformatics a Textbook. Weinheim: Wiley-VCH.

Geerts, H., Dacks, P.A., Devanarayan, V., Haas, M., Khachaturian, Z., et al., 2016. Big data to smart data in Alzheimer's disease: The brain health modeling initiative to foster actionable knowledge. Alzheimers Dement. 12, 1014–1021.

Goodman, J., 1999. Chemical Applications of Molecular Modeling. London: Royal Society of Chemistry.

Goracci, L., Tortorella, S., Tiberi, P., et al., 2017. Lipostar, a comprehensive platform-neutral cheminformatics tool for lipidomics. Anal. Chem. 89, 6257–6264.

Hansch, C., Hoekman, D., Leo, A., Weininger, D., Selassie, C., 2002. Chembioinformatics: Comparative QSAR at the interface between chemistry and biology. Chem. Rev. 102, 783–812.

Hastings, J., Adams, N., Ennis, M., Hull, D., Steinbeck, C., 2011. Chemical ontologies: What are they, what are they for and what are the challenges. J. Cheminf. 3 (Suppl 1), O4.

Helma, C.H., Kramer, S., De Raedt, L., 2002. The molecular feature miner MOLFEA. In: M.G. Hicks, M.C., Kettner, C. (Eds.), Molecular Informatics: Confronting Complexity, Proceedings of the Beilstein-Institut Workshop, Bozen, pp. 1–15.

Hinchliffe, A., 2003. Molecular Modelling for Beginners. Chichester: Wiley.

Hodgson, J., 2001. ADMET–Turning Chemicals Into Drugs. Nat. Biotechnol. 19, 722–726.

Holtje, H.-D., Sippl, W., Rognan, D., Folkers, G., 2003. Molecular Modeling. Weinheim: Wiley-VCH.

Hopkins, A.L., Keseru, G.M., Leeson, P.D., Rees, D.C., Reynolds, C.H., 2014. The role of ligand efficiency metrics in drug discovery. Nat Rev Drug Discov. 13, 105–121.

Horvath, D., Mao, B., Gozalbes, R., Barbosa, F., Rogalski, S.L., 2005. limitations Strengths and of pharmacophore-based virtual screening. In: T.I. Oprea (Ed.), Chemoinformatics in Drug Discovery. In: Mannhold, R., Kubinyi, H., Timmerman H. (Eds.), vol. 23, Methods and Principles in Medicinal Chemistry. Weinheim: Wiley-VCH, pp. 117–140.

Jager, W., Krebs, H.-J. (Eds.), 2003. Mathematics-Key Technology for the Future. Berlin: Springer.

Keil, M., Borosch, T., Exner, T.E., Brinkman, J., 2003. Computer visualization of molecular models tools for man-machine communication in molecular science. In: Gasteiger, J. (Ed.), Handbook of Chemoinformatics From Data to Knowledge. Weinheim: Wiley-VCH, pp. 320–344.

Keseru, G., Kolossvary, I., 1999. Molecular Mechanics and Conformational Analysis in Drug Design. Oxford: Blackwell Publishing.

Kimura, S.R., Hu, H.P., Ruvinsky, A.M., Sherman, W., Favia, A.D., 2017. Deciphering cryptic binding sites on proteins by mixed-solvent molecular dynamics. J. Chem. Inf. Model. 57, 388–1401.

Kirkpatrick, P., 2005. Computational chemistry: Docking on trial. Nat. Rev. Drug Discov. 4, 813.

Kitchen, D.B., Decornez, H., Furr, J.R., Bajorath, J., 2004. Docking and scoring in virtual screening for drug discovery: Methods and applications. Nat. Rev. Drug Discov. 3, 935–949.

Kochev, N., Monev, V., Bangov, I., 2004. Searching chemical structures. In: Gasteiger, J., Engel, T. (Eds.), Chemoinformatics a Textbook. Weinheim: Wiley-VCH, pp. 291–318.

Kowalski, B.R., Bender, C.F., 1975. Solving chemical problems with pattern recognition. Naturwissenschaften 62, 10–14.

Kubinyi, H., 2006. Privileged structures and analogue-based drug discovery. In: Fischer, J., Ganellin, C.R. (Eds.), Analogue-Based Drug Discovery. Weinheim: Wiley-VCH, pp. 53–65.

Leach, A.R., Gillet, V.J., 2003. An Introduction to Chemoinformatics. Dordrecht: Kluwer.

Leeson, P.D., St-Gallay, S.A., 2011. The influence of the 'organizational factor' on compound quality in drug discovery. Nat. Rev. Drug Discovery 10, 749–765.

Lipinski, C., Hopkins, A., 2004. Navigating chemical space for biology and medicine. Nature 432, 855–861.

Lipinski, C.A., Lombardo, F., Dominy, B.W., Feeney, P.J., 1997. Experimental and computational approaches to estimate solubility and permeability in drug discovery and development settings. Adv. Drug Delivery Rev. 23, 3–25.

Magdziarz, T., Mitusinska, K., Goldowska, S., *et al.*, 2017. AQUA-DUCT: A ligands tracking tool. Bioinformatics 33, 2045–2046.

Maheshwari, A., 2014. Data Analytics Made Accessible, Kindle edition Amazon.

Mallipeddi, P.L., Kumar, G., White, S.W., Webb, T.R., 2014. Recent advances in computer-aided drug design as applied to anti-influenza drug discovery. Curr. Top. Med. Chem. 14, 875–889.

Morris, P.J.T., 2002. From Classical to Modern Chemistry. London: Royal Society of Chemistry.

Motherwell, S., 2004. Chemoinformatics and crystallography. The Cambridge structural database. In: Noordik, J.H. (Ed.), Cheminformatics Developments, History, Reviews and Current Research. Amsterdam: IOS Press, pp. 37–68.

Myshkin, E., Wang, B.J., 2003. Chemometrical classification of Ephrin ligands and Eph Kinases using GRID/CPCA approach. J. Chem. Inf. Comput. Sci. 43, 1004–1010.

Nicklaus, M.C., 2003. Pharmacophore and drug discovery. In: Gasteiger, J. (Ed.), Handbook of Chemoinformatics From Data to Knowledge. Weinheim: Wiley-VCH, pp. 1687–1711.

Noordik, J.H., 2004. Cheminformatics Developments. Amsterdam: IOS Press.

Oprea, T.I., 2003. Chemoinformatics and the quest for leads in drug discovery. In: Gasteiger, J. (Ed.), Handbook of Chemoinformatics From Data to Knowledge., pp. 1509–1531.

Oprea, T., 2002. Current trends in lead discovery. Are we looking for the appropriate properties? J. Comput. -Aided Mol. Des. 16, 325–334.

Oprea, T., 2004. 3D-QSAR modeling in drug design. In: Tolleneare, J., De Winter, H., Langenaeker, W., Bultinck, P. (Eds.), Computational Medicinal Chemistry for Drug Discovery. New York: Marcel Dekker, pp. 571–616.

Ostrem, J.M., Shokat, K.M., 2016. Direct small-molecule inhibitors of KRAS: From structural insights to mechanism-based design. Nat. Rev. Drug Discov. 15, 771–785.

Ott, M.A., 2004. Chemoinformatics and organic chemistry. Computer assisted synthetic analysis. In: Noordik, J.H. (Ed.), Cheminformatics Developments, History, Reviews and Current Research. Amsterdam: IOS Press, pp. 83–110.

Pierce, K.M., Wood, L.F., Wright, B.W., Synovec, R.E.A., 2005. Comprehensive two-dimensional retention time alignment algorithm to enhance chemometric analysis of comprehensive two-dimensional separation data. Anal. Chem. 77, 7735–7743.

Polanski, J., 2003. Molecular shape analysis. In: Gasteiger, J. (Ed.), Handbook of Chemoinformatics From Data to Knowledge. Weinheim: Wiley-VCH, pp. 302–319.

Polanski, J., 2009a. Receptor dependent multidimensional QSAR for modeling drug-receptor interactions. Curr. Med. Chem. 16, 3243–3257.

Polanski, J., 2009b. Chemoinformatics. In: Walczak, B., Tauler, R., Brown, S. (Eds.), Comprehensive Chemometrics vol. 4. Amsterdam: Elsevier, pp. 459–505.

Polanski, J., 2017. Big data in structure-property studies – From definitions to models. In: Leszczynski, J., Roy, K. (Eds.), Advances in QSAR Modeling With Applications in Pharmaceutical, Chemical, Food, Agricultural, and Environmental Sciences. Berlin/Heidelberg: Springer.

Polanski, J., Gasteiger, J., 2016. Computer representation of chemical compounds. In: Leszczynski, J., Puzyn, T. (Eds.), Handbook of Computational Chemistry. Dordrecht: Springer.

Polanski, J., Gieleciak, R., Bak, A., Magdziarz, T., 2006. Robust QSAR modeling. J. Chem. Inf. Model. 46, 2310–2318.

Polanski, J., Kucia, U., Duszkiewicz, R., *et al.*, 2016. Molecular descriptor data explain market prices of a large commercial chemical compound library. Sci. Rep. 6, 28521.

Polanski, J., Tkocz, A., 2017. Between descriptors and properties: Understanding the ligand efficiency trends for G protein-coupled receptor and kinase structure-activity data sets. J. Chem. Inf. Model. 57, 1321–1329.

Pytela, O., Kulhanek, J., Ludwig, M., 1994. Chemometrical analysis of substituent effects. IV. Additivity of substituent effects in dissociation of 3,5-Disubstituted benzoic acids in organic solvents. Collect. Czech. Chem. Commun. 59, 1637–1644.

Rapaport, C., 2004. The art of molecular dynamics simulation. Cambridge: Cambridge University Press.

Reymond, J.-L., Van Deursen, R., Blum, L.C., Ruddigkeit, L., 2010. Chemical space as a source for new drugs. Med. Chem. Commun. 1, 30–38.

Richards, W.G., 2002. Virtual screening using grid computing: The screensaver project. Nat. Rev. Drug Discov. 1, 551–555.

Rodriguez-Barrios, F., Gago, F., 2004. Chemometrical identification of mutations in Hiv-1 reverse transcriptase conferring resistance or enhanced sensitivity to Arylsulfonylbenzonitriles. J. Am. Chem. Soc. 126, 2718–2719.

Rucker, C., Meringer, M., 2002. How many organic compounds are graph-theoretically nonplanar? MATCH Commun. Math. Comput. Chem. 45, 153–172.

Sadowski, J., Gasteiger, J., 1993. From atoms and bonds to three-dimensional atomic coordinates: Automatic model Builders. Chem. Rev. 93, 2567–2581.

Scannell, J.W., Blanckley, A., Boldon, H., Warrington, B., 2012. Diagnosing the decline in pharmaceutical R&D efficiency. Nat. Rev. Drug Discovery 11, 191–200.

Schneider, G., Fechner, U., 2005. Computer-based de novo design of drug-like molecules. Nat. Rev. Drug Discov. 4, 649–663.

Schneider, P., Schneider, G., 2017. Privileged structures revisited. Angew. Chem. Int. Ed. Engl. Available at: https://doi.org/10.1002/anie.201702816.

Shen, M., Beguin, C., Golbraikh, A., *et al.*, 2004. Application of predictive QSAR models to database mining: Identification and experimental validation of novel anticonvulsant compounds. J. Med. Chem. 47, 2356–2364.

Smith, A., Heckelman, P.E., O'Neil, M.J., Budavari, S. (Eds.), 2001. The Merck Index an Encyclopedia of Chemicals, Drugs, & Biologicals, thirteenth ed. Whitehouse Station: Merck & Co. Inc.

Smit, W.A., Caple, R., Bochkov, A.F., 1998. Organic synthesis the science behind the Art. London: Royal Society of Chemistry.

Steinbeck, C.H., 2003. Computer-assisted structure elucidation. In: Gasteiger, J. (Ed.), Handbook of Chemoinformatics From Data to Knowledge. Weinheim: Wiley-VCH, pp. 1378–1406.

Swinney, D.C., Anthony, J., 2011. How were new medicines discovered? Nat. Rev. Drug Discovery 10, 507–519.

Szlezak, N., Evers, M., Wang, J., Perez, L., 2014. The role of big data and advanced analytics in drug discovery, development, and commercialization. Clin. Pharmacol. Ther. 95, 492–4955.

Szymkuc, S., Gajewska, E.P., Klucznik, T., Molga, K., Dittwald, P., 2016. Computer-assisted synthetic planning: The end of the beginning. Angew. Chem. Int. Ed. Engl. 55, 5904–5937.

Todd, M.H., 2005. Computer-aided organic synthesis. Chem. Soc. Rev. 34, 247–266.

Todeschini, R., Consonni, V., 2000. Handbook of Molecular Descriptors. Weinheim: Wiley-VCH.

Tsiolaki, P.L., Nastou, K.C., Hamodrakas, S.J., Iconomidou, V.A., 2017. Mining databases for protein aggregation: A review. Amyloid. 24, 143–152.

Van de Waterbeemd, H., Gifford, E., 2003. ADMET in silico modelling: Towards prediction paradise? Nat. Rev. Drug Discov. 2, 192–204.

Wang, Y., Bryant, S.H., Cheng, T., *et al.*, 2017. PubChem bioassay statistics. Nucleic Acids Res. 45, D955–D963.

Warren, G.L., Andrews, C.W., Capelli, A.-M., *et al.*, 2006. Critical assessment of Docking programs and scoring functions. J. Med. Chem. 49, 5912–5931.

Weininger, D., 2003. SMILES – A language for molecules and reactions. In: Gasteiger, J. (Ed.), Handbook of Chemoinformatics From Data to Knowledge. Weinheim: Wiley-VCH, pp. 80–102.

Willet, P.A., 2003. History of Chemoinformatics. In: Gasteiger, J. (Ed.), Handbook of Chemoinformatics From Data to Knowledge. Weinheim: Wiley-VCH, pp. 6–20.

Wisniewski, J., 2003. Chemical nomenclature and structure representation: Algorithmic generation and conversion. In: Gasteiger, J. (Ed.), Handbook of Chemoinformatics From Data to Knowledge. Weinheim: Wiley-VCH, pp. 51–79.

Wold, S., 1995. Chemometrics; What do we mean with it, and what do we want from it? Chemometr. Intell. Lab. Syst. 30, 109–115.

Yang, C., Tarkhov, A., Marusczyk, J., *et al.*, 2015. New publicly available chemical query language, CSRML, to support chemotype representations for application to data mining and modeling. J. Chem. Inf. Model. 2015 (55), 510–528.

Further Reading

Gasteiger, J. (Ed.), 2003. Handbook of Chemoinformatics From Data to Knowledge. Weinheim: Wiley-VCH.
Gasteiger, J., Engel T. (Eds.), 2003. Chemoinformatics a Textbook, Weinheim: Wiley-VCH.
Heller, S.R., 1997. (Ed.), The Beilstein Online Database – Implementation, Content, and Retrieval ACS Symposium Series 436, American Chemical Society, Washington, DC, 1990. 13. A.D. McNaught, A. Wilkinson, Compendium of Chemical Terminology (The Gold Book), second ed., Blackwell Scientific Publications, Oxford, UK.
Kubinyi, H. (Ed.), 2008. QSAR: Hansch Analysis and Related Approaches. Weinheim, New York: VCH.
Leszczynski, J., Puzyn, T. (Eds.), 2016. Handbook of Computational Chemistry. Dordrecht: Springer.
Polanski, J., 2009. Chemoinformatics. In: Walczak, B., Tauler, R., Brown, S. (Eds.), Comprehensive Chemometrics. Amsterdam: Elsevier, pp. 459–505.
Polanski, J., 2017. Big data in structure-property studies-from definitions to models. In: Leszczynski, J., Roy, K. (Eds.), Advances in QSAR Modeling With Applications in Pharmaceutical, Chemical, Food, Agricultural, and Environmental Sciences. Berlin/Heidelberg: Springer.
Polanski, J., Gasteiger, J., 2016. Computer representation of chemical compounds. In: Leszczynski, J., Puzyn, T. (Eds.), Handbook of Computational Chemistry. Dordrecht: Springer.
Roy, K. (Ed.), 2015. Quantitative Structure-Activity Relationships in Drug Design, Predictive Toxicology and Risk Assesment. Hershey PA: Medical Information Science Reference.
Schweitzer, S.O., 2007. Pharmaceutical Economics and Policy. Oxford: University Press.
Smith, M.B., March, J., 2001. March's Advanced Organic Chemistry Reactions Mechanisms, and Structure. New York: Wiley.
Varnek, A. (Ed.), 2017. Tutorials in Chemoinformatics. Hoboken, Chichester: Wiley.
Walczak, B., Tauler, R., Brown, S. (Eds.), 2009. Comprehensive Chemometrics, Vols. 1–4. Amsterdam: Elsevier.

Relevant Websites

www.acdlabs.com
    ACD Labs.
www.beilstein-institut.de
    Beilstein-Institut.
www.chematica.net
    CHEMATICA.
www.cas.org/expertise/cascontent
    CAS.
www.daylight.com.
    DAYLIGHT.
lhasa.harvard.edu
    LHASA.
http://www.ch.ic.ac.uk/local/organic/mod/
    Molecular Modeling for Organic Chemistry.
http://www.organic-chemistry.org
    Organic Chemistry Portal.

Fingerprints and Pharmacophores

Daniela Schuster, Paracelsus Medical University of Salzburg, Salzburg, Austria

Introduction

In drug discovery and development as well as related fields such as toxicology or environmental chemical research, molecular similarity assessment is a central theme for various reasons. Comparing compounds and their properties may lead to information on the compound's physicochemical properties and even biological activities. Indeed, about 30% of compounds similar to an active compound have been reported to be active themselves (Martin et al., 2002). Very often, researchers look out for similar compounds to identify related molecules, for example for studying structure-activity relationships. However, in many cases also dissimilar compounds are sought, for example when novel chemical scaffolds are acquired for a compound collection or when intellectual property opportunities are evaluated (Maggiora et al., 2014).

Historically, similarity analysis was long based on fragment comparison or restricted to structurally very similar compound series. However, approaches describing the structures more completely and globally, thereby also allowing for more structural diversity within the studied data set, were needed (Cramer et al., 1974). Nowadays, many similarity calculations and metrics are available (Cereto-Massagué et al., 2015; Maggiora et al., 2014); however, similarity assessment is still subjective and there is so far no consensus on the 'best' methodology.

This encyclopaedia entry will focus on two widely used similarity assessment methods: fingerprints and pharmacophore models. They represent global and local representations of the chemical structures involved in the comparison and therefore serve as examples to discuss this aspect as well.

Background and Fundamentals

Fingerprints are bit-string representations of molecular structure and properties. The bit-string encodes the absence (0) or presence (1) of a specific property, which can be a molecular structure or fragment. Depending on the method transforming the molecule into a bit-string, several types of fingerprints are available. While most of them are merely based on the 2D structure of the compounds, a few can also store 3D information. Fingerprints usually consider the whole molecule for generating the bit-string and are therefore classified as so-called global descriptors. They are straightforward to compute, and so fingerprint-based comparisons are among the fastest similarity calculations available.

Historically, the concept of a pharmacophore as the essential part of a drug determining its binding to a receptor was first presented by Paul Ehrlich (Ehrlich, 1898). The concept, its understanding and application evolved over the next decades (Güner and Bowen, 2014), and is nowadays an established and fundamental approach to understanding and predicting activities of small organic molecules. A pharmacophore model consists of the chemical and spatial arrangement of chemical functionalities, which are essential for the biological activity of a compound (Wermuth et al., 1998). It contains so-called features like hydrogen bond acceptors/donors, hydrophobic areas, aromatic rings, positively or negatively ionizable groups or metal chelators. According to this definition, most features directly translate into target-ligand interactions essential for biological activity. Because only those crucial interaction points are considered when comparing compounds, pharmacophore models are classified as so-called local descriptors.

Approaches

The main fingerprint approaches include substructure keys-based fingerprints, topological fingerprints and circular fingerprints (Cereto-Massagué et al., 2015). They differ in the way they are translating structural information into the final bit string. First, in the substructure keys-based fingerprints each bit stands for a chemical substructure, for example a carboxyl group, primary amine or phenyl. The individual fingerprints belonging into this group have a pre-set bit length and definitions at which position which substructure is encoded (**Fig. 1**).

Topological fingerprints analyze all the fragments of a molecule. Starting from each atom, a usually linear path up to a predefined number of bonds is followed. Each fragment produced in this procedure corresponds to a position in the bit string (**Fig. 2**).

The third class, the circular fingerprints, records the environment of each atom up to a determined radius (**Fig. 3**). These fingerprints are widely used for full structure similarity virtual screening.

Apart from directly describing a chemical structure, fingerprints can also represent other properties of molecules such as functional classes (functional class fingerprints, FCFP), pharmacophores (pharmacophore fingerprints, PF), reactions (reaction fingerprints, RF) and others.

Once the bit-strings resulting from the individual approaches are calculated, the similarities of the compared molecules need to be quantified. In principle, several similarity coefficients are available to choose from, for example the Tanimoto, Cosine,

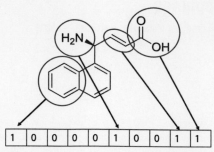

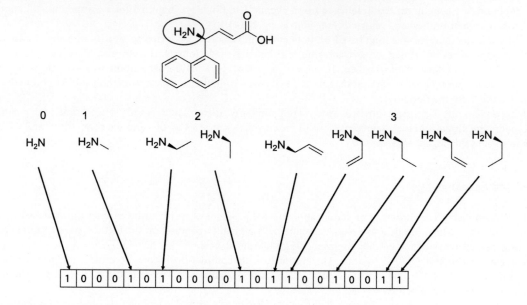

Fig. 1 Substructure keys-based fingerprints translate the presence or absence of chemical substructures into a bit string.

Fig. 2 The concept of topological fingerprints exemplified on the primary amino group of the molecule. Depending on the pre-set number of bonds, several linear pathway fragments are produced, which are then translated into a bit string.

Hamming, Russell-Rao or Forbes coefficients (Willet, 2006). In the last decade, the Tanimoto coefficient (TC) has evolved as the dominating similarity function. It is calculated according to Eq. (1)

$$TC = \frac{c}{a + b - c} \tag{1}$$

where the two compared molecules have a and b bit-sets in their fragment bit-strings and c of these bits are found in both of the bit-strings. The TC has values between 0 and 1, sometimes also expressed in percent. A value of 0 means no similarity, while 1 is calculated for identical molecules. A TC of ≥ 0.85 is considered to describe two quite similar compounds (Martin *et al.*, 2002).

Also pharmacophore models can be seen as a kind of fingerprint for a molecule. Each substructure of a molecule can be translated into a pharmacophore feature, such as for example a hydrogen bond acceptor, and a fingerprint can be created featuring the presence or absence of these features. However and more often, 3D pharmacophore models are created and used to identify chemical functionalities essential for biological activity and to search large 3D molecular databases for fitting, putatively active compounds.

In the ligand-based approach, active molecules are first transformed into 3D conformational models. These consist of representative sets of energetically favorable 3D structures for known active molecules. Those 3D structures are then superimposed in a way that equal chemical functionalities overlay. Most algorithms start with the most rigid active molecules, for which this 3D alignment is rather straightforward, and then proceed with more and more flexible molecules, which are aligned onto the more rigid ones. Finally, pharmacophore features are placed in the 3D space where chemical functionalities are common between the active molecules, which is called the common features or active analogue approach (**Fig. 4**).

There is usually a need for spatial restriction of the ligands to prevent large molecules from being reported as hits, which are likely too spacious to fit into the binding site. Technically, so-called exclusion volumes – forbidden areas for the ligands – are placed in the space surrounding the locations where the active molecules fit. If available, inactive molecules fitting the features of the pharmacophore can be used to strategically place exclusion volumes outside the active molecules' volumes.

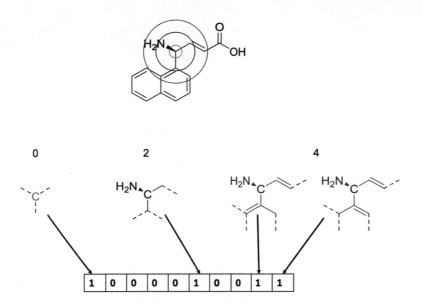

Fig. 3 Circular fingerprints start at a central atom and then consider the molecular environment in a pre-set diameter of e.g., two or four bonds, respectively. The thereby generated fragments are then translated into a bit string.

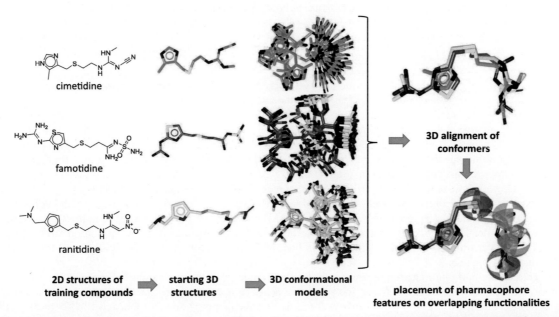

Fig. 4 Ligand-based pharmacophore model generation exemplified on three histamine A_2 receptor antagonists. Starting from a 2D structure, corresponding 3D structures and conformational models are calculated for the training ligands. The conformers are then aligned in a way that equal chemical functionalities overlay. Finally, pharmacophore features are placed on these common functionality locations. In this example, the aromatic ring feature is shown in blue, hydrogen bond donor points as green, and hydrogen bond acceptor points in red.

If there is a high quality data set available, the development of quantitative pharmacophore models aimed to predict a compound's potency is also feasible. The main limitation in this respect is the data quality. For model generation, at least 16 molecules spanning four orders of magnitude of activity, all tested in the same assay, and preferably in the same lab are required. Ideally, many more ligands from the same test are available for quantitative model validation afterwards. This stringent requirements often hamper the generation of quantitative pharmacophore models that are especially useful for ligand optimization.

In the structure-based pharmacophore modeling approach, the 3D structure of the target is known – ideally complexed to a bioactive molecule. This 3D structure can be derived experimentally by X-ray crystallography or NMR studies. Alternatively,

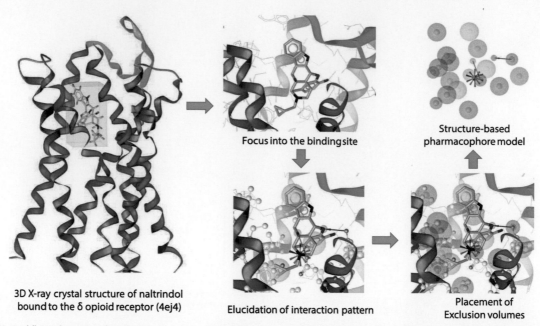

Fig. 5 The workflow of structure-based pharmacophore model generation exemplified on the δ opioid receptor antagonist naltrindole co-crystallized with its target (Protein Data Bank entry 4ej4). Starting from the 3D complex, the focus is put on just the direct environment of the ligand binding site. The ligand's surrounding amino acids are analyzed for possible direct interactions with the ligand based on their chemical functionalities, distances and contact angles. At interacting positions, pharmacophore features (yellow – hydrophobic, blue star – positively ionizable, red arrow – hydrogen bond acceptor) are placed. Exclusion volumes (grey) are placed on amino acid side chains forming the binding site. Finally, all information on the receptor and the ligand are deleted and the interaction pattern with the exclusion volumes remains as an abstract representation of binding requirements. This is the starting point for model validation, optimization and use in virtual screening.

theoretical target models, so-called homology models, with a docked active molecule can serve as basis for the model. In a first step, all observed target-ligand interactions are mapped onto the respective parts of the ligand (**Fig. 5**).

Such interaction models usually consist of far more chemical features than are necessary for biological activity. So how can the modeler find out which are the more important features? If the data situation is good, information can come from mutational analysis and other 3D target-ligand complexes. Amino acid mutants leading to a change of ligand affinity can help to identify important anchoring points for the molecule. Additionally, interactions observed in several target-ligand complexes most likely play an important role for activity modulation (Sun *et al.*, 2011). Furthermore, interaction maps from molecular dynamics simulations can help to prioritize specific interactions (Moorthy *et al.*, 2016). If such data is not available, a so-called test data set containing active and inactive compounds can help. The initial interaction model is systematically altered and applied to screen the test set. Any alteration that leads to a better retrieval of active compounds and no increase of found inactive compounds is a step towards a more general and representative model for modulators of the respective target. Furthermore, exclusion volumes can be strategically placed on amino acid residues lining the binding site (Vuorinen *et al.*, 2014). This strategy can also be followed to optimize ligand-based models.

Shape-based searches and pharmacophore features are frequently combined. Many pharmacophore modeling tools include a shape 'feature', which is derived from the shape and volume of active compounds fitting the model. The shape is included as a spatial restriction of the size of fitting hits. Also some shape-based screening programs like Openeyes ROCS can include pharmacophore-like features, which are termed color features. Although they may not be directly comparable with true physico-chemical property-based pharmacophore features, they share several functionalities such as 'hydrogen bond acceptor', 'hydrogen bond donor', 'negative charge', and some more. Another shape-pharmacophore-combining screening algorithm is the ultrafast shape recognition with CREDO atom types (USRCAT, Schreyer and Blundell, 2012).

Protein binding site-based pharmacophore models can be used for model generation without prior knowledge of active ligands. A so-called grid is calculated on the binding site surface and analyzed for potential interaction points for ligand binding (**Fig. 6**). According to available interacting amino acid residues, locations for pharmacophore features are suggested, where the modeler can place the respective features.

Similarly, pharmacophore-based docking has been reported. One of these approaches is PharmDock (Hu and Lill, 2014). It can be used for pose prediction and compound ranking like docking programs, thereby being much faster than common docking algorithms. The pharmacophore-guided docking has been shown to be superior over unbiased docking, which is not surprising. Unbiased docking lacks information on the really important interactions in the binding site, while pharmacophore-guided docking includes this information, which is added in the workflow development. However, docking with constraints forcing a docking ligand to establish certain interactions may eventually lead to similar results as the pharmacophore-guided approach. Like

Fig. 6 Visualized grid for positively ionizable ligand functionalities (blue) surrounding Asp128 in the δ opioid receptor binding site (4ej4). The empty binding site is filled with small dots representing possible locations for respective pharmacophore features. The individual dots can be selected for placing the features and manually constructing a model filling the binding site with complementary functionalities.

all other docking approaches, the inaccuracies of the available scoring functions in predicting binding affinities limit the use of pharmacophore-guided docking in compound optimization.

Illustrative Examples and Case Studies

Fingerprint-based similarity searches are widely used in cheminformatics. Large bioactivity databases like ChEMBL (Bento *et al.*, 2014) and PubChem (Wang *et al.*, 2017a) as well as chemical databases such as the SciFinder database operated by Chemical Abstracts Service have implemented fingerprints to search for similar molecules with defined Tanimoto coefficient range. A broad similarity search in various, huge databases using different fingerprints is implemented in the SwissSimilarity platform (Zoete *et al.*, 2016). For database analysis and library design, fingerprint methods are used to group similar compounds and evaluate databases for their structural diversity. They are also used to represent whole compound libraries for comparison amongst each other (Fernández de Gortari *et al.*, 2017). Furthermore they are successfully used for comparing chemical structures with compounds of known bioactivities or ensembles thereof, for example the ECFP4-based similarity ensemble approach (SEA) prediction platform (Keiser *et al.*, 2007). This tool has been successfully used for identifying off-target effects for a variety of approved drugs (Keiser *et al.*, 2009). For example amitriptyline has been identified as a nanomolar serotonin transporter inhibitor in this study, which has been previously unknown. The SEA platform has also been used to find a target for the natural product miconidin acetate (Zatelli *et al.*, 2016). In studies comparing several virtual screening approaches for their predictive power, SEA proved to be very successful in finding new active compounds (Kaserer *et al.*, 2015, 2016). However, the correct classification of inactive compounds turned out to be a challenge, possibly because the structure-activity database in the background is a limited source of information on inactive compounds. Another example for a fingerprint-based profiler is the SwissTargetPrediction platform, which uses 2D fingerprints besides 3D similarity measures for assessing ligand similarities and associations to putative targets (Gfeller *et al.*, 2014). Other fingerprint-based activity prediction tools include SuperPred (Nickel *et al.*, 2014), MOST (Huang *et al.*, 2017) and the polypharmacology browser, which even combines 10 fingerprints for similarity assessment (Awale and Reymond, 2017).

Pharmacophore models are established as useful tools to discover novel active compounds for specific targets, predict mechanism-based side effects or calculate a bioactivity profile for a molecule. For example, Markt *et al.* (2009) used two ligand-based, shape-restricted pharmacophore models for the discovery of novel cannabinoid receptor 2 (CB$_2$) ligands. From a training set of five potent CB$_2$ ligands, common feature pharmacophore models were generated and merged with one of the training compound's shape, respectively. A test set of 15 active CB$_2$ ligands and over 67,000 decoys (putative inactive compounds) were used to determine the enrichment of active compounds in a virtual screening setting. The two best performing models were selected and employed to virtually screen commercial compound databases. From initially over 920,000 compounds in the databases, nearly 30,000 fitted into either or both models. Further filtering using physicochemical properties, structural similarity analysis and visual inspection led to the selection of 14 hits for biological evaluation. Out of these 14 compounds, three showed K$_I$ values ≤2 μM, which corresponds to a success rate of over 20% true positive hits.

A classical side effect modeling study was performed by Kratz *et al.* (2014). They assembled a data set of human ether-a-go-go-related (hERG) potassium channel blockers. In the course of this study, not just one but a whole set of models was created for hERG blockers. This is because one pharmacophore model is unable to predict all active molecules for a target. Usually, this

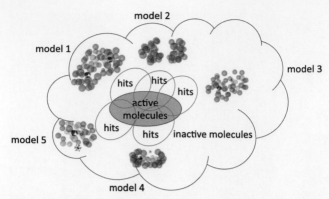

Fig. 7 The parallel use of several complementary pharmacophore models allows for a more complete retrieval of active compounds. The cloud incorporates a database of active (rose) and inactive (white) compounds. Each model used for screening derives a hit list of fitting compounds (green circles). While a single model only finds a fraction of true active compounds, the parallel use of models finding mostly different active compounds improves the overall retrieval of active molecules.

doesn't pose a problem because in a "cherry picking" scenario, the scientists are interested to find some hits for further development. In side effect studies however, the project aims to identify all potentially harmful compounds. Therefore, several models have to be generated for an adverse effect target not to miss the majority of active hits (**Fig. 7**).

The model set by Kratz *et al.* was applied to databases containing synthetic and natural products and successfully identified 13 previously unknown hERG blockers from 50 experimentally tested virtual hits.

Sometimes also the use of several pharmacophore modeling tools is beneficial because they may find complementary active hits. Temml *et al.* (2014) generated structure-based pharmacophore models from one cyclooxygenase-2 – inhibitor X-ray crystal structure using the two pharmacophore modeling programs LigandScout and Catalyst (implemented in Discovery Studio). The LigandScout model found 53 virtual hits and the Discovery Studio model 74 hits, respectively. Surprisingly, not a single hit was present in both hit lists. However, when 10 virtual hits from every hit list were biologically tested, each found 4 active compounds, respectively. So although the resulting hits of the virtual screening were completely different, each of the programs retrieved active hits.

The concept of using several models can be further expanded by applying several models for a broad set of targets. This concept is termed pharmacophore-based parallel screening or pharmacophore-based target fishing. The openly available online platform PharmMapper offers over 53,000 structure-based pharmacophore models for activity predictions (Wang *et al.*, 2017b). In comparative studies on virtual screening tools, PharmMapper performed well for cherry picking with 11 active compounds out of 19 hits predicted for cyclooxygenase (Kaserer *et al.*, 2015) and one active ligand out of two predicted ones for the peroxisome proliferator-activated receptor γ (Kaserer *et al.*, 2016), respectively. However, many active molecules from these studies were not found by this platform. Additionally, several companies offer pharmacophore databases along with their software such as Biovia Discovery Studio (PharmaDB; see Relevant Websites section) and inte:ligand (pharmacophoreDB; see Relevant Websites section). Examples from the academia include target fishing for compounds of the herbal remedy *Ruta graveolens* (Rollinger *et al.*, 2010) and the natural product leoligin, the main lignan from *Leontopodium alpinum* (Duwensee *et al.*, 2011).

Results and Discussion

Fingerprint-based methods and pharmacophore models are established and important tools in cheminformatics and virtual screening. They are likewise used in industry and in the academic field. A wide variety of fingerprints are available and used for comparing molecules for diverse purposes. In library analysis, they are important to identify the chemical diversity of a compound collection, which is regarded as beneficial for drug discovery libraries. Also when planning to purchase new compounds, fingerprint methods can report if the compounds to purchase have sufficient dissimilarity to the substances that are already on stock.

Because of their speed, fingerprint-based methods are very popular for activity predictions, especially in the field of target fishing, where query compounds are compared against millions of compounds with known activities. More and more platforms based on fingerprints are published each year and most of them are publicly available. However, success stories reporting a prospective use of these tools are rare and therefore their predictive power still needs to be critically evaluated and compared to other tools. At some point one gets the impression that developing new tools may not be meaningful unless the existing platforms have been properly validated and their strengths and weaknesses have been revealed. First validation studies suggest that the predictive power in the identification of active compounds is satisfying but still highly depending on the activity database in the background. For the same reason, the correct prediction of inactive compounds seems to remain a challenge, but will improve with the growing amount of structure-activity data. A combination with other tools that are more powerful in this respect may be a good strategy to overcome this current drawback.

In comparison, success studies reporting prospective pharmacophore-based applications are easy to spot in the scientific literature as exemplified above. Pharmacophore models are used as standalone virtual screening tools or used in combination with physicochemical filtering, structural clustering, shape-based screening or docking or a combination thereof. Nowadays, pharmacophore modeling software like many other virtual screening tools is convenient to use, with an easy to navigate graphical user interface and sometimes even 'wizards' that guide through the model generation process. This easy-to-use mentality sometimes distracts from the actually very sophisticated process of model development. At the beginning, a data set for model generation and theoretical validation is to be assembled. This data need to be of high quality and be derived from suitable assays. Essentially, if a model should reflect binding to a specific protein target, also data reporting the direct binding to this target needs to be used for the modeling. Data from cell-based assays, where also other events than direct ligand binding influence activity, are not suitable. Model generation is technically not a challenge any more, which makes a rigorous validation protocol crucial. Theoretical model optimization and validation using independent data sets must be followed by a prospective, experimental evaluation of virtual hits. The validation experiments must also comply with the rules for the data set. For instance, watching cells die is not a proof for kinase inhibition or any other specific mechanism of action. Cell-free, target-based binding assays are required to prove the pharmacophore hypothesis. Models lacking high quality experimental validation are of very limited value and journals should carefully consider if they are worth publishing at all. Of course, favorable hits should also be active in cell-based systems, but this is only the second step after prospective model validation.

Excitingly, many in silico tools are nowadays openly available to the public, either as online platforms operated on a remote server or as stand-alone programs that can be downloaded and locally installed on the workstation (Pirhadi *et al.*, 2016). This way, those tools are freely accessible to interested scientists all over the world for a more targeted planning of experimental work.

Future Directions

Both fingerprint-based methods and pharmacophore modeling have been successfully employed in a variety of research projects and are important, established tools in the analysis and discovery of active compounds. However, there is also room for improvement for both approaches.

Regarding fingerprint-based target fishing, the community currently seems to be struggling with quantitative target ranking. How shall hundreds of predicted targets be judged by the user? Which ones are the most probable ones and should be prioritized for experimental evaluation? There is currently a need to introduce a measure of significance for these tools, which enable an estimation of the chance that a virtually predicted activity is indeed probable. While there are first attempts in the field (Vogt and Bajorath, 2017), it will be a long way to go until these more sophisticated measures are implemented into the similarity-based target fishing tools. Furthermore, new fingerprint algorithms are reported regularly and their optimal application in similarity screening need to be determined.

The pharmacophore modeling branch currently suffers from severe drawbacks in feature definition and screening algorithms that rely on different concepts (Temml *et al.*, 2014; Wolber *et al.*, 2008). This leads to incompatibilities between different software or – even worse – sometimes between different versions of the same software. It will be a challenge to bring feature definitions to an up-to-date state including halogen bonds and sulfur bonds as well as a better consideration of internal hydrogen bonds in model fitting. Furthermore, hydrophobic areas are currently represented as globular spheres in the programs, which doesn't reflect reality. Adjusting the hydrophobic contacts feature to more accurate geometries may also improve quantitative scoring of hit molecules. Finally, incorporating protein-ligand-contact information from molecular dynamics simulations in the development of pharmacophore models will lead to a better understanding of ligand binding and more predictive models in the future.

Closing Remarks

Fingerprints and pharmacophore models are indispensable methods in drug research. As in every cheminformatics field, research in the fingerprint and pharmacophore area is very active and there is constant progress in the improvement of the available tools and development of new approaches. Nevertheless, critical validation studies and especially benchmarking and prospective case studies are essential to identify the power and also the pitfalls of new developments. It is also important to see and apply these tools in context with other property calculation and virtual screening tools to make the most out of them.

Acknowledgement

I thank Philipp Schuster for help in the preparation of this manuscript and Inte:Ligand GmbH for providing an academic license for the generation of pharmacophore-related figures. This work was supported by the Austrian Science Fund project P26782.

See also: Binding Site Comparison – Software and Applications. Chemical Similarity and Substructure Searches. Chemoinformatics: From Chemical Art to Chemistry in Silico. Drug Repurposing and Multi-Target Therapies. In Silico Identification of Novel Inhibitors. Introduction of Docking-Based Virtual Screening Workflow Using Desktop Personal Computer. Natural Language Processing Approaches in Bioinformatics. Pharmacophore Development. Population Analysis of Pharmacogenetic Polymorphisms. Prediction of Protein-Protein Interactions: Looking Through the Kaleidoscope. Protein Structural Bioinformatics: An Overview. Rational Structure-Based Drug Design. Small Molecule Drug Design. Structure-Based Drug Design Workflow

References

Awale, M., Reymond, J.L., 2017. The polypharmacology browser: A web-based multi-fingerprint target prediction tool using ChEMBL bioactivity data. Journal of Cheminformatics 9, 11.

Bento, A.P., Gaulton, A., Hersey, A., et al., 2014. The ChEMBL bioactivity database: An update. Nucleic Acids Research 42, D1083–D1090.

Cereto-Massagué, A., Ojeda, M.J., Valls, C., et al., 2015. Molecular fingerprint similarity search in virtual screening. Methods 71, 58–63.

Cramer III, R.D., Redl, G., Berkoff, C.E., 1974. Substructural analysis. A novel approach to the problem of drug design. Journal of Medicinal Chemistry 17, 533–535.

Duwensee, K., Schwaiger, S., Tancevski, I., et al., 2011. Leoligin, the major lignan from Edelweiss, activates cholesteryl ester transfer protein. Atherosclerosis 219, 109–115.

Ehrlich, P., 1898. Über die Constitution des Diphtheriegiftes. Deutsche Medizinische Wochenschrift 24, 597–600.

Fernández de Gortari, E., García-Jacas, C.R., Martinez-Mayorga, K., Medina-Franco, J.L., 2017. Database fingerprint (DFP): An approach to represent molecular databases. Journal of Cheminformatics 9, 9.

Gfeller, D., Grosdidier, A., Wirth, M., et al., 2014. SwissTargetPrediction: A web server for target prediction of bioactive small molecules. Nucleic Acids Research 42, W32–W38.

Güner, O.F., Bowen, J.P., 2014. Setting the record straight: The origin of the pharmacophore concept. Journal of Chemical Information and Modeling 54, 1269–1283.

Hu, B., Lill, M.A., 2014. PharmDock: A pharmacophore-based docking program. Journal of Cheminformatics 6, 14.

Huang, T., Mi, H., Lin, C.Y., et al., 2017. MOST: Most similar ligand based approach to target prediction. BMC Bioinformatics 18, 165.

Kaserer, T., Obermoser, V., Weninger, A., et al., 2016. Evaluation of selected 3D virtual screening tools for the prospective identification of peroxisome proliferator-activated receptor (PPAR) γ partial agonists. European Journal of Medicinal Chemistry 124, 49–62.

Kaserer, T., Temml, V., Kutil, Z., et al., 2015. Prospective performance evaluation of selected common virtual screening tools. Cast study: Cyclooxygenase (COX) 1 and 2. European Journal of Medicinal Chemistry 96, 445–457.

Keiser, M.J., Roth, B.L., Armbruster, B.N., et al., 2007. Relating protein pharmacology by ligand chemistry. Nature Biotechnology 25, 197–206.

Keiser, M.J., Setola, V., Irwin, J.J., et al., 2009. Predicting new molecular targets for known drugs. Nature 462, 175–182.

Kratz, J.M., Schuster, D., Edtbauer, M., et al., 2014. Experimentally validated hERG pharmacophore models as cardiotoxicity prediction tools. Journal of Chemical Information and Modeling 54, 2887–2901.

Maggiora, G., Vogt, M., Stumpfe, D., Bajorath, J., 2014. Molecular similarity in medicinal chemistry. Journal of Medicinal Chemistry 57, 3186–3204.

Markt, P., Feldmann, C., Rollinger, J.M., et al., 2009. Discovery of novel CB_2 receptor ligands by a pharmacophore-based virtual screening workflow. Journal of Medicinal Chemistry 52, 369–378.

Martin, Y.C., Kofron, J.L., Traphagen, L.M., 2002. Do structurally similar molecules have similar biological activity? Journal of Medicinal Chemistry 45, 4350–4358.

Moorthy, N.S.H.N., Souousa, S.F., Ramos, M.J., Fernandes, P.A., 2016. Molecular dynamic simulations and structure-based pharmacophore model development for farnesyltransferase inhibitors discovery. Journal of Enzyme Inhibition and Medicinal Chemistry 31, 1428–1442.

Nickel, J., Gohlke, B.O., Erehman, J., et al., 2014. SuperPred: Update on drug classification and target prediction. Nucleic Acids Research 42, W26–W31.

Pirhadi, S., Sunseri, J., Koes, D.R., 2016. Open source molecular modeling. Journal of Molecular Graphics and Modeling 69, 127–143.

Rollinger, J.M., Schuster, D., Danzl, B., et al., 2010. In silico target fishing for rationalized ligand discovery exemplified on constituents of Ruta graveolens. Planta Medica 75, 195–204.

Schreyer, A.M., Blundell, T., 2012. USRCAT: Real-time ultrafast shape recognition with pharmacophoric constraints. Journal of Cheminformatics 4, 27.

Sun, H.P., Zhu, J., Chen, F.H., You, Q.D., 2011. Structure-based pharmacophore modeling from multicomplex: A comprehensive pharmacophore generation of protein kinase CK2 and virtual screening based on it for novel inhibitors. Molecular Informatics 30, 579–592.

Temml, V., Kaserer, T., Kutil, Z., et al., 2014. Pharmacophore modeling for COX-1 and -2 inhibitors with LigandScout in comparison with Discovery Studio. Future Medicinal Chemistry 6, 1869–1881.

Vogt, M., Bajorath, J., 2017. Modeling Tanimoto similarity value distributions and predicting search results. Molecular Informatics 36, 1600131.

Vuorinen, A., Nashev, L.G., Oderrmatt, A., Rollinger, J.M., Schuster, D., 2014. Pharmacophore model refinement for 11β-hydroxysteroid dehydrogenase inhibitors: Search for modulators of intracellular glucocorticoid concentrations. Molecular Informatics 33, 15–25.

Wang, Y., Bryant, S.H., Cheng, T., et al., 2017a. PubChem BioAssay: 2017 update. Nucleic Acids Research 45, D955–D963.

Wang, X., Shen, Y., Wang, S., et al., 2017b. PharmMapper 2017 update: A web server for potential drug target identification with a comprehensive target pharmacophore database. Nucleic Acids Research 45, W356–W360.

Wermuth, C.G., Ganellin, C.R., Lindberg, P., Mitscher, L.A., 1998. Glossary of terms used in medicinal chemistry. Pure & Applied Chemistry 70, 1129–1143.

Willet, P., 2006. Similarity-based virtual screening using 2D fingerprints. Drug Discovery Today 11, 1046–1053.

Wolber, G., Seidel, T., Bendix, F., Langer, T., 2008. Molecule-pharmacophore superpositioning and pattern matching in computational drug design. Drug Discovery Today 13, 23–29.

Zatelli, G.A., Temml, V., Kutil, Z., et al., 2016. Miconidin acetate and primin as potent 5-lipoxygenase inhibitors from Brazilian Eugenia hiemalis (myrtaceae). Planta Medica Letters 3, e17–e19.

Zoete, V., Daina, A., Bovigny, C., Michielin, O., 2016. SwissSimilarity: A web tool for low to ultra high throughput ligand-based virtual screening. Journal of Chemical Information and Modeling 56, 1399–1404.

Further Reading

Akram, M., Kaserer, T., Schuster, D., 2017. Pharmacophore modeling and pharmacophore-based virtual screening. In: Cavasotto, C.N. (Ed.), In Silico Drug Discovery and Design: Theory, Methods, Challenges, and Applications. Boca Raton, FL, USA: CRC Press, pp. 123–153.

Maggiora, G.M., Shanmugasundaram, V., 2011. Molecular similarity measures. In: Bajorath, J. (Ed.), Springer Protocols: Chemoinformatics and Computational Chemical Biology. New York: Humana Press.

Muegge, I., Mukherjee, P., 2016. An overview of molecular fingerprint similarity search in virtual screening. Expert Opinion on Drug Discovery 11, 137–148.

Phamacophores and Pharmacophore Searches. Wiley VCH, Thierry Langer. Available at: https://www.wiley.com/en-at/Pharmacophores + and + Pharmacophore + Searches-p-9783527608720.

Rogers, D., Hahn, M., 2010. Extended-connectivity fingerprints. Journal of Chemical Information and Modeling 50, 742–754.

Willett, P., Barnard, J., Downs, G., 1998. Chemical similarity searching. Journal of Chemical Information and Computer Sciences 38, 983–996.

Relevant Websites

http://accelrys.com/products/collaborative-science/biovia-discovery-studio/pharmacophore-and-ligand-based-design.html
  BIOVIA Discovery Studio I Pharmacophore and Ligand-Based Design.
http://www.inteligand.com/pharmdb/
  Inteligand.
www.click2drug.org
  List of programs, tools and software for molecular modeling, analysis and drug discovery.
http://lilab.ecust.edu.cn/pharmmapper/
  PharmMapper.
http://gdbtools.unibe.ch:8080/PPB/
  Polypharmacology browser.
http://sea16.ucsf.bkslab.org/
  SEA activity profiler.
http://prediction.charite.de/
  SuperPred.
http://www.swisstargetprediction.ch/
  SwissTargetPrediction.

Biographical Sketch

Daniela Schuster studied pharmacy at the University of Innsbruck, Austria, and graduated in 2003. In her PhD time, she worked in an e-learning project at the University of Graz, Austria. She also worked as senior application scientist for Inte:Ligand in Vienna. As the first Erika Cremer habilitation fellow and one of the first Ingeborg Hochmair professors at the University of Innsbruck, she led the Computer-Aided Molecular Design Group at the Institute of Pharmacy/Pharmaceutical Chemistry. In 2018, she accepted a position as professor for Pharmaceutical and Medicinal Chemistry at the Paracelsus Medical University in Salzburg, Austria.

Public Chemical Databases

Sunghwan Kim, National Institutes of Health, Bethesda, MD, United States

Nomenclature

ADMET	Absorption, distribution, metabolism, excretion, and toxicity
ADR	Adverse drug reaction
BPS	British Pharmacological Society
CPIC	Clinical Pharmacogenomics Implementation Consortium
CSSP	ChemSpider SyntheticPages
CTD	Comparative Toxicogenomics Database
CYP450	Cytochrome P450
DOI	Digital object identifier
EBI	European Bioinformatics Institute
EMBL	European Molecular Biology Laboratory
FDA	Food and Drug Administration
GC–MS	Gas chromatography-mass spectrometry
GO	Gene ontology
GPCR	G protein-coupled receptor
GRAC	Guide to receptors and channels
GtoPdb	Guide to PHARMACOLOGY
hERG	Human ether-a-go-go related gene
HMDB	Human Metabolome Database
HSDB	Hazardous Substances Data Bank
HTS	High-throughput screening
IC_{50}	Half-maximal inhibitory concentration
InChI	International Chemical Identifier
INN	International non-proprietary name
IUPHAR	International Union of Basic and Clinical Pharmacology
K_d	Dissociation constant
K_i	Inhibitory constant
MESH	Medical subject headings
MS	Mass spectrometry
NC-IUPHAR	IUPHAR Committee on Receptor Nomenclature and Drug Classification
NER	Named entity recognition
NHR	Nuclear hormone receptor
NIH	U.S. National Institutes of Health
NMR	Nuclear magnetic resonance
NLP	Natural language processing
PD	Pharmacodynamics
PDB	Protein Data Bank
PGRN	Pharmacogenomics Research Network
PK	Pharmacokinetics
SIDER	Side effect resource
SRP	Scientific review panel

Introduction

For the past decade, the amount of chemical data available in the public domain has been rapidly increasing (Cheng *et al.*, 2014; Hu and Bajorath, 2012; Nicola *et al.*, 2012). This led to the establishment of many databases that collect, curate, and disseminate publicly available chemical information with the aim of providing easier access to and better utility of these data (Akhondi *et al.*, 2012; Hersey *et al.*, 2015; Lipinski *et al.*, 2015; Ma *et al.*, 2012; Muresan *et al.*, 2011; Nicola *et al.*, 2012; Tiikkainen and Franke, 2012; Tiikkainen *et al.*, 2013; Williams, 2008a,b; Williams *et al.*, 2012). Importantly, these databases annotate chemical data with information from other scientific domains (such as genetics, genomics, pharmacology, toxicology, medicine, bioinformatics, and systems biology), assisting biomedical researchers in taking advantage of these public data for the discovery of new medications. To maximize the usefulness of the public chemical data, scientists should be aware of which databases have the information they need, and know how to exploit their full potential.

This paper provides a brief overview of some important chemical databases available in the public domain (see **Fig. 1**). Their uniform resource locators (URLs) are listed in **Fig. 1** as well as in the Relevant Websites section at the end of this article. These databases vary in data contents, size, and quality (see **Figs. 1** and **2**) (Akhondi *et al.*, 2012; Hersey *et al.*, 2015; Lipinski *et al.*, 2015; Muresan *et al.*, 2011; Tiikkainen and Franke, 2012; Tiikkainen *et al.*, 2013; Williams *et al.*, 2012). Some of them provide a wide variety of information on millions of molecules, aggregated from hundreds of data sources. Others contain specialized information for a few thousand molecules, manually curated from peer-reviewed scientific articles. Because data exchange between these public databases is a very common practice, there are some overlaps in the data contents, but each database does have unique contents valuable to the scientific community. While they provide various tools and services for searching, browsing, visualization, download, programmatic access, etc., the primary focus of this article is on their data contents, to provide some guidance on what database to use to get a particular kind of information.

- **PubChem (https://pubchem.ncbi.nlm.nih.gov)**
  Provides comprehensive information on >93 million unique compounds, collected from >500 data sources, along with experimentally determined bioactivity data.

- **ChemSpider (http://www.chemspider.com)**
  Contains information for >59 million compounds, collected from ~500 data sources.

- **ChEMBL (https://www.ebi.ac.uk/chembl/)**
  Contains information on bioactive drug-like small molecules and their bioactivities, abstracted from literature.

- **BindingDB (https://www.bindingdb.org)**
  Contains experimental binding affinities, focusing on the interactions of proteins considered to be drug-targets with drug-like small molecules.

- **PDBbind (http://www.pdbbind.org.cn)**
  Provides experimental binding affinity data for the biomolecular complexes archived in PDB.

- **Binding MOAD (http://www.bindingmoad.org)**
  Provides experimental binding affinity data for the high-quality structures of protein-ligand complexes derived from PDB.

- **DrugBank (https://www.drugbank.ca)**
  Provides comprehensive information for FDA-approved and investigational drugs.

- **IUPHAR/BPS Guide To PHARMACOLOGY (http://www.guidetopharmacology.org)**
  Contains structures of small molecule ligands, peptides, and antibodies, with their affinities at protein targets.

- **PharmGKB (Pharmacogenomics Knowledge Base) (https://www.pharmgkb.org)**
  Provides curated information about the impact of genetic variation on drug response.

- **SIDER (Side Effect Resource) (http://sideeffects.embl.de)**
  Provides information on adverse drug reactions (ADRs) for drugs, extracted from drug labels and package inserts.

- **Comparative Toxicogenomics Database (CTD) (https://ctdbase.org)**
  Provides information on interactions between chemicals and genes/proteins and their relationship to diseases.

- **Hazardous Substances Data Bank (HSDB) (https://www.toxnet.nlm.nih.gov/newtoxnet/hsdb.htm)**
  Provides toxicological information for more than 5,800 potentially hazardous chemicals.

- **Chemical Entities of Biological Interest (ChEBI) (https://www.ebi.ac.uk/chebi/)**
  A freely available dictionary of molecular entities, focused on small chemical compounds.

- **Human Metabolome Database (HMDB) (http://www.hmdb.ca)**
  Contains detailed information about small molecule metabolites found in the human body.

- **UniChem (https://www.ebi.ac.uk/unichem/)**
  A free compound identifier mapping service, containing more than 145 million chemical structures stored in >30 databases.

Fig. 1 List of public chemical databases discussed in this paper.

Large-Scale Data Aggregators

PubChem

PubChem (Kim, 2016; Kim *et al.*, 2016a; Wang *et al.*, 2017) is a public chemical information resource, developed and maintained by the National Center for Biotechnology Information (NCBI) at the National Library of Medicine (NLM), an institute within the U.S. National institutes of Health (NIH). It collects chemical substance descriptions and their biological activities from more than 500 data sources and disseminates these data to the public free of charge. Since the launch in 2004 as a component of the NIH Molecular Libraries Roadmap Initiatives, PubChem has been a key information resource for biomedical research communities in many areas such as cheminformatics, chemical biology, medicinal chemistry, and drug discovery.

PubChem contains various types of chemical information, including 2-D and 3-D structures, chemical and physical properties, bioactivity data, pharmacology, toxicology, drug target, metabolism, safety and handling, relevant patents and scientific papers, etc. While the majority of PubChem's records are about small molecules, it also contains information on a broad range of chemical

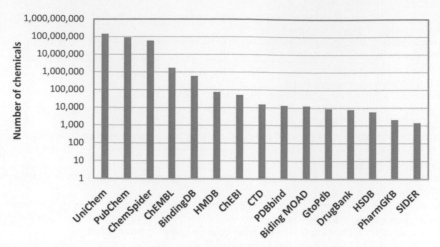

Fig. 2 Number of chemical structures contained in the databases discussed in the present paper (as of June 2017).

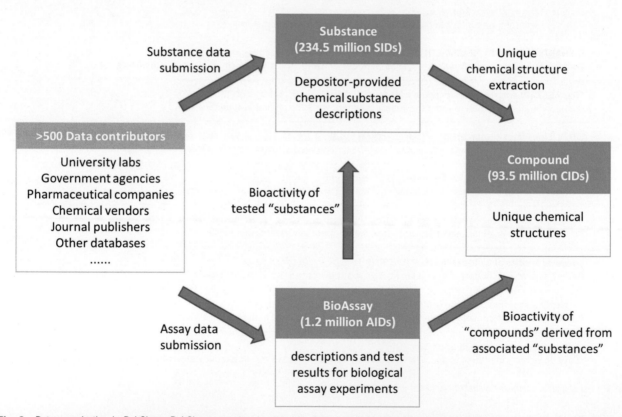

Fig. 3 Data organization in PubChem. PubChem organizes its data into three inter-linked databases called Substance, Compound, and BioAssay. SID, CID, and AID are record identifiers used in the Substance, Compound, and BioAssay databases, respectively.

entities, including siRNAs, miRNAs, carbohydrates, lipids, peptides, chemically modified macromolecules, and many others. These data are provided by various contributors, including government agencies, university labs, pharmaceutical companies, chemical vendors, publishers and a number of chemical biology resources. Most of the chemical databases discussed in this paper also contribute their data to PubChem.

As shown in **Fig. 3**, PubChem organizes its data into three inter-linked databases: Substance, Compound, and BioAssay (Kim *et al.*, 2016a) (see the Relevant Websites section). The Substance database contains chemical substance descriptions submitted by individual data depositors. Unique chemical structures are extracted from the Substance database and stored in the Compound database. The BioAssay database contains descriptions and results of biological assay experiments performed on chemical

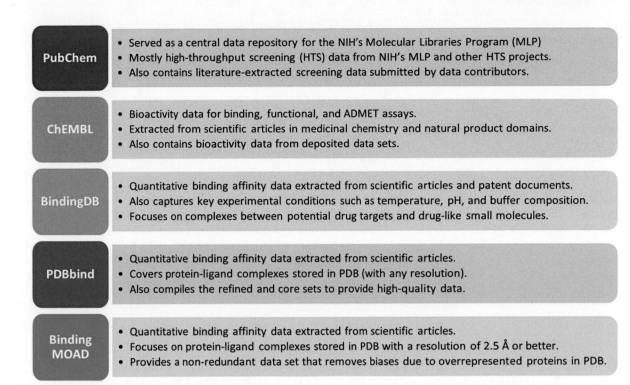

PubChem
- Served as a central data repository for the NIH's Molecular Libraries Program (MLP)
- Mostly high-throughput screening (HTS) data from NIH's MLP and other HTS projects.
- Also contains literature-extracted screening data submitted by data contributors.

ChEMBL
- Bioactivity data for binding, functional, and ADMET assays.
- Extracted from scientific articles in medicinal chemistry and natural product domains.
- Also contains bioactivity data from deposited data sets.

BindingDB
- Quantitative binding affinity data extracted from scientific articles and patent documents.
- Also captures key experimental conditions such as temperature, pH, and buffer composition.
- Focuses on complexes between potential drug targets and drug-like small molecules.

PDBbind
- Quantitative binding affinity data extracted from scientific articles.
- Covers protein-ligand complexes stored in PDB (with any resolution).
- Also compiles the refined and core sets to provide high-quality data.

Binding MOAD
- Quantitative binding affinity data extracted from scientific articles.
- Focuses on protein-ligand complexes stored in PDB with a resolution of 2.5 Å or better.
- Provides a non-redundant data set that removes biases due to overrepresented proteins in PDB.

Fig. 4 Comparison of PubChem and other databases that provide bioactivity data of small molecules.

substances. The records in the Substance, Compound, and BioAssay databases are called substances, compounds, and bioassays, respectively. Similarly, the record accessions used for the respective PubChem databases are the Substance ID (SID), Compound ID (CID), and Assay ID (AID). Currently, PubChem contains more than 234 million depositor-provided substances, 93 million unique compounds, and 233 million bioactivity test results from 1.25 million bioassays, covering more than 10,000 protein target sequences.

Although this article describes PubChem under Section Large-Scale Data Aggregators, it should be emphasized that PubChem contains the largest amount of bioactivity data available in the public domain. These data are primarily generated from high-throughput screening (HTS) experiments, because PubChem served as a central repository for the now-concluded NIH's Molecular Libraries Program (MLP). However, it also contains a substantial amount of high-quality bioactivity data extracted from research articles and patent documents, thanks to data contributions by many bioactivity databases, including ChEMBL, BindingDB, and PDBbind (to be discussed later in the present paper) (Kim *et al.*, 2016b). In **Fig. 4**, the PubChem BioAssay database is compared with other bioactivity databases discussed in the present article.

ChemSpider

Another example of large-scale chemical data aggregators in the public domain is ChemSpider, owned by the Royal Society of Chemistry (RSC) (Williams and Tkachenko, 2014; Pence and Williams, 2010). ChemSpider provides information for ∼59 million unique chemical structures integrated from ∼500 data sources. To improve the accuracy of the data, it uses a crowdsourcing approach, in which registered users can enter information as well as annotate and curate the records. Especially, users can upload various spectral data for chemicals, including infrared, Raman, nuclear magnetic resonance (NMR), and mass spectra, and ChemSpider displays these spectra in an interactive visualization widget that allows zooming and expansion.

ChemSpider extends the crowdsourcing capabilities to chemical reactions and syntheses, by providing a database service called ChemSpider SyntheticPages (CSSP; see the Relevant Websites section). CSSP is essentially a publishing platform through which a short article about a chemical synthesis procedure is submitted by the user and published after rapid review by one or two editorial board members (typically within 48 h). Once published, it can be commented on by the community. All published articles are assigned digital object identifiers (DOI), making them searchable and citable.

Bioactivity Databases

ChEMBL

ChEMBL is a large-scale bioactivity database developed by the European Bioinformatics Institute (EBI), a part of the European Molecular Biology Laboratory (EMBL) (Gaulton *et al.*, 2012, 2017; Bento *et al.*, 2014). The majority of the bioactivity data in

ChEMBL are manually extracted from the full text of peer-reviewed scientific articles published in a variety of journals in medicinal chemistry and natural product domain. From each publication, the details of the compounds tested, the assays performed and any target information for these assays are extracted. These data are further curated and standardized to maximize their quality and utility across a wide range of chemical biology and drug-discovery research problems.

ChEMBL regularly updates its data, with new releases every 3–4 months. The current release of ChEMBL (release 23) provides information extracted from >67,000 publications and ~50 deposited data sets. In total, it contains >14 million bioactivity values for >1.7 million unique compounds from >1.3 million assays. These assays cover more than 11,000 targets (including ~4300 human protein targets).

While ChEMBL's primary focus is on the bioactivity data extracted from the medicinal chemistry literature, it also collects bioactivity data and annotations from various sources. For example, ChEMBL contains a subset of PubChem's bioactivity data (for confirmatory assays with dose-response endpoints). In addition, bioactivity data extracted by BindingDB from patent documents are integrated into ChEMBL (see Section BindingDB below). Recently, ChEMBL's data scope has expanded to several new areas, including data sets from neglected disease screening, crop protection data for agrochemicals, and drug metabolism and disposition data (Gaulton *et al.*, 2017).

It is worth mentioning more about data exchange between ChEMBL and PubChem (Kim, 2016; Gaulton *et al.*, 2012; Kim *et al.*, 2016a). ChEMBL focuses on compiling bioactivity data for drug-like small molecules that can perturb a biological system in a dose-response way. With that said, the majority of PubChem's bioassay data are not suitable for ChEMBL's data collection, because they were generated from HTS experiments that measured bioactivity of a large number of molecules typically at a "single" concentration. However, a small portion of PubChem assays (confirmatory and panel assays with dose-response endpoints) are within the scope of ChEMBL's data coverage, and have been loaded into ChEMBL. In turn, all ChEMBL assays derived from literature are integrated into PubChem.

BindingDB

BindingDB is a database of experimental protein-ligand affinity data, focusing primarily on proteins considered to be drug targets and their interactions with drug-like small molecules (Gilson *et al.*, 2016; Liu *et al.*, 2007). It currently contains more than 1.3 million binding data between ~7000 protein targets and ~600,000 small molecules. Many of these data are manually curated from scientific journals in the chemical biology and biochemistry domains. BindingDB captures not only the quantitative affinity data (e.g., IC_{50}, K_i, and K_d) but also important experimental conditions such as the temperature, pH, and buffer composition. In addition, it extracts quantitative binding data from patent documents. This patent-derived bioactivity data set is a valuable resource, considering that most of them are not reported in journal articles and that curation efforts in other databases (e.g., SureChEMBL; Papadatos *et al.*, 2016) usually focuses on chemical "mentions" in patent documents without capturing quantitative binding affinity values.

BindingDB also contains affinity data collected from other public databases, such as PubChem (Wang *et al.*, 2017) and ChEMBL (Gaulton *et al.*, 2017), while BindingDB's data are integrated into them in turn. Because BindingDB aims to collect binding affinity data for protein-ligand complexes, it imports ChEMBL's binding assay data but excludes those from functional and ADMET assays. Similarly, PubChem's confirmatory assays with binding affinity data are integrated into BindingDB, whereas PubChem's primary screening data are not imported in BindingDB because they are mostly HTS data generated at a single concentration.

PDBbind

The PDBbind database (Liu *et al.*, 2015) provides experimentally measured binding data for the biomolecular complexes archived in the Protein Data Bank (PDB) (Berman *et al.*, 2003). While PDBbind was originally created at the University of Michigan in 2004 (Wang *et al.*, 2004), it has been maintained and further developed since 2007 at the Shanghai Institute of Organic Chemistry. The biomolecular complexes covered by PDBbind include: (1) complexes between proteins and small molecule ligands; (2) complexes between nucleic acids and small molecule ligands; (3) complexes between two proteins; and (4) complexes between protein and nucleic acid.

High-quality data sets are required for developing and validating molecular docking and scoring functions for structure-based drug design. However, the data quality in PDBbind varies in terms of the resolution of the protein-ligand structures as well as the accuracy of binding affinities. To address this issue, PDBbind provides the "refined set," which consists of high-quality protein-ligand complexes selected by applying a set of rules that ensure the quality of the complex structures and binding data as well as the biological/chemical nature of the complex (Li *et al.*, 2014).

Whereas the refined set contains high-quality data, it is not appropriate to use as a standard benchmark set for evaluating docking/scoring methods, because it suffers from a substantial sample redundancy. For example, ~10% of the complexes in the refined set are those with HIV-1 protease (Liu *et al.*, 2015). Therefore, the refined set was further reduced to the "core set" through a systematic, non-redundant sampling after grouping proteins into clusters with a 90% sequence identity cut off (Li *et al.*, 2014). Relative to these refined and core sets, the entire PDBbind data set is called the "general" set.

PDBbind updates its data annually to keep up with the growth of PDB, along with updated refined and core sets. Currently, PDBbind (v.2016) contains the binding data of 16,179 biomolecular complexes, including 13,308 protein-ligand, 118 nucleic acid-ligand, 777 protein-nucleic acid, and 1976 protein-protein complexes. The refined set of this release contains 4057 protein-ligand complexes, which are further reduced to the core set of 290 complexes in 58 protein clusters.

Binding MOAD

Binding MOAD provides high-quality structures of protein-ligand complexes derived from PDB (with a resolution of 2.5 Å or better) and, where available, their experimental binding affinity data (K_d, K_i, and IC_{50}) (Ahmed *et al.*, 2015; Benson *et al.*, 2008; Hu *et al.*, 2005). The data in Binding MOAD are validated through manual curations of the references cited within the PDB structure file. The proteins are clustered into different families based on 90% sequence identity, and one representative protein is selected from each family (typically, the one complexed with the tightest-binding ligand). This allows BindingDB to provide a non-redundant data set that removes biases from proteins heavily represented in the PDB. The current version of Binding MOAD (released in 2014) contains 25,769 complexes between 12,440 distinct ligands and 7599 distinct protein families, along with 9142 binding affinities.

Binding MOAD and PDBbind have similar goals to each other in the sense that both aim to provide binding data for biomolecular complexes archived in PDB, which is a key difference of the two databases from BindingDB. All three databases provide binding affinity information for protein-ligand complexes, but only a small fraction of protein-ligand affinity data in BindingDB have corresponding crystal structures in PDB. On the other hand, Binding MOAD and PDBbind do have some notable differences from each other. For example, whereas Binding MOAD considers only high-quality structures with a resolution of 2.5 Å or better, PDBbind covers all complexes regardless of the resolution. Instead, PDBbind compiles the refined and core sets to provide high-quality data (see Section PDBbind above). In addition, BindingDB focuses on protein-ligand complexes, but PDBbind has a broader coverage, including protein-ligand complexes as well as other types of complexes (e.g., protein-protein complexes).

Databases of drug information

DrugBank

DrugBank offers comprehensive information on more than 8000 small-molecule drugs and biotech drugs (Law *et al.*, 2014; Knox *et al.*, 2011; Wishart *et al.*, 2006, 2008). While the initial release of DrugBank provided information on selected U.S. Food and Drug Administration (FDA)-approved drugs and their targets, it has been expanding in the depth and breadth of its data and the current release (DrugBank 5.0) contains ~2000 FDA-approved (human and veterinary) drugs, ~1000 investigational drugs (in phase I, II, and III clinical trials), and ~5000 experimental drugs (experimentally shown to bind to specific proteins in humans as well as known bacterial, viral, and fungal pathogens). It also covers hundreds of withdrawn drugs, illicit drugs, and nutraceuticals. For each drug, DrugBank compiles a wide range of information, including their targets, metabolic enzymes, transporters, carriers, drug–drug and drug–food interactions, pharmacogenomic and pharmacoeconomic data, and many others. Because most of these data are manually curated by domain experts and biocurators, DrugBank serves as the source of reference drug data for many chemical databases such as PubChem, PharmGKB, and ChEBI.

Notably, DrugBank has a rich collection of drug metabolism information, including 1200 drug metabolites and 1300 drug metabolism reactions. In addition, a large amount of ADMET (absorption, distribution, metabolism, excretion and toxicity) data are recently added to DrugBank, including Caco2 cell permeability, blood–brain barrier permeability, human intestinal absorption levels, P-glycoprotein activity, CYP450 substrate preferences, renal transport activity, carcinogenicity, Ames test activity, human Ether-a-go-go Related Gene (hERG) activity, etc. DrugBank also provides reference mass spectrometry (MS) and NMR spectra of pure drugs and their metabolites, along with specialized tools for visualizing and searching these spectral data. These spectral data and tools are very useful for identification and/or quantification of compounds in biological matrices.

IUPHAR/BPS Guide to PHARMACOLOGY

The Guide to PHARMACOLOGY (GtoPdb) is a pharmacological information portal that provides expert-curated information on molecular interaction between drug targets and their selective ligands (Southan *et al.*, 2016; Pawson *et al.*, 2014). It was created from collaborative efforts between the International Union of Basic and Clinical Pharmacology (IUPHAR) and the British Pharmacological Society (BPS), by curating and integrating data from the IUPHAR-DB and the published BPS "Guide to Receptors and Channels" (GRAC) compendium.

The early versions of GtoPdb covered the G protein-coupled receptors (GPCRs), ion channels, and nuclear hormone receptors (NHRs) because many drug targets belong to these protein classes. However, the scope of GtoPdb has been expanded to other protein classes, including catalytic receptors, transporters, proteases, kinases, enzymes, etc., with the aim of capturing the likely targets of future medicines. While focused on human proteins, GtoPdb also contains information on mouse and rat orthologs because rodent binding data are most commonly encountered in scientific articles. For each target, GtoPdb provides selective ligands and their quantitative binding data (e.g., K_i, IC_{50}, or K_d), manually curated from scientific research articles. These ligands may be endogenous (e.g., metabolites, hormones, neurotransmitters, and cytokines) or exogenous (e.g., drugs, research leads, toxins, and probes). Currently, GtoPdb covers 2813 targets and 8900 ligands.

GtoPdb uses a unique curation strategy based on the collaboration between in-house curators and the IUPHAR Committee on Receptor Nomenclature and Drug Classification (NC-IUPHAR), which has a network of hundreds of domain experts organized into 60 subcommittees specialized in individual target families. These subcommittees guide target and ligand annotation in GtoPdb, by identifying key pharmacological properties of each target and its quantitative bioactivity data with important ligands. They also monitor de-orphanization of receptors (i.e., identifying new endogenous ligands). In addition, GtoPdb incorporates nomenclature recommendations from NC-IUPHAR. It also adopts and disseminates NC-IUPHAR-derived standards and terminology in quantitative pharmacology.

Pharmacogenomics Knowledge Base (PharmGKB)

The Pharmacogenomics Knowledge Base (PharmGKB) offers information on the impact of genetic variations on drug responses (Whirl-Carrillo et al., 2012; Hernandez-Boussard et al., 2008; Hewett et al., 2002). It contains the relationships between drugs, diseases/phenotypes and genes involved in pharmacodynamics (PD) and pharmacokinetics (PK), extracted from literature using manual curation and natural language processing (NLP) techniques. Importantly, PharmGKB extracts from scientific publications the reported associations between gene variants (single-polymorphism or haplotype) and drug phenotypes, along with key study parameters such as the study size, population ethnicity, and statistics (e.g., p-values and odd ratios). In addition, it collects information on drug-centered pathways that contain the genes involved in the PD/PK of a particular drug. These pieces of information are used to create Very Important Pharmacogene (VIP) summaries that provide a concise overview of critical genes involved in drug response, with links to the literature, gene variant and haplotype details, and relevant drugs.

In PharmGKB, multiple annotations derived from different publications may exist for a given association between a single genetic variant and a drug phenotype. PharmGKB combines these annotations into a "clinical annotation" for that variant-drug-phenotype association, which essentially provides a genotype-based summary of the clinical impact of a genomic variant based on the information in PharmGKB. Each clinical annotation is scored for the strength of supporting evidence on a scale of 1A to 4, with 1A being the strongest evidence.

PharmGKB plays an important role in clinical implementation of pharmacogenomics knowledge. With the Pharmacogenomics Research Network (PGRN), PharmGKB participates in the Clinical Pharmacogenomics Implementation Consortium (CPIC), which publishes genotype-based drug guidelines to help clinicians optimize drug therapy using available genetic test results (Caudle et al., 2014; Relling and Klein, 2011) (see the Relevant Websites section). Upon publication in a scientific journal after peer-review, these guidelines are simultaneously posted to PharmGKB with supplemental data and updates.

SIDER

In the drug discovery and development process, it is critical to identify potential undesired side effects of drug candidates as early as possible in order to reduce the late-stage attrition rate. Therefore, many attempts have been made to predict potential adverse drug reactions (ADRs) and elucidate the underlying biological mechanisms of the ADRs. In addition, several databases have been developed to provide information on ADRs and related drug targets (Juan-Blanco et al., 2015; Cai et al., 2015). One of them is the "Side Effect Resource" (SIDER) database, which currently contains 140,064 drug-ADR pairs between 1430 drugs and 5880 ADRs (Kuhn et al., 2010, 2016). These data are extracted from drug labels and package inserts (including the Structured Product Labels from U.S. FDA) through a dictionary-based named entity recognition (NER) approach. SIDER also uses NLP to extract drug indications from the package inserts, which are used to identify false positives in the ADRs detected through NER.

Databases of chemical toxicity

Comparative Toxicogenomics Database (CTD)

The Comparative Toxicogenomics database (CTD) is a toxicogenomics resource that provides information on interactions between chemicals and genes/proteins and their relationships to diseases (Davis et al., 2013, 2015, 2017). The core contents of CTD are chemical-gene, chemical-disease, and gene-disease interactions, extracted from scientific articles and manually curated by professional biocurators using controlled vocabularies and structured notation. This information is integrated with data from selected external resources, including Gene Ontology (GO) (Carbon et al., 2017), Reactome (Fabregat et al., 2016), NCBI Taxonomy and Gene, BioGRID (Chatr-Aryamontri et al., 2017), and Medical Subject Headings (MeSH).

While the primary focus of CTD has been on environmental chemicals, it also contains a substantial amount of curated data from 88,000 articles that describe the toxic actions of pharmaceuticals on cardiovascular, neurological, hepatic and renal systems, thanks to the collaboration with Pfizer (Davis et al., 2013).

In addition to curated interactions among the chemical-gene-disease triads, CTD also provides inferred relationships between them. For example, if a chemical has a curated interaction with a gene and if that gene has a curated association with a disease, the chemical is inferred to have a relationship with the disease and a putative link between them is created. This helps users to generate testable hypotheses. To assist users in navigating and prioritizing these inferences, CTD computes a ranking score for each inferred relationship, based on the topology of a local network, in which chemicals, genes, and diseases are represented as nodes connected by edges corresponding to the curated interactions (King *et al.*, 2012). Integration with external annotations yields additional inferred relationships.

CTD currently contains more than 1.7 million manually curated interactions (>1.5 million chemical-gene interactions, >34 thousand gene-disease associations, and >200 thousand chemical-disease associations) for 14,975 chemicals, 43,537 genes, and 6416 diseases extracted from 119,720 peer-reviewed scientific papers. It also has 20 million inferred gene-disease relationships and 1.9 million inferred chemical-disease relationships. In total, CTD offers more than 31 million toxicogenomic connections useful for analysis and hypothesis development.

Hazardous Substances Data Bank (HSDB)

The Hazardous Substances Data Bank (HSDB) at the NLM provides toxicological information for more than 5800 potentially hazardous chemicals, including heavy metal compounds, pollutants, herbicides, insecticides, radionuclides, solvents, pharmaceuticals, dietary supplements, venoms, and nanomaterials (Fonger *et al.*, 2014). Chemicals for HSDB record creation and update are nominated by NLM staff, scientific and regulatory agencies, advisory groups, and the public. These candidate chemicals are evaluated and selected by the HSDB chemical selection team, based on a number of considerations, including the level of toxicity, human and environmental exposure, the amount of production and use, and related factors such as regulatory status in the United States and other countries.

For the selected compounds, HSDB compiles various kinds of information, including (but not limited to) human exposure, industrial hygiene, emergency handling procedures, environmental fate, and regulatory requirements. These data are collected from a core set of books, government documents, technical reports, and journal articles. The collected information is peer-reviewed by the Scientific Review Panel (SRP), which consists of experts in major subject areas within the HSDB's scope.

The data contents in HSDB are labeled with one of three data quality tags: "PEER REVIEWED", "QC REVIEWED", and "UNREVIEWED". The data peer-reviewed by the SRP are tagged as "PEER REVIEWED". The "QC REVIEWED" tag means that a quality control review has been conducted by a HSDB senior reviewer or SRP subcommittee while a full peer review by SRP is pending. The "UNREVIEWED" tag indicates that the data has not been evaluated for scientific accuracy. Because this tag is used in the internal record building process, public HSDB data do not have the "UNREVIEWED" tag unless there is a special circumstance.

Databases in other categories

Human Metabolome Database (HMDB)

The Human Metabolome Database (HMDB) is a major metabolomics database, which offers comprehensive information on human metabolites, including their abundance, associated proteins, and disease-related properties (Wishart *et al.*, 2007, 2009, 2013). Currently, HMDB (version 3.6) contains detailed information on 52,658 metabolites, which can be classified into two groups: detected metabolites and expected metabolites. The detected metabolites are further divided into two groups: (1) detected and quantified and (2) detected but not quantified. The expected metabolites are those which could be or will be detected in the future, based on the current knowledge of human biochemistry along with human food and drug consumption patterns. Metabolites are further classified as being endogenous, microbial, drug derived, a toxin/pollutant, food derived, or different combinations of these.

The HMDB provides information on the experimentally determined concentrations of metabolites in various biofluids and/or tissues (e.g., cerebrospinal fluid, serum, urine, saliva) and tissues (e.g., prostate). Disease association and abnormal concentrations were manually curated from literature. In addition, it has reference NMR, tandem mass spectrometry (MS/MS), and gas chromatography-mass spectrometry (GC–MS) spectra. MS and NMR spectra were collected, assigned, and/or annotated by the HMDB curation team. Hundreds of additional annotated/assigned MS and NMR reference spectra were obtained from the BioMagResBank (Ulrich *et al.*, 2008), METLIN (Smith *et al.*, 2005; Tautenhahn *et al.*, 2012), and MassBank (Horai *et al.*, 2010). These spectral data are available for download as image files or in widely used spectral data exchange formats, such as JCAMP-DX (Davies and Lampen, 1993; Lampen *et al.*, 1994), nmrML, and mzML (Deutsch, 2008; Martens *et al.*, 2011), etc, which capture relevant spectral features, spectral collection conditions, assignments, and chemical structure information.

Chemical Entities of Biological Interest (ChEBI)

Chemical Entities of Biological Interest (ChEBI) at the EMBL-EBI is a free online dictionary of "small" molecular entities (Hastings *et al.*, 2013, 2016; De Matos *et al.*, 2010). The term "molecular entity" refers to "any constitutionally or isotopically distinct atom, molecule, ion, ion pair, radical, radical ion, complex, conformer, etc., identifiable as a separately distinguishable entity" (Degtyarenko *et al.*, 2008). While ChEBI focuses on small chemical compounds (either naturally-occurring or synthetic) that affect biological processes of living organisms, it excludes molecules directly encoded by the genome (e.g., nucleic acids, proteins and peptides derived from proteins by cleavage) as a rule.

The current release of ChEBI (Release 152) contains more than 50,000 fully curated entries with another ~50,000 entries that have not been processed yet. For each annotated entry, a wide range of manually curated data items are provided, including chemical structures and chemical structure representations (International Chemical Identifier (InChI) (Heller *et al.*, 2013, 2015), InChIKey, and simplified molecular-input line-entry system (SMILES) strings (Weininger, 1988, 1990; Weininger *et al.*, 1989)), chemical names (IUPAC names, International Non-proprietary Names (INNs), brand names, and other synonyms), and literature citations. In addition, each entry is extensively cross-referenced to related records in external databases.

An important feature of ChEBI is that each entry in ChEBI is classified within the "ChEBI ontology". It consists of two main subontologies: (1) a chemical entity ontology in which chemical entities are classified based on shared structural features and (2) a role ontology in which classifies entities based on their activities in biological or chemical systems or their use in applications. Through this ontological classification, ChEBI specifies the relationship between molecular entities (or classes of entities).

UniChem

Chemical information available in the public domain is scattered across many different databases. Identifying equivalent chemical entities from these resources is not a trivial task because individual resources adopt different data models, release schedules, structure standardization rules, chemical nomenclatures, etc. UniChem is a free compound identifier mapping service developed by EMBL-EBI to address this data integration issue (Chambers et al., 2013, 2014; Hersey et al., 2015). It contains standard InChIs (Heller et al., 2013, 2015) for more than 145 million chemical structures stored in >30 databases, along with pointers between these structures and chemical identifiers from all the individual databases. Therefore, UniChem is very useful to cross-reference equivalent chemical records between different databases.

Originally, UniChem produced mappings between chemicals in different sources on the basis of complete identity between their standard InChIs (i.e., only if their standard InChIs are completely identical) (Chambers et al., 2013). However, this approach appeared too narrow and stringent for some purposes because stereoisomers, isotopes, and salts of otherwise identical molecules could not be related. For this reason, UniChem introduced a service called "Connectivity Search", in which the equivalence between molecules can be determined at different levels of structural specification (Chambers et al., 2014). This service exploits the layered structural representation of standard InChIs, by matching molecules on the basis of the complete identity of their connectivity layers and then comparing the remaining layers to check the stereochemical and isotopic differences. It also supports mixtures and salts, by using standard InChI sublayers.

UniChem is a highly automated system. When new data are loaded into one source, all links between this source and all the others are automatically calculated and immediately available. UniChem also provides users with a RESTful web service interface for programmatic access to its data.

Closing Remarks

In this article, several public chemical databases useful for cheminformatics and bioinformatics research have been described. Some of these databases like PubChem and ChemSpider contain a very large number of molecules (91 millions and 59 millions, respectively) and provide a wide variety of information, including chemical, physical, and spectral properties, pharmacology, toxicology, metabolism, safety and handling, environmental health and many others. Essentially, these databases are data aggregators, which collect and integrate information from various sources.

Other databases usually offer specialized information for a small number of compounds relevant to narrow focus areas. For example, ChEMBL, BindingDB, PDBbind, and Binding MOAD extract bioactivity data from scientific articles and/or patent documents. DrugBank contains manually curated information for FDA-approved and experimental drugs. GtoPdb provides information on the interaction between drug targets and their ligands, and PharmGKB focuses on pharmacogenomics aspects of drug response. SIDER collects information on the side effect of drugs. CTD and HSDB are specialized in toxicological information, and HMDB compiles metabolomics information. ChEBI provides ontology-based classifications of chemicals and UniChem deals with cross-reference information among different databases.

Although these databases often show substantial overlaps with each other in terms of the compound coverage as well as the type of information they provide, each database does have unique data contents, which can complement the information in other databases. While data exchange between different databases has become a very common practice, this also has raised serious concerns over quality control of public-domain chemistry data. For example, the lack of a standard in chemical structure representation has made individual database developers adopt different chemical structure standardization rules that suit their needs. As a result, it is not trivial to figure out which compound in one database is the same as a compound in another database, often leading to incorrect mapping between compounds from different databases.

Users of public chemical databases should be aware of potential data quality issues (Southan et al., 2013; Kramer et al., 2012; Hu and Bajorath, 2012; Muresan et al., 2011; Tiikkainen et al., 2013; Williams et al., 2012; Williams and Ekins, 2011; Hersey et al., 2015; Akhondi et al., 2012; Fourches et al., 2010; Tiikkainen and Franke, 2012; Williams, 2008b). Therefore, some important papers about the data quality issues are listed in the Further Reading section. In addition, this section contains papers about (1) programmatic access to selected databases (Kim et al., 2015; Davies et al., 2015; Swainston et al., 2016); (2) Resource Description Framework (RDF)-formatted chemistry data for data exchange, sharing, and integration (Willighagen et al., 2013; Jupp et al., 2014; Fu et al., 2015); and (3) commonly used chemical structure representations (Warr, 2011; Heller et al., 2015; Weininger, 1988).

Acknowledgements

This work was supported by the Intramural Research Program of the National Institutes of Health, National Library of Medicine. The author thanks Evan Bolton at PubChem and Bradley Otterson at the NIH Library Editing Service for reviewing this manuscript.

See also: Bioinformatics Data Models, Representation and Storage. Biological Database Searching. Chemoinformatics: From Chemical Art to Chemistry in Silico. Data Storage and Representation. Drug Repurposing and Multi-Target Therapies. In Silico Identification of Novel Inhibitors. Information Retrieval in Life Sciences. Introduction of Docking-Based Virtual Screening Workflow Using Desktop Personal Computer. Introduction of Docking-Based Virtual Screening Workflow Using Desktop Personal Computer. Natural Language Processing Approaches in Bioinformatics. Rational Structure-Based Drug Design. Small Molecule Drug Design

---

References

Ahmed, A., Smith, R.D., Clark, J.J., Dunbar, J.B., Carlson, H.A., 2015. Recent improvements to Binding MOAD: A resource for protein-ligand binding affinities and structures. Nucleic Acids Research 43, D465–D469.

Akhondi, S.A., Kors, J.A., Muresan, S., 2012. Consistency of systematic chemical identifiers within and between small-molecule databases. Journal of Cheminformatics 4, 35.

Benson, M.L., Smith, R.D., Khazanov, N.A., et al., 2008. Binding MOAD, a high-quality protein-ligand database. Nucleic Acids Research 36, D674–D678.

Bento, A.P., Gaulton, A., Hersey, A., et al., 2014. The ChEMBL bioactivity database: An update. Nucleic Acids Research 42, D1083–D1090.

Berman, H., Henrick, K., Nakamura, H., 2003. Announcing the worldwide Protein Data Bank. Nature Structural Biology 10, 980.

Cai, M.C., Xu, Q., Pan, Y.J., et al., 2015. ADReCS: An ontology database for aiding standardization and hierarchical classification of adverse drug reaction terms. Nucleic Acids Research 43, D907–D913.

Carbon, S., Dietze, H., Lewis, S.E., et al., 2017. Expansion of the gene ontology knowledge base and resources. Nucleic Acids Research 45, D331–D338.

Caudle, K.E., Klein, T.E., Hoffman, J.M., et al., 2014. Incorporation of pharmacogenomics into routine clinical practice: The Clinical Pharmacogenetics Implementation Consortium (CPIC) guideline development process. Current Drug Metabolism 15, 209–217.

Chambers, J., Davies, M., Gaulton, A., et al., 2013. UniChem: A unified chemical structure cross-referencing and identifier tracking system. Journal of Cheminformatics 5, 3.

Chambers, J., Davies, M., Gaulton, A., et al., 2014. UniChem: Extension of InChI-based compound mapping to salt, connectivity and stereochemistry layers. Journal of Cheminformatics 6, 43.

Chatr-Aryamontri, A., Oughtred, R., Boucher, L., et al., 2017. The BioGRID interaction database: 2017 update. Nucleic Acids Research 45, D369–D379.

Cheng, T., Pan, Y., Hao, M., Wang, Y., Bryant, S.H., 2014. PubChem applications in drug discovery: A bibliometric analysis. Drug Discovery Today 19, 1751–1756.

Davies, A.N., Lampen, P., 1993. Jcamp-DX for NMR. Applied Spectroscopy 47, 1093–1099.

Davies, M., Nowotka, M., Papadatos, G., et al., 2015. ChEMBL web services: Streamlining access to drug discovery data and utilities. Nucleic Acids Research 43, W612–W620.

Davis, A.P., Grondin, C.J., Johnson, R.J., et al., 2017. The comparative toxicogenomics database: Update 2017. Nucleic Acids Research 45, D972–D978.

Davis, A.P., Grondin, C.J., Lennon-Hopkins, K., et al., 2015. The comparative toxicogenomics database's 10th year anniversary: Update 2015. Nucleic Acids Research 43, D914–D920.

Davis, A.P., Murphy, C.G., Johnson, R., et al., 2013. The comparative toxicogenomics database: Update 2013. Nucleic Acids Research 41, D1104–D1114.

Davis, A.P., Wiegers, T.C., Roberts, P.M., et al., 2013. A CTD-Pfizer collaboration: Manual curation of 88,000 scientific articles text mined for drug-disease and drug-phenotype interactions. Database (Oxford) 2013, bat080.

Degtyarenko, K., De Matos, P., Ennis, M., et al., 2008. ChEBI: A database and ontology for chemical entities of biological interest. Nucleic Acids Research 36, D344–D350.

De Matos, P., Alcantara, R., Dekker, A., et al., 2010. Chemical entities of biological interest: An update. Nucleic Acids Research 38, D249–D254.

Deutsch, E., 2008. MzML: A single, unifying data format for mass spectrometer output. Proteomics 8, 2776–2777.

Fabregat, A., Sidiropoulos, K., Garapati, P., et al., 2016. The reactome pathway knowledgebase. Nucleic Acids Research 44, D481–D487.

Fonger, G.C., Hakkinen, P., Jordan, S., Publicker, S., 2014. The national library of medicine's (NLM) hazardous substances data bank (HSDB): Background, recent enhancements and future plans. Toxicology 325, 209–216.

Fourches, D., Muratov, E., Tropsha, A., 2010. Trust, but verify: On the importance of chemical structure curation in cheminformatics and QSAR modeling research. Journal of Chemical Information and Modeling 50, 1189–1204.

Fu, G., Batchelor, C., Dumontier, M., et al., 2015. PubChemRDF: Towards the semantic annotation of PubChem compound and substance databases. Journal of Cheminformatics 7, 34.

Gaulton, A., Bellis, L.J., Bento, A.P., et al., 2012. ChEMBL: A large-scale bioactivity database for drug discovery. Nucleic Acids Research 40, D1100–D1107.

Gaulton, A., Hersey, A., Nowotka, M., et al., 2017. The ChEMBL database in 2017. Nucleic Acids Research 45, D945–D954.

Gilson, M.K., Liu, T.Q., Baitaluk, M., et al., 2016. BindingDB in 2015: A public database for medicinal chemistry, computational chemistry and systems pharmacology. Nucleic Acids Research 44, D1045–D1053.

Hastings, J., De Matos, P., Dekker, A., et al., 2013. The ChEBI reference database and ontology for biologically relevant chemistry: Enhancements for 2013. Nucleic Acids Research 41, D456–D463.

Hastings, J., Owen, G., Dekker, A., et al., 2016. ChEBI in 2016: Improved services and an expanding collection of metabolites. Nucleic Acids Research 44, D1214–D1219.

Heller, S., Mcnaught, A., Pletnev, I., Stein, S., Tchekhovskoi, D., 2015. InChI, the IUPAC international chemical identifier. Journal of Cheminformatics 7, 23.

Heller, S., Mcnaught, A., Stein, S., Tchekhovskoi, D., Pletnev, I., 2013. InChI – The worldwide chemical structure identifier standard. Journal of Cheminformatics 5, 7.

Hernandez-Boussard, T., Whirl-Carrillo, M., Hebert, J.M., et al., 2008. The pharmacogenetics and pharmacogenomics knowledge base: Accentuating the knowledge. Nucleic Acids Research 36, D913–D918.

Hersey, A., Chambers, J., Bellis, L., et al., 2015. Chemical databases: Curation or integration by user-defined equivalence? Drug Discovery Today: Technologies 14, 17–24.

Hewett, M., Oliver, D.E., Rubin, D.L., et al., 2002. PharmGKB: The Pharmacogenetics Knowledge Base. Nucleic Acids Research 30, 163–165.

Horai, H., Arita, M., Kanaya, S., et al., 2010. MassBank: A public repository for sharing mass spectral data for life sciences. Journal of Mass Spectrometry 45, 703–714.

Hu, L.G., Benson, M.L., Smith, R.D., Lerner, M.G., Carlson, H.A., 2005. Binding MOAD (Mother of all databases). Proteins-Structure Function and Bioinformatics 60, 333–340.

Hu, Y., Bajorath, J., 2012. Growth of ligand-target interaction data in ChEMBL is associated with increasing and activity measurement-dependent compound promiscuity. Journal of Chemical Information and Modeling 52, 2550–2558.

Juan-Blanco, T., Duran-Frigola, M., Aloy, P., 2015. IntSide: A web server for the chemical and biological examination of drug side effects. Bioinformatics 31, 612–613.

Jupp, S., Malone, J., Bolleman, J., et al., 2014. The EBI RDF platform: Linked open data for the life sciences. Bioinformatics 30, 1338–1339.

Kim, S., 2016. Getting the most out of PubChem for virtual screening. Expert Opinion on Drug Discovery 11, 843–855.

Kim, S., Thiessen, P.A., Bolton, E.E., Bryant, S.H., 2015. PUG-SOAP and PUG-rest: Web services for programmatic access to chemical information in PubChem. Nucleic Acids Research 43, W605–W611.

Kim, S., Thiessen, P.A., Bolton, E.E., et al., 2016a. PubChem Substance and Compound databases. Nucleic Acids Research 44, D1202–D1213.

Kim, S., Thiessen, P.A., Cheng, T., et al., 2016b. Literature information in PubChem: Associations between PubChem records and scientific articles. Journal of Cheminformatics 8, 32.

King, B.L., Davis, A.P., Rosenstein, M.C., Wiegers, T.C., Mattingly, C.J., 2012. Ranking transitive chemical-disease inferences using local network topology in the comparative toxicogenomics database. Plos One 7, e46524.

Knox, C., Law, V., Jewison, T., et al., 2011. DrugBank 3.0: A comprehensive resource for 'Omics' research on drugs. Nucleic Acids Research 39, D1035–D1041.

Kramer, C., Kalliokoski, T., Gedeck, P., Vulpetti, A., 2012. The experimental uncertainty of heterogeneous public K-i data. Journal of Medicinal Chemistry 55, 5165–5173.

Kuhn, M., Campillos, M., Letunic, I., Jensen, L.J., Bork, P., 2010. A side effect resource to capture phenotypic effects of drugs. Molecular Systems Biology 6, 343.

Kuhn, M., Letunic, I., Jensen, L.J., Bork, P., 2016. The SIDER database of drugs and side effects. Nucleic Acids Research 44, D1075–D1079.

Lampen, P., Hillig, H., Davies, A.N., Linscheid, M., 1994. JCAMP-DX for mass-spectrometry. Applied Spectroscopy 48, 1545–1552.

Law, V., Knox, C., Djoumbou, Y., et al., 2014. DrugBank 4.0: Shedding new light on drug metabolism. Nucleic Acids Research 42, D1091–D1097.

Lipinski, C.A., Litterman, N.K., Southan, C., et al., 2015. Parallel worlds of public and commercial bioactive chemistry data. Journal of Medicinal Chemistry 58, 2068–2076.

Liu, T.Q., Lin, Y.M., Wen, X., Jorissen, R.N., Gilson, M.K., 2007. BindingDB: A web-accessible database of experimentally determined protein-ligand binding affinities. Nucleic Acids Research 35, D198–D201.

Liu, Z.H., Li, Y., Han, L., et al., 2015. PDB-wide collection of binding data: Current status of the PDBbind database. Bioinformatics 31, 405–412.

Li, Y., Liu, Z.H., Li, J., et al., 2014. Comparative assessment of scoring functions on an updated benchmark: 1. Compilation of the test set. Journal of Chemical Information and Modeling 54, 1700–1716.

Martens, L., Chambers, M., Sturm, M., et al., 2011. MzML-a community standard for mass spectrometry data. Molecular & Cellular Proteomics 10.R110.000133.

Ma, X.H., Zhu, F., Liu, X., et al., 2012. Virtual screening methods as tools for drug lead discovery from large chemical libraries. Current Medicinal Chemistry 19, 5562–5571.

Muresan, S., Petrov, P., Southan, C., et al., 2011. Making every SAR point count: The development of chemistry connect for the large-scale integration of structure and bioactivity data. Drug Discovery Today 16, 1019–1030.

Nicola, G., Liu, T.Q., Gilson, M.K., 2012. Public domain databases for medicinal chemistry. Journal of Medicinal Chemistry 55, 6987–7002.

Papadatos, G., Davies, M., Dedman, N., et al., 2016. SureChEMBL: A large-scale, chemically annotated patent document database. Nucleic Acids Research 44, D1220–D1228.

Pawson, A.J., Sharman, J.L., Benson, H.E., et al., 2014. The IUPHAR/BPS guide to PHARMACOLOGY: An expert-driven knowledgebase of drug targets and their ligands. Nucleic Acids Research 42, D1098–D1106.

Pence, H.E., Williams, A., 2010. ChemSpider: An online chemical information resource. Journal of Chemical Education 87, 1123–1124.

Relling, M.V., Klein, T.E., 2011. CPIC: Clinical pharmacogenetics implementation consortium of the pharmacogenomics research network. Clinical Pharmacology & Therapeutics 89, 464–467.

Smith, C.A., O'maille, G., Want, E.J., et al., 2005. METLIN – A metabolite mass spectral database. Therapeutic Drug Monitoring 27, 747–751.

Southan, C., Sharman, J.L., Benson, H.E., et al., 2016. The IUPHAR/BPS guide to PHARMACOLOGY in 2016: towards curated quantitative interactions between 1300 protein targets and 6000 ligands. Nucleic Acids Research 44, D1054–D1068.

Southan, C., Williams, A.J., Ekins, S., 2013. Challenges and recommendations for obtaining chemical structures of industry-provided repurposing candidates. Drug Discovery Today 18, 58–70.

Swainston, N., Hastings, J., Dekker, A., et al., 2016. LibChEBI: An API for accessing the ChEBI database. Journal of Cheminformatics 8, 11.

Tautenhahn, R., Cho, K., Uritboonthai, W., et al., 2012. An accelerated workflow for untargeted metabolomics using the METLIN database. Nature Biotechnology 30, 826–828.

Tiikkainen, P., Bellis, L., Light, Y., Franke, L., 2013. Estimating error rates in bioactivity databases. Journal of Chemical Information and Modeling 53, 2499–2505.

Tiikkainen, P., Franke, L., 2012. Analysis of commercial and public bioactivity databases. Journal of Chemical Information and Modeling 52, 319–326.

Ulrich, E.L., Akutsu, H., Doreleijers, J.F., et al., 2008. BioMagResBank. Nucleic Acids Research 36, D402–D408.

Wang, R.X., Fang, X.L., Lu, Y.P., Wang, S.M., 2004. The PDBbind database: Collection of binding affinities for protein-ligand complexes with known three-dimensional structures. Journal of Medicinal Chemistry 47, 2977–2980.

Wang, Y., Bryant, S.H., Cheng, T., et al., 2017. PubChem BioAssay: 2017 update. Nucleic Acids Research 45, D955–D963.

Warr, W.A., 2011. Representation of chemical structures. Wiley Interdisciplinary Reviews–Computational Molecular Science 1, 557–579.

Weininger, D., 1988. Smiles, a chemical language and information-system. 1. Introduction to methodology and encoding rules. Journal of Chemical Information and Computer Sciences 28, 31–36.

Weininger, D., 1990. Smiles. 3. Depict - graphical depiction of chemical structures. Journal of Chemical Information and Computer Sciences 30, 237–243.

Weininger, D., Weininger, A., Weininger, J.L., 1989. Smiles. 2. Algorithm for generation of unique smiles notation. Journal of Chemical Information and Computer Sciences 29, 97–101.

Whirl-Carrillo, M., Mcdonagh, E.M., Hebert, J.M., et al., 2012. Pharmacogenomics knowledge for personalized medicine. Clinical Pharmacology & Therapeutics 92, 414–417.

Williams, A.J., Ekins, S., 2011. A quality alert and call for improved curation of public chemistry databases. Drug Discovery Today 16, 747–750.

Williams, A.J., Ekins, S., Tkachenko, V., 2012. Towards a gold standard: Regarding quality in public domain chemistry databases and approaches to improving the situation. Drug Discovery Today 17, 685–701.

Williams, A., Tkachenko, V., 2014. The royal society of chemistry and the delivery of chemistry data repositories for the community. Journal of Computer-Aided Molecular Design 28, 1023–1030.

Williams, A.J., 2008a. A perspective of publicly accessible/open-access chemistry databases. Drug Discovery Today 13, 495–501.

Williams, A.J., 2008b. Public chemical compound databases. Current Opinion in Drug Discovery & Development 11, 393–404.

Willighagen, E.L., Waagmeester, A., Spjuth, O., et al., 2013. The ChEMBL database as linked open data. Journal of Cheminformatics 5, 23.

Wishart, D.S., Jewison, T., Guo, A.C., et al., 2013. HMDB 3.0 – The Human Metabolome Database in 2013. Nucleic Acids Research 41, D801–D807.

Wishart, D.S., Knox, C., Guo, A.C., et al., 2006. DrugBank: A comprehensive resource for in silico drug discovery and exploration. Nucleic Acids Research 34, D668–D672.

Wishart, D.S., Knox, C., Guo, A.C., et al., 2008. DrugBank: A knowledgebase for drugs, drug actions and drug targets. Nucleic Acids Research 36, D901–D906.

Wishart, D.S., Knox, C., Guo, A.C., et al., 2009. HMDB: A knowledgebase for the human metabolome. Nucleic Acids Research 37, D603–D610.

Wishart, D.S., Tzur, D., Knox, C., et al., 2007. HMDB: The human metabolome database. Nucleic Acids Research 35, D521–D526.

Further Reading

Akhondi, S.A., Kors, J.A., Muresan, S., 2012. Consistency of systematic chemical identifiers within and between small-molecule databases. Journal of Cheminformatics 4, 35.

Davies, M., Nowotka, M., Papadatos, G., et al., 2015. ChEMBL web services: Streamlining access to drug discovery data and utilities. Nucleic Acids Research 43, W612–W620.

Fu, G., Batchelor, C., Dumontier, M., et al., 2015. PubChemRDF: Towards the semantic annotation of PubChem compound and substance databases. Journal of Cheminformatics 7, 34.

Heller, S., Mcnaught, A., Pletnev, I., Stein, S., Tchekhovskoi, D., 2015. InChI, the IUPAC international chemical identifier. Journal of Cheminformatics 7, 23.

Hersey, A., Chambers, J., Bellis, L., et al., 2015. Chemical databases: Curation or integration by user-defined equivalence? Drug Discovery Today: Technologies 14, 17–24.

Jupp, S., Malone, J., Bolleman, J., *et al.*, 2014. The EBI RDF platform: Linked open data for the life sciences. Bioinformatics 30, 1338–1339.

Kim, S., Thiessen, P.A., Bolton, E.E., Bryant, S.H., 2015. PUG-SOAP and PUG-rest: Web services for programmatic access to chemical information in PubChem. Nucleic Acids Research 43, W605–W611.

Swainston, N., Hastings, J., Dekker, A., *et al.*, 2016. libChEBI: An API for accessing the ChEBI database. Journal of Cheminformatics 8, 11.

Tiikkainen, P., Bellis, L., Light, Y., Franke, L., 2013. Estimating error rates in bioactivity databases. Journal of Chemical Information and Modeling 53, 2499–2505.

Warr, W.A., 2011. Representation of chemical structures. Wiley Interdisciplinary Reviews-Computational Molecular Science 1, 557–579.

Weininger, D., 1988. Smiles, a chemical language and information-system. 1. Introduction to methodology and encoding rules. Journal of Chemical Information and Computer Sciences 28, 31–36.

Williams, A.J., Ekins, S., 2011. A quality alert and call for improved curation of public chemistry databases. Drug Discovery Today 16, 747–750.

Williams, A.J., Ekins, S., Tkachenko, V., 2012. Towards a gold standard: Regarding quality in public domain chemistry databases and approaches to improving the situation. Drug Discovery Today 17, 685–701.

Willighagen, E.L., Waagmeester, A., Spjuth, O., *et al.*, 2013. The ChEMBL database as linked open data. Journal of Cheminformatics 5, 23.

Relevant Websites

http://www.BindingMOAD.org
    Binding MOAD.
https://www.bindingdb.org
    BindingDB.
https://www.ebi.ac.uk/chembl
    ChEMBL.
https://www.ebi.ac.uk/chebi/
    Chemical Entities of Biological Interest (ChEBI).
http://www.chemspider.com
    ChemSpider.
http://cssp.chemspider.com
    ChemSpider SyntheticPages.
https://ctdbase.org
    Comparative Toxicogenomics Database (CTD).
https://cpicpgx.org
    Clinical Pharmacogenomics Implementation Consortium (CPIC).
http://www.drugbank.ca
    DrugBank.
http://www.guidetopharmacology.org
    Guide To PHARMACOLOGY.
https://www.toxnet.nlm.nih.gov/newtoxnet/hsdb.htm
    Hazardous Substances Data Bank (HSDB).
http://www.hmdb.ca
    Human Metabolome Database (HMDB).
http://www.pdbbind.org.cn
    PDBbind.
https://www.pharmgkb.org
    Pharmacogenomics Knowledge Base (PharmGKB).
https://www.pgrn.org
    Pharmacogenomics Research Network (PGRN).
https://pubchem.ncbi.nlm.nih.gov
    PubChem.
https://www.ncbi.nlm.nih.gov/pcassay/
    PubChem BioAssay.
https://www.ncbi.nlm.nih.gov/pccompound/
    PubChem Compound.
https://www.ncbi.nlm.nih.gov/pcsubstance/
    PubChem Substance.
http://sideeffects.embl.de
    SIDER.
https://www.ebi.ac.uk/unichem
    UniChem.

Biographical Sketch

Sunghwan Kim is a Staff Scientist at the National Center for Biotechnology Information (NCBI), National Library of Medicine (NLM), U.S. National Institutes of Health (NIH). As a computational chemist and cheminformatician, he is actively involved in the PubChem project, which develops and maintains a small-molecule database called PubChem. Specifically, his research has been focused on building and improving "PubChem3D," which is PubChem's chemical information resource derived from 3-dimensional (3-D) molecular structures. He holds a MSc in Inorganic Chemistry (from Hanyang University, South Korea) and a PhD in Physical Chemistry (from the University of Georgia at Athens).

Chemical Similarity and Substructure Searches

Nils M Kriege and Lina Humbeck, TU, Dortmund, Germany
Oliver Koch, Technical University of Dortmund, Dortmund, Germany

Introduction

A fundamental concept in cheminformatics is the assumption that chemically similar molecules share similar bioactivities. Therefore, comparison of molecules and the determination of similarities between molecules is an important and often applied task for the identification and development of bioactive molecules (Humbeck and Koch, 2017). Many different methods and concepts have been developed to solve this essential task and it is crucial to understand and know what they are based on as similarity is not a rigid, globally defined or fixed property but rather reflects the perspective of the investigator. As with so many things in life, similarity is in the eye of the beholder.

Firstly, the molecules have to be represented. This can be achieved by a descriptor, a fingerprint or a structural formula. A descriptor is a number like the molecular masses of two molecules, a fingerprint is a vector representation and a structural formula is a molecular graph. The latter one is an intuitive as well as exact representation where vertices represent atoms and edges represent bonds. Secondly, one can discriminate between different types of searches. For example there are cases where one wants to find the exact same structure, only molecules which contain the query molecule as a substructure or molecules sharing the same fragment. In other cases only molecules that are somehow similar are of interest.

These similarity based concepts can be further divided into similarity based on substructures, pharmacophores, fingerprints, shapes, biological activities and so on. Two molecules that share a similar molecular surface (shape) are expected to be similar. A fingerprint could be a bit representation of the presence or absence of fragments and two molecules with similar fingerprints are also expected to be similar. A pharmacophore is a three dimensional arrangement of molecular features that is important for the interaction with the protein target (Guner, 2002). The three dimensional molecular graph of a query molecule is compared to this spatial arrangement and the molecule fulfills this pharmacophore, if its feature arrangement based on its graph is also similar. The search for a largest possible substructure which is present in two molecules is called MCS (Maximum Common Substructure) search. Nevertheless, often different kinds of representations and searches are used together.

In the following some specific concepts are presented to get an idea of the importance of chemical similarity. Quantitative Structure-Property Relationships (QSPR) and Quantitative Structure-Activity Relationships (QSAR) are two such concepts. One of the first QSPR analyses was done in 1947 and included the Wiener Index as description of molecular properties to predict the boiling points of alkane molecules (Wiener, 1947). A more specialized property – the bioactivity – is analyzed using corresponding QSAR (Leach and Gillet, 2007). As described, it might be valuable to compare the shapes of molecules. Therefore, tools like ROCS (Hawkins et al., 2007) or ShaEP (Vainio et al., 2009) can be used. One can also group molecules regarding their target proteins (Keiser et al., 2007) or correlate descriptors with certain properties to guide synthesis to optimized molecules.

Another often discussed approach are privileged scaffolds, which are scaffolds – meaning same core substructures of different molecules – that show activities on different protein families. Hence, they are interesting starting points for drug design projects and additionally close to the medicinal chemists view of molecules. A tool which enables a scaffold focused view of molecular databases is Scaffold Hunter (Schäfer et al., 2017). One also has to be aware of substructures that interfere with assay systems termed PAINS (Pan Assay INterferece Substances) (Baell et al., 2013) or that are unspecific destabilizers of proteins, e.g. aggregators (Irwin et al., 2015).

As mentioned above the topic of similarity and comparability is closely related to the challenge to properly represent the molecule. Often molecules are represented as vectors which make them applicable to a bunch of methods, e.g., QSAR approaches. However, this representation loses information, e.g., regarding the connectivity and structure of molecules. In contrast a graph based representation is closer to the intuitive chemical understanding of molecules. Unfortunately, efficient algorithms are needed to keep the increased calculation times small. In this article we introduce some graph based algorithms to compare molecules and discuss their applications.

Background and Fundamentals

By the middle of the nineteenth century the growing knowledge about chemical atoms and how they combine to form larger structures led to meaningful diagrams, which can be seen as early applications of *graph theory* (Biggs et al., 1986). This section introduces into graph theory as well as molecular graph representation and summarizes graph theoretical fundamentals using modern terms and notation.

Graph Theory

A *graph* $G = (V, E)$ consists of a finite set $V(G) = V$ of *vertices* and a finite set $E(G) = E$ of *edges*, where each edge connects two distinct vertices. We denote an edge connecting a vertex u and a vertex v by uv or vu, where both refers to the same edge. A *path* of length n is

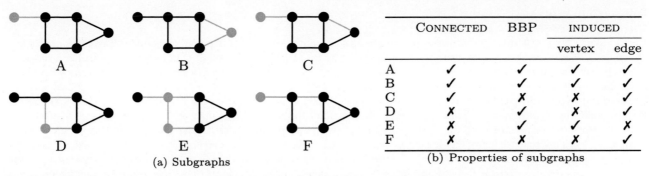

	CONNECTED	BBP	INDUCED	
			vertex	edge
A	✓	✓	✓	✓
B	✓	✓	✓	✓
C	✓	✗	✗	✓
D	✗	✓	✗	✓
E	✗	✓	✓	✗
F	✗	✗	✗	✓

(a) Subgraphs   (b) Properties of subgraphs

Fig. 1 Examples of subgraphs and their properties, where gray vertices and edges are not contained in the subgraph.

a sequence of vertices $(v_0,...,v_n)$ such that $v_i v_{i+1} \in E$ for $0 \leq i < n$. A graph is *connected* if at least one path between any two vertices exists and is *disconnected* otherwise. A *cycle* is a path of length at least 3 with no repeated vertices except $v_0 = v_n$. A graph $G' = (V', E')$ is a *subgraph* of a graph $G = (V, E)$, written $G' \subseteq G$, if $V' \subseteq V$ and $E' \subseteq E$. A subgraph $G' \subseteq G$ is said to *span* G if $V(G') = V(G)$. Let $V' \subseteq V$ and $E' = \{uv \in E \mid u, v \in V'\}$, then $G' = (V', E')$ is said to be a *(vertex-)induced* subgraph of G and is denoted by $G[V']$. Let $E' \subseteq E$ and V' be the set of vertices that appear as endpoint of at least one edge in E', then $G' = (V', E')$ is said to be an *edge-induced* subgraph of G and denoted by $G[E']$. An edge is a *bridge* if it is not contained in any cycle. A *block* is a maximal connected subgraph that does not contain a bridge. A subgraph $G' \subseteq G$ is *block and bridge preserving* (BBP) if (i) each bridge in G' is a bridge in G, (ii) any two edges in different blocks in G' are in different blocks in G. **Fig. 1** shows examples of different subgraphs.

An *isomorphism* between two graphs G and H is a bijection $\psi : V(G) \to V(H)$ such that $uv \in E(G) \Leftrightarrow \psi(u)\psi(v) \in E(H)$ for all $u, v \in V$ (G). Two graphs G and H are said to be *isomorphic*, written $G \simeq H$, if an isomorphism between G and H exists. A *subgraph isomorphism* from a graph G to a graph H is an isomorphism between G and a subgraph $H' \subseteq H$. A *common subgraph isomorphism* between G and H is an isomorphism between subgraphs $G' \subseteq G$ and $H' \subseteq H$.

A graph is called *complete* if every pair of distinct vertices is adjacent. Given a graph $G = (V, E)$, a *clique* is a subset of vertices $C \subseteq V$ such that $G[C]$ is a complete graph. A clique C is said to be *maximum* if there is no clique C' with $|C| < |C'|$. A *tree* is a connected graph containing no cycles. The following concepts intuitively allow to measure how "tree-like" a graph is. A graph G is called a *k-almost tree* if $|E(B)| < |V(B)| + k$ is satisfied for every block B in G. A graph is said to be a *partial k-trees* if it has a tree decomposition of width at most k (see, e.g., Brandstadt *et al.*, 1999, for details). The partial 2-trees are also called *series-parallel graphs*. A graph is *planar* if it admits a drawing in the plane such that no two edges cross; it is called *outerplanar* if it admits such a drawing with every vertex lying on the boundary of the unbounded region.

Graph Representations of Molecules

The *molecular graph* of a chemical compound is the graph, where the vertices represent the atoms and the edges the bonds of the molecule. The vertices of a molecular graph are typically annotated by the element symbols of the atoms they represent and the edges by the bond types. Additional information like a more differentiated atom type classification or 3D coordinates may be used also. An isomorphism between molecular graphs typically has to preserve such labels. Hydrogen atoms may or may not be represented explicitly. Molecular graphs exhibit several characteristic properties.

- The distribution of vertex and edge labels is non-uniform, e.g., the majority of vertices represent carbon atoms and is labeled by 'C'.
- The maximum degree of molecular graphs is bounded by a small constant (≤ 4 with very few exceptions) as a result of the atom valency.
- The vast majority of molecular graphs are planar, and most are even outerplanar (Horváth *et al.*, 2010). Here, we mean planar in a graph theoretical sense, while otherwise a molecule typically is considered planar if all atoms are on the same plane w.r.t. their 3D arrangement.
- Molecular graphs typically are tree-like. Yamaguchi *et al.* (2003) computed the tree width of 9712 molecular graphs derived from the LIGAND database (Goto *et al.*, 2002) and found that there is only one graph with tree width 4, while all other have tree width at most 3. Actually, 19.4% of the graphs are trees and 95% are series-parallel graphs.

These properties can be exploited in order to obtain algorithms, which solve subgraph isomorphism problems efficiently for molecular graphs.

Apart from this natural representation so-called *reduced graph* models have been developed, which represent groups of connected atoms by a single vertex (see Birchall and Gillet, 2011, and references therein). Reduced graphs typically contain only a fraction of the vertices and edges of molecular graphs, but may still contain cycles. Simplifying the representation further, Rarey and Dixon (1998) proposed to represent molecules by trees, which encode their hydrophobic fragments and functional groups.

Substructure Search

Given a data set of molecules and a query structure, the *substructure search problem* is to determine the subset of molecules that contain the query structure. Hence, this task requires to decide for each molecular graph G in the data set whether a subgraph isomorphism from the query graph Q to G exists. The subgraph isomorphism problem is well-studied and algorithms have been developed in cheminformatics since the 1950s (Barnard, 1993). Experimental evaluations in the 1980s suggested that the backtracking algorithm by Ullmann (1976) achieves a superior performance to these early algorithms (see Willett, 1999). Since then backtracking algorithms were heavily applied in practice and have been further enhanced. These approaches extend a partial mapping from the vertices of Q to the vertices of G step-by-step and perform backtracking if the current mapping cannot be completed. Most approaches maintain for every vertex of Q a set of eligible candidate vertices of G, which is filtered whenever the mapping is extended. Moreover, the ordering, in which the mapping is extended is a crucial component of this type of algorithm and has an important effect on the running time. The early algorithm by Ullmann (1976) uses elaborate candidate filtering referred to as *refinement*. Later computational less demanding filtering techniques in combination with certain vertex orderings were shown to be more efficient for practical instances. Cordella *et al.* (2004) proposed the VF2 algorithm and experimentally showed it to outperform Ullmann's algorithm on instances from pattern recognition. More recently, it was shown that straightforward backtracking approaches similar to VF2, which do not manage candidate sets explicitly, but exploit the low degree and diverse vertex labels of molecular graphs, are highly efficient for molecular graphs in practice (Kriege, 2009; Klein *et al.*, 2011; Ehrlich and Rarey, 2012). Moreover, Ullmann (2011) proposed an improved version of his algorithm relying more on search and less on a demanding refinement procedure.

Although the subgraph isomorphism problem is NP-complete for general graphs, the above mentioned backtracking algorithms with exponential worst-case running time are convenient for molecular graphs. Therefore, polynomial-time algorithms, which exist for many graph classes (Marx and Pilipczuk, 2014) (including almost all molecular graphs) are of little practical relevance. However, since chemical databases contain millions of molecules and must be able to answer a large number of substructure queries within a short response time, indexing techniques have been developed. In a preprocessing step an index data structure is build, which then typically allows to process a query as follows. First, in the filtering step the existence of a subgraph isomorphism is ruled out for most of the molecules. Then, only the remaining candidates are tested by a subgraph isomorphism algorithm in the subsequent verification step. We refer the reader to Barnard (1993), and references therein for more details on implementations in chemical information systems and to Klein *et al.* (2011), and references therein for more recent approaches studied in the broader field of graph indexing and mining.

Maximum Common Subgraph Problems

The identification of substructures which two molecules share is related to the maximum common subgraph problem. This problem is independent of the used graph model and is defined as follows. Given two graphs, determine the maximum size of a common subgraph isomorphism between them. This leads to several problem variants, which differ w.r.t. the definition of a size function and the types of allowed subgraphs. The considered subgraphs may be (i) required to be connected, or (ii) allowed to be disconnected. Moreover, they may be required to preserve ring systems, i.e., to be block and bridge preserving, and may be required to be vertex- or edge- induced. This already leads to twelve possible variants, which are summarized in **Table 1** together with an adequate size function and a name commonly used (although the terminology in the literature is far from being consistent).

All the above mentioned variants of the maximum common subgraph problem are NP-hard in general graphs. It is suspected that no polynomial-time algorithms for such problems exist. However, several special cases for restricted graph classes are known that allow to obtain polynomial-time algorithms. There is a complex interplay between graph classes, the properties of the desired common subgraph and the complexity of the problem, which is yet not fully understood. **Table 2** summarizes known results that are particularly relevant for molecular graphs.

We review classical techniques in Sections "Reduction to the Clique problem" and "Direct branch-and-bound approaches" and summarize polynomial-time algorithms of practical relevance in Section "Polynomial-time algorithms for restricted graph classes".

Algorithms

As a consequence of the practical relevance of the problem maximum common subgraph algorithms have been studied intensively in cheminformatics and other disciplines like pattern recognition (Conte *et al.*, 2004). Since maximum common subgraph

Table 1   Variants of the maximum common subgraph problem with commonly used size functions. In addition, each variant can be restricted to common subgraphs that are connected or block and bridge preserving

Name: Maximum...	Abbrev.	Size	Induced	Connected	Preserve rings
Common Subgraph	MCS	$|V(G)| + |V(E)|$	–	MCCS	BBP-MC(C)S
Common Induced Subgraph	MCIS	$|V(G)|$	vertex	MCCIS	BBP-MC(C)IS
Common Edge Subgraph	MCES	$|E(G)|$	edge	MCCES	BBP-MC(C)ES

Table 2   Summary on complexity results for the maximum common induced subgraph problem and its variants. Note that the graph classes are related as follows: trees $\subset$ outerplanar graphs $\subset$ series-parallel graphs $\subset$ partial k-trees; and trees $\subset$ k-almost trees $\subset$ partial k-trees (Bodlaender, 1986); (Δ-bounded – the maximum degree is bounded by a constant; * – the result was originally shown for the edge-induced variant)

Input graph class	Constraint	Complexity
Trees	–	NP -hard (Brandenburg, 2000)
	connected	P (Matula, 1978; Droschinsky *et al.*, 2016)
***k*-almost trees**	connected	NP-hard for any $k \geq 1$ (Akutsu, 1993)
Δ-bounded	connected	P (Akutsu, 1993)
Outerplanar graphs	connected	NP-hard (Syslo, 1982)
	biconnected	P (Droschinsky *et al.*, 2017)
	connected, BBP	P (Droschinsky *et al.*, 2017)
Δ-bounded	connected	P (Akutsu and Tamura, 2013)*
Series-parallel graphs	biconnected	P (Kriege and Mutzel, 2014)
	connected, BBP	P (Kriege *et al.*, 2014, 2018)
Partialk *k*-trees	*j*-connected, $j < k$	NP-hard for any $k \geq 1$ (Brandenburg, 2000)
	k-connected	Open for $k > 2$
Δ-bounded, $k=11$	connected	NP-hard (Akutsu and Tamura, 2012b)
Δ-bounded	connected	Open for $k \in \{2,\dots,10\}$

problems are NP-hard unless the input graphs are heavily restricted, general purpose algorithms commonly used in practice have an exponential worst-case running time. These algorithms typically rely on branch-and-bound techniques to eventually solve the problem to optimality, but still differ in their approach. We group these techniques accordingly.

Reduction to the Clique problem

The most common approach in practice is to exploit the one-to-one correspondence between common induced subgraph isomorphisms of the input graphs and cliques in their association graph (also *compatibility graph* (Koch, 2001; Durand *et al.*, 1999), *product graph* (Koch, 2001; Cazals and Karande, 2005), *derived graph* (Barrow and Burstall, 1976), or *weak modular product graph* (Hammack *et al.*, 2011)), which was described by Levi (1973) and later by Barrow and Burstall (1976). For two graphs $G = (V, E)$, $H = (V', E')$, their *association graph* is denoted by $G \nabla H = (\mathcal{V}, \mathcal{E})$, where $\mathcal{V} = V \times V'$ and $\mathcal{E}$ contains an edge connecting (u, u'), $(v, v') \in \mathcal{V}$ if $u \neq v$, $u' \neq v'$ and $uv \in E \Leftrightarrow u'v' \in E'$. Each vertex of the association graph represents a mapping of a vertex of G to a vertex of H. Two vertices in the association graph are adjacent if their corresponding vertex mappings are compatible, i.e., the involved vertices exhibit the same relation (adjacent or non-adjacent) in both graphs.

Levi (1973) has shown that two graphs G and H have a common induced subgraph with k vertices if and only if there is a clique with k vertices in $G \nabla H$. Further, there is a one-to-one correspondence between the common induced subgraph isomorphisms and cliques in the association graph. Let $C = \{(v_1, v_1'), \dots, (v_k, v_k')\} \subseteq \mathcal{V}$ be a clique in $G \nabla H$, $U = \{v_1, \dots, v_k\} \subseteq V(G)$ and $U' = \{v_1', \dots, v_k'\} \subseteq V(H)$, then $\psi(v_i) = v_i'$, for $i \in \{1, \dots, k\}$, is an isomorphism between $G[U]$ and $H[U']$. Consequently, the maximum cliques in the association graph correspond to the isomorphisms between maximum common induced subgraphs. Note that there may be different isomorphisms acting on the same sets of vertices of the two input graphs.

This relation allows to reduce the MCIS problem to the maximum clique problem. There is a multitude of algorithms devised for maximum clique detection, see the surveys by Pardalos and Xue (1994) and Bomze *et al.* (1999) for an overview. The algorithm by Bron and Kerbosch (1973) enumerates all maximal cliques and is often used in practice (see, e.g., Koch, 2001). If one is only interested in an arbitrary maximum clique or just the size of a maximum clique, branch-and-bound techniques are used to speed up the search. These essentially allow to ignore unfruitful branches in the search tree, whenever an upper bound of the possible solution size within the branch drops below the best known current solution. Consequently, these techniques benefit from heuristics to find a good solution early. One example of these algorithms is due to Wood (1997), which is employed by Stahl *et al.* (2005). Conte *et al.* (2007) performed an extensive experimental comparison including several clique detection algorithms. However, the study does not reveal clear evidence for one algorithm being preferable in practice.

In order to preserve labels of molecular graphs with this approach, it suffices to create a vertex in the association graph only if the vertex labels of the input graphs agree (and analogously for edges and edge labels). More generally, a similarity measure on the vertex and edge labels can be defined, which leads to a weighted association graph in which the clique of maximum weight corresponds a maximum common subgraph isomorphism w.r.t. the weight function (Barrow and Burstall, 1976).

Finding edge-induced common subgraphs. Nicholson *et al.* (1987) proposed a general reduction from MCES to MCIS based on the above approach, which was later applied on various occasions (Tonnelier *et al.*, 1990; Durand *et al.*, 1999; Koch, 2001). Instead of directly creating the association graph of the two input graphs, the association graph of their line graphs is created. The *line graph* $L(G)$ of a graph G is the graph with vertex set $V(L(G)) = E(G)$, in which two vertices are adjacent if and only if the two corresponding edges are adjacent in G. Obviously, two isomorphic graphs have isomorphic line graphs. A classical result states that, except for the graphs depicted in **Fig. 2**, the reverse holds as well.

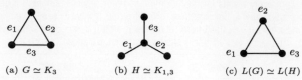

(a) $G \simeq K_3$ (b) $H \simeq K_{1,3}$ (c) $L(G) \simeq L(H)$

Fig. 2 The non-isomorphic graphs (a) and (b) have the same line graph (c).

Theorem: (Whitney, 1932). *Given two connected graphs G and H. If $L(G) \simeq L(H)$ then $G \simeq H$, unless $G \simeq K_{1,3}$ and $H \simeq K_3$ or vice versa* (The result was originally formulated without the notion of line graphs and instead directly relates graph isomorphism to edge isomorphism. An *edge isomorphism* is a bijection between edges such that any two edges are adjacent in one graph if and only if the associated edges in the other graph are adjacent).

The single exception is referred to as ΔY-*exchange* (Raymond and Willett, 2002a; Raymond *et al.*, 2002b) or *triode-triangle interchange* (Durand *et al.*, 1999).

The family of line graphs is closed under vertex removal, i.e., each induced subgraph of a line graph again is a line graph. Furthermore each vertex-induced subgraph of $L(G)$ directly corresponds to an edge-induced subgraph of G and vice versa. Based on this idea common edge subgraphs can be detected by finding cliques in $L(G) \nabla L(H)$, if ΔY-exchanges are handled adequately: For a common induced subgraph K_3 between two line graphs, it must be verified that the corresponding edge-induced subgraphs in the original graphs are either both K_3 or both $K_{1,3}$. The check can be incorporated in the clique detection algorithm, but the same idea also can be used for other exact MCIS algorithms. The transformation can be extended to cope with labeled graphs by creating labeled line graphs (Raymond *et al.*, 2002b). Tonnelier *et al.* (1990) as well as Durand *et al.* (1999) and later Koch (2001) directly define an equivalent edge association graph without using the notion of line graphs.

Finding connected common subgraphs. The above approach can be adapted to find connected common subgraphs. Tonnelier *et al.* (1990) considered the problem of enumerating all maximal connected common edge subgraph isomorphisms between two graphs G and H. A naïve approach is to enumerate all maximal cliques in the association graph $L(G) \nabla L(H)$ and to keep only the connected solutions. Koch (2001) avoids the exhaustive enumeration by directly restricting the search space. The key idea is to distinguish between two types of edges in the association graph: Edges that are due to common adjacencies are called *c-edges*, all other edges represent the absence of adjacency in both graphs and are referred to as *d-edges*. A clique C in an association graph A is called *c-clique* if $A[C]$ has a spanning subgraph induced by c-edges. Koch (2001) showed that a clique in the association graph corresponds to an isomorphism between common connected edge subgraphs if and only if it is a c-clique. Based on this observation the algorithm by Bron and Kerbosch (1973) is extended to find only c-cliques, which reduces the solution space considerably. Some flaws in the algorithm were later corrected by Cazals and Karande (2005).

Direct branch-and-bound approaches

McGregor (1982) proposed an early branch-and-bound algorithm for MCES, which has been used for classifying chemical reactions (McGregor and Willett, 1981). The approach incrementally extends an initially empty partial mapping between $V(G)$ and $V(H)$, where at each step the i-th vertex of G is mapped to a vertex of H or is omitted from the correspondence. The generated partial mappings form a tree with the empty mapping represented by the root, and children extending the mapping of their parent by an additional vertex pair. In order to reduce the search space no further extension is performed when the size of a total solution can be foreseen not to surpass the best solution found so far. By this means unfruitful branches are cut-off. Therefore, heuristics for finding good solutions early strengthen the effect. Krissinel and Henrick (2004) improved McGregor's algorithm by adding extensions in a prescribed ordering and refining the set of possible extensions, which strengthens pruning. Such branch-and-bound optimizations are an integral part of constraint programming approaches, see Section "General purpose solvers". The algorithm has been used for searching the EBI-MSD ligand database (Boutselakis *et al.*, 2003). The refinement procedure is inspired by a classical subgraph isomorphism algorithm due to Ullmann (1976). Cao *et al.* (2008) proposed a branch-and-bound approach to MCIS inspired by the VF2 subgraph isomorphism algorithm (Cordella *et al.*, 2004), which involves a new ordering and bound based on maximum bipartite matching. Moreover, the approach allows to constraint the number of connected components of the common subgraph.

General purpose solvers

Constraint programming (CP) is a general approach to solve combinatorial problems, which are stated by a set of variables allowed to take values from a predefined domain and a set of constraints which must be satisfied by a solution. CP solvers typically follow an elaborated branch-and-bound approach. Vismara and Valery (2008) formulated two explicit CP models for MCCES, one inspired by the methods detailed in Section "Reduction to the Clique problem", and an original one based on a subdivision graph. The resulting problem again corresponds to a variant of the clique problem in a product graph with additional constraints. Ndiaye and Solnon (2011) proposed improved CP models for MCIS and MCES. In a recent experimental comparison McCreesh *et al.* (2016) showed that clique-based approaches for MCIS and MCCIS remain competitive with state-of-the-art CP solvers when using modern clique detection algorithms, in particular for labeled graphs. Moreover, maximum common subgraph problems have been formulated as integer linear programs (see Manić *et al.*, 2009; Bahiense *et al.*, 2012; Piva and de Souza, 2012). These approaches support to incorporate additional side constraints easily, but are not yet competitive regarding running time with clique-based algorithms (Bahiense *et al.*, 2012).

Polynomial-time algorithms for restricted graph classes

Polynomial-time algorithms for MCCIS/MCCES in trees have been pioneered by J. Edmonds and Matula in the 1960s. They rely on solving a series of maximum bipartite matching instances (Matula, 1978). Only recently it has been shown by Droschinsky *et al.* (2016) that the approach can be implemented in cubic time. Akutsu (1993) showed that MCCES in k-almost trees is NP-hard for any $k \geq 1$, but can be solved in polynomial time in almost trees of bounded degree. Moreover, MCCIS can also be solved in polynomial time when one of the input graphs is a bounded-degree partial k-tree and the other is a connected graph with a polynomial number of possible spanning trees (Yamaguchi *et al.*, 2004). Solving BBP-MCCES in outerplanar graphs can be achieved in polynomial time (Schietgat *et al.*, 2013); the same holds for BBP-MCCIS in outerplanar (Droschinsky *et al.*, 2017) and series-parallel graphs (Kriege *et al.*, 2014, 2018; Kriege and Mutzel, 2014). By non-trivial modification of an algorithm for BBP-MCCES Akutsu and Tamura (2012a, 2013) proved that MCCES in outerplanar graphs of bounded degree is polynomial-time solvable. Moreover, Akutsu and Tamura (2012b) have shown that MCCIS and MCCES both are NP-hard in vertex-labeled partial 11-trees of bounded degree. Kann (1992) studied the approximability of maximum common subgraph problems and Abu-Khzam (2014); Abu-Khzam *et al.* (2015, 2017) their parametrized complexity.

Most of the above mentioned results focus on the theoretical complexity and are not promising for practical applications. However, the BBP-MCCIS and BBP-MCCES algorithms have been implemented and applied to molecular graphs in practice. The BBP-MCCES approach by Schietgat *et al.* (2013) runs in $\mathcal{O}(n^4)$ time (Different worst-case bounds have been provided over time in consecutive publications: Starting with $\mathcal{O}(n^7)$ (Schietgat *et al.*, 2007), then $\mathcal{O}(n^5)$ (Schietgat *et al.*, 2008; Schietgat, 2010) and finally $\mathcal{O}(n^2 \sqrt{n})$ (Schietgat *et al.*, 2013). However, it has been shown that the analysis is flawed and the algorithm allows no better bound than $\mathcal{O}(n^4)$ (Droschinsky *et al.*, 2016).) and was shown to outperform the direct backtracking algorithm by Cao *et al.* (2008) in terms of running time in practice (Schietgat, 2010; Schietgat *et al.*, 2013). The BBP-MCCIS of Droschinsky *et al.* (2017) runs in $\mathcal{O}(\Delta n^2)$ time in outerplanar graphs on n vertices with maximum degree Δ and, thus, has a quadratic time complexity in molecular graphs. Experimentally, the technique is shown to outperform the BBP-MCCES approach by Schietgat *et al.* (2013) by orders of magnitude and allows to structurally compare drug-like molecular graphs in the range of a few milliseconds.

Heuristic algorithms

Some applications have strict constraints on running time, but do not require an accurate solution. In this case heuristic algorithms are applicable, which quickly compute a possibly non-optimal solution. We refer the reader to Englert and Kovács (2015), and references therein for heuristic algorithms. Note that the branch-and-bound approaches mentioned above may benefit from heuristics which find good solution quickly.

Variations of the Maximum Common Subgraph Problem

Fuzziness in the detection of maximum common subgraphs is often mentioned as a possible improvement over the strict matching rules (see, e.g., Sheridan and Miller, 1998; Raymond *et al.*, 2003; Hattori *et al.*, 2003). The above mentioned algorithms typically already allow, or can be adapted, to take this into account. The *multiple maximum common subgraph problem* asks for the maximum common subgraph of more than two input graphs. The problem already has been studied by Varkony *et al.* (1979), later by Bayada *et al.* (1992) and more recently by Hariharan *et al.* (2011) and Dalke and Hastings (2013). When large diverse data sets are considered, their common subgraph typically consists of small disconnected fragments, which are not meaningful. Therefore, the *frequent subgraph mining problem* asks for subgraphs that are contained in at least a certain percentage of the data set. This problem and several variants have been studied extensively, see, e.g., Cheng *et al.* (2010).

Applications

Every scientist working with molecules has for a certainty applied common substructure searches based on a maximum common subgraph. In general there are two major application domains for small molecule similarity searches and structure comparisons: prediction/identification and clustering. The former relies on the assumption that the properties of similar structures are similar, which enables the prediction of a property unknown for the molecule of interest but known for a similar one. In the latter application domain a segmentation of a dataset is of interest, e.g., to investigate the diversity. A good overview about possible applications can be found in Humbeck and Koch (2017). In the following some softwares will be mentioned for identification or prediction purposes and clustering tasks. A special open source software called *small molecule subgraph detector* was developed by Rahman *et al.* (2009). It includes different algorithms and chooses the most appropriate one by analyzing the input structures. The software allows to take chemical knowledge into account and finds best solutions w.r.t. atom type match and bond sensitivity.

Identification of Similar Molecules and Prediction of Properties

Raymond *et al.* (2002b) proposed RASCAL for the calculation of graph similarities based on MCES. The approach consists of an initial screening to decide quickly if a certain minimum similarity cannot be reached. Otherwise an elaborated MCES algorithm based on the approach described in Section "Reduction to the Clique problem" is employed, which uses a tailored clique detection algorithm and exploits symmetries. The approach incorporates results of a long line of research in one MCES algorithm and employs domain-specific heuristics (Raymond *et al.*, 2002a). Similarly, Marialke *et al.* (2007) proposed *graph-based molecular alignment*, where the common subgraph not necessarily has to preserve adjacencies, but the shortest-path distances between atoms. This approach assures consistent arrangements of connected components.

Hartenfeller *et al.* (2012) developed a method which compares molecules using a graph representation to predict similar bioactivity. They collate *in silico* synthesized molecules with known bioactive molecules for the *de novo* design of molecules with similar bioactivity. Furthermore, they use three interesting approaches: A graph representation of the pharmacophoric features of the molecule, reduced graphs and a difference in feature count corrected Tanimoto similarity measure. Although they use a graph representation they do not use MCS algorithms for comparison. Algorithms based on MCS for virtual screening purposes are developed by Lešnik *et al.* (2015), Vainio *et al.* (2009) and Droschinsky *et al.* (2017). The first is called LiSiCA and uses a maximum clique algorithm to detect two and three dimensional similarities of molecular graphs. In contrast to the representation used by Hartenfeller *et al.* (2012) atom types instead of features are modeled by vertices. A product graph is generated and the maximum clique gets detected. The second approach, ShaEP, as well as LiSiCA works with a three dimensional representation, but uses electrostatical potentials and a local shape descriptor as vertex labels, which are not coincident with the heavy atoms of the molecule. Furthermore, they generate a completely connected graph and use the backtracking algorithm of Krissinel and Henrick (2004). ShaEP can also be used for the alignment of molecules.

Substructure searches are often applied to query large databases. For example PubChem supports this kind of approaches for the identification of molecules in their database that contain a molecule drawn on their webpage as substructure. This allows determining other molecules with, e.g., known bioactivity to get a first impression about the possible bioactivity profile of the query molecule.

Classification of Datasets

Due to the high complexety and long run time, clustering of datasets is often based on fingerprint based similarity. However, with increasing computational power and the development of sophisticated algorithms, molecular clustering based on graph representation arouse great interest. Schäfer and Mutzel (2017), for example, present a frequent subgraph based clustering approach termed StruClus including the identification of cluster representatives. Thus, the clustering is chemically intuitive and comprehensible. As the algorithm scales linearly with the dataset size it is capable of clustering datasets of tens of millions of molecules and therefore meets the challenges of ever growing (*in silico*) databases.

A clustering method based on MCES was proposed by Stahl *et al.* (2005) and introduced a similarity measure with penalty terms for inconsistent arrangements of the connected components of the common subgraph in the input graphs. The approach is shown to be able to produce clusterings of molecules which are in a better alignment with their target classes than previous common subgraph or fingerprint based approaches.

Another solution is given by Gardiner *et al.* (2007). They only focus on finding representatives of clusters by finding MCES using the RASCAL algorithm in a non-hierarchically preclustered dataset. The performance is improved by using reduced graphs, where functional groups are represented by a single vertex.

Results and discussion

The general suitability of graph based similarity and structure comparisons was demonstrated in many applications. Nevertheless, one has to choose an appropriate algorithm for a given task and attention should be paid to the graph representation of molecules. Vertices and edges may have different meanings and weighting may be possible. Moreover, reduced graph models allow for more efficient maximum common subgraph computations. Another critical point are the applied similarity or distance measures. Raymond and Willett (2002b) give an overview about common measures and present some intuitive similarity measures between molecules that can be defined on the basis of the size of their maximum common subgraph.

Typically maximum common subgraph algorithms are utilized as a subroutine in more complex applications like clustering. It depends on the application which type of algorithm is most suitable. If it is acceptable to trade accuracy for running time, heuristics may provide an adequate solution. Well-engineered exact algorithms have been developed, which solve the maximum common subgraph problem for most pairs of graphs within milliseconds, but are reported to take a very long time for few tough instances (Stahl *et al.*, 2005). The recent development of exact polynomial-time algorithms for restricted graph classes constitutes a step to overcome this problem. Yet further progress in this direction has to be made to obtain generally applicable algorithms.

Acknowledgement

This work was supported by the German Research Foundation (DFG), priority programme "Algorithms for Big Data" (SPP 1736). O. Koch is funded by the German Federal Ministry for Education and Research (BMBF, Medizinische Chemie in Dortmund, Grant BMBF 1316053).

See also: Algorithms for Graph and Network Analysis: Clustering and Search of Motifs in Graphs. Binding Site Comparison – Software and Applications. Biological Database Searching. Biomolecular Structures: Prediction, Identification and Analyses. Chemoinformatics: From Chemical Art to Chemistry in Silico. Drug Repurposing and Multi-Target Therapies. Fingerprints and Pharmacophores. In Silico Identification of Novel Inhibitors. Natural Language Processing Approaches in Bioinformatics. Pharmacophore Development. Protocol for Protein Structure Modelling. Public Chemical Databases. Quantitative Structure-Activity Relationship (QSAR): Modeling Approaches to Biological Applications. Rational Structure-Based Drug Design. Small Molecule Drug Design. Structure-Based Design of Peptide Inhibitors for Protein Arginine Deiminase Type IV (PAD4). Structure-Based Drug Design Workflow

References

Abu-Khzam, F.N., 2014. Maximum common induced subgraph parameterized by vertex cover. Information Processing Letters 114 (3), 99–103. ISSN 0020-0190. Available at: http://www.sciencedirect.com/science/article/pii/S0020019013002755.

Abu-Khzam, F.N., Bonnet, E., Sikora, F., 2015. On the complexity of various parameterizations of common induced subgraph isomorphism. In: Jan, K., Miller, M., Froncek, D. (Eds.), Combinatorial Algorithms: 25th International Workshop, IWOCA 2014, Duluth, MN, October 15–17, 2014, Revised Selected Papers, pp. 1–12. Springer International Publishing, Cham. ISBN 978-3-319-19315-1. Available at: https://doi.org/10.1007/978-3-319-19315.

Abu-Khzam, F.N., Bonnet, E., Sikora, F., 2017. On the complexity of various parameterizations of common induced subgraph isomorphism. Theoretical Computer Science. ISSN 0304-3975. Available at: http://www.sciencedirect.com/science/article/pii/S0304397517305480.

Akutsu, T., 1993. A polynomial time algorithm for finding a largest common subgraph of almost trees of bounded degree. IEICE Transactions on Fundamentals of Electronics, Communications and Computer Sciences E76-A (9),

Akutsu, T., Tamura, T., 2012a. A polynomial-time algorithm for computing the maximum common subgraph of outerplanar graphs of bounded degree. In: Branislav Rovan, V., Sassone Widmayer, P. (Eds.), Mathematical Foundations of Computer Science, vol. 7464 of Lecture Notes in Computer Science. Berlin/Heidelberg: Springer, pp. 76–87. ISBN 978-3-642-32588-5. Available at: https://doi.org/10.1007/978-3-642-32589-2_10.

Akutsu, T., Tamura, T., 2012b. On the complexity of the maximum common subgraph problem for partial fc-trees of bounded degree. In: Chao, K.-M., Hsu, T.-s., Lee, D.-T. (Eds.), Algorithms and Computation, vol. 7676 of Lecture Notes in Computer Science. Berlin Heidelberg: Springer, pp. 146–155. ISBN 978-3-642-35260-7. Available at: https://doi.org/10.1007/978-3-642-35261-4_18.

Akutsu, T., Tamura, T., 2013. A polynomial-time algorithm for computing the maximum common connected edge subgraph of outerplanar graphs of bounded degree. Algorithms 6 (1), 119–135. Available at: http://www.mdpi.com/1999-4893/6/1/119.

Baell, J.B., Ferrins, L., Falk, H., Nikolakopoulos, G., 2013. Pains: Relevance to tool compound discovery and fragment-based screening. Australian Journal of Chemistry 66 (12), 1483. doi:10.1071/CH13551. ISSN 0004-9425.

Bahiense, L., Manic, G., Piva, B., de Souza Cid C., 2012. The maximum common edge subgraph problem: A polyhedral investigation. Discrete Applied Mathematics 160(18), 2523–2541. ISSN 0166-218X. Available at: http://www.sciencedirect.com/science/article/pii/S0166218 × 12000340. V Latin American Algorithms, Graphs, and Optimization Symposium, Gramado, Brazil, 2009.

Barrow, H.G., Burstall, R.M., 1976. Subgraph isomorphism, matching relational structures and maximal cliques. Information Processing Letters 4 (4), 83–84.

Bayada, D.M., Simpson, R.W., Johnson, A.P., Laurenco, C., 1992. An algorithm for the multiple common subgraph problem. Journal of Chemical Information and Computer Sciences 32 (6), 680–685. Available at: http://pubs.acs.org/doi/abs/10.1021/ci00010a015.

Biggs, N.L., Lloyd, E.K., Wilson, R.J., 1986. Graph Theory 1736–1936. New York, NY: Clarendon Press, (ISBN 0-198-53916-9).

Birchall, K., Gillet, V.J., 2011. Reduced graphs and their applications in chemoinformatics. Methods in Molecular Biology 672, 197–212. Available at: https://doi.org/10.1007/978-1-60761-839-3_8.

Bodlaender, H.L., 1986. Classes of graphs with bounded treewidth. Technical Report RUU-CS-86-22, Department of Computer Science, Utrecht University.

Bomze, I.M., Budinich, M., Pardalos, P.M., Pelillo, M., 1999. The maximum clique problem. In: Du, D.-Z., Pardalos, P.M. (Eds.), Handbook of Combinatorial Optimization, vol. A. Kluwer Academic Publishers.

Boutselakis, H., Dimitropoulos, D., Fillon, J., et al., 2003. E-msd: The european bioinformatics institute macromolecular structure database. Nucleic Acids Research 31, 458–462. (ISSN 1362-4962).

Brandenburg, F.J., 2000. Subgraph isomorphism problems for k-connected partial k-trees. Unpublished Manuscript.

Brandstadt, A., Le, Van Bang, Spinrad, J.P., 1999. Graph Classes: A Survey. Philadelphia, PA: Society for Industrial and Applied Mathematics, (ISBN 0-89871-432-X).

Bron, C., Kerbosch, J., 1973. Algorithm 457: Finding all cliques of an undirected graph. Communications of the ACM 16, 575–577. ISSN 0001-0782. Available at: http://doi.acm.org/10.1145/362342.362367.

Cao, Y., Jiang, T., Girke, Thomas, 2008. A maximum common substructure-based algorithm for searching and predicting drug-like compounds. ISMB.

Cazals, F., Karande, C., 2005. An algorithm for reporting maximal c-cliques. Theoretical Computer Science 349 (3), 484–490.

Cheng, H., Yan, X., Han, J., 2010. Mining graph patterns. In: Aggarwal, C., Wang, H. (Eds.), Managing and Mining Graph Data. US, Boston, MA: Springer, pp. 365–392. ISBN 978-1-4419-6045-0. Available at: https://doi.org/10.1007/978-1-4419-6045-0_12.

Conte, D., Foggia, P., Sansone, C., Vento, M., 2004. Thirty years of graph matching in pattern recognition. International Journal of Pattern Recognition and Artificial Intelligence. Available at: https://doi.org/10.1142/S0218001404003228.

Conte, D., Foggia, P., Vento, M., 2007. Challenging complexity of maximum common subgraph detection algorithms: A performance analysis of three algorithms on a wide database of graphs. Journal of Graph Algorithms and Applications 11 (1), 99–143.

Cordella, L.P., Foggia, P., Sansone, C., Vento, M., 2004. A (sub)graph isomorphism algorithm for matching large graphs. IEEE Transactions on Pattern Analysis and Machine Intelligence 26 (10), 1367–1372. ISSN 0162-8828. Available at: https://doi.org/10.1109/TPAMI.2004.75.

Dalke, A., Hastings, J., 2013. FMCS: A novel algorithm for the multiple mcs problem. Journal of Cheminformatics 5 (1), O6. Available at: https://doi.org/10.1186/1758-2946-5-S1-O6.

Droschinsky, A., Kriege, N.M., Mutzel, P., 2016. Faster algorithms for the maximum common subtree isomorphism problem. In: Faliszewski, P., Muscholl, A., Niedermeier, R. (Eds.), Proceedings of the 41st International Symposium on Mathematical Foundations of Computer Science (MFCS 2016), vol. 58 of Leibniz International Proceedings in Informatics (LIPIcs), Dagstuhl, Germany, Schloss Dagstuhl-Leibniz-Zentrum fuer Informatik, pp. 33:1–33:14. ISBN 978-3-95977-016-3. Available at: http://drops.dagstuhl.de/opus/volltexte/2016/6447.

Droschinsky, A., Kriege, N., Mutzel, P., 2017. Finding Largest Common Substructures of Molecules in Quadratic Time. In: Steffen, B., Baier, C., van den Brand, M., et al. (Eds.), SOFSEM 2017: Theory and Practice of Computer Science. SOFSEM 2017. Lecture Notes in Computer Science. Cham, vol 10139: Springer International Publishing. ISBN 978-3-319-51963-0. Available at: https://doi.org/10.1007/978-3-319-51963-024.

Durand, P.J., Pasari, R., Baker, J.W., Tsai, C.-c., 1999. An efficient algorithm for similarity analysis of molecules. Internet Journal of Chemistry 2, 1–12.

Ehrlich, H.-C., Rarey, M., 2012. Systematic benchmark of substructure search in molecular graphs – from ullmann to vf2. Journal of Cheminformatics 4 (1), 13. Available at: https://doi.org/10.1186/1758-2946-4-13.

Englert, P., Kovács, P., 2015. Efficient heuristics for maximum common substructure search. Journal of Chemical Information and Modeling 55 (5), 941–955. Available at: https://doi.org/10.1021/acs.jcim.5b00036. PMID: 25865959.

Gardiner, E.J., Gillet, V.J., Willett, P., Cosgrove, D.A., 2007. Representing clusters using a maximum common edge substructure algorithm applied to reduced graphs and molecular graphs. Journal of Chemical Infromation and Modeling. Available at: https://doi.org/10.1021/ci600444g.

Goto, S., Okuno, Y., Hattori, M., et al., 2002. Ligand: Database of chemical compounds and reactions in biological pathways. Nucleic Acids Research 30 (1), 402–404. Available at: http://nar.oxfordjournals.org/content/ 30/1/402.abstract.

Guner, O, 2002. History and evolution of the pharmacophore concept in computer-aided drug design. Current Topics in Medicinal Chemistry 2 (12), 1321–1332. doi:10.2174/1568026023392940. (ISSN 15680266).

Hammack, R, Imrich, W, Klavzar, S, 2011. Handbook of Product Graphs. Discrete Mathematics and Its Applications. Taylor and Francis. (ISBN 9781439813041).

Hariharan, R, Janakiraman, A., Nilakantan, R., et al., 2011. Multimcs: A fast algorithm for the maximum common substructure problem on multiple molecules. Journal of Chemical Infromation and Modeling 51 (4), 788–806. Availalble at: https://doi.org/10.1021/ci100297y.

Hartenfeller, M., Zettl, H., Walter, M., et al., 2012. Dogs: reaction-driven de novo design of bioactive compounds. PLOS Computational Biology 8 (2), e1002380. doi:10.1371/journal.pcbi.1002380. (ISSN 1553-7358).

Hattori, M., Okuno, Y., Goto, S., Kanehisa, M., 2003. Heuristics for chemical compound matching. Genome Information 14, 144–153.

Hawkins, P.C.D., Skillman, A.G., Nicholls, A., 2007. Comparison of shape-matching and docking as virtual screening tools. Journal of Medicinal Chemistry 50 (1), 74–82. doi:10.1021/jm0603365. (ISSN 0022-2623).

Horváth, T., Ramon, J., Wrobel, S., 2010. Frequent subgraph mining in outerplanar graphs. Data Mining and Knowledge Discovery 21, 472–508. Available at: https://doi.org/10.1007/s10618-009-0162-1.

Humbeck, L., Koch, O., 2017. What can we learn from bioactivity data? Chemoinformatics tools and applications in chemical biology research. ACS Chemical Biology 12 (1), 23–35. doi:10.1021/acschembio.6b00706. (ISSN 1554-8937).

Irwin, J.J., Duan, D., Torosyan, H., et al., 2015. An aggregation advisor for ligand discovery. Journal of Medicinal Chemistry 58 (17), 7076–7087. doi:10.1021/acs.jmedchem. (ISSN 0022-2623).

John, M., 1993. Barnard. Substructure searching methods: Old and new. Journal of Chemical Information and Computer Sciences 33 (4), 532–538.

Kann, V., 1992. On the approximability of the maximum common subgraph problem. In: Proceedings of the 9th Annual Symposium on Theoretical Aspects of Computer Science, STACS '92, pages 377-388, London, UK, UK, Springer-Verlag. ISBN 3-540-55210-3. Available at: http://dl.acm.org/citation.cfm?Id=646508.694493.

Keiser, M.J., Roth, B.L., Armbruster, B.N., et al., 2007. Relating protein pharmacology by ligand chemistry. Nature Biotechnology 25 (2), 197–206. doi:10.1038/nbt1284. (ISSN 1087-0156).

Klein, K., Kriege, N., Mutzel, P., 2011. CT-index: Fingerprint-based graph indexing combining cycles and trees. In: IEEE Proceedings of the 27th International Conference on Data Engineering (ICDE), pp. 1115–1126, April. doi:10.1109/ICDE.2011.5767909.

Koch, I., 2001. Enumerating all connected maximal common subgraphs in two graphs. Theoretical Computer Science 250 (1-2), 1–30.

Kriege, N., 2009. Erweiterte Substruktursuche in Molekuldatenbanken und ihre Integration in Scaffold Hunter. Master's thesis, TU Dortmund.

Kriege, N., Kurpicz, F., Mutzel, P., 2014. On maximum common subgraph problems in series-parallel graphs. In: Jan, K., Miller, M., Froncek, D. (Eds.), International Workshop on Combinatorial Algorithms, IWOCA 2014, vol. 8986 of Lecture Notes in Computer Science. Springer International Publishing, pp. 200–212. ISBN 978-3-319-19314-4. Available at: https://doi.org/10.1007/978-3-319-19315-1_18.

Kriege, N., Kurpicz, F., Mutzel, P., 2018. On maximum common subgraph problems in series-parallel graphs. European Journal on Combinatorics (EJC) 68, 79–95. https://doi.org/10.1016/j.ejc.2017.07.012.

Kriege, N., Mutzel, P., 2014. Finding maximum common biconnected subgraphs in series-parallel graphs. In: Csuhaj-Varjú, E., Dietzfelbinger, M., Ésik, Zoltán (Eds.), Mathematical Foundations of Computer Science 2014, vol. 8635 of Lecture Notes in Computer Science. Berlin Heidelberg: Springer, pp. 505–516. ISBN 978-3-662-44464-1. Available at: https://doi.org/10.1007/978-3-662-44465-8_43.

Krissinel, E.B., Henrick, K., 2004. Common subgraph isomorphism detection by backtracking search. Software: Practice and Experience 34 (6), 591–607. Available at: https://doi.org/10.1002/spe.588.

Leach, A.R., Gillet, V.J., 2007. An introduction to chemoinformatics, rev ed. edition Dordrecht and London: Springer, (ISBN 1402092915).

Lešnik, S., Štular, T., Brus, B., et al., 2015. Lisica: A software for ligand-based virtual screening and its application for the discovery of butyrylcholinesterase inhibitors. Journal of Chemical Information and Modeling 55 (8), 1521–1528. doi:10.1021/acs.jcim.5b00136. (ISSN 1549-9596).

Levi, G., 1973. A note on the derivation of maximal common subgraphs of two directed or undirected graphs. Calcolo, Jan. Available at: http://www.springerlink.com/index/B37657486G578502.pdf.

Manić, G., Bahiense, L, Souza, C.D., 2009. A branch&cut algorithm for the maximum common edge subgraph problem. Electronic Notes in Discrete Mathematics, 35(0):47–52. ISSN 1571–0653. Available at: http://www.sciencedirect.com/science/article/pii/S1571065309001620. Proceedings of the Latin-American Algorithms, Graphs and Optimization Symposium (LAGOS '09).

Marialke, J., Korner, R., Tietze, S., Apostolakis, J., 2007. Graph-based molecular alignment (gma). Journal of Chemical Information and Modeling 47 (2), 591–601Available at: http://dx.doi.org/10.1021/ci600387r

Marx, D., Pilipczuk, M., 2014. Everything you always wanted to know about the parameterized complexity of Subgraph Isomorphism (but were afraid to ask). In: Mayr, E.W., Portier, N. (Eds.), Proceedings of the 31st Inter-national Symposium on Theoretical Aspects of Computer Science (STACS 2014), volume 25 of Leibniz International Proceedings in Informatics (LIPIcs), pages 542-553, Dagstuhl, Germany. Schloss Dagstuhl-Leibniz-Zentrum fuer Informatik. ISBN 978-3-93989765-1. Available at: http://drops.dagstuhl.de/opus/volltexte/2014/4486. arXiv:1307.2187.

Matula, D.W., 1978. Subtree isomorphism in $O(n^{5/2})$. In: Hell, P., Alspach, B., Miller, D.J. (Eds.), Algorithmic Aspects of Combinatorics, Vol 2 of Annals of Discrete Mathematics. Elsevier, pp. 91–106. Available at: http://www.sciencedirect.com/science/article/pii/S0167506008703248.

McCreesh, C., Ndiaye, S.N., Prosser, P., Solnon, C., 2016. Clique and Constraint Models for Maximum Common (Connected) Subgraph Problems. Cham: Springer International Publishing, ISBN 978-3-319-44953-1. Available at: https://doi.org/10.1007/978-3-319-44953-1_23.

McGregor, J.J., 1982. Backtrack search algorithms and the maximal common subgraph problem. Software: Practice and Experience 12 (1), 23–34. ISSN 1097-024X. Available at: https://doi.org/10.1002/spe.4380120103.

McGregor, J.J., Willett, P., 1981. Use of a maximum common subgraph algorithm in the automatic identification of ostensible bond changes occurring in chemical reactions. Journal of Chemical Information and Computer Sciences 21 (3), 137–140.

Ndiaye, S. Ndojh, Solnon, C., 2011. CP Models for Maximum Common Subgraph Problems. Berlin, Heidelberg: Springer, ISBN 978-3-642-23786-7. Availble at: https://doi.org/10.1007/978-3-642-23786-7_48.

Nicholson, V., Tsai, C.-C., Johnson, M., Naim, M., 1987. A subgraph isomorphism theorem for molecular graphs. In Graph Theory and Topology in Chemistry, number 51 in Stud. Physical Theoretical Chemistry. Elsevier.

Pardalos, P.M., Xue, J., 1994. The maximum clique problem. Journal of Global Optimization 4 (3), 301–328. Available at: https://doi.org/10.1007/BF01098364.

Piva, B., de Souza, C.C., 2012. Polyhedral study of the maximum common induced subgraph problem. Annals of Operations Research 199 (1), 77–102. Available at: https://doi.org/10.1007/s10479-011-1019-8.

Rahman, S.A., Bashton, M., Holliday, G.L., Schrader, R., Thornton, J.M., 2009. Small molecule subgraph detector (smsd) toolkit. J Cheminform 1 (1), 12. Available at: https://doi.org/10.1186/1758-2946-1-12.

Rarey, M., Dixon, J.S., 1998. Feature trees: A new molecular similarity measure based on tree matching. Journal of Computer-Aided Molecular Design 12, 471–490. Available at https://doi.org/10.1023/A:1008068904628.

Raymond, J.W., Blankley, J.C., Willett, P., 2003. Comparison of chemical clustering methods using graph- and fingerprint-based similarity measures. Journal of Molecular Graphics and Modelling 21 (5), 421–433. Available at: https://doi.org/10.1016/S1093-3263(02)00188-2.

Raymond, J.W., Gardiner, E.J., Willett, P., 2002a. Heuristics for similarity searching of chemical graphs using a maximum common edge subgraph algorithm. Journal of Chemical Information and Computer Sciences 42 (2), 305–316.

Raymond, J.W., Gardiner, E.J., Willett, P, 2002b. RASCAL: Calculation of graph similarity using maximum common edge subgraphs. The Computer Journal 45 (6), 631–644.

Raymond, J.W., Willett, P., 2002a. Maximum common subgraph isomorphism algorithms for the matching of chemical structures. Journal of Computer-Aided Molecular Design 16 (7), 521–533.

Raymond, J.W., Willett, P., 2002b. Effectiveness of graph-based and fingerprint-based similarity measures for virtual screening of 2d chemical structure databases. Journal of Computer-Aided Molecular Design 16, 59–71. Available at: https://doi.org/10.1023/A:1016387816342.

Schäfer, T., Kriege, N., Humbeck, L., et al., 2017. Scaffold hunter: A comprehensive visual analytics framework for drug discovery. Journal of Cheminformatics 9 (1), 1075. doi:10.1186/s13321-017-0213-3.

Schäfer, T., Mutzel, P., 2017. Struclus: Scalable structural graph set clustering with representative sampling. In: Proceedings of the 13th International Conference on Advanced Data Mining and Applications (ADMA 2017), Singapore, accepted for publication.

Schietgat, L., 2010. Graph-Based Data Mining for Biological Applications. Schietgat, Leander, 2010. Graph-Based Data Mining for Biological Applications. PhD Thesis, Informatics Section, Department of Computer Science, Faculty of Engineering, Hendrik Blockeel and Maurice Bruynooghe (supervisors). Available at: https://lirias.kuleuven. be/handle/123456789/267094.

Schietgat, L., Ramon, J., Bruynooghe, M., 2007. A polynomial-time metric for outerplanar graphs. In: Frasconi, P., Kersting, K., Koji Tsuda, (Eds.), Mining and Learning with Graphs, MLG 2007 Proceedings Firence, Italy, August 1-3, 2007, pp. 67–70.

Schietgat, L., Ramon, J., Bruynooghe, M., 2013. A polynomial-time maximum common subgraph algorithm for outerplanar graphs and its application to chemoinformatics. Annals of Mathematics and Artificial Intelligence 69 (4), 343–376. Available at: https://doi.org/10.1007/s10472-013-9335-0.

Schietgat, L., Ramon, J., Bruynooghe, M., Blockeel, H., 2008. An efficiently computable graph-based metric for the classification of small molecules. In: Jean-Francois Boulicaut, M., Berthold Horvath, T. (Eds.), Discovery Science, Vol. 5255 of Lecture Notes in Computer Science. Berlin / Heidelberg: Springer, pp. 197–209. Available at:.http://dx.doi.org/10.1007/978-3-540-88411-8_20

Sheridan, R.P., Miller, M.D., 1998. A method for visualizing recurrent topological substructures in sets of active molecules. Journal of Chemical Information and Computer Sciences 38 (5), 915–924.

Stahl, M., Mauser, H., Tsui, M., Taylor, N.R., 2005. A robust clustering method for chemical structures. Journal of Medicinal Chemistry 48 (13), 4358–4366. Available at: https://doi.org/10.1021/jm040213p.

Syslo, Maciej M., 1982. The subgraph isomorphism problem for outerplanar graphs. Theoretical Computer Science 17 (1), 91–97. Available at:.http://www.sciencedirect.com/science/article/pii/0304397582901335

Tonnelier, C., Jauffret, P., Hanser, T., Kaufmann, G., 1990. Machine learning of generic reactions: 3. An efficient algorithm for maximal common substructure determination. Tetrahedron Computer Methodology 3 (6), 351–358. Available at:.http://dx.doi.org/10.1016/0898-5529(90)90061-C

Ullmann, J.R., 1976. An algorithm for subgraph isomorphism. Journal of the. ACM 23 (1), 31–42. ISSN 0004-5411. doi: http://doi.acm.org/10.1145/321921.321925.

Ullmann, J.R., 2011. Bit-vector algorithms for binary constraint satisfaction and subgraph isomorphism. Journal of Experimental Algorithmics 15, 1.6:1.1–1.6:1.64. Available at: http://doi.acm.org/10.1145/1671970.1921702.

Vainio, M.J., Puranen, J.S., Johnson, M.S., 2009. Shaep: Molecular overlay based on shape and electrostatic potential. Journal of Chemical Information and Modeling 49 (2), 492–502. doi:10.1021/ci800315d.

Varkony, T.H., Shiloach, Y., Smith, D.H., 1979. Computer-assisted examination of chemical compounds for structural similarities. Journal of Chemical Information and Computer Sciences 19 (2), 104–111. Available at: http://pubs.acs.org/doi/abs/10.1021/ ci60018a014.

Vismara, P., Valery, B., 2008. Finding maximum common connected subgraphs using clique detection or constraint satisfaction algorithms. In: Thi, H.L., Bouvry, P., Dinh, T.P. (Eds.), Modelling, Computation and Optimization in Information Systems and Management Sciences, vol. 14 of Communications in Computer and Information Science. Berlin Heidelberg: Springer, pp. 358–368. ISBN 978-3-540-87476-8. Available at: https://doi.org/10.1007/978-3-540-87477-5_39.

Whitney, H., 1932. Congruent graphs and the connectivity of graphs. American Journal of Mathematics 54 (1), 150–168. ISSN 00029327. Available at: http://www.jstor.org/stable/2371086.

Wiener, H., 1947. Structural determination of paraffin boiling points. Journal of the American Chemical Society 69 (1), 17–20. doi:10.1021/ja01193a005. (ISSN 0002-7863).

Willett, P., 1999. Matching of chemical and biological structures using subgraph and maximal common subgraph isomorphism algorithms. The IMA Volumes in Mathematics and its Applications 108, 11–38.

Wood, D.R., 1997. An algorithm for finding a maximum clique in a graph. Operations Research Letters 21 (5), 211–217. Available at: http://www.sciencedirect.com/science/article/B6V8M-3V6083C-1/2/fc0edc68f2eca1ec3d6ab81ac824778c.

Yamaguchi, A., Aoki, K.F., Mamitsuka, H., 2003. Graph complexity of chemical compounds in biological pathways. Genome Informatics 14, 376–377.

Yamaguchi, A., Aoki, K.F., Mamitsuka, H., 2004. Finding the maximum common subgraph of a partial k-tree and a graph with a polynomially bounded number of spanning trees. Information Processing Letters 92 (2), 57–63.

Further Reading

Barnard, J.M., 1993. Substructure searching methods: Old and new. Journal of Chemical Information and Computer Sciences 33 (4), 532–538.

Chen, L., 2003. Substructure and maximal common substructure searching. In: Bultinck, P., Winter, H. De, Langenaeker, W., Tollenare, J.P. (Eds.), Computational Medicinal Chemistry for Drug Discovery. CRC Press, pp. 483–514.

Ehrlich, H.-C., Rarey, M., 2011. Maximum common subgraph isomorphism algorithms and their applications in molecular science: A review. Wiley Interdisciplinary Reviews: Computational Molecular Science 1 (1), 68–79. Available at: https://doi.org/10.1002/wcms.5.

Raymond, J.W., Gardiner, E.J., Willett, P., 2002b. RASCAL: Calculation of graph similarity using maximum common edge subgraphs. Computer Journal 45 (6), 631–644.

Raymond, John W., Willett, Peter, 2002a. Maximum common subgraph isomorphism algorithms for the matching of chemical structures. Journal of Computer-Aided Molecular Design 16 (7), 521–533.

Relevant Websites

http://insilab.org/lisica/
   In Silico Laboratory.
http://users.abo.fi/mivainio/shaep/index.php
   ShaEP.

Binding Site Comparison – Software and Applications

Christiane Ehrt and Tobias Brinkjost, TU Dortmund University, Dortmund, Germany
Oliver Koch, Technical University of Dortmund, Dortmund, Germany

Introduction

The automated comparison of binding sites can provide useful insights into enzyme function, evolutionary relationships, poly-pharmacology, off-target effects and can assist in drug repurposing. Therefore, it has become a valuable tool especially in medicinal chemistry during the process of drug discovery. Another possible application domain is the functional characterization of novel binding sites by a comparison against already annotated and characterized ligand cavities. Recent scientific publications demonstrate that binding site similarity is by no means as obvious as one might expect, i.e., cavities of functionally and structurally distinct proteins might show a high similarity (Ehrt *et al.*, 2016). Usually, the focus is on evolutionary related proteins by comparing protein structures or sequences. This strategy completely dismisses the possibility that unrelated proteins could bind similar ligands. Some very recent publications focus on the impact of local protein similarities on protein function (Garma *et al.*, 2016; Mudgal *et al.*, 2017). They underpin the general observation that a similar fold does not necessarily result in a similar function whereas a common function can be fulfilled by proteins with different folds. Therefore, the comparison of the binding site of interest against a data set of diverse binding sites might unravel some disregarded similarities. Nevertheless, it has to be stated that proteins binding to similar ligands or sharing a common function do not necessarily share a similar small molecule binding site (Barelier *et al.*, 2015).

Encouraged by a variety of useful applications, a multitude of binding site comparison tools were developed. They can be classified by the way the binding site is modeled, the comparison algorithm, or the similarity metric used for binding site similarity ranking. Binding site modeling is of special interest for the choice of the appropriate method according to the particular problem. Therefore, the first part of this article presents software with a special focus on binding site modeling. The second part is dedicated to the model principles and corresponding algorithms for binding site comparison explaining the different approaches in more detail. Finally, four examples for the impact of binding site comparison are discussed. The chosen problems illustrate that tools which are based on rather different binding site modeling, comparison, and similarity scoring concepts led interesting findings in the past. Therefore, the combination of various tools – also those not presented herein – might prove beneficial for a certain application.

A summary of all methods and their characteristics presented herein can be found in **Table 1**. The number of published tools is considerably large. Hence, only a small subset of software tools, which were successfully applied in various scientific projects and are available as standalone software, are discussed in this article. Additionally, we decided to exclusively introduce tools that allow for a comparison of user-defined datasets.

One aspect that is beyond the scope of this article is the automated identification of binding sites. It represents an important pre-processing step, for example, for the binding site annotation of novel targets. Cavity prediction is possible via elaborate binding site detection algorithms such as energy-based (SiteHound (Ghersi and Sanchez, 2009)), knowledge-based (ConCavity (Capra *et al.*, 2009)), and geometry-based (LIGSITE^{CS} (Huang and Schroeder, 2006)) methods. An overview over binding site identification software and its impact can be found elsewhere (Krone *et al.*, 2016).

In the following, we will first discuss different approaches to model small molecule binding sites with a special focus on one method per modeling approach (SMAP, FLAP, APF, SiteEngine). Second, the different model principles of those tools are explained in more detail. Afterward, we provide one application example per tool to illustrate the broad applicability of the software introduced herein. Finally, some hints are provided for successfully using binding site comparison tools.

Binding Site Modeling

The characterization of the binding site is a crucial part of the comparison process as it determines the final outcome of the study (Henrich *et al.*, 2010). The different binding site representation schemes are illustrated in **Fig. 1**. While some tools solely rely on residues and a crude abstraction of residue properties, other tools use more or less elaborate surface calculations for a projection of physicochemical properties on the binding site surface patches or solvent accessible residue atoms. Moreover, a continuous representation of pharmacophoric properties in 3D space can be applied, i.e., different pharmacophoric potentials are assigned to all 3D points. The probably highest level of abstraction is a comparison based on a description of the interactions between protein and ligand. The outcome of such methods is often highly dependent on the chemical nature of the small molecule interacting with cavity residues. In the following, one method per modeling scheme is explained in more detail to characterize the different concepts. For the remaining methods, a short summary is given.

Binding Site Modeling Based on Residues

A straightforward approach is the description of a binding site by 3D and physicochemical characteristics of the underlying residues. Nevertheless, a broad variety of modeling approaches is possible.

Table 1   Binding site comparison tools presented and discussed in this article

Software tool	Binding site representation	Model principle	Field of successful application (Ehrt et al., 2016)	Availability
Cavbse (Schmitt et al., 2001, 2002)	Residue-based	Graph	Protein–ligand interactions, Virtual screening, Evolutionary relationships	CCDC
SMAP (Xie et al., 2009)		Graph	Drug repurposing, Polypharmacology	Download
FuzCav (Weill and Rognan, 2010)		Fingerprint	Drug repurposing, Protein–ligand interactions	Upon request
SiteAlign (Schalon et al., 2008)		Fingerprint	Protein–ligand interactions	Upon request
PocketMatch (Yeturu and Chandra, 2008)		Distance lists	Function prediction, Polypharmacology, Homolog analysis	Download
TM-align (Zhang and Skolnick, 2005)		Matrix	Drug repurposing	Download
IsoMIF (Chartier and Najmanovich, 2015)	Interaction-based	Graph(grid)	Drug repurposing	Download
FLAP (Baroni et al., 2007)		Fingerprint	Prediction of compound selectivity profiles and affinities	Molecular Discovery
KRIPO (Ritschel et al., 2014)		Fingerprint	Off-target prediction	Download
TIFP (Desaphy et al., 2013)		Fingerprint	Virtual screening	Upon request
APF (Totrov, 2011)	Observer point-based	3D-points (grid)	Off-target prediction, Homolog analysis	Molsoft
ProBiS (Konc and Janezic, 2010)	Surface-based	Graph	Function prediction	Download
PSIM (Spitzer et al., 2011)		3D-points (grid)	Off-target prediction	BioPharmics LLC
Shaper (Desaphy et al., 2012)		3D points (grid)	Protein–ligand interactions	Upon request
SiteEngine (Shulman-Peleg et al., 2004)		3D points	Protein–protein interactions	Download

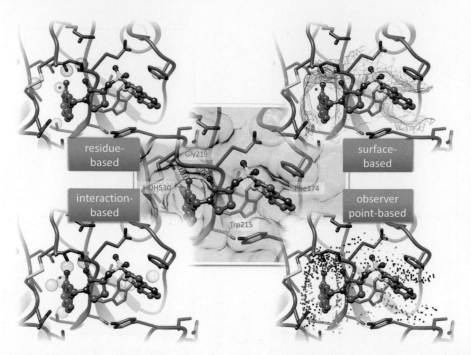

Fig. 1   Four different ways of binding site modeling that are applied by the comparison algorithms presented in this article. The chosen example shows the binding site of coagulation factor Xa (1f0r.A@pdb) in complex with the nanomolar inhibitor RPR208815 in four different representation schemes.

The representation in SMAP (Xie *et al.*, 2009) relies on the residues' Cα atoms as well as their characteristics. It is an extension of the previously developed algorithm named Sequence Order Independent Profile-Profile Alignments (SOIPPA) (Xie and Bourne, 2008). Binding sites are represented by Cα atoms and a mesh surface based on their Delaunay tessellation. Each Cα atom is assigned a normal vector that is perpendicular to the surrounding surface and a weight that accounts for chemical similarity and evolutionary relationship. The McLachlan (1971) chemical similarity matrix is utilized to represent physicochemical similarity. It is based on polar or non-polar character, size, shape, and charge of residues. A position-specific scoring matrix of the 20 amino acids which depends on the residue's sequence position expresses the evolutionary relationship. Taken together, the binding site Cα atoms, their spatial orientation, and the weights of their respective residues are modeled by a graph.

The basic idea of Cavbase (Schmitt *et al.*, 2001, 2002) is to characterize the binding site properties by means of pseudocenters representing the physicochemical properties of binding site residue moieties: hydrogen bond acceptor and donor, mixed hydrogen bond acceptor/donor, center of an aromatic ring, centers comprising π electrons, aliphatic groups, and metal ions. In FuzCav (Weill and Rognan, 2010), the Cα atom coordinates of binding sites are labeled with six pharmacophoric properties of the respective residue: hydrogen bond acceptor and donor, positive and negative ionizable, aromatic, and aliphatic. A final fingerprint representation encodes all unique pharmacophore triplets and the respective binned Cα atom distances. PocketMatch (Yeturu and Chandra, 2008) classifies binding site residues according to their physicochemical properties: aliphatic, positively and negatively polarized, aromatic, and polar. They are represented by three types of points (Cα, Cβ, the geometric center of the side chain atoms). SiteAlign (Schalon *et al.*, 2008) includes topological properties (e.g., side chain orientation, size) as well as physico-chemical properties (e.g., hydrogen bond acceptor/donor, charge, aromaticity). These characteristics are projected on a discretized 80 triangle sphere by deriving a geometrical vector from the Cα atom of each binding site residue to the sphere center. The TM-align (Zhang and Skolnick, 2005) algorithm was originally developed for protein structure alignment. Nonetheless, it was also successfully applied to compare protein binding sites. The comparison is based on the Cα coordinates and the underlying secondary structures elements based on the DSSP assignment (Kabsch and Sander, 1983). However, TM-align is the only approach presented herein which is sequence order dependent, i.e., a discontinuous alignment of binding site residues is impossible.

Binding Site Modeling Based on Interactions

These methods represent the binding site by modeling known ligand interaction patterns to find distinct relationships between pockets. However, some of them rely on the presence of structurally characterized protein-ligand complexes and are not suited to compare novel predicted binding sites against already known ones.

Fingerprints for Ligands and Proteins (FLAP) (Baroni *et al.*, 2007) utilizes GRID (Goodford, 1985) molecular interaction fields (MIFs) to analyze the protein cavity. Six probes (probes for hydrogen bond interactions, salt bridges, hydrophobic interactions, shape etc.) can be chosen to define energetically favorable and unfavorable interactions for the binding site of interest. This information is used to determine target-based pharmacophoric points using a weighted energy-based function. All possible energetically favorable arrangements of four pharmacophoric points are generated. A subsequent selection of groups of most important points can be achieved manually or automatically. Finally, a fingerprint is derived based on the distances, probe types, and chirality as four probes assume relative chiral positions in 3D space.

MIFs are also the basis for the method IsoMIF (Chartier and Najmanovich, 2015), but the GetCleft (Gaudreault *et al.*, 2015) algorithm is used to label grid points with the interaction energy for different probes. The final MIF is the set of interaction vectors at all vertex positions in the grid. Key Representation of Interaction in Pockets (KRIPO) (Ritschel *et al.*, 2014) uses intermolecular interaction features and generates 3-point pharmacophore fingerprints for ligands as well as for ligand fragments. Each triplet is encoded by the interaction properties and binned distances. Pharmacophoric properties including cation-aromatic interactions and metal coordination are also used for Interaction Fingerprint Triplets (TIFP) (Desaphy *et al.*, 2013). The interaction patterns can be represented by the protein atom coordinates, the ligand atom coordinates, or those of the geometric center of two interacting atoms.

Binding Site Modeling Based on Continuous Binding Site Properties

The method Atomic Property Fields (APF) (Totrov, 2011) treads another and unique path for modeling. It relies on a continuous binding site representation using seven atom properties: hydrogen bond acceptor and donor, lipophilicity, size, electronegativity, charge, aromaticity/hybridization. They are projected on a grid by the means of Gaussian functions. The binding site comprises all receptor atoms within a 6 Å radius of the bound ligand. The method is derived from an approach to describe the similarity of 3D property distributions of two ligands, which led to an optimized measure of chemical similarity (Totrov, 2008).

Binding Site Modeling Based on Surface Properties

The assignment of physicochemical properties in SiteEngine (Shulman-Peleg *et al.*, 2004) is similar to that of the Cavbase (Schmitt *et al.*, 2001, 2002) approach. The atoms of each binding site residue in a 4 Å radius of the binding partner are grouped according to their potential ligand interaction. A Connolly surface of the binding site is constructed and only pseudocenters with at least one surface exposed atom are retained. Pseudocenter triplets are generated and encoded with the respective side lengths and a physicochemical index.

Protein Binding Site (ProBiS) (Konc and Janezic, 2010) defines pseudocenters as realized in the Cavbase approach (Schmitt *et al.*, 2001, 2002). Solvent accessible surface atoms of a ligand binding site or a protein structure of interest are identified and labeled. Structurally related cavities can be identified based on binding sites or protein structures. In Protein SIMilarity (PSIM) (Spitzer *et al.*, 2011) the cavity is represented by surface moieties that can recognize a ligand. Distances to the molecular surface from so-called *observer points* are compared by placing them on a uniform grid. An additional weight is assigned in dependence of the minimum distance to the surface resulting in a defined number of *observer points*. Gaussian functions of the distance deviations from each *observer point* are calculated (minimum distance to any surface point to represent the shape, to a positive polarized atomic surface, and to a negative polarized atomic surface point). The binding site representation for Shaper comparisons is generated by VolSite (Desaphy *et al.*, 2012). A cube is placed at the center of mass of the bound ligand and filled with a grid. *IN* and *OUT* properties are assigned to the grid cells depending on the presence or absence of protein atoms and their degree of buriedness. The resulting site points are assigned pharmacophoric properties and a druggability prediction is performed. Druggable annotated binding sites can subsequently be compared.

Model Principles, Their Definitions, and Criteria for Similarity

The choice of an appropiate model principle is crucial for every application, so it is for the comparison of protein binding sites. The model principle defines and limits the level of abstraction, which is sometimes referred to as accuracy. It also defines the complexity of answering certain questions such as: what is the highest similarity of two given models? The complexity has a strong influence on the runtime required to answer such questions. There are three frequently used model principles: graphs, fingerprints, and points in 3D space. Fingerprints and graphs are classical data structures whereas points in 3D space are usually regarded as input data and the way they are stored and managed is referred to as their data structure (e.g., k-d trees (Bentley, 1975)). In contrast to these model principles, hashing is a function to map the input onto the output, without being a data structure such as graphs or fingerprints, and without an implicit similarity such as the root mean square deviation (RMSD) of points in 3D space.

Graphs

A graph $G = (V, E)$ consists of a set of vertices V and a set of edges E connecting these vertices. Metaphorically speaking, graphs contain objects and their relations to each other. In many of the available methods, both (vertices and edges) are labeled with additional information such as pharmacophore features at vertices and distances at edges for instance. The definition of the similarity of two given graphs is frequently based on the maximum common (induced) subgraph (MCS). A graph $G' = (V', E')$ is a subgraph of $G=(V, E)$ if $V' \subseteq V$ and $E' \subseteq E$. It is said to be (vertex-)induced by V' if $E' = (V \times V) \cap E$ holds. Let $G=(V, E)$, $G_1=(V_1, E_1)$, and $G_2=(V_2, E_2)$ be graphs. G is subgraph-isomorphic to a graph G_1 if there exists an injection $(\varphi): V \to V_1$ in such a way that $\forall u, v \in V: (u, v) \in E \Rightarrow ((\varphi)(u), (\varphi)(v)) \in E_1$. G is said to be a common subgraph of G_1 and G_2 if G is subgraph-isomorphic to G_1 and G_2. The MCS G_{max} of G_1 and G_2 is a common subgraph of G_1 and G_2 for which $|G'| \leq |G_{max}|$ holds for all common subgraphs G' of G_1 and G_2. Note that the MCS is not necessarily unique.

The determination of the MCS of two graphs can be reduced to the problem of finding the largest clique in an appropriately defined modular product graph, also known as compatibility, correspondence, or association graph and was first described by Levi (1973). A clique C in an undirected graph $G=(V, E)$ is a subset of vertices $C \subseteq V$ such that every two distinct vertices are adjacent. In a modular product graph $G_P=(V_P, E_P)$ of two labeled graphs $G_1=(V_1, E_1)$ and $G_2=(V_2, E_2)$ the set of vertices $V_P \subseteq V_1 \times V_2$ contains a vertex (v_1, v_2) with $v_1 \in V_1$ and $v_2 \in V_2$ if their labels are compatible. The set of edges E_P contains an edge connecting the vertices (u_1, u_2), $(v_1, v_2) \in V_P$ if either $e_1=(u_1, v_1) \in E_1$ and $e_2=(u_2, v_2) \in E_2$ and if their labels are compatible, or if $e_1 \notin E_1$ and $e_2 \notin E_2$. There is no general definition of the compatibility of labels.

The maximal clique detection algorithm by Bron and Kerbosch (1973) is widely used (e.g., by Cavbase, IsoMIF, ProBiS). It is a recursive backtracking algorithm that uses sets of vertices R (temporary result), P (possible extensions), X (excluded set) and finds the maximal cliques that include all of the vertices in R, some of the vertices in P, and none of the vertices in X. However, a graph $G=(V, E)$ can contain up to $3^{V/3}$ cliques (Moon and Moser, 1965) which makes their determination a demanding objective. Roughly speaking, a linear increase in the size of the input graphs may result in an exponential increase of cliques and therefore of the time required for their determination (runtime). There are optimizations available (Tomita *et al.*, 2006) to reduce the branches of the recursion tree, the recursive calls, respectively.

SMAP represents ligand binding sites as graphs modeling Cα atoms as vertices and the connections of the tessellation (mesh) as edges. The similarity of two graphs is based on determining the maximum weighted common subgraph (MWCS) which also uses an algorithm for the maximum-weight clique problem (Kumlander, 2004). This algorithm is based on the assumption that two vertices of an independent set cannot be in the same clique. An independent set is a set of vertices that are pairwise not connected by an edge (non-adjacent). The difference between a MCS and a MWCS is that the maximality is not based on the size of the subgraph but on the sum of weights reflecting the compatibility of each pair of matched vertices. Each such pair does not match in a binary manner (yes/no) but to a certain weight of compatibility. This weight is based on Gaussian density functions taking residue substitution matrices and geometrical properties into account.

Fingerprints

Fingerprints, such as molecular fingerprints (Todeschini *et al.*, 2009) or extended-connectivity fingerprints (Rogers and Hahn, 2010), are widely used in bioinformatics, chemoinformatics, and computational biology. In general, fingerprints are vectorial descriptors consisting of binary or integer values. In the binary case, each value is usually correlated to the absence (0) or occurrence (1) of a certain property in the input. In the integer case, each value represents a counter for the number of occurrences of a certain property such as pharmacophore features or distances. The latter are usually binned to discrete value ranges. The major advantage of fingerprints is the ability to determine the similarity of two given fingerprints fast. Given two fingerprints $F_1=(v_1, v_2, ..., v_n)$ and $F_2=(w_1, w_2, ..., w_m)$ with $|F_1| = n$ and $|F_2|=m$ the similarity is defined by the number of compatible values at mapping indices. If both fingerprints are of equal length $n=m$, there is a direct mapping of v_1 on w_1, v_2 on w_2 and so on. Otherwise, the fingerprints have to be aligned to obtain an indices mapping.

The definition of compatibility is trivial for binary values, but in the case of integer values, a slight deviation of values at mapping indices may still be defined as a match or compatible, respectively. A final score can be obtained using a similarity or distance metric such as the Tanimoto coefficient, which, among others, has proven to be a meaningful metric (Bajusz *et al.*, 2015). It is based on the number of compatible values C, numbers of non-zero values Z_0 in F_1 and Z_1 in F_2, and defined by: $Tanimoto(F_1, F_2)=C/(Z_0 + Z_1 - C)$. Thus, determining the similarity of two fingerprints is based on counting zero and non-zero values, which can be done by scanning both fingerprints simultaneously and once only. Thus, a linear increase in input size only results in a linear increase in runtime.

FLAP uses fingerprints in a different way. Here, a fingerprint representing a protein's binding site is an array of varying length containing several 11 integer value fingerprints. These 11 fields contain the information of one favorable arrangement of four pharmacophoric points, all pairwise distances of these points, and a value proportional to the sum of their energy. During the comparison of two protein binding sites P_1 and P_2 a fingerprint in P_1 similar to a fingerprint in P_2 is searched for. This is done by sorting the fingerprints with respect to their distances which supports the use of sophisticated search algorithms. Once a similar pair of fingerprints is found, the matching of the four pharmacophoric points of each fingerprint is used to determine the orientation of the corresponding binding sites. If these also superimpose within a certain threshold, the two protein binding sites are finally reported as similar.

Points in 3D

The comparison of clouds of points in 3D space $C_1=(p_1, p_2, ..., p_n)$ and $C_2=(q_1, q_2, ..., q_n)$ of the same size is in general an optimization problem of finding an alignment which has the lowest RMSD value: $RMSD(C_1, C_2) = \sqrt{\left(\frac{1}{n}\sum_{i=1}^{n} d(p_i, q_i)^2\right)}$ with distance function d. During the optimization routine one cloud is usually fixed in space while the other one undergoes several translations and rotations to obtain an alignment with a minimum RMSD value. If the clouds are of different size, one has to define a mapping of the points of the smaller cloud onto the ones of the larger cloud.

APF utilizes a Monte Carlo algorithm to obtain a mapping of the two binding sites to be compared. A Monte Carlo algorithm is a randomized algorithm whose output may be incorrect with a certain probability. In contrast, a deterministic algorithm is always expected to give correct answers. APF uses a pseudo-Brownian Monte Carlo algorithm to minimize local gradients. In each iteration, the algorithm calculates a random rotation of the molecule with respect to chemical meaningfulness, for example, avoid atom clashes, to find a favorable and energetically minimized orientation. Although the selection of a randomized rotation is fast, several rotations are obtained and evaluated to find a satisfactory solution which is not necessarily the optimal one.

Hashing

Hashing has a broad field of applications among which are data storage, searching, and cryptography for instance. Hashing or more precisely a hash function f maps elements of a universe U to a set of keys K: f: $U \rightarrow K$. The basic idea is that $|U|$ is usually much larger than $|K|$ or even unlimited whereas K is of fixed size. Because of this, collisions can appear: $f(u) = f(v)$ with u, $v \in U$ and $u \neq v$. Therefore, the choice of a specific hash function for the universe of the individual application is crucial and should minimize the chance of collisions. A very simple hash function is the modulo function: $f(u) := u \bmod |K|$. However, there are many different hash functions such as double hashing using two hash functions, middle-square hashing based on von Neumann (1951), and Cuckoo hashing (Pagh and Rodler, 2004). A combination of a hash function and a fingerprint-like bit vector is the Bloom filter (Bloom, 1970). It is used to test whether an element $u \in U$ is a member of a set or not. If u is added to the set the bit at the index of the fingerprint obtained from the hash function applied on u is set to 1. In a test the algorithm also uses the function on u and checks whether the corresponding bit is set to 1 and if true, the element may be in. Otherwise, it is definitely not part of the set. Usually, a Bloom filter uses multiple hash functions.

Hashing is a comparably fast technique, but the runtime strongly relies on the complexity of the function itself, in particular the underlying input (e.g., graphs, fingerprints, integers, etc.).

The hash function of SiteEngine (Shulman-Peleg et al., 2004) is based on 4-tuples consisting of the side length of triangles defined by a triplet of pseudocenters and a physicochemical index encoding the properties of the pseudocenters. It is a geometric hash function and finds almost congruent matching triangles of which each pair defines a candidate transformation. The resulting matchings are finally re-evaluated by an elaborate filtering and scoring workflow to obtain the final superposition.

Examples for Successfully Applied Binding Site Comparison Tools

The following examples will illustrate that binding site similarities are by no means easily detectable via sequence or protein structure comparison. An abstraction of the various available binding site properties is often required. This abstraction and spatial comparison led to some astonishing results in the past. Examples for unrelated proteins with respect to sequence and overall fold that share similar ligands and highly similar binding sites underline the necessity of binding site comparison. Additionally, examples for similar proteins with rather dissimilar binding sites are given that underline the importance of binding site comparison to elucidate probable off-targets and ways to enhance selectivity. The examples were chosen to illustrate that various approaches that make use of different types of binding site modeling, comparison, and similarity scoring can shed light on highly interesting relationships between otherwise unrelated proteins.

Polypharmacology

The first example illustrates the impact of binding site comparison on polypharmacology (Xie et al., 2011). The binding site comparison tool SMAP (Xie et al., 2009) was used to compare the nelfinavir binding site of the HIV protease dimer (1ohr@pdb) against 5985 human protein structures or their homologues from other organisms. The 126 most similar superimposed protein binding sites were used for molecular docking of nelfinavir. This led to 92 putative off-targets. Most of the hits were protein kinases belonging to tyrosine, cAMP-dependent, cGMP-dependent kinase families, or protein kinase C family. Exhaustive MD (molecular dynamics) simulation studies and Molecular Mechanics/Generalized Born Surface Area (MM/GBSA) free energy calculations led to the conclusion that nelfinavir interacts with EGFR, might also interact with IGF-1R, FAK, Akt2, CDK2, ARK, PKD1, and will probably not bind to FGFR, EphB4 and Abl. In a subsequent biochemical screening, the authors provided evidence for a nelfinavir-induced inhibition of EGFR and ErbB2. **Fig. 2** shows a superposition of the protease (1ohr.A@pdb) and the EGFR pocket (2j6m.A@pdb) to illustrate the similarities.

Selectivity Profiling

The second example provides a glimpse into the wide variety of possible application domains for binding site comparison. The method FLAP was applied to correlate inhibition profiles for kinase inhibitors with identified binding site features for various kinase binding sites (Sciabola et al., 2010). Based on a binding site feature similarity matrix of fourteen protein kinases partial least-squares models were created to predict pIC_{50} values. The authors of this study were able to choose a combination of probes

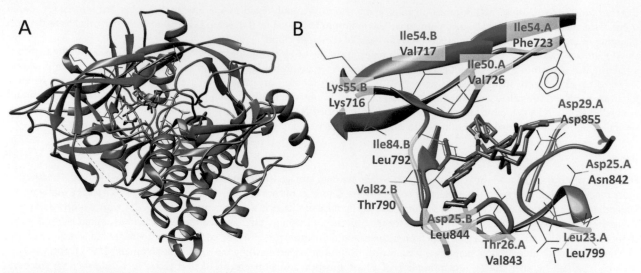

Fig. 2 Binding site similarities between the HIV protease (1ohr@pdb, green) and EGFR (2j6m.A@pdb, orange) based on a SMAP (Xie *et al.*, 2009) comparison. (A) Complete structures superimposed based on the binding site similarities. (B) A detailed view of the binding site similarities. Nelfinavir bound to HIV protease is shown in green sticks while the EGFR ligand with the PDB-ID AEE is represented by orange sticks.

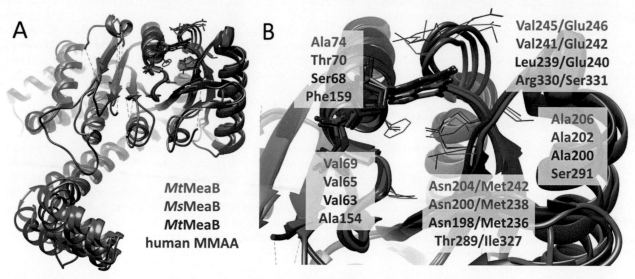

Fig. 3 Comparison between homologous proteins from different Mycobacteria species and a human related enzyme. The MeaB structures from *M. tuberculosis* (3md0@pdb), *M. smegmatis* (3nxs@pdb), *and M. thermoresistibile* (3tk1@pdb) are represented in green, orange, and purple, respectively. The structure of the human homolog MMAA is shown in pink. Reprinted and modified with permission from (Ehrt *et al.*, 2016). Copyright 2016 American Chemical Society. (A) The superposition of all structures illustrates their close relationship. (B) A detailed comparison of the underlying binding sites provided insights into possible starting points to increase the selectivity of inhibitors towards the mycobacterial enzymes.

to generate MIFs that enabled them to predict the level of inhibition of a compound on different protein kinases. Consequently, this study underlines the possible impact of binding site comparison for selectivity prediction for protein kinase inhibitors.

Off-Target Analysis

The method APF was applied to analyze differences between the binding sites of homologous structures (Edwards *et al.*, 2015). The methylmalonyl CoA mutase-associated GTPase MeaB which is an essential enzyme for the growth of pathogenic bacteria (e.g., *Mycobacterium tuberculosis*) represents an interesting drug target. The nucleotide binding sites of crystal structures from different bacteria and a structure of the human homologous protein methylmalonic aciduria associated protein A (MMAA, 2www@pdb) were compared to analyze differences between the structures. They can be exploited to circumvent a possible inhibition of the human enzyme which might result in fatal methylmalonic aciduria. **Fig. 3** shows the result of a binding site comparison between

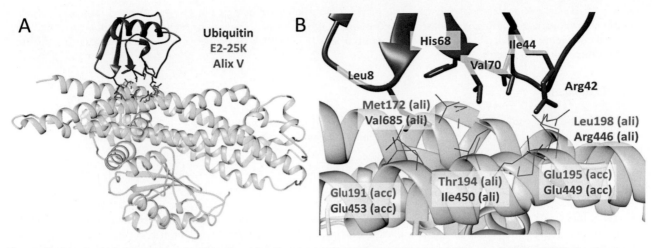

Fig. 4   Binding site similarities between the Ubiquitin-conjugating enzyme E2 K (3k9p.A@pdb) and a surface patch of the ALIX-V domain (2ojq@pdb). The template structures of the Ubiquitin-conjugation enzyme E2 K is shown in green with the bound Ubiquitin molecule in purple. The ALIX-V domain is depicted in orange. (A) Superposition of both protein structures with respect to the identified local similarity. (B) Zoomed view to illustrate identified binding site similarities with respect to the ubiquitin binding site of the Ubiquitin-conjugating enzyme. Some of the crucial similarities between both binding sites are depicted and the involved residues are labeled accordingly.

three mycobacterial MeaB structures and the human MMAA. A proper visualization of the major dissimilarities can assist the development of safe MeaB inhibitors.

Protein–Protein Interactions

The last example illustrates the applicability of binding site comparison for the analysis of protein–protein interactions. SiteEngine (Shulman-Peleg *et al.*, 2004) which is based on similarities between surface patches of binding sites was utilized to analyze potential protein–protein interactions (Keren-Kaplan *et al.*, 2013). A template ubiquitin-binding domain represented the starting point to search for new potential ubiquitin-binding domains within the PDB. Ubiquitylation plays a crucial role in trafficking, for example, of retroviral transmembrane proteins. The ALIX-V domain (2ojq@pdb) was identified as new potential ubiquitin binding domain due to its high local similarity to the ubiquitin binding site. Subsequent protein-protein docking and experimental binding affinity measurements confirmed this result. The finding was also verified in an independent study (Dowlatshahi *et al.*, 2012). The match between the template structure (3k9p@pdb) and ALIX-V is depicted in **Fig. 4**.

Remarks Concerning Binding Site Comparison

Finally, we want to provide some useful hints which might assist in applying binding site comparison tools and software for various projects.

First, proteins are no rigid constructs as commonly observed in protein crystal structures. NMR structures and MD simulations can provide useful information with respect to flexible protein regions and binding site flexibility. The knowledge of binding site dynamics can have a huge impact on computational strategies (Stank *et al.*, 2016). As shown in previous studies, it can be beneficial to use an ensemble of binding site structures to identify or compare ligand binding sites (Lanig *et al.*, 2015; Miguel *et al.*, 2015; Möller-Acuña *et al.*, 2015; Zhao *et al.*, 2012). This is especially important as the results of some comparison algorithms are very sensitive toward binding site definition. Given a large amount of crystal structures of the protein bound to different ligands, all available binding sites should be considered.

Moreover, although various methods provide some measure of significance to analyze the impact of the identified similarities, the obtained results should be visualized and rationalized. Sometimes the use of two or three tools is advisable. Different binding site comparison methods were evaluated by the means of highly distinct datasets and designed to address problems of varying application domains. Therefore, the used benchmark datasets can provide insights to appropriate tools for the aim of the study. Tools validated with respect to classification performance are promising tools to infer evolutionary relationships and protein function. Other tools were analyzed using datasets of proteins binding similar ligands. Those might be an appropriate choice for applications such as off-target prediction, polypharmacology, and drug repurposing. Nevertheless, there are examples for various tools that underline their impact on fields they were not developed for.

A further challenge is the ranking of binding site similarities. Many tools offer different similarity measures and it is often beneficial to use a scoring scheme that is most appropriate for the aim of the study. A method which models and compares a binding site based on residues might nonetheless offer scores that consider the shape overlap of two cavities or common interaction patterns. This issue was not addressed in this article and the reader is referred to the publications cited herein.

Finally, a database search cannot only be performed based on known binding sites. In cases of novel targets, it is highly recommended to predict binding sites based on geometrical, evolutionary information, or probe interaction potentials and compare them against large databases of annotated binding sites (e.g., sc-PDB (Kellenberger *et al.*, 2006)) to gain insights into potential druggable binding sites.

Acknowledgement

C.E. is funded by the Kekulé Mobility Fellowship of the Chemical Industry Fund (FCI). O.K. is funded by the German Federal Ministry for Education and Research (BMBF, Medizinische Chemie in Dortmund, TU Dortmund University, Grant BMBF 1316053).

Appendix

Connolly surface: solvent-accessible molecular surface as calculated by applying a virtual solvent molecule as probe sphere; docking: computational prediction of small molecule binding modes within predefined protein cavities; drug repurposing: investigation of new applications for existing drugs with known properties outside their original medicinal scope; Gaussian function: a continuous bell-shaped function which represents probability distributions; MD simulation: molecular dynamics simulation, a method to calculate the time-dependent behavior of molecular systems; Monte Carlo algorithm: a randomized algorithm that is, with a certain and usually small probability, allowed to provide none or wrong output; substitution matrix: a scoring matrix which contains a value for all possible substitutions (e.g., of amino acid residues) to calculate the degree of similarity between two objects; tessellation: a gapless and non-overlapping covering of shapes by repeating smaller shapes (often triangles) to represent, for example, molecular surfaces; virtual screening: *in silico* search for molecules that might modulate the activity of a protein in a desired manner.

See also: Algorithms for Graph and Network Analysis: Clustering and Search of Motifs in Graphs. Chemical Similarity and Substructure Searches. Chemoinformatics: From Chemical Art to Chemistry in Silico. Computational Tools for Structural Analysis of Proteins. Fingerprints and Pharmacophores. Natural Language Processing Approaches in Bioinformatics. Pharmacophore Development. Prediction of Protein-Binding Sites in DNA Sequences. Prediction of Protein-Protein Interactions: Looking Through the Kaleidoscope. Protein Structural Bioinformatics: An Overview. Small Molecule Drug Design

---

References

Bajusz, D., Rácz, A., Héberger, K., 2015. Why is Tanimoto index an appropriate choice for fingerprint-based similarity calculations? Journal of Cheminformatics 7, 20.

Barelier, S., Sterling, T., O'Meara, M.J., Shoichet, B.K., 2015. The recognition of identical ligands by unrelated proteins. ACS Chemical Biology 10 (12), 2772–2784.

Baroni, M., Cruciani, G., Sciabola, S., Perruccio, F., Mason, J.S., 2007. A common reference framework for analyzing/comparing proteins and ligands. Fingerprints for Ligands and Proteins (FLAP): Theory and application. Journal of Chemical Information and Modeling 47 (2), 279–294.

Bentley, J.L., 1975. Multidimensional binary search trees used for associative searching. Communications of the ACM 18 (9), 509–517.

Bloom, B.H., 1970. Space/time trade-offs in hash coding with allowable errors. Communications of the ACM 13 (7), 422–426.

Bron, C., Kerbosch, J., 1973. Algorithm 457: Finding all cliques of an undirected graph. Communications of the ACM 16 (9), 575–577.

Capra, J.A., Laskowski, R.A., Thornton, J.M., Singh, M., Funkhouser, T.A., 2009. Predicting protein ligand binding sites by combining evolutionary sequence conservation and 3D structure. PLOS Computational Biology 5 (12), e1000585.

Chartier, M., Najmanovich, R., 2015. Detection of binding site molecular interaction field similarities. Journal of Chemical Information and Modeling 55 (8), 1600–1615.

Desaphy, J., Azdimousa, K., Kellenberger, E., Rognan, D., 2012. Comparison and druggability prediction of protein-ligand binding sites from pharmacophore-annotated cavity shapes. Journal of Chemical Information and Modeling 52 (8), 2287–2299.

Desaphy, J., Raimbaud, E., Ducrot, P., Rognan, D., 2013. Encoding protein-ligand interaction patterns in fingerprints and graphs. Journal of Chemical Information and Modeling 53 (3), 623–637.

Dowlatshahi, D.P., Sandrin, V., Vivona, S., et al., 2012. ALIX is a Lys63-specific polyubiquitin binding protein that functions in retrovirus budding. Developmental Cell 23 (6), 1247–1254. doi: 10.1016/j.devcel.2012.10.023.

Edwards, T.E., Baugh, L., Bullen, J., et al., 2015. Crystal structures of Mycobacterial MeaB and MMAA-like GTPases. Journal of Structural and Functional Genomics 16 (2), 91–99.

Ehrt, C., Brinkjost, T., Koch, O., 2016. Impact of binding site comparisons on medicinal chemistry and rational molecular design. Journal of Medicinal Chemistry 59 (9), 4121–4151.

Garma, L.D., Medina, M., Juffer, A.H., 2016. Structure-based classification of FAD binding sites: A comparative study of structural alignment tools. Proteins 84 (11), 1728–1747.

Gaudreault, F., Morency, L.-P., Najmanovich, R.J., 2015. NRGsuite: A PyMOL plugin to perform docking simulations in real time using FlexAID. Bioinformatics 31 (23), 3856–3858.

Ghersi, D., Sanchez, R., 2009. EasyMIFS and SiteHound: A toolkit for the identification of ligand-binding sites in protein structures. Bioinformatics 25 (23), 3185–3186.

Goodford, P.J., 1985. A computational procedure for determining energetically favorable binding sites on biologically important macromolecules. Journal of Medicinal Chemistry 28 (7), 849–857.

Henrich, S., Salo-Ahen, O.M.H., Huang, B., et al., 2010. Computational approaches to identifying and characterizing protein binding sites for ligand design. Journal of Molecular Recognition 23 (2), 209–219.

Huang, B., Schroeder, M., 2006. LIGSITEcsc: Predicting ligand binding sites using the Connolly surface and degree of conservation. BMC Structural Biology 6, 19.

Kabsch, W., Sander, C., 1983. Dictionary of protein secondary structure: Pattern recognition of hydrogen-bonded and geometrical features. Biopolymers 22 (12), 2577–2637.

Kellenberger, E., Muller, P., Schalon, C., et al., 2006. sc-PDB: An annotated database of druggable binding sites from the Protein Data Bank. Journal of Chemical Information and Modeling 46 (2), 717–727.

Keren-Kaplan, T., Attali, I., Estrin, M., et al., 2013. Structure-based in silico identification of ubiquitin-binding domains provides insights into the ALIX-V: Ubiquitin complex and retrovirus budding. The EMBO Journal 32 (4), 538–551.

Konc, J., Janezic, D., 2010. ProBiS algorithm for detection of structurally similar protein binding sites by local structural alignment. Bioinformatics 26 (9), 1160–1168.

Krone, M., Kozlíková, B., Lindow, N., et al., 2016. Visual Analysis of Biomolecular Cavities: State of the Art. Computer Graphics Forum 35, 527–551.

Kumlander, D., 2004. A new exact algorithm for the maximum-weight clique problem based on a heuristic vertex-colouring and a backtrack search. In: Proceedings of the Fourth International Conference on Engineering Computational Technology, Lisbon, Portugal. Stirlingshire: Civil-Comp Press.

Lanig, H., Reisen, F., Whitley, D., et al., 2015. In silico adoption of an orphan nuclear receptor NR4A1. PLOS ONE 10 (8), e0135246.

Levi, G., 1973. A note on the derivation of maximal common subgraphs of two directed or undirected graphs. Calcolo 9 (4), 341–352.

McLachlan, A.D., 1971. Tests for comparing related amino-acid sequences. Cytochrome c and cytochrome c 551. Journal of Molecular Biology 61 (2), 409–424.

Miguel, A., Hsin, J., Liu, T., et al., 2015. Variations in the binding pocket of an inhibitor of the bacterial division protein FtsZ across genotypes and species. PLOS Computational Biology 11 (3), e1004117.

Möller-Acuña, P., Contreras-Riquelme, J.S., Rojas-Fuentes, C., et al., 2015. Similarities between the binding sites of SB-206553 at serotonin type 2 and alpha7 acetylcholine nicotinic receptors: Rationale for its polypharmacological profile. PLOS ONE 10 (8), e0134444.

Moon, J.W., Moser, L., 1965. On cliques in graphs. Israel Journal of Mathematics 3 (1), 23–28.

Mudgal, R., Srinivasan, N., Chandra, N., 2017. Resolving protein structure-function-binding site relationships from a binding site similarity network perspective. Proteins.

Pagh, R., Rodler, F.F., 2004. Cuckoo hashing. Journal of Algorithms 51 (2), 122–144.

Ritschel, T., Schirris, T.J., Russel, F.G., 2014. KRIPO – A structure-based pharmacophores approach explains polypharmacological effects. Journal of Cheminformatics 6 (Suppl. 1), O26.

Rogers, D., Hahn, M., 2010. Extended-connectivity fingerprints. Journal of Chemical Information and Modeling 50 (5), 742–754.

Schalon, C., Surgand, J.-S., Kellenberger, E., Rognan, D., 2008. A simple and fuzzy method to align and compare druggable ligand-binding sites. Proteins 71 (4), 1755–1778.

Schmitt, S., Hendlich, M., Klebe, G., 2001. From structure to function: A new approach to detect functional similarity among proteins independent from sequence and fold homology. Angewandte Chemie International Edition 40 (17), 3141–3144.

Schmitt, S., Kuhn, D., Klebe, G., 2002. A new method to detect related function among proteins independent of sequence and fold homology. Journal of Molecular Biology 323 (2), 387–406.

Sciabola, S., Stanton, R.V., Mills, J.E., et al., 2010. High-throughput virtual screening of proteins using GRID molecular interaction fields. Journal of Chemical Information and Modeling 50 (1), 155–169.

Shulman-Peleg, A., Nussinov, R., Wolfson, H.J., 2004. Recognition of functional sites in protein structures. Journal of Molecular Biology 339 (3), 607–633.

Spitzer, R., Cleves, A.E., Jain, A.N., 2011. Surface-based protein binding pocket similarity. Proteins 79 (9), 2746–2763.

Stank, A., Kokh, D.B., Fuller, J.C., Wade, R.C., 2016. Protein binding pocket dynamics. Accounts of Chemical Research 49 (5), 809–815.

Todeschini, R., Consonni, V., Mannhold, R., Kubinyi, H., Folkers, G., 2009. Molecular Descriptors for Chemoinformatics. Volume I: Alphabetical Listing, Volume II: Appendices. Hoboken: Wiley-VCH.

Tomita, E., Tanaka, A., Takahashi, H., 2006. The worst-case time complexity for generating all maximal cliques and computational experiments. Theoretical Computer Science 363 (1), 28–42.

Totrov, M., 2008. Atomic property fields: Generalized 3D pharmacophoric potential for automated ligand superposition, pharmacophore elucidation and 3D QSAR. Chemical Biology & Drug Design 71 (1), 15–27.

Totrov, M., 2011. Ligand binding site superposition and comparison based on atomic property fields: Identification of distant homologues, convergent evolution and PDB-wide clustering of binding sites. BMC Bioinformatics 12 (Suppl. 1), S35.

von Neumann, John, 1951. Various techniques used in connection with random digits. In: Householder, A.S., Forsythe, G.E., Germond, H.H. (Eds.), Monte Carlo Method 12. Washington, DC: US Government Printing Office, National Bureau of Standards Applied Mathematics Series, pp. 36–38.

Weill, N., Rognan, D., 2010. Alignment-free ultra-high-throughput comparison of druggable protein-ligand binding sites. Journal of Chemical Information and Modeling 50 (1), 123–135.

Xie, L., Bourne, P.E., 2008. Detecting evolutionary relationships across existing fold space, using sequence order-independent profile-profile alignments. Proceedings of the National Academy of Sciences of the United States of America 105 (14), 5441–5446.

Xie, L., Evangelidis, T., Xie, L., Bourne, P.E., 2011. Drug discovery using chemical systems biology: Weak inhibition of multiple kinases may contribute to the anti-cancer effect of nelfinavir. PLOS Computational Biology 7 (4), e1002037.

Xie, L., Xie, L., Bourne, P.E., 2009. A unified statistical model to support local sequence order independent similarity searching for ligand-binding sites and its application to genome-based drug discovery. Bioinformatics 25 (12), i305–i312.

Yeturu, K., Chandra, N., 2008. PocketMatch: A new algorithm to compare binding sites in protein structures. BMC Bioinformatics 9, 543.

Zhang, Y., Skolnick, J., 2005. TM-align: A protein structure alignment algorithm based on the TM-score. Nucleic Acids Research 33 (7), 2302–2309.

Zhao, Y., Wang, J., Wang, Y., Huang, J., 2012. A comparative analysis of protein targets of withdrawn cardiovascular drugs in human and mouse. Journal of Clinical Bioinformatics 2 (1), 10.

Further Reading

Broomhead, N.K., Soliman, M.E., 2017. Can we rely on computational predictions to correctly identify ligand binding sites on novel protein drug targets? Assessment of binding site prediction methods and a protocol for validation of predicted binding sites. Cell Biochemistry and Biophysics 75 (1), 15–23.

Brown, N., 2011. Algorithms for chemoinformatics. Wiley Interdisciplinary Reviews: Computational Molecular Science 1 (5), 716–726.

Desaphy, J., Bret, G., Rognan, D., Kellenberger, E., 2015. sc-PDB: A 3D-database of ligandable binding sites – 10 years on. Nucleic Acids Research 43 (Database issue), D399–D404.

Ehrlich, H.-C., Rarey, M., 2011. Maximum common subgraph isomorphism algorithms and their applications in molecular science: A review. Wiley Interdisciplinary Reviews: Computational Molecular Science 1 (1), 68–79.

Inhester, T., Rarey, M., 2014. Protein-ligand interaction databases: Advanced tools to mine activity data and interactions on a structural level. Wiley Interdisciplinary Reviews: Computational Molecular Science 4 (6), 562–575.

Jalencas, X., Mestres, J., 2013. Identification of similar binding sites to detect distant polypharmacology. Molecular Informatics 32 (11–12), 976–990.

Kellenberger, E., Schalon, C., Rognan, D., 2008. How to measure the similarity between protein ligand-binding sites? Current Computer Aided-Drug Design 4 (3), 209–220.

Konc, J., Janežič, D., 2014. Binding site comparison for function prediction and pharmaceutical discovery. Current Opinion in Structural Biology 25, 34–39.

Pérot, S., Sperandio, O., Miteva, M.A., Camproux, A.-C., Villoutreix, B.O., 2010. Druggable pockets and binding site centric chemical space: A paradigm shift in drug discovery. Drug Discovery Today 15 (15–16), 656–667.

Raymond, J.W., Willett, P., 2002. Maximum common subgraph isomorphism algorithms for the matching of chemical structures. Journal of Computer-Aided Molecular Design 16 (7), 521–533.

Shoemaker, B.A., Zhang, D., Tyagi, M., *et al.*, 2012. IBIS (Inferred Biomolecular Interaction Server) reports, predicts and integrates multiple types of conserved interactions for proteins. Nucleic Acids Research 40 (Database issue), D834–D840.

Volkamer, A., Rarey, M., 2014. Exploiting structural information for drug-target assessment. Future Medicinal Chemistry 6 (3), 319–331.

Xie, Z.-R., Hwang, M.-J., 2015. Methods for predicting protein-ligand binding sites. Methods in Molecular Biology 1215, 383–398.

Relevant Websites

https://www.ccdc.cam.ac.uk/
  Cavbase.
http://www.moldiscovery.com/software/flap/
  FLAP.
http://bioinfo-pharma.u-strasbg.fr/labwebsite/download.html
  FuzCav, SiteAlign, TIFP, Shaper.
http://biophys.umontreal.ca/nrg/NRG/IsoMIF.html
  IsoMIF.
http://3d-e-chem.github.io/kripodb/
  KRIPO.
http://proline.physics.iisc.ernet.in/pocketmatch/
  PocketMatch.
http://www.biopharmics.com/
  PSIM.
http://bioinfo3d.cs.tau.ac.il/SiteEngine/
  SiteEngine.
http://compsci.hunter.cuny.edu/~leixie/smap/smap.html
  SMAP.
https://zhanglab.ccmb.med.umich.edu/TM-align/
  TM-align.

Biographical Sketch

Christiane Ehrt received her MSc in Biochemistry at the Martin-Luther-University Halle-Wittenberg. Currently, she works as a PhD student at the Faculty of Chemistry and Chemical Biology at the TU Dortmund University in the group of Dr. Oliver Koch. Her interests and current focus include the identification and comparison of protein ligand binding sites, virtual screening studies for novel targets, and in this context, comparative modeling, MD simulations, and biochemical assay development. Tobias Brinkjost received his Diploma in Computer Science from the TU Dortmund University in 2012. Afterward he joined the Medicinal Chemistry group of Dr. Oliver Koch in collaboration with the Chair for Algorithm Engineering of Prof. Dr. Petra Mutzel at the TU Dortmund University for his PhD studies. His current research mainly focusses on the development of models and algorithms in the field of rational drug design. Oliver Koch studied pharmacy and computer science at the Philipps-University Marburg, Germany, where he also obtained his PhD in Pharmaceutical Chemistry with Prof. G. Klebe. After postdoctoral research at the Cambridge Crystallographic Data Center in 2008 and working in drug discovery at MSD Animal Health Innovation, he started his independent academic career in 2012 as an independent junior group leader for medicinal chemistry at the TU Dortmund University, Germany. His research interests involve the development and application of computational methods in computational molecular design and medicinal chemistry.

Quantitative Structure-Activity Relationship (QSAR): Modeling Approaches to Biological Applications

Swathik Clarancia Peter[1], Tamil Nadu Agricultural University, Coimbatore, India
Jaspreet Kaur Dhanjal[1], Vidhi Malik, and Navaneethan Radhakrishnan, Indian Institute of Technology Delhi, New Delhi, India
Mannu Jayakanthan, Tamil Nadu Agricultural University, Coimbatore, India
Durai Sundar, Indian Institute of Technology Delhi, New Delhi, India

Introduction

Quantitative structure-activity relationship (QSAR) approach relies on the basic principle of chemistry that states that the biological activity of any ligand or compound is associated with the arrangement of atoms forming the molecular structure. In other words, structurally related molecules possess similar biological activities. This structural information can be defined in terms of a series of parameters called molecular descriptors. In QSAR, the biological activity is represented as a function of these molecular descriptors as depicted in Eq. (1).

$$\text{Biological response or activity} = f(\text{molecular descriptors})\ldots \tag{1}$$

The model thus developed based on the biological activities of known ligands is used to predict the response of new compounds.

QSAR finds applicability in a wide range of fields including toxicology (Wang et al., 2014; Rochani et al., 2010), ecotoxicology (Hermens et al., 1984; Van Gestel and Ma, 1990; Escher et al., 2006), drug design and discovery (Zernov et al., 2003; Speck-Planche et al., 2012; Buolamwini and Assefa, 2002), chemical data mining (Shen et al., 2004), combinatorial library design (Ghose et al., 1999; Zhang et al., 2008) and so on.

QSAR studies therefore involve selection of active and inactive compounds with the measure of their biological activity, description and calculation of molecular descriptors, selection of appropriate features followed by construction of the mathematical model and its evaluation.

QSAR and QSPR

Quantitative structure-activity relationship (QSAR) prediction depends on the structure of molecules and atoms present in the compound. Biological activity is understood in terms of numerical values (example bioavailability, inhibitory concentration) and presence/absence of a condition (example infected/not infected, mutagenic/non mutagenic). Various QSAR studies have been carried out to understand biological properties such as pharmacokinetics (Vieira et al., 2014; Gombar and Hall, 2013), blood brain barrier penetration (BBB) (Zhang et al., 2008), carcinogenicity (Fjodorova et al., 2010; Kar and Roy, 2011), drug metabolism (Braga and Andrade, 2012; Lewis, 2000), bio-concentration (Grisoni et al., 2016; Papa et al., 2007), permeability (Gozalbes et al., 2011; Fujikawa et al., 2007), drug clearance (Manga et al., 2003; Boik and Newman, 2008), mutagenicity (Valencia et al., 2013; Barber et al., 2016), and so on. Another term associated with this approach is Quantitative structure-property relationship (QSPR). In QSPR, physiochemical properties of the chemical compounds are determined based on the molecular structure information. Physiochemical properties such as melting point (Katritzky et al., 2002; Modarresi et al., 2006), boiling point (Sola et al., 2008; Dai et al., 2013), solubility (Duchowicz and Castro, 2009; Gao et al., 2002), stability (Dioury et al., 2014; Ghasemi et al., 2010), dielectric constant (Achary, 2014; Soltanpour et al., 2016), reactivity (Toropov et al., 2004), diffusion coefficient (Mirkhani et al., 2012), thermodynamic properties (Puri et al., 2002; Duchowicz et al., 2006), hydrophobicity (Zou et al., 2016; Berinde, 2013) have been exploited to determine quantitative structure-property relationships.

History

The concept of QSAR had begun a century ago. Crum-Brown and Fraser in 1868 proposed the physiological activity of molecules based on their composition (Crum-Brown and Fraser, 1868). Then the narcotic effect of the primary alcohols respective to the molecular weight was studied by Richardson in 1869 (Richardson, 1869). Following this, studies of simple organic compounds in response to water solubility (Richet, 1893), potency variation of narcotic compounds (Meyer, 1899), study of chemical reactivity of substituted benzenes (Hammett, 1937), narcotic study based on logP and thermodynamic (Ferguson, 1939), physical organic chemistry to linear steric energy relationships (Taft, 1952, 1953a, 1953b), linear free energy relationship model by Hansch and Fujita (1964), QSAR study based on molecular fragments by Free and Wilson (1964) and substituent-based structure-activity

---

[1]Equal contribution.

relationship (Fujrra and Ban, 1971) are some of the important hallmarks in the history of QSAR. The development of 2D-QSAR began in early 1970s and the development of 3D-QSAR started in early 1980s. With the arrival of new technologies and perspectives, QSAR has now become multidimensional. More robust and accurate QSAR approaches came into practice with increased dimensionality like 4D, 5D and 6D, which has led to increased predictability, reliability and precision of the models.

Types of QSAR Methodologies

QSAR can be broadly classified in two ways (Qiao *et al.*, 2014). The first classification is based on the dimensions of the descriptors involved in the model such as 1D, 2D, 3D, 4D, 5D and so forth. Other is based on the type of biological activity predicted as a dependent variable. This includes Quantitative structure-toxicity relationship (QSTR) (Can, 2014), Quantitative structure-metabolism relationship (QSMR), Quantitative structure-reactivity relationship or Quantitative structure-retention relationship (QSRR) (Hemmateenejad *et al.*, 2009; Goryński *et al.*, 2013), Quantitative structure-permeability relationship or Quantitative structure-pharmacokinetics relationship (QSPR) (Moss *et al.*, 2002; Mayer and Van De Waterbeemd, 1985), Quantitative structure-bioavailability relationship or Quantitative structure-binding affinity relationship (QSBR) (Andrews *et al.*, 2000; Zhang *et al.*, 2006), and so forth.

QSAR models can also be grouped based on analysis of correlation-linear and non-linear (Roy and Mandal, 2008), or depending on binding nature of molecule and receptor–receptor dependent and receptor independent (Magdziarz *et al.*, 2009).

QSAR Model Construction

The quality of the QSAR model to a large extent depends on the data used for its construction. Hence, prior to the model development, it is necessary to gain insights about the data. Thorough understanding of the problem and influencing factors assists in discerning meaningful relationships. Relevant background information about the system under study either biological or chemical is to be collected via literature search. Data sets for model construction need to be chosen carefully as poor and inconsistent data would lead to corrupt model. There are several other factors like division of data into training and test data sets, molecular descriptors, statistical methods for model development that influence the quality of the QSAR model.

The schematic representation of QSAR model construction is given in **Fig. 1**. A brief description of the steps involved in QSAR model generation is discussed in the following section.

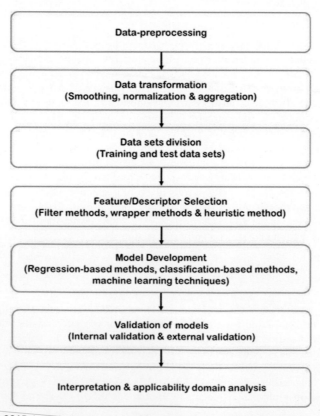

Fig. 1   Schematic representation of QSAR (quantitative structure-activity relationship) model development.

Data Pre-Processing

Pre-processing eliminates noise and redundancy in the data sets. It involves data transformation that includes smoothing, normalization and aggregation (Tomar and Agarwal, 2014), data reduction, sampling when data sets are large, noise elimination, feature selection, data cleaning, data integration when data is collected from heterogeneous sources, and discretization (Cocu *et al.*, 2008).

Training and Test Data Sets

The data is divided into training and test data sets. The training set is used to formulate the QSAR model, while the test set is used to evaluate its predictability and accuracy. Dataset is generally divided in such a way that both the sets occupy entire descriptor space. Employing appropriate data splitting techniques improves the model prediction. The various approaches available for the division of dataset include k-means clustering, based on X response, based on Y response, random selection, statistical molecular design, sphere exclusion, Kennard-Stone selection, Kohonen's self-organizing map selection, and extrapolation-oriented test set selection (Roy *et al.*, 2008).

Calculation of Molecular Descriptors

The information about the structure of molecules, defined by molecular descriptors obtained from different representations such as 2D, 3D, etc., is embedded into the QSAR model. Molecular conformations used should be correct for a better predictive model. There are different kinds of descriptors like count descriptors (0D), fingerprints (1D), topological descriptors (2D), geometrical (3D), grid based (4D) and so forth. The complexity of information and power to discriminate between similar structures as provided by different descriptors increases with dimensionality. 0D and 1D descriptors provide basic information like molecular weight and number of constituent elements that are directly derived from molecular formula. Net charge of the molecule is a 1D descriptor. Topological indices that are computed from the structural formula are 2D descriptors. These are based on the graph theory and reflect the connections in the structure. The most widely used topological descriptor is connectivity index proposed by Randić (2001); Li and Shi (2008); Kier (1985). Other topological indices are Wiener's index W (Ivanciuc, 2000; Wiener, 1947), Connectivity indices (Kier *et al.*, 1975), Kier Shape (Kier, 1985), Balaban J Index (Balaban, 1982) and Zagreb indices (Gutman and Trinajstić, 1972). The 3D descriptors are based on three-dimensional coordinates of atoms comprising the compounds. Some commonly used methods for calculating 3D descriptors are CoMFA, CoMSIA, CoMBINE, GERM, CoMMA, GRIND, WHIM, HoloQSAR and CoSA. 4D, 5D and 6D descriptors are multidimensional descriptors, which include the parameters involved in the structure and flexibility of the receptor-binding site in conjunction with ligand topology. 4D descriptors are based on reference grids and molecular dynamic simulations. Descriptors calculated using multiple conformations, orientations, protonation states and isosteriomers of the ligand constitute 5D descriptors. The solvation terms constitute 6D descriptors. Molecular descriptors can be calculated using various software, some of which are listed in **Table 1** (Damale *et al.*, 2014).

Feature Selection

Feature selection reduces the dataset horizontally. Among the large number of calculated descriptors, only few are chosen to define the model. Feature selection after descriptor calculation removes collinearity between the descriptor pairs. Selection of the most appropriate features is done using filter and wrapper methods (Goodarzi *et al.*, 2012). Filter methods involve filtering out descriptors, thereby reducing the pool size of descriptors based on inter-variable correlations. So, molecular descriptors that show inter-correlation are removed retaining only one descriptor from a pair (Roy *et al.*, 2015a). Descriptors with lowest variance are also removed. Filter methods use techniques like chi-square analysis, Shannon entropy, odds ratio, GSS coefficient (Liu, 2004), correlation based feature selection (Demel *et al.*, 2009), Fisher Score, Kolmogorov-Smirnov statistics (Guyon *et al.*, 2002), and principle component analysis. Distance based methods like Euclidean distance measures are also grouped under filter methods. Wrapper methods use regression-based approaches to select descriptors. In general, wrapper methods involve more computational power and perform better than filter methods. Recursive feature elimination (Xue *et al.*, 2004), variable selection and modeling based on the prediction (Liu *et al.*, 2003), k nearest neighbor, backward elimination, forward selection, genetic algorithm, Bayesian regularized neural network, factor analysis and combinatorial protocol are some of the commonly used wrapper methods. Hybrid methods are also being used that combine both filter and wrapper methods for selecting features (Goodarzi *et al.*, 2012).

Table 1   Software for calculation of molecular descriptors

Software	Description	Availability
ACD/LogP Freeware	logP prediction by fragment-based algorithm	Freely available
Dragon	Calculation of topological, constitutional & geometrical descriptors	Commercial
MOLGEN	Calculation of topological, constitutional & geometrical descriptors	Freely available
PaDEL Descriptors	Calculation of 2D & 3D descriptors	Freely available

Heuristic methods based on multiple linear regression are also used in selecting descriptors for QSAR models. It is fast when compared to other methods. It discards descriptors with constant values and removes descriptors whose values are not available for all the structures in the dataset thereby removing trivial descriptors. It also removes highly correlated descriptors (Liu and Long, 2009). Heuristic method has used for searching descriptor space and selection of vital descriptors in QSAR study of 1,4-dihydropyridine calcium channel antagonists (Si *et al.*, 2006), to select descriptors for multivariable linear model to predict the Percent of Applied Dose Dermally Absorbed (PADA) of the polycyclic aromatic hydrocarbons (Wang *et al.*, 2008), and descriptors selection in QSAR analysis of photosystem II electron transfer inhibitors (Karacan *et al.*, 2012).

QSAR Methods

Statistical methods are used in model construction and feature selection when there are large numbers of descriptors. They are helpful in obtaining functional endpoints. The statistical methods can be classified into regression-based approaches, classification based approaches, and machine learning techniques. QSAR modeling can be done for both linear and non-linear properties. Some of the methods for modeling linear properties include linear regression and partial regression, while artificial neural networks are being employed for modeling non-linear properties.

Models can be constructed using both supervised and unsupervised techniques. The chance effects in the unsupervised learning are less when compared to supervised learning, as it does not change to fit the model. Semi-supervised learning is advantageous over the supervised and unsupervised techniques. It considers both the labeled and unlabeled data giving better performance (Settles, 2012). Comparison of supervised learning algorithms and semi-supervised algorithms in different data sets suggested that semi-supervised learning can assist in understanding the addition of unlabeled data and hence is helpful for certain type of dataset and methods (Levatic *et al.*, 2013).

Regression based methods
Multiple linear regression
Multiple Linear Regression (MLR) method helps in establishing correlation between the independent and dependent variables. Here, the dependent variables are the biological activity or physiochemical property of the system that is being studied and the independent variables are molecular descriptors obtained from different representations. In linear regression models, the dependent variable is predicted using only one descriptor or feature. Multiple linear regression models consider more than one descriptor for the prediction of property/activity in question. The model based on the linear regression can be represented as a mathematical equation given below-

$$y = a + bx \ldots \tag{2}$$

where, y is the dependent/response variable representing the physiochemical property or biological activity, x is the independent or predictor variable accounting for the molecular descriptor, and b is the regression coefficient.

Examples of QSAR studies involving the use of MLR method include the prediction of binding affinities of H3 antagonists (Dastmalchi *et al.*, 2012) and inhibitory activity of human non-pancreatic secretory phospholipase A_2 (Singh and Verma, 2014).

Partial least squares method
Partial Least Squares (PLS), developed from the principal component regression, helps in building models predicting more than one dependent variable (Lorber *et al.*, 1987). This method is used when the number of variables are more than the number of compounds in the datasets and where the variables considered for the study are correlated (Cramer, 1993). It is applied in 3D-QSAR technique, Comparative Molecular Field Analysis (CoMFA) to reduce the number of descriptors. PLS is also used in the validation metrics of the models (Mota *et al.*, 2009). It is advantageous over other regression models (Cramer, 1993). PLS has been used in the construction of many successful QSAR models. Prediction of binding affinity of polycyclic aromatic compounds with the rat liver 2,3,7,8-tetrachlorodibenzene-p-dioxin (TCCD) receptor (Johnels *et al.*, 1989) is one such example. PLS in combination with other methods – Genetic Partial Least Squares (G/PLS), Factor analysis Partial Least Squares (FA-PLS) and Orthogonal Signal Correction Partial Least Squares (OSC-PLS) (Liu and Long, 2009) is also being used for QSAR studies.

Classification based methods
Cluster analysis
Clustering involves placing similar data into a group in a way that maximizes similarity within groups and dissimilarity between groups. It involves methods like hierarchical clustering and k-means clustering. In hierarchical clustering, clusters are grouped on the basis of the dissimilarities calculated through the distances between the objects (Euclidean distances). The k-means clustering is a non-hierarchical method. It is based on k-centroids. Some of the other classification based methods are linear discriminant analysis and logistic regression (Roy *et al.*, 2015c; Agresti, 2007; Harrell, 2001).

Machine learning techniques
Artificial neural network
Artificial Neural Network (ANN) mimics the behavior of biological neurons. The ANN has input layer, hidden layer(s) and output layer. The molecular information is fed through the input layer, which is processed by a number of processing units in

parallel and the biological activity or property is obtained as an output. The commonly used ANNs in QSAR are back propagation neural networks, probabilistic neural networks, Kohonen self-organizing maps and Bayesian regularized neural networks. Neural networks can be supervised or unsupervised in nature. The learning is supervised when the trained model is validated by a separate test set. The training set helps in fitting weight parameters and decides the number of hidden layers in the network architecture. ANN methods have been proven to be highly adaptable and are deployed in modeling non-linear systems with high variability in data sets. Superior models can be obtained from ANNs compared to traditional approaches like MLR and PLR (Shi et al., 2010). Implementing techniques like dropout in training data set reduces over-fitting of data and produces improved results when compared to conventional ANN models. However, Bayesian networks still outstand in their performance.

ANNs have been recently used in many QSAR studies. Some examples include the study of neurotrophic effects of N-p-Tolyl/phenylsulfonyl L-amino acid thioester derivatives (Luo et al., 2011) and the study of antibacterial activity of oxazolidinone derivatives (Zou and Zhou, 2007).

Support vector machine

Support Vector Machine (SVM) is a machine learning approach that uses a linear classifier to classify data into two categories. The classifier is non-probabilistic. It performs better than other 3D QSAR models. In a comparative study of 3D QSAR modeling and SVMs for predicting the activity of BRAF-V600E and HIV integrase inhibitors, SVMs outperformed 3D-QSAR models (Wesley et al., 2016). SVMs are also being used in combination with other methods like MLR, PLS and so forth for building more powerful and accurate QSAR models. In one of the QSAR reports where the anti-Alzheimer activity of triazolylthiopenes as cyclin dependent kinase 5 inhibitors was studied (Garkani-Nejad and Ghanbari, 2016), MLR was used for selecting molecular descriptors, and support vector regression and PLS were used for constructing non-linear and linear models. When these methods were compared, support vector regression outperformed other methods. SVMs are employed in dimensionality reduction through variable ranking and selection (Bi et al., 2003). SVMs also overcome the issue of over-fitting observed in artificial networks.

In a study conducted for predicting the reduction of dihydrofolate reductase by pyrimidines, SVMs outperformed three neural networks, namely radial basis function network, nearest neighbor classifier and decision tree (Burbidge et al., 2001). QSPR models are also being developed by employing a combination of SVM with other methods like Principal Component Least Square methods and so forth (Veyseh et al., 2015; Khorshidi et al., 2014).

Gene expression programming

Gene Expression Programming (GEP) is based on the genetic algorithm and genetic programming. GEP has been used in QSAR modeling for the prediction of dermal penetration (PADA, Percent of Applied Dose Dermally Absorbed) of polycyclic aromatic hydrocarbons (Wang et al., 2008), prediction of EC_{50} of anti-HIV drugs (Si et al., 2008), prediction of binding affinity of substituted 1-(3,3-diphenylpropyl)-piperidinyl amides and ureas with the chemokine receptor 5, and also for the prediction of toxicity of aromatic compounds (Shi et al., 2010). GEP proved to be better in prediction when compared to the previously discussed methods like ANNs (Shi et al., 2010), and SVMs (Si et al., 2008). Further, Improved Gene Expression Programming (IGEP) proves to be more efficient than existing methods (Fu et al., 2010).

Some of other methods deployed in QSAR studies include Monte Carlo Simulations (Kumar and Chauhan, 2017), principal component analysis (Suzuki et al., 2001), and decision trees and random forest algorithm (Polishchuk et al., 2009; Simeon et al., 2016). Still newer methods like Projection Pursuit Regression (Du et al., 2011) and Local Lazy Regression (Guha et al., 2006; Lei et al., 2010) are also implemented in QSAR models.

Software for QSAR Studies and Modeling

QSAR studies are carried out using various platforms that help in building models for predicting chemical, biological and toxicological activities. Some of these are listed in **Table 2**.

List of Databases Used in QSAR Studies

Attempt has also been made to archive the constructed QSAR models for further reference and usage. Some of these are listed in **Table 3**.

Validation of Models

Once the model is constructed, it must be validated. Validating the models avoid chance correlation of numerous descriptors used in the model and also over-fitting of data. It helps in assessing the accuracy and prediction of the model. The Organization for Economic Cooperation and Development (OECD) has put forth five principles to test the model. They are (1) a defined endpoint,

Table 2   Platforms used for QSAR modeling

Software	Description	Availability
3D-QSAR	To build 3-D QSAR models	Freely available
ACD/Tox Suite	Used for prediction of toxicity endpoints	Commercial, free web service
ADMET Predictor	Calculation of ADMET properties	Commercial
AZOrange	Machine learning platform for QSAR modeling	Freely available
BioPPsy	Prediction of pharmacokinetic properties of drug candidates by QSPR modeling	Freely available
BioTriangle	Web-based platform for calculating molecular descriptors	Freely available
BlueDesc	Molecular Descriptor Calculator	Freely available
CACTVS	Molecular Descriptor Calculator	Freely available
CAESAR	Models for developmental toxicity	Freely available
ChemDes	Descriptor and fingerprint calculation (web-based)	Freely available
CODESSA	Generate predictive QSAR models from Quantum chemical, topological & electrostatic descriptors	Commercial
CoFFer	Web-based QSAR service for the prediction of chemical compounds	Freely available
CORALSEA	Building of quantitative structure - property / activity relationships	Freely available
Derek	Rules based system with structural alerts for developmental toxicity, teratogenicity, testicular toxicity, and oesterogenecity	Commercial
DMax	Data mining tool for QSAR, virtual screening and compound screening data analysis	Freely available
DWFS	Parallel GA wrapper Feature selection (web-based)	Freely available
ECOSAR	Calculates aquatic toxicities	Freely available
EPISuite	Suite of programs for estimation of physiochemical property calculation and environmental fate	Freely available
eTOXlab	Development and validation of QSAR models	Freely available
GUSAR	Development of QSAR/QSPR models (web-based)	Freely available
HASL	Software package for 3D-QSAR	Commercial
HYBOT-PLUS	Descriptor calculation	Commercial
Leadscope	QSAR models to predict reproductive and developmental toxicity for rodent foetus	Commercial
Mathematica	Software package for ANN development	Commercial
Matlab	Software package for ANN development	Commercial
MC – 3DQSAR	Generates QSAR equations	Freely available
Molcode Toolbox	Prediction of toxicological endpoints	Commercial
MultiCASE	Models for developmental toxicity	Commercial
Neuralware	Software package for ANN development	Commercial
OECD QSAR Application Toolbox	QSAR models to fill data gaps & missing data	Freely available
PASS	Predicts biological activity using Bayesian algorithm, predicts embryotoxicity and teratogenicity	Commercial
QSARpro	Predicts activity and optimizes lead compounds using QSAR models	Commercial
SPSS	Software package for ANN development	Commercial
Statistica	Software package for ANN development	Commercial
TerraQSAR	Database compounds with structure-specific Biological activity	Freely available
T.E.S.T	Predicts toxicities of compounds by applying QSAR methodologies	Freeware
TIMES	Prediction of oestrogen, androgen and aryl hydrocarbon binding compound	Commercial
TOPKAT	Prediction of toxicological endpoints	Commercial
Toxmatch	Provides chemical similarity indices to assist in read-cross assessments & developing categories	Freely available
WEBCDK	Calculation of molecular descriptors (web-based)	Freely available
VCCL	Suite of programs for descriptor calculation, dimensionality reduction & data analysis	Freely available
VEGA-QSAR	QSAR models for regulatory purposes can be accessed and new QSAR models can be built	Freely available
NVirtualToxLab	Based on the combination of Auto flexible docking and mQSAR	Commercial

(2) an unambiguous algorithm, (3) a defined domain of applicability, (4) appropriate measures of goodness-of-fit, robustness and prediction accuracy, and (5) a mechanistic interpretation, if possible (Roy *et al.*, 2015b). Models can be validated through techniques such as internal validation, external validation, and cross validation (Veerasamy *et al.*, 2011). In internal validation, activity is predicted and parameters are estimated to analyze the precision of the prediction based on the compounds used for model construction. This is not suitable when new test set of compounds is used. But the external validation technique works well

Table 3   List of databases related to QSAR studies

Database	Description	URL
Danish QSAR database	Database of QSAR predictions	http://qsar.food.dtu.dk
ECOTOX	Online database that provides toxicity data on aquatic life, terrestrial plants & wild life	https://cfpub.epa.gov/ecotox/
EDKB (Endocrine Disruptor Knownledge Base)	Online database with predictive models to predict binding affinity of the compounds with oesterogen & androgen nuclear receptor proteins	http://edkb.fda.gov/webstart/edkb/index.html
JRC QSAR Model Database	Database of QSAR models	https://eurl-ecvam.jrc.ec.europa.eu/databases/jrc-qsar-model-database
MOE	Database of molecular data and QSAR modeling	https://www.chemcomp.com/MOE-Molecular_Operating_Environment.htm
MOLE db	Free database for molecular descriptors	http://michem.disat.unimib.it/mole_db/
QsarDB	Database of QSAR/QSPR models	https://qsardb.org/

even with the new datasets. In this case, the dataset is divided into test and training data sets. Model validation is done by test set compounds that are independent of training set (Roy *et al.*, 2015b; Veerasamy *et al.*, 2011; Roy and Kar, 2015). However, external validation is not worthy as it leaves a large portion of data set for testing (Hawkins *et al.*, 2003). Golbraikh and Tropsha (2002) proposed high value of cross-validated R^2 to be one of the criteria to have high predictive power for a QSAR model. It was emphasized that depending on q^2 for predictivity is incorrect. Instead external data set should be used for validation to have high predictive power for the model (Golbraikh and Tropsha, 2002).

Validation metrics are of two types based on type of QSAR model (Kar and Roy, 2011; Roy *et al.*, 2015b).

1. Regression-based QSAR models
2. Classification-based QSAR models

Both regression and classification based methods have their unique metrics for validation.

Validation Metrics for Regression-Based Methods

Validation metrics for regression-based models are calculated for both internal and external validation strategies.

Validation metrics for internal validation

Internal validation of QSAR models employs the use of molecules from training set to test the predictability of the model. Some of the most common methods used for internal validation of QSAR models are described here.

Least square fitting

Least square fitting is similar to linear regression and is the most commonly used validation method. It is the measure of square correlation coefficient (R^2) between the predicted and experimental value of activity. Outliers can be removed from the training data set, in order to optimize QSAR model, if difference between R^2 and R^2adj is less than 0.3 (Veerasamy *et al.*, 2011).

Chi-squared (χ^2) and root-mean squared error (RMSE)

The χ^2 and RMSE values are used to assess the predictive quality of a model. χ^2 value shows the difference between experimentally determined bioactivity values and the values predicted by the model, whereas the RMSE value is the depiction of error between the mean of experimental and predicted activity values. Even for models with large R^2 value (that is $>=7$), values of χ^2 and RMSE should be lower than 0.5 and 0.3 respectively, for good predictive ability of the model (Veerasamy *et al.*, 2011).

Cross validation

Cross validation approaches for internal validation include Leave-Group-Out (LGO), which involves leaving of a molecule or a group of molecules while creating model and evaluating the predictability of the model using the molecules left. Some of the important measures used in the internal cross validation of QSAR models are listed below.

Leave-One-Out (LOO) cross validation

In LOO cross validation, one compound is left out and the QSAR model is constructed using remaining compounds. The eliminated compound is used as a test for the predicted model. This process is repeated eliminating each of the compounds in the dataset one by one. The results so obtained from this are used for estimation of parameters involved in validation metrics. The predictability of the model is assessed by Predicted Residual Sum of Squares (PRESS) and cross-validated R^2 (Q^2) when Standard Deviation of Error of Prediction (SDEP) is obtained from PRESS (Roy and Kar, 2015; Roy *et al.*, 2015b).

Leave-Some-Out cross validation

In case of Leave-Some-Out (LSO) or Leave-Many-Out (LMO) a set of data compounds are eliminated and models are created with rest of the compounds. The left out compounds are then used to check the predictability of the model. Similar to LOO approach, LMO approach also involves repetitive cycles of elimination and model creation until each and every compound has been treated as a test set (Veerasamy *et al.*, 2011; Kar and Roy, 2011; Roy *et al.*, 2015b). On completion of all cycles of model training and testing, overall LMO-Q^2 value is calculated based on the compounds' predicted activity values. LMO methods is more consistent as compared to LOO (Veerasamy *et al.*, 2011).

Value of Q^2 is usually smaller than value of R^2. In order to avoid over-fitting of the model, the difference between R^2 and Q^2 should not exceed 0.3. Over-fitted model may very well predict the activity of compounds for training set but for new compounds predictivity is compromised (Veerasamy *et al.*, 2011).

True Q^2 and r_m^2 metrics

True Q^2 proposed by Hawkins *et al.* is used for small data sets and r_m^2 metric is calculated based on the scaled values of observed and predicted activity. Q^2 should not be treated as an ultimate proof for good predictability of models. Value of Q^2 higher than 0.5 should not be interpreted as high predictive power of QSAR models; until the ability of model to predict the activity of large number of compounds that are not used for training of model is tested (Golbraikh and Tropsha, 2002). Some of the other metrics used for internal validation are true $r_{m(LOO)}^2$ and Y-Randomization, a metric for chance correlation. The true $r_{m(LOO)}^2$ metric reveals external validation characteristics as its value is derived from the model developed after repetitive cycles of LOO. Y randomization test involves process randomization and model randomization to validate the model by permuting the response values with respect to unaltered matrix (Roy *et al.*, 2015b).

Validation metrics for external validation

The validation metrics employed in external validation are as following:

a) Predictive R^2 that can also be given as $Q^2_{(F1)}$ says about the correlation of observed and predicted data. Model is said to have good predictive power if the value of $Q^2_{(F1)}$ is greater than 0.5 (Roy *et al.*, 2015b).
b) $Q^2_{(F2)}$ and $Q^2_{(F3)}$ using the mean of test data set and training data set respectively (SchüÜRmann *et al.*, 2008). For validation of QSAR model, threshold value of 0.5 is defined for both metrics (Roy *et al.*, 2015b).
c) Golbraikh and Tropsha's criteria puts forth condition for selection of training and test data sets. For having a good predictive power, QSAR model should satisfy following conditions (Golbraikh and Tropsha, 2002; Veerasamy *et al.*, 2011):
   i. $Q^2_{training} > 0.5$
   ii. $R^2_{test} > 0.6$
   iii. $(r^2 - r_0^2)/r^2 < 0.1$ or $(r^2 - r_0'^2)/r^2 < 0.1$, where r_0^2 is R^2 of predicted *vs.* observed activities and $r_0'^2$ is R^2 of observed *vs.* predicted activities.
   iv. $0.85 <= k <= 1.15$ or $0.85 <= k' <= 1.15$, where k and k' are the slopes of regression lines through the origin.
d) Other metrics includes Root Mean Square Error of Prediction (RMSEP) to calculate prediction error of QSAR model (Roy *et al.*, 2015b); Concordance Correlation Coefficient (CCC), the most restrictive and precautionary measure, with ideal value of 1 (Chirico and Gramatica, 2011); $r_{m(rank)}^2$ which makes rank order predictions (Roy *et al.*, 2015b); and $r_{m\ (test)}^2$ to understand the relationship between observed and predicted values (Roy and Mitra, 2011; Roy *et al.*, 2015b).

Validation Metrics for Classification-Based Methods

The validation matrix employed in classification-based methods is the Wilks lambda (λ) statistics. It is used to test the significance of discriminant model function and is calculated as the ratio of within-category sum of squares to total dispersion. The value ranges between $0 < \lambda < 1$, with lower value corresponding to higher level of discrimination. Further, canonical index (Rc) is used to estimate the strength of relationship between various dependent and independent variables; Chi-square ($\chi2$) to check the quality of the classification based model; and Squared Mahalanobis distance is a measure calculated using random data points (Roy and Mitra, 2011; Roy *et al.*, 2015b).

Interpretation & Applicability Domain Analysis

The parameters or descriptors used in the model should be interpretable. Mechanistic interpretation of the built QSAR model helps in understanding the influence of descriptors in the predicted activity. Applicability domain analysis helps us to understand whether the built QSAR model can be used for any set of compounds. The applicability domain model is built on the theoretical region present in the chemical space of descriptors and activity modeled. It enables to understand the feasibility of activity or response prediction by the constructed QSAR model for a given set of compounds. So, the QSAR model prediction for a set of compounds is only reliable if the chemical space or applicability domain of those compound falls within the applicability domain of the compounds used for training the model (Roy *et al.*, 2015b). The theoretical region in the chemical space is identified or estimated by different methods. Application domain assessment is done through probability density distribution, geometrical methods, distance-based methods, ranges in descriptor space when descriptor space are used, and the range of response variable when modeled response space is used (Jaworska *et al.*, 2005).

Multidimensional QSAR

QSAR started with 0D and has evolved to 6D. Each dimension arose as a need to overcome the limitations of previous dimensions and to be more advantageous than the former. 1D QSAR calculated the molecular properties such as electronic, hydrophobic, steric and so forth. 2D QSAR considers geometric parameters, topological indices, molecular fingerprints, polar surface area but it excludes steric properties. 3D QSAR technique focuses on the spatial properties of the compound. So, the drawbacks of 2D QSAR get addressed in 3D QSAR method. The descriptor methods used in 3D QSAR are alignment-dependent and alignment-independent. The alignment dependent methods are Comparative Molecular Field Analysis (CoMFA), Comparative Molecular Similiarity Indices Analysis (CoMSIA), Comparative Binding Energy Analysis (CoMBINE), Comparative Residue Interaction Analysis (CoRIA), Hint Interaction Field Analysis (HIFA) and so forth. The alignment independent methods include Comparative Molecular Moment Analysis (CoMMA), Comparative Spectral Analysis (CoSA), and Holo-QSAR (HQSAR). 3D QSAR employs methods like Artificial Neural networks (ANN), Partial Least Squares Method (PLS), cluster analysis, and principal component analysis, and others for descriptor selection makes it more powerful than 2D QSAR. Even then, the 3D QSAR technique faces difficulties when there are large numbers of compounds in the data set, and has to compromise with the prediction accuracy. To overcome these drawbacks, 4D QSAR evolved. 4D-QSAR with fourth dimension of ensemble sampling addresses the issues of 3D QSAR. It includes descriptors for gird occupancy measures. 4D-QSAR can be applied for both, receptor independent and receptor dependent analysis (Hopfinger *et al.*, 1997). Further, 5D-QSAR evolved with the addition of new dimension to the 4D-QSAR, the new dimension being the multiple ligand topology representations. Ensembles of multiple representations deployed in 5D-QSAR makes this approach less biased when compared to 4D QSAR (Vedani and Dobler, 2002). 6D-QSAR improves the former 5D-QSAR strategy by including another dimension for solvation function that helps in analyzing different solvation models (Polanski, 2009).

QSAR in Drug Designing and Discovery

Drug design and discovery is a laborious and time-consuming process. It takes roughly 10–12 years for a molecule to be identified and approved as a drug. Most of the drugs fail during pre-clinical and clinical trials. The cost of drug discovery is very high. Determining drug candidates through QSAR studies would reduce cost of production and failure at an early stage. QSAR approaches help in identifying hits from a large library of compounds. The identified hit molecules can be purchased and studied for activity through experiments (Tang *et al.*, 2009; Montero-Torres *et al.*, 2006). The molecules with proven activity can be further optimized to design promising drug candidates. Thereby, QSAR studies avoid synthesis and testing of large number of compounds saving enormous time and cost.

Case Studies

A number of success stories reflecting the potential of QSAR in building reliable models have been reported in the literature. We have discussed here a few studies that dealt with a wide range of problems to get insights into the applications of QSAR approach.

The first example presented here is a QSAR model for designing better drugs for combating cancer. Apoptosis is an important process that can decide the fate of a cell. Malfunctioning of the process with atypical expression of B-cell lymphoma-2 (Bcl-2) anti-apoptotic proteins is a promising hallmark of cancer and in most of the cases results in resistance to chemo and radiotherapy. Multiple approaches including antisense oligonucleotides (ASOs), peptides and small molecule inhibitory compounds have been designed against these anti-apoptotic proteins. However, low cost and easy delivery makes small molecules a method of choice. To develop better compounds on the basis of available information, various QSAR models have been generated. But these models are limited to a single scaffold. In one such study, attempt was made to develop QSAR models for seven different classes of inhibitors reported in literature targeting Bcl-2 and Bcl-xL (Kanakaveti *et al.*, 2017). 453 such small molecule inhibitors with known IC_{50} (ranging from1 nm to 100 μm) were grouped into seven categories comprising of Apogossypol (89 compounds), Quinazoline thione (51 compounds), Pyrazole pyrimidine phenyl acyl (110 compounds), Quinolone (56 compounds), Thiomorpholine (42 compounds), Benzothiazole hydrazine (78 compounds) and Polyquinoline (27 compounds) depending upon the core structure. A total of 787 features (including constitutional, topological, electrostatic, geometrical and physicochemical descriptors) were tested. The most relevant descriptors were chosen and fitted using multiple linear regression analysis. Two to three parametric models were generated for all the different classes of compounds. $(n - 1)$ and $(n-10)$ leave-out cross validations were used to evaluate the performance of the generated models. The QSAR analysis resulted in models showing Pearson correlation coefficient ranging from 0.95 to 0.985. Three already known inhibitors- ABT-199, Navitoclax and Sabutoclax were tested against their generated models. The predicted IC_{50} was comparable to the reported activity. A correlation between pIC_{50} and pKi (− logKi) was delineated and also found commonalities in the activity shown by the seven families with respect to structural disparities using an approach called similarity–descriptor coupling. The study has been translated into a user-friendly webtool to predict a pan or specific inhibitor for Bcl-2 and Bcl-xL targets (Kanakaveti *et al.*, 2017).

This second example illustrates how pharmacokinetic properties of drug molecules can be predicted using QSAR modeling. Penetration through blood brain barrier (BBB) is one of the most important features that reflect the drug-likeliness of small molecular compounds. Drugs designed for central nervous system should be able to cross this barrier however penetration of BBB by peripherally acting drugs should be minimize to avoid side effects. So estimation of pharmacokinetic properties of candidate compounds *in silico* before testing them experimentally can save enormous time and money. In this study, a set of 529 organic compounds was used to correlate their chemical structures and distribution coefficients between the brain and blood using artificial neural networks. The molecular structures were defined in terms of occurrence of number of various types of fragments containing up to 10 atoms. Out of all the fragment descriptors, the most important ones were selected using the stepwise multiple linear regression. It was seen that the best model was based on the fragments comprising up to nine atoms. The model was further validated using a dataset of 2053 compounds, which was categorized as BBB+ (penetrating) and BBB- (non-penetrating). The constructed model could correctly classify 90% of BBB+ compounds from a test set, however the prediction specificity for BBB- category was low. Some of the features that have the strongest positive effects on the LogBB included hydrophobic fragments like alkyl or aromatic and presence of hydroxyl group in α- position to the carbonyl group. On the other hand, it was found that permeability decreases if the structure contains strongly polar groups like hydroxyl, carboxyl, and guanidine. This model has been integrated into a web service for predicting ADMET parameters of drugs developed in the Laboratory of Medicinal Chemistry, Department of Chemistry, Lomonosov Moscow State University (Dyabina *et al.*, 2016).

With increasing industrialization, degradation of pollutants has become a crucial need to keep the environment clean. Benzene and its derivatives are the most common chemical structure found in the nature. It is also a structural part of degradation intermediates of complex pollutants like pesticides, pharmaceuticals, surfactants or synthetic dyes. However, the chemical properties deciding their environmental fate and behavior depends on the type, number and position of functional groups present at substitution sites. Here QSAR proved to be a fast and reliable approach for testing the degradation of aromatic compounds. Thirty six congeneric single-benzene ring compounds with known biodegradability were divided into training and test sets (24 and 12 compounds, respectively). The semi-empirical quantum-chemical descriptors like dipole moment, energy of the highest occupied molecular orbital (EHOMO), energy of the lowest unoccupied molecular orbital (ELUMO), energetic difference between EHOMO and ELUMO, final heat of formation and ionization potential, and various other molecular descriptors were calculated using various software. The correlation between descriptors and biodegradability prior and at half-life was obtained using variable selection Genetic Algorithm and Multiple Linear Regression Analysis methods. The validation of best-selected models based on statistical parameters was performed using Leave Many Out and "Y-scrambling" tests. The generated models were thus used to study the key structural features that influence the biodegradability of the compound of interest and correlate to its degradation mechanisms by UV-C/H_2O_2. Molecular mass, number of C-C bonds determining the rate of saturation of benzene ring making it susceptible to cleavage into more readily degradable aliphatic compounds, electron donating/withdrawing groups, symmetry of the molecule, presence of sulfo-group, ionization potential and electrotopological states were some of the important descriptors related to the biodegradability of aromatic compounds in water. The potential of such models can be extended to more extensive purposes like risk assessment studies (Cvetnic *et al.*, 2017).

In the following example, structural factors of ionic liquids (ILs) have been correlated with the process of micelization. Critical Micelization Concentration (CMC) of any ionic liquid depends upon the types of ions it possess. Micelization affects the synthesis, purification and regeneration routes of these ionic liquids and thus is an important feature to be considered. In this study, an attempt was made to derive a qualitative relationship between structural features of ions and their effect on micelization of ILs. It was also verified whether the micelization process is governed by the constituent ions separately or they have additive effects. Literature was explored to collect experimental data of ILs with their CMC. A dataset of 59 structurally diversified IL's with the CMC ranging between the 0.098 and 902 mM was prepared. 42 compounds were used to train the model while rests were used for testing. Various molecular descriptors were generated using DRAGON software and the most appropriate ones were chosen using genetic algorithm implemented in QSARINS software. The correlation between the descriptors and CMCs was derived using Multiple Linear Regression (MLR) technique. Various statistical parameters like determination coefficient, Concordance Correlation Coefficient, Root Mean Square Error, Mean Average Error and F-value were used to evaluate the fitting of model and its significance. The leave-one-out and Y-scrambling approach were used for internal validation and investigating the robustness of the generated model. An altogether new test data set was used for external validation. Decrease in CMC was attributed to less spherical, improperly folded cations containing larger hydrophobic domain. For anions, bigger size was associated with decrease in CMC. Also the effect of cations and anions in determining the CMC was found to be independent of each other (Barycki *et al.*, 2017).

People today have become more careful about their eating habits to adopt a healthy life style. They avoid overconsumption of high-calorific food to reduce the risk of various metabolic disorders, cardiovascular diseases, obesity and diabetes. Sugars or saccharides are major contributors in this. Industry is focusing on finding out new compounds, natural or synthetic, with low calories but high sweetness. There are various QSAR models in the literature that help in extracting various structural features responsible for imparting the sweetness to the compounds, or distinguishing between sweet and non-sweet compounds. A study was carried out for virtually screening known natural compounds using QSAR modeling for identification of new sweeteners (Chéron *et al.*, 2017). A database of 316 compounds belonging to seventeen chemical families and sweetness values ranging from 0.20 to 225,000 was created. The sweetness index for all the compounds was relative to sucrose. The protonation state of compounds was adjusted according to the pH value of saliva, i.e., 6.5. Descriptors were calculated for both 2D and 3D structures of compounds using Dragon software. Random Forest and Support Vector Regression, machine learning algorithms were used to generate the models. Training set consisted of 225 molecules while the test set had 91 molecules. Leave-one-out cross-validation methods were used for both 2D and 3D QSAR models. Additional filters to avoid undesirable properties like bitterness and toxicity were applied. It was found that majority of the identified natural sweeteners were from terpene family, less than 200 molecules belonged to the category of saccharides, polyphenols or phenylpropanoids. The most potent natural sweeteners were based on saponin and stevioside scaffolds, with 1000–10,000 times more sweetness than sucrose (Chéron *et al.*, 2017).

See also: Biomolecular Structures: Prediction, Identification and Analyses. Chemoinformatics: From Chemical Art to Chemistry in Silico. Cross-Validation. Data Mining: Clustering. Drug Repurposing and Multi-Target Therapies. In Silico Identification of Novel Inhibitors. Introduction of Docking-Based Virtual Screening Workflow Using Desktop Personal Computer. Kernel Methods: Support Vector Machines. Measurements of Accuracy in Biostatistics. Molecular Dynamics and Simulation. Molecular Dynamics Simulations in Drug Discovery. Natural Language Processing Approaches in Bioinformatics. Parametric and Multivariate Methods. Protein Properties. Public Chemical Databases. Rational Structure-Based Drug Design. Regression Analysis. Small Molecule Drug Design. Structure-Based Design of Peptide Inhibitors for Protein Arginine Deiminase Type IV (PAD4). Structure-Based Drug Design Workflow

---

References

Achary, P., 2014. QSPR modelling of dielectric constants of π-conjugated organic compounds by means of the CORAL software. SAR and QSAR in Environmental Research 25, 507–526.

Agresti, A., 2007. An Introduction to Categorical Data Analysis. John Wiley.

Andrews, C.W., Bennett, L., Lawrence, X.Y., 2000. Predicting human oral bioavailability of a compound: Development of a novel quantitative structure-bioavailability relationship. Pharmaceutical Research 17, 639–644.

Balaban, A.T., 1982. Highly discriminating distance-based topological index. Chemical Physics Letters 89, 399–404.

Barber, C.E., Marshall, D.A., Mosher, D.P., *et al.*, 2016. Development of system-level performance measures for evaluation of models of care for inflammatory arthritis in Canada. The Journal of Rheumatology. 150839. [jrheum].

Barycki, M., Sosnowska, A., Puzyn, T., 2017. Which structural features stand behind micelization of ionic liquids? Quantitative structure-property relationship studies. Journal of Colloid and Interface Science 487, 475–483.

Berinde, Z., 2013. A QSPR study of hydrophobicity of phenols and 2-(aryloxy-α-acetyl)-phenoxathiin derivatives using the topological index ZEP. Creative Mathematics and Informatics 22, 33–40.

Bi, J., Bennett, K., Embrechts, M., Breneman, C., Song, M., 2003. Dimensionality reduction via sparse support vector machines. Journal of Machine Learning Research 3, 1229–1243.

Boik, J.C., Newman, R.A., 2008. Structure-activity models of oral clearance, cytotoxicity, and LD50: A screen for promising anticancer compounds. BMC Pharmacology 8, 12.

Braga, R.C., Andrade, C.H., 2012. QSAR and QM/MM approaches applied to drug metabolism prediction. Mini Reviews in Medicinal Chemistry 12, 573–582.

Buolamwini, J.K., Assefa, H., 2002. CoMFA and CoMSIA 3D QSAR and docking studies on conformationally-restrained cinnamoyl HIV-1 integrase inhibitors: Exploration of a binding mode at the active site. Journal of Medicinal Chemistry 45, 841–852.

Burbidge, R., Trotter, M., Buxton, B., Holden, S., 2001. Drug design by machine learning: Support vector machines for pharmaceutical data analysis. Computers & Chemistry 26, 5–14.

Can, A., 2014. Quantitative structure–toxicity relationship (QSTR) studies on the organophosphate insecticides. Toxicology Letters 230, 434–443.

Chéron, J.-B., Casciuc, I., Golebiowski, J., Antonczak, S., Fiorucci, S., 2017. Sweetness prediction of natural compounds. Food Chemistry 221, 1421–1425.

Chirico, N., Gramatica, P., 2011. Real external predictivity of QSAR models: How to evaluate it? Comparison of different validation criteria and proposal of using the concordance correlation coefficient. Journal of Chemical Information and Modeling 51, 2320–2335.

Cocu, A., Dumitriu, L., Craciun, M., Segal, C., 2008. A Hybrid Approach for Data Preprocessing in the QSAR Problem. Knowledge-Based Intelligent Information and Engineering Systems. Springer. pp. 565–572.

Cramer, R.D., 1993. Partial least squares (PLS): Its strengths and limitations. Perspectives in Drug Discovery and Design 1, 269–278.

Crum-Brown, A., Fraser, T., 1868. On the connection between chemical constitution and physiological action. Part 1. On the physiological action of the ammonium bases, derived from Strychia, Brucia, Thebaia, Codeia, Morphia and Nicotia. Transactions of the Royal Society of Edinburgh 25, 151–203.

Cvetnic, M., Perisic, D.J., Kovacic, M., *et al.*, 2017. Prediction of biodegradability of aromatics in water using QSAR modeling. Ecotoxicology and Environmental Safety 139, 139–149.

Dai, Y.-M., Zhu, Z.-P., Cao, Z., *et al.*, 2013. Prediction of boiling points of organic compounds by QSPR tools. Journal of Molecular Graphics and Modelling 44, 113–119.

Damale, M.G., Harke, S.N., Kalam Khan, F.A., Shinde, D.B., Sangshetti, J.N., 2014. Recent advances in multidimensional QSAR (4D-6D): A critical review. Mini Reviews in Medicinal Chemistry 14, 35–55.

Dastmalchi, S., Hamzeh-Mivehroud, M., Asadpour-Zeynali, K., 2012. Comparison of different 2D and 3D-QSAR methods on activity prediction of histamine H3 receptor antagonists. Iranian Journal of Pharmaceutical Research: IJPR 11, 97.

Demel, M.A., Janecek, A.G., Gansterer, W.N., Ecker, G.F., 2009. Comparison of contemporary feature selection algorithms: Application to the classification of ABC-transporter substrates. Molecular Informatics 28, 1087–1091.

Dioury, F., Duprat, A., Dreyfus, G.R., Ferroud, C., Cossy, J., 2014. QSPR prediction of the stability constants of gadolinium (III) complexes for magnetic resonance imaging. Journal of Chemical Information and Modeling 54, 2718–2731.

Du, H., Hu, Z., Bazzoli, A., Zhang, Y., 2011. Prediction of inhibitory activity of epidermal growth factor receptor inhibitors using grid search-projection pursuit regression method. PLOS ONE 6, e22367.

Duchowicz, P.R., Castro, E.A., 2009. QSPR studies on aqueous solubilities of drug-like compounds. International Journal of Molecular Sciences 10, 2558–2577.

Duchowicz, P.R., Castro, E.A., Fernandez, F., Pankratov, A., 2006. QSPR evaluation of thermodynamic properties of acyclic and aromatic compounds. Anales de la Asociación Química Argentina. SciELO Argentina. 31–45.

Dyabina, A., Radchenko, E., Palyulin, V., Zefirov, N., 2016. Prediction of blood-brain barrier permeability of organic compounds. Doklady Biochemistry and Biophysics. Springer. pp. 371–374.

Escher, B.I., Bramaz, N., Richter, M., Lienert, J., 2006. Comparative ecotoxicological hazard assessment of beta-blockers and their human metabolites using a mode-of-action-based test battery and a QSAR approach. Environmental Science & Technology 40, 7402–7408.

Ferguson, J., 1939. The use of chemical potentials as indices of toxicity. Proceedings of the Royal Society of London Series B, Biological Sciences 127, 387–404.

Fjodorova, N., Vračko, M., Novič, M., Roncaglioni, A., Benfenati, E., 2010. New public QSAR model for carcinogenicity. Chemistry Central Journal. Springer. p. S3.

Free, S.M., Wilson, J.W., 1964. A mathematical contribution to structure-activity studies. Journal of Medicinal Chemistry 7, 395–399.

Fujikawa, M., Nakao, K., Shimizu, R., Akamatsu, M., 2007. QSAR study on permeability of hydrophobic compounds with artificial membranes. Bioorganic & Medicinal Chemistry 15, 3756–3767.

Fujrra, T., Ban, T., 1971. Structure-activity study of phenethyiamines as substrates of biosynthetic enzymes of sympathetic enzymes of sympathetic transmitters. ZMed. CRem 14, 148–152.

Fu, W., Zhang, Y., Cheng, Z., 2010. Improved gene expression programming and its application to QSAR. In: Proceedings of the Sixth International Conference on Natural Computation (ICNC), IEEE, 4057–4061.

Gao, H., Shanmugasundaram, V., Lee, P., 2002. Estimation of aqueous solubility of organic compounds with QSPR approach. Pharmaceutical Research 19, 497–503.

Garkani-Nejad, Z., Ghanbari, A., 2016. Application of support vector machine in QSAR study of triazolyl thiophenes as cyclin dependent kinase-5 inhibitors for their anti-alzheimer activity.

Ghasemi, J.B., Ahmadi, S., Ayati, M., 2010. QSPR modeling of stability constants of the Li-hemispherands complexes using MLR: A theoretical host-guest study. Macroheterocycles 3, 234–242.

Ghose, A.K., Viswanadhan, V.N., Wendoloski, J.J., 1999. A knowledge-based approach in designing combinatorial or medicinal chemistry libraries for drug discovery. 1. A qualitative and quantitative characterization of known drug databases. Journal of Combinatorial Chemistry 1, 55–68.

Golbraikh, A., Tropsha, A., 2002. Beware of q 2!. Journal of Molecular Graphics and Modelling 20, 269–276.

Gombar, V.K., Hall, S.D., 2013. Quantitative structure–activity relationship models of clinical pharmacokinetics: Clearance and volume of distribution. Journal of Chemical Information and Modeling 53, 948–957.

Goodarzi, M., Dejaegher, B., Heyden, Y.V., 2012. Feature selection methods in QSAR studies. Journal of AOAC International 95, 636–651.

Goryński, K., Bojko, B., Nowaczyk, A., et al., 2013. Quantitative structure–retention relationships models for prediction of high performance liquid chromatography retention time of small molecules: Endogenous metabolites and banned compounds. Analytica Chimica Acta 797, 13–19.

Gozalbes, R., Jacewicz, M., Annand, R., Tsaioun, K., Pineda-Lucena, A., 2011. QSAR-based permeability model for drug-like compounds. Bioorganic & Medicinal Chemistry 19, 2615–2624.

Grisoni, F., Consonni, V., Vighi, M., Villa, S., Todeschini, R., 2016. Investigating the mechanisms of bioconcentration through QSAR classification trees. Environment International 88, 198–205.

Guha, R., Dutta, D., Jurs, P.C., Chen, T., 2006. Local lazy regression: Making use of the neighborhood to improve QSAR predictions. Journal of Chemical Information and Modeling 46, 1836–1847.

Gutman, I., Trinajstić, N., 1972. Graph theory and molecular orbitals. Total (φ)-electron energy of alternant hydrocarbons. Chemical Physics Letters 17, 535–538.

Guyon, I., Weston, J., Barnhill, S., Vapnik, V., 2002. Gene selection for cancer classification using support vector machines. Machine Learning 46, 389–422.

Hammett, L.P., 1937. The effect of structure upon the reactions of organic compounds. Benzene derivatives. Journal of the American Chemical Society 59, 96–103.

Hansch, C., Fujita, T., 1964. p-σ-π Analysis. A method for the correlation of biological activity and chemical structure. Journal of the American Chemical Society 86, 1616–1626.

Harrell, F., 2001. Regression modeling strategies. 2001. Nashville: Springer CrossRef Google Scholar.

Hawkins, D.M., Basak, S.C., Mills, D., 2003. Assessing model fit by cross-validation. Journal of Chemical Information and Computer Sciences 43, 579–586.

Hemmateenejad, B., Sanchooli, M., Mehdipour, A., 2009. Quantitative structure–reactivity relationship studies on the catalyzed Michael addition reactions. Journal of Physical Organic Chemistry 22, 613–618.

Hermens, J., Canton, H., Janssen, P., De Jong, R., 1984. Quantitative structure-activity relationships and toxicity studies of mixtures of chemicals with anaesthetic potency: Acute lethal and sublethal toxicity to Daphnia magna. Aquatic Toxicology 5, 143–154.

Hopfinger, A., Wang, S., Tokarski, J.S., et al., 1997. Construction of 3D-QSAR models using the 4D-QSAR analysis formalism. Journal of the American Chemical Society 119, 10509–10524.

Ivanciuc, O., 2000. QSAR comparative study of Wiener descriptors for weighted molecular graphs. Journal of Chemical Information and Computer Sciences 40, 1412–1422.

Jaworska, J., Nikolova-Jeliazkova, N., Aldenberg, T., 2005. QSAR applicability domain estimation by projection of the training set descriptor space: A review. ATLA-NOTTINGHAM- 33, 445.

Johnels, D., Gillner, M., Nordén, B., Toftgård, R., Gustafsson, J.Å., 1989. Quantitative structure-activity relationship (QSAR) analysis using the partial least squares (PLS) method: The binding of polycyclic aromatic hydrocarbons (PAH) to the rat liver 2, 3, 7, 8–tetrachlorodibenzo-P-dioxin (TCDD) receptor. Molecular Informatics 8, 83–89.

Kanakaveti, V., Sakthivel, R., Rayala, S., Gromiha, M.M., 2017. Importance of functional groups in predicting the activity of small molecule inhibitors for Bcl-2 and Bcl-xL. Chemical Biology & Drug Design.

Kar, S., Roy, K., 2011. Development and validation of a robust QSAR model for prediction of carcinogenicity of drugs. Indian Journal of Biochemistry & Biophysics 48 (2),

Karacan, M.S., Yakan, Ç., Yakan, M., et al., 2012. Quantitative structure–activity relationship analysis of perfluoroiso-propyldinitrobenzene derivatives known as photosystem II electron transfer inhibitors. Biochimica et Biophysica Acta (BBA)-Bioenergetics 1817, 1229–1236.

Katritzky, A.R., Lomaka, A., Petrukhin, R., et al., 2002. QSPR correlation of the melting point for pyridinium bromides, potential ionic liquids. Journal of Chemical Information and Computer Sciences 42, 71–74.

Khorshidi, N., Sarkhosh, M., Niazi, A., 2014. QSPR study of maximum absorption wavelength of various flavones by multivariate image analysis and principal components-least squares support vector machine. Journal of Scientific and Innovative Research 3, 189–202.

Kier, L.B., 1985. A shape index from molecular graphs. Molecular Informatics 4, 109–116.

Kier, L.B., Hall, L.H., Murray, W.J., Randi, M., 1975. Molecular connectivity I: Relationship to nonspecific local anesthesia. Journal of Pharmaceutical Sciences 64, 1971–1974.

Kumar, A., Chauhan, S., 2017. Monte Carlo method based QSAR modelling of natural lipase inhibitors using hybrid optimal descriptors. SAR and QSAR in Environmental Research 28, 179–197.

Lei, B., Ma, Y., Li, J., et al., 2010. Prediction of the adsorption capability onto activated carbon of a large data set of chemicals by local lazy regression method. Atmospheric Environment 44, 2954–2960.

Levatic, J., Dzeroski, S., Supek, F., Smuc, T., 2013. Semi-supervised learning for quantitative structure-activity modeling. Informatica 37, 173.

Lewis, D.F., 2000. Structural characteristics of human P450s involved in drug metabolism: Qsars and lipophilicity profiles. Toxicology 144, 197–203.

Li, X., Shi, Y., 2008. A survey on the Randic index. MATCH Communications in Mathematical and in Computer Chemistry 59, 127–156.

Liu, P., Long, W., 2009. Current mathematical methods used in QSAR/QSPR studies. International Journal of Molecular Sciences 10, 1978–1998.

Liu, S.-S., Liu, H.-L., Yin, C.-S., Wang, L.-S., 2003. VSMP: A novel variable selection and modeling method based on the prediction. Journal of Chemical Information and Computer Sciences 43, 964–969.

Liu, Y., 2004. A comparative study on feature selection methods for drug discovery. Journal of Chemical Information and Computer Sciences 44, 1823–1828.

Lorber, A., Wangen, L.E., Kowalski, B.R., 1987. A theoretical foundation for the PLS algorithm. Journal of Chemometrics 1, 19–31.

Luo, J., Hu, J., Fu, L., Liu, C., Jin, X., 2011. Use of artificial neural network for a QSAR study on neurotrophic activities of Np-tolyl/phenylsulfonyl L-amino acid thiolester derivatives. Procedia Engineering 15, 5158–5163.

Magdziarz, T., Mazur, P., Polanski, J., 2009. Receptor independent and receptor dependent CoMSA modeling with IVE-PLS: Application to CBG benchmark steroids and reductase activators. Journal of Molecular Modeling 15, 41–51.

Manga, N.N., Duffy, J.C., Rowe, P.H., Cronin, M.T., 2003. A hierarchical QSAR model for urinary excretion of drugs in humans as a predictive tool for biotransformation. Molecular Informatics 22, 263–273.

Mayer, J.M., Van De Waterbeemd, H., 1985. Development of quantitative structure-pharmacokinetic relationships. Environmental Health Perspectives 61, 295.

Meyer, H., 1899. Zur theorie der alkoholnarkose. Naunyn-Schmiedeberg's Archives of Pharmacology 42, 109–118.

Mirkhani, S.A., Gharagheizi, F., Sattari, M., 2012. A QSPR model for prediction of diffusion coefficient of non-electrolyte organic compounds in air at ambient condition. Chemosphere 86, 959–966.

Modarresi, H., Dearden, J.C., Modarress, H., 2006. QSPR correlation of melting point for drug compounds based on different sources of molecular descriptors. Journal of Chemical Information and Modeling 46, 930–936.

Montero-Torres, A., García-Sánchez, R.N., Marrero-Ponce, Y., et al., 2006. Non-stochastic quadratic fingerprints and LDA-based QSAR models in hit and lead generation through virtual screening: Theoretical and experimental assessment of a promising method for the discovery of new antimalarial compounds. European Journal of Medicinal Chemistry 41, 483–493.

Moss, G.P., Dearden, J.C., Patel, H., Cronin, M.T., 2002. Quantitative structure–permeability relationships (QSPRs) for percutaneous absorption. Toxicology in Vitro 16, 299–317.

Mota, S.G., Barros, T.F., Castilho, M.S., 2009. 2D QSAR studies on a series of bifonazole derivatives with antifungal activity. Journal of the Brazilian Chemical Society 20, 451–459.

Papa, E., Dearden, J., Gramatica, P., 2007. Linear QSAR regression models for the prediction of bioconcentration factors by physicochemical properties and structural theoretical molecular descriptors. Chemosphere 67, 351–358.

Polanski, J., 2009. Receptor dependent multidimensional QSAR for modeling drug-receptor interactions. Current Medicinal Chemistry 16, 3243–3257.

Polishchuk, P.G., Muratov, E.N., Artemenko, A.G., et al., 2009. Application of random forest approach to QSAR prediction of aquatic toxicity. Journal of Chemical Information and Modeling 49, 2481–2488.

Puri, S., Chickos, J.S., Welsh, W.J., 2002. Three-dimensional quantitative structure–property relationship (3d-qspr) models for prediction of thermodynamic properties of polychlorinated biphenyls (PCBs): Enthalpy of sublimation. Journal of Chemical Information and Computer Sciences 42, 109–116.

Qiao, L.S., Cai, Y.-L., He, Y.-S., et al., 2014. Trend of multi-scale QSAR in drug design. Asian Journal of Chemistry 26, 5917.

Randić, M., 2001. The connectivity index 25 years after. Journal of Molecular Graphics and Modelling 20, 19–35.

Richardson, B., 1869. Physiological research on alcohols. The Medical Times and Gazett. 703–706.

Richet, M., 1893. Note sur le rapport entre la toxicité et les propriétes physiques des corps. Compt Rend Soc Biol ((Paris)) 45, 775–776.

Rochani, A.K., Suma, B., Kumar, S., Jays, J., Madhavan, V., 2010. QSAR, ADME AND QSTR Studies of Some Synthesized Anti-Cancer 2-Indolinone Derivatives. International Journal of Pharma and Bio Sciences 1, 208–218.

Roy, K., Kar, S., 2015. How to judge predictive quality of classification and regression based QSAR models. In: Haq, Z.U., Madura, J. (Eds.), Frontiers of Computational Chemistry. Sharjah, UAE: Bentham Science, pp. 71–120.

Roy, K., Kar, S., Das, R.N., 2015a. A Primer on QSAR/QSPR Modeling: Fundamental Concepts. Springer.

Roy, K., Kar, S., Das, R.N., 2015b. Statistical methods in QSAR/QSPR. A Primer on QSAR/QSPR Modeling. 37–59.

Roy, K., Kar, S., Das, R.N., 2015c. Understanding the Basics of QSAR for Applications in Pharmaceutical Sciences and Risk Assessment. Academic Press.

Roy, K., Mandal, A.S., 2008. Development of linear and nonlinear predictive QSAR models and their external validation using molecular similarity principle for anti-HIV indolyl aryl sulfones. Journal of Enzyme Inhibition and Medicinal Chemistry 23, 980–995.

Roy, K., Mitra, I., 2011. On various metrics used for validation of predictive QSAR models with applications in virtual screening and focused library design. Combinatorial Chemistry & High Throughput Screening 14, 450–474.

Roy, K., Roy, P., Leonard, J., 2008. On some aspects of validation of predictive QSAR models. Chemistry Central Journal 2, P9.

SchüÜRmann, G., Ebert, R.-U., Chen, J., Wang, B., KüHne, R., 2008. External validation and prediction employing the predictive squared correlation coefficient test set activity mean vs training set activity mean. Journal of Chemical Information and Modeling 48, 2140–2145.

Settles, B., 2012. Active learning: Synthesis lectures on artificial intelligence and machine learning. Long Island, NY: Morgan & Clay Pool.

Shen, M., Béguin, C., Golbraikh, A., et al., 2004. Application of predictive QSAR models to database mining: Identification and experimental validation of novel anticonvulsant compounds. Journal of Medicinal Chemistry 47, 2356–2364.

Shi, W., Zhang, X., Shen, Q., 2010. Quantitative structure-activity relationships studies of CCR5 inhibitors and toxicity of aromatic compounds using gene expression programming. European Journal of Medicinal Chemistry 45, 49–54.

Si, H., Yuan, S., Zhang, K., et al., 2008. Quantitative structure activity relationship study on EC50 of anti-HIV drugs. Chemometrics and Intelligent Laboratory Systems 90, 15–24.

Si, H.Z., Wang, T., Zhang, K.J., De Hu, Z., Fan, B.T., 2006. QSAR study of 1, 4-dihydropyridine calcium channel antagonists based on gene expression programming. Bioorganic & Medicinal Chemistry 14, 4834–4841.

Simeon, S., Anuwongcharoen, N., Shoombuatong, W., et al., 2016. Probing the origins of human acetylcholinesterase inhibition via QSAR modeling and molecular docking. PeerJ 4, e2322.

Singh, K., Verma, N., 2014. 3-Dimensional QSAR and Molecular Docking Studies of a Series of Indole Analogues as Inhibitors of Human Non-Pancreatic Secretory Phospholipase. A2.

Sola, D., Ferri, A., Banchero, M., Manna, L., Sicardi, S., 2008. QSPR prediction of N-boiling point and critical properties of organic compounds and comparison with a group-contribution method. Fluid Phase Equilibria 263, 33–42.

Soltanpour, S., Shahbazy, M., Omidikia, N., Kompany-Zareh, M., Baharifard, M.T., 2016. A comprehensive QSPR model for dielectric constants of binary solvent mixtures. SAR and QSAR in Environmental Research 27, 165–181.

Speck-Planche, A., Kleandrova, V.V., Luan, F., Cordeiro, M.N.D., 2012. Rational drug design for anti-cancer chemotherapy: Multi-target QSAR models for the in silico discovery of anti-colorectal cancer agents. Bioorganic & Medicinal Chemistry 20, 4848–4855.

Suzuki, T., Ide, K., Ishida, M., Shapiro, S., 2001. Classification of environmental estrogens by physicochemical properties using principal component analysis and hierarchical cluster analysis. Journal of Chemical Information and Computer Sciences 41, 718–726.

Taft Jr, R.W., 1952. Polar and steric substituent constants for aliphatic and o-Benzoate groups from rates of esterification and hydrolysis of esters1. Journal of the American Chemical Society 74, 3120–3128.

Taft Jr, R.W., 1953a. The general nature of the proportionality of polar effects of substituent groups in organic chemistry. Journal of the American Chemical Society 75, 4231–4238.

Taft Jr, R.W., 1953b. Linear steric energy relationships. Journal of the American Chemical Society 75, 4538–4539.

Tang, H., Wang, X.S., Huang, X.-P., et al., 2009. Novel inhibitors of human histone deacetylase (HDAC) identified by QSAR modeling of known inhibitors, virtual screening, and experimental validation. Journal of Chemical Information and Modeling 49, 461–476.

Tomar, D., Agarwal, S., 2014. A survey on pre-processing and post-processing techniques in data mining. International Journal of Database Theory and Application 7, 99–128.

Toropov, A., Kudyshkin, V., Voropaeva, N., Ruban, I., Rashidova, S.S., 2004. QSPR modeling of the reactivity parameters of monomers in radical copolymerizations. Journal of Structural Chemistry 45, 945–950.

Valencia, A., Prous, J., Mora, O., Sadrieh, N., Valerio, L.G., 2013. A novel QSAR model of Salmonella mutagenicity and its application in the safety assessment of drug impurities. Toxicology and Applied Pharmacology 273, 427–434.

Van Gestel, C., Ma, W.-C., 1990. An approach to quantitative structure-activity relationships (QSARs) in earthworm toxicity studies. Chemosphere 21, 1023–1033.

Vedani, A., Dobler, M., 2002. 5D-QSAR: The key for simulating induced fit? Journal of Medicinal Chemistry 45, 2139–2149.

Veerasamy, R., Rajak, H., Jain, A., et al., 2011. Validation of QSAR models-strategies and importance. International Journal of Drug Design & Discovery 3, 511–519.

Veyseh, S., Hamzehali, H., Niazi, A., Ghasemi, J.B., 2015. Application of multivariate image analysis in qspr study of pKa of various acids by principal components-least squares support vector machine. Journal of the Chilean Chemical Society 60, 2985–2987.

Vieira, J.B., Braga, F.S., Lobato, C.C., et al., 2014. A QSAR, pharmacokinetic and toxicological study of new artemisinin compounds with anticancer activity. Molecules 19, 10670–10697.

Wang, D.-D., Feng, L.-L., He, G.-Y., Chen, H.-Q., 2014. QSAR studies for the acute toxicity of nitrobenzenes to the Tetrahymena pyriformis. Journal of the Serbian Chemical Society 79, 1111–1125.

Wang, T., Si, H., Chen, P., Zhang, K., Yao, X., 2008. QSAR models for the dermal penetration of polycyclic aromatic hydrocarbons based on Gene Expression Programming. Molecular Informatics 27, 913–921.

Wesley, L., Veerapaneni, S., Desai, R., et al., 2016. 3D-QSAR and SVM Prediction of BRAF-V600E and HIV integrase inhibitors: A comparative study and characterization of performance with a new expected prediction performance metric. American Journal of Biochemistry and Biotechnology 12, 253–262.

Wiener, H., 1947. Structural determination of paraffin boiling points. Journal of the American Chemical Society 69, 17–20.

Xue, Y., Li, Z.-R., Yap, C.W., et al., 2004. Effect of molecular descriptor feature selection in support vector machine classification of pharmacokinetic and toxicological properties of chemical agents. Journal of Chemical Information and Computer Sciences 44, 1630–1638.

Zernov, V.V., Balakin, K.V., Ivaschenko, A.A., Savchuk, N.P., Pletnev, I.V., 2003. Drug discovery using support vector machines. The case studies of drug-likeness, agrochemical-likeness, and enzyme inhibition predictions. Journal of Chemical Information and Computer Sciences 43, 2048–2056.

Zhang, L., Zhu, H., Oprea, T.I., Golbraikh, A., Tropsha, A., 2008. QSAR modeling of the blood–brain barrier permeability for diverse organic compounds. Pharmaceutical Research 25, 1902.

Zhang, S., Golbraikh, A., Tropsha, A., 2006. Development of quantitative structure–binding affinity relationship models based on novel geometrical chemical descriptors of the protein–ligand interfaces. Journal of Medicinal Chemistry 49, 2713–2724.

Zou, C., Zhou, L., 2007. QSAR study of oxazolidinone antibacterial agents using artificial neural networks. Molecular Simulation 33, 517–530.

Zou, J.-W., Huang, M., Huang, J.-X., Hu, G.-X., Jiang, Y.-J., 2016. Quantitative structure–hydrophobicity relationships of molecular fragments and beyond. Journal of Molecular Graphics and Modelling 64, 110–120.

Biographical Sketch

Swathik Clarancia Peter received her M.Tech. in Computational Biology from Pondicherry University, India in 2017 and B.Tech. in Bioinformatics from Tamil Nadu Agricultural University, Coimbatore, India in 2015. At present she is working as a senior research fellow in ICAR-Sugarcane Breeding Institute, Coimbatore, India. She works in the field of drug design & discovery and transcriptome data analysis.

Mannu Jayakanthan is an Assistant Professor in the Department of Plant Molecular Biology and Bioinformatics at Tamil Nadu Agricultural University, Coimbatore, India. He obtained his PhD from Pondicherry University and pursued his postdoctoral fellowship at the Centre for Cellular and Molecular Biology (CCMB) in Hyderabad, India. His research interest lies in computer-aided drug discovery.

Jaspreet Kaur received her M.Tech. in Bioinformatics from Delhi Technological University, India in 2013, and B. Tech. in Biotechnology from Amity University, India (2011). She joined Indian Institute of Technology, Delhi, India in 2014 for the Ph.D. program. She works in the field of computer aided drug designing and devising computational approaches for aiding in targeted genome editing.

Vidhi Malik received her M.Tech. in Bioinformatics from Delhi Technological University, India in 2013 and B. Tech. in Biotechnology form Sardar Vallabhbhai Patel University of Agriculture and Technology, India in 2011. She joined Indian Institute of Technology, Delhi, India in 2015 for the Ph.D. program. Her research interest lies in next generation sequence (NGS) data analysis and computer-aided drug designing.

Navaneethan Radhakrishnan received his B.Tech. in Bioinformatics from Tamil Nadu Agricultural University, India in 2016. He joined Indian Institute of Technology Delhi, India in 2016 as Junior Research Fellow in the Department of Biochemical Engineering and Biotechnology. His research interest lies in understanding the biological systems using computational approaches.

Durai Sundar is a DuPont Young Professor in the Department of Biochemical Engineering and Biotechnology at Indian Institute of Technology, Delhi. He obtained his education from Pondicherry University and Johns Hopkins University, Baltimore, United States. He is a specialist in molecular and computational biology and his current research interests are in rational design of genome editing tools and in the biological activity of natural drugs.

Pharmacophore Development

Balakumar Chandrasekaran and Nikhil Agrawal, University of KwaZulu-Natal, Durban, South Africa
Sandeep Kaushik, European Institute of Excellence on Tissue Engineering and Regenerative Medicine, Guimaraes, Portuga and University of Minho, Braga, Portugal

Introduction

Discovery of new drug molecules is an ever challenging task, even until today. It is a fact that the success rate in this area is extremely low. Typically, only one out of a million compounds finally obtain approval to be used as a drug. Starting from the process of target identification, a huge investment (approximately 2 billion dollars) and a long period (ranging 8–16 years) in research and development (R&D) activities are normally required to yield a new drug. After the post-marketing surveillance (phase 4 of the clinical trials), if the drug is found to have any adverse effects, it would be withdrawn from the market with a loss which amounts to about million dollars. Hence, the entire process starting from the drug discovery to design is a risky affair and within a short period of time, the pharmaceutical companies may need to take back the revenue out of their investment. Hence, there is an enormous pressure on pharmaceutical industries to come up with new drug molecules to treat existing diseases. Moreover, the environmental changes around the globe have heavily contributed to the evolution of drug-resistant strains; therefore, there is an urgent need for alternate/new more effective drugs.

Computer-aided drug design (CADD) tools have assisted medicinal chemists to a large extent by reducing the time, cost and number of compounds to be synthesized as leads. In this computational approach, the integrated aspects of chemistry (ligands) and biology (proteins/enzymes/receptors) are studied in order to search for new drug candidates. Currently, 20% of pharmaceutical R&D expenditure is allotted to CADD segment which helps face various challenges encountered during the R&D of new drugs. Further, the computational drug design is based on two major approaches namely, structure-based drug designing (SBDD) and ligand-based drug designing (LBDD). SBDD approach uses the structural information of the target proteins like enzymes or receptors, to identify compounds that can potentially be used as a drug (Kalva *et al.*, 2016; Agrawal *et al.*, 2016; Agrawal *et al.*, 2018). On the other hand, the ligand-based approach involves the development of 3D pharmacophore models and modeling quantitative structure-activity relationship (QSAR) or quantitative structure-property relationship (QSPR) (Balakumar *et al.*, 2017; Bheemanapalli *et al.*, 2013). This involves using the physicochemical properties of the ligand molecules for drug development. The 3D pharmacophore modeling is one of the useful drug design approaches in new drug discovery process.

Definition of a Pharmacophore: Initially, the term pharmacophore was used to refer to – "chemical groups" of a molecule that are responsible for a biological effect (Güner and Bowen, 2014). However, with time its meaning has shifted to patterns of "abstract features" that are responsible for a biological effect. The basis of modern day International Union of Pure and Applied Chemistry (IUPAC) definition of a pharmacophore is a redefinition of the above statements as given by Schueler (1960). The IUPAC definition is as follows:

> *"A pharmacophore is an ensemble of steric and electronic features that is necessary to ensure the optimal supramolecular interactions with a specific biological target structure and to trigger (or to block) its biological response."*

This is the modern accepted definition of a pharmacophore, presently (Güner, 2000; Wermuth *et al.*, 1998). Pharmacophore modeling has a wide range of applications in drug design and discovery such as 3D database search, scaffold hopping, virtual screening, ligand profiling, pose filtering, fragment designing and predicting the biological activities. Most commonly observed features in pharmacophore models are hydrogen bond donors, hydrogen bond acceptors, aromatic rings, hydrophobic sites, positively ionizable groups and negatively ionizable groups.

Historical Perspective of Pharmacophore Modeling

Paul Ehrlich, the father of chemotherapy, initially conceptualized the pharmacophores and that concept is still unperturbed for the past century (Ariens, 1966; Ehrlich, 1909, 1898). In simpler words, the bioactive molecules interact with their targets via a 3D arrangement of abstract features that define interaction types rather than specific functional groups. These interaction types could be the formation of hydrogen bonds, charged interactions, metal interactions, or hydrophobic and aromatic contacts (Qing *et al.*, 2014). Using this knowledge compounds with desired biological activities have been synthesized. The earliest success of a 2D pharmacophore model was demonstrated by synthesizing *trans*-diethylstilbestrol (an estrogenic agent) based on the pharmacophore similarities with estradiol even though the latter had a non-planar conformation (**Fig. 1**; Dodds and Lawson, 1938). Similarly, an anti-hypertensive drug called clonidine (**Fig. 2**) was found to interact with the central norepinephrine receptor through 3-point attachment (an ionic bond, hydrogen bond and stacking interaction) (Andén *et al.*, 1970). Though norepinephrine and clonidine had different chemical structures, they retained common features exhibiting functional similarities. Another important discovery in this regard was the determination of critical intramolecular distances from protonated nitrogen (N^+) to the aromatic ring (D=5.1–5.2 Å) and for the elevation of the positive charge to the aromatic plane (H=1.2–1.4 Å) (Pullman *et al.*, 1972).

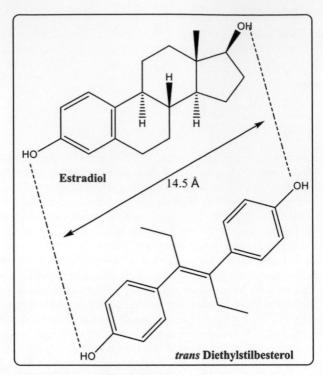

Fig. 1 Structural comparison between estradiol and *trans*-diethylstilbestrol.

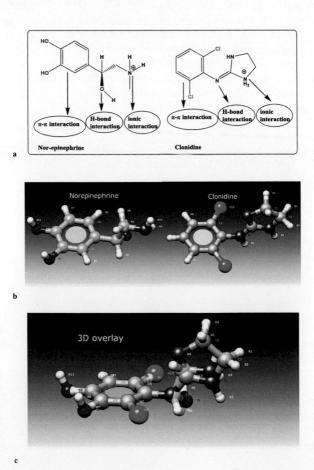

Fig. 2 (a) Clonidine presents the same type of molecular interactions as nor-epinephrine despite the difference in the orientation of the imidazolidinium ring towards the phenyl ring. (b) 3D structures of nor-epinephrine (*cyano*) and clonidine (*grey*) (c) 3D overlay of nor-epinephrine (*cyano*) and clonidine (*grey*).

Pharmacophore Modeling Approaches

Two types of pharmacophore modeling techniques are employed in drug design, namely, ligand-based pharmacophore modeling and structure-based pharmacophore modeling. This section describes the two different approaches used for pharmacophore modeling.

Ligand-Based Pharmacophore Modeling

If the target protein structure is not available, then the ligand-based pharmacophore modeling approach is employed for the design of new chemical entities (NCE). Two important steps of the ligand-based pharmacophore modeling are

1. Feature analysis in the training set molecules/ligands, and
2. Alignment of all bioactive conformations of the ligand to establish better overlay of features.

Currently, many automated software and tools (**Table 1**) for pharmacophore model generation are available for example, Hip-Hop and HypoGen (Catalyst software packaged as a part of BIOVIA Discovery Studio Dassault Systems, 2016), DISCO (Martin *et al.*, 1993), GASP (Jones *et al.*, 1995), GALAHAD (Richmond *et al.*, 2006), PHASE (Dixon *et al.*, 2006), MOE (Molecular Operating Environment MOE, 2017) and several other free academic programs (Bandyopadhyay and Agrafiotis, 2008; Baroni *et al.*, 2007; Nettles *et al.*, 2007). Each program differs in its algorithm that handles the generation and validation of pharmacophore models. There are several challenges and hurdles that one faces during pharmacophore development. For example, bioactive conformations of the compound may be unknown, then conformational flexibility remains one of the significant hurdles in developing a pharmacophore model. Another challenge is the molecular alignment of the ligand conformations in the training set.

Pharmacophore alignment: Molecular alignment can be classified into two types namely, point-based or property-based approaches (Wolber *et al.*, 2008). Atoms or chemical feature point distances are minimized in the point-based approaches whereas the property based approaches use molecular field descriptors to generate alignments. Point-based alignment algorithms have been used by a majority of the programs. The point-based method can be further categorized based on whether points represent atoms, fragments or chemical features. Each of the points can be superimposed using a least-squares fitting method. However, these assumptions become one of the greatest limitations for the all point-based methods while aligning dissimilar ligands.

On the other hand, molecular field descriptors or Gaussian functions are used for the property-based algorithms for the alignments. The stochastic proximity embedding, atomic property fields, fuzzy pattern recognition and grid-based interaction

Table 1 Some commercial tools and web servers for pharmacophore development

Software	*Brief discerption*	*Weblink*
GASP	• Uses a genetic algorithm • for pharmacophore identification. • GASP finds similarity between functional groups in different molecules and the alignment of these groups in a common geometry. • GASP is accessible through the SYBYL molecular modeling package.	https://www.certara.com/pressreleases/certara-enhances-sybyl-x-drug-design-and-discovery-software-suite/
CATALYST	• Identifies promising new molecular entities with or without target-structured data. • CTALAYST is a part of the BIOVIA Discovery Studio.	http://accelrys.com/products/collaborative-science/biovia-discovery-studio/pharmacophore-and-ligand-based-design.html
MOE pharmacophore modeling	• MOE can perform ligand, structure and fragment based pharmacophore modeling. • MOE Pharmacophore methods use a generalized molecular recognition representation and geometric constraints.	https://www.chemcomp.com/MOEPharmacophore_Discovery.htm
PHASE	• Phase uses a newly developed common pharmacophore perception algorithm. • Uses pharmacophore-based shape alignments to quickly create high-quality hypothesis from a handful to hundreds of known active ligands. • PHASE is accessible through Schrodinger molecular modeling package.	https://www.schrodinger.com/phase
PharmaGist	• PharmaGist is a freely available web server for ligand based pharmacophore identification. • PharmaGist server calculates candidate pharmacophores by multiple flexible alignments of the input ligands.	http://bioinfo3d.cs.tau.ac.il/pharma/about.html
PharmMapper	• PharmMapper freely available web-server designed to identify potential target candidates for the given probe small molecules. • PharmMapper uses semi-rigid pharmacophore mapping protocol.	http://lilab.ecust.edu.cn/pharmmapper/help.php
ZINCPharmer	• ZINCPharmer is free pharmacophore search software for screening the purchasable subset of the ZINC database. • Uses the Pharmer open source pharmacophore search technology to efficiently search a large database of fixed conformers for pharmacophore matches.	http://zincpharmer.csb.pitt.edu/

energies are the recently developed alignment methods (Bandyopadhyay and Agrafiotis, 2008; Baroni *et al.*, 2007; Nettles *et al.*, 2007; Totrov, 2008).

Appropriate selection of the training set is another crucial step in the generation of a valid pharmacophore model. When a different training set is employed for the pharmacophore generation using the same software program, it usually results in entirely different models (Hecker *et al.*, 2002; Toba *et al.*2006; Vadivelan *et al.*, 2007). For example, three different pharmacophore models were proposed for cyclin-dependent kinase 2 (CDK2) inhibitors bearing different features and location constraints using the same software program Catalyst (now a part of BIOVIA Discovery Studio Dassault Systems, 2016; Vadivelan *et al.*, 2007).

The ligand-based pharmacophore models can be developed in either qualitative (common feature-based hypothesis) or quantitative (3D-QSAR based models) manner. Below we describe each of the approach in brief.

Qualitative Models: The qualitative model is based on common feature pharmacophore generation from a set of bioactive ligands/training set (usually 2–16 ligands) of diversified structures. In the qualitative model, we assume similar binding modes for the training set with the same target protein and the 3D spatial arrangements of chemical features shared among the training set ligands is deduced. In this common feature-based hypothesis, explicit biological activity data is not required, partial matches are allowed with the preference of structural diversity among the ligands. Commonly used software like Hip-Hop, Phase, and MOE analyse ligands based on their chemical features and superimpose all the features to identify a specific conformation (Dassault Systems, 2016; Dixon *et al.*, 2006; Molecular Operating Environment MOE, 2017).

Quantitative Models: In case of the quantitative model, the biological activity data (*e.g.*, IC_{50} or K_i) are required while generating the predictive hypothesis to correlate chemical structure with biological activity thereby predicting the activity of novel compounds. The three important phases namely constructive, subtractive and optimization are involved in the generation of a quantitative pharmacophore model. Appropriate alignment is the key factor that decides the prediction power of a 3D quantitative model for virtual screening tasks. In this approach, a training dataset of 18–30 ligands depicting appropriate structural diversity with their pharmacological activity data are required (as represented by IC_{50}, EC_{50}, or K_i values of similar assay conditions). As the quality of a pharmacophore model and its predictability depends on the training data set, it is very crucial to have ligands with a wide range of biological activity in the training dataset. The activity could span over four orders of magnitude comprising categories of most active, moderately active, least active ligands and decoys. Commonly used software such as Hypogen, Phase-Macromodel and MOE are widely used for the generation of 3D-QSAR pharmacophore models (Dassault Systems, 2016; Dixon *et al.*, 2006; Molecular Operating Environment MOE, 2017). A 3D-conformational search is performed for each of the ligands within an energy range to yield multiple conformers using an appropriate algorithm. These algorithms consider coverage of the energy landscape and diversity among the conformational models of the ligands under study. Other algorithms implemented in commercial software programs are (Rubicon Daylight Chemical Information Systems, Inc.), (Omega, OpenEye Scientific Software) and stochastic proximity embedding (SPE) (Agrafiotis *et al.*, 2013).

Structure-Based Pharmacophore Modeling

The term structure in '*structure-based pharmacophore modeling*' refers to the 3D structure of a macromolecule which is usually a receptor or an enzyme with or without a bound ligand. This structure is determined through X-ray crystallography or nuclear magnetic resonance (NMR) spectroscopy techniques or may be constructed using homology modeling (Vyas *et al.*, 2012). If the protein structure is available as a potential drug target, then structure-based pharmacophore modeling will be preferred as this approach considers interaction points within the protein binding site. This method involves key aspects such as protein preparation, identification or prediction of ligand binding site, pharmacophore feature generation and selection of crucial features contributing to the pharmacological activity. In this approach, protein structure is the crucial source of information whether obtained from protein data bank (PDB) or modeled using a suitable template by homology modeling. The second most important step is the binding site prediction which is performed using different algorithms. To distinguish the binding site from other parts of the protein surface, most of these methods have exploited one or more of the following properties namely, evolutionary, geometric, energetic, statistical, and combined or knowledge-based methods (Xie and Hwang, 2015). After a careful analysis of the binding site, the programs identify the complementary chemical features of the binding-site amino acid residues and their spatial arrangements to derive a meaningful pharmacophore model. These programs initially generate interaction maps and convert them into several pharmacophoric features. This methodology yields several chemical features which must be optimized using customized tools implemented in the respective software programs. This task can be accomplished using approaches like clustering of features, removing certain unimportant features, comparing the features which complement the docked pose of ligands, and site-directed mutagenesis data. Finally, the key features responsible for contributing the desired pharmacological response will be carefully selected. For structure-based pharmacophore modeling, software programs like BIOVIA Discovery Studio (Dassault Systems, 2016), Sybyl (SYBYL7.3 Tripos Inc, 2003), MOE (Molecular Operating Environment MOE, 2017), Chemogenomics (Bredel and Jacoby, 2004), GBPM (Ortuso *et al.*, 2006), LigandScout (Wolber and Langer, 2005), Schrodinger-Maestro (Schrödinger, 2017), Pocket v2 (Chen and Lai, 2006), and Snooker (Sanders *et al.*, 2011) are commonly employed.

If the protein exists in its ligand-bound form, the key interaction points between these two can help derive at the pharmacophore model. To construct the structure-based pharmacophore model from the complex, the ligand is initially hybridized and the bonds are characterized. A widely employed program for this purpose is LigandScout (Wolber and Langer, 2005). It uses a heuristic approach along with the template-based numeric analysis of the protein-ligand complex which may have been obtained through PDB or the docking procedures. According to the protein-ligand binding information, a consensus model will be derived either by overlaying all the obtained models or by model refinement procedures.

Two major drawbacks associated with structure-based pharmacophore modeling are protein flexibility and conformational flexibility of the ligand. To overcome these drawbacks, two approaches can be used (1) deriving a model from the docked complex obtained through a flexible docking of ligands into the protein binding-site; (2) building and aligning the models simultaneously from the different snapshots of the protein or protein-ligand complexes of molecular dynamics (MD) simulations (Drwal and Griffith, 2013).

The energetically optimized structure-based pharmacophores (e-pharmacophores) can be used to screen millions of ligands as implemented in Schrodinger software (Therese *et al.*, 2014). These-pharmacophores remove the requirement of known active ligands while generating pharmacophore models and assist in identifying new regions of the active-site for drug design. In this protocol, the fragment docking (Glide XP Friesner *et al.*, 2004, 2006) data will be used to generate the hypotheses for each conformation using the e-pharmacophore script. This script maps the Glide XP energies for the top 2000 docked poses onto the ligands aligned to the receptor conformations followed by the clustering of the fragments. The software, Phase (Dixon *et al.*, 2006), will be used to generate the hypotheses from the docked poses with default features.

Role of Excluded Volumes in Pharmacophore Modeling

The so-derived pharmacophore model may possess the requisite features to bind to the target protein but yet fail to bind. One factor that may be responsible for this observation is the steric clashes between the ligand and the receptor. In such cases, the quality of the generated model can be improved by incorporating exclusion volume constraints. Ligand excluded surface represents an approximation of the regions accessible to the ligand under consideration. Receptor-based excluded volumes are represented as spheres centred on appropriate atoms in and around the binding site, with sizes dictated by the associated atomic van der Waals radii. If the ligand maps all the key features of the model, it is considered as active whereas if the functional groups of the ligand overlap the excluded volume regions, it will be inferred as pharmacologically inactive ligands. In some cases, this excluded volume constraint plays a crucial role in identifying potent ligands which match exclusively essential pharmacophoric features without occupying the exclusion volume (Schuster *et al.*, 2006).

A hard-sphere approximation is usually encountered while employing an exclusion volume in the model. In cases of inadequate sampling, this approximation may be deemed as too strict. Therefore, this approximation can be relaxed by customizing the excluded volume in the following ways

(1) Decreasing the size of the spheres,
(2) Removing spheres that are in close proximity to the bound ligand, and
(3) Tolerating a certain amount of overlap between the ligand and the excluded volumes.

In case of ligand-based pharmacophore modeling, when the receptor structure is unavailable, assigning meaningful excluded volumes is not so easy. However, this limitation may be overcome by inferring the excluded volumes from a set of known active ligands (Shrink-Wrap Method Van Drie, 1997) or inactive ligands (Murray *et al.*, 1999). Various software packages allow either manual or automated placement of excluded volumes based on a visual inspection of aligned structures and generate spheres at the desired locations.

The quality of the generated pharmacophore models can be assessed based on a scoring and ranking to determine how well the ligands map onto the pharmacophore model. A higher ranking indicates a better reliability and quality.

Validation of Pharmacophore Models

Once a pharmacophore model has been generated, it must be validated to ascertain its reliability and quality for successive applications in different molecular modeling tasks. To test whether or not our models are good enough to predict the active compounds, we perform pharmacophore validation. For the identification of a valid model, various validation parameters are analyzed through statistical analysis, the goodness of hit list (GH), receiver operating characteristic (ROC) curve, test set prediction, cost analysis (Hypogen), and Fischer's randomization test (Hypogen). This helps in differentiating the active ligand from inactive ligands for the specified drug targets. Below we discuss these analyses in brief.

Statistical Analysis

A regression analysis of the model can be performed against the test set ligands to verify a pharmacophore model. If the pharmacophore model demonstrates the correlation of $r^2 > 0.7$ between the predicted activity data and experimentally reported data, it is considered to be a valid model. To perform this task, a test set of ligands is constructed all of which act on the same target as that of training set ligand. The test set is classified as most active, moderately active, and less active based on the available range of the activity data to understand the predictability of the generated pharmacophore model.

Güner-Henry (GH) Method

The Güner-Henry method is a commonly employed validation protocol to score the hit lists and provide a quantitative "goodness of hit list" (GH) score. While using qualitative data, our aim is to identify and retrieve the maximum possible number of molecules that have the same activity as the target structure and to simultaneously minimize the number of inactive molecules. Güner and Henry introduced the GH score to evaluate the effectiveness of 3D database searches and can be applied to the evaluation of any sort of search for which qualitative bioactivity data are available (Willett, 2004). It helps in the retrieval of active ligands from a large pool of ligands that includes inactive and decoy ligands as well. For this analysis, a dataset is built using inactive ligands from the literature or from the database of decoys (DUD-E) tools (Mysinger et al., 2012). These decoys are mostly less active/inactive ligands that resemble ligands physically but are topologically dissimilar to minimize the likelihood of actual binding at the same time (Mysinger et al., 2012). The value of GH score approaching 1 indicates an ideal model (Guner et al., 2004). The following Eqs. 1–4 are used for calculation of GH score.

$$\text{Percent yield of actives } (\%\gamma) = \frac{\text{Ha}}{\text{Ht}} \times 100 \tag{1}$$

$$\text{Percent ratio of actives}(\%A) = \frac{\text{Ha}}{\text{A}} \times 100 \tag{2}$$

$$\text{Enrichment factor}(\text{EF}) = \frac{\text{Ha/Ht}}{\text{A/D}} \tag{3}$$

$$\text{Goodness of Hit} - \text{list (GH)} = \left(\frac{\text{HA}}{4\text{HtA}}\right) \times (3\text{A} + \text{Ht}) \times \left(1 - \frac{\text{Ht} - \text{Ha}}{\text{D} - \text{A}}\right) \tag{4}$$

Here, D is total number of molecules in a database, A is total number of active molecules in a database, Ht is total hits, Ha is active hits, GH is goodness of hit score.

Receiver Operating Characteristic (ROC) Curve

The ROC plot (specificity vs sensitivity) is used to determine the ability of the generated pharmacophore to distinguish between active and inactive ligands. This can be represented as an area under the curve (AUC) and other relevant parameters such as true positives (TP), true negatives (TN), false positives (FP), and false negatives (FN) (Triballeau et al., 2005). Ideally, AUC near to 1 demonstrates the perfection of the valid model. The formula to calculate the above-mentioned parameters are mentioned in the Eqs. 5–7.

$$\text{Sensitivity (Se)} = \frac{\text{TP}}{\text{TP} + \text{FN}} \tag{5}$$

$$\text{Specificity (Sp)} = \frac{\text{TN}}{\text{TN} + \text{FP}} \tag{6}$$

$$\text{Concordance} = \frac{\text{TP} + \text{TN}}{\text{TP} + \text{TN} + \text{FP} + \text{FN}} \tag{7}$$

Cost Analysis

The cost analysis method (as implemented in Hypogen, BIOVIA-DS program (Dassault Systems, 2016)) is used to analyse the statistical significance of the pharmacophore model in terms of fixed cost, null cost, and total cost. The fixed cost is the simplest model that fits all data while the null cost represents a high cost model with no features (Kandakatla and Ramakrishnan, 2014). The activity in the high cost model is determined as an average of the activity data from the training set compounds. For a valid pharmacophore model, ideally the total cost should be close to the fixed cost and the difference between the null and fixed cost value is large. A cost difference of 60 means, between 40 and 60 and below 40 means indicates a true correlation, 75%–90% and poor correlation (Kandakatla and Ramakrishnan, 2014). Similarly, configuration cost and error costs are also considered during the validation. The former refers to the complexity of the pharmacophore model space and the latter is represented by root-mean-square deviations (RMSDs) between the estimated and the experimental activities of the training set molecules. A valid pharmacophore model should comprise of good correlation, highest cost difference, and the lowest RMSD.

Fischer's Randomization Test

This test utilizes the Cat Scramble program of Catalyst/BIOVIA-DS (Dassault Systems, 2016) to create the random spreadsheets for the training set and reassign the activity values to each of the ligands to generate different pharmacophore models bearing same features. Models are generated at high (>=95%) confidence level with a cost value lesser than the original pharmacophore model. This indicates that there is a 95% chance for the model to represent a true correlation which offers strong confidence in obtaining an accurate and reasonable pharmacophore model.

Applications of Pharmacophore Modeling

Pharmacophore modeling techniques are widely used in the qualitative and quantitative analysis of ligands to predict the pharmacological activity of hypothetical molecules (Böhm *et al.*, 2004; Sun *et al.*2012). Hence, the pharmacophore modeling is a crucial and integral part of CADD approach in drug discovery projects. Once a pharmacophore model has been generated, it can be used for searching for potential ligands from a chemical database. Therefore, the pharmacophore model is used for 3D database search, scaffold hopping, virtual screening (VS), ligand profiling, pose filtering, fragment designing and predicting the pharmacological activities. **Fig. 3** depicts various applications of pharmacophore modeling in drug design and discovery stages. A brief description of the applications of pharmacophore modeling is provided in the following subsections.

3D Database Search

A validated pharmacophore model can be used for searching bioactive molecules from a chemical database with a large number of chemical structures. The *"hits"* obtained can provide good candidates for the lead optimization processes. For the purpose of design and optimization of novel lead compounds, programs such as BUILDER and LUDI (Bohm, 1992; Roe and Kuntz, 1995) are helpful to search different scaffolds that possess the essential features of the generated pharmacophore model.

Scaffold Hopping

While determining the best candidate for drug development potential, hits are identified by high-throughput or virtual screening of corporate compound collections. The manipulation of the scaffold (or the core structure) of a hit molecule is much more complicated and associated with a considerable loss of activity (Zhao, 2007). That is why, the most vital step to the success of drug development is the selection of compounds with the optimal scaffold. For this purpose, a given pharmacophore model template is used to search for a hit among ligands that share a common pharmacological activity but have different scaffolds (Yang, 2010). This is known as scaffold hopping or lead hopping. It typically starts with a set of known active compounds and tries discover structurally novel compounds by modifying the scaffold of the molecule (Böhm *et al.*, 2004). After accomplishment of an initial structure-activity relationship (SAR) study, a valid pharmacophore model is incorporated for scaffold hopping to deduce potential hits compounds (Sun *et al.*, 2012).

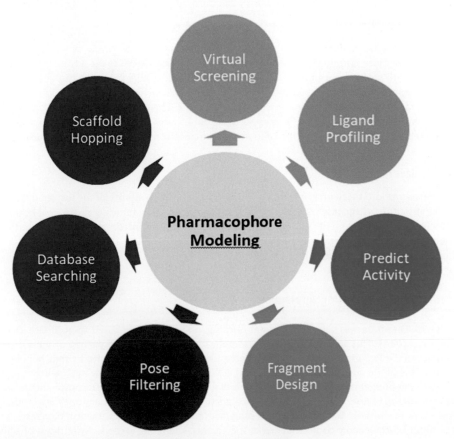

Fig. 3 Applications of pharmacophore modeling in drug design and discovery processes.

Pharmacophore-Based Virtual Screening

Pharmacophore-based virtual screening (PVS) is a subtype of ligand-based VS approach that utilizes the pharmacophoric feature perspectives in the identification of hit compounds (Balakumar *et al.*, 2017). In contrast to the structure-based VS such as docking (Bali *et al.*, 2012), the major advantage of PVS is that it is easy and less time consuming. A validated pharmacophore model serves as a template to search a large database of ligands to derive novel scaffolds bearing chemical features similar to the query pharmacophore. Handling of the conformational flexibility of database ligands and pharmacophore mapping are the two major steps in VS. The former considers multiple conformations for each ligand in the database while the latter verifies whether a query pharmacophore template is present in a given conformer or not. In case the query finds a hit, then both rigid and flexible fitting approaches will be employed. Rigid method of fitting considers the rigidity of conformations of ligands while calculating the best fit, whereas the flexible fitting permits conformational flexibility to the ligands by energy minimization, inter-feature distance and atom mapping parameters. Thus, the fit value will be assigned for each of the ligands and a higher value would represent the best mapping to the pharmacophore model and will result in the identification of highly active hit molecules.

Ligand Profiling

The structure-based pharmacophore modeling helps in the profiling of a set of ligands to predict the most likely targets for orphan bioactive ligands and anticipate adverse reactions, side effects (off-target effects) and proposing novel targets for drugs. Though, ligand-based pharmacophore modeling is also used for ligand profiling (Rognan, 2010), the application of structure-based pharmacophore modeling to profile ligands is usually preferred as it is fast, reliable and an efficient alternative to molecular docking while still representing specific ligand-protein interactions (Rella *et al.*, 2006).

Pose Filtering

The ligands that score highly as per the docking program scoring may not bind well in reality. Therefore, new methods are used where a docking program generates various ligand poses without considering the score given by the docking algorithm. Then, the poses are filtered with receptor-based pharmacophore searches. In pose filtering, the pharmacophore query can be placed into the binding pocket of the protein to filter potential ligand binding poses bearing all essential pharmacophore features to improve the probability of retrieving highly potent molecules for *in vitro* and *in vivo* experiments (Qing *et al.*, 2014).

Pharmacophore-Guided Fragment Design

It is generally easier to build up a fragment using smaller molecules than it is to reduce the size of a large one. This fragment-based approach is highly desirable because combining low molecular weight fragments offers the advantage of increased sampling of chemical space and the possibility of improved drug-like properties. Pharmacophore modeling can guide the designing of such a fragment that settles in an appropriate location in the binding site. When two or more pharmacologically active fragments are integrated, the chances of identifying the ligand pose seems to be greater. Thereby, we generate a group of fragments that may collectively yield pharmacologically active ligands. This application contributes to the lead molecule optimization stage wherein focussed hits are considered for refinement using the validated pharmacophore query (Miller and Roitberg, 2013).

Prediction of Pharmacological Activity

The 3D-QSAR based pharmacophore model has been used to predict the binding affinity of the test set ligands thereby predicting the pharmacological activity of those ligands well before an actual synthesis. Also, building pharmacophores of human ADMET-related proteins (cytochrome enzymes) aids in the identification of various pharmacokinetic properties such as ADMET (absorption, distribution, metabolism, excretion, and toxicity). Hence, pharmacophore modeling is very useful in the prediction of toxicity effects of drug-like molecules or even drugs (Rathee *et al.*, 2017).

Limitations of Pharmacophore Modeling Approaches

Unlike structure-based CADD approaches such as molecular docking, pharmacophore-based VS approach lacks reliable and good scoring metrics (Qing *et al.*, 2014). This significantly limits its applications in extending the scope of stand-alone predictions. Due to this missing element, even the ligands that may have all the requisite pharmacophoric features, may fail to exhibit *in vitro/in vivo* pharmacological activity. One of the reason for this observation is that pharmacophore modeling has inadequate consideration on the similarity with known potent ligands and overall compatibility with the target protein that complements the ligand. Hence, the structure of the identified hits may not complement the binding site of the target protein and may have other functional groups attached. Also, during a pharmacophore-based virtual screening from precomputed conformation databases, the database usually have only a limited number of low-energy conformations of the ligands. Therefore, successfully retrieval of active ligands may not be possible due to the missing conformations of the active ligand. Importantly, if a ligand has many rotatable bonds in its structure, then it is very difficult to differentiate multiple rotations during the conformation generation process. As a rule of thumb, training set ligands are selected and thereby pharmacophore model is constructed. In such cases, it results in the generation of different pharmacophore models which upon VS yield different scaffolds claiming to be active hits. For example, in most of the cases, the kinase-inhibitor driven pharmacophore models result in the identification of novel scaffolds which may not be active for the specified kinase enzyme (Vancraenenbroeck *et al.*, 2014). Also, the success of pharmacophore modeling may be limited because the project is dealing with new compounds or targets that lack sufficient molecular information (Scior *et al.*, 2012).

Another challenging aspect of the pharmacophore-based virtual screening is the yield of a higher 'false positive' rate. This means that only a small percentage of the virtual hits are really bioactive (Yang, 2010). In simple words, the virtually identified hit ligands may not be pharmacologically 'active'. There are many factors associated with this limitation such as deficiency of required hypothesis, quality, the composition of the model, and the realistic behaviour of the ligand under biological environment. To overcome this major limitation of 'false positive' models, the following factors need to be considered (1) blending expert knowledge, (2) incorporating crucial target information, (3) rigorous validation, and (4) integrating other *in silico* approaches that offer a synergistic contribution to the success of the pharmacophore modeling.

Illustrative Examples

A couple of successful application of pharmacophore modeling have been discussed in this section to illustrate their potential significance in drug design and discovery of novel ligands.

Recently, a virtual screening of National Cancer Institute (NCI) database against the IdeR (a transcription factor of *Mycobacterium tuberculosis*) DNA binding domain was performed (Rohilla *et al.*, 2017). It was followed by *in vitro* inhibition studies using Electrophoretic Mobility Shift Assay (EMSA) and 9 lead compounds were identified and subjected to the structure-based similarity search to obtain four potent compounds. Further, lead optimization through the development of energy based pharmacophore and VS of ZINC database followed by molecular docking studies. These efforts yield one highly potent IdeR inhibitor. Authors developed five point energy based pharmacophore model using e-pharmacophore script under Schrodinger Package (Schrödinger, 2017). An energy based pharmacophore extracted and weighed interactions made by an inhibitor with the protein by using the descriptor information generated by the docking. A map of critical features contributing to the interaction between inhibitor and protein was generated using Glide module (Friesner *et al.*, 2004). LigPrep module of Schrodinger package was employed to generate energy minimized conformers and stereoisomers. All the resulted conformers were docked at the active site of the protein (PDB ID:1U8R (Wisedchaisri *et al.*, 2004)) by using Glide XP-precision mode. Based on the obtained XP descriptor file, the pharmacophore hypothesis was generated and was submitted to the e-pharmacophore script to yield a pharmacophore model. Thus, the created model consists of 5 features such as two hydrogen bond acceptors, two rings and one negative ionizable group. Based on the proximity with the key residues involved in DNA binding, each pharmacophoric points were carefully selected by these researchers. Further, screening of ZINC drug-like database (~ 13 million compounds) was carried out by using Phase (Dixon *et al.*, 2006). This resulted in $\sim 110,000$ molecules with a filter of 4 out of 5 molecular features to be matched. In this study, a five-point pharmacophore model provided valuable insight into the pharmacophoric features essential for IdeR inhibition. Also, the study led to the identification of one hit molecule exhibiting potent inhibition of IdeR.

Recently, we identified a hit molecule inhibiting the human kinesin spindle protein (KSP) through ligand- and structure-based *in silico* studies (Balakumar *et al.*, 2017). We have developed ligand-based pharmacophore models based on the available KSP inhibitors using BIOVIA-Discovery Studio software package (Dassault Systems, 2016). All the generated pharmacophore models were validated against *in house* test set ligands (54 compounds). Based on the validation parameters such as enrichment factor (E) and goodness of hit score, Hypo1 was selected as a valid model that consists of one ring aromatic (R), one positively ionizable (P), one hydrophobic (H), and two hydrogen bond acceptors (A_1 and A_2). The validated model (Hypo1) was then taken for database screening (Maybridge and ChemBridge) to yield hits, which were further filtered for their drug-likeliness. To identify the ligand binding landscape, the potential hits retrieved from virtual database screening were docked to the active-site of the KSP (PDB: 4BXN Talapatra *et al.*, 2013) using CDOCKER (Wu *et al.*, 2003). The top-ranked hits obtained from molecular docking were used in molecular dynamics simulations to deduce the ligand binding affinity. This study identified MB-41570 as an inhibitor of KSP.

Conclusion and Future Perspectives

Unequivocally, the pharmacophore modeling approaches still serve as one of the most widely used and successful tools in CADD projects and medicinal chemistry. Though a considerable development in the pharmacophore modeling has happened, further improvements in pharmacophore modeling techniques are required. To enhance its success rate in real-time experiments, several aspects such as developing accurate, optimized/predictive models, better handling of ligand flexibility and efficient algorithms for ligand alignments must be considered. There could be a positive contribution in the form of the target binding-site information that can assist in developing the structure-based models more efficiently. Better algorithms can be developed in the area of fragment-based drug design involving pharmacophore fingerprint-based similarity search and generation of 3D pharmacophore queries. Another potential scope of pharmacophore modeling is the development of small molecule inhibitors of protein-protein interactions (PPIs), wherein the pharmacophoric queries will be generated by encoding the crucial interactions at the PPI interface. Hence, there are multiple new horizons to be explored through the application of pharmacophore modeling techniques in designing therapeutics.

See also: Biomolecular Structures: Prediction, Identification and Analyses. Chemoinformatics: From Chemical Art to Chemistry in Silico. Drug Repurposing and Multi-Target Therapies. Fingerprints and Pharmacophores. In Silico Identification of Novel Inhibitors. Natural Language Processing Approaches in Bioinformatics. Population Analysis of Pharmacogenetic Polymorphisms. Protein Structural Bioinformatics: An Overview. Rational Structure-Based Drug Design. Small Molecule Drug Design. Structure-Based Drug Design Workflow

References

Agrafiotis, D.K., Bandyopadhyay, D., Yang, E., 2013. Stochastic proximity embedding: A simple, fast and scalable algorithm for solving the distance geometry problem. In: Mucherino, A., *et al.* (Eds.), Distance Geometry: Theory, Methods, and Applications. New York, NY: Springer New York, pp. 291–311.

Agrawal, N., Skelton, A.A., 2016. 12-Crown-4 Ether Disrupts the Patient Brain-Derived Amyloid-β-Fibril Trimer: Insight from All-Atom Molecular Dynamics Simulations. ACS chemical neuroscience 7, 1433–1441.

Agrawal, N., Skelton, A.A., 2018. Binding of 12-crown-4 with Alzheimer's Aβ40 and Aβ42 monomers and its effect on their conformation: insight from molecular dynamics simulations. Molecular Pharmaceutics 15, 289–299.

Andén, N.E., *et al.*, 1970. Evidence for a central noradrenaline receptor stimulation by clonidine. Life Sciences 9 (9), 513–523.

Ariens, E.J., 1966. Molecular pharmacology, a basis for drug design. Fortschr Arzneimittelforsch 10, 429–529.

Balakumar, C., *et al.*, 2017. Ligand- and structure-based in silico studies to identify kinesin spindle protein (KSP) inhibitors as potential anticancer agents. Journal of Biomolecular Structure and Dynamics. 1–18.

Bali, A., Dhillon, S.K., Balakumar, C., 2012. Alkoxyphenyl methanesulfonamides: Synthesis, anti-inflammatory effect, and docking studies. Medicinal Chemistry Research 21 (10), 3053–3062.

Bandyopadhyay, D., Agrafiotis, D.K., 2008. A self-organizing algorithm for molecular alignment and pharmacophore development. Journal of Computational Chemistry 29 (6), 965–982.

Baroni, M., *et al.*, 2007. A common reference framework for analyzing/comparing proteins and ligands. Fingerprints for Ligands And Proteins (FLAP): Theory and application. Journal of Chemical Information and Modeling 47 (2), 279–294.

Bheemanapalli, L.N., Balakumar, C., Kaki, V.R., Kaur, R., Akkinepally, R.R., 2013. Pharmacophore based 3D-QSAR study of biphenyl derivatives as nonsteroidal aromatase inhibitors in JEG-3 cell lines. Medicinal Chemistry 9, 974–984.

BIOVIA, Discovery Studio Modeling Environment. San Diego: Dassault Systèmes.

Bohm, H.J., 1992. The computer program LUDI: A new method for the de novo design of enzyme inhibitors. Journal of Computer-Aided Molecular Design 6 (1), 61–78.

Böhm, H.-J., Flohr, A., Stahl, M., 2004. Scaffold hopping. Drug Discovery Today: Technologies 1 (3), 217–224.

Bredel, M., Jacoby, E., 2004. Chemogenomics: An emerging strategy for rapid target and drug discovery. Nature Reviews Genetics 5 (4), 262–275.

Chen, J., Lai, L., 2006. Pocket v.2: Further developments on receptor-based pharmacophore modeling. Journal of Chemical Information and Modeling 46 (6), 2684–2691.

Dixon, S.L., *et al.*, 2006. PHASE: A new engine for pharmacophore perception, 3D QSAR model development, and 3D database screening: 1. Methodology and preliminary results. Journal of Computer-Aided Molecular Design 20 (10–11), 647–671.

Dodds, E.C., Lawson, W., 1938. Molecular structure in relation to oestrogenic activity. Compounds without a phenanthrene nucleus. Proceedings of the Royal Society of London. Series B, Biological Sciences 125 (839), 222–232.

Drwal, M.N., Griffith, R., 2013. Combination of ligand- and structure-based methods in virtual screening. Drug Discovery Today Technology 10 (3), e395–e401.

Ehrlich, P., 1898. Ueber die Constitution des Diphtheriegiftes. Dtsch med Wochenschr 24 (38), 597–600.

Ehrlich, P., 1909. Über den jetzigen stand der Chemotherapie. Berichte Der Deutschen Chemischen Gesellschaft 42 (1), 17–47.

Friesner, R.A., *et al.*, 2004. Glide: A new approach for rapid, accurate docking and scoring. 1. Method and assessment of docking accuracy. Journal of Medicinal Chemistry 47.

Friesner, R.A., *et al.*, 2006. Extra precision glide: Docking and scoring incorporating a model of hydrophobic enclosure for protein − Ligand complexes. Journal of Medicinal Chemistry 49 (21), 6177–6196.

Güner, O.F., 2000. Pharmacophore Perception, Development and Use in Drug Design. Dr. Igor Tsigelny.

Güner, O.F., Bowen, J.P., 2014. Setting the record straight: The origin of the pharmacophore concept. Journal of Chemical Information and Modeling 54 (5), 1269–1283.

Guner, O., Clement, O., Kurogi, Y., 2004. Pharmacophore modeling and three dimensional database searching for drug design using catalyst: Recent advances. Current Medicinal Chemistry 11 (22), 2991–3005.

Hecker, E.A., *et al.*, 2002. Use of catalyst pharmacophore models for screening of large combinatorial libraries. Journal of Chemical Information and Computer Sciences 42 (5), 1204–1211.

Jones, G., Willett, P., Glen, R.C., 1995. A genetic algorithm for flexible molecular overlay and pharmacophore elucidation. Journal of Computer-Aided Molecular Design 9 (6), 532–549.

Kalva, S., Agrawal, N., Skelton, A.A., Saleena, L.M., 2016. Identification of novel selective MMP-9 inhibitors as potential anti-metastatic lead using structure-based hierarchical virtual screening and molecular dynamics simulation. Molecular BioSystems 12, 2519–2531.

Kandakatla, N., Ramakrishnan, G., 2014. Ligand based pharmacophore modeling and virtual screening Studies to design novel HDAC2 inhibitors. Advances in Bioinformatics 2014, 812148.

Martin, Y.C., *et al.*, 1993. A fast new approach to pharmacophore mapping and its application to dopaminergic and benzodiazepine agonists. Journal of Computer-Aided Molecular Design 7 (1), 83–102.

Miller 3rd, B.R., Roitberg, A.E., 2013. Design of e-pharmacophore models using compound fragments for the trans-sialidase of Trypanosoma cruzi: Screening for novel inhibitor scaffolds. Journal of Molecular Graphics and Modelling 45, 84–97.

Molecular Operating Environment (MOE), 2017. 2013.08; Chemical Computing Group ULC, 1010 Sherbooke St. West, Suite #910, Montreal, QC, Canada, H3A 2R7.

Murray, C.W., Baxter, C.A., Frenkel, A.D., 1999. The sensitivity of the results of molecular docking to induced fit effects: Application to thrombin, thermolysin and neuraminidase. Journal of Computer-Aided Molecular Design 13 (6), 547–562.

Mysinger, M.M., *et al.*, 2012. Directory of useful decoys, enhanced (DUD-E): Better ligands and decoys for better benchmarking. Journal of Medicinal Chemistry 55 (14), 6582–6594.

Nettles, J.H., *et al.*, 2007. Flexible 3D pharmacophores as descriptors of dynamic biological space. Journal of Molecular Graphics and Modelling 26 (3), 622–633.

Omega; OpenEye Scientific Software, 3600 Cerrillos Rd., Suite 1107, Santa Fe, NM 87507, USA.

Ortuso, F., Langer, T., Alcaro, S., 2006. GBPM: GRID-based pharmacophore model: Concept and application studies to protein-protein recognition. Bioinformatics 22 (12), 1449–1455.

Pullman, B., *et al.*, 1972. Quantum mechanical study of the conformational properties of phenethylamines of biochemical and medicinal interest. Journal of Medicinal Chemistry 15 (1), 17–23.

Qing, X., L, X., De Raeymaecker, J., *et al.*, 2014. Pharmacophore modeling: Advances, limitations, and current utility in drug discovery. Journal of Receptor, Ligand and Channel Research 7, 81–92.

Rathee, D., Lather, V., Dureja, H., 2017. Pharmacophore modeling and 3D QSAR studies for prediction of matrix metalloproteinases inhibitory activity of hydroxamate derivatives. Biotechnology Research and Innovation 1 (1), 112–122.

Rella, M., *et al.*, 2006. Structure-Based Pharmacophore Design and Virtual Screening for Novel Angiotensin Converting Enzyme 2 Inhibitors. Journal of Chemical Information and Modeling 46 (2), 708–716.

Richmond, N.J., *et al.*, 2006. GALAHAD: 1. pharmacophore identification by hypermolecular alignment of ligands in 3D. Journal of Computer-Aided Molecular Design 20 (9), 567–587.

Roe, D.C., Kuntz, I.D., 1995. BUILDER v.2: Improving the chemistry of a de novo design strategy. Journal of Computer-Aided Molecular Design 9 (3), 269–282.

Rognan, D., 2010. Structure-based approaches to target fishing and ligand profiling. Molecular Informatics 29 (3), 176–187.

Rohilla, A., Khare, G., Tyagi, A.K., 2017. Virtual screening, pharmacophore development and structure based similarity search to identify inhibitors against IdeR, a transcription factor of Mycobacterium tuberculosis. Scientific Reports 7 (1), 4653.

Rubicon; Daylight Chemical Information Systems, Inc., 120 Vantis, Suite 550, Aliso Viejo, CA 92656, USA.

Sanders, M.P.A., *et al.*, 2011. Snooker: A structure-based pharmacophore generation tool applied to class A GPCRs. Journal of Chemical Information and Modeling 51 (9), 2277–2292.

Schrödinger Release 2017-4: Maestro. New York, NY: Schrödinger, LLC.

Schueler, F.W., 1960. Chemobiodynamics and drug design. Journal of Pharmaceutical Sciences 50, 92.

Schuster, D., *et al.*, 2006. Pharmacophore modeling and in silico screening for new P450 19 (aromatase) inhibitors. Journal of Chemical Information and Modeling 46 (3), 1301–1311.

Scior, T., *et al.*, 2012. Recognizing pitfalls in virtual screening: A critical review. Journal of Chemical Information and Modeling 52 (4), 867–881.

Sun, H., Tawa, G., Wallqvist, A., 2012. Classification of scaffold-hopping approaches. Drug Discovery Today 17 (7), 310–324.

SYBYL7.3 Tripos Inc., 2003. St. Louis, MO, USA.; Available online: http://www.tripos.com.

Talapatra, S.K., *et al.*, 2013. Mitotic kinesin Eg5 overcomes inhibition to the phase I/II clinical candidate SB743921 by an allosteric resistance mechanism. Journal of Medicinal Chemistry 56 (16), 6317–6329.

Therese, P.J., *et al.*, 2014. Multiple e-Pharmacophore modeling, 3D-QSAR, and high-throughput virtual screening of hepatitis C virus NS5B polymerase inhibitors. Journal of Chemical Information and Modeling 54 (2), 539–552.

Toba, S., *et al.*, 2006. Using pharmacophore models to gain insight into structural binding and virtual screening: An application study with CDK2 and human DHFR. Journal of Chemical Information and Modeling 46 (2), 728–735.

Totrov, M., 2008. Atomic property fields: Generalized 3D pharmacophoric potential for automated ligand superposition, pharmacophore elucidation and 3D QSAR. Chemical Biology & Drug Design 71 (1), 15–27.

Triballeau, N., *et al.*, 2005. Virtual screening workflow development guided by the "Receiver Operating Characteristic" curve approach. Application to high-throughput docking on metabotropic glutamate receptor subtype 4. Journal of Medicinal Chemistry 48 (7), 2534–2547.

Vadivelan, S., *et al.*, 2007. Virtual screening studies to design potent CDK2-cyclin A inhibitors. Journal of Chemical Information and Modeling 47 (4), 1526–1535.

Vancraenenbroeck, R., *et al.*, 2014. In silico, in vitro and cellular analysis with a kinome-wide inhibitor panel correlates cellular LRRK2 dephosphorylation to inhibitor activity on LRRK2. Frontiers in Molecular Neuroscience 7, 51.

Van Drie, J.H., 1997. "Shrink-Wrap" Surfaces: A new method for incorporating shape into pharmacophoric 3D database searching. Journal of Chemical Information and Computer Sciences 37 (1), 38–42.

Vyas, V.K., *et al.*, 2012. Homology modeling a fast tool for drug discovery: Current perspectives. Indian Journal of Pharmaceutical Sciences 74 (1), 1–17.

Wermuth, C.G., *et al.*, 1998. Glossary of terms used in medicinal chemistry (IUPAC Recommendations 1998). Pure and Applied Chemistry. 1129.

Willett, P., 2004. Evaluation of molecular similarity and molecular diversity methods using biological activity data. In: Bajorath, J. (Ed.), Chemoinformatics: Concepts, Methods, and Tools for Drug Discovery. Humana Press.

Wisedchaisri, G., Holmes, R.K., Hol, W.G.J., 2004. Crystal structure of an IdeR–DNA complex reveals a conformational change in activated IdeR for base-specific interactions. Journal of Molecular Biology 342 (4), 1155–1169.

Wolber, G., Langer, T., 2005. LigandScout: 3-D pharmacophores derived from protein-bound ligands and their use as virtual screening filters. Journal of Chemical Information and Modeling 45 (1), 160–169.

Wolber, G., *et al.*, 2008. Molecule-pharmacophore superpositioning and pattern matching in computational drug design. Drug Discovery Today 13 (1–2), 23–29.

Wu, G., *et al.*, 2003. Detailed analysis of grid-based molecular docking: A case study of CDOCKER-A CHARMm-based MD docking algorithm. Journal of Computational Chemistry 24 (13), 1549–1562.

Xie, Z.-R., Hwang, M.-J., 2015. Methods for predicting protein–ligand binding sites. In: Kukol, A. (Ed.), Molecular Modeling of Proteins. New York, NY: Springer New York, pp. 383–398.

Yang, S.-Y., 2010. Pharmacophore modeling and applications in drug discovery: Challenges and recent advances. Drug Discovery Today 15 (11), 444–450.

Zhao, H., 2007. Scaffold selection and scaffold hopping in lead generation: A medicinal chemistry perspective. Drug Discovery Today 12 (3), 149–155.

Introduction of Docking-Based Virtual Screening Workflow Using Desktop Personal Computer

Muhammad Yusuf, Ari Hardianto, Muchtaridi Muchtaridi, and Rina F Nuwarda, Universitas Padjadjaran, Jatinangor, Indonesia
Toto Subroto, Universitas Padjadjaran, Bandung, West Java, Indonesia

Introduction

In the drug discovery process, finding new leads is an expensive and time-consuming process. The development of the combinatorial chemistry and high-throughput screening (HTS) technologies in the early 1990s has allowed enormous libraries of compounds to be synthesized and screened in a short period of time. However, many of the identified have failed in the lead optimization process due to their poor pharmacokinetic properties. Therefore, it was necessary to develop alternative strategies that might help to select appropriate series of compounds and eliminate unsuitable structures and thus could save a significant amount of time and resources. For this purpose, computational techniques such as virtual screening (VS) can be applied as an important step in the early stage of the drug discovery process (Lavecchia and Di Giovanni, 2013). While HTS aims to experimentally test a large number of compounds in the most efficient manner, VS attempts to rationalize those compounds to reduce the number for experimental testing as much as possible (Ripphausen, et al., 2011). Furthermore, the compounds evaluated do not necessarily exist, or in other words, any molecule of compound theoretically can be evaluated by VS (Lavecchia and Di Giovanni, 2013). In the VS process, a large library of compounds is filtered and reduced by a computational algorithm based on their predictive binding mode with the target, which will then be tested experimentally (Tanrikulu, et al., 2013). **Fig. 1** shows the differences between DBVS and HTS. Unlike HTS, DBVS allows us to test only compounds with high scoring, hence lowering the cost of hits discovery (Shoichet et al., 2002; Villoutreix et al., 2000). VS can be classified into two categories, namely ligand-based virtual screening (LBVS) and structure-based virtual screening (SBVS). LBVS involves the utilization of structural data from a series of known active compounds to find candidates for experimental testing. It includes several methods such as similarity searching, quantitative structure relationships (QSAR), and pharmacophore fitting. Whereas SBVS, or docking-based virtual screening (DBVS), employs the three-dimensional (3D) structure of the target macromolecule that is obtained experimentally either through X-ray crystallography or NMR or computationally through homology modeling. The candidate molecules were docked to the target macromolecule to investigate the binding mode and rank them based on their predicted binding affinity (Lavecchia and Di Giovanni, 2013).

Both LBVS and SBVS approaches are powerful technologies that promise to accelerate the pace and reduce the cost of discovering new drugs, which can be applied to VS for lead identification and optimization (Villoutreix et al., 2000; Lengauer et al., 2004; Vyas et al., 2008; Xu and Agrafiotis, 2002).

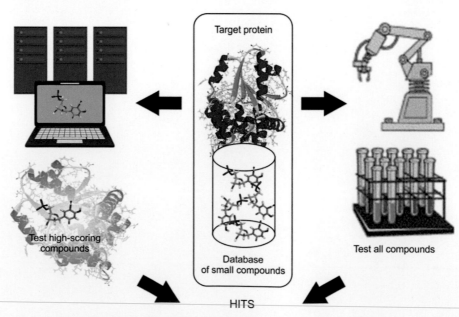

Fig. 1 DBVS selects hits compound based on the scoring function, while all compounds in HTS are being tested.

Molecular docking is a computational tool of structure-based drug design to predict protein–ligand interaction geometries and binding affinities (Cosconati *et al.*, 2010; Vyas *et al.*, 2008). In DBVS, compounds in the database are scored and ranked based on their steric and electrostatic interactions with the target site and the best compounds are tested further with biochemical assays (Singh *et al.*, 2006). The search algorithm should generate an optimum number of conformations for the ligands and of poses that should include the experimentally determined binding mode. The scoring functions commonly used in docking are force-field based, empirical-based, and knowledge-based. For this reason, there are many considerations in choosing the best compound based on the ranking in DBVS result. Several scoring functions have been developed during the past few years (Breda *et al.*, 2008; Friesner *et al.*, 2004; Jain, 2006; Kastritis and Bonvin, 2010; Kroemer, 2003; Shoichet *et al.*, 2002). There are numerous programs available for DBVS, such as Autodock (Morris *et al.*, 1996), GOLD (Jones *et al.*, 1997), DOCK (Ewing *et al.*, 2001), FlexX (Rarey *et al.*, 1996), Glide (Halgren *et al.*, 2004), LigPlot (Wallace *et al.*, 1995), and FRED (McGann, 2011). Among these programs, Autodock is the most used program for docking in scientific journals and freely available for academic work. There are some important factors to be considered in choosing docking software, including the capability for iterative refinement of docking parameter/protocol based on new results, the adaptability to additional scoring functions, pre- and/or post-docking filters, design components and results of validation studies, user learning curve, customer supports, cost, speed, user interface, input/output structural file formats, code availability, and upgrading possibility (Ghosh *et al.*, 2006).

Databases for Virtual Screening

Databases of small molecules are required for all VS projects. There are millions of compounds from structural databases of more than 32 providers gathered and stored in a single chemical database. By using the "Lipinski rule of five," around 2.1 million of compounds are categorized as "drug-like," and more than 1 million are lead like (Monge *et al.*, 2006).

However, the drug-likeness of chemicals from the database should be calculated, to avoid a clinical candidate that is too big, resulting in bad solubility and/or permeability (Verheij, 2006). When selecting compounds for screening, traditional "druglike" property cut-off values (e.g., Lipinski's "rule of five": MW < 500, LogP < 5, hydrogen acceptor < 10, and hydrogen donor < 5) (Lipinski *et al.*, 2001; Verheij, 2006) might cause problems later during the lead development stage (Verheij, 2006).

Recently, many collections of chemical library databases are available on the web. Some of the commonly used small molecule databases in VS are ZINC (free database of 3.3 million commercially available compounds), the Available Chemicals Directory (ACD; 4 million entries, not free), National Cancer Institute compound database (NCI; free database of 400,000 entries), and MDDR (MDL; Drug Data Report > 147,000 entries). Besides, a chemical library containing synthetic single compounds, combinatorial compounds, and natural product compounds (Hong, 2011; Rollinger *et al.*, 2006), and marketed drugs library (Hong, 2011; Lopez-Perez *et al.*, 2007; Quinn *et al.*, 2008) are also available. Lopez-Perez and colleagues developed NAPROC13 (Lopez-Perez *et al.*, 2007) containing ^{13}C spectral information of over 6000 natural compounds. This database allows us to identify the known compounds present in the crude extracts and provides insight into the structural elucidation of unknown compounds. Moreover, Dunkel *et al.* (2006) developed a SuperNatural database that contains 3D structures and conformers of 45,917 natural compounds and about 300 natural compounds that are similar to active ingredients of drugs. It is known that 8% (3600) of this database is identical to the marketed drugs.

Another database of natural compounds is NADI (Natural Drug Discovery), which was developed by Wahab *et al.* (2009). It is a database of Malaysian medicinal plants that aimed to be a one-stop center for in silico drug discovery from natural products. It provides structural information of around 3000 different compounds, complete with the botanical sources of plants species that could be used in VS. Compounds in the NADI database can be searched based on the name, plants, structure formula, molecular weight, substructure, and SMILES code. In addition, around 40 monographs of Malaysian herbal medicine were available. They contain an integrated information regarding the description of plants, bioactivity, efficacy and usefulness, diseases target, and references. NADI is also enriched with the target protein database, composed of 645 validated drug targets useful for rapid VS. This database was successfully used to obtain lead compounds for neuraminidase inhibitors. Ikram *et al.* (2015) screened 3000 compounds from the NADI database using AutoDock against neuraminidase, a drug target of the influenza virus. The top 100 compounds were then categorized into five plants. Interestingly, some fractions and pure compounds from the five plants showed good inhibitory activity against the neuraminidase of the influenza virus (Muchtaridi *et al.*, 2014).

Case Study

In this case study, we will do a DBVS of compounds from National Cancer Institute (NCI) database to find the potential hits as *Mycobacterium tuberculosis* (Mtb) thymidylate kinase (TMPK) inhibitor. TMPK is one of the validated drug targets for Mtb. NCI provides a diversity set consisting of about 2000 compounds representing a 140,000-chemical space. The NCI diversity set is a good choice of database for virtual screening exercise because NCI also provides real compounds for further experimental testing. The compounds, when available, can be requested for free (see "Relevant Websites section"). The procedure to obtain the vialed samples is explained in this web link provided in: "Relevant Websites section". A crystal structure of Mtb TMPK in complex with its

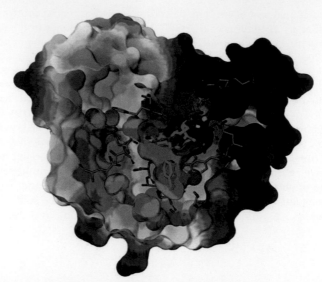

Fig. 2 Mtb TMPK in complex with TMP analog. Protein surface represents an increasing level of hydrophobicity (blue → white → brown colors). The ligand is visualized in green sticks. A green sphere, a magnesium ion, is located near the ligand. This figure is generated from PDB ID 1MRS using Biovia Discovery Studio Visualizer.

inhibitor with PDB ID 1MRS is used in this case study (**Fig. 2**). The ligand is thymidylate monophosphate (TMP) analog, possessed a nanomolar inhibition towards TMPK (Haouz *et al.*, 2003).

Programs that used in this case study are AutoDock 4.2, AutoDockTools (ADT), Raccoon, and Biovia Discovery Studio Visualizer (BDSV). ADT is a GUI of AutoDock which can be used to set up input files and docking parameters. AutoDock 4.2 and ADT is available in Windows, Linux, and Mac OS. They can be downloaded from websites provided in "Relevant Websites section" and "Relevant Websites section," respectively. BDSV is also a freeware that can be obtained through website provided in "Relevant Websites section." Moreover, automated DBVS of thousands of ligands will be executed by a bash script. In this case study, we choose one of the most popular and user-friendly Linux OS, that is, Ubuntu 16.04. Linux is a native environment of most of the computational chemistry program. Therefore, to have experience in operating DBVS under Linux should be beneficial to the beginner.

In general, DBVS using AutoDock consists of eight main steps: Ligand processing, receptor preparation, grid parameter setup, affinity maps calculation, protocol validation, docking parameter setup, virtual screening execution, and lastly screening result analysis.

Preparation of Ligand From Database

The database of ligands is usually compiled in a single "MOL2" file, not the individual file. For this reason, a tool called Raccoon (see "Relevant Websites section") can be used to prepare the PDBQT format for each of ligand, a native filetype of AutoDock. The Graphical User Interface of Raccoon is presented in **Fig. 3**.

If all the ligands are wrapped in a single MOL2 file, then we must click 'Utilities → Split a MOL2 → Open'. All the spliced MOL2 files will be generated accordingly in the same directory with its parent file. Furthermore, an automatic process of converting 2000 ligands from MOL2 format into PDBQT is accomplished by using '[+] Add ligands' button. It is noted that not all ligands will be successfully converted into the PDBQT format due to some reasons, for example, a number of torsions, atom-type parameter. Therefore, in this case study, a hassle-free PDBQT-formatted NCI Diversity Set II, namely "NCI-Diversity-AutoDock-0.2.tar.gz," can be directly downloaded from website provided in "Relevant Websites section." This file consists of 1541 ligands, a few hundred less than the original database. All ligands with AutoDock-unsupported atom type (such as selenium) and a large number of active torsions have been discarded from the list. For the sake of accuracy, AutoDock has a default maximum rotatable bonds of 32.

Preparation of Receptor and Control Ligand

(1) The receptor preparation will be done by ADT, which can be opened by typing 'adt' in the Linux terminal. The interface of ADT is presented in **Fig. 4**.

(2) A working directory of '/home/*USER*/vs' in 'File → Preferences → Set → Startup Directory' should be determined. This step will ensure all the modified files will be placed in one working directory.

Fig. 3 The interface of Raccoon.

(3) The PDB file of TMPK complex can be directly downloaded using ADT by clicking 'File → Import → Fetch from Web → 1MRS'.

(4) AutoDock uses implicit water solvent model. Therefore, in most cases, all of the crystal water should be removed by using 'Edit → Delete Water'.

(5) Most of the crystal structures in PDB lack hydrogens. All the hydrogen atoms should be added to the structure by using 'Edit → Hydrogens → Add → All Hydrogens → noBondOrder'. Furthermore, the hydrogenated PDB is saved using 'File → Save → Write PDB → 1MRS_H.pdb'.

(6) The ligand and receptor are separated into an individual file using Linux terminal. In the terminal, go to the working directory by typing:

$ cd ~/vs/source

In PDB 1MRS, the residue name for the ligand is "5HU". Therefore, the ligand can be extracted from the complex using grep command:

$ grep "5HU" 1MRS_H.pdb > 1MRS_ligand.pdb

Whereas the receptor can be extracted by typing:

$ grep "ATOM" 1MRS_H.pdb > 1MRS_receptor.pdb.

As mentioned before (**Fig. 2**), a magnesium cofactor is located near the ligand, and hence might have an important role in the active site. This ion, with residue name "Mg," can be added to the receptor file by typing:

$ grep "Mg" 1MRS_H.pdb > > 1MRS_receptor.pdb

please note the use of "> >" instead of ">"

"> >" means to add a new line(s) to the current file, while ">" means to replace the whole text with the new one. In this step, we want to add Mg as part of the receptor.

(7) Finally, the receptor and control ligand is required to be converted into PDBQT format using ADT. Here are a few steps to be done:

- Clean up the molecule windows by using 'Edit → Delete → Delete All Molecules',
- In ADT4.2 widget, select 'Ligand → Input → Open' and choose 'PDB files' in 'Files of type' menu. Select '1MRS_ligand.pdb' then click 'Open'. Save the ligand in PDBQT format by select 'Ligand → Output → Save As PDBQT', name it as '1MRS_ligand.pdbqt' then click 'Save',
- In ADT4.2 widget, select 'Grid → Macromolecule → Open', and choose 'all files' option in the 'Files of type'. Select 1MRS_receptor.pdb and click 'Open'. ADT will read its coordinates, add charges, merge nonpolar hydrogens, and assign appropriate atom types. Click 'OK' to accept the changes. A window will pop up to write the PDBQT file. Click 'Save' to write the file namely '1MRS_receptor.pdbqt',
- By default, the atomic charge of Mg will be set to 0. This value can be modified later manually using any text editor in Linux (such as gedit, nano, or geany) before docking,
- Please note to keep the receptor and control ligand loaded in the molecule window to continue with the next step.

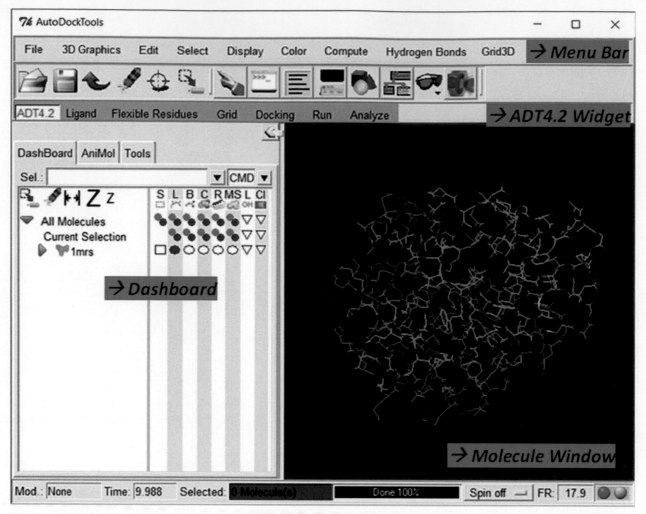

Fig. 4   Interface of AutoDockTools.

Grid Parameter Setup and Affinity Maps Calculation

In this step, we will calculate the affinity maps for each atom types of the ligand. Before that, we should determine the coordinate and the size of the ligand binding site, represented by a grid box, which is used as a target for VS.

(1) In ADT4.2 widget, specify the atom types included in the loaded 1MRS_ligand.pdbqt by using 'Grid → Set Map Types → Choose Ligand → Select Ligand'.

(2) Define the center of searching space based on the position of control ligand by using 'Grid → Grid Box → Center → Center on Ligand'. The size of grid box can be adjusted manually to cover the interacting residues (**Fig. 5**). Save the setting by clicking 'File → Close Saving Current'.

(3) Save the control ligand's grid parameter file as "control.gpf" using 'Grid → Output → Save GPF'. However, this gpf file is not the one that will be calculated using AutoGrid.

(4) It is noted that in a virtual screening setting, the size of grid box should consider the size of the compound in the database used for SBVS. This can be performed by using a python script included in the Utilities24 folder, that is, prepare_gpf4.py. This script can also be used to extract all the atom types consisted of the database. Here, "control.gpf" will be used as a reference file to create a global gpf file for the whole virtual screening process.

```
$ cd ~/vs/source
$ pythonsh ~/Utilities24/prepare_gpf4.py -r 1MRS_receptor.pdbqt
   -d ../ligand -i control.gpf -o vs.gpf -v
```

This command will generate a new gpf file covering all atom types in the database and adjusting the size of grid box based on the size of all included compound without changing the center of the grid box. It is noted that several ligands have a larger size than the original box size. In this case, it is important to consider whether to follow the new size, which increases the searching space or to stick with the previous size (if the majority of ligands covered). *Take note that the symbol "\" means to*

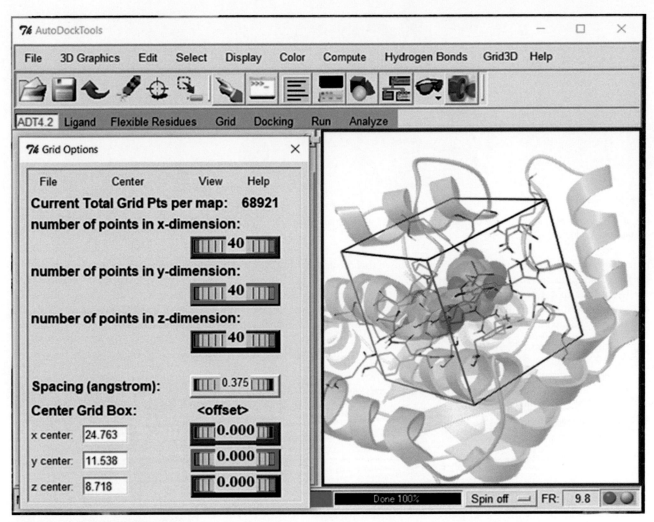

Fig. 5. Grid box setting in AutoDockTools.

continue with the next line. The symbol "\" itself does not need to be typed.

(5) Calculate the affinity maps of all atom types in the database by using AutoGrid.

```
$ autogrid4 -p vs.gpf -l vs.glg
```

Docking Parameter Setup

In general, there is no good-for-all docking parameter setting. It depends mostly on the searching problem, which is represented by the size of searching space and ligand's degree of freedom. Before preparing the docking parameter for VS, we must do one for the control ligand, which can be used as a reference setting later. AutoDock stores the docking parameter in a DPF file, which can be prepared as follows:

(1) Select the macromolecule by clicking 'Docking →Macromolecule →Set Rigid Filename' in ADT4.2 widget and choose the '1MRS_receptor.pdbqt'.
(2) Select the ligand by using 'Docking →Ligand →Choose →1MRS_ligand.pdbqt'.
(3) Select the genetic algorithm as the docking method by using 'Docking →Search Parameters →Genetic Algorithm Parameters'. Modify the number of GA runs and a maximum number of evaluation to 100 and 5,000,000, respectively.
(4) If needed, a more advanced parameter setting can be modified through this menu in the ADT4.2 widget: 'Docking →Docking Parameters'.
(5) Save the docking parameter file as "control.dpf" by using 'Docking →Output →Lamarckian GA'.

Control Docking or Validation the Docking Method

We may test our docking parameter by control docking method. It is basically a redocking of the cocrystallized ligand from the random state back into its receptor. When the docking is able to reproduce the experimental conformation and position, its

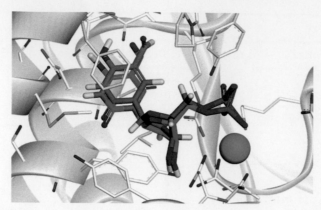

Fig. 6 The result of control docking of ligand (green colored stick) as compared to its crystal pose (gray colored stick). A green sphere is a magnesium ion located at the binding site.

parameter is considered to be used for VS. It is known that the acceptable root-mean-square-deviation (RMSD) value between the docked ligand and crystal ligand is below 2 Å.

Please note that a magnesium ion existed in the receptor file. By default, its charge was set to 0, which is not right. Before docking, we should manually edit the '1MRS_receptor.pdbqt' using any text editor by changing the atomic charge of Mg. In this session, we changed its charge from 0 to 1 (not 2) to mimic the solvent effect around the magnesium ion.

(1) In the Linux terminal, type:
$ cd ∼/vs/source
$ autodock4 -p control.dpf -l control.dlg
We can use 'tail -f control.dlg' to monitor the progress of docking process.

(2) Analyze the DLG file by using any text editor, especially under 'Clustering Histogram' section. The lowest energy docking pose is mentioned after the RMSD Table.

RMSD TABLE

Rank	Sub-Rank	Run	Binding Energy	Cluster RMSD	Reference RMSD	Grep Pattern
1	1	1	-9.28	0.00	0.97	RANKING
1	2	5	-8.07	1.02	1.08	RANKING
1	3	3	-8.06	0.93	1.31	RANKING
1	4	4	-7.96	1.77	1.70	RANKING

A successful control docking was indicated by the small RMSD value between docked pose and that of reference (0.97 Å) (**Fig. 6**). It is noted that if we did not change the atomic charge of Mg, the docking result would be worse than this. A separate docking was conducted using the Mg charge of 0, which resulted in an RMSD of 4 Å, indicating a nonreliable docking parameter.

Virtual Screening

Fig. 7 shows that our VS working directory consists of three subfolders, namely ligand, docking, and source. All ligand PDBQT are stored in ligand folder. Using a reference grid and docking parameter files, receptor PDBQT and affinity maps in the source folder, we will generate a working directory for each ligand in docking folder. Each ligand's folder will consist of PDBQT files of ligand and receptor, affinity maps, and docking parameter file. Moreover, there are Python scripts that are very useful in helping the virtual screening process, located in the folder "Utilities24" of ADT installation (by default, it is located in '/home/USER/mgltools*/ MGLToolsPckgs/AutoDockTools/Utilities24'. In this practical session, the whole directory of Utilities24 has been copied to the home directory. Hence it can be accessed at '∼/Utilities24'. (note: "∼" means /home/USER/ directory)

A map of working directories is visualized in **Fig. 7**, particularly to organize the virtual screening process in this exercise.

Bash script provided in this article might not the smartest way to do DBVS. However, it is noted that this article is provided to the beginner, especially those who want to do DBVS using their personal computer, not on the multinodes rack server. This process can be done by executing a bash script as follows: (*If you are still not familiar with a bash script, please note to press "enter" at the end of each line*)

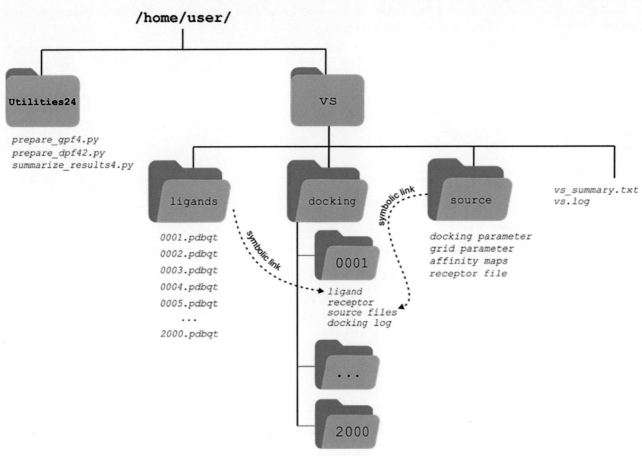

Fig. 7 Structure of directory in DBVS.

To type in the terminal	Explanation
$ cd ~/vs/docking/	Enter the working directory
$ for i in $(ls ../ligand/)	For each "i" in the result of "ls" of ligand directory
do echo $i	Display list of "i"
j=($(basename $i .pdbqt))	Declare "j" from the removal of ".pdbqt" from "i"
echo $j	Display the list of "j"
mkdir $j	Create the directory of "j"
cp ../ligand/$i $j	Copy the ligand.pdbqt to its respective folder
done	Done, iterate back until all "i" finished

At this point, each ligand's working directory consists of ligand.pdbqt. Since a docking process needs a receptor file, affinity maps, and docking parameter file, then the following script should be executed:

To type in the terminal	Explanation
$ cd ~/vs/docking/	Enter the working directory
$ for i in $(ls)	For each "i" in the result of "ls"
do echo $i	Display list of "i"
cd $i	Enter the directory of "i"
ln −s ../../source/1MRS_receptor.pdbqt.	Create a link to receptor file
ln −s ../../source/*map*.	Create a link to map files
pythonsh ~/Utilities24/prepare_dpf42.py	Prepare a docking parameter file for each ligand based on the reference file (control docking)
−l $i.pdbqt −r 1MRS_receptor.pdbqt	
−i ../../source/control.dpf -v	
cd..	Exit the directory "i"
done	Done, iterate back until all "i" finished

The script of "prepare_dpf42.py" will use the parameters in control docking "dpf" as the reference to generate a new "dpf" file for each ligand docking.

Finally, a run of VS will be executed. This script will enter each ligand's directory and do docking. After finishing with one docking, this script tells to go back to the parent directory and continue with the next ligand.

To type in the terminal	Explanation
$ cd ~/vs/docking/	Enter the working directory
$ for i in $(ls)	For each "i" in the result of "ls"
do echo $i	Display list of "i"
cd $i	Enter the directory of "i"
autodock4 -p *.dpf -l $i.dlg	Execute docking of ligand "i"
cd ..	Exit the directory of "i"
done	Done, iterate back until all "i" finished

Analyze the Virtual Screening Results

Summarize the VS results by using a Python script from AutoDockTools, namely summarize_results4.py, which located in the "Utilities24" folder.

```
$ cd ~/vs/docking
$ for i in $(ls)
do echo $i
pythonsh ~/Utilities24/summarize_results4.py -d $i −t 2.0 −B −k −e
−r 1MRS_receptor.pdbqt −o ../vs_summary.txt −a −v > > ../vs.log
done
```

Here is the detail of the options of summarize_results4.py:

-d Set the directory name that represented the docking folder of each ligand

-t RMSD tolerance between poses to cluster the docking result

-B to result in only the best docking and the largest cluster only (not all poses will be printed)

-k to report the number of hydrogen bond and its energy

-e to break down the binding energy into electrostatic, van der Waals, hydrogen bond, and desolvation energies

-r to identify the receptor file

-o output filename

-a append to output filename. This is ideal for analyzing VS results since they will be appended to a single file

-v verbose output

The resulted "vs_summary.txt" is a comma-separated value (CSV) file that can be imported to a popular worksheet program such as Microsoft Excel (as XLS file). Here is an example of an XLS file, a popular worksheet format, derived from the VS summary file.

Ligand	Lowest energy	No. of H-bond	Electrostatic energy	H-bond energy	Van der Waals energy	Desolvation energy	No. of atoms	No. of torsions	Ligand efficiency
1	− 11.42	5	− 4.94	− 2.35	− 10.34	4.82	25	5	− 0.54
2	− 8.22	5	− 0.66	− 2.17	− 10.05	3.77	23	4	− 0.41
3	− 7.32	4	− 0.72	− 1.81	− 8.71	3.33	19	3	− 0.46
4	− 11.46	3	− 5.34	− 2.56	− 11.28	5.45	28	7	− 0.50
5	− 11.17	3	− 4.60	− 1.54	− 12.48	4.98	29	6	− 0.41
Control	− 10.82	6	− 4.17	− 2.75	− 10.89	5.00	25	7	− 0.49

In principle, the calculated binding energy represents the predicted affinity of ligand to the receptor. However, many DBVS efforts resulted in a false positive result if the selection of active hits was only based on the binding energy. Docking score is not the absolute binding energy, hence we should consider the other aspects in filtering the screening result. For example, the TMPK's binding site is composed of many charged residues. Therefore, electrostatic interactions between the ligand and receptor are expected, including hydrogen bond formation. The screening result then can be sorted based on its electrostatic or hydrogen bond energy. The script of "summarize_results4.py" provides the breakdown of total binding energy into its components, such as a number of the hydrogen bond, intermolecular energy, and desolvation energy. Ligand efficiency also can be used to predict the ligand potency in binding to the receptor. It is a ratio between the binding energy and number of the atom. A higher value of ligand efficiency indicates a major contribution of a ligand's atoms to the interaction with the binding site. For this reason, the result of VS can be sorted based on many parameters, not only the binding energy.

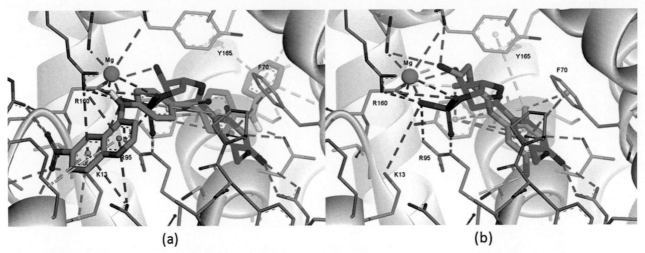

(a)   (b)

Fig. 8 Molecular interaction of (a) the best ligand (blue sticks) and (b) the ligand with similar pose with the positive control (gray sticks).

Besides the energy value, the selection of the best ligand can be assessed based on the intermolecular interaction with the receptor. For this purpose, the best conformation of each ligand can be prepared using the following script:

To type in the terminal	Explanation
$ cd ~/vs/docking/	Enter the working directory
$ for i in $(ls)	For each "i" in the result of "ls"
do echo $i	Display list of "i"
cd $i	Enter the directory of "i"
pythonsh ~/Utilities24/write_lowest_energy_ligand.py –f $i.dlg -v	Extract the ligand's conformation with the lowest energy, resulted in a "i_BE.pdbqt" file
pythonsh ~/Utilities24/pdbqt_to_pdb.py -f *BE.pdbqt	Convert the pdbqt format of "i_BE.pdbqt" into pdb file
cd ..	Exit the directory of "i"
Done	Done, iterate back until all "i" finished

Then, we collect all the best conformation of ligands into a new folder called "best_pose."

To type in the terminal	Explanation
$ cd ~/vs/docking/	Enter the working directory
$ mkdir ../best_pose	Create a new directory of "best_pose" in the upper level
$ cp */*BE.pdb ../best_pose	Copy all of the best poses of ligands into the new directory

Furthermore, we can combine the information derived from the table of results with the visual assessment. The PDB file resulting from the step above can be imported directly to BDSV (or any visualizer program). **Fig. 8(a)** shows that ligands with the best docking score sometimes do not form similar interaction with the control ligand. Near the magnesium ion, there are three positively charged residues, i.e., K13, R95, R160. While the control ligand formed a hydrogen bond, the best ligand formed Pi–cation interactions. Another ligand with moderate binding energy could form similar pose and interaction (**Fig. 8(b)**). The aromatic ring of this ligand perfectly fit the position of control ligand. Therefore, this ligand is also worth testing in further experiments.

Conclusion

The advances of docking software and computer hardware technology, and the availability of structure databases and experimental data, are strong reasons for integrating DBVS into each drug discovery project. DBVS is useful for narrowing hits candidates before in vitro experiment. However, in-depth analysis of DBVS result would enhance the successful rate of prediction. Deep understanding of the protein target should be useful to pick important parameters in selecting the best ligands for further testing, besides the docking binding energy. The protocol explained in this article is an introduction of how DBVS can be done using a personal desktop computer, which is usually found in the office or laboratory. A tutorial of DBVS in a Linux environment can be also a good start to further advance VS procedures, which mostly run in Linux.

See also: Algorithms for Structure Comparison and Analysis: Docking. Biomolecular Structures: Prediction, Identification and Analyses. Chemoinformatics: From Chemical Art to Chemistry in Silico. Drug Repurposing and Multi-Target Therapies. In Silico Identification of Novel Inhibitors. Integrative Bioinformatics. Molecular Dynamics Simulations in Drug Discovery. Natural Language Processing Approaches in Bioinformatics. Pharmacophore Development. Quantitative Structure-Activity Relationship (QSAR): Modeling Approaches to Biological Applications. Rational Structure-Based Drug Design. Small Molecule Drug Design. Structure-Based Drug Design Workflow

---

References

Breda, A., Basso, L.A., Santos, D.S., de Azevedo Jr, W.F., 2008. Virtual screening of drugs: Score functions, docking, and drug design. Current Computer-Aided Drug Designs 4, 265–272.

Cosconati, S., Forli, S., Perryman, A.L., et al., 2010. Virtual screening with autodock: Theory and practice. Expert Opinion on Drug Discovery 5 (6), 597–607. doi:10.1517/17460441.2010.484460.

Dunkel, M., Fullbeck, M., Neumann, S., Preissner, R., 2006. SuperNatural: A searchable database of available natural compounds. Nucleic Acids Research 34 (Database issue), D678–D683.

Ewing, T.J., Makino, S., Skillman, A.G., Kuntz, I.D., 2001. DOCK 4.0: Search strategies for automated molecular docking of flexible molecule databases. Journal of Computer-Aided Molecular Design 15 (5), 411–428.

Friesner, R.A., Banks, J.L., Murphy, R.B., et al., 2004. Glide: A new approach for rapid, accurate docking and scoring. 1. Method and assessment of docking accuracy. Journal of Medicinal Chemistry 47 (7), 1739–1749. doi:10.1021/jm0306430.

Ghosh, S., Nie, A., An, J., Huang, Z., 2006. Structure-based virtual screening of chemical libraries for drug discovery. Current Opinion in Chemical Biology 10 (3), 194–202.

Halgren, T.A., Murphy, R.B., Friesner, R.A., et al., 2004. Glide: A new approach for rapid, accurate docking and scoring. 2. Enrichment factors in database screening. Journal of Medicinal Chemistry 47 (7), 1750–1759. doi:10.1021/jm030644s.

Haouz, A., Vanheusden, V., et al., 2003. Enzymatic and Structural Analysis of Inhibitors Designed against Mycobacterium tuberculosis Thymidylate Kinase: New Insights into the Phosphoryl Transfer Mechanism. Journal of Biological Chemistry 278 (7), 4963–4971.

Hong, J., 2011. Role of natural product diversity in chemical biology. Current Opinion in Chemical Biology 15 (3), 350–354.

Ikram, N.K., Durrant, J.D., Muchtaridi, M., et al., 2015. A virtual screening approach for identifying plants with anti H5N1 neuraminidase activity. Journal of Chemical Information and Modeling 55 (2), 308–316. doi:10.1021/ci500405g.

Jain, A.N., 2006. Scoring functions for protein-ligand docking. Current Protein & Peptide Science 7 (5), 407–420.

Jones, G., Willett, P., Glen, R.C., Leach, A.R., Taylor, R., 1997. Development and validation of a genetic algorithm for flexible docking. Journal of Molecular Biology 267 (3), 727–748.

Kastritis, P.L., Bonvin, A.M., 2010. Are scoring functions in protein–protein docking ready to predict interactomes? Clues from a novel binding affinity benchmark. Journal of Proteome Research 9 (5), 2216–2225. doi:10.1021/pr9009854.

Kroemer, R.T., 2003. Molecular modelling probes: Docking and scoring. Biochemical Society Transactions 31 (Pt 5), 980–984. doi:10.1042/bst0310980.

Lavecchia, A., Di Giovanni, C., 2013. Virtual Screening Strategies in Drug Discovery: A Critical Review. Current Medicinal Chemistry 20 (23), 2839–2860.

Lengauer, T., Lemmen, C., Rarey, M., Zimmermann, M., 2004. Novel technologies for virtual screening. Drug Discovery Today 9 (1), 27–34.

Lipinski, C.A., Lombardo, F., Dominy, B.W., Feeney, P.J., 2001. Experimental and computational approaches to estimate solubility and permeability in drug discovery and development settings. Advanced Drug Delivery Reviews 46 (1-3), 3–26. S0169-409X(00)00129-0 [pii].

Lopez-Perez, J.L., Theron, R., del Olmo, E., Diaz, D., 2007. NAPROC-13: A database for the dereplication of natural product mixtures in bioassay-guided protocols. Bioinformatics 23 (23), 3256–3257.

McGann, M., 2011. FRED pose prediction and virtual screening accuracy. Journal of Chemical Information and Modeling 51 (3), 578–596. doi:10.1021/ci100436p.

Monge, A., Arrault, A., Marot, C., Morin-Allory, L., 2006. Managing, profiling and analyzing a library of 2.6 million compounds gathered from 32 chemical providers. Molecular Diversity 10 (3), 389–403. doi:10.1007/s11030-006-9033-5.

Morris, G.M., Goodsell, D.S., Huey, R., Olson, A.J., 1996. Distributed automated docking of flexible ligands to proteins: Parallel applications of AutoDock 2.4. Journal of Computer-Aided Molecular Design 10 (4), 293–304.

Muchtaridi, M., Bing, C.S., Abdurrahim, A.S., Wahab, H.A., 2014. Evidence of combining pharmacophore modeling-docking simulation for screening on neuraminidase inhibitors activity of natural product compounds. Asian Journal of Chemistry 14 (S.1), 59–63.

Quinn, R.J., Carroll, A.R., Pham, N.B., et al., 2008. Developing a drug-like natural product library. Journal of Natural Products 71 (3), 464–468. doi:10.1021/np070526y.

Rarey, M., Kramer, B., Lengauer, T., Klebe, G., 1996. A fast flexible docking method using an incremental construction algorithm. Journal of Molecular Biology 261 (3), 470–489.

Ripphausen, P., Nisius, B., et al., 2013. State-of-the-art in ligand-based virtual screening. Drug Discovery Today 16 (9), 372–376.

Rollinger, J.M., Langer, T., Stuppner, H., 2006. Strategies for efficient lead structure discovery from natural products. Current Medicinal Chemistry 13 (13), 1491–1507.

Shoichet, B.K., McGovern, S.L., Wei, B., Irwin, J.J., 2002. Lead discovery using molecular docking. Current Opinion in Chemical Biology 6 (4), 439–446. S1367593102003393 [pii].

Singh, S., Malik, B.K., Sharma, D.K., 2006. Molecular drug targets and structure based drug design: A holistic approach. Bioinformation 1 (8), 314–320.

Tanrikulu, Y., Krüger, B., et al., 2013. The holistic integration of virtual screening in drug discovery. Drug Discovery Today 18 (7), 358–364.

Verheij, H.J., 2006. Leadlikeness and structural diversity of synthetic screening libraries. Molecular Diversity 10 (3), 377–388. doi:10.1007/s11030-006-9040-6.

Villoutreix, B.O., Eudes, R., Miteva, M.A., 2000. Structure-based virtual ligand screening: recent success stories. Combinatorial Chemistry & High Throughput Screening 12, 1000–1016.

Vyas, V., Jain, A., Jain, A., Gupta, A., 2008. Virtual screening: A fast tool for drug design. Scientia Pharmaceutica 76, 333–360.

Wahab, H.A., Asarudin, R.M., Ahmad, S., et al. (2009). Nature based drug discovery (NADI) and its application to novel neuraminidase inhibitors identification by virtual screening, pharmacophore modelling and mapping of Malaysian medicinal plants. Trieste, Italy: Paper Presented at the Drug Design and Discovery for Developing Countries.

Wallace, A.C., Laskowski, R.A., Thornton, J.M., 1995. LIGPLOT: A program to generate schematic diagrams of protein-ligand interactions. Protein Engineering 8 (2), 127–134.

Xu, H., Agrafiotis, D.K., 2002. Retrospect and prospect of virtual screening in drug discovery. Current Topics in Medicinal Chemistry 2, 13005. 11320.

Relevant Websites

https://dtp.cancer.gov/RequestCompounds/index.xhtml
 Compound Request Form - Developmental Therapeutics Program.
https://dtp.cancer.gov/organization/dscb/obtaining/vialed.htm
 DSCB I Obtain Vialed.
http://autodock.scripps.edu/downloads
 Download Instructions - AutoDock - The Scripps Research Institute.
http://autodock.scripps.edu/resources/databases
 Databases - AutoDock - The Scripps Research Institute.
http://mgltools.scripps.edu/downloads
 Downloads - MGLTools - The Scripps Research Institute.
http://accelrys.com/resource-center/downloads/freeware/index.html
 Freeware Software - Accelrys.
http://autodock.scripps.edu/resources/raccoon
 Raccoon I AutoDock - The Scripps Research Institute.

Molecular Phylogenetics

Michael A Charleston, University of Tasmania, Hobart, TAS, Australia

Introduction

This Section begins with a basic motivation as to why (molecular) phylogenetics is important, and then introduces some of the relevant terminology. Section "Overview" continues with an overview of the phylogenetic inference process: From sequence alignment to constructing or estimating trees. Section "Common Assumptions in Molecular Phylogenetics" then explores the assumptions that are made in the name of phylogenetic inference: To do with the alignment itself, assumptions about independence, the direction of time, the nature of speciation, and similar. The last section discusses methods of finding the best tree, followed by a brief discussion on Bayesian methods and MCMC before some final comments on what (else) can possibly go wrong with molecular phylogenetic inference.

Motivation

Phylogenetic analysis is strongly motivated by our need to understand the relationships among species that have evolved from common ancestors. As – at least in theory – *everything living* has arisen from a common ancestor around 3.5–4.4 billion years ago, it is not beyond our imagination to conceive of a single over-arching "Tree of life" that describes the relationships among all extant and extinct species, from viruses to plesiosaurs. Understanding such relationships, both in terms of patterns of speciation and the times at which they occurred, is key to our ability to compare species within a framework that makes statistical *sense*. Without frameworks like phylogenetic trees (or even networks – see later) it is not possible to properly account for the shared history of species when comparing their characteristics.

But estimating phylogenetic history is hard. There are a number of reasons for this, but perhaps the most obvious is that we cannot travel backward in time to check whether what we have attempted to uncover is in fact correct. Another difficulty is that the number of phylogenetic trees is enormous. It's been long known (since at least 1870 (Schröder, 1870); more elegantly in (Cavalli-Svorza and Edwards, 1967)) that the number of possible trees for n species grows at a frightening rate: so much so that for even modest numbers of species, around 100, there are more than 3×10^{184} trees – many, many more than the estimated number (something around 10^{82}) of particles in the universe! So not only is it hard to check whether a tree is "right" (leaving aside for a moment the potential issues of assuming that evolution is tree-like), but it's also computationally intractable to search through the space of all possible trees for the best one.

Language of Phylogenetic Inference

Phylogenetics is rich in terminology, having gained it from mathematics, biology, and computer science, and it can be confusing. It is worth defining the terms we will be relying on at this point (**Box 1**).

Box 1: Terminology

Analogy: a feature of related species that is similar, but is not derived from their common ancestor (c.f., homology).

Branch: also known as an edge in a phylogenetic tree, it is equivalent to a split of the leaf nodes (taxa) into two subsets. The "same" branches can therefore occur in multiple trees, so long as they separate the taxa into the same subsets.

Character, character state: we distinguish the character, which is the particular feature such as "nucleotide" at a given site, with its state, which in this case would be A, C, G, or T.

Clade: a monophyletic group.

Graph: a collection of nodes, and pairs of nodes called edges.

Homologous: of a character, feature or site that is shared and derived from a common ancestor (c.f., analogy).

Leaf (tip) a node of a tree that is adjacent to exactly one branch (or edge); corresponding to extant taxa.

Likelihood: a method by which the probability of a given observation can be calculated, conditional on an assumed probabilistic model.

Locus; plural loci: a site or sequence within a molecular sequence.

MRCA most recent common ancestor of a set S of taxa: The node in a rooted phylogeny that is closest to the leaves, that is ancestral to all the taxa in S.

Monophyletic, monophyly: a taxon set S in a tree is monophyletic if is contains all the descendants of the most recent common ancestor of S.

Network: in phylogenetics, a graph that contains multiple paths between at least some pairs of nodes, either representing a hypothesized history of hybridization and/or horizontal gene transfer, or support for conflicting splits of the taxa.

Paraphyletic, paraphyly: a taxon set S in a tree is paraphyletic if it contains most, but not all, of the descendants of the MRCA of S.

Phylogeny: also called a phylogenetic tree, a tree whose nodes represent taxa or operational taxonomic units (OTUs) and whose branches correspond to evolutionary lineages.

 Encyclopedia of Bioinformatics and Computational Biology, Volume 2   doi:10.1016/B978-0-12-809633-8.20675-4

Root: a node of a phylogenetic tree that is specially designated as the common ancestor of all the other nodes.
Sequence: in the context of molecular phylogenetics, sequence means linear array of biological characters, e.g., nucleotides.
Site: a position in a molecular sequence, e.g., the 10th nucleotide in the gene coding for *gag* protein.
Split: a partition of the taxon set of interested into two subsets, with empty intersection and no taxa missing. Equivalent to a branch.
Taxon; plural taxa: a phylogenetically distinct biological type, e.g., species or genus.
Tip: a node corresponding to an extant taxon.
Tree: a graph in which each pair of nodes has exactly one path connecting them.

Overview

Molecular phylogenetics is concerned with the estimation of the evolutionary relationships of species or other types of taxa that we can observe today. It is a computationally and statistically challenging task, but it is central to enabling comparison of species within an appropriate statistical framework.

Molecular phylogenetics is a routine pursuit in modern biology, and while it is broadly a simple process, the details are not inconsequential. The general process is as follows: (1) Obtain a set of comparable molecular sequences across the range of taxa of interest, (2) align those sequences so individual positions in each sequence correspond to each other, (3) build or search for a tree or set of trees with methods as briefly described herein, (4) test the tree(s) for their correctness, and generally (5) use the tree(s) to answer a question about the species, e.g., when was the divergence into the major clades of birds. It is possible to skip step (2) if re-analysing an existing data set.

Data

Beginning the whole inference process is the creation and curation of suitable data.

For molecular phylogenetics, the data are sets of homologous sequences, e.g., those of particular genes, represented as strings of adenine (A), guanine (G), cytosine (C) and thymine (T) for DNA, or A, C, G, uracil (U) for RNA sequences; occasionally too these sequences are transformed into simpler encoding as purines (A, G) and pyrimidines (C, T/U). Of course it is also possible to use amino acids as characters in such sequences, or even the 64 codons that encode them (usually removing the 3 "stop" codons from consideration within a protein encoding gene).

Homologous means that the sequences correspond to each other across species in the obvious desirable way: They branched and diversified from each other at the speciation events that we're trying to recover. Sequences that are paralogous, that is, they don't have the same evolutionary history as the species in which they reside, are more problematic for phylogenetic analysis because correctly recovering the evolutionary relationships among the sequences does not directly give you the relationships among the species.

Usually sequences that are likely to show some evolutionary "signal" are chosen for phylogenetic inference: those that have some constraints so they do not evolve too quickly as to appear random with respect to each other.

Depending on the overall age of the set of taxa of interest, slower or faster regions in the homologous sequences are chosen. For example to estimate the phylogenetic relationships among a group of viruses over 100 years, the *env* gene might be used; for investigating deep historical relationships among archaea, bacteria and eukaryota, much more conserved genetic regions must be chosen such as DNA polymerases.

Aligning the Sequences

The sequences must be aligned, that is, arranged relative to each other with the possible insertion of gaps at different places, such that for each sequence, the characters (e.g., nucleotides, amino acids, purines/pyrimidines etc) at position *i* have the same evolutionary history as each other. In this sense the positions, which are generally called *sites*, are also hoped to be homologous. **Fig. 1** shows how some toy DNA sequences might evolve on a tree, showing homologous sites, insertion/deletion events, and illustrating how reconstructing the branching order might be tricky. Maybe reword here.

Alignment would be trivially easy if the only mutations that could occur were changes of state, e.g., from A to T, but sequences evolve in other ways also. In the context of sequence alignment the main other mode by which sequences can change is via deletion of sections of sequence, or insertion of new sections. Faced with two sequences that have been optimally aligned, in which there is a section in one that is "extra" (see **Fig. 1** and the table below), it is not possible to determine whether that section was inserted in one sequence, or deleted from the other one. For this reason such regions are referred to also as "indels" and the events are indel (insertion/deletion) events.

To find the optimal alignment of just two sequences is relatively straightforward: first, choose a costing function that accounts for similarities and differences between aligned sites in the two sequences, as well as gaps in either caused by insertion or deletion events during their evolution, and then use dynamic programming to search through the complete space of possible alignments in an efficient way (Waterman *et al.*, 1991). Costing the match or mismatch of characters like nucleotides, amino acids or codons is usually done by performing log-odds analyses of pairs of curated sequences that are agreed to be correctly aligned, at various levels of difference (e.g., BLOSUM (Henikoff and Henikoff, 1992), **Fig. 2**; PAM (Dayhoff *et al.*, 1978); JTT, (Jones *et al.*, 1992) etc.). The match and mismatch scores are often rounded to integer values for computational efficiency, which is generally a "safe" assumption given that the log-odds are themselves approximations, derived from hand-picked and often hand-aligned data.

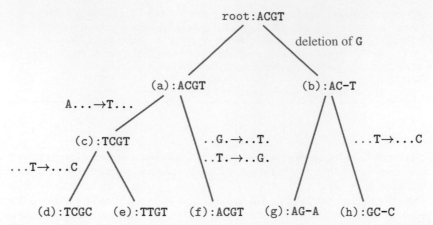

Fig. 1 A simple rooted, binary tree with some exemplar DNA sequences. All sites are homologous, with site 3 in the ancestral sequence above (b) undergoing a deletion. Site 1 shows a single change above (c), leading to support for the split (d, e l f, g, h). Site 2 is constant except for parsimony uninformative changes above (e) and (g) (which are left out, for clarity). Site 3 shows a deletion above (b), which would be indistinguishable from an insertion event above (a), and a silent change above (f) from G to T and back. Site 4 shows parallel changes above (d) and (h), leading to support for a (d, h l e, f, g) split, which is not part of the tree.

Ala	4																			
Arg	-1	5																		
Asn	-2	0	6																	
Asp	-2	-2	1	6																
Cys	0	-3	-3	-3	9															
Gln	-1	1	0	0	-3	5														
Glu	-1	0	0	2	-4	2	5													
Gly	0	-2	0	-1	-3	-2	-2	6												
His	-2	0	1	-1	-3	0	0	-2	8											
Ile	-1	-3	-3	-3	-1	-3	-3	-4	-3	4										
Leu	-1	-2	-3	-4	-1	-2	-3	-4	-3	2	4									
Lys	-1	2	0	-1	-3	1	1	-2	-1	-3	-2	5								
Met	-1	-1	-2	-3	-1	0	-2	-3	-2	1	2	-1	5							
Phe	-2	-3	-3	-3	-2	-3	-3	-3	-1	0	0	-3	0	6						
Pro	-1	-2	-2	-1	-3	-1	-1	-2	-2	-3	-3	-1	-2	-4	7					
Ser	1	-1	1	0	-1	0	0	0	-1	-2	-2	0	-1	-2	-1	4				
Thr	0	-1	0	-1	-1	-1	-1	-2	-2	-1	-1	-1	-1	-2	-1	1	5			
Trp	-3	-3	-4	-4	-2	-2	-3	-2	-2	-3	-2	-3	-1	1	-4	-3	-2	11		
Tyr	-2	-2	-2	-3	-2	-1	-2	-3	2	-1	-1	-2	-1	3	-3	-2	-2	2	7	
Val	0	-3	-3	-3	-1	-2	-2	-3	-3	3	1	-2	1	-1	-2	-2	0	-3	-1	4
	Ala	**Arg**	**Asn**	**Asp**	**Cys**	**Gln**	**Glu**	**Gly**	**His**	**Ile**	**Leu**	**Lys**	**Met**	**Phe**	**Pro**	**Ser**	**Thr**	**Trp**	**Tyr**	**Val**

Fig. 2 BLOSUM62 matrix of log-odds based amino-acid match/mismatch scores.

The most complex function to score gaps that can be accommodated in this pair-wise alignment problem is called *affine* scoring: Under this scheme there is a cost of *opening* a gap, and a cost of *extending* it. This method neatly accounts for the fact that it is much less likely for two adjacent regions to be inserted/deleted than having a single indel event. The score of a gap of length d is then $a + bd$ where a is the gap opening cost and b is the gap extension cost.

Here are two alignments of the sequences at the tips of the tree in **Fig. 1**, with the nucleotides aligned in columns:

taxon	sequence
d	TCGC
e	TTGT
f	ACGT
g	AGA
h	GCC

taxon	Alignment 1			
d	T	C	G	C
e	T	T	G	T
f	A	C	G	T
g	A	G	–	A
h	G	C	–	C
site:	1	2	3	4

taxon	Alignment 2: too "gappy"							
d	–	T	C	–	G	–	–	C
e	–	T	–	T	G	T	–	–
f	A	–	C	–	G	T	–	–
g	A	–	–	–	G	–	A	–
h	–	–	–	–	G	C	–	C
site:	1	2	3	4	5	6	7	8

$$
\text{Score} = \begin{array}{c c c c c}
A & 4 & & & \\
G & 1 & 5 & & \\
C & 0 & -1 & 4 & \\
T & 1 & -1 & 0 & 3 \\
\text{gap} & -7 & -7 & -7 & -7 \\
& A & G & C & T
\end{array}
$$

Fig. 3 Example cost matrices for nucleotide alignment. These scores are illustrative only.

In Alignment 1 a gap has been inserted at position 3 for taxa g and h, with the remaining sequences having a G at that position; the other sites all show multiple character states. Alignment 2 has a constant site at position 5, but at considerable expense elsewhere in the alignment as more gaps have been inserted than are biologically realistic. (Note that if Alignment 2 were optimal for some alignment program, the gap cost is set far too low in comparison to that for a mismatch between character states.) In general we hope to obtain alignments that show a degree of conservatism across species in the majority of sites, because these have a higher chance of each site being aligned to its homolog in other species. If the alignment for a set of sequences or parts thereof is difficult with much ambiguity, e.g., evidenced by particularly "gappy" alignments (as with Alignment 2 above), then such data may be dropped from the analysis. There are several software tools to automatically drop such sites from analysis, e.g., GBlocks (Castresana, 2000).

Multiple sequence alignment (MSA), for more than two sequences, is known to be computationally intractable: To guarantee an optimal scoring alignment even under the relatively simple affine scoring method above, requires an amount of space and time that grows exponentially with the number of sequences. Therefore for MSA the typical approach is to use one or more of a suite of well-established heuristic sequence alignment methods, such as MAFFT (Katoh et al., 2002), muscle (Edgar, 2004), k-align (Lassmann and Sonnhammer, 2005), and T-Coffee (Notredame et al., 2000) and often to then adjust their outputs "by eye". Heuristic methods do not guarantee optimality (unlike for pair-wise alignment), and may not always yield the same output. For example two separate runs of any of these programs may in principle yield two different alignments, particularly if there are any equally high-scoring options, as ties have to be resolved either arbitrarily (giving rise to unbiased but inconsistent behaviour) or deterministically (giving rise to consistent, but biased behaviour).

MSA, or often "sequence alignment" or even simply "alignment" – context notwithstanding – just like pairwise alignment, has the aim of associating individual positions along every sequence with corresponding positions in the other sequences, such that each position corresponds to the same evolutionary history.

It is in principle possible to combine MSA with tree estimation, but these methods have yet to become standard in the phylogenetics community.

While the multiple alignment problem is computationally too demanding to solve perfectly, current heuristic methods can be very fast and reliable: Using default parameters that are moderately well tuned, it is entirely possible to perform this part of the phylogenetic inference workflow quite quickly, and without inspecting the data at all. *Caveat emptor.*

For protein coding genes, more complex alignment methods exist that take into account the three-dimensional structure of the proteins.

The oldest, but largely superseded, automatic alignment program is Clustal (Higgins et al., 1992). It has default values of a gap opening cost of 10.0 and a gap extension cost of 0.1, so a gap of length 3 nucleotides has a cost of 10.3 by default.

In general the gap opening and extension costs can have different best values for different data sets (see **Fig. 3**).

Common Assumptions in Molecular Phylogenetics

In any inference process we must make assumptions. In phylogenetics these assumptions are mostly incorrect, but it turns out we can get away with making them in many situations. Some of the major ones are listed below:

The Sequences are Homologous

For many organisms, there is only one copy of any given gene, or of some other phylogenetic useful marker. However in eukaryotes (classically, hæmoglobins in primates), it is not a safe assumption that genes that encode corresponding proteins across multiple organisms are necessarily homologous: They could be paralogous, that is, their most recent common ancestor might be before the most recent common ancestor of the organisms themselves (see **Fig. 4**).

The Alignment is Correct

Partly as a result of the computational complexity of inferring a tree from an alignment, it is rarely possible to move forward with phylogenetic inference without assuming that the alignment is good enough and moving on. Therefore the usual practise is to treat the alignment as fixed, and then undertake the even harder task of inferring the phylogeny based on it.

There are some methods available that attempt to perform MSA and tree estimation concurrently – but these are very lengthy approaches, and it is not yet clear that they are an ideal solution, particularly for large data sets.

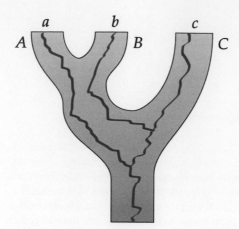

Fig. 4 The species tree here (thick pipes) has a gene tree within it that is not congruent: The two left-most genes, for *a* and *b*, are paralogous: Their last divergence is before the divergence of their host species.

Sequence Evolution is Independent and Identically Distributed

This is the most ubiquitous of assumptions made in molecular phylogenetics and is the most often realised as patently false. It states that each site in a sequence evolves in the same way, and independently of, its neighbor. A moment's thought reveals that this cannot be true: In codons first, second and third positions have different selective constraints on them. Loci that code for microRNAs (miRNAs) will have sites within them that, if they change, must change in concert else they will destroy the ability of the miRNA to fold correctly. In protein-coding genes the bulk of evolving sites correspond to amino acids that interact with others. There are many cases where concomitantly varying sequences must be a better reflection of reality than the i.i.d. assumption (Lockhart *et al.*, 1998; Wang *et al.*, 2006) but yet for practical purposes it is often still useful to make this simplification: To a great extent the huge success of molecular phylogenetics is down to the flood of phylogenetically informative data that permit the robust estimation of phylogenetic history despite this incorrect assumption.

Time is Continuous

Biologically speaking, it does not make sense that time be treated as a continuous quantity: mutations occur at discrete time points such as cell division after all. But it is an assumption that we can get away with (as we do with the i.i.d. assumption), because in most cases the evolutionary changes occur at such a gradual pace in comparison with the time periods over which we measure them, that it may as well be modelled as a stochastic process in continuous time.

The mathematical benefits are significant for moving to continuous time, as we can employ tools from stochastic modelling such as the exponentiation of rate matrices to produce probabilities of change over arbitrary time periods. Without the assumption of continuous time, we really could not do phylogenetic estimation in any statistically justifiable way, but an assumption it is and for completeness, it is included here.

Time-Reversibility

Until recently almost all models of molecular evolution have worked under the assumption that evolution does not care about the direction of time: The probability of changing from one nucleotide to another in a given time interval is always exactly the same as that probability in the reverse direction.

Clearly this is not biologically correct! As soon as one understands that there can be selection toward a particular sequence then reversibility is broken. However, this is a crucial simplification that enables the likelihood of a set of aligned sequences under a model to be calculated regardless of where the root of the tree might be: Without the assumption, we must in principle calculate the likelihood from every possible location of the root, a considerable computational burden.

Tree-Like Evolution

Possibly the most famous phylogenetic tree is that which was sketched by Charles Darwin in his book "On the origin of species" (Darwin, 1859), with the telling note, "I think." Darwin was clearly imagining the formation of new species by a process of branching, with the possibility of lineages (species) going extinct, but not in terms of them converging again.

This is an important point because, if it is true (and it mostly is), it means that *trees* are the objects that we must use to represent species' origins. If we do not adopt this idea wholly, then we must permit more complex structures, in which there could be multiple paths from a common ancestor to some descendant. In a tree, all paths are unique – this can fact be part of the definition of what a mathematical tree is: A network connecting a collection of objects, such that there is always *exactly* one path between every two such objects (see later; other places in the book). For the most part, phylogenetic inference (whether using molecular data or anything else) makes this assumption, but there is (often heated) discussion in the scientific literature as to the extend of non-tree-like evolution, caused by such mechanisms as lateral/horizontal gene transfer (LGT, HGT respectively) and hybridization.

Therefore if we think trees describe evolutionary relationships, we need to use the tools of branching processes in order to describe them. All this discussion so far is regarding a process that is moving forward with time, but there are also a wealth of tools to be employed if we look backwards. These latter tools use the concept of lineages *coalescing* as we move into the past, and they continue to yield powerful results that enable researchers to understand better the way species, and particularly lineages within species, emerge.

Another generalisation that we must consider is non-treelike evolution, corresponding to lineages converging, e.g., by hybridization or horizontal gene transfer. In phylogenetics these representations are usually referred to as *networks* (though to a mathematician, all trees are also networks).

For the moment we restrict ourselves to phylogenies as trees (but see later).

Further, we constrain our thinking to trees in which speciation events always give rise to exactly two "new" lineages – such trees are called *binary*, and differentiate between those with a chosen root and those without one (see Time and Reversibility).

Non-Tree-Like Evolution

A given site is assumed to evolve according to a tree-like pattern, only branching as they move forwards in time, but there are other units of evolution that can converge instead. For example, genes may be horizontally (or "laterally") transferred from one species to another (Bergthorsson *et al.*, 2003), and species can hybridize (Harrison and Larson, 2014). In those cases the genes may not share the same evolutionary histories, and the species phylogeny is better represented as a hybridization network.

An alternative is that due to uncertainty in the phylogenetic signal, even though the tree history may be tree-like (branching), it is desired to show a range of alternative hypotheses in the same figure. These phylogenetic networks can illustrate support for branching patterns and splits of the taxa of interest that are conflicting: They cannot both be part of a tree, but they can be shown as, for example, a "NeighborNet" (Bryant and Moulton, 2003). NeighborNets have become a standard tool in molecular phylogenetics. Their construction is based on estimates of pair-wise distances between taxa, rather than on a scoring method for a tree.

What Makes a Tree a Good Tree

Because of the vast number of possible trees, growing super-exponentially in the number of taxa involved, it is a virtual certainty that the estimated phylogeny for very large sets of taxa will be incorrect – somewhere. However, that is not to say that everything in the tree is wrong. It is helpful to think of a phylogeny as a set of non-independent hypotheses of relatedness, many of which may be correct!

For example in **Fig. 4** there are several hypotheses displayed:

1. A and B are each others' closest relatives and form a monophyletic group (a.k.a. clade);
2. A, B and C are monophyletic/form a clade;
3. B and C are paraphyletic in this tree.

A very standard method of measuring robustness of a phylogeny is via *bootstrapping*, which is a method by which sites in an alignment are sampled at uniform random and with replacement, in order to create "pseudo-replicates" of the original data set (Efron *et al.*, 1996). Each pseudo-replicate is then subjected to the same inference methods as for the original data set, and for each tree inferred on the pseudo-replicates, a note is made of which branches are in that tree. The proportion (or percentage) of times in which each branch is recovered is a measure of the phylogenetic consistency of the data.

The philosophy behind such resampling – leading to the idea of "pulling oneself up by one's bootstraps" is that, if the observed (assumed to be homologous, and independently evolving) sites were representative of a larger population of such sites, then resampling from them would be equivalent to resampling from that larger population. This is a strong assumption, so interpretation of the resulting bootstrap values must be done with care. It is technically not a measure of confidence, as such, as it is possible to gain arbitrarily large bootstrap values simply by having large data sets (Jermiin *et al.*, 2005), but in general high values (e.g., 80% or more) are taken to be indications that the data are providing strong signal in favour of a particular branch.

Choosing the Best Tree

In order to select the tree that is the best possible explanation of the evolutionary relationships among a set of species, we must be able either to construct the tree according to some algorithm, such as Neighbor-Joining (Saitou and Nei, 1987; Studier and Keppler, 1988; Gascuel and Steel, 2006), or to evaluate each tree with some kind of score function and search through the space of all possible trees for the one(s) that maximise(s) the score. Next we discuss the two main score functions for trees, and then discuss tree searching.

In order to recover a tree from molecular sequence data we must have in mind some model that effectively describes what we know about sequence evolution. Models arise from different philosophies, and in molecular phylogenetics, the two main ones are that of maximum parsimony, and of (probabilistic) likelihood. Both of these ideas require a search through tree space: the set of all possible trees that relate the set of taxa of interest, with trees being considered as adjacent with respect to some perturbation if one can be modified to match the other.

Maximum Parsimony

According to *maximum parsimony*, also referred to as Occam's Razor, it is the simplest explanation that is the best explanation of what happened. The origin of this idea is ancient, but it still has some value in modern science, provided one recognises its limitations.

Fitch (Fitch, 1971) devised an ingenious algorithm that counts the minimum number of changes of state that are required on a given tree in order to explain the sequences observed at its tips. The total Parsimony score – Also called the Parsimony *length*, is the sum of the minimum number of changes that must occur on the tree to account for the characters at the leaves of the tree, and it is this score that is to be minimised over all possible trees.

The algorithm is performed on each site in turn and the score (length) for site is accumulated into the total. (Obviously, one can speed things up by calculating it only on unique site patterns and then multiplying by the number of times it occurs, and there are other optimisations possible, but they are not illuminating here; the interested reader is directed to).

Fitch's algorithm was later shown to be a special case of a more general algorithm of Sankoff (Sankoff, 1975), which can also take into account weights for different substitutions. Sankoff's algorithm requires slightly more bookkeeping, in that one must calculate and retain the minimal total cost for every possible character state at each node, but it is beyond the scope of this brief article to go into that detail; the interested reader is directed to Sankoff (1975).

The Fitch algorithm is outlined in **Box 2** for rooted binary trees on nucleotide data.

Box 2: Algorithm for calculating the Parsimony length of a tree

Algorithm 1: `ParsimonyLength(T,A)`

> **given** T, a rooted tree with tips labelled $1,\ldots,n$
> **given** A, an aligned set of n observed sequences of length ℓ
> **let** r **be** the root node of T
> **let** p **be** the parsimony length of T
> **let** V **be** the set $V = \{v_1,\ldots,v_{n-1}\}$ of non-leaf nodes of T
> $p \leftarrow 0$
> **for each** $(i = 1,\ldots,\ell)$ **do** {
> · $p \leftarrow p + \texttt{Fitch}(r,A,i)$
> }
> **return**(p)

Algorithm 2: `Fitch(v,A,i)`

> **given** v, a node in a rooted binary tree
> **given** A, an aligned set of n observed sequences of length ℓ
> **given** i, the position of a site of interest in A
> **let** $F(v)$ **be** the character state(s) assigned to v
> **let** q **be** the Parsimony score of this site
> $q \leftarrow 0$
> **if** (v is a leaf) **then** {
>   /* $F(v)$ is our observed data */
> · $F(v) \leftarrow A_{i,v}$
> ·   **return**(0)
> } **else** {
> ·   **let** S **be** a set of states allowed (e.g., A, C, G, T)
> · $S \leftarrow F(v.leftchild) \cap F(v.rightchild)$
> ·   **if** ($S = \emptyset$) **then** {
> ·   · $q \leftarrow q + 1$
> ·   · $S \leftarrow F(v.leftchild) \cup F(v.rightchild)$
> ·   }
> · $F(v) \leftarrow S$
> }
> **return**(p)

For example, consider again the alignment that corresponds to the tree in **Fig. 1**:

taxon	sequence
d	TCGC
e	TTGT
f	ACGT
g	AG–A
h	GC–C

There are five possible states: A, C, G, T, and the "gap" state –. Consider the first site pattern, TTAAG for the five taxa d, e, f, g, h in order. For this part we use the notation (a)=X to mean node (a) has character state X. Proceeding from the leaves of the tree to the root, we first consider node (c), which has the two child nodes and character states (d) T and (e)=T. $\{T\} \cap \{T\} = \{T\}$, so we assign T to node (c). Moving up to node (a) which now has children (c)=T and (f)=A. $\{T\} \cap \{A\}$ is empty so we set (c)={A, T} and add 1 to the Parsimony length of this tree. Next is node (b), which has as its child nodes (and character states) (g)=A and (h)=G. Again we find the intersection of these two sets of character states is empty so we increment the Parsimony length and assign (b)={A, G}. The last node is the root, which now has children (a)={A, T} and (b)={A, G}. Since $\{A,T\} \cap \{A,G\} = \{A\}$ is non-empty, we assign A to the root node. The Parsimony score for this site is 2.

To calculate the Parsimony score for the complete alignment we repeat the process over the other three site patterns; and if we wish to account for different costs of different substitutions, we can keep track of all possible character states at all the nodes, and retain their minimal scores as we move "up" the tree to the root.

Maximum Likelihood

By far the most common tool used currently to infer molecular phylogenies is maximum likelihood (ML). This is a simple concept but one which has grown to comprise a very rich suite of models and methods. We begin with a toy example to illustrate the key concepts and then mention some (but not all) ways in which it is generalised.

The essential idea of maximum likelihood (ML) is that under a probabilistic model, we may calculate the probability of observing a given outcome. The given outcome we choose is our set of observed data: That is, a set of aligned sequences. The simplest possible model for illustration purposes is a two-state one, in which we have sequences of 0s and 1s, evolving according to a symmetric rate matrix

$$Q_{CF} = \begin{bmatrix} -\alpha & \alpha \\ \alpha & -\alpha \end{bmatrix}$$

This matrix is labelled "CF" as it corresponds to the Cavender-Farris model (Cavender, 1978) Rate matrices correspond to limits of instantaneous probabilities of change from one state to another; but to calculate an actual probability for a given time period, we must convert from rate matrices to probabilities.

Assuming a Poisson process for mutation events we can determine that for a rate matrix Q operating over a time period t, the probability matrix that corresponds to changes of state over that time period is given by $P=\exp(Qt)$. In the case that Q is conveniently exponentiated this leads to closed forms that can be used to calculate the probability of a given (e.g., nucleotide) state into another, conditioned on that rate matrix being correct.

The above instantaneous rate matrix Q is called doubly stochastic: Both its rows and its columns individually sum to zero. Doubly stochastic models have the property that the steady state distribution of character states is uniform: In this case, our two character states will settle over a long enough period to occur in equal proportions. In the case of DNA models that are doubly stochastic, we expect their nucleotide base frequencies to tend to $(\frac{1}{4}, \frac{1}{4}, \frac{1}{4}, \frac{1}{4})$.

The probability matrix that corresponds to this rate matrix is given by

$$Exp(Q_{CF}t) = \frac{1}{2} \begin{bmatrix} 1 + e^{-2\alpha t} & 1 - e^{-2\alpha t} \\ 1 - e^{-2\alpha t} & 1 + e^{-2\alpha t} \end{bmatrix}$$

For the next simplest model, still doubly stochastic but with four states, we have

$$Q_{JC} = \begin{bmatrix} -3\alpha & \alpha & \alpha & \alpha \\ \alpha & -3\alpha & \alpha & \alpha \\ \alpha & \alpha & -3\alpha & \alpha \\ \alpha & \alpha & \alpha & -3\alpha \end{bmatrix}$$

yielding

$$Exp(Q_{JC}t) = \begin{bmatrix} 1 + 3e^{-4\alpha t} & 1 - e^{-4\alpha t} & 1 - e^{-4\alpha t} & 1 - e^{-4\alpha t} \\ 1 - e^{-4\alpha t} & 1 + 3e^{-4\alpha t} & 1 - e^{-4\alpha t} & 1 - e^{-4\alpha t} \\ 1 - e^{-4\alpha t} & 1 - e^{-4\alpha t} & 1 + 3e^{-4\alpha t} & 1 - e^{-4\alpha t} \\ 1 - e^{-4\alpha t} & 1 - e^{-4\alpha t} & 1 - e^{-4\alpha t} & 1 + 3e^{-4\alpha t} \end{bmatrix} = \frac{1}{4} \begin{bmatrix} 1 + 3\beta & 1 - \beta & 1 - \beta & 1 - \beta \\ 1 - \beta & 1 + 3\beta & 1 - \beta & 1 - \beta \\ 1 - \beta & 1 - \beta & 1 + 3\beta & 1 - \beta \\ 1 - \beta & 1 - \beta & 1 - \beta & 1 + 3\beta \end{bmatrix}$$

where $\beta = e^{-4\alpha t}$.

The most general time-reversible model is known as GTR, the General Time-Reversible model (Tavaré, 1986). GTR parameters can be expressed simply as the equilibrium base frequencies a four-dimensional vector $\pi = (\pi_A, \pi_C, \pi_G, \pi_T)$ whose non-negative entries sum to 1, and a rate matrix

$$Q_{GTR} = \begin{bmatrix} -(\alpha\pi_G + \beta\pi_C + \gamma\pi_T) & \alpha\pi_G & \beta\pi_C & \gamma\pi_T \\ \alpha\pi_A & -(\alpha\pi_A + \delta\pi_C + \varepsilon\pi_T) & \delta\pi_C & \varepsilon\pi_T \\ \beta\pi_A & \delta\pi_G & -(\beta\pi_A + \delta\pi_G + \eta\pi_T) & \eta\pi_T \\ \gamma\pi_A & \varepsilon\pi_G & \eta\pi_C & -(\gamma\pi_A + \varepsilon\pi_G + \eta\pi_C) \end{bmatrix}$$

where a moment's inspection reveals that if $\pi_A = \pi_C = \pi_G = \pi_T = 0.25$ and $\alpha = \beta = \cdots = \eta$, we recover the Jukes-Cantor model as a special case.

The most general model possible for DNA is the General Markov Model, which is exactly as it sounds: Simply a general Markov model on four states.

Matrices like those above can yield well defined probabilities of change from one state (nucleotide) to another given a certain amount of time. This is the basis of constructing the likelihood on a complete tree, as follows: Consider a branch in a tree, with some length ℓ. We may sum over all the possible character states at the ends of this branch and, for each possible combination, calculate the probability of change from the state at one end to the state at the other. Felsenstein created an efficient "pruning algorithm" to do this calculation for a complete tree, over all possible internal character states (in the case of DNA, nucleotides) in the tree, by only keeping track of four possible states at each node, representing the maximum possible likelihood of character state assignments in its descendants (Felsenstein, 1973). The branch lengths and the model parameters (α, β etc. above) directly affect the likelihood. Thus an ML search for the best tree is a process that should in principle search over all possible sets of branch lengths, and all possible sets of parameter values: The computational task is immense.

There are hundreds of possible matrices for DNA, and more of course for amino acids or codons. The decision on which model to choose is often a difficult one, and various approaches have been devised to help. Early in the relatively short history of maximum likelihood applied to phylogenetic inference, likelihood ratio tests were used to compare nested models. The JC model above is a special case of the GTR, but they differ in the number of free parameters (JC can be considered as having just 1 parameter, and GTR 9: They differ by 8 free parameters). Twice the difference in log-likelihoods is known to be χ^2-distributed, with degrees of freedom equal to the difference in the number of their free parameters, so this method has been used successfully to guide the researcher through a decision tree in the well-cited program, ModelTest (Posada, 2008). This approach has its limitations though, and a standard scheme is now to determine for a given tree the Akaike Information Criterion (AIC) score or one of its derivations (AICc, BIC) to give for each model a simple score, and the researcher may then choose that which has the best score. This avoids the computational complexity of comparing every possible pair of nested models, and enables comparison of models that are not nested.

In general constructing a tree from an algorithm is considerably faster than searching through tree space, taking in the order of $O(n^3)$ operations over all, with n taxa. To search the entirety of tree space is prohibitive: The number of trees increases with n, the number of taxa, at a rate approaching n^n (exact numbers below); hence for all but the most moderately sized examples, we must use heuristics.

Searching Tree Space

The Fitch and Sankoff algorithms solve the "Small Parsimony Problem" – Finding the minimum number of changes that are required to explain an alignment on a tree. The related "Large Parsimony Problem" is that of finding that tree that has the best score, out of all $(2n-3)!!$ possible rooted binary trees on n taxa (though in fact, as we are regarding the substitutions as equally costly in either direction, we can treat the tree as unrooted, and there are "only" $(2n-5)!!$ of those). Similarly if the score function is likelihood we are faced with the corresponding problem: find the tree (with branch lengths and other model parameters) that has the highest likelihood of generating the data we observed.

The Large Parsimony Problem, and the tree search problem in general, is solved using local search, typically hill-climbing or simulated annealing. The starting tree is usually constructed quickly using Neighbor-Joining (NJ) or similar. The hill-climbing search operates by perturbing the current tree in some well-defined way and calculating score of the perturbed tree. Hill-climbing is relatively fast among local searches, but it has the obvious problem that if the landscape of the search space has multiple peaks (optima) then it can become stuck in one of them that is not globally optimal. There are several ways that have been developed to get around this issue, including performing multiple hill-climbing searches from different starting points, and using more sophisticated methods such as Simulated Annealing (Van Laarhoven and Aarts, 1987) and Tabu Search (Glover, 1989). Even with these more sophisticated methods it is not possible to guarantee that the best tree has been found.

Note also that our choice of score function, whether it be Maximum Parsimony or Maximum Likelihood, has to be viewed as an approximation to "best" – the assumption is that the tree that optimises the score we choose is the correct one. Many studies find that the most parsimonious tree (that with the lowest Parsimony length) is not the same as the most likely tree (that with the highest conditional probability of having generated the observed sequences). Therefore a common practise is to perform multiple tree searches with different objective functions and hope for consensus.

Time and Reversibility

The presence of a root induces a sense of direction – away from the root and toward the tips, corresponding to increasing time. Rooted trees must be a better representation of evolutionary dynamics, because time does indeed move forward.

However, most – if not all practised – methods that are used in phylogenetic inference assume that there is no inherent direction implied by any of our molecular data: For example there is no sense in which it is quicker or easier to go from A to B than it is from B to A. These models are referred to as time-reversible: The measure of evolutionary work that must be done, or the probability of a given train of sequence evolution, is independent of whether we are moving in the positive or negative time direction.

Thus, the inferred phylogeny we initially create using methods such as maximum likelihood or maximum parsimony (see later), are perforce unrooted. Since rooted phylogenies are so desirable, we must therefore also have methods to determine where the root is.

Rooting trees can be achieved in several ways, in general based on the idea that sequences have diverged from some kind of mean, and/or, at an approximately constant rate (see later). If all the sequences have evolved at approximately the same rate as each other, then we can expect them to be approximately the same distance from their common ancestor, which would, if the branches of the tree accurately reflect evolutionary time, be in the middle of the tree in a well defined way. A simple approach to rooting the tree then is to find the point on the tree that corresponds to the average of the half-way points between every pair of leaves.

Bayesian Methods and Markov Chain Monte Carlo

Thomas Bayes' Theory has become a central tool in molecular phylogenetics as it enables the use of prior information to be correctly incorporated into a likelihood estimation. In phylogenetics, Bayes' Theorem takes the form of priors on model components such as the set of possible phylogenetic trees, prior probabilities of particular sequence evolution models or their parameters, or on certain dates of divergence.

For example we might have some estimate on the probability distribution of phylogenetic trees on a taxon set: It could be that "all trees are equally likely" prior to our analysis (a model that has no biological realism it must be said), or that trees arise under a standard birth/death process; the Yule model (Yule, 1924). These priors modify the posterior probability that the data arose from particular models, and enable researchers to make much more powerful conclusions about the phylogenetic relationships under study, including estimates of population size, divergence times based on fossil calibrations, and selection.

Bayesian inference has come to the forefront of phylogenetic inference through such programs as MrBayes (Huelsenbeck and Ronquist, 2001) and BEAST (Drummond *et al.*, 2012). These programs perform Markov chain Monte Carlo analyses and enable researchers to sample from the posterior distribution of trees estimated, and inferences made based on the relative likelihoods of these trees. For example given a posterior distribution of trees, each with their posterior probability, a weighted average can be formed of the overall age of the tree.

What can go Wrong

Given that there are so many components to inferring phylogenies – and the difficulty in verifying that a given phylogenetic inference is even correct – it's no surprise that there is a veritable cornucopia of things that can go wrong. Here is a non-exhaustive list, in no particular order:

1. *Alignment error*: It is entirely possible that the alignment is wrong. Many authors attempt to avoid this possibility by removing areas of the alignment that are in particular doubt (highly "gappy" loci for example) or, less commonly, by performing analyses with different plausible alignments.
2. *Bad calibration (mutation rates)*: Good fossil records can afford useful prior probability distributions on divergence dates, but it is known that the choice of prior can have a significant effect on the accuracy of divergence time estimations in phylogenetics. Suggested ways around this problem include testing over a range of priors, more stringent screening of calibration points, and sub-sampling from the fossil calibration points to gauge sensitivity of the phylogenetic estimation.
3. *Covarion model*: Our most basic assumption is that sites evolve independently on a sequence, and we can in general get away with this. However there are cases where this assumption is violated and it has an effect on our tree estimation.
4. *Heuristics have no guarantees*: It is worth remembering that heuristic methods have no guarantee of producing an optimal solution. It's also known that there can be many cases were even our most favoured phylogenetic scoring systems (parsimony, Likelihood) may well have multiple optima for the same data set: there can easily be cases where more than one tree has an optimal likelihood or parsimony length. This not only means that heuristic search methods can become accidentally stuck in

the wrong places (globally optimal or not), but also that even if all the globally optimal solutions could be located, we would still have the troublesome task of choosing the best among equals.

5. *Incomplete burn-in for MCMC*: Markov chain Monte Carlo requires a burn-in to escape poor areas of the likelihood landscape and navigate to those areas that contribute significantly to the likelihood mass of phylogenies that could have given rise to our observed data. It is entirely possible – in fact it may be common – that burn-in proceeds in jumps, with long periods of proposed moves during which there is no major change in parameter values, followed later by location of new plateaus in "better" regions of tree space (Nylander *et al.*, 2007).

6. *It's not a tree after all*: As mentioned above, it's possible that by combining phylogenetically informative sequences from different loci, such as genes, we are combining incongruent trees: In such cases the conflicting trees should be reconciled with a single species tree; often through consensus methods.

7. *Long branch attraction*: Long branches give sequences a chance to converge towards each other simply by chance. Consider a tree in which all but two branches are short, yet the two long ones are not each other's closest relatives. Under such circumstances the two sequences at their leaves may look more like each other simply by dint of the other sequences being less diverged. This phenomenon is known as long branch attraction.

8. *Model mis-specification*: Getting the underlying model "wrong" runs the considerable risk that the tree is also incorrect (Jermiin *et al.*, 2006).

9. *Rates across sites*: individual sites may evolve under different conditions and so modelling them as though they all evolve at the same rate is an error.

10. *Saturation of sites*: The alphabet of available character states is very restricted: In DNA and RNA it is only 4 states, so accumulating many changes among those states renders sequences effectively random with respect to each other. If the steady state distribution of nucleotide frequencies is uniform (e.g., if the molecular evolution model is doubly stochastic) then the expected distance between sequences will tend to 0.75 as time increases.

11. *Software error*: Testing bioinformatics software is often complex and difficult, because testing is ideally done using an "oracle" – A ground truth against which we can test, and there are few such oracles in bioinformatics. While the software currently forming the standard set of tools for phylogenetic inference is in general quite well established and tested by its large user base, it is wise to exert caution when using new software, which may not have had such rigorous testing.

Further Reading

The interested reader is directed to some excellent texts and a wealth of publications in the scientific literature. Joe Felsenstein's "Inferring phylogenies" (Felsenstein, 2004) remains a classic work that is highly readable and yet comprehensive. Nei and Kumar's "Molecular Evolution and Phylogenetics" (Nei and Kumar, 2000) is another detailed work that goes into detail of molecular evolution in amino acids and in nucleotides, discusses phylogenetic inference in far more detail than can be achieved in this short article, and includes material on population genetics. Page and Holmes' "Molecular Evolution: A Phylogenetic Approach" (Page and Holmes, 1998) is a slightly earlier work that remains informative, beginning with a focus on genome evolution as a whole, and the how function relates to evolution. Wen-Hsiung Li's simply titled "Molecular Evolution" (Li, 1997) is from around the same time, and while it also discusses molecular phylogenetics spends a great deal more time on different mechanisms of molecular evolution. Moving into the more mathematical side, Olivier Gascuel's edited book "Mathematics of Evolution and Phylogeny" (Gascuel, 2005) touches on a variety of topics that were not covered here (as well as most that were), including the elegant Hadamard Conjugation method of phylogenetic analysis, mixture models, and even reconstruction of phylogenies from distances estimated by genome rearrangements. Mike Steel and Charles Semple have produced "Phylogenetics" (Semple and Steel, 2003) – Which is largely mathematical work suitable for the mathematician, and "Phylogeny: Discrete and Random Processes in Evolution" (Steel, 2016) is a further example of enjoyable, yet dense, mathematical writing. Yang's "Molecular Evolution: A Statistical Approach" (Yang, 2014) is a more modern piece that is rigorous and accessible. Finally, Drummond and Bouckaert's "Bayesian Evolutionary Analysis with BEAST" (Drummond and Bouckaert, 2015) is an excellent handbook for this sophisticated program, which is one of the standard tools of phylogenetic inference.

Software

Molecular phylogenetics is impossible without good software. Among the standard tools are, in no particular order, BEAST and MrBayes (Bayesian analysis), PhyML and IQTree (Maximum likelihood methods), PHYLIP, PAUP and MEGA (complete suites of methods).

See also: Algorithms for Strings and Sequences: Multiple Alignment. Stochastic Processes. Comparative and Evolutionary Genomics. Evolutionary Models. Molecular Clock. Phylogenetic Tree Rooting. Tree Evaluation and Robustness Testing. Applications of the Coalescent for the Evolutionary Analysis of Genetic Data. Natural Language Processing Approaches in Bioinformatics. Sequence Analysis. Sequence Composition. Molecular Mechanisms Responsible for Drug Resistance. Phylogenetic analysis: Early evolution of life. Inference of Horizontal Gene Transfer: Gaining Insights Into Evolution via Lateral Acquisition of Genetic Material. Gene Duplication and Speciation. Epidemiology: A Review. Identification of Homologs

References

Bergthorsson, U., Adams, K.L., Thomason, B., Palmer, J.D., 2003. Widespread horizontal transfer of mitochondrial genes in flowering plants. Nature 424, 197–201.

Bryant, D., Moulton, V., 2003. Neighbor-net: An agglomerative method for the construction of phylogenetic networks. Molecular Biology and Evolution 21 (2), 255–265.

Castresana, J., 2000. Selection of conserved blocks from multiple alignments for their use in phylogenetic analysis. Molecular Biology and Evolution 17, 540–552.

Cavalli-Svorza, L.L., Edwards, A.W.F., 1967. Phylogenetic analysis: Models and estimation procedures. American Journal of Human Genetics 19, 233–257.

Cavender, J.A., 1978. Taxonomy with confidence. Mathbio 40, 271–280.

Darwin, C., 1859. On the Origin of Species by Means of Natural Selection, or Preservation of Favoured Races in the Struggle for Life. London: John Murray.

Dayhoff, M.O., Schwartz, R.M., Orcutt, B.C., 1978. A model of evolutionary change in proteins: Matrices for detecting distant relationships. In: Dayhoff, M.O. (Ed.), Atlas of Protein Sequence and Structure, vol. 5. Washington D.C.: National Biomedical Research Foundation, pp. 345–358.

Drummond, A.J., Bouckaert, R.R., 2015. Bayesian Evolutionary Analysis with BEAST. Cambridge University Press.

Drummond, A.J., Suchard, M.A., Xie, D., Rambaut, A., 2012. Bayesian phylogenetics with BEAUti and the BEAST 1.7. Molecular Biology and Evolution 29 (8), 1969–1973.

Edgar, R.C., 2004. MUSCLE: Multiple sequence alignment with high accuracy and high throughput. Nucleic Acids Research 32 (5), 1792–1797.

Efron, B., Halloran, E., Holmes, S., 1996. Bootstrap confidence levels for phylogenetic trees. Proceedings of the National Academy of Sciences 93, 7085–7090.

Felsenstein, J., 1973. Maximum likelihood and minimum-steps methods for estimating evolutionary trees from data on discrete characters. Systematic Biology 22, 240–249.

Felsenstein, J., 2004. Inferring Phylogenies, 2nd ed. Sunderland, Massachusetts: Sinauer Associates.

Fitch, W.M., 1971. Toward defining the course of evolution: Minimum change for a specific tree topology. Systematic Zoology 20, 406–416.

Gascuel, O., 2005. Mathematics of Evolution and Phylogenetics. Oxford University Press.

Gascuel, O., Steel, M., 2006. Neighbor-Joining revealed. Molecular Biology and Evolution 23, 1997–2000.

Glover, F., 1989. Tabu search – Part 1. ORSA Journal on Computing 1, 190–206.

Harrison, R.G., Larson, E.L., 2014. Hybridization, introgression, and the nature of species boundaries. Journal of Heredity 105 (S1), 795–809.

Henikoff, S., Henikoff, J.G., 1992. Amino acid substitution matrices from protein blocks. Proceedings of the National Academy of Sciences 89, 10915–10919.

Higgins, D.G., Bleasby, A.J., Fuchs, R., 1992. CLUSTAL V: Improved software for multiple sequence alignment. Computer Applications in the Biosciences 8, 189–191.

Huelsenbeck, J.P., Ronquist, F., 2001. MRBAYES: Bayesian inference of phylogenetic trees. Bioinformatics 17 (8), 754–755. aug.

Jermiin, L.S., Ho, S.Y.W., Ababneh, F., Robinson, J., Larkum, A.W., 2006. The biasing effect of compositional heterogeneity on phylogenetic estimates may be underestimated. 61 (February 2004), pp. 1–22.

Jermiin, L.S., Poladian, L., Charleston, M.A., 2005. Is the "Big Bang" in animal evolution real? Science 310 (5756), 1910–1911.

Jones, D.T., Taylor, W.R., Thornton, J.M., 1992. The rapid generation of mutation data matrices from protein sequences. Computer Applications in the Biosciences 8 (3), 275–282.

Katoh, K., Misawa, K., Kuma, K.-i., Miyata, T., 2002. Mafft: A novel method for rapid multiple sequence alignment based on fast fourier transform. Nucleic Acids Research 30, 3059–3066.

Van Laarhoven, P.J.M., Aarts, E.H.L., 1987. Simulated Annealing: Theory and Applications. Boston: Reidel.

Lassmann, T., Sonnhammer, E.L.L., 2005. Kalign – An accurate and fast multiple sequence alignment algorithm. BMC Bioinformatics 6, 298.

Li, W.-H., 1997. Molecular Evolution. Sinauer Associates Inc.

Lockhart, P.J., Steel, M.A., Barbrook, A.C., Huson, D.H., Howe, C.J., 1998. A covariotide model describes the evolution of oxygenic photosynthesis. Molecular Biology and Evolution 15, 1183–1188.

Nei, M., Kumar, S., 2000. Molecular Evolution and Phylogenetics. Oxford: Oxford University Press.

Notredame, C., Higgins, D.G., Heringa, J., 2000. T-Coffee: A novel method for fast and accurate multiple sequence alignment. Journal of Molecular Biology 302, 205–217.

Nylander, J.A., Wilgenbusch, J.C., Warren, D.L., Swofford, D.L., 2007. AWTY (are we there yet?): A system for graphical exploration of MCMC convergence in bayesian phylogenetics. Bioinformatics 24 (4), 581–583.

Page, R.D., Holmes, E.C., 1998. Molecular Evolution: A Phylogenetic Approach. John Wiley & Sons.

Posada, D., 2008. jModelTest: Phylogenetic model averaging. Molecular Biology and Evolution 25 (7), 1253–1256.

Saitou, N., Nei, M., 1987. The Neighbor-Joining method: A new method for reconstructing phylogenetic trees. Molecular Biology and Evolution 4 (4), 406–425.

Sankoff, D., 1975. Minimal mutation trees of sequences. SIAM Journal of Applied Mathematics 28, 35–42.

Schröder, E., 1870. Vier combinatorische Probleme. Zeitschrift für Mathematik und Physik 15, 361–376.

Semple, C., Steel, M.A., 2003. Phylogenetics. vol. 24. Oxford University Press.

Steel, M., 2016. Phylogeny: Discrete and Random Processes in Evolution (CBMS-NSF Regional Conference Series). SIAM-Society for Industrial and Applied Mathematics.

Studier, J.A., Keppler, K.J., 1988. A note on the neighbor-joining algorithm of Saitou and Nei. Molecular Biology and Evolution 5, 729–731.

Tavaré, S., 1986. Some probabilistic and statistical problems in the analysis of Dna sequences. Lectures on Mathematics in the Life Sciences 17, 57–86.

Wang H.-C., Spencer M., Susko E., 2006. Testing for covarion-like evolution in protein sequences. (902).

Waterman, M.S., Joyce, J., Eggert, M., 1991. Computer alignment of sequences. In: Miyamoto, M.M., Cracraft, J. (Eds.), Phylogenetic Analysis of {DNA} Sequences. Oxford: Oxford University Press, pp. 59–72.

Yang, Z., 2014. Molecular Evolution: A Statistical Approach. Oxford University Press.

Yule, G.U., 1924. A mathematical theory of evolution based upon the conclusions of Dr. J.C. Willis, FRS. Philosophical Transactions of the Royal Society B 213, 21–87.

Evolutionary Models

David A Liberles, Temple University, Philadelphia, PA, United States
Barbara R Holland, University of Tasmania, Hobart, TAS, Australia

Introduction

In evolutionary biology, differences are observed between closely related species, both in their genotype and aspects of their phenotype. Alignments of homologous characters (those descended from a common ancestor), and corresponding phylogenetic trees that show the ancestral relationships between entries in the alignment, are commonly used to characterize the evolutionary process. In moving from sets of homologous characters to characterizations of evolution, explicit models of the evolutionary process are commonly used. These models describe the probabilities of different types of evolutionary changes allowing the most likely trajectory to be identified. The models are frequently structured as Markov processes. Evolutionary models can be applied to data at multiple levels of biological organization, from morphological change, to gene content in genomes, to gene duplication, to genetic sequence data. This article will focus on models for biological sequence evolution.

Multiple sequence alignment is one problem where evolutionary models are particularly useful. Probabilities of different types of sequence change are compared with the probabilities of an insertion or deletion having been introduced (commonly using affine gap penalties) to generate an alignment of sequence characters (DNA or protein) that derived from a common ancestor (Anisimova *et al.*, 2010). There are many methods to infer phylogenetic trees based on multiple sequence alignments, many of these methods are based on explicit models of sequence change. The phylogenetic tree shows the branching pattern of descendent sequences (from species or gene copies within a genome), with internal nodes reflecting ancestral character states. Branch lengths are estimated as well, with units of expected changes per site under the model as the most common units. Inference of sequences at internal nodes is a sub-field of its own, known as ancestral sequence (state) reconstruction. Lastly, models are also used in the search for lineage-specific positive selection, in characterizing the patterns of sequence change on particular branches of a phylogenetic tree (Benner *et al.*, 2007).

Because alignment and phylogenetic analysis both rely on models of the same evolutionary process, one strategy that has been developed, but is not currently widely used, is to simultaneously estimate both. Not only does this enable the consistency of using the same model for the evolutionary process in both steps, but it also enables accounting for alignment uncertainty in tree estimation. *Baliphy* is an example of an approach implemented in a Bayesian statistical framework that uses models to simultaneously estimate alignments and trees (Redelings and Suchard, 2005).

Markov Models

Models of sequence evolution range from phenomenological models that are designed to fit data as well as possible with as few parameters as possible, to mechanistic or generative models that are designed to model the process generating the data. The mechanistic models may not describe data as well as the phenomenological models on interpolated data, but may enable extrapolation to longer evolutionary distances in cases that might be prohibitive for phenomenological models (Liberles *et al.*, 2013). Because mechanistic models are meant to describe the process, there is the expectation that they will have predictive value on data that they were not fit to.

In order to model the evolution of molecular sequences, some key simplifying assumptions are made in almost all cases. A strong assumption that provides great benefits in terms of the tractability of models is that the evolution of different sites in an alignment is independent and happens according to a common process – this is known as the i.i.d. assumption (independent and identically distributed). The second assumption is that sequences evolve according to a Markov process. This means that the evolution of a particular character depends only on its current state rather than on any additional information about the past states.

With these assumptions in place, sequence evolution can be modeled with a continuous time Markov chain (CTMC). In the case of modeling sequence evolution for DNA, amino acids, or codons, the state space is discrete, e.g., for DNA the state space is $\{A, C, G, T\}$. The instantaneous rates at which any particular character state substitutes into any other state can be gathered into a rate matrix, Q, sometimes known as the generator matrix. If these rates are in effect for time t then the probability of a transition from state i to state j is given by the ijth entry of the matrix P, where $P = \exp(Qt)$.

Given an edge-weighted phylogenetic tree and a continuous time model given by Q it is possible to calculate the probability of observing any particular site pattern at the tips of the phylogenetic tree in extant sequences (Felsenstein, 2004). This means such models can be used as a basis for likelihood or Bayesian style statistical approaches to determine which choice of tree and edge weights give the highest probability of producing a particular sequence alignment.

A sequence alignment reflects the collection of substitutions that result from mutations. It should be noted that the term *mutation* formally refers to a mutation introduced to a population at frequency $1/N$ in a haploid population. When this mutation fixes (goes to frequency 1), this is referred to as a *substitution*. Inter-specific evolutionary models are designed to model substitution rates and processes. If sequences are evolving neutrally then the rate of mutation and substitution are the same (Kimura, 1983), but if sequences are under selection then the distinction between mutation and substitution is an important modeling consideration.

Parameterized DNA Models

Standard DNA models (see Felsenstein (2004) for a further discussion of standard models and their original references) started with the Jukes-Cantor model (Jukes and Cantor, 1969), which has a single rate parameter that describes the equal probability of exchange between all bases. From a statistical perspective, this is a natural model that characterizes the rate of change without any biological assumptions. However, from a biological perspective, this model tends to explain sequence substitution patterns poorly. The Kimura 2-parameter model (Kimura, 1980) is a standard model in the hierarchy of models that differentiates between transitions (substitutions between purines or between pyrimidines) and transversions (substitutions that convert between purines and pyrimidines). The reason this is natural is that transitions are more likely due to oxidative deamination (at least for C to U changes) and due to polymerase error, and also because of the genetic code, where 2-fold degenerate codons preserve the encoded amino acid with a transition but not a transversion.

Another introduction of model complexity that is similarly nested over the Jukes-Cantor model involved the introduction of unequal base equilibrium frequencies with a single rate parameter to produce the (Felsenstein, 1981) model. One example of the importance of unequal equilibrium frequencies is the hypothesis of temperature dependence to GC usage because of its effect on DNA melting temperatures (Groussin and Gouy, 2011). The parameters of K2P and F81 were combined in the HKY model (Hasegawa *et al.*, 1985). A further relaxation of this involves the GTR model (Lanave *et al.*, 1984; Tavaré, 1986), where each of the six inter-conversions has an independent rate parameter.

GTR is the most complex model that is time-reversible such that, given the equilibrium base-frequency distribution $\pi = (\pi_A, \pi_C, \pi_G, \pi_T)$, $\pi_i Q_{ij} = \pi_j Q_{ji}$. A further relaxation from six to twelve rate parameters gives the General Markov Model (GMM) (Barry and Hartigan, 1987), which is not time reversible. This set of standard models is summarized in **Fig. 1**. Between Jukes-Cantor and GTR for symmetrical matrices and GMM to include non-symmetrical matrices, is a hierarchy of conceivable models (Huelsenbeck *et al.*, 2004). The property of being time reversible makes models much more efficient to compute with (Felsenstein, 2004). To use non-time-reversible models, a rooted tree must be used and likelihoods calculated in a different manner (Boussau and Gouy, 2006). Despite this computational complexity, biologically one does not expect evolution to be time-reversible when the underlying processes are not in equilibrium and the changes are not neutral.

Commonly Used Additional Parameters

Additional parameters are commonly used together with phylogenetic Markov Models for DNA and for proteins. Parameters that change the equilibrium frequencies of the underlying model (particularly for empirical models described below) to fit those of the

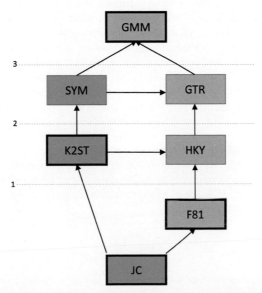

Fig. 1 An overview of common DNA models is shown. Models with equal base-frequencies are shown with a blue background, models with free base-frequencies are shown with a green background. Models with the Lie-Markov property have a bold border. Models above the first dashed line allow a transition/transversion ratio. Models above the second dashed line allow 6 different substitution rates. Models below the third dashed line are time-reversible. Arrows directed from one model to another indicate that the model is a sub-model, i.e., that it is nested within the more complicated model.

dataset (+ F) are commonly included as extra parameters. Additionally, some sites are of such (context-dependent) importance in a dataset that they show a substitution rate of zero over the period of evolution over which the data were generated. Modeling a fraction of invariant sites (+ I) with substitution rate zero is another common parameter that is added to phylogenetic Markov Models.

More generally, different sites evolve at different rates, both due to variation in the mutation process and due to differential selection on different sites. There have been two main approaches to tackling this: mixture models and partitioned models. The most common application of mixture models is to use the gamma distribution (+ G) to model rate heterogeneity across sites (Yang, 1994). A strength of this approach is that a single shape parameter can provide a continuum of models ranging from equal rates to very right-skewed rates. This is most commonly implemented using a discretization of the gamma distribution into four rate categories, although alternative mechanisms of using a gamma distribution are possible. No index parameters are included, i.e., sites are not allocated to particular rate classes, instead all four rate categories are used as a mixture process applied to all sites. Using a gamma distribution is somewhat arbitrary, a slightly more parameter-rich alternative is the free-rate model, which allows a mixture model over an arbitrary set of rates (Yang, 1995; Pagel and Meade, 2004; Mayrose et al., 2005). The other strategy for dealing with heterogeneity is to use prior knowledge of the data, e.g., gene boundaries, positions within a protein structure, and/or codon position, to partition the sequence alignment and then fit different models to each part of the partition (Bull et al., 1993).

Lie-Markov Models

Even with the extensions stated above, most phylogenetic applications of evolutionary models assume that a common process operates over all sites and in all parts of the evolutionary tree. This assumption does not appear to be biologically realistic. In particular, a prediction of such models is that base frequencies should be similar in all species under consideration, but this has been shown empirically not always to be the case (Jayaswal et al., 2011). Codon models have also been used to show that evolution can be a heterogeneous process with evidence for selection acting in some branches but not others.

If one allows the possibility that the evolutionary process may be different in different edges of the phylogeny or differ along edges then this introduces an interesting consideration about how one might want the Markov models in use to behave. If one considers that evolution happens according to some rates Q_1 for some period of time t_1 followed by some rates Q_2 for time t_2 then one can work out the transition probabilities over time $t_1 + t_2$ to be $P = \exp(Q_2 t_2)^* \exp(Q_1 t_1)$. A Markov model has the closure property if this matrix P can still be expressed as $P = \exp(Qt)$ for some Q chosen from our model. Many commonly used Markov models, including GTR and HKY, do not have this closure property. Recent work by Sumner et al. (2012) has developed a hierarchy of evolutionary models, known as Lie-Markov models, that are defined by having this closure property.

Model Selection

As discussed above, a common approach to add biological realism is to differentiate between types of nucleotide substitutions. For example, Huelsenbeck et al. (2004) consider all 203 possible partitions of relative rates for time-reversible models, reflecting all possible models of this class. Another approach is to consider that there are different classes of bases (e.g., AG and CT) and then use the resulting symmetries to find appropriate constraints on substitution rates (Fernández-Sánchez et al., 2015).

Given the wealth of possibilities for parameterized models a problem that arises is how best to choose a model to fit a particular dataset. The best inference comes when a model is appropriately complex, with all of the necessary parameters, but without extra parameters that do not contribute to data fit and/or that can cause model mis-specification (Steel, 2005). An approach, initially proposed by Posada and Crandall (2001), was to perform likelihood ratio tests of nested models (see **Fig. 1**). More recently the field has moved to prefer information criterion biased approaches such as the Akaike Information Criterion (AIC) or Bayesian Information Criterion (BIC) (Posada and Buckley, 2004). The AIC score of a model depends on both the likelihood and on the number of free parameters in the model, AIC $= -2 \ln L + 2K$ where lnL is the log likelihood and K is the number of parameters. The BIC $= -2 \ln L + K \ln n$, where n is the sample size (e.g., the length of the alignment) this penalizes additional parameters more strongly than the AIC. For both AIC and BIC, models with smaller scores are preferred. A strong advantage of both the AIC and BIC is that they do not rely on models being nested. For partitioned models the number of possible models makes it prohibitive to check the AIC or BIC of all models and it is common to take a heuristic approach to finding models that fit well instead (Lanfear et al., 2012).

Empirical Models

An alternative to using parameterized models to fit substitution data is to pre-calculate empirical substitution matrices from large datasets. This approach is commonly applied when working with amino acid as opposed to DNA data. One reason to do this is that while a maximal time-reversible DNA matrix can have up to 6 rate parameters, an equivalent matrix for proteins could have 190 rate parameters. Furthermore, some transitions between rare amino acids or those that are separated by two or three nucleotide changes in the genetic code occur too rarely to be able to estimate well from typical datasets.

The first instance of an empirical amino-acid model was produced by Dayhoff *et al.* (1978), they used closely related aligned proteins on a well-established phylogeny to estimate the rate at which particular amino acids substitute for other amino acids. Similar ideas were used by later authors although they tended to use much larger databases of curated alignments and to be more computationally intensive as they incorporated tree-estimation simultaneously with estimation of the empirical substitution parameters (e.g., JTT (Jones *et al.*, 1992), WAG (Whelan and Goldman, 2001), LG (Le and Gascuel, 2008), see also Zoller and Schneider (2013)). WAG included the innovation of maximum likelihood parameter estimation, while LG included the innovation of gamma distributed substitution rates.

A number of alternative matrices for either specific datasets, for specific attributes of protein structures, or statistically defined matrices have been created. A particular example of this is the set of empirical matrices for proteins that are partitioned by secondary structure and solvent accessibility (Koshi and Goldstein, 1995).

Codon Models

There is independent information that is encompassed in DNA models and in protein models and both sources of information can be harnessed when using codon models. Codon models can be built in three different broad types. The first type is to adopt a nucleotide substitution model such as HKY85 to describe the substitution process at the DNA level, with a genetic code matrix embedded on top of it. This genetic code matrix typically has three values, 1 for synonymous changes that involve a single nucleotide change, 0 for all changes that involve more than a single nucleotide change, and free parameter ω (dN/dS) for nonsynonymous changes that involve a single nucleotide change. This is motivated by the expectation that selective constraints on nucleotide changes that change the amino acid will be different from selective constraints on nucleotide changes that do not, leading to different rates. In the Goldman-Yang model, there are independent DNA process at each codon position, such that a different rate matrix is estimated for the 1st, 2nd, and 3rd codon positions (Goldman and Yang, 1994). In the Muse-Gaut model (Muse and Gaut, 1994), there is a single DNA level process that applies equivalently to all three positions, reflected in a common substitution matrix between them.

A second type of codon model also involves free parameters estimated directly from data, but stems from an attempt to model the underlying population genetics. In Halpern-Bruno style models (also called mutation-selection models), the probability of introducing a mutation to a population is reflected by the product of the mutation rate and the population size (Halpern and Bruno, 1998). For a mutation that is introduced, the fixation probability reflects the probability of going from frequency 1/N (1/2N in a diploid population) to fixation. This second probability depends upon the population size and the selective coefficient for the change, which in linear space is $(f'/f) - 1$, where f' is the fitness of the new amino acid and f is the fitness of the original amino acid, sampled from a vector of 20 amino acid fitnesses (19 parameters).

The third class of models is a 64×64 (61×61 when stop codons are excluded) empirical model calculated in a similar manner to amino acid substitution models. An early instance of such a model that has been constructed comes from Schneider *et al.* (2005). This class of model explicitly contains averaged data on amino acid properties that is lacking from the ω based models.

Recent work has enabled model selection across classes of models (including nucleotide, codon, and amino acid models), but has only included the first class of codon models in the set of models that are available for testing (ModelOMatic; Whelan *et al.* (2015)).

CAT Models

One class of models that is currently widely held as the state of the art for amino acid models is the CAT models (Lartillot and Philippe, 2004), based upon its ability to fit data well for the number of parameters. This model class includes a distribution of equilibrium frequencies across sites that describes the propensities for different amino acids to occur at different positions within a protein, bearing some similarity to a mixture of F81 models in protein state space. The number of categories is either defined a priori or by a Dirichlet distribution. To account for temporal heterogeneity at a site, due to epistasis (the non-independence of the evolutionary process at each site), the CAT-BP model was introduced that allows for class switching at a site (Blanquart and Lartillot, 2008).

Covarion Models

The models described so far (besides CAT-BP and some implementations of codon models) all assume that the same evolutionary process applies to all sites in an alignment on all lineages of a phylogenetic tree. While they can be applied as heterogeneous process models with discrete breaks in process at nodes of phylogenetic trees, a class of models called covarion models describes explicit rate shifting parameters (Miyamoto and Fitch, 1995). The simplest model, Tuffley-Steel, includes a parameter for shifting between a particular rate class and an invariant site (Tuffley and Steel, 1998). Galtier described a model where sites with different rate classes can shift between each other (Galtier and Jean-Marie, 2004). Roger generalized this as the General Covarion Model, with parameters to shift between rate classes and into and out of invariant states (Wang *et al.*, 2007). Overall, this modeling

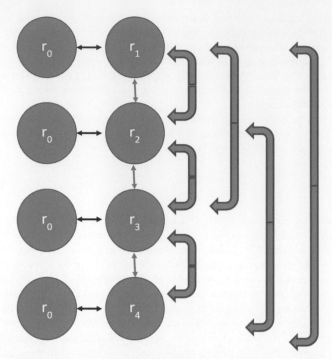

Fig. 2 An overview of covarion models is shown. r_0 states on the left reflect temporally invariant states that are accessible from states with different rates in models that incorporate this (red arrows) "Tuffley-Steel" process. Blue arrows depict the shifts between states with different rates, shown as four distinct gamma categories here. The general covarion model allows both the red and blue arrow states to occur.

framework can be applied as a general Markov process framework for temporal heterogeneity at a site. **Fig. 2** shows a summary of the modeling framework, including transitions for a site between a rate category and invariant states as well as between rate categories. Covarion processes are expected to account for the role of epistasis in enabling temporal shifts in the evolutionary processes at individual sites.

Physicochemical Models of Substitution

An alternative to using empirical matrices to characterize amino acid substitution is to use metrics of the physical chemical distance between amino acid side chains. The Grantham matrix (Grantham, 1974), the most widely used of such matrices, is calculated as a normalized RMSD between side chain atomic composition, polarity, and volume. It is most commonly used to weight amino acid differences or in metrics that examine rates of radical and conservative amino acid changes based upon Grantham matrix values. It can be extended to use as a full Markov model for phylogenetic purposes as well.

Empirical Contact Models

Another alternative to a traditional substitution matrix is to use a contact potential matrix. Miyazawa-Jernigan (1985) is a matrix that reflects the average pairwise potential of two amino acids in contact at a given distance based upon a statistical analysis of protein structures (see **Fig. 3**). This treatment enables explicit consideration of protein structure, that is not treated in other models that have been discussed. It does not consider either angular or distance information between residues that can be important in characterizing energies of interaction. To use this matrix for substitution, the difference in the sum of contacts between the original and substituted amino acids in the protein structure is used to generate an expected energetic change in the folding energy. Assuming that the strongest selective pressures are reflected in the fraction folded of a molecule based upon this energy, the fitness difference (selective pressure) associated with a change is then calculated by placing a Boltzmann distribution in a function to calculate the probability of fixation of the change.

Bastolla has generated updated contact potential matrices and most recently, a mean field model for use in ancestral sequence reconstruction (Arenas *et al.*, 2017). Robinson *et al.* (2003) introduced an early codon model with structure dependencies and novel statistics to enable computation. Kleinman *et al.* (2010) and Grahnen *et al.* (2011) have generated fuller structure-aware models for amino acid substitution that include distance and angular terms, but these methods are slow and have not performed well in characterizing individual site probabilities of substitution in actual alignments when compared with the best performing models (e.g., CAT).

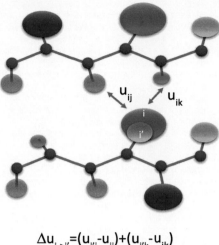

$$\Delta u_{i->i'}=(u_{i'j}-u_{ij})+(u_{i'k}-u_{ik})$$

Fig. 3 An overview of the use of a contact potential matrix for substitution is shown. Residues j and k interact with residue i. On mutation from i to i', $\Delta u_{i\rightarrow i'}$ reflects the corresponding change in potential resulting from the two underlying physical interactions in the protein structure.

Conclusions and Future Directions

Markov models are widely used in biological sequence analysis, as described in the approaches to date. DNA models are fairly good approximations to underlying biological processes, particularly for neutrally evolving sites. One ongoing area of research with DNA models is their implementation as time heterogeneous models that can account for changes in base frequencies over time. Codon models come in two main varieties, those built with a genetic code model over a DNA model and those built as empirical codon models. As with amino acid models, codon models approximate the complexity of selection on protein sequences, structures, and functions (see Chi and Liberles (2016) for a recent discussion of this). An incomplete synthesis involves incorporating structural or advanced statistical techniques to build better models for proteins. Markov models of all of these varieties are an important trajectory in bioinformatics.

Acknowledgments

We thank Peter Chi, Jeremy Sumner, Claudia Weber, and Josh Schraiber for helpful comments on both the scientific content and the general readability of this work. DAL acknowledges support from the US National Science Foundation grant DBI-1515704, to which this article will contribute to the broader impact mission.

See also: Applications of the Coalescent for the Evolutionary Analysis of Genetic Data. Gene Duplication and Speciation. Inference of Horizontal Gene Transfer: Gaining Insights Into Evolution via Lateral Acquisition of Genetic Material. Molecular Clock. Molecular Phylogenetics. Natural Language Processing Approaches in Bioinformatics. Phylogenetic analysis: Early evolution of life. Population Genetics. Stochastic Processes

References

Anisimova, M., Cannarozzi, G.M., Liberles, D.A., 2010. Finding the balance between mathematical and biological optima in multiple sequence alignment. Trends Evol. Biol. 2, e7.

Arenas, M., Weber, C.C., Liberles, D.A., Bastolla, U., 2017. ProtASR: An evolutionary framework for ancestral protein reconstruction with selection on folding stability. Syst. Biol. syw121. doi:10.1093/sysbio/syw121.

Barry, D., Hartigan, J.A., 1987. Statistical analysis of hominoid molecular evolution. Stat. Sci. 2, 191–210.

Benner, S.A., Sassi, S.O., Gaucher, E.A., 2007. Molecular paleoscience: Systems biology from the past. Adv. Enzymol. Relat. Areas Mol. Biol. 75, 1–132.

Blanquart, S., Lartillot, N., 2008. A site- and time-heterogeneous model of amino acid replacement. Mol. Biol. Evol. 25 (5), 842–858.

Boussau, B., Gouy, M., 2006. Efficient likelihood computations with nonreversible models of evolution. Syst. Biol. 55 (5), 756–768.

Bull, J.J., Huelsenbeck, J.P., Cunningham, C.W., Swofford, D.L., Waddell, P.J., 1993. Partitioning and combining data in phylogenetic analysis. Syst. Biol. 42 (3), 384–397.

Chi, P.B., Liberles, D.A., 2016. Selection on protein structure, interaction, and sequence. Protein Sci. 25 (7), 1168–1178.

Dayhoff, M.O., Schwartz, R.M., Orcutt, B.C., 1978. A model of evolutionary change in proteins. In: Dayhoff, M.O., (Ed.), Atlas of Protein Sequence and Structure National Biomedical Research Foundation, Washington DC, 5(3), 345–352.

Felsenstein, J., 1981. Evolutionary trees from DNA sequences: A maximum likelihood approach. J. Mol. Evol. 17 (6), 368–376.

Felsenstein, J., 2004. Inferring Phylogenies. Sunderland, MA: Sinauer Associates, Inc.

Fernández-Sánchez, J, Jeremy, G.S., Jarvis, P.D., Woodhams, M.D., 2015. Lie Markov models with purine/pyrimidine symmetry. J. Math. Biol. 4, 855–891.

Galtier, N., Jean-Marie, A., 2004. Markov-modulated Markov chains and the covarion process of molecular evolution. J.Comput. Biol. 11 (4), 727–733.

Goldman, N., Yang, Z., 1994. A codon-based model of nucleotide substitution for protein-coding DNA sequences. Mol. Biol. Evol. 11, 725–736.

Grahnen, J.A., Nandakumar, P., Kubelka, J., Liberles, D.A., 2011. Biophysical and structural considerations for protein sequence evolution. BMC Evol. Biol. 11, 361.

Grantham, R., 1974. Amino acid difference formula to help explain protein evolution. Science 185 (4154), 862–864.

Groussin, M., Gouy, M., 2011. Adaptation to environmental temperature is a major determinant of molecular evolutionary rates in archaea. Mol. Biol. Evol. 28 (9), 2661–2674.

Halpern, A.L., Bruno, W.J., 1998. Evolutionary distances for protein-coding sequences: Modeling site-specific residue frequencies. Mol. Biol. Evol. 15 (7), 910–917.

Hasegawa, M., Kishino, H., Yano, T., 1985. Dating of the human-ape splitting by a molecular clock of mitochondrial DNA. J. Mol. Evol. 22 (2), 160–174.

Huelsenbeck, J.P., Larget, B., Alfaro, M.E., 2004. Bayesian phylogenetic model selection using reversible jump Markov chain Monte Carlo. Mol. Biol. Evol. 21 (6), 1123–1133.

Jayaswal, V., Jermiin, L.S., Poladian, L., Robinson, J., 2011. Two stationary nonhomogeneous Markov models of nucleotide sequence evolution. Syst. Biol. 60 (1), 74–86.

Jones, D.T., Taylor, W.R., Thornton, J.M., 1992. The rapid generation of mutation data matrices from protein sequences. Comput. Appl. Biosci. 8 (3), 275–282.

Jukes, T.H., Cantor, C.R., 1969. Evolution of Protein Molecules. New York, NY: Academic Press, pp. 21–132.

Kimura, M., 1980. A simple method for estimating evolutionary rates of base substitutions through comparative studies of nucleotide sequences. J. Mol. Evol. 16 (2), 111–120.

Kimura, M., 1983. The Neutral Theory of Molecular Evolution. Cambridge: Cambridge University Press.

Kleinman, C.L., Rodrigue, N., Lartillot, N., Philippe, H., 2010. Statistical potentials for improved structurally constrained evolutionary models. Mol. Biol. Evol. 27 (7), 1546–1560.

Koshi, J.M., Goldstein, R.A., 1995. Context-dependent optimal substitution matrices. Protein Eng. 8 (7), 641–645.

Lanave, C., Preparata, G., Saccone, C., Serio, G., 1984. A new method for calculating evolutionary substitution rates. J. Mol. Evol. 20 (1), 86–93.

Lanfear, R., Calcott, B., Ho, S.Y., Guindon, S., 2012. Partitionfinder: Combined selection of partitioning schemes and substitution models for phylogenetic analyses. Mol. Biol. Evol. 29 (6), 1695–1701.

Lartillot, N., Philippe, H., 2004. A Bayesian mixture model for across-site heterogeneities in the amino-acid replacement process. Mol. Biol. Evol. 21 (6), 1095–1109.

Le, S.Q., Gascuel, O., 2008. An improved general amino acid replacement matrix. Mol. Biol. Evol. 25 (7), 1307–1320.

Liberles, D.A., Teufel, A.I., Liu, L., Stadler, T., 2013. On the need for mechanistic models in computational genomics and metagenomics. Genome Biol. Evol. 5 (10), 2008–2018.

Mayrose, I., Friedman, N., Pupko, T., 2005. A gamma mixture model better accounts for among site rate heterogeneity. Bioinformatics 21 (Suppl. 2), 151–158.

Miyamoto, M.M., Fitch, W.M., 1995. Testing the covarion hypothesis of molecular evolution. Mol. Biol. Evol. 12 (3), 503–513.

Miyazawa, S., Jernigan, R.L., 1985. Estimation of effective interresidue contact energies from protein crystal structures: Quasi-chemical approximation. Macromolecules 18, 534–552.

Muse, S.V., Gaut, B.S., 1994. A likelihood approach for comparing synonymous and nonsynonymous nucleotide substitution rates, with application to the chloroplast genome. Mol. Biol. Evol. 11 (5), 715–724.

Pagel, M., Meade, A., 2004. A phylogenetic mixture model for detecting pattern-heterogeneity in gene sequence or character-state data. Syst. Biol. 53 (4), 571–581.

Posada, D., Buckley, T.R., 2004. Model selection and model averaging in phylogenetics: Advantages of Akaike information criterion and Bayesian approaches over likelihood ratio tests. Syst. Biol. 53 (5), 793–808.

Posada, D., Crandall, K.A., 2001. Selecting the best-fit model of nucleotide substitution. Syst. Biol. 50 (4), 580–601.

Redelings, B.D., Suchard, M.A., 2005. Joint Bayesian estimation of alignment and phylogeny. Syst. Biol. 54 (3), 401–418.

Robinson, D.M., Jones, D.T., Kishino, H., Goldman, N., Thorne, J.L., 2003. Protein evolution with dependence among codons due to tertiary structure. Mol. Biol. Evol. 20 (10), 1692–1704.

Schneider, A., Cannarozzi, G.M., Gonnet, G.H., 2005. Empirical codon substitution matrix. BMC Bioinform. 6, 134.

Steel, M., 2005. Should phylogenetic models be trying to 'fit an elephant'? TrendsGenet. 21, 307–309.

Sumner, J.G., Fernández-Sánchez, J., Jarvis, P.D., 2012. Lie Markov models. J. Theor. Biol. 298, 16–31.

Tavaré, S., 1986. Some Probabilistic and Statistical Problems in the Analysis of DNA Sequences. Lectures on Mathematics in the Life Sciences 17,American Mathematical Society, pp. 57–86.

Tuffley, C., Steel, M., 1998. Modeling the covarion hypothesis of nucleotide substitution. Math. Biosci. 147 (1), 63–91.

Wang, H.C., Spencer, M., Susko, E., Roger, A.J., 2007. Testing for covarion-like evolution in protein sequences. Mol. Biol. Evol. 24 (1), 294–305.

Whelan, S., Allen, J.E., Blackburne, B.P., Talavera, D., 2015. ModelOMatic: Fast and automated model selection between RY, nucleotide, amino acid, and codon substitution models. Syst. Biol. 64 (1), 42–55.

Whelan, S., Goldman, N., 2001. A general empirical model of protein evolution derived from multiple protein families using a maximum-likelihood approach. Mol. Biol. Evol. 18 (5), 691–699.

Yang, Z., 1994. Maximum likelihood phylogenetic estimation from DNA sequences with variable rates over sites: Approximate methods. J. Mol. Evol. 39 (3), 306–314.

Yang, Z., 1995. A space-time process model for the evolution of DNA sequences. Genetics 139 (2), 993–1005.

Zoller, S., Schneider, A., 2013. Improving phylogenetic inference with a semiempirical amino acid substitution model. Mol. Biol. Evol. 30 (2), 469–479.

Molecular Clock

Cara Van Der Wal, University of Sydney, Sydney NSW, Australia and Australian Museum Research Institute, Australian Museum, Sydney NSW, Australia
Simon YW Ho, University of Sydney, Sydney NSW, Australia

Introduction

Our understanding of biological patterns and processes is greatly improved by knowledge of evolutionary timescales. The inference of these timescales has traditionally been based on evidence from the fossil record, which provides an indication of when species diverged and when key biological innovations arose. Unfortunately, the fossil record is very incomplete for many branches of the Tree of Life. Groups such as bacteria and various invertebrates have left scant traces, despite the persistence of their lineages through many millions of years. Additionally, fossil evidence is usually unable to resolve the timing of very recent events, such as demographic changes occurring within species or the transmission of pathogens across their host populations (Arbogast et al., 2002).

When the fossil record is uninformative about evolutionary and demographic timescales, we require temporal information from another source. The answer to this problem arrived in the 1960s when the idea of a 'molecular evolutionary clock' was proposed by Zuckerkandl and Pauling (1962, 1965). Molecular clocks describe the relationship between evolutionary rate and time, with the simplest clock model assuming that the rate of molecular evolution is constant across species. This allows us to use phylogenetic analyses of genetic data to estimate evolutionary and demographic timescales (**Fig. 1**). The most widely used forms of data for this purpose are nucleotide and amino acid sequences. To give estimates in terms of absolute time, however, molecular clocks need to be calibrated using independent evidence about the timing of one or more evolutionary divergences in the phylogeny.

Molecular clocks have been used to provide insights into a number of significant evolutionary events, ranging from human prehistory to the deep divergences among the kingdoms of life. They are now being applied to genome-scale data sets that comprise thousands of genes from dozens to hundreds of taxa (Tong et al., 2016). For example, evolutionary timescales have been inferred using large data sets from birds (Jarvis et al., 2014), flowering plants (Foster et al., 2017), and insects (Misof et al., 2014). The methods underpinning molecular-clock analyses continue to evolve, which is imperative if we are to make the most of the growing wealth of genomic data.

The objective of this article is to provide an overview of the molecular clock. First, we will present a brief history of the molecular-clock hypothesis and its early applications. Second, molecular-clock methods and models will be described, as will the challenges to their implementation. Following a case study on the origins of lobsters, we will discuss the role of molecular clocks in the genomic era.

Background

The Origins of the Molecular Clock

Zuckerkandl and Pauling (1962) were the first to propose the idea of a molecular clock. In their analysis of globin proteins, they assumed a constant evolutionary rate in order to estimate the timing of past gene duplications in vertebrates. They found 18 amino acid differences separating human and horse, and estimated the evolutionary rate of this protein by assuming that these two species split from each other 100–160 million years ago. This evolutionary rate was then extrapolated to the globin genes of other animals to show that humans and gorillas split from each other approximately 11 million years ago, and that different globin genes first diverged during the late Precambrian (Zuckerkandl and Pauling, 1962).

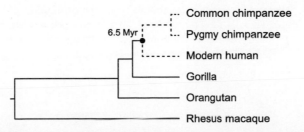

Fig. 1 Simple illustration of molecular dating. The phylogenetic tree depicts the evolutionary relationships among six primate species, inferred using genetic data from each species. Fossil evidence indicates that modern humans last shared a common ancestor with chimpanzees about 6.5 million years ago. This 'calibration' allows us to infer the rate of molecular evolution within the descendent branches (dashed lines). If this rate is assumed to remain constant throughout the tree (known as a 'strict molecular clock'), we can estimate the timing of all other divergence events.

Studies of other proteins throughout the 1960s provided further evidence of constant, clocklike evolution, leading Zuckerkandl and Pauling (1965) to put forward the term 'molecular evolutionary clock'. Promising to be a valuable tool in biological research, the power of the molecular clock was soon demonstrated when Sarich and Wilson (1967) inferred the evolutionary timescale of hominids and related primates.

The molecular-clock hypothesis relies on the idea that rates of molecular evolution, as seen in mutations in protein or DNA sequences over time, are relatively constant between lineages. The occurrence of mutations is often modelled as a Poisson process, reflecting the stochastic nature of the molecular clock. Although individual mutations do not occur at strictly regular intervals, the genetic difference between any two species will increase monotonically over time. If the evolutionary rate has been constant, then genetic divergence will be directly proportional to the time since the two species last shared a common ancestor. This relationship allows evolutionary timescales to be inferred from genetic data, provided that we know the rate of molecular evolution.

When initially proposed, the notion of a constant rate of molecular evolution created controversy. A number of researchers, including the prominent palaeontologist Simpson (1964), viewed the molecular clock with scepticism. It seemed implausible that a simple statistical model could describe a process as seemingly complex as that of molecular evolution. In the late 1960s, however, new ideas about molecular evolution would provide some support for the concept of a molecular clock.

Neutral and Nearly Neutral Theories of Molecular Evolution

Alongside the early criticisms of the molecular clock, there was growing evidence of high evolutionary rates in a range of proteins. This suggested that a substantial proportion of amino acid mutations have negligible impacts on fitness. In response to this evidence, Kimura (1968) proposed the neutral theory of molecular evolution. This theory states that many mutations do not have a measurable impact on the organism's fitness, so that their frequencies in the population vary stochastically by genetic drift rather than being driven by natural selection. These neutral mutations might represent the replacement of amino acids by other amino acids with similar biochemical properties. In protein-coding genes, some nucleotide mutations do not result in any changes to the encoded protein, so their effects on fitness are likely to be very small.

The neutral theory of molecular evolution predicts that evolutionary rates, measured on a per-generation basis, are constant across lineages (Kimura, 1968). However, generation lengths vary between species, meaning that we expect a generation-time effect: molecular evolution occurs more quickly per unit of time in species with shorter generations. For example, rodents express a higher rate of evolution than primates. Such generation-time effects have been identified in a broad range of organisms.

At the end of the 1960s, some studies found that the rate of protein evolution was independent of generation time. This contradicted the expectations of the neutral theory, leading Ohta (1972, 1973) to propose the nearly neutral theory of molecular evolution. In this model, most mutations are assumed to have a small effect on fitness. In contrast with the neutral theory, the nearly neutral theory gives a central role to the effect of population size. Natural selection is effective at removing harmful mutations in large populations, but the process of genetic drift dominates in small populations. However, there are more opportunities for mutations to arise in large populations. Ohta (1972) predicted that these two effects would offset each other, with the net effect being a constant evolutionary rate per unit of time across species.

Molecular Clocks in Practice

In the decades following the proposal of the molecular clock, genetic data have been used to estimate the evolutionary timescales of various parts of the Tree of Life. Among the most conspicuous studies were those of the deep divergences among metazoan phyla. The date estimates from these analyses pointed to a much longer timescale of evolution than indicated by the fossil record. In conflict with the idea of an explosive evolutionary radiation in the Cambrian period, the basal metazoan divergences were estimated to have occurred much deeper in the Precambrian (e.g., Runnegar (1982), Bromham et al. (1998)). These striking discrepancies between genetic and fossil-based estimates of evolutionary timescales led many to question the reliability of molecular clocks.

There has also been abundant evidence of rate variation across lineages, contradicting the simplistic assumption of clocklike evolution. These challenges to rate-homogeneous models have driven the development of 'relaxed' molecular clocks, the first of which appeared in the late 1990s. These clock models are able to relax the assumption of rate homogeneity among lineages (Sanderson, 1997; Thorne et al., 1998). There is now a wide range of relaxed-clock models, implemented in various statistical frameworks (Ho and Duchêne, 2014). Among the most commonly used in phylogenetic inference are those based on the Bayesian approach (dos Reis et al., 2016).

Approaches

Molecular-Clock Methods

The molecular clock is typically used as part of a phylogenetic analysis, in a procedure known as molecular dating. This involves inferring the evolutionary relationships among a sample of taxa, along with the evolutionary rates and time durations of the branches of the phylogenetic tree. Molecular dating comprises several key steps (**Fig. 2**): (i) selecting a data set, (ii) identifying

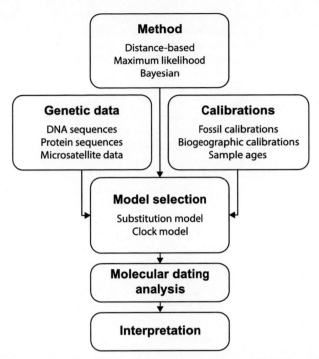

Fig. 2 Major components of a molecular dating analysis. The user begins by assembling a genetic data set, identifying suitable calibrations for the molecular clock, and selecting a method of analysis. Statistical model selection should then be performed to identify the best-fitting models of substitution and rate variation. Following the analysis of the data, the results then need to be interpreted and placed into the appropriate biological and evolutionary context.

calibrations for the molecular clock, (iii) selecting an estimation method and a model of rate variation, and (iv) running the analysis.

Choosing a data set for analysis depends on the questions that are being addressed. In most cases, nucleotide sequences are used for molecular-clock analyses because of the relative ease of obtaining these data and because of their information content. Sequences should represent the diversity of the organisms being analysed and should be sufficiently long to enable precise estimation of the amount of evolutionary change. However, genetic markers vary in their utility for molecular dating. Evolution tends to occur very slowly in functionally important genes, making them useful genetic markers for studying deep evolutionary timescales. In contrast, analyses of recent events are most effective when rapidly evolving genes are used. In animals, the mitochondrial genome typically evolves much more rapidly than the nuclear genome (Brown *et al.*, 1979). This pattern is reversed in plants: nucleotide substitutions tend to occur more rapidly in the nuclear genome than in the mitochondrial and chloroplast genomes. Mutation rates in viral genomes are higher than those in animals and plants by several orders of magnitude. In any case, analysing multiple genetic markers is now commonplace, and genome-scale data sets are seeing increasing use in molecular-clock studies (Ho, 2014).

The process of selecting calibrations, estimation methods, and models of rate variation can be informed by statistical analysis, but still involves a large degree of uncertainty. A wide range of methods are available, and these appeal to various philosophies and differ in their strengths and weaknesses (Ho and Duchêne, 2014; dos Reis *et al.*, 2016). There are also many practical factors that need to be considered, especially when analysing large data sets that comprise many sequences and/or many taxa.

Calibrating the Molecular Clock

The most critical component of molecular-clock analysis is the process of calibration. If the rate of evolution is unknown, the molecular clock must be calibrated in order for evolutionary timescales to be estimated. This involves constraining the age of at least one node in the phylogeny, allowing the average rate of its descendent branches to be inferred. The remaining nodes in the tree can then be estimated by assuming some model of rate variation across branches.

Calibrations can be chosen according to evidence from the fossil record, geological events, or past climatic events. In practice, fossil calibrations are probably the most common (Hipsley and Müller, 2014). Provided that a fossil taxon can be assigned to a lineage in the phylogeny, its age can be used to place a minimum constraint on the age of its ancestor. However, minimum age constraints need to be combined with at least one maximum age constraint in the tree. Choosing an appropriate maximum age constraint for the tree is a difficult exercise, because it relies on the assumption that a particular lineage or group did not exist before a certain time. Various criteria have been put forward for choosing appropriate fossil calibrations for the molecular clock (Parham *et al.*, 2012; Sauquet *et al.*, 2012).

In Bayesian methods, calibrations are implemented by specifying the prior distributions of the relevant node ages. For example, a node-age prior can be specified in the form of a lognormal or gamma distribution. These calibration priors can allow various sources of uncertainty to be taken into account, including errors in radiometric dating and the probability of fossil preservation (Ho and Phillips, 2009). Nevertheless, the uncertainty in the fossil evidence is rarely modelled explicitly, and the fossils themselves are only used indirectly to inform the calibration priors.

Recent developments in calibration methods have aimed to incorporate fossil evidence more comprehensively. Total-evidence dating uses combined data sets comprising genetic data from extant species and morphological data from both extant and extinct species. The morphological data are used to infer the placement of the extinct species in the tree, and the ages of these species calibrate the clock by constraining the ages of their ancestral nodes (Ronquist et al., 2012). This method removes the need to choose age constraints for nodes based on fossil evidence, by instead allowing the fossils to be included directly in the analysis. Nevertheless, the method has a number of problems that still need to be addressed (O'Reilly et al., 2015). In this regard, a key area of research is the development of more realistic models of morphological evolution.

Geological events can also be used for calibration, provided that they can be tied to evolutionary divergence events among the sampled taxa (Ho et al., 2015). These divergences can be caused by disruptions in the range of an ancestral species (vicariance), the formation of connections between previously isolated habitats (geodispersal), or dispersal across barriers and establishment of new populations (biological dispersal). Examples of these include evolutionary diversification associated with continental fragmentation or the colonization of newly formed islands (e.g., Fleischer et al. (1998)). The timing of these events can be estimated using radiometric dating methods. However, biogeographic calibrations are typically associated with large amounts of uncertainty and are sometimes regarded as unreliable (Kodandaramaiah, 2011).

In some cases, the ages of the sequences can act as calibrations (Rambaut, 2000). These tip-calibrations can be used when there has been enough time separating the sequences to allow an appreciable amount of genetic change to occur (Drummond et al., 2003). This is the case for some data sets comprising virus sequences sampled over time, or that include ancient sequences that have been radiocarbon-dated.

In the absence of primary calibrating information, age constraints can be applied to nodes in the tree based on previous molecular estimates. This practice is deprecated by some researchers, given the uncertainties in molecular dating and the potential sensitivity of date estimates to various confounding factors. However, such an approach might be the only feasible option in analyses of taxa with a poor fossil history. If secondary calibrations are employed, their uncertainty should be taken into account explicitly (Graur and Martin, 2004); this is readily done in Bayesian phylogenetic methods.

Models of Rate Variation

Evolutionary rates vary in a number of important ways, placing great importance on the act of selecting an appropriate clock model. Clock models describe how rates are expected to vary across branches in the tree and across genes in the data set. A simple way of accounting for different rates across genes is to assign a relative rate parameter to each gene. Accounting for rate variation across branches is more difficult, and various models have been developed for this purpose. One convenient way of classifying these models is to consider the number of distinct rates (k) that they allow in comparison with the number of branches (n) (**Fig. 3**; Ho and Duchêne, 2014).

The simplest model of rate variation is the strict clock, which assumes that all branches share the same rate ($k=1$). This model is often rejected in practice, with significant evidence of rate variation across branches being found for many data sets. However, the strict clock might be valid when analysing sequences from conspecific individuals or closely related species. It is also a useful null model when testing for evolutionary rate heterogeneity.

Local-clock models assume that there are several distinct evolutionary rates across the tree ($1<k<n$). These models might be appropriate when there is reason to believe that some branches in the tree are likely to share an evolutionary rate that is different from those along other branches (Yoder and Yang, 2000). Some local-clock models require the tree topology to be fixed and for distinct rates to be assigned to pre-defined sets of branches, but the Bayesian random local clock allows all of these to be inferred jointly (Drummond and Suchard, 2010).

Variants of local-clock models allow the branches sharing the same rate to be distributed throughout the tree, rather than necessarily being closely related to each other. These are sometimes known as 'discrete clocks' (Fourment and Holmes, 2014). Discrete clocks have been implemented in a range of statistical frameworks, including likelihood and Bayesian methods (e.g., Heath et al. (2012)).

The assumptions of the molecular clock can be relaxed even further, to allow a distinct rate of evolution along each branch in the tree ($k=n$). In order for these rates to be identifiable, there must be some form of constraint on their variation. The earliest relaxed-clock models assumed that rates evolve gradually throughout the tree, such that rates are autocorrelated (Sanderson, 1997, 2002; Thorne et al., 1998). In these models, evolutionary rates are assumed to be similar between adjacent branches in the tree. There are implementations of autocorrelated-rate models in likelihood methods, but most of the recent developments have occurred in a Bayesian framework.

In a second class of relaxed-clock models, known as uncorrelated relaxed clocks, the branch rates are drawn independently from an underlying probability distribution (Drummond et al., 2006; Rannala and Yang, 2007). Thus, there is no prior assumption that the rates along adjacent branches are similar to each other. A mixed relaxed clock can incorporate both autocorrelated and uncorrelated models by partitioning the rate variation between them (Lartillot et al., 2016).

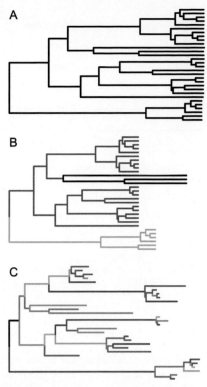

Fig. 3 The three major classes of clock model, based on the number of distinct rates that are allowed across branches in the phylogenetic tree. Branch shading indicates the relative evolutionary rate. (a) Strict clock, in which a single rate shared by all branches. (b) Local clock, in which the number of rates is much smaller than the number of branches. Rates are typically shared between adjacent branches in the tree. In this example, there are three different rates across the tree. (c) Relaxed clock, which allows a distinct evolutionary rate for each branch in the tree.

Challenges

The molecular clock has faced a variety of challenges over the past few decades, and some of the most prominent estimates of evolutionary timescales have been controversial. The most important criticisms of molecular clocks have concerned the impacts of evolutionary rate variation across timescales, along with the use of calibrations.

The notion of constant evolutionary rates has been subject to numerous criticisms. It is now widely accepted that rates of evolution vary substantially across the Tree of Life. This rate variation is due to biological factors such as differences in population size, generation length, and natural selection, as well as abiotic factors such as exposure to ultraviolet radiation (Bromham, 2009).

Part of the controversy surrounding rate variation is the use of 'universal' molecular clocks, which assume a constant rate among species within a broad taxonomic group. For example, there is widespread use of universal clocks in studies of arthropods, birds, and mammals, whereby an evolutionary rate of 1% per million years is assumed for mitochondrial DNA (e.g., Weir and Schluter, 2008). However, the reliability of assuming a standard rate across an entire taxonomic group has been regularly challenged (e.g., Nabholz et al., 2008; Nguyen and Ho, 2016). Given the abundant evidence that even closely related species can show substantial rate variation, the application of universal clocks can produce very misleading estimates of evolutionary timescales. Nevertheless, these clocks are still commonly used because there is often no other information about rates at hand.

The outcome of a molecular-clock analysis is substantially influenced by the choice of calibrations, and this has also been a major criticism of molecular dating. If calibrations are incorrectly modelled, the resulting date estimates can be highly inaccurate. When fossil calibrations are used, the selection of fossils can be subjective and their placement needs to be adequately justified. In Bayesian molecular dating, a further problem is that the calibration priors can interact with each other, producing joint priors on node times that no longer represent the fossil evidence used to specify the calibrations individually (Heled and Drummond, 2012; Warnock et al., 2012).

The uncertainty in fossil calibrations can be modelled and quantified in various ways (e.g., Marshall (2008), Wilkinson et al. (2011)), and recently developed methods allow fossil data to be incorporated in a more satisfying manner. For example, total-evidence dating and the fossilized birth-death process allow fossil taxa to be treated as part of the evolutionary process (Heath et al., 2014; Ronquist et al., 2016). Although some aspects of calibration methodology remain debated, best practice involves using multiple calibrations for molecular dating.

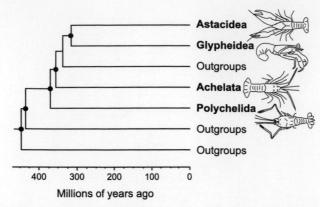

Fig. 4 Phylogeny and evolutionary timescale of lobster-like decapod crustaceans, inferred using a Bayesian phylogenetic analysis of molecular and morphological data by Bracken-Grissom *et al.* (2014). 'Outgroups' indicate groups of crustaceans that do not belong to the four infraorders of lobster-like decapods. Filled circles at internal nodes indicate the placement of fossil calibrations. Line drawings of animals show representatives of the four infraorders indicated in bold, and have been traced from images presented in the study by Bracken-Grissom *et al.*, 2014. The emergence of lobsters: phylogenetic relationships, morphological evolution and divergence time comparisons of an ancient group (Decapoda: Achelata, Astacidea, Glypheidea, Polychelida). Systematic Biology 63, 457–479.

Case Study: The Emergence of Lobsters

Lobsters are a morphologically diverse group of crustaceans belonging to the order Decapoda. They are economically and culturally important, representing a key resource for many fisheries throughout the world. Their exoskeletons increase their probability of preservation, meaning that they have a rich fossil record extending back approximately 350 million years. This record provides a unique opportunity to study the origins and diversification of major lineages within the group.

Bracken-Grissom *et al.* (2014) analysed data from 173 species to infer the phylogeny of lobster-like decapods. This sampling included 94% of extant genera and allowed the authors to resolve contentious relationships among and within groups (**Fig. 4**). Using a total-evidence approach, the authors analysed a combination of 190 morphological characters, three nuclear genes (*18S*, *28S*, and *H3*), and three mitochondrial genes (*16S*, *12S*, and *CO1*). The genetic markers included both fast- and slow-evolving genes, allowing various relationships to be resolved throughout the tree.

For each of the genes in the data set, Bracken-Grissom *et al.* (2014) identified the best-fitting model of nucleotide substitution. By inferring the tree from each gene individually, the authors checked for congruence in the phylogenetic signal across the six genes. Both maximum-likelihood and Bayesian phylogenetic methods were used to infer the evolutionary relationships among taxa. A relaxed molecular clock was used to account for rate variation across lineages. In one analysis, the clock was calibrated using the ages of 28 relevant fossils. These calibrations were implemented as exponential or uniform prior distributions for the ages of the corresponding nodes in the tree. In a second analysis, calibrations were based on a model of fossilization, speciation, and recovery (Wilkinson and Tavaré, 2009). The two approaches to calibration led to very similar sets of divergence-time estimates.

Bracken-Grissom *et al.* (2014) found that decapods arose in the Ordovician, with the most recent common ancestor of extant genera appearing in the Cretaceous. During the Jurassic, the infraorder Achelata rapidly diversified, coinciding with the early break-up of Pangaea. Southern and Northern Hemisphere crayfish were estimated to have diverged approximately 261 million years ago. In addition to these date estimates, Bracken-Grissom *et al.* (2014) provided a range of useful guidelines for incorporating fossil calibrations in molecular-clock studies.

Future Directions

Over the past decade, there have been substantial advances in DNA-sequencing technologies, leading to a dramatic increase in the data available for molecular dating. As a consequence, molecular-clock methods have had to evolve to deal with large data sets (Ho, 2014). Most of these data sets comprise either large numbers of taxa or large numbers of genes, but there is a growing trend towards genome-scale data sets that include many dozens of taxa (e.g., Jarvis *et al.* (2014), Misof *et al.* (2014)). These data sets can contain significant amounts of complex rate variation and can pose difficult computational challenges. The growing availability of genome-scale data sets has spurred the development of methods that can cope with large data sets, either by improved distribution of computational load or by relying on approximations of likelihood calculations (e.g., dos Reis and Yang (2011)). Nevertheless, current Bayesian phylogenetic methods are not capable of analysing data sets comprising large numbers of taxa and genes.

Molecular dating using genome-scale data sets must deal with the problem of incongruent gene trees. When the divergences between lineages have occurred in rapid succession, there is an increased probability of the phylogenetic signal being heterogeneous across the genome. In particular, the evolutionary histories of individual genes might not match the relationships among the species, in a phenomenon known as incomplete lineage sorting. This problem has not been satisfactorily addressed in molecular dating, but it remains an active area of research.

The growing access to genome-scale data sets raises an important question: does the precision of molecular dating improve with the amount of data? Various investigations have revealed that even with infinite data, the error in Bayesian date estimates depends on the uncertainty in the calibrations (dos Reis and Yang (2013)). Further improvements in estimates of evolutionary timescales will rely on the discovery of informative fossils and the development of better methods for integrating fossil and molecular data.

Closing Remarks

The molecular clock has been used in thousands of scientific studies, with no signs of declining relevance in the genomic era. Throughout its long history, the molecular clock has undergone substantial evolution and has proven to be a valuable tool for estimating evolutionary timescales. Applications of the molecular clock have been widespread, ranging from macroevolutionary studies of speciation and extinction to monitoring the evolutionary dynamics of modern-day pathogens.

Molecular-clock approaches have evolved from the assumption of constant evolutionary rates to the current models of rate variation across lineages. There has been an increase in efforts to incorporate fossil data into molecular dating, through total-evidence dating and the fossilized birth-death process. By taking advantage of the large amounts of genomic data being generated, molecular-clock methods will continue to play an important role in understanding the timescale of the Tree of Life.

See also: Applications of the Coalescent for the Evolutionary Analysis of Genetic Data. Evolutionary Models. Gene Duplication and Speciation. Inference of Horizontal Gene Transfer: Gaining Insights Into Evolution via Lateral Acquisition of Genetic Material. Molecular Phylogenetics. Natural Language Processing Approaches in Bioinformatics. Phylogenetic analysis: Early evolution of life. Phylogenetic Tree Rooting

References

Arbogast, B.S., Edwards, S.V., Wakeley, J., Beerli, P., Slowinski, J.B., 2002. Estimating divergence times from molecular data on phylogenetic and population genetic timescales. Annual Review of Ecology, Evolution, and Systematics 33, 707–740.

Bracken-Grissom, H.D., Ahyong, S.T., Wilkinson, R.D., *et al.*, 2014. The emergence of lobsters: Phylogenetic relationships, morphological evolution and divergence time comparisons of an ancient group (Decapoda: Achelata, Astacidea, Glypheidea, Polychelida). Systematic Biology 63, 457–479.

Bromham, L., 2009. Why do species vary in their rate of molecular evolution? Biology Letters 5, 401–404.

Bromham, L., Rambaut, A., Fortey, R., Cooper, A., Penny, D., 1998. Testing the Cambrian explosion hypothesis by using a molecular dating technique. Proceedings of the National Academy of Sciences of the USA 95, 12386–12389.

Brown, W.M., George, M., Wilson, A.C., 1979. Rapid evolution of animal mitochondrial DNA. Proceedings of the National Academy of Sciences of the USA 76, 1967–1971.

dos Reis, M., Donoghue, P.C.J., Yang, Z., 2016. Bayesian molecular clock dating of species divergences in the genomics era. Nature Reviews Genetics 17, 71–80.

dos Reis, M., Yang, Z., 2011. Approximate likelihood calculation on a phylogeny for Bayesian estimation of divergence times. Molecular Biology and Evolution 28, 2161–2172.

dos Reis, M., Yang, Z., 2013. The unbearable uncertainty of Bayesian divergence time estimation. Journal of Systematics and Evolution 51, 30–43.

Drummond, A.J., Ho, S.Y.W., Phillips, M.J., Rambaut, A., 2006. Relaxed phylogenetics and dating with confidence. PLOS Biology 4, e88.

Drummond, A.J., Pybus, O.G., Rambaut, A., Forsberg, R., Rodrigo, A.G., 2003. Measurably evolving populations. Trends in Ecology and Evolution 18, 481–488.

Drummond, A.J., Suchard, M.A., 2010. Bayesian random local clocks, or one rate to rule them all. BMC Biology 8, 114.

Fleischer, R.C., McIntosh, C.E., Tarr, C.L., 1998. Evolution on a volcanic conveyor belt: Using phylogeographic reconstructions and K-Ar-based ages of the Hawaiian Islands to estimate molecular evolutionary rates. Molecular Ecology 7, 533–545.

Foster, C.S.P., Sauquet, H., Van der Merwe, M., *et al.*, 2017. Evaluating the impact of genomic data and priors on Bayesian estimates of the angiosperm evolutionary timescale. Systematic Biology 66, 338–351.

Fourment, M., Holmes, E.C., 2014. Novel non-parametric models to estimate evolutionary rates and divergence times from heterochronous sequence data. BMC Evolutionary Biology 14, 163.

Graur, D., Martin, W., 2004. Reading the entrails of chickens: Molecular timescales of evolution and the illusion of precision. Trends in Genetics 20, 80–86.

Heath, T.A., Holder, M.T., Huelsenbeck, J.P., 2012. A Dirichlet process prior for estimating lineage-specific substitution rates. Molecular Biology and Evolution 29, 939–955.

Heath, T.A., Huelsenbeck, J.P., Stadler, T., 2014. The fossilized birth-death process for coherent calibration of divergence-time estimates. Proceedings of the National Academy of Sciences of the USA 111, 2957–2966.

Heled, J., Drummond, A.J., 2012. Calibrated tree priors for relaxed phylogenetics and divergence time estimation. Systematic Biology 61, 138–149.

Hipsley, C.A., Müller, J., 2014. Beyond fossil calibrations: Realities of molecular clock practices in evolutionary biology. Frontiers in Genetics 5, 138.

Ho, S.Y.W., 2014. The changing face of the molecular evolutionary clock. Trends in Ecology and Evolution 29, 496–503.

Ho, S.Y.W., Duchêne, S., 2014. Molecular-clock methods for estimating evolutionary rates and timescales. Molecular Ecology 23, 5947–5965.

Ho, S.Y.W., Phillips, M.J., 2009. Accounting for calibration uncertainty in phylogenetic estimation of evolutionary divergence times. Systematic Biology 58, 367–380.

Ho, S.Y.W., Tong, K.J., Foster, C.S.P., *et al.*, 2015. Biogeographic calibrations for the molecular clock. Biology Letters 11, 20150194.

Jarvis, E.D., Mirarab, S., Aberer, A.J., *et al.*, 2014. Whole-genome analyses resolve early branches in the tree of life of modern birds. Science 346, 1320–1331.

Kimura, M., 1968. Evolutionary rate at the molecular level. Nature 217, 624–626.

Kodandaramaiah, U., 2011. Tectonic calibrations in molecular dating. Current Zoology 57, 116–124.

Lartillot, N., Phillips, M.J., Ronquist, F., 2016. A mixed relaxed clock model. Philosophical Transactions of the Royal Society B 371, 20150132.

Marshall, C.R., 2008. A simple method for bracketing absolute divergence times on molecular phylogenies using multiple fossil calibration points. American Naturalist 171, 726–742.

Misof, B., Liu, S., Meusemann, K., *et al.*, 2014. Phylogenomics resolves the timing and pattern of insect evolution. Science 346, 763–767.

Nabholz, B., Glémin, S., Galtier, N., 2008. Strong variation of mitochondrial mutation rate across mammals – the longevity hypothesis. Molecular Biology and Evolution 25, 120–130.

Nguyen, J.M.T., Ho, S.Y.W., 2016. Mitochondrial rate variation among lineages of passerine birds. Journal of Avian Biology 47, 690–696.

Ohta, T., 1972. Evolutionary rate of cistrons and DNA divergence. Journal of Molecular Evolution 1, 150–157.

Ohta, T., 1973. Slightly deleterious mutant substitutions in evolution. Nature 246, 96–98.

O'Reilly, J.E., dos Reis, M., Donoghue, P.C.J., 2015. Dating tips for divergence-time estimation. Trends in Genetics 31, 637–650.

Parham, J.F., Donoghue, P.C.J., Bell, C.J., et al., 2012. Best practices for justifying fossil calibrations. Systematic Biology 61, 346–359.

Rambaut, A., 2000. Estimating the rate of molecular evolution: Incorporating non-contemporaneous sequences into maximum likelihood phylogenies. Bioinformatics 16, 395–399.

Rannala, B., Yang, Z., 2007. Inferring speciation times under an episodic molecular clock. Systematic Biology 56, 453–466.

Ronquist, F., Klopfstein, S., Vilhelmsen, L., et al., 2012. A total-evidence approach to dating with fossils, applied to the early radiation of the Hymenoptera. Systematic Biology 61, 973–999.

Ronquist, F., Lartillot, N., Phillips, M.J., 2016. Closing the gap between rocks and clocks using total-evidence dating. Philosophical Transactions of the Royal Society B 371, 20150136.

Runnegar, B., 1982. A molecular-clock date for the origin of the animal phyla. Lethaia 15, 199–205.

Sanderson, M.J., 1997. A nonparametric approach to estimating divergence times in the absence of rate constancy. Molecular Biology and Evolution 14, 1218–1231.

Sanderson, M.J., 2002. Estimating absolute rates of molecular evolution and divergence times: a penalized likelihood approach. Molecular Biology and Evolution 19, 101–109.

Sarich, V.M., Wilson, A.C., 1967. Immunological time scale for hominid evolution. Science 158, 1200–1203.

Sauquet, H., Ho, S.Y.W., Gandolfo, M.A., et al., 2012. Testing the impact of calibration on molecular divergence times using a fossil-rich group: The case of Nothofagus (Fagales). Systematic Biology 61, 289–313.

Simpson, G.G., 1964. Organisms and molecules in evolution. Science 146, 1535–1538.

Thorne, J.L., Kishino, H., Painter, I.S., 1998. Estimating the rate of evolution of the rate of molecular evolution. Molecular Biology and Evolution 15, 1647–1657.

Tong, K.J., Lo, N., Ho, S.Y.W., 2016. Reconstructing evolutionary timescales using phylogenomics. Zoological Systematics 41, 343–351.

Warnock, R.C.M., Yang, Z., Donoghue, P.C.J., 2012. Exploring uncertainty in the calibration of the molecular clock. Biology Letters 8, 156–159.

Weir, J., Schluter, D., 2008. Calibrating the avian molecular clock. Molecular Ecology 17, 2321–2328.

Wilkinson, R.D., Steiper, M.E., Soligo, C., et al., 2011. Dating primate divergences through an integrated analysis of palaeontological and molecular data. Systematic Biology 60, 16–31.

Wilkinson, R.D., Tavaré, S., 2009. Estimating primate divergence times by using conditioned birth-and-death processes. Theoretical Population Biology 75, 278–285.

Yoder, A.D., Yang, Z., 2000. Estimation of primate speciation dates using local molecular clocks. Molecular Biology and Evolution 17, 1081–1090.

Zuckerkandl, E., Pauling, L., 1962. Molecular disease, evolution and genic heterogeneity. In: Kasha, M., Pullman, B. (Eds.), Horizons in Biochemistry. New York: Academic Press, pp. 189–225.

Zuckerkandl, E., Pauling, L., 1965. Evolutionary divergence and convergence in proteins. In: Bryson, V., Vogel, H.J. (Eds.), Evolving Genes and Proteins. New York: Academic Press, pp. 97–166.

Further Reading

Bromham, L., 2011. The genome as a life-history character: Why rate of molecular evolution varies between mammal species. Philosophical Transactions of the Royal Society B 366, 2503–2513.

Donoghue, P.C.J., Yang, Z., 2016. The evolution of methods for establishing evolutionary timescales. Philosophical Transactions of the Royal Society B 371, 20160020.

dos Reis, M., Donoghue, P.C.J., Yang, Z., 2016. Bayesian molecular clock dating of species divergences in the genomics era. Nature Reviews Genetics 17, 71–80.

Heath, T.A., Huelsenbeck, J.P., Stadler, T., 2014. The fossilized birth-death process for coherent calibration of divergence-time estimates. Proceedings of the National Academy of Sciences of the USA 111, 2957–2966.

Heath, T.A., Moore, B.R., 2014. Bayesian Inference of Species Divergence Times. In: Chen, M.-H., Kuo, L., Lewis, P.O. (Eds.), Bayesian Phylogenetics: Methods, Algorithms, and Applications. Boca Raton: CRC Press, pp. 277–318.

Hipsley, C.A., Müller, J., 2014. Beyond fossil calibrations: Realities of molecular clock practices in evolutionary biology. Frontiers in Genetics 5, 138.

Ho, S.Y.W., 2014. The changing face of the molecular evolutionary clock. Trends in Ecology and Evolution 29, 496–503.

Ho, S.Y.W., Duchêne, S., 2014. Molecular-clock methods for estimating evolutionary rates and timescales. Molecular Ecology 23, 5947–5965.

Kumar, S., Hedges, S.B., 2016. Advances in time estimation methods for molecular data. Molecular Biology and Evolution 33, 863–869.

O'Reilly, J.E., dos Reis, M., Donoghue, P.C.J., 2015. Dating tips for divergence-time estimation. Trends in Genetics 31, 637–650.

Sauquet, H., 2013. A practical guide to molecular dating. Comptes Rendus Palevol 12, 355–367.

Tong, K.J., Lo, N., Ho, S.Y.W., 2016. Reconstructing evolutionary timescales using phylogenomics. Zoological Systematics 41, 343–351.

Biographical Sketch

Cara Van Der Wal is a marine evolutionary biologist interested in molecular clocks, phylogenetics, computational biology, systematics, and population genetics. She is a PhD candidate at the University of Sydney and the Australian Museum. Prior to commencing her PhD, Cara studied marine biology at the University of Technology Sydney and did research on stomatopod crustaceans at the University of Sydney.

Simon Ho is a Professor of Molecular Evolution at the University of Sydney, where he leds the Molecular Ecology, Evolution, and Phylogenetics research group. He is a computational evolutionary biologist interested in molecular clocks, evolutionary rates, phylogenetic methods, genomic evolution, and ancient DNA. He previously worked at the Australian National University and the University of Oxford, after being awarded his DPhil from the University of Oxford in 2006.

Phylogenetic Tree Rooting

Richard J Edwards, University of New South Wales, Sydney, NSW, Australia

Glossary

Additive tree   A tree in which the branch lengths are scaled according to time or the amount of evolutionary change.

Clade   A group of terminal nodes (OTUs) that share a common ancestral node (HTU).

HTU   Hypothetical Taxonomic Unit. An internal node of a phylogenetic tree.

Molecular Clock   The prediction of The Neutral Theory that, if most fixed changes are the result of neutral mutations, molecular evolution will occur at a reasonably regular "clock-like" rate, determined primarily by the neutral mutation rate.

Monophyletic   A trait (physical or genetic) that occurs within a single clade on a phylogenetic tree and thus can be explained by a single evolutionary event.

OTU   Operational Taxonomic Unit. A sequence or organism used as the terminal nodes of a phylogenetic tree.

Parsimony   The simplest explanation for an observation. The smallest number of changes needed to explain the data.

Topology   The branching order of a phylogenetic tree.

Ultrametric   Additive tree where all root-to-tip paths have an equal summed branch length.

Introduction: Phylogenetic Trees and the Role of the Root

A **phylogenetic tree** is a graphical representation of the evolutionary relationships between biological entities (**Box 1**). Phylogenetic trees can be used to model the relationships of organisms ("species trees") or, in the case of molecular phylogenetics, biological sequences ("gene trees" or "protein trees"). The methods differ in their details but the essence is the same, in which attempts are made to generate a branching structure that recapitulates the true biological relationships of the organisms or sequences in question. Because molecular phylogenetics is (a) more common (thanks to the ease of generating genetic data), and (b) inherently quantitative (due to the digital nature of the data itself and the quantitative nature of substitutional models), the rest of this chapter will be targeted at molecular phylogenetics. Unless otherwise stated, what is true for molecular phylogenies will generally also apply to species trees inferred from other types of data (e.g. morphology).

Phylogenetic trees are a special kind of dendrogram (or hierarchical clustering) in which the internal nodes are inferred to represent actual biological entities, namely common evolutionary ancestors of two or more leaves. The **role of the root** is to add direction to these relationships and clearly distinguish which internal nodes are ancestral to which tips (**Fig. 1(a)**). Relationships between sequences are captured by the **topology**, which is the branching order (**Fig. 1(b)**) and, where appropriate, **branch lengths** that indicate the amount of evolutionary change (or time) between nodes (**Fig. 2**). Note that **topology only** trees (such as **Figs. 1(a)** and **2(a)**) do not have meaningful branch lengths, which limits the rooting options (discussed below).

Phylogenetic methods can broadly be divided into three classes:

1. **Parsimony methods** seeks to minimize the number of evolutionary events required to explain the data. Parsimony methods are often biologically meaningful but only use a subset of "informative" data that can be used to split sequences into two or more groups of two or more sequences.

2. **Distance methods** convert pairwise comparisons of sequences into a distance matrix and then cluster sequences based on proximity or use an algorithm to try to minimize overall branch lengths. Distance methods use all the data but add a layer of abstraction between the actual biology and the inferred phylogeny. They tend to be very computationally efficient and thus are popular choices for non-experts.

3. **Model-based methods** (generally Maximum Likelihood or Bayesian methods) use an explicit evolutionary model to try to infer the phylogenetic tree that best fits the data. In principle, these methods are best as they are both biologically meaningful *and* use all the data. The downside is that they are dependent on the selected evolutionary model and can be very computationally intensive.

Rooting is largely independent of the method used to infer the phylogeny. Here, the primary consideration is whether branches have lengths, or only the topology (branching order) is established (**Fig. 2**).

In **additive trees**, branch lengths represent evolutionary change. For species trees, this is often an estimation of literal times of divergence. For sequence trees, branch lengths generally represent evolutionary change more directly, in terms of substitutions per site. Converting non-sequence data into a scaling numerical value with a reasonably constant rate through time is very challenging. For this reason, trees based on morphology tend to use parsimony approaches and often do not have branch lengths. Where branch lengths (and roots) are set, it is often using independent data, such as fossils. Where fossil evidence is incorporated, it is possible to have actual historical samples in the tree. This is quite different from molecular phylogenetics where, except in very rare cases (e.g. frozen viral samples), the tree is wholly inferred from contemporary data.

Box 1   Trees as graphs

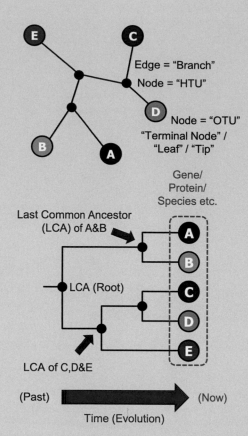

Phylogenetic trees are a special case of what is known as a graph, in which a series of **nodes** are joined by **edges**, known as **branches**. Nodes are split into terminal nodes (or "leaves" to keep with the tree analogy), also known as **Operational Taxonomic Units (OTUs)** and internal nodes, also known and **Hypothetical Taxonomic Units (HTUs)**; OTUs are generally the data on which the phylogenetic method operates, whereas HTUs are the hypothetical ancestors inferred by the method.

Unrooted trees (top figure) are "undirected graphs" in which the relationship between two connected nodes is symmetrical. Adding a **root** (bottom figure) adds directionality to the graph, such that each branch has an **ancestral node** closer to the root and a **descendant node** closer to the tips. Internal nodes now represent the **Last Common Ancestor (LCA)** of a **clade** of terminal nodes.

This chapter deals with strictly **bifurcating** (or **binary**) trees in which each HTU is connected to three branches (see Section "Midpoint Rooting"). In the case of rooted trees, this corresponds to one ancestral and two descendant branches.

Unless a **molecular clock** is assumed, substitution rates can differ throughout the tree. Many phylogenetic methods are vulnerable to large rate variations and, whilst they can compensate for an extent of variation, confidence in the phylogeny produced generally decreases as rate variation increases (see Section "Resolving Monophyly and Homoplasy").

Rooting and Evolutionary Change

Rooting a tree adds an additional internal "node" (or HTU) to a tree, which indicates the **last common ancestor** of everything in the tree (**Box 1**). In doing so, the tree is given *direction*. In addition to inferring the relative relationships of our species/sequences based on the topology, we can now look at directions of change. One key thing to remember is that placing a root does not actually change the amount of evolutionary change in the tree. Rooting a tree on a branch simply partitions up the evolutionary changes attributed to that branch into the two post-root branches (**Fig. 3**). The critical difference is that evolutionary change now has an explicit **ancestral state** and **derived state** (**Fig. 1**) (see Section "Ancestral Sequence Prediction").

Multifurcating Trees

For simplicity, this chapter is dealing with strictly **bifurcating** or **binary trees**, in which each internal node has one ancestral branch and two descendant branches, for a total of three branches (**Box 1**). Trees do not need to be strictly bifurcating and a node can have

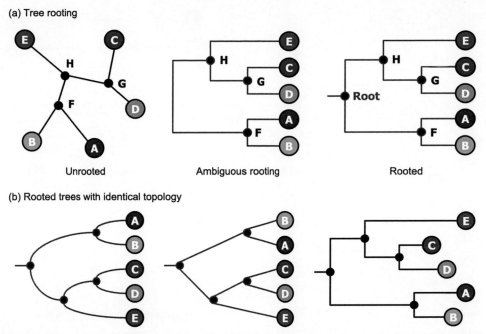

Fig. 1 Tree rooting and topology. The same phylogenetic tree of five sequences A-E with different representations. (a) Hypothetical nodes F-H represent inferred ancestors but in an unrooted tree (left) the relationship between those ancestors and other nodes is unknown. Rooting the tree (right) does not change the topology (branching order) but adds an additional ancestral node that gives directionality. In the unrooted tree, node H could be the common ancestor of (AB)E, (CD)E or (AB)(CD). In the rooted tree, node H is unambiguously the common ancestor of (CD)E. Sometimes a tree appears to be rooted but has no explicit root node marked (center). Unless otherwise stated, it is safest to assume that such a tree is unrooted. (b) Evolutionary relationships are represented by the tree topology, i.e. the branching order. This is independent of the way that branches are represented, or the vertical arrangement of terminal nodes. Despite different vertical arrangements and styles, all three trees have the same topology.

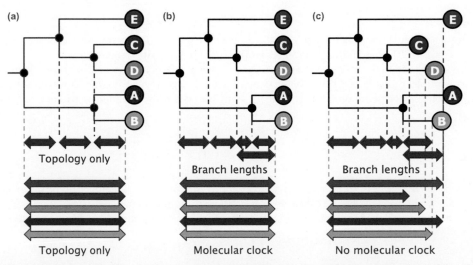

Fig. 2 Branch lengths and molecular clock. When looking at a rooted phylogenetic tree, the spacing between nodes and the root can be used to identify whether branch lengths are being shown and, if so, whether a molecular clock is in operation. (a) A topology only tree is ultrametric (all terminal nodes are the same distance from the root) and its internal nodes are typically evenly spaced at regular intervals as branches join, starting closest to the tip. (Note: vertical spacing in this representation has no meaning.) (b) If the tree is ultrametric but internal nodes are not regularly spaced then branch lengths are being used in the context of a molecular clock (a constant rate of evolution). This is usually means it is an assumption of the phylogenetic inference method but it *could* also reflect empirically clock-like evolution. Such a tree is inherently midpoint-rooted. (c) If neither terminal nor internal nodes show regular spacing, branch lengths are being used and no (universal) molecular clock is assumed. In this example, midpoint rooting is used.

more than two descendant branches. This is known as **multifurcation.** This could be a genuine biological phenomenon – a single ancestral population or sequence may radiate into more than two distinct lineages in parallel. Alternatively, it could simply represent a lack of confidence in tree topology; low confidence branches may be collapsed to zero length to explicitly acknowledge

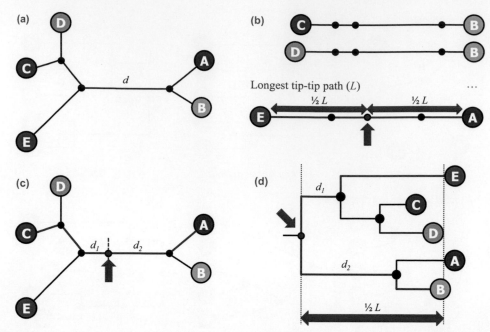

Fig. 3 Midpoint rooting. (a) Midpoint rooting starts with an unrooted tree with branch lengths. (b) Each pairwise tip-to-tip distance is calculated (only three shown) and the longest distance, L, selected to determine the root placement. The root is placed on this longest tip-to-tip path, equidistant from the two tips. (c) The evolutionary change along the rooted branched, d, is partitioned between the two new post-root branches such that $d_1 + d_2 = d$. (d) In the final tree, the two tips from the longest path, A and E, are furthest and equidistant from the root.

the lack of certainty about the branching order in that part of the tree. Unless otherwise stated, methods and considerations for rooting binary trees will also apply to multifurcating trees. (Multifurcation can be represented as a series of bifurcating branches of zero length).

Phylogenetic networks

Evolution is not always a neatly bifurcating process. Sexual reproduction, recombination and horizontal gene transfer can generate scenarios in which a given OTU has multiple ancestral HTUs. Traditional phylogenetic methods adopt a "winner takes all" approach to ancestry and it is up to the user to pre-partition the data where mixed ancestry may be an issue. **Phylogenetic network** methods will explicitly model mixed ancestry in addition to non-binary divergence. As with simpler multifurcation of descendants, rooting methods and considerations are broadly the same for a phylogenetic network and these will not be considered further in this chapter.

Rooting Methods

To correctly interpret a rooted tree, it is important to understand the assumptions and limitations of the rooting method used (see also Section "Why root a phylogenetic tree?"). There are a few different rooting methods, depending on the available data and reason for rooting (see Section "Inferring Rooting Methods"). These will be briefly summarized in this section.

Random Rooting

The simplest method for rooting a tree is to place the root on a random branch. This is clearly not a biologically meaningful rooting but repeating the process multiple times can be useful for establishing how robust the ultimate conclusions of a study are to incorrect rooting. In its simplest form, topology alone is considered. Alternatively, branches could be weighted by length, so that the root was more likely to fall in a long branch.

Midpoint Rooting

The easiest biologically meaningful rooting method is **midpoint rooting** (**Fig. 3**). Midpoint rooting needs an additive tree (i.e. scaled branch lengths, **Fig. 2**) and makes the explicit assumption that there are no major deviations in evolutionary rates (see Section "Resolving Monophyly and Homoplasy") and that sequences are evolving in a reasonably "clock-like" fashion. Under this model, it follows that the root will be equidistant from the most divergent pair of sequences in the tree: if rates are reasonably constant, the reason they are the most diverged is because they have had the longest period of independent evolution since splitting.

Midpoint rooting works by first establishing all the tip-to-tip distances by summing up the branch lengths of the shortest distance between two tips. The root is then placed in the midpoint of the longest tip-to-tip path (**Fig. 3**). If there are multiple equally-long paths, one is chosen at random, as they will all share an identical midpoint. **Ultrametric trees (Fig. 2(b))** represent a special case of midpoint rooting where the evolutionary rate is constant throughout the tree and thus all tips end up equidistant from the root.

Midpoint rooting is attractive because it does not rely on any prior knowledge and can be calculated simply from the tree itself. However, the explicit assumption of constant evolutionary rates makes it quite unreliable in many real scenarios. It is advisable to look at how "clock-like" a tree appears, i.e. how much spread there is in root-to-tip distances following midpoint rooting (**Fig. 2(c)**), before deciding whether to trust midpoint rooting.

Rate-adjusted midpoint rooting

There are two main considerations for evolutionary rates:

1. The selective constraint of the sequence in that lineage.
2. The mutation rate of the sequence in that lineage, which is affected by generation time and the number of germline cell divisions per generation.

In principle, rate variations between branches can be modelled (as part of a model-based phylogenetic method) and the branch lengths normalized for different rates to convert them into evolutionary time. The tree could then be midpoint rooted based on rate-corrected branch lengths. However, this is often done by imposing a known species tree onto the gene tree to establish the branch-specific rates, which is essentially a form of outgroup rooting (see Section "Long Branch Artifacts").

Outgroup Rooting

Outgroup rooting is the generally considered to be the Gold Standard of tree rooting. Here, prior knowledge is used to manually identify the ancestral branch in the tree. This is done by identifying a **clade** that is known to have branched off first from the rest of the tree, and then placing the root on the branch leading to that clade (**Fig. 4**). Often, the outgroup clade is a single OTU.

Selecting an outgroup

Selecting outgroup sequences is not always easy. As with all knowledge-based methods, outgroup rooting is only as good as the knowledge used. It will also make some explicit assumptions of the data, e.g. that all the sequences in the tree are orthologues (related by speciation). If a tree accidentally includes some paralogues (homologous sequences derived from duplication events),

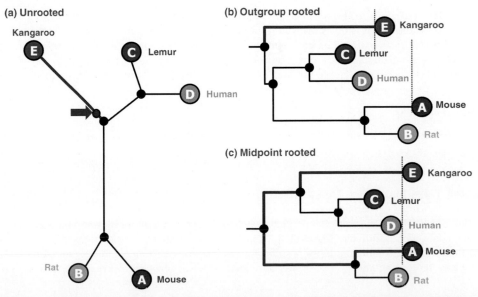

Fig. 4 Outgroup rooting. (a) Outgroup rooting identifies the OTU or clade known to have diverged first from the rest of the tree. In this case, the marsupial kangaroo diverged from the four placental mammals. (b) The root is placed at an arbitrary point on the branch leading to the outgroup. Typically, this is done to minimize the difference in the root-to-tip distances. Unlike midpoint rooting, the two most divergent tips will not necessarily be equidistant from the root. (c) In this example, the rodent lineage is evolving faster than the rest of the tree, which would result in an incorrect midpoint rooting in which primates and marsupials share a common ancestor since their divergence from rodents.

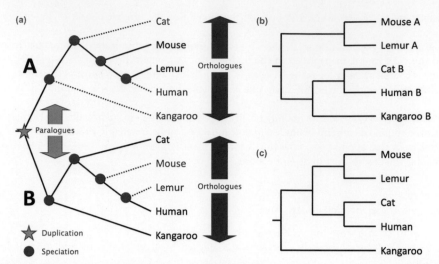

Fig. 5 Paralogue rooting artefacts. (a) True topology-only tree of a gene family with two genes A and B in five mammalian species. The root of the tree is a gene duplication event (star). Other nodes (filled circles) are speciation events. Genes related by duplication are paralogues. Genes related by speciation are orthologues. Missing data for subsequent trees is marked in red. (b) The true topology-only tree without the missing data from the original tree. Orthologues still form monophyletic clades and the root is a duplication event. (c) Using kangaroo as an outgroup but ignoring the paralogous nature of some of the sequences incorrectly roots the reduced tree and produces a confusing topology in which primates are non-monophyletic.

this assumption would be violated and the outgroup rooting could possibly go wrong (**Fig. 5**). Single, distant OTUs are more likely to be incorrectly placed in a tree, so it is usually better practice to root on an outgroup clade rather than a single sequence.

Using gene duplications as an outgroup

Although incorrect classification of proteins as orthologues or paralogues can lead to rooting issues (**Fig. 5**), known relationships between different members of a gene family can be used for outgroup rooting in the absence of known taxonomic relationships. In **Fig. 5(a)**, for example, the duplication node can be used to root the tree. In effect, gene A is acting as an outgroup for gene B and vice versa. This can be useful even when one subfamily is poorly sampled, e.g. Human B could be used to outgroup root the entire Gene A clade.

Adding outgroups

Often, there is no obvious outgroup in the set of sequences or species from which you have made your phylogeny. (Indeed, establishing the nature of relationships is often the reason for making a phylogeny from these data.) In these cases, it is useful to add one or more additional sequences from known outgroups and then use these to root the tree. For example, when generating a phylogeny of placental mammalian species or sequences, one might add one or more marsupials to root the tree (e.g. **Fig. 4**). Alternatively, one can add paralogous sequences as described above (see Section "Using gene duplications as an outgroup").

Unrooting and Re-Rooting Trees

Unrooting and re-rooting a tree is very simple. The two post-root branches are simply stuck together to (re)generate an unrooted tree (e.g. converting **Fig. 3(d)** to **(c)**). The new root can then be placed using any of the standard methods.

Recognizing Rooted Trees (and other Tree Attributes)

Correct interpretation of a phylogenetic tree relies on understanding the method used for rooting. Unfortunately, this information is not always provided. Fortunately, different tree attributes and rooting methods leave typical signs that can be used to identify the rooting method with reasonable confidence. If not possible, the safest course of action is probably to re-root the tree yourself (see Section "Unrooting and Re-Rooting Trees"). (Of course, once identified, you might not agree with the method used and want to re-root anyway).

Recognizing an Unrooted Tree

The easiest rooting method to recognize is where no rooting has been applied and the tree is in a radiation representation (**Fig. 1 (a)**, left). Sometimes unrooted trees are still drawn in a traditional "square branch" configuration (**Fig. 1(a)**, center) even though there is no actual root. This is most common – and hardest to spot – when only the tree topology is shown (**Fig. 2(a)**) but is sometimes the default output for phylogenetics programs. Ideally, a rooted tree will have an actual root node and/or branch shown (**Fig. 1(a)**, right) but this is not always the case and so trees that look rooted but lack a root are inherently ambiguous. Displaying branch lengths will often make it clear whether such a tree has been rooted; any major disparity between the root-tip distances each side of the root should be a cause for concern. If in doubt, explicitly unroot or re-root the tree (see Section "Unrooting and Re-Rooting Trees"), depending on your needs.

Inferring Rooting Methods

When interpreting a rooted phylogenetic tree, there are five basic attributes that one should look for:

1. **Topology.** This is the basic shape of the tree, e.g. its branching pattern (**Fig. 1**). All phylogenetic trees have a topology and it is the only attribute that one can guarantee. Topology-only trees are recognized by the uniformity of the branching pattern (**Fig. 2(a)**).
2. **Branch lengths.** "Additive" trees have evolutionary change information encoded in their branches, giving different branches different lengths. Additive trees can be recognized by the fact that (except for rare situations) the internal nodes are no longer spaced evenly and "line up" (**Fig. 2**). This is a pre-requisite for midpoint rooting (see Section "Inferring Rooting Methods").
3. **Molecular clock.** Most phylogenetics methods perform best when evolutionary rates are constant. Some methods go as far as to enforce this, producing what is known as an ultrametric tree. In this case, all leaves are equidistant from the root even though the internal nodes are not evenly spaced (**Fig. 2(b)**). If no molecular clock is imposed, different leaves will be different distances from the root (**Fig. 2(c)**).
4. **Midpoint rooting.** Even in the absence of other knowledge, it is possible to assess whether an additive tree is consistent with midpoint rooting. If it is then the longest root-to-leaf distance will the same for each "side" of the tree (i.e. for each of the two descendant clades that split at the root itself) (**Figs. 3(d)** and **4(c)**). On the other hand, if one side of the tree sticks out further than the other, midpoint rooting has not been used (**Fig. 4(b)**). In this case, the assumption is usually that outgroup rooting has been used but this can only be confirmed using independent knowledge. Note that for reliable trees, midpoint rooting and outgroup rooting will give the same root; consistency with midpoint rooting does not therefore rule out the use of outgroup rooting.
5. **Branch confidence.** Rooting on a particular branch can only ever be as confident as the branch itself. Methods of assessing branch confidence (typically bootstrapping or probability estimates) are beyond the scope of this chapter but will be recorded on a tree in the form of numbers associated with each branch. If the rooted branch has low support, the root itself should be considered uncertain. (Note that in most cases, the confidence estimates are made on the unrooted tree and so both descendant branches from the root will have the same confidence score.) Where confidence values are absent, branch lengths can be used as a proxy. (Note: if some scores are given and not others, missing values normally indicate low confidence.) Usually, longer branches will be more robustly supported by the data; the reason the branch is long is that a lot of data supports that division of the tree. This makes rooting particularly uncertain in scenarios of rapid early radiation, where deep branches are short. Neither branch confidence nor branch length is outright confirmation that the root (or even the topology) is correct; a phylogenetic method or alignment error might be consistently biased into producing the wrong tree. However, a low confidence and/or very short rooting branch should always make one wary of root placement. In principle, one can apply the same confidence methods to the root itself, e.g. midpoint rooting every bootstrap tree to establish how many times the root ends up in the same place. In practice, this is rarely done.

Long Branch Artifacts

Phylogenetics methods work best when:

1. Evolutionary rates are reasonably constant across the whole tree.
2. Sampling of OTUs is reasonably uniform so that there is enough data to determine relationships but not so many differences that they begin to get masked by "multiple hits" (where the same nucleotide or amino acid undergoes several substitutions *on the same branch*, masking the earlier changes).

Where sequences violate these conditions, and have a lot of divergence relative to all the other OTUs, methods can get mislead and interpret an increased evolutionary rate as an increased time of divergence. Distance methods are particularly prone to this issue but Maximum Parsimony (and even likelihood methods) can also struggle when substitutions are getting saturated in rapidly evolving lineages.

For distance methods, very long branches tend to migrate towards the midpoint root of the tree because the other OTUs cluster first. This can result in both an incorrect (but sometimes highly supported) basal placement of those branches but also an incorrect rooting if midpoint rooting is used.

For Maximum Parsimony and model-based methods, the problem is subtler. Large numbers of substitutions increase the chance of randomly shared substitutions. If these begin to exceed the genuine shared substitutions of these sequences with their more slowly evolving neighbors, the rapidly evolving lineages can begin to cluster together in a phenomenon known as "long branch attraction". Again, midpoint rooting will inherently tend to put the attracted clade near the root as the long branch lengths will also make them more likely to be part of the longest tip-to-tip path (**Fig. 3**).

Unfortunately, outgroup rooting is often performed by selecting a single sequence known to be more distantly related to all other OTUs. Assuming reasonably constant evolutionary rates, rooting on this outgroup will match midpoint rooting. This is a good thing but can be visually indistinguishable from a scenario where a single rapidly-evolving lineage has migrated to the root. This emphasizes the need to document the rooting method used. The outgroup might also "attract" other rapidly evolving sequences, which share sequence variants by chance. Where possible, outgroup rooting should be done using multiple sequences to produce a more "balanced" tree and avoid this issue.

Why Root a Phylogenetic Tree?

Most phylogenetic methods are inherently undirected. This enables the user to understand the hierarchal relationships between sequences or species but does not reveal (or predict) the ancestral state. The first question when considering the rooting of a phylogenetic tree is whether one needs to root the tree at all. Often this is not required for an analysis. This depends on how important it is to infer the direction of change (see Section "Random Rooting", **Fig. 1**).

Phylogenetic trees are often generated to provide context for certain species or sequences. In this case, we are primarily interested in which clade contains our query sequence. This is often obvious from an unrooted tree, in which case it might be considered safer not to try to establish a root. For example, in **Fig. 4**, mouse and rat are clearly most closely related, wherever the root is placed. Likewise, when trying to identify an unknown gene, its phylogenetic position within a clade might be enough to predict its identity, independent of strict ancestry: a complete unrooted version of **Fig. 5(a)** would suffice to clearly identify which mouse sequence was Gene A and which was Gene B. (Things might be less clear for the kangaroo sequences).

Ancestry becomes important when we are interested in *how* a trait evolved and want to know when it appeared. Alternatively, perhaps we know a trait appeared at a certain time and want to see what sequence changes correlate with its appearance. As with the example above, sometimes it is not actually necessary to *accurately* root the tree to get the ancestry information that one needs.

Ultrametric Phylogenetic Methods

One trivial reason to root a tree is that the chosen phylogenetic method inherently produces a rooted tree. As previously discussed, "Ultrametric" phylogenetic methods (e.g. Unweighted Pair Group Method with Arithmetic mean (UPGMA)) explicitly assume a constant rate of evolution across all branches. As such, the tree root is set as part of the phylogenetic reconstruction itself at the point where the last pair of groups are joined. In Ultrametric trees, all terminal nodes (OTUs) are equidistant from the root (**Fig. 2(a) and (b)**). Ultrametric tree rooting is therefore a special case of midpoint rooting where there are multiple longest tip-to-tip paths.

Ancestral Sequence Prediction

One of the main reasons for wanting to know the direction of branching is to establish the direction and timing of evolutionary change (**Fig. 1**). Note that it can be very challenging to correctly allocate the ancestral state of the root itself when there is disagreement between the two post-root nodes (**Fig. 6**). In general, the ancestral node will tend to converge on the node with the shortest branch length because each substitution is more likely to occur on the longer branch. This can lead to the slightly paradoxical situation where a non-zero-length branch has zero substitutions assigned to it. This, in turn, highlights some of the difficulties in interpreting distance-based trees, which tend to spread out discrete evolutionary events across multiple branches.

Resolving Monophyly and Homoplasy

Traits can either be **monophyletic** or **non-monophyletic**. Monophyletic traits are shared by all members of a clade, such as leucine (or valine) in **Fig. 6(c)**. Non-mono-phyletic traits exclude one or more members of the clade, such as leucine (but not valine) in **Fig. 6(b)**. Correct tree rooting can distinguish these scenarios (**Fig. 6**).

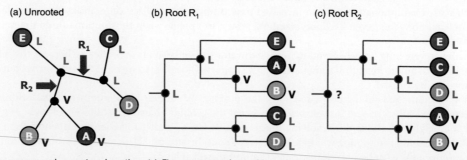

Fig. 6 Ancestral sequence assignment and rooting. (a) Five sequences have either a leucine (L) or valine (V) at a site of interest. Parsimony can be used to assign the ancestral states of internal nodes even in the absence of a root but the state of the last common ancestor and direction of change is unclear. (b) Where both post-root nodes share the same state, parsimony can be used to infer the ancestral state: in this case, leucine. (c) Where post-root nodes differ in their state, the ancestral state is unclear.

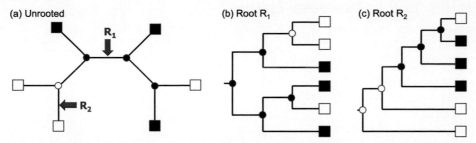

Fig. 7 Resolving homoplasy. (a) This unrooted tree shows homoplasy: the white state appears twice without shared ancestry. Rooting the tree resolves the nature of this homoplasy. (b) If the root is at R_1 the ancestral state is black and the homoplasy has arisen through (parallel) convergent evolution. (c) If the root is at R_2, the ancestral state is white and the homoplasy is the product of secondary loss, or reversion to the ancestral state.

In **Fig. 6(b)**, the leucine is **paraphyletic**, and arises from common ancestry but does not include all members of the clade. An alternative form of non-monophyly is **polyphyly**, in which the same trait appears in unrelated clades (**Fig. 7**). This is known as **homoplasy**, which is the situation where OTUs share a characteristic (either a phenotypic trait, or a specific nucleotide/residue in the same position) that is not the product of common ancestry (**homology**).

Another common reason to root a phylogenetic tree is to resolve apparent homoplasy. Correctly rooting the phylogeny can resolve whether a dispersed trait is truly homoplastic (rather than paraphyletic as in **Fig. 6(b)**). Furthermore, it can provide insight into the evolutionary history behind the homoplasy. This could be convergent (or parallel) evolution, in which the same trait independently evolves on two lineages (**Fig. 7(b)**), or secondary loss, in which a lineage reverts to the ancestral state (**Fig. 7(c)**).

Concluding Remarks

As with all analysis, there is no one-size-fits-all rule for when to root a tree and/or how carefully it needs to be done. The key thing is to make sure that the rooting is adequate for the biological question being asked – and to be wary of roots originally placed to answer a different biological question that may have different concerns and constraints.

See also: Gene Duplication and Speciation. Inference of Horizontal Gene Transfer: Gaining Insights Into Evolution via Lateral Acquisition of Genetic Material. Molecular Clock. Molecular Phylogenetics. Natural Language Processing Approaches in Bioinformatics. Phylogenetic analysis: Early evolution of life

Further Reading

Felsenstein, J., 2004. Inferring Phylogenies. Sunderland, MA: Sinauer Associates, Inc.
Page, D.M.P., Holmes, E.C., 1998. Molecular Evolution: a phylogenetic approach. Oxford: Blackwell Science Ltd.

Biographical Sketch

Rich Edwards is a Senior Lecturer in Bioinformatics in the School of Biotechnology and Biomolecular Sciences (BABS) at the University of New South Wales. Rich trained a geneticist at the University of Nottingham (UK), studying the population genetics of transposable elements in bacteria for his PhD. He moved to Dublin (Ireland) to become a full time bioinformatician in 2001, developing sequence analysis methods for the rational design of biologically active short peptides based on ancestral sequence prediction. This developed into an interest in Short Linear Motifs (SLiMs), which are short regions of proteins that mediate interactions with other proteins. Rich has developed several tools for the prediction and analysis of SLiMs, distributed in the SLiMSuite package. Rich established his lab in 2007 at the University of Southampton (UK) before moving to UNSW in 2013. Research interests in the lab stem from a fascination with molecular basis of evolutionary change and how we can harness the genetic sequence patterns left behind to make useful predictions about contemporary biological systems. Core research is divided between SLiM analysis/prediction and genomics projects. Rich has also collaborated on numerous projects involving DNA and/or protein sequence analysis.

Tree Evaluation and Robustness Testing

Mahendra Mariadassou, INRA, Paris, France
Avner Bar-Hen, CNAM, Paris, France
Hirohisa Kishino, University of Tokyo, Tokyo, Japan

Motivation

Applications of Phylogenies

Molecular phylogenetics is a lively field of research with a number of practical applications. Reconstructing large phylogenies, such as the bird (Prum *et al.*, 2015) or mammal phylogeny (Bininda-Emonds *et al.*, 2007), is of intrinsic interest to evolutionary biologists, but those phylogenies are also *the basic structures necessary to think clearly about differences between species, and to analyse those differences statistically* (Felsenstein, 2004). They arise frequently in comparative genomics, conservation issues (Bordewich *et al.*, 2008), functional prediction of genes (Eisen, 1998) and more generally are at the heart of phylogenetic comparative methods (Revell *et al.*, 2008; Pennell and Harmon, 2013). Most, if not all, applications of phylogenetics have in common that they rely on accurate phylogenetic trees and it is crucial to validate the tree as different trees can lead to vastly different conclusions concerning the origin and evolution of a trait (Geneva *et al.*, 2015).

With the advent of molecular data and increased formalism of the field (Gascuel, 2005), modern phylogenetic reconstruction is now essentially a statistical inference problem. Many popular reconstruction softwares such as PhyML (Guindon and Gascuel, 2003), RAxML (Stamatakis, 2006), FastTree (Price *et al.*, 2010) or MrBayes (Ronquist and Huelsenbeck, 2003) produce a statistical estimate of the tree and we therefore frame validation in a statistical framework. We first discuss the different sources of inaccuracies in the reconstructed tree (Section Sources of Error) and distinguish between natural variability ($\simeq$ variance) and modeling errors ($\simeq$ bias). We then briefly describe and discuss popular support values (Section Robustness) aimed at validating a tree. In addition to support values, the variability observed in a forest of trees, can be summarized in order to produce robust tree estimates (Section Consensus Methods). Finally, we end this article by reviewing promising developments in the field of continuous distance and highlight their use for validation (Section Detection of Conicting Phylogenetic Signals).

Validating the Tree

A tree is a complex object that encodes the evolutionary relationship of a set of species. Many inference methods return a single focal tree and validation is most often concerned with it. There are two families of validation methods based on (i) the scale at which validation is performed and (ii) whether the tree is considered by itself or with respect to other trees.

- The first family is *local* and grounded on the observation that a tree is uniquely determined by its branches (Buneman, 1971). A tree can thus be validated by computing a *support value* for each of its branch. Support values tell us which parts of the tree are *reliable*, in a yet to be defined way.
- The second family is *global* and considers the focal tree as a single object rather than a collection of branches. It compares it to a set of alternative trees using statistical tests (Shimodaira and Hasegawa, 1999). The tests tell us whether the tree is strictly better (i.e. a better fit to the molecular data) than the alternatives.

The two approaches have a different focus but are complimentary. In particular, global tests can be used to compute local support values.

Robust Estimate

Most softwares return the best tree for a given criteria (e.g. likelihood for PhyML and RAxML). Support values tell us whether that tree is reliable and tests tell us whether it's much better than second best or other alternatives.

However, in the presence of outlier data, the superiority of the best tree may lie exclusively in a few data points: slight changes in the molecular data may dramatically change the best tree, with deep clades moving from one position in the tree to another (Bar-Hen *et al.*, 2008). Since molecular data are inherently noisy, it is interesting to produce *robust* trees that nearly, but not completely, optimize the criteria while being resilient to small changes in the molecular data.

A straightforward way to build a robust estimate is to start from a forest of *good* trees and summarize them in some way to build a *consensus* tree. The forest can consist of trees that are only slightly worse than the best tree (e.g. bayesian consensus) or that are inferred from slightly perturbed data (e.g. bootstrap consensus).

However, more recently, the statistical perspective that observed variability between collections of trees is interesting of itself has gained traction. This has spawned a number of modern methods that utilize the elegant mathematics of tree space to permit new data analysis methods and associated new insights in a number of different contexts.

Sources of Error

Genome-scale analyses indicate that different genes yield conflicting phylogenies. Broadly speaking, these conflicts arise from two main sources: (i) the inability of traditional reconstruction methods to deal with the complexity of molecular evolution (methodological sources) and/or (ii) genuine biological events such as lateral gene transfer (LGT), incomplete lineage sorting and others that lead to different evolutionary histories along the genome (biological sources).

Methodological factors affecting phylogenetic reconstruction include the choice of optimality criterion, limited data availability, taxon sampling and specific assumptions in the modeling of sequence evolution. Biological processes such as the natural selection, small population size, etc may also cause the gene tree to differ from the species tree. The large number of potential explanations for observed incongruences in molecular phylogenetics makes decisions of how to handle them quite difficult (Rokas *et al.*, 2003b).

Biological Sources

Biological source of errors are very diverse and among other we may cite contamination, frameshift events, incorrect annotations, erroneous chimerical sequences, wrong orthology assessment, horizontal gene transfer, gene conversion, incomplete lineage sorting or hybridization, etc. (see Philippe *et al.*, 2017 for a survey).

With the improvement of sequencing technologies, the amount and quality of molecular data used in phylogenetics has drastically increased to the point where sequence quality is superseded by other concerns. For example, inclusion of non-orthologous sequences in a study can have drastic consequences on the final results (Laurin-Lemay *et al.*, 2012; Philippe *et al.*, 2011b). It can, for example, arise through undetected contamination of the genetic material during the sampling or sequencing steps (cross-contamination). It can also arise when paralogs are misidentified as orthologs, a common occurrence given the high frequency of gene/genome duplication, gene conversion and gene loss.

Finally, correct alignments are of crucial importance in phylogenetics as repeatedly pointed out (Morrison and Ellis, 1997; Ogden and Rosenberg, 2006; Talavera and Castresana, 2007; Wong *et al.*, 2008). Yet, due to the lack of tractable models of sequence evolution in the presence of insertion and deletion events (indels), the criteria optimized by alignment software are mostly *ad hoc* and based on the simplistic assumptions that homologous characters should be similar and that indels are rare events.

Modeling Errors

Every model is only a rough sketch of the complex reality of molecular evolution. First and foremost, the assumption of independent and identical evolutionary forces across sites is certainly not true in reality (Adrian, 1986; Yu and Jeffrey, 2006). What happens at one site depends both on its genomic context and on its interaction with other sites in the secondary and ternary structures of the molecule. Some relaxation allow sites to evolve at different sites (Goldman and Yang, 1994). Formally, each site has its own rate, modeled as a random effect drawn from a gamma distribution. Felsenstein and Churchill (1996) extended this approach to allow the rate at one site to depend on neighbouring sites rates using a hidden Markov model. This model captures dependence on the local genomic context (primary structure) and others have also been developed for dependence induced by proximity of sites in the secondary and ternary structure (Robinson *et al.*, 2003) but they are too complex for routine analyses.

For protein evolution, the rate matrices used to model evolution (e.g. JTT, PAM, LG, etc.) are derived by averaging over patterns observed in thousands of sites. However it is clear (Halpern and Bruno, 1998; Parisi and Echave, 2001; Susko *et al.*, 2002) that observed amino acid frequencies strongly deviate from the one expected under the JTT matrix. Lartillot and Philippe (2004) implemented a Bayesian mixture model which allows the inference of site specific rate matrices. These studies show that relaxing the assumption that all sites evolve according to the same rate matrices is crucial for accurate phylogenetic inference. However, such parameter-rich pose problems of their own, which can be difficult to diagnose (see Rannala (2002), for a discussion of over-parameterization in the context of Bayesian phylogenetic inference).

Additionally, many models assume stationarity, homogeneity and reversibility of sequence evolution. This is at odds with observations in archea, where evolution exhibit a trend towards some nucleotides over very long time scales (Boussau and Gouy, 2006). In bacteria, codon usage bias (ref codon usage bias) can lead to non reversible evolution and different underlying stationarity state frequencies in different parts of the tree. If distantly related lineages begin to display similar state frequencies, these lineages can be artefactually grouped together when using evolution model that do not explicity allow it (see for example Foster, 2004; Jermiin *et al.*, 2004). Finally, phylogenetic trees are probably not the adequate tool to analyse and represent non-vertical transmission and reticulated evolution (Huson and Bryant, 2005).

Robustness

As discussed in Section Motivation, support values are a popular way to validate a focal tree. We present here the most popular ones before describing other methods to validate a tree or fortify it.

Support Values

Bootstrap

Bootstrap values (Felsenstein, 1985) are probably the most popular and easiest to understand support values. Bootstrap involves resampling with replacement from one's molecular data with to create fictional datasets, called *bootstrap replicates*, of the same size. Specifically, the molecular data is typically organized as a multiple sequence alignment (MSA) of s species $\times n$ characters. Since most models assume independent characters, we generate a replicate by sampling n characters, with replacement, from the original MSA and do this B times. Note that in each replicate, some characters are sampled more than once and some left out entirely. The B replicates are used to estimate a forest of B bootstrap trees (one per replicate). Finally the bootstrap value (BP) of a branch of the original tree is its frequency of occurrence in the forest. The process is illustrated in **Fig. 1**.

Intuitively, the variation obtained by resampling n sites from the original data should be the same as the variation obtained by sampling n new characters. Bootstrap values capture, among other, the *sampling* variability induced by short MSA. When n increases, so does BP in general and it is quite common to achieve very high values for all branches when working on genome-scale alignments (Rokas *et al.*, 2003a).

BP provides a guide for the amount of support a branch has: branches with high BP occur more often and are more reliable than those with low BP. Although it might be tempting to interpret BP as the probability that a branch is present in the (unknown) true tree, this is not the case in general. Zharkikh and Li (1992) showed in a simple case that BP is biased and underestimates that probability. Using simulation studies, Hillis and Bull (1993) showed that BP values as small as 70% could react highly supported branches. Many studies (Felsenstein and Kishino, 1993; Efron *et al.*, 1996; Susko, 2008, 2010) examined the theoretical properties of bootstrap values and concluded that they are indeed biased. This bias is partly induced by the peculiar geometry of tree space (see Billera *et al.*, 2001; Susko (2010); Section Detection of Conicting Phylogenetic Signals).

The final limitation of bootstrap values, shared with other support values based on resampling techniques, is their high computational cost: the budget required to compute B bootstrap trees is B-times higher than the budget for the original tree. Clever implementations can substantially reduce that cost (Stamatakis, 2014) but it remains prohibitive for very large trees.

Posterior probabilities

Posterior Probabilities (PP) are mostly used in a Bayesian framework and similar in spirit bootstrap values. The main difference lies in the forest of trees used to compute support values. Bayesian procedures estimate the posterior distribution of trees. In practice, the distribution is too complex to fully explore and software produce a Monte Carlo Markov Chain (MCMC) sample from the posterior distribution (Yang and Rannala, 1997). The PP of a branch is computed, just like BP, as the probability of occurrence

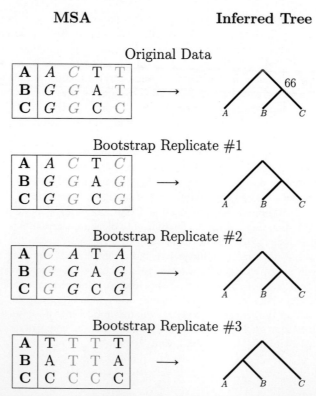

Fig. 1 Principle of the bootstrap for phylogenies. Each character is identified by its color and style. Characters are sampled with replacement to produce bootstrap replicates, which are then used to infer phylogenies. The split *A*|*BC* appears in 2 out of 3 bootstrap trees and therefore has a bootstrap value of BP=2/3 or 66%.

of that branch in the MCMC sample. MCMC trees constitute a set of highly likely trees for the original dataset. *PP* are easier to interpret than *BP* as they approximate directly the probability that a branch is present in the true tree, given the original data. Furthermore, since MCMC trees are a natural byproduct of the bayesian estimation procedure, there is almost no overhead in computing *PP*.

Unfortunately, *PP* are not immune to bias. Empirical studies found that *PP* are generally higher than *BP* (Anisimova *et al.*, 2011) and sometimes even overconfident, with the "star-tree paradox" (Yang, 2007) being the perfect example of overconfidence. Yang (2007) showed that when the actual tree is a 3-species star tree, so that all 3 potential inner branches are wrong, the bayesian method picks at random whose *PP* goes to 100% when sequence length goes to infinity, whereas one could expect the *PP* to fluctuate around 33%.

Intuitively, *PP* are higher than *BP* because they cover fewer sources of variability. Unlike bootstrap trees, MCMC trees all originate from the same dataset. *PP* are quite good at capturing the lack of phylogenetic signal in the original MSA but not the impact of a few inuential characters. For example, outlier characters with a strong effect on the tree or a tiny majority of character that favor one inner branch over another will affect all MCMC trees consistently. By contrast, they will be included in some bootstrap replicates but left out from others leading to more variation among bootstrap trees than among MCMC trees. Finally, in genome-scale context where inaccuracies are more likely to arise from modeling errors than from sampling variability, *PP* are uniformly high and as uninformative as *BP* (Philippe *et al.*, 2011a; Kumar *et al.*, 2012).

Likelihood-based support values

Both *BP* and *PP* quantify the agreement between a focal tree and forest of trees. Likelihood-based supports are fast alternatives that bypass the need for a forest and deal exclusively with the focal tree (Anisimova and Gascuel, 2006).

For any inner branch in the focal tree, there are 3 NNI configurations around that branch: the focal one T_1 and two alternatives T_2, T_3 (see **Fig. 2**). If we note $\ell_i = \log Pr(D \| T_i)$ the likelihood of the data under tree i and assume that T_1 is the maximum-likelihood tree, we have $\ell_1 \geq \max(\ell_2, \ell_3)$. Likelihood-based supports values essentially test whether $\delta = \ell_1 - \max(\ell_2, \ell_3)$ is significantly larger than 0.

The most popular support values are:

- the approximate Likelihood Ratio Tests (aLRT) values which evaluates the statistics δ and compares it to $0.5\chi_0^2 + 0.5\chi_1^2$ to compute a p-value. The p-value is then converted into a support value between 1/8 and 1. A branch with high δ will have high support.
- the SH-corrected aLRT (SH-aLRT) values are based on the same idea but use the non-parametric (Shimodaira and Hasegawa, 1999) procedure to compute the p-value of δ.
- Finally approximate bayes (aBayes) is an approximation of the posterior probability of tree T_i computed as:

$$Pr(T_i|D) = \frac{Pr(T_i)Pr(D|T_i)}{\sum_{j-1}^{3} Pr(T_j)Pr(D|T_j)}$$

with a flat prior $Pr(T_1) = Pr(T_2) = Pr(T_3)$.

All likelihood-based supports (aBayes, aLRT, SH-aLRT) amount to testing if T_1 is significantly better than T_2 and T_3. By focusing on one branch at the time rather than questioning the whole tree, likelihood-based supports are less conservative than *BP* and *PP*. They can also recycle likelihood computed while estimating the focal tree and are therefore much faster to compute than standard *BP*. Finally, they proved to be accurate in simulations studies (Anisimova *et al.*, 2011). They are the default support values in PhyML (Guindon and Gascuel, 2003).

Outliers in the Data

The aforementioned support values aggregate all variations in the data set and are unable to pinpoint variation due to outliers. The nature of resampling techniques is to use the empirical distribution as a surrogate for the true distribution. However, the empirical distribution may be polluted by outliers, defined here as "entry in the data set that are anomalous with respect to the behavior seen in the majority of the other entries in the data set" (Barnett and Lewis, 1994). This is a common occurrence in multi-locus studies where some characters can evolve according to one a tree, and others according to another tree (Degnan and Rosenberg, 2009). In

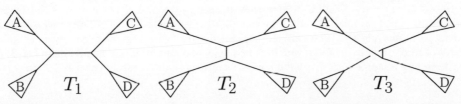

Fig. 2 The maximum likelihood tree (T_1, left) and its two NNI-alternatives (T_2, middle and T_3 right) corresponding to different resolutions of the inner branch. Subtrees are sketched as triangles.

that case, a single phylogeny is not a good fit to all the characters and (Swofford *et al.*, 1996) argued that it should be interesting to pinpoint where the phylogeny is not a good fit of the molecular data. Restricting the analyses to congruent characters usually leads to higher support values (Bar-Hen *et al.*, 2008).

Several approaches have been developed to identify outlier characters. Many studies (Rodríguez-Ezpeleta *et al.*, 2007; Burleigh and Mathews, 2004) advocate removing fast-evolving characters which are a well-known cause of misleading phylogenetic signal and long branch attraction (LBA) where distantly related taxa are grouped together in the tree due to parallel or convergent evolution (Felsenstein, 1978). Lopez *et al.* (1999) also suggest to investigate and remove characters with high rate variations (i.e. fast-evolving in some parts of the tree, slow-evolving in others). However both methods assume that good topologies are available to accurately estimate rates, leading to a circularity problem.

Bar-Hen *et al.* (2008) adapted instead inuence functions (Hampel, 1974) to phylogenetics in order to assess the impact of a single site on the likelihood. The main idea consists in removing one character at a time, to create *jackknife* replicates, and to infer a tree on each replicate. Jackknife trees are used to find influential characters whose removal most affect the tree likelihood. Bar-Hen *et al.* (2008) report that inuential sites have a strong impact on the topology and correspond mostly to fast evolving sites. All approaches found that removing outliers leads to more stable phylogenies but none is available as a routine in popular softwares.

Taxon Sampling

In phylogenomics studies, it is common to have conicting trees with support values higher than 95% for all inner branches (Rydin and Källersjö, 2002). This correspond to setups where the estimated tree has a very small estimation variance and differences between inferred trees result mostly from bias and modeling errors. In particular, Swofford *et al.* (1996) argues that adequate taxon sampling is one of the primary factors for accurate phylogenetic estimates, on par with enough sequence data. For example, dense taxon sampling can reduce the impact of LBA by splitting long branches. Similarly, Holland *et al.* (2003) and Shavit *et al.* (2007) showed that the inclusion of an outgroup to the analysis may disrupt the ingroup phylogeny. When there are only a few taxa, but many characters, phylogenetic analysis can produce high support values (*BP, PP, etc.*) for incorrect or misleading phylogenies (Rokas *et al.*, 2003a; Rokas and Carroll, 2005; Heath *et al.*, 2008).

Analysis of sensitivity to taxon inclusion should be a part of careful and thorough phylogenetic analysis (Heath *et al.*, 2008). Mariadassou *et al.* (2012) defined a the Taxon Inuence Index (TII) to assess the inuence of each taxon on the phylogeny. Using any inference method, we define T^* to be the tree inferred from the complete MSA. Let T_k be a smaller tree, inferred from the alignment deprived of taxon k and T_k the tree obtained by pruning taxon k from T^* The TII is the distance between trees T_k and T_k^*, such that

$$TII(k) = d(T_k, T_k^*)$$

They found that most taxa have small TII(k) and little inuence on the topology whereas a few are highly inuential *rogue taxa* and alter the phylogeny in clades even loosely related to their placement in the tree. Aberer *et al.* (2013) use a different approach to find rogue taxa, they start from a forest of trees (e.g. bootstrap trees) and search for a small set of taxa whose pruning increases the agreement between trees in the forest. The method is implemented in the webservice RogueNarok. The rationale in both cases is that reliable trees over smaller taxa sets are preferable over uncertain trees of larger taxa sets. Both methods find that pruning rogue taxa improves accuracy and results in more stable phylogenies with higher support values.

Consensus Methods

Bootstrap, jackknife and bayesian estimation naturally produce a forest of trees with the same species set. But different trees can also be estimated by using different methods or different sources of data. One way to summarize the forest is to *project* it on a focal tree to compute support values (see Section Robustness). Testing Alternatively, one can bypass the focal tree altogether and combine all trees in the forest to get a single tree. That is the purpose of consensus trees methods.

Consensus Trees

Consensus trees are trees that summarize a forest of trees with the same species set. We present here only the *strict* consensus, the *majority rule* consensus and the *extended majority rule* consensus but there are many other consensus (see Bryant (2003) for an extensive survey). The different notions are best understood on an example. Consider the forest featured in **Fig. 3** with 40 copies of trees T_1, 48 of tree T_2 and 12 of tree T_3. Each tree is completely defined by the bipartitions it induces (or clades for rooted trees). For example, T_1 induces the partitions *AB|CDEF*, *ABCD|EF* and *ABEF|CD* (or clades *AB, CD, EF* and *CDEF* if considered as rooted), in addition to all trivial partitions *A|BCDFE*, *B|ACDEF*, etc. not shown in the figure. Our three consensus methods scan the forest to build a list of all partitions occurring in the forest with their frequency of occurrence (middle column of **Fig. 3**). They then select a subset of partitions according to some rules and build a consensus tree from that subset only.

Strict consensus

The *strict consensus* tree (Rohlf, 1982) only uses partitions that appear in all trees, i.e. with a 100% occurence frequency. The strict consensus is fully compatible with all trees in the forest. However, it is less resolved than any tree and usually too strict. In our

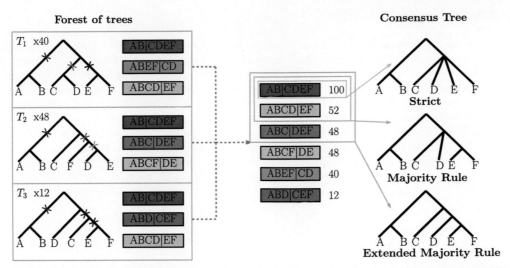

Forest of trees

Consensus Tree

Fig. 3 Left: a forest of 100 trees, corresponding to 3 topologies. Each colored cross correspond to a non trivial bipartition. Middle: Set of all bipartitions, with their occurence frequencies, found in the forest. Right: Different consensus made up by increasing large sets of partitions.

example, T_1 and T_2 and T_3 only differ in the position of D: if we removed D from all trees, they would be identical. We could therefore expect a branch separating EF from ABC. However, the set $CDEF$ is completely unresolved in the strict consensus.

Majority-rule consensus

The *majority-rule* consensus tree (Margush and McMorris, 1981) relaxes the condition that a bipartition must appear in all trees to be included in the consensus. Instead, it must only appear in *most* trees, i.e. have a occurrence frequency higher than 50%. Although not obvious, all such partitions are pairwise-compatible and can be used to build a proper tree (Buneman, 1971). The majority-rule tree is more resolved than the strict consensus one but is not compatible with all trees in the forest. In our example, the partition $ABCD|EF$ seen in the majority-rule consensus is in conflict with the partition $ABCF|DE$ present in T_2.

Extended majority-rule consensus

The *extended majority-rule* consensus (Felsenstein, 2005), also called greedy consensus, relaxes the occurrence frequency condition even further. The consensus is build by sequentially adding partition one at a time, in decreasing order of occurrence and only if compatible with previously included partitions, until the tree is fully resolved or no more partitions can be added. Since all partitions with frequency higher than 50% are compatible, they are part of the selection and the greedy consensus is thus a refinement of the majority-rule consensus. In our example, after including partitions $ABCD|EF$ and $AB|CDEF$, we can add either $ABC|DEF$ or $ABCF|DE$. The latter is in conflict with $ABCD|EF$ and thus not included whereas the former is compatible with both partitions and thus included. After addition of $ABC|DEF$, the greedy consensus is fully resolved. Note that the greedy consensus is different from all trees in the forest.

Branch lengths in consensus trees

The strict, majority-rule and extended majority-rule consensus trees only use *topologies* and produce consensus topologies. One way to add branch lengths to that consensus is to take, for each branch, its average length over trees where it is present. This is the approach used in MrBayes (Ronquist and Huelsenbeck, 2003) when building a consensus tree from the sample of posterior trees.

Distance-Based Consensus

The previous consensus methods are easy to understand and implement. Some also have a theoretical grounding that leads to generalizations. Barthélemy and McMorris (1986) showed that the majority-rule consensus is the *center* of the forest in the sense that it minimizes the sum of Robinson-Foulds distances (Robinson and Foulds, 1979) to all trees in the forest. It is thus the forest *median tree*. One could minimize total squared distance to all trees in the forest to find the *mean tree*.

The Robinson-Foulds is but one of many distances between trees (see St. John (2017) for an excellent review) and one can define other consensus similarly as the forest mean or median tree for some distance between tree. Although seducing, this approach suffered in the past from two shortcomings that severely limit its use. First, it was not obvious that the mean, or median, tree is well-defined or unique for some distances (Billera *et al.*, 2001). Second, even when the mean was well-defined, there was no routine to compute it in practice. Recent progress in the field (Miller *et al.*, 2015a) have made it easier to compute sophisticated distances and use them to validate trees.

Detection of Conflicting Phylogenetic Signals

The very idea of a consensus tree relies on the existante of a *center* around which trees are distributed. However, trees that are far away from the consensus can be the sign of either high variability or conflicting signals. While high variability implies weak phylogenetic signal, conflicting signals implies incongruence in the underlying data. Most distances, such as RF, are discrete and lead to very coarse information about the tree distribution. By contrast, continuous distances are a natural way to characterize variance and distinguish high variability from conflicting signals. Moreover continuous distance lend themselves nicely to other standard aspects of inferential statistics such as confidence sets and hypothesis testing.

The geometric model of Billera *et al.* (2001) allows one to compare phylogenetic trees with the same taxa set (of size *m*) in a quantitative way. This space has a natural metric, giving a way of measuring *continuous* distances between phylogenies and providing some procedures for averaging or combining several trees with the same leaves. This geometry also shows which trees appear within a fixed distance of a given tree and enables one to build confidence convex hulls from a set of trees. It also provides a justification for calling outliers in a collection of otherwise congruent trees and limit the consensus to those congruent tree.

Tree Space Definition

The distance $d(T,T')$ between two trees T and T' accounts for differences with respect to both their tree topologies (branching structure) and branch lengths. The space is constructed by representing each of the $(2m-3)!!$ possible tree topologies by a single non-negative Euclidean orthant of dimension $m-3$ (the largest possible number of internal branches). The orthants are then "glued together" along appropriate axes. Specifically, nearest neighbor interchange (NNI) topologies lie in adjacent non-negative orthants along the boundary corresponding to the collapse of the relevant NNI edge.

For two trees with different topologies, the BHV distance is the length of the shortest path that link them in the treespace. The length of any path can be computed by calculating the Euclidean distance of the path restricted to each orthant that it passes though, and summing these lengths. The shortest path is called a geodesic, and will pass from one orthant to the next orthant through lower-dimensional boundaries corresponding to less resolved trees. Since the space is nonpositively curved, the geodesics are unique (**Fig. 4**).

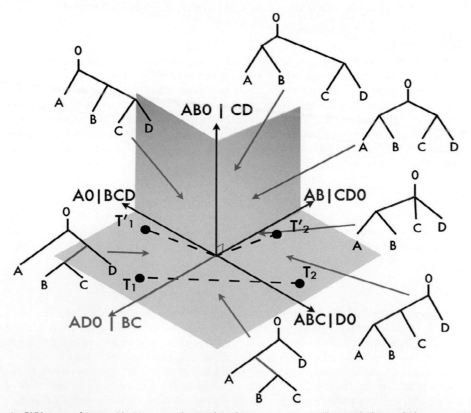

Fig. 4   Orthants in the BHV space of trees with 4 leaves and examples of two geodesics. As illustrated, the geodesics are not identical and therefore depend on both the topology and the actual branch lengths. Reproduced from Brown, D.G., Owen, M., 2017. Mean and variance of phylogenetic trees. arXiv preprint arXiv:1708.00294.

In Euclidean space, the Fréchet mean is the point minimizing the sum of the squared distances to the sample points, and is equivalent to the coordinate-wise average of the sample points. Similarly, the Fréchet mean tree is the tree minimizing the sum of squared BHV distances to a set of tree and that sum is the variance which quantifies how spread out the forest is Miller *et al.* (2015b) and Brown and Owen (2017). Billera *et al.* (2001) showed that mean is unique and is not necessarily a refinement of the majority-rule consensus.

Use of BHV Distance

BHV distances can be used in much the same way as euclidian distance and the standard tools of multivariate statistics can be used to analyse the forest. For example, Barden *et al.* (2014) proved a Central Limit Theorem in the BHV space that can be used to either detect weakly and strongly supported edges in the mean tree or to test hypotheses on some portion of the tree. One of their main tool is the so-called log-map which projects the BHV metric space to the usual euclidian space and allows one to visualize the forest as a scatter plot to gain some understanding of the estimate uncertainty.

In the same spirit, Willis (2016) use the distance matrix to estimate the principal directions of variability *via* principal components analysis (PCA). The axes of the $\mathbb{R}^m$ ellipsoid indicate the relative directions of precision, and the ellipsoid can be shrunk to be wholly contained in the same orthant as a focal tree $\hat{T}$. This gives an unambiguous indication of the relative confidence in the edges of $\hat{T}$. Note that the procedure is also explicit about the trees contained in the confidence set for a given confidence level α. de Vienne *et al.* (2012) use a similar approach based on a different mapping of the trees to a euclidian space and use multiple co-inertia analysis (MCOA) instead of PCA to detect the principal directions of variability.

Finally Weyenberg *et al.* (2014) propose a non-parametric estimator of the distribution that generated the forest $T_1, \ldots, T_n$. This estimator can be viewed as a refined version of histogram-based estimation of a density. The kernel function, is a non-negative function defined on pairs of trees, which measures how similar two trees are. Kernel density estimation use the fact that points close to sample points tend to have higher likelihood than distant outlier points. The ultimate goal is to detect outlier trees, T_j, which are not actually drawn from the true distribution.

Conclusions

Nowadays efficient algorithms permit to compute means of continuous distances in a reasonable time. Moreover a huge amount of work has been done to derive mathematical properties of theses distances. Therefore it is possible to go from a theoretical work to applications. For example, Kendall and Colijn (2016) use tree mapping to show that 3 genes of Ebolavirus have markedly different phylogenies. This result is quite typical in genomic-scales analysis where different loci have different evolution history. This is a natural consequence of the gene trees/species tree problem whereby the evolutionary history of a gene can be different from the species tree (Degnan and Rosenberg, 2009). This has spanned the entire research field of reconciliation devoted to the reconstruction of species tree from sets of gene tree.

Distances between trees can help cluster trees and/or loci into congruent groups with the intent of performing one inference per group to infer more robust phylogenetic estimates. In an era of cheap and abundant sequences, tree validation is as much a matter of computing support values as a matter of validating and making sure there are not too many conflicting signals in the molecular data (taxa sct, gene set) used for tree reconstruction.

See also: Comparative and Evolutionary Genomics. Evolutionary Models. Gene Duplication and Speciation. Inference of Horizontal Gene Transfer: Gaining Insights Into Evolution via Lateral Acquisition of Genetic Material. Introduction to the Non-Parametric Bootstrap. Molecular Clock. Molecular Phylogenetics. Natural Language Processing Approaches in Bioinformatics. Phylogenetic analysis: Early evolution of life. Phylogenetic Tree Rooting

---

References

Aberer, A.J., Krompass, D., Stamatakis, A., 2013. Pruning rogue taxa improves phylogenetic accuracy: An efficient algorithm and webservice. Syst Biol 62 (1), 162–166. doi:10.1093/sysbio/sys078.

Adrian, P.B., 1986. Cpg-rich islands and the function of dna methylation. Nature 321 (6067), 209–213.

Anisimova, M., Gascuel, O., 2006. Approximate likelihood-ratio test for branches: A fast, accurate, and powerful alternative. Syst Biol 55 (4), 539–552. doi:10.1080/10635150600755453.

Anisimova, M., Gil, M., Dufayard, J.-F., Dessimoz, C., Gascuel, O., 2011. Survey of branch support methods demonstrates accuracy, power, and robustness of fast likelihood-based approximation schemes. Syst Biol 60 (5), 685–699. doi:10.1093/sysbio/syr041.

Barden, D., Le, H., Owen, M., 2014. Limiting behaviour of fréchet means in the space of phylogenetic trees. Ann Inst Stat Math. 1–31.

Bar-Hen, A., Mariadassou, M., Poursat, M.-A., Van-denkoornhuyse, P., 2008. Influence function for robust phylogenetic reconstructions. Mol Biol Evol 25 (5), 869–873. doi:10.1093/molbev/msn030.

Barnett, V., Lewis, T., 1994. Outliers in Statistical Data. John Wiley & Sons.

Barthélemy, J.-P., McMorris, F.R., 1986. The median procedure for n-trees. J Classif 3 (2), 329–334. doi:10.1007/BF01894194.

Billera, L.J., Holmes, S.P., Vogtmann, K., 2001. Geometry of the space of phylogenetic trees. Adv Appl Math 27, 733–767. Available at: http://www.math.cornell.edu/~billera/papers/treespace.pdf.

Bininda-Emonds, O.R.P., Cardillo, M., Jones, K.E., *et al.*, 2007. The delayed rise of present-day mammals. Nature 446, 507–512. doi:10.1038/nature05634.

Bordewich, M., Rodrigo, A.G., Semple, C., 2008. Selecting taxa to save or sequence: Desirable criteria and a greedy solution. Syst Biol 57 (6), 825–834. doi:10.1080/10635150802552831.

Boussau, B., Gouy, M., 2006. Efficient likelihood computations with nonreversible models of evolution. Systematic Biology 55 (5), 756–768.

Brown, D.G., Owen, M., 2017. Mean and variance of phylogenetic trees. arXiv preprint. arXiv:1708.00294.

Bryant, D., 2003. A classification of consensus methods for phylogenetics. Bioconsensus 61, 163–183.

Buneman, P., 1971. Mathematics the Archeological and Historical Sciences, Chapter: The Recovery of Trees from Measures of Dissimilarity. Edinburgh University Press. (ISBN: 9780852242131).

Burleigh, J.G., Mathews, S., 2004. Phylogenetic signal in nucleotide data from seed plants: Implications for resolving the seed plant tree of life. Am J Bot 91 (10), 1599–1613. doi:10.3732/ajb.91.10.1599. Available at: http://www.amjbot.org/content/91/10/1599.abstract.

Degnan, J.H., Rosenberg, N.A., 2009. Gene tree discordance, phylogenetic inference and the multispecies coalescent. Trends Ecol Evol 24 (6), 332–340. doi:10.1016/j.tree.2009.01.009.

de Vienne, D.M., Ollier, S., Aguileta, G., 2012. Phylo-mcoa: A fast and efficient method to detect outlier genes and species in phylogenomics using multiple co-inertia analysis. Mol Biol Evol 29 (6), 1587–1598.

Efron, B., Halloran, E., Holmes, S., 1996. Bootstrap confidence levels for phylogenetic trees. Proc Natl Acad Sci USA 93 (14), 7085–7090. Available at: http://www.pnas.org/cgi/content/full/93/23/13429.

Eisen, J.A., 1998. Phylogenomics: Improving functional predictions for uncharacterized genes by evolutionary analysis. Genome Res 8 (3), 163–167.

Felsenstein, J., 1978. Cases in which parsimony or compatibility methods will be positively misleading. Syst Zool 27 (4), 401–410. Available at: http://www.molecularevolution.org/si/resources/references/files/Felsenstein_1978.pdf..

Felsenstein, J., 1985. Confidence limits on phylogenies: An approach using the bootstrap. Evolution 39 (4), 783–791. doi:10.2307/2408678. Available at: http://links.jstor.org/sici?sici=0014-3820(198507)39:4%3C783:CLOPAA%3E2.0.CO;2-L.

Felsenstein, J., 2005. Phylip (phylogeny inference package) version 3.6. Distributed by the author.

Felsenstein, J., Kishino, H., 1993. Is there something wrong with the bootstrap on phylogenies? A reply to hillis and bull. Syst Biol 42 (2), 193–200. doi:10.2307/2992541. Available at: http://links.jstor.org/sici?sici=1063-5157(199306)42%3A2%3C193%3AITSWwt%3E2.0.CO%3B2-Y.

Felsenstein, J., 2004. Inferring Phylogenies. Sinauer Associates.

Felsenstein, J., Churchill, G.A., 1996. A hidden markov model approach to variation among sites in rate of evolution. Mol Biol Evol 13 (1), 93–104.

Foster, P.G., 2004. Modeling compositional heterogeneity. Syst Biol 53 (3), 485–495.

Gascuel, O., 2005. Mathematics of Evolution and Phylogeny. Oxford University Press.

Geneva, A.J., Hilton, J., Noll, S., Glor, R.E., 2015. Multilocus phylogenetic analyses of hispaniolan and bahamian trunk anoles (distichus species group). Mol Phylogenet Evol 87, 105–117.

Goldman, N., Yang, Z., 1994. A codon-based model of nucleotide substitution for protein-coding dna sequences. Mol Biol Evol 11 (5), 725–736.

Guindon, S., Gascuel, O., 2003. A simple, fast, and accurate algorithm to estimate large phylogenies by maximum likelihood. Syst Biol 52 (5), 696–704. doi:10.1080/10635150390235520. Available at: http://www.informaworld.com/smpp/ftinterface~content=a713850337~fulltext=713240930.

Halpern, A.L., Bruno, W.J., 1998. Evolutionary distances for protein-coding sequences: Modeling site-specific residue frequencies. Mol Biol Evol 15 (7), 910–917.

Hampel, F.R., 1974. The influence curve and its role in robust estimation. J Am Stat Assoc 69, 383–393.

Heath, T.A., Hedtke, S.M., Hillis, D.M., 2008. Taxon sampling and the accuracy of phylogenetic analyses. J Mol Evol 46, 239–257. Available at: http://www.plantsystematics.com/qikan/epaper/hb_zhaiyao.asp.

Hillis, D.M., Bull, J.J., 1993. An empirical test of bootstrapping as a method for assessing confidence in phylogenetic analysis. Syst Biol 42 (2), 182–192. doi:10.2307/2992540.

Holland, B.R., Penny, D., Hendy, M.D., 2003. Outgroup misplacement and phylogenetic inaccuracy under a molecular clock-a simulation study. Syst Biol 52 (2), 229–238. doi:10.1080/10635150390192771. Available at: http://www.informaworld.com/smpp/content~content=a713850188~db=all~order=page.

Huson, D.H., Bryant, D., 2005. Application of phylogenetic networks in evolutionary studies. Mol Biol Evol 23 (2), 254–267.

Jermiin, L.S., Ho, S.Y.W., Ababneh, F., Robinson, J., Larkum, A.W.D., 2004. The biasing effect of compositional heterogeneity on phylogenetic estimates may be underestimated. Syst Biol 53 (4), 638–643.

Kendall, M., Colijn, C., 2016. Mapping phylogenetic trees to reveal distinct patterns of evolution. Mol Biol Evol 33 (10), 2735–2743. doi:10.1093/molbev/msw124.

Kumar, S., Filipski, A.J., Battistuzzi, F.U., Kosakovsky Pond, S.L., Tamura, K., 2012. Statistics and truth in phylogenomics. Mol Biol Evol 29 (2), 457–472. doi:10.1093/molbev/msr202.

Lartillot, N., Philippe, H., 2004. A bayesian mixture model for across-site heterogeneities in the amino-acid replacement process. Mol Biol Evol 21 (6), 1095–1109.

Laurin-Lemay, S., Brinkmann, H., Philippe, H., 2012. Origin of land plants revisited in the light of sequence contamination and missing data. Curr Biol 22 (15), R593–R594.

Lopez, P., Forterre, P., Philippe, H., 1999. The root of the tree of life in the light of the co-varion model. J Mol Evol 49 (4), 496–508. Available at: http://www.springerlink.com/content/hwle29fxv74ra6xy/.

Margush, T., McMorris, F.R., 1981. Consensus n-trees. Bull Math Biol 43 (2), 239–244. doi:10.1007/BF02459446.

Mariadassou, M., Bar-Hen, A., Kishino, H., 2012. Taxon influence index: Assessing taxon-induced incongruities in phylogenetic inference. Syst Biol 61 (2), 337–345. doi:10.1093/sysbio/syr129.

Miller, E., Owen, M., Provan, J.S., 2015a. Polyhedral computational geometry for averaging metric phylogenetic trees. Adv Appl Math 68, 51–91. Available at: https://doi.org/10.1016/j.aam.2015.04.002; http://www.sciencedirect.com/science/article/pii/S0196885815000470.

Miller, E., Owen, M., Provan, J.S., 2015b. Polyhedral computational geometry for averaging metric phylogenetic trees. Adv Appl Math 68, 51–91.

Morrison, D.A., Ellis, J.T., 1997. Effects of nucleotide sequence alignment on phylogeny estimation: A case study of 18s rDNAs of apicomplexa. Mol Biol Evol 14 (4), 428–441.

Ogden, T.H., Rosenberg, M.S., 2006. Multiple sequence alignment accuracy and phylogenetic inference. Syst Biol 55 (2), 314–328.

Parisi, G., Echave, J., 2001. Structural constraints and emergence of sequence patterns in protein evolution. Mol Biol Evol 18 (5), 750–756.

Pennell, M.W., Harmon, L.J., 2013. An integrative view of phylogenetic comparative methods: Connections to population genetics, community ecology, and paleobiology. Ann N.Y. Acad Sci 1289, 90–105. doi:10.1111/nyas.12157.

Philippe, H., Brinkmann, H., Lavrov, D.V., *et al.*, 2011a. Resolving difficult phylogenetic questions: Why more sequences are not enough. PLOS Biol 9 (3), e1000602. doi:10.1371/journal.pbio.1000602.

Philippe, H., Brinkmann, H., Lavrov, D.V., *et al.*, 2011b. Resolving difficult phylogenetic questions: Why more sequences are not enough. PLOS Biology 9 (3), e1000602.

Philippe, H., De Vienne, D.M., Ranwez, V., *et al.*, 2017. Pitfalls in supermatrix phylogenomics. Eur J Taxon 283, 1–25.

Price, M.N., Dehal, P.S., Arkin, A.P., 2010. Fasttree 2-approximately maximum-likelihood trees for large alignments. PLoS One 5 (3), e9490. doi:10.1371/journal.pone.0009490.

Prum, R.O., Berv, J.S., Dornburg, A., *et al.*, 2015. A comprehensive phylogeny of birds (Aves) using targeted next-generation DNA sequencing. Nature 526, 569–573. doi:10.1038/nature15697.

Rannala, B., 2002. Identifiability of parameters in mcmc bayesian inference of phylogeny. Syst Biol 51 (5), 754–760.

Revell, L.J., Harmon, L.J., Collar, D.C., 2008. Phylogenetic signal, evolutionary process, and rate. Syst Biol 57 (4), 591–601. doi:10.1080/10635150802302427.

Robinson, D.F., Foulds, L.R., 1979. Comparison of weighted labelled trees. Lectures Note in Mathematics, vol. 748. Berlin: Springer-Verlag, pp. 119–126.

Robinson, D.M., Jones, D.T., Kishino, H., Goldman, N., Thorne, J.L., 2003. Protein evolution with dependence among codons due to tertiary structure. Mol Biol Evol 20 (10), 1692–1704. doi:10.1093/molbev/ msg184.

Rodríguez-Ezpeleta, N., Brinkmann, H., Roure, B., *et al.*, 2007. Detecting and overcoming systematic errors in genome-scale phylogenies. Systematic Biology 56 (3), 389–399. doi:10.1080/10635150701397643.

Rohlf, F.J., 1982. Consensus indices for comparing classifications. Mathematical Biosciences 59 (1), 131–144. Available at: https://doi.org/10.1016/0025-5564(82)90112-2; http://www.sciencedirect.com/science/article/pii/0025556482901122.

Rokas, A., Carroll, S.B., 2005. More genes or more taxa? The relative contribution of gene number and taxon number to phylogenetic accuracy. Mol Biol Evol 22 (5), 1337–1344. doi:10.1093/molbev/msi121.

Rokas, A., Williams, B.L., King, N., Carroll, S.B., 2003a. Genome-scale approaches to resolving incongruence in molecular phylogenies. Nature 425 (6960), 798–804. doi:10.1038/nature02053.

Rokas, A., Williams, B.L., King, N., Carroll, S.B., 2003b. Genome-scale approaches to resolving incongruence in molecular phylogenies. Nature 425 (6960), 798.

Ronquist, F., Huelsenbeck, J.P., 2003. Mrbayes 3: Bayesian phylogenetic inference under mixed models. Bioinformatics 19 (12), 1572–1574. Available at: http://bioinformatics. oxfordjournals.org/cgi/reprint/19/12/1572.

Rydin, C., Kallersjo, M., 2002. Taxon sampling and seed plant phylogeny. Cladis-tics 18 (5), 485–513. doi:https://doi.org/10.1016/S0748-3007(02)00104-4. Available at: http://www.sciencedirect.com/science/article/pii/S0748300702001044. ISSN: 0748-3007

Shavit, L., Penny, D., Hendy, M.D., Holland, B.R., 2007. The problem of rooting rapid radiations. Mol Biol Evol 24 (11), 2400–2411. doi:10.1093/ molbev/msm178.

Shimodaira, H., Hasegawa, M., 1999. Multiple comparisons of log-likelihoods with applications to phylogenetic inference. Mol Biol Evol 16 (8), 1114–1116.

St. John, K., 2017. Review paper: The shape of phylogenetic treespace. Syst Biol 66 (1), e83–e94. doi:10.1093/sysbio/syw025.

Stamatakis, A., 2006. Raxml-vi-hpc: Maximum likelihood-based phylogenetic analyses with thousands of taxa and mixed models. Bioinformatics 22 (21), 2688–2690. doi:10.1093/bioinformatics/btl446.

Stamatakis, A., 2014. Raxml version 8: A tool for phylogenetic analysis and post-analysis of large phylogenies. Bioinformatics 30 (9), 1312–1313. doi:10.1093/bioinformatics/btu033.

Susko, E., 2008. On the distributions of bootstrap support and posterior distributions for a star tree. Syst Biol 57 (4), 602–612. doi:10.1080/10635150802302468.

Susko, E., 2010. First-order correct bootstrap support adjustments for splits that allow hypothesis testing when using maximum likelihood estimation. Mol Biol Evol. doi:10.1093/molbev/msq048.

Susko, E., Inagaki, Y., Field, C., Holder, M.E., Roger, A.J., 2002. Testing for differences in rates-across-sites distributions in phylogenetic subtrees. Mol Bio Evol 19 (9), 1514–1523.

Swofford, D.L., Olsen, G.J., Waddell, P.J., Hillis, D.M., 1996. Phylogenetic inference. In: Hillis, D.M., Moritz, C., Mable, B.K. (Eds.), Molecular Systematics, second ed. Sunderland, MA: Sinauer Associates, Inc., pp. 407–514.

Talavera, G., Castresana, J., 2007. Improvement of phylogenies after removing divergent and ambiguously aligned blocks from protein sequence alignments. Syst Biol 56 (4), 564–577.

Weyenberg, G., Huggins, P.M., Schardl, C.L., Howe, D.K., Yoshida, R., 2014. Kdetrees: Non-parametric estimation of phylogenetic tree distributions. Bioinformatics 30 (16), 2280–2287.

Willis, A., 2016. Confidence sets for phylogenetic trees. arXiv preprint. arXiv:1607.08288.

Wong, K.M., Suchard, M.A., Huelsenbeck, J.P., 2008. Alignment uncertainty and genomic analysis. Science 319 (5862), 473–476.

Yang, Z., 2007. Fair-balance paradox, star-tree paradox, and bayesian phylogenetics. Mol Biol Evol 24 (8), 1639–1655. doi:10.1093/molbev/msm081.

Yang, Z., Rannala, B., 1997. Bayesian phylogenetic inference using DNA sequences: A markov chain monte carlo method. Mol Biol Evol 14 (7), 717–724. Available at: http://mbe.oxfordjournals.org/cgi/reprint/14/7/717.

Yu, J., Jeffrey, L., 2006. Thorne. Dependence among sites in RNA evolution. Mol Biol Evol 23 (8), 1525–1537. doi:10.1093/molbev/msl015.

Zharkikh, A., Li, W.H., 1992. Statistical properties of bootstrap estimation of phylogenetic variability from nucleotide sequences. i. four taxa with a molecular clock. Mol Biol Evol 9 (6), 1119–1147. Available at: http://mbe.oxfordjournals.org/cgi/reprint/9/6/1119.

Further Reading

Bryant, D., 2003. A classification of consensus methods for phylogenetics. Bioconsensus 61, 163–183.

Degnan, J.H., Rosenberg, N.A., 2009. Gene tree discordance, phylogenetic inference and the multispecies coalescent. Trends Ecol Evol 24 (6), 332–340. doi:10.1016/j.tree.2009.01.009.

Felsenstein, J., 2004. Inferring Phylogenies. Sinauer Associates.

Philippe, H., De Vienne, D.M., Ranwez, V., *et al.*, 2017. Pitfalls in supermatrix phylogenomics. Eur J Taxon 283, 1–25.

St. John, K., 2017. Review paper: The shape of phylogenetic treespace. Syst Biol 66 (1), e83–e94. doi:10.1093/sysbio/syw025.

Susko, E., 2010. First-order correct bootstrap support adjustments for splits that allow hypothesis testing when using maximum likelihood estimation. Mol Biol Evol. doi:10.1093/molbev/msq048.

Yang, Z., 2007. Fair-balance paradox, star-tree paradox, and Bayesian phylogenetics. Mol Biol Evol 24 (8), 1639–1655. doi:10.1093/molbev/msm081.

Applications of the Coalescent for the Evolutionary Analysis of Genetic Data

Miguel Arenas, University of Vigo, Vigo, Spain

Introduction

The coalescent (Kingman, 1982) is a stochastic model to simulate the evolutionary history of a sample of a large population. Instead of simulating the entire population and then sampling, the coalescent simulates the history of the sample to its most recent common ancestor (MRCA). The population size is considered in the coalescent but only to determine the probability of coalescent events, not to simulate the population history (Section "The Coalescent Theory and its Extensions). As a consequence, coalescent simulations are generally rapid, which is convenient for a variety of applications.

First coalescent approaches were rapidly extended to mimic diverse fundamental evolutionary processes such as recombination (Hudson and Kaplan, 1988; Kaplan *et al.*, 1991; Hudson, 1990), population structure with migration (Hudson, 1998; Kaplan *et al.*, 1991; Hudson, 1990) and selection (Hudson and Kaplan, 1988, 1995; Kaplan *et al.*, 1988, 1991; Hudson, 1990) (Section "The Coalescent Theory and its Extensions"), with goals but also with limitations (Section "Goals and Limitations of the Coalescent"). All these extensions were implemented into a variety of evolutionary frameworks (Section "Coalescent Simulators). The coalescent can be applied to address diverse evolutionary questions (Section "Applications of the Coalescent"). For example, it can be used for hypothesis testing, validation of analytical methods, selection of evolutionary models and estimation of evolutionary parameters.

This article revises the current state of the coalescent with a special focus on its applications. First, it provides a brief introduction on the fundamental insights of the standard coalescent and its extensions (some complementary reviews are provided). Next it describes goals and limitations of the coalescent, currently available coalescent simulators and applications in population genetics and evolution. Finally, the article provides suggestions to design a coalescent-based experiment, including some practical examples.

The Coalescent Theory and Its Extensions

The standard coalescent only requires sample size (k) and effective population size (N) to simulate the evolutionary history of a sample. This simulation is based on a backwards in time process, where the sample is evolved from the present to the past (**Fig. 1**). The probability that two individuals descend from a common ancestor of the population is $1/2N$. Thus, the probability of a coalescent event between two lineages of a sample of k individuals can be obtained by calculating, $1/2N \times k(k-1)/2 = k(k-1)/4N$, because there are $k(k-1)/2$ pairs of individuals in a sample of k individuals (note that if k is 2 the result is again $1/2N$). **Fig. 1** presents the probabilities of coalescent events in an illustrative example. Next, considering that the time for a coalescent event follows an exponential distribution (for $k=2$, $P(t+1) = (1-(1/2N))^t \times (1/2N) \approx (1/2N)e^{-t/2N}$), the expectation of this time is

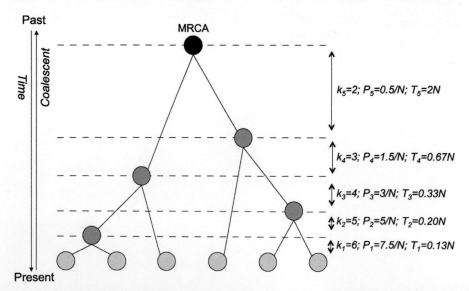

Fig. 1 Illustrative example of the standard coalescent. Simulation of a coalescent tree from a sample of six individuals (circles in clear grey) and population size *N*. Coalescent events generate internal nodes (circles in dark grey) that are simulated going backwards in time until reaching only one active lineage, the MRCA (circle in black). For each period of time before a coalescent event the following information is shown: number of active lineages *k*, probability of a coalescence *P* and expected time for a coalescence *T*.

$E(T_k)=4N/k(k-1)$. These expressions require $4N/k(k-1)\gg1$ that only works when the sample size is much smaller than the population size, which is one of the most important assumptions of the coalescent. The time to the MRCA (TMRCA) can be obtained by summing coalescence times from k to 2 individuals, $E(T_{MRCA})=E(T_k)+E(T_{k-1})+E(T_{k-2})+\ldots+E(T_{k=2})=4N(1-1/k)$. Interestingly, note that the last coalescent event (when $k=2$) requires $2N$ generations, which is almost half of the time required to reach the MRCA. Further details about the mathematical expressions of the coalescent can be found in the following reviews (Nordborg, 2007; Neuhauser and Tavaré, 2001; Hein et al., 2005). It follows with an overview on the most relevant extensions of the standard coalescent.

The Coalescent With Variable Population Size

Fluctuations in population size through time are common and, consequently, one of the first extensions of the coalescent was oriented to consider this aspect. As noted above, the probability or expected time for a coalescent event depends on the population size. The larger is the population size, the smaller is the probability of a coalescent event or the longer is the expected time for a coalescent event, which leads to longer branch lengths (**Fig. 2(B)**). Thus, if the population size increased over time (it decreased if we go back in time) longer branches are expected to appear near present (coalescences mainly appear in the past) leading to a star-like tree (**Fig. 2(A)**). By contrast, if population size decreases through time, longer branches are expected to appear in the past, leading to trees with a comb-like shape (**Fig. 2(C)**).

The variation of the population size through time can be systematic or irregular. A systematic variation can be modeled with only a few parameters (i.e., the commonly used population growth rate g where $N_{(t)}=N_{(o)}\times e^{-gt}$) that can be easily incorporated to the coalescent. However, the population size could also vary in a different manner at different periods of time (i.e., as it occurs many times in virus populations (Lemey et al., 2006)). Since these demographic periods may not be coincident with times between coalescent events, the implementation of this scenario is slightly more complex (details in Hein et al. (2005)). An interesting scenario is a population bottleneck (an important decline in population size during a period of time). During a bottleneck the population size can be very small and, consequently, the rate of coalescence will increase reducing the TMRCA (Hein et al., 2005).

The Coalescent With Population Structure and Migration

Sometimes individuals may only mate with others that present particular features (i.e., proximity or phenotype). Under these situations, one can devise a model where a population structure defined as a set of subpopulations (or demes) presents mating (coalescence) only between individuals of the same deme but with possible migration of individuals between demes (**Fig. 3**). If the sample comes from different demes and there is not a connection between them (i.e., lack of a demes tree and/or migration between demes) the coalescent cannot be applied because a MRCA common to all demes cannot be reached: migration or demes coalescence is required to coalesce all the lineages. Migration rate can affect both topology and branch lengths. The smaller is the migration rate, the longer is the expected TMRCA (Hein et al., 2005). Nevertheless, migration can be modeled as a process independent of coalescence and recombination and, consequently, it can be easily considered to determine the probability or expected time of coalescent and migration events (Hein et al., 2005; Notohara, 1990). For mathematical details about probabilities and expected times of migration and coalescence events the reader is referred to Hein et al. (2005). The choice of a receiving deme in a migration event depends on the specified subdivision (migration) model and migration rates (**Fig. 4(A)**). The most used migration model in the coalescent is the island model where all demes can exchange individuals with all demes (Hudson, 1998). The stepping stone migration model (Kimura and Weiss, 1964), where demes only exchange migrants with neighboring demes, is also well-established. Another useful model is the continent-island model that considers migration between a central deme and peripheral demes (Wright, 1931). The coalescent can also be structured considering a demes tree (Nielsen and Wakeley, 2001; **Fig. 4(B)**). Under this scenario, demes share a common ancestor and migration is not mandatory to reach the MRCA.

The Coalescent With Recombination

Recombination is a fundamental evolutionary force for most organisms, especially virus and bacteria (Perez-Losada et al., 2015). Recombination events generate an ancestral recombination graph (ARG) (Arenas, 2013b; Griffiths and Marjoram, 1997, 1996)

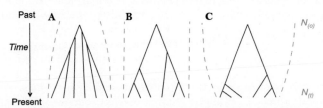

Fig. 2 Illustrative example of the influence of the population size on the coalescent process. The variation of the population size through time is shown with grey dashed lines. (B) Coalescent tree simulated with constant population size ($N_{(t)}=N_{(o)}$). (A) Coalescent tree simulated with a positive population growth rate (population size increases with time, $N_{(t)}>N_{(o)}$). (C) Coalescent tree simulated with a negative population growth rate (population size decreases with time, $N_{(t)}<N_{(o)}$).

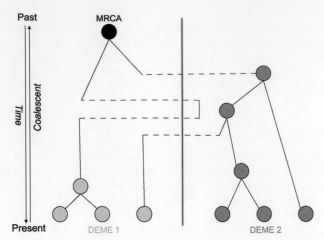

Fig. 3 Illustrative example of the coalescent with population structure and migration. Simulation of a coalescent tree from a sample of six individuals (circles at the present) that belong to two different demes (deme 1 in clear grey and deme 2 in dark grey). Coalescent events generate internal nodes (circles). Migration events are depicted with dashed lines. The process occurs going backwards in time until reaching the MRCA (circle in black).

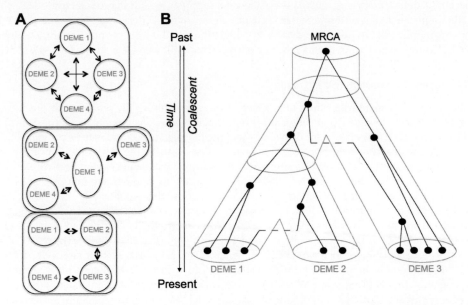

Fig. 4 Migration models and the coalescent within a demes tree. (A) Migration or subdivision models: Island model (top), continent-island model (middle), stepping stone model (bottom). (B) Simulation of a coalescent tree within a demes tree with migration. Migration events are depicted with dashed lines. The process occurs going backwards in time until reaching the MRCA.

where genomic fragments can present different evolutionary histories (**Fig. 5**) and that should be carefully considered for inferring phylogenetic trees (Schierup and Hein, 2000a; Arenas and Posada, 2010b; Mallo *et al.*, 2016; Posada and Crandall, 2002). The number of simulated recombination events is based on the population recombination rate $\rho = 4Nrl$, where r is the recombination rate per generation per site and l is the sequence length. Importantly, a recombination breakpoint can only occur in sites that did not reach their MRCA yet. The expected number of recombination events in a panmictic population with constant population size and a sample of k individuals is defined as, $E(number\ of\ recombination\ events) = \rho \sum_{i=1}^{k-1} \frac{1}{i}$.

It is also possible to simulate the coalescent with heterogeneous recombination along the sequences. There, background (homogeneous) and hotspot (heterogeneous) recombination rates can be applied to simulate hotspot location, heterogeneity and precision (Wiuf and Posada, 2003).

An special coalescent process with recombination is the coalescent simulation of intra-codon recombination (Arenas and Posada, 2010a). Substitution models of codon evolution require complete codons to operate and a codon broken due to a recombination event can be problematic because can lead to branches assigned to incomplete codons. This problem was solved by Arenas and Posada (2010a) with a coalescent algorithm where ancestral material that does not belong to the sample but that is required to complete broken codons (called as pseudo-ancestral material) can be considered in the coalescent, leading to slightly longer ARGs but allowing codon models to operate (Arenas and Posada, 2012).

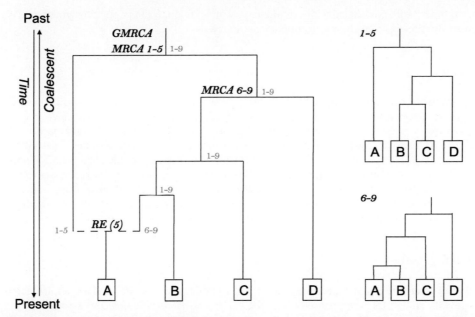

Fig. 5 Illustrative example of an ancestral recombination graph with a tree for each recombinant partition. Ancestral recombination graph for a sample of 4 sequences (A, B, C, D) with 9 sites (i.e., nucleotides). A recombination event (RE, dashed line) occurs with breakpoint at position 5, leading to the recombinant partitions 1–5 and 6–9. Each recombinant partition presents a particular coalescent tree (depicted on the right).

The Coalescent With Selection

Previous coalescent approaches assume neutral evolution. However, in the real world, some lineages may present a higher fitness (i.e., better adapted to the environment) than others. Therefore, the coalescent was also extended to mimic selection through two main approaches.

The first one is the simulation of the ancestral selection graph (ASG) (Krone and Neuhauser, 1997; Neuhauser and Krone, 1997). The ASG presents the genealogy of the lineages (or alleles) into a graph based on coalescence and branching events, similar to the ARG (**Fig. 5**), where removing branches (i.e., alleles with low fitness) can lead to a tree. The ASG is especially useful to simulate evolutionary histories under weak selection (moderate or strong selection in the ASG may not be computationally tractable). It was later extended to model ancestry conditioned on the frequencies of the selected alleles (Slade and Neuhauser, 2003; Slade, 2000a,b) and even to population subdivision and migration (Slade and Wakeley, 2005).

The other approach is the conditional structured coalescent (Kaplan et al., 1988; Hudson and Kaplan, 1988) where historical allele frequencies are used to generate classes or subpopulations based on their levels of selection. There is not selection within each subpopulation but each subpopulation may present specific values for parameters such as population size or population growth rate. The methods for the coalescent with population structure and migration (see above) can thus be applied to this approach. A migration event can be considered as a mutation or a recombination event (i.e., a beneficial mutation or recombination event could move an allele from low to high selection − or the opposite −, thus moving it from a particular subpopulation to another one). Some examples of this approach involving balancing selection, background selection and selective sweeps are detailed described in Nordborg (2007).

Goals and Limitations of the Coalescent

In addition to the coalescent, some other approaches have been developed to simulate evolutionary histories in population genetics (Arenas, 2012; Hoban et al., 2012). One of the most relevant is the forward in time approach, which simulates the evolutionary history of the entire population from the past to the present (Peng et al., 2012; Carvajal-Rodriguez, 2010). Simulations generated with the forward in time approach consider the ancestral information of the entire population and are useful to study diverse complex evolutionary processes such as mating systems and complex migration (i.e., sex biased dispersal or long-distance dispersal) (e.g., Alves et al., 2016; Rasteiro et al., 2012; Peng and Amos, 2008) or complex selection (Peng et al., 2007) that cannot be studied with the coalescent yet. However, forward in time simulations usually require extensive computational costs and memory capacity to simulate the entire population. In this concern, coalescent simulations are generally much more efficient than forward in time simulations. The coalescent is convenient for studies that require a large number of simulations such as Bayesian (Beaumont and Rannala, 2004) and approximation Bayesian computation (ABC) (Arenas, 2015; Beaumont, 2010) approaches.

A problem in the coalescent can occur with the simulation of the ARG (and, similarly, the ASG) under very high population recombination rates (i.e., $\rho > 1000$). Under this situation the number of active lineages may dramatically increase going backwards in time leading to computer problems related with out of memory. In addition, if the population growth rate is too negative, the TMRCA can be almost infinite, again leading to computational problems.

Coalescent Simulators

A number of freely available coalescent simulators have been developed (**Table 1**). Most of available coalescent simulators implement demographics, population structure and recombination. However, selection is implemented in only a few coalescent programs mainly due to its complexity (see above) in comparison with forward in time approaches (Peng *et al.*, 2012; Carvajal-Rodriguez, 2010). Current trends of coalescent frameworks include the adaptation of the coalescent to simulate specific genetic markers such as codons (Arenas and Posada, 2010a, 2007), proteins (Arenas *et al.*, 2013) or whole-genome sequences (e.g., Arenas and Posada, 2014b; Kelleher *et al.*, 2016). Another trend is to develop extensions of the coalescent to mimic the evolution of specific organisms such as bacteria (De Maio and Wilson, 2017; Brown *et al.*, 2016) and cancer (Zhao *et al.*, 2014; Nicolas *et al.*, 2007). In all, the development of coalescent simulators is a very active area of research in population genetics with a continuous emergence of more efficient, and with more capabilities, frameworks.

Applications of the Coalescent

The coalescent provides clear and rapid representations of real evolutionary processes and its implementation in computer frameworks is, in general, straightforward. Consequently, the coalescent has been applied to a variety of purposes, the most relevant are presented below.

Hypothesis Testing

The coalescent has been widely used for testing hypothesis such as the effects between evolutionary parameters or the influence of evolutionary parameters on phylogenetic reconstructions. It follows with some illustrative examples.

Computer simulations of the coalescent with recombination (Arenas, 2013a), followed by the simulation of codon evolution upon the simulated coalescent histories (Arenas and Posada, 2012), were applied in several studies to analyze the influence of ignored recombination on the estimation of selection (the nonsynonymous (dN) to synonymous (dS) substitution ratio, dN/dS or ω) (Arenas and Posada, 2010a, 2014a; Anisimova *et al.*, 2003; Shriner *et al.*, 2003). These studies found that ignored recombination biases the estimation of dN/dS by generating false positively selected sites. As a consequence, those studies concluded that recombination must be considered for the estimation of dN/dS. That can be performed by the following methodology: (i) detection of recombination breakpoints, (ii) inference of a phylogenetic tree for each recombination partition and, (iii) estimation of dN/dS per site (or per recombination partition) accounting for the corresponding phylogenetic tree. This methodology was already applied in several studies (e.g., Perez-Losada *et al.*, 2009; Perez-Losada *et al.*, 2011).

Another application of coalescent simulations was the analysis of the influence of recombination on ancestral sequence reconstruction (ASR). Arenas and Posada (2010b) performed coalescent simulations under different levels of recombination and simulated DNA sequence evolution over those coalescent histories to generate sequences for every tip and internal node. Next, they performed ASR from the simulated sample and a previously inferred ML or Bayesian phylogenetic tree, to infer the ancestral sequences. Finally, they compared the simulated (true) ancestral sequences with the inferred ancestral sequences to evaluate the impact of recombination on ASR. They found that recombination seriously bias ASR, especially under high levels of genetic diversity (Arenas and Posada, 2010b). In order to reduce this bias, they suggested that ASR should be performed for each recombinant partition independently (see below).

The influence of recombination on the estimation of dN/dS and on ASR is derived from the influence of recombination on phylogenetic tree reconstruction. Recombination can bias phylogenetic tree inferences because each recombinant partition can present an evolutionary history that differs from the evolutionary history of other recombinant partitions. This was investigated by Schierup and Hein (2000a) with coalescent simulations based on the Hudson's algorithm of the coalescent with recombination (Hudson, 1983). They found that if recombination is present but ignored, derived phylogenetic trees can be biased with incorrect topologies and branch lengths (Schierup and Hein, 2000a). Moreover, ignoring recombination can result in loss of molecular clock, apparent homoplasies and incorrect substitution rate heterogeneity (Schierup and Hein, 2000a,b). In addition, phylogenetic incongruence in empirical data was observed and attributed to ignored recombination (Worobey and Holmes, 1999; Feil *et al.*, 2001). Consequently, under the presence of recombination one should either infer a phylogenetic tree for each recombinant partition (recombinant partitions can be identified with a framework for the detection of recombination breakpoints (e.g., Martin *et al.*, 2015; Kosakovsky Pond *et al.*, 2006) (Arenas and Posada, 2010b) or infer a phylogenetic recombination network (Arenas, 2013b; Griffiths and Marjoram, 1997; Huson and Bryant, 2006; Huson, 1998). Additionally, a variety of analytical evolutionary frameworks (i.e., *BEAST* (Drummond *et al.*, 2012)) are based on the coalescent without recombination and can therefore generate biases from data that suffered recombination events. Detecting recombination partitions and analyzing them separately is once again recommended (e.g., Perez-Losada *et al.*, 2007).

Table 1   Main coalescent simulators and their capabilities

Program	Evolutionary scenarios	Additional capabilities	Genetic data	OS	Reference
FastSimBac	Dm, Pm, Re	Bacterial recombination / Recombination hotspots / Species/population trees	Nt	SC	De Maio and Wilson (2017)
SimBac	Re	Bacterial recombination	Nt	SC	Brown et al. (2016)
msprime	Dm, Pm, Re	Capabilities of ms and the following: / Fast and accurate simulation of huge samples	–	SC	Kelleher et al. (2016)
scrm	Ls, Dm, Pm, Re	Species/population trees	–	SC	Staab et al. (2015)
CoalEvol	Ls, Dm, Pm, Re	Capabilities of NetRecodon and the following: / Recombination hotspots / Species/population trees	Nt, Cd, Aa	SC, Linux, Mac	Arenas and Posada (2014b)
SGWE	Ls, Dm, Pm, Re	Capabilities of CoalEvol and the following: Different models of evolution for different partitions	Nt, Cd, Aa	SC, Linux, Mac	Arenas and Posada (2014b)
Cosi2	Dm, Pm, Re, Se	Recombination hotspots Gene conversion	SNP	SC	Shlyakhter et al. (2014)
ProteinEvolver	Ls, Dm, Pm, Re	Capabilities of NetRecodon and the following: / Recombination hotspots	Nt, Cd, Aa	SC, Linux, Mac	Arenas et al. (2013)
GUMS	Dm, Pm	Species/population trees	–	Linux, Mac	Heled et al. (2013)
FTEC	Dm, Pm, Re		SNP	SC	Reppell et al. (2012)
Fastsimcoal	Dm, Pm, Re	Capabilities of SIMCOAL2 and the following: / Demographic periods / Species/population trees Admixture	Nt, STR, SNP	Linux, Mac, Win	Excoffier and Foll (2011)
NetRecodon	Ls, Dm, Pm, Re	Capabilities of Recodon and the following: / Intracodon recombination Printed ARG[a]	Nt, Cd	All	Arenas and Posada (2010a)
msms	Dm, Pm, Re, Se	Capabilities of ms and the following: Selection (single diploid locus)	SNP	All	Ewing and Hermisson (2010)
mbs	Dm, Pm, Re, Se	Capabilities of ms and the following: / Selection (biallelic site) / Recombination hot-spots	SNP, Nt	SC	Teshima and Innan (2009)
MaCS	Dm, Pm, Re	Species/population trees	–	SC	Chen et al. (2009)
SimMLST	Dm, Re	Multi-locus sequence typing data	Nt	SC, Linux, Win	Didelot et al. (2009)
msHOT	Dm, Pm, Re	Capabilities of ms and the following: Multiple crossover hotspots and multiple gene conversion hotspots	SNP	SC	Hellenthal and Stephens (2007)

(Continued)

Table 1   Continued

Program	Evolutionary scenarios	Additional capabilities	Genetic data	OS	Reference
GENOME	Dm, Pm, Re	Recombination hotspots Species/population trees	SNP	SC, Linux, Win	Liang et al. (2007)
mlcoalsim	Dm, Pm, Re	Capabilities of ms and the following: Recombination includes hot-spots	Nt	SC, Mac, Win	Ramos-Onsins and Mitchell-Olds (2007)
Recodon	Dm, Pm, Re	Demographics according to a population growth rate and demographic periods	Nt, Cd	All	Arenas and Posada (2007)
Serial Simcoal	Ls, Dm, Pm	Capabilities of SIMCOAL and the following: Longitudinal sampling Species/population trees	Nt, RFLP, STR	SC, Mac, Win	Anderson et al. (2005)
CoaSim	Dm, Pm, Re	Gene conversion Species/population trees	SNP, STR	SC, Linux	Mailund et al. (2005)
Cosi	Dm, Pm, Re	Recombination hotspots	SNP	SC	Schaffner et al. (2005)
Simcoal2	Dm, Pm, Re	Gene conversion Species/population trees Capabilities of SIMCOAL and the following: Recombination hotspots	Nt, RFLP, STR, SNP	SC, Win	Laval and Excoffier (2004)
SelSim	Re, Se		Nt, STR, SNP	SC, Linux, Win	Spencer and Coop (2004)
SNPsim	Dm, Re	Recombination hotspots	SNP	SC	Posada and Wiuf (2003)
CodonRecSim	Re	–	Cd	SC, Win	Anisimova et al. (2003)
ms	Dm, Pm, Re	–	SNP	SC	Hudson (2002)
Simcoal	Dm, Pm	–	Nt, RFLP, STR	SC, Win	Excoffier et al. (2000)
Treevolve	Dm, Pm, Re	–	Nt	SC, Mac	Grassly et al. (1999)

^aThe ARG can be analyzed/visualized with NetTest (Arenas et al., 2010).

Note. The simulators are presented by publication date, from the present to the past. The table shows: (i) implemented evolutionary scenarios (Ls, Dm, Pm, Re and Se, refers to longitudinal sampling, demographics, population structure and migration, recombination and selection, respectively); (ii) additional capabilities to the cited evolutionary scenarios; (iii) type of simulated genetic data (in addition to the simulated coalescent tree/s) like sequences of SNPs, microsatellites (STR), nucleotides (Nt), codons (Cd) and amino acids (Aa); (iv) operative system (OS) considering the availability of binary files (Linux, Macintosh, Windows) and/or source code (SC); (v) references. This list is not exhaustive but is meant to highlight the diversity of the field.

Another example is the evaluation of the criteria for selecting substitution models of evolution. Luo *et al.* (2010) simulated coalescent evolutionary histories and used those trees to simulate DNA sequence evolution under different substitution models of evolution. Next, they estimated the best fitting substitution model of evolution for the simulated data under different statistical criteria such as Bayesian information criterion, decision theory, Akaike information criterion and hierarchical likelihood-ratio test. They found that Bayesian information and decision theory performed better for identifying the correct substitution model and, therefore, these criteria should be preferred in substitution model selection.

Validation of Analytical Methods

Any analytical framework must be properly validated. In population genetics, the "true" underlined evolutionary process of a dataset is usually unknown. Therefore, the evaluation of analytical frameworks is frequently assessed with computer simulations where the "true" (simulated) evolutionary process is known. The coalescent has been widely used for the evaluation of analytical evolutionary frameworks. Some illustrative examples are described below.

It was remarkable in the previous subsection that genetic recombination should be considered in the evolutionary analysis of genetic data. Because of that, and other aspects about the important role of recombination in the evolution of many organisms (e.g., Arenas *et al.*, 2017; Castelhano *et al.*; 2017, Arenas *et al.*, 2016; Perez-Losada *et al.*, 2015), a variety of frameworks were developed to detect and quantify recombination and to infer phylogenetic recombination frameworks. Many of these frameworks were validated with the coalescent extended with recombination. The following are some examples: (i) Marttinen *et al.* (2012) with a method to detect homologous recombination events (validated comparing the number of simulated and predicted recombination events); (ii) Westesson and Holmes (2009) with a method to identify recombination breakpoints (validated comparing simulated and detected recombinant regions); (iii) Dialdestoro *et al.* (2016) with a method to estimate recombination rate and other population genetics parameters such as population size (validated comparing the distance between simulated and estimated recombination rates); (iv) White *et al.* (2013) and Sun *et al.* (2011) evaluating some of the existing recombination detection tests (validated comparing simulated and estimated recombination rate); (v) Lopes *et al.* (2014) with a method to coestimate recombination rate with other parameters of molecular evolution (also validated comparing simulated and estimated recombination rate).

Coalescent simulations are also frequently applied to evaluate reconstruction methods of phylogenetic trees (e.g., Mccormack *et al.*, 2009; Chen *et al.*, 2011) and networks (e.g., Wen *et al.*, 2016; Javed *et al.*, 2011; Yu *et al.*, 2014; Yu and Nakhleh, 2015; Hassanzadeh *et al.*, 2012; Wang *et al.*, 2013) where the topology and branch lengths of simulated and reconstructed phylogenies are compared.

Coalescent simulations have also been applied to validate other analytical methods in population genetics such as the estimation of pairwise distance between sequences (Domazet-Loso and Haubold, 2009) or the estimation of population size and migration rate (Beerli, 2006).

Estimation of Evolutionary Parameters and Model Selection

A variety of computer programs based on the coalescent can estimate evolutionary parameters. Some of the most relevant are, (i) *BEAST*, estimation of variable population size, substitution rate and other evolutionary parameters, also phylogenetic tree reconstruction; (ii) *GENETREE* (Bahlo and Griffiths, 2000), estimation of population growth, mutation and migration rates; (iii) *IMa* (Hey and Nielsen, 2007), estimation of population size, migration rate and divergence time; (iv) *LAMARC* (Kuhner, 2006), estimation of population growth, migration and recombination rates; (v) *MIGRATE* (Beerli and Felsenstein, 2001), estimation of population size and migration rates.

The cited programs estimate evolutionary parameters through a likelihood function. However, under complex models of evolution a likelihood function could not be mathematically designed or could be computationally intractable. In these situations, the ABC approach provides a very useful alternative because it does not require the likelihood function. Moreover, ABC can outperform maximum-likelihood (ML) methods based on approximate models (Arenas *et al.*, 2015; Lopes *et al.*, 2014). Briefly, ABC provides model selection and parameters estimation by the following pipeline: (i) Design of prior distributions for parameters, these prior distributions should contain the real value. (ii) Extensive computer simulations after sampling from the prior distributions. (iii) Design and computation of summary statistics for simulated and real data. (iv) Selection of summary statistics from simulated data that are closer to the summary statistics from real data according to a given tolerance. (v) With the retained summary statistics from simulated data and summary statistics from real data, ABC applies a rejection (Pritchard *et al.*, 1999) or a multiple regression approach (Beaumont *et al.*, 2002) to obtain the posterior distributions for the studied models or parameters. For further details about the ABC approach the reader is referred to the following reviews (Beaumont, 2010; Arenas, 2015; Csillery *et al.*, 2010; Sunnaker *et al.*, 2013; Bertorelle *et al.*, 2010; Lopes and Beaumont, 2010). ABC usually requires a number of computer simulations that should be longer enough (from thousands to millions) for properly covering the prior distributions. Since the coalescent provides a rapid computation, most of ABC studies have applied the coalescent to simulate datasets (e.g., Arenas *et al.*, 2015; Lopes *et al.*, 2014; Alves *et al.*, 2016; Wilson *et al.*, 2009; Fan and Kubatko, 2011; Li and Jakobsson, 2012). Because of this important application of the coalescent, many coalescent simulators incorporate adaptations to facilitate the design of

ABC analyses (i.e., simulating from predefined prior distributions) (e.g., Arenas and Posada, 2014b; Excoffier and Foll, 2011; Pavlidis *et al.*, 2010).

Designing a Coalescent-Based Experiment

Studies involving coalescent simulations may require a previous design. The researcher has to choose a coalescent tool to simulate an evolutionary scenario and obtain a desired type of genetic data. A few aspects should be considered to design the study.

1. Evolutionary scenario. The currently available coalescent simulators implement a large variety of models to mimic diverse evolutionary scenarios (see Section "Coalescent Simulators" and **Table 1**). The following sentences provide suggestions from the experience of the author. The simulation of basic scenarios such as a constant population size can be easily performed with the *ms* program. This program is also useful for simulating basic (homogeneous) recombination. For scenarios with population structure and migration, the simulators *SimCoal2* and *fastsimcoal* provide a variety of parameters (i.e., a matrix of migration rates). Complex recombination (i.e., hotspots) can be simulated with *SNPsim*, which implements a wide variety of models of heterogeneous recombination. Scenarios of selection can be simulated with *mbs* and *msms*.
2. Desired output data. Any coalescent framework simulates a coalescent tree but may not simulate genetic data. Here there are two options. (i) Apply a coalescent framework that internally simulates genetic data (i.e., *CoalEvol* simulates nucleotide, codon and protein sequences under a variety of substitution models of evolution); (ii) Apply the chosen coalescent framework to simulate the coalescent history and then simulate sequence evolution (Yang, 2006) upon such a history with frameworks such as *INDELible* (Fletcher and Yang, 2009), *CoalEvol*, *SeqGen* (Rambaut and Grassly, 1997), *Pyvolve* (Spielman and Wilke, 2015) or *Phylosim* (Sipos *et al.*, 2011), see for a review (Arenas, 2012).

Illustrative Practical Examples

Here five practical examples to simulate a particular type of genetic data under a particular evolutionary scenario are presented.

1. *Codon sequences under recombination.* Probably the easiest procedure to perform this simulation is with the programs *NetRecodon* or *CoalEvol*. These programs implement the simulation of the ARG and the subsequent simulation of coding sequence evolution (under a variety of codon substitution models) over the ARG. Another possibility is to apply the program *ms* to obtain the ARG (a coalescent tree for each recombinant partition) and then apply *INDELible* (or a similar framework, see above) to simulate codon evolution upon the coalescent trees. The advantage from using the first procedure is that *NetRecodon* and *CoalEvol* implement the simulation of both intercodon and intracodon recombination, which is much more realistic than consider only intercodon recombination in the second procedure.
2. *Protein sequences under a population structure with migration.* Again, the program *CoalEvol* implements the simulation of population structure with migration and the simulation of protein sequences over the previously simulated tree. However, *CoalEvol* implements a limited number of migration parameters (i.e., same migration rate between all demes). As an alternative, one could apply *fastsimcoal*, which includes a variety of parameters to better mimic the population structure with migration, and then simulate protein sequences with *INDELible* (or similar framework).
3. *Protein sequences under selection.* First, coalescent trees under selection can be simulated with *mbs* or *msms* and then, protein sequences can be simulated over those trees with *INDELible* (or similar framework). For simulations under complex models of selection one could explore forward in time simulation frameworks (Peng *et al.*, 2012).
4. *Nucleotide sequences under heterogeneous recombination (hotspots).* I would recommend simulate the coalescent trees (ARG) with *SNPsim* (although there are other advanced simulators such as *Cosi2*, *msHOT*, *GENOME* and *SimCoal2*) and next simulate nucleotide sequences with *INDELible* (or similar framework). Another methodology is to perform both the coalescent simulation and sequence simulation with *CoalEvol*, which implements the heterogeneous recombination models of *SNPsim*.
5. *Codon sequences with longitudinal sampling accounting for variation of the population size and population structure.* Longitudinal sampling, variable population size through time and population structure can be simulated with *Serial SimCoal*, *scrm* and *CoalEvol*. *CoalEvol* also allows the simulation of codon sequence evolution. In case of *Serial SimCoal* and *scrm*, codon sequence evolution can be simulated with *INDELible* (or similar framework).

Concluding Remarks

The coalescent is a stochastic statistical process established in population genetics for more than 30 years. Instead of reducing its popularity through time, the coalescent is progressively more famous, used and incorporated into new frameworks (usually published in important journals of the discipline because of their utility, as presented in this study) to both simulate and analyze genetic data in an evolutionary perspective.

A number of coalescent simulators exist implementing fundamental evolutionary processes of the real world such as demographics, recombination, population structure and selection. Some coalescent simulators are also oriented to mimic the evolution of specific organisms such as bacteria or cancer. However, further work is required to accommodate more complex evolutionary models, especially complex selection, as done in the forward in time simulation approach.

Some famous analytical tools (i.e., *BEAST* and *LAMARC*) are based on the coalescent and consequently, users of those tools should have in mind the limitations of the underlined coalescent approach (i.e., the cited programs ignore selection). It is worth highlighting the incorporation of the coalescent into the ABC approach, which is a methodology of increasing application in population genetics and ecology to analyze genetic data under complex models of evolution. Instead of being forced to only consider the limited models implemented in ML or Bayesian tools, ABC allows the user to apply any model that can be simulated. In this concern, the coalescent provides fast simulations that are convenient to achieve the extensive number of simulations usually required in ABC.

The future of the coalescent is likely to be highly promising because of their important applications. However, we still need to develop more extensions of the coalescent, and subsequent implementation in simulation frameworks, to better fit with real evolutionary processes.

Acknowledgements

MA was supported by the Grant "Ramón y Cajal" RYC-2015-18241 from the Spanish Government.

See also: Evolutionary Models. Gene Duplication and Speciation. Genetics and Population Analysis. Inference of Horizontal Gene Transfer: Gaining Insights Into Evolution via Lateral Acquisition of Genetic Material. Molecular Clock. Molecular Phylogenetics. Natural Language Processing Approaches in Bioinformatics. Phylogenetic analysis: Early evolution of life. Phylogenetic Tree Rooting. Population Genetics

References

Alves, I., Arenas, M., Currat, M., et al., 2016. Long-distance dispersal shaped patterns of human genetic diversity in Eurasia. Mol Biol Evol 33, 946–958.

Anderson, C.N., Ramakrishnan, U., Chan, Y.L., Hadly, E.A., 2005. Serial SimCoal: A population genetics model for data from multiple populations and points in time. Bioinformatics 21, 1733–1734.

Anisimova, M., Nielsen, R., Yang, Z., 2003. Effect of recombination on the accuracy of the likelihood method for detecting positive selection at amino acid sites. Genetics 164, 1229–1236.

Arenas, M., 2012. Simulation of molecular data under diverse evolutionary scenarios. PLOS Comput Biol 8, e1002495.

Arenas, M., 2013a. Computer programs and methodologies for the simulation of DNA sequence data with recombination. Front Genet 4, 9.

Arenas, M., 2013b. The importance and application of the ancestral recombination graph. Front Genet 4, 206.

Arenas, M., 2015. Advances in computer simulation of genome evolution: Toward more realistic evolutionary genomics analysis by approximate Bayesian computation. J Mol Evol 80, 189–192.

Arenas, M., Araujo, N.M., Branco, C., et al., 2017. Mutation and recombination in pathogen evolution: Relevance, methods and controversies. Infect Genet Evol.

Arenas, M., Dos Santos, H.G., Posada, D., Bastolla, U., 2013. Protein evolution along phylogenetic histories under structurally constrained substitution models. Bioinformatics 29, 3020–3028.

Arenas, M., Lopes, J.S., Beaumont, M.A., Posada, D., 2015. CodABC: A computational framework to coestimate recombination, substitution, and molecular adaptation rates by approximate Bayesian computation. Mol Biol Evol 32, 1109–1112.

Arenas, M., Lorenzo-Redondo, R., Lopez-Galindez, C., 2016. Influence of mutation and recombination on HIV-1 in vitro fitness recovery. Mol Phylogenet Evol 94, 264–270.

Arenas, M., Patricio, M., Posada, D., Valiente, G., 2010. Characterization of phylogenetic networks with NetTest. BMC Bioinformatics 11, 268.

Arenas, M., Posada, D., 2007. Recodon: Coalescent simulation of coding DNA sequences with recombination, migration and demography. BMC Bioinformatics 8, 458.

Arenas, M., Posada, D., 2010a. Coalescent simulation of intracodon recombination. Genetics 184, 429–437.

Arenas, M., Posada, D., 2010b. The effect of recombination on the reconstruction of ancestral sequences. Genetics 184, 1133–1139.

Arenas, M., Posada, D., 2012. Simulation of coding sequence evolution. In: Cannarozzi, G.M., Schneider, A. (Eds.), Codon Evolution. Oxford: Oxford University Press.

Arenas, M., Posada, D., 2014a. The influence of recombination on the estimation of selection from coding sequence alignments. In: Fares, M.A. (Ed.), Natural Selection: Methods and Applications. Boca Raton: CRC Press/Taylor & Francis.

Arenas, M., Posada, D., 2014b. Simulation of genome-wide evolution under heterogeneous substitution models and complex multispecies coalescent histories. Mol Biol Evol 31, 1295–1301.

Bahlo, M., Griffiths, R.C., 2000. Inference from gene trees in a subdivided population. Theor Popul Biol 57, 79–95.

Beaumont, M.A., 2010. Approximate Bayesian computation in evolution and ecology. Annu Rev Ecol Evol Syst 41, 379–405.

Beaumont, M.A., Rannala, B., 2004. The Bayesian revolution in genetics. Nat Rev Genet 5, 251–261.

Beaumont, M.A., Zhang, W., Balding, D.J., 2002. Approximate Bayesian computation in population genetics. Genetics 162, 2025–2035.

Beerli, P., 2006. Comparison of Bayesian and maximum-likelihood inference of population genetic parameters. Bioinformatics 22, 341–345.

Beerli, P., Felsenstein, J., 2001. Maximum likelihood estimation of a migration matrix and efective population sizes in *n* subpopulations by using a coalescent approach. Proc Natl Acad Sci USA 98, 4563–4568.

Bertorelle, G., Benazzo, A., Mona, S., 2010. ABC as a flexible framework to estimate demography over space and time: Some cons, many pros. Mol Ecol 19, 2609–2625.

Brown, T., Didelot, X., Wilson, D.J., De Maio, N., 2016. SimBac: Simulation of whole bacterial genomes with homologous recombination. Microb Genom 2.

Carvajal-Rodriguez, A., 2010. Simulation of genes and genomes forward in time. Curr Genomics 11, 58–61.

Castelhano, N., Araujo, N.M., Arenas, M., 2017. Heterogeneous recombination among Hepatitis B virus genotypes. Infect Genet Evol 54, 486–490.

Chen, G.K., Marjoram, P., Wall, J.D., 2009. Fast and flexible simulation of DNA sequence data. Genome Res 19, 136–142.

Chen, S.C., Rosenberg, M.S., Lindsay, B.G., 2011. MixtureTree: A program for constructing phylogeny. BMC Bioinformatics 12, 111.

Csillery, K., Blum, M.G.B., Gaggiotti, O.E., Francois, O., 2010. Approximate Bayesian computation (ABC) in practice. Trends Ecol Evol 25, 410–418.

De Maio, N., Wilson, D.J., 2017. The bacterial sequential markov coalescent. Genetics 206, 333–343.

Dialdestoro, K., Sibbesen, J.A., Maretty, L., *et al.*, 2016. Coalescent inference using serially sampled, high-throughput sequencing data from intrahost HIV infection. Genetics 202, 1449–1472.

Didelot, X., Lawson, D., Falush, D., 2009. SimMLST: Simulation of multi-locus sequence typing data under a neutral model. Bioinformatics 25, 1442–1444.

Domazet-Loso, M., Haubold, B., 2009. Efficient estimation of pairwise distances between genomes. Bioinformatics 25, 3221–3227.

Drummond, A.J., Suchard, M.A., Xie, D., Rambaut, A., 2012. Bayesian phylogenetics with BEAUti and the BEAST 1.7. Mol Biol Evol 29, 1969–1973.

Ewing, G., Hermisson, J., 2010. MSMS: A coalescent simulation program including recombination, demographic structure and selection at a single locus. Bioinformatics 26, 2064–2065.

Excoffier, L., Foll, M., 2011. fastsimcoal: A continuous-time coalescent simulator of genomic diversity under arbitrarily complex evolutionary scenarios. Bioinformatics 27, 1332–1334.

Excoffier, L., Novembre, J., Schneider, S., 2000. SIMCOAL: A general coalescent program for the simulation of molecular data in interconnected populations with arbitrary demography. J Hered 91, 506–509.

Fan, H.H., Kubatko, L.S., 2011. Estimating species trees using approximate Bayesian computation. Mol Phylogenet Evol 59, 354–363.

Feil, E.J., Holmes, E.C., Bessen, D.E., *et al.*, 2001. Recombination within natural populations of pathogenic bacteria: Short-term empirical estimates and long-term phylogenetic consequences. Proc Natl Acad Sci USA 98, 182–187.

Fletcher, W., Yang, Z., 2009. INDELible: A flexible simulator of biological sequence evolution. Mol Biol Evol 26, 1879–1888.

Grassly, N.C., Harvey, P.H., Holmes, E.C., 1999. Population dynamics of HIV-1 inferred from gene sequences. Genetics 151, 427–438.

Griffiths, R.C., Marjoram, P., 1996. Ancestral inference from samples of DNA sequences with recombination. J Comput Biol 3, 479–502.

Griffiths, R.C., Marjoram, P., 1997. An ancestral recombination graph. In: Donelly, P., Tavaré, S. (Eds.), Progress in population genetics and human evolution. Berlin: Springer-Verlag.

Hassanzadeh, R., Eslahchi, C., Sung, W.K., 2012. Constructing phylogenetic supernetworks based on simulated annealing. Mol Phylogenet Evol 63, 738–744.

Hein, J., Schierup, M., Wiuf, C., 2005. Gene genealogies, variation and evolution: A primer in coalescent theory. Oxford University Press.

Heled, J., Bryant, D., Drummond, A.J., 2013. Simulating gene trees under the multispecies coalescent and time-dependent migration. BMC Evol Biol 13, 44.

Hellenthal, G., Stephens, M., 2007. msHOT: Modifying Hudson's ms simulator to incorporate crossover and gene conversion hotspots. Bioinformatics 23, 520–521.

Hey, J., Nielsen, R., 2007. Integration within the Felsenstein equation for improved Markov chain Monte Carlo methods in population genetics. Proc Natl Acad Sci USA 104, 2785–2790.

Hoban, S., Bertorelle, G., Gaggiotti, O.E., 2012. Computer simulations: Tools for population and evolutionary genetics. Nat Rev Genet 13, 110–122.

Huson, D.H., 1998. SplitsTree: Analyzing and visualizing evolutionary data. Bioinformatics 14, 68–73.

Huson, D.H., Bryant, D., 2006. Application of phylogenetic networks in evolutionary studies. Mol Biol Evol 23, 254–267.

Hudson, R.R., 1983. Properties of a neutral allele model with intragenic recombination. Theor Popul Biol 23, 183–201.

Hudson, R.R., 1990. Gene genealogies and the coalescent process. Oxford Surv Evol Biol 7, 1–44.

Hudson, R.R., 1998. Island models and the coalescent process. Mol Ecol 7, 413–418.

Hudson, R.R., 2002. Generating samples under a Wright-Fisher neutral model of genetic variation. Bioinformatics 18, 337–338.

Hudson, R.R., Kaplan, N.L., 1988. The coalescent process in models with selection and recombination. Genetics 120, 831–840.

Hudson, R.R., Kaplan, N.L., 1995. The coalescent process and background selection. Philos Trans R Soc Lond B Biol Sci 349, 19–23.

Javed, A., Pybus, M., Mele, M., *et al.*, 2011. IRiS: Construction of ARG networks at genomic scales. Bioinformatics 27, 2448–2450.

Kaplan, N.L., Darden, T., Hudson, R.R., 1988. The coalescent process in models with selection. Genetics 120, 819–829.

Kaplan, N.L., Hudson, R.R., Iizuka, M., 1991. The coalescent process in models with selection, recombination and geographic subdivision. Genet Res Camb 57, 83–91.

Kelleher, J., Etheridge, A.M., Mcvean, G., 2016. Efficient coalescent simulation and genealogical analysis for large sample sizes. PLOS Comput Biol 12, e1004842.

Kimura, M., Weiss, G.H., 1964. The stepping stone model of population structure and the decrease of genetic correlation with distance. Genetics 49, 561–576.

Kingman, J.F.C., 1982. The coalescent. Stoch Process Appl 13, 235–248.

Kosakovsky Pond, S.L., Posada, D., Gravenor, M.B., Woelk, C.H., Frost, S.D., 2006. GARD: A genetic algorithm for recombination detection. Bioinformatics 22, 3096–3098.

Krone, S.M., Neuhauser, C., 1997. Ancestral processes with selection. Theor Popul Biol 51, 210–237.

Kuhner, M.K., 2006. LAMARC 2.0: Maximum likelihood and Bayesian estimation of population parameters. Bioinformatics 22, 768–770.

Laval, G., Excoffier, L., 2004. SIMCOAL 2.0: A program to simulate genomic diversity over large recombining regions in a subdivided population with a complex history. Bioinformatics 20, 2485–2487.

Lemey, P., Rambaut, A., Pybus, O.G., 2006. HIV evolutionary dynamics within and among hosts. AIDS Rev 8, 125–140.

Li, S., Jakobsson, M., 2012. Estimating demographic parameters from large-scale population genomic data using approximate Bayesian computation. BMC Genet 13, 22.

Liang, L., Zollner, S., Abecasis, G.R., 2007. GENOME: A rapid coalescent-based whole genome simulator. Bioinformatics 23, 1565–1567.

Lopes, J.S., Arenas, M., Posada, D., Beaumont, M.A., 2014. Coestimation of recombination, substitution and molecular adaptation rates by approximate Bayesian computation. Heredity 112, 255–264.

Lopes, J.S., Beaumont, M.A., 2010. ABC: A useful Bayesian tool for the analysis of population data. Infect Genet Evol 10, 826–833.

Luo, A., Qiao, H., Zhang, Y., *et al.*, 2010. Performance of criteria for selecting evolutionary models in phylogenetics: A comprehensive study based on simulated datasets. BMC Evol Biol 10, 242.

Mailund, T., Schierup, M.H., Pedersen, C.N., *et al.*, 2005. CoaSim: A flexible environment for simulating genetic data under coalescent models. BMC Bioinformatics 6, 252.

Mallo, D., Sánchez-Cobos, A., Arenas, M., 2016. Diverse considerations for successful phylogenetic tree reconstruction: Impacts from model misspecification, recombination, homoplasy, and pattern recognition. In: Elloumi, M., Iliopoulos, C., Wang, J., Zomaya, A. (Eds.), Pattern Recognition in Computational Molecular Biology. John Wiley & Sons, Inc.

Martin, D.P., Murrell, B., Golden, M., Khoosal, A., Muhire, B., 2015. RDP4: Detection and analysis of recombination patterns in virus genomes. Virus Evol 1, vev003.

Marttinen, P., Hanage, W.P., Croucher, N.J., *et al.*, 2012. Detection of recombination events in bacterial genomes from large population samples. Nucleic Acids Res 40, e6.

Mccormack, J.E., Huang, H., Knowles, L.L., 2009. Maximum likelihood estimates of species trees: How accuracy of phylogenetic inference depends upon the divergence history and sampling design. Syst Biol 58, 501–508.

Neuhauser, C., Krone, S.M., 1997. The genealogy of samples in models with selection. Genetics 145, 519–534.

Neuhauser, C., Tavaré, S., 2001. The coalescent. Encyclopedia of Genetics. New York: Academic Press.

Nicolas, P., Kim, K.M., Shibata, D., Tavare, S., 2007. The stem cell population of the human colon crypt: Analysis via methylation patterns. PLOS Comput Biol 3, e28.

Nielsen, R., Wakeley, J., 2001. Distinguishing migration from isolation: A Markov chain Monte Carlo approach. Genetics 158, 885–896.

Nordborg, M., 2007. Coalescent theory. In: Balding, D.J., Bishop, M., Cannings, C. (Eds.), Handbook of Statistical Genetics, Third ed. Chichester, UK: John Wiley & Sons, Ltd..

Notohara, M., 1990. The coalescent and the genealogical process in geographically structured population. J Math Biol 29, 59–75.

Pavlidis, P., Laurent, S., Stephan, W., 2010. msABC: A modification of Hudson's ms to facilitate multi-locus ABC analysis. Mol Ecol Resour 10, 723–727.

Peng, B., Amos, C.I., 2008. Forward-time simulations of non-random mating populations using simuPOP. Bioinformatics 24, 1408–1409.

Peng, B., Amos, C.I., Kimmel, M., 2007. Forward-time simulations of human populations with complex diseases. PLOS Genet 3, e47.

Peng, B., Kimmel, M., Amos, C.I., 2012. Forward-time population genetics simulations: Methods, implementation, and applications. Hoboken, NJ: John Wiley & Sons.

Perez-Losada, M., Arenas, M., Galan, J.C., Palero, F., Gonzalez-Candelas, F., 2015. Recombination in viruses: Mechanisms, methods of study, and evolutionary consequences. Infect Genet Evol 30C, 296–307.

Perez-Losada, M., Crandall, K.A., Zenilman, J., Viscidi, R.P., 2007. Temporal trends in gonococcal population genetics in a high prevalence urban community. Infect Genet Evol 7, 271–278.

Perez-Losada, M., Jobes, D.V., Sinangil, F., *et al.*, 2011. Phylodynamics of HIV-1 from a phase III AIDS vaccine trial in Bangkok, Thailand. PLOS One 6, e16902.

Perez-Losada, M., Posada, D., Arenas, M., *et al.*, 2009. Ethnic differences in the adaptation rate of HIV gp120 from a vaccine trial. Retrovirology 6, 67.

Posada, D., Crandall, K.A., 2002. The effect of recombination on the accuracy of phylogeny estimation. J Mol Evol 54, 396–402.

Posada, D., Wiuf, C., 2003. Simulating haplotype blocks in the human genome. Bioinformatics 19, 289–290.

Pritchard, J.K., Seielstad, M.T., Perez-Lezaun, A., Feldman, M.W., 1999. Population growth of human Y chromosomes: A study of Y chromosome microsatellites. Mol Biol Evol 16, 1791–1798.

Rambaut, A., Grassly, N.C., 1997. Seq-Gen: An application for the Monte Carlo simulation of DNA sequence evolution along phylogenetic trees. Comput. Appl. Biosciences 13, 235–238.

Ramos-Onsins, S.E., Mitchell-Olds, T., 2007. Mlcoalsim: Multilocus coalescent simulations. Evol Bioinform Online 3, 41–44.

Rasteiro, R., Bouttier, P.A., Sousa, V.C., Chikhi, L., 2012. Investigating sex-biased migration during the neolithic transition in Europe, using an explicit spatial simulation framework. Proc Biol Sci 279, 2409–2416.

Reppell, M., Boehnke, M., Zollner, S., 2012. FTEC: A coalescent simulator for modeling faster than exponential growth. Bioinformatics 28, 1282–1283.

Schaffner, S.F., Foo, C., Gabriel, S., et al., 2005. Calibrating a coalescent simulation of human genome sequence variation. Genome Res 15, 1576–1583.

Schierup, M.H., Hein, J., 2000a. Consequences of recombination on traditional phylogenetic analysis. Genetics 156, 879–891.

Schierup, M.H., Hein, J., 2000b. Recombination and the molecular clock. Mol Biol Evol 17, 1578–1579.

Shlyakhter, I., Sabeti, P.C., Schaffner, S.F., 2014. Cosi2: An efficient simulator of exact and approximate coalescent with selection. Bioinformatics 30, 3427–3429.

Shriner, D., Nickle, D.C., Jensen, M.A., Mullins, J.I., 2003. Potential impact of recombination on sitewise approaches for detecting positive natural selection. Genet Res 81, 115–121.

Sipos, B., Massingham, T., Jordan, G.E., Goldman, N., 2011. PhyloSim — Monte Carlo simulation of sequence evolution in the R statistical computing environment. BMC Bioinform 12, 104.

Slade, P.F., 2000a. Most recent common ancestor probability distributions in gene genealogies under selection. Theor Popul Biol 58, 291–305.

Slade, P.F., 2000b. Simulation of selected genealogies. Theor Popul Biol 57, 35–49.

Slade, P.F., Neuhauser, C., 2003. Nonneutral genealogical structure and algorithmic enhancements of the ancestral selection graph. Comment Theor Biol 8, 255–277.

Slade, P.F., Wakeley, J., 2005. The structured ancestral selection graph and the many-demes limit. Genetics 169, 1117–1131.

Spencer, C.C., Coop, G., 2004. SelSim: A program to simulate population genetic data with natural selection and recombination. Bioinformatics 20, 3673–3675.

Spielman, S.J., Wilke, C.O., 2015. Pyvolve: A flexible python module for simulating sequences along phylogenies. PLOS One 10, e0139047.

Staab, P.R., Zhu, S., Metzler, D., Lunter, G., 2015. SCRM: Efficiently simulating long sequences using the approximated coalescent with recombination. Bioinformatics 31, 1680–1682.

Sun, S., Evans, B.J., Golding, G.B., 2011. "Patchy-tachy" leads to false positives for recombination. Mol Biol Evol 28, 2549–2559.

Sunnaker, M., Busetto, A.G., Numminen, E., et al., 2013. Approximate Bayesian computation. PLOS Comput Biol 9, e1002803.

Teshima, K.M., Innan, H., 2009. mbs: Modifying Hudson's ms software to generate samples of DNA sequences with a biallelic site under selection. BMC Bioinform 10, 166.

Wang, J., Guo, M., Liu, X., et al., 2013. LNETWORK: An efficient and effective method for constructing phylogenetic networks. Bioinformatics 29, 2269–2276.

Wen, D., Yu, Y., Nakhleh, L., 2016. Bayesian inference of reticulate phylogenies under the multispecies network coalescent. PLOS Genet 12, e1006006.

Westesson, O., Holmes, I., 2009. Accurate detection of recombinant breakpoints in whole-genome alignments. PLOS Comput Biol 5, e1000318.

White, D.J., Bryant, D., Gemmell, N.J., 2013. How good are indirect tests at detecting recombination in human mtDNA? G3 (Bethesda) 3, 1095–1104.

Wilson, D.J., Gabriel, E., Leatherbarrow, A.J., et al., 2009. Rapid evolution and the importance of recombination to the gastroenteric pathogen Campylobacter jejuni. Mol Biol Evol 26, 385–397.

Wiuf, C., Posada, D., 2003. A coalescent model of recombination hotspots. Genetics 164, 407–417.

Worobey, M., Holmes, E.C., 1999. Evolutionary aspects of recombination in RNA viruses. J Gen Virol 80, 2535–2543.

Wright, S., 1931. Evolution in Mendelian populations. Genetics 16, 97–159.

Yang, Z., 2006. Computational Molecular Evolution. Oxford, England: Oxford University Press.

Yu, Y., Dong, J., Liu, K.J., Nakhleh, L., 2014. Maximum likelihood inference of reticulate evolutionary histories. Proc Natl Acad Sci USA 111, 16448–16453.

Yu, Y., Nakhleh, L., 2015. A maximum pseudo-likelihood approach for phylogenetic networks. BMC Genomics 16 (Suppl 10), S10.

Zhao, H., Wei, Q., Fu, Y.-X., 2014. Coalescent analysis of modeling mutation process in colorectal cancer. Cancer Res 66, 370.

Further Reading

Arenas, M., 2012. Simulation of molecular data under diverse evolutionary scenarios. PLOS Comput Biol 8, e1002495.

Arenas, M., 2015. Advances in computer simulation of genome evolution: Toward more realistic evolutionary genomics analysis by approximate bayesian computation. J Mol Evol 80, 189–192.

Beaumont, M.A., 2010. Approximate Bayesian computation in evolution and ecology. Annu Rev Ecol Evol Syst 41, 379–405.

Hein, J., Schierup, M., Wiuf, C., 2005. Gene Genealogies, Variation and Evolution: A Primer in Coalescent Theory. Oxford University Press.

Hudson, R.R., 1990. Gene genealogies and the coalescent process. Oxford Surv Evol Biol 7, 1–44.

Hudson, R.R., 1998. Island models and the coalescent process. Mol Ecol 7, 413–418.

Kaplan, N.L., Hudson, R.R., Iizuka, M., 1991. The coalescent process in models with selection, recombination and geographic subdivision. Genet Res Camb 57, 83–91.

Kingman, J.F.C., 1982. The coalescent. Stochas Process Appl 13, 235–248.

Mallo, D., Sánchez-Cobos, A., Arenas, M., 2016. Diverse considerations for successful phylogenetic tree reconstruction: Impacts from model misspecification, recombination, homoplasy, and pattern recognition. In: Elloumi, M., Iliopoulos, C., Wang, J., Zomaya, A. (Eds.), Pattern Recognition in Computational Molecular Biology. John Wiley & Sons, Inc.

Neuhauser, C., Tavaré, S., 2001. The coalescent. Encyclopedia of Genetics. New York: Academic Press.

Nordborg, M., 2007. Coalescent theory. In: Balding, D.J., Bishop, M., Cannings, C. (Eds.), Handbook of Statistical Genetics, third ed. Chichester, UK: John Wiley & Sons, Ltd.

Notohara, M., 1990. The coalescent and the genealogical process in geographically structured population. J Math Biol 29, 59–75.

Peng, B., Kimmel, M., Amos, C.I., 2012. Forward-Time Population Genetics Simulations: Methods, Implementation, and Applications. Hoboken, NJ: John Wiley & Sons.

Yang, Z., 2006. Computational Molecular Evolution. Oxford, England: Oxford University Press.

Relevant Website

http://www.coalescent.dk/
Educational tools for understanding the coalescent simulation.

Biographical Sketch

Miguel Arenas is a principal investigator at the Department of Biochemistry, Genetics and Immunology of the University of Vigo. His research interests include the development of models, methods and frameworks to simulate and analyze the evolution of genetic material. He is also interested in the understanding of the evolution of organisms at both molecular and population levels.

Population Genetics

Conrad J Burden, Australian National University, Canberra, ACT, Australia

Nomenclature

$A1, A2,$	Allele types.
$B(.,.)$	The beta function.
$f(x; p, t)$	Solution of the forward or backward Kolmogorov equation corresponding to the initial condition $X(0)=p$.
h	Degree of dominance.
$I_1(\cdot)$	Modified bessel function of the first kind of order 1.
K	In multi-allelic models, the number of allele types.
m_0	Current population size for the coalescent; Initial population size in the Bienaymé-Galton-Watson model.
M	Haploid population size, or for diploid populations equivalent haploid population size $2N$.
n_{sample}	The size of a randomly chosen sample of individuals in the population.
N	Diploid population size.
p_{ij}	In discrete-time models, Markov transition matrix from $Y(\tau)=i$ to $Y(\tau+1)=j$.
q_{ab}	In the diffusion limit, the rate of mutations from allele A_a to allele A_b.
Q	In the diffusion limit with multiple alleles, the instantaneous rate matrix, whose elements are q_{ab}.
s	Selection coefficient.
S	The proportion of genomic sites which are segregating.
$\mathcal{S}$	The $(K-1)$-dimensional simplex.
t	Continuum time in the diffusion limit.
T_i	The inter-coalescence time.
u_{ab}	In discrete-time models, the probability of allele A_a mutating to allele A_b in one generation.

V_α	The random number of offspring of individual α in the Cannings and Bienaymé-Galton-Watson models.
$X(t)$	In the diffusion limit, relative proportion of A_1 alleles in the population at time t.
$Y(\tau)$	In discrete-time models, the number of individuals in the population of allele-type A_1 at timestep τ.
$\mathbf{Y}(\tau)$	In discrete-time models with multiple alleles, the number of individuals in the population of allele-type $A_1,...,A_K$ at timestep τ.
α	The relative mutation bias in a 2-allele model.
γ	In the diffusion limit, the scaled selection coefficient.
$\Gamma(\cdot)$	The gamma function.
$\delta(\cdot)$	Dirac delta function. See 14.
θ	In the diffusion limit, the total mutation rate.
$\hat{\theta}_W$	The Watterson estimator.
λ	Expected number of offspring per individual in the Bienaymé-Galton-Watson model.
π_i	Probability that allele A_1 fixes given $Y(0)=i$.
σ^2	Variance of the number of offspring per individual in the Bienaymé-Galton-Watson model.
τ	Discrete time steps in Wright-Fisher and Bienaymé-Galton-Watson models.
τ_C	The coalescent time in generations of two randomly chosen individuals in a population.
$\bar{\tau}(i)$	Expected time for allele A_1 to fix conditional on $Y(0)=i$.
$\bar{\tau}^*(i)$	Expected time for allele A_1 to fix conditional on A_1 fixing and $Y(0)=i$.
Ω_Z	The range of a random variable Z.

Introduction

If we were to zoom in on a multiple alignment of one of our chromosomes across the entire human population we might see something like **Fig. 1**. Assuming it is not an X or Y, every individual has two copies of the chromosome. At first sight the sequence appears to be identical across the population. However we would soon notice certain isolated points called *single nucleotide polymorphisms* (SNPs) which have been highlighted by vertical bars in the figure. At most SNPs two possible letters are observed, for instance A and T in the left hand SNP, and C and T in the right hand SNP in **Fig. 1**. The two possible letters are an example of what are referred to as *alleles*. Very occasionally three-allele SNPs occur (Cao *et al.*, 2015), while four-allele SNPs are extremely rare indeed (Phillips *et al.*, 2015). Sequencing of the human genome has revealed in excess of 10 million SNPs (1000 Genomes Project Consortium, 2010; Shen *et al.*, 2013), at an average separation of about 300 bases, or, to put it another way roughly 1 in 300 genomic sites on average is observed to be a SNP. Furthermore, as the number of individuals sequenced increases, so does the reported density of SNPs increase as more sites are revealed to be SNPs. SNPs occur at higher density in non-protein-coding regions than in coding regions. SNPs are the most common type of sequence variation, estimated to account for 90% of all sequence variation.

Before proceeding, a word about terminology is in order. The word allele introduced above is generically used in population genetics to mean any one of a number of alternate forms of a genetic locus. For instance, if a locus contains n SNPs with two

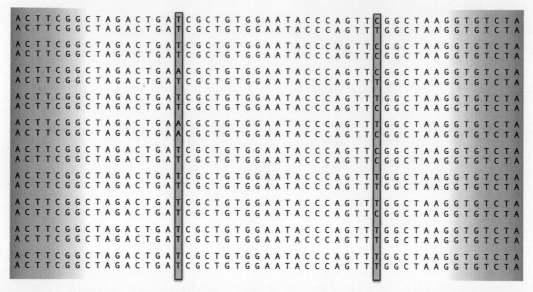

Fig. 1 A small portion of a multiple genomic alignment across the entire population showing a sample of ten individuals. Only one strand of the double helix is shown. The sequences are grouped into pairs to indicate that each individual has two copies of each chromosome. The single nucleotide polymorphisms, or SNPs, are highlighted.

alternate bases observed at each SNP, the 2^n possible patterns may be referred to as 2^n alleles. Many older texts use the word *gene* as a synonym for allele. We avoid this use here to avoid confusion with the more common use of the word in genomics, namely a protein-coding sequence of nucleotides. A more extensive discussion of various subclasses of allele can be found in the text by Gillespie (2004).

In addition to SNPs there are other genomic differences between individuals, such as the presence of copy number variations (CNVs). These occur when an extended region which may be anywhere from a kilobase to several megabases is repeated a number of times, the number of repeats varying from one individual to another. This variation accounts for roughly 5% to 10% of the human genome. Other forms of genetic variation include insertions and deletions of a single letter or of extended regions, and reversals of extended regions. Without genetic variations, people of a given sex would all look very similar to one another; in other words, we would all be clones.

As a general rule, each instance of genetic variation is caused by a copying error or mutation occurring in the germline in a single individual, which is passed on to future generations. Here the word *germline* refers to the lineage of cells which, as cells split, are passed on to the next generation. In the case of sexual reproduction, the germline comprises those sperm or eggs which contribute to the next generation and the lineage of cells from which they are descended, all the way back to the initial cell from which the individual developed, called the zygote. The vast majority of the estimated 10^{13} to 10^{14} cells in a human are of course not part of the germline. Population genetics is concerned with genetic differences across and within populations and the dynamics of how errors occurring in the germlines of single individuals can in time propagate through an entire population to create the phenomenon we know as evolution.

In this article we will concentrate principally on SNPs and ignore other forms of genetic variation. Mathematical modelling of the propagation of SNPs through a population takes into account two processes: *genetic drift* and *selection*. Genetic drift is conceptually the more difficult of the two. It refers to the diffusion process by which an allele already present at a given SNP will become more or less prevalent in the population mainly as a result of variability in the number of offspring of each individual (Masel, 2011). In the next section we will look at the simplest a model of genetic drift, the Wright-Fisher model, which is traditionally used as a starting point from which to develop more detailed and realistic models of population genetics.

Selection refers to the force driving an allele frequency in one direction or the other as a result of the fitness conferred by the mutation from which the SNP arose. Many mutations are harmful and may for instance make it unlikely an individual will survive long enough to produce offspring. For such mutations the frequency of occurrences within the population of the mutant allele will be forced downwards over generations. Occasionally a mutation will occur which confers an advantage which makes the individual more fit to produce offspring, thus driving the allele frequency upwards over generations. Mutations which confer neither an advantage or a disadvantage, such as mutations in a redundant part of the genome, are called *neutral*. Selection is modelled mathematically by augmenting existing models of genetic drift. Mathematical models which take into account only drift and mutations and not selection are referred to as models of neutral evolution.

Naïvely one may think that for evolution to proceed from point mutations in single individuals in a large population, propagation of mutations must be assisted by positive selection: one would expect the probability of a random mutation in a single individual to drift through the entire population must be very small indeed. Surprisingly, we will see that this is not the case:

nucleotide substitutions can occur throughout an entire population through neutral evolution without the aid of positive selection, driven by nothing other than genetic drift.

The Wright-Fisher Model

The simplest and possibly the earliest serious mathematical model of population genetics, known as the Wright-Fisher (WF) model, is attributed to Fisher (1930) and Wright (1931), neither of whom published a clear definition of the model per se, though each did publish a number of consequences of the model. To begin with we will look at the simplest form, namely the two-allele WF model with genetic drift but not mutations or selection. We start with a number of assumptions:

Non-overlapping generations: Assume that evolution occurs over discrete timesteps $\tau = 0, 1, 2,...$, so that the entire population is replaced by a new population each timestep.
Fixed population: Assume the population size N remains fixed, independent of τ.
Diploid population: Assume each individual has two copies of the genome.
Monoecious population: Assume each individual carries both male and female reproductive organs, so that mating can occur between two individuals or within one individual to produce offspring.
Random mating: Assume each individual is the offspring of two randomly and independently chosen parents from the previous generation. Since the population is monoecious, both parents may be the same individual.

This list of assumptions is highly restrictive, but does not rule out all species of life on earth. Annual plants complete their life cycle from germination to production of seed in one year and then die. Many trees, such as oaks, cedars and figs, are monoecious. In particular, sunflowers are both monoecious and annuals, and it would not be totally impossible to contrive a situation in which a sunflower crop's size is held more or less constant. In spite of these restrictions, its mathematical simplicity makes the WF model a very popular starting point for much of the published work in population genetics, even though the assumptions are clearly violated for most populations. Often one will encounter the concept of "effective population size", for which there seems to be no universally accepted unambiguous definition. Loosely speaking, it means the value to which the parameter N in an 'ideal' model with some or all of the above assumptions should be set in order to give correct results, depending on what characteristics one wants to calculate, when applied to a population for which the assumptions may not necessarily be appropriate. See Sections 1.6 and 3.7 of Ewens (2004) for a detailed discussion of this point.

Given the assumptions, it is more convenient mathematically to think in terms of a population of $M = 2N$ copies of the genome, rather than a population of N diploid individuals. Equivalently, one may think of the model as applying to a haploid population with the following assumptions:

Non-overlapping generations: Assume that evolution occurs over discrete timesteps $\tau = 0, 1, 2,...$, so that the entire population is replaced by a new population each timestep.
Fixed population: Assume the population size M remains fixed, independent of τ.
Haploid population: Assume each individual has one copy of the genome.
Random parents: Assume each individual is the offspring of one randomly and independently chosen parent.

Male ants, for instance, are haploid because they develop from unfertilised haploid eggs. Another example is mitochondrial DNA, which is normally only inherited from the mother in sexually reproducing species. The female portion of a population is therefore haploid with respect its mitochondrial genome.

Now consider a given locus within a genome at which there are two distinct alleles, A_1 and A_2. Define the random variable $Y(\tau)$, whose range is $\Omega_{Y(\tau)} = \{0, 1,...,M\}$, to be the number of copies of allele A_1 within the (effective haploid) population of size $M = 2N$. Under the assumption of fixed total population size and an assumption that every individual chromosome at timestep τ is equally likely to be the progenitor of this SNP in a given individual chromosome at timestep $\tau + 1$, the number of A_1 alleles in the population at time $\tau + 1$, given the number of A_1 alleles in the previous generation τ, follows a binomial distribution:

$$\text{Prob } (Y(\tau + 1) = j | Y(\tau) = i) = p_{ij} \qquad (1)$$

where

$$p_{ij} = \binom{M}{j} \left(\frac{i}{M}\right)^j \left(1 - \frac{i}{M}\right)^{M-j}, \quad i, j = 0, 1, ..., M, \qquad (2)$$

since each new offspring inherits either allele A_1 with probability i/M, or allele A_2 with probability $1 - i/M$. Eqs. (1) and (2) define the WF model without mutations or selection, that is, a model in which the only effect driving the dynamics is genetic drift.

This system is an example of a finite-state Markov chain, with states labelled $i = 0, 1, ..., M$ and transition matrix p_{ij}. Eq. (2) implies that

$$p_{0j} = \begin{cases} 1 & \text{if } j = 0 \\ 0 & \text{if } j = 1, ..., M \end{cases}, \qquad p_{Mj} = \begin{cases} 0 & \text{if } j = 0, ..., M - 1 \\ 1 & \text{if } j = M. \end{cases} \qquad (3)$$

Thus $Y(\tau)=0$ and $Y(\tau)=M$ are absorbing states of the Markov chain. These two states correspond respectively to the cases where the entire population has allele A_2 or A_1 at the site in question, and no further change is allowed by the model. If $Y(\tau)=M$ eventuates, we say that A_1 has become *fixed* in the population, and when $Y(\tau)=0$ we say A_2 has become fixed. If $Y(\tau)\in\{1,2,\ldots,M-1\}$, neither allele is yet fixed, and the site in question is said to be a *segregating site*.

Probability of Fixation

So far we have not built into the model any mechanism for a site to become a segregating site in the first place. We will come to that later when we deal with mutations in the section Including Mutations in the Wright-Fisher Model. For the time being, consider the initial condition $Y(0)=i$ for some $i=0,\ldots,M$ and ask the question: What is the probability that allele A_1 becomes fixed? We have

$$\begin{aligned}
&\text{Prob } (A_1 \text{ becomes fixed}|Y(0)=i)\\
&= \text{Prob } (Y(\infty)=M|Y(0)=i)\\
&= \sum_{j=1}^{M} \text{Prob } (Y(\infty)=M|Y(1)=j) \text{ Prob } (Y(1)=j|Y(0)=i).
\end{aligned} \tag{4}$$

We also have boundary conditions

$$\text{Prob}(Y(\infty)=M|Y(0)=0)=0, \qquad \text{Prob}(Y(\infty)=M|Y(0)=M)=1, \tag{5}$$

since 0 and M are absorbing states. Defining

$$\pi_i = \text{Prob } (Y(\infty)=M|Y(\tau)=i), \tag{6}$$

which we note is independent of τ since a Markov chain has no memory of prior events, Eqs. (4) and (5) become

$$\pi_i = \sum_{j=0}^{M} p_{ij}\pi_j, \qquad \pi_0=0, \qquad \pi_M=1. \tag{7}$$

One easily checks that the solution is

$$\text{Prob}(A_1 \text{ becomes fixed}|Y(0)=i)=\pi_i=\frac{i}{M}. \tag{8}$$

Similarly, replacing the boundary conditions with $\pi_0=1$, $\pi_M=0$ gives

$$\text{Prob}(A_2 \text{ becomes fixed}|Y(0)=i)=1-\pi_i=1-\frac{i}{M}, \tag{9}$$

implying that either one allele or the other must become fixed (Or, more precisely, one allele or the other becomes fixed 'almost surely', which is statistical jargon for 'there exist trajectories in state space which never become fixed, but the probability of such a trajectory occurring is zero'.).

To interpret this result, consider a mutation which occurs in one member of the population at $\tau=0$. Intuitively one might expect that, for large populations, it is highly unlikely a mutation could become fixed by pure chance without the aid of natural selection. The above calculation shows that the WF model predicts a $1/M$ probability that a mutation in a single individual will become fixed in the population due to random genetic drift only. For instance, **Fig. 2** shows a numerical simulation of 300 trajectories of the WF model with a population $M=100$, and, as predicted, in roughly 1 in 100 cases the mutation has become fixed (One case has not fixed on either allele yet, but A_1 looks like dying out after coming very close to fixing). On the other hand, the rate at which such mutations occur anywhere in the genome is proportional to the population size M, so we arrive at the surprising result that, even without natural selection, the model predicts that random mutations can fix in a population at a rate which is approximately independent of population size.

Expected Time to Fixation

A second question suggested by **Fig. 2** is: What is the expected time for an allele to fix in the WF model? Consider first the unconditional problem of fixation in either of the two absorbing states. Define $\bar{\tau}(i)$ to be the expected number of time steps before fixation at either 0 or M given the initial conditions $Y(0)=i$. For either of the cases $i=0$ or $i=M$ the answer is trivial, while for any other value of i we can write a difference equation based on a sum of conditional expectation values labelled by the first step,

$$\bar{\tau}(i) = \sum_{j=0}^{M} p_{ij}\bar{\tau}(j) + 1, \quad i=1,\ldots,M-1, \tag{10}$$

$$\bar{\tau}(0) = \bar{\tau}(M) = 0. \tag{11}$$

This equation does not admit a simple analytic solution. However, assuming M is large, we can obtain a good approximation to the solution by taking a continuum limit in time

$$t = \frac{\tau}{M}, \quad \delta t = \frac{1}{M}, \tag{12}$$

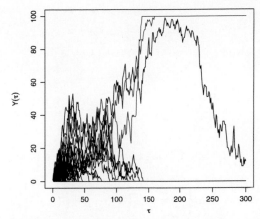

Fig. 2 Simulation of the WF model: 300 independent trajectories for a population of $M=100$ copies of the genome, with a mutant allele introduced into one member of the population at $\tau=0$. Trajectories for which the mutant allele has become fixed are shown in red.

while at the same time defining a rescaled continuous random variable

$$X(t) = \frac{1}{M}Y(\tau), \quad \Omega_{X(t)} = [0, 1], \tag{13}$$

equal to the fraction of the population with allele A_1. In Eq. (10) we make the substitutions

$$x = \frac{i}{M}, \quad u = \frac{j}{M}, \quad \bar{t}(x) = \frac{1}{M}\bar{\tau}(i), \tag{14}$$

and set

$$\delta X = X(\delta t) - X(0) = \frac{1}{M}(Y(1) - Y(0)), \tag{15}$$

to obtain

$$\bar{t}(x) = \int_0^1 \text{Prob}(X(\delta t) = u|X(0) = x)\bar{t}(u)du + \frac{1}{M}$$

$$= \int_0^1 \text{Prob}(X(\delta t) = u|X(0) = x)$$

$$\times \left\{ \bar{t}(x) + (u-x)\bar{t}'(x) + \frac{1}{2}(u-x)^2\bar{t}''(x) + \ldots \right\}du + \delta t$$

$$= \left\{ \bar{t}(x) + E(\delta X|X(0) = x)\bar{t}'(x) + \frac{1}{2}E(\delta X^2|X(0) = x)\bar{t}''(x) + \ldots \right\} + \delta t. \tag{16}$$

Eq. (16) holds for any model scaled according to Eqs. (12) and (13) irrespective of the form of the transition matrix p_{ij}. For the WF model in particular, $Y(1)|(Y(0)=i) \sim \text{Bin}(M, i/M)$, so

$$E(\delta X|X(0) = x) = \frac{1}{M}E(Y(1) - Y(0)|Y(0) = i)$$

$$= \frac{1}{M}\{E(Y(1)|Y(0) = i) - i\}$$

$$= \frac{1}{M}\left(M \times \frac{i}{M} - i\right)$$

$$= 0, \tag{17}$$

and

$$E(\delta X^2|X(0) = x) = \text{Var}(\delta X|X(0) = x) + E(\delta X|X(0) = x)^2$$

$$= \frac{1}{M^2}\text{Var}(Y(1) - Y(0)|Y(0) = i) + 0$$

$$= \frac{1}{M^2}\left\{M \times \frac{i}{M} \times \left(1 - \frac{i}{M}\right)\right\}$$

$$= x(1-x)\delta t. \tag{18}$$

Similarly one can show

$$E(\delta X^k | X(0) = x) = o(\delta t), \qquad \text{for } k = 3, 4, \dots \text{ as } \delta t \to 0. \tag{19}$$

Substituting back into Eq. (16) and tidying up gives the differential equation

$$\frac{1}{2} x(1 - x)\bar{t}''(x) + 1 = 0, \tag{20}$$

and from Eq. (11) we have the boundary conditions

$$\bar{t}(0) = \bar{t}(1) = 0. \tag{21}$$

One easily checks that the solution is

$$\bar{t}(x) = -2\left\{ x \log x + (1 - x)\log(1 - x) \right\} \tag{22}$$

or

$$\bar{\tau}(i) \approx -2M\left\{ \frac{i}{M} \log \frac{i}{M} + \left(1 - \frac{i}{M}\right)\log\left(1 - \frac{i}{M}\right) \right\}. \tag{23}$$

For instance, a SNP with both alleles equally populated could be expected to fix in $\bar{\tau}(M/2) \approx 1.39M$ generations.

A more pertinent question relevant to evolution might be to ask the expected time to fixation for trajectories such as those highlighted in red in **Fig. 2**, where a newly arrived allele A_1 has managed to fix. To answer this question, consider a new random variable

$$Y^*(\tau) = Y(\tau) | (Y(\infty) = M), \qquad \Omega_{Y^*} = 1, \dots, M, \tag{24}$$

equal to the number of individuals in the population with allele A_1, conditioned on the restricted set of trajectories for which A_1 fixes. The corresponding Markov transition matrix is

$$
\begin{aligned}
p^*_{ij} &= \text{Prob}\left(Y(\tau + 1) = j | Y(\tau) = i, Y(\infty) = M\right) \\
&= \frac{\text{Prob}\left(Y(\tau + 1) = j, Y(\infty) = M\right) | Y(\tau) = i)}{\text{Prob}\left(Y(\infty) = M\right) | Y(\tau) = i)} \\
&= \frac{\text{Prob}\left(Y(\tau + 1) = j | Y(\tau) = i\right)\text{Prob}\left(Y(\infty) = M | Y(\tau + 1) = j\right)}{\text{Prob}\left(Y(\infty) = M\right) | Y(\tau) = i)} \\
&= \frac{p_{ij}\pi_j}{\pi_i}, \qquad i, j = 1, \dots, M
\end{aligned} \tag{25}
$$

where, in the third line, we have used the Markovian property that the trajectory $\tau \to \tau + 1$ is independent of the trajectory $\tau + 1 \to \infty$.

For the case of the WF model, it is straightforward to check that Eqs. (2), (8), and (25) imply

$$p^*_{ij} = \binom{M - 1}{j - 1}\left(\frac{i}{M}\right)^{j-1}\left(1 - \frac{i}{M}\right)^{M-j}, \quad i, j = 1, \dots, M, \tag{26}$$

and that consequently the random variable $Y^*(\tau + 1) - 1$ conditioned on $Y^*(\tau) = i$ is binomial:

$$Y^*(\tau + 1) | (Y^*(\tau) = i) \sim \text{Bin}(M - 1, i/M) + 1. \tag{27}$$

Following the same line of reasoning as for the unrestricted case, we find that Eq. (16) also applies to an analogous set of starred variables defined by

$$X^*(t) = \frac{1}{M}Y^*(\tau), \quad \delta X^* = \frac{1}{M}(Y^*(1) - Y^*(0)), \quad \bar{t}^*(x) = \frac{1}{M}\bar{\tau}^*(i). \tag{28}$$

However, calculations analogous to Eqs. (17) and (18) now give

$$E(\delta X^* | X^*(0) = x) = (1 - x)\,\delta t, \tag{29}$$

$$E\left(\delta X^{*2} | X^*(0) = x\right) = x(1 - x)\delta t + O(\delta t^2), \tag{30}$$

leading to the differential equation

$$(1 - x)\bar{t}^{*'}(x) + \tfrac{1}{2}x(1 - x)\bar{t}^{*''}(x) + 1 = 0, \tag{31}$$

for the expected time to fixation conditional on A_1 fixing, $\bar{t}^*(x)$. The solution should also satisfy the boundary condition

$$\bar{t}^*(1) = 0, \tag{32}$$

and the symmetry condition

$$
\begin{aligned}
\bar{t}(x) &= \text{Prob}\left(X(\infty) = 1\right)\bar{t}^*(x) + \text{Prob}\left(X(\infty) = 0\right)\bar{t}^{**}(x) \\
&= x\bar{t}^*(x) + (1-x)\bar{t}^*(1-x),
\end{aligned} \tag{33}
$$

where $\bar{t}(x)$ is given by Eq. (22) and $\bar{t}^{**}(x)$ is the expected time to fixation conditional on A_2 fixing. The required solution is

$$\bar{t}^*(x) = -\frac{2}{x}(1 - x)\log(1 - x), \tag{34}$$

or

$$\bar{\tau}^*(i) = -\frac{2M^2}{i}\left(1 - \frac{i}{M}\right)\log\left(1 - \frac{i}{M}\right). \tag{35}$$

If a mutation is introduced into a single member of a large population at $\tau = 0$ and the mutation fixes, Eq. (35) says that the expected time to fix is approximately $\bar{\tau}^*(1) \approx 2M(1 + O(1/M))$. The red trajectories in **Fig. 2** are broadly in agreement with this result.

Including Mutations in the Wright-Fisher Model

So far mutations have only been included in the model as an ad hoc specification of initial conditions. In this section we introduce mutations dynamically in a way which can be generalised to incorporate instantaneous rate matrices for multiple alleles in the section Multiple Alleles.

Consider again a locus in which only two alleles are present, or equivalently, assume a site in a simplified model genome in which the genomic alphabet only consists of two letters, A_1 and A_2. Returning to the assumptions of the WF model, we add one more assumption:

Random mutation: Following the germline over one generation, a locus occupied by an A_1 allele at the initial time of inheritance has a probability u_{12} of mutating to A_2 before reproductive maturity, and a genomic site initially occupied by an A_2 allele has a probability u_{21} of mutating to A_1 before reproductive maturity.

By 'reproductive maturity' we mean the point at which the genomic information is passed onto the next generation. In principle any values in the range $0 \le u_{12}, u_{21} \le 1$ are allowed by the mathematics, though in practice the mutation rates per generation are typically very small.

Now interpret the random variable $Y(\tau)$ as the number of copies of allele A_1 within the population at birth. Conditioning on the event $Y(\tau) = i$ for some $i = 0,...,M$, the following four events are possible in any one member of the population:

E1: The locus inherits allele A_1 and does not mutate;
E2: The locus inherits allele A_1 and mutates to A_2;
E3: The locus inherits allele A_2 and does not mutate;
E4: The locus inherits allele A_2 and mutates to A_1.

Then the probability the locus will be occupied by allele A_1 at maturity, given that $Y(\tau) = i$ at birth, is

$$\psi(i) = \text{Prob}(E_1) + \text{Prob}(E_4) = \frac{i}{M}(1 - u_{12}) + \left(1 - \frac{i}{M}\right)u_{21} \tag{36}$$

and the probability it will be occupied by allele A_2 is

$$1 - \psi(i) = \text{Prob}(E_2) + \text{Prob}(E_3) = \frac{i}{M}u_{12} + \left(1 - \frac{i}{M}\right)(1 - u_{21}) \tag{37}$$

Again we see that $Y(\tau + 1)|Y(\tau) = i$ is a binomial random variable, but the Markov transition matrix defined by Eq. (1) is modified from Eq. (2) to

$$p_{ij} = \binom{M}{j}\psi(i)^j(1 - \psi(i))^{M-j}, \quad i, j = 0, 1, ..., M, \tag{38}$$

where $\psi(i)$ is defined above.

One can check that if u_{12} and u_{21} are both non-zero, the Markov chain has no absorbing states; if $u_{12} = 0$ and $u_{21} > 0$, $Y(\tau) = M$ is the only absorbing state, implying that A_1 must fix (almost surely); and if $u_{21} = 0$ and $u_{12} > 0$, $Y(\tau) = 0$ is the only absorbing state, implying that A_2 must fix (almost surely).

Fig. 3 shows a simulation of a trajectory of this model with parameter values $M = 100$, $u_{12} = 0.0005$ and $u_{21} = 0.0003$ chosen to demonstrate the behaviour. In most real-world scenarios populations are much higher (in studies of humans effective population sizes of $\sim 10^5$ are often used) and mutation rates much lower (human genomic mutation rates at a neutral site may typically be $\sim 10^{-8}$ per individual per generation). Nevertheless, we note the important features that most of the time a genomic site is not segregating (i.e. $Y(\tau) = 0$ or M), and the transition between the two almost-fixed states $Y(\tau) = 0$ and M occurs over a time roughly the same order of magnitude as the fixation time $\bar{\tau}^*(1) \approx 2M$ obtained for the simpler model of Section Expected Time to Fixation. A similar effect is explained by Vogl and Bergman (2015) in the context of the 2-allele Moran model as a strong dominance of genetic drift over mutations for polymorphic sites.

The rate at which new alleles fix throughout a population is a called the *substitution rate*. For the current WF model it can be estimated as follows. For a given site which is currently non-segregating of type A_1, the number of individuals out of a population of size M who mutate to type A_2 in a given generation is Poisson with parameter $u_{12}M$. Thus the probability that at least one mutation occurs is $1 - e^{-u_{12}M} \approx u_{12}M$ for $u_{12}M << 1$. Thus the expected time between 'attempted' mutations in a population of size

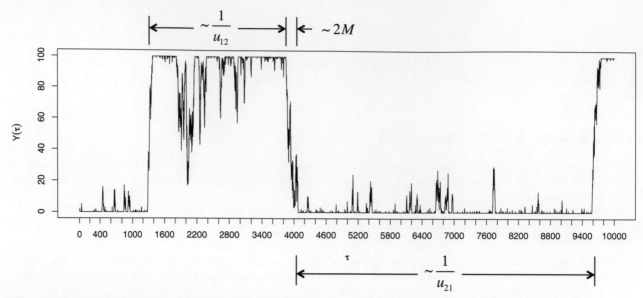

Fig. 3 Simulation of a single trajectory of the WF model with mutations. The parameters are: population of $M=100$ copies of the genome, mutation rate A_1 to A_2 of $u_{12}=0.0005$ per generation and A_2 to A_1 of $u_{21}=0.0003$ per generation. $Y(\tau)$ is the number of A_1 alleles in the population. As described in the text, the expected substitution time before an A_2 allele fixes given that a site is currently fixed at A_1, is approximately $1/u_{12}$; the expected A_2 to A_1 substitution time is approximately $1/u_{21}$, and the expected time the substitution takes to occur is $2M$.

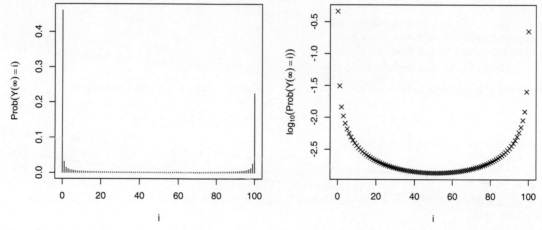

Fig. 4 Stationary distribution of WF model with mutations. Parameters are the same as for **Fig. 3**. The right-hand panel is the same plot on a logarithmic scale.

M is $1/(u_{12}M)$. Of those attempted mutations, Eqs. (8) and (9) tell us that a fraction $1/M$ will fix throughout the population, assuming fixation is driven mainly by drift for low mutation rates. Thus the expected time to the next time allele A_2 fixes throughout the population at the site in question is $M \times 1/(u_{12}M) = 1/u_{12}$, and the population's substitution rate for A_1 to A_2 is u_{12}. Similarly the substitution rate for A_2 to A_1 is u_{21}. In summary, for the WF model with low mutation rates, the population's substitution rate matches the genomic mutation rate. To put it another way, if we consider the genomic alphabet to consist of only 2 letters instead of 4, the instantaneous rate matrix appropriate to, say, maximum likelihood estimates of phylogenetic trees, is

$$U = \begin{pmatrix} -u_{12} & u_{12} \\ u_{21} & -u_{21} \end{pmatrix} \tag{39}$$

provided time from one generation to the next is used as the unit of time.

Another way to interpret the two-allele WF model with mutations is to consider the stationary distribution of the Markov transition matrix Eq. (38), namely $\text{Prob}(Y(\infty)=i)$. The numerically determined distribution is plotted in **Fig. 4** for the same set of parameters as **Fig. 3**. Recall that the model can be thought of as a toy model of a genomic site corresponding to a world in which the genomic alphabet has only two possible nucleotides, A_1 and A_2. If we consider the genome to be a set of independent genomic sites which have been evolving according to the WF model for a considerable time, this distribution can be thought of as the distribution of allele frequencies throughout the genome. Most genomic sites are not segregating, but contribute to the spike in the left hand plot at $i=100$ if the site is occupied by nucleotide A_1 or the spike at $i=0$ if the site is occupied by nucleotide A_2. The remaining vertical bars in the plot correspond to the segregating sites, or SNPs, the height of the bar being the proportion of sites in a large genome at which the allele A_1 occurs in a fraction i/M of the population. This distribution is called the *site frequency spectrum* (SFS). The SFS is important because it is directly observable from sequencing data, and can, in principle, enable estimation of mutation rates under the assumption of a given population genetics model. In a later section we show how to solve for the SFS for WF type models in the continuum limit $M \to \infty$.

Before moving on, we make the observation in passing that the model we have considered is of course a very crude approximation to the observed SFS of any real genome, even if the assumptions of the Wright-Fisher Model hold and we restrict ourselves to parts of the genome undergoing neutral evolution, that is, evolution without selection. One important problem is that genetic drift is an extremely slow process for large populations, calling into question any assumption of stationarity (Gillespie, 2004, pp. 25–26). Another problem is that sites are not independent to the extent that mutation rates are known to depend on neighbouring bases (Aggarwala and Voight, 2016; Zhu *et al.*, 2017). Models which include neighbouring-base context dependence are very difficult to make any progress with analytically, because the neighbouring bases themselves mutate on the same timescale. Consequently, analytic results for population genetics models which include context dependence are almost completely absent from the scientific literature.

Including Selection in the Wright-Fisher Model

Return now to the set of assumptions set out in the section. The Wright-Fisher Model leading to the WF model for a diploid population. Note that each individual in the population can be any one of three *genotypes*: A_1A_1, A_1A_2 and A_2A_2. Suppose we add one extra assumption:

Selection: Assume that the probability an individual survives to maturity and therefore has the potential to procreate is dependent on the genotype of the individual. This probability is known as the *fitness* of the genotype.

A common convention is to specify the ratios of relative fitnesses of the three genotypes, without explicit regard to normalisation. In order to specify three relative fitnesses, two parameters are needed. Here we will adopt the convention used in Equation (1.2b) of Ewens (2004) and Section 5.3 of Etheridge (2011) that the relative fitnesses of the three genotypes are in the proportions

$$1+s : 1+hs : 1 \tag{40}$$

where s is called the *selection coefficient* and h is called the *degree of dominance*: if h is close to 1, A_1 is dominant to A_2 in fitness; if h is close to zero, A_2 is dominant to A_1 in fitness; and if $h = \frac{1}{2}$ there is no dominance in fitness.

If, at the beginning of generation τ, the number of A_1 alleles in a population of size $M = 2N$ is $Y(\tau) = i \in \{0, \ldots M\}$, then the assumption of random mating implies that the relative proportion of new-born genotypes A_1A_1, A_1A_2 and A_2A_2 is

$$i^2 : 2i(M-i) : (M-i)^2. \tag{41}$$

This set of relative proportions is known as *Hardy-Weinberg equilibrium* (HWE) (Hardy, 1908; Stern, 1943). The relative proportion of surviving mature genotypes in the population is then

$$i^2(1+s) : 2i(M-i)(1+hs) : (M-i)^2 \tag{42}$$

The number of A_1 alleles present in the newborn population at time $\tau+1$, given that there were i type-A_1 alleles in the newborn population at time τ, that is $Y(\tau+1)|(Y(\tau)=i)$, will once again be a binomial random variable with Markov transition matrix of the form

$$p_{ij} = \binom{M}{j} \eta(i)^j (1-\eta(i))^{M-j}, \quad i,j = 0,1,\ldots,M. \tag{43}$$

Here $\eta(i)$ is the probability that a given newborn allele will be A_1. Note that random mating of the previous mature generation ensures that the $\eta(i)$ are inherited from genotypes of the previous mature generation who are in the relative ratios

$$\eta(i)^2 : 2\eta(i)(1-\eta(i)) : (1-\eta(i))^2. \tag{44}$$

Comparing Eqs. (42) and (44) we have

$$\eta(i)^2 = \frac{i^2(1+s)}{w}, \qquad 2\eta(i)(1-\eta(i)) = \frac{2i(M-i)(1+hs)}{w}, \tag{45}$$

where $w = i^2(1+s) + 2i(M-i)(1+hs) + (M-i)^2$. Thus

$$\eta(i) = \eta(i)^2 + \eta(i)(1-\eta(i))$$

$$= \frac{i^2(1+s) + i(M-i)(1+hs)}{i^2(1+s) + 2i(M-i)(1+hs) + (M-i)^2}. \tag{46}$$

Thus, Eqs. (43) and (46) define the Markov transition matrix of the 2-allele WF model for a randomly-mating diploid population with selection, but no mutation.

Including selection in the WF model for a haploid population is somewhat simpler. In this case there is no distinction between allele types and genotypes. Without loss of generality we can define relative fitnesses of the two allele types A_1 and A_2 in terms of a single parameter s as

$$1 + \tfrac{1}{2}s : 1. \tag{47}$$

If at time step τ the number of newborn A_1 alleles is $Y(\tau) = i$ and the number of newborn A_2 alleles is $M - i$, then the relative probabilities of an individual inheriting an A_1 or A_2 allele are

$$i(1 + \tfrac{1}{2}s) : M - i \tag{48}$$

implying that the probability of a newborn individual at time step $\tau + 1$ being A_1 is

$$\eta(i) = \frac{i(1 + \tfrac{1}{2}s)}{i(1 + \tfrac{1}{2}s) + M - i}. \tag{49}$$

Eqs. (43) and (49) define the Markov transition matrix of the 2-allele WF model for a haploid population with selection, but no mutation.

Diffusion Limit and the Forward Kolmogorov Equation

Although no analytic solution exists for the SFS shown in **Fig. 4** corresponding to the stationary distribution of the WF model with mutations, Eq. (38), or for the analogous stationary distribution of the WF model with selection, Eq (43), very close approximations can be found in the form of exact stationary distributions of a continuum model constructed by taking the so-called *diffusion limit* $M \to \infty$. This limit involves constructing a partial differential equation known variously as the *forward Komogorov equation* or *Fokker-Planck equation*, which describes the time evolution of a probability density over allele frequencies.

More generally, the Fokker-Planck equation was published by Adriaan Fokker in 1914 and Max Planck in 1917 in the context of studying the velocity distribution of particles undergoing Brownian motion. The forward Kolmogorov equation is equivalent and was published by Andrei Kolmogorov in 1931 in the context of a more general study of the continuous-time diffusion limit of Markov processes. The influence of the diffusion limit and the forward Kolmogorov equation in population genetics is generally attributed to Kimura (1964) who obtained diffusion limit solutions to a number of models including the neutral WF model (Kimura, 1955a) and the WF model with selection (Kimura, 1955b).

Appendix A contains a derivation of the forward Kolmogorov equation for a generic population genetics model based on a discrete time finite state Markov chain. The derivation given is for a genomic site at which two alleles A_1 and A_2 are possible in a population of fixed size M. We assume also that we are given a time-independent Markov transition matrix

$$p_{ij} = \text{Prob } (Y(\tau + 1) = j | Y(\tau) = i), \qquad i, j = 0, \dots M, \tag{50}$$

where $Y(\tau)$ are the number of A_1 alleles present in the population at discrete times $\tau = 0, 1, 2, \dots,$

With the intention of taking the limit $M \to \infty$, consider the substitutions which proved useful above for calculating fixation times:

$$t = \frac{\tau}{M}, \quad \delta t = \frac{1}{M}, \quad X(t) = \frac{1}{M} Y(\tau). \tag{51}$$

After taking the limit $M \to \infty$, t becomes a continuous time variable over the interval $[0, \infty)$ and the fraction $X(t)$ of the population with allele A_1 becomes a continuous random variable with range $\Omega_X = [0, 1]$. The aim of the forward Kolmogorov equation is to describe the evolution of the probability density $f(x; p, t)$ corresponding to $X(t)$, conditional on an initial condition $X(0) = p$, that is,

$$f(x; p, t)dx = \text{Prob } (x \leq X(t) < x + dx | X(0) = p), \qquad 0 \leq x, p \leq 1, \quad t > 0. \tag{52}$$

In order to obtain a sensible continuum limit from Eq. (51), there is an extra requirement that the moments of the random variable $\delta X(t) = X(t + \delta t) - X(t)$ must be of the form

$$E[\delta X(t) | X(t) = u] = a(u)\delta t + o(\delta t)$$

$$E[(\delta X(t))^2 | X(t) = u] = b(u)\delta t + o(\delta t)$$

$$E[(\delta X(t))^k | X(t) = u] = o(\delta t), \ k = 3, 4, \dots, \tag{53}$$

as $\delta t \to 0$ for finite, well behaved functions $a(u)$ and $b(u)$ defined for $0 \leq u \leq 1$. We have already seen this condition satisfied for the WF model by Eqs. (17) to (19) and the WF model conditioned on fixing allele A_1 by Eqs. (29) and (30), and below we will also see it satisfied for WF model with mutations and selection. For some population genetics models such as the Moran model described later in the chapter, the continuum limit is defined differently to Eq. (51), in which case Eq. (53) must be adjusted accordingly.

With this machinery in place it is shown in Appendix A that the forward Kolmogorov equation takes the form

$$\frac{\partial f(x; p, t)}{\partial t} = -\frac{\partial}{\partial x}[a(x)f(x; p, t)] + \frac{1}{2}\frac{\partial^2}{\partial x^2}[b(x)f(x; p, t)]. \tag{54}$$

For any specific population genetics model, the functions $a(x)$ and $b(x)$ are determined from the Markov transition matrix p_{ij}.

The Continuum Wright-Fisher Model With Mutations

We now return to the WF model with 2-way mutations derived earlier. Including Mutations in the Wright-Fisher Model. To determine the functions $a(x)$ and $b(x)$ in the forward Kolmogorov equation, we need to apply the continuum limit defined by Eq. (51). Because the mutation rates u_{12} and u_{21} are defined as probabilities of mutation per generation, and the generation time $\delta t = 1/M$ becomes infinitesimal in the limit $M \to \infty$, rescaled mutation rates with respect to the continuum time t are needed. Accordingly we set

$$q_{12} = Mu_{12}, \qquad q_{21} = Mu_{21}. \tag{55}$$

Note from the argument leading to Eq. (39) that when $q_{12}, q_{21} << 1$, the matrix

$$Q = \begin{pmatrix} -q_{12} & q_{12} \\ q_{21} & -q_{21} \end{pmatrix}, \tag{56}$$

is also the instantaneous rate matrix with respect to the continuum time t for allele substitution throughout the population.

For the transition matrix Eq. (38),

$$E[\delta X(t)|X(t) = x] = \frac{1}{M}E[Y(\tau + 1) - Y(\tau)|Y(\tau) = i]$$

$$= \frac{1}{M}(M\psi(i) - i)$$

$$= -\frac{i}{M}u_{12} + \left(1 - \frac{i}{M}\right)u_{21}$$

$$= [-q_{12}x + q_{21}(1-x)]\delta t. \tag{57}$$

Comparing with Eq. (53) then gives

$$a(x) = -q_{12}x + q_{21}(1 - x), \tag{58}$$

while a similar calculation using the variance gives

$$b(x) = x(1 - x). \tag{59}$$

We are now in a position to determine the stationary distribution of the continuum limit of the WF model with two-way mutations. Substituting $a(x)$ and $b(x)$ into Eq. (54), setting $\partial f/\partial t = 0$ and integrating once gives the ordinary differential equation

$$[q_{12}x - q_{21}(1 - x)]f(x) + \frac{1}{2}\frac{d}{dx}[x(1 - x)f(x)] = \text{const.} \tag{60}$$

It is possible to show that the constant of integration on the right hand side corresponds to a flux of probability across the boundaries at $x=0$ and $x=1$, which must be zero. Then it is straightforward to check that the function

$$f(x) = \frac{1}{B(2q_{21}, 2q_{12})}x^{2q_{21}-1}(1 - x)^{2q_{12}-1} \quad 0 \leq x \leq 1, \tag{61}$$

is a solution to the ordinary differential equation and is thus the required stationary distribution of the forward Kolmogorov equation. Here the prefactor, which is written in terms of the beta function, whose definition is (Abramowitz et al., 1972)

$$B(z_1, z_2) = \int_0^1 x^{z_1-1}(1 - x)^{z_2-1}dx, \quad z_1, z_2 > 0, \tag{62}$$

ensures that the probability density function $f(x)$ has the correct normalisation, i.e., $\int_0^1 f(x)dx = 1$. This distribution is the well known beta distribution, and was first identified as the stationary distribution in the presence of two-way mutations by Wright (1931), without explicit reference to the forward Kolmogorov equation.

Fig. 5 shows the SFS for the example in Fig. 4 reproduced with the stationary solution to the continuum theory, Eq. (61), superimposed. The horizontal and vertical axes have been rescaled by factors of $1/M$ and M respectively in order to make the comparison and the parameters $2q_{12}=0.05$ and $2q_{21}=0.03$ calculated from Eq. (55). To compare with the SFS at $i=0$ and M,

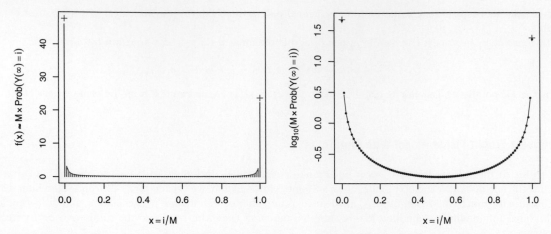

Fig. 5 The SFS computed in **Fig. 4** for a WF model with two-way mutation for a finite population M, rescaled to the continuum parameter $x = i/M$ (black), and the continuum stationary solution Eq. (61) to the forward Kolmogorov equation with corresponding parameters $2q_{12} = 0.05$ and $2q_{21} = 0.03$ (red). The red plus signs at $x = 0$ and 1 are computed from Eqs. (63) and (64).

contributions from the singularities in $f(x)$ at $x = 0$ and $x = 1$ are estimated as

$$\text{Prob}(Y(\infty) = 0) \approx \int_0^{1/M} f(x) dx$$

$$\approx \frac{1}{B(2q_{21}, 2q_{12})} \int_0^{1/M} x^{2q_{21}-1} dx$$

$$= \frac{M^{-2q_{21}}}{2q_{21} B(2q_{21}, 2q_{12})}, \tag{63}$$

and

$$\text{Prob}(Y(\infty) = 1) \approx \int_{1-1/M}^1 f(x) dx$$

$$\approx \frac{1}{B(2q_{21}, 2q_{12})} \int_{1-1/M}^1 (1 - x)^{2q_{12}-1} dx$$

$$= \frac{M^{-2q_{12}}}{2q_{12} B(2q_{21}, 2q_{12})}. \tag{64}$$

Clearly, the agreement in **Fig. 5** is excellent even at this moderate value of M, indicating that the continuum limit $M \to \infty$ is reached rapidly.

In biologically relevant situations it often happens that $q_{12}, q_{21} << 1$. In this case a useful approximation to the stationary distribution is

$$f(x) \approx \frac{2q_{12} q_{21}}{q_{12} + q_{21}} \left(\frac{1}{x} + \frac{1}{1-x} \right), \tag{65}$$

valid for $q_{21} \log x << 1$ and $q_{12} \log(1 - x) << 1$, with the terminal spikes approximated by

$$\text{Prob}(Y(\infty) = 0) \approx \frac{q_{12}}{q_{12} + q_{21}} (1 - 2q_{21} \log M),$$

$$\text{Prob}(Y(\infty) = 1) \approx \frac{q_{21}}{q_{12} + q_{21}} (1 - 2q_{12} \log M). \tag{66}$$

These approximations are a consequence of the property of the beta function that

$$\frac{1}{B(\varepsilon, \eta)} = \frac{\varepsilon \eta}{\varepsilon + \eta} (1 + O(\varepsilon) + O(\eta)) \quad as \ \varepsilon, \eta \to 0+. \tag{67}$$

The function Eq. (65) is easily seen to be a solution to Eq. (60) with q_{12} and q_{21} set to zero. The physical import of this approximation is therefore that the dynamics is driven by drift, and not mutation, over most of the range of x, and that the role of mutations is to set the normalisation of Eq. (60) via the influence of boundary effects at x close to 0 and 1. This interpretation is described in the context of the Moran model with mutations and selection by Vogl and Bergman (2015) who introduce the terminology *boundary-mutation model* to refer to the low mutation rate approximation.

The Continuum Wright-Fisher Model With Mutations and Selection

Analogous to the continuous time mutation rates Eq. (55), we define a continuous-time selection coefficient

$$\gamma = Ms, \tag{68}$$

where s was defined by Eqs. (40) and (47) for diploid and haploid populations respectively. A calculation analogous to Eq. (57) using the transition matrix for WF with selection, Eqs. (43), (46) and (49), yields the first order derivative coefficient

$$a(x) = \begin{cases} \gamma x(1-x)[x + h(1-2x)] & \text{(diploid)}, \\ \dfrac{1}{2}\gamma x(1-x) & \text{(haploid)}, \end{cases} \tag{69}$$

in the forward Kolmogorov equation, while the second order derivative coefficient $b(x)$ in Eq. (59) remains unchanged.

Note that forward Kolmogorov equation for the diploid case of no-dominance, $h = \frac{1}{2}$, is identical to haploid equation. If both mutations and selection are included, the no-dominance diploid or haploid forward Kolmogorov equation is

$$\frac{\partial f(x; p, t)}{\partial t} = -\frac{\partial}{\partial x}\left[\left(\tfrac{1}{2}\gamma x(1-x) - q_{12}x + q_{21}(1-x)f(x; p, t)\right)\right]$$

$$+ \frac{1}{2}\frac{\partial^2}{\partial x^2}[x(1-x)f(x; p, t)]. \tag{70}$$

The stationary solution to this equation is

$$f(x) = Cx^{2q_{21}-1}(1-x)^{2q_{12}-1}e^{\gamma x}, \tag{71}$$

where the constant C is chosen to ensure $\int_0^1 f(x)\, dx = 1$.

A complete time dependent solution to Eq. (70) for the case of selection but no mutation (i.e. $q_{12} = q_{21} = 0$) is derived in Section 8.6 of Crow and Kimura (1970). The solution is highly complex, however it is relatively easy to derive the result that, given an initial relative frequency p, allele A_1 will fix with probability (Waxman, 2011; Ewens, 2004, Sections 4.3 and 5.3)

$$\text{Prob}(X(\infty) = 1 | X(0) = p) = \frac{1 - e^{-\gamma p}}{1 - e^{-\gamma}}. \tag{72}$$

Estimation of Mutation Rates

We return now to the stationary solution to the diffusion limit WF model with neutral mutations, namely Eq. (61), and address the problem of estimating mutation rates from sampled SNP data. In the following it will prove useful to use an alternate parameterisation similar to that used by Vogl (2014), and set

$$\theta = q_{12} + q_{21}, \qquad \alpha = \frac{q_{12}}{q_{12} + q_{21}}. \tag{73}$$

The parameter θ is the scaled total mutation rate and the parameter α is a measure of the relative bias between A_1 to A_2 mutations and A_2 to A_1 mutations. In terms of the newly defined parameters, Eq. (61) becomes

$$f(x) = \frac{1}{B(2(1-\alpha)\theta, 2\alpha\theta)} x^{2(1-\alpha)\theta - 1}(1-x)^{2\alpha\theta - 1}, \quad 0 \le x \le 1. \tag{74}$$

Generally one expects an effective population to be considerably larger than that of the toy examples considered in the numerical simulations Figs. 2 to 5, and if stationarity is assumed it should be possible in principle to estimate the continuum limit parameters such as θ and α from an observed SFS. However it is important to realise that the distribution $f(x)$ of the SFS is not observed directly, simply because a census of the genomic makeup an entire population is not practical. Instead, biologists will sequence the genomes of a sample of the population, which is ideally randomly chosen.

Maximum Likelihood Estimator

Suppose we assume a sample of n_{sample} individual chromosomes across n_{site} independent genomic sites, each assumed chosen to be a site not susceptible to selective pressures. We also assume the genome to have evolved so as to represent the stationary state of the simplified model in which the genome consists only of letters A_1 and A_2 from a 2-letter alphabet, and that all the assumptions leading to the neutral WF model with two-way mutations hold.

In this case the data will consist of a set of i.i.d. random numbers $U_1, U_2, \ldots, U_{n_{\text{site}}}$ of type A_1 alleles, each taking a value in the set $\{0, 1, \ldots, n_{\text{sample}}\}$, observed at the n_{site} genomic sites. At any given site j, conditional on an (unobserved) underlying population fraction x of A_1-type alleles at that site, each U_j will be a binomial random variable:

$$U_j | (X = x) \sim \text{bin}(n_{\text{sample}}, x), \tag{75}$$

where X has the beta distribution, Eq. (74). Thus the unconditional distribution of each U_j is

$$
\begin{aligned}
\mathrm{Prob}(U_j = u) &= \int_0^1 \mathrm{Prob}\,(U_j = u | X = x) f(x)\ dx \\
&= \int_0^1 \binom{n_{\mathrm{sample}}}{u} x^u (1-x)^{n_{\mathrm{sample}}-u} \times \frac{x^{2(1-\alpha)\theta-1}(1-x)^{2\alpha\theta-1}}{B(2(1-\alpha)\theta, 2\alpha\theta)}\ dx \\
&= \binom{n_{\mathrm{sample}}}{u} \frac{1}{B(2(1-\alpha)\theta, 2\alpha\theta)} \\
&\quad \times \int_0^1 x^{u+2(1-\alpha)\theta-1}(1-x)^{n_{\mathrm{sample}}-u+2\alpha\theta-1}\ dx \\
&= \binom{n_{\mathrm{sample}}}{u} \frac{B(u+2(1-\alpha)\theta, n_{\mathrm{sample}}-u+2\alpha\theta)}{B(2(1-\alpha)\theta, 2\alpha\theta)},
\end{aligned}
\tag{76}
$$

for $u = 0, \ldots, n_{\mathrm{sample}}$. Eq. (76) is the probability function of the beta-binomial distribution (Johnson *et al.*, 1993). This is a 3-parameter family of discrete distributions whose generic probability mass function takes the form

$$
P_U(u; n, a, b) = \binom{n}{k} \frac{B(u+a, n-u+b)}{B(a, b)}, \quad u = 0, \ldots, n,
\tag{77}
$$

where n is a non-negative integer and a and b are positive real numbers.

Given the dataset described above, the parameters θ and α can be estimated numerically via a maximum likelihood calculation. In this case one would maximize the log likelihood

$$
L(\theta, \alpha | u_1, \ldots, u_{n_{\mathrm{site}}}) = \sum_{j=1}^{n_{\mathrm{site}}} \log P_U(u_j; n_{\mathrm{sample}}, 2(1-\alpha)\theta, 2\alpha\theta)),
\tag{78}
$$

with respect to θ and α, where $u_1, \ldots, u_{n_{\mathrm{site}}}$ are the observed counts of type-A_1 alleles at each site. Vogl (2014) has developed an expectation-maximisation algorithm for performing this maximisation.

Relationship to the Watterson Estimator

A commonly encountered estimator of the scaled total mutation rate θ, used when $\theta \ll 1$ and $\alpha = \frac{1}{2}$, is the *Watterson estimator* (Watterson, 1975). In the language of our current analysis it states that, if n_{sample} genomes are sampled, and the proportion of sites which are observed to be segregating is S, then

$$
\hat{\theta}_W = \frac{S}{\sum_{i=1}^{n_{\mathrm{sample}}-1}(1/i)},
\tag{79}
$$

is an unbiased estimate of θ. A site is said to be segregating if more than one allele is observed at that site (i.e. it is observed to be a SNP on the basis of a sample of size n_{sample}). An demonstration that the Watterson estimator is unbiased will be given from the point of view of the coalescent in the section The Kingman Coalescent and the Watterson Estimator.

By carrying out an expansion of the beta-binomial distribution in the small parameter θ, Vogl (2014) has obtained maximum likelihood estimators which generalise Waterson's result to the case $\alpha \neq \frac{1}{2}$. Vogl's estimators are

$$
\hat{\vartheta}_V = \frac{S}{\sum_{i=1}^{n_{\mathrm{sample}}-1}(1/i)}, \quad \hat{\alpha}_V = S_2 + \tfrac{1}{2}S,
\tag{80}
$$

where S is the proportion of sites that are segregating and S_2 is the proportion of sites that are non-segregating with the observed allele being A_2. Here $\hat{\vartheta}_V$ is an unbiased, maximum likelihood estimator of the quantity

$$
\vartheta = 4\alpha(1-\alpha)\theta,
\tag{81}
$$

which reduces to θ when $\alpha = \frac{1}{2}$.

Below we give a brief proof that $\hat{\vartheta}_V$ is unbiased. Note first that S is an unbiased estimator of the probability that a site chosen at random from a very large number of independent sites is observed to be segregating. Equivalently, $E[S]$ is the probability that the random variable U_j in the distribution Eq. (76) is neither 1 or n_{sample}. Thus to demonstrate that $\hat{\vartheta}_V$ is indeed an unbiased estimator of ϑ it suffices to show that

$$
1 - \mathrm{Prob}\,(U_j = 0) - \mathrm{Prob}\,(U_j = n_{\mathrm{sample}}) = 4\alpha(1-\alpha)\theta \sum_{i=1}^{n_{\mathrm{sample}}-1} \frac{1}{i}, \quad \text{as } \theta \to 0,
\tag{82}
$$

where U_j has the beta-binomial distribution Eq. (76).

To obtain the small θ approximation, we start with some properties of the beta and related functions (Abramowitz *et al.*, 1972). The beta function can be written in terms of gamma functions as

$$B(z_1, z_2) = \frac{\Gamma(z_1)\Gamma(z_2)}{\Gamma(z_1 + z_2)}. \tag{83}$$

For ε small and $n = 1, 2, \ldots,$

$$\Gamma(\varepsilon) = \frac{1}{\varepsilon} - \gamma + O(\varepsilon), \qquad \Gamma(n + \varepsilon) = (1 + \varepsilon\psi(n))\Gamma(n) + O(\varepsilon^2), \tag{84}$$

where γ is Euler's constant and $\psi(z) = d\log\Gamma(z)/dz = \Gamma'(z)/\Gamma(z)$ is called the digamma function. It satisfies

$$\psi(1) = -\gamma, \qquad \psi(n) = -\gamma + \sum_{i=1}^{n-1} \frac{1}{i}, \quad n = 1, 2, \ldots. \tag{85}$$

Armed with these properties, we have from Eq. (76)

$$\text{Prob}(U_j = 0) = \binom{n_{\text{sample}}}{0} \frac{B(2(1-\alpha)\theta, n_{\text{sample}} + 2\alpha\theta)}{B(2(1-\alpha)\theta, 2\alpha\theta)}$$

$$= \frac{\Gamma(n_{\text{sample}} + 2\alpha\theta)}{\Gamma(n_{\text{sample}} + 2\theta)} \cdot \frac{\Gamma(2\theta)}{\Gamma(2\alpha\theta)}$$

$$= \frac{1 + 2\alpha\theta\psi(n_{\text{sample}})}{1 + 2\theta\psi(n_{\text{sample}})} \cdot \frac{1/(2\theta) - \gamma}{1/(2\alpha\theta) - \gamma} + O(\theta^2)$$

$$= \alpha\left[1 - 2(1-\alpha)\theta\psi(n_{\text{sample}}) - 2(1-\alpha)\theta\gamma\right] + O(\theta^2)$$

$$= \alpha - 2\alpha(1-\alpha)\theta \sum_{i=1}^{n_{\text{sample}}-1} \frac{1}{i} + O(\theta^2). \tag{86}$$

Similarly,

$$\text{Prob}(U_j = n_{\text{sample}}) = 1 - \alpha - 2\alpha(1-\alpha)\theta \sum_{i=1}^{n_{\text{sample}}-1} \frac{1}{i} + O(\theta^2), \tag{87}$$

and Eq. (82) follows.

Multiple Alleles

We now consider extending the number of alleles from two to any number K. For instance, at any genomic site, the possible nucleotides are $\{A,C,G,T\}$, giving $K=4$ possible alleles.

In general we will label the alleles $A_1, \ldots, A_K$, and consider a model with a fixed population size of M haploid or $M=2N$ diploid individuals. Define the random variables $Y_a(\tau)$ to be the number individuals of type A_a in the population at birth at the discrete times $\tau = 0, 1, 2, \ldots$, with the constraint that $\sum_{a=1}^{K} Y_a(\tau) = M$. The transition matrix Eq. (38) of the 2-allele WF model generalises to a matrix of probabilities of transitioning from a set of allele frequencies $\mathbf{i} = (i_1, \ldots, i_K)$ to a set of allele frequencies $\mathbf{j} = (j_1, \ldots, j_K)$ given by the multinomial distribution

$$p_{\mathbf{ij}} = \text{Prob}(\mathbf{Y}(\tau + 1) = \mathbf{j} | \mathbf{Y}(\tau) = \mathbf{i})$$

$$= \begin{cases} \dfrac{M!}{\prod_{a=1}^{K} j_a!} \displaystyle\prod_{a=1}^{K} \psi_a(\mathbf{i})^{j_a} & \text{if } \displaystyle\sum_{a=1}^{K} i_a = \sum_{a=1}^{K} j_a = M, \\ 0 & \text{otherwise} \end{cases} \tag{88}$$

where $\psi_a(\mathbf{i})$ is the probability of an individual inheriting allele type A_a given the previous generation's initial allele frequencies $\mathbf{i}$. This transition matrix defines a finite state Markov chain with a state space of dimension $\binom{M + K - 1}{K - 1}$, corresponding to the sites of a simplicial lattice in K dimensions. An example of the lattice is shown in **Fig. 6** for the case $K=4$ and $M=30$.

Consider now the neutral WF model with K alleles. Suppose that u_{ab}, where $1 \leq a$, $b \leq K$, $u_{ab} \geq 0$ and $\sum_{b=1}^{K} u_{ab} = 1$, is the probability of an individual mutating from allele type A_a to type A_b in a single timestep. Generalising Eqs. (36) and (37) we have

$$\psi_a(\mathbf{i}) = \sum_{b=1}^{K} \frac{i_b}{M} u_{ba}. \tag{89}$$

The diffusion limit is obtained by defining random variables $X_a(t) = Y_a(\tau)/M$ equal to the relative proportion of type-A_a alleles within the population at continuous time $t = \tau/M$. The limit $M \to \infty$ and $u_{ab} \to 0$ for $a \neq b$ is taken in such a way that the $K \times K$

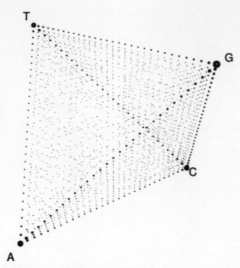

Fig. 6 Stationary distribution of allele frequencies for the multi-allele neutral WF model defined by Eqs. (88) and (89) for a haploid population of size $M=30$ and scaled rate matrix Q (see Eq. (90)) given by Eq. (95). The corners labelled A, C, G and T correspond to allele frequencies $\mathbf{i}=(N,0,0,0)$, $(0,N,0,0)$, $(0,0,N,0)$ and $(0,0,0,N)$ respectively, and the volume of the sphere at each coordinate point is proportional to the probability mass function. Reproduced from Burden, C.J., Tang, Y., 2016. An approximate stationary solution for multi-allele neutral diffusion with low mutation rates. Theoretical Population Biology 112, 22–32.

instantaneous rate matrix Q, whose elements are defined by

$$q_{ab} = M(u_{ab} - \delta_{ab}), \tag{90}$$

remains finite. Here δ_{ab} is the Kronecker delta, equal to 1 if $a=b$ and 0 otherwise. Note that Eq. (90) is the generalisation of Eq. (55), and that Q is an instantaneous rate matrix (Isaev, 2004, Section 5.4) with the properties

$$q_{ab} \geq 0 \text{ for } a \neq b, \qquad \sum_{b=1}^{K} q_{ab} = 0. \tag{91}$$

Using an argument similar to that illustrated by **Fig. 3**, the rate matrix Q can be identified as the whole-population substitution matrix relative to the continuum time t.

The diffusion limit gives the forward Kolmogorov equation

$$\frac{\partial f}{\partial t} = -\sum_{a=1}^{K-1} \frac{\partial}{\partial x_a} \sum_{b=1}^{K} x_b q_{ba} f + \frac{1}{2} \sum_{a,b=1}^{K-1} \frac{\partial^2}{\partial x_a \partial x_b} \{(\delta_{ab} x_a - x_a x_b) f\}, \tag{92}$$

for the density function $f(x_1,\ldots x_{K-1};t)$ of the vector of continuous random variables $X_1(t),\ldots X_{K-1}(t)$. The function f is defined over the simplex

$$S = \left\{ (x_1,\ldots,x_{K-1}) : x_1,\ldots,x_{K-1} \geq 0, \sum_{a=1}^{K-1} x_a \leq 1 \right\}. \tag{93}$$

For notational convenience, we have defined $x_K = 1 - \sum_{a=1}^{K-1} x_a$ in Eq. (92), and also in Eq. (94) below. For details of the derivation of Eq. (92) see Lemma 4.1 of Etheridge (2011) or Eq. (5.125) of Ewens (2004).

Stationary Distribution of the Multi-Allele Neutral WF Model

The stationary distribution $f(x_1,\ldots x_{K-1})$ to the forward Kolmogorov equation is obtained by setting $\partial f / \partial t$ to zero on the left hand side of Eq. (92). The problem of finding an analytic stationary solution for an arbitrary rate matrix Q is as yet unsolved, though an exact analytic solution is known for a special case of parent-independent rate matrices in which the elements of the rate matrix take the form $q_{ab} = q_b$ (independent of a). In this case the solution is the multi-dimensional generalisation of the beta distribution, Eq. (61), namely the Dirichlet distribution (Wright, 1969; Tier and Keller, 1978; Griffiths, 1979):

$$f(x_1,\ldots x_{K-1}) = \frac{\Gamma\left(2\sum_{a=1}^{K} q_a\right)}{\prod_{a=1}^{K} \Gamma(2q_a)} \prod_{a=1}^{K} x_a^{2q_a - 1}. \tag{94}$$

Regrettably, there is no biological reason to consider parent-independent rate matrices, and in fact, most realistic models in phylogenetics are based on rate matrices such as the Hasegawa-Kishino-Yano (HKY) rate matrix (Hasegawa *et al.*, 1985), which are not parent-independent. On the other hand, an accurate approximate solution is known in the biologically relevant limit of low scaled mutation rates, $q_{ab} \ll 1$ where $a \neq b$, for otherwise arbitrary rate matrices (Burden and Tang, 2016).

To understand the idea behind Burden and Tang's solution, consider for instance the numerical stationary solution to the discrete Wright-Fisher defined by Eqs. (88) to (90) shown in **Fig. 6**. For the purposes of illustration the solution has been simulated using an HKY matrix

$$Q = \begin{pmatrix} -0.11 & 0.03 & 0.06 & 0.02 \\ 0.02 & -0.09 & 0.03 & 0.04 \\ 0.04 & 0.03 & -0.09 & 0.02 \\ 0.02 & 0.06 & 0.03 & -0.11 \end{pmatrix} \tag{95}$$

and a small population $M = 30$ in order to render the simulation numerically tractable. The mutation rates are unrealistically high by at least two orders of magnitude to enable the distribution to be visible over the entire simplex on the scale of the plot. Nevertheless, the distribution is dominated by the corners of the tetrahedron, indicating that the majority of genomic sites are not SNPs. Most of the remaining support of the distribution lies on the edges of the tetrahedron, which correspond to 2-allele SNPs. The interiors of the four faces, corresponding to 3-allele SNPs, and the interior volume of the tetrahedron, corresponding to 4-allele SNPs, account for only a small fraction of the total probability. This distribution is consistent with the observed occurrence of SNP multiplicities described at the beginning of the Introduction.

Exploiting this property of the low mutation rate limit, one can argue that the stationary distribution is concentrated close to the edges of the simplex S and can be represented accurately as a set of line densities defined along those edges. Suppose we label the corners of S by the allele prevalent within the population at that corner, so the corner A_a corresponds to the co-ordinate $(x_1, \ldots, x_K) = e_a$, where e_a is the K-dimensional vector with 1 in the ath position and zeroes elsewhere, as illustrated in **Fig. 6** for the 4-letter genomic alphabet. On the edge joining corner A_a to corner A_b a line density $f_{ab}(x)$ is defined for each pair of indices a and b, with the convention adopted that the argument x is the relative proportion of type-a alleles, and $1 - x$ is the relative proportion of type-b alleles. The relative proportion of the remaining $K - 2$ alleles along this edge is taken to be zero.

The line densities $f_{ab}(x)$ are solutions of an effective 2-allele model and satisfy a stationary forward Kolmogorov equation of the form of Eq. (60) with q_{12} and q_{21} replaced by q_{ab} and q_{ba} respectively. However, there is one major difference, namely that the constant on the right hand side can be a non-zero flux of probability Φ_{ab} flowing through the corners of the simplex S. Burden and Tang show that the presence of this flux modifies the approximate solution Eq. (65) to the form

$$f_{ab}(x) \approx C_{ab}\left(\frac{1}{x} + \frac{1}{1-x}\right) - \Phi_{ab}\left(\frac{1}{x} - \frac{1}{1-x}\right), \tag{96}$$

where the constants C_{ab} and Φ_{ab} are determined from rate matrix Q via the formulae

$$C_{ab} = \pi_a q_{ab} + \pi_b q_{ba}, \qquad \Phi_{ab} = \pi_a q_{ab} - \pi_b q_{ba}. \tag{97}$$

Here $\pi^T = (\pi_1 \ldots \pi_K)$ is the left eigenvector of Q satisfying

$$\pi_a \geq 0, \qquad \sum_{a=1}^{K} \pi_a = 1, \qquad \sum_{a=1}^{K} \pi_a q_{ab} = 0. \tag{98}$$

A sufficient condition for a unique π^T to exist is that $q_{ab} > 0$ for all $a \neq b$, which is expected to include any biologically realistic model. For computational reasons it is common in the phylogenetics literature to assume models in which rate matrices are reversible, that is, matrices satisfying the constraint $\pi_a q_{ab} = \pi_b q_{ba}$ (Lanave et al., 1984; Tavare, 1986). We note that the introduction of non-zero values of the fluxes Φ_{ab} is equivalent to invoking non-reversible rate matrices.

The approximation Eq. (96) loses accuracy as the non-integrable singularities at $x = 0$ and 1 are approached, that is, in the vicinity of the corners of S. However, one can show that for a population of size M, the value of the stationary distribution at the corners of the simplex corresponding to the discrete problem defined by Eqs. (88) and (89) is approximately

$$P(Y(\infty) = e_a) \approx \pi_a - \sum_{b \neq a} C_{ab} \log M. \tag{99}$$

The approximate stationary solution defined by Eqs. (96) and (99) reduces to the $K = 2$-allele approximate solution, Eqs. (65) and (66) on observing that, for a 2×2 rate matrix Q,

$$(\pi_1, \pi_2) = \frac{1}{q_{12} + q_{21}}(q_{21}, q_{12}), \quad C_{12} = C_{21} = \frac{2q_{12}q_{21}}{q_{12} + q_{21}}, \quad \Phi_{12} = \Phi_{21} = 0. \tag{100}$$

Fig. 7 shows the numerically determined stationary solution for a population of size $M = 30$ and the rate 4×4 matrix

$$Q^{GRM} = \begin{pmatrix} -5.375 & 3.125 & 2.125 & 0.1250 \\ 0.833 & -17.500 & 0.583 & 16.083 \\ 15.000 & 0.500 & -21.000 & 5.500 \\ 4.600 & 15.900 & 0.100 & -20.600 \end{pmatrix} \times 10^{-4}, \tag{101}$$

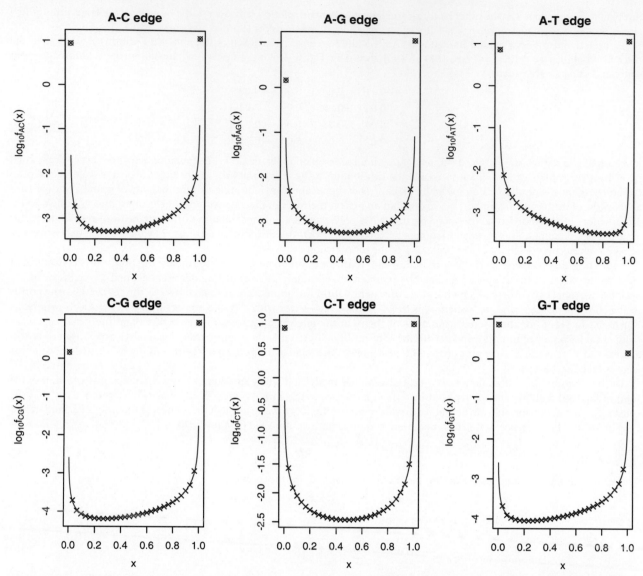

Fig. 7 Simulation of the neutral Wright-Fisher model with $K=4$ alleles and a population size $M=30$ and the rate matrix Q defined by Eq. (101). Blue crosses are the numerically determined stationary distribution along each edge of the simplex over which the site frequency distribution is defined. For any 2 alleles, A_a and A_b, the parameter x is the relative proportion of A_a alleles and $1-x$ is the relative proportion of A_b alleles. Superimposed in red are the theoretical line densities $f_{ab}(x)$, Eq. (96) plotted on a logarithmic scale. The red circles are the theoretical probabilities at the simplex corners, Eq. (99). The probabilities of the stationary distribution and the probabilities in Eq. (99) have been multiplied by a factor M for comparison with the continuum line densities.

plotted along each of the six edges of the simplex S, together with the approximate solution Eqs. (96) and (99). This matrix is taken from Eq. (53) of Burden and Tang (2017), and was chosen reproduce approximately the time-reversible matrix in Table 5 of Lanave *et al.* (1984) estimated from the rat-mouse phylogeny, with an arbitrary non-reversible part added.

Zeng (2010) has exploited the fact that stationary distributions of multi-allelic WF models are concentrated on the corners and edges of the simplex S to demonstrate that it is feasible to estimate all parameters of an evolutionary rate matrix Q from site-frequency data via numerical solution of the multi-allelic discrete WF model. Burden and Tang (2017) have extended the procedure to take advantage of the approximate analytic solution described above. Zeng's approach has the advantage the he is able to estimate selection parameters as well as mutation rates. On the other hand, Burden and Tang's approach has the advantages that the likelihood function takes a relatively simple analytic form entailing efficient numerical calculation for a given observed site-frequency data set, and that it provides a statistical test of the hypothesis, commonly assumed without biological justification, that the rate matrix is reversible.

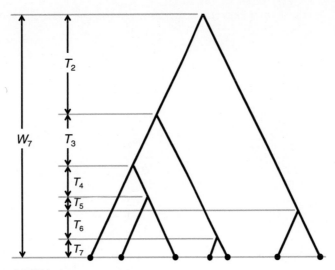

Fig. 8 A coalescent tree of $n_{\text{sample}}=7$ individuals.

The Kingman Coalescent and the Watterson Estimator

The *coalescent* of a sample of individuals in a population is the ancestral lineage of the sample traced backwards in time to their most recent common ancestor. Use of the coalescent as a tool of analysis in population genetics was developed by a number of authors in the 1970s and early 1980s, though the definitive treatment of the subject is due to Kingman (1982, 2000). Here we give a brief description of the coalescent in the context of the haploid Wright-Fisher model, making use primarily of the property that each individual is the offspring of one randomly and independently chosen parent from the previous generation.

Define τ_C to be the number of generations counted backwards in time to the most recent common ancestor of two randomly chosen members of a population of size M. Clearly we have that $\text{Prob}(\tau_C = 1)=1/M$ and $\text{Prob}(\tau_C \neq 1)=1-1/M$. Given that generations are independent, the event "$\tau_C = \tau$" corresponds to $\tau - 1$ generations for which the ancestral line traced backwards does not coalesce, followed by one generation for which the ancestral line does coalesce, thus

$$\text{Prob}\,(\tau_C = \tau) = \left(1 - \frac{1}{M}\right)^{\tau-1} \frac{1}{M}. \tag{102}$$

Now consider the continuum limit $M \to \infty$ defined with a continuous time $t=\tau/M$, $\delta t=1/M$, which proved useful in calculations of the fixation time and for deriving the forward Kolmogorov Equation. Defining the continuous time to coalescence of a sample of size two as $T_2=\tau_C/M$, Eq. (102) implies

$$\text{Prob}\,(t \leq T_2 < t + \delta t) = e^{-t}\delta t + O\!\left(\delta t^2\right). \tag{103}$$

Thus T_2 is an exponential random variable with rate 1. Moreover, we have that the probability that the next coalescent event backwards in time between two given individuals whose ancestry has not coalesced is

$$\text{Prob}\,(t \leq T_2 < t + \delta t | T_2 > t) = \delta t + O\!\left(\delta t^2\right). \tag{104}$$

Next we consider the coalescent of a sample of n_{sample} randomly chosen individuals, as shown in **Fig. 8**. Define T_j to be the time interval between the coalescence event which reduces the ancestry to j individuals, and the coalescence event which reduces the ancestry to $j-1$ individuals. At a point when the ancestry consists of j individuals there are $j(j-1)/2$ possibilities for the next pairwise coalescent event backwards in time. Thus from Eq. (104),

$$\text{Prob}\,(t \leq T_j < t + \delta t | T_j > t) = \frac{j(j-1)}{2}\delta t + O\!\left(\delta t^2\right), \tag{105}$$

which implies that the T_j are independent exponential random variables with respective rates $j(j-1)/2$, and that

$$E(T_j) = \frac{2}{j(j-1)}, \qquad \text{Var}(T_j) = \frac{4}{j^2(j-1)^2}. \tag{106}$$

Set

$$W_{n_{\text{sample}}} = \sum_{j=2}^{n_{\text{sample}}} T_j, \tag{107}$$

to be the time back to the most recent common ancestor of the entire sample. The mean and variance of $W_{n_{\text{sample}}}$ are

$$E\left(W_{n_{\text{sample}}}\right) = 2 \sum_{j=2}^{n_{\text{sample}}} \frac{1}{j(j-1)}$$

$$= 2 \sum_{j=2}^{n_{\text{sample}}} \left(\frac{1}{j-1} - \frac{1}{j}\right)$$

$$= 2\left(1 - \frac{1}{n_{\text{sample}}}\right)$$

$$= 2 + O\left(n_{\text{sample}}^{-1}\right), \quad \text{as } n_{\text{sample}} \to \infty, \tag{108}$$

and

$$\text{Var}\left(W_{n_{\text{sample}}}\right) = 4 \sum_{j=2}^{n_{\text{sample}}} \left(\frac{1}{j-1} - \frac{1}{j}\right)^2$$

$$= 8 \sum_{j=1}^{n_{\text{sample}}-1} \frac{1}{j^2} - 12 + \frac{8}{n_{\text{sample}}} + \frac{4}{n_{\text{sample}}^2}$$

$$= \frac{4\pi^2}{3} - 12 + O\left(n_{\text{sample}}^{-1}\right), \quad \text{as } n_{\text{sample}} \to \infty. \tag{109}$$

Note that half of the contribution to $E\left(W_{n_{\text{sample}}}\right)$ comes from the time taken for the coalescence of the earliest two ancestors, namely T_2. To put it another way, the longest arms of the tree tend to be those for which there is a small number of individuals in the ancestry. Furthermore, as $n_{\text{sample}} \to \infty$ the expected coalescence time of the entire population is $2M(1 + O(1/M))$ generations, which is consistent with the time $\bar{\tau}^*(1)$ obtained in the section Expected time to Fixation for a mutation in a single individual to fix throughout the entire population.

Recall the Watterson estimator $\hat{\theta}_W$, Eq. (79), for the total mutation rate θ in neutral evolution, calculated in terms of the proportion S of segregating sites in the genomes of a sample of individuals. The total length of the branches of the coalescent tree can be used to show that $\hat{\theta}_W$ is an unbiased estimator of θ as follows. From **Fig. 8**, the total tree length is

$$L = \sum_{j=2}^{n_{\text{sample}}} jT_j. \tag{110}$$

If mutations are rare over the timescale of the figure and $\frac{1}{2}\theta$ is the average mutation rate per site (see Eq. (73)), then the proportion of segregating sites, conditional on the total length L of the tree, is a Poisson random variable with mean $\frac{1}{2}\theta L$. Then using the law of total expectation,

$$E(S) = E(E(S|L))$$

$$= \tfrac{1}{2}\theta E(L)$$

$$= \tfrac{1}{2}\theta \sum_{j=2}^{n_{\text{sample}}} jE(T_j)$$

$$= \theta \sum_{j=1}^{n_{\text{sample}}-1} \frac{1}{j}. \tag{111}$$

It is then immediately obvious from Eq. (79) that $E(\hat{\theta}_W) = \theta$ and so $\hat{\theta}_W$ is an unbiased estimate of θ.

The Coalescent for Varying Populations

Slatkin and Hudson (1991) have generalised the coalescent for $n_{\text{sample}} = 2$ individuals to the case of time-varying population sizes. Suppose we have a deterministically imposed varying population of size $M(\tau)$ where τ is the number of generations counted backwards in time from the present population, $M(0) = m_0$. Eq. (102) generalises to

$$\text{Prob}(\tau_C = \tau) = \prod_{\ell=1}^{\tau-1} \left(1 - \frac{1}{M(\ell)}\right) \frac{1}{M(\tau)}. \tag{112}$$

We define a continuous-time limit by setting

$$t = \frac{\tau}{m_0}, \quad \delta t = \frac{1}{m_0}, \quad T_2 = \frac{\tau_C}{m_0}, \quad v(t) = \frac{M(\tau)}{m_0}, \tag{113}$$

and taking the limit $m_0 \to \infty$. Rearranging Eq. (112) we have

$$\log(M(\tau)\mathrm{Prob}(\tau_c = \tau)) = \sum_{\ell=1}^{\tau-1} \log\left(1 - \frac{1}{M(\ell)}\right)$$

$$= -\sum_{\ell=1}^{\tau-1}\left[\frac{1}{m_0 v(m_0 \ell)} + O\left(\frac{1}{m_0^2}\right)\right]. \tag{114}$$

Thus

$$\log\left(\frac{v(t)}{\delta t}\mathrm{Prob}\left(t \le T_2 < t + \delta t\right)\right) = -\int_0^t \frac{du}{v(u)} + O(\delta t), \tag{115}$$

which rearranges to

$$\mathrm{Prob}(t \le T_2 < t + \delta t) = \frac{1}{v(t)}\exp\left(-\int_0^t \frac{du}{v(u)}\right)\delta t + O(\delta t^2). \tag{116}$$

The density function for the coalescent time T_2 is then

$$f_{T_2}(t) = \frac{1}{v(t)}\exp\left(-\int_0^t \frac{du}{v(u)}\right). \tag{117}$$

For the case of exponential population growth,

$$v(t) = e^{-\alpha t}, \tag{118}$$

one easily obtains

$$f_{T_2}(t) = e^{\alpha t}\exp\left(\frac{1 - e^{\alpha t}}{\alpha}\right), \qquad 0 \le t < \infty. \tag{119}$$

The mean time to coalescence is

$$E(T_2) = \frac{e^{1/\alpha}}{\alpha}E_1\left(\frac{1}{\alpha}\right), \tag{120}$$

where $E_1(z) = \int_z^\infty e^{-t}/t\, dt$ is the exponential integral (Abramowitz et al.,1972, Eq. (5.5.1)). Slatkin and Hudson (1991) note that for $\alpha \gg 1$, the distribution is narrow and concentrated around a mean $E(T_2) \approx (\log\alpha - \gamma)e^{1/\alpha}/\alpha$ where $\gamma \approx 0.577$ is the Euler-Mascheroni constant.

Note that for a time-varying population the problem of determining the coalescent for $n_{\mathrm{sample}} \ge 3$ individuals is nontrivial as the inter-coalescence times T_j illustrated in **Fig. 8** are not mutually independent: There is no simple analog of Eq. (105) as restarting the clock at the last coalescence event requires knowing the population size at that event, which in turn depends on $T_{j+1} + \cdots + T_{n_{\mathrm{sample}}}$. For further analysis of this problem, including an explicit formula for the joint distribution of $T_2, \ldots, T_{n_{\mathrm{sample}}}$ see Griffiths and Tavare (1994) and Chen and Chen (2013).

Alternatives to Wright-Fisher

So far we have concentrated only on the Wright-Fisher model. Below is a survey of some of the more important alternative models of population genetics.

The Moran Model

In common with the WF model, the *Moran model* (Moran, 1958) assumes a fixed population size and discrete time steps $\tau = 0,1,2,\ldots$. The main difference, however, is that the in the Moran model generations are allowed to overlap.

Consider the simplest form of the model, namely a haploid population of size M, each individual carrying one of two alleles, A_1 and A_2, which are inherited without mutations. At each time step one randomly chosen member of the population reproduces, with the offspring carrying the same allele as the parent, and one randomly chosen individual, possibly the same individual, dies. With the convention that the number of individuals carrying allele A_1 at time step τ is $Y(\tau)$ the model defines a finite state Markov chain with transition probabilities

$$
\begin{aligned}
p_{ij} &= \mathrm{Prob}\left(Y(\tau+1) = j | Y(\tau) = i\right) \\
&= \begin{cases} \dfrac{i(M-i)}{M^2} & \text{if } j = i \pm 1, \\[2mm] \dfrac{i^2 + (M-i)^2}{M^2} & \text{if } j = i, \\[2mm] 0 & \text{otherwise} \end{cases}
\end{aligned} \tag{121}
$$

for $i, j = 0, \ldots, M$. As for the WF model, it is easy to check that Eq. (3) holds, so that $Y(\tau) = 0$ and $Y(\tau) = M$ are absorbing states of the Markov chain. Also Eqs. (8) and (9) hold, and so if $Y(0) = i$, either A_1 or A_2 must fix almost surely, with probabilities i/M and $1 - i/M$ respectively.

The most straightforward way to include mutations dynamically in Eq. (121) is through the *decoupled Moran model*, which was introduced by Baake and Bialowons (2008) and Etheridge and Griffiths (2009) for the more general K-allele case. For 2 alleles the model is defined as

$$p_{ij} = \begin{cases} \dfrac{i(M-i)}{M^2} + u_{12}\dfrac{i}{M} & \text{if } j = i - 1, \\[2mm] \dfrac{i^2 + (M-i)^2}{M^2} - u_{12}\dfrac{i}{M} - u_{21}\dfrac{M-i}{M} & \text{if } j = i, \\[2mm] \dfrac{i(M-i)}{M^2} + u_{21}\dfrac{M-i}{M} & \text{if } j = i + 1, \\[2mm] 0 & \text{otherwise} \end{cases} \tag{122}$$

where u_{12} and u_{21} are mutation rates per unit of the time-step τ.

To obtain a sensible diffusion limit it is necessary to use a different scaling for the continuum time axis from that used for the WF model. It turns out that, if we define

$$t = \frac{2\tau}{M^2}, \quad \delta t = \frac{2}{M^2}, \quad X(t) = \frac{1}{M}Y(\tau), \quad q_{12} = \tfrac{1}{2}Mu_{12}, \quad q_{21} = \tfrac{1}{2}Mu_{21}. \tag{123}$$

then we again arrive at Eqs. (58) and (59), and hence recover identical forward and backward Kolmogorov equations as those for the WF model with mutations.

The Cannings Model

The WF and Moran models are particular cases of a class of models known collectively as the *Cannings model* (Cannings, 1974). Again consider a population of fixed size M and discrete generations $\tau = 0, 1, 2, \ldots$. Define a random vector

$$\mathbf{V}(\tau) = (V_1(\tau), \ldots, V_M(\tau)), \tag{124}$$

equal to the (non-negative integer) number of offspring produced by the M individuals at time τ. The $\mathbf{V}(\tau)$ are assumed to be i.i.d. for different τ, and to share their distribution with a generic random variable V satisfying firstly that

$$\sum_{\alpha=1}^{M} V_\alpha = M, \tag{125}$$

and secondly that

$$\text{Prob } (V_1 = v_1, \ldots, V_M = v_M) = \text{Prob } (V_1 = v_{\pi(1)}, \ldots, V_M = v_{\pi(M)}), \tag{126}$$

where $\pi(\cdot)$ is any permutation of the indices $1, \ldots, M$. The second requirement is known as the exchangeability property, and the model is sometimes referred to as the Cannings exchangeable model. Note that Eqs. (125) and (126) imply that $E(V_\alpha) = 1$.

Suppose now that the population partitions into two alleles types, A_1 and A_2, which are inherited without mutation. Let $Y(\tau)$ be the number of A_1 alleles and $M - Y(\tau)$ be the number of A_2 alleles present at time τ, so by appropriate labelling of the indices we can write

$$Y(\tau + 1) = \sum_{\alpha=1}^{Y(\tau)} V_\alpha(\tau). \tag{127}$$

The 2-allele WF model without mutations is the particular case

$$\mathbf{V}^{\text{WF}} \sim \text{multinomial}\left(M; \frac{1}{M}, \ldots, \frac{1}{M}\right), \tag{128}$$

and the Moran model is the particular case

$$\mathbf{V}^{\text{Moran}} \sim \text{uniform}\{(k_1, \ldots, k_M) : (k_{\pi(1)}, \ldots, k_{\pi(M)}) = (2, 0, 1, \ldots, 1)\}. \tag{129}$$

Other specific cases are a trivial case in which each individual has exactly 1 offspring, for which neither allele ever fixes unless $Y(0) = 0$ or M, and another trivial case in which one randomly chosen individual has all M offspring, in which the allele of the chosen individual fixes in 1 generation. In general, a Cannings model can be extended to include multiple alleles and may include neutral mutations, but not selection, as that would violate exchangeability.

The coalescent generalises easily to any Cannings model (Etheridge, 2011, Section 2.2). Recall that in deriving the coalescent for a WF model we defined τ_C to be the number of generations counted backwards in time to the most recent common ancestor of two randomly chosen individuals within the population. The probability that the two chosen individuals are siblings from the same parent α is $v_\alpha(v_\alpha - 1)/(M(M-1))$, where v_α is the number of children produced by α. Defining c_M to be the probability that τ_C

is one generation, it follows that

$$c_M = \text{Prob}\,(\tau_C = 1)$$

$$= \sum_{v_1,\ldots,v_M} \text{Prob}(\tau_C = 1 | \mathbf{V}(-1) = (v_1, \ldots, v_M)) \times \text{Prob}(\mathbf{V}(-1) = (v_1, \ldots, v_M))$$

$$= \sum_{v_1,\ldots,v_M} \sum_{\alpha=1}^{M} \frac{v_\alpha(v_\alpha - 1)}{M(M-1)} \,\text{Prob}\,(\mathbf{V}(-1) = (v_1, \ldots, v_M))$$

$$= \frac{1}{M(M-1)} \sum_{\alpha=1}^{M} E[V_\alpha(V_\alpha - 1)]$$

$$= \frac{1}{M-1} E[V_1(V_1 - 1)]$$

$$= \frac{1}{M-1} \,\text{Var}\,(V_1), \tag{130}$$

where we have used the exchangeability property in the second last line and the fact that $E(V_1) = 1$ in the last line. Eq. (102) then generalises to

$$\text{Prob}(\tau_C = \tau) = (1 - c_M)^{\tau-1} c_M. \tag{131}$$

The transition to a continuum time limit depends on the scaling behaviour of $\text{Var}(V_1)$ as $M \to \infty$. For instance, Eq. (51) defines an appropriate time scale for WF, while Eq. (123) defines an appropriate time scale for the Moran model. Once an appropriate coalescence time scale is established, the results in the section The Kingman Coalescent and the Watterson Estimator will in general carry through with appropriate modifications.

Branching Models

Branching models have their origins in the unpublished work of Bienaymé dating from 1845 (see Heyde and Seneta, 1972), and the work of Watson and Galton (1875) on the fate of patrilineally inherited family surnames. The Bienaymé-Galton-Watson (BGW) branching model differs from Cannings models in that the total population size is not fixed, or even specified deterministically, but varies stochastically.

Consider a population of $M(\tau)$ haploid individuals which are assumed to reproduce in discrete, non-overlapping generations $\tau = 0,1,2,\ldots$. The population is partitioned into K allele types such that the number of copies of type k within the population is $Y_k(\tau)$. Thus $M(\tau) = \sum_{k=1}^{K} Y_i(\tau)$. In the simplest version of the model described here the alleles are assumed to be neutral with respect to selection and no mutation between allele types occurs. The central tenet of the BGW model is an assumption that the (non-negative integer) number of offspring per individual in any generation is given by a set of i.i.d. random variables $V_\alpha^{(k)}$, $\alpha = 1,\ldots,Y_k(\tau)$, whose common distribution is denoted by a generic non-negative integer valued random variable V with mean and variance

$$E(V) = \lambda, \quad \text{Var}(V) = \sigma^2, \tag{132}$$

and finite moments to all higher orders. Thus

$$Y_k(\tau + 1) = \sum_{\alpha=1}^{Y_k(\tau)} V_\alpha^{(k)}, \tag{133}$$

and if the $Y_k(0)$ are mutually independent, then $Y_k(\tau)$ are mutually independent at all subsequent times τ. The standard formula for the mean of the sum of a random number of i.i.d. random variables (Ewens and Grant, 2005, p90) gives $E(Y_k(\tau + 1)) = \lambda E(Y_k(\tau))$. Given initial conditions

$$Y_k(0) = i_k, \qquad M(0) = m_0 = \sum_{k=0}^{K} i_k, \tag{134}$$

it follows that

$$E(Y_k(\tau)) = i_k \lambda^\tau, \qquad E(M(\tau)) = m_0 \lambda^\tau. \tag{135}$$

Consider now the fate of any one allele, the number of copies of which we will denote generically as $Y(\tau)$. If the probability generating functions of V and $Y(\tau)$ are respectively

$$G_V(\xi) = E(\xi^V), \qquad G_{Y(\tau)}(\xi) = E\left(\xi^{Y(\tau)}\right), \tag{136}$$

then the formula for the probability generating function for the sum of a random number of i.i.d. random variables (Johnson et al., 1993, p344) gives

$$G_{Y(\tau+1)}(\xi) = G_{Y(\tau)}(G_V(\xi)). \tag{137}$$

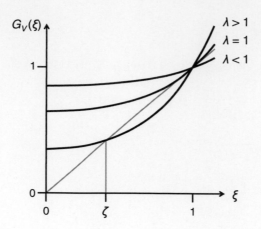

Fig. 9 The stable fixed point of $\zeta = G_V(\zeta)$.

Suppose we start with an initial population of just one individual, in which case $Y(0) = 1$ and $G_{Y(0)}(\xi) = \xi$. Applying the above iterative formula gives

$$
\begin{aligned}
G_{Y(1)}(\xi) &= G_{Y(0)}(G_V(\xi)) = G_V(\xi), \\
G_{Y(2)}(\xi) &= G_{Y(1)}(G_V(\xi)) = G_V(G_V(\xi)), \\
&\vdots \\
G_{Y(\tau)}(\xi) &= \underbrace{G_V(G_V(...G_V(\xi)...))}_{\tau \text{ times}} = G_V(G_{Y(\tau-1)}(\xi)).
\end{aligned}
\tag{138}
$$

Also, since

$$
G_{Y(\tau)}(\xi) = \sum_{y=0}^{\infty} \xi^y \, \mathrm{Prob}(Y(\tau) = y) = \mathrm{Prob}(Y(\tau) = 0) + O(\xi),
\tag{139}
$$

the probability of extinction at or before generation τ is

$$
\mathrm{Prob}\,(Y(\tau) = 0) = G_{Y(\tau)}(0).
\tag{140}
$$

Thus the probability of eventual extinction, $\zeta = \mathrm{Prob}(Y(\infty) = 0) = G_{Y(\infty)}(0)$, is, by Eq. (138), the stable fixed point of the iterative equation

$$
\zeta = G_V(\zeta).
\tag{141}
$$

Noting that

$$
G_V(1) = 1, \quad G_V'(1) = \lambda, \quad \text{and} \quad G_V''(1) = \sigma^2 + \lambda^2 - \lambda > 0,
\tag{142}
$$

since $\lambda^2 - \lambda = E(V(V-1)) \geq 0$ if V only takes values $0,1,2,...$, the graph of $G_V(\xi)$ passes through $(1,1)$ with slope λ and is convex, as shown in **Fig. 9**. It follows that

$$
\zeta \begin{cases} <1 & \text{if } \lambda > 1, \\ =1 & \text{if } \lambda \leq 1. \end{cases}
\tag{143}
$$

Moreover, since lineages are independent, if the initial abundance of an allele is $Y(0) = i$, then the probability of eventual extinction is

$$
\mathrm{Prob}(Y(\infty) = 0 | Y(0) = i) = \begin{cases} \zeta^i & \text{if } \lambda > 1, \\ 1 & \text{if } \lambda \leq 1. \end{cases}
\tag{144}
$$

The upshot of this is that each allele will almost surely eventually become extinct in the case of sub-critical growth, $\lambda < 1$, or critical growth, $\lambda = 1$, and that there is a finite probability $1 - \zeta^i$ that an allele will never become extinct in the case of supercritical growth, $\lambda > 1$. Furthermore, in the presence of 2 alleles with initial abundances i_1 and i_2 and supercritical growth, there is a probability $(1 - \zeta^{i_1})(1 - \zeta^{i_2})$ that the population will persist indefinitely with neither allele ever fixing throughout the population.

The Continuum Bienaymé-Galton-Watson Branching Model

The diffusion approximation of a BGW branching process via a forward Kolmogorov diffusion equation has been analysed in detail by Feller (1951a) and is summarised in the texts by Bailey (1964) and Cox and Miller (1978). For an analysis in the context of genetic drift, see Burden and Simon (2016).

For each allele type A_k the continuum limit is obtained by defining

$$t = \frac{\tau}{m_0}, \quad \delta t = \frac{1}{m_0}, \quad X_k(t) = \frac{Y_k(\tau)}{m_0}, \quad x_{k0} = \frac{i_k}{m_0}, \tag{145}$$

where m_0 and i_k are the initial conditions specified in Eq. (134). The simultaneous limits $m_0 \to \infty$, $\lambda \to 1$ are then taken in such a way that

$$\alpha = m_0 \log \lambda, \tag{146}$$

and σ^2 remain fixed. Setting $\delta X_k(t) = (Y_k(\tau+1) - Y_k(\tau))/m_0$, one obtains

$$E(\delta X_k(t)|X_k(t) = x) = \frac{1}{m_0}(\lambda - 1)x m_0$$

$$= \alpha x \, \delta t + o(\delta t), \tag{147}$$

and

$$E(\delta X_k(t)^2|X_k(t) = x) = \text{Var}\,(\delta X_k(t)|X_k(t) = x) + E(\delta X_k(t)|X_k(t) = x)^2$$

$$= \frac{1}{m_0^2}\sigma^2 x m_0 + O\left(\frac{1}{m_0^2}\right)$$

$$= \sigma^2 x \, \delta t + o(\delta t). \tag{148}$$

Comparing with the general form Eqs. (53) and (54) yields the forward Kolmogorov equation for the density function $f_{X_k}(x,t)$,

$$\frac{\partial f_{X_k}}{\partial t} = -\alpha \frac{\partial}{\partial x}(x f_{X_k}) + \frac{1}{2}\sigma^2 \frac{\partial^2}{\partial x^2}(x f_{X_k}). \tag{149}$$

The process defined by this forward Kolmogorov equation is often referred to as a *Feller diffusion*.

The solution obtained by Feller (1951a,b) corresponding to the initial condition $f_{X_k}(x,0) = \delta(x - x_{k0})$ is

$$f_{X_k}(x,t) = \delta(x)p_0(t)$$

$$+\tilde{\kappa}(t)\left(\frac{x_{k0}e^{-\alpha\tau}}{x}\right)^{\frac{1}{2}}\exp\{-\tilde{\kappa}(t)(x_{k0} + xe^{-\alpha\tau})\}I_1\left(2\tilde{\kappa}(t)(x_{k0}xe^{-\alpha\tau})^{\frac{1}{2}}\right) \tag{150}$$

where $\delta(x)$ is the Dirac delta function defined in Appendix A, $I_1(\,\cdot\,)$ is the modified Bessel function of order 1,

$$\tilde{\kappa}(\tau) = \begin{cases} \dfrac{2\alpha}{\sigma^2(1 - e^{-\alpha t})} & \text{if } \alpha \neq 0 \\[2mm] \dfrac{2}{\sigma^2 t} & \text{if } \alpha = 0, \end{cases} \tag{151}$$

and

$$p_0(t) = e^{-\tilde{\kappa}x_{k0}}. \tag{152}$$

An outline of Feller's derivation of this solution using the Laplace transform of $f_{X_k}(x,t)$ is given on pages 235 and 250 of Cox and Miller (1978).

The delta-function term represents a point mass at zero indicating the finite probability $p_0(t)$ that the allele becomes extinct at or before time t. The probability of eventual extinction is

$$\lim_{t \to \infty} p_0(t) = \begin{cases} 1 & \text{if } \alpha \leq 0, \\ e^{-2\alpha x_{k0}/\sigma^2} & \text{if } \alpha > 0. \end{cases} \tag{153}$$

It can be checked that this is consistent with the corresponding probability of extinction for the discrete case as follows. From Eqs. (141) and (142),

$$\zeta = 1 + \lambda(\zeta - 1) + \tfrac{1}{2}(\sigma^2 + \lambda^2 - \lambda)(\zeta - 1)^2 + O((\zeta - 1)^3). \tag{154}$$

Solving for the stable root shown in **Fig. 9**, and noting that ζ is close to 1 for λ close to 1, we have either $\zeta = 1$ if $\lambda \leq 1$, or

$$\zeta = 1 + \frac{2(1 - \lambda)}{\sigma^2} + O((\lambda - 1)^2)$$

$$= 1 - \frac{2\log\lambda}{\sigma^2} + O((\log\lambda)^2)$$

$$= 1 - \frac{2\alpha}{m_0\sigma^2} + O\left(\frac{1}{m_0^2}\right), \tag{155}$$

if $\lambda > 1$. In the continuum limit, and noting that $i_k = m_0 x_{k0}$, Eq. (144) gives the probability of extinction for the $\lambda > 1$ case as

$$\lim_{m_0 \to \infty} \left(1 - \frac{2\alpha}{m_0 \sigma^2}\right)^{m_0 x_{k0}} = e^{-2\alpha x_{k0}/\sigma^2}, \tag{156}$$

agreeing with the $\alpha > 0$ case of Eq. (153). The $\alpha \leq 0$ case clearly corresponds to the $\lambda \leq 1$ stable solution.

For the critical value $\alpha = 0$ and $K = 2$ alleles, Burden and Simon, (2016) have shown that the diffusion limit of the BGW model has similar, but not identical, distributions of fixation times to the neutral WF and Cannings models without mutations. More specifically, they demonstrate that, given an initial population in which a fraction x_0 are of allele-type A_1 and a fraction $1 - x_0$ are of allele-type A_2, the probability that A_1 fixes is x_0, and the probability that A_2 fixes is $1 - x_0$, in agreement with Eqs. (8) and (9) for the WF model. Furthermore, if $T^*(x_0)$ is defined to be the time for A_1 to fix, conditional on A_1 fixing from an initial population x_0, then

$$E(T^*(x_0)) = -\frac{2(1 - x_0)}{x_0 \sigma^2} \log (1 - x_0). \tag{157}$$

When $\sigma^2 = 1$ this result agrees with the fixation time for the WF model, Eq. (34). On the other hand, it can be shown that the variance of $T^*(x_0)$ differs from that of the Wright-Fisher model. Results for fixation probabilities can be extended to the case of supercritical growth, $\alpha > 0$, with complications due to the fact that there is a finite probability that neither allele ever fixes.

Closing Remarks

The intention behind this article is to provide a solid mathematical background to many of the models underpinning current research in population genetics. No survey of this length could possibly cover all aspects of mathematical population genetics.

Some of the basic topics we have not covered include infinitely-many alleles models (Kimura and Crow, 1964), tests for neutrality versus selection such as Tajima's D statistic (Tajima, 1989), and interactions between multiple loci, which is also known as epistasis. Also not covered are non-random mating, recombination, and selective sweeps, that is, loss of genetic variation in the genomic vicinity of a beneficial mutation. The basic form of the WF and more general exchangeable models can be extended to study spatially structured populations and gene flow, including the transfer of alleles from one population to another due to migration. From the mathematical point of view, the diffusion limit of any model accessible to the forward Kolmogorov equation can equally be analysed in terms of stochastic differential equations.

The coalescent has had a major influence on the population genetics literature as improvements in genotyping technologies and the development of several software packages for simulating genealogies have become available, and perhaps deserves an extensive review in its own right.

From the point of view of population sizes, loss of genetic diversity through population bottlenecks and the founder effect are also of interest. Finally, the Bienaymé-Galton-Watson branching model can be adapted in a number of ways. Two examples are the stabilization of population size by introducing logistic growth (Lambert *et al.*, 2005), and the introduction of multiple alleles via multitype branching processes (Mode, 1971).

Appendix A   Derivation of the Forward Kolmogorov Equation

In order to facilitate the derivation of the forward Komogorov equation we first make a short digression to introduce a mathematical abstraction known as the Dirac delta function. The Dirac delta function (or just 'delta function' for short) is an example of a construct known to pure mathematicians as a generalised function. It was introduced by the physicist Paul Dirac (1930) as a convenient notation for studying quantum mechanics and was subsequently formalised rigorously by the mathematician Laurent Schwartz. However, to have a practical working understanding of the delta function it is not necessary to delve into the arcane pure mathematics of generalised functions. Here we take a simple but adequate heuristic approach.

The Dirac delta function, $\delta(x)$ is equal to zero everywhere on the real number line except at $x = 0$, where it is envisaged as an infinite spike, the area beneath which is 1. Thus

$$\delta(x) = \begin{cases} \infty & \text{if } x = 0, \\ 0 & \text{if } x \neq 0, \end{cases} \tag{A.1}$$

and

$$\int_{-\varepsilon}^{\varepsilon} \delta(x)\, dx = 1, \tag{A.2}$$

for any $\varepsilon > 0$. The delta function can be represented as a limit of the normal distribution density functions centred about zero as the variance tends to 0 (see **Fig. 10**). An important property of the delta function is that, for any function $f(x)$ which is continuous at $x = a$,

$$\int_{-\infty}^{\infty} f(x)\delta(x - a)\, dx = f(a). \tag{A.3}$$

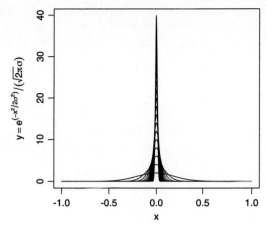

Fig. 10 Representation of the Dirac delta function as the limit of a normal distribution centred about zero as its variance σ^2 tends to zero.

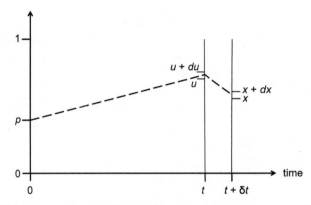

Fig. 11 Derivation of the forward Kolmogorov equation: decomposing $f(x, p, t+\delta t)dx$ into a sum over paths through intermediate events $u \leq X(t) < u + du$ at time t.

By formally integrating by parts we also obtain meaning for derivatives of the delta function, namely

$$\int_{-\infty}^{\infty} f(x)\delta'(x-a)\,dx = -f'(a), \qquad \int_{-\infty}^{\infty} f(x)\delta''(x-a)\,dx = f''(a). \tag{A.4}$$

We will derive the forward Kolmogorov equation for a generic population genetics model based on a discrete time finite state Markov chain. As previously, consider a genomic site at which two alleles A_1 and A_2 are possible in a population of fixed size M, and let the random variable $Y(\tau)$ be the number of A_1 alleles present at timestep $\tau = 0,1,2,\ldots$ labelling non-overlapping generations. The range of $Y(\tau)$ is $\Omega_{Y(\tau)} = \{0,\ldots,M\}$. Assume also that we are given a time-independent Markov transition matrix

$$p_{ij} = \text{Prob}(Y(\tau+1)=j|Y(\tau)=i), \qquad i,j = 0,\ldots M. \tag{A.5}$$

With the intention of taking the limit $M \to \infty$, make the substitutions:

$$t = \frac{\tau}{M}, \quad \delta t = \frac{1}{M}, \quad X(t) = \frac{1}{M}Y(\tau),$$

$$\delta X(t) = X(t+\delta t) - X(t) = \frac{1}{M}[Y(\tau+1) - Y(\tau)]. \tag{A.6}$$

After taking the limit $M \to \infty$, t becomes a continuous time variable over the interval $[0,\infty)$ and the fraction $X(t)$ of the population with allele A_1 becomes a continuous random variable with range $\Omega_X = [0,1]$. Now define the probability density $f(x;p,t)$ corresponding to $X(t)$, conditional on the initial condition $X(0)=p$, that is,

$$f(x;p,t)\,dx = \text{Prob}\,(x \leq X(t) < x+dx|X(0)=p), \qquad 0 \leq x,p \leq 1, \quad t > 0. \tag{A.7}$$

Our aim is to find a partial differential equation governing the evolution of $f(x; p, t)$.

The derivation of the forward Kolmogorov equation begins with decomposing $\text{Prob}(x \leq X(t+\delta t) < x + dx \mid X(0) = p)$ into a sum over intermediate events $u \leq X(t) < u + du$, which we write formally as

$$\text{Prob}\,(x \leq X(t+\delta t) < x + dx \mid X(0) = p)$$

$$= \sum_{u \in [0,1]} \text{Prob}\,(x \leq X(t+\delta t) < x + dx \mid u \leq X(t) < u + du)$$

$$\times \text{Prob}\,(u \leq X(t) < u + du \mid X(0) = p). \tag{A.8}$$

A path through one such intermediate state is shown in **Fig. 11**. At this point one should think of du and dx as being infinitesimal, and δt as being finite but small, in which case we can use Eq. (A.7) to rewrite the formal sum over u as

$$f(x; p, t + \delta t) = \int_0^1 du\, f(x; u, \delta t) f(u; p, t). \tag{A.9}$$

In order to make sense of the last equation we introduce one more assumption, namely that the moments of the random variable $\delta X(t)$ defined by Eq. (A.6) are of the form

$$E[\delta X(t) \mid X(t) = u] = a(u)\delta t + o(\delta t)$$

$$E[(\delta X(t))^2 \mid X(t) = u] = b(u)\delta t + o(\delta t)$$

$$E[(\delta X(t))^k \mid X(t) = u] = o(\delta t), \quad k = 3, 4, \ldots \tag{A.10}$$

as $\delta t \to 0$ for finite, well behaved functions $a(u)$ and $b(u)$ defined for $0 \leq u \leq 1$. This assumption essentially means that the density function $f(x; u, \delta t)$ is a narrow distribution centred about a mean approximately equal to $u + a(u)\delta t$ with variance approximately equal to $b(u)\delta t$ when δt is small. In other words, the point x in **Fig. 11** is unlikely to stray further than a distance of $O(\sqrt{\delta t})$ from the point u. We claim we can therefore represent $f(x; u, \delta t)$ in terms of the Dirac delta function as

$$f(x; u, \delta t) = \delta(x - u) - a(u)\delta'(x - u)\delta t + \tfrac{1}{2}b(u)\delta''(x - u)\delta t + o(\delta t). \tag{A.11}$$

One can check using the properties of the delta function that this is consistent with Eq. (A.10). For instance,

$$E[\delta X(t) \mid X(t) = u] = E[X(t+\delta t) - X(t) \mid X(t) = u]$$

$$= \int_0^1 (x - u) f(x; u, \delta t)\, dx$$

$$= \int_0^1 (x - u)\left[\delta(x - u) - a(u)\delta'(x - u)\delta t + \tfrac{1}{2}b(u)\delta''(x - u)\delta t\right] dx + o(\delta t)$$

$$= 0 + \int_0^1 \left[\frac{\partial}{\partial x}(x - u)\right] a(u)\delta(x - u)\delta t\, dx$$

$$\quad + \frac{1}{2}\int_0^1 \left[\frac{\partial^2}{\partial x^2}(x - u)\right] b(u)\delta(x - u)\delta t\, dx + o(\delta t)$$

$$= 0 + \int_0^1 a(u)\delta(x - u)\delta t\, dx + 0 + o(\delta t)$$

$$= a(u)\delta t + o(\delta t). \tag{A.12}$$

Similarly one can confirm the remaining two expectation values.

Returning to Eq. (A.9) and expanding the left hand side to first order in δt, we have

$$f(x; p, t) + \frac{\partial f(x; p, t)}{\partial t}\delta t$$

$$= \int_0^1 f(u; p, t)\left[\delta(x - u) - a(u)\delta'(x - u)\delta t + \tfrac{1}{2}b(u)\delta''(x - u)\delta t\right] du + o(\delta t)$$

$$= f(x; p, t) - \int_0^1 \frac{\partial}{\partial u}[f(u; p, t)a(u)]\delta(x - u)\delta t\, du$$

$$\quad + \frac{1}{2}\int_0^1 \frac{\partial^2}{\partial u^2}[f(u; p, t)b(u)]\delta(x - u)\delta t\, du + o(\delta t)$$

$$= f(x; p, t) + \left(-\frac{\partial}{\partial x}[f(x; p, t)a(x)] + \frac{1}{2}\frac{\partial^2}{\partial x^2}[f(x; p, t)b(x)]\right)\delta t + o(\delta t). \tag{A.13}$$

Equating the coefficient of δt on both sides finally gives the forward Kolmogorov equation

$$\frac{\partial f(x;p,t)}{\partial t} = -\frac{\partial}{\partial x}[a(x)f(x;p,t)] + \frac{1}{2}\frac{\partial^2}{\partial x^2}[b(x)f(x;p,t)]. \tag{A.14}$$

For any specific population genetics model, the functions $a(x)$ and $b(x)$ are determined from the Markov transition matrix p_{ij}. One can perform a similar calculation by partitioning the path from $X(0)=p$ to $X(t+\delta t)=x$ into an interval 0 to δt and an interval from δt to $t+\delta t$ to obtain the *backward Kolmogorov equation* (Ewens, 2004, p138)

$$\frac{\partial f(x;p,t)}{\partial t} = a(p)\frac{\partial}{\partial p}f(x;p,t) + \frac{1}{2}b(p)\frac{\partial^2}{\partial p^2}f(x;p,t). \tag{A.15}$$

The backward equation yields the same solution as the forward equation, the difference being that in the forward equation $f(x;p,t)$ is treated as a function with independent variables x and t, and p treated as a parameter, whereas in the backward equation $f(x;p,t)$ is treated as a function with independent variables p and t, and x treated as a parameter.

See also: Algorithms for Strings and Sequences: Multiple Alignment. Comparative Epigenomics. Epidemiology: A Review. Genetics and Population Analysis. Molecular Phylogenetics. Natural Language Processing Approaches in Bioinformatics

---

References

Abramowitz, M., Stegun, I.A., et al., 1972. Handbook of Mathematical Functions With Formulas, Graphs, and Mathematical Tables. New York, NY: Dover.

Aggarwala, V., Voight, B.F., 2016. An expanded sequence context model broadly explains variability in polymorphism levels across the human genome. Nature Genetics 48 (4), 349.

Baake, E., Bialowons, R., 2008. Ancestral processes with selection: Branching and Moran models. In: Miekisz, J. (Ed.), Stochastic Models in Biological Sciences, vol. 80. Banach Center Publications, Institute of Mathematics, Polish Academy of Sciences, pp. 33–52.

Bailey, N.T.J., 1964. The Elements of Stochastic Processes With Applications to the Natural Sciences. New York, NY: Wiley.

Burden, C.J., Tang, Y., 2016. An approximate stationary solution for multi-allele neutral diffusion with low mutation rates. Theoretical Population Biology 112, 22–32.

Burden, C.J., Simon, H., 2016. Genetic drift in populations governed by a Galton–Watson branching process. Theoretical Population Biology 109, 63–74.

Burden, C.J., Tang, Y., 2017. Rate matrix estimation from site frequency data. Theoretical Population Biology 113, 23–33.

Cannings, C., 1974. The latent roots of certain markov chains arising in genetics: A new approach, i. haploid models. Advances in Applied Probability. 260–290.

Cao, M., Shi, J., Wang, J., et al., 2015. Analysis of human triallelic SNPs by next-generation sequencing. Annals of Human Genetics 79 (4), 275–281.

Chen, H., Chen, K., 2013. Asymptotic distributions of coalescence times and ancestral lineage numbers for populations with temporally varying size. Genetics 194 (3), 721–736.

Cox, D.R., Miller, H.D., 1978. The Theory of Stochastic Processes. London: Chapman and Hall.

Crow, J.F., Kimura, M., 1970. An Introduction to Population Genetics Theory. New York, NY: Evanston, IL: London: Harper & Row, Publishers.

Dirac, P.A.M., 1930. The Principles of Quantum Mechanics, first ed. Oxford: Oxford University Press.

Etheridge, A., 2011. Some Mathematical Models From Population Genetics. Ecole D'Ete de Probabilites de Saint-Flour XXXIX-2009, Volume 2012 of Lecture Notes in Mathematics. Berlin; Heidelberg: Springer.

Etheridge, A., Griffiths, R., 2009. A coalescent dual process in a Moran model with genic selection. Theoretical Population Biology 75 (4), 320–330.

Ewens, W.J., 2004. Mathematical Population Genetics, second ed. New York, NY: Springer.

Ewens, W.J., Grant, G.R., 2005. Statistical Methods in Bioinformatics: An introduction, second ed. New York, NY: Springer.

Feller, W., 1951a. Diffusion processes in genetics. Proceedings of the Second Berkeley Symposium in Mathematical Statistics and Probability 227, 246.

Feller, W., 1951b. Two singular diffusion problems. Annals of Mathematics 54 (1), 173–182.

Fisher, R.A., 1930. The Genetical Theory of Natural Selection. Oxford University Press.

1000 Genomes Project Consortium, 2010. A map of human genome variation from population scale sequencing. Nature 467 (7319), 1061.

Gillespie, J.H., 2004. Population Genetics: A Concise Guide, second ed. Baltimore, MD: Johns Hopkins University Press.

Griffiths, R., 1979. A transition density expansion for a multi-allele diffusion model. Advances in Applied Probability. 310–325.

Griffiths, R.C., Tavare, S., 1994. Sampling theory for neutral alleles in a varying environment. Philosophical Transactions of the Royal Society B: Biological Sciences 344 (1310), 403–410.

Hardy, G.H., 1908. Mendelian proportions in a mixed population. Science 28 (706), 49–50.

Hasegawa, M., Kishino, H., Yano, T.-A., 1985. Dating of the human-ape splitting by a molecular clock of mitochondrial DNA. Journal of Molecular Evolution 22 (2), 160–174.

Heyde, C., Seneta, E., 1972. Studies in the history of probability and statistics. xxxi. The simple branching process, a turning point test and a fundamental inequality: A historical note on I. J. Bienayme. Biometrika 59 (3), 680–683.

Isaev, A., 2004. Introduction to Mathematical Methods in Bioinformatics. Berlin; Heidelberg: Springer-Verlag.

Johnson, N., Kotz, S., Kemp, A., 1993. Univariate discrete distributions, Wiley Series in Probability and Mathematical Statistics: Probability and Mathematical Statistics, second ed. New York: John Wiley & Sons.

Kimura, M., 1955a. Solution of a process of random genetic drift with a continuous model. Proceedings of the National Academy of Sciences of the United States of America 41 (3), 144–150.

Kimura, M., 1955b. Stochastic processes and distribution of gene frequencies under natural selection. Cold Spring Harbor Symposium Quantitative Biology 20, 33–53.

Kimura, M., 1964. Diffusion models in population genetics. Journal of Applied Probability 1 (2), 177–232.

Kimura, M., Crow, J.F., 1964. The number of alleles that can be maintained in a finite population. Genetics 49 (4), 725.

Kingman, J.F., 1982. On the genealogy of large populations. Journal of Applied Probability 19, 27–43.

Kingman, J.F., 2000. Origins of the coalescent. 1974–1982. Genetics 156 (4), 1461–1463.

Lambert, A., et al., 2005. The branching process with logistic growth. The Annals of Applied Probability 15 (2), 1506–1535.

Lanave, C., Preparata, G., Sacone, C., Serio, G., 1984. A new method for calculating evolutionary substitution rates. Journal of Molecular Evolution 20 (1), 86–93.

Masel, J., 2011. Genetic drift. Current Biology 21 (20), R837–R838.

Mode, C. J., 1971. Multitype Branching Processes: Theory and Applications Volume 34 of Modern Analytic and Computational Methods in Science and Mathematics New York, NY: American Elsevier Pub. Co.

Moran, P.A.P., 1958. Random processes in genetics. Mathematical Proceedings of the Cambridge Philosophical Society 54 (1), 60–71.

Phillips, C., Amigo, J., Carracedo, A., Lareu, M., 2015. Tetra-allelic SNPs: Informative forensic markers compiled from public whole-genome sequence data. Forensic Science International: Genetics 19, 100–106.

Shen, H., Li, J., Zhang, J., et al., 2013. Comprehensive characterization of human genome variation by high coverage whole-genome sequencing of forty four caucasians. PLOS ONE 8 (4), e59494.

Slatkin, M., Hudson, R.R., 1991. Pairwise comparisons of mitochondrial DNA sequences in stable and exponentially growing populations. Genetics 129 (2), 555–562.

Stern, C., 1943. The Hardy–Weinberg law. Science 97 (2510), 137–138.

Tajima, F., 1989. Statistical method for testing the neutral mutation hypothesis by dna polymorphism. Genetics 123 (3), 585–595.

Tavare, S., 1986. Some probabilistic and statistical problems in the analysis of DNA sequences. Lectures on Mathematics in the Life Sciences 17, 57–86.

Tier, C., Keller, J.B., 1978. A tri-allelic diffusion model with selection. SIAM Journal on Applied Mathematics 35 (3), 521–535.

Vogl, C., 2014. Estimating the scaled mutation rate and mutation bias with site frequency data. Theoretical Population Biology 98, 19–27.

Vogl, C., Bergman, J., 2015. Inference of directional selection and mutation parameters assuming equilibrium. Theoretical Population Biology 106, 7182.

Watson, H.W., Galton, F., 1875. On the probability of the extinction of families. The Journal of the Anthropological Institute of Great Britain and Ireland 4, 138–144.

Watterson, G., 1975. On the number of segregating sites in genetical models without recombination. Theoretical Population Biology 7 (2), 256–276.

Waxman, D., 2011. A unified treatment of the probability of fixation when population size and the strength of selection change over time. Genetics 188 (4), 907–913.

Wright, S., 1931. Evolution in mendelian populations. Genetics 16 (2), 97159.

Wright, S., 1969. Evolution and the Genetics of Populations: The Theory of Gene Frequencies. vol. 2. Chicago, IL: University of Chicago Press.

Zeng, K., 2010. A simple multiallele model and its application to identifying preferred-unpreferred codons using polymorphism data. Molecular Biology and Evolution 27 (6), 1327–1337.

Zhu, Y., Neeman, T., Yap, V.B., Huttley, G.A., 2017. Statistical methods for identifying sequence motifs affecting point mutations. Genetics 205 (2), 843–856.

Further Reading

Handbook of statistical genetics. In: Balding, D.J., Bishop, M., Cannings, C. (Eds.), Part 5: Population Genetics 2. Chichester: John Wiley & Sons.

Durrett, R., 2008. Probability Models for DNA Sequence Evolution. New York, NY: Springer Science & Business Media.

Haccou, P., Jagers, P., Vatutin, V.A., 2005. Branching Processes: Variation, Growth, and Extinction of Populations. Cambridge University Press. No. 5 in Cambridge Studies in Adaptive Dynamics.

Hartl, D., Clark, A., 2007. Principles of Population Genetics, fourth ed. Sunderland, MA: Sinauer.

Lambert, A., 2008. Population dynamics and random genealogies. Stochastic Models 24 (S1), 45–163.

Mode, C.J., Sleeman, C.K., 2012. Stochastic Processes in Genetics and Evolution: Computer Experiments in the Quantification of Mutation and Selection. Singapore: World Scientific.

Rosenberg, N.A., Nordborg, M., 2002. Genealogical trees, coalescent theory and the analysis of genetic polymorphisms. Nature Reviews Genetics 3 (5), 380–390.

Tavare, S., 2004. Part I: Ancestral inference in population genetics. In: Picard, J. (Ed.), Lectures on Probability Theory and Statistics: Ecole d'Ete de Probabilites de Saint-Flour XXXI − 2001. Berlin; Heidelberg: Springer, pp. 1–188.

Wakeley, J., 2009. Coalescent Theory: An Introduction. Greenwood Village, CO: Roberts & Company Publishers.

Yuan, X., Miller, D.J., Zhang, J., Herrington, D., Wang, Y., 2012. An overview of population genetic data simulation. Journal of Computational Biology 19 (1), 42–54.

Relevant Websites

http://www.radford.edu/~rsheehy/Gen_flash/popgen/
    Population Genetics Simulation Program, Radford University.
http://www.mabs.at/teaching/
    Teaching Resources of the Mathematics and BioSciences Group, University of Vienna, Including a Tutorial on Population Genetics by P. Pfaffelhuber et al.
http://evolution.gs.washington.edu/pgbook/
    Theoretical Evolutionary Genetics, Draft of Text by Joe Felsenstein, University of Washington.
http://evolution.genetics.washington.edu/geccobw.pdf
    Tutorial on Theoretical Population Genetics by Joe Felsenstein, University of Washington.

Biographical Sketch

Conrad Burden is an associate professor in the Mathematical Sciences Institute at the Australian National University lecturing in bioinformatics and biological modelling. His research interests include mathematical population genetics, alignment-free sequence comparison, the analysis of high-throughput sequencing data, and the physico-chemical modelling of microarrays. Before making the move into bioinformatics in 2003 he spent 20 years as a theoretical physicist followed by a short period working in the IT industry. He graduated from the University of Queensland with first class honours in Applied Mathematics in 1979 and The Australian National University with a PhD in Theoretical Physics in 1983. Between 1983 and 1988 he held postdoctoral fellowships in theoretical physics at the Weizmann Institute of Science, Glasgow University, The Australian National University and Flinders University. Between 1988 and 1999 he held a research position in the Department of Theoretical Physics, Australian National University, working mainly in quantum field theory and subatomic particle physics. Between 1999 and 2002 he was a programmer/developer in the IT industry. He is a Fellow of the Australian Institute of Physics.

Computational Systems Biology

Sucheendra K Palaniappan, Ayako Yachie-Kinoshita, and Samik Ghosh, The Systems Biology Institute, Tokyo, Japan and University of Toronto, Toronto, ON, Canada

Introduction

Rapid advancements in novel and high throughput technologies to observe and record different biological players and their interactions in multi-dimensions have opened interesting "Big Data" challenges. Using novel computational methods to use all this data to convert them into actionable insights is crucial in the biomedical domain. This is even more relevant in systems biology based approaches where the motivation is to use computational techniques to connect and reconcile data from different spatiotemporal scales to gain a systems level understanding of biological systems.

Extensive literature is available in the community highlighting the variety of tools, algorithms and databases available for different aspects of analysis in systems biology (Ghosh et al., 2011) – from Data and Knowledge management, deep curation, in silico simulation and model analysis, to physiological modeling, molecular interaction modeling (Kröger and Bry, 2003; Brazma et al., 2006). Successful application of these methodologies in modeling cell cycle dynamics, such as a computational model that explained the effects of over 120 knockout mutations on cell cycle dynamics in yeast (Tyson et al., 2001; Novak and Tyson, 1993), have been demonstrated. Significant progress has also been made in the analysis of signaling pathways – for example, in understanding the dynamics of mitogen-activated protein kinase (MAPK) signaling (Chen et al., 2004; Aoki et al., 2011) – and in cancer drug discovery applications, in which a reagent that was developed using model-based computational analysis is now in clinical trials (Schoeberl et al., 2009, 2010).

Current tools and techniques in computational systems biology has demonstrated their usage in various application areas, as illustrated schematically in **Fig. 1**. At the same time, the shift in paradigm both in experimental techniques which generate vast quantities of data, and in powerful data analytics, modeling and visualization methodologies, entail the empowerment of computational systems biology models and methodologies which leverage developments in machine learning, big data management and analysis, large scale modeling and simulations, to name a few. This topic article endeavors to provide some key areas of modeling and methodologies – highlighting new directions and developments, to enable computational systems biology to address the new challenges in biology and medicine.

The rest of the paper provides a brief overview of some of the key model and methodologies – including, Pathway Curation, Network analysis, Modeling and Simulation and Data Analytics. We conclude with a perspective on the promises and perils for application of novel models and methodologies in deeper understanding of basic biology as well as drug discovery and healthcare.

Pathway Curation

Biological systems can be thought of as composition of different biological players whose concerted actions lead to the biological function. Advances in high throughput technologies for recording these interactions have led to increased data which

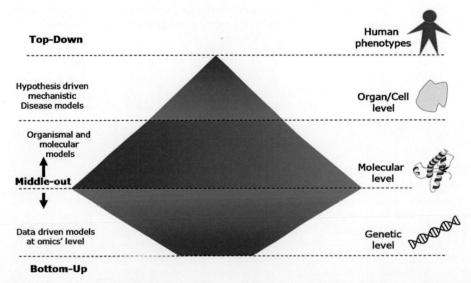

Fig. 1 Models and methodologies in computational systems biology. Taken from Hird, N., Ghosh, S., Kitano, H., 2016. Digital health revolution: Perfect storm or perfect opportunity for pharmaceutical R&D? Drug Discovery Today 21 (6), 900–911. Available at: http://doi.org/10.1016/j.drudis.2016.01.010.

has led to increased understanding about these molecules and their associated interactions. With this one can now more objectively reason about how these biological players work in unison at the systems level to perform various biological functions. Systematically recording and understanding these interactions is important and for this we construct and use Pathway maps. Pathway maps can be thought of as a formal representation of the collection of interactions among various biological players. They represent important insights pertaining to the flow of information within a biological system. They could either encode chemical reactions which are involved in the production or breakdown of different metabolites (metabolic pathways), or regulatory interaction between genes and proteins (gene regulatory networks) or record interactions that occur in response to external/internal stimuli (signaling pathways). Studying, analyzing and simulating these pathways can lead to crucial insights leading to designing effective cures for diseases (Kitano, 2002). One of the important undertakings in systems biology is creating formal (and machine readable) representations of these pathways and constitutes the task of Pathway curation (Kitano *et al.*, 2005). Pathways are usually encoded using standards such as SBML (Hucka *et al.*, 2003) and BioPAX (Demir *et al.*, 2010). Numerous pathway databases such as KEGG (Kanehisa and Goto, 2000) and Reactome (Joshi-Tope *et al.*, 2005) store pathway interactions. Attempts to construct detailed process level maps (such as the mTOR pathway map (Caron *et al.*, 2010) and TLR pathway map (Oda and Kitano, 2006)) and disease level pathway maps (such as Alzheimer's (Mizuno *et al.*, 2012) and Parkinson disease maps (Fujita *et al.*, 2014)) have been undertaken.

Typically, pathways are constructed and curated by expert pathway curators who manually read relevant scientific literature in detail, comprehend and select the relevant information about molecular interactions relevant in context and represent it as a pathway diagram. This process is facilitated by several graphical tools such as CellDesigner (Funahashi *et al.*, 2008).

The primary issues with manual approach to pathway curation is time consuming, importantly that the scale at which basic research is progressing, it is often hard to keep pace. Many times, the interpretation of the details is left to the judgment of the curator, hence leading to considerable inter-curator variability. Next, as new research finding are reported, these pathway maps need to be updated or augmented and doing this manually is almost impossible. Developing (semi) automated pathway curation methods using text mining and natural language processing (NLP) for has been an active field of research (Ananiadou *et al.*, 2010; Chowdhury and Sarkar, 2015; Cohen, 2015; Valenzuela-Escarcega *et al.*, 2015;) and is among important goals of large funded research efforts such as the recent DARPA's Big Mechanism Project (Cohen, 2015).

The task of automated pathway curation is essentially that of event extraction where given a text, the task of the automated system will be to first identify different biologically relevant players (using Named entity recognition), identifying the events (or reactions) and mapping the identified biological players to the corresponding reactions. There have been numerous tools such as TEES (Bjorne and Salakoski, 2011), Reach framework (Valenzuela-Escarcega *et al.*, 2015) and more recently the INDRA (Gyori *et al.*, 2017) which specifically tackle this problem. All these methods try to solve the problem of assembling individual reactions types and evaluating their performance. While this is useful, there has been limited acceptance of these techniques in the mainstream curation community. This is primarily because current extraction methods look at the atomic task of extracting at reaction level from text, while ignoring the aspect of threading these pieces together to reconstruct a holistic picture of the pathway (network). In fact, when a human curator is building a pathway map, he is thinking in terms of a network as the ultimate objective which is not well captured by the current NLP techniques.

There have been recent efforts to study and bridge this gap recently (Spranger *et al.*, 2016, 2015). These studies highlight some crucial aspects of improvement for current event extraction systems such as species normalization (improvements needed in unambiguous disambiguation of biological players), Complex identification, looking at a holistic view of the pathway outcome rather than focus only on reaction inference and finally for NLP systems to understand the level of abstraction at which a human interprets the publication and represents it on the pathway map.

Network Based Approaches

Once a pathway map (essentially a network) is obtained after curation, multiple analysis tasks can be carried out. The most straight forward is to purely consider the pathway as a network and performing static analysis on it.

Controllability/Network Motif Analysis

Controllability analysis focuses on analyzing a complex pathway network with the goal of providing insights about the key genes/proteins (called drivers) required to control the network, whose perturbation has a profound impact on the state of the network (Liu *et al.*, 2011). Network controllability has been successfully used in the context of large biological network to characterize how biological networks have special properties which confer then unique capabilities (Müller and Schuppert, 2011; Kawakami *et al.*, 2016). This can be useful for designing therapeutic interventions where identifying the drivers of a network and characterizing their relative importance in changing healthy to disease state is important for targeting them.

Since the pathway is essentially a network/graph, analyzing the structure of the graph can reveal unique substructures within the network that are expressed in majority and confer specific properties to the network. These substructures are often called network motifs and they are believed to serve as building blocks of the network. These network motifs and the impact they have on the dynamic features and biological functions of the overall network is well studied. In fact, it is believed that these motifs play a crucial part in the regulation and robustness of the network (Shen-Orr *et al.*, 2002; Novák and Tyson, 2008).

Network Inference

While pathway curation concerns with organizing the current knowledge about pathways one the other interesting facet is to discover new previously unknown links in the pathway network. Again, thanks to the high through put technologies, we can measure thousands of biomolecules concurrently over time. The data that is produced from these experiments can be used to discover novel pathway links or study how pathways and their links get affected in diseases/healthy states. The field of network inference deals with inferring these causal relationships between biological players from experimental data. The process is challenging as the network of molecular interactions are complex have numerous control and regulatory mechanisms governing the interactions and the experimental measurements are currently sparse and noisy.

There has been a lot of effort in developing and using efficient algorithms to infer these networks (Hase *et al.*, 2013; Sachs *et al.*, 2009, 2005). Considerable effort needs to also be put into producing accurate networks even with noisy observation data and will continue to be a major challenge in this field.

Modeling and Simulation

Pathway maps such as those discussed the pathway curation section provide a static picture of the events transpiring in a cell, however the dynamics of their interaction in time and space is what confers cellular function and behavior. For this, modeling pathway systems using mathematical formulations and consequently analyzing these models using techniques such as simulations is crucial.

Usually, for modeling studies, one starts from curated pathway maps which statically encode our understand of the mechanisms. A part of the map which is relevant to the current study is isolated after which suitable modeling formalism is chosen to mathematically encode interactions or hypotheses. These mathematical formalisms have kinetic parameters which need to be tuned so that these models faithfully replicate the observed dynamics and forms a large part of the model calibration and validation efforts. Once a model is calibrated, it faithfully captures the dynamics of the underlying system and can now be used for further analysis. These models once calibrated, can be used to predict behavior of the system under perturbations which is crucial to assess robustness of the system. Additionally, they can potentially assist in designing better experiments whereby one can evaluate potential hypotheses on the models first (which have considerably less cost) before validating them using wet-lab experiments. This section attempts to provide a broad overview of field.

There are many formalisms for modeling biological systems. The choice of the framework is usually dictated by the biological process under study, the questions for which we seek to gain insights, the experimental data that is available to us for calibrating and analysis of the model and the abstract level of detail at which we seek to gain insights. Models can be built to either give qualitative (Simao *et al.*, 2005; Fisher *et al.*, 2007; Schaub *et al.*, 2007) insights in case the nature of data is limited to qualitative observations or quantitative (Hua *et al.*, 2006; Janes and Yaffe, 2006). On the other hand, model formulation can be deterministic, stochastic or a hybrid between the two paradigms. Common methods of mathematical modeling include Ordinary differential equations (Aldridge *et al.*, 2006), Partial differential equations (Leung, 1989), Boolean networks (Raeymaekers, 2002), Petri nets (Ruths *et al.*, 2008) or Rule-based approaches (Danos *et al.*, 2008) to name a few.

Deterministic vs Stochastic Approaches to Modeling

The most common approach of modeling biological systems is using deterministic kinetics. Deterministic models such as Ordinary differential equations ignore random fluctuations in molecule numbers and other sources of noise. Consequently, given an initial state of the system its dynamics is uniquely determined by the underlying kinetics.

The most widely used deterministic model is Ordinary Differential Equation (ODEs). They capture the concentration changes of different biological players through the reactions they take part in. The concentration of every biological player over time is assumed to be governed by a set of coupled differential equations. The formulation of these equations is dictated by kinetic laws that govern each reaction (Aldridge *et al.*, 2006). Formulating ODEs relevant to biological systems requires one to have a detailed knowledge about the mechanisms of the reactions, rate constants. May times, restricted classes of these models are used by making several simplifying assumptions on the model formalism. Examples include the piecewise − multi affine models (Batt *et al.*, 2008), qualitative differential equations (Trelease *et al.*, 1999).

Stochasticity is important in the context of biological systems especially when considering biological players present in very low particle numbers (such as molecules participating in transcription, translation or their regulation). At these low numbers the effect of randomness in the system is not ignorable and has major implications on function (McAdams and Arkin, 1997). Additionally, cell-to-cell variability can occur due to random events in the cell which can change the order and number of interactions (Elowitz *et al.*, 2002; Spencer *et al.*, 2009). The most popular method for modeling stochastic systems include Chemical Master Equation (Gillespie, 1977) and Rule based formalisms (Danos *et al.*, 2008) to name a few.

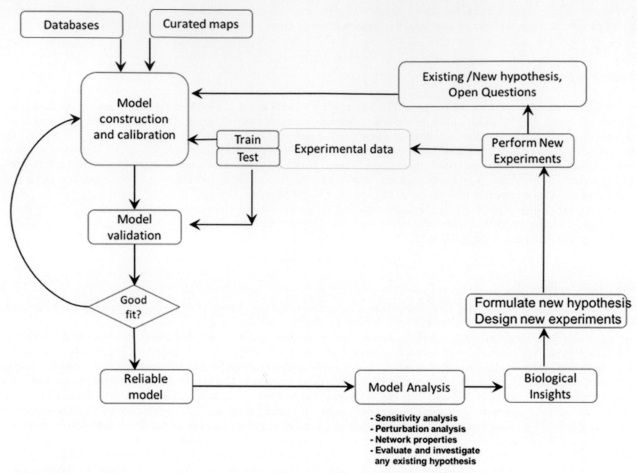

Fig. 2 Model Building steps. adapted from Palaniappan, S.K., 2013. Probabilistic verification and analysis of biopathway dynamics. (Dissertation).

Model Construction

Now we will very briefly discuss the main steps involved in building a reliable computational model as described in **Fig. 2**. Having decided the formalism and the scope of the modeling exercise, we first build the connections/structure of the model which incorporates all our current understanding of the pathway, this can also incorporate any hypotheses that we have. Depending on the data we have at our disposal and the questions to be asked of it, we choose a suitable modeling formalism to model the pathway.

Model Calibration and Tuning

Model calibration and tuning (often referred to as parameter estimation), deals with estimating unknown parameters (such as kinetic rate constants, initial concentrations of unmeasured players) of the model (depending on the chosen formalism). The final goal is to tune the model so that it can explain the given experimental observations. Usually, the experimental data is divided into two sets, one is used for estimating the parameters of model and the other is used as a validation or test set to check for the robustness of estimated parameters or check for overfitting. Model calibration is essentially a mathematical optimization problem where the goal is to minimize an objective function. The objective function is a goodness of fit measure which measures how far away is the model prediction from experimental data. Parameter estimation is a resource intensive task since evaluating the objective function for each parameter combination involves repeatedly simulating the model. After calibration, the model does through a round of model validation where the model output is evaluated for goodness of fit with the test data (which has not been seen by the model yet). If the fit is reasonably good, the conclusion is that the model is not over fitted and has a reasonable predictive power, it can then be used for further analysis tasks, else more calibrations rounds may be necessary or more data may be necessary to train the model.

Model Analysis

Once we have a model that is calibrated, it can be now be used for various model analysis which will help us harness the full potential of the model. Multiple analysis tasks can be performed on the model. For instance, bifurcation analysis gives insights on the effect of parameters on qualitative behavior of the system. Bifurcation points point to areas along the parameter space where there is a switch in the system behavior and give important insights about how the system switched between states (say diseased and healthy state). It has been used in the context of biological systems for robustness analysis (Morohashi *et al.*, 2002).

Sensitivity analysis (Schoeberl *et al.*, 2008; Rodriguez-Fernandez and Banga, 2008; Rodriguez-Fernandez *et al.*, 2012) on the other hand quantitatively studies the effect of the changes in kinetic parameters or initial concentrations of model players on the dynamic behavior the model. Sensitivity analysis plays an important role in robustness analysis, optimal experimental design, model reduction and drug target selection. In industrial biology and synthetic biology applications they are used for yield improvement where perturbations to the system can lead to target dynamic profile of select biological players.

As we discussed in previous sections, modeling is essentially an iterative process where, as new data or evidence is obtained or when new hypothesis in the mechanisms needs to be incorporated, we will need to re-estimate some parameters or add new links to the model. Additionally, with every change that is done to the model there is a need to verify and validate these models. This ensures that the new model is consistent with what is known already (nothing that is known has changed during the change) and that it explains the new data or hypothesis that has been newly incorporated. Often as the scale of the modeling exercise increases, ensuring that models are in accordance with the current knowledge is a non-trivial task. Manual analysis and validation is error prone.

Automated methods for analysis and verification of these models using Formal methods such as Model checking offers an attractive alternative (Fisher and Henzinger (2007)). The basic idea is to formalize our knowledge about the system behavior either in qualitative or quantitative terms using a specification language called temporal logics. These formal statements are then automatically analyzed against the Model using efficient algorithms to check if the system conforms to them. We can now automatically reason about system behavior and establish the extent to which a given model is consistent with observations of interest. There is a lot of interest in using formal methods for analyzing the dynamics of models in systems biology (Monteiro *et al.*, 2008; Clarke *et al.*, 2008; Hlavacek, 2009). In fact, it has been suggested model checking could serve as yardstick to check the reliability of models in public domain. Formal methods have also been used successfully for model calibration and sensitivity analysis tasks also (Barnat *et al.*, 2012; Calzone *et al.*, 2006; Palaniappan *et al.*, 2013). While there has been considerable effort on Model checking for systems biology, its adoption into mainstream modeling workflows are still limited owing to limited availability of usable tools which can seamlessly connect into a modeler's workflow.

Data Analytics

While the early approaches in computational systems biology focused primarily on modeling of pathways, biochemical reactions and their dynamic simulation in space and time, combining these techniques with data-driven approaches have gained increasing traction in the community. Enrichment analysis based on biological processes, pathways and functions have formed the cornerstone of bioinformatics and systems biology analysis pipelines (Ghosh *et al.*, 2011; Glass *et al.*, 2014).

With the availability of omics-scale data, novel data analysis workflows for gene expression analysis, proteomics and metabolomics analysis have been successfully developed to study the dynamics of those systems (Glass *et al.*, 2014; Khatri *et al.*, 2012; Caron *et al.*, 2017). A recent trend in data analytics for systems biology is the development of multi-omics and trans-omics analysis – methodologies to connect multi-layered data across the omics scale to reconstruct interaction networks linking genetic factors with phenotypes and environmental factors (Yugi *et al.*, 2016).

The rapid rise in machine learning techniques, particularly the class of feature learning convolution neural network models (CNN), known popularly as deep learning architectures, have opened a range of powerful approaches for data analytics in biology (Webb, 2018). Wide-ranging applications of deep learning models have been developed in cheminformatics, translational biomarker and drug discovery, proteomics, to healthcare and clinical informatics (comprehensive reference available at [deeplearning-biology]). Particular success in genomics field including the DeepVariant model for variant calling, enhancer predictions, and population genomics (Poplin *et al.*, 2016).

New horizons in data analytics are opening rapidly and have the potential for far-reaching impact on systems level study of complex biology of health and disease states. The ability to connect the different models and methodologies in a flexible platform (Ghosh *et al.*, 2011), will play a pivotal role in the next generation of computational systems biology applications as we posit next in the conclusion section.

Conclusion

The paper outlines new directions of models and methodologies in computational systems biology ranging from text-mining guided molecular pathway curation to data-driven approaches in network analysis, inference, dynamic modeling and simulation studies.

While early work in systems biology focused on top-down, hypothesis-driven approaches (refer **Fig. 1**) to model relatively well-defined processes, with the rise of high-throughput omics scale data together with advances in machine learning based approaches, the recent focus has weighted more on data-driven approaches (refer **Fig. 1**), especially for the applications in medicine and healthcare. Data-driven approaches allow the ability to harness large volumes of data and obtain insights by extracting patterns. Contrary to hypothesis-driven approaches, these techniques do not always require deep knowledge of the underlying biological process which was studied in the experiment and generated the data used in the models. On the other hand, hypothesis-driven approaches embed the biology mechanism in the modeling approach and thus provide better *explainability* of output results compared to pure data-driven approaches, albeit at the cost of injecting a *hypothesis bias* as well as compromise on the model scope and time to integrate the process information in the models. As we move towards systems medicine, it becomes apparent that a balanced approach which combines the different models and methodologies in a middle-out manner will be critically needed for computational systems biology to empower the future of medicine – the ability to analyze large-scale data to obtain novel insights together with the ability to explain the mechanism of those insights at molecular levels.

See also: Biological and Medical Ontologies: GO and GOA. Biological and Medical Ontologies: GO and GOA. Biological and Medical Ontologies: Systems Biology Ontology (SBO). Cell Modeling and Simulation. Cluster Analysis of Biological Networks. Computational Approaches for Modeling Signal Transduction Networks. Computational Lipidomics. Computational Systems Biology Applications. Coupling Cell Division to Metabolic Pathways Through Transcription. Data Formats for Systems Biology and Quantitative Modeling. Experimental Platforms for Extracting Biological Data: Mass Spectrometry, Microarray, Next Generation Sequencing. Functional Enrichment Analysis Methods. Integrative Bioinformatics. Multi-Scale Modelling in Biology. Natural Language Processing Approaches in Bioinformatics. Network Inference and Reconstruction in Bioinformatics. Network Models. Networks in Biology. Pathway Informatics. Protein-Protein Interactions: An Overview. Quantitative Modelling Approaches. Standards and Models for Biological Data: FGED and HUPO. Standards and Models for Biological Data: SBML. Studies of Body Systems. Visualization of Biological Pathways

---

References

Aldridge, B.B., Burke, J.M., Lauffenburger, D.A., Sorger, P.K., 2006. Physicochemical modelling of cell signalling pathways. Nature Cell Biology 8, 1195–1203.

Ananiadou, S., Pyysalo, S., Tsujii, J., Kell, D., 2010. Event extraction for systems biology by text mining the literature. Trends in Biotechnology 28, 381–390.

Aoki, K., Yamada, M., Kunida, K., Yasuda, S., Matsuda, M., 2011. Processive phosphorylation of ERK MAP kinase in mammalian cells. Proceedings of the National Academy of Sciences of the United States of America 108, 12675–12680.

Barnat, J., Brim, L., Krejci, A., et al., 2012. On parameter synthesis by parallel model checking. IEEE/ACM Transactions on Computational Biology and Bioinformatics 9 (3), 693–705.

Batt, G., Belta, C., Weiss, R., 2008. Temporal logic analysis of gene networks under parameter uncertainty. IEEE Transactions on Circuits and Systems I (Special Issue on Systems Biology) 53, 215–229.

Bjorne, J., Salakoski, T., 2011. Generalizing biomedical event extraction. In: Proceedings of the BioNLP Shared Task Workshop, pp. 183–191. (Association for Computational Linguistics).

Brazma, A., Krestyaninova, M., Sarkans, U., 2006. Standards for systems biology. Nature Rev. Genet. 7, 593–605.

Calzone, L., Chabrier-Rivier, N., Fages, F., Soliman, S., 2006. Machine learning biochemical networks from temporal logic properties. Transactions on Computational Systems Biology VI. 68–94.

Caron, E., Ghosh, S., Matsuoka, Y., et al., 2010. A comprehensive map of the mtor signaling network. Molecular Systems Biology 6.

Caron, E., Roncagalli, R., Hase, T., et al., 2017. Precise Temporal Profiling of Signaling Complexes in Primary Cells Using SWATH Mass Spectrometry. Cell Reports 18, 3219–3226.

Chen, K.C., et al., 2004. Integrative analysis of cell cycle control in budding yeast. Molecular Biology of the Cell 15, 3841–3862.

Chowdhury, S., Sarkar, R.R., 2015. Comparison of human cell signaling pathway databases—evolution, drawbacks and challenges. Database 2015, bau126.

Clarke, E., Faeder, J., Langmead, C., et al., 2008. Statistical model checking in BioLab: Applications to the automated analysis of T-Cell receptor signaling pathway. Computational Methods in Systems Biology. 231–250.

Cohen, P., 2015. Darpa's big mechanism program. Physical Biology 12 (045008), Available at: http://stacks.iop.org/1478-3975/12/i=4/a=045008.

Danos, V., Feret, J., Fontana, W., Harmer, R., Krivine, J., 2008. Rule-based modeling of signal cellular signalling. In: Proceedings of the International Conference on Concurrency Theory, pp. 17–41.

Demir, E., Cary, M.P., Paley, S., et al., 2010. The biopax community standard for pathway data sharing. Nature Biotechnology 28 (9), 935–942.

Elowitz, M.B., Levine, A.J., Siggia, E.D., Swain, P.S., 2002. Stochastic gene expression in a single cell. Science Signalling 297 (5584), 1183.

Fisher, J., Piterman, N., Hajnal, A., Henzinger, T.A., 2007. Predictive modeling of signaling crosstalk during c. elegans vulval development. PLOS Computational Biology 3 (5), e92.

Fujita, K., Ostaszewski, M., Matsuoka, Y., et al., 2014. Integrating pathways of parkinson's disease in a molecular interaction map. Molecular Neurobiology 49, 88–102.

Funahashi, A., Mat- suoka, Y., Jouraku, A., et al., 2008. Celldesigner 3.5: A versatile modeling tool for biochemical net-works. In: Proceedings of the IEEE, 96, pp. 1254–1265.

Ghosh, S., Matsuoka, Y., Asai, Y., Hsin, K.Y., Kitano, H., 2011. Software for systems biology: From tools to integrated platforms. Nature Reviews Genetics. doi:10.1038/nrg3096.

Gillespie, D.T., 1977. Exact stochastic simulation of coupled chemical reactions. The Journal of Physical Chemistry 81 (25), 2340–2361.

Glass, K., et al., 2014. Annotation enrichment analysis: An alternative method for evaluating the functional properties of gene sets, Kimberly Glass & Michelle Girvan. Scientific Reports 4, 4191. doi:10.1038/srep04191.

Gyori, B.M., Bachman, J.A., Subramanian, K., et al., 2017. From word models to executable models of signaling networks using automated assembly. Molecular Systems Biology 13 (11), 954.

Hase, T., Ghosh, S., Yamanaka, R., Kitano, H., 2013. Harnessing diversity towards the reconstructing of large scale gene regulatory networks. PLOS Computational Biology 9 (11), e1003361.

Hlavacek, W.S., 2009. How to deal with large models. Molecular Systems Biology 5, 240.

Hua, F., Hautaniemi, S., Yokoo, R., Lauffenburger, D.A., 2006. Integrated mechanistic and data-driven modelling for multivariate analysis of signalling pathways. Journal of the Royal Society Interface 3 (9), 515–526.

Hucka, M., Finney, A., Sauro, H.M., et al., 2003. The systems biology markup language (sbml): A medium for representation and exchange of biochemical network models. Bioinformatics 19 (4), 524–531.

Janes, K.A., Yaffe, M.B., 2006. Data-driven modelling of signal-transduction networks. Nature Reviews Molecular Cell Biology 7 (11), 820–828.

Joshi-Tope, G., Gillespie, M., Vastrik, I., et al., 2005. Reactome: A knowledgebase of biological pathways. Nucleic Acids Research 33, D428–D432.

Kanehisa, M., Goto, S., 2000. Kegg: Kyoto encyclopedia of genes and genomes. Nucleic Acids Research 28, 27–30.

Kawakami, E., Singh, V.K., Matsubara, K., et al., 2016. Network analyses based on comprehensive molecular interaction maps reveal robust control structures in yeast stress response pathways. Npj Systems Biology and Applications 2, 15018. Available at: https://doi.org/10.1038/npjsba.2015.18.

Khatri, P., Sirota, M., Butte, A.J., 2012. , Ten years of pathway analysis: Current approaches and outstanding challenges. PLOS Computational Biology 8 (2), e1002375. Available at: https://doi.org/10.1371/journal.pcbi.1002375.

Kitano, H., 2002. Computational systems biology. Nature 420 (6912), 206–210.

Kitano, H., Funahashi, A., Matsuoka, Y., Oda, K., 2005. Using process diagrams for the graphical representation of biological networks. Nature Biotechnology 23 (8), 961–966.

Kröger, P., Bry, F., 2003. A computational biology database digest: Data, data analysis, and data management. Distributed and Parallel Databases 13, 7–42.

Leung, A.W., 1989. Systems of nonlinear partial differential equations: Applications to biology and engineering. In: Proceedings of the Mathematics and its Applications, Kluwer Academic Publishers.

Liu, Y.-Y., Slotine, J.-J., Barabási, A.-L., 2011. Controllability of complex networks. Nature 473 (7346), 167–173. doi:10.1038/nature10011.

McAdams, H.H., Arkin, A., 1997. Stochastic mechanisms in gene expression. Proceedings of the National Academy of Sciences of the United States of America 94 (3), 814–819.

Mizuno, S., Iijima, R., Ogishima, S., et al., 2012. Alzpathway: A comprehensive map of signaling pathways of alzheimer's disease. BMC Systems Biology 6, 52.

Monteiro, P.T., Ropers, D., Mateescu, R., Freitas, A.T., de Jong, H., 2008. Temporal logic patterns for querying dynamic models of cellular interaction networks. Bioinformatics 24 (16), i227–i233.

Morohashi, M., Winn, A.E., Borisuk, M.T., et al., 2002. Robustness as a measure of plausibility in models of biochemical networks. Journal of Theoretical Biology 216 (1), 19–30.

Müller, F.-J., Schuppert, A., 2011. Few inputs can reprogram biological networks. Nature 478 (7369), E4-E4. doi:10.1038/nature10543.

Novak, B., Tyson, J.J., 1993. Numerical analysis of a comprehensive model of M-phase control in Xenopus oocyte extracts and intact embryos. Journal of Cell Science 106, 1153–1168.

Novák, B., Tyson, J.J., 2008. Design principles of biochemical oscillators. Nature Reviews Molecular Cell Biology 9, 981–991.

Oda, K., Kitano, H., 2006. A comprehensive map of the toll-like receptor signaling network. Molecular Systems Biology 2.

Palaniappan, S.K., Sucheendra K., et al., 2013. Statistical model checking based calibration and analysis of bio-pathway models. In: Proceedings of the International Conference on Computational Methods in Systems Biology, Berlin, Heidelberg: Springer.

Palaniappan, S.K., 2013. Probabilistic verification and analysis of biopathway dynamics. (Dissertation).

Poplin, R., Newburger, D., Dijamco, J., et al., 2016. Creating a universal SNP and small indel variant caller with deep neural networks. Available at: https://doi.org/10.1101/092890.

Raeymaekers, L., 2002. Dynamics of boolean networks controlled by biologically meaningful functions. Journal of Theoretical Biology 218 (3), 331–341. PMID: 12381434.

Rodriguez-Fernandez, M., Banga, J.R., Doyle III, F.J., 2012. Novel global sensitivity analysis methodology accounting for the crucial role of the distribution of input parameters: Application to systems biology models. International Journal of Robust and Nonlinear Control 22, 1082–1102.

Rodriguez-Fernandez, M., Banga, J.R., 2008. Global sensitivity analysis of a biochemical pathway model. In: Proccedings of the2nd International Workshop on Practical Applications of Computational Biology and Bioinformatics,pp. 233–242.

Ruths, D., Muller, M., Tseng, J.T., Nakhleh, L., Ram, P.T., 2008. The signaling Petri net-based simulator: A non-parametric strategy for characterizing the dynamics of cell-specific signaling networks. PLOS Computational Biology 4 (2), 1–15.

Sachs, K., Itani, S., Carlisle, J., et al., 2009. Learning signaling network structures with sparsely distributed data. Journal of Computational Biology 16 (2), 201–212.

Sachs, K., Perez, O., Pe'er, D., Lauffenburger, D.A., Nolan, G.P., 2005. Causal protein-signaling networks derived from multiparameter single-cell data. Science 308 (5721), 523–529.

Schaub, M.A., Henzinger, T.A., Fisher, J., 2007. Qualitative networks: A symbolic approach to analyze biological signaling networks. BMC Systems Biology 1 (1), 4.

Schoeberl, B., et al., 2009. Therapeutically targeting ErbB3: A key node in ligand-induced activation of the ErbB receptor-PI3K axis. Science Signaling 2, ra31.

Schoeberl, B., et al., 2010. An ErbB3 antibody, MM-121, is active in cancers with ligand-dependent activation. Cancer Research 70, 2485–2494.

Schoeberl, B., Eichler-Jonsson, C., Gilles, E.D., Muller, G., 2008. Computational modeling of the dynamics of the map kinase cascade activated by surface and internalized EGF receptors. Nature Biotechnology 20 (4), 370–375.

Shen-Orr, S.S., Milo, R., Mangan, S., Alon, U., 2002. Network motifs in the transcriptional regulation network of Escherichia coli. Nature Genetics 31, 64–68.

Simao, E., Remy, E., Thieffry, D., Chaouiya., C., 2005. Qualitative modelling of regulated metabolic pathways: Application to the tryptophan biosynthesis in E. coli. Bioinformatics 21 (suppl. 2), ii190–ii196.

Spencer, S.L., Gaudet, S., Albeck, J.G., Burke, J.M., Sorger, P.K., 2009. Non-genetic origins of cell-to-cell variability in trail-induced apoptosis. Nature 459 (7245), 428–432.

Spranger, M., Palaniappan, S.K., Ghosh, S., 2015. Extracting biological pathway models from NLP event representations. In: Proceedings of the ACL 2015 Workshop on Biomedical Natural Language Processing (BioNLP'15).

Spranger, M., Palaniappan, S.K., Ghosh, S., 2016. Measuring the state of the art of automated pathway curation using graph algorithms – A case study of the mTOR pathway. BioNLP, Berlin, Germany.

Trelease, R.B., Henderson, R.A., Park, J.B., 1999. A qualitative process system for modeling nf-κb and ap-1 gene regulation in immune cell biology research. Artificial Intelligence in Medicine 17 (3), 303–321.

Tyson, J.J., Chen, K., Novak, B., 2001. Network dynamics and cell physiology. Nature Reviews Molecular Cell Biology 2, 908–916.

Valenzuela-Escarcega, M., Hahn-Powell, G., Hicks, T., Surdeanu, M., 2015. A domain-independent rule-based framework for event extraction. In: Proceedings of the 53rd Annual Meeting of the Association for Computational Linguistics (ACL- IJCNLP), pp. 127–132. ACL.

Webb, S., 2018. Deep learning for biology. In: Proceedings of the Nature Technology Feature, Available at: https://www.nature.com/articles/d41586-018-02174-z#correction-0.

Yugi, K., et al., 2016. Trans-omics: How to reconstruct biochemical networks across multiple 'Omic' layers. Trends in Biotechnology 34 (4), 276–290.

Relevant Website

https://github.com/hussius/deeplearning-biology
Deeplearning-Biology.

Pathway Informatics

Sarita Poonia and Smriti Chawla, Indraprastha Institute of Information Technology, Delhi, India
Sandeep Kaushik, European Institute of Excellence on Tissue Engineering and Regenerative Medicine, Guimaraes, Portugal and University of Minho, Braga, Portugal
Debarka Sengupta, Indraprastha Institute of Information Technology, Delhi, India

Introduction

Traditional biology focuses on structure and function of important biomolecules such as genes and proteins. But this scientific reductionist approach hinders understanding of their interregulation and its role in execution of the various biophysical processes (Joyner and Pedersen, 2011). With the advancement of technologies and exponential growth in data availability, scientists are now probing into the higher order structures that underlie biological activities. This holistic approach has been widening our knowledge about the functionally interdependent molecular circuitry that orchestrates life (Chan and Loscalzo, 2012).

Pathway informatics is a flourishing interdisciplinary research area that is progressively growing and evolving through a blend of computational, experimental, and theoretical approaches (Karp, 2003). This field focuses on understanding how various types of biomolecules interact with each other to form context specific, complex networks, which we often refer to as biological pathways.

To understand the dynamics of biological systems, various computational approaches and pathway databases have been developed, which help in pathway representation, simulation, visualization for quantitative and qualitative analysis (Cary *et al.*, 2005), thus, helping in better understand cellular processes. In this article, we will discuss different biological pathways, their corresponding databases, pathway analysis, and computational approaches for network inference.

Broad Categories of Molecular Pathways

Metabolic Pathways

A metabolic pathway is a series of interlinked biochemical reactions catalyzed by enzymes. Three major classes of molecules in higher order organisms are proteins, lipids, and carbohydrates. Together, these are considered as the building blocks of life. Metabolic pathways that involve synthesis and storage of these biomolecules are referred to as the anabolic pathways. On the other hand, pathways responsible for breakdown of these biomolecules to generate energy are referred to as the catabolic pathways (Ward, 2016).

Regulatory Pathways

Gene regulatory pathways are responsible for turning genes on or off, which in turn determines availability and concentrations of the related protein molecules (Ma and Zhao, 2013). In eukaryotes, genes, by default, are in the off state when they are tightly bound to histones. But eukaryotic genes manage to escape this silencing due to modifications in histones. While histone acetylation makes DNA open and accessible to DNA binding protein, methylation, coupled with histone modification obscure DNA, which in turn lead to gene silencing. Once DNA is open, it is accessible to transcription factors (TFs), which can either transcribe genes or repress them (Hoopes, 2008). Apart from TFs, a number of long and short noncoding RNAs also participate in the ultra-complex gene regulatory network primarily through post transcriptional modifications (Sengupta and Bandyopadhyay, 2011, 2013; Holoch and Moazed, 2015).

Signaling Pathways

Signal transduction is a cascade of biochemical reactions that take place in a cell when a signal molecule such as hormone or biomolecule binds to a receptor on the cell membrane to perform specific biological processes (Pawson, 1995). Cells use a large number of singaling pathways to regulate biological activities. These signals pass information either through protein–protein interactions or via diffusible elements called referred as secondary messengers (Berridge, 2014). Signaling pathway can be of two types: Extracellular and intracellular. The entire signaling process is regulated by feedback pathways (Berg *et al.*, 2002). Any deviation in the signal transduction pathways could cause dysregulation of biological processes and can lead to diseases such as cancer, diabetes, etc. (Liu *et al.*, 2008). Examples of extracellular stimuli are cyclic AMP singling pathway, voltage-operated channels (VOCs), receptor-operated channels (ROCs), etc. Endoplasmic reticulum (ER) stress singling, metabolic messengers are the examples of intracellular stimuli (Berridge, 2014).

Drug Related Pathways

Drugs are xenobiotic, small, foreign molecules. To deal with drugs the human body attempts a number of responses. Some drugs are excreted from the body without involving metabolism. But most drugs need structural modification to facilitate excretion,

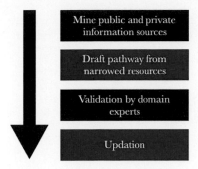

Fig. 1 Pathway construction by curation.

called drug metabolism. There are numerous pathways through which a drug is metabolized or biotransformed. Chemical processes to metabolize drugs are oxidation, hydrolysis, reduction, hydration, condensation, and conjugation. Study of these pathways as the route of metabolism of a drug can help to assess if a drug shows any pharmacological, clinical relevance, and toxicological activity. The primary site for drug metabolism is the liver (Wilkinson, 2005).

Pathway Construction: Data Curation Based Approaches

The curation process of biological networks and pathways involves manual or computer aided curation of information from various structured and unstructured sources, followed by their assembly in a centralized repository. Major steps involved in pathway building are as follows. First step in pathway construction is mining of relevant data from available public and private data sources about biological entities and their interactions. The pathway draft thus obtained is further enriched while accounting for cell, species, and disease type annotations. This specific annotated pathway is then reviewed by domain experts and iteratively updated before finalization (Viswanathan *et al.*, 2008) (see **Fig. 1** for the steps).

Increasingly, natural language processing based methods are taking over manual curation processes. Such approaches are capable of automatic extraction of event–entity relationships from literature (Ananiadou *et al.*, 2010). This accelerated curation process reduces time and labor in accumulating experimentally validated relationships, which, in turn, are used in pathway construction (Yu *et al.*, 2013).

Among several methods, one of the simplest and embracive approach is cooccurrence based literature mining (Li *et al.*, 2006). For example, if breast cancer and *BRCA1* gene cooccur in a sentence, a relationship between them is assumed. However, this system is prone to errors and may not accurately capture all true relationships that may be latently explained through a number of experimental results.

The second approach is rule-based systems, which exploits some frequently observed linguistic motifs, often used in citing biologically pertinent facts. Such systems either use hard-coded patterns or semantic analysis to find about classes of biologically nontrivial facts. Related to this, a commonly used approach is to use inferential statistics or machine learning to classify sentences and documents into classes of interests (Cohen and Hunter, 2008).

A more sophisticated approach is to infer biological association networks (BANs) through Medscan based sentence parsing. Medscan is capable of extracting relations between proteins and various cellular processes/diseases. If information regarding the direction of relation, mechanism of action and effect on a target is present in the same sentence, Medscan can easily understand and interpret the same. Also, this algorithm can identify if a same relationship is portrayed differently by different authors, thereby reduces redundancy.

Pathway Databases

To keep up with the expansion of our knowledge about biological pathways, systematic and periodic archival of the pathway related information is essential. Below is a list of the key pathway related databases, heavily referred to in biomedical research.

Databases for metabolic pathway

Metabolic pathway databases generally contain detailed information about pathways connected as series of biochemical reactions. One of the main database for metabolic pathways is MetaCyc (Caspi *et al.*, 2013). The MetaCyc is an intensively curated, nonredundant, and comprehensive database for both primary and secondary metabolic pathways. The statistical details of this database are shown in **Table 1**. It contains valuable information about metabolites, reactions, and enzymatic data that have been experimentally validated through extensive literature scoping from all domains of life.

Another popular database in this domain is Kyoto Encyclopedia of Genes and Genomes (KEGG), which is well-integrated reference knowledge base for molecular interaction, reactions, and network relations for metabolism (Kanehisa and Goto, 2000). Relevant statistics for KEGG are shown in **Table 2**.

Table 1   MetaCyC statistics

Feature	Statistics
Pathways	2,527
Reactions	14,347
Enzymes	11,547
Chemical compounds	14,003
Organisms	2,883
Citations	54,196

Table 2   KEGG statistics

Feature	Pathway information	Statistics
KEGG pathway	Pathway maps	519 (542,037)
KEGG genes	Genes in KEGG organisms and other categories	24,338,758
KEGG compound	Metabolites and other small molecules	18,111
KEGG reaction	Biochemical reactions	10,668
KEGG disease	Human diseases	1,910

Table 3   TRANSFAC statistics for year 2017

Feature	Statistics
Factors	23,800
miRNAs	1,279
DNA sites	49,945
mRNA sites	21,558
Genes	82,975
ChIP TFBS	27,243,261
References	36,911

Human Metabolome Database (HMDB) is an open source containing the most comprehensive curated information about human metabolites and human metabolism data (Wishart *et al.*, 2007). It features over 2180 endogenous metabolites and their details gathered from extensive literature scoping. HMDB also contains metabolite concentration data collected from mass spectra (MS) and Nuclear Magnetic resonance (NMR) data analysis. The HMDB is composed of various tools for browsing, extensive searching, and querying.

Gene regulation databases

Gene regulation databases tend to cover information on relationships between TFs and genes regulated by them. These databases have broad coverage of organisms and some of the databases tend to incorporate DNA binding data from high-throughput assays such as chromatin immunoprecipitation (ChIP). Such protein–DNA binding information only indicates that TFs bind to promoters of genes to regulate them but it doesn't provide functional consequences of such binding (Cary *et al.*, 2005). The most prominent gene regulation databases is TRANSFAC. TRANSFAC is a manually curated database for eukaryotic TFs and their genomic binding sites (Wingender *et al.*, 2000). It is an encyclopedia for transcriptional regulation and is an apt tool for extensive genomic analysis to predict potential transcription factor binding sites (TFBSs). Computationally TFBSs are predicted using basic positional weight matrices, which are involved motifs based on the detected nucleotide patterns in a set of TFBSs for the corresponding TF (Mathelier and Wasserman, 2013). More detailed statistics of TRANSFAC are shown in **Table 3**. The basic structure of TRANSFAC comprises of two domains, one is involved in documenting TFBSs in promoters and enhancers and another domain is involved in describing TFs (Matys *et al.*, 2006).

TRUSST (transcriptional regulatory relationships unraveled by sentence-based text-mining) is a publicly available manually curated database for human TFs target interactions. Approximately 20 million Medline abstracts have been subjected to sentence-based text-mining approach for manual curation of regulatory networks. TRUSST currently covers 8015 interactions amongst 748 TF genes and 1975 non-TF genes (Han *et al.*, 2015).

Table 4 Signaling pathway databases

Database	URL	Description
Biocarta	http://www.biocarta.com	Database of graphs and annotations about signal pathways
KEGG	http://www.genome.ad.jp/kegg	Manually drawn pathway maps of interaction and reaction network
Reactome	http://www.reactome.org	Curated resource of core reactions and pathways
PID	http://pid.nci.nih.gov	Cellular processes and biomolecular interactions assembled into reliable human signaling pathways
STKE	http://stke.sciencemag.org	Information on the components of cellular signaling pathways and their relations to one another, organized into pathways
AfCS	http://www.signaling-gateway.org	A database of cell signaling and protein interaction in pathways and complex reactions
AMAZE	http://www.amaze.ulb.ac.be	A database to manage, analyze, represent and annotate cell signaling information
BIND	http://www.bind.ca	A data platform binding public and patented sequence, interaction, and related information
DOQCS	http://doqcs.ncbs.res.in	A repository of signaling pathways models, biochemical reaction schemes, rate constants, concentrations, and annotations on the models
SigPath	http://sigpath.org	A database having pathways of dynamic signaling

Signaling pathway databases

Signaling pathway databases contain detailed information about cell signaling molecules and their interactions. Some important signaling pathway databases are discussed here.

BioCarta: BioCarta is the interactive signaling pathway database formed by a group of engineers and scientists, having detailed information of individual molecule. Information provided in BioCarta is not updated so it might be outdated now. Pathways figures are available in the website provided in "see Section Relevant Website" (Nishimura, 2001). KEGG: KEGG is a collection of manually drawn pathway maps. This database is easy to download but less comprehensive than BioCarta (Kanehisa and Goto, 2000). STKE: STKE is a freely accessible database that has general data of cells and some special signals in cells. STKE offers data access through machine-generated interactive pathway diagrams. The information in database is ordered into generalized and specific pathways (Gough, 2002). AfCS: AfCS database provides signaling pathway map and interactions based on molecule interaction (Gilman et al., 2002). AMAZE: AMAZE database provides an object-oriented platform, gene regulation entries, cell signaling pathways, and interaction information (Lemer et al., 2004). DOQCS and sigPath are quantitative signaling pathway databases. DOQCS includes rate constants, reaction schemes, concentrations, as well as annotations on the models (Sivakumaran et al., 2003; Campagne et al., 2004).

Some important signaling pathways databases, URLs, and a short description are given in **Table 4**.

Drug pathway databases

Drug pathway databases contain detailed drug data, drug target information, pharmacological information about drugs, mechanism of action, and metabolism of drugs. Some important signaling pathway databases are discussed here.

DrugBank: DrugBank database is a unique resource of cheminformatics and bioinformatics that combines detailed drug data with complete information on drug targets. Latest updated version (version 5.0.9, released 2017-10-02) of Drugbank contains 10,500 drug entries in which 1737 are approved small molecule drugs, 103 nutraceuticals, 870 approved biotech drugs, and 5023 experimental drugs (Law et al., 2013). SMPDB (Small Molecule Pathway DataBase): This is a visual, interactive database that contains more than 30,000 human small molecule pathways. Majority of these pathways are only found in SMPDB. Each small molecule in SMPDB is hyperlinked to complete information in DrugBank and HMDB (Frolkis et al., 2009). KEGG: KEGG is an integrated database resource that contains 16 main pathway databases. The health information category of KEGG includes DRUG, DISEASE, ENVIRON, and DGROUP databases for drug and disease information (Kanehisa and Goto, 2000). Transformer: Transformer is a comprehensive database providing information on transport and transformation of drugs, phase-I and phase-II reactions, alimentary, and traditional Chinese medicine compounds (Hoffmann et al., 2013). Drug-Path: Drug-Path is a database generated by KEGG pathway consisting of drug-induced pathway enrichment analysis for drug-induced downregulated and upregulated genes that are coming from datasets of drug-induced gene expression in Connectivity Map. Pipeline of this database relies on information provided by pharmaceutical partners. This provides user-friendly interfaces to download, retrieve, and visualize drug-induced pathway data. In addition, deregulated genes by a given drug are highlighted. The DrugPath database contains information on 2081 drugs, 751 companies, and 722 targets (Zeng et al., 2015).

Other databases

WikiPathways

WikiPathways is a collaborative platform and freely available manually curated public resource for capturing and publicizing models of biological pathways for data curation, visualization, and analysis (Kutmon et al., 2015). It provides you with the facilities of zoomable pathway viewer, provision for pathway annotations, and convenient hyperlinks to pathways and their components, thus enabling biologists to capture their productive, intuitive mental models of biological pathways without any hassle. The most observable feature of WikiPathways is the pathway page. Each pathway has its own dedicated page providing information about the specific biological mechanism and pathway diagram, along with description and hyperlinks to genes, proteins, or metabolites for detailed information (Kelder *et al.*, 2011). It was launched in 2008 and since then it has seen a massive increase in its content, which is continuously evolving and benefiting the scientific community in several ways (Kutmon *et al.*, 2015; Kelder *et al.*, 2011). Initially, it started out with 500 pathways spanning across 6 species. But today it contains over 2300 pathways spanning over 25 different species. The human pathway coverage is the largest and most active collection by species. This database is quite flexible to use as there are no restrictions on pathway models, therefore accepting any pathway that researchers ep continue to grow, helping researchers across the globe and capturing each and every biological pathway of interest and publicizing it in as many useful ways as possible (Kutmon *et al.*, 2015).

Small molecule pathway database

SMPDB is a freely available manually curated comprehensive, interactive, and visual database intended to carry out clinical omics studies with ease, with specific focus on clinical metabolomics (Frolkis *et al.*, 2009). SMPDB covers over 600 human pathways of small molecule metabolism or processes central to small molecules and nearly 75% of these pathways are not found in other databases such as KEGG, Reactome, and WikiPathways (Jewison *et al.*, 2013). Pathways covered in this database are metabolic pathways, small molecule disease pathways, small molecule drug pathways, and small molecule signaling pathways.

It is the only pathway database that embraces substantial numbers of metabolic disease and drug pathways. SMPDB keeps on adding new information, describing each pathway in the milieu of human physiology and human biochemistry. Also, each drug or metabolite in SMPDB is hyperlinked, which provides detailed information about that particular molecule. In its pathway diagrams, it provides valuable and useful graphical content, which includes the depiction of the organelles, cofactors, and other important cellular features. SMPDB possesses a number of useful, user friendly, and unique features such as use of thumbnail images to facilitate easy pathway viewing and browsing and scrollable table for displaying pathways and pathways synopsis. Although SMPDB work is still in progress, new pathways are constantly being added and existing ones are continuously being improved. Therefore, SMBPD is a useful addition to a collection of already existing pathway databases (Frolkis *et al.*, 2009).

Reactome pathway database

Reactome is an open source of manually curated and peer-reviewed pathway database of human pathways, reactions, and processes. The Reactome data model streamlines the concept of reaction by taking into consideration transformations of different biological entities such as proteins, nucleic acids, and macromolecular complexes. These transformations could be the transport of biological entities from one compartment to another and formation of complexes. Thus, streamlining of Reactome reaction model allows capturing a range of biological processes that span across signaling, metabolic, transcriptional regulation, and apoptosis in a single integrated platform in a computationally navigated format. Reactome is a complete and comprehensive resource of human pathways for facilitating basic research, genomic analysis, pathway modeling, and analysis (Croft *et al.*, 2010).

Pathway Construction: Quantitative Approaches

For obvious reasons curation based approaches lack comprehensiveness. Use of computational and mathematical models is therefore inevitable. Quantitative approaches to model biological pathways can be broadly split into three categories:

a. Network-based methods
b. Mathematical modeling

Below we elaborate on the above mentioned categories.

Network-Based Methods

Network-based methods use graph theoretical representation of molecules and their interactions. Nodes in such graphs denote molecules, whereas edges denote their regulatory relationships. Probabilistic graph models are a commonly used technique for modeling molecular interdependencies as observed from genome wide molecular profiling data. Gene expression data is commonly used for such modeling.

Gene coexpression networks (GCN) are progressively being used to model the coexpression relationships among genes. Within a GCN, genes are represented by nodes and edges exists between pair of genes when their expression profiles are significantly correlated across a set of expression-measurement samples (Ficklin *et al.*, 2017). Coexpression analysis network identifies set of genes that have tendency to display a synchronized expression pattern across a group of samples. Coexpression network construction involves use of different measures for covariability. Pearson's/Spearman's correlation coefficients, mutual information, etc., are popular among these. Coexpression network construction is often followed by identification of community of coexpressed genes using different clustering techniques (van Dam *et al.*, 2017).

Mathematical Modeling

Mathematical modeling of biological processes arises with deep knowledge of complex cellular systems (Wang *et al.*, 2012). Mathematical models learn and analyze the principal network and then convert the reactions and entities of underlying network into matrix form. To study biological pathways, complexity of networks, and different type of biochemical reactions, a number of mathematical methods have been developed (Hou *et al.*, 2015). Below we provide an excerpt of these methods for modeling signaling, regulatory, and metabolic pathways respectively.

Signaling pathways: Signaling pathways are studied in either top-down or bottom-up fashion. Top-down approaches work on omics datasets to infer downstream biological processes. On the other hand, bottom-up approaches are used to model the interaction between the biological components such as proteins, genes, and metabolites to learn about the dynamic behavior of the cell. The most widely used bottom-up approach is continuous dynamic modeling, which needs adequate kinetic parameters (synthesis and degradation rates) and mechanistic details. Discrete dynamic modeling, for example, Petri nets, Boolean network models, and multivalued logical models yield qualitative dynamic illustration of a system behavior that does not need kinetic parameters (Liu and Betterton, 2012; Wang *et al.*, 2012).

Regulatory network: Understanding the dynamism of regulatory networks is critical in understanding the mechanism of diseases that occur due to dysregulation of these regulatory networks (Karlebach and Shamir, 2008). Differential equations can be used to represent these regulatory networks, in which interactions among proteins, mRNAs are defined in terms of rate equations. As the system evolves, these rate equations specify the levels of each mRNA and protein as a function of other components. Differential equation based models use time and/or space dependent variables (mRNA and protein concentration, production rate, degradation rate).

These models can be divided into two groups:

a. Ordinary differential equations (ODE), which involve a single variable such as time.
b. Partial differential equations (PDE), which depend on more than one variable such as time and space.

Finding analytical solutions for ODEs is hard. Approximate solutions can be found by various approximate numerical methods (Ay and Arnosti, 2011).

Differential equations have been used in modeling regulatory networks in various organisms:

- In bacteriophage T7 developmental cycle (Endy *et al.*, 2000)
- Tryptophan synthesis in *E. coli* (Santillán and Mackey, 2001)
- In *E. coli* lac operon induction (Carrier and Keasling, 1999)
- Cell division in *Xenopus* (Novak and Tyson, 1993)
- Circadian rhythms in *Drosophila* (Leloup and Goldbeter, 1998)
- In lytic and lysogenic cycle of bacteriophage (McAdams and Shapiro, 1995)

Metabolic pathway: Mathematical interpretations of protein and gene expression, developmental biology, and enzyme kinetics are required to understand the control system that underlies metabolic flux. To design a mathematical model for metabolic pathways, dynamic mathematical models of metabolic networks and stoichiometric models and flux analysis are used. Dynamic models are designed to predict how metabolism will respond to genetics as well as environmental manipulations. Stoichiometric models and flux analysis are largely used and descriptive models, which gives the complete description of metabolism. The former is a purely descriptive tool that yields detailed quantitative snapshots of metabolism that could be employed to gain insight into physiology. The latter class of tools (e.g., dynamic models) move from the realm of description to potential prediction as to how metabolism will respond to manipulation, either environmental or genetic. Flux balance analysis (FBA) is a control based modeling method that can be used to produce computational models by integrating transcriptomic data into metabolic network regenerations. Metabolic flux analysis (MFA) is a well-known method to study metabolic flux in microorganisms but it is still a challenge to adapt these methods for a complex eukaryotic system. MFA has been used to study how various factors (genetic or environmental) and conditions can affect cellular metabolism in fungi. In mammalian cells, MFA is also used to in medical studies, toxicological research, and cell culture techniques.

A flowchart for pathway modeling is given in **Fig. 2**.

Some Common Applications of Pathways

Pathway analysis methods possess a wide range of applications in biomedical research. They help researchers to figure out important biomolecules, which are key in understanding the phenomena under study. Pathways also help in pathology specific determination of biological function of candidate genes, of which many have therapeutic value. Another application of pathway analysis is that it helps in the determination of functional similarities and dissimilarities between samples based on differential molecular abundance. In summary, pathway analysis is an indispensable tool for data driven sectors in biomedical and cell sciences.

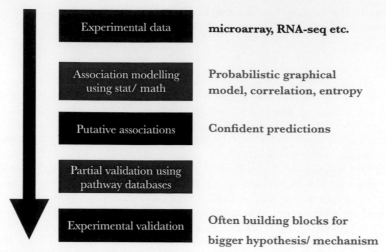

Fig. 2 Template for quantitative approaches for pathway construction.

Overrepresentation Analysis

Overrepresentation analysis (ORA) or functional enrichment analysis approach is routinely used in functional genomic studies to identify overrepresented pathways with a list of interesting or susceptible genes, genes that are likely to increase the likelihood of developing particular disease by using traditional statistical tests such as Fisher's exact test for contingency table (Jin *et al.*, 2014). ORA involves creating an input gene list based on certain criteria, which is followed by counting the number of genes in the list of susceptible genes that hit the gene set annotated by Gene Ontology, KEGG, BioCarta, and Reactome. This is followed by assessing the significance of overlaps using traditional statistical tests and finding out relevant pathways based on their p values. Although it is a widely used method there are certain limitations, in that it does not take into account complex gene–gene interactions and secondly it assumes that each gene is of equal importance, which is not the case in biology (Dong *et al.*, 2016). It is not suitable to use ORA when genes to be analyzed are very large as it is based on a stringent threshold. Its statistical power is low when GWAS data is used for identification of significant genes (Jin *et al.*, 2014).

Gene Set-Based Scoring

Gene set-based scoring methods are an extension of ORA but they don't assume that each individual gene is of equal importance; rather they rank these genes by some statistical criteria or *p* values. They are different from ORA in two ways. Firstly, they remove criteria of using an arbitrary threshold as used by ORA and thus consider all of the genes in an experiment. Secondly, they permute phenotype class labels to assess the significance and they allow preserving of gene–gene correlation, thus resulting in a more accurate null model. Statistical tests such as Wilcoxon rank sum or Kolmogorov–Smirnov can be used to know the overall impact of a gene set on biological phenotypes (Jin *et al.*, 2014).

One of the applications of gene set-based scoring is in gene set enrichment analysis (GSEA), a powerful method for inferring results of gene expression data at a level of gene sets. This method is quite powerful since it focuses on gene sets, i.e., genes sharing common biological functions chromosomal location or regulation. Gene sets are defined based on prior biological knowledge such as published biochemical pathway data or other coexpression experimental data. Ranking of genes is based on the correlation between their expression and class distinction can be done by using any suitable metric. The aim of GSEA is to determine whether the majority of genes from a gene set falls towards the top or bottom of the list and in which case the gene set can be correlated with the disease phenotypes. Main steps involved in GSEA are:

a. Calculation of enrichment score (ES): ES represents degree to which gene set is overrepresented at the extremes (top or bottom) of the entire ranked list. This score corresponds to weighted Kolmogorov–Smirnov like statistics.
b. Estimation of significance level of ES: Statistical significance is estimated by empirical phenotypic based permutation test procedures in order to generate a null distribution for the ES.
c. Adjustment for multiple hypothesis testing: When a large number of gene sets are being analyzed at one time, ES for each gene set is normalized, and false discovery rate is calculated (Subramanian *et al.*, 2005).

Network Topology-Based Analysis

Topological analysis is involved in identification of global qualitative properties of the biological system. Most methods in topological analysis are developed for analyzing gene expression data, and they virtually can be extended to other data types (e.g., GWAS) fairly easily. They combine conventionally measured molecular data and also the structural information of the pathway provided by biological databases. Classical graph theory is used by one approach to identify several motifs, which are groups of interacting biological entities capable of processing information that occurs repeatedly in a pathway and are represented as a directed graph. Boolean network analysis can be used to identify feedback loops, i.e., impact of negative or positive regulations of each interaction. Bayesian networks can also be used for inferring indirect relationships and studying complex cellular networks from quantitative experimental data (Viswanathan *et al.*, 2008). Two algorithms based on topological based analysis are:

CliPPER: CliPPER is an empirical two-step method based on Gaussian graphical models that tries to identify signal paths within significantly altered pathways. First step in CliPPER is testing of the whole pathway. Under specific conditions, strength of molecular interactions within a pathway could be transformed (Martini *et al.*, 2012). The P value of the test defining whether two graphical Gaussian models (cases and control) are homoscedastic as the weight is collected to compute the relevance of each path (Jin *et al.*, 2014). Second step involves identification within these pathways the signal paths having the greatest association with a specific phenotype (Martini *et al.*, 2012).

Signaling Pathway Impact Analysis (SPIA): SPIA is topology-based pathway analysis method that combines two types of evidence, classical enrichment analysis and actual perturbation on a given pathway (Tarca *et al.*, 2008). It is involved in the capturing of several features of data such as changes in gene expression, the pathway enrichment, and the topology of signaling pathways (Jin *et al.*, 2014). In order to assess the significance of the observed total pathway perturbation, a bootstrap procedure is used (Tarca *et al.*, 2008). This method models signaling pathways as a graph where nodes are represented by genes and edges represent interactions (Jin *et al.*, 2014). Furthermore, it defines a gene-level statistics and pathway level statistics, which is called perturbation factor (PF). The PF for a gene is defined as a sum of its measured change in expression and a linear function of the PFs of all genes in a pathway. PF in case of pathway is defined as a sum of PFs of all genes in a pathway (Tarca *et al.*, 2008).

Conclusion

Pathway analysis has emerged as an indispensable tool in functional genomics studies. It is routinely being used in generating biological hypothesis testing on the premise of quantitative molecular investigations. With the accumulation of underanalyzed genomic data deluge an unprecedented opportunity has been created to accelerate pathway research. The role of big consortia is felt in managing goal oriented crowd-sourced projects for expanding and crystallizing our knowledge about biological pathways.

Also, for a better interoperability of multiple software systems and data types, pathway databases must adopt common data standards and formats as it is highly desirable to combine data from different pathway databases. This will allow better coverage of all the reactions and biological entities involved in a pathway (Bauer-Mehren *et al.*, 2009).

In summary, advancement in the field of pathway informatics together with next-generation sequencing technologies presents an unparalleled opportunity for fully exploiting the power of pathway database and pathway analysis resources, tools, and software for understanding cell and disease biology better.

See also: Algorithms for Graph and Network Analysis: Clustering and Search of Motifs in Graphs. Algorithms for Graph and Network Analysis: Graph Alignment. Algorithms for Graph and Network Analysis: Graph Indexes/Descriptors. Algorithms for Graph and Network Analysis: Traversing/Searching/Sampling Graphs. Alignment of Protein-Protein Interaction Networks. Biological and Medical Ontologies: GO and GOA. Biological Pathway Analysis. Biological Pathway Data Formats and Standards. Biological Pathways. Cell Modeling and Simulation. Cluster Analysis of Biological Networks. Community Detection in Biological Networks. Computational Approaches for Modeling Signal Transduction Networks. Computational Systems Biology Applications. Computational Systems Biology. Data Mining: Mining Frequent Patterns, Associations Rules, and Correlations. Functional Enrichment Analysis Methods. Gene Regulatory Network Review. Gene-Gene Interactions: An Essential Component to Modeling Complexity for Precision Medicine. Graph Algorithms. Graph Isomorphism. Graph Theory and Definitions. Graphlets and Motifs in Biological Networks. Investigating Metabolic Pathways and Networks. Metabolic Models. Metabolic Profiling. Metabolome Analysis. Natural Language Processing Approaches in Bioinformatics. Network Centralities and Node Ranking. Network Inference and Reconstruction in Bioinformatics. Network Models. Network Properties. Network Topology. Network-Based Analysis for Biological Discovery. Network-Based Analysis of Host-Pathogen Interactions. Networks in Biology. Prediction of Protein-Protein Interactions: Looking Through the Kaleidoscope. Protein–Protein Interaction Databases. Protein-Protein Interactions: An Overview. Quantitative Modelling Approaches. Transcriptome Analysis. Two Decades of Biological Pathway Databases: Results and Challenges. Visualization of Biological Pathways. Visualization of Biomedical Networks

---

References

Ananiadou, S., Pyysalo, S., Tsujii, J.I., Kell, D.B., 2010. Event extraction for systems biology by text mining the literature. Trends in Biotechnology 28 (7), 381–390.

Ay, A., Arnosti, D.N., 2011. Mathematical modeling of gene expression: A guide for the perplexed biologist. Critical Reviews in Biochemistry and Molecular Biology 46 (2), 137–151.

Bauer-Mehren, A., Furlong, L.I., Sanz, F., 2009. Pathway databases and tools for their exploitation: Benefits, current limitations and challenges. Molecular Systems Biology 5 (1), 290.

Berg, J.M., Tymoczko, J.L., Stryer, L., 2002. Signal-transduction pathways: An introduction to information metabolism. Biochemistry. 395–424.

Berridge, M.J., 2014. Module 7: Cellular processes. Cell Signalling Biology 6, csb0001007.

Campagne, F., Neves, S., Chang, C.W., et al., 2004. Quantitative information management for the biochemical computation of cellular networks. Science STKE. 2004 (248), pl11.
Carrier, T.A., Keasling, J.D., 1999. Investigating autocatalytic gene expression systems through mechanistic modeling. Journal of Theoretical Biology 201 (1), 25–36.
Cary, M.P., Bader, G.D., Sander, C., 2005. Pathway information for systems biology. FEBS Letters 579 (8), 1815–1820.
Caspi, R., Altman, T., Billington, R., et al., 2013. The MetaCyc database of metabolic pathways and enzymes and the BioCyc collection of pathway/genome databases. Nucleic Acids Research 42 (D1), D459–D471.
Chan, S.Y., Loscalzo, J., 2012. The emerging paradigm of network medicine in the study of human disease. Circulation Research 111 (3), 359–374.
Cohen, K.B., Hunter, L., 2008. Getting started in text mining. PLOS Computational Biology 4 (1), e20.
Croft, D., O'Kelly, G., Wu, G., et al., 2010. Reactome: A database of reactions, pathways and biological processes. Nucleic Acids Research 39 (Suppl. 1), D691–D697.
Dong, X., Hao, Y., Wang, X., Tian, W., 2016. LEGO: A novel method for gene set over-representation analysis by incorporating network-based gene weights. Scientific Reports 6.
Endy, D., You, L., Yin, J., Molineux, I.J., 2000. Computation, prediction, and experimental tests of fitness for bacteriophage T7 mutants with permuted genomes. Proceedings of the National Academy of Sciences 97 (10), 5375–5380.
Ficklin, S.P., Dunwoodie, L.J., Poehlman, W.L., et al., 2017. Discovering condition-specific gene co-expression patterns using gaussian mixture models: A cancer case study. Scientific Reports 7 (1), 8617.
Frolkis, A., Knox, C., Lim, E., et al., 2009. SMPDB: The small molecule pathway database. Nucleic Acids Research 38 (Suppl. 1), D480–D487.
Gilman, A.G., Simon, M.I., Bourne, H.R., et al., 2002. Overview of the alliance for cellular signaling. Nature 420 (6916), 703–706.
Gough, N.R., 2002. Science's signal transduction knowledge environment. Annals of the New York Academy of Sciences 971 (1), 585–587.
Han, H., Shim, H., Shin, D., et al., 2015. TRRUST: A reference database of human transcriptional regulatory interactions. Scientific Reports 5, 11432.
Hoffmann, M.F., Preissner, S.C., Nickel, J., et al., 2013. The transformer database: Biotransformation of xenobiotics. Nucleic Acids Research 42 (D1), D1113–D1117.
Holoch, D., Moazed, D., 2015. RNA-mediated epigenetic regulation of gene expression. Nature Reviews Genetics 16 (2), 71–84.
Hoopes, L., 2008. Introduction to the gene expression and regulation topic room. National Education 1 (1), 160.
Hou, J., Acharya, L., Zhu, D., Cheng, J., 2015. An overview of bioinformatics methods for modeling biological pathways in yeast. Briefings in Functional Genomics 15 (2), 95–108.
Jewison, T., Su, Y., Disfany, F.M., et al., 2013. SMPDB 2.0: Big improvements to the small molecule pathway database. Nucleic Acids Research 42 (D1), D478–D484.
Jin, L., Zuo, X.Y., Su, W.Y., et al., 2014. Pathway-based analysis tools for complex diseases: A review. Genomics, Proteomics & Bioinformatics 12 (5), 210–220.
Joyner, M.J., Pedersen, B.K., 2011. Ten questions about systems biology. The Journal of Physiology 589 (5), 1017–1030.
Kanehisa, M., Goto, S., 2000. KEGG: Kyoto encyclopedia of genes and genomes. Nucleic Acids Research 28 (1), 27–30.
Karlebach, G., Shamir, R., 2008. Modelling and analysis of gene regulatory networks. Nature Reviews Molecular Cell Biology 9 (10), 770–780.
Karp, P.D., 2003. Pathway bioinformatics. In: CSB, p. 27.
Kelder, T., van Iersel, M.P., Hanspers, K., et al., 2011. WikiPathways: Building research communities on biological pathways. Nucleic Acids Research 40 (D1), D1301–D1307.
Kutmon, M., Riutta, A., Nunes, N., et al., 2015. WikiPathways: Capturing the full diversity of pathway knowledge. Nucleic Acids Research 44 (D1), D488–D494.
Law, V., Knox, C., Djoumbou, Y., et al., 2013. DrugBank 4.0: Shedding new light on drug metabolism. Nucleic Acids Research 42 (D1), D1091–D1097.
Leloup, J.C., Goldbeter, A., 1998. A model for circadian rhythms in Drosophila incorporating the formation of a complex between the PER and TIM proteins. Journal of Biological Rhythms 13 (1), 70–87.
Lerner, C., Antezana, E., Couche, F., et al., 2004. The aMAZE LightBench: A web interface to a relational database of cellular processes. Nucleic Acids Research 32 (Suppl. 1), D443–D448.
Li, S., Wu, L., Zhang, Z., 2006. Constructing biological networks through combined literature mining and microarray analysis: A LMMA approach. Bioinformatics 22 (17), 2143–2150.
Liu, W., Li, D., Zhu, Y., He, F., 2008. Bioinformatics analyses for signal transduction networks. Science in China Series C: Life Sciences 51 (11), 994–1002.
Liu, X., Betterton, M.D. (Eds.), 2012. Computational Modeling of Signaling Networks. Humana Press.
Ma, H., Zhao, H., 2013. Drug target inference through pathway analysis of genomics data. Advanced Drug Delivery Reviews 65 (7), 966–972.
Martini, P., Sales, G., Massa, M.S., Chiogna, M., Romualdi, C., 2012. Along signal paths: An empirical gene set approach exploiting pathway topology. Nucleic Acids Research 41 (1), e19.
Mathelier, A., Wasserman, W.W., 2013. The next generation of transcription factor binding site prediction. PLOS Computational Biology 9 (9), e1003214.
Matys, V., Kel-Margoulis, O.V., Fricke, E., et al., 2006. TRANSFAC® and its module TRANSCompel®: Transcriptional gene regulation in eukaryotes. Nucleic Acids Research 34 (Suppl. 1), D108–D110.
McAdams, H.H., Shapiro, L., 1995. Circuit simulation of genetic networks. Science 269 (5224), 650–656.
Nishimura, D., 2001. BioCarta. Biotech Software & Internet Report: The Computer Software Journal for Scientific 2 (3), 117–120.
Novak, B., Tyson, J.J., 1993. Modeling the cell division cycle: M-phase trigger, oscillations, and size control. Journal of Theoretical Biology 165 (1), 101–134.
Pawson, T., 1995. Protein modules and signalling networks. Nature 373 (6515), 573–580.
Santillán, M., Mackey, M.C., 2001. Dynamic regulation of the tryptophan operon: A modeling study and comparison with experimental data. Proceedings of the National Academy of Sciences 98 (4), 1364–1369.
Sengupta, D., Bandyopadhyay, S., 2011. Participation of microRNAs in human interactome: Extraction of microRNA–microRNA regulations. Molecular Biosystems 7 (6), 1966–1973.
Sengupta, D., Bandyopadhyay, S., 2013. Topological patterns in microRNA–gene regulatory network: Studies in colorectal and breast cancer. Molecular Biosystems 9 (6), 1360–1371.
Sivakumaran, S., Hariharaputran, S., Mishra, J., Bhalla, U.S., 2003. The database of quantitative cellular signaling: Management and analysis of chemical kinetic models of signaling networks. Bioinformatics 19 (3), 408–415.
Subramanian, A., Tamayo, P., Mootha, V.K., et al., 2005. Gene set enrichment analysis: A knowledge-based approach for interpreting genome-wide expression profiles. Proceedings of the National Academy of Sciences 102 (43), 15545–15550.
Tarca, A.L., Draghici, S., Khatri, P., et al., 2008. A novel signaling pathway impact analysis. Bioinformatics 25 (1), 75–82.
van Dam, S., Võsa, U., van der Graaf, A., Franke, L., de Magalhães, J.P., 2017. Gene co-expression analysis for functional classification and gene–disease predictions. Briefings in Bioinformatics. bbw139.
Viswanathan, G.A., Seto, J., Patil, S., Nudelman, G., Sealfon, S.C., 2008. Getting started in biological pathway construction and analysis. PLOS Computational Biology 4 (2), e16.
Wang, R.S., Saadatpour, A., Albert, R., 2012. Boolean modeling in systems biology: An overview of methodology and applications. Physical Biology 9 (5), 055001.
Ward, C., 2016. Metabolic pathways [internet]. 2016 Jan 19; Diapedia 5105765817 rev. no. 25.
Wilkinson, G.R., 2005. Drug metabolism and variability among patients in drug response. New England Journal of Medicine 352 (21), 2211–2221.
Wingender, E., Chen, X., Hehl, R., et al., 2000. TRANSFAC: An integrated system for gene expression regulation. Nucleic Acids Research 28 (1), 316–319.
Wishart, D.S., Tzur, D., Knox, C., et al., 2007. HMDB: The human metabolome database. Nucleic Acids Research 35 (Suppl. 1), D521–D526.
Yu, D., Kim, M., Xiao, G., Hwang, T.H., 2013. Review of biological network data and its applications. Genomics & Informatics 11 (4), 200–210.
Zeng, H., Qiu, C., Cui, Q., 2015. Drug-Path: A database for drug-induced pathways. Database 2015.

Relevant Website

http://cgap.nci.nih.gov/Pathways/BioCarta_Pathways
    BioCarta_Pathways.

Network Inference and Reconstruction in Bioinformatics

Paolo Tieri, CNR National Research Council, IAC Institute for Applied Computing "Mauro Picone", Rome, Italy
Lorenzo Farina, Department of Computer, Control and Management Engineering "A. Ruberti", Sapienza University of Rome, Italy
and CNR National Research Council, IASI Institute of Systems Analysis and Computer Science "Antonio Ruberti", Rome, Italy
Manuela Petti, Department of Computer, Control and Management Engineering "A. Ruberti", Sapienza University of Rome, Italy and
Neuroelectrical Imaging and BCI Lab, Fondazione Santa Lucia IRCSS, Rome, Italy
Laura Astolfi, Department of Computer, Control and Management Engineering "A. Ruberti", Sapienza University of Rome, Italy and
Neuroelectrical Imaging and BCI Lab, Fondazione Santa Lucia IRCSS, Rome, Italy
Paola Paci, CNR National Research Council, IASI Institute of Systems Analysis and Computer Science "Antonio Ruberti", Rome,
Italy
Filippo Castiglione, National Research Council of Italy, Rome, Italy

Introduction

Biological networks inference, also referred to as 'reverse engineering', is the scientific process of using (low- and high-throughput) experimental data, statistical and computational techniques to reconstruct how the elements of the biological network (genes, proteins, signaling molecules, cells) interact and operate as a system.

The paradigm of 'network' is crucial for the sciences of complexity, as well as for systems biology. This concept represents a potent instrument for the description and analysis of complex systems, their elements and their connections, able to detect underlying topologies, structures, architectures as well as functions emerging from the arrangement of their components. Such methodology has been fruitfully employed for the representation and for the analysis of numerous different systems in various fields of study, e.g., engineering and technology, social studies and, of course, life sciences.

The power of the network approach basically resides in the capability to catch specific features of any type of systems, e.g., steady and/or physically connected systems (such as the physical internet cabling, or power grids) as well as dynamic, virtual and non-cabled (such as social networks, air traffic networks, or gene/protein interactions).

In bioinformatics, such interdisciplinary approach allowed to undertake, probably for the first time, rigorous mathematical analysis of larger portions of, or even whole biological systems.

A critical step is therefore the process of reconstruction of such networks starting from accessible data. Here, making no claim to completeness, we briefly outline inference and reconstruction approaches of some among the most relevant types of biological networks in molecular biology and neuroscience, and some of the most promising methodologies applied in the recent field of network medicine.

Molecular Networks Inference

Protein–Protein Interaction Networks

Protein-protein interactions (PPIs) are highly specific physical contacts, stable or transient in time, between two or more proteins. PPIs involve electrostatic forces between the so-called protein surfaces, i.e. the 'exposed' regions of the three-dimensional structures of folded proteins. PPIs are essential to almost every process in a biological system, e.g., in signal transduction, cell metabolism, membrane transport, muscle contraction, among others, and consequently their mapping and understanding is critical for the comprehension of cellular physiology in normal and as well as in pathological conditions.

The representation of interconnected webs of PPIs at the cellular level, or, in other words, the union of all considered proteins and the interactions among them, are generally referred to as protein interaction networks (PINs). Such networks are usually represented with undirected graphs and undergo mathematical and computational analysis in order to study cell processes, functions and organization at the system level.

PINs reconstruction is carried using several experimental (in vivo and in vitro) and computational methods, involving specific and different pros and cons (Rao *et al.*, 2014; Cafarelli *et al.*, 2017).

Among the most widely known experimental approaches, Yeast 2 hybrid (Y2H) is an *in vivo* method, suitable for both transient and stable interactions. It is implemented by screening the protein of interest against a library of potential interacting protein partners through the activation of a reporter gene (usually the gene *Gal4* in the yeast *Saccharomyces cerevisiae*). Y2H is affordable, fast and low-tech, but recently it has been criticized as not completely reliable in terms of yielding high numbers of false positive identifications. Y2H is the only *in vivo* method based on the detection of real physical interactions, being the other *in vivo* one, namely synthetic lethality, based on functional interactions, i.e. using deletions or mutations in genes to observe protein interactions.

Among the most used *in vitro* techniques, tandem affinity purification-mass spectroscopy (TAP-MS) is based on labelling the protein under scrutiny on its chromosomal locus with a designed protein tag, and on a two-step purification method followed by mass spectroscopic analysis. It is usually considered suitable for stable interactions only (due to the purification steps), simple to implement and quantitatively reliable. Other approaches include protein microarrays, affinity chromatography, coimmunoprecipitation, X-ray crystallography or NMR spectroscopy, among others.

Among the *in silico* methods, the ortholog-based sequence approach is a commonly used technique: it is grounded on the homologous nature of the query protein in the annotated protein databases using pairwise local sequence algorithm. Other in silico approaches are based on protein sequence (as the summentioned ortholog-based one) or protein structure, on gene fusion or gene expression, on chromosomal proximity, on phylogenetic tree or gene ontology.

Methods for the assessment of PPIs, among other approaches, can be implemented computationally e.g., by using specific reliability indexes such as 'expression profile reliability' (EPR) index (used to assess whole datasets by comparing the RNA expression profiles for the proteins whose interactions are found in the screen with expression profiles for known interacting and non-interacting pairs of proteins) or the paralogous verification method (PVM, used to score individual interactions), which judges an interaction likely if the putatively interacting pair has paralogs that also interact. PPI validation can be also implemented by using known macromolecular complexes that provide defined and unbiased set of protein interactions (Przulj, 2011). It is foreseen that prospective PIN reconstruction attempts will focus on the dynamic and quantitative properties, surpassing current stationary representations (Cafarelli *et al.*, 2017).

Gene Regulatory Networks

Gene regulatory networks (GRNs, also known as transcriptional regulatory networks) are networks of causal interactions among transcription factors and downstream genes, and are usually represented with directed graphs and inferred by gene expression data. Links between elements of a GRN represent biochemical process such as a reaction, transformation, interaction, activation or inhibition. GRNs are a crucial area of systems biology research, providing understanding of the inward mechanisms of a biological system.

Many methods to infer GRNs are based on mutual information (MI) (e.g., MI with background, maximal MI, conditional MI, three-way MI), a probabilistic concept and dimensionless quantity that states how much the knowledge of a random variable tells about another one: in other words, mutual information can be thought as the decrease in uncertainty of a random variable once given knowledge of another, so that mutual information equal to zero between two random variables means that the variables are completely independent (nothing can be said about a variable given the other), and increases of mutual information denote decreases in such uncertainty.

Inferred GRNs represent causal biochemical interactions between genes, RNA and proteins which links aim to correspond to real physical, directed, and quantitatively determined interaction events between such molecules that can potentially lead to the discover of, for example, crucial functional relationships between RNA expression and chemotherapeutic susceptibility (Butte *et al.*, 2000). Recently, data from single-cell gene expression have become mature and have been approached using partial information decomposition to detect putative functional associations and to formulate systematic hypotheses (Chan *et al.*, 2017; Aibar *et al.*, 2017).

The validation of GRNs implies the definition of 'gold standards' (i.e. sets of interactions validated with a given level of certainty) and the use statistical thresholds to quantitatively evaluate the adherence of the GRN to such standards, or perturbation analysis of the biological (sub)system to observe and measure perturbation effects and subsequently validate the network under scrutiny (Chai *et al.*, 2014; Hecker *et al.*, 2009; Emmert-Streib *et al.*, 2014; Veiga *et al.*, 2010; Lee and Tzou, 2009; Bansal *et al.*, 2007).

Gene Co-Expression Networks

Gene co-expression networks (GCNs) are transcript–transcript association networks, generally reported as undirected graphs, where genes are connected when an appreciable co-expression association between them exists.

GCNs are built from gene expression data by calculating co-expression values in terms of pairwise gene similarity score and choosing a significance threshold. Debates on normalization methods, co-expression correlation (e.g., based on Pearson's or Spearman's correlation measures, among others) and significance and relevance are still alive and ongoing. Graphical Gaussian Models (also said 'concentration graph' or 'covariance selection' models) are also popular in the field of GCNs, the key idea behind them being the use of partial correlations, able to discriminate between direct and indirect interactions, as a degree of independence of any pair of genes (Schafer and Strimmer, 2005). Other strategies for GCN inference include edge removal based on gene triplets analysis (e.g., ARACNE) (Margolin *et al.*, 2006), regression methods and Bayesian networks (aimed to discover the best gene set predictor of a target gene expression).

GCNs are a potent approach to gather biologically relevant information, e.g., for the identification of genes not yet associated with explicit biological questions, and for accelerating the interpretation of molecular mechanisms at the root of significant biological processes. Trends in this field include, among other approaches, the combination of co-expression analysis with other omics techniques, such as metabolomics, for estimating the coordinated behavior between gene expression and metabolites, as well as for assessing metabolite-regulated genetic networks (Serin *et al.*, 2016; Usadel *et al.*, 2009).

Metabolic Networks

Metabolism, apart from obviously being the set of life-sustaining biochemical reactions in every organisms, is also among the major contributors to many human diseases, such as cardiovascular diseases, obesity, diabetes and cancer, to cite some. Metabolic network reconstruction is commonly referred to as the annotation process of genes and metabolites for the determination of the metabolic network's elements, relationships, structure and dynamics. It is generally possible to infer the enzymatic function of individual proteins, or to reconstruct larger (or whole) metabolic networks in their entirety.

Originally introduced at the beginning of 1900, the first successful attempts to quantitatively reconstruct small metabolic networks, involving linear sequences of few reactions, date back to the Eighties of the past century. Later on, coordinated efforts led to the reconstruction of the global human metabolic network, based on recent genome annotation and on a large amount of bibliomic data (Mo *et al.*, 2007). Today, the main aim is usually the reconstruction of genome-scale metabolic networks, and techniques such as metabolic flux analysis (MFA) and its enhancements (e.g., isotopically nonstationary metabolic flux analysis), and flux balance analysis (FBA) has become widely used for the simultaneous predictions of fluxes of multiple reactions. The methods are generally based on stable isotope-labeled substrates and measuring time courses of intracellular metabolites labels. FBA is able to generate flux predictions without prior information of enzyme-kinetic parameters, based on optimization principles that theoretically should have shaped efficiency (and fluxes) of metabolic networks. Integrated computational approaches coupling metabolic flux analysis with mass spectrometry approaches have been also recently used. From the computational side, single enzyme function prediction is carried out by using machine learning (e.g., when the enzyme does not show significant similarity to existing proteins from which predictions can be made) or the so-called 'annotation transfer' approaches (based on DNA sequences or reference databases or orthologs, to predict the occurrence of enzymes). Comparative pathway prediction techniques generally use existing functional annotations to check for the presence of new reactions, while explorative pathway prediction approaches, i.e. without the use of existing annotations, can be graph-theoretic (e.g., by weighting paths of metabolite connectivity) or constraint-based (e.g., elementary mode analysis), or both (Pitkanen *et al.*, 2010; Nikoloski *et al.*, 2015).

Signaling Networks

Signaling networks are cascades of molecular interactions and chemical modifications to carry stimuli (e.g., signaling molecules, hormones, pathogens, nutrients) sensed by cell membrane receptors down to the nucleus to coordinate initiate proper metabolic and genetic responses. The processes of identifying signaling networks constituents and of network reconstruction have been often carried out by using gene knockout techniques, which have been effective in describing cascades in a linear manner. Nevertheless, the complexity and intricate nature of such networks require new methods that can infer aspects and features of signaling processes from high-throughput omic data in a faster and systemic way. In this perspective, efficient algorithms are necessary to analyze and detect relations among numerous data points, and in general such inference problems can be reported to the definition of suitable optimal connected subgraphs of a network originally defined by the available data, which can result in being computationally intractable. Some of the approaches employ Steiner tree approaches (e.g., based on the determination of the shortest total lengths of paths of interacting proteins), or linear programming and maximum-likelihood (e.g., based on tagging proteins as activators or repressors to explain the maximum number of observed gene knockout); other use probabilistic network approaches, such as network flow optimization (e.g., based on Bayesian weighting schemes for underlying PPIs coupled with other omic data), or network propagation (e.g., based on a gene prioritization function that scores the strength-of-association of proteins with a given disease), or information flow analysis (a computational approach that identifies proteins dominant in the communication of biological information across the network) (Tuncbag *et al.*, 2013; Ourfali *et al.*, 2007; Missiuro *et al.*, 2009; Huang and Fraenkel, 2009; Aldridge *et al.*, 2006; Papin *et al.*, 2005; Ritz *et al.*, 2016; Molinelli *et al.*, 2013; Hyduke and Palsson, 2010)

Assessment and Validation of Inferred Molecular Networks

The methods to reconstruct biological networks as well as inferred networks need to undergo rigorous validation and assessment processes based on theoretical points, on *in silico* experiments, and on 'wet lab' data. Known issues about assessment approaches include the volume of information accessible related to the network entities, the general sparseness of biological networks as well as their topology, and, often neglected, the lack of experimentally verified non-interacting pairs. Some general considerations arise, applicable to the different types of networks and of methods. As a general procedure, network inference methods have been often assessed using receiver operating characteristic (ROC) curves and precision-recall (PR) curves: being them differently sensitive to the ratio between (true/false/actual/predicted) positives and negatives among the tested pairs, ROC curves seem to show and advantage in comparing different classification methods (Schrynemackers *et al.*, 2013). Another general consideration regards the relevance of the lack of experimental support for non-interacting pairs in most biological networks, especially in the field of PPIs, lack that can heavily impinge upon the assessment of false positives, issue that is recently receiving more attention (Blohm *et al.*, 2014; Smialowski *et al.*, 2010; Launay *et al.*, 2017; Srivastava *et al.*, 2016). Some specific measures have been used to assess inferential validity for GRNs, such as the Hamming distance (e.g., to measure the network topology closeness), or steady-state mass difference for network dynamic behavior similarity (Qian and Dougherty, 2013). As a general consideration, the challenging nature of network inference tasks should be highlighted, as a remarkable amount of the evaluated algorithms seems to not accomplish better results than random (Lopes and Bontempi, 2013; De Smet and Marchal, 2010), but it is also recognized that despite inferred networks might not be error-free at the level of mechanistic description, they can still be fruitfully used to comprehend emergent functions and behavior of biological systems (Stolovitzky *et al.*, 2007).

Networks in Neuroscience

In neuroscience, network science can be applied to the patterns obtained by studying the brain anatomy and activity at different spatial (molecules, neurons, circuits, whole brain, social behavior) and temporal scales (milliseconds, minutes, hour, days, years). Examples include networks like those described above (PPI, genetic regulatory networks…) and networks consisting of anatomical or functional connections among brain areas. The concept of functional brain connectivity is crucial to understand how communication between cortical regions is organized: brain areas interacting with a functional connection are not necessarily connected by a direct physical link; on the other hand, an anatomical link does not necessarily imply that a functional connection has established at any time point.

Anatomical connectivity can be tracked by diffusion weighted imaging techniques such as Diffusion Tensor Imaging (DTI): this technique shows anatomical connections of neurons by identifying nerve fiber pathways in the brain.

Brain functional connectivity can be estimated from a wide range of biomedical signals and with different neuroimaging techniques – functional magnetic resonance imaging (fMRI), electroencephalography (EEG), magnetoencephalography (MEG) positron emission tomography (PET). In the following a description of two of the main approaches is provided.

- **Functional magnetic resonance imaging (fMRI)** is a specific magnetic resonance imaging procedure to measure brain activity by detecting associated changes in blood flow: specifically, brain activity is measured through low frequency blood oxygenation level dependent (BOLD) signal in the brain. fMRI technique is characterized by high spatial resolution, but by poor time resolution.
- **Electroencephalography** is a method to record electrical activity of the brain. Scalp EEG data have many advantages: high temporal resolution, non-invasiveness and low-cost. On the other hand, it is characterized by low spatial resolution. An EEG setup can be used in a wide range of pathological conditions and can be easily adapted to different clinical settings. To overcome the limitation of the low spatial resolution, in the past decades, the inverse EEG solutions has received considerable attention: this approach provides estimation of source distributions within the brain that correspond to the scalp-recorded signals. Different methods (e.g., minimum norm estimates (MNE), low resolution electrical tomography (LORETA), beam-forming approaches) have been proposed to solve the EEG inverse problem (Grech et al., 2008). Once the inverse problem for EEG source localization is solved, connectivity estimation can be performed in the source space.

Three classes of methods are commonly used to estimate functional connectivity:

- **Model-based functional connectivity approaches**. Seed based connectivity is a hypothesis-driven method based on a priori decision regarding the regions of interest (ROI). It is usually adopted with fMRI data. *Methods based on Granger causality* (Granger, 1969) also belong to this class. They are defined both in time and in frequency domain, and being based on bivariate or multivariate autoregressive models they allow to reconstruct the direction of information flows. Examples of Granger Causality-based methods are: i) *Partial Directed Coherence* (Baccalá and Sameshima, 2001); ii) *Directed Transfer Function* (Kamiński and Blinowska, 1991); iii) *time reversed Granger Causality* (Haufe, Nikulin, and Nolte, 2012).
- **Model-free functional connectivity approaches**. These methods are especially useful in the analysis of spatially distributed functional connectivity networks. An example is the *imaginary coherence* (Nolte et al., 2004). Coherency between two EEG-channels is a measure of the linear relationship of the two at a specific frequency. The idea is to isolate the imaginary part of the coherency which reflects interactions. This method does not return the information about interaction direction and it has been proposed as a possible solution to the interpretation of observed interactions between time series when they are affected by volume conduction.
- ***Biologically Inspired models***. *Dynamic Causal Modelling* (DCM, (Friston et al., 2003)) belongs to this class. DCM is a general framework for inferring processes and mechanisms at the neuronal level from measurements of brain activity with different techniques (fMRI, EEG/MEG, etc.): it combines a model of the hidden neuronal dynamics with a forward model that translates neuronal states into predicted measurements. This method returns directed networks.

In the last decades, several studies demonstrated that by combining neuroimaging techniques (functional MRI, EEG, etc.) with network science, it is possible to characterize the topological properties of human brain networks. Nowadays, network characterization of structural or functional connectivity data is increasing (Bassett and Sporns, 2017; Bullmore and Sporns, 2009) and rests on several important motivations. Indeed, complex network analysis promises to reliably quantify brain networks with a small number of neurobiologically meaningful and easily computable measures (Pichiorri et al., 2017; Petti et al., 2016; Achard and Bullmore, 2007; Toppi et al., 2012).

Network Medicine

Until recently, the investigation of disease etiology, diagnosis and treatment, has been based on a conventional reductionist approach. This tenet argues that critical biological factors work in a simple linear mechanism to control disease pathobiology. Rather, they are nearly always the result of multiple pathobiological pathways that interact through an interconnected network: a disease is rarely a direct consequence of an abnormality in a single gene or molecular component (Chan and LoScalzo, 2012). For example, complex diseases like hypertension, atherosclerosis or cancers of various sorts, have extraordinary complex biological phenomena that underlie them.

Today, big data, genomics, and quantitative *in silico* methodologies integration, have the potential to push forward the frontiers of medicine in an unprecedented way. Clinicians, diagnosticians and therapists have long strived to determine single molecular traits that lead to diseases. What they had in mind was the idea that a single "golden bullet" drug might provide a cure. But, this reductionist approach, largely ignored the essential complexity of human diseases. Indeed, a large body of evidence that is now emerging from new genomic technologies, points out directly to the cause of disease as "perturbations" within the "interactome", i.e. the comprehensive network map of molecular components and their interactions. The interactome integrates all physical interactions within a cell, from protein-protein, regulatory protein-DNA and metabolic interactions. Currently, more than 140,000 interactions between more than 13,000 proteins are known.

Consequently, a paradigm shift is needed towards the development of temporal and spatial multi-level models, from molecular machineries to single cells, whole organism and individuals, including the environment, to reveal the underlying links among components. This new type of medical paradigm is called "Network Medicine". Rather than trying to understand pathogenesis into a reductionist framework, network medicine entwines the many facets of disease in many different types of networks: from the physical interactions acting in a cell to the information flow through biological components. The human genome sequencing using high-throughput next-generation devices is being deeply affecting current conceptions of biomedical and clinical research. More recently, entering the era of personal whole-genome sequencing, 38 million genetic variants some of which are rare mutations and thus may be associated with large size effect.

How to use this big data and information deluge to generate better understanding of disease and find appropriate drug targets? An entirely new perspective is required as a shift from a reductionist to a holistic and data-driven research: looking at networks without a specific biomedical bias in mind and let the data speak by themselves, so to formulate new hypothesis to be further validated by experimentalist and so on, moving within a virtuous circle of shared knowledge. Most importantly, the representation of complex systems as networks is of paramount importance for visualizing the interactome underlying structure, revealing new functional roles, and proposing new and fresh interpretations of data. Networks can be obtained from any sort of information: known protein-protein interactions, gene expression profiles, functional annotation, etc. In fact, the fundamental tenet of network medicine is to look at diseases as perturbations within the 'interactome', i.e. the overall network map of molecular components and their interactions. Recently, has been shown the genes associated with a disease are localized in specific neighborhoods, or 'disease modules', within such an interactome. The overall ambition of this initiative is to both developing a global understanding of how interactome perturbations result in disease traits, and to translate computational insights into concrete clinical applications, such as new drugs and therapies or diagnostic tools.

A true inter-disciplinary environment for network medicine must be based on the accurate preparation the interactions among experts of different fields, especially from those trained in medical/biological sciences with those experts in data analysis, bioinformatics, statistical modelling, and network algorithms. The gap between the biological and the informational mindset can be daunting and might impair from the beginning the development of shared concepts. However, the network medicine setting will certainly facilitate communications across disciplines given the immediate and intuitive understanding of the "network" concept, a metaphor that can be used by molecular biologists to visualize their knowledge in a structured way ready to be translated into an algorithm on the available data, as medical or biological goals are defined. Barabási *et al.* (2011) proposed the following research pipeline:

for any specific disease, the identification and validation of disease modules consists of several steps:

- Interactome reconstruction merges the most up-to-date information on protein–protein interactions, co-complex memberships, regulatory interactions and metabolic network maps in the tissue and cell line of interest.
- Disease gene (seed) identification collects the known disease-associated genes obtained from linkage analysis, genome-wide association studies or other sources, which serve as the seed of the disease module.
- In disease module identification, the seed genes are placed on the interactome, with the aim of identifying a subnetwork that contains most of the disease-associated components, exploiting both the functional and topological modularity of the network.
- Pathway identification can be used in instances in which the number of components contained in the ascertained disease module is so large that it cannot serve as a tractable starting point for further experimental work.
- During validation disease modules are tested for their functional and dynamic homogeneity.

Disease Gene Prediction Methodologies

A classification of the disease gene prediction methodologies can be performed following different criteria (Piro and Di Cunto, 2012):

- Type of evidence: different data sources can be exploited to predict disease-genes (sequence data, protein-protein interactions, gene expression...).
- Level of knowledge: disease-gene prediction methods can be classified based on the use of prior knowledge for example about causal genes for the disease phenotype under investigation or about affected tissues/organs.
- Type of prediction: prediction methods can return a prioritization of the candidate genes (a ranking according to their likelihood of being involved in a disorder) or a selection of best candidates. In the first case the goal is to rank the disease gene as high as possible in the prioritized candidate list, while candidate selection methods provide a candidates' set (as small as possible).
- Scope of application: based on this criterion, we can distinguish between disease-centered methods and undifferentiated approaches. In the first case, candidates are evaluated in relation to a specific disease (or disorders class), while methods belonging to the second class provide an overall evaluation of the involvement of the candidates in some disease in general.

More recently, the main classification was performed based on network analysis strategy:

- local methods, algorithms based on the search for direct neighbors of disease genes;

- global methods: methods that model the information flow in the cell to assess the proximity and connectivity between known disease genes and candidate genes.

Here we focus on disease-gene prediction methods that are based on the analysis of the topological properties of protein-protein interaction (PPI) networks and we describe with more detail some of the most used and newest approaches at the state of the art.

Local Methods

- **Direct Interaction (DI).** Oti *et al.* (2006) proposed a procedure to detect disease-genes in known disease loci by exploring the degree to which proteins linked to known disease-genes are also associated with the same phenotype. The algorithm performs the following steps:
  - identification of neighboring proteins for each known disease protein;
  - identification of chromosomal locations of the genes coding for these interacting proteins using gene location data from the Ensembl database;
  - check of the chromosomal locations: each interacting protein gene located within one or more disease loci (of the same disease) was considered a candidate gene prediction;
  - predictions counting: if a candidate gene lay within multiple (overlapping) loci of the same disease, each of them was counted as a separate prediction.
- **GenePANDA (Gene Prioritizing Approach using Network Distance Analysis).** In the last year another algorithm of this class has been proposed: GenePANDA (Gene Prioritizing Approach using Network Distance Analysis) (Yin *et al.*, 2017). The novelty of GenePANDA is the introduction of adjusted network distance that is derived by considering not only the direct network distance between two genes, but also their respective mean network distances to all other genes in the network. This method uses the STRING network (Franceschini *et al.*, 2013) and the Genetic Association Database (GAD) as the resource of disease data, and it consists of three steps:
  - adjusted network distance computation: adjusted network distance is computed considering not only the direct network distance between two genes, but also their respective mean network distances to all other genes in the network. Given the genes a and b, their adjusted network distance is defined as:

$$D_{ab}^{adj} = \frac{D_{ab}}{\sqrt{\mu_a \times \mu_b}} \tag{1}$$

where D_{ab} is the raw network distance between gene a and gene b, μ_a and μ_b the mean raw network distances for a and b respectively, defined as:

$$\mu_a = \frac{\sum_{j=1}^{N} D_{aj}}{N} \tag{2}$$

with N the total number of genes in the network, and D_{aj} the raw network distance between a and b;
  - *disease-specific gene weighting* with the hypothesis that a candidate disease gene should have stronger functional interaction with known disease genes than with random genes in the network. Based on this hypothesis, a disease-specific gene weight was defined as:

$$w_i = \frac{\sum_{j=1}^{N} D_{ab}^{adj}}{N} - \frac{\sum_{j=1}^{K} D_{ab}^{adj}}{K} \tag{3}$$

where N is the total number of genes in the network, K is the total number of disease genes. Higher weight indicating higher probability to be a candidate disease gene.
  - score conversion: for a given disease, the disease-specific gene weights can be compared with each other related to the same disease, but they cannot be directly compared across diseases. To overcome this limitation, a score conversion procedure was introduced converting the weights into probabilities.

Global Methods

- **Random Walk with Restart (RWR).** Kohler *et al.* (2008) proposed the random walk analysis for definition of similarity in PPI networks demonstrating that global network-similarity measures (random walk and diffusion kernel) are greatly superior to local distance measures (direct links or shortest paths) in the disease-genes prioritization problem. They performed a random walk on the PPI network, starting at the known disease genes, and rank candidate genes by the steady state probabilities induced by the walk. In the RWR procedure, a random walk starts at one node in the set S of known disease genes (seed genes): the random walker can move to a randomly selected neighbor or it can restart at one of the proteins in S. For each restart, the probability of restarting at a specific protein is r. Formally, RWR is defined as:

$$p^{t+1} = (1-r)Wp^t + rp^0 \tag{4}$$

where W is the column-normalized adjacency matrix of the graph and p^t is a vector of probability of being at node i at time step t. In Kohler *et al.* (2008), the authors set the initial probability vector p0 with equal probabilities to the nodes of S, with the sum of the probabilities equal to 1 (disease genes with equal probability).

- **PRINCE (PRIoritizatioN and Complex Elucidation).** To compute the association between disease genes and candidate proteins, Vanunu *et al.* (2010) proposed a network propagation algorithm based on formulating constraints on the prioritization function that relate to its smoothness over the network and usage of prior information. PRINCE receives as input a disease-disease similarity measure and a network of protein-protein interactions. The algorithm infers a strength-of-association scoring function that models an information pump that originates at the seed sets: this function is smooth over the network (adjacent nodes are assigned with similar values) and respects the prior knowledge on causal genes for the same disease or similar ones.

- **DA DA.** In Erten *et al.* (2011), the authors proposed DA DA (Degree-Aware Algorithms for Network-Based Disease Gene Prioritization), a free available suite implemented in Matlab. This algorithm applies statistical adjustment methods to correct for degree bias in information flow based disease gene prioritization with the aim to limit the prediction errors due to the incompleteness and the noisy network of PPI networks. Despite methods based on random walks or network propagation as the two above described (Kohler *et al*, V1.36., 2008; Vanunu *et al.*2010), are less affected by this problem by considering multiple alternate paths and whole topology of PPI networks, DADA outperforms these methods.

- **ProDiGe.** The algorithm proposed in Mordelet and Vert (2011) implements a novel machine learning strategy based on learning from positive and unlabeled examples: it assumes that a set of gene-disease associations is already known to infer new ones. The main difference from the existing approaches, is the definition of a scoring function using both the set P of known disease genes (positive examples) and the set U of candidate genes (unlabeled examples), formulating the problem of disease genes prediction as an instance of the problem known as learning from positive and unlabeled examples (PU learning). Moreover, ProDiGe is characterized by other two properties: i) it exploits information about known disease genes across diseases; ii) it allows heterogeneous data integration (sequence features, expression levels in different conditions…).

- **DIAMOnD (DIseAse MOdule Detection).** In Ghiassian *et al.* (2015), the authors showed that disease associated proteins do not reside within locally dense communities and that connectivity significance is the most predictive quantity. Based on these results, they developed a novel algorithm (DIAMOnD) that performs a systematic analysis of the network properties exploiting the connectivity significance instead of connection density. To identify disease modules, DIAMOnD algorithm performs the following steps:

  i. computation of the connectivity significance for all proteins with at least one connection to any of the seed proteins:

$$p - value(k, k_s) = \sum_{k_i=k_s}^{k} p(k, k_i) \tag{5}$$

where s_0 is the relatively small number of seed proteins associated with a particular disease and $p(k,ks)$ is the probability that a protein with k links has exactly ks $(ks<k)$ links to seed proteins given by the hypergeometric distribution:

$$p(k, k_s) = \frac{\dbinom{s_0}{k_s}\dbinom{N-s_0}{k-k_s}}{\dbinom{N}{k}} \tag{6}$$

where N is the number of proteins;

  ii. ranking of the protein according to their respective p-values;

  iii. the protein with the highest rank (i.e. lowest p-value) is added to the set of seed nodes, increasing their number;

  iv. steps (i)-(iii) are iterated with the expanded set of seed proteins, pulling in one protein at a time into the growing disease module.

The procedure (i)-(iv) can be continued until the entire network is agglomerated. The order in which the proteins are being pulled in to the module reflects their topological relevance to the disease, resulting in a ranking of all proteins.

Conclusion and Perspectives

Among the obstacles and future perspectives concerning network inference, the discovery of 'actual' causal regulatory networks in the gargantuan space of possible networks is among the most formidable ones, just considering that the number of possible arrangements, i.e. different networks, of a relatively 'trivial' gene regulatory network containing as little as 50 nodes outclasses the estimated number of atoms of the observable universe ($\sim 10^{80}$). In such a challenge, it can be easily anticipated that previous knowledge, in terms of background, exploitable information, including already extracted causal connections and subnetworks, will likely impact the quality and quantity of successive attempts (or network 'updates') of network reconstruction in the different fields. Optimization methods, algorithmic progresses, innovative analytics and well-founded statistical inference as well as efficient approaches for integrating omic data and information will drive advancements in network reconstruction that should eventually speed up biological discovery, by helping in the avoidance of inaccurate or far from optimal network configurations.

Lately, artificial neural networks -deep learning- approaches are receiving more attention and resources in computational biology, thanks to their ability to capitalize on the accessibility and convenience of high-throughput omic data: better performance and higher-level characteristics can be obtained over conventional modelling methodology, and they could potentially deliver crucial understanding about the organization of biological networks.

The last decade has witnessed the relentless growth of biological molecular data (the so-called "omics" revolution) driven by highly efficient biotechnological devices and methods. Also, neuroscientists recently started collecting multi-subject, multi-modal human neuroimaging datasets: speaking more in general, 'big data' is the new reality also in biology, medicine and neuroscience. This growth is still ongoing and even increasing in speed and quality of outcomes. However, despite initial enthusiasms, the challenge of its management and use for the understanding of diseases has become over time a daunting task. In other words, the goal of turning lead into gold, i.e. transform data into knowledge, seems to step far away every single day. As a matter of fact, experimental measurements on almost every molecular entity acting in a cell are now currently available on databases spread all over the world, and this "heap" of data is incredibly huge and ever-increasing. The inevitable consequence of such wild expansion is that now it is practically impossible to put all the pieces together for two basic, simple reasons: (1) there are many missing ones, and (2) they are virtually infinite, so that the current 'data factory' is unlikely to stop working or even slow down. Now, it is time for asking questions and provide frameworks for real inter-disciplinary studies.

Network representations are invaluable for this purpose, as just shown. A network is the ideal metaphor for representing relationships among data, finding "patterns" for biomarker studies and, most importantly, any kind of expertise can find the opportunity to fully express itself using a common language. In fact, the network metaphor provides plenty of words and concepts that can be exploited by biologists and clinicians to represent their knowledge, and by bioinformaticians or computational modellers to derive algorithms and find answers using computation and mathematics. In other words, we envisage a new way of conceiving a "inter-disciplinary team" where biology and computation can flourish together (Korcsmaros et al., 2017). The key point is that researchers should leave behind grand visions of "universal principles" and start an "earthly" work of untangling one by one the long list of problems that ask for (at least partial) answers in a peer-to-peer inter-disciplinary setting where the network metaphor provides the language to go beyond the conceptual/computational barriers.

See also: Algorithms for Graph and Network Analysis: Graph Alignment. Algorithms for Graph and Network Analysis: Graph Indexes/ Descriptors. Algorithms for Graph and Network Analysis: Traversing/Searching/Sampling Graphs. Computational Systems Biology Applications. Computational Systems Biology. Data Mining: Mining Frequent Patterns, Associations Rules, and Correlations. Gene Regulatory Network Review. Graphlets and Motifs in Biological Networks. Natural Language Processing Approaches in Bioinformatics. Network Models. Network Properties. Network-Based Analysis for Biological Discovery. Network-Based Analysis of Host-Pathogen Interactions. Networks in Biology. Pathway Informatics

References

Achard, S., Bullmore, E., 2007. Efficiency and cost of economical brain functional networks. PLOS Comput. Biol. 3, e17.

Aibar, S., Gonzalez-Blas, C.B., Moerman, T., et al., 2017. SCENIC: Single-cell regulatory network inference and clustering. Nat. Methods 14, 1083–1086.

Aldridge, B.B., Burke, J.M., Lauffenburger, D.A., Sorger, P.K., 2006. Physicochemical modelling of cell signalling pathways. Nat. Cell Biol. 8, 1195–1203.

Baccalá, Luiz A., Sameshima, Koichi, 2001. Partial Directed Coherence: A New Concept in Neural Structure Determination. Biological Cybernetics 84 (6), 463–474. doi:10.1007/PL00007990.

Bansal, M., Belcastro, V., Ambesi-Impiombato, A., Bernardo, D.I., 2007. How to infer gene networks from expression profiles. Mol. Syst. Biol. 3, 78.

Barabási, A.L., Gulbahce, N., Loscalzo, J., 2011. Network medicine: A network-based approach to human disease. Nat. Rev. Genet. 12, 56–68.

Bassett, D.S., Sporns, O., 2017. Network neuroscience. Nat. Neurosci. 20, 353–364.

Blohm, P., Frishman, G., Smialowski, P., et al., 2014. Negatome 2.0: A database of non-interacting proteins derived by literature mining, manual annotation and protein structure analysis. Nucleic Acids Res. 42, D396–D400.

Bullmore, E., Sporns, O., 2009. Complex brain networks: Graph theoretical analysis of structural and functional systems. Nat. Rev. Neurosci. 10, 186–198.

Butte, A.J., Tamayo, P., Slonim, D., Golub, T.R., Kohane, I.S., 2000. Discovering functional relationships between RNA expression and chemotherapeutic susceptibility using relevance networks. Proc. Natl. Acad. Sci. USA 97, 12182–12186.

Cafarelli, T.M., Desbuleux, A., Wang, Y., et al., 2017. Mapping, modeling, and characterization of protein–protein interactions on a proteomic scale. Curr. Opin. Struct. Biol. 44, 201–210.

Chai, L.E., Loh, S.K., Low, S.T., et al., 2014. A review on the computational approaches for gene regulatory network construction. Comput. Biol. Med. 48, 55–65.

Chan, S.Y., LoScalzo, J., 2012. The emerging paradigm of network medicine in the study of human disease. Circulation Research. 111, 359–6374

Chan, T.E., Stumpf, M.P.H., Babtie, A.C., 2017. Gene regulatory network inference from single-cell data using multivariate information measures. Cell Syst. 5, 251–267. (e3).

De Smet, R., Marchal, K., 2010. Advantages and limitations of current network inference methods. Nat. Rev. Microbiol. 8, 717–729.

Emmert-Streib, F., Dehmer, M., Haibe-Kains, B., 2014. Gene regulatory networks and their applications: Understanding biological and medical problems in terms of networks. Front. Cell Dev. Biol. 2, 38.

Erten, S., Bebek, G., Ewing, R.M., Koyuturk, M., 2011. DADA: Degree-aware Algorithms for network-based disease gene prioritization. BioData Min. 4, 19.

Franceschini, A., Szklarczyk, D., Frankild, S., et al., 2013. STRING v9.1: Protein-protein interaction networks, with increased coverage and integration. Nucleic Acids Res. 41, D808–D815.

Friston, K.J., Harrison, L., Penny, W., 2003. Dynamic causal modelling. Neuroimage 19, 1273–1302.

Ghiassian, S.D., Menche, J., Barabási, A.L., 2015. A DIseAse MOdule Detection (DIAMOnD) algorithm derived from a systematic analysis of connectivity patterns of disease proteins in the human interactome. PLOS Comput. Biol. 11, e1004120.

Granger, C.W.J., 1969. Investigating Causal Relations by Econometric Models and Cross-Spectral Methods. Econometrica 37 (3), 424–438. doi:10.2307/1912791.

Grech, R., Cassar, T., Muscat, J., et al., 2008. Review on solving the inverse problem in EEG source analysis. J. Neuroeng. Rehabil. 5, 25.

Haufe, Stefan, Vadim, V. Nikulin, Nolte, Guido, 2012. Alleviating the Influence of Weak Data Asymmetries on Granger-Causal Analyses. In: Latent Variable Analysis and Signal Separation. Berlin, Heidelberg: Springer, pp. 25–33. https://doi.org/10.1007/978-3-642-28551-6_4.

Hecker, M., Lambeck, S., Toepfer, S., Van Someren, E., Guthke, R., 2009. Gene regulatory network inference: Data integration in dynamic models-a review. Biosystems 96, 86–103.

Huang, S.S., Fraenkel, E., 2009. Integrating proteomic, transcriptional, and interactome data reveals hidden components of signaling and regulatory networks. Sci. Signal. 2, ra40.

Hyduke, D.R., Palsson, B.O., 2010. Towards genome-scale signalling network reconstructions. Nat. Rev. Genet. 11, 297–307.

Kaminski, M.J., Blinowska, K.J., 1991. A New Method of the Description of the Information Flow in the Brain Structures. Biological Cybernetics 65 (3), 203–210.

Kohler, S., Bauer, S., Horn, D., Robinson, P.N., 2008. Walking the interactome for prioritization of candidate disease genes. Am. J. Hum. Genet. 82, 949–958.

Korcsmaros, T., Schneider, M.V., Superti-Furga, G., 2017. Next generation of network medicine: Interdisciplinary signaling approaches. Integr. Biol. (Camb.) 9, 97–108.

Launay, G., Ceres, N., Martin, J., 2017. Non-interacting proteins may resemble interacting proteins: Prevalence and implications. Sci. Rep. 7, 40419.

Lee, W.P., Tzou, W.S., 2009. Computational methods for discovering gene networks from expression data. Brief Bioinform. 10, 408–423.

Lopes, M., Bontempi, G., 2013. Experimental assessment of static and dynamic algorithms for gene regulation inference from time series expression data. Front. Genet. 4, 303.

Margolin, A.A., Nemenman, I., Basso, K., et al., 2006. ARACNE: An algorithm for the reconstruction of gene regulatory networks in a mammalian cellular context. BMC Bioinform. 7 (Suppl 1), S7.

Missiuro, P.V., Liu, K., Zou, L., et al., 2009. Information flow analysis of interactome networks. PLOS Comput. Biol. 5, e1000350.

Molinelli, E.J., Korkut, A., Wang, W., et al., 2013. Perturbation biology: Inferring signaling networks in cellular systems. PLOS Comput. Biol. 9, e1003290.

Mo, M.L., Jamshidi, N., Palsson, B.O., 2007. A genome-scale, constraint-based approach to systems biology of human metabolism. Mol. Biosyst. 3, 598–603.

Mordelet, F., Vert, J.P., 2011. ProDiGe: Prioritization Of Disease Genes with multitask machine learning from positive and unlabeled examples. BMC Bioinform. 12, 389.

Nikoloski, Z., Perez-Storey, R., Sweetlove, L.J., 2015. Inference and prediction of metabolic network fluxes. Plant Physiol. 169, 1443–1455.

Nolte, G., Bai, O., Wheaton, L., et al., 2004. Identifying true brain interaction from EEG data using the imaginary part of coherency. Clin. Neurophysiol. 115, 2292–2307.

Oti, M., Snel, B., Huynen, M.A., Brunner, H.G., 2006. Predicting disease genes using protein-protein interactions. J. Med. Genet. 43 (8), 691–698.

Ourfali, O., Shlomi, T., Ideker, T., Ruppin, E., Sharan, R., 2007. SPINE: A framework for signaling-regulatory pathway inference from cause-effect experiments. Bioinformatics 23, i359–i366.

Papin, J.A., Hunter, T., Palsson, B.O., Subramaniam, S., 2005. Reconstruction of cellular signalling networks and analysis of their properties. Nat. Rev. Mol. Cell Biol. 6, 99–111.

Petti, M., Toppi, J., Babiloni, F., et al., 2016. EEG resting-state brain topological reorganization as a function of age. Comput. Intell. Neurosci. 2016, 6243694.

Pichiorri, F., Petti, M., Caschera, S., et al., 2017. An EEG index of sensorimotor interhemispheric coupling after unilateral stroke: Clinical and neurophysiological study. Eur. J. Neurosci.

Piro, R.M., Di Cunto, F., 2012. Computational approaches to disease-gene prediction: Rationale, classification and successes. FEBS J. 279, 678–696.

Pitkanen, E., Rousu, J., Ukkonen, E., 2010. Computational methods for metabolic reconstruction. Curr. Opin. Biotechnol. 21, 70–77.

Przulj, N., 2011. Protein-protein interactions: Making sense of networks via graph-theoretic modeling. Bioessays 33, 115–123.

Qian, X., Dougherty, E.R., 2013. Validation of gene regulatory network inference based on controllability. Front. Genet. 4, 272.

Rao, V.S., Srinivas, K., Sujini, G.N., Kumar, G.N., 2014. Protein-protein interaction detection: Methods and analysis. Int. J. Proteomics 2014, 147648.

Ritz, A., Poirel, C.L., Tegge, A.N., et al., 2016. Pathways on demand: Automated reconstruction of human signaling networks. NPJ Syst. Biol. Appl. 2, 16002.

Schafer, J., Strimmer, K., 2005. An empirical bayes approach to inferring large-scale gene association networks. Bioinformatics 21, 754–764.

Schrynemackers, M., Kuffner, R., Geurts, P., 2013. On protocols and measures for the validation of supervised methods for the inference of biological networks. Front. Genet. 4, 262.

Serin, E.A., Nijveen, H., Hilhorst, H.W., Ligterink, W., 2016. Learning from co-expression networks: Possibilities and challenges. Front. Plant Sci. 7, 444.

Smialowski, P., Pagel, P., Wong, P., et al., 2010. The negatome database: A reference set of non-interacting protein pairs. Nucleic Acids Res. 38, D540–D544.

Srivastava, A., Mazzocco, G., Kel, A., Wyrwicz, L.S., Plewczynski, D., 2016. Detecting reliable non interacting proteins (NIPs) significantly enhancing the computational prediction of protein-protein interactions using machine learning methods. Mol. Biosyst. 12, 778–785.

Stolovitzky, G., Monroe, D., Califano, A., 2007. Dialogue on reverse-engineering assessment and methods: The DREAM of high-throughput pathway inference. Ann. NY Acad. Sci. 1115, 1–22.

Toppi, J., De Vico Fallani, F., Vecchiato, G., et al., 2012. How the statistical validation of functional connectivity patterns can prevent erroneous definition of small-world properties of a brain connectivity network. Comput. Math. Methods Med. 2012, 130985.

Tuncbag, N., Braunstein, A., Pagnani, A., et al., 2013. Simultaneous reconstruction of multiple signaling pathways via the prize-collecting steiner forest problem. J. Comput. Biol. 20, 124–136.

Usadel, B., Obayashi, T., Mutwil, M., et al., 2009. Co-expression tools for plant biology: Opportunities for hypothesis generation and caveats. Plant Cell Environ. 32, 1633–1651.

Vanunu, O., Magger, O., Ruppin, E., Shlomi, T., Sharan, R., 2010. Associating genes and protein complexes with disease via network propagation. PLOS Comput. Biol. 6, e1000641.

Veiga, D.F., Dutta, B., Balazsi, G., 2010. Network inference and network response identification: Moving genome-scale data to the next level of biological discovery. Mol. Biosyst. 6, 469–480.

Yin, T., Chen, S., Wu, X., Tian, W., 2017. GenePANDA – A novel network-based gene prioritizing tool for complex diseases. Sci. Rep. 7, 43258.

Further Reading

Barabási, A.L., Oltvai, Z.N., 2004. Network biology: Understanding the cell's functional organization. Nat. Rev. Genet. 5 (2), 101–113. PMID: 14735121.

Bassett, D.S., Sporns, O., 2017. Network neuroscience. Nat. Neurosci. 20 (3), 353–364. doi:10.1038/nn.4502. PMID: 28230844.

Csermely, P., Korcsmáros, T., Kiss, H.J., London, G., Nussinov, R., 2013. Structure and dynamics of molecular networks: A novel paradigm of drug discovery: A comprehensive review. Pharmacol. Ther. 138 (3), 333–408. doi:10.1016/j.pharmthera.2013.01.016. PMID: 23384594.

Le Novère, N., 2015. Quantitative and logic modelling of molecular and gene networks. Nat. Rev. Genet. 16 (3), 146–158. doi:10.1038/nrg3885. PMID: 25645874.

Loscalzo, Joseph, Barabási, Albert-László, Silverman, Edwin K., 2017. Network Medicine: Complex Systems in Human Disease and Therapeutics. Harvard University Press.

Graphlets and Motifs in Biological Networks

Laurentino Quiroga Moreno, Birkbeck College, University of London, United Kingdom

Introduction

Large-scale biological networks are being generated at a growing rate. Advances in experimental biotechnologies are yielding molecular interaction networks of various types, including protein-protein interaction (PPI), epistatic genetic interaction (GI), metabolic reaction, and transcriptional regulation data, among others (Snider *et al.*, 2015). As each data set in isolation carries only limited information about biology and as the data sets complement each other (Przulj and Malod-Dognin, 2016), extracting new biological information hidden in the wiring patterns of these complex interconnected "big data" has received much attention in the scientific community and industry. These data are often modelled as *networks* (also called *graphs*), in which biomolecules are modelled as *nodes* (also called *vertices*) and their interactions are modelled as *edges* (also called *links*). Hence, biological systems have been modelled and analysed by using networks (Barabasi and Oltavi, 2004).

However, analysing large networks is mostly computationally intractable. Computational complexity theory studies hardness of computational problems and many of the problems underlying analytics of large networked data fall into the category of computationally intractable problems, so called NP-hard and NP-complete problems (Bondy and Murty, 1976). In short, computational problems roughly fall into the two main categories: polynomial time solvable problems (P) and non-deterministic polynomial time hard (NP-hard). Class P contains problems for which we can compute exact solutions in the time that is polynomial of the size of the input data. These problems are called computationally *easy*. In contrast, NP-hard problems cannot be solved exactly on large scale data, so we must solve them approximately, or heuristically. Some intractable problems include network comparison and alignment, which are hard due to NP-completeness (special type of NP-hard problems) of the underlying sub-graph isomorphism problem (Cook, 1971).

Proteins are key to the functioning of the cell and hence PPI networks play important roles in the functioning of cells. Similarities in the wiring of proteins within a PPI network have been shown to also mean similarity in biological function. Similarities in wiring patterns of proteins in PPI networks can also be used to guide network alignment algorithms, which analogous to sequence alignments, find and align similarly wired and similarly functioning proteins over PPI networks of different species. This, in turn, is used to transfer annotation from model organisms to human. Network motifs and graphlets have been used to address all these issues.

Network Motifs and Graphlets

For the above reasons, many approximate methods to analyse the structure of biological networks have been proposed. They can roughly and historically be divided into *local* and *global network properties*. Simple examples include the *degree* of a node (the number of edges that the node participates in), the *degree distribution* (the distribution of degrees over all nodes of a network), the *clustering coefficient* of a node (the number of edges between the neighbours of a node as a percentage of the maximum possible number of edges between them), the *average clustering coefficient* of the network over all its nodes etc. (Newman, 2010). More detailed descriptors of network structure include methods based on *network motifs* and *graphlets*, which are the subject of this article.

Network Motifs

Network motifs have been introduced by the group of Uri Alon (Milo *et al.*, 2002). They are defined as small patterns of interconnections occurring in complex networks at numbers that are significantly higher than those in randomized networks (Milo *et al.*, 2002). They were used to study the transcriptional regulation networks of well-studied microorganisms (Alon and Mangan, 2003; Mangan *et al.*, 2003), as well as of higher order organisms (Charney *et al.*, 2017; Datta *et al.*, 2017). It was shown that these networks appear to be made up of a small set of recurring regulation patterns, captured by network motifs.

Example motifs include positive and negative autoregulation, positive and negative cascades, positive and negative feedback loops, feedforward loops (FFLs), single input modules, and combinations of these, illustrated in **Fig. 1** (Shoval and Alon, 2010). They were linked with biological function (illustrated in **Fig. 1**). Since the same network motifs have been found in diverse organisms from bacteria to humans, it has been suggested that they serve as basic building blocks of transcription networks. Network motifs in other biological networks also seem to exist, including signalling (Ptacek *et al.*, 2005; Itzkovitz *et al.*, 2005; Ma'ayan *et al.*, 2005) and neuronal networks (Sporns and Kotter, 2004; Sakata *et al.*, 2005). Also, they were used to analyse integrated networks of transcriptional regulation and protein-protein interactions (Yeger-Lotem *et al.*, 2004). For a review, see Alon (2007).

"Temporal motifs" were introduced to study the temporal changes in local network structure over time-evolving networks. For example, static network motifs described above were counted in each network snapshot over time and then their counts compared across the snapshots (Braha and Bar-Yam, 2009). This approach ignored any motif relationships between different network

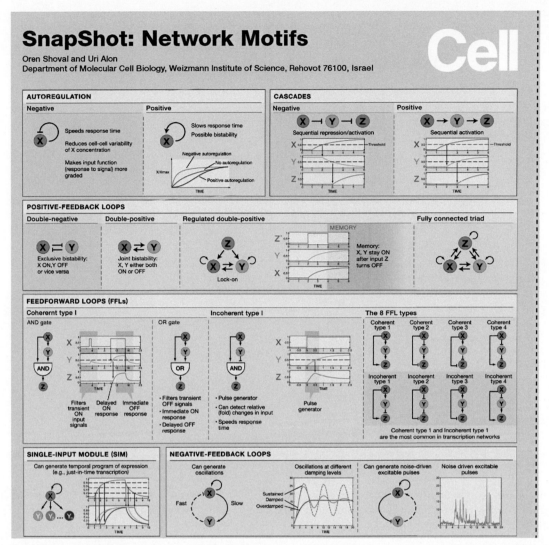

Fig. 1 An illustration of different types of network motifs and their functions. Taken from Shoval, O., Alon, U. 2010. SnapShot: Network motifs. Cell 143, 326.

snapshots. Hence, static network motifs were extended into several notions of temporal motifs (Chechik *et al.*, 2008; Kovanen *et al.*, 2011). Intuitively, temporal motifs are classes of sequences of similar network event sequences, where the similarity refers both to topology and to the temporal order of the events.

To understand the difference in the definition of motifs and graphlets, we need to introduce the following simple graph theoretic terms. A *partial subgraph* is a subgraph of a larger network in which once we pick the nodes that form the subgraph, we can pick any subset of edges between the chosen nodes of the larger network. An *induced subgraph* is a subgraph in which we must pick *all* the edges between the chosen nodes of the larger network to form the subgraph. Both of these subgraph definitions are illustrated in **Fig. 2**. Network motifs are partial subgraphs that are significantly overrepresented in the data compared to a chosen random graph model that is assumed to fit well the data.

However, as network comparison is computational intractable, determining a well-fitting network model is hard, as it involves comparing the data network with model networks. A random network model that is usually used as well-fitting is that of a random graph constructed to have the same degree distribution as the data network, while edges are drawn at random. Also, note that motifs are partial subgraphs, while characterizing the structure of any graph class is based on induced subgraphs (Andreas Brandstädt, 1999). An anti-motif is defined similar to a network motif, but it needs to be under-represented in the data compared to a random network model.

Hence, analyses of biological networks involving network motifs have been criticised, since the definitions of motifs and anti-motifs heavily depend on the choice of a random graph (network) model (Artzy-Randrap *et al.*, 2004; Milo *et al.*, 2004), since they are partial subgraphs (Przulj *et al.*, 2004) and since they can exhibit a whole range of dynamic behaviours (Ingram *et al.*, 2006). Furthermore, it is computationally hard to identify network motifs (and the same holds for graphlets), as the number of possible sub-graphs increases exponentially with the network and motif size (node and edge counts). Hence, various algorithms for their

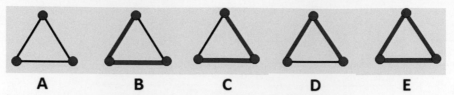

Fig. 2 Partial *vs* induced subgraphs. In the fully connected network with 3 nodes illustrated in panel A, if we pick all 3 of its nodes to make connected subgraphs, there exist three partial subgraphs corresponding to 3-node paths, illustrated in panels B, C and D, but only one induced subgraph, a triangle, illustrated in panel E.

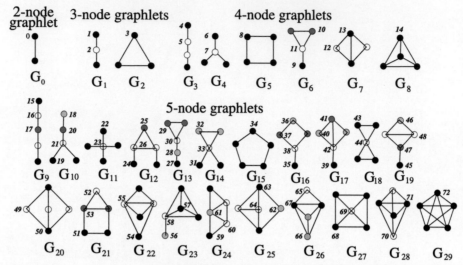

Fig. 3 All 2- to 5-node graphlets, G_0 to G_{29}, along with their symmetry groups (automorphism orbits), numbered from 0 to 72. Taken from Przulj, N., 2006. Biological Network Comparision Using Graphlet Degree. Oxford University Press, pp. e177–e183.

detection have been proposed, including those involving parallel and cloud computing (Kim *et al.*, 2013). A survey paper from 2011 discusses the biological significance of network motifs, the motivation for finding them, as well as the strategies for solving various aspects of this problem (Wong *et al.*, 2011).

Graphlets

As a response to the above mentioned drawbacks in the definition of network motifs, *graphlets* were introduced (Przulj *et al.*, 2004). Graphlets are small, connected, induced subgraphs of large networks (Przulj *et al.*, 2004). They can appear in the data network at any frequency and hence their definition does not depend on any assumed random graph model that fits well the data. Also, they are induced, so they are a suitable tool for designing various algorithms for mining domain relevant new information from the structures (topologies) of biological and other network data. All 2- to 5-node undirected graphlets are illustrated in **Fig. 3**.

As systems-level molecular network data present an opportunity to discover new biological information, several data analytics algorithms included graphlets as the underlying methodology aimed at uncovering the additional meaning. For instance, *graphlet frequency distribution* is introduced to be superior to the *degree distribution* (which is the first in the spectrum of 73 graphlet degree distributions), the *clustering coefficient* (the measure of "cliquishness" of the network) and *network diameter* (which measures how "far spread" the nodes of the network are) by imposing a large number of similarity constrains on the networks being compared (Przulj, 2006). It does so by generalizing the degree distribution (which measures the number of nodes touching k edges) into a spectrum of 73 distributions measuring the number of nodes touching k graphlets at a particular orbit for each value of k (the 73 orbits for 2- to 5-node graphlets are numbered in **Fig. 3**). Basically, it compares sequences of numbers over 73 pairs to compare two networks. It can be classified into both types of network comparison heuristics, global and local, as it does comparison globally over the entire network, but uses local network features in the comparisons.

As biological networks are globally incomplete, but locally more complete (due to biases in biotechnologies and experimental design focusing on heavily exploring the parts of networks relevant for human disease), focusing more on local network properties provides a better measure of network structure for these particular data (Przulj, 2006). This was demonstrated on 14 real PPI networks and the four network models (Przulj, 2006). Also, it was shown for the first time that *geometric random graphs* are fitting PPI networks better than other previously used models (Przulj *et al.*, 2004). In geometric random graphs, nodes correspond to points randomly distributed in a metric space and edges are drawn between the points if the corresponding points are close enough (within a given radius) in space. This model was later refined to mimic network growth inspired by gene duplication and mutation events (Przulj *et al.*, 2010).

A more sophisticated heuristic measure of distance between networks is called *Graphlet Correlation Distance (GCD)*, which exploits correlations between graphlet orbits (symmetries in graphlets, see **Fig. 3**) over all nodes in a network, to compress the topology of a network of any size and density into an 11 by 11 matrix of numbers between -1 and 1, the so called Graphlet Correlation Matrix (GCM). It compares GCMs of different networks to find the topological similarity between the networks (Yaveroglu *et al.*, 2014). The superiority of GCD over other measures was demonstrated not only in accuracy, but also in noise tolerance and computational efficiency (Yaveroglu *et al.*, 2015). GCD demonstrated relationships between unrelated networks, such as, Facebook, metabolic, and protein structure networks. Biological networks were also compared by the méthod involving differential graphlet communities (Wong *et al.*, 2015).

The constant production of large scale, dynamic and directed correlated data has demanded development of more sophisticated methods to analyse them and gain additional biological insights. Analysing directed networks by using graphlets was proposed, since their above described undirected counterparts were successful in analysing undirected network data. Graphlets and their degrees were extended to directed data, as well as the GCD measure (Sarajlic *et al.*, 2016). It was demonstrated that these directed versions of graphlet-based tools are more sensitive in comparing directed networks than if the undirected version was applied to directed data. Additionally, it was necessary to extend the conical correlation analysis framework to facilitate uncovering the relationships between directed network wirings and their annotations. The method was applied to metabolic networks (Sarajlic *et al.*, 2016).

Algorithms for Graphlet Identification

However, identifying graphlets is computationally demanding as one needs to search through exponentially many possible subgraphs, so heuristic approaches have been sought (Przulj *et al.*, 2006; Wang, *et al.*, 2016). For instance, by using geometric graphs as a well-fitting model of PPI networks, a heuristic exploits the property of geometric graphs being sparser at the "boundary" and denser in the "core" of the network to quickly estimate graphlet counts in geometric graphs and also in PPI networks, since they are fit well by geometric graphs. This heuristic is called Targeted Node Processing (TNP), since it targets counting of graphlets in these sparse, network boundary neighbourhoods and from them estimates the counts in the entire network. The same paper proposes another heuristic called Neighbourhood Local Search (NLS) that is based on a more standard sampling search procedure of randomly sampling nodes and counting graphlets around them to produce the estimates (Przulj *et al.*, 2006). Both heuristics work well for high-confidence PPI networks, which have power-law degree distribution, as well as for geometric random graphs which do not, indicating that the behaviour of the heuristics is dictated by the local rather than global structure of these networks. A combinatorial method was also proposed, using a system of equations that connect counts of orbits in graphlets to allow for computing all orbit counts by enumerating a single one, hence reducing practical time complexity in sparse graphs by an order of magnitude compared to the exhaustive enumeration-based algorithms (Hocevar and Demsar, 2014).

Graphlets Link Network Structure and Biological Function

Since few protein structures have been resolved, predicting structure and function of uncharacterized proteins has been attempted by using various data sets. The traditional techniques were based on homology, but with the availability of systems-level PPI data that aim to capture the entire proteome, the questions arose of whether protein function could also be predicted from PPI network data. In this direction, a graphlet-based method demonstrated that the local structure around a protein in a PPI network and its biological function are related (Milenkovic and Przulj, 2008). The method uses a vector of graphlet degrees, called a *signature* of the protein in the PPI network, and computes the similarities of signatures of proteins across all protein pairs. It demonstrated that when the local wiring patterns around proteins are similar and proteins are grouped based on these similarities, the grouped proteins tend to perform the same biological functions, belong to the same subcellular compartments and have the same tissue expression. These observations opened up research directions for experimental design and for disease protein prediction from proteome-scale network data.

Also, new disease genes were identified from the analysis of the topology of PPI networks. It was assumed that neighbours of disease genes in the PPI network are also involved in the same or similar diseases (Ideker and Sharan, 2008). Simple network properties of disease genes were used in order to identify new gene-disease, or drug to drug-target associations. For instance, it has been demonstrated that cancer genes have larger connectivities and centralities compared with non-cancer genes (Jonsson and Bates, 2006). However, the relationship between disease genes and their network degree (number of neighbours they have in the network) needed to be analysed in more detail, because many diseases genes code for proteins that are not hubs in PPI networks (Goh *et al.*, 2007). Hence, demonstrating a relationship between cancer genes and their wirings in the PPI network required more sophisticated methodology.

Since it has been demonstrated that proteins with similar wirings in PPI networks can have the same functions (Milenkovic and Przulj, 2008), the new challenge was to establish if cancer genes also exhibit similarity of their topological graphlet degree signatures. Indeed, when signatures were used to predict new cancer genes purely from the topology of PPI networks, 80% of the predicted new cancer genes were validated in the literature (Milenkovic *et al.*, 2009). Furthermore, the biological validations with siRNA screens confirmed some of the predictions, in particular cancer-related negative regulators of melanogenesis (Milenkovic *et al.,* 2009). The results were more encouraging than expected, as the predictions were obtained only from PPI network topology, demonstrating their importance as a source of new biological information. Also, the study showed evidence that the structure of the PPI network around cancer genes is different from the structure around non-cancer genes.

Furthermore, graphlets were used to show that similarity of wiring of proteins and the link of their wiring with biological function is conserved over species even as distant as yeast and human (Davis *et al.,* 2014). It was noted that the variability of topology-function relationships required more investigation to make functional predictions and do annotation transfers, because the topology-function relationship may differ between different species and between different protein functions. To address this issue, a statistical framework was built by using conical correlation analysis (CCA). In this direction, the wirings around proteins in a PPI network are represented by graphlet degree vectors and correlated with Gene Ontology (GO) annotations (Ashburner *et al.,* 2000) by using CCA. As a result, statistically significant topology-function relationships were obtained between yeast and human. Functions that have conserved topology in PPI networks of these species were discovered, which were called "topologically orthologous" functions (Davis *et al.,* 2014).

A question of how much functional information could be obtained from sequence and how much from PPI network topology, and whether the two data types as sources of biological information corroborate, or complement each other has also been addressed by using graphlets (Memisevic *et al.,* 2010). Complementarity of PPI network and sequence information in homologous proteins has been examined, as PPI network topology and protein sequence may provide details into different slices of biological information (Memisevic *et al.,* 2010). The goal was to ensure that no biological information is lost by considering only sequence as a relevant source of biological information. First, it was examined if the information about homologues captured by the PPI network topology by using graphlet degree vectors (signatures) differs and to what extent from the information captured by their sequences. The similarity of topology around homologous proteins in the PPI network was statistically significantly higher than in non-homologous proteins. Second, none of the data sources, network topology and sequence, seemed to capture homology information entirely. However, it was demonstrated that both data sources provide relevant and complementary functional information, concluding that topological neighbourhoods of a protein in the PPI network could complement sequence-based information to identify homologous proteins.

Graphlets were used to analyse other biological networks as well. In a brain functional network, brain regions are modelled as nodes and if they are simultaneously active while the brain performs a cognitive task as measured by electrocorticographic signals, then the corresponding nodes are linked by an edge. It was established that these brain connectivity networks have a small-world topology (Basset and Bullmore, 2006). However, novel graph theoretic techniques analysed brain function networks to allow us to identify structural changes in the brain's wiring as a result various stimuli, or cognitive tasks. On brain functional networks of six epileptic patients, by using graphlet-based network similarity measures, it was found that during resting, these networks are differently wired than during performing cognitive tasks (Kuchaiev *et al.,* 2009).

In addition, analysing temporal real-world networks is a challenge. There are various options on how to model a temporal network, for example, as single aggregate statistic networks, or as a series of time-specific snapshots. Depending on the modelling, there can exist limitations in extracting information from them, because valuable temporal information could be lost if a static analysis is implemented. They were studied by considering evolution of their global network properties (Leskovec *et al.,* 2005; Nicosia *et al.,* 2012). Also, graphlets were used to study temporal networks by registering intermediate snapshot relationships (Hulovatyy *et al.,* 2015). This use of graphlets was proposed as an alternative to the one based on temporal motifs described above to alleviate the above noted limitations of finding the well-fitting null model that's required to define motifs, which is a complex task. This method was applied to age-specific molecular networks to better understand the processes of human aging.

Graphlets in Network Aligners

Another bioinformatics application in which graphlets are used is that of network alignment. Just as sequence alignment algorithms have revolutionized our understanding of biology, aligning systems-scale molecular networks will have similar ground-breaking impacts. However, as described above, aligning networks is computationally intractable due to NP-completeness of the underlying subgraph isomorphism problem. Hence, heuristic algorithms for network alignment have been proposed (Kelley *et al.,* 2004). One of the first algorithms to use purely network topology without sequence to align pairs of PPI networks is based on a seed-and-extend approach where nodes with the most similar graphlet-based topological signatures in the PPI network are aligned and then the alignment was extended around these seed nodes in a principled way (Kuchaiev *et al.,* 2010). It was applied to biological networks to produce the most comprehensive network alignments at the time, exposing surprisingly large regions of PPI network topological similarity even in species as distant as yeast and human. It produced new insight into phylogeny and biological function. As there are many possible alignments, the question of which one is optimal was addressed by using the Hungarian algorithm for solving the bipartite matching problem, where the bipartitions are nodes (proteins) of PPI network being aligned (e.g., yeast and human) and the edges across represent graphlet degree vector similarity scores between the nodes (Milenkovic *et al.,* 2010). The latest in the series of graplhet-based aligners is L-GRAAL, which combines the statistics of graphlets with Lagrangian relaxation to produce the state-of-the art alignments (Malod-Dognin and Przulj, 2015).

Another application of graphlets-based network aligners is in structural biology. Alignments of protein structures aim to deliver the most accurate mappings of equivalent residues and elucidate protein function. Aligning protein structures has attracted considerable research efforts, since there is no algorithm that yields satisfactory results and since the existing algorithms are not applicable to large scale data. Hence, proteins have been modelled as networks, in which nodes are residues and an edge exists if the corresponding residues are close enough, usually within 5 Å. As graphlets were used to design sensitive topological similarity measures between nodes and networks, the question was whether they could also be used to align protein structures. A method called GR-align was introduced for alignment of protein structures, which uses a maximum Contact Map Overlap (CMO) that is suitable for database searches (Malod-Dognin and Przulj, 2014). GR-align is more efficient in terms of speed than other similar methods. Additionally, its similarity scores agree with the structural classification of proteins. The method was applied on the Astral-40 dataset. Apart from producing better scoring than other structure alignment heuristics, it also produced results with higher level of agreement with the structural classifications proteins. GR-align also allowed for transferring more information across proteins (Malod-Dognin and Przulj, 2014).

Despite these efforts, aligning pairs of large networks still represents a challenge. Aligning multiple networks is even more challenging, as performance of such aligners decreases with increase in the number of networks being aligned. Another challenge to be met by a multiple network aligner is to deliver clusters of nodes (genes) that are evolutionarily and functionally conserved across all aligned networks, i.e., to produce both topologically and biologically meaningful alignments. The current state-of-the-art multiple network alignment algorithm, called Fuse, works in two steps (Gligorijevic et al., 2015). First, it generates node (protein) similarity scores across multiple PPI networks by using penalized (graph-regularized) non-negative matrix tri-factorization (NMTF) for fusing information from wiring patterns of all aligned protein-protein interaction (PPI) networks and sequences similarities between their proteins. Second, it identifies clusters of aligned proteins over all networks, for which it proposes a new heuristic for addressing an NP-hard problem of k-partite matching for k > 2. Fuse outperforms others aligners and produces a larger number of biologically consistent clusters that cover all aligned protein-protein interaction (PPI) networks.

Concluding Remarks

Despite significant advances in methods based on network motifs and graphlets for analysing and extracting new biological and medial information from large-scale molecular networks, we are far from comprehensively understanding biological systems such as cells, tissues, or organs. Roles of molecular networks in health and disease are still to be uncovered and once uncovered, they need to be altered by therapeutics to improve patient treatment and care. Each individual network type gives insight only into one slice of biological information. Graplhet- and motif-based methods need to be complemented by new sophisticated graph theoretic methods capable of capturing multi-scale organization of biological systems. Furthermore, these new methods and formalisms need to be integrated within data fusion frameworks, so that we would get a data-driven, full understanding of a biological system by mining all available data collectively.

See also: Algorithms for Graph and Network Analysis: Clustering and Search of Motifs in Graphs. Algorithms for Graph and Network Analysis: Graph Alignment. Algorithms for Graph and Network Analysis: Graph Indexes/Descriptors. Algorithms for Graph and Network Analysis: Traversing/Searching/Sampling Graphs. Alignment of Protein-Protein Interaction Networks. Cluster Analysis of Biological Networks. Community Detection in Biological Networks. Computational Systems Biology Applications. Computational Systems Biology. Gene Regulatory Network Review. Graph Algorithms. Graph Isomorphism. Graph Theory and Definitions. Natural Language Processing Approaches in Bioinformatics. Network Inference and Reconstruction in Bioinformatics. Network Models. Network Properties. Network Topology. Network-Based Analysis for Biological Discovery. Networks in Biology. Pathway Informatics

---

References

Alon, U., 2007. Network motifs: Theory and experimental approaches. Nature Reviews Genetics. (8), 450–461.

Alon, U., Mangan, S., 2003. Structure and function of the feed-forward loop network motif. Proceedings of the National Academy of Sciences of the United States of America 100 (21), 11980–11985.

Andreas Brandstädt, V.B., 1999. GRAPH Classes: A Survey. SIAM.

Artzy-Randrap, Y., Fleishman, S., Ben-Tal, N., Stone, L., 2004. COMMENT on "Network motifs simple blocks of complex networks" and "superfamilies of evolved and designed networks". Science.

Ashburner, et al., 2000. Gene ontology: Tool for the unification of biology. Nature Genetics 25 (1), 25–29.

Barabasi, A., Oltavi, Z., 2004. NETWORK biology: Understanding the cell's functional organization. Nature Reviews Genetics 5, 101–113.

Basset, D., Bullmore, E., 2006. Small-world brain networks. The Neuroscientist 12 (6), 512–523.

Bondy, J., Murty, U., 1976. Graph Theory With Applications. Macmillan.

Braha, D., Bar-Yam, Y., 2009. Time-dependent complex networks: Dynamic centrality, dynamic motifs, and cycles of social interactions. Adaptive Networks. Berlin: Springer, pp. 39–50.

Charney, R.M., Paraiso, K.D., Blitz, I.L., Cho, K.W., 2017. A Gene Regulatory Program Controlling Early Xenopus Mesendoderm Formation: Network Conservation and Motifs, 66. Elsevier. pp. 12–24.

Chechik, G., al, e., 2008. Activity motifs reveal principles of timing in transcriptional control of the yeast metabolic network. Nature Biotechnology 26, 1251–1259.

Cook, S., 1971. THE complexity of theorem-proving procedures. In: Proceedings of the Third Annual ACM Symposium on Theory of Computing. ACM, pp. 151–158.

Datta, R., et al., 2017. A feed-forward relay between bicoid and orthodenticle regulates the timing of embryonic patterning in Drosophila. bioRxiv. 1–46.

Davis, D., Yaveroglu, O.N., Malod-Dognin, N., Stojmirovic, A., Przulj, N., 2014. Topology–Function Conservation in Protein–Protein Interactions Networks. Oxford University Press. pp. 1–7.

Gligorijevic, V., Malod-Dognin, N., Przulj, N., 2015. FUSE: Multiple Network Alignment via DAT Fusion. Oxford University Press. pp. 1195–1203.

Goh, K., Cusick, M., Valle, D., et al., 2007. The human disease network. National Academy of Sciences. 8685–8690.

Hocevar, T., Demsar, J., 2014. A combinatorial approach to graphlet counting. Bioinformatics. 559–565.

Hulovatyy, Y., Chen, H., Milenkovic, T., 2015. Exploring the Structure and Function of Temporal Networks With Dynamic Graphlets. Oxford University Press. pp. i171–i180.

Ideker, T., Sharan, R., 2008. Protein networks in disease. Genome Research 18, 644–652.

Ingram, P., Stumpf, M., Stark, J., 2006. Network motifs: Structure does not determine function. BMC Genomics 7 (1), 108.

Itzkovitz, S., et al., 2005. Coarse-graining and self-dissimilarity of complex networks. Physical Review E 71, 016127.

Jonsson, P., Bates, P., 2006. Global topological features of cancer proteins in the human ineteractome. Bioinformatics 22, 2291–2297.

Kelley, B., Yuan, B., Lewitter, F., et al., 2004. PathBLAST: A tool for alignment of protein interaction networks. Nucleic Acids Research. W83–W88.

Kim, W., Diko, M., Rawson, K., 2013. Network motif detection: Algorithms, parallel and cloud computing, and related tools. IEEE Xplore 18 (5), 469–489.

Kovanen, L., et al., 2011. Temporal motifs in time-dependent networks. Journal of Statistical Mechanics: Theory and Experiment. P11005.

Kuchaiev, O., Milenkovik, T., Memisevic, V., Hayes, W., Przulj, N., 2010. Topological network alignment uncovers biological function and phylogeny. Journal of the Royal Society. 1341–1354.

Kuchaiev, O., Wang, P., Nanedic, Z., Przulj, N., 2009. STRUCTURE of brain functional networks. In: Proceedings of the 31st Annual International Conference of the IEEE Engineering in Medicine and Biology Society (EMBC'09), pp. 1–5. Minneapolis, Minnesota: IEEE.

Leskovec, J., et al., 2005. Graphs over time: Densification laws, shrinking diameters and possible explanations. In: Proceedings of the 11th ACM SIGKDD International Conference on Knowledge Discovery and Data Mining, ACM Press. pp. 177–187.

Ma'ayan, A., et al., 2005. Formation of regulatory patterns during signal propagation in a mammalian cellular network. Science 309, 1078–1083.

Malod-Dognin, N., Przulj, N., 2015. L-GRAAL: Lagrangian Graphlet-Based Network Aligner. Oxford University Press. pp. 2182–2189.

Malod-Dognin, N., Przulj, N., 2014. GR-ALIGN: Fast and Flexible Alignment of Protein 3D Structures Using Graphlet Degree Similarity. Oxford University Press. pp. 1259–1265.

Mangan, S., Zaslaver, A., Alon, U., 2003. The coherent feedforward loop serves as a sign-sensitive delay element in transcription networks. Jounal of Molecular Biology 334 (2), 197–204.

Memisevic, V., Milenkovic, T., Przulj, N., 2010. Complementary of network and sequence information in homologous proteins. Journal if Integrative Bioinformatics. 1–15.

Milenkovic, T., Ng, W. Leong, Hayes, W., Przulj, N., 2010. OPTIMAL network alignment with graphlet degree vectors. Cancer Informatics. 121–137.

Milenkovic, T., Memisevic, V., Ganesan, A., Przulj, N., 2009b. Systems-levels cancer gene identification from protein interaction network topology applied to melanogenesis-related functional genomics data. Journal of the Royal Society. 1–15.

Milenkovic, T., Przulj, N., 2008. Uncovering biological network function via graphlet degree signatures. Cancer Informatics. 257–273.

Milo, R., Itzkovitz, S., Kashtan, N., Levitt, R., Alon, U., 2004. Response to comment on "Network motifs: Simple building blocks of complex network" and "Superfamilies of evolved and designed networks". Science. 2.

Milo, R., Shen-Orr, S., Itzkovitz, S., et al., 2002. Network motifs: Simple building blocks of complex networks. Science. 824–827.

Newman, M., 2010. Networks: An Introduction. Oxford University Press.

Nicosia, V., et al., 2012. Components in time-varying graphs. Chaos 22, 3101.

Przulj, N., Corneil, D.G., Jurisica, I., 2004a. Modeling interactome: Scale-free or Geometric? Bioinformatics. 3508–3515.

Przulj, N., Corneil, D., Jurisica, I., 2004b. Modeling interactome: Scale-free or geometric? Bioinformatics. 3508–3515.

Przulj, N., Corneil, D., Jurisica, I., 2006. Efficient Estimation of Graphlet Frequency Distributions in Protein–Protein Interaction Networks. Oxford University Press. pp. 974–980.

Przulj, N., Kuchaiev, O., Stefanovic, A., Hayes, W., 2010. Geometric evolutionary dynamics of protein interaction networks. In: Proceedings of the 2010 Pacific Symposium on Biocomputing (PSB). Big Island, Hawaii.

Przulj, N., Malod-Dognin, N., 2016. Network analytics in the age of big data. Science. 123–124.

Przulj, N., 2006. Biological Network Comparision Using Graphlet Degree. Oxford University Press. pp. e177–e183.

Ptacek, J., et al., 2005. Global analysis of protein phosphorylation in yeast. Nature 438, 679–684.

Sakata, S., Komatsu, Y., Yamamori, T., 2005. Local design principles of mammalian cortical networks. Neuroscience Research 51, 309–315.

Sarajlic, A., Malod-Dognin, N., Yaveroglu, O.N., Przulj, N., 2016. Graplet-based characterization of directed networks. Scientific Reports. 1–14.

Shoval, O., Alon, U., 2010. SnapShot: Network motifs. Cell 143, 326.

Snider, J., Kotlyar, M., Saraon, P., et al., 2015. Fundamentals of protein interaction network mapping. Molecular System Biology. 11–848.

Sporns, O., Kotter, R., 2004. Motifs in brain networks. PLOS Biology 2, e369.

Wang, P., Zhang, X., Li, Z., et al., 2016. A fast sampling method of exploring graphlet degrees of large directed. arXiv 1604, 08691.

Wong, E., Baur, B., Quader, S., Huang, C.-H., 2011. Biological Network Motif Detection: Principles and Practice. pp. 202–215.

Wong, S.W., Cercone, N., Jurisica, I., 2015. Comparative network analysis via. Proteomics. 608–617.

Yaveroglu, O.N., Malod-Dognin, N., Davis, D., et al., 2014. Revealing the hidden language of complex networks. Scientific Reports. 1–9.

Yaveroglu, O.N., Milenkovic, T., Przulj, N., 2015. Proper Evaluation of Alignment-Free Network Comparison Methods. Oxford University Press. pp. 2697–2704.

Yeger-Lotem, E., et al., 2004. Network motifs in integrated cellular networks of transcription-regulation and protein-protein interaction. Proceedings of the National Academy of Sciences of the United States of America 101 (16), 5934–5939.

Biographical Sketch

Laurentino Quoroga Moreno is a Colombian engineer. He moved from his home country to the UK in 2007 to study English and enrol into post-graduate studies at King's College London supported by a scholarship from Colombia. In 2012, he obtained a postgraduate certificate in Engineering with Business Management from King's College London. He worked in private and public sectors for many years. In 2016, he changed disciplines and entered an MSc program in Bioinformatics with Systems Biology at Birkbeck College, University of London. Currently, he is carrying out his research project at University College London in the department of Computer Science on disease-disease relationships from systems-level molecular network data.

Protein-Protein Interactions: An Overview

Edson L Folador, Federal University of Paraíba, João Pessoa, Brazil
Sandeep Tiwari, Federal University of Minas Gerais, Belo Horizonte, Brazil
Camila E Da Paz Barbosa, Federal University of Paraíba, João Pessoa, Brazil
Syed B Jamal, Federal University of Minas Gerais, Belo Horizonte, Brazil
Marco Da Costa Schulze, Federal University of Paraíba, João Pessoa, Brazil
Debmalya Barh, Institute of Integrative Omics and Applied Biotechnology, Purba Medinipur, India
Vasco Azevedo, Federal University of Minas Gerais, Belo Horizonte, Brazil

Introduction

Biological Interaction

Proteins are organic macromolecules which have the ability to act together with several other molecules, including other protein (Moraes, 2013). They are normally identified by their individual actions as catalysts for reactions in metabolic pathways, signaling molecules in intracellular regulation cascade, defense proteins such as antibodies, nutrient or storage proteins, motile or contractile proteins (Jeong et al., 2001; Phizicky and Fields, 1995). But the proteins are also identified by structural actions carried out by the interaction by forming complexes such as proteins of the cytoskeleton, signal transduction, cell-to-cell communication, transcription, replication and membrane transport (Keskin et al., 2016). Regardless of the activities to be structural, metabolic or regulatory, proteins are important in maintaining homeostasis of the organism through the interactions that form each other and with other molecules, are essential for any living organism.

For some proteins become active and perform its function, it is required that binds to other molecules, thus forming monomers, dimers, trimers, polymers, etc. A monomer is defined as a molecule capable of binding to other molecules. When two monomers (molecules/proteins) are bound are called dimers, and if the dimer is formed by the same protein is called homodimers. A dimer sample can be observed with a binder in the structure of thymidylate synthase (TS) (PDB 5X69), an enzyme of the transferase class with function to catalyze the synthesis monophosphate thymidine (dTMP) from uridine monophosphate (dUMP), essential in DNA synthesis and directly linked to tumor proliferation rate (Chen et al., 2017). Trimers are joints of three monomers, for example the Dynamic HIV Envelope Trimer Apex complexed with neutralizing HIV antibodies (PDB: 5V8L) (Lee et al., 2017) and when it is formed by the junction of the same molecules is termed homotrimer. The polymers contain a large number of monomers linked such as retinal protein for cell-cell adhesion (retinoschisina – RS1) an octamer consisting of eight monomers (PDB: 3JD6). When they are composed of the same molecule form a homopolymer such as the Ebola virus membrane-associated matrix protein (vP40) formed by four antiparallel homodimers (PDB: 1H2D).

The biological interactions maintain proper functioning of the body and occur at all times to ensure the production or non-production of biomolecules involved in metabolism. Proteins are key elements in the interactions processes and they are responsible for modulating catabolism and anabolism processes with synthesis or degradation of essential molecules, respectively.

The biological interactions can be defined as a mutual operation of several components of an organism to ensure proper sequence of events that together ensure the development of the individual. With the premise that all living beings have evolved from a common ancestor we can assume that, in addition to genes and proteins, the interactions are conserved.

The biological interactions occur only within a single organism or cell, they also occur between different host-pathogen interactions of bacteria *Mycobacterium tuberculosis* and Homo sapiens and the knowledge of this protein-protein interaction network (PIN) is an important tool for understanding how the bacteria interact with their host and expresses its virulence (Huo et al., 2015). The knowledge about the proteins and how they interacts in a PIN enables a systemic understanding of the organism, cell or metabolic pathway, and may able since the discovery of new therapies to treat diseases until the knowledge about the metabolism of the microorganisms to be used in industrial, pharmacological, agricultural or livestock bioprocesses. Estimates indicate on average 30% of the organisms' proteins are identified in the interaction networks, with PINs being a large area of study and with many possibilities still open (Hao et al., 2016).

For an interaction to exist, it is necessary to have a fine space-time coordination, that is, both partner proteins must coexist in the same cellular location at the same time (Hao et al., 2016). This leads to the thought that for an interaction to occur, it is not enough just to have the respective coding genes, the proteins need to be transcribed and translated at the same time and in the same place. This fact has an even greater relevance in eukaryotic, because at least the interacting proteins must be expressed or transported to the same tissue. In time, transmembrane or secreted proteins are able to interact with proteins outside the cell environment such as occurs in the host-pathogen interactions. These characteristics exert direct influence on the methods of identification and prediction of interaction and, if not considered, can lead to false positive or false negative identification.

In order to understand a PIN is necessary to know the types and characteristics of each protein interaction.

Interactions Types

Interactions between proteins can be classified as physical (structural) or functional (regulatory). An interaction is physical when there is physical contact between two proteins and it is functional when there is an activation or inhibition event between the biomolecules, often mediated by intermediary cellular processes as in a metabolic pathway.

Interactions may also be classified according to duration time as transient or persistent. The transient interactions occur for a short period of time when a protein binds to the other only at a time and, after performing its function are separated. Persistent interactions tend to be more stable and occur when proteins remain bound for a longer period or until they are degraded (Nooren and Thornton, 2003). In time, the proteins may classified according to the amount of interaction. Proteins with many interactions are termed hubs and play a central role in PIN by connecting several other proteins, probably assuming several functions and participating in several metabolic pathways.

These types of interactions are not exclusive and may occur simultaneously. Due to the temporary nature, transient interactions are more difficult to identify and study. Physical interactions are important targets in the development of a new class of drugs with the potential to inhibit or stabilize interactions.

Interaction Representation

The representation for PINs view is based on graph theory, coming from the exact area. An interaction is symbolized by nodes (representing proteins) and by vertices linking these nodes (representing the interaction), as can be observed in **Fig. 1**.

In biological networks, typically the interactions are non-directional, indicating there is no source or destination for interaction and can be stated without loss of information that both the protein 'A' interacts with the protein 'B' as protein 'B' interacts with 'A'.

A directional interaction is represented by an arrow (>) at the end of vertex indicating the direction of interaction, thus having source and destination in the proteins of the interaction. Two proteins form an interaction, and several interactions, peer-to-peer, form a complex Protein Interaction Network (PIN), which may contain hundreds or thousands of interactions.

Protein-Protein Interaction Network (PIN)

The PINs, also known as interactome, are important tools for the systemic study of an organism, since they represent a set of protein interactions whose characteristics are susceptible of analyzes inherited from the graph theory. It is possible to observe in a PIN, since interactions between proteins in a complex or in a metabolic pathway until a set of pathways, comprising both organisms showing low complexity as those with high complexity (Szklarczyk et al., 2011). By analyzing the characteristics, a PIN provides elements for a great variety of studies, such as (i) understanding the organism studied at a systemic level, (ii) visualizing and complementing metabolic pathways, (iii) presuming protein functions and assisting the annotation process of genes and hypothetical proteins, (iv) identify new targets for drugs and (v) suggest new therapies to treat diseases (Pavlopoulos et al., 2011).

Biological Network Characteristics

A PIN such as the World Wide Web, as Facebook ® or personal network is a biological network. The biological networks in general, whether of people or proteins, present similar characteristics but different from non-biological networks. Using analysis of graph theory these characteristics can be identified and applied for a better understanding of the network aiming at biological purposes.

Degree: is a measure that indicates the number of edges possess a node. In PIN is a particular characteristic of each protein, indicates with how many other proteins a protein interacts. It is a feature that generally can change the topology of a PIN and have important biological applications. A protein that has a high degree of interaction, which interacts with many other proteins, is called a *hub*. The hub proteins by interacting with many proteins form large tightly connected groups and tend to participate in several metabolic pathways or biological processes, assuming diverse functions.

The hub proteins have important cellular function, indicating that if they are removed from the PIN, the organism may not perform its basic functions, justified by the disruption that the removal of a hub node causes in the network, probably affecting several metabolic pathways (Pavlopoulos et al., 2011; Jeong et al., 2001). *In silico* studies suggest that the hub proteins are under high evolutionary pressure, because (i) they are more conserved among organisms, (ii) they have a lower mutation rate in their sequences, they also have a conserved evolution, and (iii) they are correlated with proteins essential (Dilucca et al., 2017). Just because they are conserved and under selective pressure, it indicates the hub proteins have some singular significance for the organisms.

However, how many interactions a protein must have to be considered a hub? This answer is not given in absolute numbers because the amount of interactions identified in a PIN is dependent on the methodology used, ie in the same organism, different

Fig. 1   Nodes represent the proteins (A and B) and the vertex is the interaction between the nodes.

methodologies identify amount of different proteins and interactions. This is a problem even to replicate or validate experiments. For this reason the hub proteins are sorted by relative values, such as 15% (Folador *et al.*, 2016) or 20% (Delprato, 2012) of the protein with a higher degree of interaction in the PIN. There is no consensus on the exact value to define hub proteins and the choice of one or another value depends on how much you want the identification is more or less sensitive.

When a hub protein interacts with its partner proteins simultaneously it is named the Party hub. Party hub generally interact physically with their partners and form complexes performing distinct function, such as proteins composing polymers or homopolymers. Complementarily, a protein hub that interacts with different partners at different time or place is named date hub. Date hubs are proteins that participate in functional interactions, usually assuming role of regulation in signaling cascades or participating enzymatically in the stages of different metabolic pathways (Hao *et al.*, 2016).

The interactions around hub proteins influence the topology of a PIN and can form two network models, assortative or dissortative. When hub proteins are more connected to other hub proteins, the network is named assortative and, contrary, when connecting to proteins having low interaction degree is named dissortative (Pavlopoulos *et al.*, 2011). Usually biological networks are dissortatives with hub proteins interacting more with proteins having low interaction degree than protein hub (Newman, 2003). These features contribute to the specificity of biological networks and differentiation from non-biological or random networks, and may used as a first layer to in-silico validation by analysis of the interaction degree distribution for PPIs.

Scale-free distribution: It is a topological measure of interaction networks. Biological networks have a scale-free distribution to interaction degree of each node with tendency to power-law (**Fig. 2** left), different from random networks showing normal distribution (**Fig. 2** right).

The PIN having a scale-free distribution indicate has a large number of proteins with few interactions and a small number of proteins with a large number of interactions (Barabási and Albert, 1999). This fact is in agreement with the presence of hub proteins in PPI concentrating most interactions while other proteins have few interactions. This characteristic indicates biological interactions do not occur randomly or by chance, reinforcing the evolutionary pressure occurs in the interactions, evidencing the importance of precise interactions in the maintenance of life.

Small-world-effect: It is a topological measure of interaction networks. The effect of small world (or shortest path) indicates that any two nodes are separated by only a few edges (Barabasi and Oltvai, 2004), this explains for example that anyone in the world is connected to another on average by 5 or 6 people. In biological networks, this means that the disturbance of any protein will propagate in the network because a few proteins separate all other proteins, suggesting the interactions within a cell are closely related. Certainly, hub proteins contribute for this effect by keeping a strongly connected PIN, reducing the distance between the proteins. Without the hub nodes, the distance between all proteins would probably the same indicating both paths and random network.

Clustering coefficient: in graph theory, is a measure indicating how much the nodes tend to cluster in a network (Pavlopoulos *et al.*, 2011). Clusters are densely interconnected sets of nodes and vertices and the clustering coefficient is used as a feature of biological networks. Generally, PINs present a high clustering coefficient in relation to random networks, forming cohesive groups of interacting proteins acting on the same metabolic pathway. While in PPI the clustering coefficient reaches values greater than 0,4 on random networks is generally less than 0.01 with $p < 0.05$ (Rezende *et al.*, 2012; Folador *et al.*, 2016). Since there are several metabolic pathways in an organism, naturally there are several groups of proteins interacting more intensively in these pathways, giving the PPI a higher clustering coefficient.

Betweenness centrality: is a network centrality measure that calculates the proximity between the vertices in a graph, determining, among the shortest paths to cross the network, how many pass through the node. The higher betweenness value more shortest paths pass through node and enables cross the network side to side in a few steps. In PPI are proteins involved in different biological processes or metabolic pathways, connecting to several groups of proteins without necessarily be a protein hub.

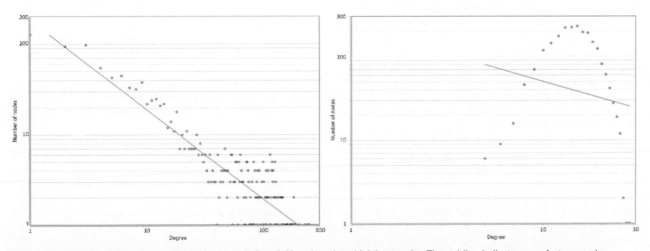

Fig. 2 Sample of interaction degree distribution in scale-free (left) and random (right) networks. The red line indicates a perfect power-law.

However, some hub proteins, because they have many interactions and connect several modules, are also characteristic of betweenness (Pavlopoulos *et al.*, 2011).

The topological measures are features that should be present in PPI generated by experimental or computational methods, and are useful as a first layer to validate the predicted PINs by computational methods. Derived from graph theory, there are several other metrics that can be applied to PPI to extract some biological significance, however without validations described. Having a PPI these and other measures can be easily calculated, e.g., using Network Analyzer plugin (Assenov *et al.*, 2008).

Methods

To investigate protein interactions, both experimental methods in the laboratory (*in vitro* or *in vivo*) and computational (*in silico*) are used. An interaction can be identified by experimental methods or predicted by computational methods, considering biological premises. In the case of PINs, the methods can be classified into (i) low throughput when used to characterize only a few interactions or (ii) high throughput when used to characterize a large number of interactions. Each method class has advantages and disadvantages. Generally, low-throughput methods are more reliable in identifying the interaction while high-throughput methods are more sensitive and identify the entire interaction network.

Experimental Methods

Experimental methods for identifying protein-protein interactions are performed by in vitro or in vivo techniques. The challenge of the test methods is to identify interactions in their natural state, just they occur in a living cell, without changing properties of proteins or the environment where they are, so as not induce or suppress interactions that naturally would not occur.

Co-immunoprecipitation (Co-IP): Co-IP is able to identify interaction between proteins or even protein complexes by using beads attached to an antibody specific for a target protein, whose interactions we wish to know. The antibodies bound to beads and the cell media are arranged together, establishing the interactions: the target protein interacts with its natural partners and interacts with the antibody. By centrifuging process the complex formed by the bead, antibody, target protein and the partner interaction are pelleted and isolated. The supernatant proteins are discarded and the pelleted proteins can be identified by mass spectrometry or other technique (Lin and Lai, 2017).

Tandem Affinity Purification (TAP): Similar to Co-IP TAP is able to investigate the interactions as they occur in the cell environment and has low contamination rate, although a high-throughput method has an accuracy similar to low-throughput methods. The method consists of attaching the TAP tag to a bait protein, arranged together with other proteins in concentrations similar to that in vivo, allowing interactions are established and protein complexes formed. TAP has affinity to the purification column and retains the complex formed with the bait protein in the washing process, when discarded molecules not interacting in the complex. In another purification step, the protein complex is separated from TAP and isolated, allowing the characterization of proteins by mass spectrometry (MS-TAP) (Puig *et al.*, 2001). Due consecutive purification step the method have a deficiency: proteins participating weak or transient interactions may not detected by turning off from TAP complex.

Yeast two-hybrid (Y2H): The two-hybrid method is capable of identifying physical interactions of proteins. Y2H is a method that uses transcription machinery to express a reporter gene. The transcription factor is activated only when the proteins (bait and prey) interact, and the reporter gene is transcribed. Thus, the interaction is perceived by the expression of the reporter gene or its phenotype. Bait is a target protein anchored in the DNA-binding domain which in turn binds to the promoter region Upstream Activation Sequence (UAS) of DNA. Prey are the Bait partner proteins, anchored to the transcriptional activation domain (TA), which will be interrogated. When Bait and Prey interact the transcription process is activated and the gene reporter expresses (Ito *et al.*, 2001). This process can be reproduced on a large scale and investigate the entire interactome.

Bimolecular fluorescence complementation (BiFC): is a technology used to identify physical protein interactions. It is based on the use of a fragmented fluorescent protein (e.g., enhanced yellow EYFP) in two non-fluorescent peptides, respectively the N- and C-terminal peptides, each one attached to a specific know target protein by a linker. When these two target proteins expressed on the live cell interacts, the N- and C-terminal peptides became near and interact too, reformulating to the native three-dimensional structure da fluorescent reporter protein (e.g., EYFP). The emitted fluorescent signal can detected within the cell by an inverted fluorescence microscope making possible identify the location of interacting proteins and known the interaction type by the signal intensity (Wang *et al.*, 2017).

Cross-linking coupled with mass spectrometry (XL-MS): The cross-link are markers connecting a polymer chain to another (e.g., protein-protein), usually developed by the use of reagents or cell engineering and can modify the characteristics of the investigated protein or its natural environment. Several methodologies researching *in-vivo* interactions use different cross-link to identify the interaction. The cross-linking form covalent bonds and stabilizes the interaction between proteins, making it possible to identify even the weak or transient interactions by mass spectrometry (Holding, 2015).

Laser UV Cross-linKing (LUCK): The major difference of the LUCK method is to use as cross-link the aromatic residues (Phe, Trp, Tyr). These residues are naturally present in most proteins, especially at binding sites, establishing strong bonds, having the advantage of not modifying the characteristics of the investigated protein or its natural environment. When irradiated with UV light these aromatic residues produce radicals that react and generate covalent bonds with near residues, making it possible to identify even the transient interactions. The LUCK has already been used both to identify the formation of stable dimers of GAPDH in HeLa Cell (Itri *et al.*, 2016) as to detect the protein interaction between GAPDH and alpha-enolase in living cells (Itri *et al.*, 2017).

Fluorescence resonance energy transfer (FRET): is a method that uses different fluorescent proteins (FP) to identify interactions, having potential for unveils temporal and spatial information on proteins in living cells. Both FP are named donor and acceptor respectively, which when exited by a specific wavelength range emit in response a different specific wavelength range. Thus, considering a donor with excitation and emission peaks at X and Ynm respectively and a receptor with excitation and emission peaks at Y and Znm respectively, the interaction is identified when perceive peaks at Znm when excited at Xnm. Justified by energy transfer from donor to receiver when the signal emitted by the donor excites the near receiver.

Several other methods are described in the literature taking advantage of the physicochemical property for label proteins and permits identify interactions by diverse techniques, including sequencing transcritos as bPPI-seq method (Zhang *et al.*, 2017).

Computational Methods

The computational methods, also known as in silico, uses computational tools to solve biological problems also considered a more viable alternative due to its low cost when compared with experimental methods. Another advantage of in silico approaches is the high capacity to generate and analyze data inexpensively. Due to the rapid development in computing field is increasingly feasible the use of more powerful techniques to study biological networks.

The use of previously unviable or non-existent methodologies, such as the use of high-performance computing, has been increasingly frequent, as well as the emergence of new and more effective methodologies (Keskin *et al.*, 2016). Aided by the amount of information and biological data available added to the evolution of computers and algorithms have emerged methods for PPI prediction based on diverse approaches as binding strength between molecules, sequence alignment, phylogenetic trees, text mining, artificial intelligence and even using mixed approach.

Three-Dimensional Structure-Based Interaction Prediction

Bond strength methods use three-dimensional structures of two proteins to find a contact region (binding site) and measure the affinity between these two structures. In order, to reduce the computing power required for docking, usually the protein structures are considered a rigid body, differing from biological reality but enabling the prediction be performed quickly to a pair of proteins. Conversely, molecular dynamics considers the protein as a flexible body and besides calculating the affinity between molecules also calculates the angles of each atom taking into account the possible movements that the proteins can present recursively, consequently increasing the computational requirements. To achieve a few milliseconds with molecular dynamics, it takes days of processing, depending on the computational power available (Keskin *et al.*, 2016).

These methods are considered low throughput mainly for two reasons: (i) the high computational power and time required especially when using molecular dynamics technique, or (ii) a lack of three-dimensional structures determined experimentally to be used with docking or dynamic molecular techniques compared to the amount of available sequences.

The processing time can be considerably reduced when the docking or molecular dynamics programs are able to use the resources of the graphic processing units (GPUs) that, in addition to being accessible, are currently in the order of 5000 processing units, exceeding the amount of central processing unit (CPU) available on good and expensive servers. To take best advantage of both techniques, docking is usually performed to bind two proteins and then runs the molecular dynamics to optimize the interacting partner proteins in their less energy state.

A software widely used for molecular dynamics is GROMACS (Abraham *et al.*, 2015) while docking can be performed with HEX (Ritchie, 2003), MegaDock (Ohue *et al.*, 2014) and HADDOCK (De Vries *et al.*, 2010), SwarmDock Server (Torchala *et al.*, 2013), ZDOCK Server (Pierce *et al.*, 2014), Multi-LZerD (Esquivel-Rodríguez *et al.*, 2012), LightDock, ClusPro (Jiménez-García *et al.*, 2017; Kozakov *et al.*, 2017), among others.

In order to encourage the development of new algorithms and methods as well as to improve existing ones, the Critical Assessment of PRedected Interactions (CAPRI) was created in 2000. The CAPRI is an open event that occurs in competition format to evaluate the algorithms in blind predictions of the structure of protein-protein complexes. Prior to publication, the organizers make available to competitors proteins offered by crystallographers as targets, without prior knowledge of the complex. The competitors perform the complex structure prediction by docking the individual components. Afterwards, the predicted models are submitted to evaluators compare the geometry and interaction site against the experimental models, scoring the best predictors.

Multiple Sequence Alignment (MSA)-Based Interaction Prediction

With the popularity of DNA sequencers, many genome projects were undertaken, and various organisms had their genome sequenced and annotated. Many genomes as their amino acid or nucleotide sequences are available in several public databases, and the GenBank, EMBL and DDBJ are the largest repositories of such information. In contrast to the low number of three-dimensional structures, the number of sequences in the public databases is extremely higher, favoring the scientific community that uses these sequences in their researches, including the prediction of interaction. A class of method uses multiple sequence alignment (MSA) to predict protein-protein interaction and each is based on a biological premise that generally aims to identify conservation among the organisms' sequences using alignment software. They are:

Phylogenetic profile-based prediction (co-occurrence): The method is based on the biological premise that if two genes remain conserved when comparing several organisms, the respective protein translated from these genes will interact. This assumption becomes even stronger when both conserved genes are not present in one or more organisms compared, demonstrating selective pressure for both remain or be removed together from genome. The uncertainty in using this method is to identify how phylogenetically distant should be the organisms compared therefore, by selecting close organisms makes it difficult to identify whether proteins are conserved because they are under evolutionary pressure or simply because there has not been enough time to diverge. The more distant organisms are, the more reliable the prediction will be, but fewer interactions will be identified (Valencia and Pazos, 2002).

Neighborhood genes-based prediction: This method is an evolution of the previous one, considers the same premise and presents the same doubt, but considers that the genes, besides being conserved, must be neighbors. This method prioritizes to identify conservation of closer genes taking into account they tend to evolve in similar periods, like operons transcribed and translated at the same time, probably generating protein structures interacting in a same function or metabolic pathway (Valencia and Pazos, 2002).

Phylogenetic tree-based prediction (Mirrortree): The method is based on the biological premise that if two proteins interact they undergo similar evolutionary pressures and co-evolve in an organism (Zhou and Jakobsson, 2013). Thus, when comparing the same protein in several organisms, the respective interacting proteins, will present a similar phylogenetic tree, one mirrored in the other, representing similar evolution in the different organisms. However, proteins evolving differently may also interact and be disregarded by this method. To identify an interaction phylogenies of various proteins are made and, the phylogenetic trees presenting similar topology reveals the respective proteins participating in the interaction (Pellegrini *et al.*, 1999). This can be made feasible by converting pairs of phylogenetic trees into distance matrix and calculating the correlation coefficient. Matrices with high correlation coefficient indicates a phylogenetic tree with similar topology.

Gene fusion-based method: The biological premise underlying this method is the existence of two genes (ga, gb) in an organism with functional domains homologous to a single gene in another phylogenetically close organism (gc), probably the proteins encoded by the two individual genes (pa and pb) interact physically to play their role, explained by an evolutionary process that led to the fusion of genes and translation of a single protein into another organism (pc). The homology between genes can be identified via sequence alignment and this assumption becomes even stronger when one perceives in multiple alignment a group of organisms presenting the two individual genes and another group presenting the fused genes (Zahiri *et al.*, 2013).

Interolog Mapping

This method is based on the biological premise that if it is known the interaction between two proteins in a particular organism and exist both orthologous proteins in another organism, they probably also interact. It is a high throughput method by allowing mapping known available interactions from public databases to an organism of interest using sequence alignment. The mapping of the interactions is performed in two steps: (i) identification of the orthologous genes or proteins by aligning the sequences from public databases against the sequences of your organism and (ii) composition of the orthologous pairs of interacting proteins based on interactions from public database. Although simple, due to the large volume of sequences and interactions in some databases, in the order of millions, it may require considerable computational resource for the prediction process. The confidence of the predicted interaction will depend on two factors: (i) that truly homologous proteins are mapped and, (ii) how reliable is the known interaction from public databases. Despite being an effective and easy to use method, it is able to map the interactions already determined in other organisms, without the ability to predict new interactions not yet reported.

Text Mining

The text mining is a computational technique used to obtain important information recorded in texts such as abstracts or scientific articles, having commercial, governmental and scientific applications. Briefly, if an interaction is described in a text, it is theoretically possible to be identified using different algorithms. In summary, a described interaction theoretically can be identified using different techniques and text mining algorithms (Zahiri *et al.*, 2013). The identification of interactions is performed in four steps: (i) download of library to be used in the research, in the case summaries and articles containing information about proteins or genes; (Ii) identifying entities (genes or proteins) of interest; (Iii) identification and verification of relationship between the entities in sentences and paragraphs, in the case indicating interaction; and (iv) extraction of entities to form the interaction pair. As interolog mapping this is a high throughput method able to identify interactions previously described, without the ability to identify new interactions. To reduce the sampling range and increase reliability in prediction, the literature is usually retrieved from a specific species, limiting the method to identify specific interactions. Another limiting factor is how the entities are described (e.g., the gene name, the protein name, codes from different databases), if untreated, can limit the method does not to identify all possible interactions and generate a lot of false-negative. However, it can be a very useful method, especially when it is desired to know the relation of the interactions, usually identified by action verbs in sentences (e.g., bind, interact, regulate), making it more informative. Interactions identified experimentally and published, when predicted by text-mining method and curated, in many cases have experimental equivalence, since it was first identified by an experimental method.

Machine Learning

Machine learning is an Artificial Intelligence subfield of the Computing area. The algorithms form models and patterns from sets of gold standard data where the different characteristics that determine an interaction are already known. The algorithms are executed in two steps: (i) learning, where the characteristics are observed and identified in a set of gold standard data whose interactions are known aiming to establish a pattern e; (Ii) prediction, where the characteristics are checked on a real data set to determine whether an interaction exists or not, or even establish a degree of confidence in the interaction. In PPI can be used diversified characteristics such as genetic or protein sequence, physicochemical data, three-dimensional structures, cellular location, binding sites, conserved regions or even the combination of these characteristics. Prediction confidence is dependent on learning, ie if the algorithm is able to learn and determine all the relevant characteristics for an interaction to exist, it will be able to infer whether an interaction is possible or not based on these characteristics. Due to the large number of interactions available, experimentally determined or not, it is possible to create a set of interacting proteins to train the algorithms (Dick *et al.*, 2017). For an efficient training, it is important to highlight and imperative the algorithm learns which characteristics determine an interaction, as well as the characteristics determining not interaction, that is, for the learning, besides a positive control set a negative control set is necessary. This is plausible since there are cured databases of proteins that do not interact as The Negatome Database (Blohm *et al.*, 2013).

Abstractly, we can break any prediction method down into.

Sample description: how each sample can be described to the method. Not everything is accepted by all methods. For instance, in simple linear regression, each sample can only be described by two numbers (x, y). *Model:* every method assumes a certain structure between the samples. For example, in linear regression, that assumption is that the set of samples can be somewhat accurately represented by a straight line. Clustering, on the other hand, assumes "related" samples are also "close" in the sample space. *Parameters:* variables of that prediction method. *Training:* is the process of finding the best parameters that fit the samples, for some interpretation of "best", how to find the right (or best) parameters. Training might be "supervised", where previously classified data is used to find the appropriate parameters, or it might be "unsupervised", when no such premade classification exists. *Prediction:* how predictions are made. Different methods accept different kinds of questions and output their predictions in different ways. Questions may include: "Given a sample with incomplete data, what best fits the blanks?" or "How well does this new sample fit?". Predictions might be in the form of an yes/no answer, a real number, a label, or a mix of any of these.

These categories will be used to describe the prediction methods above.

Linear regression

Sample description: two (simple linear regression) or more numbers (multivariate linear regression). Linear regression assumes certain statistical properties over the training samples. Those properties can vary depending on which regression is used out of many variants, but may include assumptions such as weak exogeneity or homoscedasticity. *Model:* samples are represented as a straight line that is approximately close to all samples. *Parameters:* a and b where $y = xa + b$. In simple linear regression, a and b are real numbers. In multivariate linear regression, a and b are m/cross n matrices, where m is the number of independent variables and n the number of independent variables. *Training:* always supervised and accepting only positive examples. The number of published and used estimators are too great to list here, but includes the Ordinary Least Squares method, the Maximum-Likelihood Estimation, the Bayesian Linear Regression, and the Theil-Sen Estimator. *Prediction:* given how the training phase yields a line, the distance to that line is a natural metric to evaluate how fit any new sample is. New samples can be taken from the line. *Examples:* Logistic regression (a generalization of linear regression for discrete dependent variables) was evaluated against other five methods with 16 feature categories used for sample description, including: gene expression, GO molecular function, GO biological process, GO component, and essentiality (Qi *et al.*, 2006). Logistic regression was used to create a confidence score for experimental data on protein-protein interactions. The confidence score was calculated based on the results of Luminescence-based Mammalian IntERactome (LUMIER), Mammalian Protein-Protein Interaction Trap (MAPPIT), Yeast-2-Hybrid (Y2H), Yellow Fluorescent Protein (YFP), Protein Complementation Assay (PCA), and a modified version of the Nucleic Acid Programmable Protein Array (wNAPPA) assays (Braun *et al.*, 2009).

k-means clustering

Sample description: each sample is described by a real vector with a fixed dimension. *Model:* the sample space is partitioned into k clusters and each sample belongs to a single cluster. Sample variance is minimized within a single cluster (Bishop, 2006). *Parameters:* the centroid of each cluster. *Training:* supervised, unsupervised, or a mixture of both. In the worst case (unsupervised), finding the absolute best centroids is NP-hard, requiring training to run for a time proportional to the exponential of the number of samples. Because of this, heuristics are usually used, giving "good enough" centroids most of the time. *Prediction:* k-means clustering naturally classifies samples in one of k categories. The centroid also serves as the archetype of each cluster. *Examples:* Mohamed and colleagues (Mohamed *et al.*, 2010) used k-means clustering where each sample, describing a protein-protein interaction, was represented by a vector with 27 dimensions describing features such as GO molecular function, GO biological process, GO component, and sequence similarity. K-means clustering was used as a method for dimension reduction, where each sample was the result of multiple liquid chromatography–mass spectrometry (LC-MS) alignments, each giving noisy indications of protein-protein interactions (Rinner *et al.*, 2007).

Hierarchical clustering

Sample description: anything with a meaningful distance function. *Model:* a hierarchy of clusters. Each cluster may encompass samples and other clusters, but each cluster can be encompassed by at most a single parent cluster. *Parameters:* edges and internal nodes, forming a tree. *Training:* supervised, unsupervised, or a mixture of both. Two strategies are usually used: agglomerative, where samples are grouped into clusters, or divisive, where a cluster encompassing all samples is recursively subdivided. There are many ways to decide whether to agglutinate or split, called the linkage criteria, including Complete-Linkage Clustering, Single-Linkage Clustering, and Minimum Energy Clustering. *Prediction:* the training result cannot be directly used for prediction. Instead, any new sample is added to the initial sample set and clusters are regenerated. *Examples:* Brohée and colleagues (Brohee and Van Helden, 2006) evaluates different clustering algorithms for protein-protein interaction networks. General properties of protein-protein interaction networks are noted based on their hierarchical clusters (Yook *et al.*, 2004).

Support vector machine (SVM)

Sample description: each sample is described by a real vector with a fixed dimension. *Model:* samples separated into two areas by a hyperplane, optionally put in a transformed space. *Parameters:* a hyperplane, the minimum separation between the hyperplane and training samples (soft margin), and an optional transformation function (Haykin, 1994). *Training:* supervised only. Given a set of positive and negative samples, the parameters of usual transformation functions and the hyperplane can be found by solving a quadratic optimization problem, which is NP-hard and takes exponential time relative to the number of samples. Iterative alternatives include Sub-Gradient Descent and Coordinate Descent. *Prediction:* any sample can be classified as positive or negative by applying the transformation and checking whether the transformed point lies in the positive or negative half-space. Extensions allow for a broader classification than only positive/negative. *Examples:* Guo and colleagues (Guo *et al.*, 2008) uses an SVM to model and predict protein interactions based on many physicochemical properties of their amino acids: hydrophobicity, hydrophicility, volumes of side chains of amino acids, polarity, polarizability, solvent-accessible surface area, and net charge index of side chains of amino acids. Koike and colleagues uses a SVM to predict the interaction sites of proteins based on their sequence of amino acids (Koike and Takagi, 2004).

Hidden markov model (HMM)

Sample description: a sequence of symbols in a fixed alphabet. The alphabet need not be ordered (Bishop, 2006). *Model:* an HMM is composed by a set of distinct states, a set of normalized (summing to 1) probabilities for the state to change from any state to any other state (including itself), and a set of normalized probabilities that an observation occurs given a state. Alternatively, an "observation" can be seen as a letter in an appropriate alphabet that the model "emits" one at a time based on its current state and a random number. *Parameters:* transition and observation probabilities. *Training:* supervised or unsupervised. Given a set of states and possible observations (or emissions), the expectation–maximization algorithm is usually used to, iteratively, find the set of probabilities. Finding the "best" probabilities is possible but there is no known efficient algorithm for that as the problem is NP-hard. *Prediction:* a trained HMM can score sequences based on their similarity to sequences used in training. HMMs can also run in reverse, giving sequences that are likely similar to the ones used in training. *Examples:* In (Wojcik and Schächter, 2001), HMMs are used to predict protein-protein interaction (PPI) maps given a PPI map of another organism. Eddy (Eddy, 1998) gives a review of Profile Hidden Markov Models (pHMM) which can perform sequence alignment based on similarity to certain sequences.

Artificial neural network (ANN)

Sample description: any type of data with a fixed size. Depending on how the network is built, variably-sized data can be used as well (Haykin, 1994). *Model:* a set of artificial neurons forming a network, where each one combines, transforms, and filter incoming data, passing the result forward. *Parameters:* network topology, which can be premade, and parameters for each artificial neuron. *Training:* supervised and unsupervised. Training procedure is highly dependent on network type and topology, but most used are Gradient Descent, Simulated Annealing, Evolutionary Algorithms, Particle Swarm Optimization, and others. Training an ANN usually takes considerable computer resources and time. *Prediction:* any kind, depending on topology and how it was trained. *Examples:* An ANN is used to predict interaction sites between proteins given a sequence of aminoacids (Ofran and Rost, 2003). Protein are given a predicted function with an ANN. The authors also suggest that predicted function can be used to validate protein-protein interaction databases that might contain false negatives (Jensen *et al.*, 2002).

As each predictive computational method considers different biological premise, it can of course disregard biological factors relevant to identification of interaction, thus, a good practice, is to use predictive computational methods together in order to guarantee greater sensitivity in the PIN predictions. This strategy besides useful to add information to the PIN is useful to validate interactions, because an interaction identified by more than one method reduces prediction by chance and increases the probability of being true interaction. Similarly, when possible, using both experimental and computational methods can also aggregate information to the PIN.

Different methods predicted PINs for various organisms whose interactions usually are deposited in public databases.

Public Databases

Aiming to make public and share data interactions for the scientific community, several databases were designed, containing interactions identified in different organisms by experimental and computational methods. With the wide variety of organisms and methods for PPI prediction also came the need to distinguish these public databases reliability of interactions. Thus, arose in September 2005 the International Molecular Exchange Consortium (IMEx) as an international collaboration having as participants several public databases joining efforts to validate protein-protein interactions through curation, ensuring a detailed and highly reliable set of non-redundant interactions to be made publicly available (Orchard *et al.*, 2012).

Public databases store and provides physical or functional interaction data experimentally characterized or predicted computationally or even cured, some store interaction of various organisms while others store specific organism interaction or even specialized in biological process or functionality (**Table 1**).

The classes of positive and negative interactions are important biologically but also computationally. Besides being useful at systems biology level for better understand an organism, they serve as training data for machine learning algorithms, still useful as the gold standard to validate computational predictive methods (Dick *et al.*, 2017). More details of methods and databases can be observed in the work of Rao and colleagues (Rao *et al.*, 2014).

The Minimum Information About a Molecular Interaction Experiment (Mimix)

Recently, to publish a PIN article, magazines began to suggest the interactions identified in the experiments were deposited in a public database, a great initiative to provide and share information with the scientific community. However, with the diversity of databases, it was also realized the need to standardize the information concerning the PIN experiments that would be deposited aiming to facilitate the comparison, exchange and verification between these databases. Thus, was developed a guide advising how to describe a molecular interaction experiment and which information it are important (Orchard *et al.*, 2007), available on the HUPO Proteomics Standards Initiative – "see Relevant Website section" (access in 2017-10-26).

Thus, interactions from the most diverse experiments, studied and applied for biological purposes are deposited in the various public databases.

PIN Applications

The PINs have various biological applications such as providing subsidies for characterization of hypothetical proteins; understand systematically an organism, process or metabolic pathway; identification of key proteins or even suggest potential targets for developing new drugs.

Protein Annotation

The no knowledge about the function of certain proteins prevents the understanding of an organism in relation to its development, proliferation, virulence and biological processes thus, annotation of hypothetical proteins is essential for a better

Table 1   Protein-protein interaction public databases

Name	Organisms	Interaction type (Fisica/funcional)	Protein and interaction amount	Method type (Experimetal, computacional)	Curated	Link
String v 10.5	2031	Both	P: 9,6 million I: 1380 billions	Both	No	https://string-db.org/
DIP	834	Both	P: 28,826 I: 81,762	Ex	Yes (IMEX)	http://dip.doe-mbi.ucla.edu/dip/Main.cgi
Intact v 4.2.10	Organisms model	Both	P: 93,720 I: 786,143 2017/09	Text-mining	Yes (IMEX)	https://www.ebi.ac.uk/intact/
MINT	611	Both	P: 25,530 I: 125,464	Text-mining	Yes (IMEX)	http://mint.bio.uniroma2.it/
I2D	Five model organisms and human	Both	P: N/A I: 1,279,157	Co	Yes (IMEX)	http://ophid.utoronto.ca/ophidv2.204/index.jsp
MAtrixDB	Extracellular matrix proteins, proteoglycans and polysaccharides interactions	N/A	P: 14,902 I:15,018	Ex	Yes (IMEX)	http://matrixdb.univ-lyon1.fr/
InnateDB v 5.4	innate immune response of humans, mice and bovines to microbial infection	Both	P: 25,110 Ic: 367,503 Ip: 462,421	Ex	Yes (IMEX)	http://www.Innatedb.com/
Negatome v 2.0	non-interacting protein	Phy	P: N/A I: 30,756	Co	Yes	http://mips.helmholtz-muenchen.de/proj/ppi/negatome/

understanding of the organism. Generally, protein annotation is performed by sequence alignment and homology identification against proteins from public databases, but even with the large number of sequences not all proteins have their function annotated, requiring other approaches.

Currently, proteins having similar or complementary functions are known to be in close proximity on genome and tend to be transcribed together such as operons. Thus, it is expected these proteins are located in a cluster on the PIN and are closely connected among themselves (Pellegrini *et al.*, 2004).

As PIN provides knowledge of the various proteins working together in biological processes, it is sometimes possible to deduce the function of a hypothetical protein based on the function of its interaction partners in the cluster (Marcotte *et al.*, 1999). This method was applied in *Eriocheir sinensis* and based on PIN modularity feature the functions of 677 proteins were annotated (Hao *et al.*, 2015).

System Biology

The identification and annotation of IPP is important to understand how proteins form complexes to perform tasks inside the cell, subsidizing to understand by the set of interactions, how the organism works systematically. A metabolic pathway can be defined as a set of actions or interactions between genes and their products that results in the formation or change of some component of the system, essential for the correct functioning of a biological system. Thus, the identification of molecules and pathways in which they are involved is essential in the studies seeking to discover mechanisms involved in phenotypes and diseases (Hu *et al.*, 2017).

The PIN use is becoming increasingly common and essential for helping to elucidate biological processes clarifying the molecular origins of many diseases. As interactions are extremely disturbed by diseases such as cancer, heart disease and neurodegenerative diseases, the data generated by PIN provides identifying reasons to study them in living cells, enabling more accurate results and near reality (Itri *et al.*, 2017).

Based on gene modules constructed from human PINs to measure the relationship between pathways, a total of 2143 KEGG pathway pairs with close connections were identified, whose results were consistent with available evidence. The method was applied to explain the potential pathway targets that may be important in the etiology and development of Parkinson's disease (Hu *et al.*, 2017). The PIN usage in *Borrelia burgdoferi* allowed to discover two proteins with important functions related to infection and the biology of the organism, having different binding sites the interaction between proteins controls the infectivity of the organism (Thakur *et al.*, 2017).

PINs are important tools enabling better understanding of an organism mainly when added to other layers of functional and experimental data such as cell location or gene expression.

Essential Proteins

For optimal functioning the organism needs proteins guaranteeing primary resources for survival, and the lack or diminution of these proteins may be deleterious (Winzeler *et al.*, 1999). High-interaction proteins (hubs) are usually related to the essential genes in organisms, having important cellular functions form densely connected clusters whose removal causes a great rupture in the PIN and the organism may not perform its basic functions (Pavlopoulos *et al.*, 2011).

Recently *in silico* experiment showed the hub proteins are correlated with essential genes, they are highly conserved among species and possess low mutation rate, probably due to strong evolutionary pressure receiving (Dilucca *et al.*, 2017). With these features, the hub proteins are potential targets for drugs, offering new possibilities to research effective drugs for pathogenic organisms (Becker and Palsson, 2005).

Another in silico work identified 181 hub protein in nine *Corynebacterium pseudotuberculosis* strains, 180 with proven essentiality in other organisms. Because they were highly conserved, only 41 had no homology to the host and were considered promising targets for drugs (Folador *et al.*, 2016).

Host-Pathogen Interaction

The PINs still provide the possibility to investigate host-pathogen interactions and can be applied to understand the relationship between pathogen and host at systems biology level as suggesting proteins or interactions as potential targets for drug. Aspects related to virulence such as adhesion, escape of the immune system, proliferation are often dependent on the host-pathogen interaction and the knowledge of these mechanisms, even partially, can be determinant to understand the infection and to guide new experiments.

In addition to intracellular PIN, should be considered spatio-temporal aspects of host-pathogen PIN prediction, that is, the proteins participants of interaction must coexist and must be in cellular location favoring interaction, being potential participants in host-pathogen interaction the membrane, transmembrane or secreted proteins. The identification of host-pathogen interaction mainly promotes research into new interaction-inhibitors drug preventing infection be harmful to the host.

The host-pathogen PINs were used to identify interactions among *Mycobacterium tuberculosis* and *Homo sapiens* (Huo *et al.*, 2015). Using the SVM method was predicted the PIN of viruses with *Homo sapiens* as interactions of both hepatitis C virus and papillomavirus against *humans* obtaining 80% accuracy (Cui *et al.*, 2012).

In the study between *Nipah* virus and *human* 101 interactions were identified, 88 never found previously, being the main targets the miRNA processing machinery and the preprocessing factor of mRNA PRP19, including proteins capable of altering the p53 gene expression (Martinez-Gil *et al.*, 2017). Many targets for existing drugs are contained in a small number of families as guanine nucleotide-binding protein (G-protein) – acute receptor, nuclear hormone receptor, ionic channel kinase, and protease.

New Drug Classes

Many targets for existing drugs are contained in a small number of families as guanine nucleotide-binding protein (G-protein) – acute receptor, nuclear hormone receptor, ionic channel kinase, and protease (Groom and Hopkins, 2002; Santos *et al.*, 2017). Discovering new drugs has become a difficult task as known targets in the human proteome are increasingly limited being PINs the future targets for new drug discovery (Shin *et al.*, 2017). To overcome the difficulty to find new targets for drugs, researchers expect PIN become drug targets for diseases not yet having effective treatments such as Parkinson's and Alzheimer's disease (Milroy *et al.*, 2014). It is estimated in the human interactome there are between 130,000 and 650,000 interactions being drug targets only a small percentage (Venkatesan *et al.*, 2009; Rognan, 2015). Only in recent years more than 40 interactions were targeted and many inhibitors are in clinical trials phase (Arkin *et al.*, 2014), however, more than 90% of the interactions in humans have not yet been deciphered, having wide scope for research and improvement of predictive methods (Shin *et al.*, 2017).

Classical drugs tend to inhibit proteins that are sometimes essential in other biological processes and can cause adversities by not acting on an exclusive component (Petrakis and Andrade-Navarro, 2016). Due to the great structural and conformational variability the interactions emerge as an important target for specific drug development that act by stabilizing or inhibiting a distinct interaction sites of an exclusive component (Bultinck *et al.*, 2012). An example is the use of docking to identify inhibitory molecules of the Aurora Kinase-TPX2 PPI and prevent the growth of malignant tumors such as colon and stomach cancer, preventing the multiplication of cancerous cells (Cole *et al.*, 2017). Research combating cancer identified the molecules NVP-CGM097 and NVP-HDM201 with affinity to MDM2 and with potential p53-MDM2 PPI inhibitor, already in phase 1 clinical development (Holzer, 2017). A successful example is the SMPPIIs inhibitors acting on bromodomaine, four in clinical trials for cancer (I-BET762, CPI-0610, Ten-010 and OTX15) for the potential of binding to the bromodomain hydrophobic core and inhibiting the interaction between the two parts of bromodomo-histone (Arkin *et al.*, 2014).

The NusB-NusE interactions, fundamental to the formation of stable complexes required for RNA transcription, have antibiotic activity and are potential emerging antibacterial targets (Cossar *et al.*, 2017). Another example is 14-3-3s family proteins, abundant in the central nervous system, bind to serine and threonine regions when phosphorylated, are related to certain diseases and the use as a target for drug showed therapeutic effects (Kaplan and Fournier, 2017).

However, the identification of molecules inhibiting or stabilizing interactions presents challenges, mainly because they do not present traditional drugs characteristics as the Lipinski's five rules. The PPI interfaces are distinct from traditional targets, both geometrically when because the interactions are largely dominated by hydrophobic atoms (Shin *et al.*, 2017), being the natural compounds products potent candidates in the drug design targeting protein interface (Jin *et al.*, 2017).

Conclusions

Besides the work explained in this article, there are other public databases containing PINs from a diversity of sources, specialized in some specific organism. Similarly, there are other methods for identification or prediction of PINs and following technological developments, others are likely to be developed. Regardless of whether it is experimental or computational, low-throughput methods present greater scientific reproducibility than high-throughput methods, identifying large scale interactions. The low reproducibility of high-throughput experimental methods is probably due to the inability to reproduce a physico-chemical cellular environment similar to that occurring in vivo, especially with respect to variables that are not easily controlled as the space and time the interactions are carried out. For high-throughput computational methods, the low reproducibility is mainly motivated by the different and specific biological approach of each method, which use different logic, algorithm or data input, consequently generating different results. This shows that we still have a lot to learn and develop until we get PINs similar to those in nature.

PINs are applied for various purposes ranging from understanding an organism, understanding a metabolic pathway or process systematically, identifying proteins or interactions with potential use as a target for drugs or even understanding the relationship between pathogen and host. However, in order to obtain better results, it is not enough to only have PINs, it is necessary to aggregate layers of information from the different omics to the network, making possible a more complete interpretation within a biological context. Although, there are tools for visualization, formatting and analysis of PINs, the biological interpretation is usually done by a specialist researcher, with a broad knowledge of the biological context for which the network was built or knowledge about the organism. A PIN is not usually the final object of a search, but rather a tool that helps researcher in the organization, visualization, and integration of data to generate knowledge. Combining the researcher's expertise and knowledge, a PIN can serve to raise new hypotheses and to direct more assertive experiments in the laboratory.

See also: Alignment of Protein-Protein Interaction Networks. Community Detection in Biological Networks. Computational Approaches for Modeling Signal Transduction Networks. Computational Systems Biology. Experimental Platforms for Extracting Biological Data: Mass Spectrometry, Microarray, Next Generation Sequencing. Functional Genomics. Natural Language Processing Approaches in Bioinformatics. Network Inference and Reconstruction in Bioinformatics. Network Models. Network Properties. Network-Based Analysis of Host-Pathogen Interactions. Networks in Biology. Pathway Informatics. Prediction of Protein Localization. Prediction of Protein-Protein Interactions: Looking Through the Kaleidoscope. Protein Post-Translational Modification Prediction. Protein Properties. Protein-Peptide Interactions in Regulatory Events. Protein–Protein Interaction Databases. Protein–Protein Interaction Databases. The Evolution of Protein Family Databases

References

Abraham, M.J., Murtola, T., Schulz, R., *et al.*, 2015. GROMACS: High performance molecular simulations through multi-level parallelism from laptops to supercomputers. SoftwareX 1, 19–25.

Arkin, M.R., Tang, Y., Wells, J.A., 2014. Small-molecule inhibitors of protein-protein interactions: Progressing toward the reality. Chemistry & Biology 21, 1102–1114.

Assenov, Y., Ramírez, F., Schelhorn, S.-E., Lengauer, T., Albrecht, M., 2008. Computing topological parameters of biological networks. Bioinformatics 24, 282–284.

Barabási, A.-L., Albert, R., 1999. Emergence of scaling in random networks. Science 286, 509–512.

Barabasi, A.-L., Oltvai, Z.N., 2004. Network biology: Understanding the cell's functional organization. Nature Reviews Genetics 5, 101.

Becker, S.A., Palsson, B.Ø., 2005. Genome-scale reconstruction of the metabolic network in Staphylococcus aureus N315: An initial draft to the two-dimensional annotation. BMC Microbiology 5, 8.

Bishop, C.M., 2006. Pattern Recognition and Machine Learning. springer.

Blohm, P., Frishman, G., Smialowski, P., *et al.*, 2013. Negatome 2.0: A database of non-interacting proteins derived by literature mining, manual annotation and protein structure analysis. Nucleic Acids Research 42, D396–D400.

Braun, P., Tasan, M., Dreze, M., *et al.*, 2009. An experimentally derived confidence score for binary protein-protein interactions. Nature Methods 6, 91–97.

Brohee, S., Van Helden, J., 2006. Evaluation of clustering algorithms for protein-protein interaction networks. BMC Bioinformatics 7, 488.

Bultinck, J., Lievens, S., Tavernier, J., 2012. Protein-protein interactions: Network analysis and applications in drug discovery. Current Pharmaceutical Design 18, 4619–4629.

Chen, D., Jansson, A., Sim, D., Larsson, A., Nordlund, P., 2017. Structural analyses of human thymidylate synthase reveal a site that may control conformational switching between active and inactive states. Journal of Biological Chemistry 292, 13449–13458.

Cole, D.J., Janecek, M., Stokes, J.E., *et al.*, 2017. Computationally-guided optimization of small-molecule inhibitors of the Aurora A kinase–TPX2 protein–protein interaction. Chemical Communications 53, 9372–9375.

Cossar, P.J., Abdel-Hamid, M.K., Ma, C., *et al.*, 2017. Small-molecule inhibitors of the NusB–NusE protein–protein interaction with antibiotic activity. ACS Omega 2, 3839–3857.

Cui, G., Fang, C., Han, K., 2012. Prediction of protein-protein interactions between viruses and human by an SVM model. BMC Bioinformatics 13, S5.

Delprato, A., 2012. Topological and functional properties of the small GTPases protein interaction network. PLOS ONE 7, e44882.

De Vries, S.J., Van Dijk, M., Bonvin, A.M., 2010. The HADDOCK web server for data-driven biomolecular docking. Nature Protocols 5, 883.

Dick, K., Dehne, F., Golshani, A., Green, J.R., 2017. Positome: A method for improving protein interaction quality and prediction accuracy. In: Proceeding of 2017 IEEE Conference on Computational Intelligence in Bioinformatics and Computational Biology (CIBCB), pp. 1–8. IEEE.

Dilucca, M., Cimini, G., Giansanti, A., 2017. Topological transition in bacterial protein-protein interaction networks ruled by gene conservation, essentiality and function. arXiv preprint arXiv:1708.02299.

Eddy, S.R., 1998. Profile hidden Markov models. Bioinformatics (Oxford) 14, 755–763.

Esquivel-Rodríguez, J., Yang, Y.D., Kihara, D., 2012. Multi-LZerD: Multiple protein docking for asymmetric complexes. Proteins: Structure, Function, and Bioinformatics 80, 1818–1833.

Folador, E.L., De Carvalho, P.V.S.D., Silva, W.M., *et al.*, 2016. In silico identification of essential proteins in Corynebacterium pseudotuberculosis based on protein-protein interaction networks. BMC Systems Biology 10, 103.

Groom, C.R., Hopkins, A.L., 2002. Protein kinase drugs–optimism doesn't wait on facts. Drug Discovery Today 7, 801–802.

Guo, Y., Yu, L., Wen, Z., Li, M., 2008. Using support vector machine combined with auto covariance to predict protein–protein interactions from protein sequences. Nucleic Acids Research 36, 3025–3030.

Hao, T., Peng, W., Wang, Q., Wang, B., Sun, J., 2016. Reconstruction and application of protein–protein interaction network. International Journal of Molecular Sciences 17, 907.

Hao, T., Yu, A., Wang, B., Liu, A., Sun, J., 2015. Function annotation of proteins in eriocheir sinensis based on the protein-protein interaction network. In: The Proceedings of the Third International Conference on Communications, Signal Processing, and Systems, pp. 831–837. Springer.

Haykin, S., 1994. Neural Networks: A Comprehensive Foundation. Prentice Hall PTR.

Holding, A.N., 2015. XL-MS: Protein cross-linking coupled with mass spectrometry. Methods 89, 54–63.

Holzer, P., 2017. Discovery of potent and selective p53-MDM2 protein-protein interaction inhibitors as anticancer drugs. Chimia 71, 716.

Huo, T., Liu, W., Guo, Y., *et al.*, 2015. Prediction of host-pathogen protein interactions between Mycobacterium tuberculosis and Homo sapiens using sequence motifs. BMC Bioinformatics 16, 100.

Hu, Y., Yang, Y., Fang, Z., *et al.*, 2017. Detecting pathway relationship in the context of human protein-protein interaction network and its application to Parkinson's disease. Methods.

Ito, T., Chiba, T., Ozawa, R., *et al.*, 2001. A comprehensive two-hybrid analysis to explore the yeast protein interactome. Proceedings of the National Academy of Sciences 98, 4569–4574.

Itri, F., Monti, D.M., Chino, M., *et al.*, 2017. Identification of novel direct protein-protein interactions by irradiating living cells with femtosecond UV laser pulses. Biochemical and Biophysical Research Communications.

Itri, F., Monti, D.M., Della Ventura, B., *et al.*, 2016. Femtosecond UV-laser pulses to unveil protein–protein interactions in living cells. Cellular and Molecular Life Sciences 73, 637–648.

Jensen, L.J., Gupta, R., Blom, N., *et al.*, 2002. Prediction of human protein function from post-translational modifications and localization features. Journal of Molecular Biology 319, 1257–1265.

Jeong, H., Mason, S.P., Barabasi, A.-L., Oltvai, Z.N., 2001. Lethality and centrality in protein networks. arXiv preprint cond-mat/0105306.

Jiménez-García, B., Roel-Touris, J., Romero-Durana, M., *et al.*, 2017. LightDock: A new multi-scale approach to protein–protein docking. Bioinformatics.

Jin, X., Lee, K., Kim, N.H., *et al.*, 2017. Natural products used as a chemical library for protein–protein interaction targeted drug discovery. Journal of Molecular Graphics and Modelling.

Kaplan, A., Fournier, A.E., 2017. Targeting 14-3-3 adaptor protein-protein interactions to stimulate central nervous system repair. Neural Regeneration Research 12, 1040.

Keskin, O., Tuncbag, N., Gursoy, A., 2016. Predicting protein–protein interactions from the molecular to the proteome level. Chemical Reviews 116, 4884–4909.

Koike, A., Takagi, T., 2004. Prediction of protein–protein interaction sites using support vector machines. Protein Engineering Design and Selection 17, 165–173.

Kozakov, D., Hall, D.R., Xia, B., et al., 2017. The ClusPro web server for protein-protein docking. Nature Protocols 12, 255–278.

Lee, J.H., Andrabi, R., Su, C.-Y., et al., 2017. A broadly neutralizing antibody targets the dynamic HIV envelope trimer apex via a long, rigidified, and anionic β-hairpin structure. Immunity 46, 690–702.

Lin, J.-S., Lai, E.-M., 2017. Protein–protein interactions: Co-immunoprecipitation. In: Bacterial Protein Secretion Systems. Springer.

Marcotte, E.M., Pellegrini, M., Ng, H.-L., et al., 1999. Detecting protein function and protein-protein interactions from genome sequences. Science 285, 751–753.

Martinez-Gil, L., Vera-Velasco, N.M., Mingarro, I., 2017. Exploring the Human-Nipah virus protein-protein interactions. Journal of Virology, JVI. doi:10.1128/JVI.01461-17.

Milroy, L.-G., Grossmann, T.N., Hennig, S., Brunsveld, L., Ottmann, C., 2014. Modulators of protein–protein interactions. Chemical Reviews 114, 4695–4748.

Mohamed, T.P., Carbonell, J.G., Ganapathiraju, M.K., 2010. Active learning for human protein-protein interaction prediction. BMC Bioinformatics 11, S57.

Moraes, C., 2013. Série em biologia celular e molecular: Métodos experimentais no estudo de proteínas. Rio de Janeiro: Fundação Oswaldo Cruz–Fiocruz.

Newman, M.E., 2003. Mixing patterns in networks. Physical Review E 67, 026126.

Nooren, I.M., Thornton, J.M., 2003. Diversity of protein–protein interactions. The EMBO Journal 22, 3486–3492.

Ofran, Y., Rost, B., 2003. Predicted protein–protein interaction sites from local sequence information. FEBS Letters 544, 236–239.

Ohue, M., Shimoda, T., Suzuki, S., et al., 2014. MEGADOCK 4.0: An ultra–high-performance protein–protein docking software for heterogeneous supercomputers. Bioinformatics 30, 3281–3283.

Orchard, S., Kerrien, S., Abbani, S., et al., 2012. Protein interaction data curation: The International Molecular Exchange (IMEx) consortium. Nature Methods 9, 345–350.

Orchard, S., Salwinski, L., Kerrien, S., et al., 2007. The minimum information required for reporting a molecular interaction experiment (MIMIx). Nature Biotechnology 25, 894–898.

Pavlopoulos, G.A., Secrier, M., Moschopoulos, C.N., et al., 2011. Using graph theory to analyze biological networks. BioData Mining 4, 10.

Pellegrini, M., Haynor, D., Johnson, J.M., 2004. Protein interaction networks. Expert Review of Proteomics 1, 239–249.

Pellegrini, M., Marcotte, E.M., Thompson, M.J., Eisenberg, D., Yeates, T.O., 1999. Assigning protein functions by comparative genome analysis: Protein phylogenetic profiles. Proceedings of the National Academy of Sciences 96, 4285–4288.

Petrakis, S., Andrade-Navarro, M.A., 2016. Protein interaction networks in health and disease. Frontiers in Genetics 7.

Phizicky, E.M., Fields, S., 1995. Protein-protein interactions: Methods for detection and analysis. Microbiological Reviews 59, 94–123.

Pierce, B.G., Wiehe, K., Hwang, H., et al., 2014. ZDOCK server: Interactive docking prediction of protein–protein complexes and symmetric multimers. Bioinformatics 30, 1771–1773.

Puig, O., Caspary, F., Rigaut, G., et al., 2001. The tandem affinity purification (TAP) method: A general procedure of protein complex purification. Methods 24, 218–229.

Qi, Y., Bar-Joseph,, Z., Klein-Seetharaman, J., 2006. Evaluation of different biological data and computational classification methods for use in protein interaction prediction. Proteins: Structure, Function, and Bioinformatics 63, 490–500.

Rao, V.S., Srinivas, K., Sujini, G., Kumar, G., 2014. Protein-protein interaction detection: methods and analysis. International Journal of Proteomics 2014.

Rezende, A.M., Folador, E.L., Resende, D.D.M., Ruiz, J.C., 2012. Computational prediction of protein-protein interactions in Leishmania predicted proteomes. PLOS ONE 7, e51304.

Rinner, O., Mueller, L.N., Hubálek, M., et al., 2007. An integrated mass spectrometric and computational framework for the analysis of protein interaction networks. Nature Biotechnology 25, 345–352.

Ritchie, D.W., 2003. Evaluation of protein docking predictions using Hex 3.1 in CAPRI rounds 1 and 2. Proteins: Structure, Function, and Bioinformatics 52, 98–106.

Rognan, D., 2015. Rational design of protein–protein interaction inhibitors. MedChemComm 6, 51–60.

Santos, R., Ursu, O., Gaulton, A., et al., 2017. A comprehensive map of molecular drug targets. Nature Reviews Drug Discovery 16, 19–34.

Shin, W.-H., Christoffer, C.W., Kihara, D., 2017. In silico structure-based approaches to discover protein-protein interaction-targeting drugs. Methods.

Szklarczyk, D., Franceschini, A., Kuhn, M., et al., 2011. The STRING database in 2011: Functional interaction networks of proteins, globally integrated and scored. Nucleic Acids Research 39, D561–D568.

Thakur, M., Sharma, K., Chao, K., et al., 2017. A protein-protein interaction dictates Borrelial infectivity. Scientific Reports. 7.

Torchala, M., Moal, I.H., Chaleil, R.A., Fernandez-Recio, J., Bates, P.A., 2013. SwarmDock: A server for flexible protein–protein docking. Bioinformatics 29, 807–809.

Valencia, A., Pazos, F., 2002. Computational methods for the prediction of protein interactions. Current Opinion in Structural Biology 12, 368–373.

Venkatesan, K., Rual, J.-F., Vazquez, A., et al., 2009. An empirical framework for binary interactome mapping. Nature Methods 6, 83–90.

Wang, S., Ding, M., Chen, X., Chang, L., Sun, Y., 2017. Development of bimolecular fluorescence complementation using rsEGFP2 for detection and super-resolution imaging of protein-protein interactions in live cells. Biomedical Optics Express 8, 3119–3131.

Winzeler, E.A., Shoemaker, D.D., Astromoff, A., et al., 1999. Functional characterization of the S. cerevisiae genome by gene deletion and parallel analysis. Science 285, 901–906.

Wojcik, J., Schächter, V., 2001. Protein-protein interaction map inference using interacting domain profile pairs. Bioinformatics 17, S296–S305.

Yook, S.H., Oltvai, Z.N., Barabási, A.L., 2004. Functional and topological characterization of protein interaction networks. Proteomics 4, 928–942.

Zahiri, J., Hannon Bozorgmehr, J., Masoudi-Nejad, A., 2013. Computational prediction of protein–protein interaction networks: Algorithms and resources. Current Genomics 14, 397–414.

Zhang, Y., Ku, W.L., Liu, S., et al., 2017. Genome-wide identification of histone H2A and histone variant H2A. Z-interacting proteins by bPPI-seq. Cell Research.

Zhou, H., Jakobsson, E., 2013. Predicting protein-protein interaction by the mirrortree method: Possibilities and limitations. PLOS ONE 8, e81100.

Relevant Website

http://www.psidev.info/mimix
    MIMIx.

Prediction of Protein-Protein Interactions: Looking Through the Kaleidoscope

Anna Laddach, Sun Sook Chung, and Franca Fraternali, King's College, London, UK

Glossary

Asymmetric unit The smallest protein unit that can, by duplication and symmetry operations, generate a multimeric Protein Data Bank (PDB) structure.

Biological unit The functional protein unit of a PDB structure.

B-factor Also called temperature value, is associated with the spread of the measured electron density of an atom. It describes the uncertainty associated with the position of an atom in a crystallographic structure.

Interface core Amino acid residues which localise to the centre of a protein-protein interface and are fully buried upon interface formation.

Interface rim Amino acid residues which localise to the periphery of a protein-protein interface and are not fully buried upon complex formation.

Homologs Genes/proteins which share a common ancestor. A threshold sequence identity of 30% is normally used to detect homologs.

Obligate interaction A permanent interaction between proteins which cannot form stable structures on their own.

Orthologs Homologs which have arisen through speciation and perform similar functions in different species.

Paralogs Homologs which have arisen through a gene duplication event and have (generally) evolved to perform distinct functions.

Permanent interaction An interaction which is usually very stable and only exists in its complexed form.

Protein core The interior of a folded protein structure which is not exposed to solvent.

Protein interaction interface Regions of separate proteins which directly contact one another on complex formation.

Protein surface The surface of a protein structure which is exposed to solvent.

Solvent accessible surface area (SASA) The surface area, of a protein 30 component, which is exposed to solvent. Normally calculated at the residue level.

Transient interaction An interaction which can be formed briefly in the cell and in a reversible manner. Can be further classified as strong transient or weak transient based on measured interaction affinity.

Introduction

The genomic revolution of the past few decades has pushed technology to deliver efficient and precise tools for the fast and accurate mapping of genes in the genomes of thousands of species (Koepfli *et al.*, 2015). As a consequence, our knowledge of the human gene atlas is quite exhaustive (Telenti *et al.*, 2016). But knowledge without understanding has limited use. What one would aim for is to master the functional relationships between these genes, and their complex communication wired in a cellular context. The gene products, proteins, play an important role in this communication and in the networking capability of the cell; in this sense, they have been rightly defined as the workhorse of the cell. Therefore one of the next challenges is an exhaustive discovery of protein-protein interactions (PPIs) and the molecular aspects which play a role in their functional interplay. However, mapping the landscape of this multidimensional space of interactions is challenging, and techniques used to date vary in both complexity and efficacy. Recent renewed efforts, both in the theoretical and experimental fields, have been promoted by a number of studies which have challenged PPI detection techniques, such as Yeast Two-hybrid (Y2H) and Mass Spectrometry, to deliver much more accurate datasets for binary PPI maps (Fessenden, 2017). Data collected by the Human Proteome Project, since 2010, have recently (Feb 2016) achieved a coverage of 82% of this proteome, identifying altogether more than 90,000 unique protein interactions (Deutsch *et al.*, 2015). In spite of this progress, there is little overlap in the interactome uncovered by different experimental techniques (Fernandes *et al.*, 2010; Drew *et al.*, 2017). We are therefore obliged to look at these detailed snapshots as through a kaleido-scope and store, catalogue and assemble the unique pictures they display. From these, one can generate accurately annotated sub-sets of proteins and testable hypotheses, which can help in assigning confidence to functional modular units that play a role in the cellular machinery (Fessenden, 2017; Chung *et al.*, 2015).

Unfortunately, the large scale experimental probing of protein-protein interactions is both time-consuming and costly. However, computational methods can play a large role in filling the gap between gene knowledge and the functional communication. Moreover the results of such computational predictions can be validated by targeted experiments, thereby making more efficient use of experimental resources (Havugimana *et al.*, 2012; Carlin *et al.*, 2011).

Computational approaches to the study of PPIs include automated literature curation to generate large databases, such as STRING (Szklarczyk *et al.*, 2015) and the exploitation of sequence similarity/evolutionary relationships in order to both predict and extend functional knowledge of PPIs. Moreover further features, including the physicochemical properties of the interacting protomers, can be incorporated in methods to predict protein-protein interactions. Protein structural information, deposited in the PDB, has recently been exploited to add reliability to the collected interactions. Here, the structural coverage of the network has been extended through the large-scale homology modelling of binary interaction complexes (Mosca *et al.*, 2013). Recent work using machine learning techniques and Bayesian approaches (Garzon *et al.*, 2016) has considerably progressed the field of protein-protein interaction prediction combining multiple scores, such as those based on protein abundance, expression profiles and partner redundancy and coarse grained structural modelling, along with the previously described features. Structural information on single protomers is also being used in docking procedures to predict and score binary complexes. These methods have considerably improved in the recent years (Xue *et al.*, 2015; Gromiha *et al.*, 2017) but are still dependent of some prior information on the binding features.

As the protein-protein interaction space covered by amalgamating data from 90 experimental investigations is sparse, it is becoming clear that effective data integration from multiple sources with different granularity, including computational analyses, can play a determinant role in expanding the known interactome and laying the foundation for improved predictions (Moal *et al.*, 2017). In this article we will give an overview of current methods for determination and prediction of PPIs, with a particular focus on predicting interfaces of interaction complexes and their structural properties.

Experimental Techniques and Data resources

Computational analyses of PPIs can extract useful information for the interpretation and prediction of properties that characterise PPIs. It is therefore important to be aware of the experimental source used for data determination, as this can impact on its underlying quality and accuracy. Moreover, an interaction can be considered more reliable if its existence is supported by multiple experimental techniques. This section briefly describes the most commonly used experimental methods for the detection of protein-protein interactions, high-lighting the advantages and limitations of the different techniques, and recent advanced data resources, which have constructed protein-protein interaction networks from large-scale human proteome studies. These have stimulated the creation of integrative databases which make the results accessible to users.

Yeast Two-Hybrid (Y2H)

The yeast two-hybrid method (Two-hybrid screening) is the most widely used assay for detecting binary protein-protein interactions. It has been developed, refined and adapted to large-scale studies (Durfee *et al.*, 1993; Vidal *et al.*, 1996) since first introduced (Fields and Song, 1989). Two protein domains are tagged as a DNA binding domain (DBD) (bait protein) and an activation domain (AD) (prey protein) both of which are required for the transcription of a reporter gene and thus their interactions are inferred by expression of this reporter gene (Ito *et al.*, 2001). The Y2H method and its applications have been largely exploited to provide high-throughput protein-protein interaction (PPI) information, however some technical limitations persist. These are mainly caused by the lack of specificity between bait and prey, which may cause false-positive and false-negative interaction detection (Rao *et al.*, 2014; Westermarck *et al.*, 2013).

Affinity Purification Coupled with Mass Spectrometry (AP-MS)

Affinity Purification coupled with mass spectrometry (AP-MS) can identify protein complexes by directly purifying the protein of interest from cell lysates. This is followed by identification of the protein or proteins present by mass spectrometry analysis (Bantscheff *et al.*, 2007; Gingras *et al.*, 2007). Recent biotechno-logical advances (Konermann *et al.*, 2014; Yates, 2013) have overcome earlier problems such as specificity (segregation of over-expressed proteins or epitope-tagged proteins into abnormal cellular compartments) and co-purification of contaminating proteins (Rao *et al.*, 2014; Westermarck *et al.*, 2013).

Co-Fractionation Mass Spectrometry (CF-MS)

While the AP-MS method has been used to detect labelled bait proteins and the proteins with which they associate, it is limited to the identification of proteins for which antibodies are available or reliable epitope-tagged expression can be achieved. A Co-fractionation Mass Spectrometry (CF-MS) method has been developed to overcome such limitations, and establish comprehensive, less biased, protein associations of endogenous cellular proteins (Havugimana *et al.*, 2012; Kristensen *et al.*, 2012). The method involves preparing cellular lysates under mild, non-denaturing conditions. Soluble protein complexes in these lysates are then separated by several biochemical fractionation procedures, such as high-performance ion exchange chromatography, size-exclusion chromatography or isoelectro focussing. Fractions from each biochemical purification method are collected, trypsinized and the proteins present are identified by mass spectrometry analyses of each fraction. The components of each protein complex that co-fractionate are calculated by statistical and computational methods (Havugimana *et al.*, 2012).

3D Structures: X-ray, NMR, Cryo-EM

Accurate atomic detail of single and protein assemblies is regularly stored in the Protein Data Bank (PDB) (Berman *et al.*, 2000), the largest archive of structural data for biological macromolecules. The number of entries has increased rapidly due to progress in the techniques of X-ray crystallography, Nuclear Magnetic Resonance Spectrometry (NMR) and Cryo-electron microscopy (cryo-EM) (Callaway, 2015) (60,332 homo-/hetero-mers among 121,046 protein structures, May 2017), however coverage of the human proteome and interactome is still incomplete (see **Table 1**).

PDB entries annotate possible assemblies of the characterised proteins. The asymmetric unit refers to the smallest protein unit that can, by duplication and symmetry operations, generate the multimeric structure. This result may actually not be related to the biological role of complex. On the other hand, the biological assembly is annotated as such if it has been shown, or predicted, to be the functional unit (Krissinel and Henrick, 2007). Thus, to understand the functional role of a structure, it is preferable to refer the biological assembly, whenever possible.

PDB entries annotate possible assemblies of the characterised proteins. The asymmetric unit refers to the smallest protein unit that can, by duplication and symmetry operations, generate the multimeric structure. This result may actually not be related to the biological role of complex. On the other hand, the biological assembly is annotated as such if it has been shown, or predicted, to be the functional unit (Krissinel and Henrick, 2007). Thus, to understand the functional role of a structure, it is preferable to refer the biological assembly, whenever possible.

Large-scale Human Proteome Studies

There have been an increasing number of high-throughput PPI studies published recently, which made use of the methods described in Sections Yeast Two-hybrid (Y2H), Affinity Purification Coupled with Mass Spectrometry (AP-MS), Co-fractionation Mass Spectrometry (CF-MS). For human protein networks, four accurate large-scale studies from different research communities, which applied different technologies such as Y2H (Rolland et al.,2014), AP-MS (Hein et al., 2015; Huttlin et al., 2015) and CF-MS (Wan et al., 2015), have been published. In spite of the fact that each dataset covers three to seven thousand proteins, with considerable overlap in protein content (30% to 68%), their overlap in protein interactions is still very limited (3% to 6%) (Drew et al., 2017).

Public Databases

The need for reliable and public protein interaction data sources due to the progress in the technologies used to detect protein interaction information, has stimulated the development of a number of databases in recent years (Chatr-Aryamontri et al., 2015; Orchard et al., 2014; Szklarczyk et al., 2015). These 180 rely on a collaborative effort throughout the International Molecular Exchange (IMEx) consortium (Orchard et al., 2012) which develops solid data formats and defines curation rules to improve data integrity. Not only integrative databases, but also databases specifically tailored for structural annotations on protein-protein interaction networks, have been developed (Mosca et al., 2013; Lu et al., 2016) (**Table 2**).

Table 1   Structural coverage of the human protein-protein interaction network; statistics taken from Interactome3D

	Total	With structure	With model	Without structure
Proteins	14190	1843 (13.0%)	7600 (53.6%)	4747 (33.5%)
Interactions	95152	5757 (6.1%)	5978 (6.3%)	83417 (87.7%)

Source: Mosca, R., Ceol, A., Aloy, P., 2013. Interactome3D: Adding structural details to protein networks. Nat. Methods 10, 47–53. doi:10.1038/nmeth.2289.

Table 2   Examples of public PPI databases

Name	Key features	Public website
IntAct (Orchard et al., 2014)	Protein interaction data repository curated from the literature and direct experimental results Updated monthly	http://www.ebi.ac.uk/intact/
BioGRID (Chatr-Aryamontri et al., 2015)	Curated set of physical and genetic interactions including chemical associations and post-translational modifications Updated monthly	https://thebiogrid.org/
STRING (Szklarczyk et al., 2015)	Known and predicted protein-protein interactions Main sources: High-throughput lab experiments, genomic context prediction, Co-expression, automated textmining, other knowledge databases	https://string-db.org
Interactome3D (Mosca et al., 2013)	Structural annotation of protein-protein interaction networks from the available three-dimensional structures and modeling	http://interactome3d.irbbarcelona.org/
PinSnps (Lu et al., 2016)	Genetic variant data mapped to a 3D integrated protein-protein interaction network	http://fraternalilab.kcl.ac.uk/PinSnps/
PrePPI (Garzon et al., 2016)	Large scale database of interactions predicted by combining structural and non-structural methods	https://honiglab.c2b2.columbia.edu/PrePPI/

Features of Protein-Protein Interactions

The wealth of experimental data, described in Section Experimental Techniques and Data Resources, in addition to sequencing data, has been mined to uncover features which are characteristic of protein-protein interactions. These features are the key ingredients which have been exploited to predict both new interactions and interaction interfaces. Here we describe these features, followed by an account of how different approaches use such features in Section Computational Approaches to Predicting Protein-Protein Interactions.

Homology and Conservation

Since the late 80s, homology modelling has been used to infer the structure of proteins which lack experimentally resolved structures (Chothia and Lesk, 1986; Blundell et al., 1988). This has played an important role in expanding the structural coverage of the proteome, as known protein sequences currently greatly outnumber experimentally solved structures (as described in Section 3D Structures: X-ray, NMR, Cryo-EM).

Importantly, this approach can be extended to expand the structural coverage of known or putative protein-protein interactions (see Fig. 1(a)); however, the coverage of known interactions by experimentally resolved binary complexes is even smaller (see Section 3D Structures: X-ray, NMR, Cryo-EM). The proviso must also be taken that, although interaction interfaces are generally conserved, a number of homologs have been observed to interact using different interfaces (Aloy et al., 2003; Xue et al., 2011) (see Fig. 1(c)). This is particularly the case when proteins are paralogs (proteins which have arisen through gene duplication which generally perform distinct functions (Pearson, 2013) rather than orthologs. The exact degree of interface conservation also appears to depend on a number of parameters associated with the nature of the interaction, e.g. binding sites of obligate interactions tend to be highly conserved and those of transient interactions less conserved (Xue et al., 2011) (see Fig. 2 for a summary of interaction types). It has also been shown that homologs tend to interact using the same interface, even if the interaction partner is different (Esmaielbeiki et al., 2016) (see Fig. 1(b)). However, partner-specific interfaces have been shown to demonstrate greater conserva tion than generic interfaces, even with respect to those occurring in transient interactions (Xue et al., 2011).

Homology-based approaches are rooted in the knowledge that, during evolution, constraints arising from protein structure and function impact on the conservation of residues (and therefore sequence identity). A classic example is that residues which reside in the core of a protein tend to be more conserved than surface residues. This is because such residues often play a crucial role in maintaining the fold of the protein and thus mutations which localise to this region are generally selected against (Echave et al., 2016). Protein interactions represent another type of constraint, both structural and functional. A protein must maintain a structure which allows it to sterically and physicochemically interact with its partners; the affinity and kinetics of an interaction must also be suitably maintained so

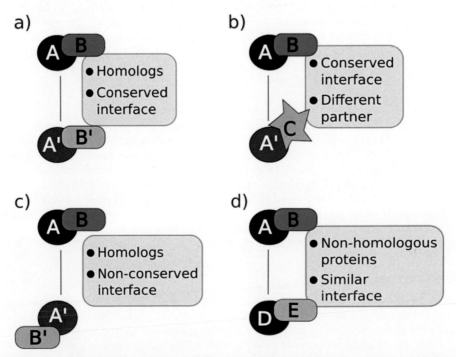

Fig. 1 Scenarios exploring the relationship between interface similarity and homology. Letters refer to protomers which form binary interaction complexes and the prime symbol is used to denote homology. (a) Homologs A′ and B′ use the same interface as A and B, (b) homolog A′ uses the same interface to interact with a protein which is unrelated to B, (c) homologs A′ and B′ interact via a different interface, (d) non-homologous proteins D and E use similar interface due to similar structural properties).

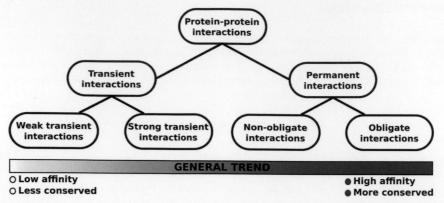

Fig. 2 Schematic showing interaction types. Obligate interactions refer to permanent complexes composed of protomers that are not stable on their own (see Glossary of terms).

that biological function is not disrupted. Indeed, it has been shown that surface residues which are involved in protein interaction interfaces are more highly conserved than non-interacting surface residues (Choi *et al.*, 2009; Ashkenazy *et al.*, 2016). An important observation, recently exploited in the prediction of protein interaction interfaces, is that when mutants do localise to an interface, this can result in the selection of compensatory mutants on opposing interface. This phenomenon, which is essentially due to co-evolution of the residues at the interface, allows functions abrogated by the initial mutation to be "rescued" by the compensatory mutation on the opposite side of the interface (de Juan *et al.*, 2013; Bitbol *et al.*, 2016).

Notably, it has also been shown that convergent evolution can take place. Here non-homologous proteins develop similar interaction interfaces (from a steric and physicochemical perspective) in order to interact with the same interaction partner (see **Fig. 1(d)**). An example of this is the mimicry of host protein interfaces by viral proteins in order to hijack host cell functions (Zhao *et al.*, 2011).

Protein Domains

In the vast majority of cases interactions do not involve entire proteins but rather specific regions, termed domains, which interact with one another (Raghavachari *et al.*, 2008). Therefore knowledge of a proteins underlying domain-architecture can aid in the prediction of both interactions and interaction interfaces, as discussed in Section Structure-based predictors.

Domains can be described as units within a protein which have a distinct structure or function, can fold independently, and can evolve with a degree of independence from one another. A number of resources define protein domains, including CATH (Sillitoe *et al.*, 2015), a structure-based classification system, and PFAM, which makes use of evolutionary information to infer domain boundaries (Finn *et al.*, 2014). Correct domain definitions are particularl crucial when dealing with large multi-domain proteins, such as titin (Laddach *et al.*, 2017).

Protein Recognition Mechanisms

Three different models exist to describe the nature of protein protein interactions (see **Fig. 3**). The lock and key model, originally proposed by Emil Fischer, describes interactions which are rigid in nature (Kastritis and Bonvin, 2013a; Fischer, 1894). Here both interaction interfaces are complementary in shape and negligible conformational changes take place on binding. In contrast the induced fit model involves conformational change on binding, allowing proteins with varying degrees of shape complementarity in the unbound state to interact (Kastritis and Bonvin, 2013a; Koshland, 1958). A subset of interactions which follow this model involve intrinsically disordered regions. Here an increase in order upon binding is associated with a decrease in entropy and results in weak, transient interactions (Perkins *et al.*, 2010). A third model, termed conformational selection, has been proposed. As proteins are dynamic in structure, each interaction partner will explore a number of conformational states, of which only particular states which will be able to interact (Kastritis and Bonvin, 2013a; Ma *et al.*, 1999; Bosshard, 2001; Boehr *et al.*, 2009).

Physicochemical Properties of Interaction Interface Regions

A number of features make important contributions to the properties of protein-protein interaction interfaces. These include the physicochemical properties of amino acids, such as charge and hydrophobicity, in addition to solvent accessibility, hydrogen bonds, secondary structural elements, steric properties (i.e., shape complementarity) and flexibility (Gromiha and Yugandhar, 2017; Keskin *et al.*, 2016). Protein interaction interfaces can be further divided into 280 core and rim regions (see **Fig. 4**). It has been shown that interface core regions, although surface exposed before binding, are enriched in hydrophobic residues and have amino acid compositions similar to the buried interior of a protein. In contrast, the rim regions differ less from other surface regions, and maintain a degree of solvent exposure upon binding. Hydrophilic and charged residues are usually enriched in such regions (Chakrabarti and Janin, 2002; Guharoy and Chakrabarti, 2005).

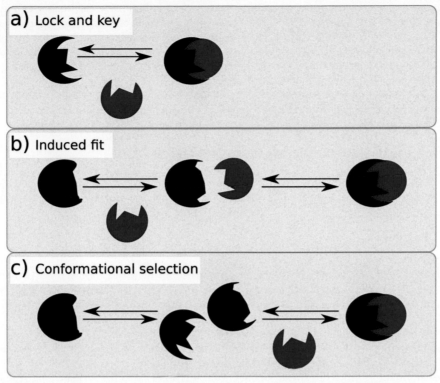

Fig. 3 Schematic showing protein mechanisms through with protein A (blue) recognises its binding parter, protein B (red): (a) lock and key, (b) induced fit, (c) conformational selection. Protein B can also undergo conformational changes, however here it is represented as a rigid body for simplicity. Adapted from Kastritis, P., Bonvin, A., 2013a. On the binding affinity of macromolecular interactions: Daring to ask why proteins interact. J R Soc Interface 10, 20120835. doi:10.1098/rsif.2012.0835.

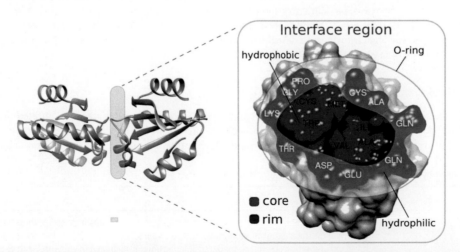

Fig. 4 Schematic showing homodimeric thioredoxin interface (PDB:3zzx) – core (red) and rim regions (blue) can clearly be distinguished. Note the core is composed of hydrophobic amino acids and the rim is composed of hydrophilic amino acids. The localisation of the interface is depicted on the left side in a ribbon representation of the homodimeric complex (Reproduced from Campos-Acevedo, A.A., Rudino-Pinera, E., 2014. Crystallographic studies evidencing the high energy tolerance to disrupting the interface disulfide bond of thioredoxin 1 from white leg shrimp Litopenaeus vannamei. Molecules 19, 21113-21126. doi:10.3390/molecules191221113).

Distinctions also exist between the properties of obligate and transient interaction interfaces. Hydrophobic interactions make a larger contribution to obligate interfaces whereas transient interfaces rely more on the formation of salt bridges (Keskin *et al.*, 2016). Phosphorylation sites also frequently play a role in transient interactions (Keskin *et al.*, 2016). Obligate interactions have been coined hairy by Kufareva *et al.* (2007), who found such interactions to rely on amino acid side chains. This contrasts with their findings on transient interactions, where protein backbones were determined to play a greater role. Although rare, disulphide bonds have been shown to play an important role in stabilising a number small complexes (Acuner Ozbabacan *et al.*, 2011).

It has been shown that only a small number of residues make large contributions to the binding affinity of protein-protein interactions (Bogan and Thorn, 1998; Moreira *et al.*, 2007). These are termed hotspot residues and tend to cluster within the core of interfaces (Keskin *et al.*, 2005); furthermore, they have been shown to be shielded from bulk solvent by the side chains of neighbouring residues (Li and Liu, 2009). Such shielding residues have been termed the O-ring due to the shape they form around core residues. Arginine, tryptophan and tyrosine have been found to be both most enriched in hotspots and to account for the greatest proportion of hotspot residues (Bogan and Thorn, 1998).

Finally, it is important to note that solvent water molecules can play an important role in protein interactions, in particular as they can fill cavities in the interaction complex, and be involved in bridging hydrogen bonds (Janin, 1999).

Computational Approaches to Predicting Protein-Protein Interactions

In this section we overview different approaches used in the computational prediction of protein-protein interactions. We then focus techniques which add a three-dimensional structural element to the interaction network, namely computational interface prediction and protein-protein docking.

Overview

The computational prediction of protein-interactions can be approached in numerous ways and with multiple levels of granularity. At the coarsest level a binary prediction of whether two proteins interact can be performed. One factor to consider in evaluating the feasibility of an interaction *in vivo*, is the cellular compartment in which the interacting protein operate and their likely co-occurence. A task of much finer granularity is the prediction of the exact atomistic nature of protein-protein interactions; this enables protein interaction interfaces to be inferred.

A recently developed database which exemplifies the use of multiple methods to predict protein-protein interactions is PrePPI (Garzon *et al.*, 2016). Approaches used by PrePPI are summarised in **Table 3**. For a more general review see Keskin *et al.* (2016).

Given the wealth of information which can already be obtained from PrePPI and other databases (e.g STRING (Szklarczyk *et al.*, 2015)), the rest of this article will focus on methods which enable the prediction of interaction interfaces (i.e. exactly which residues contact one another in a complex). This information has several important applications, for example in the rational design of drugs which target protein-protein interactions (Jin *et al.*, 2014; Mosca *et al.*, 2015; Duran-Frigola *et al.*, 2017; Kategaya *et al.*, 2017), and in the interpretation of genetic variants and expression data. Here 3D structural data can be used to assess the impact of a genetic variant on the binding affinity of an interaction, and thus its edgetic impact on the network. Variants which localise to a protein core may disrupt a protein's structure, effectively removing a node from the network, whereas those which localise to interaction interfaces may affect specific interactions (edges in the network) (Sahni *et al.*, 2015; Engin *et al.*, 2016; Yi *et al.*, 2017). Therefore residue level information describing protein interaction interfaces is essential for the edgetic modelling of disease-associated variants (Mosca *et al.*, 2015).

Protein-Protein Interface Prediction

Interface predictors make use of known features of protein-protein interactions interfaces, as outlined in Section Affinity Purification Coupled with Mass Spectrometry (AP-MS), in order to predict which residues are involved in such interfaces. A distinction must be made between partner-specific and non-partner-specific predictors. Partner-specific predictors

Table 3   Methods used by PrePPI in the prediction of protein-protein interactions

Method	Description
Molecular Modelling	From the structures or models of a complex A:B, A and B are superimposed onto experimentally determined structures of A' in complex with B'. Interaction probability is then evaluated based on structural properties of the modelled interface and structural similarity of A and B with their homologs.
Phylogenetic profile	If orthologs of A and B are frequently present in the same species they are deemed more likely to interact.
Gene ontology	If proteins A and B have similar functions they are deemed more likely to interact.
Orthology	If orthologs of A and B are known to interact in other species they are deemed likely to interact.
Expression profile	If A and B demonstrate similar expression profiles they are deemed likely to interact. This can be extended to the expression profiles of orthologs.
Partner redundancy	If protein A interacts with a number n of proteins which are structurally similar to B, the likelihood of B interacting with A increases with n.
Protein peptide	If A contains a peptide (sequence motif) known to interact with a particular domain type present in B, A and B are considered likely to interact.

Source: Garzon, J.I., Deng, L., Murray, D., Shapira, S., Petrey, D., Honig, B., 2016. A computational interactome and functional annotation for the human pro-710 teome. eLife 5. doi:10.7554/eLife.18715.

(e.g., PS-HomPPI (Xue *et al.*, 2011), EV-complex (Hopf *et al.*, 2014)) take as input two query proteins, A and B, which are known to interact, and predict which residues constitute their interaction interface. Non-partner specific predictors (e.g. NPS-HomPPI (Xue *et al.*, 2011), SPPIDER (Porollo and Meller, 2007)) take a single query protein, A, and predict the residues that are likely to be involved in interactions with any interaction partner.

The following sections outline different categories of interface predictors and representative methods for each category are summarised in **Table 4**.

Sequence-based predictors

Features which can be derived from sequence alone, such as the physicochemical properties of amino acids, hydrophobicity and interface propensity can be used to predict whether residues are involved in protein-protein interactions. Properties are generally calculated using a sliding window or subsequence of a certain number of residues (Xue *et al.*, 2011; Rydevik *et al.*, 1991; Ofran and Rost, 2003). Some sequence-based predictors, such as PSIVER (Murakami and Mizuguchi, 2010) and ISIS (Ofran and Rost, 2007) also use predicted structural information, including predicted SASA and secondary structure; this has been shown to improve their performance (Esmaielbeiki *et al.*, 2016). SeRenDIP, a recently developed random forest based predictor, goes one step further by incor-porating predicted dynamics (predicted backbone flexibility) in its predictions (Hou *et al.*, 2017).

As interface residues are more conserved than non-interface surface residues (as detailed in Section Homology and Con-servation), evolutionary information derived from multiple sequence alignments can be used in the prediction of interface residues. A more sophisticated use of evolutionary information to predict protein interaction interfaces is used by co-evolutionary methods; these make use of the theory, outlined in Section Homology and Conservation, that compensatory mutations on opposing interfaces should be selected for during evolution. A key challenge in these methods has been distinguishing between the direct coupling of residues (as they are in direct contact) and indirect coupling of residues (residue A and B covary as they are both in contact with residue C). A number of approaches, which rely on the creation of a global statistical model, have been used to address this challenge. These include direct coupling analysis, protein sparse inverse covariance and Bayesian network-based approaches (de Juan *et al.*, 2013).

Sequence-based methods are generally less accurate than structure-based methods, but are able to obtain a much greater coverage of the proteome due to the higher availability of sequence information (Gallet *et al.*, 2000). Coevo-lutionary methods have improved dramatically over recent years (Esmaielbeiki *et al.*, 2016), but are limited to predicting interfaces where both partners are known, and a large number of homologous sequences are available. This is often a hindrance when protein partners are exclusive to a particular taxonomic branch (Hopf *et al.*, 2014).

Structure-based predictors

A number of predictors use structural features, such as solvent accessible surface area (SASA), secondary structure, geometry and B-factor, directly (Esmaielbeiki *et al.*, 2016; Porollo and Meller, 2007). The performance of predictors which already use sequence-based features is greatly improved by this; particularly as, through the use of SASA, buried residues in the protein core can be immediately ruled out (Xue *et al.*, 2011). Other methods map features on to the protein structure and search for 3D patches of residues whose properties are consistent with being involved in protein-protein interaction interfaces (Porollo and Meller, 2007) (this can be seen as a 3D version of the sliding window approach outlined in Section Sequence-Based Predictors). Although not specific to protein interaction interfaces, ConSurf notably uses this method to facilitate the identification of functional sites (Ashkenazy *et al.*, 2010).

Table 4   Representative methods for the prediction of protein-protein interaction interfaces

Name	Key features	Model type	Public website
PSIVER (Murakami and Mizuguchi, 2010)	sequence-based predictor	Naï've Bayes	https://omictools.com/psiver-tool
SeRenDIP (Hou *et al.*, 2017)	sequence-based predictor	random forests	http://www.ibi.vu.nl/programs/serendipwww/
EVcomplex (Hopf *et al.*, 2014)	co-evolutionary method	maximum entropy	https://evcomplex.hms.harvard.edu/
SPPIDER (Porollo and Meller, 2007)	structural and sequence-based features used	neural network	http://sppider.cchmc.org/
ISPRED4 (Zhao and Gong, 2017)	structural and sequence-based features used	SVM with grammar based correction	https://ispred4.biocomp.unibo.it/ispred/
HomPPI (Xue *et al.*, 2011)	structural template-based method	regression	http://ailab1.ist.psu.edu/PSHOMPPIv1.2/http://ailab1.ist.psu.edu/NPSHOMPPI/
PriSE (Jordan *et al.*, 2012)	structural neighbourhood method	empirical scoring system	http://ailab1.ist.psu.edu/prise/index.py
PRISM (Tuncbag *et al.*, 2011)	structural alignment of interface regions	empirical scoring system	http://cosbi.ku.edu.tr/prism/

Partner specific structural predictors can take into consideration features of both structures, for example shape complementarity and electrostatics of the interface regions. Such predictions are thought to be more reliable if binding follows the lock and key model (see Section Physicochemical Properties of Interaction Interface Regions) as shape complementarity of unbound structures will be less complete if binding follows the induced fit model (Xue *et al.*, 2011). Furthermore, this strategy is most successful if available monomeric structures are similar in conformation to that in which the partners bind.

Template-based predictors of interfaces search for homologous proteins in structurally resolved complexes. Here the homologous complex may involve ho-mologs of both proteins A and B, or it may only involve a homolog of one of the interaction partners. In the case that only one interacting partner is involved, the observation that proteins often use the same interface to interact with different partners, as outlined in Section Homology and Conservation, can be used to make a less confident prediction of the interaction interface. Where homologous complexes for partner specific interactions are available, these methods tend to give the best results. As outlined in Section Protein Domains, interactions may only occur between specific domains of two interacting proteins. Therefore, where no global templates are available to model an interaction, templates which are homologous at the domain level, if available, can be used; this is the approach which has been taken by Interac-tome3D (Mosca *et al.*, 2013), which uses the 3DID database (Mosca *et al.*, 2014) as a resource to identify interacting domain-domain templates. Some methods, such as HomPPI (Xue *et al.*, 2011), incorporate information from multiple structural templates, where available, in order to improve performance.

Structural neighbourhood methods also use template-based prediction methods. For these methods the protein of interest must have an experimental structure, despite the absence of a structure where it is present with its interaction partner. Here, rather than homologs, templates are searched for where the sub-units are structurally similar to the query protein. Methods such as PredUs (Zhang *et al.*, 2011) use global similarity of the query structure to the template whereas PrISE (Jordan *et al.*, 2012) also takes interface similarity specifically into account.

Combining features and methods

Information from different features can be combined to make predictions in multiple ways. Empirical scoring functions can be used; however, these functions can suffer if knowledge of the system (in this case the parameters which determine whether proteins interact) is incomplete. Such scoring functions have the disadvantage that their form must be predetermined and are generally linear in structure. Machine learning techniques have the advantage that they can make use of non-linear combinations of features and can, to a greater degree, infer their functional form from the data (Wo´jcikowski *et al.*, 2017). A number of different algorithms including Support Vector Machines (SVMs) (Sriwastava *et al.*, 2015; Zhao and Gong, 2017), neural networks (Porollo and Meller, 2007) and random forests (Hou *et al.*, 2017) can be used to classify residues as interacting or non-interacting. Researchers are beginning to explore the application of deep learning methods to the prediction of protein-protein interaction interfaces (Zhao and Gong, 2017), however these techniques require vast amounts of training data.

A number of methods, for predicting protein interaction interfaces, incor-porate several distinct steps. For example, PRISM (Baspinar *et al.*, 2014), although primarily a template based method, guides structural alignments through the evolutionary prediction of hotspot residues. Succeeding structural alignment, further docking (see Section Docking for more information) and refinement steps are performed.

As more predictors become available, their results can be used as features to train meta-predictors, which generally achieve better performance than each individual predictor (Qin and Zhou, 2007; de Vries and Bonvin, 2011). Here, care must be taken that testing data used to evaluate the meta predictor does not overlap with the training data of its component predictors.

Docking

Given two protomers or proteins with structures, docking is a computational method to predict 3D structural protein-protein interaction complexes of target proteins (Rodrigues and Bonvin, 2014). It has evolved rapidly in parallel to the increase of publicly available molecular structures (Soni and Madhusudhan, 2017) and advances in computational technology (Schlick *et al.*, 2011). The Critical Assessment of Predicted Interactions (CAPRI) (Janin, 2005) is dedicated to the evaluation of protein-protein docking and organised by the research community to target unpublished crystal or NMR structural complexes. It has encouraged parti-cipations from different institutes worldwide and has become an important event in the field of protein structure prediction. Several representative computational tools and participants are listed in **Table 5**.

The key elements of docking are 3D structures (experimental or modelled) of target proteins, a sampling procedure to generate the conformational landscape of the predicted structural interactions, and a scoring function to compare and assess the predicted models (Rodrigues and Bonvin, 2014). Mainly, three types of scoring functions are used to evaluate binding affinity of protein interaction models: force-field scoring (calculating van der Waals (VDW) energy, electrostatic energy and bond stretching/bending/torsional forces); empirical scoring (integrating different weighted energy scores selectively to fit empirical binding data; and knowledge-based scoring (calculating statistical potential of the protein complex based on experimentally solved atomic structures) (Huang *et al.*, 2010). However, sampling and scoring are still challenging because of the incomplete quantitative information available regarding the impact of conformational change and molecular flexibility on binding affinity, as well as difficulties associated with certain protein types (e.g. lipid or membrane binding proteins) and multi-component assemblies (Kastritis and

Table 5   Examples of docking methods

Name	Key features	Public website
ATTRACT (deVries *et al.*, 2015)	Energy minimization in rotational and translational degrees of freedom	http://attract.ph.tum.de/services/ATTRACT/attract.html
ClusPro (Kozakov *et al.*, 2017)	Rigid-body docking (FFT), cluster retained conformations and refine by CHARMM minimization	https://cluspro.bu.edu
HADDOCK (Dominguez *et al.*, 2003)	Rigid body energy minimisation, semi-flexible refinement in torsion angle space, final refinement in explicit solvent refinement	http://milou.science.uu.nl/services/HADDOCK2.2/
PATCHDOCK (Schneidman-Duhovny *et al.*, 2005)	Molecular Shape representation, surface patch matching, filtering and scoring	https://bioinfo3d.cs.tau.ac.il/PatchDock/
RosettaDock (Lyskov and Gray, 2008)	Rigid-body position by random perturbation and selection based on Gaussian distribution, energy minimisation orientation, side-chain conformation optimisation	http://rosie.graylab.jhu.edu/

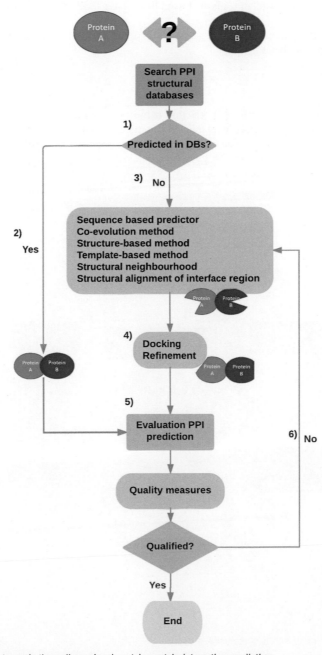

Fig. 5   A flowchart showing steps towards three dimensional protein-protein interaction prediction.

Bonvin, 2013b).

A Step by Step Guide to the Prediction of Three-Dimensional Protein-Protein Interactions

Here we outline how methods and resources, introduced throughout this article, can be applied to a query, consisting of two proteins (A and B), in order to obtain a 3D binary interaction complex model. Proposed steps are outlined below and illustrated in **Fig. 5**.

1) Given proteins A and B, for which we want to predict interactions, search PPI databases with structural annotations (e.g., Interactome3D (Mosca *et al.*, 2013), PrePPI (Garzon *et al.*, 2016)) and identify whether there are available models.

2) If there are predicted models of the complex, we can evaluate them as described in point 5) below.

3) If no models are available, different approaches can be applied. Models can be built using information derived from sequence based predictors and/or structure-based predictors. See Section Protein-Protein Interface Prediction and **Table 4**.

4) Apply docking or other refinement methods in order to improve predictions. See Sections Combining Features and Methods and Docking.

5) Evaluate models based on some quality measurement of prediction methods or scoring functions described in Docking. In addition, general protein structure assessment tools can be applied such as MolProbity (Chen *et al.*, 2010) or QMEAN (Benkert *et al.*, 2008).

6) Re-apply prediction methods if the qualities are not satisfied. Make use of other listed approaches, these can be applied and adapted to improve PPI model predictions.

Future Directions

This article has focussed on the prediction of binary interaction complexes, and more specifically on the prediction of protein-protein interfaces. However most biological complexes that play a crucial functional role in the cell, are made up of multiple protein components. Template based methods can be used to model such assemblies, but these are limited by the lack of available

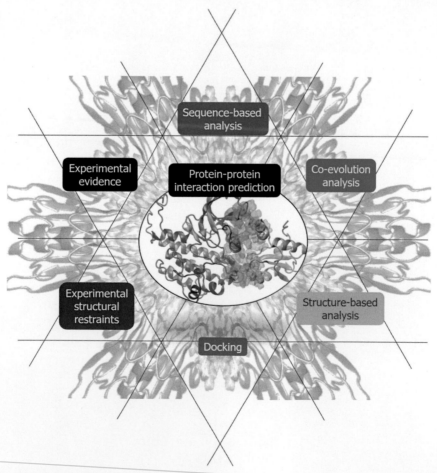

Fig. 6   A kaleidoscopic schematic for protein-protein interaction prediction.

experimentally resolved multicomponent structures. These are still particularly challenging to address by modern techniques. Notably, recent advances in the cryo-EM methodology allow for the structural determination of large complexes at near atomic resolution (Chlanda and Krijnse Locker, 2017; Orlov et al., 2017; Lengyel et al., 2014). Additionally, this technique can be complemented by data from crystallography and NMR experiments, as smaller complexes or single component structures can be projected onto EM density maps and help in the accurate reconstruction of a complex. In light of this, the PDB has launched a prototype framework for depositing hybrid models which result from a combination of experimental techniques (Burley et al., 2017). This will allow for efficient use of integrated data in prediction methods.

A challenge will be in the extension and development of predictors to make efficient use of available multimeric information and the associated level of resolution. The Swiss-Model developers have started to address this quest with the release of a new pipeline, which enables the automated modelling of oligomeric complexes (Bertoni et al., 2017). In parallel, a handful of docking programs are starting to address the automated docking of multiple components, although the procedures still rely mainly the pairwise docking of all the complex components (Soni and Madhusudhan, 2017).

As more experimental data are released, it becomes imperative to incorporate these into both computational predictors and docking methods. A number of computational procedures are already being developed to accomplish the efficient merging of restraints from data derived from different sources (this data may be varied in both nature and accuracy) (Politis and Schmidt, 2017; Schmidt et al., 2017; Tamo et al., 2017). Low resolution data, for example from small angle x-ray scattering (SAXS), EM, Ion Mobility Mass Spectrom-etry, and even data from site-directed mutagenesis experiments, can provide a conditional universe in which predictions can take place. One example which uses this approach is the docking software HADDOCK (Rodrigues and Bonvin, 2014), which already allows the integration of data from multiple sources, in the form of restraints.

Throughout this article we have seen how kaleidoscopic snapshots of the protein interaction universe (**Fig. 6**) can be expanded and integrated through the use of computational methods. However, only through the continued development and integration of both experimental and computational methods, it is conceivable that we may one day achieve a complete and accurate multidimensional map of the protein interaction landscape.

Acknowledgements

The authors thank Joseph Chi-fung Ng for his critical reading of the manuscript. This research was supported by the British Heart Foundation (to FF and AL), Bloodwise (to FF and SSC) and the Medical Research Council (MR/L01257X/1 to FF). AL is funded by the British Heart Foundation (RE/13/2/30182) and SSC is funded by a Bloodwise Gordon Piller PhD Studentship.

See also: Algorithms for Structure Comparison and Analysis: Docking. Alignment of Protein-Protein Interaction Networks. Artificial Intelligence and Machine Learning in Bioinformatics. Data Mining: Classification and Prediction. Machine Learning in Bioinformatics. Natural Language Processing Approaches in Bioinformatics. Protein-Peptide Interactions in Regulatory Events. Protein–Protein Interaction Databases. Protein–Protein Interaction Databases. Protein-Protein Interactions: An Overview. Supervised Learning: Classification. Visualization of Biomedical Networks

References

Acuner Ozbabacan, S., Engin, H., Gursoy, A., Keskin, O., 2011. Transient protein–protein interactions. Protein Eng. Des. Sel. 24, 635–648. doi:10.1093/ protein/gzr025.

Aloy, P., Ceulemans, H., Stark, A., Russell, R., 2003. The relationship between sequence and interaction divergence in proteins. J. Mol. Biol. 332, 989–998.

Ashkenazy, H., Abadi, S., Martz, E., et al., 2016. ConSurf 2016: An improved methodology to estimate and visualize evolutionary conservation in macromolecules. Nucleic Acids Res. 44, W344–W350. doi:10.1093/nar/gkw408.

Ashkenazy, H., Erez, E., Martz, E., Pupko, T., Ben-Tal, N., 2010. ConSurf 2010: Calculating evolutionary conservation in sequence and structure of proteins and nucleic acids. Nucleic Acids Res. 38, W529–W533. doi:10.1093/nar/gkq399.

Bantscheff, M., Schirle, M., Sweetman, G., Rick, J., Kuster, B., 2007. Quantitative mass spectrometry in proteomics: A critical review. Anal. Bioanal. Chem. 389, 1017–1031. doi:10.1007/s00216-007-1486-6. Available at: https://www.ncbi.nlm.nih.gov/pubmed/17668192.

Baspinar, A., Cukuroglu, E., Nussinov, R., Keskin, O., Gursoy, A., 2014. PRISM: A web server and repository for prediction of protein–protein inter actions and modeling their 3D complexes. Nucleic Acids Res. 42, W285–W289. doi:10.1093/nar/gku397.

Benkert, P., Tosatto, S.C., Schomburg, D., 2008. Qmean: A comprehensive scoring function for model quality assessment. Proteins: Struct., Funct., Bioinform. 71, 261–277.

Berman, H.M., Westbrook, J., Feng, Z., et al., 2000. The protein data bank. Nucleic Acids Res. 28, 235–242. Available at: https://www.ncbi.nlm.nih.gov/pubmed/10592235.

Bertoni, M., Kiefer, F., Biasini, M., Bordoli, L., Schwede, T., 2017. Modeling protein quaternary structure of homo- and hetero-oligomers beyond binary interactions by homology. Sci. Rep. 7, 10480. doi:10.1038/ s41598-017-09654-8.

Bitbol, A., Dwyer, R., Colwell, L., Wingreen, N., 2016. Inferring interaction partners from protein sequences. Proc. Natl. Acad. Sci. USA 113, 12180–12185. doi:10.1073/pnas.1606762113.

Blundell, T., Carney, D., Gardner, S., et al., 1988. 18th Sir Hans Krebs lecture. Knowledge-based protein modelling and design. Eur. J. Biochem. 172, 513–520.

Boehr, D., Nussinov, R., Wright, P., 2009. The role of dynamic conformational ensembles in biomolecular recognition. Nat. Chem. Biol. 5, 789–796. doi:10.1038/ nchembio.232.

Bogan, A., Thorn, K., 1998. Anatomy of hot spots in protein interfaces. J. Mol. Biol. 280, 1–9. doi:10.1006/jmbi.1998.1843.

Bosshard, H., 2001. Molecular recognition by induced fit: How fit is the concept? News Physiol. Sci. 16, 171–173.

Burley, S., Kurisu, G., Markley, J., et al., 2017. PDB-Dev: A prototype system for depositing integrative/hybrid structural models. Structure 25, 1317–1318. doi:10.1016/j.str.2017.08.001.

Callaway, E., 2015. The revolution will not be crystallized: A new method sweeps through structural biology. Nature 525, 172–174. doi:10.1038/525172a. Available at: https://www.ncbi.nlm.nih.gov/pubmed/26354465.

Carlin, L.M., Evans, R., Milewicz, H., et al., 2011. A targeted siRNA screen identifies regulators of Cdc42 activity at the natural killer cell immunological synapse. Sci. Signal. 4, ra81. doi:10.1126/scisignal.2001729.

Chakrabarti, P., Janin, J., 2002. Dissecting protein–protein recognition sites. Proteins 47, 334–343.

Chatr-Aryamontri, A., Breitkreutz, B.J., Oughtred, R., et al., 2015. The biogrid interaction database: 2015 Update. Nucleic Acids Res. 43, D470–D478. doi:10.1093/nar/gku1204. Available at: https://www.ncbi.nlm.nih.gov/pubmed/25428363, https://www.ncbi.nlm.nihgov/pmc/articles/PMC4383984, http://nar.oxfordjournals.org/cgi/pmidlookup?view=long&pmid=25428363.

Chen, V.B., Arendall, W.B., Headd, J.J., et al., 2010. Mol-probity: All-atom structure validation for macromolecular crystallography. Acta Crystallogr. Sect. D: Biol. Crystallogr. 66, 12–21.

Chlanda, P., Krijnse Locker, J., 2017. The sleeping beauty kissed awake: New methods in electron microscopy to study cellular membranes. Biochem. J. 474, 41–1053. doi:10.1042/BCJ20160990.

Choi, Y., Yang, J., Choi, Y., Ryu, S., Kim, S., 2009. Evolutionary conservation in multiple faces of protein interaction. Proteins 77, 14–25. doi:10.1002/ prot.22410.

Chothia, C., Lesk, A., 1986. The relation between the divergence of sequence and structure in proteins. EMBO J. 5, 823–826.

Chung, S.S., Pandini, A., Annibale, A., et al., 2015. Bridging topological and functional information in protein interaction networks by short loops profiling. Sci. Rep. 5, 8540. doi:10.1038/ srep08540.

Deutsch, E., Sun, Z., Campbell, D., et al., 2015. State of the Human Proteome in 2014/2015 As Viewed through PeptideAtlas: Enhancing accuracy and coverage through the AtlasProphet. J. Proteome Res. 14, 3461–3473. doi:10.1021/acs.jproteome.5b00500

de Vries, S., Bonvin, A., 2011. CPORT: A consensus interface predictor and its performance in prediction-driven docking with HADDOCK. PLOS One 6, e17695. doi:10.1371/ journal.pone.0017695.

deVries, S., Schindler, C., ChauvotdeBeauchne, I., Zacharias, M., 2015. A web interface for easy flexible protein–protein docking with attract. Biophys. J. 108, 462–465. Available at: http://www.sciencedirect.com/science/article/pii/S0006349514047602, doi: https://doi.org/10.1016/j.bpj.2014.12.015.

Dominguez, C., Boelens, R., Bonvin, A.M., 2003. Haddock: A protein–protein docking approach based on biochemical or biophysical information. J. Am. Chem. Soc. 125, 1731–1737. doi:10.1021/ja026939x. Available at: https://www.ncbi.nlm.nih.gov/pubmed/12580598.

Drew, K., Lee, C., Huizar, R.L., et al., 2017. Integration of over 9,000 mass spectrometry experiments builds a global map of human protein complexes. Mol. Syst. Biol. 13. doi:10.15252/msb.20167490. Available at: http://msb.embopress.org/content/13/6/932.full.pdf.

Duran-Frigola, M., Siragusa, L., Ruppin, E., et al., 2017. Detecting similar binding pockets to enable systems polypharmacology. PLOS Comput. Biol. 13, e1005522. doi:10.1371/journal.pcbi.1005522.

Durfee, T., Becherer, K., Chen, P.L., et al., 1993. The retinoblastoma protein associates with the protein phosphatase type 1 catalytic subunit. Genes Dev. 7, 555–569. Available at: https://www.ncbi.nlm.nih.gov/pubmed/8384581.

Echave, J., Spielman, S., Wilke, C., 2016. Causes of evolutionary rate variation among protein sites. Nat. Rev. Genet. 17, 109–121. doi:10.1038/nrg.2015.18.

Engin, H., Kreisberg, J., Carter, H., 2016. Structure-based analysis reveals cancer missense mutations target protein interaction interfaces. PLOS One 11, e0152929. doi:10.1371/journal.pone.0152929.

Esmaielbeiki, R., Krawczyk, K., Knapp, B., Nebel, J., Deane, C., 2016. Progress and challenges in predicting protein interfaces. Brief. Bioinform. 17, 117–131. doi:10.1093/bib/bbv027.

Fernandes, L., Annibale, A., Kleinjung, J., Coolen, A., Fraternali, F., 2010. Protein networks reveal detection bias and species consistency when analysed by information-theoretic methods. PLOS One 5, e12083. doi:10.1371/journal. pone.0012083.

Fessenden, M., 2017. Protein maps chart the causes of disease. Nature 549, 293–295. doi:10.1038/549293a.

Fields, S., Song, O., 1989. A novel genetic system to detect protein–protein interactions. Nature 340, 245–246. doi:10.1038/340245a0. Available at: https://www.ncbi.nlm.nih.gov/pubmed/2547163.

Finn, R., Bateman, A., Clements, J., et al., 2014. Pfam: The protein families database. Nucleic Acids Res. 42, D222–D230. doi:10.1093/nar/gkt1223.

Fischer, E., 1894. Einfluss der configuration auf die wirkung der enzyme. Eur. J. Inorg. Chem. 27, 2985–2993.

Gallet, X., Charloteaux, B., Thomas, A., Brasseur, R., 2000. A fast method to predict protein interaction sites from sequences. J. Mol. Biol. 302, 917–926. doi:10.1006/ jmbi.2000.4092.

Garzon, J.I., Deng, L., Murray, D., et al., 2016. A computational interactome and functional annotation for the human proteome. eLife 5. doi:10.7554/eLife.18715.

Gingras, A.C., Gstaiger, M., Raught, B., Aebersold, R., 2007. Analysis of protein complexes using mass spectrometry. Nat. Rev. Mol. Cell Biol. 8, 645–654. doi:10.1038/ nrm2208. Available at: https://www.ncbi.nlm.nih.gov/pubmed/17593931.

Gromiha, M.M., Yugandhar, K., 2017. Integrating computational methods and experimental data for understanding the recognition mechanism and binding affinity of protein–protein complexes. Prog. Biophys. Mol. Biol. 128, 33–38. doi:10.1016/j.pbiomolbio.2017.01.001.

Gromiha, M., Yugandhar, K., Jemimah, S., 2017. Protein–protein interactions: Scoring schemes and binding affinity. Curr. Opin. Struct. Biol. 44, 31–38. doi:10.1016/j. sbi.2016.10.016.

Guharoy, M., Chakrabarti, P., 2005. Conservation and relative importance of residues across protein–protein interfaces. Proc. Natl. Acad. Sci. USA 102, 15447–15452. doi:10.1073/pnas.0505425102.

Havugimana, P.C., Hart, G.T., Nepusz, T., et al., 2012. A census of human soluble protein complexes. Cell 150, 1068–1081. doi:10.1016/j.cell.2012.08.011.

Hein, M.Y., Hubner, N.C., Poser, I., et al., 2015. A human interactome in three quantitative dimensions organized by stoichiometries and abundances. Cell 163, 712–723. doi:10.1016/j.cell.2015.09.053. Available at: https://www.ncbi.nlm.nih.gov/pubmed/26496610, https://linkinghub.elsevier.com/retrieve/pii/S0092-8674(15)01270-2.

Hopf, T.A., Scharfe, C.P., Rodrigues, J.P., et al., 2014. Sequence co-evolution gives 3D contacts and structures of protein complexes. eLife 3. doi:10.7554/eLife.03430.

Hou, Q., De Geest, P.F.G., Vranken, W.F., Heringa, J., Feenstra, K.A., 2017. Seeing the trees through the forest: Sequence-based homo- and heteromeric protein–protein interaction sites prediction using random forest. Bioinformatics 33, 1479–1487. doi:10.1093/bioinformatics/btx005.

Huang, S.Y., Grinter, S.Z., Zou, X., 2010. Scoring functions and their evaluation methods for protein–ligand docking: Recent advances and future directions. Phys. Chem. Chem. Phys/. 12, 12899–12908.

Huttlin, E.L., Ting, L., Bruckner, R.J., et al., 2015. The bioplex network: A systematic exploration of the human interactome. Cell 162, 425–440. doi:10.1016/j.cell.2015.06.043. Available at: https://www.ncbi.nlm.nih.gov/pubmed/26186194, https://www.ncbi.nlm.nih.gov/pmc/articles/PMC4617211, https://linkinghub.elsevier.com/retrieve/pii/S0092-8674 (15)00768-0.

Ito, T., Chiba, T., Ozawa, R., et al., 2001. A comprehensive two-hybrid analysis to explore the yeast protein interactome. Proc. Natl. Acad. Sci. USA 98, 4569–4574. doi:10.1073/pnas.061034498. Available at: https://www.ncbi.nlm.nih.gov/pubmed/11283351.

Janin, J., 1999. Wet and dry interfaces: The role of solvent in protein–protein and protein–DNA recognition. Structure 7, R277–R279.

Janin, J., 2005. Assessing predictions of protein–protein interaction: The capri experiment. Protein Sci. 14, 278–283. doi:10.1110/ps.041081905. Available at: https://www.ncbi.nlm.nih.gov/pubmed/15659362.

Jin, L., Wang, W., Fang, G., 2014. Targeting protein–protein interaction by small molecules. Annu. Rev. Pharmacol. Toxicol. 54, 435–456. doi:10.1146/ annurev-pharmtox-011613-140028.

Jordan, R., El-Manzalawy, Y., Dobbs, D., Honavar, V., 2012. Predicting protein–protein interface residues using local surface structural similarity. BMC Bioinform. 13, 41. doi:10.1186/1471-2105-13-41.775de.

Juan, D., Pazos, F., Valencia, A., 2013. Emerging methods in protein co-evolution. Nat. Rev. Genet. 14, 249–261. doi:10.1038/nrg3414.

Kastritis, P., Bonvin, A., 2013a. On the binding affinity of macromolecular interactions: Daring to ask why proteins interact. J. R. Soc. Interface 10, 20120835. doi:10.1098/rsif.2012.0835.780.

Kastritis, P.L., Bonvin, A.M., 2013b. Molecular origins of binding affinity: Seeking the archimedean point. Curr. Opin. Struct. Biol. 23, 868–877. doi:10.1016/j.sbi.2013.07.001. Available at: https://www.ncbi.nlm.nih.gov/pubmed/23876790.

Kategaya, L., Di Lello, P., Rouge, L., et al., 2017. USP7 small-molecule inhibitors interfere with ubiquitin binding. Nature. doi:10.1038/nature24006.

Keskin, O., Ma, B., Nussinov, R., 2005. Hot regions in protein–protein interactions: The organization and contribution of structurally conserved hot spot residues. J. Mol. Biol. 345, 1281–1294. doi:10.1016/j.jmb.2004.10.077.

Keskin, O., Tuncbag, N., Gursoy, A., 2016. Predicting protein–protein interactions from the molecular to the proteome level. Chem. Rev. 116, 4884–4909. doi:10.1021/acs.chemrev.5b00683.

Koepfli, K., Paten, B., O'Brien, S., 2015. The Genome 10K Project: A way forward. Annu. Rev. Anim. Biosci. 3, 57–111. doi:10.1146/ annurev-animal-090414-014900.

Konermann, L., Vahidi, S., Sowole, M.A., 2014. Mass spectrometry methods for studying structure and dynamics of biological macromolecules. Anal. Chem. 86, 213–232. doi:10.1021/ac4039306. Available at: https://www.ncbi.nlm.nih.gov/pubmed/24304427.

Koshland, D., 1958. Application of a theory of enzyme specificity to protein synthesis. Proc. Natl. Acad. Sci. USA 44, 98–104.

Kozakov, D., Hall, D.R., Xia, B., et al., 2017. The cluspro web server for protein–protein docking. Nat. Protoc. 12, 255–278. doi:10.1038/nprot.2016.169. Available at: https://www.ncbi.nlm.nih.gov/pubmed/28079879.

Krissinel, E., Henrick, K., 2007. Inference of macromolecular assemblies from crystalline state. J. Mol. Biol. 372, 774–797. doi:10.1016/j.jmb.2007.05.022. Available at: http://www.ncbi.nlm.nih.gov/pubmed/17681537.

Kristensen, A.R., Gsponer, J., Foster, L.J., 2012. A high-throughput approach for measuring temporal changes in the interactome. Nat. Methods 9, 907. doi:10.1038/nmeth.2131379. Available at: https://www.ncbi.nlm.nih.gov/pubmed/22863883.

Kufareva, I., Budagyan, L., Raush, E., Totrov, M., Abagyan, R., 2007. PIER: Protein interface recognition for structural proteomics. Proteins 67, 400–417. doi:10.1002/prot.21233.

Laddach, A., Gautel, M., Fraternali, F., 2017. TITINdb-a computational tool to assess titins role as a disease gene. Bioinformatics. btx424. doi:10.1093/ bioinformatics/btx424.

Lengyel, J., Hnath, E., Storms, M., Wohlfarth, T., 2014. Towards an integrative structural biology approach: Combining Cryo-TEM, X-ray crystallography, and NMR. J. Struct. Funct. Genomics 15, 117–124. doi:10.1007/s10969-014-9179-9.

Li, J., Liu, Q., 2009. 'Double water exclusion': A hypothesis refining the O-ring theory for the hot spots at protein interfaces. Bioinformatics 25, 743–750. doi:10.1093/bioinformatics/btp058.

Lu, H.C., Herrera Braga, J., Fraternali, F., 2016. Pinsnps: Structural and functional analysis of snps in the context of protein interaction networks. Bioinformatics 32, 2534–2536.

Lyskov, S., Gray, J.J., 2008. The rosettadock server for local protein–protein docking. Nucleic Acids Res. 36, W233–W238. doi:10.1093/nar/gkn216. Available at: https://www.ncbi.nlm.nih.gov/pubmed/18442991.

Ma, B., Kumar, S., Tsai, C., Nussinov, R., 1999. Folding funnels and binding mechanisms. Protein Eng. 12, 713–720.

Moal, I., Barradas-Bautista, D., Jimenez-Garcla, B., et al., 2017. IRaPPA: Information retrieval based integration of biophysical models for protein assembly selection. Bioinformatics 33, 1806–1813. doi:10.1093/ bioinformatics/btx068.

Moreira, I., Fernandes, P., Ramos, M., 2007. Hot spots – A review of the protein–protein interface determinant amino-acid residues. Proteins 68, 80312. doi:10.1002/prot.21396.

Mosca, R., Ceol, A., Aloy, P., 2013. Interactome3D: Adding structural details to protein networks. Nat. Methods 10, 47–53. doi:10.1038/nmeth.2289.

Mosca, R., Ceol, A., Stein, A., Olivella, R., Aloy, P., 2014. 3did: A catalog of domain-based interactions of known three-dimensional structure. Nucleic Acids Res. 42, D374–D379. doi:10.1093/nar/gkt887.

Mosca, R., Tenorio-Laranga, J., Olivella, R., et al., 2015. dSysMap: Exploring the edgetic role of disease mutations. Nat. Methods 12, 167–168. doi:10.1038/nmeth.3289.

Murakami, Y., Mizuguchi, K., 2010. Applying the Naive Bayes classifier with kernel density estimation to the prediction of protein–protein interaction sites. Bioinformatics 26, 1841–1848. doi:10.1093/bioinformatics/btq302.

Ofran, Y., Rost, B., 2003. Predicted protein-protein interaction sites from local sequence information. FEBS Lett. 544, 236–239.

Ofran, Y., Rost, B., 2007. ISIS: Interaction sites identified from sequence. Bioinformatics 23, e13–e16. doi:pmid17237081

Orchard, S., Ammari, M., Aranda, B., et al., 2014. The mintact project – Intact as a common curation platform for 11 molecular interaction databases. Nucleic Acids Res. 42, D358–D363. doi:10.1093/nar/gkt1115. Available at: https://www.ncbi.nlm.nih.gov/pubmed/24234451, https://www.ncbi.nlm.nih.gov/pmc/articles/PMC3965093, http://nar.oxfordjournals.org/cgi/pmidlookup?view=long&pmid=24234451.

Orchard, S., Kerrien, S., Abbani, S., et al., 2012. Protein interaction data curation: The international molecular exchange (imex) consortium. Nat. Methods 9, 345–350. doi:10.1038/nmeth.1931. Available at: https://www.ncbi.nlm.nih.gov/pubmed/22453911, https://www.ncbi.nlm.nih.gov/pmc/articles/PMC3703241.

Orlov, I., Myasnikov, A., Andronov, L., et al., 2017. The integrative role of cryo electron microscopy in molecular and cellular structural biology. Biol. Cell 109, 81–93. doi:10.1111/boc.201600042.

Pearson, W.R., 2013. An introduction to sequence similarity ("homology") searching. Curr. Protoc. Bioinform. 1. doi:10.1002/0471250953.bi0301s42. Chapter 3, Unit3.

Perkins, J., Diboun, I., Dessailly, B., Lees, J., Orengo, C., 2010. Transient protein–protein interactions: Structural, functional, and network properties. Structure 18, 1233–1243. doi:10.1016/j.str.2010.08.007.

Politis, A., Schmidt, C., 2017. Structural characterisation of medically relevant protein assemblies by integrating mass spectrometry with computational modelling. J Proteomics. doi:10.1016/j.jprot.2017.04.019.

Porollo, A., Meller, J., 2007. Prediction-based fingerprints of protein–protein interactions. Proteins 66, 630–645. doi:10.1002/prot.21248.

Qin, S., Zhou, H., 2007. meta-PPISP: A meta web server for protein–protein interaction site prediction. Bioinformatics 23, 3386–3387. doi:10.1093/ bioinformatics/btm434.

Raghavachari, B., Tasneem, A., Przytycka, T., Jothi, R., 2008. DOMINE: A database of protein domain interactions. Nucleic Acids Res. 36, D656–D661. doi:10.1093/nar/gkm761.

Rao, V.S., Srinivas, K., Sujini, G.N., Kumar, G.N., 2014. Protein-protein interaction detection: Methods and analysis. Int. J. Proteomics 2014, 147648. Available at: https://www.ncbi.nlm.nih.gov/pubmed/24693427, doi:10.1155/2014/147648.

Rodrigues, J., Bonvin, A., 2014. Integrative computational modeling of protein interactions. FEBS J. 281, 1988–2003. doi:10.1111/febs.12771.

Rolland, T., Taan, M., Charloteaux, B., et al., 2014. A proteome-scale map of the human interactome network. Cell 159, 1212–1226. doi:10.1016/j.cell.2014.10.050. Available at: https://www.ncbi.nlm.nih.gov/pubmed/25416956, https://www.ncbi.nlm.nih.gov/pmc/articles/PMC4266588, https://linkinghub.elsevier.com/retrieve/pii/S0092-8674(14)01422-6.

Rydevik, B., Pedowitz, R., Hargens, A., et al., 1991. Effects of acute, graded compression on spinal nerve root function and structure: An experimental Study of the Pig Cauda Equina. Spine (Phila. PA 1976) 16, 487–493.

Sahni, N., Yi, S., Taipale, M., et al., 2015. Widespread macromolecular interaction perturbations in human genetic disorders. Cell 161, 647–660. doi:10.1016/j.cell.2015.04.013.

Schlick, T., Collepardo-Guevara, R., Halvorsen, L.A., Jung, S., Xiao, X., 2011. Biomolecularmodeling and simulation: A field coming of age. Q. Rev. Biophys. 44, 191–228. doi:10.1017/S0033583510000284. Available at: https://www.ncbi.nlm.nih.gov/pubmed/21226976.

Schmidt, C., Macpherson, J.A., Lau, A.M., et al., 2017. Surface accessibility and dynamics of macromolecular assemblies probed by covalent labeling mass spectrometry and integrative modeling. Anal. Chem. 89, 1459–1468. doi:10.1021/acs.analchem.6b02875.

Schneidman-Duhovny, D., Inbar, Y., Nussinov, R., Wolfson, H.J., 2005. Patchdock and symmdock: Servers for rigid and symmetric docking. Nucleic Acids Research 33, W363–W367. doi:10.1093/nar/gki481. Available at: https://doi.org/10.1093/nar/gki481.

Sillitoe, I., Dawson, N., Thornton, J., Orengo, C., 2015. The history of the CATH structural classification of protein domains. Biochimie 119, 209–217. doi:10.1016/j.biochi.2015.08.004.

Soni, N., Madhusudhan, M., 2017. Computational modeling of protein assemblies. Curr. Opin. Struct. Biol. 44, 179–189. doi:10.1016/j.sbi.2017.04.006.

Sriwastava, B.K., Basu, S., Maulik, U., 2015. Protein–protein interaction site prediction in *Homo sapiens* and *E. coli* using an interaction-affinity based membership function in fuzzy SVM. J. Biosci. 40, 809–818. doi:10.1007/s12038-015-9564-y.

Szklarczyk, D., Franceschini, A., Wyder, S., *et al.*, 2015. STRING v10: Protein–protein interaction networks, integrated over the tree of life. Nucleic Acids Res. 43, D447–D452. doi:10.1093/nar/gku1003.

Tamo, G., Maesani, A., Trager, S., *et al.*, 2017. Disentangling constraints using viability evolution principles in integrative modeling of macromolecular assemblies. Sci. Rep. 7, 235. doi:10.1038/s41598-017-00266-w.

Telenti, A., Pierce, L., Biggs, W., *et al.*, 2016. Deep sequencing of 10,000 human genomes. Proc. Natl. Acad. Sci. USA 113, 11901–11906. doi:10.1073/pnas.1613365113.

Tuncbag, N., Gursoy, A., Nussinov, R., Keskin, O., 2011. Predicting protein–protein interactions on a proteome scale by matching evolutionary and structural similarities at interfaces using PRISM. Nat. Protoc. 6, 1341–1354. doi:10.1038/nprot.2011.367.

Vidal, M., Brachmann, R.K., Fattaey, A., Harlow, E., Boeke, J.D., 1996. Reverse two-hybrid and one-hybrid systems to detect dissociation of protein–protein and DNA–protein interactions. Proc. Natl. Acad. Sci. USA 93, 10315–10320. Available at: https://www.ncbi.nlm.nih.gov/pubmed/8816797.

Wan, C., Borgeson, B., Phanse, S., *et al.*, 2015. Panorama of ancient metazoan macromolecular complexes. Nature 525, 339–344. doi:10.1038/nature14877. Available at: https://www.ncbi.nlm.nih.gov/pubmed/26344197, https://www.ncbi.nlm.nih.gov/pmc/articles/PMC5036527.

Westermarck, J., Ivaska, J., Ivaska, G.L., 2013 Identification of protein interactions involved in cellular signaling. Mol.Cell Proteomics 12, 1752–1763. Available at: https://www.ncbi.nlm.nih.gov/pubmed/23481661,doi:10.1074/mcp.R113.027771.

Wo´jcikowski, M., Ballester, P., Siedlecki, P., 2017. Performance of machine-learning scoring functions in structure-based virtual screening. Sci. Rep. 7, 46710. doi:10.1038/srep46710.

Xue, L., Dobbs, D., Bonvin, A., Honavar, V., 2015. Computational prediction of protein interfaces: A review of data driven methods. FEBS Lett. 589, 3516–3526. doi:10.1016/j.febslet.2015.10.003.

Xue, L., Dobbs, D., Honavar, V., 2011. HomPPI: A class of sequence homology based protein–protein interface prediction methods. BMC Bioinform. 12, 244. doi:10.1186/1471-2105-12-244.

Yates, J.R., 2013. The revolution and evolution of shotgun proteomics for large-scale proteome analysis. J. Am. Chem. Soc. 135, 1629–1640. doi:10.1021/ja3094313. Available at: https://www.ncbi.nlm.nih.gov/pubmed/23294060.

Yi, S., Lin, S., Li, Y., *et al.*, 2017. Functional variomics and network perturbation: Connecting genotype to phenotype in cancer. Nat. Rev. Genet. 18, 395–410. doi:10.1038/nrg.2017.8.

Zhang, Q., Deng, L., Fisher, M., *et al.*, 2011. Pre-dUs: A web server for predicting protein interfaces using structural neighbors. Nucleic Acids Res. 39, W283–W287. doi:10.1093/nar/gkr311.

Zhao, N., Pang, B., Shyu, C., Korkin, D., 2011. Structural similarity and classification of protein interaction interfaces. PLOS One 6, e19554. doi:10.1371/journal.pone.0019554.

Zhao, Z., Gong, X., 2017. Protein–protein interaction interface residue pair prediction based on deep learning architecture. IEEE/ACM Trans. Comput. Biol. Bioinform. doi:10.1109/TCBB.2017.2706682.

Protein–Protein Interaction Databases

Ashwini Patil, The University of Tokyo, Tokyo, Japan

Introduction

The function of a protein in a cell depends on its interaction partners in the form of other proteins, nucleic acids and small molecules. These interactions change with time and cellular state allowing proteins to perform their functions. The knowledge of these interactions can be used to make interaction networks that can be analyzed to identify functional modules of proteins and their interconnections to ultimately understand cellular function (Cafarelli *et al.*, 2017). The identification of protein-protein interactions (PPIs) is thus an important means of understanding biological processes.

The detection of interactions between proteins began on a large scale in the early 2000s, prior to which PPIs were identified mainly as those between proteins of interest or those associated with a specific pathway or function. Starting in the year 2000, several studies reported a large number of protein-protein interactions identified using high-throughput screening methods (Uetz *et al.*, 2000; Ito *et al.*, 2001; Ho *et al.*, 2002; Giot *et al.*, 2003; Li *et al.*, 2004; Rual *et al.*, 2005; Stelzl *et al.*, 2005). Over the years, the number of interaction datasets has increased exponentially. This lead to the need for resources that would combine the large number of interactions, annotate them and make them interoperable. This need is fulfilled by PPI databases which organize, curate, integrate and annotate PPI data and make it available for downstream analysis.

All PPI databases collect protein-protein interactions either directly from literature, through direct deposition by authors, or from other databases. Databases use different criteria to filter, annotate and group the collected interactions and provide distinct user interfaces to allow users to access the PPI data. In some cases, methods to analyze this data are also provided. Some databases are limited to one or a few species. An important service provided by the databases is the assessment of the PPIs for reliability. It is known that PPIs identified in-vitro by existing methods have a large number of false-positives i.e. interactions that show incorrect binding or are not observed in-vivo (Von Mering *et al.*, 2002). This is now changing with increasing confirmation of the identified interactions using validation assays (Luck *et al.*, 2017). However, several databases now assign scores to interactions that help users identify those interactions that are more likely to exist in the cell. Due to limitations of interaction detection techniques, the number of interactions experimentally identified is much smaller than those existing in the cell. Therefore, a few databases also provide predicted PPIs, either as predicted interactions between proteins or functional associations identified by genetic interactions.

Thus, there is a wide variety of PPI databases available. As the number of PPIs detected increases and the databases become more specialized, it is becoming increasingly difficult to keep abreast of the various databases and their offerings. Therefore, it is important for users to understand these databases to help identify the ones that will be most appropriate in a specific study. This article provides an overview of the current major PPI databases. The article first describes the different types of interactions identified and stored in PPI databases. This is followed by a brief description of several databases, their special features and the methods they use for PPI quality assessment. Next, a description of the recent development of community-wide standards and protocols that make it easier to access database content for in-depth analyses is provided. Finally, a discussion of the strengths and weaknesses of the current PPI databases is presented and suggestions are made for their improvement.

Types of Interactions

PPIs can be broadly classified into two groups – direct or physical, and indirect or non-physical interactions. These are further divided into sub-groups (**Table 1**). These groups are frequently used to classify PPIs in databases.

Physical Interactions

Physical interactions are those in which the interactors physically bind to each other. These may be interactions between proteins or proteins and other molecules. Physical protein-protein interactions can further be divided into two categories depending on the type of method used to identify them. Binary interactions are direct binding events that are detected between two proteins. Several

Table 1   Classification of interactions in PPI databases

Class	Definition
Physical/direct	Direct binding event between two proteins
Non-physical/Indirect	Functional association – genetic or predicted interaction
Small-scale	Detected in small experiment with 100 or less interactions
High-throughput	Detected in proteome-scale experiment
Binary	Interaction between two proteins
Complex	Association of multiple proteins; direct interactions unknown

methods can be used to detect binary interactions, such as, yeast two hybrid (Y2H) systems (Ito *et al.*, 2001; Giot *et al.*, 2003; Li *et al.*, 2004; Rolland *et al.*, 2014), protein arrays (Yazaki *et al.*, 2016), fluorescence resonance energy transfer (FRET), and structure determination methods like X-ray crystallography and NMR Spectroscopy. PPIs can also be identified in the form of protein complexes, or associations, using methods like affinity purification followed by mass spectroscopy (AP-MS) (Hein *et al.*, 2015; Huttlin *et al.*, 2015, 2017) and co-fractionation followed by mass spectroscopy (CoFrac-MS) (Wan *et al.*, 2015). Of these methods, Y2H, AP-MS and CoFrac-MS are amenable to high-throughput experiments and have been used to identify PPIs on the proteomic scale. Methods identifying protein complexes do not provide information about direct binding events in the complex. Hence binary interactions are often predicted from protein complex data before they are stored in databases. Databases use one of two models to convert protein complex data into binary interactions – spoke model or matrix model (Bader and Hogue, 2002). The spoke model assumes that the bait protein, i.e. the protein that is tagged and used to elute other proteins that associate with it, directly binds to all the other associated proteins. The matrix model assumes that all proteins eluted together interact with each other. The spoke model is more conservative and results in fewer false positive interactions (Bader and Hogue, 2002) and hence, is the model of choice in several databases. However, neither model gives the true binding states, often resulting in the prediction of incorrect binary interactions.

An additional classification that is common for PPIs is that based on the size of the experiment in which they are detected. PPIs that are detected in small, focused studies, typically identifying less than 100 interactions, are often denoted as small-scale (Patil *et al.*, 2011). On the other hand, interactions identified in proteome-wide studies are deemed as high-throughput. The cutoff interaction count separating small-scale from high-throughput experiments varies with databases. It has long been argued that interactions identified in high-throughput experiments have larger number of false positives or non-physiological interactions (Von Mering *et al.*, 2002; Wodak *et al.*, 2013). This makes scoring interactions for reliability very important. It has also been observed that interactions from the same species identified by different methods are complementary and have very little overlap (Von Mering *et al.*, 2002; Wodak *et al.*, 2013). It has been proposed that this may be the result of the large number of potential interactions in the cell all of which cannot be identified by a single method, and the low sensitivity of the methods, rather than the high rate of false interactions identified by these methods (Wodak *et al.*, 2013; Luck *et al.*, 2017).

Non-Physical Interactions

Indirect or non-physical interactions are functional associations between proteins and/or other molecules that may or may not be direct binding events. Indirect associations of proteins may be identified as genetic interactions using synthetic lethal arrays (Costanzo *et al.*, 2010). They can also be predicted interactions or associations (Skrabanek *et al.*, 2008). Sequence homology is often used to predict protein associations such that homologs of interacting proteins in the same or different species are also predicted to interact. This is a method that is often used to transfer interactions between closely related species. For example, interactions identified in mouse can be transferred to human through orthologous proteins (Brown and Jurisica, 2005). Structural methods can also be used to predict PPIs through binding surface prediction and docking of two proteins with known structures (Mosca *et al.*, 2013). Genomic context approaches are another commonly used method of PPI prediction. These include methods based on (1) identifying pairs of genes that are in close proximity or fused in a few species, (2) the coordinated presence or absence of genes in different genomes and, the presence of conserved regions and (3) correlated mutations in two protein/gene sequences indicating a possible role in binding (Skrabanek *et al.*, 2008). Finally, co-occurring terms in scientific abstracts obtained through text-mining can also be used to predict functional associations between proteins (Papanikolaou *et al.*, 2015).

Protein–Protein Interaction Databases

Protein-protein interaction databases are divided into two categories: (1) primary databases, and (2) derived or consolidated databases.

Primary Databases

Primary databases collect and store interactions directly from literature or through author submissions (**Table 2**). They may contain data from small-scale focused studies or large-scale high-throughput experiments. Following are some of the major primary databases storing PPI data:

1. IntAct: A database containing molecular interactions curated from literature or obtained by direct submission from authors (Orchard *et al.*, 2014). Most of the interactions in IntAct are physical associations from high-throughput experiments, and the species with the highest representation is human. It is one of the few PPI databases with extensive annotations and proteins mapped to multiple database identifiers.
2. BioGRID: A primary database that collects interactions from literature, mainly from high-throughput experiments, and stores them after curation (Chatr-Aryamontri *et al.*, 2017). The interactions in BioGRID are classified into physical or genetic interactions. BioGRID also stores data about post-translational modifications in proteins. It contains the largest number of curated interactions in yeast. All interactions in BioGRID are in binary format i.e. complex associations have been converted to binary format using the spoke model in which the bait protein interacts with each prey protein.

Table 2   Primary protein-protein interaction databases

Database	Interactions	PPI type	Associations	Scoring	Focus	Reference
IntAct	Protein-molecule	Experimental	All	Yes	All	Orchard *et al.* (2014)
BioGRID	Protein-molecule	Experimental	All	No	All	Chatr-Aryamontri *et al.* (2017)
MINT	Protein-molecule	Experimental	All	Yes	All	Ceol *et al.* (2010)
DIP	Protein-protein	Experimental	All	No	All	Salwinski *et al.* (2004)
HPRD	Protein-protein	Experimental	Physical	No	Human	Prasad *et al.* (2009)
MatrixDB	Protein-protein	Experimental	All	No	All	Launay *et al.* (2015)
CORUM	Protein-protein	Experimental	Physical	No	Mammals	Ruepp *et al.* (2010)
MIPS	Protein-protein	Experimental	Physical	No	Mammals	Pagel *et al.* (2005)

Interactions: Protein-protein – interactions exclusively between proteins; Protein-molecule – interactions between proteins and proteins, DNA, RNA, drugs or other small molecules.
PPI type: Experimental – experimentally identified interactions; Predicted – predicted associations.
Associations: All – physical and indirect associations between interactors; Physical – only interactions involving physical binding.
Scoring: Presence of confidence score assessing interaction reliability.
Focus: Specific area of focus of the database eg. Species or biological process specific.

3. MINT: MINT collects and curates PPI data on a smaller scale than IntAct and BioGRID (Ceol *et al.*, 2010). This data is now hosted by the IntAct database. However, MINT still contains some interactions that have not been transferred to IntAct.
4. DIP: DIP is a much smaller database compared to IntAct and BioGRID. It stores curated protein-protein interactions from literature (Salwinski *et al.*, 2004).
5. HPRD: The Human Protein Reference Database (HPRD) is one of the most comprehensive resources of physical protein-protein interactions in humans (Prasad *et al.*, 2009). The data is manually curated from literature, primarily from small-scale experiments. Interactions in HPRD are in binary format and come with isoform information. HPRD data available for download is not as detailed as that provided by some of the other primary databases.
6. MatrixDB: This is a database that curates and stores interactions of extracellular matrix proteins obtained from literature (Launay *et al.*, 2015).
7. CORUM: A database containing manually annotated protein complexes from mammals. This database contains small-scale interactions collected and manually curated from literature (Ruepp *et al.*, 2010). These interactions are taken only from small-scale focused experiments and are, therefore, considered to be more reliable than those present in larger databases which contain a greater number of interactions from high-throughput experiments. The data from CORUM is also frequently used as a gold standard, or known true, dataset in PPI quality evaluation.
8. MIPS: MIPS is a database of manually curated protein-protein interactions in mammals from literature (Pagel *et al.*, 2005). MIPS is also used as a gold standard dataset for binary protein-protein interactions.

Primary databases perform a commendable task of collecting, curating and making the PPI data available. But several issues need to be addressed. At present, PPI data is distributed across multiple primary databases. As a result, getting a complete set of interactions for network or pathway analysis requires combining data from multiple sources. The adoption of standards by several primary databases has significantly eased this process. However, combining data from multiple databases and annotating them is still non-trivial for two reasons. Firstly, databases use different interactor identifiers requiring the mapping of identifiers from multiple databases onto a common identifier before merging. Secondly, the presence of interactions using obsolete or incorrect identifiers makes merging challenging. The depth of curation of data varies with each database. Thirdly, some databases are better curated and have more annotations than others. As a result, integrating information from databases requires time and resources, and derived or consolidated databases help resolve these issues.

Derived Databases

Derived databases take PPI data from primary interaction databases, integrate it, filter it and annotate it (**Table 3**). Many of the derived databases also provide an additional level of curation. The derived databases try to alleviate the problems users face during integration of multiple PPI datasets by mapping all the interactor identifiers from multiple databases to a common identifier followed by combining the interactions into a unique set. After integration, interactions may be selected by the databases for one or more species or a specific biological system. Along with integration and selection, several databases also provide confidence scores to assess the quality of the interactions. Some databases also provide predicted interactions and interactions with annotations of 3D structure. As a result, derived databases are often the ones from which users get their interaction data for further analysis. Following are some of the major derived PPI databases.

1. STRING: STRING is a database of functional associations (Szklarczyk *et al.*, 2015). It is one of the largest derived databases. It collects physical and indirect PPIs from multiple primary databases. It also contains predicted functional associations between proteins. All interactions in STRING are scored using multiple evidences for their reliability. The data from STRING is now available for network analysis through a plugin in the software, Cytoscape (Cline *et al.*, 2007).

Table 3   Derived protein-protein interaction databases. Column descriptions are same as those in **Table 2**

Database	Interactions	PPI type	Associations	Scoring	Focus	References
STRING	Molecular	Experimental + Predicted	All	Yes	All	Szklarczyk et al. (2015)
HitPredict	Proteins	Experimental	Physical	Yes	All	Lopez et al. (2015)
iRefWeb	Proteins	Experimental	All	Yes	All	Turner et al. (2010)
mentha	Proteins	Experimental	Physical	Yes	All	Calderone et al. (2013)
InnateDB	Molecular	Experimental + Predicted	All	No	Mammals (immunity)	Breuer et al. (2013)
IID	Proteins	Experimental + Predicted	All	No	6 organisms	Kotlyar et al. (2016)
HPIDB	Proteins	Experimental + Predicted	All	No	Human	Ammari et al. (2016)
ConsensusPathDB	Molecular	Experimental	All	No	3 organisms	Kamburov et al. (2013)
HAPPI	Proteins	Experimental + Predicted	All	Yes	Human	Chen et al. (2017)
HIPPIE	Proteins	Experimental	All	Yes	Human	Alanis-Lobato et al. (2017)
DroID	Molecular	Experimental + Predicted	All	Yes	Drosophila	Murali et al. (2011)
Interactome3D	Proteins	Experimental	Physical	No	All (3D structure)	Mosca et al. (2013)

2. HitPredict: HitPredict is a derived database that includes physical protein-protein interactions from multiple primary databases (Lopez et al., 2015). HitPredict is a deeply curated database using several automatic and manual checks to assess the quality of the PPI data from primary sources. Extensive checks are performed to confirm the validity of the interacting proteins and significant effort is put into assigning the correct identifier to proteins. Interactions of proteins that do not map to valid identifiers, are not in the same species, or that do not have valid experimental annotations are not included. HitPredict includes interactions from organisms that have at least 10 high-quality interactions. HitPredict also provides a reliability score for PPIs.

3. iRefWeb: This is a derived database that integrates PPI data from multiple sources (Turner et al., 2010). It includes physical as well as indirect interactions. iRefWeb provides detailed information about each interaction in the form of the number of primary databases it was obtained from and the annotation of the interaction in each source database. It also provides extensive mapping to multiple protein and gene identifiers including the various isoforms. iRefWeb provides a reliability score for each interaction.

4. mentha: This is a derived database that automatically collects and integrates data published by multiple primary databases using a web service (Calderone et al., 2013). mentha depends on the source databases to provide the curation and the correct interactor identifiers. mentha also provides a reliability score. It provides some network analysis tools to extract subnetworks of genes and identify the possible paths between two set of genes.

5. InnateDB: InnateDB is a database of associations between mammalian genes and proteins that are specific to the innate immune response (Breuer et al., 2013). Interactions are limited to those in human, cow and mouse. It includes a small set of curated interactions, a set of interactions validated from publications and a set of predicted interactions. Interactions are physical as well as indirect associations between proteins, DNA and RNA. InnateDB provides tools for pathway, gene ontology and network analyses. The interactions in InnateDB are not scored.

6. IID: The Integrated Interaction Database combines experimentally identified interactions from primary databases and includes predicted PPIs in yeast, worm, fly, rat, mouse and human (Kotlyar et al., 2016). It also provides tissue specific PPIs. IID does not provide reliability scoring for interactions.

7. HPIDB: HPIDB provides interactions between proteins from pathogens and their host organisms (Ammari et al., 2016). PPIs are collected from primary and some derived databases without reliability scoring.

8. ConsensusPathDB: Physical and indirect interactions from human, yeast and mouse are collected from primary sources and curated (Kamburov et al., 2013). Network analysis and gene set enrichment analysis is provided. This database also includes information about protein-drug interactions. A confidence score is provided for each interaction.

9. HAPPI: HAPPI is a database of experimentally identified and predicted human protein-protein interactions collected from primary databases (Chen et al., 2017). All interactions are curated and scored.

10. HIPPIE: A human protein-protein interaction database with confidence scoring and network analysis tools (Alanis-Lobato et al., 2017). Interactions are collected from multiple PPI databases.

11. DroID: A database of physical and genetic interactions between proteins, DNA and RNA from the fly, Drosophila melanogaster (Murali et al., 2011). It provides extensive links to fly specific databases such as FlyBase (Gramates et al., 2017). It also includes normalized gene expression values and expression correlation values. Orthologous interactions predicted from known human, worm and yeast interactions are also part of this database.

12. Interactome3D: A derived database that collects PPIs from primary source databases and provides 3D structural annotations to the interacting proteins (Mosca et al., 2013). 3D structures are predicted for the complete or partial protein complex, when unknown, along with the binding mode.

Quality Assessment and Scoring

Most of the experimentally identified PPIs in databases are detected using high-throughput experimental methods on a proteome-wide scale and potentially contain spurious interactions (Von Mering et al., 2002; Wodak et al., 2013). In spite of the increasing accuracy of these methods, assessing the existing interaction data for reliability is important before its use in downstream analyses. As described earlier, several PPI databases now provide a reliability score for the interactions that they contain.

Experimentally detected interactions can be scored in two ways. Firstly, they can be experimentally verified either by confirming a small set of the identified interactions or by assessing the accuracy of the experimental method to estimate the fraction of high-confidence interactions among those detected (Luck et al., 2017). With the increase in experimental validation and assessment, most primary and derived databases provide information about author assigned confidence for interactions and experimental datasets. Secondly, interactions can be assessed computationally to identify the probability of the association or interaction being true and occurring in the cell.

The accuracy of the interaction detected is calculated using information about the type of association detected (physical or indirect), the method used to detect the interaction (Y2H, AP-MS), the type of experiment (small-scale or high-throughput) and the number of publications that support the interaction. Several databases including IntAct, MINT, mentha, HAPPI, HIPPIE and HitPredict incorporate these experimental details into the reliability score they calculate.

To assess the probability of an interaction occurring in vivo, the features of the interacting proteins or the presence of homologous interactions may be used. HitPredict uses three characteristics to calculate an annotation score for an interaction – the presence of a homologous interaction, the same gene ontology terms associated with the interacting proteins and the presence of domains that are observed to interact in 3D structural data in the interacting proteins (Patil and Nakamura, 2005). The annotation score is combined with the experiment score, calculated using information about the experimental method, to give the final confidence score in HitPredict (Lopez et al., 2015). STRING calculates the probability of two random pairs of proteins binding and includes a score based on homology and co-occurrence to calculate the final score (Szklarczyk et al., 2015). The reliability score in mentha is based on the number of publications supporting an interaction and the type of method used to detect the method (Calderone et al., 2013). ConsensusPathDB integrates interaction assessment scores calculated using graph-based topology, literature evidence, pathway co-occurrence of the interacting proteins and the functional term similarities (Kamburov et al., 2013). iRefWeb calculates the reliability score using the number of publications supporting the interaction, the method used to identify the interaction and the presence of the same interaction in other species (Turner et al., 2010). HIPPIE uses a semi-automated scoring procedure based on experimental evidence of the interaction (Alanis-Lobato et al., 2017). HAPPI uses a combination of heuristic scores for the source and detection method to calculate a final score (Chen et al., 2017).

With each database adopting a different method to assess the interaction quality, it is not surprising that the high-confidence interactions identified by each database are different (Wodak et al., 2013). Another reason for differences in the interactions deemed as high-confidence between databases is the different gold standard interactions that are used by these methods to evaluate their performance and set a threshold score that differentiates between the high and low confidence interactions. The gold standard datasets can be often biased and error-prone and tend to sometimes result in incorrect threshold scores. Therefore, it must be noted that while interaction scores are important for assessing the quality of interactions, they must be checked across multiple databases for the interactions of interest and used with caution when filtering interactions.

Community Standards

Given the large number of PPI databases and the increasing number of interactions in these databases, inter-operability is an issue that needs to be addressed. With each database having its own data model, terminology, filtering and curation criteria, combining data from multiple databases and mapping the proteins to common identifiers has become tedious and time-consuming. Community standards have been introduced by the Human Proteome Organisation Proteomics Standards Initiative – Molecular Interactions (HUPO PSI-MI) to address this issue (Orchard and Hermjakob, 2008).

PSI-MI has developed common data formats for sharing interaction data in the form of XML (PSI-MI XML) and tab-delimited (MITAB) files. Controlled vocabulary has been developed to describe interaction properties such as the molecular features of the interactors (protein, DNA, etc.), the type of interaction (direct, genetic, etc.) and the interaction detection method (biophysical, biochemical, etc.). The controlled vocabulary includes a standardized term to describe the interaction feature and an accession number that can be used to programmatically search and filter interactions based on their features (Orchard, 2012). HUPO PSI-MI also provides a standardized method for calculating the interaction score in the form of a PSIScore (Villaveces et al., 2015). Most of the databases now provide downloadable files in these file formats and several follow the use of controlled vocabularies.

It is also possible to programmatically access the data within PPI databases so that it can be combined and analyzed. The PSI Common Query InterfaCe (PSICQUIC) is a query interface that allows users to programmatically access interaction databases and get data in PSI-MI XML or MITAB formats (Orchard, 2012). Depending on the database compliance, PSICQUIC also allows users to query for proteins using multiple identifiers. The user can query the most recent version of multiple databases that provide this web service. The PSICQUIC View at EMBL-EBI provides information about the databases that implement this service and their current status. Some databases like IntAct, BioGrid, MINT DIP, InnateDB, STRING and mentha are available through PSICQUIC.

Along with the definition of standardized formats and the availability of a standardized query interface, it is also important to specify common standards for data quality and curation. The International Molecular Exchange (IMEx) consortium has established common curation rules for PPI data. IntAct, MINT, DIP, MatrixDB and InnateDB are some of the members of the consortium. Member databases are committed to curating the data from published papers through a central database and web service, IMEx Central. The data is curated to standards specified by the IMEx consortium. Data curation is divided between the members of the consortium to prevent overlapping curation efforts. The IMEx consortium provides users with a non-redundant set of interactions through the PSICQUIC service (Orchard *et al.*, 2012).

Future Directions

With the improvement in interaction detection technology and the rapidly increasing number of PPIs being detected, the role of databases in collecting, organizing and curating these interactions has become crucial. The existing PPI databases are doing an excellent job of making the interactions available to the scientific community. The established standards and web services have eased the access to interactions and their integration for downstream network analyses. It has now become possible to perform large-scale network analysis studies and to combine other omics and disease-related data in the context of the network to identify activated cellular pathways and novel functions of proteins.

Some problems associated with the data in PPI databases are the result of the properties of the techniques used to identify these interactions. For instance, there is a known functional bias in the interaction data as a result of the techniques used to detect the PPIs (Yu *et al.*, 2008). For example, Y2H data is enriched in interactions involving signal transduction while AP-MS interactions are enriched for proteins functioning in translation. This may be the result of Y2H being more suited to identify transient interactions while AP-MS being better for the identification of protein complexes. Current high-throughput methods also provide no information about the affinity of interactions, protein abundance, post-translational modifications of the proteins, tissue-specific expression patterns of genes, and the isoform participating in the interaction. All these are important properties of proteins and are known to impact the interaction landscape in the cell.

An estimate of the size of interaction universe in organisms shows that many new PPIs are yet to be discovered (Vidal and Fields, 2014). As the amount of protein-protein interaction data increases and the quality of these interactions improves, resources have to evolve to keep up with these changes. Therefore, existing databases will have to scale and improve their integration, curation and annotation methods. Firstly, integrating interaction data with other omics data needs to become easier through the use of extensive cross-referencing and identifier mapping by all databases. Secondly, the depth of curation needs to be improved in primary as well as derived databases. Obsolete identifiers, incorrect methods, interaction types, evidence types and other annotations are some of the errors that can creep into PPI data. Automatic, or semi-automatic, methods are necessary to correct these errors and reduce the task of manual curators. The scoring schemes used by different databases can also lead to different high confidence datasets and hence, greater uniformity in scoring of interactions is recommended. Finally, methods to analyze and visualize the interaction information are also important and need to be integrated into the databases. Basic network analyses and function enrichment analyses methods will dramatically increase the utility of PPI databases.

Despite these shortcomings, protein-protein interaction databases are critical for the future of large-scale biological data analyses. The next several years will be important as these databases develop new methods to deal with the large amount of data being generated as they make it easily accessible and available for analysis.

See also: Alignment of Protein-Protein Interaction Networks. Bioinformatics Data Models, Representation and Storage. Biological Database Searching. Biological Pathway Analysis. Community Detection in Biological Networks. Data Storage and Representation. Graphlets and Motifs in Biological Networks. Natural Language Processing Approaches in Bioinformatics. Network Inference and Reconstruction in Bioinformatics. Pathway Informatics. Prediction of Protein-Protein Interactions: Looking Through the Kaleidoscope. Protein Functional Annotation. Protein–Protein Interaction Databases. Protein-Protein Interactions: An Overview. Proteomics Data Representation and Databases. Two Decades of Biological Pathway Databases: Results and Challenges. Visualization of Biomedical Networks

---

References

Alanis-Lobato, G., Andrade-Navarro, M.A., Schaefer, M.H., 2017. HIPPIE v2.0: Enhancing meaningfulness and reliability of protein–protein interaction networks. Nucleic Acids Res 45, D408–D414.
Ammari, M.G., Gresham, C.R., Mccarthy, F.M., Nanduri, B., 2016. HPIDB 2.0: A curated database for host–pathogen interactions. Database: J Biol Databases Curation 2016, baw103.
Bader, G.D., Hogue, C.W., 2002. Analyzing yeast protein–protein interaction data obtained from different sources. Nat Biotechnol 20, 991–997.
Breuer, K., Foroushani, A.K., Laird, M.R., *et al.*, 2013. InnateDB: Systems biology of innate immunity and beyond – Recent updates and continuing curation. Nucleic Acids Res 41, D1228–D1233.
Brown, K.R., Jurisica, I., 2005. Online predicted human interaction database. Bioinformatics 21, 2076–2082.
Cafarelli, T.M., Desbuleux, A., Wang, Y., *et al.*, 2017. Mapping, modeling, and characterization of protein–protein interactions on a proteomic scale. Curr Opin Struct Biol 44, 201–210.
Calderone, A., Castagnoli, L., Cesareni, G., 2013. Mentha: A resource for browsing integrated protein-interaction networks. Nat Methods 10, 690–691.
Ceol, A., Chatr Aryamontri, A., Licata, L., *et al.*, 2010. MINT, the molecular interaction database: 2009 Update. Nucleic Acids Res 38, D532–D539.
Chatr-Aryamontri, A., Oughtred, R., Boucher, L., *et al.*, 2017. The BioGRID interaction database: 2017 Update. Nucleic Acids Res 45, D369–D379.
Chen, J.Y., Pandey, R., Nguyen, T.M., 2017. HAPPI-2: A comprehensive and high-quality map of human annotated and predicted protein interactions. BMC Genom 18, 182.
Cline, M.S., Smoot, M., Cerami, E., *et al.*, 2007. Integration of biological networks and gene expression data using cytoscape. Nat Protoc 2, 2366–2382.
Costanzo, M., Baryshnikova, A., Bellay, J., *et al.*, 2010. The genetic landscape of a cell. Science 327, 425–431.

Giot, L., Bader, J.S., Brouwer, C., *et al.*, 2003. A protein interaction map of *Drosophila melanogaster*. Science 302, 1727–1736.

Gramates, L.S., Marygold, S.J., Santos, G.D., *et al.*, 2017. FlyBase at 25: Looking to the future. Nucleic Acids Res 45, D663–D671.

Hein, M.Y., Hubner, N.C., Poser, I., *et al.*, 2015. A human interactome in three quantitative dimensions organized by stoichiometries and abundances. Cell 163, 712–723.

Ho, Y., Gruhler, A., Heilbut, A., *et al.*, 2002. Systematic identification of protein complexes in *Saccharomyces cerevisiae* by mass spectrometry. Nature 415, 180–183.

Huttlin, E.L., Bruckner, R.J., Paulo, J.A., *et al.*, 2017. Architecture of the human interactome defines protein communities and disease networks. Nature 545, 505–509.

Huttlin, E.L., Ting, L., Bruckner, R.J., *et al.*, 2015. The BioPlex network: A systematic exploration of the human interactome. Cell 162, 425–440.

Ito, T., Chiba, T., Ozawa, R., *et al.*, 2001. A comprehensive two-hybrid analysis to explore the yeast protein interactome. Proc Natl Acad Sci USA 98, 4569–4574.

Kamburov, A., Stelzl, U., Lehrach, H., Herwig, R., 2013. The ConsensusPathDB interaction database: 2013 Update. Nucleic Acids Res 41, D793–D800.

Kotlyar, M., Pastrello, C., Sheahan, N., Jurisica, I., 2016. Integrated interactions database: Tissue-specific view of the human and model organism interactomes. Nucleic Acids Res 44, D536–D541.

Launay, G., Salza, R., Multedo, D., *et al.*, 2015. MatrixDB, the extracellular matrix interaction database: Updated content, a new navigator and expanded functionalities. Nucleic Acids Res 43, D321–D327.

Li, S., Armstrong, C.M., Bertin, N., *et al.*, 2004. A map of the interactome network of the metazoan *C. elegans*. Science 303, 540–543.

Lopez, Y., Nakai, K., Patil, A., 2015. HitPredict version 4: Comprehensive reliability scoring of physical protein-protein interactions from more than 100 species. Database (Oxford). 2015:bav117.

Luck, K., Sheynkman, G.M., Zhang, I., Vidal, M., 2017. Proteome-scale human interactomics. Trends Biochem Sci 42, 342–354.

Mosca, R., Ceol, A., Aloy, P., 2013. Interactome3D: Adding structural details to protein networks. Nat Meth 10, 47–53.

Murali, T., Pacifico, S., Yu, J., *et al.*, 2011. DroID 2011: A comprehensive, integrated resource for protein, transcription factor, RNA and gene interactions for *Drosophila*. Nucleic Acids Res 39, D736–D743.

Orchard, S., 2012. Molecular interaction databases. Proteomics 12, 1656–1662.

Orchard, S., Ammari, M., Aranda, B., *et al.*, 2014. The MIntAct project – IntAct as a common curation platform for 11 molecular interaction databases. Nucleic Acids Res 42, D358–D363.

Orchard, S., Hermjakob, H., 2008. The HUPO proteomics standards initiative – Easing communication and minimizing data loss in a changing world. Brief Bioinform 9, 166–173.

Orchard, S., Kerrien, S., Abbani, S., *et al.*, 2012. Protein interaction data curation: The International Molecular Exchange (IMEx) consortium. Nat Methods 9, 345–350.

Pagel, P., Kovac, S., Oesterheld, M., *et al.*, 2005. The MIPS mammalian protein–protein interaction database. Bioinformatics 21, 832–834.

Papanikolaou, N., Pavlopoulos, G.A., Theodosiou, T., Iliopoulos, I., 2015. Protein–protein interaction predictions using text mining methods. Methods 74, 47–53.

Patil, A., Nakai, K., Nakamura, H., 2011. HitPredict: A database of quality assessed protein–protein interactions in nine species. Nucleic Acids Res 39, D744–D749.

Patil, A., Nakamura, H., 2005. Filtering high-throughput protein–protein interaction data using a combination of genomic features. BMC Bioinform 6, 100.

Prasad, T.S., Kandasamy, K., Pandey, A., 2009. Human protein reference database and human proteinpedia as discovery tools for systems biology. Methods Mol Biol 577, 67–79.

Rolland, T., Tasan, M., Charloteaux, B., *et al.*, 2014. A proteome-scale map of the human interactome network. Cell 159, 1212–1226.

Rual, J.F., Venkatesan, K., Hao, T., *et al.*, 2005. Towards a proteome-scale map of the human protein-protein interaction network. Nature 437, 1173–1178.

Ruepp, A., Waegele, B., Lechner, M., *et al.*, 2010. CORUM: The comprehensive resource of mammalian protein complexes – 2009. Nucleic Acids Res 38, D497–D501.

Salwinski, L., Miller, C.S., Smith, A.J., *et al.*, 2004. The database of interacting proteins: 2004 Update. Nucleic Acids Res 32, D449–D451.

Skrabanek, L., Saini, H.K., Bader, G.D., Enright, A.J., 2008. Computational prediction of protein–protein interactions. Mol Biotechnol 38, 1–17.

Stelzl, U., Worm, U., Lalowski, M., *et al.*, 2005. A human protein–protein interaction network: A resource for annotating the proteome. Cell 122, 957–968.

Szklarczyk, D., Franceschini, A., Wyder, S., *et al.*, 2015. STRING v10: Protein–protein interaction networks, integrated over the tree of life. Nucleic Acids Res 43, D447–D452.

Turner, B., Razick, S., Turinsky, A.L., *et al.*, 2010. iRefWeb: Interactive analysis of consolidated protein interaction data and their supporting evidence. Database 2010, baq023.

Uetz, P., Giot, L., Cagney, G., *et al.*, 2000. A comprehensive analysis of protein–protein interactions in Saccharomyces cerevisiae. Nature 403, 623–627.

Vidal, M., Fields, S., 2014. The yeast two-hybrid assay: Still finding connections after 25 years. Nat Methods 11, 1203–1206.

Villaveces, J.M., Jimenez, R.C., Porras, P., *et al.*, 2015. Merging and scoring molecular interactions utilising existing community standards: Tools, use-cases and a case study. Database (Oxford). p. 2015.

Von Mering, C., Krause, R., Snel, B., *et al.*, 2002. Comparative assessment of large-scale data sets of protein–protein interactions. Nature 417, 399–403.

Wan, C., Borgeson, B., Phanse, S., *et al.*, 2015. Panorama of ancient metazoan macromolecular complexes. Nature 525, 339–344.

Wodak, S.J., Vlasblom, J., Turinsky, A.L., Pu, S., 2013. Protein–protein interaction networks: The puzzling riches. Curr Opin Struct Biol 23, 941–953.

Yazaki, J., Galli, M., Kim, A.Y., *et al.*, 2016. Mapping transcription factor interactome networks using HaloTag protein arrays. Proc Natl Acad Sci USA 113, E4238–E4247.

Yu, H., Braun, P., Yildirim, M.A., *et al.*, 2008. High-quality binary protein interaction map of the yeast interactome network. Science 322, 104–110.

Relevant Website

http://www.ebi.ac.uk/Tools/webservices/psicquic/view/main.xhtml
  EMBL-EBI.

Computational Approaches for Modeling Signal Transduction Networks

Zhongming Zhao, The University of Texas Health Science Center at Houston, Houston, TX, United States
Junfeng Xia, Anhui University, Hefei, China

Introduction

Signal transduction is an essential biological process in cellular systems. It involves numerous biological functions in cell, and its disruption may lead to various diseases, phenotypes and drug treatment outcomes. Through dynamic and refined signal transduction, molecules connect and interact in a predefined fashion in signaling pathways so that the living organisms fulfill overall function in a harmonious way. Aberrant signal transduction in the pathways may cause dysregulation of biological processes, resulting in diseases such as cancer (Sever and Brugge, 2015). This article first introduces the basic concept about the signaling transduction, biological pathways, and molecular networks. Then, we describe the signaling transduction networks (STNs) in complex diseases by using cancer and its mitogen-activated protein kinase (MAPK) pathway as a specific example. MAPK pathway plays critical roles in multiple processes associated with many types of cancer and other diseases. Next, we introduce essential structural motifs in STNs, including feedback and feed forward loops. Finally, we review the databases and analysis tools that have been used in STNs and highlight some applications of computational modeling of STNs to complex disease studies.

Basic Concept in Signal Transduction Networks

Complex cellular activities such as cell growth and division, cell death, cell fate, and cell motility are generally regulated through the refined response of cells to stimuli in their environment. The transportation of an extracellular cue through cell membrane to the nucleus and intracellular signal passing between organelles or from organelles to other cellular components in response to cell stress or metabolic status are all referred to signal transduction (Liu *et al.*, 2008). Signal transduction allows cells to respond to the cues from either neighboring cells or even distant cells and to change their behavior accordingly. Cells constantly and concordantly receive a variety of signals that will affect one or multiple intersections within a signaling transduction pathway, as well as multiple signaling transduction pathways. Although numerous cell signaling transduction events and related molecular mechanisms have been uncovered, a deep understanding of cellular signaling transduction has still remained a main challenge in biological and biomedical research. This is mainly because there are thousands of small molecules, often with different forms or isoforms, that interact in a dynamic and complex fashion (e.g., spatiotemporal dimensions). Such a complex, dynamic feature is a major limit to apply the currently available statistical or computational modeling. However, deep deciphering of the complexity of cell signaling is ultimately essential for investigators to understand the normal biological functions as well as the pathophysiology of diseases. Such knowledge builds the foundation on the development of enhanced therapeutic strategies for many diseases, including cancer (Csermely *et al.*, 2016). So far, many molecular therapeutic approaches in cancer treatment have been based on the actionable mutations in kinase domains, which are the critical components in signaling pathways (Cheng *et al.*, 2014).

Extensive signaling studies have reconstructed a great number of carefully curated canonical signaling pathways, including the Jak-STAT, NF-κB, Wnt, Notch, Ras/ERK, and G protein-coupled receptor pathways (Housden and Perrimon, 2014). These pathways form a complex network called signal transduction network (STN) (Kolch *et al.*, 2015) that allows cells to integrate multiple environmental signals, and respond with output signals specifically tailored to the environmental change or condition. In these STNs or subnetworks, nodes are primarily the molecules such as signaling proteins and microRNAs, while edges (links between STN nodes) are activating or inhibitory physical connections of participating signaling proteins, microRNAs and other regulators. The regulation involves in binding at the critical sites of the molecules, enzymatic reaction (e.g., phosphorylation and dephosphorylation), and allosteric regulation, among others (Csermely *et al.*, 2016).

The Raf–MEK–ERK MAPK Signaling Pathway in Cancer

There are many signaling pathways in our knowledgebase. For example, one of the popularly used pathway databases, Kyoto Encyclopedia of Genes and Genomes (KEGG) database, currently curates a total of 320 human pathways (Kanehisa *et al.*, 2016). Illustration of each pathway and how it works is beyond the scope of this article. Here, we take signaling pathway in cancer as one example. Tumorigenesis has been perceived as dysfunction of STNs that regulate molecular communications and cellular processes (Kolch *et al.*, 2015). During cancer initiation and development, the STN undergoes systematic changes in terms of expression level of nodes, as well as in the sign (inhibition or activation) and weight (strength) of their edges. Dysregulation of cellular signal transduction pathways underlies most well-defined features in cancer cells (Sever and Brugge, 2015). We performed literature survey using keywords "signal transduction", "network", and "cancer". As shown in **Fig. 1**, it clearly shows the research in signaling pathways in cancer has been growing rapidly and consistently. We expect the number of cancer STN studies to increase substantially in the near future, considering that more –omics data and better pathway and network tools will be released in the near

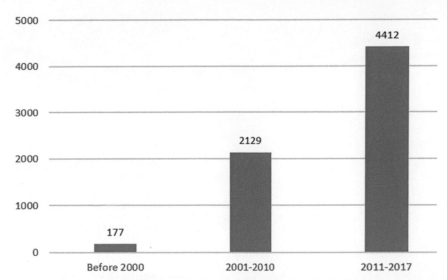

Fig. 1 Number of publications (*y*-axis) from PubMed title/abstract search using keywords "signal transduction" AND "network" AND "cancer". This simple search is not to have an exhaustive list but shows a strong trend of the growing studies in cancer by using signaling transduction network approaches.

future. To cover all of the signaling molecules involved and their myriad contributions to cancer would be too much in this article. We therefore focus on mitogen-activated protein kinase (MAPK or MAP kinase) signaling pathway, which attracts most attention and research interest due to its central role in regulating cell proliferation and survival and significant impact in various types of cancer (Roberts and Der, 2007; Cheng *et al.*, 2013).

In cancer, especially in melanoma (Cheng *et al.*, 2013), MAPK signaling cascades are critical and constitutively activated by a variety of mechanisms. These activation mechanisms, as well as the key genetic alterations detected in the pathway, have enabled the investigators to successfully develop kinase inhibitors and applied to the treatment of cancer, namely kinase or pathway based, molecularly targeted therapies. MAPK signaling pathway is one of key cellular mechanisms involving in cancer through promoting cell proliferation, survival, invasion, and tumor angiogenesis. This pathway consists of four widely studied MAP kinases, and these cascades sequentially transfer proliferative signals from the cell surface receptors into the nucleus via a series of protein phosphorylation events. Each MAPK cascade is comprised of three protein kinases that act as a signaling relay: MAP kinase kinase kinase (MAPKKK, MAP3K, or MEKK), MAP kinase kinase (MAPKK, MAP2K, or MEK), and MAPK. The four alternative terminal MAP kinases are the extracellular signal-regulated kinase 1 and 2 (ERK1/2), the c-Jun amino-terminal kinases (JNK12/3), p38 kinases, and ERK5. Among these four cascades, RAS-RAF-MEK-ERK1/2 pathway (**Fig. 2**) has attracted most research interest due to its central involvement in the regulation of cell proliferation and cell survival (Roberts and Der, 2007). In response to binding of various activators such as growth factors and hormones, receptor tyrosine kinases (RTKs; e.g., KIT) form a dimer, which causes the activation of Ras to GTP-bound state, and facilitates the recruitment of Raf (e.g., Braf) from the cytosol to cell membrane where it becomes an activated form. The adaptor protein growth factor receptor-bound protein 2 (Grb2), which comprises of Src homology 2 domain, promotes the binding of guanine nucleotide exchange factor son of sevenless (SOS), thereby leading to the formation of active Ras-GTP. Once being activated, Raf kinases phosphorylate and activate Mek proteins (Mek1 and Mek2), which subsequently lead to the phosphorylation and activation of Erk (Erk1 and Erk2). Finally, phosphorylated Erk1/2 proteins enter nucleus and regulate the activities of a large number of substrates – an estimate of over 160 proteins including ELK-1, Fos, Myc, and others – to control multiple cellular processes (Roberts and Der, 2007). The cascade of kinase reaction from RAF to MEK to ERK is a classic example of a signaling pathway. This pathway also represents an excellent example to illustrate feedback regulation (see below) and signal amplification.

Structural Motifs in STN

In a highly clustered STN, essential modules can be detected by identifying recurring patterns of highly interlinked groups of nodes (Liu *et al.*, 2008). A number of structural motifs exist in STNs, such as feed forward loops (FFLs) (Mangan and Alon, 2003) and feedback loops (FBLs) (Brandman and Meyer, 2008). These motifs enable a STN to respond selectively to a particular stimulus at the dynamic signal level, and therefore, result in precision action as the response.

The FFL (Mangan and Alon, 2003) is composed of three signaling molecules (e.g., transcription factors or microRNAs): a regulator (X) that regulates a molecule Y and targets a molecule Z, and Z is jointly regulated by X and Y. Because each of the three regulatory interactions in the FFL can be either positive (activation) or negative (repression), there are eight possible structural types of FFL (**Fig. 3**). Four of these FFLs are termed "coherent", where the sign of the direct signaling path (from X to Z) is the same

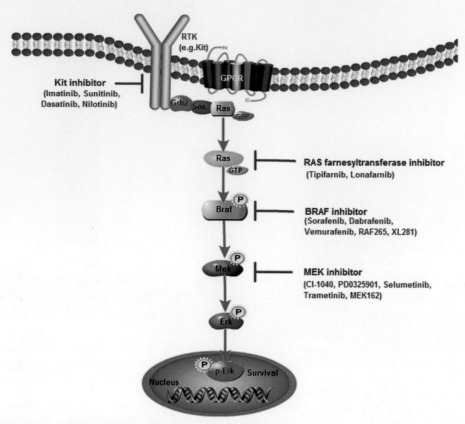

Fig. 2 Schematic diagram of the MAPK signaling pathway and selected inhibitors.

as the overall sign of the indirect signaling path (from X through Y to Z). The remaining four FFL configurations are termed "incoherent" where the signs of the direct and indirect signaling paths are opposite. In complex disease research field, the FFL approach was first applied to schizophrenia in 2010, where 32 FFLs among the compiled schizophrenia-related transcription factors (TFs), miRNAs and genes were identified (Guo et al., 2010). This approach has been frequently applied to cancer and other disease or phenotype since then, and a database for TF-miRNA-gene modules has been developed.

The FBL (Brandman and Meyer, 2008) is another common network motif that connect output signals back to their inputs. In this case, a downstream signaling molecule of a pathway activates (positive feedback) or represses (negative feedback) an upstream signaling molecule (**Fig. 3**). Positive FBLs can be utilized to control the response of a signaling pathway to different stimulation durations by locking the system into an on or off state following a transient signal. On the contrary, negative FBLs can cause a change in the dynamics of signaling because in that scenario, an activation of the signaling pathway can subsequently lead to a repression of the same pathway. And this may have several different effects depending on the kinetics of the FBL. FBL as a motif unit has been widely used in biomedical and biological research, such as the kinase cascade introduced in the subsection 'The Raf–MEK–ERK MAPK signaling pathway in cancer'.

STN Databases and Enrichment Analysis Tools

There have been numerous studies of STNs or the subcomponents of a STN. The rich information has been available from literatures, databases and other resources. Accordingly, knowledgebase on STNs has become one of the important efforts to summarize the previous findings. Such knowledgebase can provide us with a systematic view of the biological processes and assist us for future studies of disease mechanisms (Liu et al., 2008). To meet various needs, researchers have established many databases through careful curation and storage of such data, as well as allowing users to access and visualize the biological pathways and molecular interactions (**Table 1**). Among these databases, KEGG database (Kanehisa et al., 2016) is one of the popular, manually curated pathway resources. It includes the knowledge in the molecular interaction, reaction and relation networks, and pharmacological information. KEGG consists of seven types of network context: metabolism, genetic information processing, environmental information processing, cellular processes, organismal systems, human diseases, and drug development. The KEGG pathway information is machine-readable via KEGG Markup Language. KEGG includes many canonical signaling transduction pathways. It was initially publicly available but later the access requires fee-based subscription. Reactome (Fabregat et al., 2014) is another popular, well curated, and peer reviewed pathway database. The current version (V62) comprises 11,302 reactions (e.g.,

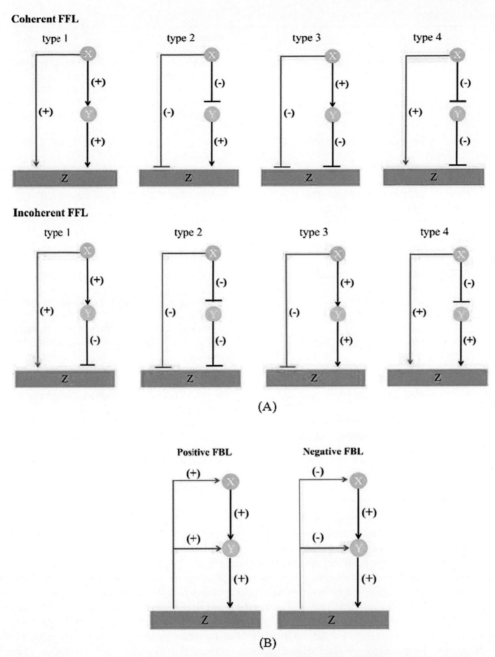

Fig. 3 Motifs of feed forward loop (FFL) and feedback loop (FBL) in signal transduction network. (A) The eight types of FFLs can be classified into four coherent and four incoherent FFLs. In coherent FFLs, the sign of the direct path from signaling molecule X to output Z is the same as the overall sign of the indirect path through signaling molecule Y. Incoherent FFLs have inconsistent or opposite signs for the two paths. (B) The two types of FBLs: a downstream molecule activates (positive FBL) or represses (negative FBL) an upstream molecule.

phosphorylation, acetylation, etc.) organized into 2176 human pathways, involved in 10,878 proteins and 1768 small molecules. These data were extracted from 27,526 publications. For machine readability, Reactome adopted the Systems Biology Markup Language (SBML) format for systematic access and analysis. In addition to those self-curated pathway databases, other databases assemble different sources of pathways. For example, the Pathway Commons database (Cerami *et al.*, 2011) contains extensive collection of pathway and interaction data from partner databases such as KEGG, Reactome, and MSigDB (the Molecular Signatures Database) (Subramanian *et al.*, 2005) and is represented in the Biological Pathway Exchange (BioPAX) standard. WikiPathways (Kutmon *et al.*, 2016) is another integrated database of biological pathways maintained by and for the scientific community. The advantage of WikiPathways lies in the scalable, community-based curation and unrestricted pathway model through accepting any pathway that researchers find useful in their work. **Table 1** provides a list of pathway databases, including the description and access URLs. Each of these pathway databases include signaling transduction pathways.

Table 1   Summary of databases and tools for signaling transduction networks

Name	URLs	Description
KEGG	www.genome.jp/kegg/pathway.html	A collection of manually pathway maps representing our knowledge on the molecular interaction, reaction and relation networks
Reactome	www.reactome.org	A free, open-source, curated and peer reviewed pathway database
Gene Ontology	www.geneontology.org	A computational representation of gene products in terms of their associated biological processes, cellular components and molecular functions
Pathway Commons	www.pathwaycommons.org	An integrated resource from public pathway and interactions databases
NCI PID	https://pid.nci.nih.gov	Biomolecular interactions and cellular processes assembled into authoritative human signaling pathways
BioCarta	https://cgap.nci.nih.gov/Pathways/ BioCarta_Pathways	Online maps of metabolic and signaling pathways
WikiPathways	www.wikipathways.org	A database of biological pathways maintained by and for the scientific community
UCSD Signaling Gateway	www.signalinggateway.org	A database providing essential information on the proteins involved in cell signaling
MSigDB	https://software.broadinstitute.org/ gsea/msigdb	A collection of annotated gene sets for gene set enrichment analysis
NetPath	www.netpath.org	A manually curated resource of signal transduction pathways in humans
SignaLink	www.signalink.org	A signaling pathway resource with multi-layered regulatory networks
Pascal	www2.unil.ch/cbg/index.php? title=Pascal	A tool for gene scoring and pathway analysis from GWAS results
DAVID	https://david.ncifcrf.gov	A comprehensive set of functional annotation tools to understand biological meaning behind large list of genes
webGestalt	www.webgestalt.org	A suite of tools for functional enrichment analysis in various biological contexts
Ingenuity	www.qiagenbioinformatics.com/ products/ingenuity-pathway-analysis	A comprehensive analysis and search tool of global molecular network based on manual curation
ToppGene	https://toppgene.cchmc.org	A portal for gene list enrichment analysis and candidate gene prioritization based on functional annotations and protein interactions network

Based on the prior knowledge (pathway databases), many tools have been developed to interpret gene lists derived from the high-throughput experiments (Wadi *et al.*, 2016). These pathway analysis tools use various strategies to aggregate genes across sets of the related genes (**Table 1**). The most common approach used for gene set enrichment of pathway function is based on binary enrichment tests, which rely on a threshold parameter to determine which genes are significantly associated with the disease or trait. One disadvantage of such an approach is that potential contributions of the weakly associated genes that just missed the threshold would be lost and there is no clear rule on the threshold setting. To address this problem, Lamparter *et al.* (2016) presented a powerful tool called Pascal (Pathway scoring algorithm) for computing pathway scores. Pascal integrates individual gene scores without the need for a tunable threshold parameter to dichotomize gene scores for binary membership enrichment analysis. As a result, it shows better discovery of confirmed pathways compared with other popular methods. However, challenges and limitations exist in all the pathway enrichment methods currently available. These challenges include the multiple test correction, redundancy of the genes shared by different pathways, appropriate statistical tests, incomplete pathway data being available to test, and gene length bias on measuring the signals from the genomic data, among others (Jia *et al.*, 2011; Wang *et al.*, 2010).

Computational Modeling of STNs and Its Applications

In this section, we present several examples of the computational modeling of STNs.

Genetic Mutations Impact STNs

Somatic mutations may cause abnormal cell growth, resulting in tumorigenesis. These mutations confer a selective advantage to the cancer cells, which together with changes in the microenvironment, promote cancer growth and progression. Analyses of massive amounts of cancer genomics data generated from several large-scale cancer genome sequencing projects such as The

Cancer Genome Atlas (TCGA) and the International Cancer Genome Consortium (ICGC) have discovered more than 68 million simple somatic mutations and many other types of somatic mutations (as of December 11, 2017). However, most of these mutations are passenger mutations. Typically, only two to eight in a cancer cell are driver mutations that cause progression of the cancer (Vogelstein *et al.*, 2013). These may be single-nucleotide variants, small insertions or deletions, copy number variations, or structure variants.

Based on these mutation data, we can connect the genetic mutations in cancer cells within STNs that control processes associated with tumorigenesis and place them in the context of distortions of wider STNs that fuel cancer progression. Genetic mutations can cause abnormal expression of proteins involved in a STN (e.g., gene amplification) or change the individual contributions of a specific STN by producing mutant proteins whose activities are dysregulated (e.g., point mutations, truncations, and fusions). Alternatively, deletions in genes may inactivate negative regulators, which may cause the absence of the specific nodes within a STN. Moreover, multiple mutations can occur within a cell simultaneously and among cells of a specific tumor type. Recently, Zhao *et al.* (2017) proposed a computational oncoproteomics approach, namely kinome-wide network module for cancer pharmacogenomics (KNMPx), for identifying actionable mutations that rewired STNs by incorporating the somatic missense mutations into the protein phosphorylation sites. Specifically, 746,631 missense mutations in 4997 tumor samples across 16 major cancer types/subtypes from TCGA were integrated into over 170,000 carefully curated non-redundant phosphorylation sites covering 18,610 proteins. Forty-seven mutated proteins (e.g., ERBB2, TP53, and CTNNB1) had enriched missense mutations at their phosphorylation sites in pan-cancer analysis. In addition, tissue-specific kinase-substrate interaction modules altered by somatic mutations that were identified in cancer genomes were significantly associated with patient survival. Interestingly, the authors found that cell lines could highly reproduce oncogenic phosphorylation site mutations identified in primary tumors, supporting the confidence in their associations with sensitivity/resistance of inhibitors targeting EGF, MAPK, PI3K, mTOR, and Wnt signaling pathways. According to the results, the KNMPx approach is powerful for identifying oncogenic alterations in signaling molecules via rewiring phosphorylation-related STNs and for predicting drug sensitivity or resistance in the era of precision oncology.

Pan-Cancer Analysis of Dysregulated Transcription Factor – microRNA FFLs in STNs

Transcription factor and microRNA (miRNA) can mutually regulate each other and jointly regulate their shared target genes. This is an ideal example for FFLs in cellular system. Since 2010, TF-miRNA FFLs have been widely used to identify the cancer-associated genes or miRNAs in many tumor types. As one example, Jiang *et al.* examined dysregulated FFLs in pan-cancer (13 cancer types) (Jiang *et al.*, 2016). They identified 26 pan-cancer FFLs, each of which was found to be dysregulated in at least five tumor types using TCGA data. They found these pan-cancer FFLs could communicate with each other and form functionally consistent subnetworks, such as epithelial to mesenchymal transition-related subnetwork. Many proteins and miRNAs in each subnetwork belong to the same protein family and miRNA family, respectively. Importantly, cancer-associated genes and drug targets were enriched in these pan-cancer FFLs, in which the genes and miRNAs also tended to be hubs and bottlenecks. In addition, they identified potential anticancer indications for existing drugs with novel mechanism of action. Collectively, their study demonstrated the potential of pan-cancer FFLs as critical regulatory units that are useful for elucidating pathogenesis of cancer and developing anticancer drugs.

Decipher STNs for Drug Effect in Complex Disease

A drug exerts its effects typically through a STN cascade, which is non-linear and involves intertwined networks of multiple signaling pathways. Construction of such a STN may help the investigators to identify novel drug targets and further understand drug action. Sun *et al.* developed a novel computational framework, the Drug-specific Signaling Pathway Network (DSPathNet), to tackle this issue (Sun *et al.*, 2015). The DSPathNet amalgamates the prior drug knowledge and drug-induced gene expression via random walk algorithms. Using the drug metformin as an example, they illustrated this framework and obtained one metformin-specific SPNetwork containing 477 nodes and 1366 edges. The further examination of this subnetwork using one type 2 diabetes (T2D) genome-wide association study (GWAS) dataset, three cancer GWAS datasets, and one GWAS dataset of cancer patients with T2D on metformin indicated that the metformin network was significantly enriched with disease genes for both T2D and cancer, and that the network also included genes that may be associated with metformin-associated cancer survival. The authors also generated a subnetwork to highlight the molecules' crosstalk between T2D and cancer. The follow-up network analyses and literature mining revealed that some valuable insights into the mode of metformin action, which will facilitate our understanding of the molecular mechanisms underlying drug treatments, disease pathogenesis, and identification of novel drug targets and repurposed drugs. As many drugs may target signaling molecules, or through metabolism that links to signaling pathways, novel computational approaches for deciphering STNs will be much needed.

Applications of STNs in Other Fields

Besides cancer, STNs have been frequently used to study the molecular regulation in specific cellular conditions, cell types, epigenetic regulation, or in other complex disease or phenotype. For example, Bernabò *et al.* have applied STN approach to investigate the cell signal transduction, which is a complex phenomenon and has a central role in cell surviving and adaptation, in non-medical systems. The authors presented two applications of STNs for cell signaling. One is to study the architecture of signaling systems for acquisition of a complex cellular function (i.e., process of activation of spermatozoa) and the other is to find the organization of specific signaling systems that are active in specific cells and/or tissues (i.e., the endocannabinoid system) (Bernabò *et al.*, 2014). In both the applications, the authors found that the STNs follow a scale free and small world topology. Such network characteristics will allow the investigators to identify molecules (e.g., hubs) or modules that play key role in signaling system. STNs approaches have also been applied to study human pluripotent stem cells. As an example, Fgf/MAPK, TGFβ/SMAD2,3 and insulin/PI3K signaling pathways are found to be required for maintenance of the stem cell state (Dalton, 2013). Moreover, DNA methylation data has been integrated with other genomic data into STNs for exploring epigenetic regulation patterns for complex disease like bone mineral density (BMD) (Zhang *et al.*, 2015). In synthetic biology field, the progress of STN approaches in synthetic signal transduction implementation has been systematically reviewed in diverse biological contexts including bacteria, yeast, vertebrate, and plant cells (Hansen and Benenson, 2016), and various systems STN concepts have been implemented into metabolic engineering and synthetic biology (He *et al.*, 2016).

Closing Remarks

Complex disease like cancer is highly heterogeneous – it involves dynamic regulation and interactions of STNs at the cellular level. It is often tissue-specific and development stage specific. Identification of STN changes that are critical for cell function, development, or drug binding will substantially improve our knowledge and have potential for healthcare. As we have entered the big data era, we expect more novel, systematic studies will be conducted, leading to not only knowledge discovery in signaling pathways and networks, but also more disease mutations and drug-targetable sites. Advanced statistical and computational methods and tools will be further needed to meet such a strong demand in future. For example, novel tools that can effectively integrate the genomic and genetic data in a spatiotemporal fashion into the STNs will power the investigators to measure the genetic or biological changes in high resolution and more reflective on the cellular condition.

Acknowledgements

Dr. Zhao was partially supported by National Institutes of Health grants [R01LM012806 and R21CA196508]. The funders had no role in study design, data collection and analysis, decision to publish, or preparation of the manuscript.

See also: Cell Modeling and Simulation. Computational Systems Biology Applications. Computational Systems Biology. Functional Genomics. Graphlets and Motifs in Biological Networks. Natural Language Processing Approaches in Bioinformatics. Network Inference and Reconstruction in Bioinformatics. Network Models. Network-Based Analysis for Biological Discovery. Networks in Biology. Pathway Informatics. Prediction of Protein-Protein Interactions: Looking Through the Kaleidoscope. Protein Post-Translational Modification Prediction. Protein-Protein Interactions: An Overview. Visualization of Biomedical Networks

---

References

Bernabò, N., Barboni, B., Maccarrone, M., 2014. The biological networks in studying cell signal transduction complexity: The examples of sperm capacitation and of endocannabinoid system. Computational and Structural Biotechnology Journal 11, 11–21.

Brandman, O., Meyer, T., 2008. Feedback loops shape cellular signals in space and time. Science 322, 390–395.

Cerami, E., Gross, B., Demir, E., *et al.*, 2011. Pathway Commons, a web resource for biological pathway data. Nucleic Acids Research 39, 685–690.

Cheng, F., Jia, P., Wang, Q., Zhao, Z., 2014. Quantitative network mapping of the human kinome interactome reveals new clues for rational kinase inhibitor discovery and individualized cancer therapy. Oncotarget 5, 3697–3710.

Cheng, Y., Zhang, G., Li, G., 2013. Targeting MAPK pathway in melanoma therapy. Cancer and Metastasis Reviews 32, 567–584.

Csermely, P., Korcsmaros, T., Nussinov, R., 2016. Intracellular and intercellular signaling networks in cancer initiation, development and precision anti-cancer therapy: Ras acts as contextual signaling hub. Seminars in Cell & Developmental Biology 58, 55–59.

Dalton, S., 2013. Signaling networks in human pluripotent stem cells. Current Opinion in Cell Biology 25, 241–246.

Fabregat, A., Sidiropoulos, K., Garapati, P., *et al.*, 2014. The reactome pathway knowledgebase. Nucleic Acids Research 42, 472–477.

Guo, A.Y., Sun, J., Jia, P., Zhao, Z., 2010. A novel microRNA and transcription factor mediated regulatory network in schizophrenia. BMC Systems Biology 4, 10.

Hansen, J., Benenson, Y., 2016. Synthetic biology of cell signaling. Natural Computing 15, 5–13.

He, F., Murabito, E., Westerhoff, H.V., 2016. Synthetic biology and regulatory networks: Where metabolic systems biology meets control engineering. Journal of the Royal Society Interface 13, 20151046.

Housden, B.E., Perrimon, N., 2014. Spatial and temporal organization of signaling pathways. Trends in Biochemical Sciences 39, 457–464.

Jia, P., Wang, L., Meltzer, H.Y., Zhao, Z., 2011. Pathway-based analysis of GWAS datasets: Effective but caution required. International Journal of Neuropsychopharmacology 14 (4), 567–572.

Jiang, W., Mitra, R., Lin, C., *et al.*, 2016. Systematic dissection of dysregulated transcription factor-miRNA feed-forward loops across tumor types. Briefings in Bioinformatics 17, 996–1008.

Kanehisa, M., Sato, Y., Kawashima, M., Furumichi, M., Tanabe, M., 2016. KEGG as a reference resource for gene and protein annotation. Nucleic Acids Research 44, 457–462.

Kolch, W., Halasz, M., Granovskaya, M., Kholodenko, B.N., 2015. The dynamic control of signal transduction networks in cancer cells. Nature Reviews Cancer 15, 515–527.

Kutmon, M., Riutta, A., Nunes, N., *et al.*, 2016. WikiPathways: Capturing the full diversity of pathway knowledge. Nucleic Acids Research 44, 488–494.

Lamparter, D., Marbach, D., Rueedi, R., Kutalik, Z., Bergmann, S., 2016. Fast and rigorous computation of gene and pathway scores from SNP-based summary statistics. PLOS Computational Biology 12 (1), e1004714.

Liu, W., Li, D., Zhu, Y., He, F., 2008. Bioinformatics analyses for signal transduction networks. Science China-life Sciences 51, 994–1002.

Mangan, S., Alon, U., 2003. Structure and function of the feed-forward loop network motif. Proceedings of the National Academy of Sciences of the United States of America 100, 11980–11985.

Roberts, P.J., Der, C.J., 2007. Targeting the Raf-MEK-ERK mitogen-activated protein kinase cascade for the treatment of cancer. Oncogene 26, 3291–3310.

Sever, R., Brugge, J.S., 2015. Signal transduction in cancer. Cold Spring Harbor Perspectives in Medicine 5, a006098.

Subramanian, A., Tamayo, P., Mootha, V.K., *et al.*, 2005. Gene set enrichment analysis: A knowledge-based approach for interpreting genome-wide expression profiles. Proceedings of the National Academy of Sciences of the United States of America 102, 15545–15550.

Sun, J., Zhao, M., Jia, P., *et al.*, 2015. Deciphering signaling pathway networks to understand the molecular mechanisms of metformin action. PLOS Computational Biology 11, e1004202.

Vogelstein, B., Papadopoulos, N., Velculescu, V.E., *et al.*, 2013. Cancer genome landscapes. Science 339, 1546–1558.

Wadi, L., Meyer, M., Weiser, J., Stein, L., Reimand, J., 2016. Impact of outdated gene annotations on pathway enrichment analysis. Nature Methods 13, 705–706.

Wang, K., Li, M., Hakonarson, H., 2010. Analysing biological pathways in genome-wide association studies. Nature Reviews Genetics 11, 843–854.

Zhang, J.G., Tan, L.J., Xu, C., *et al.*, 2015. Integrative analysis of transcriptomic and epigenomic data to reveal regulation patterns for BMD variation. PLOS ONE 10, e0138524.

Zhao, J., Cheng, F., Zhao, Z., 2017. Tissue-specific signaling networks rewired by major somatic mutations in human cancer revealed by proteome-wide discovery. Cancer Research 7, 2810–2821.

Further Reading

Alon, U., 2007. Network motifs: Theory and experimental approaches. Nature Reviews Genetics 8, 450–461.

Inui, M., Martello, G., Piccolo, S., 2010. MicroRNA control of signal transduction. Nature Reviews Molecular Cell Biology 11 (4), 252–263.

Logue, J.S., Morrison, D.K., 2012. Complexity in the signaling network: Insights from the use of targeted inhibitors in cancer therapy. Genes & Development 26, 641–650.

Prasasya, R.D., Tian, D., Kreeger, P.K., 2011. Analysis of cancer signaling networks by systems biology to develop therapies. Seminars in Cancer Biology 21, 200–206.

Cell Modeling and Simulation

Ayako Yachie-Kinoshita, The Systems Biology Institute, Tokyo, Japan and University of Toronto, Toronto, ON, Canada
Kazunari Kaizu, RIKEN Quantitative Biology Center (QBiC), Osaka, Japan

Introduction

Since the 21st century has begun, the so-called post-genome era, the biggest open questions are how the phenotype is generated from the genotype, and how evolution is achieved from the phenotype. Towards this ultimate goal, systems biology has emerged as a conceptual approach to understand life as an evolvable and robust system (Kitano, 2002). Mathematical modeling and simulation of the biological systems have played the central role in the context of systems-level understanding and predictability of living organisms (**Fig. 1**).

As with modeling of other systems, the abstraction of the cell properties is the first key step in cell modeling. This step requires the definition of boundaries in the focused cellular system.

The cell contains highly integrated subsystems in itself, although it is often called a building unit for the structure and function of living organisms. The complexity lays in space and time. Moreover, although the cell has independent replicability, its activity is dependent not only internal state but also extrinsic perturbations, that is, its surroundings, which are orchestrated towards robust biological functions. It is thus difficult to define physical and functional boundaries in the cellular system, as is clear from the fact that the cellular behavior observed in vitro is hardly reproduced in vitro. Under these circumstances, cell simulation often requires drastic abstraction of the system to keep the balance of desired trade-offs with predictability.

The demands for cell simulations are also largely regulated by how the technology in molecular biology has evolved. The electronic kinetics emerged in the 1950s and drove the first mathematical modeling of cellular phenotypes reported by Hodgkin–Huxley (Hodgkin and Huxley, 1952), focusing on the dynamics of heart activity by calculating the ensemble of voltage-dependent channels using nonlinear ordinary differential equations, which is still a basis for studying the behavior of ion channels. The protein purification led to massive studies on enzymes in the 1970s that initiated metabolic simulations based on enzyme kinetics (Garfinkel et al., 1970; Heinrich et al., 1978; Tyson and Othmer, 1978). From the late 1990s to the present, with genomics as a start, comprehensive measurements of each biological layer have emerged, including metabolomics, proteomics, and epigenomics, and have changed our view of biological systems towards data-driven science (Hey et al., 2009). In recent years, it has become possible to observe cell behaviors at the single cell resolution (Kalisky et al., 2011; Xia et al., 2013). This has made us aware of the significance of cell–cell heterogeneity at the population level (Iwamoto et al., 2016), inter- and intracellular spatio-temporal dynamics considering diffusion and reaction kinetics (Takahashi et al., 2005), and inherent biological noise (Mettetal et al., 2006).

Simulation models where essential components in the system are appropriately defined and assembled can predict the system's response towards given perturbations. The hypothesis derived from such simulation then becomes valuable for experimental validations. The simulation–experiment cycle through mathematical models can minimize costly or impossible experiments such as sensitivity analysis, impact estimation of unknown factors, assessment of systems threshold, or finding necessary and sufficient minimum components for the focused event.

Approaches in Cell Simulation

Due to the complexity in both cellular systems and the demands for cell simulation, a variety of approaches and methods have been considerable, depending on the feature of the focused system including cell type and phenomena, the accessibility to the system, and the availability of the data (**Fig. 2**).

Dynamic Modeling and Static Modeling

The first decision branch on cell modeling is the definition of time in the simulation. Dynamic models typified by ordinary differential equations (ODEs) and partial differential equations (PDEs) determine the time development of the cellular system based on the precedent states and response to the environmental changes.

On the other hand, static cell models represent the analysis of the static diagrams of model components, including their attributes and relationships between components. For example, for metabolic systems, topological pathway analysis such as elementary modes (Schuster et al., 2000) and extreme pathways (Schilling et al., 2000) predict essential chain of processes by using stoichiometric relationships and mass conservation law. Metabolic flux analysis is also a stoichiometry-based static model to predict gross error and flux distribution of the metabolic system by introducing flux measurements into the pathway topology (Zupke and Stephanopoulos, 1994; Schmidt et al., 1997). Metabolic control analysis is the static modeling to predict rate or pool controlling steps from the pathway diagram (Fell, 1992). In a broader sense, the static modeling can be expanded to include the generation of comprehensive networks of physical and functional interactions between cellular components by certain network

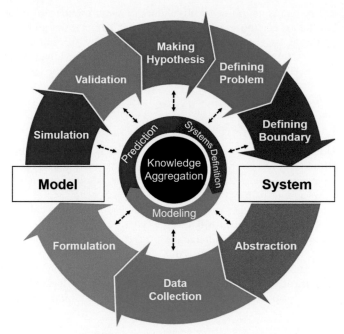

Fig. 1 A roadmap in predictive biology based on cell modeling and simulation. Once we define the cell system, the system is modeled through mathematical formulation based on data. The model is then used to predict the focused cellular phenomena to be validated. The prediction generates, strengthens, or refines the hypothesis on cell system, which drives another cycle of the road map.

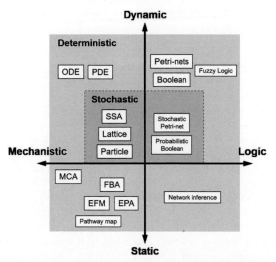

Fig. 2 Simulation approaches of cell systems. Cell modeling is starting from the selection of the suitable mathematical framework from a variety of simulation approaches displayed along several axes. Horizontal axis and vertical axis sort out simulation techniques by mechanistic or logic, dynamic or static, respectively. In addition, dynamic approaches are also split into two groups depending on stochasticity (deterministic or stochastic). Abbreviation used: ODE (ordinary differential equation), PDE (partial differential equation), FBA (flux balance analysis), SSA (stochastic simulation algorithm), EPA (extreme pathway analysis), MCA (metabolic control analysis), EFM (elementary flux mode).

inference approaches (Marbach *et al.*, 2012), or manual curation of the molecular pathway maps (Oda *et al.*, 2005; Oda and Kitano, 2006; Kaizu *et al.*, 2010; Matsuoka *et al.*, 2013).

Deterministic Simulation and Stochastic Simulation

One of the major aspects segregating approaches of cell simulation is the consideration of the stochasticity in the processes. Deterministic behavior can be defined as a consecutive chain of precedent events where no randomness is affecting the next states. In this sense, deterministic cellular models reproduce the same predictions for a given initial condition. This is applicable in case of well-stirred systems, in other words, the systems where cell behavior can be assumed as the behavior of the cell population where the cell belongs.

However, in case the target cell system is assuming small volumes or small numbers of molecules, this definition of deterministic behavior is not applicable (Rao et al., 2002), and stochasticity of the process should be introduced, such as the Gillespie algorithm (Gillespie, 1977). For example, each transcription process is governed by a specific transcription factor whose number is as low as several dozen (Elowitz et al., 2002), and the deterministic modeling can no longer reflect the behavior of cell systems. Moreover, recent emergence of single cell tracking techniques shed light on the stochastic fluctuations of molecules, which exert large impact on significant processes (Gillespie, 2007).

Mechanism-Based Modeling and Logic-Based Modeling

Modeling of the cell system considering biochemical properties for each process, for example, biochemical rules, requires a large amount of mechanistic and quantitative information. This shortage of data on modeling has never been solved despite the rapid spread of high-throughput technologies, as the mechanism-based modeling requires data restricted to influential range of the focused variables while the high-throughput data typically provide more macroscopic view.

Logic-based models are the alternative approaches to overcome the issue around data availability where relationships between variables are described without molecular details, such as Boolean models, fuzzy logic models, and Petri-net models (Kauffman, 1969; Reddy et al., 1996; Hofestadt and Thelen, 1998; Aldridge et al., 2009). Logic-based models assume that one variable respond to the others in a gate (switch-like) fashion (Watterson et al., 2008; Morris et al., 2010; Wynn et al., 2012; Le Novere, 2015). This mathematical simplification is analogous to a steep Hill function with high cooperativity while dose-response features can't be represented. In contrast, it may be feasible to some extent, for example, to apply logic-based modeling to predict the gene expression patterns, with the generally applicable fact that gene expression is a bimodality distributed at the single cell level (Shalek et al. 2013).

Simulation Models in Biological Layers

Simulation models are focused on specific phenotype and thus generally focused on specific biological layers. Due to the characteristics of each layer, frequently applied approaches are varied.

Metabolism

The nature of metabolism is the well-regulated mass transfer reaction chains of small molecules called metabolites, whose number is enough large in most cases to assume well-stirred systems. Due to these characteristics of metabolic pathways, dynamic, deterministic and mechanistic approach has been dominantly applied empowered by the massive studies of enzyme kinetics. The case example of modeling of red blood cell metabolism and its application is discussed in the following section.

On the other hand, apart from the dynamic kinetic models, cellular metabolic systems have long played a central role in the tight connectivity between mathematics and wet-lab biological experiments, specifically in the area of systems control theory in metabolic engineering. Indeed, the distinguished scientists in metabolic engineering and their studies took the stage when systems biology emerged and formed as a field (see the subsection "Dynamic Modeling and Static Modeling" in the section "Approaches in Cell Simulation"). Under the mass conservation law and the steady-state assumption, which is relatively fair in metabolism given the robust metabolic homeostasis, it is possible to calculate the flux distribution of the given metabolic system, called flux balance analysis (FBA; Orth et al., 2010). Further assumptions that the cell metabolism is optimized or can be engineered towards the objective cellular functions introduce optimization techniques into FBA to examine the ideal space of metabolic states (i.e., flux distribution patterns), which is called constraint-based modeling (O'Brien et al., 2015; Oberhardt et al., 2009). We now have information from genomes in most model organisms and their annotations with high quality, and thus have a comprehensive topology comprising metabolic reactions at whole-genome scale (Bao and Eddy, 2003). In addition to the bottom-up reconstruction of the model, top-down integration of high-throughput measurements as a further constraint has led the increasingly relevant predictions on cell metabolism (Bordbar et al., 2014). For example, the constraint-models of a genome-scale metabolic pathway have been applied to engineer a strain of *Escherichia coli* producing a valuable chemical, such as L-threonine and 1,4-Butanediol (Lee et al., 2007; Yim et al., 2011), and those of human metabolism have predicted tissue-specific significance of disease gene activity (Shlomi et al., 2008; Jerby et al., 2010).

Signaling

In contrast to metabolic pathways where small molecules and their mass transfer processes are the major variables and processes, the central components of signaling pathways are proteins and the sets of chemical chain reactions based on their states, which is called a signaling cascade. In addition, the key signaling cascades often contain physical information transfer such as nuclear translocations, as the signal propagation should occur in a local and restricted environment. This nature of cellular signaling has called modelers' attention toward spatiotemporal approaches, as described in the section "Spatial Simulation Techniques."

The cellular response to biological signals is significantly varied in contrast to the limited number of known signaling molecules and their cascades. Nevertheless the biological phenomena as a final outcome of the signaling cascade are decoded and often digitized, such as cell fate decision. Due to this discreteness and the dynamic variations of the systems output, the consideration of stochasticity of each process is important in modeling of signaling systems. Remarkably, the models have predicted the significance of stochasticity in small subcellular volume for the propagation of information through certain signaling cascade (Bhalla, 2004a,b, Fujii *et al.*, 2017).

Logic-based modeling approaches have also been applied to simulate signaling cascades (Morris *et al.*, 2010) while the detailed mechanistic and kinetic models of signaling systems have given us significant insights on both fundamental topology and hidden dynamics of the signaling cascade (Kholodenko, 2006). For example, using the mitogen-activated protein kinase signaling model, it has been revealed that the frequently observed signaling cascade arrangement has unexpected consequences for the dynamics that can give an ultrasensitive response to input signal (Huang and Ferrell, 1996). It has been also revealed by the kinetic model that transiently high and sustained concentration of the upstream input signal make different outputs of extracellular-signal-regulated kinase signaling cascade (Sasagawa *et al.*, 2005). This indicated that cell systems encode the distinct dynamics of signals to specific proteins using the signaling cascade to keep the variations of cellular response with minimum set of molecules.

Gene Regulatory Networks

A gene regulatory network (GRN) is a collection of regulatory relationships between transcription factors (TFs) and TF-binding sites of specific mRNA to govern certain expression levels of mRNA and their resulted proteins. Up to 10% of the ORF-coding genomes are coding TFs (Levine and Tjian, 2003) each of which are highly interconnected in the GRN. The GRN itself is often recognized as a static model and frequently described as a large "graph" consisting of hundreds of TFs where nodes represent TFs and edges between nodes represent regulatory relationships between them. The edges in the GRN graph are assessed experimentally by the detection of protein–DNA interactions through chromatin immunoprecipitation techniques (Lee *et al.*, 2002; Ren *et al.*, 2000; Robertson *et al.*, 2007; Johnson *et al.*, 2007; Mikkelsen *et al.*, 2007) or protein–protein interactions (Fields and Song, 1989; Chien *et al.*, 1991). The computational approaches have also been massively developed (Bansal *et al.*, 2007; Ideker and Krogan, 2012; Marbach *et al.*, 2012) to infer GRNs.

Apart from static modeling, dynamic approaches to model GRNs have considerable difficulties including the time delay between transcription and translation, and the shortage of information on the essential topology to represent the focused phenotype and the kinetics. In this context, logic-based models have frequently been applied to simulate and understand the transient states of GRNs (Chai *et al.*, 2014). A recent attempt identified a minimal set of nodes (genes) and their interactions sufficient to explain the experimentally observed cell behavior, by automated reasoning using Boolean models with possible interactions in the GRN in mouse embryonic stem cells (Dunn *et al.*, 2014).

Nevertheless, cell fate transitions occur when cells switch from one GRN to another. Boundaries between these cell states are influenced both by intrinsic fluctuation expression (Eldar and Elowitz, 2010, Singer *et al.*, 2014) and by extrinsic variations in the cellular microenvironment. Moreover, cell–cell heterogeneity of GRN states is commonly observed that can yield stochastic single cell state transitions coalescing into a stable cell population (Glauche *et al.*, 2010; Herberg *et al.*, 2014). To tackle these fundamental properties of GRNs while keeping predictive power, the novel simulation framework has been developed based on asynchronous Boolean simulation followed by graph-theory analysis of the GRN state-transitions (Yachie-Kinoshita *et al.*, 2017). The simulation quantitatively depicts the dynamic changes in cellular GRN states as a function of signaling inputs and their consequential feedback loops within the GRN. The outputs from the simulation provide predictions of the probability of transitions between distinct single cell states, population-average expression frequencies, and the emergence of stable subpopulations in response to different input conditions. The simulation framework was then applied to a relatively large GRN (around 30 nodes in the graph) in mouse embryonic stem cells and five major signaling pathways around pluripotency. As a result, the model predicted a novel combination of signaling inputs that drives mESCs to a specific cell fate.

Application of Predictive Kinetic Model

Red blood cell metabolism and its enzymology have been well studied over the last four decades due to its availability as human cell samples and relative simplicity where the mature cells lack the dynamism from gene expression and the individual cells are physically and functionally compartmentalized via the cell membrane. Although the essential biochemical pathways in red blood cells to maintain the cell state are established earlier and the kinetic information corresponding to each enzymatic reaction in these pathways is available, the quantitative and physiological role of the metabolism is still an open question, because the nature of the cellular function is the complex dynamics of all metabolites.

The ODE-based model of the comprehensive red blood cell metabolism representing detailed enzyme kinetics was then constructed on the E-Cell simulation platform (Takahashi *et al.*, 2004) The model comprises nearly 100 reactions and around 50 metabolites each of which belongs to glycolysis, pentose phosphate pathway, purine salvage pathway, membrane transports, ion leak processes, ATP (adenosine triphosphate)-dependent $Na+/K+$ pump process, or binding reactions between metabolites (Kinoshita *et al.*, 2007a,b).

The initial role of this model was to predict acute hypoxic response of red blood cell metabolism to understand the cellular states in microcirculation. To this end, the hemoglobin allostery in binding and releasing oxygen was modeled in ODE formalism, as well as the binding reactions between key energy metabolites and distinct forms of hemoglobin with and without oxygen.

In the simulation of hypoxia, the oxygen pressure was set from 100 mmHg (arterial level) to 30 mmHg (venous level), then the predicted dynamics in metabolite concentrations were compared with the experimental data from metabolome analysis. The predicted hypoxia response from the model was not in agreement with the experimental data from metabolome analysis; rather no remarkable dynamics was predicted upon state transition of hemoglobins, in spite of that the model successfully reproduce the metabolic states upon other environmental or genetic changes of enzymes.

Given the discrepancy between the prediction and the validation, the model was then refined and expanded to include the other biochemical events by literature curation focusing on connectivity between hemoglobins and metabolism. It is known that both hemoglobin forms with and without oxygen bind to a membrane anion transporter protein called band 3 in distinct association constants. Separately, it is also proved that three enzymes catalyzing intermediate steps of glycolysis also bind to band 3 membrane protein. The model refinement step was carried out by the curation and implementation of these binding processes with distinct kinetic parameters, that is, binding constants.

The refined model considering band 3 membrane protein successfully predicted the measured dynamics. These results from the simulation–experiment cycle indicated the significance of this protein for the metabolic alteration during hypoxia as well as the hypothesized nature of hypoxic response of metabolism where the hemoglobin state transition is the dominant factor of acute changes of red blood cell metabolism upon hypoxia. Moreover, from the analysis of the model, it was predicted that this hypoxic response is necessary to maintain metabolic resources (ATP and 2,3-bisphosphoglycerate (2,3-BPG)) in red blood cells during hypoxia. These hypotheses were confirmed by the biochemical experiment using carbon monoxide-treated hemoglobins, which can mimic the oxygen-containing form of hemoglobin without oxygen molecule.

This precise and predictive model of red blood cell metabolism has further been applied to predict the residual content of metabolic resources in practical long-term storage (Nishino *et al.*, 2009; Nishino *et al.*, 2013). The kinetic parameters of the model are to be altered by temperature, time, and the storage solution. In the modeling studies, these unknown parameters were estimated by real number genetic algorithm to fit with experimental measurements of time course chances in key metabolite concentrations in preserved red blood cells. As a result of the parameter calibration, the model successfully reproduced the dynamics of intermediate metabolites measured in metabolome analysis. The simulation analysis of the model predicted the fates and roles of additive metabolites in the solution for red blood cell storage, which showed somewhat of a trade-off and interlocking relationships between the benefits and possible side effects.

The predictive models and the validation experiments in this way can help us to generate testable hypothesis through the model refinement cycles and give insights to design cellular behaviors (**Fig. 1**).

Spatial Simulation Techniques

The rapid progress in spatial simulation techniques together with the significant developments in advanced single-molecule measurement such as TIRF (total internal reflection fluorescence) microscopy, and FRET (fluorescence resonance energy transfer) (Liu *et al.*, 2015) enables various applications to study significant cellular phenomena, which are hardly assumed to be spatially uniform.

A conventional way of spatial representation of the cell system is compartmental modeling, where an intracellular space is partitioned into compartments, such as cytoplasm, nucleus, and membrane (**Fig. 3(a)**). Each compartment is assumed to be well-stirred and the molecules are distributed uniformly in it. In the compartmental modeling framework, the systems dynamics, that is, time-dependent changes of molecular concentrations are either deterministically formulated as ODEs or a chemical master equation for the kinetic reactions, or stochastically formulated as a stochastic simulation algorithm typed by the Gillespie method to be numerically calculated for the chemical system taking into account its intrinsic fluctuation.

On the other hand, the mesh model is widely applied to take account of spatial heterogeneity of molecular distribution in the cell (**Fig. 3(b)**). In the mesh models, intracellular space is divided into small mesh or subvolumes (0.1–1 μm, in general). Molecules diffuse between neighboring subvolumes, and react in each subvolume. In the deterministic formalization, the chemical reaction–diffusion system is represented as PDEs, and is solved by finite element method, a widely used and common approach in various scientific fields. Various implementations of stochastic algorithms have been proposed for the mesh model with Gillespie method-based stochasticity (see **Table 1** for the mesoscopic scope). The mesh models with multiple molecules in each voxel are also known as a mesoscopic lattice model. For example, distinct spatial patterns of synaptic response have been studied by mesh modeling of dendrites and spines (Blackwell, 2013).

Furthermore, along with the improvement in spatio-temporal resolutions of measurements, single molecule-level biophysical approaches have emerged where chemical reactions and diffusion processes of individual molecules are represented rather than the number or concentration of each molecule. As contrasted with mesoscopic lattice method, this type of modeling approach where each molecule occupies a molecular-scale voxel (typically few nanometers) is called the microscopic lattice method (**Fig. 3(c)**). In the microscopic lattice models, each molecule diffuses across a lattice, and reacts with another molecule at the collision (Arjunan and Tomita, 2010; Arjunan and Takahashi, 2017). The successful models with this type of approach have shown good agreement with experimental observations even with single-particle tracking (SPT) microscopy (Watabe *et al.*, 2015; Lindén *et al.*, 2016).

Yet another modeling approach for spatio-temporal simulation is the particle method based on Brownian dynamics, which treats molecules explicitly as particles, enabling a nanoscale simulation in continuous space (**Fig. 3(d)**). The Brownian dynamics (BD) method

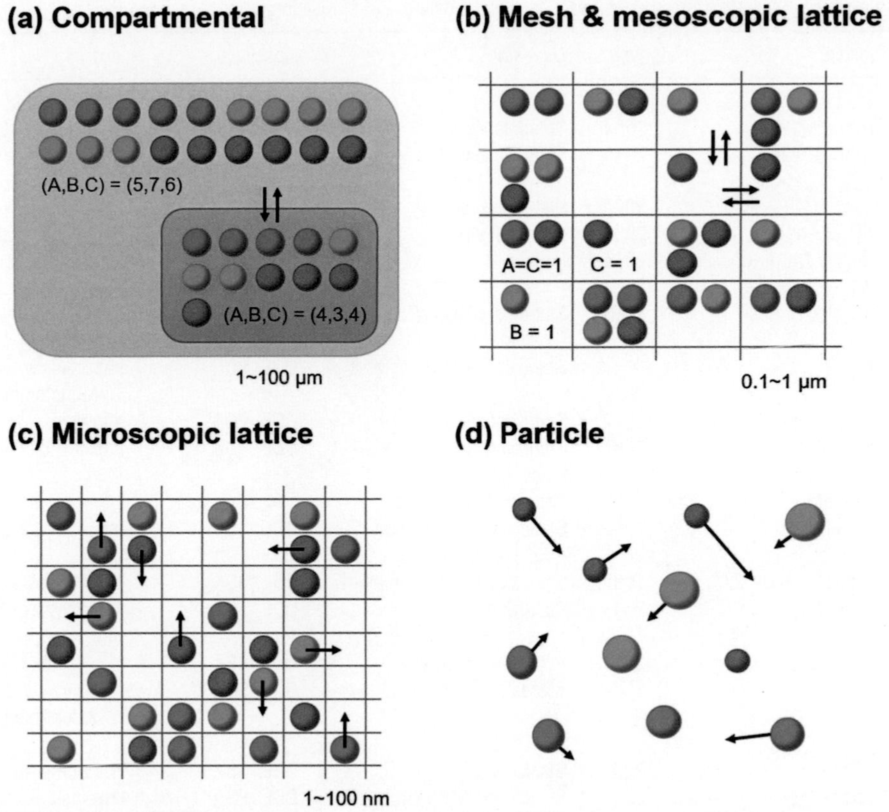

(a) Compartmental

(A,B,C) = (5,7,6)

(A,B,C) = (4,3,4)

1~100 μm

(b) Mesh & mesoscopic lattice

A=C=1   C = 1

B = 1

0.1~1 μm

(c) Microscopic lattice

1~100 nm

(d) Particle

Fig. 3 Spatial representations and simulation approaches. (a) The compartmental model is based on the numbers or concentrations of molecules in well-stirred compartments. The spatial resolution is comparable to the size of cell, nucleus, and organelles (about 1–100 μm). (b) The mesh and mesoscopic lattice models discretize a space into small volumes or areas. Each subvolume can contain multiple molecules in it. Molecules diffuse between and react in subvolumes. The length scale is smaller than a cell but larger than a molecule (0.1–1 μm). (c) The microscopic lattice model consists of small voxels that have the same size with molecules (1–100 nm). A molecule occupies a voxel at the location, and other molecules cannot penetrate there. (d) In the particle model, a molecule is explicitly represented as a particle or a set of particles in continuous space.

generally takes significantly long time to simulate the models and is assumed not applicable to the practical simulations of cell behavior. However, recently developed algorithms with remarkable speed-up have pushed the field towards cellular scale simulation with particle model in several minutes (Schöneberg *et al.*, 2014). As one of the representative particle simulations, Green's function reaction dynamics (GFRD) has accomplished fast and exact simulations of the many-body problem by decomposing it into one- and two-body problems while applying the analytical solution in an event-driven manner (Van Zon and Ten Wolde, 2005; Opplestrup *et al.*, 2006). In the context of the particle simulations at the single-molecule level, various representations of molecules can be applied (Schöneberg *et al.*, 2014). The particle model describes the molecule as a point particle (Andrews, 2017; Plimpton and Slepoy, 2005; Kerr *et al.*, 2008), as a sphere (Takahashi *et al.*, 2010, Klann *et al.*, 2011), or as a collection of spheres resembling molecular structure (Schöneberg and Noé, 2013; Michalski and Loew, 2016; Vijaykumar *et al.*, 2017). For example, GFRD-based particle modeling of ligand-receptor reactions predicted sensing efficiency in bacterial chemotaxis against biological noise (Kaizu *et al.*, 2014).

Tools for Cell Modeling and Simulation

Here we provide an up-to-date list of software and tools for cellular modeling and simulation (**Table 1**). Due to increasing need for integration of multiple models and approaches, extensibility has become a key factor in tool selection. In this context, the license information for each software package or tool has also been provided in the comprehensive list.

Discussion

Given the emergence of high-throughput and high-resolution experimental techniques as well as the demand for predictive science in cell biology, "whole-cell simulation" is now set as an ultimate but intended goal of systems biology (Tomita *et al.*, 1999; Karr *et al.*, 2012). Whole-cell modeling is a set of challenges in various aspects including modeling approaches, software connectivity, and big data handling (Karr *et al.*, 2015; Goldberg *et al.*, 2018).

Table 1   An up-to-date list of available tools and software for cell modeling and simulation

Name	License	Year	Usability	Approach	Access
SBW	BSD License	2002	Built-in?	ODE	http://sbw.sourceforge.net/
Cell Illustrator	Proprietary?	2004	Standalone/GUI	ODE	http://www.cellillustrator.com/home
DBSolveOptimum	Proprietary?	2010	Standalone/GUI	ODE	http://insysbio.com/en/software/db-solve-optimum
JWS Online		2002	Web-based/GUI	ODE	http://jjj.biochem.sun.ac.za/
SloppyCell	BSD License	2007	Library (Python)	ODE	http://sloppycell.sourceforge.net/
PySCeS	Creative Commons Zero v1.0 Universal	2004	Python	ODE	http://pysces.sourceforge.net/
JSim (physiome)	BSD-like License	2013	Standalone/GUI/Web-based (Java)	ODE, PDE	http://www.physiome.org/jsim/
MATLAB (Mathworks Simulink)	Proprietary		Standalone/GUI/Scriptable	ODE, PDE	https://www.mathworks.com/products/matlab.html
PhysioDesigner/FLINT	MIT License	2012	Standalone/GUI	ODE, PDE	http://www.physiodesigner.org/
MOOSE	GPL v3	2007	Standalon e/GU I/Python-scriptable	ODE, PDE	https://moose.ncbs.res.in/
CADLIVE	Proprietary?	2003	Standalone/GUI	ODE, S-system	http://www.cadlive.jp/
E-Cell3	GPL v2	1996	Standalone/GUI/Python-Scriptable	ODE, SSA	http://www.e-cell.org/
COPASI (Gepasi)	Artistic License 2.0	1993	Standalone/GUI/Scriptable?	ODE, SSA	http://copasi.org/
CellDesigner (with COPASI, SOSlib)	Proprietary ?	2003	Standalone/GUI	ODE, SSA	http://www.celldesigner.org/
PySB	BSD 2-clause License	2008	Python-scriptable	ODE, SSA	http://pysb.org/
BioNetGen (RuleBender, BioLab, NFSim)		2004	Standalone/GUI	ODE, SSA	http://www.csb.pitt.edu/Faculty/Faeder/?page_id=409
KaSim (Kappa, KaSa)	LGPL v3	2003	Standalone	ODE, SSA	http://dev.executableknowledge.org/
tellurium (libroadrunner, antimony, phrasedml, libsbml,	Apache 2.0	2014	Standalone/GUI/Library (Python)	ODE, SSA	http://tellurium.analogmachine.org/
BIOCHAM	GPL v2	2003	Web-based/Standalone(Java)/GUI	ODE, SSA, Boolean	https://lifeware.inria.fr/biocham/
BioUML	Open-Source?	2004	Standalone/GUI	ODE, SSA, FBA	http://wiki.biouml.org/index.php/Landing
iBioSim	Apache 2.0	2009	Standalone/GUI	ODE, SSA, FBA	www.async.ece.utah.edu/ibiosim
E-Cell4	GPL v2	2005	Python-Scriptable	ODE, SSA, Meso/Microscop ic, Particle	http://www.e-cell.org/ecell4/
The Virtual Cell	MIT License	1997	Web-based/Standalone (Java)	PDE	http://vcell.org/
FreeFEM	LGPL v2	1987	Scriptable? (own language)	PDE	http://www.freefem.org/
openCOBRA	LGPL/GPL v2	2007	Scriptable (MATLAB, Python Julia)	FBA	https://opencobra.github.io/
BetaWB (BlenX)	Proprietary?	2008	Standalone	SSA	https://sites.google.com/site/aromanel/software/bwb
Bio-PEPA (Bio-PEPA Workbench)	GPL v2	2008	Plugin(Eclipse)/Standalone	SSA	http://homepages.inf.ed.ac.uk/stg/software/biopepa/bpwb.html
SPiM	Microsoft License Agreement (Open-Source, but noncommercial use)	2004	Standalone/GUI	SSA	https://www.microsoft.com/en-us/research/project/stochastic-pi-machine/
BoolNet	Artistic License 2.0	2009	Library (R-lang)	Boolean	http://sysbio.uni-ulm.de/?Software:BoolNet
BooleanNet	MIT License	2007	Python-scriptable	Boolean	https://pypi.python.org/pypi/BooleanNet/1.2.8
ViSiBooL	Open-source?	2017	Standalone (Java)	Boolean	http://sysbio.uni-ulm.de/?Software:ViSiBooL
GINsim	GPL v3	2006	Standalone/GUI (Java)	Boolean	http://ginsim.org/
ePNK	GPL	2012	Eclipse Plugin	Petri nets	http://www.imm.dtu.dk/~ekki/projects/ePNK/index.shtml
WoPeD	LGPL v3	2003	Standalone	Petri nets	http://woped.dhbw-karlsruhe.de/woped/

Table 1   Continued

Name	License	Year	Usability	Approach	Access
MONALISA	Artistic License 2.0	2013	Standalone/GUI (Java)	Petri nets	http://www.bioinformatik.uni-frankfurt.de/tools/monalisa/
GreatSPN	Proprietary, but free for educational use?	1995	Standalone/GUI (Java)	Petri nets	http://www.di.unito.it/~greatspn/index.html
Snoopy	Proprietary, but free for non-commercial use?	2003	Standalone/GUI	Petri nets	http://www-dssz.informatik.tu-cottbus.de/DSSZ/Software/Snoopy
MCell (CellBlender)	GPL v2	1996	Standalone/GUI	Agent-based	http://mcell.org/
FLAME	GNU LGPL?	2006	Standalone	Agent-based	http://flame.ac.uk/
REPAST	BSD-like License	2004	Standalone (GUI)	Agent-based	https://repast.github.io/
URDME (StochSS)	GPL v3	2008	Scriptable (Python/MATLAB)	Mesoscopic	http://www.stochss.org/
MesoRD	GPL v2	2005	Standalone/GUI	Mesoscopic	http://mesord.sourceforge.net/
STEPS	GPL v2	2009	Python-scriptable	Mesoscopic	http://steps.sourceforge.net/STEPS/default.php
Lattice Microbes (pyLM)	University of Illinois Open Source License	2013	Python-scriptable	Mesoscopic	http://www.scs.illinois.edu/schulten/lm/
Simmune	Proprietary	2006	Standalone/GUI	Mesoscopic	https://www.niaid.nih.gov/research/simmune-project
CompuCell3D	Open-Source?	2012	Standalone/Python-Scriptable/GUI	Mesoscopic	http://www.compucell3d.org/
Spatiocyte	GPL v2	2009	Standalone/GUI/Python-scriptable	Microscopic	http://spatiocyte.org/
GridCell	Proprietary, but free for academic use	2007	Standalone/GUI/Web-based	Microscopic	http://www.isip.ece.mcgill.ca/research/gridcell/start
Smoldyn	Mostly LGPL	2003	Standalone/GUI	Particle	http://www.smoldyn.org/
GFRD	GPL v2	2005	Standalone	Particle	http://gfrd.org/
ReaDDy	BSD/3-clause	2011	Python-scriptable	Particle	http://www.readdy-project.org/
ChemCell (Pizza.py)	GPL v2	2005	Standalone/Python-scriptable	Particle	http://chemcell.sandia.gov/
SpringSalad	Open-Source?	2016	Standalone	Particle	http://vcell.org/springsalad-2
CDS	Open-Source?	2010	Standalone/GUI	Particle	https://med.uth.edu/nba/cds/
SRSim	GPL v2	2010	Standalone	Particle	http://www.biosys.uni-jena.de/Members/Gerd+Gruenert/SRSim.html
dReal (dReach)	GPL v3	2012	Standalone	Model Checking	http://dreal.github.io/dReach/
Pathway Logic	GPL v2	2006	Standalone/GUI	Model Checking	http://pl.csl.sri.com/
PRISM	GPL v2	2000	Standalone	Model Checking	http://www.prismmodelchecker.org/
Neuron	The 3-Clause BSD License?	1989	GUI/Python-Scriptable	Nerve Equations	https://www.neuron.yale.edu/neuron/

For example, as discussed in the section of "Simulation Models in Biological Layers," each biological layer has distinct scales of time, space, and quantity, all of which are orchestrated into one cell system. The integration of distinct biological scales so far cannot be done with a single simulation algorithm. In this context, for example, the E-Cell System originally launched in 1996 (Tomita *et al.*, 1997; Tomita *et al.*, 1999) endeavors to model and simulate the distinct spatio-temporal scales at the same time by developing a mathematical framework to combine multi-algorithms and multi-timescales (Takahashi *et al.*, 2004; The E-Cell System version 4).

As in the up-to-date list in **Table 1**, we now have a variety of tools and software, each of which has different focus on applicable biological layers, simulation algorithms, and usabilities. Under these circumstances, the need for unified platform to support multi-tools has become more apparent. From the "soft power" point of view, a standard format of simulation model would increase exchangeability among simulation tools and reproducibility and reusability of the models are secured. The Systems Biology Markup Language (SBML) has been developed and now popularized in the scientific community (Hucka *et al.*, 2003; Hucka *et al.*, 2015; Waltemath *et al.*, 2016). The movement of packaging software for cell modeling and simulation has also come out such as BioNetGen (Blinov *et al.*, 2004) and PySB for Python-based programmatic model construction (Lopez *et al.*, 2013).

The scalability of the simulation models towards whole-cell simulation is largely dependent on (generally experimental) data availability and its accessibility. The complete computational workflow including data collection, data mining, network inference, modeling, simulation, model analysis, and model validation has to be executed to complete the systems biology roadmap and that should be connected with the "big pond" of the experimental data. The complete workflow to integrate whole available knowledge from independent small sciences needs the novel computational framework and connectivity platform (Kitano *et al.*, 2011). Recent advances in the development of such platforms include Garuda (Ghosh *et al.*, 2011), Galaxy (Goecks *et al.*, 2010), GenomeSpace (Qu *et al.*, 2016), and Taverna (Oinn *et al.*, 2004).

All these bottom-up and top-down efforts contributing to whole-cell modeling have opened up opportunities for the systems-level understanding of cells to assess the mechanism of how cell behavior can advance the evolvability of life.

See also: Biological and Medical Ontologies: GO and GOA. Computational Approaches for Modeling Signal Transduction Networks. Computational Systems Biology Applications. Computational Systems Biology. Coupling Cell Division to Metabolic Pathways Through Transcription. Data Formats for Systems Biology and Quantitative Modeling. Gene Regulatory Network Review. Graphlets and Motifs in Biological Networks. Integrative Analysis of Multi-Omics Data. Integrative Bioinformatics. Molecular Dynamics and Simulation. Multi-Scale Modelling in Biology. Natural Language Processing Approaches in Bioinformatics. Network Inference and Reconstruction in Bioinformatics. Pathway Informatics. Prediction of Protein Localization. Prediction of Protein-Protein Interactions: Looking Through the Kaleidoscope. Protein-Protein Interactions: An Overview. Quantitative Modelling Approaches. Visualization of Biomedical Networks

References

Aldridge, B.B., Saez-Rodriguez, J., Muhlich, J.L., Sorger, P.K., Lauffenburger, D.A., 2009. Fuzzy logic analysis of kinase pathway crosstalk in TNF/EGF/insulin-induced signaling. PLOS Computational Biology 5, e1000340.

Andrews, S.S., 2017. Smoldyn: Particle-based simulation with rule-based modeling, improved molecular interaction and a library interface. Bioinformatics 33, 710–717.

Arjunan, S.N.V. & Takahashi, K., 2017. Multi-algorithm particle simulations with spatiocyte. In: Methods in Molecular Biology, vol. 1611, New York, NY: Humana Press, pp. 219–236.

Arjunan, S.N.V., Tomita, M., 2010. A new multicompartmental reaction-diffusion modeling method links transient membrane attachment of E. coli MinE to E-ring formation. Systems and Synthetic Biology 4, 35–53.

Bansal, M., Belcastro, V., Ambesi-Impiombato, A., Di Bernardo, D., 2007. How to infer gene networks from expression profiles. Molecular Systems Biology 3, 78.

Bao, Z., Eddy, S.R., 2003. Automated de novo identification of repeat sequence families in sequenced genomes. Genome Research 13, 1269–1276.

Bhalla, U.S., 2004a. Signaling in small subcellular volumes. I. Stochastic and diffusion effects on individual pathways. Biophysical Journal 87, 733–744.

Bhalla, U.S., 2004b. Signaling in small subcellular volumes. II. Stochastic and diffusion effects on synaptic network properties. Biophysical Journal 87, 745–753.

Blackwell, K.T., 2013. Approaches and tools for modeling signaling pathways and calcium dynamics in neurons. Journal of Neuroscience Methods 220, 131–140.

Blinov, M.L., Faeder, J.R., Goldstein, B., Hlavacek, W.S., 2004. BioNetGen: Software for rule-based modeling of signal transduction based on the interactions of molecular domains. Bioinformatics 20, 3289–3291.

Bordbar, A., Monk, J.M., King, Z.A., Palsson, B.O., 2014. Constraint-based models predict metabolic and associated cellular functions. Nature Reviews Genetics 15, 107–120.

Chai, L.E., et al., 2014. A review on the computational approaches for gene regulatory network construction. Computers in Biology and Medicine 48, 55–65.

Chien, C.T., Bartel, P.L., Sternglanz, R., Fields, S., 1991. The two-hybrid system: A method to identify and clone genes for proteins that interact with a protein of interest. Proceedings of the National Academy of Sciences of the United States of America 88, 9578–9582.

Dunn, S.J., Martello, G., Yordanov, B., Emmott, S., Smith, A.G., 2014. Defining an essential transcription factor program for naïve pluripotency. Science (80-) 344, 1156–1160.

Eldar, A., Elowitz, M.B., 2010. Functional roles for noise in genetic circuits. Nature 467, 167–173.

Elowitz, M.B., Levine, A.J., Siggia, E.D., Swain, P.S., 2002. Stochastic gene expression in a single cell. Science (80-) 297, 1183–1186.

Fell, D.A., 1992. Metabolic control analysis: A survey of its theoretical and experimental development. Biochemical Journal 286, 313–330.

Fields, S., Song, O., 1989. A novel genetic system to detect protein–protein interactions. Nature 340, 245–246.

Fujii, M., Ohashi, K., Karasawa, Y., Hikichi, M., Kuroda, S., 2017. Small-volume effect enables robust, sensitive, and efficient information transfer in the spine. Biophysical Journal 112, 813–826.

Garfinkel, D., Garfinkel, L., Pring, M., Green, S.B., Chance, B., 1970. Computer applications to biochemical kinetics. Annual Review of Biochemistry 39, 473–498.

Ghosh, S., Matsuoka, Y., Asai, Y., Hsin, K.Y., Kitano, H., 2011. Software for systems biology: From tools to integrated platforms. Nature Reviews Genetics 12, 821–832.

Gillespie, D.T., 2007. Stochastic simulation of chemical kinetics. Annual Review of Physical Chemistry 58, 35–55.

Gillespie, D.T., 1977. Exact stochastic simulation of coupled chemical reactions. The Journal of Physical Chemistry 81, 2340–2361.

Glauche, I., Herberg, M., Roeder, I., 2010. Nanog variability and pluripotency regulation of embryonic stem cells – Insights from a mathematical model analysis. PLOS ONE 5, e11238.

Goecks, J., et al., 2010. Galaxy: A comprehensive approach for supporting accessible, reproducible, and transparent computational research in the life sciences. Genome Biology 11, R86.

Goldberg, A.P., et al., 2018. Emerging whole-cell modeling principles and methods. Current Opinion in Biotechnology 51, 97–102.

Heinrich, R., Rapoport, S.M., Rapoport, T.A., 1978. Metabolic regulation and mathematical models. Progress in Biophysics & Molecular Biology 32, 1–82.

Herberg, M., Kalkan, T., Glauche, I., Smith, A., Roeder, I., 2014. A model-based analysis of culture-dependent phenotypes of mESCs. PLOS ONE 9, e92496.

Hey T., Tansley S., and Tolle K., 2009, The Fourth Paradigm: Data-Intensive Scientific Discovery, ISBN 978-0-9825442-0-4.

Hodgkin, A.L., Huxley, A.F., 1952. A quantitative description of membrane current and its application to conduction and excitation in nerve. The Journal of Physiology 117, 500–544. (https://doi.org/10.1113/jphysiol.1952.sp004764).

Hofestadt, R., Thelen, S., 1998. Quantitative modeling of biochemical networks. In Silico Biology 1, 39–53.

Huang, C.Y., Ferrell, J.E., 1996. Ultrasensitivity in the mitogen-activated protein kinase cascade. Proceedings of the National Academy of Sciences of the United States of America 93, 10078–10083.

Hucka, M., et al., 2003. The systems biology markup language (SBML): A medium for representation and exchange of biochemical network models. Bioinformatics 19, 524–531.

Hucka, M., et al., 2015. Promoting coordinated development of community-based information standards for modeling in biology: The COMBINE initiative. Frontiers in Bioengineering and Biotechnology 3, 19.

Ideker, T., Krogan, N.J., 2012. Differential network biology. Molecular Systems Biology 8, 565.

Iwamoto, K., Shindo, Y., Takahashi, K., 2016. Modeling cellular noise underlying heterogeneous cell responses in the epidermal growth factor signaling pathway. PLOS Computational Biology 12, e1005222.

Jerby, L., Shlomi, T., Ruppin, E., 2010. Computational reconstruction of tissue-specific metabolic models: Application to human liver metabolism. Molecular Systems Biology 6, 401.

Johnson, D.S., Mortazavi, A., Myers, R.M., Wold, B., 2007. Genome-wide mapping of in vivo protein–DNA interactions. Science (80-) 316, 1497–1502.

Kaizu, K., et al., 2014. The berg–purcell limit revisited. Biophysical Journal 106, 976–985.

Kaizu, K., et al., 2010. A comprehensive molecular interaction map of the budding yeast cell cycle. Molecular Systems Biology 6, 415.

Kalisky, T., Blainey, P., Quake, S.R., 2011. Genomic analysis at the single-cell level. Annual Review of Genetics 45, 431–445.

Karr, J.R., et al., 2012. A whole-cell computational model predicts phenotype from genotype. Cell 150, 389–401.

Karr, J.R., Takahashi, K., Funahashi, A., 2015. The principles of whole-cell modeling. Current Opinion in Microbiology 27, 18–24.

Kauffman, S., 1969. Homeostasis and differentiation in random genetic control networks. Nature 224, 177–178.

Kerr, R.A., et al., 2008. Fast Monte Carlo simulation methods for biological reaction–diffusion systems in solution and on surfaces. SIAM Journal on Scientific Computing 30, 3126–3149.

Kholodenko, B.N., 2006. Cell-signalling dynamics in time and space. Nature Reviews Molecular Cell Biology 7, 165–176.

Kinoshita, A., *et al.*, 2007a. Roles of hemoglobin allostery in hypoxia-induced metabolic alterations in erythrocytes: Simulation and its verification by metabolome analysis. Journal of Biological Chemistry 282, 10731–10741.

Kinoshita, A., Nakayama, Y., Kitayama, T., Tomita, M., 2007b. Simulation study of methemoglobin reduction in erythrocytes: Differential contributions of two pathways to tolerance to oxidative stress. The FEBS Journal 274, 1449–1458.

Kitano, H., 2002. Systems biology: A brief overview. Science (80-) 295, 1662–1664.

Kitano, H., Ghosh, S., Matsuoka, Y., 2011. Social engineering for virtual 'big science' in systems biology. Nature Chemical Biology 7, 323–326.

Klann, M.T., Lapin, A., Reuss, M., 2011. Agent-based simulation of reactions in the crowded and structured intracellular environment: Influence of mobility and location of the reactants. BMC Systems Biology 5, 71.

Lee, K.H., Park, J.H., Kim, T.Y., Kim, H.U., Lee, S.Y., 2007. Systems metabolic engineering of Escherichia coli for L-threonine production. Molecular Systems Biology 3, 149.

Lee, T.I., *et al.*, 2002. Transcriptional regulatory networks in Saccharomyces cerevisiae. Science (80-) 298, 799–804.

Levine, M., Tjian, R., 2003. Transcription regulation and animal diversity. Nature 424, 147–151.

Lindén, M., Ćurić, V., Boucharin, A., Fange, D., Elf, J., 2016. Simulated single molecule microscopy with SMeagol. Bioinformatics 32, 2394–2395.

Liu, Z., Lavis, L.D., Betzig, E., 2015. Imaging live-cell dynamics and structure at the single-molecule level. Molecular Cell 58, 644.

Lopez, C.F., Muhlich, J.L., Bachman, J.A., Sorger, P.K., 2013. Programming biological models in Python using PySB. Molecular Systems Biology 9, 646.

Marbach, D., *et al.*, 2012. Wisdom of crowds for robust gene network inference. Nature Methods 9, 796–804.

Matsuoka, Y., *et al.*, 2013. A comprehensive map of the influenza A virus replication cycle. BMC Systems Biology 7, 97.

Mettetal, J.T., Muzzey, D., Pedraza, J.M., Ozbudak, E.M., van Oudenaarden, A., 2006. Predicting stochastic gene expression dynamics in single cells. Proceedings of the National Academy of Sciences of the United States of America 103, 7304–7309.

Michalski, P.J., Loew, L.M., 2016. SpringSaLaD: A spatial, particle-based biochemical simulation platform with excluded volume. Biophysical Journal 110, 523–529.

Mikkelsen, T.S., *et al.*, 2007. Genome-wide maps of chromatin state in pluripotent and lineage-committed cells. Nature 448, 553–560.

Morris, M.K., Saez-Rodriguez, J., Sorger, P.K., Lauffenburger, D.A., 2010. Logic-based models for the analysis of cell signaling networks. Biochemistry 49, 3216–3224.

Nishino, T., *et al.*, 2009. In silico modeling and metabolome analysis of long-stored erythrocytes to improve blood storage methods. Journal of Biotechnology 144, 212–223.

Nishino, T., *et al.*, 2013. Dynamic simulation and metabolome analysis of long-term erythrocyte storage in adenine–guanosine solution. PLOS ONE 8, e71060.

Le Novere, N., 2015. Quantitative and logic modelling of molecular and gene networks. Nature Reviews Genetics 16, 146–158.

Oberhardt, M.A., Palsson, B., Papin, J.A., 2009. Applications of genome-scale metabolic reconstructions. Molecular Systems Biology 5, 320.

Oda, K., Kitano, H., 2006. A comprehensive map of the toll-like receptor signaling network. Molecular Systems Biology 2, 2006.0015.

Oda, K., Matsuoka, Y., Funahashi, A., Kitano, H., 2005. A comprehensive pathway map of epidermal growth factor receptor signaling. Molecular Systems Biology 1, E1–E17.

Oinn, T., *et al.*, 2004. Taverna: A tool for the composition and enactment of bioinformatics workflows. Bioinformatics, 20. pp. 3045–3054.

Opplestrup, T., Bulatov, V.V., Gilmer, G.H., Kalos, M.H., Sadigh, B., 2006. First-passage Monte Carlo algorithm: Diffusion without all the hops. Physical Review Letters 97, 230602.

Orth, J.D., Thiele, I., Palsson, B.O., 2010. What is flux balance analysis? Nature Biotechnology 28, 245–248.

O'Brien, E.J., Monk, J.M., Palsson, B.O., 2015. Using genome-scale models to predict biological capabilities. Cell 161, 971–987.

Plimpton, S.J., Slepoy, A., 2005. Microbial cell modeling via reacting diffusive particles. Journal of Physics: Conference Series 16, 305–309.

Qu, K., *et al.*, 2016. Integrative genomic analysis by interoperation of bioinformatics tools in GenomeSpace. Nature Methods 13, 245–247.

Rao, C.V., Wolf, D.M., Arkin, A.P., 2002. Control, exploitation and tolerance of intracellular noise. Nature 420, 231–237.

Reddy, V.N., Liebman, M.N., Mavrovouniotis, M.L., 1996. Qualitative analysis of biochemical reaction systems. Computers in Biology and Medicine 26, 9–24.

Ren, B., *et al.*, 2000. Genome-wide location and function of DNA binding proteins. Science (80-) 290, 2306–2309.

Robertson, G., *et al.*, 2007. Genome-wide profiles of STAT1 DNA association using chromatin immunoprecipitation and massively parallel sequencing. Nature Methods 4, 651–657.

Sasagawa, S., Ozaki, Y.I., Fujita, K., Kuroda, S., 2005. Prediction and validation of the distinct dynamics of transient and sustained ERK activation. Nature Cell Biology 7, 365–373.

Schilling, C.H., Letscher, D., Palsson, B.O., 2000. Theory for the systemic definition of metabolic pathways and their use in interpreting metabolic function from a pathway-oriented perspective. Journal of Theoretical Biology 203, 229–248.

Schmidt, K., Carlsen, M., Nielsen, J., Villadsen, J., 1997. Modeling isotopomer distributions in biochemical networks using isotopomer mapping matrices. Biotechnology and Bioengineering 55, 831–840.

Schöneberg, J., Noé, F., 2013. ReaDDy – A software for particle-based reaction–diffusion dynamics in crowded cellular environments. PLOS ONE 8, e74261.

Schöneberg, J., Ullrich, A., Noë, F., 2014. Simulation tools for particle-based reaction–diffusion dynamics in continuous space. BMC Biophysics 7, 11.

Schuster, S., Fell, D.A., Dandekar, T., 2000. A general definition of metabolic pathways useful for systematic organization and analysis of complex metabolic networks. Nature Biotechnology 18, 326–332.

Shalek, A.K., *et al.*, 2013. Single-cell transcriptomics reveals bimodality in expression and splicing in immune cells. Nature 498, 236–240.

Shlomi, T., Cabili, M.N., Herrgård, M.J., Palsson, B.Ø., Ruppin, E., 2008. Network-based prediction of human tissue-specific metabolism. Nature Biotechnology 26, 1003–1010.

Singer, Z.S., *et al.*, 2014. Dynamic heterogeneity and DNA methylation in embryonic stem cells. Molecular Cell 55, 319–331.

Takahashi, K., Kaizu, K., Hu, B., Tomita, M., 2004. A multi-algorithm, multi-timescale method for cell simulation. Bioinformatics 20, 538–546.

Takahashi, K., Tanase-Nicola, S., ten Wolde, P.R., 2010. Spatio-temporal correlations can drastically change the response of a MAPK pathway. Proceedings of the National Academy of Sciences of the United States of America 107, 2473–2478.

Takahashi, K., Vel Arjunan, S.N., Tomita, M., 2005. Space in systems biology of signaling pathways – Towards intracellular molecular crowding in silico. FEBS Letters 579, 1783–1788.

The E-Cell System version 4: An integrated environment for modeling, simulation and analysis of cell systems with multi-algorithms, multi-timescales and multi-spatial-representations. doi:10.5281/zenodo.1119017.

Tomita, M., *et al.*, 1997. E-cell: Software environment for whole cell simulation. Genome Informatics. Workshop on Genome Informatics 8, 147–155.

Tomita, M., *et al.*, 1999. E-cell: Software environment for whole-cell simulation. Bioinformatics 15, 72–84.

Tyson, J., Othmer, H.G., 1978. The dynamics of feedback control circuits in biochemical pathways. Progress in Theoretical Biology 5, 2–62.

Vijaykumar, A., Ouldridge, T.E., Ten Wolde, P.R., Bolhuis, P.G., 2017. Multiscale simulations of anisotropic particles combining molecular dynamics and Green's function reaction dynamics. Journal of Chemical Physics 146, 114106.

Waltemath, D., *et al.*, 2016. Toward community standards and software for whole-cell modeling. IEEE Transactions on Biomedical Engineering 63, 2007–2014.

Watabe, M., *et al.*, 2015. A computational framework for bioimaging simulation. PLOS ONE 10, e0130089.

Watterson, S., Marshall, S., Ghazal, P., 2008. Logic models of pathway biology. Drug Discovery Today 13, 447–456.

Wynn, M.L., Consul, N., Merajver, S.D., Schnell, S., 2012. Logic-based models in systems biology: A predictive and parameter-free network analysis method. Integrative Biology 4, 1323.

Xia, T., Li, N., Fang, X., 2013. Single-molecule fluorescence imaging in living cells. Annual Review of Physical Chemistry 64, 459–480.

Yachie-Kinoshita, A., *et al.*, 2017. Modeling signaling-dependent pluripotent cell states with Boolean logic can predict cell fate transitions. bioRxiv 14, e7952.

Yim, H., *et al.*, 2011. Metabolic engineering of Escherichia coli for direct production of 1,4-butanediol. Nature Chemical Biology 7, 445–452.

Van Zon, J.S., Ten Wolde, P.R., 2005. Simulating biochemical networks at the particle level and in time and space: Green's function reaction dynamics. Physical Review Letters 94, 128103.

Zupke, C., Stephanopoulos, G., 1994. Modeling of isotope distributions and intracellular fluxes in metabolic networks using atom mapping matrices. Biotechnology Progress 10, 489–498.

Quantitative Modelling Approaches

Filippo Castiglione, National Research Council of Italy, Rome, Italy
Emiliano Mancini, University of Amsterdam, Amsterdam, Netherlands
Marco Pedicini, Roma Tre University, Rome, Italy
Abdul Salam Jarrah, American University of Sharjah, Sharjah, UAE

Glossary

Complexity science A field of study aiming at defining methods and tools to assess and predict the dynamics of a complex system using a holistic approach.

Complex system A hierarchical self-organized system with a large number of components characterized by non-linear connections. In such systems order is an emergent property of its internal microscopic dynamics.

Computation All steps necessary for a digital computer to solve a problem. A computation transforms the input, i.e., the definition of a problem in mathematical terms, into outputs, i.e., the solution of the problem in terms of values of some of the problem's variables.

Dynamics Area of classical mechanics concerned with the study of forces and torques and their effect on motion. More in general, the dynamics of a system describes the trajectory of the constituting variables in a proper mathematical space.

Mathematical modelling Description of a natural systems by means of the mathematical language that is then suitable for transformations and other mathematical treatments to find non-trivial relations among system elements.

Networks Graphical description of the relations among entities that is then viable to mathematical treatments such as those of graph theory.

Simulation computer recreation of the dynamics of natural or artificial systems.

Stochasticity The presence of something that is not determinisitically determined. In mathematics, it is used to indicate the use of a random element in the description of a system.

Introduction

The recent wide availability of experimental high-throughput omics data (genomics, proteomics, metabolomics, etc.) is fostering the development of theoretical biology. This data is analysed by established bioinformatics tools that have been developed in the past decades as well as by novel approaches devised at a great pace by the community. However, the results of such analysis are in some cases inconclusive and further investigation is necessary to make sense of it. This is when mathematical, statistical and computational approaches such as modelling and simulation come into play. This article briefly describes few but well-representative methods that have been used to this purpose in the last couple of decades.

Quantitative modelling of biological phenomena comprise all methods borrowed by other quantitative or *exact* disciplines such as physics, mathematics, computer science or engineering, that are nowadays used to tame the biological complexity. Given the enormous diversity found in biological systems, it is in general difficult to obtain quantitative estimation of parameters of interest and often the results are dependent on the modelling assumptions. Therefore, the methods range between the two extremes of being completely qualitative to perfectly quantitative. Qualitative methods provide an approximate explanation of what is going on in the studied system without being able to estimate values precisely or give clear indications about how to influence the system (Saadatpour and Albert, 2016). On the other hand quantitative methods provide answers to specific questions with numbers, that is, with something that is readily useful in real practice without the need of further (complex) transformations or assumptions (Mogilner *et al.*, 2006).

The usefulness of computational models in systems biology has been pointed out in many recent literature review articles (Bartocci and Lió, 2016; Fisher and Henzinger, 2007; Ji *et al.*, 2017; Machado *et al.*, 2011; Materi and Wishart, 2007). This article, however, aims at providing an informal and not exhaustive introduction to the main computational modelling approaches adopted in systems biology. Besides the classical differential equation approach, this paper presents other computational dynamic approaches such as Petri nets, Boolean networks, cellular automata, agent based models, and the statistical methods of Bayesian networks and formal systems such as process algebras.

Background/Fundamentals

Mathematical modelling has a long tradition and is central to most disciplines of science and engineering. It characterizes the scientific effort in describing nature in quantitative terms. Newton's work which was published in the 17th century, is probably the most striking example of application of the concepts of calculus to describe a physical system (i.e., the motion of planets in terms of differential equations). A model is, today as then, a mathematical description of the elements of a process and the ways in which they influence each other. The mathematical description, sometimes amenable of symbolic transformations, leads to a formulation of the "solution" which reveals aspects of the structure and dynamics of the system that are not plainly recognisable at a first sight. Mathematical (now computational alike) modelling is an iterative process which goes from the real-life phenomenon to the virtual mathematical object called the "model" that produces "results" in terms of how it behaves in hypothetical scenarios.

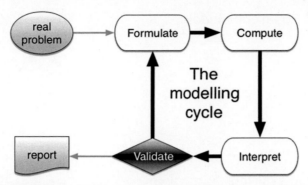

Fig. 1 Mathematical or computational modelling is an iterative process which involves the translation of the most significant characteristics of the real problem into a mathematical formalism amenable of computation (or analytical treatment) to produce results that are then interpreted to be translated back to reality. Those need then to be compared to the real system to be validated so to show the correctness and even usefulness of the model. If this is not the case, the model is improved and the whole cycle is repeated.

To be scientifically sound, these results should be compared to actual data that were produced by the real-life phenomenon and were not used to build or infer the model yet This comparison, which is called model validation, usually identifies shortcomings or problems in the model, advocating some model adjustment, hence the iterative scheme described in **Fig. 1**.

This practice is not different when the system under consideration comprise life. On one side, we have a biological system we wish to understand and on the other side we have a model corresponding to our understanding of the real system that allows us to perform analysis and testing of hypothetical scenarios. This is for instance the case of a model describing the pathogenesis of a disease permitting to test different therapeutic interventions thus enabling the search for the most successful treatment options.

Models can be broadly classified according to their structure. A first important distinction to be made relates to whether or not time is one of the variables of the model. A *dynamic* model accounts for time-dependent changes in the state of the system whereas *static* models only consider the system at equilibrium. Static models are often used to describe the invariable relations among object such as, for instance, proteins, metabolites or genes. A subfield of systems biology called network biology deals with such descriptions in terms of graphs or networks. Dynamical models allow for "forecasting" hypothetical scenarios and are quite valuable, for instance, in finding optimal clinical interventions in modelling of disease progress.

Another distinction concerns the involvement of the concept of randomness among the elements of the model. A model is *deterministic* if the current state of the model is uniquely determined by the model parameters and its previous states. That is, given an initial state, a deterministic model always follows the same "trajectory". In contrast, the trajectories of a *stochastic* model may vary depending on the randomness present, in other words, by the variance of the stochastic variables that are therefore not described by unique values, but rather by probability distributions.

A third important classification criterion is the structure of the space from which the variables of the model take on their values. A variable is called *discrete* if it takes on values from a countable set (finite such as the set $\{0,1\}$ or infinite such as the set of integers), otherwise, it is called *continuous*., that is the case when its values can be arbitrarily close real numbers. If all variables are discrete then the model is said to be discrete. On the other hand, if all variables stand for continuous values then the model is continuous. Finally, hybrid models contain both continuous and discrete components. For instance a model of bacterial growth could represent with $n(t,x)$ the number of bacteria at time t in a certain volume at position x with time and space continuous but state n discrete.

In the following sections, we briefly illustrate different mathematical or computational modelling paradigms classifiable according to the mentioned criteria above. We leave to the reader the leisure of recognising each of them. The paragraphs also report examples of application of the methodologies in the biomedical field with the hope to convey the message of the convenience of quantitative reasoning in these fields. References of relevant works are provided and further references for interested readers are provided at the end of the article.

Differential Equations

Ordinary Differential Equations

To our knowledge one of the earliest work in the literature that used differential equations in biology was the influential and controversial paper by Daniel Bernoulli which was published in 1760. Using a simple model of differential equations, Bernoulli was able to show that the long-term benefits of inoculation overweight the risk of immediate death as it may lead to increasing the life expectancy. For more information about the original Bernoulli's model see Chapter 4 of Bacaër (2011), and, for recent work on Bernoulli's model, see Dietz and Heesterbeek (2002).

For a given biological system of interacting species (substances, chemicals, …) where their concentrations are functions of only one variable (time, age, or cell size,…), an ordinary differential equation (ODE) model of the system is basically a collection of equations, one for each species, describing the rate of change in the concentration of the species, with respect to the variable, in terms of the concentrations of other species in the system, see **Fig. 2**, which presents the well-known Lotka-Volterra or predator-prey model of two interacting quantities x (prey) and y (predator) with the initial concentrations being x_0 and y_0, respectively. This

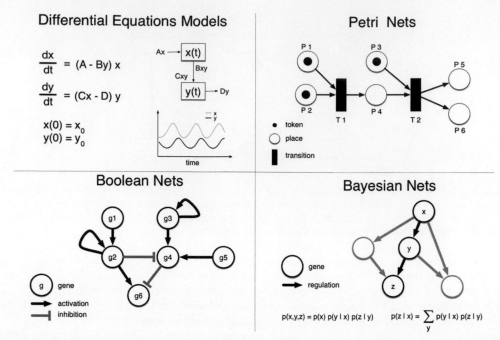

Fig. 2 Differential equations models exemplified in a predator-prey relation. The Petri-nets modelling paradigm is very generic and apt to model concurrency not just in biological systems. It consists in tokens moving among places through transitions epitomising conditions. Boolean networks have been conceived to describe gene regulation but are general enough to be applied to other field of science well. Bayesian networks are statistical methods which allow to estimate the strength (probabilistically speaking) of a relation among two objects such as genes, give a set of data.

model has four parameters, namely A, B, C, and D. For more information about predator-prey models, see Chapter 13 of Bacaër (2011). A similar but more specific example of ODE system applied to clinic is those in Dell'Acqua and Castiglione (2009) and Jarrah *et al.* (2014) about modelling muscular dystrophy.

Values of the parameters of a model are determined based on the available data and using any of several parameter estimation methods such as Genetic Algorithm, Particle Swarm, Simulated Annealing, Steepest Descent, to name few (Moles *et al.*, 2003) and multiple shooting method for stochastic systems (Peifer and Timmer, 2007). For biochemical networks, the parameters could be assumed to be of one of two forms depending on the underlying kinetics: mass action or Hill function which includes Michaelis-Menten as a special case. While Mass Action (polynomial form) is appropriate when the reaction rate among species is proportional to the probability of the collision of the particles of the species, the Hill function in general is used in enzyme kinetics as well as protein's activation or inhibition.

Regardless of the optimization method, parameter estimation is probably the hardest step in the ODE approach for modelling biological systems (Chou and Voit, 2009) mainly due to the large number of species being modelled, their nonlinear interactions, and the fact that available quantitative data are limited. Yet the ODE approach is probably the most popular among all, mainly due to its solid and old mathematical foundation as well as its huge success in modelling a variety of biological systems at all levels; such as the famous population dynamic model predator-prey, Hodgkin–Huxley model of excitable neuron (Hodgkin and Huxley, 1952), and gene regulatory network of yeast cell cycle (Tyson and Novák, 2015) to name a few.

There are many different variations of the ODE approach, in particular, stochastic ODE (Bachar *et al.*, 2013), delay ODE (Parmar *et al.*, 2015). Also the S-systems (Savageau, 1988) is a special class of ODE models that assumes a specific structure of the system.

There are several general ODE solvers as well as specialized standalone software that can be used for parameter estimations and solving ODE. The standalone software CellDesigner (Matsuoka *et al.*, 2014) and COPASI (Hoops *et al.*, 2006) can be used for the modelling and simulation of biochemical networks and both have most of the mentioned methods and features above.

One of the main limitations of the ODE approach is the fact that it assumes that the species are well-mixed, and hence it cannot be used to properly model a system that is space-depended. This can be easily resolved when using partial differential equations.

Partial Differential Equations

This approach enables modelling interacting species as functions of more than one variable. In particular, a partial differential equation (PDE) represents the change in the concentration of a substance with respect to not only time, as most ODE models, but also to other variables such as place (spatial coordinates), age and size of a cell.

Alan Turing developed a simple PDE model of two diffusible and interacting chemicals, and hence the name reaction-diffusion model, and he used the model to explain the pattern formation in the developing animal embryo (Turing, 1952). Since then, the reaction-diffusion model has been the basis of most PDE methods for the PDE modelling and simulation of complex biological patterns (Kondo and Miura, 2010). For a comprehensive treatment of the subject, see Murray (2003).

Modelling methods using the PDE approach have been developed to study various kinds of biological systems at all levels; such as ion singling (Ramay *et al.*, 2010), chemotaxis-driven migration of T-cells toward infection sites (Vroomans *et al.*, 2012), and climate change (Goosse *et al.*, 2010) as well as weather prediction (Steppeler *et al.*, 2003).

Similar to the ODE approach, solving PDE equations is not easy and is usually done using numerical techniques (Tadmor, 2012). Furthermore, methods for the parameter estimations of PDE models have recently been developed using different approaches such as least squares method (Muller and Timmer, 2002) and Bayesian methods (Xun *et al.*, 2013).

Difference Equations

In *Liber Abaci*, which was published in 1202, Fibonacci investigated how fast rabbits could breed. Assuming ideal environment, he developed a model for the number of rabbits using a difference equation and produced the famous Fibonacci's sequence: 1, 1, 2, 3, 5, 8, 13, 21, …. It is worth mentioning that Fibonacci's sequence has appeared in other areas within as well as outside biology. Fast forward to the 20th century, difference equations models of various biological systems are ubiquitous, in particular, the difference equations approach provides a discrete version of population biology in general (Cull *et al.*, 2005). Also, finite difference equations are used to develop numerical methods for solving ODE and PDE systems (Sewell, 2006). Furthermore, difference equations approach provide a generalization as well as a mathematical framework for cellular automata (Itoh and Chua, 2009) and other finite discrete methods. See the classical textbook (Elaydi, 2005) for more information on the theory of difference equations, and Allman and Rhodes (2004) and Cull *et al.* (2005) for examples of how the difference equations approach is used to model several biological systems such logistic growth and phylogenetic trees. Difference equations are best for studying discrete-time, population dynamics with stochastic features (Witten and de la Torre, 1984).

Petri Nets

Petri nets (PN) are directed bipartite graphs with nodes belonging to one of the two sets called states (also called places) and transitions (see **Fig. 2**). They were conceived by Carl Adam Petri in 1962 to describe chemical reactions (Petri and Reisig, 2008). Indeed, the states indicate substances and the transitions the reactions. Nodes identifying states are filled with tokens to indicate that the relative substance is present (alternatively, that the substance is present in sufficient concentration to allow the reaction to take place). When a transition has more than one incoming arcs (called *take* arcs) linked to corresponding states, those states all have to contain sufficient tokens to allow the transition to fire, meaning that the reaction takes place and the token is moved ahead in the states linked downward the exit arcs (called *give* arcs) (see **Fig. 2**). In this manner, the states code for conditions that need to be satisfied for a transition (or more generically an action) to take place. The flexibility of this modelling formalism allow its use to describe concurrent process in distributed systems. It has indeed been extensively used in computer science to describe distributed processing. In systems biology, Petri nets and its variants, developed to remove constraints such as the use of a discrete number of tokens (i.e., the coloured PN), or the use of deterministic transitions (i.e., the stochastic PN), have been employed to model biochemical networks (Chaouiya, 2007; Goss and Peccoud, 1998; Srivastava *et al.*, 2001) such as metabolic processes (Simao *et al.*, 2005) and cell signalling (Li *et al.*, 2007). The description of a biological process such as those at the molecular level just mentioned, is quite straightforward with PN. Moreover, given the intrinsic account for the temporal scale, such formalism is quite appealing for biologist that can analyse, mainly in a qualitative fashion but in principle also in a quantitative manner, complex biological phenomena by answering what-if questions such as "what will happen if the production of a molecular compound is inhibited?"

Boolean Networks

Boolean networks (BN) have been introduced by Kauffman in 1969 to model gene regulation (Kauffman, 1969, 1993). They represent genes as nodes of a graph with two types of arcs, activation or inhibition, denoting respectively an activation of, say, gene $g2$ subsequent to the activation of gene $g1$, and inhibition, indicating the capacity of an express gene to block the transcription of another gene (see **Fig. 2**). The model describes the temporal evolution, carried out in discrete steps, of the active/inactive states of the set of genes. At each time step the state of each gene is determined by a Boolean logical rule which is a function of the state of the genes which impinge upon it (i.e., its regulators). For instance, for the network in **Fig. 2**, the Boolean formula for node g4 could be $g4(t+1) = g3(t)$ *AND* $g5(t)$ *AND NOT* $g2(t)$.

The state of the genes can be updated all at once (synchronous update) or one at a time with a specified or unspecified order (asynchronous update). The two evolutions are not equivalent with the latter being much more complex to compute than the former. The space of the possible trajectories of the set of gene-expression grows exponentially with the number of genes thus its exploration to determine the dynamical properties of the BN becomes quickly unfeasible. On the other hand, the analysis of network properties such as its robustness or the steady-states of its dynamics which identify specific biological functions, are exactly the reward of using such models (Li *et al.*, 2004).

Boolean networks are generally inferred from time-series of gene expression data (D'haeseleer *et al.*, 2000) or are constructed by expert manual curation from literature data (Pedicini *et al.*, 2010). Boolean networks have also been employed to model signalling pathways (Saez-Rodriguez *et al.*, 2007). For these biological applications the limitation of the two-values logic has been removed in the multi-valued extension (Schaub *et al.*, 2007). Moreover, to account for the presence of noise and uncertainty, the stochastic extensions of probabilistic Boolean networks were introduced (Shmulevich *et al.*, 2002).

Bayesian Networks

A Bayesian network is a statistical model describing a set of variables and their conditional dependencies via a directed acyclic graph (see **Fig. 2**). The nodes of the network represent random variables whereas the edges stand for dependencies between couple of variables. The value of each variable is determined by a probabilistic function of the variables associated to the nodes than impinges upon it.

Bayesian networks were introduced by the work of Pearl (Pearl, 1988) as generic probabilistic methods to infer relationships among variables given a set of data. In fact, based on the formula from Bayes, there exist learning methods to infer the probability parameters also in case of incomplete data. Data coming from gene regulation activity provide a natural example of applicability of Bayesian networks. In this case the nodes-variables are the expression levels of the genes and the probabilistic relationships identified by the edges are estimated by means of the inference algorithms (Friedman, 2004). Data from signalling networks is amenable to this representation and analysis (Sachs *et al.*, 2005). The disadvantage due to the inability to model feedback loops present in biological networks has been removed in an extension called dynamic Bayesian networks (Husmeier, 2003; Kim *et al.*, 2003; Zou and Conzen, 2005).

Microscopic Ab-initio Models

Models in this category are usually discrete (both in time and space) dynamic models that aim to simulate/address biological problems for which spatial properties strongly affects the resulting dynamics. Discrete models are better suited to capture the discrete nature of individual cells than continuous models, especially when dealing with complex systems involving many variables such as sets of cells in different stages of their life cycle or distinguishable by the affinity of their receptors for different antigens. Cell-based models such as cellular automata, cellular Potts and agent-based models are able to represent the behaviour and interactions of each individual cell in a biological system. In these models, the macroscopic dynamics of a system is an emergent property of the microscopic interactions and for this reason they are considered bottom-up modelling approaches.

The Ising model (1925) is the simplest theoretical model of ferromagnetism used to study ferromagnetic phase transitions and is considered by many the precursor of cellular automata and agent based models. In the Ising model each magnetic dipole moment of atomic spins is represented as a discrete variable characterized by one of two states ($+1$ or -1) and arranged on a lattice. The interactions between spins are local and each spin is allowed to change synchronously based on the spin state of its neighbours.

Cellular Automata

Cellular automata (1940s) are a general computational model characterized by discrete space, state, and time (Ilachinski, 2001; Toffoli and Margolus, 1987; Wolfram, 2002) inspired in the first half of the 20th century by the work of John von Neumann (von Neumann, 1966) and Stanislaw Ulam (Ulam, 1952). A cellular automata model is a discrete dynamic system defined by a grid (a finite metric space), a finite set of possible cells states, a function describing the neighbourhood of a cell (the set of cells influencing the one under consideration) and a transition function determining the state of a cell in the next time step depending on the state of the cells in its neighbourhood (see **Fig. 3**).

Classical cellular automata are synchronous (all cells update their state simultaneously thus the grid state is independent from the order of update of the grid), local, and homogeneous. Many modifications of the classical model exist: asynchronous automata, heterogeneous automata, non-uniform cellular automata, stochastic automata and many others such as movable cellular automata (Psakhie *et al.*, 1995), Brownian cellular automata (Lee and Peper, 2010).

Cellular automata have been extensively used as a modelling paradigm in biology (Ermentrout and Edelstein-Keshet, 1993). Some applications include the study of idiotypic B cell proliferation in the human immune system (De Boer *et al.*, 1992) or the dynamics of infectious diseases (Zorzenon dos Santos and Coutinho, 2001).

Cellular Potts models, also known as Glazier Graner-Hogeweg models (Graner and Glazier, 1992; Glazier and Graner, 1993), are an evolution of cellular automata in which each cell occupies more than one lattice site or pixel. For this reason, each cell has a shape which varies over time as a function of a Hamiltonian representing the adhesion energy of the cell and other mechanical constraints such as surface-area-to-volume ratio. One of the most known framework for cellular Potts model is CompuCell3D (Chaturvedi *et al.*, 2005; Popławski *et al.*, 2008) which has been used in several different applications, from anatomical and pathological conditions of tissues and organs to models of single-species and multi-species biofilm growth. In this hybrid framework, the mechanical properties of the cell membrane can be integrated with continuous models of reaction-diffusion dynamics and models of intracellular signalling dynamics achieving a complete and realistic multi-scale model of a complex biological system.

Agent-Based Models

Agent-based models (ABM) are an evolution of the simpler cellular automata model in which each agent is an autonomous decision-making entity interacting on the basis of a given set of rules which are not necessarily identical (Bonabeau, 2002; Gilbert, 2008). Strictly speaking, ABMs are not discrete in time and space as they are exemplary of hybrid systems where discrete time variables are combined with continuous state-space ones. Nevertheless, in their applications to biology and medicine and with the purpose to speed up execution on a computer, agents are usually placed on a structure (a two dimensional or three-dimensional lattice representing real-world spatial environment such as a lymph node or a complex network representing social contacts or sexual interactions) that defines the couples or groups of agents that are going to interact at a given time. Agent-based models have

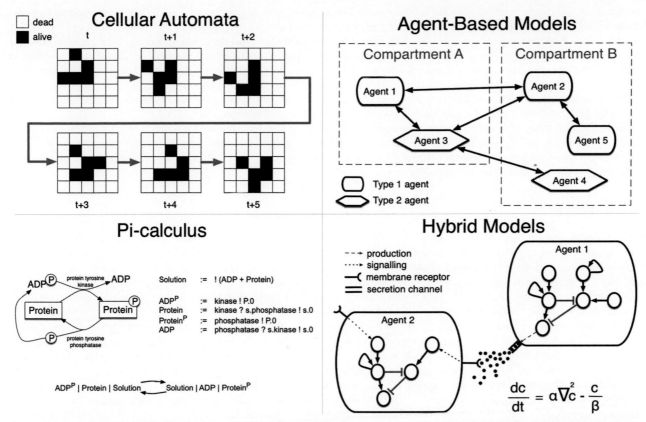

Fig. 3 Cellular automata and agent-based models represent entities individually. These evolve according to defined rules whose complexity range from rather simple (often Boolean) functions in cellular automata to sophisticated algorithms for the agents' behaviour in agent-based models. Furthermore, as an example, one can construct complex hybrid models embedding Boolean networks in agents representing biological cells to set up the rule for its differentiation. Pi-calculus is a different approach used to formally describe a complex system such as a chemical reactor and to conduct (in the case of stochastic calculi, e.g., stochastic k-calculus) simulations much like in the previously described methodologies.

been used to study many different problems. For instance, they have been used to simulate HIV or gonorrhoea epidemics where agents represent individuals part of a sexual contact network (Mei *et al.*, 2011) or to describe the population dynamics of ecosystems in which each agent is an animal or a plant. They have also being used to model the intracellular dynamics phenomena in which agents are viral particles moving within the cytoplasm of an infected cell. Finally, they have been extensively used to describe the dynamics of the immune system. In this case the agents represent all the entities involved in the immune response to a pathogen such as immune cells, epithelial cells, bacteria, viral particles etc (Castiglione and Celada, 2015).

Several frameworks can be used to implement agent based models: NETLOGO, FLAME, MASON, JASON and REPAST are some of the most versatile and supported frameworks.

Process Calculi

The design of a computational implementation of any biological system starts from the knowledge of the living system under consideration and by abstracting its main characteristics. This task is fulfilled by means of some formal method such as those describing the syntax of a computer language. The most striking feature of biological phenomena that is difficult to capture in symbolic formal methods is concurrency. Formal systems, therefore, need to be able to fully denote such biological complexity or otherwise turn out inaccurate and eventually useless.

Process calculi or process Algebras (PA) are formal systems devised to easily express parallel computations and their properties (like Church's lambda-calculus introduced to express Turing computable arithmetic functions). The most successful system was Milner's pi-calculus introduced to express interactions between concurrent processes (Milner, 1989). PAs share some similarity with formalism used in the description of chemical reactions (see **Fig. 3** for the example of the phosphorylation/dephosphorylation reaction expressed in pi-calculus). This was precisely established with the introduction of the Chemical Abstract Machine or CHAM (Berry and Boudol, 1992), where the states of the parallel machine are chemical solutions and molecules can interact according to reaction rules.

Biological phenomena involve millions of moving particles in any single cell while millions of cells form tissues which transport substances, transform energy use signals, etc. Notably, biological systems process information necessary to the organization of life by the same mechanisms they use to processes any other element, like the nutrients. Process algebras are considered suitable formalisms to model these kinds of systems because the transmitted information is processed *concurrently* and *asynchronously*. Moreover, part of essential transactions which happen in biological systems are inherently computational and therefore they can be expressed with formal languages used to describe concurrent processes.

While representing biological systems with PAs, several researchers realised that there was a missing dimension: quantitative aspects were neglected in the description of the interaction rules. The stochastic pi-calculus (Priami, 1995) was an enriched Milner's formalism with a quantitative context semantics giving access to the world of simulations of biochemical processes (Regev *et al.*, 2001; Priami *et al.*, 2001). Further developments were obtained by deriving new formal systems by adding missing features to PAs: bio-ambients (Cardelli and Gordon, 2000) and brane-calculi (Cardelli, 2005) allowing compartments, the kappa-calculus (Danos and Laneve, 2004) with the specific purpose of representing protein interactions.

The prominent aspect to remark regarding the convenience for using this approach in describing biological processes is that it enables, for instance, the syntactic and semantic comparison (hence prompt translation of the properties of the first to the second system) between two concurrent systems once formalised with a PA or to build incrementally complex process from a multitude of elementary processes being either independent or interrelated.

Closing Remarks

Computational methods are used to make sense of the vast amount of data produced by experimental biology high throughput technologies and, importantly, to "fill the dots" where data is incomplete in revealing the underlying structure or in describing the temporal evolution of the system under study. Mathematical or computational modelling methods are applied and even developed anew to analyse such data, to dig information out of it, to discover structures, to generalize conclusions, to provide scenarios to choose from, etc. The accuracy of these computational approaches ranges from low (that is when we call the approach qualitative) to high (i.e., that is quantitative) depending on the data availability.

The methodologies range from classical differential equations to more unconventional agent-based simulation, passing through process algebra and statistical approaches. These models have shown to be promising if not even concretely useful in some cases, revealing biological aspects of the phenomena under consideration which were impossible to see just by inspecting the data. Vice versa, biology is providing a number of challenging problems to appraise the methods themselves, encouraging computational scientists to extend and improve existing methodologies (Machado *et al.*, 2011).

One of the greatest challenge in this respect is the construction of multi-scale models embracing the wide range of time and length scales of biological phenomena. Whereas few modelling techniques though theoretical able to handle different scales (e.g., process algebras or Petri nets), their practical use is impaired by the combinatorial explosion of variants eventually leading to computational intractable problems. In this case the obvious trick is to combine two or (rarely) more approaches that have their strength in different time/length scale into a unified hybrid multi-scale system (Fisher and Henzinger, 2007; Bortolussi and Policriti, 2008; Fromentin *et al.*, 2010). This is for instance the case of agent-based models of cellular systems where each agent/cell is equipped with a system of ordinary differential equations or a Boolean network to describe the regulatory network among genes driving its behaviour (see **Fig. 3**).

Computational biology is a flourishing field of applied mathematics and computer science. Only in the past two decades the collective effort of computational scientists combined with the expertise of field biologist have produced a number of open-software that offer the model techniques described in this brief article to analyse experimental data and/or to simulate hypothetical scenarios. Review papers such as Bartocci and Lió (2016) offer a nice view of what is available at the time of that publication. There is no reason to believe that such vigorous effort in the systems biology community will grow smaller, but rather just the opposite, new tools will be available soon, either as improvements of existing methodologies or able to exploit and combine different formalisms and modelling paradigms to provide powerful platforms for the analysis of the ever increasing experimental and diagnostic data.

Acknowledgement

Abdul Jarrah was supported by the National Science Foundation under Grant No. DMS-1440140 in the Fall 2018 while the author was in residence at the Mathematical Sciences Research Institute in Berkeley, California, USA. Filippo Castiglione acknowledge partial support from the Institute for Advanced Studies of the University of Amsterdam. Emiliano Mancini acknowledges partial support from the SimCity project of the Dutch NWO, eScience agency under contract C.2324.0293.

See also: Cell Modeling and Simulation. Computational Systems Biology Applications. Computational Systems Biology. Multi-Scale Modelling in Biology. Natural Language Processing Approaches in Bioinformatics. Pathway Informatics

References

Allman, E., Rhodes, J., 2004. Mathematical Models in Biology: An Introduction. New York: Cambridge University Press.
Bacaër, N., 2011. A Short History of Mathematical Population Dynamics. London: Springer-Verlag.
Bachar, M., Batzel, J., Ditlevsen, S., 2013. Stochastic Biomathematical Models: With Applications to Neuronal Modeling. Heidelberg: Springer.
Bartocci, E., Lió, P., 2016. Computational modeling, formal analysis, and tools for systems biology. PLOS Computational Biology 12 (1), e1004591. doi:10.1371/journal.pcbi.1004591.
Berry, G., Boudol, G., 1992. The chemical abstract machine. Theoretical Computer Science 96, 217–248.

Bonabeau, E., 2002. Agent-based modeling: Methods and techniques for simulating human systems. Proceedings of the National Academy of Sciences of the United States of America 99, 7280–7287.

Bortolussi, L., Policriti, A., 2008. Hybrid systems and biology. In: Bernardo, Marco, Degano, Pierpaolo, Zavattaro, Gianluigi (Eds.), Formal Methods for Computational Systems Biology. vol. 5016 of LNCS. Springer, pp. 424–448.

Cardelli, L., Gordon, A.D., 2000. Mobile ambients. Theoretical Computer Science 240 (1), 177–213.

Cardelli, L., 2005. Brane Calculi. In: Danos, V., Schachter, V. (Eds.), Computational Methods in Systems Biology (CMSB 2004). Lecute Notes in Computer Science 3082. Berlin, Heidelberg: Springer, pp. 257–278.

Castiglione, F., Celada, F., 2015. Immune System Modeling and Simulation. Boca Raton, FL, USA: CRC Press.

Chaouiya, C., 2007. Petri net modelling of biological networks. Briefings in Bioinformatics 8, 210–219.

Chaturvedi, R., Huang, C., Kazmierczak, B., et al., 2005. On multiscale approaches to three-dimensional modeling of morphogenesis. Journal of the Royal Society Interface 2, 237–253.

Chou, C., Voit, E.O., 2009. Recent developments in parameter estimation and structure identification of biochemical and genomic systems. Mathematical Biosciences 219, 57–83.

Cull, P., Flahive, M., Robson, R., 2005. Difference Equations: From Rabbits to Chaos. New York: Springer.

Danos, V., Laneve, C., 2004. Formal molecular biology. Theoretical Computer Science 325 (1), 69–110.

De Boer, R.J., Segel, L.A., Perelson, A.S., 1992. Pattern formation in one- and two-dimensional shape-space models of the immune system. Journal of Theoretical Biology 155 (3), 295–333.

Dell'Acqua, G., Castiglione, F., 2009. Stability and phase transitions in a mathematical model of Duchenne muscular dystrophy. Journal of Theoretical Biology 260 (2), 283–289. doi:10.1016/j.jtbi.2009.05.037.

Dietz, K., Heesterbeek, J.A.P., 2002. Daniel Bernoulli's epidemiological model revisited. Mathematical Biosciences 180, 1–21.

D'haeseleer, P., Liang, S., Somogyi, R., 2000. Genetic network inference: From co- expression clustering to reverse engineering. Bioinformatics 16 (8), 707–726.

Elaydi, S., 2005. An Introduction to Difference Equations. New York: Springer.

Ermentrout, G.B., Edelstein-Keshet, L., 1993. Cellular automata approaches to biological modeling. Journal of Theoretical Biology 160 (1), 97–133.

Fisher, J., Henzinger, T.A., 2007. Executable cell biology. Nature Biotechnology 25 (11), 1239–1249. doi:10.1038/nbt1356.

Friedman, N., 2004. Inferring cellular networks using probabilistic graphical models. Science 303 (5659), 799–805.

Fromentin, J., Eveillard, D., Roux, O., 2010. Hybrid modeling of biological networks: Mixing temporal and qualitative biological properties. BMC Systems Biology 4, 79. doi:10.1186/1752-0509-4-79. (PMID: 20525331).

Gilbert, N., 2008. Agent-Based Models. Los Angeles: Sage Publications.

Glazier, J.A., Graner, F., 1993. Simulation of the differential adhesion driven rearrangement of biological cells. Physical Review E 47, 2128–2154.

Goosse, H., Barriat, P.Y., Lefebvre, W., Loutre, M.F., Zunz, V., 2010. Chapter 3: Modelling the Climate System of Introduction to Climate Dynamics and Climate Modeling. Online textbook available at http://www.climate.be/textbook.

Goss, P.J., Peccoud, J., 1998. Quantitative modeling of stochastic systems in molecular biology by using stochastic Petri nets. Proceedings of the National Academy of Sciences of the United States of America 95, 6750–6755.

Graner, F., Glazier, J.A., 1992. Simulation of biological cell sorting using a two-dimensional extended Potts model. Physical Review Letters 69 (13), 785–790.

Hodgkin, A.L., Huxley, A.F., 1952. A quantitative description of membrane current and its application to conduction and excitation in nerve. The Journal of Physiology 117, 500–544.

Hoops, S., Sahle, S., Gauges, R., et al., 2006. COPASI: A complex pathway simulator. Bioinformatics 22, 3067–3074.

Husmeier, D., 2003. Sensitivity and specificity of inferring genetic regulatory interactions from microarray experiments with dynamic Bayesian networks. Bioinformatics 19 (17), 2271–2282.

Ilachinski, A., 2001. Cellular Automata: A Discrete Universe. River Edge, NJ, USA: World Scientific Publishing Company.

Itoh, M., Chua, L., 2009. Difference equations for cellular automata. International Journal of Bifurcation and Chaos 19, 805–830.

Jarrah, A.S., Castiglione, F., Evans, N.P., Grange, R.W., Laubenbacher, R., 2014. A mathematical model of skeletal muscle disease and immune response in the mdx mouse. BioMed Research International. 871810. doi:10.1155/2014/871810.

Ji, Z., Yan, K., Li, W., Hu, H., Zhu, X., 2017. Mathematical and computational modeling in complex biological systems. BioMed Research International. 5958321. doi:10.1155/2017/5958321.

Kauffman, S., 1969. Metabolic stability and epigenesis in randomly constructed genetic nets. Journal of Theoretical Biology 22, 437–467.

Kauffman, S.A., 1993. The Origins of Order: Self-organization and Selection in Evolution. New York: Oxford University Press.

Kim, S., Imoto, S., Miyano, S., 2003. Inferring gene networks from time series microarray data using dynamic Bayesian networks. Briefings in Bioinformatics 4 (3), 228–235.

Kondo, S., Miura, T., 2010. Reaction-diffusion model as a framework for understanding biological pattern formation. Science 329, 1616–1620.

Lee, J., Peper, F., 2010. Efficient computation in Brownian cellular automata. In: Peper, F., Umeo, H., Matsui, N., Isokawa, T. (Eds.), Natural Computing. Proceedings in Information and Communications Technology 2. Tokyo: Springer.

Li, C., Ge, Q.W., Nakata, M., Matsuno, H., Miyano, S., 2007. Modelling and simulation of signal transductions in an apoptosis pathway by using timed Petri nets. Journal of Biosciences 32, 113–127.

Li, F., Long, T., Lu, Y., Ouyang, Q., Tang, C., 2004. The yeast cell-cycle network is robustly designed. Proceedings of the National Academy of Sciences of the United States of America 101 (14), 4781–4786.

Machado, D., Costa, R.S., Rocha, M., et al., 2011. Modeling formalisms in systems biology. AMB Express 1, 45. doi:10.1186/2191-0855-1-45.

Materi, W., Wishart, D.S., 2007. Computational systems biology in drug discovery and development: Methods and applications. Drug Discovery Today 12 (7–8), 295–303. doi:10.1016/j.drudis.2007.02.013.

Matsuoka, Y., Funahashi, A., Ghosh, S., Kitano, H., 2014. Modeling and simulation using CellDesigner. In: Miyamoto-Sato, E., Ohashi, H., Sasaki, H., Nishikawa, J., Yanagawa, H. (Eds.), Transcription Factor Regulatory Networks. Methods in Molecular Biology (Methods and Protocols), vol. 1164. New York: Humana Press.

Mei, S., Quax, R., van de Vijver, D., Zhu, Y., Sloot, P.M.A., 2011. Increasing risk behaviour can outweigh the benefits of antiretroviral drug treatment on the HIV incidence among men-having-sex-with-men in Amsterdam. BMC Infectious Diseases 11, 118.

Milner, R., 1989. Communication and Concurrency. Upper Saddle River, NJ, USA: Prentice Hall, Inc., (ISBN:0-13-115007-3).

Mogilner, A., Wollman, R., Marshall, W.F., 2006. Quantitative modeling in cell biology: What is it good for? Developmental Cell 11, 279–287.

Moles, C.G., Mendes, P., Banga, J.R., 2003. Parameter estimation in biochemical pathways: A comparison of global optimization methods. Genome Research 13, 2467–2474.

Muller, T., Timmer, J., 2002. Fitting parameters in partial differential equations from partially observed noisy data. Physical Review 171, 1–7.

Murray, J., 2003. Mathematical Biology. Berlin: Springer-Verlag.

Parmar, K., Blyuss, K.B., Kyrychko, Y.N., Hogan, S.J., 2015. Time-delayed models of gene regulatory networks. Computational and Mathematical Methods in Medicine 2015, 16. doi:10.1155/2015/347273. (Article ID 347273).

Pearl, J., 1988. Probabilistic Reasoning in Intelligent Systems: Networks of Plausible Inference. San Francisco: Morgan Kaufmann Publishers Inc.

Pedicini, M., Barrenas, F., Clancy, et al., 2010. Combining network modeling and gene expression microarray analysis to explore the dynamics of Th1 and Th2 cell regulation. PLOS Computational Biology 6 (12), e1001032. doi:10.1371/journal.pcbi.1001032.

Pcifer, M., Timmer, J., 2007. Deterministic inference for stochastic systems using multiple shooting and a linear noise approximation for the transition probabilities. IET System Biology 1, 78–88.

Petri, C.A., Reisig, W., 2008. Petri net. Scholarpedia 3 (4), 6477. Available at: http://www.scholarpedia.org/article/Petri_net.

Popławski, N.J., Shirinifard, A., Swat, M., Glazier, J.A., 2008. Simulation of single-species bacterial-biofilm growth using the Glazier-Graner-Hogeweg model and the CompuCell3D modeling environment. Mathematical Biosciences and Engineering 5 (2), 355.

Priami, C., 1995. Stochastic pi-calculus. Computer Journal 38 (7), 578–589.

Priami, C., Regev, A., Shapiro, E., Silverman, W., 2001. Application of a stochastic name-passing calculus to representation and simulation of molecular processes. Information Processing Letters 80 (1), 25–31.

Psakhie, S.G., Horie, Y., Korostelev, S.Y., et al., 1995. Method of movable cellular automata as a tool for simulation within the framework of mesomechanics. Russian Physics Journal 38 (11), 1157–1168.

Ramay, H., Jafri, M.S., Lederer, W.J., Sobie, E.A., 2010. Predicting local SR Ca^{2+} dynamics during Ca^{2+} wave propagation in ventricular myocytes. Biophysical Journal 98, 2515–2523.

Regev, A., Silverman, W., Shapiro, E., 2001. Representation and simulation of biochemical processes using the π-calculus process algebra. In: Altman, R.B., Dunker, A.K., Hunter, L., Klein, T.E. (Eds.), Pacific Symposium on Biocomputing 6. Singapore: World Scientific Press, pp. 459–470.

Saadatpour, A., Albert, R., 2016. A comparative study of qualitative and quantitative dynamic models of biological regulatory networks. EPJ Nonlinear Biomedical Physics 4, 5. doi:10.1140/epjnbp/s40366-016-0031-y.

Saez-Rodriguez, J., Simeoni, L., Lindquist, J., et al., 2007. A logical model provides insights into T cell receptor signaling. PLOS Computational Biology 3 (8), e163.

Sachs, K., Perez, O., Pe'er, D., Lauffenburger, D., Nolan, G., 2005. Causal protein-signaling networks derived from multi-parameter single-cell data. Science 308 (5721), 523–529.

Savageau, M.A., 1988. Introduction to S-systems and the underlying power-law formalism. Mathematical and Computer Modelling 11, 546–551.

Schaub, M.A., Henzinger, T.A., Fisher, J., 2007. Qualitative networks: A symbolic approach to analyze biological signaling networks. BMC Systems Biology 1, 4.

Sewell, G., 2006. The Numerical Solution of Ordinary and Partial Differential Equations. Chicester, United Kingdom: John Wiley & Sons.

Shmulevich, I., Dougherty, E.R., Zhang, W., 2002. From Boolean to probabilistic Boolean networks as models of genetic regulatory networks. Proceedings of the IEEE 90 (11), 1778–1792.

Simao, E., Remy, E., Thieffry, D., Chaouiya, C., 2005. Qualitative modelling of regulated metabolic pathways: Application to the tryptophan biosynthesis in E. coli. Bioinformatics 21 (2), 190–196.

Srivastava, R., Peterson, M.S., Bentley, W.E., 2001. Stochastic kinetic analysis of the Escherichia coli stress circuit using sigma(32)-targeted antisense. Biotechnology and Bioengineering 75, 120–129.

Steppeler, J., Hess, R., Schattler, U., Bonaventura, L., 2003. Review of numerical methods for nonhydrostatic weather prediction models. Meteorology and Atmospheric Physics 82 (1–4), 287–301.

Tadmor, E., 2012. A review of numerical methods for nonlinear partial differential equations. Bulletin of the American Mathematical Society 49, 507–554.

Toffoli, T., Margolus, N., 1987. Cellular Automata Machines: A New Environment for Modeling. Cambridge, MA, USA: MIT Press.

Turing, A., 1952. The chemical basis of morphogenesis. Philosophical Transactions of the Royal Society of London Series B Biological Sciences 237, 37–72.

Tyson, J., Novák, B., 2015. Models in biology: Lessons from modeling regulation of the eukaryotic cell cycle. BMC Biology 13, 46 doi:10.1186/s12915-015-0158-9.

Ulam, S., 1952. Random processes and transformations. In: Proceedings of the International Congress of Mathematicians, pp. 264–275. Rhode Island: American Mathematical Society.

von Neumann, J., 1966. The Theory of Self-reproducing Automata. Urbana, IL: University of Illinois Press.

Vroomans, R.M.A., Marée, A.F.M., de Boer, R.J., Beltman, J.B., 2012. Chemotactic migration of T cells toward dendritic cells promotes the detection of rare antigens. PLOS Computational Biology 8, e1002763.

Witten, M., de la Torre, D., 1984. Biological populations obeying difference equations: The effects of stochastic perturbation. Journal of Theoretical Biology 111, 493–507.

Wolfram, S., 2002. A New Kind of Science. Champaign, IL: Wolfram Media, Inc.

Xun, X., Cao, J., Mallick, B., Carroll, R.J., Maity, A., 2013. Parameter estimation of partial differential equation models. Journal of the American Statistical Association 108 (503), doi:10.1080/01621459.2013.794730.

Zorzenon dos Santos, R.M., Coutinho, S., 2001. Dynamics of HIV infection: A cellular automata approach. Physical Review Letters 87, 168102.

Zou, M., Conzen, S., 2005. A new dynamic Bayesian network (DBN) approach for identifying gene regulatory networks from time course microarray data. Bioinformatics 21 (1), 71–79.

Further Reading

Castiglione, F., Celada, F., 2015. Immune System Modeling and Simulation. Boca Raton: CRC Press.

Ciocchetta, F., Hillston, J., 2009. Bio-PEPA: A framework for the modelling and analysis of biological systems. Theoretical Computer Science 410 (33–34), 3065–3084.

Deutsch, A., Dormann, S., 2005. Cellular Automaton Modeling of Biological Pattern Formation: Characterization, Applications, and Analysis. Boston: Birkhäuser.

Elaydi, S., 2005. An Introduction to Difference Equations. New York: Springer.

Ellner, S.P., Guckenheimer, J., 2006. Dynamic Models in Biology. New Jersey: Princeton University Press.

Gilbert, N., 2008. Agent-Based Models. Los Angeles: Sage Publications.

Ilachinski, A., 2001. Cellular Automata: A Discrete Universe. River Edge, NJ, USA: World Scientific Publishing Company.

Kauffman, S.A., 1995. At Home in the Universe: The Search for Laws of Self-organization and Complexity. Oxford: Oxford University Press.

Laneve, C., Tarissan, F., 2007. Simple calculus for proteins and cells. Electronic Notes in Theoretical Computer Science 171, 139–154.

Milner, R., 1999. Communicating and Mobile Systems: The π-Calculus. Cambridge: Cambridge University Press.

Murray, J., 2003. Mathematical Biology. Berlin: Springer-Verlag.

Biographical Sketch

Filippo Castiglione is researcher at the Istituto per le Applicazioni del Calcolo of the National Research Council of Italy and adjunct professor of Computational Biology at the Department of Mathematics and Physics of Roma Tre University. He graduated in computer science at the University of Milan, Italy, and got a PhD in Scientific Computing at the University of Cologne, Germany. He has been PostDoc at the Institute for Medical Bio-Mathematics in Tel Aviv, Israel and visiting research fellow at the IBM - T.J. Watson Research Center, Yorktown Heights (NY), at the Department of Molecular Biology at Princeton University and at the Department of Cell Biology, Harvard Medical School, Boston. Filippo has published one book and about 100 reviewed research papers among journals, books and conferences proceedings. He is the main author of the C-ImmSim agent-based simulation model of the immune system. Filippo has been involved as principal investigator in two EU-funded projects of the ICT for Health of the 6th Framework Programme (FP6) and has coordinate a project of the FP7. His research interests range from the study of complex systems to the modeling of biological systems, machine learning and high-performance computing.

Emiliano Mancini is programme developer on Health Systems Complexity at the Institute for Advanced Study in Amsterdam, where he coordinates the research in various academic disciplines for interdisciplinary research projects. In this role, he also engages private companies and stakeholders for the valorisation and utilization of these academic projects as well as collaborate on the strategic plan for the growth of the Institute. Emiliano obtained his Master degree in Physics at University of Rome Tor Vergata and a PhD in Computational Science at University of Amsterdam simulating infectious diseases (HIV and H1N1) using agent-based modeling. As a postdoc Emiliano has worked between University of Amsterdam and the Nanyang Technological University in Singapore on modeling the dynamics of several complex biological systems. He worked as a consultant or team member of several start-up companies and has been involved in the development of a clinical decision support platform for HIV treatment and on a non-invasive continuous glucose meter funded by SMART-MIT. Emiliano was also responsible for the organization and supervision of several experiments on crowd dynamics for the Kumbh Mela Experiment: an Indo-Dutch project to study the Kumbh Mela, the largest religious event in the world.

Marco Pedicini is professor of Computer Science at the Department of Mathematics and Physics of Roma Tre University. He received his Ph.D. in Mathematics (Mathematical Logic and Theoretical Computer Science) from the University of Paris 7 in 1998. He was researcher at Italian National Research Council (CNR). Since his PhD thesis, Marco has developed ideas in TCS and his interests include logic in computer science, cryptography, computer security, parallel and distributed computing, computational methods for systems biology, computational number theory, and more recently quantum computing. Although his whole activity could be classifiable in TCS it finds in interdisciplinary research its own specificity. Interactions with other disciplines are the key to evaluate his scientific activity: in fact, he developed a very special kind of expertise while coping deep theoretical results with practical issues in advanced applications in collaboration with colleagues of different disciplines. Marco has published 30 scientific papers, many of them in leading international journals and selective peer reviewed conferences. In order to witness a broad area in Marco interests, these papers not only appeared on top conferences and journals in logic for computer science but also on top journals covering application of computer science to biology and medicine.

Abdul Salam Jarrah is a Professor of Mathematics and a faculty member of the Master of Science in Biomedical Engineering Programme at the American University of Sharjah (AUS) in the United Arab Emirates. Before joining the AUS in 2009, he was a senior research scientist at the Virginia Bioinformatics Institute and Affiliate Assistant Professor at the Department of Mathematics in Virginia Tech, USA. Prior to that, he was as an Assistant Professor of Applied Mathematics at East Tennessee State University, USA. In addition, Abdul was a member of the Mathematical Sciences Research Institute at Berkeley, USA from August 2016 to January 2018. He received his PhD in Mathematics from New Mexico State University in 2002. His current research interests span many areas of mathematics including discrete dynamical systems, discrete homology theory of graphs, computational algebra, and their applications in biology, especially the modeling and simulation of biological systems. He has made several contributions to these fields through many publications in international journals.

Data Formats for Systems Biology and Quantitative Modeling

Martin Golebiewski, Heidelberg Institute for Theoretical Studies (HITS gGmbH), Heidelberg, Germany

Standards for Computational Systems Biology

Standards support communities with a basis for mutual understanding and information exchange and thereby shape our everyday life. Common standards are indispensable for collaborative work. An example is the Baltimore fire in 1904, when more than 1500 buildings were destroyed even though fire departments from Washington, Philadelphia, New York, and other towns provided aid, as they found their hose couplings did not match the city's fire hydrants due to a lack of standardization. This reduced the impact of help significantly and led afterwards to a national standard for hose couplings and fire hydrants in the USA (Seck and Evans, 2004). A similar breakdown of collaborative work due to a lack of standardization, however, could happen in any field that requires interfacing between different parts, especially in research fields such as in the modern life sciences with their increasing flood and complexity of data and all the vast variety of technologies that have to be brought together like in a jigsaw puzzle.

Standardization of data and their documentation is all the more important for highly interdisciplinary and collaborative fields such as systems biology and systems medicine, where computational models are created that can assist with understanding, describing, analyzing, and prediction of biological systems and their functions. Over the last 20 years computational modeling has grown from a niche activity in systems biology into a standard tool in the life sciences, as documented by the increasing number of published models stored in public model repositories (Li et al., 2010; Yu et al., 2011). Model-driven design, testing, analysis, and optimization of biological systems are performed computationally by integrating a variety of heterogeneous data together within mathematical frameworks in order to understand and predict the complex and dynamic, often nonlinear molecular interactions within cells, as well as the interplay of cells within tissues, organs, organisms, and beyond that even the interaction of different organisms in whole biological ecosystems. Because of the interdisciplinary approach in systems biology these datasets are coming from different sources, derived from different experimental setups, and correspond to various technologies applied to collect the data.

A computational model in systems biology, thus, typically relies on a collection of varied data, which can include, for example, data from genomics, transcriptomics, proteomics, or metabolomics experimental setups, reaction kinetics information, and imaging data. This requires a consistent structuring and description of data and computer models with their underlying experimental or modeling processes, comprising the standardized description of applied methods, biological material and workflows for data processing, analysis, exchange, and integration (e.g., into computational models), as well as of the setup, handling, and simulation of the resulting models. Hence, standards for formatting and describing experimental data, applied workflows, and computer models have become important, especially for data integration across the biological scales for multiscale approaches. Such a consistent documentation based on standards ensures that the data and corresponding metadata (data describing the data and its context), as well as models, methods, and visualizations are structured and described in a "FAIR" manner: Findable, Accessible, Interoperable, and Reusable (Wilkinson et al., 2016).

Systems biology standards are not static but develop and evolve with the progress of science and technology. A major challenge is to harmonize the standardization efforts in the different fields that refer to different approaches and technologies and to make the corresponding standards interoperable. The standards used for encoding and annotating models and related data are designed by experts with an understanding of what key information will comprise the outcome of an experiment, and how this information is best structured and formatted. The development of such standards for systems biology typically occur at a grassroots level driven by the corresponding scientific communities themselves with their need for exchanging data and models. The COMBINE (see "Relevant Websites section") network ("COmputational Modeling in BIology NEtwork" (Hucka et al., 2015b; Myers et al., 2017)), for example, is a consortium of groups involved in the development of open community standards and formats used in computational modeling in biology. It was formed in 2009 following the observation that many standardization efforts shared similar goals and sometimes even involved the same individuals. The network helps foster greater interaction and awareness of the activities in different standards' development, which encourages the federated projects to develop standards that are more likely to be interoperable and less likely to overlap substantially than if the efforts proceeded separately. Building on the experience of mature standards, which already have stable specifications, software support, user bases, and community governance, it also supports emerging efforts aimed at filling gaps or addressing new needs in the overall interoperability landscape. The COMBINE initiative published the first collection of systems and synthetic biology standards as a special issue of the *Journal of Integrative Bioinformatics* (*JIB*) in 2015 (Schreiber et al., 2015). Since then a regular special issue of *JIB* serves as both an overview of existing standards as well as an update of the current state of standards in the domain (Schreiber et al., 2016, 2018).

Given the diversity of research topics in computational biology, and the different facets of modeling, multiple specific standards, with particular purposes, are needed – also beyond the COMBINE standards – in order to formalize and serialize the data (Stromback et al., 2007; Klipp et al., 2007). In the following, several standards for specific tasks in systems biology and related fields will be introduced, in order to provide examples for typical workflows and the standards that can be applied therein.

Standards for Data Used for Model Construction

The first step in generating a model is collating the datasets that need to be integrated into the model. This task heavily relies on the datasets being formatted and annotated correctly. Reporting and annotation checklists, or "minimum information guidelines" as they are often called, for different types of data derived from a big variety of laboratory technologies are readily cataloged for the biosciences as part of MIBBI (Minimum Information for Biological and Biomedical Investigations (Taylor *et al.*, 2008)). Such guidelines are widely accepted for some data types, for example, for the description of microarray experiments (MIAME, Minimum Information About a Microarray Experiment (Brazma *et al.*, 2001)), genome sequences (MIGS, Minimum Information about a Genome Sequence (Field *et al.*, 2008)), or proteomics experiments (MIAPE, Minimum Information About a Proteomics Experiment (Taylor *et al.*, 2007)). The FAIRsharing community makes a wide range of minimum information checklists available for researchers in the MIBBI FAIRsharing collection (see "Relevant Websites section"). Finding comprehensive lists about formatting standards for different types of data can be difficult, and not all formatting standards or annotation checklists are still maintained, or fully usable. To reduce the time required to identify standards that are maintained and usable, FAIRDOM (see "Relevant Websites section") maintains a FAIRsharing collection of formatting standards, and checklists that are known to be actively used within the systems biology community (see "Relevant Websites section"). This collection is based on a community survey commissioned by Infrastructure for Systems Biology Europe (ISBE) (see "Relevant Websites section") in 2015 (Stanford *et al.*, 2015).

Experimentally obtained kinetic data that describe the dynamics of biochemical reactions (e.g., metabolic or signaling reactions within a cellular network), for instance, can be obtained from databases such as SABIO-RK (Wittig *et al.*, 2012) (see "Relevant Websites section"). This manually curated database contains data that are either extracted by hand from the scientific literature (Wittig *et al.*, 2014) or directly submitted from laboratory experiments (Swainston *et al.*, 2010). The database content includes kinetic parameters in relation to the reactions and biological sources, as well as experimental conditions under which they were obtained, and is fully annotated to other resources, making the data easily accessible via web interface or web services for integration into the computational models. Through export of the data (together with its annotations and SABIO-RK identifiers for tracing back to the original dataset) in standardized data exchange formats like SBML (Systems Biology Markup Language) (Hucka *et al.*, 2003) this allows the direct data integration into models.

Standards for Structuring Models of Biological Systems

Model representation formats standardize the structured encoding of biological models. Examples mainly used for models consisting of molecular entities and describing their interactions and dynamical interplay include:

- SBML (Systems Biology Markup Language) as standardized interchange format for computer models of biological processes (Hucka *et al.*, 2003; Hucka *et al.*, 2018).
- CellML, a standard format to store and exchange reusable, modular computer-based mathematical models (Cuellar *et al.*, 2003).
- NeuroML (Neural Open Markup Language), which allows standardization of model descriptions in computational neuroscience (Gleeson *et al.*, 2010).

All three are based on XML (Extensible Markup Language) (see "Relevant Websites section"), a markup language that defines a set of rules for encoding documents in a format that is both human-readable and machine-readable. These formats represent the structure of a model (e.g., the biochemical network) and enable annotation of the model to better convey a model's intention. Here the focus will be on two of the most widely used standards, SBML and CellML.

SBML (Systems Biology Markup Language)

The Systems Biology Markup Language (SBML) (Hucka *et al.*, 2003) is a machine-readable representation format for computational models in systems biology. This format began its development in 2000 as a way of encoding models for the exchange between modeling software platforms and ensuring longevity of the models beyond the software used to create it. The evolution of SBML proceeds in stages in which each "Level" is an attempt to achieve a consistent language at a certain level of complexity. The current version of SBML is Level 3 Version 2 (Hucka *et al.*, 2018). SBML Level 3 is modular, with the core usable in its own right and Level 3 packages being additional "layers" that add features to the core. By itself, core SBML Level 3 is well suited to representing such things as classical metabolic models and cell signaling models, involving well-mixed substances and spatially homogeneous compartments where they are located. Other model types can also be expressed using SBML's core constructs, but SBML Level 3 packages that extend the core and are optional in their use, add more natural support for such types and additional model features, such as visualizations (Gauges *et al.*, 2015), constraint-based models (Olivier and Bergmann, 2015; Olivier and Bergmann, 2018), hierarchical model composition (Smith *et al.*, 2015), or grouping of elements (Hucka and Smith, 2016).

SBML models are decomposed into explicitly labeled constituent elements. A valid model may consist of various user-defined elements, for example, substances, products and modifiers involved in processes, or compartments where they are located. Models are not cast directly into a specific form such as differential equations. Many SBML models represent biochemical reaction networks by reactants, called *species* in SBML, that participate in *reactions*, as a substrate, a product, or a modifier (e.g., a *catalyst* of an enzyme catalyzed reaction or an *inhibitor* or *activator*) and that are part of a *compartment*, for example, a cellular substructure or a defined cell. Mathematical descriptions of the reactions within the model, such as the kinetic rate equations or kinetic constants, are formed using *functions*, *units*, *parameters*, *assignments*, and *kinetic definitions*. The mathematical system can be enhanced using *constraints*, *rules*, and *events*, which can mimic biological phenomena such as feed-source constraints, fixed substrate ratios if needed (e.g., ATP, ADP, AMP balances), nutrient pulse experiments, and more. Every major element carries an identifier, and MathML (see "Relevant Websites section") is used to encode mathematical descriptions of reactions.

In 2010 the SBML community developed an annotation scheme (Hucka *et al.*, 2015a; Hucka *et al.*, 2018) based on the Resource Description Format (RDF (Lassila and Swick, 1999)) of the guideline standard "Minimum information requested in the annotation of biochemical models" (MIRIAM (Le Novere *et al.*, 2005)) and also based on unambiguous identifiers retrieved from the corresponding registry (Juty *et al.*, 2012). This annotation framework enables users to describe their models, and include standard identifiers, so that models are more reusable and exchangeable. SBML is supported by hundreds of different software tools for reading, writing, simulating, editing, and otherwise working with models encoded in the standard, as well as by many databases and other resources (see "Relevant Websites section").

CellML

CellML (Cuellar *et al.*, 2003) is a description language to define models of cellular and subcellular processes. At its core, CellML defines lightweight XML constructs that group mathematical relationships within modules. The variables used in mathematics are defined within each module and connections between variables in different modules can be specified. It began development in 1998, completely independently of SBML, in order to support the construction and exchange of cardiac cell models. The first stable specification for CellML 1.0 was released in 2001, at the time of writing the current stable version is CellML 1.1 (Cuellar *et al.*, 2003), but a public draft version 2.0 has been available since May 2017.

Primarily used for describing systems of differential algebraic equations, CellML provides the syntactic constructs to describe mathematical equations in a modular framework called components. One of the key defining features of CellML is its ability to support component-based modeling, allowing models to import other models, or subparts of models, therefore strongly encouraging their reuse (Lloyd *et al.*, 2004; Wimalaratne *et al.*, 2009a) and facilitating a modularized modeling approach. A CellML model typically consists of components, which may contain variables and mathematics that describe the behavior of that component. CellML provides means to reuse and group components into hierarchical structures. Similar to SBML, all entities (elements) carry an identifier and mathematical definitions are encoded using MathML. The mathematical model is considered to be the primary data and biological context is provided by annotating the variables and equations with metadata using Resource Description Format (RDF (Lassila and Swick, 1999)). All numerical values and variables used in a CellML document are required to unambiguously define their physical units that can be defined document-wide, or specifically for certain components. Due to the requirement for physical units, numerical quantities can vary between modules and software is expected to convert units automatically. A CellML metadata standard allows for biological and biophysical annotation of the models (Beard *et al.*, 2009; Wimalaratne *et al.*, 2009b). A wide range of software supports CellML (Garny *et al.*, 2008; Wimalaratne *et al.*, 2009a) and includes software for modeling, visualization, simulation, validation, and conversion. All the software tools can be viewed on the CellML website (see "Relevant Websites section").

Both CellML and SBML facilitate semantic annotations (e.g., links to controlled vocabularies and ontologies), which further describe the biological meaning of single certain elements (Courtot *et al.*, 2011). Controlled vocabularies and ontologies are detailed later in this article. Neither CellML nor SBML encode what is done with a model nor the results of doing something with it (e.g., simulating it) – these are aspects addressed by other standards such as SED-ML (see below).

Standard for Visualizing Models: The Systems Biology Graphical Notation

When it comes to visualizing networks, each researcher has their own ideas of how the different biological components should be illustrated. The style of diagram can be affected by many factors, including the researchers' backgrounds, and what elements of the diagram they want to emphasize for their work. This creates bias, and makes creating useful visuals challenging; particularly because each researcher also interprets these figures differently (Kitano *et al.*, 2005). In the same way that electronic circuit diagrams have been standardized in electrical engineering for many decades, consistency and uniformity can be brought to the graphical expression of network diagrams in biology by defining common standards for visualization. The Systems Biology Graphical Notation (SBGN (Le Novere *et al.*, 2009)) standardizes the visual notation used to depict biological networks and processes. The use of such a standard visual notation is vital to ensure that diagrams are unambiguous and consistent. SBGN was first developed in 2006, and is comprised of three languages, which allow biological networks to be viewed from different perspectives and at different levels of detail:

(1)   The process description (PD) variant visualizes biochemical interactions as processes over time (Moodie *et al.*, 2015). As the process map extends from left to right, time evolves, allowing the exact order of reactions to be represented. It is normal in this diagram for the same entity to appear multiple times. PD can describe each process in a network in great detail (e.g., biochemical reaction, binding/unbinding of proteins, and the like) and is useful to represent chemical kinetics models. However, some biological phenomena entail a combinatorial explosion of possible interrelated states, making them extremely difficult to depict at this level of detail (**Fig. 1**).

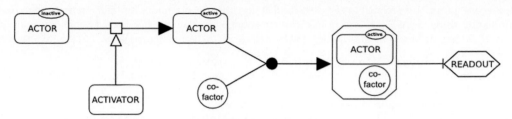

Fig. 1   Example of a SBGN process description map. provided by Nicolas Le Novère, UK.

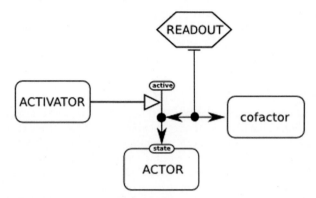

Fig. 2   Example of a SBGN entity relationship map. provided by Nicolas Le Novère, UK.

Fig. 3   Example of a SBGN activity flow map. provided by Nicolas Le Novère, UK.

(2)   The entity relationship (ER) variant visualizes all interactions and relationships between entities, for a given entity (Sorokin *et al.*, 2015). Thereby ER abstracts away the notion of time and focuses on depicting only the relationships between elements, independent of each other. It is useful to represent rule-based models, among other things (**Fig. 2**).

(3)   The activity flow (AF) variant visualizes the flow of information between biological entities (Mi *et al.*, 2015). So, AF maps focus on the influences between elements rather than the actual processes and are especially useful for representing qualitative models (**Fig. 3**).

In each instance, a standardized set of entities, and relations visualize the model in a so-called map. A standardized markup language (SBGN-ML) can be used to independently encode the map (van Iersel *et al.*, 2012). SBGN can be used to visualize data and models in SBML and CellML formats, among others.

Standard for Recording Simulation Environments and Setups: The Simulation Experiment Description Markup Language

Consistent and standardized descriptions and formatting of models are not the only requirements for improved reproducible modeling results. Published models are often validated using a set of simulations, and results published from the model are the outcome of a set of virtual experiments. The number of options available to simulate models in different ways, however, can range from diverse algorithms that can run, for example, time courses, steady-states, metabolic control analysis, parameter sensitivities and others, to fixing of various cellular concentrations as precondition, or introducing stepwise changes of variables at specific time-points. This information about a simulation setup can be captured in a method description, but mistakes can be made in recording what steps were taken, and in reimplementing the steps. The Simulation Experiment Description Markup Language (SED-ML (Waltemath *et al.*, 2011b)) was designed to record such descriptive information necessary to rerun a model, such that it can be exported from one simulation tool, and imported into another. These standardized descriptions ensure that virtual experiments, when applied to a computational model, reproduce a given result. So, SED-ML allows the exact simulation setup to be configured and rerun. The first official version was released in 2011. Similarly to SBML, SED-ML evolves in Levels and Versions. The standard is not specific to any simulation software, or modeling format. It is widely used to exchange simulation experiments in computational biology (Olivier and Bergmann, 2015).

SED-ML files record information related to the checklist that describes the minimum information required to understand and reproduce a simulation study, MIASE (Minimum Information About a Simulation Experiment (Waltemath *et al.*, 2011a)). This typically consists of five major blocks of information (Waltemath *et al.*, 2011a; Bergmann *et al.*, 2015):

(1) Reference to the models being used in the simulation;
(2) Descriptions of modifications applied to the model before simulation (e.g., the initialization of the variables);
(3) Descriptions of the simulation steps, including the configuration of the software tool or numerical algorithm;
(4) Descriptions of the post-processing of result data after simulation; and
(5) Specifications of the results, including definition of plots and numerical reports.

Libraries to read and write SED-ML are provided by the community and some software tools already consume and export SED-ML files (see "Relevant Websites section"), for example, COPASI (Hoops *et al.*, 2006) or JWS Online (Olivier and Snoep, 2004; Snoep and Olivier, 2003), or Tellurium (Choi *et al.*, 2016). SED-ML elements can also be linked to semantic annotations (Courtot *et al.*, 2011). Simulation Experiment Description Markup Language files can be linked to model descriptions in other formats, notably SBML or CellML, to ensure reproducibility of experiments presented in scientific publications. Such links can, for example, be instantiated via the provision of files in a COMBINE archive (Bergmann *et al.*, 2014), or through provision via public model repositories such as BioModels Database (Li *et al.*, 2010) or the Physiome Repository (Yu *et al.*, 2011).

Recording Modeling Results in a Standardized Way

Similarly to models and simulation setups, communicating the results for in silico modeling experiments for exchange, validation, and reuse can be difficult. Results of modeling in the life sciences typically include numerical values, which may also be turned into figures. Numerical results are usually tables or matrices, encoded in CSV (comma-separated values) (Shafranovich, 2005). The flexibility of CSV files means that software and tooling that produces the data can structure and format the data in many different orientations. However, the same data are often structured and labeled differently by each software. This variation hampers ease of exchange, validation, and reuse. Early, generic approaches to providing structure and schema definitions to CSV-like files includes fielded text (Klink, 2016), which is XML-based annotations that describe the header fields of a CSV file. Whilst useful, the amount of life science-specific information that can be used to describe the data is limited. To improve the level of annotation, and to ensure consistent formatting of results data, the Systems Biology Results Mark-up Language (SBRML (Dada *et al.*, 2010)) was developed in 2010. It is specialized for encoding simulation results obtained by running an SBML model. Semantic annotations can be included, with SBRML able to include terms from any ontology that may be relevant to the results. Developing software to analyze the results is also improved by the consistent data structure that can be expected, improving validation and reuse of the data. SBRML is currently being used as a basis to develop a more general results exchanging format NuML (Numerical Markup Language) (see "Relevant Websites section"). The associated library libNUML provides software support for reading, writing, and manipulating data in NuML format on all operating systems (see "Relevant Websites section").

Metadata, Controlled Vocabularies, and Ontologies for Adding Semantic Information

Sustainable model reuse requires a basic understanding of (1) the biological background, (2) the modeled system, and (3) possible parameterizations under different conditions (Scharm and Waltemath, 2016). This knowledge can be transferred to end-users and computers by using metadata in the form of semantic annotations. Metadata is data about data – it clarifies the intended semantics of the biological data, improving understanding of their scope and validity. Through this it improves the shareability and interoperability of the model (Gennari *et al.*, 2011) – especially computationally. Machine-readable annotations can automate exchange, validation, reuse, composition of models and to convert machine-readable code into human-readable formats (Finney *et al.*, 2006; Misirli *et al.*, 2016; Swainston and Mendes, 2009; Rodriguez *et al.*, 2016). The semantic layer can also be exploited to convert machine-readable code into human-readable formats, such as PDF documents (Dräger *et al.*, 2009; Shen *et al.*, 2010) or visualizations (Junker *et al.*, 2012; Shen *et al.*, 2010) to aid human comprehension.

The metadata can comprise of many things, from free text descriptions, through to inclusion of more formal representations of knowledge like controlled vocabularies and ontologies (Rosse and Mejino, 2003; Bard and Rhee, 2004). Controlled vocabularies are an organized arrangement of precise terms and definitions for a given research domain. They work similarly to a taxonomy system or hierarchical classifications, defining given objects by is_a relationships e.g., ethanol is_a primary alcohol. Ontologies are similar to controlled vocabularies, but define more comprehensive relationships between objects (e.g., part_of, has_role). Along with other descriptions, these comprise the information that would be used to semantically annotate data and models, promoting their reuse (Misirli *et al.*, 2016). Terms from controlled vocabularies and ontologies are linked to the entities of a project using semantic technologies, such as the Resource Description Framework (see "Relevant Websites section") (Lassila and Swick, 1999). There is software available for embedding semantic annotations within spreadsheets, improving semantic adoption among laboratory scientists (Wolstencroft *et al.*, 2011; Maguire *et al.*, 2013).

Especially important ontologies in the domain of modeling in systems biology are (see "Relevant Websites section"):

- Systems Biology Ontology (SBO) (Courtot *et al.*, 2011) designed for models in the domain of systems biology,
- Kinetic Simulation Algorithm Ontology (KiSAO) (Courtot *et al.*, 2011) designed for specifying simulation algorithms and their parameterizations,

- TErminology for the Description of DYnamics (TEDDY) (Courtot *et al.*, 2011) designed for dynamical behaviors and results,
- Just Enough Results Model (JERM) (Wolstencroft *et al.*, 2013) designed to be a minimal descriptor of key information required for ensuring reproducibility of systems biology experiments.

There are also a number of other ontologies and controlled vocabularies that are commonly used within computational systems biology for describing the content and environmental context of models. These include for example, the Gene Ontology (Ashburner *et al.*, 2000), which provides information on genes and their molecular function; the NCBI Taxonomy (Federhen, 2012), which provides information about the nomenclature of organisms; the Protein Ontology (Natale *et al.*, 2014), which provides information about proteins; ChEBI (Degtyarenko *et al.*, 2008), which provides information about chemical compounds of biological interest. In addition, using the vCard ontology (Iannella and McKinney, 2014) it is possible to relate to people and organizations. Using the COMODI ontology it is also possible to encode knowledge about differences between computational models and versions thereof (Scharm *et al.*, 2016).

There are a number of guidelines and checklists available in order to support researchers in annotating their scientific results. For example, the initiative "Minimum Information for Biological and Biomedical Investigations" (MIBBI (Taylor *et al.*, 2008)) provides a general checklist for results in the domain of the life sciences. More specifically, the guideline on "Minimum information requested in the annotation of biochemical models" (MIRIAM (Le Novere *et al.*, 2005)) provides a checklist specifically entailed for computational models. According to MIRIAM, the description of a computational model requires:

- A valid implementation in an appropriate language,
- An initial parameterization,
- Proper metadata about the model provenance (creators, contributors, and creation and modification dates, terms of distribution, etc.) and references to corresponding publications or similar documentations,
- The reproducibility of the references publication/documentation.

Similarly, the guideline on "Minimum Information About a Simulation Experiment" (MIASE, (Waltemath *et al.*, 2011a)) provides a checklist for simulation descriptions of modeling projects. According to MIASE, a simulation experiment needs to specify:

- A comprehensive model including equations and parameterizations,
- A simulation description including precise description of the simulation steps,
- Anything that is necessary to obtain described results.

Survey and Synopsis of Modeling Formats: The NormSys Registry

We have seen that in systems biology, the consistent structuring and description of data and computer models with their underlying experimental or modeling processes is only possible by applying interoperable standards for formatting and describing data, workflows, and models, as well as the metadata describing the interconnection between all these. However, with hundreds of available standards in the field, it is cumbersome for the modelers as potential users of these standards to get an overview about the available formats and supporting standards, as well as their potential fields of application.

To survey some of the most common freely available standards for model description and exchange used in systems biology and related fields, the NormSys registry (see "Relevant Websites section") was first released in October 2015. This publicly accessible platform provides a single access point for consistent information about these model-exchange formats and aims at listing them in a synopsis to bundle detailed and coherent information about their major features, as well as their similarities, relationships, and differences. The standards are classified by potential fields of application to provide the users guidance on the application of certain standards for a given task (e.g., a certain modeling approach that needs to be implemented) and representative use-cases give concrete model examples for these tasks (**Fig. 4**). Moreover, possible translation options from one standard to the other are illustrated, as well as interfacing options between the analyzed modeling standards. The standards covered by the first release of the NormSys registry include:

- **SBML** (Systems Biology Markup Language) as standardized interchange format for computer models of biological processes;
- **SBGN** (Systems Biology Graphical Notation) as standard for the graphical description of biological networks and pathways;
- **CellML** as standard for the exchange of mathematical models in biology and physiology;
- **FieldML** as declarative standard language for building hierarchical models represented by generalized mathematical fields;
- **SED-ML** (Simulation Experiment Description Markup Language) as standardized format for encoding simulation setups, to ensure exchangeability and reproducibility of simulation experiments;
- **SBOL** (Synthetic Biology Open Language) to represent genetic designs through a standardized vocabulary and standardized formats;
- **NeuroML** as a standardized description language that provides a common data format for defining and exchanging descriptions of neuronal cell and network models; and
- **PharmML** (Pharmacometrics Markup Language) as exchange format for encoding of models, associated tasks, and their annotation in pharmacometrics.

The NormSys registry also comprises an extension for validation and certification of computer models of biological systems that are described in the standard formats. This NormSys model validator, which was released at the end of 2016, includes both a formal syntax

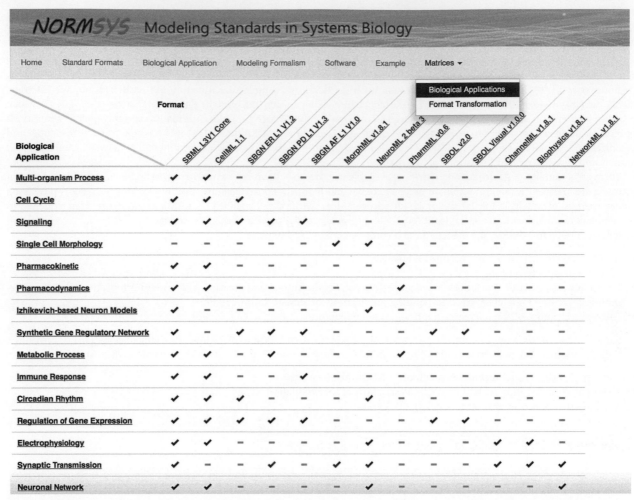

Fig. 4 The NormSys registry (http://normsys.h-its.org) provides details about modeling standards in systems biology and indicates potential fields of biological application.

check (XML-based) and a semantic check of the models and the integrity and consistency of their content, their entities, and their annotations. For syntax validation, a generic framework was implemented that makes use of the XML schema information on the corresponding standards stored in the NormSys registry. The consistency and validity of model entities and their corresponding annotations are checked for semantic validation. This semantic annotation test of an uploaded model is performed by a compliance check according to the MIRIAM guidelines (Le Novere *et al.*, 2005), which provide a checklist catalog for model annotations in biology ("Minimum information requested in the annotation of biochemical models").

Acknowledgment

The author received funding support for this work from the German Federal Ministry for Economic Affairs and Energy (BMWi) via the NormSys project (grant FKZ 01FS14019), from the German Federal Ministry of Education and Research (BMBF) as part of the Liver Systems Medicine Network (LiSyM, FKZ 031L0056), as well as from the Klaus Tschira Foundation (KTS).

See also: Bioinformatics Data Models, Representation and Storage. Biological and Medical Ontologies: GO and GOA. Cell Modeling and Simulation. Computational Systems Biology Applications. Computational Systems Biology. Data Cleaning. Data Integration and Transformation. Data Storage and Representation. Information Retrieval in Life Sciences. Integrative Analysis of Multi-Omics Data. Integrative Bioinformatics of Transcriptome: Databases, Tools and Pipelines. Multi-Scale Modelling in Biology. Natural Language Processing Approaches in Bioinformatics. Pre-Processing: A Data Preparation Step. Quantitative Modelling Approaches. Standards and Models for Biological Data: BioPax. Standards and Models for Biological Data: Common Formats. Standards and Models for Biological Data: SBML. Studies of Body Systems

References

Ashburner, M., Ball, C.A., Blake, J.A., *et al.*, 2000. Gene ontology: Tool for the unification of biology. The Gene Ontology Consortium. Nature Genetics 25 (1), 25–29. doi:10.1038/75556.

Bard, J.B., Rhee, S.Y., 2004. Ontologies in biology: Design, applications and future challenges. Nature Reviews Genetics 5 (3), 213–222. doi:10.1038/nrg1295.

Beard, D.A., Britten, R., Cooling, M.T., *et al.*, 2009. CellML metadata standards, associated tools and repositories. Philosophical Transactions Series A, Mathematical, Physical, and Engineering Sciences 367 (1895), 1845–1867. doi:10.1098/rsta.2008.0310.

Bergmann, F.T., Adams, R., Moodie, S., *et al.*, 2014. One file to share them all: Using the COMBINE archive and the OMEX format to share all information about a modeling project. BMC Bioinformatics 15, 369. doi:10.1186/s12859-014-0369-z.

Bergmann, F.T., Cooper, J., Le Novere, N., Nickerson, D., Waltemath, D., 2015. Simulation experiment description markup language (SED-ML) level 1 version 2. Journal of Integrative Bioinformatics 12 (2), 119–212. doi:10.1515/jib-2015-262.

Brazma, A., Hingamp, P., Quackenbush, J., *et al.*, 2001. Minimum information about a microarray experiment (MIAME)-toward standards for microarray data. Nature Genetics 29 (4), 365–371. doi:10.1038/ng1201-365.

Choi, K., Medley, J.K., Cannistra, C., *et al.*, 2016. Tellurium: A python based modeling and reproducibility platform for systems biology. bioRxiv. doi:10.1101/054601.

Courtot, M., Juty, N., Knupfer, C., *et al.*, 2011. Controlled vocabularies and semantics in systems biology. Molecular Systems Biology 7, 543. doi:10.1038/msb.2011.77.

Cuellar, A.A., Lloyd, C.M., Nielsen, P.F., *et al.*, 2003. An overview of CellML 1.1, a biological model description language. Simulation 79 (12), 740–747.

Dada, J.O., Spasic, I., Paton, N.W., Mendes, P., 2010. SBRML: A markup language for associating systems biology data with models. Bioinformatics 26 (7), 932–938. doi:10.1093/bioinformatics/btq069.

Degtyarenko, K., de Matos, P., Ennis, M., *et al.*, 2008. ChEBI: A database and ontology for chemical entities of biological interest. Nucleic Acids Research 36 (Database issue), D344–D350. doi:10.1093/nar/gkm791.

Dräger, A., Planatscher, H., Motsou Wouamba, D., *et al.*, 2009. SBML2L(A)T(E)X: Conversion of SBML files into human-readable reports. Bioinformatics 25 (11), 1455–1456. doi:10.1093/bioinformatics/btp170.

Federhen, S., 2012. The NCBI Taxonomy database. Nucleic Acids Research 40 (Database issue), D136–D143. doi:10.1093/nar/gkr1178.

Field, D., Garrity, G., Gray, T., *et al.*, 2008. The minimum information about a genome sequence (MIGS) specification. Nature Biotechnology 26 (5), 541–547. doi:10.1038/nbt1360.

Finney, A., Hucka, M., Bornstein, B.J., Keating, S.M., Shapiro, B.E., 2006. Software infrastructure for effective communication and reuse of computational models. In: Szallasi, Z., Stelling, J., Periwal, V. (Eds.), System Modeling in Cell Biology: From Concepts to Nuts and Bolts. Cambridge, MA: MIT Press, pp. 355–378.

Garny, A., Nickerson, D.P., Cooper, J., *et al.*, 2008. CellML and associated tools and techniques. Philosophical Transactions Series A, Mathematical, Physical, and Engineering Sciences 366 (1878), 3017–3043. doi:10.1098/rsta.2008.0094.

Gauges, R., Rost, U., Sahle, S., Wengler, K., Bergmann, F.T., 2015. The systems biology markup language (SBML) level 3 package: Layout, version 1 core. Journal of Integrative Bioinformatics 12 (2), 550–602. doi:10.1515/jib-2015-267.

Gennari, J.H., Neal, M.L., Galdzicki, M., Cook, D.L., 2011. Multiple ontologies in action: Composite annotations for biosimulation models. Journal of Biomedical Informatics 44 (1), 146–154. doi:10.1016/j.jbi.2010.06.007.

Gleeson, P., Crook, S., Cannon, R.C., *et al.*, 2010. NeuroML: A language for describing data driven models of neurons and networks with a high degree of biological detail. PLOS Computational Biology 6 (6), e1000815. doi:10.1371/journal.pcbi.1000815.

Hoops, S., Sahle, S., Gauges, R., *et al.*, 2006. COPASI – A complex pathway simulator. Bioinformatics 22 (24), 3067–3074. doi:10.1093/bioinformatics/btl485.

Hucka, M., Bergmann, F.T., Dräger, A., *et al.*, 2018. The systems biology markup language (SBML): Language specification for level 3 version 2 core. Journal of Integrative Bioinformatics 15, 20170081. doi:10.1515/jib-2017-0081.

Hucka, M., Bergmann, F.T., Hoops, S., *et al.*, 2015a. The systems biology markup language (SBML): Language specification for level 3 version 1 core. Journal of Integrative Bioinformatics 12 (2), 382–549. doi:10.1515/jib-2015-266.

Hucka, M., Finney, A., Sauro, H.M., *et al.*, 2003. The systems biology markup language (SBML): A medium for representation and exchange of biochemical network models. Bioinformatics 19 (4), 524–531.

Hucka, M., Nickerson, D.P., Bader, G.D., *et al.*, 2015b. Promoting coordinated development of community-based information standards for modeling in biology: The COMBINE initiative. Frontiers in Bioengineering and Biotechnology 3, 19. doi:10.3389/fbioe.2015.00019.

Hucka, M., Smith, L.P., 2016. SBML level 3 package: Groups, version 1 release 1. Journal of Integrative Bioinformatics 13 (3), 8–29. doi:10.1515/jib-2016-290.

Iannella, R., McKinney, J., 2014. vCard ontology – For describing people and organizations. W3c. https://www.w3.org/TR/vcard-rdf/ (accessed 03.04.17).

Junker, A., Rohn, H., Czauderna, T., *et al.*, 2012. Creating interactive, web-based and data-enriched maps with the systems biology graphical notation. Nature Protocols 7 (3), 579–593. doi:10.1038/nprot.2012.002.

Juty, N., Le Novere, N., Laibe, C., 2012. Identifiers.org and MIRIAM registry: Community resources to provide persistent identification. Nucleic Acids Research 40 (Database issue), D580–D586. doi:10.1093/nar/gkr1097.

Kitano, H., Funahashi, A., Matsuoka, Y., Oda, K., 2005. Using process diagrams for the graphical representation of biological networks. Nature Biotechnology 23 (8), 961–966. doi:10.1038/nbt1111.

Klink P. (2016) FieldedText. Available at: http://www.fieldedtext.org/ (accessed 01.05.17).

Klipp, E., Liebermeister, W., Helbig, A., Kowald, A., Schaber, J., 2007. Systems biology standards – The community speaks. Nature Biotechnology 25 (4), 390–391. doi:10.1038/nbt0407-390.

Lassila O., Swick R.R., 1999. Resource Description Framework (RDF) model and syntax specification. Technical report. World Wide Web Consortium.

Li, C., Donizelli, M., Rodriguez, N., *et al.*, 2010. BioModels Database: An enhanced, curated and annotated resource for published quantitative kinetic models. BMC Systems Biololgy 4, 92. doi:10.1186/1752-0509-4-92.

Lloyd, C.M., Halstead, M.D., Nielsen, P.F., 2004. CellML: Its future, present and past. Progress in Biophysics & Molecular Biology 85 (2–3), 433–450. doi:10.1016/j.pbiomolbio.2004.01.004.

Maguire, E., Gonzalez-Beltran, A., Whetzel, P.L., Sansone, S.A, Rocca-Serra, P., 2013. OntoMaton: A bioportal powered ontology widget for google spreadsheets. Bioinformatics 29 (4), 525–527. doi:10.1093/bioinformatics/bts718.

Misirli, G., Cavaliere, M., Waites, W., *et al.*, 2016. Annotation of rule-based models with formal semantics to enable creation, analysis, reuse and visualization. Bioinformatics 32 (6), 908–917. doi:10.1093/bioinformatics/btv660.

Mi, H., Schreiber, F., Moodie, S., *et al.*, 2015. Systems biology graphical notation: Activity flow language level 1 version 1.2. Journal of Integrative Bioinformatics 12 (2), 340–381. doi:10.1515/jib-2015-265.

Moodie, S., Le Novere, N., Demir, E., Mi, H., Villeger, A., 2015. Systems biology graphical notation: Process description language level 1 version 1.3. Journal of Integrative Bioinformatics 12 (2), 213–280. doi:10.1515/jib-2015-263.

Myers C.J., Bader G.D., Gleeson P., *et al*, 2017. A brief history of COMBINE, 2017 Winter Simulation Conference (WSC), pp. 884–895. Las Vegas, NV, USA. doi: 10.1109/WSC.2017.8247840.

Natale, D.A., Arighi, C.N., Blake, J.A., *et al.*, 2014. Protein ontology: A controlled structured network of protein entities. Nucleic Acids Research 42 (Database issue), D415–D421. doi:10.1093/nar/gkt1173.

Le Novere, N., Finney, A., Hucka, M., *et al.*, 2005. Minimum information requested in the annotation of biochemical models (MIRIAM). Nature Biotechnology 23 (12), 1509–1515. doi:10.1038/nbt1156.

Le Novere, N., Hucka, M., Mi, H., *et al.*, 2009. The systems biology graphical notation. Nature Biotechnology 27 (8), 735–741. doi:10.1038/nbt.1558.

Olivier, B.G., Bergmann, F.T., 2015. The Systems Biology Markup Language (SBML) level 3 package: Flux balance constraints. Journal of Integrative Bioinformatics 12 (2), 660–690. doi:10.1515/jib-2015-269.

Olivier, B.G., Bergmann, F.T., 2018. SBML level 3 package: Flux balance constraints version 2. Journal of Integrative Bioinformatics 15 (1), 20170082. doi:10.1515/jib-2017-0082.

Olivier, B.G., Snoep, J.L., 2004. Web-based kinetic modelling using JWS online. Bioinformatics 20 (13), 2143–2144. doi:10.1093/bioinformatics/bth200.

Rodriguez, N., Pettit, J.B., Dalle Pezze, P., Li, L., *et al.*, 2016. The systems biology format converter. BMC Bioinformatics 17, 154. doi:10.1186/s12859-016-1000-2.

Rosse, C., Mejino Jr., J.L., 2003. A reference ontology for biomedical informatics: The foundational model of anatomy. Journal of Biomedical Informatics 36 (6), 478–500. doi:10.1016/j.jbi.2003.11.007.

Scharm, M., Waltemath, D., 2016. A fully featured COMBINE archive of a simulation study on syncytial mitotic cycles in Drosophila embryos. F1000 Research 5, 2421. doi:10.12688/f1000research.9379.1.

Scharm, M., Waltemath, D., Mendes, P., Wolkenhauer, O., 2016. COMODI: An ontology to characterise differences in versions of computational models in biology. Journal of Biomedical Semantics 7 (1), 46. doi:10.1186/s13326-016-0080-2.

Schreiber, F., Bader, G.D., Gleeson, P., *et al.*, 2016. Specifications of standards in systems and synthetic biology: Status and developments in 2016. Journal of Integrative Bioinformatics 13, 289. doi:10.2390/biecoll-jib-2016-289.

Schreiber, F., Bader, G., Gleeson, P., *et al.*, 2018. Specifications of standards in systems and synthetic biology: Status and developments in 2017. Journal of Integrative Bioinformatics 15 (1), 20180013 https://doi.org/10.1515/jib-2018-0013.

Schreiber, F., Bader, G.D., Golebiewski, M., *et al.*, 2015. Specifications of standards in systems and synthetic biology. Journal of Integrative Bioinformatics 12, 258. doi:10.2390/biecoll-jib-2015-258.

SeckM.D., Evans, D.D., 2004. Major U.S. cities using national standard fire hydrants, one century after the great Baltimore fire, vol. 7158, pp. 7–9. Gaithersburg: National Institute of Standards and Technology, NISTIR.

ShafranovichY., 2005. Common format and MIME type for comma-separated values (CSV) Files. IETF p. 1 (RFC 4180). The Internet Society. doi:10.17487/RFC4180.

Shen, S.Y., Bergmann, F., Sauro, H.M., 2010. SBML2TikZ: Supporting the SBML render extension in LaTeX. Bioinformatics 26 (21), 2794–2795. doi:10.1093/bioinformatics/btq512.

Smith, L.P., Hucka, M., Hoops, S., *et al.*, 2015. SBML level 3 package: Hierarchical model composition, version 1 release 3. Journal of Integrative Bioinformatics 12 (2), 603–659. doi:10.1515/jib-2015-268.

Snoep, J.L., Olivier, B.G., 2003. JWS online cellular systems modelling and microbiology. Microbiology 149 (Pt 11), 3045–3047. doi:10.1099/mic.0.C0124-0.

Sorokin, A., Le Novere, N., Luna, A., *et al.*, 2015. Systems biology graphical notation: Entity relationship language level 1 version 2. Journal of Integrative Bioinformatics 12 (2), 281–339. doi:10.1515/jib-2015-264.

Stanford, N.J., Wolstencroft, K., Golebiewski, M., *et al.*, 2015. The evolution of standards and data management practices in systems biology. Molecular Systems Biology 11, 851. doi:10.15252/msb.20156053.

Stromback, L., Hall, D., Lambrix, P., 2007. A review of standards for data exchange within systems biology. Proteomics 7 (6), 857–867. doi:10.1002/pmic.200600438.

Swainston, N., Golebiewski, M., Messiha, H.L., *et al.*, 2010. Enzyme kinetics informatics: From instrument to browser. FEBS Journal 277, 3769–3779. doi:10.1111/j.1742-4658.2010.07778.x.

Swainston, N., Mendes, P., 2009. libAnnotationSBML: A library for exploiting SBML annotations. Bioinformatics 25 (17), 2292–2293. doi:10.1093/bioinformatics/btp392.

Taylor, C.F., Field, D., Sansone, S.A., *et al.*, 2008. Promoting coherent minimum reporting guidelines for biological and biomedical investigations: The MIBBI project. Nature Biotechnology 26 (8), 889–896. doi:10.1038/nbt.1411.

Taylor, C.F., Paton, N.W., Lilley, K.S., *et al.*, 2007. The minimum information about a proteomics experiment (MIAPE). Nature Biotechnology 25 (8), 887–893. doi:10.1038/nbt1329.

van Iersel, M.P., Villeger, A.C., Czauderna, T., *et al.*, 2012. Software support for SBGN maps: SBGN-ML and LibSBGN. Bioinformatics 28 (15), 2016–2021. doi:10.1093/bioinformatics/bts270.

Waltemath, D., Adams, R., Beard, D.A., *et al.*, 2011a. Minimum information about a simulation experiment (MIASE). PLOS Computational Biology 7 (4), e1001122. doi:10.1371/journal.pcbi.1001122.

Waltemath, D., Adams, R., Bergmann, F.T., *et al.*, 2011b. Reproducible computational biology experiments with SED-ML – The simulation experiment description markup language. BMC Systems Biology 5, 198. doi:10.1186/1752-0509-5-198.

Wilkinson, M.D., Dumontier, M., Aalbersberg, I.J., *et al.*, 2016. Comment: The FAIR guiding principles for scientific data management and stewardship. Scientific Data 3. doi:10.1038/sdata.2016.18.

Wimalaratne, S.M., Halstead, M.D., Lloyd, C.M., *et al.*, 2009a. Facilitating modularity and reuse: Guidelines for structuring CellML 1.1 models by isolating common biophysical concepts. Experimental Physiology 94 (5), 472–485. doi:10.1113/expphysiol.2008.045161.

Wimalaratne, S.M., Halstead, M.D., Lloyd, C.M., Crampin, E.J., Nielsen, P.F., 2009b. Biophysical annotation and representation of CellML models. Bioinformatics 25 (17), 2263–2270. doi:10.1093/bioinformatics/btp391.

Wittig, U., Kania, R., Golebiewski, M., *et al.*, 2012. SABIO-RK – Database for biochemical reaction kinetics. Nucleic Acids Research 40, D790–D796. doi:10.1093/nar/gkr1046.

Wittig, U., Rey, M., Kania, R., *et al.*, 2014. Challenges for an enzymatic reaction kinetics database. FEBS Journal 281, 572–582. doi:10.1111/febs.12562.

Wolstencroft, K., Owen, S., Horridge, M., *et al.*, 2011. RightField: Embedding ontology annotation in spreadsheets. Bioinformatics 27 (14), 2021–2022. doi:10.1093/bioinformatics/btr312.

Wolstencroft K., Owen S., Krebs O., *et al.*, 2013. Semantic Data and Models Sharing in Systems Biology: The Just Enough Results Model and the SEEK Platform. In: Alani H., Kagal L., Fokoue A *et al.* (Eds.), The Semantic Web – ISWC 2013. Lecture Notes in Computer Science, vol. 8219, 212-227. Berlin, Heidelberg: Springer. doi:10.1007/978-3-642-41338-4_14

Yu, T., Lloyd, C.M., Nickerson, D.P., *et al.*, 2011. The physiome model repository 2. Bioinformatics 27 (5), 743–744. doi:10.1093/bioinformatics/btq723.

Relevant Websites

http://co.mbine.org/
   COMBINE.
https://www.cellml.org/tools
   CellML tools
   CellML.
https://www.w3.org/XML/
   Extensible Markup Language (XML) ı World Wide Web Consortium (W3C).

https://fairsharing.org/collection/MIBBI
>   FAIRsharing MIBBI Collection.

https://fair-dom.org/
>   FAIRDOM.

https://fairsharing.org/collection/FAIRDOM
>   FAIRsharing Collection: FAIRDOM Community Standards.

http://project.isbe.eu/
>   ISBE Project Website.

http://jermontology.org/
>   JERM Ontology.

http://co.mbine.org/standards/kisao
>   Kinetic Simulation Algorithm Ontology ǀ COMBINE.

https://www.w3.org/Math/
>   MathML ǀ World Wide Web Consortium (W3C).

https://github.com/numl/numl
>   NuML · GitHub.

https://github.com/NuML/NuML/tree/master/libnuml
>   NuML/libnuml at master · NuML/NuML · GitHub.

http://normsys.h-its.org
>   Normsys Standard Registry ǀ HITS gGmbH.

https://www.w3.org/TR/2014/REC-rdf11-concepts-20140225/
>   RDF 1.1 Concepts and Abstract Syntax ǀ World Wide Web Consortium (W3C).

http://sabiork.h-its.org
>   SABIO-RK Biochemical Reaction Kinetics Database ǀ HITS gGmbH.

http://sbml.org/SBML_Software_Guide/SBML_Software_Matrix
>   SBML Software Guide/SBML Software Matrix ǀ Systems Biology Markup Language (SBML).

https://sed-ml.github.io/showcase.html
>   SED-ML Tools and Libraries.

https://www.ebi.ac.uk/sbo/main/
>   SBO (Systems Biology Ontology) ǀ EMBL-EBI.

http://co.mbine.org/standards/teddy/
>   TEDDY-Ontology ǀ COMBINE.

Computational Lipidomics

Josch K Pauling, Humboldt University of Berlin, Berlin, Germany

Introduction

Lipidomics, the study of the lipidome, is a rather new member of the omics family (Wenk, 2005; Dennis, 2009). A first European initiative with a focus on lipidomics started in 2005/2006 (van Meer *et al.*, 2007). The term *cellular lipidome* refers to the full lipid complement of a cell (van Meer, 2005). It first appeared in a publication by Han and Gross (2003) where the authors describe a mass spectrometry-based method to detect and quantify all lipids in a cell extract, the lipidome. Even though lipids had been known and measured for over three centuries, the ability to now measure the entire lipidome was a great advancement from traditional approaches that focused on efficiently extracting and measuring a single targeted lipid class or species (Hsu and Turk, 2000a,b, 2002, 2003, 2008; Liebisch *et al.*, 2006; Deems *et al.*, 2007; Krank *et al.*, 2007) often in a specific biochemical or biomedical context. Lipids display a great diversity in chemical structure and have distinct chemical properties (van Meer, 2005) and thus measuring lipidomes is a complex task that requires suitable extraction protocols (Folch *et al.*, 1957; Bligh and Dyer, 1959), ionization (Fenn *et al.*, 1989; Ho *et al.*, 2003), detection and quantification techniques (Sandhoff *et al.*, 1999; Ivanova *et al.*, 2007; Almeida *et al.*, 2015), algorithms to automatically read and process mass spectra (Niemelä *et al.*, 2009; Husen *et al.*, 2013; Herzog *et al.*, 2013; Marella *et al.*, 2015; Kochen *et al.*, 2016; Peng and Ahrends, 2016) (explained throughout this article), nomenclature rules (IUPAC-IUB (CBN), 1967, 1977; IUPAC-IUB (JCBN), 1987, 1989a,b; Moss, 1996; Liebisch *et al.*, 2013), databases to store information (Fahy *et al.*, 2005, 2007, 2009; Wishart *et al.*, 2007; Sud *et al.*, 2007; Schmelzer *et al.*, 2007; Foster *et al.*, 2013; Aimo *et al.*, 2015), and software to link and mine this information (Hadadi *et al.*, 2014). Hence, lipidomics research integrates a wide range of scientific disciplines such as biochemistry, mass spectrometry, bioinformatics, computational biology, biomedicine and biophysics.

Lipids and Functions

Cellular lipids are biomolecules with a very high structural and functional diversity (Gross and Han, 2011). Glycerophospholipids, for example, consist of a hydrophilic head group and two hydrophobic hydrocarbon tails. Together with sphingolipids and cholesterol they can create lipid bilayers (Edidin, 2003; van Meer *et al.*, 2008) that separate two aqueous environments from each other. Within a cell these lipid bilayers act as biological membranes creating cellular compartments. A membrane is the site of metabolic exchange between the organelle and the extra-organelle environment, for example, the cytosol with membrane proteins managing the actual exchange and transport across the membrane. The composition of lipids determines the specific membrane properties such as structure and fluidity. The regulation of membrane-lipid composition is very rigid and slight alterations in it can impair membrane functionality.

Another, and widely known, function of lipids is the storage and provision of cellular energy. Glycerolipids, especially triacylglycerols (TAGs), are storage lipids that serve as carriers of fatty-acids. They consist of a glycerol backbone where fatty-acyl side chains are attached. A TAG consists of three fatty-acyl side chains, which is the maximal number that can be directly attached to a glycerol. Through enzymatic activity these fatty-acyl side chains can be released from the glycerol providing non-esterified fatty-acids (NEFAs), which can serve as a source of ATP when oxidized. There are many more lipids and functions but glycerophospholipids, specifically phosphatidylcholine (PC 34:1 as shown in **Fig. 1**) will be used in following examples.

Computational Lipidomics

The lipidomics workflow starts with extracting lipids from a cell or tissue sample. Protocols such as the one proposed by Folch *et al.* (1957) or by Bligh and Dyer (1959) have been widely used for decades and the corresponding publications have received

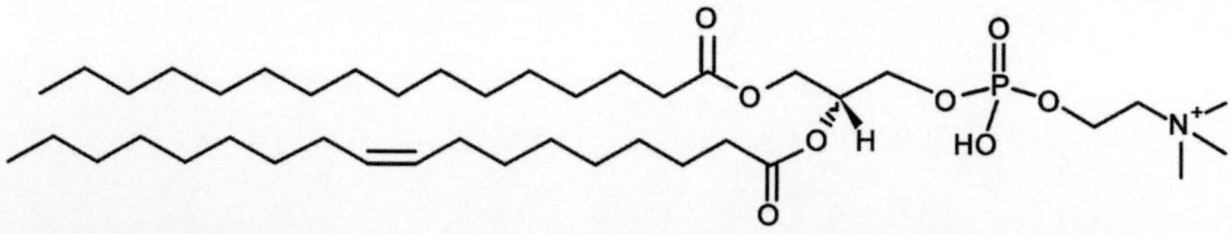

Fig. 1 Chemical structure of a phosphatidylcholine (PC). Together, both fatty-acyls consist of 34 carbon atoms and one double bond (PC 34:1). A shorthand notation of this particular species is PC 16:0/18:1 indicating the position-specific distribution of fatty-acyls. A fatty-acyl with 16 carbon atoms is bound to the stereospecific numbering (sn)-1 position of the glycerol. A fatty-acyl with 18 carbon atoms and one double bond is bound to the sn-2 position of the glycerol. Given that the depicted double bond is of the *cis*-type (Z according to the E-Z-scheme) the full name is 1-hexadecanoyl-2-(9Z-octadecenoyl)-*sn*-glycero-3-phosphocholine according to IUPAC nomenclature rules.

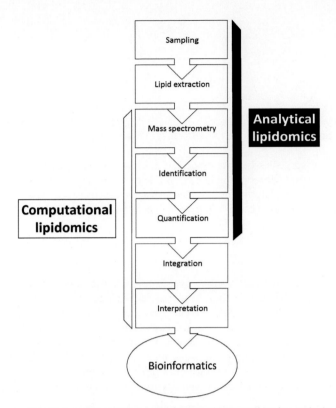

Fig. 2   The lipidomics workflow from a sample to an interpretation in the context of integrated omics analyses. This scheme is described in this article.

over 50,000 and over 40,000 citations, respectively. After extraction, a feasible mix of exogenous lipid standards is selected and added to the extract (Moore *et al.*, 2007). Finally, the sample extract is injected into a mass spectrometer for measurement. There are various mass spectrometric methods but the basic decision is whether or not to apply a chromatographic separation prior to injection, for example, liquid chromatography (LC/MS) or gas chromatography (GC/MS) (McDonald *et al.*, 2007; Sullards *et al.*, 2007; Cruz-Hernandez and Destaillats, 2012). The alternative is to inject the sample extract directly into the electrospray ionization (ESI) device of a mass spectrometer. This is called "shotgun lipidomics" (Han, 2003; Han and Gross, 2005; Schwudke *et al.*, 2007, 2011; Schuhmann *et al.*, 2011, 2012).

The path from generating a set of mass spectra to algorithms mining this data and meta data by querying multiple databases is a very long and tedious one (**Fig. 2**). The first few steps are very technical and require detailed background knowledge of lipid biochemistry and mass spectrometry. Mass spectra representing a lipidome are generated and a readout is stored. Subsequently, the data must be analyzed and prepared in a way that meta data is quickly accessible. This includes quality control mechanisms and statistical analysis. At last there are classical bioinformatics tasks such as database development, data mining and linking the data to other omics data sets, for example, to study health and disease. At this point, the applied bioinformatics methods are not specifically tailored to deal with lipidome data and may therefore be more generally summarized as "lipid bioinformatics". However, the earlier steps are highly specific to mass spectra of lipidomes and may thus rather be termed "computational lipidomics" which is in the focus of this article.

Mass Spectral Analysis and Peak Detection

High throughput, high resolution lipid experiments may become quite complex when measuring various different tissue samples under various conditions at different time points with biological and technical replicates. This may result in thousands of mass spectra contained in a single data set. Experts can analyze a mass spectrum quickly by looking at it but this method is not feasible for more than about one hundred mass spectra. Automated routines are necessary to analyze such a data set in an adequate time frame. The first step in such a routine is to identify peaks that each resemble lipid molecules of a specific mass per charge ratio (m/z) in a long list of data points which comprise two values (1) m/z value and (2) intensity. Ideally, these data point lists, typically one list per mass spectrum, are kept in computer memory. However, mass spectrometer vendors keep the format in which mass spectra are stored secret and instead provide software tools to convert them to regular tab- or comma-separated text files. Reading and storing the lists in memory is followed by a peak detection heuristic that maps groups of data points to one peak and determines this peak's centroid m/z and intensity tuple. It is important to note that this is not always a straight forward process as

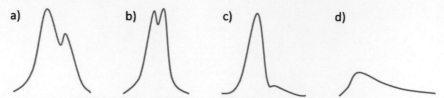

Fig. 3 Various schemata of peak appearances in a mass spectrum. (a) through (c) may depict two or more convoluted peaks. The higher the resolution of the mass spectrometer the more molecules are fully resolved as singular peaks.

shown in **Fig. 3**. In most of the depicted peak schemata the peak is not resembling a clean Gaussian distribution of intensities over the peak width and in some cases one may observe multiple peaks that could not be fully resolved. The challenge is to deconvolute overlapping peaks and to estimate the m/z and intensity values for each peak. The apex value of a fitted Gaussian bell curve is a good estimator to start with but more sophisticated solutions are needed when this method fails.

Identification and Quantification of Lipids

Assuming that m/z values and intensities have been assigned to each peak, the m/z value can be utilized to annotate each peak with the name of the lipid it resembles. This identification process is vastly dependent on the accuracy of the assigned m/z value and the calibration of the mass spectrometer at the time of measurement. A shift in m/z values may result in a false mapping to a lipid with a slightly higher or lower m/z value. It is therefore essential in quality control to become aware of and correct for such shifts that may occur over the time of a measurement and can only be detected by measuring control samples in between the actual samples. Additionally, there exist lipids with the same atomic composition but yet different chemical structure. These are isomeric molecules and it is not possible to distinguish one from the other just by the peak m/z value.

After mapping every peak to a lipid species the intensity value is utilized to quantify the full complement of lipids in the sample of interest. Normalization is key to guarantee comparability between several mass spectra and experiments. Prior to injection into the mass spectrometer, a mix of internal standards (non-endogenous lipid species) of specific and exact amounts is added to each sample extract to obtain a relationship between the peak signal and the actual concentration in the sample. Lipid class intensities can then be normalized to the respective lipid standard. Doing so ensures that intensity values from different mass spectra become directly comparable. At the same time, differences in ionization efficiencies between lipid classes are corrected.

In a final step after normalization all detected and identified lipids including their intensities are reassembled to formulate the lipidome of a particular sample. To acquire a complete data set, this procedure is done for each sample measured as a part of an experiment. In case one fills an entire 96-well plate with samples, controls, and blanks this is an enormous effort that requires sophisticated computational support.

The Next Dimension

The previous section provided insights into the computational challenges to solve when measuring a lipidome based on intact lipid molecules (abbreviated as MS or MS1). This measurement is not sufficient to reveal the particular fatty-acyl side chains on one of those lipid species or to resolve isomeric lipids. All one can infer from the m/z value is the atomic composition of the whole molecule but not its structural arrangement. In lipid research it is often of interest to track fatty-acid metabolism and the side chains of membrane glycerophospholipids. To achieve this information it is necessary to break a lipid into smaller fragments, for example, via collision-induced dissociation (CID) (Sleno and Volmer, 2004; Wells and McLuckey, 2005). With this technique (abbreviated as MS2, one fragmentation step) it is possible to dissociate the fatty-acyl side chains and the head group from the glycerol backbone and afterwards measure the m/z values of these fragments. This allows to infer the particular fatty-acyls. **Fig. 4** guides through this process. Finally, fragments may be fragmented again to reveal even further structural detail (abbreviated MS3, MS4, etc., or simply MSn, where n denotes the number of fragmentation steps).

The lipid molecule of interest, a PC 34:1, has a m/z value of 760.58 Da. Multiple peaks are shown each representing fragments with lower m/z values. The dissociated part of the molecule did not have a charge and is therefore not represented in the mass spectrum. This is called a neutral loss (NL) while the observable charged fragment is called a product ion. The neutral loss can occur in two slightly different ways leaving the two peaks at 478 and 496 Da. The corresponding compounds are shown above. However, according to the mass difference between the complete lipid molecule and the fragments one can infer that in both cases a fatty-acyl with 18 carbon atoms and one double bond is the missing part of the fragment. Hence, the presence of the sub-species PC 16:0–18:1 can be confirmed. The same procedure can be applied to all peaks in the mass spectrum. Doing so additionally shows the neutral losses of different fatty-acyls (red annotation) revealing the presence of another sub-species PC 16:1–18:0 in which the double bond is found on the shorter fatty-acyl. Fragmentation by collisional dissociation is a powerful tool to detect lipid molecules with a high level of structural detail. However, it too has its limitations since some structural details such as double bond type and position and the position of hydroxyl group modifications cannot directly be detected.

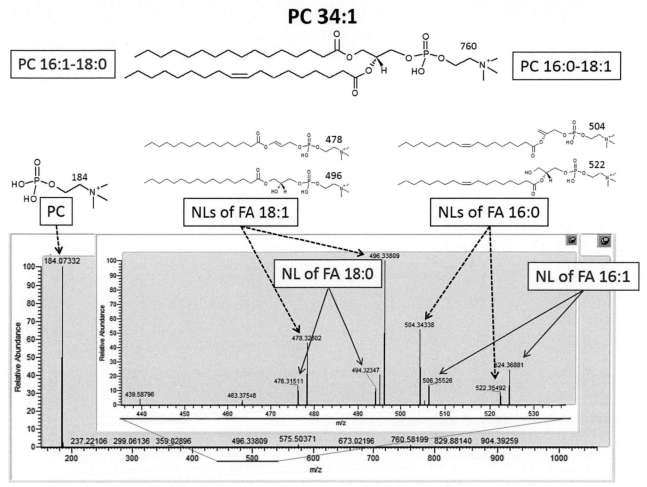

Fig. 4 A raw mass spectrum of fragmented PC 34:1 as it is generated by a mass spectrometer. The spectrum in front depicts a zoom in the *m/z* range between 440 and 540 Da as indicated by the horizontal, orange lines. The spectrum in the back shows the entire mass range and the prominent peak at 184, a phosphocholine. The chemical structure of this fragment is shown above. The other peaks strongly suggest that there are two detected sub-species (labels in black and red). The corresponding fragments to each peak are shown above. From the evidence one can deduce that there is not only the black labelled molecule PC 16:0–18:1 but also the red labelled PC 16:1–18:0. These are molecules with different chemical structure but with the same atomic composition and *m/z* value. For simplicity, the structure of the red labelled PC 16:1–18:0 is not shown. On top of the figure only the black labelled PC 16:0–18:1 is shown. Please also note that the sn-position of the fatty-acyl side chains cannot be deduced from a peak's *m/z* value. Therefore, the annotation "PC 16:0-18:1" is used instead of the position-specific annotation "PC 16:0/18:1" despite the position-specific graphical structure of a PC 16:0/18:1.

When fragmentation is utilized in an experiment the number of mass spectra is significantly increased from hundreds (one mass spectrum per well on a 96-well plate, double injections) to possibly several thousand. The challenge is to automate the process described in **Fig. 4** and to accurately infer the various fatty-acyl side chain compositions for a particular lipid species. **Fig. 5** depicts the difference between a MS measurement and a MS2 (or MS*n*) measurement visualized as a schematic bar plot.

With the progression and advancement of today's analytical capacities from smaller scale lipid studies to large scale high-throuput scans of full lipidomes and from whole molecule identification to a higher degree of structural detail it will continue to be a vital task to co-develop computational support routines. To develop sophisticated solutions it is essential to further expand on interdisciplinary expertise in the lipid-oriented research community. In comparison with other omics technology the range of lipid-centric informatics solutions is currently small and it is therefore not surprising that the few existing ones may not be usable free of charge. It is advisable to acquire a solid basic understanding of how computational biology support has helped to fuel the advancement of the omics landscape. Lipidomics and proteomics are largely mass spectrometry-based disciplines. Thus, computational lipidomics can profit from translational efforts. This, however, requires at least basic expertise in proteomics, lipidomics, and computational biology as well as bioinformatics and potentially additional disciplines. In today's biomedical research, there is already a need for position-specific structural characterization of lipid molecules on a lipidome-wide scale within a multi-omics setting. This underlines the importance of the emerging field of computational lipidomics research.

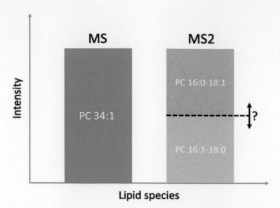

Fig. 5 The levels of detail gained from either a full MS scan or a MS2 up to MSn fragmentation analysis. The stacked bar on the right visualizes the sub-sepcies composition. The exact quantity of detected sub-species cannot accurately be determined solely based on fragment spectra and must be approximated.

See also: Computational Systems Biology. Experimental Platforms for Extracting Biological Data: Mass Spectrometry, Microarray, Next Generation Sequencing. Natural Language Processing Approaches in Bioinformatics

References

Aimo, L., Liechti, R., Hyka-Nouspikel, N., *et al.*, 2015. The swisslipids knowledgebase for lipid biology. Bioinformatics 31, 2860–2866.

Almeida, R., Pauling, J.K., Sokol, E., Hannibal-Bach, H.K., Ejsing, C.S., 2015. Comprehensive lipidome analysis by shotgun lipidomics on a hybrid quadrupole-orbitrap-linear ion trap mass spectrometer. Journal of the American Society for Mass Spectrometry 26, 133–148.

Bligh, E.G., Dyer, W.J., 1959. A rapid method of total lipid extraction and purification. Canadian Journal of Biochemistry and Physiology 37 (8), 911–917.

Cruz-Hernandez, C., Destaillats, F., 2012. Analysis of lipids by gas chromatography. Gas chromatography. 529–549. doi:10.1016/B978-0-12-385540-4.00023-7a.

Deems, R., Buczynski, M.W., Bowers-Gentry, R., Harkewicz, R., Dennis, E.A., 2007. Detection and quantitation of eicosanoids via high performance liquid chromatography-electrospray ionization-mass spectrometry. Methods in Enzymology 432, 59–82.

Dennis, E.A., 2009. Lipidomics joins the omics evolution. Proceedings of the National Academy of Sciences of the United States of America 106 (7), 2089–2090.

Edidin, M., 2003. Lipids on the frontier: A century of cell-membrane bilayers. Nature Reviews. Molecular Cell Biology 4, 414–418.

Fahy, E., Subramaniam, S., Brown, H.A., *et al.*, 2005. A comprehensive classification system for lipids. Journal of Lipid Research 46, 839–861.

Fahy, E., Subramaniam, S., Murphy, R.C., *et al.*, 2009. Update of the lipid maps comprehensive classification system for lipids. Journal of Lipid Research 50 (Suppl.), S9–S14.

Fahy, E., Sud, M., Cotter, D., Subramaniam, S., 2007. Lipid maps online tools for lipid research. Nucleic Acids Research 35, W606–W612.

Fenn, J.B., Mann, M., Meng, C.K., Wong, S.F., Whitehouse, C.M., 1989. Electrospray ionization for mass spectrometry of large biomolecules. Science 246, 64–71.

Folch, J., Lees, M., Sloane Stanley, G.H., 1957. A simple method for the isolation and purification of total lipides from animal tissues. The Journal of Biological Chemistry 226, 497–509.

Foster, J.M., Moreno, P., Fabregat, A., *et al.*, 2013. Lipidhome: A database of theoretical lipids optimized for high throughput mass spectrometry lipidomics. PLOS ONE 8, e61951.

Gross, R.W., Han, X., 2011. Lipidomics at the interface of structure and function in systems biology. Chemistry & Biology 18, 284–291.

Hadadi, N., Cher Soh, K., Seijo, M., *et al.*, 2014. A computational framework for integration of lipidomics data into metabolic pathways. Metabolic Engineering 23, 1–8.

Han, X., Gross, R.W., 2003. Global analyses of cellular lipidomes directly from crude extracts of biological samples by ESI mass spectrometry: A bridge to lipidomics. The Journal of Lipid Research 44 (6), 1071–1079.

Han, X., Gross, R.W., 2005. Shotgun lipidomics: Multidimensional MS analysis of cellular lipidomes. Expert Review of Proteomics 2, 253–264.

Herzog, R., Schwudke, D., Shevchenko, A., 2013. Lipidxplorer: Software for quantitative shotgun lipidomics compatible with multiple mass spectrometry platforms. Current protocols in bioinformatics 43 (14), 12.1–14.1230.

Ho, C.S., Lam, C.W.K., Chan, M.H.M., *et al.*, 2003. Electrospray ionisation mass spectrometry: Principles and clinical applications. The Clinical Biochemist. Reviews 24, 3–12.

Hsu, F.F., Turk, J., 2000a. Charge-driven fragmentation processes in diacyl glycerophosphatidic acids upon low-energy collisional activation. A mechanistic proposal. Journal of the American Society for Mass Spectrometry 11, 797–803.

Hsu, F.F., Turk, J., 2000b. Charge-remote and charge-driven fragmentation processes in diacyl glycerophosphoethanolamine upon low-energy collisional activation: A mechanistic proposal. Journal of the American Society for Mass Spectrometry 11, 892–899.

Hsu, F.-F., Turk, J., 2002. Characterization of ceramides by low energy collisional-activated dissociation tandem mass spectrometry with negative-ion electrospray ionization. Journal of the American Society for Mass Spectrometry 13, 558–570.

Hsu, F.-F., Turk, J., 2003. Electrospray ionization/tandem quadrupole mass spectrometric studies on phosphatidylcholines: The fragmentation processes. Journal of the American Society for Mass Spectrometry 14, 352–363.

Hsu, F.-F., Turk, J., 2008. Structural characterization of unsaturated glycerophospholipids by multiple-stage linear ion-trap mass spectrometry with electrospray ionization. Journal of the American Society for Mass Spectrometry 19, 1681–1691.

Husen, P., Tarasov, K., Katafiasz, M., *et al.*, 2013. Analysis of lipid experiments (alex): A software framework for analysis of high-resolution shotgun lipidomics data. PLOS ONE 8, e79736.

IUPAC-IUB commission on biochemical nomenclature (IUPAC-IUB (CBN)), 1967. The nomenclature of lipids. European Journal of Biochemistry 2, 127–131.

IUPAC-IUB commission on biochemical nomenclature (IUPAC-IUB (CBN)), 1977. The nomenclature of lipids recommendations (1976). Lipids 12, 455–468.

IUPAC-IUB joint commission on biochemical nomenclature (IUPAC-IUB (JCBN)), 1987. Prenol nomenclature recommendations 1986. European Journal of Biochemistry 167, 181–184.

IUPAC-IUB joint commission on biochemical nomenclature (IUPAC-IUB (JCBN)), 1989a. The nomenclature of steroids recommendations 1989. European Journal of Biochemistry 186, 429–458.

IUPAC-IUB Joint Commission on Biochemical Nomenclature (IUPAC-IUB (JCBN)), 1989b. Guidelines on eicosanoid nomenclature. Eicosanoids 2, 65–68.

Ivanova, P.T., Milne, S.B., Byrne, M.O., Xiang, Y., Brown, H.A., 2007. Glycerophospholipid identification and quantitation by electrospray ionization mass spectrometry. Methods in Enzymology 432, 21–57.

Kochen, M.A., Chambers, M.C., Holman, J.D., *et al.*, 2016. Greazy: Open-source software for automated phospholipid tandem mass spectrometry identification. Analytical Chemistry 88, 5733–5741.

Krank, J., Murphy, R.C., Barkley, R.M., Duchoslav, E., McAnoy, A., 2007. Qualitative analysis and quantitative assessment of changes in neutral glycerol lipid molecular species within cells. Methods in Enzymology 432, 1–20.

Liebisch, G., Binder, M., Schifferer, R., *et al.*, 1761. High throughput quantification of cholesterol and cholesteryl ester by electrospray ionization tandem mass spectrometry (esi-ms/ms). Biochimica et Biophysica Acta 121–128, 2006.

Liebisch, G., Vizcaino, J.A., Köfeler, H., *et al.*, 2013. Shorthand notation for lipid structures derived from mass spectrometry. Journal of Lipid Research 54, 1523–1530.

Marella, C., Torda, A.E., Schwudke, D., 2015. The lux score: A metric for lipidome homology. PLOS Computational Biology 11, e1004511.

McDonald, J.G., Thompson, B.M., McCrum, E.C., Russell, D.W., 2007. Extraction and analysis of sterols in biological matrices by high performance liquid chromatography electrospray ionization mass spectrometry. Methods in Enzymology 432, 145–170.

Moore, J.D., Caufield, W.V., Shaw, W.A., 2007. Quantitation and standardization of lipid internal standards for mass spectroscopy. Methods in Enzymology 432, 351–367.

Moss, G.P., 1996. Basic terminology of stereochemistry (IUPAC recommendations 1996). Pure and Applied Chemistry 68 (12).

Niemelä, P.S., Castillo, S., Sysi-Aho, M., Oresic, M., 2009. Bioinformatics and computational methods for lipidomics. Journal of Chromatography. B, Analytical Technologies in the Biomedical and Life Sciences 877, 2855–2862.

Peng, B., Ahrends, R., 2016. Adaptation of skyline for targeted lipidomics. Journal of Proteome Research 15, 291–301.

Sandhoff, R., Brügger, B., Jeckel, D., Lehmann, W.D., Wieland, F.T., 1999. Determination of cholesterol at the low picomole level by nano-electrospray ionization tandem mass spectrometry. Journal of Lipid Research 40, 126–132.

Schmelzer, K., Fahy, E., Subramaniam, S., Dennis, E.A., 2007. The lipid maps initiative in lipidomics. Methods in Enzymology 432, 171–183.

Schuhmann, K., Almeida, R., Baumert, M., *et al.*, 2012. Shotgun lipidomics on a ltq orbitrap mass spectrometer by successive switching between acquisition polarity modes. Journal of Mass Spectrometry: JMS 47, 96–104.

Schuhmann, K., Herzog, R., Schwudke, D., *et al.*, 2011. Bottom-up shotgun lipidomics by higher energy collisional dissociation on ltq orbitrap mass spectrometers. Analytical Chemistry 83, 5480–5487.

Schwudke, D., Liebisch, G., Herzog, R., Schmitz, G., Shevchenko, A., 2007. Shotgun lipidomics by tandem mass spectrometry under data-dependent acquisition control. Methods in Enzymology 433, 175–191.

Schwudke, D., Schuhmann, K., Herzog, R., Bornstein, S.R., Shevchenko, A., 2011. Shotgun lipidomics on high resolution mass spectrometers. Cold Spring Harbor Perspectives in Biology 3, a004614.

Sleno, L., Volmer, D.A., 2004. Ion activation methods for tandem mass spectrometry. Journal of Mass Spectrometry: JMS 39, 1091–1112.

Sud, M., Fahy, E., Cotter, D., *et al.*, 2007. LMSD: Lipid maps structure database. Nucleic Acids Research 35, D527–D532.

Sullards, M.C., Allegood, J.C., Kelly, S., *et al.*, 2007. Structure-specific, quantitative methods for analysis of sphingolipids by liquid chromatography-tandem mass spectrometry: "Inside-out" sphingolipidomics. Methods in Enzymology 432, 83–115.

van Meer, G., 2005. Cellular lipidomics. The EMBO Journal 24, 3159–3165.

van Meer, G., Leeflang, B.R., Liebisch, G., Schmitz, G., Goi, F.M., 2007. The european lipidomics initiative: Enabling technologies. Methods in Enzymology 432, 213–232.

van Meer, G., Voelker, D.R., Feigenson, G.W., 2008. Membrane lipids: Where they are and how they behave. Nature Reviews. Molecular Cell Biology 9, 112–124.

Wells, J.M., McLuckey, S.A., 2005. Collision-induced dissociation (cid) of peptides and proteins. Methods in Enzymology 402, 148–185.

Wenk, M.R., 2005. The emerging field of lipidomics. Nature Reviews. Drug Discovery 4, 594–610.

Wishart, D.S., Tzur, D., Knox, C., *et al.*, 2007. Hmdb: The human metabolome database. Nucleic Acids Research 35, D521–D526.

Multi-Scale Modelling in Biology

Socrates Dokos, University of New South Wales, Sydney, NSW, Australia

Introduction

Multi-scale modelling aims at developing computational models of physical systems characterised by phenomena spanning multiple spatial or temporal scales. Non-biological examples include modelling whole-material properties from microstructural and molecular-level features (Horstemeyer, 2009), chemical engineering processes from molecular-level interactions (Salciccioli *et al.*, 2011; Zhao *et al.*, 2012), nuclear reactor design (Mahadevan *et al.*, 2014) and astrophysics (van Elteren *et al.*, 2014). In biology and physiology, multi-scale models are often used to integrate biological processes operating from the molecular and sub-cellular scale through to cell and tissue function, up to whole-organ and organ systems (Hunter and Nielsen, 2005; Southern *et al.*, 2008; Qu *et al.*, 2011). Examples include biomechanics (Viceconti, 2012), electrophysiology (Winslow *et al.*, 2000), pulmonary gas exchange (Donovan, 2011), tissue patterning (Thorne *et al.*, 2007), tumour growth (Zangooei and Habibi, 2017), biochemical signalling pathways (Walker *et al.*, 2008) and systems biology in general (Dada and Mendes, 2011). An overview of the typical spatial scales employed in multi-scale biological models is shown in **Fig. 1**.

Multi-Scale Modelling Approaches

Approaches to formulating multi-scale models and their computational implementation will depend on the choice of spatial scales adopted, as well as the particular physics being coupled.

Model Formulation

Multi-scale models are generally constructed from one or more sub-component models, each characterised by a finer spatial and/or temporal scale than the overall model. Integration of these into a single course scale model involves some degree of macroscopic approximation, loosely termed *homogenisation* (across space) or *averaging* (across time) (Matthies, 2006). These methods include model reduction and course-graining (Gorban *et al.*, 2006), continuous field approximation (Bonilla *et al.*, 2014), compartmental and lumped-parameter modelling (Quarteroni *et al.*, 2016) and cellular automata approaches (Ermentrout and Edelstein-Keshet, 1993).

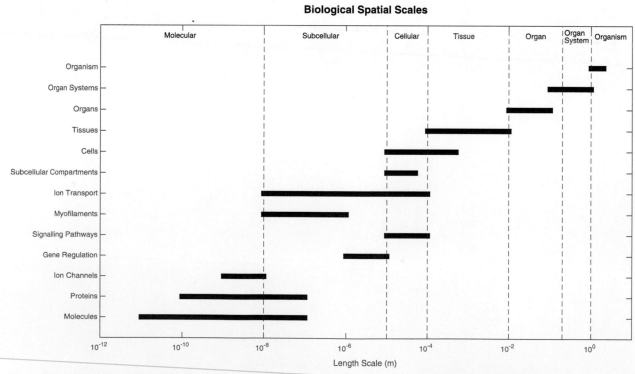

Fig. 1 Typical biological spatial scales used in multiple-scale modelling. Larger scales are shown typical for human physiology.

Course-graining methods approximate spatial heterogeneities by discretising the model's spatial domain into course regions, assigning spatially-averaged properties of the original system to each. In molecular dynamics simulations for example, instead of representing the kinetics of every atom in a system, groups of atoms are clustered together into single mesoscale entities, to provide a simplified, low-resolution approximation (Saunders and Voth, 2013). In contrast, *model reduction* seeks to lower the total degrees of freedom in a system of equations by projecting state variables onto lower-dimensional subspaces whilst preserving overall model behaviour. A simple example is Fitzhugh's (1961) reduction of the Hodgkin and Huxley (1952) equations of neural electrical activity from four state-variables to two, subsequently known as the Fitzhugh-Nagumo model.

In biology, many multi-scale models are typically formulated using *continuous field approximation*, in which reaction-diffusion equations are used to represent macroscopic properties of the system. For example, to model the propagation of electrical activity in excitable tissues such as the heart, the complex network of constituent interconnected cells can be represented using a simplified continuum approximation of cell membrane potential at each point in the tissue (Pullan *et al.*, 2005) (see also section "Modelling Spatial Propagation of Electrical Activation in Heart Tissue" below). Another approach to multi-scale model formulation is the use of *lumped-parameter* or *compartmental* descriptions to represent separable model components such as, for example, modelling of insulin-glucose kinetics using only two discrete intercellular and blood plasma compartments (Tolić *et al.*, 2000). Finally, *cellular-automata* models have also been used to represent multi-scale biological systems, in which simplified rule-based descriptions are employed to characterise biological function of constituent unit components. An example is the cardiac arrhythmia model of Mitchell *et al.* (1992), who used cellular automata to characterise single-cell electrical activity, modelling electrical conduction at the whole-tissue level.

Computational Implementation

To solve multi-scale biological models, commercial multiphysics finite-element numerical software such as COMSOL Multiphysics (COMSOL AB, Switzerland) (Dokos, 2017), or SIMULIA (Dassault Systèmes Simulia, USA) (Baillargeon *et al.*, 2014), are increasingly being employed, along with a range of specialised open-source software platforms including OpenCMISS (see "Relevant Websites section"), Chaste (see "Relevant Websites section"), CHeart (see "Relevant Websites section") and Continuity (see "Relevant Websites section"): the latter platforms utilised within multi-scale biological modelling frameworks such as the Physiome Project (see "Relevant Websites section") (Hunter *et al.*, 2008). Increasingly important to multi-scale biological model development are mark-up language specifications for biological/physiological systems including SBML (see "Relevant Websites section"), CellML (see "Relevant Websites section") and FieldML (see "Relevant Websites section"). Similar multi-scale computational frameworks have been described for chemical process engineering applications (Yang, 2013). An advantage of such frameworks is that, at least in principle, models of component subsystems can be readily swapped in and out from existing model libraries, providing a powerful means of formulating multi-scale biological models (Cooling *et al.*, 2010).

Case Study – Modelling Cardiac Function

An interesting example of multi-scale modelling applied to biology is in the area of cardiac electromechanics, whereby ion channel dynamics in individual cardiomyocytes determines the pattern of electrical activation of the whole heart and its associated contractile performance (Land *et al.*, 2012). Such models could have important applications for a range of patient-specific cardiovascular therapies in future, including development of antiarrhythmic drugs for long-QT syndrome (Hunter and Smith, 2016). A multi-scale model of the heart will incorporate a geometric description of the heart and its chambers, coupled to spatially-heterogeneous models of ion channel kinetics, local wall microstructure and active force generation, all linked to a lumped-parameter description of the circulation system acting as a load on the heart, as shown in **Fig. 2**.

Single-Cell Ionic Models

The basic modelling unit of multi-scale cardiac electromechanics is the single-cell ionic model of cardiac electric activity. Ion flow through voltage-gated transmembrane pores can be experimentally determined from patch-clamp data, yielding whole-cell membrane current macroscopic approximations expressed using either Hodgkin and Huxley (1952) formalism, or as Markov-type models (Hille, 2001). For example, in the human ventricular myocyte model of ten Tusscher *et al.* (2004), the rapid potassium ionic current I_{Kr} can be expressed in Hodgkin-Huxley form as

$$I_{Kr} = G_{Kr} \sqrt{\frac{[K]_o}{5.4}} x_{r1} x_{r2} (V_m - E_K)$$

where V_m is the membrane potential, G_{Kr} is the maximum membrane conductance of I_{Kr} channels, E_K is the equilibrium potential for potassium, $[K]_o$ is the extracellular potassium ion concentration, and x_{r1}, x_{r2} are gating variables, each governed by a first-order ordinary differential equation of the form

$$\frac{dx_{r1}}{dt} = \alpha_{r1}(1 - x_{r1}) - \beta_{r1} x_{r1}$$

where α_{r1}, β_{r1} are rate coefficients that are empirical functions of the membrane potential V_m. The same ionic current can be expressed also in Markov form, whereby several channel states are defined as either open, closed or inactive, and transition rates

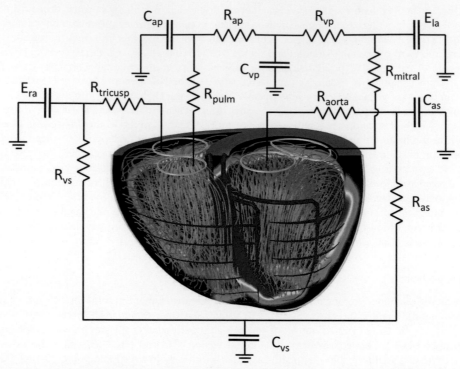

Fig. 2 Multiscale model of cardiac electromechanics, according to Bakir *et al.*, 2017 illustrating multi-physics coupling between cardiac electrical activation (activated regions shown in red, quiescent regions in blue), blood dynamics (shown as velocity streamlines within the left and right ventricular cavities), and lumped-parameter models of systemic and pulmonary circulations. Also shown are the tree-like structures corresponding to the rapidly-conducting cardiac conduction system.

between these states are given as a function of membrane potential V_m (Bett *et al.*, 2011). If the fraction of open states is given by O, then

$$I_{Kr} = G_{Kr}O(V_m - E_K)$$

Summing together all the ion currents present in the cell to form the total ionic current I_{tot}, leads to the space-clamped equation for membrane potential:

$$\frac{dV_m}{dt} = -\frac{I_{tot}}{C_m}$$

where C_m is the membrane capacitance per unit area. This formulation allows for the reconstruction of a cardiac action potential waveform, given an external stimulus of sufficient magnitude (added to I_{tot} in the above).

Several single-cell cardiac ionic models are available from the CellML online repository (see "Relevant Websites section") (Lloyd *et al.*, 2004), which can be used to specify different expressions for I_{tot} in various regions of the heart. Many of these ionic models also include intracellular compartments for simulating calcium ion transients or other ions, second messengers and metabolites.

Modelling Spatial Propagation of Electrical Activation in Heart Tissue

Once I_{tot} has been specified for each region of the heart, propagation of electrical activation across the myocardium can be simulated using the macroscopic continuum approximation of excitable tissues known as the *monodomain formulation* (Cloherty *et al.*, 2006):

$$\beta\left(C_m\frac{\partial V_m}{\partial t} + i_{ion}\right) = \nabla \cdot (\sigma\nabla V_m)$$

where β is the tissue cell membrane surface-to-volume ratio, σ is the tissue conductivity, and ∇ is the *nabla* or *del* spatial partial derivative operator, given in 3D by

$$\nabla \equiv \begin{pmatrix} \dfrac{\partial}{\partial x} \\ \dfrac{\partial}{\partial y} \\ \dfrac{\partial}{\partial z} \end{pmatrix}$$

If the myocardium is *isotropic*, that is its physical properties are not dependent on spatial orientation of the tissue, then σ will be a scalar. In the general case however, myocardial tissue is anisotropic, both in its electrical and mechanical properties, with electrical conductivity dependent on local fine microstructure of the tissue. This microstructure consists of cardiac muscle *fibres* and *fibre sheets*, representing layers of bundled fibres. An orthogonal set of local coordinate axes can then be defined along the fibre, sheet, and_normal-to-sheet_ directions, with electrical conductivity greatest along the fibre axis, followed by the sheet, then the normal-to-sheet axis (Hooks *et al.*, 2007). Defining the unit vectors of these local fibre, sheet and normal axes with respect to the global coordinate system as $\hat{\mathbf{f}}$, $\hat{\mathbf{s}}$, $\hat{\mathbf{n}}$, we can define the conductivity *tensor* for use with the monodomain equation as

$$\sigma = \sigma_f \hat{\mathbf{f}} \otimes \hat{\mathbf{f}} + \sigma_s \hat{\mathbf{s}} \otimes \hat{\mathbf{s}} + \sigma_n \hat{\mathbf{n}} \otimes \hat{\mathbf{n}}$$

where σ_f, σ_s and σ_n denote the scalar conductivities along the local fibre, sheet and normal-to-sheet orientations respectively, and $\otimes$ denotes the tensor outer product. Note that the orientation of the local microstructural axes will vary with spatial position in the heart.

Modelling Passive and Active Mechanical Properties of the Myocardium

In additional to electrical function, passive mechanical properties of the myocardium will also depend on the local microstructure (Dokos *et al.*, 2002). These properties can be expressed using a continuum mechanics approximation of the tissue, typically though use of a *hyperelastic* constitutive law, such as that of Holzapfel *et al.* (2000):

$$W = \frac{c_0}{2}\left(\bar{I}_1 - 3\right) + \frac{c_{f1}}{2c_{f2}}\left[e^{c_{f2}\left(\bar{I}_f - 1\right)^2} - 1\right]$$

where W denotes a scalar *strain energy* function, $\bar{I}_1$ is the first invariant of the Cauchy-Green deformation tensor, c_0, c_{f1} and c_{f2} are material parameters of the tissue, and $\bar{I}_f$ denotes the stretch along the fibre direction $\hat{\mathbf{f}}$. Detailed step-by-step implementation of this material description for myocardium using COMSOL Multiphysics software is given in Dokos (2017).

Many cardiac single-cell ionic models also contain descriptions of cytosolic calcium transients, which can be used to model calcium ion interaction with contractile proteins to generate active force (Hunter *et al.*, 1998). Other models simply link active force generation to the membrane potential itself (Nash and Panfilov, 2004). Either way, these models can be used to generate active tension T_a, which can then be incorporated into the passive strain energy function to determine the 2nd-Piola Kirchhoff stress tensor components S_{ij} within the myocardium according to

$$S_{ij} = \frac{\partial W}{\partial E_{ij}} + T_a \hat{\mathbf{f}} \otimes \hat{\mathbf{f}} + \gamma T_a \hat{\mathbf{s}} \otimes \hat{\mathbf{s}} + \gamma T_a \hat{\mathbf{n}} \otimes \hat{\mathbf{n}}$$

where E_{ij} are the components of the Green strain tensor, and γ represents the fractional component of active tension in the cross-fibre directions due to fibre branching. Passive constitutive laws coupled with active force generation allows structural deformation of the myocardium to be determined (Holzapfel, 2000).

Modelling Fluid Dynamics and Circulatory Blood Flow

Blood flow within the ventricles can be modelled using incompressible Navier-Stokes equations (Fung, 1997):

$$\rho\left(\frac{\partial \mathbf{u}}{\partial t} + \mathbf{u} \cdot \nabla \mathbf{u}\right) = -\nabla p + \mu \nabla^2 \mathbf{u}$$

$$\nabla \cdot \mathbf{u} = 0$$

where $\mathbf{u}$ is the fluid velocity field and p is its pressure, ρ is the fluid density and μ its viscosity. Boundary conditions for the fluid equations include a moving wall condition on the endocardial surface, whereby the fluid velocity is set to equal the velocity of the moving wall determined from the structural mechanics deformation. Because of the deforming fluid domain, a moving mesh approach must also be implemented, known as the Arbitrary Lagrangian-Eulerian (ALE) method (Donea *et al.*, 2004). In turn, the fluid pressure p can be used as an endocardial boundary load for the wall deformation mechanics. In practice however, the pressure will be nearly-constant across the surface at each time instant. As a result, most multi-scale models of cardiac electromechanics omit the Navier-Stokes equations, determining the endocardial fluid pressure load from the circulatory load model instead.

As shown in **Fig. 2**, the circulatory load on the heart can also be approximated by hydraulic circuits, analogous to electric circuits where the analogue of current is fluid flow rate, and the analogue of voltage is fluid pressure. Also shown are capacitive elements, representing the compliance of arteries and veins due to their elastic distensibility. These circuit elements are lumped parameter representations of circulatory components including tissue beds, large arteries and the lungs.

Multi-Scale Overview of Cardiac Electromechanics

From the above descriptions, it is evident that the multi-physics model of cardiac electromechanics combines several spatial scales into a single, albeit complex, model formulation. Beginning with the membrane ionic channels, whose properties can be modelled at the molecular level using molecular dynamics, a macroscopic representation of their function is adopted in the form of Hodgkin-Huxley formalism or Markov-state models. These descriptions of membrane currents are then incorporated into single-cell electrophysiology models, which can additionally include expressions for sub-cellular ionic concentrations, metabolites, contractile processes and signalling pathways. The cell models can then be combined into the monodomain macroscopic formulation of tissue electrical activity to solve for the propagating electrical activation wavefront. Similar macroscopic approximations are also made for the passive and active structural mechanics, incorporating features of the local fine tissue microstructure. Finally, the entire circulation, which sets the loading conditions on the heart, is approximated as a lumped-parameter description, representing various components of the vasculature.

Conclusion

Biological systems are inherently complex, characterised by multiple phenomena operating at various spatial and temporal scales from the molecular and sub-cellular, through to whole-cells and tissues, up to the whole-organ level and beyond. Multi-scale modelling approaches are well-suited to the study of biological systems, integrating knowledge of the various subsystems into a coherent understanding of whole-system properties and function. In future, with ever-increasing computational power becoming available, multi-scale models will undoubtedly yield fundamental insights into numerous complex biological phenomena, even playing a key role in drug design and personalised therapies.

See also: Cell Modeling and Simulation. Computational Approaches for Modeling Signal Transduction Networks. Computational Systems Biology Applications. Computational Systems Biology. Data Formats for Systems Biology and Quantitative Modeling. Integrative Bioinformatics. Molecular Dynamics and Simulation. Natural Language Processing Approaches in Bioinformatics. Pathway Informatics. Quantitative Modelling Approaches. Studies of Body Systems

References

Baillargeon, B., Rebelo, N., Fox, D.D., Taylor, R.L., Kuhl, E., 2014. The living heart project: A robust and integrative simulator for human heart function. European Journal of Mechanics A/Solids 48, 38–47.

Bakir, A.A., Al, A.A., Lovell, N.H. Dokos, S., 2017. Dokos A generic cardiac biventricular fluid-electromechanics model. In: Proceedings of the 39th Annual International Conference of the IEEE Engineering in Medicine and Biology Society (EMBC), pp. 3680–3683.

Bett, G.C.L., Zhou, Q., Rasmusson, R.L., 2011. Models of HERG gating. Biophysical Journal 101, 631–642.

Bonilla, L.L., Capasso, V., Alvaro, M., Carretero, M., 2014. Hybrid modeling of tumor-induced angiogenesis. Physical Review E 90, 062716.

Cloherty, S.L., Dokos, S., Lovell, N.H., 2006. Modeling of electrical activity in cardiac tissue. In: Akay, M. (Ed.), Wiley Encyclopedia of Biomedical Engineering, vol. 2. Hoboken, NJ: John Wiley and Sons, pp. 1216–1226.

Cooling, M.T., Rouilly, V., Misirli, G., et al., 2010. Standard virtual biological parts: A repository of modular modeling components for synthetic biology. Bioinformatics 26, 925–931.

Dada, J.O., Mendes, P., 2011. Multi-scale modelling and simulation in systems biology. Integrative Biology 3, 86–96.

Dokos, S., 2017. Modelling Organs, Tissues, Cells and Devices Using Matlab and Comsol Multiphysics. Berlin: Springer-Verlag.

Dokos, S., Smaill, B.H., Young, A.A., LeGrice, I.J., 2002. Shear properties of passive ventricular myocardium. American Journal of Physiology: Heart and Circulatory Physiology 283, H2650–H2659.

Donea, J., Huerta, A., Ponthot, J.-Ph., Rodríguez-Ferran, A., 2004. Arbitrary Lagrangian-Eulerian methods. In: Stein, E., de Borst, R., Hughes, T.J.R. (Eds.), Encyclopedia of Computational Mechanics. Vol. 1: Fundamentals. New York: Wiley, pp. 413–437.

Donovan, G.M., 2011. Multiscale mathematical models of airway constriction and disease. Pulmonary Pharmacology and Therapeutics 24, 533–539.

van Elteren, A., Pelupessy, I., Portegies Zwart, S., 2014. Multi-scale and multi-domain computational astrophysics. Philosophical Transactions of the Royal Society A 372, 20130385.

Ermentrout, G.B., Edelstein-Keshet, L., 1993. Cellular automata approaches to biological modeling. Journal of Theoretical Biology 160, 97–133.

Fitzhugh, R., 1961. Impulses and physiological states in theoretical models of nerve membrane. Biophysical Journal 1, 445–466.

Fung, Y.C., 1997. Biomechanics: Circulation, second ed. New York: Springer.

Gorban, A.N., Kazantsis, N.K., Kevrekides, I.G., Öttinger, H.C., Theodoropoulos, C., 2006. Model Reduction and Coarse-Graining Approaches for Multiscale Phenomena. Berlin: Springer.

Hille, B., 2001. Ion Channels Of Excitable Membranes, third ed. Sunderland, MA: Sinauer Associates.

Hodgkin, A.L., Huxley, A.F., 1952. A quantitative description of membrane current and its application to conduction and excitation in nerve. Journal of Physiology 117, 500–544.

Holzapfel, G.A., 2000. Nonlinear Solid Mechanics – A Continuum Approach for Engineering. UK: Wiley-Blackwell.

Holzapfel, G.A., Gasser, T.C., Ogden, R.W., 2000. A new constitutive framework for arterial wall mechanics and a comparative study of material models. Journal of Elasticity and the Physical Science of Solids 61, 1–48.

Hooks, D.A., Trew, M.L., Caldwell, B.J., et al., 2007. Laminar arrangement of ventricular myocytes influences electrical behavior of the heart. Circulation Research 101, e103–e112.

Horstemeyer, M.F., 2009. Multiscale modeling: A review. In: Leszczynski, J., Shukla, M.K. (Eds.), Practical Aspects of Computational Chemistry. New York: Springer, pp. 87–135.

Hunter, P.J., Crampin, E.J., Nielsen, P.M.F., 2008. Bioinformatics, multiscale modelling and the IUPS physiome project. Briefings in Bioinformatics 9, 333–343.

Hunter, P.J., McCulloch, A.D., ter Keurs, H., 1998. Modelling the mechanical properties of cardiac muscle. Progress in Biophysics and Molecular Biology 69, 289–331.

Hunter, P., Nielsen, P., 2005. A strategy for integrative computational physiology. Physiology (Bethesda) 20, 316–325.

Hunter, P.J., Smith, N.P., 2016. The cardiac physiome project. Journal of Physiology 594, 6815–6816.

Land, S., Niederer, S.A., Smith, P., 2012. Efficient computational methods for strongly coupled cardiac electromechanics. IEEE Transactions on Biomedical Engineering 59, 1219–1228.

Lloyd, C.M., Halstead, M.D.B., Nielsen, P.F., 2004. CellML: Its future, present and past. Progress in Biophysics and Molecular Biology 85, 433–450.

Mahadevan, V.S., Merzari, E., Tautges, T., et al., 2014. High-resolution coupled physics solvers for analysing fine-scale nuclear reactor design problems. Philosophical Transactions of the Royal Society A 372, 20130381.

Matthies, K., 2006. Exponential estimates in averaging and homogenisation. In: Mielke, A. (Ed.), Analysis, Modeling and Simulation of Multiscale Problems. Berlin: Springer, pp. 1–19.

Mitchell, R.H., Bailey, A.H., Anderson, J., 1992. Cellular automaton model of ventricular fibrillation. IEEE Transactions on Biomedical Engineering 39, 253–259.

Nash, M.P., Panfilov, A.V., 2004. Electromechanical model of excitable tissue to study reentrant cardiac arrhythmias. Progress in Biophysics and Molecular Biology 85, 501–522.

Pullan, A.J., Buist, M.L., Cheng, L.K., 2005. Mathematically Modelling the Electrical Activity of the Heart: From Cell to Body Surface and Back Again. Singapore: World Scientific Publishing.

Quarteroni, A., Veneziani, A., Vergara, C., 2016. Geometric multiscale modeling of the cardiovascular system, between theory and practice. Computer Methods in Applied Mechanics and Engineering 302, 193–252.

Qu, Z., Garfinkel, A., Weiss, J.N., Nivala, M., 2011. Multi-scale modeling in biology: How to bridge the gaps between scales? Progress in Biophysics and Molecular Biology 107, 21–31.

Salciccioli, M., Stamatakis, M., Caratzoulas, S., Vlachos, D.G., 2011. A review of multiscale modeling of metal-catalyzed reactions: Mechanism development for complexity and emergent behavior. Chemical Engineering Science 66, 4319–4355.

Saunders, M.G., Voth, G.A., 2013. Course-graining methods for computational biology. Annual Reviews of Biophysics 42, 73–93.

Southern, J., Pitt-Francis, J., Whiteley, J., et al., 2008. Multi-scale computational modelling in biology and physiology. Progress in Biophysics and Molecular Biology 96, 60–89.

Thorne, B.C., Bailey, A.M., Peirce, S.M., 2007. Combining experiments with multi-cell agent-based modeling to study biological tissue patterning. Briefings in Bioinformatics 8, 245–257.

Tolić, I.M., Mosekilde, E., Sturis, J., 2000. Modeling the insulin-glucose feedback system: The significance of pulsatile insulin secretion. Journal of Theoretical Biology 207, 361–375.

ten Tusscher, K.H.W.J., Noble, D., Noble, P.J., Panfilov, A.V., 2004. A model for human ventricular tissue. American Journal of Physiology: Heart and Circulatory Physiology 286, H1573–H1589.

Viceconti, M., 2012. Multiscale Modelling of the Skeletal System. Cambridge: Cambridge University Press.

Walker, D.C., Georgopoulos, N.T., Southgate, J., 2008. From pathway to population – A multiscale model of juxtacrine EGFR-MAPK signalling. BMC Systems Biology 2, 102.

Winslow, R.L., Scollan, D.F., Holmes, A., et al., 2000. Electrophysiological modeling of cardiac ventricular function: From cell to organ. Annual Review of Biomedical Engineering 2, 119–155.

Yang, A., 2013. On the common conceptual and computational frameworks for multiscale modeling. Industrial and Engineering Chemistry Research 52, 11451–11462.

Zangooei, M.H., Habibi, J., 2017. Hybrid multiscale modeling and prediction of cancer cell behavior. PLOS ONE 12, e0183810.

Zhao, Y., Jiang, C., Yang, A., 2012. Towards computer-aided multiscale modelling: An overarching methodology and support of conceptual modelling. Computers and Chemical Engineering 36, 10–21.

Relevant Websites

https://www.cellml.org
　　cellML.
http://www.cs.ox.ac.uk/chaste
　　Chaste.
http://cheart.co.uk
　　CHeart.
http://continuity.ucsd.edu
　　CONTINUITY 6.
http://opencmiss.org
　　OpenCMISS.
http://physiomeproject.org
　　Physiome Project.
http://physiomeproject.org/software/fieldml
　　Physiome Project.
http://sbml.org
　　SBML.org.

Computational Immunogenetics

Marta Gómez Perosanz, Complutense University of Madrid, Madrid, Spain
Giulia Russo, University of Catania, Catania, Italy
Jose Luis Sanchez-Trincado Lopez, Complutense University of Madrid, Madrid, Spain
Marzio Pennisi, University of Catania, Catania, Italy
Pedro A Reche, Complutense University of Madrid, Madrid, Spain
Adrian Shepherd, University of London, London, United Kingdom
Francesco Pappalardo, University of Catania, Catania, Italy

Introduction

The extraordinary availability of large sets of data produced in biotechnology through the advancement of information technology is promoting the amplification of the knowledge of biological systems. These innovations have swung the pendulum in the way biomedical research, development and applications are done. Clinical data complement biological data, enabling detailed descriptions of various healthy and diseased states, progression and responses to therapies. The amount of data representing various biological states, processes, and their time dependencies enable the study of biological systems at various levels of organization, from molecule to organism, and even at population levels. Multiple sources of data support a rapidly growing body of biomedical knowledge; however, our ability to analyse and interpret these data lags far behind data generation and storage capacity.

Mathematical and computational models are increasingly used to help interpret biomedical data produced by high-throughput genomics and proteomics projects. Advanced applications of computer models that support the simulation of biological processes are used to generate hypotheses and plan experiments. Appropriately interfaced with biomedical databases, computational models are necessary for rapid access to and sharing of knowledge through data mining and knowledge discovery approaches.

The immune system (IS) represents one of the most fascinating systems from many points of view, including biology, physics, computer science and mathematics. Together with the central nervous system, it is probably the most complex, adaptive and highly distributed system in life sciences. It constitutes the fundamental defence mechanism of the vertebrate animals, including human beings, from invading from the pathogens and harmful foreign substances. Following the evolution of multicellular organisms, the immune system developed various defence mechanisms and abilities against all kinds of pathogens, like viruses, bacteria, fungi and other parasites.

Innate Immune System

The "evolutionary-older" mechanisms belong to the class of "innate immunity" defences, and include physical barriers, soluble mediators that directly inhibit foreign micro-organisms and specialized cells able to phagocytose and kill micro-organisms.

The simplest physical barrier is represented by the skin which gives a very effective protection from infection. The mildly acidic pH of skin also inhibits bacterial proliferation. Moreover, the blood coagulation system stops the bleeding, and creates a protective clot over wounds, while the internal mucosae inhibits foreign invaders from entering in the organism. The very acidic pH of stomach juices is also able of effectively sterilize ingested materials. Even the mechanisms that regulate the host temperature (i.e. fever) represent a defence mechanism against the proliferation of some pathogens.

Many soluble molecules such as defensins (natural antibiotics), lysozyme (an enzyme that lyses the wall of some bacteria), the complement (a system of proteins activated in cascade in the cell membrane of certain pathogens that result in their lysis), opsonins (molecules that coat pathogens and facilitate phagocytosis) and some ancient cytokines such as interferons can inhibit viral infections and also kill bacteria.

Innate immunity is also composed by many cellular-driven mechanisms that include the cellular migration toward invaders and phagocytosis, followed by intracellular destruction of the ingested micro-organism. Extracellular killing is also possible if phagocytes release substances that can cause lysis. In mammals such actions are carried by granulocytes and macrophages. The former contains intracellular granules with lytic substances, whereas the latter emigrate to practically all tissues and organs to carry out their phagocyte functions, also important in the turnover of aging cell components.

Even if their evolution is relatively recent and goes in parallel with the evolution of lymphocytes, natural killer (NK) cells are commonly considered as non-phagocytic innate immunity cells, and are able to kill virus-infected cells.

In innate immunity, the recognition of pathogens is done thanks to the presence of pathogen-associated molecular patterns expressed by micro-organisms. Cells belonging to innate immunity hold receptors, globally referred to as "pattern-recognition receptors", that are able to recognize and match with such molecular patterns. These receptors include the family of Toll-like receptors (TLR), mannose receptors (MR) and seven-transmembrane spanning receptors (TM7). TLR recognize bacterial and viral nucleic acids, flagellin, bacterial peptides, lipopolysaccharide (LPS) and other bacterial components. MR receptors bind carbohydrate moieties of several pathogens, such as bacteria, fungi, parasites, and viruses. Finally, TM7 receptors are activated by bacterial peptides or by endogenous chemokines.

Adaptive Immune System

The evolutionary process brought to the developing of newer adaptive and specific responses that couple the classical innate (or natural) immune responses. Adaptive immunity is in fact evolutionary recent. Starting from sharks, the evolution of adaptive immunity followed those of vertebrates, contributing to their evolutionary success and their long life spans. It should be said that adaptive immunity must be seen as a complement of innate immunity, as the two systems strongly interact and cooperate to eradicate pathogens.

Lymphocytes represent the principal effectors of adaptive immunity. Lymphocytes circulate in the blood, tissues, lymphatic vessels and reside in lymphoid organs, which include the thymus, the spleen and lymph nodes. Why are lymphocyte functions so different from innate immunity cells? The answer can be reassumed as follows: specificity, memory and immune tolerance. The lymphocyte population is composed by millions of clones specialized (thanks to their clonotypic antigen receptors) in recognizing a specific antigenic sequence. Only the clones that express the receptors capable of recognizing a specific micro-organism are activated and thus proliferate upon infection, while the others remain inactive. A unique random DNA rearrangement procedure is responsible to bring, from a relatively small pool of DNA sequences, billions of different receptors (a theoretical estimate is of the order of 10^{20}). The immunological meaning of memory represent the capability of lymphocytes to "remember", thanks to specialized memory cells, the first encounter with a given antigen (primary immune response) in order to respond more promptly and more efficiently to subsequent encounters (secondary immune response). Immune tolerance is a selective mechanism used by the immune system to exclude potentially autologous (self) lymphocytes clones through their destruction or their inactivation.

Lymphocytes populations can be distinguished in many ways, starting from the different types of antigen receptors or different specific functions. The most fundamental distinction is between T and B cells. B cells are the main effectors of the so-called "humoral response" and use immunoglobulins(Ig)-like membrane antigen receptors. After recognition and stimulation by the antigen, activated B cells differentiate into plasma cells which secrete a soluble form of the receptor, called antibody. The help of CD4 T-cells (helper T-cells) is a fundamental requirement to mount an appropriate response of B-cells in producing mature antibodies. Each antibody molecule is a dimer of a heavy and a light chain, and each one has a variable and a constant part. The basic antibody molecule is Y-shaped with two independent antigen-binding sites (at the ends of the diagonal segments of the Y). The constant stem of the Y mediates the so-called effector functions of the antibody. We can distinguish among various classes of antibody types. IgM are the first antibodies produced during the primary immune response. IgG are the main class of antibodies released in blood during secondary immune responses. In humans there are four different IgG types (IgG1, IgG2, IgG3 and IgG4). IgA are specialized for functioning in secretions like milk, tears, saliva and intestinal. IgE are best known in conjunction with allergies.

Antibodies bind and inactivate their cognate antigens, and with different mechanisms the promote immunity against foreign pathogens. Antibodies binding to cell membranes may activate the complement system that can lead to cell lysis. Moreover, leukocytes that express surface receptors for the constant stem of antibodies can act with phagocytosis (opsonization) and cell lysis (antibody-dependent cell-mediated cytotoxicity, ADCC) of antibody-bound cells. In this way, macrophages or NK cells acquire the antigen specificity of adaptive immunity. Finally, insoluble antigen-antibody complexes (immunocomplexes) can be rapidly removed from the host.

T cells instead represent the main effectors of the "cell-mediated response". There are many important differences between B and T cells. First, antibodies are at the same time the receptor and the effector molecule of B cells, while the T cell receptor (TCR) is a membrane receptor that activates a series of effector actions mediated by other molecules. Moreover, antibodies can bind any conceivable molecular species such as proteins, lipids, sugars and small organic molecules while T cells receptors are specialized in recognizing only small peptides on the surface of cells. Furthermore, antibodies recognize the antigen in its free native form, whereas TCRs recognize only specific peptides present on the host cell membrane bound to a cellular protein called Major Histocompatibility Complex (MHC), not the whole antigen structure.

There are various T cell populations. Cytotoxic T cells (Tc, also referred to as CTL) directly kill cells expressing the antigen, helper T cells (Th) promote the activities of B and Tc cells, while regulatory T cells (Treg) modulate the immune responses. The Th population can be further distinguished in Th1 and Th2 sub-populations, according to the cytokines they secrete. The former mostly release gamma-interferon and other cytokines to stimulate immune responses against virus and intracellular bacteria, whereas the latter release interleukin 4 (IL-4) against parasites.

Adaptive immune response involves a complex cooperation between both innate and adaptive cells. In general, immune responses start on the periphery of the body where antigen presenting cells (i.e., cells that commonly are able to process and endocytose antigenic sequences or proteins) capture an antigen and move through lymphatic system towards lymph nodes. Here they present the antigenic peptides exposed on their membrane in conjunction with MHC molecules to Th cells and secrete interleukins, such as Intereleukin 12 (IL-12), to promote Th responses. Th cells with specialized TCR for the target antigen are activated by both recognition and co-stimulatory molecules.

Activated Th cells proliferate and secrete other cytokines activating antibody production by B cells and replication of Tc cells. Antibodies, Th and Tc cells reach the periphery where they encounter the pathogen. Antibodies bind to pathogens promoting, for example, phagocytosis by macrophages, activation of the complement system and the ADCC in conjunction with NK cells, respectively. Th cells secrete multiple cytokines with diverse functions such as attraction of other leukocytes in the place of infection, inhibition of viral replication and stimulation of haemathopoiesis to increase the number of leukocytes. Tc cells recognize and directly kill infected cells expressing peptide-MHC complexes on the cell membrane.

As previously stated, lymphocytes receptors are obtained by random rearrangements of a relatively small number of DNA molecules. This random rearrangement sometimes implies the risk of generating receptors which can bring self-reactive responses (autoimmunity). The way the immune system uses to avoid autoimmunity is to generate new cells clones and to destroy potentially harmful clones thereafter. After initial development in the bone marrow, T cell precursors migrate to the thymus. Positive selection occurs in cortex of the thymus over double positive (CD8 + CD4 +) positive thymocytes, selecting those that are capable of engaging with self peptide-MHC complexes. During positive selection, thymocytes commit to recognize either MHC class I or MHC class II molecules, developing into single positive (SP) cells that express either CD8 or CD4. Those cells can not engage with peptide-MHC complex die by apoptosis (death by neglect). Up to 95% of the cells die by neglect. Subsequently, SP cells migrate to the medulla of the thymus and in the region between the cortex and the medulla they undergo the process of negative selection. SP cells that recognized self-peptide/MHC complexes with high affinity undergo apoptosis. Positive selection guarantee MHC restriction while negative selection provides a mechanism of central tolerance.

Elimination of self-reactive T clones in the thymus (central tolerance) is not 100% efficient. Self-reactive clones that surpass the central tolerance are rendered harmless by a mechanism called "peripheral tolerance". T cell activation depends from antigen presentation of the MHC-peptide complex by APC. Only these cells are able to release the appropriate costimulatory molecules required to completely activate T clones. In lack of these costimulatory molecules, the interaction between a self-reactive T cell and a parenchymal cell brings to the inhibition of the T cell (anergy).

Sometimes a breakdown of the tolerance mechanisms may entitle the appearing of autoimmune diseases. Auto-reactive cells can be activated by foreign micro-organisms that express antigens similar to autologous molecules (antigen mimicry) or as a consequence of the presence of cytokines that have been released to activate other clones, but have the collateral effect to counterbalance the lack of costimulatory molecules needed for anergy.

Allergy is due to a pathological immune response (hypersensitivity) to specific antigens (allergens) contained in food, dust, pollens, animal components, chemical products and drugs.

Recognition of the allergen elicits a Th2 response leading to IgE antibody production and capture by receptors on the surface of mast cells and basophil granulocytes (sensitization phase). Allergen binding then triggers the release from cells of various mediators, such as histamine, prostaglandins and cytokines, which cause the allergic reaction.

The type of reaction is typically related to the mode of entry of allergens, e.g. inhaled pollens cause hay fever with sneezing and cough, while food allergens cause gastrointestinal reactions.

The most severe type of allergic reaction, anaphylactic shock, may result from allergen injection, for example bee stings.

A lack of immune response can be caused either by heritable genetic defects (primary immunodeficiency) or by extrinsic causes as, for example, the HIV infection or drug treatments (secondary immunodeficiencies). The fact that all immunodeficient individuals experience a significant increase in sensitivity to chronic and lethal infections is a clear illustration of the defensive role of the immune system toward microorganisms.

The immune system is capable of responding against anything that is considered as non-self. Transplantation of cells, organs or tissues between unrelated individuals elicits in the recipient a very strong immune response directed against the graft. The cause of tissue incompatibilities within the same species is due to the fact that individuals express unique antigens (alloantigens). Either recipient antibodies or by T cells recognizing donor MHC antigens can be cause of rejection. The solution to this problem is given by immunosuppressive drugs like cyclosporin that block T cell activation and response.

In immunology, the use of mathematical and computational methodologies revealed to be a very important support tool not only for the understanding of the immune system behaviour when dealing with pathogens, tumours and auto-immunity disorders, but also for testing, predicting and optimizing possible treatments for them.

Modelling the Immune System

A model is a mapping from a real-world domain to a mathematical/computational domain; thus, it highlights some of the essential believed properties while ignoring believed unessential ones. Mathematical modelling permits the integration of biological and clinical data at various levels and, in doing so, has the potential to provide insights into complex diseases. Modelling necessitates the statement of explicit hypotheses, a process which improves our understanding of the biological system and can uncover critical points where understanding is still poor. Subsequent simulations can reveal hidden patterns and/ or counter-intuitive mechanisms in complex systems. Theoretical thinking and mathematical modelling help generate new hypotheses that can be tested in the laboratory. Mathematical and computational models are usually designed to improve our understanding of phenomena, to validate or reject hypotheses, to shorten production costs, or to find possible points of intervention in order to drive the behaviour of a system towards given goals.

Of course, such models usually represent coarse sketch of the real biological scenario, however they often demonstrated able to capture the complexity of the biological problem, being able to reproduce the most known important immunological features and to bring out some unknown hidden features.

Models help in finding possible therapeutic targets for many diseases, in optimize existing treatments, in shorten, both in terms of costs and time, the pharmaceutical research and discovery pipeline. Furthermore, a technological revolution is nowadays beginning, as models are starting to be personalized on specific individual data. Patient specific models will allow to use such techniques to reproduce both the immunological background and future of the individual, and to personalize therapeutic and prophylactic interventions directly on him, achieving better efficacy and minimizing the risk of undesirable effects. This will drive towards what we call a "personalized medicine" scenario.

A good model should be: (1) relevant, capturing the essential properties of the phenomenon; (2) computable, driving computational knowledge into mathematical representation; (3) understandable, offering a conceptual framework for thinking about the scientific domain; and (4) extensible, allowing the inclusion of additional real properties in the same mathematical scheme.

In the framework of the modelling of immune system dynamics, relevant means that the model should be able to capture the essential properties of the system, namely its organization and dynamic behaviour; computable refers to the model's ability to simulate both the dynamic behaviour and the evolution and interactions of system entities; understandable means that the model must reproduce concepts and ideas of immunology while opening new computational possibilities for understanding immune system interactions; finally, extensible refers to the possibility to include new immunology concepts and knowledge with limited effort using the same mathematical and computational framework.

Immunoinformatics represents a multidisciplinary field dealing with experimental immunology and computer science. In particular, it takes advantage of computational approaches that help to understand the overwhelming and complex dynamics of immune system and to face with the enormous amount of data in immunological context.

Recently, immunoinformatics has grown significantly in scientific field and it potentially has the cards to become a discipline of outstanding value. Immunoinformatics involves several subareas that include immunogenomics, immunoproteomics, epitope prediction, in silico vaccination and systems biology approaches applied to immunology. Whereas immunoinformatics aims to develop mathematical and computational methods to study the dynamics of cellular and molecular entities during the immune response (Pappalardo *et al.*, 2015), the immunological bioinformatics focuses on proposing methods to investigate big genomic and proteomic immunological-related datasets and predict new knowledge mainly by statistical inference and machine learning algorithms.

The glut of data produced by high-throughput instrumentation, notably genomics, transcriptomics, epigenetics, and proteomics methods, requires computational tools for acquisition, storage, and analysis of immunological data. The exploitation of such a huge amount of immunological data usually requires its conversion into computational problems, their solution using mathematical and computational approaches, and then the translation of the obtained results into immunologically meaningful interpretations.

Immunogenetics Modelling Approaches

Antibody Modelling

An antibody (Ab) is a protein with a shape depicting a large Y. It is a protein that is released largely by plasma cells. It's a component of the humoral immunity able to make pathogens not harmful. Specifically, the antibody recognizes the antigen expressed by the pathogen or on the cell surface. The Ab binding region is built by a paratope that is specific for one particular epitope on an antigen. As soon as the Ab binds specifically the antigen, the pathogen or the target cell can be killed directly by the Ab or signalled to other specific immune cells for killing.

The specific function of an antibody (Ab), i.e. where it binds and with what affinity, is intimately related to its structure. However, the number of solved Ab structures is tiny compared to the number of known sequences. Given that precisely solving the structure of an Ab (notably using X-ray crystallography) is typically both challenging and time consuming, modelling the structure computationally is a highly-desirable option.

At the time of writing, there are over 2600 Ab structures in the Protein Data Bank (PDB) (see Relevant Website section), of which around 1700 are bound to some kind of ligand (such as protein, peptide and glycan). Even Abs from different species are sufficiently similar that producing homology models of Ab structures with good overall similarity (measured, for example, using the root-mean-square deviation (RMSD) between the α carbons of model and target) is relatively straightforward using standard homology modelling tools, such as Modeller (Webb and Sali, 2014). But low overall RMSD is rarely the end goal of Ab modelling – what matters for nearly all practical applications is that key regions of an Ab are modelled with high accuracy, notably those parts of the Ab responsible for binding to an Ag. In most cases this requires multiple Complementarity-determining regions (CDRs) – which, depending on the definitions used, consist exclusively or predominantly of loops – to be modelled. CDRs are part of the variable chains in immunoglobulins (antibodies) and T cell receptors, generated by B-cells and T-cells respectively. There are three types of CDRs, CDR1, CDR2 and CDR3. Hypervariable regions associated with both immunoglobulins and T cell receptors are found in CDRs. In general, loop modelling is a considerable challenge and becomes increasingly intractable beyond a length of around 8 amino acids (Fiser *et al.*, 2000). CDR3 loops are usually longer than 10 residues and comprise some of V, all of diversity (D, heavy chains only) and joining (J) regions.

Thankfully there are a number of dedicated Ab modelling tools that exploit knowledge of canonical CDR conformations to produce accurate models of most Ab CDRs, at least when conditions are favourable. Example are Pigs (Marcatili *et al.*, 2014), RosettaAntibody (Weitzner *et al.*, 2017). Modelling and docking of antibody structures with Rosetta, and ABodyBuilder (Leem *et al.*, 2016). A typical strategy for such a tool is to choose appropriate template structures from the PDB (often separate templates for the heavy and light chains), graft canonical CDR loops onto the templates (typically good canonical CDR matching will be found for all but CDR-H3), and model the CDR-H3 (either from a template, or *ab initio*).

The accuracy with which tools can generate models of unknown structures has been evaluated via two Antibody Modelling Assessment initiatives, AMA-I (Almagro *et al.*, 2011) and AMA-II (Almagro *et al.*, 2014), at which research groups were given the task of modelling Ab variable domains of unpublished high-resolution structures. The outcome of AMA-II was positive in several important respects. Accurate models within an RMSD of 1.5 Å are possible with most software for most CDRs and most structures, and higher accuracy is often achievable when someone skilled at modelling is prepared to spend time manually checking and tweaking the model. On the other hand, performance is often poor with Abs from under-represented species and models of the most variable CDR, CDR-H3, are often poor. Given that CDR-H3 is often the most important CDR for Ag binding, this represents a significant problem and an important research topic in its own right (Marks and Deane, 2017).

Even if one has a method capable of generating highly accurate models of unbound Abs, it is not clear how useful that is if one wishes to understand where that Ab binds – even if one has (as is often the case) a solved structure of its known, or likely, Ag. This introduces another challenging problem, that of Ab docking. Protein-protein docking in general has been the subject of the blinded prediction evaluation initiative, CAPRI, since 2002 (Gray et al., 2003). Effective docking commonly involves two key components: the generation of lots of models (decoys) that widely sample the potential "binding space"; and the effective scoring of these models to identify the best decoy. Ab-Ag docking has its own characteristics that make it somewhat different from the more general case – in some respects easier (we know in advance that Abs bind Ags mainly via their CDRs), in other respects potentially more challenging (the CDRs are flexible), and in further respects simply different (Ab contact residues tend to be of different types to those of proteins in general (Brenke et al., 2012). This has led to the development of Ab-specific approaches to docking e.g. SNUGDOCK (Sircar and Gray, 2010), although there appears ample scope for further improvement. In summary, the obstacles to generating a good model of a bound Ab should not be underestimated. However, when circumstances are favourable (which partly depends on the availability of good structural templates), and particularly when an experienced modeller is involved, success is certainly possible.

One context in which there is strong potential to build high quality models of bound Abs is where the model differs only slightly (e.g. by as little as a single residue) from that of a solved structure. Such analyses can be used to infer the impact of mutations to the epitope (e.g. in the context of antigenic escape by an evolving virus) or to the paratope (e.g. where the goal is to engineer a higher-affinity Ab). There are several general approaches for analysing the impact of mutations on the binding between two proteins that can be readily applied to the binding of Ab to Ag. These include online tools such as ANCHOR (Meireles et al., 2010) and mCSM (Pires et al., 2014) that predict changes in the binding affinity between proteins. A much more complex and time-consuming option, but one that has proved remarkably accurate in recent study focusing on broad-spectrum Abs binding to the stalk of influenza A (Lees et al., 2017), is molecular dynamics (MD). MD involves the simulation of molecular forces between atoms and molecules, and can now be performed using relatively modest computational resources through the exploiting of cheap and fast graphical processing units (GPUs). It seems reasonable to assume that MD will make an increasingly important future contribution to the modelling of Abs and Ab/Ag complexes.

Prediction of T- and B-Cell Epitopes

B and T-cells recognize portions within their cognate antigens known as epitopes. The identification of the precise location of these epitopes in antigens is called epitope mapping and is relevant for understanding disease etiology, immune monitoring, developing diagnosis assays and for designing epitope-based vaccines. However, epitope mapping is hampered by the need for experimental testing on large arrays of potential epitope candidates. Luckily, researchers have developed in silico prediction methods that dramatically reduce the burden associated with epitope mapping by decreasing the list of potential epitope candidates. Here, we will review some of the most relevant tools for epitope prediction, with special focus on those that are available for free online use.

T- cell epitope prediction

T-cell epitopes are peptide antigens displayed on the surface of antigen-presenting cells (APCs) and bound to major histocompatibility complex (MHC) molecules. MHC molecules fall in two classes, class I (MHC I) and II (MHC II), that are recognized by two distinct sets of T-cells, CD8 T and CD4 T-cells, respectively (**Fig. 1**). Subsequently, there are two types of T-cell epitopes, CD8 and CD4.

CD8 T-cells become cytotoxic T lymphocyes (CTL) through a priming process that requires the recognition of peptide antigens presented by MHC I molecules on the cell surface of dendritic cells (DC). Thereby, CD8 T-cell epitopes are often known as CTL epitopes. Primed CTLs kill target T-cells displaying their cognate antigens bound to MHC I molecules, thus playing a key role in the clearance of infected and tumoral cells. On the other hand, DC-primed CD4 T-cells become helper (Th) or regulatory (Treg) T-cells that control the immune response through the production of cytokines and cell-to-cell contact.

T-cell epitope mapping consists of isolating the shortest amino acid sequence of an antigen that is able to stimulate either CD4 or CD8 T-cells (Ahmed and Maeurer, 2009). This capacity to stimulate T-cells is called immunogenicity and it is tested in assays requiring synthetic peptides derived from antigens (Ahmad Eweida and El-Sayed, 2016; Malherbe, 2009). There are many distinct peptides within antigens and T-cell prediction methods aim to identify those that are immunogenic. T-cell epitope immuno-genicity is contingent on three basic steps: i) antigen processing, ii) peptide binding to MHC molecules and iii) recognition by cognate TCR. Of these three events, MHC-peptide binding is the most selective one at determining T-cell epitopes (Lafuente and Reche, 2009; Jensen, 2007). Therefore, prediction of peptide-MHC binding is the main basis to anticipate T-cell epitopes.

Prediction of peptide-MHC binding

MHC I and MHC II molecules have similar 3D-structures with bound peptides sitting in a groove delineated by two a-helices overlying a floor comprised of eight antiparallel b-stranded sheets (**Fig. 2**). However, there are also key differences between MHC I

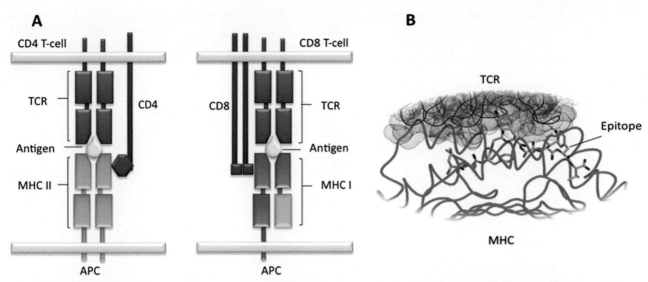

Fig. 1 T-cell epitope recognition (A) Schematic view of T-cell recognition. T cells recognize via T cell receptors (TCR) peptide antigens bound to MHC molecules displayed in the cell surface of antigen presenting cells. CD8 T-cells express CD8 co-receptor that binds to MHC I, while CD4 T-cells express the CD4 co-receptor, which binds to MHC II. (B) Close up of representative 3D-structure of a peptide-MHC I complex recognized by a TCR. MHC I molecule is depicted as green ribbons and the T cell epitope, peptide, in sticks. The TCR α and β chains are shown as blue and green ribbons, respectively, covered by atom surface mesh matching the color of the TCR chains. TCR recognition of peptide-MHC II complexes is similar to that of peptide-MHC I complexes.

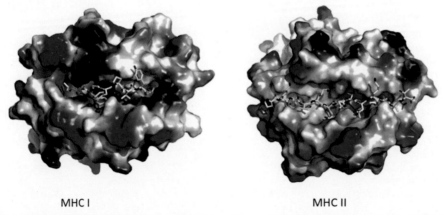

MHC I   MHC II

Fig. 2 Binding groove of MHC molecules. Figure depicts the molecular surface as seen by the TCR of representative MHC I and II molecules. The binding groove of MHC I molecules is closed while is open in MHC II molecules. As a result, MHC I molecules bind short peptides (8–10 amino acids), while MHC II molecules bind longer peptides (9–22 amino acids).

and II binding grooves that condition peptide-binding predictions. The peptide binding cleft of MHC I molecules is closed as it is made by a single a chain. As a result, MHC I molecules can only bind short peptides ranging from 9 to 11 amino acids, whose N- and C-terminal ends remain pinned to conserved residues of the MHC I molecule through a network of hydrogen bonds (Stern and Wiley, 1994; Madden, 1995). The MHC I peptide binding groove also contains deep binding pockets with tight physico-chemical preferences that facilitate binding predictions. There is a complication however. Peptides that have different sizes and bind to the same MHC I molecule often use alternative binding pockets (Madden *et al.*, 1993). Therefore, methods predicting peptide-MHC I binding aim for a fixed peptide length; yet it is generally preferable to predict peptides with nine residues (9-mers) as most MHCI-peptide ligands have that size. In contrast, the peptide-binding groove of MHC II molecules is open, allowing the N- and C-terminal ends of a peptide to extend beyond the binding (Stern and Wiley, 1994; Madden, 1995). As a result, MHC II-bound peptides vary widely in length (9–22 residues), although only a core of nine residues (peptide binding core) sits into the MHC II binding groove. Therefore, peptide-MHC II binding prediction methods generally target to identify these peptide-binding cores. MHCII molecules binding pockets are also shallower and less demanding than those of MHC I molecules. Therefore, peptide-binding prediction to MHC II molecules is less reliable than that of MHC I molecules.

Given the relevance of the problem, there are myriad of methods to predict peptide-MHC binding (**Table 1**). They can be divided in two main categories: data-driven and structure-based methods. Structure-based approaches generally rely on modelling the peptide-MHC structure followed by evaluation of the interaction through methods such as molecular dynamics simulations (Lafuente and Reche, 2009; Desai and Kulkarni-Kale, 2014; Khan *et al.*, 2011; Khan and Ranganathan, 2011, Khan *et al.*, 2010;

Table 1   Selected T cell epitope prediction tools available online for free public use

Tool	URL	Method[a]	MHC Class	S	T	P	Ref.
EpiDOCK	http://epidock.ddg-pharmfac.net	SB	II	–	–	–	Atanasova *et al.* (2013)
MotifScan	https://www.hiv.lanl.gov/content/immunology/motif_scan/motif_scan	SM	I and II	X	–	–	(–)
Rankpep	http://imed.med.ucm.es/Tools/rankpep.html	MM	I and II	–	–	X	Reche *et al.* (2004)
SYFPEITHI	http://www.syfpeithi.de/	MM	I and II	–	–	–	Rammensee *et al.* (1999)
MAPPP	http://www.mpiib-berlin.mpg.de/MAPPP/	MM	I	X	–	X	Hakenberg *et al.* (2003)
PREDIVAC	http://predivac.biosci.uq.edu.au/	MM	II	–	–	–	Oyarzun *et al.* (2013)
PEPVAC	http://imed.med.ucm.es/PEPVAC/	MM	I	X	–	X	Reche and Reinherz (2005)
EPISOPT	http://bio.med.ucm.es/episopt.html	MM	I	X	–	–	Molero-Abraham *et al.* 2013)
Vaxign	http://www.violinet.org/vaxign/	MM	I and II	–	–	–	He *et al.* (2010)
MHCPred	http://www.ddg-pharmfac.net/mhcpred/MHCPred/	QSAR	I and II	–	–	–	Guan *et al.* (2003)
EpiTOP	http://www.pharmfac.net/EpiTOP	QSAR	II	–	–	–	Dimitrov *et al.* (2010)
BIMAS	https://www-bimas.cit.nih.gov/molbio/hla_bind/	QAM	I				Parker *et al.* (1994)
TEPITOPE	http://datamining-iip.fudan.edu.cn/service/TEPITOPEpan/TEPITOPEpan.html	QAM	II	–	–	–	Sturniolo *et al.* (1999)
Propred	http://www.imtech.res.in/raghava/propred/	QAM	II	X	–	–	Singh and Raghava (2001)
Propred-1	http://www.imtech.res.in/raghava/propred1/	QAM	I	X	–	X	Singh and Raghava (2003)
EpiJen	http://www.ddg-pharmfac.net/epijen/EpiJen/EpiJen.htm	QAM	I	–	X	X	Doytchinova *et al.* (2006)
IEDB-MHCI	http://tools.immuneepitope.org/mhci/	Combined	I	–	–	–	Zhang *et al.* (2008)
IEDB-MHCII	http://tools.immuneepitope.org/mhcii/	Combined	II	–	–	–	Zhang *et al.* (2008)
MULTIPRED2	http://cvc.dfci.harvard.edu/multipred2/index.php	ANN	I and II	X	–	–	Zhang *et al.* (2011)
MHC2PRED	http://www.imtech.res.in/raghava/mhc2pred/index.html	SVM	II	–	–	–	Bhasin and Raghava (2004a,b)
NetMHC	http://www.cbs.dtu.dk/services/NetMHC/	ANN	I	–	–	–	Nielsen *et al.* (2003)
NetMHCII	http://www.cbs.dtu.dk/services/NetMHCII/	ANN	II	–	–	–	Nielsen *et al.* (2007)
NetMHCpan	http://www.cbs.dtu.dk/services/NetMHCpan/	ANN	I	–	–	–	Nielsen *et al.* (2007)
NetMHCIIpan	http://www.cbs.dtu.dk/services/NetMHCIIpan/	ANN	II	–	–	–	Nielsen *et al.* (2008)
nHLApred	http://www.imtech.res.in/raghava/nhlapred/	ANN	I	–	–	X	Bhasin and Raghava (2007)
SVMHC	http://abi.inf.uni-tuebingen.de/Services/SVMHC/	SVM	I and II	–	–	–	Donnes and Elofsson (2002)
SVRMHC	http://us.accurascience.com/SVRMHCdb/	SVM	I and II	–	–	–	Liu *et al.* (2006)
NetCTL	http://www.cbs.dtu.dk/services/NetCTL/	ANN	I	X	X	X	Larsen *et al.* (2005)
WAPP	https://abi.inf.uni-tuebingen.de/Services/WAPP/index_html	SVM	I	–	X	X	Donnes and Kohlbacher (2005)

[a]Method used for prediction of peptide-MHC binding. Keys for methods: SM: sequence-motif; SB: structure-based; MM: Motrif-matrix; QAM: quantitative affinity matrix; SVM: support vector machine; ANN: artificial neural network; QSAR: Quantitative structure–activity relationship model; Combined: Tool uses different methods including ANN and QAM, selecting the more appropriated for each distinct MHC molecule. The table also indicate whether the tool predict supertypes (S), TAP binding (T) and proteosomal cleavage (P); marked with an X.

Tong and Ranganathan, 2013). Structure-based methods have the great advantage of not needing experimental data. However, they are seldom used for they are computationally intensive and could exhibit a lower predictive performance than data-driven methods (Patronov and Doytchinova, 2013).

Data-driven methods for peptide-MHC binding prediction are based on peptide sequences that are known to bind to MHC molecules. These peptides sequences are generally available in specialized epitope databases such as IEDB (Vita *et al.*, 2015), EPIMHC (Molero-Abraham *et al.*, 2014), Antijen (Toseland *et al.*, 2005) and others (Singh and Mishra, 2016). Both MHC I and II binding peptides contain frequently occurring amino acids at particular peptides positions, known as anchor residues. Thereby, prediction of peptide-MHC binding was first approached using sequence motifs (SM) reflecting amino acid preferences of MHC molecules at anchor positions (D'Amaro *et al.*, 1995). However, it was soon shown that non-anchor residues also contribute to the capacity of a peptide to bind to a given MHC molecule (Bouvier and Wiley, 1994; Ruppert *et al.*, 1993). Subsequently, researchers developed motif-matrices (MM) that could evaluate the contribution of each and all peptide positions to the binding with the MHC molecule (Nielsen *et al.*, 2004; Rammensee *et al.*, 1999; Reche *et al.*, 2002; Reche and Reinherz, 2007). The most sophisticated form of motif-matrices consists of profiles (Reche *et al.*, 2002, 2004; Reche and Reinherz, 2007) that are similar to those used for detecting sequence homology (Gribskov and Veretnik, 1996). Motif-matrices are often confused with quantitative affinity matrices (QAMs) since both produce peptide scores. However, MMs are derived without taking in consideration values of binding affinities and therefore resulting peptide scores are not suited to address binding affinity. In contrast, QAMs are trained on peptides and corresponding binding affinities, and aim to predict binding affinity. The first method based on QAMs was developed by Parker *et al.* (1994) **(Table 1)**. Subsequently, various approaches were developed to obtain QAMs from peptide affinity data and predict peptide binding to MHC I and II molecules (Bui *et al.*, 2005; Nielsen *et al.*, 2007; Peters and Sette, 2005; Sturniolo *et al.*, 1999).

QAMs and motif-matrices assume an independent contribution of peptide-side chains to the binding. This assumption is well supported by experimental data but there is also evidence that neighbouring peptide residues interfere with others (Peters *et al.*, 2003). To account for those interferences, researchers introduced quantitative structure activity relationship (QSAR) additive models wherein the binding affinity of peptides to MHC is computed as the sum of amino acid contributions at each position plus the contribution of adjacent side chains interactions (Guan *et al.*, 2003). However, machine learning (ML) is the most popular and robust approach introduced to deal with non-linearity of peptide-MHC binding data (Lafuente and Reche, 2009). Researchers have used ML for two distinct problems: discrimination of MHC binders from non-binders and prediction of binding affinity of peptides MHC molecules.

For discrimination models, ML algorithms are trained on data sets consisting of peptides that either bind or do not bind to MHC molecules. Relevant examples of ML-based discrimination models are those based on artificial neural networks (ANNs) (Milik *et al.*, 1998; Brusic *et al.*, 1998), support vector machines (SVMs) (Donnes and Kohlbacher, 2005; Bhasin and Raghava, 2004a,b; Jacob and Vert, 2008), decision trees (DTs) (Zhu *et al.*, 2006; Savoie *et al.*, 1999) and Hidden Markov models (HMMs), that can also cope with non-linear data, have also been used to discriminate peptides binding to MHC molecules. However, unlike other ML algorithms, they have to be trained only on positive data. Three types of HMMs have been used to predict MHC-peptide binding: fully connected HMMs (Mamitsuka, 1998), structure-optimized HMMs (Zhang *et al.*, 2006) and profile HMMs (Zhang *et al.*, 2006; Lacerda *et al.*, 2010). Of those, only fully connected HMMs (fcHMMs) and structure-optimized HMMs (soHMMs) can model non-linear data, recognizing different patterns in the peptide binders. In fact, profile HMMs that are derived from sets of ungapped alignments (the case for peptides binding to MHC), are nearly identical to profile matrices (Durbin *et al.*, 1998) **(Table 1)**.

With regard to predicting binding affinity, ML algorithms are trained on datasets consisting of peptides with known affinity to MHC molecules. Both SVMs and ANNs have been used for such purpose. SVMs were first applied to predict peptide binding affinity to MHCI molecules (Liu *et al.*, 2006) and later to MHC II molecules (Jandrlic, 2016) **(Table 1)**. Likewise, ANNs were also applied first to the prediction of peptide binding to MHC I (Buus *et al.*, 2003; Nielsen *et al.*, 2003) and latter to MHC II molecules (Nielsen and Lund, 2009) **(Table 1)**. Benchmarking of peptide-MHC binding prediction methods appears to indicate that those based on ANNs are superior to those based on QAMs and MMs. However, the differences between the distinct methods are marginal and vary for different MHC molecules (Yu *et al.*, 2002). Moreover, it has been shown that performance of peptide-MHC predictions is improved by combining several methods and providing consensus predictions (Wang *et al.*, 2008).

A major complication for predicting T-cell epitopes through peptide-MHC binding models is MHC polymorphism. In the human, MHC molecules are known as human leukocytes antigens (HLAs), and there are hundreds of allelic variants of class I (HLA I) and class II (HLA II) molecules. These HLA allelic variants bind distinct sets of peptides (Reche and Reinherz, 2003) and hence require specific models for predicting peptide-MHC binding. However, peptide-binding data is only available for a minority of HLA molecules. To overcome this limitation, some researchers have developed pan-MHC specific methods by training ANNs on input data combining MHC residues that contact the peptide with peptide-binding affinity that are capable of predicting peptide-binding affinities to uncharacterized HLA alleles (Nielsen *et al.*, 2007, 2008).

HLA polymorphism also hampers the development of T-cell epitope-based vaccines with wide population coverage as HLA variants are also expressed at vastly variable frequencies in different ethnic groups (Terasaki, 2007). Interestingly, different HLA molecules can also bind similar sets of peptides (Sette and Sidney, 1998, 1999) and researchers have devised methods to cluster them in groups, known as HLA supertypes, consisting of HLA alleles with similar peptide-binding specificities (Doytchinova and Flower, 2005; Greenbaum *et al.*, 2011; Lund *et al.*, 2004). The HLA-A2, HLA-A3, and HLA-B7 are relevant examples of supertypes; 88% of the population expresses at least an allele included in these supertypes (Reche and Reinherz, 2007; Sette and Sidney, 1998, 1999). Identification of promiscuous peptide-binding to HLA supertypes enables the development of T-cell epitope vaccines with high population coverage using a limited number of peptides. Currently, several web-based methods allow the prediction of promiscuous peptide-binding to HLA supertypes for epitope vaccine design including MULTIPRED (Zhang *et al.*, 2011) and PEPVAC (Reche and

Reinherz, 2005) (see **Table 1**). A method to identify promiscuous peptide-binding beyond HLA supertypes was developed and implemented by Molero-Abraham *et al.* (2013) with the name of EPISOPT. EPISOPT predicts HLA I presentation profiles of individual peptides regardless of supertypes and identifies epitope combinations providing a wider population protection coverage.

Prediction of antigen processing and integration with peptide-MHC binding predictions

Antigen processing shapes the peptide repertoire available for MHC binding and is a limiting step determining T-cell epitope immunogenicity (Zhong *et al.*, 2003). Subsequently, computational modelling of the antigen processing pathway provides a mean to enhance T-cell epitope predictions. Antigen presentation by MHC I and II molecules proceed by two different pathways. MHC II presents peptide antigens derived from endocytosed antigens that are degraded and loaded onto the MHC II molecule in endosomal compartments (Blum *et al.*, 2013). Class II antigen degradation is poorly understood and there is lack of good prediction algorithms yet (Hoze *et al.*, 2013). In contrast, MHC I molecules present peptides derive mainly from antigens degraded in the cytosol. The resulting peptides antigens are then transported to the endoplasmic reticulum by TAP where they are loaded onto nascent MHC I molecules (Blum *et al.*, 2013) (**Fig. 3**). Prior to loading, peptides often undergo trimming by ERAAP N-terminal amino peptidases (Hammer *et al.*, 2006).

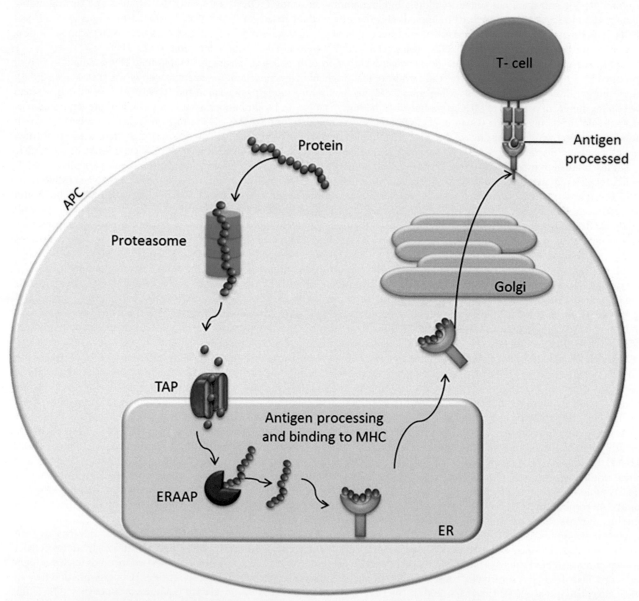

Fig. 3 Class I antigen processing. Figure depicts the major step involved in antigen presentation by MHC I molecules. Proteins are degraded by the proteasome and peptide fragments transported to the endoplasmic reticulum (ER) by TAP before being presented by MHC molecules. TAP transport peptides ranging from 8 to 16 amino acids. Long peptides often become suitable for binding to MCH I molecules after trimming at the N-terminus by ERAAP.

Proteosomal cleavage and peptide-binding to TAP have been studied in detail and there are computational methods to that predict both processes. Proteosomal cleavage prediction models have been derived from peptide fragments generated *in vitro* by human constitutive proteasomes (Nussbaum *et al.*, 2001; Holzhutter *et al.*, 1999) and from sets of MHC I restricted ligands mapped onto their source proteins (Nielsen *et al.*, 2005; Diez-Rivero *et al.*, 2010a,b; Bhasin and Raghava, 2005). On the other hand, TAP binding prediction methods have been developed by training different algorithms on peptides of known affinity to TAP (Bhasin and Raghava, 2004a,b; Diez-Rivero *et al.*, 2010a,b; Daniel *et al.*, 1998; Brusic *et al.*, 1999). Combination of proteosomal cleavage and peptide-binding to TAP with peptide-MHC binding predictions increases T-cell epitope predictive rate in comparison to just peptide-binding to MHC I (Donnes and Kohlbacher, 2005; Tenzer *et al.*, 2005; Bhasin and Raghava, 2004a,b; Doytchinova *et al.*, 2006; Larsen *et al.*, 2005). Subsequently, researchers have developed resources to predict CD8 T-cell epitopes through multistep approaches integrating proteosomal cleavage, TAP transport and peptide-binding to MHC molecules (Reche *et al.*, 2004; Donnes and Kohlbacher, 2005; Doytchinova *et al.*, 2006; Larsen *et al.*, 2005; Schubert *et al.*, 2015, 2016; Rammensee *et al.*, 1995) (**Table 1**).

T-cell epitope prediction concluding remarks

Current peptide-MHC binding predictions methods can readily anticipate T-cell epitopes from primary sequences. However, while it is possible to verify experimental binding data for most peptides predicted to bind to MHC molecules, only ~10% of those are shown to be immunogenic (able to elicit a T-cell response) (Zhong *et al.*, 2003). This scenario remains true when using T-cell epitope predictions methods that combine models for all the steps that determine T-cell epitope immunogenicity. Such a low T-cell epitope discovery rate is due to the fact that we do not have adequate models for antigen processing. In order to accelerate epitope identification and translational vaccine research, we must improve epitope immunogenicity prediction and define rationales and platforms for prioritizing protein antigens in epitope prediction and vaccine design (Flower *et al.*, 2010; Diez-Rivero and Reche, 2012; Molero-Abraham *et al.*, 2015). An alternative approach to overcome the lack of immunogenicity of predicted T-cell epitopes is to combine legacy experimentation, consisting of experimentally defined epitopes, with immunoinformatics predictions. This strategy was first conceived to assemble CD8 T-cell epitope vaccines (Reche *et al.*, 2006; Molero-Abraham *et al.*, 2013) and later extended to CD4 T-cell epitope vaccines (Sheikh *et al.*, 2016). Key criteria for epitope inclusion/selection are conservation and binding to multiple MHC molecules for maximum population protection coverage.

B-cell epitope prediction

B-cells recognize solvent exposed antigens through antigen receptors, B-cell receptor (BCR), consisting of membrane-bound immunoglobulins (**Fig. 4**). Upon activation, B-cells differentiate and secrete soluble forms of the immunoglobulin, also known as antibodies. A B-cell epitope or antigenic determinant is the antigen portion binding to the immunoglobulin or antibody. Antigens recognized by B-cells can be of different chemical nature but most of them are proteins and here will focus on them.

Any solvent exposed region in the antigen can be subject of recognition by antibodies. Nonetheless, B-cell epitopes can be divided in two main groups: linear and conformational (**Fig. 5**). Linear B-cell epitopes consist of sequential residues, peptides, whereas conformational B-cell epitopes consist of patches of solvent exposed atoms from residues that are not necessarily sequential (**Fig. 5**). Therefore, linear and conformational B-cell epitopes are also known as continuous and discontinuous B-cell epitopes, respectively. Antibodies recognizing linear B-cell epitopes can recognize denatured antigens, while denaturing the antigen results in loss of recognition for conformational B-cell epitopes. Most B-cell epitopes (approximately a 90%) are conformational and in fact only a minority of native antigens contains linear B-cell epitopes (Van Regenmortel, 2009).

B-cell epitopes are determined by structural studies that require solving the 3D-structure of antigen-antibody complexes using X-ray crystallography, nucleic magnetic resonance (NMR) or electron microscopy (EM). Alternatively, B-cell epitopes can be identified by functional assays in which the antigen is mutated and the interaction antibody–antigen is evaluated and by screening

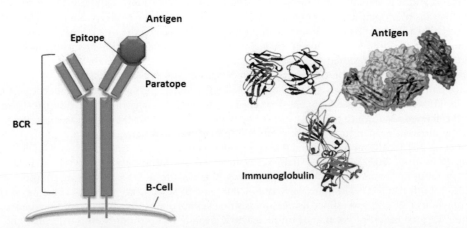

Fig. 4 Antigen-antibody interaction. B-cell epitopes are solvent exposed portions of the antigen that bind to immunoglobins and antibodies. The portion of the antibody recognizing the epitope is known as paratope.

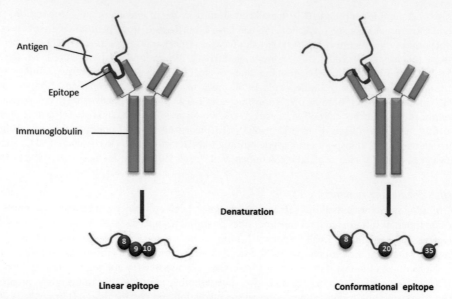

Fig. 5 Linear and conformational B-cell epitopes. Linear B-cell epitopes are composed of sequential/continuous residues, while conformational B-cell epitopes contain scattered/discontinuous residues along the sequence.

of peptide libraries for antibody binding (Potocnakova *et al.*, 2016). B-cell epitope prediction aims to facilitate B-cell epitope identification with the practical purpose of replacing the antigen for antibody production or for carrying structure-function studies.

Prediction of linear B-cell epitopes

For Linear B-cell epitopes are peptides they can easily be used to replace antigens for antibody production. Therefore, despite being a minority, prediction of linear B-cell epitopes has received major attention. Linear B-cell epitopes are predicted from the primary sequence of antigens using sequence-based methods. Early computational methods for the prediction of B-cell epitopes were based on simple amino acid propensities scales depicting physico-chemical features of B-cell epitopes. For example, Hoop and Wood applied residue hydrophilicity calculations for B-cell epitope prediction (Hopp and Woods, 1981, 1983) on the assumption that hydrophilic regions are predominantly located on the protein surface and are potentially antigenic. We know now however that protein surfaces contain roughly the same number of hydrophilic and hydrophobic residues (Lins *et al.*, 2003). Other amino acid propensity scales introduced for B-cell epitope prediction are based on flexibility (Karplus and Schulz, 1985), surface accessibility (Emini Hughes *et al.*, 1985) and ß-turn propensity (Pellequer *et al.*, 1993). Current available bioinformatics tools to predict linear B-cell epitopes using propensity scales include PREDITOP (Pellequer and Westhof, 1993) and PEOPLE (Alix, 1999) (**Table 2**). PREDITOP (Pellequer and Westhof, 1993) uses a multi-parametric algorithm based on hydrophilicity, accessibility, flexibility and secondary structure properties of the amino acids. PEOPLE (Alix, 1999) uses the same parameters and in addition include the assessment of ß-turns. A related method to predict B-cell epitopes was introduced by Kolaskar and Tongaonkar (Kolaskar and Tongaonkar, 1990) consisting of a simple antigenicity scale derived from physicochemical properties and frequencies of amino acids in experimentally determined B-cell epitopes. This index is perhaps the most popular antigenic scale for B-cell epitope prediction and it is actually implemented by GCG (Womble, 2000) and EMBOSS (Rice *et al.*, 2000) packages.

Comparative evaluations of propensity scales carried out on a dataset of 85 linear B-cell epitopes showed that most propensity scales predicted between 50%–70% of B-cell epitopes, with the ß-turn scale reaching the best values (Pellequer *et al.*, 1993, 1991). It has also been shown that combining the different scales does not appear to improve the predictions (Pellequer and Westhof, 1993; Odorico and Pellequer, 2003). Moreover, Blythe and Flower (Blythe and Flower, 2005) demonstrated that single-scale amino acid propensity scales are not reliable to predict epitope location.

Poor performance amino acid scales for prediction of linear B-cell epitopes prompted the introduction of machine learning (ML)-based methods (**Table 3**). These methods are developed by training ML algorithms to distinguish experimental B-cell epitopes from non-B-cell epitopes. Prior to training B-cell epitopes are translated into features vectors capturing selected properties, such as those given by different propensity scales. Relevant examples of B-cell epitope prediction methods based on ML include BepiPred (Jespersen *et al.*, 2017), ABCpred (Saha and Raghava, 2006), LBtope (Singh *et al.*, 2013), BCPREDS (El-Manzalawy *et al.*, 2008) and SVMtrip (Yao *et al.*, 2012). Datasets, training features and algorithms used for developing these methods differ. BepiPred is based on random forests trained on B-cell epitopes obtained from 3D-structures of antigen-antibody complexes (Jespersen *et al.*, 2017). Both BCPREDS (El-Manzalawy *et al.*, 2008) and SVMtrip (Yao *et al.*, 2012) are based on support vector machines (SVM) but while BCPREDS was trained using various string kernels that eliminate the need for representing the sequence into fix length feature vectors, SMVtrip was trained on fix length tripeptide composition vectors. ABCpred and LBtope methods consist of artificial neural networks (ANN) trained on similar postive data, B-cell epitopes, but differ on the negative data, non B-cell epitopes. Negative data used for training ABCpred consisted on random peptides while negative data used for LBtope

Table 2 Selected B-cell epitope prediction methods available for free online use

Tool	Method	Server (URL)	Ref.
Linear B cell epitope			
PEOPLE	Propensity scale method	http://www.iedb.org/	Alix (1999)
BepiPred	ML (DT)	http://www.cbs.dtu.dk/ services/BepiPred/	Jespersen *et al.* (2017)
ABCpred	ML (ANN)	http://www.imtech.res.in/ raghava/abcpred/	Saha and Raghava (2006)
LBtope	ML (ANN)	http://www.imtech.res.in/ raghava/lbtope/	Singh *et al.* (2013)
BCPREDS	ML (SVM)	http://ailab.ist.psu.edu/ bcpred/	El-Manzalawy *et al.* (2008)
SVMtrip	ML (SVM)	http://sysbio.unl.edu/ SVMTriP/prediction.php	Yao *et al.* (2012)
Conformational B-cell epitope			
CEP	Structure-based method (solvent accessibility)	http://bioinfo.ernet.in/cep.htm	Kulkarni-Kale *et al.* (2005)
DiscoTope	Structure-based method (surface accessibility and propensity amino acid score)	http://tools.iedb.org/ discotope/	Haste Andersen *et al.* (2006)
ElliPro	Structure-based method (geometrical properties)	http://tools.iedb.org/ellipro/	Ponomarenko *et al.* (2008)
PEPITO	Structure-based method (physicochemical properties and geometrical structure)	http://pepito.proteomics.ics. uci.edu/	Sweredoski and Baldi (2008)
SEPPA	Structure-based method (physicochemical properties and geometrical structure)	http://lifecenter.sgst.cn/seppa/	Sun *et al.* (2009)
EPITOPIA	Structure-based method (ML- Naïve bayes)	http://epitopia.tau.ac.il/	Rubinstein *et al.* (2009)
EPSVR	Structure-based method (ML-SVR)	http://sysbio.unl.edu/EPSVR/	Liang *et al.* (2010)
EPIPRED	Structure-based method (ASEP, Docking)	http://opig.stats.ox.ac.uk/ webapps/sabdab-sabpred/ EpiPred.php	Krawczyk *et al.* (2014)
PEASE	Structure-based method (ASEP, ML)	http://www.ofranlab.org/ PEASE	Sela-Culang *et al.* (2015a)
MIMOX	Mimotope	http://immunet.cn/mimox/ helps.html	Huang *et al.* (2006)
PEPITOPE	Mimotope	http://pepitope.tau.ac.il/	Mayrose *et al.* (2007)
EpiSearch	Mimotope	http://curie.utmb.edu/ episearch.html	Negi and Braun (2009)
MIMOPRO	Mimotope	http://informatics.nenu.edu. cn/MimoPro	Chen *et al.* (2011)
CBTOPE	Sequence based (SVM)	http://www.imtech.res.in/ raghava/cbtope/submit.php	Ansari and Raghava (2010)

consisted of experimentally validated non B-cell epitopes form IEDB (Vita *et al.*, 2015). In general, B-cell epitope prediction methods based on ML-algorithm are reported to outperform those based on amino acid propensity scales. However, some authors have reported that ML algorithms show little improvement over single scale-based methods (Greenbaum *et al.*, 2007).

Prediction of conformational B-cell epitopes

Most B-cell epitopes are conformational and yet, prediction of conformational B-cell epitopes has lagged behind that of linear B-cell epitopes. There are two main practical reasons for that. First of all, prediction of conformational B-cell epitopes requires in general knowing the protein three-dimensional (3D)-structure and this information is only available for a fraction of the proteins (Levitt, 2009). Secondly, isolating conformational B-cell epitopes from their protein context for selective antibody production is a difficult task that requires suitable scaffold for epitope grafting (Levitt, 2009). Thereby, prediction of conformational B-cell prediction is currently of little relevance for epitope-vaccine design and antibody-based technologies. Nonetheless, prediction of conformational B-cell epitopes is relevant for carrying structure-function studies involving antibody-antigen interactions.

Table 3   A summary list of the best-known and recent databases relevant in immunology research

Databases	Description	References
AAgAtlas http://aagatlas.ncpsb.org	Human autoantigen database	Wang *et al.* (2017)
ALPSbase http://www.niaid.nih.gov/topics/alps/Pages/default.aspx	Autoimmune lymphoproliferative syndrome database	Puck (2005)
AntigenDB http://www.imtech.res.in/raghava/antigend	Sequence, structure and other data on pathogen antigens	Ansari *et al.* (2010)
AntiJen http://www.ddg-pharmfac.net/antijen/AntiJen/antijenhomepage.htm	Quantitative binding data for peptides and proteins of immunological interest	Toseland *et al.* (2005)
BCIpep http://www.imtech.res.in/raghava/bcipep/	Database of experimentally determined B-cell epitopes of antigenic proteins	Saha *et al.* (2005)
bNAber http://bnaber.org/	A database of broadly neutralizing HIV − 1 antibodies	Eroshkin *et al.* (2014)
dbMHC http://www.ncbi.nlm.nih.gov/gv/mhc/	dbMHC provides access to HLA sequences, tools to support genetic testing of HLA loci, HLA allele and haplotype frequencies of over 90 populations worldwide	Helmberg *et al.* (2004)
DIGIT http://circe.med.uniroma1.it/digit	Database of immunoglobulin variable domain sequences annotated with the type of antigen, the germline sequences and pairing information between light and heavy chains	Chailyan *et al.* (2012)
EPIMHC http://imed.med.ucm.es/epimhc/	A curated database of MHC-binding peptides for customized computational vaccinology	Reche *et al.* (2005)
Epitome http://rostlab.org/services/epitome/	Database of all known antigenic residues and the antibodies that interact with them	Schlessinger *et al.* (2006)
GPX-Macrophage Expression Atlas http://gpxmea.gti.ed.ac.uk/	An online resource for expression based studies of a range of macrophage cell types following treatment with pathogens and immune modulators	Grimes *et al.* (2005)
HaptenDB http://www.imtech.res.in/raghava/haptendb	Database of hapten molecules, small molecule not immunogenic by itself, that can react with antibodies of appropriate specificity	Singh *et al.* (2006)
HPTAA http://www.bioinfo.org.cn/hptaa/	Database of potential tumor-associated antigens that uses expression data from various expression platforms, including carefully chosen publicly available microarray expression data, GEO SAGE data and Unigene expression data	Wang *et al.* (2006)
IEDB http://www.iedb.org/	The Immune Epitope Database (IEDB) provides a catalog of experimentally characterised B and T cell epitopes, as well as data on MHC binding and MHC ligand elution experiments	Vita *et al.* (2015)
EDB − 3D http://www.immuneepitope.org/bb_structure.php)	Structural data within the Immune Epitope Database	Ponomarenko *et al.* (2011)
IL2Rgbase http://research.nhgri.nih.gov/scid/	X-linked severe combined immunodeficiency mutations	Puck (1996)
IMGT http://www.imgt.org/	The international ImMunoGeneTics information system (iMGT) is a high-quality integrated knowledge resource specialized in IG, T cell receptors and MHC of human and other vertebrate species and related proteins of the immune system (RPI)	Lefranc *et al.* (2005)
IMGT/GENE-DB http://www.imgt.org/IMGT_GENE-DB/GENElect?livret=0/	IMGT/GENE-DB is part of IMGT database and allows a search of IG and TR gene entries by locus, group, subgroup, based on the classification axiom of IMGT-ONTOLOGY	Giudicelli and Lefranc (1999)
IMGT/HLA http://www.ebi.ac.uk/imgt/hla/	Database of recognized HLA alleles and their sequences for performing sequence alignments	Robinson *et al.* (2000)
IMGT/LIGM-DB http://www.imgt.org/ligmdb/	IMGT/LIGM-DB is a database of immunoglobulin and T cell receptor nucleotide sequences	Giudicelli *et al.* (2006)
IMGT/mAb-DB http://www.imgt.org	Database for therapeutic monoclonal antibodies	Lefranc *et al.* (2015)
ImmuNet http://immunet.princeton.edu	Database of immunological networks and functional relationships between molecular entities, able to predict disease-associated genes	Gorenshteyn *et al.* (2015)
InnateDB http://www.innatedb.com/	Database of genes, proteins, experimentally-verified interactions and signalling pathways involved in the innate immune response of humans, mice and bovines to microbial infection	Lynn *et al.* (2008)
Interferon Stimulated Gene Database http://www.lerner.ccf.org/labs/williams/xchip-html.cgi	A database of interferon stimulated genes (ISGs)	Samarajiwa *et al.* (2009)
IPD - Immuno Polymorphism Database http://www.ebi.ac.uk/ipd/	Databases of polymorphic genes in the immune system	Robinson *et al.* (2003)

Table 3 Continued

Databases	Description	References
IPD-ESTDAB http://www.ebi.ac.uk/ipd/estdab/	IPD-ESTDAB is a database of immunologically characterised melanoma cell lines	Robinson *et al.* (2003)
IPD-HPA - Human Platelet Antigens http://www.ebi.ac.uk/ipd/hpa/	IPD-HPA is a database of alloantigens expressed only on platelets	Robinson *et al.* (2003)
IPD-KIR - Killer-cell Immunoglobulin-like Receptors http://www.ebi.ac.uk/ipd/kir/	IPD-KIR contains the allelic sequences of Killer-cell Immunoglobulin-like Receptors	Robinson *et al.* (2003)
IPD-MHC http://www.ebi.ac.uk/ipd/mhc/	IPD-MHC is a database of sequences of the major histocompatibility complex of different species	Robinson *et al.* (2003)
MHCBN http://www.imtech.res.in/raghava/mhcbn/	MHCBN is a comprehensive database comprising over 23000 peptids sequences, whose binding affinity with MHC or TAP molecules has been assayed experimentally	Bhasin *et al.* (2003)
MPID-T2 http://biolinfo.org/mpid-t2/	MPID-T2 is a highly curated database for sequence-structure-function information on MHC-peptide interactions	Kangueane *et al.* (2001)
MUGEN Mouse Database http://www.mugen-noe.org/database/	Murine models of immune processes and immunological diseases	Aidinis *et al.* (2008)
Protegen http://www.violinet.org/protegen/	Protective antigen database and analysis system	Yang *et al.* (2011)
SAbDab http://opig.stats.ox.ac.uk/webapps/sabdab	Structural Antibody Database	Dunbar *et al.* (2014)
SYFPEITHI http://www.syfpeithi.de/	A database of MHC ligands and motifs	Rammensee *et al.* (1995)
SuperHapten http://bioinformatics.charite.de/superhapten/	Database integrating information from literature and web resources about haptens	Günther *et al.* (2007)
VBASE2 http://www.vbase2.org/	Database of germ-line V genes from the immunoglobulin loci of human and mouse	Retter *et al.* (2005)

There are several methods available to predict conformational B-cell epitopes (**Table 2**). The first to be introduced was CEP (Kulkarni-Kale *et al.*, 2005) which relied almost entirely on predicting patches of solvent exposed residues. It followed DiscoTope (Haste Andersen *et al.*, 2006), which in addiction to solvent accessibility, considered amino acid statistics and spatial information to predict conformational B-cell epitopes. An independent evaluation of these two methods using a benchmark dataset of 59 conformational epitopes revealed that they did not exceed 40% of precision and 46% of recall (Ponomarenko and Bourne, 2007). Subsequently, more methods were developed like ElliPro (Ponomarenko *et al.*, 2008), that aims to identify protruding regions in antigen surfaces, and PEPITO (Sweredoski and Baldi, 2008) and SEPPA (Sun *et al.*, 2009) that manage to combine single physicochemical properties of amino acids and geometrical structure properties. The reported Area Under the Curve (AUC) of these methods is around 0.7, which is indicative of poor discrimination capacity yet better than random. However, in an independent evaluation SEPPA reached an AUC of 0.62 while all the mentioned methods had AUC around 0.5 (Xu *et al.*, 2010). ML has also been applied to predict conformational B-cell epitopes in 3D-structures. Relevant examples include EPITOPIA (Rubinstein *et al.*, 2009) and EPSVR (Liang *et al.*, 2010) which are based on naïve bayes and support vector regression, respectively, train on features vectors combining different scores. The reported AUC of these two methods is around 0.6.

The above methods for conformational B-cell epitope prediction identify generic antigenic regions regardless of antibodies that are ignored. However, there are also methods for antibody-specific epitope prediction (Sela-Culang *et al.*, 2015b). This approach was pioneered by Soga *et al.* (2010) who defined an antibody-specific epitope propensity (ASEP) index after analysing the interfaces of antigen-antibody 3D-structures. Using this index, they developed a novel method of predicting epitope residues for individual antibodies that worked by narrowing down candidate epitope residues predicted by conventional methods. More recently Krawczyk *et al.* (2014) developed EpiPred, a method that uses a docking like approach to matchup antibody-antigen structures and thus identify epitope regions on the antigen. A similar approach is used by PEASE (Sela-Culang *et al.*, 2015a) only that the method utilizes the sequence of the antibody and the 3D-structure of the antigen. Briefly, for each pair of antibody sequence and antigen structure, PEASE uses a machine-learning model trained on properties from 120 antibody-antigen complexes to identify pair combination of residues from CDRs of the antibody and the antigen that are likely to interact.

Another method to identify conformational B-cell epitopes in a protein with a known 3D structure is through mimotope-based methods. Mimotopes are peptides selected from randomized peptide libraries for their ability to bind to an antibody raised against a native antigen. Mimotope-based methods require as input antibody affinity-selected peptides and the 3D structure of antigen. Examples of bioinformatics tools for conformational B-cell epitope prediction using mimotopes include MIMOX (Huang *et al.*, 2006), PEPITOPE (Mayrose *et al.*, 2007), EPISEARCH (Negi and Braun, 2009), MIMOPRO (Chen *et al.*, 2011) and PEPMAPPER (Chen *et al.*, 2012) (**Table 2**).

Methods for conformational B-cell epitope prediction require the 3D-structure of the antigen. Exceptionally, however, Ansari and Raghava (2010) developed a method (CBTOPE) for the identification of conformational B-cell epitope form the primary sequence of the antigen. CBTOPE is based on SVM trained on physicochemical and sequence-derived features of conformational B-cell epitopes. CBTOPE reported accuracy was 86.6% in cross-validation experiments.

B-cell epitope prediction concluding remarks

Prediction B-cell epitopes is clearly relevant for epitope-vaccine design and for developing pharmaceutical and biotech applications based on antibody-antigen interactions. As a result, there are many methods available to predict both conformational and liner B-cell epitopes. However, practical application of B-cell epitope prediction has seldom arrived for different reasons. First of all, prediction of B-cell epitopes is still unreliable for both linear and conformational B-cell epitopes. Secondly, linear B-cell epitopes do usually elicit antibodies that do not cross-react with native antigens. Third, the great majority of B-cell epitopes are conformational and yet predicting conformational epitopes has few applications, as they cannot be isolated from their protein context. Under this scenario, it is key to improve current methods for B-cell epitope prediction and more importantly to develop approaches and platforms for epitope grafting onto suitable scaffolds capable of replacing the native antigen.

Modelling Signalling Pathways in Immunology

The study of immunology includes also the study of immune signalling networks (Dower and Qwarnstrom, 2003). Immunological networks oscillate from the macro to the micro level and within this range the breadth of computational tools co-adjuvate both in vitro and in vivo studies (Kidd *et al.*, 2014).

Several signalling pathway components, such as kinases, phosphatases, adaptor molecules and molecular scaffolds are essential to orchestrate extracellular signals with distinct cellular function. In immune context, cellular signalling leads to the activation of different cells owing specific immune activities (Harwood and Batista, 2010; Kawai and Akira, 2011). Specific ligands bind to a specific receptor on the cellular membrane of an immune system cell to trigger biochemical reactions and signal transduction pathways. Cytokines are secreted by immune cells in response to cellular signalling; they bind to specific membrane receptors to propagate the signal through second messengers, such as tyrosine kinase complexes, with the aim to modify cellular activity and regulate gene expression (Turnera *et al.*, 2014). Interleukins represent the largest class of cytokines and are secreted by a specific leukocyte that indirectly acts on other ones recruited as signalling ligands.

Understanding and targeting these networks may be the key for the development of novel immunomodulatory therapies and the treatment of immune dysfunction and dysregulation disorders. Graph theory (Biggs *et al.*, 1986) and dynamical modelling (Banks *et al.*, 2017) represent two major computational approaches used to study signalling networks (Janes and Lauffenburger, 2013). Both approaches result useful: specifically, network analysis and graph theory improve the understanding of signalling network architecture and organization, and predict the information-processing capabilities of the model through statistical methods, whereas dynamical modelling is helpful to determine how the entire system varies in time and space upon receiving specific and simultaneous stimuli, such as in cross-talk phenomena. Dynamic modelling approaches can be divided in two classes: deterministic and stochastic models. Deterministic models usually use ordinary differential equations (ODEs) based systems of biochemical interactions. Stochastic models use Gillespie's algorithm as the standard approach for computations (Gillespie, 2007).

A new computational approach could be represented by a combination of methodologies consisting of agent-based technique implemented with signalling metabolic pathway analysis. A combination model like this, could be considered exemplary in precision medicine due to the fact that these models own the potential to create therapeutic protocols tailored to a patient's specific biology, needs and health conditions. Indeed, it could be used to better understand the pathogenic mechanisms of autoimmune disease such as multiple sclerosis disorder and to suggest predictive information capabilities. The latter concerns the effectiveness of drugs in typical sclerosis multiple's patient scenario, according to his/her own genetic risk factors, anamnesis, average number and typologies of relapses. To facilitate the design and the simulation of the dynamics of immune system related signalling pathways, specific software like Complex Pathways Simulator (COPASI), may help to this aim (Hoops *et al.*, 2006).

Tools for Modelling Cellular Immune System Dynamics

During the years, many models and tools have been developed to mimic the immune system functions at the cellular scale. However, all the adopted strategies can be grouped in two main approaches: the top-down and the bottom-up approaches.

The first approach is represented by so-called top-down approach. The top down approach works by looking down from above to the problem and estimates the behaviour at the macroscopic level. This is usually obtained by the use of Differential Equations (DE). The DE-based models are all population-based, and the spatiality and topology which both depend on individual interactions are, in general, ignored. Historically, the first models that have been developed for representing the immune system function at the cellular scale make use of such an approach thanks to the possibility to borrow the well-established mathematical techniques widely used for engineering, mathematics and physics. The simplest models based on the top-down approach make use of Ordinary Differential Equations (ODE). For very simple ODE models it is possible to take advantage of years of mathematical studies and methods that can bring out analytical solutions, steady states and asymptotic behaviour. However, as the biological description grows, they become almost intractable and can be only solved numerically. The application of some extension, such as the use of Partial Differential Equations (PDE) can bring some spatial description, whereas the use of stochastic differential equations (SDE) and delayed differential equations (DDE) can allow to reproduce stochastic effects and temporal delays, respectively. However, all of these usually entitle instability problems, a higher computational effort and numerical simulations only. Another problem of differential equations is given by their formalism. The mathematical "language" used is not commonly known to biologists, making the communication with mathematicians/computer scientists even hard, and damaging the interdisciplinary collaboration. Despite of that, models based on differential equations have been successfully used for years bringing out important results and helping in understanding, verifying and validating the basic concepts of immunology.

Many equation-based models have been presented in the literature. In the field of tumor immunology, a simple model for effector cells and tumor cells has been defined by Kuznetsov and co-workers (Kuznetsov et al., 1994). In their model, a threshold above which there is uncontrollable tumor growth is predicted. Below such threshold, the disease can be attenuated with periodic exacerbations occurring every 3–4 months. DeLisi and Rescigno (1977) and Adam (1996) take also into account, besides of IS populations and tumor cells, the external stimulation of immune cells, showing that such a stimulation may increase survival. On the other hand, it has been possible to show that, in some particular scenario, an increase in the number of effector cells may increase the chance of tumor survival. Nani and Ŏguztöreli (1994) include adoptive cellular immunotherapy in their model against tumor growth. This model takes into account stochastic effects (SDE) on the IS-cancer interactions, showing that success of treatment is dependent on the initial tumor burden. Kirschner and co-workers (Kirschner and Webb, 1997) used several differential equations to demonstrate the progression HIV infection by including some important characteristics of the pathology as well as the effects of a combined drug therapy. The model by Nowak et al. (1991) indicated that virus diversity is able to speed up the infection progress of HIV.

Gullo et al. (2015) developed an ODE model for predicting and improving the expansion of human cord blood CD133 + hematopoietic stem/progenitor cells. In this model, no pathology is reproduced. The combined effects (survival, duplication and differentiation) of different cytokines combinations on progenitor stem cells are instead predicted to suggest the best cytokine combination that maximizes duplication, while minimizing differentiation towards other cell populations. Another example of the immune system – tumor interaction under the actions of an external stimulus can be found in the work by Bianca et al. (2012). In their model the authors describe with sufficient degree of detail both the humoral and cellular response of the immune system to the tumor associated antigens and the recognition process between B cells, T cells and antigen presenting cells. The control of the tumor cells growth occurs through the definition of different vaccine protocols with good agreement with in vivo experiments on transgenic mice.

The bottom-up approach works at the opposite side by observing the entities behaviour individually, and thus representing the system at a lower, microscopic level. With the bottom-up approach, unexpected complex global properties can "emerge" as the sum of the rather simple individual behaviours (emergent properties). This approach usually requires greater computational power in order to simulate a significant number of entities, and for this reason it has been pushed only in the last years following the gains in computational performance achieved in the computer world. Cellular automata (CA) and (Multi)Agent-based methods are the most used bottom-up approaches.

CA are composed by a set of identical elements, called cells, each of one occupying a node of a regular and discrete spatial network (a lattice of n-tuples of integer number). The possible states of a cell are discrete and belong to a finite set of states. CA evolve using discrete time steps, and all cells change their states according to a local rule δ (or transition function) homogeneously applied at every step which depends on cell state itself and on the state of the cells in the neighbourhood. The CA-based approach was adopted to reproduce the IS behaviour manly to study their ability to self-adapt and regulate. In this field, the papers by Santos (1999) and Hershberg et al. (2001) showed how, with CA, it was possible to study the HIV-immune dynamics in physical space and Shape Space, respectively. Grilo et al. (1999) combined of Genetic Algorithms and CA to represent the dynamical changes that may occur in the virus - the immune cells equilibrium. In Beauchemin et al. (2005), Beauchemin and co-workers present a simple two-dimensional CA model of influenza infection. CA have been also used to reproduce the IS-Tumor interactions (Mallet and De Pillis, 2006). Other CA-based models for the IS can be found in Bandini (1996), Hu and Ruan (2003) and Perelson (2002).

Agent based models can be seen as an extension of CA. ABM suppose the presence of individuals (agents) that are usually placed on a simulation space (i.e. a lattice). Such agents can be of heterogeneous nature, can have internal properties (i.e. lifetime, internal state and energy), and can act and take decisions (i.e. move, interact with other agents in their neighbourhood, modify their internal state or die) individually or as a result with the interaction with other agents. Models based on CA and ABM have various advantages. By definition, they can be stochastic and include both delays and a spatial description. Furthermore, they allow a richer description of the biological aspects. As a consequence of that, approximations are usually more biological in character than mathematical. Nonlinearities are managed easily and it is always possible to add complexity or modify the model without introducing any new difficulties in solving it. Such methods are also intrinsically numerically stable thanks to the fact that most of the variables representing the attributes of the entities are integers and very few floating point operations are required.

Guo and co-workers validated the three stages of HIV infection by using a multi-agent model (Guo et al., 2005). Perrin and co-workers emphasized the diversity and mutation of HIV virus, which are important characteristics that influence the immune response latency (Perrin et al., 2006). Jacob et al. (2004) presented a swarm-based, three-dimensional model for the human immune system, innate response and adaptive response. The model takes on a strengthening reaction to the previous encountering pathogen, giving a simple representation of immune memory mechanisms.

C-IMMSIM and SIMTRIPLEX are ABM models developed in the C programming language (Bernaschi and Castiglione, 2001; Palladini et al., 2010), with focus on improved efficiency and simulation size and complexity. These simulators, initially developed for HIV and mammary carcinoma, respectively, mimic the receptor interactions by using binary strings. Many mechanism of the immune system (i.e., memory, specificity, tolerance, homeostasis etc.) are here represented. In these simulators, the IS response is designed and coded to allow simulations considering millions of cells with a very high degree of complexity.

The same computational framework developed with SimTriplex has been also used to predict the effects of candidate vaccine adjuvants derived by citrus in boosting the immune system activation (Pappalardo *et al.*, 2016), to investigate the induced immune system response against B16-melanoma (Pappalardo *et al.*, 2011), and, when combined in a hybrid fashion with a lattice Boltzmann (an equation based method), to model the avascular tumor growth under nutrient diffusion, also including the relative immune response (Alemani *et al.*, 2012). SIMMUNE investigates how context adaptive behaviour of the IS might emerge from local cell-cell and cell-molecule interactions (see Relevant Websites section). It is based on molecule interactions on a cells surface. Cells do not have states but behaviours that depend on rules based on cellular response to external stimuli. CYCELLS (Warrender *et al.*, 2006), designed for studying intercellular interactions, uses a hybrid model that represents molecular concentrations continuously and cells discretely. Each type of molecular signal (e.g. cytokines) is represented by a decay and diffusion rate.

Finally, ABM specific development tools have also been used to model the Immune system behaviour, as in the model by Pennisi *et al.* (2013). In their work the authors use an ABM development framework named NetLogo to reproduce the complex cross regulation mechanisms that occur between CD8 T cells and T regulatory cells and show how the appearing of relapsing-remitting multiple sclerosis can occur in predisposed patients that also present breakdown of such regulatory functions.

In the last few years also Petri Nets (PN) have been applied to catch up the immune system dynamics. PN are graphical modelling tool initially developed to reproduce the behaviour of concurrent distributed systems. PN include some graphical elements such as places (represented by empty circles) to represent the possible states of a system, transitions (represented by boxes) to represent possible changes in the states of the system, and arrows to connect places to transitions and vice-versa. Inside places we can find the tokens (small black spheres) to quantify the occurrences or the number of elements in given state. The happening of an event is described by the fact that tokens are taken from input places and consumed by transitions, that can then produce some new tokens in the output places according to well-defined rules. Many analytical methodologies for studying the properties of a PN model have been also presented. A brief summary of PN related methodologies can be found in Murata (1989).

PN have been increasingly applied in other fields of engineering, computer science, chemistry and, lastly, in biology for modelling signalling pathways. Their application to model the immune system function is relatively new and shows how it is possible to apply a particular extension of PN (the Fuzzy continuous PN) for the modelling of the Immune system response (Park *et al.*, 2006). However, such an approach presented some flaws; among them, the biggest one was represented by the fact that it did not allow to distinguish among different kind of cells that have different internal states, receptors, or a different position in the simulation space.

Recently, Pennisi and co-workers presented a novel methodological approach based on Coloured Petri Nets (CPN) to model the immune system function (Pennisi *et al.*, 2016). CPN associate "colours" to tokens. Colours can describe token specific properties and/or internal states (i.e., life-time, position, receptor) and then it is possible to discern among tokens with different internal states. Through a simple example model of the humoral immune system response that was able to represent some of the most complex features of the adaptive IS response like memory and specificity, such a methodology demonstrated able to allow a good level of granularity in the description of cells behaviour, without losing the possibility to have some sort of qualitative analysis, and thus positioning itself somewhat in the middle between top-down and bottom-up approaches.

Immunological Databases

Over recent years database resources have become a fundamental tool for immunological research, due to the growing explosion of extensive information from advanced biotechnology that produce large amounts of biological data. The access to immuno-logical databases is increasingly widespread in immunological research for extracting existing information, planning experiments and analysing experimental results. Immunological data include biological sequences, antigenic and epitope structures, immu-nogenetic properties and specificity of immune interactions. Immunological databases include both bioinformatic resources in which immunological data are stored and computational methodologies suitable for the extraction and the analysis of infor-mation data. These tools have become increasingly sophisticate to allow a speed up of the discovery process, a quick identification of sequences of interest and an achievement of significant bibliographic, taxonomic and feature information.

Repertoire Sequencing

The advent of comparatively cheap Next Generation Sequencing (NGS) is transforming many areas of biology, and immunology is no exception. NGS is a term used to describe a number of high-throughput sequencing technologies that typically generate large numbers of fairly short reads spanning part of a longer nucleic acid sequences. There are many potential applications in this field, ranging from the identification of polymorphisms in immune-related genes, to understanding how rapidly-mutating viruses evolve within a single individual in response to host immune pressure. But many of the most interesting and challenging applications involve characterizing the properties of T-cell and/or B-cell repertoires by sequencing the genomic DNA or messenger RNA that encodes individual TCRs and Abs, a technique known as Rep-seq. The discussion here focuses on Ab repertoire sequencing in order that the additional complexities associated with somatic hypermutation can be discussed, but in other respects the similarities with T-cell repertoire sequencing are considerable.

Ideally, we would like to capture the whole repertoire – paired, full-length sequences of Abs – at sufficiently frequent intervals to elucidate the dynamic properties of interest (potential applications are considered below), but there are important practical limitations. Firstly, there are limits to what proportion of the repertoire is present in a biological sample – only around 2% of the full human repertoire is found in peripheral blood, the commonest type of sample (much larger proportions are present in lymph nodes and spleens). Secondly, the choice of sequencing platform affects both read length and depth-of-coverage (i.e. the number reads spanning a given base in the target sequence). Here it is worth considering the specific challenges posed by Ab repertoires as targets for NGS.

Standard RNA-seq with paired-end read lengths of around 100 (2×50) or 150 (2×75) nucleotides is the norm for genomic studies. These short sequences are then assembled, often using a reference genome, into full-length sequences. However, the pattern of variability within a typical set of Ab sequences – with highly variable regions (the CDRs) separated by less variable regions (the framing regions) – means that naïve approaches to sequence assembly risk generating large numbers of chimeric artefacts, i.e., sequences that do not existing *in vivo*. For this reason, Rep-seq is generally undertaken using reads that are long enough to span all, or most, of the region of interest – which, for the heavy chain, spans all three CDRs and, for some applications, the part of the constant region that determines Ab isotype. In practice, many different read-lengths have been used with 600 (2×300) nucleotides (Illumina MiSeq) arguably the most popular current option. Although no assembly is required, this length is still shorter than one would choose, as it means that the 5′ end of each read starts within a comparatively variable region of the Ab sequence. To address this problem, it is typically necessary to use a large set of V-gene-specific primers.

Given the high diversity of Ab sequences, a further challenge arises from the combination of sequencing depth constraints and sequencing error rates (approximately 2.5×10^6 reads per sample and 10^{-4} per base respectively with Illumina MiSeq). The number of reads is much lower than the actual repertoire diversity; in practice, this means that it is impossible to distinguish between a true sequence polymorphism observed only once and a sequence error. This problem can be partially solved by adding a molecular barcode to each sequence prior to PCR amplification; sequences containing errors that share the same barcode (and hence should be identical) can then be combined to form the correct consensus sequence.

One additional, and crucial, limitation is that standard sequencing platforms fail to retain information about the pairing of heavy and light Ab chains. New technologies are emerging that address this issue using microfluidic single-cell technology. More generally, it seems reasonable to assume that technologies better suited to the specific requirements of Rep-seq will emerge in the fairly near future and give an additional boost to this emerging field.

Bioinformatics plays a crucial role in the analysis of Rep-seq data. The aim of such analyses is often to probe the dynamic properties of the repertoire (e.g. to characterize the polyclonality of a response to a specific challenge as it changes over time), and also the fine details of an immune response (e.g. to elucidate the maturation pathway of a particular Ab of interest). Rep-seq analytics is an active area of research, with new tools regularly developed. A maintained list is available at OMICtools. The key steps that are undertaken in many Rep-seq analyses comprise (1) Preliminary sequence processing, including: NGS sequence quality control; primer masking; handling barcodes (if present); and the assembly of pair-end reads; (2) Junction analysis: the assignment of V, J and D genes (typically in descending order of confidence); and the identification of CDRs; (3) Clonal family inference, i.e. inferring which sequences arose from the same V(D)J rearrangement and (4) Maturation pathway reconstruction, i.e. inferring the unmutated ancestor and intermediates sequences for an Ab of interest.

Additionally, Rep-seq data is often integrated with structural and/or functional information about individual Abs, information that has been derived using a range of techniques including X-ray crystallography, hybridoma technology, or tandem mass spectrometry. Rep-seq of B- and T-cell repertoires has a wide range of applications. It is being used to characterise immune responses to different challenges, including vaccination (Galson *et al.*, 2014), infection (e.g. HIV Zhu *et al.*, 2013), West Nile virus (Tsioris *et al.*, 2015), and herpesviruses (Klarenbeek *et al.*, 2012), and to cancer (van Heijst *et al.*, 2013). It is also being used to help us gain a better understanding of the properties of 'normal' repertoires, such as the degree of similarity/dissimilarity between the repertoires of individuals (including twin studies (Zvyagin *et al.*, 2014), and to elucidate the impact of ageing (Rubelt *et al.*, 2012)). Rep-seq can also help us gain insights into the properties of immune dysfunction (e.g. acute Systemic Lupus Erythematosus (Tipton *et al.*, 2015) and rheumatoid arthritis (Tan *et al.*, 2014).

Conclusions

Computational immunogenetics involves the development and application of bioinformatics methods, mathematical models and statistical techniques for the study of immune system biology. The immune system composition is highly heterogenic and all the entities involved in its functionality play both centralized and de-centralized roles. The metaphor that depicts the immune system like an orchestra is tight-fitting as it really shows a clear image on the machinery that governs the immune function. Tens of different cell types and thousands of interconnected molecular pathways and signals act in real time to carry on the defence against pathogens and tumours. Systems approaches can be used to predict how the immune system will respond to a particular infection or therapeutic strategies, helping in the comprehension on how to act on an immunotherapy to enhance their effect and to reduce possible side-effects. Moreover, computational approaches are increasingly vital to understand the implications of the wealth of gene expression and epigenomics data being gathered from immune cells. The availability of dozens of immune database really helps in organizing and deciphering the huge amount of data that is collected day by day from the availability of new bio-technological approaches. Multi-scale methodologies are needed to tether molecular, cellular and organism levels, in order to gain a global vision on how the things really go. There are already such methodologies but they still suffer of complex issues related to the communications among them i.e., the scales speak different languages in terms of time and space. Computational immunology is also required in the new challenge in the field of computational biomedicine: in silico trials. In silico trial platforms allow the simulation of the relevant individual human physiology and physiopathology in patients. Virtual populations of individuals aim to study the effects of treatments, allowing the simulation of the action of the vaccination strategies and predicting the treatments outcomes in order to have a personalized medicine approach. In the near future, in silico trial could predict, explore and inform of the reasons for failure should the vaccinations strategies against a determined pathology under testing found not efficient, which will suggest possible improvements, which can be rapidly explored on in silico platforms. Another possible benefit of the in silico trial platform could be the reduction of human testing: by providing a reliable prediction of the phase III outcomes on the basis of the data collected a phase II clinical trial, it will increase the confidence in investing in a phase III trial to demonstrate the efficacy in term of reduced recurrence and to drastically reduce the numerosity of the enrolled patients to obtain enough statistical power.

See also: Computing for Bioinformatics. Extraction of Immune Epitope Information. Immunoglobulin Clonotype and Ontogeny Inference. Immunoinformatics Databases. Natural Language Processing Approaches in Bioinformatics. Vaccine Target Discovery

References

Adam, J.A., 1996. Effects of vascularization on lymphocyte/tumor cell dynamics: Qalitative features. Mathematical and Computer Modelling 23, 1–10.

Ahmad, T.A., Eweida, A.E., El-Sayed, L.H., 2016. T-cell epitope mapping for the design of powerful vaccines. Vaccine Reports 6, 13–22.

Ahmed, R.K., Maeurer, M.J., 2009. T-cell epitope mapping. Methods in Molecular Biology 524, 427–438.

Aidinis, V., Chandras, C., Manoloukos, M., *et al.*, 2008. MUGEN mouse database; Animal models of human immunological diseases. Nucleic Acids Research 36, D1048–D1054.

Alemani, D., Pappalardo, F., Pennisi, M., Motta, S., Brusic, V., 2012. Combining cellular automata and lattice Boltzmann method to model multiscale avascular tumor growth coupled with nutrient diffusion and immune competition. Journal of Immunological Methods 376, 55–68.

Alix, A.J., 1999. Predictive estimation of protein linear epitopes by using the program people. Vaccine 18, 311–314.

Almagro, J.C., Beavers, M.P., Hernandez-Guzman, F., *et al.*, 2011. Antibody modeling assessment. Proteins: Structure, Function, and Bioinformatics 79, 3050–3066.

Almagro, J.C., Teplyakov, A., Luo, J., *et al.*, 2014. Second antibody modeling assessment (AMA-II). Proteins: Structure, Function, and Bioinformatics 82, 1553–1562.

Ansari, H.R., Flower, D.R., Raghava, G.P., 2010. AntigenDB: An immunoinformatics database of pathogen antigens. Nucleic Acids Research 38, D847–D853.

Ansari, H.R., Raghava, G.P., 2010. Identification of conformational B-cell Epitopes in an antigen from its primary sequence. Immunome Research 6, 6.

Atanasova, M., Patronov, A., Dimitrov, I., Flower, D.R., Doytchinova, I., 2013. EpiDOCK: A molecular docking-based tool for MHC class II binding prediction. Protein Engineering, Design and Selection 26, 631–634.

Bandini, S., 1996. Hyper-cellular automata for the simulation of complex biological systems: A model for the immune system, special issue on advances in mathematical modeling of biological processes. International Journal of Applied Science and Computation 3, 1076–5131.

Banks, H.T., Hu, S., Rosenberg, E., 2017. A dynamical modeling approach for analysis of longitudinal clinical trials in the presence of missing endpoints. Applied Mathematics Letters 63, 109–117.

Beauchemin, C., Samuel, J., Tuszynski, J., 2005. A simple cellular automaton model for influenza A viral infections. Journal of Theoretical Biology 232, 223–234.

Bernaschi, M., Castiglione, F., 2001. Design and implementation of an immune system simulator. Computation in Biology and Medicine 31, 303–331.

Bhasin, M., Raghava, G.P.S., 2004a. Analysis and prediction of affinity of TAP binding peptides using cascade SVM. Protein Science 13, 596–607.

Bhasin, M., Raghava, G.P.S., 2004b. SVM based method for predicting HLADRB1*0401 binding peptides in an antigen sequence. Bioinformatics 20, 421–423.

Bhasin, M., Raghava, G.P.S., 2005. Pcleavage: An SVM based method for prediction of constitutive proteasome and immunoproteasome cleavage sites in antigenic sequences. Nucleic Acids Research 33, W202–W207.

Bhasin, M., Raghava, G.P., 2007. A hybrid approach for predicting promiscuous MHC class I restricted T cell epitopes. Journal of Biosciences 32, 31–42.

Bhasin, M., Singh, H., Raghava, G.P.S., 2003. MHCBN: A comprehensive database of MHC binding and non-binding peptides. Bioinformatics 19, 665–666.

Bianca, C., Chiacchio, F., Pappalardo, F., Pennisi, M., 2012. Mathematical modeling of the immune system recognition to mammary carcinoma antigen. BMC Bioinformatics 13, S21.

Biggs, N., Lloyd, E.K., Wilson, R.J., 1986. Graph Theory. Oxford: Clarendon.

Blum, J.S., Wearsch, P.A., Cresswell, P., 2013. Pathways of antigen processing. Annual Review of Immunology 31, 443–473.

Blythe, M.J., Flower, D.R., 2005. Benchmarking B cell epitope prediction: Underperformance of existing methods. Protein Science 14, 246–248.

Bouvier, M., Wiley, D.C., 1994. Importance of peptide amino and carboxyl termini to the stability of MHC class I molecules. Science 265, 398–402.

Brenke, R., Hall, D.R., Chuang, G.Y., *et al.*, 2012. Application of asymmetric statistical potentials to antibody–protein docking. Bioinformatics 28, 2608–2614.

Brusic, V., Rudy, G., Honeyman, G., Hammer, J., Harrison, L., 1998. Prediction of MHC class II-binding peptides using an evolutionary algorithm and artificial neural network. Bioinformatics 14, 121–130.

Brusic, V., van Endert, P., Zeleznikow, J., *et al.*, 1999. A neural network model approach to the study of human TAP transporter. In Silico Biology 1, 109–121.

Bui, H.H., Sidney, J., Peters, B., *et al.*, 2005. Automated generation and evaluation of specific MHC binding predictive tools: Arb matrix applications. Immunogenetics 57, 304–314.

Buus, S., Lauemoller, S.L., Worning, P., *et al.*, 2003. Sensitive quantitative predictions of peptide-MHC binding by a 'Query by Committee' artificial neural network approach. Tissue Antigens 62, 378–384.

Chailyan, A., Tramontano, A., Marcatili, P., 2012. A database of immunoglobulins with integrated tools: Digit. Nucleic Acids Research 40, D1230–D1234.

Chen, W., Guo, W.W., Huang, Y., Ma, Z., 2012. PepMapper: A collaborative web tool for mapping epitopes from affinity-selected peptides. PLoS One 7, e37869.

Chen, W.H., Sun, P.P., Lu, Y., *et al.*, 2011. MimoPro: A more efficient Web-based tool for epitope prediction using phage display libraries. BMC Bioinformatics 12, 199.

D'Amaro, J., Houbiers, J.G., Drijfhout, J.W., *et al.*, 1995. A computer program for predicting possible cytotoxic lymphocyte epitopes based on HLA class I peptide-binding motifs. Human Immunology 43, 13–18.

Daniel, S., Brusic, v., Caillat-Zucman, S., *et al.*, 1998. Relationship between peptide selectivities of human transporters associated with antigen processing and HLA class I molecules. The Journal of Immunology 161, 617–624.

DeLisi, C., Rescigno, A., 1977. Immune surveillance and neoplasia-I: A minimal mathematical model. Bulletin of Mathematical Biology 39, 201–221.

Desai, D.V., Kulkarni-Kale, U., 2014. T-cell epitope prediction methods: An overview. Methods In Molecular Biology 1184, 333–364.

Diez-Rivero, C.M., Chenlo, B., Zuluaga, P., Reche, P.A., 2010a. Quantitative modeling of peptide binding to TAP using support vector machine. Proteins 78, 63–72.

Diez-Rivero, C.M., Lafuente, E.M., Reche, P.A., 2010b. Computational analysis and modeling of cleavage by the immunoproteasome and the constitutive proteasome. BMC Bioinformatics 11, 479.

Diez-Rivero, C.M., Reche, P.A., 2012. CD8 T cell epitope distribution in viruses reveals patterns of protein biosynthesis. PLOS One 7, e43674.

Dimitrov, I., Garnev, P., Flower, D.R., Doytchinova, I., 2010. EpiTOP– a proteochemometric tool for MHC class II binding prediction. Bioinformatics 26, 2066–2068.

Donnes, P., Elofsson, A., 2002. Prediction of MHC class I binding peptides, using SVMHC. BMC Bioinformatics 3, 25.

Donnes, P., Kohlbacher, O., 2005. Integrated modelling of the major events in the MHC class I antigen processing pathway. Protein Science 14, 2132–2140.

Dower, S.K., Qwarnstrom, E.E., 2003. Signalling networks, inflammation and innate immunity. Biochemical Society Transactions 31, 1462–1471.

Doytchinova, I.A., Flower, D.R., 2005. In silico identification of supertypes for class II MHCs. The Journal of Immunology 174, 7085–7095.

Doytchinova, I.A., Guan, P., Flower, D.R., 2006. EpiJen: A server for multistep T cell epitope prediction. BMC Bioinformatics 7, 131.

Dunbar, J., Krawczyk, K., Leem, J., et al., 2014. SAbDab: The structural antibody database. Nucleic Acids Research 42, D1140–D1146.

Durbin, R., Eddy, S., Krogh, A., Mitchison, G., 1998. Biological sequence analysis: Probabilistic models of proteins and nucleic acids. Cambridge: Cambridge University Press.

El-Manzalawy, Y., Dobbs, D., Honavar, V., 2008. Predicting linear B-cell epitopes using string kernels. Journal of Molecular Recognition 21, 243–255.

Emini, E.A., Hughes, J.V., Perlow, D.S., Boger, J., 1985. Induction of hepatitis A virus-neutralizing antibody by a virus-specific synthetic peptide. Journal of Virology 55, 836–839.

Eroshkin, A.M., LeBlanc, A., Weekes, D., et al., 2014. bNAber: Database of broadly neutralizing HIV antibodies. Nucleic Acids Research 42, D1133–D1139.

Fiser, A., Do, R.K.G., Šali, A., 2000. Modeling of loops in protein structures. Protein Science 9, 1753–1773.

Flower, D.R., Macdonald, I.K., Ramakrishnan, K., Davies, M.N., Doytchinova, I.A., 2010. Computer aided selection of candidate vaccine antigens. Immunome Research 6, S1.

Galson, J.D., Pollard, A.J., Trück, J., Kelly, D.F., 2014. Studying the antibody repertoire after vaccination: Practical applications. Trends in immunology 35, 319–331.

Gillespie, D.T., 2007. Stochastic simulation of chemical kinetics. Annual Review of Physical Chemistry 58, 35–55.

Giudicelli, V., Ginestoux, C., Folch, G., et al., 2006. IMGT/LIGM-DB, the IMGT®comprehensive database of immunoglobulin and T cell receptor nucleotide sequences. Nucleic Acids Research 34, D781–D784.

Giudicelli, V., Lefranc, M.P., 1999. Ontology for immunogenetics: The IMGT-Ontology. Bioinformatics 15, 1047–1054.

Gorenshteyn, D., Zaslavsky, E., Fribourg, M., et al., 2015. Interactive big data resource to elucidate human immune pathways and diseases. Immunity 43, 605–614.

Gray, J.J., Moughon, S.E., Kortemme, T., et al., 2003. Protein–protein docking predictions for the CAPRI experiment. Proteins: Structure, Function, and Bioinformatics 52, 118–122.

Greenbaum, J.A., Andersen, P.H., Blythe, M., et al., 2007. Towards a consensus on datasets and evaluation metrics for developing B-cell epitope prediction tools. Journal of Molecular Recognition 20, 75–82.

Greenbaum, J., Sidney, J., Chung, J., et al., 2011. Functional classification of class II human leukocyte antigen (HLA) molecules reveals seven different supertypes and a surprising degree of repertoire sharing across supertypes. Immunogenetics 63, 325–335.

Gribskov, M., Veretnik, S., 1996. Identification of sequence pattern with profile analysis. Methods in Enzymology 266, 198–212.

Grilo, A., Caetano, A., Rosa, A., 1999. Immune system simulation through a complex adaptive system model. In: Dasgupta, D., Nino, F. (Eds.), Proceedings of the 3rd Workshop on Genetic Algorithms and Artificial Life (GAAL99), pp. 1–2. Lisbon: CRC Press.

Grimes, G.R., Moodie, S., Beattie, J.S., et al., 2005. GPX-Macrophage Expression Atlas: A database for expression profiles of macrophages challenged with a variety of pro-inflammatory, anti-inflammatory, benign and pathogen insults. BMC Genomics 6, 178.

Guan, P., Doytchinova, A., Zygouri, C., Flower, D.R., 2003. MHCPred: A server for quantitative prediction of peptide-MHC binding. Nucleic Acids Research 31, 3621–3624.

Gullo, F., van der Garde, M., Russo, G., et al., 2015. Computational modeling of the expansion of human cord blood CD133 + hematopoietic stem/progenitor cells with different cytokine combinations. Bioinformatics 31, 2514–2522.

Guo, Z., Han, H.K., Tay, J.C., 2005. Sufficiency verification of HIV-1 pathogenesis based on multi-agent simulation. In: Beyer, H., O'Reilly, U., Arnold, D., et al. (Eds.), Proceedings of the ACM Genetic and Evolutionary Computation Conference 2005 (GECCO'05), pp. 305–312. Washington: ACM Press.

Günther, S., Hempel, D., Dunkel, M., Rother, K., Preissner, R., 2007. SuperHapten: A comprehensive database for small immunogenic compounds. Nucleic Acids Research 35, D906–D910.

Hakenberg, J., Nussbaum, A.K., Schild, H., et al., 2003. MAPPP: MHC class I antigenic peptide processing prediction. Appl Bioinformatics 2, 155–158.

Hammer, G.E., Gonzalez, F., Champsaur, M., Cado, D., Shastri, N., 2006. The aminopeptidase ERAAP shapes the peptide repertoire displayed by major histocompatibility complex class I molecules. Nature Immunology 7, 103–112.

Harwood, N.E., Batista, F.D., 2010. Early events in B cell activation. Annual Review of Immunology 28, 185–210.

Haste Andersen, P., Nielsen, M., Lund, O., 2006. Prediction of residues in discontinuous B-cell epitopes using protein 3D structures. Protein Science 15, 2558–2567.

Helmberg, W., Dunivin, R., Feolo, M., 2004. The sequencing-based typing tool of dbMHC: Typing highly polymorphic gene sequences. Nucleic Acids Research 32, W173–W175.

Hershberg, U., Louzoun, Y., Atlan, H., Solomon, S., 2001. HIV time hierarchy: Winning the war while, losing all the battles. Physica A 289, 178–190.

He, Y., Xiang, Z., Mobley, H.L., 2010. Vaxign: The first web-based vaccine design program for reverse vaccinology and applications for vaccine development. Journal of Biomedicene and Biotechnology 2010, 297505.

Holzhutter, H.G., Frommel, C., Kloetzel, P.M., 1999. A theoretical approach towards the identification of cleavage-determining amino acid motifs of the 20 S proteasome. Journal of Molecular Biology 286, 1251–1265.

Hoops, S., Sahle, S., Gauges, et al., 2006. COPASI: A COmplex PAthway SImulator. Bioinformatics 22, 3067–3074.

Hopp, T.P., Woods, K.R., 1981. Prediction of protein antigenic determinants from amino acid sequences. Proceedings of the National Academy of Sciences of the USA 78, 3824–3828.

Hopp, T.P., Woods, K.R., 1983. A computer program for predicting protein antigenic determinants. Molecular Immunology 20, 483–489.

Hoze, E., Tsaban, L., Maman, Y., Louzoun, Y., 2013. Predictor for the effect of amino acid composition on CD4 + T cell epitopes preprocessing. Journal of Immunological Methods 391, 163–173.

Huang, J., Gutteridge, A., Honda, W., Kanehisa, M., 2006. MIMOX: A web tool for phage display based epitope mapping. BMC Bioinformatics 7, 451. [Immunogenetics 41, 178–228].

Hu, R., Ruan, X., 2003. A simple cellular automaton model for tumor-immunity system. In: Proceedings of IEEE International Conference of Robotics, Intelligent Systems and Signal Processing, pp. 1031–1035. Changsha, Hunan, China: IEEE Press.

Jacob, C., Litorco, J., Lee, L., 2004. Immunity through swarms: Agent-based simulations of the human immune system. Lecture Notes in Computer Science 3239, 400–412.

Jacob, L., Vert, J.P., 2008. Efficient peptide-MHC-I binding prediction for alleles with few known binders. Bioinformatics 24, 358–366.

Jandrlic, D.R., 2016. SVM and SVR-based MHC-binding prediction using a mathematical presentation of peptide sequences. Computational Biology and Chemistry 65, 117–127.

Janes, K.A., Lauffenburger, D.A., 2013. Models of signalling networks – what cell biologists can gain from them and give to them. Journal of Cell Science 126, 1913–1921.

Jensen, P.E., 2007. Recent advances in antigen processing and presentation. Natural Immunology 8, 1041–1048.

Jespersen, M.C., Peters, B., Nielsen, M., Marcatili, P., 2017. BepiPred-2.0: Improving sequence-based B-cell epitope prediction using conformational epitopes. Nucleic Acids Research 2. doi:10.1093/nar/gkx346.

Kangueane, P., Sakharkar, M.K., Kolatkar, P.R., Ren, E.C., 2001. Towards the MHC-Peptide combinatorics. Human Immunology 62, 539–556.

Karplus, P.A., Schulz, G.E., 1985. Prediction of chain flexibility in proteins: A tool for the selection of peptide antigen. Naturwissenschaften 72, 212–213.

Kawai, T., Akira, S., 2011. Toll-like receptors and their crosstalk with other innate receptors in infection and immunity. Immunity 34, 637–650.

Khan, J.M., Cheruku, H.R., Tong, J.C., Ranganathan, S., 2011. MPID-T2: A database for sequence-structure-function analyses of pMHC and TR/pMHC structures. Bioinformatics 27, 1192–1193.

Khan, J.M., Ranganathan, S., 2011. Understanding TR binding to pMHC complexes: How does a TR scan many pMHC complexes yet preferentially bind to one. PLoS ONE 6 (2), e17194.

Khan, J.M., Tong, J.C., Ranganathan, S., 2010. Structural Immunoinformatics: Understanding MHC-Peptide-TR binding. In: Davies, N., Ranganathan, S., Flower, D.R. (Eds.), Bioinformatics for Immunomics. New York: Springer, pp. 77–93.

Kidd, B.A., Peters, L.A., Schadt, E.E., Dudley, J.T., 2014. Unifying immunology with informatics and multiscale biology. Nature Immunology 15, 118–127.

Kirschner, D.E., Webb, G.F., 1997. A mathematical model of combined drug therapy of HIV infection. Journal of Theoretical Medicine 1, 25–34.

Klarenbeek, P.L., Remmerswaal, E.B.M., ten Berge, I.J.M., et al., 2012. Deep sequencing of antiviral T-cell responses to HCMV and EBV in humans reveals a stable repertoire that is maintained for many years. PLoS Pathogens 8, e1002889.

Kolaskar, A.S., Tongaonkar, P.C., 1990. A semi-empirical method for prediction of antigenic determinants on protein antigens. FEBS Letters 276, 172–174.

Krawczyk, K., Liu, X., Baker, T., et al., 2014. Improving B-cell epitope prediction and its application to global antibody-antigen docking. Bioinformatics 30, 2288–2294.

Kulkarni-Kale, U., Bhosle, S., Kolaskar, A.S., 2005. CEP: A conformational epitope prediction server. Nucleic Acids Research 33, W168–W171.

Kuznetsov, V.A., Makalkin, I.A., Taylor, M.A., Perelson, M.A., 1994. Non-linear dynamics of immunogenic tumors: Parameter estimation and global bifurcation analysis. Bulletin of Mathematical Biology 56, 295–321.

Lacerda, M., Scheffler, K., Seoighe, C., 2010. Epitope discovery with phylogenetic hidden Markov models. Molecular Biology and Evolution 27, 1212–1220.

Lafuente, E.M., Reche, P.A., 2009. Prediction of MHC-peptide binding: A systematic and comprehensive overview. Current Pharmaceuticals Design 15, 3209–3220.

Larsen, M.V., Lundegaard, C., Lamberth, K., et al., 2005. An integrative approach to CTL epitope prediction: A combined algorithm integrating MHC class I binding, TAP transport efficiency, and proteasomal cleavage predictions. European Journal of Immunology 35, 2295–2303.

Leem, J., Dunbar, J., Georges, G., Shi, J., Deane, C.M., et al., 2016. ABodyBuilder: Automated antibody structure prediction with data–driven accuracy estimation. MAbs 8, 1259–1268.

Lees, W.D., Stejskal, L., Moss, D.S., Shepherd, A.J., 2017. Investigating substitutions in antibody–antigen complexes Using Molecular Dynamics: A case study with Broad-spectrum, influenza a antibodies. Frontiers In Immunology 8, 143.

Lefranc, M.-P., Giudicelli, V., Duroux, P., et al., 2015. IMGT®, the international ImMunoGeneTics information system® 25 years on. Nucleic Acids Research 43, D413–D422.

Lefranc, M.P., Giudicelli, V., Kaas, Q., et al., 2005. IMGT, the international ImMunoGeneTics information system®. Nucleic Acids Research 33, D593–D597.

Levitt, M., 2009. Nature of the protein universe. Proceedings of the National Academy of Sciences of the USA 106, 11079–11084.

Liang, S., Zheng, D., Standley, D.M., et al., 2010. EPSVR and EPMeta: Prediction of antigenic epitopes using support vector regression and multiple server results. BMC Bioinformatics 11, 381.

Lins, L., Thomas, A., Brasseur, R., 2003. Analysis of accessible surface of residues in proteins. Protein Science 12, 1406–1417.

Liu, W., Meng, X., Xu, Q., Flower, D.R., Li, T., 2006. Quantitative prediction of mouse class I MHC peptide binding affinity using support vector machine regression (SVR) models. BMC Bioinformatics 7, 182.

Lund, O., Nielsen, M., Kesmir, C., et al., 2004. Definition of supertypes for HLA molecules using clustering of specificity matrices. Immunogenetics 55, 797–810.

Lynn, D.J., Winsor, G.L., Chan, C., et al., 2008. InnateDB: Facilitating systems-level analyses of the mammalian innate immune response. Molecular Systems Biology 4, 218.

Madden, D.R., 1995. The three-dimensional structure of peptide-MHC complexes. Annual Review of Immunology 13, 587–622.

Madden, D.R., Garboczi, D.N., Wiley, D.C., 1993. The antigenic identity of peptide-MHC complexes: A comparison of the conformations of five viral peptides presented by HLA-A2. Cell 75, 693–708.

Malherbe, L., 2009. T-cell epitope mapping. Annals of Allergy Asthma and Immunology 103, 76–79.

Mallet, D., De Pillis, L., 2006. A cellular automata model of tumor immune system interactions. Journal of Theoretical Biology 239, 334–350.

Mamitsuka, H., 1998. Predicting peptides that bind to MHC molecules using supervised learning of hidden Markov models. Proteins 33, 460–474.

Marcatili, P., Olimpieri, P.P., Chailyan, A., Tramontano, A., 2014. Antibody modeling using the prediction of ImmunoGlobulin structure (PIGS) web server. Nature Protocols 9, 2771–2783.

Marks, C., Deane, C.M., 2017. Antibody H3 structure prediction. Computational and Structural Biotechnology Journal 15, 222–231.

Mayrose, I., Penn, O., Erez, E., et al., 2007. Pepitope: Epitope mapping from affinity-selected peptides. Bioinformatics 23, 3244–3246.

Meireles, L.M.C., Dömling, A.S., Camacho, C.J., 2010. ANCHOR: A web server and database for analysis of protein–protein interaction binding pockets for drug discovery. Nucleic Acids Research 38, W407–W411.

Milik, M., Sauer, D., Brunmark, A.P., et al., 1998. Application of an artificial neural network to predict specific class I MHC binding peptide sequences. Natural Biotechnology 16, 753–756.

Molero-Abraham, M., Glutting, J.P., Flower, D.R., Lafuente, E.M., Reche, P.A., 2015. EPIPOX: Immunoinformatic characterization of the shared T-Cell epitome between Variola virus and related pathogenic Orthopoxviruses. Journal of Immunology Research 2015, 738020.

Molero-Abraham, M., Lafuente, E.M., Flower, D.R., Reche, P.A., 2013. Selection of conserved epitopes from hepatitis C virus for pan-populational stimulation of T-cell responses. Clinical and Developmental Immunology 2013, 601943.

Molero-Abraham, M., Lafuente, E.M., Reche, P., 2014. Customized predictions of peptide-MHC binding and T-cell epitopes using EPIMHC. Methods In Molecular Biology 1184, 319–332.

Murata, T., 1989. Petri Nets: Properties, analysis and applications. Proceedings of the IEEE 77, 541–580.

Nani, F.K., Öguztöreli, M.N., 1994. Modelling and simulation of Rosenberg- type adoptive cellular immunotherapy. IMA Journal of Mathematics Applied in Medicine & Biology 11, 107–147.

Negi, S.S., Braun, W., 2009. Automated detection of conformational epitopes using phage display Peptide sequences. Bioinformatics and Biology Insights 3, 71–81.

Nielsen, M., Lund, O., 2009. NN-align. An artificial neural network-based alignment algorithm for MHC class II peptide binding prediction. BMC Bioinformatics 10, 296.

Nielsen, M., Lundegaard, C., Blicher, T., et al., 2007. NetMHCpan, a method for quantitative predictions of peptide binding to any HLA-A and -B locus protein of known sequence. PLoS One 2 (8), e796.

Nielsen, M., Lundegaard, C., Blicher, T., et al., 2008. Quantitative predictions of peptide binding to any HLA-DR molecule of known sequence: Netmhciipan. PLoS Computational Biology 4, e1000107.

Nielsen, M., Lundegaard, C., Lund, O., 2007. Prediction of MHC class II binding affinity using SMM-align, a novel stabilization matrix alignment method. BMC Bioinformatics 8, 238.

Nielsen, M., Lundegaard, C., Lund, O., Kesmir, C., 2005. The role of the proteasome in generating cytotoxic T-cell epitopes: Insights obtained from improved predictions of proteasomal cleavage. Immunogenetics 57, 33–41.

Nielsen, M., Lundegaard, C., Worning, P., et al., 2003. Reliable prediction of T-cell epitopes using neural networks with novel sequence representations. Protein Science 12, 1007–1017.

Nielsen, M., Lundegaard, C., Worning, P., et al., 2004. Improved prediction of MHC class I and class II epitopes using a novel Gibbs sampling approach. Bioinformatics 20, 1388–1397.

Nowak, M.A., Anderson, R.M., Mclean, A.R., et al., 1991. Antigenic diversity thresholds and the development of AIDS. Science 254, 963–969.

Nussbaum, A.K., Kuttler, C., Hadeler, K.P., Rammensee, H.G., Schild, H., 2001. PAProC: A prediction algorithm for proteasomal cleavages available on the WWW. Immunogenetics 53, 87–94.

Odorico, M., Pellequer, J.L., 2003. BEPITOPE: Predicting the location of continuous epitopes and patterns in proteins. Journal of Molecular Recognition 16, 20–22.

Oyarzun, P., Ellis, J.J., Boden, M., Kobe, B., 2013. PREDIVAC: CD4+ T-cell epitope prediction for vaccine design that covers 95% of HLA class II DR protein diversity. BMC Bioinformatics 14, 52.

Palladini, A., Nicoletti, G., Pappalardo, F., Murgo, A., Grosso, V., et al., 2010. In silico modeling and in vivo efficacy of cancer preventive vaccinations. Cancer Research 70, 7755–7763.

Pappalardo, F., Fichera, E., Paparone, N., Lombardo, A., Pennisi, M., et al., 2016. A computational model to predict the immune system activation by citrus derived vaccine adjuvants. Bioinformatics 32, 2672–2680.

Pappalardo, F., Flower, D., Russo, G., Pennisi, M., Motta, S., 2015. Computational modelling approaches to vaccinology. Pharmacological Research 92, 40–45.

Pappalardo, F., Forero, I.M., Pennisi, M., Palazon, A., Melero, I., et al., 2011. SimB16: Modeling induced immune system response against B16-melanoma. PLoS ONE 6.

Parker, K.C., Bednarek, M.A., Coligan, J.E., 1994. Scheme for ranking potential HLA-A2 binding peptides based on independent binding of individual peptide side chains. Journal of Immunology 152, 163–175.

Park, I., Na, D., Lee, D., Lee, K.H., 2006. Fuzzy continuous Petri Net-based approach for modeling immune systems. Lecture Notes In Computer Science 3931, 278–285.

Patronov, A., Doytchinova, I., 2013. T-cell epitope vaccine design by immunoinformatics. Open Biology 3, 120139.

Pellequer, J.L., Westhof, E., 1993. PREDITOP: A program for antigenicity prediction. Journal of Molecular Graphics 11, 204–210.

Pellequer, J.L., Westhof, E., Van Regenmortel, M.H., 1991. Predicting location of continuous epitopes in proteins from their primary structures. Methods In Enzymology 203, 176–201.

Pellequer, J.L., Westhof, E., Van Regenmortel, M.H., 1993. Correlation between the location of antigenic sites and the prediction of turns in proteins. Immunology Letters 36, 83–99.

Pennisi, M., Cavalieri, S., Motta, S., Pappalardo, F., 2016. A methodological approach for using High-Level Petri Nets to model the adaptive immune system response. BMC Bioinformatics 16, 91–105.

Pennisi, M., Rajput, A.-M., Toldo, L., Pappalardo, F., 2013. Agent based modeling of Treg-Teff cross regulation in relapsing-remitting multiple sclerosis. BMC Bioinformatics 14.

Perelson, A.S., 2002. Modelling viral and immune system dynamics. Nature Reviews Immunology 2, 28–36.

Perrin, D., Ruskin, H.J., Burns, J., Crane, M., 2006. An agent-based approach to immune modelling. Lecture Notes in Computer Science 3980, 612–621.

Peters, B., Sette, A., 2005. Generating quantitative models describing the sequence specificity of biological processes with the stabilized matrix method. BMC Bioinformatics 6, 132.

Peters, B., Tong, W., Sidney, J., Sette, A., Weng, Z., 2003. Examining the independent binding assumption for binding of peptide epitopes to MHC-I molecules. Bioinformatics 19, 1765–1772.

Pires, D.E., Ascher, D.B., Blundell, T.L., 2014. mCSM: Predicting the effects of mutations in proteins using graph-based signatures. Bioinformatics 30, 335–342.

Ponomarenko, J.V., Bourne, P.E., 2007. Antibody-protein interactions: Benchmark datasets and prediction tools evaluation. BMC Structural Biology 7, 64.

Ponomarenko, J., Bui, H.H., Li, W., et al., 2008. ElliPro: A new structure-based tool for the prediction of antibody epitopes. BMC Bioinformatics 9, 514.

Ponomarenko, J., Papangelopoulos, N., Zajonc, D.M., et al., 2011. IEDB-3D: Structural data within the immune epitope database. Nucleic Acids Research 39, D1164–D1170.

Potocnakova, L., Bhide, M., Pulzova, L.B., 2016. An introduction to B-Cell epitope mapping and in silico epitope prediction. Journal of Immunology Research 2016, 6760830.

Puck, J.M., 1996. IL2RGbase: A database of gamma c-chain defects causing human X-SCID. Immunology Today 17, 507–511.

Puck J.M., 2005. ALPSbase: Database of mutation causing human ALPS. Available online at: http://research.nhgri.nih.gov/alps/.

Rammensee, H., Bachmann, J., Emmerich, N.P., Bachor, O.A., Stevanovic, S., 1999. SYFPEITHI: Database for MHC ligands and peptide motifs. Immunogenetics 50, 213–219.

Rammensee, H.G., Friede, T., Stevanoviic, S., 1995. MHC ligands and peptide motifs: First listing. Immunogenetics 41, 178–228.

Reche, P.A., Glutting, J.P., Reinherz, E.L., 2002. Prediction of MHC class I binding peptides using profile motifs. Human Immunology 63, 701–709.

Reche, P.A., Glutting, J.P., Zhang, H., Reinherz, E.L., 2004. Enhancement to the RANKPEP resource for the prediction of peptide binding to MHC molecules using profiles. Immunogenetics 56, 405–419.

Reche, P.A., Keskin, D.B., Hussey, R.E., et al., 2006. Elicitation from virus-naive individuals of cytotoxic T lymphocytes directed against conserved HIV-1 epitopes. Medical Immunology 5, 1.

Reche, P.A., Reinherz, E.L., 2003. Sequence variability analysis of human class I and class II MHC molecules: Functional and structural correlates of amino acid polymorphisms. Journal of Molecular Biology 331, 623–641.

Reche, P.A., Reinherz, E.L., 2005. PEPVAC: A web server for multi-epitope vaccine development based on the prediction of supertypic MHC ligands. Nucleic Acids Research 33, W138–W142.

Reche, P., Reinherz, E.L., 2007. Definition of MHC supertypes through clustering of MHC peptide-binding repertoires. Methods In Molecular Biology 409, 163–173.

Reche, P.A., Zhang, H., Glutting, J.-P., Reinherz, E.L., 2005. EPIMHC: A curated database of MHC-binding peptides for customized computational vaccinology. Bioinformatics 21, 2140–2141.

Retter, I., Althaus, H.H., Münch, R., Müller, W., 2005. VBASE2, an integrative V gene database. Nucleic Acids Research 33, D671–D674.

Rice, P., Longden, I., Bleasby, A., 2000. EMBOSS: The European molecular biology open software suite. Trends in Genetics 16, 276–277.

Robinson, J., Malik, A., Parham, P., Bodmer, J.G., Marsh, S.G.E., 2000. IMGT/HLA database – a sequence database for the human major histocompatibility complex. Tissue Antigens 55, 280–287.

Robinson, J., Waller, M.J., Parham, P., et al., 2003. IMGT/HLA and IMGT/MHC: Sequence databases for the study of the major histocompatibility complex. Nucleic Acids Research 31, 311–314.

Rubelt, F., Sievert, V., Knaust, F., et al., 2012. Onset of immune senescence defined by unbiased pyrosequencing of human immunoglobulin mRNA repertoires. PLoS One 7, e49774.

Rubinstein, N.D., Mayrose, I., Martz, E., Pupko, T., 2009. Epitopia: A web-server for predicting B-cell epitopes. BMC Bioinformatics 10, 287.

Ruppert, J., Sidney, J., Celis, E., et al., 1993. Prominent role of secondary anchor residues in peptide binding to HLA-A2.1 molecules. Cell 74, 929–937.

Saha, S., Bhasin, M., Raghava, G.P., 2005. Bcipep: A database of B-cell epitopes. BMC Genomics 6, 79.

Saha, S., Raghava, G.P., 2006. Prediction of continuous B-cell epitopes in an antigen using recurrent neural network. Proteins 65, 40–48.

Samarajiwa, S.A., Forster, S., Auchetti, K., Hertzog, P.J., 2009. INTERFEROME: The database of interferon regulated genes. Nucleic Acids Research 37, D852–D857.

Santos, R., 1999. Immune responses: Getting close to experimental results with cellular automata models. Singapore: World Scientific Publishing Company.

Savoie, C.J., Kamikawaji, N., Sasazuki, T., Kuhara, S., 1999. Use of BONSAI decision trees for the identification of potential MHC class I peptide epitope motifs. In: Pacific Symposium on Biocomputing, pp. 182-9.

Schlessinger, A., Ofran, Y., Yachdav, G., Rost, B., 2006. Epitome: Database of structure-inferred antigenic epitopes. Nucleic Acids Research 34, D777–D780.

Schubert, B., Brachvogel, H.P., Jurges, C., Kohlbacher, O., 2015. EpiToolKit – a web-based workbench for vaccine design. Bioinformatics 31, 2211–2213.

Schubert, B., Walzer, M., Brachvogel, H.P., et al., 2016. FRED 2: An immunoinformatics framework for Python. Bioinformatics 32, 2044–2046.

Sela-Culang, I., Ashkenazi, S., Peters, B., Ofran, Y., 2015a. PFASE: Predicting B-cell epitopes utilizing antibody sequence. Bioinformatics 31, 1313–1315.

Sela-Culang, I., Ofran, Y., Peters, B., 2015b. Antibody specific epitope prediction-emergence of a new paradigm. Current Opinion in Virology 11, 98–102.

Sette, A., Sidney, J., 1998. HLA supertypes and supermotifs: A functional perspective on HLA polymorphism. Current Opinion in Immunology 10, 478–482.

Sette, A., Sidney, J., 1999. Nine major HLA class I supertypes account for the vast preponderance of HLA-A and -B polymorphism. Immunogenetics 50, 201–212.

Sheikh, Q.M., Gatherer, D., Reche, P.A., Flower, D.R., 2016. Towards the knowledge-based design of universal influenza epitope ensemble vaccines. Bioinformatics 32, 3233–3239.

Singh, H., Ansari, H.R., Raghava, G.P., 2013. Improved method for linear B-cell epitope prediction using antigen's primary sequence. PLoS One 8, e62216.

Singh, S.P., Mishra, B.N., 2016. Major histocompatibility complex linked databases and prediction tools for designing vaccines. Human Immunology 77, 295–306.

Singh, H., Raghava, G.P., 2001. ProPred: Prediction of HLA-DR binding sites. Bioinformatics 17 (1236-7), 2001.

Singh, H., Raghava, G.P., 2003. ProPred1: Prediction of promiscuous MHC Class-I binding sites. Bioinformatics 19, 1009–1014.

Singh, M.K., Srivastava, S., Raghava, G.P., Varshney, G.C., 2006. HaptenDB: A comprehensive database of haptens, carrier proteins and anti-hapten antibodies. Bioinformatics 22, 253–255.

Sircar, A., Gray, J.J., 2010. SnugDock: Paratope structural optimization during antibody-antigen docking compensates for errors in antibody homology models. PLoS Computational Biology 6, e1000644.

Soga, S., Kuroda, D., Shirai, H., et al., 2010. Use of amino acid composition to predict epitope residues of individual antibodies. Protein Engineering, Design and Selection 23, 441–448.

Stern, L.J., Wiley, D.C., 1994. Antigenic peptide binding by class I and class II histocompatibility proteins. Structure 2, 245–251.

Sturniolo, T., Bono, E., Ding, J., et al., 1999. Generation of tissue-specific and promiscuous HLA ligand databases using DNA microarrays and virtual HLA class II matrices. Nature Biotechnology 17, 555–561.

Sun, J., Wu, D., Xu, T., et al., 2009. SEPPA: A computational server for spatial epitope prediction of protein antigens. Nucleic Acids Research, 37. pp. W612–W616.

Sweredoski, M.J., Baldi, P., 2008. PEPITO: Improved discontinuous B-cell epitope prediction using multiple distance thresholds and half sphere exposure. Bioinformatics 24, 1459–1460.

Tan, Y.C., Kongpachith, S., Blum, L.K., et al., 2014. Barcode-enabled sequencing of plasmablast antibody repertoires in rheumatoid arthritis. Arthritis & rheumatology 66, 2706–2715.

Tenzer, S., Peters, B., Bulik, S., et al., 2005. Modeling the MHC class I pathway by combining predictions of proteasomal cleavage, TAP transport and MHC class I binding. Cellular and Molecular Life Sciences 62, 1025–1037.

Terasaki, P.I., 2007. A brief history of HLA. Immunologic Research 38, 139–148.

Tipton, C.M., Fucile, C.F., Darce, J., et al., 2015. Diversity, cellular origin and autoreactivity of antibody-secreting cell expansions in acute Systemic Lupus Erythematosus. Nature immunology 16, 755.

Tong, J.C., Ranganathan, S., 2013. Computer-aided vaccine design. Woodhead Publishing Series in Biomedicine, 23. Cambridge: Woodhead, pp. 1–164.

Toseland, C.P., Clayton, D.J., McSparron, H., et al., 2005. AntiJen: A quantitative immunology database integrating functional, thermodynamic, kinetic, biophysical, and cellular data. Immunome Research 1, 4.

Tsioris, K., Gupta, N.T., Ogunniyi, A.O., et al., 2015. Neutralizing antibodies against West Nile virus identified directly from human B cells by single-cell analysis and next generation sequencing. Integrative Biology 7, 1587–1597.

Turnera, M.D., Nedjaib, B., Hursta, T., Penningtonc, D.J., 2014. Cytokines and chemokines: At the crossroads of cell signalling and inflammatory disease. Molecular Cell 11, 2563–2582.

van Heijst, J.W., Ceberio, I., Lipuma, L.B., et al., 2013. Quantitative assessment of T-cell repertoire recovery after hematopoietic stem cell transplantation. Nature medicine 19, 372.

Van Regenmortel, M.H., 2009. What is a B-cell epitope? Methods in Molecular Biology 524, 3–20.

Vita, R., Overton, J.A., Greenbaum, J.A., et al., 2015. The immune epitope database (IEDB) 3.0. Nucleic Acids Research 43, D405–D412.

Wang, D., Yang, L., Zhang, P., et al., 2017. AAgAtlas 1.0: A human autoantigen database. Nucleic Acids Research 45, D769–D776.

Wang, P., Sidney, J., Dow, C., et al., 2008. A systematic assessment of MHC class II peptide binding predictions and evaluation of a consensus approach. PLoS Computational Biology 4, e1000048.

Wang, X., Zhao, H., Xu, Q., et al., 2006. HPtaa database-potential target genes for clinical diagnosis and immunotherapy of human carcinoma. Nucleic Acids Research 34, D607–D612.

Warrender, C., Forrest, S., Koster, F., 2006. Modeling intercellular interactions in early Mycobaterium infection. Bulletin of Mathematical Biology 68, 2233–2261.

Webb, B., Sali, A., 2014. Protein structure modeling with MODELLER. Protein Structure Prediction 1137, 1–15.

Weitzner, B.D., Jeliazkov, J.R., Lyskov, S., et al., 2017. Modeling and docking of antibody structures with Rosetta. Nature Protocols 12, 401–416.

Womble, D.D., 2000. GCG: The Wisconsin Package of sequence analysis programs. Methods In Molecula Biology 132, 3–22.

Xu, X., Sun, J., Liu, Q., et al., 2010. Evaluation of spatial epitope computational tools based on experimentally-confirmed dataset for protein antigens. Chinese Science Bulletin 55, 5.

Yang, B., Sayers, S., Xiang, Z., He, Y., 2011. Protegen: A web-based protective antigen database and analysis system. Nucleic Acids Research 39, D1073–D1078.

Yao, B., Zhang, L., Liang, S., Zhang, C., 2012. SVMTriP: A method to predict antigenic epitopes using support vector machine to integrate tri-peptide similarity and propensity. PLoS One 7, e45152.

Yu, K., Petrovsky, N., Schonbach, C., Koh, J.Y., Brusic, V., 2002. Methods for prediction of peptide binding to MHC molecules: A comparative study. Molecular Medicine 8, 137–148.

Zhang, C., Bickis, M.G., Wu, F.X., Kusalik, A.J., 2006. Optimally-connected hidden markov models for predicting MHC-binding peptides. Journal of Bioinformatics and Computational Biology 4, 959–980.

Zhang, G.L., DeLuca, D.S., Keskin, D.B., et al., 2011. MULTIPRED2: Acomputational system for large-scale identification of peptides predicted to bind to HLA supertypes and alleles. Journal of Immunological Methods 374, 53–61.

Zhang, Q., Wang, P., Kim, Y., et al., 2008. Immune epitope database analysis resource (IEDB-AR). Nucleic Acids Research 36, W513–W518.

Zhong, W., Reche, P.A., Lai, C.C., Reinhold, B., Reinherz, E.L., 2003. Genome-wide characterization of a viral cytotoxic T lymphocyte epitope repertoire. The Journal of Biological Chemistry 278, 45135–45144.

Zhu, J., Wu, X., Zhang, B., et al., 2013. De novo identification of VRC01 class HIV-1-neutralizing antibodies by next-generation sequencing of B-cell transcripts. Proceedings of the National Academy of Sciences 110, E4088–E4097.

Zhu, S., Udaka, K., Sidney, J., et al., 2006. Improving MHC binding peptide prediction by Incorporating binding data of auxiliary MHC molecules. Bioinformatics 22, 1648–1655.

Zvyagin, I.V., Pogorelyy, M.V., Ivanova, M.E., et al., 2014. Distinctive properties of identical twins' TCR repertoires revealed by high-throughput sequencing. Proceedings of the National Academy of Sciences of the United States of America 111, 5980–5985.

Relevant Websites

https://www.rcsb.org/pdb/
  RCSB, Protein Data Bank.
https://www.niaid.nih.gov/research/simmune-project
  National Insitute of Allergy and Infectious Diseases.

Biographical Sketch

Marta has a bachelor's Degree in Biology (2015) and a master Degree in Biomedical Investigation (2016) both from Universidad Complutense of Madrid. She has two years of experience in the Department of Clinical Microbiology and Infectious Diseases of the Hospital General Universitario Gregorio Marañon of Madrid. After completing her master degree Marta joined the Immumedicine Group at the School of Medicine of the Complutense University where she is pursuing her PhD, which is based on computer-assisted design of epitope-based vaccines.

Giulia Russo received the MSc degree in Pharmacy from the University of Catania, in 2014. In 2015, she started her research activities at the Department of Drug Sciences, University of Catania, in the field of computational modeling in oncology and immunology in Prof. Francesco Pappalardo's research group. Her current re-search interests include signalling pathways analysis, systems biology and computational biology in oncology. In 2015, she won the competition for the PhD program in biomedical and biotechnological sciences, with a particular interest in computational approaches in systems biomedicine.

Jose Luis has a Bachelor's Degree in Biotechnology (2015), from the Pablo de Olavide University, followed by a Master's Degree in Advanced Genetics, from the Autonomous University of Barcelona (2016). Then, he worked as research assistant in Severo Ochoa's Center for Molecular Biology to work on Huntington's disease. Jose Luis joined the Immunomedine group early in 2017 to complete his PhD in T-cell immunology.

Marzio Pennisi is currently a post-doc researcher at the University of Catania, Italy. He earned his MSc. and PhD. degrees from University of Catania, in 2006 and 2010 respectively. His research topics include stochastic optimization techniques, evolutionary algorithms, constrained and unconstrained optimization and computational and mathematical methods for the simulation of the immune system. Up to now, he published more than 60 reviewed research papers. He serves the scientific community as editorial board member, reviewer and program committee member for various journals and conferences in the area of computer science, computational biology and bioinformatics.

Pedro A Reche is Chemist (1990) with a Ph.D. in Molecular Biology/Biochemistry (1995) from the University of Granada, Spain. He obtained post-doctoral training at the Department of Biochemistry of the University of Cambridge (England) (1995-1998) and the DNAX Research Institute, California, USA (1998-2001). From 2001 to 2006, he worked at the Dana-Farber Cancer Institute, Boston, MA, USA, directing the Bioinformatics cores of the Molecular Immunology Foundation and the Cancer Vaccine Center. He also held the appointment of Instructor of Harvard Medical School. Late 2006, he joined the University Complutense to establish the Immunomedicine Group. Pedro is a multidisplinary scientist with relevant contributions on the fields of Biochemistry, Molecular Biology, Bioinformatics and Immunology. He has published over 60 articles in peer review journals that have received over 4700 citations (h-index = 28) and he is in the Editorial Board of 5 international journals. Pedro's main research lines are A) prediction of immunogenicity, B) discovery of therapeutic molecules and C) study of adaptive immune conditioning by epithelial cells.

Dr Adrian Shepherd was awarded a PhD in Neural Computing at UCL in 1995, then joined the research group of Prof. Janet Thornton as a Post-Doctoral Research Fellow to design machine learning classifiers for protein structure prediction. He joined Birkbeck, University of London as a lecturer in 2002, and was a founding member of the European ImmunoGrid Consortium. Now a Reader in Computational Biology at the Institute of Structural and Molecular Biology, Birkbeck, his research focuses on human adaptive immune response using a range of computational methods - from Next Generation Sequencing of immune repertoires (Rep-seq) to molecular dynamics simulations of antibody binding. His group collaborates with clinicians, virologists and other wet-lab research scientists to address medical problems that include the design of more effective vaccines and the stratification of patient responses to therapy. He is currently working on the adaptive immune response to several pathogens (influenza A virus, hepatitis C virus and Bacillus anthracis), to cancer (hepatocellular carcinoma and breast cancer), and to therapeutic proteins (notably replacement Factor VIII used in the treatment of hemophilia A).

Francesco Pappalardo is Associate Professor of Computer Science at the University of Catania, Italy and Visiting Professor at Metropolitan College of Boston, USA. He did research in computer operating system security. He was visiting researcher at Dana-Farber Cancer Institute in Boston (USA) and at the Molecular Immunogenetics Labs, IMGT in Montpellier (France). His major research area is on computational modelling in systems biology and medicine, with a special focus on the immune system. Up to now, Francesco Pappalardo published more than 100 reviewed research papers. He serves the scientific community as editorial board member, reviewer and program committee member for major journals and conferences in the area of computer science and computational biology.

Immunoinformatics Databases

Christine LP Eng and Tin Wee Tan, National University of Singapore, Singapore
Joo Chuan Tong, National University of Singapore, Singapore and Institute of High Performance Computing, Singapore

Introduction

Immunological databases play important roles in knowledge dissemination and contribute directly to our better understanding of the mechanisms underlying the defence of human body. Without appropriate data, effective bioinformatics tools cannot be built that allow for meaningful interpretations of immunological outcomes. As of May 2017, 36 immunological databases have been described in the NAR Molecular Biology Database Collection (Galperin *et al.*, 2017). Immune sequence databases provide access to experimentally characterized immune cell sequences that are implicated in infections and autoimmunity. Immune epitope databases are a set of specialist databases that stores binding data of immune molecules. Such information is commonly used to inform the design of subunit vaccines. Here, the most important databases are reviewed.

The Immune System

The human body has developed a complex immune system against the constant threat by pathogens from the environment. The first line of defence is helmed by the innate immune system, which is a general, non-specific immunity against infection that occurs rapidly upon infection (Kimbrell and Beutler, 2001). The cells from the innate immune system recognize and respond to pathogens generically, as they have different types of receptors to recognize common features of pathogens. Hence, the effects of immunity to the host are only short-lived.

In contrast, the adaptive immune system is much more specialized, capable of a more effective elimination of the infection in the event that the pathogens escape innate immunity. The adaptive immune system, consisting of B-cells and T-cells, is shaped to recognize specific infections, resulting in long-lasting immunity (Murphy *et al.*, 2008). The lymphocytes of the adaptive immune system have a single receptor type recognizing specific chemical structures, which leads to a great diversity from the huge repertoire of immunoglobulins and T-cell receptors.

What is an Epitope?

An epitope is a localized region on the surface of an antigen that is recognized by the immune system, specifically the B- or T-cells. B-cell epitopes are antigenic determinants that interact with B-cell receptors. There are two types of B-cell epitopes – linear and non-linear. About 10% of B-cell epitopes are continuous, consisting of a linear stretch of amino acids along the polypeptide chain. Most B-cell epitopes, though, are discontinuous in nature, where distant residues are brought into spatial proximity by protein folding. Studies have shown that not all residues within a B-cell epitope are functionally important for binding, and the specificity could be reduced or eliminated by single-site amino acid substitution.

T-cell epitopes are short linear peptides that bind to T-cell receptors in association with the major histocompatibility complex (MHC) molecules. Two classes of T cells are available: 1) CD8 + T cytotoxic (Tc) cells, which recognize peptides displayed by MHC class I molecules, and 2) CD4 + T helper (Th) cells, which recognize peptides in association with MHC class II molecules (#49). Tc cells release cytotoxins which are responsible for cell lysis, and granzymes which induces apoptosis. Th1 cells produce interferon γ (IFN-γ) and tumour necrosis factor β (TNF-β) and are involved in delayed-type hypersensitivity (DTH) reactions. By contrast, Th2 cells produce interleukin 4 (IL-4), IL-5, IL-10 and IL-13, which are responsible for strong antibody responses, including the activation and recruitment of IgE antibody-producing B-cells, mast cells, eosinophils, and the inhibition of several macrophage functions. T-cell epitopes presented by MHC class I molecules are typically peptides about 8–10 amino acids long, whereas MHC class II molecules binds longer peptides of about 13–20 residues.

Major Databases

In the following section, we review the major databases that are available for immunoinformatics research. This ranges from comprehensive databases such as IEDB providing epitope data and structure information, to databases specialising in genes or MHC binding information, as well as databases focusing on sequence/structure/function with emphasis on interaction information.

IEDB

The Immune Epitope Database and Analysis Resource (IEDB; see Relevant Websites section) is a central resource for immunoinformatics, with an extensive collection of B and T cell epitopes characterized experimentally in humans, non-human primates, mice, as well as other species (Fleri *et al.*, 2017). Funded by the National Institute of Allergy and Infectious Diseases (NIAID), the database was developed in 2004 at the La Jolla Institute for Allergy and Immunology with the aim of assisting researchers in the development of novel diagnostics, therapeutics and vaccines (**Fig. 1**).

The database contains information on peptidic and non-peptidic epitopes, as well as T cell, B cell and MHC ligand assays, which were curated from peer-reviewed literature and researcher submitted data. Each record consists of the sequence, source organism, source antigen, and references linked to additional information which include author, title and abstract of the literature. As of June 2017, IEDB includes over 298,000 peptidic epitopes and 2500 non-peptidic epitopes from over 3600 source organisms, based on approximately 319,000 T cell assays, 393,000 B cell assays, and 601,000 MHC ligand assays. This represents over 99% of all peptidic epitope information available publicly.

IMGT

The international ImMunoGeneTics information system® (IMGT®; see Relevant Websites section) is an integrated knowledge resource for immunoglobulins, major histocompatibility, and T-cell receptors of human and other vertebrates (Lefranc *et al.*, 2015). Created at 1989 by Marie-Paule Lefranc, IMGT® manages the diverse and complex genes and proteins of immunoglobulin and T-cell receptors, as well as the major histocompatibility proteins polymorphism. It comprises seven databases in total: (**Fig. 2**).

1. IMGT/LIGM-DB is a comprehensive database of fully-annotated nucleotide sequences of immunoglobulin and T-cell receptor from humans and other vertebrates (Giudicelli *et al.*, 2006). Each sequence in the database include information on sequence identification, classification of gene and allele, constitutive and specific motif description, numbering of the codon and amino acid, and the sequence obtaining information. As of June 2017, there are over 178,000 immunoglobulin and T-cell receptor sequences from 351 species.

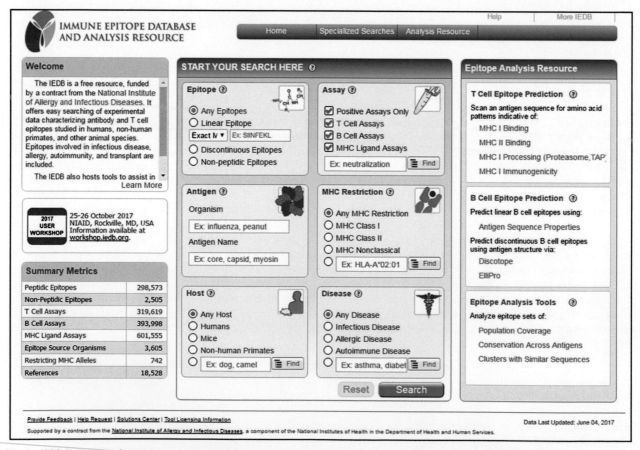

Fig. 1   IEDB homepage where users can start accessing data by performing queries based on epitopes, assays, antigens, MHC restriction type, hosts and diseases. IEDB. Available at: http://www.iedb.org.

WELCOME!
to the IMGT Home page

THE INTERNATIONAL IMMUNOGENETICS INFORMATION SYSTEM®

http://www.imgt.org

- IMGT®
- References and News
- Contacts & Legal notices

IMGT®, the international ImMunoGeneTics information system® http://www.imgt.org, is the global reference in immunogenetics and immunoinformatics, created in 1989 by Marie-Paule Lefranc (Université de Montpellier and CNRS). IMGT® is a high-quality integrated knowledge resource specialized in the immunoglobulins (IG) or antibodies, T cell receptors (TR), major histocompatibility (MH) of human and other vertebrate species, and in the immunoglobulin superfamily (IgSF), MH superfamily (MhSF) and related proteins of the immune system (RPI) of vertebrates and invertebrates. IMGT® provides a common access to sequence, genome and structure Immunogenetics data, based on the concepts of IMGT-ONTOLOGY and on the IMGT Scientific chart rules. IMGT® works in close collaboration with EBI (Europe), DDBJ (Japan) and NCBI (USA). IMGT® consists of sequence databases, genome database, structure database, and monoclonal antibodies database, **Web resources** and **interactive tools.**

IMGT founder and director: Marie-Paule Lefranc (Marie-Paule.Lefranc@igh.cnrs.fr), Université de Montpellier, CNRS, LIGM, IGH, SFR, Montpellier (France)

The 2015 IMGT® Customer Satisfaction Survey ⬇
The Quality Management System of IMGT® Montpellier France has been approved by Lloyd's Register Quality Assurance France SAS to the following Quality Management System Standard: ISO 9001:2008

IMGT databases

IMGT/LIGM-DB (doc) LIGM, Montpellier, France
Nucleotide sequences of IG and TR from 351 species **(178 929 entries)**
IMGT/MH-DB ANRI, BPRC, hosted at EBI
Sequences of the human MH (HLA)
IMGT/PRIMER-DB (doc) LIGM, Montpellier, France
Oligonucleotides (primers) of IG and TR from 11 species **(1 864 entries)**
IMGT/CLL-DB (bylaws) LIGM, Montpellier, France
IG sequences from CLL, an initiative of the IMGT/CLL-DB group

IMGT/GENE-DB (doc) LIGM, Montpellier, France
International nomenclature for IG and TR genes from human, mouse, rat and rabbit **(4 147 genes, 5 858 alleles)**

IMGT/3Dstructure-DB and IMGT/2Dstructure-DB (doc) LIGM, Montpellier, France
3D structures (IMGT Colliers de Perles) of IG antibodies, TR, MH and RPI **(4 808 entries)**
Source: PDB, INN, Kabat

IMGT/mAb-DB (doc) LIGM, Montpellier, France
Monoclonal antibodies (IG, mAb), fusion proteins for immune applications (FPIA), composite proteins for clinical applications (CPCA), and related proteins (RPI) of therapeutic interest **(722 entries)**

IMGT Web resources

IMGT Repertoire (IG and TR, MH and RPI)
IMGT Scientific chart (Sequence and 3D structure identification and description, Numbering, Nomenclature, Representation rules)
IMGT Index (FactsBook, IMGT-ONTOLOGY, Sequence submission, Taxonomy...)
IMGT Bloc-notes (Interesting links, PubMed, Meeting announcements, Postdoctoral positions and jobs, Messages, Search engines...)
IMGT Education (IMGT Lexique, Aide-mémoire, Tutorials, Questions and answers, Enseignements...)
IMGT Posters and diaporama
The IMGT Medical page
The IMGT Veterinary page
The IMGT Biotechnology page
The IMGT Immunoinformatics page

Fig. 2 The IMGT homepage with links to the seven databases. IMGT®. Available at: http://www.imgt.org.

2. IMGT/MH-DB specializes in human major histocompatibility complex sequences (Robinson *et al.*, 2015). Established as a locus-specific database for the allelic sequences of the HLA genes, the database also includes official sequences specified by the World Health Organization Nomenclature Committee For Factors of the HLA System. As of June 2017, there are 12,351 HLA class I alleles, 4404 HLA class II alleles, and 178 non-HLA alleles in the database.

3. IMGT/PRIMER-DB is a database of oligonucleotides (primer) for immunoglobulin and T-cell receptor. The primers can be used for antibody single chain Fragment variable (scFy), combinatorial library design, microarray technologies and phage display. The database comprises 1864 primers from 11 species.

4. IMGT/CLL-DB is a database specializing in primary immunoglobulin sequences associated with clinical and biological data of Chronic Lymphocytic Leukemia (CLL) patients. There are current restrictions to the database for members of the IMGT/CLL-DB group.

5. IMGT/GENE-DB is a comprehensive database for immunoglobulin and T-cell receptor genes from human, mouse, rat and rabbit (Giudicelli *et al.*, 2005). The database is the international reference for the gene nomenclature of immunoglobulin and T-cell receptor, displaying gene data related to genome, allelic polymorphisms, gene expression, protein and structures. As of June 2017, the database contains 4147 genes and 5858 alleles from 24 species.

6. IMGT/3Dstructure-DB is a resource on annotated structural data of immunoglobulin, T-cell receptor, MHC and related proteins of the immune system (Ehrenmann *et al.*, 2010). With 4766 entries currently, each record provides information on the sequence, 2D structures and 3D structures.

7. IMGT/mAB-DB is a monoclonal antibodies database, providing resource on immunoglobulins or monoclonal antibodies with clinical indications, as well as fusion proteins for immune applications. As of June 2017, the database contains 722 records.

SYFPEITHI

One of the first publicly available database for MHC ligands and peptide motifs, SYFPEITHI (see Relevant Websites section) is a database of known peptide sequences bound to MHC class I and II molecules (Rammensee *et al.*, 1995). It was developed in 1999

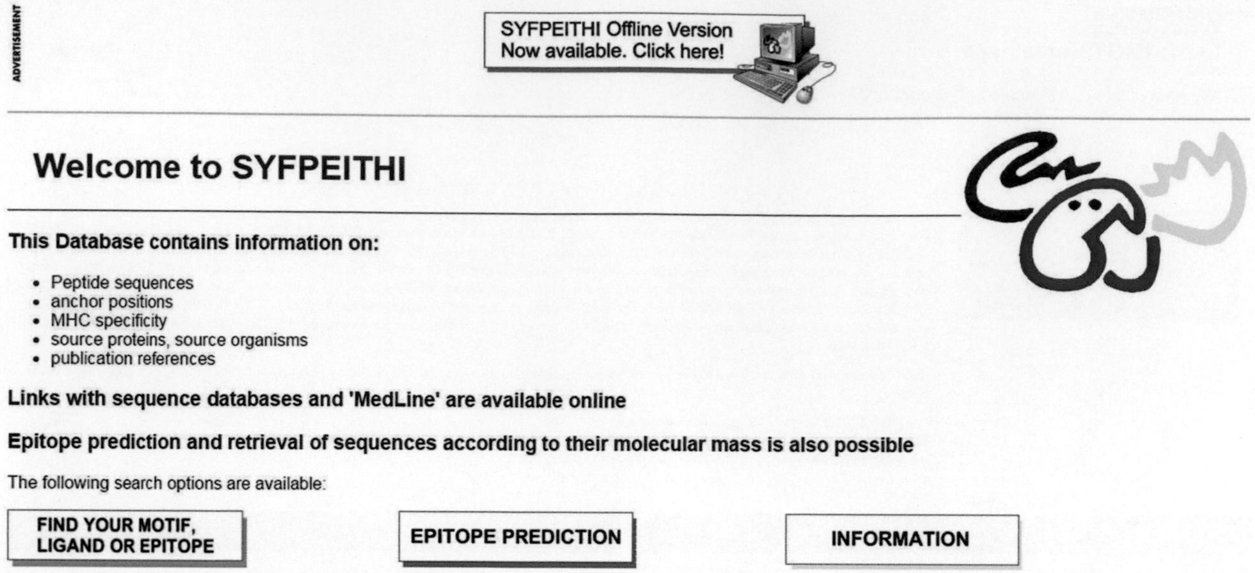

Welcome to SYFPEITHI

This Database contains information on:

- Peptide sequences
- anchor positions
- MHC specificity
- source proteins, source organisms
- publication references

Links with sequence databases and 'MedLine' are available online

Epitope prediction and retrieval of sequences according to their molecular mass is also possible

The following search options are available:

| FIND YOUR MOTIF, LIGAND OR EPITOPE | EPITOPE PREDICTION | INFORMATION |

Fig. 3   Homepage of SYFPEITHI, with links to access the database. SYFPEITHI. Available at: http://www.syfpeithi.de.

by Hans-Georg Rammensee's group at the Institute of Cell Biology, University of Tübingen. With over 7000 MHC ligands, motifs and T-cell epitopes, the database aims to facilitate the search for peptides and aiding in the prediction of T-cell epitopes. Manually curated from published literature, each record provides information on the source and reference (**Fig. 3**).

AntiJen

AntiJen (see Relevant Websites section) is a large compilation of quantitative binding data (McSparron *et al.*, 2003). Developed by Darren Flower's group at the Edward Jenner Institute for Vaccine Research, the database was developed with the aim of improving computational vaccinology for building tools to accurately predict epitopes. The database includes MHC ligand molecules and kinetics, T cell epitopes, Transmembrane Peptide Transporter and B cell epitopes, protein complexes and protein-protein interactions, as well as data peptide library, diffusion coefficient and copy numbers. With over 24,000 entries based on experimentally determined data, AntiJen is among the largest immune epitope databases available (**Fig. 4**).

MHCBN

MHCBN (see Relevant Websites section) is a comprehensive database on MHC binding and non-binding peptides curated from existing databases and published literature (Lata *et al.*, 2009). The database was developed by Gagendra Raghava's group at the Institute of Microbial Technology in India. With over 25,800 peptide sequences, the database provides sequence and structure data on the source proteins as well as the MHC molecules (**Fig. 5**).

MPID-T2

The MHC-Peptide Interaction Database-TR version 2 (MPID-T2; see Relevant Websites section) is a new generation database with information on the sequence-structure-function of T-cell receptor/peptide/MHC interactions (Tong *et al.*, 2006). First developed at National University of Singapore in 2006 by Shoba Ranganathan's group, and subsequently at the Macquarie University in 2010, the MPID-T2 contains information on all known structures of T-cell receptor/peptide/MHC and peptide/MHC complexes, emphasizing on the structural characterization. Protein interaction information such as hydrogen bonds, non-hydrogen bonds, solvent accessibility, contact residues, interface area, gap index and gap volume is also included. Manually curated and populated with data from the Protein Data Bank (PDB), the database contains 415 records from humans (282), murine (127), rat (3), chicken (2), and monkey (1), which spans across 56 alleles. The database aims to facilitate the development of tools to predict peptide binding to MHC alleles (**Fig. 6**).

BEID

The B-cell Epitope Interaction Database (BEID) is a repository of sequence-structure-function information on interactions of immunoglobulin-antigen (Tong *et al.*, 2008). The database contains 164 antigens, 126 immunoglobulins and 189 immunoglobulin-antigen complexes extracted from PDB where they were manually verified, classified, and then analysed for the intermolecular

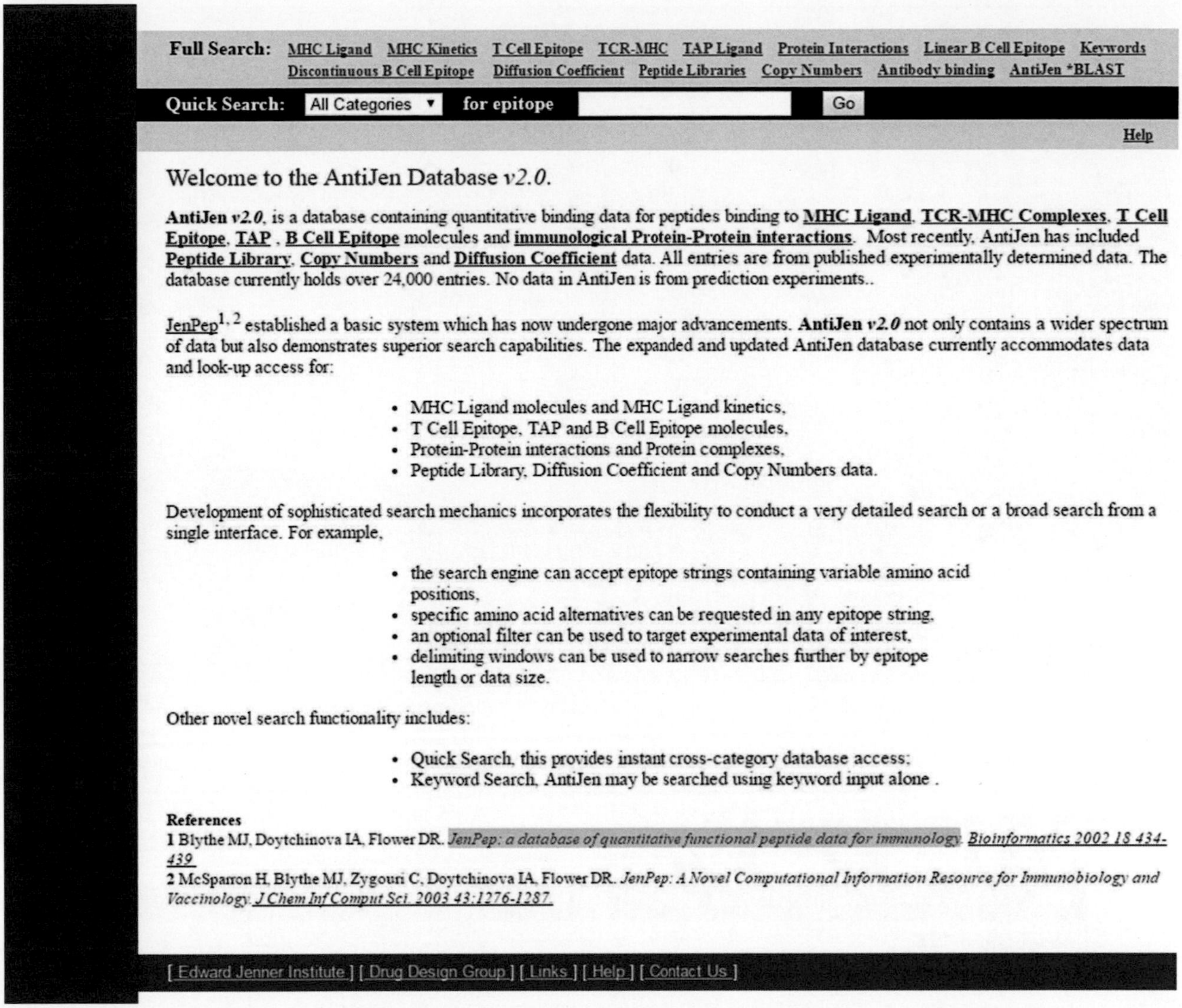

Fig. 4 Homepage of AntiJen with quick search functions for epitopes. AntiJen. AVailable at: http://www.ddg-pharmfac.net/antijen/.

interactions. Each record contains interaction information on gap index, gap volume, interface area, solvent accessibility, contact residues, hydrogen bonds and non-hydrogen bonds.

CED

The Conformational Epitope Database (CED; see Relevant Websites section) is a specialist database on conformational B cell epitopes (Huang and Honda, 2006). Manually curated from peer-reviewed literature, each entry contains information on the location and composition of the epitope, immunological property, source antigen and corresponding immunoglobulin. As of June 2017, the database contains 225 entries from protein antigens, nucleic acids, glycans and lipids, derived from various sources.

DFRMLI

The Dana-Farber Repository for Machine Learning in Immunology (DFRMLI; see Relevant Websites section) is a database specializing in standardized datasets for the application of machine learning in immunology (Zhang et al., 2011). Processed specifically for the development of machine learning tools for use in epitope prediction, the repository provides standard datasets of experimentally validated binding affinities of HLA-binding peptides mapped onto a common scale.

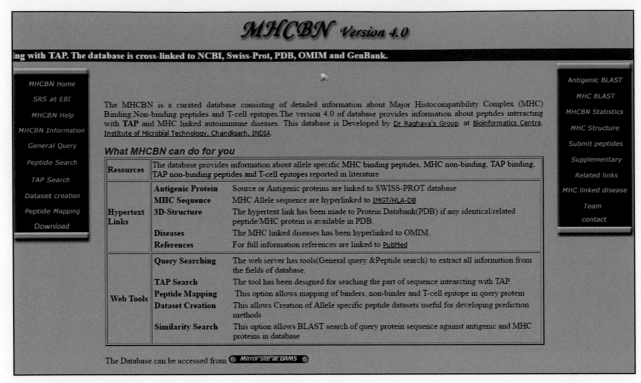

Fig. 5   MHCBN homepage with information and links to peptide query. MHCBN. Available at: http://imtech.res.in/raghava/mhcbn.

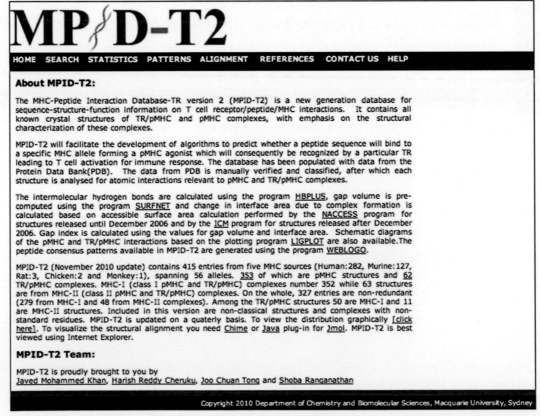

Fig. 6   MPID-T2 homepage with information and links to access the database. MHCBN. Available at: http://biolinfo.org/mpid-t2.

IPD

The Immuno Polymorphism Database (IPD; see Relevant Websites section) is a compilation of specialist databases associated with the study of gene polymorphisms in the immune system (Robinson *et al.*, 2013). It was developed in 2003 from a collaboration between the European Bioinformatics Institute (EBI) and the HLA Informatics Group of the Anthony Nolan Research Institute. The IPD comprises four databases: (**Fig. 7**).

1. IPD-KIR provides a repository on human Killer-cell Immunoglobulin-like Receptors (KIR) sequences. There are over 600 alleles in the database, coding for more than 320 unique sequences.
2. IPD-MHC is a centralised database of MHC sequences from different species. Manually curated by field experts, the database now contains over 4000 alleles from 47 species of non-human primates, as well as other species including canines, felines, cattle, teleost fish, rats, sheep, and swine.

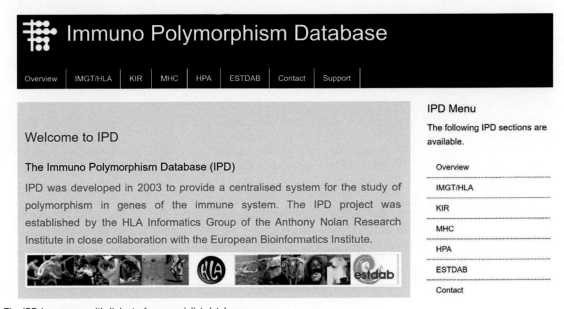

Fig. 7   The IPD homepage with links to four specialist databases.

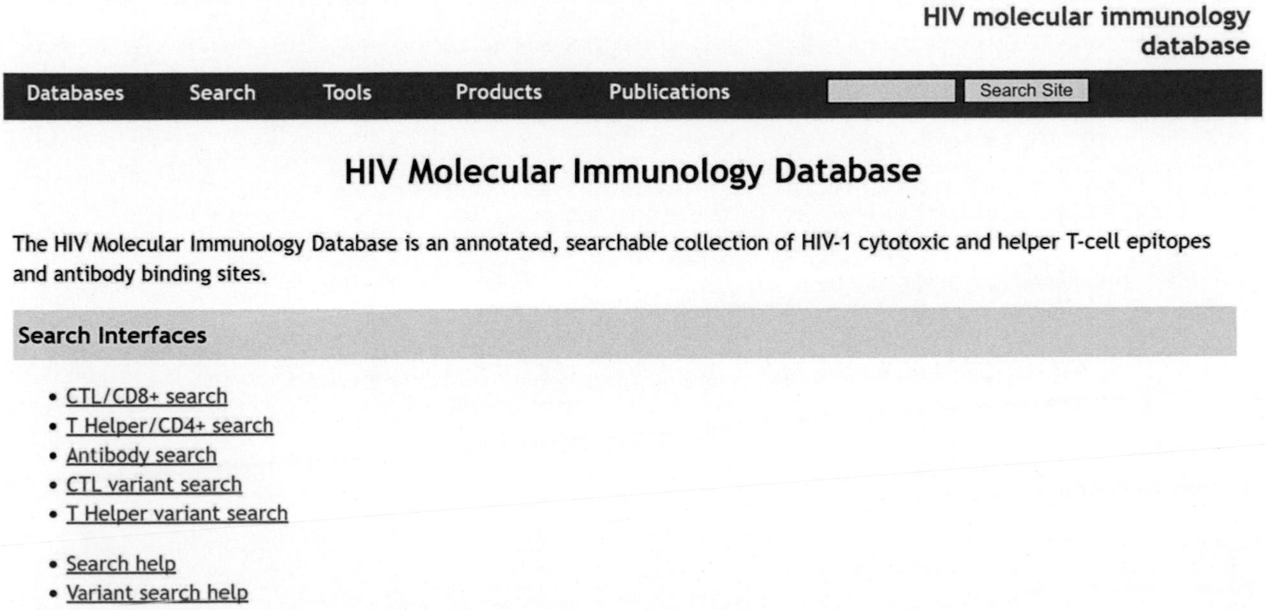

HIV molecular immunology database

Databases   Search   Tools   Products   Publications   Search Site

HIV Molecular Immunology Database

The HIV Molecular Immunology Database is an annotated, searchable collection of HIV-1 cytotoxic and helper T-cell epitopes and antibody binding sites.

Search Interfaces

- CTL/CD8+ search
- T Helper/CD4+ search
- Antibody search
- CTL variant search
- T Helper variant search

- Search help
- Variant search help

Fig. 8   Homepage of HIV Molecular Immunology Database with links to search interfaces.

3. IPD-HPA details data on the human platelet antigens (HPA). The database provides information on the HPA as well as additional background information.
4. IPD-ESTDAB is the European Searchable Tumour Line Database and Cell Bank for immunologically characterized melanoma cell lines. The database provides a search facility for HLA typed, immunologically characterized tumour cells.

HIV Molecular Immunology Database

The HIV Molecular Immunology Database (see Relevant Websites section) is an annotated database of HIV-1 cytotoxic and helper T-cell epitopes, and immunoglobulin binding sites. The database is funded by the Division of AIDS, National Institute of Allergy and Infectious Diseases and developed by the Los Alamos National Laboratory. The data is extracted from HIV immunology literature, and includes information on immunoglobulin sequence, escape mutations, cross-reactivity, T-cell receptor usage, functional domains overlapping with epitopes, immune response among others. As of June 2017, there are 905 HIV-1 cytotoxic T cell epitopes, 1023 HIV-1 helper T-cell epitopes and 1448 immunoglobulin binding sites in the database (**Fig. 8**).

Discussion

The immunoinformatics field has matured over the past decade, with the creation of many high-quality immunological databases available publicly. Many of the databases discussed above contain well-curated and comprehensive data on MHC ligands and immune epitopes, including both naturally processed peptides as well as synthetic peptides available on IEDB. Some of the databases such as IEDB, AntiJen, MHCBN, MPID-T2 and BEID also include information on the structure and protein-protein interactions which provides much more detail to study individual binding interactions. On the other hand, there are also specialized databases such as DFRMLI in addition to IEDB and IMGT, which provide training datasets with large number of MHC-peptide binding data to facilitate the development of machine learning approaches for epitope prediction. While there are a multitude of databases available, the challenge remains to update the database with new information regularly so that more tools would be able to utilize the rich resource to advance the field of immunoinformatics.

See also: Bioinformatics Data Models, Representation and Storage. Biological Database Searching. Computational Immunogenetics. Data Storage and Representation. Extraction of Immune Epitope Information. Immunoglobulin Clonotype and Ontogeny Inference. Information Retrieval in Life Sciences. Natural Language Processing Approaches in Bioinformatics. Vaccine Target Discovery

---

References

Ehrenmann, F., Kaas, Q., Lefranc, M.P., 2010. IMGT/3Dstructure-DB and IMGT/DomainGapAlign: A database and a tool for immunoglobulins or antibodies, T cell receptors, MHC, IgSF and MhcSF. Nucleic Acids Res. 38 (Database Issue), D301–D307.
Fleri, W., et al., 2017. The immune epitope database and analysis resource in epitope discovery and synthetic vaccine design. Front. Immunol. 8, 278.
Galperin, M.Y., Fernandez-Suarez, X.M., Rigden, D.J., 2017. The 24th annual nucleic acids research database issue: A look back and upcoming changes. Nucleic Acids Res. 45 (9), 5627.
Giudicelli, V., et al., 2006. IMGT/LIGM-DB, the IMGT comprehensive database of immunoglobulin and T cell receptor nucleotide sequences. Nucleic Acids Res. 34 (Database Issue), D781–D784.
Giudicelli, V., Chaume, D., Lefranc, M.P., 2005. IMGT/GENE-DB: A comprehensive database for human and mouse immunoglobulin and T cell receptor genes. Nucleic Acids Res. 33 (Database Issue), D256–D261.
Huang, J., Honda, W., 2006. CED: A conformational epitope database. BMC Immunol. 7, 7.
Kimbrell, D.A., Beutler, B., 2001. The evolution and genetics of innate immunity. Nat. Rev. Genet. 2 (4), 256–267.
Lata, S., Bhasin, M., Raghava, G.P., 2009. MHCBN 4.0: A database of MHC/TAP binding peptides and T-cell epitopes. BMC Res. Notes 2, 61.
Lefranc, M.P., et al., 2015. IMGT(R), the international ImMunoGeneTics information system(R) 25 years on. Nucleic Acids Res. 43 (Database Issue), D413–D422.
McSparron, H., et al., 2003. JenPep: A novel computational information resource for immunobiology and vaccinology. J. Chem. Inf. Comput. Sci. 43 (4), 1276–1287.
Murphy, K.P., et al., 2008. Janeway's Immunobiology. New York, NY: Garland Pub.
Rammensee, H.G., Friede, T., Stevanoviic, S., 1995. MHC ligands and peptide motifs: First listing. Immunogenetics 41 (4), 178–228.
Robinson, J., et al., 2013. IPD – The Immuno polymorphism database. Nucleic Acids Res. 41 (Database Issue), D1234–D1240.
Robinson, J., et al., 2015. The IPD and IMGT/HLA database: Allele variant databases. Nucleic Acids Res 43 (Database issue), D423–D431.
Tong, J.C., et al., 2006. MPID-T: Database for sequence–structure–function information on T-cell receptor/peptide/MHC interactions. Appl. Bioinform 5 (2), 111–114.
Tong, J.C., et al., 2008. BEID: Database for sequence–structure–function information on antigen–antibody interactions. Bioinformation 3 (2), 58–60.
Zhang, G.L., et al., 2011. Dana-Farber repository for machine learning in immunology. J. Immunol. Methods 374 (1–2), 18–25.

Relevant Websites

http://www.ddg-pharmfac.net/antijen/
    AntiJen.
http://immunet.cn/ced
    CED.
http://bio.dfci.harvard.edu/DFRMLI
    DFRMLI.

http://www.hiv.lanl.gov/content/immunology
    HIV Molecular Immunology Database.
http://www.iedb.org
    IDEB.
http://www.imgt.org
    IMGT.
http://www.ebi.ac.uk/ipd
    IPD.
http://imtech.res.in/raghava/mhcbn
    MHCBN.
http://biolinfo.org/mpid-t2
    MPID-T2.
http://www.syfpeithi.de
    SYFPEITHI.

Study of Human Antibody Responses From Analysis of Immunoglobulin Gene Sequences

Katherine JL Jackson, Garvan Institute of Medical Research, Darlinghurst, NSW, Australia

Introduction

Antibodies are a crucial component of the immune system and the genes that encode antibodies are subject to more manipulation and modification than any other human genes. The many processes that occur during antibody ontogeny are captured within the sequence of the rearranged immunoglobulin genes. The study of the sequences of these genes therefore provides a window into not only the antigen response but also into the underlying genetics of the person who produced the response. Sequencing technologies that capture millions of immune receptor gene transcripts offer a means for high resolution analysis of the antibody response in contexts ranging from responses to infection (Wu *et al.*, 2011; Parameswaran *et al.*, 2013; Liao *et al.*, 2013) and vaccination (Wang *et al.*, 2015; Jackson *et al.*, 2014; Galson *et al.*, 2015; Ellebedy *et al.*, 2016; Truck *et al.*, 2015), to autoimmunity and immune deficiency (Roskin *et al.*, 2015), in allergy (Hoh *et al.*, 2016; Wang *et al.*, 2014; Levin *et al.*, 2016) and in lymphoid cancers (Boyd *et al.*, 2009; Faham *et al.*, 2012).

Each antibody protein in the human is comprised of four polypeptide chains – two identical heavy chains encoded by an immunoglobulin heavy chain (IGH) gene and two identical light chains encoded by either an immunoglobulin kappa (IGK) or lambda light chain (IGL) gene (**Fig. 1**) (Tonegawa, 1983). Each polypeptide chain is broadly divided into two regions; the variable region and the constant region. The variable portion of immunoglobulin genes are formed by a series of genomic rearrangements within each B cell's genome at the heavy (chromosome 14 (Croce *et al.*, 1979)) and light chain loci (kappa: chromosome 2, lambda: chromosome 22 (McBride *et al.*, 1982)) that bring together several gene segments to form an immunoglobulin gene (Tonegawa, 1983) (**Fig. 1(A)**). The resulting gene is often termed a 'rearranged' immunoglobulin gene in order to differentiate it from the unrearranged or 'germline' gene segments within the immunoglobulin gene loci that are the building blocks from which functional antibodies are created. The process of genomic recombination allows for a vast diversity of antibody proteins to be generated from a relatively small number of variable (IGHV), diversity (IGHD) and joining (IGHJ) gene segments for the heavy chain and variable (IGKV/IGLV) and joining (IGKJ/IGLJ) gene segments for the light chains. The same V(D)J recombination mechanism also forms the T cell receptor (TCR) gene repertoire (Davis, 1990).

The variable region of an immunoglobulin protein provides the antigen specificity. It can be further sub-divided into four structurally important framework regions (FR1, 2, 3, 4) and three diverse complementarity determining regions (CDR1, 2, 3) that tend to contact the antigen (Chothia and Lesk, 1987). Immunoglobulin protein function is dictated by the constant region, encoded by the heavy chain constant region (IGHC) gene, which determines the isotype or subclass of the immunoglobulin (Schroeder and Cavacini, 2010). IGHC exons are associated with the variable region at the level of mRNA transcription and each isotype class and subclass has varying in its functional capabilities. Prior to antigen exposure, naïve B cells express their VDJ rearrangement in association with both IgM and IgD through alternative mRNA splicing (**Fig. 1(D)**). Following antigen encounter, further genomic changes, termed class switch recombination (CSR), alter the associated constant region by deletion of IGHC genes from the genome and allowing expression of IgG subclasses (IgG3, IgG1, IgG2, IgG4), IgA subclasses (IgA1, IgA2), or IgE (**Fig. 1(D)**). The constant region also determines if paired heavy and light chains are secreted from the B cell as antibodies, or if they are membrane bound and act as B cell receptors (BCRs). This is based on whether the constant region includes a terminal secretory (antibodies) or trans-membrane domain (BCRs). Light chains also have constant regions encoded by IGKC or IGKL genes but these do not undergo CSR.

In addition to CSR, following encounter with an antigen to which the BCR can bind, immunoglobulin genes will also undergo affinity maturation (**Fig. 1(E)**) (Eisen, 2014). This is driven by a mutation process termed somatic hypermutation (SHM) because the mutations are introduced within the V(D)J gene rearrangement at a rate of 10^5 to 10^6 above the normal background somatic mutation rate (Weigert *et al.*, 1970) (Bernard *et al.*, 1978). The point mutations introduced within the V(D)J may be positively selected when they improve the interaction between the antigen and the immunoglobulin and negatively selected when they are deleterious. CSR and affinity maturation occur in the context of the B cell clonal expansion whereby the initial B cell that expressed an immunoglobulin with some specificity for antigen divides and each daughter cell inherits the immunoglobulin rearrangement, but may acquire further SHM and undergo further class switch, to build a B cell clonal lineage.

Diversification of Immunoglobulin Genes

The diversity of the human antibody repertoire within a single person is predicted to be as high as 10^{11} unique antibodies (Glanville *et al.*, 2009). Given that a human genome includes fewer than 20,000 protein-coding genes (Ezkurdia *et al.*, 2014), it is not possible for a single gene for each unique antibody to be maintained within the germline genome, rather a diverse immunoglobulin repertoire is generated through processes that involve rearrangements and modifications to the genomes of single immune cells (Tonegawa, 1983). Immunoglobulin diversification processes can be broadly divided into 'combinatorial' and

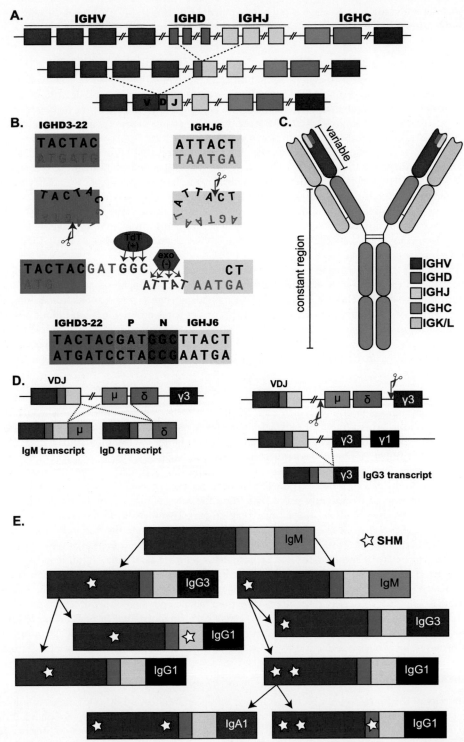

Fig. 1 Rearrangement of immunoglobulin genes and the generation of diversity in humans. (A) The human IGH locus is comprised of sets of IGHV, IGHD and IGHJ gene segments. A rearranged immunoglobulin gene is generated by genomic recombination events that first join single IGHD and IGHJ gene segments, followed by joining to a single IGHV gene segment. (B) Joining of gene segments is imprecise as hairpin structures that form following RAG cleavage at RSS sites are opened asymmetrically and the joints are acted on by exonucleases that remove nucleotides from coding ends and TdT which incorporates untemplated nucleotides, leading to diverse junctions. (C) The IGH and light chain polypeptide chains are joined by disulfide bonds (pink) to form an immunoglobulin protein which is broadly divided into variable and constant regions. (D) Rearranged immunoglobulin genes are associated with IGHC exons at the mRNA level, naïve B cells can express both IgM and IgD through alternative splicing. Genomic recombination events can alter the IGHC locus leading to expression of downstream isotypes classes and subclasses. (E) Rearranged genes undergo somatic hypermutation and class switch recombination as part of the B cell clonal expansion to give rise to B cell lineages.

'junctional' which generate the vastly diverse naïve B cell compartment (**Fig. 1(A)** and **(B)**). Naïve B cells are those that are yet to experience foreign antigen interactions. Diversification subsequent to antigen encounter is SHM driven.

Combinatorial Diversity

IGH are formed via the joining of three distinct gene segments; IGHV, IGHD and IGHJ. Sets of each type of gene segments are present within the IGH locus on chromosome 14. IGH gene segments are both polygenic and polymorphic. Within the human population each gene may exist as a number of allelic variants with any single individual carrying just one (homozygous) or two (heterozygous) of the possible alleles (Watson *et al.*, 2013). More than two variants of some genes, for example IGHV1–69, may be carried by a single person due copy number variant haplotypes (Watson *et al.*, 2013).

Functional IGH gene segments are flanked by recombination signal sequences (RSS) (Tonegawa, 1983). Each RSS is composed of a conserved heptamer (6 base pairs) and nonamer (9 base pair) sequence separated by a variable spacer region of either 12 ± 1 (IGHD segments) or 23 ± 1 (IGHV and IGHJ segments) base pairs (bp) in length. Only genes with different spacer lengths can efficiently recombine and this 12/23 rule promotes the preferential formation IGHD: IGHJ and IGHV: IGHDJ joins. The formation of the rearranged immunoglobulin variable gene that will eventually be transcribed as the IGH begins with the pairing of single IGHD and IGHJ segments from the approximately 25 IGHD and 6 IGHJ segments within an individuals' IGH locus. Subsequently, the IGHD: IGHJ will be joined to one of between 45 to 50 functional IGHV segments (Boyd *et al.*, 2010; Kidd *et al.*, 2012; Watson *et al.*, 2013) to give the rearranged IGH variable region gene.

The recombination of select gene segments to generate the final gene sequence provides for approximately 22,500 unique IGH rearrangements as the IGHDs can be used in all three reading frames (RFs). The light chain provides for approximately 200 different IGKs by recombining 34 to 40 IGKV to 5 IGKJ and 165 lambda light chains combinations from joining between 29 to 33 IGLV and 5 IGLJ. The pairing of heavy and light chains therefore provides for just over 8 million unique antibodies.

A number of biases in the joining of particular segments to each other have been observed, for example, a tendency for 3` IGHD genes to preferentially pair with 5` IGHJ genes (Kidd *et al.*, 2016; Volpe and Kepler, 2008; Souto-Carneiro *et al.*, 2005). This may be the result of RSS efficiency or other mechanisms such as chromatin structure that impact gene segment accessibility.

Junctional Diversity

The approximately 8 million immunoglobulins formed from combinatorial diversity accounts for only a fraction of the predicted diversity of the immunoglobulin repertoire within a single individual's repertoire. Further diversity stems from the imprecise nature of the gene segment joining with nucleotides added and removed during the recombination process (**Fig, 1(B)**) (Alt and Baltimore, 1982; Lafaille *et al.*, 1989). Joining of two gene coding segments begins with the recruitment of the lymphocyte specific Recombination Activation Gene (RAG) complex consisting of RAG-1 and RAG-2 proteins to the RSS elements (van Gent *et al.*, 1995). The RAG complex introduces a nick between the RSS and the adjacent gene coding segment that is converted a double-stranded break with the formation of a covalently sealed hairpin on the gene coding side of the break (McBlane *et al.*, 1995).

The hairpin loops at the gene segment ends are opened by a complex formed between the nuclease Artemis and the catalytic subunit of the DNA-dependent kinase (DNA-PK$_{CS}$) (Ma *et al.*, 2002). Artemis preferentially cleaves the hairpin loop at a position 3′ to the tip generating a single stranded overhang (**Fig. 1(B)**). The single stranded overhang is an inverted repeat of the sequence of the end of the gene segment and this sequence may be incorporated into the rearranged gene as palindromic (P) nucleotides. The overhangs of the open hair pins are processed in two ways - exonuclease trimming and non-template encoded (N) addition (Lafaille *et al.*, 1989). Exonuclease trimming may remove all P nucleotides, plus additional nucleotides from the coding sequence beyond the original break. The nuclease responsible is unknown. N nucleotide addition arises from the incorporation of free dNTPs to the 3′ end of DNA by the short isoform of nuclear enzyme terminal deoxynucleotidyl transferase (TdT) within the joint (Benedict *et al.*, 2000). These regions are termed the N-regions (N1 at the IGHV to IGHD join and N2 at the IGHD to IGHJ join). TdT preferentially incorporates G. Following processing, the joint between the two gene segment ends is resolved by non-homologous DNA end joining (Lieber, 1999).

Somatic Hypermutation

SHM expands the diversity of the immunoglobulin repertoire and tailors antibodies and BCRs for higher affinity interactions with antigen (**Fig. 1(E)**). SHM takes place within specialized micro-environments within secondary lymphoid tissue call germline centers (GC) (Bannard and Cyster, 2017). Within the GC, B cells undergo rapid clonal expansion, during which the B cell's immunoglobulin genes mutate at the rate of approximately 10^{-3} changes per nucleotide per cell division. Point mutations are introduced into the domain that starts approximately 150 bp downstream from the IGHV promoter and extends 2kpbs downstream with a frequency that decays exponentially from the IGHV promoter (Rada and Milstein, 2001).

Substitutions are targeted to intrinsic mutational hotspots (Jolly *et al.*, 1996). Activation induced cytosine deaminase (AID) is specifically expressed in activated B cells and undertakes targeted deamination of deoxycytidine residues (Muramatsu *et al.*, 2000). Deamination of C by AID changes C:G base pairs to U:G mispairs and is targeted to WRC motifs with a major intrinsic hotspot being overlapping WRC motifs on the coding and non-coding strands – WGCW – which suggests that AID binds as an oligomer (Ta *et al.*, 2003). The U:G mispair can be resolved in a number of ways. Replication using the mismatching template can produce transition mutations where C/G are replaced by A/T. Alternatively, the U can be removed by uracil-DNA glycolsylase such as UNG2 creating an apyrimidinic side (AP) which can be repaired by base excision repair (BER) (Teng and Papavasiliou, 2007). As BER results in untemplated replication opposite the AP site any dNTP can be incorporated at the position leading to either reversion, transition or transversion mutations. The nucleotide inserted is dependent on the DNA polymerase involved in the BER as polymerases are biased in the nucleotides that they insert.

Targeting of mutations to A:T pairs may arise through several proposed mechanism (Teng and Papavasiliou, 2007). For example, the binding of MSH2-MSH6 at U:G mispairs can recruit exonuclease 1 where it can create a gap that when filled by an error prone polymerase leads to mutations at A:T pairs. Alternatively, during short patch repair of the AID U:G lesion, U may be inappropriately incorporated due to dNTP pool imbalances creating A:U mispairs in the region surrounding U:G lesion and the resolution of the A:U mispair may introduce mutations at A:T pairs.

Sequencing of Immunoglobulin Gene Repertoires

Each immunoglobulin transcript captures details of the gene segments that formed the rearranged gene, the processing the that gene segments underwent during recombination, and, if the B cell has encountered antigen, the mutations that have accumulated during affinity maturation, along with the isotype subclass that the VDJ is associated with. Amplicons from tens to hundreds of thousands of B cells may be sequenced from a single peripheral blood sample to capture the different members of clonal expansions. Furthermore, as VDJ rearrangement is an intra-chromosomal event, analysis of low mutation rearrangements can also inform on the underlying immune genotype and haplotype of the individual (Boyd et al., 2010) (Kidd et al., 2012). Repeated longitudinal sampling from a single subject over the course of an immune response, such as infection or vaccination, allows tracking of the origin and fate of B cell clones over the course of the response.

Several approaches for deep sequencing of immune receptor repertoires have been published (Jackson et al., 2014; Galson et al., 2015; Ellebedy et al., 2016). The common themes to these protocols are amplification from cDNA to capture VDJ and isotype information and multiplexing of samples using nucleotide sequence barcodes to permit single libraries to be generated from many different samples. Some protocols incorporate unique molecular identifiers (UMI) at the cDNA synthesis step to tag each RNA molecule with unique identifier before PCR amplification to allow counting of single transcripts and for error correction (Turcha-ninova et al., 2016). In addition to cDNA based protocols, the relationship between single B cells and amplicons may be maintained by genomic DNA amplification in replicate to study clonal expansions at the sacrifice of isotype information (Boyd et al., 2009).

Full length VDJ amplified from leader primers, or near-full length from FR1 primers, with enough CH1 exon to discriminate IgG and IgA subclasses can be captured using paired end libraries sequences on the Illumina MiSeq platform with 600 cycle kits for in excess of 20 million reads. Longer read, lower throughput technologies, such as PacBio and Oxford Nanopore have potential to offer ability to capture full length VDJ plus constant region amplicons. Usually amplifications are from bulk cell populations, either total peripheral blood mononuclear cells or sorted cell populations from fluorescent or magnetic cell sorting to enrich to particular phenotypic cell subsets (eg. plasmablasts, memory B cells, activated B cells, or naïve B cells), however, immunoglobulin genes may also be studied from single B cells, to capture paired heavy and light chain gene transcripts (Busse et al., 2014). Attempts have also been made to reconstruct rearranged immunoglobulin genes from single cell RNA transcriptome sequencing methods to permit linking of paired heavy and light chain transcripts to overall gene transcription within a cell (Rizzetto et al., 2017).

Analysis of Immunoglobulin Gene Sequences

Partitioning of Rearranged Immunoglobulin Genes

The most basic challenge of the study of immune receptor genes through sequencing datasets is the partitioning of rearranged genes into their various components such as IGHV, IGHD and IGHJ segments and identification of nucleotide loss and addition of the IGHV-IGHD and IGHD-IGHJ joints to determine of SHM spectra.

A number of tools exist for this purpose (**Table 1**). The earliest contributions to the field treated the partitioning problem as three separate alignment problems; one for each gene segment. These include IMGT/V-QUEST (Brochet et al., 2008), which appears to be based on a dynamic programming algorithm with a single recursive rule as alignment scores are re-created by a simple derivation of the Smith-Waterman local sequence alignment algorithm, and IgBLAST (Ye et al., 2013) based on the Basic Local Alignment Search Tool (BLAST) algorithm (Camacho et al., 2009). Contemporary tools have attempted to gain speed advantages to the alignment-approach by using modified kmer-chaining based methods (MiXCR (Bolotin et al., 2015)), dynamic programming algorithms constrained by conserved motifs (HTJoinSolver (Russ et al., 2015)) and fast-tag-searching algorithms (LymAnalyzer (Yu et al., 2016)). IMGT/V-QUEST and IgBLAST remain perhaps the most widely used germline gene annotation tools for rearranged gene sequences. Only MiXCR identifies the associated isotype from the IGHC gene, otherwise this is undertaken a customized post-processing step using BLAST (Camacho et al., 2009) or string matching.

The variety of processes that underlie the gene rearrangement however complicate the partitioning, particularly the discrimination of the components within the CDR3 that spans from the IGHV end, across the V-D junction, IGHD, D-J junction, through to the IGHJ start. Tools specific for the CDR3 were therefore developed. The first wave of such tools included IMGT/Junction Analysis (IMGT/JA (Yousfi Monod et al., 2004)) and JOINSOLVER (Souto-Carneiro et al., 2004). IMGT/JA uses a five-stage process to identify the P nucleotides, the N nucleotides, the IGHD, to consider the processing of IGHV and IGHJ ends, and to

Table 1   Example of immune receptor tools and pipelines available for processing and analysis of rearranged immune gene repertoire datasets

Name	Website	Purpose	Type
ALPHABETR	https://github.com/edwardslee/alphabetr	R package for inference of TCR alpha-beta pairing	Analysis tool
BASELINE	http://selection.med.yale.edu/baseline/	Antigen selection	Analysis tool
IgPhyML	https://github.com/kbhoehn/IgPhyML	Analysis of immunoglobulin phylogeny	Analysis tool
MaxSnippetModel	https://github.com/jostmey/MaxSnippetModel	Statistical classifier for immune repertoires	Analysis tool
RDI	http://bitbucket.org/cbolen1/rdicore	Repertoire dissimilarity metric	Analysis tool
VDJSeq-Solver	http://eda.polito.it/VDJSeq-Solver/	Identification of clonal populations from neoplastic datasets	Analysis tool
Decombinator	https://github.com/innate2adaptive/Decombinator	TCR, string matching algorithm	Annotation
HTJoinSolver	https://dcb.cit.nih.gov/HTJoinSolver/	Partitioning by dynamic programming constrained by conserved motifs	Annotation
IgBLAST	https://www.ncbi.nlm.nih.gov/igblast/	BLAST-based partitioning of BCR and TCR via web-interface or stand-alone	Annotation
iHMMune-align	http://ihmmune.web.cse.unsw.edu.au/home.htm	HMM partitioning of IGH	Annotation
IMGT/V-QUEST	http://www.imgt.org/IMGT_vquest/vquest	Partitioning tool using IMGT reference datasets via web-interface	Annotation
JOINSOLVER	https://joinsolver.niaid.nih.gov/	IGH partitioning with a focus on the confident determination of the IGHD within the CDR3	Annotation
partis	https://github.com/psathyrella/partis/	HMM partitioning tool	Annotation
SoDA, SoDA2	No longer supported, succeeded to cloanalyst	Probabilistic partitioning of IGH	Annotation
TCRbiter	https://github.com/AstraZeneca-NGS/tcrbiter	TCR annotation tool	Annotation
TCRklass	http://sourceforge.net/projects/tcrklass	K-string based algorithm for TCR analysis	Annotation
VDJsolver	http://www.cbs.dtu.dk/services/VDJsolver/	IGH partitioning tool using Maximum Likelihood for model fitting	Annotation
IGoR	https://bitbucket.org/qmarcou/igor	Probabilistic alignment using a sparse Expectation-Maximization algorithm, three modes – statistics learning, sequence analysis and sequence generation.	Annotation & Simulation
IgDiscover	https://github.com/NBISweden/IgDiscover/	Allele discovery	Germline inference
IMPre	https://github.com/zhangwei2015/IMPre	Gene segment prediction	Germline inference
ARGalaxy	https://bioinf-galaxian.erasmusmc.nl/argalaxy/	Antibody analysis within the Galaxy platform	GUI
ARResT/Interrogate	https://github.com/InfspiredBAT/ARResT.Interrogate	Interactive IG/TR analysis, currently down	GUI
ClonoCalc & ClonoPlot	https://bitbucket.org/ClonoSuite/clonocalc-plot	GUI front end for MiXCR with visualization tools	GUI
IGGalaxy	http://bioinformatics.erasmusmc.nl/wiki/index.php/Immunoglobulin_Galaxy	Galaxy service for detection and quantification, uses IMGT and IgBLAST for alignment	GUI
VDJServer	https://vdjserver.org/	Web-based repertoire analysis and storage	GUI
Vidjil	http://www.vidjil.org/	Interactive analysis of HTS via web-based interface with IMGT/V-QUEST, IgBLAST or MiXCR processing	GUI
Migmap	https://github.com/mikessh/migmap	IgBLAST wrapper	Other
VDJMLpy	https://vdjserver.org/vdjml/	File format for aligned sequence data	Other
AbMining Toolbox	https://sourceforge.net/projects/abmining/	Toolkit focusing on CDR3 analysis of antibody libraries	Pipeline
Cloanalyst	http://www.bu.edu/computationalimmunology/research/software/	Bayesian based approaches to gene annotation and unmuted ancestor inference	Pipeline
IgSCUEAL	https://github.com/spond/IgSCUEAL	Pipeline for B cell transcript analysis based on genetic algorithm for viral sub-typing	Pipeline
Immcantation	http://immcantation.readthedocs.io/	Pipeline, uses either IMGT or IgBLAST for align, many features including allelic inference	Pipeline
ImmuneDB	http://immunedb.com/	Storage and analysis of B- and T-cell sequence data, pipeline with custom alignment algorithm	Pipeline

Tool	URL	Description	Type
ImmunedeveRsity	https://bitbucket.org/ImmunedeveRsity/immunediversity	Manipulation and processing of HTS reads to identify VDJ usage and clonal origin, IgBLAST	Pipeline
IMonitor	https://github.com/zhangwei2015/IMonitor	TCR/BCR pipeline using BLAST for alignment	Pipeline
IMSEQ	http://www.imtools.org/	Pipeline focused on clonotype inference	Pipeline
LymAnalyzer	https://sourceforge.net/projects/lymanalyzer	Allele discovery, fast-tag-search algorithm for alignment of dastsets	Pipeline
MiXCR	https://milaboratory.com/software/mixcr/	Pipeline for BCR and TCR analysis with focus on clonotype building	Pipeline
RTCR	http://uubram.github.io/RTCR/	TCR analysis pipeline	Pipeline
SONAR	https://github.com/scharch/SONAR	Pipeline for analysis of repertoire datasets	Pipeline
VDJFasta	https://sourceforge.net/projects/vdjfasta/	BCR analysis tools including partitioning	Pipeline
clonotypeR	http://clonotyper.branchable.com	R package for the identification and analysis of clonotypes	Post-analysis
IMEX	http://bioinformatics.fh-hagenberg.at/immunexplorer/	Post-processing of IMGT/HighV-QUEST output	Post-analysis
LymphoSeq	https://bioconductor.org/packages/release/bioc/html/LymphoSeq.html	Post-analysis of Adapative Biotechnologies ImmunoSEQ output	Post-analysis
tcrR	http://imminfo.github.io/tcr/	Post processing of MiXCR, ImmunoSEQ or MiGEC	Post-analysis
TRIgS	https://github.com/williamdlees/TRIgS	Clonal lineage analysis of processed datasets	Post-analysis
VDJtools	https://github.com/mikessh/vdjtools	Toolkit for the post-analysis of datasets	Post-analysis
VDJviz	https://github.com/antigenomics/vdjviz	Visualization and analysis of datasets	Post-analysis
IgRepertoireConstructor	http://yana-safonova.github.io/ig_repertoire_constructor/	Antibody repertoire construction for MS/MS applications	Simulation
IgSimulator	http://yana-safonova.github.io/ig_simulator/	Simulation of antibody repertoires	Simulation
repgenHMM	https://bitbucket.org/yuvalel/repgenhmm	HMM for generating synthetic rearrangements	Simulation
sciReptor	https://github.com/b-cell-immunology/sciReptor	Single cell immunoglobulin repertoire analysis	Single cell
TraCeR	https://github.com/teichlab/tracer	Analysis of TCRs from single cell transcriptomes	Single cell
VDJPuzzle	https://bitbucket.org/kirbyvisp/vdjpuzzle2	TCR and BCR reconstruction from scRNA-seq data	Single cell

Note: The 'type' column provides a general classification for the tool; 'analysis' tools focus on performing targeted analysis on processed datasets, 'annotation' tools focus on the partitioning of rearranged sequences into their components, 'germline inference' tools attempt to define the genotypes from rearranged datasets, 'GUIs' provide easy-to-use interfaces for tools or pipelines, 'pipelines' are end-to-end solutions starting from raw data, 'post-analysis' tools offer general post-processing of annotated datasets, and 'single-cell' are tools are for the reconstruction of genes from single cell datasets.

allow for somatic point mutations within the ends by placing predefined maximum values on the number of mutations and by excluding the possibility of consecutive mutations (Yousfi Monod *et al.*, 2004)).

The location of the IGHD amongst the N-additions means that care must be taken to delineate N additions from germline gene contributions. JOINSOLVER considers the likelihood that N-addition may give rise to sequences that mimic germline IGHDs (Souto-Carneiro *et al.*, 2004) by defining the maximum number of consecutive matches that differentiate between a true germline IGHD gene and N-additions. The minimum number consecutive matches were determine using a Monte Carlo simulation whereby large simulated datasets representative of N additions are used to calculate the frequency at which IGHD sequence motifs occur to establish probabilities. For appropriate probabilities to be calculated the simulated N additions should reflect real N-additions and models for simulation of N additions have been proposed (Jackson *et al.*, 2007).

Partitioning approaches also consider the problem of immunoglobulin partitioning in the context of the biological rearrangement process. Rather than a series of local alignments against reference sets of germline genes, models of rearrangement are created, and mathematical techniques are used to find the combination of germline genes, mutation and nucleotide loss and addition that best explain a rearranged sequence in the context of the model. SoDA for example utilized a 3D alignment algorithm for simultaneous alignments of all genes segments and to account for nucleotide loss and addition (Volpe *et al.*, 2006). JointML uses a maximum likelihood method to determine the best fit to a model of IGH rearrangement that considers all possible combinations of germline genes and N and P additions (Ohm-Laursen *et al.*, 2006), and iHMMune-align uses the veterbi algorithm to find the best path through a hidden markov model (HMM) which consists of a series of probabilistically defined states whose relationship to each other are described by transitions that are themselves also associated with probabilities (Gaeta *et al.*, 2007). These utilities were developed in the pre-deep sequencing era and generally do not scale well to very large repertoire datasets.

Contemporary tools that use model-based approaches include partis (Ralph and Matsen, 2016a), IGoR (Marcou *et al.*, 2017) and IgSCUEAL (Frost *et al.*, 2015). The partis package uses a novel HMM factorization strategy across large datasets to build a parameter rich model for partitioning of individual sequences into their germline gene and N addition contributions. It overcomes the limitations of previous HMM-bases approaches by using flexible categorical distributions rather than probability distributions, implementing a new HMM compiler (ham), and by performing inferences on collections of HMMs rather than a single all-encompassing HMM (Ralph and Matsen, 2016a). IgSCUEAL attempts to annotate IGHV and IGHJ genes within rearrangements using a phylogenetic-based approach that combines a genetic algorithm with maximum likelihood fitting of evolutionary models, but IGHD identification is performed using pair-wise alignment (Frost *et al.*, 2015). IGoR uses a Sparse Expectation-Maximization algorithm to learn statistics of rearrangements for use in annotating rearranged genes in a probabilistic manner and can also generate simulated repertoire datasets based on the learned statistics (Marcou *et al.*, 2017).

Gene Reference Datasets and Allelic Inference

The major reference set utilized in the analysis of rearranged immunoglobulin genes is the IMGT reference set (Lefranc and Lefranc, 2001). Alternative collections of human immunoglobulin gene segments include NCBI (Ye *et al.*, 2013) and VBASE2 (Retter *et al.*, 2005). As increasingly diverse human populations are studied it has become apparent that existing reference datasets do not capture all polymorphisms that exist in the human population (Watson *et al.*, 2013; Boyd *et al.*, 2010; Kidd *et al.*, 2012; Watson *et al.*, 2015; Wang *et al.*, 2011; Scheepers *et al.*, 2015).

Several tools for the inference of novel gene segment polymorphisms within repertoire datasets have been developed including IgDiscover (Corcoran *et al.*, 2016), TIgGER (Gadala-Maria *et al.*, 2015), LymAnalyzer (Yu *et al.*, 2016) and IMPre (Zhang *et al.*, 2016). These tools leverage the ability of rearranged immunoglobulin genes to be used to infer set of available gene segments within the IGH locus of an individual (Boyd *et al.*, 2010). IgDiscover applies an iterative clustering, consensus building and filtering algorithm to IgM antibody libraries to define dataset specific germline gene segment sets *de novo* (Corcoran *et al.*, 2016). TIgGER utilizes datasets that have previously been aligned against a reference set, such as the IMGT reference set (Lefranc and Lefranc, 2001), and applies regression-based methods to detect distinct patterns of SHM the distinguish polymorphic positions from positions targeted by SHM (Gadala-Maria *et al.*, 2015). LymAnalyzer is a toolkit that aligns to the IMGT reference set using a fast-tag-search alignment and applies a two-step heuristic approach, adapted from (Boyd *et al.*, 2010), to the detection of novel alleles in the aligned data, testing for enrichment of a mismatch to the reference in multiple distinct rearrangements and requiring the difference to be present in at least 10% of the alignments to a reference gene (Yu *et al.*, 2016). IMPre utilizes a custom algorithm combining k-mer seed clustering, multiway tree-based assembly and optimization by filtering, merging and error correction, for *de novo* inference of germline gene segments from rearranged repertoire datasets (Zhang *et al.*, 2016).

Analysis suggests that parallel use haplotype inference, as originally proposed by (Kidd *et al.*, 2012), in conjunction with novel allele inference, provides for high sensitivity and specificity in repertoire analysis (Kirik *et al.*, 2017). The use of dataset specific reference sets does however restrict the options with respect to partitioning tools to those, such as IgBLAST (Ye *et al.*, 2013), that permit the user defined datasets, or to the application of re-assignment methods such as within TIgGER (Gadala-Maria *et al.*, 2015). Attempts to define additional allelic variants using other data sources have also been undertaken. For example, AlleleMiner (Yu *et al.*, 2017) attempts to recover immunoglobulin gene segments alleles from genomic sequence data by mining genomic variant calls from the short-read sequencing datasets such as the 1000 Genomes Project data. There are many challenges in building reference datasets from this type of single nucleotide polymorphism (SNP) data and caution must be exercised especially with respect to the mapping of the SNP data to the reference prior to inference (Watson *et al.*, 2017).

Analysis of Clonal Lineages

Once a dataset of high quality annotated gene rearrangements has been gathered it is desirable to understand the lineage structure within the dataset, or across datasets in the case of longitudinal or multiple subset or multiple site sampling from a single subject. Historically, this has been approached as a sequence clustering problem aimed at grouping highly similar immunoglobulin gene rearrangements together under the assumption that their similarity has resulted from their sharing the same initial B cell. This often involves clustering the CDR3 sequences of rearranged IGH that share the same IGHV and IGHJ segments and have equal CDR3 lengths at an identity threshold that allows from some SHM within the CDR3 but which shouldn't bring together CDR3s that were generated by distinct VDJ recombination events.

Several software packages implement sequence-based clustering for clone grouping, for example, IMSEQ's clone building using shared CDR3 sequences (Kuchenbecker *et al.*, 2015) and the clone clustering within the ImmuneDB (Rosenfeld *et al.*, 2017). The clone clustering implemented by ImmuneDB, which iterates over unique reads in descending size order, grouping smaller copy reads with higher count reads where they differ by less than 15% at the CDR3 amino acid level, was recently applied to B cells sampled from a number of different anatomical sites to develop an atlas of B-cell clonal distribution (Meng *et al.*, 2017). An attempt to improve on the results of sequence-based clustering approached is offered by the likelihood-based clonal inference using multi-HMMs implemented as part of partis (Ralph and Matsen, 2016b).

Detection of Antigen Selection

Antigen selection refers to testing whether or not the mutational spectra of immunoglobulin genes have been influenced by enrichment of mutations predicted to improve interaction with antigen. One package for testing if mutational distributions have been skewed by antigen selection is BASELINe (Yaari *et al.*, 2012). BASELINe provides a framework for quantifying the strength of antigen selection on individual sequences or within clonal lineages using Bayesian estimates of replacement frequency and has the capacity to aggregate probability density functions over multiple sequences in order to compare sets of sequences.

Additional approaches for studying antigen selection have been described, but not implemented as tools. Approaches are largely based on detecting an enrichment of mutations that have led to amino acid substitutions within the CDRs, and the avoidance of such changes in the FRs, and differ in their hypothesis testing statistics. The first of such methods was proposed by (Shlomchik *et al.*, 1987) and tested replacement (amino acid change) and silent (no amino acid change) in CDRs and FRs against a random model of mutation using the binomial distribution. This was later modified by (Chang and Casali, 1994) in an attempt to better account for the differences in codon usage between CDRs and FRs as CDRs are more intrinsically biased toward replacement mutations owing to their codon usage and (Lossos *et al.*, 2000) suggested the use of a multinomial, rather than binomial, model to determine significance. The failure to account for mutational hotspots and biases in mutational outcomes was suggested to lead to inaccuracies in these approaches (Bose and Sinha, 2005). This led to the proposed method by (Dahlke *et al.*, 2006) which extended prior literature to model hotspots using trinucleotides mutability scoring based on observed CDR and FR sequences to compare the accumulation of replacement mutations in the CDR in the context of the overall level of IGHV mutation.

Comparing Immunoglobulin Repertoires

The ability to compare repertoires, either from different samples from a single individual, or between different individuals generally relies upon comparing various metrics calculated after repertoire partitioning and annotation of clonal relationships and SHM spectra. For example, the frequency at which different gene segments are utilized, the frequency of somatic hypermutation, the isotype subclass utilization or characteristics and distributions of the clonal expansions. Frameworks for statistical comparisons of overall repertoires have been historically lacking. Two recently published frameworks to compare repertoires are the repertoire dissimilarity index (RDI) (Bolen *et al.*, 2017) and the MaxSnippetModel (Ostmeyer *et al.*, 2017). RDI is a non-parametric method to quantitatively compare repertoires by average variance in gene segment utilization through a combination of bootstrapped sub-sampling and simulation (Bolen *et al.*, 2017). In comparison, the MaxSnippetModel focusses on determining differences in CDR3 sub-motif composition between repertoires from different disease contexts by attempting to statistically classify sequences by representing CDR3 AA 'snippets' (overlapping sub-strings of fixed length) sequences as vector-based representation of AA properties using Atchely factors (Ostmeyer *et al.*, 2017). A TensorFlow implementation is used to score the snippet Atchley factors using logistic modelling to define the maximum score for each sample before model fitting using gradient optimization for classification.

Analysis Pipelines for Repertoire Data Analysis

A variety of immunoglobulin repertoire datasets are available from public repositories. A sampling of public immunoglobulin repertoire sequence datasets is provided in **Table 2**. There is no a single approach to the analysis of repertoire data. The approach taken will often depend upon the source of the data sequenced, the sequencing technology used, the computational skills of the user, and the research question they are seeking to. Many toolkits that implement end-to-end pipelines for analysis are now available with extensive use-case documentation, for example MiXCR (see Relevant Websites section), Immcantation (see Relevant Websites section) or ImmuneDB (see Relevant Websites section).

Table 2   Examples of publically available human immunoglobulin repertoire datasets

Accession	Accession type	Technology	Description
PRJNA324093	BioProject	Illumina MiSeq	Longitudinal vaccine response with different cell subsets
PRJNA337970	BioProject	Illumina MiSeq	B cell subsets from lifelong malaria exposures
PRJNA330659	BioProject	Illumina HiSeq	Pre- and post-treatment samples for childhood B-ALL cohort
PRJNA309577	BioProject	Illumina HiSeq	5′ RACE derived BCR and TCR repertoire
PRJNA291102	BioProject	Illumina HiSeq	Comparison of methodologies for repertoire amplifications
PRJNA277373	BioProject	Illumina MiSeq	Amplification from total PBMCs of a healthy donor
PRJNA260985	BioProject	Illumina MiSeq	IGH sequencing from B cells and plasmablasts in a variety of conditions including autoimmunity
PRJNA260905	BioProject	Illumina HiSeq	Immunosuppression in context of transplantation

In the interest of highlighting the general components of repertoire data analysis pipelines an example of the basic processing steps from raw Illumina MiSeq paired end sequenced library of cDNA amplicons generated from primers positioned in the IGHV and IGHC to post-processed data using stand-alone tools would be as follows:

- **Pre-processing**: Merging of read pairs (eg. PEAR (Zhang *et al.*, 2014), de-multiplexing of barcodes and extraction of UMIs (eg. UMI-tools (Smith *et al.*, 2017)) and read trimming to remove primers and adaptors and quality filtering (eg. trimmomatic (Bolger *et al.*, 2014))
- **Partitioning**: Partitioning of merged read pairs using stand-alone instance of IgBLAST (Ye *et al.*, 2013) using either the IMGT reference sets of IGHV, IGHD and IGHJ segment or customized germline datasets.
- **Isotype calling**: Using either script-based string matching (eg, perl or python) or BLAST alignment (Camacho *et al.*, 2009) to determine which isotype subclass is associated with each VDJ based on the expected CH1 exon sequence governed by the IGHC primer location.
- **Parsing**: Scripting to parse key features from the IgBLAST output such as gene segment calls, mutations, N/P nucleotides and CDR/FRs and to integrate the isotype and other metadata.
- **Clone building**: CDR3 sequence subset based on the subject, IGHV, IGHJ and CDR3 length may be clustered using tools such as cd-hit (Fu *et al.*, 2012) or uclust (Edgar, 2010) to infer clonally related sequences within an individual response. Similar clustering methods can be used across responses from different subject to define convergent IGH signatures.
- **Post-processing**: Custom databases may be used for storage to serve as a basis for analysis or custom file formats generated. Scripting languages can be used to develop analysis to explore the features of interest within the dataset.

Current Challenges and Conclusions

Like any field in rapid change, the immune receptor repertoire sequencing field has lacked standardization as it has moved into the era of deep sequencing, however, this gap is beginning to be addressed (Yaari and Kleinstein, 2015; Breden *et al.*, 2017). Challenges exist around the development and implementation of frameworks for the standardization of data submission, data formats for tool input and output and in the incorporation of new data types into traditional resources.

The integration of repertoire sequencing datasets with other 'omics data derived from the same subject is becoming increasingly popular. For example, annotation of deeply sequenced repertoire data for antigen specificity through the use of monoclonal antibodies (Jackson *et al.*, 2014), or matching of transcript datasets to proteomic (Adamson *et al.*, 2017). Tools that provide researchers with a unified analysis of their various data types will become increasingly important as these gaps are currently filled by custom, in-house pipelines that link often disparate tools and toolkits to undertake analysis. Leveraging deep sequenced dataset for deep and machine learning applications is also expected to become increasingly informative as the variety, number and depth of datasets continues to increase.

See also: Computational Immunogenetics. Extraction of Immune Epitope Information. Immunoglobulin Clonotype and Ontogeny Inference. Immunoinformatics Databases. Natural Language Processing Approaches in Bioinformatics. Sequence Analysis. Sequence Composition. Vaccine Target Discovery

References

Adamson, P.J., Al Kindi, M.A., Wang, J.J., *et al.*, 2017. Proteomic analysis of influenza haemagglutinin-specific antibodies following vaccination reveals convergent immunoglobulin variable region signatures. Vaccine 35, 5576–5580.
Alt, F.W., Baltimore, D., 1982. Joining of immunoglobulin heavy chain gene segments: Implications from a chromosome with evidence of three D–JH fusions. Proc Natl Acad Sci USA 79, 4118–4122.
Bannard, O., Cyster, J.G., 2017. Germinal centers: Programmed for affinity maturation and antibody diversification. Curr Opin Immunol 45, 21–30.
Benedict, C.L., Gilfillan, S., Thai, T.H., Kearney, J.F., 2000. Terminal deoxynucleotidyl transferase and repertoire development. Immunol Rev 175, 150–157.

Bernard, O., Hozumi, N., Tonegawa, S., 1978. Sequences of mouse immunoglobulin light chain genes before and after somatic changes. Cell 15, 1133–1144.

Bolen, C.R., Rubelt, F., vander Heiden, J.A., Davis, M.M., 2017. The Repertoire Dissimilarity Index as a method to compare lymphocyte receptor repertoires. BMC Bioinformatics 18, 155.

Bolger, A.M., Lohse, M., Usadel, B., 2014. Trimmomatic: A flexible trimmer for Illumina sequence data. Bioinformatics 30, 2114–2120.

Bolotin, D.A., Poslavsky, S., Mitrophanov, I., et al., 2015. MiXCR: Software for comprehensive adaptive immunity profiling. Nat Methods 12, 380–381.

Bose, B., Sinha, S., 2005. Problems in using statistical analysis of replacement and silent mutations in antibody genes for determining antigen-driven affinity selection. Immunology 116, 172–183.

Boyd, S.D., Gaeta, B.A., Jackson, K.J., et al., 2010. Individual variation in the germline Ig gene repertoire inferred from variable region gene rearrangements. J Immunol 184, 6986–6992.

Boyd, S.D., Marshall, E.L., Merker, J.D., et al., 2009. Measurement and clinical monitoring of human lymphocyte clonality by massively parallel VDJ pyrosequencing. Sci Transl Med 1, 12ra23.

Breden, F., Luning Prak, E.T., Peters, B., et al., 2017. Perspective: Reproducibility and reuse of adaptive immune receptor repertoire data. Front Immunol.

Brochet, X., Lefranc, M.P., Giudicelli, V., 2008. IMGT/V-QUEST: The highly customized and integrated system for IG and TR standardized V–J and V–D–J sequence analysis. Nucleic Acids Res 36, W503–W508.

Busse, C.E., Czogiel, I., Braun, P., Arndt, P.F., Wardemann, H., 2014. Single-cell based high-throughput sequencing of full-length immunoglobulin heavy and light chain genes. Eur J Immunol 44, 597–603.

Camacho, C., Coulouris, G., Avagyan, V., et al., 2009. BLAST+: Architecture and applications. BMC Bioinform 10, 421.

Chang, B., Casali, P., 1994. The CDR1 sequences of a major proportion of human germline Ig VH genes are inherently susceptible to amino acid replacement. Immunol Today 15, 367–373.

Chothia, C., Lesk, A.M., 1987. Canonical structures for the hypervariable regions of immunoglobulins. J Mol Biol 196, 901–917.

Corcoran, M.M., Phad, G.E., Vazquez Bernat, N., et al., 2016. Production of individualized V gene databases reveals high levels of immunoglobulin genetic diversity. Nat Commun 7, 13642.

Croce, C.M., Shander, M., Martinis, J., et al., 1979. Chromosomal location of the genes for human immunoglobulin heavy chains. Proc Natl Acad Sci USA 76, 3416–3419.

Dahlke, I., Nott, D.J., Ruhno, J., Sewell, W.A., Collins, A.M., 2006. Antigen selection in the IgE response of allergic and nonallergic individuals. J Allergy Clin Immunol 117, 1477–1483.

Davis, M.M., 1990. T cell receptor gene diversity and selection. Annu Rev Biochem 59, 475–496.

Edgar, R.C., 2010. Search and clustering orders of magnitude faster than BLAST. Bioinformatics 26, 2460–2461.

Eisen, H.N., 2014. Affinity enhancement of antibodies: How low-affinity antibodies produced early in immune responses are followed by high-affinity antibodies later and in memory B-cell responses. Cancer Immunol Res 2, 381–392.

Ellebedy, A.H., Jackson, K.J., Kissick, H.T., et al., 2016. Defining antigen-specific plasmablast and memory B cell subsets in human blood after viral infection or vaccination. Nat Immunol 17, 1226–1234.

Ezkurdia, I., Juan, D., Rodriguez, J.M., et al., 2014. Multiple evidence strands suggest that there may be as few as 19,000 human protein-coding genes. Hum Mol Genet 23, 5866–5878.

Faham, M., Zheng, J., Moorhead, M., et al., 2012. Deep-sequencing approach for minimal residual disease detection in acute lymphoblastic leukemia. Blood 120, 5173–5180.

Frost, S.D., Murrell, B., Hossain, A.S., Silverman, G.J., Pond, S.L., 2015. Assigning and visualizing germline genes in antibody repertoires. Philos Trans R Soc Lond B Biol Sci. 370.

Fu, L., Niu, B., Zhu, Z., Wu, S., Li, W., 2012. CD-HIT: Accelerated for clustering the next-generation sequencing data. Bioinformatics 28, 3150–3152.

Gadala-Maria, D., Yaari, G., Uduman, M., Kleinstein, S.H., 2015. Automated analysis of high-throughput B-cell sequencing data reveals a high frequency of novel immunoglobulin V gene segment alleles. Proc Natl Acad Sci USA 112, E862–E870.

Gaeta, B.A., Malming, H.R., Jackson, K.J., et al., 2007. iHMMune-align: Hidden Markov model-based alignment and identification of germline genes in rearranged immunoglobulin gene sequences. Bioinformatics 23, 1580–1587.

Galson, J.D., Clutterbuck, E.A., Truck, J., et al., 2015. BCR repertoire sequencing: Different patterns of B-cell activation after two Meningococcal vaccines. Immunol Cell Biol 93, 885–895.

Glanville, J., Zhai, W., Berka, J., et al., 2009. Precise determination of the diversity of a combinatorial antibody library gives insight into the human immunoglobulin repertoire. Proc Natl Acad Sci USA 106, 20216–20221.

Hoh, R.A., Joshi, S.A., Liu, Y., et al., 2016. Single B-cell deconvolution of peanut-specific antibody responses in allergic patients. J Allergy Clin Immunol 137, 157–167.

Jackson, K.J., Gaeta, B.A., Collins, A.M., 2007. Identifying highly mutated IGHD genes in the junctions of rearranged human immunoglobulin heavy chain genes. J Immunol Methods 324, 26–37.

Jackson, K.J., Liu, Y., Roskin, K.M., et al., 2014. Human responses to influenza vaccination show seroconversion signatures and convergent antibody rearrangements. Cell Host Microbe 16, 105–114.

Jolly, C.J., Wagner, S.D., Rada, C., et al., 1996. The targeting of somatic hypermutation. Semin Immunol 8, 159–168.

Kidd, M.J., Chen, Z., Wang, Y., et al., 2012. The inference of phased haplotypes for the immunoglobulin H chain V region gene loci by analysis of VDJ gene rearrangements. J Immunol 188, 1333–1340.

Kidd, M.J., Jackson, K.J., Boyd, S.D., Collins, A.M., 2016. DJ pairing during VDJ recombination shows positional biases that vary among individuals with differing IGHD locus immunogenotypes. J Immunol 196, 1158–1164.

Kirik, U., Greiff, L., Levander, F., Ohlin, M., 2017. Parallel antibody germline gene and haplotype analyses support the validity of immunoglobulin germline gene inference and discovery. Mol Immunol 87, 12–22.

Kuchenbecker, L., Nienen, M., Hecht, J., et al., 2015. IMSEQ – A fast and error aware approach to immunogenetic sequence analysis. Bioinformatics 31, 2963–2971.

Lafaille, J.J., Decloux, A., Bonneville, M., Takagaki, Y., Tonegawa, S., 1989. Junctional sequences of T cell receptor gamma delta genes: Implications for gamma delta T cell lineages and for a novel intermediate of V-(D)-J joining. Cell 59, 859–870.

Lefranc, M.-P., Lefranc, G., 2001. The Immunoglobulin Factsbook. San Diego: Academic Press.

Levin, M., King, J.J., Glanville, J., et al., 2016. Persistence and evolution of allergen-specific IgE repertoires during subcutaneous specific immunotherapy. J Allergy Clin Immunol 137, 1535–1544.

Liao, H.X., Lynch, R., Zhou, T., et al., 2013. Co-evolution of a broadly neutralizing HIV-1 antibody and founder virus. Nature 496, 469–476.

Lieber, M.R., 1999. The biochemistry and biological significance of nonhomologous DNA end joining: An essential repair process in multicellular eukaryotes. Genes Cells 4, 77–85.

Lossos, I.S., Tibshirani, R., Narasimhan, B., Levy, R., 2000. The inference of antigen selection on Ig genes. J Immunol 165, 5122–5126.

Ma, Y., Pannicke, U., Schwarz, K., Lieber, M.R., 2002. Hairpin opening and overhang processing by an Artemis/DNA-dependent protein kinase complex in nonhomologous end joining and V(D)J recombination. Cell 108, 781–794.

Marcou, Q., Mora, T. & Walczak, A.M. 2017. IGoR: A Tool For High-Throughput Immune Repertoire Analysis. bioRxiv.

McBlane, J.F., van Gent, D.C., Ramsden, D.A., et al., 1995. Cleavage at a V(D)J recombination signal requires only RAG1 and RAG2 proteins and occurs in two steps. Cell 83, 387–395.

McBride, O.W., Hieter, P.A., Hollis, G.F., et al., 1982. Chromosomal location of human kappa and lambda immunoglobulin light chain constant region genes. J Exp Med 155, 1480–1490.

Meng, W., Zhang, B., Schwartz, G.W., et al., 2017. An atlas of B-cell clonal distribution in the human body. Nat Biotechnol 35, 879–884.

Muramatsu, M., Kinoshita, K., Fagarasan, S., et al., 2000. Class switch recombination and hypermutation require activation-induced cytidine deaminase (AID), a potential RNA editing enzyme. Cell 102, 553–563.

Ohm-Laursen, L., Nielsen, M., Larsen, S.R., Barington, T., 2006. No evidence for the use of DIR, D-D fusions, chromosome 15 open reading frames or VH replacement in the peripheral repertoire was found on application of an improved algorithm, JointML, to 6329 human immunoglobulin H rearrangements. Immunology 119, 265–277.

Ostmeyer, J., Christley, S., Rounds, W.H., et al., 2017. Statistical classifiers for diagnosing disease from immune repertoires: A case study using multiple sclerosis. BMC Bioinformatics 18, 401.

Parameswaran, P., Liu, Y., Roskin, K.M., et al., 2013. Convergent antibody signatures in human dengue. Cell Host Microbe 13, 691–700.

Rada, C., Milstein, C., 2001. The intrinsic hypermutability of antibody heavy and light chain genes decays exponentially. EMBO J 20, 4570–4576.

Ralph, D.K., Matsen, F.A.T., 2016a. Consistency of VDJ rearrangement and substitution parameters enables accurate B cell receptor sequence annotation. PLOS Comput Biol 12, e1004409.

Ralph, D.K., Matsen, F.A.T., 2016b. Likelihood-based inference of B cell clonal families. PLOS Comput Biol 12, e1005086.

Retter, I., Althaus, H.H., Munch, R., Muller, W., 2005. VBASE2, an integrative V gene database. Nucleic Acids Res 33, D671–D674.

Rizzetto, S., Koppstein, D.N., Samir, J., et al., 2017. B-cell receptor reconstruction from single-cell RNA-seq with VDJPuzzle. bioRxiv.

Rosenfeld, A.M., Meng, W., Luning Prak, E.T., Hershberg, U., 2017. ImmuneDB: A system for the analysis and exploration of high-throughput adaptive immune receptor sequencing data. Bioinformatics 33, 292–293.

Roskin, K.M., Simchoni, N., Liu, Y., et al., 2015. IgH sequences in common variable immune deficiency reveal altered B cell development and selection. Sci Transl Med 7, 302ra135.

Russ, D.E., Ho, K.Y., Longo, N.S., 2015. HTJoinSolver: Human immunoglobulin VDJ partitioning using approximate dynamic programming constrained by conserved motifs. BMC Bioinform 16, 170.

Scheepers, C., Shrestha, R.K., Lambson, B.E., et al., 2015. Ability to develop broadly neutralizing HIV-1 antibodies is not restricted by the germline Ig gene repertoire. J Immunol 194, 4371–4378.

Schroeder Jr., H.W., Cavacini, L., 2010. Structure and function of immunoglobulins. J Allergy Clin Immunol 125, S41–S52.

Shlomchik, M.J., Aucoin, A.H., Pisetsky, D.S., Weigert, M.G., 1987. Structure and function of anti-DNA autoantibodies derived from a single autoimmune mouse. Proc Natl Acad Sci USA 84, 9150–9154.

Smith, T., Heger, A., Sudbery, I., 2017. UMI-tools: Modeling sequencing errors in Unique Molecular Identifiers to improve quantification accuracy. Genome Res 27, 491–499.

Souto-Carneiro, M.M., Longo, N.S., Russ, D.E., Sun, H.W., Lipsky, P.E., 2004. Characterization of the human Ig heavy chain antigen binding complementarity determining region 3 using a newly developed software algorithm, JOINSOLVER. J Immunol 172, 6790–6802.

Souto-Carneiro, M.M., Sims, G.P., Girschik, H., Lee, J., Lipsky, P.E., 2005. Developmental changes in the human heavy chain CDR3. J Immunol 175, 7425–7436.

Ta, V.T., Nagaoka, H., Catalan, N., et al., 2003. AID mutant analyses indicate requirement for class-switch-specific cofactors. Nat Immunol 4, 843–848.

Teng, G., Papavasiliou, F.N., 2007. Immunoglobulin somatic hypermutation. Annu Rev Genet 41, 107–120.

Tonegawa, S., 1983. Somatic generation of antibody diversity. Nature 302, 575–581.

Truck, J., Ramasamy, M.N., Galson, J.D., et al., 2015. Identification of antigen-specific B cell receptor sequences using public repertoire analysis. J Immunol 194, 252–261.

Turchaninova, M.A., Davydov, A., Britanova, O.V., et al., 2016. High-quality full-length immunoglobulin profiling with unique molecular barcoding. Nat Protoc 11, 1599–1616.

van Gent, D.C., McBlane, J.F., Ramsden, D.A., et al., 1995. Initiation of V(D)J recombination in a cell-free system. Cell 81, 925–934.

Volpe, J.M., Cowell, L.G., Kepler, T.B., 2006. SoDA: Implementation of a 3D alignment algorithm for inference of antigen receptor recombinations. Bioinformatics 22, 438–444.

Volpe, J.M., Kepler, T.B., 2008. Large-scale analysis of human heavy chain V(D)J recombination patterns. Immunome Res 4, 3.

Wang, C., Liu, Y., Cavanagh, M.M., et al., 2015. B-cell repertoire responses to varicella-zoster vaccination in human identical twins. Proc Natl Acad Sci USA 112, 500–505.

Wang, Y., Jackson, K.J., Davies, J., et al., 2014. IgE-associated IGHV genes from venom and peanut allergic individuals lack mutational evidence of antigen selection. PLOS One 9, e89730.

Wang, Y., Jackson, K.J., Gaeta, B., et al., 2011. Genomic screening by 454 pyrosequencing identifies a new human IGHV gene and sixteen other new IGHV allelic variants. Immunogenetics 63, 259–265.

Watson, C.T., Matsen, F.A.T., Jackson, K.J.L., et al., 2017. Comment on "A database of human immune receptor alleles recovered from population sequencing data". J Immunol 198, 3371–3373.

Watson, C.T., Steinberg, K.M., Graves, T.A., et al., 2015. Sequencing of the human IG light chain loci from a hydatidiform mole BAC library reveals locus-specific signatures of genetic diversity. Genes Immun 16, 24–34.

Watson, C.T., Steinberg, K.M., Huddleston, J., et al., 2013. Complete haplotype sequence of the human immunoglobulin heavy-chain variable, diversity, and joining genes and characterization of allelic and copy-number variation. Am J Hum Genet 92, 530–546.

Weigert, M.G., Cesari, I.M., Yonkovich, S.J., Cohn, M., 1970. Variability in the lambda light chain sequences of mouse antibody. Nature 228, 1045–1047.

Wu, X., Zhou, T., Zhu, J., et al., 2011. Focused evolution of HIV-1 neutralizing antibodies revealed by structures and deep sequencing. Science 333, 1593–1602.

Yaari, G., Kleinstein, S.H., 2015. Practical guidelines for B-cell receptor repertoire sequencing analysis. Genome Med 7, 121.

Yaari, G., Uduman, M., Kleinstein, S.H., 2012. Quantifying selection in high-throughput Immunoglobulin sequencing data sets. Nucleic Acids Res 40, e134.

Ye, J., Ma, N., Madden, T.L., Ostell, J.M., 2013. IgBLAST: An immunoglobulin variable domain sequence analysis tool. Nucleic Acids Res 41, W34–W40.

Yousfi Monod, M., Giudicelli, V., Chaume, D., Lefranc, M.P., 2004. IMGT/Junction Analysis: The first tool for the analysis of the immunoglobulin and T cell receptor complex V–J and V–D–J Junctions. Bioinformatics 20 (Suppl 1), i379–i385.

Yu, Y., Ceredig, R., Seoighe, C., 2016. LymAnalyzer: A tool for comprehensive analysis of next generation sequencing data of T cell receptors and immunoglobulins. Nucleic Acids Res 44, e31.

Yu, Y., Ceredig, R., Seoighe, C., 2017. A database of human immune receptor alleles recovered from population sequencing data. J Immunol 198, 2202–2210.

Zhang, J., Kobert, K., Flouri, T., Stamatakis, A., 2014. PEAR: A fast and accurate Illumina Paired-End reAd mergeR. Bioinformatics, 30, 614–620.

Zhang, W., Wang, I.M., Wang, C., et al., 2016. IMPre: An accurate and efficient software for prediction of T- and B-cell receptor germline genes and alleles from rearranged repertoire data. Front Immunol 7, 457.

Relevant Websites

http://immcantation.readthedocs.io/
Immcacntation.
http://immunedb.com/
ImmuneDB.
http://mixcr.readthedocs.io/
MiXCR.

Biographical Sketch

Dr Katherine J.L. Jackson: Dr Jackson is currently a Senior Research Officer at the Garvan Institute of Medical Research in Sydney, Australia. She completed her graduate studies at the University of New South Wales in Sydney, Australia on the detection of antigen selection signatures in immunoglobulin sequence datasets and has completed postdoctoral training at both the University of New South Wales and Stanford University, CA, USA. Her research applies immune repertoire sequencing methodologies to the study a variety of human immune responses including primary and secondary responses to vaccination and infection, allergy and autoimmunity, while also using exploring the immunogenetics of both humans and mice.

Epitope Predictions

Roman Kogay, Nazarbayev University, Astana, Republic of Kazakhstan
Christian Schönbach, Nazarbayev University, Astana, Kazakhstan

Abbreviations

ACO	Ant colony optimization	**MCC**	Matthews correlation coefficient
ANN	Artificial neural network	**SVM**	Support vector machine
A$_{ROC}$	Area under receiver operating characteristic curve	**SVR**	Support vector regression
		WEKA	Waikato environment for knowledge analysis

Introduction

Hallmarks of the adaptive immune system are discrimination of self from non-self recognized as antigenic determinants or epitopes through concerted activation of antibody and cell-mediated primary or memory responses. A complex network of molecular and cellular interactions that includes antigen presenting cells, T cell and B cells regulates the steps from antigen processing, epitope presentation and recognition to memory recall (Litman *et al.*, 2010; Frank, 2002). One key question in epitope prediction is whether a predicted epitope will be immunogenic. Indeed theoretical estimates for MHC class I restricted T cell epitopes indicate that only 1 in 2000 peptides of non-self antigens may induce a dominant CTL response (Yewdell and Bennink, 1999).

T Cell Epitopes

Antigen presenting cells process proteins into peptides that if recognized by T cells are called T cell epitopes. Two distinct pathways facilitate the processing of exogenous and endogenous (self and foreign) proteins into peptides which were comprehensively reviewed by Blum *et al.* (2013). Most peptides generated by proteolysis through 26S proteasome are transported by TAP (transporter associated with antigen processing) into the endoplasmatic reticulum where they bind to MHC class I molecules. If the affinity of the peptides to MHC class I molecules is sufficiently high, stable peptide-MHC-I complexes are transported through the Golgi apparatus to the cell surface where they are recognized by T-cell receptors (TCR) of CD8$^+$ T cells (**Fig. 1**). In contrast, MHC class II molecules usually bind in the endosome to peptides derived from lysosomal proteolysis of exogeneous proteins trafficked by phago- and endocyotsis. Peptide-MHC II complexes are then transported in endosomal vesicles to the cell surface where they are recognized by TCR expressed on CD4$^+$ T cells (**Fig. 2**). Although the binding of the peptides to MHC is crucial in

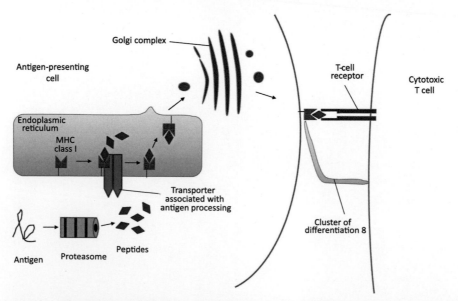

Fig. 1 Processing, presentation and recognition of MHC class I-restricted T cell epitopes. Modified from Fig 17.21 (a) in Karp, G., 2008. Cell and Molecular Biology. Concepts and Experiments, fifth ed. Asia: John Wiley & Sons (Asia) Pte. Ltd.

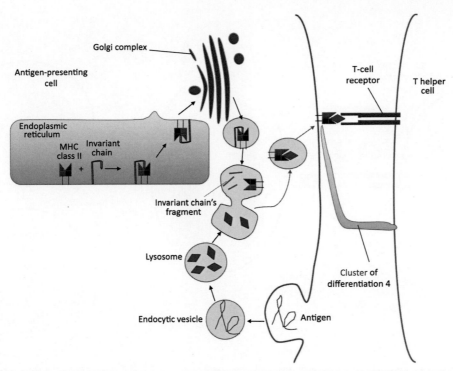

Fig. 2 Processing, presentation and recognition of MHC class II restricted T cell epitopes. Modified from Fig 17.21 (a) in Karp, G., 2008. Cell and Molecular Biology. Concepts and Experiments, fifth ed. Asia: John Wiley & Sons (Asia) Pte. Ltd.

defining whether peptides may become epitopes, cross-presentation of peptides generated from phagocytosed exogenous proteins by MHC class I or endogenous proteins processed in the lysosome by MHC class II complicates matters in epitope prediction-assisted vaccine design. Cross-priming of naïve $CD8^+$ T cells mediated by cross-presentation is just one example where the choice of epitopes in a vaccine will affect primary and secondary T-cell responses (Grotzke *et al.*, 2017). Since epitope-based vaccines are meant to mimic the natural protective immunity that activates the functions of both T and B cells (**Fig. 3**) we need to consider also B cell epitopes.

B Cell Epitopes

B cell epitopes do not require processing and are predominantly of conformational or discontinuous nature (Barlow *et al.*, 1986; Greenbaum *et al.*, 2007). The B cell epitope is a specific 3D surface area of an antigen that is recognized by the paratope, the antigen-binding part of the antibody (Sela-Culang *et al.*, 2013). Antibodies represent the secreted form of B-cell receptors (hence B cell epitopes) which are produced by mature B cells, the plasma cells (**Fig. 3**). The antibodies produced by plasma cells are the result of a somatic hypermutation process of the complementary determining regions (CDRs) that increases the antibody affinity to the antigen (affinity maturation) (Schroeder and Cavacini, 2010; Neu and Wilson, 2016). Unlike T cell epitope predictions the conformational nature of the B cell epitopes and conformational diversity of antibodies themselves, in addition to conformational changes induced in both antibodies and epitopes when they bind to each other, pose an additional challenge in identifying immunogenic candidate B cell epitopes. Unsurprisingly, advances in predicting B cell epitopes are less pronounced than for T cell epitopes although the problem of conformational epitopes has been known since 1986 when the first crystal structure of a lysozyme-antibody complex was published (Barlow *et al.*, 1986).

Background

Immunogenicity of Epitopes

In both T and B cell epitope predictions of pathogen-derived proteins the goal is to identify potential immunogenic epitopes that induce a protective immune response. In case of T cell epitopes, peptides that are predicted to bind with high affinity (good binders) to MHC molecules are indeed very often immunogenic. Yet the MHC binding affinity remains a surrogate that does not provide clues about the type of T cell response an epitope may trigger. Co-stimulatory signals mediated by expression of various members of the cluster of differentiation (CD) family on both antigen presenting cells and T cells will determine which signaling pathway downstream of the TCR-peptide-MHC complex are activated, and hence whether the epitope triggers an effector T cell response or a suppressive, regulatory T cell response. Thus future assessments of predicted epitopes towards immunogenicity,

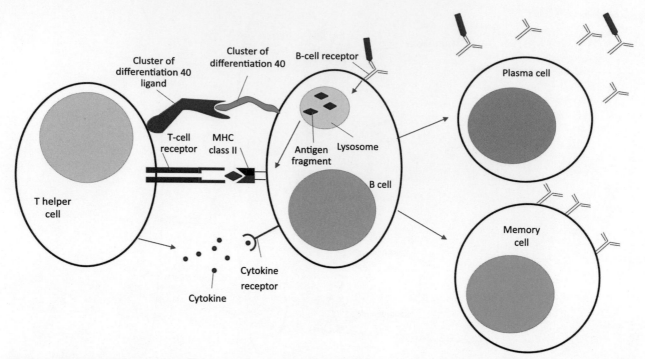

Fig. 3 Interaction of activated T helper cells activates B cells which differentiate into memory and plasma cells. The latter secrete antigen-specific antibodies. Modified from Fig 17.10 (a) in Karp, G., 2008. Cell and Molecular Biology. Concepts and Experiments, fifth ed. Asia: John Wiley & Sons (Asia) Pte. Ltd.

particularly in context of human vaccine development, should take into account the data derived from systems biology approaches (Querec *et al.*, 2009) and biomarker assays (Rappuoli and Aderem, 2011) in the context, dynamics and quality of a protective immune response to further boost the efficacy of data-driven predictions.

Overview of Epitope Predictions

Epitope predictions utilize statistical, machine learning or structural models derived from experimental data as reviewed by Soria-Guerra *et al.* (2015). Early epitope prediction methods BIMAS (Parker *et al.*, 1994) and EpiMatrix (Schafer *et al.*, 1998) were based on quantitative matrices (QM) or binding motifs (SYFPEITHI) (Rammensee *et al.*, 1999) that capture simple linear relationships of binding and non-binding data. Both methods are limited by the number of available experimental binding and non-binding data for a specific MHC allotype to derive a binding motif based on amino acids observed at peptide residue positions or to score a peptide's binding potential using position-dependent coefficients of amino acids. The limitations of quantitative matrices including overfitting, are balanced in practice by their easy implementation and use combined with an acceptable accuracy for selected MHC allotypes (Schönbach *et al.*, 2002; De Groot *et al.*, 2002).

The continuous growth of data enabled the application of more sophisticated methods utilizing artificial neural networks (ANN) (Adams and Koziol, 1995), hidden Markov models (HMM) ranging from fully connected HMMs to profile HMMs (Mamitsuka, 1998; Brusic *et al.*, 2002; Zhang *et al.*, 2006; Larsen *et al.*, 2006), and support vector machines (SVM) (Zhang *et al.*, 2007; Jacob and Vert, 2007; Chen *et al.*, 2007) that are adept in extracting and processing complex nonlinear relationships in peptide-MHC binding or B cell epitope data to derive predictive models. The predictive power of these models largely depends on the quantity and quality of (unbiased) annotated data. Immune Epitope Database (IEDB) (Vita *et al.*, 2014), a manually curated database of experimentally characterized immune epitopes has been instrumental in providing since 2005 high-quality data for training and testing datasets. Successful application of these data-driven methods that reduced experimental time and costs spurned further improvements towards increasing the accuracy of CTL epitope predictions by combining different individual prediction-based models followed by integrating top-predicting scores to derive a consensus score (consensus methods) (Moutaftsi *et al.*, 2006). Similarly, the integration of predictions for distinct processes such as proteasomal processing, TAP binding and peptide-MHC binding while decreasing the dependency on MHC allotype-specific binding data as implemented in NetCTL-pan (Stranzl *et al.*, 2010) improved the predictive power.

Structure-based methods (Patronov and Doytchinova, 2013) yield high-quality quantitative predictions of peptide-MHC binding affinities based on free Gibbs energy calculations that take into account the interactions and distances between atoms of peptide and MHC amino acid residues. Threading of peptides predicted to bind to MHC is based on the computationally intensive free energy calculations of individual interactions (Logean and Rognan, 2002) or pairwise interactions using an energy potential matrix (Schueler-Furman *et al.*, 2000). Quantitative structure-activity relationship (QSAR) methods that are employed in drug discovery, assuming similar molecular structures result in similar activities, analyze differences in free energy and structures (Doytchinova and Flower, 2002). The limited number of available high quality structures restricts the applicability of structure-based methods particularly for B cell epitope predictions.

The performance of prediction methods is evaluated by applying Pearson correlation, Matthews correlation coefficients or receiver operating characteristic (ROC) analyses which includes positive-predictive value (PPV), negative-predictive value (NPV), area under the receiver operating characteristic curve (A_{ROC}), sensitivity (SE), specificity (SP) or root mean squared deviations (RSMD) of distances in Å for structure predictions (Yang and Yu, 2009; Hattotuwagama *et al.*, 2007). SE provides the ratio of correctly predicted real positives, whereas SP for example, correctly predicted true negatives indicate the quality of predictions. PPV and NPV provide an overall success rate as proportions of true positives (negatives) of all positive (negative) predicted, whereas the A_{ROC} value reflects the overall quality of predictions within SE and SP value ranges. The A_{ROC} value is a robust measure as long as the number of positives and negatives in a test set are not extremely lop-sided. When no independent test data sets are available leave-one-out cross-validation (LOOC) (Sammut and Webb, 2010) is often used. LOOC may indicate an acceptable performance of a tool that becomes inacceptable when applied to a new data set.

Approaches

T Cell Epitope Prediction

T cell epitope predictions can be divided into integrated, MHC-pan and allotype-specific binding, TAP binding, proteasomal cleavage, and a few miscellaneous approaches which are represented by 38 publicly available, mostly web accessible tools shown in **Table 1**. Each tool is summarized by its name, URL, applicable species, and salient features including limitations, performance and methods applied to provide a convenient desk reference for choosing a T cell epitope prediction tool.

Independent of the approach categories the majority of tools rely on the application of QM, ANN, SVM, consensus or hybrid methods. An ANN is a network of interconnected nodes whose connections carry numeric data that are trained by weighting the connections (Honeyman *et al.*, 1998). Prior to using an ANN for peptide sequences the sequences must be transformed to numeric descriptors to input into the network layers. Popular ANN-based prediction tools include MULTIPRED2 (Zhang *et al.*, 2011a), NetMHC 4.0 (Andreatta and Nielsen, 2015) and NetMHCpan 3.0 (Nielsen and Andreatta, 2016) for MHC class I and NetMHCII 2.2 (Nielsen and Lund, 2009) and NetMHCIIpan 3.1 (Andreatta *et al.*, 2015) for MHC class II peptide binding.

SVMs comprise machine learning algorithms that classify complex data by placing it into a multidimensional space and constructing a hyperplane through an appropriate kernel function. Representative tools using SVM are TAP binding predictor TAPreg (Diez-Rivero *et al.*, 2010), MHC class II peptide binding predictor MHC2SKpan (Guo *et al.*, 2013) and T cell epitope immunogenicity predictor Repitope (Ogishi and Yotsuyanagi, 2017).

Several MHC-pan (PSSMHCpan (Liu *et al.*, 2017) and TEPITOPEpan (Zhang *et al.*, 2012)) and allotype-specific tools (Pick-Pocket 1.1 (Zhang *et al.*, 2009), HLaffy (Mukherjee *et al.*, 2016)) and PREDIVAC 2.0 (Oyarzún *et al.*, 2013) use QM-based methods that assign coefficients for each amino acid in a peptide frame and discriminate potential binders from non-binders through the derived peptide score. The only QM-based tool that successfully integrated proteasomal processing, TAP, MHC class I and II binding prediction is TEpredict (Antonets and Maksyutov, 2010) whereas ANN- and QM-based NetCTLpan 1.1 (Stranzl *et al.*, 2010) do not cover MHC class II.

Integrated approaches that are based on the consensus of independent methods tend to outperform single methods. NetMHCcons 1.1 which couples NetMHC 3.4, NetMHCpan 2.8 and PickPocket 1.1 methods is a leading example (Karosiene *et al.*, 2012).

The only MHC class II allelle-specific candidate epitope prediction tools that utilize structure-based methods (Yao *et al.*, 2013) are EpiDOCK (Atanasova *et al.*, 2013) and EpiTOP (Dimitrov *et al.*, 2010a; Dimitrov *et al.*, 2010b). EpiDOCK uses docking score-based quantitative matrices to evaluate the binding potential of overlapping nonamer peptides generated from an input sequence to five HLA class II DPA/DBP, six DQA/DQB and 12 DRB1 proteins. EpiTOP employs QSAR-derived quantitative matrices that encode both peptide and HLA protein pocket interactions derived from 12 HLA-DRB1 allotypes to predict peptide-HLA-DRB1 binding.

Two tools, categorized under miscellaneous approaches use hybrid SVM and motif methods to identify potential T-helper cell epitopes by predicting cytokine-inducing peptides. IL4Pred (Dhanda *et al.*, 2013a) and IFNepitope (Dhanda *et al.*, 2013b) predict interferon gamma- and interleukin 4-inducing MHC class II peptides, respectively.

B Cell Epitope Prediction

The conformational and linear B cell epitope prediction approaches comprising 21 web accessible tools are summarized according to their characteristics including performance, limitations and methods in **Table 2**.

Representatives of conformational epitope prediction approaches using structure-based methods are Discotope 2.0 (Kringelum *et al.*, 2012) and BEpro (Sweredoski and Baldi, 2008b). Discotope 2.0 uses amino acid residue propensity scores within a defined sphere whereas BEPro applies an amino acid propensity scale in combination with side chain orientation and solvent accessibility properties.

Table 1   T cell epitope prediction approaches

Name and URL	Species	Functionality and Performance	Limitations	Method
Integrated approaches				
NetCTLpan 1.1 http://www.cbs.dtu.dk/services/NetCTLpan/	Hosa, Patr, Mamu, Susc, Mumu, Gogo, Bota	Integrates prediction of proteasomal cleavage, TAP and MHC class I binding affinity; A_{ROC}: 0.920–0.977 (depending on data); (Stranzl et al., 2010).	Predictions for 8–11mer; at most 5000 sequences per submission and each sequence must be <20,000 amino acids; maximum 20 MHC allotypes per submission.	ANN and matrix (matrix for TAP).
NetTepi 1.0 http://www.cbs.dtu.dk/services/NetTepi/	Hosa	Integrates peptide-MHC binding affinity, peptide-MHC stability and TCR propensity; prediction values are weighted sum or % rank; sorted by combined, affinity, stability, TCR propensity scores; $AUC_{0.1}$: 0.9285–0.9305 (depending on the combined model) (Trolle and Nielsen, 2014).	13 HLA allotypes; 8–14mer peptides; at most 5000 sequences per submission and each sequence must be <20,000 amino acids; T cell propensity tool was validated only on 9mer peptides.	ANN and matrix.
NetMHCcons 1.1 http://www.cbs.dtu.dk/services/NetMHCcons/	Hosa, Patr, Mamu, Susc, Mumu, Gogo, Bota	Integrates methods of NetMHC, NetMHCpan and Pickpocket; predictions for 8–15mer peptides; prediction values in nM IC_{50} and % Rank; sorting by predicted binding affinity; PCC: 0.23–0.72 (depending on the group of allotypes) (Karosiene et al., 2012).	101 MHC I allotypes; at most 5000 sequences per submission and each sequence must be <20,000 amino acids; maximum 20 MHC allotypes per submission.	Consensus method: ANN, pan-specific ANN, and matrix.
RANKPEP http://imed.med.ucm.es/Tools/rankpep.html	Hosa, Mumu	Cleavage prediction; immunodominance filter; molecular weight filter; predictions can be made from MSA; >80% of CD8$^+$ T cell epitopes are among top 2% of scoring peptides; 3–10% threshold was required to predict 80% of CD4$^+$ T cells epitopes; (Reche et al., 2004).	102 MHC I and 80 MHC II allotypes; predictions for MHC I binder are limited 8–11mer peptide; matrices are limited by the quality of sequences.	Matrix and SVM (Immuno-dominance).
NetCTL 1.2 http://www.cbs.dtu.dk/services/NetCTL/	Hosa	Integrates HLA class I binding, C terminal cleavage, TAP transport efficiency; (depending on the threshold); A_{ROC}: 0.941, sensitivity: 72% (among 5% top-scoring peptide) (Larsen et al., 2007).	12 HLA supertypes; predicts only 9mer epitopes.	ANN (HLA binding and C terminal cleavage) and matrix (TAP transport efficiency).

Tool / URL	Species	Description	Notes	Method
ProPred1 http://crdd.osdd.net/raghava/propred1/	Hosa; Mumu	Prediction of the standard proteasome and immunoproteasome cleavage sites; identification of MHC binders with cleavage sites at C terminus; output displayed in different formats; allows subsequence analysis; accuracy: 38–80% for HLA-A*0201, 70–80% for H2-Kb (depends on the threshold); (Singh and Raghava, 2003).	47 MHC I allotypes; only 9mer peptides are predicted; over-generalized – the matrices were obtained for enolase-I protein.	Matrix.
nHLAPred http://crdd.osdd.net/raghava/nhlapred/	Hosa, Mumu	Average accuracy for hybrid approach: 92.8%; prediction of the standard proteasome and immunoproteasome cleavage sites; identification of MHC binders with cleavage sites at C terminus; (Bhasin and Raghava, 2007).	67 MHC I allotypes; only nonamer peptides.	ANN and matrix or matrix only.
MULTIPRED 2.0 http://cvc.dfci.harvard.edu/multipred2/	Hosa	Allows input of pre-calculated viral proteomes; employs NetMHCpan and NetMHCIIpan as predictive engines; heatmaps are generated to visualize results; (Zhang et al., 2011).	26 HLA class I and II supertypes; 8–11 mer peptides for HLA I and 9mer peptides for HLA II.	ANN.
TepiTool (Pipeline) http://tools.iecb.org/tepitool/	Hosa, Patr, Mamu, Susc, Mumu, Gogo, Bota and others	Integrates several predictive tools; predicts MHC I and MHC II binders; allow to select most frequent allotypes; different types of filtering, such as percentile rank, absolute rank, based on IC_{50} (Paul et al., 2016).	Predicts 8–14mer epitopes.	Includes but not limited to consensus, ANN, and matrix methods.
MHCPred 2.0 http://www.ddg-pharmfac.net/mhcpred/MHCPred/	Hosa, Mumu	Two models of prediction: amino acids contribution; amino acids + their interactions; anchor positions (maximum 4) can be chosen; predicts affinity to TAP; (Guan et al., 2006).	17 MHC class I and II allotypes; sequences are limited to 1000 residues; only plain format is supported.	Matrix.
FRED 2.0 (Pipeline) http://fred-2.github.io/	N/A	HLA typing, epitope prediction, epitope selection, and epitope assembly; implemented in Python; provides unified access to different epitope prediction tools and databases; (Schubert et al., 2016).	Difficult to execute for unfamiliar users; requires separate installation of several external tools from Center for Biological Sequence Analysis, Technical University of Denmark.	Python-based framework.
TEpredict http://tepredict.sourceforge.net/downloads.html	N/A	Predicts MHC I and MHC II binders, proteasomal and immunoproteasomal processing, peptide and TAP binding;	Considers only 9mer peptides.	Matrix.

(Continued)

Table 1 Continued

Name and URL	Species	Functionality and Performance	Limitations	Method
SVMHC (works only via Epitoolkit pipeline) https://abi.inf.uni-tuebingen.de/Services/SVMHC	Hosa, Mumu	sensitivity: 50–80%; specificity: 75–99%; (Antonets and Maksyutov, 2010). Allows accessions/database identifiers as input; analysis of single amino acid polymorphism; different output views; (Dönnes and Kohlbacher, 2006).	Predictions for 8–10 mer peptides; 26 MHC I allotypes from MHCPEP, 24 MHC I allotypes from SYFPEITHI and 51 MHC II allotypes; only one sequence per submission; thresholds cannot be set by users.	SVM (MHC I) and matrices (MHC II).
PickPocket 1.1 http://www.cbs.dtu.dk/services/PickPocket/	Hosa, Patr, Mamu, Susc, Mumu, Gogo, Bota	Robust when data is scarce and when the similarity to MHC molecules with characterized binding specificities is low; PCC: 0.26–0.6 (depending on the evaluated set) (Zhang et al., 2009).	8–12mer peptides; >150 MHC I and II allotypes; at most 5000 sequences per submission and each sequence must be <20,000 amino acids; maximum 20 MHC per submission.	Matrix.
MetaMHCpan http://datamining-iip.fudan.edu.cn/MetaMHCpan/index.php/pages/view/info	Hosa, Mumu	Meta-server with different predictive methods (Xu et al., 2016).	8–11mer peptides for MHC I and 9–25mer peptides for MHC II; 41 MHC I and >600 MHC II allotypess.	Matrix, SVM, and multiple instance learning method.
MHC I pan approaches				
NetMHCpan 3.0 http://www.cbs.dtu.dk/services/NetMHCpan/	Hosa, Patr, Mamu, Susc, Mumu, Gogo, Bota	91% of ligands are recovered at a rank threshold of 2% with a specificity of 98%; A_{ROC} : 0.83–0.89; (depending on the epitope length); (Nielsen and Andreatta, 2016).	172 MHC I molecules; 8–14mer peptides; at most 5000 sequences per submission and each sequence must be <20,000 amino acids; maximum 20 MHC allotypes per submission.	ANN.
PSSMHCpan https://github.com/BGI2016/PSSMHCpan	Hosa	A_{ROC}: 0.94, accuracy: 85%; able to predict neoantigens; 8–25 mer peptide length prediction; (Liu et al., 2017).	87 HLA class I allotypes.	Matrix.
MHC I allotype-specific approaches				
NetMHC 4.0 http://www.cbs.dtu.dk/services/NetMHC/	Hosa, Patr, Mamu, Susc, Mumu, Bota	A_{ROC}:0.882–0.895; (depending on the peptide epitope length); (Andreatta and Nielsen, 2015).	122 MHC I allotypes; 8–13 mer peptides; at most 5000 sequences per submission and each sequence must be <20,000 amino acids;	ANN.

Tool / URL	Species	Description	Specifications	Method
HLAffy http://proline.biochem.iisc.ernet.in/HLAffy/?tab=1	Hosa	Estimates peptide affinity for HLA class I; accuracy: 92% and correlation: 0.85, for IEDB dataset accuracy: 82.5%; provides a histogram view; representative molecular models with favorable peptide-HLA residue interactions are available (Mukherjee et al., 2016).	maximum 20 MHC allotypes per submission. Only for 9mer peptides.	Matrix.
MHC II pan approaches				
NetMHCIIpan 3.1 http://www.cbs.dtu.dk/services/NetMHCIIpan/	Hosa, Mumu	Prediction values are given as IC_{50} (inhibitory concentration) and as % ranks; A_{ROC}: 0.80–0.90 for the binding between peptide and MHC; (Andreatta et al., 2015).	4 MHC II isotypes; at most 5000 sequences per submission and each sequence must be <20,000 amino acids; maximum 20 MHC per submission.	ANN.
TEPITOPEpan http://datamining-iip.fudan.edu.cn/service/TEPITOPEpan/TEPITOPEpan.html	Hosa	Shows binding cores and percentile ranks; prediction for 9–25mer peptides; average A_{ROC}: 0.717–0.833 (depending on the dataset); predicts over 700 HLA-DR allotypes; (Zhang et al., 2012).	Limited to HLA-DR.	Matrix.
MHC2SKpan http://datamining-iip.fudan.edu.cn/service/MHC2SKpan/info.html	Hosa	Predicts MHC class II peptide binding; A_{ROC}: 0.734–0.843 (depending on the dataset); (Guo et al., 2013).	9–25mer peptides; DRB only.	SVM.
PREDIVAC – 2.0 http://predivac.biosci.uq.edu.au/	Hosa	A_{ROC}: 0.842–0.872 for HLA II binding; A_{ROC}: 0.749 for CD4-T-cell epitopes; 883 HLA II allotypes; target population epitope predictions (Oyarzún et al., 2013).	DRB only.	Matrix.
MHC II allotype-specific approaches				
NetMHCII 2.2 http://www.cbs.dtu.dk/services/NetMHCII/	Hosa, Mumu	Predictions are given as % Rank and in nM IC_{50} values; A_{ROC}: 0.68–0.82 (depending on the dataset and method) (Nielsen and Lund, 2009).	28 MHC II allotypes; at most 5000 sequences per submission, each sequence must be <20,000 amino acids.	ANN.
EpiDOCK http://epidock.ddg-pharmfac.net/	Hosa	Identifies 90% true binders and 76% true non-binders; overall accuracy: 83% (Atanasova et al., 2013).	23 HLA II allotypes; only 9mer peptides are predicted; no sorting mechanism.	Matrix.
EpiTOP http://www.pharmfac.net/EpiTOP/	Hosa		12 HLA-DRB1 allotypes; only 9mer peptides are predicted;	Matrix.

(Continued)

Table 1   Continued

Name and URL	Species	Functionality and Performance	Limitations	Method
		Identifies 89% of known epitopes within top 20% of predicted binders (Dimitrov et al., 2010a,b).	allotype selection did not work at the time of testing.	
MHC2Pred http://crdd.osdd.net/raghava/mhc2pred/	Hosa, Mumu	Accuracy: >78%; (Lata et al., 2007).	42 MHC II allotypes; only 9mer peptides are predicted; inconvenient output format without sorting options.	SVM.
Propred http://crdd.osdd.net/raghava/propred/	Hosa	Output displayed in different formats; (Singh and Raghava, 2001).	51 HLA-DR allotypes.	Matrix.
HLA-DR4Pred http://crdd.osdd.net/raghava/hladr4pred/	Hosa	Accuracy: 86% and 78% for SVM and ANN, respectively. (Bhasin and Raghava, 2004b)	Only for HLA-DRB1*0401.	SVM and ANN.
Proteasomal cleavage approaches				
Pcleavage http://crdd.osdd.net/raghava/pcleavage/	N/A	Predicts proteasome cleavage sites; MCCs: 0.54 and 0.43 for in vitro and MHC ligand data respectively (Bhasin and Raghava, 2005).	The C-terminal of MHC ligands represents only a subset of cleavages that occur in vivo.	SVM.
NetChop 3.1 http://www.cbs.dtu.dk/services/NetChop/	Hosa	Predicts cleavage sites of human proteasome; two different methods: C-terminus and 20S; specificity: 48%, sensitivity: 81% MCC: 0.31 (Nielsen et al., 2005).	At most 100 sequences and 100,000 amino acids per submission.	ANN.
TAP binding approaches				
TAPPred http://crdd.osdd.net/raghava/tappred/	Hosa	Predicts TAP-binding peptides; correlation coefficients: 0.88 (Cascade SVM) and 0.8 (SVM); cascade SVM uses features of amino acids along with sequence whereas SVM uses only sequence (Bhasin et al., 2007).	N/A	SVM or Cascade SVM.
TAPreg http://imed.med.ucm.es/Tools/tapreg/	N/A	Predicts affinity of TAP binding ligands; maximum PCC: 0.89 ± 0.03 (Diez-Rivero et al., 2010).	Data training was done on 9mer peptides only.	SVM.
Miscellaneous T cell epitope related prediction approaches				
Repitope https://github.com/masato-ogishi/Repitope	Hosa	Epitope immunogenicity prediction; accuracy: 70–80% (Ogishi and Yotsuyanagi, 2017).	Retrospective observational study; simplified assumptions biophysical-chemical nature of the TCR-peptide-MHC	SVM.

Tool	Species	Description	Method
IL4pred http://crdd.osdd.net/raghava/il4pred/	N/A	Predicts interleukin 4-inducing MHC class II binders; maximum accuracy: 75.76% and MCC: 0.51 (hybrid method) (Dhanda et al., 2013a).	interactions; limited to TCR-V beta. Training set consists of only 8–22mer peptides created without species considerations. Motif, SVM or hybrid.
IFNepitope http://crdd.osdd.net/raghava/ifnepitope/	N/A	Prediction and design of interferon-γ inducing MHC class II binding peptides; maximum prediction accuracy of hybrid approach: 81.3% (MCC: 0.57) – 82.1% (MCC: 0.62) (depends on dataset) (Dhanda et al., 2013b).	Analysis of positions-specific preference of residues was done only for a small number of peptides. Motif, SVM or hybrid.
CTLPred http://crdd.osdd.net/raghava/ctlpred/	N/A	Predicts CTL epitopes; accuracy for QM, ANN and SVM: 70.0, 72.2% and 75.2% respectively; has combined and consensus prediction approaches; (Bhasin and Raghava, 2004a).	Training data contains only 9mer epitopes; small blind dataset (63 epitopes for 2 subgroups). ANN, SVM and matrix.
MMBPred http://crdd.osdd.net/raghava/mmbpred/	Hosa, Mamu, Mumu	Prediction of mutated promiscuous binders e.g. increase/decrease of peptide-MHC binding; analyzes position and type of mutations (Bhasin and Raghava, 2003).	67 MHC I allotypes; only 9mer peptides are predicted; promiscuous binding prediction was not supported at the time of testing. Matrix.

Abbreviations: ANN: artificial neural network, A_{ROC} or AUC: area under receiver operating characteristic curve, Bota: *Bos taurus*, Gogo: Gorilla gorilla, Hosa: *Homo sapiens*, HLA: human leukocyte antigen, Mamu: *Macaca mulatta*, MCC: Matthews correlation coefficient, MHC: major histocompatibility complex, MSA: multiple sequence alignment, Mumu: *Mus musculus*, Patr: *Pan troglodytes*, PCC: Pearson correlation coefficient, SVM: support vector machine, Susc: *Sus scrofa*, TAP: transporter associated with antigen processing, and QM: quantitative matrix.

Table 2   B cell epitope prediction approaches

Discontinuous/conformational epitope prediction approaches

Name and URL	Functionality and Performance	Limitations	Method
SEPPA 2.0 http://lifecenter.sgst.cn/seppa2/	Subcellular localization of an antigen and host species are considered; A_{ROC}: 0.785–0.823 (depending on the host and localization of antigen protein); Jmol is used for visualization (Qi et al., 2014).	Antigens with less than <25 residues were not considered.	Optimized logistic regression algorithm.
SEPIa https://github.com/SEPIaTool/SEPIa	A_{ROC}: 0.65; prediction is based on the amino acid sequence; amino acid features and sequence-based features are taken into account (Dalkas and Rooman, 2017).	Only window of 9 residues is used to test whether the middle residue is an epitope or not.	Gaussian Naïve Bayes and Random Forest algorithms based on 13 features.
DiscoTope 2.0 http://www.cbs.dtu.dk/services/DiscoTope/	A_{ROC}: 0.824 and 0.727 (for training and independent data sets) (Kringelum et al., 2012).	Trained on a dataset with the total number of residues per epitope 9–22 and with the longest sequential stretch 3–12 residues per epitope.	Definition of the spatial neighborhood to sum propensity scores and half-sphere exposure as a surface measure.
EPSVR http://sysbio.unl.edu/EPSVR/	A_{ROC}: 0.597; evaluates 6 different features: residue epitope propensity, conservation score, side chain energy score, contact number, surface planarity score and secondary structure composition (Liang et al., 2010).	Multiple epitopes in one antigen were not considered.	SVR with six attributes.
EPMeta http://sysbio.unl.edu/EPMeta/	A_{ROC}: 0.638 (Liang et al., 2010).	Works only on Linux OS.	Consensus method for EPSVR, EPCES, Epitopia, SEPPA, BEpro and Discotope 1.2.
BEpro http://pepito.proteomics.ics.uci.edu/index.html	A_{ROC}: 0.754 and 0.683 (for Discotope and Epitome datasets respectively); Jmol is used for visualization (Sweredoski and Baldi, 2008b).	Propensity scale scores averaged over a window of exactly 9 residues; only PDB file format is supported.	Linear combination of amino-acid propensity scale, side chain orientation and solvent accessibility information using half sphere exposure values.
CBTOPE http://crdd.osdd.net/raghava/cbtope/	Accuracy: 86.59%, A_{ROC}: 0.90, MCC: 0.73; prediction is based on the amino acid sequence using different window length patterns, standard binary and physicochemical profiles of pattern (Ansari and Raghava, 2010).	Lacks 3D structural output.	SVM.
EPCES http://sysbio.unl.edu/EPCES/	Sensitivity: 47.8%, Specificity: 69.5%, A_{ROC}: 0.632; evaluates residue epitope propensity, conservation score, side chain energy score, contact number, surface planarity score and secondary structure composition (Liang et al., 2009).	Low accuracy for the bound structures.	Consensus scoring utilizing six different scoring functions.

Method	Performance / details	Features / notes	Approach
Bpredictor https://code.google.com/archive/p/my-project-bpredictor/downloads	A_{ROC}: 0.633 and 0.654 for bound and unbound datasets respectively; impact of interior residues and different contributions of adjacent residues were considered (Zhang et al., 2011a,b,c).	Window size: 3–15 residues.	Random-forest algorithm with distance based feature.
EpiSearch http://curie.utmb.edu/episearch.html	In most cases covers >50% of experimentally validated residues; Jmol is used for visualization (Negi and Braun, 2009).	In addition to the 3D structure of the antigen the input requires the set of mimotopes (up to 12 mer).	Patch analysis that identifies cluster of residues on the surface of antigen with similar physicochemical properties as in mimotopes.
PEP – 3D-Search http://kyc.nenu.edu.cn/Pep3DSearch/	MCC: 0.176, sensitivity: 36.4%, Precision: 69.5% (Huang et al., 2008).	In addition to the 3D structure of the antigen the input requires the set of mimotopes; compatible only with Windows OS.	Mimotope-based analysis via ACO algorithm.
CEP http://196.1.114.49/cgi-bin/cep.pl	Accuracy: 75%; Jmol is used for visualization (Kulkarni-Kale et al., 2005).	Simplified approach with few attributes.	Algorithm uses accessibility of residues and spatial distance cut-off.
Linear B cell epitope prediction approaches			
LBtope http://crdd.osdd.net/raghava/lbtope/	Able to identify mutation(s) in peptide and convert it/them to the epitope; accuracy: 81% (54–86% depending on model and dataset); mutation tool is available to design better epitope or for de-immunization purpose (Singh et al., 2013).	5–30mer epitope prediction.	SVM, using diverse features; binary profile, di-peptide composition and amino acid pair profile.
COBEpro http://scratch.proteomics.ics.uci.edu/	A_{ROC}: 0.606–0.829 (depending on the dataset); Predicts epitopes of any length (Sweredoski and Baldi, 2008a).	Prediction is limited to sequences of 1500 amino acids length.	SVM for the similarity measure based on the total number of identical substrings.
ABCpred http://crdd.osdd.net/raghava/abcpred/	Sensitivity: 67.14%, specificity: 64.71%, accuracy: 65.93%; has overlapping filter (Saha and Raghava, 2006).	10, 14, 16, 18, 20mer predicted epitope length in plain text format; developed with a small dataset derived from database BCIPEP (Saha et al., 2005)	ANN.
SVMTriP http://sysbio.unl.edu/SVMTriP/	Precision: 54.1–57.1%, sensitivity: 68.5–80.1%, A_{ROC}: 0.674–0.702 (depending on the length of epitope); prediction combines tri-peptide similarity and propensity scores (Yao et al., 2012).	10, 14, 16, 18, 20mer predicted epitope length; only one sequence per time in FASTA format; waiting time at the time of testing was 10–30 min.	SVM.
IgPred http://crdd.osdd.net/raghava/igpred/	Predicts B-cell epitopes that can induce a specific class of antibody; MCC: 0.44 (IgG), 0.7 (IgE), 0.45 (IgA), accuracy is around 80%; epitope mapping and motif scan functions are available (Gupta et al., 2013).	4–20 mer epitopes.	SVM and WEKA. package tools.
Bcepred http://crdd.osdd.net/raghava/bceprec/		Only plain text format is supported; developed with a small dataset derived	Evaluation of different physicochemical scales by different method for each.

(Continued)

Table 2   Continued

Name and URL	Functionality and Performance	Limitations	Method
	Prediction accuracy for various properties: 52.92–57.53%; highest accuracy via combination of properties: 58.70% (Saha and Raghava, 2004).	from database BCIPEP (Saha et al., 2005)	
Linear and discontinuous/conformational epitope prediction approaches			
Epitopia http://epitopia.tau.ac.il/	A_{ROC}: 0.6 (conformational) and 0.59 (linear); visualization through Jmol and RasMol ; calculates immunogenicity for each solvent accessible residue for 3D structure input and for every amino acid for sequence input (Rubinstein et al., 2009).	Immunogenicity of each residue is determined by analysis of physicochemical properties of three flanking residues.	Naïve Bayes classifier.
ElliPro http://tools.iedb.org/ellipro/	A_{ROC}: 0.732, sensitivity: 60.1%; Jmol and MODELLER program are available for visualization and prediction 3D structure (Ponomarenko et al., 2008).	Generalize all proteins as ellipsoids.	Modified Thornton's method with residue (pI value) clustering
BepiPred 2.0 http://www.cbs.dtu.dk/services/BepiPred/index.php	A_{ROC} for structural epitopes: 0.62, A_{ROC} for linear epitopes: 0.574 (Jespersen et al., 2017).	5–25mer epitope predictions; at most 50 sequences (max. 300,000 residues per submission, <6000 residues/sequence	Random Forest algorithm

Implementations of machine learning algorithms for conformational epitope predictions are Epitopia (Rubinstein *et al.*, 2009), EPSVR (Liang *et al.*, 2010) and Bpredictor (Zhang *et al.*, 2011c). These tools utilize naïve Bayesian classifiers, SVM and random forest algorithm, respectively. A consensus method (Liang *et al.*, 2009) was developed for EPMeta (Liang *et al.*, 2010). EPMeta combines EPSVR, EPCES, Epitopia, SEPPA, BEpro and Discotope 1.2 predictions to predict surface residues as an epitope if two or more single tools have voted for it. None of the methods applied increased the predictive performances which range between A_{ROC} of 0.597 and 0.824, depending on the datasets tested, above moderate levels.

Sequence-based methods rely on amino acid sequence-based features to evaluate the probability of each residue to be a part of a conformational epitope (Sun *et al.*, 2013). CBTOPE is a SVM-based tool that applies standard binary and physico-chemical profiles of patterns (Ansari and Raghava, 2010). SEPIa employs naïve Bayesian classifier and random forest ensemble methods to classify residues using 13 different features (Dalkas and Rooman, 2017). Although sequence-based methods can be a solution for conformational epitope predictions the performance improvement of complex methods as implemented in SEPIa is minor compared to simpler methods.

Mimotopes are peptides selected from random libraries for their ability to bind to an antibody directed against a specific antigen. Since the mimicry relies on similarities in physicochemical properties and spatial organization (Moreau *et al.*, 2006) mimotope analysis methods applied to conformational epitope predictions require both mimotopes and a 3D structure of the target antigen as an input. The mimotopes are mapped to the surface of the antigen to identify the best sequence alignment, and predict potential epitope regions (Sun *et al.*, 2016). For example PEP-3D-Search searches for matching paths on an antigen surface with respect to the query mimotopes using an ant colony optimization algorithm (Huang *et al.*, 2008).

Linear epitope prediction approaches utilize SVM, ANN and random forest machine learning methods to evaluate multiple amino acid residue properties (Potocnakova *et al.*, 2016). The machine learning approaches have substituted older, poorly performing methods that used only amino acid propensity scores (Blythe and Flower, 2005). The features assessed include for example hydrophilicity, flexibility, turns, solvent accessibility and amino acid pair antigenicity scale (Yasser and Honavar, 2010). SVMTriP (Yao *et al.*, 2012), LBtope (Singh *et al.*, 2013) and COBEpro (Sweredoski and Baldi, 2008a) are representatives of SVM-based tools. ABCpred (Saha and Raghava, 2006) and Bepipred 2.0 (Jespersen *et al.*, 2017) are ANN and random forest-based tools, respectively.

A few tools allow the prediction of both conformational and linear B cell epitopes. ElliPro (Ponomarenko *et al.*, 2008) is based on the implementation of three different algorithms that treat a protein as an ellipsoid shape (Taylor *et al.*, 1983), derive a residue protrusion index using a modified Thornton method (Thornton *et al.*, 1986), and cluster neighboring residues according to their pI (isoelectric point) values. When tested on a conformational epitope dataset constructed from antibody-protein complex 3D structures ElliPro's performance was moderate (A_{ROC} 0.732). Naïve Bayes classifier-based Epitopia (Rubinstein *et al.*, 2009) performs slightly inferior for conformational (A_{ROC} 0.60) and linear (A_{ROC} 0.59) epitopes. Similarly modest performances were reported for BepiPred 2.0 with A_{ROC} 0.62 for conformational and A_{ROC} 0.574 for linear epitopes (Jespersen *et al.*, 2017).

MHC and Epitope Databases

Sustained public accessibility of curated high-quality data on HLA allele sequences, experimentally derived epitope and non-epitope data has enabled the development and improvement of epitope prediction tools. IMGT® (the international ImMuno-GeneTics information system®) established in 1989 became the global reference in immunogenetics and immunoinformatics for immunoglobulins or antibodies, T-cell receptors, human and vertebrate major histocompatibility, immunoglobulin superfamily and related proteins of the immune system of vertebrates and invertebrates (Lefranc *et al.*, 2014).

SYFPEITHI, one of the oldest databases in the field had been utilized together with IEDB (Vita *et al.*, 2014) to develop T cell epitope prediction tools. SYFPEITHY includes more than 7000 peptides that bind to MHC class I and II molecules (Rammensee *et al.*, 1999). With increasing utility of IEDB and integration of epitope data SYFPEITHY became static in 2012.

IEDB stores data on more than 300,000 peptide epitopes including epitope information derived from antigen-antibody complex of PDB, and more than 2500 non-peptide epitopes. The epitope data is not restricted to human and mouse but includes also chimpanzee, macaque, cow and swine. Integrated epitope prediction tools for linear and discontinuous B cell epitopes such as ElliPro (Ponomarenko *et al.*, 2008) or the pipeline TepiTool pipeline (Paul *et al.*, 2016) for vaccine, diagnostic, therapeutic candidate epitope discovery render IEDB by far the most user-friendly and effective resource. Yet a prediction tool that identifies in one process candidate B and T cell epitopes still awaits its implementation.

Illustrative Examples or Case Studies

The applications of epitope predictions are as multifarious as immunoinformatics the field that rose from the beginnings in theoretical immunology to make an impact on vaccine research and development. A representative example is reverse vaccinology (Sette and Rappuoli, 2010). This strategy allows the rapid design of novel vaccines based on pathogen genome information used in epitope predictions. Guttierez and collaborators have demonstrated successfully the design of and effective epitope-based vaccine against swine Influenza A virus through MHC class I and II restricted epitope prediction using PigMatrix (Gutiérrez *et al.*, 2016). The vaccinated pigs responded to virus re-stimulation, showing that the epitope-based vaccination gave rise to T cells that were cross-reactive in vitro with epitopes present in the whole virus.

Another important application of epitope predictions are personalized therapeutic vaccines with the aim to treat epithelial cancer and melanoma as reviewed by Bobisse *et al.* (Bobisse *et al.*, 2016). Typically, epitope predictions of mutated antigens (neonantigens) involves NetMHC or NetMHCpan to identify high-affinity candidate neo-epitopes for *in vitro* validation before stimulating and expanding a patient's tumor infiltrating lymphocytes *ex vivo* for use in adoptive cell therapy. Several clinical trials are ongoing for example a phase I study on a personalized melanoma neoantigen cancer vaccine that started in 2013 (see relevant websites). Although it is too early to pass judgement on the success or failure of the approach even negative data, provided they are published and shared, can help to improve predictions, particularly prediction methods for class II restricted epitopes which perform less accurately than for class I.

In a recent report by Anagnostou *et al.* (2017) epitope prediction had an essential role to elucidate the mechanism of resistance to immune checkpoint blockade drugs in cancer patients. The study of neoantigen evolution during immune checkpoint blockade in non-small cell lung cancer required to assess the immunogenicity of somatic mutations in tumors using wild-type and mutated peptides. NetMHCpan was used to predict the class I binding potential of each peptide, and NetCTLpan to evaluate antigen processing to classify epitopes and non-epitopes. Interestingly, candidate mutation-associated neoantigens with high HLA binding affinity disappeared, probably driven by immune-mediated elimination of cancer cells, thereby depriving patients the means to mount an effective functional immune response.

A lesser known application area of epitope prediction is graft-versus-host disease (GVHD) in organ transplantation research. Monitoring GVHD onset and prevention depends is critically depended on the knowledge of minor histocompatbility antigen (MiHA) match/mismatch between donor and recipient. Van Bergen and collaborators used NetCTLpan to evaluate amino acid polymorphisms among 19 MiHAs for the potential to be targeted by alloreactive CD8 T cells in GVHD patients (van Bergen *et al.*, 2017). The approach yielded 13 new MiHA T cell epitopes.

Numerous other case studies ranging from allergen cross-reactivity, infectious diseases, autoimmunity to therapeutic proteins could be listed here that demonstrate the applicability of current epitope prediction methods in basic, applied and clinical research accompanied by savings in time and cost.

Results and discussion

One of the challenges in epitope prediction is the high polymorphism of MHC genes. Immuno Polymorphism Database IPD-HLA/IMGT (release 3.29.0, 2017) contains 12,544 HLA class I alleles and 4622 HLA class II alleles (Robinson *et al.*, 2014). In IEDB only 272 (2.17%) HLA class I and 247 (5.34%) HLA class II allotypes are associated with experimental peptide binding or T cell epitope data. Therefore allotype-specific prediction methods (e.g. QM-based methods) are not able to deal with MHC allotypes that are uncharacterized with regard to peptide binding. The first method that satisfactorily addressed the issue for HLA-A and -B allotypes was NetMHCpan (Nielsen *et al.*, 2007; Zhang *et al.*, 2011a,b,c) by combining all HLA sequence and peptide data as input into an ANN to derive general features between HLA sequences and peptides and make inferences on binding affinities. Although the initial method has been refined and overcame the problem of variation in peptide length (Lundegaard *et al.* 2008) and potential multiple binding frames for MHC class II (NetMHCIIpan) it has been of limited use for various non-human primate, bovine and swine MHC where dissimilarities among MHC sequences exceeded the threshold for reasonably accurate epitope predictions. Considering the effects of MHC sequence and peptide data diversity and redundancy on the accuracy of epitope predictions (Kim *et al.*, 2014) Mattsson *et al.* proposed in 2016 an MHC-pan method improvement that relies on MHC class I similarity redundancy reduction rather than peptide similarity by using pseudo-sequences derived from the binding cleft amino acid environment of each MHC class I molecule in the input space (Mattsson *et al.*, 2016). In principle, the method could be extended to all MHC class II allotypes opening the possibility to integrate OptiType HLA-typing from next-generation sequencing data (Szolek *et al.*, 2014) with epitope prediction for patient-specific therapeutic vaccine or in transplantation.

Despite the limitations of QM-based methods compared to ANN-based methods, they still yield robust results as demonstrated by De Groot and collaborators who have applied a genome-to-vaccine approach to design (not produce) an HLA class I and II epitope-based vaccine for avian H7N9 influenza using EpiMatrix in 20 h (De Groot *et al.*, 2013). The integration of EpiMatrix and JanusMatrix (Moise *et al.*, 2013), a prediction tool which allows to evaluate potential undesired cross-reactivity of predicted epitopes with human proteins or commensal gut bacteria that may trigger autoimmune responses into iVax (Moise *et al.*, 2015) improves the selection of epitope candidates to be tested for a vaccine. Unfortunately iVax is not an unrestricted-access tool which limits its spread and use, and may trigger future improvements of unrestricted accessible ANN-based tools emulating the JanusMatrix concept.

The performance of B cell epitope predictions is limited by the lack of a sufficiently accurate propensity scales for sequence-based methods and the scarcity of structural data of antigen-antibody complexes. For instance the training data sets of EPSVR and Discotope 2.0 comprised only 48 and 75 complexes, respectively. Therefore it is not too surprising that experimental validations of predicted B cell epitopes result in an average accuracy of 60% (Bergmann-Leitner *et al.*, 2013). Since NMR or X-ray crystallographic antigen-antibody and unbound structural data will increase only slowly, the development of new hybrid and meta-learning methods incorporating qualitative and quantitative features extracted from data generated by experimental methods of lower cost and effort such as phage-display library screening and "quality of antibody response" workflows that generate data of antibody binding from surface plasmon resonance (Davidoff *et al.*, 2015) and hydrogen deuterium exchange data combined with mass spectrometry (Yang *et al.*, 2016) are likely to improve the performance of B cell epitope predictions in the near future.

Epitope prediction tools are an integral part of the reverse vaccinology approach, yet a straightforward performance assessment similar to Critical Assessment of Structure Prediction (CASP) (Moult *et al.*, 2014) has not been conducted so far. CASP has helped to raise the accuracy of homology models and their acceptance as *bona fide* structural information source. In 2006 a NIAD B cell epitope prediction tool workshop (Greenbaum *et al.*, 2007) recommended the creation of annotated datasets and the development of methods and metrics to assess the prediction tools. IEDB has been very successful in assembling richly annotated T and B cell epitope data including negative non-epitope data. The latter were found to be biased towards short non-B-cell epitopes which negatively affected the performance of B cell epitope predictions when included in testing data sets (Rahman *et al.*, 2016) Yet, we are still lacking a community-based consensus procedure and minimal standards to assess the prediction tools. The establishment of a Critical Assessment of Epitope Predictions (CAEP) initiative is overdue to guide the improvement of predictions tools.

Future Directions

The success of epitope predictions particularly for HLA class I T cell epitopes has been largely driven by data and their level of integration with the current state of knowledge of immunological processes and host-pathogen interactions. Knowledge gaps in the details of immunological processes that define relations and differentiation dynamics of effector and memory T cell pools, and central and peripheral tolerance represent just two bottlenecks among others that require many more experimental data to advance predictive methods. At present we can identify immunogenic epitope candidates, but not optimal protective ones. For example, more work and data on molecular structures, TCR and BCR repertoire data are needed to predict cross-reactive and -neutralizing epitopes that will not be reduced in efficacy nor trigger autoimmunity when a pathogen is encountered two or more times.

Rational vaccine design and evaluation resembles in its temporal and mechanistic order of steps a biological pathway with epitope prediction positioned fairly upstream. Efforts to improve connectivity and feedback among individual components including epitope prediction may come to fruition on proof-of-concept level within in the next five to ten year if concerted and standardized generation and collection of longitudinal data downstream of epitope predictions on gene and protein expression levels, protein-protein interactions and pathways, pathogen and antigen variation is funded and performed. The Human Vaccines Project (Koff *et al.*, 2014) a nonprofit public-private project that aims to elucidate the molecular cellular principles of vaccine-induced immunity is a promising high-impact initiative to enhance rational vaccine design. Translation into effective and affordable preventive vaccines for complex targets such as tuberculosis or HIV and therapeutic cancer vaccines may take longer because research progress on the effects and impact of natural genome and transcriptome variations in populations, and impact of environmental factors for example metals and chemicals, on the development, functioning and ageing of the immune system is slower. Areas deserving more attention include the development of algorithms to predict non-protein epitopes e.g. carbohydrates that are not unimportant for vaccines.

Closing Remarks

In the hierarchy of epitope prediction performance HLA class I binding predictions occupy the top position. Basically we can predict with current systems for any classical HLA class I allotype T cell epitope candidates. Second ranked are MHC class I-restricted T cell epitope predictions for chimpanzee, macaque, cow and swine. Therefore the development and actual use of epitope-based veterinary vaccines might precede the ones for human vaccines. Third ranked are predictions of HLA class II DR- and DQ-restricted epitope candidates. There is still room for significant improvements with regard to correctly identifying the core binding residues and increasing the number of experimental data for less studied allotypes, especially for HLA-DP. At the bottom of the performance hierarchy are B cell epitope predictions. This fact is expected to incentivize the development of both new experimental and data-driven computational methods that are anticipated to lift the accuracy and efficacy of B cell epitope predictions to acceptable levels until 2025.

Acknowledgement

R.K. acknowledges the award of an IRCMS Internship by International Research Center for Medical Sciences, Kumamoto University. C.S. acknowledges the support for the work on epitope predictions by Kumamoto University, International Research Center for Medical Sciences (#005–5700101122).

See also: Data Mining: Classification and Prediction. Data Mining: Prediction Methods. Extraction of Immune Epitope Information. Immunoinformatics Databases. Natural Language Processing Approaches in Bioinformatics. Prediction of Protein-Protein Interactions: Looking Through the Kaleidoscope. Protein Post-Translational Modification Prediction. Protein Properties. Secondary Structure Prediction. Sequence Analysis. Sequence Composition. Supervised Learning: Classification. Vaccine Target Discovery

References

Adams, H.-P., Koziol, J.A., 1995. Prediction of binding to MHC class I molecules. Journal of Immunological Methods 185, 181–190.

Anagnostou, V., Smith, K.N., Forde, P.M., et al., 2017. Evolution of neoantigen landscape during immune checkpoint blockade in non-small cell lung cancer. Cancer Discovery 7, 264–276.

Andreatta, M., Karosiene, E., Rasmussen, M., et al., 2015. Accurate pan-specific prediction of peptide-MHC class II binding affinity with improved binding core identification. Immunogenetics 67, 641–650.

Andreatta, M., Nielsen, M., 2015. Gapped sequence alignment using artificial neural networks: Application to the MHC class I system. Bioinformatics 32, 511–517.

Ansari, H.R., Raghava, G.P., 2010. Identification of conformational B-cell epitopes in an antigen from its primary sequence. Immunome Research 6, 6.

Antonets, D., Maksyutov, A., 2010. TEpredict: Software for T-cell epitope prediction. Molecular Biology 44, 119–127.

Atanasova, M., Patronov, A., Dimitrov, I., Flower, D.R., Doytchinova, I., 2013. EpiDOCK: A molecular docking-based tool for MHC class II binding prediction. Protein Engineering, Design & Selection 26, 631–634.

Barlow, D., Edwards, M., Thornton, J., 1986. Continuous and discontinuous protein antigenic determinants. Nature 322, 747–748.

Bergmann-Leitner, E.S., Chaudhury, S., Steers, N.J., et al., 2013. Computational and experimental validation of B and T-cell epitopes of the in vivo immune response to a novel malarial antigen. PloS One 8, e71610.

Bhasin, M., Lata, S., Raghava, G., 2007. TAPPred prediction of TAP-binding peptides in antigens. Immunoinformatics: Predicting Immunogenicity In Silico. 381–386.

Bhasin, M., Raghava, G., 2003. Prediction of promiscuous and high-affinity mutated MHC binders. Hybridoma and Hybridomics 22, 229–234.

Bhasin, M., Raghava, G., 2004a. Prediction of CTL epitopes using QM, SVM and ANN techniques. Vaccine 22, 3195–3204.

Bhasin, M., Raghava, G., 2004b. SVM based method for predicting HLA-DRB1* 0401 binding peptides in an antigen sequence. Bioinformatics 20, 421–423.

Bhasin, M., Raghava, G., 2005. Pcleavage: An SVM based method for prediction of constitutive proteasome and immunoproteasome cleavage sites in antigenic sequences. Nucleic Acids Research 33, W202–W207.

Bhasin, M., Raghava, G., 2007. A hybrid approach for predicting promiscuous MHC class I restricted T cell epitopes. Journal of Biosciences 32, 31–42.

Blum, J.S., Wearsch, P.A., Creswell, P., 2013. Pathways of antigen processing. Annual Review of immunology 31, 443–473.

Blythe, M.J., Flower, D.R., 2005. Benchmarking B cell epitope prediction: Underperformance of existing methods. Protein Science 14, 246–248.

Bobisse, S., Foukas, P.G., Coukos, G., Harari, A., 2016. Neoantigen-based cancer immunotherapy. Annals of Translational Medicine 4, 262.

Brusic, V., Petrovsky, N., Zhang, G., Bajic, V.B., 2002. Prediction of promiscuous peptides that bind HLA class I molecules. Immunology and Cell Biology 80, 280.

Chen, J., Liu, H., Yang, J., Chou, K.-C., 2007. Prediction of linear B-cell epitopes using amino acid pair antigenicity scale. Amino acids 33, 423–428.

Dalkas, G.A., Rooman, M., 2017. SEPIa, a knowledge-driven algorithm for predicting conformational B-cell epitopes from the amino acid sequence. BMC Bioinformatics 18, 95.

Davidoff, S.N., Ditto, N.T., Brooks, A.E., Eckman, J., Brooks, B.D., 2015. Surface plasmon resonance for therapeutic antibody characterization. Label-Free Biosensor Methods in Drug Discovery. 35–76.

De Groot, A.S., Einck, L., Moise, L., et al., 2013. Making vaccines "on demand" A potential solution for emerging pathogens and biodefense? Human Vaccines and Immunotherapeutics 9, 1877–1884.

De Groot, A.S., Sbai, H., Saint Aubin, C., et al., 2002. Immuno-informatics: Mining genomes for vaccine components. Immunology and Cell Biology 80, 255.

Dhanda, S.K., Gupta, S., Vir, P., Raghava, G., 2013a. Prediction of IL4 inducing peptides. Clinical and Developmental Immunology 2013, 263952.

Dhanda, S.K., Vir, P., Raghava, G.P., 2013b. Designing of interferon-gamma inducing MHC class-II binders. Biology Direct 8, 30.

Diez-Rivero, C.M., Chenlo, B., Zuluaga, P., Reche, P.A., 2010. Quantitative modeling of peptide binding to TAP using support vector machine. Proteins: Structure, Function, and Bioinformatics 78, 63–72.

Dimitrov, I., Garnev, P., Flower, D.R., Doytchinova, I., 2010a. EpiTOP – a proteochemometric tool for MHC class II binding prediction. Bioinformatics 26, 2066–2068.

Dimitrov, I., Garnev, P., Flower, D.R., Doytchinova, I., 2010b. Peptide binding to the HLA-DRB1 supertype: A proteochemometrics analysis. European Journal of Medicinal Chemistry 45, 236–243.

Dönnes, P., Kohlbacher, O., 2006. SVMHC: A server for prediction of MHC-binding peptides. Nucleic Acids Research 34, W194–W197.

Doytchinova, I.A., Flower, D.R., 2002. Physicochemical explanation of peptide binding to HLA-A* 0201 major histocompatibility complex: A three-dimensional quantitative structure-activity relationship study. Proteins: Structure, Function, and Bioinformatics 48, 505–518.

Frank, S.A., 2002. Immunology and Evolution of Infectious Disease. Princeton, NJ: Princeton University Press.

Greenbaum, J.A., Andersen, P.H., Blythe, M., et al., 2007. Towards a consensus on datasets and evaluation metrics for developing B-cell epitope prediction tools. Journal of Molecular Recognition 20, 75–82.

Grotzke, J.E., Sengupta, D., Lu, Q., Cresswell, P., 2017. The ongoing saga of the mechanism (s) of MHC class I-restricted cross-presentation. Current Opinion in Immunology 46, 89–96.

Guan, P., Hattotuwagama, C.K., Doytchinova, I.A., Flower, D.R., 2006. MHCPred 2.0. Applied Bioinformatics 5, 55–61.

Guo, L., Luo, C., Zhu, S., 2013. MHC2SKpan: A novel kernel based approach for pan-specific MHC class II peptide binding prediction. BMC Genomics 14, S11.

Gupta, S., Ansari, H.R., Gautam, A., Raghava, G.P., 2013. Identification of B-cell epitopes in an antigen for inducing specific class of antibodies. Biology Direct 8, 27.

Gutiérrez, A.H., Loving, C., Moise, L., et al., 2016. In vivo validation of predicted and conserved T cell epitopes in a swine influenza model. PlOS One 11, e0159237.

Hattotuwagama, C.K., Doytchinova, I.A., Flower, D.R., 2007. Toward the prediction of class I and II mouse major histocompatibility complex-peptide-binding affinity: In silico bioinformatic step-by-step guide using quantitative structure-activity relationships. Immunoinformatics: Predicting Immunogenicity In Silico. 227–245.

Honeyman, M.C., Brusic, V., Stone, N.L., Harrison, L.C., 1998. Neural network-based prediction of candidate T-cell epitopes. Nature Biotechnology 16, 966–969.

Huang, Y.X., Bao, Y.L., Guo, S.Y., et al., 2008. Pep-3D-Search: A method for B-cell epitope prediction based on mimotope analysis. BMC Bioinformatics 9, 538.

Jacob, L., Vert, J.-P., 2007. Efficient peptide-MHC-I binding prediction for alleles with few known binders. Bioinformatics 24, 358–366.

Jespersen, M.C., Peters, B., Nielsen, M., Marcatili, P., 2017. BepiPred-2.0: Improving sequence-based B-cell epitope prediction using conformational epitopes. Nucleic Acids Research 45, W24–W29.

Karosiene, E., Lundegaard, C., Lund, O., Nielsen, M., 2012. NetMHCcons: A consensus method for the major histocompatibility complex class I predictions. Immunogenetics 64, 177–186.

Kim, Y., Sidney, J., Buus, S., et al., 2014. Dataset size and composition impact the reliability of performance benchmarks for peptide-MHC binding predictions. BMC Bioinformatics 15, 241.

Koff, W.C., Gust, I.D., Plotkin, S.A., 2014. Toward a human vaccines project. Nature Immunology 15, 589–592.

Kringelum, J.V., Lundegaard, C., Lund, O., Nielsen, M., 2012. Reliable B cell epitope predictions: Impacts of method development and improved benchmarking. PLOS Computational Biology 8, e1002829.

Kulkarni-Kale, U., Bhosle, S., Kolaskar, A.S., 2005. CEP: A conformational epitope prediction server. Nucleic Acids Research 33, W168–W171.

Larsen, J.E., Lund, O., Nielsen, M., 2006. Improved method for predicting linear B-cell epitopes. Immunome Research 2, 2.

Larsen, M.V., Lundegaard, C., Lamberth, K., et al., 2007. Large-scale validation of methods for cytotoxic T-lymphocyte epitope prediction. BMC Bioinformatics 8, 424.

Lata, S., Bhasin, M., Raghava, G.P., 2007. Application of machine learning techniques in predicting MHC binders. Methods in Molecular Biology 409, 201–215.

Lefranc, M.-P., Giudicelli, V., Duroux, P., et al., 2014. IMGT®, the international ImMunoGeneTics information system® 25 years on. Nucleic Acids Research 43, D413–D422.

Liang, S., Zheng, D., Standley, D.M., *et al.*, 2010. EPSVR and EPMeta: Prediction of antigenic epitopes using support vector regression and multiple server results. BMC Bioinformatics 11, 381.

Liang, S., Zheng, D., Zhang, C., Zacharias, M., 2009. Prediction of antigenic epitopes on protein surfaces by consensus scoring. BMC Bioinformatics 10, 302.

Litman, G.W., Rast, J.P., Fugmann, S.D., 2010. The origins of vertebrate adaptive immunity. Nature Reviews. Immunology 10, 543.

Liu, G., Li, D., Li, Z., *et al.*, 2017. PSSMHCpan: A novel PSSM-based software for predicting class I peptide-HLA binding affinity. Giga Science 6, 1–11.

Logean, A., Rognan, D., 2002. Recovery of known T-cell epitopes by computational scanning of a viral genome. Journal of Computer-aided Molecular Design 16, 229–243.

Lundegaard, C., Lund, O., Nielsen, M., 2008. Accurate approximation method for prediction of class I MHC affinities for peptides of length 8, 10 and 11 using prediction tools trained on 9mers. Bioinformatics 24, 1397–1398.

Mamitsuka, H., 1998. Predicting peptides that bind to MHC molecules using supervised learning of hidden Markov models. Proteins Structure Function and Genetics 33, 460–474.

Mattsson, A.H., Kringelum, J.V., Garde, C., Nielsen, M., 2016. Improved pan-specific prediction of MHC class I peptide binding using a novel receptor clustering data partitioning strategy. HLA 88, 287–292.

Moise, L., Gutierrez, A., Kibria, F., *et al.*, 2015. iVAX: An integrated toolkit for the selection and optimization of antigens and the design of epitope-driven vaccines. Human Vaccines and Immunotherapeutics 11, 2312–2321.

Moise, L., Gutierrez, A.H., Bailey-Kellogg, C., *et al.*, 2013. The two-faced T cell epitope: Examining the host-microbe interface with JanusMatrix. Human Vaccines and Immunotherapeutics 9, 1577–1586.

Moreau, V., Granier, C., Villard, S., Laune, D., Molina, F., 2006. Discontinuous epitope prediction based on mimotope analysis. Bioinformatics 22, 1088–1095.

Moult, J., Fidelis, K., Kryshtafovych, A., Schwede, T., Tramontano, A., 2014. Critical assessment of methods of protein structure prediction (CASP) – round x. Proteins: Structure, Function, and Bioinformatics 82, 1–6.

Moutaftsi, M., Peters, B., Pasquetto, V., *et al.*, 2006. A consensus epitope prediction approach identifies the breadth of murine TCD8 + -cell responses to vaccinia virus. Nature Biotechnology 24, 817.

Mukherjee, S., Bhattacharyya, C., Chandra, N., 2016. HLaffy: Estimating peptide affinities for Class-1 HLA molecules by learning position-specific pair potentials. Bioinformatics 32, 2297–2305.

Negi, S.S., Braun, W., 2009. Automated detection of conformational epitopes using phage display peptide sequences. Bioinformatics and Biology Insights 3, 71.

Neu, K.E., Wilson, P.C., 2016. Taking the broad view on B cell affinity maturation. Immunity 44, 518–520.

Nielsen, M., Andreatta, M., 2016. NetMHCpan-3.0; improved prediction of binding to MHC class I molecules integrating information from multiple receptor and peptide length datasets. Genome Medicine 8, 33.

Nielsen, M., Lund, O., 2009. NN-align. An artificial neural network-based alignment algorithm for MHC class II peptide binding prediction. BMC Bioinformatics 10, 296.

Nielsen, M., Lundegaard, C., Blicher, T., *et al.*, 2007. NetMHCpan, a method for quantitative predictions of peptide binding to any HLA-A and-B locus protein of known sequence. PlOS One 2, e796.

Nielsen, M., Lundegaard, C., Lund, O., Keşmir, C., 2005. The role of the proteasome in generating cytotoxic T-cell epitopes: Insights obtained from improved predictions of proteasomal cleavage. Immunogenetics 57, 33–41.

Ogishi, M., Yotsuyanagi, H., 2017. Epitope immunogenicity prediction through repertoire-wide TCR-peptide contact profiles. *bioRxiv* 155317.

Oyarzún, P., Ellis, J.J., Bodén, M., Kobe, B., 2013. PREDIVAC: CD4 + T-cell epitope prediction for vaccine design that covers 95% of HLA class II DR protein diversity. BMC Bioinformatics 14, 52.

Parker, K.C., Bednarek, M.A., Coligan, J.E., 1994. Scheme for ranking potential HLA-A2 binding peptides based on independent binding of individual peptide side-chains. The Journal of Immunology 152, 163–175.

Patronov, A., Doytchinova, I., 2013. T-cell epitope vaccine design by immunoinformatics. Open Biology 3, 120139.

Paul, S., Sidney, J., Sette, A., Peters, B., 2016. TepiTool: A pipeline for computational prediction of T cell epitope candidates. Current Protocols in Immunology 114, 18.19.1.

Ponomarenko, J., Bui, H.-H., l i, W, *et al.*, 2008. ElliPro: A new structure-based tool for the prediction of antibody epitopes. BMC Bioinformatics 9, 514.

Potocnakova, L., Bhide, M., Pulzova, L.B., 2016. An Introduction to B-cell epitope mapping and in silico epitope prediction. Journal of Immunology Research 2016, 6760830.

Qi, T., Qiu, T., Zhang, Q., *et al.*, 2014. SEPPA 2.0 – more refined server to predict spatial epitope considering species of immune host and subcellular localization of protein antigen. Nucleic Acids Research 42, W59–W63.

Querec, T.D., Akondy, R.S., Lee, E.K., *et al.*, 2009. Systems biology approach predicts immunogenicity of the yellow fever vaccine in humans. Nature Immunology 10, 116–125.

Rahman, K.S., Chowdhury, E.U., Sachse, K., Kaltenboeck, B., 2016. Inadequate reference datasets biased toward short non-epitopes confound B-cell epitope prediction. The Journal of Biological Chemistry 29, 14585–14599.

Rammensee, H.-G., Bachmann, J., Emeerich, N.P.N., Bachor, O.A., Stevanović, S., 1999. SYFPEITHI: Database for MHC ligands and peptide motifs. Immunogenetics 50, 213–219.

Rappuoli, R., Aderem, A., 2011. A 2020 vision for vaccines against HIV, tuberculosis and malaria. Nature 473, 463.

Reche, P.A., Glutting, J.-P., Zhang, H., Reinherz, E.L., 2004. Enhancement to the RANKPEP resource for the prediction of peptide binding to MHC molecules using profiles. Immunogenetics 56, 405–419.

Robinson, J., Halliwell, J.A., Hayhurst, J.D., *et al.*, 2014. The IPD and IMGT/HLA database: Allele variant databases. Nucleic Acids Research 43, D423–D431.

Rubinstein, N.D., Mayrose, I., Martz, E., Pupko, T., 2009. Epitopia: A web-server for predicting B-cell epitopes. BMC Bioinformatics 10, 287.

Saha, S., Bhasin, M., Raghava, G.P., 2005. Bcipep: A database of B-cell epitopes. BMC Genomics 6, 79.

Saha, S., Raghava, G., 2006. Prediction of continuous B-cell epitopes in an antigen using recurrent neural network. Proteins: Structure, Function, and Bioinformatics 65, 40–48.

Saha, S., Raghava, G.P.S., 2004. BcePred: Prediction of continuous B-Cell epitopes in antigenic sequences using physico-chemical Properties. In: Nicosia, G., Cutello, V., Bentley, P.J., Timmis, J. (Eds.), ICARIS. Berlin: Springer, pp. 197–204.

Sammut, C., Webb, G.I., 2010. Leave-One-Out Cross-Validation. In: Sammut, C., Webb, G.I. (Eds.), Encyclopedia of Machine Learning, 2nd edn. Boston: Springer US.

Schafer, J.R.A., Jesdale, B.M., George, J.A., Kouttab, N.M., De Groot, A.S., 1998. Prediction of well-conserved HIV-1 ligands using a matrix-based algorithm, EpiMatrix. Vaccine 16, 1880–1884.

Schönbach, C., Kun, Y., Brusic, V., 2002. Large-scale computational identification of HIV T-cell epitopes. Immunology and Cell Biology 80, 300.

Schroeder, H.W., Cavacini, L., 2010. Structure and function of immunoglobulins. Journal of Allergy and Clinical Immunology 125, S41–S52.

Schubert, B., Walzer, M., Brachvogel, H.-P., *et al.*, 2016. FRED 2: An immunoinformatics framework for Python. Bioinformatics 32, 2044–2046.

Schueler-Furman, O., Altuvia, Y., Sette, A., Margalit, H., 2000. Structure-based prediction of binding peptides to MHC class I molecules: Application to a broad range of MHC alleles. Protein Science 9, 1838–1846.

Sela-Culang, I., Kunik, V., Ofran, Y., 2013. The structural basis of antibody-antigen recognition. Frontiers in Immunology 4, 302.

Sette, A., Rappuoli, R., 2010. Reverse vaccinology: Developing vaccines in the era of genomics. Immunity 33, 530–541.

Singh, H., Ansari, H.R., Raghava, G.P., 2013. Improved method for linear B-cell epitope prediction using antigen's primary sequence. PLOS One 8, e62216.

Singh, H., Raghava, G., 2001. ProPred: Prediction of HLA-DR binding sites. Bioinformatics 17, 1236–1237.

Singh, H., Raghava, G., 2003. ProPred1: Prediction of promiscuous MHC Class-I binding sites. Bioinformatics 19, 1009–1014.

Soria-Guerra, R.E., Nieto-Gomez, R., Govea-Alonso, D.O., Rosales-Mendoza, S., 2015. An overview of bioinformatics tools for epitope prediction: Implications on vaccine development. Journal of Biomedical Informatics 53, 405–414.

Stranzl, T., Larsen, M.V., Lundegaard, C., Nielsen, M., 2010. NetCTLpan: Pan-specific MHC class I pathway epitope predictions. Immunogenetics 62, 357–368.

Sun, P., Ju, H., Liu, Z., et al., 2013. Bioinformatics resources and tools for conformational B-cell epitope prediction. Computational and Mathematical Methods in Medicine. 2013), 943636.

Sun, P., Qi, J., Zhao, Y., et al., 2016. A novel conformational B-cell epitope prediction method based on mimotope and patch analysis. Journal of Theoretical Biology 394, 102–108.

Sweredoski, M.J., Baldi, P., 2008a. COBEpro: A novel system for predicting continuous B-cell epitopes. Protein Engineering, Design and Selection 22, 113–120.

Sweredoski, M.J., Baldi, P., 2008b. PEPITO: Improved discontinuous B-cell epitope prediction using multiple distance thresholds and half sphere exposure. Bioinformatics 24, 1459–1460.

Szolek, A., Schubert, B., Mohr, C., et al., 2014. OptiType: Precision HLA typing from next-generation sequencing data. Bioinformatics 30, 3310–3316.

Taylor, W., Thornton, J.T., Turnell, W., 1983. An ellipsoidal approximation of protein shape. Journal of Molecular Graphics 1, 30–38.

Thornton, J., Edwards, M., Taylor, W., Barlow, D., 1986. Location of 'continuous' antigenic determinants in the protruding regions of proteins. The EMBO Journal 5, 409.

Trolle, T., Nielsen, M., 2014. NetTepi: An integrated method for the prediction of T cell epitopes. Immunogenetics 66, 449–456.

van Bergen, C.A., Van Luxemburg-Heijs, S.A., De Wreede, L.C., et al., 2017. Selective graft-versus-leukemia depends on magnitude and diversity of the alloreactive T cell response. The Journal of Clinical Investigation 127, 517.

Vita, R., Overton, J.A., Greenbaum, J.A., et al., 2014. The immune epitope database (IEDB) 3.0. Nucleic Acids Research 43, D405–D412.

Xu, Y., Luo, C., Mamitsuka, H., Zhu, S., 2016. MetaMHCpan, a meta approach for pan-specific MHC peptide binding prediction. Vaccine Design: Methods and Protocols, Volume 2: Vaccines for Veterinary Diseases. 753–760.

Yang, D., Frego, L., Lasaro, M., et al., 2016. Efficient qualitative and quantitative determination of antigen-induced immune responses. The Journal of Biological Chemistry 291, 16361–16374.

Yang, X., Yu, X., 2009. An introduction to epitope prediction methods and software. Reviews in Medical Virology 19, 77–96.

Yao, B., Zheng, D., Liang, S., Zhang, C., 2013. Conformational B-cell epitope prediction on antigen protein structures: A review of current algorithms and comparison with common binding site prediction methods. PLOS One 8, e62249.

Yao, B., Zhang, L., Liang, S., Zhang, C., 2012. SVMTriP: A method to predict antigenic epitopes using support vector machine to integrate tri-peptide similarity and propensity. PIOS One 7, e45152.

Yasser, E.-M., Honavar, V., 2010. Recent advances in B-cell epitope prediction methods. Immunome Research 6, S2.

Yewdell, J.W., Bennink, J.R., 1999. Immunodominance in major histocompatibility complex class I-restricted T lymphocyte responses. Annual Review of Immunology 17, 51–88.

Zhang, C., Bickis, M.G., Wu, F.-X., Kusalik, A.J., 2006. Optimally-connected hidden markov models for predicting MHC-binding peptides. Journal of Bioinformatics and Computational Biology 4, 959–980.

Zhang, G.L., Bozic, I., Kwoh, C.K., August, J.T., Brusic, V., 2007. Prediction of supertype-specific HLA class I binding peptides using support vector machines. Journal of Immunological Methods 320, 143–154.

Zhang, G.L., Deluca, D.S., Keskin, D.B., et al., 2011a. MULTIPRED2: A computational system for large-scale identification of peptides predicted to bind to HLA supertypes and alleles. Journal of Immunological Methods 374, 53–61.

Zhang, H., Lund, O., Nielsen, M., 2009. The PickPocket method for predicting binding specificities for receptors based on receptor pocket similarities: Application to MHC-peptide binding. Bioinformatics 25, 1293–1299.

Zhang, L., Chen, Y., Wong, H.-S., et al., 2012. TEPITOPEpan: Extending TEPITOPE for peptide binding prediction covering over 700 HLA-DR molecules. PLOS One 7, e30483.

Zhang, L., Udaka, K., Mamitsuka, H., Zhu, S., 2011b. Toward more accurate pan-specific MHC-peptide binding prediction: A review of current methods and tools. Briefings in Bioinformatics 13, 350–364.

Zhang, W., Xiong, Y., Zhao, M., et al., 2011c. Prediction of conformational B-cell epitopes from 3D structures by random forests with a distance-based feature. BMC Bioinformatics 12, 341.

Further Reading

Belden, O.S., Baker, S.C., Baker, B.M., 2015. Citizens unite for computational immunology!. Trends in Immunology 36, 385–387.

Brusic, V., Gottardo, R., Kleinstein, S.H., Davis, M.M., 2014. Computational resources for high-dimensional immune analysis from the Human Immunology Project Consortium. Nature Biotechnology 32, 146–148.

De, R.K., Tomar, N., 2014. Immunoinformatics. [eds.] In: Walker, (Ed.), Methods in Molecular Biology, second ed., 1184. New York: Humana Press.

De Gregorio, E., Rappuoli, R., 2014. From empiricism to rational design: A personal perspective of the evolution of vaccine development. Nature Reviews. Immunology 14, 505.

Ditto, N.T., Brooks, B.D., 2016. The emerging role of biosensor-based epitope binning and mapping in antibody-based drug discovery. Expert Opinion on Drug Discovery 11, 925–937.

Fleri, W., Paul, S., Dhanda, S.K., et al., 2017. The immune epitope database and analysis resource in epitope discovery and synthetic vaccine design. Frontiers in Immunology 8, 278.

He, L., Zhu, J., 2015. Computational tools for epitope vaccine design and evaluation. Current Opinion in Virology 11, 103–112.

Liljeroos, L., Malito, E., Ferlenghi, I., Bottomley, M.J., 2015. Structural and computational biology in the design of immunogenic vaccine antigens. Journal of Immunology Research 2015, 156241.

Scheuermann, R.H., Sinkovits, R.S., Schenkelberg, T., Koff, W.C., 2017. A bioinformatics roadmap for the human vaccines project. Expert Review of Vaccines 16, 535–544.

Sette, A., Peters, B., 2007. Immune epitope mapping in the post-genomic era: Lessons for vaccine development. Current Opinion in Immunology 19, 106–110.

Relevant Websites

https://clinicaltrials.gov/ct2/show/NCT01970358
    A phase I study with a personalized neoantigen cancer vaccine in melanoma.
https://www.immunespace.org/
    Enabling integrative modelling of human immunological data. The Human Immunology Project Consortium.
http://www.iedb.org/
    Immune Epitope Database Analysis Resource.
http://www.imgt.org
    IMGT®, the international ImMunoGeneTics information system®.
http://www.humanvaccinesproject.org
    The Human Vaccines Project.

Biographical Sketch

Roman Kogay is an undergraduate student at School of Science and Technology, Department of Biology, Nazarbayev University. His research interests include application of bioinformatics in metabolomics, genomics, proteomics and biomedical sciences. After graduation he hopes to pursue a doctoral degree.

Christian Schönbach is Professor at Graduate School of Medical Sciences, International Research Center for Medical Sciences, Kumamoto University. Prior to joining Kumamoto University he was Professor at School of Science and Technology, Department of Biology, Nazarbayev University (2013–2016). His research and teaching interests revolve around bioinformatics, genomics and immunology. Since 2010 he is serving the bioinformatics community of Asia-Pacific Bioinformatics Network (APBioNet) in various leadership roles.

Immunoglobulin Clonotype and Ontogeny Inference

Pazit Polak, Ramit Mehr, and Gur Yaari, Bar-Ilan University, Ramt Gan, Israel

Introduction

The adaptive immune response functions mainly through B and T lymphocytes. B lymphocytes express immunoglobulins (Igs, aka antibodies), which specifically bind antigens on surfaces of pathogens and neutralize them. Specificity towards such a large pool of threats is achieved by the diversity and dynamics of Ig repertoires. The DNA encoding for Igs comes from random selection of the variable (V), diversity (D), and joining (J) genes, through somatic rearrangement. In addition to the combinatorial diversity coming from the choice of V, D, and J genes, further diversity is added at the boundaries where the segments are joined, by imprecise excision and addition of random nucleotides, resulting in a staggering receptor diversity of up to 10^{11} different B cell clones in each human (see **Fig. 1**) (Boyd et al., 2010, 2009; Murphy and Weaver, 2016; Weinstein et al., 2009).

Following initial antigen recognition, B cells further undergo cycles of affinity maturation, during which their rearranged Ig gene loci are mutated in a process called somatic hypermutation (SHM), and cells with increased affinity to the antigen are selected for clonal expansion. This results in improved affinity of the Igs to the antigens, and the whole process is referred to as affinity maturation. The outcome of affinity maturation is short- or long-lived plasma cells secreting Igs, and short- or long-lived memory B cells.

Thus, the Ig repertoire of an individual stores information about current and past threats that the body has encountered. In addition to studying the development of the adaptive immune system (Pogorelyy et al., 2017; Rechavi and Somech, 2017; Rechavi et al., 2015) and other fundamental processes underlying the immune system in healthy individuals (Arnaout et al., 2011; Boyd et al., 2010, 2009; Wu et al., 2010b), investigation of the repertoire has the potential to reveal dysregulation in abnormal conditions such as insufficient, overly active or misdirected immune responses (von Büdingen et al., 2012; Cameron et al., 2009; Lehmann-Horn et al., 2013; Palanichamy et al., 2014; Singh et al., 2013; Snir et al., 2015; Stern et al., 2014; Zuckerman et al., 2010a), as well as infectious diseases (Laserson et al., 2014; Parameswaran et al., 2013; Sok et al., 2013; Tsioris et al., 2015), allergy (Patil et al., 2015; Wu et al., 2014), cancer (Fridman et al., 2012; Galon et al., 2006; Glanville et al., 2011; Jiang et al., 2015; Katoh et al., 2017; Lossos et al., 2000; Yahalom et al., 2013), and aging (Ademokun et al., 2011; Dunn-Walters and Ademokun, 2010; Dunn-Walters et al., 2003; Wu et al., 2012).

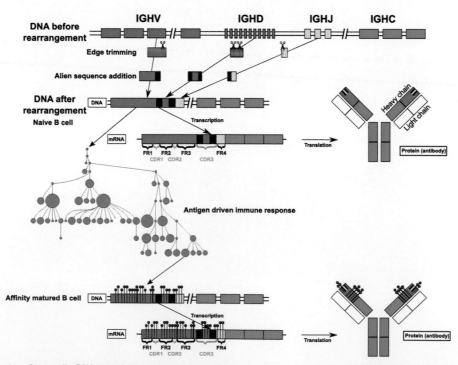

Fig. 1 B cell maturation. Stem cells DNA contains many potential genes for the V, D and J regions. During B cell development, one V gene, one D gene and one J gene are recombined to form the variable region of the Ig. Further diversity is added at the boundaries between the segments, by trimming and template-less addition of nucleotides. After encountering a pathogen, activated B cells undergo an affinity maturation process, resulting in mutated high-affinity Igs.

Ig repertoires are also starting to be utilized in several clinical settings including the detection of minimal residual disease in leukemia (Logan *et al.*, 2011) and analysis of tumor-infiltrating B cells for cancer prognosis (Zhang *et al.*, 2017). A snapshot of the immune system could serve as a natural bio-sensor for diagnostic and prognostic purposes. Indeed, recent studies have found that common Ig sequences can be found in unrelated individuals, for example, following Dengue (Parameswaran *et al.*, 2013) or Zika virus infection (Magnani *et al.*, 2017), in multiple sclerosis (Cameron *et al.*, 2009), and rheumatoid arthritis (Tak *et al.*, 2017), and therefore may be of diagnostic value. Repertoire analysis could also allow for diagnosis of diseases in early stages, and personalized medical treatment. Moreover, analysis of Ig repertoire sequential snapshots can teach us about the dynamics of the arms race between the immune system and the threat.

The diverse repertoire of B lymphocytes is constantly changing. Ig diversification endows the system with the ability to recognize any biological molecule or pathogen. However, the huge diversity and dynamic nature of Ig repertoires make their study challenging. Recent developments in high throughput sequencing (HTS) enable researchers to obtain large numbers of sequences from multiple samples simultaneously and increase the detection power (Ademokun *et al.*, 2011; Boyd *et al.*, 2010; Campbell *et al.*, 2008; Prabakaran *et al.*, 2012; Scheid *et al.*, 2009). Here, we describe protocols for construction Ig libraries for HTS, analysis methods, clonotype inference, i.e. classification of Igs into clones based on their V(D)J sequences, ontogeny inference, i.e. lineage trees of Ig evolution during antigen-driven affinity selection, and applications of clonotype and ontogeny inference.

Sequencing Approaches – Pros and Cons

Dramatic improvements in HTS technologies now enable large-scale characterization of Ig repertoires (Benichou *et al.*, 2012; Georgiou *et al.*, 2014). HTS platforms such as MiSeq by Illumina are being exploited by most researchers in the field (Arnaout *et al.*, 2011; Boyd *et al.*, 2010, 2009; von Büdingen *et al.*, 2012; Jiang *et al.*, 2011; Krause *et al.*, 2011; Weinstein *et al.*, 2009; Wu *et al.*, 2010b). MiSeq can generate 10 to 20 million paired-end 300 base-pair reads in a single run with relatively few errors (Loman *et al.*, 2012) and is expected to be extended to 400 base-pair reads in the near future. Other HTS platforms include Ion torrent, Pacbio, and minION (Goodwin *et al.*, 2016).

The choice of library preparation protocol involves several decisions that influence the experimental setup, the bioinformatic pipeline, and the type of information that can be extracted from the data. For example, single-cell sequencing is becoming more and more widespread in many biological applications, and is expected to spread also into Ig repertoire sequencing (Briggs *et al.*, 2017; DeKosky *et al.*, 2014). However, current technologies for Ig single cell sequencing are dramatically more expensive and complicated, and as a consequence far less popular. Other aspects that need to be taken into consideration when choosing a library preparation protocol are whether to use DNA or RNA as starting material, which sets of primers should be used, and if and how to include unique molecular identifiers. The pros and cons of choosing between DNA vs. RNA are summarized in **Table 1**.

UMI

A relatively recent major improvement to Ig repertoire library preparation protocols is the inclusion of unique molecular identifiers (UMIs), i.e. random sequences of 10–15 nucleotides that are added to the tail of the reverse transcription primer(s) (Jiang *et al.*, 2013; Mamedov *et al.*, 2013; Shiroguchi *et al.*, 2012; Shugay *et al.*, 2014). These barcodes tag each individual molecule with a specific barcode, so each amplicon can be traced to a consensus sequence, providing a means to distinguish real somatic mutations

Table 1   Pros and Cons for using DNA vs. RNA

	DNA	RNA
Number of molecules	(+) One DNA molecule encoding for one type of functional Ig in each cell, and thus the sequencing results reflect cell type distribution to some extent.	(−) Many RNA molecules in each cell, the number is highly variable between cells, thus the sequencing results reflect mRNA distribution which can be skewed due to Ig secreting plasmablasts.
Material stability	(+) Much more stable, enables working with difficult samples, e.g., formalin preserved.	(−) Less stable. Even if stored in − 80°C, RNA can degrade after a few months.
Number of steps required for library preparation	(+) Ready for PCR.	(−) Requires an additional reverse transcription step.
Isotype information	(−) Lost, since the intron between the constant and variable regions is too large for PCR.	(+) One can use primers corresponding to the 5′ edge of the constant region to obtain isotype information with minimal addition to the length of each molecule in the library.
Published protocols	(−) Few. None includes UMI, which makes PCR and sequencing error as well as bias correction impossible.	(+) Many, including UMI, so error and bias correction are possible.

from PCR amplification and sequencing-dependent errors. UMIs also help in correcting PCR amplification biases stemming from varied primer efficiencies. Although theoretically possible, no protocol for using UMIs on Ig DNAs has been published yet.

The Choice of Primer Sets

The high diversity of the Ig encoding region poses a challenge for reverse transcription or amplification during library preparation. In humans, for example, there are 9 possible constant genes for IgM, IgD, IgE, IgA1, IgA2, IgG1, IgG2, IgG3, IgG4, or 6 possible J genes, and there is no single primer that can capture all constant or J genes at the 3′ end. At the 5′ end, the challenge is even greater, with more than 200 possibilities for functional V alleles in the heavy chain (Lefranc *et al.*, 2009). To cover these sequences there is a need for a minimum of ∼40 primers. Extension of the transcript into the 5′UTR reduces the minimum number of necessary primers to 12, still a considerable and bias-prone number for a single PCR reaction. Moreover, due to SHM, even if the Ig sequence originated from a sequence that was included in the primer set, the mutated Ig sequence, including its 5′ UTR, can be very different and hence will not be amplified. A method to circumvent the need for so many primers is to perform the reverse transcription with primers only for the J or constant regions, using a 5′ RACE polymerase. This polymerase incorporates a sequence of several free cytosine (C) nucleotides at the end of each newly synthesized molecule, followed by template switching with any primer that contains a sequence of 3 RNA guanine (rG) nucleotides at the 3′ end. As a result, 5′ RACE eliminates the potential amplification bias and can be used to amplify highly mutated sequences and sequences of species with limited information about the diversity of their Ig loci. To date, 5′ RACE can only be used when the starting material is RNA.

Single Cell Sequencing

Each Ig molecule comprises two identical light chains and two identical heavy chains, linked by disulfide bonds. Most current experimental protocols sequence only the heavy chain gene, since this gene contains most of the diversity of the Ig molecule. In some cases, heavy and light chains are sequenced separately from the same sample, but high-throughput pairing of light and heavy chain sequences originating from a single lymphocyte remains a highly non-trivial challenge (Busse *et al.*, 2014; Dekosky *et al.*, 2013, 2014; Tan *et al.*, 2014; Zhu *et al.*, 2013). Currently, the best experimental approach for retaining the pairing of heavy and light chains is performing single cell barcoding (Briggs *et al.*, 2017) or linkage PCR (Dekosky *et al.*, 2013; McDaniel *et al.*, 2016) within emulsion droplets. Although promising, currently these methods are much more costly than bulk sequencing, typically resulting in a reduced number of sequenced cells per experiment. Another approach is to combinatorially identify and pair heavy and light chains based on their frequencies in the B cell repertoire (Zhu *et al.*, 2013), i.e. the most abundant VL with the most abundant VH, and so forth. However, this approach is noisy for large data sets and works well only for very abundant sequences.

Library Preparation

B cell isolation or enrichment can be performed, for example, by FACS sorting or ficoll gradient, respectively. Cells are lysed and RNA or DNA is extracted. Then the variable region of the Ig gene is amplified by reverse transcription (in the case of RNA) and PCR, using primers as described above (see "The choice of primer sets" subsection). The primers include overhanging extensions with UMIs, sample barcodes, annealing sites for the sequencing primers, and adaptors for the sequencer. Usually, two PCR steps are required, due to the length of overhang extensions on the primers, and sometimes the need to perform nested PCR to increase specificity. Another possibility to add the above extensions is by ligation. After verification of product size and quality, e.g. by tapestation or bioanalyzer analysis, the concentration is adjusted to fit the sequencer requirements, and the library is ready for sequencing (**Fig. 2**).

A specific problem for Ig repertoire HTS is the highly similar nature of the Ig sequences, which leads to failure of the Illumina sequencer to distinguish neighboring clusters. This can be alleviated by spiking the desired library with a control phiX virus library, and/or replacing each primer with a mixture of 4 primers, where the body of the primer is identical except for 0–3 extra random nucleotides that are added at the beginning of it.

Ig Repertoire Sequence Data Pre-Processing, Annotation and Clonal Assignment

The theoretical diversity of Igs is immense. Igs are combined of a light and a heavy chain, of varying lengths. For example, a typical heavy chain is between 330 and 390 base pairs, so naively, one could think that there are 4^{330}-4^{390} possible combinations for this sequence. Structural constraints reduce this number dramatically, but still the estimated diversity of the heavy chain only is ∼10^9 (Murphy and Weaver, 2016).

Pre-Processing

Many non-Ig HTS applications assume that the sequenced amplicons can be mapped and compared to a reference genome (Michaeli *et al.*, 2012). Due to the somatic rearrangement discussed above, a B cell clone creates its unique Ig sequence, with no

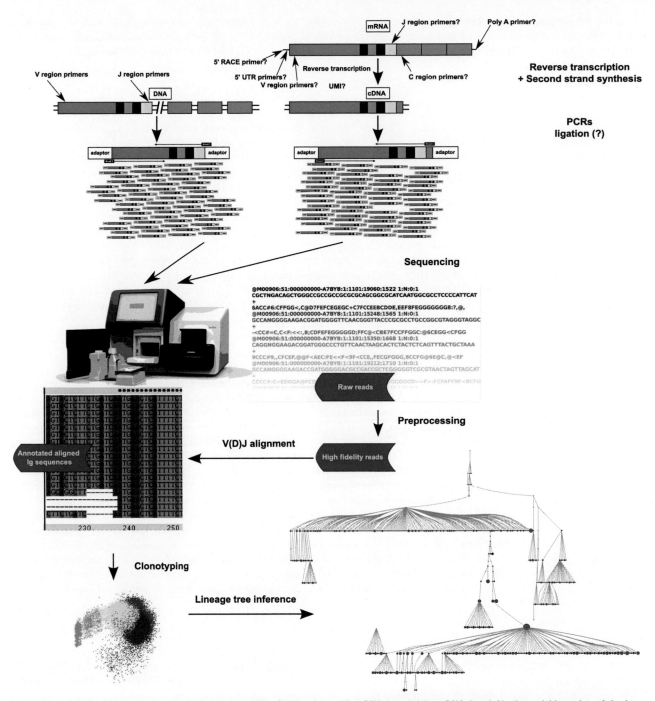

Fig. 2 The process of Ig repertoire sequencing and analysis. Starting from either RNA (top right) or DNA (top left), the variable region of the Ig gene is enriched and amplified, and adaptors are added for HTS. The raw sequencing reads undergo preprocessing to filter out irrelevant and low-quality sequences, and the high-fidelity reads are aligned to the known reference genes, annotated and clonotyped, followed by lineage tree construction.

complete reference in the genome, and further diversifies it through SHM. Thus, tailored tools for preprocessing Ig repertoires sequencing data had to be developed by the computational immunology community (see, e.g., Georgiou *et al.*, 2014; Vander Heiden *et al.*, 2014; Michaeli *et al.*, 2012, 2013; Moorhouse *et al.*, 2014; Schaller *et al.*, 2015; Wardemann and Busse, 2017; Yaari and Kleinstein, 2015). First, Ig repertoire sequencing data require a large amount of pre-processing work to clean out the data and prepare them for more specific downstream analysis related to clonal identification and mutation analysis. Specific computational toolkits tailored for Ig repertoires sequencing purposes perform pre-processing of raw sequencing reads and produce error-corrected, sorted and annotated sequence sets, along with a wealth of quality control metrics (Bolotin *et al.*, 2015; Vander Heiden *et al.*, 2014; Michaeli *et al.*, 2012). These tools include, for example: a. Removal of low-quality reads. b. Removal of reads where the primer could not be identified or had a poor alignment score. c. Identification of sets of sequences with identical UMIs. These are collapsed into one consensus sequence per set. d. Assembly of the two consensus paired-end reads into a complete Ig sequence. e. Removal of sequences that do not appear in a single sample with at least two independent molecular IDs (see also Stern *et al.*, 2014).

Annotation

Ig repertoire analysis benefits from adequate annotations. For example, each Ig sequence is derived from a combination of V, D, and J germline genes, which could be inferred from the Ig sequence using different tools (Bolotin *et al.*, 2015; Brochet *et al.*, 2008; Gaëta *et al.*, 2007; Munshaw and Kepler, 2010; Ralph and Matsen, 2016; Ye *et al.*, 2013). These VDJ assignment tools return for each Ig an aligned germline sequence, constructed from the inferred V, D, and J genes, taking into account the possibilities for trimming the edges of these genes and inserting alien nucleotides in between them (N and P additions) ((Murphy and Weaver, 2016), **Fig. 1**). Germline sequence annotations reduce the dimensionality of the sequence space, and enable a more comprehensible and interpretable analysis. Furthermore, and most importantly, somatic mutations can be identified from the germline sequence, and be used for phylogenetic tree construction, selection quantification, and building somatic mutation models. To identify the V and J genes, methods based on sequence similarity are commonly used (Lefranc *et al.*, 2009; Ye *et al.*, 2013). The inference of the D gene and its flanking sequences is a more difficult task, due to the stochastic processes underlying VDJ recombination and the relatively short lengths of the D genes. Several statistical frameworks to cope with this stochastic process exist, including HMM (Gaëta *et al.*, 2007; Munshaw and Kepler, 2010) and an E-M iterative process that first estimates the probabilities for each event from the whole dataset, and then aligns the sequence to the inferred germline (Marcou *et al.*, 2017).

Germline Repertoire Inference

Another important point in VDJ assignment is the ability to a priori set the V, D, and J, germline repertoires. Studying polymorphisms in the Ig loci in human populations is an active field. Recent studies suggest that there are complex events of deletion and duplication in these loci, yielding an undiscovered wealth of unknown alleles (Kidd *et al.*, 2012; Luo *et al.*, 2016; Wang *et al.*, 2008; Watson and Breden, 2012; Watson *et al.*, 2013). IMGT has a central repository for V, D, and J allele sequences; to add new sequences to this repertoire there is a need for direct DNA sequencing of non B cells. In recent years, several independent methods to detect novel polymorphisms from Ig repertoires have been developed (Corcoran *et al.*, 2016; Gadala-Maria *et al.*, 2015), and there is an ongoing effort to define a set of rules that will allow recognition by the community of these inferred alleles (Brenden *et al.*, 2017). Comparing Ig sequences to all known alleles can result in erroneous gene assignments, because alleles might be missing in current germline gene datasets, and since highly mutated sequences may be mis-assigned. To reduce this risk, a genotype and/or a haplotype step that generates a personalized pool of possible alleles can be performed. In a genotype step, all sampled sequences from an individual are combined, and a decision is made about the set of alleles that the individual actually has. After creating such a genotype, VDJ assignment is repeated with the restricted set of available germline genes. Such a step can reduce the number of mis-assigned sequences dramatically (Gadala-Maria *et al.*, 2015). To haplotype a repertoire, one can use J (or D) gene heterozygosity and analyze the frequency of recombined J (or D) alleles with V alleles (Kidd *et al.*, 2012). After inferring a haplotype for an individual, VDJ assignment should be repeated to ensure that alleles are co-assigned to a sequence only if they are present on the same chromosome. After the VDJ assignment step, each sequence is annotated with its inferred germline genes, trimmed and added sequences, CDR3 length, charge, etc.

Clonal Assignment

One of the leading themes behind our understanding of adaptive immunity is clonal selection, and the idea that T and B cell repertoires are comprised of groups of cells of varying sizes, each of which derived from a common ancestor (Hodgkin *et al.*, 2007). While in T cells it is straightforward to identify these clonally-related sequences, as they are identical in all the cells belonging to the same clone, in B cells, due to SHM, inferring these groups from Ig repertoire sequencing data is a statistically non-trivial task (Chen *et al.*, 2010; Hershberg and Luning Prak, 2015). Since the number of sequences in a typical Ig repertoire sequencing experiment is in the order of millions, clustering these sequences requires tailored approaches from machine learning and statistics. Clones are usually inferred based on their heavy chain sequences, as light chain diversity alone is too low for this task (Saada *et al.*, 2007). When heavy and light chain sequences are paired (Dekosky *et al.*, 2013, 2014), the combined information can be used for clonal assignment.

Clonal assignment is normally performed in two steps. First, sequences are grouped based on their V-J gene calls and junction length annotations. Second, a sequence-based distance measure is used to cluster the sequences inside each group. Due to the nature of SHM, the common distance metric is based on nucleotide similarity, and in most cases is applied to the CDR3 part of the sequence. In principle, it is possible to cluster the sequences based on sequence similarity throughout the whole variable region, and also allow for insertions and deletions that can occur during SHM (Kepler *et al.*, 2014a). However, in most studies these are not considered. It is also possible to use a distance metric that is based on a mutability model that takes into account hot and cold spots of SHM (Elhanati *et al.*, 2015; Yaari *et al.*, 2013).

After constructing a distance matrix, clustering can be performed using approaches such as imposing a threshold on a hierarchical tree using either single, complete or average linkage. To determine the adequate threshold, a "distance-to-nearest" plot (Glanville *et al.*, 2011) can be utilized. Since biological clones are restricted to a single individual, inferred clones that span multiple individuals can be used to estimate specificity for clonal assignment. With these assumptions, it was shown that "hamming distance" using single linkage clustering performs better in terms of specificity and sensitivity (Gupta *et al.*, 2015; Yermanos *et al.*, 2017). Alternatives to the distance matrix clustering include cutting lineage trees (see below) to create subtrees that can be interpreted as clones (Liberman *et al.*, 2013), and maximum likelihood approaches (Kepler *et al.*, 2014a; Wu *et al.*, 2011). Initial V(D)J assignments can be refined after clonal assignments as all sequences stemmed from the same cell have the same germline (Kepler, 2013; Stern *et al.*, 2014). Although V(D)J assignments are not required for clonal assignment (Giraud *et al.*, 2014), utilizing these annotations makes it possible to cluster much larger data sets with standard clustering methods.

Grouping similar sequences can also serve other purposes. Similar receptor sequences could indicate similarity in function. For example, the similarity of TCRs was recently shown to help in predicting epitopes (Dash *et al.*, 2017; Glanville *et al.*, 2017). Functionally similar Ig sequences across individuals indicate "public" immune responses (Glanville *et al.*, 2011). These sequences can also indicate convergent evolution where multiple clones produce independently similar Ig sequences at the amino acid level, but not necessarily using the same genes (Jackson *et al.*, 2014; Parameswaran *et al.*, 2013). For these applications, sequence similarity at the amino acid level can be used for clustering.

Repertoire Analyses

Common Repertoire Measures and Coping With Inherent Biases

Ig repertoires are characterized by many features that allow researchers to reduce their high dimensionality and describe them in biologically meaningful terms. Examples include: V-D-J gene usage distribution, isotype usage distribution, various measures of repertoire diversity and similarity, CDR3 length distribution, amino acid composition, and mutation patterns. The interpretation of these distributions and measures depends on various factors, such as the sequenced material (mRNA or DNA), whether a UMI approach was taken, and whether bulk or single cell sequencing was done (see **Table 1**). For example, amplification biases can dramatically influence the results if each sequence is counted independently and UMIs are not utilized. Even if UMIs are utilized, the measured distribution will describe the mRNA distribution, which can be very misleading as some cells (plasmablasts) have orders of magnitude higher numbers of mRNA molecules than B cells. A common way to confront these biases is to construct the distributions from the set of unique sequences. However, this approach still cannot overcome the experimental artifacts, such as transcription and amplification errors. Yet, a repetitive important task in Ig repertoire analysis is to calculate the above feature distributions and conclude from them something about repertoire generation and dynamics. For example, using the D-J genes distribution to infer the inherent constraints and preferences of the V-D-J recombination machinery. To do this, researchers commonly pick one sequence from every clone and compute the distributions from these representatives. On the other hand, taking only one representative from every clone leaves most of the data unanalyzed. To exploit more information from the data, it is necessary to also integrate the construction of lineage trees into the analysis, as discussed below.

In a healthy repertoire, large clones are rare, and we usually sample only one or a few cells per clone (Hershberg and Luning Prak, 2015). Even during an acute infection, only a few responding clones grow and may constitute at most a few percent of the repertoire. In contrast, if the clonal size distribution contains one or more highly frequent clones, we may suspect the presence of an autoimmune disease such as Sjogren's syndrome (Hershberg *et al.*, 2014) systemic lupus erythematosus following rituximab therapy (Sfikakis *et al.*, 2009), or a B cell malignancy such as chronic lymphocytic leukemia (Logan *et al.*, 2011). Positive selection for independent B cell clones often results in shared amino acid sequence motifs that bind to particular antigens. Such public CDR3 motifs have been documented in many chronic disorders ranging from viral (Jackson *et al.*, 2014; Wrammert *et al.*, 2011) or bacterial (Scott *et al.*, 1989; Silverman and Lucas, 1991) infections and autoimmunity to cancer.

Repertoire Diversity

Repertoires may differ from one another in many ways. Quantifying these differences is a research field of its own, and includes many possibilities to assess diversity and similarity of repertoires. One potent approach to measure diversity is the Hill diversity index (Hill, 1973). This index can be applied to any feature distribution (e.g. V gene usage), and usually is applied to characterize the clone size distribution. It is defined as

$$qD = \left(\sum_{i=1}^{i=n} p_i^q \right)^{(1/(1-q))}$$

where the sum is over all clones and p_i is the fraction of clone i out of the entire population. This diversity index converges to all of the known diversity measurements for different values of the parameter q. For example, for q = 0, D is the number of species (a.k.a. richness); for q = ∞, D is the reciprocal of the fraction of the largest clone; for q = $-\infty$, D is the reciprocal of the fraction of the smallest clone; for q = 1, D is the exponent of Shannon's entropy, and higher values of q (2, 3 etc.) give a higher weight to the larger clones. Studying the shape of this curve is much more informative than measuring a single diversity score along with an "evenness" score (Hill, 1973). When comparing diversity profiles between samples, it is advised to subsample (with replacement) the larger repertoire to the size of the smaller one and only then to compare the two (Gupta *et al.*, 2015).

Lineage Tree Analysis & Lessons Learned From Tree Structure

During affinity maturation, Ig loci accumulate mutations via SHM. An important step in Ig repertoire analysis is the inference of each clone's pedigree. These pedigrees are referred to as lineage trees, and methods to infer them are usually borrowed from the field of phylogeny, despite different assumptions underlying each case. For example, in B cell lineages, the trees are rooted by the rearranged, pre-mutation sequence. This sequence is not necessarily the most recent common ancestor of the tree, but is used to construct the tree.

Lineage Tree Construction

The phylogenetic methods most commonly used for analysis of Ig repertoire sequencing data are maximum likelihood, maximum parsimony, neighbor joining (Barak *et al.*, 2008) and Bayesian inference (Yermanos *et al.*, 2017). These methods utilize a distance metric to construct the trees. The most common distance metrics are the Hamming, Kimura substitution model (Kimura, 1968), and Levenshtein distances. Since SHM preferences are based on neighboring nucleotides, a more adequate distance metric should take into account these neighboring nucleotides. A first step to integrate neighboring bases effects in constructing a B cell lineage tree was taken recently (Hoehn *et al.*, 2017). It was evaluated using trees constructed from HIV patients (Wu *et al.*, 2015), and showed better performance. A method based on natural SHM metrics, such as the S5F (Yaari *et al.*, 2013) or 7mers (Elhanati *et al.*, 2015) is still lacking. Given a distance metric, the differences in performance between the methods are mild (Yermanos *et al.*, 2017). In order to use the outputs of lineage tree construction algorithms, there is a need in parsing the resulting trees and fit them to Ig repertoires (Barak *et al.*, 2008; Vander Heiden *et al.*, 2015), as most phylogenetic algorithms assume that all observed sequences are at the leaves of the tree. Also, in B cell lineages the germline sequence should be upstream of the root of the tree.

Lessons Learned From Lineage Trees

Merely constructing Ig lineage trees sheds light on fundamental processes underlying the immune system in health and disease, and may aid in vaccine and drug design. We present here a few examples, which are by no means an exhaustive set. Tabibian-Keissar *et al.* (Tabibian-Keissar *et al.*, 2008) used lineage trees to give the first direct evidence of trafficking between human colon and adjacent lymph nodes, Bergqvist *et al.* (Bergqvist *et al.*, 2013) used trees to show the spreading and synchronization of IgA responses throughout the murine gut, and Pabst *et al.* (Pabst *et al.*, 2015) used trees of sequences from serial mouse gut biopsy samples to show the development of the murine gut immune system. Hazanov *et al.* (Hazanov *et al.*, 2015) have used lineage tree structures to elucidate the relationships between switched memory, IgM memory, naïve and IgD-CD27 – (double-negative, "DN") B cells in humans. Meng *et al.* (Meng *et al.*, 2017), with the aim of enhancing our understanding of the distribution of B cell clones in the human body, sequenced and analyzed over 38 million B cells from eight anatomic compartments. They found two major networks of large clones, one in the blood, bone marrow, spleen and lung, and another in the GI tract, with little overlap between the networks. The GI tract contained large clones with the highest levels of somatic hypermutation.

SHM and class switching

Horns *et al.* (Horns *et al.*, 2016) developed a way to reconstruct the B cell proliferation process and thereby trace the lineage of individual B cells, using somatic hypermutations as a molecular clock. They revealed that closely related B cells often switch to the same class, but lose coherence as somatic mutations accumulate. They also found that naïve classes (IgM or IgD) accumulated more mutations before undergoing class switch recombination to activated classes (IgG, IgA, or IgE), in comparison with class switch recombination between activated classes. These findings lay a foundation for developing techniques to direct Ig class switching. Kepler *et al.* (Kepler *et al.*, 2014b) analyzed paired heavy-light lineage trees of an influenza-infected individual, and showed how affinity maturation occurs stepwise with time within clones, increasing 1000-fold from the unmutated ancestor to the highest affinity observed Igs.

Aging

Jiang *et al.* (Jiang *et al.*, 2013) and the Dunn-Walters lab (Ademokun *et al.*, 2011; Tabibian-Keissar *et al.*, 2016; Wu *et al.*, 2012) determined the lineage structure of the Ig repertoire before and after influenza vaccination, and observed age related changes. Elderly subjects had fewer lineages than other age groups, both before and after vaccination, and influenza vaccination resulted in expansion of far fewer B cell lineages and a reduced B cell clonal diversity in the elderly compared to the younger age groups. The elderly had a higher percentage of somatic mutations, likely since their clonal expansions draw upon a pool of B cells having more somatic mutations to begin with.

Infection and response dynamics

Highly effective, broadly neutralizing Igs towards HIV have unusual characteristics (Burton *et al.*, 1994; Calarese *et al.*, 2003; Haynes *et al.*, 2005; Huang *et al.*, 2004; Klein *et al.*, 2013; Pancera *et al.*, 2010; Pejchal *et al.*, 2010; Scheid *et al.*, 2011; Walker *et al.*,

2009, 2011; Wu *et al.*, 2010a, 2011; Zhou *et al.*, 2007) and usually take years of chronic infection to develop (Gray *et al.*, 2011; Wu *et al.*, 2006). The best studied broadly neutralizing Ig lineages are VRC01, VRC26, and CH103 (Doria-Rose *et al.*, 2014; Liao *et al.*, 2013; Wu *et al.*, 2015). Wu *et al.* (Wu *et al.*, 2015) investigated VRC01, which targets gp120, the site of CD4 engagement on HIV-1. They identified and described Ig lineage characteristics over the course of 15 years in a single patient. Most Igs differ from the germline by ~5%, but the VRC01 lineage is over 30% different. Over the course of the study, the authors showed using lineage trees that VRC01 Igs continuously evolved, with a rate of ~2 substitutions per 100 nucleotides per year, comparable to that of HIV-1 evolution. Liao *et al* (Liao *et al.*, 2013) followed the CH103 Ig lineage, also targeting gp120, in a single donor over 3 years. They showed the evolution of the lineage up to a mature broadly neutralizing Ig that could neutralize ~55% of HIV-1 isolates. This lineage is less mutated than other CD4-binding Igs, and may be first detectable as early as 14 weeks after infection. Therefore, the CD4-binding site may be a vaccine target. Doria-Rose *et al.* (Doria-Rose *et al.*, 2014) delineated longitudinal interactions between the development of 12 related broadly neutralizing Igs and HIV-1 within a single donor over 4 years. They found that the unmutated ancestor of the lineage emerged between weeks 30–38 post-infection, bound and neutralized the virus weakly and did not bind any other HIV strains. Later variants demonstrated increasing affinity towards several strains of the virus. Subsequent affinity maturation focused within CDR H3 and allowed for progressively greater binding and neutralization. This work defined the molecular requirements and genetic pathways leading to virus neutralization, providing a template for their vaccine elicitation. Sok *et al.* (Sok *et al.*, 2013) developed a novel method called ImmuniTree, which is an alternative approach to conventional phylogenetic analyses and is designed specifically to model Ig somatic hypermutations. They used this method to study the PGT121-134 broadly neutralizing Ig lineage, which is also characterized by high levels of somatic mutations and is among the most potent lineages described to date. They discovered Igs in the lineage with half the mutation level of PGT121-134, capable of neutralizing 40%–80% of PGT121-134 sensitive viruses. Such Igs will likely be easier to induce through vaccination.

Autoimmune diseases

Autoimmune responses also generate large clones, and various insights can be gleaned from analysis of the corresponding lineage trees. Stern *et al.* (Stern *et al.*, 2014) and Palanichamy *et al.* (Palanichamy *et al.*, 2014) analyzed B cell repertoires from the brain and lymph nodes of multiple sclerosis patients, and showed that B cells populating the multiple sclerosis brain mature in the draining cervical lymph nodes, and that the traffic between the periphery and the brain is an ongoing process, and not a one-time event as was previously thought. This provides insight into the trafficking of B cells in MS. Therefore, modulation of specific B cell subsets could provide therapeutic benefit in MS patients. Snir *et al.* (Di Niro *et al.*, 2016; Snir *et al.*, 2015) compared B cell clones in the gut and blood of celiac patients, and found that the generation of plasma cells producing autoantibodies likely occurs outside the gut. Such knowledge helps to shed light on the molecular mechanisms causing the disease, and may facilitate the development of immune-based therapies.

Mutation Analyses Relying on Lineage Tree Structure

Lineage tree graphical properties can be used by themselves to estimate the relative strength of selection, mutation and initial competitive advantage of clones (Shahaf *et al.*, 2008), and thus to characterize a specific immune response. However, the most useful feature of the trees is the mutations they are constructed from. Analyzing mutation patterns in Ig repertoires can shed light on the underlying biology of B cell adaptation, and can be used to estimate affinity dependent selection. For any analysis of mutation patterns, reliable lineage trees should be used to identify independent events of mutations. If one ignores the lineage structure of a B cell clone, mutations can be counted multiple times as an artificially larger number of independent events. This may lead to severe biases in mutation analyses. Integrating lineage tree structure in the analysis should thus be used to count each mutation event only once. In addition, tree structure enables more correct identification of each mutation (as each sequence is compared to its most recent identifiable ancestor, and not to the root), and the identification of reversion mutations.

The patterns of mutations carry the fingerprints of the SHM machinery – from AID targeting sites (Chandra *et al.*, 2015) to the various pathways to deamination site resolution (Zanotti and Gearhart, 2016). In most of our studies, whether the focus was aging, chronic or autoimmune diseases or B cell malignancies, we have found no differences from controls in mutation statistics such as targeting motifs, transition: transversion ratios, the fraction of mutations from and to each nucleotide, etc. A noteworthy exception was seen in studying ectopic GC from myasthenia gravis patient thymi. Using tree-based mutation analysis, Zuckerman at al. have predicted changes in the SHM mechanism (Zuckerman *et al.*, 2010b), a prediction which was verified by gene expression analysis.

Selection analysis has many applications, including identification of potentially high affinity sequences, understanding how different genetic manipulations impact affinity maturation, and investigating whether disease processes are antigen-driven. It is common to look for an increased frequency of non-synonymous mutations as evidence of antigen-driven positive selection, and a decreased frequency of non-synonymous mutations as evidence of negative selection (Hershberg *et al.*, 2008; Uduman *et al.*, 2011; Yaari *et al.*, 2012). Expectations are calculated either using a uniform targeting model in which all mutations occur with the same rate (Lossos *et al.*, 2000; Shlomchik *et al.*, 1987), or using a mutability model such as the S5F (Yaari *et al.*, 2013).

When quantifying selection in Ig sequences, mutations can be calculated in several time scales. An Ig sequence can be compared to the germline sequence, to the most recent common ancestor, or to the sequence that is one branch upstream in the tree. These different comparisons yield different selection estimations. Lineage trees were used to show how selection is stronger along the tree vs. close to its leaves (Uduman *et al.*, 2014). It was also shown that memory cells from past infections experienced stronger selection compared to current clonal expansion (Yaari *et al.*, 2015).

Summary and Challenges for the Future

The analysis of TCR and BCR clonal repertoires differs from other bioinformatic practices due to the lack of reference genes, and with the amounts of data generated by HTS, is more complex. On the other hand, as shown by the examples shown above, it has an enormous potential for generating new insights into the structure and activity of the immune system at every stage of life and when faced with any challenge. We eagerly anticipate the day in which knowing a person's BCR and TCR "haplotypes" and having a good sample of their repertoire would give us as much knowledge of their immune status as the rest of their genome can tell about their health status and potential.

Before we reach that day, however, there are several challenges to overcome. First, we do not have a full understanding of the structure and contents of human BCR and TCR genomic loci; better knowledge of these genomic regions, in particular the identification of larger numbers of alleles, will accumulate as the repertoires from more and varied population groups are sequenced. Second, for sequencing of such magnitude to be possible, we need ways to generate longer, more reliable reads at far lower cost. Third, in order to understand the function of the BCRs and TCRs sequenced, we need a reliable, easy and low-cost protocol for sequencing paired heavy and light chains or beta and alpha chains. While emulsion-based methods are much less work-intensive and have higher throughputs than well plate-based methods, paired sequencing is still basically a single-cell method of analysis and hence its current output is not as high as single-chain sequencing. Fourth, in order to compare and evaluate studies and enable meta-analysis, we must standardize protocols and data format between labs. Research in the field will be facilitated by data and protocol sharing and adherence to meticulous standards (see Relevant Website Section). Finally, we should devote considerable efforts to the continued invention of new and creative ways to look at the data collected. We need ways to obtain a deeper view of the true repertoire, as even the new methods only sample a small portion of it.

Acknowledgement

This research was supported by the Israel Science Foundation (grant No. 832/16).

See also: Computational Immunogenetics. Epitope Predictions. Extraction of Immune Epitope Information. Immunoinformatics Databases. Natural Language Processing Approaches in Bioinformatics. Vaccine Target Discovery

References

Ademokun, A., Wu, Y.-C., Martin, V., *et al.*, 2011. Vaccination-induced changes in human B-cell repertoire and pneumococcal IgM and IgA antibody at different ages. Aging Cell 10, 922–930.

Arnaout, R., Lee, W., Cahill, P., *et al.*, 2011. High-resolution description of antibody heavy-chain repertoires in humans. PLOS ONE 6, e22365.

Barak, M., Zuckerman, N.S., Edelman, H., Unger, R., Mehr, R., 2008. IgTree: Creating Immunoglobulin variable region gene lineage trees. J. Immunol. Methods 338, 67–74.

Benichou, J., Ben-Hamo, R., Louzoun, Y., Efroni, S., 2012. Rep-Seq: Uncovering the immunological repertoire through next-generation sequencing. Immunology 135, 183–191.

Bergqvist, P., Stensson, A., Hazanov, L., *et al.*, 2013. Re-utili zation of germinal centers in multiple Peyer's patches results in highly synchronized, oligoclonal, and affinity-matured gut IgA responses. Mucosal Immunol. 6, 122–135.

Bolotin, D.A., Poslavsky, S., Mitrophanov, I., *et al.*, 2015. MiXCR: Software for comprehensive adaptive immunity profiling. Nat. Methods 12, 380–381.

Boyd, S.D., Gaëta, B.A., Jackson, K.J., *et al.*, 2010. Individual variation in the germline Ig gene repertoire inferred from variable region gene rearrangements. J. Immunol. 184, 6986–6992.

Boyd, S.D.S., Marshall, E.L., Merker, J.D., *et al.*, 2009. Measurement and clinical monitoring of human lymphocyte clonality by massively parallel VDJ pyrosequencing. Sci. Transl. Med. 1, 12ra23.

Breden, F., Luning Prak, E.T., Peters, B., *et al.*, 2017. Reproducibility and reuse of adaptive immune receptor repertoire data. Front. Immunol. 8, Article 1418.

Briggs, A.W., Goldfless, S.J., Timberlake, S., *et al.*, 2017. Tumor-infiltrating immune repertoires captured by single-cell barcoding in emulsion. BioRxiv.

Brochet, X., Lefranc, M.-P., Giudicelli, V., 2008. IMGT/V-QUEST: The highly customized and integrated system for IG and TR standardized V-J and V-D-J sequence analysis. Nucleic Acids Res. 36, W503–W508.

Burton, D.R., Pyati, J., Koduri, R., *et al.*, 1994. Efficient neutralization of primary isolates of HIV-1 by a recombinant human monoclonal antibody. Science 266, 1024–1027.

Busse, C.E., Czogiel, I., Braun, P., Arndt, P.F., Wardemann, H., 2014. Single-cell based high-throughput sequencing of full-length immunoglobulin heavy and light chain genes. Eur. J. Immunol. 44, 597–603.

Calarese, D.A., Scanlan, C.N., Zwick, M.B., *et al.*, 2003. Antibody domain exchange is an immunological solution to carbohydrate cluster recognition. Science 300, 2065–2071.

Cameron, E.M., Spencer, S., Lazarini, J., *et al.*, 2009. Potential of a unique antibody gene signature to predict conversion to clinically definite multiple sclerosis. J. Neuroimmunol. 213, 123–130.

Campbell, P.J., Pleasance, E.D., Stephens, P.J., *et al.*, 2008. Subclonal phylogenetic structures in cancer revealed by ultra-deep sequencing. Proc. Natl. Acad. Sci. USA. 105, 13081–13086.

Chandra, V., Bortnick, A., Murre, C., 2015. AID targeting: Old mysteries and new challenges. Trends Immunol. 36, 527–535.

Chen, Z., Collins, A.M., Wang, Y., Gaëta, B.A., 2010. Clustering-based identification of clonally-related immunoglobulin gene sequence sets. Immunome Res. 6, S4.

Corcoran, M.M., Phad, G.E., Vázquez Bernat, N., *et al.*, 2016. Production of individualized V gene databases reveals high levels of immunoglobulin genetic diversity. Nat. Commun. 7, 13642.

Dash, P., Fiore-Gartland, A.J., Hertz, T., *et al.*, 2017. Quantifiable predictive features define epitope-specific T cell receptor repertoires. Nature 547, 89–93.

Dekosky, B.J., Ippolito, G.C., Deschner, R.P., *et al.*, 2013. High-throughput sequencing of the paired human immunoglobulin heavy and light chain repertoire. Nat. Biotechnol. 31, 166–169.

DeKosky, B.J., Kojima, T., Rodin, A., *et al.*, 2014. In-depth determination and analysis of the human paired heavy- and light-chain antibody repertoire. Nat. Med. 21, 86–91.

Di Niro, R., Snir, O., Kaukinen, K., *et al.*, 2016. Responsive population dynamics and wide seeding into the duodenal lamina propria of transglutaminase-2-specific plasma cells in celiac disease. Mucosal Immunol. 9, 254–264.

Doria-Rose, N.A., Schramm, C.A., Gorman, J., *et al.*, 2014a. Developmental pathway for potent V1V2-directed HIV-neutralizing antibodies. Nature 509, 55–62.

Dunn-Walters, D.K., Ademokun, A.A., 2010. B cell repertoire and ageing. Curr. Opin. Immunol. 22, 514–520.

Dunn-Walters, D.K., Banerjee, M., Mehr, R., 2003. Effects of age on antibody affinity maturation. Biochem. Soc. Trans. 31, 447–448.

Elhanati, Y., Sethna, Z., Marcou, Q., *et al.*, 2015. Inferring processes underlying B-cell repertoire diversity. Philos. Trans. R. Soc. B Biol. Sci 370, 20140243.

Fridman, W.H., Pagès, F., Sautès-Fridman, C., Galon, J., 2012. The immune contexture in human tumours: Impact on clinical outcome. Nat. Rev. Cancer 12, 298–306.

Gadala-Maria, D., Yaari, G., Uduman, M., Kleinstein, S.H., 2015. Automated analysis of high-throughput B-cell sequencing data reveals a high frequency of novel immunoglobulin V gene segment alleles. Proc. Natl. Acad. Sci. USA. 112, E862–E870.

Gaëta, B.A., Malming, H.R., Jackson, K.J.L., *et al.*, 2007. iHMMune-align: Hidden Markov model-based alignment and identification of germline genes in rearranged immunoglobulin gene sequences. Bioinformatics 23, 1580–1587.

Galon, J., Costes, A., Sanchez-Cabo, F., *et al.*, 2006. Type, density, and location of immune cells within human colorectal tumors predict clinical outcome. Science 313, 1960–1964.

Georgiou, G., Ippolito, G.C., Beausang, J., *et al.*, 2014. The promise and challenge of high-throughput sequencing of the antibody repertoire. Nat. Biotechnol. 32, 158–168.

Giraud, M., Salson, M., Duez, M., *et al.*, 2014. Fast multiclonal clusterization of V(D)J recombinations from high-throughput sequencing. BMC Genomics 15, 409.

Glanville, J., Kuo, T.C., Büdingen, H.-C., *et al.*, 2011. Naive antibody gene-segment frequencies are heritable and unaltered by chronic lymphocyte ablation. Proc. Natl. Acad. Sci. USA. 108, 20066–20071.

Glanville, J., Huang, H., Nau, A., *et al.*, 2017. Identifying specificity groups in the T cell receptor repertoire. Nature 547, 94–98.

Goodwin, S., McPherson, J.D., McCombie, W.R., 2016. Coming of age: Ten years of next-generation sequencing technologies. Nat. Rev. Genet. 17, 333–351.

Gray, E.S., Madiga, M.C., Hermanus, T., *et al.*, 2011. The neutralization breadth of HIV-1 develops incrementally over four years and is associated with CD4+ T cell decline and high viral load during acute infection. J. Virol. 85, 4828–4840.

Gupta, N.T., Vander Heiden, J.A., Uduman, M., *et al.*, 2015. Change-O: A toolkit for analyzing large-scale B cell immunoglobulin repertoire sequencing data: Table 1. Bioinformatics 31, 3356–3358.

Haynes, B.F., Fleming, J., St Clair, E.W., *et al.*, 2005. Cardiolipin polyspecific autoreactivity in two broadly neutralizing HIV-1 antibodies. Science 308, 1906–1908.

Hazanov, L., Mehr, R., Wu, Y.-C.B., Dunn-Walters, D.K., 2015. Lineage tree analysis of high throughput immunoglobulin sequencing clarifies B cell maturation pathways In: 2015 International Workshop on Artificial Immune Systems (AIS). pp. 1–6. IEEE.

Hershberg, U., Luning Prak, E.T., 2015. The analysis of clonal expansions in normal and autoimmune B cell repertoires. Philos. Trans. R. Soc. Lond. B. Biol. Sci. 370.

Hershberg, U., Meng, W., Zhang, B., *et al.*, 2014. Persistence and selection of an expanded B cell clone in the setting of rituximab therapy for Sjogren's syndrome. Arthritis Res. Ther. 16, R51.

Hershberg, U., Uduman, M., Shlomchik, M.J., Kleinstein, S.H., 2008. Improved methods for detecting selection by mutation analysis of Ig V region sequences. Int Immunol 20, 683–694.

Hill, M.O., 1973. Diversity and evenness: A unifying notation and its consequences. Ecology 54, 427.

Hodgkin, P.D., Heath, W.R., Baxter, A.G., 2007. The clonal selection theory: 50 years since the revolution. Nat. Immunol. 8, 1019–1026.

Hoehn, K.B., Lunter, G., Pybus, O.G., 2017. A phylogenetic codon substitution model for antibody lineages. Genetics 206, 417–427.

Horns, F., Vollmers, C., Croote, D., *et al.*, 2016. Lineage tracing of human B cells reveals the in vivo landscape of human antibody class switching. Elife 5.

Huang, C., Venturi, M., Majeed, S., *et al.*, 2004. Structural basis of tyrosine sulfation and VH-gene usage in antibodies that recognize the HIV type 1 coreceptor-binding site on gp120. Proc. Natl. Acad. Sci. USA. 101, 2706–2711.

Jackson, K.J.L.L., Liu, Y., Roskin, K.M., *et al.*, 2014. Human responses to influenza vaccination show seroconversion signatures and convergent antibody rearrangements. Cell Host Microbe 16, 105–114.

Jiang, N., He, J., Weinstein, J.A., *et al.*, 2013. Lineage structure of the human antibody repertoire in response to influenza vaccination. Sci. Transl. Med. 5, 171ra19.

Jiang, N., Weinstein, J.A., Penland, L., *et al.*, 2011. Determinism and stochasticity during maturation of the zebrafish antibody repertoire. Proc. Natl. Acad. Sci. USA. 108, 5348–5353.

Jiang, Y., Nie, K., Redmond, D., *et al.*, 2015. VDJ-Seq: Deep sequencing analysis of rearranged immunoglobulin heavy chain gene to reveal clonal evolution patterns of B cell lymphoma. J. Vis. Exp. e53215.

Katoh, H., Komura, D., Konishi, H., *et al.*, 2017. Immunogenetic profiling for gastric cancers identifies sulfated glycosaminoglycans as major and functional B cell antigens in human malignancies. Cell Rep. 20, 1073–1087.

Kepler, T.B., 2013. Reconstructing a B-cell clonal lineage I. Statistical inference of unobserved ancestors. F000Research 2, 103.

Kepler, T.B., Liao, H.-X., Alam, S.M., *et al.*, 2014a. Immunoglobulin gene insertions and deletions in the affinity maturation of HIV-1 broadly reactive neutralizing antibodies. Cell Host Microbe 16, 304–313.

Kepler, T.B., Munshaw, S., Wiehe, K., *et al.*, 2014b. Reconstructing a B-cell clonal lineage. II. Mutation, selection, and affinity maturation. Front. Immunol. 5, 170.

Kidd, M.J., Chen, Z., Wang, Y., *et al.*, 2012. The inference of phased haplotypes for the immunoglobulin H chain V region gene loci by analysis of VDJ gene rearrangements. J. Immunol. 188, 1333–1340.

Kimura, M., 1968. Evolutionary rate at the molecular level. Nature 217, 624–626.

Klein, F., Diskin, R., Scheid, J.F., *et al.*, 2013. Somatic mutations of the immunoglobulin framework are generally required for broad and potent HIV-1 neutralization. Cell 153, 126–138.

Krause, J.C., Tsibane, T., Tumpey, T.M., *et al.*, 2011. Epitope-specific human influenza antibody repertoires diversify by B cell intraclonal sequence divergence and interclonal convergence. J. Immunol. 187, 3704–3711.

Laserson, U., Vigneault, F., Gadala-Maria, D., *et al.*, 2014. High-resolution antibody dynamics of vaccine-induced immune responses. Proc. Natl. Acad. Sci. 111, 4928–4933.

Lefranc, M.-P., Giudicelli, V., Ginestoux, C., *et al.*, 2009. IMGT, the international ImMunoGeneTics information system. Nucleic Acids Res. 37, D1006–D1012.

Lehmann-Horn, K., Kronsbein, H.C., Weber, M.S., 2013. Targeting B cells in the treatment of multiple sclerosis: Recent advances and remaining challenges. Ther. Adv. Neurol. Disord 6, 161–173.

Liao, H.-X., Lynch, R., Zhou, T., *et al.*, 2013. Co-evolution of a broadly neutralizing HIV-1 antibody and founder virus. Nature 496, 469–476.

Liberman, G., Benichou, J., Tsaban, L., Glanville, J., Louzoun, Y., 2013. Multi step selection in Ig H chains is initially focused on CDR3 and then on other CDR regions. Front. Immunol. 4, 274.

Logan, A.C., Gao, H., Wang, C., *et al.*, 2011. High-throughput VDJ sequencing for quantification of minimal residual disease in chronic lymphocytic leukemia and immune reconstitution assessment. Proc. Natl. Acad. Sci. USA. 108, 21194–21199.

Loman, N.J., Misra, R.V., Dallman, T.J., *et al.*, 2012. Performance comparison of benchtop high-throughput sequencing platforms. Nat. Biotechnol. 30, 434–439.

Lossos, I.S., Okada, C.Y., Tibshirani, R., *et al.*, 2000. Molecular analysis of immunoglobulin genes in diffuse large B-cell lymphomas. Blood 95, 1797–1803.

Luo, S., Yu, J.A., Song, Y.S., 2016. Genotyping allelic and copy number variation in the immunoglobulin heavy chain locus.

Magnani, D.M., Silveira, C.G.T., Rosen, B.C., et al., 2017. A human inferred germline antibody binds to an immunodominant epitope and neutralizes Zika virus. PLoS Negl. Trop. Dis. 11, e0005655.

Mamedov, I.Z., Britanova, O.V., Zvyagin, I.V., et al., 2013. Preparing unbiased T-Cell receptor and antibody cDNA libraries for the deep next generation sequencing profiling. Front. Immunol. 4, 456.

Marcou, Q., Mora, T., Walczak, A.M., 2017. IGoR: A tool for high-throughput immune repertoire analysis.

McDaniel, J.R., DeKosky, B.J., Tanno, H., Ellington, A.D., Georgiou, G., 2016. Ultra-high-throughput sequencing of the immune receptor repertoire from millions of lymphocytes. Nat. Protoc. 11, 429–442.

Meng, W., Zhang, B., Schwartz, G.W., et al., 2017. An atlas of B-cell clonal distribution in the human body. Nat. Biotechnol. 35, 879–884.

Michaeli, M., Barak, M., Hazanov, L., Noga, H., Mehr, R., 2013. Automated analysis of immunoglobulin genes from high-throughput sequencing: Life without a template. J. Clin. Bioinforma. 3, 15.

Michaeli, M., Noga, H., Tabibian-Keissar, H., Barshack, I., Mehr, R., 2012. Automated cleaning and pre-processing of immunoglobulin gene sequences from high-throughput sequencing. Front. Immunol. 3, 386.

Moorhouse, M.J., van Zessen, D., IJspeert, H., et al., 2014. ImmunoGlobulin galaxy (IGGalaxy) for simple determination and quantitation of immunoglobulin heavy chain rearrangements from NGS. BMC Immunol. 15, 59.

Munshaw, S., Kepler, T.B., 2010. SoDA2: A Hidden Markov Model approach for identification of immunoglobulin rearrangements. Bioinformatics 26, 867–872.

Murphy, K., Weaver, C., 2016. Janeway's Immunobiology. Garland Science.

Pabst, O., Hazanov, H., Mehr, R., 2015. Old questions, new tools: Does next-generation sequencing hold the key to unraveling intestinal B-cell responses? Mucosal Immunol. 8, 29–37.

Palanichamy, A., Apeltsin, L., Kuo, T.C., et al., 2014. Immunoglobulin class-switched B cells form an active immune axis between CNS and periphery in multiple sclerosis. Sci. Transl. Med. 6, 248ra106.

Pancera, M., McLellan, J.S., Wu, X., et al., 2010. Crystal structure of PG16 and chimeric dissection with somatically related PG9: Structure-function analysis of two quaternary-specific antibodies that effectively neutralize HIV-1. J. Virol. 84, 8098–8110.

Parameswaran, P., Liu, Y., Roskin, K.M., et al., 2013. Convergent antibody signatures in human dengue. Cell Host Microbe 13, 691–700.

Patil, S.U., Ogunniyi, A.O., Calatroni, A., et al., 2015. Peanut oral immunotherapy transiently expands circulating Ara h 2-specific B cells with a homologous repertoire in unrelated subjects. J. Allergy Clin. Immunol. 136, 125–134. [e12].

Pejchal, R., Walker, L.M., Stanfield, R.L., et al., 2010. Structure and function of broadly reactive antibody PG16 reveal an H3 subdomain that mediates potent neutralization of HIV-1. Proc. Natl. Acad. Sci. USA. 107, 11483–11488.

Pogorelyy, M.V., Elhanati, Y., Marcou, Q., et al., 2017. Persisting fetal clonotypes influence the structure and overlap of adult human T cell receptor repertoires. PLOS Comput. Biol. 13, e1005572.

Prabakaran, P., Chen, W., Singarayan, M.G., et al., 2012. Expressed antibody repertoires in human cord blood cells: 454 sequencing and IMGT/HighV-QUEST analysis of germline gene usage, junctional diversity, and somatic mutations. Immunogenetics 64, 337–350.

Ralph, D.K., Matsen, F.A., 2016. Consistency of VDJ rearrangement and substitution parameters enables accurate B cell receptor sequence annotation. PLOS Comput. Biol. 12, e1004409.

Rechavi, E., Lev, A., Lee, Y.N., et al., 2015. Timely and spatially regulated maturation of B and T cell repertoire during human fetal development. Sci. Transl. Med. 7, 276ra25.

Rechavi, E., Somech, R., 2017. Survival of the fetus: Fetal B and T cell receptor repertoire development. Semin. Immunopathol.

Saada, R., Weinberger, M., Shahaf, G., Mehr, R., 2007. Models for antigen receptor gene rearrangement: CDR3 length. Immunol. Cell Biol. 85, 323–332.

Schaller, S., Weinberger, J., Jimenez-Heredia, R., et al., 2015. ImmunExplorer (IMEX): A software framework for diversity and clonality analyses of immunoglobulins and T cell receptors on the basis of IMGT/HighV-QUEST preprocessed NGS data. BMC Bioinformatics 16, 252.

Scheid, J.F., Mouquet, H., Feldhahn, N., et al., 2009. Broad diversity of neutralizing antibodies isolated from memory B cells in HIV-infected individuals. Nature 458, 636–640.

Scheid, J.F., Mouquet, H., Ueberheide, B., et al., 2011. Sequence and structural convergence of broad and potent HIV antibodies that mimic CD4 binding. Science 333, 1633–1637.

Scott, M.G., Tarrand, J.J., Crimmins, D.L., et al., 1989. Clonal characterization of the human IgG antibody repertoire to Haemophilus influenzae type b polysaccharide. II. IgG antibodies contain VH genes from a single VH family and VL genes from at least four VL families. J. Immunol. 143, 293–298.

Sfikakis, P.P., Karali, V., Lilakos, K., Georgiou, G., Panayiotidis, P., 2009. Clonal expansion of B-cells in human systemic lupus erythematosus: Evidence from studies before and after therapeutic B-cell depletion. Clin. Immunol. 132, 19–31.

Shahaf, G., Barak, M., Zuckerman, N.S., et al., 2008. Antigen-driven selection in germinal centers as reflected by the shape characteristics of immunoglobulin gene lineage trees: A large-scale simulation study. J. Theor. Biol. 255, 210–222.

Shiroguchi, K., Jia, T.Z., Sims, P.A., Xie, X.S., 2012. Digital RNA sequencing minimizes sequence-dependent bias and amplification noise with optimized single-molecule barcodes. Proc. Natl. Acad. Sci. USA. 109, 1347–1352.

Shlomchik, M.J., Marshak-Rothstein, A., Wolfowicz, C.B., Rothstein, T.L., Weigert, M.G., 1987. The role of clonal selection and somatic mutation in autoimmunity. Nature 328, 805–811.

Shugay, M., Britanova, O.V., Merzlyak, E.M., et al., 2014. Towards error-free profiling of immune repertoires. Nat. Methods.

Silverman, G.J., Lucas, A.H., 1991. Variable region diversity in human circulating antibodies specific for the capsular polysaccharide of Haemophilus influenzae type b. Preferential usage of two types of VH3 heavy chains. J. Clin. Invest. 88, 911–920.

Singh, V., Stoop, M.P., Stingl, C., et al., 2013. Cerebrospinal-fluid-derived immunoglobulin G of different multiple sclerosis patients shares mutated sequences in complementarity determining regions. Mol. Cell. Proteomics 12, 3924–3934.

Snir, O., Mesin, L., Gidoni, M., et al., 2015. Analysis of celiac disease autoreactive gut plasma cells and their corresponding memory compartment in peripheral blood using high-throughput sequencing. J. Immunol. 194, 5703–5713.

Sok, D., Laserson, U., Laserson, J., et al., 2013. The effects of somatic hypermutation on neutralization and binding in the PGT121 family of broadly neutralizing HIV antibodies. PLoS Pathog. 9, e1003754.

Stern, J.N.H., Yaari, G., Vander Heiden, J.A., et al., 2014. B cells populating the multiple sclerosis brain mature in the draining cervical lymph nodes. Sci. Transl. Med. 6, 248ra107.

Tabibian-Keissar, H., Hazanov, L., Schiby, G., et al., 2016. Aging affects B-cell antigen receptor repertoire diversity in primary and secondary lymphoid tissues. Eur. J. Immunol. 46, 480–492.

Tabibian-Keissar, H., Zuckerman, N.S., Barak, M., et al., 2008. B-cell clonal diversification and gut-lymph node trafficking in ulcerative colitis revealed using lineage tree analysis. Eur. J. Immunol. 38, 2600–2609.

Tak, P.P., Doorenspleet, M.E., de Hair, M.J.H., et al., 2017. Dominant B cell receptor clones in peripheral blood predict onset of arthritis in individuals at risk for rheumatoid arthritis. Ann. Rheum. Dis. (annrheumdis-2017-211351).

Tan, Y.-C., Blum, L.K., Kongpachith, S., et al., 2014. High-throughput sequencing of natively paired antibody chains provides evidence for original antigenic sin shaping the antibody response to influenza vaccination. Clin. Immunol. 151, 55–65.

Tsioris, K., Gupta, N.T., Ogunniyi, A.O., et al., 2015. Neutralizing antibodies against West Nile virus identified directly from human B cells by single-cell analysis and next generation sequencing. Integr. Biol. 7, 1587–1597.

Uduman, M., Shlomchik, M.J., Vigneault, F., Church, G.M., Kleinstein, S.H., 2014. Integrating B cell lineage information into statistical tests for detecting selection in Ig sequences. J. Immunol. 192, 867–874.

Uduman, M., Yaari, G., Hershberg, U., et al., 2011. Detecting selection in immunoglobulin sequences. Nucleic Acids Res. 39, W499–W504.

Vander Heiden, J.A., Yaari, G., Uduman, M., et al., 2014. pRESTO: A toolkit for processing high-throughput sequencing raw reads of lymphocyte receptor repertoires. Bioinformatics 30, 1930–1932.

Vander Heiden, J., Gupta, N., Marquez, S., et al., 2015. alakazam: Immunoglobulin clonal lineage and diversity analysis.

von Büdingen, H.-C., Kuo, T.C., Sirota, M., et al., 2012. B cell exchange across the blood-brain barrier in multiple sclerosis. J. Clin. Invest. 122, 4533–4543.

Walker, L.M., Huber, M., Doores, K.J., et al., 2011. Broad neutralization coverage of HIV by multiple highly potent antibodies. Nature 477, 466–470.

Walker, L.M., Phogat, S.K., Chan-Hui, P.-Y., et al., 2009. Broad and potent neutralizing antibodies from an African donor reveal a new HIV-1 vaccine target. Science 326, 285–289.

Wang, Y., Jackson, K.J.L., Sewell, W.A., Collins, A.M., 2008. Many human immunoglobulin heavy-chain IGHV gene polymorphisms have been reported in error. Immunol. Cell Biol. 86, 111–115.

Wardemann, H., Busse, C.E., 2017. Novel approaches to analyze immunoglobulin repertoires. Trends Immunol. 38, 471–482.

Watson, C.T., Breden, F., 2012. The immunoglobulin heavy chain locus: Genetic variation, missing data, and implications for human disease. Genes Immun. 13, 363–373.

Watson, C.T., Steinberg, K.M., Huddleston, J., et al., 2013. Complete haplotype sequence of the human immunoglobulin heavy-chain variable, diversity, and joining genes and characterization of allelic and copy-number variation. Am. J. Hum. Genet. 92, 530–546.

Weinstein, J.A., Jiang, N., White, R.A., Fisher, D.S., Quake, S.R., 2009. High-throughput sequencing of the zebrafish antibody repertoire. Science 324, 807–810.

Wrammert, J., Koutsonanos, D., Li, G.-M., et al., 2011. Broadly cross-reactive antibodies dominate the human B cell response against 2009 pandemic H1N1 influenza virus infection. J. Exp. Med. 208, 181–193.

Wu, L., Yang, Z.-Y., Xu, L., et al., 2006. Cross-clade recognition and neutralization by the V3 region from clade C human immunodeficiency virus-1 envelope. Vaccine 24, 4995–5002.

Wu, X., Yang, Z.-Y., Li, Y., et al., 2010a. Rational design of envelope identifies broadly neutralizing human monoclonal antibodies to HIV-1. Science 329, 856–861.

Wu, X., Zhang, Z., Schramm, C.A., et al., 2015a. Maturation and diversity of the VRC01-antibody lineage over 15 years of chronic HIV-1 infection. Cell 161, 470–485.

Wu, X., Zhou, T., Zhu, J., et al., 2011a. Focused evolution of HIV-1 neutralizing antibodies revealed by structures and deep sequencing. Science 333, 1593–1602.

Wu, Y.-C.B., James, L.K., Vander Heiden, J.A., et al., 2014. Influence of seasonal exposure to grass pollen on local and peripheral blood IgE repertoires in patients with allergic rhinitis. J. Allergy Clin. Immunol. 134, 604–612.

Wu, Y.-C.B., Kipling, D., Dunn-Walters, D.K., 2012. Age-related changes in human peripheral blood IGH repertoire following vaccination. Front. Immunol. 3, 193.

Wu, Y.-C., Kipling, D., Leong, H.S., et al., 2010b. High-throughput immunoglobulin repertoire analysis distinguishes between human IgM memory and switched memory B-cell populations. Blood 116, 1070–1078.

Yaari, G., Benichou, J.I.C., Vander Heiden, J.A., Kleinstein, S.H., Louzoun, Y., 2015. The mutation patterns in B-cell immunoglobulin receptors reflect the influence of selection acting at multiple time-scales. Philos. Trans. R. Soc. B Biol. Sci. 370, 20140242.

Yaari, G., Kleinstein, S.H., 2015. Practical guidelines for B-cell receptor repertoire sequencing analysis. Genome Med. 7, 121.

Yaari, G., Uduman, M., Kleinstein, S.H., 2012. Quantifying selection in high-throughput immunoglobulin sequencing data sets. Nucleic Acids Res 40, e134.

Yaari, G., Vander Heiden, J.A., Uduman, M., et al., 2013. Models of somatic hypermutation targeting and substitution based on synonymous mutations from high-throughput immunoglobulin sequencing data. Front. Immunol. 4.

Yahalom, G., Weiss, D., Novikov, I., et al., 2013. An antibody-based blood test utilizing a panel of biomarkers as a new method for improved breast cancer diagnosis. Biomark. Cancer 5, 71–80.

Ye, J., Ma, N., Madden, T.L., Ostell, J.M., 2013. IgBLAST: An immunoglobulin variable domain sequence analysis tool. Nucleic Acids Res. 41, W34–W40.

Yermanos, A., Greiff, V., Krautler, N.J., et al., 2017. Comparison of methods for phylogenetic B-cell lineage inference using time-resolved antibody repertoire simulations (AbSim). Bioinformatics.

Zanotti, K.J., Gearhart, P.J., 2016. Antibody diversification caused by disrupted mismatch repair and promiscuous DNA polymerases. DNA Repair (Amst) 38, 110–116.

Zhang, W., Feng, Q., Wang, C., et al., 2017. Characterization of the B cell receptor repertoire in the intestinal mucosa and of tumor-infiltrating lymphocytes in colorectal adenoma and carcinoma. J. Immunol. 198, 3719–3728.

Zhou, T., Xu, L., Dey, B., et al., 2007. Structural definition of a conserved neutralization epitope on HIV-1 gp120. Nature 445, 732–737.

Zhu, J., Ofek, G., Yang, Y., et al., 2013. Mining the antibodyome for HIV-1-neutralizing antibodies with next-generation sequencing and phylogenetic pairing of heavy/light chains. Proc. Natl. Acad. Sci. USA. 110, 6470–6475.

Zuckerman, N.S., Hazanov, H., Barak, M., et al., 2010a. Somatic hypermutation and antigen-driven selection of B cells are altered in autoimmune diseases. J. Autoimmun. 35, 325–335.

Zuckerman, N.S., Howard, W.A., Bismuth, J., et al., 2010b. Ectopic GC in the thymus of myasthenia gravis patients show characteristics of normal GC. Eur. J. Immunol. 40, 1150–1161.

Relevant Website

http://airr.irmacs.sfu.ca/
AIRR Community.

Quantitative Immunology by Data Analysis Using Mathematical Models

Shoya Iwanami, Kyushu University, Fukuoka, Japan
Shingo Iwami, Kyushu University, Fukuoka, Japan and Japan Science and Technology Agency, Kawaguchi, Japan

Introduction

Immunology covers many areas of research, such as the production mechanisms of diverse antibodies, the formulation and maintenance of the T-cell repertoire, the development and maturation of lymphocytes, discrimination of self and non-self, and the interactions between immune cells and viruses or cancer cells (**Fig. 1**). By investigating these topics from both basic and clinical viewpoints, we can understand pathology and develop treatment modalities for allergies, autoimmune diseases and immunodeficiency (Kishimoto, 2005). The broad-ranging topics in immunology are underpinned by the fundamental processes of immune systems, which dictate how cells, antibodies, and pathogens interact and the consequent effects of these interactions (Kitano, 2002). Owing to recently developed experimental techniques, we can clearly visualize the immune response process and its associated changes of signal transductions and gene expressions (Orkin and Zon, 2008; Notta *et al.*, 2016; Gewirtz *et al.*, 2001; Fehniger *et al.*, 1999). Immunologists can also exploit the large body of accumulated datasets. For example, bioinformatics and statistics have revealed several parts of homeostatic mechanisms in living bodies (Martinez *et al.*, 2006). To apply the knowledge obtained from cell or animal experiments at the clinical stage (e.g., in drug development and drug treatment strategies), we must quantitatively understand immune system dynamics such as antibody production changes and cell cycling (Perelson, 2002; Perelson and Nelson, 1999; Koizumi *et al.*, 2017). Mathematical models have become increasingly popular in experimental and clinical data analysis, and are now regarded as cutting-edge tools for quantitative interpretation in immunology (Perelson and Nelson, 1999; Perelson, 2002; Michor *et al.*, 2005; Bernitz *et al.*, 2016; De Boer and Perelson, 2013). In this paper, we introduce and discuss several important data analysis studies that quantify immune-cell differentiation, lymphocyte homeostasis, and viral infection using mathematical models.

Fundamentals

In response to potentially harmful internal and external stimuli, the immune system initiates many interactions of cells, antigens, antibodies and/or cytokines. These mechanisms can be described by "population dynamics" models (Hirsch *et al.*, 2012), which temporally evolve the concentration of cells, antigens, antibodies or cytokines throughout their interactions (Eftimie *et al.*, 2011). In general, cell populations undergo four processes: "birth", "death", "immigration" and "emigration". In populations of immune cells, these four processes correspond to cell proliferation, death, differentiation from other populations, and differentiation into other populations, respectively (see **Fig. 1**). These processes are generally formulated as follows:

$$dN_i(t)/dt = (p_i - a_i - (\Sigma)^m_{j=1, j \neq i} d_{ij}) N_i(t) + (\Sigma)^m_{j=1, j \neq i} h_{ji} N_j(t) \tag{1}$$

where $N_i(t)$ is the number of cells in population i (i = 1, 2,..., m) at time t. The parameters p_i and a_i are the proliferation and death rates of cell population i, respectively, d_{ij} is the emigration rate from population i to population j, and h_{ji} is the immigration rate from population j to population i. If cells in population i migrate into population j by differentiation, then d_{ij} and h_{ji} are equal. If cells of population i are produced from cells in population j, d_{ij} is 0 and h_{ji} is positive. The coefficients might be constants or functions of $N_h(t)$ (h = 1, 2,..., m) to describe interactions with other populations. This generalized model can be tailored to

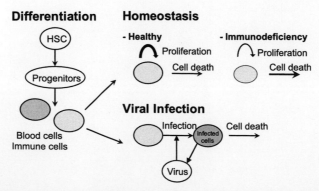

Fig. 1 Schematic of the immune system. Immune cells are derived from HSCs and the immune system is homeostatically maintained in a healthy state that protects against pathogens. Homeostasis is considered to fail in the disease state. The arrows correspond to interactions between populations of cells or antigens. The interactions among populations can be incorporated into mathematical models of dynamic population changes.

model the interactions and differentiation mechanisms of cell populations of interest to the researcher. It can also include hypothetical but experimentally testable interactions. Moreover, using statistical criteria such as Akaike's Information Criterion (AIC) and the likelihood ratio test, we can select the most suitable model among several hypothetical models, discussing empirical possibilities and guiding the next steps in experimental studies (Akaike, 1974). The generalized model can clearly define the focal areas of a study, enhancing our understanding of immune systems rather more than complex models. The following sections exemplify some of the novel insights gained from the generalized model.

Applications

Owing to recent technological developments, the snapshot comparisons in classical researches can be supplemented by novel researches that quantitatively characterize biological behaviors from the dynamical changes in time-course data. In the immunology field, mathematical modeling is increasingly used for describing and quantifying the time-course data of clinical practices and animal/cell experiments.

Cell Differentiation

Cell differentiation is the process by which dividing cells change their functional or phenotypical type. All cells presumably derive from stem cells and obtain their functions as they mature. Cellular composition is often modeled as a hierarchical scheme with stem cells at the top of the hierarchy. Mathematical models of cell differentiation consider the proliferation, death, and differentiation of an appropriately distinguished cell population.

For instance, the cells in a stem cell population differentiate into multiple cells but maintain their number by self-renewal. A progenitor cell population derives from stem cells and differentiates into a restricted lineage, whereas a matured cell population derives from progenitor cells, performs some functions, and dies a natural death. Considering these populations as cell differentiation, Michor et al. (2005) proposed a set of ordinary differential equations (ODEs) describing the population dynamics of four cell populations (hematopoietic stem cells, progenitors, differentiated cells and terminally differentiated cells), which explains the clinical data of leukemia. In successful cancer therapy, the number of leukemic cells exhibits a biphasic exponential decline. By fitting the model to the clinical data, the authors revealed that this behavior corresponds to the turnover rates of differentiated and progenitor cells. Furthermore, their study implied that cancer therapy strongly inhibits the production of differentiated leukemic cells, but does not deplete leukemic stem cells (Michor et al., 2005). Their mathematical modeling provided valuable quantitative insights.

Abundant datasets are also available for hematopoietic cell systems, for which the cell hierarchy is well characterized (Orkin and Zon, 2008; Dingli et al., 2007). In hematopoietic stem cell research, the number of divisions of stem cells can be measured by labeling techniques. Bernitz et al. (2016) distinguished the numbers of different hematopoietic stem cells by the numbers of cell divisions, which were determined from the intensities of fluorescent protein incorporated in the stem cells. Fitting a mathematical model to the data, they estimated the rate at which the hematopoietic stem cells leaned toward a myeloid lineage during differentiation. Thus, the mathematical modeling of cell differentiation quantifies the parameters that cannot be obtained from conventional experimental data.

Lymphoid Cell Turnover

Humans maintain a stable internal environment by special mechanisms called homeostatic mechanisms, which maintain the number of cells in the body within a certain range. Consequently, the number of human lymphocytes such as T lymphocytes is usually unchanged. Dysregulation of homeostasis leads to disease conditions. Abnormal cells that increase their numbers by uncontrolled cell division become malignant cancer cells. Immune cells can become depleted or malfunctioning (leading to immune deficiency) or unusually activated to attack the body's own tissues (causing autoimmune diseases). Such disease states cannot be understood by merely comparing the numbers of cells in healthy and diseased states.

An analog of pyrimidine, 5-Bromo-2'-deoxyuridine (BrdU), can be incorporated into DNA during cell division. Under BrdU administration, dividing cells become labeled and can be detected by a BrdU antibody for measuring the dynamics of lymphocytes (Gratzner, 1982). Since the 2000s, immunology studies have combined classical BrdU labeling with mathematical modeling (De Boer and Perelson, 2013). Interestingly, BrdU labeling and mathematical modeling can together quantify the proliferation or death rates of cells, and detect differences in the BrdU-labeled lymphocyte dynamics. For example, Mohri et al. (1998) compared the time-course BrdU labeling data from rhesus macaques infected with simian immunodeficiency virus (SIV) and their non-infected counterparts. They found that SIV infection dramatically increases the turnover of T lymphocytes. Their assumption – a constant total number of T lymphocytes – was relaxed in a later mathematical model (Bonhoeffer et al., 2000). Similar approaches have quantified the turnovers of other lymphocyte populations such as B cells, T cells, and NK cells (De Boer et al., 2003b). De Boer et al. (2003a) proposed a new index, the "average turnover rate", defined as the cellular death rate averaged over all subpopulations in the model (i.e., $\bar{\delta} = (\Sigma)_i{}^m (\alpha_i \delta_i)$ ($i = 1, \ldots, m$), where $\bar{\delta}$, δ_i and α_i are the average death rate, a death rate of subpopulation i and a fraction of subpopulation i among whole cell population, respectively, and m is the number of subpopulations.). This value is estimated similarly by different models. In summary, mathematical modeling complements conventional BrdU labeling experiments to enhance our quantitative understanding of lymphocyte dynamics.

Virus Dynamics

The representative virus of immunology studies is human immunodeficiency virus type I (HIV-1), which causes acquired immune deficiency syndrome (AIDS). Viral infection behaviors, called "virus dynamics", have been mathematically studied since the 1990s (Ho *et al.*, 1995; Wei *et al.*, 1995). Mathematical models of virus dynamics have analyzed many kinds of *in vivo* experimental and clinical data, and have elucidated various aspects of HIV biology, especially, the average lifespan of virus-producing cells (Simon and Ho, 2003; Perelson, 2002). Many recent studies have combined mathematical modeling with cell culture (i.e., *in vitro*) experiments (Beauchemin *et al.*, 2017; Ito *et al.*, 2017; Iwami *et al.*, 2015; Iwanami *et al.*, 2017; Mahgoub *et al.*, 2018; Ikeda *et al.*, 2015). In cell culture experiments, one can measure several different kinds of time-course experimental data, enabling robust parameter estimation for the mathematical model. For example, Ikeda *et al.* (2015) coupled mathematical modeling with cell culture experiments to quantify the antiviral effects of interferon (IFN) on HIV-1 infection. They deduced that IFN acts by inhibiting *de novo* infection rather than virus production. Thus, quantitative estimates can help to resolve unanswered questions in immunology.

Analysis and Assessment

Mathematical Modeling of Hematopoietic Stem Cell Differentiation

The type and number of the target cells can be measured in flow cytometers. The functional types of immune cells are easily distinguished by markers attached on the cells. Therefore, the differentiation rates or pathways can be theoretically estimated by analyzing the count data of cell phenotypes in mathematical models. In simple terms, differentiation occurs when either or both of the two daughter cells change(s) after cell division. To mathematically describe cell differentiation, we set the number of daughter cells as twice the number of mother cells, then immigrate some of the daughters into the next cell population at a certain rate, maintaining the others in their current precursor population. For example, Thomas-Vaslin *et al.* (2008) described the differentiation process from CD4 and CD8 double-negative T lymphocytes in the thymus to CD4 or CD8 single-positive naïve T lymphocytes in the spleen by ODEs. Fitting their mathematical model to the measured number of T lymphocytes in each cell-differentiation stage, they estimated the turnover rate of each stage and the death rate of T lymphocytes in the thymus. Schlub *et al.*(2009) theoretically explained the formulation mechanism of central- memory and effector-memory T lymphocyte repertoires. Here, the memory cells compose the immunological memory. Buchholz *et al.* (2013) mathematically identified conceivable differentiation models (from naïve T cells to central memory precursors, effector memory precursors, or effector cells) by fitting them to time-course data. The best-fit model was selected by ranking the AICs of the models. These studies are expected to elucidate the homeostasis mechanism of the immune system against various stimuli.

For a deeper understanding of immune-system homeostasis, we need to elucidate the differentiation of hematopoietic stem cells (HSCs), which sustain the production of immune cells. Although HSCs are known to differentiate into two major lineages (the lymphoid and myeloid lineages), how these lineages branch during HSC differentiation and how aging HSCs maintain their homeostasis and functions are not fully understood. In fact, diseases related to the immune system are sometimes considered to arise from disorders of aging HSCs. Crude markers and functional heterogeneity of HSCs have been reported (Yamamoto *et al.*, 2013). Therefore, by combining mathematical modeling with flow-cytometer measurements of HSC numbers, we might elucidate the mechanism of HSC differentiation. HSC differentiation is among the most fundamental processes in immune-cell differentiation, and HSC research has recently adopted mathematical modeling for quantitatively understanding the differentiation phenomena. Assuming that HSCs expand through four symmetric divisions, the increasing HSC numbers during aging can be described by the following ODEs (**Fig. 2**):

$$dx_0(t)/dt = -k_0 x_0(t)$$
$$dx_1(t)/dt = 2k_0 x_0(t) - k_1 x_1(t)$$
$$dx_2(t)/dt = 2k_1 x_1(t) - k_2 x_2(t)$$
$$dx_3(t)/dt = 2k_2 x_2(t) - k_3 x_3(t)$$
$$dx_4(t)/dt = 2k_3 x_3(t) - k_4 x_4(t)$$

(2)

Here, $x_n(t)$ is the number of HSCs that have divided n times (n=0, 1, 2, 3, 4) and $1/k_n$ is the expected time interval to the next division for a cell that has previously divided n times. After fitting the model to the measured number of HSCs in young mice, the division rate of murine HSCs that had divided n times was best described by $k_n = k\beta^{2n}$ $(0 < \beta < 1)$ (Bernitz *et al.*, 2016). Note that by using phenotypes (which roughly distinguish HSCs from other cells), and by tracking the intensity of fluorescent protein (which decreases by half after each cell division), we can measure the number of cells corresponding to each of the n division events.

As HSCs age, their differentiation ability leans toward the myeloid lineage (Notta *et al.*, 2016), and the fraction of cells expressing CD41 (a marker of increased myeloid differentiation) is lower in young mice than in adult mice (Bernitz *et al.*, 2016). Considering the bias toward the myeloid lineage, defined by CD41 expression, the HSC dynamics can be described by the following system of ODEs (**Fig. 2**):

4 step symmetric division model

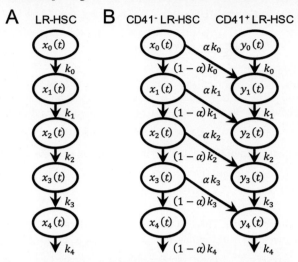

Fig. 2 Division scheme of HSCs with aging. Mathematical model describing the dynamics of aging HSCs. (A) $x_n(t)$ denotes the number of HSCs that have divided n times (n = 0, 1, 2, 3, 4) at time t, and $1/k_n$ is the expected time interval to the next division for a cell that has previously divided n times. (B) $x_n(t)$ and $y_n(t)$ denote the number of CD41$^-$ and CD41$^+$ HSCs, respectively, that have divided n times at time t. A fraction α of the divided cells in the CD41$^-$ HSC population migrate into the CD41$^+$ HSC population.

$$
\begin{aligned}
dx_0(t)/dt &= -k_0 x_0(t) \\
dx_1(t)/dt &= 2(1-\alpha)k_0 x_0(t) - k_1 x_1(t) \\
dx_2(t)/dt &= 2(1-\alpha)k_1 x_1(t) - k_2 x_2(t) \\
dx_3(t)/dt &= 2(1-\alpha)k_2 x_2(t) - k_3 x_3(t) \\
dx_4(t)/dt &= 2(1-\alpha)k_3 x_3(t) - k_4 x_4(t) \\
dy_0(t)/dt &= -k_0 y_0(t) \\
dy_1(t)/dt &= 2k_0 y_0(t) + 2\alpha k_0 x_0(t) - k_1 y_1(t) \\
dy_2(t)/dt &= 2k_1 y_1(t) + 2\alpha k_1 x_1(t) - k_2 y_2(t) \\
dy_3(t)/dt &= 2k_2 y_2(t) + 2\alpha k_2 x_2(t) - k_3 y_3(t) \\
dy_4(t)/dt &= 2k_3 y_3(t) + 2\alpha k_3 x_3(t) - k_4 y_4(t)
\end{aligned}
\tag{3}
$$

Here, $x_n(t)$ and $y_n(t)$ are the number of HSCs not expressing and expressing CD41, respectively. The cells have divided n times (n = 0, 1, 2, 3, 4). In this model, the CD41$^-$ HSCs gain CD41 expression at a rate α with each division. Analyzing HSC count data, α was estimated as 0.12, meaning that the cells gain CD41 expression after every ten divisions (Bernitz *et al.*, 2016). In this way, the differentiation rate of HSCs was quantitatively deduced from detailed cell count data and the mathematical model.

Basic Model of BrdU Labeling

Lymphocyte dynamics are commonly quantified by two DNA labeling methods, the BrdU labeling technique introduced in Section "Applications", and deuterium labeling, which can be provided as heavy water or deuterated glucose. Under deuterium administration, some of the hydrogen atoms of the DNA that was newly synthesized during cell division are replaced with deuterium atoms, which are detectable by mass spectrometry. Deuterium glucose is non-toxic to humans, so is available for quantifying the dynamics of human lymphocytes (Macallan *et al.*, 2003, 2005; Mohri *et al.*, 2001; Ribeiro *et al.*, 2002). Both methods label the DNA of dividing cells. DNA can also be labeled by an intracellular fluorescent dye called carboxy-fluorescein diacetate succinimidyl ester (CSFE) (Lyons, 2000). After measuring the fluorescence intensity of the CFSE-labeled cells, one can determine the distribution of the number of cell divisions undergone by each cell, because (as mentioned above) the fluorescence intensity halves after each cell division. CFSE-labeling data have been analyzed by several ODE models (Revy *et al.*, 2001; Luzyanina *et al.*, 2007), delay equations based on the cell cycle (De Boer and Perelson, 2005) and stochastic processes (Hawkins *et al.*, 2007).

In this subsection, we discuss mathematical models for analyzing BrdU labeling data. The fraction of cells incorporating BrdU is measured by detecting the BrdU antibodies attached to the cells. During BrdU administration, all unlabeled cells become two labeled daughter cells after a cell division, and all labeled cells divide into two labeled daughter cells. After the BrdU administration, the unlabeled cells divide into two unlabeled cells but the labeled cells divide into two labeled daughter cells with diluted

BrdU. Let us assume that a cell inflowing from the outside is labeled during the BrdU administration but is not labeled after the administration is withdrawn (**Fig. 3**; De Boer *et al.*, 2003a,b; Kaur *et al.*, 2008). The lymphocyte dynamics during the BrdU administration are then described by the following ODEs:

$$dL(t)/dt = s + 2pU(t) + (p-d)L(t)$$
$$dU(t)/dt = -(p+d)U(t)$$

(4)

Here, $L(t)$ and $U(t)$ are the fractions of BrdU-labeled and unlabeled cells among the activated lymphocytes, respectively. s is a constant rate of inflowing cells per unit time, and p and d are the proliferation and death rate constants of the lymphocytes, respectively. As the number of total activated cells, $A(t) = L(t) + U(t)$, usually remains constant, we obtain:

$$A(t) = A = s/(p-d)$$

(5)

The lymphocyte dynamics after the BrdU administration are described by the following ODEs:

$$dL(t)/dt = (p-d)L(t)$$
$$dU(t)/dt = s + (p-d)U(t)$$

(6)

Let R be the number of unactivated lymphocytes. The total number of lymphocytes is then $T = A + R$, and the fractions of BrdU-labeled lymphocytes during and after the BrdU administration ($f_L(t) = L(t)/T$) can be analytically solved as follows:

$$f_L(t) = 100\alpha\left(1 - e^{-(d+p)t}\right)\left(t \leq T_{end}\right)$$
$$f_L(t) = 100\alpha\left(1 - e^{-(d+p)T_{end}}\right)e^{(p-d)(t-T_{end})}\left(t > T_{end}\right)$$

(7)

Here T_{end} represents the end of the BrdU administration, and α is the maximum fraction of BrdU-labeled lymphocytes if the BrdU is sufficiently administrated (i.e., $\alpha = A/T$) (De Boer *et al.*, 2003a,b). Fitting Eq. (7) to the BrdU-labeled experimental data and comparing the estimated parameters α, p and d under different conditions (e.g., health and disease), one can detect the affected process in terms of parameter values. In the macaque study of (Mohri *et al.*, 1998), the replacement rates of CD3 + CD4 + / − lymphocytes (defined as the difference between the death and proliferation rates of the lymphocytes) were approximately two- to three-fold higher in SIV-infected macaques than in uninfected macaques, indicating that SIV infection is associated with heightened lymphocyte turnover (Mohri *et al.*, 1998). Quantified cell dynamics, i.e., the rates of cell division and cell death, can effectively reveal a pathophysiology or the effect of a treatment, and have been reported in many studies (De Boer *et al.*, 2003a,b; De Boer and Perelson, 2013; Kaur *et al.*, 2008; Asquith *et al.*, 2007).

Modeling the Antiviral Effect of IFN

Model (8) is a classical but well-parameterized mathematical model of virus infection, which has been especially applied in cell culture (see **Fig. 4**; Iwami *et al.*, 2012b; Perelson, 2002; Perelson and Nelson, 1999):

$$dT(t)/dt = -\beta T(t)V(t)$$
$$dI(t)/dt = \beta T(t)V(t) - \delta I(t)$$
$$dV(t)/dt = pI(t) - cV(t)$$

(8)

Here, $T(t)$ and $I(t)$ are the numbers of target (i.e., uninfected) cells and infected cells, respectively, and $V(t)$ is the viral load at time t. β is the rate constant of the infection. Infected cells die at rate δ per unit time. Viruses are produced from infected cells at rate p per unit time, and degrade at rate c per unit time. To quantify the antiviral effect of IFN on HIV-1-infected cells in culture, Ikeda *et al.* (2015) proposed the following extended mathematical model of target cell growth:

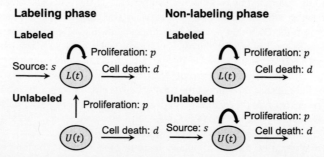

Fig. 3 BrdU-labeling scheme during and after administration. BrdU-labeling dynamics of a leukocyte population. L(t) and U(t) denote the numbers of labeled and unlabeled cells, respectively, at time t. In the labeling phase, unlabeled cells become labeled after cell proliferation and outside cells flow into the population of labeled cells. In the non-labeling phase, outside cells flow into the population of unlabeled cells.

$$dT(t)/dt = gT(t)(1 - (T(t) + I(t))/T_{max}) - (1 - \eta)\beta T(t)V(t)$$
$$dI(t)/dt = (1 - \eta)\beta T(t)V(t) - \delta I(t) \tag{9}$$
$$dV(t)/dt = (1 - \varepsilon)pI(t) - cV(t)$$

In Model (9), IFN is assumed to inhibit *de novo* infection and virus production at rates η and ε, respectively ($0 < \eta$, $\varepsilon < 1$). By fitting Eq. (9) to the experimental cell-culture data in the presence and absence of IFN, Ikeda *et al.*, (2015) found that IFN is at least two-fold more effective against *de novo* infection (η) than against virus production (ε). As discussed in landmark papers (Neumann *et al.*, 1998; Dixit *et al.*, 2004), mathematical modeling can extract the unknown action mechanisms of antiviral drugs from experimental and clinical data.

Illustrative Examples

All models in Section "Analysis and Assessment" were derived from the generalized mathematical model Eq. (1) introduced in Section "Fundamentals".

In Model (2), the $- k_n x_n(t)$ terms denote emigrations from cell populations that have divided n times (nth-population cells) to populations of cells that have divided $n + 1$ times ([n + 1]th- population cells) by cell division (i.e., k_n corresponds to d_{ij}, $i = x_n$, $j = x_{n+1}$). Immigrations from the nth population to the $(n + 1)$th population are described by $2k_n x_n(t)$, because cells divide into two daughter cells (i.e., k_n corresponds to d_{ji}, where $i = x_{n+1}$, $j = x_n$, and $2k_n$ corresponds to h_{ji}, where $i = x_{n+1}$, $j = x_n$). In Model (3), the CD41$^-$ cells emigrated from the nth population divide into two daughter cells, which then immigrate to the $(n + 1)$th populations of CD41$^-$ cells at a rate of $1 - \alpha$, or to the $(n + 1)$th population of CD41$^+$ cells while gaining CD41 expression at rate α. In this case, $2(1 - \alpha)k_n$ corresponds to h_{ji}, where $i = x_{n+1}$, $j = x_n$, and $2\alpha k_n$ corresponds to h_{ji}, where $i = x_{n+1}$, $j = y_n$. The emigrated CD41$^+$ cells of the nth population divide into two daughter cells and immigrate to the $(n + 1)$th population of CD41$^+$ cells. In models (2) and (3), the proliferation and death processes are omitted because both daughter cells generated by the cell division migrate to the next population and HSCs are not assumed to die within the time scale of the experiment.

In Model (4), the emigrated unlabeled cells ($- pU(t)$), which divide into two daughter cells and become labeled by BrdU, immigrate into populations of BrdU-labeled cells ($2pU(t)$) during the BrdU administration (i.e., p corresponds to d_{ij}, where $i = U$, $j = L$, and 2p corresponds to h_{ji}, where $i = L$, $j = U$). Outside cells immigrating into a population of BrdU-labeled cells are described by s, because all such cells are assumed to be labeled by BrdU. The terms $- dL(t)$ and $- dU(t)$ denote a cell death event among the BrdU-labeled and unlabeled cells, respectively (i.e., d corresponds to a_i, where $i = L$, U). In Model (6), the BrdU-labeled and -unlabeled cells proliferate and die at rates p and d after BrdU administration, respectively (i.e., p corresponds to p_i, with $i = L$, U, and d corresponds to a_i, with $i = L$, U). Outside immigration into a population of unlabeled cells is described by s because all such cells are assumed to be unlabeled (i.e., s corresponds to d_{ji}, where $i = L$ and $j = $ outside).

In Model (8), the term $\beta T(t) V(t)$ describes a migration event from the target cells into the virally infected cells (i.e., a new viral infection of a target cell). This migration rate $\beta V(t)$ assumes that the target cells and viruses are well-mixed and that all target cells encounter all viruses at the same frequency (i.e., $\beta V(t)$ corresponds to d_{ij}, where $i = T$, $j = I$). The infected cells die by cytopathogenesis at rate δ (i.e., δ corresponds to a_i, with $i = I$). Immigration into the virus population is denoted by $pI(t)$, which describes the newly produced viruses from infected cells (i.e., p corresponds to h_{ji}, where $i = V$, $j = I$). In Model (9), the cell-growth term g T(t) $(1 - (T(t) + I(t))/T_{max})$ is a form of the "logistic equation" (i.e., $(1 - (T(t) + I(t))/T_{max})$ corresponds to p_i, with $i = T$). Migrations from the target cells into the infected cells by infection, and immigrations into the virus population by viral production in infected cells, are inhibited at rates $(1 - \eta)$ and $(1 - \varepsilon)$ by the antiviral effects of IFN, respectively. In this case, $(1 - \eta)\beta V(t)$ corresponds to d_{ij}, where $i = T$, $j = I$, and $(1 - \varepsilon)p$ corresponds to h_{ji}, where $i = V$, $j = I$.

Results and Discussion

Mathematical modeling of experimental systems is important for obtaining quantitative results. However, the estimated parameters and the interpreted results of a data analysis depend on the experimental system and/or formulation of the mathematical model. Therefore, the assumptions underlying these models should be biologically reasonable. When several mathematical models can conceivably describe the analysis result, immunologists must base their choice on the precise assumptions underlying the different models (De Boer and Perelson, 2013). Moreover, when developing a mathematical model based on the generalized model (1), the modeler should include only those cell populations and interactions with a sound scientific or immunological basis.

For example, the cell population dynamics in Model (3) are distinguished by the number of cell divisions undergone by each cell, and by the presence or absence of CD41 expression (Ito *et al.*, 2017; Bernitz *et al.*, 2016). On the one hand, this mathematical model sufficiently estimates the rate at which CD41$^-$ HSCs gain CD41 expression from the cell-division number distribution in the presence or absence of CD41 expression, which was experimentally determined. On the other hand, Model (4) considers two subpopulations of the activated lymphocytes (BrdU-labeled and BrdU-unlabeled). Mathematical models of BrdU-labeling, regardless of their formulation, are known to robustly estimate "average turnover" (De Boer *et al.*, 2003a,b). Thus, under clear assumptions, a theoretical study is advantageous for understanding the relationships between different mathematical models. Viral infection models of target cells, infected cells and viruses are also derived from the generalized model (1). An example is Model (9), which was established for both clinical and experimental data analysis (Perelson, 2002). Other model branched from this fundamental model including the non-infectious virus population was fitted to the measured concentration of target cells,

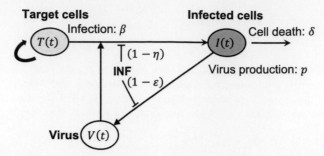

Fig. 4 Infection scheme including the effect of INF. Dynamics of HIV-1 infection, showing the anti-viral effect of IFN on *de novo* infection and virus production. T(t) and I(t) denote the number of target (uninfected) cells and infected cells at time t, respectively. V(t) denotes the viral load at time t. In this model, target cells are infected with viruses at rate β and infected cells produce viruses at rate p. The inhibition rates of infection and virus production by IFN are η and ε, respectively.

concentration of infected cells, total viral load, and infectious viral load. From the fittings, the production efficacy of infectious virus was estimated (Iwami *et al.*, 2012a; Iwanami *et al.*, 2017).

In summary, the simple generalized model (1) can provide novel immunological insights. A deep and mutual understanding between theoreticians and experimentalists can establish a strong collaboration of mathematical models and experimental data.

Future Directions

Various experimental methods for cell counting and data analysis have been established in the last 20 years. The coupling of experimental methods with mathematical models can quantitatively reveal the dynamics of different cell types, as discussed herein. This approach should be followed by mathematical models and computer simulations that reveal the complex, hierarchical effective interactions among immune cells that maintain homeostasis. Recall that a failure of this balanced system leads to a diseased state. For example, osteoimmunology is an interdisciplinary field that tries to link immunological interactions to the mechanisms of bone formation and bone resorption (Takayanagi, 2007). Osteoblasts and osteoclasts interact through cytokines represented by RANKL, a member of the tumor necrosis factor family of cytokines. RANKL also works in the immune system. These interactions maintain the balance of bone formation and bone resorption. When this balanced interaction fails, bone diseases such as osteoporosis and fibrodysplasia ossificans progressiva will develop. The generalized model (i.e., Eq. (1)) and its derived version are appropriate and suitable for elucidating the mechanisms of these complex immune-related diseases.

Closing Remarks

Immunological research has greatly benefitted from experimental techniques in molecular and cell biology, which have rapidly advanced in recent years. However, many of these experimental techniques elucidate only one aspect of immunological phenomena. In tandem with rigorous experimental work, mathematical modeling offers an excellent opportunity for comprehensively analyzing immunological events. At one time, modeling work was essentially ignored by experimental immunologists, but in the last 20 years, it has become an important tool in biological research. In fact, almost all experimental biology groups are now collaborating with a theoretical scientist. Mathematical modeling provides quantitative insights that cannot be obtained by experimental and clinical studies alone. The examples in the present article were selected to highlight these strengths.

Acknowledgment

This work was supported in part by the JST PRESTO and CREST program (to S.I.), the Japan Society for the Promotion of Science (JSPS) KAKENHI Grant Numbers 16H04845, 16K13777, 15KT0107 and 26287025 (to S.I.), a Grant-in-Aid for Scientific Research on Innovative Areas from the Ministry of Education, Culture, Science, Sports, and Technology (MEXT) of Japan 16H06429, 16K21723, and 17H05819 (to S.I.), J-PRIDE 17fm0208006h0001, 17fm0208019h0101, 17fm0208014h0001 (to S.I.), the Program on the Innovative Development and the Application of New Drugs for Hepatitis B 17fk0310114j0001, AMED (to S.I.), the Mitsui Life Social Welfare Foundation (to S.I.), the Shin-Nihon of Advanced Medical Research (to S.I.), a GSK Japan Research Grant 2016 (to S.I.), the Mochida Memorial Foundation for Medical and Pharmaceutical Research (to S.I.), the Suzuken Memorial Foundation (to S.I.), the SEI Group CSR Foundation (to S.I.), the Life Science Foundation of Japan (to S.I.), the SECOM Science and Technology Foundation (to S.I.), the Center for Clinical and Translational Research of Kyushu University Hospital (to S.I.), the Kyushu University-initiated venture business seed development program (to S.I.), and the Japan Prize Foundation (to S.I.). We thank Leonie Pipe, PhD, from Edanz Group (see "Relevant Website section") for editing a draft of this manuscript.

See also: Bioinformatics Approaches for Studying Alternative Splicing. Chromatin: A Semi-Structured Polymer. Comparative and Evolutionary Genomics. Computational Immunogenetics. Exome Sequencing Data Analysis. Functional Genomics. Gene Mapping. Genome Alignment. Genome Annotation. Genome Annotation: Perspective From Bacterial Genomes. Genome Databases and Browsers. Genome Informatics. Genome-Wide Association Studies. Hidden Markov Models. Integrative Bioinformatics of Transcriptome: Databases, Tools and Pipelines. Learning Chromatin Interaction Using Hi-C Datasets. Linkage Disequilibrium. Natural Language Processing Approaches in Bioinformatics. Next Generation Sequencing Data Analysis. Nucleosome Positioning. Phylogenetic Footprinting. Repeat in Genomes: How and Why You Should Consider Them in Genome Analyses?. Whole Genome Sequencing Analysis

References

Akaike, H., 1974. A new look at the statistical model identification. IEEE Trans. Autom. Control 19 (6), 716–723.

Asquith, B., Zhang, Y., Mosley, A.J., et al., 2007. In vivo T lymphocyte dynamics in humans and the impact of human T-lymphotropic virus 1 infection. Proc. Natl. Acad. Sci. USA 104 (19), 8035–8040.

Beauchemin, C.A., Miura, T., Iwami, S., 2017. Duration of SHIV production by infected cells is not exponentially distributed: Implications for estimates of infection parameters and antiviral efficacy. Sci. Rep. 7, 42765.

Bernitz, J.M., Kim, H.S., Macarthur, B., Sieburg, H., Moore, K., 2016. Hematopoietic stem cells count and remember self-renewal divisions. Cell 167, 1296–1309. e10.

Bonhoeffer, S., Mohri, H., Ho, D., Perelson, A.S., 2000. Quantification of cell turnover kinetics using 5-bromo-2 '-deoxyuridine. J. Immunol. 164, 5049–5054.

Buchholz, V.R., Flossdorf, M., Hensel, I., et al., 2013. Disparate individual fates compose robust CD8+ T cell immunity. Science 340, 630–635.

De Boer, R.J., Mohri, H., Ho, D.D., Perelson, A.S., 2003a. Estimating average cellular turnover from 5-bromo-2'-deoxyuridine (BrdU) measurements. Proc. Biol. Sci. 270, 849–858.

De Boer, R.J., Mohri, H., Ho, D.D., Perelson, A.S., 2003b. Turnover rates of B cells, T cells, and NK cells in simian immunodeficiency virus-infected and uninfected rhesus macaques. J. Immunol. 170, 2479–2487.

De Boer, R.J., Perelson, A.S., 2005. Estimating division and death rates from CFSE data. J. Comput. Appl. Math. 184 (1), 140–164.

De Boer, R.J., Perelson, A.S., 2013. Quantifying T lymphocyte turnover. J. Theor. Biol. 327, 45–87.

Dingli, D., Traulsen, A., Pacheco, J.M., 2007. Compartmental architecture and dynamics of hematopoiesis. PLOS ONE 2, e345.

Dixit, N.M., Layden-Almer, J.E., Layden, T.J., Perelson, A.S., 2004. Modelling how ribavirin improves interferon response rates in hepatitis C virus infection. Nature 432, 922–924.

Eftimie, R., Bramson, J.L., Earn, D.J., 2011. Interactions between the immune system and cancer: A brief review of non-spatial mathematical models. Bull. Math. Biol. 73, 2–32.

Fehniger, T.A., Shah, M.H., Turner, M.J., et al., 1999. Differential cytokine and chemokine gene expression by human NK cells following activation with IL-18 or IL-15 in combination with IL-12: Implications for the innate immune response. J. Immunol. 162 (8), 4511–4520.

Gewirtz, A.T., Navas, T.A., Lyons, S., Godowski, P.J., Madara, J.L., 2001. Cutting edge: Bacterial flagellin activates basolaterally expressed TLR5 to induce epithelial proinflammatory gene expression. J. Immunol. 167, 1882–1885.

Gratzner, H.G., 1982. Monoclonal antibody to 5-bromo- and 5-iododeoxyuridine: A new reagent for detection of DNA replication. Science 218, 474–475.

Hawkins, E.D., Turner, M.L., Dowling, M.R., Van Gend, C., Hodgkin, P.D., 2007. A model of immune regulation as a consequence of randomized lymphocyte division and death times. Proc. Natl. Acad. Sci. USA 104, 5032–5037.

Hirsch, M.W., Smale, S., Devaney, R.L., 2012. Differential Equations, Dynamical Systems, and an Introduction to Chaos. Academic Press.

Ho, D.D., Neumann, A.U., Perelson, A.S., et al., 1995. Rapid turnover of plasma virions and CD4 lymphocytes in HIV-1 infection. Nature 373, 123–126.

Ikeda, H., Godinho-Santos, A., Rato, S., et al., 2015. Quantifying the antiviral effect of IFN on HIV-1 Replication in Cell Culture. Sci. Rep. 5, 11761.

Ito, Y., Remion, A., Tauzin, A., et al., 2017. Number of infection events per cell during HIV-1 cell-free infection. Sci. Rep. 7, 6559.

Iwami, S., Holder, B.P., Beauchemin, C.A., et al., 2012a. Quantification system for the viral dynamics of a highly pathogenic simian/human immunodeficiency virus based on an in vitro experiment and a mathematical model. Retrovirology 9, 18.

Iwami, S., Sato, K., De Boer, R.J., et al., 2012b. Identifying viral parameters from in vitro cell cultures. Front. Microbiol. 3, 319.

Iwami, S., Takeuchi, J.S., Nakaoka, S., et al., 2015. Cell-to-cell infection by HIV contributes over half of virus infection. elife. 4.

Iwanami, S., Kakizoe, Y., Morita, S., et al., 2017. A highly pathogenic simian/human immunodeficiency virus effectively produces infectious virions compared with a less pathogenic virus in cell culture. Theor. Biol. Med. Model. 14, 9.

Kaur, A., Di Mascio, M., Barabasz, A., et al., 2008. Dynamics of T- and B-lymphocyte turnover in a natural host of simian immunodeficiency virus. J. Virol. 82, 1084–1093.

Kishimoto, T., 2005. Interleukin-6: From basic science to medicine – 40 years in immunology. Annu. Rev. Immunol. 23, 1–21.

Kitano, H., 2002. Systems biology: A brief overview. Science 295, 1662–1664.

Koizumi, Y., Ohashi, H., Nakajima, S., et al., 2017. Quantifying antiviral activity optimizes drug combinations against hepatitis C virus infection. Proc. Natl. Acad. Sci. USA 114, 1922–1927.

Luzyanina, T., Mrusek, S., Edwards, J.T., et al., 2007. Computational analysis of CFSE proliferation assay. J. Math. Biol. 54, 57–89.

Lyons, A.B., 2000. Analysing cell division in vivo and in vitro using flow cytometric measurement of CFSE dye dilution. J. Immunol. Methods 243, 147–154.

Macallan, D.C., Asquith, B., Irvine, A.J., et al., 2003. Measurement and modeling of human T cell kinetics. Eur. J. Immunol. 33, 2316–2326.

Macallan, D.C., Wallace, D.L., Zhang, Y., et al., 2005. B-cell kinetics in humans: Rapid turnover of peripheral blood memory cells. Blood 105, 3633–3640.

Mahgoub, M., Yasunaga, J.I., Iwami, S., et al., 2018. Sporadic on/off switching of HTLV-1 Tax expression is crucial to maintain the whole population of virus-induced leukemic cells. Proc. Natl. Acad. Sci. USA 115, E1269–E1278.

Martinez, F.O., Gordon, S., Locati, M., Mantovani, A., 2006. Transcriptional profiling of the human monocyte-to-macrophage differentiation and polarization: New molecules and patterns of gene expression. J. Immunol. 177, 7303–7311.

Michor, F., Hughes, T.P., Iwasa, Y., et al., 2005. Dynamics of chronic myeloid leukaemia. Nature 435, 1267–1270.

Mohri, H., Bonhoeffer, S., Monard, S., Perelson, A.S., Ho, D.D., 1998. Rapid turnover of T lymphocytes in SIV-infected rhesus macaques. Science 279, 1223–1227.

Mohri, H., Perelson, A.S., Tung, K., et al., 2001. Increased turnover of T lymphocytes in HIV-1 infection and its reduction by antiretroviral therapy. J. Exp. Med. 194, 1277–1287.

Neumann, A.U., Lam, N.P., Dahari, H., et al., 1998. Hepatitis C viral dynamics in vivo and the antiviral efficacy of interferon-alpha therapy. Science 282, 103–107.

Notta, F., Zandl, S., Takayama, N., et al., 2016. Distinct routes of lineage development reshape the human blood hierarchy across ontogeny. Science 351, aab2116.

Orkin, S.H., Zon, L.I., 2008. Hematopoiesis: An evolving paradigm for stem cell biology. Cell 132, 631–644.

Perelson, A.S., 2002. Modelling viral and immune system dynamics. Nat. Rev. Immunol. 2, 28–36.

Perelson, A.S., Nelson, P.W., 1999. Mathematical analysis of HIV-1 dynamics in vivo. SIAM Rev. 41, 3–44.

Revy, P., Sospedra, M., Barbour, B., Trautmann, A., 2001. Functional antigen-independent synapses formed between T cells and dendritic cells. Nat. Immunol. 2, 925–931.

Ribeiro, R.M., Mohri, H., Ho, D.D., Perelson, A.S., 2002. In vivo dynamics of T cell activation, proliferation, and death in HIV-1 infection: Why are CD4+ but not CD8+ T cells depleted? Proc. Natl. Acad. Sci. USA 99, 15572–15577.

Schlub, T.E., Venturi, V., Kedzierska, K., *et al.*, 2009. Division-linked differentiation can account for CD8(+) T-cell phenotype in vivo. Eur. J. Immunol. 39, 67–77.

Simon, V., Ho, D.D., 2003. HIV-1 dynamics in vivo: Implications for therapy. Nat. Rev. Microbiol. 1, 181–190.

Takayanagi, H., 2007. Osteoimmunology: Shared mechanisms and crosstalk between the immune and bone systems. Nat. Rev. Immunol. 7, 292–304.

Thomas-Vaslin, V., Altes, H.K., De Boer, R.J., Klatzmann, D., 2008. Comprehensive assessment and mathematical modeling of T cell population dynamics and homeostasis. J. Immunol. 180, 2240–2250.

Wei, X., Ghosh, S.K., Taylor, M.E., *et al.*, 1995. Viral dynamics in human immunodeficiency virus type 1 infection. Nature 373, 117–122.

Yamamoto, R., Morita, Y., Ooehara, J., *et al.*, 2013. Clonal analysis unveils self-renewing lineage-restricted progenitors generated directly from hematopoietic stem cells. Cell 154, 1112–1126.

Relevant Website

www.edanzediting.com/ac
   Edanz Editing - Expert English Editing.

Biographical Sketch

Shoya Iwanami I am a graduate student in the mathematical biology laboratory, Department of Biology, in Kyushu University, Japan. I studied biology as an undergraduate student and joined the mathematical biology laboratory in 2015. My research interest is quantifying the biological characteristics of immunology, virus dynamics and stem cells using mathematical models.

Shingo Iwami I am an Associate Professor in the mathematical biology laboratory, Department of Biology, in Kyushu University, Japan. I have been working on theoretical analyses of virus infection *in vivo* and *in vitro* since 2007, covering various infections. My main interest is mathematical modeling and data analysis based on mechanistic models of virus infection dynamics.

Bioimage Informatics

Sorayya Malek, University of Malaya, Kuala Lumpur, Malaysia
Mogeeb Mosleh, University of Taiz, Taiz, Yemen
Sarinder K Dhillon and Pozi Milow, University of Malaya, Kuala Lumpur, Malaysia

Introduction

Bioimaging techniques were developed, which produce images from biological and medical specimens of entire organism, to single molecules and organs (Swedlow, 2012). Common biological imaging methods include magnetic resonance imaging (MRI) computed tomography (CT), X-rays, and positron emission tomography images. These methods are used especially in medical imaging for clinical purposes for vascular, tumor, anatomical, and metabolic imaging and they have a lower resolution. Confocal or two-photon laser scanning microscopy (Pawley, 2006), scanning or transmission electron microscopy, and atomic force microscopy (Keller and Stelzer, 2010) are imaging methods that have higher resolutions commonly used for visualization of cell structures, mapping cell surface, and discerning protein structure (Schmidt et al., 2013). Both the medical and the high resolution imaging techniques are important sources for bioimaging techniques. Methods based on light microscopy are also important to understand the cell structure and function, and which lie between medical imaging and high resolution techniques. Imaging techniques such as photoactivation localization microscopy, fluorescence photoactivation localization microscopy, and stochastic optical reconstruction microscopy provide locations and localization uncertainties of individual molecules accurately up to 20 nm (Betzig et al., 2006; Hess et al., 2006; Rust et al., 2006). Advancement in imaging technologies generates quantitative measurements from images that enhance knowledge and understanding of biological phenomena.

Advances in the field of microscopy in terms of three dimensional (3D) images provides better understanding of biological images (Schmid et al., 2010; Mayer et al., 2012; Guo et al., 2013; Allen et al., 2015). In four-dimensional (4D) imaging, time is the added dimension in the existing spatial dimensions that allows studying biological phenomena in four dimensions. 4D imaging is important in the field of medical imaging such as tumor detection (Handels et al., 2007), changes in tissues intensities (Gorbunova et al., 2012) and cell invasion (Kelley et al., 2017).

Image processing can be used to determine whether an image included specific information or contains some specific objects, features, or activities. The identification task is solved automatically by using a digital image processing field with less human effort. The surge of complex biological and biomedical images leads to the existence of numerous image data analysis and informatics methods to extract quantitative information via segmentation and motion tracking, comparison, search, and management of the biological information of the images (Kvilekval et al., 2010). Bioimage informatics techniques translate image data into valuable biological information (Peng, 2008). Bioimage informatics allows us to discover answers to biological problems from biological and medical specimens of the entire organism, single molecules and organs that would require higher cost to solve (Peng, 2008; Myers, 2012). Bioimage informatics research is heading towards automated invention of models of biological systems and machine learning methods (Murphy, 2014). Obara et al. (2012) established an image analysis method to identify and illustrate huge compound fungal networks using a graph representation that can be used for predicting the physiological performance of the fungal system. Puniyani and Xing (2013) used images of *Drosophila* embryo using a supervised learning method to understand gene interaction networks. Liu et al. (2012) developed a phenotype recognition model using zebrafish embryos' development response towards HTS toxicity. Aging is studied using phenotypes described in *C. elegans* upon different levels of mitochondrial alteration to develop an automated high-content strategy to identify new potential pro-longevity interventions.

Coelho et al. (2015) developed an automated quantification and identification of neutrophil extracellular traps (NETs) using supervised machine learning method that is important in controlling bacterial pathogens. Deep convolutional neural networks have been used for used for automated segmentation of bacterial and mammalian cells and identification of wild animal species in camera-trap images (Sadanandan et al., 2017; Gomez and Salazar, 2016; Villa et al., 2017). Perre et al. (2016) applied image analysis techniques to automatically identify invasive fruit fly species based on wing and aculeus images.

Mosleh et al. (2016) developed a program for automated cephalometric analysis to identify important landmarks point in the human skull that can also be applied to species identification. Computerizing cephalometric is important to lessen the time required for orthodontic procedures by providing X-ray enhancement, consistent measurements, presurgical simulation. Cephalometric analysis can be extended to include species classification.

Automated image recognition has been used to identify various species of algae and harmful species of algae to access environmental condition in aquatic habitat (Mosleh et al., 2012; Verikas et al., 2012; Luo et al., 2011; Dimitrovski et al., 2012; Coltelli et al., 2014). Identification of flatworms is important for preservation of marine life was conducted by Kalafi et al. (2016) using images of hard parts of the haptoral organs of flatworms with machine learning approach for classification. Nguyen et al. (2016) proposed a method SLIC-MMED (simple linear iterative clustering on multichannel microscopy with edge detection) that is able to segment muscle fibers and can be applied to stem cell regeneration and cell lineage tracing. Savkare and Narote (2012) and Rosado et al. (2016) used image processing approach to extract red blood cell and classify them as normal or malaria infected cell. Intarapanich et al. (2016) developed an object-based extended depths of field method to obtain a clear focus image of the foreground properties of the image which constitutes important features. This can be applied in medical diagnostic applications such as diagnosis of malaria infection that can considerably lessen image processing time while preserving significant image features.

Quantitative information resulting from microscopy bioimages is important to enhance understanding of biological processes. With the advancement in imaging technology, image processing techniques are in dire need. There is vast amount of studies conducted on basic image processing, however there very is little information on image processing methods that involves bioimage informatics. Furthermore the absence of high-quality precise data sets for assessment of various common bioimage analysis methods is the main obstacle. A high-quality, well-defined data set for bioimage analysis consists of benchmarked data with well-defined verified and validated structure that is the gold standard for bioimage analysis. A benchmarked dataset reduces analysis and evaluation time and is important for assessing performance, consistency, and accuracy of methods in bioimage analysis (Gelasca *et al.*, 2009).

The important procedures of bioimage informatics include bioimage acquisition, visualization, analysis, and management. Bioimage informatics also covers creation and understanding of biological information generated from biomages using image processing, data mining, machine learning, and pattern recognition methods.

In this article, image acquisition and processing, image segmentation, feature extraction, feature selection, classification registration, and annotation are discussed. Examples and an appropriate case study are provided in each section.

Image Acquisition and Preprocessing

Digital images are produced by image equipment and the process of transferring the analogue image into digital images using imaging devices is image acquisition. The main unit of digital images is pixel which represent the values correspond to the light intensity in one or more spectral bands. Image acquisition is a crucial stage in bioimage informatics related research and automated systems (Coltelli *et al.*, 2014). Captured microscopy images usually suffer from problems such as varying in colors, darkness, resolution, contrast, and brightness that is related to light source, lens, and method of capturing. Microscope images also include unavoidable scum existing beside target cells, and some holes or small objects that strongly affect image quality. Images may contain unwanted areas, shadows, and appear blurry. The preprocessing of captured images is a preparation and treatment process used to enhance the features of images to produce clearer details, remove noise, remove intelligibility of images, and improve the overall appearance of images. Image enhancement is defined as the conversion of an image quality to create a clear image and ensure the accuracy of the process of segmentation and feature extraction. Some common techniques used in image enhancement include image conversion to grayscale, filtering, histogram conversion, and color composition. Kho *et al.* (2017) converted the RGB value of a leaf image to HSV value to remove shadow and to ensure clear image contrast for image segmentation and feature extraction as depicted in **Fig. 1**.

Differences in light or brightness can be reduced using histogram equalization method and contrast enhancement in images can be carried out by stretching the histogram of digital image (Gonzalez and Woods, 2007). Histogram equalization can be applied to enhance the contrast of the color image intensity before the image is converted into a gray scale image. The frequency occurrence of pixel intensities given by the histogram image intensity was adjusted to increase image contrast. Histogram equalization is one type of gray scale conversion; it used to convert the histogram of the original image into an equalized histogram. An accumulated histogram was calculated from original image and then divided into a number of equal regions. The corresponding gray scale in each region was assigned to a converted gray scale. The effect of histogram equalization was the enhancement of image parts that have more frequency variation, whereas parts of an image with less frequency were neglected. **Fig. 2** shows a comparison between the original and converted images after histogram equalization was applied, and shows a comparison between the histogram of the original image with the histogram of a produced algae image.

Images acquisition condition such as parameter and equipment should be consistent and images should be of high-quality with clear details and less noise. Applying resampling method, reducing sensor noise and contrast enhancement can increase the quality of image being captured. Images captured using different equipment should be labeled with scales. Problem related to using different equipment in medical images may produce images that are not alike in appearance. Images can contain noise such

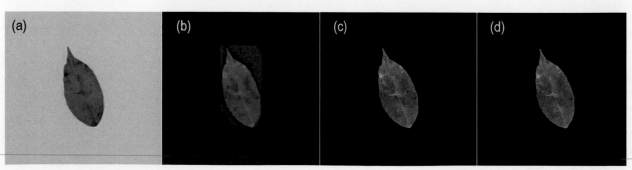

Fig. 1 Shadow removing image preprocessing. Thin streak of shadow can still be seen in (c), which is then removed in (d).

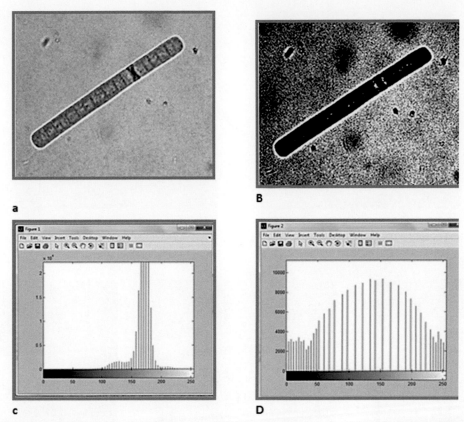

Fig. 2 Examples: Histogram equalization application process as (a) Original image of *Oscillatoria sp.*, (b) Result image after equalization, (c) Accumulated histogram of original image, (d) Accumulated histogram for grayscale image.

as Gaussian film grain, nonisotropic, speckle and periodic noise, and salt and pepper noise from imaging equipment that can reduce the image quality especially salt and pepper noise in medical images (Sánchez *et al.*, 2012). Noise is added into an image at the time of image acquisition or image capturing. Noise reduction is necessary to preserve the quality of image. Noise reduction filters are comprised of linear, nonlinear, adaptive, median, and fuzzy filters (Mythili and Kavitha, 2011). The main purpose of the filters is to produce a high-quality image by sharpening, edge highlighting, and contrast improvement. High-quality images are important for image processing especially medical images. Some of these developed filters to enhance image quality are median filter, a nonlinear filter used mostly to reduce image noise in better way than other methods such as convolution approach (Lim, 1990). **Fig. 3** illustrates application of median filter to algae species. Median filters have been applied by Kho *et al.* (2017) to remove noise from a leaf image and Barry and Williams (2011) for fungal network extraction.

The Kalman filter an adaptive filtering approach for color noise based on the correlative method (Xinding *et al.*, 1996). The Gaussian filter is an image-blurring filter that uses a normal distribution, also called Gaussian distribution, for calculating the transformation to apply to each pixel in the image. It is used widely to filter medical images to remove fine detail and noise (Yue *et al.*, 2006). Gaussian filters are also used to smooth the input image (Yue *et al.*, 2006). Gaussian filters can separate the holes from the noise in gray image region filters to smooth gray images for contrast enhancement. Gaussian filters are widely used for detection of edges and peaks in images. Unsharp filters are used to filter low spatial frequencies from the image data to enhance the image. An X-ray or clinical medical image generally suffers from low contrast quality and degradations that vary from one region to another. Unsharp filtering or masking is an edge-enhancement technique by isolating the edges in an image, amplifying them, and then adding them back into the image. The unsharp filter and Gaussian filter are the most common filters used for medical image filtering (Chalazonitis *et al.*, 2003).

In X-rays of hard and soft tissue in a cephalometric system image quality influences the accuracy of locating landmarks points automatically. High-quality images are important to identify landmark points automatically as manual identification is not accurate (Mosleh *et al.*, 2016). X-Cephalometric images are the basis of cephalometric measurement and analysis, which have been widely applied in the fields of orthodontics and provide a scientific basis for making the diagnosis, treatment, and accurate and effective prediction. The cephalometric system analysis requires good quality X-rays. If the X-ray is poorly exposed, the lines are fragmented. The critical lines defining the landmarks may not be tracked properly. The quality of images directly influence accuracy of locating landmarks. The original image must be filtered to ensure the accuracy of locations and measurements. During sampling and transmission, images are often degraded by noises that originated from a multiplicity of sources. These noises are colorful and their variances are not known beforehand. The original images must be filtered to ensure the accuracy of location and measurement. Mosleh *et al.* (2016) applied unsharp and Gaussian filters to enhance image properties for feature extraction. Unsharp filters in this

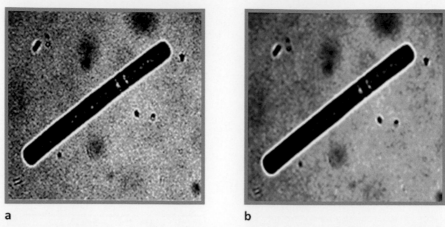

Fig. 3 Median filter application (a) *Oscillatoria sp.* image in gray scale before the process (b) *Oscillatoria sp.* after median filter was applied.

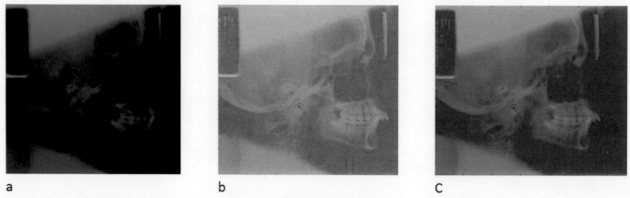

Fig. 4 Illustrates application filter for image enhancement (a) original image (b) unsharp filter, (c) Gaussian filter.

study are used for sharpening of X-ray images in the presence of low noise levels via an adaptive filtering algorithm. Adaptive filtering using these kernels can be performed by filtering the image with each kernel, in obtaining the output image with sharp filter. Gaussian filters are designed to give no overshoot to a step function input while maximizing the rise and fall time. This behavior is closely connected to the fact that the Gaussian filter has the minimum possible group delay. The gradient operators in a Gaussian filter enhance image edges and at the same time enhance noise. Emphasizing the fine details of a radiographic image while suppressing noise is possible by employing a median filter prior to applying a gradient operator. Approximations to the pill box blur and Gaussian low pass. The idea of Gaussian smoothing is to use this 2-D distribution as a point-spread function, and this is achieved by convolution. Before the convolution is performed a discrete approximation to the Gaussian function was produced. In theory, the Gaussian distribution is nonzero everywhere, which would require an infinitely large convolution kernel, but in practice it is effectively zero when it is more than three standard deviations from the mean (Davies, 1990).

Gaussian filter is generated with some parameters as a 2-D array because the X-ray image is a 2-D array. Input is the unfiltered X-ray image and the output is the filtered X-ray image. Both implemented filters can work in real time at system runtime.

Fig. 4(a) illustrates X-ray images without and with filtering. The low intensity contrast that is almost hidden in the background in **Fig. 4(a)** can be distinguished clearly in **Fig. 4(b)** and **Fig. 4(c)**. The image in **Fig. 4(a)** is considered a slightly overexposed cephalometric radiograph. The contrast of the bone anatomical structures is quite satisfactory, but soft tissue tends to mix with the background in some zones. After the application of unsharp filter as shown in **Fig. 4(b)** or Gaussian filter as shown in **Fig. 4(c)**, the bone contrast is greatly improved, and anatomical structures become more visible.

Image Segmentation

Microscopic images normally contain different objects, therefore image processing is required to isolate the image into different object components. Detection processes for specific image points or regions are performed to determine the object for further processing. This process of isolating and dividing the image into a subimage, where each image contains at least one object, is called segmentation. Segmentation is an essential phase in image processing, which is carried out after image processing and before feature extraction and classification (Haralick and Shapiro, 1992). The image segmentation process is used to isolate individual objects in captured images. The segmentation process is important for quantification and analysis of bioimages. There are numerous image segmentation methods such as the thresholding based method, clustering based method, edge based method, region growing based method, watershed method contour based method, and model based method (Basavaprasad and Ravi, 2014). The thresholding based segmentation method is a common method for image segmentation that is applied to the image pixels. This method can be categorized into otsu, global, and local thresholding techniques. Rosada *et al.* (2016) reviewed numerous methods involved in automated malaria cell segmentation where the thresholding based method is most commonly applied. In bioimage segmentation the otsu based method has been applied in identifying *Drosophila* egg stages (Jia *et al.*, 2016), identification of herbs leaves and plant disease identification (Isnanto *et al.*, 2016; Akhtar *et al.*, 2013).

Edge based segmentation is another common method and is implemented using filters such as Canny's, Sobels, Laplacian, Robert, and gradient filters (Gonzalez and Woods, 2007). A Canny edge detector is considered as the most powerful edge detector for image segmentation (Canny, 1987). It is used to identify discontinuities in an image intensity value or the edge of the image and it has been used widely in species identification (Kalafi *et al.*, 2016; Mosleh *et al.*, 2012; Murat *et al.*, 2017). Both Canny and Sobel have been applied in automatic identification of algae due to substantial edges and shapes of the species (Santhi *et al.*, 2013).

Thresholding based method together with edge based method have been applied in Munisami *et al.* (2015) for plant leaf recognition and brain tumor detection (Manasa *et al.*, 2016). Identification of bioimage segments is often more effective when making use of boundaries and shape information extracted by segmentation methods. The Grabcut algorithm (Rother *et al.*, 2004) is a segmentation technique used in automated identification of species systems (Hernández-Serna and Jiménez-Segura, 2014) to remove background. In this technique, hard segmentation made by iterative graph-cut optimization is combined with border matting to get rid of mixed and blurred pixels on the boundaries of an object.

Overlapping objects in microscopic images makes object detection difficult. There have been various methods to separate overlapping objects including watershed algorithm, edge detection method, morphological erosion, active contour method, and others. The application of the edge detection method together with morphological erosion and dilation are implemented in

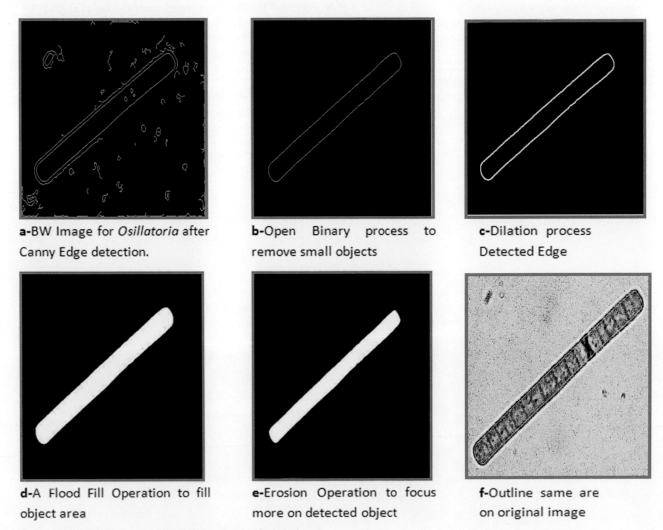

a-BW Image for *Osillatoria* after Canny Edge detection.

b-Open Binary process to remove small objects

c-Dilation process Detected Edge

d-A Flood Fill Operation to fill object area

e-Erosion Operation to focus more on detected object

f-Outline same are on original image

Fig. 5   Image samples for morphological operation on *Oscillatoria sp.*

Mosleh *et al.* (2012) in automatic algae detection. In morphological operations, the value of each pixel in the output image was compared with the corresponding pixel in the input image of its neighbors. **Fig. 5** illustrates application of morphological operation and Canny edge detection.

The edge detection method does not perform well in instances where the overlapping objects are intensely textured or there is no noticeable intensity variance between the overlapping. The active contour and snake methods (Kass *et al.*, 1988) require high computer processing capacity and it not suitable for large objects. It has been used by Chopin *et al.* (2016) to determine in identification of wheat leaves and Sabeena Beevi *et al.*, (2017) in identification of mitotic nuclei from microscopy images of breast histopathology slides.

The watershed method is considered as an effective and efficient method to separate overlapping objects (Savkare and Narote, 2012). Watershed method segmentation is also used to detect and extract a graph representation of complex curvilinear networks and overcomes a critical bottleneck in biological network analysis (Obara *et al.*, 2012). Segmentation of globular object can be challenging due to object irregularity. This requires additional step such as hole filling before applying watershed segmentation method (Long *et al.*, 2007). Yang and Ahuja (2014) used local density clustering to identify pixels that matter before the watershed algorithm is used to separate the overlapping objects in blob or granular object recognition for cell/nuclei segmentation. Automatic segmentation of cell nuclei in microscopic images is highly desirable, however it is difficult due to huge intensity inhomogeneities in the cell nuclei and the background. Unsupervised methods are popular for cell nuclei segmentation thresholding, *k*-means clustering, watershed, active contour, level set and graph-based models. Thresholding, *k*-means, and watershed are applicable when there is great contrast between object and background and it is not suitable with images with inhomogeneities.

Other techniques such as *k*-means clustering, which attempts to minimize the variance of intensity values within a predefined number of clusters, and Gaussian mixture models (GMM) (Permuter *et al.*, 2006), which use probability distributions to classify pixels as foreground or background. GMM has been used in the brain segmentation of magnetic resonance images (MRI) due to ease of use and high contrast between different tissue classes of the brain (Song *et al.*, 2014). Combination of *k*-means clustering and otsu method was applied in identification of plant leaf disease after RGB color transformation on leaf. *K*-means clustering was used to divide a leaf image into segments to identify disease where the leaf is infected by more than one disease (Al-Hiary *et al.*, 2011).

Chopin *et al.* (2016) applied hybrid segmentation technique for whole plant such as wheat image segmentation comprising the leaf tips and twists. Initial segmentation was done using GMM, *K*-mean, and multidimensional histogram thresholding and active contour model by Kass *et al.* (1988), and applied to enhance the segmentation after identification of control points in the image.

Before feature extraction is applied, images, especially species or plant images, need to be aligned. Mosleh *et al.* (2012) applied techniques for image objects alignments. This method also has been adopted by Kho *et al.* (2017) in *ficus* plant leaf identification. This technique performs automatic rotation for objects to be aligned with the horizontal axis, which increases the classification accuracy. A small routine is developed to obtain the angle of inclination for an object automatically. This routine is then used for rotating image objects to be aligned horizontally as shown in **Fig. 6**.

The angle of inclination is calculated automatically by obtaining the longest path between each two points on the object boundary. Identified point $P1$ $(X1,Y1)$, $P2(X2,Y2)$, and Origin point $P(0,0)$ are used to determine the angle of inclination using the following equation:

$$\theta = tan^{-1}(m1 - m2)/(1 + m1*m2) \tag{1}$$

where $m1$ and $m2$ are the slopes of lines that form the angle that is obtained using the following equations:

$$m1 = (Y2 - Y1)/(X2 - X1) \tag{2}$$

$$m2 = (Y1 - Y0)/(X1 - X0) \tag{3}$$

This routine is designed to align the rotated shape into horizontal lines, which will ease the feature extraction process, and improve the accuracy and performance of the recognition process by ensuring that all the extracted features are calculated with similar positioning of the object coordinates.

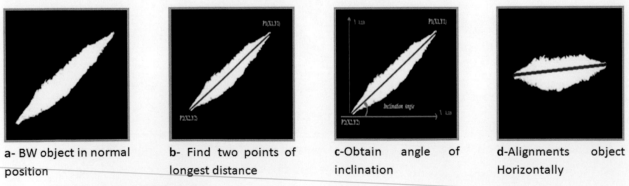

a- BW object in normal position

b- Find two points of longest distance

c- Obtain angle of inclination

d- Alignments object Horizontally

Fig. 6 Sample steps of auto-orientation method.

Feature Extraction

Features extraction and selection is a crucial step as features extracted from bioimages will be used for bioimage identification. Selection of significant features from bioimages is important as the number of features extracted can be large and this can lead to heavy computational effort. Feature selection allows an optimized number of features being selected. The success of bioimage feature extraction and selection techniques is determined by the quality of the extracted data and the type of classifiers used in bioimage identification and recognition (Chora's, 2007; Kiranyaz et al., 2011).

Feature extraction is a process to obtain useful information from the segmented bioimages that can be used to classify or identify them. Feature extraction is necessary for successful bioimage classification. Features used in bioimage classification normally are a combination of several types of features (Song et al., 2016). There are three distinct features: Texture, shape, and color (Islam et al., 2008; Pin Tian, 2013).

Color is the most significant feature for image classification (Seyyid, 2015; Chora's, 2007). Color histograms, coherence vector, moments, correlogram, scalable color descriptor, dominant color descriptor, and structure color descriptor are various methods to extract color features in an image that are based on mean, skewness, and standard deviation of an image pixel. Color histogram is the most common method to detect color distribution in an image that meets basic requirements meanwhile color moment is the simplest method to implement (Pin Tian, 2013; Seyyid, 2015). In automated identification of zebrafish embryo development in response to toxicity, color features such as scalable color descriptor, color layout descriptor, and color histogram with texture features gave successful results (Liu et al., 2012). Color variance is an important feature in skin cancer detection (Bhuiyan et al., 2013). Chopin et al. (2016) used color of leaf together with the direction and magnitude of leaf edges in whole plant identification.

Texture features are significant as well however it is used with other features such as color and shape to enhance its effectiveness. Texture features are important in digital image identification especially in plant leaf identification (Brilhador et al., 2013; Jamil et al., 2015). Combination of texture and shape feature produced better identification in plant leaf identification than using combination of texture, shape and color (Jamil et al., 2015). Texture features should be used as an input when color or shape features are not distinctive enough on the plant species images (Camargo and Smith 2009). Textures features are categorized into statistical and structural features such as Haralick, Gabor, and wavelet and Fourier transform, co-occurrence matrices, shift-invariant principal component analysis, and Tamura features (Chora's, 2007). The Gabor filters compared to gray level co-occurrence matrix for image segmentation and classification can be utilized as is or converted into feature vectors (Kumar et al., 2015). Al-Hiary et al. (2011) used texture features derived from co-occurrence matrix with color features to identify plant disease. Camargo et al. (2009) used color features together with shape and texture features to enhance identification of visual symptoms of plant disease. Akhtar et al. (2013) combined Haralick texture features (Haralick et al., 1973), discrete cosine transform (DCT), and discrete wavelet transform (DWT) features for automatic identification of plant disease.

Combination of Haralick texture features, Pearson correlation between channel intensities, and SURF features (Bay et al., 2008) was used to automatically identify NETs images (Coelho et al., 2015). Texture feature based on DWT is important for bioimage classification of cancerous cells. However in breast cancer cell diagnosis complex daubechies wavelet transform (CDWT) performed better than DWT with color features (Niwas et al., 2013). In identification mitotic nuclei from images of breast histopathology shape and textures features are used (Beevi et al., 2014). Morphological changes of *Caenorhabditis elegans* (*C. elegans*) is used for studying aging due to the small size, transparent body, well-characterized cell types and lineages. An automatic image processing method for both normal and abnormal structures is contrasted using geometric, intensity, and texture features that describe the properties of nuclei in *C. elegans* (Zhao et al., 2017).

Shape feature is important in bioimage object identification. The shape feature extraction method can be categorized into contour based and region-based techniques. Shape features typically comprises of geometric features such as aspect ratio, circularity and irregularity. Advance method for shape features is feature moment such as Zernike moment (ZM) a region-based descriptor. Detailed descriptions of shape based features are explained in Mingqiang et al. (2008). Shape, size scale-invariant feature transform (SIFT) are used in identification of cell nuclei in microscopic images (Song et al., 2013). Kalafi et al. (2016) extracted various shape features from anchors and bars of monogeneans such as Euler number, perimeter, area, density, and bounding box. Automated stage identification of the *Drosophila* egg chambers includes morphological shape features such as egg chamber size, oocyte size, egg chamber ratio and distribution of follicle cell (Jia et al., 2016).

Shape feature is the most common feature used in plant leaf identification (Jamil et al., 2015). Spatial interrelation shape feature such as convex hull, morphological shape feature of plant, distance features, and color histogram have been used in plant leaf recognition (Munisami et al., 2015). Convex hull allows segmentation of a complex object into a polygon without holes that are convex in nature and shape is very essential in object recognition (Jayaram and Fleyeh 2016). An herbal plant leaf identification system (Isnanto et al., 2016) is based on the shape of the herbal plants' leaf region-based invariant feature extraction or Hu's seven moment invariants. Hu's seven moment invariants are indifferent against translation, scaling, and rotation, as well as in mirror position of the herbal leaf. A combination of shape features using SIFT, color using color moment, and texture feature using segmentation-based fractal texture analysis (SFTA) in herbal plant identification gave better results compared to using a single feature (Jamil et al., 2015). A combination of shape features, texture, and color features have also been applied successfully in classification of plant leaves and algae recognition (Kho et al., 2017; Mosleh et al., 2012).

Below are examples of shape feature extraction based on Mosleh et al. (2012). In this study, a specific feature set was selected to obtain the essential characteristics for the selected algae. Feature extraction process is implemented to extract some parameters

Fig. 7 Example of area extraction on algae species.

from both binary and color images of algae including shape index, area, perimeter, minor and major axes, centroid, equivalent perimeters, bounded box, and Fourier spectrum with principal combination analysis (PCA). Other features were derived from the main feature to obtain the relation between shape parameters such as that between area and parameters.

(1) *Area*

The area represents the actual number of white pixels in the selected region. It is calculated by using the number of white or "1" pixels inside the image object as shows in **Fig. 7**. Area is used as a parameter in classifying operations because it gives an indication of the size of objects.

(2) *Perimeter*

The perimeter of the object is the summation of the distance between each adjoining pair of pixels around the object border, as shown in red pixel in **Fig. 8**.

(3) *Major and minor axes*

Major and minor axes are extracted where two points are identified automatically by calculating the maximum distance between given points in the objects vector. The major axis represents the line segment connecting between the base points in the X axis, whereas the minor axis represents the maximum width which is perpendicular to the major axis. The major axis can be used as object length and minor axis as object width as represented on **Fig. 9**.

(4) *Object width factor*

Object width factor is a customized shape feature developed for algae feature extraction that can be applied for any plant images. It is calculated by slicing across the major axis and parallel to the minor axis, the feature points are normalized into a number of vertical strips, and for each strip, the ratio of strip length to the object width is calculated using the following equation: $R_c = W_c/L$, where R_c is the ratio at column c, W_c is the width of object at column c, and L is the object length as shown in **Fig. 10**. This feature is used to differentiate between some algae that is identical in width and length, and have some essential variations in width at different positions such as *Navicula*, which has similar value of width as *Oscillatoria* at the middle position; however values of width vary in different positions as illustrated in **Fig. 10**.

(5) *Equivalent diameter*

Equivalent diameter is the feature that determines the diameter of a circle with the same area of a segmented image. It is computed as:

$$EquDi = Sqrt(4*Area/pi)$$

(6) *Euler number*

This feature is suitable to identify the number of holes inside the image object. The Euler function segments the image before it is filled by holes. The Euler number is equal to the number of objects in the region minus the number of holes in those objects. **Fig. 11** shows one example of using Euler number to improve classification method of algae and plant leaf recognition.

Fig. 8 Perimeter feature extraction.

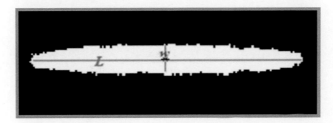

Fig. 9 Major and minor axes.

a- Sample for Slicing *Navicula*   **b**-Sample of Slicing *Oscillatoria*

Fig. 10 Slicing process on *Navicula* and *Oscillatoria*.

(7) *Bounding box*

Bounding box is identified as the smallest rectangle that contains the region. The box has four feature parameters, including the left corner point of the rectangle ($X1$, $Y1$), length, and width of bounding rectangle. **Fig. 12** shows an example for the bounding box object. The rectangle bounding the object is used to compute the extent parameter, which represents the proportion of pixels in the bounding box within the region. The results are obtained by dividing the object area with the area of the bounding box.

(8) *Centroid*

The centroid is the center of mass for the segmented region, and helps us determine the object center coordinates. The elements of the centroid are two points, the x- and y-coordinates, of the center of mass. **Fig. 13** represents the centroid of an object as small blue star in the center of the object.

The centroid as parameter does not provide a distinguishing feature for selected algae however it can be used to derive a new feature to enhance the classifications method. The centroid coordinate is used to extract the length and slope of the line that connects between the centroid coordinate points and the left corner point extracted from the bounding box; the line is shown also in **Fig. 13** in blue.

(9) *Derived features*

Selected feature ratios are used for improving the classifying process such as the ratio of the area and perimeter (perimeter/ area), ratio of perimeter and object length (perimeter/length), ratio of perimeter and object width (perimeter/width), and ratio of minor to major axes (minor/major).

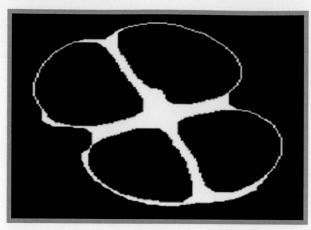

Fig. 11   Euler number for *Chroococcus sp.*

Fig. 12   Bounding box feature example.

Fig. 13   Illustrates centroid and slope lines.

(10) *Shape index*

Extracted shape index is one of the customized features constructed in this study. Algal taxonomy based on algal shapes is used to develop a new function that can categorize algal shape into three different types: Circular, spiral, and irregular. This function is designed to identify algal shape type and provided us with a new feature called shape index, which is equal to "1" if the shape is circular, "0.5" if the shape is cylindrical, "0" if the shape is spiral, and " − 1" if the shape is irregular. This classifier is used to improve accuracy rate and optimize the time of recognition process (Abdullah, 2013). The results obtained from shape extraction will be included as one of the input parameters for algal classification using MLP. The shape index feature can be used as an early stage classifier for the algal taxonomy area. The measurement of the shape index is obtained by evaluating the roundness or cylinder of the objects as described in the following steps:

- Comparing the longest and shortest diameters for the object.
- Comparing the area with the formula $\pi \times r^2$.
- Comparing the perimeter with the formula $2\pi \times r$.
- Eccentricity is also used to improve the results of shape index measurements. Eccentricity refers to the ratio of the distance between the foci of the ellipse bounded the object and its major axis length.

(11) *Entropy of gray image*

Entropy is defined as the statistical measurement of randomness that can be used to characterize the texture of the input image. Entropy is used to extract texture features on the gray scale image. During segmentation processes, the detected object is used to obtain the outline of both color and gray images. Entropy is implemented in our system to obtain one output feature from the outlined gray image.

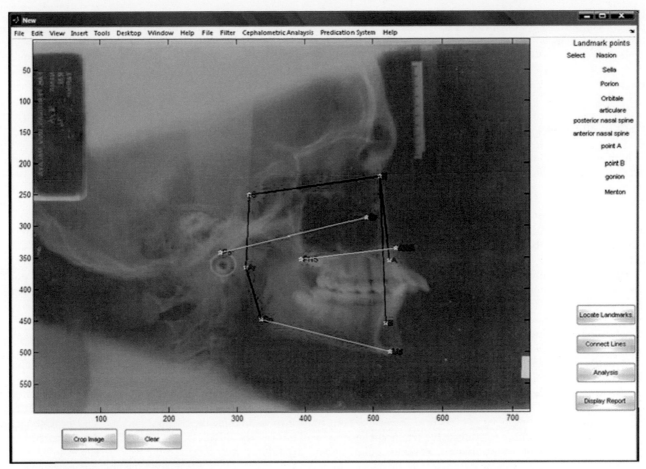

Fig. 14 Illustrates customized features extraction.

Morphological shape features and Euclidean distance are extracted from a graph representation of complex curvilinear or fungal networks. The approach is generic and can be applied to a wide range of biomedical images (Obara *et al.*, 2012). To further improve efficiency of feature descriptors for particular problem areas, tailored features have been constructed manually for feature extraction from bioimages (Song *et al.*, 2016). Some of the customized shape features for algae and plant leaf detection by Mosleh *et al.* (2012) have been discussed such as object width factor and shape index. Mosleh *et al.* (2016) extracted morphological or shape features for a cephalometric identification system. Tailored feature extraction methods have been constructed to extract line, line angles, and line length. Two points in a binary image are connected if a path can be found between them along which the characteristic function remains constant. Since every selected landmark point has a position in a coordinate system (*X, Y*), it can be used as parameters to create a line between two selected points as shown in **Fig. 14**. To perform this operation is a new matrix, which consists of all points between the two points where each generated point is represented by the shortest path in this matrix. Angle measurements are essential for any cephalometric measurement process. The angular measurements module is designed to calculate the size of these angles. Angle is a result of intersection between two lines in a plane, or via connecting all three points. In the two-dimensional space, the angle θ between two lines is the slope of a straight line in the *XY* plane represents changes of the ordinate *Y* of the point of the line per a unit-change of the abscissa *X* of the point. It requires that the line is not vertical. However, in this model angles are considered dimensionless, since they are defined as the ratio of lengths, the unit of measurement is radian. All angle measurements are converted to degree scale instead of radian scale because the degree scale is more standardized and easy to understand for orthodontics specialists. The line measurement of the distance between two points or pixels is constructed using trigonometry given by Pythagoras' theorem, the linear measurements are converted from pixel to inch. Cephalometric line measurements are important because they give information about the skeletal relation. Image resolution and screen resolution is used to calculate the linear measurements in inches. By considering the factor scales, all image types, whatever their resolution, can be analyzed and measured. Scale factor is calculated in our system to minimize the errors of measurement.

Some bioimage software such as BIOimage Classification and Annotation Tool (BIOCAT) and CP-CHARM are image based classification algorithms that are able to extract features without performing segmentation. CP-CHARM algorithm is based on WND-CHARM (Weighted Neighbor Distances using a Compound Hierarchy of Algorithms Representing Morphology) (Orlov *et al.*, 2008). CP-CHARM algorithm is capable of extracting morphological features of an image minus the necessities of segmentation. Features extracted from CP-CHARM are as follows: Moments features, texture features (Haralick, Tamura), edge features, high contrast features, Gabor features, polynomial decompositions (Fourier, Haar wavelet, and discrete Chebyshev transforms), pixel statistics (histograms, moments), texture measurements, and image levels (transforms and compound transforms) (Uhlmann *et al.*, 2016). CP-CHARM is easier to use and it is an effective substitute to more comprehensive and difficult tools such as BIOCAT (Zhou *et al.*, 2013). Features extraction in BIOCAT comprises of 14 different features such as texture features using DTW, morphological features that comprises of Hu moments, ZMs, Hessian features for identifying tubular features, Gaussian derivatives, and Laplacian features and object statistics (Zhou *et al.*, 2013).

Feature Selection

Feature selection is an important method to construct a classification model that produces higher accuracy rates in bioimage recognition or classification. Feature selection eliminates features that are uninformative, noisy, or redundant, which leads to a dataset that is expressed with important information that is essential for better classification with controlled or little loss of information.

The importance of feature selection method is as follows: It can ovoid overfitting of the classification model, provide better understanding into the underlying processes that generated the data, improve model performance, and lower processing time. Feature selection method can be classified into the following three methods: Filter, wrapper, and embedded methods (George and Raj, 2011).

Filter method uses ranking of features based on univariate metric. The top ranking features are utilized in model construction whereas the remaining low ranking features are eliminated. The method is only dependent on training data for feature selection, therefore the results from filter method does not affect model classification. Filter methods evaluate the significance of features by considering solely inherent properties of the data. Filter methods computational requirement are low and are independent of the classification algorithm. Therefore feature selection needs to be executed only once before carrying out classification (Saeys *et al.*, 2007). Brilhador *et al.* (2013) applied correlation feature selection (CFS) method for plant image classification with the Image-CLEF dataset. CFS is an automatic filter based method that identifies features that have no correlation with each other but are greatly correlated with a classification problem (Hall, 1999).

The wrapper method applies a learning algorithm to rank features based on their importance to classification model performance. It is a simple and convenient method of feature selection applicable irrespective of the selected learning algorithm. The learning algorithm is considered as black box. The wrapper procedure involves in using the classification model performance of a particular learning algorithm to evaluate the significant importance of the features. Specific criteria needs to be determined such as on search procedures of features subsets, assessment of m-learning algorithm performance and the type of features to be used. Performance evaluation of the learning classification algorithm is carried out using a cross validation method. Some examples of common classification models are decision trees, naive Bayes, and genetic algorithm (GA). Yusof *et al.* (2013) applied kernel based GA for feature selection from tropical wood texture features. The application of GA produced better results compared with a classification model that did not apply any feature selection. A disadvantage of the wrapper method is the high computational requirements. However this problem can be solved using effective search approaches, which do not sacrifice classification or predictive performance. Such approach involves greedy search strategies; forward selection and backward elimination are computationally beneficial and robust against overfitting. In forward selection, model performance is estimated by adding unselected features to the features subsets. Features that improve model performance are selected. Forward selection procedure is terminated when the expected performance of adding any feature is less than the performance of the feature set already selected. Whereas backward elimination model performance is accessed using all available features and gradually eliminates the least significant ones. The wrapper method is easy to use but embedded methods that incorporate feature selection in the training process might be more effective as there is no need for data splitting into the training and validation set. An example of embedded method is decision trees such as classification and regression trees (CART), which is equipped with a feature selection mechanism (Breiman *et al.*, 1984). The feature selection in an embedded method is built into the classification method. Embedded methods are explicit to a particular learning algorithm. Embedded methods comprise the interaction with the classification model. Examples of embedded methods are random forests (RFs), support vector machine (SVM) and artificial neural network (ANN) (Guyon and Elisseeff, 2003; Saeys *et al.*, 2007). The RF variable importance method has been applied in feature selection of plant leaf; variables with higher mean decrease accuracy are kept and retrained for better classification accuracy (Kumar *et al.*, 2015).

An alternative set of methods are the ones that construct new features from existing features that possibly could lead to loss of information. The methods are linear discriminant analysis (LDA) and principal component analysis (PCA) (Lewis *et al.*, 2012). Landmark localization uses PCA to extract landmarks on the surface of the human face using shape and texture information from 2D data human facial images (Guo *et al.*, 2013). Mosleh *et al.* (2012) in his study applied PCA to extract feature texture for algae classification. CP-CHARM bioimage classification software also utilizes PCA as its feature selection algorithm (Uhlmann *et al.*, 2016). Clustering based algorithms also have been used for feature selection. The most common algorithms comprise *K*-means and hierarchical clustering. Clustering is an unsupervised learning method. Kalafi *et al.* (2016) applied LDA for feature selection in fully automated classification of monogeneans at the species level. Huh *et al.* (2009) applied extension of LDA method for yeast image classification. However this method only works well when the numbers of classes are large as it tends to over reduce features. Stepwise discriminant analysis has been proven to work well in identifying protein in human cancer cells (Xu *et al.*, 2014). Bioimage classification software BIOCAT, comprises of the Fisher's linear discriminant feature selection method.

Classification

Machine learning and data mining techniques have the ability to accurately and successfully analyze substantial amounts of image data. Many studies have demonstrated the benefits of applying machine learning classification techniques to classify images using features extracted from them. Some of the common classification approaches are SVM, LDA, the K-nearest neighbor (KNN) classifier, ANNs, RF and decision tree classifier, naive Bayes (Abbas et al., 2015). These classification methods for cell nuclei classification normally classify images based on variance in intensities of foreground and background histograms. The classification model performance is subjected to the variances in intensity between image background and foreground. Features such as local Fourier transform, spatial information, shape, and morphological features are normally used for cell nuclei classification (Song et al., 2013).

SVM classifier separates data points of different classes to achieve largest margin between the classes. The SVM classifies the data by finding an optimal hyperplane that is the one with the largest margin between the classes. Margin is defined as the distance from the decision surface to the support vectors (Vapnik and Vapnik, 1998). RF is a nonparametric approach that builds an assembled model of decision trees during the model training. The output is the mode of the individual trees (Breiman, 2001). ANNs are a machine learning method that processes information by adopting the way in which the neurons of human brains work (Daliakopoulos et al., 2005), which consists of a set of nodes that imitate the neuron and carries activation signals of different strength. ANN is a black box as it does not provide any insights on the structure of the function being approximated. There are two approaches to ANN development, which are supervised and unsupervised learning algorithm. The KNN approach comprises of input samples of k-nearest neighbors in the training set in the feature space and assigns the labels of classes that are most similar among its k-nearest neighbors. Performance of KNN is largely reliant on the value of k and the applied distance metric such as Euclidean distances or Manhattan distance. The CART algorithm uses tree structure where the roots are at the top with the leaves at the bottom (Kuhn and Johnson, 2013). A node represents a condition of an attribute, each branch signifies the result of a condition, and each leaf node holds a class label. CART involves three phases: Creation of an initial tree from examples, pruning the tree to remove branches with little statistical validity, and processing the pruned tree to improve its understandability (Alpaydin, 2004; Kotsiantis, 2007). Naive Bayes is a linear classification model that uses Bayes theorem to analyze feature probabilities and assumes feature independence. This approach is particularly useful when the dataset is large and contains many different features.

There is no one particular classification method that is superior and outperforms other available classification methods. Every method has its own advantages and disadvantages. SVM using radial basis function (RBF) classifier is superior in performance and is tolerant to irrelevant and highly correlated features as compared to decision trees, ANN, and KNN classifiers (Alpaydin, 2004; Kotsiantis, 2007).

SVM is suitable when data is not linearly separable. However the SVM algorithm is slower due to the optimization of the hyperplane parameters, the cost parameter C and γ. The selection of these parameters' value involves a grid search. The cost parameter provides an adjustment between training error and model complexity. Higher value of C indicates higher cost for nonseparable samples (Alpaydin, 2004; Kotsiantis, 2007). The SVM model depends on mapping data to a higher dimensional space by using the function of a kernel, the maximum-margin hyperplane will be selected in order to separate training data. Hence SVM improves the accuracy via optimization of the space separation. Thus, it produces a better result compared to the other classification models. ANN and SVM models are considered good at fitting functions and recognizing patterns in various datasets. Both ANN and SVM can approximate practically all types of nonlinear functions (Desai et al., 2008; Gulati et al., 2010). However, there are some limitations such as that standardized coefficients for variables may not be straightforwardly calculated and presented, and it is considered as a "black box" approach. The complete insight into the internal workings of the model or information for evaluating the interaction of inputs is unknown (Dayhoff and DeLeo, 2001). RF compared to ANN and SVM is relatively easier to use and RF provides insight to the variable importance.

Nuclear abnormality is a hallmark of progeria in humans. Analysis of age-dependent nuclear morphological changes in C. elegans is of great value to aging research. Classification of C. elegans images was made using several machine learning methods such as linear SVM, RF, KNN, ANN, and DT. The data used in this study is imbalanced where the data inhibits unequal amount of data for each class. In order to improve classification performance when imbalanced data set is involved each class is assigned weight based on number of total sample over number of samples in each class. Optimum number of classifiers are identified using the K-fold method and the RF algorithm outperformed all other machine learning algorithms including SVM. The reason behind this is the choice of SVM algorithm in this study, that is, linear SVM. Performance classification of SVM-linear is faster compared to SVM-RBF. However SVM-RBF demonstrates higher classification accuracy and lower misclassification compared to SVM-linear. SVM-linear is a suitable choice for variable selection during iterative training phase and SVM-RBF for all cell types' classification (Abbas et al., 2014; Mao and Tao, 2017). SVM has been used as a classifier to assess the existence of malaria cell P. falciparum trophozoites and white blood cells in Giemsa stained thick blood smears. Images were obtained using a smartphone and manually annotated. The results were 80.5% sensitivity for malaria cell and 98.2% sensitivity for white blood cell (Rosado et al., 2016).

Automated plant identification using machine learning methods can be applied to classify plants into appropriate taxonomies. Brilhador et al. (2013) compared KNN, ANN, SVM, RF, and naïve Bayes for plant leaf image classification using all features and selected features. SVM outperformed all other classifiers without and with using feature selection method. RF and ANN performance was slightly lower than SVM. Kho et al. (2017) applied SVM and ANN classifier for automated ficus leaf identification that resulted in both models having similar performance. K-means method was used for diseased plant leaf image clustering and ANN for classification resulted in accuracy of 94% (Al-Hiary et al., 2011).

Plant leaf identification of 32 different species using convex hull, morphological, color histogram, and distance map features using KNN as classifier achieved an accuracy of 87.3% (Munisami et al., 2015). Automated plant disease using Otsu segmentation method was carried out using four different types of classifiers, that is KNN with k value of 1, ANN, naïve Bayesian, linear SVM, and decision tree. SVM classifier produced the highest accuracy of 94.45% using discrete cosine and wavelet transform and 90% accuracy using texture features alone. Previous studies in plant disease identification (Camargo and Smith, 2009; Al-Hiary et al., 2011) using SVM with texture, shape, histogram of frequency features using thresholding segmentation method, and ANN with K-means clustering using texture features produced a slightly lower classification result of 94% (Akhtar et al., 2013).

In a species identification system successful results have been obtained using ANN and KNN as a classifier. A species identification system using photographic images of fish, plant, and butterfly species uses geometry, morphology, and texture feature and ANN classifier for image recognition. ANN achieved accuracy of about 90% for all species (Hernández-Serna and Jiménez-Segura, 2014). DAISY is a generic species identification system for insect groups using ANN, SVM, and a plastic self-organizing map as classifiers (ONeill, 2007). ANN and shape features were used to classify the skulls, sex, and region of species of *Suncus murinus*. Skull classification achieved 100% and the other was about 80%. This indicates that shape alone is sufficient for species identification (Abu et al., 2016). KNN was applied to classify the monogenean specimens and achieved classification accuracy of 90% using morphological shape features (Kalafi et al., 2016).

In the medical domain the SVM classifier achieved higher accuracy as well compared to other methods. Comparative study on automated identification of ophthalmic images using local binary pattern and SVM as classifier and wavelet transformation and SVM attained accuracy of 87%. Combination of color and texture feature using SVMs or kNN classifier identification of ophthalmic images can be used for mobile devices application due to ease of implementation and faster processing time (Wang et al., 2017). Early detection of skin cancer melanoma using morphological features of pixel intensity extracted from diseased layer of skin was classified using SVM classifier, which achieved 92% (Li et al., 2013). Identification of tumor location from head and neck MRI images data is vital but very complicated. SVM classifier is used with discrete sine transform and DCT achieve better performance compared to the existing techniques (Gouid et al., 2014). Cervical cell nuclei classification using morphometric and Haralick texture features with PCA for dimension reduction was conducted using linear method, KNN, Mahalanobis, and SVM. SVM classifier with PCA achieved highest classification accuracy (Lorenzo-Ginori et al., 2013).

Application of feature selection and different combination features improves classifier performance in bioimage classification. Feature selection algorithm effect on the performance accuracy of plant leaf classification was studied using KNN, CART, RF, and J48. Gabor texture features and RF feature selection method was used to select significant features in plant leaf identification. RF model outperformed with and without feature selection; meanwhile CART was the worst performing classification model. Classification accuracy increased using selected features compared to using a full set of features for all classification algorithm (Kumar et al., 2015). Adaboost classifiers boosting algorithm for binary classification was used with a combination of shape features using SIFT, color feature using color moment, and texture feature using SFTA in herbal plant identification. Single texture feature and combination of texture and shape feature produced almost similar classification results of 92% compared to using all three features together of 94% or using either shape or color feature alone (Jamil et al., 2015). SVM with RBF kernel classifier using feature reduction method for yeast image classification for 20 classes achieved 87% and classification without feature selection is 95% (Huh et al., 2009)

Deep learning applications have been gaining popularity in bioimage informatics. The deep learning method uses a convolutional neural network (CNN), a special kind of ANN. Individual neurons execute a convolution of a kernel with an input image and generate a filtered output image or a feature map. The input image can be made out of a few channels, and each layer in the neural net holds as many channels of feature maps. The feature maps in the last layer are considered the final features learned by the network and are used for classification. The main difference of CNN from conventional feature based classification methods are features, and the weights of the kernels, are learnt by the CNN algorithm. CNN performance was compared with conventional machine learning classifiers such as SVM, RF, and LDA. CNN outperformed with high accuracy rate of 97%. CNN as a features learning algorithm is as good as conventional features and supervised learning algorithm (Dürr and Sick, 2016). Related literature on CNN application include classification of red blood cells in sickle cell anemia (Xu et al., 2017), breast cancer histopathological image classification (Spanhol et al., 2016), and automatic plant type identification. CNN has been compared with SVM and yielded higher accuracy (Yalcin and Razavi, 2016).

Bioimage software for image analysis and classification uses various classification methods and is easy to use. Area, shape, intensity, texture, and advanced features such as Haralick, Gabor, and Zernike features are the most commonly used features for cell identification in different bioimage software packages (Abbas et al., 2014). CP-CHARM uses LDA as classification and PCA for feature dimension reduction (Uhlmann et al., 2016). WND-CHARM uses thresholding and weighting as feature selection algorithm and weighted nearest neighbor for classification (Orlov et al., 2008). CellProfiler is a Matlab-based package for analysis of biological images using joint boosting classifier limited to two classes of classification (Lamprecht et al., 2007). To enhance CellProfiler capabilities, enhanced CellClassifier (Misselwitz et al., 2010) uses an images analyst by CellProfiler for multiclass classification using SVM. BIOCAT software is used for automatic classification and annotation of 2D and 3D and regions of interest on individual biological images. It is mainly used for cell biology and neuroscience. Feature selection is carried out using Fisher's criterion and supports classifiers such as SVM, KNN, naïve Bayes, DT, and RF (Zhou et al., 2013). FARSIGHT is software comprised of modules of bioimage analysis and classification that uses supervised spectral clustering. CellCognition (Held et al., 2010) uses a hidden Markov model algorithm as classifier. CellXpress is based on SVM and Ilastik (Sommer et al., 2011) uses RF as it classifier. SVM linear and SVM RBF have proven to be superior in classification task compared to joint boosting and LDA classifier. Abbas et al. (2014) suggests that the SVM based classifier should be incorporated in bioimage analysis tools' fast and efficient analysis of high-throughput data.

Image Registration and Annotation

Image registration is defined as a process that overlays two or more images from various imaging equipment or sensors taken at different times and angles, or from the same scene to geometrically align the images for analysis (Zitová and Flusser, 2003). Objects taken in different imaging equipment are not alike. Existence of calcification can be identified efficiently by using computed tomography images (CT scan). In MRI uneven intravascular calcification plaque and blood disorders can create image artifacts. Application of using similar image scanning method for example MRI on different scanning equipment with different parameter can also lead to differences in acquired image. The image registration method can be applied to resolve the issues from differences in images acquired due to equipment and parameter settings (Chen *et al.*, 2017).

The basic procedure in the image registration method comprises feature detection, feature alignment from the captured image and the reference image, mapping functions parameter estimation, and image resampling and transformation. Image registration methods are unique to each problem and there is no common technique of image registration that is applicable to all registration tasks due to images' geometric radiometric distortions and noise disturbance, data characteristic, and accuracy threshold level.

Vascular registration of retinal and MRA methods proposed by Chen *et al.* (2017) comprises area and feature based registration methods. The area based method comprises of cross correlation matching and mutual information methods and correspondences estimation from the original image. The feature based method consists of a description of invariance obtained from feature extraction such as correlation coefficient, closed region, geometric features, and coherent point drift. This method proposed by the author can be applied to other tubular structures.

There is a huge amount of images of proteins with cancerous tissues, however due to absence of clear and precise annotations, images from cancerous tissues are not being utilized for developing supervised models for cancer classification (Uhlen *et al.*, 2010). Annotation of biological images has conventionally been done manually by a domain expert and it is very time consuming. Automated annotation *Drosophila* gene expression pattern images and identification of the images have been carried out using the bags-of-words (BoW) method. Images are denoted as BoW based on visual features extracted from the image to produce a visual codebook about the image. Feature of an image based on the regions of an image is extracted using SIFT descriptor for image codebook generation. The spatial BoW method on the FlyExpress image database is used to generate a histogram for each image and frequency of words appearing in an image. The BoW trails the location of visual words appearing on the image. The spatial BoW is an enhancement of the BoW method with spatial properties of an image with better accuracy compared to using nonspatial BoW or nonannotated images (Yuan *et al.*, 2012). Software such as the BIOCAT is used for automated bioimage annotation. The software is able to perform automated image recognition, classification and annotation of images and region of interests with application to cell biology and neuroscience (Zhou *et al.*, 2013).

Closing Remarks

This work discusses general aspects of bioimage informatics covering cell, organism, species, and medical images. Techniques discussed in this work can be applied to various biological images.

See also: Data Mining: Classification and Prediction. Experimental Platforms for Extracting Biological Data: Mass Spectrometry, Microarray, Next Generation Sequencing. Information Retrieval in Life Sciences. Integrative Bioinformatics. Integrative Bioinformatics of Transcriptome: Databases, Tools and Pipelines. Natural Language Processing Approaches in Bioinformatics. Proteome Informatics

---

References

Abbas, M.M., Mohie-Eldin, M.M., El-Manzalawy, Y., 2015. Assessing the Effects of Data Selection and Representation on the Development of Reliable E. coli Sigma 70 Promoter Region Predictors. Plos One 10 (3), doi:10.1371/journal.pone.0119721.

Abbas, S., Dijkstra, T.M., Heskes, T., 2014. A comparative study of cell classifiers for image-based high-throughput screening. BMC Bioinformatics 15, 342. doi:10.1186/1471-2105-15-342.

Abdullah, H.M., 2013. A preliminary study on an automated freshwater algae recognition and classification system.

Abu, A., Leow, L.K., Ramli, R., Omar, H., 2016. Classification of Suncus murinus species complex (Soricidae: Crocidurinae) in peninsular Malaysia using image analysis and machine learning approaches. BMC Bioinformatics 17 (Suppl. 19), S505. doi:10.1186/s12859-016-1362-5.

Akhtar, A., Khanum, A., Khan, S.A., Shaukat, A. (2013). Automated Plant Disease Analysis (APDA): Performance comparison of machine learning techniques. In: Proceedings of the 2013 11th International Conference on Frontiers of Information Technology. doi:10.1109/fit.2013.19.

Al-Hiary, H., Bani-Ahmad, S., Reyalat, M., Braik, M., ALRahamneh, Z., 2011. Fast and accurate detection and classification of plant diseases. International Journal of Computer Applications 17, 31–38.

Allen, M.J., Kanteti, R., Riehm, J.J., El-Hashani, E., Salgia, R., 2015. Whole-animal mounts of Caenorhabditis elegans for 3D imaging using atomic force microscopy. Nanomedicine: Nanotechnology, Biology and Medicine 11, 1971–1974. doi:10.1016/j.nano.2015.07.014.

Alpaydin, E., 2004. Introduction to Machine Learning (Adaptive Computation and Machine Learning). The MIT Press. ISBN: 026201243.

Barry, D., Williams, G., 2011. Microscopic characterisation of filamentous microbes: Towards fully automated morphological quantification through image analysis. Journal of Microscopy 244 (1), 1–20. doi:10.1111/j.1365-2818.2011.03506.x.

Basavaprasad, B., Ravi, M., 2014. A comparative study on classification of image segmentation methods with a focus on graph based techniques. International Journal of Research in Engineering and Technology 03, 310–315. doi:10.15623/ijret.2014.0315060.

Bay, H., *et al.*, 2008. Speeded-up robust features (SURF). Comput. Vis. Image Understand. 110, 346–359.

Beevi, K.S., Nair, M.S., Bindu, G.R., 2014. Automatic segmentation and classification of mitotic cell nuclei in histopathology images based on Active Contour Model. In: Proceedings of the 2014 International Conference on Contemporary Computing and Informatics (IC3I). doi:10.1109/ic3i.2014.7019762.

Betzig, E., Patterson, G.H., Sougrat, R., *et al.*, 2006. Imaging intracellular fluorescent proteins at nanometer resolution. Science 313, 1642–1645. doi:10.1126/science.1127344.

Bhuiyan, M A.M., Azad, I., Uddin, M.U., 2013. Image processing for skin cancer features extraction. International Journal of Scientific & Engineering Research 4 (2),

Breiman, L., Friedman, J., Stone, C.J., Olshen, R.A., 1984. Classification and Regression Trees. Taylor & Francis.

Breiman, L., 2001. Random forests. Machine Learning 45, 5–32.

Brilhador, A., Colonhezi, T.P., Bugatti, P.H., Lopes, F.M., 2013. Combining texture and shape descriptors for bioimages classification: A case of study in ImageCLEF dataset. Progress in Pattern Recognition, Image Analysis, Computer Vision, and Applications, Lecture Notes in Computer Science. 431–438. doi:10.1007/978-3-642-41822-8_54.

Camargo, A., Smith, J.S., 2009. Image pattern classification for the identification of disease causing agents in plants. Computers and Electronics in Agriculture 66, 121–125.

Canny, J., 1987. A Computational Approach to Edge Detection. Readings in Computer Vision. 184–203. doi:10.1016/b978-0-08-051581-6.50024-6.

Chalazonitis, A.N., Koumarianos, D., Tzovara, J., Chronopoulos, P., 2003. How to optimize radiological images captured from digital cameras, using the Adobe Photoshop 6.0 program. Journal of Digital Imaging 16, 216–229. doi:10.1007/s10278-003-1651-1.

Chen, L., Lian, Y., Guo, Y., *et al.*, 2017. A vascular image registration method based on network structure and circuit simulation. BMC Bioinformatics 18 (1), doi:10.1186/s12859-017-1649-1.

Chopin, J., Laga, H., Miklavcic, S.J., 2016. A hybrid approach for improving image segmentation: Application to phenotyping of wheat leaves. PLOS ONE 11, e0168496. doi:10.1371/journal.pone.0168496.

Chora's, R.S., 2007. Image feature extraction techniques and their applications for CBIR and biometrics systems. International Journal of Biology and Biomedical Engineering 1 (1), 6–16.

Coelho, L.P., Pato, C., Friães, A., *et al.*, 2015. Automatic determination of NET (neutrophil extracellular traps) coverage in fluorescent microscopy images. Bioinformatics 31, 2364–2370. doi:10.1093/bioinformatics/btv156.

Coltelli, P., Barsanti, L., Evangelista, V., Frassanito, A.M., Gualtieri, P., 2014. Water monitoring: Automated and real time identification and classification of algae using digital microscopy. Environ. Sci.: Processes Impacts 16 (11), 2656–2665. doi:10.1039/c4em00451e.

Daliakopoulos, I.N., Coulibaly, P., Tsanis, I.K., 2005. Groundwater level forecasting using artificial neural networks. Journal of Hydrology 309, 229–240.

Davies, E.R., 1990. Machine Vision: Theory, Algorithms and Practicalities. London: Academic Press, pp. 42–44.

Dayhoff, J.E., DeLeo, J.M., 2001. Artificial neural networks. Cancer 91, 1615–1635.

Desai, K.M., Survase, S.A., Saudagar, P.S., Lele, S., Singhal, R.S., 2008. Comparison of artificial neural network (ANN) and response surface methodology (RSM) in fermentation media optimization: Case study of fermentative production of scleroglucan. Biochemical Engineering Journal 41 (3), 266–273. doi:10.1016/j.bej.2008.05.009.

Dimitrovski, I., Kocev, D., Loskovska, S., Džeroski, S., 2012. Hierarchical classification of diatom images using ensembles of predictive clustering trees. Ecological Informatics 7 (1), 19–29. doi:10.1016/j.ecoinf.2011.09.001.

Dürr, O., Sick, B., 2016. Single-cell phenotype classification using deep convolutional neural networks. Journal of Biomolecular Screening 21, 998–1003. doi:10.1177/1087057116631284.

Gelasca, E.D., Obara, B., Fedorov, D., Kvilekval, K., Manjunath, B., 2009. A biosegmentation benchmark for evaluation of bioimage analysis methods. BMC Bioinformatics 10, 368. doi:10.1186/1471-2105-10-368.

George, G.V.S., Raj, V.C., 2011. Review on feature selection techniques and the impact of SVM for cancer classification using gene expression profile. CoRR. *abs/1109.1062*.

Gomez, A., Diez, G., Salazar, A., Diaz, A., 2016. Animal Identification in Low Quality Camera-Trap Images Using Very Deep Convolutional Neural Networks and Confidence Thresholds. In: Bebis, G., *et al.* (Eds.), Advances in Visual Computing. ISVC 2016. Lecture Notes in Computer Science, vol 10072. Cham: Springer.

Gonzalez, R., Woods, R., 2007. Digital Image Processing, third ed. Prentice Hall.

Gorbunova, V., Sporring, J., Lo, P., *et al.*, 2012. Mass preserving image registration for lung CT. Medical Image Analysis 16, 786–795. doi:10.1016/j.media.2011.11.001.

Gouid, G., Nasser, A., Mostafa, M., El-Hennawi, D., 2014. Automatic identification of Head and Neck Swellings in MRI images using support vector machines based on cepstral analysis. In: Proceedings of the 2014 IEEE Conference on Computational Intelligence in Bioinformatics and Computational Biology. doi:10.1109/cibcb.2014.6845504.

Gulati, T., Chakrabarti, M., Singh, A., Duvuuri, M., Banerjee, R., 2010. Comparative study of response surface methodology, artificial neural network and genetic algorithms for optimization of soybean hydration. Food Technol Biotechnol 48 (1), 11–18.

Guo, J., Mei, X., Tang, K., 2013. Automatic landmark annotation and dense correspondence registration for 3D human facial images. BMC Bioinformatics 14, 232. doi:10.1186/1471-2105-14-232.

Guyon, I., Elisseeff, A., 2003. An introduction to variable and feature selection. Journal of Machine Learning Research 3, 1157–1182.

Hall, M.A., 1999. Correlation-based feature selection for machine learning.

Handels, H., Werner, R., Schmidt, R., *et al.*, 2007. 4D medical image computing and visualization of lung tumor mobility in spatio-temporal CT image data. International Journal of Medical Informatics 76 (Suppl. 3), S433–S439. doi:10.1016/j.ijmedinf.2007.05.003.

Haralick, R.M., Shapiro, L.G., 1992. Computer and robot vision. Reading, MA: Addison-Wesley.

Haralick, R.M., Shanmugam, K., *et al.*, 1973. Textural features for image classification. IEEE Transactions on Systems, Man, and Cybernetics 3, 610–621.

Held, M., Schmitz, M.H., Fischer, B., *et al.*, 2010. CellCognition: Time-resolved phenotype annotation in high-throughput live cell imaging. Nature Methods 7, 747–754. doi:10.1038/nmeth.1486.

Hernández-Serna, A., Jiménez-Segura, L.F., 2014. Automatic identification of species with neural networks. PeerJ 2, e563. doi:10.7717/peerj.563.

Hess, S.T., Girirajan, T.P.K., Mason, M.D., 2006. Ultra-high resolution imaging by fluorescence photoactivation localization microscopy. Biophysical Journal 91, 4258–4272. doi:10.1529/biophysj.106.091116.

Huh, S., Lee, D., Murphy, R.F., 2009. Efficient framework for automated classification of subcellular patterns in budding yeast. Cytometry Part A 75A, 934–940. doi:10.1002/cyto.a.20793.

Intarapanich, A., Kaewkamnerd, S., Pannarut, M., Shaw, P.J., Tongsima, S., 2016. Fast processing of microscopic images using object based extended depth of field. BMC Bioinformatics 17 (Suppl. 19), 516. doi:10.1186/s12859-016-1373-2.

Islam, M.M., D. Zhang, G. Lu, 2008. A geometric method to compute directionality features for texture images. In: Presented at the IEEE International Conference of Multimedia and Expo (ICME), IEEE. doi:10.1109/ICME.2008.4607736.

Isnanto, R.R., Zahra, A.A., Julietta, P., 2016. Pattern recognition on herbs leaves using region-based invariants feature extraction. In: Proceedings of the 2016 3rd International Conference on Information Technology, Computer, and Electrical Engineering (ICITACEE). doi:10.1109/icitacee.2016.7892491.

Jamil, N., Hussin, N.A., Nordin, S., Awang, K., 2015. Automatic plant identification: Is shape the key feature? Procedia Computer Science 76, 436–442. doi:10.1016/j.procs.2015.12.287.

Jayaram, M.A., Fleyeh, H., 2016. Convex hulls in image processing: A scoping review. American Journal of Intelligent Systems 6, 48–58. doi:10.5923/j.ajis.20160602.03.

Jia, D., Xu, Q., Xie, Q., Mio, W., Deng, W., 2016. Automatic stage identification of Drosophila egg chamber based on DAPI images. Scientific Reports 6. doi:10.1038/srep18850.

Kalafi, E.Y., Boon, T.W., Town, C., Dhillon, S.K., 2016. Automated identification of Monogeneans using digital image processing and k-nearest neighbour approaches. BMC Bioinformatics 17 (Suppl. 19), S511. doi:10.1186/s12859-016-1376-z.

Kass, M., Witkin, A., Terzopoulos, D., 1988. Snakes: Active contour models. International Journal of Computer Vision 1 (4), 321–331. doi:10.1007/BF00133570.

Keller, P.J., Stelzer, E.H.K., 2010. Digital scanned laser light sheet fluorescence microscopy. Cold Spring Harb Protocots 2010 (5), pdb.top78.

Kelley, L.C., Wang, Z., Hagedorn, E.J., et al., 2017. Live-cell confocal microscopy and quantitative 4D image analysis of anchor-cell invasion through the basement membrane in Caenorhabditis elegans. Nature Protocols 12, 2081–2096. doi:10.1038/nprot.2017.093.

Kho, S.J., Manickam, S., Malek, S., Mosleh, M.A., Dhillon, S.K., 2017. Automated plant identification using artificial neural network and support vector machine. Frontiers in Life Science 10 (1), 98–107. doi:10.1080/21553769.2017.1412361. org.

Kiranyaz, S., Ince, T., Pulkkinen, J., et al., 2011. Classification and retrieval on macroinvertebrate image databases. Computers in Biology and Medicine 41, 463–472.

Kotsiantis, S.B., 2007. Supervised machine learning: A review of classification techniques. Informatica. 31 (3), 249–268.

Kuhn, M., Johnson, K., 2013. Applied Predictive Modeling. vol. 26. Springer.

Kumar, A., Patidar, V., Khazanchi, D., Saini, P., 2015. Role of feature selection on leaf image classification. Journal of Data Analysis and Information Processing 03, 175–183. doi:10.4236/jdaip.2015.34018.

Kvilekval, K., et al., 2010. Bisque: A platform for bioimage analysis and management. Bioinformatics 26, 544–552.

Lamprecht, M., Sabatini, D., Carpenter, A., 2007. CellProfiler™: Free, versatile software for automated biological image analysis. BioTechniques 42, 71–75. doi:10.2144/000112257.

Lewis, J.M., Van Der Maaten, L., De Sa,.V.R., 2012. A behavioral investigation of dimensionality reduction. In: Proceedings of the 34th Annual Conference of the Cognitive Science Society, pp. 671–676.

Lim, J.S., 1990. Two-dimensional signal and image processing. Englewood Cliffs, NJ: Prentice Hall, p. 710.

Liu, R., Lin, S., Rallo, R., et al., 2012. Automated phenotype recognition for zebrafish embryo based in vivo high throughput toxicity screening of engineered nano-materials. PLOS ONE 7 (4), e35014. doi:10.1371/journal.pone.0035014.

Li, L., Zhang, Q., Ding, Y., et al., 2013. A computer-aided spectroscopic system for early diagnosis of melanoma. In: Proceedings of the 2013 IEEE 25th International Conference on Tools with Artificial Intelligence. doi:10.1109/ictai.2013.31.

Long, F., Peng, H., Myers, E., 2007. Automatic Segmentation of Nuclei In 3D Microscopy Images of C.elegans. 2007 4th IEEE International Symposium on Biomedical Imaging: From Nano to Macro. doi:10.1109/isbi.2007.356907.

Lorenzo-Ginori, J.V., Curbelo-Jardines, W., López-Cabrera, J.D., Huergo-Suárez, S.B., 2013. Cervical cell classification using features related to morphometry and texture of nuclei. In: Ruiz-Shulcloper, J., Sanniti di Baja, G. (Eds.), Progress in Pattern Recognition, Image Analysis, Computer Vision, and Applications. CIARP 2013. Lecture Notes in Computer Science, vol 8259. Berlin, Heidelberg: Springer.

Luo, Q., Gao, Y., Luo, J., et al., 2011. Automatic Identification of Diatoms with Circular Shape using Texture Analysis. Journal of Software 6 (3), doi:10.4304/jsw.6.3.428-435.

Mao, H., Tao, L., 2017. Segmentation and classification of two-channel C. elegans nucleus-labeled fluorescence images. BMC Bioinformatic 18, 412. doi:10.1186/s12859-017-1817-3.

Mayer, J., Swoger, J., Ozga, A.J., Stein, J.V., Sharpe, J., 2012. Quantitative measurements in 3-dimensional datasets of mouse lymph nodes resolve organ-wide functional dependencies. Computational and Mathematical Methods in Medicine 2012. Article ID 128431, 8 pages.

Myers, G., 2012. Why bioimage informatics matters. Nature Methods 9 (7), 659–660. doi:10.1038/nmeth.2024.

Mingqiang, Y., Kidiyo, K., Joseph, R., 2008. A survey of shape feature extraction techniques. Pattern Recognition Techniques, Technology and Applications. doi:10.5772/6237.

Misselwitz, B., Strittmatter, G., Periaswamy, B., et al., 2010. Enhanced CellClassifier: A multi-class classification tool for microscopy images. BMC Bioinformatics 11, 30. doi:10.1186/1471-2105-11-30.

Mosleh, M.A.A., Baba, M.S., Sorayya, M., Almaktari, R.A., 2016. Ceph-X: Development and evaluation of 2D cephalometric system. BMC Bioinformatics 17 (Suppl. 19), S499. doi:10.1186/s12859-016-1370-5.

Mosleh, M.A., Mansor, H., Malek, S., Milow, P., Salleh, A., 2012. A preliminary study on automated freshwater algae recognition and classification system. BMC Bioinformatics 13, S25. doi:10.1186/1471-2105-13-25.

Munisami, T., Ramsurn, M., Kishnah, S., Pudaruth, S., 2015. Plant leaf recognition using shape features and colour histogram with K-nearest neighbour classifiers. Procedia Computer Science 58, 740–747. doi:10.1016/j.procs.2015.08.095.

Murphy, R.F., 2014. A new era in bioimage informatics. Bioinformatics 30, 1353. doi:10.1093/bioinformatics/btu158.

Murat, M., Chang, S., Abu, A., Yap, H.J., Yong, K., 2017. Automated classification of tropical shrub species: A hybrid of leaf shape and machine learning approach. PeerJ. 5. doi:10.7717/peerj.3792.

Mythili, C., Kavitha, V., 2011. Efficient Technique for Color Image Noise Reduction. The research bulletin of Jordan. ACM.

Nguyen B.P., Heemskerk H., So P.T.C., Tucker-Kellogg, L., 2016. Superpixel-based segmentation of muscle fibers in multi-channel microscopy. BMC Systems Biol. doi:10.1186/s12918-016-0372-2. In: Proceedings of the 16th International Conference on Bioinformatics (InCoB 2017): Available at: http://incob.apbionet.org/incob17 (accessed 18.11.16).

Niwas, S.I., Palanisamy, P., Sujathan, K., Bengtsson, E., 2013. Analysis of nuclei textures of fine needle aspirated cytology images for breast cancer diagnosis using Complex Daubechies wavelets. Signal Processing 93, 2828–2837. doi:10.1016/j.sigpro.2012.06.029.

Obara, B., Grau, V., Fricker, M.D., 2012. A bioimage informatics approach to automatically extract complex fungal networks. Bioinformatics 28, 2374–2381. doi:10.1093/bioinformatics/bts364.

Oneill, M., 2007. Daisy. Systematics Association Special Volumes Automated Taxon Identification in Systematics. 101–114. doi:10.1201/9781420008074.ch7.

Orlov, N., Shamir, L., Macura, T., et al., 2008. WND-CHARM: Multi-purpose image classification using compound image transforms. Pattern Recognition Letters 29, 1684–1693. doi:10.1016/j.patrec.2008.04.013.

Pawley, J. (Ed.), 2006. Handbook Of Biological Confocal Microscopy. doi:10.1007/978-0-387-45524-2.

Peng, H., 2008. Bioimage informatics: A new area of engineering biology. Bioinformatics 24, 1827–1836.

Permuter, H., Francos, J., Jermyn, I., 2006. A study of Gaussian mixture models of color and texture features for image classification and segmentation. Pattern Recognition 39, 695–706.

Perre, P., Faria, F.A., Jorge, L.R., et al., 2016. Toward an automated identification of anastrepha fruit flies in the fraterculus group (Diptera, Tephritidae). Neotropical Entomology 45, 554–558. doi:10.1007/s13744-016-0403-0.

Pin Tian, D., 2013. A Review on Image Feature Extraction and Representation Techniques. International Journal of Multimedia and Ubiquitous Engineering 8 (4),

Puniyani, K., Xing, E.P., 2013. GINI: From ISH Images to Gene Interaction Networks. PLoS Computational Biology 9 (10), doi:10.1371/journal.pcbi.1003227.

Rosado, L., Correia Da Costa, J.M., Elias, D., Cardoso, J.S., 2016. A review of automatic malaria parasites detection and segmentation in microscopic images. Anti-Infective Agents 14, 11–22. doi:10.2174/2211352514011160302121107.

Rother, C., Kolmogorov, V., Blake, A., 2004. GrabCut. ACM SIGGRAPH 2004 Papers on - SIGGRAPH 04. doi:10.1145/1186562.1015720.

Rust, M.J., Bates, M., Zhuang, X., 2006. Sub-diffraction-limit imaging by stochastic optical reconstruction microscopy (STORM). Nature Methods 3, 793–796. doi:10.1038/nmeth929.

Sabeena Beevi, K., Madhu, S. Nair, Bindu, G.R., 2017. A Multi-Classifier System for Automatic Mitosis Detection in Breast Histopathology Images using Deep Belief Networks. IEEE Journal of Translational Engineering in Health and Medicine 5 (1), 1–11.

Sadanandan, S.K., Ranefall, P., Guyader, S.L., Wählby, C., 2017. Automated training of deep convolutional neural networks for cell segmentation. Scientific Reports 7 (1), 7860. doi:10.1038/s41598-017-07599-6.

Saeys, Y., Inza, I., Larranaga, P., 2007. A review of feature selection techniques in bioinformatics. Bioinformatics 23, 2507–2517. doi:10.1093/bioinformatics/btm344.

Santhi, N., Pradeepa, C., Subashini, P., Kalaiselvi, S., 2013. Automatic Identification of Algal Community from Microscopic Images. Bioinformatics and Biology Insights 7. doi:10.4137/bbi.s12844.

Sánchez, M.G., Vidal, V., Verdú, G., Mayo, P., Rodenas, F., 2012. Medical image restoration with different types of noise. In: Proceedings of the 34th Annual International Conference of the IEEE Engineering in Medicine and Biology Society, pp. 4382–4385.

Savkare, S., Narote, S., 2012. Automatic System for Classification of Erythrocytes Infected with Malaria and Identification of Parasites Life Stage. Procedia Technology 6, 405–410. doi:10.1016/j.protcy.2012.10.048.

Schmidt, T., Dürr, J., Keuper, M., et al., 2013. Variational attenuation correction in two-view confocal microscopy. BMC Bioinformatics 14, 366. doi:10.1186/1471-2105-14-366.

Schmid, B., Schindelin, J., Cardona, A., Longair, M., Heisenberg, M., 2010. A high-level 3D visualization API for Java and ImageJ. BMC Bioinformatics 11, 274.

Seyyid, A.M., 2015. A comparative study of feature extraction methods in images classification. International Journal of Image, Graphics and Signal Processing 3, 16–23. doi:10.5815/ijigsp.2015.03.03.

Sommer, C., Straehle, C., Kothe, U., Hamprecht, F.A. 2011. Ilastik: Interactive learning and segmentation toolkit. In: Proceedings of the 2011 IEEE International Symposium on Biomedical Imaging: From Nano to Macro. doi:10.1109/isbi.2011.5872394.

Song, Y., Cai, W., Huang, H., et al., 2016. Bioimage classification with subcategory discriminant transform of high dimensional visual descriptors. BMC Bioinformatics 17 (1), 465. doi:10.1186/s12859-016-1318-9.

Song, Y., Cai, W., Huang, H., et al., 2013. Region-based progressive localization of cell nuclei in microscopic images with data adaptive modeling. BMC Bioinformatics 14, 173. doi:10.1186/1471-2105-14-173.

Song, Y., Ji, Z., Sun, Q., 2014. An extension Gaussian mixture model for brain MRI segmentation. In: Proceedings of the 2014 36th Annual International Conference of the IEEE Engineering in Medicine and Biology Society. doi:10.1109/embc.2014.6944676.

Spanhol, F.A., Oliveira, L.S., Petitjean, C., Heutte, L., 2016. Breast cancer histopathological image classification using Convolutional Neural Networks. In: Proceedings of the 2016 International Joint Conference on Neural Networks (IJCNN). doi:10.1109/ijcnn.2016.7727519.

Swedlow, J.R., 2012. Innovation in biological microscopy: Current status and future directions. Bioessays 34, 333–340.

Uhlen, M., et al., 2010. Towards a knowledge-based human protein atlas. Nature Biotechnology 28, 1248–1250.

Uhlmann, V., Singh, S., Carpenter, A.E., 2016. CP-CHARM: Segmentation-free image classification made accessible. BMC Bioinformatics 17, 51. doi:10.1186/s12859-016-0895-y.

Vapnik, V.N., Vapnik, V., 1998. Statistical Learning Theory. vol. 1. New York: Wiley.

Verikas, A., Gelzinis, A., Bacauskiene, M., et al., 2012. Automated image analysis- and soft computing-based detection of the invasive dinoflagellate Prorocentrum minimum (Pavillard) Schiller. Expert Systems with Applications 39 (5), 6069–6077. doi:10.1016/j.eswa.2011.12.006.

Villa, A.G., Salazar, A., Vargas, F., 2017. Towards automatic wild animal monitoring: Identification of animal species in camera-trap images using very deep convolutional neural networks. Ecological Informatics 41, 24–32. doi:10.1016/j.ecoinf.2017.07.004.

Wang, L., Zhang, K., Liu, X., et al., 2017. Comparative analysis of image classification methods for automatic diagnosis of ophthalmic images. Scientific Reports 7, 41545. doi:10.1038/srep41545.

Xu, M., Papageorgiou, D.P., Abidi, S.Z., et al., 2017. A deep convolutional neural network for classification of red blood cells in sickle cell anemia. PLOS Computational Biology 13 (10), doi:10.1371/journal.pcbi.1005746.

Xu, Y., Yang, F., Zhang, Y., Shen, H., 2014. Bioimaging-based detection of mislocalized proteins in human cancers by semi-supervised learning. Bioinformatics 31, 1111–1119. doi:10.1093/bioinformatics/btu772.

Xinding, S., Zhenming, X., Changsheng, X., Yan, W. (n.d.). Adaptive Kalman filtering approach of color noise in cephalometric image. Proceedings of Third International Conference on Signal Processing (ICSP96). doi:10.1109/icsigp.1996.567341.

Yang, H., Ahuja, N., 2014. Automatic segmentation of granular objects in images: Combining local density clustering and gradient-barrier watershed. Pattern Recognition 47 (6), 2266–2279. doi:10.1016/j.patcog.2013.11.004.

Yalcin, H., Razavi, S., 2016. Plant classification using convolutional neural networks. In: Proceedings of the 2016 Fifth International Conference on Agro-Geoinformatics (Agro-Geoinformatics). doi:10.1109/agro-geoinformatics.2016.7577698.

Yuan, L., Woodard, A., Ji, S., et al., 2012. Learning sparse representations for fruit-fly gene expression pattern image annotation and retrieval. BMC Bioinformatics 13, 107. doi:10.1186/1471-2105-13-107.

Yue, Y., Croitoru, M., Bidani, A., Zwischenberger, J., Clark, J., 2006. Nonlinear multiscale wavelet diffusion for speckle suppression and edge enhancement in ultrasound images. IEEE Transactions on Medical Imaging 25 (3), 297–311. doi:10.1109/tmi.2005.862737.

Yusof, R., Khalid, M., Khairuddin, A.S., 2013. Application of kernel-genetic algorithm as nonlinear feature selection in tropical wood species recognition system. Computers and Electronics in Agriculture 93, 68–77. doi:10.1016/j.compag.2013.01.007.

Zhao, M., An, J., Li, H., et al., 2017. Segmentation and classification of two-channel C. elegans nucleus-labeled fluorescence images. BMC Bioinformatics 18, 412. doi:10.1186/s12859-017-1817-3.

Zhou, J., Lamichhane, S., Sterne, G., Ye, B., Peng, H., 2013. Biocat: A pattern recognition platform for customizable biological image classification and annotation. BMC Bioinformatics 14, 291.

Zitová, B., Flusser, J., 2003. Image registration methods: A survey. Image and Vision Computing 21, 977–1000. doi:10.1016/s0262-8856(03)00137-9.

Bioimage Databases

Arpah Abu and Sarinder K Dhillon, University of Malaya, Kuala Lumpur, Malaysia

Introduction

In biology, image data can be classified as high-resolution microscopy data, medical imaging data, observed effects in cell, high content screening, molecular imaging and biodiversity (organisms) data. These data are generated via advance experimental and field work as well as from high throughput analysis using the imaging technology. Besides images of organisms, images of protein structures and structural models of proteins and RNA molecules are also available for browsing, querying as well as analysis. Besides the raw source of images obtained, the annotations that describe the related details are valuable source of information. However, till date, very little has been reported in the development of bioimage databases, particularly for storing and pooling the vast amount of scientific image data both at molecular or cellular level. Nevertheless, a substantial amount of work has been done in documenting digital pictures of organisms as in terms of technology it is much simple and affordable.

Generally, bioimage databases are developed for storage and retrieval purposes. The data formats used by these databases are specific to the type of images stored. Retrieval of data from image databases can be done in two ways; the first method accepts only text query to retrieve images (or the content of the database) and the other method accepts image query which referred as content-based image retrieval. In this article, existing work done using both of these approaches are presented. The problems concerning bioimage database are discussed and finally we present a viable solution to develop bioimage databases.

Bioimage Databases Using Text-Based Query

In this section, some of the important and widely used bioimage databases are presented. These databases are text query based, which means the input in the search function is string rather than an image.

Flybase (see "Relevant Websites section")

FlyBase (McQuilton et al., 2012) is an image database of Drosophila genes and genomes. One of the query tools provided in this system is ImageBrowse (see **Fig. 1**) for browsing the images based on the organ system, life-cycle stages, major tagma, germ layer and all species images.

The images in this database were collected from the researchers' publications such as journal articles and books. The retrieved images are listed in unranked order along with short description (see **Fig. 2**).

Planetary Biodiversity Inventory – PBI (see "Relevant Websites section")

The Planetary Biodiversity Inventory – PBI (Gregorev et al., 2003) for Plant Bugs provides a resource about the global plant bug subfamilies Orthotylinae and Phylinae (Insecta: Heteroptera: Miridae) to various users such as systematists, insect ecologists, conservation biologists, the general public, and students via the Internet. In the latest version 6, all users have access to a systematic catalog of plant bug taxa with more than 10,000 species, a digital library of ~30,000 pages, locality database, and updated Methods and News and Outputs. Furthermore, users can view taxon treatments and distribution maps though discoverlife.org by using the search box provided on the page as well as through a dedicated web page, provided in "Relevant Websites section". Through these portals specimen records are available for approximately 4000 species and more than 250,000 specimens. 56 papers have so far been published with at least 15 additional papers in preparation or in press.

The Plant Bug PBI was made possible with the financial support from the National Science Foundation (NSF) from September 1, 2003 to August 31, 2011.

The images along with taxon information and distribution can be retrieved using Image Database interface in **Fig. 3**. Users can enter the relevant bug name such as ant, bee and wasp and the retrieved images are listed in unranked order.

Specimen Image Database – SID (see "Relevant Websites section")

Specimen Image Database – SID (Simon and Vince, 2011) is a searchable database of high-resolution images for phylogenetic and biodiversity research. This database is intended as a reference collection of named specimens and a resource for comparative morphological research. Each image is accompanied by fully searchable annotation, and can be browsed, searched or downloaded. Public users as well can register in this database and the registered users can add, annotate or label the images. Currently, this database is devoted to the insect order Phthiraptera (lice) and contains 7650 images of 440 taxa.

Fig. 1   ImageBrowse in FlyBase.

Key features of SID (see **Fig. 4**) include web upload/download of images, bulk and single image annotation via web forms, extensive browse and search options by text query, web service facility, web utility to label specific image features, taxonomy served and validated independently by the Glasgow Taxonomy Name Server, alias addresses for images by accession number and freeware which is allows anyone to set up the database and serve their own images. The retrieved images are listed in unranked order with the taxon information, host as well as image properties.

Orchidaceae of Central Africa (see "Relevant Websites section")

Orchidaceae of central Africa (Droissart *et al.*, 2012) is an image database on orchids of Central Africa. It includes scientific names, distribution data, photos, interactive identification keys, web links, and references. The images presented in this database are from intensive fieldwork carried out by the Université Libre de Bruxelles, the Université de Yaoundé I, the Missouri Botanical Garden and the Institut de Recherche pour le Développement in several countries of Central Africa. This information will promote land conservation and sound natural resource management in tropical Africa. The information presented in this database may be suitable for academic, educational, and general purpose.

The database currently comprises information on about 300 illustrated orchid species. The images can be retrieved by using text query based on the taxon name; or browsing the entire database or with few options such as taxa by genus, country or author as shown in **Fig. 5**. The retrieved images are listed in unranked order with species and authors name.

MonoDb (see "Relevant Websites section")

MonoDb (Andy and James, 2012) is another biodiversity database that provides image gallery as one of the features in the database. MonoDb is a web-host for the parasite monogenea. As mentioned in this website, the purpose of this website is to help children, adults, experts and non-experts to learn more about this fascinating group of animals. Images in this database can be

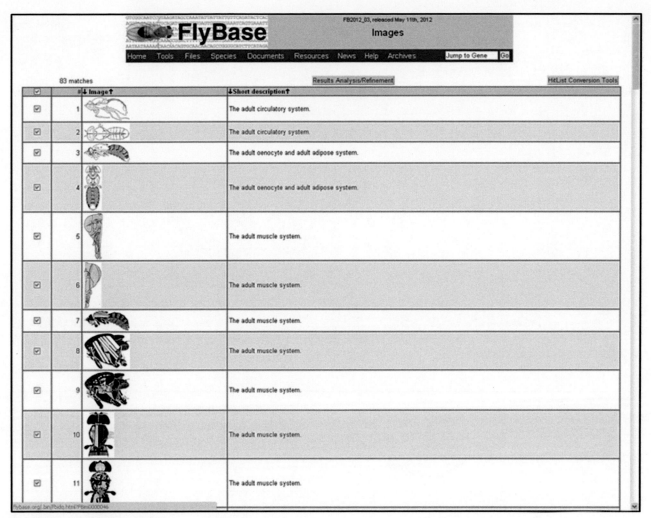

Fig. 2 Retrieval results from FlyBase.

retrieved by browsing the entire images provided in the database. The images are listed randomly and no information attached to the images (see **Fig. 6**).

Cell Image Library – CIL-CCDB (see "Relevant Websites section")

CIL-CCDB (Orloff *et al.*, 2013) is a searchable database and archive of cellular images. As a repository for microscopy data, it accepts all forms of cell imaging from light and electron microscopy, including multi-dimensional images, Z- and time stacks in a broad variety of raw-data formats, as well as movies and animations. A CIL-CCDB was developed as a resource and a tool for the cell biology community such as for researchers, educators and the general public. Images in this database can be retrieved by browsing the entire images provided in the database and by using the simple or advance search features as shown in **Fig. 7**.

ModBase: Database of Comparative Protein Structure Models (see "Relevant Websites section")

MODBASE (Ursula *et al.*, 2014) is a queryable database of annotated protein structure models. The models are derived by ModPipe, an automated modelling pipeline relying on the programs PSI-BLAST and MODELLER. The database also includes the fold assignments and alignments on which the models were based. ModBase was developed by Sali Lab. **Fig. 8** shows the main home page of ModBase. The data was organized into datasets in which can be used either available to the public, to the academicians or to specific users.

SWISS-MODEL Repository (see "Relevant Websites section")

The SWISS-MODEL Repository (Bienert *et al.*, 2017) as shown in **Fig. 9** is a database of annotated 3D protein structure models generated by the SWISS-MODEL homology-modelling pipeline. It was developed by Swiss Institute of Bioinformatics. The aim of

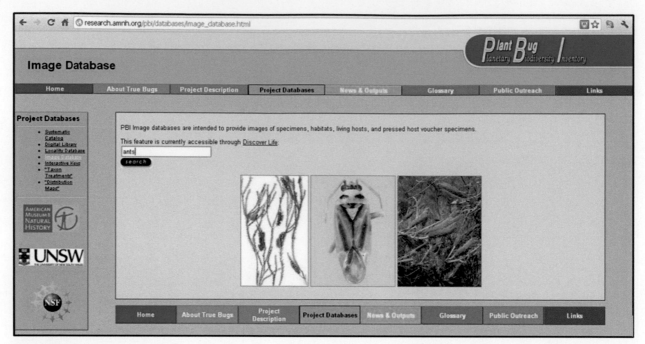

Fig. 3   PBI – Image Database interface.

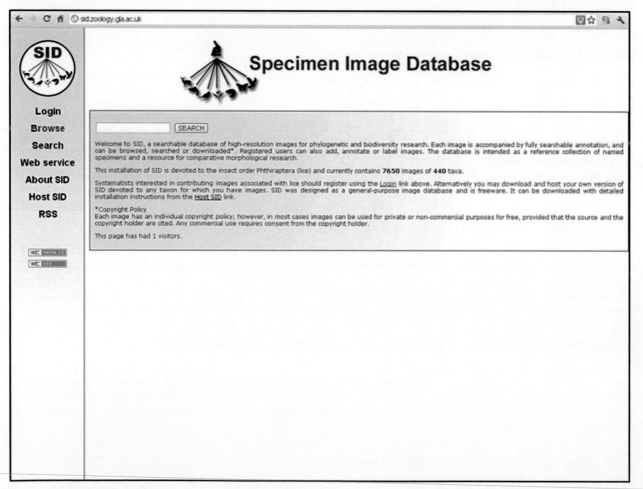

Fig. 4   SID – Search page interface.

Fig. 5 Orchidaceae of central Africa – Search page interface.

Fig. 6 Monogenean images in MonoDb.

the SWISS-MODEL Repository is to provide access to an up-to-date collection of annotated 3D protein models generated by automated homology modelling for relevant model organisms and experimental structure information for all sequences in UniProtKB. Regular updates ensure that target coverage is complete, that models are built using the most recent sequence and template structure databases, and that improvements in the underlying modelling pipeline are fully utilized. It also allows users to assess the quality of the models using the latest QMEAN results. If a sequence has not been modeled, the user can build models interactively via the SWISS-MODEL workspace.

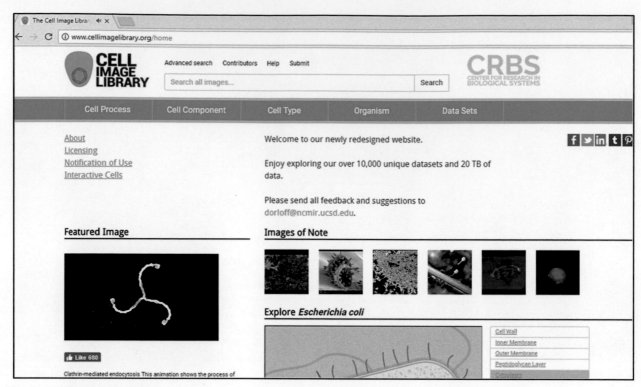

Fig. 7 Home page of CIL-CCDB.

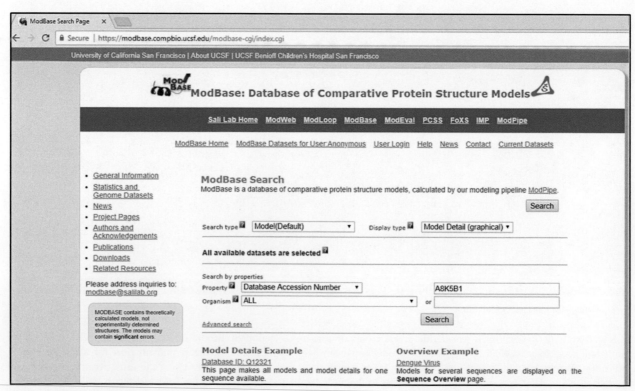

Fig. 8 ModBase home page.

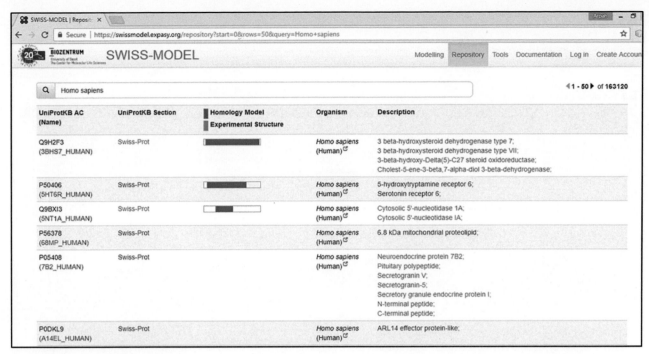

Fig. 9 Search page in SWISS MODEL.

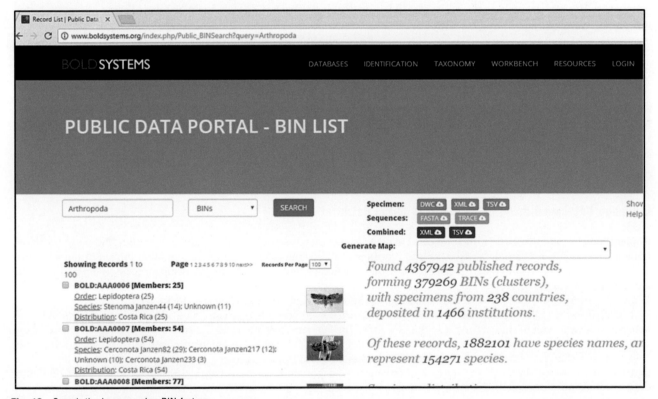

Fig. 10 Search the images using BIN feature.

BOLD SYSTEMS (see "Relevant Websites section")

BOLD is a cloud-based data storage and analysis platform developed at the Centre for Biodiversity Genomics in Canada. It consists of four main modules, a data portal, an educational portal, a registry of BINs (putative species), and a data collection and analysis

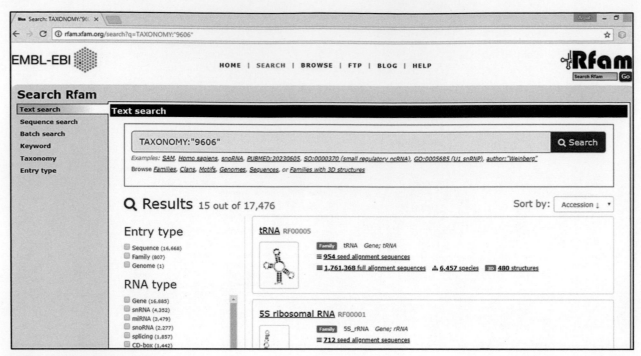

Fig. 11   Rfam searching page.

workbench. The BIN database as shown in **Fig. 10** is a searchable database of Barcode Index Numbers (BINs), sequence clusters that closely approximate species. The retrieved images are displayed with their description that give meaning to the images.

Rfam (see "Relevant Websites section")

The Rfam (Kalvari *et al.*, 2017) database is a collection of RNA families, each represented by multiple sequence alignments, consensus secondary structures and covariance models (CMs). The Rfam database is curated and maintained at the European Bioinformatics Institute in Cambridge, UK. The resource is collaboration with researchers from Sean Eddy lab at Harvard University, USA. The secondary structures images are retrieved as results with annotated information using a text-based query as shown in **Fig. 11**.

Bioimage Databases Using Content-Based Image Retrieval

Content-based Image Retrieval (CBIR) is a concept used in image databases. The work presented in Section "Bioimage Databases Using Text-Based Query" are databases with text-based retrieval capabilities. Hence the functionality is limited in the sense that only text query can be used to retrieve images, unlike CBIR database systems, whereby the query input can be an image. Due to the escalation of biological data, new ways have been introduced to store and analyze data. The need of efficient management of the vast visual information has introduced new techniques and tools for content-based image retrieval approach.

Fig. 12 shows a typical architecture of CBIR system depicted from Torres and Falcao (2006). The goal of this approach is to search and retrieve a set of similar images to the user query. The interface layer allows user to send a query image. The images from image database are then assigned as training set images. Both query and training set images features (such as shape, texture and color) are extracted and formed the feature vectors in the feature space. The similarity comparison (such as Euclidean distance and Mahalanobis distance) between the query and training set images are then measured, and the classifier (such as minimum distance, maximum distance and k-nearest classifier) is used to classify the retrieved images. The results are then returned to the user through user interface. As for the results, the retrieved images must be accurate, relevant and related to the user query.

Rui *et al.* (1999), Smeulders *et al.* (2000), Feng *et al.* (2003) and Torres and Falcao (2006) introduced some fundamental techniques as well as technical achievements in the field of CBIR.

As mentioned previously, biologists produce a vast of digital images. From these images, it can be used for identification as well as for teaching and research. Wang *et al.* (2012) presents a work on butterfly family identification using CBIR, while Sheikh *et al.* (2011) developed CBIR system for various types of marine life images. A content-based pattern analysis system for a biological specimen collection was developed by Joyita and Gardner (2007) and a CBIR system for fish taxonomy research was also

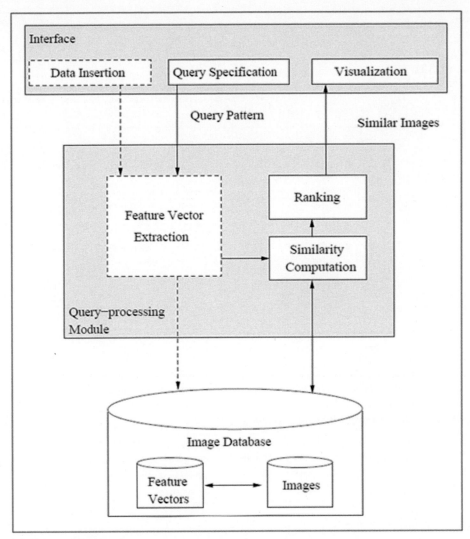

Fig. 12 A typical architecture of CBIR system.

developed by Chen *et al.* (2005). As stated in Wang *et al.* (2012), CBIR is applied because of its capacity for mass processing and operability.

On the other hand, because of the heterogeneous data, complexity of the biology images as well as the images descriptions are often ignored to attach to together with the images, there are few works such as mentioned in EKEY (2012), Torres *et al.* (2004) and Murthy *et al.* (2009) to enhance the CBIR capability

The CBIR approach has been applied in medical for teaching, research and diagnostics on diseases. The benefits and future directions have been discussed in Müller *et al.* (2004). In Kak and Pavlopoulou (2002), CBIR is used to automate retrieval from large medical image databases and presented solutions to some of them in the specific context of HRCT images of lung and liver. Scott and Chi-Ren (2007) presents a knowledge-driven multidimensional indexing structure for biomedical media database retrieval. While in El-Naqa *et al.* (2004) and Rosa *et al.* (2008), they use CBIR approach for digital mammographic masses. As to improve the retrieval efficiency, some of the works such as mentioned in Demner-Fushman *et al.* (2009), Hsu *et al.* (2009) and You *et al.* (2011), they have combining the metadata approach into CBIR approach.

The CBIR approach has been used as well in several applications such as face identification, digital libraries, historical research, medical and geology. It is probably the most useful application in biology. In this article, some of these applications are presented as in this section, focusing on medical and biology applications.

Google Image Search (see "Relevant Websites section")

Google Image Search is a Google's CBIR system. Query specification follows the query by example using external images. However, it does not work on all images. Metadata query function is also provided.

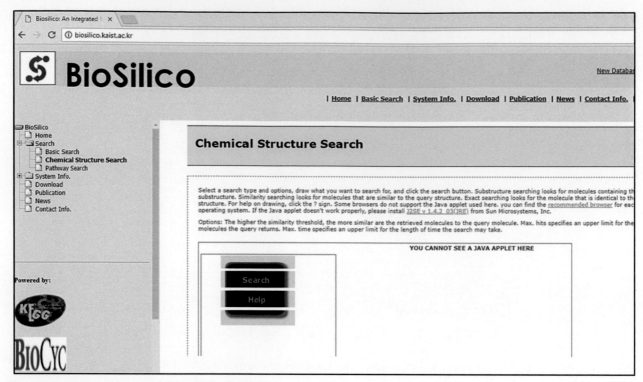

Fig. 13   Chemical structure search using CBIR in BioSilico.

Macroglossa Visual Search (see "Relevant Websites section")

Macroglossa is a visual search engine based on the comparison of images, coming from an Italian Group. For image retrieval, query specification follows the query by example using external images or query by category such animals, biological, panoramic, artistic or botanical. Macroglossa supports all popular image extensions such jpeg, png, bmp, gif and video formats such avi, mov, mp4, m4v, 3gp, wmv, mpeg.

BioSilico (see "Relevant Websites section")

BioSilico is a web-based database system that facilitates the search and analysis of metabolic pathways. Heterogeneous metabolic databases including LIGAND, ENZYME, EcoCyc and MetaCyc are integrated in a systematic way, thereby allowing users to efficiently retrieve the relevant information on enzymes, biochemical compounds and reactions. In addition, it provides well-designed view pages for more detailed summary information as shown in **Fig. 13**.

CATH (see "Relevant Websites section")

The CATH (Sillitoe *et al.*, 2015) database is a free, publicly available online resource that provides information on the evolutionary relationships of protein domains. It was created in the mid-1990s by Professor Christine Orengo and colleagues, and continues to be developed by the Orengo group at University College London. The CATH database is a hierarchical domain classification of protein structures in the Protein Data Bank. Protein structures are classified using a combination of automated and manual procedures. There are four major levels in this hierarchy which are Class, Architecture, Topology, and Homologous superfamily. For any given structure classified in the database, CATH gives you information on the structure and function of that protein. The evolutionary relationships involving the structure of interest and other proteins in the database can also be determined. These tasks can be performed by searching using Text or ID, Sequence or Structure in PDB image format as shown in **Fig. 14**.

IDR – Image Data Resource (see "Relevant Websites section")

IDR (Williams *et al.*, 2017) is online, public data repository seeks to store, integrate and serve image datasets from published scientific studies. It was developed by University of Dundee & Open Microscopy Environment. IDR allow the community to search, view, mine and even process and analyze large, complex, multidimensional life sciences image data as shown in **Fig. 15**. Sharing data promotes the validation of experimental methods and scientific conclusions, the comparison with new data obtained by the global scientific community and enables data reuse by developers of new analysis and processing tools.

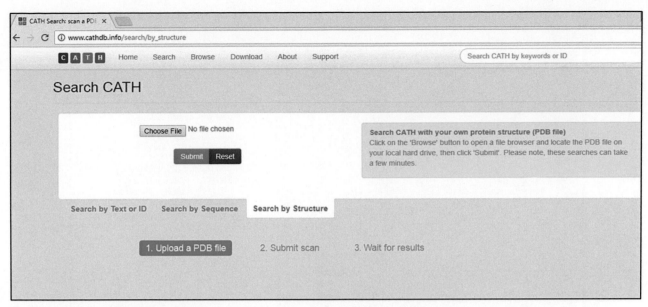

Fig. 14 CATH image database using image query.

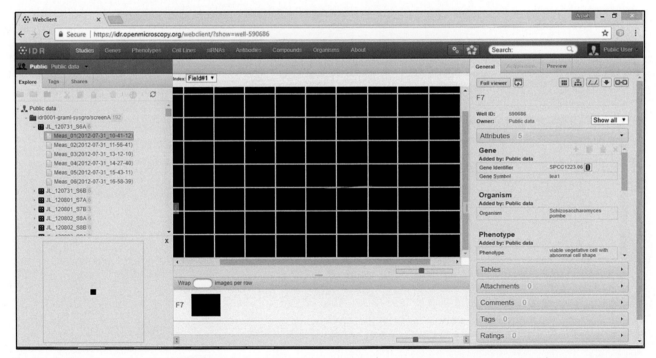

Fig. 15 Retrieved images from fungi dataset.

BioDIG (see "Relevant Websites section")

BioDIG (Oberlin *et al.*, 2013) a generic set a tool that allows linking of image data to genomic data. BioDIG features the following: rapid construction of web-based workbenches, community-based annotation, user management and web services. By using BioDIG to create websites, researchers and curators can rapidly annotate a large number of images with genomic information. The BioDIG includes an image module, a genome module and a user management module as shown in **Fig. 16**.

RNA 3D Motif Atlas (see "Relevant Websites section")

RNA 3D Motif Atlas (Petrov *et al.*, 2013) is a comprehensive and representative collection of internal and hairpin loop RNA 3D motifs extracted from the Representative Sets of RNA 3D structures. It was developed by BGSU RNA Structural Bioinformatics at

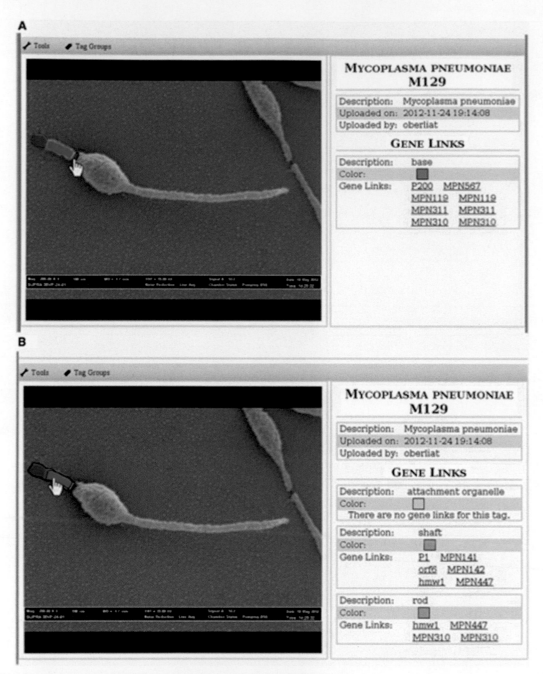

Fig. 16 The *Mycoplasma pneumonia* image annotated with genome data in BioDIG interface.

Bowling Green State University. The 3D images of RNA motif are retrieved using browsing featured motifs which are Kink-turn (as shown in **Fig. 17**), C-loop, Sarcin, Triple sheared, Double sheared T-loop, and GNRA.

Discussion on Bioimage Databases

Many issues have been discussed and suggested (Rui *et al.*, 1999; Smeulders *et al.*, 2000; Shandilya and Singhai, 2010) to improve the typical Content-based Image Retrieval (CBIR) based image databases such as involve user interaction, integration of multi-disciplines approach, relevance feedback and reducing semantic gap. The main purpose of these improvements is to enhance the efficiency of image retrieval.

Most previous works focused on image representation (Krishnapuram *et al.*, 2004; Wei *et al.*, 2006; Lamard *et al.*, 2007; Sergyan, 2008), classifier algorithm (Xin and Jin, 2004; Duan *et al.*, 2005; Liu *et al.*, 2008), the use of image database (Kak and Pavlopoulou, 2002), and relevance feedback (Stejić *et al.*, 2003; Zhang *et al.*, 2003; Ortega-Binderberger and Mehrotra, 2004;

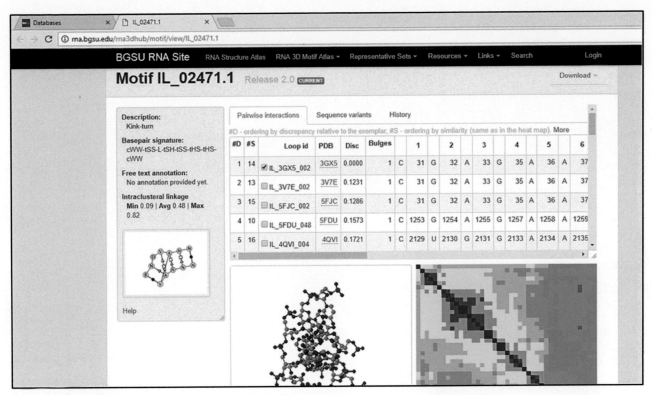

Fig. 17 Retrieved RNA 3D motifs using Kink-turn featured motif.

Wang and Ma, 2005; Wei and Li, 2006) to enhance the CBIR system. However, as mentioned in Liu *et al.* (2007), research focus has been shifted into reducing the semantic gap and identified five major categories of the-state-of-the-art techniques in narrowing down the semantic gap i.e., (i) using object ontology to define high-level concepts; (ii) using machine learning methods to associate low-level features with query concepts; (iii) using relevance feedback to learn users' intention; (iv) generating semantic template to support high-level image retrieval; (v) fusing the evidences from HTML text and the visual content of images for WWW image retrieval.

On the other hand, as stated in Torres *et al.* (2004), the implementation of CBIR systems raises several research challenges such as (i) new tool for annotating needs to be developed to deal with the semantic gap presented in images and their textual descriptions; (ii) automatic tool for extracting semantic features from images; (iii) development of new data fusion algorithm to support text-based and content-based retrieval when combining information of different heterogeneous formats; (iv) text mining techniques to be combined with content-based descriptions; and (v) investigating user interfaces for annotating, browsing and searching based on image content.

Related works focusing on reducing the semantic gap between the visual (low-level) features and the richness of human semantic (high-level features) is in progress. Lin *et al.* (2007) proposes the integration of textual and visual information for cross-language image retrieval; Zhang *et al.* (2011) presents automatic image tagging automatically assigns image with semantic keyword called tag, which significantly facilitates image search and organization; Aye and Thein (2012) presents a retrieval framework can support various types of queries and can accept multimedia examples and metadata-based document; and Lee and Wang (2012) presents a utilization of text- and photo- types of location information with a novel approach of information fusion that exploit effective image annotation and location based text-mining approaches to enhance identification of geographic location and spatial cognition.

Generally, there are two approaches to image retrieval which are textual-based that retrieves images based on human-annotated metadata, and content-based that analyzes the actual image data (Avril, 2005). Many digital libraries have annotation capabilities for image digital libraries such as Imense Image Search Portal (imense, 2007), Incogna Image Search (Incogna, 2012), and Anaktisi (Chatzichristofis *et al.*, 2010). Yet, in biology, very few are providing the capabilities to integrate text- and content-based annotation and retrieval of parts of images such as EKEY (EKEY, 2012), SuperIDR (Murthy *et al.*, 2009), and teaching tool for parasitology (Kozievitch *et al.*, 2010). EKEY is a web-based system that provides taxonomic classification, dichotomous key, text-based search and combination of shape and text-based search which taking into account fish shape outlines and textual terms. For the SuperIDR, instead of provides same features as EKEY, it enables the user interaction to add content, support for working with specific parts of images, performing content-based image annotation and retrieval and has pen-input capabilities which mimicking free-hand drawing and writing on paper. While in terms of database system, the relational database architecture was used for text annotation. Both systems are used in the Ichthyology domain. As an alternative approach to teach, compare and learn concept about parasites in general, research group in Kozievitch *et al.* (2010) adapted SuperIDR.

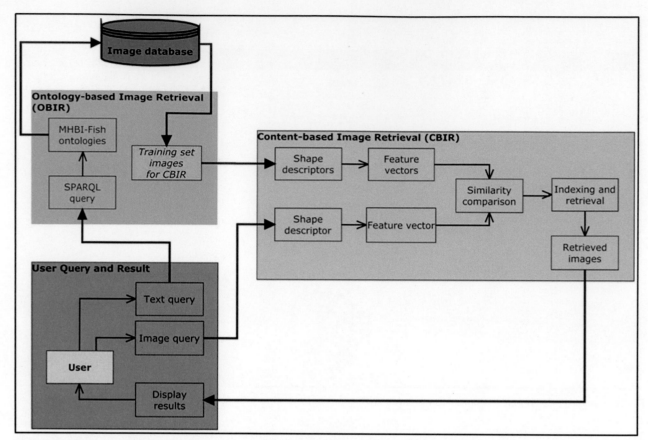

Fig. 18   The Model procedural flow.

To summarize, based on the review done, several main points are derived:-

(1) Most of the image databases provide only text query option to retrieve the images from the database.
(2) None of the existing systems uses ontology-based image annotation and retrieval to perform image pre-classification. The nearest are the EKEY and SuperIDR whereby each image is annotated with certain parameters such as species name. Thus, the user can customize their search according to these parameters to find the nearest match to the query image.
(3) Biology data is heterogeneous, containing complex images and terminology to describe the data is always evolving overtime. Thus, graph data is suitable approach for text data modelling.
(4) Both text- and content-based image retrieval approaches have their own advantages. Yet both approaches as well have a same limitation, which is, may retrieve irrelevant images.

Finally, in order to improve the efficiency of image retrieval in CBIR approach, one of the solutions to reduce the semantic gap limitation is by combining text-based image retrieval into CBIR. By using this approach, it will narrow down the most relevant images to be used for training set images. There are few aspects that are important to consider when developing the integrating text- and content-based image retrieval model, i.e., textual data representation, image annotations, query specification and the expected output of the retrieval process.

Suitable and proper vocabularies are needed in order to annotate the image because the images will be retrieved based on the vocabularies. The annotated data is required to represent in meaningful, dynamic and flexible manner so that any inclusion new vocabulary in future will be able to accommodate without changing the whole data structure. As for query specification, both textual and image query are needed. The last aspect is the output of the retrieval process, which is crucial in determining whether the retrieval process works well and in an efficient manner. Thus, to achieve this, the most relevance images must be retrieved with their annotation.

Proposed Solution for Bioimage Databases Using Ontology and CBIR Approaches

The details of the proposed solution in relation to the identified problems are presented here. The proposed solution aims to integrated both textual retrieval and image retrieval from image databases. The proposed image retrieval using ontology and CBIR approaches is shown in **Fig. 18**. This architecture is proposed by Arpah *et al.* (2013).

In this model, the Ontology Based Information Retrieval (OBIR) layer determines the training set for Content Based Image Retrieval (CBIR). Ontology-based image retrieval is used as technique to reduce the training images for the CBIR layer by eliminating the irrelevant images using the text-based query in OBIR layer. This technique is also referred as data reduction usually used in data pre-processing to obtain a reduced representation of the dataset, which is smaller in quantity, yet closely maintains the integrity of the original data.

Conclusion

In this article, bioimage databases are presented in two classified categories which are text-based retrieval and image-based retrieval or more commonly known as content-based image retrieval. Based on the reviews presented in this article, we conclude that bioimage databases have not been able to completely harness the power of computer vision and image analytics in building the data archives. As such, a substantial amount of work in this field is required to build scientific databases, using ontologies for text queries and content based image retrieval methods for image queries. The methodology presented by Arpah *et al.* (2013) and the initiative started by Williams *et al.* (2017) are good example of research to be modeled in building bioimage databases. Finally, for the advancement of science, particularly biology, access to a wide range of image data is necessary to conduct transformative research.

> *See also*: Bioinformatics Data Models, Representation and Storage. Biological Database Searching. Data Storage and Representation. Information Retrieval in Life Sciences. Integrative Bioinformatics. Integrative Bioinformatics of Transcriptome: Databases, Tools and Pipelines. Natural Language Processing Approaches in Bioinformatics

References

Andy, S., James, B., 2012, 2009. MonoDb homepage. Retrieved August 2011, from: http://www.monodb.org/index.php.

Abu, A., Lim, L.H.S., Sidhu, A.S., Dhillon, S.K., 2013. Biodiversity image retrieval framework for monogeneans. Systematics and Biodiversity 11 (1), 19–33.

Avril, S., 2005. Ontology-based image annotation and retrieval. Master of Science, University of Helsinki. Retrieved from: http://www.cs.helsinki.fi/u/astyrman/gradu.pdf.

Aye, K.N., Thein, N.L., 2012. Efficient indexing and searching framework for unstructured data. In: Zeng, Z.L.Y. (Ed.), Proceedings of the Fourth International Conference on Machine Vision, vol. 8349.

Bienert, S., Waterhouse, A., de Beer, T.A., *et al.*, 2017. The SWISS-MODEL repository - new features and functionality. Nucleic Acids Research 45 (D1), D313–D319.

Chatzichristofis, S.A., Zagoris, K., Boutalis, Y.S., Papamarkos, N., 2010. Accurate image retrieval based on compact composite descriptors and relevance feedback information. International Journal of Pattern Recognition and Artificial Intelligence 24 (2), 207–244.

Chen, Y., Henry, L., Bart, J., Teng, F., 2005. A content-based image retrieval system for fish taxonomy. In: Paper Presented at the Proceedings of the 7th ACM SIGMM International Workshop on Multimedia information retrieval, Hilton, Singapore.

Demner-Fushman, D., Antani, S., Simpson, M., Thoma, G.R., 2009. Annotation and retrieval of clinically relevant images. International Journal of Medical Informatics 78 (12), e59–e67. doi:10.1016/j.ijmedinf.2009.05.003.

Droissart, V., Simo, M., Sonké, B., Geerinck, D., Stévart, T., 2012. Orchidaceae of Central Africa. Retrieved August 2011, from: http://www.orchid-africa.net/.

Duan, L., Gao, W., Zeng, W., Zhao, D., 2005. Adaptive relevance feedback based on Bayesian inference for image retrieval. Signal Processing 85 (2), 395–399. doi:10.1016/j.sigpro.2004.10.006.

El-Naqa, I., Yongyi, Y., Galatsanos, N.P., Nishikawa, R.M., Wernick, M.N., 2004. A similarity learning approach to content-based image retrieval: Application to digital mammography. IEEE Transactions on Medical Imaging 23 (10), 1233–1244. doi:10.1109/tmi.2004.834601.

EKEY. (2012). EKEY - The Electronic Key for Identifying Freshwater Fishes Retrieved July, 2009, from http://digitalcorpora.org/corp/nps/files/govdocs1/054/054359.html.

Feng, D., Siu, W.C., Zhang, H.J., 2003. Multimedia Information Retrieval and Management: Technological Fundamentals and Applications. Springer.

Gregorev, N., Huber, B., Shah, M.R.A.V., *et al.*, 2003. Plant bug: Planetary biodiversity inventory, 2011, from: http://www.monodb.org/index.php.

Hsu, W., Antani, S., Long, L.R., Neve, L., Thoma, G.R., 2009. SPIRS: A Web-based image retrieval system for large biomedical databases. International Journal of Medical Informatics 78 (Suppl. 1), S13–S24. doi:10.1016/j.ijmedinf.2008.09.006.

imense, 2007. Imense Image Search Portal. Retrieved March 2010, from: http://imense.com/.

Incogna, 2012. Incogna image search. Retrieved August 2011, from: http://www.incogna.com/.

Kalvari, I., Argasinska, J., Quinones-Olvera, N., *et al.*, 2017. Rfam 13.0: Shifting to a genome-centric resource for non-coding RNA families. Nucleic Acids Research. doi:10.1093/nar/gkx1038.

Mallik, J., Samal, A., Gardner, S.L., 2007. A content based pattern analysis system for a biological specimen collection. In: Proceedings of the Seventh IEEE International Conference on Data Mining Workshops. doi: 10.1109/ICDMW.2007.3.

Kak, A., Pavlopoulou, C., 2002. Content-based image retrieval from large medical databases. In: Paper Presented at Proceedings of the First International Symposium on 3D Data Processing Visualization and Transmission.

Kozievitch, N.P., Torres, R.D., Andrade, F., *et al.*, 2010. A teaching tool for parasitology: Enhancing learning with annotation and image retrieval. In: Lalmas, M., Jose, J., Rauber, A., Sebastiani, F., Frommholz, I. (Eds.), Research and Advanced Technology for Digital Libraries 6273. Berlin: Springer-Verlag Berlin, pp. 466–469.

Krishnapuram, R., Medasani, S., Sung-Hwan, J., Young-Sik, C., Balasubramaniam, R., 2004. Content-based image retrieval based on a fuzzy approach. IEEE Transactions on Knowledge and Data Engineering 16 (10), 1185–1199. doi:10.1109/tkde.2004.53.

Lamard, M., Cazuguel, G., Quellec, G., *et al.*, 22–26 August 2007. Content based image retrieval based on wavelet transform coefficients distribution. In: Paper Presented at the 29th Annual International Conference of the IEEE Engineering in Medicine and Biology Society, EMBS 2007.

Lee, C.-H., Wang, S.-H., 2012. An information fusion approach to integrate image annotation and text mining methods for geographic knowledge discovery. Expert Systems with Applications 39 (10), 8954–8967. doi:10.1016/j.eswa.2012.02.028.

Lin, W.-C., Chang, Y.-C., Chen, H.-H., 2007. Integrating textual and visual information for cross-language image retrieval: A trans-media dictionary approach. Information Processing & Management 43 (2), 488–502. doi:10.1016/j.ipm.2006.07.015.

Liu, Y., Zhang, D., Lu, G., Ma, W.-Y., 2007. A survey of content-based image retrieval with high-level semantics. Pattern Recognition 40 (1), 262–282. doi:10.1016/j.patcog.2006.04.045.

Liu, R., Wang, Y., Baba, T., Masumoto, D., Nagata, S., 2008. SVM-based active feedback in image retrieval using clustering and unlabeled data. Pattern Recognition, 41(8), 2645-2655. doi:10.1016/j.patcog.2008.01.023.

McQuilton, P., Pierre, S.E.S., Thurmond, J., 2012.FlyBase 101 – The basics of navigating FlyBase. Nucleic Acids Research 40 (Database issue), D706–D714. doi:10.1093/nar/gkr1030.

Müller, H., Michoux, N., Bandon, D., Geissbuhler, A., 2004. A review of content-based image retrieval systems in medical applications – Clinical benefits and future directions. International Journal of Medical Informatics 73 (1), 1–23. doi:10.1016/j.ijmedinf.2003.11.024.

Murthy, U., Fox, E.A., Chen, Y., et al., 2009. Superimposed Image description and retrieval for fish species identification. In: Agnosti, M.B.J.K.S.P.C.T.G. (Ed.), Proceedings of the Research and Advanced Technology for Digital Libraries, vol. 5714, pp. 285–296.

Oberlin, A.T., Jurkovic, D.A., Balish, M.F., Friedberg, I., 2013. Biological database of images and genomes: Tools for community annotations linking image and genomic information. Database: The Journal of Biological Databases and Curation 2013, bat016. doi:10.1093/database/bat016.

Orloff, D.N., Iwasa, J.H., Martone, M.E., Ellisman, M.H., Kane, C.M., 2013. The cell: An image library-CCDB: A curated repository of microscopy data. Nucleic Acids Research 41 (Database issue), D1241–D1250. doi:10.1093/nar/gks1257.

Ortega-Binderberger, M., Mehrotra, S., 2004. Relevance feedback techniques in the MARS image retrieval system. Multimedia Systems 9 (6), 535–547. doi:10.1007/s00530-003-0126-z.

Petrov, A.I., Zirbel, C.L., & Leontis, N.B., 2013. Automated classification of RNA 3D motifs and the RNA 3D Motif Atlas. RNA, 19(10), 1327–1340. http://doi.org/10.1261/rna.039438.113.

Rosa, N.A., Felipe, J.C., Traina, A.J.M., et al., 20–25 August 2008. Using relevance feedback to reduce the semantic gap in content-based image retrieval of mammographic masses. In: Paper Presented at the 30th Annual International Conference of the IEEE Engineering in Medicine and Biology Society, EMBS 2008.

Rui, Y., Huang, T.S., Chang, S.-F., 1999. Image retrieval: Current techniques, promising directions, and open issues. Journal of Visual Communication and Image Representation 10, 39–62.

Scott, G., Chi-Ren, S., 2007. Knowledge-driven multidimensional indexing structure for biomedical media database retrieval. IEEE Transactions on Information Technology in Biomedicine 11 (3), 320–331. doi:10.1109/titb.2006.880551.

Sergyan, S., 21–22 January 2008. Color histogram features based image classification in content-based image retrieval systems. In: Paper Presented at the 6th International Symposium on Applied Machine Intelligence and Informatics, SAMI 2008.

Shandilya, S.K., Singhai, N., 2010. A survey on: Content based image retrieval systems. International Journal of Computer Applications 4 (2), 22–26.

Sheikh, A.R., Lye, M.H., Mansor, S., Fauzi, M.F.A., Anuar, F.M., 14–17 June 2011. A content based image retrieval system for marine life images. In: Paper Presented at the IEEE 15th International Symposium on the Consumer Electronics, ISCE.

Sillitoe, I., Lewis, T.E., Cuff, A.L., et al., 2015. CATH: Comprehensive structural and functional annotations for genome sequences. Nucleic Acids Research 43 (Database issue), D376–D381. doi:10.1093/nar/gku947.

Simon, R., Vince, S., 2011. SID: Specimen image database. Retrieved August 2011, from: http://sid.zoology.gla.ac.uk/.

Smeulders, A.W.M., Worring, M., Santini, S., Gupta, A., Jain, R., 2000. Content-based image retrieval at the end of the early years. IEEE Transactions on Pattern Analysis and Machine Intelligence 22 (12), 1349–1380. doi:10.1109/34.895972.

Stejić, Z., Takama, Y., Hirota, K., 2003. Genetic algorithm-based relevance feedback for image retrieval using local similarity patterns. Information Processing and Management 39 (1), 1–23. doi:10.1016/s0306-4573(02)00024-9.

Torres, R.D.S., Falcao, A.X., 2006. Content-based image retrieval: Theory and applications. Revista de Informática Teórica e Aplicada 13, 161–185.

Torres, R.D., Medeiros, C.B., Dividino, R.Q., et al., 2004. Using digital library components for biodiversity systems.

Ursula, P., Webb, B.M., Qiang Dong, G., et al., 2014. MODBASE, a database of annotated comparative protein structure models and associated resources. Nucleic Acids Research 42, D336–D346.

Wang, J., Ji, L., Liang, A., Yuan, D., 2012. The identification of butterfly families using content-based image retrieval. Biosystems Engineering 111 (1), 24–32. doi:10.1016/j.biosystemseng.2011.10.003.

Wang, D., Ma, X., 2005. A hybrid image retrieval system with user's relevance feedback using neurocomputing. Informatica 29 (3), 271–280.

Wei, J., Guihua, E., Qionghai, D., Jinwei, G., 2006. Similarity-based online feature selection in content-based image retrieval. IEEE Transactions on Image Processing 15 (3), 702–712. doi:10.1109/tip.2005.863105.

Wei, C.-H., Li, C.-T., 2006. Calcification descriptor and relevance feedback learning algorithms for content-based mammogram retrieval. In: Astley, S., Brady, M., Rose, C., Zwiggelaar, R., (Eds.), Proceedings of International Workshop on Digital Mammography Digital Mammography, vol. 4046, pp. 307–314. Heidelberg: Springer Berlin.

Williams, E., Moore, J., Li, S.W., et al., 2017. Image data resource: A bioimage data integration and publication platform. Nature Methods 14, 775. doi:10.1038/nmeth.4326.

Xin, J., Jin, J.S., 2004. Relevance feedback for content-based image retrieval using Bayesian network. In: Paper Presented at the Proceedings of the Pan-Sydney Area Workshop on Visual Information Processing. Available at: http://dl.acm.org/citation.cfm?id=1082137.

You, D., Antani, S., Demner-Fushman, D., et al., 2011. Automatic identification of ROI in figure images toward improving hybrid (text and image) biomedical document retrieval. In: Agam, G., ViardGaudin, C. (Eds.), Document Recognition and Retrieval Xviii 7874. Bellingham: Spie-Int Soc Optical Engineering.

Zhang, H., Chen, Z., Li, M., Su, Z., 2003. Relevance feedback and learning in content-based image search. World Wide Web 6 (2), 131–155. doi:10.1023/a:1023618504691.

Zhang, X.M., Huang, Z., Shen, H.T., Li, Z.J., 2011. Probabilistic image tagging with tags expanded by text-based search. In: Yu, J.X., Kim, M.H., Unland, R. (Eds.), Database Systems for Advanced Applications, Pt I 6587. Berlin: Springer-Verlag Berlin, pp. 269–283.

Relevant Websites

http://www.monodb.org/index.php
  A web-host for the monogenea.
http://biosilico.kaist.ac.kr/
  Biosilico: An Integrated Metabolic Database System.
http://www.cathdb.info/search/by_structure
  CATH Search.
http://rfam.xfam.org/
  EMBL-EBI.
http://sid.zoology.gla.ac.uk/
  Florida State University, Tallahassee, USA.
http://flybase.org/
  FlyBase Homepage.

http://biodig.org
    Friedberg Lab, Miami University, Oxford OH, USA.
http://images.google.com/
    Google Images.
http://research.amnh.org/pbi/heteropteraspeciespage/
    Heteroptera Species Pages - Our Research.
https://idr.openmicroscopy.org
    Image Data Resource: IDR.
https://modbase.compbio.ucsf.edu
    ModBase Search Page.
http://www.macroglossa.com/
    MVE Photography.
http://www.orchid-africa.net/
    Orchids of Central Africa.
https://research.amnh.org/pbi/
    Plant Bug :: Planetary Biodiversity Inventory.
http://rna.bgsu.edu/rna3dhub/motifs
    RNA 3D Motif Atlas - BGSU RNA Structural Bioinformatics.
https://swissmodel.expasy.org/repository
    SWISS-MODEL - Repository.
http://www.boldsystems.org/index.php
    SWISS-MODEL.
http://www.cellimagelibrary.org
    The Cell Image Library.

Segmentation Techniques for Bioimages

Saowaluck Kaewkamnerd and Apichart Intarapanich, National Electronics and Computer Technology Center, Pathum Thani, Thailand
Sissades Tongsima, National Center for Genetic Engineering and Biotechnology, Pathum Thani, Thailand

Introduction

Bioimage is a combination of the two words "Biology" and "image". Biology means the study of life and living organisms, including their physical and chemical structure, function, development and evolution (Murphy and Davidson, 2001) and thus, Bioimages broadly imply biologically related images. These images can be taken from large varieties of organisms of any size, ranging from whole organisms down to a single molecule level. There are many techniques to obtain these images, that can be categorized according to sizes into three levels, namely molecular, microscopic and whole organism levels. First, images from the molecular level include scanning or transmission electron microscopy (Tsien, 2003). For the microscopic level, Bioimages include those from confocal or two-photon laser scanning microscopy (Tsien, 2003). The last category is composed of those Bioimages of parts or organs of organisms, obtained from X-ray, magnetic resonance, ultrasound, endoscopy, thermography, positron emission tomography (PET), Single-photon emission computed tomography (SPECT) and inferred (Jia *et al.*, 2016) and others. These Bioimages require domain expert visual instruction to interpret them. However, these manual interpretation steps are very tedious, error prone and time consuming. To overcome the limitation, the use of computer-aided systems such as image processing techniques, has been introduced. Thus, components/objects from Bioimages must be automatically extracted and converted into a format that any image-analysis algorithm can efficiently processed.

Bioimage Segmentation: Definition and Challenges?

Humans understand and identify specific parts of an image because of our recognition through memorization and retrieval of said information. To be able to do this using computer, automatic identification of useful components in an image is needed. Image segmentation is a process to partition an image into objects, which can be either meaningful or non-meaningful and therefore, it is a vital process to identify objects of interest. Bioimage segmentation plays an important role in many branches of life sciences such as, biotechnology, precision agricultures, pharmacology, advanced medicine and microbiology. These domains require large number of images to be processed. For more efficiency and automation, computer-based systems are required to identify objects or region of interest before passing them to other stages of image processing, such as feature extraction, image classification and image recognition (Gonzalez and Woods, 2008). Even though research on image segmentation has been conducted for more than 40 years (Zhang, 2009), there are no single solution for all kinds of images. Such a challenge majorly stems from a wide spectrum of image qualities. For example, in microscopic image analysis, the difficulties in segmentation are from inconsistent stained color, over-exposed or under-exposed to microscope light, poor contrast, varieties of cell types, cell overlapping and unwanted objects (artifacts) in the image (Win and Choomchuay, 2017). Another example of confounding signals is found in an analysis of intravascular ultrasound images whose poor qualities come from the presence of calcifications, shadows, transducer reflection, speckle noises and bifurcations and intensity-variant of structure (Jodas *et al.*, 2003). Customized segmentation algorithms are yet to be devised for some specific problems (Gupta *et al.*, 2017).

Segmentation Techniques: Review

In image analysis, the segmentation step is carried out after some pre-processing processes such as noise reduction, color conversion and enhancement processes to suppress unwilling distortions, transform data or enhance features for further processing. Segmentation techniques can be further classified into seven classes based on their methodologies (**Fig. 1**), including thresholding segmentation, region-based segmentation, edge-based segmentation, contour-based segmentation, clustering-based segmentation, recognition-based segmentation and computational intelligence-based segmentation (Patil and Deore, 2013; Kaur and Kaur, 2014; Anjna and Kaur, 2017). The underlying theoretical foundations and limitations of the methods in these classes are discussed in the following subsections.

Thresholding Segmentation

Segmentation algorithms from this class share a common scheme that is using a pixel intensity as a threshold to partition an image into two partitions. For any pixel p located at a coordinate (x,y), the pixel p is assigned to one partition if its intensity I, at the coordinate (x,y) is greater than or equal to the threshold value Th, i.e., $I(x,y) \geq Th$. Otherwise the pixel is assigned to the other partition (Kaur and Kaur, 2014). The performance of this kind of thresholding protocol relies on its threshold value, which can be determined manually or automatically.

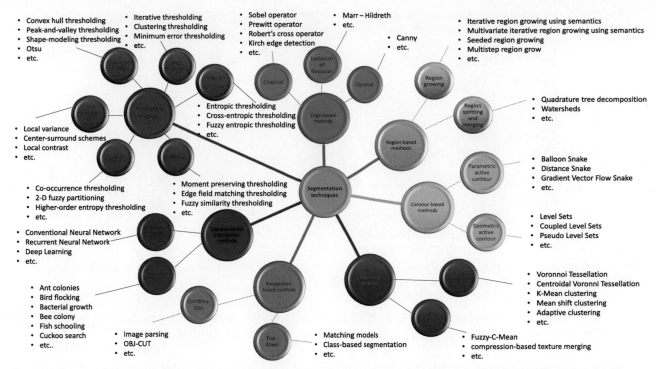

- Convex hull thresholding
- Peak-and-valley thresholding
- Shape-modeling thresholding
- Otsu
- etc.

- Iterative thresholding
- Clustering thresholding
- Minimum error thresholding
- etc.

- Sobel operator
- Prewitt operator
- Robert's cross operator
- Kirch edge detection
- etc.

- Marr – Hildreth
- etc.

- Canny
- etc.

- Iterative region growing using semantics
- Multivariate iterative region growing using semantics
- Seeded region growing
- Multistep region grow
- etc.

- Local variance
- Center-surround schemes
- Local contrast
- etc.

- Entropic thresholding
- Cross-entropic thresholding
- Fuzzy entropic thresholding
  etc.

- Quadrature tree decomposition
- Watersheds
- etc.

- Co-occurrence thresholding
- 2-D fuzzy partitioning
- Higher-order entropy thresholding
- etc.

- Moment preserving thresholding
- Edge field matching thresholding
- Fuzzy similarity thresholding
- etc.

- Balloon Snake
- Distance Snake
- Gradient Vector Flow Snake
- etc.

- Conventional Neural Network
- Recurrent Neural Network
- Deep Learning
- etc.

- Level Sets
- Coupled Level Sets
- Pseudo Level Sets
- etc.

- Ant colonies
- Bird flocking
- Bacterial growth
- Bee colony
- Fish schooling
- Cuckoo search
- etc..

- Voronnoi Tessellation
- Centroidal Voronni Tessellation
- K-Mean clustering
- Mean shift clustering
- Adaptive clustering
- etc.

- Image parsing
- OBJ-CUT
- etc.

- Matching models
- Class-based segmentation
- etc.

- Fuzzy-C-Mean
- compression-based texture merging
- etc.

Fig. 1 Seven main classes of segmentation methods, that are classified based upon their core techniques used to partition input images: (1) thresholding, (2) region-based, (3) edge-based, (4) contour-based, (5) clustering-based, (6) recognition-based, and (7) computational intelligence-based classes.

The automatic thresholding techniques may be further divided into six sub-categories based on distinguishable characteristic at the pixel level such as histogram shape information, measurement space clustering, histogram entropy information, image attribute information, spatial information and local characteristics (Zhang, 2001). The first category uses the information of histogram shape such as peaks, valleys and curvature for thresholding. Examples of shape thresholding are Otsu, Convex hull thresholding, Peak-and-valley thresholding, Shape-modeling thresholding and others (Sezgin and Sankur, 2004). The main idea of the Otsu method, named after the inventor, Otsu (1979) is to find an optimum threshold for separating an image into two classes (object and background). This Otsu technique operates by iterating over all possible threshold values and selects the threshold that minimizes their intra-class variances or maximizes their inter-class variances. For the second sub-category, measurement of space clustering, the number of clusters is always set to two, namely the object cluster and the background cluster. The thresholding value, also called the discriminator, is used to create a mask (0 or 1 corresponding to each pixel from the original image) that can separate objects from the background. Examples of the measurement of space clustering category include Iterative thresholding, Clustering thresholding, Minimum error thresholding (Sezgin and Sankur, 2004). The entropy-based thresholding sub-category utilizes the concept of the average level of information, known as the entropy as a criteria to look for the most suitable thresholding value that best preserves the information containing in the original image. A method, called maximum entropy thresholding, was introduced by Kapur *et al.* (1985). In their approach, the threshold is selected if the sum of entropies of the object and the background reaches the maximum value. Another example of entropy-based is the cross-entropy technique (Kapur *et al.*, 1985) when the threshold value is selected by minimizing the information difference between the input and output images. Object attribute-based segmentation algorithms commonly employ similarity of attributes such as shape compactness, gray-level moments and texture between input and output images (Tsai, 1985; Pal and Rosenfeld, 1988). Thresholding methods, belonging to the spatial sub-category, uses higher-order probability distribution of gray value of pixel and its neighborhood such as context probabilities, correlation functions and co-occurrence probabilities (Cheng and Chen, 1999). Finally, those local characteristic segmentation algorithms make use of local statistics like range and variance of range and variance of a pixel and those from other pixels around it. More examples of the segmentation techniques based on thresholding are presented in Sezgin and Sankur (2004).

The above thresholding-based image segmentation techniques are computationally fast and very easy to implement. They are commonly adopted and implemented by many real-time applications (Dass *et al.*, 2012; Anjna and Kaur, 2017; Patil and Deore, 2013). However, these simple segmentation techniques are not good when processing images that have uneven illumination background with broad and flat valleys. Using a threshold to segment poor images with high degree of noise interferences results in poor segmentation outputs. Furthermore, these algorithms do not offer spatial characteristic of image. To demonstrate, how thresholding-based segmentation performs, Fiji (Schindelin *et al.*, 2012), which is a nice image processing software that bundles all popular image processing algorithms in ImageJ (Schneider *et al.*, 2012) was used. To segment an MRI image using thresholding

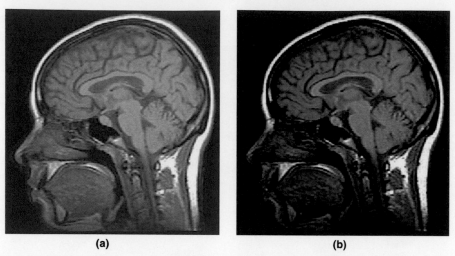

(a)   (b)

Fig. 2 An example of a brain Magnetic Resonance Imaging (MRI) image. (a) The original MRI scanned image and (b) the segmented image with a threshold set to 30, where regions highlighted in red represent detected background.

(**Fig. 2**), choose Fiji: image → adjust → threshold from the main menu. A pop-up window displaying the histogram profile of the selected image will be shown where the user can adjust the thresholding parameter using a slide bar.

Edge-Based Segmentation

Edge-based segmentation methods detect "discontinuity" properties such as gray change, color distinctness and texture variety that can demarcate between different regions. Particularly, edges and region boundaries are closely related because lines around an object create region boundaries. This kind of edge detection scheme is commonly used as a pre-processing step to enhance other segmentation techniques. There are different types of edges that are considered in algorithms from this edge-based segmentation class, namely ramp edge (gradually change in intensity), step edge (an abrupt change in intensity), roof edge (not instantaneous over a short distance) and spike edge (a quick change in intensity values) (Anjna and Kaur, 2017). Krishnan *et al.* (2017) classified edge-based segmentation according to three operators used to detect the changes: classical operator, Laplacian of Gaussian operator and optimal operator. A classical operator is the first order derivation with examples such as Sobel operator (Gonzalez and Woods, 2008), Prewitt operator (Prewitt, 1970), Robert's cross operator (Bansal *et al.*, 2012), Kirsch edge detection (Kirsch, 1971) etc. Laplacian of Gaussian operator uses to find the second derivation. Finally the optimal operator uses a multi-stage algorithm to detect a wide range of edge information, e.g., Canny edge detection (Canny, 1986). **Fig. 3** shows a resulting segmented fingerprint image (right panel) using ImageJ by choosing ImageJ: Plugins → Canny Edge Detector from ImageJ menu.

Edge-based methods are simple to implement and yield good results in most cases. However, for those images which contain too much noises too many traces, resulting edges identified by an edge detection method are usually discontinued and very fragmented. Therefore an edge linking algorithm to join these fragments and construct separable boundaries is needed.

Region-Based Segmentation

Region-based segmentation methods partition an image using a concept of region similarity which must be predefined. To form a region, a picture element (pixel) that is an elementary building block for bitmap (raster) images (Kaur and Kaur, 2014) is compared to the neighboring pixels. If the similarity criteria is met, these pixels will incrementally be formed as a region. These region-based segmentation methods can be categorized into two groups, (1) region growing group and (2) region splitting and merging group (Kaur and Kaur, 2014). The region growing group incrementally expands a boundary of a region by merging those similar pixels together. Examples of algorithms from the region growing group include Seeded Region Growing, Theoretical Criteria, and Multistep Region Growing (Vantaram and Saber, 2012). A recent review by Adamu and Ding (2015) further classifies these region growing segmentation techniques into Seeded Region Growing (SRG) and Unseeded Region Growing (UsRG). For SRG, the growing process defines seed points or seed pixels whose properties can be used as inclusion criteria for expanding a region. The neighboring pixels that satisfy the inclusion criteria join the existing seed pixels. This inclusion selection is done iteratively until all pixels are examined. Clearly, segmentation results from the SRG algorithms are very much subjective and dependent upon the selected seeds. Unlike the SRG protocol, UsRG techniques do not require preselected explicit seeds. UsRG algorithms arbitrarily define one single region covering some pixels that will be used to grow the region boundary iteratively by recruiting more neighboring pixels. When encountering a pixel whose property greatly differs from the current growing region, that pixel will start a new region, which in turn start growing accordingly.

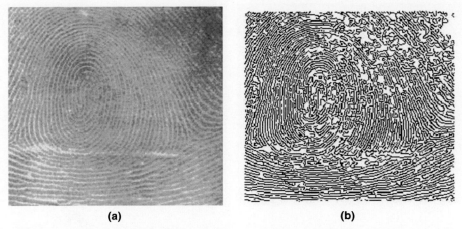

Fig. 3 A fingerprint image contains mostly traces of fingerprints presenting a challenge in segmentation based on classical edge-detection since detected edges may not be connected to form good boundaries for partitioning. (a) The original fingerprint image and (b) The segmented image using Canny edge detection, where contour traces show potential edges extracted from the original fingerprint.

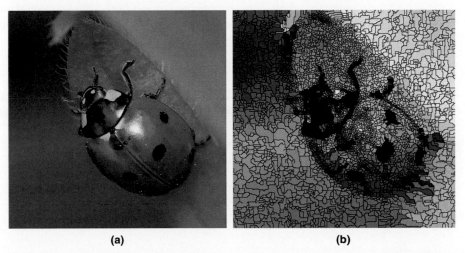

Fig. 4 Using Watershed to partition a clear region boundary image of a ladybug produces over-segmented result: (a) The original image of a ladybug and (b) the over-segmented result after using Watershed.

For the region splitting and merging class, algorithms in this class use divide-and-conquer to recursively partition (split) the whole image and then quickly merge those small yet similar regions to create bigger regions. In (Gorte, 1996), the quadtree segmentation algorithm was used by setting the whole image as a root of the quadtree. The splitting commences if all pixels in this whole image are not homogeneous. The splitting process create four child squares, each of which is carried out recursively using the aforementioned splitting procedure. Splitting terminate if all the pixels in the region are homogeneous (being sufficiently similar according to the criterion) or having only one pixel in it. The merging of similar regions grows from the leaves of this quadtree toward the root. As an example, Watershed is an efficient region-based splitting method (Courprie and Bertrand, 1997). The term "Watershed" refers to a geological watershed, which separates adjacent drainage basins. An input image is treated like a topographic map, whose pixels' brightness represent the heights of ridges. Examples of Watershed-based algorithms used to partition regions are watershed by flooding, watershed by topographic distance, watershed by the drop of water principle, inter-pixel watershed and topological watershed (Courprie and Bertrand, 1997).

Advantages of region-based methods are having high robustness, preserving the boundary of the segmented regions, producing coherent regions (regions with high degree of pixel uniformity). Nevertheless, objects with multiple disconnected regions present greater challenges, requiring human intervention for selecting input seed pixels. Furthermore, the resulting segmentations by the region-based methods are highly dependent on the choice of input seeds for seeded region growing. Watershed-based methods are likely to produce over-segmented results. As an example, **Fig. 4** presents a ladybug image with Watershed-based segmentation algorithm, provided in Fiji to partition the image. To operate this, from Fiji main menu choose: Fiji: Plugins → MorphoLibJ → Segmentation → Classic Watershed. The resulting image displays too many regions that Watershed attempted to segment.

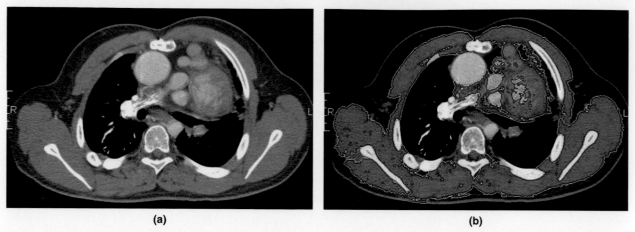

(a)   **(b)**

Fig. 5   A CT scan image does not have explicit edge to be detected (a). Using contour-based segmentation, important objects in this image can be detected as shown in the right panel (b).

Contour-Based Segmentation

A contour-based method is used for tracking boundaries (boundary tracing) in an image. First, a closed region (closed contour line) is designated as the starting boundary. This starting boundary is iteratively modified by either shink/expansion operations; thus this changing boundary is also called an active contour. There are two types of active contour models, (1) the parametric active contour model such as Balloon Snake, Distance Snake and Gradient Vector Flow Snake and (2) the geometric active contour model such as level sets, coupled level-set, pseudo level-set etc. (Baswaraj *et al.*, 2012; Vantaram and Saber, 2012). Particularly, Snake (Kass *et al.*, 1988) moves a contour based on an energy minimization while level-set (Osher and Sethian, 1988) moves a contour according to a level of a function. A contour changes its location and shape based on "a predefined force function" that pulls the contours toward features of interest. In Snake, a set of snake points must be selected from an image. Those snake points will be moved to the next location by determining the local minimum energy until the points stop at the object boundary. In Level set, the algorithm partitions several objects in an image by implicit curve propagation. The curve is moved as the evolution of the level set function. This method helps estimate the geometric properties of the evolution structure (Bhaidasna and Mehta, 2013). Dynamic contour concept provides a clever way to detect objects from the input image with continuous boundaries, stability and irrelevancy with topology. A good segmentation relies on the initial locations (snake points) and having only one object in the image. **Fig. 5** presents a CT scan image that does not have clear boundaries. By applying level-set from Fiji by choosing: Plugins → Segmentation → Level Sets. Significant objects can be detected by means of contour-based segmentation.

Clustering-Based Segmentation

Clustering is a method to group data into clusters based on their similarities (Dehariya *et al.*, 2010). The idea of clustering-based segmentation methods is to segment an image by assigning pixels with similar characteristics to a *k* number of groups (clusters) when *k* is given (Sharma and Suji, 2016). A membership function is used to define if a pixel belongs to the cluster. Such a function is described using a membership matrix with numbers between 0 and 1 representing a degree of membership of a pixel to a cluster. The pixel assignment continues until the maximum separation between clusters exceeds a given threshold.

There are two main types of clustering, (1) hard clustering such as Voronnoi Tessellation, Centroidal Voronni Tessellation, K-Mean clustering, Mean shift clustering, and Adaptive clustering (Vantaram and Saber, 2012) and (2) soft clustering such as Fuzzy-C-Mean (Kaur and Kaur, 2014). Hard clustering partitions an image into a given number of clusters with a restricted constraint that only one pixel belongs to one cluster. In this case, the hard clustering membership functions with the entry value of either 1 or 0 representing a membership status (being a member or not). K-Means clustering is a well-known hard clustering technique that first estimates a center for each of the *K* clusters, then each pixel will be assigned to those *K* clusters based on a similarity criterion such as distance, connectivity, intensity between pixel and cluster. The pixel assignment process attempts to minimize the inter-cluster similarity (between different clusters) and maximizing the intra-cluster similarity (within the group). Unlike hard clustering, soft clustering makes use of a flexible membership function and allows a pixel to be assigned to more than one cluster with varying degrees of membership ranging from 0 to 1. This type of flexibility is more natural than that of hard clustering. The Fuzzy-C-Mean algorithm (FCM) belongs to the soft clustering class (Christ and Parvathi, 2011). FCM is more flexible but its computational time could be expensive. The computational time of FCM can be dramatically reduced by means of using an intensity histogram of an image instead of accessing at the pixel level (Kaur and Kaur, 2014).

In summary, these clustering-based segmentation algorithms are quite robust to use in most images. However, all clustering-based segmentation methods require a predefined number of clusters. Most clustering techniques are largely sensitive to noises and other image artifacts that could interfere with the clustering assignment. **Fig. 6** presents a segmentation result of a flower image

Fig. 6 K-mean clustering-based segmentation algorithm was used to detect different flowers. (a) A flower image with big (yellow) and small (white) flowers with leaves among them (b) the segmented image where both type of flowers are clearly separated.

that use K-mean clustering when $K=4$ to identify these flowers. The analysis was performed using ImageJ: Plugins → Segmentation → k-means Clustering.

Recognition-Based Segmentation

The aforementioned image segmentation techniques segment an image by grouping pixels that have something in common, like common colors, textures, intensities etc. Once the image is segmented using these techniques, a whole object or region may be oversegmented. Thus these segmented regions must be combined so that true objects can finally be extracted. The process of using fundamental image segmentation to identify uniquely primitive regions and then merge them to form the objects and background is called the bottom-up method. Bottom-up is quite useful for most images and generally produce good segmentation results; however, there are some complex images containing objects that cannot be well partitioned/detected by the former segmentation algorithms. For example, a segmentation algorithm may partition background from real object(s) but toward the final step, similar colors/textures of background and the objects may be merged. This happens because of the lack of meaningful information about the objects of interest. Human utilizes knowledge about each object in order to differentiate an object from the background or other objects. Top-down image segmentation (Borenstein and Ullman, 2002) imitates human recognition process by including a matching model to do the job by utilizing some characteristics of objects with known shapes to guide its segmentation process.

Another variant of image segmentation algorithm is the combination of both bottom-up and top-down approach (Ferrari *et al.*, 2006; Wang *et al.*, 2007; Levin and Weiss, 2009). The approach by Wang *et al.* presents an algorithm that combines both top-down recognition and bottom-up image segmentation. It consists of two steps, (1) hypothesis generation and (2) verification. In the first step, a set of object locations and figure-ground masks, called hypotheses, are constructed. Then the verification step uses these object hypotheses to compute a set of feasible segmented objects where the false positive objects are subject to be eliminated using False Positive Pruning (FPP) (Wang *et al.*, 2007). Advantages of these recognition-based segmentation algorithms include the unique ability to handle images with extensive clutter of objects, dominant occlusion, and large scale and viewpoint changes (Levin and Weiss, 2009). In particular, these algorithms can be used to track an object in an image so that no matter how the original image rotates or even shifted, the object will always be recognized. **Fig. 7** shows how a red flower with a specific shape and color is tracked using a recognition-based segmentation algorithm from Fiji. From the main menu of Fiji: Plugins → Feature Extraction → Extract SIFT Correspondences.

Computational Intelligence-Based Segmentation

Human vision can efficiently recognize or identify an object from its background even if the object is not clearly or completely standing out. Clearly, gathering and using primitive information from an image to help partition is not enough. To improve this computational processing, researchers employ computational intelligence including Neural Network and other nature-inspired algorithms. Examples of Neural Network are Conventional Neural Network, Concurrent Neural Network, Deep Learning (Zheng *et al.*, 2015). The Deep Neural Networks was adopted to help detect brain tumor of various shapes, sizes and contrasts from magnetic resonance images (Havaei *et al.*, 2017). The adoption of Deep Neural Network in image segmentation creates a novel Convolutional Neural Networks (CNN) architecture that concurrently utilizes both local and global contextual features. Nature-inspired algorithms efficiently identify the best solution from a large number of finite solutions by mimicking how nature solves a combinatorial optimization problem and comes up with a good solution. Examples of these nature-inspired optimization algorithms include swarm intelligence, ant colony optimization (ACO), bird flocking, bacterial growth, bee colony, fish schooling, cuckoo search and others (Passino, 2002; Sharma *et al.*, 2015). A recent nature-inspired segmentation algorithm is the combination of ACO and genetic algorithm (Rahebi *et al.*, 2010). This work by Rahebi *et al.* is a probabilistic technique mimicking ant foraging behavior to choose an

Fig. 7 An image with many red flowers scattered and buried in the background. To track a particular object (a flower in a small window in the left panel), a recognition-based segmentation technique is used to perform the task.

optimal path by following ant pheromone to bring food back to their nest quickly. ACO segmentation models an image as a graph where nodes represent pixels. The algorithm generates artificial ants to store pixel intensities and construct a pheromone matrix that keeps track of paths (graph edges) that ants often visit. Nature-inspired algorithms can detect meaningful objects that may be incomplete or partly overlapped with other background. However, these algorithms can be very time consuming as learning stage is required to provide sufficient intelligence to these algorithms. **Fig. 8** presents an image containing different fishes. Using "Trainable Weka Segmentation" in Fiji: Plugins → Segmentation → Trainable Weka Segmentation, fishes can be detected.

Segmentation Techniques: Case study

In order to demonstrate the basic idea of each technique, a Giemsa-stained blood smear image with varying image conditions, namely in-focus, with noise and with uneven illumination, are used (**Fig. 9**) This blood smear image (**Fig. 9(a)**) was obtained from Kaewkamnerd *et al.* (2012) containing white blood cells (big black dots) and malaria parasites (small dots). Random noises, dots whose sizes are smaller than those of malaria parasite in the original image, were added to the original image (**Fig. 9(b)**). Finally the illumination scale is adjusted so that the image has illumination gradient (**Fig. 9(c)**). These three images are used as inputs for different segmentation algorithms. The image processing tool, named Fiji, is used for comparison purposes.

Thresholding-Based Method

Thresholding segmentation in Fiji was used to process the above three images in **Fig. 9**. Thresholding is able to segment all of the white blood cells (marked by blue region around each cell) as well as most of malaria parasites (**Fig. 10(a)**). With the present of noises, the segmentation result (**Fig. 10(b)**) reveals that only few malaria parasites are identified; random noises clearly interfere with thresholding segmentation. When illumination is uneven, thresholding segmentation could neither detect both white blood cells nor malaria parasites in the area affected with dark illumination. Particularly, the whole area with dark illumination was partitioned into one big object region (blue shaded area in **Fig. 10(c)**).

Edge-Based Method

For edge-based segmentation, the Canny edge-based method in Fiji was used. Similar to thresholding, Canny was able to segment the original in-focus blood smear image (**Fig. 11(a)**). For the noisy image, Canny identified more objects (**Fig. 11(b)**), via its edge detection scheme, than that from in-focus original image (**Fig. 11(a)**) as well as from the uneven illumination image (**Fig. 11(c)**). Noises introduce discontinuities of pixel intensity that provide suitable environment for edge-based segmentation to make a decision. However, most objects detected by Canny are not connected, causing difficulties to further post-process this image.

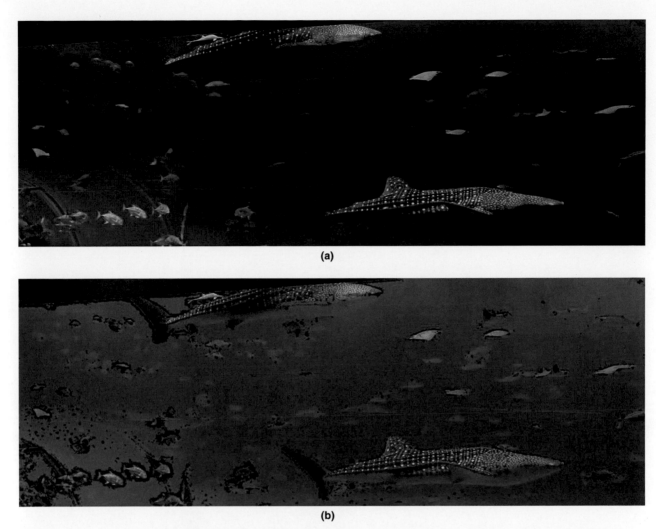

Fig. 8 An aquarium image has different kinds of fishes that may blend in with the blue background (a), presenting a challenge in image segmentation. Using a computational intelligence, fish objects can be detected (b).

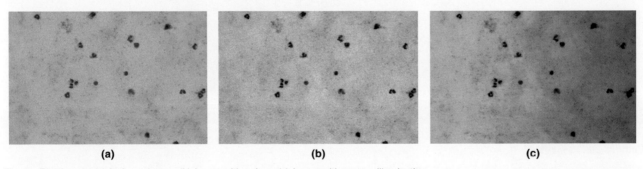

Fig. 9 Test Images (a) In-focus image. (b) Image with noises. (c) Image with uneven illumination.

Region-Based Method

The Watershed region-based method in Fiji was used to perform segmentation on the blood smear images in **Fig. 9**. However, Watershed over-segmented all of the three input images. **Fig. 12(a)** shows that the segmentation of white blood cells includes some of the malaria parasites around them. The light gray area of the background was clearly over-segmented. Such over-segmented effects appear more in the noisy image (**Fig. 12(b)**) which might have been influenced by those small noisy dots added to the original image. **Fig. 12(c)** also reveal the over-segmented result on uneven illumination blood smear image. Watershed is still over sensitive to changes of pixel illumination.

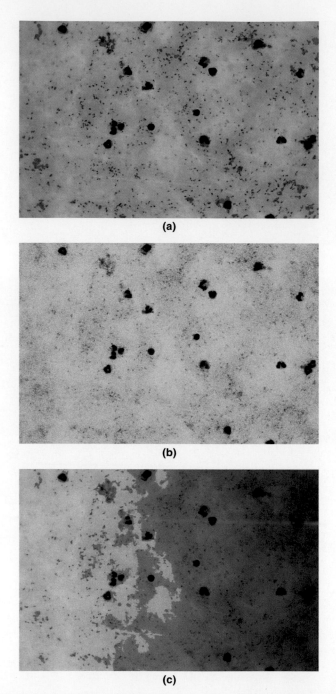

Fig. 10   Result images using thresholding method (a) In-focus image. (b) Image with noise. (c) Image with uneven illumination.

Contour-Based Method

In this experiment, the Level-set contour-based segmentation was used. To demonstrate the region growing in Level-set, we assigned the initial seeds to the centers of those white blood cells in three images. Level set was able to correctly detect boundaries of all white blood cells from the original blood smear image (**Fig. 13(a)**). However, those noisy dots affect the performance of the region growing process (**Fig. 13(b)**) where fewer white blood cells were correctly traced. Uneven illumination did not interfere with the object detection (**Fig. 13(c)**).

Clustering-Based Method

K-Mean clustering in ImageJ was used in this experiment. K-Mean clustering identifies regions in the in-focus image into two clusters (K=2), namely white blood cells and malaria parasites. All white blood cells are identified while only some of malaria

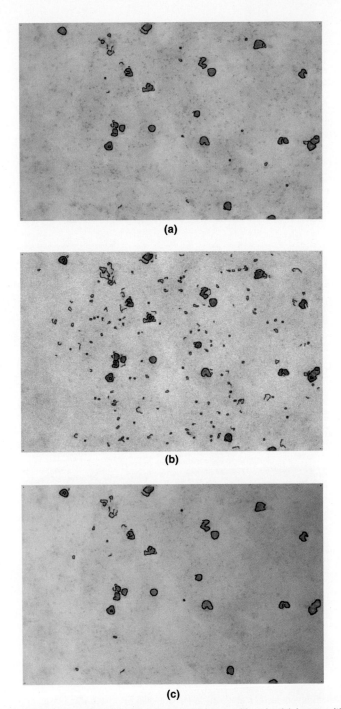

Fig. 11 Result images using Canny edge-based method (a) In-focus image. (b) Image with noise.(c) Image with uneven illumination.

parasites are detected (**Fig. 14(a)**). Random noises caused K-Mean clustering to be erratic and not able to properly segment the image. Only white blood cells were detected while a few of parasites were marked (**Fig. 14(b)**). Uneven illumination poses more problem to K-Mean clustering where the darker illumination was detected as one big cluster (**Fig. 14(c)**). K-Mean clustering was able to properly segment the area with lighter illumination in **Fig. 14(c)**.

Recognition-Based Method

The scale-invariant feature transform (SIFT) in Fiji (JavaSIFT) was used to identify objects in the test images. SIFT attempted to extract and store key features of objects of interest from the input image; it then use these features to recognize the objects, e.g., tracking them. To demonstrate image tracking feature, a 15 degree rotated image of **Fig. 9** was used as a source image

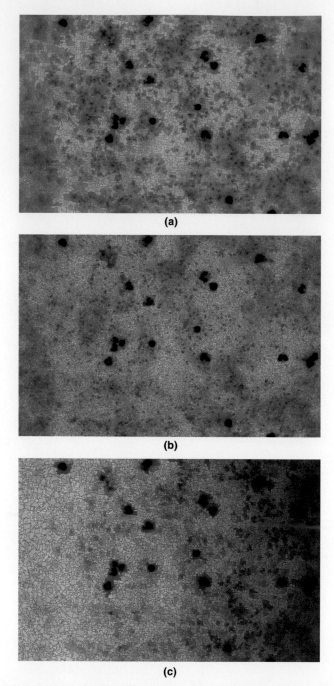

(a)

(b)

(c)

Fig. 12   Result images using Watershed region-based method (a) In-focus image. (b) Image with noise. (c) Image with uneven illumination.

(see **Fig. 15(a)**); then SIFT extracted object features for future looking up. Using **Fig. 9(a–c)** as the input SIFT tracked those objects from **Fig. 15(a)** and produced the tracked results in **Fig. 15(b–d)**, respectively. Yellow marks on these resulting figures reveal the tracked objects. SIFT is able to track (recognize) the objects using the stored key features on all variant of blood smear images.

Computational Intelligence Method

The trainable Weka segmentation with Bayesian classifier in Fiji is used in to analyze the previous malaria images. Prior to performing image segmentation, we assigned two groups of the training set, namely white blood cell and malaria parasite groups. After training, Bayesian classifier was able to properly segment the original blood smear image, where majority of white blood cells and parasites were correctly segmented (**Fig. 16(a)**). When encountering with random noise signals, however,

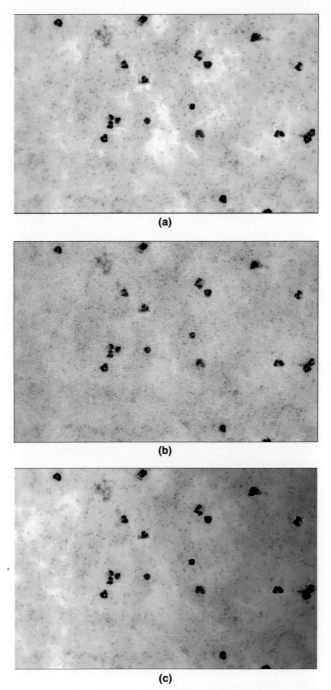

Fig. 13   Result images using Level set contour-based method (a) In-focus image. (b) Image with noise. (c) Image with uneven illumination.

the number of detected malaria parasites slightly dropped (**Fig. 16(b)**). The Bayesian classifier was able to largely detect the objects from both groups, except the region toward the rightmost end of the uneven illumination image that has the darkest shade (**Fig. 16(c)**).

Segmentation Techniques: Performance Criteria

Performance of an image processing algorithm can either be subjective (judging by feeling) evaluation or objective (judging by some measurable quality values) evaluation. Subjective evaluation is commonly conducted via human visual inspection; thus, it may be different from one evaluator to the other. In image segmentation, the performance of a segmentation method can be done

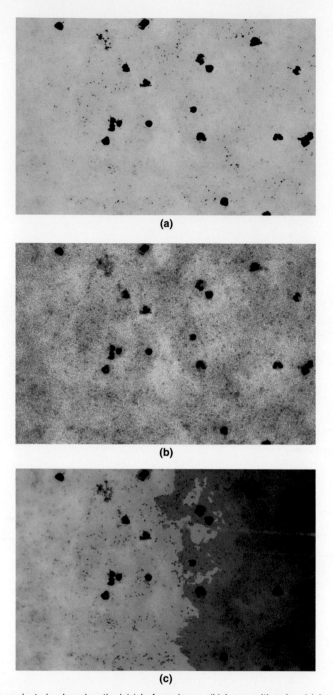

Fig. 14 Result images using K-Mean clustering-based method (a) In-focus image. (b) Image with noise. (c) Image with uneven illumination.

by visually comparing the original image to the segmented image. For objective evaluation, various mathematical measurement criteria are adopted for examples, misclassification error, edge mismatch, relative foreground area error, modified Hausdorff distance and region nonuniformity (Sezgin and Sankur, 2004). Misclassification error estimates the percentage of the misclassification between object and background. Edge mismatch calculates differences of the edge map between the original image and the corresponding segmented image. Relative foreground area error compares object properties in segmented image with the original one. The modified Hausdorff distance assesses the similarity of shape of segmented regions and compares with the ground-truth shapes. For the region nonuniformity measurement and comparison, the uniformity of objects and background from a segmented image are used to calculate variance of both segmented region. Since only objective evaluation may not be enough to effectively and correctly appraise a segmentation algorithm, subjective evaluation is also a common practice. Both objective and subjective evaluation.

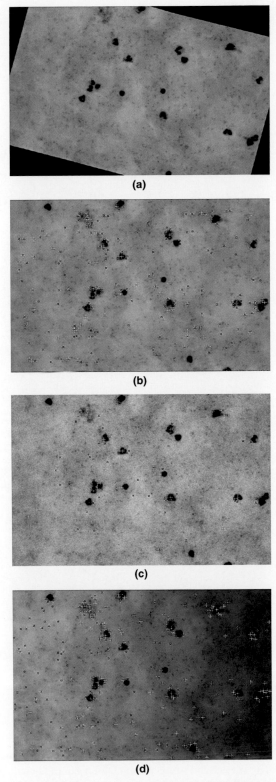

Fig. 15 Result images using SIFT recognition-based method (a) 15 degree rotated image for SIFT feature extraction, (b) tracked objects on In-focus image, (c) tracked objects on noisy image, and (d) tracked objects from the uneven illumination image.

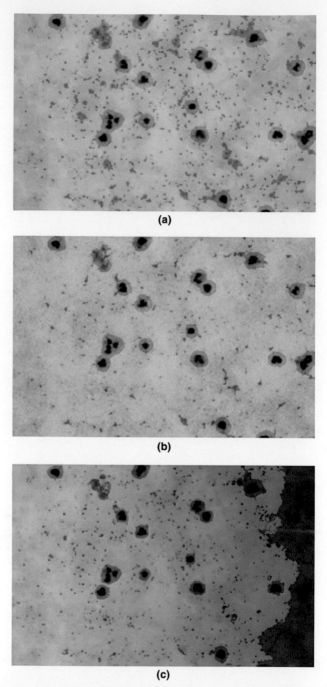

Fig. 16   Result images using Natural inspired-method (a) In-focus image. (b) Image with noise. (c) Image with uneven illumination.

Conclusion

The ultimate goal of image segmentation is to efficiently and effectively partition an image into meaningful objects or regions. Segmentation techniques have been widely studied and some of them have been implemented as tools at different levels, namely, lower level, middle level and high level. In this article, the segmentation techniques are categorized into seven classes, thresholding, edge-based, region-based, contour-based, clustering, recognition-based and computation-intelligence segmentations. Each class is evaluated based on three aspects, input constraints, output of methods and performance (**Table 1**).

Based on **Table 1**, Thresholding methods, Edge-based methods, Region-based methods, Contour-based methods and clustering methods are not required prior knowledge; however, these methods are very sensitive and my fail when dealing with noisy image and uneven illumination image while the recognition methods and computational intelligence methods could perform better in both images but require a prior knowledge from user. The last two require high complexity and more stability comparing to the others. The contour-

Table 1   The evaluation of segmentation methods

Methods	Input constraints			Output of methods				Performance	
	Noise	Uneven illumination	Requiring prior knowledge	Edge	Region	Contour	Object of interest	High complexity	Stability
Thresholding	✗	✗	✗	✗	✓	✗	✗	✗	✗
Edge-based	✗	✗	✗	✓	✗	✗	✗	✗	✗
Region-based	✗	✗	✗	✗	✓	✗	✗	✗	✗
Contour-based	✗	✗	✗	✗	✓	✓	✗	✗	✓
Clustering	✗	✗	✗	✗	✓	✗	✗	✗	✓
Recognition	✓	✓	✓	✗	✓	✗	✓	✓	✓
Computational Intelligence	✓	✓	✓	✗	✓	✗	✓	✓	✓

Notes: ✗ means a method does not have an ability and ✓ means a method do have an ability.
Noises: unwanted interruptions which causes a number of pixels in the image to change in color and/or brightness.
Prior knowledge: addition data added to the system to help it identify, recognize, classify, differentiate or etc.
Uneven illumination: the level of brightness is not the same throughout the image.
Edge: where there is a sudden change of intensity between pixels. (most edges do not connect).
Region: an area of pixels which have similar properties.
Contour: boundary of regions which could be formed by a set of connected edges.
Object of interest: a meaningful object or region.
High complexity: computational complexity.
Stability: ability of a technique to remain unchanged under unexpected condition.

based method is stable as it can perform well on the uneven illumination image. Although there are many segmentation techniques have been proposed, they are not able to efficiently handle all possible image conditions as shown in the case study. Many factors such as image preparation process, acquisition process and complexity of an image itself present greater challenges for researchers to explore.

Acknowledgement

We would like to thank Napat Kaewkamnerd for preparing and processing those malaria images using various algorithms in Fiji presented in the case study section.

See also: Natural Language Processing Approaches in Bioinformatics

References

Adamu, M.J., Ding, X., 2015. A relative study on image segmentation methods. International Journal of Science and Research 6, 1975–1981.
Anjna, E., Kaur, R., 2017. Review of image segmentation technique. International Journal of Advanced Research in Computer Science 8, 36–39.
Bansal, B., Saini, J.S., Bansal, V., Kaur, G., 2012. Comparison of various edge detection techniques. Journal of Information and Operations Management 3, 103–106.
Baswaraj, D., Govardhan, A., Premchand, P., 2012. Active contours and image segmentation the current state of the art. Global Journal of Computer Science and Technology Graphics & Vision 12, 1–13.
Bhaidasna, Z.C., Mehta, S., 2013. A review on level set method for image segmentation. International Journal of Computer Applications 63, 20–22.
Borenstein, E., Ullman, S., 2002. Class-specific, top-down segmentation. In: Proceedings of the 7th Proceeding of the European Conference on Computer Vision-Part II, pp. 109–122. London: Springer-verlag.
Canny, J., 1986. A computational approach to edge detection. IEEE Transaction on Pattern Analysis and Machine Intelligence 6, 679–698.
Cheng, H.D., Chen, Y.H., 1999. Fuzzy partition of two-dimensional histogram and its application to thresholding. Pattern Recognition 32, 825–843.
Christ, M.C.J., Parvathi, R.M.S., 2011. Fuzzy c-means algorithm for medical image segmentation. In: Proceedings of the 3rd International Conference on Electronics Computer Technology, pp. 33–36. Kanyakumari: IEEE.
Courprie, M., Bertrand, G., 1997. Topological grayscale watershed transformation. SPIE Vision Geometry 3168, 136–146.
Dass, R., Priyanka, Devi, S., 2012. Image segmentation techniques. The International Journal of Electronics & Communication Technology 3, 66–70.
Dehariya, V.K., Shrivaastava, S.K., Jain, R.C., 2010. Clustering of image data asset using K-Means and Fuzzy K-Means algorithms. In: International Conference on Computational Intelligence and Communication Networks. pp. 386–391. Bhopal, IEEE.
Ferrari, V., Tuytelaars, T., Gool, L.V., 2006. Simultaneous object recognition and segmentation from single or multiple model views. International Journal of Computer Vision 67, 1–26.
Gonzalez, R.C., Woods, R.E., 2008. Digital Image Processing, third ed. Upper Saddle River: Pearson Prentice Hall.
Gorte, B.G.H., 1996. Multi-spectral quadtree based image segmentation. International Archives of Photogrammetry and Remote Sensing 31, 251–356.
Gupta, H., Schmitter, D., Uhlmann, V., Unser, M., 2017. General surface energy for spinal cord and aorta segmentation. In: Proceedings of the 14th International Symposium on Biomedical Imaging, pp. 319–322. Melbourne: IEEE.
Havaei, M., Davy, A., Warde-Farley, D., et al., 2017. Brain tumor segmentation with deep neural networks. Medical Image Analysis 35, 18–31.
Jia, G., Heymsfield, S.B., Zhou, J., Yand, G., Takayama, Y., 2016. Quantitative biomedical imaging: Techniques and clinical applications. BioMed Research International 2016, 1–2.
Jodas, D.S., Pereira, A.S., Tavares, R.S., 2003. Automatic segmentation of the lumen region in intravascular images of the coronary artery. Medical Image Analysis 40, 60–79.

Kaewkamnerd, S., Uthaipibul, C., Intarapanich, A., *et al.*, 2012. An automatic device for detection and classification of malaria parasite species in thick blood film. BMC Bioinformatics 13, 1–10.

Kapur, J.N., Sahoo, P.K., Wong, A.K., 1985. A new method for gray-level picture thresholding using the entropy of the histogram. Computer Vision Graphics and Image Processing 29, 273–285.

Kass, M., Witkin, A., Terzopoulos, D., 1988. Snakes, active contour model. International Journal of Computer Vision. 321–331.

Kaur, D., Kaur, Y., 2014. Various image segmentation techniques: A review. International Journal of Computer Science and Mobile Computing 3, 809–814.

Kirsch, R., 1971. Computer determination of the constituent structure of biological images. Computers and Biomedical Research 4, 315–328.

Krishnan, K.B., Ranga, S.P., Guptha, N., 2017. A survey on different edge detection techniques for image segmentation. Indian Journal of Science and Technology 10, 1–8.

Levin, A., Weiss, Y., 2009. Learning to combine bottom-up and top-down segmentation. International Journal of Computer Vision 81, 105–118.

Murphy, D.B., Davidson, M.W., 2001. Fundamentals of light microscopy and electronic imaging, second ed. Singapore: Wiley-Blackwell.

Osher, S., Sethian, J.A., 1988. Fronts propagating with curvature-dependent speed: Algorithms based on Hamilton–Jacobi formulations. Journal of Computation Physics 79, 12–49.

Otsu, N., 1979. A threshold selection method from gray-level histogram. IEEE Transaction on System Man Cyberent SMC-8. 62–66.

Pal, S.K., Rosenfeld, A., 1988. Image enhancement and thresholding by optimization of fuzzy compactness. Pattern Recognition Letter 7, 77–86.

Passino, K.M., 2002. Biomimicry of bacterial foraging for distributed optimization and control. IEEE Control Systems Magazine 22, 52–67.

Patil, D.D., Deore, S.G., 2013. Medical image segmentation. International Journal of Computer Science and Mobile Computing 2, 22–27.

Prewitt, J.M.S., 1970. Object enhancement and extraction. In: Rosenfeld, A. (Ed.), Picture Processing and Psychopictorics. New York, NY: Academic Press, pp. 75–149.

Rahebi, J., Elmi, Z., Farzamnia, A., Shayan, K., 2010. Digital image edge detection using an ant colony optimization based on genetic algorithm. In: Conference on Cybernetics and Intelligent Systems, pp. 145–149. Singapore: IEEE.

Schindelin, J., Arganda-Carreras, I., Frise, E., *et al.*, 2012. Fiji: An open-source platform for biological-image analysis. Nature Methods 7, 676–682.

Schneider, C.A., Rasband, W.S., Eliceiri, K.W., 2012. NIH Image to ImageJ: 25 years of image analysis. Nature Methods 7, 671–675.

Sezgin, M., Sankur, B., 2004. Survey over image thresholding techniques and quantitative performance evaluation. Journal of Electronic Imaging 13, 146–165.

Sharma, A., Chaturvedi, R., Dwivedi, U.K., 2015. Recent trends and techniques in image segmentation using particle Swarm optimization-a survey. International Journal of Scientific and Research Publication 5, 1–6.

Sharma, P., Suji, J., 2016. A review on image segmentation with its clustering techniques. International Journal of Signal Processing, Image Processing and Pattern Recognition 9, 209–218.

Tsai, W.H., 1985. Moment-preserving thresholding: A new approach. Graphics Models Image Process 19, 377–393.

Tsien, R.Y., 2003. Imagining imaging's future. Nature Review Molecular Cell Biology 4, SS16–SS21.

Vantaram, S.R., Saber, E., 2012. Survey of contemporary trends in color image segmentation. Journal of Electronic Imaging 4, 1–28.

Wang, L., Shi, J., Song, G., Shen, I., 2007. Object detection combining recognition and segmentation. In: Proceedings of the 8th Asian Conference on Computer Vision – Volume Part I, pp. 189–199. Heidelberg: Springer-Verlag.

Win, Y.K., Choomchuay, S., 2017. Automated segmentation of cell nucli in cytology pleural fluid images using OTSU thresholding. In: International Conference on Digital Arts, Media and Technology, pp. 1–5.

Zhang, Y.J., 2001. An overview of image and video segmentation in the last 40 years. In: Proceedings of the 6th International Symposium on Signal Processing and Its Applications, pp. 144–51. IGI Global Disseminator of Knowledge.

Zhang, Y.J., 2009. Image segmentation in the last 40 years. In: Khosrow-Pour, M. (Ed.), Encyclopedia of Information Science and Technology, second ed. New York, NY: Information Science Reference, pp. 1818–1823.

Zheng, S., Jayasumana, S., Romera-Paredes, B., *et al.*, 2015. Conditional random fields as recurrent neural networks. In: Proceeding of the IEEE International Conference on Computer Vision, pp. 1529–1537. Washington, DC: IEEE Computer Society.

Relevant Websites

http://bioimages.vanderbilt.edu
  Bioimage.
https://fiji.sc
  Fiji.
https://imagej.net/Welcome
  ImageJ.
https://en.wikipedia.org/wiki/Medical_imaging
  Medical Imaging.

Biographical Sketch

Saowaluck Kaewkamnerd is currently a principal researcher and the head of Biomedical Signal Processing Laboratory at National Electronics and Computer Technology Center (NECTEC), National Science and Technology Development Agency (NSTDA), Thailand. Dr. Kaewkamnerd received her Ph.D. in Electrical Engineering from the University of Texas at Arlington, United States, in 2001. She has experienced in conducting research in biomedical image processing as well as other signal processing related projects.

Apichart Intarapanich is currently a researcher in Biomedical Signal Processing Laboratory at National Electronics and Computer Technology Center (NECTEC), National Science and Technology Development Agency (NSTDA), Thailand. He obtained his doctoral degree in Electrical Engineering from the University of Calgary, Canada in 2005. His specialized skills include Signal Processing for Telecommunication Section, implementing a real-time microphone array system for sound quality improvement, antenna design for WLAN, smart Antenna System development and PHS handset development.

Sissades Tongsima received his Ph.D. in Computer Science and Engineering from the University of Notre Dame, United States in 1999. He worked at the National Electronics and Computer Technology Center (NECTEC), National Science and Technology Development Agency (NSTDA) until 2003. Since then he has moved to the National Center for Genetic Engineering and Biotechnology (BIOTEC), NSTDA as a senior researcher. Currently, he is now a principal researcher and head of Bioinformatics Laboratory, Genome Technology Research Unit at BIOTEC. His research interests include – omics Big data analysis as well as utilizing digital signal processing in his research. He serves as an executive committee of Asia Pacific Bioinformatic Network (APBioNET) and an associate editor of the journal of Human Genetics.

Translational and Disease Bioinformatics

Sandeep Kaushik, European Institute of Excellence on Tissue Engineering and Regenerative Medicine, Guimaraes, Portugal and University of Minho, Braga, Portugal
Chakrawarti Prasun and Deepak Sharma, Indian Institute of Technology Roorkee, India

Introduction

The completion of the human genome project and a huge amount of data generated from the health care system have presented us with tremendous opportunities to understand and analyze our biological system in a greater detail. The data generated during clinical practice/research is used in finding its correlation with genomic data to obtain biologically significant insights. Due to the enormous size of data, the knowledge of information technology is required to store, integrate and analyze it. Analysis of complex biological data using knowledge of computer science, statistics and mathematics has resulted in the birth of a new field of life sciences, 'bioinformatics'.

Translational bioinformatics is an essential player in bench-to-bedside research, i.e., transformation of data into diagnostics, prognostics and therapeutics (Kann, 2010; Yan, 2011). It is the development of storage-related, analytic and interpretive methods by applying informatics methodology on voluminous biomedical and genomic data. It is particularly useful to find specific knowledge and medical devices that can be utilized by various stakeholders like biomedical scientists, clinicians and patients for improving patient care system, real time health monitoring and rational drug discovery. To understand translational bioinformatics, it is imperative to know the various types of data as well as the numerous tools and techniques to integrate and interpret these data (Altman, 2012). In this article, we describe various components of translational bioinformatics (**Fig. 1**).

Types of Datasets

The data utilized in translational bioinformatics can be broadly classified into biological and clinical data. An effective integration of these large heterogeneous datasets from classified areas, for instance molecular biology and clinical practice, provides clinically significant information (Sorani *et al.*, 2010).

Biological Data

The biological data encompasses molecular biology databases, which includes systematic arrangement of genomics, proteomics, metabolomics, microarray gene expression and phylogenetic data (Baxevanis, 2001). It can be further categorized into sequence, structure, functional or gene ontology database. Sequence databases include nucleic acid and protein sequences. Structure database encompasses RNA and protein structure. Functional database includes the function of gene and gene products. Functional or gene ontology is classified further based on molecular function, cellular component and biological process. If the functional database relates to physiological role of gene products, it is termed as molecular function database; if it determines the subcellular structures, it is called cellular component database; and if it determines the biological pathway in combination of other gene products, it is known as biological process database (Ashburner *et al.*, 2000). Gene ontology in conjunction with anatomical data

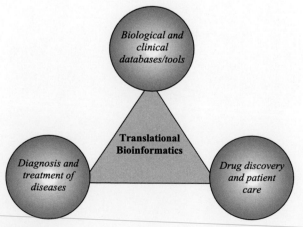

Fig. 1 Components of translational bioinformatics.

connects the parent and children phenotype. Similarly, biological process and disease progression, which are genetically determined, can be studied, and prophylactic action can be taken to slow down or avoid disease conditions from occurring.

The prominent biological databases implicated in the field of translational bioinformatics encompass data across metabolic/signaling pathways, protein-protein interactions, gene expression, polymorphisms, enzyme databases as well as drug related databases, which include drug structure, indication, pharmacological information, target structure/binding information and drug toxicity/adverse drug reactions (ADRs) data. Several databases and tools have been developed on metabolic/signaling pathways analysis, amongst them human metabolome database (HMDB), Kyoto Encyclopedia of Genes and Genomes (KEGG), Pathguide, pathDIP and XTalkDB are the most prominent and with distinct focus (**Table 1**).

HMDB provides information of human metabolite/metabolism, its detailed description, synonyms, structure, physico-chemical properties and reference NMR/MS spectra along with disease associations, pathway information, enzyme data, gene sequence data, SNP (single-nucleotide polymorphism) and mutation data (Wishart *et al.*, 2018). On the other hand, KEGG is a database and tool for analysis of gene function for genomes of 24 species including human, by linking the genomic information with cellular processes [such as metabolism, membrane transport, signal transduction and cell cycle] (Kanehisa and Goto, 2000). It has three databases, GENES database for genomic information, PATHWAY database for graphical representations of cellular processes and LIGAND database for the information about chemical compounds, enzyme molecules and enzymatic reactions. In contrast, Pathguide is a meta-database on

Table 1 Prominent biological and clinical databases/tools

Website	Key feature(s)	CI^a	$Access^b$	URL
BindingDB	Database of protein-ligand binding affinities	85.2	W	https://www.bindingdb.org/bind/index.jsp
BRENDA	Comprehensive enzyme information database	43.6	W	https://www.brenda-enzymes.org/
CAGE	Cap-analysis gene expression (CAGE) analysis database	27.7	W	https://cage-seq.com/
CTD	Illuminates how chemicals affect human health	22.7	W	http://ctdbase.org/
DrugBank	Comprehensive drug target database	174.6	W	https://www.drugbank.ca/
DynamProt	For tracking the levels and location of endogenous proteins in human cells	1.6	W	http://www.weizmann.ac.il/mcb/UriAlon/DynamProt/
Gene expression barcode	Barcode algorithm to estimate expressed and unexpressed genes in a microarray hybridization	20.9	W	http://barcode.luhs.org/
GeneWeaver	Cross-species data and gene entity integration database for scalable hierarchical analysis	12.4	W	https://geneweaver.org/
HIT	Herb Ingredients' Targets database	14.6	W	http://lifecenter.sgst.cn/hit/
HMDB	Human metabolite/metabolism database	142.7	W	http://www.hmdb.ca/
KEGG	Knowledge base for systematic analysis of gene functions, linking genomic information with higher order functional information	553.2	W	http://www.genome.jp/kegg/
LigAsite	Database of biologically relevant binding sites in proteins with known apo-structures	6.8	W	http://ligasite.org/
MIMIC	Repository of physiologic signals and vital signs, captured from patient monitors, and comprehensive clinical data	87.8	W	https://physionet.org/mimic2/
MINT	Experimentally verified protein-protein interactions database	71.0	W	http://mint.bio.uniroma2.it/
PANTHER	Tool to decipher the function of uncharacterized genes based on their evolutionary relationships	49.3	W	http://www.pantherdb.org/
pathDIP	Database of signaling cascades, comprising core pathways	3.3	W	http://ophid.utoronto.ca/pathdip/
Pathguide	Meta-database of biological pathways	31.3	W	http://www.pathguide.org/
PolyDoms	Whole genome database for the identification of non-synonymous coding SNPs	5.8	W	https://polydoms.cchmc.org/polydoms/
SIDER	Repository of marketed medicines and their recorded ADRs	60.0	W	http://sideeffects.embl.de/
SuperDRUG2	Database of approved/marketed drugs with chemical structures, dosage, biological targets, physicochemical properties, side-effects and pharmacokinetic data	$-^c$	W	http://cheminfo.charite.de/superdrug2/
T3DB	Toxin and toxin target database	15.2	W	http://www.t3db.ca/
TTD	Therapeutic protein and nucleic acid targets database	19.2	W	http://bidd.nus.edu.sg/group/cjttd/
WeGET	Tool to find mammalian genes that strongly co-express with a human query gene set	3.8	W	http://weget.cmbi.umcn.nl/
XTalkDB	Database of signaling pathway crosstalk	8.4	W	http://www.xtalkdb.org/home
ZINC	Database of commercially-available compounds for virtual screening	215.9	W	http://zinc.docking.org/

[a]CI, citation index (number of citations per year).
[b]W, Web based.
[c]-, no citations.

metabolic pathways, signaling pathways, transcription factor targets, gene regulatory networks, genetic interactions, protein-compound interactions and protein-protein interactions (Bader *et al.*, 2006). It is an overview of more than 190 web-accessible biological pathway and network databases, maintained by diverse groups in different locations. Recently, a resource pathDIP was developed that integrates core pathways (signaling/metabolic pathway databases) to predict biologically relevant protein-pathway associations (Rahmati *et al.*, 2017). It annotates 17070 protein-coding genes with 4678 pathways. Concurrently, a distinct tool XTalkDB was established for the analysis of signaling pathways and their crosstalk (Sam *et al.*, 2017). In addition, it covers the molecular components (e.g., proteins, hormones, microRNAs) that mediate crosstalk between a pair of pathways and the species and tissue.

Out of several protein-protein interactions databases, the most used are PANTHER (Protein ANalysis THrough Evolutionary Relationships), MINT (Molecular INTeraction Database) and DynamProt (Dynamics Proteomics) (**Table 1**). One of the first one to be developed was PANTHER that provides the function of uncharacterized genes (from any organism) based on their evolutionary relationships to genes with known functions (Mi *et al.*, 2005). It contains statistical models (Hidden Markov Models, or HMMs) to find out proteins (and their genes) family and subfamily that allows more precise evolutionary information. Subsequently, MINT database was established that catalogs physical interactions (obtained from large scale, genome wide experiment) between proteins stored in structured format (Chatr-aryamontri *et al.*, 2007). It also includes an integrated feature, HomoMINT for interactions between human proteins with orthologous proteins in model organisms. Another useful resource is DynamProt which is compendium of endogenously tagged human proteins expressed from its endogenous chromosomal location under its natural regulation (Frenkel-Morgenstern *et al.*, 2010). It illustrates the protein dynamics and localizations in individual living human cells following drug administration which is crucial for understanding effect of drug on cell.

Gene expression databases include CAGE, GeneWeaver, Gene Expression Barcode and WeGET (**Table 1**). CAGE (Cap analysis gene expression) is a highly sensitive and precise means of gene expression technique to produce high-throughput CAGE library (Kodzius *et al.*, 2006). It is primarily used to locate exact transcription start sites in the genome and allows investigation of promoter structure necessary for gene expression. It contributes to genome annotation, gene discovery and expression profiling. In comparison, GeneWeaver is a platform for integration of cross-species data and gene entity as well as scalable hierarchical analysis of investigator's experimental data with a community-built and curated data archive of gene sets and gene networks (Baker *et al.*, 2012). It provides means for data driven comparison of biological, behavioral and disease concepts. Gene Expression Barcode is the first database to provide absolute measures of expression for most annotated genes (McCall *et al.*, 2014). The expression of genes across tissues and cell types will help in generating hypothesis for gene function and clinical predictions using gene expression signatures. Recently, WeGET (Weighted Gene Expression Tool and database) was developed to predict new genes of a molecular system by correlating gene expression (Szklarczyk *et al.*, 2016). It ranks new candidate genes based on their weighted co-expression with that system and predict novel genes that co-express with a custom query set. On the other hand, BRENDA is one of the most comprehensive enzyme repositories (Schomburg *et al.*, 2004). It provides functional and molecular information of enzymes which is crucial for research of enzyme mechanisms, metabolic pathways and, furthermore, for medicinal diagnostics and pharmaceutical research. Additionally, PolyDoms is a whole genome database for polymorphism analysis and identification of non-synonymous coding SNPs (nsSNPs) with the potential to impact disease (Jegga *et al.*, 2007). It also predicts structural and functional impacts of all nsSNPs and identifies potentially harmful nsSNPs among multiple genes associated with specific diseases, anatomies, mammalian phenotypes, gene ontologies, pathways or protein domains.

The drug related databases cover drug structure database such as ZINC, DrugBank and SuperDRUG2, adverse drug reaction or toxicological databases such as CTD (Comparative Toxicogenomics Database), T3DB (Toxin and Toxin Target Database) and SIDER (Side Effect Resource) as well as drug target databases like TTD (Therapeutic Target Database) and HIT (Herb Ingredients' Targets). ZINC contains over 35 million molecules and provides their 3D structures for virtual screening against the specified target (Irwin and Shoichet, 2005) while DrugBank contains prescribing and pharmacological information including kinetics, dynamics, mechanism, target, interactions and metabolism pathway along with chemical, spectroscopic, patent information and 3D structure of a drug (Wishart *et al.*, 2008). In a similar vein, SuperDRUG2 provides indications, drug targets, side-effects, physicochemical properties, drug-drug interactions, pharmacological, chemical and regulatormarketed drug/marketed drugs (Siramshetty *et al.*, 2018). In contrast, CTD provides the information about the effects of environmental chemicals on human health (Davis *et al.*, 2009). It also curates chemical – Gene interactions, chemical – Disease relationships and gene – Disease relationships to construct chemical – Gene – Disease networks by data integration. Similarly, T3DB is a database of toxic exposome (environmental exposures of human to an acutely toxic compounds) (Wishart *et al.*, 2015). It contains detailed information of toxic substances that is, chemical properties, descriptions, targets, toxic effects, toxicity thresholds, sequences (for both targets and toxins) and mechanisms. Recently, SIDER database was developed that hosts drug's side effects or adverse drug reactions [ADRs] (Kuhn *et al.*, 2016). TTD is a repository of known therapeutic protein and nucleic acid targets (Chen *et al.*, 2002). It covers sequence, 3D structure, function, nomenclature, drug/ligand binding properties and drug effects. Another distinct useful resource is HIT which is a database of herbal ingredients with protein target information (Ye *et al.*, 2011).

Some important binding site information databases are BindingDB and LigAsite. BindingDB is a resource of binding affinity of protein-ligand complexes (Liu *et al.*, 2007). In comparison, LigAsite is a repository of biologically relevant binding site in proteins for which at least one apo- and one holo-structure (ligand unbound and bound structure, respectively) are available (Dessailly *et al.*, 2008).

It is noteworthy to mention that the popular/useful general tools and databases of DDBJ, NCBI, EMBL, ExPASy as well as CLUSTALW and PDB have not been covered as they are beyond the focus/scope of this article.

Clinical Data

The data collected by physicians while treating patients and during clinical trials are termed as clinical databases (Blum, 1982). It includes comprehensive data pertaining to drug effects, electronic health records (EHRs), drug-drug interactions data, patients' medical histories and demographic data, pathology, diagnostic data and data of concomitant medications. The various databases where such clinical data are stored are MIMIC, ClinicalTrials.gov, EMA Clinical Data and Clinical Data Repository.

MIMIC (Multiparameter Intelligent Monitoring in Intensive Care) is a database of physiologic signals and vital signs captured from patient monitors (Saeed *et al.*, 2011). Clinical Trials (see "Relevant Websites section") is a resource of clinical trials and their outcomes, provided by the U.S. National Library of Medicine. EMA Clinical Data (see "Relevant Websites section") web portal hosts clinical data published under the European Medicines Agency. Clinical Data Repository (CDR) is a real time database that houses data from a variety of clinical sources such as 'UVa clinical data repository' (see "Relevant Websites section"), containing clinical information on over 1 million patients and 5 million clinical encounters, the clinical data repository ALLOY from Liaison (see "Relevant Websites section") and Inovalon's clinical data repository (see "Relevant Websites section").

Integration and Analysis of Data for Diagnosis and Treatment

After the collection of voluminous data on numerous diseases, the next crucial step is to store it systematically so that it can be easily retrieved/processed and analyzed. Volume, Variety and Velocity (3V's) are formidable challenges in such big data management (Sagiroglu and Sinanc, 2013). To manage these challenges, numerous tools and techniques have been developed. Big data could be structured (stored as arrays, files, records, tables or trees), unstructured (either in text-format for instance PowerPoint presentations, word documents, PDFs or non-text format like images, audios, videos) or semi structured [mixed] (Raghupathi and Raghupathi, 2014). Many approaches and technologies for storage and processing have been developed for better management and quick retrieval/response. Notable among these are Distributed File System (a server based application that allows remote data access of structured data to several clients simultaneously e.g., Apache Hadoop), NoSQL (for storage and retrieval of non-tabular data), Cloud Computing (for storage on a network of remote servers hosted on internet rather than a personal computer or local server) and Massive Parallel Processing (MPP, coordinated processing of a program by different processors working on different parts of the program) (Katal *et al.*, 2013).

Apart from storage and processing techniques, powerful and highly analytical tools are required to visualize, analyze and integrate the vast data. Some of the most popular tools toward this context include EHDViz (for data visualization), Theano (for deep learning), PreGel, Apache Giraph and Gephi (for graph processing) as well as Apache Mahout, Spark MLlib and Weka (for machine learning) (Shameer *et al.*, 2017). Machine and deep learning techniques are used to process and analyze variety of data with high velocity. Machine learning is a part of artificial intelligence (AI) that trains the computer to perform the desired task automatically and more efficiently and accurately. There are different machine learning techniques like supervised learning (viz. Naïve Bayes Classifiers, Decision Tree, Support Vector Machine), unsupervised learning, reinforcement learning, deep learning, association rule learning and numeric prediction (Tan and Gilbert, 2003). Towards this end, several web servers and programs have been developed. For instance, an algorithm like Decision Support Systems (DSS), a program which facilitates complex decision making and problem solving by compiling, either graphically or as written report, massive reams of data to provide analytical result (Somek and Hercigonja-Szekeres, 2017). KDSS (Knowledge-based Decision Support System) is an example of DSS using a knowledge based coding to provide a novel workflow for protein complex extraction (Fiannaca *et al.*, 2013).

Cardiovascular diseases, central nervous system (CNS) diseases, respiratory diseases, cancer, diabetes, tuberculosis and viral diseases (like HIV, hepatitis, dengue, influenza/flu, measles) are the major cause of deaths globally (WHO report 2017). Amongst these, computational tools and databases implicated in cancer (Blekherman *et al.*, 2011), tuberculosis (Sharma *et al.*, 2015) and viral diseases (Sharma and Surolia, 2011) have already been compiled/described in great details. Prominent tools for cardiovascular, CNS and respiratory diseases as well as diabetes research are listed in **Table 2**. It also enlists useful tools that are not specific to any particular set of diseases.

CADgene, T-HOD or CardioSignal are gene databases associated with cardiovascular diseases (**Table 2**). CADgene is a comprehensive resource and integration tool for coronary artery disease genes (Liu *et al.*, 2011). It provides detailed information about the size of case-control, population, SNP, odds ratio and P-value, for each gene. T-HOD is a gene database for hypertension, obesity and diabetes (Dai *et al.*, 2013). It has a gene/disease identification system and a disease-gene relation extraction system to affirm the association of genes with these three diseases. CardioSignal is a repository of transcriptional regulation involved in heart development and cardiac hypertrophy (Zhen *et al.*, 2007). On the other hand, PubAngioGen is a comprehensive database for exploring the connection between angiogenesis and diseases at multi-levels including protein-protein interaction, drug-target, disease-gene and signaling pathways (Li *et al.*, 2015). In contrast, miRHrt is a database of microRNA function in heart development and by computational analysis it illustrates the correlation of differentially expressed miRNAs with cellular functions and heart development (Liu *et al.*, 2010). CARFMAP is an interactive cardiac fibroblast pathway map derived from biomedical literature (Nim *et al.*, 2015). It enables identification of new genes that are relevant to cardiac research and differentially regulated in high-throughput assays. Another useful tool is ECGSYN (see "Relevant Websites section") that generates a synthesized ECG signal with user-settable mean heart rate, number of beats, sampling frequency, waveform morphology (P, Q, R, S, and T timing, amplitude and duration), standard deviation of the RR interval, and LF/HF ratio (a measure of the relative contributions of the low

Table 2 Disease-specific tools and databases

Disease	Website	Key feature(s)	CI[a]	Access[b]	URL
Cardiovascular diseases	CADgene	Coronary artery disease (CAD) gene database	9.5	W	http://www.bioguo.org/CADgene/about_us.php
	CardioSignal	Repository of transcriptional regulation in cardiac development and hypertrophy	0.8	W	http://www.cardiosignal.org/index.html
	CARFMAP	For analyses of cardiac fibroblasts	1.6	W	http://visionet.erc.monash.edu.au/CARFMAP/
	ECGSYN	ECG waveform generator with user-settable parameters	-[c]	W	https://www.physionet.org/physiotools/ecgsyn/
	miRHrt	Database of miRNAs function in heart development	2.2	W	http://cistrome.org/cr/miRHrt_for_loading/
	PubAngioGen	Resource dedicated to angiogenesis	1.4	W	http://www.megabionet.org/aspd/
	T-HOD	Gene database for hypertension, obesity and diabetes	3.0	W	http://bws.iis.sinica.edu.tw/THOD
Central Nervous System (CNS) diseases	ADHDgene	Genetic resource of Attention Deficit Hyperactivity Disorder	8.2	W	http://adhd.psych.ac.cn
	AlzGene	Database of Alzheimer disease susceptibility genes	137.0	W	http://www.alzgene.org
	AutDB	Repository of genes implicated in autism susceptibility	23.1	W	http://autism.mindspec.org/autdb/Welcome.do
	EpilepsyGene	Catalogs genes and mutations related to epilepsy	5.6	W	http://122.228.158.106/EpilepsyGene
	G2Cdb	Database for linking synaptic genes to cognition and disease	5.4	W	http://www.genes2cognition.org
	NSDNA	Nervous System Disease NcRNAome Atlas	2.4	W	http://www.bio-bigdata.net/nsdna/
	PDGene	Database to provide genetic association results in Parkinson's disease	62.5	W	http://www.pdgene.org
	SZGR	Schizophrenia Gene Resource	10.1	W	https://bioinfo.uth.edu/SZGR/
Diabetes	Islet Regulome Browser	Pancreatic islet genomic database	-	W	http://gattaca.imppc.org/isletregulome/home
	T1DBase	Molecular genetics database for T1D susceptibility and pathogenesis	3.3	W	http://T1DBase.org
	T2D Knowledge Portal	Data/tools to promote understanding and treatment of T2D and its complications	-	W	http://www.type2diabetesgenetics.org
	T2D-Db	Database of the molecular components involved in the T2D pathogenesis	3.7	W	http://t2ddb.ibab.ac.in
Respiratory diseases	IGDB.NSCLC	Repository of non-small cell lung cancer genes and microRNAs	2.4	W	http://igdb.nsclc.ibms.sinica.edu.tw/
	LGEA	Resource for lung gene expression analysis	14.0	W	https://research.cchmc.org/pbge/lunggens/mainportal.html
	MyMpn	*Mycoplasma pneumoniae* database	6.3	W	http://mympn.crg.eu/
	SEGEL	For visualization of smoking effects on human lung gene expression	0.4	W	http://www.chengfeng.info/smoking_database.html
General tools	DisGeNET	Database to provide associations between genes and diseases, disorders and clinical or abnormal human phenotypes	75.5	W	http://www.disgenet.org/web/DisGeNET/menu/home
	DistiLD	Disease-gene associations database	3.1	W	http://distild.jensenlab.org
	Imprinted Gene Catalogue	Imprinted gene and parent-of-origin effect database	7.3	W	http://igc.otago.ac.nz/home.html
	LncRNADisease	Resource for long- non-coding RNA-associated diseases	66.4	W	http://cmbi.bjmu.edu.cn/lncrnadisease
	Organ System Heterogeneity DB	Repository of phenotypic effects of human diseases and drugs	0.7	W	http://mips.helmholtz-muenchen.de/Organ_System_Heterogeneity/

[a]CI, citation index (number of citations per year).
[b]W, Web based.
[c]-, no citations.

and high frequency components of the RR time series to total heart rate variability). Additionally, a method for simulating atrial fibrillation has been developed using ECGSYN (Healey *et al.*, 2004). The simulator uses a timing operator to switch from generating normal ECG morphologies to atrial fibrillation.

Several genomic databases for central nervous system diseases include AlzGene, PDGene, ADHDGene, EpilepsyGene, AutDB or SZGR (**Table 2**). AlzGene is a resource of Alzheimer disease susceptibility genes with statistically significant allelic summary

(Bertram *et al.*, 2007). It also performs systematic meta-analyses for each polymorphism with available genotype data. Similarly, PDGene is a meta-analyses database of genes associated with Parkinson's disease (Lill *et al.*, 2012). ADHDGene is a genetic resource and analysis platform for Attention Deficit Hyperactivity Disorder [ADHD] (Zhang *et al.*, 2012). EpilepsyGene is a repository of genes and mutations related to epilepsy (Ran *et al.*, 2015). It also includes functional annotation, gene prioritization, functional analysis of prioritized genes, and overlap analysis focusing on the comorbidity. In contrast, AutDB is a well classified (as per genetic variation of candidates identified from genetic association studies, rare single gene mutations and genes linked to syndromic autism) database of genes connected to Autism Spectrum Disorders [ASD] (Basu *et al.*, 2009). It covers gene annotation for their relevance to autism, along with an in-depth view of their molecular functions. Schizophrenia Gene Resource (SZGR) catalogs genetic data from association studies, linkage scans, gene expression, literature, gene ontology (GO) annotations, gene networks, cellular and regulatory networks across all mycobacterial spe, as well as microRNAs and their target sites (Jia *et al.*, 2010). Another useful tool is Genes to Cognition database (G2Cdb) that houses sets of genes and biochemically isolated (from mammalian synapse) or experimentally elucidated proteins (Croning *et al.*, 2009). It links genes, proteins, (patho)physiology, anatomy and behavior across species. Recently, a manually curated database Nervous System Disease NcRNAome Atlas (NSDNA) was established that provides comprehensive experimentally supported associations about nervous system diseases (NSDs) and noncoding RNAs (ncRNAs) (Wang *et al.*, 2017).

The databases/tools dedicated for diabetic research and study are T1DBase, T2D-Db, Type 2 Diabetes Knowledge Portal and Islet Regulome Browser (**Table 2**). T1DBase is a web server of molecular genetics to study susceptibility and pathogenesis of type-1 diabetes (T1D) (Smink *et al.*, 2005). It covers annotated genome sequence for human, rat and mouse; information on genetically identified T1D susceptibility regions in human, rat and mouse, and genetic linkage and association studies pertaining to T1D; the Beta Cell Gene Expression Bank, which reports expression levels of genes in beta cells under various conditions, and annotations of gene function in beta cells; data on gene expression in a variety of tissues and organs; and biological networks across all mycobacterial spe from KEGG and BioCarta. It also contains tools like GBrowse (genome browser), site-wide context dependent search, Connect-the-Dots (for connecting gene and other identifiers from multiple data sources), Cytoscape (for visualizing and analyzing biological networks), and the GESTALT workbench (for genome annotation). On the other hand, T2D-Db (Type 2 diabetes database) provides integrated data on molecular components involved in the pathogenesis of T2D of human, mouse and rat (Agrawal *et al.*, 2008). It also contains information on candidate genes, SNPs (Single Nucleotide Polymorphism) in candidate genes or candidate regions, Genome-Wide Association Studies (GWAS), tissue specific gene expression patterns, EST (Expressed Sequence Tag) data, expression information from microarray data, pathways, protein-protein interactions and disease associated risk factors. Similarly, 'Type 2 Diabetes Knowledge Portal' is a repository of DNA sequence, functional and epigenomic information, and clinical data on T2D and its macro- and microvascular complications, and creating analytic tools to analyze these data (see "Relevant Websites section"). Islet Regulome Browser is a tool for exploring pancreatic islet epigenomic and transcriptomic data and provides interactive access to GWAS variants, different classes of regulatory elements, together with enhancer clusters, stretch-enhancers and transcription factor binding sites in pancreatic progenitors and adult human pancreatic islets (Mularoni *et al.*, 2017).

The databases involved in research of respiratory diseases are IGDB.NSCLC, LGEA, MyMpn and SEGEL (**Table 2**). IGDB.NSCLC is an integrated genomic resource for non-small cell lung cancer (Kao *et al.*, 2012). It covers aberrant expressed genes and microRNAs, somatic mutations and experimental evidence and clinical information of non-small cell lung cancer patients. In contrast, LGEA (Lung Gene Expression Analysis) is a gene expression database for lung diseases (Du *et al.*, 2017). It provides different tools such as LungGENS (for single cell transcriptomes analysis), LungSortedCells (for sorting lung cell population), LungDTC (for development time course analysis) and LungDisease (for disease analysis). MyMpn is an online resource for studying the human pathogen *Mycoplasma pneumonia* and hosts data obtained from gene expression profiling experiments, gene essentiality studies, protein abundance profiling, protein complex analysis, metabolic reactions and network modeling, cell growth experiments, comparative genomics and 3D tomography (Wodke *et al.*, 2015). Concurrently, a distinct tool SEGEL (Smoking Effects on Gene Expression of Lung) for visualizing smoking effects on human lung gene expression was developed (Xu *et al.*, 2015). It integrates expression microarray data sets from trachea epithelial cells, large airway epithelial cells, small airway epithelial cells and alveolar macrophages.

In addition to disease-specific tools, there are several general tools to integrate and analyze association between gene and disease progression such as DisGeNet, DistiLD, Imprinted Gene Catalogue, LncRNADisease, and 'Organ System Heterogeneity DB' (**Table 2**). DisGeNet integrates expert-curated databases with text-mined data, covers information on Mendelian and complex diseases and includes data from animal disease models (Pinero *et al.*, 2015). It features a score based on the supporting evidence to prioritize gene-disease associations. In comparison, DistiLD focuses on disease-associated SNPs and genes in their chromosomal context (Palleja *et al.*, 2012). Another useful tool is Imprinted Gene Catalogue that helps in examining parental origin trends for different types of spontaneous mutations (Glaser *et al.*, 2006). LncRNADisease is a database for long-non-coding RNA-associated disease diagnosis, treatment and prognosis (Chen *et al.*, 2013). In addition, 'Organ System Heterogeneity DB' displays distribution of phenotypic effects across organ systems along with the heterogeneity value (Mannil *et al.*, 2015).

Strategies and Scope of Translational Bioinformatics

Translational bioinformatics assists in bringing laboratory work to clinical world as well as helps in drug discovery, clinical research, surveillance of approved drugs and patient-specific therapy regimes.

Drug Discovery

Drug molecule development

The important aspects of drug molecule development include:

Molecular variation

The efficacy and safety of any new compound is first evaluated in an animal model under preclinical studies. We find that several of these new compounds show efficacy in animal models, but they fail in humans and vice-versa. The reason of this efficacy variation is molecular variation between humans and animals (Shanks *et al.*, 2009; Buchan *et al.*, 2011). The two models might have distinct molecular pathologies. Translational bioinformatics helps to analyze the molecular variation in data obtained from the animal models and humans, and helps to take decision whether further development should proceed or not (Cox *et al.*, 2011). The drugs which are efficacious during preclinical phase and get rejected in late clinical phase, increase the cost of drug development. Furthermore, the safety of drugs could be an issue, as molecule may be safe in animal but can be life threatening in case of human beings (Ioannidis, 2012). The cost of drug development has been increasing steadily due to drug failure, the major cause of which is lack of clinical efficacy as well as toxicity. As per the executive summary prepared by FDA in October 2017, the estimated cost of failed clinical trials ranges from $800 million to $1.4 billion. Translational bioinformatics can help in pre-analysis and ensure cost effectiveness and safe drug development.

Similarly considering the reverse case, several molecules which might be life saving for human beings, are discarded due to their failure in animal models, and thus are not studied in humans. So, translational bioinformatics may help in identifying drugs, which may be efficacious in humans.

Target identification/rational drug discovery

The discovery of drugs based on the knowledge of the target is termed as rational drug discovery. There are several inventive approaches to find out and characterize the target. The techniques of transcriptional bioinformatics, metabolic and signaling pathway analysis, knowledge of interacting proteins and information on enzyme activity are used as a protocol for target identification/rational drug discovery. Numerous bioinformatics tools/web servers **(Table 1)** are frequently used to find target structure and binding site information that can translate data into drugs.

Clinical research

Clinical research comprises of pharmacogenomics for optimal treatment and biomarker efficacy.

Pharmacogenomics for optimal treatment

The study of how genes affect individual's response to the drug is termed as pharmacogenomics, which could be studied as a part of precision medicine (Thorn *et al.*, 2010). The available genomic and phenomic data studied in relation with Electronic Medical Records (EMRs) and pharmacology help to evaluate safe and effective medicine and transplant for individuals.

Characterization of patient's genome by high throughput sequence analysis and comparative analysis with standard database gives an idea whether any gene is responsible for disease expression (Sarkar *et al.*, 2011). The inhibitor of that gene will only be effective in the patient who is affected due to overexpression of it, not by any other cause. The knowledge of pharmacogenomics helps in the selection of people with desired genetic composition for optimal treatment, with minimum side effects.

Biomarker (tracking) efficacy

A biomarker is a biomolecule that provides early diagnosis of a disease or its state of progression (Xia *et al.*, 2013). It is utilized to quantify drug efficacy, and identify drug targets and mechanism. Biomarkers that monitor specific physiological or pharmacological phenomenon can be used to select multiple therapeutic targets of a drug molecule (Kusumegi *et al.*, 2004).

Post-approval surveillance

The safety of drugs and repurposing drugs for new ailments are important considerations once drugs get approved for marketing.

Drug safety monitoring

To monitor safety of drugs, adverse drug reactions (ADRs) data are prepared. ADRs data can be easily obtained by transforming the big data (refers to electronic health records, administrative or health claims data, disease and drug monitoring registries) for pharmacovigilance system (Harpaz *et al.*, 2016). Sometimes, this data helps in finding new indications for existing drugs. A beneficial side effect could be considered as an additional therapeutic effect of the drug. For example, minoxidil, which is a potassium channel potentiator, is used as a antihypertensive drug, with the side effect of hypertrichosis (excessive growth of hair). This is considered as an additional therapeutic benefit and is used to control alopecia (hair loss or baldness).

Some ADRs are also observed in persons with specific genetic make-up. ADR databases help to identify the correct set of patients for a particular drug molecule. In addition, the analysis of these ADR database gives an idea about extra effects of the drugs and the pharmacological mechanism of drug action (Harpaz *et al.*, 2016; Abernethy *et al.*, 2011).

Drug repurposing

The reuse of already approved drugs for new diseases is termed as drug repositioning or repurposing. Since drug development and approval are costly and time consuming, this is an efficient way to bypass the early development phase and proceed directly with clinical trials for possible new diseases.

Translational bioinformatics strategies for repurposing involve finding new targets for approved drug molecules, by protein-protein, and drug-protein interactions, pathway analysis and similarity in gene expression during the progression of different diseases that may involve same molecular mechanism. Furthermore, high throughput screening of a drug's chemical structure against the various receptors allows finding new targets.

Patient-Specific Treatment Regimes

These include personalized medicine and real-time health monitoring systems.

Personalized medicine and pharmacogenomics

An individual genetic/molecular variation predicts disease susceptibility and responses to therapeutic agents. Biological data of people of various demographic conditions enable us to understand differential treatment approach required by individuals of different types of genes, lifestyle and environments (Reynolds, 2012). The data related to medical history of the patient is a key requirement for personalized medication. Apart from diagnostic reports, metagenomic analysis (environmental analysis) and genetic testing of patients are utilized to compare and analyze the class of medication that must be prescribed to the patient. The term 'personalized medicine' is used for group of people having similar gene type, lifestyle and environments, providing a basis for similar treatment strategies.

Although the term 'precision medicine' is relatively new, the concept has been a part of healthcare for more than a century. For instance, specific blood group of people can be transfused with only specific types of blood due to their antigen, which is result of their genetic composition. Presently, precision medicine is being intensively studied and, if future, could lead into new era of medicine.

Real-time health monitoring

To get the real time data of the individual medical aspects, various real time health monitoring wearable or implantable devices are available, for e.g., Fitbit Surge (heart rate, activity, sleep, and caloric burn), iHealth BP5 (blood pressure), iHealth glucometer (Blood glucose), Muse Headband (EEG), Scanadu Scanaflo (urine analysis), Zephyr BioPatch (respiratory rate, ECG), Empatica E3 (photoplethysmograph, electrodermal activity). The data provided by these devices are known as Electronic Health Records (EHRs) or Electronic Medical Records (EMRs), which are integrated to clinical repositories or biobank by machine learning technology and predictive analysis offered by various tools and programs such as (i) eMERGE (see "Relevant Websites section"), a tool to combine DNA biorepositories with EMR systems for large scale, high-throughput genetic research in support of implementing genomics (Gottesman et al., 2013a); (ii) CLIPMERGE PGx, a program for personalized medicine through EHRs and genomics-pharmacogenomics (Gottesman et al., 2013b); (iii) i2b2 (see "Relevant Websites section"), integrating biology and bedside, a translational engine of patient's clinical data (Takai-Igarashi et al., 2011).

EMRs are standard medical and clinical data gathered and stored electronically by health care provider. Analysis of these data with respect to available database helps to understand and identify the transition of the healthy state of the person to the diseased state, and the individual's disease progression (Shameer et al., 2017). Most prominent databases linked with EMRs are (i) BioMe BioBank (see "Relevant Websites section"), a database of blood samples and health information, (ii) HarmoniMD (see "Relevant Websites section"), a repository of electronic clinical documents and cloud based EHR system, and (iii) HealthVault (see "Relevant Websites section"), a platform to store personal health information and generate insight from it.

Drug Interactions

Drug-drug interaction (DDI) can occur when two or more drugs are co-administered to a patient. It can lead to changed systemic exposure, resulting in variations in drug response of the co-administered drugs. DDIs generally occur due to inhibition of the metabolism for one drug by the other. It leads to a rise in plasma concentration of the drug whose metabolism is inhibited. The therapeutic index of the drug that has increased concentration in plasma could be decreased, leading to adverse reaction or increase in toxicity. The drug efficacy databases and DDI databases, play crucial roles during concomitant medication and avoid risk of adverse impact on the patient (Duke et al., 2012). Evaluation of potential drug interactions also provides the idea about physiological pathways of its kinetics and dynamics.

Computational Tools Implicated in Drug Discovery

Multitude of biological and clinical databases have led to the development of tools for drug discovery (**Table 3**). It includes data integration software, virtual screening or structure activity relationship tool, docking, modeling or interaction visualization tool, disease progression or signaling pathway analysis tools.

Table 3 Drug discovery tools and web servers

Website	Key feature(s)	CI[a]	Access[b]	URL
CAVER analyst	For calculation, analysis and real time visualization of static or dynamic protein tunnel interaction	20.2	W	http://www.caver.cz/index.php
ConsensusPathDB	Tool to integrate protein-protein, genetic, metabolic, signaling, gene regulatory and drug-target and biochemical pathways interactions	24.1	W	http://cpdb.molgen.mpg.de/
Drug2Gene	For identifying gene targeting compound	5.4	W	http://www.drug2gene.com/
MLViS	Machine learning-based virtual screening tool	1.1	W	http://www.biosoft.hacettepe.edu.tr/MLViS/
Molsoft	Tools for cheminformatics, QSAR, mutation analysis and sequence analysis	-[c]	W	https://www.molsoft.com/index.html
MycoRegDB	For delineating regulatory/pathways in mycobacteria and in turn identify new drug targets for tuberculosis	1.8	W	http://compbio.iitr.ac.in/mycoregdb/
NNScore	For calculating receptor-ligand scoring function based on neural network	9.0	W	http://www.nbcr.net/software/nnscore/
PharmGKB	To find impact of genetic variation on drug response	21.4	W	https://www.pharmgkb.org/
PROMISCUOUS	For drug repositioning by integrating drug, target and signaling pathway	20.6	W	http://bioinformatics.charite.de/promiscuous/
SuperTarget	Web server to analyze drug-target interactions	32.0	W	http://insilico.charite.de/supertarget/index.php?site=home

[a]CI, citation index (number of citations per year).
[b]W, Web based.
[c]-, no citations.

ConsensusPathDB, PharmGKB and PROMISCUOUS are database integration and analysis tools that integrate -omic (genomic, proteomic or metabolomics) data with signaling or biochemical pathway data to identify drug targets. ConsensusPathDB integrates and provides visualization of diverse functional interactions (protein – Protein interactions, biochemical reactions, gene regulatory interactions) and more than 1700 metabolic/biochemical pathways of human and provides visualization of interaction networks and substructures derived from the integrated interaction network (Kamburov et al., 2009). On the other hand, MycoRegDB delineates regulatory pathways/networks across all mycobacterial species (Sharma et al., 2009; Sharma and Surolia, 2013, unpublished data). This in turn will pave way for identification of new drugs for tuberculosis. In contrast, PharmGKB provides pathway and pharmacogene summary by integrating genotype, molecular as well as clinical knowledge and helps in investigating how genetic variation affects drug response (Thorn et al., 2013). PROMISCUOUS is a database for understanding and prediction of off-target effects of drugs and allows a rational approach for drug-repositioning (von Eichborn et al., 2011). It measures structural similarity for drugs and connects it to protein-protein interactions to establish and analyze networks responsible for its known side-effects. Based on this network-based approach, it helps in drug-repositioning. Another useful tool is MLViS that classifies molecules as drug-like and nondrug-like based on various machine-learning methods, including discriminant, tree-based, kernel-based, ensemble and other algorithms (Korkmaz et al., 2015). Subsequently, it helps in performing virtual screening of drug like molecules against the target to find out active drug compounds. NNScore is a tool for scoring function (score is used during virtual screening) based on neural network for the characterization of protein-ligand complex (Durrant and McCammon, 2010). A distinct tool SuperTarget has also been developed that integrates drug-related information like medical indication, adverse drug effects, drug metabolization, pathways and Gene Ontology to find out drug-target relations (Gunther et al., 2008). 'CAVER Analyst' is a graphic tool for interactive real-time visualization and analysis of tunnels and channels in static and dynamic protein structures (Kozlikova et al., 2014). Molsoft contains a quantitative structure activity relationship (QSAR) tool, which helps in drug designing by providing the quantitative information about functional group's position/type present in drug molecule (see "Relevant Websites section"). In addition, Drug2Gene is a database that combines the compound/drug-gene/protein information and integrates it to identify compounds targeting a given gene product or for finding all known targets of a drug (Roider et al., 2014).

Discussion

Translational bioinformatics has opened new doors for churning out precious information from the existing knowledge of human health, disease progression mechanism, prophylactic conditions and human variation. The current challenges in healthcare are cost, ADRs, efficacy variance, late diagnosis and high rate of organ failures. These challenges can be addressed in the patient if the right medication can be administered at an appropriate time. Against this backdrop, it becomes important to understand the genetic makeup of an individual as well as the impact of environmental and demographic factors to provide right medication in sync with the requirement. The biological database helps us in this endeavor, provided we have scientific base for finding gene interactors and gene(s) responsible for disease progression and its molecular mechanism. Various relevant tools help us to understand the connection between genotype and phenotype via proteomic and metabolomics information.

On the other hand, the structure-function relationship provides information for drug designing, pharmacophore modeling and QSAR studies, thereby leading to drug discovery. Numerous molecules are now available in market or under development owing to this rational method of drug discovery rather than the traditional methods of study. It has not only reduced the cost of pharmaceutical industry significantly but also provided motivation for new findings or synthesizing analogs of higher efficacy and least toxicity. In addition, translational bioinformatics has proved the significance and utility of biomarkers by studying systemic interactions among genes, metabolites, tissue system, drugs and intermediate molecules, including environmental factors. Bio-markers information helps in timely diagnosis, and thereby ensures better treatment outcomes.

Similarly, the clinical information about patient's conditions, disease progression, adverse drug reactions (ADRs), pathological, diagnostics and wellness reports provide records for real time health monitoring by comparing and analyzing with the help of relevant tools and applications. It provides information about disease progression that helps to take required treatment at an appropriate time. Furthermore, the integration of these data help in the prediction of the rate of occurrence of any disease based on pollution condition or type, socioeconomic and geographic distribution. This helps to suggest further precaution or prophylactic action to the group of individuals. The total environmental exposure of human is termed as 'exposome', which indicates the disease susceptibility (e.g., lung cancer, skin cancer, asthma, allergy, baldness and deficient to specific micronutrient) based on exposure to the ambient environmental conditions (e.g., air-quality, smoke, radiations and specific nutrient chelators in food). Some devices are also there to measure exposome along with EMRs.

Summary

In the new era of data-driven healthcare evolution, translational bioinformatics is a crucial approach for drug discovery and development process. The innovative translational and computational analysis approaches are making a great impact on discovery, preclinical, clinical and post launch of a drug. The computational study saves millions in cost of drug development as compared to classical method. It gives rational approach, which reduces the chance of failure as compared to blind testing. It facilitates a better drug designing of improved efficacy since structural data of receptors gives fair idea about pharmacophore modeling and quantitative structure activity relationship (QSAR) studies. Additionally, the sequence data helps in generating model of unknown structures with the help of several tools, which are used to study its interaction with drug molecules. It also enables to understand and decipher the mechanism of drug molecules by analyzing the functional data and laboratory verification. The development of various databases in health care system is a huge resource for advancement and evolution of safe and efficacious medicinal system. As discussed above, the clinical data independently or in integration with biological data help to find out new drug or improved safety prescriptions. In addition, proper selection of patients during trial improves design of the trial and its success rate. Personalized medicine and real time health monitoring is a new era in the medical history, wherein huge efforts are being made. It has a great role in preventing life threatening disease like cancer where proper drug selection is still a challenge. Translational bioinformatics is an emerging field with lot of opportunities in medical advancement with development of algorithms, tools, programs and methods in order to find out significant information from available databases. Hopefully in the near future, translational bioinformatics with utilization of artificial intelligence will revolutionize medical landscape.

See also: Biological and Medical Ontologies: Disease Ontology (DO). Biological and Medical Ontologies: Human Phenotype Ontology (HPO). Clinical Proteomics. Comparative Transcriptomics Analysis. Disease Biomarker Discovery. Experimental Platforms for Extracting Biological Data: Mass Spectrometry, Microarray, Next Generation Sequencing. Identification and Extraction of Biomarker Information. Information Retrieval in Life Sciences. Integrative Analysis of Multi-Omics Data. Integrative Bioinformatics. Integrative Bioinformatics of Transcriptome: Databases, Tools and Pipelines. Natural Language Processing Approaches in Bioinformatics. Ontology in Bioinformatics. Predicting Non-Synonymous Single Nucleotide Variants Pathogenic Effects in Human Diseases

References

Abernethy, D.R., Woodcock, J., Lesko, L.J., 2011. Pharmacological mechanism-based drug safety assessment and prediction. Clin. Pharmacol. Ther. 89 (6), 793–797.

Agrawal, S., *et al.*, 2008. T2D-Db: An integrated platform to study the molecular basis of Type 2 diabetes. BMC Genom. 9, 320.

Altman, R.B., 2012. Translational bioinformatics: Linking the molecular world to the clinical world. Clin. Pharmacol. Ther. 91 (6), 994–1000.

Ashburner, M., *et al.*, 2000. Gene ontology: Tool for the unification of biology. Nat. Genet. 25 (1), 25–29.

Bader, G.D., Cary, M.P., Sander, C., 2006. Pathguide: A pathway resource list. Nucleic Acids Res. 34 (Database issue), D504–D506.

Baker, E.J., *et al.*, 2012. GeneWeaver: A web-based system for integrative functional genomics. Nucleic Acids Res. 40 (Database issue), D1067–D1076.

Basu, S.N., Kollu, R., Banerjee-Basu, S., 2009. AutDB: A gene reference resource for autism research. Nucleic Acids Res. 37 (Database issue), D832–D836.

Baxevanis, A.D., 2001. The molecular biology database collection: An updated compilation of biological database resources. Nucleic Acids Res. 29 (1), 1–10.

Bertram, L., *et al.*, 2007. Systematic meta-analyses of Alzheimer disease genetic association studies: The AlzGene database. Nat. Genet. 39 (1), 17–23.

Blekherman, G., *et al.*, 2011. Bioinformatics tools for cancer metabolomics. Metabolomics 7 (3), 329–343.

Blum, R.L., 1982. Discovery, confirmation, and incorporation of causal relationships from a large time-oriented clinical data base: The RX project. Comput. Biomed. Res. 15 (2), 164–187.

Buchan, N.S., *et al.*, 2011. The role of translational bioinformatics in drug discovery. Drug Discov. Today 16 (9–10), 426–434.

Chatr-aryamontri, A., *et al.*, 2007. MINT: The Molecular INTeraction database. Nucleic Acids Res. 35 (Database issue), D572–D574.

Chen, G., *et al.*, 2013. LncRNADisease: A database for long-non-coding RNA-associated diseases. Nucleic Acids Res. 41 (Database issue), D983–D986.

Chen, X., Ji, Z.L., Chen, Y.Z., 2002. TTD: Therapeutic target database. Nucleic Acids Res. 30 (1), 412–415.

Cox, B., *et al.*, 2011. Translational analysis of mouse and human placental protein and mRNA reveals distinct molecular pathologies in human preeclampsia. Mol. Cell. Proteomics 10 (12), doi:10.1074/mcp.M111.012526.

Croning, M.D., *et al.*, 2009. G2Cdb: The genes to cognition database. Nucleic Acids Res. 37 (Database issue), D846–D851.

Dai, H.J., *et al.*, 2013. T-HOD: A literature-based candidate gene database for hypertension, obesity and diabetes. Database (Oxford) 2013, bas061.

Davis, A.P., *et al.*, 2009. Comparative toxicogenomics database: A knowledgebase and discovery tool for chemical-gene-disease networks. Nucleic Acids Res. 37 (Database issue), D786–D792.

Dessailly, B.H., *et al.*, 2008. LigASite – A database of biologically relevant binding sites in proteins with known apo-structures. Nucleic Acids Res. 36, D667–D673.

Duke, J.D., *et al.*, 2012. Literature based drug interaction prediction with clinical assessment using electronic medical records: Novel myopathy associated drug interactions. PLOS Comput. Biol. 8 (8), e1002614.

Durrant, J.D., McCammon, J.A., 2010. NNScore: A neural-network-based scoring function for the characterization of protein-ligand complexes. J. Chem. Inf. Model. 50 (10), 1865–1871.

Du, Y., *et al.*, 2017. Lung Gene Expression Analysis (LGEA): An integrative web portal for comprehensive gene expression data analysis in lung development. Thorax 72 (5), 481–484.

Fiannaca, A., *et al.*, 2013. A knowledge-based decision support system in bioinformatics: An application to protein complex extraction. BMC Bioinform. 14 (Suppl 1), S5.

Frenkel-Morgenstern, M., *et al.*, 2010. Dynamic proteomics: A database for dynamics and localizations of endogenous fluorescently-tagged proteins in living human cells. Nucleic Acids Res. 38 (Database issue), D508–D512.

Glaser, R.L., Ramsay, J.P., Morison, I.M., 2006. The imprinted gene and parent-of-origin effect database now includes parental origin of de novo mutations. Nucleic Acids Res. 34 (Database issue), D29–D31.

Gottesman, O., *et al.*, 2013a. The Electronic Medical Records and Genomics (eMERGE) network: Past, present, and future. Genet. Med. 15 (10), 761–771.

Gottesman, O., *et al.*, 2013b. The CLIPMERGE PGx program: Clinical implementation of personalized medicine through electronic health records and genomics-pharmacogenomics. Clin. Pharmacol. Ther. 94 (2), 214–217.

Grover, G., Sharma, D., unpublished data.

Gunther, S., *et al.*, 2008. SuperTarget and matador: Resources for exploring drug-target relationships. Nucleic Acids Res. 36 (Database issue), D919–D922.

Harpaz, R., DuMochel, W., Shah, N.H., 2016. Big data and adverse drug reaction detection. Clin. Pharmacol. Ther. 99 (3), 268–270.

Healey, J., *et al.*, 2004. An open-source method for simulating atrial fibrillation using ECGSYN. Comput. Cardiol. 31, 425–427.

Ioannidis, J.P., 2012. Extrapolating from animals to humans. Sci. Transl. Med. 4 (151), 151ps15.

Irwin, J.J., Shoichet, B.K., 2005. ZINC – A free database of commercially available compounds for virtual screening. J. Chem. Inf. Model 45 (1), 177–182.

Jegga, A.G., *et al.*, 2007. PolyDoms: A whole genome database for the identification of non-synonymous coding SNPs with the potential to impact disease. Nucleic Acids Res. 35 (Database issue), D700–D706.

Jia, P., *et al.*, 2010. SZGR: A comprehensive schizophrenia gene resource. Mol. Psychiatry 15 (5), 453–462.

Kamburov, A., *et al.*, 2009. ConsensusPathDB – A database for integrating human functional interaction networks. Nucleic Acids Res. 37 (Database issue), D623–D628.

Kanehisa, M., Goto, S., 2000. KEGG: Kyoto encyclopedia of genes and genomes. Nucleic Acids Res. 28 (1), 27–30.

Kann, M.G., 2010. Advances in translational bioinformatics: Computational approaches for the hunting of disease genes. Brief. Bioinform. 11 (1), 96–110.

Kao, S., *et al.*, 2012. IGDB.NSCLC: Integrated genomic database of non-small cell lung cancer. Nucleic Acids Res. 40 (Database issue), D972–D977.

Katal, A., Wazid, M., Goudar, R.H., 2013. Big data: Issues, challenges, tools and good practices. In: Proceedings of the 2013 Sixth International Conference on Contemporary Computing (Ic3), pp. 404–409.

Kodzius, R., *et al.*, 2006. CAGE: Cap analysis of gene expression. Nat. Methods 3 (3), 211–222.

Korkmaz, S., Zararsiz, G., Goksuluk, D., 2015. MLViS: A web tool for machine learning-based virtual screening in early-phase of drug discovery and development. PLOS ONE 10 (4), e0124600.

Kozlikova, B., *et al.*, 2014. CAVER Analyst 1.0: Graphic tool for interactive visualization and analysis of tunnels and channels in protein structures. Bioinformatics 30 (18), 2684–2685.

Kuhn, M., *et al.*, 2016. The SIDER database of drugs and side effects. Nucleic Acids Res. 44 (D1), D1075–D1079.

Kusumegi, T., *et al.*, 2004. BMP7/ActRIIB regulates estrogen-dependent apoptosis: New biomarkers for environmental estrogens. J. Biochem. Mol. Toxicol. 18 (1), 1–11.

Lill, C.M., *et al.*, 2012. Comprehensive research synopsis and systematic meta-analyses in Parkinson's disease genetics: The PDGene database. PLOS Genet. 8 (3), e1002548.

Li, P., *et al.*, 2015. PubAngioGen: A database and knowledge for angiogenesis and related diseases. Nucleic Acids Res. 43 (Database issue), D963–D967.

Liu, G., *et al.*, 2010. Computational analysis of microRNA function in heart development. Acta Biochim. Biophys. Sin. (Shanghai) 42 (9), 662–670.

Liu, H., *et al.*, 2011. CADgene: A comprehensive database for coronary artery disease genes. Nucleic Acids Res. 39 (Database issue), D991–D996.

Liu, T., *et al.*, 2007. BindingDB: A web-accessible database of experimentally determined protein-ligand binding affinities. Nucleic Acids Res. 35 (Database issue), D198–D201.

Mannil, D., *et al.*, 2015. Organ system heterogeneity DB: A database for the visualization of phenotypes at the organ system level. Nucleic Acids Res. 43 (Database issue), D900–D906.

McCall, M.N., *et al.*, 2014. The gene expression barcode 3.0: Improved data processing and mining tools. Nucleic Acids Res. 42 (Database issue), D938–D943.

Mi, H., *et al.*, 2005. The PANTHER database of protein families, subfamilies, functions and pathways. Nucleic Acids Res. 33 (Database issue), D284–D288.

Mularoni, L., Ramos-Rodriguez, M., Pasquali, L., 2017. The pancreatic islet regulome browser. Front. Genet. 8, 13.

Nim, H.T., *et al.*, 2015. CARFMAP: A curated pathway map of cardiac fibroblasts. PLOS ONE 10 (12), e0143274.

Palleja, A., *et al.*, 2012. DistiLD database: Diseases and traits in linkage disequilibrium blocks. Nucleic Acids Res. 40 (Database issue), D1036–D1040.

Pinero, J., *et al.*, 2015. DisGeNET: A discovery platform for the dynamical exploration of human diseases and their genes. Database (Oxford) 2015, bav028.

Raghupathi, W., Raghupathi, V., 2014. Big data analytics in healthcare: Promise and potential. Health Inf. Sci. Syst. 2, 3.

Rahmati, S., *et al.*, 2017. pathDIP: An annotated resource for known and predicted human gene-pathway associations and pathway enrichment analysis. Nucleic Acids Res. 45 (D1), D419–D426.

Ran, X., *et al.*, 2015. EpilepsyGene: A genetic resource for genes and mutations related to epilepsy. Nucleic Acids Res. 43 (Database issue), D893–D899.

Reynolds, K.S., 2012. Achieving the promise of personalized medicine. Clin. Pharmacol. Ther. 92 (4), 401–405.

Roider, H.G., *et al.*, 2014. Drug2Gene: An exhaustive resource to explore effectively the drug-target relation network. BMC Bioinform. 15, 68.

Saeed, M., *et al.*, 2011. Multiparameter intelligent monitoring in intensive care II: A public-access intensive care unit database. Crit. Care Med. 39 (5), 952–960.

Sagiroglu, S., Sinanc, D., 2013. Big data: A review. In: Proceedings of the 2013 International Conference on Collaboration Technologies and Systems (Cts), pp. 42–47.

Sam, S.A., *et al.*, 2017. XTalkDB: A database of signaling pathway crosstalk. Nucleic Acids Res. 45 (D1), D432–D439.

Sarkar, I.N., *et al.*, 2011. Translational bioinformatics: Linking knowledge across biological and clinical realms. J. Am. Med. Inform. Assoc. 18 (4), 354–357.

Schomburg, I., *et al.*, 2004. BRENDA, the enzyme database: Updates and major new developments. Nucleic Acids Res. 32 (Database issue), D431–D433.

Sharma, D., Mohanty, D., Surolia, A., 2009. RegAnalyst: a web interface for the analysis of regulatory motifs, networks and pathways. Nucleic Acids Res. 37, W193–W201.

Sharma, D., Surolia, A., 2013. Pathway targeting, antimycobacterial drug design. Encyclopedia of Systems Biology. Springer. pp. 1656–1659.

Shameer, K., *et al.*, 2017. Translational bioinformatics in the era of real-time biomedical, health care and wellness data streams. Brief. Bioinform. 18 (1), 105–124.

Shanks, N., Greek, R., Greek, J., 2009. Are animal models predictive for humans? Philos. Ethics Humanit. Med. 4, 2.

Sharma, D., Priyadarshini, P., Vrati, S., 2015. Unraveling the web of viroinformatics: Computational tools and databases in virus research. J. Virol. 89 (3), 1489–1501.

Sharma, D., Surolia, A., 2011. Computational tools to study and understand the intricate biology of mycobacteria. Tuberculosis (Edinb.) 91 (3), 273–276.

Siramshetty, V.B., et al., 2018. SuperDRUG2: A one stop resource for approved/marketed drugs. Nucleic Acids Res. 46 (D1), D1137–D1143.

Smink, L.J., et al., 2005. T1DBase, a community web-based resource for type 1 diabetes research. Nucleic Acids Res. 33 (Database issue), D544–D549.

Somek, M., Hercigonja-Szekeres, M., 2017. Decision support systems in health care - Velocity of Apriori algorithm. Stud. Health Technol. Inform. 244, 53–57.

Sorani, M.D., et al., 2010. Clinical and biological data integration for biomarker discovery. Drug Dis. Today 15 (17–18), 741–748.

Szklarczyk, R., et al., 2016. WeGET: Predicting new genes for molecular systems by weighted co-expression. Nucleic Acids Res. 44 (D1), D567–D573.

Takai-Igarashi, T., et al., 2011. On experiences of i2b2 (Informatics for integrating biology and the bedside) database with Japanese clinical patients' data. Bioinformation 6 (2), 86–90.

Tan, A.C., Gilbert, D., 2003. Ensemble machine learning on gene expression data for cancer classification. Appl. Bioinform. 2 (3 Suppl), S75–S83.

Thorn, C.F., Klein, T.E., Altman, R.B., 2010. Pharmacogenomics and bioinformatics: PharmGKB. Pharmacogenomics 11 (4), 501–505.

Thorn, C.F., Klein, T.E., Altman, R.B., 2013. PharmGKB: The pharmacogenomics knowledge base. Methods Mol. Biol. 1015, 311–320.

von Eichborn, J., et al., 2011. PROMISCUOUS: A database for network-based drug-repositioning. Nucleic Acids Res. 39 (Database issue), D1060–D1066.

Wang, J., et al., 2017. NSDNA: A manually curated database of experimentally supported ncRNAs associated with nervous system diseases. Nucleic Acids Res. 45 (D1), D902–D907.

Wishart, D.S., et al., 2008. DrugBank: A knowledgebase for drugs, drug actions and drug targets. Nucleic Acids Res. 36 (Database issue), D901–D906.

Wishart, D., et al., 2015. T3DB: The toxic exposome database. Nucleic Acids Res. 43 (Database issue), D928–D934.

Wishart, D.S., et al., 2018. HMDB 4.0: The human metabolome database for 2018. Nucleic Acids Res. 46 (D1), D608–D617.

Wodke, J.A., et al., 2015. MyMpn: A database for the systems biology model organism Mycoplasma pneumoniae. Nucleic Acids Res. 43 (Database issue), D618–D623.

Xia, J., et al., 2013. Translational biomarker discovery in clinical metabolomics: An introductory tutorial. Metabolomics 9 (2), 280–299.

Xu, Y., et al., 2015. SEGEL: A web server for visualization of smoking effects on human lung gene expression. PLOS ONE 10 (5), e0128326.

Yan, Q., 2011. Toward the integration of personalized and systems medicine: Challenges, opportunities and approaches. Personalized Med. 8 (1), 1–4.

Ye, H., et al., 2011. HIT: Linking herbal active ingredients to targets. Nucleic Acids Res. 39 (Database issue), D1055–D1059.

Zhang, L., et al., 2012. ADHDgene: A genetic database for attention deficit hyperactivity disorder. Nucleic Acids Res. 40 (Database issue), D1003–D1009.

Zhen, Y., et al., 2007. CardioSignal: A database of transcriptional regulation in cardiac development and hypertrophy. Int. J. Cardiol. 116 (3), 338–347.

Relevant Websites

http://www.type2diabetesgenetics.org/
   ACCELERATING MEDICINES PARTNERSHIP (AMP).
http://icahn.mssm.edu/research/ipm/programs/biome-biobank
   BioMe.
https://clinicaldata.ema.europa.eu/web/cdp/home
   Clinical data.
https://www.physionet.org/physiotools/ecgsyn/
   ECGSYN.
https://emerge.mc.vanderbilt.edu/
   eMERGE.
http://about.harmonimd.com/
   HarmoniMD.
https://international.healthvault.com/in/en
   HealthVault.
https://www.i2b2.org/
   i2b2.
http://www.inovalon.com/howwehelp/cdr
   Inovalon.
https://www.liaison.com/healthcare-data-management/solutions/clinical-data-repository/
   LIAISON.
https://clinicaltrials.gov/
   NIH.
https://www.molsoft.com/
   Search ResultsMolsoft L.L.C.
http://data.hsl.virginia.edu/clinical-data-repository
   UVA Clinical Data Repository.

Translational Bioinformatics Databases

Onkar Singh, Institute of Information Science, Academia Sinica, Taipei, Taiwan, Republic of China and National Yang-Ming University, Taipei, Taiwan, Republic of China
Nai-Wen Chang, National Taiwan University, Taipei city, Taiwan, Republic of China and National Yang-Ming University, Taipei, Taiwan, Republic of China
Hong-Jie Dai, National Taitung University, Taipei, Taiwan, Republic of China
Jitendra Jonnagaddala, University of New South Wales, Sydney, NSW, Australia

Introduction

Translational bioinformatics (TBI) is an emerging field with great potential to assist clinicians to achieve success in the advancement of personalized medicine. Translational bioinformatics (TBI) is very young field at the horizon of TBI comprises molecular bioinformatics, clinical informatics, genetics, biostatics, health informatics and. The convergence of these fields into one makes TBI a prime component of biomedical research. The American Medical Informatics Association defined TBI as "the development of storage, analytic, and interpretive methods to optimize the transformation of increasingly voluminous biomedical data, and genomic data, into proactive, predictive, preventive, and participatory health" (see "Relevant Website section"). TBI integrates biological findings with the clinical information by using advanced computing environments and methodologies to bridge the gap between bench and bedside. TBI mainly focus on development of novel techniques for the integration of biological and clinical data and applying bioinformatics to translate the biological observations to practical application in a clinical setting. After the completion of the human genome project, clinicians acknowledge the potential use of the data generated, that it might give valuable information to know more about the basic mechanism of a human body. This accelerated the use of genome sequencing techniques resulting in the generation of a massive data that helps characterize the transcriptome, the proteome, and the metabolome (Altman, 2012). TBI not only created new knowledge but also gave new insights into underlying genetic mechanism of a disease (Londin and Barash, 2015). The four main foci of bioinformatics are clinical genomics, genomic medicine, pharmacogenomics, and genetic epidemiology (**Fig. 1**). These form the foundation of practices used by TBI to bridge the gap of research discoveries to clinical applications and their respective databases (Sarkar *et al.*, 2011).

In order to provide precision medicine-based therapies, heterogeneous data from clinical evaluations, bio-specimens and experimental investigations to characterize an individual patient's disease progression need to be integrated. This requires systematic collection, standardized storage, and analysis of data (Jonnagaddala *et al.*, 2016, 2014). This integrated information often resides in several public and private databases. These comprehensive databases have become increasingly important among the researchers for hypothesis generation, validation, and verification in many areas such as drug discovery and clinical research (Zou *et al.*, 2015). The TBI rely on the four fundamental pillars of bioinformatics i.e., clinical genomics, genomic medicine, pharmacogenomics, and genetic epidemiology. These emerging fields laid down the foundation of approaches used in TBI to

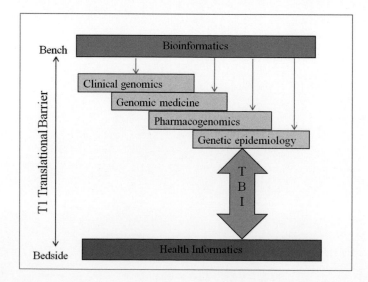

Fig. 1 Schematic illustration of four main areas of bioinformatics that laid the foundation of the approaches used by TBI to bridge the gap between bench and bedside.

bridge the gap between bench and bedside. This article discusses various TBI databases. These databases have already made a significant impact worldwide and continue to assist clinicians and other health care providers to implement personalized medicine. We also briefly discuss about the use of TBI databases in clinical genomics and in drug discovery and repurposing.

Translational Bioinformatics Databases

Over the decade, several TBI databases are developed to store biological findings, high-throughput data; scientific studies, published literature, and computational analysis. Most of the data stored in these databases contain evolving information which can give new insights to the researchers and help to facilitate the advancement of existing research. Here in this article we present few databases which are frequently used by diverse groups including from biologists to bioinformaticians.

Clinical Genomics Databases

Genomics is a study of the function, evolution, structure, and inheritance of the entire set of genetic material of an organism. Genomics involves the study of genes at all levels (e.g., DNA, mRNA, protein, cell, and tissue). The genomic studies become very efficient after the development of advanced technologies to facilitate high throughput sequencing and functional analysis. These technologies generate huge amounts of data that consist of DNA and protein sequence, data generated from various functional analysis and genetic mapping data. Most data are freely available for the public to access. Some of the popular genomic databases are: Clinical Genomic Database (CGD) (Solomon et al., 2013) is a valuable resource of known genetic conditions contains the data of single gene alteration. COXPRESdb (Obayashi et al., 2012) provides co-expression relationship for several animal species. COXPRESdb is a resource to discover a function of new gene candidates or to identify functional modules in metabolic and signaling pathways. Another popular database is "Database of genetic variants" (DGV) (MacDonald et al., 2014) provides comprehensive information of structural variant in the human genome (healthy controls). dbVar is a publicly accessible, and freely available database that is developed and maintained by the National Center for Biotechnology Information (NCBI). The dbVar database is slightly different from DGV, as it also contains structural variation information for human but also includes clinically relevant structural variation data from ClinVar. The human genome database (GDB) is a public repository mainly focus on human genes, clones, polymorphism, and STSs (Letovsky et al., 1998).

Genomic Medicine Databases

Genomic medicine is an emerging medical field which necessitates the use of an individual's genomic information to advance preventive treatment strategies and tailor-made treatments. Genomic medicine is adequate to make a mark on the personal healthcare by understanding the underlying mechanism of a disease. Here we will briefly describe several web resources that fall under the genomic medicine category. Cancer genome interpreter (CGI) is a web interface mainly developed to identify validated oncogenic alterations and to anticipate cancer driver genes among mutations of unknown significance. This CGI also have a comprehensive catalogue of validated oncogenic alterations, biomarker of drug response, cancer genes, and cancer types (Tamborero et al., 2018). ClinVar is a freely available archive which addresses the relationship of human variability and ascertained health condition with detailed interpretation knowledge (Landrum et al., 2016). The database of curated mutation (DoCM) in cancer is an open resource with the collection of biological important cancer variants (Ainscough et al., 2016). OncoKB, a precision-based oncology knowledge base provides evidenced based information of individual somatic mutation and structural alterations (Chakravarty et al., 2017). Like OncoKB, Precision Medicine Knowledge Base (PMKB) is also precision-based knowledge base valuable resource for cancer variants and represented the interpretations in structured manner. This knowledge base is freely accessible which allows the user to submit interpretations as well as modify existing information (Huang et al., 2017). Personalized cancer therapy (PCT) is a knowledge base for precision oncology that was developed to suggest a potential therapy to patients and clinicians based on particular tumor biomarkers.

Pharmacogenomics Databases

The pharmacogenomics is a combination of pharmacology and genomics where genomic information is used to examine the individual response to a drug agent. Pharmacogenomic research and discovery strongly assist health practitioners and clinicians in determining the correct drugs and accurate dosage for each individual, which may prevent adverse drug reaction and helps in the advancement of drug therapy (Zhang et al., 2015). In this section, we will describe most commonly used pharmacogenomic database. The PharmacoDB (Smirnov et al., 2018) is the largest collection of cancer pharmacogenomic studies. This database allows researchers to explore the information of a particular drug or cell line across publicly available datasets and compare the dose-response data for a specific cell line – drug pair from any of the studies included in the database. The pharmacogenomic knowledgebase (PharmGKB) is a curated collection of genetic variants, drug labels, genes, drugs and their associations and relationships (Whirl-Carrillo et al., 2012). The Clinical Pharmacogenetics Implementation Consortium (CPIC) is an international consortium formed in 2009 with the aim to address the implementation of genetic test results into an actionable form to make decision for affected drugs. This web resource is the collection of updated, detailed gene/drug clinical practice guidelines. The

DrugBank is a freely available resource containing drug and drug target related information. This database is full of both bioinformatics and chemo-informatics resources. The SCAN database is the collection of genetic and genomic data with multiple algorithms that can be used in data mining. This database consists two categories of SNPs: (1) Physical-based annotation and (2) functional annotation. CTDBase curates scientific data that describes relationships among chemicals/drugs, genes/proteins, diseases, taxa, phenotypes, GO annotations, pathways, and interaction modules. The primary goal of CTDBase is to enhance the understanding of the effects of environmental chemicals on human health on the genetic level, a field called toxicogenomic (Davis et al., 2017). The drug gene interaction database (DGIdb) provides information of candidate genes or drugs against the known and potentially druggable genome (Cotto et al., 2018).

Genetic Epidemiology Databases

Genetic factors or genes have all detailed information of our body; it has a crucial role in maintaining health and causing disease. Genetic epidemiology is the study of the role of genetic factors in our body for the determination of inherited disease among families. Such study can lead us to understand the underlying mechanism of genetic factors. In this topic we will discuss several genetic epidemiology databases include neuropsychiatric disorders, infectious diseases, breast cancer, kidney disease, cardiovascular disease, diabetes and aging. Neuropsychiatric disorder de novo mutation database (NPdenovo) isa huge collection of new (de novo) mutations causing mental disorders (autism spectrum disorder, intellectual disability, epileptic encephalopathy, schizophrenia and unaffected siblings). All information is identified by using whole-exome sequencing (Li et al., 2016). BRCA share is a largest database containing clinical BRCA gene variants responsible for breast cancer (Beroud et al., 2016). T2D@ZJU is comprehensive knowledge base contained information of type 2 database. All information are categorized in three types retrieved from pathways database, Protein-Protein Interaction (PPI) database and research articles (Yang et al., 2013). Type 2 diabetes genetic association database (T2DGADB) is developed to provide valuable information to the clinicians about the genetic risk factors, which plays vital role in the development of Type 2 diabetes (Lim et al., 2010). The clinical database for kidney disease (CDKD) includes physiological data of kidney patients (Singh et al., 2012). CardioGenBase, a comprehensive database for major cardiovascular disease (MCVDs) and causing genes. Allgene/protein information are collected from MCVD based literatures. This database provides valuable information to the clinicians to unveil novel disease mechanism (Alexandar et al., 2015). Alzforum is an information resource aiming to help researchers to accelerate discovery and development of diagnostics and treatments for Alzheimer's disease and related disorders. This

Table 1   A list of few currently available TBI databases. The databases are categorized into four different categories as discussed in this article

Database	URL
Clinical Genomic Database	
Clinical Genomic Database	https://research.nhgri.nih.gov/CGD/
COXPRESdb	http://coxpresdb.jp/
Database of Genomic Variants	http://dgv.tcag.ca/dgv/app/home
dbVar	https://www.ncbi.nlm.nih.gov/dbvar/
GDB	http://www.gdb.org
Genomic Medicine Databases	
CGI	https://www.cancergenomeinterpreter.org/home
ClinVar	https://www.ncbi.nlm.nih.gov/clinvar/
DoCM	http://docm.info/
OncoKB	http://oncokb.org/#/
PMKB	https://pmkb.weill.cornell.edu/
PCT	https://pct.mdanderson.org
Pharmacogenomics Databases	
PharmGKB	https://www.pharmgkb.org/
CPIC	https://cpicpgx.org/
Drug Bank	https://www.drugbank.ca/
SCAN	http://www.scandb.org/
CTDBase	http://ctdbase.org/
DGidb	http://www.dgidb.org/
Genetic Epidemiology Database	
NPdenovo	http://www.wzgenomics.cn/NPdenovo/
BRCA share	http://umd.be/BRCA1/
T2D@ZJU	http://tcm.zju.edu.cn/t2d
CardioGenBase	http://www.CardioGenBase.com/
CDKD	http://www.cdkd.org/
Alzforum	https://www.alzforum.org/

information resource has several databases dedicated to Alzheimer's disease. A list of few currently available TBI databases are presented in **Table 1**.

Recent Advances

Drug discovery is a process of designing potential medicines to stop or reverse the effects of the disease. This process starts after the diagnosis and characterization of symptoms affecting human life (Xia, 2017). Drug discovery needs a large investment, enormous time and effort. The average time frame to develop a drug is 10–17 years. Despite this, Food and Drug Administration (FDA) approved a very small number of drugs every year (Jadamba and Shin, 2016). Drug repurposing is an approach that addresses these issues to some extent. It takes advantage of the fact that approved drugs and many left alone compounds have already been tested in humans and detailed information of clinical and pharmacokinetic data is available (Astin *et al.*, 2017). The applications of TBI approaches are on a high demand in the pharmaceutical industry especially in drug discovery and drug repositioning aspects. By using the new knowledge created in the TBI databases, pharmaceuticals are addressing the identification and development of potential drugs (Buchan *et al.*, 2011). Several sophisticated computational tools are employed to analyze these databases that help in decision making in several aspects of drug discovery and development.

Conclusion

In summary, in this article we discussed the role of TBI databases and also presented mostly commonly used databases by diverse group of users, including but not limited to clinicians, biologists, clinical researchers and bioinformaticians. We provided an overview and some insights on how these TBI databases are used. The TBI databases are valuable resources available to generate, test or validate new hypothesis shiftily in a short amount of period. TBI databases continue to consistently contribute to the discovery and development of novel therapeutics and treatments. We believe in future there will be more TBI databases of high quality which integrates and brings together knowledge already acquired in the current TBI databases.

Acknowledgments

This article was prepared as part of the Translational Cancer research network (TCRN) research programs. TCRN is funded by Cancer Institute of New South Wales and Prince of Wales Clinical School, UNSW Medicine.

See also: Bioinformatics Data Models, Representation and Storage. Biological Database Searching. Data Storage and Representation. Genome Annotation. Genome Databases and Browsers. Information Retrieval in Life Sciences. Integrative Bioinformatics. Integrative Bioinformatics of Transcriptome: Databases, Tools and Pipelines. Models for Computable Phenotyping. Natural Language Processing Approaches in Bioinformatics. Networks in Biology. Standards and Models for Biological Data: FGED and HUPO

References

Ainscough, B.J., *et al.*, 2016. DoCM: A database of curated mutations in cancer. Nature Methods 13 (10), 806–807.

Alexandar, V., *et al.*, 2015. CardioGenBase: A literature based multi-omics database for major cardiovascular diseases. PLOS ONE 10 (12), e0143188.

Altman, R.B., 2012. Translational bioinformatics: Linking the molecular world to the clinical world. Clinical Pharmacology and Therapeutics 91 (6), 994–1000.

Astin, J.W., *et al.*, 2017. Chapter 2 – Innate immune cells and bacterial infection in zebrafish. In: Detrich, H.W., Westerfield, M., Zon, L.I. (Eds.), Methods in Cell Biology. Academic Press, pp. 31–60.

Beroud, C., *et al.*, 2016. BRCA share: A collection of clinical BRCA Gene variants. Human Mutation 37 (12), 1318–1328.

Buchan, N.S., *et al.*, 2011. The role of translational bioinformatics in drug discovery. Drug Discovery Today 16 (9), 426–434.

Chakravarty, D., *et al.*, 2017. OncoKB: A precision oncology knowledge base. JCO Precision Oncology 1, 1–16.

Cotto, K.C., *et al.*, 2018. DGIdb 3.0: A redesign and expansion of the drug–gene interaction database. Nucleic Acids Research 46 (D1), D1068–D1073.

Davis, A.P., *et al.*, 2017. The comparative toxicogenomics database: Update 2017. Nucleic Acids Research 45 (D1), D972–D978.

Huang, L., *et al.*, 2017. The cancer precision medicine knowledge base for structured clinical-grade mutations and interpretations. Journal of the American Medical Informatics Association 24 (3), 513–519.

Jadamba, E., Shin, M., 2016. A systematic framework for drug repositioning from integrated omics and drug phenotype profiles using pathway-drug network. BioMed Research International 2016, 7147039.

Jonnagaddala, J., *et al.*, 2014. Data sharing challenges and recommendations for human biorepositories: A systematic literature review. The International Technology Management Review 4 (2), 68–77.

Jonnagaddala, J., *et al.*, 2016. Integration and analysis of heterogeneous or translational. Nursing Informatics 2016, 387.

Landrum, M.J., *et al.*, 2016. ClinVar: Public archive of interpretations of clinically relevant variants. Nucleic Acids Research 44 (D1), D862–D868.

Letovsky, S.I., *et al.*, 1998. GDB: The human genome database. Nucleic Acids Research 26 (1), 94–99.

Li, J., *et al.*, 2016. Genes with de novo mutations are shared by four neuropsychiatric disorders discovered from NPdenovo database. Molecular Psychiatry 21 (2), 290–297.

Lim, J.E., *et al.*, 2010. Type 2 diabetes genetic association database manually curated for the study design and odds ratio. BMC Medical Informatics and Decision Making 10 (1), 76.

Londin, E.R., Barash, C.I., 2015. What is translational bioinformatics? Applied & Translational Genomics 6, 1–2.

MacDonald, J.R., *et al.*, 2014. The database of genomic variants: A curated collection of structural variation in the human genome. Nucleic Acids Research 42 (Database issue), D986–D992.

Obayashi, T., *et al.*, 2012. COXPRESdb: A database of comparative gene coexpression networks of eleven species for mammals. Nucleic Acids Research 41 (D1), D1014–D1020.

Sarkar, I.N., *et al.*, 2011. Translational bioinformatics: Linking knowledge across biological and clinical realms. Journal of the American Medical Informatics Association 18 (4), 354–357.

Singh, S.K., *et al.*, 2012. CDKD: A clinical database of kidney diseases. BMC Nephrology 13, 23.

Smirnov, P., *et al.*, 2018. PharmacoDB: An integrative database for mining in vitro anticancer drug screening studies. Nucleic Acids Research 46 (Database issue), D994–D1002.

Solomon, B.D., *et al.*, 2013. Clinical genomic database. Proceedings of the National Academy of Sciences of the United States of America 110 (24), 9851–9855.

Tamborero, D., *et al.*, 2018. Cancer genome interpreter annotates the biological and clinical relevance of tumor alterations. Genome Medicine 10 (1), 25.

Whirl-Carrillo, M., *et al.*, 2012. Pharmacogenomics knowledge for personalized medicine. Clinical Pharmacology and Therapeutics 92 (4), 414–417.

Xia, X., 2017. Bioinformatics and drug discovery. Current Topics in Medicinal Chemistry 17 (15), 1709–1726.

Yang, Z., *et al.*, 2013. T2D@ZJU: A knowledgebase integrating heterogeneous connections associated with type 2 diabetes mellitus. Database 2013, bat052.

Zhang, G., *et al.*, 2015. Web resources for pharmacogenomics. Genomics, Proteomics & Bioinformatics 13 (1), 51–54.

Zou, D., *et al.*, 2015. Biological databases for human research. Genomics, Proteomics & Bioinformatics 13 (1), 55–63.

Relevant Website

http://www.amia.org/applicationsinforatcs/translational-bioinformatics
American Medical Informatics Association (AMIA).

Electronic Health Record Integration

Hong Yung Yip, Nur A Taib, Haris A Khan, and Sarinder K Dhillon, University of Malaya, Kuala Lumpur, Malaysia

Introduction

An Electronic Medical Record (EMR) is a computerized system that contains a patient's health record generated in a medical practice. Its primary purpose is to integrate health care information to improve quality of care (Gunter and Terry, 2005). EMR systems are designed to store data accurately and to capture and reflect the state of a patient across time which includes records such as patient personal statistics and demographics, past medical history, diagnosis information such as laboratory test results, and surgery related information. The improved access of readily available healthcare information provides a well-coordinated, collaborative, and multidisciplinary care that contribute to better quality of healthcare delivery (Cutler, 2010). Traditionally, EMR data is designed and stored in relational database management system (RDBMS), since it offers a powerful declarative query language known as Structured Query Language (SQL) which is capable of handling complex queries. The limitations of the current implementation of EMR using relational model are presented in later section.

Data visualization is visual representation of data which is indispensable in order to convey message, facilitate understanding and discovery, make sense of complex data which assists users in analyzing and reasoning evidence (Aparicio and Costa, 2015). In the case of a clinical studies of EMR data, data visualization allows one to quickly identify areas requiring attention or improvement, identify relationships and patterns of a disease outbreak, pinpoint emerging trends, etc. (SAS, 2016).

Linked data is one of the use cases in visualization model. It is a method of publishing structured data so that it can be interlinked and become informative through semantic queries (**Fig. 1**). This technology has revolutionized the era of semantic web and defined a new way of how information can be traversed (Bizer *et al.*, 2009).

The primary objective of this study is to propose a novel way of visualizing and linking EMR data by developing a NoSQL graph database using Neo4j. With the same set of EMR data, this study also compares and contrasts the performance, in terms of query runtime and storage requirement between relational (Microsoft SQL Server) and NoSQL databases (Neo4j and MongoDB document-oriented database). Other criteria such as scalability, availability, consistency, data models and open source availability are also subjectively measured. Additionally, this study evaluates the differences between MS SQL Server and Neo4j in terms of schema design, ease of programming language, maturity and level of support, flexibility, and security to demonstrate how graph databases can provide a viable data storage and management alternative applicable to EMR systems.

Electronic Medical Record (EMR)

A system that handles operations and stores EMR records is referred to as an EMR system. Some of the benefits of an EMR system include efficient information sharing function to improve the value and quality of integrated health care to support development of health policies, medical education and advanced research (van der Linden *et al.*, 2009). However, despite its benefits, there are many obstacles and challenges in relation to EMR systems such as standardization of vocabulary, security, privacy and data quality. Nonetheless, Orfanidis *et al.* (2004) claimed that the one of the most critical issues with the current EMR systems is the ever expanding size of healthcare data. The exponential and uncoordinated growth of information systems in healthcare industry in terms of size and heterogeneity of data has resulted in no standardised data that makes, linking, retrieval and analysis difficult. Moreover, most EMRs are not customizable enough to cater to the information needs of researchers from different backgrounds. Fortunately, data visualization or visual analytics holds the potential to address the current information overload that is becoming more and more prevalent.

Data Visualization and Linked Data

Data visualization is the science of analytic reasoning which involves the study of the visual representation of data. It has revolutionized data sciences by enabling mapping, and visualization to allow better inferences of complicated data (Huang *et al.*, 2015). Proper visualization provides an alternative approach to illustrate potential data interconnectedness and relationships. Data visualization, despite having applied in various domains of healthcare (Huang *et al.*, 2015), remain absent from EMR data visualization using NoSQL graph model.

One of many use cases in visualization model involves Linked Data, which uses uniform resource identifiers (URIs), hypertext transfer protocol (HTTP) and resource description framework (RDF/JSON) technologies. Its importance include large scale integration of data on the Web (Bizer *et al.*, 2009). One of the best example of a large Linked Dataset is DBPedia, which utilizes the content in RDF format from Wikipedia to link to other datasets on the Web (W3C, 2015).

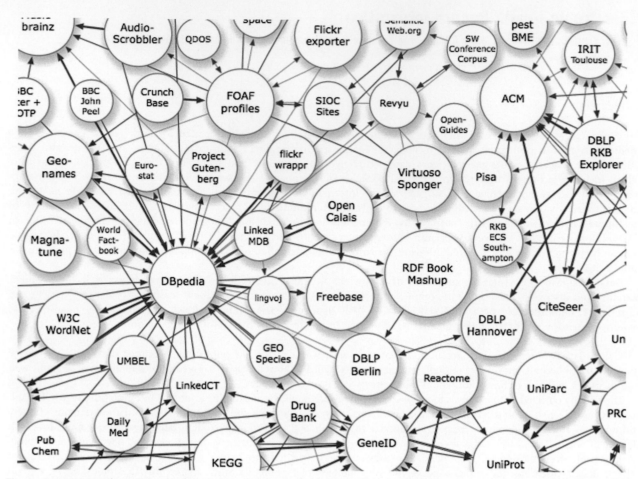

Fig. 1   Part of Linking Open Data project cloud diagram. Reproduced from Linkeddata.org., 2009. What is linked data? Retrieved November 15, 2016, from http://linkeddata.org.

The current implementation of EMR systems in hospitals often distributes and stores patient records across multiple databases, which causes undesirable data redundancy and isolation. Linked Data systems such as NoSQL databases can be used to offer a dynamic, graph-based model for recording and storing data that is interchangeable.

NoSQL (Not Only SQL) Databases

With the exponential increase in size and heterogeneity of data in modern information systems, a need has risen to cater to more complex and vast data structures compared to the previous applications of the traditional relational databases. This has led to the development of "NoSQL" databases. NoSQL architecture is designed to overcome the limited scalability, flexibility, performance, availability and infrastructure cost issues associated with relational databases (Stonebraker, 2009). They depend on horizontal scalability, method to increase performance of a system by increasing the number of nodes, to provide more computing capacity, as opposed to increasing the computer power of an individual node (Abramova and Bernardino, 2013).

Types of NoSQL databases

Though there are numerous types of NoSQL databases, they can be grouped in categories as explained below (**Table 1**):

BASE (Basically available, soft state, eventually consistent)

Unlike relational databases, NoSQL databases does not possess all strong ACID properties due to the its scattered architecture that lacks neither a coordinator nor a master node. Instead they based on Basically Available, Soft state and Eventually consistent (BASE principal). The principal is based on the fact that a system should be able to continue functioning despite the scenario that a single node fails in a distributed architecture (Bailis et al., 2013). Based on CAP Theorem, **Figs. 2** and **3** illustrate the strengths of relational data models in delivering consistency but at a lesser availability, whereas NoSQL key-value, column-oriented, tabular and document-oriented data models are capable of delivering either consistency of data storage or availability of data for retrieval.

Table 1   Categories of NoSQL databases (Abramova and Bernardino, 2013)

Category	Description
Key-value store	Data are stored as key-value pairs or hashes. The unique keys are utilized to index and retrieve the information stored in values. Examples are Redis, Azure Table Storage and DynamoDB.
Document store	Document store is similar to key-value store, however, their values are typically documents in formats such as eXtensible Markup Language (XML) or JSON. This category offers great performance and horizontal scalability. The internal storage is similar to relational databases but more flexible due to the lack of schema. It is also great for storage since it does not store empty or null values.
Column-family	Data are stored in columns. Each key is associated with one or more columns. This category is suitable for data mining and analytic systems due to high scalability. Examples are Cassandra and HBase.
Graph database	Data are stored as graph in the form of nodes (objects) and edges (relationships). The primary goal is to visualize the relationships between nodes. Examples are Neo4j and InfoGrid.

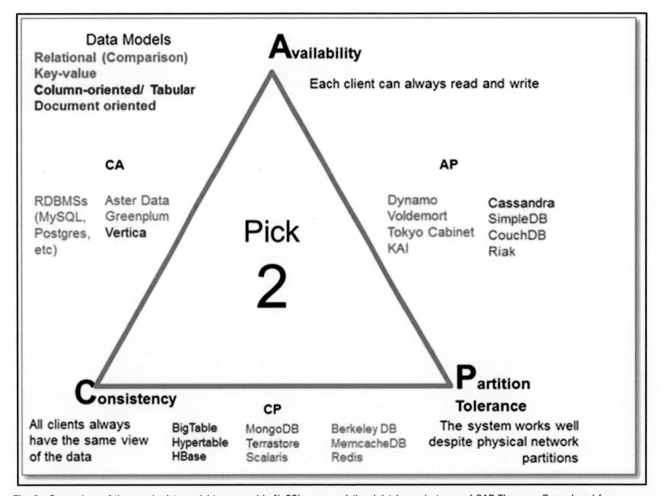

Fig. 2 Comparison of three main data model types used in NoSQL versus relational databases in terms of CAP Theorem. Reproduced from Ercan, M.Z., 2014. Evaluation of NoSQL databases for EHR systems. In: Proceedings of the 25th Australasian Conference on Information Systems, pp. 1–10. Auckland, New Zealand.

Potential benefits of NoSQL databases for EMR systems

Table 2 describes the main requirements of an EMR system and how newer NoSQL database system features can address these requirements to develop a distributed healthcare system.

NoSQL graph database

Graph databases enable queries that allow semantic web technology to function with the data defined and represented by nodes, edges and properties. They data retrieval from complex data structures easily as compared to relational systems (Neo4j, 2016a,b,c,d,e). Contrary to relational databases that operate with SQL language, the difficulty with graph

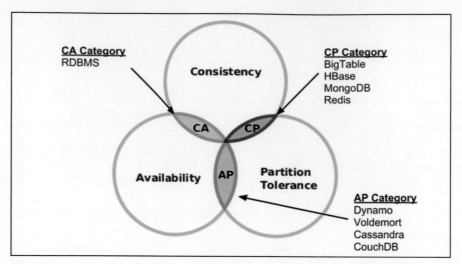

Fig. 3   Three categories of CAP Theorem. Reproduced from Ercan, M.Z., 2014. Evaluation of NoSQL databases for EHR systems. In: Proceedings of the 25th Australasian Conference on Information Systems, pp. 1–10. Auckland, New Zealand.

Table 2   Comparison of EMR requirements and corresponding NoSQL database features (Ercan, 2014)

EMR requirement	NoSQL database feature
Size of healthcare data expanded over time and eventually became a bottleneck for traditional EMR systems.	NoSQL databases are based on horizontal scalability which permits effortless and automatic scaling.
Increasing heterogeneity of healthcare data such as free-text notes, images and other complex data that are unstructured or semi-structured require new storage alternatives.	Flexible data models or schemas powered by NoSQL databases allow complex data to be stored easily.
Healthcare records should always be accessible for undisruptive and continuity of healthcare services.	NoSQL databases provide high availability due to its distributed nature and data replication.
Healthcare records are normally added, not updated.	Eventual consistency of NoSQL database architecture provides an acceptable EMR use cases.
Sharing of healthcare data requires concurrent access to EMR systems from multiple locations which requires a high-performance system to respond to data access and retrieval request in a timely manner.	Contrary to relational databases, NoSQL distributed databases offer higher performance that is capable of spanning data across server nodes, racks, or even multiple data centers with no single point of failure.
Implementation and infrastructure costs are high.	Most NoSQL database systems are open-source and can run on inexpensive commodity hardware architectures.

database is that no query language has ever been standardized. Most graph based applications make used of application programming interfaces (APIs) for querying, though there are some query languages available like SPARQL and Gremlin (Wood, 2012).

Property graph model

Graph databases employ nodes, edges, and properties for data representation (**Fig. 4**). A node can hold any number of attributes or properties or key-value-pairs. Edges represent graphs or relationships, which are the lines that connect nodes to other nodes to describe relationship between them (**Fig. 5**).

Neo4j is the implementation chosen to represent graph database. Sponsored by Neo Technology, it is an open-source NoSQL graph database implemented in Java and Scala for all non-commercial uses (Vicknair *et al.*, 2010). Neo4j is an embedded, disk-based, fully transactional Java persistence engine that stores data structured in graph rather than in tables. It implements the property graph model with native graph storage and provides full database characteristics including ACID transaction compliance, cluster support, and runtime failover, making it suitable for the development of EMR database (Neo4j, 2016a,b,c,d,e).

Database and Database Management Systems (DBMS)

A database by definition, is a set of data and definition of its organization. It is an electronic collection of schemas, tables, queries, reports, views and other objects. A general-purpose DBMS provides functionalities to define, create, query, update and administer databases. DBMSs are often classified according to the database model that they support such as relational and

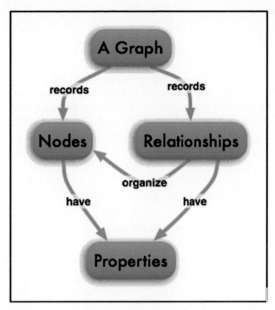

Fig. 4 The building blocks of the property graph model. Reproduced from Neo4j, 2016a,b,c,d,e. What is a graph database? Retrieved from Neo4j: https://neo4j.com/developer/graph-database/.

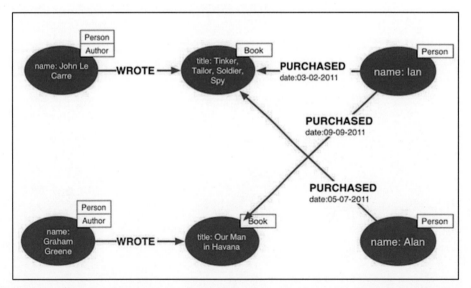

Fig. 5 Labeled property graph data model. Reproduced from Neo4j, 2016a,b,c,d,e. What is a graph database? Retrieved from Neo4j: https://neo4j.com/developer/graph-database/.

Table 3   Types of database languages (Beynon-Davies, 2003)

Language	Description
Data definition language (DDL)	Defines data types and their relationships.
Data manipulation language (DML)	Performs transactions such as insert, update or delete records.
Query language	Enables searching for information and computing derived information.

non-relational (NoSQL). Generally, a database is not backwards compatible through a range of different DBMSs. However, DBMS can be standardized to allow compatibility. Databases can be manipulated using one or more of the following languages (**Table 3**).

Case Study of a EMR Graph Model

EMR Document Database Development

The EMR NoSQL document-oriented database was developed using MongoDB. To convert and export the existing EMR relational database to MongoDB, a third party application, MongoDB SQL Server Importer (SQL2Mongo) (Tuldoklambat, 2012) was used to achieve such task. SQL2Mongo required the specification of the source destination and the target database, and then automatically converts SQL data types directly to JSON data types as JSON output. Subsequently, a database transaction was performed in MongoChef (MongoDB GUI) to ensure the imported database and tables were consistent with those created in MS SQL Server. A sample query was also executed to verify the database integrity and to ensure data reproducibility of both MS SQL Server and MongoDB.

EMR Graph Database Development

Contrary to MongoDB, to import data from MS SQL Server into Neo4j, it is essential to transform a relational database schema and model it as a graph prior implementation. Since Neo4j graph database revolves around property graph model, a graph schema (**Figs. 6** and **7**) was derived from the EMR relational model with the following guidelines (Neo4j, 2016a,b,c,d,e):

- Each *entity table* is a *label* on nodes
- Each *row* in the entity table is a *node*
- *Columns* on the entity table become node *properties*

Once the graph schema was established, EMR data and their appropriate tables were extracted and exported from MS SQL Server using the built-in SSMS data exporter interface in CSV format. Since Neo4j is a Shell-based application without GUI unlike MS SQL Server, all actions were performed using its native language, Cypher (Neo4j, 2016a,b,c,d,e). Cypher's LOAD CSV command was used to load the contents of the CSV file into Neo4j to generate a graph structure according to the derived graph schema (**Figs. 6** and **7**).

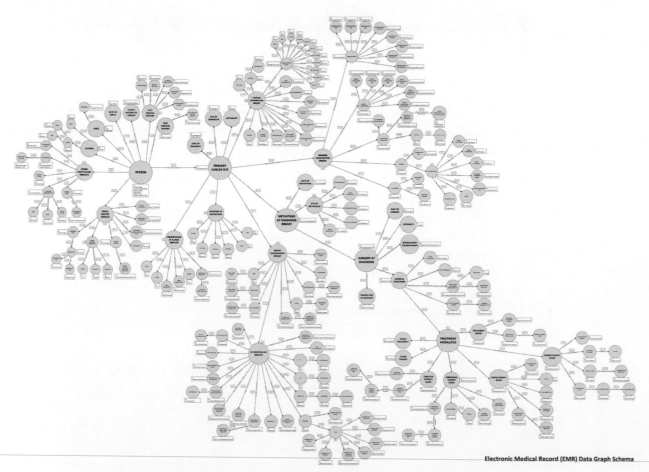

Electronic Medical Record (EMR) Data Graph Schema

Fig. 6 Full graph schema of EMR database. A snapshot view of the red outlined graph schema is shown in **Fig. 7**.

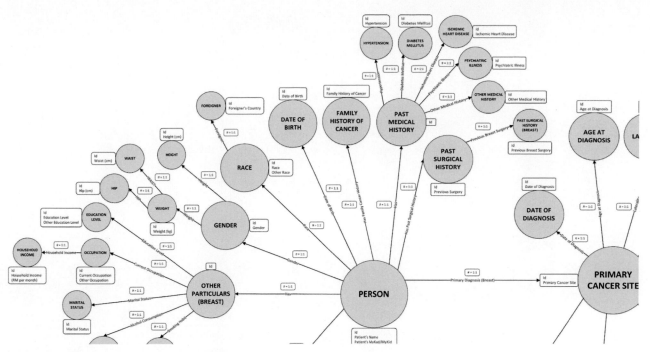

Fig. 7 Closer view of graph schema of EMR database.

The resulting graph with a total of 231 unique labels and 190 relationship types with 22,869 nodes, 22,869 relationships and 23,680 properties was generated.

Neo4j EMR Data Visualization

A single EMR record was fully visualized in Neo4j with all its adjacent relationships defined to confer semantics of EMR data. The full visualization (**Fig. 8(a–c)**) includes EMR data ranging from:

- Patient personal identifying information
- Patient family history of cancer
- Past medical and surgical history
- History of presenting illness
- Clinical examination information such as cancer location, size and description
- Imaging examination information such as mammogram, breast ultrasound, CT scan and bone scan
- Biopsy examination such as core biopsy and excision biopsy of tumor tissue
- Histopathology
- Diagnosis and metastasis information
- Surgical relation information
- Follow up therapy treatment records

Fig. 9 illustrates a simple visualization of how patient demographics information can be linked with past medical history to allow meaningful data representation.

Discussion

Data Visualization and Linked Data

In relational databases such as MS SQL Server, EMR data is retrieved by the users according to a set of defined criteria. In such cases, the results will be limited to only what the user intended to query. EMR data contain extensive amount of information about a patient and metadata on how healthcare is delivered. However, Graph databases such as Neo4j provide visual representations to facilitate effective visualization renders to assist users in analyzing (Aparicio and Costa, 2015). According to **Fig. 8(a–c)**, EMR data were well presented as a graph consisting of nodes interconnected with relationships to confer semantics for sense-making and data analysis. **Fig. 9** illustrated a simple visualization of how patient demographics data were linked with past medical history in Neo4j.

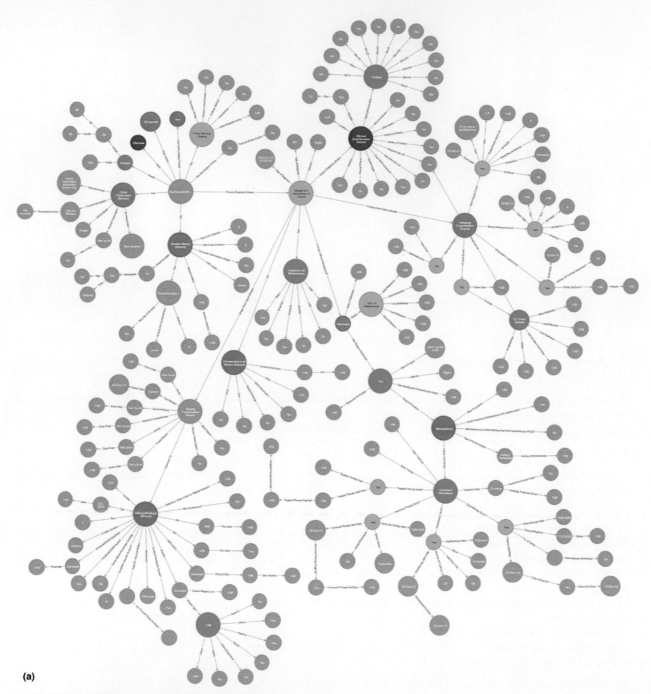

(a)

Fig. 8 (a) Full visualization of a single EMR record in Neo4j.

Relational Versus NoSQL Databases

Performance of MS SQL server versus MongoDB

According to the results obtained, MongoDB was 4 times faster than MS SQL Server with an average execution time of 1ms to 4ms respectively in performing simple queries 1–4. Results were in agreement with comparative study conducted by Parker *et al.* (2013).

For the more complex query 5, involving multiple object types and nested queries, MS SQL Server had a runtime of 6 ms while MongoDB had a runtime of 11.75 ms. MongoDB was 2 times slower than MS SQL Server. MS SQL Server excels in join operations of relatively modest amount of structured data and SQL comes with native aggregate function like DATEDIFF(). In the present study, EMR records/ documents (Date of surgery) from one collection was first unwind to be copied/ joined to the second collection (Date of diagnosis), and subsequently underwent subtraction and division operations to calculate the time taken to surgery in days. This slows down the performance by a considerable amount. Unlike using SQL, MongoDB uses reference to locate

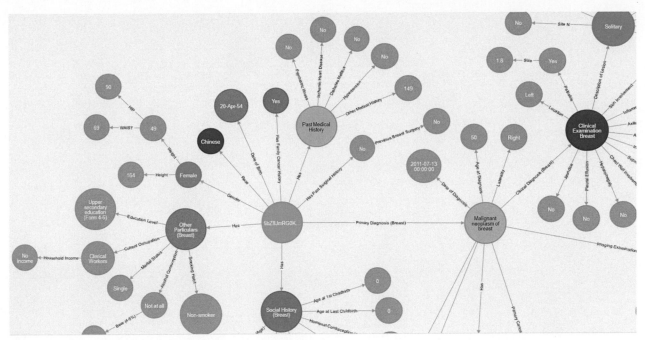

(b)

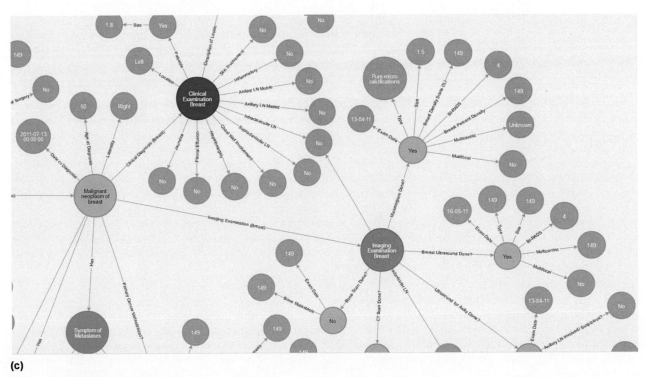

(c)

Fig. 8 (*Continued*) (b) The relationship between adjacent nodes in a record in Neo4j conferring semantics of EMR data. (c) Visualization of a record in Neo4j, showing the relationships between clinical symptoms and test carried out.

data in memory, additional decisions and aggregate functions such as lookup were required to determine how to implement or join relationships between collections, hence the slower performance (Parker *et al.*, 2013).

Performance of MS SQL server versus Neo4j

According to out results, MS SQL Server was 2 times faster on average compared to Neo4j with a mean execution time of 4.5–8.5 ms in performing simple queries 1–4. Results were in agreement with study by Vicknair *et al.* (2010). With regard to performance, Neo4j appears to have lower optimization features around querying for null values since it does not store empty or null values

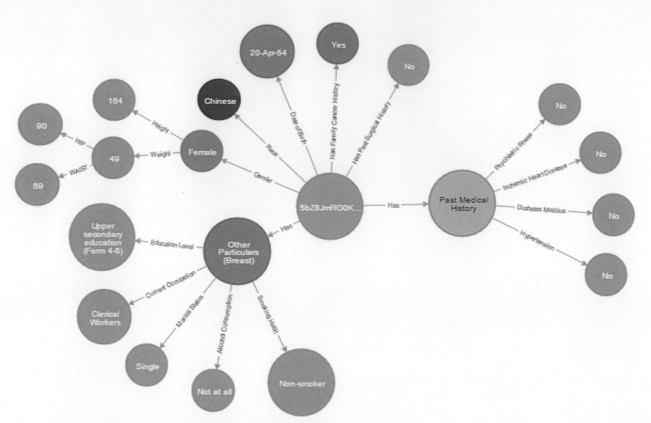

Fig. 9 Simple visualization of patient demographics linked with past medical history in Neo4j.

natively (Neo4j, 2015). Query 1–4 involve conditions using CASE statement with SUM aggregate function to search through each and every record (total 99) to query for the total number of breast cancer cases and surgeries performed respectively. However, the Patient records retrieved in this case study were not fully populated and contain some empty values, hence the slower performance from Neo4j.

Since query test 1–4 were performed on default configurations, theoretically speaking, options such as memory allocation can be adjusted, tuned and optimized to improve input/ouput (I/O) and read/write performance. According to Vicknair *et al.* (2010), relational database performed better than Neo4j at small scale, but scaling upward dramatically shifted the search times in favor of Neo4j.

For the more complex query 5 which involves date and time notions, MS SQL Server had a runtime of 6 ms, whereas the test was not applicable to Neo4j as Neo4j currently does not have a native date/time/timestamp value type (Neo4j, 2015). Therefore, aggregate function to determine the date/time difference was not feasible in Neo4j. Neo4j is considerably new with only seven years of production. It is still undergoing active development.

Storage requirement of MS SQL server versus MongoDB

Database storage comprises of the internal (physical) level in the database architecture as well all the information needed such as metadata to construct the conceptual and external level. According to the results obtained, with the identical set of EMR data, MongoDB was 51 times more space-efficient than MS SQL Server, occupying only 0.50 MB in disk space compared to 25.50 MB in MS SQL Server. This can be attributed to differences in storage engine and the way the data is stored in MS SQL Server and MongoDB (Nguyen, 2010).

The storage engine manages the nature of how the data is stored in a database. MS SQL Server uses SYBASE storage engine. It is a full-featured engine but is generally slower due to its ACID compliance for reliability (VCU, 2013). Most of the data are stored in disk which explains the higher storage requirement. Read and write speed can also be bottlenecked by the performance of disk I/O. On the contrary, MongoDB comes with two supported storage engines: MMAPv1 (Memory Mapped Version 1) engine and WiredTiger storage engine (MongoDB, 2016a,b,c,d,e,f). MMAPv1 is MongoDB's original storage engine based on memory mapped files. It prioritizes memory usage over disk as its cache to read and write documents. In addition, WiredTiger is a newer storage engine that scales on modern multi-CPU architectures. It offers document level concurrency control with native compression to minimize on-disk overhead and I/O, which significantly reduces the amount of storage required by up to 80%, unlocking new infrastructure efficiency and cost savings (MongoDB, 2016a,b,c,d,e,f). Hence, the observed lower disk space consumption by MongoDB compared to MS SQL Server.

Storage requirement of MS SQL server versus Neo4j

According to the results obtained, Neo4j was 1.35 times more space-efficient than MS SQL Server, consuming only 18.87 MB to 25.50 MB in disk space. Likewise, the observed difference can be attributed to the variation in storage engine and the way how the data is stored.

Neo4j storage engine is powered by Neo Technology, which behaves similarly to MongoDB, where it prioritizes memory usage rather than disk with newer compression engine to read and write data. However, Neo4j is actually a fully transactional database with ACID compliance (Neo4j, 2016a,b,c,d,e), which explains the higher disk space usage compared to MongoDB.

Performance and storage scalability

The requirement of scalability for EMR systems is considered restrictive as most of the current systems are based on relational databases which are not designed to scale, which calls for a need of newer scalable database systems to maximize output at the expense of lower storage requirement (Jin *et al.*, 2011). Based on the objective tests (performance and storage size) conducted in this study, it was noticeable that NoSQL databases such as MongoDB and Neo4j performed better with a lower disk space requirement compared to MS SQL Server.

To scale is to either increase hardware capacity or increase the amount of data stored per capacity. For an example based on results in, 25.5 MB was required to store 99 records in MS SQL Server, whereas 99 records occupied only 0.50 MB and 18.87 MB in MongoDB and Neo4j respectively. Therefore, with an effective similar storage size of 25.5 MB, MongoDB and Neo4j can theoretically store up to 5049 and 135 records, a 51 and 1.35 time increased in the amount of data stored per same capacity respectively due to newer storage and compression technology.

On the contrary, NoSQL database systems are based on shared-nothing approach where servers have their own dedicated resources such as RAM, processor or storage, allowing scaling horizontally (Cattell, 2011).

Availability

Schmitt and Majchrzak (2012) suggested that the nature of EMR data holds critical patient records which therefore, requires the prerequisites of having both high availability and distributed data management to allow the system to function even a single point of failure within it. Based on CAP Theorem, NoSQL database systems trade consistency for availability and introduce eventual consistency concept (Dede *et al.*, 2013). In addition, data replications allow failovers to enable "always-on" database systems which is crucial to the day-to-day operations in hospital settings.

Consistency

NoSQL is not ACID compliant, instead it adopts BASE principles. In NoSQL databases, data read by clients immediately after being updated may be outdated as not all nodes have been updated instantly. Due to this though there is no guarantee of consistency at any instance of time, it eventually will be consistent at some point in time (Vogels, 2009). Bailis and Ghodsi (2013) showed that the inconsistency is less than a second and therefore, eventual consistency model is claimed to be adequate in most use cases. Due to this, weaker consistency models can therefore be applied using NoSQL databases in healthcare environment without the fear of causing any major setback in terms of data consistency (Frank *et al.*, 2014).

Data models

The need to store heterogeneous medical data has increased since using relational databases has proven inefficient and difficult to handle and manage such data (Jin *et al.*, 2011). NoSQL databases support various types of data structures which allow flexible modelling of data, contrary to relational databases which support only structured model.

Relational Versus Graph Databases

Schema design

Schema refers to the organization of data as a blueprint of how the database is constructed prior implementation. A schema design in relational model adheres to the notion of theory in predicate calculus to define tables, fields, relationships, views, and indexes (Rybiński, 1987). Hence, relational databases are limited to homogenous data structure and extremely ineffective for heterogeneous, unstructured and frequently changing data.

In a clinical setting, data collected from patient can be in the form of texts or images. According to relational databases, data structure must be homogenous. On the contrary, a new label can be easily defined and linked to store both text and image records since graph databases embrace relationships over data types. Therefore, in terms of handling multiple variants of data, graph databases like Neo4j are more suitable for the task as relational schema is more rigid and less flexible.

Language

Relational databases like MS SQL Server use SQL to query and manage relational data. It is based upon relational algebra and tuple relational calculus. It consists of a DDL, DML and Data Control Language (DCL). SQL is a declarative language which includes data insert, query, update and delete, schema creation and modification, and data access control. It is sub-divided into several language elements (**Table 4**).

Table 4   SQL language elements (ISO.org, 2011)

Element	Description
Clauses	Constituent components of statements and queries.
Expressions	Produce either scalar values, or tables consisting of columns and rows of data.
Predicates	Specify conditions that can be evaluated to SQL three-valued logic (3 VL) (true/false/unknown) or Boolean truth values that are used to change program flow and limit the effects of statements and queries.
Queries	Retrieve data based on specific criteria.
Statements	Control transactions, program flow, connections, sessions, or diagnostics.

Neo4j uses Cypher language which is a declarative graph query language that allows for expressive and efficient querying and updating of the graph store. It is a relatively simple yet powerful. Unlike SELECT statement, Cypher uses MATCH and WHERE. WHERE is used to add additional constraints to patterns. CREATE and DELETE are used to create and delete nodes and relationships. SET and REMOVE are used to set values to properties and add labels to nodes (Neo4j, 2016a,b,c,d,e).

Maturity and level of support

Relational database has been around for decades and is one of the most used systems. They are used in almost every commercial and academic pursuits and have spawned several commercial ventures such as Oracle and Microsoft. Relational databases benefit from having a unified language, SQL. Since SQL does not differ greatly between implementations, support for one implementation is generally applicable to other implementations. Graph databases on the other hand, have less market penetration in general. Since most of the graph database systems are open source, their commercial ventures are usually smaller and lack a unified language. Neo4j however, has a reasonable amount of support. Hence in terms of maturity and level of support, relational databases have a better edge.

Flexibility

Both relational and graph databases perform equally well in the environments for which they were designed. Flexibility measures how well they perform outside of those environments (Vicknair *et al.*, 2010).

MS SQL Server is a large server application designed to perform in a large-scale multi-user environment. While it is optimized, it may be too heavy weighted for small applications users as there some overhead for features that are not necessarily beneficial for the small user.

Neo4j is small, light-weight, and efficient. It performs well at server scale and does not impose the overhead often associated with server applications (Vicknair *et al.*, 2010). Unlike MS SQL Server which only works on Microsoft ecosystems like Windows, as a Java component, Neo4j supports cross-platform deployment, meaning it can be run on Windows, Linux or Mac OS (Neo4j, 2016a,b,c,d,e). Therefore in terms of flexibility, Neo4j graph database is the clear winner.

Security

Most relational databases in general such as MS SQL Server have extensive built-in multi-user support. Different privileges can be assigned to individuals or groups to grant different user views and actions. On the contrary, many graph databases including Neo4j lack much support for multi-user environments. It forces all user management to be handled at the application level. By extension, Neo4j assumes a trusted environment and does not have any built-in security support (Vicknair *et al.*, 2010). Relational databases such MS SQL Server are better at ACL-based security.

Limitations of Current Study and Proposed Future Work

The present study illustrated a full scale visualization of EMR data in Neo4j and demonstrated a simple use case of how records can be linked among EMR data. Future work can look into importing and integrating open data in RDF/JSON format with current EMR graph database system, Neo4j to study and assess the proposed idea of Linked Data initiative.

Future work shall look into assessing the scalability of NoSQL databases in terms of performance and storage in comparison to a relational database in various configurations and scenarios by using different size of EMR data. In addition, EMR data can be distributed among different nodes to evaluate the performance and horizontal scalability of NoSQL database systems in a distributed environment compared to a single system using relational database. Future work can also look into assessing availability of NoSQL databases by deploying and maintaining a number of replications to simulate failovers in the event of power or systems failure.

Conclusion

Electronic Medical Records store and manage huge amounts of clinical data. New approaches are needed in order to achieve higher performance and to store data efficiently than is possible with traditional relational databases. In this study, we demonstrated that NoSQL databases (MongoDB and Neo4j), particularly the graph model is a powerful tool to visualize EMR data for sense-making and data analysis. Graph model allows one to identify make predictive inferences which are beneficial in qualitative research. Graph model stores data in JSON format, which can be easily linked with open data to create new relationships and extend information as linked data uses a universal interchangeable RDF/JSON format to describe things. Nonetheless, the capacity of graph model and linked data are underutilized due to the lack of adoption in healthcare domain.

This study also demonstrated that NoSQL databases such as MongoDB and Neo4j performed better with a lower disk space consumption compared to relational MS SQL server. Although we cannot conclude the clear superiority of NoSQL databases over relational database, we however provide enough reasons such as generic suitability and technical and financial reliefs for large scale data-intensive applications that is provided by NoSQL databases. Given the importance of EMR systems for healthcare delivery and overall health systems, distributed NoSQL databases arguable has a clear potential to dominate the development of EMR applications due to the scalability, high availability, and flexibility it provides.

Overall, this study demonstrated how graph database system such as Neo4j is the way forward due to its powerful visualization model and potential for linked data. Although it fell short in terms of maturity, level of support and security compared to relational databases, its flexibility in handling heterogeneous data with dynamic schema, SQL-like Cypher programming language and advantages as a NoSQL database provide a better data storage and management alternative viable to hospital EMR systems.

See also: Biological and Medical Ontologies: Disease Ontology (DO). Biological and Medical Ontologies: Human Phenotype Ontology (HPO). Information Retrieval in Life Sciences. Integrative Bioinformatics. Integrative Bioinformatics of Transcriptome: Databases, Tools and Pipelines. Models for Computable Phenotyping. Natural Language Processing Approaches in Bioinformatics. Ontology: Introduction

---

References

Abramova, V., Bernardino, J., 2013. NoSQL databases. In: Proceedings of the International C* Conference on Computer Science and Software Engineering – C3S2E '13, pp. 14–22. New York, New York, USA: ACM Press. Available at: https://doi.org/10.1145/2494444.2494447.

Aparicio, M., Costa, C.J., 2015. Data visualization. Communication Design Quarterly Review 3 (1), 7–11. Available at: http://doi.org/10.1145/2721882.2721883.

Bailis, P., Fekete, A., Ghodsi, A., Stoica, I., 2013. HAT, not CAP: Towards highly available transactions. In: Proceedings of the 14th USENIX Conference on Hot Topics in Operating Systems, pp. 24–24. Santa Ana Pueblo, New Mexcio: USENIX Association.

Bailis, P., Ghodsi, A., 2013. Eventual consistency today: Limitations, extensions, and beyond. Queue 11 (3), 20.

Beynon-Davies, P., 2003. Database Systems, third ed. Palgrave Macmillan.

Bizer, C., Heath, T., Berners-Lee, T., 2009. Linked data – The story so far. International Journal on Semantic Web and Information Systems 5 (3), 1–22. Available at: http://doi.org/10.4018/jswis.2009081901.

Cattell, R., 2011. Scalable SQL and NoSQL data stores. ACM SIGMOD Record 39 (4), 12. Available at: http://doi.org/10.1145/1978915.1978919.

Cutler, D., 2010. Analysis & commentary. How health care reform must bend the cost curve. Health Affairs 29 (6), 1131–1135. Available at: http://doi.org/10.1377/hlthaff.2010.0416.

Dede, E., Govindaraju, M., Gunter, D., Canon, R.S., Ramakrishnan, L., 2013. Performance evaluation of a MongoDB and hadoop platform for scientific data analysis. In: Proceedings of the 4th ACM workshop on Scientific Cloud Computing – Science Cloud '13, p. 13. New York, New York, USA: ACM Press. Available at: https://doi.org/10.1145/2465848.2465849.

Frank, L., Pedersen, R.U., Frank, C.H., Larsson, N.J., 2014. The CAP theorem versus databases with relaxed ACID properties. In: Proceedings of the 8th International Conference on Ubiquitous Information Management and Communication – ICUIMC '14, pp. 1–7. New York, New York, USA: ACM Press. Available at: https://doi.org/10.1145/2557977.2557981.

Gunter, T.D., Terry, N.P., 2005. The emergence of national electronic health record architectures in the United States and Australia: Models, costs, and questions. Journal of Medical Internet Research 7 (1), e3. Available at: http://doi.org/10.2196/jmir.7.1.e3.

Huang, C.-W., Lu, R., Iqbal, U., *et al.*, 2015. A richly interactive exploratory data analysis and visualization tool using electronic medical records. BMC Medical Informatics and Decision Making 15 (1), 92. Available at: http://doi.org/10.1186/s12911-015-0218-7.

ISO.org, 2011. ISO/IEC 9075-2:2011. Retrieved from information technology: http://www.iso.org/iso/iso_catalogue/catalogue_tc/catalogue_detail.htm?csnumber=53682.

Yang, J., Tang, D., Zheng, X., 2011. Research on the distributed electronic medical records storage model. In: Proceedings of 2011 IEEE International Symposium on IT in Medicine and Education, pp. 288–292. IEEE. Available to: https://doi.org/10.1109/ITiME.2011.6132041.

MongoDB, 2016a. Documents. Retrieved from Documentation: https://docs.mongodb.com/manual/core/document/.

MongoDB, 2016b. Replication. Retrieved from Documentation: https://docs.mongodb.com/manual/replication/.

MongoDB, 2016c. Sharding. Retrieved from Documentation: https://docs.mongodb.com/manual/sharding/.

MongoDB, 2016d. Storage engines. Retrieved from Storage: https://docs.mongodb.com/v3.2/core/storage-engines/.

MongoDB, 2016e. What is NoSQL? Retrieved from NoSQL Databases Explained: https://www.mongodb.com/nosql-explained.

MongoDB, 2016f. WiredTiger storage engine. Retrieved from Storage Engines: https://docs.mongodb.com/v3.2/core/wiredtiger/.

Neo4j, 2015. Neo4j: Real-world performance experience with a graph model. Retrieved from Neo4j: https://neo4j.com/blog/neo4j-real-world-performance/.

Neo4j, 2016a. From relational to Neo4j. Retrieved from What is Neo4j?: https://neo4j.com/developer/graph-db-vs-rdbms/.

Neo4j, 2016b. From SQL to cypher – A hands-on guide. Retrieved from Cypher Query Language: https://neo4j.com/developer/guide-sql-to-cypher/.

Neo4j, 2016c. Graph database use cases. Retrieved November 15, 2016, from: https://neo4j.com/use-cases/.

Neo4j, 2016d. Neo4j release candidate. Retrieved from Products Download: https://neo4j.com/download/other-releases/.

Neo4j, 2016e. What is a graph database? Retrieved from Neo4j: https://neo4j.com/developer/graph-database/.

Nguyen, L., 2010. SQL server database engine basics. Retrieved from SQL Server: http://sqlmag.com/sql-server/sql-server-database-engine-basics.

Orfanidis, L., Bamidis, P.D., Eaglestone, B., 2004. Data quality issues in electronic health records: an adaptation framework for the Greek health system. Health informatics journal 10 (1), 23–36.

Parker, Z., Poe, S., Vrbsky, S.V., 2013. Comparing NoSQL MongoDB to an SQL DB. In: Proceedings of the 51st ACM Southeast Conference on – ACMSE '13, p. 1. New York, New York, USA: ACM Press. Available at: https://doi.org/10.1145/2498328.2500047.

Rybiński, H., 1987. On first-order-logic databases. ACM Transactions on Database Systems 12 (3), 325–349. Available at: http://doi.org/10.1145/27629.27630.

SAS, 2016. Data visualization: What it is and why it matters. Retrieved November 15, 2016, from: http://www.sas.com/en_us/insights/big-data/data-visualization.html.

Schmitt, O., Majchrzak, T.A., 2012. Using document-based databases for medical information systems in unreliable environments. In: Proceedings of the 9th International Conference on Information Systems for Crisis Response and Management (ISCRAM). Vancouver, CA.

Stonebraker, M., 2009. SQL databases vs NoSQL databases. Retrieved November 15, 2016, from: http://cacm.acm.org/blogs/blog-cacm/50678-the-nosql-discussion-has-nothing-to-do-with-sql/fulltext.

Tuldoklambat, 2012. MongoDB SQL server importer. Retrieved from CodePlex: https://sql2mongo.codeplex.com/.

van der Linden, H., Kalra, D., Hasman, A., Talmon, J., 2009. Inter-organizational future proof EHR systems. International Journal of Medical Informatics 78 (3), 141–160. Available at: http://doi.org/10.1016/j.ijmedinf.2008.06.013.

VCU, 2013. A comparision of MySQL vs microsoft SQL server. Retrieved from Virginia Commonwealth University: http://www.people.vcu.edu/~agnew/Misc/MySQL-MS-SQL.HTML.

Vicknair, C., Macias, M., Zhao, Z., et al., 2010. A comparison of a graph database and a relational database. In: Proceedings of the 48th Annual Southeast Regional Conference on – ACM SE '10, p. 1. New York, New York, USA: ACM Press. Available at: https://doi.org/10.1145/1900008.1900067.

Vogels, W., 2009. Eventually consistent. Communications of the ACM 52 (1), 40. Available at: http://doi.org/10.1145/1435417.1435432.

W3C, 2015. Linked data. Retrieved November 15, 2016, from: https://www.w3.org/standards/semanticweb/data.

Wood, P.T., 2012. Query languages for graph databases. ACM SIGMOD Record 41 (1), 50. Available at: http://doi.org/10.1145/2206869.2206879.

Genome Databases and Browsers for Cancer

Madhu Goyal, University of Technology Sydney, Sydney, NSW, Australia

Introduction

Researchers over the last two to three decades supporting the fact that different type of cancers are fundamentally a disease of the genome. Genes are being studied and identified that underlie the different types of cancer. A better understanding of these genes is one of the most pressing needs in basic cancer research. Huge amount of the genome sequence data has become available in the last two decades for researchers to do research on cancer. This explosion of data has also provided opportunity for researchers to identify patterns among data and to develop understanding the way cancer is caused by changes in the DNA, RNA, and proteins of a cell that drive over growth of cells. This big data has also witnessed the development of the systematic study of the cancer genome and sophisticated sequencing technologies. The knowledge of cancer genome databases is very important to guide the development of more effective approaches for reducing cancer morbidity and its mortality. The identification of the genomic variations that arise in cancer can also help researchers decipher cancer development and improve upon the diagnosis and treatment of cancers.

Cancer genomics data (Chin et al., 2011) is complex and heterogeneous just like cancer itself. But a comprehensive analysis of the cancer genome still remains an overwhelming task. This is partly due to the limitations in current technologies to visualize, integrate, compare and analyze cancer genomics data. Such limitations prevent investigators from truly appreciating the breadth and depth of these genomics and epigenomes resources. Thus careful statistical and algorithmic considerations are required to integrate the information provided by the large volume and variety of data alongside clinical annotations. The specific conclusions extracted from data must be presented in a coherent system for display and analysis should be accessible to the scientific and medical communities.

Genomes databases (Schattner, 2008; Yang et al., 2015) for Cancer contain the information of DNA sequence and gene expression differences between tumor cells and normal host cells. Genome database mean a data repository that includes all or most of the genomic DNA sequence data of one more organisms. Human genome database includes "annotations" that either describes the features of the DNA sequence itself or other biological properties of humans. The multiple fundamental tasks are involved in building a genome database, which include identification of the locations of the genes within the genome sequence, aligning transcript data into the genomic sequence, sequencing the genomic DNA and assembling the fragments of DNA sequence onto the length of chromosomes.

A genome database typically also includes a web based interface referred to as 'genome bowser'. Genome browsers have been created to allow the simultaneous display of multiple annotations within a graphical interface. They provide the ability to search for markers and sequences, to extract annotations for specific regions or for the whole genome and to act as a central starting point for genomic research Genome browsers allow researchers to navigate the genome in an comparable way to navigating the internet e.g., with Internet Explorer, Mozilla Firefox or Chrome. Nowadays the amount of available genomic data is vast or big and different open source databases aim to make these data accessible to all researchers. The number and variety of annotations has increased dramatically and thus there is huge requirement of a detailed view of many aspects of the genomes. In this paper several popular cancer genomics data repositories and browser applications are described (**Table 1**).

Various genome databases and browsers have provided different analysis tools, which have proven to be successful at facilitating a better understanding of the genomes. The overwhelming amount of cancer genomics data from multiple different technical platforms has provided increasingly opportunities to perform data integration, exploration, and analytics, especially for scientists without a computational background. These cancer genomics resources have specifically lowered the barriers of access to the complex data sets and thereby accelerate the translation of genomic data into new biological insights, therapies, and clinical trials. They have offered researchers comprehensive repositories, contextual information, and curated data as a method of handling the exponentially growing amount of sequence data. Although these resources have been useful and have solved many issues, but they will continue to face new types and ever-growing amounts of data that will exacerbate many more challenges.

Systems and Applications

NCI'S Genomic Data Commons (GDC)

The NCI's Genomic Data Commons (GDC) provides the cancer research community with an integrated data repository that enables data sharing across different cancer genomic studies. The National Cancer Institute (NCI)'s Genomic Data Commons was previously managed by the Cancer Genomics Hub (CGHub), which was established in August 2011 to provide a repository to promote the sharing of cancer data, collaboration between cancer researchers, and to generate effective treatments for the Cancer. CGHub rapidly grew to be the largest database of cancer genomes in the world, storing more than 2.5 petabytes of data and serving downloads of nearly 3 petabytes per month. The Cancer Genomics Hub mission project was completed in 2016. The GDC supports several cancer genome programs at the NCI's Center for Cancer Genomics (CCG), including The Cancer Genome Atlas (TCGA) and Therapeutically Applicable Research to Generate Effective Treatments (TARGET).

Table 1　Key resources for cancer genomics research and study

Name	Website	Key features
NCI's genomic data commons (GDC)	https://gdc.nci.nih.gov/	Big data repository
Cancer genome atlas (TCGA)	https://cancergenome.nih.gov/	Big data repository
GDC's data analysis, visualization, and exploration (DAVE) tools	https://gdc.cancer.gov/analyze-data/gdc-dave-tools	Web interface for exploring and analyzing GDC's cancer genomic data
ICGC data portal (DCC)	http://icgc.org/	Huge data knowledgebase of somatic mutations, abnormal expression of genes, epigenetic modifications.
cBio cancer genomics portal	http://cbioportal.org	Open-access resource of cancer genomics data sets
COSMIC	http://cancer.sanger.ac.uk/cosmic	Largest somatic mutation database
Cancer cell line encyclopedia	https://portals.broadinstitute.org/ccle	Analysis and visualization of DNA copy number, mRNA expression, mutation data of cancer cell lines
FireHose	http://gdac.broadinstitute.org/	Systemize the analyses from the cancer genome atlas TCGA
FireBrowsem	http://firebrowse.org	Explore FireHose cancer data with graphical tools.
TumorPortal	http://tumorportal.org/	Analyzing and visualizing genes, cancers, DNA mutations and annotations
Integrative genomics viewer (IGV)	http://www.igv.org/	High-performance visualization tool for interactive exploration of large, integrated genomic datasets
Genome analysis toolkit	https://software.broadinstitute.org/gatk/	Industry standard for identifying SNPs, DNA and RNAseq data
Hail	https://hail.is/	Open-source, scalable framework for exploring and analyzing genomic data
European genome-phenome archive (EGA)	https://www.ebi.ac.uk/ega/about	Huge repository for all types of sequence and genotype experiments and data
Network of cancer genes (NCG)	http://ncg.kcl.ac.uk/index.php	Web based tool to study duplicability, orthology and evolutionary appearance of cancer genes
MutaGene	https://www.ncbi.nlm.nih.gov/research/mutagene/	Web tool for computational exploration of DNA context-dependent mutational patterns
UCSC Xena	http://xena.ucsc.edu/	Cancer genomics browser for viewing, analyzing, visualizing the public data hubs of cancer genomics and clinical data
canEvolve	http://www.canevolve.org/	Analysis of functional genomics which includes gene, mRNA, microRNA and protein expression, genome variations and protein–protein interactions
MethyCancer	http://methycancer.psych.ac.cn/	Hosts integrated data of DNA methylation, mutation and cancer information from public resources
SomamiR 2.0	http://compbio.uthsc.edu/SomamiR	Database of cancer somatic mutations in microRNAs (miRNA)
canSAR	https://cansar.icr.ac.uk/	Open source, multidisciplinary, cancer database developed to make provision for cancer translational research and drug discovery
NONCODE	http://www.bioinfo.org/noncode/	Database of cancer somatic mutations in microRNAs (miRNA)
China cancer genome database (CCGD)	https://db.cngb.org/cancer/	Large comprehensive genome database of Chinese cancer patients

The NCI's Center for Cancer Genomics (CCG) was established to command the NCI effort in generating, cataloging and unification of datasets of alterations seen in human tumors. It also aims to be at the forefront of supporting the development of analytical tools and computation approaches for improving the understanding of large-scale multidimensional data. The CCG supports several comprehensive cancer genome research programs including The Cancer Genome Atlas (TCGA) and the Office of Cancer Genomics (OCG). OCG includes two programs i.e., Therapeutically Applicable Research to Generate Effective Treatments (TARGET) program and the Cancer Genome Characterization Initiative (CGCI). TCGA, TARGET, CGCI, and other CCG programs have provided a complete characterization of genomic changes in several human cancers; however these characterizations are maintained in separate repositories, in diverse formats, and with different data management infrastructures. NCI established the GDC to unify these efforts and to provide the cancer research community with a data service supporting the receipt, quality control, integration, storage, and redistribution of standardized cancer genomic data sets derived from various legacy and active NCI programs.

Cancer genome atlas (TCGA)

The Cancer Genome Atlas (TCGA) is the result of collaboration between the National Cancer Institute (NCI) and National Human Genome Research Institute (NHGRI). It has created a genomic data analysis pipeline that can effectively collect, select, and analyze human tissues for genomic alterations on a very large scale. It has also generated complete multi-dimensional maps of the key genomic changes in 33 types of cancer. The TCGA dataset of about 2.5 petabytes describing tumor tissue of more than 11,000 patients is publically available and has been used widely by the research community. The data have contributed to more than a thousand studies of cancer by independent researchers and to the TCGA research network publications. The success of this national network of research and technology teams serves as a model for future projects and exemplifies the tremendous power of teamwork in science.

GDC's data analysis, visualization, and exploration (DAVE) tools

DAVE is a web interface for intuitively exploring and analyzing GDC's cancer genomic data. It provides an unprecedented level of flexibility in exploring the data by selecting patients with particular altered genes or other relevant biological and clinical features. Researchers can navigate from project cohorts to individual patients, to specific genes and mutations of interest. The tool can generate specialized graphs to help researchers visualize the genes in which the most somatic mutations are observed. Users can also plot patient survival curves and identify the molecular consequence of a mutation on the resultant protein. All cases of the project can be plotted and visualize the top 50 mutated genes affected by high impact mutations in the GDC's OncoGrid.

ICGC Data Portal (DCC)

The primary goal of International Cancer Genome Consortium (ICGC) has been to catalogue and coordinate a large number of research projects that have the common aim of interpreting the genomic changes present in many forms of cancers. ICGC has generated huge data knowledgebase of genomic abnormalities (somatic mutations, abnormal expression of genes, epigenetic modifications) in tumors from 50 different cancer types and/or subtypes across the globe and make the data available to the entire research community. The ICGC's Data Portal provides tools for visualizing, querying and downloading the data released quarterly by the consortium's member projects. The ICGC facilitates communication among the members and provides a forum for coordination with the objective of maximizing efficiency among the scientists working to understand, treat, and prevent the different cancers.

cBio Cancer Genomics Portal

The cBioPortal for Cancer Genomics was originally developed at Memorial Sloan Kettering Cancer Center (MSK). The public cBioPortal site is hosted by the Center for Molecular Oncology at MSK. The cBioPortal software is now available under an open source license via GitHub. The cBio Cancer Genomics Portal (Gao et al., 2013) is an open-access resource for exploration of multidimensional cancer genomics data sets. It currently provides access to data from more than 160 cancer studies. The portal provides graphical summaries of gene-level data from multiple platforms, network visualization, survival analysis and also software programmatic access. The cBio Cancer Genomics Portal significantly lowers the barriers between complex genomic data and cancer researchers who want rapid, intuitive, and high-quality access to molecular profiles and clinical attributes from large-scale cancer genomics projects.

COSMIC

COSMIC (Forbes et al., 2017), the Catalogue of Somatic Mutations in Cancer, is the world's largest and high resolution resource for exploring the genetics of human cancer. COSMIC is a database that collects these somatic mutation data from a variety of public sources into one standard repository containing all forms of human cancer, from the most frequent cancers in lung, breast and colon, to extremely rare forms of blood cancer. Currently, COSMIC contains 1335 disease descriptions across more than 5000 detailed classifications. Its public website has been custom-built to make the many annotations in it which can be easily explored in user-friendly graphical ways whilst also providing large tabulated data sets. The front page of website offers multiple ways to explore the database e.g., 'Resources', 'Tools', and a range of pages describing the database content and details on its accessibility.

Cancer Program Resource Gateway

The central component of the Broad Institute's Cancer Program is Cancer Program Resource Gateway, which addresses unanswered questions of cancer genomics through platforms, datasets and resources. The institute was founded in 2004 due to the need that arose from the Human Genome Project to successfully decipher the entire human genetic code. This prestigious and famous institute is owned by MIT and Harvard and the main goal is to improve human health by using genomics to advance understanding and treatment of human diseases. The cancer program resource gateway has two strands i.e. data and tools.

Some of the popular resources are:

Cancer cell line encyclopedia

The CCLE (Cancer Cell Line Encyclopedia) project is collaboration between the Broad Institute, and the Novartis Institutes for Biomedical Research and its Genomics Institute of the Novartis Research Foundation. Its main objective is to conduct a detailed genetic and pharmacologic characterization of a large panel of human cancer models. The Cancer Cell Line Encyclopedia (CCLE) project is an effort to conduct a detailed genetic characterization of a large panel of human cancer cell lines. The CCLE provides public access analysis and visualization of DNA copy number, mRNA expression, mutation data and more, for 1000 cancer cell lines.

FireHose

In order to systemize the analyses from The Cancer Genome Atlas TCGA and to study the remaining diseases FireHose was developed. GDAC Firehose now sits atop ~55 terabytes of analysis-ready TCGA data and reliably executes thousands of pipelines per month.

FireBrowse

FireBrowse is a simple way to explore cancer data, backed by a powerful computational infrastructure, application programming interface (API) and graphical tools. It sits above the TCGA GDAC Firehose one of the deepest and most integrated *open* cancer datasets in the world with over 80 K sample aliquots from 11,000 + cancer patients, spanning 38 unique disease cohorts.

TumorPortal

TumorPortal is for analyzing and visualizing genes, cancers, DNA mutations and annotations. In TumorPortal datasets can be explored by tumor types as well as genes.

Integrative genomics viewer (IGV)

The Integrative Genomics Viewer (IGV) is a high-performance visualization tool for interactive exploration of large, integrated genomic datasets. It supports a wide variety of data types, including array-based and next-generation sequence data, and genomic annotations. Although IGV is often used to view genomic data from public sources, its primary emphasis is to support researchers who wish to visualize and explore their own data sets or those from colleagues.

Genome analysis toolkit

The GATK is the industry standard for identifying SNPs, DNA and RNAseq data. It also has the capability to find somatic variation, to tackle copy number (CNV) and structural variation (SV). GATK also includes many utilities to perform related tasks such as processing and quality control of high-throughput sequencing data. And although it was originally developed for human genetics, the GATK has since evolved to handle genome data from any organism

Hail

Hail is an open-source, scalable framework for exploring and analyzing genomic data. It can generate variant annotations like call rate, Hardy-Weinberg equilibrium p-value, and population-specific allele count. Hail can also generate new annotations from existing ones as well as genotypes, and use these to filter samples, variants, and genotypes.

European Genome-Phenome Archive

The European Genome-phenome Archive (EGA) is designed to be a repository for all types of sequence and genotype experiments, including case-control, and family studies. It includes SNP and CNV genotypes from array based methods and genotyping done with re-sequencing methods. EGA serve as a permanent archive that will archive several levels of data including the raw data (which could, for example, be re-analysed in the future by other algorithms) as well as the genotype calls provided by the submitters. Data at EGA is collected from individuals whose consent agreements to authorise data release only for specific research use to bona fide researchers. The EGA accepted data types include raw data formats from the array-based and new sequencing platforms as well as phenotype files describing study samples. EGA offers a range of tools for the file and meta-data upload like Java EgaCryptor, which encrypts and verifies all the accepted file types, and the EGA Webin portal for the metadata submission.

Network of Cancer Genes (NCG)

The network of cancer genes (NCG) is a web based tool to duplicability, orthology and evolutionary appearance of cancer genes. It introduces a more robust procedure to extract manually curated genes from published cancer mutational screenings. NCG allows to interpreting cancer sequencing by determining if a gene has already been reported as a driver gene in a given cancer type. This helps interpreting the mutational landscape of a tumor sample and to identify whether the genes mutated have a known role in cancer or if they are likely to be passengers. NCG provides new approaches to target cancer genes i.e., the systems-level properties reported for each cancer gene and allow identifying new targets and strategies to target a mutated gene.

MutaGene

MutaGene (Goncearenco *et al.*, 2017) is an online computational framework, which explores DNA context-dependent mutational patterns and underlying somatic cancer mutagenesis. It explores mutational profiles of cancer samples, identifies the combinations of underlying mutagenic processes including those related to infidelity of DNA replication and repair machinery, and various other endogenous and exogenous mutagenic factors. It also calculate expected mutability for each DNA and protein site using background mutational models.

UCSC Xena

UCSC Xena (Goldman *et al.*, 2015) is a cancer genomics browser for viewing, analyzing, visualizing the public data hubs of cancer genomics and clinical data. It is as web-based application that is developed in response to a crucial demand in the cancer research field for integrative visualization of large, complex genomic datasets arising from different technology platforms. It hosts many public databases such as TCGA, ICGC, TARGET, GTEx, CCLE, and others. These databases are normalized before use so that they can be combined, linked, filtered, explored and downloaded. Xena is very flexible with data and can load most types of genomic or phenotype data into it. Researchers can explore the relationship between genomic alterations and phenotypes by visualizing various genomic data alongside clinical and phenotypic features, such as age, subtype classifications and genomic biomarkers. It can also generate Kaplan Meier plot (KM plot) to test whether a genomic or phenotype variable significantly affects survival. Xena can group samples by any data (cell lines, xenografts, organoids, or patients) and then use these groups to compare expression or any other genomic data.

canEvolve

CanEvolve (Samur *et al.*, 2013) supports different type of analysis of information extracted from 90 cancer genomics studies comprising of more than 10,000 patients. The portal stores functional genomics and other large-scale data on cancer which includes gene, mRNA, microRNA and protein expression, genome variations and protein–protein interactions. The portal also provides stored knowledge in database (canEvolve web portal is implemented using mySQL open source system) as well as generate analysis results from oncogenomic profiles in response to user queries. It allows visualization of knowledge and analysis results in an appropriate manner and let the user download query results and related information from the portal. The querying for primary analysis includes differential gene and miRNA expression as well as changes in gene copy number measured with SNP microarrays. Finally, canEvolve provides query functionalities to fulfill most frequent analysis requirements of cancer researchers towards generating novel biological hypotheses.

MethyCancer

DNA methylation plays a vital role in the development of cancer and is associated with oncogene activation and chromosomal instability. MethyCancer (He *et al.*, 2008) hosts integrated data of DNA methylation, mutation and cancer information from public resources, and the CpG Island (CGI) clones derived from large-scale sequencing. The database also has graphical Methy-View, which shows DNA methylation in context of genomics and genetics data. It thus facilitates the research in cancer to understand genetic mechanisms that make dramatic changes in gene expression of tumor cells.

SomamiR

Genetic and somatic mutations or miRNA–mRNA interactions have been associated with various cancers. The miRNAs are small non-coding RNAs, known for their role as post-transcriptional regulators of protein-coding mRNAs. SomamiR 2.0 (Bhattacharya and Cui, 2016) is a database of cancer somatic mutations in microRNAs (miRNA). It helps researchers study the interactions between miRNAs and competing endogenous RNAs (ceRNA) including mRNAs, circular RNAs (circRNA) and long noncoding RNAs (lncRNA). It also has webserver miR2GO which is integrated with the database to provide a seamless pipeline for assessing functional impacts of somatic mutations in miRNA seed regions.

canSAR

canSAR (Tym *et al.*, 2016) is open source, multidisciplinary, cancer database developed to make provision for cancer translational research and drug discovery. It integrates comprehensive multidisciplinary knowledge i.e., annotation for genes, pharmacological, drug and chemical data with structural biology to enable target validation and drug discovery. It also applies machine learning approaches to provide drug-discovery useful predictions. It is known globally as the key resource to aid target selection and prioritization of drug discovery for cancer. canSAR thus provides unique views on genes and proteins, drugs, 3D structures, protein interaction networks, cancer cell lines and cancer clinical trials.

NONCODE

Mounting evidence shows that non-coding RNAs play key roles in various biological processes and in disease etiology. The recently reduced cost of RNA sequencing has produced an explosion of newly identified data, and as a result there has been an explosive rise in the number of newly identified non-coding RNAs. NONCODE (Yi *et al.*, 2015) is an integrated database of cancer somatic mutations in microRNAs (miRNA)from several species such as human and mouse. NONCODE provides a subset searching interface, literature support, other database support and long-read sequencing method support. The quality controls also include selection of exon numbers, the lengths of the transcripts and prediction tools support. The web interface presents the subset according to the conditions users chose and allow users to download the data.

China Cancer Genome Database (CCGD)

The China Cancer Genome Database (CCGD) is built by China National GeneBank as a first, large comprehensive genome database in China. The CCGD Data Portal stores, catalogs, and accesses the cancer related data, and provides a platform for researchers to download data sets. The types of data include raw data, clean data, alignment data, function analysis data, phenotype data. This database collects and stores cancer genome data and shares with all cancer research centers in order to promote the development of China cancer genome research. The database utilizes these data or research mechanism of the most frequently occurred cancers among Chinese population and largely promotes research of the early diagnosis and cure of cancer in China.

Conclusions

Advancement in software and information technologies has enabled software developers and bioinformatics researchers to integrate different publicly-accessible genetics data types from a large variety of sources. The coordinated efforts of many organizations have also led to the development of large-scale cancer genomics though which complete catalogs of the genomic alterations in specific cancer types can be obtained. Overall there are four different categories of genome resources discussed in this article. The first category represents very large repositories of genomics data such as TCGA, ICGC, COSMIC, EGA, cBio portal, China Cancer database with their own browsing tools. The second category resources offer tools for data analysis, data integration, and visualization (e.g., UCSC Xena, FireBrowse, TumorPortal, IGV, GATK, Hail, NCG, MutaGene). The third class includes databases that focus on inferring the association of specific biological features to cancer (e.g., MethyCancer, SomamiR, and NONCODE). Finally, databases such as CanSAR support the application of genomics to drug discovery. There are other useful resources for cancer research, but are not included in this paper due to the limitation of space and scope.

Although the methods for analyzing cancer genomes have improved at a swift rate, but still many challenges remain. Big Data management and advanced computational methods should be developed and integrated with tools to establish the clinical relevance of cancer genomic discovery. Collecting accurate clinical information on tumor samples remains an important and challenging task, but one that is necessary for the interpretation of genomic findings in depth. However, most available samples are associated with incomplete annotation or lack appropriate consent to permit the linkage of clinical information. There is thus a huge scope for personalized cancer genomic information, which once widely available and successfully implemented, holds considerable promise for improving the lives of many patients with cancer. Lastly, but importantly, the growing number of genome databases, analysis tools, and other resources available on the web has made it an unnerving task for researchers to use these resources effectively.

See also: Bioinformatics Data Models, Representation and Storage. Biological Database Searching. Data Storage and Representation. Genome Annotation. Genome Informatics. Integrative Analysis of Multi-Omics Data. Integrative Bioinformatics. Natural Language Processing Approaches in Bioinformatics. Next Generation Sequencing Data Analysis. Predicting Non-Synonymous Single Nucleotide Variants Pathogenic Effects in Human Diseases. Quantitative Immunology by Data Analysis Using Mathematical Models. Standards and Models for Biological Data: FGED and HUPO. Whole Genome Sequencing Analysis

---

References

Bhattacharya, A., Cui, Y., 2016. SomamiR 2.0: A database of cancer somatic mutations altering microRNA–ceRNA interactions. Nucleic Acids Research 44 (D1), D1005–D1010. PMID: 26578591.
Chin, L., Hahn, W.C., Getz, G., Meyerson, M., 2011. Making sense of cancer genomic data. Genes & Development 25, 534–555.
Forbes, S.A., Beare, D., Boutselakis, H., *et al.*, 2017. COSMIC: Somatic cancer genetics at high-resolution. Nucleic Acids Research 45 (Issue D1), D777–D783.
Gao, J., Aksoy, B.A., Dogrusoz, U., *et al.*, 2013. Integrative analysis of complex cancer genomics and clinical profiles using the cBioPortal. Science Signaling 6 (269), pl1.
Goldman, M., Craft, B., Swatloski, T., *et al.*, 2015. The UCSC cancer genomics browser: Update 2015. Nucleic Acids Research 43 (Issue D1), D812–D817.
Goncearenco, A., Rager, L.S., Li, M., *et al.*, 2017. Exploring background mutational processes to decipher cancer genetic heterogeneity. Nucleic Acids Research 45 (Issue W1), W514–W522.
He, X., Chang, S., Zhang, J., *et al.*, 2008. MethyCancer: The database of human DNA methylation and cancer. Nucleic Acids Research 36, D836–D841.
Samur, M.K., Yan, Z., Wang, X., *et al.*, 2013. CanEvolve: A web portal for integrative oncogenomics. PLOS ONE 8, e56228.
Schattner, P., 2008. Genomes, Browsers and Databases: Data-Mining Tools for Integrated Genomic Databases. Cambridge University Press.
Tym, J.E., Mitsopoulos, C., Coker, E.A., *et al.*, 2016. canSAR: An updated cancer research and drug discovery knowledgebase. Nucleic Acids Research 44 (D1), D938–D943.
Yang, Y., Dong, X., Xie, B., *et al.*, 2015. Databases and web tools for cancer genomics study. Genomics, Proteomics & Bioinformatics 13, 46–50.
Yi, Z., Hui, L., Fang, S., *et al.*, 2015. NONCODE 2016: An informative and valuable data source of long non-coding RNAs. Nucleic Acids Research 44 (D1), D203–D208.

Text Mining Resources for Bioinformatics

Sandeep Kaushik, European Institute of Excellence on Tissue Engineering and Regenerative Medicine, Guimaraes, Portugal and University of Minho, Braga, Portugal
Priyanka Baloni and Charu K Midha, Institute for Systems Biology, Seattle, WA, United States

Introduction

With the digital revolution, there has been an explosion in the amount of information available, which in turn has transformed the traditional ways of data analyses. Mining important information from terabytes of data is the need of the hour. Data mining uses sophisticated algorithms and statistical measurements to extract useful information from existing data. In modern day computing, the main types of data are text, numbers and multimedia. Text mining is the process of analyzing a collection of unstructured data in the form of text and deriving useful information or generating a new hypothesis. There is a wide scope and application of this technology in risk management, cyber-crime prevention (Berry and Kogan, 2010; Witten *et al.*, 2016; Kontostathis *et al.*, 2010), fraud detection (Joudaki *et al.*, 2015), biomedical research (Rebholz-Schuhmann *et al.*, 2012), to name a few.

Text mining for bioinformatics has an overwhelming potential for translational research. There are thousands of research articles published every year showcasing the discoveries made in the scientific field, making it difficult for the researchers to follow them in great details. Due to the massive information generated, there is a need for computational tools to parse and analyze the information to make the learning process simpler. Text mining, as the term suggests, is the use of an automated method for extracting meaningful data from the wealth of information available in literature and other resources. The process of text mining comprises mainly of information retrieval, information extraction, knowledge discovery and hypothesis generation. The information extracted in the process is generally stored in database for future reference.

There are various tools and resources that are used by bioinformaticians for retrieving and extracting information from unformatted data. Some of the extensively used retrieval engines for biomedical research are PubMed, PubChem, RefMED, UK PubMed Central (Walport and Kiley, 2006), WorldWideScience, MedlinePlus, GoPubMed and Google Scholar (Ananiadou *et al.*, 2006). In order to extract information, some of the commonly used resources are iHOP, Textpresso, Open Biomedical Annotator, InAct, MedScan, Reactome as well as KEGG (Rebholz-Schuhmann *et al.*, 2012). There is a constant effort in building curated databases that contain information retrieved and extracted for specific purposes. STRING, STITCH, SIDER, HPID, HPRD, Bio-Caster, PharmGKB are few databases that have been curated and extensively used by researchers to conclude valuable inferences (Rebholz-Schuhmann *et al.*, 2012; Zhu *et al.*, 2013). **Fig. 1** depicts the process of text mining involving converting unformatted text to machine-interpretable structured format.

Text mining is computationally intensive and is performed using technologies such as natural language processing, knowledge management, information extraction, machine learning as well as pattern recognition. The present text mining tools have virtually explored only the tip of the iceberg and there is still lots to be explored. The field of text mining has a wide scope for research development and has immensely assisted researchers in the advancement of biomedical research. Although, there are challenges associated with this field (discussed later), with the development of powerful algorithms and techniques, this field has received great impetus. This article discusses the fundamental concepts and various text mining resources developed for researchers.

Is Text Mining Ready to Deliver?

Information mining is a process that involves a clear understanding of various stages involved in parsing information. Some of these stages not only require bioinformatic techniques like pattern evaluation from databases and automatic detection, but scientific knowledge, analyst creativity as well as basic knowledge of the field. A better understanding of the biological process or problems help us to structure text mining projects, therefore, they are closer to accurate analysis rather than driven by chance or individual's insight.

History of Text Mining

Text mining techniques have evolved over many years with a view of expansion and improving methodologies by learning from previous attempts. Text mining was commenced in the earlier times to catalog the literature in the form of books in libraries. Soon, it was deviated to data retrieval using Natural Language Processing methods (abbreviated as NLP – it is an attempt to understand the modelling of the natural human language using computers). Indeed, text mining has evolved with time owing to improvised computational techniques. Since 2001, information on the internet has an expanded from over 100 terabytes to 1500 terabytes in 2009, which is nearly 40% explosion in the information generation and storage. This exponential growth of information is mainly composed of a vast number of personal, public and corporate pages, news, electronic books, scientific data repository, databases *etc.* The Library of Congress, one of the largest library in the world contains 17 million catalog records for books, serials, manuscripts, maps, music, recordings, images, and electronic resources in its collections and is still growing exponentially. Although, this field is still in its developmental stage, it has already contributed to many scientific discoveries.

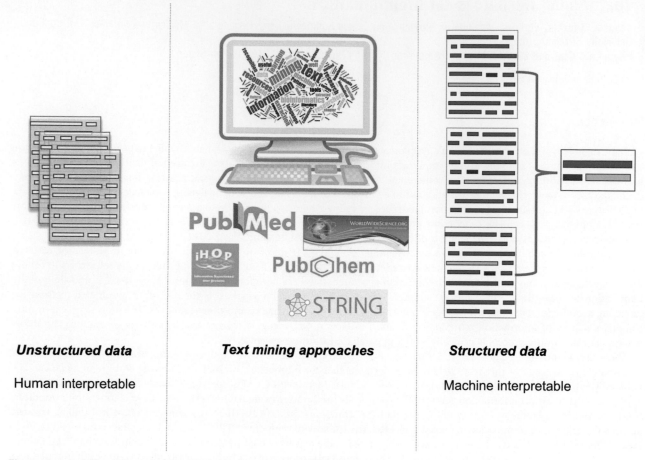

Unstructured data

Human interpretable

Text mining approaches

Structured data

Machine interpretable

Fig. 1 Deconvolution of the unstructured data into machine-interpretable format using various text mining approaches.

Trends in Text Mining

In earlier days, library catalog involved text summarization and classification which was contributed by Thomas Hyde from University of Oxford for the Bodleian Library in 1674. Later in 1876, Melvil Dewey introduced the index card for library card catalog at Yale University. Text mining has not sprouted in one instance, rather several technological advancements, extensive data generations of different types and applications have contributed to the present modern data retrieving systems.

Later, by the year 1898, the text processing included summarization of the information for generating abstracts, majorly in Science stream (Luhn, 1958). It was as a joint collaboration between the Physical Society of London and Institute of Electrical Engineers. Luhn used computers to generate the document abstracts in 1958 by exploiting the word frequency analysis on an early IBM 701 vacuum tube computer (Luhn, 1958). It gave relative significance measurement of sentences which were combined according to their linear distances between the words to produce a metric of significance sentence contributing to form an abstract of a document. This automated procedure very well adapted in library sciences. Electronic versions of the scientific abstracts had become available since 1967 onwards.

As the technology advanced from huge room size computers to personal computers and revolution in digitization, document cataloging has improved with more access to other existing documents. Science Citation Indexing (SCI) is a citation index created by Eugene Garfield in 1964 from Institute for Scientific Information and is currently owned by Clarivate Analytics after Thomson Reuters. It's enhanced version Science Citation Index Expanded includes more than 8500 journals, across 150 disciplines (Garfield, 2006, 2007, 1964).

The searching mechanism was advanced by text manipulations, indexing, and development of Natural Language Processing (NLP). To access the information from the repository, two processes – Information Retrieval and Information extraction are used. Eventually, computational power in the 1960s invented NLP applications, clustering, information extraction and modern text mining methods. All these methods are briefly described in the next section.

Since 2000, major innovations and methods in text mining include modern information extraction engines (based on tokenization); categorization of text with Machine Learning techniques, Support Vector Machines, kernel-based learning methods, Neural Networks, automation, linking and statistical approaches (Miner, 2012). A glimpse of few of these methods are discussed in the next section.

Background

Mining information from big data repositories involves strategies and techniques that can process data in fast and reliable manner. In order to improvise scientific inferences and narrow down the search parameters, it is important to carry out pattern recognition, relation extraction, knowledge discovery as well as hypothesis generation. One of the major challenges of retrieving data from scientific databases demands the ability to think how such fundamental concepts apply to specific biological problems (Sharma and Srivastava, 2016). This section highlights the main concepts underlying textmining approaches.

Concepts

Text mining
The term denotes a process aimed at analyzing unstructured text extracted from various text resources, detecting lexical patterns useful for deriving previously unknown information (Berry and Castellanos, 2008; Sebastiani, 2002).

Machine learning (ML)
It is a method to train computers to make and improve predictions based on given data without being explicitly programmed. Some of the widely used machine learning approaches are Hidden Markov Models (HMM), Support Vector Machines (SVMs), Conditional Random Fields (CRFs) as well as Maximum Entropy (ME) (Sharma and Srivastava, 2016).

Natural language processing (NLP)
It relies on machine learning algorithms to analyze and understand the context in human language that is easily interpretable by computers. NLP is used for text mining, translation, relationship extraction as well as speech recognition. This method involves processing of words in the text and relies on statistical techniques. Some of the open source libraries for NLP are Natural Language Toolkit (NLTK), Apache OpenNLP, Stanford NLP and MALLET (see "Relevant Websites section").

Information retrieval
Information retrieval (IR) deals with searching for information as well as recovery of textual information from a collection of resources. The desired information is often posed as a search query, which in turn recovers those articles from a repository that are most relevant and matches to the given input. Google scholar, PubMed, UK PubMed Central [06] as well as Elsevier's Science Direct are some of the widely-used document search and retrieval engines for research or publications. Also, resources like PubMed QUEST are very helpful in the rapid searching of PubMed for publications of interest (PubMed, 1946).

Information extraction
Information extraction (IE) is the task of automatically extracting meaningful semantic structures using different strategies and identify relations between concepts. Open Biomedical Annotator, iHOP, Textpresso, CoPub are examples of resources used for information extraction (Rebholz-Schuhmann et al., 2012).

Pattern recognition
It is a branch of machine learning and involves recognizing patterns in data and classify the input data into different classes based on certain features.

Document summarization
This tool takes the documents into consideration and returns a concise summary of the text that highlights key points in the longer text.

Importance of Text Mining

From Experiments to Articles to Models

The experimentalist designs and carry out the lab work (experiments) to obtain the results followed by its processing and interpretation. The significant outcomes are published in scientific journals that follow specific journal format comprising of Title, Abstract, Keywords, Text body, References, Tables and Figures.

National Center for Biotechnology Information (NCBI) developed PubMed database (see "Relevant Websites section"), which is an information retrieval system at National Library of Medicine (NLM). This database is mainly dedicated to life sciences literature. Access to the literature is through NCBI Entrez retrieval system which is a text-based search and retrieval method. It can have a search query by author name, the title of the papers, or the keywords.

Scientific community submits their citations electronically from different journals into PubMed. Presently, PubMed has more than 27 million citations for biomedical literature from MEDLINE, life science journals, and online books. Citations may include links to full-text content from PubMed Central and publisher websites. Among these published articles, few are indexed with

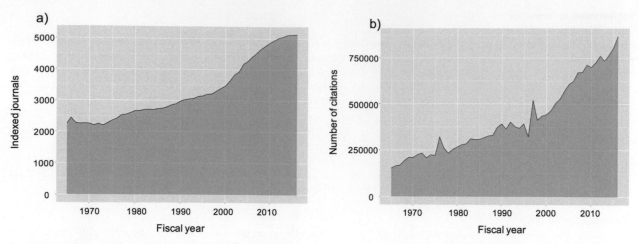

Fig. 2 The growth in indexed journals in PubMed (a) and recorded citations from 1965 to 2016 (b) is represented in plots. (https://www.nlm.nih. gov/bsd/index_stats_comp.html).

Table 1   A table illustrating the most used tools for text mining (Biological networks, Protein–Protein interactions, NMR/Crystal structure)

Database name [Reference]	Name of database	URL
NCBI-PubMed (Hanauer and Chinnaiyan, 2006)	Published biomedical literature	https://www.ncbi.nlm.nih.gov/pubmed
DIP (Salwinski et al., 2004)	Database of interacting proteins	http://dip.doe-mbi.ucla.edu/dip/Main.cgi
HPID (Han et al., 2004)	Human protein interaction database	http://wilab.inha.ac.kr/hpid/webforms/intro.aspx
MINT (Licata et al., 2011)	Molecular interaction database	http://mint.bio.uniroma2.it/
STRING (Szklarczyk et al., 2017)	Search tool for the retrieval of interacting genes	https://string-db.org/
PDB (Berman et al., 2002)	Protein data bank	https://www.rcsb.org/pdb/home/home.do
UniProt (UniProt Consortium, 2011)	The universal protein resource	http://www.uniprot.org/

MeSH (expanded as Medical Subject Headings) term publication types to GenBank Accession numbers. MeSH is NLM curated medical vocabulary resource (see "Relevant Websites section"). MeSH, indexes and catalogues the terminology belonging to biomedical information (MEDLINE/PubMed and other NLM databases) in hierarchical order. The growth in the number of indexed journals in PubMed and recorded citations from 1965 to 2016 is represented in **Fig. 2**.

As discussed in concepts above, Google has developed Google Scholar search engine targeting academic and research users. It's a scholarly literature that includes peer-reviewed papers, theses, books, abstracts, reports. On passing query it returns results based on full text, authors, publications types/journals and number of citations.

Biomedical literature has characteristics features in terms of usage of domain-specific terminologies (technical terms); words possessing ambiguity in their meaning e.g., Drosophila has genes named archipelago, capicua or ebony; new terminology and names, low frequency of certain words, typographical variants etc. to keep in mind while searching (**Table 1**).

Case Studies of Use of Text Mining in Bioinformatics/Biomedical Research

Accessing information from database: PlasmoDB as an example

Among the biological databases available for the scientific community, several dedicated databases exist belonging to a specific class, organism, characterization etc. For example – Eukaryotic Pathogen Database Resources, EuPathDB (see "Relevant websites section") (**Fig. 3**) is a portal provided by Bioinformatics Resource Center for accessing genomic-scale datasets from wide eukaryotic microbes listed in **Table 2**.

This article will cover one of the databases from EuPathDB, called Plasmodium Genomics Resource, PlasmoDB (see "Relevant Websites section") in detail (Aurrecoechea et al., 2008).

Of the five species of Plasmodium- *Plasmodium vivax, P. falciparum, P. ovale, P. malariae* and *P. Knowlesi*, mostly the first four species cause human malaria and occasionally by *P. Knowlesi* which causes animal malaria. Among these *P. falciparum* is the most lethal. PlasmoDB comprises of all the updated information on each and every gene of these species/strains uploaded and submitted by the experimentalists. It has linked up the information to other relevant databases and summarized results in a single window.

The home page of PlasmoDB shows wide range of search options such as: text-based, gene models (exon count/gene type), annotations, curation and identifiers (gene IDs), genomic location, taxonomy (organism), sequence analysis (BLAST, protein

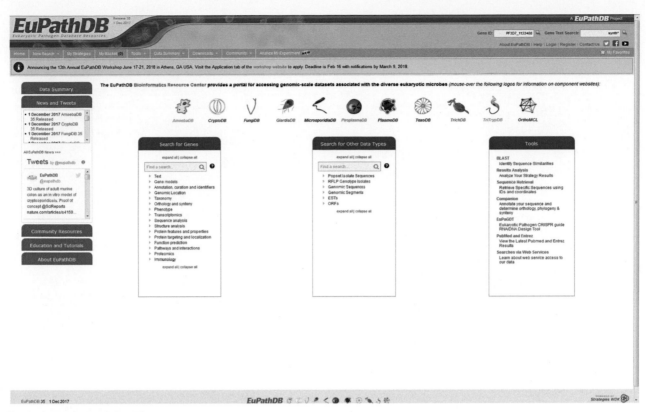

Fig. 3   Screenshot of EuPathDB.

Table 2   List of databases included in the EuPathDB consortium

Database	Url	Specification
AmoebaDB	http://amoebadb.org/amoeba/	Acanthamoeba, Entamoeba and Naegleria
CryptoDB	http://cryptodb.org/cryptodb/	Chromera, Cryptosporidium, Gregarina and Vitrella
FungiDB	http://fungidb.org/fungidb/	Agaricomycetes, Blastocladiomycetes, Chytridiomycetes, Eurotiomycetes, Leotiomycetes, Oomycetes, Pneumocystidomycetes, Pucciniomycetes, Saccharomycetes, Schizosaccharomyces, Sordariomycetes, Tremellomycetes, Ustilaginomycetes, Zygomycetes
GiardialDB	http://giardiadb.org/giardiadb/	Giardia and Spironucleus
MicrosporidiaDB	http://microsporidiadb.org/micro/	Annacaliia, Edhazardia, Encephalitozoon, Enterocytozoon, Hamiltosporidium, Nematocida, Nosema, Spraguea, Trachipleistophora, Vavraia, Vittaforma
PiroplasmaDB	http://piroplasmadb.org/piro/	Babesia and Theileria
PlasmoDB	http://plasmodb.org/plasmo/	Plasmodium
ToxoDB	http://toxodb.org/toxo/	Eimeria, Hammondia, Neospora, Sarcocystis, Toxoplasma
TrichDB	http://trichdb.org/trichdb/	Trichomonas
TriTrypDB	http://tritrypdb.org/tritrypdb/	Crithidiam Endotrypanum, Leishmania, Trypanosoma
OrthoMCL	http://orthomcl.org/orthomcl/	Ortholog Groups of Protein Sequences

motif pattern, TF Binding Site Evidence), structure analysis (PDB 3D structures, predicted 3D structures, protein secondary structures), and micro-array expression based searches.

On looking in detail the results obtained after searching tyrosyl- tRNA synthetases by its Gene ID: **PF11_0181**, opens a page with a brief summary of the gene at the left side (**Fig. 4**): its name, type as in coding/non-coding, chromosome no. and its location, species and strain name.

Right panel reveals the experimental information, gene model- exons and transcripts with their length in pictorially; annotation, curations and identifiers – listing its previous IDs and alias notes that are specific to annotator or user comments; Link outs-external links to different databases and tools for the gene of interest (such as Entrez Gene, Gene DB, Malaria Literature Database, Ontology Based pattern Identification, PlasmoDraft, PubMed, UniProt); genomic locations on the chromosome; associated literatures to this gene/protein with their PubMed ID, DOI links, titles and authors; taxonomic classification- to which

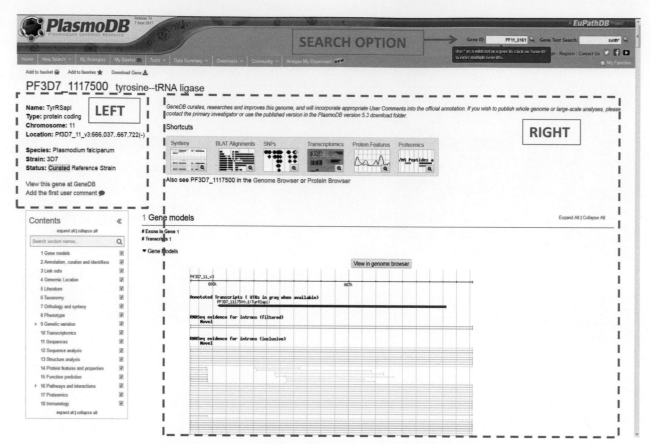

Fig. 4   Screenshot of the PlasmoDB with tyrosine-tRNA ligase as a hit.

superkingdom, phylum, class, order, family, genus and species it belongs; Orthologous and paralogous list of proteins from different species; Synteny of this gene and its domains in pictorial form, phenotypic information, and genetic variation; transcriptomics expression information; sequences -protein/mRNA/genomic sequence and their lengths, structural information-structural homologous PDB IDs from different organisms, and structural information coverage of the protein; protein family classifications, signal peptides presence and its length and relevant BLASTP hits and low complexity regions, secondary structure predictions (Helix or Beta strands); functional classification, gene ontology classification, pathways involved and protein-protein interactions, proteomic- Mass Spectrometry based expression information at different stages of *Plasmodium falciparum* life cycle.

In addition, it allows to perform the multiple sequence alignment (ClustalW) or Multi-FASTA of the nucleotide sequence with any other set of sequences through external links. To conclude, this DB platform provides the best practices to retrieve all or any information of a gene/protein belonging to Plasmodium species.

Personalized medicine and text mining

Text mining has played a pivotal role in the development of diverse fields. The field of personalized medicine has immensely benefitted from development in text mining approaches. As shown in **Fig. 5(a)**, the number of publications containing the term 'text mining' in title or abstract when queried in PubMed has increased greatly since 2001. We can forecast the number of publications for text mining from the trend of previous years. With rapid advances in genomics and healthcare, it has become possible to tailor the best medical intervention to an individual. With the knowledge of underlying genotypic variations responsible for a particular phenotype, we can improve predictions in turn leading to prevention and cure of disease. The field of personalized medicine is gradually becoming mainstream practice. With the options of genome sequencing and various health monitoring trackers, the field of personalized medicine has gained momentum. The field of text mining has immensely helped in deriving the relationships between various factors such as genes, proteins, metabolites and others. As can be seen in **Fig. 5(b)**, the number of publications on personalized medicine is increasing every year.

Identifying the disease-gene relationship is a major challenge in this field. In many cases, ML-based methods are used to identify the genetic mutations described in biomedical literature related to particular disorder or disease. In one such study, the authors retrieved biomedical literature related to target disease and extracted information for mutation and disease (Singhal *et al.*, 2016). They used several features such as statistical, distance and sentiment features for a ML-based method to predict the relationship between disease and gene mutation. The advantage of using this approach is generalizability and scalability as it is feasible to use the model to extract relationship for any disease using larger biomedical literature knowledgebase. Some of the

a)

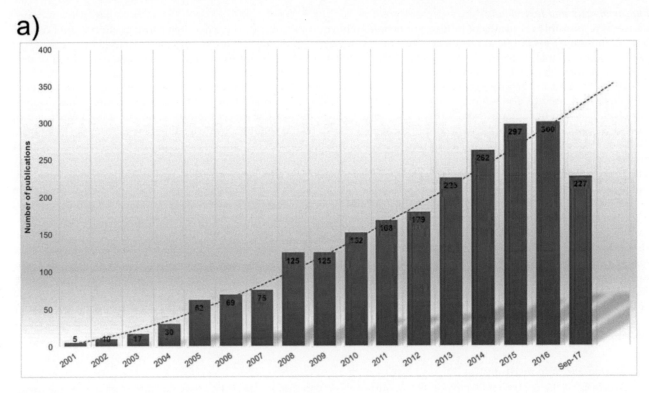

b)

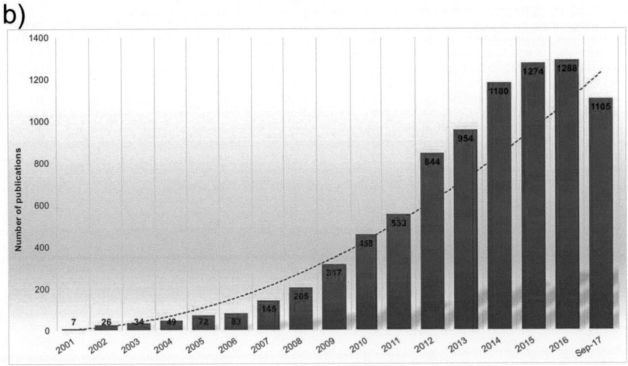

Fig. 5 Plots representing an increase in the number of publications using (a) 'text mining' and (b) 'personalized medicine' as query term in PubMed. The data from 2001 to September 2017 is plotted.

areas of improvement for text mining in personalized medicine is the availability of highly curated gene-mutation database, a crowdsourcing effort to curate existing databases, use of electronic health records and clinical trials data in addition to biomedical literature as well as robust methods for such analysis. Text mining and curation have facilitated biomedical research, development of knowledge base for precision cancer medicine being one of the relevant examples.

Cancer models and text mining

Cancer is responsible for deaths of millions of people every year In the world. There has been constant effort to find cure for cancer, and numerous publications feature cancer research each year. Biomedical text mining has been used by researchers in mining cancer-related genes and proteins and their relationship to various forms of cancer. Automatic recognition of relationship between cancer and genes or protein terms associated with the disease is one of the important aspects of bioinformatics. In order to identify genes related to prostate cancer and their relationship, prostate cancer-related abstracts from MEDLINE were used to develop maximum entropy-based system (Chun et al., 2006). Other than identifying genes involved in cancer, it is also important to verify them. In one such study, the prostate cancer biomarkers obtained from experimental studies were validated by employing text mining approach with OMIM (Online Mendelian Inheritance in Man) (Deng et al., 2006). One of the primary goals of cancer research is developing strategies for early detection of cancer so that it is useful for prevention and controlling the disease. Using biomedical text mining techniques, cancer risk assessment can be evaluated by using terms from literature as features, and developing classifiers for the purpose of automatic identification from text. Databases such as PubMeth, MeInfoText, and others are used for obtaining information such as genes and their association in cancer methylation (Ongenaert et al., 2007; Fang et al., 2011). Patient's clinical records also serves as a good source to extract relevant information. There has been great progress in creating automated systems for constructing models from free-text pathology reports. MedTAS/P is useful in mapping the pathology reports with cancer model (Coden et al., 2009). This system also efficiently captures the details of grading and staging cancer. The clinical significance of such an NLP-based system is the ease in cancer practice management by sub-classifying cancer patients based on the grade or stage of cancer. This practice is slowly entering mainstream medicine and will be helpful to clinicians in deciding the course of treatment in near future. Some of the well-documented text mining systems used for clinical purposes are MedLEE (Medical Language Extraction and Encoding) system and cTAKES (clinical Text Analysis and Knowledge Extraction System) (Zhu et al., 2013). Thus, to study the complex mechanisms of cancer, the need of the hour is text mining from hierarchical network view and discover new knowledge using this systems biomedicine perspective.

Text mining and drug discovery

Drug discovery is the process of identifying new candidate molecules that are selected by screening hits, their affinity, selectivity, potency, stability as well as bioavailability. The field of drug discovery has undergone a paradigm shift from using traditional methods to identify active ingredient to employing computational methods for such discovery (Bull et al., 2000). Publicly-available chemical databases such as PubChem, Chemical Entities of Biological Interest (ChEBI), ChemExpr, ChemBank, Side Effect Resource (SIDER), ChemSpider, Therapeutic Target Database (TTD), and DrugBank provides information of chemical entities that can be explored for therapeutic purposes (Papanikolaou et al., 2016). Text mining approach has been applied to extract information from DrugBank. DrugQuest is an information retrieval and extraction tool that uses textual information from DrugBank and clusters these records. Once the user provides query, the DrugBank records are selected on the basis of the fields such as the mechanism of action, pharmacodynamics, and indication. Named Entity Recognition techniques have been used for storing and parsing information from DrugBank. Tagging services such as Reflect and BeCAS (Papanikolaou et al., 2016) help in minimizing the ambiguity in the gene, protein and chemical terms and are useful in resolving the issue of multiple synonyms while searching the text. The retrieved documents from DrugBank are clustered algorithmically on the basis of the information contained in them. DrugQuest efficiently summarizes the results of the analysis and also provides user the option of visual representation of the results. Although DrugQuest has the limitation of 5000 textual records per analysis, it takes only a few seconds to process and compute the output (Papanikolaou et al., 2016). **Fig. 6** shows the output obtained when 'isoniazid' and 'methotrexate' are given as query in DrugQuest.

Thus, text mining approaches are useful in mining chemical repositories to find relationship between chemical entities and leverage this information for drug repositioning. It will be even more useful if chemical information can be extracted simultaneously from various repositories such as PubChem, ChemExpr, ChEBI, and SIDER. This kind of concerted effort will boost the development of this field and provide information that was unknown till now.

Discussion

There is a surge of scientific data that is generated at an exponential rate and limited methods for comprehensive analysis of data. There is an urgent need for the development of searching methods to minimize the gap between powerful storehouses and fetching useful information in an efficient manner. User-specific problems have to be addressed primarily and in more structured way. Thus, an understanding of these facts is crucial for data scientists, recruiters, investors, or the end user scholars. A better and clear understanding of process and stages involved helps in our rationale and systematic thinking leading to less errors and their omissions.

There are challenges involved in searching methodology like human formulated questions using natural language – "What are the molecular functions of Glycogenin?", alternative scientific terms, new terminologies derivations etc. Inferences derived using text mining resources can be of immense value to the scientific community as a whole. Indeed, there are evidences, claiming that data-oriented decisions and big data technologies improve scientific performances.

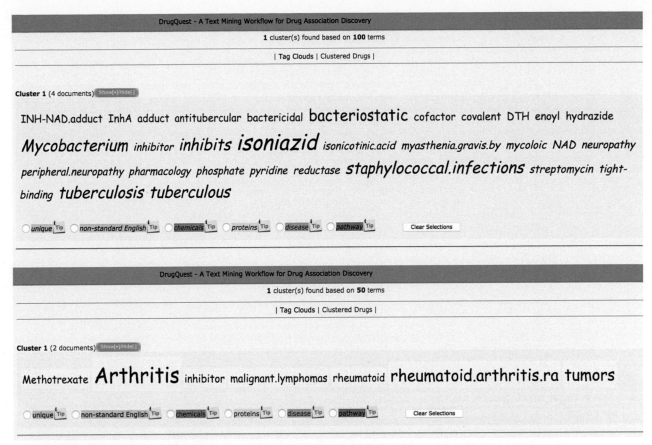

Fig. 6 The clustered output from DrugQuest for terms such as 'isoniazid' and 'methotrexate', respectively. Reproduced from Zhu, F., Patumcharoenpol, P., Zhang, C., *et al.*, 2013. Biomedical text mining and its applications in cancer research. Journal of biomedical informatics, 46 (2), 200–211.

Closing Remarks

Here we encourage the scholar to read listed references -published papers and books, relevant to the topic for advance information. Explore the information from given links of different bioinformatics databases and text mining tools. This will enable the learner to acquire the knowledge that is available while implementing it all together.

Acknowledgement

Sandeep Kaushik, Priyanka Baloni and Charu K Midha, designed the workflow and wrote the article.

See also: Bayes' Theorem and Naive Bayes Classifier. Data-Information-Concept Continuum From a Text Mining Perspective. Integrative Bioinformatics. Natural Language Processing Approaches in Bioinformatics. Text Mining Applications. Text Mining Basics in Bioinformatics. Text Mining for Bioinformatics Using Biomedical Literature

References

Ananiadou, S., Kell, D.B., Tsujii, J.I., 2006. Text mining and its potential applications in systems biology. Trends in Biotechnology 24 (12), 571–579.
Aurrecoechea, C., Brestelli, J., Brunk, B.P., *et al.*, 2008. PlasmoDB: A functional genomic database for malaria parasites. Nucleic Acids Research 37 (Suppl. 1), D539–D543.
Berman, H.M., Battistuz, T., Bhat, T.N., *et al.*, 2002. The protein data bank. Acta Crystallographica Section D: Biological Crystallography 58 (6), 899–907.
Berry, M.W., Castellanos, M., 2008. Survey of Text Mining II. vol. 6. New York: Springer.
Berry, M.W., Kogan, J. (Eds.), 2010. Text Mining: Applications and Theory. John Wiley & Sons.
Bull, A.T., Ward, A.C., Goodfellow, M., 2000. Search and discovery strategies for biotechnology: The paradigm shift. Microbiology and Molecular Biology Reviews 64 (3), 573–606.

Chun, H.W., Tsuruoka, Y., Kim, J.D., *et al.*, 2006. Automatic recognition of topic-classified relations between prostate cancer and genes using MEDLINE abstracts. BMC Bioinformatics 7 (3), S4.

Coden, A., Savova, G., Sominsky, I., *et al.*, 2009. Automatically extracting cancer disease characteristics from pathology reports into a Disease Knowledge Representation Model. Journal of Biomedical Informatics 42 (5), 937–949.

Deng, X., Geng, H., Bastola, D.R., Ali, H.H., 2006. Link test – A statistical method for finding prostate cancer biomarkers. Computational Biology and Chemistry 30 (6), 425–433.

Fang, Y.C., Lai, P.T., Dai, H.J., Hsu, W.L., 2011. MeInfoText 2.0: Gene methylation and cancer relation extraction from biomedical literature. BMC Bioinformatics 12 (1), 471.

Garfield, E., 1964. Science Citation Index – A new dimension in indexing. Science 144 (3619), 649–654.

Garfield, E., 2006. Citation indexes for science. A new dimension in documentation through association of ideas. International Journal of Epidemiology 35 (5), 1123–1127.

Garfield, E., 2007. The evolution of the science citation index. International Microbiology 10 (1), 65.

Hanauer, D.A., Chinnaiyan, A.M., 2006. PubMed QUEST: The PubMed Query Search Tool. An informatics tool to aid cancer centers and cancer investigators in searching the PubMed databases. Cancer Informatics 2, 79.

Han, K., Park, B., Kim, H., Hong, J., Park, J., 2004. HPID: The human protein interaction database. Bioinformatics 20 (15), 2466–2470.

Joudaki, H., Rashidian, A., Minaei-Bidgoli, B., *et al.*, 2015. Using data mining to detect health care fraud and abuse: A review of literature. Global Journal of Health Science 7 (1), 194.

Kontostathis, A., Edwards, L., Leatherman, A., 2010. Text mining and cybercrime. In: Text Mining: Applications and Theory. Chichester, UK: John Wiley & Sons, Ltd.

Licata, L., Briganti, L., Peluso, D., *et al.*, 2011. MINT, the molecular interaction database: 2012 update. Nucleic Acids Research 40 (D1), D857–D861.

Luhn, H.P., 1958. The automatic creation of literature abstracts. IBM Journal of Research and Development 2 (2), 159–165.

Miner, G., 2012. Practical Text Mining and Statistical Analysis for Non-Structured Text Data Applications. Academic Press.

Ongenaert, M., Van Neste, L., De Meyer, T., *et al.*, 2007. PubMeth: A cancer methylation database combining text-mining and expert annotation. Nucleic Acids Research 36 (Suppl. 1), D842–D846.

Papanikolaou, N., Pavlopoulos, G.A., Theodosiou, T., Vizirianakis, I.S., Iliopoulos, I., 2016. DrugQuest – A text mining workflow for drug association discovery. BMC Bioinformatics 17 (5), 182.

PubMed, 1946. Bethesda (MD): National Library of Medicine (US). (cited 2017 Dec 21).

Rebholz-Schuhmann, D., Oellrich, A., Hoehndorf, R., 2012. Text-mining solutions for biomedical research: Enabling integrative biology. Nature Reviews Genetics 13 (12), 829–839.

Salwinski, L., Miller, C.S., Smith, A.J., *et al.*, 2004. The database of interacting proteins: 2004 update. Nucleic Acids Research 32 (Suppl. 1), D449–D451.

Sebastiani, F., 2002. Machine learning in automated text categorization. ACM Computing Surveys (CSUR) 34 (1), 1–47.

Sharma, S., Srivastava, S.K., 2016. Review on text mining algorithms. International Journal of Computer Applications 134 (8).

Singhal, A., Simmons, M., Lu, Z., 2016. Text mining for precision medicine: Automating disease-mutation relationship extraction from biomedical literature. Journal of the American Medical Informatics Association 23 (4), 766–772.

Szklarczyk, D., Morris, J.H., Cook, H., *et al.*, 2017. The STRING database in 2017: Quality-controlled protein–protein association networks, made broadly accessible. Nucleic Acids Research 45 (D1), D362–D368.

UniProt Consortium, 2011. Reorganizing the protein space at the Universal Protein Resource (UniProt). Nucleic Acids Research. gkr981.

Walport, M., Kiley, R., 2006. Open access, UK PubMed central and the wellcome trust. Journal of the Royal Society of Medicine 99 (9), 438–439.

Witten, I.H., Frank, E., Hall, M.A., Pal, C.J., 2016. Data Mining: Practical Machine Learning Tools and Techniques. Morgan Kaufmann.

Zhu, F., Patumcharoenpol, P., Zhang, C., *et al.*, 2013. Biomedical text mining and its applications in cancer research. Journal of Biomedical Informatics 46 (2), 200–211.

Relevant Websites

http://eupathdb.org/eupathdb/
  EuPathDB.
https://www.nlm.nih.gov/mesh/
  MeSH Browser.
http://www.phontron.com/nlptools.php
  Natural Language Processing Tools.
https://www.ncbi.nlm.nih.gov/pubmed/
  PubMed
  NCBI
  NIH.
http://plasmodb.org/plasmo/
  PlasmoDB.

Preclinical: Drug Target Identification and Validation in Human

Meena K Sakharkar and Karthic Rajamanickam, University of Saskatchewan, Saskatoon, SK, Canada
Chidambaram S Babu, JSS University, Mysuru, India
Jitender Madan, Chandigarh College of Pharmacy Landran, Mohali, India
Ramesh Chandra, University of Delhi, Delhi, India
Jian Yang, University of Saskatchewan, Saskatoon, SK, Canada

Introduction

Target identification is the first step in drug discovery. A target is an entity to which an endogenous ligand or a drug binds resulting in a change in its behavior or function. Enzymes, receptors and transport proteins along with DNA and RNA are the main targets for drugs at the molecular level. The pre-clinical drug development process is based on the hypothesis that modulation of the target(s) of interest will result in a therapeutic effect in a disease state. The human genome data is an excellent resource to understand the genetic factors in human disease, and one of its prime goals was to pave the way for new strategies for disease diagnosis, treatment and prevention; as it has the potential to reveal targets involved in a specific disease. However, not many new molecular entities have been discovered since the sequencing of the human genome. The paucity of translatable results can be attributed to several factors. Lack of accurate information on gene architecture and gene annotation which is essential for drug discovery as this allows insight into splice variants and drug cross-reactivity. Moreover, potential targets provided by the genome projects are not endowed with elaborate background knowledge because of our inability to infer the function of most of the DNA in the genome. Many of the computationally derived annotations in the databases are either minimal or incorrect (apart from a carefully manually-curated database such as Uniprot) (Zhao *et al.*, 2007). Also, as the annotation of genes is provided by multiple public resources using different methodologies, the resultant information may be similar but not always identical (Kangueane *et al.*, 2015). This is further confounded by our limitation to infer the number of splice variants or protein-protein interactions from gene sequences alone, thereby limiting our analyses of genome and in consequence, our estimation of the total number of targets. Lack of information on protein structure, protein function and binding of a specific drug to the target of interest adds another dimension to this puzzle. In conclusion, the increased number of targets and the lack of functional knowledge about them are generating a bottleneck in the target validation process (Ofran *et al.*, 2005). Despite the relatively limited knowledge about the complex relationship between chemical space and genomic space, drug development strategies have been influenced profoundly by the wealth of potential targets offered by genome projects (Finan *et al.*, 2017). This article highlights some of the key aspects in human pre-clinical drug target identification and validation.

Target Identification *in Silico*

Over the past two decades, several computational (*in silico*) methods have been developed and applied to generate pharmacology hypotheses. These *in silico* methods are primarily used alongside *in vitro* data to create the in vivo model for pre-clinical testing and optimization of novel molecules with the potential to bind to a target, and the ADMET (absorption, distribution, metabolism, excretion and toxicity) properties in addition to the physicochemical characterization.

Drugs Databases and Drug Target Databases

The first step towards drug discovery is the correct identification and validation of the drug target and its interaction with the drug. The key drug databases and drug target databases are:

Drug databases

The **PubChem** database contains about 35 million compounds. It is estimated that <7000 compounds have the information on their corresponding target proteins (Chen *et al.*, 2016). It has been reported that the known chemical space probably contains on the order of 100 million molecules and the estimate for Lipinski virtual chemical space is around 10^{60} compounds or a more modest 10^{20}–10^{24} molecules if the combination of known fragments are considered (Ertl, 2003). **ChEMBL** contains 1.6 million distinct compounds, 14 million bioactivities, 11K biological targets, and other related data organized in 72 tables manually collected from the published literature. These data are very useful for drug discovery and include binding, functional and ADMET (i.e., assessment of *in vivo* absorption, distribution, metabolism, excretion and toxicity properties) information for a larger number of drug-like bioactive compounds (Nowotka *et al.*, 2017). **ZINC** contains over 120 million compounds for ligand discovery and virtual screening and allows for investigators to seek chemical matter for their biological targets (Sterling and Irwin, 2015). **DCDB** (Drug Combination Database) offers data on drug combinations. Its current version comprises of 1363 approved or investigational drug combinations, including 237 unsuccessful drug combinations, involving 904 individual drugs, from >6000 references

(Liu *et al.*, 2014). The information about drugs and their targets is manually annotated based on the literature and relevant databases such as Drugbank (Wishart *et al.*, 2017), PubChem, UniProt and Drugs.com.

Thus, even with today's computational resource, the entire chemical space cannot be exhaustively enumerated computationally and prioritisation and targeted selection is essential for virtual screening.

Drug target databases

The DrugBank database is a richly annotated bioinformatics and cheminformatics resource that combines detailed drug data (e.g., chemical, pharmacological and pharmaceutical) with comprehensive target information (e.g., sequence, structure and pathway) (Wishart *et al.*, 2008). Therapeutic target database (TTD) provides the information about known and explored therapeutic protein and nucleic acid targets, the targeted diseases, pathway information and corresponding drugs directed at each of these targets (Chen *et al.*, 2002). Recently, the information of 1755 biomarkers for 365 disease conditions and 210 drug scaffolds for 714 drugs and leads has been added into this database (Qin *et al.*, 2014). PharmGKB is another pharmacogenomics knowledge resource that encompasses clinical information including dosing guidelines and drug labels, potentially clinically actionable gene-drug associations and genotype-phenotype relationships. PharmGKB also collects curates and disseminates knowledge about the impact of human genetic variation on drug responses (Thorn *et al.*, 2013).

Computational Approaches in Drug Target Identification

In order to understand the cellular processes taking place in response to a drug, it is critical to elucidate its molecular targets. This has tremendous implications for disease prevention and treatment. Here, it is important to mention that the incorrect identification of targets is a significant factor in drug failures and the low drug approval rate during drug development stemming from an incomplete knowledge on the underlying physiology of the target and incorrect biological hypothesis (Vasaikar *et al.*, 2016). Specifically, in line with this, the association of a drug with additional targets beyond its direct ones, may give rise to hazardous side effects, precluding further drug development and usage. 82% of Food and Drug Administration (FDA) approved drugs have an assigned mechanism-of-action (Overington *et al.*, 2006). However, at least half of the drugs with unassigned mechanism of action date from the pre-molecular era and have no targets defined as their action was explored against whole tissues, rarely on isolated proteins, and target identities were only inferred from tissue-based responses (Gregori-Puigjane *et al.*, 2012). Even though the total number of predicted drug targets of pharmacological interest is estimated in the range of 6000–8000 in the human genome, drugs have been approved for only a fraction of these targets (Landry and Gies, 2008; Plewczynski and Rychlewski, 2009). This is because a large number of these putative drug targets remain to be validated. Druggability is the presence of protein folds (quaternary structures) that favor interactions with drug-like chemical compounds (Hajduk *et al.*, 2005). The ability of a protein to bind a small molecule with the appropriate chemical properties and binding affinity might make it druggable, but does not necessarily make it a potential drug target. This is true, though, for proteins that are linked to human diseases (Hopkins and Groom, 2002). In this regard, Hopkins and Groom have highlighted that ~10% of the entire human genome is involved in disease onset or progression, resulting in ~3000 potential targets suitable for therapeutic intervention (Perumal *et al.*, 2009; Sakharkar and Sakharkar, 2007; Sakharkar *et al.*, 2007). An attempt to identify the number of druggable disease genes using DGIdb database (see "Relevant Websites section"), which is a compendium of 3860 druggable genes from BaderLabGenes, CarisMolecularIntelligence, FoundationOneGenes, GO, GuideToPharmacologyGenes, HopkinsGroom, MskImpact, RussLampel, and dGene resources and the DisGeNet platform which reports on the gene disease associations derived from expert curated databases and text mined data (see "Relevant Websites section") shows that there are 1645 druggable disease genes which are associated with 923 diseases. Here, it is important to mention that DisGeNet provides information on 26,522 disease gene associations reporting 7878 disease genes and 6761 unique diseases. Taking genes involved in breast cancer from BCDB database and mapping them onto the 1645 druggable genes, we found 471 genes that are involved in breast cancer and are druggable (**Fig. 1**).

Once the disease mechanism is understood and the target and lead compounds are identified, chemoinformatic tools present a tremendous potential to help decipher the interactions between protein targets and the ligands before expensive and time-consuming experiments. Towards this end, we have earlier shown the coffee component HHQ as a putative ligand for PPAR gamma and its implications in breast cancer in vitro (BMC Genomics) (**Fig. 2**) (permission applied).

However, one should keep in mind that the identification of novel drugs and their targets is still an extremely difficult goal owing to the limitation in mapping drugs and targets in chemical space and genomic space into a unified space, called pharmacological space.

Approaches for Target Validation *in Silico, in Vitro* and *in Vivo*

Validation is a crucial step in the drug discovery process. The only way to be completely certain that a protein is instrumental in a given disease is to test the idea in clinical trials. With the evolution of sequence targeted therapeutics like CRISPR and RNAi silencing, gene expression levels can be raised or lowered at transcriptional, translational or post-translational stages. The effects of these can be evaluated in preclinical models generated by the gene editing of experimental animals and human cells (2017). Additionally, the role of Genome-wide association studies (GWAS) of disease predisposition, metabolism and gene expression provides data on

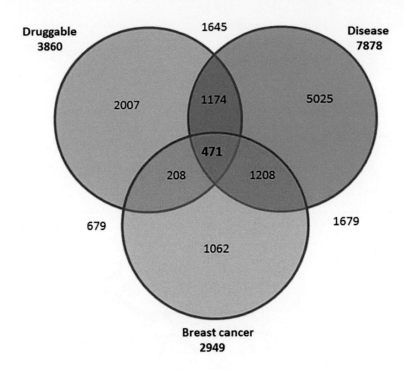

Fig. 1 Schematic representation of number of genes involved in breast cancer as provided by breast cancer data base (BCDB). Out of 1645 druggable genes, 471 genes were involved in breast cancer.

functioning of protein-targeting drugs. GWAS also use the regulatory variation in human genomes to guide the therapeutic future of targeted gene regulation (2017). Thus, the protein based concept of druggability needs to be expanded and redefined. Several resources and techniques help validate the predicted drug-target interaction. Some of them are briefly described below:

Gene and protein expression profiling datasets: Gene expression profiling through transcriptomics (RNA profiling) and proteomics (protein profiling) has contributed significantly to our efforts in understanding complex biological processes. Technologies such as microarrays, quantitative real-time polymerase chain reactions, and next-generation sequencing have helped decipher the targeted receptor(s), pathway or network through the identification of downregulated or upregulated pattern(s) of the intended drug target (s). **Connectivity Map (CMAP)** is a publicly accessible gene expression database that contains over 1.5 million gene expression profiles from ~5000 small-molecule compounds, and ~3000 genetic reagents, tested in multiple cell types (see "Relevant Websites section"). CMAP collates, compares and links all changes in gene expression ("signatures") arising from a disease or gene modulation (knockdown or overexpression of a gene), or treatment with a small molecule (Lamb *et al.*, 2006). Differential expressions that show highly similar or dissimilar, expression signatures are termed "connected". Similar physiological effects on the cells imply that the transcriptional effects are related. GEO is a database of gene expression profiles derived from over 100 organisms and more than a billion individual gene expression measurements addressing several biological issues such as disease states and stages of development. It is a public repository that freely distributes this information that is submitted by the scientific community. It can be explored, queried, and visualized using user-friendly Web-based tools. **DeSigN** is a web-based tool for predicting drug efficacy against cancer cell lines using gene expression patterns (Lee *et al.*, 2017). It can use gene expression analyses to identify candidate drugs using an input gene signature. Identifying distinct common pathways/gene modules that are shared by pathophysiological processes of multiple diseases is made easier by the integration of these diverse datasets.

Phenotypic Screening

Phenotypic screening is based on the alteration(s) of metabolically active systems such as tissues or whole animals produced at physiological or molecular structural levels upon exposure to a chemical entity. Hits are identified in phenotypic screening and followed by the leads thereof by structural activity relationship assays. Phenotypic screenings are based on the system suitability, the stimulus and the end-point readout rather than the pathological and clinical endpoints (Vincent *et al.*, 2015). These ensure robustness, repeatability and cost-effectiveness of the procedure. Phenotypic screenings are carried out using test systems which include cell lines, situ organ/tissue preparations and animal models. The animal models used in phenotypic screening include *C. elegans*, Danio *rerio* (*Zebrafish*), X. laevis, and *Drosophila melanogaster*. In disease models, phenotypic screening involves the determination of changes in structural morphology and biochemical characteristics and effects of compounds (drugs) on reversal of the changes. The screening of compounds on the neurite growth and amyloid-*β*, neurofibrillary tangles in Alzheimer diseases

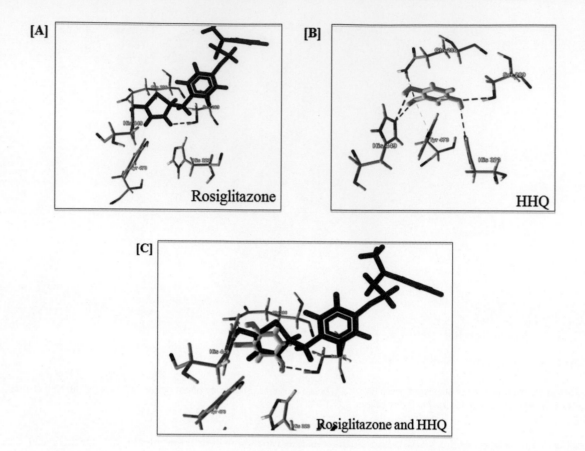

Fig. 2 HHQ docked in the ligand binding domain of PPARγ protein crystal structure for PDB solved in conjunction with Rosiglitazone (PDBID: 2PRG). [A] Represents hydrogen bonds (in black dotted line) observed for Rosiglitazone (in red) with the active site residues in the ligand binding domain of PPARγ (2PRG). [B] Represents hydrogen bonds (in black dotted line) observed for HHQ (in yellow) with the active site residues in the ligand binding domain of 2PRG. [C] Represents superposition of the best conformation of Rosiglitazone (in red) and HHQ (in yellow) in the ligand binding domain of 2PRG. Rosiglitazone is a known ligand of PPARγ (published in Shashni, B., Sharma, K., Singh, R., *et al.*, 2013. Coffee component hydroxyl hydroquinone (HHQ) as a putative ligand for PPAR gamma and implications in breast cancer. BMC Genomics 14 (Suppl. 5), S6).

and measurements is one of the best examples for phenotypic screening in which both the cellular morphology and the biochemical changes and effects of drugs are studied in parallel.

One of the key limitations of phenotypic screening is that it usually provides very little information on the possible targets of the compounds. Moreover, compounds with a poor ADMET may not be active in primary screens. These factors demand cell-based assays or transgenic/knockout models to identify the target proteins and provide some information on the possible mechanism of action of the compounds/drugs.

Transgenic organisms

Genetically modified animals are widely in research to create models of human diseases. These animal models are created based on a molecular understanding of genomic/proteomic alterations in human diseases. However, the intensity of up- or down-regulation, progress and intracellular accumulation may vary phenotypically from the human disease. Animal models are genetically modified to over-express (transgenic), remove ("knock-out"), or replace ("knock-in") specific genes. These animals are widely used to understand the pathology and investigate the effects of hits/leads for their efficacy (Bolon, 2004). Furthermore, the use of genetically modified models provides information on target identification and helps establish the precise mechanism efficacy or toxicity of the new chemical entities or drugs. For example, in the transgenic mouse model of Parkinsonism, the animals were modified to over-express human α-synuclein, a major protein that aggregates in neuronal cells in Parkinson's disease (PD). This animal model mimics PD pathology and simulates it closer to the human disease. Further, the animals which over-expressed human α-synuclein showed decreased dopamine turn over in turn motor deficits. Thus, the model serves as a target for the phenotypic screening of drug candidates having effects on α-synucelin. Transgenic models also help to rule out the down- or upstream signalling pathways, which are further helpful in understanding how the disease progresses and its possible therapeutic options.

Imaging

Imaging techniques have often been used to study gross anatomy. Read outs usually show structural deformities. With advancements in technology, it became feasible to image physiological and biochemical events in metabolically active systems such as cell lines and animal models. This is called functional imaging. The live imaging of specific proteins or target proteins of interest was made possible with the help of the development of molecular probes, this is often called molecular imaging. The main advantages of molecular imaging are that it is repetitive, non-invasive, and the effects are quantifiable at cellular and sub-cellular levels. These imaging techniques use various energy forms to penetrate or interact with the endogenous proteins or fluorescence tagged proteins in in-vivo systems (**Fig. 3**).

With the help of sophisticated imaging techniques, the identification of marker proteins and effects of drugs on the target proteins is performed in phenotypic screening. This helps to screen huge numbers of compounds at fast rates and more economically in drug discovery processes. Molecular imaging techniques are also being used to study the pharmacokinetics of drug following its administration. On the other hand, in vivo imaging techniques serve as an important tool for studying the pathophysiology and efficacy of drug on the reversal processes in an evidence based manner. As mentioned earlier, the use of molecular imaging became is a pivotal method in preclinical oncology research, including the xenograft and orthotopic models. For example, in a mouse lymphoma xenograft model, the effect of temozolomide on tumor progression and the prolongation of survival was studied. The tumor cells were transfected with the luciferase gene, enabling the visualization of lymphoma cells using bioluminescence (BL) imaging. The tumor growths between untreated and treated groups were studied with reference to the luciferase imaging (Kadoch *et al.*, 2009).

Biomarkers: Biomarker assays not only assist in patient selection and the design of clinical trials but also act as indicators of drug efficacy, toxicity and disease progression (Ganesalingam and Bowser, 2010).

This substantially increases their importance in drug discovery processes. These assays are initiated in cell lines and primary blood cells or isolated tissue cells and are used to validate in vitro modulation of target by the drug and the effect of drugs on cells that is routine cellular screening. Typically, these assays are initiated in human cell lines, primary blood cells or isolated tissue cells. Biomarker assays are also used to identify markers of drug sensitivity that are used in selection of patient populations. Biomarker assays for pharmacokinetic/pharmacodynamic (PK/PD) models have also been developed, profiling molecules prior to testing in longer term disease models as initial proofs of concept. PK/PD models can also assist in dose-to-man scaling predictions for use in clinical trials. The biomarker assays developed during the in vitro discovery phases are frequently used as efficacy or toxicity endpoints in the clinic. Clinical trials, particularly in oncology, are frequently designed around these biomarkers.

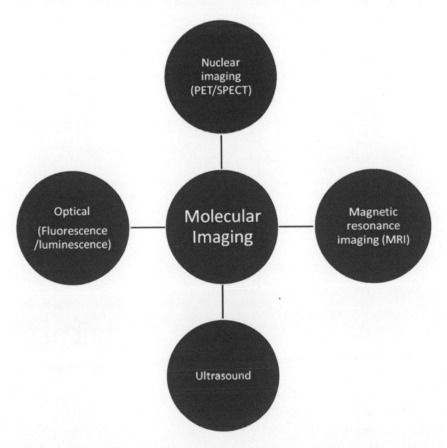

Fig. 3 Armamentarium of molecular imaging techniques employed to quantify target proteins. Nuclear imaging, magnetic resonance imaging, ultrasound imaging, and optical imaging techniques have been categorized as repetitive and non-invasive molecular imaging techniques.

Conclusions

Target identification and validation is costly and labour intensive. Failures in clinical trials are likely to have multifactorial reasons. However, selecting the most biologically plausible molecular targets that are relevant to the disease state is a critical first step to improve the probability of success. Towards this end, it is key to get a complete picture of the target(s), their role in a biological process and how they can be modulated to achieve the required output.

See also: Chemoinformatics: From Chemical Art to Chemistry in Silico. Comparative Epigenomics. Functional Enrichment Analysis. Gene Prioritization Using Semantic Similarity. Integrative Bioinformatics. Introduction of Docking-Based Virtual Screening Workflow Using Desktop Personal Computer. Molecular Mechanisms Responsible for Drug Resistance. Natural Language Processing Approaches in Bioinformatics. Rational Structure-Based Drug Design. Structure-Based Drug Design Workflow. Transmembrane Domain Prediction

References

Bolon, B., 2004. Genetically engineered animals in drug discovery and development: A maturing resource for toxicological research. Basic and Clinical Pharmacology and Toxicology 95, 154–161.

Chen, X., Ji, Z.L., Chen, Y.Z., 2002. TTD: Therapeutic target database. Nucleic Acids Research 30, 412–415.

Chen, X., Yan, C.C., Zhang, X., et al., 2016. Drug-target interaction prediction: Databases, web servers and computational models. Briefings in Bioinformatics 17, 696–712.

Ertl, P., 2003. Cheminformatics analysis of organic substituents: Identification of the most common substituents, calculation of substituent properties, and automatic identification of drug-like bioisosteric groups. Journal of Chemical Information and Modeling 43, 374–380.

Finan, C., Gaulton, A., Kruger, F.A., et al., 2017. The druggable genome and support for target identification and validation in drug development. Science Translational Medicine 9.

Ganesalingam, J., Bowser, R., 2010. The application of biomarkers in clinical trials for motor neuron disease. Biomarkers in Medicine 4, 281–297.

Gregori-PuigJane, E., Setola, V., Hert, J., et al., 2012. Identifying mechanism-of-action targets for drugs and probes. Proceedings of National Academy of Science 109, 11178–11183.

Hajduk, P.J., Huth, J.R., Tse, C., 2005. Predicting protein druggability. Drug Discovery Today 10, 1675–1682.

Hopkins, A.L., Groom, C.R., 2002. The druggable genome. Nature Reviews Drug Discovery 1, 727–730.

Kadoch, C., Dinca, E.B., Voicu, R., et al., 2009. Pathologic correlates of primary central nervous system lymphoma defined in an orthotopic xenograft model. Clinical Cancer Research 15, 1989–1997.

Kangueane, P., Sowmya, G., Anupriya, S., et al., 2015. Short peptide vaccine design and development: Promises and challenges. In: Global Virology I-Identifying and Investigating Viral Diseases. Springer.

Lamb, J., Crawford, E.D., Peck, D., et al., 2006. The Connectivity map: Using gene-expression signatures to connect small molecules, genes, and disease. Science 313, 1929–1935.

Landry, Y., Gies, J.P., 2008. Drugs and their molecular targets: An updated overview. Fundamental and Clinical Pharmacology 22, 1–18.

Lee, B.K.B., Tiong, K.H., Chang, J.K., et al., 2017. DeSigN: Connecting gene expression with therapeutics for drug repurposing and development. BMC Genomics 18, 934.

Liu, Y., Wei, Q., Yu, G., et al., 2014. DCDB 2.0: A major update of the drug combination database. Database, Oxford 2014, bau124.

Nowotka, M.M., Gaulton, A., Mendez, D., et al., 2017. Using ChEMBL web services for building applications and data processing workflows relevant to drug discovery. Expert Opinion on Drug Discovery 12, 757–767.

Ofran, Y., Punta, M., Schneider, R., Rost, B., 2005. Beyond annotation transfer by homology: Novel protein-function prediction methods to assist drug discovery. Drug Discovery Today 10, 1475–1482.

Overington, J.P., Al-Lazikani, B., Hopkins, A.L., 2006. How many drug targets are there? Nature Reviews Drug Discovery 5, 993–996.

Perumal, D., Lim, C.S., Sakharkar, M.K., 2009. A comparative study of metabolic network topology between a pathogenic and a non-pathogenic bacterium for potential drug target identification. Summit on Translational Bioinformatics 2009, 100–104.

Plewczynski, D., Rychlewski, L., 2009. Meta-basic estimates the size of druggable human genome. Journal of Molecular Modeling 15, 695–699.

Qin, C., Zhang, C., Zhu, F., et al., 2014. Therapeutic target database update 2014: A resource for targeted therapeutics. Nucleic Acids Research 42, D1118–D1123.

Sakharkar, M.K., Sakharkar, K.R., 2007. Targetability of human disease genes. Current Drug Discovery Technology 4, 48–58.

Sakharkar, M.K., Sakharkar, K.R., Pervaiz, S., 2007. Druggability of human disease genes. International Journal of Biochemistry and Cell Biology 39, 1156–1164.

Shashni, B., Sharma, K., Singh, R., et al., 2013. Coffee component hydroxyl hydroquinone (HHQ) as a putative ligand for PPAR gamma and implications in breast cancer. BMC Genomics 14 (Suppl. 5), S6.

Sterling, T., Irwin, J.J., 2015. ZINC 15 – Ligand discovery for everyone. Journal of Chemical Information and Modeling 55, 2324–2337.

Thorn, C.F., Klein, T.E., Altman, R.B., 2013. PharmGKB: The pharmacogenomics knowledge base. Methods in Molecular Biology 1015, 311–320.

Vasaikar, S., Bhatia, P., Bhatia, P.G., Chu Yaiw, K., 2016. Complementary approaches to existing target based drug discovery for identifying novel drug targets. Biomedicines 4, 27.

Vincent, F., Loria, P., Pregel, M., et al., 2015. Developing predictive assays: The phenotypic screening "rule of 3". Science Translational Medicine 7, 293ps15.

Wishart, D.S., Feunang, Y.D., Guo, A.C., et al., 2017. DrugBank 5.0: A major update to the DrugBank database for 2018. Nucleic Acids Research.

Wishart, D.S., Knox, C., Guo, A.C., et al., 2008. DrugBank: A knowledgebase for drugs, drug actions and drug targets. Nucleic Acids Research 36, D901–D906.

Zhao, B., Sakharkar, K.R., Lim, C.S., Kangueane, P., Sakharkar, M.K., 2007. MHC-Peptide binding prediction for epitope based vaccine design. International Journal of Integrative Biology 1, 127–140.

Relevant Websites

https://clue.io/cmap
    CMap.
http://www.disgenet.org/web/DisGeNET/menu/downloads
    DisGeNET.
http://dgidb.org
    DGIdb.

Biomedical Text Mining

Hagit Shatkay, University of Delaware, Newark, DE, United States

1 Introduction

The vast majority of biomedical knowledge is conveyed and shared by means of written text, ranging from publications in journals and conferences, summary abstracts such as those stored in PubMed (PubMed, 2016), notes in electronic health records, drug labels, or short descriptive entries in biomedical databases (Chatr-Aryamontri *et al.*, 2015; Dowell *et al.*, 2009; Eppig *et al.*, 2015; The UniProt Consortium, 2012; Van Auken *et al.*, 2012). As text is typically provided in the form of natural language sentences, searching for and gleaning information conveyed within requires much human labor or computational tools that can sift through text and identify relevant words or passages. Developing and applying such tools for processing text and finding relevant information is typically referred to as Text Mining. This is a broad area of activity and research, involving several disciplines. A key discipline is natural language processing (NLP), and specifically information extraction (IE). It focuses on obtaining structured information from natural-language text, by identifying certain entities (e.g., genes, proteins, or diseases) and relations between them (e.g., phosphorylation, activation, repression) mentioned within the text. The entities and the relationships are typically converted into a standard form, such as ontological terms or numerical identifiers, and stored in a database. Ontological terms are terms listed within a controlled and carefully maintained vocabulary that denotes entities pertaining to a certain domain, along with relationships among these entities. The Gene Ontology (GO) is the most widely used ontology in the biological domain (Gene Ontology, 2016; The Gene Ontology Consortium, 2000), while the Medical Subject Headings (MeSH) is one of several controlled vocabularies used in medicine, covering organs, symptoms, diseases, and other conditions (MeSH, 2016).

Another major text mining task is the identification of publications or text passages that are relevant to individual users or user-communities. Examples include papers discussing gene expression in the mouse, which are of interest for curation by the Mouse Genome Database (Eppig *et al.*, 2015), or papers indicating side effects of a drug – which may be of interest to individual physicians. The latter may also be of interest for curation by the FDA (The United States Food and Drug Administration). This area of text analysis is known as information retrieval.

Another area related to information retrieval is that of text classification. It can be viewed as another way of identifying documents relevant to specific tasks; however, rather than viewing the goal as that of satisfying the different needs of an individual end user, the objective is to partition a collection of articles into individual subcategories based on topics of interest. For instance, one may partition documents into those that are relevant to breast cancer, those that pertain to prostate cancer, and those that are not relevant to any type of cancer. Each relevance-category is called a class, and the objective is to label each document by its appropriate class. Clearly, there are many ways to classify documents into different categories, and as such, text classification is applicable in a variety of contexts, and used for a wide range of text-related tasks.

Throughout this module we provide background about these different text-related tasks, along with examples of their application within the biomedical domain.

2 Natural Language Processing and Information Extraction

As noted above, biomedically-relevant text is available from many sources and repositories, ranging from sentences and comments stored as annotations, to curated genes, proteins or processes, to complete publications stored in publishers databases and in

PubMed/PubMed-central (2016). To effectively seek information within text, computational methods that perform NLP are often used (Jurafsky and Martin, 2009; Manning and Schütze, 1999).

2.1 Basic Concepts and Challenges in Natural Language Processing

Consider an application that tries to automatically seek information about a specific protein within text. Basic processing of the text's natural language requires first identifying orthographical features including word breaks, word-capitalization, and punctuation marks, which help identify words, phrases, or tokens within sentences (Tokens are character-sequences that are treated as a single unit, a word, or a term; the task of identifying them is known as tokenization). **Fig. 1** shows an example of a sentence, along with the components comprising it. The example demonstrates the value of orthographic information for identifying specific entities of interest; the names of genes/proteins like *FSP27* (Fat Specific Protein 27) and *PLIN1* (prepilin 1) are typically shown as a short sequence of capital letters followed by a number.

Notably, orthographic rules are not hard-and-fast; while they serve as useful cues for identifying word boundaries in general, as well as genes and proteins in the biological domain, capitalization is often used for a variety of other purposes, indicating abbreviations, acronyms, or the start of a sentence. Similarly, hyphens, dashes, periods and other delimiters can serve a variety of purposes. The multiple roles such features play within a sentence make the identification of components a nontrivial task.

Another indicator of a word's meaning is its morphology, that is, its structure as a composition of smaller common units. For instance, the words interaction, interacts, and interacting all share the same root form interact and the same semantics, where the differences are relegated to the suffix. Conversely, certain suffixes can be used to convey a common semantic meaning, for example, the suffix "ase" often denotes an enzyme (e.g., kinase, ligase). Thus, morphological analysis including suffix stripping or suffix detection can be useful in identifying words that denote certain entity types or convey a similar meaning – such as enzymes, diseases, drugs, or interactions among these entities.

The meaning of a phrase or a sentence depends on relationships among words and the role of each word within a broader context. Breaking the sentence into its components – known as parsing, and identifying the role of words – known as part of speech (POS) tagging, are both fundamental steps in NLP. **Fig. 1** illustrates the syntactic parsing of an example sentence, and the assignment of POS tags to its components: *FSP27* and *PLIN1* are both nouns, interact is a verb, in and with are prepositions. The set of POS labels used in practice is much more comprehensive (see for instance the *Penn Treebank corpora* (Marcus *et al.*, 1993), that include tens of thousands manually annotated sentences, employing tens of POS different tags).

Similar to the dependency of each word's meaning on its neighboring words, the meaning of a sentence as a whole also depends on surrounding sentences. That is, the meaning of any sequence of words or sentences is context-dependent. Discourse analysis (Afantenos *et al.*, 2010; Jurafsky and Martin, 2009; Conrath *et al.*, 2014; Dascalu, 2014) is an area within NLP focusing on the meaning conveyed within broader text regions through relations among multiple sentences.

Seeking the actual meaning conveyed in the text through NLP is a complex and challenging task. A key obstacle to understanding natural language, by a computer or by a human being, is ambiguity. Ambiguity manifests itself in every level of the language. Orthographical cues have multiple interpretations where capitalization can indicate a beginning of a sentence, a proper noun, or an acronym. At the word level, a single word may convey multiple meanings (polysemy, e.g., the word fly may serve as a noun denoting an animal or a verb denoting the flight action); at the sentence-level, syntactic ambiguity occurs where a sentence may be parsed in more than one way conveying different intentions. Much work in NLP is dedicated to overcoming ambiguity, providing accurate tools for identifying terms within text, tagging and parsing sentences, and identifying types of discourse (see e. g., (Jurafsky and Martin, 2009; Manning and Schütze, 1999; Shatkay and Craven, 2012) for details and examples).

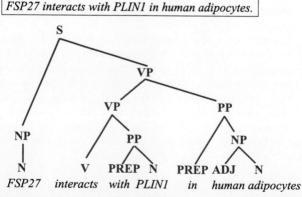

Fig. 1 A sentence (S) taken from a biomedical abstract (Grahn et al., 2014), along with its parse tree that shows parts-of-speech tags. Phrase tags, NP, VP, and PP correspond to Noun-, Verb-, and Prepositional-phrase respectively, while terminal tags – one layer above the bottom of the tree – indicate nouns (N), verbs (V), prepositions (PREP), and adjectives (ADJ). The leaves of the tree are the terminals, i.e., the words comprising the sentence.

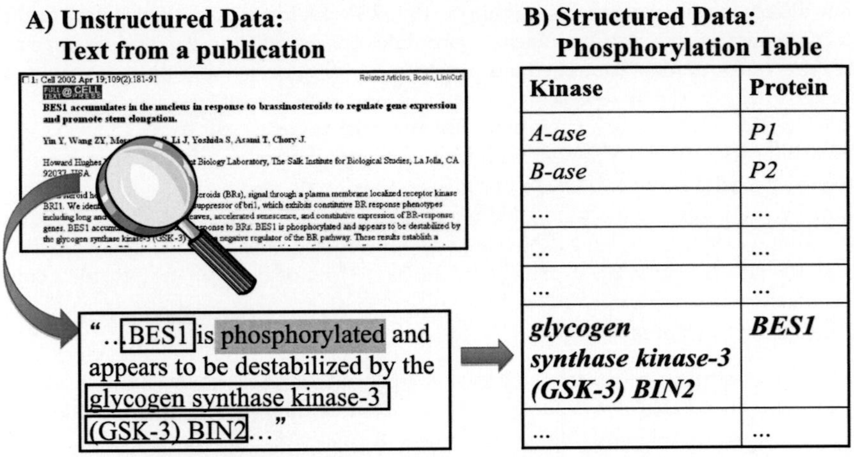

A) Unstructured Data: Text from a publication

B) Structured Data: Phosphorylation Table

Kinase	Protein
A-ase	*P1*
B-ase	*P2*
...	...
...	...
...	...
glycogen synthase kinase-3 (GSK-3) BIN2	***BES1***
...	...

Fig. 2 (A) A biomedical paper (Yin et al., 2002) – shown as free, unstructured text (top), and a sentence within it discussing phosphorylation (bottom). (B) A structured phosphorylation table, showing a list of proteins and the kinase that phosphorylates each. The entry shown toward the bottom of the table is populated through information extraction methods applied to the sentence shown in A (bottom, left). The names of the identified protein and kinase are shown in dark frames, while the verb indicating phosphorylation is shown in gray.

The accuracy of NLP tools often depends on the domain in which they have been developed and employed (Ferraro *et al.*, 2013; McClosky *et al.*, 2010). Thus, as a basis for building NLP tools in the biological domain, specialized NLP resources providing annotations and tagging of biomedical sentences have been developed (Tateisi *et al.*, 2005; Thompson *et al.*, 2017) along with specialized tools for language analysis, for example (Cardie, 1997; Ferraro *et al.*, 2013; Kang *et al.*, 2011; Smith *et al.*, 2004).

2.2 Information Extraction: From Unstructured Text to Structured Data

Two specific tasks that are often addressed within NLP and are of special interest within the biomedical domain are named entity recognition and relation extraction. The former means identifying mentions of relevant biological entities like genes, proteins, chemical compounds, drugs, diseases and many others; the latter refers to detecting relationships among these entities, such as reduction in drug efficacy due to a gene's activity, or interaction between two proteins or between two drugs. Gleaning such information from unstructured text and placing it in a database or within another structured format is known as IE (Cardie, 1997; Cowie and Lehnert, 1996).

Fig. 2 shows an illustration of IE applied to a text abstract. The task is the identification and extraction of phosphorylation relations between a kinase and a protein. The Protein and the Kinase entities are both identified, along with a verb that indicates a phosphorylation relationship. The entities are then harvested and placed in an entry of a structured phosphorylation table.

Named entity recognition is clearly needed for relation extraction (as illustrated in **Fig. 2**). It is also an important step toward indexing documents based on their contents so that documents discussing particular topics can be easily found, as discussed in Section 3.1. As such, much work in the field of biomedical text mining focuses on named entity recognition. It ranges from developing tools for gene- or protein-name extraction (Fluck *et al.*, 2007; Leser and Hakenberg, 2005; Settles, 2005; Tanabe and Wilbur, 2002) to chemical- drug- or disease-identification (Batista-Navarro *et al.*, 2015; Krallinger *et al.*, 2015; Kuhn *et al.*, 2016; Leaman *et al.*, 2013; Lowe and Sayle, 2015; Savova *et al.*, 2010). These methods rely on matching word-patterns to dictionaries containing curated entity-names, as well as on rules and machine learning methods, which are designed to distinguish named-entities from other words and terms. Extraction work is also being conducted in order to identify relations and statements pertaining to drug-interaction or other relationships among biological entities, for example (Kolchinsky *et al.*, 2010; Kuhn *et al.*, 2016; Savova *et al.*, 2010; Comeau *et al.*, 2014; Howe *et al.*, 2016).

Notably, extraction efforts aim to identify explicit names and statements within text that contains relevant information. Identifying the publications and text passages that are indeed relevant to a certain subject matter is a task known as information retrieval, which is discussed next.

3 Information Retrieval and Text Classification

Information retrieval aims to identify a set of relevant documents in a large body of text. Commonly used search engines, like Google or PubMed, perform information retrieval known as ad hoc retrieval. The basic search mechanism in this case relies on a query specified by the user, where the query consists of a search term or a Boolean combination of terms. The search engine scans the large document collection and retrieves all documents that satisfy the query. The documents are then displayed in an order that is based on certain criteria such as temporal order, relevance, importance, or the number of citations.

Boolean term-combinations can be a cumbersome way to express a query, often resulting in many false-positives and false-negatives. A complementary form of ad hoc retrieval relies on similarity queries, within what is known as the vector space

(Manning and Raghavan, 2008; van Rijsbergen, 1979; Salton, 1989; Sparck-Jones *et al.*, 2000). Under this framework, a set of terms or words – which can even be a complete document – constitutes the query. The latter is represented as an algebraic vector in high-dimensional space; the vector is then compared against the text collection by employing a vector-similarity measure. The documents retrieved are those whose vector representation is most similar to that of the query vector.

Another aspect of information retrieval is text classification (e.g., Sebastiani, 2002; Yang and Liu, 1999). Rather than issue queries against the whole collection, the documents are split into different categories, where each category corresponds to a topic of interest. For instance, the categories may be defined a priori, where a collection of medical publications can be partitioned based on the disease discussed; here documents discussing kidney disease form a separate class from those discussing pneumonia. Categories may also be automatically uncovered by the categorization system, a process known as clustering. Throughout the rest of this section we discuss several mechanisms supporting the different types of queries, and provide examples for their use.

3.1 Boolean Queries and Index Structures

The basic mechanism that enables most information retrieval engines and Boolean queries in particular, is the index. As noted earlier, any document is initially processed by breaking it into its constituent terms, which can be individual words (also called unigrams), pairs of consecutive words (bigrams), longer sequences of consecutive words (n-grams), or grammatical phrases. An index is a data structure mapping each of the terms to all its occurrences within documents in the text collection. An illustration of an index structure is shown in **Fig. 3**. A Boolean query is executed by scanning through the index structure for the query terms and identifying the documents that satisfy the required Boolean combination of term occurrences (Manning and Raghavan, 2008; Witten *et al.*, 1999).

The basic index structure contains terms occurring in the document collection. However, other terms denoting fundamental concepts can be associated with the documents and included in the index. These may include, for instance, terms from the National Library of Medicine's MeSH controlled vocabulary (MeSH, 2016) assigned by human curators to documents, or terms from the GO (Gene Ontology, 2016; The Gene Ontology Consortium, 2000) that indicate properties of the proteins mentioned in the publication.

Boolean queries, expressed as terms connected by Boolean operators, (e.g., "BRCA1 AND Cancer") are supported in a straightforward way by index structures as shown in **Fig. 3**, albeit in their basic form can be ineffective for satisfying domain-specific information needs. Their limitations stem primarily from the inherent ambiguity of natural language as discussed in Section 1. Firstly, many terms occur often in a very large number of documents, thus the set of documents satisfying a simple Boolean query is too large for a user to practically scan through. Secondly, a query term typically has multiple meanings, varying by context (e.g., expression has a different meaning in the context of "gene expression" vs. "facial expression," while fly in the context of "fruit fly" carries a different meaning than in the context "birds fly"). Thus a large portion of documents that satisfy the literal query may end up being irrelevant to the user's actual need. Thirdly, documents that are relevant but use different words from the ones mentioned in the query (e.g., synonyms of the query terms) to denote the same idea, are not retrieved.

The vector space model, discussed next, offers an alternative mechanism, relying less on explicit or specific words and more on the whole context and the semantics denoted by words and phrases.

3.2 Similarity Queries: Documents in Vector Space

Naturally, we think about and visualize documents as sequences of words, terms, or sentences. As a step toward applying computational methods to text, we abstract away from our own use of documents, and view both documents and queries as

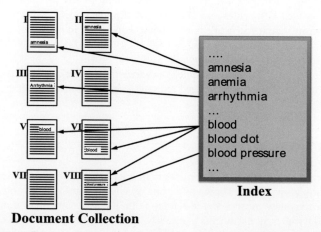

Fig. 3 An example of an index structure. The term entries in the index (right) are sorted alphabetically, and each term entry references the documents in which the term appears. Such a structure supports fast query processing, as documents containing the query terms can be quickly identified and accessed.

formal mathematical entities represented as algebraic vectors of weights, where each weight reflects the significance of an individual term. That is, every document d is represented as an N-dimensional vector of term-weights $\langle w_{t_1}^q, \cdots, w_{t_N}^q \rangle$, where $t_1, \ldots, t_N$ are the terms in the vocabulary and $w_{t_i}^q$ is a weight reflecting the significance of term t_i within the document d or the set of documents. A query q is likewise represented as a vector $\langle w_{t_1}^q, \cdots, w_{t_N}^q \rangle$.

The retrieval task thus amounts to finding the document-vectors most similar to the query vector. Such queries are supported, for instance, by PubMed (PubMed, 2016) – where one can click on the link "Similar articles" next to a retrieved article and obtain links to all articles that share similar words or contents. Similarity-based retrieval methods vary across the specific choice of terms included in the vocabulary, the weighting scheme used to assign weights to these terms, and the similarity measure employed for comparing the vectors.

A simple weighting scheme is to assign a binary value (0 or 1) to each term, corresponding to its presence or absence in the document. Other weighting strategies are often applied, guided by several intuitive principles (Sparck-Jones *et al.*, 2000): (1) A term occurring frequently within a document, is typically important in the document's context, and its weight should reflect this. (Local Term Frequency); (2) A term occurring in a very large number of documents typically does not characterize any specific document, and as such is less significant (Inverse Document Frequency); (3) A term occurring a few times in a short document is more significant to the document than if it occurs the same number of times in a long document (Inverse Document Length).

Weighting schemes that employ these principles thus include Local term frequency – the number of times the term occurs in the document, and variations on what is known as TF × IDF (Term Frequency × Inverse Document Frequency); the latter is a product of some function of the Local term frequency and the inverse of the number of documents within the collection that contain the term. Detailed weighting scheme strategies are discussed in the information retrieval literature (Manning and Raghavan, 2008; Salton, 1989; Sparck-Jones *et al.*, 2000; Witten *et al.*, 1999; Wilbur and Yang, 1996), where (Wilbur and Yang, 1996) is particularly relevant to the biomedical literature.

Once documents and queries are represented in vector space, vector-similarity is calculated to assess similarity between pairs of documents or between a query and each document in the collection. A commonly used vector-similarity measure is the cosine coefficient, which is the cosine of the angle between two vectors. For vectors $V_1 = \langle w_1^1, \ldots, w_N^1 \rangle$ and $V_2 = \langle w_1^2, \ldots, w_N^2 \rangle$, it is calculated as: $\left(\sum_{i=1}^{N} w_i^1 \cdot w_i^2 \right) / (\|V_1\| \cdot \|V_2\|)$, where $(\|V_1\| \cdot \|V_2\|)$ are the norms of the vectors V_1, V_2, respectively.

Several approaches, discussed next, take advantage of the vector model, using it to group together articles that share a common topic or theme. Some of these approaches use it directly, while others utilize a probabilistic interpretation of the weight-vector.

3.3 Text Categorization: Classification, Clusters, and Topic Models

One way to obtain documents that are relevant to a specific area of interest is to pre-categorize documents by such areas. For instance, as illustrated in **Fig. 4**, documents can be grouped according to the disease or other medical subjects of interest discussed in them. When a user looks for documents relevant to such a subject or area, all documents that are categorized as belonging to this area can be retrieved and returned.

A variety of machine learning methods are applied to address this task (Cohen and Singer, 1999; Dumais *et al.*, 1998; Joachims, 1998; Lewis, 1998; Lewis *et al.*, 2004; Sebastiani, 2002; Xu *et al.*, 2016; Yang and Liu, 1999). We distinguish between two types of categorization approaches: supervised categorization, known as classification, and unsupervised, known as clustering. In classification, the set of class labels is fixed and a set of pre-categorized documents, regarded as training examples, is available. A classifier is learned from the training examples, where the goal is to correctly assign class labels to documents outside the training set. Clustering, on the other hand, aims to identify subsets (clusters) of interrelated documents within the dataset, while separating unrelated documents into disjoint clusters – without having any predefined labels or pre-categorized training examples. Much of the categorization work done in the biomedical domain is concerned with supervised learning, i.e., classification. The methods that are often used in practice include decision trees (Mitchel, 1997; Quinlan, 1993), their extension to Random Forests (Breiman, 2001), naïve Bayes (Mitchel, 1997; Lewis, 1998; Sohn *et al.*, 2008), support vector machines (SVMs) (Joachims, 1998; Vapnik, 1995) and logistic regression (Schütze *et al.*, 1995; Vlachos and Craven, 2010).

Text categorization is based on the idea of similarity, where documents that are similar according to a certain similarity criterion are grouped together, while those that are dissimilar are placed in separate categories. In Section 3.2 we have discussed the vector representation for documents, along with the cosine similarity measure. This measure can be applied both in the

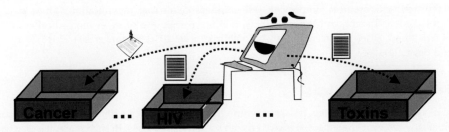

Fig. 4 Automatic text categorization. Documents are processed and assigned to a specific category based on their contents.

context of query-answering and for categorization. Another interpretation of the vector model views each weight w_i^j of a term t_i in a document d^j as reflecting the probability of the term to occur in the document. Under such probabilistic interpretation, the information retrieval task amounts to finding documents that with high probability satisfy the information need expressed by the query. Probabilistic models of retrieval have been introduced by Van Rijsbergen and others several decades ago (van Rijsbergen, 1977; Sparck-Jones et al., 2000), and were further developed and formalized as language models (Ponte and Croft, 1998) and, more recently, as topic models (Hofmann, 1999; Shatkay and Wilbur, 2000; Blei et al., 2003; Blei, 2012). Under a simple language model, we view a document as though it was produced by sampling terms from a language, where the language is modeled as a multinomial distribution over terms. Within the language-model framework, similarity between documents is expressed as their probability to have been generated by the same language model. A theme or a topic model (Shatkay and Wilbur, 2000; Blei et al., 2003) extends the basic model, by having documents viewed as sampled from a mixture of several multinomial distributions, where the mixture defines a combination of subjects that are discussed within the same context. Some of the early work on theme models was developed specifically for analyzing biomedical literature, aiming to support gene-expression interpretation and enrichment-analysis for certain genomic functions (Shatkay and Craven, 2012; Shatkay and Wilbur, 2000). A more in-depth discussion of probabilistic, language and topic models can be found in the literature cited above as well as in more comprehensive books (van Rijsbergen, 1979; Manning and Raghavan, 2008; Shatkay and Craven, 2012).

Within the biomedical domain, much work on text categorization is concerned with supporting biomedical data-curation for public databases. For example, organism-specific databases including the Mouse Genome Informatics databases maintained by the Jackson Lab (Dowell et al., 2009; Eppig et al., 2015), WormBase (Van Auken et al., 2012) and others, store information about gene expression or the effects of genetic mutations in the respective organisms. The information is obtained by curators who scan the relevant literature. Automatically identifying the relevant body of literature is a text-categorization task: a set of documents relevant to a specific database forms a class, while irrelevant documents form another class. Curators for databases that store information about proteins, their properties, and their interactions, such as BioGrid (Chatr-Aryamontri et al., 2015), IntAct (Kerrien et al., 2012) or UniProt (The UniProt Consortium, 2012) face a similar challenge. Such categorization of documents by relevance, is also known as triage, and has been the focus of work and shared tasks within the biomedical text mining community (e.g., Hersh et al., 2006; Hirschman et al., 2012).

4 Biomedical Applications, Shared Tasks, and Current/Future Challenges

We have already touched on several text-based methods applicable in biomedicine. This section provides a brief overview of text-related tasks that have been addressed and text-based methods that have been applied in biology and medicine – both early and recent.

4.1 Information Retrieval and Extraction Tasks

In terms of information retrieval, PubMed (2016) is the most comprehensive and widely used biomedical text-retrieval system. It supports Boolean queries, similarity queries, as well as refinement of the retrieval task utilizing pre-classification of the articles by species (e.g., Human, Other Animals), publication type (e.g., Review, Clinical Trial), and additional categories of common interest. Other large broadly-used biomedical data resources, such as the protein database UniProt (The UniProt Consortium, 2012), support text-based Boolean queries, which provide access to annotated entries based on the text description and GO terms associated with these entries.

Much other information-retrieval work involves text categorization. For example, identifying articles discussing mouse genomics is a text-categorization task supporting mouse-related curation. This task was one of the first text-related shared tasks, as further discussed below (Hersh et al., 2006; Hersh, 2008). Other examples of text categorization include identifying documents and sentences discussing protein-interaction or protein-characterization (Denroche et al., 2010; Donaldson et al., 2003; Kolchinsky et al., 2010; Krallinger et al., 2011; Xu et al., 2009), adverse drug reaction (Duda et al., 2005; Sarker and Gonzalez, 2015), and many others.

In terms of fine-grained extraction and NLP, a lot of effort has been dedicated to identifying bio-entities – genes, proteins, drugs, chemicals and diseases, and on finding statements that link between such entities. The very first applications of text mining in biology (Leek, 1997; Blaschke et al., 1999; Craven and Kumlien, 1999), specifically looked to identify protein names within sentences searching for statements of protein–protein interactions (Blaschke et al., 1999), and for locations of proteins within the cell and location of protein-coding genes on the chromosome (Craven and Kumlien, 1999; Leek, 1997). This pioneering work was followed by much research and progress, improving gene- and protein-name identification (Fluck et al., 2007; Leser and Hakenberg, 2005; Settles, 2005; Tanabe and Wilbur, 2002), and the recognition of other biomedical entities and relationships (Comeau et al., 2014; Friedman, 2009; Hunter et al., 2008; Kolchinsky et al., 2010; Krallinger et al., 2015; Kuhn et al., 2016; Leaman et al., 2013; Lowe and Sayle, 2015).

4.2 Community Challenges and Shared Tasks

The growing interest in systems that utilize text resources within biomedicine, along with the importance of assessing their utility toward meeting biologically-relevant challenges, has given rise to several shared tasks. In such shared tasks, a text-related challenge is proposed to the community by the challenge organizers, and teams come together to address the challenge. A gold standard consisting of expert-annotated document sets is made available to the community, and the systems that are being developed are trained and tested over the provided datasets. The latter are typically formed by a well-recognized organization that maintains and curates large biomedical data resources, while the annotators are typically biologists, database-curators or physicians who have much expertise and experience in the relevant area. Performance measures that enable quantitative evaluation of the systems are also decided on as part of the challenge definition. While standard performance measures – accuracy, precision, recall, and variations over them are often used (see e.g., (Shatkay and Craven, 2012), Chapter 5.1), certain tasks require developing specific performance evaluation metrics.

The first shared task on biological text mining was introduced as part of the KDD-cup 2002 (Yeh *et al.*, 2002), where the challenge involved identifying evidence for gene expression in the fruit fly. This event was followed by several series of long-term organized tasks. One series, TREC-Genomics (Hersh *et al.*, 2006; Hersh, 2008; Roberts *et al.*, 2009) was introduced as part of the well-established text retrieval conferences (TREC) sponsored by the U.S. National Institutes of Standards and Technology (NIST) (TREC, 2016). It focused primarily on ad hoc retrieval (that is, addressing a variety of user queries), particularly searching for documents discussing specific aspects of genes and proteins (e.g., gene function, role in disease, expression within certain tissue types). TREC-Genomics was followed by additional TREC medical/clinical tracks for addressing retrieval from electronic health records and support clinical decision-making (Simpson *et al.*, 2014; Voorhees and Hersh, 2012).

Another important series of shared tasks, BioCreative (BioCreative, 2016; Hirschman *et al.*, 2005), focuses primarily on biological text mining in support of database curation. Aiming to facilitate Critical Assessment of IE in Biology, BioCreative has been running regularly since 2004, presenting challenges ranging from identifying named-entities such as genes or chemicals (Krallinger *et al.*, 2015), through finding evidence for protein–protein interactions (Krallinger *et al.*, 2011) to creating effective interfaces for textual annotation by curators (Wang *et al.*, 2016; Hirschman *et al.*, 2012). As high-quality ground-truth text annotations are essential for both training text mining systems and for assessing their performance throughout the shared challenges, using crowd-sourcing for obtaining reliable annotations is an important recent direction that is being studied as part of biological text mining (Hirschman *et al.*, 2016). Additional recent area-specific shared tasks include the i2b2 (Informatics for Integrating Biology and the Bedside) NLP tasks (i2b2, 2016), BioNLP challenges (BioNLP Shared Task, 2016) primarily affiliated with the Association for Computational Linguistics, and BioASQ (BioASQ, 2016; Tsatsaronis *et al.*, 2015) targeting question-answering and extraction of certain types of information from text records or the literature. Additional information about such shared tasks can be found recently published survey on the topic (Huang and Lu, 2016).

4.3 Knowledge Discovery Through Text and Future Directions

The basic form of automated text mining revolves around finding within the text facts and assertions that have already been uncovered and published. However, a highly useful and less conventional way of leveraging published text is by identifying indirect links that expose hitherto unknown information, giving rise to new knowledge.

Swanson's pioneering work (Swanson, 1986; Swanson, 1990; Swanson *et al.*, 2001) introduced the idea of uncovering new links between concepts or entities, by following indirect connections among them. A canonical example illustrating the idea was his showing that Raynaud's syndrome can be treated by increasing the intake of fish oil (Swanson, 1986). To do so he pointed out two separate recurring statements in the literature: (1) fish oil can reduce blood viscosity and (2) Raynaud's syndrome is characterized by increased blood viscosity. While no previous publication stated these two facts together, Swanson established the connection, noting the overlapping concept blood viscosity occurring in both the context of fish oil and the context of Raynaud's syndrome. He thus proposed that fish oil may be used for treating Raynaud's syndrome. The idea is illustrated in **Fig. 5**. Following Swanson's uncovering the putative connection between fish oil and Raynaud's treatment, an independent clinical study

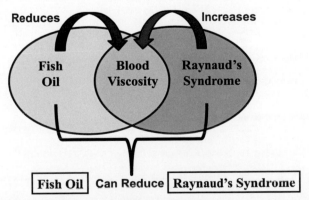

Fig. 5 Example of a relationship discovery by connecting two indirect link. Swanson established a hitherto unknown relationship between fish oil intake and reduction in Raynaud's Syndrome, by noting that the literature concerning Raynaud's syndrome (shown in pink on the right) reports an increase in blood viscosity, while the literature about fish oil (blue, left) reports reduction in blood viscosity. The gray area in the middle indicates the common concept, blood viscosity, which establishes the connection between the other two terms.

corroborated this link (DiGiacomo *et al.*, 1989). The framework has been implemented in Swanson's Arrowsmith discovery system (Swanson *et al.*, 2001), and further extended by others in later years toward literature-based knowledge discovery (Lee *et al.*, 2007; Srinivasan, 2004; Srinivasan and Libbus, 2004).

Another way of uncovering new knowledge is by treating text as raw data, while searching for common motifs and patterns within it (Shatkay *et al.*, 2015; Shatkay *et al.*, 2000). This use of text is similar to the way in which homology and recurring motifs are scanned for on genomic and proteomic sequences, aiming to reveal relationships among such biological entities. Like genomic sequences, the text is viewed as a sequence of symbols, words, terms, phrases, or grammatical objects, and the statistical properties of their distribution across documents are used to expose similarities and connections among different text passages. Unlike most biological data, text components have readily understandable semantics, making results of such statistical pattern analysis easy to interpret and to validate. Early work establishing this direction included the utilization of statistical theme models in the gene-literature in order to identify functional relations among genes (Shatkay *et al.*, 2000). Similar ideas were employed for identifying sets of related genes or proteins by grouping together text passages discussing them (Chagoyen *et al.*, 2006; Renner and Aszódi, 2000).

A distinct but similar line of work directly employs text-based features associated with biological entities in order to predict biological properties such as protein homology, function, or subcellular location (Shatkay *et al.*, 2015; Chang *et al.*, 2001; Nair and Rost, 2002; Brady and Shatkay, 2008). Moreover, the integration of both text and sequence data into a unified prediction framework has shown very effective in improving prediction accuracy (Briesemeister *et al.*, 2009; Shatkay *et al.*, 2007). Thus, a direction that shows much promise in utilizing text is its integration as a source of data with other forms of biological data obtained from expression experiments, sequencing, or mutation analysis.

Another source of data that is currently being studied in conjunction with text is image information, where figures accompanying the text within medical records and publications are being examined as a source of additional knowledge (Ahmed *et al.*, 2010; Bockhorst *et al.*, 2012; Cohen *et al.*, 2003; Demner-Fushman *et al.*, 2012; Kalpathy-Cramer *et al.*, 2014; Ma *et al.*, 2015; Shatkay *et al.*, 2006; Xu *et al.*, 2008). This new direction that takes advantage of figures in addition to text holds much promise, both for supporting reliable curation of existing knowledge and for identifying significant pieces of information within the published literature.

See also: Bayes' Theorem and Naive Bayes Classifier. Data Mining: Mining Frequent Patterns, Associations Rules, and Correlations. Data-Information-Concept Continuum From a Text Mining Perspective. Knowledge Discovery in Databases. Natural Language Processing Approaches in Bioinformatics. Text Mining Applications. Text Mining Basics in Bioinformatics. Text Mining for Bioinformatics Using Biomedical Literature

---

References

Afantenos, S., Denis, P., Muller, P., Danlos, L., 2010. Learning recursive segments for discourse parsing. In: Proceedings of 7th Language Resources and Evaluation Conference (LREC'10), pp. 3578–3584.

Ahmed, A., Arnold, A., Coelho, L.P., *et al.*, 2010. Structured literature image finder: Parsing text and figures in biomedical literature. Web Semantics: Science, Services and Agents on the World Wide Web 8 (2), 151–154.

Batista-Navarro, R., Rak, R., Ananiadou, S., 2015. Optimizing chemical named entity recognition with pre-processing analytics, knowledge-rich features and heuristics. Journal of Cheminformatics 7 (Suppl 1), S6.

BioASQ, 2016. Available at: http://bioasq.org/

BioCreative, 2016. BioCreative: Critical assessment of information extraction in biology. Available at: http://www.biocreative.org/

BioNLP Shared Task, 2016. Available at: http://www.bionlp-st.org

Blaschke, C., Andrade, M., Ouzounis, O., Valencia, A., 1999. Automatic extraction of biological information from scientific text: Protein–protein interactions. In: Proceedings of the 7th International Conference on Intelligent Systems for Molecular Biology (ISMB'99), aaAI Press, pp. 60–67.

Blei, D.M., Ng, A.Y., Jordan, M.I., Lafferty, J., 2003. Latent Dirichlet allocation. Journal of Machine Learning Research 3, 993–1022.

Blei, D.M., 2012. Probabilistic topic models. Communications of the ACM 55 (4), 77–84.

Bockhorst, J.P., Conroy, J.M., Agrawal, S., O'Leary, D.P., Yu, H., 2012. Beyond captions: Linking figures with abstract sentences in biomedical articles. PLOS ONE 7 (7), 1–15.

Brady, S., Shatkay, H., 2008. EpiLoc: A (working) text-based system for predicting protein subcellular location. In: Proceedings of the Pacific Symposium on Biocomputing, pp. 604–615.

Breiman, L., 2001. Random Forests. Machine Learning 45 (1), 5–32.

Briesemeister, S., Blum, T., Brady, S., *et al.*, 2009. Sherloc2: A high-accuracy hybrid method for predicting subcellular localization of proteins. Journal of Proteome Research 8 (11), 5363–5366.

Cardie, C., 1997. Empirical methods in information extraction. AI Magazine 18 (4), 65–80.

Chagoyen, M., Carmona-Saez, P., Shatkay, H., Caraz, J.M., Pascual-Montano, A., 2006. Discovering semantic features in the literature: A foundation for building functional associations. BMC Bioinformatics 7, 41.

Chang, J.T., Raychaudhuri, S., Altman, R.B., 2001. Including biological literature improves homology search. In: Proceedings of the Pacific Symposium on Biocomputing, pp. 374–383.

Chatr-Aryamontri, A., Breitkreutz, B.J., Oughtred, R., *et al.*, 2015. The BioGRID interaction database: 2015 update. Nucleic Acids Research 43 (Database Issue), D470–D478. Available at: http://thebiogrid.org/

Cohen, W., Kou, Z., Murphy, R.F. 2003. Extracting information from text and images for location proteomics. In: Proceedings of the 3rd ACM SIGKDD Workshop on Data Mining in Bioinformatics (BIOKDD'03), pp. 2–9.

Cohen, W.W., Singer, Y., 1999. Context-sensitive learning methods for text categorization. ACM Transactions on Information Systems 17 (2), 141–173.

Comeau, D.C., Liu, H., Doğan, R.I., Wilbur, W.J., 2014. Natural Language processing pipelines to annotate BioC Collections with an Application to the NCBI Disease Corpus. Database. http://dx.doi.org/10.1093/database/bau056

Conrath, J., Afantenos, S., Asher, N., Muller, P., 2014. Unsupervised extraction of semantic relations using discourse cues. In: Proceedings of the International Conference on Computational Linguistics (COLING'14), pp. 2184–2194.

Cowie, J., Lehnert, W., 1996. Information extraction. Communications of the ACM 39 (1), 80–91.

Craven, M., Kumlien, J., 1999. Constructing biological knowledge bases by extracting information from text sources. In: Proceedings of the 7th International Conference on Intelligent Systems for Molecular Biology (ISMB'99), AAAI Press, pp. 77–86.

Dascalu, M., 2014. Computational discourse analysis. Analyzing Discourse and Text Complexity for Learning and Collaborating: Series on Studies in Computational Intelligence (534). Switzerland: Springer, pp. 53–77. (Chapter 4).

Demner-Fushman, D., Antani, S., Thoma, G.R., 2012. Design and development of a multimodal biomedical information retrieval system. Journal of Computing Science and Engineering 6 (2), 168–177.

Denroche, R., Madupu, R., Yooseph, S., Sutton, G., Shatkay, H., 2010. Toward computer-assisted text curation: Classification is easy (choosing training data can be hard...). In: Blaschke, C., Shatkay, H. (Eds.), Linking Literature, Information, and Knowledge for Biology. Berlin, Heidelberg: Springer, pp. 33–42.

DiGiacomo, R., Kremer, J., Shah, D., 1989. Fish-oil dietary supplementation in patients with Raynaud's phenomenon: A double-blind, controlled, prospective study. The American Journal of Medicine 86 (2), 158–164.

Donaldson, I., Martin, J., Wolting, C., et al., 2003. PreBIND and Textomy – mining the biomedical literature for protein-protein interactions using a support vector machine. BMC Bioinformatics 4, 11.

Dowell, K., McAndrews-Hill, M., Hill, D., Drabkin, H., Blake, J., 2009. Integrating text mining into the MGI biocuration workflow. Database. http://dx.doi.org/10.1093/database/bap019

Duda, S., Aliferis, C., Miller, R., Statnikov, A., Johnson, K., 2005. Extracting drug–drug interaction articles from MEDLINE to improve the content of drug databases. In: Proceedings of the AMIA Annual Symposium, p. 216.

Dumais, S.T., Platt, J., Heckerman, D., Sahami M., 1998. Inductive learning algorithms and representations for text categorization. In: Proceedings of the ACM International Conference on Information and Knowledge Management (CIKM), pp. 148–155.

Eppig, J.T., Blake, J.A., Bult, C.J., Kadin, J.A., Richardson, J.E., and The Mouse Genome Database Group, 2015. The mouse genome database (MGD): Facilitating mouse as a model for human biology and disease. Nucleic Acids Research 43 (Database Issue), D726–D736.

Ferraro, J.P., Daumé, H., DuVall, S.L., et al., 2013. Improving performance of natural language processing part-of-speech tagging on clinical narratives through domain adaptation. Journal of the American Medical Informatics Association 20 (5), 931–939.

Fluck, J., Mevissen, H.T., Dach, H., Oster, M., Hofmann-Apitius, M., 2007. ProMiner: Recognition of human gene and protein names using regularly updated dictionaries. In: Proceedings of Second BioCreative Challenge Evaluation Workshop, pp. 149–151.

Friedman, C. 2009. Discovering novel adverse drug events using natural language processing and mining of the electronic health record. In: Proceedings of the 12th Conference on Artificial Intelligence in Medicine (AIME), pp. 1–5.

Gene Ontology, 2016. Gene ontology consortium. Available at: www.geneontology.org

Grahn, T.H.M., Kaur, R., Yin, J., et al., 2014. Fat-specific protein 27 (FSP27) interacts with Adipose Triglyceride Lipase (ATGL) to regulate lipolysis and insulin sensitivity in human adipocytes. Journal of Biological Chemistry 289 (17), 12029–12039.

Hersh, W., 2008. Information Retrieval: A Health and Biomedical Perspective. New York, NY: Springer-Verlag, (Chapter 9.2).

Hersh, W.R., Cohen, A., Yang, J., et al., 2006. TREC 2005 genomics track overview. In: Proceedings of the 14th Text Retrieval Conference – TREC'05, NIST Special Publication, pp. 14–25.

Hirschman, L., Burns, G.A., Krallinger, M., et al., 2012. Text mining for biocuration workflow. Database. http://dx.doi.org/10.1093/database/bas020

Hirschman, L., Fort, K., Boue, S., et al., 2016. Crowdsourcing and curation: Perspectives from biology and natural language processing. Database. http://dx.doi.org/10.1093/database/baw115

Hirschman, L., Yeh, A., Blaschke, C., Valencia, A., 2005. Overview of BioCreAtIvE: Critical assessment of information extraction for biology. BMC Bioinformatics 6 (Suppl 1), S1.

Hofmann, T., 1999. Probabilistic latent semantic indexing. In: Proceedings of the 22nd International Conference on Research and Development in Information Retrieval, (SIGIR'99), pp. 50–57.

Howe, K.L., Bolt, B.J., Cain, S., et al., 2016. WormBase 2016: Expanding to enable Helminth genomic research. Nucleic Acids Research 44 (Database Issue), D774–D780.

Huang, C.C., Lu, Z., 2016. Community challenges in biomedical text mining over 10 years: Success, failure and the future. Briefings in Bioinformatics 17 (1), 132–144.

Hunter, L., Lu, Z., Firby, J., et al., 2008. OpenDMAP: An open source, ontology-driven concept analysis engine with applications to capturing knowledge regarding protein transport, protein interactions and cell-type specific gene expression. BMC Bioinformatics 9, 78.

i2b2, 2016. Informatics for integrating biology & the bedside. Available at: https://www.i2b2.org/NLP/

Joachims, T., 1998. Text categorization with support vector machines: Learning with many relevant features. In: Proceedings of the Tenth European Conference on Machine Learning, pp. 137–142.

Jurafsky, D., Martin, J., 2009. Speech and Language Processing: An Introduction to Natural Language Processing, Speech Recognition, and Computational Linguistics, second ed. Englewood Cliffs, NJ: Prentice-Hall.

Kalpathy-Cramer, J., de Herrera, A.G.S., Demner-Fushman, D., et al., 2014. Evaluating performance of biomedical image retrieval systems – An overview of the medical image retrieval task at imageCLEF 2004–2014. Comp. Medical Imaging and Graphics 39, 55–61.

Kang, N., van Mulligan, E.M., Kors, J.A., 2011. Comparing and combining chunkers of biomedical text. Journal of Biomedical Informatics 44 (2), 354–360.

Kerrien, S., Aranda, B., Breuza, L., et al., 2012. The IntAct molecular interaction database in 2010. Nucleic Acids Research 40 (D1), D841–D846. Available at: http://www.ebi.ac.uk/intact/

Kolchinsky, A., Abi-Haidar, A., Kaur, J., Hamed, A.A., Rocha, L.M., 2010. Classification of protein–protein interaction full-text documents using text and citation network features. IEEE/ACM Transactions on Computational Biology and Bioinformatics 7 (3), 400–411.

Krallinger, M., Leitner, F., Rabal, O., et al., 2015. CHEMDNER: The drugs and chemical names extraction challenge. Journal of ChemInformatics 7 (1), 1.

Krallinger, M., Vazquez, M., Leitner, F., et al., 2011. The protein–protein interaction tasks of BioCreative III: Classification/ranking of articles and linking bio-ontology concepts to full text. BMC Bioinformatics 12 (Suppl 8), S3.

Kuhn, M., Letunic, I., Jensen, L.J., Bork, P., 2016. The SIDER database of drugs and side effects. Nucleic Acids Research 44 (D1), D1075–D1079.

Leaman, R., Doğan, R.I., Lu, Z., 2013. DNorm: Disease name normalization with pairwise learning to rank. Bioinformatics 29 (22), 2909–2917.

Leek, T., 1997. Information extraction using hidden Markov models. Master's Thesis, Department of Computer Science and Engineering, University of California.

Lee, W.J., Raschid, L., Srinivasan, P., et al., 2007. Using annotations from controlled vocabularies to find meaningful associations. In: Proceedings of the Workshop on Data Integration in the Life Sciences, Lecture Notes in Computer Science, Springer, pp. 247–263.

Leser, U., Hakenberg, J., 2005. What makes a gene name? Named entity recognition in the biomedical literature. Briefings in Bioinformatics 6 (4), 357–369.

Lewis, D.D., Yang, Y., Rose, T., Li, F., 2004. RCV1: A new benchmark collection for text categorization research. Journal of Machine Learning Research 5, 361–397.

Lewis, D.D., 1998. Naïve (Bayes) at forty: The independence assumption in information retrieval. In: Proceedings of the 10th European Conference on Machine Learning (ECML'98), pp. 4–15.

Lowe, D.M., Sayle, R.A., 2015. LeadMine: A grammar and dictionary driven approach to entity recognition. Journal of ChemInformatics 7 (1), S5.

Manning, C., Raghavan, P., 2008. Introduction to Information Retrieval. New York, NY: Cambridge University Press.

Manning, C., Schütze, H., 1999. Foundations of Statistical Natural Language Processing. Cambridge, MA: MIT Press.

Marcus, M.P., Santorini, B., Marcinkiewicz, M.A., 1993. Building a large annotated corpus of english: The Penn Treebank. Computational Linguistics 19 (2), 313–330.

Ma K., Jeong H., Rohith M.V., et al. 2015. Utilizing image-based features in biomedical document classification. In: Proceedings of the International Conference on Image Processing (ICIP'15), pp. 4451–4455.

McClosky, D., Charniak, E., Johnson M., 2010. Automatic domain adaptation for parsing. In: Proceedings of Human Language Technologies: The Annual Conference of the North American Chapter of the Association for Computational Linguistics (ACL'10), pp. 28–36.

MeSH, 2016. Medical Subject Headings. Available at: https://www.nlm.nih.gov/mesh/

Mitchel, T., 1997. Machine Learning. New York, NY: Mcgraw-Hill.

Nair, R., Rost, B., 2002. Inferring sub-cellular localization through automated lexical analysis. Bioinformatics 18 (Suppl. 1), S78–S86.

Ponte, J.M., Croft, W.B., 1998. A language modeling approach to information retrieval. In: Proceedings of the 21st International Conference on Research and Development in Information Retrieval (SIGIR'98), pp. 275–281.

PubMed, 2016. Available at: https://www.ncbi.nlm.nih.gov/pubmed/ (accessed Oct 2016).

Quinlan, J., 1993. C4.5: Programs for Machine Learning. San Francisco, CA: Morgan Kaufmann.

Renner, A., Aszodi, A., 2000. High-throughput functional annotation of novel gene products using document clustering. In :Proceedings of the Pacific Symposium on Biocomputing, pp. 54–65.

Roberts, P., Cohen, A.M., Hersh, W.R., 2009. Tasks, topics and relevance judging for the TREC genomics track: Five years of experience evaluating biomedical text information retrieval systems. Information Retrieval 12 (1), 81–97.

Salton, G., 1989. Automatic Text Processing. Boston, MA: Addison-Wesley.

Sarker, A., Gonzalez, G., 2015. Portable automatic text classification for adverse drug reaction detection via multi-corpus training. Journal of Biomedical Informatics 53, 196–207.

Savova, G.K., Masanz, J.J., Ogren, P.V., et al., 2010. Mayo clinical text analysis and knowledge extraction system (cTAKES): Architecture, component evaluation and applications. Journal of the American Medical Informatics Association 17 (5), 507–513.

Schütze, H., Hull, D.A., Pedersen, J.O., 1995. A comparison of classifiers and document representations for the routing problem. In: Proceedings of the 18th International Conference on Research and Development in Information Retrieval (SIGIR'95), ACM, pp. 229–237.

Sebastiani, F., 2002. Machine learning in automated text categorization. ACM Computing Surveys 34 (1), 1–47.

Settles, B., 2005. ABNER: An open source tool for automatically tagging genes, proteins, and other entity names in text. Bioinformatics 21 (14), 3191–3192.

Shatkay, H, Brady, S., Wong, A., 2015. Text as data: Using text-based features for proteins representation and for computational prediction of their characteristics. Methods 74, 54–64. Special Issue on Text Mining of Biomedical Literature.

Shatkay, H., Chen, N., Blostein, D., 2006. Integrating image data into biomedical text categorization. Bioinformatics 22 (14), e446–e453.

Shatkay, H., Craven, M., 2012. Mining the Biomedical Literature. Cambridge, MA: MIT Press.

Shatkay, H., Edwards, S., Wilbur, W.J., Boguski, M., 2000. Genes, themes and microarrays: Using information retrieval for large scale gene analysis. In: Proceedings of the 8th International Conference on Intelligent Systems for Molecular Biology, AAAI Press, pp. 317–328.

Shatkay, H., Höglund, A., Brady, S., et al., 2007. Sherloc: High-accuracy prediction of protein subcellular localization by integrating text and protein sequence data. Bioinformatics 23 (11), 1410–1417.

Shatkay, H., Wilbur, W.J., 2000. Finding themes in MEDLINE documents: Probabilistic similarity search. In: Proceedings of the IEEE Conference on Advances in Digital Libraries, pp. 183–192.

Simpson, M.S., Voorhees, E., Hersh, W., 2014. Overview of the TREC 2014 clinical decision support track. In: Proceedings of the 23rd Text Retrieval Conference – TREC'14, NIST Special Publication.

Smith, L., Rindflesch, T., Wilbur, W.J., 2004. MedPost: A part-of-speech tagger for bioMedical text. Bioinformatics 20 (14), 2320–2321.

Sohn, S., Kim, W., Comeau, D.C., Wilbur, W.J., 2008. Optimal training sets for Bayesian prediction of MeSH assignment. Journal of the American Medical Informatics Association 15 (4), 546–553.

Sparck-Jones, K., Walker, S., Robertson, S., 2000. A probabilistic model of information retrieval: Development and status. Information Processing and Management 36 (6), 779–840.

Srinivasan, P., 2004. Text mining: Generating hypotheses from MEDLINE. Journal of the American Society for Information Science (JASIS) 55 (5), 396–413.

Srinivasan, P., Libbus, B., 2004. Mining MEDLINE for implicit links between dietary substances and diseases. Bioinformatics 20 (Suppl. 1), i290–i296.

Swanson, D.R., 1986. Fish-oil, Raynaud's syndrome and undiscovered public knowledge. Perspectives in Biology and Medicine 30 (1), 7–18.

Swanson, D.R., 1990. Somatomedin C and arginine: Implicit connections between mutually isolated literatures. Perspectives in Biology and Medicine 33 (2), 157–186.

Swanson, D.R., Smalheiser, N.R., Bookstein, A., 2001. Information discovery from complementary literatures: Categorizing viruses as potential weapons. Journal of the American Society for Information Science and Technology 52 (10), 797–812.

Tanabe, L., Wilbur, W.J., 2002. Tagging gene and protein names in full text articles. In: Proceedings of the ACL-02 Workshop on Natural Language Processing in the Biomedical Domain, vol. 3, pp. 9–13.

Tateisi, Y., Yakushiji, A., Ohta, T., Tsujii, J., 2005. Syntax annotation for the GENIAcorpus. In: Dale, R., Wong, K.F., Su, J., Kwong, O.Y. (Eds.), Natural Language Processing – IJCNLP 2005. Springer, pp. 222–227.

TREC. 2016. Text retrieval conference. Available at: http://trec.nist.gov

The Gene Ontology Consortium, 2000. Gene ontology: Tool for the unification of biology. Nature Genetics 25, 25–29.

The UniProt Consortium, 2012. Reorganizing the protein space at the Universal Protein Resource (UniProt). Nucleic Acids Research 40, D71–D75.

Thompson, P., Ananiadou, S., Tsujii, J., 2017. The GENIA corpus: Annotation levels and applications. In: Ide N., Pustejovsky J.(Eds.), Handbook of Linguistic Annotation, Springer, pp. 1421–1432.

Tsatsaronis, G., Balikas, G., Malakasiotis, P., et al., 2015. An overview of the BIOASQ large-scale biomedical semantic indexing and question answering competition. BMC Bioinformatics 16, 138.

Van Auken, K., Fey, P., Berardini, T.Z., et al., 2012. Text mining in the biocuration workflow: Applications for literature curation at wormbase, dicty base and TAIR. Database. http://dx.doi.org/10.1093/database/bas040

van Rijsbergen, C.J., 1977. A theoretical basis for the use of co-occurrence data in information retrieval. Journal of Documentation 33 (2), 106–119.

van Rijsbergen, C.J., 1979. Information Retrieval. London: Butterworth.

Vapnik, V., 1995. The Nature of Statistical Learning Theory. Berlin, Heidelberg: Springer-Verlag.

Vlachos, A., Craven, M., 2010. Detecting speculative language using syntactic dependencies and logistic regression. In: Proceedings of the Conference on Computational Natural Language Learning, pp. 18–25.

Voorhees, E., Hersh, W., 2012. Overview of the TREC 2012 medical records track. In: Proceedings of the 21st Text Retrieval Conference – TREC'12, NIST Special Publication.

Wang, Q., Abdul, S.S., Almeida, L., et al., 2016. Overview of the interactive task in BioCreative V. Database. http://dx.doi.org/10.1093/database/baw119

Wilbur, W.J., Yang, Y., 1996. An analysis of statistical term strength and its use in the indexing and retrieval of molecular biology text. Computers in Biology and Medicine 26 (3), 209–222.

Witten, I.H, Moffat, A., Bell, T.C., 1999. Managing Gigabytes: Compressing and Indexing Documents and Images, second ed. San Francisco, CA: Morgan-Kaufmann.

Xu, G., Niu, Z., Uetz P., et al., 2009. Semi-supervised learning of text classification on bacterial protein–protein interaction documents. In: Proceedings of the International Joint Conference on Bioinformatics, Systems Biology and Intelligent Computing (IJCBS'09), pp. 263–270.

Xu, R., Yang, Y., Liu, H., Hsi, A., 2016. Cross-lingual text classification via model translation with limited dictionaries. In: Proceedings of the 25th ACM International Conference on Information and Knowledge Management (CIKM'16), pp. 95–104.

Xu, S., McCusker, J., Krauthammer, M., 2008. Yale Image Finder (YIF): A new search engine for retrieving biomedical images. Bioinformatics 24 (17), 1968–1970.

Yang, Y., Liu, X., 1999. A re-examination of text categorization methods. In: Proceedings of the 22nd International Conference on Research and Development in Information Retrieval (SIGIR'99), pp. 42–49.

Yeh, A., Hirschman, L., Morgan, A., 2002. Background and overview for KDD Cup 2002 Task 1: Information extraction from biomedical articles. SIGKDD Explorations 4 (2), 87–89.

Yin, Y., Wang, Z.Y., Mora-Garcia, S., et al., 2002. BES1 accumulates in the nucleus in response to brassinosteroids to regulate gene expression and promote stem elongation. Cell 109 (2), 181–191.

Further Reading

Cohen, K.B., Demner-Fushman, D., 2014. Biomedical Natural Language Processing. Amsterdam: John Benjamin Publishing Company.

Hersh, W., 2009. Information Retrieval: A Health and Biomedical Perspective. New York, NY: Springer.

Manning, C., Raghavan, P., 2008. Introduction to Information Retrieval. New York, NY: Cambridge University Press.

Manning, C., Schütze, H., 1999. Foundations of Statistical Natural Language Processing. Cambridge, MA: MIT Press.

Mitchel, T., 1997. Machine Learning. Mcgraw-Hill.

Przybyła, P., Shardlow, M., Aubin, S., et al., 2016. Text mining resources for the life sciences. Database. http://dx.doi.org/10.1093/database/baw145

Shatkay, H., Craven, M., 2012. Mining the Biomedical Literature. Cambridge, MA: MIT Press.

Relevant Websites

http://bioasq.org/
 BioASQ.
http://www.biocreative.org/
 Biocreative.
http://bionlp.org/
 BioNLP.
http://www.geneontology.org/
 Gene Ontology Consortium.
https://www.ncbi.nlm.nih.gov/pubmed/
 NCBI.

Biographical Sketch

Hagit Shatkay is an Associate Professor and Director of the Computational Biomedicine and Machine Learning Lab at the Department of Computer and Information Sciences, University of Delaware. She has a PhD in Computer Science from Brown University, and an MSc and BSc in Computer Science from the Hebrew University of Jerusalem. Her research is focused on the development and use of machine learning and data-mining methods for addressing data-intensive problems in biology and medicine. She is an active member of the biomedical informatics and bio-text research communities, has presented many invited talks and international tutorials, and has authored, with Mark Craven, the book *Mining the Biomedical Literature* (MIT Press, 2012). Among many organizational and editorial roles, she is Section Editor for Knowledge Based Analysis at BMC Bioinformatics, associate editor in several journals, a board member of the International Society for Computational Biology (ISCB), has been an Area Chair of the Text Mining area at the International Conference on Intelligent Systems for Molecular Biology (ISMB) since 2008, co-organizer of the BioLINK Special Interest Group on linking biology and literature at ISMB, on the advisory board of BioASQ since its inception (2012), the steering committee of for TREC-Genomics 2005–2008, and others.

Biodiversity Databases and Tools

Lina AM Zalani and Kho S Jye, University of Malaya, Kuala Lumpur, Malaysia
Shaarmini Balakrishnan, Manipal International University, Nilai, Negeri Sembilan, Malaysia
Sarinder K Dhillon, University of Malaya, Kuala Lumpur, Malaysia

Background

Biological database is a large, organized library of life sciences information, collected from scientific experiments, published literature, high-throughput experiment technology, and computational analyses. It is associated with the usage of computerized software to store, update, query, and retrieve information within the system (Kumar, 2005). Database provides a convenient method to store enormous amount of data and allows biologists or researchers to collect the data, organize them systematically, analyze the data, and share the information with other researchers efficiently.

Biological data is often comprised of complicated data. There are different types of biological data, such as DNA sequence, protein sequence, protein structure, biodiversity, and many other specialized fields. Development of a database should acknowledge the complexity of the database content the capability of the technology in managing the complex data. In this article, we focus on biodiversity databases and tools, focusing on important projects done in this field. We also discuss the importance of biodiversity databases and the future direction anticipated inline with the advancement of technology and ICT.

Biodiversity Databases

Recent developments in information and communication technology have given birth to new experiences in the integration, analysis and visualization of biodiversity information, leading to the development of biodiversity databases (Canhos et al., 2004). It includes application of information technology to the management, algorithmic exploration, analysis and interpretation of primary data regarding life, particularly at the species level of organization (Soberon and Peterson, 2004).

Biodiversity databases deals with managing information of unnamed taxa that are produced by environmental samples or sequencing of samples, and also cover the computational problems specific to the names of biological entities. Algorithms are developed to cope with those problems. For example, Ch'ng (2009) developed a segmentation algorithm for entity interaction, which increases the efficiency of real-time simulation that models the biotic interactions of large population datasets. Besides, biodiversity databases involve development of the syntax and semantics for publishing and integrating biodiversity information, while Darwin Core archive is the data standard developed to improve interoperability (Wieczorek et al., 2012).

Existing Biodiversity Databases

Antweb

AntWeb (2018) focuses on ants, and it features as an image database. Being the leading online database of specimen records, images, and natural history information on ants, it aims to publish to the scientific community the high-quality images of global ant species. AntWeb focuses on data at specimen level and the corresponding images of the specimens. besides distribution maps. It accepts contributions on natural history information and field images that are linked directly to taxonomic names. Its content are also provided to the Global Biodiversity Information Facility (gbif.org), the Encyclopedia of Life (EOL.org) and Wikipedia (see "Relevant Websites section").

Fishbase

Fishbase (see "Relevant Websites section") (FishBase.org, 2017) is a global text and image database on finfishes. It contains a wide range of data on all species currently known worldwide. This includes information on the taxonomy, biology, trophic ecology, life history, uses, and historical data reaching back to 250 years. The detailed information in the database, which includes online analytical and graphical tools, cater to the needs of stakeholders, consists of researchers, scientists, policy makers, fisheries managers, donor, conservationists, teachers, and students. FishBase aims to provide for sustainable fisheries management, biodiversity conservation, and environmental protection.

Georgian Biodiversity Database

The Georgian Biodiversity Database (see "Relevant Websites section") (Georgian Biodiversity Database, 2015) focuses on the biological diversity of Georgia, and to a certain extent, the Caucasus ecoregion. This database can be categorized as a text and image database, but mostly as an image database. This resource aims to introduce its biological diversity to the worldwide scientific community and also to any users alike. The Georgian Biodiversity database contains species list from terrestrial and freshwater ecosystems of Georgia, categorized according to the taxonomic hierarchical system. It also contains brief information on the individual species and higher-order taxa, including verbal descriptions of key ecological features and range, systematic remarks, bibliographies, images, and distribution maps.

Species + Database

Species + (see "Relevant Websites section") (Species +, 2017) is a text database developed by UN Environment World Conservation Monitoring Centre (See Relevant Websites section) and the Convention on International Trade in Endangered Species of Wild Fauna and Flora (CITES) Secretariat (cites.org). It is designed as a centralized data warehouse for accessing key information on species such as Legal, Names, Distribution, References, Documents, and many others.

Freshwater Ecoregions of the World, FEOW

The main focus of this database is Freshwater biodiversity and it is known as a text and image database. FEOW (see "Relevant Websites section") (FEOW.org, 2015) provides a new global biogeographic regionalization of this planet's freshwater biodiversity. It covers virtually all freshwater habitats on Earth, and is the first-ever ecoregion map. Together with associated species data, it is a useful tool for underpinning global and regional conservation planning efforts, particularly to identify outstanding and imperiled freshwater systems. It also serves as a logical framework for a large-scale conservation strategies, and a global-scale knowledge base for increasing freshwater biogeographic literacy.

Bacterial Diversity Metadatabase, Bacdive

The Bacterial Diversity Metadatabase is a Metadatabase that provides strain-linked information about bacterial and archaeal biodiversity. It features a text database. BacDive (see "Relevant Websites section") (BacDive, 2017) represents a collection of organism-linked information covering the multifarious aspects of bacterial and archaeal biodiversity. Its content covers on taxonomy, morphology, physiology, sampling and environmental conditions, and molecular biology too. BacDive allows simple search at the portal entrance for a quick access to strain related information. As for TAXplorer, it can be used to select a strain by browsing through taxonomy.

Scalenet

ScaleNet (see "Relevant Websites section") (ScaleNet, 2016) is a text database on scale insects (superfamily Coccoidea). It contains information about scale insects, their taxonomic diversity, summary of nomenclatural history, geographic distribution, ecological associations, and economic importance and a complete bibliography. To date it contains data from 23,708 references pertaining to 8187 valid species names.

Naturdata

Naturdata (see "Relevant Websites section") (Naturdata.com, 2017) is a text and image database on Portuguese taxa. It contains checklist of Portuguese species with a page for each species pertaining to the taxonomy, synonyms, vernacular names, images, videos and distribution of the species. Until 2009, it was not possible to know the count and the type of species existed in Portugal. Some groups, such as the chordates, insects, arachnids were well inventoried, but in all, these groups represented a minimal percentage of Portugal's biodiversity (with current information, about 5%). Beyond 2009, Naturdata was created as a data warehouse to collect, process and share information on Portugal's biodiversity.

Pan-European Species-Directories Infrastructure, PESI

The Pan-European Species-directories Infrastructure, PESI (see "Relevant Websites section") (PESI, 2017) features European taxa and functions as a text and image database. PESI is an authoritative taxonomic checklist of European species, including higher taxonomy, synonyms, vernacular names, and European distribution. It is Europe's e-infrastructure for taxonomic information on species occurring in Europe. The data warehouse contains nearly 450,000 scientific names, 240,000 valid (sub) species names, and 140,000 vernacular names in 89 languages. Besides taxonomic information, PESI gathers relevant information on species, such as images, literature, distributions, and conservation status. PESI also offers its services to those who are interested to build their proprietary species applications.

Wikispecies

WikiSpecies (see "Relevant Websites section") (WikiSpecies, 2017) focuses on all forms life. This database is both a text and image database, with free species directory on Animalia, Plantae, Fungi, Bacteria, Archaea, Protista, and all other forms of life. Currently, there are about 549,468 articles in the database. Interestingly, it is able to do taxon navigation, and includes higher taxonomy, synonyms, vernacular names, and references.

All Catfish Species Inventory, ACSI

All Catfish Species Inventory (see "Relevant Websites section") (ACSI, 2017) is a text and image database on catfish, categorized by the genera. This inventory that also estimates the total species count, aids in the discovery, description and dissemination of knowledge of all catfish species by the world consortium of taxonomists and systematists. The aim of ACSI is to discover around 1750 new species of catfish and describe between 2300 and 4600 new species of freshwater fishes, which will then result to a complete taxonomy of Siluriformes and later in the completed taxonomy of Otophysi, the clade containing over two-thirds of all freshwater fishes. Besides a completed taxonomy of catfishes with up-to-date identification guides, ACSI will include atlases, catalogues and checklists of species, phylogenetic studies of higher-level relationships among catfishes, and data analytics. It is expected to produce large samples of freshwater fishes from poorly collected regions to be added to permanent collections in US. and foreign institutions, and enhanced international communication among fish taxonomists. The ACSI is expected to serve as a channel of communication among global taxonomists on topics pertaining to research, educational and outreach opportunities.

Arctos

Arctos (see "Relevant Websites section") (Arctos, 2017) is a text and image database that features specimen holdings of several natural history museums, agencies, and accessible private collections. It consists of vertebrates, invertebrates, parasites, vascular and non-vascular plants, many with images and extensive usage data. It is both a community and a collection management information system. It provides fundamental research infrastructure for biodiversity data, and is intended for curators, collection managers, investigators, educators, and anyone interested in natural and cultural history. Besides serving individuals, Arctos is a consortium of museums that collaborate to serve data on over 3 million records from natural and cultural history collections. It is a collaborative effort to share data vocabulary and standards while curating and improving the quality of shared data such as agents and taxonomy. Arctos on its own is an information system that integrates data from a myriad of disciplines which are anthropology, botany, entomology, ethnology, herpetology, ichthyology, mammalogy, ornithology, paleontology, and parasitology. The content of Arctos includes specimen records, observations, tissues, endoparasites and ectoparasites, stomach contents, field notes and other documents, and media such as images, audio recordings, and video. Arctos displays all that is known about a museum record, and provides solutions to managing and integrating collections data with object tracking (barcodes, RFID tags), transactions (loans, borrows, accessions, permits), geospatial information (collecting events, coordinates), agents (people, organizations), and usage (publications, projects, citations).

ACB Database

The ASEAN Centre for Biodiversity (see "Relevant Websites section") (ACB, 2015), is a text and image database consists of information on amphibians, birds, butterflies, dragonflies, edible plants, freshwater fishes, mammals, plants, reptiles and Malesian mosses of Southeast Asia. Established in 2005, besides serving as a data store , ACB facilitates cooperation and coordination among the ten ASEAN Member States (AMS) on the conservation and sustainable use of biological diversity, and the fair and equitable sharing of benefits arising from the use of such natural treasures. ACB implements its programs and projects through five components, which are Programme Development and Implementation, Capacity Building, Biodiversity Information Management, Communication and Public Affairs, and Organizational Management and Resource Mobilization.

Vertnet

While most biodiversity databases are built as date warehouses, VertNet (see "Relevant Websites section") (VertNet.org, 2016) uses the distributed database concept. It contains text and image specimens on vertebrates. VertNet is built upon conventional VertNET networks such FishNet, MaNIS, HerpNET, ORNIS by combining them into a single interface for the community to discover, capture, and publish biodiversity data online. It is also an engine for training current and future professionals to use and build upon best practices in data quality, curation, research, and data publishing in taxonomy.

Database of Plant Biodiversity of West Bengal, WBPBDIVDB

The Database of Plant Biodiversity of West Bengal (see "Relevant Websites section") (WBPBDIVDB, 2017) is a plant biodiversity of West Bengal, and it functions as a text and image database. The database covers the richness of floral diversity of West Bengal from Terai, Duars, Darjeeling, the eastern Himalayan region and in the mangrove forests of Sundarbans. West Bengal is an important biodiversity spot in India as it contributes around 12% of the total angiosperm diversity found all over India while the southern deltaic parts of West Bengal owns more than 60 species of the Sundarban Mangrove ecosystem, which outnumbers the total mangrove diversity in India. There are also information on the IUCN status of Plants in West Bengal, showing Family, RDB status, distribution sites and average altitudes. The Genera included are algae, bryophyte, dicotyledon, fungi, gymnosperm, lichen, monocotyledon and pteridophyta. It also shows the species and ecological statistics, by displaying total species and frequency of occurrences.

Reptiles Database

The Reptiles Database (see "Relevant Websites section") (The Reptiles Database, 2017) is a text and image database on reptiles, containing taxonomic information, names, and photos on about 10,000 species and 2800 subspecies. This database provides a catalogue of all living reptile species and their classification. It covers all living snakes, lizards, turtles, amphisbaenians, tuataras, and crocodiles, emphasizing on taxonomic data, particularly names and synonyms, distribution and type data, and literature references. It has no commercial interest and therefore depends on contributions from volunteers. Most of the data comes from published sources, which are curated by editors.

Inaturalist

iNaturalist (see "Relevant Websites section") (iNaturalist.org, 2017) is an online social network of nature enthusiasts sharing biodiversity knowledge. In 2014, the California Academy of Sciences acquired iNaturalist, and now serves as the home for this endeavor. iNaturalist records a life based on number of observations by the community, which is then noted by scientists and land managers, hence serves as a crowdsourced species identification system and an organism occurrence recording tool. The community in iNaturalist records observations, get help with identifications, collaborate with others to collect similar information for a common purpose, or access the observational data collected by other iNaturalist users. It is a unique database as the scientifically valuable biodiversity data is generated by the user's personal encounters and experiences. Scientists and taxonomic experts can provide enormous value to the community by helping to identify observations.

Integrated Biodiversity Information System, IBIS

The Integrated Biodiversity Information System (see "Relevant Websites section") (IBIS, 2011) is a text and image database focusing on Australian plants. It provides taxonomic information, collection details and photographs, incorporating the integration, retrieval and dissemination of biodiversity information via collections management of nomenclatural and taxonomic indices, information delivery, web services and federated database support. IBIS links the data held in the various collections of the Australian National Botanic Gardens, the Australian National Herbarium, the Australian Plant Image Index and the Australian Plant Name Index. IBIS is built upon open source solutions such as Apache, Perl and Java using the Oracle RDBMS for the application and SQL engine. The content uses global naming conventions such as HISPID, TDWG, and BioCase for information management and delivery.

Natureserve Explorer

NatureServe (see "Relevant Websites section") (NatureServe Explorer, 2017) is a database that houses about 70,000 plants, animals, and ecosystems of the United States and Canada, with an in-depth coverage for rare and endangered species. NatureServe easily locates information on scientific and common names, conservation status, distribution maps and images for thousands of species, life histories and conservation needs.

Virtual Tour Database System of Rimba Ilmu Botanic Garden, University of Malaya

This system is a virtual reality (VR) application on plants in Rimba Ilmu Botanical Garden, University of Malaya (UM) (see "Relevant Websites section") (Rimba Ilmu Botanic Garden, 2016), Kuala Lumpur, Malaysia. In this application, the VR technology was used to develop an innovative VR database system which includes visual analytics features. The interactive virtual tour functions side by side the biodiversity database, as it involves an online interactive website to be developed so that users have a virtual tour of a botanic garden and are able to select distinct hotspots throughout the garden. A biodiversity database that consists of the plant species details and information is linked to this interactive portal. The information on the plant species of Rimba Ilmu Botanic Garden such as taxonomy and locality, are stored in a plant database. Geographical representation data are also available in the portal. Visualizing multiple hotspots in a botanical garden using the concept of virtual tour is a novel method of visualizing biodiversity. There are seven hotspots which were selected according to significant plants within the hotspot. Overall, it is a portal on Rimba Ilmu that links to a plant species database and the virtual tours around the botanic garden. The application is a useful tool for biodiversity researchers, non-specialists and also for educating and engaging the public. This system is based on the effort started by Sarinder et al. (2007) in digitizing the local biodiversity using diverse computational tools and technology including building ontologies in biodiversity.

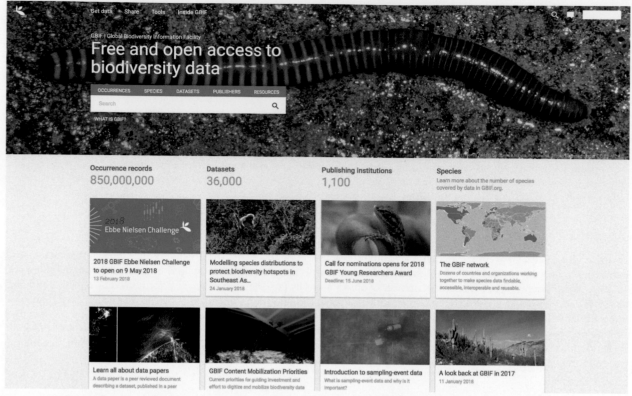

Fig. 1   The GBIF website. *Source*: GBIF.org, 2017. GBIF Home Page. Available from: http://gbif.org (27.02.18).

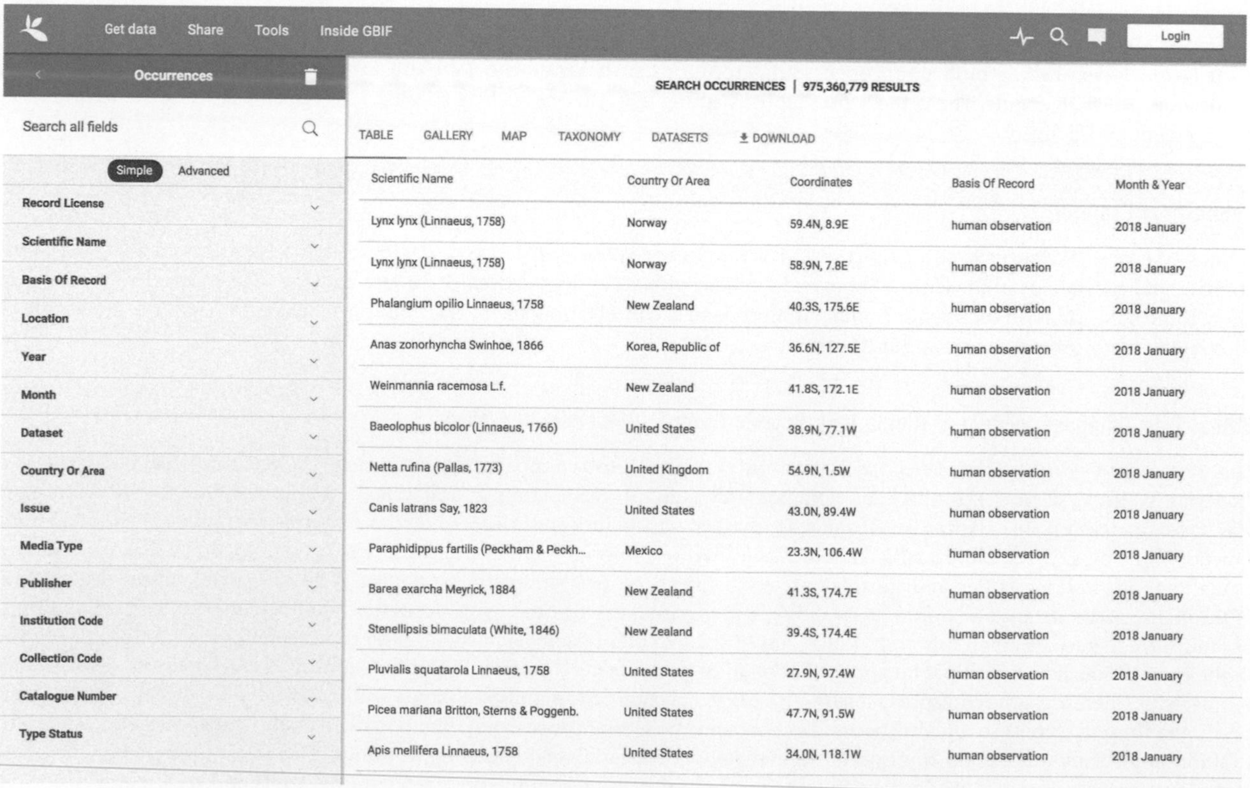

Fig. 2   Occurrences information in GBIF. *Source*: GBIF.org, 2017. GBIF. Available from: http://gbif.org (27.02.18).

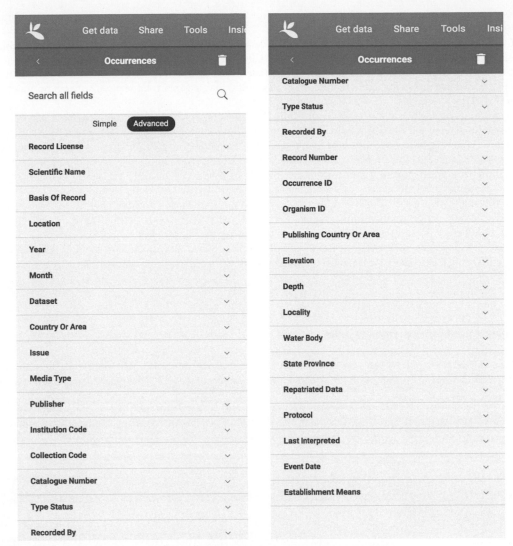

Fig. 3 Searchable fields in GBIF. *Source*: GBIF.org, 2017. GBIF. Available from: http://gbif.org (27.02.18).

The next section presents two very important and large biodiversity databases and tools widely used by scientists; the Global Biodiversity Information Facility (GBIF) and the Encyclopedia of Life (EoL) database. These databases are discussed as case studies in the next section.

Case Studies on Biodiversity Databases

This case study describes two databases, GBIF database and EOL database, which are some of the best examples of existing biodiversity databases and tools that are available on the web. It gives a description of the research and biodiversity needs motivated by a set of technologies with front end portals providing internal and external users with a comprehensive list of capabilities.

GBIF: Free and Open Access to Biodiversity Data

The Global Biodiversity Information Facility, GBIF (see "Relevant Websites section") (GBIF.org, 2017) is an open-data research infrastructure which is funded by governments around the globe, facilitating access more to over 377 million records from more than 400 data publishers (Ariño, 2010). GBIF arose from a 1999 recommendation by the Biodiversity Informatics Subgroup of the Organization for Economic Cooperation and Development's Megascience Forum. One of its report concluded that "An international mechanism is needed to make biodiversity data and information accessible worldwide", stressing the fact that this mechanism could produce many economic and social benefits and enable sustainable development by providing sound scientific evidence. Following this mechanism, in 2001, GBIF was officially established through Memorandum of Understanding between participating governments.

The aim of GBIF is to provide easy access to text and image data on all types of life on Earth. The coordinator with its network of participating countries and organizations provide data owners around the globe with common standards and open-source tools that allows sharing of information such as the location and date of the recorded species. GBIF data is derived from multiple sources, including from museum specimens collected in the past and geotagged smartphone photos shared by amateur naturalists in recent dates (**Fig. 1**).

The GBIF network draws all the sources together using the Darwin Core standard, which forms the basis of millions of species occurrence records. An assessment of data count resulted in 267 million occurrences records (**Fig. 2**) accessible through the GBIF network in which 62% belonged to Kingdom Animalia, followed by Kingdom Plantae (23%), Fungi (1.55%), Protozoa (0.67%), and Bacteria (0.59%) (Gaiji *et al.*, 2003). Every year, hundreds of peer-reviewed publications and policy papers are written based on datasets published by GBIF, covering topics from the impacts of climate change and the spread of pests, to priorities for conservation and protected areas, food security and human health (see "Relevant Websites section").

The role of GBIF is to provide a discovery window on the posted statistics. Such a position requires reconciling, decoding and publishing the crucial key attributes: taxonomic, temporal and geospatial (Gaiji *et al.*, 2003). The searchable fields in GBIF is presented in **Fig. 3** while an example of species page is presented in **Fig. 4**. **Fig. 5** shows example of georeferenced records.

GBIF's ultimate plan is to serve the needs of GBIF's global community and infrastructure to the next decade by empowering global network, enhancing biodiversity information infrastructure, filling data gaps, improving data quality, and also by delivering relevant data.

EOL: Free, Online Collaborative Encyclopedia

The Encyclopedia of Life (EOL) (see "Relevant Websites section") (EOL.org, 2017) is a free, online collaborative encyclopedia intended to document all of the 1.9 million living species known to science (**Fig. 1**). It gives global access to knowledge about life on Earth (**Fig. 6**).

Under the umbrella of EOL, data previously scattered in books, journals, databases, websites, specimen collections, reports are now collated and made available to the public, in the form of articles, media, maps, data in a single website with a collection of application programming interfaces (APIs) (see "Relevant Websites section"). An example of a species page is shown in **Fig. 7**

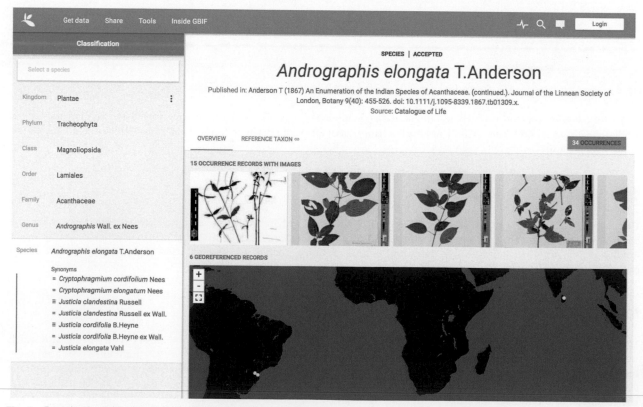

Fig. 4 Example of species page in GBIF. *Source*: GBIF.org, 2017. GBIF. Available from: http://gbif.org (7.02.18).

Fig. 5 GBIF's georeferenced records. *Source*: GBIF.org, 2017. GBIF. Available from: http://gbif.org (27.02.18).

while **Fig. 8** shows an example of textual and image content in the EOL website (see "Relevant Websites section"). The type of taxonomy content is shown in **Fig. 9**.

EOL is supported by some very important institutions, which are Atlas of Living Australia, La Comisión Nacional para el Conocimiento y Uso de la Biodiversidad (CONABIO), Harvard University, Marine Biological Laboratory, New Library of Alexandria, and also Smithsonian Institution's National Museum of Natural History. In addition, EOL collaborates with global bioinformatics projects, like the Biodiversity Heritage Library (BHL), Barcode of Life (BOLD), Catalogue of Life (COL), and Global Biodiversity Information Facility (GBIF) (see "Relevant Websites section").

Essentially, the Encyclopedia of Life serves the global community and by collaborating with its Global Partners, EOL seeks to remove the barriers of geography and language to support international professional and citizen scientists interested in biodiversity and related topics (see "Relevant Websites section"). More than just a database, EOL offers tools and resources that help users harness and apply the information found on EOL.

EOL's mission is to increase awareness and understanding of living nature through a service engine that gathers, generates, and shares knowledge in an open, freely accessible and trusted digital resource (see "Relevant Websites section"). EOL has achieved its mission as it is currently one of the biggest and widely accessible data store on biodiversity. Over time, the content in EOL will be richer as more partners joins this initiative and the current partners stay active in depositing data into the portal.

Importance of Digitizing Biodiversity Information

The collection of biodiversity information plays a critical role in understanding biodiversity and ecosystem. The significance of biodiversity databases are highlighted in this section.

Sebastes chrysomelas
Rockfish

EOL News   see more

Biodiversity Heritage Library User Survey
The Biodiversity Heritage Library (BHL), is an international consortium of natural history
and botanical libraries that cooperate to digitize and make freely available the biodiversity
literature held in... more
JULY 20, 2017 13:40

Moth Week
The Sixth Annual National Moth Week will be taking place throughout the United States
and around the world July 22 - 30, 2017. Learn more and find an event near you
at: http://nationalmothweek.org
JULY 14, 2017 13:27

New EOL Education and Learning Website
EOL's Learning and Education group has a new website! Our goal is to help make the
wealth of biodiversity information on EOL accessible through free tools, resources and

Community Activity   see more

Kento Furui added an unknown common name in
an unknown language to "Veniliornis".
ABOUT 3 HOURS AGO
reply

Daniel Hartley added an unknown item to the
collection "Flora Lisboa e Vale do Tejo".
ABOUT 11 HOURS AGO
reply

C. Michael Hogan marked the classification from
"Species 2000 & ITIS Catalogue of Life: April 2013"
as preferred for "Prosopis velutina Wooton".
ABOUT 14 HOURS AGO

Fig. 6   EOL website. *Source*: EOL.org, 2017. Encyclopedia of Life Home Page. Available from: http://www.eol.org (27.02.18).

Biodiversity and Environmental Change

Biodiversity databases and tools allows documentation of the biological diversity of life and demonstration of changes in the environment that have taken place through time (Baird, 2010). The presence or absence of a species as well as their frequency changes in a geographic region over the course of time are documented by examining the records. If a gradual decline of an important species is observed, it may serve as an indicator whether a recovery program should be carried out to restore the population of the species. The collections that are well documented and deposited can offer proof that the program was successful, provide evidence that changes has taken place, and can help to effectively monitor rare, threatened and endangered species.

Invasive Alien Species and Biosecurity

Other than monitoring endangering species, biodiversity informatics is capable of revealing the invasive alien species that are threat for biodiversity. Increasing travel, trade, and tourism associated with globalization and expansion of the human population have facilitated intentional and unintentional movement of species beyond natural biogeographical barriers (Forest Research Institute Malaysia, 2011). These species are known as alien species and many of them have become invasive. Invasive alien species are known to cause biodiversity loss and lead to a nation's economic loss. Thus, it is critical to differentiate native from alien species, and respond rapidly to these threat, either to remove the species from a region, or suppress their population growth if unable to remove it.

Besides revealing invasive alien species, biodiversity informatics allows prediction of potential of a non-native species to be invasive alien species. Faulkner *et al.* (2014) has created and applied a simple, rapid methodology for developing invasive species watch lists. The watch lists are created using three predictors of three invasion success: History of invasion, environmental suitability and propagule pressure. The authors claim that building the list may be an important step in developing biosecurity scheme, especially for resource poor regions.

Public Health and Wildlife Disease

There are many infections that manifest themselves in human and non-human populations, such as H1N1 Influenza, West Nile Virus, Lyme disease, tuberculosis, Ebola Virus, and SARS. For these and other zoonoses, approximately 70% of new important disease affecting human health are believed to have a wild animal source, and can have profound impacts on human health and economy (Blancou *et al.*, 2005).

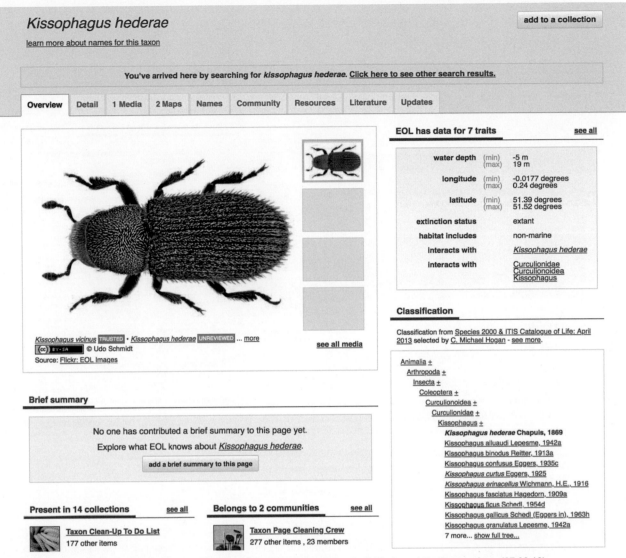

Fig. 7 EOL's species page. *Source*: EOL.org, 2017. Encyclopedia of Life website. Available from: http://www.eol.org (27.02.18).

Using the occurrence data from GBIF, Pigott *et al.* (2014) mapped the area potentially at risk from outbreaks of Ebola virus, based on the environmental niche of bat species believed to act as reservoir hosts of the disease. The authors determined the national population at risk, and claimed that this would be a strong rationale for improving, prioritizing, and stratifying surveillance for EVD outbreaks and diagnostic capacity in these countries.

Economics and Regulatory Framework

The Economics of Ecosystems and Biodiversity (TEEB) (see "Relevant Websites section") is a global initiative focused on "making nature's values visible". The objective of TEEB is to introduce the importance of putting economic value to biodiversity, particularly for policies and decision making at all levels. It follows a structured approach to valuation of ecosystem in economic terms in order to recognize the benefits of biodiversity protection, which is inline with the objectives of the Convention of Biological Diversity. While economic impact studies on biodiversity is still new, sooner or later it is going to change the way humans perceive biodiversity.

Access and Benefit Sharing

The obvious significance of digitization of biodiversity information is dissemination of information globally online. Before digitization, the information is held by institutes in developed countries, whereas scientists may do their field work at

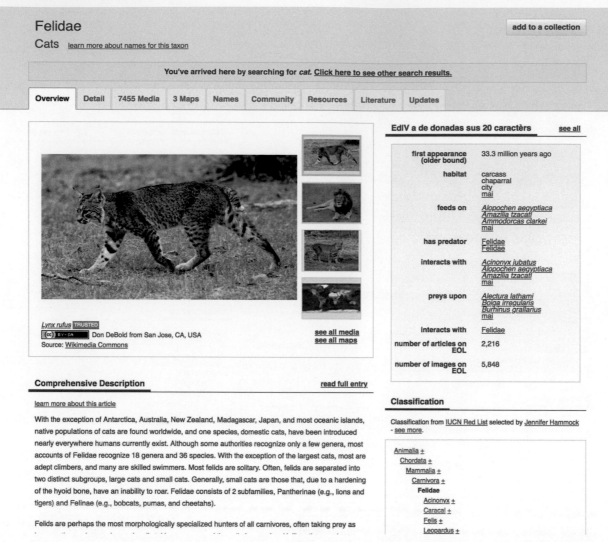

Fig. 8   Textual and image content in EOL. *Source*: EOL.org, 2017. Encyclopedia of Life website. Available from: http://www.eol.org (27.02.18).

underdeveloped country. Digitization allows researcher to access to the information remotely via the web. Besides, digitization allows sharing of information among researchers. The information may be integrated and analyzed to contribute to their study.

Future Directions and Closing Remarks

The production of biodiversity data has accelerated enormously in the past few years, and this has led to the creation of databases and tools to seek deeper understanding of the myriad of values it offers. While every biodiversity and conservation entity in the world is trying to collate and store their data in a digital form, without sharing it in a central repository, the actual value cannot be perceived. The engagement programs started by world class organizations like GBIF and EOL have introduced novel methodologies in managing, accessing and visualizing primary and secondary biodiversity datasets. It is hoped that, with the integration of biodiversity data globally, new discoveries in science is produced, besides creating a new economy in this field.

Apart from curated databases, we strongly belief that species identification, which is currently dependent on expertise of a few individuals, can be automated using technologies in pattern recognition and semantics, for both image and text retrieval. Future efforts in biodiversity should focus on areas such as (1) to what extent can biodiversity be linked using the current technology; (2) how does biodiversity fits into Big Data Analytics; and (3) to what extent automated systems can help in returning accurate species identifications with annotations? With these findings, we should aim to build integrated biodiversity platforms, focusing on semantic search and big data analytics, along with high throughput automated systems capable of accurate species identifications in seconds. This potential system can then be used as a benchmark for indexing millions of identified species.

The actual value of a species can only be known if sufficient data about the species is linked, in a manner when new insights can be discovered easily, through the technology of linked data. This is similar to the social media data linking people such as Facebook, Twitter, LinkedIn, ResearchGate, ResearcherID, WeChat and Hangouts. A similar idea could be implemented for specimens or species.

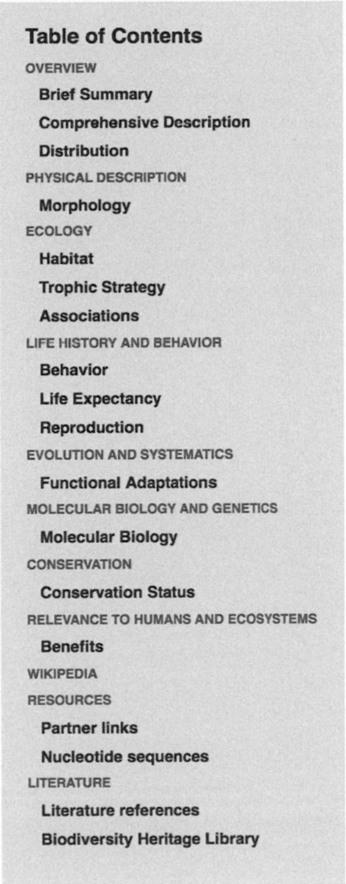

Table of Contents

OVERVIEW
  Brief Summary
  Comprehensive Description
  Distribution
PHYSICAL DESCRIPTION
  Morphology
ECOLOGY
  Habitat
  Trophic Strategy
  Associations
LIFE HISTORY AND BEHAVIOR
  Behavior
  Life Expectancy
  Reproduction
EVOLUTION AND SYSTEMATICS
  Functional Adaptations
MOLECULAR BIOLOGY AND GENETICS
  Molecular Biology
CONSERVATION
  Conservation Status
RELEVANCE TO HUMANS AND ECOSYSTEMS
  Benefits
WIKIPEDIA
RESOURCES
  Partner links
  Nucleotide sequences
LITERATURE
  Literature references
  Biodiversity Heritage Library

Fig. 9 Taxonomy data content in EOL. *Source*: EOL.org, 2017. Encyclopedia of Life website. Available from: http://www.eol.org (27.02.18).

Technology such as semantic web and ontologies, big data analytics, cloud computing and data science are controlling the ICT age and many new discoveries are anticipated. We strongly believe application of these technologies in biodiversity studies will help scientists in future investigation, inquiry, reasoning and analysis of biodiversity data. Moreover, students, policymakers and stakeholders will gain insight to biodiversity data captured and interpreted precisely.

See also: Bioinformatics Data Models, Representation and Storage. Biological Database Searching. Challenges in Creating Online Biodiversity Repositories with Taxonomic Classification. Data Storage and Representation. Ecosystem Monitoring Through Predictive Modeling. Information Retrieval in Life Sciences. Large Scale Ecological Modeling With Viruses: A Review. Mapping the Environmental Microbiome. Molecular Phylogenetics. Natural Language Processing Approaches in Bioinformatics. Population Genetics

References

ACB, 2015. ASEAN Centre for Biodiversity (ACB) Home Page. Available from: https://aseanbiodiversity.org (06.09.17).
ACSI, 2017. All Catfish Species Inventory (ACSI) Home Page. Available from: http://silurus.acnatsci.org (06.09.17).
AntWeb, 2018. Available from http://www.antweb.org. (25.04.18).
Arctos, 2017. Arctos Home Page. Available from: https://arctosdb.org (06.09.17).
Ariño, A.H., 2010. Approaches to estimating the universe of natural history collections data. Biodiversity Informatics 7, 81. Available from: https://journals.ku.edu/jbi/article/view/3991.
BacDive, 2017. The Bacterial Diversity Metadatabase Home Page. Available from: https://bacdive.dsmz.de (23.08.17).
Baird, R., 2010. Leveraging the fullest potential of scientific collections through digitization. Biodiversity Informatics 7, 130–136.
Blancou, J., Chomel, B.B., Belotto, A., Meslin, F.X., 2005. Emerging or re-emerging bacterial zoonoses: Factors of emergence, surveillance and control. Veterinary Research 36, 507–522. doi:10.1051/vetres:2005008.
Canhos, V.P., Souza, S., Giovannia, R., Canhos, D.A.L., 2004. Global biodiversity informatics: Setting the scene for "new world" of ecological modeling. Biodiversity Informatics 1, 1–13.
Ch'ng, E., 2009. An efficient segmentation algorithm for entity interaction. Biodiversity Informatics 6, 5–17.
EOL.org, 2017. Encyclopedia of Life Home Page. Available from: http://www.eol.org (31.12.17).
Faulkner, K.T., Robertson, M.P., Rouget, M., Wilson, J.R.U., 2014. A simple, rapid methodology for developing invasive species watch lists. Biological Conservation 179, 25–32. doi:10.1016/j.biocon.2014.08.014.
FEOW.org, 2015. Freshwater Ecoregions of the World (FEOW) Home Page. Available from: http://www.feow.org (14.08.17).

FishBase.org, 2017. FishBase Search Page. Available from: http://www.fishbase.org (01.01.18).

Gaiji, S., Chavan, V., Arino, A.H., 2013. Content assessment of the primary biodiversity data published through GBIF network: Status, challenges and potentials. Biodiversity Informatics 8, 94–172. Available from: https://www.jcel-pub.org/jbi/article/view/4124/4201.

GBIF.org, 2017, GBIF Home Page. Available from: http://gbif.org (31.12.17).

Georgian Biodiversity Database, 2015, Georgian Biodiversity Database Home Page. Available from: http://www.biodiversity-georgia.net (14.08.17).

IBIS, 2011. IBIS (Integrated Biodiversity Information System) Introduction Page. Available from: http://www.anbg.gov.au/ibis (07.09.17).

iNaturalist.org, 2017. iNaturalist Home Page. Available from: http://www.inaturalist.org (07.09.17).

Kumar, S., 2005. Molecular biology database/biological database/bioinformatics database – An overview. Available from: http://bioinformaticsweb.net/

Naturdata.com, 2017. Naturdata Home Page. Available from: http://naturdata.com (23.08.17).

NatureServe Explorer, 2017. NatureServe Explorer Home Page. Available from: http://explorer.natureserve.org (07.09.17).

PESI, 2017. Pan-European Species-directories Infrastructure (PESI) Home Page. Available from: http://www.eu-nomen.eu/portal (23.08.17).

Pigott, D.M., Golding, N., Mylne, A., et al., 2014. Mapping the zoonotic niche of Ebola virus disease in Africa. eLife 3, e04395. doi:10.7554/eLife.04395.

Rimba Ilmu Botanic Garden, 2016. Rimba Ilmu Botanic Garden Home Page. Available from: http://rimba.um.edu.my (28.02.18).

Sarinder, K.K.S., Lim, L.H.S., Dimyati, K., Merican, A.F., 2007. An indigenous integration system for biodiversity databases. Online Journal of Bioinformatics 8 (1), 56–60.

ScaleNet, 2016. ScaleNet Home Page. Available from: http://scalenet.info (23.08.17).

Soberon, J., Peterson, A.T., 2004. Biodiversity informatics: Managing and applying primary biodiversity data. Philosophical Transactions of the Royal Society B: Biological Sciences 359, 689–698. doi:10.1098/rstb.2003.143.

Species +, 2017. Species + Home Page. Available from: https://speciesplus.net (14.08.17).

The Reptiles Database, 2017. The Reptiles Database Home Page. Available from: http://www.reptile-database.org (07.09.17).

VertNet.org, 2016. VertNet Home Page. Available from: http://www.vertnet.org/index.html (07.09.17).

WBPBDIVDB, 2017. Database of Plant Biodiversity of West Bengal Introduction Page. Available from: http://thebiome.in/databases/wbpbdivdb_intro.php (07.09.17).

Wieczorek, J., Bloom, D., Guralnick, R., Blum, S., et al., 2012. Darwin core: An evolving community-developed biodiversity data standard. PLOS ONE 7 (1), e29715. doi:10.1371/journal.pone.0029715.

WikiSpecies, 2017. WikiSpecies Main Page. Available from: https://species.wikimedia.org/wiki/Main_Page (23.08.17).

Relevant Websites

http://silurus.acnatsci.org/
All Catfish Species Inventory.

http://www.antweb.org
Antweb.org - AntWeb.

https://arctosdb.org/
Arctos.

https://aseanbiodiversity.org/
ASEAN Centre for Biodiversity.

https://bacdive.dsmz.de/
BacDive.

http://eol.org/
Encyclopedia of Life.

http://www.eu-nomen.eu/portal/
EU-nomen.

http://www.feow.org/
FEOW.

http://www.fishbase.org
FishBase.

https://www.gbif.org/
GBIF.

http://www.biodiversity-georgia.net
Georgian Biodiversity Database.

http://www.anbg.gov.au/ibis/
IBIS.

http://www.inaturalist.org/
iNaturalist.

http://naturdata.com/
NATURDATA.

http://explorer.natureserve.org/
NatureServe Explorer.

http://rimba.um.edu.my
Rimba Ilmu.

http://scalenet.info/
Scale Net.

https://speciesplus.net/
Species +.

http://www.teebweb.org/
TEEB.

http://thebiome.in/databases/wbpbdivdb_intro.php
The Biome.

http://www.reptile-database.org/
THE REPTILE DATABASE.

https://www.unep-wcmc.org/
Unep-WCMC.

http://www.vertnet.org/index.html
   VertNet.
https://www.wikipedia.org/
   Wikipedia.
https://species.wikimedia.org/wiki/Main_Page
   Wikispecies.

Challenges in Creating Online Biodiversity Repositories With Taxonomic Classification

Mohd NM Ali, Amy Y Then Hui, and Sarinder K Dhillon, University of Malaya, Kuala Lumpur, Malaysia

Introduction

Presently many biodiversity databases with species information annotations are available online. Although these databases tend to focus on specific taxa, accurate and updated taxonomic classification remain challenging given that new species are being discovered from time to time while extant species may become or have gone extinct. According to the United Nations Environment Programme, scientists estimate that 150–200 species of plants, insects, birds, and mammals become extinct every 24 h and there are around 15% of mammal species and 11% of bird species that are classified as threatened with extinction (Vidal, 2010). On the other hand, new species are being described every three days over last decade (Tobias, 2010). This situation depicts the challenges taxonomists and biologists have to face in order to keep up with updated, accurate information on biodiversity profiles and classification. Most of the online taxonomic databases or portals contain information on authoritative source(s), namely who is managing the data, how it is curated and credibility of the data. Most of these authoritative sources are fellow scientists or researchers, who form a group that manages the taxonomic information shared to the public.

Taxonomy is a field of classifying living things through defining and naming based on shared characteristics. Over time, there are varying definitions of taxonomy within the scientific community, but the core of its practice is revolved around the conception, naming, and classification of groups of organisms. The Linnaean Taxonomy, a system for categorizing organisms and naming living things through a binomial nomenclature, is regarded as the origin of the practice. This system is named after Carl Linnaeus, a Swedish botanist and the father of taxonomy, who introduced the science of naming organisms (either living or extinct) through the usage of two descriptors, namely the genus that the organism belong to, followed by the species modifier. For example, a Chinook salmon or better known as king salmon, has the scientific name of *Oncorhynchus tschawytscha* where *Oncorhynchus* is the genus and *tschawytscha* is the species modifier.

In modern biological classification, taxonomic hierarchy is used to place organisms in increasingly exclusive groupings or ranks, starting with the broadest rank of domain, followed by kingdom, phylum, class, order, family, genus, and species. Within these categorizations, there are more specific groups for them known as subdivisions, which further clarify their classification ranks. For example, superclass, subclass, infraclass, or parvclass form taxonomic subdivisions of the class rank.

Taxonomic classification is one of the main branches of biology, practiced by experts known as taxonomists. The roles of trained taxonomists are clear, which include describing new species and conducting taxonomic revisions to described species as needed. Such exercises are not easy as prior knowledge of the taxa of interest is required and formal taxonomic training on the subject of study is crucial. Incorrect identification of biological entities will introduce misleading information and errors in subsequent work pertaining to that particular information, e.g., worldwide distribution of a species. As such, competent taxonomic work is fundamentally important for the understanding of biology and conservation of biodiversity.

Taxonomy is separated into several sub-branches, each with their unique approach of classifying organisms. Some of the popular sub-branches are alpha taxonomy, beta taxonomy, cytotaxonomy, chemotaxonomy, and cladistics taxonomy. Alpha taxonomy and beta taxonomy both deal with external morphology; however the latter includes internal features such as anatomical characteristics and organ studies. Cytotaxonomy deals with plant classification through the use of cytology (cell) studies, which greatly aid plant classification. Chemotaxonomy also focuses on plants, but utilizes information on chemical constituents of the plants for classification. A widely used taxonomic approach in the modern age is the cladistic taxonomy which groups organisms into clades based on shared, derived characteristics from the most recent common ancestor.

Taxonomy Related Issues

It is evident that even in the age of technology, where information can be accessed and shared anywhere in the world via the internet, dedicated online databases have proven to be useful for the scientific community to publish their findings. The use of an online database is a necessity for taxonomic research; however there are a number of issues related to biodiversity conservation and data management which will be discussed. The first issue, which is quite alarming, is insufficiency of taxonomists or taxonomic research capacity. This is essentially a global issue affecting biology in general. A study showed that although systematic research as a whole has been on the increase, traditional taxonomists, a unique breed of scientists with specific skill sets that are difficult to train and replace, are dwindling in numbers (Lee and Palci, 2015). Another study on the other hand showed that the number of taxonomists has actually been increasing; however the number of species described per taxonomist has not seen decline for poorly studied taxa such as beetles and parasitic wasps (Bacher, 2012). This suggests that much remains to be done for new species discovery and description. One recent study pointed out that while taxonomy is still very much alive, taxonomists are struggling to obtain sufficient funding and appropriate recognition for their field (Bik, 2017).

The second issue, which is linked to the previous, is about species unknown to science going extinct before they are discovered and described. It is commonly accepted in the scientific community that at the current rate of extinction, many species will die before they are managed to be described by science (Dirzo and Raven, 2003; Gaston and May, 1992). In a research done in the Cape Floristic Region (one of the smallest and popular biodiversity hotspots in South Africa), it was reported that there is a finite number of species that may be processed and identified by a taxonomist during his/her working life (Treurnicht *et al.*, 2017). The same research further claimed that the current state of technology and infrastructure does not support an increase in individual taxonomic output; improper operational concepts practiced by taxonomists may also affect a taxonomist's output.

The third issue is regarding the correct identification of species by biologists in general and by taxonomists, which is fundamental for any taxonomy-based studies. The field of taxonomy is thwarted by inconsistent taxonomy literature and vague species descriptions containing incorrect identifications due to work by amateur taxonomists, or inadequate access to original works on the species. Also as mentioned earlier, there are various sub-branches of taxonomy. The use of integrated taxonomic methods, i.e., other branches in conjunction with traditional morphological methods, has led to taxonomic revisions for many known species; this indicates that accurate classification of species remains an issue. However, in modern taxonomy practice, traditional morphology tends to be overlooked in favor of genetic studies. Decoupling of traditional and modern taxonomy affects the type of data shared on online databases, which usually have groups or consortium of sorts regulating these information. Given the present age of Big Data, it is impossible to review and regulate all taxonomic information needed for correct species identification. This is an issue which both taxonomists and data scientists need to tackle in order to provide reliable data for and to facilitate accurate classification of living things.

Impact of the Issues to Online Taxonomy Databases

The issues raised above will create an impact on online taxonomy databases, particularly pertaining to users' expectation to have updated content available to them whenever needed. The decline in the number of expert taxonomists in the near future will impede progress in taxonomic research (Lim and Gibson, 2010). This situation is keenly felt in developing countries, which are often areas of high biodiversity and where not only the number of taxonomists is low but funding for basic taxonomic research is scarce; the net result is often inadequate data quality and analysis and poor data management.

Another impact is bias in research funding. Traditional taxonomy, a fundamental branch of biology, is not perceived as popular as molecular studies, the latter usually preferred by funding bodies. For any research community, funding availability is critical to the progress of the research. In this day and age, dissemination of updated information is done primarily via online means; insufficient funds will also impact dedicated curation of online taxonomic data. Creative funding opportunities must be sought to alleviate the biodiversity crisis, and to promote progress of research on of online taxonomy databases.

Benefits and Risks of Putting Data Online

The necessity for scientific research, taxonomy notwithstanding, to be merged with computational technology is evident (Bik, 2017); this includes making research information publicly available and updated. One benefit of having research data online is to increase visibility of the research, to make them accessible for research continuity and to avoid loss of invaluable data. Even though 'old data' might not have a direct or immediate value, such data might generate new findings after recalibration. Another benefit for putting research data online is to provide public documentation of a topic of research even though it has not gone through a peer-reviewed process. This minimizes the possibility of research hijacking or another party claiming authority over the same research topic as their own. Besides sharing research data online, it is possible to link the data to other information sources via data sharing portals. This enables new analysis and research while increasing the value of the previous research. Current online databases have already started this practice where taxonomic data are linked to genetic databases pertaining to a taxonomy rank. This is discussed further in the next section.

However, there are valid reasons why scientists may not prefer to share their data online. The human factor is perhaps the main stumbling block to online publishing. Publishing taxonomy data exposes information into the public domain. Unlike peer reviewed publications where researchers get scientific credit via the citation process (an important academic performance measure), taxonomists publishing raw data tend to remain anonymous while the web data curators governing the information are known. Another risk is plagiarism. Scientists fear sharing data openly, especially from research in the initial stages, as unethical researchers might illegally use these data and represent them as their original work. Data published online is prone to security threats. There is a risk of data being hacked and manipulated. Data owners who are not computationally savvy or bogged down by other research work may not be able to manage the online data content, which involves creating a change log, and notifying users about the change. Finally, there is a risk of inappropriate use of data whereby they are used to conduct inappropriate or unethical research.

Comparison Between Databases

Taxonomic classification data contain information on scientific names, taxon ranks, morphological features, distributions, and other related information; all these information provided in species profile databases, species lineage or classification information are valuable. The lineage or classification data in most databases are usually linked to sequence data repositories, such as GenBank (see "Relevant Websites section"). Online taxonomic databases have clearly defined ways to manage taxonomic information. Here we discuss some of the popular databases, with emphasis on fish taxonomy.

FishBase (see "Relevant Websites section") is a portal managed by the FishBase Consortium group and is one of the trusted online resources on fish. This portal holds a huge amount of information on fish but here we are concerned only with taxonomic classification. FishBase displays species classification data on individual species profile page, such as common names, climate, distribution, and size parameters; the portal presents relevant information specific to the species without covering the species lineage or phylogenetic tree. Nevertheless, FishBase provides a link to the Deep Fin Classification for every species stored in its database.

Another popular taxonomic database is the NCBI Taxonomy Browser managed by the NCBI group (see "Relevant Websites section"). It provides some of the most comprehensive information for a species. In addition to displaying a full lineage of any species of interest, it also provides links to its mitochondrial genetic code, nucleotide sequences and protein sequences. The coverage is elaborated with external information sources which include origin of the name, author, abbreviated taxonomy, and biotic interactions.

FishBase is a database made to handle solely fish data while the NCBI Taxon is developed for classification of living organisms. As such there is a difference in the way data are displayed and organized. FishBase displays information related to the fish species, leaving out information about the lineage of the species. NCBI Taxon, on the other hand may increase the coverage by elaborating on the genetic information of the classification for interested users; when any doubt arises, they may challenge the findings. The possibilities of using these data to produce new research avenues are endless and it is up to the scientific community on how to use it.

Next, we discuss how Wikipedia (see "Relevant Websites section") handles species information and compare it to more credible sources such as FishBase and NCBI Taxon discussed above. Wikipedia, although not necessarily an authoritative source, is often a preferred choice among students or even researchers wanting to get a quick overview on any topic. Data in Wikipedia are generated mostly by voluntary contributors; thus, available taxonomic information provided on each species page may vary depending on the popularity and extent of knowledge of the species. Nevertheless, there are some standards set by Wikipedia for publishing species information such as full scientific names, scientific classification, conservation status, binomial name, synonyms (if available), and references used to write the species article. Information on species classification, however, is basic and linked to a Wikipedia article page, not another authoritative source. **Table 1** presents the comparison between popular online taxonomic databases, including those discussed above.

Ontology and Taxonomic Classification

With the birth of a new Web commonly referred to as the Semantic Web, it is imperative for scientists to embrace the world of data sharing and Big Data. This shift in scientific operation and paradigm requires that the wealth of scientific data, in all manners of format and complexity, be captured and mapped correctly especially in the context of species classification. A preferred way to do this for species data is in the form of an ontology, a formal naming convention that defines any domain of interest and handles information about the types, properties, and interrelationships of these entities.

Ontology is different from a database in terms of their data structure and the way to query data. The most important difference is the triple data structure of an ontology, which creates relationships between a class and properties. As such there are differences in the purpose of creating an online database versus an ontology. The latter is created primarily to give definition to a certain domain; an example is the Gene Ontology that is created specifically to describe genes. It is generally advisable to create lower level domain ontologies with a defined focus, such as gene sequences and skeletal structure, since it is easier to handle and limit the amount of information and to reuse it for other ontologies or databases. Creation of ontologies for large-scale domains such as human or flowers would result in huge amount of data that need to be managed which may contain information that are vague.

In 2017, we published a Fish Ontology (FISHO) (see "Relevant Websites section"), with the aim of performing automated identification of fish using taxonomy (Ali *et al.*, 2017). The FISHO consists of 652 classes (terms), and 27 object properties (relationships) (can be viewed at "Relevant Websites section"). There are 10 main classes which act as the core classes covering fish-related and non-related terms within the FISHO structure. The FISHO provides terms related to fish and infers species related information based on data that are fed to the ontology. The current version of the FISHO is able to classify jawless fish, early jawed fish and living fossil fish. The FISHO contains 253 classes dedicated to fish studies and 38 classes related to fish sampling processes. **Table 2** shows the adoption of the terms in the FISHO using 3 sources which include an authoritative book on fish (Helfman *et al.*, 2009), the Vertebrate Taxonomy Ontology (Midford *et al.*, 2013), and the NCBI Taxon (Federhen, 2016) (**Fig. 1**).

The FISHO is created using Protégé (Protégé, 2018), an open source ontology editor provided by the Stanford Center for Biomedical Informatics Research at the Stanford University School of Medicine. This software contains all the tools needed to develop the FISHO with many supporting features that can assist in the development and visualization of ontology. One of the feature provided by Protégé is the inferring capabilities. In **Fig. 2**, we show the results of the inferring process within the FISHO. Inferred results of a sample specimen with certain parameters are shown with bright orange color. Once it is inferred, more results can be gained from the inferred results; as an example, Specimen5 was recognized not only as a whale in Infer Result A, the FISHO also inferred that it is not a fish, but a mammal. For Sample2, it is recognized as a longtail carpet shark after the inferring process, and furthermore recognized as an early jawed fish in Infer Result B.

Table 1   Differences between online databases in handling and displaying taxonomic classification information

Database name	Purpose	Classification tree	Link to taxonomic information source	Related Data complementing the taxonomy data	External Links related to the taxonomy data
FishBase (www.fishbase.org)	Fish profile	Partial	Yes	Names, Synonyms, Climates, Size, Distribution	Minimal. (Catalog of Fishes, Deep Fin Classification)
NCBI taxon (www.ncbi.nlm.nih.gov/Taxonomy/taxonomyhome.html)	Taxonomy	Complete including ranks subdivisions	Yes	Taxon Id, Inherited Blast Name, Genetic Code, Mitochondrial Genetic Code, Entrez records	Moderate. (Encyclopedia of Life, Global Biotic Interaction, Dryad Digital Repositories, Arctos Specimen Database)
Wikipedia (www.wikipedia.org)	General knowledge	Partial	Yes	Description, History (if available), Elaborated Information (based on necessity and availability), conservation status, binomial name, synonym	Minimal. (IUCN Redlist, and random links, depending on the author)
WoRMS (www.marinespecies.org)	Marine species registry	Partial	Yes	Status, Ranks, Parents, Original name, Synonyms, Vernacular names, Environment, and Distribution	Comprehensive. (Barcode of Life, Biodiversity Heritage Library, Encyclopedia of Life, and Fishbase to name a few)
The IUCN red list of threatened species (www.iucnredlist.org)	Red list of threatened species	Full	Yes	Names, synonyms, assessment information, population, usage, threats level, and conservation actions	None.
Barcode of life data system (www.boldsystems.org)	Generation & application of DNA barcode data	Full	No	Sample ID, Museum ID, Collection data, Sequencing records (sequence ID, GenBank Accession, Genome, Locus, Nucleotides, Amino Acids, and Illustrative Barcode)	Minimal. (Genbank)

Table 2   Term adoption in the Fish Ontology

Example of terms	Sources			Implementations in the Fish Ontology
	Helfman (2009)	Vertebrate Taxonomy Ontology (VTO)	NCBI Taxon	
Furcacaudiformes (order)	Classified as Subclass of Thelodonti (superclass)	Classified as subclass of Agnatha (class)	Not classified	Follows and reuses the VTO terms
Jawless fish	Contains species and information for jawless fish species	No classes and annotations found, but related species are classified	No classes and annotations found, but related species are classified	Follows Helfman (2009) for labeling
Lobe finned fish	Classified as Actinopterygii (page 4)	No classes and annotations found, but related species are classified	Classified as Coelacanthiformes	Follow Helfman (2009) for classification and labeling
Gobiidae (family)	Listed and classified as family	Listed and classified as family	Listed and classified as family	Follows and reuses the VTO terms
Oxudercinae (subfamily)	Not listed or classified	Not listed or classified	Classified as a subclass of Gobiidae (family)	Follows and reuses the VTO classification up to the lowest existing taxonomic terms covered (Family Gobiidae). Adopts NCBI Taxon terms for subfamily Oxudercinae onwards

The FISHO was reviewed by several experts, one of them being a museum curator who raised concerns over taxonomic classification within the ontology. From a cladistics standpoint, mammals are derived from fish and the inclusion of mammals (and other tetrapods) under Sarcopterygii (an ancient fish group with lobed fins) allows it to be a monophyletic group. However, the main point of creating the ontology is not primarily for taxonomic classification, but to create an automated fish recognition using taxonomy. Presently in the FISHO, the class Mammalia is not placed within the general fish classification structure but under the Chordata phylum. Inclusion of the Mammalia class in the FISHO is due to the existence of some aquatic mammals, such as

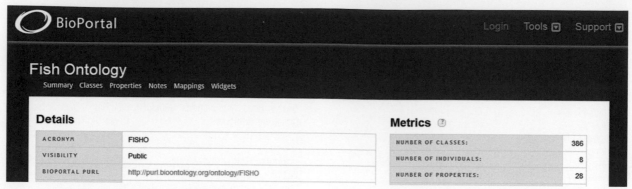

Fig. 1 The Fish Ontology (FISHO) hosted at Bioportal.

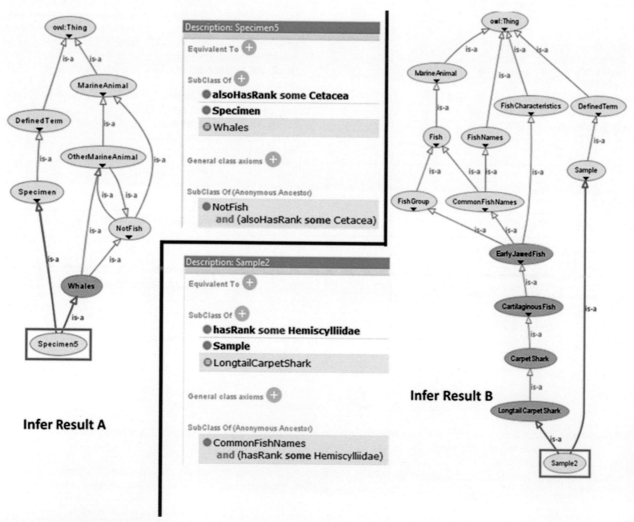

Fig. 2 Inferring process in the FISHO.

whales and dolphins, which may be erroneously classified as fish. Under the ideal FISHO classification (in our opinion), whales would not be recognized as a fish but a mammal. As such, the consequence of explicitly classifying mammals under Sarcopterygii in the FISHO is such that a whale would then be recognized as both a mammal and a fish. Further explanation of this situation is shown in **Fig. 3**, where we show the specific use case of other sources compared to the FISHO.

Another comment from the reviewer highlighted the confusion about jawless fishes in the FISHO. From a phylogenetic point of view, Agnatha (jawless fish group that includes lampreys, hagfish and ostracoderms) is not a monophyletic group. However, our

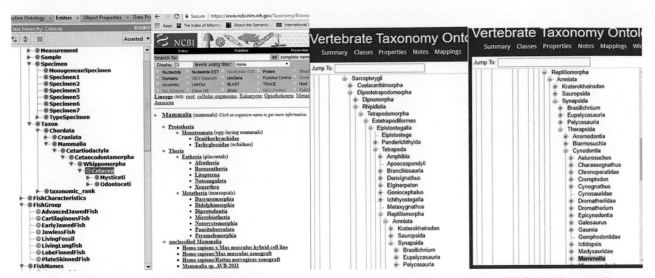

Fig. 3 Different usage of other classification sources compared to the FISHO. The leftmost figure shows how the FISHO classify mammals compared to the NCBI Taxon and the Vertebrate Taxonomy Ontology (VTO).

purpose of including jawless fishes in the FISHO was to capture it as one type of fish grouping or characteristic. For example, a hagfish would be characterized as (or having the attribute of) a jawless fish in the FISHO. This grouping term is not intended as a form of taxonomic classification (i.e., this term is not a class under taxonomic rank, but under Fish Group in the FISHO). The nature of the ontology is such that broad inclusion of relevant fish-related terms such as 'jawless fish' would improve the inference capacity of the FISHO reasoner, thus allowing for relevant and comprehensive search automation and classification of fish.

With these two examples of the challenges faced during the development of the FISHO, one solution that may be adopted to further upgrade the FISHO is to create 2 types of classification, one based on traditional morphology-based taxonomic classification and the other based on modern of phylogenetic taxonomy. Through this approach, the ontology might have the ability to recognize fish specimens using both species and DNA characteristics. This solution may also be applied to other databases or recognition softwares having similar issues. Finally, we propose that future development of computational systems should utilize a variety of species specific data types to speed up the taxonomic classification processes.

Conclusions

An enduring and elusive challenge in biodiversity classification is to determine the numbers of species inhabiting the earth. It had been estimated that there are around 8.7 million species on land and sea, but many are still awaiting description (Mora *et al.*, 2011). This staggering number demonstrates the amount of effort that has been and still needs to be invested to properly classify and capture this biodiversity. As such, the pressure is on for taxonomists to expedite the process and for taxonomic databases to manage the voluminous amount of data accurately and with timely updates.

Scientific data, which refer to taxonomic information in this context, have limited usefulness if they are not read or discovered by others. It has been more than 200 years now since Linnaeus founded taxonomy. His research, and that of his counterparts, had resulted in many natural history drawings, identification keys, species descriptions, and monographs. However, such important records are almost untouched in this digital age. The way forward is to integrate technology with these traditional taxonomic knowledge base. Availability of such data in the online medium will promote secondary research using modern data science techniques, which may give birth to new knowledge and facilitate reproducible research. With the help of semantic data and ontology, digitized taxonomic resources could be made immediately accessible to the public and facilitate seamless integration of new taxonomic information.

See also: Bioinformatics Data Models, Representation and Storage. Data Storage and Representation. Ecosystem Monitoring Through Predictive Modeling. Information Retrieval in Life Sciences. Large Scale Ecological Modeling With Viruses: A Review. Mapping the Environmental Microbiome. Molecular Phylogenetics. Natural Language Processing Approaches in Bioinformatics. Population Genetics

References

Ali, N.M., Khan, H.A., Then, A.Y.-H., *et al.*, 2017. Fish Ontology framework for taxonomy-based fish recognition. PeerJ 5, e3811.
Bacher, S., 2012. Still not enough taxonomists: Reply to Joppa *et al.* Trends in Ecology & Evolution 27 (2), 65–66.
Bik, H.M., 2017. Let's rise up to unite taxonomy and technology. PLOS Biology 15 (8), e2002231. Available at: https://doi.org/10.1371/journal.pbio.2002231.

Dirzo, R., Raven, P.H., 2003. Global state of biodiversity and loss. Annual Review of Environment and Resources 28 (1), 137–167.

Federhen, S., 2016. NCBI organismal classification – An ontology representation of the NCBI organismal taxonomy. Retrieved from http://www.obofoundry.org/ontology/ncbitaxon.html.

Gaston, K.J., May, R.M., 1992. Taxonomy of taxonomists. Nature 356 (6367), 281–282. https://doi.org/10.1038/356281a0.

Helfman, G.S., Collette, B.B., Facey, D.E., Bowen, B.W., 2009. The Diversity of Fishes: Biology, Evolution, and Ecology. vol. 2. Atlantic: John Wiley & Sons.

Lim, L.H.S., Gibson, D.I., 2010. Taxonomy, taxonomists & biodiversity. In: Manurung, R., Zaliha, C.A., Fasihuddin, B.A., kuek, C. (Eds.), Biodiversity-Biotechnology: Gateway to Discoveries, Sustainable Utilization and Wealth Creation. Kuching, Sarawak, Malaysia, pp. 33–43.

Lee, M.S., Palci, A., 2015. Morphological phylogenetics in the genomic age. Current Biology 25 (19), R922–R929.

Midford, P., Dececchi, T., Balhoff, J., et al., 2013. The vertebrate taxonomy ontology: A framework for reasoning across model organism and species phenotypes. Journal of Biomedical Semantics 4 (1), 34.

Mora, C., Tittensor, D.P., Adl, S., Simpson, A.G.B., Worm, B., 2011. How many species are there on earth and in the ocean? PLOS Biology 9 (8), e1001127..

Protégé. (2018). Stanford Center for Biomedical Informatics Research. Retrieved from http://protege.stanford.edu/.

Tobias, J., 2010. Amazon alive. In: World Wildlife Fund. Retrieved from https://www.worldwildlife.org/stories/amazon-alive.

Treurnicht, M., Colville, J.F., Joppa, L.N., Huyser, O., Manning, J., 2017. Counting complete? Finalising the plant inventory of a global biodiversity hotspot. PeerJ 5, e2984.

Vidal, J., 2010. Protect nature for world economic security, warns UN biodiversity chief. The Guardian. Retrieved from https://www.theguardian.com/environment/2010/aug/16/nature-economic-security.

Relevant Websites

www.worldwildlife.org/stories/amazon-alive
    Amazon Alive.
www.boldsystems.org
    Barcode of Life Data System.
https://bioportal.bioontology.org/ontologies/FISHO
    Fish Ontology.
www.fishbase.org
    FishBase.
www.ncbi.nlm.nih.gov/genbank
    GenBank.
https://mohdnajib1985.github.io/FOWebPage/
    mohdnajib1985 · GitHub.
www.ncbi.nlm.nih.gov/Taxonomy/taxonomyhome.html
    NCBI Taxonomy Browser.
www.theguardian.com
    The Guardian.
www.iucnredlist.org
    The IUCN Red List of Threatened Species.
www.obofoundry.org/ontology/vto.html
    Vertebrate Taxonomy Ontology (VTO).
www.wikipedia.org
    Wikipedia.
www.marinespecies.org
    WoRMS.

Ecological Networks

Kazuhiro Takemoto and Midori Iida, Kyushu Institute of Technology, Fukuoka, Japan

Introduction

In the natural world, many species complexly interact with one another via various types of relationships (e.g., commensalism, amensalism, competition, mutualism, prey-predator relationship), and they compose ecosystems (Allesina and Tang, 2012). It is important to understand the function and stability against environmental perturbations (e.g., climate change) of ecosystems in the context of biodiversity maintenance and environmental assessment (Bascompte, 2010; Evans et al., 2013; Pocock et al., 2012). In computational biology, up until now, ecological networks have been mainly examined using theory (mathematical models such as Lotka–Volterra equations (Jansen and Kokkoris, 2003) describing the time evolutions of species populations). For example, May's pioneering theoretical works on the relationship between ecosystem complexity and stability (May, 1976; May, 1972) are interesting because ecosystem stability is related to biodiversity maintenance.

More recently, ecosystems have also been studied from a data analysis perspective. The development of field observation techniques and the improvement of infrastructure such as databases have increased the ecological data availability and have enabled large-scale data analyses of real-world ecosystems. In this context, *network science* plays an important role. Network science in ecology is called *network ecology*. In this article, we will overview the network ecology topic and the approaches to investigate ecological networks.

Network Ecology

Network science is a research area in which complex networks are studied, and it originates from graph theory (Barabási, 2013). Networks describe the relationships among elements, and are, thus, simple and powerful tools for describing complicated systems. The concept of networks is universal and can be applied to a wide range of fields (e.g., mathematics, computer science, economy, sociology, chemistry, biology). In recent years, considerable data on interaction has been accumulated. Thus, networks have become quite important for understanding real-world systems and extracting knowledge of complex systems. For example, network science has been applied to biology. The biomolecules of the living organisms such as proteins and metabolites undergo several interactions and chemical reactions which lead to the occurrence of various life phenomena. To understand the biological processes, it is important to obtain an understanding of the networks; thus, network biology (Barabási and Oltvai, 2004) and network medicine (Barabasi et al., 2011) have already attracted attention. Ecological communities are also represented as networks or graphs (so-called *ecological networks* in which nodes and edges correspond to species and interspecific interactions, respectively). As the availability of ecological data has increased, network science has also been applied to ecology (Proulx et al., 2005).

Ecological networks are often classified according to interaction types (e.g., mutualism and prey–predator relationship). The representative ecological networks are *food webs* and *mutualistic networks*.

Food webs indicate who eats whom, and they have discussed in ecology for a very long time (Kondoh et al., 2010; Thompson et al., 2012). Food webs are represented as networks in which nodes and edges correspond to organisms and trophic links (prey–predator relationships). It should be noted that food webs are generally represented as unipartite directed networks (**Fig. 1(A)**) because of predator–prey relationships which are direction-oriented. Food webs are often also called *antagonistic networks* and *trophic networks*. However, plant–herbivore food webs are represented as bipartite networks (Muto-Fujita et al., 2017) defined as graphs having two different node sets.

Mutualistic networks are generally mentioned in the context of plant–animal mutualism. The representative mutualistic networks are *pollination networks* (plant–pollinator relationships) and *seed-dispersal networks* (the relationships between plants and animals that transport the plant seeds). The mutualistic networks are represented as bipartite networks (**Fig. 1(B)**) because mutualistic links are only found between two types of organisms (i.e., plants and animals) (Bascompte et al., 2003; Olesen et al., 2007).

Approaches

Searching for Non-Random Structural Patterns

Which factors determinate ecosystem stability is a long-standing question in ecology (see also Section Structure–Stability Relationship). Earlier theoretical studies (e.g., May, 1976, 1972) indicated that more complex ecosystems (i.e., larger and/or denser ecological networks) are less stable. However, this prediction is inconsistent with the fact that real-world ecosystems are complex. This is known as May's paradox. Resolving the paradox is a central topic in theoretical ecology. The paradox is expected to result from an assumption, in particular, the studies that considered that ecological networks are random for simplicity.

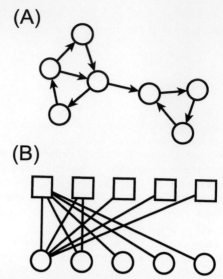

Fig. 1 A network representation of ecological networks. (A) A food web represented as a directed network. Nodes and directed edges correspond to species and prey–predator relationships, respectively. (B) A mutualistic network represented as a bipartite network. Square nodes and circle nodes are plants and animals. Edges are mutualistic relationships (pollination between plants and animals).

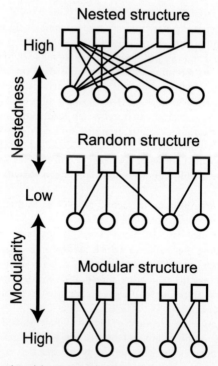

Fig. 2 Schematic diagram of nested structure and modular structure in bipartite networks.

However, is this assumption really sufficient in real-world networks? Network analyses of real-world ecological networks have found that the networks are not random. Specifically, real-world ecological networks are known to display two non-random structural patterns (**Fig. 2**). One structural pattern is the nested architecture (*nestedness*) (Bascompte *et al.*, 2003), a hierarchical structure in which the interaction pairs of a specialist species are included in those of another (generalist) species. In other words, the interaction partners of a specialist species are the subset of those of another generalist species. Another structural pattern is the modular structure (*modularity*) (Olesen *et al.*, 2007), a compartmentalized structure in which a number of dense sub-networks (modules) are weakly interconnected. Despite the correlation between them, these two structural patterns can provide complementary information on how interactions are organized in communities (Fortuna *et al.*, 2010). The degree of nestedness and modularity differ between food webs and mutualistic networks, in particular, the modularity of mutualistic networks is typically lower than that of food webs, whereas the nestedness of mutualistic networks is generally higher than that of food webs (Bascompte *et al.*, 2003; Thébault and Fontaine, 2010). However, food web subnetworks are significantly nested (Kondoh *et al.*, 2010).

Structure–Stability Relationship

Non-random structural patterns (e.g., nestedness and modularity) are expected to influence ecosystem stability. In fact, nestedness plays important roles in increasing ecosystem stability in mutualistic networks (Bastolla *et al.*, 2009; Saavedra *et al.*, 2016, 2011), and it emerges as a result of an optimization principle aimed at maximizing species abundance (Suweis *et al.*, 2013). Modularity is a particularly important property because it is related to the robustness (Hartwell *et al.*, 1999). Thus, modularity is a significant property of biomolecular networks such as signaling networks (Takemoto and Kihara, 2013) and metabolic networks (Takemoto and Borjigin, 2011; Takemoto and Oosawa, 2012). According to these previous studies, modularity is expected to affect ecosystem stability. In fact, modularity can have moderate stabilizing effects, while anti-modularity can greatly destabilize ecological networks (Grilli *et al.*, 2016). Moreover, both nestedness and modularity influence ecosystem stability (Thébault and Fontaine, 2010). The contributions of nestedness and modularity to ecosystem stability differ between food webs and mutualistic networks. For example, increasing nestedness and/or decreasing modularity enhance the stability of mutualistic networks but reduce the stability of food webs. However, note that several recent studies have cast doubt on the importance of nestedness. For example, nested architecture can be more easily acquired than previously thought (Takemoto and Arita, 2010). A study based on numerical simulation indicated that biodiversity in mutualistic communities is described by the number of mutualistic partners a species has (i.e., node degree) rather than nestedness (James *et al.*, 2012). A theoretical study (Feng and Takemoto, 2014) supports this conclusion.

Effects of Environmental Factors on Ecological Networks

In the context of the structure–stability relationship, the effects of environmental or external factors on ecological networks are also important. Given that environmental factors can be sources of perturbation (e.g., rainfall, seasonal variation of climate), it would be expected that ecological networks have an optimal structure that maximizes the ecosystem stability against such perturbations. Macroecological studies are useful for evaluating the effects of environments on ecological networks (Trøjelsgaard and Olesen, 2013), and they have enabled by the combined use of global-scale environmental data (see Section Impact of Global Warming on Ecological Networks for details). Such studies have demonstrated the climate change effects and human impacts on ecological network structure/ecosystem stability.

In the context of the structure–stability relationship, environmental factors influence the nestedness and modularity. Climatic parameters are linked to nestedness and modularity in mutualistic networks and food webs. For example, nestedness in pollination networks decreased with annual precipitation (Trøjelsgaard and Olesen, 2013), whereas modularity in seed-dispersal networks and food webs increased with temperature seasonality (Schleuning *et al.*, 2014), and precipitation seasonality (Takemoto *et al.*, 2014), respectively. The type of climatic seasonality influencing the network structure differs among ecosystems. For example, network properties were mainly affected by rainfall seasonality in freshwater ecosystems but primarily by temperature seasonality in terrestrial ecosystems (Takemoto *et al.*, 2014).

The effects of climate change and human activities are more interesting in the context of biodiversity maintenance and environmental assessment. For instance, modularity and nestedness in pollination networks correlated with the historical rate of warming (Dalsgaard *et al.*, 2013) and human impacts (e.g., human population density, land use change, infrastructure development, and so forth) (Takemoto and Kajihara, 2016). Modularity declined and nestedness increased in seed-dispersal networks in response to human impacts (Sebastián-González *et al.*, 2015) and warming velocity (Takemoto and Kajihara, 2016). Food web nestedness increased and modularity declined in response to the global warming (Takemoto and Kajihara, 2016).

Finding Keystone Species

It is also important to characterize local properties (i.e., the characteristics of each node in a complex network) in addition to the global features of ecological networks such as non-random structural patterns. Specifically, finding important (keystone) species in ecological networks is challenging (Jordan, 2009). In this context, *Centrality* analysis is useful which is an important concept in network analysis because it helps in finding central (important) nodes in complex networks (Takemoto and Oosawa, 2012). It plays an important role in network biology and network medicine (Vidal *et al.*, 2011). In general, it is expected that hub species (i.e., species that interact with many other species) are important. However, recent studies encourage a reconsideration of the importance of hubs. For example, hubs can be classified into 2 types (Han *et al.*, 2004): party hubs that coordinate a specific functional modules and date hubs that play a role in intermediates between different specific functional modules. Moreover, they found that the effect of hub removals on complex networks is different between party hubs and date hubs. In particular, the removal of date hubs leads to a more immediate collapse of the networks than that by the removal of party hubs. This finding implies the importance of hubs bridging between different network modules. Therefore, module decomposition or community detection is also useful. In particular, a functional cartography method (Guimerà and Amaral, 2005), revealing patterns of intra- and inter-module connections in complex networks, and the concept of bottlenecks (Yu *et al.*, 2007) is more useful to find important nodes. The bottlenecks are not always hubs. Importantly, the method can also evaluate the importance of inconspicuous (i.e., low-degree) nodes. In fact, a study examined pollination networks using this method, and it concluded that species serving as hubs and connectors should receive high conservation priorities (Olesen *et al.*, 2007). To detect important nodes based on the concept of bottlenecks, a number of centrality measures have been proposed (Takemoto and Oosawa, 2012). For example, an application of the PageRank algorithm for a website ranking of ecological networks is interesting. Such a centrality analysis is also useful for finding the importance of species for coextinction (Allesina and Pascual, 2009).

Illustrative Examples

In this section, we present some illustrative examples of network ecology approaches along with available databases and analysis tools.

Databases and Network Analysis Tools

Databases on ecological networks

Several databases on ecological networks are available, for example, 359 food webs are available in the GlobalWeb database (Thompson *et al.*, 2012; see Relevant Websites section). The Interaction Web DataBase (see Relevant Websites section) includes a number of ecological networks types, such as 27 food webs, 42 pollination networks, 12 seed-dispersal networks, 4 plant––herbivore networks, 4 plant–ant networks, 7 host–parasite networks, and 2 anemone–fish networks. The Web-of-Life Database (see Relevant Websites section) provides the data on a number of ecological networks: 143 pollination networks, 34 seed-dispersal networks, 34 host–parasite networks, 4 plant–herbivore networks, and 4 plant–ant networks. The Web-of-Life Database is useful for macro-ecological studies because the locations (i.e., latitude and longitude) of observation sites of the networks are also available in the database. Note that there are data duplications among the databases and the statistics as of October 4, 2017.

Network analysis tools

A number of network analysis tools are available to accelerate ecological network studies, in particular, the packages of the R-software (see Relevant Websites section). For example, the package *igraph* (igraph.org) is used for general network analysis (i.e. centrality analysis, community detection). A tutorial of *igraph* is available online at "Network Analysis and Visualization with R and igraph" link (see Relevant Websites section). The package *vegan* (see Relevant Websites section) and *bipartite* (see Relevant Websites section) are useful for analyzing mutualistic networks, represented as bipartite networks. The *nested* and *computeModules* functions in the *bipartite* package are used to calculate nestedness and modularity of bipartite networks, respectively. The functional cartography method is also available.

Environmental data

To explore the relationship between environments and ecological networks, environmental data at observation sites of ecological networks are needed. Several useful databases are available for this purpose. For example, the WorldClim database (Hijmans *et al.*, 2005; see Relevant Websites section) provides the gridded climate data with a spatial resolution of about 1 km^2. These data include annual mean temperature, temperature seasonality, annual precipitation, and rainfall seasonality (see Relevant Websites section). The data can be easily obtained using the R-package *raster*. In addition, the WorldClim database provides the past climate data (e.g., last glacial maximum climate conditions) estimated by the paleoclimate modeling intercomparison project. Climate change velocity is estimated by a comparison between the current and past climate (Dalsgaard *et al.*, 2013; Takemoto and Kajihara, 2016).

For human activities, the human footprint (HF) score from the global human footprint dataset compiled by the last of the wild project is available (Sanderson *et al.*, 2002). The HF score is provided with a spatial resolution of 1–km grid cells. According to the descriptions in the last of the wild database (see Relevant Websites section), HF scores are calculated by normalizing the human influence index defined as the sum of eight human activity-related variables: human population density, human land use change and infrastructure (built-up areas, nighttime lights, land use/land cover), and human access (coastlines, roads, railroads, navigable rivers).

The National Aeronautics and Space Act (NASA) Earth Data website (see Relevant Websites section) may be useful for obtaining other global environmental data.

Evaluating Significance of a Structural Pattern

As mentioned in Section Searching for Non-Random Structural Patterns, it is important to find non-random structural patterns of ecological networks. A comparison with randomized model networks was performed to evaluate the statistical significance of an observed network measure or structural pattern. Randomized networks are generated from a real-world network using edge-switching algorithms (Milo *et al.*, 2002) and configuration models (Catanzaro *et al.*, 2005; see also Takemoto and Oosawa, 2012). The significance of the network measure X was evaluated based on the Z-score: $Z_X = (X_{real} - X_{rand})/SD_{rand}$, where X_{real} is the

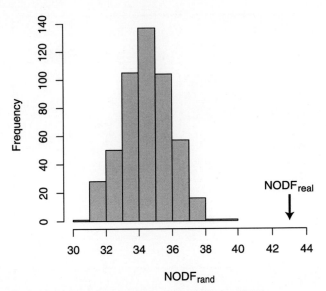

Fig. 3 The distribution of NODF$_{rand}$ obtained from 500 randomized networks compared to NODF$_{real}$.

network measure of a real-world network, $\underline{X}_{rand}$ is the average value of the network measure, and SD$_{rand}$ is the standard deviation obtained from a large number of randomized model networks. Moreover, Z-scores are also used in the context of standardization (i.e., to allow comparisons among matrices).

As an example, we indicated the significance of nestedness in a pollination network (Memmott, 1999) (**Fig. 3**). Although several definitions of nestedness are proposed, we used the Nestedness metric based on Overlap and Decreasing Fill (NODF) metrics for evaluating nestedness (Almeida-Neto *et al.*, 2008). The NODF metrics is related to the proportion of shared interactions between species pairs over a bipartite network, and it ranges from 0 to 100. A null model based on configuration models (Bascompte *et al.*, 2003) was used to generate 500 randomized networks.

Comparison of Structural Patterns Between Different Types of Ecological Networks

Are structural patterns different between different types of ecological networks? Using a dataset (Takemoto and Kajihara, 2016), we compared nestedness and modularity among food webs, pollination networks and seed-dispersal networks (**Fig. 4**). We used the standardized NODF (Z_{NODF}) and standardized M (Z_M) for evaluating nestedness and modularity, respectively. M is a well-used modularity score, and it is defined as the fraction of edges that lie within, rather than between, modules relative to that expected by chance (Fortunato, 2010). M ranges from 0 to 1; a network with a higher M indicates a higher modular structure. We used the BIPARTMOD software (Guimerà *et al.*, 2007) (see Relevant Websites section) to calculate the modularity (M) of directed networks and bipartite networks because the BIPARTMOD software finds the maximum M based on simulated annealing to minimize the resolution-limit problem in community detection (Fortunato, 2010; Fortunato and Barthélemy, 2007).

Impact of Global Warming on Ecological Networks

As mentioned in Section Effects of Environmental Factors on Ecological Networks, it is hypothesized that non-random striatal patterns are associated with environmental perturbations, given the structure–stability relationship. To test this hypothesis, using a dataset (Takemoto and Kajihara, 2016), we examined the relationship between warming velocity and nestedness/modularity in food webs (**Fig. 5**). We used the standardized NODF and standardized M to evaluate nestedness and modularity, respectively. Warming velocity (temperature change) is defined as the temporal climate gradient divided by the spatial climate gradient, with the temporal gradient in turn defined as the absolute difference between current and the last glacial maximum (LGM) climate conditions (see Takemoto and Kajihara, 2016 for details).

Results and Discussion

In the pollination network, the observed NODF score (42.8) was significantly higher than the average NODF score (34.5) obtained from the randomized networks (**Fig. 3**; $Z_{NODF} = 5.8$, $p = 7.2 \times 10^{-9}$ using the Z-test). This result indicated that the pollination network was nested than expected by chance. The significance of nestedness was generally concluded in pollination networks and seed-dispersal networks (i.e., mutualistic networks) (**Fig. 4(A)**). However, the significance of structural patterns was different between food webs and mutualistic networks. Food webs were less nested than expected by chance, while mutualistic networks

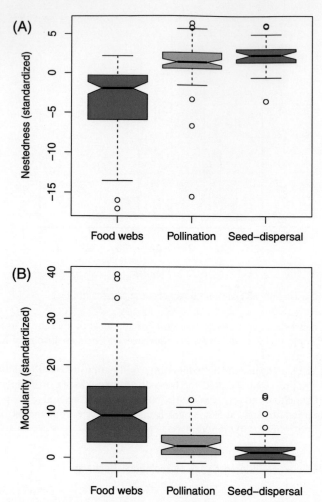

Fig. 4 Differences in nestedness (A) and modularity (B) among food webs, pollination networks, and seed-dispersal networks.

(pollination and seed-dispersal networks) are highly nested (**Fig. 4(A)**; $p < 2.2 \times 10^{-16}$ using the Kruskal–Wallis rank sum test). On the other hand, food webs were highly modularized (compartmentalized) while mutualistic networks were less modularized (**Fig. 4(B)**; $p = 8.2 \times 10^{-13}$ using the Kruskal–Wallis rank sum test). This result may be due to the fact that the contributions of nestedness and modularity to ecosystem stability differ between mutualistic networks and food webs, in that increasing nestedness and/or decreasing modularity enhance the stability of mutualistic networks but reduce the stability of food webs (Thébault and Fontaine, 2010).

A positive correlation between warming velocity and nestedness was observed (Spearman's rank correlation coefficient $r_s = 0.44$, $p = 2.7 \times 10^{-7}$). On the other hand, modularity was negatively correlated with warming velocity ($r_s = -0.30$, $p = 6.1 \times 10^{-4}$) (**Fig. 5**). Increasing nestedness and/or decreasing modularity reduced ecosystem stability in food webs (Thébault and Fontaine, 2010); as such, this result suggested that food-web stability decreases in response to environmental changes. However, more careful examinations are required. In macroecological studies, we need to remove any inherent spatial auto-correlation. In this context, a spatial eigenvector mapping modeling approach is useful (Takemoto and Kajihara, 2016).

Future Directions

Network ecology has scientific and societal impacts; however, it remains controversial.

Weighted Network Analysis

Ideally, ecological networks should be represented as weighted networks because of interaction weights, in particular, a different conclusion may be derived from comparisons between weighted and binary networks. For example, nestedness is statistically significant in binary networks but not in weighted networks (Staniczenko et al., 2013). Temperature seasonality was correlated with the weighted version of modularity, but not with the binary version of modularity (Schleuning et al., 2014). However, many

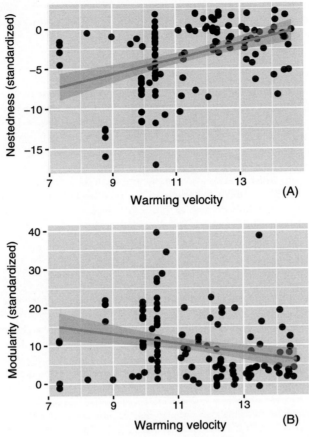

Fig. 5 Scatter plots of a network parameter versus warming velocity in food webs. Nestedness (A) and modularity (B).

of ecological network studies have considered binary networks (i.e., presence: 1, or absence: 0, of a given link) because the datasets include numerous binary data, and many software programs for network analysis generally assume binary networks. This is also because the definition of interaction weight is not uniform throughout the ecological network datasets. The interaction weight assigned to a given species pair is based on the number of contacts they share. However, the weight need to be corrected (or normalized) for factors such as sampling effort (e.g., observation area and observation time) and species abundance. Such normalization methods differ among studies. Thus, database normalizations are needed (see Section Construction of Highly Normalized Databases).

Multiplex Network Analysis

A mixture of interaction types is essential for more realistic interactions although ecological networks have been generally classified according to the interaction types. In particular, the mixture of interaction types may contribute to increasing ecosystem stability (Allesina and Tang, 2012; Mougi and Kondoh, 2012). Network measures for multiplex networks should be considered in the future to evaluate ecological network structure under more realistic conditions. Multiplex networks consist of a set of nodes connected by different link types, and have been studied in the context of social network analysis (Hoang and Antoncic, 2003). Recently, multiplex ecological network analyses have also begun (Baggio *et al.*, 2016; Pilosof *et al.*, 2017).

Construction of Highly Normalized Databases

More highly normalized databases should be constructed. In particular, information on ecological networks (e.g., species taxonomy, ecosystem type, observed site, observed area, and time) should be summarized. For example, it is possible that sampling effort affects network parameters (e.g., nestedness and modularity) when considering the species–area relationship (McGuinness, 1984) which states that the number of observed species increases with the observed area. However, the relevant information on sampling effort is not often obtained because it is not always clearly delineated in the literature and databases. In addition, the effects of phylogenetic signals should be examined because several studies have reported that phylogenetic signals are weak in ecological networks (Rezende *et al.*, 2007; Schleuning *et al.*, 2014). However, ecological networks partly consist of species whose descriptions are unknown (i.e., they are expressed as species A and species B) or ambiguous. Moreover, a restricted understanding of interspecific reactions (i.e., missing links) is a more serious limitation. To avoid these limitations, larger-scale and more highly normalized databases should be constructed, and it is especially important that data on weighted networks be expanded. In this context, data sharing (Parr and Cummings, 2005) is also important.

Linking to Dynamics

The relationship between network structure and dynamics (e.g., ecosystem stability) remains unclear. Alternative approaches are needed. Recently, the novel analytical framework revealed the universal relationship between the system's dynamics (coextinctions or cascading failures in ecological networks, in particular) and network topology (Gao et al., 2016). In particular, the authors indicated that a single universal resilience function and parameter can characterize the behavior of different networks. Stochastic models for coextinctions in ecological networks are also useful (Vieira and Almeida-Neto, 2014). Such models are capable of simulating extinction cascades far more complex than those observed in network topology-based models. For example, the models produce complex extinction cascades in which species losses in one trophic level may lead to indirect additional species losses at the same trophic level, and they predicted extinction cascades to be more likely in highly connected ecological communities. However, the validity of these approaches is still debatable; thus, more careful examinations are required. The development of such novel network-based analytical frameworks remains an important topic.

Closing Remarks

This article has described the current understanding of ecological networks revealed through data analysis and theoretical studies. Networks are useful for extracting knowledge of complex ecosystems. Real-world ecological networks are non-random, and such non-random structural patterns are different among the ecological network types. Moreover, the non-random structural patterns influence ecosystems stability, and they are linked to environmental changes. Network ecology approaches enhance our understanding of ecosystem stability and the effects of environmental change on ecosystems. Considering that ecological networks have not been fully understood yet, network ecology is a research field with high future growth potential.

Acknowledgement

KT was partly supported by a Grant-in-Aid for Young Scientists (A) from the Japan Society for the Promotion of Science (no. 17H04703).

See also: Challenges in Creating Online Biodiversity Repositories with Taxonomic Classification. Community Detection in Biological Networks. Ecosystem Monitoring Through Predictive Modeling. Large Scale Ecological Modeling With Viruses: A Review. Mapping the Environmental Microbiome. Natural Language Processing Approaches in Bioinformatics. Network Models. Network-Based Analysis for Biological Discovery. Transcriptome Analysis

References

Allesina, S., Pascual, M., 2009. Googling food webs: Can an eigenvector measure species' importance for coextinctions? PLOS Comput. Biol. 5, e1000494. doi:10.1371/journal.pcbi.1000494.

Allesina, S., Tang, S., 2012. Stability criteria for complex ecosystems. Nature 483, 205–208. doi:10.1038/nature10832.

Almeida-Neto, M., Guimarães, P., Guimarães, P.R., Loyola, R.D., Ulrich, W., 2008. A consistent metric for nestedness analysis in ecological systems: Reconciling concept and measurement. Oikos 117, 1227–1239. doi:10.1111/j.0030-1299.2008.16644.x.

Baggio, J.A., BurnSilver, S.B., Arenas, A., et al., 2016. Multiplex social ecological network analysis reveals how social changes affect community robustness more than resource depletion. Proc. Natl. Acad. Sci. USA 113, 13708–13713. doi:10.1073/pnas.1604401113.

Barabási, A.-L., 2013. Network science. Philos. Trans. R. Soc. A 371, 20120375. doi:10.1098/rsta.2012.0375.

Barabási, A.-L., Oltvai, Z.N., 2004. Network biology: Understanding the cell's functional organization. Nat. Rev. Genet. 5, 101–113. doi:10.1038/nrg1272.

Barabasi, A.L., Gulbahce, N., Loscalzo, J., 2011. Network medicine: A network-based approach to human disease. Nat. Rev. Genet. 12, 56–68. doi:10.1038/nrg2918.

Bascompte, J., 2010. Structure and dynamics of ecological networks. Science 329, 765–766. doi:10.1126/science.1194255.

Bascompte, J., Jordano, P., Melián, C.J., Olesen, J.M., 2003. The nested assembly of plant-animal mutualistic networks. Proc. Natl. Acad. Sci. USA 100, 9383–9387. doi:10.1073/pnas.1633576100.

Bastolla, U., Fortuna, M.A., Pascual-García, A., et al., 2009. The architecture of mutualistic networks minimizes competition and increases biodiversity. Nature 458, 1018–1020. doi:10.1038/nature07950.

Catanzaro, M., Boguñá, M., Pastor-Satorras, R., 2005. Generation of uncorrelated random scale-free networks. Phys. Rev. E 71, 27103. doi:10.1103/PhysRevE.71.027103.

Dalsgaard, B., Trøjelsgaard, K., Martín González, A.M., et al., 2013. Historical climate-change influences modularity and nestedness of pollination networks. Ecography (Cop.) 36, 1331–1340. doi:10.1111/j.1600-0587.2013.00201.x.

Evans, D.M., Pocock, M.J.O., Memmott, J., 2013. The robustness of a network of ecological networks to habitat loss. Ecol. Lett. doi:10.1111/ele.12117. [n/a-n/a].

Feng, W., Takemoto, K., 2014. Heterogeneity in ecological mutualistic networks dominantly determines community stability. Sci. Rep. 4, 5912. doi:10.1038/srep05912.

Fortuna, M.A., Stouffer, D.B., Olesen, J.M., *et al.*, 2010. Nestedness versus modularity in ecological networks: Two sides of the same coin? J. Anim. Ecol. 79, 811–817. doi:10.1111/j.1365-2656.2010.01688.x.

Fortunato, S., 2010. Community detection in graphs. Phys. Rep. 486, 75–174. doi:10.1016/j.physrep.2009.11.002.

Fortunato, S., Barthélemy, M., 2007. Resolution limit in community detection. Proc. Natl. Acad. Sci. USA 104, 36–41. doi:10.1073/pnas.0605965104.

Gao, J., Barzel, B., Barabási, A.-L., 2016. Universal resilience patterns in complex networks. Nature 530, 307–312. doi:10.1038/nature16948.

Grilli, J., Rogers, T., Allesina, S., 2016. Modularity and stability in ecological communities. Nat. Commun. 7, 1–10. doi:10.1038/NCOMMS12031.

Guimerà, R., Amaral, L.A.N., 2005. Functional cartography of complex metabolic networks. Nature 433, 895–900. doi:10.1038/nature03288.

Guimerà, R., Sales-Pardo, M., Amaral, L., 2007. Module identification in bipartite and directed networks. Phys. Rev. E 76, 36102. doi:10.1103/PhysRevE.76.036102.

Han, J.-D.J., Bertin, N., Hao, T., *et al.*, 2004. Evidence for dynamically organized modularity in the yeast protein–protein interaction network. Nature 430, 88–93. doi:10.1038/nature02555.

Hartwell, L.H., Hopfield, J.J., Leibler, S., Murray, A.W., 1999. From molecular to modular cell biology. Nature 402, C47–C52. doi:10.1038/35011540.

Hijmans, R.J., Cameron, S.E., Parra, J.L., Jones, P.G., Jarvis, A., 2005. Very high resolution interpolated climate surfaces for global land areas. Int. J. Climatol. 25, 1965–1978. doi:10.1002/joc.1276.

Hoang, H., Antoncic, B., 2003. Network-based research in entrepreneurship. J. Bus. Ventur. 18, 165–187. doi:10.1016/S0883-9026(02)00081-2.

James, A., Pitchford, J.W., Plank, M.J., 2012. Disentangling nestedness from models of ecological complexity. Nature 487, 227–230. doi:10.1038/nature11214.

Jansen, V.A.A., Kokkoris, G.D., 2003. Complexity and stability revisited. Ecol. Lett. 6, 498–502. doi:10.1046/j.1461-0248.2003.00464.x.

Jordan, F., 2009. Keystone species and food webs. Philos. Trans. R. Soc. B Biol. Sci. 364, 1733–1741. doi:10.1098/rstb.2008.0335.

Kondoh, M., Kato, S., Sakato, Y., 2010. Food webs are built up with nested subwebs. Ecology 91, 3123–3130.

May, R.M., 1972. Will a large complex system be stable? Nature 238, 413–414. doi:10.1038/238413a0.

May, R.M., 1976. Thresholds and breakpoints in ecosystms with a multiplicity of stable states. Nature 260, 471–477. doi:10.1038/269471a0.

McGuinness, K.A., 1984. Species-area curves. Biol. Rev. 59, 423–440. doi:10.1111/j.1469-185X.1984.tb00711.x.

Memmott, J., 1999. The structure of a plant-pollinator food web. Ecol. Lett. 2, 276–280. doi:10.1046/j.1461-0248.1999.00087.x.

Milo, R., Shen-Orr, S., Itzkovitz, S., *et al.*, 2002. Network motifs: Simple building blocks of complex networks. Science 298, 824–827. doi:10.1126/science.298.5594.824.

Mougi, A., Kondoh, M., 2012. Diversity of interaction types and ecological community stability. Science 337, 349–351. doi:10.1126/science.1220529.

Muto-Fujita, A., Takemoto, K., Kanaya, S., *et al.*, 2017. Data integration aids understanding of butterfly–host plant networks. Sci. Rep. 7, 43368. doi:10.1038/srep43368.

Olesen, J.M., Bascompte, J., Dupont, Y.L., Jordano, P., 2007. The modularity of pollination networks. Proc. Natl. Acad. Sci. USA 104, 19891–19896. doi:10.1073/pnas.0706375104.

Parr, C., Cummings, M., 2005. Data sharing in ecology and evolution. Trends Ecol. Evol. 20, 362–363. doi:10.1016/j.tree.2005.04.023.

Pilosof, S., Porter, M.A., Pascual, M., Kéfi, S., 2017. The multilayer nature of ecological networks. Nat. Ecol. Evol. 1, 101. doi:10.1038/s41559-017-0101.

Pocock, M.J.O., Evans, D.M., Memmott, J., 2012. The robustness and restoration of a network of ecological networks. Science 335, 973–977. doi:10.1126/science.1214915.

Proulx, S.R., Promislow, D.E.L., Phillips, P.C., 2005. Network thinking in ecology and evolution. Trends Ecol. Evol. 20, 345–353. doi:10.1016/j.tree.2005.04.004.

Rezende, E.L., Jordano, P., Bascompte, J., 2007. Effects of phenotypic complementarity and phylogeny on the nested structure of mutualistic networks. Oikos 116, 1919–1929. doi:10.1111/j.2007.0030-1299.16029.x.

Saavedra, S., Rohr, R.P., Olesen, J.M., Bascompte, J., 2016. Nested species interactions promote feasibility over stability during the assembly of a pollinator community. Ecol. Evol. 6, 997–1007. doi:10.1002/ece3.1930.

Saavedra, S., Stouffer, D.B., Uzzi, B., Bascompte, J., 2011. Strong contributors to network persistence are the most vulnerable to extinction. Nature 478, 233–235. doi:10.1038/nature10433.

Sanderson, E.W., Jaiteh, M., Levy, M.A., *et al.*, 2002. The human footprint and the last of the wild. Bioscience 52, 891–904. doi:10.1641/0006-3568(2002)052[0891:THFATL]2.0.CO;2.

Schleuning, M., Ingmann, L., Strauß, R., *et al.*, 2014. Ecological, historical and evolutionary determinants of modularity in weighted seed-dispersal networks. Ecol. Lett. doi:10.1111/ele.12245. [n/a-n/a].

Sebastián-González, E., Dalsgaard, B., Sandel, B., Guimarães, P.R., 2015. Macroecological trends in nestedness and modularity of seed-dispersal networks: Human impact matters. Glob. Ecol. Biogeogr. 24, 293–303. doi:10.1111/geb.12270.

Staniczenko, P.P.A., Kopp, J.C., Allesina, S., 2013. The ghost of nestedness in ecological networks. Nat. Commun. 4, 1391. doi:10.1038/ncomms2422.

Suweis, S., Simini, F., Banavar, J.R., Maritan, A., 2013. Emergence of structural and dynamical properties of ecological mutualistic networks. Nature 500, 449–452. doi:10.1038/nature12438.

Takemoto, K., Arita, M., 2010. Nested structure acquired through simple evolutionary process. J. Theor. Biol. 264, 782–786. doi:10.1016/j.jtbi.2010.03.029.

Takemoto, K., Borjigin, S., 2011. Metabolic network modularity in Archaea depends on growth conditions. PLOS ONE 6, e25874. doi:10.1371/journal.pone.0025874.

Takemoto, K., Kajihara, K., 2016. Human impacts and climate change influence nestedness and modularity in food-web and mutualistic networks. PLOS ONE 11, e0157929. doi:10.1371/journal.pone.0157929.

Takemoto, K., Kanamaru, S., Feng, W., 2014. Climatic seasonality may affect ecological network structure: Food webs and mutualistic networks. Biosystems 121, 29–37. doi:10.1016/j.biosystems.2014.06.002.

Takemoto, K., Kihara, K., 2013. Modular organization of cancer signaling networks is associated with patient survivability. Biosystems 113, 149–154. doi:10.1016/j.biosystems.2013.06.003.

Takemoto, K., Oosawa, C., 2012. Introduction to complex networks: Measures, statistical properties, and models. Stat. Mach. Learn. Approaches Netw. Anal. 45–75. doi:10.1002/9781118346990.ch2.

Thébault, E., Fontaine, C., 2010. Stability of ecological communities and the architecture of mutualistic and trophic networks. Science 329, 853–856. doi:10.1126/science.1188321.

Thompson, R.M., Brose, U., Dunne, J.A., *et al.*, 2012. Food webs: Reconciling the structure and function of biodiversity. Trends Ecol. Evol. 27, 689–697. doi:10.1016/j.tree.2012.08.005.

Trøjelsgaard, K., Olesen, J.M., 2013. Macroecology of pollination networks. Glob. Ecol. Biogeogr. 22, 149–162. doi:10.1111/j.1466-8238.2012.00777.x.

Vidal, M., Cusick, M.E., Barabási, A.-L., 2011. Interactome Networks and Human Disease. Cell 144, 986–998. doi:10.1016/j.cell.2011.02.016.

Vieira, M.C., Almeida-Neto, M., 2014. A simple stochastic model for complex coextinctions in mutualistic networks: Robustness decreases with connectance. Ecol. Lett. 18, 144–152. doi:10.1111/ele.12394.

Yu, H., Kim, P.M., Sprecher, E., Trifonov, V., Gerstein, M., 2007. The importance of bottlenecks in protein networks: Correlation with gene essentiality and expression dynamics. PLOS Comput. Biol. 3, e59. doi:10.1371/journal.pcbi.0030059.

Further Reading

Takemoto, K., Oosawa, C., 2012. Introduction to complex networks: Measures, statistical properties, and models. Stat. Mach. Learn. Approaches Netw. Anal. 45–75. doi:10.1002/9781118346990.ch2.

Hastings, A., Gross, L. (Eds.), 2012. Encyclopedia of Theoretical Ecology. University of California Press. Available at: http://www.jstor.org/stable/10.1525/j.ctt1pp0s7.

Relevant Websites

CRAN.R-project.org/package=bipartite
  Bipartite: Visualising Bipartite Networks and Calculating Some (Ecological) Indices
www.globalwebdb.com
  GlobalWeb.
https://www.globalwebdb.com
  GlobalWeb: An Online Collection of Food Webs.
www.nceas.ucsb.edu/interactionweb
  Interaction Web DataBase.
earthdata.nasa.gov
  NASA.
http://kateto.net/networks-r-igraph
  Network Analysis and Visualization With R and Igraph.
http://networksciencebook.com
  Network Science.
seeslab.info/downloads/bipartite-modularity
  SEES:lab.
sedac.ciesin.columbia.edu/data/collection/wildareas-v2/methods
  Socioeconomic Data and Applications Center (SEDAC)
www.R-project.org
  The R Project for Statistical Computing.
CRAN.R-project.org/package=vegan
  Vegan: Community Ecology Package
www.web-of-life.es
  Web of Life.
http://www.web-of-life.es
  Web of Life: Ecological Networks Database.
www.worldclim.org
  WorldClim
  Global Climate Data.
www.worldclim.org/bioclim
  WorldClim
  Global Climate Data: Bioclimatic Variables.

Biographical Sketch

Kazuhiro Takemoto received his doctoral degree in Informatics from Kyoto University in 2008 after earning his bachelor's and master's degrees in Computer Science from Kyushu Institute of Technology (Kyutech) in 2004 and 2006, respectively. After serving as a JSPS research fellow (2007–2009), a postdoctoral fellow at University of Tokyo (2009), Japan Science Technology Agency PRESTO researcher (2009–2013) and assistant professor at Kyutech (2012–2015), he is currently an associate professor at Kyutech (2015–). His research interests are network science, computational and integrative biology.

Midori Iida received her doctoral degree in Science from Ehime University in 2013 after earning her bachelor's and master's degrees in the university in 2008 and 2010, respectively. She is currently a postdoctoral fellow at Kyushu Institute of Technology (2013–). Her research interests are ecotoxicology, aquatic life science, and bioinformatics.

Dedicated Bioinformatics Analysis Hardware

Bertil Schmidt and Andreas Hildebrandt, Institut für Informatik, Mainz, Germany

Introduction

Recent years have seen a tremendous increase in the volume of data generated in the life sciences, especially propelled by the rapid progress of next-generation sequencing (NGS) technologies. For example, the sequence read archive (SRA) from NCBI which stores sequence data obtained from NGS technology now contains well over 10^{16} base-pairs (bps) of DNA sequence data and doubles in size roughly every 18 months. It is estimated that between one hundred million and two billion human genomes could be sequenced within the next decade (Stephens *et al.*, 2015) which will have a major impact on many areas of life in general with many applications in precision and personalized medicine as well as drug discovery (Schmidt and Hildebrandt, 2017).

This review provides an overview of the usage of GPUs, FPGAs, and dedicated hardware in bioinformatics. We provide some background information in Section Background/Fundamentals. In Section Applications we discuss a number of typical bioinformatics applications that can benefit from massive parallelism while Section Illustrative Examples focuses on two illustrative examples. We finish with a discussion, upcoming trends, and closing remarks in Sections Discussion, Future Directions, and Closing Remarks, respectively.

Bioinformatics is a wide field. We focus on sequence-based and structure-based applications. Further popular areas such as the usage of GPU/FPGA computing in systems biology, genome-wide analysis, or mass spectrometer data processing are outside the scope of our review.

Background/Fundamentals

Inherent to sequence-based approaches is a very common abstraction in bioinformatics – the representation of biopolymers (proteins, DNA, and RNA) as one-dimensional strings or sequences of characters. While this abstraction works surprisingly well for a wide variety of application scenarios, more detailed insight into bio-molecular function often requires more complex representations.

In structural or structure-based bioinformatics, molecules are represented as collections of atoms with properties such as element type, partial charge, covalent bonds, radius, and three-dimensional position and velocity in classical or semi-classical models, or by models of their quantum mechanical wave functions in quantum chemistry.

Even though sequences and structures are different molecular abstractions, both sequence-based and structure-based bioinformatics have the need for efficient implementations of core algorithms in common, albeit generally for different reasons (while sequence-based bioinformatics needs to scale to exploding experimental data set sizes, structure-based approaches generate tremendous amounts of floating-point computations). The traditional approach to implement bioinformatics applications is based on standard microprocessors (CPUs). Unfortunately, the compute performance provided by CPUs is often insufficient to meet current and future requirements. For example, it is projected that variant calling from NGS data alone requires around two trillion CPU hours by 2025 (Stephens *et al.*, 2015). Thus, bioinformatics data analysis shows the increased importance of computational techniques and the usage of high performance computing (HPC) infrastructures (Schatz, 2015).

Technology relevant for HPC applications comes in different variants, each with their own characteristics. Accelerator technologies such as Field Programmable Gate Arrays (FPGAs) and Graphics Processing Units (GPUs) can be an attractive option compared to large-scale compute clusters and clouds, especially in terms of price-performance ratio and energy-efficiency.

GPUs can provide around one order-of-magnitude higher peak performance compared to CPUs through massive fine-grained parallelism at a highly competitive price-performance ratio (Owens *et al.*, 2008). Using programming languages such as CUDA or OpenCL they can be used for general-purpose applications. While applications in structural bioinformat-ics are typically floating point-based and highly compute-intensive, sequence analysis algorithms are usually integer-based and more data-intensive. In order to realize an efficient GPU-parallelization, both careful algorithm design and optimized implementation are required.

Field programmable gate arrays (FPGAs) are programmable hardware chips mainly consisting of configurable logic gates and memory blocks (Compton and Hauck, 2002). FPGA configurations are generally specified using a hardware description languages such as VHDL or Verilog. FPGAs are particularly attractive for applications in cryptography and sequence analysis where relatively simple highly regular operations are performed on short integers. Thus, using FPGAs for analyzing large-scale sequence data offers a great opportunity for algorithm acceleration.

Finally, if monetary cost is not an issue, dedicated hardware can be designed for the problem at hand in the form of so-called application-specific integrated circuits (ASICs). These hand-tailored processors can achieve unparalleled efficiency and performance, but suffer from inflexibility with respect to alternative workloads.

Applications

BLAST

The most popular software to search a library of sequences with a query sequence is BLAST. BLAST is a family programs of which BLASTN (Nucleotide-nucleotide BLAST) and BLASTP (Protein-protein BLAST) are most widely used. The heuristics used consist of a pipeline of several stages, which have different characteristics (e.g., two-hit method in BLASTP vs. word-matching in BLASTN) but also contain common components.

Acceleration of BLAST on FPGAs and GPUs has received much attention. Mercury BLASTP (Jacob et al., 2008) and Mercury BLASTN (Lancaster et al., 2009) implement all stages of NCBI BLASTP and NCBI BLASTN. They achieve speedups of around one order-of-magnitude on a system with two Xilinx Virtex-II 6000-6 FPGAs compared to NCBI BLAST executed on two Opteron CPUs. The more recent CAAD BLAST (Mahram and Herbordt, 2015) reports a 5x speedup on a single Virtex-6 FPGA over a fully parallel implementation of the reference code on a modern multi-core CPU. Wienbrandt (2014) presents a BLASTP implementation that scales to 128 Spartan3-5000 FPGAs. Timelogic's Tera-BLAST (2017) is a commercial FPGA implementation of various BLAST algorithms. GPU-BLASTP (Vouzis and Sahinidis, 2010), CUDA-BLASTP (Liu et al., 2011a,b), and cuBLASTP (Zhang et al., 2015) are implementations of BLASTP on CUDA-enabled GPUs. G-BLASTN (Zhao and Chu, 2014) and HS-BLASTN (Chen et al., 2015) report massively parallel GPU accelerations of BLASTN and MegaBLAST. The recent H-BLAST (Ye et al., 2017) achieves speedups ranging between 4 and 10 on one K20x GPU over the sequential NCBI-BLASTP.

NGS Read Mapping

The *mapping* of NGS reads to a reference genome sequence is usually one of the first steps in many NGS data processing pipelines. Most existing read mappers (or aligners) are based on a seed-and-extend approach where short exact matches (seeds) are first identified using an index data structure. These seeds are then verified whether they can be extended to a full alignment. Approaches to accelerate read mapping can be categorized by the utilized indexing data structure.

Early GPU approaches including CUSHAW (Liu et al., 2012) and SOAP3-dp (Luo et al., 2013) are based on the Burrows Wheeler transform (BWT) with FM-index. This approach has a low memory footprint and thus can fit into the limited GPU global memory. Arioc (Wilton et al., 2015) and PEANUT (Koster and Rahmann, 2014) are examples of more recent GPU-accelerated read mapper based on hash tables. nvBowtie is a GPU-accelerated of the popular Bowtie2 software built on top of the NVBio (NVIDIA, see Relevant Websites section) library achieving a speedup of up to 8 on a K80 GPU compared to Bowtie2 running with 20 CPU threads. BowMapCL (Nogueira et al., 2016) is based on OpenCL instead of CUDA in order to support multiple heterogeneous accelerators reporting speedups between 2 and 7.5 on a single GeForce GTX Ti GPU compared to multi-threaded Bowtie and BWA. Chen et al. (2013) proposed a hash-based short read mapper for FPGAs. Houtgast et al. (2015) implemented an FPGA-accelerated as well as a GPU-accelerated (Houtgast et al., 2017) version of BWA-MEM, which achieve roughly the same performance. Arram et al. (2017) leverage FPGAs to accelerated Bowtie2 and achieve a 28x speedup on a Maxeler MPC-X2000 node consisting of eight Altera Stratix-V FPGAs compared to multi-threaded Bowtie2 running with 16 CPU threads. Fernandez et al. (2015) reported a 12-fold speed gain on a Conway HC-2ex system compared to Bowtie running eight CPU threads.

Variant Calling

Variant calling is a an important operation performed on the alignments returned by mapping NGS reads to a reference genome. GSNP (Lu et al., 2011) is a GPU-accelerated version of SOAPsnp reporting speedups of around 40. Luo et al. (2014) leverages the power of GPUs to implement a pipeline consisting of read mapping, re-alignment and variant calling called BALSA. It is able to process 90 samples from the 1000 genomes project in around 3 days on a cluster with five GTX680 GPUs. The commercial DRAGEN platform provides FPGA-accelerated implementations of genome and transcriptome NGS analysis pipelines. Miller et al. (2015) report whole genome sequencing (WGS) on the DRAGEN system within 26-h time from blood sample to provisional diagnosis.

Denovo Assembly and Error Correction

While approaches for accelerating denovo genome assembly have been limited, the time-consuming pre-processing step for correcting errors in NGS reads has received considerable more attention. CUDA-EC (Shi et al., 2010) was the first approach to speedup error correction (EC) on GPUs based on Bloom filters. DecGPU (Liu et al., 2011a,b) improved this approach to GPU clusters. nvLighter is a GPU-accelerated re-engineering of the Lighter error corrector using the NVBio library (NVIDIA, see Relevant Websites section). Ramachandran et al. (2015) perform FPGA-accelerated EC achieving speedups of around 40x on a Stratix V FPGA compared to the CPU-based BLESS algorithm.

Multiple Sequence Alignment

Progressive alignment is a widely used but time-consuming approach for computing multiple sequence alignments (MSAs). MSA-CUDA (Liu et al., 2009b) was the first GPU-implementation of all three stages of the ClustalW pipeline. CUDA ClustalW (Hung et al., 2015) is an efficient GPU implementation of ClustalW v2 on multiple GPUs. G-MSA (Blazewicz et al., 2013) accelerates the

T-Coffee algorithm to achieve around two orders-of-magnitude speedup. QuickProbs (Gudys and Deorowicz, 2014) is a variant of the highly accurate MSAProbs (Liu *et al.*, 2010b) algorithm suited for GPUs. A reconfigurable computing approach proposed by Oliver *et al.* (2005b) accelerates the first stage of ClustalW by computing pairwise alignments with a linear systolic array. Lloyd and Snell (2011) mapped the third phase onto FPGAs reporting a speedup of up to 150 versus a single core. Mahram and Herbordt (2012) accelerate ClustalW through pipelined prefiltering on a FPGA.

Molecular Force Fields

A fundamental abstraction in structural bioinformatics is the notion of a molecular force field. In principle, simulating molecular behaviour at atomic resolution would require modelling quantum mechanics, as non-classical effects are crucial at such length scales. Such models, however, have very large computational demands. Molecular force fields address this problem by prescribing collections of interatomic interaction energies that would lead to a behaviour compatible with quantum mechanics and experimental evidence. For example, a quadratic energy component can be used to model covalent bonds, as long as the expected length deviations are not too large. More realistic models for larger deviations can be obtained using higher order polynomials or exponential models, albeit at greater computational cost.

Force fields contain two different classes of interaction: *bonded* and *non-bonded*. Bonded interactions occur only between atoms sharing a covalent bond, while non-bonded ones (typically van-der-Waals interactions and electrostatics) can occur between any pair of atoms. Since the number of bonds per atom is bounded by a small number, the number of bonded interactions is linear in the number of atoms in the system. Non-bonded interactions, on the other hand, grow quadratically and, hence, typically dominate computational cost of force field evaluations.

Depending on the system under consideration and the required accuracy, more sophisticated types of force fields may be needed, such as polarizable (Baker, 2015) or reactive (Farah *et al.*, 2012). On the other hand, force fields may be coarse-grained over collections of atoms to save on computational cost (Barnoud and Monticelli, 2015).

Molecular Docking

Molecules exert influence on one another through physical interactions. To regulate a system as complex as a living cell, or even a multi-cellular organism, at the molecular level requires a high degree of specificity: a molecule intended to activate or suppress a particular target can not be allowed to blindly interact with any other protein in the vicinity. The physical origin of this specificity is the short range and relatively weak strength of the pair-wise interactions: to exert a noticeable influence, many pairs of atoms on the surfaces of both molecules need to come into close contact. This implies a degree of geometric compatibility. To use Emil Fischer's famous image, one molecule fits like a key into the other molecule's lock (Fischer, 1894), or more appropriately using Koshland's induced fit-model (Koshland, 1958), like a hand into a glove that is slightly too small, modifying its shape along the way.

Predicting whether a given molecule could bind to an active site of another, and with what binding strength, is of great interest to many fields of science and industry, such as drug design. In principle, the structure and binding strength of such molecular complexes could be studied using molecular dynamics (MD), given suitable force fields. The computational effort required for this task, though, is currently out of reach outside of dedicated MD hardware such as Anton 2. However, in many cases it might not be necessary to study the full dynamics of molecular complexes to predict molecular binding. Empirically, we often find that one geometric arrangement of such a complex has significantly lower energy than others and, hence, dominates the partition function of the system (Rarey *et al.*, 1996). Predicting this structure and approximating its binding energy is the task of docking algorithms.

In essence, docking methods are optimization procedures for some scoring function approximating the binding strength with respect to the molecular degrees of freedom. In rigid docking, only six degrees of freedom are optimized: the translation and rotation of one of the partners. In reality, though, molecules are non-rigid and can undergo flexible rearrangements upon binding, which greatly increases the dimensionality of the optimization problem as well as the number of local minima.

Most work on the GPU-based acceleration of flexible docking algorithms was devoted to porting heuristic global optimization approaches, such as genetic algorithms, simulated annealing, or differential evolution, or to vectorize the interaction energy calculations required for these methods (Korb *et al.*, 2011). Autodock (Morris *et al.*, 2009), for instance, has seen several user-contributed ports to the GPU, some of which are quite popular. In addition, Autodock has been ported to FPGAs (Pechan and Feher, 2011), yielding significant speed-up compared to a CPU-implementation, but being more or less on par with a GPU variant.

Illustrative Examples

Pairwise Sequence Alignment

For pairwise sequence similarity computation, the Smith-Waterman (SW) algorithm, the Needleman Wunsch (NW) algorithm, and their variants are widely used. They compute the optimal local, global, or semi-global alignment of two sequences under a given scoring scheme by means of dynamic programming (DP). The associated time complexity proportional to the product of sequence lengths makes this method time-consuming in many application scenarios. Consequently, parallelization approaches have been proposed for a variety of hardware platforms. We compare the achieved performance in terms of billion cell updates per second (GCUPS).

Early solutions based on dedicated hardware were already proposed in the 1980s. Lipton and Lopresti (1985) proposed one of the first designs based on a systolic array implemented on an ASIC. Systolic array approaches are typically based on a linear array of simple processing elements (PEs) working in lock-step. The DP matrix is typically calculated using a *wavefront* scheme, whereby in every clock cycle a minor diagonal of the DP matrix can be calculated in parallel.

Consider two sequences S_1 and S_2 of length l_1 and l_2. Assuming there are l_1 PEs each storing one character of S_1, then we can compute the whole DP matrix in $l_1 + l_2$ time steps instead of $\mathcal{O}(l_1 \cdot l_2)$ on a single processor. If the number of PEs is less than the sequence length, a suitable partitioning scheme needs to be implemented.

FPGA-based solutions typically implement the systolic array approach on reconfigurable hardware instead of an ASIC. An early solution for DNA sequences using linear gap scoring by Hoang and Lopresti (1992) used a SPLASH board consisting of 32 Xilinx XC3090 FPGA chips to implement a linear of 248 PEs to achieve 0.27 GCUPS. Oliver et al. (2005a) proposed a systolic array of 252 PEs to scan a database of protein sequences with a query sequence using affine gap penalties and achieved 5.8 GCUPS on a single Virtex II XC2V6000 FPGA. Zhang et al. (2007) further improved the PE design to achieve 25.6 GCUPS on an Altera Stratix II FPGA. Benkrid et al. (2009) improved the flexibility with respect to supporting different sequence types and scoring schemes by implementing a parameterized SW algorithm FPGA-based accelerator. Recent work by Xia et al. (2017) extends the design from a score-only alignment computation to performing backtracking on a linear systolic array to achieve a performance of 106 GCUPS using 512 PEs on a Xilinx XC7VX1140T FPGA.

SW has also been implemented on a number of commercial FPGA systems. Wienbrandt (2014) reports 6020 GCUPS for aligning DNA sequences on the SciEngines RIVYERA S6-LX150 platform equipped with 128 FPGAs of type Xilinx Spartan6-LX150. Vermij (2011) achieves a performance of 460 GCUPS on the Convey HC1 system composed of four XC5VLX330 FPGA chips for SW without backtracking. Another commercial solution is from Timelogic (see Relevant Websites Section).

The first implementation of SW using CUDA was proposed by Manavski and Valle (2008) who achieved a performance of up to 3.6 GCUPS on a GeForce 8800 GTX. CUDASW++ 1.0 (Liu et al., 2009a) considerably improved performance to 16 GCUPS on a GeForce GTX 295 by a more careful consideration of the GPU memory hierarchy and the introduction of the inter-task parallelization approach where each CUDA thread computes a distinct alignment of a protein sequence database search task. The successor versions CUDASW++ 2.0 (Liu et al., 2010a) and CUDASW++ 3.0 (Liu et al., 2010a) further improved performance to 29.7 GCUPS on a GeForce GTX 295 and 185.6 GCUPS on a dual-GPU GeForce GTX 690, respectively. SWhybrid (Lan et al., 2017) is currently fastest CUDA implementation of SW-based protein database search achieving close to 300 GCUPS on a GeForce GTX 1080.

CUDAlign focuses on the computation of a single pairwise alignment of long (chromosome-length) DNA sequences. CUDAlign 1.0 (Sandes and Melo, 2010) calculates the DP matrix with the wavefront method and achieved a performance of 20.4 GCUPS on a GeForce GTX 280. CUDAlign 2.1 (Sandes and Melo, 2013) incorporates the retrieval of optimal local alignments in linear space with a performance 58.2 GCUPS on a GeForce 560Ti. The latest version (Sandes et al., 2016) achieves 10,370 GCUPS on a cluster with 384 GPUs for computing exact chromosome-wide DNA alignments.

Molecular Dynamics

Today, molecular dynamics (MD) simulations clearly belong to the most important computational techniques to study molecular properties (Karplus and McCammon, 2002). Starting from an initial configuration (i.e., position and velocity vectors for every atom in the system), a chemical parametrization (i.e., information about the chemical type for each atom) and a molecular force field, it solves for the classical equations of motion of the involved particles, with suitable modifications to achieve, e.g., constant temperature or pressure (Rapaport, 2004).

Integrating the equations of motion requires very small time steps, leading to large numbers of iterations to simulate biologically relevant time scales. Since system sizes can become quite large and since the number of non-bonded interaction evaluations grows quadratically, computation of such pairwise interaction energies typically dominates MD running times. Using increasingly long simulations of increasingly large systems with increasingly complex force fields hence leads to a growing demand for computational resources. Consequently, MD codes made use of GPU accelerators almost immediately after NVIDIA introduced CUDA in 2007 (Stone, 2007; Liu et al., 2008). Today, all practically relevant MD implementations are GPU-accelerated, and in many cases scale well to multi-GPU setups.

A fully featured MD application is a complex, often modular, system with components that cover the whole MD pipeline from input preparation up to trajectory analysis. Porting the entirety of this pipeline to the GPU would hardly be suitable. Hence, modern MD codes provide GPU-accelerated implementations for time-critical parts of the pipeline that vectorize well and typically cover the majority of the computational work. Most vectorization efforts focus on the computation of non-bonded interactions, which are further decomposed into short-range and long-range interactions depending on the distance between the two interacting atoms.

Considering the functional form of the van-der-Waals interaction, which very quickly saturates to zero with distance, it is often possible to neglect all van-der-Waals terms between atoms at a distance beyond a user-defined threshold (optionally, the terms can be smoothly pulled to zero in a small region just below the threshold using a switch-off function) and hence treat these as short-ranged interactions only. For electrostatics, though, this approximation is often unsuitable. The most popular method to handle these interactions computationally is the so-called *particle mesh Ewald* (PME) summation (Darden et al., 1993), which separates the sum of all pairwise electrostatic interactions into a short-ranged and a long-ranged part as described above. The first part converges quickly in real space and can hence be directly computed. To parallelize the second part, three different strategies can be employed (Rovigatti *et al.*, 2015):

1. Atom-decomposition, where each compute unit receives a block of atoms and computes all interactions of each atom in the block,
2. Force-decomposition, where all potentially interacting atom pairs are distributed over compute units, and
3. Space-decomposition, where a compute unit calculates the interactions of atoms within a certain spatial domain.

Atom-decomposition is currently the most popular approach on the GPU by far, implemented, e.g., in GROMACS, NAMD, or HOOMD-blue, even though force-decomposition might be preferable in some cases (Rovigatti *et al.*, 2015).

The second term of the above decomposition (long-ranged contributions to non-bonded interactions) converges slowly in real space, but quickly in reciprocal- or Fourier space. For periodic boundary conditions, this suggests to use fast Fourier transform algorithms on charge- and potential grids to convert them to Fourier space where suitable cut-offs enable efficient computation of the interaction contributions. While PME reduces computational effort on long-range interaction from $\mathcal{O}(n^2)$ to $\mathcal{O}(n \log(n))$, it is harder to parallelize than pairwise interaction summation due to non-negligible communications overhead. Hence, many MD packages, such as GROMACS (Pall and Hess, 2013; Pall *et al.*, 2015) or YASARA (Krieger and Vriend, 2015) use GPU acceleration only for the short-range contributions while keeping the rest (including PME) on the CPU which, incidentally, allows to keep CPU and GPU busy simultaneously. Other packages such as AMBER PMEMD (Götz *et al.*, 2012; Salomon-Ferrer *et al.*, 2013), ACEMD (Harvey *et al.*, 2009a,b), or HOOMD-blue (Anderson *et al.*, 2008; Glaser *et al.*, 2015) report that more than 90% of their computational workload is offloaded to the GPU.

When porting MD codes to accelerators, a further challenging complication arises from inherent numerical instabilities (e.g., through cancellation effects in summing over many small interactions). As shown by Colberg and Höfling (2011), single precision floating point calculations are insufficient even in simple situations using only Lennard-Jones potentials. While modern GPUs can handle double precision floating point computations reasonably well, albeit at the cost of greatly reduced performance. Hence, modern MD code often uses what is known as *mixed* precision or SPDP-mode (Götz *et al.*, 2012), where only sensitive operations are computed in double precision. Alternatively, fixed-point integer arithmetic or even extended precision can be used as a replacement for double precision in so-called SPFP-mode (Le Grand *et al.*, 2013) or SPXP-mode (Thall, 2006).

All this effort seems to be justified in practice. According to a white paper by NVIDIA, running at least part of the code on the GPU typically leads to speed-ups of 3x to 8x when compared to multi-threaded CPU implementations.

Considering the enormous compute requirements of MD simulations, it is not surprising that other kinds of accelerator hardware besides commodity graphics cards have been and are still being actively investigated. FPGA-based solutions such as Yang *et al.* (2007), Khan *et al.* (2013), Waidyasooriya *et al.* (2016) have repeatedly proven to yield significantly better performance than CPU-only code, but seem to suffer from sub-optimal floating point performance, in particular for those parts where double precision is required. The importance of MD as an indispensable tool for molecular studies even justifies the enormous cost of designing special-purpose MD ASIC solutions. RIKEN's MDGRAPE-4 system (Ohmura *et al.*, 2014) consists of 512 specially designed systems-on-chip (SoC), each of which has 64 pipelines for the computation of non-bonded interactions, running at 0.8 GHz and 65 Ten-silica Xtensa LX configurable processor cores with single-precision floating point units, clocked at 0.6 GHz, for the remaining calculations. Each SoC has a peak performance of 51.2 billion interaction computations per second. Optical transmitters and receivers are used for internode connections.

D.E. Shaw Research's (DESRES) Anton-architecture (Shaw *et al.*, 2008, 2014) in its second iteration is the first platform that was able to simulate several microseconds per day on systems composed of millions of atoms. All parts of the simulation pipeline are handled by the ASICs without resorting to general purpose commodity host processors. To fully leverage the system, DESRES also developed new MD algorithms adapted to the strenghts of the system. The Anton 2 systems in operation consist of 512 ASIC nodes, each of which dedicates about one quarter of its die area to the computation of pair-wise interactions and additionally contains 66 general purpose programmable cores. Each ASIC is clocked at 1.65 GHz.

As demonstrated by the Anton system, the ability to routinely study protein behaviour at the micro- or even millisecond range is truly transformative, both for understanding general principles of biochemistry and protein organization and for dedicated biological or pharmaceutical applications (Chung *et al.*, 2015; Hu *et al.*, 2015; Pan *et al.*, 2017).

Discussion

In sequence-based bioinformatics the usage of GPU/FPGA/dedicated hardware is driven by the enormous progress of NGS technologies. Typical algorithms often rely on combinations of integer-intensive computations (such as dynamic programming for alignments) or big data techniques (such as large-scale index data structures). As many applications involving a single genome have already been parallelized, we are starting to see first acceleration approaches for analyzing *metagenomic* NGS data. MEGAHIT (Li *et al.*, 2016) assembles large and complex metagenomic datasets using a parallel algorithm for constructing succinct de Bruijn Graphs on GPUs while (Kobus *et al.*, 2017) perform metagenomic read classification on GPUs.

On the other hand, structural bioinformatics models biomolecules as three-dimensional objects with non-trivial and often dynamic geometry. It relies on concepts from computational physics, such as interaction energies and force fields, and from computational geometry, such as geometric queries and comparisons. Such models are typically based on floating-point computations, often even at double precision. With increasing system sizes, simulation time steps, and model accuracy, the required amount of floating point operations grows immensely. Hence, structural bioinformatics quickly adopted accelerator technologies. Here, we have discussed some of the typical application scenarios focusing on MD and docking.

Compared to MD, using accelerators for docking has received considerably less attention. In fact, not all docking methods seem to map well to GPUs or FPGAs at all. One area that does profit significantly from GPU acceleration is rigid docking, which today is typically only considered suitable as part of a larger pipeline including flexibilization steps. Many rigid docking algorithms are based on computing correlations through multiplication in Fourier space, as pioneered by Katchalski-Katzir et al. (1992). FFT computation can easily be offloaded to the GPU, which has been used to speed up docking based on spherical harmonics (Ritchie and Venkatraman, 2010) or in the classical Katchalski-Katzir sense (Ohue et al., 2014). In addition, the code has been ported to Intel's MIC architecture, where it was found to perform less well than on GPUs (Ohue et al., 2014).

Future Directions

Programming of massively parallel accelerators can still be cumbersome. Implementing hand-optimized algorithms from scratch using languages like CUDA can be time-consuming since it requires a deep understanding of GPU architectures. However, this might change with the availability of highly optimized libraries and the advancements of pragma-based languages such as OpenACC. Furthermore, a major hurdle in taking up FPGA-based systems can be their programming complexity, leading to long software development cycles. Furthermore, application software is often not compatible across different FPGA generations. However, the recent development of higher level programming languages such as OpenCL for FPGAs and the availability of associated libraries offer an opportunity to write more portable yet efficient code.

Bioinformatics is becoming increasingly data-intensive which also implies an increased application of *deep learning* techniques (Ching et al., 2017). Central to these techniques is large performance gains on parallel hardware such as GPUs or FPGAs. A recent example is *DeepVariant* (DePristo and Poplin, 2017) a tool for variant calling from NGS data that trains a highly complex neural network from MSAs.

However, increasingly complex neural networks and data sets require an even higher performance which motivates the recent development of ASICs (such as Google's Tensor Processing Unit (TPU)) or special functional components (such as the integration of Tensor cores in Volta GPUs). Thus, we can expect the regular usage of such hardware in bioinformatics analysis in the near future.

Closing Remarks

In this article we have surveyed the usage of parallel accelerator solutions for sequence-based and structure-based bioinformatics. State-of-the-art tools based on GPUs, FPGAs, or even ASICs can provide significant speedups compared to traditional CPU-based solutions. Since biological data sets continue to increase rapidly this approach provides a promising future direction to address the *big data deluge* challenge in the 'omics' sciences.

See also: Computing for Bioinformatics. Infrastructure for High-Performance Computing: Grids and Grid Computing. Infrastructures for High-Performance Computing: Cloud Computing. Infrastructures for High-Performance Computing: Cloud Infrastructures. Models and Languages for High-Performance Computing. Natural Language Processing Approaches in Bioinformatics. Parallel Architectures for Bioinformatics

---

References

Anderson, J.A., Lorenz, C.D., Travesset, A., 2008. General purpose molecular dynamics simulations fully implemented on graphics processing units. Journal of Computational Physics 227 (10), 5342–5359.

Arram, J., et al., 2017. Leveraging FPGAs for accelerating short read alignment. IEEE/ACM Transactions on Computational Biology and Bioinformatics 14 (3), 668–677.

Baker, C.M., 2015. Polarizable force fields for molecular dynamics simulations of biomolecules. Wiley Interdisciplinary Reviews: Computational Molecular Science 5 (2), 241–254.

Barnoud, J., Monticelli, L., 2015. Coarse-grained force fields for molecular simulations. Methods in Molecular Biology 1215, 125–149.

Benkrid, K., Liu, Y., Benkrid, A., 2009. A highly parameterized and efficient FPGA-based skeleton for pairwise biological sequence alignment. IEEE Transactions on VLSI 17 (4), 561–570.

Blazewicz, J., et al., 2013. G-MSA – A GPU-based, fast and accurate algorithm for multiple sequence alignment. Journal of Parallel and Distributed Computing 73 (1), 32–41.

Chen, Y., et al., 2015. High speed BLASTN: An accelerated MegaBLAST search tool. Nucleic Acids Research 43 (16), 7762–7768.

Chen, Y., Schmidt, B., Maskell, D., 2013. A hybrid short read mapping accelerator. BMC Bioinformatics 14 (1), 67.

Ching, T., et al., 2017. Opportunities and obstacles for deep learning in biology and medicine. bioRxiv. 142760.

Chung, H.S., et al., 2015. Structural origin of slow diffusion in protein folding. Science 349 (6255), 1504–1510.

Colberg, P.H., Höfling, F., 2011. Highly accelerated simulations of glassy dynamics using GPUs: Caveats on limited floating-point precision. Computer Physics Communications 182 (5), 1120–1129.

Compton, K., Hauck, S., 2002. Reconfigurable computing: A survey of systems and software. ACM Computing Surveys 34 (2), 171–210.

Darden, T., York, D., Pedersen, L., 1993. Particle mesh Ewald: An N log(N) method for Ewald sums in large systems. The Journal of Chemical Physics 98 (12), 10089–10092.

DePristo, M., Poplin, R., 2017. DeepVariant: Highly accurate genomes with deep neural networks. Available at: https://research.googleblog.com/2017/12/deepvariant-highly-accurate-genomes.html

Farah, K., Müller-Plathe, F., Böhm, M.C., 2012. Classical reactive molecular dynamics implementations: State of the art. ChemPhysChem 13 (5), 1127–1151.

Fernandez, E.B., et al., 2015. FHAST: Fpga-based acceleration of Bowtie in hardware. IEEE/ACM Transactions on Computational Biology and Bioinformatics 12 (5), 973–981.

Fischer, E., 1894. Einfluss der Konfiguration auf die Wirkung der Enzyme. Berichte Der Deutschen Chemischen Gesellschaft 27 (3), 2985–2993.

Glaser, J., et al., 2015. Strong scaling of general-purpose molecular dynamics simulations on GPUs. Computer Physics Communications 192, 97–107.

Götz, A.W., Williamson, M.J., Xu, D., et al., 2012. Routine microsecond molecular dynamics simulations with AMBER on GPUs. Journal of Chemical Theory and Computation 8 (5), 1542–1555.

Gudy, A., Deorowicz, S., 2014. QuickProbs – A fast multiple sequence alignment algorithm designed for graphics processors. PLOS ONE 9 (2), e88901.

Harvey, M.J., De Fabritiis, G., 2009a. An implementation of the smooth particle-mesh Ewald (PME) method on GPU hardware. Journal of Chemical Theory and Computation 5, 2371–2377.

Harvey, M.J., Giupponi, G., De Fabritiis, G., 2009b. ACEMD: Accelerating biomolecular dynamics in the microsecond time scale. Journal of Chemical Theory and Computation 5 (6), 1632–1639.

Hoang, D.Z., Lopresti, D., 1992. FPGA implementation of systolic sequence alignment. International Workshop on Field Programmable Logic and Applications. 183–191.

Houtgast, E.J., et al., 2017. An efficient GPU-accelerated implementation of genomic short read mapping with BWA-MEM. ACM SIGARCH Computer Architecture News 44 (4), 38–43.

Houtgast, E.J., et al., 2015. An FPGA-based systolic array to accelerate the BWA-MEM genomic mapping algorithm. In: 2015 International Conference on Embedded Computer Systems: Architectures, Modeling, and Simulation (SAMOS), IEEE.

Hung, C., et al., 2015. CUDA ClustalW: An efficient parallel algorithm for progressive multiple sequence alignment on Multi-GPUs. Computational Biology and Chemistry 58, 62–68.

Hu, X., et al., 2015. The dynamics of single protein molecules is non-equilibrium and self-similar over thirteen decades in time. Nature Physics 12 (2), 171–174.

Jacob, A., et al., 2008. Mercury BLASTP: Accelerating protein sequence alignment. ACM Transactions on Reconfigurable Technology and Systems 1 (2), 9.

Karplus, M., McCammon, J.A., 2002. Molecular dynamics simulations of biomolecules. Nature Structural Biology 9 (9), 646–652.

Katchalski-Katzir, E., et al., 1992. Molecular surface recognition: Determination of geometric fit between proteins and their ligands by correlation techniques. Proceedings of the National Academy of Sciences of the United States of America 89 (6), 2195–2199.

Khan, M.A., Chiu, M., Herbordt, M.C., 2013. FPGA-accelerated molecular dynamics. In: Benkrid, K., Vanderbauwhede, W. (Eds.), High-Performance Computing Using FPGAs. New York, NY: Springer, p. 105135.

Kobus, R., et al., 2017. Accelerating metagenomic read classification on CUDA-enabled GPUs. BMC Bioinformatics 18 (1), 11.

Korb, O., Stutzle, Exner, T.E., 2011. Accelerating molecular docking calculations using graphics processing units. Journal of Chemical Information and Modeling 51 (4), 865–876.

Koshland, D.E., 1958. Application of a theory of enzyme specificity to protein synthesis. Proceedings of the National Academy of Sciences of the United States of America 44 (2), 98–104.

Koster, J., Rahmann, S., 2014. Massively parallel read mapping on GPUs with the q-group index and PEANUT. PeerJ 2, e606.

Krieger, E., Vriend, G., 2015. New ways to boost molecular dynamics simulations. Journal of Computational Chemistry 36 (13), 996–1007.

Lancaster, J., Buhler, J., Chamberlain, R.D., 2009. Acceleration of ungapped extension in Mercury BLAST. Microprocessors and Microsystems 33 (4), 281–289.

Lan, H., Liu, W., Liu, Y., Schmidt, B., 2017. SWhybrid: A hybrid-parallel framework for large-scale protein sequence database search. IEEE IPDPS 2017, 42–51.

Li, D., et al., 2016. MEGAHIT v1.0: A fast and scalable metagenome assembler driven by advanced methodologies and community practices. Methods 102, 3–11.

Le Grand, S., Gtz, A.W., Walker, R.C., 2013. SPFP: Speed without compromise – A mixed precision model for GPU accelerated molecular dynamics simulations. Computer Physics Communications 184 (2), 374–380.

Lipton, R.J., Lopresti, D., 1985. A systolic array for rapid string comparison. In: Proceedings of the Chapel Hill Conference on VLSI. 363–376.

Liu, Y., Schmidt, B., Maskell, D., 2010a. CUDASW ++ 2.0: Enhanced Smith-Waterman protein database search on CUDA-enabled GPUs based on SIMT and virtualized SIMD abstractions. BMC Research Notes 3 (1), 93.

Liu, W., et al., 2008. Accelerating molecular dynamics simulations using Graphics Processing Units with CUDA. Computer Physics Communications 179 (9), 634–641.

Liu, W., Schmidt, B., Müller-Wittig, K.W., 2011a. CUDA-BLASTP: Accelerating BLASTP on CUDA-enabled graphics hardware. IEEE/ACM Transactions on Computational Biology and Bioinformatics 8 (6), 1678–1684.

Liu, Y., Maskell, D., Schmidt, B., 2009a. CUDASW ++ : Optimizing Smith-Waterman sequence database searches for CUDA-enabled graphics processing units. BMC Research Notes 2 (1), 73.

Liu, Y., Schmidt, B., Maskell, D., 2009b. MSA-CUDA: Multiple sequence alignment on graphics processing units with CUDA. In: Proceedings of the 20th IEEE International Conference Application-specific Systems, Architectures and Processors.

Liu, Y., Schmidt, B., Maskell, D., 2010b. MSAProbs: Multiple sequence alignment based on pair hidden Markov models and partition function posterior probabilities. Bioinformatics 26 (16), 1958–1964.

Liu, Y., Schmidt, B., Maskell, D., 2011b. DecGPU: Distributed error correction on massively parallel graphics processing units using CUDA and MPI. BMC Bioinformatics 12 (1), 85.

Liu, Y., Schmidt, B., Maskell, D., 2012. CUSHAW: A CUDA compatible short read aligner to large genomes based on the BurrowsWheeler transform. Bioinformatics 28 (14), 1830–1837.

Lloyd, S., Snell, Q.O., 2011. Accelerated large-scale multiple sequence alignment. BMC Bioinformatics 12 (1), 466.

Lu, M., et al., 2011. GSNP: A DNA single-nucleotide polymorphism detection system with GPU acceleration. In: 2011 International Conference on Parallel Processing (ICPP), IEEE.

Luo, R., et al., 2013. SOAP3-dp: Fast, accurate and sensitive GPU-based short read aligner. PLOS ONE 8 (5), e65632.

Luo, R., et al., 2014. BALSA: Integrated secondary analysis for whole-genome and whole-exome sequencing, accelerated by GPU. PeerJ 2, e421.

Mahram, A., Herbordt, M.C., 2012. FMSA: FPGA-accelerated ClustalW-based multiple sequence alignment through pipelined prefiltering. In: Proceedings of the 20th Annual International Symposium on Field-Programmable Custom Computing Machines (FCCM), IEEE.

Mahram, A., Herbordt, M.C., 2015. NCBI BLASTP on high-performance reconfigurable computing systems. ACM Transactions on Reconfigurable Technology and Systems 7 (4), 33.

Manavski, S., Valle, G., 2008. CUDA compatible GPU cards as efficient hardware accelerators for Smith Waterman sequence alignment. BMC Bioinformatics 9 (S2),

Miller, N.A., et al., 2015. A 26-h system of highly sensitive whole genome sequencing for emergency management of genetic diseases. Genome Medicine 7 (1), 100.

Morris, G.M., et al., 2009. AutoDock4 and AutoDockTools4: Automated docking with selective receptor flexibility. Journal of Computational Chemistry 30 (16), 2785–2791.

Nogueira, D., Tomas, P., Roma, N., 2016. BowMapCL: Burrows-wheeler mapping on multiple heterogeneous accelerators. IEEE/ACM Transactions on Computational Biology and Bioinformatics 13 (5), 926–938.

Ohmura, I., et al., 2014. MDGRAPE-4: A special-purpose computer system for molecular dynamics simulations. Philosophical Transactions Series A, Mathematical, Physical, and Engineering Sciences 372 (2012),

Owens, J.D., et al., 2008. GPU computing. Proceedings of the IEEE 96 (5), 879–899.

Ohue, M., et al., 2014. MEGADOCK 4.0: An ultra-high-performance protein-protein docking software for heterogeneous supercomputers. Bioinformatics 30 (22), 3281–3283.

Oliver, T., et al., 2005a. Using reconfigurable hardware to accelerate multiple sequence alignment with ClustalW. Bioinformatics 21 (16), 3431–3432.

Oliver, T.F., Schmidt, B., Maskell, D.L., 2005b. Reconfigurable architectures for bio-sequence database scanning on FPGAs. IEEE Transactions on Circuits and Systems II 52 (12), 851–855.

Pall, S., Abraham, M.J., Kutzner, C., Hess, B., Lindahl, E., 2015. Tackling Exascale Software Challenges in Molecular Dynamics Simulations With GROMACS. Springer. pp. 3–27.

Pall, S., Hess, B., 2013. A flexible algorithm for calculating pair interactions on SIMD architectures. Computer Physics Communications 184 (12), 2641–2650.

Pan, A.C., et al., 2017. Quantitative characterization of the binding and unbinding of millimolar drug fragments with molecular dynamics simulations. Journal of Chemical Theory and Computation 13 (7), 3372–3377.

Pechan, I., Feher, B., 2011. Molecular docking on FPGA and GPU platforms. In: 2011 Proceedings of the 21st International Conference on Field Programmable Logic and Applications, pp. 474–477. IEEE.

Ramachandran, A., et al., 2015. FPGA accelerated DNA error correction. In: Design, Automation & Test in Europe Conference & Exhibition (DATE), IEEE.

Rapaport, D.C., 2004. The Art of Molecular Dynamics Simulation. Cambridge University Press.

Rarey, M., et al., 1996. A fast flexible docking method using an incremental construction algorithm. Journal of Molecular Biology 261 (3), 470–489.

Ritchie, D.W., Venkatraman, V., 2010. Ultra-fast FFT protein docking on graphics processors. Bioinformatics 26 (19), 2398–2405.

Rovigatti, L., et al., 2015. A comparison between parallelization approaches in molecular dynamics simulations on GPUs. Journal of Computational Chemistry 36 (1), 1–8.

Salomon-Ferrer, R., Gtz, A.W., Poole, D., Le Grand, S., Walker, R.C., 2013. Routine microsecond molecular dynamics simulations with AMBER on GPUs. 2. explicit solvent particle mesh ewald. Journal of Chemical Theory and Computation 9 (9), 3878–3888.

Sandes, E.F.O., Melo, A., 2010. CUDAlign: Using GPU to accelerate the comparison of megabase genomic sequences. ACM SIGPLAN Notices 45 (5), 137–146.

Sandes, E.F.O., Melo, A., 2013. Retrieving Smith-Waterman alignments with optimizations for megabase biological sequences using GPU. IEEE Transactions on Parallel and Distributed Systems 24 (5), 1009–1021.

Sandes, E.F.O., Miranda, G., Ayguade, E., et al., 2016. CUDAlign 4.0: Incremental speculative traceback for exact chromosome-wide alignment in GPU clusters. IEEE Transactions on Parallel and Distributed Systems 27 (10), 2838–2850.

Schmidt, B., Hildebrandt, A., 2017. Next-generation seqeuncing: Big data meets high performance computing. Drug Discovery Today 22 (4), 712–717.

Schatz, M.C., 2015. Biological data sciences in genome research. Genome Research 25 (10), 1417–1422.

Shaw, D.E., et al., 2014. Anton 2: Raising the bar for performance and programmability in a special-purpose molecular dynamics supercomputer. In: SC14: International Conference for High Performance Computing, Networking, Storage and Analysis, pp. 41–53. IEEE.

Shi, H., et al., 2010. A parallel algorithm for error correction in high-throughput short-read data on CUDA-enabled graphics hardware. Journal of Computational Biology 17 (4), 603–615.

Stephens, Z.D., et al., 2015. Big Data: Astronomical or genomical? PLOS Biology 13, e1002195.

Stone, J.E., Phillips, J.C., Freddolino, P.L., et al., 2007. Accelerating molecular modeling applications with graphics processors. Journal of Computational Chemistry 28 (16), 2618–2640.

Thall, A., 2006. Extended-precision floating-point numbers for GPU computation. ACM SIGGRAPH 2006 Research Posters, pp. 1–12.

Vermij, E., 2011. Genetic sequence alignment on a supercomputing platform. MS Thesis, TU Delft, Netherlands.

Vouzis, P.D., Sahinidis, N.V., 2010. GPU-BLAST: Using graphics processors to accelerate protein sequence alignment. Bioinformatics 27 (2), 182–188.

Waidyasooriya, H.M., Hariyama, M., Kasahara, K., 2016. Architecture of an FPGA accelerator for molecular dynamics simulation using OpenCL. In: 2016 IEEE/ACIS Proceedings of the 15th International Conference on Computer and Information Science (ICIS), p. 15.

Wienbrandt, L., 2014. The FPGA-based high-performance computer RIVY-ERA for applications in bioinformatics. Conference on Computability in Europe. 383–392.

Wilton, R., et al., 2015. Arioc: High-throughput read alignment with GPU-accelerated exploration of the seed-and-extend search space. PeerJ 3, e808.

Xia, F., et al., 2017. FPGASW: Accelerating large-scale Smith–Waterman sequence alignment application with backtracking on FPGA linear systolic array. Interdisciplinary Sciences: Computational Life Sciences. 1–13.

Yang, X., Mou, S., Dou, Y., 2007. FPGA-accelerated molecular dynamics simulations: An overview. Reconfigurable Computing: Architectures, Tools and Applications. 293–301.

Ye, W., et al., 2017. H-BLAST: A fast protein sequence alignment toolkit on heterogeneous computers with GPUs. Bioinformatics 33 (8), 1130–1138.

Zhang, J., Wang, H., Feng, W., 2015. cublastp: Fine-grained parallelization of protein sequence search on cpu + gpu. IEEE/ACM Transactions on Computational Biology and Bioinformatics.

Zhang, P., Tan, G., Gau, G.R., 2007. Implementation of the Smith-Waterman algorithm on a reconfigurable supercomputing platform. In: Proceedings of the 1st international workshop on high-performance reconfigurable computing technology and applications, pp. 39–48.

Zhao, K., Chu, X., 2014. G-BLASTN: Accelerating nucleotide alignment by graphics processors. Bioinformatics 30 (10), 1384–1391.

Further Reading

Anderson, J.A., Lorenz, C.D., Travesset, A., 2008. General purpose molecular dynamics simulations fully implemented on graphics processing units. Journal of Computational Physics 227 (10), 5342–5359.

Compton, K., Hauck, S., 2002. Reconfigurable computing: A survey of systems and software. ACM Computing Surveys 34 (2), 171–210.

Le Grand, S., Gtz, A.W., Walker, R.C., 2013. SPFP: Speed without compromise – A mixed precision model for GPU accelerated molecular dynamics simulations. Computer Physics Communications 184 (2), 374–380.

Liu, Y., Wirawan, A., Schmidt, B., 2013. CUDASW ++ 3.0: Accelerating Smith-Waterman protein database search by coupling CPU and GPU SIMD instructions. BMC Bioinformatics 14, 117.

Miller, N.A., et al., 2015. A 26-h system of highly sensitive whole genome sequencing for emergency management of genetic diseases. Genome Medicine 7 (1), 100.

Oliver, T.F., Schmidt, B., Maskell, D.L., 2005b. Reconfigurable architectures for bio-sequence database scanning on FPGAs. IEEE Transactions on Circuits and Systems II 52 (12), 851–855.

Owens, J.D., et al., 2008. GPU computing. Proceedings of the IEEE 96 (5), 879–899.

Schatz, M.C., 2015. Biological data sciences in genome research. Genome Research 25 (10), 1417–1422.

Schmidt, B., Hildebrandt, A., 2017. Next-generation seqeuncing: Big data meets high performance computing. Drug Discovery Today 22 (4), 712–717.

Shaw, D., et al., 2008. Anton, a special-purpose machine for molecular dynamics simulation. Communications of the ACM 51 (7), 91.

Stephens, Z.D., et al., 2015. Big Data: Astronomical or genomical? PLOS Biology 13, e1002195.

Stone, J.E., Phillips, J.C., Freddolino, P.L., et al., 2007. Accelerating molecular modeling applications with graphics processors. Journal of Computational Chemistry 28 (16), 2618–2640.

Relevant Websites

https://developer.nvidia.com/nvbio
    NVIDIA.
http://www.timelogic.com/catalog/757
    TimeLogic.
http://www.timelogic.com/catalog/758/decyphersw
    TimeLogic.

Biographical Sketch

Bertil Schmidt is tenured Full Professor and Chair for Parallel and Distributed Architectures at the University of Mainz, Germany. Prior to that he was a faculty member at Nanyang Technological University (Singapore) and at University of New South Wales (UNSW). His research group has designed a variety of algorithms and tools for Bioinformatics mainly focusing on the analysis of large-scale sequence and read datasets. For his research work, he has received a GPU Research Center award, GPU Education Center Award, CUDA Academic Partnership award, CUDA Professor Partnership award and Best Paper Awards at IEEE ASAP 2015 and IEEE ASAP 2009.

Andreas Hildebrandt is tenured Full Professor and Chair for Bioinformatics and Software Engineering at Johannes Gutenberg University Mainz in Germany. Previously, he was the head of the independent junior research group on protein–protein interactions and computational proteomics at the Center for Bioinformatics of Saarland University. His main research interests include structural bioinformatics, computational proteomics and computational systems biology, where his group develops and applies novel techniques and program packages. He is also one of the core developers of the BALL library.

Computational Pipelines and Workflows in Bioinformatics

Jeremy Leipzig, Drexel University, Philadelphia, PA, United States

Background and Fundamentals

Bioinformatic analyses aim to extract biological insights from raw data. Tenable approaches to such goals must be composed of discrete ordered steps, performed either in serial or in parallel. These steps must be communicated to other scientists, and generalized to accept both new biological samples and parameters. In many cases these steps, or *transformations*, have traditionally been performed by compiled software designed for Unix-compatible operating systems on distributed grid or cluster hardware at large academic and commercial institutions. As scientific computing has matured and borrowed from the greater technology community, this approach is evolving to further abstract out both hardware and operating infrastructure using portable, scalable, and expressive software.

Modern bioinformatics has been largely dominated by sequence analysis, whose processing and organizational demands has catalyzed the growth of pipelines and workflows in a field that has witnessed an explosion in throughput over the last two decades – from capillary gel electrophoresis systems (ABI 3700) capable of sequencing 300 kbp per day in 1999 (Mullikin and McMurragy, 1999) to a sequencing by synthesis (SBS) machines capable of sequencing 16 whole human genomes (49.6 bbp; Illumina NovaSeq 6000) at a depth of 30x – a 165,333% increase. However, virtually every -omics subfield, as well as other life sciences and physical sciences, deal with the same need to formalize ad-hoc processes for the purposes of robustness, growth, transparency, and reproducibility.

The terms "pipelines" and "workflows" are used somewhat interchangeably, although conventionally a pipeline is a narrower description of a file transformation process, while a workflow can refer to any set of processes that encompass multiple goals and sometimes manual steps, even those in the laboratory.

Several disparate and changing elements – tools, pipeline frameworks, data and metadata standards, programming languages, notebooks, computational infrastructure – make up the "workflow landscape". In particular three developments – the cloud, containerization, and semantic encoding – are rapidly changing how modern bioinformatic workflows are created. These are continue to evolve with the aim of improved scalability, accessibility, transparency, and perhaps most importantly, reproducibility.

The process of converting existing analyses into workflows catalyzes the modularization of steps common to many analyses, such that individual steps or entire pipelines can be reused for similar experiments with minimal modifications. A tangentially related aspect of this modularization is the formal introduction of common steps in machine learning workflows such as feature engineering, training, validation, testing, and optimization. The efficient dispatch of processing necessary for high-throughput analysis, processing thousands of samples and millions of data points, is an evolving focus of modern pipeline frameworks, and will be discussed in terms of scalability and computational runtime as well as in development time and handholding.

Scalability in workflows has been tackled from two sides of the high performance computing – one side the maturing use of cloud platforms to quickly deploy jobs on rented on-demand virtualized computing infrastructure using a variety of traditional cluster batch queueing systems, cloud-aware batch schedulers, container management systems, and serverless computing services. The other side is the development of distributed computing environments for big data, such as Apache Spark which can leverage large clusters while presenting a seamless interactive programming session. While these efforts are not incompatible, cloud computing has engaged the evolution of workflows and workflow frameworks more than Spark by allowing existing executables and algorithms to scale without reimplementation, basically running unaware of the workflow that is driving them.

The primary goal of a workflow, regardless of its formal structure, is "narrow-sense reproducibility" – the automated replication of results given a set of input data and parameters. A secondary or auxiliary goal is broad-sense reproducibility (aka "reviewable research," Stodden *et al.*, 2013), the transparent communication of a scientific experiment or train of thought which encompasses both the intent of an analysis and its underlying source code. Workflows connect the versioning, dependencies, and literate programming, to reproducibly produce results from data and metadata (Sandve *et al.*, 2013) in the narrow repeatable sense. Workflows will also play larger roles in broader senses of reproducibility – confirmable, generalizable, reviewable, auditable, verifiable and validatable – though these goals are dependent on semantic representations of workflows and integration with the larger scientific process, namely publishing (Hrynaszkiewicz *et al.*, 2014). Ideally, future article reviewers will swap out both individual tools and data in a workflow attached to a submission to test its underlying scientific validity.

In the information sciences community, the concept of reproducibility is often discussed in terms of "provenance" (Kanwal *et al.*, 2017; Gil *et al.*, 2007). Provenance is defined by the W3C as "a record that describes the people, institutions, entities, and activities involved in producing, influencing, or delivering a piece of data or a thing" (Moreau *et al.*, 2013). In workflow circles provenance refers to the origins of both input data, tools, results, and intermediates.

An interesting review by Cohen-Boulakia *et al.* (2017) examines workflow systems in the context of *reproducibility-friendliness*, using real-world use cases as a means to explore the needs of researchers in terms of reproducibility, and the strengths and limitations of pipeline frameworks. The workflow specification, provenance tracking, and dependency management are compared in terms of scientific demands related to reproducibility. These include rapid integration of new tools, tool version tracking, workflow execution provenance trace, and workflow segment reuse. This group also distinguishes "levels of reproducibility" from the increasingly broad senses of repeat, replicate, and reproduce, each introducing progressively more variation from the initial *in silico* experiment. The article concludes that more tools are needed to assist in semantic annotation of workflows and to take advantage of the provenance being generated in terms of both validation and discovery; many workflows are not modular enough to work in an interoperable manner (accepting intermediates from other workflows, for example). Other areas highlighted as having potential for improvement are integration with publications, and the development of more user-friendly and abstracted workbenches.

Accessibility and shareability of workflows is key to realizing the vision of enabling "open analysis" or "open research" – transmission of an entire workflow to another individual to be evaluated or reused without significant technical hurdles. This goes hand-in-hand with the goal of analytical democratization, allowing biologists and other non-specialists to build and use pipelines with their own data. In the life sciences, many of these goals have been formalized in manifests such as the FAIR principles (Wilkinson *et al.*, 2016) and are being actively pursued by various international consortiums such as GA4GH (The Global Alliance for Genomics and Health*, 2016), as well as government-funded efforts such as the NIH's Big Data to Knowledge (BD2K) program, and NCI's Cancer Genomics Cloud (Davis-Dusenbery, 2015). These initiatives and other like them seek to build shared workflow platforms on the massive data coordination that has taken place over the past two decades.

Systems and Applications

Toolkits

Bioinformatics tools that cater to novel experimental research methods tend to be composed of simple tools aggregated into "toolkits" with common data formats and syntax conventions that focus on a domain. To maintain flexibility, tools within toolkits can be used "a la carte", enabling tools from other toolkits or user-developed tools to be swapped in, and can take on a variety of software language implementations. Popular toolkits include the Genomic Analysis Toolkit for variant analysis (McKenna *et al.*, 2010) as of version 3.6 composed mostly of Java .jar files, the Tuxedo suite, comprised of tools including Bowtie, TopHat, and Cufflinks, compiled executables for DNA-seq and RNA-Seq analysis (Trapnell *et al.*, 2012). Bioconductor, which is currently composed of 1383 R packages, supports many interrelated tools organized by "Views" and "Common Workflows" such as Differential Expression and Flow Cytometry, with an emphasis on recipes and use-case driven documentation (Gentleman *et al.*, 2004). Bioconductor is supported by a backbone of key classes and methods such as GenomicRanges, SummarizedExperiment (Huber *et al.*, 2015), and VariantAnnotation (Obenchain *et al.*, 2014).

Ready-made pipelines

Ready-made pipelines such as bcbio-nextgen (Guimera, 2012), omicspipe (Fisch *et al.*, 2015) and Churchill (Kelly *et al.*, 2015) are tightly bound collections of existing third-party tools. Such pipelines are typically driven by large configuration files and are designed to provide optimal settings and high performance for typical workflows, however they are not extensible in the sense that user-developed tools cannot be easily integrated which can limit their application in experimental settings. Some ready-made pipelines, such as recount2 (Collado-Torres *et al.*, 2017), are composed of purely of libraries used within programming environments such as R, which offer the advantage of remaining in an interactive session for the duration of a pipeline, although this limits scalability.

Some pipelines are built on existing pipeline frameworks (discussed below). These include Knime4Bio (Lindenbaum *et al.*, 2011) and KNIME4NGS (Hastreiter *et al.*, 2017), based on the Konstanz Information Miner (KNIME). META-Pipe (Robertsen *et al.*, 2016) and deepTools (Ramírez *et al.*, 2014) are popular toolkits built for use within Galaxy server workbench.

Plummer and Twin (2015) compared three popular ready-made pipelines for processing 16s rRNA gene sequences to study bacterial phylogeny and taxonomy – MG-RAST (Meyer *et al.*, 2008), QIIME (Caporaso *et al.*, 2010), and Mothur (Schloss *et al.*, 2009). This study found large differences in usability and performance, as well as tertiary measures such as user support, visualizations, and overall robustness in terms of being able to stably accommodate a range of experimental scenarios and varying input quality as encountered in the field. Robustness is arguably the primary criterion that determines the longevity of a pipeline in the scientific community, and together with reproducibility are the two most important abstract properties for judging pipeline quality (Sztromwasser, 2014).

Pipeline frameworks

Pipeline frameworks, also known as workflow management systems (WMS), or scientific workflow management systems (SWfMS), provide structured means of expressing file transformations that are common to bioinformatics analyses. These provide several key advantages over their predecessors in terms of robustness, reusability, extensibility and scalability (Goble and De Roure, 2009).

Until recently, most basic bioinformatic pipelines were developed using scripting languages, either in a Unix shell or a high-level scripting language such as Perl or Python. In such a scheme, variables are used to represent files or parameters and loops to perform iterative processing. An alternative to used the build automation tool, Make (Stallman and McGrath, 2002, described below), which provides a means of representing a dependency graph in terms of filename suffix wildcard recipes. Parameterization and dependency graphs, the descendents of scripts and Makefiles, have been universal components of all modern pipeline frameworks.

The dependency graphs described above, more precisely understood as directed acyclic graphs (DAGs), can represent both serial and parallel tasks and file dependency order. DAGs serve as the conceptual logic underpinning all workflows and are useful as a visual representation as well as a functional map. The means by which DAGs are created distinguishes modern bioinformatic pipeline frameworks. DAGs are created either by the workflow itself, inferred from filename extensions or explicit inline task dependencies, or by the user, who may use a formal markup such as XML, a domain-specific shorthand, a full-featured language, or using a graphical user interface. Regardless of their origins, DAGs enable the internal workflow engine to determine which unfinished dependencies are needed to complete a task and which can be run in parallel.

A review by Leipzig categorizes frameworks using three criteria: using an implicit or explicit syntax, using a configuration, convention or class-based design paradigm and offering a command line or workbench interface (Leipzig, 2016). While these categories are broad and do not address myriad novel features developed by individual projects, the taxonomy is useful for understanding the major types of pipeline frameworks.

Implicit and explicit DSLs

Implicit frameworks can trace their lineage to the Make, a compile build tool designed in the 1970s as a domain-specific language. Make offers a set syntax consisting of symbols which relate files with different filename suffixes into implicit wildcard pattern rules, which define how these files can be transformed into one another. As such, Make can produce any file for which a rule chain exists. Make is highly dependent on file modification dates in order to assess necessary tasks. Crucial to the success of Make and other implicit frameworks is *reentrancy* – the ability to stop and resume the progress of a pipeline at any stage without penalty. Files that are already completed are not remade, but new samples introduced as input are processed. Reentrancy is indispensable during pipeline development, when errors are frequent, but is also useful for quickly recovering from unexpected events in production, as well as bypassing redundant tasks as projects that grow both in scale and scope. Implicit frameworks such as Snakemake (Köster and Rahmann, 2012) and Nextflow (Di Tommaso *et al.*, 2017) augment the Make approach with support for full featured programming languages, Python and Groovy respectively.

Snakemake remains true to the conventions introduced by Make, in that pipelines are composed of explicit rules with target inputs (usually lists of files), and implicit wildcard rules which describe file transformations. While Python functions can be used as inputs, allowing databases or other actors to be sources of filenames, the general approach is file-centric. This reliance on file naming can necessitate a number of lookup functions when there is no clear suffix chain to associate inputs and outputs, as may happen when processing files named with universally unique identifiers (UUIDs). Nextflow introduces *channels* which hold file types shared between tasks, obviating the need to carefully tag intermediates with complex file suffix name schemes. Inputs and outputs in Nextflow are typed values, offering greater flexibility in terms of holding results in memory while still retaining reentrancy through caching.

Rather than rely on inputs, outputs, and generalized rules, *explicit frameworks* link *tasks* in a set order specified by the user. This approach is more intuitive to some whose conception of a pipeline does not deviate from a set script, but limits the flexibility of the pipeline to produce target files not initially anticipated by the user.

Configuration frameworks

We consider domain-specific languages (DSL) to use a "conventional" design paradigm, whereby the DAG is formed by following syntax conventions formed by the DSL and the underlying language, with an aim to reduce the overall typing load on the developer. A more structured approach is to demand a configuration file formatted in a markup language such as XML or JSON, with little or no embedded code.

Configuration frameworks are always explicit. In addition, they also strongly distinguish workflows from provenance. The respective terms assigned are "workflow templates" and "workflow runs" – a workflow run having all variables assigned and serving as a log of steps that are performed in the creation of an output.

Of course, workflow configurations can themselves be generated by scripts, various wizards both graphical and command line, or APIs, but they must be validated and loaded using strict parsers. This leaves little room for inline on-the-fly conditional logic based on input itself, often called "dynamic branching", for example skipping corrupt files or switching sensitivity/specificity parameters.

Established configuration frameworks such as Pegasus (Deelman *et al.*, 2005), are often tightly coupled to an existing execution engine, such as Condor (Thain *et al.*, 2005), which enables efficient job scheduling but at the expense of portability. Modern configuration frameworks, such as Cromwell from the Broad Institute (Voss *et al.*, 2017), support a number of execution engines and schedulers, both for in-house clusters and the cloud.

Workbenches

Workbenches typically offer a graphical canvas that allows users to connect nodes (representing tasks) with edges (representing parallel or serial data flow). Open source server workbenches can be installed locally or in the cloud. With over 1500 citations, Galaxy (Goecks *et al.*, 2010) is arguably the most popular bioinformatics workbench. Galaxy is designed to be run on a local cluster or in the cloud. Knime (Berthold *et al.*, 2009), Taverna (Wolstencroft *et al.*, 2013), Vistrails (Callahan *et al.*, 2006), and Kepler (Ludäscher *et al.*, 2006) can be installed on a desktop computer but offer server and cloud-based execution. Commercial cloud-based workbenches such as DNANexus, SevenBridges, and Illumina's BaseSpace are remotely managed, Software-as-a-Service (SaaS) installations, which provide private accounts and abstract the provisioning of underlying cloud platforms for usability. Developments in object storage such as Amazon Web Services (AWS) Simple Storage Service (S3) buckets have allowed researchers some autonomy in how their data is owned and managed.

Some *collaborative analysis platforms* combine the mechanics of canvas workbenches with an existing suite of reference data, highly customized reports, and comparative samples, typically focused to maximize relevant results in a specific domain. One example is Cavatica, a joint project of Seven Bridges Genomics and Children's Hospital of Philadelphia to accelerate pediatric cancer research by encouraging data sharing with the open use of standards-compliant tools and workflows and visualizations developed for the cBioPortal (Gao *et al.*, 2013).

For high throughput analyses there is usually a need to automate data management away from the graphical user interface using local command line dispatch. This is normally supported using APIs. Projects such as BioBlend (Sloggett *et al.*, 2013) provide tools to remotely manage CloudMan (Afgan *et al.*, 2010) cloud instances which host Galaxy workflows. Other platforms such as Arvados (Amstutz *et al.*, 2016) and Cyverse (Devisetty *et al.*, 2016) provide support for API-driven workflows and a large data collection that can be browsed using dashboard tools.

Common workflow language

Because of the highly linked nature of these operations, workbenches require strongly typed tool definitions and workflow descriptions to perform in a robust manner. The sheer number of bioinformatic tools and frameworks warranted a level of standardization in order to reduce redundancy and enhance portability of tools and workflows. The proliferation of pipeline frameworks, each with its own idiosyncratic syntax for defining tool inputs, outputs, and parameters, presents frustrations for both users and developers. An effort spearheaded by developers at commercial cloud workbench platforms has developed a standard format for defining both workflows and tool definitions that can be used by any configuration-based framework. The Common Workflow Language (CWL) is a multi-vendor consortium that has developed a configuration-based standard for interfacing with command-line software, describing workflows, and adding semantic annotation to these elements.

Two important technologies being used by CWL are JSON-LD, a linked human-readable data format that provides semantic encoding to elements such as tools, inputs, and outputs, and Docker (see below), a containerization layer that provides isolation and a consistent environment for tools. This effort has spawned an entirely new schema validation system supporting JSON-LD (SALAD), the cwl-viewer for visualization, and a number of APIs for producing CWL programmatically (**Figs 1** and **2**).

Class-based frameworks

Class-based frameworks dispense with the symbolism of domain-specific language clauses and rely on pure-language annotators and methods to introduce workflow operations into existing code.

A number of class-based frameworks have been developed in e-commerce settings, and these tend to provide better support than scientific pipeline frameworks for event-triggered executions, live monitoring, and database operations common to real-time applications. On such framework, Luigi, is similar in spirit to implicit DSLs, but its syntax is implemented as Python classes and methods. Airflow is a pure-python DAG generator – its implementation is closer to a configuration file API than a means to annotate existing code.

Toil is a pure-Python framework developed to take full advantage of cloud infrastructure, featuring a "pluggable backend APIs for machine provisioning, job scheduling and file management" to enable execution on heterogeneous computing environments and sophisticated "caching and streaming" to reduce file system contention (Vivian *et al.*, 2017). Toil received a great deal of attention after a group at UCSC reported it the framework to process an RNA-Seq run of 200,000 samples performed over 90 h on Amazon Web Services, utilizing roughly 32,000 compute cores. The group was able to reduce the cost of analysis by using a combination of Toil optimizations and kmer-based transcript quantification tools. While Toil can ingest CWL, it also includes a number of methods to leverage the Spark-based ADAM toolset developed by the Big Data Genomics (Paten *et al.*, 2015).

Future Directions

Workflows do not exist in a vacuum but are an integral part of the scientific publishing cycle. A major driving factor in workflow standards and innovation has been responding to developments in the cloud, containerization, notebooks, and the semantic web.

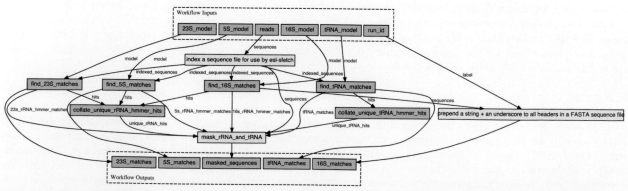

Fig. 1   A workflow using rRNA selector as viewed through the cwl-viewer. Reproduced from Robinson, M., Soiland-Reyes, S., Crusoe, M.R., Goble, C., 2017. CWL viewer: The common workflow language viewer. F1000Research 6, 1075. Available at: http://dx.doi.org/10.7490/f1000research.1114375.1 (accessed 25.06.17); Lee, J.-H., Yi, H., Chun, J., 2011. rRNASelector: A computer program for selecting ribosomal RNA encoding sequences from metagenomic and metatranscriptomic shotgun libraries. Journal of Microbiology 49 (4), 689–691.

```
inputs:
  reads:
    type: File
    format: edam:format_1929  # FASTA
…
outputs:
  16S_matches:
    type: File
    outputSource: prepend_header_for_QIIME/labeled_sequences
…
  index_reads:
    run: ../tools/esl-sfetch-index.cwl
    in:
      sequences: reads
    out: [ sequences_with_index ]
…
$namespaces:
  edam: http://edamontology.org/
  s: http://schema.org/
$schemas:
  - http://edamontology.org/EDAM_1.16.owl
  - https://schema.org/docs/schema_org_rdfa.html
```

Fig. 2 Snippets of the CWL code show input, output semantically-defined inputs and outputs, and a step in the workflow. Reproduced from Crusoe, M., 2017. Available at: https://github.com/ProteinsWebTeam/ebi-metagenomics-cwl/blob/master/workflows/rna-selector.cwl.

The cloud

While the cloud can be considered a rented timeshare of local computing infrastructure, the cloud has introduced many new opportunities for scalability and the sharing of workflows than was previously possible. Several technologies that are relatively rare in academic computing environments have also become de rigueur among workflows, namely object storage. Workflows must accommodate staging files from object store to local volumes. This would be highly impractical without frameworks.

Mixed-use schedulers

Apache Mesos and Apache YARN are cluster management platforms that are designed to conduct job scheduling across distributed heterogeneous networks. While these projects have different origins, they both compete with traditional batch engines such as SGE, LSF, and SLURM.

Cloud-native batch execution engines

Proprietary cloud execution engines such as Microsoft's Azure Batch and Amazon's AWS Batch assume job scheduling and auto-scaling as an integrated feature of the Platform-As-A-Service (PAAS) model. Because they are designed from the ground up to work in the cloud, these engines are equipped to better profile existing CPU, memory, and cost-structures to complete jobs than batch schedulers designed for traditional grids, and they require less setup time. As of this writing Nextflow has already introduced support for AWS Batch, which suggests widespread support for cloud-native engines is likely in the near future.

Serverless computing

Some cloud providers offer a class of services aimed at users who wish to dispense with all of the setup and overhead associated with servers. Serverless platforms such as AWS Lambda, Microsoft Azure Functions, and Google Cloud Functions allow simple code to run without the need to configure and deploy servers. While this "microservices" approach is quite familiar to software developers in e-commerce, rapid adoption in bioinformatics will likely coalesce around specific high-load query cases. Because serverless functions lack the basic process exit status logic of traditional DSL-based frameworks, process status checkpoints need to be manually built to work with file-centric sequence analysis programs.

Dependency management and containerization

The installation and communication of a corpus of versioned compatible tools is a major point of frustration for developers and a concern for reproducible research. Because workflows consist of many tools chained together, the management and harmony of these dependencies is of paramount importance to workflow sharing. Two developments, universal package management and containerization, have addressed this problem using different but potentially compatible approaches.

Universal package management

A package manager that originated from Python, Conda provides an elegant solution by offering a number of dependencies – including Python and R packages and compiled binary executables – in a single versioned requirements file. Conda packages specify other Conda dependencies such as linked libraries, cross-platform binaries that are distributed within the Conda ecosystem through channels. One such channel is Bioconda, which contains over 2700 bioinformatics-specific packages submitted by a wide group of developers and validated using an automated continuous testing and integration scheme.

Containerization

Docker is an open source project that provides a lightweight and configurable operating system virtualization layer called a container. Containers allow software with exact dependencies and requirements to run isolated from the external underlying operating system. This approach ensures software operates consistently and identically on different operating systems, both locally and in the cloud. By isolating programs from each other and the OS, Docker guarantees the replicable execution of scientific software within a workflow regardless of conflicting dependencies or versions with other software in that workflow or software installed by the user (Boettiger, 2015), with no appreciable loss of performance (Di Tommaso *et al.*, 2015).

The Docker architecture utilizes configured scripts called Dockerfiles that are processed to build "images", ready-to-run files used to instantiate containers. Dockerfiles can themselves be specified with base images, enabling a certain level of templating. The process of building images is far more time consuming than instantiating containers, so it is common to cache these images in a service such as Quay (quay.io). The proliferation of bioinformatics Docker containers has been aided by repositories such as Dockstore (O'Connor *et al.*, 2017) and bioShaDock (Moreews *et al.*, 2015).

In terms of distributing the underlying software behind workflows, dependency management and containers have significant overlap in terms of the goal of providing consistent software across platforms. Conda attempts to do this by providing compiled software and a compatible ecosystem within custom environments. While Conda provides environments and cross-platform dependency management, it does not provide process isolation and depends on an compatible cohort of programs to coexist within an environment, often making exact version specifications impossible. Docker isolates individual programs to eliminate conflict from each other and the underlying operating system. Both approaches are less than ideal for scientific computing, but do provide working solutions, especially when used synergistically. Dockerfiles consist mostly of dependency installation commands and a few exposed ports and paths. Some Docker images use Conda as their primary dependency installation step, and there are now projects, such as Mulled, a project within the BioContainers repository (da Veiga Leprevost *et al.*, 2017), that auto-create Docker images based on Conda packages.

The potential for Docker containers to provide a missing technical link in terms of reproducible computational research in bioinformatics has been explored by projects including evaluation of de novo assemblers (Belmann *et al.*, 2015). Beaulieu-Jones *et al.* introduced the application of Docker containers using the Drone configuration pipeline engine, allowing the automated evaluation of code changes into the analytical pipeline, which the authors call "continuous analysis" (Beaulieu-Jones and Greene, 2017). While containers can be run in a similar fashion to any command-line software, some pipeline frameworks, such as Nextflow and the CWL-enabled workbenches, offer first-class support for Docker containers.

Containers present an atomic unit of doing work. Docker Swarm and Kubernetes are frameworks that orchestrate hundreds of Docker containers, such as Docker Swarm and Kubernetes, and also offer some level of queueing and resource management, but they compete in a crowded field of existing schedulers.

Workflow sharing

Provided a data set that fits an existing approach, workbenches provide a smooth path for workflow reuse. myExperiment collates workflows from Taverna, Galaxy, and other frameworks into a searchable portal (Goble *et al.*, 2010).

Research Objects aims to bundle the entire stack of data and analysis into a cohesive stack, with a research object bundle (see "Relevant Website" section) and associated workflow ontology (see "Relevant Website" section) which together with a manifest allow researchers to understand an entire analysis from raw data to results (Bechhofer *et al.*, 2010). The motivations and rationale for packaging research objects is described in the highly-cited "Why Linked Data is Not Enough for Scientists" (Bechhofer *et al.*, 2013), which explains why bundling linked data with workflows, results, and publications is preferable to linked data without context.

Notebooks

Notebooks are web-based interactive programming environments. Like Sweave (Leisch, 2002) or knitr (Xie, 2014) literate reports, notebooks support a mix of code and markup, but they are a significant advance in that they allow code to be evaluated in live chunks which retain state. This allows for an easier transition from the type of organic "conversations" typical of analyses in a statistical environment such as R to an automated reproducible report. However, the transition from data preparation steps to pipeline to a notebook is not graceful and currently involves handoffs in the form of intermediate files to store metadata.

The Jupyter notebook is a polyglot derivation of iPython (Perez and Granger, 2007) which supports over 40 languages. The JSON-based .ipynb format recognized by Jupyter been the de facto standard for distributing analyses in the data science community and is gaining acceptance within bioinformatics for rapid dissemination of reproducible analyses (Wang and Ma'ayan, 2016). A static viewer, nbviewer, is available for reading these files in a non-interactive fashion, but few canned solutions exist for instantiating live sessions. A proof of concept demonstrating live notebooks was published in Nature (Shen, 2014) using tmpnb, a Docker-driven temporary notebook engine. An effort from the Freeman neuroscience lab, mybinder.org, accepts Conda requirements files for dependency management and leverages Docker, the Kubernetes container orchestration tool, and Google Cloud Platform to host live repository-driven notebooks.

While they provide an interactive analysis session, notebooks themselves are not integrated development environments (IDEs), and most lack features such as code completion and debugging that IDEs provide. NextflowWorkbench is a project that augments authoring Nextflow pipelines (Kurs *et al.*, 2016) and Dockerfiles using supported JetBrains IDEs.

Although they lack reentrancy frameworks and native support for parallelization, in many ways notebooks are more appropriate for the final steps of analyses than typical workflows frameworks, especially for analysis that rely on statistical tests conducted in scripting languages or statistical environments. Some notebooks such as Beaker are polyglot – supporting multiple languages simultaneously. Collaborative notebook servers such as CoCalc (previous SageMathCloud), aim to provide live editing for highly collaborative research, but self-hosted collaborative notebooks are still in development. Most cloud workbenches do not yet offer significant support for notebooks and integrating notebooks and workflows is a major area of research (Grüning *et al.*, 2017).

Semantically encoded workflows

Extending semantic encoding to workflows has long been envisioned as a way to make them understandable, discoverable, and more reproducible. For example, semantic encoding may allow researchers to find workflows that use a certain reference data set, a combination of similar tools, or even suggest a biological interaction. All of these scenarios are dependent on semantically encoded linked data. They are also highly dependent on linking data between all the various components and stages of a workflow – input data, tools, pipelines, statistical reports and publication. Fortunately, the theoretical (ontologies) and technical (RDF/OWL) foundations for this endeavor have been established by efforts to create the Semantic Web.

An ontology is a set of terms that provides meaning to data. It builds on controlled vocabularies, by defining formal relationships between concepts. Ontologies have been used extensively to describe biological concepts. Building on ontologies is the concept of linked data, which leverages remote but unambiguous addresses to point to items of data. The combinations of ontologies and linked data forms the theoretical foundation for the semantic web, a set of technologies that allow computers to gain insight into the meaning of data, giving different services an ability to both locate and relate linked data using common web Uniform Resource Identifiers (URIs). The underpinning of the semantic web is the Resource Description Framework (RDF), which allows entities to be related to each other and to attributes using triplets in the form of entity: attribute:value. The relationships themselves are defined by a standard called Web Ontology Language (OWL), in which domain-specific ontologies, including those specific to science, can be expressed.

While RDF and OWL were designed to give computers the ability to transverse the web intelligently, these technologies have been used extensively to describe biological systems in implementations such as Uniprot (Jain *et al.*, 2009), EBI (Jupp *et al.*, 2014), DisGeNet (Queralt-Rosinach *et al.*, 2016), KEGG (Kanehisa and Goto, 2000), Reactome (Joshi-Tope *et al.*, 2005), OpenLifeData (González *et al.*, 2014) and others. A popular compendium of linked biological data resources is Bio2RDF (Callahan *et al.*, 2013).

Linked data can be queried using the SPARQL language. Cytoscape and ChEMBL were early proponents of this strategy. Providing SPARQL endpoints has become especially desirable for complex data such as that hosted by the Cancer Genomics Cloud. Tools such as SADI have been extended to Galaxy to enable complex data queries within workflows (Aranguren *et al.*, 2014). For queries on workflows themselves, Taverna and myExperiment support SPARQL endpoints, as well as the Research Object Hub. This means, in theory, one can use a single query for workflows that rely on certain data, use certain tools, and were published within a certain time period by certain analysts. It is also been posited that semantic representations can be used a high-level standard to translate workflows into other frameworks (Garijo *et al.*, 2014).

A semantic workflow platform, Workflow Instance Generation and Specialization (WINGS) is a project that allows semantically encoded workflows, which are then executed using existing pipeline frameworks (Gil *et al.*, 2011). A key feature of WINGS is the ability to semantically enforce outputs, allowing technical and scientific-based sanity checks to be integrated into workflows.

WINGS is built on existing semantic web standards and workflow-specific ontologies, including PROV, a generic semantic model for describing provenance, consisting of data models (PROV-DM) and ontology (PROV-O) capable of representing typical data lineage (Missier *et al.*, 2013). The Open Provenance Model (OPM) extends PROV for use in workflows. A metadata ontology called Open Provenance Model for Workflows (OPMW) extends OPM, PROV and P-Plan in order to encompass a large range of actors, intents, and concepts. Motivations include the ability to enforce "analytical validity" (Zheng *et al.*, 2015) to encode expert knowledge within workflows.

In configuration frameworks such as workbenches there is a clear distinction between workflow templates and workflow runs, and these are reflected respectively in the ontologies wfdesc and wfprov developed by the wf4ever consortium (Corcho *et al.*, 2012), supported as exports from Taverna via TavernaPROV (Soiland-Reyes, 2016) and CWL-based frameworks, and implemented in Research Objects. CWL itself offers a level of semantic encoding using JSON-LD which should be able to leverage existing ontologies, although existing examples have been mostly focused on file formats.

An important ontology describing the components of workflows is provided by EDAM (Ison *et al.*, 2013), an OWL ontology that can describe a diverse set of tools, datasets, and operations common to bioinformatic workflows. At the analysis level, statistical ontologies such as OBCS (Zheng *et al.*, 2016) may be useful for promoting reproducibility and discovery.

Challenges and Opportunities for Pipelines and Workflows

Unruly provenance

Because workflow performance is highly dependent on both parameters and high-performance computing configurations, this should ideally be encoded into provenance metadata, and some attempts have been made to standardize this in a vendor-agnostic fashion (Santana-Perez *et al.*, 2014). Despite extensive work toward the semantic markup of workflows, simply attaching workflow runs to data may not suffice to provide a transparent provenance for data contained within results. Efforts such as LabelFlow (Alper *et al.*, 2014) and PoeM (Gaignard *et al.*, 2016) aim at providing higher-level context by identifying common usage motifs and combining domain-specific ontologies. Another problem with complex provenance is the crowding of workflow visualizations, making them incomprehensible, suggesting some lightweight annotations are needed to distinguish major and minor steps (Missier *et al.*, 2008). Binding experimental metadata to results is an under-examined step in reproducible computational research, and has been identified as a key element in the sustainability of highly variable data such as metagenomics where sample preparation and *in silico* choices have a large impact on results (Hoopen *et al.*, 2017). It is possible blockchain technology may also be useful in this area (Neisse *et al.*, 2017).

Workflow decay

The development of large multi-institutional data repositories that characterize "big science" and remote web services that support both remote data usage and the vision of "bringing the tools to the data" make the cloud an appealing replacement for local computing resources (Stein, 2010). This dependence on data and services hosted by others, however, introduces the threat of "workflow decay" (De Roure *et al.*, 2011) that requires extensive provenance tracking to freeze inputs and tools in order to ensure reproducibility at a later date.

Integration of the web

Due to the increase in the size of datasets combined with limitations of networks, web browsers, and web application frameworks, web applications as the main interface for high-throughput analyses have been less than ideal. This prompted Nucleic Acids Research to add a "Stand-Alone Programs for High-Throughput Data Analysis" to their popular web-server issue. However, there is a large gap between pipelines and notebooks, which are the primary tools used by bioinformaticians and GUI-driven interactive analysis which allows researchers to explore data in real-time. The latter has splintered into smaller "single-page" applications, which have been made popular due to easy-to-use web application frameworks such as RShiny which have spawned many convenience utilities (Mattiello *et al.*, 2016; Zhang and Taylor, 2017; Fouillet *et al.*, 2017). The interactive nature of these sites often demand technologies such as AJAX which dispense with the advantages of stable URLs and linked data enjoyed by earlier generations of bioinformatics web applications. Web applications may also disrupt reproducibility by introducing manual and unrecorded steps into an analysis.

Cloud-based workbenches offer some solutions to this problem, because of their scalability and built-in faculties for queues and the ability to spawn runs locally using APIs. But these tools tend to drop off at the point of target file generation, and even report generation and sharing is not a strong suit for workbenches. There is clearly a great deal to be done to bridge the gap from pipelines to open, linked analyses.

A related challenge, in particular for cloud-based commercial workbenches, is providing support for highly customized or interactive web reports as reporting mechanisms. Web-based QC reporting tools such as MultiQC (Ewels *et al.*, 2016) and literature programming outputs such as knitr/R Markdown pose challenges both in terms of security and navigation within hosted infrastructure.

Notebook and Spark integration

Integrating notebooks into a workflow context, so that it can access the workflow and the notebook components can be reused, parameterized, and queried, remains an open challenge. Ideally statistical reports are tied intimately to their respective workflows (i.e., metadata-driven analysis), so changes such as addition of samples are done in one sample table, and these changes are automatically propagated through the pipeline to a notebook or static report. The approach to this propagation differs significantly between DSLs and configuration-based frameworks, the former being more likely to use variables and functions to drive the pipeline and the report. Breaking notebook sessions into file-centric state engines of DSLs, much less the verbose markup of workbenches and other configuration frameworks, completely destroys their powerful interactive and exploratory nature. This approach is precisely the opposite direction of that taken by the Spark community, which intends to bring executable components into distributed data structures that can be used in live notebook sessions. The YesWorkflow project allow developers to use ontology-based lightweight tags to annotate popular scripting languages with provenance metadata (McPhillips *et al.*, 2015). A non-coincidentally named project, noWorkflow, reconstructs provenance through profiling Python code (Murta *et al.*, 2014). These can be used together to generate semi-automated annotations (Belhajjame *et al.*, 2016; Dey *et al.*, 2015). A similar project, Data Narratives (Gil and Garijo, 2017), offers human-readable annotations of workflows.

Apache Spark is a fast, in-memory data processing engine, released as an Scala API but with support for a number of scripting languages. Spark provides a consistent API to perform data analysis without leaving a notebook session and hides many of the inner workings of the distributed memory. The basic abstraction in Spark that allows distributed data to be manipulated as though it were centralized is the Resilient Distributed Dataset (RDD). RDDs themselves implement a DAG to describe changes and provenance, including their dependence on other RDDs. Therefore, Spark itself subsumes some of the functions of pipeline frameworks, provided an entire bioinformatics analysis can be implemented in Spark. Because Spark is not itself a DSL or configured DAG evaluated at runtime, the reentrancy provided by Spark through caching RDDs is not automatic but must be anticipated by the user. Though Spark can certainly behave in concert with workflows – there are extensive example of ADAM in Toil workflows – these are largely still file-centric in nature, using Spark as a distributed replacement for an compiled executable. Spark jobs are deployed on Apache YARN, and this is supported by Cromwell and Toil. However it does not appear there is yet a framework that leverages RDDs as in-memory inputs and outputs, and supports files within the same workflow context.

Limitations of technologies borrowed from elsewhere

Many technologies for bioinformatic pipelines and workflows have been adopted from other areas of scientific computing encompassing the physical and mathematical sciences, but also computer science, data science, information science, e-commerce, finance, and media. As a result, the fit of these tools and infrastructure is often not ideal for bioinformatics. Containers are an excellent example of this phenomenon.

Docker itself offers no dependency management and its approach is often contrary to the natural flow of an analysis – Docker does not enable the easy configured addition of new R packages within an existing environment, for example. Docker was originally designed to address concerns of rapidly deploying e-commerce apps, which rely heavily on databases and web application servers. These web installations do not resemble a typical executable file-transformation pipeline common to scientific computing, and so frameworks for multi-container installations (e.g., Docker Compose) or orchestration (e.g., Kubernetes) are not geared for the start/stop nature of bioinformatic workflows. Docker is also not particularly good at accepting runtime parameters. The Docker daemon requires administrative (either sudo or Docker-group) to run containers, making Docker unfeasible for some shared compute cluster environments (Silver, 2017). A related project developed at Lawrence Berkeley National Laboratory, Singularity, allows more user-centric computing environments composed of many containers to be created and distributed without these administrative privileges.

Continued need for incentives

Despite the interest in semantic encoding from workflow developers, widespread use of semantic markup on workflows remain elusive. There is also no common API for biological file format validation, which would be low-hanging fruit for encouraging semantically defined outputs. Although considerable effort has been put into WINGS, it is built to leverage Pegasus as an execution engine. However, Pegasus is not yet CWL compliant, and the syntax used for semantic enforcement, Apache Jena, is quite verbose. The move toward standardization suggests a good candidate semantic integration moving forward is through the Common Workflow Language.

Almost 3000 workflows are available in myExperiment, although submissions have lagged in the NGS era (only 10 results were obtained for the search term "RNA-Seq", for instance). It seems likely the introduction of CWL-based workflows will re-energize this repository, but a reinvention of the publishing structure to encourage open workflows as an essential prerequisite to the peer-review process would be a more effective catalyst.

Closing Remarks

Pipelines and workflows are the primary river between scientific data and publications. With the growth of high-throughput data, this river has been made wider by the standardization of workflow languages, deeper by the recognition for the semantic encoding, and is flowing faster thanks to technical advances in cloud computing and containerization. One of the main challenges to future workflow development is keeping the key components of *in silico* scientific processes – data, tools, notebooks – fluid enough to support groundbreaking work and advances from the greater computing community while maintaining standards for reproducible research.

See also: Cloud-Based Bioinformatics Platforms. Cloud-Based Bioinformatics Tools. Cloud-Based Molecular Modeling Systems. Cloud-Based Molecular Modeling Systems. Computational Pipelines and Workflows in Bioinformatics. Computing for Bioinformatics. Constructing Computational Pipelines. DNA Barcoding: Bioinformatics Workflows for Beginners. Exome Sequencing Data Analysis. Functional Enrichment Analysis. Genome Annotation: Perspective From Bacterial Genomes. Infrastructure for High-Performance Computing: Grids and Grid Computing. Infrastructures for High-Performance Computing: Cloud Computing. Integrative Analysis of Multi-Omics Data. Integrative Bioinformatics of Transcriptome: Databases, Tools and Pipelines. MapReduce in Computational Biology via Hadoop and Spark. Natural Language Processing Approaches in Bioinformatics. Network-Based Analysis for Biological Discovery. Next Generation Sequencing Data Analysis. Pipeline of High Throughput Sequencing. Prediction of Coding and Non-Coding RNA. Profiling the Gut Microbiome: Practice and Potential. Protocol for Protein Structure Modelling. Standards and Models for Biological Data: Common Formats. Vaccine Target Discovery. Whole Genome Sequencing Analysis

References

Afgan, E., *et al.*, 2010. Galaxy CloudMan: Delivering cloud compute clusters. BMC Bioinformatics 11 (Suppl. 12), S4.

Alper, P., *et al.*, 2014. LabelFlow: Exploiting workflow provenance to surface scientific data provenance. In: Provenance and Annotation of Data and Processes. Lecture Notes in Computer Science. International Provenance and Annotation Workshop. Cham: Springer, pp. 84–96.

Amstutz, P., *et al.*, 2016. Using the common workflow language (CWL) to run portable workflows with Arvados and toil. F1000Research 5.Available at: http://dx.doi.org/10.7490/f1000research.1112523.1 (accessed 21.08.17)..

Aranguren, M.E., González, A.R., Wilkinson, M.D., 2014. Executing SADI services in galaxy. Journal of Biomedical Semantics 5 (1), 42.

Beaulieu-Jones, B.K., Greene, C.S., 2017. Reproducibility of computational workflows is automated using continuous analysis. Nature Biotechnology 35 (4), 342–346.

Bechhofer, S. *et al.*, 2010. Research objects: Towards exchange and reuse of digital knowledge. In: The Future of the Web for Collaborative Science (FWCS 2010). Available at: https://eprints.soton.ac.uk/268555/ (accessed 10.08.17).

Bechhofer, S., *et al.*, 2013. Why linked data is not enough for scientists. Future Generations Computer Systems: FGCS 29 (2), 599–611.

Belhajjame, K. *et al.*, 2016. Yin & Yang: Demonstrating complementary provenance from noWorkflow & YesWorkflow. In: Provenance and Annotation of Data and Processes: Proceedings of the 6th International Provenance and Annotation Workshop, IPAW 2016, McLean, VA, USA, June 7–8, 2016, Springer, p. 161.

Belmann, P., *et al.*, 2015. Bioboxes: Standardised containers for interchangeable bioinformatics software. GigaScience 4, 47.

Berthold, M.R., *et al.*, 2009. KNIME – The Konstanz information miner: Version 2.0 and beyond. SIGKDD Explorations Newsletter 11 (1), 26–31.

Boettiger, C., 2015. An introduction to Docker for reproducible research. ACM SIGOPS Operating Systems Review 49 (1), 71–79.

Callahan, A., *et al.*, 2013. Bio2RDF release 2: Improved coverage, interoperability and provenance of life science linked data. The Semantic Web: Semantics and Big Data. Berlin; Heidelberg: Springer, pp. 200–212. Extended Semantic Web Conference.

Callahan, S.P. *et al.*, 2006. VisTrails: Visualization meets data management. In: Proceedings of the 2006 ACM SIGMOD International Conference on Management of Data, SIGMOD '06. New York, NY: ACM, pp. 745–747.

Caporaso, J.G., *et al.*, 2010. QIIME allows analysis of high-throughput community sequencing data. Nature methods 7 (5), 335–336.

Cohen-Boulakia, S., *et al.*, 2017. Scientific workflows for computational reproducibility in the life sciences: Status, challenges and opportunities. Future Generations Computer Systems: FGCS. Available at: http://dx.doi.org/10.1016/j.future.2017.01.012.

Collado-Torres, L., *et al.*, 2017. Reproducible RNA-seq analysis using recount2. Nature Biotechnology 35 (4), 319–321.

Corcho, O., *et al.*, 2012. Workflow-centric research objects: First class citizens in scholarly discourse. In: Proceedings of Workshop on the Semantic Publishing. Proceedings of the 9th Extended Semantic Web Conference Hersonissos. Facultad de Informática (UPM), p. 12.

da Veiga Leprevost, F., *et al.*, 2017. BioContainers: An open-source and community-driven framework for software standardization. Bioinformatics 33 (16), 2580–2582.

Davis-Dusenbery, B.N., 2015. Petabyte-scale cancer genomics in the cloud. Cancer Genetics 208 (6), 360.

Deelman, E., *et al.*, 2005. Pegasus: A framework for mapping complex scientific workflows onto distributed systems. Scientific Programming 13 (3), 219–237.

De Roure, D., *et al.*, 2011. Towards the preservation of scientific workflows. In: Proceedings of the 8th International Conference on Preservation of Digital Objects (iPRES 2011). ACM. Available at: http://www.amiga.iaa.csic.es/FCKeditor/UserFiles/File/wfpreservev.pdf.

Devisetty, U.K., *et al.*, 2016. Bringing your tools to CyVerse Discovery Environment using Docker. F1000Research 5, 1442.

Dey, S., *et al.*, 2015. Linking prospective and retrospective provenance in scripts. Theory and Practice of Provenance (TaPP). Available at: https://www.usenix.org/system/files/tapp15-dey.pdf..

Di Tommaso, P., *et al.*, 2015. The impact of Docker containers on the performance of genomic pipelines. PeerJ 3, e1273.

Di Tommaso, P., *et al.*, 2017. Nextflow enables reproducible computational workflows. Nature Biotechnology 35 (4), 316–319.

Ewels, P., *et al.*, 2016. MultiQC: Summarize analysis results for multiple tools and samples in a single report. Bioinformatics 32 (19), 3047–3048.

Fisch, K.M., *et al.*, 2015. Omics pipe: A community-based framework for reproducible multi-omics data analysis. Bioinformatics 31 (11), 1724–1728.

Fouillet, A., *et al.*, 2017. User-friendly Rshiny web applications for supporting syndromic surveillance analysis. Online Journal of Public Health Informatics 9 (1), Available at: http://journals.uic.edu/ojs/index.php/ojphi/article/view/7628 (accessed 22.06.17)..

Gaignard, A., Skaf-Molli, H., Bihouée, A., 2016. From scientific workflow patterns to 5-star linked open data. In: Proceedings of the 8th USENIX Conference on Theory and Practice of Provenance, USENIX Association, pp. 44–48.

Gao, J., *et al.*, 2013. Integrative analysis of complex cancer genomics and clinical profiles using the cBioPortal. Science Signaling 6 (269), l1.

Garijo, D., Gil, Y., Corcho, O., 2014. Towards workflow ecosystems through semantic and standard representations. In: Proceedings of the 9th Workshop on Workflows in Support of Large-Scale Science, WORKS '14. Piscataway, NJ: IEEE Press, pp. 94–104.

Gentleman, R.C., *et al.*, 2004. Bioconductor: Open software development for computational biology and bioinformatics. Genome Biology 5 (10), R80.

Gil, Y., *et al.*, 2007. Examining the challenges of scientific workflows. Computer 40 (12), 24–32.

Gil, Y., *et al.*, 2011. Wings: Intelligent workflow-based design of computational experiments. IEEE Intelligent Systems 26 (1), 62–72.

Gil, Y., Garijo, D., 2017. Towards automating data narratives. In: Proceedings of the 22nd International Conference on Intelligent User Interfaces, ACM, pp. 565–576.

Goble, C., De Roure, D., 2009. The impact of workflow tools on data-centric research. Available at: https://eprints.soton.ac.uk/267336/1/workflows-submitted.pdf.

Goble, C.A., *et al.*, 2010. myExperiment: A repository and social network for the sharing of bioinformatics workflows. Nucleic Acids Research 38 (Web Server Issue), W677–W682.

Goecks, J., *et al.*, 2010. Galaxy: A comprehensive approach for supporting accessible, reproducible, and transparent computational research in the life sciences. Genome Biology 11 (8), R86.

González, A.R., *et al.*, 2014. Automatically exposing OpenLifeData via SADI semantic Web Services. Journal of Biomedical Semantics 5, 46.

Grüning, B.A., *et al.*, 2017. Jupyter and galaxy: Easing entry barriers into complex data analyses for biomedical researchers. PLOS Computational Biology 13 (5), e1005425.

Guimera, R.V., 2012. Bcbio-nextgen: Automated, distributed next-gen sequencing pipeline. EMBnet. Journal 17 (B), 30.

Hastreiter, M., *et al.*, 2017. KNIME4NGS: A comprehensive toolbox for next generation sequencing analysis. Bioinformatics 33 (10), 1565–1567.

Hoopen, P.T., *et al.*, 2017. The metagenomic data life-cycle: Standards and best practices. *GigaScience*. Available at: https://academic.oup.com/gigascience/article-abstract/doi/10.1093/gigascience/gix047/3869082/The-metagenomic-data-lifecycle-standards-and-best (accessed 21.06.17).

Hrynaszkiewicz, I., Li, P., Edmunds, S., 2014. Open science and the role of publishers in reproducible research. Implementing Reproducible Research. 383–410. ISBN: 1826028603.

Huber, W., *et al.*, 2015. Orchestrating high-throughput genomic analysis with Bioconductor. Nature methods 12 (2), 115–121.

Ison, J., *et al.*, 2013. EDAM: An ontology of bioinformatics operations, types of data and identifiers, topics and formats. Bioinformatics 29 (10), 1325–1332.

Jain, E., *et al.*, 2009. Infrastructure for the life sciences: Design and implementation of the UniProt website. BMC Bioinformatics 10, 136.

Joshi-Tope, G., *et al.*, 2005. Reactome: A knowledgebase of biological pathways. Nucleic Acids Research 33 (Database Issue), D428–D432.

Jupp, S., *et al.*, 2014. The EBI RDF platform: Linked open data for the life sciences. Bioinformatics 30 (9), 1338–1339.

Kanehisa, M., Goto, S., 2000. KEGG: Kyoto encyclopedia of genes and genomes. Nucleic Acids Research 28 (1), 27–30.

Kanwal, S., *et al.*, 2017. Investigating reproducibility and tracking provenance – A genomic workflow case study. BMC Bioinformatics 18 (1), 337.

Kelly, B.J., *et al.*, 2015. Churchill: An ultra-fast, deterministic, highly scalable and balanced parallelization strategy for the discovery of human genetic variation in clinical and population-scale genomics. Genome Biology 16 (1), 6.

Köster, J., Rahmann, S., 2012. Snakemake – A scalable bioinformatics workflow engine. Bioinformatics 28 (19), 2520–2522.

Kurs, J.P., Simi, M., Campagne, F., 2016. NextflowWorkbench: Reproducible and reusable workflows for beginners and experts. bioRxiv. 041236. Available at: http://biorxiv.org/content/early/2016/02/24/041236.abstract (accessed 25.06.17)..

Leipzig, J., 2016. A review of bioinformatic pipeline frameworks. Briefings in Bioinformatics. Available at: http://dx.doi.org/10.1093/bib/bbw020..

Leisch, F., 2002. Sweave: Dynamic generation of statistical reports using literate data analysis. In: Härdle, P.D.W., Rönz, P.D.B. (Eds.), Compstat. Heidelberg: Physica-Verlag, pp. 575–580.

Lindenbaum, P., *et al.*, 2011. Knime4Bio: A set of custom nodes for the interpretation of next-generation sequencing data with KNIME. Bioinformatics 27 (22), 3200–3201.

Ludäscher, B., *et al.*, 2006. Scientific workflow management and the Kepler system: Research Articles. Concurrency and Computation: Practice & Experience 18 (10), 1039–1065.

Mattiello, F., *et al.*, 2016. A web application for sample size and power calculation in case-control microbiome studies. Bioinformatics 32 (13), 2038–2040.

McKenna, A., *et al.*, 2010. The genome analysis toolkit: A MapReduce framework for analyzing next-generation DNA sequencing data. Genome Research 20 (9), 1297–1303.

McPhillips, T. *et al.*, 2015. YesWorkflow: A user-oriented, language-independent tool for recovering workflow information from scripts. arXiv [cs.SE]. Available at: http://arxiv.org/abs/1502.02403.

Meyer, F., *et al.*, 2008. The metagenomics RAST server – A public resource for the automatic phylogenetic and functional analysis of metagenomes. BMC Bioinformatics 9 (1), 386.

Missier, P., *et al.*, 2008. Data lineage model for taverna workflows with lightweight annotation requirements. Provenance and Annotation of Data and Processes. Lecture Notes in Computer Science. International Provenance and Annotation Workshop. Berlin; Heidelberg: Springer, pp. 17–30.

Missier, P., Belhajjame, K., Cheney, J., 2013. The W3C PROV family of specifications for modelling provenance metadata. In: Proceedings of the 16th International Conference on Extending Database Technology, EDBT '13. New York, NY: ACM, pp. 773–776.

Moreau, L., *et al.*, 2013. PROV-DM: The PROV data model. Retrieved July 30, 2013.

Moreews, F., *et al.*, 2015. BioShaDock: A community driven bioinformatics shared Docker-based tools registry. F1000Research 4, 1443.

Mullikin, J.C., McMurragy, A.A., 1999. Techview: DNA sequencing. Sequencing the genome, fast. Science 283 (5409), 1867–1869.

Murta, L., *et al.*, 2014. noWorkflow: Capturing and analyzing provenance of scripts. In: Provenance and Annotation of Data and Processes. Lecture Notes in Computer Science. International Provenance and Annotation Workshop. Cham: Springer, pp. 71–83.

Neisse, R., Steri, G., Nai-Fovino, I., 2017. A blockchain-based approach for data accountability and provenance tracking. arXiv [cs.CR]. Available at: http://arxiv.org/abs/1706.04507.

Obenchain, V., *et al.*, 2014. VariantAnnotation: A bioconductor package for exploration and annotation of genetic variants. Bioinformatics 30 (14), 2076–2078.

O'Connor, B.D., *et al.*, 2017. The dockstore: Enabling modular, community-focused sharing of Docker-based genomics tools and workflows. F1000Research 6, 52.

Paten, B., *et al.*, 2015. The NIH BD2K center for big data in translational genomics. Journal of the American Medical Informatics Association: JAMIA 22 (6), 1143–1147.

Perez, F., Granger, B.E., 2007. IPython: A system for interactive scientific computing. Computing in Science Engineering 9 (3), 21–29.

Plummer, E., Twin, J., 2015. A comparison of three bioinformatics pipelines for the analysis of preterm gut microbiota using 16S rRNA gene sequencing data. Journal of Proteomics & Bioinformatics 8 (12), Available at: https://www.omicsonline.org/open-access/a-comparison-of-three-bioinformatics-pipelines-for-the-analysis-ofpreterm-gut-microbiota-using-16s-rrna-gene-sequencing-data-jpb-1000381.php?aid=65142..

Queralt-Rosinach, N., *et al.*, 2016. DisGeNET-RDF: Harnessing the innovative power of the Semantic Web to explore the genetic basis of diseases. Bioinformatics 32 (14), 2236–2238.

Ramírez, F., *et al.*, 2014. deepTools: A flexible platform for exploring deep-sequencing data. Nucleic Acids Research 42 (Web Server Issue), W187–W191.

Robertsen, E.M., *et al.*, 2016. META-pipe – Pipeline annotation, analysis and visualization of marine metagenomic sequence data. arXiv [cs.DC]. Available at: http://arxiv.org/abs/1604.04103.

Sandve, G.K., *et al.*, 2013. Ten simple rules for reproducible computational research. PLOS Computational Biology 9 (10), e1003285.

Santana-Perez, I., *et al.*, 2014. A semantic-based approach to attain reproducibility of computational environments in scientific workflows: A case study. In: Euro-Par 2014: Parallel Processing Workshops. Lecture Notes in Computer Science. European Conference on Parallel Processing. Cham: Springer, pp. 452–463.

Schloss, P.D., *et al.*, 2009. Introducing mothur: Open-source, platform-independent, community-supported software for describing and comparing microbial communities. Applied and Environmental Microbiology 75 (23), 7537–7541.

Shen, H., 2014. Interactive notebooks: Sharing the code. Nature 515 (7525), 151–152.

Silver, A., 2017. Software simplified. Nature 546 (7656), 173–174.

Sloggett, C., Goonasekera, N., Afgan, E., 2013. BioBlend: Automating pipeline analyses within Galaxy and CloudMan. Bioinformatics 29 (13), 1685–1686.

Soiland-Reyes, S., 2016. 2016-provweek-tavernaprov, Github. Available at: https://github.com/stain/2016-provweek-tavernaprov (accessed 16.08.17).

Stallman, R., McGrath, R., 2002. GNU make: A program for directing recompilation: GNU Make Version 3.79.1, Free Software Foundation.

Stein, L.D., 2010. The case for cloud computing in genome informatics. Genome Biology 11 (5), 207.

Stodden, V., Borwein, J., Bailey, D.H., 2013. Setting the default to reproducible incomputational science research. SIAM News 46, 4–6.

Sztromwasser, P., 2014. Throughput and robustness of bioinformatics pipelines for genome-scale data analysis. Available at: http://bora.uib.no/bitstream/handle/1956/7906/dr-thesis-2014-Pawe%C5%82-Sztromwasser.pdf?Sequence=3.

Thain, D., Tannenbaum, T., Livny, M., 2005. Distributed computing in practice: The Condor experience. Concurrency and Computation: Practice & Experience 17 (2–4), 323–356.

2016.A federated ecosystem for sharing genomic, clinical data. Science 352 (6291), 1278–1280.

Trapnell, C., *et al.*, 2012. Differential gene and transcript expression analysis of RNA-seq experiments with TopHat and Cufflinks. Nature Protocols 7 (3), 562–578.

Vivian, J., *et al.*, 2017. Toil enables reproducible, open source, big biomedical data analyses. Nature Biotechnology 35 (4), 314–316.

Voss, K., Van der Auwera, G., Gentry, J., 2017. Full-stack genomics pipelining with GATK4 + WDL + Cromwell. F1000Research 6. Available at: http://dx.doi.org/10.7490/f1000research.1114634.1 (accessed 21.08.17)..

Wang, Z., Ma'ayan, A., 2016. An open RNA-Seq data analysis pipeline tutorial with an example of reprocessing data from a recent Zika virus study. F1000Research 5, 1574.

Wilkinson, M.D., *et al.*, 2016. The FAIR guiding principles for scientific data management and stewardship. Scientific Data 3, 160018.

Wolstencroft, K., *et al.*, 2013. The Taverna workflow suite: Designing and executing workflows of Web Services on the desktop, web or in the cloud. Nucleic Acids Research 41 (Web Server Issue), W557–W561.

Xie, Y., 2014. Knitr: A comprehensive tool for reproducible research in R. Implementing Reproducible Research 1, 20.

Zhang, Z., Taylor, D., 2017. GSA-Genie: A web application for gene set analysis. bioRxiv 125443. Available at: http://biorxiv.org/content/early/2017/04/07/125443.abstract (accessed 22.06.17)..

Zheng, C.L., *et al.*, 2015. Use of semantic workflows to enhance transparency and reproducibility in clinical omics. Genome Medicine 7, 73.

Zheng, J., *et al.*, 2016. The ontology of biological and clinical statistics (OBCS) for standardized and reproducible statistical analysis. Journal of Biomedical Semantics 7 (1), 53.

Relevant Websites

https://w3id.org/ro/
   Wf4Ever.
http://wf4ever.github.io/ro/
   Wf4Ever.

Biographical Sketch

Jeremy Leipzig is a bioinformatics software developer at Cytovas LLC in Philadelphia. He received his MS in Computer Science from North Carolina State University. He is pursuing a PhD in Information Studies at Drexel University.